MW01347063

Climate Change Reconsidered

© 2009, Science and Environmental Policy Project and
Center for the Study of Carbon Dioxide and Global Change

Published by THE HEARTLAND INSTITUTE
19 South LaSalle Street #903
Chicago, Illinois 60603 U.S.A.
phone +1 (312) 377-4000
fax +1 (312) 377-5000
www.heartland.org

All rights reserved, including the right to reproduce
this book or portions thereof in any form.

Opinions expressed are solely those of the authors.
Nothing in this report should be construed as reflecting the views of
the Science and Environmental Policy Project,
Center for the Study of Carbon Dioxide and Global Change,
or The Heartland Institute,
or as an attempt to influence pending legislation.

Additional copies of this book are available from the
Science and Environmental Policy Project, The Heartland Institute, and
Center for the Study of Carbon Dioxide and Global Change at the following prices:

1-10 copies	$154 per copy
11-50 copies	$123 per copy
51-100 copies	$98 per copy
101 or more	$79 per copy

Please use the following citation for this report:

Craig Idso and S. Fred Singer, *Climate Change Reconsidered: 2009 Report of the Nongovernmental Panel on Climate Change (NIPCC)*, Chicago, IL: The Heartland Institute, 2009.

Printed in the United States of America
ISBN-13 – 978-1-934791-28-8
ISBN-10 – 1-934791-28-8

June 2009

Climate Change Reconsidered

Lead Authors

Craig Idso (USA), S. Fred Singer (USA)

Contributors and Reviewers

Warren Anderson (USA), J. Scott Armstrong (USA), Dennis Avery (USA),
Franco Battaglia (Italy), Robert Carter (Australia), Piers Corbyn (UK),
Richard Courtney (UK), Joseph d'Aleo (USA), Don Easterbrook (USA),
Fred Goldberg (Sweden), Vincent Gray (New Zealand), William Gray (USA),
Kesten Green (Australia), Kenneth Haapala (USA), David Hagen (USA),
Klaus Heiss (Austria), Zbigniew Jaworowski (Poland), Olavi Karner (Estonia),
Richard Alan Keen (USA), Madhav Khandekar (Canada), William Kininmonth (Australia),
Hans Labohm (Netherlands), Anthony Lupo (USA), Howard Maccabee (USA),
H. Michael Mogil (USA), Christopher Monckton (UK), Lubos Motl (Czech Republic),
Stephen Murgatroyd (Canada), Nicola Scafetta (USA), Harrison Schmitt (USA),
Tom Segalstad (Norway), George Taylor (USA), Dick Thoenes (Netherlands),
Anton Uriarte (Spain), Gerd Weber (Germany)

Editors

Joseph L. Bast (USA), Diane Carol Bast (USA)

2009 Report of the Nongovernmental International
Panel on Climate Change (NIPCC)

Published for the Nongovernmental International Panel on Climate Change (NIPCC)

The Heartland INSTITUTE

Preface

Before facing major surgery, wouldn't you want a second opinion?

When a nation faces an important decision that risks its economic future, or perhaps the fate of the ecology, it should do the same. It is a time-honored tradition in science to set up a "Team B," which examines the same original evidence but may reach a different conclusion. The Nongovernmental International Panel on Climate Change (NIPCC) was set up to examine the same climate data used by the United Nations-sponsored Intergovernmental Panel on Climate Change (IPCC).

In 2007, the IPCC released to the public its three-volume Fourth Assessment Report titled *Climate Change 2007* (IPCC-AR4, 2007). Its constituent documents were said by the IPCC to comprise "the most comprehensive and up-to-date reports available on the subject," and to constitute "the standard reference for all concerned with climate change in academia, government and industry worldwide." But are these characterizations correct?

On the most important issue, the IPCC's claim that "most of the observed increase in global average temperatures since the mid-twentieth century is *very likely* due to the observed increase in anthropogenic greenhouse gas concentrations [emphasis in the original]," NIPCC reaches the opposite conclusion—namely, that natural causes are very likely to be the dominant cause. Note: We do not say anthropogenic greenhouse gases (GHG) cannot produce some warming or has not in the past. Our conclusion is that the evidence shows they are not playing a *substantial* role.

Almost as importantly, on the question of what effects the present and future warming might have on human health and the natural environment, the IPCC says global warming will "increase the number of people suffering from death, disease and injury from heatwaves, floods, storms, fires and droughts." The NIPCC again reaches the opposite conclusion: A warmer world will be a safer and healthier world for humans and wildlife alike. Once again, we do not say global warming won't occur or have any effects (positive or negative) on human health and wildlife.

Rather, our conclusion is that the evidence shows the *net* effect of continued warming and rising carbon dioxide concentrations in the atmosphere will be beneficial to humans, plants, and wildlife.

We have reviewed the materials presented in the first two volumes of the Fourth Assessment—*The Physical Science Basis* and *Impacts, Adaptation and Vulnerability*—and we find them to be highly selective and controversial with regard to making future projections of climate change and discerning a significant human-induced influence on current and past climatic trends. Although the IPCC claims to be unbiased and to have based AR4 on the best available science, such is not the case. In many instances conclusions have been seriously exaggerated, relevant facts have been distorted, and key scientific studies have been omitted or ignored.

We present support for this thesis in the body of this volume, where we describe and reference thousands of peer-reviewed scientific journal articles that document scientific or historical facts that contradict the IPCC's central claims, that global warming is man-made and that its effects will be catastrophic. Some of this research became available after the AR4's self-imposed deadline of May 2006, but much of it was in the scientific record that was available to, and should have been familiar to, the IPCC's editors.

Below, we first sketch the history of the IPCC and NIPCC, which helps explain why two scientific bodies could study the same data and come to very different conclusions. We then explain the list of 31,478 American scientists that appears in Appendix 4, and end by expressing what we hoped to achieve by producing this report.

A Brief History of the IPCC

The rise in environmental consciousness since the 1970s has focused on a succession of 'calamities': cancer epidemics from chemicals, extinction of birds and other species by pesticides, the depletion of the ozone layer by supersonic transports and later by freons, the death of forests ('Waldsterben') because of acid rain, and finally, global warming, the "mother of all environmental scares" (according to the late Aaron Wildavsky).

The IPCC can trace its roots to World Earth Day in 1970, the Stockholm Conference in 1971-72, and the Villach Conferences in 1980 and 1985. In July 1986, the United Nations Environment Program (UNEP) and the World Meteorological Organization (WMO) established the Intergovernmental Panel on Climate Change (IPCC) as an organ of the United Nations.

The IPCC's key personnel and lead authors were appointed by governments, and its Summaries for Policymakers (SPM) have been subject to approval by member governments of the UN. The scientists involved with the IPCC are almost all supported by government contracts, which pay not only for their research but for their IPCC activities. Most travel to and hotel accommodations at exotic locations for the drafting authors is paid with government funds.

The history of the IPCC has been described in several publications. What is not emphasized, however, is the fact that it was an activist enterprise from the very beginning. Its agenda was to justify control of the emission of greenhouse gases, especially carbon dioxide. Consequently, its scientific reports have focused solely on evidence that might point toward human-induced climate change. The role of the IPCC "is to assess on a comprehensive, objective, open and transparent basis the latest scientific, technical and socio-economic literature produced worldwide relevant to the understanding of *the risk of human-induced climate change,* its observed and projected impacts and options for adaptation and mitigation" [emphasis added] (IPCC 2008).

The IPCC's three chief ideologues have been (the late) Professor Bert Bolin, a meteorologist at Stockholm University; Dr. Robert Watson, an atmospheric chemist at NASA, later at the World Bank, and now chief scientist at the UK Department of Environment, Food and Rural Affairs; and Dr. John Houghton, an atmospheric radiation physicist at Oxford University, later head of the UK Met Office as Sir John Houghton.

Watson had chaired a self-appointed group to find evidence for a human effect on stratospheric ozone and was instrumental in pushing for the 1987 Montreal Protocol to control the emission of chlorofluorocarbons (CFCs). Using the blueprint of

the Montreal Protocol, environmental lawyer David Doniger of the Natural Resources Defense Council then laid out a plan to achieve the same kind of control mechanism for greenhouse gases, a plan that eventually was adopted as the Kyoto Protocol.

From the very beginning, the IPCC was a political rather than scientific entity, with its leading scientists reflecting the positions of their governments or seeking to induce their governments to adopt the IPCC position. In particular, a small group of activists wrote the all-important Summary for Policymakers (SPM) for each of the four IPCC reports (McKitrick *et al.,* 2007).

While we are often told about the thousands of scientists on whose work the Assessment reports are based, the vast majority of these scientists had no direct influence on the conclusions expressed by the IPCC. Those policy summaries were produced by an inner core of scientists, and the SPMs were revised and agreed to, line-by-line, by representatives of member governments. This obviously is not how real scientific research is reviewed and published.

These SPMs turn out, in all cases, to be highly selective summaries of the voluminous science reports—typically 800 or more pages, with no indexes (except, finally, the Fourth Assessment Report released in 2007), and essentially unreadable except by dedicated scientists.

The IPCC's First Assessment Report (IPCC-FAR, 1990) concluded that the observed temperature changes were "broadly consistent" with greenhouse models. Without much analysis, it gave the "climate sensitivity" of a 1.5 to 4.5° C rise for a doubling of greenhouse gases. The IPCC-FAR led to the adoption of the Global Climate Treaty at the 1992 Earth Summit in Rio de Janeiro.

The FAR drew a critical response (SEPP, 1992). FAR and the IPCC's style of work also were criticized in two editorials in *Nature* (Anonymous, 1994, Maddox, 1991).

The IPCC's Second Assessment Report (IPCC-SAR, 1995) was completed in 1995 and published in 1996. Its SPM contained the memorable conclusion, "the balance of evidence suggests a discernible human influence on global climate." The SAR was again heavily criticized, this time for having undergone significant changes in the body of the report to make it 'conform' to the SPM—*after* it was finally approved by the scientists involved in writing the report. Not only was the report altered, but a key graph was also doctored to suggest a human influence. The evidence presented to support the SPM conclusion turned out to be completely spurious.

There is voluminous material available about these text changes, including a *Wall Street Journal* editorial article by Dr. Frederick Seitz (Seitz, 1996). This led to heated discussions between supporters of the IPCC and those who were aware of the altered text and graph, including an exchange of letters in the *Bulletin of the American Meteorological Society* (Singer *et al.,* 1997).

SAR also provoked the 1996 publication of the Leipzig Declaration by SEPP, which was signed by some 100 climate scientists. A booklet titled *The Scientific Case Against the Global Climate Treaty* followed in September 1997 and was translated into several languages. (SEPP, 1997. All these are available online at www.sepp.org.) In spite of its obvious shortcomings, the IPCC report provided the underpinning for the Kyoto Protocol, which was adopted in December 1997. The background is described in detail in the booklet *Climate Policy— From Rio to Kyoto,* published by the Hoover Institution (Singer, 2000).

The Third Assessment Report of the IPCC (IPCC-TAR 2001) was noteworthy for its use of spurious scientific papers to back up its SPM claim of "new and stronger evidence" of anthropogenic global warming. One of these was the so-called "hockey-stick" paper, an analysis of proxy data, which claimed the twentieth century was the warmest in the past 1,000 years. The paper was later found to contain basic errors in its statistical analysis (McIntyre and McKitrick, 2003, 2005; Wegman *et al.,* 2006). The IPCC also supported a paper that claimed pre-1940 warming was of human origin and caused by greenhouse gases. This work, too, contained fundamental errors in its statistical analysis. The SEPP response to TAR was a 2002 booklet, *The Kyoto Protocol is Not Backed by Science* (SEPP, 2002).

The Fourth Assessment Report of the IPCC (IPCC-AR4 2007) was published in 2007; the SPM of Working Group I was released in February; and the full report from this Working Group was released in May—after it had been changed, once again, to "conform" to the Summary. It is significant that AR4 no longer makes use of the hockey-stick paper or the paper claiming pre-1940 human-caused warming. Once again controversy ensued, however, this time when the IPCC refused to publicly share comments submitted by peer-reviewers, then sent all the reviewers' comments in hard copy to a library that was closed for renovation, and then finally, but only under pressure, posted them online. Inspection of those comments revealed that the authors had rejected more than half of all the reviewers' comments in the crucial chapter attributing recent warming to human activities.

AR4 concluded that "most of the observed increase in global average temperatures since the mid-20th century is *very likely* due to the observed increase in anthropogenic greenhouse gas concentrations" (emphasis in the original). However, as the present report will show, it ignored available evidence against a human contribution to current warming and the substantial research of the past few years on the effects of solar activity on climate change.

Why have IPCC reports been marred by controversy and so frequently contradicted by subsequent research? Certainly its agenda to find evidence of a human role in climate change is a major reason; its organization as a government entity beholden to political agendas is another major reason; and the large professional and financial rewards that go to scientists and bureaucrats who are willing to bend scientific facts to match those agendas is yet a third major reason.

Another reason for the IPCC's unreliability is the naive acceptance by policymakers of "peer-reviewed" literature as necessarily authoritative. It has become the case that refereeing standards for many climate-change papers are inadequate, often because of the use of an "invisible college" of reviewers of like inclination to a paper's authors (Wegman *et al.,* 2006). Policy should be set upon a background of demonstrable science, not upon simple (and often mistaken) assertions that, because a paper was refereed, its conclusions must be accepted.

Nongovernmental International Panel on Climate Change (NIPCC)

When new errors and outright falsehoods were observed in the initial drafts of AR4, SEPP set up a "Team B" to produce an independent evaluation of the available scientific evidence. While the initial organization took place at a meeting in Milan in 2003, Team B was activated after the AR4 SPM appeared in February 2007. It changed its name to the Nongovernmental International Panel on Climate Change (NIPCC) and organized an international climate workshop in Vienna in April 2007.

The present report stems from the Vienna workshop and subsequent research and contributions by a larger group of international scholars. For a list of those contributors, see page ii. Craig Idso then made a major contribution to the report by tapping the extensive collection of reviews of scientific research he helped collect and write, which is available on the Web site of the Center for the Study of Carbon Dioxide and Global Change (www.CO2science.org). A Summary for Policymakers, edited by S. Fred Singer, was published by The Heartland Institute in 2008 under the title *Nature, Not Human Activity, Rules the Planet* (Singer, 2008). Since the summary was completed prior to a major expansion and completion of the full NIPCC report, the two documents now stand on their own as independent scholarly works and substantially agree.

What was our motivation? It wasn't financial self-interest: Except for a foundation grant late in the process to enable Craig Idso to devote the many hours necessary to assemble and help edit the final product, no grants or contributions were provided or promised in return for producing this report. It wasn't political: No government agency commissioned or authorized our efforts, and we do not advise or support the candidacies of any politicians or candidates for public office.

We donated our time and best efforts to produce this report out of concern that the IPCC was provoking an irrational fear of anthropogenic global warming based on incomplete and faulty science. Global warming hype has led to demands for unrealistic efficiency standards for cars, the construction of uneconomic wind and solar energy stations, the establishment of large production facilities for uneconomic biofuels such as ethanol from corn, requirements that electric companies purchase expensive power from so-called "renewable" energy sources, and plans to sequester, at considerable expense, carbon dioxide emitted from power plants. While there is nothing wrong with initiatives to increase energy efficiency or diversify energy sources, they cannot be justified as a realistic means to control climate. Neither does science justify policies that try to hide the huge cost of greenhouse gas controls, such as cap and trade, a "clean development mechanism," carbon offsets, and similar schemes that enrich a few at the expense of the rest of us.

Seeing science clearly misused to shape public policies that have the potential to inflict severe economic harm, particularly on low-income groups,

we choose to speak up for science at a time when too few people outside the scientific community know what is happening, and too few scientists who know the truth have the will or the platforms to speak out against the IPCC.

NIPCC is what its name suggests: an international panel of *nongovernment* scientists and scholars who have come together to understand the causes and consequences of climate change. Because we are not predisposed to believe climate change is caused by human greenhouse gas emissions, we are able to look at evidence the IPCC ignores. Because we do not work for any governments, we are not biased toward the assumption that greater government activity is necessary.

The Petition Project

Attached as Appendix 4 to this report is a description of "The Petition Project" and a directory of the 31,478 American scientists who have signed the following statement:

> We urge the United States government to reject the global warming agreement that was written in Kyoto, Japan in December, 1997, and any other similar proposals. The proposed limits on greenhouse gases would harm the environment, hinder the advance of science and technology, and damage the health and welfare of mankind.
>
> There is no convincing scientific evidence that human release of carbon dioxide, methane, or other greenhouse gases is causing or will, in the foreseeable future, cause catastrophic heating of the Earth's atmosphere and disruption of the Earth's climate. Moreover, there is substantial scientific evidence that increases in atmospheric carbon dioxide produce many beneficial effects upon the natural plant and animal environments of the Earth.

This is a remarkably strong statement of dissent from the perspective advanced by the IPCC, and it is similar to the perspective represented by the NIPCC and the current report. The fact that more than *ten times* as many scientists have signed it as are alleged to have "participated" in some way or another in the research, writing, and review of IPCC AR4 is very significant. These scientists, who include among their number 9,029 individuals with Ph.D.s, actually *endorse* the statement that appears above. By contrast, fewer than 100 of the scientists (and nonscientists) who are listed in the appendices to the IPCC AR4 actually participated in the writing of the all-important Summary for Policymakers or the editing of the final report to comply with the summary, and therefore could be said to endorse the main findings of that report. Consequently, we cannot say for sure whether more than 100 scientists in the entire world actually endorse the most important claims that appear in the IPCC AR4 report.

We will not make the same mistake as the IPCC. We do not claim the 31,478 scientists whose names appear at the end of this report endorse all of the findings and conclusions of this report. As the authors of the petition say (in an introduction to the directory of signers in Appendix 4), "signatories to the petition have signed just the petition—which speaks for itself." We append the list of their names to this report with the permission of the persons who maintain the list to demonstrate unequivocally the broad support within the scientific community for the general perspective expressed in this report, and to highlight one of the most telling differences between the NIPCC and the IPCC.

For more information about The Petition Project, including the text of the letter endorsing it written by the late Dr. Frederick Seitz, past president of the National Academy of Sciences and president emeritus of Rockefeller University, please turn to Appendix 4 or visit the project's Web site at www.petitionproject.org.

Looking Ahead

The public's fear of anthropogenic global warming, despite almost hysterical coverage of the issue by the mainstream media, seems to have hit a ceiling and is falling. Only 34 percent of Americans polled (Rasmussen Reports, 2009) believe humans are causing global warming. A declining number even believe the Earth is experiencing a warming trend (Pew Research Center, 2008). A poll of 12,000 people in 11 countries, commissioned by the financial institution HSBC and environmental advocacy groups, found only one in five respondents—20 percent—said they would be willing to spend any extra money to reduce climate change, down from 28 percent a year earlier (O'Neil, 2008).

While the present report makes it clear that the scientific debate is tilting away from global warming alarmism, we are pleased to see the political debate also is not over. Global warming "skeptics" in the policy arena include Vaclav Klaus, president of the Czech Republic and 2009 president of the Council of the European Union; Helmut Schmidt, former German chancellor; and Lord Nigel Lawson, former United Kingdom chancellor of the exchequer. There is some evidence that policymakers world-wide are reconsidering the wisdom of efforts to legislate reductions in greenhouse gas (GHG) emissions.

We regret that many advocates in the debate have chosen to give up debating the science and focus almost exclusively on questioning the motives of "skeptics," name-calling, and *ad hominem* attacks. We view this as a sign of desperation on their part, and a sign that the debate has shifted toward climate realism.

We hope the present study will help bring reason and balance back into the debate over climate change, and by doing so perhaps save the peoples of the world from the burden of paying for wasteful, unnecessary energy and environmental policies. We stand ready to defend the analysis and conclusion in the study that follows, and to give further advice to policymakers who are open-minded on this most important topic.

S. Fred Singer, Ph.D.
President, Science and Environmental Policy Project
Professor Emeritus of Environmental Science,
    University of Virginia
www.sepp.org

Craig D. Idso, Ph.D.
Chairman, Center for the Study of Carbon Dioxide and Global Change
www.co2science.org

Acknowledgments: The editors thank Joseph and Diane Bast of The Heartland Institute for their editorial skill and R. Warren Anderson for his technical assistance.
www.heartland.org

References

Anonymous 1994. IPCC's ritual on global warming. *Nature* **371**: 269.

IPCC-AR4 2007. *Climate Change 2007: The Physical Science Basis. Contribution of Working Group I to the Fourth Assessment Report of the Intergovernmental Panel on Climate Change.* Cambridge University Press.

IPCC-FAR 1990. *Scientific Assessment of Climate Change. Contribution of Working Group I to the First Assessment Report of the Intergovernmental Panel on Climate Change.* Cambridge University Press.

IPCC-SAR 1996. *Climate Change 1995: The Science of Climate Change. Contribution of Working Group I to the Second Assessment Report of the Intergovernmental Panel on Climate Change,* Cambridge University Press.

IPCC-TAR 2001. *Climate Change 2001: The Scientific Basis. Contribution of Working Group I to the Third Assessment Report of the Intergovernmental Panel on Climate Change.* Cambridge University Press.

Maddox J. 1991. Making global warming public property. *Nature* **349**: 189.

McIntyre, S. and McKitrick, R. 2003. Corrections to Mann et al. (1998) proxy data base and northern hemisphere average temperature series. *Energy & Environment* **14**: 751-777.

McIntyre, S. and McKitrick, R. 2005. Hockey sticks, principal components and spurious significance. *Geophysical Research Letters* 32 L03710.

McKitrick, R. 2007. *Independent Summary for Policymakers IPCC Fourth Assessment Report.* Ed. Fraser Institute. Vancouver, BC.

O'Neil, P. 2008. Efforts to support global climate-change falls: Poll. Canwest News Service, 27 Nov.

Pew Research Center 2008. A deeper partisan divide over global warming, summary of findings. 8 May. http://people-press.org

Rasmussen Reports 2009. Energy Update. April 17. http://www.rasmussenreports.com/

Seitz, F. 1996. A major deception on global warming. *The Wall Street Journal,* 12 June.

SEPP 1992. *The Greenhouse Debate Continued: An Analysis and Critique of the IPCC Climate Assessment.* ICS Press, San Francisco, CA.

SEPP 1997. *The Scientific Case Against the Global Climate Treaty.* www.sepp.org/publications/GWbooklet/GW.html [Also available in German, French, and Spanish].

Singer, S.F. 1997, 1999. *Hot Talk Cold Science.* The Independent Institute, Oakland CA.

Singer, S.F. 2008. *Nature, Not Human Activity, Rules the Climate.* The Heartland Institute, Chicago, IL.

Wegman, E., Scott, D.W. and Said, Y. 2006. Ad Hoc Committee Report to Chairman of the House Committee on Energy & Commerce and to the Chairman of the House sub-committee on Oversight & Investigations on the Hockey-stick Global Climate Reconstructions. US House of Representatives, Washington DC.

Table of Contents

Executive Summary

The Fourth Assessment Report of the Intergovernmental Panel on Climate Change's Working Group-1 (Science) (IPCC-AR4 2007), released in 2007, is a major research effort by a group of dedicated specialists in many topics related to climate change. It forms a valuable compendium of the current state of the science, enhanced by having an index which had been lacking in previous IPCC reports. AR4 also permits access to the numerous critical comments submitted by expert reviewers, another first for the IPCC.

While AR4 is an impressive document, it is far from being a reliable reference work on some of the most important aspects of climate change science and policy. It is marred by errors and misstatements, ignores scientific data that were available but were inconsistent with the authors' pre-conceived conclusions, and has already been contradicted in important parts by research published since May 2006, the IPCC's cut-off date.

In general, the IPCC fails to consider important scientific issues, several of which would upset its major conclusion—that "most of the observed increase in global average temperatures since the mid-20th century is *very likely* due to the observed increase in anthropogenic greenhouse gas concentrations [emphasis in the original]." The IPCC defines "very likely" as at least 90 percent certain. They do not explain how they derive this number. The IPCC also does not define the word "most," nor do they provide any explanation.

The IPCC does not apply generally accepted methodologies to determine what fraction of current warming is natural, or how much is caused by the rise in greenhouse gases (GHG). A comparison of "fingerprints" from best available observations with the results of state-of-the-art GHG models leads to the conclusion that the (human-caused) GHG contribution is minor. This fingerprint evidence, though available, was ignored by the IPCC.

The IPCC continues to undervalue the overwhelming evidence that, on decadal and century-long time scales, the Sun and associated atmospheric cloud effects are responsible for much of past climate change. It is therefore highly likely that the Sun is also a major cause of twentieth-century warming, with anthropogenic GHG making only a minor contribution. In addition, the IPCC ignores, or addresses imperfectly, other science issues that call for discussion and explanation.

These errors and omissions are documented in the present report by the Nongovernmental International Panel on Climate Change (NIPCC). The report is divided into nine chapters that are briefly summarized here, and then more fully described in the remainder of this summary.

Chapter 1 describes the limitations of the IPCC's attempt to forecast future climate conditions by using computer climate models. The IPCC violates many of the rules and procedures required for scientific forecasting, making its "projections" of little use to policymakers. As sophisticated as today's state-of-the-art models are, they suffer deficiencies and shortcomings that could alter even the very *sign* (plus or minus, warming or cooling) of earth's projected temperature response to rising atmospheric CO_2 concentrations. If the global climate models on which the IPCC relies are not validated or reliable, most of the rest of the AR4, while it makes for fascinating reading, is irrelevant to the public policy debate over what should be done to stop or slow the arrival of global warming.

Chapter 2 describes feedback factors that reduce the earth's temperature sensitivity to changes in atmospheric CO_2. Scientific studies suggest the model-derived temperature sensitivity of the earth for a doubling of the pre-industrial CO_2 level is much lower than the IPCC's estimate. Corrected feedbacks in the climate system reduce climate sensitivity to values that are an order of magnitude smaller than what the IPCC employs.

Chapter 3 reviews empirical data on past temperatures. We find no support for the IPCC's claim that climate observations during the twentieth century are either unprecedented or provide evidence of an anthropogenic effect on climate. We reveal the

1

methodological errors of the "hockey stick" diagram of Mann *et al.,* evidence for the existence of a global Medieval Warm Period, flaws in the surface-based temperature record of more modern times, evidence from highly accurate satellite data that there has been no net warming over the past 29 years, and evidence that the distribution of modern warming does not bear the "fingerprint" of an anthropogenic effect.

Chapter 4 reviews observational data on glacier melting, sea ice area, variation in precipitation, and sea level rise. We find no evidence of trends that could be attributed to the supposedly anthropogenic global warming of the twentieth century.

Chapter 5 summarizes the research of a growing number of scientists who say variations in solar activity, not greenhouse gases, are the true driver of climate change. We describe the evidence of a solar-climate link and how these scientists have grappled with the problem of finding a specific mechanism that translates small changes in solar activity into larger climate effects. We summarize how they may have found the answer in the relationships between the sun, cosmic rays and reflecting clouds.

Chapter 6 investigates and debunks the widespread fears that global warming might cause more extreme weather. The IPCC claims global warming will cause (or already is causing) more droughts, floods, hurricanes, storms, storm surges, heat waves, and wildfires. We find little or no support in the peer-reviewed literature for these predictions and considerable evidence to support an opposite prediction: That weather would be *less* extreme in a warmer world.

Chapter 7 examines the biological effects of rising CO_2 concentrations and warmer temperatures. This is the largely unreported side of the global warming debate, perhaps because it is unequivocally good news. Rising CO_2 levels increase plant growth and make plants more resistant to drought and pests. It is a boon to the world's forests and prairies, as well as to farmers and ranchers and the growing populations of the developing world.

Chapter 8 examines the IPCC's claim that CO_2-induced increases in air temperature will cause unprecedented plant and animal extinctions, both on land and in the world's oceans. We find there little real-world evidence in support of such claims and an abundance of counter evidence that suggests ecosystem biodiversity will *increase* in a warmer and CO_2-enriched world.

Chapter 9 challenges the IPCC's claim that CO_2-induced global warming is harmful to human health.

The IPCC blames high-temperature events for increasing the number of cardiovascular-related deaths, enhancing respiratory problems, and fueling a more rapid and widespread distribution of deadly infectious diseases, such as malaria, dengue and yellow fever. However, a thorough examination of the peer-reviewed scientific literature reveals that further global warming would likely do just the opposite and actually reduce the number of lives lost to extreme thermal conditions. We also explain how CO_2-induced global warming would help feed a growing global population without major encroachment on natural ecosystems, and how increasing production of biofuels (a strategy recommended by the IPCC) damages the environment and raises the price of food.

The research summarized in this report is only a small portion of what is available in the peer-reviewed scientific literature. To assist readers who want to explore information not contained between the covers of this volume, we have included Internet hyperlinks to the large and continuously updated databases maintained by the Center for the Study of Carbon Dioxide and Global Change at www.co2science.org.

Key Findings by Chapter

Chapter 1. Global Climate Models and Their Limitations

- The IPCC places great confidence in the ability of general circulation models (GCMs) to simulate future climate and attribute observed climate change to anthropogenic emissions of greenhouse gases.

- The forecasts in the Fourth Assessment Report were not the outcome of validated scientific procedures. In effect, they are the opinions of scientists transformed by mathematics and obscured by complex writing. The IPCC's claim that it is making "projections" rather than "forecasts" is not a plausible defense.

- Today's state-of-the-art climate models fail to accurately simulate the physics of earth's radiative energy balance, resulting in uncertainties "as large as, or larger than, the doubled CO_2 forcing."

- A long list of major model imperfections prevents models from properly modeling cloud formation and cloud-radiation interactions, resulting in large differences between model predictions and observations.

- Computer models have failed to simulate even the correct sign of observed precipitation anomalies, such as the summer monsoon rainfall over the Indian region. Yet it is understood that precipitation plays a major role in climate change.

Chapter 2. Feedback Factors and Radiative Forcing

- Scientific research suggests the model-derived temperature sensitivity of the earth accepted by the IPCC is too large. Corrected feedbacks in the climate system could reduce climate sensitivity to values that are an order of magnitude smaller.

- Scientists may have discovered a connection between cloud creation and sea surface temperature in the tropics that creates a "thermostat-like control" that automatically vents excess heat into space. If confirmed, this could totally compensate for the warming influence of all anthropogenic CO_2 emissions experienced to date, as well as all those that are anticipated to occur in the future.

- The IPCC dramatically underestimates the total cooling effect of aerosols. Studies have found their radiative effect is comparable to or larger than the temperature forcing caused by all the increase in greenhouse gas concentrations recorded since pre-industrial times.

- Higher temperatures are known to increase emissions of dimethyl sulfide (DMS) from the world's oceans, which increases the albedo of marine stratus clouds, which has a cooling effect.

- Iodocompounds—created by marine algae—function as cloud condensation nuclei, which help create new clouds that reflect more incoming solar radiation back to space and thereby cool the planet.

- As the air's CO_2 content—and possibly its temperature—continues to rise, plants emit greater amounts of carbonyl sulfide gas, which eventually makes it way into the stratosphere, where it is transformed into solar-radiation-reflecting sulfate aerosol particles, which have a cooling effect.

- As CO_2 enrichment enhances biological growth, atmospheric levels of biosols rise, many of which function as cloud condensation nuclei. Increased cloudiness diffuses light, which stimulates plant growth and transfers more fixed carbon into plant and soil storage reservoirs.

- Since agriculture accounts for almost half of nitrous oxide (N_2O) emissions in some countries, there is concern that enhanced plant growth due to CO_2 enrichment might increase the amount and warming effect of this greenhouse gas. But field research shows that N_2O emissions fall as CO_2 concentrations and temperatures rise, indicating this is actually another negative climate feedback.

- Methane (CH_4) is a potent greenhouse gas. An enhanced CO_2 environment has been shown to have "neither positive nor negative consequences" on atmospheric methane concentrations. Higher temperatures have been shown to result in reduced methane release from peatbeds. Methane emissions from cattle have been reduced considerably by altering diet, immunization, and genetic selection.

Chapter 3. Observations: Temperature Records

- The IPCC claims to find evidence in temperature records that the warming of the twentieth century was "unprecedented" and more rapid than during any previous period in the past 1,300 years. But the evidence it cites, including the "hockey-stick" representation of earth's temperature record by Mann et al., has been discredited and contradicted by many independent scholars.

- A corrected temperature record shows temperatures around the world were warmer during the Medieval Warm Period of approximately 1,000 years ago than they are today, and have averaged 2-3°F warmer than today's temperatures over the past 10,000 years.

- Evidence of a global Medieval Warm Period is extensive and irrefutable. Scientists working with a variety of independent methodologies have found it in proxy records from Africa, Antarctica, the Arctic, Asia, Europe, North America, and South America.

- The IPCC cites as evidence of modern global warming data from surface-based recording stations yielding a 1905-2005 temperature increase of 0.74°C +/- 0.18°C. But this temperature record is known to be positively biased by insufficient corrections for the non-greenhouse-gas-induced urban heat island (UHI) effect. It may be impossible to make proper corrections for this deficiency, as the UHI of even small towns dwarfs any concomitant augmented greenhouse effect that may be present.

- Highly accurate satellite data, adjusted for orbit drift and other factors, show a much more modest warming trend in the last two decades of the twentieth century and a dramatic decline in the warming trend in the first decade of the twenty-first century.

- The "fingerprint" or pattern of warming observed in the twentieth century differs from the pattern predicted by global climate models designed to simulate CO_2-induced global warming. Evidence reported by the U.S. Climate Change Science Program (CCSP) is unequivocal: All greenhouse models show an increasing warming trend with altitude in the tropics, peaking around 10 km at roughly twice the surface value. However, the temperature data from balloons give the opposite result: no increasing warming, but rather a slight cooling with altitude.

- Temperature records in Greenland and other Arctic areas reveal that temperatures reached a maximum around 1930 and have decreased in recent decades. Longer-term studies depict oscillatory cooling since the Climatic Optimum of the mid-Holocene (~9000-5000 years BP), when it was perhaps 2.5° C warmer than it is now.

- The average temperature history of Antarctica provides no evidence of twentieth century warming. While the Antarctic peninsula shows recent warming, several research teams have documented a cooling trend for the interior of the continent since the 1970s.

Chapter 4. Observations: Glaciers, Sea Ice, Precipitation, and Sea Level

- Glaciers around the world are continuously advancing and retreating, with a general pattern of retreat since the end of the Little Ice Age. There is no evidence of a increased rate of melting overall since CO_2 levels rose above their pre-industrial levels, suggesting CO_2 is not responsible for glaciers melting.

- Sea ice area and extent have continued to increase around Antarctica over the past few decades. Evidence shows that much of the reported thinning of Arctic sea ice that occurred in the 1990s was a natural consequence of changes in ice dynamics caused by an atmospheric regime shift, of which there have been several in decades past and will likely be several in the decades to come, totally irrespective of past or future changes in the air's CO_2 content. The Arctic appears to have recovered from its 2007 decline.

- Global studies of precipitation trends show no net increase and no consistent trend with CO_2, contradicting climate model predictions that warming should cause increased precipitation. Research on Africa, the Arctic, Asia, Europe, and North and South America all find no evidence of a significant impact on precipitation that could be attributed to anthropogenic global warming.

- The cumulative discharge of the world's rivers remained statistically unchanged between 1951 and 2000, a finding that contradicts computer forecasts that a warmer world would cause large changes in global streamflow characteristics. Droughts and floods have been found to be less frequent and severe during the Current Warm Period than during past periods when temperatures were even higher than they are today.

- The results of several research studies argue strongly against claims that CO_2-induced global warming would cause catastrophic disintegration of the Greenland and Antarctic Ice Sheets. In fact, in the case of Antarctica, they suggest just the opposite—i.e., that CO_2-induced global warming would tend to buffer the world against such an outcome.

- The mean rate of global sea level rise has not accelerated over the recent past. The determinants of sea level are poorly understood due to considerable uncertainty associated with a number of basic parameters that are related to the water balance of the world's oceans and the meltwater contribution of Greenland and Antarctica. Until these uncertainties are satisfactorily resolved, we cannot be confident that short-lived changes in global temperature produce corresponding changes in sea level.

Chapter 5. Solar Variability and Climate Cycles

- The IPCC claims the radiative forcing due to changes in the solar output since 1750 is +0.12 Wm^{-2}, an order of magnitude smaller than its estimated net anthropogenic forcing of +1.66 Wm^{-2}. A large body of research suggests that the IPCC has got it backwards, that it is the sun's influence that is responsible for the lion's share of climate change during the past century and beyond.

- The total energy output of the sun changes by only 0.1 percent during the course of the solar cycle, although larger changes may be possible over periods of centuries. On the other hand, the ultraviolet radiation from the sun can change by several percent over the solar cycle – as indeed noted by observing changes in stratospheric ozone. The largest changes, however, occur in the intensity of the solar wind and interplanetary magnetic field.

- Reconstructions of ancient climates reveal a close correlation between solar magnetic activity and solar irradiance (or brightness), on the one hand, and temperatures on earth, on the other. Those correlations are much closer than the relationship between carbon dioxide and temperature.

- Cosmic rays could provide the mechanism by which changes in solar activity affect climate. During periods of greater solar magnetic activity, greater shielding of the earth occurs, resulting in less cosmic rays penetrating to the lower atmosphere, resulting in fewer cloud condensation nuclei being produced, resulting in fewer and less reflective low-level clouds occurring, which leads to more solar radiation being absorbed by the surface of the earth, resulting (finally) in increasing near-surface air temperatures and global warming.

- Strong correlations between solar variability and precipitation, droughts, floods, and monsoons have all been documented in locations around the world. Once again, these correlations are much stronger than any relationship between these weather phenomena and CO_2.

- The role of solar activity in causing climate change is so complex that most theories of solar forcing must be considered to be as yet unproven. But it would also be appropriate for climate scientists to admit the same about the role of rising atmospheric CO_2 concentrations in driving recent global warming.

Chapter 6. Observations: Extreme Weather

- The IPCC predicts that a warmer planet will lead to more extreme weather, characterized by more frequent and severe episodes of drought, flooding, cyclones, precipitation variability, storms, snow, storm surges, temperature variability, and wildfires. But has the last century – during which the IPCC claims the world experienced more rapid warming than any time in the past two millennia – experienced significant trends in any of these extreme weather events?

- Droughts have not become more extreme or erratic in response to global warming. Real-world evidence from Africa, Asia, and other continents find no trend toward more frequent or more severe droughts. In most cases, the worst droughts in recorded meteorological history were much milder than droughts that occurred periodically during much colder times.

- Floods were more frequent and more severe during the Little Ice Age than they have been during the Current Warm Period. Flooding in Asia, Europe, and North America has tended to be less frequent and less severe during the twentieth century.

- The IPCC says "it is likely that future tropical cyclones (typhoons and hurricanes) will become more intense, with larger peak wind speeds and more heavy precipitation associated with ongoing increase of tropical sea surface temperatures." But despite the supposedly "unprecedented" warming of the twentieth century, there has been no increase in the intensity or frequency of tropical cyclones globally or in any of the specific oceans.

- A number of real-world observations demonstrate that El Niño-Southern Oscillation (ENSO) conditions during the latter part of the twentieth century were not unprecedented in terms of their frequency or magnitude. Long-term records suggest that when the earth was significantly warmer than it is currently, ENSO events were substantially reduced or perhaps even absent.

- There is no support for the model-based projection that precipitation in a warming world becomes more variable and intense. In fact, some observational data suggest just the opposite, and provide support for the proposition that precipitation responds more to cyclical variations in solar activity.

- As the earth has warmed over the past 150 years, during its recovery from the global chill of the Little Ice Age, there has been no significant increase in either the frequency or intensity of stormy weather.

- Between 1950 and 2002, during which time the air's CO_2 concentration rose by 20 percent, there was no net change in either the mean onset date or duration of snow cover for the continent of North America. There appears to have been a downward trend in blizzards.

- Storm surges have not increased in either frequency or magnitude as CO_2 concentrations in the atmosphere have risen. In the majority of cases investigated, they have tended to decrease.

- Air temperature variability almost always *decreases* when mean air temperature *rises*, be it in cases of temperature change over tens of thousands of years or over mere decades, or even between individual cooler and warmer years when different ENSO states are considered. The claim that global warming will lead to more extremes of climate and weather, including more extremes of temperature itself, is not supported by real-world data.

- Although one can readily identify specific parts of the planet that have experienced both significant increases and decreases in land area burned by wildfires over the last two to three decades of the twentieth century, for the globe as a whole there was no relationship between global warming and total area burned over this period.

Chapter 7. Biological Effects of Carbon Dioxide Enhancement

- A 300-ppm increase in the air's CO_2 content typically raises the productivity of most herbaceous plants by about one-third; and this positive response occurs in plants that utilize all three of the major biochemical pathways (C_3, C_4, CAM) of photosynthesis. For woody plants, the response is even greater. The productivity benefits of CO_2 enrichment are also experienced by aquatic plants, including freshwater algae and macrophytes, and marine microalgae and macroalgae.

- The amount of carbon plants gain per unit of water lost—or water-use efficiency—typically rises as the CO_2 content of the air rises, greatly increasing their ability to withstand drought. In addition, the CO_2-induced percentage increase in plant biomass production is often greater under water-stressed conditions than it is when plants are well watered.

- Atmospheric CO_2 enrichment helps ameliorate the detrimental effects of several environmental stresses on plant growth and development, including high soil salinity, high air temperature, low light intensity and low levels of soil fertility. Elevated levels of CO_2 have additionally been demonstrated to reduce the severity of low temperature stress, oxidative stress, and the stress of herbivory. In fact, the percentage growth enhancement produced by an increase in the air's CO_2 concentration is often even greater under stressful and resource-limited conditions than it is when growing conditions are ideal.

- As the air's CO_2 content continues to rise, plants will likely exhibit enhanced rates of photosynthesis and biomass production that will not be diminished by any global warming that might occur concurrently. In fact, if the ambient air temperature rises, the growth-promoting effects of atmospheric CO_2 enrichment will likely also rise, becoming more and more robust.

- The ongoing rise in the air's CO_2 content likely will not favor the growth of weedy species over that of crops and native plants.

- The growth of plants is generally not only enhanced by CO_2-induced increases in net photosynthesis during the light period of the day, it is also enhanced by CO_2-induced decreases in respiration during the dark period.

- The ongoing rise in the air's CO_2 content, as well as any degree of warming that might possibly accompany it, will not materially alter the rate of decomposition of the world's soil organic matter and will probably enhance biological carbon sequestration. Continued increases in the air's CO_2 concentration and temperature will not result in massive losses of carbon from earth's peatlands. To the contrary, these environmental changes—if they persist—would likely work together to enhance carbon capture.

- Other biological effects of CO_2 enhancement include enhanced plant nitrogen-use efficiency, longer residence time of carbon in the soil, and increased populations of earthworms and soil nematodes.

- The aerial fertilization effect of the ongoing rise in the air's CO_2 concentration (which greatly enhances vegetative productivity) and its anti-transpiration effect (which enhances plant water-use efficiency and enables plants to grow in areas that were once too dry for them) are stimulating plant growth across the globe in places that previously were too dry or otherwise unfavorable for plant growth, leading to a significant greening of the Earth.

- Elevated CO_2 reduces, and nearly always overrides, the negative effects of ozone pollution on plant photosynthesis, growth and yield. It also reduces atmospheric concentrations of isoprene, a highly reactive non-methane hydrocarbon that is emitted in copious quantities by vegetation and is responsible for the production of vast amounts of tropospheric ozone.

Chapter 8. Species Extinction

- The IPCC claims "new evidence suggests that climate-driven extinctions and range retractions are already widespread" and the "projected impacts on biodiversity are significant and of key relevance, since global losses in biodiversity are irreversible (very high confidence)." These claims are not supported by scientific research.

- The world's species have proven to be remarkably resilient to climate change. Most wild species are at least one million years old, which means they have all been through hundreds of climate cycles involving temperature changes on par with or greater than those experienced in the twentieth century.

- The four known causes of extinctions are huge asteroids striking the planet, human hunting, human agriculture, and the introduction of alien species (e.g., lamprey eels in the Great Lakes and pigs in Hawaii). None of these causes are connected with either global temperatures or atmospheric CO_2 concentrations.

- Real-world data collected by the United Nations Environmental Program (UNEP) show the rate of extinctions at the end of the twentieth century was the lowest since the sixteenth century—despite 150 years of rising world temperatures, growing populations, and industrialization. Many, and probably most, of the world's species benefited from rising temperatures in the twentieth century.

- As long as the atmosphere's CO_2 concentration rises in tandem with its temperature, most plants will not need to migrate toward cooler conditions, as their physiology will change in ways that make them better adapted to warmer conditions. Plants will likely spread poleward in latitude and upward in elevation at the cold-limited boundaries of their ranges, thanks to longer growing seasons and less frost, while their heat-limited boundaries will probably remain pretty much as they are now or shift only slightly.

- Land animals also tend to migrate poleward and upward, to areas where cold temperatures prevented them from going in the past. They follow earth's plants, while the heat-limited boundaries of their ranges are often little affected, allowing them to also expand their ranges.

- The persistence of coral reefs through geologic time—when temperatures were as much as 10°-15°C warmer than at present, and atmospheric CO_2 concentrations were two to seven times higher than they are currently—provides substantive evidence that these marine entities can successfully adapt to a dramatically changing global environment.

- The 18- to 59-cm warming-induced sea-level rise that is predicted for the coming century by the IPCC falls well within the range (2 to 6 mm per year) of typical coral vertical extension rates, which exhibited a modal value of 7 to 8 mm per year during the Holocene and can be more than double that value in certain branching corals. Rising sea levels should therefore present no difficulties for coral reefs.

- The rising CO_2 content of the atmosphere may induce very small changes in the well-buffered ocean chemistry (pH) that could slightly reduce coral calcification rates; but potential positive effects of hydrospheric CO_2 enrichment may more than compensate for this modest negative phenomenon. Real-world observations indicate that elevated CO_2 and elevated temperatures are having a positive effect on most corals.

- Polar bears have survived changes in climate that exceed those that occurred during the twentieth century or are forecast by the IPCC's computer models.

- Most populations of polar bears are growing, not shrinking, and the biggest influence on polar bear populations is not temperature but hunting by humans, which historically has taken a large toll on polar bear populations.

- Forecasts of dwindling polar bear populations assume trends in sea ice and temperature that are counterfactual, rely on unvalidated computer climate models that are known to be unreliable, and violate most of the principles of scientific forecasting.

Chapter 9. Human Health Effects

- The IPCC alleges that "climate change currently contributes to the global burden of disease and premature deaths" and will "increase malnutrition and consequent disorders." In fact, the overwhelming weight of evidence shows that higher temperatures and rising CO_2 levels have played an indispensible role in making it possible to feed a growing global population without encroaching on natural ecosystems.

- Global warming reduces the incidence of cardiovascular disease related to low temperatures and wintry weather by a much greater degree than it increases the incidence of cardiovascular disease associated with high temperatures and summer heat waves.

- Mortality due to respiratory diseases decrease as temperatures rise and as temperature variability declines.

- Claims that malaria and tick-borne diseases are spreading or will spread across the globe as a result of CO_2-induced warming are not supported in the scientific literature.

- Total heat-related mortality rates have been shown to be lower in warmer climates and to be unaffected by rising temperatures during the twentieth century.

- The historical increase in the air's CO_2 content has improved human nutrition by raising crop

yields during the past 150 years on the order of 70 percent for wheat, 28 percent for cereals, 33 percent for fruits and melons, 62 percent for legumes, 67 percent for root and tuber crops, and 51 percent for vegetables.

- The quality of plant food in the CO_2-enriched world of the future, in terms of its protein and antioxidant (vitamin) contents, will be no lower and probably will be higher than in the past.

- There is evidence that some medicinal substances in plants will be present in significantly greater concentrations, and certainly in greater absolute amounts, than they are currently.

- The historical increase of the air's CO_2 content has probably helped lengthen human lifespans since the advent of the Industrial Revolution, and its continued upward trend will likely provide more of the same benefit.

- Higher levels of CO_2 in the air help to advance all three parts of a strategy to resolve the tension between the need to feed a growing population and the desire to preserve natural ecosystems: increasing crop yield per unit of land area, increasing crop yield per unit of nutrients applied, and increasing crop yield per unit of water used.

- Biofuels for transportation (chiefly ethanol, biodiesel, and methanol) are being used in growing quantities in the belief that they provide environmental benefits. In fact, those benefits are very dubious. By some measures, "the net effect of biofuels production ... is to increase CO_2 emissions for decades or centuries relative to the emissions caused by fossil fuel use."

- Biofuels compete with livestock growers and food processors for corn, soybeans, and other feedstocks, leading to higher food prices. Rising food prices in 2008 led to food riots in several developing countries. The production of biofuels also consumes enormous quantities of water compared with the production of gasoline.

- There can be little doubt that ethanol mandates and subsidies have made both food and energy more, not less, expensive and therefore less available to a growing population. The extensive damage to natural ecosystems already caused by this poor policy decision, and the much greater destruction yet to come, are a high price to pay for refusing to understand and utilize the true science of climate change.

1

Global Climate Models and Their Limitations

Introduction

Because the earth-ocean-atmosphere system is so vast and complex, it is impossible to conduct a small-scale experiment that reveals how the world's climate will change as the air's greenhouse gas (GHG) concentrations continue to rise. As a result, scientists try to forecast the effect of rising GHG by looking backwards at climate history to see how the climate responded to previous "forcings" of a similar kind, or by creating computer models that define a "virtual" earth-ocean-atmosphere system and run scenarios or "story lines" based on assumptions about future events.

The Intergovernmental Panel on Climate Change (IPCC) places great confidence in the ability of general circulation models (GCMs) to simulate future climate and attribute observed climate change to anthropogenic emissions of greenhouse gases. It says "climate models are based on well-established physical principles and have been demonstrated to reproduce observed features of recent climate ... and past climate changes ... There is considerable confidence that Atmosphere-Ocean General Circulation Models (AOGCMs) provide credible quantitative estimates of future climate change, particularly at continental and larger scales" (IPCC, 2007-I, p. 591).

To be of any validity, GCMs must incorporate all of the many physical, chemical, and biological processes that influence climate in the real world, and they must do so correctly. A review of the scientific literature reveals numerous deficiencies and shortcomings in today's state-of-the-art models, some of which deficiencies could even alter the sign of projected climate change. In this chapter, we first ask if computer models are capable *in principle* of producing reliable forecasts and then examine three areas of model inadequacies: radiation, clouds, and precipitation.

References

IPCC. 2007-I. *Climate Change 2007: The Physical Science Basis. Contribution of Working Group I to the Fourth Assessment Report of the Intergovernmental Panel on Climate Change.* Solomon, S., D. Qin, M. Manning, Z. Chen, M. Marquis, K.B. Averyt, M. Tignor and H.L. Miller. (Eds.) Cambridge University Press, Cambridge, UK.

1.1. Models and Forecasting

J. Scott Armstrong, professor, The Wharton School, University of Pennsylvania and a leading figure in the discipline of professional forecasting, has pointed out that forecasting is a practice and discipline in its own right, with its own institute (International Institute of Forecasters, founded in 1981), peer-reviewed journal (*International Journal of Forecasting*), and an extensive body of research that has been compiled into a set of scientific procedures, currently numbering 140, that must be used to make reliable forecasts (*Principles of Forecasting: A Handbook for Researchers and Practitioners,* by J. Scott Armstrong, Kluwer Academic Publishers, 2001).

According to Armstrong, when physicists, biologists, and other scientists who do not know the rules of forecasting attempt to make climate predictions based on their training and expertise, their forecasts are no more reliable than those made by nonexperts, even when they are communicated through complex computer models (Armstrong, 2001). In other words, forecasts by scientists, even large numbers of very distinguished scientists, are not necessarily *scientific* forecasts. In support of his position, Armstrong and a colleague cite research by Philip E. Tetlock (2005), a psychologist and professor of organizational behavior at the University of California, Berkeley, who "recruited 288 people whose professions included 'commenting or offering advice on political and economic trends.' He asked them to forecast the probability that various situations would or would not occur, picking areas (geographic and substantive) within and outside their areas of expertise. By 2003, he had accumulated more than 82,000 forecasts. The experts barely if at all outperformed non-experts and neither group did well against simple rules" (Green and Armstrong, 2007). The failure of expert opinion to lead to reliable forecasts has been confirmed in scores of empirical studies (Armstrong, 2006; Craig *et al.,* 2002; Cerf and Navasky, 1998; Ascher, 1978) and illustrated in historical examples of incorrect forecasts made by leading experts (Cerf and Navasky, 1998).

In 2007, Armstrong and Kesten C. Green of Monash University conducted a "forecasting audit" of the IPCC Fourth Assessment Report (Green and Armstrong, 2007). The authors' search of the contribution of Working Group I to the IPCC "found no references … to the primary sources of information on forecasting methods" and "the forecasting procedures that were described [in sufficient detail to be evaluated] violated 72 principles. Many of the violations were, by themselves, critical."

One principle of scientific forecasting Green and Armstrong say the IPCC violated is "Principle 1.3 Make sure forecasts are independent of politics." The two authors write, "this principle refers to keeping the forecasting process separate from the planning process. The term 'politics' is used in the broad sense of the exercise of power." Citing David Henderson (Henderson, 2007), a former head of economics and

statistics at the Organization for Economic Cooperation and Development (OECD), they say "the IPCC process is directed by non-scientists who have policy objectives and who believe that anthropogenic global warming is real and danger." They conclude:

> The forecasts in the Report were not the outcome of scientific procedures. In effect, they were the opinions of scientists transformed by mathematics and obscured by complex writing. Research on forecasting has shown that experts' predictions are not useful in situations involving uncertainty and complexity. We have been unable to identify any scientific forecasts of global warming. Claims that the Earth will get warmer have no more credence than saying that it will get colder.

Scientists working in fields characterized by complexity and uncertainty are apt to confuse the output of *models*—which are nothing more than a statement of how the modeler believes a part of the world works—with real-world trends and forecasts (Bryson, 1993). Computer climate modelers certainly fall into this trap, and they have been severely criticized for failing to notice that their models fail to replicate real-world phenomena by many scientists, including Balling (2005), Christy (2005), Essex and McKitrick (2007), Frauenfeld (2005), Michaels (2000, 2005, 2009), Pilkey and Pilkey-Jarvis (2007), Posmentier and Soon (2005), and Spencer (2008).

Canadian science writer Lawrence Solomon (2008) interviewed many of the world's leading scientists active in scientific fields relevant to climate change and asked them for their views on the reliability of the computer models used by the IPCC to detect and forecast global warming. Their answers showed a high level of skepticism:

- Prof. Freeman Dyson, professor of physics at the Institute for Advanced Study at Princeton University, one of the world's most eminent physicists, said the models used to justify global warming alarmism are "full of fudge factors" and "do not begin to describe the real world."

- Dr. Zbigniew Jaworowski, chairman of the Scientific Council of the Central Laboratory for Radiological Protection in Warsaw and former chair of the United Nations Scientific Committee on the Effects of Atomic Radiation, a world-renowned expert on the use of ancient ice cores for climate research, said the U.N. "based its global-warming hypothesis on arbitrary assumptions and these assumptions, it is now clear, are false."

- Dr. Richard Lindzen, a professor of meteorology at M.I.T. and member of the National Research Council Board on Atmospheric Sciences and Climate, said the IPCC is "trumpeting catastrophes that couldn't happen even if the models were right."

- Prof. Hendrik Tennekes, director of research at the Royal Netherlands Meteorological Institute, said "there exists no sound theoretical framework for climate predictability studies" used for global warming forecasts.

- Dr. Richard Tol, principal researcher at the Institute for Environmental Studies at Vrije Universiteit and adjunct professor at the Center for Integrated Study of the Human Dimensions of Global Change at Carnegie Mellon University, said the IPCC's Fourth Assessment Report is "preposterous ... alarmist and incompetent."

- Dr. Antonino Zichichi, emeritus professor of physics at the University of Bologna, former president of the European Physical Society, and one of the world's foremost physicists, said global warming models are "incoherent and invalid."

Princeton's Freeman Dyson has written elsewhere, "I have studied the climate models and I know what they can do. The models solve the equations of fluid dynamics, and they do a very good job of describing the fluid motions of the atmosphere and the oceans. They do a very poor job of describing the clouds, the dust, the chemistry, and the biology of fields and farms and forests. They do not begin to describe the real world that we live in" (Dyson, 2007).

Many of the scientists cited above observe that computer models can be "tweaked" to reconstruct climate histories after the fact, as the IPCC points out in the passage quoted at the beginning of this chapter. But this provides no assurance that the new model will do a better job forecasting *future* climates, and indeed points to how unreliable the models are. Individual climate models often have widely differing assumptions about basic climate mechanisms but are then "tweaked" to produce similar forecasts. This is nothing like how real scientific forecasting is done.

Kevin Trenberth, a lead author along with Philip D. Jones of chapter 3 of the Working Group I contribution to the IPCC's Fourth Assessment Report, replied to some of these scathing criticisms on the blog of the science journal *Nature*. He argued that "the IPCC does not make forecasts" but "instead proffers 'what if' projections of future climate that correspond to certain emissions scenarios," and then hopes these "projections" will "guide policy and decision makers" (Trenberth, 2007). He says "there are no such predictions [in the IPCC reports] although the projections given by the Intergovernmental Panel on Climate Change (IPCC) are often treated as such. The distinction is important."

This defense is hardly satisfactory. As Green and Armstrong point out, "the word 'forecast' and its derivatives occurred 37 times, and 'predict' and its derivatives occurred 90 times in the body of Chapter 8" of the Working Group I report, and a survey of climate scientists conducted by those same authors found "most of our respondents (29 of whom were IPCC authors or reviewers) nominated the IPCC report as the most credible source of forecasts (not 'scenarios' or 'projections') of global average temperature." They conclude that "the IPCC does provide forecasts." We agree, and add that those forecasts are unscientific and therefore likely to be wrong.

Additional information on this topic, including reviews of climate model inadequacies not discussed here, can be found at http://www.co2science.org/subject/m/subject_m.php under the heading Models of Climate.

References

Armstrong, J.S. 2001. *Principles of Forecasting – A Handbook for Researchers and Practitioners*. Kluwer Academic Publishers, Norwell, MA.

Armstrong, J.S. 2006. Findings from evidence-based forecasting: Methods for reducing forecast error. *International Journal of Forecasting* **22**: 583-598.

Ascher, W. 1978. *Forecasting: An Appraisal for Policy Makers and Planners*. Johns Hopkins University Press. Baltimore, MD.

Balling, R.C. 2005. Observational surface temperature records versus model predictions. In Michaels, P.J. (Ed.) *Shattered Consensus: The True State of Global Warming*. Rowman & Littlefield. Lanham, MD. 50-71.

Bryson, R.A. 1993. Environment, environmentalists, and global change: A skeptic's evaluation. *New Literary History:* **24**: 783-795.

Cerf, C. and Navasky, V. 1998. *The Experts Speak.* Johns Hopkins University Press. Baltimore, MD.

Christy, J. 2005. Temperature changes in the bulk atmosphere: beyond the IPCC. In Michaels, P.J. (Ed.) *Shattered Consensus: The True State of Global Warming.* Rowman & Littlefield. Lanham, MD. 72-105.

Craig, P.P., Gadgil, A., and Koomey, J.G. 2002. What can history teach us? A retrospective examination of long-term energy forecasts for the United States. *Annual Review of Energy and the Environment* **27**: 83-118.

Dyson, F. 2007. Heretical thoughts about science and society. *Edge: The Third Culture.* August.

Essex, C. and McKitrick, R. 2002. *Taken by Storm. The Troubled Science, Policy and Politics of Global Warming.* Key Porter Books. Toronto, Canada.

Frauenfeld, O.W. 2005. Predictive skill of the El Niño-Southern Oscillation and related atmospheric teleconnections. In Michaels, P.J. (Ed.) *Shattered Consensus: The True State of Global Warming.* Rowman & Littlefield. Lanham, MD. 149-182.

Green, K.C. and Armstrong, J.S. 2007. Global warming: forecasts by scientists versus scientific forecasts. *Energy Environ.* **18**: 997–1021.

Henderson, D. 2007. Governments and climate change issues: The case for rethinking. *World Economics* **8**: 183-228.

Michaels, P.J. 2009. *Climate of Extremes: Global Warming Science They Don't Want You to Know.* Cato Institute. Washington, DC.

Michaels, P.J. 2005. *Meltdown: The Predictable Distortion of Global Warming by Scientists, Politicians and the Media.* Cato Institute, Washington, DC.

Michaels, P.J. 2000. *Satanic Gases: Clearing the Air About Global Warming.* Cato Institute. Washington, DC.

Pilkey, O.H. and Pilkey-Jarvis, L. 2007. *Useless Arithmetic.* Columbia University Press, New York.

Posmentier, E.S. and Soon, W. 2005. Limitations of computer predictions of the effects of carbon dioxide on global temperature. In Michaels, P.J. (Ed.) *Shattered Consensus: The True State of Global Warming.* Rowman & Littlefield. Lanham, MD. 241-281.

Solomon, L. 2008. *The Deniers: The World Renowned Scientists Who Stood Up Against Global Warming Hysteria, Political Persecution, and Fraud**And those who are too fearful to do so.* Richard Vigilante Books. Minneapolis, MN.

Spencer, R. 2008. *Climate Confusion: How Global Warming Hysteria Leads to Bad Science, Pandering Politicians and Misguided Policies that Hurt the Poor.* Encounter Books. New York, NY.

Tetlock, P.E. 2005. *Expert Political Judgment—How Good Is It? How Can We Know?* Princeton University Press, Princeton, NJ.

Trenberth, K.E. 2007. Global warming and forecasts of climate change. *Nature* blog. http://blogs.nature.com/climatefeedback/2007/07/global_warming_and_forecasts_o.html. Last accessed May 6, 2009.

1.2. Radiation

One problem facing GCMs is how to accurately simulate the physics of earth's radiative energy balance. Of this task, Harries (2000) says "progress is excellent, on-going research is fascinating, but we have still a great deal to understand about the physics of climate."

Harries says "we must exercise great caution over the true depth of our understanding, and our ability to forecast future climate trends." As an example, he states that our knowledge of high cirrus clouds is very poor, noting that "we could easily have uncertainties of many tens of Wm^{-2} in our description of the radiative effect of such clouds, and how these properties may change under climate forcing." This state of affairs is disconcerting in light of the fact that the radiative effect of a doubling of the air's CO_2 content is in the lower single-digit range of Wm^{-2}, and, to quote Harries, "uncertainties as large as, or larger than, the doubled CO_2 forcing could easily exist in our modeling of future climate trends, due to uncertainties in the feedback processes." Because of the vast complexity of the subject, Harries says "even if [our] understanding were perfect, our ability to describe the system sufficiently well in even the largest computer models is a problem."

A related problem is illustrated by the work of Zender (1999), who characterized the spectral, vertical, regional and seasonal atmospheric heating caused by the oxygen collision pairs $O_2 \cdot O_2$ and $O_2 \cdot N_2$, which had earlier been discovered to absorb a small but significant fraction of the globally incident solar radiation. In addition, water vapor demers (a double molecule of H_2O) shows strong absorption bands in the near-infrared of the solar spectrum. Zender revealed that these molecular collisions lead to the absorption of about 1 Wm^{-2} of solar radiation, globally and annually averaged. This discovery, in Zender's words, "alters the long-standing view that H_2O, O_3, O_2, CO_2 and NO_2 are the only significant gaseous solar absorbers in earth's atmosphere," and he suggests that the phenomenon "should therefore be included in ... large-scale atmospheric models used to simulate climate and climate change."

In another revealing study, Wild (1999) compared the observed amount of solar radiation absorbed in the atmosphere over equatorial Africa with what was predicted by three GCMs and found the model predictions were much too small. Indeed, regional and seasonal model underestimation biases were as high as 30 Wm^{-2}, primarily because the models failed to properly account for spatial and temporal variations in atmospheric aerosol concentrations. In addition, Wild found the models likely underestimated the amount of solar radiation absorbed by water vapor and clouds.

Similar large model underestimations were discovered by Wild and Ohmura (1999), who analyzed a comprehensive observational dataset consisting of solar radiation fluxes measured at 720 sites across the earth's surface and corresponding top-of-the-atmosphere locations to assess the true amount of solar radiation absorbed within the atmosphere. These results were compared with estimates of solar radiation absorption derived from four GCMs and, again, it was shown that "GCM atmospheres are generally too transparent for solar radiation," as they produce a rather substantial mean error close to 20 percent below actual observations.

Another solar-related deficiency of GCMs is their failure to properly account for solar-driven variations in earth-atmosphere processes that operate over a range of timescales extending from the 11-year solar cycle to century- and millennial-scale cycles (see Section 4.11, Solar Influence on Climate). Although the absolute solar flux variations associated with these phenomena are rather small, there are a number of "multiplier effects" that may significantly amplify their impacts.

According to Chambers et al. (1999), most of the many nonlinear responses to solar activity variability are inadequately represented in the global climate models used by the IPCC to predict future greenhouse gas-induced global warming, while at the same time other amplifier effects are used to model past glacial/interglacial cycles and even the hypothesized CO_2-induced warming of the future, where CO_2 is not the major cause of the predicted temperature increase but rather an initial perturber of the climate system

that according to the IPCC sets other more powerful forces in motion that produce the bulk of the ultimate warming. There appears to be a double standard within the climate modeling community that may best be described as an inherent reluctance to deal even-handedly with different aspects of climate change. When multiplier effects suit their purposes, they use them; but when they don't suit their purposes, they don't use them.

Ghan et al. (2001) warn that "present-day radiative forcing by anthropogenic greenhouse gases is estimated to be 2.1 to 2.8 Wm^{-2}; the direct forcing by anthropogenic aerosols is estimated to be -0.3 to -1.5 Wm^{-2}, while the indirect forcing by anthropogenic aerosols is estimated to be 0 to -1.5 Wm^{-2}," so that "estimates of the total global mean present-day anthropogenic forcing range from 3 Wm^{-2} to -1 Wm^{-2}," which implies a climate change somewhere between a modest warming and a slight cooling. They conclude that "the great uncertainty in the radiative forcing must be reduced if the observed climate record is to be reconciled with model predictions and if estimates of future climate change are to be useful in formulating emission policies."

Pursuit of this goal, Ghan et al. say, requires achieving "profound reductions in the uncertainties of direct and indirect forcing by anthropogenic aerosols," which is what they set out to do in their analysis of the situation, which consisted of "a combination of process studies designed to improve understanding of the key processes involved in the forcing, closure experiments designed to evaluate that understanding, and integrated models that treat all of the necessary processes together and estimate the forcing." At the conclusion of this laborious set of operations, Ghan et al. came up with some numbers that considerably reduce the range of uncertainty in the "total global mean present-day anthropogenic forcing," but that still implied a set of climate changes stretching from a small cooling to a modest warming. They also provided a long list of other things that must be done in order to obtain a more definitive result, after which they acknowledged that even this list "is hardly complete." In fact, they conclude, "one could easily add the usual list of uncertainties in the representation of clouds, etc." Consequently, the bottom line, in their words, is that "much remains to be done before the estimates are reliable enough to base energy policy decisions upon."

Also studying the aerosol-induced radiative forcing of climate were Vogelmann et al. (2003), who report that "mineral aerosols have complex, highly varied optical properties that, for equal loadings, can cause differences in the surface IR flux between 7 and 25 Wm^{-2} (Sokolik *et al.*, 1998)." They say "only a few large-scale climate models currently consider aerosol IR effects (e.g., Tegen *et al.*, 1996; Jacobson, 2001) despite their potentially large forcing." Because of these facts, and in an attempt to persuade climate modelers to rectify the situation, Vogelmann *et al.* used high-resolution spectra to calculate the surface IR radiative forcing created by aerosols encountered in the outflow of air from northeastern Asia, based on measurements made by the Marine-Atmospheric Emitted Radiance Interferometer aboard the NOAA Ship *Ronald H. Brown* during the Aerosol Characterization Experiment-Asia. In doing so, they determined, in their words, that "daytime surface IR forcings are often a few Wm^{-2} and can reach almost 10 Wm^{-2} for large aerosol loadings," which values they say "are comparable to or larger than the 1 to 2 Wm^{-2} change in the globally averaged surface IR forcing caused by greenhouse gas increases since pre-industrial times." In a massive understatement of fact, the researchers concluded that their results "highlight the importance of aerosol IR forcing which should be included in climate model simulations." If a forcing of this magnitude is not included in current state-of-the-art climate models, what other major forcings are they ignoring?

Two papers published one year earlier and appearing in the same issue of *Science* (Chen *et al.*, 2002; Wielicki *et al.*, 2002) revealed what Hartmann (2002) called a pair of "tropical surprises." The first of the seminal discoveries was the common finding of both groups of researchers that the amount of thermal radiation emitted to space at the top of the tropical atmosphere increased by about 4 Wm^{-2} between the 1980s and the 1990s. The second was that the amount of reflected sunlight decreased by 1 to 2 Wm^{-2} over the same period, with the net result that more total radiant energy exited the tropics in the latter decade. In addition, the measured thermal radiative energy loss at the top of the tropical atmosphere was of the same magnitude as the thermal radiative energy gain that is generally predicted to result from an instantaneous doubling of the air's CO_2 content. Yet as Hartmann notes, "only very small changes in average tropical surface temperature were observed during this time." How did this occur?

The change in solar radiation reception was driven by reductions in cloud cover, which allowed more solar radiation to reach the surface of the earth's tropical region and warm it. These changes were

produced by what Chen *et al.* determined to be "a decadal-time-scale strengthening of the tropical Hadley and Walker circulations." Another helping-hand was likely provided by the past quarter-century's slowdown in the meridional overturning circulation of the upper 100 to 400 meters of the tropical Pacific Ocean (McPhaden and Zhang, 2002), which circulation slowdown also promotes tropical sea surface warming by reducing the rate-of-supply of relatively colder water to the region of equatorial upwelling.

These observations provide several new phenomena for the models to replicate as a test of their ability to properly represent the real world. In the words of McPhaden and Zhang, the time-varying meridional overturning circulation of the upper Pacific Ocean provides "an important dynamical constraint for model studies that attempt to simulate recent observed decadal changes in the Pacific."

In an eye-opening application of this principle, Wielicki *et al.* (2002) tested the ability of four state-of-the-art climate models and one weather assimilation model to reproduce the observed decadal changes in top-of-the-atmosphere thermal and solar radiative energy fluxes that occurred over the past two decades. No significant decadal variability was exhibited by *any* of the models; and they *all* failed to reproduce even the cyclical seasonal change in tropical albedo. The administrators of the test kindly concluded that "the missing variability in the models highlights the critical need to improve cloud modeling in the tropics so that prediction of tropical climate on interannual and decadal time scales can be improved." Hartmann was considerably more candid in his scoring of the test, saying flatly that the results indicated "the models are deficient." Expanding on this assessment, he noted that "if the energy budget can vary substantially in the absence of obvious forcing," as it did over the past two decades, "then the climate of earth has modes of variability that are not yet fully understood and cannot yet be accurately represented in climate models."

Also concentrating on the tropics, Bellon *et al.* (2003) note that "observed tropical sea-surface temperatures (SSTs) exhibit a maximum around 30°C," and that "this maximum appears to be robust on various timescales, from intraseasonal to millennial." Hence, they say, "identifying the stabilizing feedback(s) that help(s) maintain this threshold is essential in order to understand how the tropical climate reacts to an external perturbation," which knowledge is needed for understanding how the global climate reacts to perturbations such as those produced by solar variability and the ongoing rise in the air's CO_2 content. This contention is further substantiated by the study of Pierrehumbert (1995), which "clearly demonstrates," in the words of Bellon et al., "that the tropical climate is not determined locally, but globally." Also, they note that Pierrehumbert's work demonstrates that interactions between moist and dry regions are an essential part of tropical climate stability, which points to the "adaptive infrared iris" concept of Lindzen et al. (2001), which is reported in Section 1.2.

Noting that previous box models of tropical climate have shown it to be rather sensitive to the relative areas of moist and dry regions of the tropics, Bellon et al. analyzed various feedbacks associated with this sensitivity in a four-box model of the tropical climate "to show how they modulate the response of the tropical temperature to a radiative perturbation." In addition, they investigated the influence of the model's surface-wind parameterization in an attempt to shed further light on the nature of the underlying feedbacks that help define the global climate system that is responsible for the tropical climate observations of constrained maximum sea surface temperatures (SSTs).

Bellon et al.'s work, as they describe it, "suggests the presence of an important and as-yet-unexplored feedback in earth's tropical climate, that could contribute to maintain the 'lid' on tropical SSTs." They say the demonstrated "dependence of the surface wind on the large-scale circulation has an important effect on the sensitivity of the tropical system," specifically stating that "this dependence reduces significantly the SST sensitivity to radiative perturbations by enhancing the evaporation feedback," which injects more heat into the atmosphere and allows the atmospheric circulation to export more energy to the subtropical free troposphere, where it can be radiated to space by water vapor.

This literature review makes clear that the case is not closed on either the source or the significance of the maximum "allowable" SSTs of tropical regions. Neither, consequently, is the case closed on the degree to which the planet may warm in response to continued increases in the atmospheric concentrations of carbon dioxide and other greenhouse gases, in stark contrast to what is suggested by the climate models promoted by the IPCC.

In conclusion, there are a number of major inadequacies in the ways the earth's radiative energy balance is treated in contemporary general circulation models of the atmosphere, as well as numerous other telling inadequacies stemming from the non-treatment of pertinent phenomena that are nowhere to be found in the models. IPCC-inspired predictions of catastrophic climatic changes due to continued anthropogenic CO_2 emissions are beyond what can be soundly supported by the current state of the climate modeling enterprise.

Additional information on this topic, including reviews of newer publications as they become available, can be found at http://www.co2science.org/subject/m/inadeqradiation.php.

References

Bellon, G., Le Treut, H. and Ghil, M. 2003. Large-scale and evaporation-wind feedbacks in a box model of the tropical climate. *Geophysical Research Letters* **30**: 10.1029/2003GL017895.

Chambers, F.M., Ogle, M.I. and Blackford, J.J. 1999. Palaeoenvironmental evidence for solar forcing of Holocene climate: linkages to solar science. *Progress in Physical Geography* **23**: 181-204.

Chen, J., Carlson, B.E. and Del Genio, A.D. 2002. Evidence for strengthening of the tropical general circulation in the 1990s. *Science* **295**: 838-841.

Ghan, S.J., Easter, R.C., Chapman, E.G., Abdul-Razzak, H., Zhang, Y., Leung, L.R., Laulainen, N.S., Saylor, R.D. and Zaveri, R.A. 2001. A physically based estimate of radiative forcing by anthropogenic sulfate aerosol. *Journal of Geophysical Research* **106**: 5279-5293.

Harries, J.E. 2000. Physics of the earth's radiative energy balance. *Contemporary Physics* **41**: 309-322.

Hartmann, D.L. 2002. Tropical surprises. *Science* **295**: 811-812.

Jacobson, M.Z. 2001. Global direct radiative forcing due to multicomponent anthropogenic and natural aerosols. *Journal of Geophysical Research* **106**: 1551-1568.

Lindzen, R.S., Chou, M.-D. and Hou, A.Y. 2001. Does the earth have an adaptive infrared iris? *Bulletin of the American Meteorological Society* **82**: 417-432.

McPhaden, M.J. and Zhang, D. 2002. Slowdown of the meridional overturning circulation in the upper Pacific Ocean. *Nature* **415**: 603-608.

Pierrehumbert, R.T. 1995. Thermostats, radiator fins, and the local runaway greenhouse. *Journal of the Atmospheric Sciences* **52**: 1784-1806.

Sokolik, I.N., Toon, O.B. and Bergstrom, R.W. 1998. Modeling the radiative characteristics of airborne mineral aerosols at infrared wavelengths. *Journal of Geophysical Research* **103**: 8813-8826.

Tegen, I., Lacis, A.A. and Fung, I. 1996. The influence on climate forcing of mineral aerosols from disturbed soils. *Nature* **380**: 419-422.

Vogelmann, A.M., Flatau, P.J., Szczodrak, M., Markowicz, K.M. and Minnett, P.J. 2003. Observations of large aerosol infrared forcing at the surface. *Geophysical Research Letters* **30**: 10.1029/2002GL016829.

Wielicki, B.A., Wong, T., Allan, R.P., Slingo, A., Kiehl, J.T., Soden, B.J., Gordon, C.T., Miller, A.J., Yang, S.-K., Randall, D.A., Robertson, F., Susskind, J. and Jacobowitz, H. 2002. Evidence for large decadal variability in the tropical mean radiative energy budget. *Science* **295**: 841-844.

Wild, M. 1999. Discrepancies between model-calculated and observed shortwave atmospheric absorption in areas with high aerosol loadings. *Journal of Geophysical Research* **104**: 27,361-27,371.

Wild, M. and Ohmura, A. 1999. The role of clouds and the cloud-free atmosphere in the problem of underestimated absorption of solar radiation in GCM atmospheres. *Physics and Chemistry of the Earth* **24B**: 261-268.

Zender, C.S. 1999. Global climatology of abundance and solar absorption of oxygen collision complexes. *Journal of Geophysical Research* **104**: 24,471-24,484.

1.3. Clouds

Correctly parameterizing the influence of clouds on climate is an elusive goal that the creators of atmospheric general circulation models (GCMs) have yet to achieve. One reason for their lack of success has to do with model resolution on vertical and horizontal space scales. Lack of adequate resolution forces modelers to parameterize the ensemble large-scale effects of processes that occur on smaller scales than their models are capable of handling. This is particularly true of physical processes such as cloud formation and cloud-radiation interactions. Several studies suggest that older model parameterizations did not succeed in this regard (Groisman *et al.*, 2000), and subsequent studies suggest they still are not succeeding.

Lane *et al.* (2000) evaluated the sensitivities of the cloud-radiation parameterizations utilized in contemporary GCMs to changes in vertical model resolution, varying the latter from 16 to 60 layers in increments of four and comparing the results to observed values. This effort revealed that cloud fraction varied by approximately 10 percent over the range of resolutions tested, which corresponded to about 20 percent of the observed cloud cover fraction. Similarly, outgoing longwave radiation varied by 10 to 20 Wm^{-2} as model vertical resolution was varied, amounting to approximately 5 to 10 percent of observed values, while incoming solar radiation experienced similar significant variations across the range of resolutions tested. The model results did not converge, even at a resolution of 60 layers.

In an analysis of the multiple roles played by cloud microphysical processes in determining tropical climate, Grabowski (2000) found much the same thing, noting there were serious problems related to the degree to which computer models failed to correctly incorporate cloud microphysics. These observations led him to conclude that "it is unlikely that traditional convection parameterizations can be used to address this fundamental question in an effective way." He also became convinced that "classical convection parameterizations do not include realistic elements of cloud physics and they represent interactions among cloud physics, radiative processes, and surface processes within a very limited scope." Consequently, he says, "model results must be treated as qualitative rather than quantitative."

Reaching rather similar conclusions were Gordon *et al.* (2000), who determined that many GCMs of the late 1990s tended to under-predict the presence of subtropical marine stratocumulus clouds and failed to simulate the seasonal cycle of clouds. These deficiencies are extremely important because these particular clouds exert a major cooling influence on the surface temperatures of the sea below them. In the situation investigated by Gordon and his colleagues, the removal of the low clouds, as occurred in the normal application of their model, led to sea surface temperature increases on the order of 5.5°C.

Further condemnation of turn-of-the-century model treatments of clouds came from Harries (2000), previously cited in Section 1.1, who wrote that our knowledge of high cirrus clouds is very poor and that "we could easily have uncertainties of many tens of Wm^{-2} in our description of the radiative effect of such clouds, and how these properties may change under climate forcing."

Moving into the twenty-first century, Lindzen *et al.* (2001) analyzed cloud cover and sea surface temperature (SST) data over a large portion of the Pacific Ocean, finding a strong inverse relationship between upper-level cloud area and mean SST, such that the area of cirrus cloud coverage normalized by a measure of the area of cumulus coverage decreased by about 22 percent for each degree C increase in cloudy region SST. Essentially, as the researchers described it, "the cloudy-moist region appears to act as an infrared adaptive iris that opens up and closes down the regions free of upper-level clouds, which more effectively permit infrared cooling, in such a manner as to resist changes in tropical surface temperature." The sensitivity of this negative feedback was calculated by Lindzen et al. to be substantial. In fact, they estimated it would "more than cancel all the positive feedbacks in the more sensitive current climate models" that were being used to predict the consequences of projected increases in atmospheric CO_2 concentration.

Lindzen's challenge to what had become climatic political correctness could not go uncontested, and Hartmann and Michelsen (2002) quickly claimed the correlation noted by Lindzen et al. resulted from variations in subtropical clouds that are not physically connected to deep convection near the equator, and that it was thus "unreasonable to interpret these changes as evidence that deep tropical convective anvils contract in response to SST increases." Fu et al. (2002) also chipped away at the adaptive infrared iris concept, arguing that "the contribution of tropical high clouds to the feedback process would be small since the radiative forcing over the tropical high cloud region is near zero and not strongly positive," while also claiming to show that water vapor and low cloud effects were overestimated by Lindzen et al. by at least 60 percent and 33 percent, respectively. As a result, they obtained a feedback factor in the range of -0.15 to -0.51, compared to Lindzen et al.'s much larger negative feedback factor of -0.45 to -1.03.

In a contemporaneously published reply to this critique, Chou et al. (2002) stated that Fu et al.'s approach of specifying longwave emission and cloud albedos "appears to be inappropriate for studying the iris effect," and that since "thin cirrus are widespread in the tropics and ... low boundary clouds are optically thick, the cloud albedo calculated by [Fu et al.] is too large for cirrus clouds and too small for boundary layer clouds," so that "the near-zero contrast in cloud albedos derived by [Fu et al.] has the effect of underestimating the iris effect." In the end, however, Chou et al. agreed that Lindzen et al. "may indeed have overestimated the iris effect somewhat, though hardly by as much as that suggested by [Fu et al.]."

Although there has thus been some convergence in the two opposing views of the subject, the debate over the reality and/or magnitude of the adaptive infrared iris effect continues. It is amazing that some political leaders proclaim the debate over global warming is "over" when some of the meteorological community's best minds continue to clash over the nature and magnitude of a phenomenon that could entirely offset the effects of anthropogenic CO_2 emissions.

Grassl (2000), in a review of the then-current status of the climate-modeling enterprise two years before the infrared iris effect debate emerged, noted that changes in many climate-related phenomena, including cloud optical and precipitation properties caused by changes in the spectrum of cloud condensation nuclei, were insufficiently well known to provide useful insights into future conditions. His advice in the light of this knowledge gap was that "we must continuously evaluate and improve the GCMs we use," although he was forced to acknowledge that contemporary climate model results were already being "used by many decision-makers, including governments."

Although some may think that what we currently know about the subject is sufficient for predictive purposes, a host of questions posed by Grassl—for which we still lack definitive answers—demonstrates that this assumption is erroneous. As but a single example, Charlson et al. (1987) described a negative feedback process that links biologically-produced dimethyl sulfide (DMS) in the oceans with climate. (See Section 2.3 for a more complete discussion.) The basic tenet of this hypothesis is that the global radiation balance is significantly influenced by the albedo of marine stratus clouds, and that the albedo of these clouds is a function of cloud droplet concentration, which is dependent upon the availability of condensation nuclei that have their origin in the flux of DMS from the world's oceans to the atmosphere.

Acknowledging that the roles played by DMS oxidation products within the context described above are indeed "diverse and complex" and in many instances "not well understood," Ayers and Gillett (2000) summarized empirical evidence supporting Charlson et al.'s hypothesis that was derived from data collected at Cape Grim, Tasmania, and from reports of other pertinent studies in the peer-reviewed scientific literature. According to their findings, the "major links in the feedback chain proposed by Charlson et al. (1987) have a sound physical basis," and there is "compelling observational evidence to suggest that DMS and its atmospheric products participate significantly in processes of climate regulation and reactive atmospheric chemistry in the remote marine boundary layer of the Southern Hemisphere."

The empirical evidence analyzed by Ayers and Gillett highlights an important suite of negative feedback processes that act in opposition to model-predicted CO_2-induced global warming over the world's oceans; and these processes are not fully incorporated into even the very best of the current crop of climate models, nor are analogous phenomena that occur over land included in them, such as those discussed by Idso (1990). (See also, in this regard, Section 2.7 of this report.)

Further to this point, O'Dowd et al. (2004) measured size-resolved physical and chemical properties of aerosols found in northeast Atlantic marine air arriving at the Mace Head Atmospheric Research station on the west coast of Ireland during phytoplanktonic blooms at various times of the year. In doing so, they found that in the winter, when biological activity was at its lowest, the organic fraction of the submicrometer aerosol mass was about 15 percent. During the spring through autumn, however, when biological activity was high, they found that "the organic fraction dominates and contributes 63 percent to the submicrometer aerosol mass (about 45 percent is water-insoluble and about 18 percent water-soluble)." Based on these findings, they performed model simulations that indicated that the marine-derived organic matter "can enhance the cloud droplet concentration by 15 percent to more than 100 percent and is therefore an important component of the aerosol-cloud-climate feedback system involving marine biota."

As for the significance of their findings, O'Dowd et al. state that their data "completely change the picture of what influences marine cloud condensation nuclei given that water-soluble organic carbon, water-insoluble organic carbon and surface-active properties, all of which influence the cloud condensation nuclei activation potential, are typically not parameterized in current climate models," or as they say in another place in their paper, "an important source of organic matter from the ocean is omitted from current climate-modeling predictions and should be taken into account."

Another perspective on the cloud-climate conundrum is provided by Randall et al. (2003), who state at the outset of their review of the subject that "the representation of cloud processes in global atmospheric models has been recognized for decades as the source of much of the uncertainty surrounding predictions of climate variability." They report, however, that "despite the best efforts of [the climate modeling] community ... the problem remains largely unsolved." What is more, they say, "at the current rate of progress, cloud parameterization deficiencies will continue to plague us for many more decades into the future."

"Clouds are complicated," Randall et al. declare, as they begin to describe what they call the "appalling complexity" of the cloud parameterization situation. They state that "our understanding of the interactions of the hot towers [of cumulus convection] with the global circulation is still in a fairly primitive state," and not knowing all that much about what goes up, it's not surprising we also don't know much about what comes down, as they report that "downdrafts are either not parameterized or crudely parameterized in large-scale models."

With respect to stratiform clouds, the situation is no better, as their parameterizations are described by Randall et al. as "very rough caricatures of reality." As for interactions between convective and stratiform clouds, during the 1970s and '80s, Randall et al. report that "cumulus parameterizations were extensively tested against observations without even accounting for the effects of the attendant stratiform clouds." Even at the time of their study, they had to report that the concept of detrainment was "somewhat murky" and the conditions that trigger detrainment were "imperfectly understood." "At this time," as they put it, "no existing GCM includes a satisfactory parameterization of the effects of mesoscale cloud circulations."

Randall et al. additionally say that "the large-scale effects of microphysics, turbulence, and radiation should be parameterized as closely coupled processes acting in concert," but they report that only a few GCMs have even attempted to do so. Why? Because, as they continue, "the cloud parameterization problem is overwhelmingly complicated," and "cloud parameterization developers," as they call them, are still "struggling to identify the most important processes on the basis of woefully incomplete observations." To drive this point home, they say "there is little question why the cloud parameterization problem is taking a long time to solve: It is very, very hard." In fact, the four scientists conclude that "a sober assessment suggests that with current approaches the cloud parameterization problem will not be 'solved' in any of our lifetimes."

To show that the basis for this conclusion is robust, and cannot be said to rest on the less-than-enthusiastic remarks of a handful of exasperated climate modelers, we report the results of additional studies of the subject that were published *subsequent* to the analysis of Randall *et al.*, and which therefore could have readily refuted their assessment of the situation if they felt that such was appropriate.

Siebesma *et al.* (2004) report that "simulations with nine large-scale models [were] carried out for June/July/August 1998 and the quality of the results [was] assessed along a cross-section in the subtropical and tropical North Pacific ranging from (235°E, 35°N) to (187.5°E, 1°S)," in order to "document the performance quality of state-of-the-art GCMs in modeling the first-order characteristics of subtropical and tropical cloud systems." The main conclusions of this study, according to Siebesma *et al.*, were that "(1) almost all models strongly underpredicted both cloud cover and cloud amount in the stratocumulus regions while (2) the situation is opposite in the trade-wind region and the tropics where cloud cover and cloud amount are overpredicted by most models." In fact, they report that "these deficiencies result in an overprediction of the downwelling surface short-wave radiation of typically 60 Wm^{-2} in the stratocumulus regimes and a similar underprediction of 60 Wm^{-2} in the trade-wind regions and in the intertropical convergence zone (ITCZ)," which discrepancies are to be compared with a radiative forcing of only a couple of Wm^{-2} for a 300 ppm increase in the atmosphere's CO_2 concentration. In addition, they state that "similar biases for the short-wave radiation were found at the top of the atmosphere, while discrepancies in the outgoing long-wave radiation are most pronounced in the ITCZ."

The 17 scientists who wrote Siebesma *et al.*, hailing from nine different countries, also found "the representation of clouds in general-circulation models remains one of the most important *as yet unresolved* [our italics] issues in atmospheric modeling." This is partially due, they continue, "to the overwhelming variety of clouds observed in the atmosphere, but even more so due to the large number of physical processes governing cloud formation and evolution as well as the great complexity of their interactions." Hence, they conclude that through repeated critical evaluations of the type they conducted, "the scientific community will be forced to develop further physically sound parameterizations that *ultimately*

[our italics] result in models that are capable of simulating our climate system with increasing realism."

In an effort to assess the status of state-of-the-art climate models in simulating cloud-related processes, Zhang *et al.* (2005) compared basic cloud climatologies derived from 10 atmospheric GCMs with satellite measurements obtained from the International Satellite Cloud Climatology Project (ISCCP) and the Clouds and Earth's Radiant Energy System (CERES) program. ISCCP data were available from 1983 to 2001, while data from the CERES program were available for the winter months of 2001 and 2002 and for the summer months of 2000 and 2001. The purpose of their analysis was two-fold: (1) to assess the current status of climate models in simulating clouds so that future progress can be measured more objectively, and (2) to reveal serious deficiencies in the models so as to improve them.

The work of 20 climate modelers involved in this exercise reveals a huge list of major model imperfections. First, Zhang *et al.* report a four-fold difference in high clouds among the models, and that the majority of the models simulated only 30 to 40 percent of the observed middle clouds, with some models simulating less than a quarter of observed middle clouds. For low clouds, they report that half the models underestimated them, such that the grand mean of low clouds from all models was only 70 to 80 percent of what was observed. Furthermore, when stratified in optical thickness ranges, the majority of the models simulated optically thick clouds more than twice as frequently as was found to be the case in the satellite observations, while the grand mean of all models simulated about 80 percent of optically intermediate clouds and 60 percent of optically thin clouds. And in the case of *individual* cloud types, the group of researchers reports that "differences of seasonal amplitudes among the models and satellite measurements can reach several hundred percent." As a result of these and other observations, Zhang *et al.* conclude that "much more needs to be done to fully understand the physical causes of model cloud biases presented here and to improve the models."

L'Ecuyer and Stephens (2007) used multi-sensor observations of visible, infrared, and microwave radiance obtained from the Tropical Rainfall Measuring Mission satellite for the period from January 1998 through December 1999, in order to evaluate the sensitivity of atmospheric heating—and the factors that modify it—to changes in east-west sea surface temperature gradients associated with the strong 1998 El Niño event in the tropical Pacific, as expressed by the simulations of nine general circulation models of the atmosphere that were utilized in the IPCC's most recent Fourth Assessment Report. This protocol, in their words, "provides a natural example of a short-term climate change scenario in which clouds, precipitation, and regional energy budgets in the east and west Pacific are observed to respond to the eastward migration of warm sea surface temperatures."

Results indicated that "a majority of the models examined do not reproduce the apparent westward transport of energy in the equatorial Pacific during the 1998 El Niño event." They also found that "the intermodel variability in the responses of precipitation, total heating, and vertical motion is often larger than the intrinsic ENSO signal itself, implying an inherent lack of predictive capability in the ensemble with regard to the response of the mean zonal atmospheric circulation in the tropical Pacific to ENSO." In addition, they reported that "many models also misrepresent the radiative impacts of clouds in both regions [the east and west Pacific], implying errors in total cloudiness, cloud thickness, and the relative frequency of occurrence of high and low clouds." As a result of these much-less-than-adequate findings, the two researchers from Colorado State University's Department of Atmospheric Science conclude that "deficiencies remain in the representation of relationships between radiation, clouds, and precipitation in current climate models," and they say that these deficiencies "cannot be ignored when interpreting their predictions of future climate."

In another recent paper, this one published in the *Journal of the Atmospheric Sciences*, Zhou *et al.* (2007) state that "clouds and precipitation play key roles in linking the earth's energy cycle and water cycles," noting that "the sensitivity of deep convective cloud systems and their associated precipitation efficiency in response to climate change are key factors in predicting the future climate." They also report that cloud resolving models or CRMs "have become one of the primary tools to develop the physical parameterizations of moist and other subgrid-scale processes in global circulation and climate models," and that CRMs could someday be used in place of traditional cloud parameterizations in such models.

In this regard, the authors note that "CRMs still need parameterizations on scales smaller than their grid resolutions and have many known and unknown deficiencies." To help stimulate progress in these areas, the nine scientists compared the cloud and precipitation properties observed from the Clouds and the Earth's Radiant Energy System (CERES) and Tropical Rainfall Measuring Mission (TRMM) instruments against simulations obtained from the three-dimensional Goddard Cumulus Ensemble (GCE) model during the South China Sea Monsoon Experiment (SCSMEX) field campaign of 18 May-18 June 1998.

The authors report that: (1) "the GCE rainfall spectrum includes a greater proportion of heavy rains than PR (Precipitation Radar) or TMI (TRMM Microwave Imager) observations"; (2) "the GCE model produces excessive condensed water loading in the column, especially the amount of graupel as indicated by both TMI and PR observations"; (3) "the model also cannot simulate the bright band and the sharp decrease of radar reflectivity above the freezing level in stratiform rain as seen from PR"; (4) "the model has much higher domain-averaged OLR (outgoing longwave radiation) due to smaller total cloud fraction"; (5) "the model has a more skewed distribution of OLR and effective cloud top than CERES observations, indicating that the model's cloud field is insufficient in area extent"; (6) "the GCE is ... not very efficient in stratiform rain conditions because of the large amounts of slowly falling snow and graupel that are simulated"; and finally, and in summation, (7) "large differences between model and observations exist in the rain spectrum and the vertical hydrometeor profiles that contribute to the associated cloud field."

Even more recently, a study by Spencer and Braswell (2008) observed that "our understanding of how sensitive the climate system is to radiative perturbations has been limited by large uncertainties regarding how clouds and other elements of the climate system feed back to surface temperature change (e.g., Webster and Stephens, 1984; Cess *et al.*, 1990; Senior and Mitchell, 1993; Stephens, 2005; Soden and Held, 2006; Spencer *et al.*, 2007)." The two scientists from the Earth System Science Center at the University of Alabama in Huntsville, Alabama then point out that computer models typically assume that if the *causes* of internal sources of variability (X terms) are uncorrelated to surface temperature changes, then they will not affect the accuracy of regressions used to estimate the relationship between radiative flux changes and surface temperature (T). But "while it is true that the processes that *cause* the X terms are, by [Forster and Gregory (2006)] definition, uncorrelated to T, the *response* of T to those forcings cannot be uncorrelated to T – for the simple reason that it is a radiative forcing that causes changes in T [italics in the original]." They ask "to what degree could nonfeedback sources of radiative flux variability contaminate feedback estimates?"

Spencer and Braswell use a "very simple time-dependent model of temperature deviations away from an equilibrium state" to estimate the effects of "daily random fluctuations in an unknown nonfeedback radiative source term N, such as those one might expect from stochastic variations in low cloud cover." Repeated runs of the model found the diagnosed feedback departed from the true, expected feedback value of the radiative forcing, with the difference increasing as the amount of nonfeedback radiative flux noise was increased. "It is significant," the authors write, "that all model errors for runs consistent with satellite-observed variability are in the direction of positive feedback, raising the possibility that current observational estimates of cloud feedback are biased in the positive direction." In other words, as the authors say in their abstract, "current observational diagnoses of cloud feedback – and possibly other feedbacks – could be significantly biased in the positive direction."

In light of these findings, it is clear that CRMs still have a long way to go before they are ready to properly assess the roles of various types of clouds and forms of precipitation in the future evolution of earth's climate in response to variations in anthropogenic and background forcings. This evaluation is not meant to denigrate the CRMs, it is merely done to indicate that the climate modeling enterprise is not yet at the stage where faith should be placed in what it currently suggests about earth's climatic response to the ongoing rise in the air's CO_2 content.

The hope of the climate-modeling community of tomorrow resides, according to Randall *et al.*, in something called "cloud system-resolving models" or CSRMs, which can be compared with single-column models or SCMs that can be "surgically extracted from their host GCMs." These advanced models, as they describe them, "have resolutions fine enough to represent individual cloud elements, and space-time domains large enough to encompass many clouds over many cloud lifetimes." Of course, these improvements mean that "the computational cost of running a CSRM is hundreds or thousands of times greater than that of running an SCM." Nevertheless, in a few more decades, according to Randall *et al.*, "it

will become possible to use such global CSRMs to perform century-scale climate simulations, relevant to such problems as anthropogenic climate change."

A few more decades, however, is a little long to wait to address an issue that nations of the world are confronting now. Hence, Randall *et al.* say that an approach that could be used very soon (to possibly determine whether or not there even is a problem) is to "run a CSRM as a 'superparameterization' inside a GCM," which configuration they call a "super-GCM." Not wanting to be accused of impeding scientific progress, we say "go for it," but only with the proviso that the IPCC should admit it is truly needed in order to obtain a definitive answer to the question of CO_2-induced "anthropogenic climate change." In other words, the scientific debate over the causes and processes of global warming is still ongoing and there is no scientific case for governments to regulate greenhouse gas emissions in an expensive and likely futile attempt to alter the course of future climate.

We believe, with Randall *et al.*, that our knowledge of many aspects of earth's climate system is sadly deficient. Climate models currently do not provide a reliable scientific basis for implementing programs designed to restrict anthropogenic CO_2 emissions. The cloud parameterization problem by itself is so complex that no one can validly claim that humanity's continued utilization of fossil-fuel energy will result in massive counter-productive climatic changes. There is no justification for that conclusion in reliable theoretical models.

Additional information on this topic, including reviews of newer publications as they become available, can be found at http://www.co2science.org/subject/m/inadeqclouds.php.

References

Ayers, G.P. and Gillett, R.W. 2000. DMS and its oxidation products in the remote marine atmosphere: implications for climate and atmospheric chemistry. *Journal of Sea Research* **43**: 275-286.

Cess, R.D. *et al.* 1990. Intercomparison and interpretation of climate feedback processes in 19 atmospheric general circulation models. *Journal of Geophysical Research* **95**: 16601-16615.

Charlson, R.J., Lovelock, J.E., Andrea, M.O. and Warren, S.G. 1987. Oceanic phytoplankton, atmospheric sulfur, cloud albedo and climate. *Nature* **326**: 655-661.

Chou, M.-D., Lindzen, R.S. and Hou, A.Y. 2002. Reply to: "Tropical cirrus and water vapor: an effective Earth infrared iris feedback?" *Atmospheric Chemistry and Physics* 2: 99-101.

Forster, P.M., and Taylor, K.E. 2006. Climate forcings and climate sensitivities diagnosed from coupled climate model integrations, *Journal of Climate* 19: 6181-6194.

Fu, Q., Baker, M. and Hartmann, D.L. 2002. Tropical cirrus and water vapor: an effective Earth infrared iris feedback? *Atmospheric Chemistry and Physics* 2: 31-37.

Gordon, C.T., Rosati, A. and Gudgel, R. 2000. Tropical sensitivity of a coupled model to specified ISCCP low clouds. *Journal of Climate* 13: 2239-2260.

Grabowski, W.W. 2000. Cloud microphysics and the tropical climate: Cloud-resolving model perspective. *Journal of Climate* 13: 2306-2322.

Grassl, H. 2000. Status and improvements of coupled general circulation models. *Science* 288: 1991-1997.

Groisman, P.Ya., Bradley, R.S. and Sun, B. 2000. The relationship of cloud cover to near-surface temperature and humidity: Comparison of GCM simulations with empirical data. *Journal of Climate* 13: 1858-1878.

Harries, J.E. 2000. Physics of the earth's radiative energy balance. *Contemporary Physics* 41: 309-322.

Hartmann, D.L. and Michelsen, M.L. 2002. No evidence for IRIS. *Bulletin of the American Meteorological Society* 83: 249-254.

Idso, S.B. 1990. A role for soil microbes in moderating the carbon dioxide greenhouse effect? *Soil Science* 149: 179-180.

Lane, D.E., Somerville, R.C.J. and Iacobellis, S.F. 2000. Sensitivity of cloud and radiation parameterizations to changes in vertical resolution. *Journal of Climate* 13: 915-922.

L'Ecuyer, T.S. and Stephens, G.L. 2007. The tropical atmospheric energy budget from the TRMM perspective. Part II: Evaluating GCM representations of the sensitivity of regional energy and water cycles to the 1998-99 ENSO cycle. *Journal of Climate* 20: 4548-4571.

Lindzen, R.S., Chou, M.-D. and Hou, A.Y. 2001. Does the earth have an adaptive infrared iris? *Bulletin of the American Meteorological Society* 82: 417-432.

O'Dowd, C.D., Facchini, M.C., Cavalli, F., Ceburnis, D., Mircea, M., Decesari, S., Fuzzi, S., Yoon, Y.J. and Putaud, J.-P. 2004. Biogenically driven organic contribution to marine aerosol. *Nature* 431: 676-680.

Randall, D., Khairoutdinov, M., Arakawa, A. and Grabowski, W. 2003. Breaking the cloud parameterization deadlock. *Bulletin of the American Meteorological Society* 84: 1547-1564.

Senior, C.A. and Mitchell, J.F.B. 1993. CO_2 and climate: The impact of cloud parameterization. *Journal of Climate* 6: 393-418.

Siebesma, A.P., Jakob, C., Lenderink, G., Neggers, R.A.J., Teixeira, J., van Meijgaard, E., Calvo, J., Chlond, A., Grenier, H., Jones, C., Kohler, M., Kitagawa, H., Marquet, P., Lock, A.P., Muller, F., Olmeda, D. and Severijns, C. 2004. Cloud representation in general-circulation models over the northern Pacific Ocean: A EUROCS intercomparison study. *Quarterly Journal of the Royal Meteorological Society* 130: 3245-3267.

Soden, B.J. and Held, I.M. 2006. An assessment of climate feedbacks in coupled ocean-atmosphere models. *Journal of Climate* 19: 3354-3360.

Spencer, R.W. and Braswell, W.D. 2008. Potential biases in feedback diagnosis from observational data: A simple model demonstration. *Journal of Climate 21: 5624-5628.*

Spencer, R.W., Braswell, W.D., Christy, J.R. and Hnilo, J. 2007. Cloud and radiation budget changes associated with tropical intraseasonal oscillations. *Geophysical Research Letters* 34: L15707, doi:10.1029/2007GLO296998.

Stephens, G.L. 2005. Clouds feedbacks in the climate system: A critical review. *Journal of Climate* 18: 237-273.

Webster, P.J. and Stephens, G.L. 1984. Cloud-radiation feedback and the climate problem. In Houghton, J. (Ed.) *The Global Climate.* Cambridge University Press. 63-78.

Zhang, M.H., Lin, W.Y., Klein, S.A., Bacmeister, J.T., Bony, S., Cederwall, R.T., Del Genio, A.D., Hack, J.J., Loeb, N.G., Lohmann, U., Minnis, P., Musat, I., Pincus, R., Stier, P., Suarez, M.J., Webb, M.J., Wu, J.B., Xie, S.C., Yao, M.-S. and Yang, J.H. 2005. Comparing clouds and their seasonal variations in 10 atmospheric general circulation models with satellite measurements. *Journal of Geophysical Research* 110: D15S02, doi:10.1029/2004JD005021.

Zhou, Y.P., Tao, W.-K., Hou, A.Y., Olson, W.S., Shie, C.-L., Lau, K.-M., Chou, M.-D., Lin, X. and Grecu, M. 2007. Use of high-resolution satellite observations to evaluate cloud and precipitation statistics from cloud-resolving model simulations. Part I: South China Sea monsoon experiment. *Journal of the Atmospheric Sciences* 64: 4309-4329.

1.4. Precipitation

One of the predictions of atmospheric general circulation models (GCMs) is that the planet's hydrologic cycle will intensify as the world warms, leading to an increase in the frequency and intensity of extreme precipitation events. In an early review of the subject, Walsh and Pittock (1998) reported "there is some evidence from climate model studies that, in a warmer climate, rainfall events will be more intense," and that "there is considerable evidence that the frequency of extreme rainfall events may increase in the tropics." Upon further study, however, they were forced to conclude that "because of the insufficient resolution of climate models and their generally crude representation of sub-grid scale and convective processes, little confidence can be placed in any definite predictions of such effects."

Two years later, Lebel *et al.* (2000) compared rainfall simulations produced by a GCM with real-world observations from West Africa for the period 1960-1990. Their analysis revealed that the model output was affected by a number of temporal and spatial biases that led to significant differences between observed and modeled data. The simulated rainfall totals, for example, were significantly greater than what was typically observed, exceeding real-world values by 25 percent during the dry season and 75 percent during the rainy season. In addition, the seasonal cycle of precipitation was not well simulated, as the researchers found that the simulated rainy season began too early and that the increase in precipitation was not rapid enough. Shortcomings were also evident in the GCM's inability to accurately simulate convective rainfall events, as it typically predicted too much precipitation. Furthermore, it was found that "interannual variability [was] seriously disturbed in the GCM as compared to what it [was] in the observations." As for why the GCM performed so poorly in these several respects, Lebel *et al.* gave two main reasons: parameterization of rainfall processes in the GCM was much too simple, and spatial resolution was much too coarse.

Three years later, Woodhouse (2003) generated a tree-ring-based history of snow water equivalent (SWE) characteristic of the first day of April for each year of the period 1569-1999 for the drainage basin of the Gunnison River of western Colorado, USA. Then, because "an understanding of the long-term characteristics of snowpack variability is useful for guiding expectations for future variability," as she phrased it, she analyzed the reconstructed SWE data in such a way as to determine if there was anything unusual about the SWE record of the twentieth century, which the IPCC claims experienced a warming that was unprecedented over the past two millennia.

Woodhouse found "the twentieth century is notable for several periods that lack extreme years." Specifically, she determined that "the twentieth century is notable for several periods that contain few or no extreme years, for both low and high SWE extremes," and she reports that "the twentieth century also contains the lowest percent of extreme low SWE years." These results are in direct contradiction of what GCMs typically predict should occur in response to global warming. Their failure in this regard is especially damning because it occurred during a period of global warming that is said to have been the most significant of the past 20 centuries.

Two years later, and as a result of the fact that the 2004 summer monsoon season of India experienced a 13 percent precipitation deficit that was not predicted by any of the empirical or dynamical models regularly used in making rainfall forecasts, Gadgil *et al.* (2005) performed a historical analysis of the models forecast skill over the period 1932-2004. Despite model advancements and an ever-improving understanding of monsoon variability, they found the models' skill in forecasting the Indian monsoon's characteristics had not improved since the very first versions of the models were applied to the task some seven decades earlier. The empirical models Gadgil *et al.* evaluated generated large differences between monsoon rainfall measurements and model predictions. In addition, the models often failed to correctly predict even the *sign* of the precipitation anomaly, frequently predicting excess rainfall when drought occurred and drought when excess rainfall was received.

The dynamical models fared even worse. In comparing observed monsoon rainfall totals with simulated values obtained from 20 state-of-the-art GCMs and a supposedly superior coupled atmosphere-ocean model, Gadgil *et al.* reported that not a single one of those many models was able "to simulate correctly the interannual variation of the summer monsoon rainfall over the Indian region." And as with the empirical models, the dynamical models also frequently failed to correctly capture even the *sign* of the observed rainfall anomalies. In addition, the researchers report that Brankovic and Molteni (2004) attempted to model the Indian monsoon with a much higher-resolution GCM, but its output also proved to be "not realistic."

Lau *et al.* (2006) considered the Sahel drought of the 1970s-'90s to provide "an ideal test bed for evaluating the capability of CGCMs [coupled general circulation models] in simulating long-term drought, and the veracity of the models' representation of coupled atmosphere-ocean-land processes and their interactions." They chose to "explore the roles of sea surface temperature coupling and land surface processes in producing the Sahel drought in CGCMs that participated in the twentieth-century coupled climate simulations of the Intergovernmental Panel on Climate Change [IPCC] Assessment Report 4," in which the 19 CGCMs "are driven by combinations of realistic prescribed external forcing, including anthropogenic increase in greenhouse gases and sulfate aerosols, long-term variation in solar radiation, and volcanic eruptions."

In performing this analysis, the climate scientists found, in their words, that "only eight models produce a reasonable Sahel drought signal, seven models produce excessive rainfall over [the] Sahel during the observed drought period, and four models show no significant deviation from normal." In addition, they report that "even the model with the highest skill for the Sahel drought could only simulate the increasing trend of severe drought events but not the magnitude, nor the beginning time and duration of the events." All 19 of the CGCMs employed in the IPCC's Fourth Assessment Report, in other words, failed to adequately simulate the basic characteristics of "one of the most pronounced signals of climate change" of the past century—as defined by its start date, severity and duration."

Wentz *et al.* (2007), in a study published in *Science*, noted that the Coupled Model Intercomparison Project, as well as various climate modeling analyses, predicted an increase in precipitation on the order of 1 to 3 percent per °C of surface global warming. They decided to see what had happened in the real world in this regard over the prior 19 years (1987-2006) of supposedly unprecedented global warming, when data from the Global Historical Climatology Network and satellite measurements of the lower troposphere indicated there had been a global temperature rise on the order of 0.20°C per decade.

Using satellite observations obtained from the Special Sensor Microwave Imager (SSM/I), the four Remote Sensing Systems scientists derived precipitation trends for the world's oceans over this period, and using data obtained from the Global Precipitation Climatology Project that were acquired from both satellite and rain gauge measurements, they derived precipitation trends for earth's continents.

Appropriately combining the results of these two endeavors, they derived a real-world increase in precipitation on the order of 7 percent per °C of surface global warming, which is somewhere between 2.3 and 7.0 times *larger* than what is predicted by state-of-the-art climate models.

How was this huge discrepancy to be resolved? Wentz *et al.* concluded that the only way to bring the two results into harmony was for there to have been a 19-year decline in global wind speeds. But when looking at the past 19 years of SSM/I wind retrievals, they found just the opposite, an *increase* in global wind speeds. In quantitative terms, the two results were about as opposite as they could possibly be, as they report that "when averaged over the tropics from 30°S to 30°N, the winds increased by 0.04 m s^{-1} (0.6 percent) decade^{-1}, and over all oceans the increase was 0.08 m s^{-1} (1.0 percent) decade^{-1}," while global coupled ocean-atmosphere models or GCMs, in their words, "predict that the 1987-to-2006 warming should have been accompanied by a decrease in winds on the order of 0.8 percent decade^{-1}."

In discussing these results, Wentz *et al.* say "the reason for the discrepancy between the observational data and the GCMs is not clear." They also observe that this dramatic difference between the real world of nature and the virtual world of climate modeling "has enormous impact" and the questions raised by the discrepancy "are far from being settled."

Allan and Soden (2007) quantified trends in precipitation within ascending and descending branches of the planet's tropical circulation and compared their results with simulations of the present day and projections of future changes provided by up to 16 state-of-the-art climate models. The precipitation data for this analysis came from the Global Precipitation Climatology Project (GPCP) of Adler *et al.* (2003) and the Climate Prediction Center Merged Analysis of Precipitation (CMAP) data of Xie and Arkin (1998) for the period 1979-2006, while for the period 1987-2006 they came from the monthly mean intercalibrated Version 6 Special Sensor Microwave Imager (SSM/I) precipitation data described by Wentz *et al.* (2007).

The researchers reported "an emerging signal of rising precipitation trends in the ascending regions and decreasing trends in the descending regions are detected in the observational datasets," but that "these trends are substantially larger in magnitude than present-day simulations and projections into the 21st century," especially in the case of the descending regions. More specifically, for the tropics "the GPCP trend is about 2-3 times larger than the model ensemble mean trend, consistent with previous findings (Wentz *et al.*, 2007) and also supported by the analysis of Yu and Weller (2007)," who additionally contend that "observed increases of evaporation over the ocean are substantially greater than those simulated by climate models." What is more, Allan and Soden note that "observed precipitation changes over land also appear larger than model simulations over the 20th century (Zhang *et al.*, 2007)."

Noting that the difference between the models and real-world measurements "has important implications for future predictions of climate change," Allan and Soden say "the discrepancy cannot be explained by changes in the reanalysis fields used to subsample the observations but instead must relate to errors in the satellite data or in the model parameterizations." This same dilemma was also faced by Wentz *et al.* (2007); and they too stated that the resolution of the issue "has enormous impact" and likewise concluded that the questions raised by the discrepancy "are far from being settled."

Lin (2007) states that "a good simulation of tropical mean climate by the climate models is a prerequisite for their good simulations/predictions of tropical variabilities and global teleconnections," but "unfortunately, the tropical mean climate has not been well simulated by the coupled general circulation models (CGCMs) used for climate predictions and projections," noting that "most of the CGCMs produce a double-intertropical convergence zone (ITCZ) pattern," and acknowledging that "a synthetic view of the double-ITCZ problem is still elusive."

To explore the nature of this problem in greater depth, and in hope of making some progress in resolving it, Lin analyzed tropical mean climate simulations of the 20-year period 1979-99 provided by 22 IPCC Fourth Assessment Report CGCMs, together with concurrent Atmospheric Model Intercomparison Project (AMIP) runs from 12 of them. This work revealed, in Lin's words, that "most of the current state-of-the-art CGCMs have some degree of the double-ITCZ problem, which is characterized by excessive precipitation over much of the Tropics (e.g., Northern Hemisphere ITCZ, Southern Hemisphere SPCZ [South Pacific Convergence Zone], Maritime Continent, and equatorial Indian Ocean), and often associated with insufficient precipitation over the equatorial Pacific," as well as "overly strong trade winds, excessive LHF [latent heat flux], and insufficient SWF [shortwave flux], leading to significant cold SST (sea surface temperature) bias in much of the tropical oceans."

The authors further note that "most of the models also simulate insufficient latitudinal asymmetry in precipitation and SST over the eastern Pacific and Atlantic Oceans," and further, that "the AMIP runs also produce excessive precipitation over much of the Tropics including the equatorial Pacific, which also leads to overly strong trade winds, excessive LHF, and insufficient SWF," which suggests that "the excessive tropical precipitation is an intrinsic error of the atmospheric models." And if that is not enough, Lin adds that "over the eastern Pacific stratus region, most of the models produce insufficient stratus-SST feedback associated with insufficient sensitivity of stratus cloud amount to SST."

With the solutions to all of these long-standing problems continuing to remain "elusive," and with Lin suggesting that the sought-for solutions are in fact prerequisites for "good simulations/predictions" of future climate, there is significant reason to conclude that current state-of-the-art CGCM predictions of CO_2-induced global warming should not be considered reliable.

In conclusion, in spite of the billions of dollars spent by the United States alone on developing and improving climate models, the models' ability to correctly simulate even the largest and most regionally-important of earth's atmospheric phenomena—the tropical Indian monsoon—hasn't improved at all. The scientific literature is filled with studies documenting the inability of even the most advanced GCMs to accurately model radiation, clouds, and precipitation. Failure to model any one of these elements would be grounds for rejecting claims that the IPCC provides the evidence needed to justify regulation of anthropogenic greenhouse gas emissions.

Additional information on this topic, including reviews of newer publications as they become available, can be found at http://www.co2science.org/subject/p/precipmodelinadeq.php.

References

Adler, R.F., Huffman, G.J., Chang, A., Ferraro, R., Xie, P., Janowiak, J., Rudolf, B., Schneider, U. Curtis, S., Bolvin, D., Gruber, A., Susskind, J. Arkin, P. and Nelkin, E. 2003. The version-2 Global Precipitation Climatology Project (GPCP) monthly precipitation analysis (1979-present). *Journal of Hydrometeorology* **4**: 1147-1167.

Allan, R.P. and Soden, B.J. 2007. Large discrepancy between observed and simulated precipitation trends in the ascending and descending branches of the tropical circulation. *Geophysical Research Letters* **34**: 10.1029/2007GL031460.

Brankovic, C. and Molteni, F. 2004. Seasonal climate and variability of the ECMWF ERA-40 model. *Climate Dynamics* **22**: 139-155.

Gadgil, S., Rajeevan, M. and Nanjundiah, R. 2005. Monsoon prediction—Why yet another failure? *Current Science* **88**: 1389-1400.

Lau, K.M., Shen, S.S.P., Kim, K.-M. and Wang, H. 2006. A multimodel study of the twentieth-century simulations of Sahel drought from the 1970s to 1990s. *Journal of Geophysical Research* **111**: 10.1029/2005JD006281.

Lebel, T., Delclaux, F., Le Barbé, L. and Polcher, J. 2000. From GCM scales to hydrological scales: rainfall variability in West Africa. *Stochastic Environmental Research and Risk Assessment* **14**: 275-295.

Lin, J.-L. 2007. The double-ITCZ problem in IPCC AR4 coupled GCMs: Ocean-atmosphere feedback analysis. *Journal of Climate* **20**: 4497-4525.

Walsh, K. and Pittock, A.B. 1998. Potential changes in tropical storms, hurricanes, and extreme rainfall events as a result of climate change. *Climatic Change* **39**: 199-213.

Wentz, F.J., Ricciardulli, L., Hilburn, K. and Mears, C. 2007. How much more rain will global warming bring? *Science* **317**: 233-235.

Woodhouse, C.A. 2003. A 431-yr reconstruction of western Colorado snowpack from tree rings. *Journal of Climate* **16**: 1551-1561.

Xie, P. and Arkin, P.A. 1998. Global monthly precipitation estimates from satellite-observed outgoing longwave radiation. *Journal of Climate* **11**: 137-164.

Yu, L. and Weller, R.A. 2007. Objectively analyzed air-sea heat fluxes for the global ice-free oceans (1981-2005). *Bulletin of the American Meteorological Society* **88**: 527-539.

Zhang, X., Zwiers, F.W., Hegerl, G.C., Lambert, F.H., Gillett, N.P., Solomon, S., Stott, P.A. and Nozawa, T. 2007. Detection of human influence on twentieth-century precipitation trends. *Nature* **448**: 461-465.

2

Feedback Factors and Radiative Forcing

Introduction

According to the Intergovernmental Panel on Climate Change (IPCC), "the combined radiative forcing due to increases in carbon dioxide, methane, and nitrous oxide is +2.30 [+2.07 to +2.35] W m^{-2}, and its rate of increase during the industrial era is *very likely* to have been unprecedented in more than 10,000 years [italics in the original]" (IPCC, 2007-I, p. 3). The IPCC calculates that this sensitivity of earth's climate system to greenhouse gases (GHG) means that if CO_2 concentrations were to double, the rise in global average surface temperature "is *likely* to be in the range 2°C to 4.5°C with a best estimate of about 3°C, and is *very unlikely* to be less than 1.5°C [italics in the original]" (Ibid., p. 12).

Many scientific studies suggest this model-derived sensitivity is too large and feedbacks in the climate system reduce it to values that are an order of magnitude smaller. This chapter reviews those feedbacks most often mentioned in the scientific literature, some of which have the ability to totally offset the radiative forcing expected from the rise in atmospheric CO_2.

Additional information on this topic, including reviews of feedback factors not discussed here, can be found at http://www.co2science.org/subject/f/subject_f.php under the heading Feedback Factors.

References

IPCC. 2007-I. *Climate Change 2007: The Physical Science Basis. Contribution of Working Group I to the Fourth Assessment Report of the Intergovernmental Panel on Climate Change.* Solomon, S., D. Qin, M. Manning, Z. Chen, M. Marquis, K.B. Averyt, M. Tignor and H.L. Miller. (Eds.) Cambridge University Press, Cambridge, UK.

2.1. Clouds

Based on data obtained from the Tropical Ocean Global Atmosphere—Coupled Ocean-Atmosphere Response Experiment, Sud *et al.* (1999) demonstrated that deep convection in the tropics acts as a thermostat to keep sea surface temperature (SST) oscillating between approximately 28° and 30°C. Their analysis suggests that as SSTs reach 28°-29°C, the cloud-base airmass is charged with the moist static energy needed for clouds to reach the upper troposphere, at which point the cloud cover reduces the amount of solar radiation received at the surface of the sea, while cool and dry downdrafts promote ocean surface cooling by increasing sensible and latent heat fluxes there. This "thermostat-like control," as Sud *et al.* describe it, tends "to ventilate the tropical ocean efficiently and help contain the SST between 28°-30°C." The phenomenon would also be expected to prevent SSTs from rising any higher in response to enhanced CO_2-induced radiative forcing.

Lindzen et al. (2001) used upper-level cloudiness data obtained from the Japanese Geostationary Meteorological Satellite and SST data obtained from the National Centers for Environmental Prediction to derive a strong inverse relationship between upper-level cloud area and the mean SST of cloudy regions of the eastern part of the western Pacific (30°S-30°N; 130°E-170°W), such that the area of cirrus cloud coverage normalized by a measure of the area of cumulus coverage decreases about 22 percent per degree C increase in the SST of the cloudy region. In describing this phenomenon, Lindzen et al. say "the cloudy-moist region appears to act as an infrared adaptive iris that opens up and closes down the regions free of upper-level clouds, which more effectively permit infrared cooling, in such a manner as to resist changes in tropical surface temperature."

The findings of Lindzen et al. were subsequently criticized by Hartmann and Michelsen (2002) and Fu et al. (2002), and then Fu et al. were rebutted by Chou et al. (2002), an exchange that is summarized in Section 1.2 of this report. The debate over the infrared adaptive iris still rages in the scientific community, but Lindzen and his colleagues are not the only scientists who believe the cooling effect of clouds has been underestimated.

Croke et al. (1999) used land-based observations of cloud cover for three regions of the United States (coastal southwest, coastal northeast, and southern plains) to demonstrate that, over the period 1900-1987, cloud cover had a high correlation with global air temperature, with mean cloud cover rising from an initial value of 35 percent to a final value of 47 percent as the mean global air temperature rose by 0.5°C.

Herman et al. (2001) used Total Ozone Mapping Spectrometer 380-nm reflectivity data to determine changes in radiation reflected back to space over the period 1979 to 1992, finding that "when the 11.3-year solar-cycle and ENSO effects are removed from the time series, the zonally averaged annual linear-fit trends show that there have been increases in reflectivity (cloudiness) poleward of 40°N and 30°S, with some smaller but significant changes occurring in the equatorial and lower middle latitudes." The overall long-term effect was an increase in radiation reflected back to space of 2.8 Wm^{-2} per decade, which represents a large cloud-induced cooling influence.

Rosenfeld (2000) used satellite data obtained from the Tropical Rainfall Measuring Mission to look for terrestrial analogues of the cloud trails that form in the wakes of ships at sea as a consequence of their emissions of particulates that redistribute cloud-water into larger numbers of smaller droplets that do not rain out of the atmosphere as readily as they would in the absence of this phenomenon. Visualizations produced from the mission data clearly revealed the existence of enhanced cloud trails downwind of urban and industrial complexes in Turkey, Canada, and Australia, to which Rosenfeld gave the name *pollution tracks* in view of their similarity to *ship tracks*. Rosenfeld also demonstrated that the clouds comprising these pollution tracks were composed of droplets of reduced size that did indeed suppress precipitation by inhibiting further coalescence and ice precipitation formation. As Toon (2000) noted in a commentary on this study, these smaller droplets will not "rain out" as quickly and will therefore last longer and cover more of the earth, both of which effects tend to cool the globe.

In summation, as the earth warms, the atmosphere has a tendency to become more cloudy, which exerts a natural brake upon the rising temperature. Many of man's aerosol-producing activities tend to do the same thing. Hence, there appear to be a number of cloud-mediated processes that help the planet "keep its cool."

Additional information on this topic, including reviews of newer publications as they become available, can be found at http://www.co2science.org/subject/f/feedbackcloud.php.

References

Chou, M.-D., Lindzen, R.S. and Hou, A.Y. 2002. Reply to: "Tropical cirrus and water vapor: an effective Earth infrared iris feedback?" *Atmospheric Chemistry and Physics* **2**: 99-101.

Croke, M.S., Cess, R.D. and Hameed, S. 1999. Regional cloud cover change associated with global climate change: Case studies for three regions of the United States. *Journal of Climate* **12**: 2128-2134.

Fu, Q., Baker, M. and Hartmann, D.L. 2002. Tropical cirrus and water vapor: an effective Earth infrared iris feedback? *Atmospheric Chemistry and Physics* **2**: 31-37.

Hartmann, D.L. and Michelsen, M.L. 2002. No evidence for IRIS. *Bulletin of the American Meteorological Society* **83**: 249-254.

Herman, J.R., Larko, D., Celarier, E. and Ziemke, J. 2001. Changes in the Earth's UV reflectivity from the surface, clouds, and aerosols. *Journal of Geophysical Research* **106**: 5353-5368.

Lindzen, R.S., Chou, M.-D. and Hou, A.Y. 2001. Does the earth have an adaptive infrared iris? *Bulletin of the American Meteorological Society* **82**: 417-432.

Rosenfeld, D. 2000. Suppression of rain and snow by urban and industrial air pollution. *Science* **287**: 1793-1796.

Sud, Y.C., Walker, G.K. and Lau, K.-M. 1999. Mechanisms regulating sea-surface temperatures and deep convection in the tropics. *Geophysical Research Letters* **26**: 1019-1022.

Toon, O.W. 2000. How pollution suppresses rain. *Science* **287**: 1763-1765.

2.2. Carbonyl Sulfide

Some time ago, Idso (1990) suggested that the volatilization of reduced sulfur gases from earth's soils may be just as important as dimethyl sulfide (DMS) emissions from the world's oceans in enhancing cloud albedo and thereby cooling the planet and providing a natural brake on the tendency for anthropogenically enhanced greenhouse gases to drive global warming. (See Section 2.7.) On the basis of experiments that showed soil DMS emissions to be positively correlated with soil organic matter content, and noting that additions of organic matter to soils tend to increase the amount of sulfur gases they emit, Idso hypothesized that because atmospheric CO_2 enrichment augments plant growth and, as a result, vegetative inputs of organic matter to earth's soils, this phenomenon should produce an impetus for cooling, even in the absence of the surface warming that sets in motion the chain of events that produce the oceanic DMS-induced negative feedback that tends to cool the planet.

Two years later, Idso (1992) expanded this concept to include carbonyl sulfide (OCS), another biologically produced sulfur gas that is emitted from soils, noting that it too is likely to be emitted in increasingly greater quantities as earth's vegetation responds to the aerial fertilization effect of the ongoing rise in the air's CO_2 content, while pointing out that OCS is relatively inert in the troposphere, but that it eventually makes its way into the stratosphere, where it is transformed into solar-radiation-reflecting sulfate aerosol particles. He consequently concluded that the CO_2-induced augmentation of soil OCS emissions constitutes a mechanism that can cool the

planet's surface (1) in the absence of an impetus for warming, (2) without producing additional clouds or (3) making them any brighter.

What have we subsequently learned about biologically mediated increases in carbonyl sulfide emissions? One important thing is that the OCS-induced cooling mechanism also operates at sea, just as the DMS-induced cooling mechanism does, and that it too possesses a warming-induced component in addition to its CO_2-induced component.

In a study contemporary with that of Idso (1992), ocean-surface OCS concentrations were demonstrated by Andreae and Ferek (1992) to be highly correlated with surface-water primary productivity. So strong is this correlation, in fact, that Erickson and Eaton (1993) developed an empirical model for computing ocean-surface OCS concentrations based solely on surface-water chlorophyll concentrations and values of incoming solar radiation. It has also been learned that an even greater portion of naturally produced OCS is created in the atmosphere, where carbon disulfide and dimethyl sulfide—also largely of oceanic origin (Aydin *et al.*, 2002)—undergo photochemical oxidation (Khalil and Rasmussen, 1984; Barnes *et al.*, 1994). Hence, the majority of the tropospheric burden of OCS is ultimately dependent upon photosynthetic activity occurring near the surface of the world's oceans.

This is important because the tropospheric OCS concentration has risen by approximately 30 percent since the 1600s, from a mean value of 373 ppt over the period 1616-1694 to something on the order of 485 ppt today. This is a sizeable increase; and Aydin *et al.* (2002) note that only a fourth of it can be attributed to anthropogenic sources. Consequently, the rest of the observed OCS increase must have had a natural origin, a large portion of which must have ultimately been derived from the products and byproducts of marine photosynthetic activity, which must have increased substantially over the past three centuries. A solid case can be made for the proposition that both the increase in atmospheric CO_2 concentration and the increase in temperature experienced over this period were the driving forces for the concomitant increase in tropospheric OCS concentration and its likely subsequent transport to the stratosphere, where it could exert a cooling influence on the earth and that may have kept the warming of the globe considerably below what it might otherwise have been in the absence of this chain of events.

Another fascinating aspect of this multifaceted global "biothermostat" was revealed in a laboratory study of samples of the lichen *Ramalina menziesii*, which were collected from an open oak woodland in central California, USA, by Kuhn and Kesselmeier (2000). They found that when the lichens were optimally hydrated, they absorbed OCS from the air at a rate that gradually doubled as air temperature rose from approximately 3° to 25°C, whereupon their rate of OCS absorption began a precipitous decline that led to zero OCS absorption at 35°C.

The first portion of this response can be explained by the fact that most terrestrial plants prefer much warmer temperatures than a mere 3°C, so that as their surroundings warm and they grow better, they extract more OCS from the atmosphere in an attempt to promote even more warming and grow better still. At the point where warming becomes a detriment to them, however, they reverse this course of action and begin to rapidly reduce their rates of OCS absorption in an attempt to forestall warming-induced death. And since the consumption of OCS by lichens is under the physiological control of carbonic anhydrase—which is the key enzyme for OCS uptake in all higher plants, algae, and soil organisms—we could expect this phenomenon to be generally operative over most of the earth. Hence, this thermoregulatory function of the biosphere may well be powerful enough to define an upper limit above which the surface air temperature of the planet may be restricted from rising, even when changes in other forcing factors, such as increases in greenhouse gas concentrations, produce an impetus for it to do so.

Clearly, this multifaceted phenomenon is extremely complex, with different biological entities tending to both increase and decrease atmospheric OCS concentrations at one and the same time, while periodically reversing directions in this regard in response to climate changes that push the temperatures of their respective environments either above or below the various thermal optima at which they function best. This being the case, there is obviously much more we need to learn about the many plant physiological mechanisms that may be involved.

State-of-the-art climate models totally neglect the biological processes we have described here. Until we fully understand the ultimate impact of the OCS cycle on climate, and then incorporate them into the climate models, we cannot be certain how much of the warming experienced during the twentieth century, if any, can be attributed to anthropogenic causes.

Additional information on this topic, including reviews of newer publications as they become available, can be found at http://www.co2science.org/subject/c/carbonylsulfide.php.

References

Andreae, M.O. and Ferek, R.J. 1992. Photochemical production of carbonyl sulfide in seawater and its emission to the atmosphere. *Global Biogeochemical Cycles* **6**: 175-183.

Aydin, M., De Bruyn, W.J. and Saltzman, E.S. 2002. Preindustrial atmospheric carbonyl sulfide (OCS) from an Antarctic ice core. *Geophysical Research Letters* **29**: 10.1029/2002GL014796.

Barnes, I., Becker, K.H. and Petroescu, I. 1994. The tropospheric oxidation of DMS: a new source of OCS. *Geophysical Research Letters* **21**: 2389-2392.

Erickson III, D.J. and Eaton, B.E. 1993. Global biogeochemical cycling estimates with CZCS satellite data and general circulation models. *Geophysical Research Letters* **20**: 683-686.

Idso, S.B. 1990. A role for soil microbes in moderating the carbon dioxide greenhouse effect? *Soil Science* **149**: 179-180.

Idso, S.B. 1992. The DMS-cloud albedo feedback effect: Greatly underestimated? *Climatic Change* **21**: 429-433.

Khalil, M.A.K. and Rasmussen, R.A. 1984. Global sources, lifetimes, and mass balances of carbonyl sulfide (OCS) and carbon disulfide (CS_2) in the earth's atmosphere. *Atmospheric Environment* **18**: 1805-1813.

Kuhn, U. and Kesselmeier, J. 2000. Environmental variables controlling the uptake of carbonyl sulfide by lichens. *Journal of Geophysical Research* **105**: 26,783-26,792.

2.3. Diffuse Light

The next negative feedback phenomenon is diffused light. It operates through a chain of five linkages, triggered by the incremental enhancement of the atmosphere's greenhouse effect that is produced by an increase in the air's CO_2 content. The first linkages is the proven propensity for higher levels of atmospheric CO_2 to enhance vegetative productivity, which phenomena are themselves powerful negative feedback mechanisms of the type we envision. Greater CO_2-enhanced photosynthetic rates, for example, enable plants to remove considerably more CO_2 from the air than they do under current conditions, while CO_2-induced increases in plant water use efficiency allow plants to grow where it was previously too dry for them. (See Chapter 7 for extensive documentation of this phenomenon.) This establishes a potential for more CO_2 to be removed from the atmosphere by increasing the abundance of earth's plants and increasing their robustness.

The second linkage of the feedback loop is the ability of plants to emit gases to the atmosphere that are ultimately converted into "biosols," i.e., aerosols that owe their existence to the biological activities of earth's vegetation, many of which function as cloud condensation nuclei. It takes little imagination to realize that since the existence of these atmospheric particles is dependent upon the physiological activities of plants and their associated soil biota, the CO_2-induced presence of more, and more-highly-productive, plants will lead to the production of more of these cloud-mediating particles, which can then result in more clouds which reflect sunlight and act to cool the planet.

The third linkage is the observed propensity for increases in aerosols and cloud particles to enhance the amount of diffuse solar radiation reaching the earth's surface. The fourth linkage is the ability of enhanced diffuse lighting to reduce the volume of shade within vegetative canopies. The fifth linkage is the tendency for less internal canopy shading to enhance whole-canopy photosynthesis, which finally produces the end result: a greater biological extraction of CO_2 from the air and the subsequent sequestration of its carbon, compliments of the intensified diffuse-light-driven increase in total canopy photosynthesis and subsequent transfers of the extra fixed carbon to plant and soil storage reservoirs.

How significant is this multi-link process? Roderick *et al.* (2001) provide a good estimate based on the utilization of a unique "natural experiment," a technique that has been used extensively by Idso (1998) to evaluate the climatic sensitivity of the entire planet. Specifically, Roderick and his colleagues considered the volcanic eruption of Mt. Pinatubo in June 1991, which ejected enough gases and fine materials into the atmosphere to produce sufficient aerosol particles to greatly increase the diffuse component of the solar radiation reaching the surface of the earth from that point in time through much of 1993, while only slightly reducing the receipt of total solar radiation. Based on a set of lengthy calculations, they concluded that the Mt. Pinatubo eruption may

well have resulted in the removal of an extra 2.5 Gt of carbon from the atmosphere due to its diffuse-light-enhancing stimulation of terrestrial vegetation in the year following the eruption, which would have reduced the ongoing rise in the air's CO_2 concentration that year by about 1.2 ppm.

Interestingly, this reduction is about the magnitude of the real-world perturbation that was actually observed (Sarmiento, 1993). What makes this observation even more impressive is the fact that the CO_2 reduction was coincident with an El Niño event; because, in the words of Roderick et al., "previous and subsequent such events have been associated with *increases* in atmospheric CO_2." In addition, the observed reduction in total solar radiation received at the earth's surface during this period would have had a tendency to reduce the amount of photosynthetically active radiation incident upon earth's plants, which would also have had a tendency to cause the air's CO_2 content to rise, as it would tend to lessen global photosynthetic activity.

Significant support for the new negative feedback phenomenon was swift in coming, as the very next year a team of 33 researchers published the results of a comprehensive study (Law *et al.*, 2002) that compared seasonal and annual values of CO_2 and water vapor exchange across sites in forests, grasslands, crops and tundra—which are part of an international network called FLUXNET—investigating the responses of these exchanges to variations in a number of environmental factors, including direct and diffuse solar radiation. The researchers reported that "net carbon uptake (net ecosystem exchange, the net of photosynthesis and respiration) was greater under diffuse than under direct radiation conditions," and in discussing this finding, which is the centerpiece of the negative feedback phenomenon we describe, they noted that "cloud-cover results in a greater proportion of diffuse radiation and constitutes a higher fraction of light penetrating to lower depths of the canopy (Oechel and Lawrence, 1985)." More importantly, they also reported that "Goulden *et al.* (1997), Fitzjarrald *et al.* (1995), and Sakai *et al.* (1996) showed that net carbon uptake was consistently higher during cloudy periods in a boreal coniferous forest than during sunny periods with the same PPFD [photosynthetic photon flux density]." In fact, they wrote that "Hollinger *et al.* (1994) found that daily net CO_2 uptake was greater on cloudy days, even though total PPFD was 21-45 percent lower on cloudy days than on clear days."

One year later, Gu *et al.* (2003) reported that they "used two independent and direct methods to examine the photosynthetic response of a northern hardwood forest (Harvard Forest, 42.5°N, 72.2°W) to changes in diffuse radiation caused by Mount Pinatubo's volcanic aerosols," finding that in the eruption year of 1991, "around noontime in the mid-growing season, the gross photosynthetic rate under the perturbed cloudless solar radiation regime was 23, 8, and 4 percent higher than that under the normal cloudless solar radiation regime in 1992, 1993, and 1994, respectively," and that "integrated over a day, the enhancement for canopy gross photosynthesis by the volcanic aerosols was 21 percent in 1992, 6 percent in 1993 and 3 percent in 1994." Commenting on the significance of these observations, Gu *et al.* noted that "because of substantial increases in diffuse radiation world-wide after the eruption and strong positive effects of diffuse radiation for a variety of vegetation types, it is likely that our findings at Harvard Forest represent a global phenomenon."

In the preceding paragraph, we highlighted the fact that the diffuse-light-induced photosynthetic enhancement observed by Gu *et al.*, in addition to likely being global in scope, was caused by volcanic aerosols acting under cloudless conditions. Our reason for calling attention to these two facts is to clearly distinguish this phenomenon from a closely related one that is also described by Gu *et al.*; i.e., the propensity for the extra diffuse light created by increased cloud cover to further enhance photosynthesis, even though the total flux of solar radiation received at the earth's surface may be reduced under such conditions. Based on still more real-world data, for example, Gu *et al.* note that "Harvard Forest photosynthesis also increases with cloud cover, with a peak at about 50 percent cloud cover."

Although very impressive, in all of the situations discussed above the source of the enhanced atmospheric aerosol concentration was a singular significant event—specifically, a massive volcanic eruption—but what we really need to know is what happens under more normal conditions. This was the new and important question addressed the following year in the study of Niyogi *et al.* (2004): "Can we detect the effect of relatively routine aerosol variability on field measurements of CO_2 fluxes, and if so, how does the variability in aerosol loading affect CO_2 fluxes over different landscapes?"

To answer this question, the group of 16 researchers used CO_2 flux data from the AmeriFlux network (Baldocchi *et al.*, 2001) together with cloud-free aerosol optical depth data from the NASA Robotic Network (AERONET; Holben *et al.*, 2001) to assess the effect of aerosol loading on the net assimilation of CO_2 by three types of vegetation: trees (broadleaf deciduous forest and mixed forest), crops (winter wheat, soybeans, and corn), and grasslands. Their work revealed that an aerosol-induced increase in diffuse radiative-flux fraction [DRF = ratio of diffuse (Rd) to total or global (Rg) solar irradiance] increased the net CO_2 assimilation of trees and crops, making them larger carbon sinks, but that it decreased the net CO_2 assimilation of grasslands, making them smaller carbon sinks.

How significant were the effects observed by Niyogi *et al.*? For a summer mid-range Rg flux of 500 Wm^{-2}, going from the set of all DRF values between 0.0 and 0.4 to the set of all DRF values between 0.6 and 1.0 resulted in an approximate 50 percent increase in net CO_2 assimilation by a broadleaf deciduous forest located in Tennessee, USA. Averaged over the entire daylight period, they further determined that the shift from the lower to the higher set of DRF values "enhances photosynthetic fluxes by about 30 percent at this study site." Similar results were obtained for the mixed forest and the conglomerate of crops studied. Hence, they concluded that natural variability among commonly present aerosols can "routinely influence surface irradiance and hence the terrestrial CO_2 flux and regional carbon cycle." And for these types of land-cover (forests and agricultural crops), that influence is to significantly increase the assimilation of CO_2 from the atmosphere.

In the case of grasslands, however, the effect was found to be just the opposite, with greater aerosol loading of the atmosphere leading to less CO_2 assimilation, due most likely, in the estimation of Niyogi *et al.*, to grasslands' significantly different canopy architecture. With respect to the planet as a whole, however, the net effect is decidedly positive, as earth's trees are the primary planetary players in the sequestration of carbon. Post *et al.* (1990), for example, noted that woody plants account for approximately 75 percent of terrestrial photosynthesis, which comprises about 90 percent of the global total (Sellers and McCarthy, 1990); those numbers make earth's trees and shrubs responsible for fully two-thirds (0.75 x 90 percent = 67.5 percent) of the planet's net primary production.

What is especially exciting about these real-world observations is that much of the commonly-present aerosol burden of the atmosphere is plant-derived.

Hence, it can be appreciated that earth's woody plants are themselves responsible for emitting to the air that which ultimately enhances their own photosynthetic prowess. In other words, earth's trees significantly control their own destiny; i.e., they alter the atmospheric environment in a way that directly enhances their opportunities for greater growth.

Society helps too, in this regard, for as we pump ever more CO_2 into the atmosphere, the globe's woody plants quickly respond to its aerial fertilization effect, becoming ever more productive, which leads to even more plant-derived aerosols being released to the atmosphere, which stimulates this positive feedback cycle to a still greater degree. Stated another way, earth's trees use some of the CO_2 emitted to the atmosphere by society to alter the aerial environment so as to enable them to remove even more CO_2 from the air. The end result is that earth's trees and humanity are working hand-in-hand to significantly increase the productivity of the biosphere. This is happening in spite of all other insults to the environment that work in opposition to enhanced biological activity.

In light of these several observations, it is clear that the historical and still-ongoing CO_2-induced increase in atmospheric biosols should have had, and should be continuing to have, a significant cooling effect on the planet that exerts itself by both slowing the rate of rise of the air's CO_2 content and reducing the receipt of solar radiation at the earth's surface. Neither of these effects is fully and adequately included in any general circulation model of the atmosphere of which we are aware.

Additional information on this topic, including reviews of newer publications as they become available, can be found at http://www.co2science.org/subject/f/feedbackdiffuse.php.

References

Baldocchi, D., Falge, E., Gu, L.H., Olson, R., Hollinger, D., Running, S., Anthoni, P., Bernhofer, C., Davis, K., Evans, R., Fuentes, J., Goldstein, A., Katul, G., Law, B., Lee, X.H., Malhi, Y., Meyers, T., Munger, W., Oechel, W., Paw U, K.T., Pilegaard, K., Schmid, H.P., Valentini, R., Verma, S., Vesala, T., Wilson, K. and Wofsy, S. 2001. FLUXNET: A new tool to study the temporal and spatial variability of ecosystem-scale carbon dioxide, water vapor, and energy flux densities. *Bulletin of the American Meteorological Society* **82**: 2415-2434.

Fitzjarrald, D.R., Moore, K.E., Sakai, R.K. and Freedman, J.M. 1995. Assessing the impact of cloud cover on carbon uptake in the northern boreal forest. In: Proceedings of the American Geophysical Union Meeting, Spring 1995, *EOS Supplement*, p. S125.

Goulden, M.L., Daube, B.C., Fan, S.-M., Sutton, D.J., Bazzaz, A., Munger, J.W. and Wofsy, S.C. 1997. Physiological responses of a black spruce forest to weather. *Journal of Geophysical Research* **102**: 28,987-28,996.

Gu, L., Baldocchi, D.D., Wofsy, S.C., Munger, J.W., Michalsky, J.J., Urbanski, S.P. and Boden, T.A. 2003. Response of a deciduous forest to the Mount Pinatubo eruption: Enhanced photosynthesis. *Science* **299**: 2035-2038.

Holben, B.N., Tanré, D., Smirnov, A., Eck, T.F., Slutsker, I., Abuhassan, N., Newcomb, W.W., Schafer, J.S., Chatenet, B., Lavenu, F., Kaufman, Y.J., Castle, J.V., Setzer, A., Markham, B., Clark, D., Frouin, R., Halthore, R., Karneli, A., O'Neill, N.T., Pietras, C., Pinker, R.T., Voss, K. and Zibordi, G. 2001. An emerging ground-based aerosol climatology: Aerosol Optical Depth from AERONET. *Journal of Geophysical Research* **106**: 12,067-12,097.

Hollinger, D.Y., Kelliher, F.M., Byers, J.N. and Hunt, J.E. 1994. Carbon dioxide exchange between an undisturbed old-growth temperate forest and the atmosphere. *Ecology* **75**: 134-150.

Idso, S.B. 1998. CO_2-induced global warming: a skeptic's view of potential climate change. *Climate Research* **10**: 69-82.

Law, B.E., Falge, E., Gu, L., Baldocchi, D.D., Bakwin, P., Berbigier, P., Davis, K., Dolman, A.J., Falk, M., Fuentes, J.D., Goldstein, A., Granier, A., Grelle, A., Hollinger, D., Janssens, I.A., Jarvis, P., Jensen, N.O., Katul, G., Mahli, Y., Matteucci, G., Meyers, T., Monson, R., Munger, W., Oechel, W., Olson, R., Pilegaard, K., Paw U, K.T., Thorgeirsson, H., Valentini, R., Verma, S., Vesala, T., Wilson, K. and Wofsy, S. 2002. Environmental controls over carbon dioxide and water vapor exchange of terrestrial vegetation. *Agricultural and Forest Meteorology* **113**: 97-120.

Niyogi, D., Chang, H.-I., Saxena, V.K., Holt, T., Alapaty, K., Booker, F., Chen, F., Davis, K.J., Holben, B., Matsui, T., Meyers, T., Oechel, W.C., Pielke Sr., R.A., Wells, R., Wilson, K. and Xue, Y. 2004. Direct observations of the effects of aerosol loading on net ecosystem CO_2 exchanges over different landscapes. *Geophysical Research Letters* **31**: 10.1029/2004GL020915.

Oechel, W.C. and Lawrence, W.T. 1985. Tiaga. In: Chabot, B.F. and Mooney, H.A. (Eds.) *Physiological Ecology of North American Plant Communities*. Chapman & Hall, New York, NY, pp. 66-94.

Post, W.M., Peng, T.-H., Emanuel, W.R., King, A.W., Dale, V.H. and DeAngelis, D.L. 1990. The global carbon cycle. *American Scientist* **78**: 310-326.

Roderick, M.L., Farquhar, G.D., Berry, S.L. and Noble, I.R. 2001. On the direct effect of clouds and atmospheric particles on the productivity and structure of vegetation. *Oecologia* **129**: 21-30.

Sakai, R.K., Fitzjarrald, D.R., Moore, K.E. and Freedman, J.M. 1996. How do forest surface fluxes depend on fluctuating light level? In: *Proceedings of the 22nd Conference on Agricultural and Forest Meteorology with Symposium on Fire and Forest Meteorology*, Vol. 22, American Meteorological Society, pp. 90-93.

Sarmiento, J.L. 1993. Atmospheric CO_2 stalled. *Nature* **365**: 697-698.

Sellers, P. and McCarthy, J.J. 1990. Planet Earth, Part III, Biosphere. *EOS: Transactions of the American Geophysical Union* **71**: 1883-1884.

2.4. Iodocompounds

The climatic significance of iodinated compounds or iodocompounds was first described in the pages of *Nature* by O'Dowd *et al.* (2002). As related by Kolb (2002) in an accompanying perspective on their work, the 10-member research team discovered "a previously unrecognized source of aerosol particles" by unraveling "a photochemical phenomenon that occurs in sea air and produces aerosol particles composed largely of iodine oxides." Specifically, the team used a smog chamber operated under coastal atmospheric conditions to demonstrate, as they report, that "new particles can form from condensable iodine-containing vapors, which are the photolysis products of biogenic iodocarbons emitted from marine algae." With the help of aerosol formation models, they also demonstrated that concentrations of condensable iodine-containing vapors over the open ocean "are sufficient to influence marine particle formation."

The significance of this work is that the aerosol particles O'Dowd *et al.* discovered can function as cloud condensation nuclei (CCN), helping to create new clouds that reflect more incoming solar radiation back to space and thereby cool the planet (a negative feedback). With respect to the negative feedback nature of this phenomenon, O'Dowd *et al.* cite the work of Laturnus *et al.* (2000), which demonstrates that emissions of iodocarbons from marine biota "can increase by up to 5 times as a result of changes in

environmental conditions associated with global change." Therefore, as O'Dowd *et al.* continue, "increasing the source rate of condensable iodine vapors will result in an increase in marine aerosol and CCN concentrations of the order of 20—60 percent." Furthermore, they note that "changes in cloud albedo resulting from changes in CCN concentrations of this magnitude can lead to an increase in global radiative forcing similar in magnitude, but opposite in sign, to the forcing induced by greenhouse gases."

Four years later, Smythe-Wright *et al.* (2006) measured trace gas and pigment concentrations in seawater, while identifying and enumerating picophytoprokaryotes during two ship cruises in the Atlantic Ocean and one in the Indian Ocean, where they focused "on methyl iodide production and the importance of a biologically related source." In doing so, they encountered methyl iodide concentrations as great as 45 pmol per liter in the top 150 meters of the oceanic water column that correlated well with the abundance of *Prochlorococcus*, which they report "can account for >80 percent of the variability in the methyl iodide concentrations." They add that they "have confirmed the release of methyl iodide by this species in laboratory culture experiments."

Extrapolating their findings to the globe as a whole, the six researchers "estimate the global ocean flux of iodine [I] to the marine boundary layer from this single source to be 5.3×10^{11} g I year^{-1}," which they say "is a large fraction of the total estimated global flux of iodine (10^{11}-10^{12} g I year^{-1})." This observation is extremely important, because volatile iodinated compounds, in Smythe-Wright *et al.*'s words, "play a part in the formation of new particles and cloud condensation nuclei (CCN)," and because "an increase in the production of iodocompounds and the subsequent production of CCN would potentially result in a net cooling of the earth system and hence in a negative climate feedback mechanism, mitigating global warming." More specifically, they suggest that "as ocean waters become warmer and more stratified, nutrient concentrations will fall and there will likely be a regime shift away from microalgae toward *Prochlorococcus*," such that "colonization within the <50° latitude band will result in a ~15 percent increase in the release of iodine to the atmosphere," with consequent "important implications for global climate change," which, as previously noted, tend to counteract global warming.

Most recently, as part of the Third Pelagic Ecosystem CO_2 Enrichment Study, Wingenter *et al.* (2007) investigated the effects of atmospheric CO_2

enrichment on marine microorganisms in nine marine mesocosms maintained within two-meter-diameter polyethylene bags submerged to a depth of 10 meters in a fjord at the Large-Scale Facilities of the Biological Station of the University of Bergen in Espegrend, Norway. Three of these mesocosms were maintained at ambient levels of CO_2 (~375 ppm or base CO_2), three were maintained at levels expected to prevail at the end of the current century (760 ppm or $2xCO_2$), and three were maintained at levels predicted for the middle of the next century (1150 ppm or $3xCO_2$). During the 25 days of this experiment, the researchers followed the development and subsequent decline of an induced bloom of the coccolithophorid *Emiliania huxleyi*, carefully measuring several physical, chemical, and biological parameters along the way. This work revealed that the iodocarbon chloroiodomethane (CH_2CII) experienced its peak concentration about six to 10 days after the coccolithophorid's chlorophyll-a maximum, and that its estimated abundance was 46 percent higher in the $2xCO_2$ mesocosms and 131 percent higher in the $3xCO_2$ mesocosms.

The international team of scientists concluded that the differences in the CH_2CII concentrations "may be viewed as a result of changes to the ecosystems as a whole brought on by the CO_2 perturbations." And because emissions of various iodocarbons have been found to lead to an enhancement of cloud condensation nuclei in the marine atmosphere, as demonstrated by O'Dowd *et al.* (2002) and Jimenez *et al.* (2003), it can be appreciated that the CO_2-induced stimulation of the marine emissions of these substances provides a natural brake on the tendency for global warming to occur as a consequence of any forcing, as iodocarbons lead to the creation of more highly reflective clouds over greater areas of the world's oceans.

In conclusion, as Wingenter *et al.* sum things up, the processes described above "may help contribute to the homeostasis of the planet." And the finding of O'Dowd *et al.* that changes in cloud albedo "associated with global change" can lead to an increase in global radiative forcing that is "similar in magnitude, but opposite in sign, to the forcing induced by greenhouse gases," suggests that CO_2-induced increases in marine iodocarbon emissions likely contribute to maintaining that homeostasis.

Additional information on this topic, including reviews of newer publications as they become available, can be found at http://www.co2science.org/subject/f/feedbackiodo.php.

References

Jimenez, J.L., Bahreini, R., Cocker III, D.R., Zhuang, H., Varutbangkul, V., Flagan, R.C., Seinfeld, J.H., O'Dowd, C.D. and Hoffmann, T. 2003. New particle formation from photooxidation of diiodomethane (CH_2I_2). *Journal of Geophysical Research* 108: 10.1029/2002JD002452.

Kolb, C.E. 2002. Iodine's air of importance. *Nature* **417**: 597-598.

Laturnus, F., Giese, B., Wiencke, C. and Adams, F.C. 2000. Low-molecular-weight organoiodine and organobromine compounds released by polar macroalgae— The influence of abiotic factors. *Fresenius' Journal of Analytical Chemistry* **368**: 297-302.

O'Dowd, C.D., Jimenez, J.L., Bahreini, R., Flagan, R.C., Seinfeld, J.H., Hameri, K., Pirjola, L., Kulmala, M., Jennings, S.G. and Hoffmann, T. 2002. Marine aerosol formation from biogenic iodine emissions. *Nature* **417**: 632-636.

Smythe-Wright, D., Boswell, S.M., Breithaupt, P., Davidson, R.D., Dimmer, C.H. and Eiras Diaz, L.B. 2006. Methyl iodide production in the ocean: Implications for climate change. *Global Biogeochemical Cycles* **20**: 10.1029/2005GB002642.

Wingenter, O.W., Haase, K.B., Zeigler, M., Blake, D.R., Rowland, F.S., Sive, B.C., Paulino, A., Thyrhaug, R., Larsen, A., Schulz, K., Meyerhofer, M. and Riebesell, U. 2007. Unexpected consequences of increasing CO_2 and ocean acidity on marine production of DMS and CH_2CII: Potential climate impacts. *Geophysical Research Letters* **34**: 10.1029/2006GL028139.

2.5. Nitrous Oxide

One of the main sources of nitrous oxide (N_2O) is agriculture, which accounts for almost half of N_2O emissions in some countries (Pipatti, 1997). With N_2O originating from microbial N cycling in soil— mostly from aerobic nitrification or from anaerobic denitrification (Firestone and Davidson, 1989)—there is a concern that CO_2-induced increases in carbon input to soil, together with increasing N input from other sources, will increase substrate availability for denitrifying bacteria and may result in higher N_2O emissions from agricultural soils as the air's CO_2 content continues to rise.

In a study designed to investigate this possibility, Kettunen *et al.* (2007a) grew mixed stands of timothy (*Phleum pratense*) and red clover (*Trifolium pratense*) in sandy-loam-filled mesocosms at low and moderate soil nitrogen levels within greenhouses maintained at either 360 or 720 ppm CO_2, while measuring harvestable biomass production and N_2O evolution from the mesocosm soils over the course of three crop cuttings. This work revealed that the total harvestable biomass production of *P. pratense* was enhanced by the experimental doubling of the air's CO_2 concentration by 21 percent and 26 percent, respectively, in the low and moderate soil N treatments, while corresponding biomass enhancements for *T. pratense* were 22 percent and 18 percent. In addition, the researchers found that after emergence of the mixed stand and during vegetative growth before the first harvest and N fertilization, N_2O fluxes were higher under ambient CO_2 in both the low and moderate soil N treatments. In fact, it was not until the water table had been raised and extra fertilization given after the first harvest that the elevated CO_2 seemed to increase N_2O fluxes. The four Finnish researchers thus concluded that the mixed stand of *P. pratense* and *T. pratense* was "able to utilize the increased supply of atmospheric CO_2 for enhanced biomass production without a simultaneous increase in the N_2O fluxes," thereby raising "the possibility of maintaining N_2O emissions at their current level, while still enhancing the yield production [via the aerial fertilization effect of elevated CO_2] even under low N fertilizer additions."

In a similar study, Kettunen *et al.* (2007b) grew timothy (*Phleum pratense*) in monoculture within sandy-soil-filled mesocosms located within greenhouses maintained at atmospheric CO_2 concentrations of either 360 or 720 ppm for a period of 3.5 months at moderate (standard), low (half-standard), and high (1.5 times standard) soil N supply, while they measured the evolution of N_2O from the mesocosms, vegetative net CO_2 exchange, and final above- and below-ground biomass production over the course of three harvests. In this experiment the elevated CO_2 concentration increased the net CO_2 exchange of the ecosystems (which phenomenon was primarily driven by CO_2-induced increases in photosynthesis) by about 30 percent, 46 percent and 34 percent at the low, moderate, and high soil N levels, respectively, while it increased the above-ground biomass of the crop by about 8 percent, 14 percent, and 8 percent at the low, moderate and high soil N levels, and its below-ground biomass by 28 percent, 27 percent, and 41 percent at the same respective soil N levels. And once again, Kettunen *et al.* report that "an explicit increase in N_2O fluxes due

to the elevated atmospheric CO_2 concentration was not found."

Welzmiller *et al.* (2008) measured N_2O and denitrification emission rates in a C_4 sorghum [*Sorghum bicolor* (L.) Moench] production system with ample and limited flood irrigation rates under Free-Air CO_2 Enrichment (seasonal mean = 579 ppm) and control (seasonal mean = 396 ppm) CO_2 during the 1998 and 1999 summer growing seasons at the experimental FACE site near Maricopa, Arizona (USA). The study found "elevated CO_2 did not result in increased N_2O or N-gas emissions with either ample or limited irrigation," which findings they describe as being "consistent with findings for unirrigated western U.S. ecosystems reported by Billings *et al.* (2002) for Mojave Desert soils and by Mosier *et al.* (2002) for Colorado shortgrass steppe."

In discussing the implications of their findings, Welzmiller *et al.* say their results suggest that "as CO_2 concentrations increase, there will not be major increases in denitrification in C_4 cropping environments such as irrigated sorghum in the desert southwestern United States," which further suggests there will not be an increased impetus for global warming due to this phenomenon.

In a different type of study—driven by the possibility that the climate of the Amazon Basin may gradually become drier due to a warming-induced increase in the frequency and/or intensity of El Niño events that have historically brought severe drought to the region—Davidson *et al.* (2004) devised an experiment to determine the consequences of the drying of the soil of an Amazonian moist tropical forest for the net surface-to-air fluxes of both N_2O and methane (CH_4). This they did in the Tapajos National Forest near Santarem, Brazil, by modifying a one-hectare plot of land covered by mature evergreen trees so as to dramatically reduce the amount of rain that reached the forest floor (throughfall), while maintaining an otherwise similar one-hectare plot of land as a control for comparison.

Prior to making this modification, the three researchers measured the gas exchange characteristics of the two plots for a period of 18 months; then, after initiating the throughfall-exclusion treatment, they continued their measurements for an additional three years. This work revealed that the "drier soil conditions caused by throughfall exclusion inhibited N_2O and CH_4 production and promoted CH_4 consumption." In fact, they report that "the exclusion manipulation lowered annual N_2O emissions by >40 percent and increased rates of consumption of atmospheric CH_4 by a factor of >4," which results they attributed to the "direct effect of soil aeration on denitrification, methanogenesis, and methanotrophy."

Consequently, if global warming would indeed increase the frequency and/or intensity of El Niño events as some claim it will, the results of this study suggest that the anticipated drying of the Amazon Basin would initiate a strong negative feedback via (1) large drying-induced reductions in the evolution of both N_2O and CH_4 from its soils, and (2) a huge drying-induced increase in the consumption of CH_4 by its soils. Although Davidson et al. envisage a more extreme second phase response "in which drought-induced plant mortality is followed by increased mineralization of C and N substrates from dead fine roots and by increased foraging of termites on dead coarse roots" (an extreme response that would be expected to increase N_2O and CH_4 emissions), we note that the projected rise in the air's CO_2 content would likely prohibit such a thing from ever occurring, due to the documented tendency for atmospheric CO_2 enrichment to greatly increase the water use efficiency of essentially all plants, which would enable the forest to continue to flourish under significantly drier conditions than those of the present.

In summation, it would appear that concerns about additional global warming arising from enhanced N_2O emissions from agricultural soils in a CO_2-enriched atmosphere of the future are not well founded.

Additional information on this topic, including reviews of newer publications as they become available, can be found at http://www.co2science.org/subject/n/nitrousoxide.php

References

Billings, S.A., Schaeffer, S.M. and Evans, R.D. 2002. Trace N gas losses and mineralization in Mojave Desert soils exposed to elevated CO_2. *Soil Biology and Biochemistry* **34**: 1777-1784.

Davidson, E.A., Ishida, F.Y. and Nepstad, D.C. 2004. Effects of an experimental drought on soil emissions of carbon dioxide, methane, nitrous oxide, and nitric oxide in a moist tropical forest. *Global Change Biology* **10**: 718-730.

Firestone, M.K. and Davidson, E.A. 1989. Microbiological basis of NO and N_2O production and consumption in soil. In: Andreae, M.O. and Schimel, D.S. (Eds.) *Exchange of Trace Gases Between Terrestrial Ecosystems and the Atmosphere*. Wiley, Chichester, pp. 7-21.

Kettunen, R., Saarnio, S., Martikainen, P.J. and Silvola, J. 2007a. Can a mixed stand of N_2-fixing and non-fixing plants restrict N_2O emissions with increasing CO_2 concentration? *Soil Biology & Biochemistry* **39**: 2538-2546.

Kettunen, R., Saarnio, S. and Silvola, J. 2007b. N_2O fluxes and CO_2 exchange at different N doses under elevated CO_2 concentration in boreal agricultural mineral soil under *Phleum pratense*. *Nutrient Cycling in Agroecosystems* **78**: 197-209.

Mosier, A.R., Morgan, J.A., King, J.Y., LeCain, D. and Milchunas, D.G. 2002. Soil-atmosphere exchange of CH_4, CO_2, NO_X, and N_2O in the Colorado shortgrass steppe under elevated CO_2. *Plant and Soil* **240**: 201-211.

Pipatti, R. 1997. Suomen metaani-ja dityppioksidipäästöjen rajoittamisen mahdollisuudet ja kustannustehokkuus. VTT tiedotteita. 1835, Espoo, 62 pp.

Welzmiller, J.T., Matthias, A.D., White, S. and Thompson, T.L. 2008. Elevated carbon dioxide and irrigation effects on soil nitrogen gas exchange in irrigated sorghum. *Soil Science Society of America Journal* **72**: 393-401.

2.6. Methane

What impact do global warming, the ongoing rise in the air's carbon dioxide (CO_2) content and a number of other contemporary environmental trends have on the atmosphere's methane (CH_4) concentration? The implications of this question are huge because methane is a more powerful greenhouse gas, molecule for molecule, than is carbon dioxide. Its atmospheric concentration is determined by the difference between how much CH_4 goes into the air (emissions) and how much comes out of it (extractions) over the same time period. There are significant forces at play that will likely produce a large negative feedback toward the future warming potential of this powerful greenhouse gas, nearly all of which forces are ignored by the IPCC.

2.6.1. Extraction

Early indications that atmospheric CO_2 enrichment might significantly reduce methane emissions associated with the production of rice were provided by Schrope et al. (1999), who studied batches of rice growing in large vats filled with topsoil and placed within greenhouse tunnels maintained at atmospheric CO_2 concentrations of 350 and 700 ppm, each of which tunnels was further subdivided into four sections that provided temperature treatments ranging from ambient to as much as 5°C above ambient. As would be expected, doubling the air's CO_2 content significantly enhanced rice biomass production in this system, increasing it by up to 35 percent above-ground and by up to 83 percent below-ground. However, in a truly unanticipated development, methane emissions from the rice grown at 700 ppm CO_2 were found to be 10 to 45 times less than emissions from the plants grown at 350 ppm. As Schrope et al. describe it, "the results of this study did not support our hypothesis that an effect of both increased carbon dioxide and temperature would be an increase in methane emissions." Indeed, they report that "both increased carbon dioxide and increased temperatures were observed to produce decreased methane emissions," except for the first 2°C increase above ambient, which produced a slight increase in methane evolution from the plant-soil system.

In checking for potential problems with their experiment, Schrope et al. could find none. They thus stated that their results "unequivocally support the conclusion that, during this study, methane emissions from *Oryza sativa* [rice] plants grown under conditions of elevated CO_2 were dramatically reduced relative to plants gown in comparable conditions under ambient levels of CO_2," and to be doubly sure of this fact, they went on to replicate their experiment in a second year of sampling and obtained essentially the same results. Four years later, however, a study of the same phenomenon by a different set of scientists yielded a different result in a different set of circumstances.

Inubushi et al. (2003) grew a different cultivar of rice in 1999 and 2000 in paddy culture at Shizukuishi, Iwate, Japan in a FACE study where the air's CO_2 concentration was increased 200 ppm above ambient. They found that the extra CO_2 "significantly increased the CH_4 [methane] emissions by 38 percent in 1999 and 51 percent in 2000," which phenomenon they attributed to "accelerated CH_4 production as a result of increased root exudates and root autolysis products and to the increased plant-mediated CH_4 emission because of the higher rice tiller numbers under FACE conditions." With such a dramatically different result from that of Schrope et al., many more studies likely will be required to determine which of these results is the more typical of rice culture around the world.

A somewhat related study was conducted by Kruger and Frenzel (2003), who note that "rice paddies contribute approximately 10-13 percent to the global CH_4 emission (Neue, 1997; Crutzen and Lelieveld, 2001)," and that "during the next 30 years rice production has to be increased by at least 60 percent to meet the demands of the growing human population (Cassman et al., 1998)." Because of these facts they further note that "increasing amounts of fertilizer will have to be applied to maximize yields [and] there is ongoing discussion on the possible effects of fertilization on CH_4 emissions."

To help promote that discussion, Kruger and Frenzel investigated the effects of N-fertilizer (urea) on CH_4 emission, production, and oxidation in rice culture in laboratory, microcosm and field experiments they conducted at the Italian Rice Research Institute in northern Italy. They report that in some prior studies "N-fertilisation stimulated CH_4 emissions (Cicerone and Shetter, 1981; Banik et al., 1996; Singh et al., 1996)," while "methanogenesis and CH_4 emission was found to be inhibited in others (Cai et al., 1997; Schutz et al., 1989; Lindau et al., 1990)," similar to the polarized findings of Schrope et al. and Inubushi et al. with respect to the effects of elevated CO_2 on methane emissions. In the mean, therefore, there may well be little to no change in overall CH_4 emissions from rice fields in response to both elevated CO_2 and increased N-fertilization. With respect to their own study, for example, Kruger and Frenzel say that "combining our field, microcosm and laboratory experiments we conclude that any agricultural praxis improving the N-supply to the rice plants will also be favourable for the CH_4 oxidising bacteria," noting that "N-fertilisation had only a transient influence and was counter-balanced in the field by an elevated CH_4 production." The implication of these findings is well articulated in the concluding sentence of their paper: "neither positive nor negative consequences for the overall global warming potential could be found."

Another agricultural source of methane is the fermentation of feed in the rumen of cattle and sheep. Fievez et al. (2003) studied the effects of various types and levels of fish-oil feed additives on this process by means of both *in vitro* and *in vivo* experiments with sheep, observing a maximal 80 percent decline in the ruminants' production of methane when using fish-oil additives containing n-3-eicosapentanoic acid. With respect to cattle, Boadi et al. (2004) report that existing mitigation strategies for reducing CH_4 emissions from dairy cows include the addition of ionophores and fats to their food, as well as the use of high-quality forages and grains in their diet, while newer mitigation strategies include "the addition of probiotics, acetogens, bacteriocins, archaeal viruses, organic acids, [and] plant extracts (e.g., essential oils) to the diet, as well as immunization, and genetic selection of cows." To this end, they provide a table of 20 such strategies, where the average maximum potential CH_4 reduction that may result from the implementation of each strategy is 30 percent or more.

With as many as 20 different mitigation strategies from which to choose, each one of which (on average) has the potential to reduce CH_4 emissions from dairy cows by as much as a third, it would appear there is a tremendous potential to dramatically curtail the amount of CH_4 released to the atmosphere by these ruminants and, by implication, the host of other ruminants that mankind raises and uses for various purposes around the world. Such high-efficiency approaches to reducing the strength of the atmosphere's greenhouse effect, while not reducing the biological benefits of elevated atmospheric CO_2 concentrations in the process, should be at the top of any program designed to achieve that difficult (but still highly questionable) objective.

In view of these several observations, we can be cautiously optimistic about our agricultural intervention capabilities and their capacity to help stem the tide of earth's historically rising atmospheric methane concentration, which could take a huge bite out of methane-induced global warming. But do methane emissions from natural vegetation respond in a similar way?

We have already discussed the results of Davidson et al. (2004) in our Nitrous Oxide section, which results suggest that a global warming-induced drying of the Amazon Basin would initiate a strong negative feedback to warming via (1) large drying-induced reductions in the evolution of N_2O and CH_4 from its soils and (2) a huge drying-induced increase in the consumption of CH_4 by its soils. In a contemporaneous study, Strack et al. (2004) also reported that climate models predict increases in evapotranspiration that could lead to drying in a warming world and a subsequent lowering of water tables in high northern latitudes. This prediction cries out for an analysis of how lowered water tables will affect peatland emissions of CH_4.

In a theoretical study of the subject, Roulet et al. (1992) calculated that for a decline of 14 cm in the water tables of northern Canadian peatlands, due to

climate-model-derived increases in temperature (3°C) and precipitation (1mm/day) predicted for a doubling of the air's CO_2 content, CH_4 emissions would decline by 74-81 percent. In an attempt to obtain some experimental data on the subject, at various times over the period 2001-2003 Strack et al. measured CH_4 fluxes to the atmosphere at different locations that varied in depth-to-water table within natural portions of a poor fen in central Quebec, Canada, as well as within control portions of the fen that had been drained eight years earlier.

At the conclusion of their study, the Canadian scientists reported that "methane emissions and storage were lower in the drained fen." The greatest reductions (up to 97 percent) were measured at the higher locations, while at the lower locations there was little change in CH_4 flux. Averaged over all locations, they determined that the "growing season CH_4 emissions at the drained site were 55 percent lower than the control site," indicative of the fact that the biosphere appears to be organized to resist warming influences that could push it into a thermal regime that might otherwise prove detrimental to its health.

In another experimental study, Garnet et al. (2005) grew seedlings of three emergent aquatic macrophytes (*Orontium aquaticum* L., *Peltandra virginica* L., and *Juncus effusus* L.) plus one coniferous tree (*Taxodium distichum* L.), all of which are native to eastern North America, in a five-to-one mixture of well-fertilized mineral soil and peat moss in pots submerged in water in tubs located within controlled environment chambers for a period of eight weeks. Concomitantly, they measured the amount of CH_4 emitted by the plant foliage, along with net CO_2 assimilation rate and stomatal conductance, which were made to vary by changing the CO_2 concentration of the air surrounding the plants and the density of the photosynthetic photon flux impinging on them.

Methane emissions from the four wetland species increased linearly with increases in both stomatal conductance and net CO_2 assimilation rate; but the researchers found that changes in stomatal conductance affected foliage methane flux "three times more than equivalent changes in net CO_2 assimilation," making stomatal conductance the more significant of the two CH_4 emission-controllers. In addition, they note that evidence of stomatal control of CH_4 emission has also been reported for *Typha latifolia* (Knapp and Yavitt, 1995) and *Carex* (Morrissey et al., 1993), two other important wetland plants. Hence, since atmospheric CO_2 enrichment leads to approximately equivalent—but oppositely directed—changes in foliar net CO_2 assimilation (which is increased) and stomatal conductance (which is reduced) in most herbaceous plants (which are the type that comprise most wetlands), it can be appreciated that the ongoing rise in the air's CO_2 content should be acting to *reduce* methane emissions from earth's wetland vegetation, because of the three-times-greater negative CH_4 emission impact of the decrease in stomatal conductance compared to the positive CH_4 emission impact of the equivalent increase in net CO_2 assimilation.

According to Prinn *et al.* (1992), one of the major means by which methane is removed from the atmosphere is via oxidation by methanotrophic bacteria in the aerobic zones of soils, the magnitude of which phenomenon is believed to be equivalent to the annual input of methane to the atmosphere (Watson *et al.*, 1992). This soil sink for methane appears to be ubiquitous, as methane uptake has been observed in soils of tundra (Whalen and Reeburgh, 1990), boreal forests (Whalen *et al.*, 1992), temperate forests (Steudler *et al.*, 1989; Yavitt *et al.*, 1990), grasslands (Mosier *et al.*, 1997), arable lands (Jensen and Olsen, 1998), tropical forests (Keller, 1986; Singh *et al.*, 1997), and deserts (Striegl *et al.*, 1992), with forest soils—especially boreal and temperate forest upland soils (Whalen and Reeburgh, 1996)—appearing to be the most efficient in this regard (Le Mer and Roger, 2001).

In an attempt to learn more about this subject, Tamai *et al.* (2003) studied methane uptake rates by the soils of three Japanese cypress plantations composed of 30- to 40-year-old trees. Through all seasons of the year, they found that methane was absorbed by the soils of all three sites, being positively correlated with temperature, as has also been observed in several other studies (Peterjohn *et al.*, 1994; Dobbie and Smith, 1996); Prieme and Christensen, 1997; Saari *et al.*, 1998). Methane absorption was additionally—and even more strongly—positively correlated with the C/N ratio of the cypress plantations' soil organic matter. Based on these results, it can be appreciated that any global warming, CO_2-induced or natural, would produce two biologically mediated negative feedbacks to counter the increase in temperature: (1) a warming-induced increase in methane uptake from the atmosphere that is experienced by essentially all soils, and (2) an increase in soil methane uptake from the atmosphere that is produced by the increase in plant-litter C/N

ratio that typically results from atmospheric CO_2 enrichment.

Another study that deals with this topic is that of Menyailo and Hungate (2003), who assessed the influence of six boreal forest species—spruce, birch, Scots pine, aspen, larch, and Arolla pine—on soil CH_4 consumption in the Siberian artificial afforestation experiment, in which the six common boreal tree species had been grown under common garden conditions for the past 30 years under the watchful eye of the staff of the Laboratory of Soil Science of the Institute of Forest, Siberian Branch of the Russian Academy of Sciences (Menyailo *et al.*, 2002). They determined, in their words, that "soils under hardwood species (aspen and birch) consumed CH_4 at higher rates than soils under coniferous species and grassland." Under low soil moisture conditions, for example, the soils under the two hardwood species consumed 35 percent more CH_4 than the soils under the four conifers; under high soil moisture conditions they consumed 65 percent more. As for the implications of these findings, Pastor and Post (1988) have suggested, in the words of Menyailo and Hungate, that "changes in temperature and precipitation resulting from increasing atmospheric CO_2 concentrations will cause a northward migration of the hardwood-conifer forest border in North America." Consequently, if such a shifting of species does indeed occur, it will likely lead to an increase in methane consumption by soils and a reduction in methane-induced global warming potential, thereby providing yet another biologically mediated negative feedback factor that has yet to be incorporated into models of global climate change.

Last, we note that increases in the air's CO_2 concentration will likely lead to a net reduction in vegetative isoprene emissions, which, as explained in Section 7.7.1. under the heading of Isoprene, should also lead to a significant removal of methane from the atmosphere. Hence, as the air's CO_2 content—and possibly its temperature—continues to rise, we can expect to see a significant increase in the rate of methane removal from earth's atmosphere, which should help to reduce the potential for further global warming.

Additional information on this topic, including reviews of newer publications as they become available, can be found at http://www.co2science.org/subject/m/methaneextract.php, http://www.co2science.org/subject/m/methaneag.php, and http://www.co2science.org/subject/m/methagnatural.php.

References

Banik, A., Sen, M. and Sen, S.P. 1996. Effects of inorganic fertilizers and micronutrients on methane production from wetland rice (*Oryza sativa* L.). *Biology and Fertility of Soils* **21**: 319-322.

Boadi, D., Benchaar, C., Chiquette, J. and Masse, D. 2004. Mitigation strategies to reduce enteric methane emissions from dairy cows: Update review. *Canadian Journal of Animal Science* **84**: 319-335.

Cai, Z., Xing, G., Yan, X., Xu, H., Tsuruta, H., Yogi, K. and Minami, K. 1997. Methane and nitrous oxide emissions from rice paddy fields as affected by nitrogen fertilizers and water management. *Plant and Soil* **196**: 7-14.

Cassman, K.G., Peng, S., Olk, D.C., Ladha, J.K., Reichardt, W., Doberman, A. and Singh, U. 1998. Opportunities for increased nitrogen-use efficiency from improved resource management in irrigated rice systems. *Field Crops Research* **56**: 7-39.

Cicerone, R.J. and Shetter, J.D. 1981. Sources of atmospheric methane. Measurements in rice paddies and a discussion. *Journal of Geophysical Research* **86**: 7203-7209.

Crutzen, P.J. and Lelieveld, J. 2001. Human impacts on atmospheric chemistry. *Annual Review of Earth and Planetary Sciences* **29**: 17-45.

Davidson, E.A., Ishida, F.Y. and Nepstad, D.C. 2004. Effects of an experimental drought on soil emissions of carbon dioxide, methane, nitrous oxide, and nitric oxide in a moist tropical forest. *Global Change Biology* **10**: 718-730.

Dobbie, K.E. and Smith, K.A. 1996. Comparison of CH_4 oxidation rates in woodland, arable and set aside soils. *Soil Biology & Biochemistry* **28**: 1357-1365.

Fievez, V., Dohme, F., Danneels, M., Raes, K. and Demeyer, D. 2003. Fish oils as potent rumen methane inhibitors and associated effects on rumen fermentation in vitro and in vivo. *Animal Feed Science and Technology* **104**: 41-58.

Garnet, K.N., Megonigal, J.P., Litchfield, C. and Taylor Jr., G.E. 2005. Physiological control of leaf methane emission from wetland plants. *Aquatic Botany* **81**: 141-155.

Inubushi, K., Cheng, W., Aonuma, S., Hoque, M.M., Kobayashi, K., Miura, S., Kim, H.Y. and Okada, M. 2003. Effects of free-air CO_2 enrichment (FACE) on CH_4 emission from a rice paddy field. *Global Change Biology* **9**: 1458-1464.

Jensen, S. and Olsen, R.A. 1998. Atmospheric methane consumption in adjacent arable and forest soil systems. *Soil Biology & Biochemistry* **30**: 1187-1193.

Keller, M. 1986. Emissions of N_2O, CH_4, and CO_2 from tropical forest soils. *Journal of Geophysical Research* **91**: 11,791-11,802.

Knapp, A.K. and Yavitt, J.B. 1995. Gas exchange characteristics of *Typha latifolia* L. from nine sites across North America. *Aquatic Botany* **49**: 203-215.

Kruger, M. and Frenzel, P. 2003. Effects of N-fertilisation on CH_4 oxidation and production, and consequences for CH_4 emissions from microcosms and rice fields. *Global Change Biology* **9**: 773-784.

Le Mer, J. and Roger, P. 2001. Production, oxidation, emission and consumption of methane by soils: a review. *European Journal of Soil Biology* **37**: 25-50.

Lindau, C.W., DeLaune, R.D., Patrick Jr., W.H. *et al.* 1990. Fertilizer effects on dinitrogen, nitrous oxide, and methane emission from lowland rice. *Soil Science Society of America Journal* **54**: 1789-1794.

Menyailo, O.V. and Hungate, B.A. 2003. Interactive effects of tree species and soil moisture on methane consumption. *Soil Biology & Biochemistry* **35**: 625-628.

Menyailo, O.V., Hungate, B.A. and Zech, W. 2002. Tree species mediated soil chemical changes in a Siberian artificial afforestation experiment. *Plant and Soil* **242**: 171-182.

Morrissey, L.A., Zobel, D. and Livingston, G.P. 1993. Significance of stomatal control of methane release from *Carex*-dominated wetlands. *Chemosphere* **26**: 339-356.

Mosier, A.R., Parton, W.J., Valentine, D.W., Ojima, D.S., Schimel, D.S. and Heinemeyer, O. 1997. CH_4 and N_2O fluxes in the Colorado shortgrass steppe. 2. Long-term impact of land use change. *Global Biogeochemical Cycles* **11**: 29-42.

Nepstad, D.C., Verissimo, A., Alencar, A., Nobre, C., Lima, E., Lefebvre, P., Schlesinger, P., Potter, C., Moutinho, P., Mendoza, E., Cochrane, M. and Brooks, V. 1999. Large-scale impoverishment of Amazonian forests by logging and fire. *Nature* **398**: 505-508.

Neue, H.U. 1997. Fluxes of methane from rice fields and potential for mitigation. *Soil Use and Management* **13**: 258-267.

Pastor, J. and Post, W.M. 1988. Response of northern forests to CO_2-induced climate change. *Nature* **334**: 55-58.

Peterjohn, W.T., Melillo, J.M., Steudler, P.A. and Newkirk, K.M. 1994. Responses of trace gas fluxes and N availability to experimentally elevated soil temperatures. *Ecological Applications* **4**: 617-625.

Prieme, A. and Christensen, S. 1997. Seasonal and spatial variation of methane oxidation in a Danish spruce forest. *Soil Biology & Biochemistry* **29**: 1165-1172.

Prinn, R., Cunnold, D., Simmonds, P., Alyea, F., Boldi, R., Crawford, A., Fraser, P., Gutzler, D., Hartley, D., Rosen, R. and Rasmussen, R. 1992. Global average concentration and trend for hydroxyl radicals deduced from ALE/GAGE trichloroethane (methyl chloroform) data for 1978-1990. *Journal of Geophysical Research* **97**: 2445-2461.

Roulet, N., Moore, T., Bubier, J. and Lafleur, P. 1992. Northern fens: Methane flux and climatic change. *Tellus Series B* **44**: 100-105.

Saari, A., Heiskanen, J., Martikainen, P.J. 1998. Effect of the organic horizon on methane oxidation and uptake in soil of a boreal Scots pine forest. *FEMS Microbiology Ecology* **26**: 245-255.

Schrope, M.K., Chanton, J.P., Allen, L.H. and Baker, J.T. 1999. Effect of CO_2 enrichment and elevated temperature on methane emissions from rice, *Oryza sativa*. *Global Change Biology* **5**: 587-599.

Schutz, H., Holzapfel-Pschorrn, A., Conrad, R. *et al.* 1989. A 3-year continuous record on the influence of daytime, season, and fertilizer treatment on methane emission rates from an Italian rice paddy. *Journal of Geophysical Research* **94**: 16405-16416.

Singh, J.S., Singh, S., Raghubanshi, A.S. *et al.* 1996. Methane flux from rice/wheat agroecosystem as affected by crop phenology, fertilization and water level. *Plant and Soil* **183**: 323-327.

Singh, J.S., Singh, S., Raghubanshi, A.S., Singh, S., Kashyap, A.K. and Reddy, V.S. 1997. Effect of soil nitrogen, carbon and moisture on methane uptake by dry tropical forest soils. *Plant and Soil* **196**: 115-121.

Steudler, P.A., Bowden, R.D., Meillo, J.M. and Aber, J.D. 1989. Influence of nitrogen fertilization on CH_4 uptake in temperate forest soils. *Nature* **341**: 314-316.

Strack, M., Waddington, J.M. and Tuittila, E.-S. 2004. Effect of water table drawdown on northern peatland methane dynamics: Implications for climate change. *Global Biogeochemical Cycles* **18**: 10.1029/2003GB002209.

Striegl, R.G., McConnaughey, T.A., Thorstensen, D.C., Weeks, E.P. and Woodward, J.C. 1992. Consumption of atmospheric methane by desert soils. *Nature* **357**: 145-147.

Tamai, N., Takenaka, C., Ishizuka, S. and Tezuka, T. 2003. Methane flux and regulatory variables in soils of three equal-aged Japanese cypress (Chamaecyparis obtusa) forests in central Japan. *Soil Biology & Biochemistry* **35**: 633-641.

Watson, R.T., Meira Filho, L.G., Sanhueza, E. and Janetos, A. 1992. Sources and sinks. In: Houghton, J.T., Callander, B.A. and Varney, S.K. (Eds.), *Climate Change 1992: The Supplementary Report to The IPCC Scientific Assessment*, Cambridge University Press, Cambridge, UK, pp. 25-46.

Whalen, S.C. and Reeburgh, W.S. 1990. Consumption of atmospheric methane by tundra soils. *Nature* **346**: 160-162.

Whalen, S.C. and Reeburgh, W.S. 1996. Moisture and temperature sensitivity of CH_4 oxidation in boreal soils. *Soil Biology & Biochemistry* **28**: 1271-1281.

Whalen, S.C., Reeburgh, W.S. and Barber, V.A. 1992. Oxidation of methane in boreal forest soils: a comparison of seven measures. *Biogeochemistry* **16**: 181-211.

Yavitt, J.B., Downey, D.M., Lang, D.E. and Sextone, A.J. 1990. CH_4 consumption in two temperate forest soils. *Biogeochemistry* **9**: 39-52.

2.6.2. Concentrations

In Section 2.6.1, we reported on several real-world phenomena that can act to reduce or extract methane (CH_4) from the atmosphere, most of which feedbacks are enhanced as the air's CO_2 concentration rises. That those feedbacks may already be operating and having a significant impact on global methane concentrations is illustrated in a discussion of observed atmospheric methane trends.

We begin with Figure 2.6.2.1, the graph of real-world data from Simpson *et al.* (2002), which clearly shows a linear-trend decline in CH_4 growth rates since the mid-1980s. The authors contended it was "premature to believe" the rate of growth was falling, even though their own data bore witness against them.

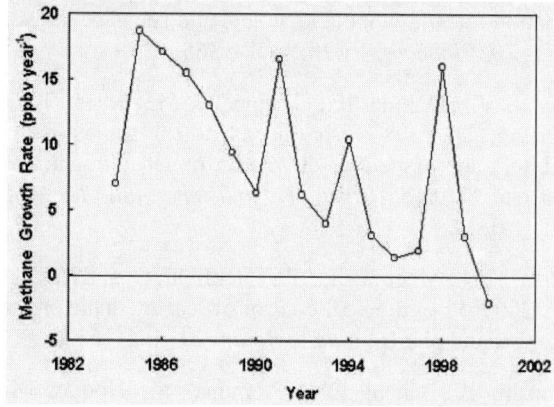

Figure 2.6.2.1. Global tropospheric methane (CH_4) growth rate vs. time. Adapted from Simpson *et al.* (2002).

The first of the 1990s' large CH_4 spikes is widely recognized as having been caused by the eruption of Mt. Pinatubo in June 1991 (Bekki *et al.*, 1994;

Dlugokencky *et al.*, 1996; Lowe *et al.*, 1997), while the last and most dramatic of the spikes has been linked to the remarkably strong El Niño of 1997-98 (Dlugokencky *et al.*, 2001). As noted earlier, Dlugokencky *et al.* (1998), Francey *et al.* (1999), and Lassey *et al.* (2000) have all suggested that the annual rate-of-rise of the atmosphere's CH_4 concentration is indeed declining and leading to a cessation of growth in the atmospheric burden of methane.

Dlugokencky *et al.* (2003) revisited the subject with an additional two years' of data. Based on measurements from 43 globally distributed remote boundary-layer sites that were obtained by means of the methods of Dlugokencky *et al.* (1994), they defined an evenly spaced matrix of surface CH_4 mole fractions as a function of time and latitude, from which they calculated global CH_4 concentration averages for the years 1984-2002. We have extracted the results from their graphical presentation and re-plotted them as shown in Figure 2.6.2.2.

With respect to these data, Dlugokencky *et al.* note that the globally averaged atmospheric methane concentration "was constant at ~1751 ppb from 1999 through 2002," which suggests, in their words, that "during this 4-year period the global methane budget has been at steady state." They caution, however, that "our understanding is still not sufficient to tell if the prolonged pause in CH_4 increase is temporary or permanent." We agree. However, we feel confident in suggesting that if the recent pause in CH_4 increase is indeed temporary, it will likely be followed by a decrease in CH_4 concentration, since that would be the next logical step in the observed progression from significant, to much smaller, to no yearly CH_4 increase.

Khalil *et al.* (2007) essentially "put the nails in the coffin" of the idea that rising atmospheric CH_4 concentrations pose any further global warming threat at all. In their study, the three Oregon (USA) researchers combined two huge atmospheric methane datasets to produce the unified dataset depicted in Figure 2.6.2.3.

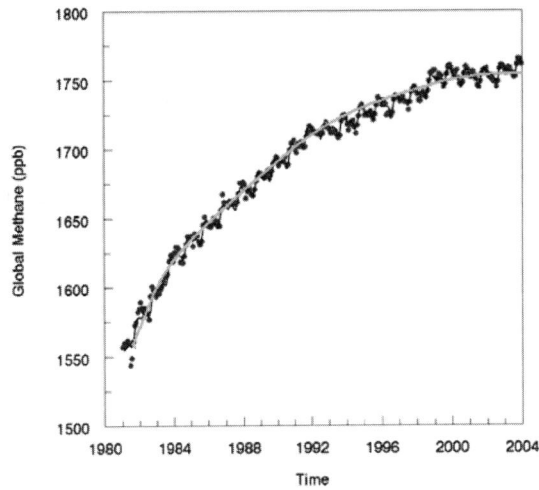

Figure 2.6.2.2. Global methane (CH_4) concentration. Adapted from Khalil *et al.* (2007).

In viewing this graph, to which we have added the smooth line, it is clear that the rate of methane increase in the atmosphere has dropped dramatically over time. As Khalil *et al.* describe it, "the trend has been decreasing for the last two decades until the present when it has reached near zero," and they go on to say that "it is questionable whether human activities can cause methane concentrations to increase greatly in the future."

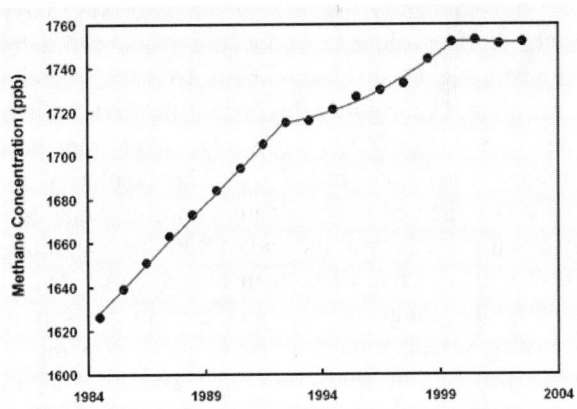

Figure 2.6.2.3. Global tropospheric methane (CH_4) concentration vs. time. Adapted from Dlugokencky *et al.* (2003).

One year later, Schnell and Dlugokencky (2008) provided an update through 2007 of atmospheric methane concentrations as determined from weekly discrete samples collected on a regular basis since 1983 at the NOAA/ESRL Mauna Loa Observatory.

Our adaptation of the graphical rendition of the data provided by the authors is presented in Figure 2.6.2.4.

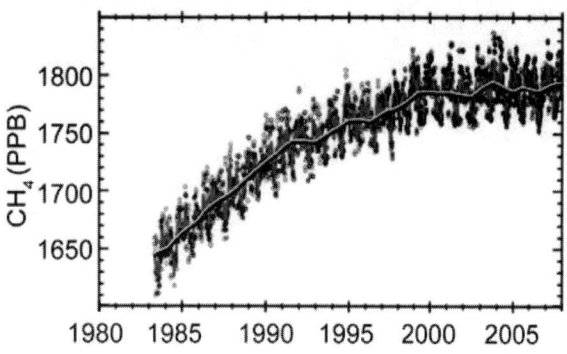

Figure 2.6.2.4. Trace gas mole fractions of methane (CH₄) as measured at Mauna Loa, Hawaii. Adapted from Schnell and Dlugokencky (2008).

In commenting on the data contained in the figure above, Schnell and Dlugokencky state that "atmospheric CH₄ has remained nearly constant since the late 1990s." This is a most important finding, because, as they also note, "methane's contribution to anthropogenic radiative forcing, including direct and indirect effects, is about 0.7 Wm⁻², about half that of CO₂." In addition, they say that "the increase in methane since the preindustrial era is responsible for approximately one-half the estimated increase in background tropospheric O₃ during that time."

Most recently, Rigby *et al.* (2008) analyzed methane data obtained from the Advanced Global Atmospheric Gases Experiment (AGAGE) and the Australian Commonwealth Scientific and Industrial Research Organization (CSIRO) over the period January 1997 to April 2008. The results of their analysis indicated that methane concentrations "show renewed growth from the end of 2006 or beginning of 2007 until the most recent measurements," with the record-long range of methane growth rates mostly hovering about zero, but sometimes dropping five parts per billion (ppb) per year into the negative range, while rising near the end of the record to mean positive values of 8 and 12 ppb per year for the two measurement networks.

Although some people might be alarmed by these findings, as well as by the US, UK, and Australian researchers' concluding statement that the methane growth rate during 2007 "was significantly elevated at all AGAGE and CSIRO sites simultaneously for the first time in almost a decade," there is also reassurance in the recent findings. We note, for example, that near the end of 1998 and the beginning of 1999, both networks measured even larger methane growth rate increases of approximately 13 ppb per year, before dropping back to zero at the beginning of the new millennium. And we note that the most current displayed data from the two networks indicate the beginning of what could well be another downward trend.

Additional reassurance in this regard comes from the work of Simpson *et al.* (2002), the findings of whom we reproduced previously in Figure 2.6.2.1. As can be seen there, even greater methane growth rates than those observed by Rigby *et al.* occurred in still earlier years. Hence, these periodic one-year-long upward spikes in methane growth rate must be the result of some normal phenomenon, the identity of which has yet to be determined.

In light of these finding, it can be appreciated that over the past decade there have been essentially no increases in methane emissions to the atmosphere, and that the leveling out of the atmosphere's methane concentration—the exact causes of which, in the words of Schnell and Dlugokencky, "are still unclear"—has resulted in a one-third reduction in the combined radiative forcing that would otherwise have been produced by a continuation of the prior rates-of-rise of the concentrations of the two atmospheric greenhouse gases.

Additional information on this topic, including reviews of newer publications as they become available, can be found at http://www.co2science.org/subject/m/methaneatmos.php.

References

Bekki, S., Law, K.S. and Pyle, J.A. 1994. Effect of ozone depletion on atmospheric CH₄ and CO concentrations. *Nature* **371**: 595-597.

Dlugokencky, E.J., Dutton, E.G., Novelli, P.C., Tans, P.P., Masarie, K.A., Lantz, K.O. and Madronich, S. 1996. Changes in CH₄ and CO growth rates after the eruption of Mt. Pinatubo and their link with changes in tropical tropospheric UV flux. *Geophysical Research Letters* **23**: 2761-2764.

Dlugokencky, E.J., Houweling, S., Bruhwiler, L., Masarie, K.A., Lang, P.M., Miller, J.B. and Tans, P.P. 2003. Atmospheric methane levels off: Temporary pause or a new steady-state? *Geophysical Research Letters* **30**: 10.1029/2003GL018126.

Dlugokencky, E.J., Masarie, K.A., Lang, P.M. and Tans, P.P. 1998. Continuing decline in the growth rate of the atmospheric methane burden. *Nature* **393**: 447-450.

Dlugokencky, E.J., Steele, L.P., Lang, P.M. and Masarie, K.A. 1994. The growth rate and distribution of atmospheric methane. *Journal of Geophysical Research* **99**: 17,021-17,043.

Dlugokencky, E.J., Walter, B.P., Masarie, K.A., Lang, P.M. and Kasischke, E.S. 2001. Measurements of an anomalous global methane increase during 1998. *Geophysical Research Letters* **28**: 499-502.

Ehhalt, D.H. and Prather, M. 2001. Atmospheric chemistry and greenhouse gases. In: *Climate Change 2001: The Scientific Basis*, Cambridge University Press, New York, NY, USA, pp. 245-287.

Francey, R.J., Manning, M.R., Allison, C.E., Coram, S.A., Etheridge, D.M., Langenfelds, R.L., Lowe, D.C. and Steele, L.P. 1999. A history of $\delta^{13}C$ in atmospheric CH_4 from the Cape Grim Air Archive and Antarctic firn air. *Journal of Geophysical Research* **104**: 23,631-23,643.

Khalil, M.A.K., Butenhoff, C.L. and Rasmussen, R.A. 2007. Atmospheric methane: Trends and cycles of sources and sinks. *Environmental Science & Technology* **10.1021**/es061791t.

Lassey, K.R., Lowe, D.C. and Manning, M.R. 2000. The trend in atmospheric methane $\delta^{13}C$ and implications for constraints on the global methane budget. *Global Biogeochemical Cycles* **14**: 41-49.

Lowe, D.C., Manning, M.R., Brailsford, G.W. and Bromley, A.M. 1997. The 1991-1992 atmospheric methane anomaly: Southern hemisphere ^{13}C decrease and growth rate fluctuations. *Geophysical Research Letters* **24**: 857-860.

Rigby, M., Prinn, R.G., Fraser, P.J., Simmonds, P.G., Langenfelds, R.L., Huang, J., Cunnold, D.M., Steele, L.P., Krummel, P.B., Weiss, R.F., O'Doherty, S., Salameh, P.K., Wang, H.J., Harth, C.M., Muhle, J. and Porter, L.W. 2008. Renewed growth of atmospheric methane. *Geophysical Research Letters* **35**: 10.1029/2008GL036037.

Schnell, R.C. and Dlugokencky, E. 2008. Methane. In: Levinson, D.H. and Lawrimore, J.H. (Eds.) *State of the Climate in 2007*. Special Supplement to the *Bulletin of the American Meteorological Society* **89**: S27.

Simpson, I.J., Blake, D.R. and Rowland, F.S. 2002. Implications of the recent fluctuations in the growth rate of tropospheric methane. *Geophysical Research Letters* **29**: 10.1029/2001GL014521.

2.7. Dimethyl Sulfide

More than two decades ago, Charlson *et al.* (1987) discussed the plausibility of a multi-stage negative feedback process, whereby warming-induced increases in the emission of dimethyl sulfide (DMS) from the world's oceans tend to counteract any initial impetus for warming. The basic tenet of their hypothesis was that the global radiation balance is significantly influenced by the albedo of marine stratus clouds (the greater the cloud albedo, the less the input of solar radiation to the earth's surface). The albedo of these clouds, in turn, is known to be a function of cloud droplet concentration (the more and smaller the cloud droplets, the greater the cloud albedo and the reflection of solar radiation), which is dependent upon the availability of cloud condensation nuclei on which the droplets form (the more cloud condensation nuclei, the more and smaller the cloud droplets). And in completing the negative feedback loop, Charlson *et al.* noted that the cloud condensation nuclei concentration often depends upon the flux of biologically produced DMS from the world's oceans (the higher the sea surface temperature, the greater the sea-to-air flux of DMS).

Since the publication of Charlson *et al.*'s initial hypothesis, much empirical evidence has been gathered in support of its several tenets. One review, for example, states that "major links in the feedback chain proposed by Charlson *et al.* (1987) have a sound physical basis," and that there is "compelling observational evidence to suggest that DMS and its atmospheric products participate significantly in processes of climate regulation and reactive atmospheric chemistry in the remote marine boundary layer of the Southern Hemisphere" (Ayers and Gillett, 2000).

But just how strong is the negative feedback phenomenon proposed by Charlson *et al.*? Is it powerful enough to counter the threat of greenhouse gas-induced global warming? According to the findings of Sciare *et al.* (2000), it may well be able to do just that.

In examining 10 years of DMS data from Amsterdam Island in the southern Indian Ocean, these researchers found that a sea surface temperature increase of only 1°C was sufficient to increase the atmospheric DMS concentration by as much as 50 percent. This finding suggests that the degree of warming typically predicted to accompany a doubling of the air's CO_2 content would increase the atmosphere's DMS concentration by a factor of three or more, providing what they call a "very important" negative feedback that could potentially offset the original impetus for warming.

Other research has shown that this same chain of events can be set in motion by means of phenomena not discussed in Charlson *et al.*'s original hypothesis. Simo and Pedros-Alio (1999), for example, discovered that the depth of the surface mixing-layer has a substantial influence on DMS yield in the short term, via a number of photo-induced (and thereby mixing-depth mediated) influences on several complex physiological phenomena, as do longer-term seasonal variations in vertical mixing, via their influence on seasonal planktonic succession scenarios and food-web structure.

More directly supportive of Charlson *et al.*'s hypothesis was the study of Kouvarakis and Mihalopoulos (2002), who measured seasonal variations of gaseous DMS and its oxidation products—non-sea-salt sulfate ($nss-SO_4^{2-}$) and methanesulfonic acid (MSA)—at a remote coastal location in the Eastern Mediterranean Sea from May 1997 through October 1999, as well as the diurnal variation of DMS during two intensive measurement campaigns conducted in September 1997. In the seasonal investigation, DMS concentrations tracked sea surface temperature (SST) almost perfectly, going from a low of 0.87 nmol m^{-3} in the winter to a high of 3.74 nmol m^{-3} in the summer. Such was also the case in the diurnal studies: DMS concentrations were lowest when it was coldest (just before sunrise), rose rapidly as it warmed thereafter to about 1100, after which they dipped slightly and then experienced a further rise to the time of maximum temperature at 2000, whereupon a decline in both temperature and DMS concentration set in that continued until just before sunrise. Consequently, because concentrations of DMS and its oxidation products (MSA and $nss-SO_4^{2-}$) rise dramatically in response to both diurnal and seasonal increases in SST, there is every reason to believe that the same negative feedback phenomenon would operate in the case of the long-term warming that could arise from increasing greenhouse gas concentrations, and that it could substantially mute the climatic impacts of those gases.

Also of note in this regard, Baboukas *et al.* (2002) report the results of nine years of measurements of methanesulfonate (MS-), an exclusive oxidation product of DMS, in rainwater at Amsterdam Island. Their data, too, revealed "a well distinguished seasonal variation with higher values in summer, in line with the seasonal variation of its gaseous precursor (DMS)," which, in their words, "further confirms the findings of Sciare *et al.* (2000)." In addition, the MS- anomalies in the rainwater were found to be closely related to SST anomalies; and Baboukas *et al.* say this observation provides even more support for "the existence of a positive ocean-atmosphere feedback on the biogenic sulfur cycle above the Austral Ocean, one of the most important DMS sources of the world."

In a newer study of this phenomenon, Toole and Siegel (2004) note that it has been shown to operate as described above in the 15 percent of the world's oceans "consisting primarily of high latitude, continental shelf, and equatorial upwelling regions," where DMS may be accurately predicted as a function of the ratio of the amount of surface chlorophyll derived from satellite observations to the depth of the climatological mixed layer, which they refer to as the "bloom-forced regime." For the other 85 percent of the world's marine waters, they demonstrate that modeled surface DMS concentrations are independent of chlorophyll and are a function of the mixed layer depth alone, which they call the "stress-forced regime." So how does the warming-induced DMS negative feedback cycle operate in these waters?

For oligotrophic regimes, Toole and Siegel find that "DMS biological production rates are negatively or insignificantly correlated with phytoplankton and bacterial indices for abundance and productivity while more than 82 percent of the variability is explained by UVR(325) [ultraviolet radiation at 325 nm]." This relationship, in their words, is "consistent with recent laboratory results (e.g., Sunda *et al.*, 2002)," who demonstrated that intracellular DMS concentration and its biological precursors (particulate and dissolved dimethylsulfoniopropionate) "dramatically increase under conditions of acute oxidative stress such as exposure to high levels of UVR," which "are a function of mixed layer depth."

These results—which Toole and Siegel confirmed via an analysis of the Dacey *et al.* (1998) 1992-1994 organic sulfur time-series that was sampled in concert with the U.S. JGOFS Bermuda Atlantic Time-Series Study (Steinberg *et al.*, 2001)—suggest, in their words, "the potential of a global change-DMS-climate feedback." Specifically, they say that "UVR doses will increase as a result of observed decreases in stratospheric ozone and the shoaling of ocean mixed layers as a result of global warming (e.g., Boyd and Doney, 2002)," and that "in response, open-ocean phytoplankton communities should increase their DMS production and ventilation to the atmosphere, increasing cloud condensing nuclei, and potentially playing out a coupled global change-DMS-climate feedback."

This second DMS-induced negative-feedback cycle, which operates over 85 percent of the world's marine waters and complements the first DMS-induced negative-feedback cycle, which operates over the other 15 percent, is another manifestation of the capacity of earth's biosphere to regulate its affairs in such a way as to maintain climatic conditions over the vast majority of the planet's surface within bounds conducive to the continued existence of life, in all its variety and richness. In addition, it has been suggested that a DMS-induced negative climate feedback phenomenon also operates over the terrestrial surface of the globe, where the volatilization of reduced sulfur gases from soils may be just as important as marine DMS emissions in enhancing cloud albedo (Idso, 1990).

On the basis of experiments that showed soil DMS emissions to be positively correlated with soil organic matter content, for example, and noting that additions of organic matter to a soil tend to increase the amount of sulfur gases emitted therefrom, Idso (1990) hypothesized that because atmospheric CO_2 is an effective aerial fertilizer, augmenting its atmospheric concentration and thereby increasing vegetative inputs of organic matter to earth's soils should also produce an impetus for cooling, even in the absence of surface warming.

Nevertheless, and in spite of the overwhelming empirical evidence for both land- and ocean-based DMS-driven negative feedbacks to global warming, the effects of these processes have not been fully incorporated into today's state-of-the-art climate models. Hence, the warming they predict in response to future anthropogenic CO_2 emissions must be considerably larger than what could actually occur in the real world. It is very possible these biologically driven phenomena could entirely compensate for the warming influence of all greenhouse gas emissions experienced to date, as well as all those anticipated to occur in the future.

Additional information on this topic, including reviews of newer publications as they become available, can be found at http://www.co2science.org/subject/d/dms.php.

References

Ayers, G.P. and Gillett, R.W. 2000. DMS and its oxidation products in the remote marine atmosphere: implications for climate and atmospheric chemistry. *Journal of Sea Research* 43: 275-286.

Baboukas, E., Sciare, J. and Mihalopoulos, N. 2002. Interannual variability of methanesulfonate in rainwater at Amsterdam Island (Southern Indian Ocean). *Atmospheric Environment* 36: 5131-5139.

Boyd, P.W. and Doney, S.C. 2002. Modeling regional responses by marine pelagic ecosystems to global climate change. *Geophysical Research Letters* 29: 10.1029/2001GL014130.

Charlson, R.J., Lovelock, J.E., Andrea, M.O. and Warren, S.G. 1987. Oceanic phytoplankton, atmospheric sulfur, cloud albedo and climate. *Nature* 326: 655-661.

Dacey, J.W.H., Howse, F.A., Michaels, A.F. and Wakeham, S.G. 1998. Temporal variability of dimethylsulfide and dimethylsulfoniopropionate in the Sargasso Sea. *Deep Sea Research* 45: 2085-2104.

Idso, S.B. 1990. A role for soil microbes in moderating the carbon dioxide greenhouse effect? *Soil Science* 149: 179-180.

Kouvarakis, G. and Mihalopoulos, N. 2002. Seasonal variation of dimethylsulfide in the gas phase and of methanesulfonate and non-sea-salt sulfate in the aerosols phase in the Eastern Mediterranean atmosphere. *Atmospheric Environment* 36: 929-938.

Sciare, J., Mihalopoulos, N. and Dentener, F.J. 2000. Interannual variability of atmospheric dimethylsulfide in the southern Indian Ocean. *Journal of Geophysical Research* 105: 26,369-26,377.

Simo, R. and Pedros-Alio, C. 1999. Role of vertical mixing in controlling the oceanic production of dimethyl sulphide. *Nature* 402: 396-399.

Steinberg, D.K., Carlson, C.A., Bates, N.R., Johnson, R.J., Michaels, A.F. and Knap, A.H. 2001. Overview of the US JGOFS Bermuda Atlantic Time-series Study (BATS): a decade-scale look at ocean biology and biogeochemistry. *Deep Sea Research Part II: Topical Studies in Oceanography* 48: 1405-1447.

Sunda, W., Kieber, D.J., Kiene, R.P. and Huntsman, S. 2002. An antioxidant function for DMSP and DMS in marine algae. *Nature* 418: 317-320.

Toole, D.A. and Siegel, D.A. 2004. Light-driven cycling of dimethylsulfide (DMS) in the Sargasso Sea: Closing the loop. *Geophysical Research Letters* 31: 10.1029/2004GL019581.

2.8. Aerosols

2.8.1. Total Aerosol Effect

The IPCC estimates the net effect of all aerosols is to produce a cooling effect, with a total direct radiative forcing of -0.5 Wm⁻² and an additional indirect cloud albedo forcing of -0.7 Wm⁻² (IPCC, 2007-I, p. 4). However, the scientific literature indicates these estimates are too low. Many studies suggest the radiative forcing of aerosols may be as large as, or larger than, the radiative forcing due to atmospheric CO₂.

Vogelmann *et al.* (2003) report that "mineral aerosols have complex, highly varied optical properties that, for equal loadings, can cause differences in the surface IR flux between 7 and 25 Wm⁻² (Sokolik *et al.*, 1998)," and "only a few large-scale climate models currently consider aerosol IR [infrared] effects (e.g., Tegen *et al.*, 1996; Jacobson, 2001) despite their potentially large forcing." In an attempt to persuade climate modelers to rectify this situation, they used high-resolution spectra to obtain the IR radiative forcing at the earth's surface for aerosols encountered in the outflow from northeastern Asia, based on measurements made by the Marine-Atmospheric Emitted Radiance Interferometer from the NOAA Ship *Ronald H. Brown* during the Aerosol Characterization Experiment-Asia. As a result of this work, the scientists determined that "daytime surface IR forcings are often a few Wm⁻² and can reach almost 10 Wm⁻² for large aerosol loadings." These values, in their words, "are comparable to or larger than the 1 to 2 Wm⁻² change in the globally averaged surface IR forcing caused by greenhouse gas increases since pre-industrial times" and "highlight the importance of aerosol IR forcing which should be included in climate model simulations."

Chou *et al.* (2002) analyzed aerosol optical properties retrieved from the satellite-mounted Sea-viewing Wide Field-of-view Sensor (SeaWiFS) and used them in conjunction with a radiative transfer model of the planet's atmosphere to calculate the climatic effects of aerosols over earth's major oceans. In general, this effort revealed that "aerosols reduce the annual-mean net downward solar flux by 5.4 Wm⁻² at the top of the atmosphere, and by 5.9 Wm⁻² at the surface." During the large Indonesian fires of September-December 1997, however, the radiative impetus for cooling at the top of the atmosphere was more than 10 Wm⁻², while it was more than 25 Wm⁻² at the surface of the sea in the vicinity of Indonesia.

These latter results are similar to those obtained earlier by Wild (1999), who used a comprehensive set of collocated surface and satellite observations to calculate the amount of solar radiation absorbed in the atmosphere over equatorial Africa and compared the results with the predictions of three general circulation models of the atmosphere. This work revealed that the climate models did not properly account for spatial and temporal variations in atmospheric aerosol concentrations, leading them to predict regional and seasonal values of solar radiation absorption in the atmosphere with underestimation biases of up to 30 Wm⁻². By way of comparison, as noted by Vogelmann *et al.*, the globally averaged surface IR forcing caused by greenhouse gas increases since pre-industrial times is 1 to 2 Wm⁻².

Aerosol uncertainties and the problems they generate figure prominently in a study by Anderson *et al.* (2003), who note there are two different ways by which the aerosol forcing of climate may be computed. The first is forward calculation, which is based, in their words, on "knowledge of the pertinent aerosol physics and chemistry." The second approach is inverse calculation, based on "the total forcing required to match climate model simulations with observed temperature changes."

The first approach utilizes known physical and chemical laws and assumes nothing about the outcome of the calculation. The second approach, in considerable contrast, is based on matching residuals, where the aerosol forcing is computed from what is required to match the calculated change in temperature with the observed change over some period of time. Consequently, in the words of Anderson *et al.*, "to the extent that climate models rely on the results of inverse calculations, the possibility of circular reasoning arises."

So which approach do climate models typically employ? "Unfortunately," according to Anderson *et al.*, "virtually all climate model studies that have included anthropogenic aerosol forcing as a driver of climate change have used only aerosol forcing values that are consistent with the inverse approach."

How significant is this choice? Anderson *et al.* report that the negative forcing of anthropogenic aerosols derived by forward calculation is "considerably greater" than that derived by inverse calculation; so much so, in fact, that if forward calculation is employed, the results "differ greatly" and "even the sign of the total forcing is in question," which implies that "natural variability (that is, variability not forced by anthropogenic emissions) is much larger than climate models currently indicate." The bottom line, in the words of Anderson *et al.*, is that "inferences about the causes of surface warming over the industrial period and about climate sensitivity may therefore be in error."

Schwartz (2004) also addressed the subject of uncertainty as it applies to the role of aerosols in climate models. Noting that the National Research Council (1979) concluded that "climate sensitivity [to CO_2 doubling] is likely to be in the range 1.5-4.5°C" and that "remarkably, despite some two decades of intervening work, neither the central value nor the uncertainty range has changed," Schwartz opined that this continuing uncertainty "precludes meaningful model evaluation by comparison with observed global temperature change or empirical determination of climate sensitivity," and that it "raises questions regarding claims of having reproduced observed large-scale changes in surface temperature over the 20th century."

Schwartz thus contends that climate model predictions of CO_2-induced global warming "are limited at present by uncertainty in radiative forcing of climate change over the industrial period, which is dominated by uncertainty in forcing by aerosols," and that if this situation is not improved, "it is likely that in another 20 years it will still not be possible to specify the climate sensitivity with [an] uncertainty range appreciably narrower than it is at present." Indeed, he says "the need for reducing the uncertainty from its present estimated value by at least a factor of 3 and perhaps a factor of 10 or more seems inescapable if the uncertainty in climate sensitivity is to be reduced to an extent where it becomes useful for formulating policy to deal with global change," which surely suggests that even the best climate models of the day are wholly inadequate for this purpose.

Coming to much the same conclusion was the study of Jaenicke *et al.* (2007), who reviewed the status of research being conducted on biological materials in the atmosphere, which they denominate primary biological atmospheric particles or PBAPs. Originally, these particles were restricted to culture-forming units, including pollen, bacteria, mold and viruses, but they also include fragments of living and dead organisms and plant debris, human and animal epithelial cells, broken hair filaments, parts of insects, shed feather fractions, etc., which they lump together under the category of "dead biological matter."

With respect to the meteorological and climatic relevance of these particles, they note that many PBAPs, including "decaying vegetation, marine plankton and bacteria are excellent ice nuclei," and "one can easily imagine the [IR] influence on cloud cover, climate forcing and feedback and global precipitation distribution."

In describing their own measurements and those of others, which they say "have now been carried out at several geographical locations covering all seasons of the year and many characteristic environments," Jaenicke *et al.* report that "by number and volume, the PBAP fraction is ~20 percent of the total aerosol, and appears rather constant during the year." In addition, they write that "the impression prevails that the biological material, whether produced directly or shed during the seasons, sits on surfaces, ready to be lifted again in resuspension."

In a brief summation of their findings, the German researchers say "the overall conclusion can only be that PBAPs are a major fraction of atmospheric aerosols, and are comparable to sea salt over the oceans and mineral particles over the continents," and, consequently, that "the biosphere must be a major source for directly injected biological particles, and those particles should be taken into account in understanding and modeling atmospheric processes." However, they note that "the IPCC-Report of 2007 does not even mention these particles," and that "this disregard of the biological particles requires a new attitude."

We agree. Over much of the planet's surface, the radiative cooling influence of atmospheric aerosols (many of which are produced by anthropogenic activities) must prevail, suggesting a probable net anthropogenic-induced climatic signal that must be very close to zero and incapable of producing what the IPCC refers to as the "unprecedented" warming of the twentieth century. Either the air temperature record they rely on is in error or the warming, if real, is due to something other than anthropogenic CO_2 emissions.

Our review of important aerosol studies continues below with a separate discussion of four important aerosol categories: (1) Biological (Aquatic), (2) Biological (Terrestrial), (3) Non-Biological (Anthropogenic), and (4) Non-Biological (Natural). Additional information on this topic, including reviews of aerosols not discussed here, can be found at http://www.co2science.org/subject/a/subject_a. php under the heading Aerosols.

References

Anderson, T.L., Charlson, R.J., Schwartz, S.E., Knutti, R., Boucher, O., Rodhe, H. and Heintzenberg, J. 2003. Climate forcing by aerosols—a hazy picture. *Science* **300**: 1103-1104.

Chou, M-D., Chan, P-K. and Wang, M. 2002. Aerosol radiative forcing derived from SeaWiFS-retrieved aerosol optical properties. *Journal of the Atmospheric Sciences* **59**: 748-757.

IPCC. 2007-I. *Climate Change 2007: The Physical Science Basis. Contribution of Working Group I to the Fourth Assessment Report of the Intergovernmental Panel on Climate Change.* Solomon, S., Qin, D., Manning, M., Chen, Z., Marquis, M., Averyt, K.B., Tignor, M. and H.L. Miller. (Eds.) Cambridge University Press, Cambridge, UK.

Jacobson, M.Z. 2001. Global direct radiative forcing due to multicomponent anthropogenic and natural aerosols. *Journal of Geophysical Research* **106**: 1551-1568.

Jaenicke, R., Matthias-Maser, S. and Gruber, S. 2007. Omnipresence of biological material in the atmosphere. *Environmental Chemistry* **4**: 217-220.

National Research Council. 1979. *Carbon Dioxide and Climate: A Scientific Assessment.* National Academy of Sciences, Washington, DC, USA.

Schwartz, S.E. 2004. Uncertainty requirements in radiative forcing of climate. *Journal of the Air & Waste Management Association* **54**: 1351-1359.

Sokolik, I.N., Toon, O.B. and Bergstrom, R.W. 1998. Modeling the radiative characteristics of airborne mineral aerosols at infrared wavelengths. *Journal of Geophysical Research* **103**: 8813-8826.

Tegen, I., Lacis, A.A. and Fung, I. 1996. The influence on climate forcing of mineral aerosols from disturbed soils. *Nature* **380**: 419-422.

Vogelmann, A.M., Flatau, P.J., Szczodrak, M., Markowicz, K.M. and Minnett, P.J. 2003. *Geophysical Research Letters* **30**: 10.1029/2002GL016829.

Wild, M. 1999. Discrepancies between model-calculated and observed shortwave atmospheric absorption in areas with high aerosol loadings. *Journal of Geophysical Research* **104**: 27,361-27,371.

2.8.2. Biological (Aquatic)

Charlson *et al.* (1987) described a multi-stage negative feedback phenomenon, several components of which have been verified by subsequent scientific studies, that links biology with climate change. The process begins with an initial impetus for warming that stimulates primary production in marine phytoplankton. This enhanced process leads to the production of more copious quantities of dimethylsulphoniopropionate, which leads in turn to the evolution of greater amounts of dimethyl sulphide, or DMS, in the surface waters of the world's oceans. Larger quantities of DMS diffuse into the atmosphere, where the gas is oxidized, leading to the creation of greater amounts of acidic aerosols that function as cloud condensation nuclei. This phenomenon then leads to the creation of more and brighter clouds that reflect more incoming solar radiation back to space, thereby providing a cooling influence that counters the initial impetus for warming.

Several recent studies have shed additional light on this complex hypothesis. Simo and Pedros-Alio (1999) used satellite imagery and *in situ* experiments to study the production of DMS by enzymatic cleavage of dimethylsulphoniopropionate in the North Atlantic Ocean about 400 km south of Iceland, finding that the depth of the surface mixing-layer has a substantial influence on DMS yield in the short term, as do seasonal variations in vertical mixing in the longer term, which observations led them to conclude that "climate-controlled mixing controls DMS production over vast regions of the ocean."

Hopke *et al.* (1999) analyzed weekly concentrations of 24 different airborne particulates measured at the northernmost manned site in the world—Alert, Northwest Territories, Canada—from 1980 to 1991. They found concentrations of biogenic sulfur, including sulfate and methane sulfonate, were low in winter but high in summer, and that the year-to-year variability in the strength of the biogenic sulfur signal was strongly correlated with the mean temperature of the Northern Hemisphere. "This result," the authors say, "suggests that as the temperature rises, there is increased biogenic production of the reduced sulfur precursor compounds that are oxidized in the atmosphere to sulfate and methane sulfonate and could be evidence of a negative feedback mechanism in the global climate system."

Ayers and Gillett (2000) summarized relevant empirical evidence collected at Cape Grim, Tasmania, along with pertinent evidence reported in many peer-reviewed scientific papers on this subject. They conclude that "major links in the feedback chain proposed by Charlson *et al.* (1987) have a sound physical basis." More specifically, they noted there is "compelling observational evidence to suggest that DMS and its atmospheric products participate significantly in processes of climate regulation and reactive atmospheric chemistry in the remote marine boundary layer of the Southern Hemisphere."

Sciare *et al.* (2000) made continuous measurements of atmospheric DMS concentration over the 10-year period 1990-1999 at Amsterdam Island in the southern Indian Ocean. Their study revealed "a clear seasonal variation with a factor of 20 in amplitude between its maximum in January (austral summer) and minimum in July-August (austral winter)." In addition, they found DMS anomalies to be "closely related to sea surface temperature anomalies, clearly indicating a link between DMS and climate changes." They found that a temperature increase of only 1°C was sufficient to increase the atmospheric DMS concentration by as much as 50 percent on a monthly basis, noting that "this is the first time that a direct link between SSTs [sea surface temperatures] and atmospheric DMS is established for a large oceanic area."

Another pertinent study was conducted by Kouvarakis and Mihalopoulos (2002), who investigated the seasonal variations of gaseous DMS and its oxidation products—non-sea-salt sulfate (nss-SO_4^{2-}) and methanesulfonic acid (MSA)—at a remote coastal location in the Eastern Mediterranean Sea from May 1997 through October 1999, as well as the diurnal variation of DMS during two intensive measurement campaigns in September 1997. In the seasonal investigation, DMS concentrations tracked sea surface temperature (SST) almost perfectly, going from a low of 0.87 nmol m^{-3} in the winter to a high of 3.74 nmol m^{-3} in the summer. Such was also the case in the diurnal study: DMS concentrations were lowest just before sunrise, rose rapidly thereafter to about 1100, were followed by a little dip and then a further rise to 2000, whereupon a decline set in that continued until just before sunrise. MSA concentrations exhibited a similar seasonal variation to that displayed by DMS, ranging from a wintertime low of 0.04 nmol m^{-3} to a summertime high of 0.99 nmol m^{-3}. The same was also true of aerosol nss-SO_4^{2-} which varied from 0.6 to 123.9 nmol m^{-3} in going from winter to summer.

A related study of methanesulfonate (MS$^-$) in rainwater at Amsterdam Island, by Baboukas *et al.* (2002), in the authors' words, "further confirms the findings of Sciare *et al.* (2000)." For more about that

study and a newer study by Toole and Siegel (2004), see Section 2.7 of this report.

As time passes, more studies confirm the Charlson *et al.* hypothesis that as marine phytoplankton are exposed to rising temperatures, they give off greater quantities of gases that lead to the production of greater quantities of cloud condensation nuclei, which create more and brighter clouds, that reflect more incoming solar radiation back to space, and thereby either reverse, stop, or slow the warming that initiated this negative feedback phenomenon. The normal hour-to hour, day-to-day, and season-to-season behaviors of the phytoplanktonic inhabitants of earth's marine ecosystems seem to be effectively combating extreme environmental temperature changes.

Additional information on this topic, including reviews of newer publications as they become available, can be found at http://www.co2science.org/subject/a/aerosolsbioaqua.php.

References

Ayers, G.P. and Gillett, R.W. 2000. DMS and its oxidation products in the remote marine atmosphere: implications for climate and atmospheric chemistry. *Journal of Sea Research* **43**: 275-286.

Baboukas, E., Sciare, J. and Mihalopoulos, N. 2002. Interannual variability of methanesulfonate in rainwater at Amsterdam Island (Southern Indian Ocean). *Atmospheric Environment* **36**: 5131-5139

Charlson, R.J., Lovelock, J.E., Andrea, M.O. and Warren, S.G. 1987. Oceanic phytoplankton, atmospheric sulfur, cloud albedo and climate. *Nature* **326**: 655-661.

Hopke, P.K., Xie, Y. and Paatero, P. 1999. Mixed multiway analysis of airborne particle composition data. *Journal of Chemometrics* **13**: 343-352.

Kouvarakis, G. and Mihalopoulos, N. 2002. Seasonal variation of dimethylsulfide in the gas phase and of methanesulfonate and non-sea-salt sulfate in the aerosols phase in the Eastern Mediterranean atmosphere. *Atmospheric Environment* **36**: 929-938.

O'Dowd, C.D., Facchini, M.C., Cavalli, F., Ceburnis, D., Mircea, M., Decesari, S., Fuzzi, S., Yoon, Y.J. and Putaud, J.-P. 2004. Biogenically driven organic contribution to marine aerosol. *Nature* **431**: 676-680.

Sciare, J., Mihalopoulos, N. and Dentener, F.J. 2000. Interannual variability of atmospheric dimethylsulfide in the southern Indian Ocean. *Journal of Geophysical Research* **105**: 26,369-26,377.

Simo, R. and Pedros-Alio, C. 1999. Role of vertical mixing in controlling the oceanic production of dimethyl sulphide. *Nature* **402**: 396-399.

Toole, D.A. and Siegel, D.A. 2004. Light-driven cycling of dimethylsulfide (DMS) in the Sargasso Sea: Closing the loop. *Geophysical Research Letters* **31**: 10.1029/2004 GL019581.

2.8.3. Biological (Terrestrial)

Just as marine phytoplankton respond to rising temperatures by giving off gases that ultimately lead to less global warming, so too do terrestrial plants. What is more, earth's terrestrial plants have a tendency to operate in this manner more effectively as the air's CO_2 content rises.

A good introduction to this subject is provided by the review paper of Peñuelas and Llusia (2003), who say biogenic volatile organic compounds (BVOCs) constitute "one of nature's biodiversity treasures." Comprised of isoprene, terpenes, alkanes, alkenes, alcohols, esters, carbonyls, and acids, this diverse group of substances is produced by a variety of processes occurring in many plant tissues. Some of the functions of these substances, according to the two scientists, include acting as "deterrents against pathogens and herbivores, or to aid wound sealing after damage (Pichersky and Gershenzon, 2002)." They also say BVOCs provide a means "to attract pollinators and herbivore predators, and to communicate with other plants and organisms (Peñuelas *et al.*, 1995; Shulaev *et al.*, 1997)."

Of particular importance within the context of global climate change, in the opinion of Peñuelas and Llusia, is the growing realization that "isoprene and monoterpenes, which constitute a major fraction of BVOCs, might confer protection against high temperatures" by acting "as scavengers of reactive oxygen species produced [within plants] under high temperatures." If this is indeed the case, it can be appreciated that with respect to the claimed ill effects of CO_2-induced global warming on earth's vegetation, there are likely to be two strong ameliorative phenomena that act to protect the planet's plants: (1) the aerial fertilization effect of atmospheric CO_2 enrichment, which is typically more strongly expressed at higher temperatures, and (2) the tendency for rising air temperatures and CO_2 concentrations to spur the production of higher concentrations of heat-stress-reducing BVOCs. With respect to temperature, Peñuelas and Llusia calculate

that "global warming over the past 30 years could have increased the BVOC global emissions by approximately 10 percent, and a further 2-3°C rise in the mean global temperature ... could increase BVOC global emissions by an additional 30-45 percent."

There may also be other phenomena that favor earth's plants within this context. Peñuelas and Llusia note that "the increased release of nitrogen into the biosphere by man probably also enhances BVOC emissions by increasing the level of carbon fixation and the activity of the responsible enzymes (Litvak *et al.*, 1996)." The conversion of abandoned agricultural lands to forests and the implementation of planned reforestation projects should help the rest of the biosphere too, since Peñuelas and Llusia report that additional numbers of "*Populus*, *Eucalyptus* or *Pinus*, which are major emitters, might greatly increase BVOC emissions."

Most intriguing of all, perhaps, is how increased BVOC emissions might impact climate change. Peñuelas and Llusia say that "BVOCs generate large quantities of organic aerosols that could affect climate significantly by forming cloud condensation nuclei." As a result, they say "there should be a net cooling of the Earth's surface during the day because of radiation interception," noting that Shallcross and Monks (2000) "have suggested that one of the reasons plants emit the aerosol isoprene might be to cool the surroundings in addition to any physiological or evaporative effects that might cool the plant directly."

Not all experiments have reported increases in plant BVOC emissions with increasing atmospheric CO_2 concentrations, one example being Constable *et al.* (1999), who found no effect of elevated CO_2 on monoterpene emissions from Ponderosa pine and Douglas fir trees. Some studies, in fact, have reported *decreases* in BVOC emissions, such as those of Vuorinen *et al.* (2004), who worked with cabbage plants, and Loreto *et al.* (2001), who studied monoterpene emissions from oak seedlings.

On the other hand, Staudt *et al.* (2001) observed CO_2-induced *increases* in BVOC emissions in the identical species of oak studied by Vuorinen *et al.* An explanation for this wide range of results comes from Baraldi *et al.* (2004), who—after exposing sections of a southern California chaparral ecosystem to atmospheric CO_2 concentrations ranging from 250 to 750 ppm in 100-ppm increments for a period of four years—concluded that "BVOC emission can remain nearly constant as rising CO_2 reduces emission per unit leaf area while stimulating biomass growth and leaf area per unit ground area." In most of the cases investigated, however, BVOC emissions tend to increase with atmospheric CO_2 enrichment; and the increases are often large.

Jasoni *et al.* (2003) who grew onions from seed for 30 days in individual cylindrical flow-through growth chambers under controlled environmental conditions at atmospheric CO_2 concentrations of either 400 or 1,000 ppm. At the end of the study, the plants in the CO_2-enriched chambers had 40 percent more biomass than the plants grown in ambient air, and their photosynthetic rates were 22 percent greater. In addition, the CO_2-enriched plants exhibited *17-fold* and *38-fold* increases in emissions of the BVOC hydrocarbons 2-undecanone and 2-tridecanone, respectively, which Jasoni *et al.* make a point of noting, "confer insect resistance against a major agricultural pest, spider mites." More generally, they conclude that "plants grown under elevated CO_2 will accumulate excess carbon and that at least a portion of this excess carbon is funneled into an increased production of BVOCs," which have many positive implications in the realms of both biology and climate, as noted above.

Raisanen *et al.* (2008) conducted an experiment designed to see to what extent a doubling of the air's CO_2 content and a 2°—6°C increase in air temperature might impact the emission of monoterpenes from 20-year-old Scots pine (*Pinus sylvestris* L.) seedlings. They studied the two phenomena (and their interaction) within closed-top chambers built on a naturally seeded stand of the trees in eastern Finland that had been exposed to the four treatments—ambient CO_2 and ambient temperature, ambient temperature and elevated CO_2, ambient CO_2 and elevated temperature, elevated temperature and elevated CO_2—for the prior five years.

Over the five-month growing season of May-September, the three Finnish researchers found that total monoterpene emissions in the elevated-CO_2-only treatment were 5 percent greater than those in the ambient CO_2, ambient temperature treatment, and that emissions in the elevated-temperature-only treatment were 9 percent less than those in ambient air. In the presence of both elevated CO_2 *and* elevated temperature, however, there was an increase of fully 126 percent in the total amount of monoterpenes emitted over the growing season, which led the authors to conclude, "the amount of monoterpenes released by Scots pines into the atmosphere during a growing season will increase substantially in the predicted future climate."

A number of studies suggest that the phenomena discussed in the preceding paragraphs do indeed operate in the real world. Kavouras *et al.* (1998), for example, measured a number of atmospheric gases and particles in a eucalyptus forest in Portugal and analyzed their observations to see if there was any evidence of biologically produced gases being converted to particles that could function as cloud condensation nuclei. Their work demonstrated that certain hydrocarbons emitted by vegetation (isoprene and terpenes, in particular) do indeed experience gas-to-particle transformations. In fact, aerosols (or *bio*sols) produced from two of these organic acids (*cis-* and *trans*-pinonic acid) comprised as much as 40 percent of the fine particle atmospheric mass during daytime hours.

A similar study was conducted by O'Dowd *et al.* (2002), who measured aerosol electrical-mobility size-distributions before and during the initial stage of an atmospheric nucleation event over a boreal forest in Finland. Simultaneously, organic vapor growth rate measurements were made of particles that nucleated into organic cloud-droplets in the flow-tube cloud chamber of a modified condensation-particle counter. This work demonstrated, in their words, that newly formed aerosol particles over forested areas "are composed primarily of organic species, such as cis-pinonic acid and pinonic acid, produced by oxidation of terpenes in organic vapours released from the canopy."

Commenting on this finding, O'Dowd *et al.* note that "aerosol particles produced over forested areas may affect climate by acting as nuclei for cloud condensation," but they say there remain numerous uncertainties involving complex feedback processes "that must be determined if we are to predict future changes in global climate."

Shifting from trees to a much smaller plant, Kuhn and Kesselmeier (2000) collected lichens from an open oak woodland in central California, USA, and studied their uptake of carbonyl sulfide (OCS) in a dynamic cuvette system under controlled conditions in the laboratory. When optimally hydrated, OCS was absorbed from the atmosphere by the lichens at a rate that gradually doubled as air temperature rose from approximately 3° to 25°C, whereupon the rate of OCS absorption dropped precipitously, reaching a value of zero at 35°C. Why is this significant?

OCS is the most stable and abundant reduced sulfur gas in the atmosphere and is thus a major player in determining earth's radiation budget. After making its way into the stratosphere, it can be photodissociated, as well as oxidized, to form SO_2, which is typically converted to sulfate aerosol particles that are highly reflective of incoming solar radiation and, therefore, have the capacity to significantly cool the earth as more and more of them collect above the tropopause. This being the case, biologically modulated COS concentrations may play a role in keeping earth's surface air temperature within bounds conducive to the continued existence of life, exactly what is implied by the observations of Kuhn and Kesselmeier.

Once air temperature rises above 25°C, the rate of removal of OCS from the air by this particular species of lichen declines dramatically. When this happens, more OCS remains in the air, which increases the potential for more OCS to make its way into the stratosphere, where it can be converted into sulfate aerosol particles that can reflect more incoming solar radiation back to space and thereby cool the earth. Since the consumption of OCS by lichens is under the physiological control of carbonic anhydrase—which is the key enzyme for OCS uptake in all higher plants, algae, and soil organisms—we could expect this phenomenon to be generally operative throughout much of the plant kingdom. This biological "thermostat" may well be powerful enough to define an upper limit above which the surface air temperature of the planet may be restricted from rising, even when changes in other forcing factors, such as greenhouse gases, produce an impetus for it to do so. For more about OCS, see Section 2.2 of this report.

Although BVOCs emitted from terrestrial plants both small and large are important to earth's climate, trees tend to dominate in this regard. Recent research suggests yet another way in which their response to atmospheric CO_2 enrichment may provide an effective counterbalance to the greenhouse properties of CO_2. The phenomenon begins with the propensity for CO_2-induced increases in BVOCs, together with the cloud particles they spawn, to enhance the amount of diffuse solar radiation reaching the earth's surface (Suraqui et al., 1974; Abakumova et al., 1996), which is followed by the ability of enhanced diffuse lighting to reduce the volume of shade within vegetative canopies (Roderick et al., 2001), which is followed by the tendency for less internal canopy shading to enhance whole-canopy photosynthesis (Healey et al., 1998), which finally produces the end result: a greater photosynthetic extraction of CO_2 from the air and the subsequent reduction of the strength of the atmosphere's greenhouse effect.

The significance of this process is described and documented at some length in Section 2.3 of this report. For example, Roderick et al. concluded that the Mt. Pinatubo eruption—a unique natural experiment to evaluate the overall climatic sensitivity of the planet—may well have resulted in the removal of an extra 2.5 Gt of carbon from the atmosphere due to its diffuse-light-enhancing stimulation of terrestrial photosynthesis in the year following the eruption. Additional real-world evidence for the existence of this phenomenon was provided by Gu et al. (2003), Law et al. (2002), Farquhar and Roderick (2003), Reichenau and Esser (2003), and Niyogi et al. (2004).

One final beneficial effect of CO_2-induced increases in BVOC emissions is the propensity of BVOCs to destroy tropospheric ozone, as documented by Goldstein et al. (2004). Earth's vegetation is responsible for the production of vast amounts of ozone (O_3) (Chameides et al., 1988; Harley et al., 1999), but it is also responsible for destroying a lot of O_3. With respect to the latter phenomenon, Goldstein et al. mention three major routes by which O_3 exits the air near the earth's surface: leaf stomatal uptake, surface deposition, and within-canopy gas-phase chemical reactions with BVOCs.

The first of these exit routes, according to Goldstein et al., accounts for 30 percent to 90 percent of total ecosystem O_3 uptake from the atmosphere (that is, O_3 destruction), while the remainder has typically been attributed to deposition on non-stomatal surfaces. However, they note that "Kurpius and Goldstein (2003) recently showed that the non-stomatal flux [from the atmosphere to oblivion] increased exponentially as a function of temperature at a coniferous forest site," and that "the exponential increase with temperature was consistent with the temperature dependence of monoterpene emissions from the same ecosystem, suggesting O_3 was lost via gas phase reactions with biogenically emitted terpenes before they could escape the forest canopy."

In a study designed to take the next step towards turning the implication of this observation into something stronger than a mere suggestion, Schade and Goldstein (2003) demonstrated that forest thinning dramatically enhances monoterpene emissions. In the current study, Goldstein et al. take another important step towards clarifying the issue by measuring the effect of forest thinning on O_3 destruction in an attempt to see if it is enhanced in parallel fashion to the thinning-induced increase in monoterpene emissions.

In a ponderosa pine plantation in the Sierra Nevada Mountains of California, USA, a management procedure to improve forest health and optimize tree growth was initiated on May 11, 2000 and continued through June 15, 2000. This procedure involved the use of a *masticator* to mechanically "chew up" smaller unwanted trees and leave their debris on site, which reduced plantation green leaf biomass by just over half. Simultaneously, monoterpene mixing ratios and fluxes were measured hourly within the plantation canopy, while total ecosystem O_3 destruction was "partitioned to differentiate loss due to gas-phase chemistry from stomatal uptake and deposition."

Goldstein *et al.* report that both the destruction of ozone due to gas-phase chemistry and emissions of monoterpenes increased dramatically with the onset of thinning, and that these phenomena continued in phase with each other thereafter. Hence, they "infer that the massive increase of O_3 flux [from the atmosphere to oblivion] during and following mastication is driven by loss of O_3 through chemical reactions with unmeasured terpenes or closely related BVOCs whose emissions were enhanced due to wounding [by the masticator]." Indeed, they say that "considered together, these observations provide a conclusive picture that the chemical loss of O_3 is due to reactions with BVOCs emitted in a similar manner as terpenes," and that "we can conceive no other possible explanation for this behavior other than chemical O_3 destruction in and above the forest canopy by reactions with BVOCs."

Goldstein *et al.* say their results "suggest that total reactive terpene emissions might be roughly a factor of 10 higher than the typically measured and modeled monoterpene emissions, making them larger than isoprene emissions on a global scale." If this proves to be the case, it will be a most important finding, for it would mean that vegetative emissions of terpenes, which lead to the destruction of ozone, are significantly greater than vegetative emissions of isoprene, which lead to the creation of ozone (Poisson et al., 2000). In addition, there is substantial evidence to suggest that the ongoing rise in the air's CO_2 content may well lead to an overall reduction in vegetative isoprene emissions, while at the same time enhancing vegetative productivity, which may well lead to an overall increase in vegetative terpene emissions. As a result, there is reason to believe that the ongoing rise in the air's CO_2 content will help to reduce the ongoing rise in the air's O_3 concentration, which should be a boon to the entire biosphere.

In conclusion, a wealth of real-world evidence is beginning to suggest that both rising air temperatures and CO_2 concentrations significantly increase desirable vegetative BVOC emissions, particularly from trees, which constitute the most prominent photosynthetic force on the planet, and that this phenomenon has a large number of extremely important and highly beneficial biospheric consequences.

These findings further demonstrate that the biology of the earth influences the climate of the earth. Specifically, they reveal a direct connection between the metabolic activity of trees and the propensity for the atmosphere to produce clouds, the metabolic activity of lichens and the presence of sulfate aerosol particles in the atmosphere that reflect incoming solar radiation, and the increased presence of BVOCs caused by rising CO_2 and an increase in diffuse solar radiation, which leads to increased photosynthetic extraction of CO_2 from the air. In each case, the relationship is one that is self-protecting of the biosphere. This being the case, we wonder how anyone can presume to decide what should or should not be done about anthropogenic CO_2 emissions.

Additional information on this topic, including reviews of newer publications as they become available, can be found at http://www.co2science.org/subject/a/aerosolsterr.php.

References

Abakumova, G.M., Feigelson, E.M., Russak, V. and Stadnik, V.V. 1996. Evaluation of long-term changes in radiation, cloudiness, and surface temperature on the territory of the former Soviet Union. *Journal of Climatology* **9**: 1319-1327.

Baldocchi, D., Falge, E., Gu, L.H., Olson, R., Hollinger, D., Running, S., Anthoni, P., Bernhofer, C., Davis, K., Evans, R., Fuentes, J., Goldstein, A., Katul, G., Law, B., Lee, X.H., Malhi, Y., Meyers, T., Munger, W., Oechel, W., Paw U, K.T., Pilegaard, K., Schmid, H.P., Valentini, R., Verma, S., Vesala, T., Wilson, K. and Wofsy, S. 2001. FLUXNET: A new tool to study the temporal and spatial variability of ecosystem-scale carbon dioxide, water vapor, and energy flux densities. *Bulletin of the American Meteorological Society* **82**: 2415-2434.

Baraldi, R., Rapparini, F., Oechel, W.C., Hastings, S.J., Bryant, P., Cheng, Y. and Miglietta, F. 2004. Monoterpene emission responses to elevated CO_2 in a Mediterranean-type ecosystem. *New Phytologist* **161**: 17-21.

Chameides, W.L., Lindsay, R.W., Richardson, J. and Kiang, C.S. 1988. The role of biogenic hydrocarbons in urban photochemical smog: Atlanta as a case study. *Science* **241**: 1473-1475.

Constable, J.V.H., Litvak, M.E., Greenberg, J.P. and Monson, R.K. 1999. Monoterpene emission from coniferous trees in response to elevated CO2 concentration and climate warming. *Global Change Biology* **5**: 255-267.

Farquhar, G.D. and Roderick, M.L. 2003. Pinatubo, diffuse light, and the carbon cycle. *Science* **299**: 1997-1998.

Goldstein, A.H., McKay, M., Kurpius, M.R., Schade, G.W., Lee, A., Holzinger, R. and Rasmussen, R.A. 2004. Forest thinning experiment confirms ozone deposition to forest canopy is dominated by reaction with biogenic VOCs. *Geophysical Research Letters* **31**: 10.1029/2004GL021259.

Gu, L., Baldocchi, D.D., Wofsy, S.C., Munger, J.W., Michalsky, J.J., Urbanski, S.P. and Boden, T.A. 2003. Response of a deciduous forest to the Mount Pinatubo eruption: Enhanced photosynthesis. *Science* **299**: 2035-2038.

Harley, P.C., Monson, R.K. and Lerdau, M.T. 1999. Ecological and evolutionary aspects of isoprene emission from plants. *Oecologia* **118**: 109-123.

Healey, K.D., Rickert, K.G., Hammer, G.L. and Bange, M.P. 1998. Radiation use efficiency increases when the diffuse component of incident radiation is enhanced under shade. *Australian Journal of Agricultural Research* **49**: 665-672.

Holben, B.N., Tanré, D., Smirnov, A., Eck, T.F., Slutsker, I., Abuhassan, N., Newcomb, W.W., Schafer, J.S., Chatenet, B., Lavenu, F., Kaufman, Y.J., Castle, J.V., Setzer, A., Markham, B., Clark, D., Frouin, R., Halthore, R., Karneli, A., O'Neill, N.T., Pietras, C., Pinker, R.T., Voss, K. and Zibordi, G. 2001. An emerging ground-based aerosol climatology: Aerosol Optical Depth from AERONET. *Journal of Geophysical Research* **106**: 12,067-12,097.

Idso, S.B. 1998. CO2-induced global warming: a skeptic's view of potential climate change. *Climate Research* **10**: 69-82.

Jasoni, R., Kane, C., Green, C., Peffley, E., Tissue, D., Thompson, L., Payton, P. and Pare, P.W. 2003. Altered leaf and root emissions from onion (*Allium cepa* L.) grown under elevated CO_2 conditions. *Environmental and Experimental Botany* **51**: 273-280.

Kavouras, I.G., Mihalopoulos, N. and Stephanou, E.G. 1998. Formation of atmospheric particles from organic acids produced by forests. *Nature* **395**: 683-686.

Kuhn, U. and Kesselmeier, J. 2000. Environmental variables controlling the uptake of carbonyl sulfide by lichens. *Journal of Geophysical Research* **105**: 26,783-26,792.

Kurpius, M.R. and Goldstein, A.H. 2003. Gas-phase chemistry dominates O_3 loss to a forest, implying a source of aerosols and hydroxyl radicals to the atmosphere. *Geophysical Research Letters* **30**: 10.1029/2002GL016785.

Law, B.E., Falge, E., Gu, L., Baldocchi, D.D., Bakwin, P., Berbigier, P., Davis, K., Dolman, A.J., Falk, M., Fuentes, J.D., Goldstein, A., Granier, A., Grelle, A., Hollinger, D., Janssens, I.A., Jarvis, P., Jensen, N.O., Katul, G., Mahli, Y., Matteucci, G., Meyers, T., Monson, R., Munger, W., Oechel, W., Olson, R., Pilegaard, K., Paw U, K.T., Thorgeirsson, H., Valentini, R., Verma, S., Vesala, T., Wilson, K. and Wofsy, S. 2002. Environmental controls over carbon dioxide and water vapor exchange of terrestrial vegetation. *Agricultural and Forest Meteorology* **113**: 97-120.

Litvak, M.E., Loreto, F., Harley, P.C., Sharkey, T.D. and Monson, R.K. 1996. The response of isoprene emission rate and photosynthetic rate to photon flux and nitrogen supply in aspen and white oak trees. *Plant, Cell and Environment* **19**: 549-559.

Loreto, F., Fischbach, R.J., Schnitzler, J.P., Ciccioli, P., Brancaleoni, E., Calfapietra, C. and Seufert, G. 2001. Monoterpene emission and monoterpene synthase activities in the Mediterranean evergreen oak *Quercus ilex* L. grown at elevated CO2 concentrations. *Global Change Biology* **7**: 709-717.

Niyogi, D., Chang, H.-I., Saxena, V.K., Holt, T., Alapaty, K., Booker, F., Chen, F., Davis, K.J., Holben, B., Matsui, T., Meyers, T., Oechel, W.C., Pielke Sr., R.A., Wells, R., Wilson, K. and Xue, Y. 2004. Direct observations of the effects of aerosol loading on net ecosystem CO_2 exchanges over different landscapes. *Geophysical Research Letters* **31**: 10.1029/2004GL020915.

O'Dowd, C.D., Aalto, P., Hameri, K., Kulmala, M. and Hoffmann, T. 2002. Atmospheric particles from organic vapours. *Nature* **416**: 497-498.

Peñuelas, J. and Llusia, J. 2003. BVOCs: plant defense against climate warming? *Trends in Plant Science* **8**: 105-109.

Peñuelas, J., Llusia, J. and Estiarte, M. 1995. Terpenoids: a plant language. *Trends in Ecology and Evolution* **10**: 289.

Pichersky, E. and Gershenzon, J. 2002. The formation and function of plant volatiles: perfumes for pollinator attraction and defense. *Current Opinion in Plant Biology* **5**: 237-243.

Poisson, N., Kanakidou, M. and Crutzen, P.J. 2000. Impact of non-methane hydrocarbons on tropospheric chemistry and the oxidizing power of the global troposphere: 3-

dimensional modeling results. *Journal of Atmospheric Chemistry* **36**: 157-230.

Raisanen, T., Ryyppo, A. and Kellomaki, S. 2008. Effects of elevated CO_2 and temperature on monoterpene emission of Scots pine (*Pinus sylvestris* L.). *Atmospheric Environment* **42**: 4160-4171.

Reichenau, T.G. and Esser, G. 2003. Is interannual fluctuation of atmospheric CO_2 dominated by combined effects of ENSO and volcanic aerosols? *Global Biogeochemical Cycles* **17**: 10.1029/2002GB002025.

Roderick, M.L., Farquhar, G.D., Berry, S.L. and Noble, I.R. 2001. On the direct effect of clouds and atmospheric particles on the productivity and structure of vegetation. *Oecologia* **129**: 21-30.

Sarmiento, J.L. 1993. Atmospheric CO_2 stalled. *Nature* **365**: 697-698.

Schade, G.W. and Goldstein, A.H. 2003. Increase of monoterpene emissions from a pine plantation as a result of mechanical disturbances. *Geophysical Research Letters* **30**: 10.1029/2002GL016138.

Shallcross, D.E. and Monks, P.S. 2000. A role for isoprene in biosphere-climate-chemistry feedbacks. *Atmospheric Environment* **34**: 1659-1660.

Shulaev, V., Silverman, P. and Raskin, I. 1997. Airborne signaling by methyl salicylate in plant pathogen resistance. *Nature* **385**: 718-721.

Stanhill, G. and Cohen, S. 2001. Global dimming: a review of the evidence for a widespread and significant reduction in global radiation with discussion of its probable causes and possible agricultural consequences. *Agricultural and Forest Meteorology* **107**: 255-278.

Staudt, M., Joffre, R., Rambal, S. and Kesselmeier, J. 2001. Effect of elevated CO_2 on monoterpene emission of young *Quercus ilex* trees and its relation to structural and ecophysiological parameters. *Tree Physiology* **21**: 437-445.

Suraqui, S., Tabor, H., Klein, W.H. and Goldberg, B. 1974. Solar radiation changes at Mt. St. Katherine after forty years. *Solar Energy* **16**: 155-158.

Vuorinen, T., Reddy, G.V.P., Nerg, A.-M. and Holopainen, J.K. 2004. Monoterpene and herbivore-induced emissions from cabbage plants grown at elevated atmospheric CO_2 concentration. *Atmospheric Environment* **38**: 675-682.

2.8.4. Non-Biological (Anthropogenic)

There are several ways the activities of humanity lead to the creation of aerosols that have the potential to alter earth's radiation balance and affect its climate.

Contrails created in the wake of emissions from jet aircraft are one example. Minnis *et al.* (2004) have calculated that nearly all of the surface warming observed over the United States between 1975 and 1994 (0.54°C) may well be explained by aircraft-induced increases in cirrus cloud coverage over that period. If true, this result would imply that little to none of the observed U.S. warming over that period could be attributed to the concomitant increase in the air's CO_2 content.

Ship tracks, or bright streaks that form in layers of marine stratus clouds, are another example. They are created by emissions from ocean-going vessels; these persistent and highly reflective linear patches of low-level clouds generally tend to cool the planet (Ferek *et al.*, 1998; Schreier *et al.*, 2006). Averaged over the surface of the earth both day and night and over the year, Capaldo *et al.* (1999) calculated that this phenomenon creates a mean negative radiative forcing of -0.16 Wm^{-2} in the Northern Hemisphere and -0.06 Wm^{-2} in the Southern Hemisphere, which values are to be compared to the much larger positive radiative forcing of approximately 4 Wm^{-2} due to a 300 ppm increase in the atmosphere's CO_2 concentration.

In some cases, the atmosphere over the sea also carries a considerable burden of anthropogenically produced aerosols from terrestrial sites. In recent years, attention to this topic has centered on highly polluted air from south and southeast Asia that makes its way over the northern Indian Ocean during the dry monsoon season. There has been much discussion about the impact of this phenomenon on regional climates. Norris (2001) looked at cloud cover as the ultimate arbiter of the various competing hypotheses, finding that daytime low-level oceanic cloud cover increased substantially over the last half of the past century in both the Northern and Southern Hemispheres at essentially all hours of the day. This finding is indicative of a pervasive net cooling effect.

Aerosol-generating human activities also have a significant impact on local, as well as more wide-ranging, climatic phenomena over land. Sahai (1998) found that although suburban areas of Nagpur, India had warmed over recent decades, the central part of the city had cooled, especially during the day, because of "increasing concentrations of suspended particulate matter." Likewise, outside of, but adjacent to, industrial complexes in the Po Valley of Italy, Facchini *et al.* (1999) found that water vapor was more likely to form on aerosols that had been altered by human-produced organic solutes, and that this phenomenon led to the creation of more numerous and more highly reflective cloud droplets that had a tendency to cool the surface below them.

In a similar vein, Rosenfeld (2000) studied pollution tracks downwind of urban/industrial complexes in Turkey, Canada and Australia. His findings indicated that the clouds comprising these pollution tracks were composed of small droplets that suppressed precipitation by inhibiting further coalescence and ice precipitation formation. In commenting on this research, Toon (2000) pointed out that when clouds are composed of smaller droplets, they will not "rain out" as quickly and will therefore last longer and cover more of the earth, both of which effects tend to cool the globe.

In reviewing these and other advances in the field of anthropogenic aerosol impacts on clouds, Charlson *et al.* (2001) note that droplet clouds "are the most important factor controlling the albedo (reflectivity) and hence the temperature of our planet." They say man-made aerosols "have a strong influence on cloud albedo, with a global mean forcing estimated to be of the same order (but opposite in sign) as that of greenhouse gases," and "both the forcing [of this man-induced impetus for cooling] and its magnitude may be even larger than anticipated." They rightly warn that lack of inclusion of the consequences of these important phenomena in climate change deliberations "poses additional uncertainty beyond that already recognized by the Intergovernmental Panel on Climate Change, making the largest uncertainty in estimating climate forcing even larger."

Another assessment of the issue was provided by Ghan *et al.* (2001), who studied both the positive radiative forcings of greenhouse gases and the negative radiative forcings of anthropogenic aerosols and reported that current best estimates of "the total global mean present-day anthropogenic forcing range from 3 Wm^{-2} to -1 Wm^{-2}," which represents everything from a modest warming to a slight cooling. After performing their own analysis of the problem, they reduced the magnitude of this range somewhat but the end result still stretched from a small cooling influence to a modest impetus for warming. "Clearly," they concluded, "the great uncertainty in the radiative forcing must be reduced if the observed climate record is to be reconciled with model predictions and if estimates of future climate change are to be useful in formulating emission policies."

Another pertinent observation comes from Stanhill and Cohen (2001), who reviewed numerous solar radiation measurement programs around the world to see if there had been any trend in the mean amount of solar radiation falling on the surface of the earth over the past half-century. They determined there was a significant 50-year downward trend in this parameter that "has globally averaged 0.51 ± 0.05 Wm^{-2} per year, equivalent to a reduction of 2.7 percent per decade, [which] now totals 20 Wm^{-2}." They also concluded that the most probable explanation for this observation "is that increases in man-made aerosols and other air pollutants have changed the optical properties of the atmosphere, in particular those of clouds."

Although this surface-cooling influence is huge, it falls right in the mid-range of a similar solar radiative perturbation documented by Satheesh and Ramanathan (2000) in their study of the effects of human-induced pollution over the tropical northern Indian Ocean, where they determined that "mean clear-sky solar radiative heating for the winters of 1998 and 1999 decreased at the ocean surface by 12 to 30 Wm^{-2}." Hence, the decline in solar radiation reception discovered by Stanhill and Cohen could well be real. And if it is, it represents a *tremendous* counter-influence to the enhanced greenhouse effect produced by the contemporaneous increase in atmospheric CO_2 concentration.

In a more recent study, Ruckstuhl *et al.* (2008) presented "observational evidence of a strong decline in aerosol optical depth over mainland Europe during the last two decades of rapid warming"—when air temperatures rose by about 1°C after 1980—via analyses of "aerosol optical depth measurements from six specific locations and surface irradiance measurements from a large number of radiation sites in Northern Germany and Switzerland."

In consequence of the observed decline in aerosol concentration of up to 60 percent, the authors state there was "a statistically significant increase of solar irradiance under cloud-free skies since the 1980s." The value of the direct aerosol effect of this radiative forcing was approximately 0.84 Wm^{-2}; and when combined with the concomitant cloud-induced radiative forcing of about 0.16 Wm^{-2}, it led to a total radiative surface climate forcing over mainland Europe of about 1 Wm^{-2} that "most probably strongly contributed to the recent rapid warming in Europe." Cleaning up significantly polluted skies, it seems, can provide an even greater impetus for climate warming than does the carbon dioxide that is concurrently emitted to them, as has apparently been the case over mainland Europe for the past quarter-century.

Anthropogenic aerosols plainly have a major effect on climate. The evidence is dear contrails created by emissions from jet aircraft, ship tracks created by ocean-going vessels, and air pollution from terrestrial sources all have effects on temperatures that rival or exceed the likely effect of rising CO_2 levels. With the progress that has been made in recent years in reducing air pollution in developed countries, it is possible the lion's share of the warming has likely been produced by the removal from the atmosphere of *true* air pollutants.

Additional information on this topic, including reviews of newer publications as they become available, can be found at http://www.co2science.org/subject/a/aerononbioanthro.php.

References

Capaldo, K., Corbett, J.J., Kasibhatla, P., Fischbeck, P. and Pandis, S.N. 1999. Effects of ship emissions on sulphur cycling and radiative climate forcing over the ocean. *Nature* **400**: 743-746.

Charlson, R.J., Seinfeld, J.H., Nenes, A., Kulmala, M., Laaksonen, A. and Facchini, M.C. 2001. Reshaping the theory of cloud formation. *Science* **292**: 2025-2026.

Facchini, M.C., Mircea, M., Fuzzi, S. and Charlson, R.J. 1999. Cloud albedo enhancement by surface-active organic solutes in growing droplets. *Nature* **401**: 257-259.

Ferek, R.J., Hegg, D.A., Hobbs, P.V., Durkee, P. and Nielsen, K. 1998. Measurements of ship-induced tracks in clouds off the Washington coast. *Journal of Geophysical Research* **103**: 23,199-23,206.

Ghan, S.J., Easter, R.C., Chapman, E.G., Abdul-Razzak, H., Zhang, Y., Leung, L.R., Laulainen, N.S., Saylor, R.D. and Zaveri, R.A. 2001. A physically based estimate of radiative forcing by anthropogenic sulfate aerosol. *Journal of Geophysical Research* **106**: 5279-5293.

Minnis, P., Ayers, J.K., Palikonda, R. and Phan, D. 2004. Contrails, cirrus trends, and climate. *Journal of Climate* **17**: 1671-1685.

Norris, J.R. 2001. Has northern Indian Ocean cloud cover changed due to increasing anthropogenic aerosol? *Geophysical Research Letters* **28**: 3271-3274.

Rosenfeld, D. 2000. Suppression of rain and snow by urban and industrial air pollution. *Science* **287**: 1793-1796.

Ruckstuhl, C., Philipona, R., Behrens, K., Coen, M.C., Durr, B., Heimo, A., Matzler, C., Nyeki, S., Ohmura, A., Vuilleumier, L., Weller, M., Wehrli, C. and Zelenka, A. 2008. Aerosol and cloud effects on solar brightening and the recent rapid warming. *Geophysical Research Letters* **35**: 10.1029/2008GL034228.

Sahai, A.K. 1998. Climate change: a case study over India. *Theoretical and Applied Climatology* **61**: 9-18.

Satheesh, S.K. and Ramanathan, V. 2000. Large differences in tropical aerosol forcing at the top of the atmosphere and Earth's surface. *Nature* **405**: 60-63.

Schreier, M., Kokhanovsky, A.A., Eyring, V., Bugliaro, L., Mannstein, H., Mayer, B., Bovensmann, H. and Burrows, J.P. 2006. Impact of ship emissions on the microphysical, optical and radiative properties of marine stratus: a case study. *Atmospheric Chemistry and Physics* **6**: 4925-4942.

Stanhill, G. and Cohen, S. 2001. Global dimming: a review of the evidence for a widespread and significant reduction in global radiation with discussion of its probable causes and possible agricultural consequences. *Agricultural and Forest Meteorology* **107**: 255-278.

Toon, O.W. 2000. How pollution suppresses rain. *Science* **287**: 1763-1765.

2.8.5. Non-Biological (Natural)

We conclude our section on aerosols with a brief discussion of a non-biological, naturally produced aerosol—dust. Dust is about as natural and ubiquitous a substance as there is. One might think we would have a pretty good handle on what it does to earth's climate as it is moved about by the planet's ever-active atmosphere. But such is not the case, as was made strikingly clear by Sokolik (1999), who with the help of nine colleagues summarized the sentiments of a number of scientists who have devoted their lives to studying the subject.

Sokolik notes state-of-the-art climate models "rely heavily on oversimplified parameterizations" of many important dust-related phenomena, "while ignoring others." As a result, the group concludes. "the magnitude and even the sign of dust net direct radiative forcing of climate remains unclear."

According to Sokolik, there are a number of unanswered questions about airborne dust, including: (1) How does one quantify dust emission rates from both natural and anthropogenic (disturbed) sources with required levels of temporal and spatial resolution? (2) How does one accurately determine the composition, size, and shape of dust particles from ground-based and aircraft measurements? (3) How does one adequately measure and model light absorption by mineral particles? (4) How does one link the ever-evolving optical, chemical, and physical properties of dust to its life cycle in the air? (5) How does one model complex multi-layered aerosol stratification in the dust-laden atmosphere? (6) How does one quantify airborne dust properties from satellite observations?

In discussing these questions, Sokolik makes some interesting observations, noting that: (1) what is currently known (or believed to be known) about dust emissions "is largely from micro-scale experiments and theoretical studies," (2) new global data sets are needed to provide "missing information" on input parameters (such as soil type, surface roughness, and soil moisture) required to model dust emission rates, (3) improvements in methods used to determine some of these parameters are also "sorely needed," (4) how to adequately measure light absorption by mineral particles is still an "outstanding problem," and (5) it "remains unknown how well these measurements represent the light absorption by aerosol particles suspended in the atmosphere."

It is easy to understand why Sokolik says "a challenge remains in relating dust climatology and the processes controlling the evolution of dust at all relevant spatial/temporal scales needed for chemistry and climate models," for until this challenge is met, we will but "see through a glass, darkly," especially when it comes to trying to discern the effects of airborne dust on earth's climate.

Vogelmann et al. (2003) reiterate that "mineral aerosols have complex, highly varied optical properties that, for equal loadings, can cause differences in the surface IR flux [of] between 7 and 25 Wm^{-2} (Sokolik et al., 1998)," while at the same time acknowledging that "only a few large-scale climate models currently consider aerosol IR effects (e.g., Tegen et al., 1996; Jacobson, 2001) despite their potentially large forcing."

Vogelmann et al. "use[d] high-resolution spectra to obtain the IR radiative forcing at the surface for aerosols encountered in the outflow from northeastern Asia," based on measurements made by the Marine-Atmospheric Emitted Radiance Interferometer aboard the NOAA Ship Ronald H. Brown during the Aerosol Characterization Experiment-Asia. This work led them to conclude that "daytime surface IR forcings are often a few Wm^{-2} and can reach almost 10 Wm^{-2} for large aerosol loadings," which values, in their words, "are comparable to or larger than the 1 to 2 Wm^{-2} change in the globally averaged surface IR forcing caused by greenhouse gas increases since pre-industrial times." And in a massive understatement of

fact, Vogelmann et al. say that these results "highlight the importance of aerosol IR forcing which should be included in climate model simulations."

Another aspect of the dust-climate connection centers on the African Sahel, which has figured prominently in discussions of climate change ever since it began to experience extended drought conditions in the late 1960s and early '70s. Initial studies of the drought attributed it to anthropogenic factors such as overgrazing of the region's fragile grasses, which tends to increase surface albedo, which was envisioned to reduce precipitation, resulting in a further reduction in the region's vegetative cover, and so on (Otterman, 1974; Charney, 1975). This scenario, however, was challenged by Jackson and Idso (1975) and Idso (1977) on the basis of empirical observations; while Lamb (1978) and Folland et al. (1986) attributed the drought to large-scale atmospheric circulation changes triggered by multidecadal variations in sea surface temperature.

Building on the insights provided by these latter investigations, Giannini et al. (2003) presented evidence based on an ensemble of integrations with a general circulation model of the atmosphere—forced only by the observed record of sea surface temperature—which suggested that the "variability of rainfall in the Sahel results from the response of the African summer monsoon to oceanic forcing amplified by land-atmosphere interaction." The success of this analysis led them to conclude that "the recent drying trend in the semi-arid Sahel is attributed to warmer-than-average low-latitude waters around Africa, which, by favoring the establishment of deep convection over the ocean, weaken the continental convergence associated with the monsoon and engender widespread drought from Senegal to Ethiopia." They further concluded that "the secular change in Sahel rainfall during the past century was not a direct consequence of regional environmental change, anthropogenic in nature or otherwise."

In a companion article, Prospero and Lamb (2003) report that measurements made from 1965 to 1998 in the Barbados trade winds show large interannual changes in the concentration of dust of African origin that are highly anticorrelated with the prior year's rainfall in the Soudano-Sahel. They say the 2001 IPCC report "assumes that natural dust sources have been effectively constant over the past several hundred years and that all variability is attributable to human land-use impacts." But "there is little firm evidence to support either of these assumptions," they say, and their findings demonstrate why: The IPCC assumptions are wrong.

Clearly, much remains to be learned about the climatic impacts of dust before anyone can place any confidence in the climatic projections of the IPCC. Additional information on this topic, including reviews of newer publications as they become available, can be found at http://www.co2science.org/subject/a/aerononbio nat.php.

References

Charney, J.G. 1975. Dynamics of desert and drought in the Sahel. *Quarterly Journal of the Royal Meteorological Society* **101**: 193-202.

Folland, C.K., Palmer, T.N. and Parker, D.E. 1986. Sahel rainfall and worldwide sea temperatures, 1901-85. *Nature* **320**: 602-607.

Giannini, A., Saravanan, R. and Chang, P. 2003. Oceanic forcing of Sahel rainfall on interannual to interdecadal time scales. *Science* **302**: 1027-1030.

Houghton, J.T., Ding, Y., Griggs, D.J., Noguer, M., van der Linden, P.J., Xiaosu, D., Maskell, K. and Johnson, C.A. (Eds.). 2001. *Climate Change 2001: The Scientific Basis.* Cambridge University Press, Cambridge, UK. (Contribution of Working Group 1 to the Third Assessment Report of the Intergovernmental Panel on Climate Change.)

Idso, S.B. 1977. A note on some recently proposed mechanisms of genesis of deserts. *Quarterly Journal of the Royal Meteorological Society* **103**: 369-370.

Jackson, R.D. and Idso, S.B. 1975. Surface albedo and desertification. *Science* **189**: 1012-1013.

Jacobson, M.Z. 2001. Global direct radiative forcing due to multicomponent anthropogenic and natural aerosols. *Journal of Geophysical Research* **106**: 1551-1568.

Lamb, P.J. 1978. Large-scale tropical Atlantic surface circulation patterns associated with sub-Saharan weather anomalies. *Tellus* **30**: 240-251.

Otterman, J. 1974. Baring high-albedo soils by overgrazing: a hypothesized desertification mechanism. *Science* **186**: 531-533.

Prospero, J.M. and Lamb, P.J. 2003. African droughts and dust transport to the Caribbean: climate change implications. *Science* **302**: 1024-1027.

Sokolik, I.N. 1999. Challenges add up in quantifying radiative impact of mineral dust. *EOS: Transactions, American Geophysical Union* **80**: 578.

Sokolik, I.N., Toon, O.B. and Bergstrom, R.W. 1998. Modeling the radiative characteristics of airborne mineral aerosols at infrared wavelengths. *Journal of Geophysical Research* **103**: 8813-8826.

Tegen, I., Lacis, A.A. and Fung, I. 1996. The influence on climate forcing of mineral aerosols from disturbed soils. *Nature* **380**: 419-422.

Vogelmann, A.M., Flatau, P.J., Szczodrak, M., Markowicz, K.M. and Minnett, P.J. 2003. *Geophysical Research Letters* **30**: 10.1029/2002GL016829.

[this page intentionally blank]

3

Observations: Temperature Records

Introduction

The Intergovernmental Panel on Climate Change (IPCC) claims to have found evidence in paeloclimatic data that higher levels of atmospheric CO_2 can cause or amplify an increase in global temperatures (IPCC, 2007-I, Chapter 6). The IPCC further claims to have evidence of an anthropogenic effect on climate in the earth's temperature history during the past century (Chapters 3, 9), in the pattern (or "fingerprint") of more recent warming (Chapter 9, Section 9.4.1.4), in data from land-based temperature stations and satellites (Chapter 3), and in the temperature records of the Artic region and Antarctica where models predict anthropogenic global warming should be detected first (Chapter 11, Section 8). In this chapter, we critically examine the data used to support each of these claims, starting with the relationship between CO_2 and temperature in ancient climates.

References

IPCC. 2007-I. *Climate Change 2007: The Physical Science Basis. Contribution of Working Group I to the Fourth Assessment Report of the Intergovernmental Panel on Climate Change.* Solomon, S., D. Qin, M. Manning, Z. Chen, M. Marquis, K.B. Averyt, M. Tignor and H.L. Miller. (Eds.) Cambridge University Press, Cambridge, UK.

3.1. Paeloclimatic Data

Rothman (2002) derived a 500-million-year history of the air's CO_2 content based on considerations related to the chemical weathering of rocks, volcanic and metamorphic degassing, and the burial of organic carbon, along with considerations related to the isotopic content of organic carbon and strontium in marine sedimentary rocks. The results of this analysis suggest that over the majority of the half-billion-year record, earth's atmospheric CO_2 concentration fluctuated between values that were two to four times greater than those of today at a dominant period on the order of 100 million years. Over the last 175 million years, however, the data depict a long-term decline in the air's CO_2 content.

Rothman reports that the CO_2 history "exhibits no systematic correspondence with the geologic record of climatic variations at tectonic time scales." A visual examination of Rothman's plot of CO_2 and concomitant major cold and warm periods indicates the three most striking peaks in the air's CO_2 concentration occur either totally or partially within periods of time when earth's climate was relatively cool.

A more detailed look at the most recent 50 million years of earth's thermal and CO_2 history was prepared by Pagani *et al.* (2005). They found about 43 million years ago, the atmosphere's CO_2 concentration was approximately 1400 ppm and the oxygen isotope ratio (a proxy for temperature) was

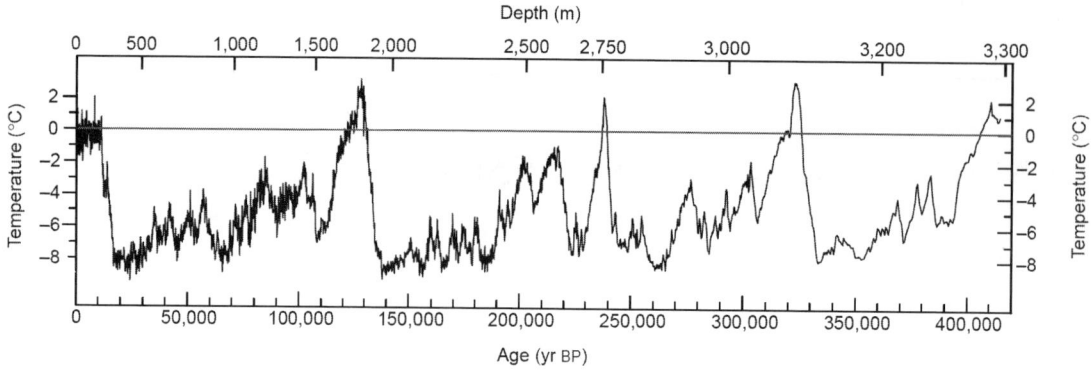

Figure 3.1. Temperature history derived by Petit *et al.* (1999) from an ice core extracted from the Russian Vostok drilling station in East Antarctica.

about 1.0 per mil. Then, over the next ten million years, the air's CO_2 concentration experienced three huge oscillations on the order of 1000 ppm from peak to valley. In the first two oscillations, temperature did not appear to respond at all to the change in CO_2, exhibiting an uninterrupted slow decline. Following the third rise in CO_2, however, temperatures seemed to respond, but in the direction opposite to what the greenhouse theory of global warming predicts, as the rise in CO_2 was followed by the sharpest drop in temperature of the entire record.

Following this large drop in temperature between 34 and 33 million years before present (Ma BP), the oxygen isotope ratio hovered around a value of 2.7 per mil from about 33 to 26 Ma BP, indicating little change in temperature over that period. The corresponding CO_2 concentration, on the other hand, experienced about a 500 ppm increase around 32 Ma BP, after which it dropped 1,000 ppm over the next two million years, only to rise again by a few hundred ppm, refuting – three times – the CO_2-induced global warming hypothesis. Next, around 26 Ma BP, the oxygen isotope ratio dropped to about 1.4 per mil (implying a significant *rise* in temperature), during which time the air's CO_2 content declined. From 24 Ma BP to the end of the record at 5 Ma BP, there were relatively small variations in atmospheric CO_2 content but relatively large variations in oxygen isotope values, both up and down. All of these many observations, according to Pagani *et al.* (2005), "argue for a decoupling between global climate and CO_2."

Moving closer to the modern era, Fischer *et al.* (1999) examined trends of atmospheric CO_2 and air temperature derived from Antarctic ice core data that extended back in time a quarter of a million years.

Over this period, the three most dramatic warming events experienced on earth were the terminations of the last three ice ages; and for each of these climatic transitions, earth's air temperature always rose well in advance of the increase in atmospheric CO_2. In fact, the air's CO_2 content did not begin to rise until 400 to 1,000 years after the planet began to warm.

Another research team, Petit *et al.* (1999), studied the beginnings rather than the ends of glacial ages. They discovered that during all glacial inceptions of the past half million years, temperature always dropped well before the decline in the air's CO_2 concentration. They said their data indicate that "the CO_2 decrease lags the temperature decrease by several thousand years." Petit *et al.* also found the current interglacial is the coolest of the five most recent such periods. In fact, the peak temperatures of the four interglacials that preceded it were, on average, more than 2°C warmer than that of the one in which we currently live. (See Figure 3.1.)

Figure 3.1 tells us three things about the current warm period. First, temperatures of the last decades of the twentieth century were "unprecedented" or "unusual" only because they were *cooler* than during past interglacial peaks. Second, the current temperature of the globe cannot be taken as evidence of an anthropogenic effect since it was warmer during parts of all preceding interglacials for which we have good proxy temperature data. And third, the higher temperatures of the past four interglacials cannot be attributed to higher CO_2 concentrations caused by some non-human influence because atmospheric CO_2 concentrations during all four prior interglacials never rose above approximately 290 ppm, whereas the air's CO_2 concentration today stands at nearly 380 ppm.

Likewise, Mudelsee (2001) determined that

variations in atmospheric CO_2 concentration lagged behind variations in air temperature by 1,300 to 5,000 years over the past 420,000 years. During certain climatic transitions characterized by rapid warmings of several degrees Centigrade, which were followed by slower coolings that returned the climate to essentially full glacial conditions, Staufer *et al.* (1998) observed the atmospheric CO_2 concentration derived from ice core records typically varied by less than 10 ppm. They, too, considered the CO_2 perturbations to have been caused by the changes in climate, rather than vice versa.

Other studies have also demonstrated this reverse coupling of atmospheric CO_2 and temperature (e.g., Cheddadi *et al.*, 1998; Gagan *et al.*, 1998; Raymo *et al.*, 1998), where temperature is the independent variable that appears to induce changes in CO_2. Steig (1999) noted cases between 7,000 and 5,000 years ago when atmospheric CO_2 concentrations increased by just over 10 ppm at a time when temperatures in both hemispheres cooled.

Caillon *et al.* (2003) measured the isotopic composition of argon – specifically, $\delta^{40}Ar$, which they argue "can be taken as a climate proxy, thus providing constraints about the timing of CO_2 and climate change" – in air bubbles in the Vostok ice core over the period that comprises what is called Glacial Termination III, which occurred about 240,000 years ago. The results of their tedious but meticulous analysis led them to conclude that "the CO_2 increase lagged Antarctic deglacial warming by 800 ± 200 years." This finding, in their words, "confirms that CO_2 is not the forcing that initially drives the climatic system during a deglaciation."

Indermuhle *et al.* (1999) determined that after the termination of the last great ice age, the CO_2 content of the air gradually rose by approximately 25 ppm in almost linear fashion between 8,200 and 1,200 years ago, over a period of time that saw a slow but steady *decline* in global air temperature. On the other hand, when working with a high-resolution temperature and atmospheric CO_2 record spanning the period 60 to 20 thousand years ago, Indermuhle et al. (2000) discovered four distinct periods when temperatures rose by approximately 2°C and CO_2 rose by about 20 ppm. However, one of the statistical tests they performed on the data suggested that the shifts in the air's CO_2 content during these intervals *followed* the shifts in air temperature by approximately 900 years; while a second statistical test yielded a mean CO_2 lag time of 1,200 years.

Another pertinent study is that of Siegenthaler *et al.* (2005), who analyzed CO_2 and proxy temperature (δD, the ratio of deuterium to hydrogen) data derived from an ice core in Antarctica. Results of their analysis revealed a coupling of Antarctic temperature and CO_2 in which they obtained the best correlation between CO_2 and temperature "for a lag of CO_2 of 1900 years." Specifically, over the course of glacial terminations V to VII, they indicate that "the highest correlation of CO_2 and deuterium, with use of a 20-ky window for each termination, yields a lag of CO_2 to deuterium of 800, 1600, and 2800 years, respectively." In addition, they note that "this value is consistent with estimates based on data from the past four glacial cycles," citing in this regard the work of Fischer *et al.* (1999), Monnin *et al.* (2001) and Caillon *et al.* (2003).

These observations seem to undermine the IPCC's claims that the CO_2 produced by the burning of fossil fuels will lead to catastrophic global warming. Nevertheless, Siegenthaler *et al.* stubbornly state that the new findings "do not cast doubt ... on the importance of CO_2 as a key amplification factor of the large observed temperature variations of glacial cycles." The previously cited Caillon *et al.* also avoid the seemingly clear implication of their own findings, that CO2 doesn't *cause* global warming. We find such disclaimers disingenuous.

When temperature is found to lead CO_2 by thousands of years, during both glacial terminations and inceptions (Genthon *et al.*, 1987; Fischer *et al.*, 1999; Petit *et al.*, 1999; Indermuhle *et al.*, 2000; Monnin *et al.*, 2001; Mudelsee, 2001; Caillon *et al.*, 2003), it is extremely likely that CO_2 plays only a minor role in enhancing temperature changes that are induced by something else. Compared with the mean conditions of the preceding four interglacials, there is currently 90 ppm more CO_2 in the air and yet it is currently more than 2°C colder than it was then. There is no way these real-world observations can be construed to suggest that a significant increase in atmospheric CO_2 would necessarily lead to *any* global warming, much less the catastrophic type that is predicted by the IPCC.

References

Caillon, N., Severinghaus, J.P., Jouzel, J., Barnola, J.-M., Kang, J. and Lipenkov, V.Y. 2003. Timing of atmospheric CO_2 and Antarctic temperature changes across Termination III. *Science* **299**: 1728-1731.

Cheddadi, R., Lamb, H.F., Guiot, J. and van der Kaars, S. 1998. Holocene climatic change in Morocco: a quantitative reconstruction from pollen data. *Climate Dynamics* **14**: 883-890.

Fischer, H., Wahlen, M., Smith, J., Mastroianni, D. and Deck, B. 1999. Ice core records of atmospheric CO_2 around the last three glacial terminations. *Science* **283**: 1712-1714.

Gagan, M.K., Ayliffe, L.K., Hopley, D., Cali, J.A., Mortimer, G.E., Chappell, J., McCulloch, M.T. and Head, M.J. 1998. Temperature and surface-ocean water balance of the mid-Holocene tropical western Pacific. *Science* **279**: 1014-1017.

Genthon, C., Barnola, J.M., Raynaud, D., Lorius, C., Jouzel, J., Barkov, N.I., Korotkevich, Y.S. and Kotlyakov, V.M. 1987. Vostok ice core: Climatic response to CO_2 and orbital forcing changes over the last climatic cycle. *Nature* **329**: 414-418.

Indermuhle, A., Monnin, E., Stauffer, B. and Stocker, T.F. 2000. Atmospheric CO_2 concentration from 60 to 20 kyr BP from the Taylor Dome ice core, Antarctica. *Geophysical Research Letters* **27**: 735-738.

Indermuhle, A., Stocker, T.F., Joos, F., Fischer, H., Smith, H.J., Wahllen, M., Deck, B., Mastroianni, D., Tschumi, J., Blunier, T., Meyer, R. and Stauffer, B. 1999. Holocene carbon-cycle dynamics based on CO_2 trapped in ice at Taylor Dome, Antarctica. *Nature* **398**: 121-126.

Monnin, E., Indermühle, A., Dällenbach, A., Flückiger, J, Stauffer, B., Stocker, T.F., Raynaud, D. and Barnola, J.-M. 2001. Atmospheric CO_2 concentrations over the last glacial termination. *Nature* **291**: 112-114.

Mudelsee, M. 2001. The phase relations among atmospheric CO_2 content, temperature and global ice volume over the past 420 ka. *Quaternary Science Reviews* **20**: 583-589.

Pagani, M., Authur, M.A. and Freeman, K.H. 1999. Miocene evolution of atmospheric carbon dioxide. *Paleoceanography* **14**: 273-292.

Pagani, M., Zachos, J.C., Freeman, K.H., Tipple, B. and Bohaty, S. 2005. Marked decline in atmospheric carbon dioxide concentrations during the Paleogene. *Science* **309**: 600-603.

Pearson, P.N. and Palmer, M.R. 1999. Middle Eocene seawater pH and atmospheric carbon dioxide concentrations. *Science* **284**: 1824-1826.

Pearson, P.N. and Palmer, M.R. 2000. Atmospheric carbon dioxide concentrations over the past 60 million years. *Nature* **406**: 695-699.

Petit, J.R., Jouzel, J., Raynaud, D., Barkov, N.I., Barnola, J.-M., Basile, I., Bender, M., Chappellaz, J., Davis, M., Delaygue, G., Delmotte, M., Kotlyakov, V.M., Legrand, M., Lipenkov, V.Y., Lorius, C., Pepin, L., Ritz, C., Saltzman, E. and Stievenard, M. 1999. Climate and atmospheric history of the past 420,000 years from the Vostok ice core, Antarctica. *Nature* **399**: 429-436.

Raymo, M.E., Ganley, K., Carter, S., Oppo, D.W. and McManus, J. 1998. Millennial-scale climate instability during the early Pleistocene epoch. *Nature* **392**: 699-702.

Rothman, D.H. 2002. Atmospheric carbon dioxide levels for the last 500 million years. *Proceedings of the National Academy of Sciences USA* **99**: 4167-4171.

Royer, D.L., Wing, S.L., Beerling, D.J., Jolley, D.W., Koch, P.L., Hickey, L.J. and Berner, R.A. 2001. Paleobotanical evidence for near present-day levels of atmospheric CO_2 during part of the Tertiary. *Science* **292**: 2310-2313.

Siegenthaler, U., Stocker, T., Monnin, E., Luthi, D., Schwander, J., Stauffer, B., Raynaud, D., Barnola, J.-M., Fischer, H., Masson-Delmotte, V. and Jouzel, J. 2005. Stable carbon cycle-climate relationship during the late Pleistocene. *Science* **310**: 1313-1317.

Staufer, B., Blunier, T., Dallenbach, A., Indermuhle, A., Schwander, J., Stocker, T.F., Tschumi, J., Chappellaz, J., Raynaud, D., Hammer, C.U. and Clausen, H.B. 1998. Atmospheric CO_2 concentration and millennial-scale climate change during the last glacial period. *Nature* **392**: 59-62.

Steig, E.J. 1999. Mid-Holocene climate change. *Science* **286**: 1485-1487.

3.2. Past 1,000 Years

The IPCC claims "average Northern Hemisphere temperatures during the second half of the 20th century were *very likely* higher than during any other 50-year period in the past 500 years and *likely* the highest in at least the past 1,300 years [italics in the original]" (IPCC, 2007-I, p. 9). Later in that report, the IPCC says "the warming observed after 1980 is unprecedented compared to the levels measured in the previous 280 years" (p. 466) and "it is likely that the 20th century was the warmest in at least the past 1.3 kyr. Considering the recent instrumental and longer proxy evidence together, it is very likely that average NH [Northern Hemisphere] temperatures during the second half of the 20th century were higher than for any other 50-year period in the last 500 years" (p. 474).

The notions that the warming of the second half of the twentieth century was "unprecedented" and that temperatures during the twentieth century were "the warmest in at least the past 1.3 kyr" will be questioned and tested again and again in the present report. We start here with an examination of the work of Mann *et al.* (1998, 1999, 2004) and Mann and Jones (2003), which captured the attention of the world in the early years of the twenty-first century and upon which the IPCC still relies heavily for its conclusions. We then present a thorough examination of temperature records around the world to test the IPCC's claim that there was no Medieval Warm Period during which temperatures exceeded those of the twentieth century, starting with data from Africa and then from Antarctica, the Arctic, Asia, Europe, North America, and finally South America. We return to Antarctica and the Arctic at the end of this chapter to discuss more recent temperature trends.

References

IPCC. 2007-I. *Climate Change 2007: The Physical Science Basis. Contribution of Working Group I to the Fourth Assessment Report of the Intergovernmental Panel on Climate Change.* Solomon, S., D. Qin, M. Manning, Z. Chen, M. Marquis, K.B. Averyt, M. Tignor and H.L. Miller. (Eds.) Cambridge University Press, Cambridge, UK.

Mann, M.E., Bradley, R.S. and Hughes, M.K. 1998. Global-scale temperature patterns and climate forcing over the past six centuries. *Nature* **392**: 779-787.

Mann, M.E., Bradley, R.S. and Hughes, M.K. 1999. Northern Hemisphere temperatures during the past millennium: Inferences, uncertainties, and limitations. *Geophysical Research Letters* **26**: 759-762.

Mann, M.E. and Jones, P.D. 2003. Global surface temperatures over the past two millennia. *Geophysical Research Letters* **30**: 10.1029/2003GL017814.

Mann *et al.* 2004. Corrigendum: Global-scale temperature patterns and climate forcing over the past six centuries. *Nature* **430**: 105.

3.2.1. The Hockey Stick

One of the most famous pieces of "evidence" for anthropogenic global warming (AGW) brought forth in recent years was the "hockey stick" diagram of Michael Mann and colleagues (Mann *et al.*, 1998; Mann *et al.*, 1999; Mann and Jones, 2003). (See Figure 3.2.1.) Because the graph played such a big role in mobilizing concern over global warming in the years since it was first released, and since the IPCC continues to rely upon and defend it in its latest report (see IPCC, 2007-I, pp. 466-471), we devote some space here to explaining its unusual origins and subsequent rejection by much of the scientific community.

The hockey stick graph first appeared in a 1998 study led by Michael Mann, a young Ph.D. from the University of Massachusetts (Mann *et al.*, 1998). Mann and his colleagues used several temperature proxies (but primarily tree rings) as a basis for assessing past temperature changes from 1000 to 1980. They then grafted the surface temperature record of the twentieth century onto the pre-1980 proxy record. The effect was visually dramatic. (See Figure 3.2.1.) Gone were the difficult-to-explain Medieval Warming and the awkward Little Ice Age. Mann gave us nine hundred years of stable global temperatures—until about 1910. Then the twentieth century's temperatures seem to rocket upward out of control.

The Mann study gave the Clinton administration the quick answer it wanted to the argument that natural climate variations exceed whatever effect human activity might have had in the twentieth century by claiming, quite simply, that even the very biggest past historic changes in temperatures *simply never happened.* The Clinton administration featured it as the first visual in the *U.S. National Assessment of the Potential Consequences of Climate Variability and Change* (later published as *Climate Change Impacts on the United States: The Potential Consequences of Climate Variability and Change* (National Assessment Synthesis Team, 2001)). Mann was named an IPCC lead author and his graph was prominently displayed in the IPCC's Third Assessment Report (IPCC-TAR, 2001), and it subsequently appeared in Al Gore's movie, "An Inconvenient Truth." Mann was named an editor of *The Journal of Climate*, a major professional journal, signaling the new order of things to the rest of his profession.

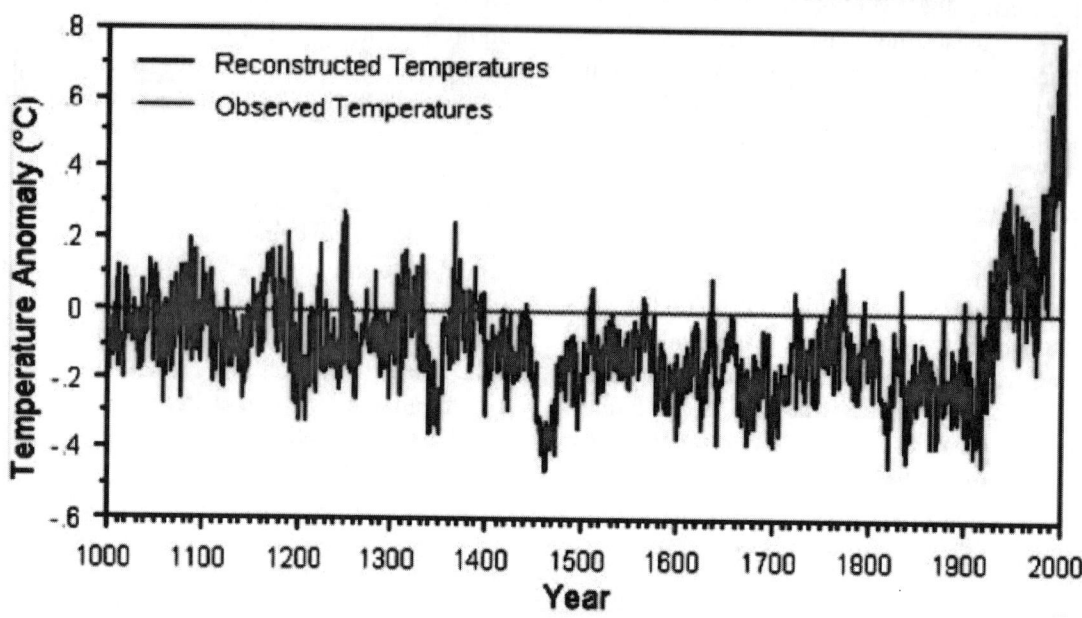

Figure 3.2.1. The 'hockey stick' temperature graph was used by the IPCC to argue that the twentieth century was unusually warm (IPCC-TAR 2001, p. 3).

The "hockey stick" graph was severely critiqued by two Canadian nonscientists who were well trained in statistics—metals expert Stephen McIntyre of Toronto and economist Ross McKitrick from Canada's University of Guelph (McIntyre and McKitrick, 2003, 2005). McIntyre and McKitrick requested the original study data from Mann. It was provided—haltingly and incompletely—indicating that no one else had previously requested the data for a peer review in connection with the original publication in *Nature*. They found the data did not produce the claimed results "due to collation errors, unjustifiable truncation or extrapolation of source data, obsolete data, geographical location errors, incorrect calculation of principal components and other quality control defects."

In their exchanges with the Mann research team, McIntyre and McKitrick learned that the Mann studies give by far the heaviest weight to tree-ring data from 14 sites in California's Sierra Nevada Mountains. At those sites, ancient, slow-growing, high-elevation bristlecone pine trees (which can live 5,000 years) showed a strong twentieth century growth spurt. The growth ring data from those trees were collected and presented in a 1993 paper by Donald Graybill and Sherwood Idso. Significantly, that paper was titled "Detecting the Aerial

Fertilization Effect of Atmospheric CO_2 Enrichment in Tree Ring Chronologies" (Graybill and Idso, 1993). Graybill and Idso specifically pointed out in their study that neither local nor regional temperature changes could account for the twentieth century growth spurt in those already-mature trees. But CO_2 acts like fertilizer for trees and plants and also increases their water-use efficiency. All trees with more CO_2 in their atmosphere are very likely to grow more rapidly. Trees like the high-altitude bristlecone pines, on the margins of both moisture and fertility, are likely to exhibit very strong responses to CO_2 enrichment—which was the point of the Graybill and Idso study.

McIntyre and McKitrick demonstrated that removing the bristlecone pine tree data eliminates the distinctive rise at the end of the "hockey stick." Mann and his coauthors could hardly have escaped knowing the CO_2 reality, since it was clearly presented in the title of the study from which they derived their most heavily weighted data sites. Using corrected and updated source data, McIntyre and McKitrick recalculated the Northern Hemisphere temperature index for the period 1400–1980 using Mann's own methodology. This was published in *Energy & Environment,* with the data refereed by the World Data Center for Paleoclimatology (McIntyre and

McKitrick, 2003). "The major finding is that the [warming] in the early 15th century exceed[s] any [warming] in the 20th century," report McIntyre and McKitrick. In other words, the Mann study was fundamentally wrong.

Mann and his team were forced to publish a correction in *Science* admitting to errors in their published proxy data, but they still claimed that "none of these errors affect our previously published results" (Mann *et al.*, 2004). That claim, too, was contradicted by later work by McIntyre and McKitrick (2005), by statistics expert Edward Wegman (Wegman *et al.*, 2006), and by a National Academy of Sciences report (NAS, 2006). The NAS skipped lightly over the errors of the hockey-stick analysis and concluded it showed only that the twentieth century was the warmest in 400 years, but this conclusion is hardly surprising, since the Little Ice Age was near its nadir 400 years ago, with temperatures at their lowest. It was the claim that temperatures in the second half of the twentieth century were the highest in the last millennium that properly generated the most attention.

Where does the IPCC stand today regarding the "hockey stick"? Surprisingly, it still defends and relies on it. It appears in a series of graphs on page 467. Critiques by Soon and Baliunas (2003) and McIntyre and McKitrick are reported briefly but both are dismissed, the first because "their qualitative approach precluded any quantitative summary of the evidence at precise times," and the latter by citing a defense of Mann by Wahl and Ammann (2006) "who show the impact on the amplitude of the final reconstruction is very small (~0.05°C)" (IPCC, 2007-I, p. 466). The Medieval Warm Period appears only in quotes in the index and body of the IPCC 2007-I report. In the glossary (Annex I), it is defined as "an interval between AD 1000 and 1300 in which some Northern Hemisphere regions were warmer than during the Little Ice Age that followed" (p. 949). In the text it is referred to as "the so-called 'Medieval Warm Period.'" In a boxed discussion of "Hemispheric Temperatures in the 'Medieval Warm Period,'" it says "medieval warmth was heterogeneous in terms of its precise timing and regional expression" and "the warmest period prior to the 20th century very likely occurred between 950 and 1100, but temperatures were probably between 0.1°C and 0.2°C below the 1961 to 1990 mean and significantly below the level shown by instrumental data after 1980" (p. 469).

One can disprove the IPCC's claim by demonstrating that about 1,000 years ago, there was a world-wide Medieval Warm Period (MWP) when global temperatures were equally as high as or higher than they were over the latter part of the twentieth century, despite there being approximately 25 percent less CO_2 in the atmosphere than there is today. This real-world fact conclusively demonstrates there is nothing unnatural about the planet's current temperature, and that whatever warming occurred during the twentieth century was likely caused by the recurrence of whatever cyclical phenomena created the equal or even greater warmth of the MWP.

The degree of warming and climatic influence during the MWP varied from region to region and, hence, its consequences were manifested in several ways. But that it occurred and was a global phenomenon is certain; there are literally hundreds of peer-reviewed scientific articles that bear witness to this truth.

The Center for the Study of Carbon Dioxide and Global Change has analyzed more than 200 peer-reviewed research papers produced by more than 660 individual scientists working in 385 separate institutions from 40 different countries that comment on the MWP. Figure 3.2.2 illustrates the spatial distribution of these studies. Squares denote studies where the scientists who conducted the work provided quantitative data that enable one to determine the degree by which the peak temperature of the MWP differed from the peak temperature of the Current Warm Period (CWP). Circles denote studies where the scientists who conducted the work provided qualitative data that enable one to determine which of the two periods was warmer, but not by how much. Triangles denote studies where the MWP was evident in the study's data, but the data did not provide a means by which the warmth of the MWP could be compared with that of the CWP. The third category includes studies that are based on data related to parameters other than temperature, such as precipitation. As can be seen from the figure, evidence of the MWP has been uncovered at locations throughout the world, revealing the truly global nature of this phenomenon.

A second question often posed with respect to the MWP is: When did it occur? A histogram of the timeframe (start year to end year) associated with the MWP of the studies plotted in Figure 3.2.2 is shown in Figure 3.2.3. The peak timeframe of all studies occurs around 1050 AD, within a more generalized 800 to 1300 AD warm era.

69

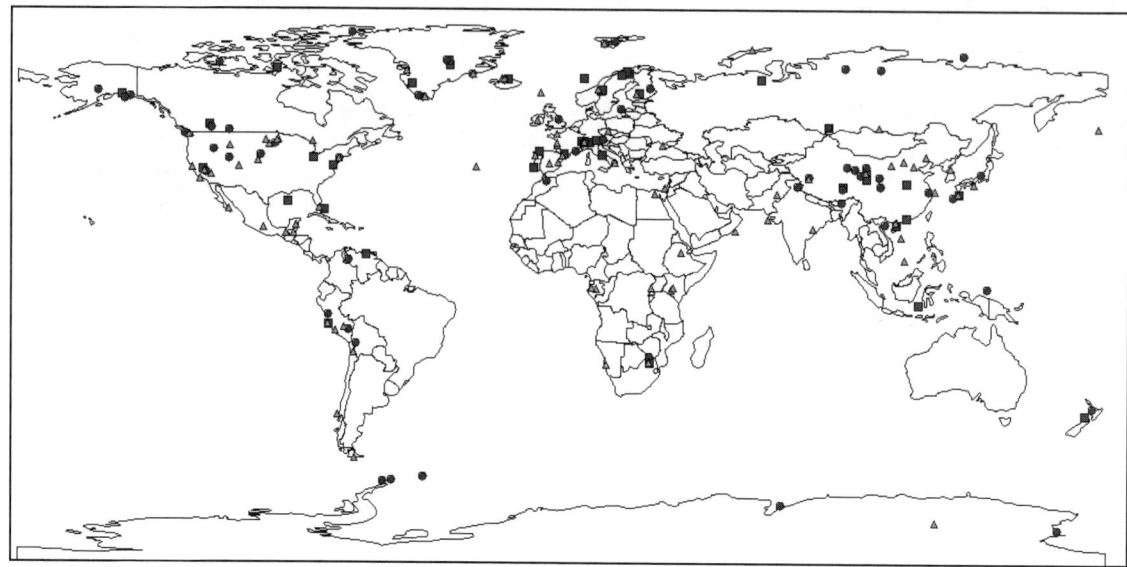

Figure 3.2.2. Plot of the locations of proxy climate studies for which (a) quantitative determinations of the temperature difference between the MWP and CWP can be made (squares), (b) qualitative determinations of the temperature difference between the MWP and CWP can be made (circles), and (c) neither quantitative nor qualitative determinations can be made, with the studies simply indicating that the Medieval Warm Period did indeed occur in the studied region (triangles).

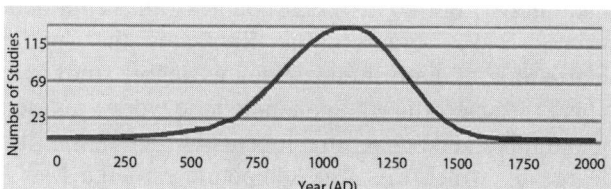

Figure 3.2.3. Histogram showing the timeframe associated with all MWP studies plotted in Figure 3.2.2.

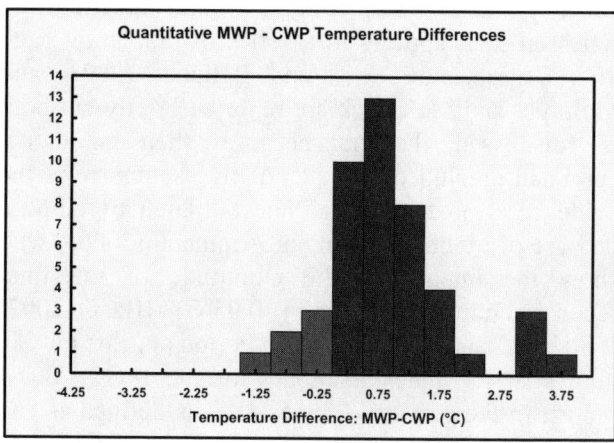

Figure 3.2.4. The distribution, in 0.5°C increments, of studies that allow one to identify the degree by which peak Medieval Warm Period temperatures either exceeded or fell short of peak Current Warm Period temperatures.

With respect to how warm it was during this period, we have plotted the frequency distribution of all MWP-CWP temperature differentials from all quantitative studies (squares) shown in Figure 3.2.2 to create Figure 3.2.4. This figure reveals there are a few studies in which the MWP was determined to have been cooler than the CWP, but the vast majority of the temperature differentials are positive, indicating the MWP was warmer than the CWP. The average of all such differentials is 1.01°C, while the median is 0.90°C.

We can further generalize the superior warmth of the MWP by analyzing the qualitative studies in Figure 3.2.2, which we have done in Figure 3.2.5. Here we have plotted the number of studies in Figure 3.2.2 in which the MWP was warmer than, cooler than, or about the same as, the CWP, based upon data

presented by the authors of the original works. The vast majority of studies indicates the MWP was warmer than the CWP.

It is often claimed that temperatures over the latter part of the twentieth century were higher than those experienced at any other time over the past one to two millennia. Based upon the synthesis of real-world data presented here (and hereafter), however, that claim is seen to be false.

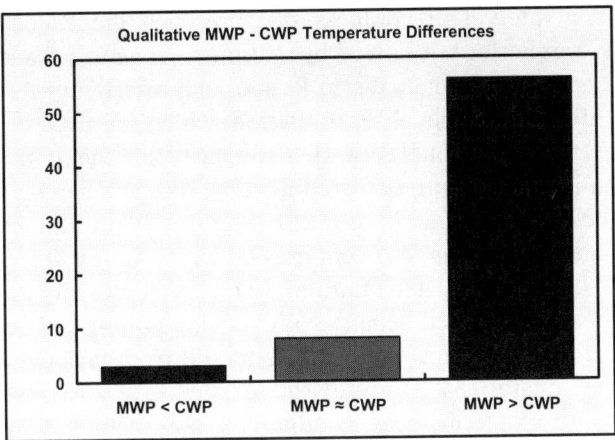

Figure 3.2.5. The distribution of studies that allow one to determine whether peak Medieval Warm Period temperatures were warmer than, equivalent to, or cooler than peak Current Warm Period temperatures.

In the rest of this section, we highlight the results of studies from regions across the globe that show the existence of a Medieval Warm Period. Additional information on this topic, including reviews on the Medieval Warm Period not discussed here, can be found at http://www.co2science.org/subject/m/ subject_m.php under the heading Medieval Warm Period.

References

Graybill, D.A. and Idso, S.B. 1993. Detecting the aerial fertilization effect of atmospheric CO2 enrichment in tree ring chronologies. *Global Biogeochemical Cycles* 7:81–95.

IPCC. 2007-I. *Climate Change 2007: The Physical Science Basis. Contribution of Working Group I to the Fourth Assessment Report of the Intergovernmental Panel on Climate Change.* Solomon, S., D. Qin, M. Manning, Z. Chen, M. Marquis, K.B. Averyt, M. Tignor and H.L. Miller. (Eds.) Cambridge University Press, Cambridge, UK.

IPCC-TAR 2001. *Climate Change 2001: The Scientific Basis. Contribution of Working Group I to the Third Assessment Report of the Intergovernmental Panel on Climate Change.* Cambridge University Press.

McIntyre, S. and McKitrick, R. 2003. Corrections to Mann *et al.* (1998) proxy data base and northern hemisphere average temperature series. *Energy & Environment* **14**: 751-777.

McIntyre, S. and McKitrick, R. 2005. Hockey sticks, principal components and spurious significance. *Geophysical Research Letters* 32 L03710.

Mann, M.E., Bradley, R.S. and Hughes, M.K. 1998. Global-scale temperature patterns and climate forcing over the past six centuries. *Nature* **392**: 779-787.

Mann, M.E., Bradley, R.S. and Hughes, M.K. 1999. Northern Hemisphere temperatures during the past millennium: Inferences, uncertainties, and limitations. *Geophysical Research Letters* **26**: 759-762.

Mann, M.E. and Jones, P.D. 2003. Global surface temperatures over the past two millennia. *Geophysical Research Letters* **30**: 10.1029/2003GL017814.

Mann *et al.* 2004. Corrigendum: Global-scale temperature patterns and climate forcing over the past six centuries. *Nature* **430**: 105.

NAS 2006. *Surface Temperature Reconstructions for the Last 2,000 Years.* National Academy Press, Washington, DC.

National Assessment Synthesis Team. 2001. *Climate Change Impacts on the United States: The Potential Consequences of Climate Variability and Change,* Report for the U.S. Global Change Research Program. Cambridge University Press, Cambridge UK.

Soon, W. and Baliunas, S. 2003. Proxy climatic and environmental changes of the past 1000 years. *Climate Research* **23** (2): 89-110.

Wahl, E.R. and Ammann, C.M. 2007. Robustness of the Mann, Bradley, Hughes reconstruction of Northern Hemisphere surface temperatures: Examination of criticisms based on the nature and processing of proxy climate evidence. *Climate Change* **85**: 33-69.

Wegman, E., Scott, D.W. and Said, Y. 2006. Ad Hoc Committee Report to Chairman of the House Committee on Energy & Commerce and to the Chairman of the House sub-committee on Oversight & Investigations on the Hockey-stick Global Climate Reconstructions. US House of Representatives, Washington, DC. Available at http://energycommerce.house.gov/108/home/07142006 Wegman Report.pdf.

3.2.2. Africa

Based on the temperature and water needs of the crops that were cultivated by the first agropastoralists of southern Africa, Huffman (1996) constructed a climate history of the region based on archaeological evidence acquired from various Iron Age settlements. In the course of completing this project, dated relic evidence of the presence of cultivated sorghum and millets was considered by Huffman to be so strong as to essentially prove that the climate of the subcontinent-wide region must have been warmer and wetter than it is today from approximately AD 900-1300, for these crops cannot be grown in this part of southern Africa under current climatic conditions, which are much too cool and dry.

Other evidence for this conclusion comes from Tyson *et al.* (2000), who obtained a quasi-decadal record of oxygen and carbon-stable isotope data from a well-dated stalagmite of Cold Air Cave in the Makapansgat Valley (30 km southwest of Pietersburg, South Africa), which they augmented with five-year-resolution temperature data that they reconstructed from color variations in banded growth-layer laminations of the stalagmite that were derived from a relationship calibrated against actual air temperatures obtained from a surrounding 49-station climatological network over the period 1981-1995, which had a correlation of +0.78 that was significant at the 99 percent confidence level. This record revealed the existence of a significantly warmer-than-present period that began prior to AD 1000 and lasted to about AD 1300. Tyson *et al.* report that the "maximum warming at Makapansgat at around 1250 produced conditions up to 3-4°C hotter than those of the present."

In a similar study, Holmgren *et al.* (2001) derived a 3,000-year temperature record for South Africa that revealed several multi-century warm and cold periods. They found a dramatic warming at approximately AD 900, when temperatures reached a level that was 2.5°C higher than that prevailing at the time of their analysis of the data.

Lamb *et al.* (2003) provided strong evidence for the hydrologic fingerprint of the Medieval Warm Period in Central Kenya in a study of pollen data obtained from a sediment core taken from Crescent Island Crater, which is a sub-basin of Lake Naivasha. Of particular interest in this regard is the strong similarity between their results and those of Verschuren *et al.* (2000). The most striking of these correspondences occurred over the period AD 980 to 1200, when lake-level was at an 1,100-year low and woody taxa were significantly underrepresented in the pollen assemblage.

Holmgren *et al.* (2003) developed a 25,000-year temperature history from a stalagmite retrieved from Makapansgat Valley's Cold Air Cave based on $\delta^{18}O$ and $\delta^{13}C$ measurements dated by ^{14}C and high-precision thermal ionization mass spectrometry using the $^{230}Th/^{234}U$ method. This work revealed, in the words of the nine researchers (together with our interspersed notes), that "cooling is evident from ~6 to 2.5ka [thousand years before present, during the

long interval of coolness that preceded the Roman Warm Period], followed by warming between 1.5 and 2.5 ka [the Roman Warm Period] and briefly at ~AD 1200 [the Medieval Warm Period, which followed the Dark Ages Cold Period]," after which "maximum Holocene cooling occurred at AD 1700 [the depth of the Little Ice Age]." They also note that "the Little Ice Age covered the four centuries between AD 1500 and 1800 and at its maximum at AD 1700 represents the most pronounced negative $\delta^{18}O$ deviation in the entire record." This new temperature record from far below the equator (24°S) reveals the existence of all of the major millennial-scale oscillations of climate that are evident in data collected from regions surrounding the North Atlantic Ocean.

Two years later, Kondrashov *et al.* (2005) applied advanced spectral methods to fill data gaps and locate interannual and interdecadal periodicities in historical records of annual low- and high-water levels on the Nile River over the 1,300-year period AD 622-1922. In doing so, several statistically significant periodicities were noted, including cycles at 256, 64, 19, 12, 7, 4.2 and 2.2 years. With respect to the causes of these cycles, the three researchers say that the 4.2- and 2.2-year oscillations are likely due to El Niño-Southern Oscillation variations, that the 7-year cycle may be related to North Atlantic influences, and that the longer-period oscillations could be due to astronomical forcings. They also note that the annual-scale resolution of their results provides a "sharper and more reliable determination of climatic-regime transitions" in tropical east Africa, including the documentation of fairly abrupt shifts in river flow at the beginning and end of the Medieval Warm Period.

Ngomanda *et al.* (2007) derived high-resolution (<40 years) paleoenvironmental reconstructions for the past 1,500 years based on pollen and carbon isotope data obtained from sediment cores retrieved from Lakes Kamalete and Nguene in the lowland rainforest of Gabon. The nine researchers state that after a sharp rise at ~1200 cal yr BP, "A/H [aquatic/hygrophytic] pollen ratios showed intermediate values and varied strongly from 1150 to 870 cal yr BP, suggesting decadal-scale fluctuations in the water balance during the 'Medieval Warm Period'." Thereafter, lower A/H pollen ratios "characterized the interval from ~500 to 300 cal yr BP, indicating lower water levels during the 'Little Ice Age'." In addition, they report that "all inferred lake-level low stands, notably between 500 and 300 cal yr BP, are associated with decreases in the score of the TRFO [Tropical Rainforest] biome."

In discussing their findings, Ngomanda *et al.* state that "the positive co-variation between lake level and rainforest cover changes may indicate a direct vegetational response to regional precipitation variability," noting that "evergreen rainforest expansion occurs during wet intervals, with contraction during periods of drought." It appears that in this part of Western Equatorial Africa, the Little Ice Age was a time of low precipitation, low lake levels, and low evergreen rainforest presence, while much the opposite was the case during the Medieval Warm Period, when fluctuating wet-dry conditions led to fluctuating lake levels and a greater evergreen rainforest presence.

Placing these findings within a broader temporal context, Ngomanda *et al.* additionally note that "rainforest environments during the late Holocene in western equatorial Africa are characterized by successive millennial-scale changes according to pollen (Elenga *et al.*, 1994, 1996; Reynaud-Farrera *et al.*, 1996; Maley and Brenac, 1998; Vincens *et al.*, 1998), diatom (Nguetsop *et al.*, 2004), geochemical (Delegue *et al.*, 2001; Giresse *et al.*, 1994), and sedimentological data (Giresse *et al.*, 2005; Wirrmann *et al.*, 2001)," and that "these changes were essentially driven by natural climatic variability (Vincens *et al.*, 1999; Elenga *et al.*, 2004)."

Esper *et al.* (2007) used *Cedrus atlantica* ring-width data "to reconstruct long-term changes in the Palmer Drought Severity Index (PDSI) over the past 953 years in Morocco, Northwest Africa." They report "the long-term PDSI reconstruction indicates generally drier conditions before ~1350, a transition period until ~1450, and generally wetter conditions until the 1970s," after which there were "dry conditions since the 1980s." In addition, they determined that "the driest 20-year period reconstructed is 1237-1256 (PDSI = -4.2)," adding that "1981-2000 conditions are in line with this historical extreme (-3.9)." Also of significance, the six researchers note that "millennium-long temperature reconstructions from Europe (Buntgen *et al.*, 2006) and the Northern Hemisphere (Esper *et al.*, 2002) indicate that Moroccan drought changes are broadly coherent with well-documented temperature fluctuations including warmth during medieval times, cold in the Little Ice Age, and recent anthropogenic warming," which latter coherency would tend to suggest that the peak warmth of the Medieval Warm Period was at least as great as that of the last two decades of the twentieth century throughout the entire Northern Hemisphere; and, if the coherency is strictly

interpreted, it suggests that the warmth of the MWP was likely even greater than that of the late twentieth century.

In light of these research findings, it appears that (1) the Medieval Warm Period did occur over wide reaches of Africa, and (2) the Medieval Warm Period was probably more extreme in Africa than has been the Current Warm Period to this point in time.

Additional information on this topic, including reviews of newer publications as they become available, can be found at http://www.co2science.org/subject/a/africamwp.php.

References

Buntgen, U., Frank, D.C., Nievergelt, D. and Esper, J. 2006. Summer temperature variations in the European Alps, A.D. 755-2004. *Journal of Climate* **19**: 5606-5623.

Delegue, A.M., Fuhr, M., Schwartz, D., Mariotti, A. and Nasi, R. 2001. Recent origin of large part of the forest cover in the Gabon coastal area based on stable carbon isotope data. *Oecologia* **129**: 106-113.

Elenga, H., Maley, J., Vincens, A. and Farrera, I. 2004. Palaeoenvironments, palaeoclimates and landscape development in Central Equatorial Africa: A review of major terrestrial key sites covering the last 25 kyrs. In: Battarbee, R.W., Gasse, F. and Stickley, C.E. (Eds.) *Past Climate Variability through Europe and Africa.* Springer, pp. 181-196.

Elenga, H., Schwartz, D. and Vincens, A. 1994. Pollen evidence of Late Quaternary vegetation and inferred climate changes in Congo. *Palaeogeography, Palaeoclimatology, Palaeoecology* **109**: 345-356.

Elenga, H., Schwartz, D., Vincens, A., Bertraux, J., de Namur, C., Martin, L., Wirrmann, D. and Servant, M. 1996. Diagramme pollinique holocene du Lac Kitina (Congo): mise en evidence de changements paleobotaniques et paleoclimatiques dans le massif forestier du Mayombe. *Compte-Rendu de l'Academie des Sciences, Paris, serie* **2a**: 345-356.

Esper, J., Cook, E.R. and Schweingruber, F.H. 2002. Low-frequency signals in long tree-ring chronologies for reconstructing past temperature variability. *Science* **295**: 2250-2253.

Esper, J., Frank, D., Buntgen, U., Verstege, A., Luterbacher, J. and Xoplaki, E. 2007. Long-term drought severity variations in Morocco. *Geophysical Research Letters* **34**: 10.1029/2007GL030844.

Giresse, P., Maley, J. and Brenac, P. 1994. Late Quaternary palaeoenvironments in Lake Barombi Mbo (West

Cameroon) deduced from pollen and carbon isotopes of organic matter. *Palaeogeography, Palaeoclimatology, Palaeoecology* **107**: 65-78.

Giresse, P., Maley, J. and Kossoni, A. 2005. Sedimentary environmental changes and millennial climatic variability in a tropical shallow lake (Lake Ossa, Cameroon) during the Holocene. *Palaeogeography, Palaeoclimatology, Palaeoecology* **218**: 257-285.

Holmgren, K., Lee-Thorp, J.A., Cooper, G.R.J., Lundblad, K., Partridge, T.C., Scott, L., Sithaldeen, R., Talma, A.S. and Tyson, P.D. 2003. Persistent millennial-scale climatic variability over the past 25,000 years in Southern Africa. *Quaternary Science Reviews* **22**: 2311-2326.

Holmgren, K., Tyson, P.D., Moberg, A. and Svanered, O. 2001. A preliminary 3000-year regional temperature reconstruction for South Africa. *South African Journal of Science* **97**: 49-51.

Huffman, T.N. 1996. Archaeological evidence for climatic change during the last 2000 years in southern Africa. *Quaternary International* **33**: 55-60.

Kondrashov, D., Feliks, Y. and Ghil, M. 2005. Oscillatory modes of extended Nile River records (A.D. 622-1922). *Geophysical Research Letters* **32**: doi:10.1029/2004 GL022156.

Lamb, H., Darbyshire, I. and Verschuren, D. 2003. Vegetation response to rainfall variation and human impact in central Kenya during the past 1100 years. *The Holocene* **13**: 285-292.

Maley, J. and Brenac, P. 1998. Vegetation dynamics, paleoenvironments and climatic changes in the forests of western Cameroon during the last 28,000 years B.P. *Review of Palaeobotany and Palynology* **99**: 157-187.

Ngomanda, A., Jolly, D., Bentaleb, I., Chepstow-Lusty, A., Makaya, M., Maley, J., Fontugne, M., Oslisly, R. and Rabenkogo, N. 2007. Lowland rainforest response to hydrological changes during the last 1500 years in Gabon, Western Equatorial Africa. *Quaternary Research* **67**: 411-425.

Nguetsop, V.F., Servant-Vildary, S. and Servant, M. 2004. Late Holocene climatic changes in west Africa, a high resolution diatom record from equatorial Cameroon. *Quaternary Science Reviews* **23**: 591-609.

Reynaud-Farrera, I., Maley, J. and Wirrmann, D. 1996. Vegetation et climat dans les forets du Sud-Ouest Cameroun depuis 4770 ans B.P.: analyse pollinique des sediments du Lac Ossa. *Compte-Rendu de l'Academie des Sciences, Paris, serie* 2a **322**: 749-755.

Tyson, P.D., Karlén, W., Holmgren, K. and Heiss, G.A. 2000. The Little Ice Age and medieval warming in South Africa. *South African Journal of Science* **96**: 121-126.

Verschuren, D., Laird, K.R. and Cumming, B.F. 2000. Rainfall and drought in equatorial east Africa during the past 1,100 years. *Nature* **403**: 410-414.

Vincens, A., Schwartz, D., Bertaux, J., Elenga, H. and de Namur, C. 1998. Late Holocene climatic changes in Western Equatorial Africa inferred from pollen from Lake Sinnda, Southern Congo. *Quaternary Research* **50**: 34-45.

Vincens, A., Schwartz, D., Elenga, H., Reynaud-Farrera, I., Alexandre, A., Bertaux, J., Mariotti, A., Martin, L., Meunier, J.-D., Nguetsop, F., Servant, M., Servant-Vildary, S. and Wirrmann, D. 1999. Forest response to climate changes in Atlantic Equatorial Africa during the last 4000 years BP and inheritance on the modern landscapes. *Journal of Biogeography* **26**: 879-885.

Wirrmann, D., Bertaux, J. and Kossoni, A. 2001. Late Holocene paleoclimatic changes in Western Central Africa inferred from mineral abundance in dated sediments from Lake Ossa (Southwest Cameroon). *Quaternary Research* **56**: 275-287.

3.2.3. Antarctica

Hemer and Harris (2003) extracted a sediment core from beneath the Amery Ice Shelf, East Antarctica, at a point that is currently about 80 km landward of the location of its present edge. In analyzing the core's characteristics over the past 5,700 ^{14}C years, the two scientists observed a peak in absolute diatom abundance in general, and the abundance of *Fragilariopsis curta* in particular—which parameters, in their words, "are associated with increased proximity to an area of primary production, such as the sea-ice zone"—at about 750 ^{14}C yr B.P., which puts the time of maximum Ice Shelf retreat in close proximity to the historical time frame of the Medieval Warm Period.

Khim *et al.* (2002) likewise analyzed a sediment core removed from the eastern Bransfield Basin just off the northern tip of the Antarctic Peninsula, including grain size, total organic carbon content, magnetic susceptibility, biogenic silica content, ^{210}Pb geochronology, and radiocarbon (^{14}C) age, all of which data clearly depicted, in their words, the presence of the "Little Ice Age and Medieval Warm period, together with preceding climatic events of similar intensity and duration."

Hall and Denton (2002) mapped the distribution and elevation of surficial deposits along the southern Scott Coast of Antarctica in the vicinity of the Wilson Piedmont Glacier, which runs parallel to the coast of the western Ross Sea from McMurdo Sound north to

Granite Harbor. The chronology of the raised beaches they studied was determined from more than 60 ^{14}C dates of incorporated organic materials they had previously collected from hand-dug excavations (Hall and Denton, 1999); the record the dates helped define demonstrated that near the end of the Medieval Warm Period, "as late as 890 ^{14}C yr BP," as Hall and Denton describe it, "the Wilson Piedmont Glacier was still less extensive than it is now," demonstrating that the climate of that period was in all likelihood considerably warmer than it is currently.

Noon *et al.* (2003) used oxygen isotopes preserved in authigenic carbonate retrieved from freshwater sediments of Sombre Lake on Signy Island (60°43'S, 45°38'W) in the Southern Ocean to construct a 7,000-year history of that region's climate. This work revealed that the general trend of temperature at the study site has been downward. Of most interest to us, however, is the millennial-scale oscillation of climate that is apparent in much of the record. This climate cycle is such that approximately 2,000 years ago, after a thousand-year gap in the data, Signy Island experienced the relative warmth of the last vestiges of the Roman Warm Period, as delineated by McDermott *et al.* (2001) on the basis of a high-resolution speleothem δ^{18}O record from southwest Ireland. Then comes the Dark Ages Cold period, which is also contemporaneous with what McDermott *et al.* observe in the Northern Hemisphere, after which the Medieval Warm Period appears at the same point in time and persists for the same length of time that it does in the vicinity of Ireland, whereupon the Little Ice Age sets in just as it does in the Northern Hemisphere. Finally, there is an indication of late twentieth century warming, but with still a long way to go before conditions comparable to those of the Medieval Warm Period are achieved.

Two years later, Castellano *et al.* (2005) derived a detailed history of Holocene volcanism from the sulfate record of the first 360 meters of the Dome Concordia ice core that covered the period 0-11.5 kyr BP, after which they compared their results for the past millennium with similar results obtained from eight other Antarctic ice cores. Before doing so, however, they normalized the results at each site by dividing its several volcanic-induced sulfate deposition values by the value produced at that site by the AD 1816 Tambora eruption, in order to reduce deposition differences among sites that might have been induced by differences in local site characteristics. This work revealed that most volcanic events in the early last millennium (AD 1000-1500)

exhibited greater among-site variability in normalized sulphate deposition than was observed thereafter.

Citing Budner and Cole-Dai (2003) in noting that "the Antarctic polar vortex is involved in the distribution of stratospheric volcanic aerosols over the continent," Castellano *et al.* say that assuming the intensity and persistence of the polar vortex in both the troposphere and stratosphere "affect the penetration of air masses to inland Antarctica, isolating the continental area during cold periods and facilitating the advection of peripheral air masses during warm periods (Krinner and Genthon, 1998), we support the hypothesis that the pattern of volcanic deposition intensity and geographical variability [higher values at coastal sites] could reflect a warmer climate of Antarctica in the early last millennium," and that "the re-establishment of colder conditions, starting in about AD 1500, reduced the variability of volcanic depositions."

Describing this phenomenon in terms of what it implies, Castellano *et al.* say "this warm/cold step could be like a Medieval Climate Optimum-like to Little Ice Age-like transition." They additionally cite Goosse *et al.* (2004) as reporting evidence from Antarctic ice-core δD and δ^{18}O data "in support of a Medieval Warming-like period in the Southern Hemisphere, delayed by about 150 years with respect to Northern Hemisphere Medieval Warming." The researchers conclude by postulating that "changes in the extent and intra-Antarctic variability of volcanic depositional fluxes may have been consequences of the establishment of a Medieval Warming-like period that lasted until about AD 1500."

A year later, Hall *et al.* (2006) collected skin and hair (and even some whole-body mummified remains) from Holocene raised-beach excavations at various locations along Antarctica's Victoria Land Coast, which they identified by both visual inspection and DNA analysis as coming from southern elephant seals, and which they analyzed for age by radiocarbon dating. By these means they obtained data from 14 different locations within their study region—which they describe as being "well south" of the seals' current "core sub-Antarctic breeding and molting grounds"—that indicate that the period of time they denominate the Seal Optimum began about 600 BC and ended about AD1400, the latter of which dates they describe as being "broadly contemporaneous with the onset of Little Ice Age climatic conditions in the Northern Hemisphere and with glacier advance near [Victoria Land's] Terra Nova Bay."

In describing the significance of their findings, the US, British, and Italian researchers say they are indicative of "warmer-than-present climate conditions" at the times and locations of the identified presence of the southern elephant seal, and that "if, as proposed in the literature, the [Ross] ice shelf survived this period, it would have been exposed to environments substantially warmer than present," which would have included both the Roman Warm Period and Medieval Warm Period.

More recently, Williams *et al.* (2007) presented methyl chloride (CH_3Cl) measurements of air extracted from a 300-m ice core that was obtained at the South Pole, Antarctica, covering the time period 160 BC to AD 1860. In describing what they found, the researchers say "CH_3Cl levels were elevated from 900-1300 AD by about 50 ppt relative to the previous 1000 years, coincident with the warm Medieval Climate Anomaly (MCA)," and that they "decreased to a minimum during the Little Ice Age cooling (1650-1800 AD), before rising again to the modern atmospheric level of 550 ppt." Noting that "today, more than 90% of the CH_3Cl sources and the majority of CH_3Cl sinks lie between 30°N and 30°S (Khalil and Rasmussen, 1999; Yoshida *et al.*, 2004)," they say "it is likely that climate-controlled variability in CH_3Cl reflects changes in tropical and subtropical conditions." They go on to say that "ice core CH_3Cl variability over the last two millennia suggests a positive relationship between atmospheric CH_3Cl and *global* [our italics] mean temperature."

As best we can determine from the graphical representation of their data, the peak CH_3Cl concentration measured by Williams *et al.* during the MCA is approximately 533 ppt, which is within 3 percent of its current mean value of 550 ppt and well within the range of 520 to 580 ppt that characterizes methyl chloride's current variability. Hence, we may validly conclude that the mean peak temperature of the MCA (which we refer to as the Medieval Warm Period) over the latitude range 30°N to 30°S—and possibly over the entire globe—may not have been materially different from the mean peak temperature so far attained during the Current Warm Period.

This conclusion, along with the findings of the other studies we have reviewed of the climate of Antarctica, suggests there is nothing unusual, unnatural, or unprecedented about the current level of earth's warmth.

Additional information on this topic, including reviews of newer publications as they become available, can be found at http://www.co2science.org/subject/a/antarcticmwp.php.

References

Budner, D. and Cole-Dai, J. 2003. The number and magnitude of large explosive volcanic eruptions between 904 and 1865 A.D.: Quantitative evidence from a new South Pole ice core. In: Robock, A. and Oppenheimer, C. (Eds.) *Volcanism and the Earth's Atmosphere, Geophysics Monograph Series* **139**: 165-176.

Castellano, E., Becagli, S., Hansson, M., Hutterli, M., Petit, J.R., Rampino, M.R., Severi, M., Steffensen, J.P., Traversi, R. and Udisti, R. 2005. Holocene volcanic history as recorded in the sulfate stratigraphy of the European Project for Ice Coring in Antarctica Dome C (EDC96) ice core. *Journal of Geophysical Research* **110**: 10.1029/JD005259.

Goosse, H., Masson-Delmotte, V., Renssen, H., Delmotte, M., Fichefet, T., Morgan, V., van Ommen, T., Khim, B.K. and Stenni, B. 2004. A late medieval warm period in the Southern Ocean as a delayed response to external forcing. *Geophysical Research Letters* **31**: 10.1029/2003GL019140.

Hall, B.L. and Denton, G.H. 1999. New relative sea-level curves for the southern Scott Coast, Antarctica: evidence for Holocene deglaciation of the western Ross Sea. *Journal of Quaternary Science* **14**: 641-650.

Hall, B.L. and Denton, G.H. 2002. Holocene history of the Wilson Piedmont Glacier along the southern Scott Coast, Antarctica. *The Holocene* **12**: 619-627.

Hall, B.L., Hoelzel, A.R., Baroni, C., Denton, G.H., Le Boeuf, B.J., Overturf, B. and Topf, A.L. 2006. Holocene elephant seal distribution implies warmer-than-present climate in the Ross Sea. *Proceedings of the National Academy of Sciences USA* **103**: 10,213-10,217.

Hemer, M.A. and Harris, P.T. 2003. Sediment core from beneath the Amery Ice Shelf, East Antarctica, suggests mid-Holocene ice-shelf retreat. *Geology* **31**: 127-130.

Khalil, M.A.K. and Rasmussen, R.A. 1999. Atmospheric methyl chloride. *Atmospheric Environment* **33**: 1305-1321.

Khim, B-K., Yoon, H.I., Kang, C.Y. and Bahk, J.J. 2002. Unstable climate oscillations during the Late Holocene in the Eastern Bransfield Basin, Antarctic Peninsula. *Quaternary Research* **58**: 234-245.

Krinner, G. and Genthon, C. 1998. GCM simulations of the Last Glacial Maximum surface climate of Greenland and Antarctica. *Climate Dynamics* **14**: 741-758.

McDermott, F., Mattey, D.P. and Hawkesworth, C. 2001. Centennial-scale Holocene climate variability revealed by a high-resolution speleothem δ¹⁸O record from SW Ireland. *Science* **294**: 1328-1331.

Noon, P.E., Leng, M.J. and Jones, V.J. 2003. Oxygen-isotope (δ¹⁸O) evidence of Holocene hydrological changes at Signy Island, maritime Antarctica. *The Holocene* **13**: 251-263.

Williams, M.B., Aydin, M., Tatum, C. and Saltzman, E.S. 2007. A 2000 year atmospheric history of methyl chloride from a South Pole ice core: Evidence for climate-controlled variability. *Geophysical Research Letters* **34**: 10.1029/2006GL029142.

Yoshida, Y., Wang, Y.H., Zeng, T. and Yantosea, R. 2004. A three-dimensional global model study of atmospheric methyl chloride budget and distributions. *Journal of Geophysical Research* **109**: 10.1029/2004JD004951.

3.2.4. Arctic

Dahl-Jensen *et al.* (1998) used temperature measurements from two Greenland Ice Sheet boreholes to reconstruct the temperature history of this portion of the earth over the past 50,000 years. Their data indicate that after the termination of the glacial period, temperatures steadily rose to a maximum of 2.5°C warmer than at present during the Holocene Climatic Optimum (4,000 to 7,000 years ago). The Medieval Warm Period and Little Ice Age were also documented in the record, with temperatures 1°C warmer and 0.5-0.7°C cooler than at present, respectively. After the Little Ice Age, they report that temperatures once again rose, but that they "have decreased during the last decades." These results thus clearly indicate that the Medieval Warm Period in this part of the Arctic was significantly warmer than current temperatures.

Wagner and Melles (2001) also worked on Greenland, where they extracted a 3.5-m-long sediment core from a lake (Raffels So) on an island (Raffles O) located just off Liverpool Land on the east coast of Greenland, which they analyzed for a number of properties related to the past presence of seabirds there, obtaining a 10,000-year record that tells us much about the region's climatic history. Key to the study were biogeochemical data that, in the words of the researchers, reflect "variations in seabird breeding colonies in the catchment which influence nutrient and cadmium supply to the lake."

Wagner and Melles' data reveal sharp increases in the values of the parameters they measured between about 1100 and 700 years before present (BP), indicative of the summer presence of significant numbers of seabirds during that "medieval warm period," as they describe it, which had been preceded by a several-hundred-year period (Dark Ages Cold Period) of little to no bird presence. Thereafter, their data suggest another absence of birds during what they call "a subsequent Little Ice Age," which they note was "the coldest period since the early Holocene in East Greenland."

The Raffels So data also show signs of a "resettlement of seabirds during the last 100 years, indicated by an increase of organic matter in the lake sediment and confirmed by bird observations." However, values of the most recent measurements are not as great as those obtained from the earlier Medieval Warm Period, which indicates that higher temperatures prevailed during the period from 1,100 to 700 years BP than what has been observed over the most recent hundred years.

A third relevant Greenland study was conducted by Kaplan *et al.* (2002), who derived a climatic history of the Holocene by analyzing the physical-chemical properties of sediments obtained from a small lake in southern Greenland. They determined that the interval from 6,000 to 3,000 years BP was marked by warmth and stability, but that the climate cooled thereafter until its culmination in the Little Ice Age. From 1,300-900 years BP, however, there was a partial amelioration during the Medieval Warm Period, which was associated with an approximate 1.5°C rise in temperature.

In a non-Greenland Arctic study, Jiang *et al.* (2002) analyzed diatom assemblages from a high-resolution core extracted from the seabed of the north Icelandic shelf to reconstruct a 4,600-year history of mean summer sea surface temperature at that location. Starting from a maximum value of about 8.1°C at 4,400 years BP, the climate was found to have cooled fitfully for about 1,700 years and then more consistently over the final 2,700 years of the record. The most dramatic departure from this long-term decline was centered on about 850 years BP, during the Medieval Warm Period, when the temperature rose by more than 1°C above the line describing the long-term downward trend to effect an almost complete recovery from the colder temperatures of the Dark Ages Cold Period, after which temperatures continued their descent into the Little Ice Age, ending with a final most recent value of approximately 6.3°C. These data also clearly indicate that the Medieval Warm Period in this part of the Arctic was significantly warmer than it is there now.

Moore *et al.* (2001) analyzed sediment cores from Donard Lake, Baffin Island, Canada, producing a 1,240-year record of average summer temperatures for this Arctic region. Over the entire period from AD 750-1990, temperatures averaged 2.9°C. However, anomalously warm decades with summer temperatures as high as 4°C occurred around AD 1000 and 1100, while at the beginning of the thirteenth century, Donard Lake witnessed "one of the largest climatic transitions in over a millennium," as "average summer temperatures rose rapidly by nearly 2°C from 1195-1220 AD, ending in the warmest decade in the record" with temperatures near 4.5°C.

This rapid warming of the thirteenth century was followed by a period of extended warmth that lasted until an abrupt cooling event occurred around 1375 and made the following decade one of the coldest in the record. This event signaled the onset of the Little Ice Age, which lasted for 400 years, until a gradual warming trend began about 1800, which was followed by a dramatic cooling event in 1900 that brought temperatures back to levels similar to those of the Little Ice Age. This cold regime lasted until about 1950, whereupon temperatures warmed for about two decades but then tended downwards again all the way to the end of the record in 1990. Hence, in this part of the Arctic the Medieval Warm Period was also warmer than it is there currently.

Grudd *et al.* (2002) assembled tree-ring widths from 880 living, dead, and subfossil northern Swedish pines into a continuous and precisely dated chronology covering the period 5407 BC to AD 1997. The strong association between these data and summer (June-August) mean temperatures of the last 129 years of the period then enabled them to produce a 7,400-year history of summer mean temperature for northern Swedish Lapland.

The most dependable portion of this record, based upon the number of trees that were sampled, consisted of the last two millennia, which the authors say "display features of century-timescale climatic variation known from other proxy and historical sources, including a warm 'Roman' period in the first centuries AD and a generally cold 'Dark Ages' climate from about AD 500 to about AD 900." They also note that "the warm period around AD 1000 may correspond to a so-called 'Mediaeval Warm Period,' known from a variety of historical sources and other proxy records." Lastly, they say "the climatic deterioration in the twelfth century can be regarded as the starting point of a prolonged cold period that continued to the first decade of the twentieth

century," which "Little Ice Age," in their words, is also "known from instrumental, historical and proxy records." Going back further in time, the tree-ring record displays several more of these relatively warmer and colder periods. They report that "the relatively warm conditions of the late twentieth century do not exceed those reconstructed for several earlier time intervals."

Seppa and Birks (2002) used a recently developed pollen-climate reconstruction model and a new pollen stratigraphy from Toskaljarvi—a tree-line lake in the continental sector of northern Fenoscandia (located just above 69°N latitude)—to derive quantitative estimates of annual precipitation and July mean temperature. As they describe it, their reconstructions "agree with the traditional concept of a 'Medieval Warm Period' (MWP) and 'Little Ice Age' in the North Atlantic region (Dansgaard *et al.*, 1975) and in northern Fennoscandia (Korhola *et al.*, 2000)." In addition, they report there is "a clear correlation between [their] MWP reconstruction and several records from Greenland ice cores," and that "comparisons of a smoothed July temperature record from Toskaljavri with measured borehole temperatures of the GRIP and Dye 3 ice cores (Dahl-Jensen *et al.*, 1998) and the $\delta^{18}O$ record from the Crete ice core (Dansgaard *et al.*, 1975) show the strong similarity in timing of the MWP between the records." Finally, they note that "July temperature values during the Medieval Warm Period (ca. 1400-1000 cal yr B.P.) were ca. 0.8°C higher than at present," where present means the last six decades of the twentieth century.

Noting that temperature changes in high latitudes are (1) sensitive indicators of *global* temperature changes, and that they can (2) serve as a basis for verifying climate model calculations, Naurzbaev *et al.* (2002) developed a 2,427-year proxy temperature history for the part of the Taimyr Peninsula of northern Russia that lies between 70°30' and 72°28' North latitude, based on a study of ring-widths of living and preserved larch trees, noting further that "it has been established that the main driver of tree-ring variability at the polar timber-line [where they worked] is temperature (Vaganov *et al.*, 1996; Briffa *et al.*, 1998; Schweingruber and Briffa, 1996)." In doing so, they found that "the warmest periods over the last two millennia in this region were clearly in the third [Roman Warm Period], tenth to twelfth [Medieval Warm Period] and during the twentieth [Current Warm Period] centuries."

With respect to the second of these periods, they emphasize that "the warmth of the two centuries AD 1058-1157 and 950-1049 attests to the reality of relative mediaeval warmth in this region." Their data also reveal three other important pieces of information: (1) the Roman and Medieval Warm Periods were both warmer than the Current Warm Period has been to date, (2) the "beginning of the end" of the Little Ice Age was somewhere in the vicinity of 1830, and (3) the Current Warm Period peaked somewhere in the vicinity of 1940. All of these observations are at odds with what is portrayed in the thousand-year Northern Hemispheric "hockey stick" temperature history of Mann *et al.* (1998, 1999) and its thousand-year global extension developed by Mann and Jones (2003).

Knudsen *et al.* (2004) documented climatic changes over the past 1,200 years by means of high-resolution multi-proxy studies of benthic and planktonic foraminiferal assemblages, stable isotopes, and ice-rafted debris found in three sediment cores retrieved from the North Icelandic shelf. This work revealed that "the time period between 1200 and around 7,800 cal. years BP, including the Medieval Warm Period, was characterized by relatively high bottom and surface water temperatures," after which "a general temperature decrease in the area marks the transition to ... the Little Ice Age." They also note that "minimum sea-surface temperatures were reached at around 350 cal. BP, when very cold conditions were indicated by several proxies." Thereafter, they say "a modern warming of surface waters ... is *not* [our italics] registered in the proxy data," and that "there is no clear indication of warming of water masses in the area during the last decades," even in sea surface temperatures measured over the period 1948-2002.

Grinsted *et al.* (2006) developed "a model of chemical fractionation in ice based on differing elution rates for pairs of ions ... as a proxy for summer melt (1130-1990)," based on data obtained from a 121-meter-long ice core they extracted from the highest ice field in Svalbard (Lomonosovfonna: 78°51'53"N, 17°25'30"E), which was "validated against twentieth-century instrumental records and longer historical climate proxies." This history indicated that "in the oldest part of the core (1130-1200), the washout indices [were] more than 4 times as high as those seen during the last century, indicating a high degree of runoff." In addition, they report they have performed regular snow pit studies near the ice core site since 1997 (Virkkunen, 2004) and that "the very warm 2001 summer resulted in similar loss of ions and washout ratios as the earliest part of the core." They then state that "this suggests that the Medieval Warm Period in Svalbard summer conditions [was] as warm (or warmer) as present-day, consistent with the Northern Hemisphere temperature reconstruction of Moberg *et al.* (2005)." In addition, they conclude that "the degree of summer melt was significantly larger during the period 1130-1300 than in the 1990s," which likewise suggests that a large portion of the Medieval Warm Period was significantly warmer than the peak warmth (1990s) of the Current Warm Period.

Besonen *et al.* (2008) derived thousand-year histories of varve thickness and sedimentation accumulation rate for Canada's Lower Murray Lake (81°20'N, 69°30'W), which is typically covered for about 11 months of each year by ice that reaches a thickness of 1.5 to 2 meters at the end of each winter. With respect to these parameters, they say—citing seven other studies—that "field-work on other High Arctic lakes clearly indicates that sediment transport and varve thickness are related to temperatures during the short summer season that prevails in this region, and we have no reason to think that this is not the case for Lower Murray Lake."

They found "the twelfth and thirteenth centuries were relatively warm," with their data indicating that Lower Murray Lake and its environs were often much warmer during this time period (AD 1080-1320) than they were at any point in the twentieth century, which has also been shown to be the case for Donard Lake (66.25°N, 62°W) by Moore *et al.* (2001).

The studies reviewed above indicate that the Arctic—which climate models suggest should be sensitive to greenhouse-gas-induced warming—is still not as warm as it was many centuries ago during portions of the Medieval Warm Period, when there was much less CO_2 and methane in the air than there is today. This further suggests the planet's more modest current warmth need not be the result of historical increases in these two greenhouse gases.

Additional information on this topic, including reviews of newer publications as they become available, can be found at http://www.co2science.org/subject/a/arcticmwp.php.

References

Besonen, M.R., Patridge, W., Bradley, R.S., Francus, P., Stoner, J.S. and Abbott, M.B. 2008. A record of climate over the last millennium based on varved lake sediments from the Canadian High Arctic. *The Holocene* **18**: 169-180.

Briffa, K.R., Schweingruber, F.H., Jones, P.D., Osborn, T.J., Shiyatov, S.G. and Vaganov, E.A. 1998. Reduced sensitivity of recent tree-growth to temperature at high northern latitudes. *Nature* **391**: 678-682.

Dahl-Jensen, D., Mosegaard, K., Gundestrup, N., Clow, G.D., Johnsen, S.J., Hansen, A.W. and Balling, N. 1998. Past temperatures directly from the Greenland Ice Sheet. *Science* **282**: 268-271.

Dansgaard, W., Johnsen, S.J., Gundestrup, N., Clausen, H.B. and Hammer, C.U. 1975. Climatic changes, Norsemen and modern man. *Nature* **255**: 24-28.

Grinsted, A., Moore, J.C., Pohjola, V., Martma, T. and Isaksson, E. 2006. Svalbard summer melting, continentality, and sea ice extent from the Lomonosovfonna ice core. *Journal of Geophysical Research* **111**: 10.1029/2005JD006494.

Grudd, H., Briffa, K.R., Karlén, W., Bartholin, T.S., Jones, P.D. and Kromer, B. 2002. A 7400-year tree-ring chronology in northern Swedish Lapland: natural climatic variability expressed on annual to millennial timescales. *The Holocene* **12**: 657-665.

Jiang, H., Seidenkrantz, M-S., Knudsen, K.L. and Eiriksson, J. 2002. Late-Holocene summer sea-surface temperatures based on a diatom record from the north Icelandic shelf. *The Holocene* **12**: 137-147.

Kaplan, M.R., Wolfe, A.P. and Miller, G.H. 2002. Holocene environmental variability in southern Greenland inferred from lake sediments. *Quaternary Research* **58**: 149-159.

Knudsen, K.L., Eiriksson, J., Jansen, E., Jiang, H., Rytter, F. and Gudmundsdottir, E.R. 2004. Palaeoceanographic changes off North Iceland through the last 1200 years: foraminifera, stable isotopes, diatoms and ice rafted debris. *Quaternary Science Reviews* **23**: 2231-2246.

Korhola, A., Weckstrom, J., Holmstrom, L. and Erasto, P. 2000. A quantitative Holocene climatic record from diatoms in northern Fennoscandia. *Quaternary Research* **54**: 284-294.

Mann, M.E., Bradley, R.S. and Hughes, M.K. 1998. Global-scale temperature patterns and climate forcing over the past six centuries. *Nature* **392**: 779-787.

Mann, M.E., Bradley, R.S. and Hughes, M.K. 1999. Northern Hemisphere temperatures during the past millennium: Inferences, uncertainties, and limitations. *Geophysical Research Letters* **26**: 759-762.

Mann, M.E. and Jones, P.D. 2003. Global surface temperatures over the past two millennia. *Geophysical Research Letters* **30**: 10.1029/2003GL017814.

Moberg, A., Sonechkin, D.M., Holmgren, K., Datsenko, N.M. and Karlén, W. 2005. Highly variable Northern Hemisphere temperatures reconstructed from low- and high-resolution proxy data. *Nature* **433**: 613-617.

Moore, J.J., Hughen, K.A., Miller, G.H. and Overpeck, J.T. 2001. Little Ice Age recorded in summer temperature reconstruction from varved sediments of Donard Lake, Baffin Island, Canada. *Journal of Paleolimnology* **25**: 503-517.

Naurzbaev, M.M., Vaganov, E.A., Sidorova, O.V. and Schweingruber, F.H. 2002. Summer temperatures in eastern Taimyr inferred from a 2427-year late-Holocene tree-ring chronology and earlier floating series. *The Holocene* **12**: 727-736.

Schweingruber, F.H. and Briffa, K.R. 1996. Tree-ring density network and climate reconstruction. In: Jones, P.D., Bradley, R.S. and Jouzel, J. (Eds.), *Climatic Variations and Forcing Mechanisms of the Last 2000 Years*, NATO ASI Series 141. Springer-Verlag, Berlin, Germany, pp. 43-66.

Seppa, H. and Birks, H.J.B. 2002. Holocene climate reconstructions from the Fennoscandian tree-line area based on pollen data from Toskaljavri. *Quaternary Research* **57**: 191-199.

Vaganov, E.A., Shiyatov, S.G. and Mazepa, V.S. 1996. *Dendroclimatic Study in Ural-Siberian Subarctic*. Nauka, Novosibirsk, Russia.

Virkkunen, K. 2004. *Snowpit Studies in 2001-2002 in Lomonosovfonna, Svalbard*. M.S. Thesis, University of Oulu, Oulu, Finland.

Wagner, B. and Melles, M. 2001. A Holocene seabird record from Raffles So sediments, East Greenland, in response to climatic and oceanic changes. *Boreas* **30**: 228-239.

3.2.5. Asia

3.2.5.1. China

Using a variety of climate records derived from peat, lake sediment, ice core, tree-ring and other proxy sources, Yang *et al.* (2002) identified a period of exceptional warmth throughout China between AD 800 and 1100. Yafeng *et al.* (1999) also observed a warm period between AD 970 and 1510 in $\delta^{18}O$ data obtained from the Guliya ice cap of the Qinghai-Tibet Plateau. Similarly, Hong *et al.* (2000) developed a 6,000-year $\delta^{18}O$ record from plant cellulose deposited in a peat bog in the Jilin Province (42° 20' N, 126° 22' E), within which they found evidence of "an obvious warm period represented by the high $\delta^{18}O$

from around AD 1100 to 1200 which may correspond to the Medieval Warm Epoch of Europe."

Shortly thereafter, Xu *et al.* (2002) determined from a study of plant cellulose $\delta^{18}O$ variations in cores retrieved from peat deposits at the northeastern edge of the Qinghai-Tibet Plateau that from AD 1100-1300 "the $\delta^{18}O$ of Hongyuan peat cellulose increased, consistent with that of Jinchuan peat cellulose and corresponding to the 'Medieval Warm Period'." In addition, Qian and Zhu (2002) analyzed the thickness of laminae in a stalagmite found in Shihua Cave, Beijing, from whence they inferred the existence of a relatively wet period running from approximately AD 940 to 1200.

Hong *et al.* (2000) also report that at the time of the MWP "the northern boundary of the cultivation of citrus tree (*Citrus reticulata* Blanco) and *Boehmeria nivea* (a perennial herb), both subtropical and thermophilous plants, moved gradually into the northern part of China, and it has been estimated that the annual mean temperature was 0.9-1.0°C higher than at present." Considering the climatic conditions required to successfully grow these plants, they further note that annual mean temperatures in that part of the country during the Medieval Warm Period must have been about 1.0°C higher than at present, with extreme January minimum temperatures fully 3.5°C warmer than they are today, citing De'er (1994).

Chu *et al.* (2002) studied the geochemistry of 1,400 years of dated sediments recovered from seven cores taken from three locations in Lake Huguangyan (21°9'N, 110°17'E) on the low-lying Leizhou Peninsula in the tropical region of South China, together with information about the presence of snow, sleet, frost, and frozen rivers over the past 1,000 years obtained from historical documents. They report that "recent publications based on the phenological phenomena, distribution patterns of subtropical plants and cold events (Wang and Gong, 2000; Man, 1998; Wu and Dang, 1998; Zhang, 1994) argue for a warm period from the beginning of the tenth century AD to the late thirteenth century AD," as their own data also suggest. In addition, they note there was a major dry period from AD 880-1260, and that "local historical chronicles support these data, suggesting that the climate of tropical South China was dry during the 'Mediaeval Warm Period'."

Paulsen *et al.* (2003) used high-resolution $\delta^{13}C$ and $\delta^{18}O$ data derived from a stalagmite found in Buddha Cave [33°40'N, 109°05'E] to infer changes in climate in central China for the past 1,270 years.

Among the climatic episodes evident in their data were "those corresponding to the Medieval Warm Period, Little Ice Age and 20th-century warming, lending support to the global extent of these events." In terms of timing, the dry-then-wet-then-dry-again MWP began about AD 965 and continued to approximately AD 1475.

Also working with a stalagmite, this one from Jingdong Cave about 90 km northeast of Beijing, Ma *et al.* (2003) assessed the climatic history of the past 3,000 years at 100-year intervals on the basis of $\delta^{18}O$ data, the Mg/Sr ratio, and the solid-liquid distribution coefficient of Mg. They found that between 200 and 500 years ago, "air temperature was about 1.2°C lower than that of the present," but that between 1,000 and 1,300 years ago, there was an equally aberrant but *warm* period that "corresponded to the Medieval Warm Period in Europe."

Based on 200 sets of phenological and meteorological records extracted from a number of historical sources, many of which are described by Gong and Chen (1980), Man (1990, 2004), Sheng (1990), and Wen and Wen (1996), Ge *et al.* (2003) produced a 2,000-year history of winter half-year temperature (October to April, when CO_2-induced global warming is projected to be most evident) for the region of China bounded by latitudes 27° and 40°N and longitudes 107° and 120°E. Their work revealed a significant warm epoch that lasted from the AD 570s to the 1310s, the peak warmth of which was "about 0.3-0.6°C higher than present for 30-year periods, but over 0.9°C warmer on a 10-year basis."

Bao *et al.* (2003) utilized proxy climate records (ice-core $\delta^{18}O$, peat-cellulose $\delta^{18}O$, tree-ring widths, tree-ring stable carbon isotopes, total organic carbon, lake water temperatures, glacier fluctuations, ice-core CH4, magnetic parameters, pollen assemblages, and sedimentary pigments) obtained from 20 prior studies to derive a 2,000-year temperature history of the northeastern, southern and western sections of the Tibetan Plateau. In each case, there was more than one prior 50-year period of time when the mean temperature of each region was warmer than it was over the most recent 50-year period. In the case of the northeastern sector of the plateau, all of the maximum-warmth intervals occurred during the Medieval Warm Period; in the case of the western sector, they occurred near the end of the Roman Warm Period, and in the case of the southern sector they occurred during both warm periods.

From these several studies, it is evident that for a considerable amount of time during the Medieval

Warm Period, many parts of China exhibited warmer conditions than those of modern times. Since those earlier high temperatures were caused by something other than high atmospheric CO_2 concentrations, whatever was responsible for them could be responsible for the warmth of today.

Additional information on this topic, including reviews of newer publications as they become available, can be found at http://www.co2science.org/subject/m/mwpchina.php.

References

Bao, Y., Brauning, A. and Yafeng, S. 2003. Late Holocene temperature fluctuations on the Tibetan Plateau. *Quaternary Science Reviews* **22**: 2335-2344.

Chu, G., Liu, J., Sun, Q., Lu, H., Gu, Z., Wang, W. and Liu, T. 2002. The 'Mediaeval Warm Period' drought recorded in Lake Huguangyan, tropical South China. *The Holocene* **12**: 511-516.

De'er, Z. 1994. Evidence for the existence of the medieval warm period in China. *Climatic Change* **26**: 289-297.

Ge, Q., Zheng, J., Fang, X., Man, Z., Zhang, X., Zhang, P. and Wang, W.-C. 2003. Winter half-year temperature reconstruction for the middle and lower reaches of the Yellow River and Yangtze River, China, during the past 2000 years. *The Holocene* **13**: 933-940.

Gong, G. and Chen, E. 1980. On the variation of the growing season and agriculture. *Scientia Atmospherica Sinica* **4**: 24-29.

Hong, Y.T., Jiang, H.B., Liu, T.S., Zhou, L.P., Beer, J., Li, H.D., Leng, X.T., Hong, B. and Qin, X.G. 2000. Response of climate to solar forcing recorded in a 6000-year $\delta^{18}O$ time-series of Chinese peat cellulose. *The Holocene* **10**: 1-7.

Ma, Z., Li, H., Xia, M., Ku, T., Peng, Z., Chen, Y. and Zhang, Z. 2003. Paleotemperature changes over the past 3000 years in eastern Beijing, China: A reconstruction based on Mg/Sr records in a stalagmite. *Chinese Science Bulletin* **48**: 395-400.

Man, M.Z. 1998. Climate in Tang Dynasty of China: discussion for its evidence. *Quaternary Sciences* **1**: 20-30.

Man, Z. 1990. Study on the cold/warm stages of Tang Dynasty and the characteristics of each cold/warm stage. *Historical Geography* **8**: 1-15.

Man, Z. 2004. *Climate Change in Historical Period of China*. Shandong Education Press, Ji'nan, China.

Paulsen, D.E., Li, H.-C. and Ku, T.-L. 2003. Climate variability in central China over the last 1270 years revealed by high-resolution stalagmite records. *Quaternary Science Reviews* **22**: 691-701.

Qian, W. and Zhu, Y. 2002. Little Ice Age climate near Beijing, China, inferred from historical and stalagmite records. *Quaternary Research* **57**: 109-119.

Sheng, F. 1990. A preliminary exploration of the warmth and coldness in Henan Province in the historical period. *Historical Geography* **7**: 160-170.

Wang, S.W. and Gong, D.Y. 2000. The temperature of several typical periods during the Holocene in China. *The Advance in Nature Science* **10**: 325-332.

Wen, H. and Wen, H. 1996. Winter-Half-Year Cold/Warm Change in Historical Period of China. Science Press, Beijing, China.

Wu, H.Q. and Dang, A.R. 1998. Fluctuation and characteristics of climate change in temperature of Sui-Tang times in China. *Quaternary Sciences* **1**: 31-38.

Xu, H., Hong, Y., Lin, Q., Hong, B., Jiang, H. and Zhu, Y. 2002. Temperature variations in the past 6000 years inferred from $\delta^{18}O$ of peat cellulose from Hongyuan, China. *Chinese Science Bulletin* **47**: 1578-1584.

Yafeng, S., Tandong, Y. and Bao, Y. 1999. Decadal climatic variations recorded in Guliya ice core and comparison with the historical documentary data from East China during the last 2000 years. *Science in China Series D-Earth Sciences* **42** Supp.: 91-100.

Yang, B., Braeuning, A., Johnson, K.R. and Yafeng, S. 2002. General characteristics of temperature variation in China during the last two millennia. *Geophysical Research Letters* **29**: 10.1029/2001GL014485.

Zhang, D.E. 1994. Evidence for the existence of the Medieval Warm Period in China. *Climatic Change* **26**: 287-297.

3.2.5.2. Russia

Demezhko and Shchapov (2001) studied a borehole extending to more than 5 km depth, reconstructing an 80,000-year history of ground surface temperature in the Middle Urals within the western rim of the Tagil subsidence (58°24' N, 59°44'E). The reconstructed temperature history revealed the existence of a number of climatic excursions, including, in their words, the "Medieval Warm Period with a culmination about 1000 years ago."

Further north, Hiller *et al.* (2001) analyzed subfossil wood samples from the Khibiny mountains on the Kola Peninsula of Russia (67-68°N, 33-34°E) in an effort to reconstruct the region's climate history over the past 1,500 years. They determined that between AD 1000 and 1300 the tree-line was located at least 100-140 m above its current elevation. This observation, in their words, suggests that mean summer temperatures during this "Medieval climatic optimum" were "at least 0.8°C higher than today," and that "the Medieval optimum was the most pronounced warm climate phase on the Kola Peninsula during the last 1500 years."

Additional evidence for the Medieval Warm Period in Russia comes from Naurzbaev and Vaganov (2000), who developed a 2,200-year proxy temperature record (212 BC to 1996 AD) using tree-ring data obtained from 118 trees near the upper timberline in Siberia. Based on their results, they concluded that the warming experienced in the twentieth century was "not extraordinary," and that "the warming at the border of the first and second millennia was longer in time and similar in amplitude."

Krenke and Chernavskaya (2002) present an impressive overview of what is known about the MWP within Russia, as well as throughout the world, based on historical evidence, glaciological evidence, hydrologic evidence, dendrological data, archaeological data, and palynological data. Concentrating on data wholly from within Russia, they report large differences in a number of variables between the Little Ice Age (LIA) and MWP. With respect to the annual mean temperature of northern Eurasia, they report an MWP to LIA drop on the order of 1.5°C. They also say that "the frequency of severe winters reported was increased from once in 33 years in the early period of time, which corresponds to the MWP, to once in 20 years in the LIA," additionally noting that "the abnormally severe winters [of the LIA] were associated with the spread of Arctic air masses over the entire Russian Plain." Finally, they note that the data they used to draw these conclusions were "not used in the reconstructions performed by Mann *et al.*," which perhaps explains why the Mann *et al.* temperature history of the past millennium does not depict the coolness of the LIA or the warmth of the MWP nearly as well as the more appropriately derived temperature history of Esper *et al.* (2002).

In discussing their approach to the subject of global warming detection and attribution, the Russians state that "an analysis of climate variations over 1000 years should help ... reveal natural multicentennial variations possible at present but not detectable in available 100-200-year series of instrumental records." In this endeavor, they were highly successful, stating unequivocally that "the Medieval Warm Period and the Little Ice Age existed globally."

Additional information on this topic, including reviews of newer publications as they become available, can be found at http://www.co2science.org/subject/m/mwprussia.php.

References

Demezhko, D.Yu. and Shchapov, V.A. 2001. 80,000 years ground surface temperature history inferred from the temperature-depth log measured in the superdeep hole SG-4 (the Urals, Russia). *Global and Planetary Change* **29**: 167-178.

Esper, J., Cook, E.R. and Schweingruber, F.H. 2002. Low-frequency signals in long tree-ring chronologies for reconstructing past temperature variability. *Science* **295**: 2250-2253.

Hiller, A., Boettger, T. and Kremenetski, C. 2001. Medieval climatic warming recorded by radiocarbon dated alpine tree-line shift on the Kola Peninsula, Russia. *The Holocene* **11**: 491-497.

Krenke, A.N. and Chernavskaya, M.M. 2002. Climate changes in the preinstrumental period of the last millennium and their manifestations over the Russian Plain. *Isvestiya, Atmospheric and Oceanic Physics* **38**: S59-S79.

Mann, M.E., Bradley, R.S. and Hughes, M.K. 1998. Global-scale temperature patterns and climate forcing over the past six centuries. *Nature* **392**: 779-787.

Mann, M.E., Bradley, R.S. and Hughes, M.K. 1999. Northern Hemisphere temperatures during the past millennium: Inferences, uncertainties, and limitations. *Geophysical Research Letters* **26**: 759-762.

Naurzbaev, M.M. and Vaganov, E.A. 2000. Variation of early summer and annual temperature in east Taymir and Putoran (Siberia) over the last two millennia inferred from tree rings. *Journal of Geophysical Research* **105**: 7317-7326.

3.2.5.3. Other Asia Locations

In addition to China and Russia, the Medieval Warm Period (MWP) has been identified in several other parts of Asia.

Schilman *et al.* (2001) analyzed foraminiferal oxygen and carbon isotopes, together with the physical and geochemical properties of sediments, contained in two cores extracted from the bed of the southeastern Mediterranean Sea off the coast of Israel, where they found evidence for the MWP centered on AD 1200. In discussing their findings, they note there is an abundance of other evidence for the existence of the MWP in the Eastern Mediterranean as well, including, in their words, "high Saharan lake levels (Schoell, 1978; Nicholson, 1980), high Dead Sea levels (Issar *et al.*, 1989, 1991; Issar, 1990, 1998; Issar and Makover-Levin, 1996), and high levels of the Sea of Galilee (Frumkin *et al.*, 1991; Issar and Makover-Levin, 1996)," in addition to "a precipitation maximum at the Nile headwaters (Bell and Menzel, 1972; Hassan, 1981; Ambrose and DeNiro, 1989) and in the northeastern Arabian Sea (von Rad *et al.*, 1999)."

Further to the east, Kar *et al.* (2002) explored the nature of climate change preserved in the sediment profile of an outwash plain two to three km from the snout of the Gangotri Glacier in the Uttarkashi district of Uttranchal, Western Himalaya. Between 2,000 and 1,700 years ago, their data reveal the existence of a relatively cool climate. Then, from 1,700 to 850 years ago, there was what they call an "amelioration of climate," during the transition from the depth of the Dark Ages Cold Period to the midst of the Medieval Warm Period. Subsequent to that time, Kar *et al.*'s data indicate the climate "became much cooler," indicative of its transition to Little Ice Age conditions, while during the last 200 years there has been a rather steady warming, as shown by Esper *et al.* (2002a) to have been characteristic of the entire Northern Hemisphere.

At a pair of other Asian locations, Esper *et al.* (2002b) used more than 200,000 ring-width measurements obtained from 384 trees at 20 individual sites ranging from the lower to upper timberline in the Northwest Karakorum of Pakistan (35-37°N, 74-76°E) and the Southern Tien Shan of Kirghizia (40°10'N, 72°35'E) to reconstruct regional patterns of climatic variations in Western Central Asia since AD 618. According to their analysis, the Medieval Warm Period was already firmly established and growing even warmer by the early seventh century; and between AD 900 and 1000, tree growth was exceptionally rapid, at rates they say "cannot be observed during any other period of the last millennium."

Between AD 1000 and 1200, however, growing conditions deteriorated; and at about 1500, minimum tree ring-widths were reached that persisted well into the seventeenth century. Towards the end of the twentieth century, ring-widths increased once again; but Esper *et al.* (2002b) report that "the twentieth-century trend does not approach the AD 1000 maximum." In fact, there is almost no comparison between the two periods, with the Medieval Warm Period being much more conducive to good tree growth than the Current Warm Period. As the authors describe the situation, "growing conditions in the twentieth century exceed the long-term average, but the amplitude of this trend is not comparable to the conditions around AD 1000."

The latest contribution to Asian temperature reconstruction is the study of Esper *et al.* (2003), who processed several extremely long juniper ring-width chronologies for the Alai Range of the western Tien Shan in Kirghizia in such a way as to preserve multi-centennial growth trends that are typically "lost during the processes of tree ring data standardization and chronology building (Cook and Kairiukstis, 1990; Fritts, 1976)." In doing so, they used two techniques that maintain low frequency signals: long-term mean standardization (LTM) and regional curve standardization (RCS), as well as the more conventional spline standardization (SPL) technique that obscures (actually *removes*) long-term trends.

Carried back in time a full thousand years, the SPL chronologies depict significant inter-decadal variations but no longer-term trends. The LTM and RCS chronologies, on the other hand, show long-term decreasing trends from the start of the record until about AD 1600, broad minima from 1600 to 1800, and long-term increasing trends from about 1800 to the present. As a result, in the words of Esper *et al.* (2003), "the main feature of the LTM and RCS Alai Range chronologies is a multi-centennial wave with high values towards both ends."

This grand result has essentially the same form as the Northern Hemisphere extratropic temperature history of Esper *et al.* (2002a), which is vastly different from the hockey stick temperature history of Mann *et al.* (1998, 1999) and Mann and Jones (2003), in that it depicts the existence of both the Little Ice Age and preceding Medieval Warm Period, which are nowhere to be found in the Mann reconstructions. In addition, the new result—especially the LTM chronology, which has a much smaller variance than the RCS chronology—depicts several periods in the first half of the last millennium that were warmer than any part of the last century. These periods include much of the latter half of the Medieval Warm Period and a good part of the first half of the fifteenth century, which has also been found to have been warmer than it is currently by McIntyre and McKitrick (2003) and by Loehle (2004).

In commenting on their important findings, Esper *et al.* (2003) remark that "if the tree ring reconstruction had been developed using 'standard' detrending procedures only, it would have been limited to inter-decadal scale variation and would have missed some of the common low frequency signal." We would also remark, with respect to the upward trend of their data since 1800, that a good portion of that trend may have been due to the aerial fertilization effect of the concomitantly increasing atmospheric CO_2 content, which is known to greatly stimulate the growth of trees. Properly accounting for this very real effect would make the warmer-than-present temperatures of the first half of the past millennium even warmer, relative to those of the past century, than they appear to be in Esper *et al.*'s LTM and RCS reconstructions.

Additional information on this topic, including reviews of newer publications as they become available, can be found at http://www.co2science.org/subject/a/asiamwp.php.

References

Ambrose, S.H. and DeNiro, M.J. 1989. Climate and habitat reconstruction using stable carbon and nitrogen isotope ratios of collagen in prehistoric herbivore teeth from Kenya. *Quaternary Research* **31**: 407-422.

Bell, B. and Menzel, D.H. 1972. Toward the observation and interpretation of solar phenomena. AFCRL F19628-69-C-0077 and AFCRL-TR-74-0357, Air Force Cambridge Research Laboratories, Bedford, MA, pp. 8-12.

Cook, E.R. and Kairiukstis, L.A. 1990. *Methods of Dendrochronology: Applications in the Environmental Sciences*. Kluwer, Dordrecht, The Netherlands.

Esper, J., Cook, E.R. and Schweingruber, F.H. 2002a. Low-frequency signals in long tree-ring chronologies and the reconstruction of past temperature variability. *Science* **295**: 2250-2253.

Esper, J., Schweingruber, F.H. and Winiger, M. 2002b. 1300 years of climatic history for Western Central Asia inferred from tree-rings. *The Holocene* **12**: 267-277.

Esper, J., Shiyatov, S.G., Mazepa, V.S., Wilson, R.J.S., Graybill, D.A. and Funkhouser, G. 2003. Temperature-

sensitive Tien Shan tree ring chronologies show multi-centennial growth trends. *Climate Dynamics* **21**: 699-706.

Fritts, H.C. 1976. *Tree Rings and Climate*. Academic Press, London, UK.

Frumkin, A., Magaritz, M., Carmi, I. and Zak, I. 1991. The Holocene climatic record of the salt caves of Mount Sedom, Israel. *Holocene* **1**: 191-200.

Hassan, F.A. 1981. Historical Nile floods and their implications for climatic change. *Science* **212**: 1142-1145.

Issar, A.S. 1990. *Water Shall Flow from the Rock*. Springer, Heidelberg, Germany.

Issar, A.S. 1998. Climate change and history during the Holocene in the eastern Mediterranean region. In: Issar, A.S. and Brown, N. (Eds.) *Water, Environment and Society in Times of Climate Change*, Kluwer Academic Publishers, Dordrecht, The Netherlands, pp. 113-128.

Issar, A.S. and Makover-Levin, D. 1996. Climate changes during the Holocene in the Mediterranean region. In: Angelakis, A.A. and Issar, A.S. (Eds.) *Diachronic Climatic Impacts on Water Resources with Emphasis on the Mediterranean Region*, NATO ASI Series, Vol. I, 36, Springer, Heidelberg, Germany, pp. 55-75.

Issar, A.S., Tsoar, H. and Levin, D. 1989. Climatic changes in Israel during historical times and their impact on hydrological, pedological and socio-economic systems. In: Leinen, M. and Sarnthein, M. (Eds.), *Paleoclimatology and Paleometeorology: Modern and Past Patterns of Global Atmospheric Transport*, Kluwer Academic Publishers, Dordrecht, The Netherlands, pp. 535-541.

Issar, A.S., Govrin, Y., Geyh, M.A., Wakshal, E. and Wolf, M. 1991. Climate changes during the Upper Holocene in Israel. *Israel Journal of Earth-Science* **40**: 219-223.

Kar, R., Ranhotra, P.S., Bhattacharyya, A. and Sekar B. 2002. Vegetation *vis-à-vis* climate and glacial fluctuations of the Gangotri Glacier since the last 2000 years. *Current Science* **82**: 347-351.

Loehle, C. 2004. Climate change: detection and attribution of trends from long-term geologic data. *Ecological Modelling* **171**: 433-450.

Mann, M.E., Bradley, R.S. and Hughes, M.K. 1998. Global-scale temperature patterns and climate forcing over the past six centuries. *Nature* **392**: 779-787.

Mann, M.E., Bradley, R.S. and Hughes, M.K. 1999. Northern Hemisphere temperatures during the past millennium: Inferences, uncertainties, and limitations. *Geophysical Research Letters* **26**: 759-762.

Mann, M.E. and Jones, P.D. 2003. Global surface temperatures over the past two millennia. *Geophysical Research Letters* **30**: 10.1029/2003GL017814.

McIntyre, S. and McKitrick, R. 2003. Corrections to the Mann *et al.* (1998) proxy data base and Northern Hemispheric average temperature series. *Energy and Environment* **14**: 751-771.

Nicholson, S.E. 1980. Saharan climates in historic times. In: Williams, M.A.J. and Faure, H. (Eds.) *The Sahara and the Nile*, Balkema, Rotterdam, The Netherlands, pp. 173-200.

Schilman, B., Bar-Matthews, M., Almogi-Labin, A. and Luz, B. 2001. Global climate instability reflected by Eastern Mediterranean marine records during the late Holocene. *Palaeogeography, Palaeoclimatology, Palaeoecology* **176**: 157-176.

Schoell, M. 1978. Oxygen isotope analysis on authigenic carbonates from Lake Van sediments and their possible bearing on the climate of the past 10,000 years. In: Degens, E.T. (Ed.) *The Geology of Lake Van, Kurtman*. The Mineral Research and Exploration Institute of Turkey, Ankara, Turkey, pp. 92-97.

von Rad, U., Schulz, H., Riech, V., den Dulk, M., Berner, U. and Sirocko, F. 1999. Multiple monsoon-controlled breakdown of oxygen-minimum conditions during the past 30,000 years documented in laminated sediments off Pakistan. *Palaeogeography, Palaeoclimatology, Palaeoecology* **152**: 129-161.

3.2.6. Europe

Based on analyses of subfossil wood samples from the Khibiny mountains on the Kola Peninsula of Russia, Hiller *et al.* (2001) were able to reconstruct a 1,500-year history of alpine tree-line elevation. This record indicates that between AD 1000 and 1300, the tree-line there was located at least 100 to 140 meters above its current location. The researchers state that this fact implies a mean summer temperature that was "at least 0.8°C higher than today."

Moving from land to water, in a study of a well-dated sediment core from the Bornholm Basin in the southwestern Baltic Sea, Andren *et al.* (2000) found evidence for a period of high primary production at approximately AD 1050. Many of the diatoms of that period were warm water species that the scientists say "cannot be found in the present Baltic Sea." This balmy period, they report, "corresponds to the time when the Vikings succeeded in colonizing Iceland and Greenland." The warmth ended rather abruptly, however, at about AD 1200, when they note there was "a major decrease in warm water taxa in the diatom assemblage and an increase in cold water taxa," which latter diatoms are characteristic of what they

call the Recent Baltic Sea Stage that prevails to this day.

In another marine study, Voronina *et al.* (2001) analyzed dinoflagellate cyst assemblages in two sediment cores retrieved from the southeastern Barents Sea, one spanning a period of 8,300 years and one spanning a period of 4,400 years. The longer of the two cores indicated a warm interval from about 8,000 to 3,000 years before present, followed by cooling pulses coincident with lowered salinity and extended ice cover in the vicinity of 5,000, 3,500, and 2,500 years ago. The shorter core additionally revealed cooling pulses at tentative dates of 1,400, 300, and 100 years before present. For the bulk of the past 4,400 years, however, ice cover lasted only two to three months per year, as opposed to the modern mean of 4.3 months per year. In addition, August temperatures ranged between 6° and 8°C, significantly warmer than the present mean of 4.6°C.

Moving back towards land, Mikalsen *et al.* (2001) made detailed measurements of a number of properties of sedimentary material extracted from the bottom of a fjord on the west coast of Norway, deriving a relative temperature history of the region that spanned the last five millennia. This record revealed the existence of a period stretching from A.D. 1330 to 1600 that, in their words, "had the highest bottom-water temperatures in Sulafjorden during the last 5000 years."

In eastern Norway, Nesje *et al.* (2001) analyzed a sediment core obtained from Lake Atnsjoen, deriving a 4,500-year record of river flooding. They observed "a period of little flood activity around the Medieval period (AD 1000-1400)," which was followed by "a period of the most extensive flood activity in the Atnsjoen catchment." This flooding, in their words, resulted from the "post-Medieval climate deterioration characterized by lower air temperature, thicker and more long-lasting snow cover, and more frequent storms associated with the 'Little Ice Age'."

Working in both Norway and Scotland, Brooks and Birks (2001) studied midges, the larval-stage head capsules of which are well preserved in lake sediments and are, in their words, "widely recognized as powerful biological proxies for inferring past climate change." Applying this technique to sediments derived from a lake in the Cairngorms region of the Scottish Highlands, they determined that temperatures there peaked at about 11°C during what they refer to as the "Little Climatic Optimum"— which we typically call the Medieval Warm Period—

"before cooling by about 1.5°C which may coincide with the 'Little Ice Age'."

These results, according to Brooks and Birks, "are in good agreement with a chironomid stratigraphy from Finse, western Norway (Velle, 1998)," where summer temperatures were "about 0.4°C warmer than the present day" during the Medieval Warm Period. This latter observation also appears to hold for the Scottish site, since the upper sample of the lake sediment core from that region, which was collected in 1993, "reconstructs the modern temperature at about 10.5°C," which is 0.5°C less than the 11°C value the authors found for the Medieval Warm Period.

Moving to Switzerland, Filippi et al. (1999) analyzed a sediment core extracted from Lake Neuchatel in the western Swiss Lowlands. During this same transition from the Medieval Warm Period (MWP) to the Little Ice Age (LIA), they detected a drop of approximately 1.5°C in mean annual air temperature. To give some context to this finding, they say that "the warming during the 20th century does not seem to have fully compensated the cooling at the MWP-LIA transition." And to make the message even more clear, they add that during the Medieval Warm Period, the mean annual air temperature was "on average higher than at present."

In Ireland, in a cave in the southwestern part of the country, McDermott et al. (2001) derived a $\delta^{18}O$ record from a stalagmite that provided evidence for climatic variations that are "broadly consistent with a Medieval Warm Period at ~1000 ± 200 years ago and a two-stage Little Ice Age." Also evident in the data were the $\delta^{18}O$ signatures of the earlier Roman Warm Period and Dark Ages Cold Period that comprised the preceding millennial-scale cycle of climate in that region.

In another study of three stalagmites found in a cave in northwest Germany, Niggemann et al. (2003) discovered that the climate records they contained "resemble records from an Irish stalagmite (McDermott et al., 1999)," specifically noting that their own records provide evidence for the existence of the Little Ice Age, the Medieval Warm Period and the Roman Warm Period, which evidence also implies the existence of what McDermott et al. (2001) call the Dark Ages Cold Period that separated the Medieval and Roman Warm Periods, as well as the existence of the unnamed cold period that preceded the Roman Warm Period.

Bodri and Cermak (1999) derived individual ground surface temperature histories from the temperature-depth logs of 98 separate boreholes drilled in the Czech Republic. From these data they detected "the existence of a medieval warm epoch lasting from A.D. 1100-1300," which they describe as "one of the warmest postglacial times. Noting that this spectacular warm period was followed by the Little Ice Age, they went on to suggest that "the observed recent warming may thus be easily a natural return of climate from the previous colder conditions back to a 'normal'."

Filippi et al. (1999) share similar views, as is demonstrated by their citing of Keigwin (1996) to the effect that "sea surface temperature (SST) reconstructions show that SST was ca. 1°C cooler than today about 400 years ago and ca. 1°C warmer than today during the MWP." Citing Bond et al. (1997), they further note that the MWP and LIA are merely the most recent manifestations of "a pervasive millennial-scale coupled atmosphere-ocean climate oscillation," which, we might add, has absolutely nothing to do with variations in the air's CO_2 content.

Lastly, we report the findings of Berglund (2003), who identified several periods of expansion and decline of human cultures in northwest Europe and compared them with a history of reconstructed climate "based on insolation, glacier activity, lake and sea levels, bog growth, tree line, and tree growth." In doing so, he determined there was a positive correlation between human impact/land-use and climate change. Specifically, in the latter part of the record, where both cultural and climate changes were best defined, there was, in his words, a great "retreat of agriculture" centered on about AD 500, which led to "reforestation in large areas of central Europe and Scandinavia." He additionally notes that "this period was one of rapid cooling indicated from tree-ring data (Eronen et al., 1999) as well as sea surface temperatures based on diatom stratigraphy in [the] Norwegian Sea (Jansen and Koc, 2000), which can be correlated with Bond's event 1 in the North Atlantic sediments (Bond et al., 1997)."

Next came what Berglund calls a "boom period" that covered "several centuries from AD 700 to 1100." This interval of time proved to be "a favourable period for agriculture in marginal areas of Northwest Europe, leading into the so-called Medieval Warm Epoch," when "the climate was warm and dry, with high treelines, glacier retreat, and reduced lake catchment erosion." This period "lasted until around AD 1200, when there was a gradual change to cool/moist climate, the beginning of the

Little Ice Age ... with severe consequences for the agrarian society."

The story from Europe seems quite clear. There was a several-hundred-year period in the first part of the last millennium that was significantly warmer than it is currently. In addition, there is reason to believe the planet may be on a natural climate trajectory that is taking it back to a state reminiscent of the Medieval Warm Period. There is nothing we can do about this natural cycle except, as is implied by the study of Berglund (2003), reap the benefits.

Additional information on this topic, including reviews of newer publications as they become available, can be found at http://www.co2science.org/subject/e/europemwp.php.

References

Andren, E., Andren, T. and Sohlenius, G. 2000. The Holocene history of the southwestern Baltic Sea as reflected in a sediment core from the Bornholm Basin. *Boreas* **29**: 233-250.

Berglund, B.E. 2003. Human impact and climate changes—synchronous events and a causal link? *Quaternary International* **105**: 7-12.

Bodri, L. and Cermak, V. 1999. Climate change of the last millennium inferred from borehole temperatures: Regional patterns of climatic changes in the Czech Republic—Part III. *Global and Planetary Change* **21**: 225-235.

Bond, G., Showers, W., Cheseby, M., Lotti, R., Almasi, P., deMenocal, P., Priori, P., Cullen, H., Hajdes, I. and Bonani, G. 1997. A pervasive millennial-scale climate cycle in the North Atlantic: The Holocene and late glacial record. *Science* **278**: 1257-1266.

Brooks, S.J. and Birks, H.J.B. 2001. Chironomid-inferred air temperatures from Lateglacial and Holocene sites in north-west Europe: progress and problems. *Quaternary Science Reviews* **20**: 1723-1741.

Eronen, M., Hyvarinen, H. and Zetterberg, P. 1999. Holocene humidity changes in northern Finnish Lapland inferred from lake sediments and submerged Scots pines dated by tree-rings. *The Holocene* **9**: 569-580.

Filippi, M.L., Lambert, P., Hunziker, J., Kubler, B. and Bernasconi, S. 1999. Climatic and anthropogenic influence on the stable isotope record from bulk carbonates and ostracodes in Lake Neuchatel, Switzerland, during the last two millennia. *Journal of Paleolimnology* **21**: 19-34.

Hiller, A., Boettger, T. and Kremenetski, C. 2001. Medieval climatic warming recorded by radiocarbon dated alpine tree-line shift on the Kola Peninsula, Russia. *The Holocene* **11**: 491-497.

Jansen, E. and Koc, N. 2000. Century to decadal scale records of Norwegian sea surface temperature variations of the past 2 millennia. *PAGES Newsletter* **8**(1): 13-14.

Keigwin, L.D. 1996. The Little Ice Age and Medieval Warm Period in the Sargasso Sea. *Science* **174**: 1504-1508.

McDermott, F., Frisia, S., Huang, Y., Longinelli, A., Spiro, S., Heaton, T.H.E., Hawkesworth, C., Borsato, A., Keppens, E., Fairchild, I., van Borgh, C., Verheyden, S. and Selmo, E. 1999. Holocene climate variability in Europe: evidence from delta18O, textural and extension-rate variations in speleothems. *Quaternary Science Reviews* **18**: 1021-1038.

McDermott, F., Mattey, D.P. and Hawkesworth, C. 2001. Centennial-scale Holocene climate variability revealed by a high-resolution speleothem $\delta^{18}O$ record from SW Ireland. *Science* **294**: 1328-1331.

Mikalsen, G., Sejrup, H.P. and Aarseth, I. 2001. Late-Holocene changes in ocean circulation and climate: foraminiferal and isotopic evidence from Sulafjord, western Norway. *The Holocene* **11**: 437-446.

Nesje, A., Dahl, S.O., Matthews, J.A. and Berrisford, M.S. 2001. A ~ 4500-yr record of river floods obtained from a sediment core in Lake Atnsjoen, eastern Norway. *Journal of Paleolimnology* **25**: 329-342.

Niggemann, S., Mangini, A., Richter, D.K. and Wurth, G. 2003. A paleoclimate record of the last 17,600 years in stalagmites from the B7 cave, Sauerland, Germany. *Quaternary Science Reviews* **22**: 555-567.

Velle, G. 1998. A paleoecological study of chironomids (Insecta: Diptera) with special reference to climate. M.Sc. Thesis, University of Bergen.

Voronina, E., Polyak, L., De Vernal, A. and Peyron, O. 2001. Holocene variations of sea-surface conditions in the southeastern Barents Sea, reconstructed from dinoflagellate cyst assemblages. *Journal of Quaternary Science* **16**: 717-726.

3.2.7. North America

Arseneault and Payette (1997) analyzed tree-ring and growth-form sequences obtained from more than 300 spruce remains buried in a presently treeless peatland in northern Quebec to produce a proxy record of climate for this region of the continent between 690 and 1591 AD. Perhaps the most outstanding feature of this history was the warm period it revealed between 860 and 1000 AD. Based on the fact that the northernmost twentieth century location of the forest tree-line is presently 130 km south of their study site, the scientists concluded that the "Medieval Warm Period was approximately 1°C warmer than the 20th century."

Shifting to the other side of the continent, Calkin *et al.* (2001) carefully reviewed what they termed "the most current and comprehensive research of Holocene glaciation" along the northernmost Gulf of Alaska between the Kenai Peninsula and Yakutat Bay, where they too detected a Medieval Warm Period that lasted for "at least a few centuries prior to A.D. 1200." Also identifying the Medieval Warm Period, as well as other major warm and cold periods of the millennial-scale climatic oscillation that is responsible for them, was Campbell (2002), who analyzed the grain sizes of sediment cores obtained from Pine Lake, Alberta, Canada (52°N, 113.5°W) to provide a non-vegetation-based high-resolution record of climate variability for this part of North America over the past 4,000 years. Periods of both increasing and decreasing grain size (related to moisture availability) were noted throughout the 4,000-year record at decadal, centennial, and millennial time scales. The most predominant departures were several-centuries-long epochs that corresponded to the Little Ice Age (about AD 1500-1900), the Medieval Warm Period (about AD 700-1300), the Dark Ages Cold Period (about BC 100 to AD 700), and the Roman Warm Period (about BC 900-100).

Laird *et al.* (2003) studied diatom assemblages in sediment cores taken from three Canadian and three United States lakes situated within the northern prairies of North America, finding that "shifts in drought conditions on decadal through multicentennial scales have prevailed in this region for at least the last two millennia." In Canada, major shifts occurred near the beginning of the Medieval Warm Period, while in the United States they occurred near its end. In giving some context to these findings, the authors state that "distinct patterns of abrupt change in the Northern Hemisphere are common at or near the termination of the Medieval Warm Period (*ca.* A.D. 800-1300) and the onset of the Little Ice Age (*ca.* A.D. 1300-1850)." They also note that "millennial-scale shifts over at least the past 5,500 years, between sustained periods of wetter and drier conditions, occurring approximately every 1,220 years, have been reported from western Canada (Cumming *et al.*, 2002)," and that "the striking correspondence of these shifts to large changes in fire frequencies, inferred from two sites several hundreds

of kilometers to the southwest in the mountain hemlock zone of southern British Columbia (Hallett *et al.*, 2003), suggests that these millennial-scale dynamics are linked and operate over wide spatial scales."

In an effort to determine whether these climate-driven millennial-scale cycles are present in the terrestrial pollen record of North America, Viau *et al.* (2002) analyzed a set of 3,076 ^{14}C dates from the North American Pollen Database used to date sequences in more than 700 pollen diagrams across North America. Results of their statistical analyses indicated there were nine millennial-scale oscillations during the past 14,000 years in which continent-wide synchronous vegetation changes with a periodicity of roughly 1,650 years were recorded in the pollen records. The most recent of the vegetation transitions was centered at approximately 600 years BP (before present). This event, in the words of the authors, "culminat[ed] in the Little Ice Age, with maximum cooling 300 years ago." Prior to that event, a major transition that began approximately 1,600 years BP represents the climatic amelioration that "culminat[ed] in the maximum warming of the Medieval Warm Period 1000 years ago."

And so it goes, on back through the Holocene and into the preceding late glacial period, with the times of all major pollen transitions being "consistent," in the words of the authors of the study, "with ice and marine records." Viau *et al.* additionally note that "the large-scale nature of these transitions and the fact that they are found in different proxies confirms the hypothesis that Holocene and late glacial climate variations of millennial-scale were abrupt transitions between climatic regimes as the atmosphere-ocean system reorganized in response to some forcing." They go on to say that "although several mechanisms for such *natural* [our italics] forcing have been advanced, recent evidence points to a potential solar forcing (Bond *et al.*, 2001) associated with ocean-atmosphere feedbacks acting as global teleconnections agents." Furthermore, they note that "these transitions are identifiable across North America and presumably the world."

Additional evidence for the solar forcing of these millennial-scale climate changes is provided by Shindell *et al.* (2001), who used a version of the Goddard Institute for Space Studies GCM to estimate climatic differences between the period of the Maunder Minimum in solar irradiance (mid-1600s to early 1700s) and a century later, when solar output was relatively high for several decades. Their results compared so well with historical and proxy climate data that they concluded, in their words, that "colder winter temperatures over the Northern Hemispheric continents during portions of the 15th through the 17th centuries (sometimes called the Little Ice Age) and warmer temperatures during the 12th through 14th centuries (the putative Medieval Warm Period) may have been influenced by long-term solar variations."

Rounding out our mini-review of the Medieval Warm Period in North America are two papers dealing with the climatic history of the Chesapeake Bay region of the United States. The first, by Brush (2001), consists of an analysis of sediment cores obtained from the Bay's tributaries, marshes and main stem that covers the past millennium, in which it is reported that "the Medieval Climatic Anomaly and the Little Ice Age are recorded in Chesapeake sediments by terrestrial indicators of dry conditions for 200 years, beginning about 1000 years ago, followed by increases in wet indicators from about 800 to 400 years ago."

Willard et al. (2003) studied the same region for the period 2,300 years BP to the present, via an investigation of fossil dinoflagellate cysts and pollen from sediment cores. Their efforts revealed that "several dry periods ranging from decades to centuries in duration are evident in Chesapeake Bay records." The first of these periods of lower-than-average precipitation, which spanned the period 200 BC-AD 300, occurred during the latter part of the Roman Warm Period, as delineated by McDermott et al. (2001) on the basis of a high-resolution speleothem $\delta^{18}O$ record from southwest Ireland. The next such dry period (~AD 800-1200), in the words of the authors, "corresponds to the 'Medieval Warm Period', which has been documented as drier than average by tree-ring (Stahle and Cleaveland, 1994) and pollen (Willard et al., 2001) records from the southeastern USA."

Willard et al. go on to say that "mid-Atlantic dry periods generally correspond to central and southwestern USA 'megadroughts', described by Woodhouse and Overpeck (1998) as major droughts of decadal or more duration that probably exceeded twentieth-century droughts in severity." They further indicate that "droughts in the late sixteenth century that lasted several decades, and those in the 'Medieval Warm Period' and between ~AD 50 and AD 350 spanning a century or more have been indicated by Great Plains tree-ring (Stahle et al., 1985; Stahle and Cleaveland, 1994), lacustrine diatom and ostracode

(Fritz et al., 2000; Laird et al., 1996a, 1996b) and detrital clastic records (Dean, 1997)."

It is evident that the Medieval Warm Period has left its mark throughout North America in the form of either warm temperature anomalies or periods of relative dryness.

Additional information on this topic, including reviews of newer publications as they become available, can be found at http://www.co2science.org/subject/n/northamericamwp.php.

References

Arseneault, D. and Payette, S. 1997. Reconstruction of millennial forest dynamics from tree remains in a subarctic tree line peatland. *Ecology* **78**: 1873-1883.

Bond, G., Kromer, B., Beer, J., Muscheler, R., Evans, M.N., Showers, W., Hoffmann, S., Lotti-Bond, R., Hajdas, I. and Bonani, G. 2001. Persistent solar influence on North Atlantic climate during the Holocene. *Science* **294**: 2130-2136.

Brush, G.S. 2001. Natural and anthropogenic changes in Chesapeake Bay during the last 1000 years. *Human and Ecological Risk Assessment* **7**: 1283-1296.

Calkin, P.E., Wiles, G.C. and Barclay, D.J. 2001. Holocene coastal glaciation of Alaska. *Quaternary Science Reviews* **20**: 449-461.

Campbell, C. 2002. Late Holocene lake sedimentology and climate change in southern Alberta, Canada. *Quaternary Research* **49**: 96-101.

Cumming, B.F., Laird, K.R., Bennett, J.R., Smol, J.P. and Salomon, A.K. 2002. Persistent millennial-scale shifts in moisture regimes in western Canada during the past six millennia. *Proceedings of the National Academy of Sciences USA* **99**: 16,117-16,121.

Dean, W.E. 1997. Rates, timing, and cyclicity of Holocene eolian activity in north-central United States: evidence from varved lake sediments. *Geology* **25**: 331-334.

Fritz, S.C., Ito, E., Yu, Z., Laird, K.R. and Engstrom, D.R. 2000. Hydrologic variation in the northern Great Plains during the last two millennia. *Quaternary Research* **53**: 175-184.

Hallett, D.J., Lepofsky, D.S., Mathewes, R.W. and Lertzman, K.P. 2003. 11,000 years of fire history and climate in the mountain hemlock rain forests of southwestern British Columbia based on sedimentary charcoal. *Canadian Journal of Forest Research* **33**: 292-312.

Laird, K.R., Fritz, S.C., Grimm, E.C. and Mueller, P.G. 1996a. Century-scale paleoclimatic reconstruction from Moon Lake, a closed-basin lake in the northern Great Plains. *Limnology and Oceanography* **41**: 890-902.

Laird, K.R., Fritz, S.C., Maasch, K.A. and Cumming, B.F. 1996b. Greater drought intensity and frequency before AD 1200 in the Northern Great Plains, USA. *Nature* **384**: 552-554.

Laird, K.R., Cumming, B.F., Wunsam, S., Rusak, J.A., Oglesby, R.J., Fritz, S.C. and Leavitt, P.R. 2003. Lake sediments record large-scale shifts in moisture regimes across the northern prairies of North America during the past two millennia. *Proceedings of the National Academy of Sciences USA* **100**: 2483-2488.

McDermott, F., Mattey, D.P. and Hawkesworth, C. 2001. Centennial-scale Holocene climate variability revealed by a high-resolution speleothem $\delta^{18}O$ record from SW Ireland. *Science* **294**: 1328-1331.

Shindell, D.T., Schmidt, G.A., Mann, M.E., Rind, D. and Waple, A. 2001. Solar forcing of regional climate change during the Maunder Minimum. *Science* **294**: 2149-2152.

Stahle, D.W. and Cleaveland, M.K. 1994. Tree-ring reconstructed rainfall over the southeastern U.S.A. during the Medieval Warm Period and Little Ice Age. *Climatic Change* **26**: 199-212.

Stahle, D.W., Cleaveland, M.K. and Hehr, J.G. 1985. A 450-year drought reconstruction for Arkansas, United States. *Nature* **316**: 530-532.

Viau, A.E., Gajewski, K., Fines, P., Atkinson, D.E. and Sawada, M.C. 2002. Widespread evidence of 1500 yr climate variability in North America during the past 14,000 yr. *Geology* **30**: 455-458.

Willard, D.A., Cronin, T.M. and Verardo, S. 2003. Late-Holocene climate and ecosystem history from Chesapeake Bay sediment cores, USA. *The Holocene* **13**: 201-214.

Willard, D.A., Weimer, L.M. and Holmes, C.W. 2001. The Florida Everglades ecosystem, climatic and anthropogenic impacts over the last two millennia. *Bulletins of American Paleontology* **361**: 41-55.

Woodhouse, C.A. and Overpeck, J.T. 1998. 2000 years of drought variability in the Central United States. *Bulletin of the American Meteorological Society* **79**: 2693-2714.

3.2.8. South America

In Argentina, Cioccale (1999) assembled what was known at the time about the climatic history of the central region of that country over the past 1,400 years, highlighting a climatic "improvement" that began some 400 years before the start of the last millennium, which ultimately came to be characterized by "a marked increase of environmental suitability, under a relatively homogeneous climate." As a result of this climatic amelioration that marked the transition of the region from the Dark Ages Cold Period to the Medieval Warm Period, Cioccale says "the population located in the lower valleys ascended to higher areas in the Andes," where they remained until around AD 1320, when the transition to the stressful and extreme climate of the Little Ice Age began.

Down at the southern tip of the country in Tierra del Fuego, Mauquoy *et al.* (2004) inferred similar changes in temperature and/or precipitation from plant macrofossils, pollen, fungal spores, testate amebae, and humification associated with peat monoliths collected from the Valle de Andorra. These new chronologies were compared with other chronologies of pertinent data from both the Southern and Northern Hemispheres in an analysis that indicated there was evidence for a period of warming-induced drier conditions from AD 960-1020, which, in their words, "seems to correspond to the Medieval Warm Period (MWP, as derived in the Northern Hemisphere)." They note that "this interval compares well to the date range of AD 950-1045 based on Northern Hemisphere extratropical tree-ring data (Esper *et al.*, 2002)," and they conclude that this correspondence "shows that the MWP was possibly synchronous in both hemispheres, as suggested by Villalba (1994)."

In Chile, Jenny *et al.* (2002) studied geochemical, sedimentological, and diatom-assemblage data derived from sediment cores extracted from one of the largest natural lakes (Laguna Aculeo) in the central part of the country. From 200 BC, when the record began, until AD 200, conditions there were primarily dry, during the latter stages of the Roman Warm Period. Subsequently, from AD 200-700, with a slight respite in the central hundred years of that period, there was a high frequency of flood events, during the Dark Ages Cold Period. Then came a several-hundred-year period of less flooding that was coeval with the Medieval Warm Period. This more benign period was then followed by another period of frequent flooding from 1300-1700 that was coincident with the Little Ice Age, after which flooding picked up again after 1850.

In Peru, Chepstow-Lusty *et al.* (1998) derived a 4,000-year climate history from a study of pollen in sediment cores obtained from a recently in-filled lake in the Patacancha Valley near Marcacocha. Their data indicated a several-century decline in pollen content after AD 100, as the Roman Warm Period gave way to the Dark Ages Cold Period. However, a "more optimum climate," as they describe it, with warmer temperatures and drier conditions, came into being and prevailed for several centuries after about AD 900, which was, of course, the Medieval Warm Period, which was followed by the Little Ice Age, all of which climatic periods are in nearly perfect temporal agreement with the climatic history derived by McDermott *et al.* (2001) from a study of a stalagmite recovered from a cave nearly half the world away in Ireland.

Subsequent work in this area was conducted by Chepstow-Lusty and Winfield (2000) and Chepstow-Lusty *et al.* (2003). Centered on approximately 1,000 years ago, Chepstow-Lusty and Winfield researchers identified what they describe as "the warm global climatic interval frequently referred to as the Medieval Warm Epoch." This extremely *arid* interval in this part of South America, in their opinion, may have played a significant role in the collapse of the Tiwanaku civilization further south, where a contemporaneous prolonged drought occurred in and around the area of Lake Titicaca (Binford *et al.*, 1997; Abbott *et al.*, 1997).

Near the start of this extended dry period, which had gradually established itself between about AD 700 and 1000, Chepstow-Lusty and Winfield report that "temperatures were beginning to increase after a sustained cold period that had precluded agricultural activity at these altitudes." This earlier colder and wetter interval was coeval with the Dark Ages Cold Period of the North Atlantic region, which in the Peruvian Andes had held sway for a good portion of the millennium preceding AD 1000, as revealed by a series of climatic records developed from sediment cores extracted from yet other lakes in the Central Peruvian Andes (Hansen *et al.*, 1994) and by proxy evidence of concomitant Peruvian glacial expansion (Wright, 1984; Seltzer and Hastorf, 1990).

Preceding the Dark Ages Cold Period in both parts of the world was what in the North Atlantic region is called the Roman Warm Period. This well-defined climatic epoch is also strikingly evident in the pollen records of Chepstow-Lusty *et al.* (2003), straddling the BC/AD calendar break with one to two hundred years of relative warmth and significant aridity on both sides of it.

Returning to the Medieval Warm Period and proceeding towards the present, the data of

Chepstow-Lusty *et al.* (2003) reveal the occurrence of the Little Ice Age, which in the Central Peruvian Andes was characterized by relative coolness and wetness. These characteristics of that climatic interval are also evident in ice cores retrieved from the Quelccaya ice cap in southern Peru, the summit of which extends 5,670 meters above mean sea level (Thompson *et al.*, 1986, 1988). Finally, both the Quelccaya ice core data and the Marcacocha pollen data reveal the transition to the drier Current Warm Period that occurred over the past 100-plus years.

In harmony with these several findings are the related observations of Rein *et al.* (2004), who derived a high-resolution flood record of the entire Holocene from an analysis of the sediments in a 20-meter core retrieved from a sheltered basin situated on the edge of the Peruvian shelf about 80 km west of Lima. These investigators found a major Holocene anomaly in the flux of lithic components from the continent onto the Peruvian shelf during the Medieval period. Specifically, they report that "lithic concentrations were very low for about 450 years during the Medieval climatic anomaly from A.D. 800 to 1250." In fact, they state that "all known terrestrial deposits of El Niño mega-floods (Magillian and Goldstein, 2001; Wells, 1990) precede or follow the medieval anomaly in our marine records and none of the El Niño mega-floods known from the continent date within the marine anomaly." In addition, they report that "this precipitation anomaly also occurred in other high-resolution records throughout the ENSO domain," citing 11 other references in support of this statement.

Consequently, because heavy winter rainfalls along and off coastal Peru occur only during times of maximum El Niño strength, and because El Niños are typically more prevalent and stronger during cooler as opposed to warmer periods (see El Niño (Relationship to Global Warming) in Chapter 5), the lack of strong El Niños from A.D. 800 to 1250 suggests that this period was truly a Medieval Warm Period; and the significance of this observation was not lost on Rein *et al.* In the introduction to their paper, they note that "discrepancies exist between the Mann curve and alternative time series for the Medieval period." Most notably, to use their words, "the global Mann curve has no temperature optimum, whereas the Esper *et al.* (2002) reconstruction shows northern hemisphere temperatures almost as high as those of the 20th century" during the Medieval period. As a result, in the final sentence of their paper they suggest that "the occurrence of a Medieval climatic anomaly (A.D.

800-1250) with persistently weak El Niños may therefore assist the interpretation of some of the regional discrepancies in thermal reconstructions of Medieval times," which is a polite way of suggesting that the Mann *et al.* (1998, 1999) hockey stick temperature history is deficient in not depicting the presence of a true Medieval Warm Period.

In Venezuala, Haug *et al.* (2001) found a temperature/precipitation relationship that was different from that of the rest of the continent. In examining the titanium and iron concentrations of an ocean sediment core taken from the Cariaco Basin on the country's northern shelf, they determined that the concentrations of these elements were lower during the Younger Dryas cold period between 12.6 and 11.5 thousand years ago, corresponding to a weakened hydrologic cycle with less precipitation and runoff, while during the warmth of the Holocene Optimum of 10.5 to 5.4 thousand years ago, titanium and iron concentrations remained at or near their highest values, suggesting wet conditions and an enhanced hydrologic cycle. Closer to the present, higher precipitation was also noted during the Medieval Warm Period from 1.05 to 0.7 thousand years ago, followed by drier conditions associated with the Little Ice Age between 550 and 200 years ago.

In an update of this study, Haug *et al.* (2003) developed a hydrologic history of pertinent portions of the record that yielded "roughly bi-monthly resolution and clear resolution of the annual signal." This record revealed that "before about 150 A.D.," which according to the climate history of McDermott *et al.* corresponds to the latter portion of the Roman Warm Period (RWP), Mayan civilization had flourished. However, during the transition to the Dark Ages Cold Period (DACP), which was accompanied by a slow but long decline in precipitation, Haug *et al.* report that "the first documented historical crisis hit the lowlands, which led to the 'Pre-Classic abandonment' (Webster, 2002) of major cities."

This crisis occurred during the first intense multi-year drought of the RWP-to-DACP transition, which was centered on about the year 250 A.D. Although the drought was devastating to the Maya, Haug *et al.* report that when it was over, "populations recovered, cities were reoccupied, and Maya culture blossomed in the following centuries during the so-called Classic period." Ultimately, however, there came a time of reckoning, between about 750 and 950 A.D., during what Haug *et al.* determined was the driest interval of the entire Dark Ages Cold Period, when they report that "the Maya experienced a demographic disaster as

profound as any other in human history," in response to a number of other intense multi-year droughts. During this Terminal Classic Collapse, as it is called, Haug *et al.* say that "many of the densely populated urban centers were abandoned permanently, and Classic Maya civilization came to an end."

In assessing the significance of these several observations near the end of their paper, Haug *et al.* conclude that the latter droughts "were the most severe to affect this region in the first millennium A.D." Although some of these spectacular droughts were "brief," lasting "only" between three and nine years, Haug *et al.* report "they occurred during an extended period of reduced overall precipitation that may have already pushed the Maya system to the verge of collapse."

Although the Mayan civilization thus faded away, Haug *et al.*'s data soon thereafter depict the development of the Medieval Warm Period, when the Vikings established their historic settlement on Greenland. Then comes the Little Ice Age, which just as quickly led to the Vikings' demise in that part of the world. This distinctive cold interval of the planet's millennial-scale climatic oscillation also must have led to hard times for the people of Mesoamerica and northern tropical South America; according to the data of Haug *et al.*, the Little Ice Age produced the lowest precipitation regime (of several hundred years' duration) of the last two millennia in that part of the world.

In conclusion, it is difficult to believe that the strong synchronicity of the century-long Northern Hemispheric and South American warm and cold periods described above was coincidental. It is much more realistic to believe it was the result of a millennial-scale oscillation of climate that is global in scope and driven by some regularly varying forcing factor. Although one can argue about the identity of that forcing factor and the means by which it exerts its influence, one thing should be clear: It is not the atmosphere's CO_2 concentration, which has exhibited a significant in-phase variation with global temperature change only over the Little Ice Age to Current Warm Period transition. This being the case, it should be clear that the climatic amelioration of the past century or more has had little or nothing to do with the concomitant rise in the air's CO_2 content but *everything* to do with the influential forcing factor that has governed the millennial-scale oscillation of earth's climate as far back in time as we have been able to detect it.

Additional information on this topic, including reviews of newer publications as they become available, can be found at http://www.co2science.org/subject/s/southamericamwp.php.

References

Abbott, M.B., Binford, M.W., Brenner, M. and Kelts, K.R. 1997. A 3500 ^{14}C yr high resolution record of water-level changes in Lake Titicaca. *Quaternary Research* **47**: 169-180.

Binford, M.W., Kolata, A.L, Brenner, M., Janusek, J.W., Seddon, M.T., Abbott, M. and Curtis. J.H. 1997. Climate variation and the rise and fall of an Andean civilization. *Quaternary Research* **47**: 235-248.

Chepstow-Lusty, A.J., Bennett, K.D., Fjeldsa, J., Kendall, A., Galiano, W. and Herrera, A.T. 1998. Tracing 4,000 years of environmental history in the Cuzco Area, Peru, from the pollen record. *Mountain Research and Development* **18**: 159-172.

Chepstow-Lusty, A., Frogley, M.R., Bauer, B.S., Bush, M.B. and Herrera, A.T. 2003. A late Holocene record of arid events from the Cuzco region, Peru. *Journal of Quaternary Science* **18**: 491-502.

Chepstow-Lusty, A. and Winfield, M. 2000. Inca agroforestry: Lessons from the past. *Ambio* **29**: 322-328.

Cioccale, M.A. 1999. Climatic fluctuations in the Central Region of Argentina in the last 1000 years. *Quaternary International* **62**: 35-47.

Esper, J., Cook, E.R. and Schweingruber, F.H. 2002. Low-frequency signals in long tree-ring chronologies for reconstructing past temperature variability. *Science* **295**: 2250-2253.

Hansen, B.C.S., Seltzer, G.O. and Wright Jr., H.E. 1994. Late Quaternary vegetational change in the central Peruvian Andes. *Palaeogeography, Palaeoclimatology, Palaeoecology* **109**: 263-285.

Haug, G.H., Gunther, D., Peterson, L.C., Sigman, D.M., Hughen, K.A. and Aeschlimann, B. 2003. Climate and the collapse of Maya civilization. *Science* **299**: 1731-1735.

Haug, G.H., Hughen, K.A., Sigman, D.M., Peterson, L.C. and Rohl, U. 2001. Southward migration of the intertropical convergence zone through the Holocene. *Science* **293**: 1304-1308.

Jenny, B., Valero-Garces, B.L., Urrutia, R., Kelts, K., Veit, H., Appleby, P.G. and Geyh, M. 2002. Moisture changes and fluctuations of the Westerlies in Mediterranean Central Chile during the last 2000 years: The Laguna Aculeo record (33°50'S). *Quaternary International* **87**: 3-18.

Magillian, F.J. and Goldstein, P.S. 2001. El Niño floods and culture change: A late Holocene flood history for the Rio Moquegua, southern Peru. *Geology* **29**: 431-434.

Mann, M.E., Bradley, R.S. and Hughes, M.K. 1998. Global-scale temperature patterns and climate forcing over the past six centuries. *Nature* **392**: 779-787.

Mann, M.E., Bradley, R.S. and Hughes, M.K. 1999. Northern Hemisphere temperatures during the past millennium: Inferences, uncertainties, and limitations. *Geophysical Research Letters* **26**: 759-762.

Mauquoy, D., Blaauw, M., van Geel, B., Borromei, A., Quattrocchio, M., Chambers, F.M. and Possnert, G. 2004. Late Holocene climatic changes in Tierra del Fuego based on multiproxy analyses of peat deposits. *Quaternary Research* **61**: 148-158.

McDermott, F., Mattey, D.P. and Hawkesworth, C. 2001. Centennial-scale Holocene climate variability revealed by a high-resolution speleothem $\delta^{18}O$ record from SW Ireland. *Science* **294**: 1328-1331.

Rein, B., Luckge, A. and Sirocko, F. 2004. A major Holocene ENSO anomaly during the Medieval period. *Geophysical Research Letters* **31**: 10.1029/2004GL020161.

Seltzer, G. and Hastorf, C. 1990. Climatic change and its effect on Prehispanic agriculture in the central Peruvian Andes. *Journal of Field Archaeology* **17**: 397-414.

Thompson, L.G., Mosley-Thompson, E., Dansgaard, W. and Grootes, P.M. 1986. The Little Ice Age as recorded in the stratigraphy of the tropical Quelccaya ice cap. *Science* **234**: 361-364.

Thompson, L.G., Davis, M.E., Mosley-Thompson, E. and Liu, K.-B. 1988. Pre-Incan agricultural activity recorded in dust layers in two tropical ice cores. *Nature* **307**: 763-765.

Villalba, R. 1994. Tree-ring and glacial evidence for the Medieval Warm Epoch and the 'Little Ice Age' in southern South America. *Climatic Change* **26**: 183-197.

Webster, D. 2002. *The Fall of the Ancient Maya*. Thames and Hudson, London, UK.

Wells, L.E. 1990. Holocene history of the El Niño phenomenon as recorded in flood sediments of northern coastal Peru. *Geology* **18**: 1134-1137.

Wright Jr., H.E. 1984. Late glacial and Late Holocene moraines in the Cerros Cuchpanga, central Peru. *Quaternary Research* **21**: 275-285.

3.3. Urban Heat Islands

How accurate are the surface temperature records cited by the IPCC as showing unprecedented millennial warmth over the past couple decades? The IPCC considers them very accurate and nearly free of any contaminating influence, yielding a 1905-2005 increase of 0.74°C ± 0.18°C (IPCC, 2007-I, p. 237). Warming in many growing cities, on the other hand, may have been a full order of magnitude greater. Since nearly all near-surface air temperature records of this period have been obtained from sensors located in population centers that have experienced significant growth, it is essential that urban heat island (UHI) effects be removed from all original temperature records when attempting to accurately assess what has truly happened in the natural non-urban environment.

The IPCC dismisses this concern, saying the UHI is "an order of magnitude smaller than decadal and longer time-scale trends" (p. 244) and "UHI effects are real but local, and have a negligible influence (less than 0.006° C per decade over land and zero over the oceans) on these [observed temperature] values" (p. 5). On this extremely important matter, the IPCC is simply wrong, as the rest of this section demonstrates.

References

IPCC. 2007-I. *Climate Change 2007: The Physical Science Basis. Contribution of Working Group I to the Fourth Assessment Report of the Intergovernmental Panel on Climate Change.* Solomon, S., D. Qin, M. Manning, Z. Chen, M. Marquis, K.B. Averyt, M. Tignor and H.L. Miller. (Eds.) Cambridge University Press, Cambridge, UK.

3.3.1. Global

Hegerl *et al.* (2001) describe UHI-induced temperature perversions as one of three types of systematic error in the surface air temperature record whose magnitude "cannot be assessed at present." Nevertheless, they go on to do just that, claiming "it has been estimated that temperature trends over rural stations only are very similar to trends using all station data, suggesting that the effect of urbanization on estimates of global-scale signals should be small." This statement is patently false.

De Laat and Maurellis (2004) used a global dataset developed by Van Aardenne *et al.* (2001), which reveals the spatial distribution of various levels of industrial activity over the planet as quantified by the intensity of anthropogenic CO_2 emissions to divide the surface of the earth into non-industrial and industrial sectors of various intensity levels, after which they plotted the 1979-2001 temperature trends (°C/decade) of the different sectors using data from both the surface and the lower and middle troposphere.

The two scientists report that "measurements of surface and lower tropospheric temperature change give a very different picture from climate model predictions and show strong observational evidence that the degree of industrialization is correlated with surface temperature increases as well as lower tropospheric temperature changes." Specifically, they find that the surface and lower tropospheric warming trends of all industrial regions are greater than the mean warming trend of the earth's non-industrial regions, and that the difference in warming rate between the two types of land use grows ever larger as the degree of industrialization increases.

In discussing the implications of their findings, De Laat and Maurellis say "areas with larger temperature trends (corresponding to higher CO_2 emissions) cover a considerable part of the globe," which implies that "the 'real' global mean surface temperature trend is very likely to be considerably smaller than the temperature trend in the CRU [Hadley Center/Climate Research Unit] data," since the temperature measurements that comprise that data base "are often conducted in the vicinity of human (industrial) activity." These observations, in their words, "suggest a hitherto-overlooked driver of local surface temperature increases, which is linked to the degree of industrialization" and "lends strong support to other indications that surface processes (possibly changes in land-use or the urban heat effect) are crucial players in observed surface temperature changes (Kalnay and Cai, 2003; Gallo *et al.*, 1996, 1999)." They conclude that "the observed surface temperature changes might be a result of local surface heating processes and not related to radiative greenhouse gas forcing."

A similar study was conducted by McKitrick and Michaels (2004), who calculated 1979-2000 linear trends of monthly mean near-surface air temperature for 218 stations in 93 countries, based upon data they obtained from the Goddard Institute of Space Studies (GISS), after which they regressed the results against indicators of local economic activity—such as income, gross domestic product growth rates, and coal use—to see if there was any evidence of these socioeconomic factors affecting the supposedly "pristine as possible" temperature data. Then, they repeated the process using the gridded surface air temperature data of the IPCC.

The two scientists report that the spatial pattern of trends they derived from the GISS data was "significantly correlated with non-climatic factors, including economic activity and sociopolitical characteristics." Likewise, with respect to the IPCC data, they say "very similar correlations appear, despite previous attempts to remove non-climatic effects." These "socioeconomic effects," in the words of McKitrick and Michaels, "add up to a net warming bias," although they say "precise estimation of its magnitude will require further work."

We can get a good feel for the magnitude of the "socioeconomic effect" in some *past* work, such as that of Oke (1973), who measured the urban heat island strength of 10 settlements in the St. Lawrence Lowlands of Canada that had populations ranging from approximately 1,000 to 2,000,000 people, after which he compared his results with those obtained for a number of other cities in North America, as well as some in Europe. Over the population range studied, Oke found that the magnitude of the urban heat island was linearly correlated with the logarithm of population; this relationship indicated that at the lowest population value encountered, i.e., 1,000 inhabitants, there was an urban heat island effect of 2° to 2.5°C, which warming is more than twice as great as the increase in mean global air temperature believed to have occurred since the end of the Little Ice Age. It should be abundantly clear there is ample opportunity for large errors to occur in thermometer-derived surface air temperature histories of the twentieth century, and that error is probably best described as a large and growing warming bias.

That this urban heat island-induced error has indeed corrupted data bases that are claimed to be immune from it is suggested by the work of Hegerl and Wallace (2000), who attempted to determine if trends in recognizable atmospheric modes of variability could account for all or part of the observed trend in surface-troposphere temperature differential, i.e., lapse rate, which has been driven by the upward-inclined trend in surface-derived temperatures and the nearly level trend in satellite-derived tropospheric temperatures over the last two decades of the twentieth century. After doing

everything they could conceive of doing, they had to conclude that "modes of variability that affect surface temperature cannot explain trends in the observed lapse rate," and that "no mechanism with clear spatial or time structure can be found that accounts for that trend." In addition, they had to acknowledge that "all attempts to explain all or a significant part of the observed lapse rate trend by modes of climate variability with structured patterns from observations have failed," and that "an approach applying model data to isolate such a pattern has also failed." Nor could they find any evidence "that interdecadal variations in radiative forcing, such as might be caused by volcanic eruptions, variations in solar output, or stratospheric ozone depletion alone, offer a compelling explanation." Hence, the two scientists ultimately concluded that "there remains a gap in our fundamental understanding of the processes that cause the lapse rate to vary on interdecadal timescales."

On the other hand, the reason why no meteorological or climatic explanation could be found for the ever-increasing difference between the surface- and satellite-derived temperature trends of the past 20-plus years may be that one of the temperature records is incorrect. Faced with this possibility, one would logically want to determine which of the records is likely to be erroneous and then assess the consequences of that determination. Although this task may seem daunting, it is really not that difficult. One reason why is the good correspondence Hegerl and Wallace found to exist between the satellite and radiosonde temperature trends, which leaves little reason for doubting the veracity of the satellite results, since this comparison essentially amounts to an *in situ* validation of the satellite record. A second important reason comes from the realization that it would be extremely easy for a spurious warming of 0.12°C per decade to be introduced into the surface air temperature trend as a consequence of the worldwide intensification of the urban heat island effect that was likely driven by the world population increase that occurred in most of the places where surface air temperature measurements were made over the last two decades of the twentieth century.

It appears almost certain that surface-based temperature histories of the globe contain a significant warming bias introduced by insufficient corrections for the *non*-greenhouse-gas-induced urban heat island effect. Furthermore, it may well be next to impossible to make proper corrections for this deficiency, as the urban heat island of even small towns *dwarfs* any concomitant augmented greenhouse effect that may be present.

References

De Laat, A.T.J. and Maurellis, A.N. 2004. Industrial CO_2 emissions as a proxy for anthropogenic influence on lower tropospheric temperature trends. *Geophysical Research Letters* **31**: 10.1029/2003GL019024.

Gallo, K.P., Easterling, D.R. and Peterson, T.C. 1996. The influence of land use/land cover on climatological values of the diurnal temperature range. *Journal of Climate* **9**: 2941-2944.

Gallo, K.P., Owen, T.W., Easterling, D.R. and Jameson, P.F. 1999. Temperature trends of the historical climatologic network based on satellite-designated land use/land cover. *Journal of Climate* **12**: 1344-1348.

Hegerl, G.C., Jones, P.D. and Barnett, T.P. 2001. Effect of observational sampling error on the detection of anthropogenic climate change. *Journal of Climate* **14**: 198-207.

Hegerl, G.C. and Wallace, J.M. 2002. Influence of patterns of climate variability on the difference between satellite and surface temperature trends. *Journal of Climate* **15**: 2412-2428.

IPCC. 2007. Climate Change 2007: The Physical Science Basis. Contribution of Working Group I to the Fourth Assessment Report of the Intergovernmental Panel on Climate Change. Cambridge University Press.

Kalnay, E. and Cai, M. 2003. Impact of urbanization and land use change on climate. *Nature* **423**: 528-531.

McKitrick, R. and Michaels, P.J. 2004. A test of corrections for extraneous signals in gridded surface temperature data. *Climate Research* **26**: 159-173.

Oke, T.R. 1973. City size and the urban heat island. *Atmospheric Environment* **7**: 769-779.

Van Aardenne, J.A., Dentener, F.J., Olivier, J.G.J., Klein Goldewijk, C.G.M. and Lelieveld, J. 2001. A 1° x 1° resolution dataset of historical anthropogenic trace gas emissions for the period 1890-1990. *Global Biogeochemical Cycles* **15**: 909-928.

3.3.2. North America

In studying the urban heat island (UHI) of Houston, Texas, Streutker (2003) analyzed 82 sets of nighttime radiation data obtained from the split-window infrared channels of the Advanced Very High Resolution Radiometer on board the NOAA-9 satellite during March 1985 through February 1987 and from 125 sets of similar data obtained from the NOAA-14 satellite during July 1999 through June 2001. Between these two periods, it was found that the mean nighttime surface temperature of Houston rose by 0.82 ± 0.10 °C. In addition, Streutker notes that the growth of the Houston UHI, both in magnitude and spatial extent, "scales roughly with the increase in population," and that the mean rural temperature measured during the second interval was "virtually identical to the earlier interval."

This informative study demonstrates that the UHI phenomenon can sometimes be very powerful, for in just 12 years the UHI of Houston grew by more than the IPCC contends the mean surface air temperature of the planet rose over the entire past century, during which period earth's population rose by approximately 280 percent, or nearly an order of magnitude more than the 30 percent population growth experienced by Houston over the 12 years of Streutker's study.

A very different type of study was conducted by Maul and Davis (2001), who analyzed air and seawater temperature data obtained over the past century at the sites of several primary tide gauges maintained by the U.S. Coast and Geodetic Survey. Noting that each of these sites "experienced significant population growth in the last 100 years," and that "with the increase in maritime traffic and discharge of wastewater one would expect water temperatures to rise" (due to a maritime analogue of the urban heat island effect), they calculated trends for the 14 longest records and derived a mean century-long seawater warming of 0.74°C, with Boston registering a 100-year warming of 3.6°C. In addition, they report that air temperature trends at the tide gauge sites, which represent the standard urban heat island effect, were "much larger" than the seawater temperature trends.

In another different type of study, Dow and DeWalle (2000) analyzed trends in annual evaporation and Bowen ratio measurements on 51 eastern U.S. watersheds that had experienced various degrees of urbanization between 1920 and 1990. In doing so, they determined that as residential development progressively occurred on what originally were rural watersheds, watershed evaporation decreased and sensible heating of the atmosphere increased. And from relationships derived from the suite of watersheds investigated, they calculated that complete transformation from 100 percent rural to 100 percent urban characteristics resulted in a 31 percent decrease in watershed evaporation and a 13 W/m^2 increase in sensible heating of the atmosphere.

Climate modeling exercises suggest that a doubling of the air's CO_2 concentration will result in a nominal 4 W/m^2 increase in the radiative forcing of earth's surface-troposphere system, which has often been predicted to produce an approximate 4°C increase in the mean near-surface air temperature of the globe, indicative of an order-of-magnitude climate sensitivity of 1°C per W/m^2 change in radiative forcing. Thus, to a first approximation, the 13 W/m^2 increase in the sensible heating of the near-surface atmosphere produced by the total urbanization of a pristine rural watershed in the eastern United States could be expected to produce an increase of about 13°C in near-surface air temperature over the central portion of the watershed, which is consistent with maximum urban heat island effects observed in large and densely populated cities. Hence, a 10 percent rural-to-urban transformation could well produce a warming on the order of 1.3°C, and a mere 2 percent transformation could increase the near-surface air temperature by as much as a quarter of a degree Centigrade.

This powerful anthropogenic but non-greenhouse-gas-induced effect of urbanization on the energy balance of watersheds and the temperature of the boundary-layer air above them begins to express itself with the very first hint of urbanization and, hence, may be readily overlooked in studies seeking to identify a greenhouse-gas-induced global warming signal. In fact, the fledgling urban heat island effect may already be present in many temperature records that have routinely been considered "rural enough" to be devoid of all human influence.

A case in point is provided by the study of Changnon (1999), who used a series of measurements of soil temperatures obtained in a totally rural setting in central Illinois between 1889 and 1952 and a contemporary set of air temperature measurements made in an adjacent growing community (as well as similar data obtained from other nearby small towns), to evaluate the magnitude of unsuspected heat island effects that may be present in small towns and cities that are typically assumed to be free of urban-induced warming. This work revealed that soil temperature in the totally rural setting experienced an increase from the decade of 1901-1910 to the decade of 1941-1950 that amounted to 0.4°C.

This warming is 0.2°C *less* than the 0.6°C warming determined for the same time period from the entire dataset of the U.S. Historical Climatology Network, which is supposedly corrected for urban heating effects. It is also 0.2°C less than the 0.6°C warming determined for this time period by 11 benchmark stations in Illinois with the highest quality long-term temperature data, all of which are located in communities that had populations of less than 6,000 people as of 1990. And it is 0.17°C less than the 0.57°C warming derived from data obtained at the three benchmark stations closest to the site of the soil temperature measurements and with populations of less than 2,000 people.

Changnon says his findings suggest that "both sets of surface air temperature data for Illinois believed to have the best data quality with little or no urban effects may contain urban influences causing increases of 0.2°C from 1901 to 1950." He further notes—in a grand understatement—that "this could be significant because the IPCC (1995) indicated that the global mean temperature increased 0.3°C from 1890 to 1950."

DeGaetano and Allen (2002b) used data from the U.S. Historical Climatology Network to calculate trends in the occurrence of maximum and minimum temperatures greater than the 90th, 95th, and 99th percentile across the United States over the period 1960-1996. In the case of daily warm minimum temperatures, the slope of the regression line fit to the data of a plot of the annual number of 95th percentile exceedences vs. year was found to be 0.09 exceedences per year for rural stations, 0.16 for suburban stations, and 0.26 for urban stations, making the rate of increase in extreme warm minimum temperatures at urban stations nearly three times greater than the rate of increase at rural stations less affected by growing urban heat islands. Likewise, the rate of increase in the annual number of daily maximum temperature 95th percentile exceedences per year over the same time period was found to be 50 percent greater at urban stations than it was at rural stations.

Working on the Arctic Coastal Plain near the Chuckchi Sea at Barrow, Alaska—which is described by Hinkel *et al.* (2003) as "the northernmost settlement in the USA and the largest native community in the Arctic," the population of which "has grown from about 300 residents in 1900 to more than 4600 in 2000"—the four researchers installed 54 temperature-recording instruments in mid-June of 2001, half of them within the urban area and the other half distributed across approximately 150 km^2 of surrounding land, all of which provided air temperature data at hourly intervals approximately two meters above the surface of the ground. In this paper, they describe the results they obtained for the following winter. Based on urban-rural spatial averages for the entire winter period (December 2001-March 2002), they determined the urban area to be 2.2°C warmer than the rural area. During this period, the mean daily urban-rural temperature difference increased with decreasing temperature, "reaching a peak value of around 6°C in January-February." It was also determined that the daily urban-rural temperature difference increased with decreasing wind speed, such that under calm conditions (< 2 m s^{-1}) the daily urban-rural temperature difference was 3.2°C in the winter. Last of all, under simultaneous calm and cold conditions, the urban-rural temperature difference was observed to achieve hourly magnitudes exceeding 9°C.

Four years later, Hinkel and Nelson (2007) reported that for the period 1 December to 31 March of four consecutive winters, the spatially averaged temperature of the urban area of Barrow was about 2°C warmer than that of the rural area, and that it was not uncommon for the daily magnitude of the urban heat island to exceed 4°C. In fact, they say that on some days the magnitude of the urban heat island exceeded 6°C, and that values in excess of 8°C were sometimes recorded, while noting that the warmest individual site temperatures were "consistently observed in the urban core area."

These results indicate just how difficult it is to measure a background global temperature increase that is believed to have been less than 1°C over the past century (representing a warming of less than 0.1°C per decade), when the presence of a mere 4,500 people can create a winter heat island that may be two orders of magnitude greater than the signal being sought. Clearly, there is no way that temperature measurements made within the range of influence of even a small village can be adjusted to the degree of accuracy that is required to reveal the true magnitude of the pristine rural temperature change.

Moving south, we find Ziska *et al.* (2004) working within and around Baltimore, Maryland, where they characterized the gradual changes that occur in a number of environmental variables as one moves from a rural location (a farm approximately 50 km from the city center) to a suburban location (a park approximately 10 km from the city center) to an urban location (the Baltimore Science Center

approximately 0.5 km from the city center). At each of these locations, four 2 x 2 m plots were excavated to a depth of about 1.1 m, after which they were filled with identical soils, the top layers of which contained seeds of naturally occurring plants of the area. These seeds sprouted in the spring of the year, and the plants they produced were allowed to grow until they senesced in the fall, after which all of them were cut at ground level, removed, dried and weighed.

Ziska *et al.* report that along the rural-to-suburban-to-urban transect, the only consistent differences in the environmental variables they measured were a rural-to-urban increase of 21 percent in average daytime atmospheric CO_2 concentration and increases of 1.6 and 3.3°C in maximum (daytime) and minimum (nighttime) daily temperatures, respectively, which changes, in their words, are "consistent with most short-term (~50 year) global change scenarios regarding CO_2 concentration and air temperature." In addition, they determined that "productivity, determined as final above-ground biomass, and maximum plant height were positively affected by daytime and soil temperatures as well as enhanced CO_2, increasing 60 and 115% for the suburban and urban sites, respectively, relative to the rural site."

The three researchers say their results suggest that "urban environments may act as a reasonable surrogate for investigating future climatic change in vegetative communities," and those results indicate that rising air temperatures and CO_2 concentrations tend to produce dramatic increases in the productivity of the natural ecosystems typical of the greater Baltimore area and, by inference, probably those of many other areas as well.

Three years later, George *et al.* (2007) reported on five years of work at the same three transect locations, stating that "atmospheric CO_2 was consistently and significantly increased on average by 66 ppm from the rural to the urban site over the five years of the study," and that "air temperature was also consistently and significantly higher at the urban site (14.8°C) compared to the suburban (13.6°C) and rural (12.7°C) sites." And they again noted that the increases in atmospheric CO_2 and air temperature they observed "are similar to changes predicted in the short term with global climate change, therefore providing an environment suitable for studying future effects of climate change on terrestrial ecosystems," specifically noting that "urban areas are currently experiencing elevated atmospheric CO_2 and temperature levels that can significantly affect plant growth compared to rural areas."

Working further south still, LaDochy *et al.* (2007) report that "when speculating on how global warming would impact the state [of California], climate change models and assessments often assume that the influence would be uniform (Hansen *et al.*, 1998; Hayhoe *et al.*, 2004; Leung *et al.*, 2004)." Feeling a need to assess the validity of this assumption, they calculated temperature trends over the 50-year period 1950-2000 to explore the extent of warming in various sub-regions of the state, after which they attempted to evaluate the influence of human-induced changes to the landscape on the observed temperature trends and determine their significance compared to those caused by changes in atmospheric composition, such as the air's CO_2 concentration.

In pursuing this protocol, the three researchers found that "most regions showed a stronger increase in minimum temperatures than with mean and maximum temperatures," and that "areas of intensive urbanization showed the largest positive trends, while rural, non-agricultural regions showed the least warming." In fact, they report that the Northeast Interior Basins of the state actually experienced *cooling*. Large urban sites, on the other hand, exhibited rates of warming "over twice those for the state, for the mean maximum temperature, and over five times the state's mean rate for the minimum temperature." Consequently, they concluded that "if we assume that global warming affects all regions of the state, then the small increases seen in rural stations can be an estimate of this general warming pattern over land," which implies that "larger increases," such as those they observed in areas of intensive urbanization, "must then be due to local or regional surface changes."

Noting that "breezy cities on small tropical islands ... may not be exempt from the same local climate change effects and urban heat island effects seen in large continental cities," Gonzalez *et al.* (2005) describe the results of their research into this topic, which they conducted in and about San Juan, Puerto Rico. In this particular study, a NASA Learjet—carrying the Airborne Thermal and Land Applications Sensor (ATLAS) that operates in visual and infrared wavebands—flew several flight lines, both day and night, over the San Juan metropolitan area, the El Yunque National Forest east of San Juan, plus other nearby areas, obtaining surface temperatures, while strategically placed ground instruments recorded local air temperatures. This

work revealed that surface temperature differences between urbanized areas and limited vegetated areas were higher than 3°C during daytime, creating an urban heat island with "the peak of the high temperature dome exactly over the commercial area of downtown," where noontime air temperatures were as much as 3°C greater than those of surrounding rural areas. In addition, the eleven researchers report that "a recent climatological analysis of the surface [air] temperature of the city has revealed that the local temperature has been increasing over the neighboring vegetated areas at a rate of 0.06°C per year for the past 30 years."

In discussing their findings, Gonzalez *et al.* state that "the urban heat island dominates the sea breeze effects in downtown areas," and they say that "trends similar to those reported in [their] article may be expected in the future as coastal cities become more populated." Indeed, it is probable that this phenomenon has long been operative in coastal cities around the world, helping to erroneously inflate the surface air temperature record of the planet and contributing to the infamous "hockey stick" representation of this parameter that has been so highly touted by the Intergovernmental Panel on Climate Change.

One year later, Velazquez-Lozada *et al.* (2006) evaluated the thermal impacts of historical land cover and land use (LCLU) changes in San Juan, Puerto Rico over the last four decades of the twentieth century via an analysis of air temperatures measured at a height of approximately two meters above ground level within four different LCLU types (urban-coastal, rural-inland, rural-coastal and urban-inland), after which they estimated what the strength of the urban heat island might be in the year 2050, based on anticipated LCLU changes and a model predicated upon their data of the past 40 years. In doing so, their work revealed "the existence of an urban heat island in the tropical coastal city of San Juan, Puerto Rico that has been increasing at a rate of 0.06°C per year for the last 40 years." In addition, they report that predicted LCLU changes between now and 2050 will lead to an urban heat island effect "as high as 8°C for the year 2050."

Noting that a mass population migration from rural Mexico into medium- and large-sized cities took place throughout the second half of the twentieth century, Jáuregui (2005) examined the effect of this rapid urbanization on city air temperatures, analyzing the 1950-1990 minimum air temperature series of seven large cities with populations in excess of a million people and seven medium-sized cities with populations ranging from 125,000 to 700,000 people. This work indicated that temperature trends were positive at all locations, ranging from 0.02°C per decade to 0.74°C per decade. Grouped by population, the average trend for the seven large cities was 0.57°C per decade, while the average trend for the seven mid-sized cities was 0.37°C per decade. In discussing these results, Jáuregui says they "suggest that the accelerated urbanization process in recent decades may have substantially contributed to the warming of the urban air observed in large cities in Mexico."

One additional question that may arise in relation to this topic is the direct heating of near-surface air in towns and cities by the urban CO_2 dome that occurs above them. Does it contribute significantly to the urban heat island?

In a study designed to answer this question, Balling *et al.* (2002) obtained vertical profiles of atmospheric CO_2 concentration, temperature, and humidity over Phoenix, Arizona from measurements made in association with once-daily aircraft flights conducted over a 14-day period in January 2000 that extended through, and far above, the top of the city's urban CO_2 dome during the times of its maximum manifestation. They then employed a one-dimensional infrared radiation simulation model to determine the thermal impact of the urban CO_2 dome on the near-surface temperature of the city. These exercises revealed that the CO_2 concentration of the air over Phoenix dropped off rapidly with altitude, returning from a central-city surface value on the order of 600 ppm to a normal non-urban background value of approximately 378 ppm at an air pressure of 800 hPa, creating a calculated surface warming of only 0.12°C at the time of maximum CO_2-induced warming potential, which is about an order of magnitude less than the urban heat island effect of cities the size of Phoenix. In fact, the authors concluded that the warming induced by the urban CO_2 dome of Phoenix is possibly two orders of magnitude smaller than that produced by other sources of the city's urban heat island. Although the doings of man are indeed responsible for high urban air temperatures (which can sometimes rise 10°C or more above those of surrounding rural areas), these high values are not the result of a local CO_2-enhanced greenhouse effect.

Meteorologist Anthony Watts (2009), in research too new to have appeared yet in a peer-reviewed journal, discovered compelling evidence that the temperature stations used to reconstruct the U.S.

surface temperature are unreliable and systemically biased toward recording more warming over time. Watts recruited a team of more than 650 volunteers to visually inspect the temperature stations used by the National Oceanic and Atmospheric Administration (NOAA) and the National Aeronautics and Space Administration (NASA) to measure changes in temperatures in the U.S. In the researcher's words, "using the same quality standards established by NOAA, they found that 89 percent of the stations – nearly 9 of every 10 – fail to meet NOAA's own siting requirements for stations with an expected reporting error of less than 1° C. Many of them fall *far* short of that standard [italics in the original]."

Watts goes on to report finding stations "located next to the exhaust fans of air conditioning units, causing them to report much- higher-than-actual temperatures. We found stations surrounded by asphalt parking lots and located near roads, sidewalks, and buildings that absorb and radiate heat. We found 68 stations located at wastewater treatment plants, where the process of waste digestion causes temperatures to be higher than in surrounding areas."

Watts also discovered that failure to adequately account for changes in the technology used by temperature stations over time—including moving from whitewash to latex paint and from mercury thermostats to digital technology—"have further contaminated the data, once again in the direction of falsely raising temperature readings." Watts is also extremely critical of adjustments to the raw data made by both NOAA and NASA, which "far from correcting the warming biases, actually compounded the measurement errors."

The results of these several North American studies demonstrate that the impact of population growth on the urban heat island effect is very real and can be very large, overshadowing the effects of natural temperature change. This insight is not new: more than three decades ago, Oke (1973) demonstrated that towns with as few as a thousand inhabitants typically create a warming of the air within them that is more than twice as great as the increase in mean global air temperature believed to have occurred since the end of the Little Ice Age, while the urban heat islands of the great metropolises of the world create warmings that rival those that occur between full-fledged ice ages and interglacials. Extensive research conducted since then by independent scientists has confirmed Oke's finding. Due to extensive corruption of land-based temperature data from urban heat islands, the North

American temperature record cannot be cited as providing reliable data in support of the greenhouse theory of global warming.

Additional information on this topic, including reviews of newer publications as they become available, can be found at http://www.co2science.org/subject/u/uhinorthamerica.php.

References

Balling Jr., R.C., Cerveny, R.S. and Idso, C.D. 2002. Does the urban CO_2 dome of Phoenix, Arizona contribute to its heat island? *Geophysical Research Letters* **28**: 4599-4601.

Changnon, S.A. 1999. A rare long record of deep soil temperatures defines temporal temperature changes and an urban heat island. *Climatic Change* **42**: 531-538.

DeGaetano, A.T. and Allen, R.J. 2002. Trends in twentieth-century temperature extremes across the United States. *Journal of Climate* **15**: 3188-3205.

Dow, C.L. and DeWalle, D.R. 2000. Trends in evaporation and Bowen ratio on urbanizing watersheds in eastern United States. *Water Resources Research* **36**: 1835-1843.

George, K., Ziska, L.H., Bunce, J.A. and Quebedeaux, B. 2007. Elevated atmospheric CO_2 concentration and temperature across an urban-rural transect. *Atmospheric Environment* **41**: 7654-7665.

Gonzalez, J.E., Luvall, J.C., Rickman, D., Comarazamy, D., Picon, A., Harmsen, E., Parsiani, H., Vasquez, R.E., Ramirez, N., Williams, R. and Waide, R.W. 2005. Urban heat islands developing in coastal tropical cities. *EOS: Transactions, American Geophysical Union* **86**: 397,403.

Hansen, J., Sato, M., Glascoe, J. and Ruedy, R. 1998. A commonsense climatic index: Is climate change noticeable? *Proceedings of the National Academy of Sciences USA* **95**: 4113-4120.

Hayhoe, K., Cayan, D., Field, C.B., Frumhoff, P.C. *et al.* 2004. Emissions, pathways, climate change, and impacts on California. *Proceedings of the National Academy of Sciences USA* **101**: 12,422-12,427.

Hinkel, K.M. and Nelson, F.E. 2007. Anthropogenic heat island at Barrow, Alaska, during winter: 2001-2005. *Journal of Geophysical Research* **112**: 10.1029/2006JD007837.

Hinkel, K.M., Nelson, F.E., Klene, A.E. and Bell, J.H. 2003. The urban heat island in winter at Barrow, Alaska. *International Journal of Climatology* **23**: 1889-1905.

Intergovernmental Panel on Climate Change (IPCC). 1995. *Climate Change 1995, The Science of Climate Change.* Cambridge University Press, Cambridge, U.K.

Jáuregui, E. 2005. Possible impact of urbanization on the thermal climate of some large cities in Mexico. *Atmosfera* **18**: 249-252.

LaDochy, S., Medina, R. and Patzert, W. 2007. Recent California climate variability: spatial and temporal patterns in temperature trends. *Climate Research* **33**: 159-169.

Leung, L.R., Qian, Y., Bian, X., Washington, W.M., Han, J. and Roads, J.O. 2004. Mid-century ensemble regional climate change scenarios for the western United States. *Climatic Change* **62**: 75-113.

Maul, G.A. and Davis, A.M. 2001. Seawater temperature trends at USA tide gauge sites. *Geophysical Research Letters* **28**: 3935-3937.

Oke, T.R. 1973. City size and the urban heat island. *Atmospheric Environment* **7**: 769-779.

Streutker, D.R. 2003. Satellite-measured growth of the urban heat island of Houston, Texas. *Remote Sensing of Environment* **85**: 282-289.

Velazquez-Lozada, A., Gonzalez, J.E. and Winter, A. 2006. Urban heat island effect analysis for San Juan, Puerto Rico. *Atmospheric Environment* **40**: 1731-1741.

Watts, A. 2009. *Is the U.S. Temperature Record Reliable?* Chicago, IL: The Heartland Institute.

Ziska, L.H., Bunce, J.A. and Goins, E.W. 2004. Characterization of an urban-rural CO_2/temperature gradient and associated changes in initial plant productivity during secondary succession. *Oecologia* **139**: 454-458.

3.3.3. Asia

Hasanean (2001) investigated surface air temperature trends with data obtained from meteorological stations located in eight Eastern Mediterranean cities: Malta, Athens, Tripoli, Alexandria, Amman, Beirut, Jerusalem, and Latakia. The period of analysis varied from station to station according to available data, with Malta having the longest temperature record (1853-1991) and Latakia the shortest (1952-1991). Four of the cities exhibited overall warming trends and four of them cooling trends. In addition, there was an important warming around 1910 that began nearly simultaneously at all of the longer-record stations, as well as a second warming in the 1970s; but Hasanean reports that the latter warming was "not uniform, continuous or of the same order" as the warming that began about 1910, nor was it evident at all of the stations. One interpretation of this non-uniformity of temperature behavior in the 1970s is that it may have been the result of temporal differences in city urbanization histories that were accentuated about that time, which could have resulted in significantly different urban heat island trajectories at the several sites over the latter portions of their records.

In a more direct study of the urban heat island effect that was conducted in South Korea, Choi et al. (2003) compared the mean station temperatures of three groupings of cities (one comprised of four large urban stations with a mean 1995 population of 4,830,000, one comprised of six smaller urban stations with a mean 1995 population of 548,000, and one comprised of six "rural" stations with a mean 1995 population of 214,000) over the period 1968-1999. This analysis revealed, in their words, that the "temperatures of large urban stations exhibit higher urban bias than those of smaller urban stations and that the magnitude of urban bias has increased since the late 1980s." Specifically, they note that "estimates of the annual mean magnitude of urban bias range from 0.35°C for smaller urban stations to 0.50°C for large urban stations." In addition, they indicate that "none of the rural stations used for this study can represent a true non-urbanized environment." Hence, they correctly conclude that their results are underestimates of the true urban effect, and that "urban growth biases are very serious in South Korea and must be taken into account when assessing the reliability of temperature trends."

In a second study conducted in South Korea, Chung et al. (2004a) report there was an "overlapping of the rapid urbanization-industrialization period with the global warming era," and that the background climatic trends from urbanized areas might therefore be contaminated by a growing urban heat island effect. To investigate this possibility, they say "monthly averages of daily minimum, maximum, and mean temperature at 14 synoptic stations were prepared for 1951-1980 (past normal) and 1971-2000 (current normal) periods," after which "regression equations were used to determine potential effects of urbanization and to extract the net contribution of regional climate change to the apparent temperature change." Twelve of these stations were growing urban sites of various size, while two (where populations actually *decreased*) were rural, one being located inland and one on a remote island.

In terms of change over the 20 years that separated the two normal periods, Chung et al. report that in Seoul, where population increase was greatest, annual mean daily minimum temperature increased by 0.7°C, while a mere 0.1°C increase was detected at

one of the two rural sites and a 0.1°C decrease was detected at the other, for no net change in their aggregate mean value. In the case of annual mean daily maximum temperature, a 0.4°C increase was observed at Seoul and a 0.3°C increase was observed at the two rural sites. Hence, the change in the annual mean daily mean temperature was an increase of 0.15°C at the two rural sites (indicative of regional background warming of 0.075°C per decade), while the change of annual mean daily mean temperature at Seoul was an increase of 0.55°C, or 0.275°C per decade (indicative of an urban-induced warming of 0.2°C per decade in addition to the regional background warming of 0.075°C per decade). Also, corresponding results for urban areas of intermediate size defined a linear relationship that connected these two extreme results when plotted against the logarithm of population increase over the two-decade period.

In light of the significantly intensifying urban heat island effect detected in their study, Chung et al. say it is "necessary to subtract the computed urbanization effect from the observed data at urban stations in order to prepare an intended nationwide climatic atlas," noting that "rural climatological normals should be used instead of the conventional normals to simulate ecosystem responses to climatic change, because the urban area is still much smaller than natural and agricultural ecosystems in Korea."

A third study of South Korea conducted by Chung et al. (2004b) evaluated temperature changes at 10 urban and rural Korean stations over the period 1974-2002. They found "during the last 29 years, the increase in annual mean temperature was 1.5°C for Seoul and 0.6°C for the rural and seashore stations," while increases in mean January temperatures ranged from 0.8 to 2.4°C for the 10 stations. In addition, they state that "rapid industrialization of the Korean Peninsula occurred during the late 1970s and late 1980s," and that when plotted on a map, "the remarkable industrialization and expansion ... correlate with the distribution of increases in temperature." Consequently, as in the study of Chung et al. (2004a), Chung et al. (2004b) found that over the past several decades, much (and in many cases *most*) of the warming experienced in the urban areas of Korea was the result of local urban influences that were not indicative of regional background warming.

Shifting attention to China, Weng (2001) evaluated the effect of land cover changes on surface temperatures of the Zhujiang Delta (an area of slightly more than 17,000 km^2) via a series of analyses of remotely sensed Landsat Thematic Mapper data. They found that between 1989 and 1997, the area of land devoted to agriculture declined by nearly 50 percent, while urban land area increased by close to the same percentage. Then, upon normalizing the surface radiant temperature for the years 1989 and 1997, they used image differencing to produce a radiant temperature change image that they overlaid with images of urban expansion. The results indicated, in Weng's words, that "urban development between 1989 and 1997 has given rise to an average increase of 13.01°C in surface radiant temperature."

In Shanghai, Chen et al. (2003) evaluated several characteristics of that city's urban heat island, including its likely cause, based on analyses of monthly meteorological data from 1961 to 1997 at 16 stations in and around this hub of economic activity that is one of the most flourishing urban areas in all of China. Defining the urban heat island of Shanghai as the mean annual air temperature difference between urban Longhua and suburban Songjiang, Chen et al. found that its strength increased in essentially linear fashion from 1977 to 1997 by 1°C.

Commenting on this finding, Chen et al. say "the main factor causing the intensity of the heat island in Shanghai is associated with the increasing energy consumption due to economic development," noting that in 1995 the Environment Research Center of Peking University determined that the annual heating intensity due to energy consumption by human activities was approximately 25 Wm^{-2} in the urban area of Shanghai but only 0.5 Wm^{-2} in its suburbs. In addition, they point out that the 0.5°C/decade intensification of Shanghai's urban heat island is an order of magnitude greater than the 0.05°C/decade global warming of the earth over the past century, which is indicative of the fact that ongoing intensification of even strong urban heat islands cannot be discounted.

Simultaneously, Kalnay and Cai (2003) used differences between trends in directly observed surface air temperature and trends determined from the NCEP-NCAR 50-year Reanalysis (NNR) project (based on atmospheric vertical soundings derived from satellites and balloons) to estimate the impact of land-use changes on surface warming. Over undisturbed rural areas of the United States, they found that the surface- and reanalysis-derived air temperature data yielded essentially identical trends, implying that differences between the two approaches over urban areas would represent urban heat island effects. Consequently, Zhou et al. (2004) applied the

same technique over southeast China, using an improved version of reanalysis that includes newer physics, observed soil moisture forcing, and a more accurate characterization of clouds.

For the period January 1979 to December 1998, the eight scientists involved in the work derived an "estimated warming of mean surface [air] temperature of 0.05°C per decade attributable to urbanization," which they say "is much larger than previous estimates for other periods and locations, including the estimate of 0.027°C for the continental U.S. (Kalnay and Cai, 2003)." They note, however, that because their analysis "is from the winter season over a period of rapid urbanization and for a country with a much higher population density, we expect our results to give higher values than those estimated in other locations and over longer periods."

In a similar study, Frauenfeld et al. (2005) used daily surface air temperature measurements from 161 stations located throughout the Tibetan Plateau (TP) to calculate the region's mean annual temperature for each year from 1958 through 2000, while in the second approach they used 2-meter temperatures from the European Centre for Medium-Range Weather Forecasts (ECMWF) reanalysis (ERA-40), which temperatures, in their words, "are derived from rawinsonde profiles, satellite retrievals, aircraft reports, and other sources including some surface observations." This approach, according to them, results in "more temporally homogeneous fields" that provide "a better assessment of large-scale temperature variability across the plateau."

Frauenfeld et al. report that over the period 1958-2000, "time series based on aggregating all station data on the TP show a statistically significant positive trend of 0.16°C per decade," as has also been reported by Liu and Chen (2000). However, they say that "no trends are evident in the ERA-40 data for the plateau as a whole."

In discussing this discrepancy, the three scientists suggest that "a potential explanation for the difference between reanalysis and station trends is the extensive local and regional land use change that has occurred across the TP over the last 50 years." They note, for example, that "over the last 30 years, livestock numbers across the TP have increased more than 200% due to inappropriate land management practices and are now at levels that exceed the carrying capacity of the region (Du et al., 2004)." The resultant overgrazing, in their words, "has caused land degradation and desertification at an alarming rate (Zhu and Li, 2000; Zeng et al., 2003)," and they note that "in other parts of the world, land degradation due to overgrazing has been shown to cause significant local temperature increases (e.g., Balling *et al.*, 1998)."

Another point they raise is that "urbanization, which can result in 8°-11°C higher temperatures than in surrounding rural areas (e.g., Brandsma *et al.*, 2003), has also occurred extensively on the TP," noting that "construction of a gas pipeline in the 1970s and highway expansion projects in the early 1980s have resulted in a dramatic population influx from other parts of China, contributing to both urbanization and a changed landscape." In this regard, they say that "the original Tibetan section of Lhasa (i.e., the pre-1950 Lhasa) now only comprises 4% of the city, suggesting a 2400% increase in size over the last 50 years." And they add that "similar population increases have occurred at other locations across the TP," and that "even villages and small towns can exhibit a strong urban heat island effect."

In concluding their analysis of the situation, Frauenfeld *et al.* contend that "these local changes are reflected in station temperature records." We note that when the surface-generated anomalies are removed, as in the case of the ERA-40 reanalysis results they present, it is clear there has been no warming of the Tibetan Plateau since at least 1958. Likewise, we submit that the other results reported in this section imply much the same about other parts of China and greater Asia.

In conclusion, a large body of research conducted by scores of scientists working in countries around the world reveals that the twentieth century warming claimed by the IPCC, Mann *et al.* (1998, 1999), and Mann and Jones (2003) to represent mean global background conditions is likely significantly biased towards warming over the past 30 years and is therefore not a true representation of earth's recent thermal history.

Additional information on this topic, including reviews of newer publications as they become available, can be found at http://www.co2science.org/subject/u/uhiasia.php

References

Balling Jr., R.C., Klopatek, J.M., Hildebrandt, M.L., Moritz, C.K. and Watts, C.J. 1998. Impacts of land degradation on historical temperature records from the Sonoran desert. *Climatic Change* **40**: 669-681.

Brandsma, T., Konnen, G.P. and Wessels, H.R.A. 2003. Empirical estimation of the effect of urban heat advection on the temperature series of DeBilt (the Netherlands). *International Journal of Climatology* **23**: 829-845.

Chen, L., Zhu, W., Zhou, X. and Zhou, Z. 2003. Characteristics of the heat island effect in Shanghai and its possible mechanism. *Advances in Atmospheric Sciences* **20**: 991-1001.

Choi, Y., Jung, H.-S., Nam, K.-Y. and Kwon, W.-T. 2003. Adjusting urban bias in the regional mean surface temperature series of South Korea, 1968-99. *International Journal of Climatology* **23**: 577-591.

Chung, U., Choi, J. and Yun, J.I. 2004a. Urbanization effect on the observed change in mean monthly temperatures between 1951-1980 and 1971-2000. *Climatic Change* **66**: 127-136.

Chung, Y.-S., Yoon, M.-B. and Kim, H.-S. 2004b. On climate variations and changes observed in South Korea. *Climatic Change* **66**: 151-161.

Du, M., Kawashima, S., Yonemura, S., Zhang, X. and Chen, S. 2004. Mutual influence between human activities and climate change in the Tibetan Plateau during recent years. *Global and Planetary Change* **41**: 241-249.

Frauenfeld, O.W., Zhang, T. and Serreze, M.C. 2005. Climate change and variability using European Centre for Medium-Range Weather Forecasts reanalysis (ERA-40) temperatures on the Tibetan Plateau. *Journal of Geophysical Research* **110**: 10.1029/2004JD005230.

Hasanean, H.M. 2001. Fluctuations of surface air temperature in the Eastern Mediterranean. *Theoretical and Applied Climatology* **68**: 75-87.

Kalnay, E. and Cai, M. 2003. Impact of urbanization and land-use change on climate. *Nature* **423**: 528-531.

Liu, X. and Chen, B. 2000. Climatic warming in the Tibetan Plateau during recent decades. *International Journal of Climatology* **20**: 1729-1742.

Mann, M.E., Bradley, R.S. and Hughes, M.K. 1998. Global-scale temperature patterns and climate forcing over the past six centuries. *Nature* **392**: 779-787.

Mann, M.E., Bradley, R.S. and Hughes, M.K. 1999. Northern Hemisphere temperatures during the past millennium: Inferences, uncertainties, and limitations. *Geophysical Research Letters* **26**: 759-762.

Mann, M.E. and Jones, P.D. 2003. Global surface temperatures over the past two millennia. *Geophysical Research Letters* **30**: 10.1029/2003GL017814.

Weng, Q. 2001. A remote sensing-GIS evaluation of urban expansion and its impact on surface temperature in the

Zhujiang Delta, China. *International Journal of Remote Sensing* **22**: 1999-2014.

Zeng, Y., Feng, Z. and Cao, G. 2003. Land cover change and its environmental impact in the upper reaches of the Yellow River, northeast Qinghai-Tibetan Plateau. *Mountain Research and Development* **23**: 353-361.

Zhou, L., Dickinson, R.E., Tian, Y., Fang, J., Li, Q., Kaufmann, R.K., Tucker, C.J. and Myneni, R.B. 2004. Evidence for a significant urbanization effect on climate in China. *Proceedings of the National Academy of Sciences USA* **101**: 9540-9544.

Zhu, L. and Li, B. 2000. Natural hazards and environmental issues. In: Zheng, D., Zhang, Q. and Wu, S. (Eds.) *Mountain Genecology and Sustainable Development of the Tibetan Plateau*, Springer, New York, New York, USA, pp. 203-222.

3.4. Fingerprints

Is there a method that can distinguish anthropogenic global warming from natural warming? The IPCC (IPCC-SAR, 1996, p. 411; IPCC, 2007-I, p. 668) and many scientists believe the "fingerprint" method is the only reliable one. It compares the observed pattern of warming with a pattern calculated from greenhouse models. While an agreement of such fingerprints cannot prove an anthropogenic origin for warming, it would be consistent with such a conclusion. A mismatch would argue strongly against any significant contribution from greenhouse gas (GHG) forcing and support the conclusion that the observed warming is mostly of natural origin.

Climate models all predict that, if GHG is driving climate change, there will be a unique fingerprint in the form of a warming trend increasing with altitude in the tropical troposphere, the region of the atmosphere up to about 15 kilometers. (See Figure 3.4.1.) Climate changes due to solar variability or other known natural factors will not yield this pattern; only sustained greenhouse warming will do so.

The fingerprint method was first attempted in the IPCC's Second Assessment Report (SAR) (IPCC-SAR, 1996, p. 411). Its Chapter 8, titled "Detection and Attribution," attributed observed temperature changes to anthropogenic factors—greenhouse gases and aerosols. The attempted match of warming trends with altitude turned out to be spurious, since it depended entirely on a particular choice of time interval for the comparison (Michaels and Knappenberger, 1996). Similarly, an attempt to correlate the observed and calculated geographic

distribution of surface temperature trends (Santer *et al.* 1996) involved making changes on a published graph that could and did mislead readers (Singer, 1999, p. 9; Singer, 2000, pp. 15, 43-44). In spite of these shortcomings, IPCC-SAR concluded that "the balance of evidence" supported AGW.

With the availability of higher-quality temperature data, especially from balloons and satellites, and with improved GH models, it has become possible to apply the fingerprint method in a more realistic way. This was done in a report issued by the U.S. Climate Change Science Program (CCSP) in April 2006—making it readily available to the IPCC for its Fourth Assessment Report—and it permits the most realistic comparison of fingerprints (Karl *et al.*, 2006).

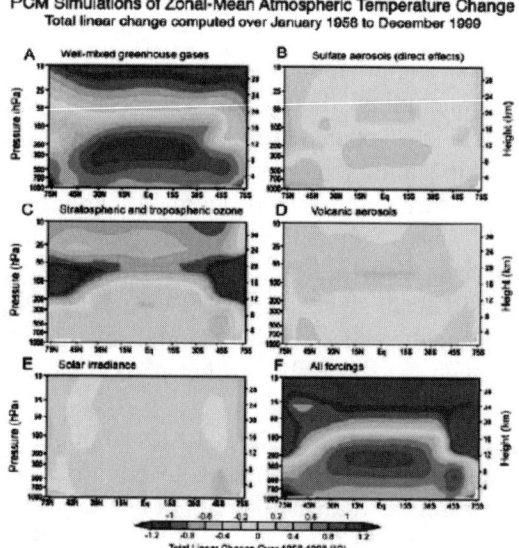

Figure 3.4.1. Model-calculated zonal mean atmospheric temperature change from 1890 to 1999 (degrees C per century) as simulated by climate models from [A] well-mixed greenhouse gases, [B] sulfate aerosols (direct effects only), [C] stratospheric and tropospheric ozone, [D] volcanic aerosols, [E] solar irradiance, and [F] all forcings (U.S. Climate Change Science Program 2006, p. 22). Note the pronounced increase in warming trend with altitude in figures A and F, which the IPCC identified as the 'fingerprint' of greenhouse forcing.

The CCSP report is an outgrowth of an NAS report "Reconciling Observations of Global Temperature Change" issued in January 2000 (NAS, 2000). That NAS report compared surface and troposphere temperature trends and concluded they cannot be reconciled. Six years later, the CCSP report expanded considerably on the NAS study. It is

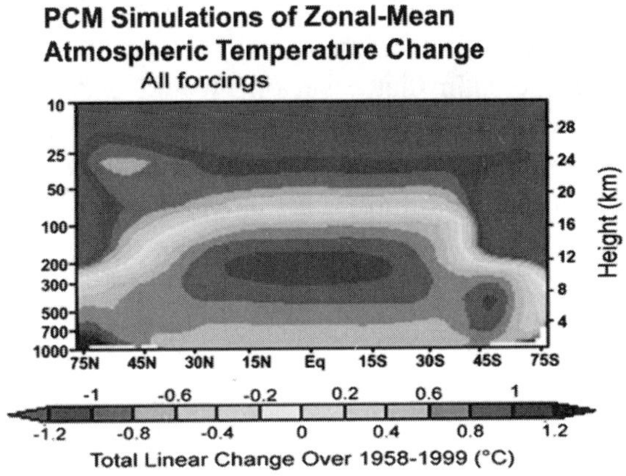

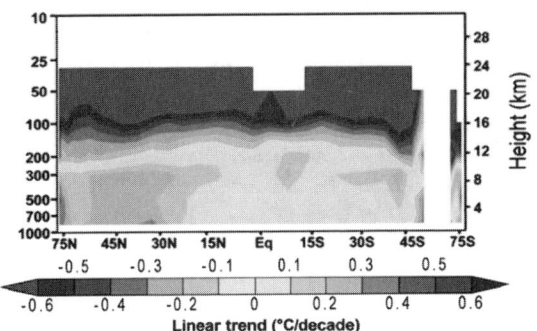

Figure 3.4.2. Greenhouse-model-predicted temperature trends versus latitude and altitude; this is figure 1.3F from CCSP 2006, p. 25. Note the increased temperature trends in the tropical mid_troposphere, in agreement also with the IPCC result (IPCC-AR4 2007, p. 675).

Figure 3.4.3. By contrast, observed temperature trends versus latitude and altitude; this is figure 5.7E from CCSP 2006, p. 116. These trends are based on the analysis of radiosonde data by the Hadley Centre and are in good agreement with the corresponding U.S. analyses. Notice the absence of increased temperature trends in the tropical mid-troposphere.

essentially a specialized report addressing the most crucial issue in the global warming debate: Is current global warming anthropogenic or natural? The CCSP result is unequivocal. While all greenhouse models show an increasing warming trend with altitude, peaking around 10 km at roughly two times the surface value, the temperature data from balloons give the opposite result: no increasing warming, but rather a slight cooling with altitude in the tropical zone. See Figures 3.4.2 and 3.4.3, taken directly from the CCSP report.

The CCSP executive summary inexplicably claims agreement between observed and calculated patterns, the opposite of what the report itself documents. It tries to dismiss the obvious disagreement shown in the body of the report by suggesting there might be something wrong with both balloon and satellite data. Unfortunately, many people do not read beyond the summary and have therefore been misled to believe the CCSP report supports anthropogenic warming. It does not.

The same information also can be expressed by plotting the difference between surface trend and troposphere trend for the models and for the data (Singer, 2001). As seen in Figure 3.4.4 and 3.4.5, the models show a histogram of negative values (i.e. surface trend less than troposphere trend) indicating that atmospheric warming will be greater than surface warming. By contrast, the data show mainly positive values for the difference in trends, demonstrating that measured warming is occurring principally on the surface and not in the atmosphere.

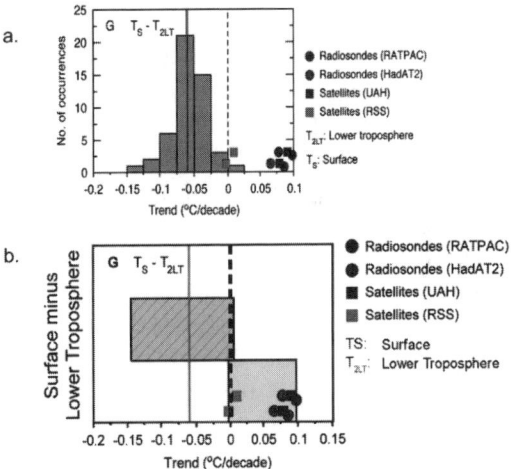

Figure 3.4.4. Another way of presenting the difference between temperature trends of surface and lower troposphere; this is figure 5.4G from CCSP 2006, p. 111. The model results show a spread of values (histogram); the data points show balloon and satellite trend values. Note the model results hardly overlap with the actual observed trends. (The apparent deviation of the RSS analysis of the satellite data is as yet unexplained.)

Figure 3.4.5. By contrast, the executive summary of the CCSP report presents the same information as Figure 3.4.4 in terms of 'range' and shows a slight overlap between modeled and observed temperature trends (Figure 4G, p. 13). However, the use of 'range' is clearly inappropriate (Douglass et al. 2007) since it gives undue weight to 'outliers.'

The same information can be expressed in yet a different way, as seen in research papers by Douglass *et al.* (2004, 2007), as shown in Figure 3.4.6. The models show an increase in temperature trend with altitude but the observations show the opposite.

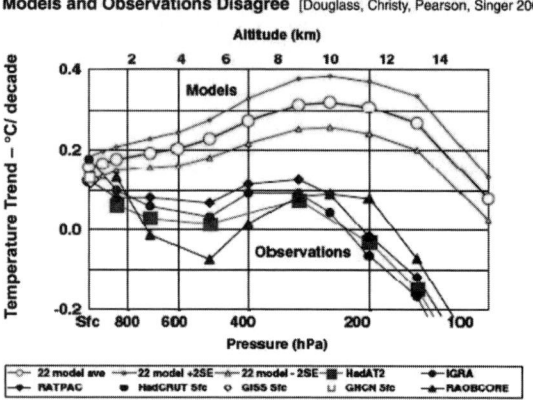

Figure 3.4.6. A more detailed view of the disparity of temperature trends is given in this plot of trends (in degrees C/decade) versus altitude in the tropics [Douglass et al. 2007]. Models show an increase in the warming trend with altitude, but balloon and satellite observations do not.

This mismatch of observed and calculated fingerprints clearly falsifies the hypothesis of anthropogenic global warming (AGW). We must conclude therefore that anthropogenic greenhouse gases can contribute only in a minor way to the current warming, which is mainly of natural origin. The IPCC seems to be aware of this contrary evidence but has tried to ignore it or wish it away. The summary for policymakers of IPCC's Fourth Assessment Report (IPCC 2007-I, p. 5) distorts the key result of the CCSP report: "New analyses of balloon-borne and satellite measurements of lower- and mid-tropospheric temperature show warming rates that are similar to those of the surface temperature record, and are consistent within their respective uncertainties, largely reconciling a discrepancy noted in the TAR." How is this possible? It is done partly by using the concept of "range" instead of the statistical distribution shown in Figure 12a. But "range" is not a robust statistical measure because it gives undue weight to "outlier" results (Figure 12b). If robust probability distributions were used they would show an exceedingly low probability of any overlap of modeled and the observed temperature trends.

If one takes greenhouse model results seriously, then the greenhouse fingerprint would suggest the true surface trend should be only 30 to 50 percent of the observed balloon/satellite trends in the troposphere. In that case, one would end up with a much-reduced surface warming trend, an insignificant AGW effect, and a minor greenhouse-induced warming in the future.

References

Douglass, D.H., Pearson, B. and Singer, S.F. 2004. Altitude dependence of atmospheric temperature trends: Climate models versus observations. *Geophysical Research Letters* **31**.

Douglass, D.H., Christy, J.R. , Pearson, B.D. and Singer, S.F. 2007. A comparison of tropical temperature trends with model predictions. *International Journal of Climatology* (Royal Meteorol Soc). DOI:10.1002/joc.1651.

IPCC. 2007-I. *Climate Change 2007: The Physical Science Basis. Contribution of Working Group I to the Fourth Assessment Report of the Intergovernmental Panel on Climate Change.* Solomon, S., D. Qin, M. Manning, Z. Chen, M. Marquis, K.B. Averyt, M. Tignor and H.L. Miller. (Eds.) Cambridge University Press, Cambridge, UK.

IPCC-SAR 1996. *Climate Change 1995: The Science of Climate Change. Contribution of Working Group I to the Second Assessment Report of the Intergovernmental Panel on Climate Change.* Cambridge University Press, Cambridge, UK.

Karl, T.R., Hassol, S.J. , Miller, C.D. and Murray, W.L. (Eds.) 2006. *Temperature Trends in the Lower Atmosphere: Steps for Understanding and Reconciling Differences.* A report by the Climate Change Science Program and Subcommittee on Global Change Research, http://www.climatescience.gov/Library/sap/sap1-1/final report/default.htm.

Michaels, P.J. and Knappenberger, P.C. 1996. Human effect on global climate? *Nature* **384**: 522-523.

NAS 2000. *Reconciling Observations of Global Temperature Change.* National Academy of Sciences. National Academy Press, Washington, DC.

Santer, B.D., *et al.* 1996. Towards the detection and attribution of an anthropogenic effect on climate. *Climate Dynamics* **12**: 79-100.

Singer, S.F. 1999. Human contribution to climate change remains questionable. Also, Reply. *EOS: Transactions, American Geophysical Union* **80**: 33, 186-187 and 372-373.

Singer, S.F. 2000. Climate policy—From Rio to Kyoto a political issue for 2000 and beyond. *Essays in Public Policy* **102**. Hoover Institution, Stanford University, Stanford, CA.

Singer, S.F. 2001. Disparity of temperature trends of atmosphere and surface. Paper presented at 12th Symposium on Global Climate Change, American Meteorological Society, Albuquerque, NM.

3.5. Satellite Data

The IPCC claims that data collected by satellite-mounted microwave sounding units (MSU) and advanced MSU measurements since 1979 reveal a warming trend of 0.12° C to 0.19° C per decade, which it says "is broadly consistent with surface temperature trends" (IPCC, 2007-I, p. 237). This would be surprising, since we indicated in the previous section that the surface-based temperature record is unreliable and biased toward a spurious warming trend. In this section we investigate the truth of the IPCC's claim in this regard and report other findings based on satellite data.

Most climate models predict that the troposphere should warm about 1.2 times more than the surface globally, and about 1.5 times more in the tropics. Although the MSUs mounted on satellites sent into orbit by NASA for the National Oceanic and Atmospheric Administration (NOAA) were not originally intended to be used to measure temperatures in the troposphere, they have been used for this purpose since 1979 and, despite some ongoing debate, are acknowledged to be a reliable source of information about temperatures in the troposphere (Christy *et al.,* 2003, 2007; Santer *et al.,* 2005). As Wentz and Schabel observed in an article in *Nature* in 1998, "the detection and measurement of small changes in the Earth's climate require extremely precise global observations of a broad spectrum of complementary physical variables. In this endeavour, satellite observations are playing an increasingly important role. As compared to conventional *in situ* observations, satellites provide daily near-global coverage with a very high statistical precision that results from averaging millions of individual observations" (Wentz and Schabel, 1998).

Four groups currently report MSU measurements: the University of Alabama in Huntsville (UAH), Remote Sensing System (RSS) (a small private weather forecasting firm led by the previously cited Frank Wentz), the University of Maryland (UMd),

and a group from NOAA whose data series begins in 1987. RSS and UAH produce estimates of temperatures for the lower troposphere (LT), mid-troposphere (MT), and lower stratosphere (LS). UMd produces estimates only for MT (Christy and Norris, 2006). New data for the UAH series is posted every month on a Web site maintained at the University of Alabama at Huntsville.

The first satellite record was produced by Roy Spencer, then with NASA and now the U.S. Science Team leader for the Advanced Microwave Scanning Radiometer flying on NASA's Aqua satellite, and John Christy, distinguished professor of atmospheric science and director of the Earth System Science Center at the University of Alabama in Huntsville. Published in *Science* in 1990 (Spencer and Christy, 1990), the article presented the first 10 years of satellite measurements of lower atmospheric temperature changes (from 1979 to 1988) and found "no obvious trend for the 10-year period." Although this finding covered too short a period of time to prove a trend, it seemed to contradict claims by some scientists at the time that a warming trend was underway. It triggered a long-running debate, which continues to this day, over the accuracy of the satellite data.

In 1997, Kevin Trenberth, of the U.S. National Center for Atmospheric Research (NCAR) and later a lead author of the IPCC's Third Assessment Report, along with coauthors challenged the reliability of the satellite data (Trenberth, 1997; Trenberth and Hurrell, 1997; Hurrell and Trenberth, 1997). Trenberth argued that Spencer and Christy had failed to properly calibrate the sensors on each new satellite as older satellites were retired and new ones launched into orbit, based on a surface-satellite comparison. Spencer and Christy, however, showed that the surface and tropospheric discrepancy was real, as it also was found in independent balloon comparisons (Christy *et al.,* 1997).

Critics of the satellite data pointed to other possible and actual errors in the satellite record, and Spencer and Christy made two adjustments based on these external criticisms for such things as orbit decay and changes in technology. One of the larger changes was made to correct for drift in local crossing time (i.e., change in the time-of-day that the measures are taken), an error discovered by Mears and Wentz (2005) and subsequently corrected by Christy and Spencer (2005). Many of the adjustments made by Christy and Spencer resulted in the satellite record showing a small warming trend of 0.123 °C between 1979 and 2005.

In 2006, a panel of the U.S. Climate Change Science Program (CCSP) attempted to reconcile differences between satellite and surface-station data. While the executive summary of the report claimed (as the IPCC does) that "this significant discrepancy [between surface station records and satellite records] no longer exists because errors in the satellite and radiosonde data have been identified and corrected," in fact significant differences in some values (especially in the important region of the tropics, as discussed in Section 3.4. Fingerprints) remained unsolved (CCSP, 2006).

Satellite data allow us to check the accuracy of the warming trend during the last three decades reported by the three combined land-surface air temperature and sea-surface temperature (SST) records used by the IPCC: CRU (from the Climate Research Unit (CRU) at the University of East Anglia, in Norwich, England), NCDC (from the National Climatic Data Center), and GISS (from NASA's Goddard Institute for Space Studies). The IPCC lists temperature trends (°C /decade) for each record for the periods 1850-2005, 1901-2005, and 1979-2005 (IPCC, 2007-I, Table 3.3, p. 248). All three surface temperature records used by the IPCC show positive trends in global temperatures during the 1979-2005 period (the period that can be checked against satellite data) of between 0.163° C/decade and 0.174 °C /decade, compared to the UAH record of only 0.123° C/decade. This means the IPCC's estimates of warming are between 33 percent and 41.5 percent more rapid than the most scientifically accurate record we have of global temperatures during this 26-year period. The IPCC claims an even higher estimate of 0.177° C/decade (44 percent higher than UAH) in a graph on page 253, a variation of which appears in the Summary for Policymakers (p. 6). Finally, we note that none of the warming rates reported in the IPCC's Table 3.3 reaches the 0.19° C that the IPCC claimed to be the upper end of the range of credible estimates, while the UAH record of 0.128° C/decade sits very close to the lowest estimate of 0.12° C.

Similarly, the IPCC's temperature records for the Northern Hemisphere are CRU's 0.234° C and NCDC's 0.245° C, approximately 17 percent to 22.5 percent higher than the UAH's record of 0.20° C. For the Southern Hemisphere, the IPCC's estimates are 0.092° C (CRU) and 0.096 °C (NCDC), or a very large 84 percent and 92 percent higher than the

UAH's 0.05° C. To say this is "broadly consistent," as the IPCC does, is not accurate. In light of the large discrepancy between satellite and surface records for the Southern Hemisphere, it is notable that Christy has been using the UAH database to detect and correct errors in the Australian radiosondes record (Christy and Norris, 2009) and the surface station record in East Africa (Christy *et al.,* 2009).

Satellite data also allow us to compare real-world temperatures to the predictions (or "scenarios") offered by those who have been predicting warming since the 1980s. Figure 3.5.1. compares the UAH and RSS temperature records "adjusted to mimic surface temperature variations for an apples to apples comparison with the model projections (factor of 1.2, CCSP SAP 1.1)" to three model projections of global surface temperature presented by NASA's James Hansen in Senate testimony in 1988 (Christy, 2009). "GISS-A 88" and "GISS-B 88," at the top of the graph, are Hansen's two "business-as-usual" model projections of temperature which assumed greenhouse gas emissions would be similar to what actually has happened. "GISS-C 88" is Hansen's temperature forecast if drastic GHG reductions were made. Obviously, real-world temperatures have failed to rise as Hansen had predicted, and indeed, global temperatures in 2009 were no higher than when Hansen testified in 1988. As Christy comments, "Even the model projection for drastic CO_2 cuts still overshot the observations. This would be considered a failed hypothesis test for the models from 1988" (Christy, 2009).

By 2008, the UAH data series indicated that global temperatures in the lower atmosphere had warmed at the slightly higher rate of about 0.14° C/decade from January 1979 through December 2007, while the RSS data series showed a warming rate of 0.17° C/decade. Recent research by Randall and Herman (2009) using data collected from a subset of weather balloon observations thought to be most reliable suggests the RSS data incorporate an improper handling of diurnal cycle effects that causes a small warming bias over global land areas, thus suggesting that the lower UAH estimate of 0.14° C/decade may have been closer to correct. Graphs showing both data sets and a third graph showing the difference between the two data sets appear in Figure 3.5.2.

Explanations exist for two of the biggest differences between the two datasets. The first is a sudden warming in RSS relative to UAH in January 1992. This feature has been found in comparison with

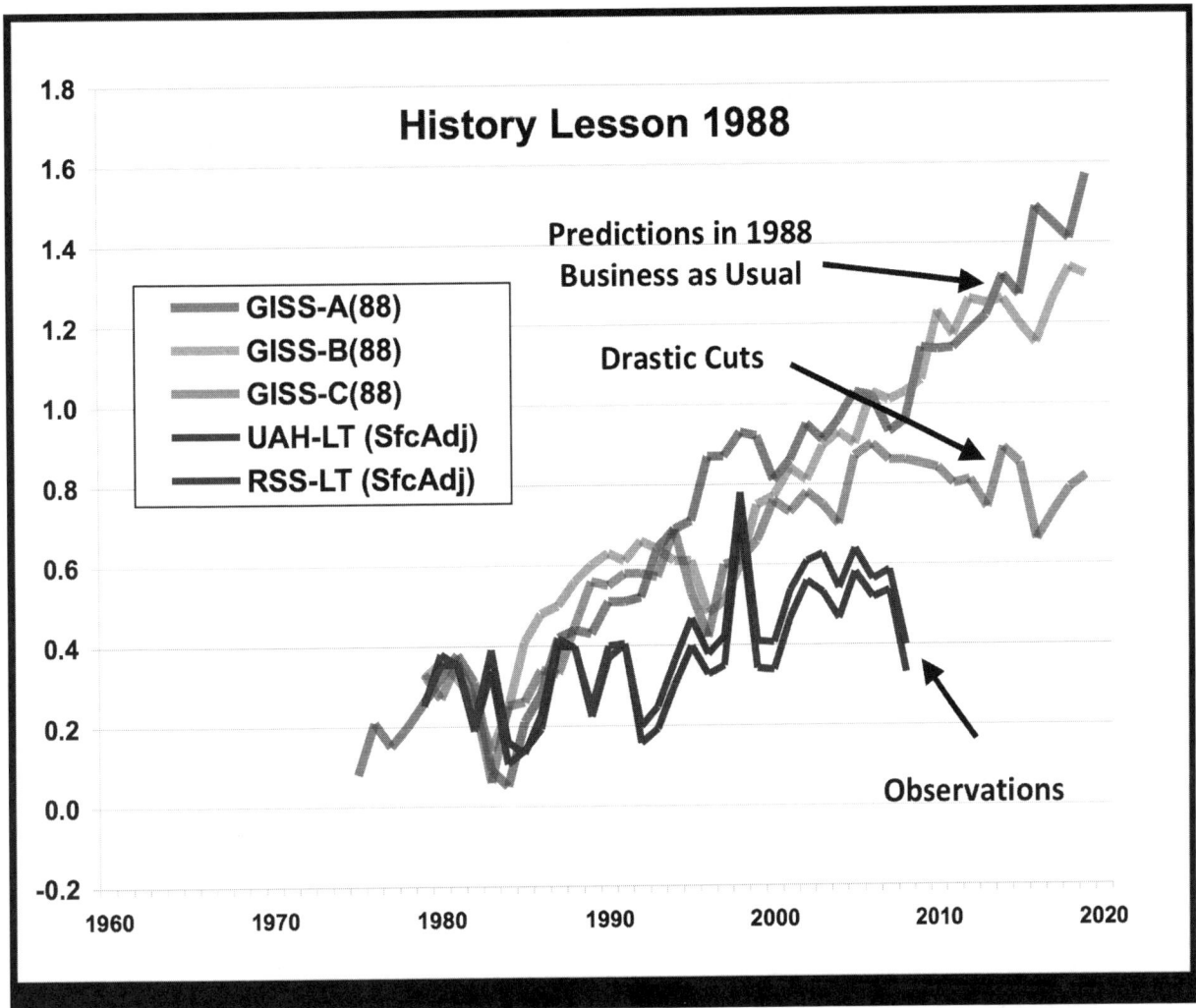

Figure 3.5.1. Actual temperature changes from UAH and RSS satellite data, adjusted to mimic surface temperatures, compared to predictions made by James Hansen to Congress in 1988. Source: Christy, J.R. 2009. Written testimony to House Ways and Means Committee. 25 February. http://waysandmeans.house.gov/media/pdf/111/ctest.pdf, last accessed May 10, 2009

every other surface and tropospheric temperature dataset, indicating RSS contains a spurious warming shift at that time (Christy *et al.*, 2007). The second feature is the relative cooling of RSS vs. UAH since 2006. This can be explained by the fact UAH uses a spacecraft (NASA's Aqua) that is not subject to orbital drifting, whereas RSS relies on NOAA-15, which is drifting into warmer diurnal times. The implication here is that RSS is overcorrecting for this spurious warming by reporting too much cooling. Overall, the shifts unique to RSS create a spurious warming in the record, which is being slowly mitigated by the more recent spurious cooling.

The graphs show that the temperature anomalies in the RSS dataset for November 2007 and December 2007 were below the 1979-1998 mean average for the first time since 2000. Both data series show the rate of warming has slowed dramatically during the past seven to 12 years. Between the end of 2007 and early 2009, global temperature anomalies fell even further, effectively returning the world to the temperatures that prevailed in the late 1980s and early 1990s. See Figure 3.5.3. below.

The new trend toward less warming has prompted some scientists to wonder if the world's climate experienced a trend break in 2001-2002, similar to ones that occurred around 1910 (ending a cooling trend), the early 1940s (ending a warming trend), and the mid-1970s (ending a cooling trend). Swanson and Anastasios (2009), writing in *Geophysical Research Letters*, say "a break in the global mean temperature trend from the consistent warming over the 1976/77–2001/02 period may have occurred." Moreover, the episodic nature of temperature changes during the

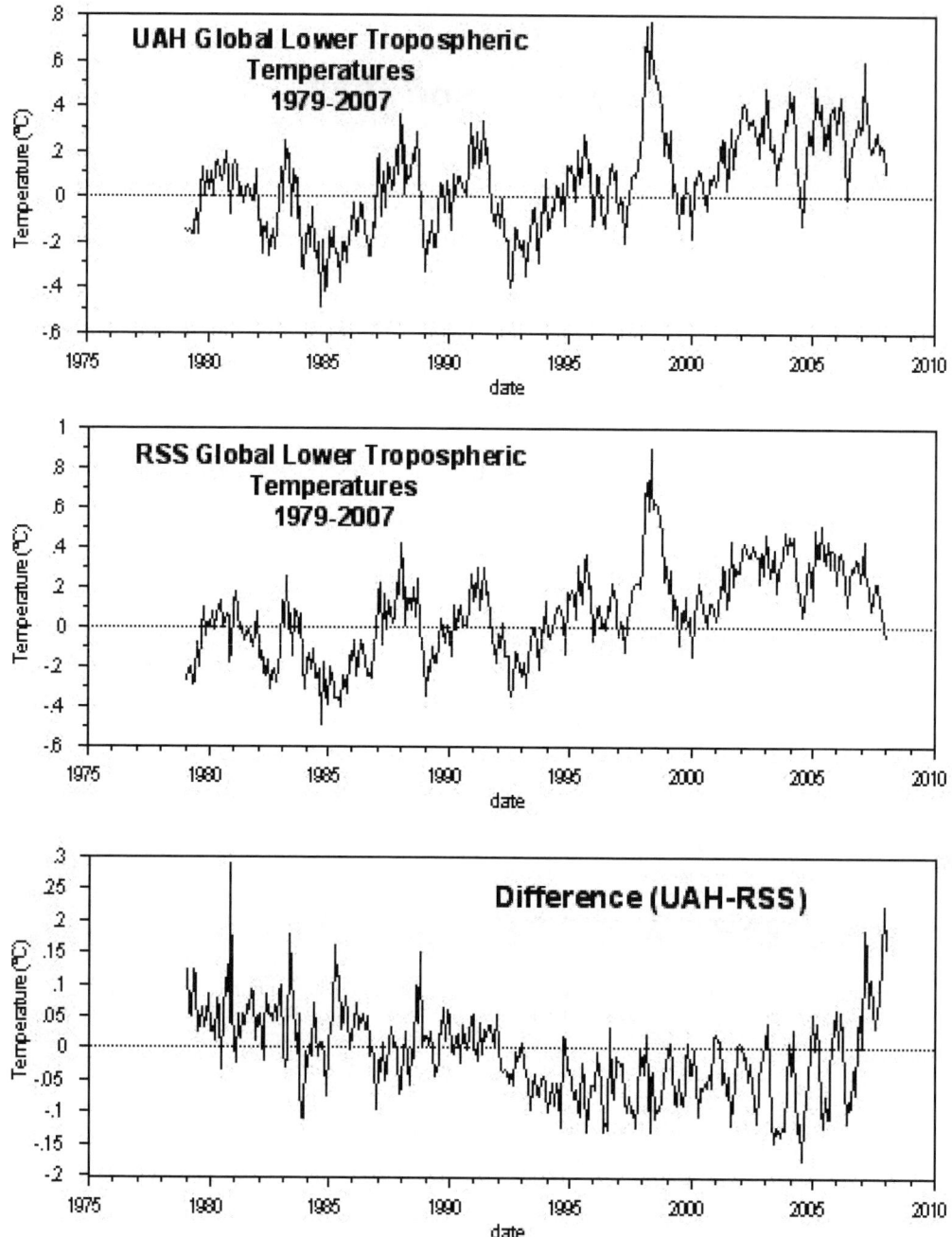

Figure 3.5.2. Global temperature anomalies from the lower troposphere. January 1979 through December 2007. (top) Data compiled at the University of Alabama, (middle) data compiled by Remote Sensing Systems, (bottom) difference between the two datasets (UAH minus RSS). Graphs were produced by Patrick Michaels using UAH and RSS data and first appeared in *World Climate* Report on February 7, 2008, http://www.worldclimatereport.com/index.php/category/temperature-history/satelliteballoons/.

past century is "difficult to reconcile with the presumed smooth evolution of anthropogenic greenhouse gas and aerosol radiative forcing with respect to time" and "suggests that an internal reorganization of the climate system may underlie such shifts."

Noel Keenlyside, a scientist with Germany's Leibniz Institute of Marine Science, writing with colleagues in a letter published in *Nature*, said "the climate of the North Atlantic region exhibits fluctuations on decadal timescales that have large societal consequences" and "these multidecadal

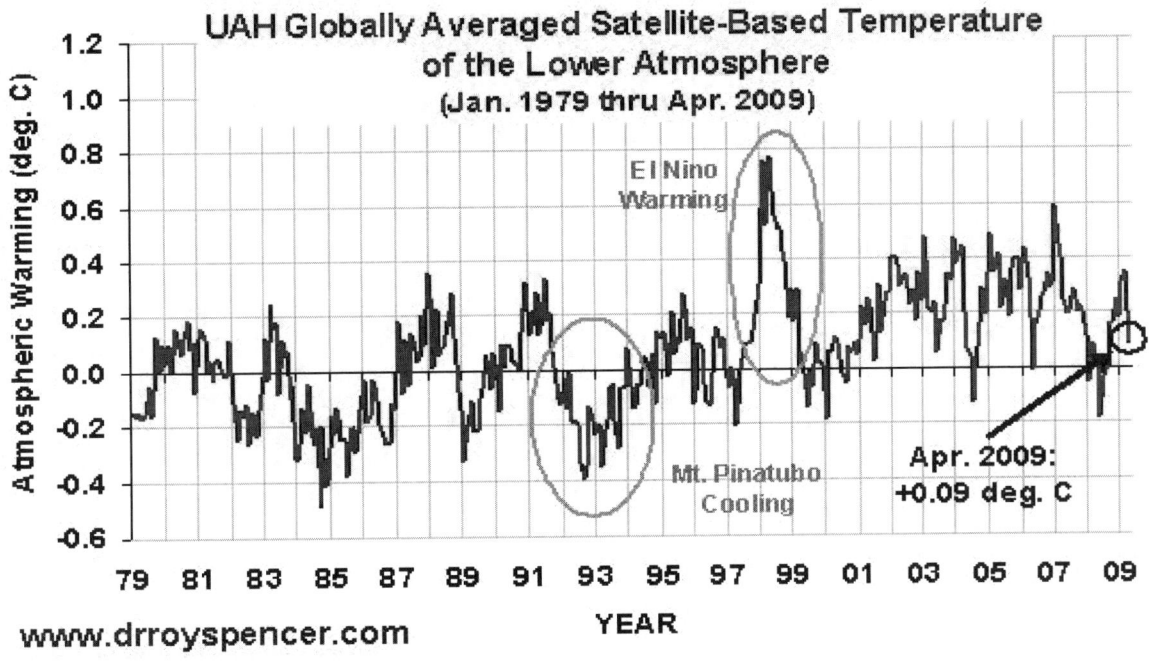

Figure 3.5.3. UAH Globally Averaged Satellite-Based Temperature of the Lower Atmosphere, January 1979 – April 2009. Source: Roy Spencer, http://www.drroyspencer.com/latest-global-temperatures/, last accessed May 10, 2009.

variations are potentially predictable if the current state of the ocean is known" (Keenlyside *et al.*, 2008). Using a database of sea-surface temperature (SST) observations, they "make the following forecast: over the next decade, the current Atlantic meridional overturning circulation will weaken to its long-term mean; moreover, North Atlantic SST and European and North American surface temperatures will cool slightly, whereas tropical Pacific SST will remain almost unchanged. Our results suggest that global surface temperature may not increase over the next decade, as natural climate variations in the North Atlantic and tropical Pacific temporarily offset the projected anthropogenic warming."

We predict more predictions of this kind as more scientists recognize, first, that estimates of past warming have been exaggerated by reliance on surface-station data that have been discredited by physical observation and by testing against superior satellite data; second, that recent temperature trends contradict past and recent forecasts by the IPCC and other prominent advocates of the theory that temperatures will steadily rise in response to increasing forcing by rising CO_2 levels in the atmosphere; and third, as attention turns to natural cycles like those modeled by Keenlyside *et al.*, as most scientists have known all along are more influential than the small effects of rising CO_2 in the atmosphere.

References

CCSP. 2006. *Temperature Trends in the Lower Atmosphere: Steps for Understanding and Reconciling Differences*. Karl, T.R., Hassol, S.J., Miller, C.D., and Murray, W.L. (Eds.) A Report by the Climate Change Science Program and the Subcommittee on Global Change Research, Washington, DC.

Christy, J.R. 2009. Written testimony to House Ways and Means Committee. 25 February. http://waysandmeans. house.gov/media/pdf/111/ctest.pdf, last accessed May 10, 2009.

Christy, J.R. and Norris, W.B. 2006. Satellite and VIZ-radiosonde intercomparisons for diagnosis of nonclimatic influences. *Journal of Atmospheric and Oceanic Technology* **23**: 1181-1194.

Christy, J.R. and Norris, W.B. 2009. Discontinuity issues with radiosondes and satellite temperatures in the Australia region 1979-2006. *Journal of Atmospheric and Oceanic Technology* **25**: OI:10.1175/2008JTECHA1126.1.

Christy, J.R., Norris, W.B. and McNider, R.T. 2009. Surface temperature variations in East Africa and possible causes. *Journal of Climate.* **22** (in press).

Christy, J.R., Norris, W.B., Spencer, R.W. and Hnilo, J.J. 2007. Tropospheric temperature change since 1979 from tropical radiosonde and satellite measurements. *Journal of Geophysical Research* **112**: doi:10.1029/2005JD0068.

Christy, J.R. and Spencer, R.W. 2005. Correcting temperature data sets. *Science* **310**: 972.

Christy, J.R., Spencer, R.W. and Braswell, D. 1997. How accurate are satellite "thermometers"? *Nature* **389**: 342-3.

Christy, J.R., Spencer, R.W., Norris, W.B., Braswell, W.D. and Parker, D.E. 2003. Error estimates of version 5.0 of MSU-AMSU bulk atmospheric temperatures. *Journal of Atmospheric and Oceanic Technology* **20**: 613-629.

Douglass, D.H., Christy, J.R., Pearson, B.D. and Singer, S.F. 2007. A comparison of tropical temperature trends with model predictions. *International Journal of Climatology.* DOI:10.1002/joc.1651.

Hurrell, J.W. and Trenberth, K.E. 1997. Spurious trends in the satellite MSU temperature record arising from merging different satellite records. *Nature* **386**: 164-167.

IPCC. 2007-I. *Climate Change 2007: The Physical Science Basis. Contribution of Working Group I to the Fourth Assessment Report of the Intergovernmental Panel on Climate Change.* Solomon, S., D. Qin, M. Manning, Z. Chen, M. Marquis, K.B. Averyt, M. Tignor and H.L. Miller. (Eds.) Cambridge University Press, Cambridge, UK.

Keenlyside, N.S., Latif, M., Jungclaus, J., Kornblueh, L. and Roeckner, E. 2008. Advancing decadal-scale climate prediction in the North Atlantic sector. *Nature* **453**: 84-88.

Mears, C.A. and Wentz, F.J. 2005. The effect of diurnal correction on satellite-derived lower tropospheric temperature. *Science* **309**: 1548-1551.

Randall, R.M. and Herman, B.M. 2008. Using limited time period trends as a means to determine attribution of discrepancies in microwave sounding unit derived tropospheric temperature time series. *Journal of Geophysical Research*: doi:10.1029/2007JD008864.

Santer, B.D. *et al.* 2005. Amplification of surface temperature trends and variability in the tropical atmosphere. *Science* **309**: 1551-1556.

Spencer, R.W., Braswell, W.D., Christy, J.R. and Hnilo, J. 2007. Cloud and radiation budget changes associated with tropical intraseasonal oscillations. *Geophysical Research Letters* **34**: L15707, doi:10.1029/2007GLO296998.

Spencer, R.W. and Christy, J.R. 1990. Precise monitoring of global temperature trends from satellites. *Science* **247**: 1558-1562.

Swanson, K.L. and Tsonis, A.A. 2009. Has the climate shifted? *Geophysical Research Letters* **36**: L06711, doi:10.1029/2008GL037022.

Trenberth, K.E. 1997. The use and abuse of climate models in climate change research. *Nature* **386**: 131-133.

Trenberth, K.E. and Hurrell, J.W. 1997. How accurate are satellite "thermometers." *Nature* **389**: 342-343.

Wentz, F.J. and Schabel, M. 1998. Effects of satellite orbital decay on MSU lower tropospheric temperature trends. *Nature* **394**: 661-664.

3.6. Arctic

The IPCC claims "average arctic temperatures increased at almost twice the global average rate in the past 100 years," though it then acknowledges that "arctic temperatures have high decadal variability, and a warm period was also observed from 1925 to 1945" (IPCC 2007-I, p. 7). Later in the report, the IPCC says "the warming over land in the Arctic north of 65°C is more than double the warming in the global mean from the 19th century to the 21st century and also from about the late 1960s to the present. In the arctic series, 2005 is the warmest year" (p. 248). But the IPCC then admits that "a few areas have cooled since 1901, most notably the northern North Atlantic near southern Greenland" (p. 252). So has the Artic really experienced the so-called unprecedented warming of the twentieth century?

References

IPCC. 2007-I. *Climate Change 2007: The Physical Science Basis. Contribution of Working Group I to the Fourth Assessment Report of the Intergovernmental Panel on Climate Change.* Solomon, S., D. Qin, M. Manning, Z. Chen, M. Marquis, K.B. Averyt, M. Tignor and H.L. Miller. (Eds.) Cambridge University Press, Cambridge, UK.

3.6.1. Greenland

Dahl-Jensen *et al.* (1998) used data from two ice sheet boreholes to reconstruct the temperature history of Greenland over the past 50,000 years. Their analysis indicated that temperatures on the Greenland Ice Sheet during the Last Glacial Maximum (about 25,000 years ago) were 23 ± 2 °C colder than at present. After the termination of the glacial period, however, temperatures increased steadily to a value that was 2.5°C *warmer* than at present, during the Climatic Optimum of 4,000 to 7,000 years ago. The Medieval Warm Period and Little Ice Age were also evident in the borehole data, with temperatures 1°C warmer and 0.5-0.7°C cooler than at present, respectively. Then, after the Little Ice Age, the scientists report "temperatures reached a maximum around 1930 AD" and that "temperatures have *decreased* [our italics] during the last decades."

The results of this study stand in stark contrast to the predictions of general circulation models of the atmosphere, which consistently suggest there should have been a significant CO_2-induced warming in high northern latitudes over the past several decades. They also depict large temperature excursions over the past 10,000 years, when the air's CO_2 content was relatively stable. Each of these observations raises serious doubts about the models' ability to correctly forecast earth's climatic response to the ongoing rise in the air's CO_2 content.

In another study of Greenland climate that included both glacial and interglacial periods, Bard (2002) reviews the concept of rapid climate change. Of this phenomenon, he writes that "it is now recognized that the ocean-atmosphere system exhibits several stable regimes under equivalent external forcings," and that "the transition from one state to another occurs very rapidly when certain climatic parameters attain threshold values." Specifically, he notes that in the models "a slight increase in the freshwater flux above the modern level F produces a decrease in the NADW [North Atlantic Deep Water] convection and a moderate cooling in the North Atlantic," but that "the system flips to another state once the flux reaches a threshold value $F + \text{delta}F$," which state has no deep convection and "is characterized by surface temperatures up to 6°C lower in and around the North Atlantic."

With respect to what has been learned from observations, Bard concentrates on the region of the North Atlantic, describing glacial-period millennial-scale episodes of dramatic warming called Dansgaard-Oeschger events (with temperature increases "of more than 10°C"), which are evident in Greenland ice core records, as well as episodes of "drastic cooling" called Heinrich events (with temperature drops "of up to about 5°C"), which are evident in sea surface temperature records derived from the study of North Atlantic deep-sea sediment cores.

In the Greenland record, according to Bard, the progression of these events is such that "the temperature warms abruptly to reach a maximum and then slowly decreases for a few centuries before reaching a threshold, after which it drops back to the cold values that prevailed before the warm event." He also reports that "models coupling the atmosphere, ocean, and ice sheets are still unable to correctly simulate that variability on all scales in both time and space," which suggests we do not fully understand the dynamics of these rapid climate changes. Bard states, "all the studies so far carried out fail to answer the crucial question: How close are we to the next bifurcation [which could cause a rapid change-of-state in earth's climate system]?" In this regard, he notes that "an intense debate continues in the modeling community about the reality of such instabilities under warm conditions," which is a particularly important point, since all dramatic warming and cooling events that have been detected to date have occurred in either full glacials or transitional periods between glacials and interglacials.

This latter real-world fact clearly suggests we are unlikely to experience any dramatic warming or cooling surprises in the near future, as long as the earth does not begin drifting towards glacial conditions, which is another reason to not be concerned about the ongoing rise in the air's CO_2 content. In fact, it suggests that allowing more CO_2 to accumulate in the atmosphere provides an "insurance policy" against abrupt climate change; interglacial warmth seems to inoculate the planet against climatic instabilities, allowing only the mild millennial-scale climatic oscillation that alternately brings the earth slightly warmer and cooler conditions typical of the Medieval Warm Period and Little Ice Age.

Focusing on the more pertinent period of the current interglacial or Holocene, we next consider a number of papers that bear upon the reality of the Medieval Warm Period and Little Ice Age: two well-known multi-century periods of significant climatic aberration. These periods of modest climatic aberration, plus the analogous warm and cool periods that preceded them (the Roman Warm Period and

Dark Ages Cold Period), provide strong evidence for the existence of a millennial-scale oscillation of climate that is unforced by changes in the air's CO_2 content, which in turn suggests that the global warming of the Little Ice Age-to-Current Warm Period transition was likely totally independent of the coincidental concomitant increase in the air's CO_2 content that accompanied the Industrial Revolution.

We begin with the study of Keigwin and Boyle (2000), who briefly reviewed what is known about the millennial-scale oscillation of earth's climate that is evident in a wealth of proxy climate data from around the world. Stating that "mounting evidence indicates that the Little Ice Age was a global event, and that its onset was synchronous within a few years in both Greenland and Antarctica," they remark that in Greenland it was characterized by a cooling of approximately 1.7°C. Likewise, in an article titled "Was the Medieval Warm Period Global?" Broecker (2001) answers yes, citing borehole temperature data that reveal the magnitude of the temperature drop over Greenland from the peak warmth of the Medieval Warm Period (800 to 1200 A.D.) to the coldest part of the Little Ice Age (1350 to 1860 A.D.) to have been approximately 2°C, and noting that as many as six thousand borehole records from all continents of the world confirm that the earth was a significantly warmer place a thousand years ago than it is today.

McDermott *et al.* (2001) derived a $\delta^{18}O$ record from a stalagmite discovered in Crag Cave in southwestern Ireland, after which they compared this record with the $\delta^{18}O$ records from the GRIP and GISP2 ice cores from Greenland. In doing so, they found evidence for "centennial-scale $\delta^{18}O$ variations that correlate with subtle $\delta^{18}O$ changes in the Greenland ice cores, indicating regionally coherent variability in the early Holocene." They additionally report that the Crag Cave data "exhibit variations that are broadly consistent with a Medieval Warm Period at ~1000 ± 200 years ago and a two-stage Little Ice Age, as reconstructed by inverse modeling of temperature profiles in the Greenland Ice Sheet." Also evident in the Crag Cave data were the $\delta^{18}O$ signatures of the earlier Roman Warm Period and Dark Ages Cold Period that comprised the prior such cycle of climate in that region. In concluding they reiterate the important fact that the coherent $\delta^{18}O$ variations in the records from both sides of the North Atlantic "indicate that many of the subtle multicentury $\delta^{18}O$ variations in the Greenland ice

cores reflect regional North Atlantic margin climate signals rather than local effects."

Another study that looked at temperature variations on both sides of the North Atlantic was that of Seppa and Birks (2002), who used a recently developed pollen-climate reconstruction model and a new pollen stratigraphy from Toskaljavri, a tree-line lake in the continental sector of northern Fenoscandia (located just above 69°N latitude), to derive quantitative estimates of annual precipitation and July mean temperature. The two scientists say their reconstructions "agree with the traditional concept of a 'Medieval Warm Period' (MWP) and 'Little Ice Age' in the North Atlantic region (Dansgaard *et al.*, 1975)." Specifically, they report there is "a clear correlation between our MWP reconstruction and several records from Greenland ice cores," and that "comparisons of a smoothed July temperature record from Toskaljavri with measured borehole temperatures of the GRIP and Dye 3 ice cores (Dahl-Jensen *et al.*, 1998) and the $\delta^{18}O$ record from the Crete ice core (Dansgaard *et al.*, 1975) show the strong similarity in timing of the MWP between the records." Last of all, they note that "July temperature values during the Medieval Warm Period (ca. 1400-1000 cal yr B.P.) were ca. 0.8°C higher than at present," where present means the last six decades of the twentieth century.

Concentrating solely on Greenland and its immediate environs are several other papers, among which is the study of Wagner and Melles (2001), who retrieved a sediment core from a lake on an island situated just off Liverpool Land on the east coast of Greenland. Analyzing it for a number of properties related to the past presence of seabirds there, they obtained a 10,000-year record that tells us much about the region's climatic history.

Key to the study were certain biogeochemical data that reflected variations in seabird breeding colonies in the catchment area of the lake. These data revealed high levels of the various parameters measured by Wagner and Melles between about 1,100 and 700 years before present (BP) that were indicative of the summer presence of significant numbers of seabirds during that "medieval warm period," as they describe it, which had been preceded by a several-hundred-year period of little to no inferred bird presence. Then, after the Medieval Warm Period, the data suggested another absence of birds during what they refer to as "a subsequent Little Ice Age," which they note was "the coldest period since the early Holocene in East Greenland." Their data also showed signs of a "resettlement of seabirds during the last 100 years, indicated by an increase of organic matter in the lake sediment and confirmed by bird observations." However, values of the most recent data were not as great as those obtained from the earlier Medieval Warm Period; and temperatures derived from two Greenland ice cores led to the same conclusion: it was warmer at various times between 1,100 to 700 years BP than it was over the twentieth century.

Kaplan et al. (2002) also worked with data obtained from a small lake, this one in southern Greenland, analyzing sediment physical-chemical properties, including magnetic susceptibility, density, water content, and biogenic silica and organic matter concentrations. They discovered that "the interval from 6000 to 3000 cal yr BP was marked by warmth and stability." Thereafter, however, the climate cooled "until its culmination during the Little Ice Age," but from 1,300-900 years BP, there was a partial amelioration of climate (the Medieval Warm Period) that was associated with an approximate 1.5°C rise in temperature.

Following another brief warming between AD 1500 and 1750, the second and more severe portion of the Little Ice Age occurred, which was in turn followed by "naturally initiated post-Little Ice Age warming since AD 1850, which is recorded throughout the Arctic." They report that Viking "colonization around the northwestern North Atlantic occurred during peak Medieval Warm Period conditions that ended in southern Greenland by AD 1100," noting that Norse movements around the region thereafter "occurred at perhaps the worst time in the last 10,000 years, in terms of the overall stability of the environment for sustained plant and animal husbandry."

We can further explore these aspects of Greenland's climatic history from three important papers that reconstructed environmental conditions in the vicinity of Igaliku Fjord, South Greenland, before, during, and after the period of Norse habitation of this and other parts of the ice-covered island's coast, beginning with the study of Lassen et al. (2004), who provide some historical background to their palaeoclimatic work by reporting that "the Norse, under Eric the Red, were able to colonize South Greenland at AD 985, according to the Icelandic Sagas, owing to the mild Medieval Warm Period climate with favorable open-ocean conditions." They also mention, in this regard, that the arrival of the gritty Norsemen was "close to the peak of Medieval warming recorded in the GISP2 ice core which was dated at AD 975 (Stuiver et al., 1995)," while we additionally note that Esper et al. (2002) independently identified the peak warmth of this period throughout North American extratropical latitudes as "occurring around 990." Hence, it would appear that the window of climatic opportunity provided by the peak warmth of the Medieval Warm Period was indeed a major factor enabling seafaring Scandinavians to establish long-enduring settlements on the coast of Greenland.

As time progressed, however, the glowing promise of the apex of Medieval warmth gave way to the debilitating reality of the depth of Little Ice Age cold. Jensen et al. (2004), for example, report that the diatom record of Igaliku Fjord "yields evidence of a relatively moist and warm climate at the beginning of settlement, which was crucial for Norse land use," but that "a regime of more extreme climatic fluctuations began soon after AD 1000, and after AD c. 1350 cooling became more severe." Lassen et al. additionally note that "historical documents on Iceland report the presence of the Norse in South Greenland for the last time in AD 1408," during what they describe as a period of "unprecedented influx of (ice-loaded) East Greenland Current water masses into the innermost parts of Igaliku Fjord." They also report that "studies of a Canadian high-Arctic ice core and nearby geothermal data (Koerner and Fisher, 1990) correspondingly show a significant temperature lowering at AD 1350-1400," when, in their words, "the Norse society in Greenland was declining and reaching its final stage probably before the end of the fifteenth century." Consequently, what the relative warmth of the Medieval Warm Period provided the Norse settlers, the relative cold of the Little Ice Age took from them: the ability to survive on Greenland.

More details of the saga of five centuries of Nordic survival at the foot of the Greenland Ice Cap are provided by the trio of papers addressing the palaeohistory of Igaliku Fjord. Based on a high-resolution record of the fjord's subsurface water-mass properties derived from analyses of benthic foraminifera, Lassen et al. conclude that stratification of the water column, with Atlantic water masses in its lower reaches, appears to have prevailed throughout the last 3,200 years, except for the Medieval Warm Period. During this period, which they describe as occurring between AD 885 and 1235, the outer part of Igaliku Fjord experienced enhanced vertical mixing (which they attribute to increased wind stress) that would have been expected to increase nutrient availability there. A similar conclusion was reached by Roncaglia and Kuijpers (2004), who found evidence of increased bottom-water ventilation between AD 960 and 1285. Hence, based on these findings, plus evidence of the presence of *Melonis barleeanus* during the Medieval Warm Period (the distribution of which is mainly controlled by the presence of partly decomposed organic matter), Lassen *et al.* conclude that surface productivity in the fjord during this interval of unusual relative warmth was "high and thus could have provided a good supply of marine food for the Norse people."

Shortly thereafter, the cooling that led to the Little Ice Age was accompanied by a gradual re-stratification of the water column, which curtailed nutrient upwelling and reduced the high level of marine productivity that had prevailed throughout the Medieval Warm Period. These linked events, according to Lassen *et al.*, "contributed to the loss of the Norse settlement in Greenland." Indeed, with deteriorating growing conditions on land and simultaneous reductions in oceanic productivity, the odds were truly stacked against the Nordic colonies, and it was only a matter of time before their fate was sealed. As Lassen *et al.* describe it, "around AD 1450, the climate further deteriorated with further increasing stratification of the water-column associated with stronger advection of (ice-loaded) East Greenland Current water masses." This development, in their words, led to an even greater "increase of the ice season and a decrease of primary production and marine food supply," which "could also have had a dramatic influence on the local seal population and thus the feeding basis for the Norse population."

The end result of these several conjoined phenomena, in the words of Lassen *et al.*, was that "climatic and hydrographic changes in the area of the Eastern Settlement were significant in the crucial period when the Norse disappeared." Also, Jensen *et al.* report that "geomorphological studies in Northeast Greenland have shown evidence of increased winter wind speed, particularly in the period between AD 1420 and 1580 (Christiansen, 1998)," noting that "this climatic deterioration coincides with reports of increased sea-ice conditions that caused difficulties in using the old sailing routes from Iceland westbound and further southward along the east coast of Greenland, forcing sailing on more southerly routes when going to Greenland (Seaver, 1996)."

In light of these observations, Jensen *et al.* state that "life conditions certainly became harsher during the 500 years of Norse colonization," and that this severe cooling-induced environmental deterioration "may very likely have hastened the disappearance of the culture." At the same time, it is also clear that the more favorable living conditions associated with the peak warmth of the Medieval Warm Period—which occurred between approximately AD 975 (Stuiver *et al.*, 1995) and AD 990 (Esper *et al.*, 2002)—were what originally enabled the Norse to successfully colonize the region. In the thousand-plus subsequent years, there has never been a sustained period of comparable warmth, nor of comparable terrestrial or marine productivity, either locally or hemispherically (and likely globally, as well), the strident protestations of Mann *et al.* (2003) notwithstanding.

Concentrating on the twentieth century, Hanna and Cappelen (2003) determined the air temperature history of coastal southern Greenland from 1958-2001, based on data from eight Danish Meteorological Institute stations in coastal and near-coastal southern Greenland, as well as the concomitant sea surface temperature (SST) history of the Labrador Sea off southwest Greenland, based on three previously published and subsequently extended SST datasets (Parker *et al.*, 1995; Rayner *et al.*, 1996; Kalnay *et al.*, 1996). The coastal temperature data showed a *cooling* of 1.29°C over the period of study, while two of the three SST databases also depicted cooling: by 0.44°C in one case and by 0.80°C in the other. Both the land-based air temperature and SST series followed similar patterns and were strongly correlated, but with no obvious lead/lag either way. In addition, it was determined that the cooling was "significantly inversely correlated with an increased phase of the North Atlantic Oscillation (NAO) over the past few decades." The two researchers say this "NAO-temperature link doesn't explain what caused the observed cooling in coastal southern Greenland but it does lend it credibility."

In referring to what they call "this important regional exception to recent 'global warming'," Hanna and Cappelen note that the "recent cooling may have significantly added to the mass balance of at least the southern half of the [Greenland] Ice Sheet." Consequently, since this part of the ice sheet is the portion that would likely be the first to experience melting in a warming world, it would appear that whatever caused the cooling has not only protected the Greenland Ice Sheet against warming-induced disintegration but actually fortified it against that possibility.

Several other studies have also reported late-twentieth century cooling on Greenland. Based on mean monthly temperatures of 37 Arctic and seven sub-Arctic stations, as well as temperature anomalies of 30 grid-boxes from the updated dataset of Jones, for example, Przybylak (2000) found that "the level of temperature in Greenland in the last 10-20 years is similar to that observed in the 19th century." Likewise, in a study that utilized satellite imagery of the Odden ice tongue (a winter ice cover that occurs in the Greenland Sea with a length of about 1,300 km and an aerial coverage of as much as 330,000 square kilometers) plus surface air temperature data from adjacent Jan Mayen Island, Comiso *et al.* (2001) determined that the ice phenomenon was "a relatively smaller feature several decades ago," due to the warmer temperatures that were prevalent at that time. In addition, they report that observational evidence from Jan Mayen Island indicates temperatures there cooled at a rate of 0.15 ± 0.03°C per decade during the past 75 years.

Taurisano *et al.* (2004) examined the temperature history of the Nuuk fjord during the last century, where their analyses of all pertinent regional data led them to conclude that "at all stations in the Nuuk fjord, both the annual mean and the average temperature of the three summer months (June, July and August) exhibit a pattern in agreement with the trends observed at other stations in south and west Greenland (Humlum 1999; Hanna and Cappelen, 2003)." As they describe it, the temperature data "show that a warming trend occurred in the Nuuk fjord during the first 50 years of the 1900s, followed by a cooling over the second part of the century, when the average annual temperatures decreased by approximately 1.5°C." Coincident with this cooling trend there was also what they describe as "a remarkable increase in the number of snowfall days (+59 days)." What is more, they report that "not only did the cooling affect the winter months, as suggested by Hannna and Cappelen (2002), but also the summer mean," noting that "the summer cooling is rather important information for glaciological studies, due to the ablation-temperature relations."

In a study of three coastal stations in southern and central Greenland that possess almost uninterrupted temperature records between 1950 and 2000, Chylek *et al.* (2004) discovered that "summer temperatures, which are most relevant to Greenland ice sheet melting rates, do not show any persistent increase during the last fifty years." In fact, working with the two stations with the longest records (both over a century in length), they determined that coastal Greenland's peak temperatures occurred between 1930 and 1940, and that the subsequent decrease in temperature was so substantial and sustained that current coastal temperatures "are about 1°C below their 1940 values." Furthermore, they note that "at the summit of the Greenland ice sheet the summer average temperature has decreased at the rate of 2.2°C per decade since the beginning of the measurements in 1987." Hence, as with the Arctic as a whole, it would appear that Greenland has not experienced any net warming over the most dramatic period of atmospheric CO_2 increase on record. In fact, it has *cooled* during this period.

At the start of the twentieth century, however, Greenland was warming, as it emerged, along with the rest of the world, from the depths of the Little Ice Age. Between 1920 and 1930, when the atmosphere's CO_2 concentration rose by a mere 3 to 4 ppm, there was a phenomenal warming at all five coastal locations for which contemporary temperature records are available. In the words of Chylek *et al.*, "average annual temperature rose between 2 and 4°C [and by as much as 6°C in the winter] in less than ten years." And this warming, as they note, "is also seen in the $^{18}O/^{16}O$ record of the Summit ice core (Steig *et al.*, 1994; Stuiver *et al.*, 1995; White *et al.*, 1997)."

In commenting on this dramatic temperature rise, which they call the "great Greenland warming of the 1920s," Chylek *et al.* conclude that "since there was no significant increase in the atmospheric greenhouse gas concentration during that time, the Greenland warming of the 1920s demonstrates that a large and rapid temperature increase can occur over Greenland, and perhaps in other regions of the Arctic, due to internal climate variability such as the NAM/NAO [Northern Annular Mode/North Atlantic Oscillation], without a significant anthropogenic influence." These facts led them to speculate that "the NAO may play a crucial role in determining local Greenland climate during the 21st century, resulting in a local climate that may defy the global climate change."

Clearly, there is no substance to the claim that Greenland provides evidence for an impending CO_2-induced warming. These many studies of the temperature history of Greenland depict long-term oscillatory cooling ever since the Climatic Optimum of the mid-Holocene, when it was perhaps 2.5°C warmer than it is now, within which cooling trend is included the Medieval Warm Period, when it was about 1°C warmer than it is currently, and the Little Ice Age, when it was 0.5 to 0.7°C cooler than now, after which temperatures rebounded to a new maximum in the 1930s, only to fall steadily thereafter.

Additional information on this topic, including reviews of newer publications as they become available, can be found at http://www.co2science.org/subject/g/greenland.php.

References

Bard, E. 2002. Climate shock: Abrupt changes over millennial time scales. *Physics Today* **55(12)**: 32-38.

Broecker, W.S. 2001. Was the Medieval Warm Period global? *Science* **291**: 1497-1499.

Christiansen, H.H. 1998. 'Little Ice Age' navigation activity in northeast Greenland. *The Holocene* **8**: 719-728.

Chylek, P., Box, J.E. and Lesins, G. 2004. Global warming and the Greenland ice sheet. *Climatic Change* **63**: 201-221.

Comiso, J.C., Wadhams, P., Pedersen, L.T. and Gersten, R.A. 2001. Seasonal and interannual variability of the Odden ice tongue and a study of environmental effects. *Journal of Geophysical Research* **106**: 9093-9116.

Dahl-Jensen, D., Mosegaard, K., Gundestrup, N., Clow, G.D., Johnsen, S.J., Hansen, A.W. and Balling, N. 1998. Past temperatures directly from the Greenland Ice Sheet. *Science* **282**: 268-271.

Dansgaard, W., Johnsen, S.J., Gundestrup, N., Clausen, H.B. and Hammer, C.U. 1975. Climatic changes, Norsemen and modern man. *Nature* **255**: 24-28.

Esper, J., Cook, E.R. and Schweingruber, F.H. 2002. Low-frequency signals in long tree-ring chronologies for reconstructing past temperature variability. *Science* **295**: 2250-2253.

Hanna, E. and Cappelen, J. 2002. Recent climate of Southern Greenland. *Weather* **57**: 320-328.

Hanna, E. and Cappelen, J. 2003. Recent cooling in coastal southern Greenland and relation with the North Atlantic Oscillation. *Geophysical Research Letters* **30**: 10.1029/2002GL015797.

Humlum, O. 1999. Late-Holocene climate in central West Greenland: meteorological data and rock-glacier isotope evidence. *The Holocene* **9**: 581-594.

Jensen, K.G., Kuijpers, A., Koc, N. and Heinemeier, J. 2004. Diatom evidence of hydrographic changes and ice conditions in Igaliku Fjord, South Greenland, during the past 1500 years. *The Holocene* **14**: 152-164.

Kalnay, E., Kanamitsu, M., Kistler, R., Collins, W., Deaven, D., Gandin, L., Iredell, M., Saha, S., White, G., Woollen, J., Zhu, Y., Chelliah, M., Ebisuzaki, W., Higgins, W., Janowiak, J., Mo, K.C., Ropelewski, C., Wang, J., Leetmaa, A., Reynolds, R., Jenne, R. and Joseph, D. 1996. The NCEP/NCAR 40-year reanalysis project. *Bulletin of the American Meteorological Society* **77**: 437-471.

Kaplan, M.R., Wolfe, A.P. and Miller, G.H. 2002. Holocene environmental variability in southern Greenland inferred from lake sediments. *Quaternary Research* **58**: 149-159.

Keigwin, L.D. and Boyle, E.A. 2000. Detecting Holocene changes in thermohaline circulation. *Proceedings of the National Academy of Sciences USA* **97**: 1343-1346.

Koerner, R.M. and Fisher, D.A. 1990. A record of Holocene summer climate from a Canadian high-Arctic ice core. *Nature* **343**: 630-631.

Lassen, S.J., Kuijpers, A., Kunzendorf, H., Hoffmann-Wieck, G., Mikkelsen, N. and Konradi, P. 2004. Late-Holocene Atlantic bottom-water variability in Igaliku Fjord, South Greenland, reconstructed from foraminifera faunas. *The Holocene* **14**: 165-171.

Mann, M., Amman, C., Bradley, R., Briffa, K., Jones, P., Osborn, T., Crowley, T., Hughes, M., Oppenheimer, M., Overpeck, J., Rutherford, S., Trenberth, K. and Wigley, T. 2003. On past temperatures and anomalous late-20th century warmth. *EOS: Transactions, American Geophysical Union* **84**: 256-257.

McDermott, F., Mattey, D.P. and Hawkesworth, C. 2001. Centennial-scale Holocene climate variability revealed by a high-resolution speleothem $\delta^{18}O$ record from SW Ireland. *Science* **294**: 1328-1331.

Parker, D.E., Folland, C.K. and Jackson, M. 1995. Marine surface temperature: Observed variations and data requirements. *Climatic Change* **31**: 559-600.

Petit, J.R., Jouzel, J., Raynaud, D., Barkov, N.I., Barnola, J.-M., Basile, I., Bender, M., Chappellaz, J., Davis, M., Delaygue, G., Delmotte, M., Kotlyakov, V.M., Legrand, M., Lipenkov, V.Y., Lorius, C., Pepin, L., Ritz, C., Saltzman, E. and Stievenard, M. 1999. Climate and atmospheric history of the past 420,000 years from the Vostok ice core, Antarctica. *Nature* **399**: 429-436.

Przybylak, R. 2000. Temporal and spatial variation of surface air temperature over the period of instrumental observations in the Arctic. *International Journal of Climatology* **20**: 587-614.

Rayner, N.A., Horton, E.B., Parker, D.E., Folland, C.K. and Hackett, R.B. 1996. Version 2.2 of the global sea-ice and sea surface temperature data set, 1903-1994. *Climate Research Technical Note 74*, Hadley Centre, U.K. Meteorological Office, Bracknell, Berkshire, UK.

Roncaglia, L. and Kuijpers A. 2004. Palynofacies analysis and organic-walled dinoflagellate cysts in late-Holocene sediments from Igaliku Fjord, South Greenland. *The Holocene* 14: 172-184.

Seaver, K.A. 1996. The Frozen Echo: Greenland and the Exploration of North America AD c. 1000-1500. Stanford University Press, Stanford, CA, USA.

Seppa, H. and Birks, H.J.B. 2002. Holocene climate reconstructions from the Fennoscandian tree-line area based on pollen data from Toskaljavri. *Quaternary Research* 57: 191-199.

Steig, E.J., Grootes, P.M. and Stuiver, M. 1994. Seasonal precipitation timing and ice core records. *Science* 266: 1885-1886.

Stuiver, M., Grootes, P.M. and Braziunas, T.F. 1995. The GISP2 $\delta^{18}O$ climate record of the past 16,500 years and the role of the sun, ocean, and volcanoes. *Quaternary Research* 44: 341-354.

Taurisano, A., Boggild, C.E. and Karlsen, H.G. 2004. A century of climate variability and climate gradients from coast to ice sheet in West Greenland. *Geografiska Annaler* 86A: 217-224.

Wagner, B. and Melles, M. 2001. A Holocene seabird record from Raffles So sediments, East Greenland, in response to climatic and oceanic changes. *Boreas* 30: 228-239.

White, J.W.C., Barlow, L.K., Fisher, D., Grootes, P.M., Jouzel, J., Johnsen, S.J., Stuiver, M. and Clausen, H.B. 1997. The climate signal in the stable isotopes of snow from Summit, Greenland: Results of comparisons with modern climate observations. *Journal of Geophysical Research* 102: 26,425-26,439.

3.6.1.2. Rest of Arctic

Overpeck *et al.* (1997) combined paleoclimatic records obtained from lake and marine sediments, trees, and glaciers to develop a 400-year history of circum-Arctic surface air temperature. From this record they determined that the most dramatic warming of the last four centuries of the past millennium (1.5°C) occurred between 1840 and 1955, over which period the air's CO_2 concentration rose from approximately 285 ppm to 313 ppm, or by 28 ppm. Then, from 1955 to the end of the record (about 1990), the mean circum-Arctic air temperature declined by 0.4°C, while the air's CO_2 concentration rose from 313 ppm to 354 ppm, or by 41 ppm. On the basis of these observations, which apply to the entire Arctic, it is not possible to assess the influence of atmospheric CO_2 on surface air temperature or even to conclude it has any effect at all. Why? Because over the first 115 years of warming, as the air's CO_2 concentration rose by an average of 0.24 ppm/year, air temperature rose by an average of 0.013°C/year; over the final 35 years of the record, when the air's CO_2 content rose at a mean rate of 1.17 ppm/year (nearly five times the rate at which it had risen in the prior period), the rate-of-rise of surface air temperature decelerated, to a mean value (0.011°C/year) that was nearly the same as the rate at which it had previously risen.

Naurzbaev and Vaganov (2000) developed a 2,200-year temperature history using tree-ring data obtained from 118 trees near the upper-timberline in Siberia for the period 212 BC to AD 1996, as well as a similar history covering the period of the Holocene Climatic Optimum (3300 to 2600 BC). They compared their results with those obtained from an analysis of isotopic oxygen data extracted from a Greenland ice core. This work revealed that fluctuations in average annual temperature derived from the Siberian record agreed well with air temperature variations reconstructed from the Greenland data, suggesting to the two researchers that "the tree ring chronology of [the Siberian] region can be used to analyze both regional peculiarities and global temperature variations in the Northern Hemisphere."

Naurzbaev and Vaganov reported that several warm and cool periods prevailed for several multi-century periods throughout the last two millennia: a cool period in the first two centuries AD, a warm period from AD 200 to 600, cooling again from 600 to 800 AD, followed by the Medieval Warm Period from about AD 850 to 1150, the cooling of the Little Ice Age from AD 1200 though 1800, followed by the recovery warming of the twentieth century. In regard to this latter temperature rise, however, the two scientists say it was "not extraordinary," and that "the warming at the border of the first and second millennia [AD 1000] was longer in time and similar in amplitude." In addition, their reconstructed temperatures for the Holocene Climatic Optimum revealed there was an even warmer period about 5,000 years ago, when temperatures averaged 3.3°C more than they did over the past two millennia.

Contemporaneously, Vaganov *et al.* (2000) also used tree-ring width as a temperature proxy, reporting temperature variations for the Asian subarctic region over the past 600 years. Their graph of these data reveals that temperatures in this region exhibited a small positive trend from the beginning of the record until about AD 1750. Thereafter, a severe cooling trend ensued, followed by a 130-year warming trend from about 1820 through 1950, after which temperatures fell once again.

In analyzing the entire record, the researchers determined that the amplitude of twentieth century warming "does not go beyond the limits of reconstructed natural temperature fluctuations in the Holocene subarctic zone." And in attempting to determine the cause or causes of the temperature fluctuations, they report finding a significant correlation with solar radiation and volcanic activity over the entire 600-year period (r = 0.32 for solar radiation, r = -0.41 for volcanic activity), which correlation improved over the shorter interval (1800-1990) of the industrial period (r = 0.68 for solar radiation, r = -0.59 for volcanic activity). It is also enlightening to note, in this regard, that in this region of the world, where climate models predict large increases in temperature as a result of the historical rise in the air's CO_2 concentration, real-world data show an actual cooling trend since around 1940. Where warming does exist in the record—between about 1820 and 1940—much of it correlates with changes in solar irradiance and volcanic activity, two factors that are free of anthropogenic influence.

One year later, Moore *et al.* (2001) analyzed sediment cores extracted from Donard Lake, Baffin Island, Canada (~66.25°N, 62°W), to produce a 1,240-year record of mean summer temperature for this region that averaged 2.9°C over the period AD 750-1990. Within this period there were several anomalously warm decades with temperatures that were as high as 4°C around AD 1000 and 1100, while at the beginning of the thirteenth century Donard Lake witnessed what they called "one of the largest climatic transitions in over a millennium," as "average summer temperatures rose rapidly by nearly 2°C from AD 1195-1220, ending in the warmest decade in the record," with temperatures near 4.5°C. This latter temperature rise was then followed by a period of extended warmth that lasted until an abrupt cooling event occurred around AD 1375, resulting in the following decade being one of the coldest in the record and signaling the onset of the Little Ice Age on Baffin Island, which lasted 400 years. At the modern end of the record, a gradual warming trend occurred over the period 1800-1900, followed by a dramatic cooling event that brought temperatures back to levels characteristic of the Little Ice Age, which chilliness lasted until about 1950. Thereafter, temperatures rose

once more throughout the 1950s and 1960s, whereupon they trended downwards toward cooler conditions to the end of the record in 1990.

Gedalof and Smith (2001) compiled a transect of six tree ring-width chronologies from stands of mountain hemlock growing near the treeline that extends from southern Oregon to the Kenai Peninsula, Alaska. Over the period of their study (AD 1599-1983), they determined that "much of the pre-instrumental record in the Pacific Northwest region of North America [was] characterized by alternating regimes of relatively warmer and cooler SST [sea surface temperature] in the North Pacific, punctuated by abrupt shifts in the mean background state," which were found to be "relatively common occurrences." They concluded that "regime shifts in the North Pacific have occurred 11 times since 1650." A significant aspect of these findings is the fact that the abrupt 1976-77 shift in this Pacific Decadal Oscillation, as it is generally called, is what was responsible for the vast majority of the past half-century's warming in Alaska, which some commentators wrongly point to as evidence of CO_2-induced global warming.

About the same time, Kasper and Allard (2001) examined soil deformations caused by ice wedges (a widespread and abundant form of ground ice in permafrost regions that can grow during colder periods and deform and crack the soil). Working near Salluit, northern Quebéc (approx. 62°N, 75.75°W), they found evidence of ice wedge activity prior to AD 140, reflecting cold climatic conditions. Between AD 140 and 1030, however, this activity decreased, reflective of warmer conditions. Then, from AD 1030 to 1500, conditions cooled; and from 1500 to 1900 ice wedge activity was at its peak, when the Little Ice Age ruled, suggesting this climatic interval exhibited the coldest conditions of the past 4,000 years. Thereafter, a warmer period prevailed, from about 1900 to 1946, which was followed by a return to cold conditions during the last five decades of the twentieth century, during which time more than 90 percent of the ice wedges studied reactivated and grew by 20-30 cm, in harmony with a reported temperature decline of 1.1°C observed at the meteorological station in Salluit.

In another study from the same year, Zeeberg and Forman (2001) analyzed twentieth century changes in glacier terminus positions on north Novaya Zemlya, a Russian island located between the Barents and Kara Seas in the Arctic Ocean, providing in the process a quantitative assessment of the effects of temperature and precipitation on glacial mass balance. This work revealed a significant and accelerated post-Little Ice Age glacial retreat in the first and second decades of the twentieth century; but by 1952, the region's glaciers had experienced between 75 to 100 percent of their net twentieth century retreat. During the next 50 years, the recession of more than half of the glaciers stopped, and many tidewater glaciers actually began to advance. These glacial stabilizations and advances were attributed by the two scientists to observed increases in precipitation and/or decreases in temperature. In the four decades since 1961, for example, weather stations at Novaya Zemlya show summer temperatures to have been 0.3° to 0.5°C colder than they were over the prior 40 years, while winter temperatures were 2.3° to 2.8°C colder than they were over the prior 40-year period. Such observations, in Zeeberg and Forman's words, are "counter to warming of the Eurasian Arctic predicted for the twenty-first century by climate models, particularly for the winter season."

Comiso *et al.* (2000) utilized satellite imagery to analyze and quantify a number of attributes of the Odden ice, including its average concentration, maximum area, and maximum extent over the period 1979-1998. They used surface air temperature data from Jan Mayen Island, located within the region of study, to infer the behavior of the phenomenon over the past 75 years.

The Odden ice tongue was found to vary in size, shape, and length of occurrence during the 20-year period, displaying a fair amount of interannual variability. Quantitatively, trend analyses revealed that the ice tongue had exhibited no statistically significant change in any of the parameters studied over the short 20-year period. However, a proxy reconstruction of the Odden ice tongue for the past 75 years revealed the ice phenomenon to have been "a relatively smaller feature several decades ago," due to the significantly warmer temperatures that prevailed at that time.

The fact that the Odden ice tongue has persisted, virtually unchanged in the mean during the past 20 years, is in direct contrast with predictions of rapid and increasing warmth in earth's polar regions as a result of CO_2-induced global warming. This observation, along with the observational evidence from Jan Mayen Island that temperatures there actually cooled at a rate of 0.15 ± 0.03°C per decade during the past 75 years, bolsters the view that there has been little to no warming in this part of the Arctic,

as well as most of its other parts, over the past seven decades.

Polyakov *et al.* (2002b) used newly available long-term Russian observations of surface air temperature from coastal stations to gain new insights into trends and variability in the Arctic environment poleward of 62°N. Throughout the 125-year history they developed, they identified "strong intrinsic variability, dominated by multi-decadal fluctuations with a timescale of 60-80 years"; they found temperature trends in the Arctic to be highly dependent on the particular time period selected for analysis. They found they could "identify periods when Arctic trends were actually smaller or *of different sign* [our italics] than Northern Hemisphere trends." Over the bulk of the twentieth century, when they say "multi-decadal variability had little net effect on computed trends," the temperature histories of the two regions were "similar," but they did "not support amplified warming in polar regions predicted by GCMs."

In a concomitant study, Naurzbaev *et al.* (2002) developed a 2,427-year proxy temperature history for the part of the Taimyr Peninsula, northern Russia, lying between 70°30' and 72°28' North latitude, based on a study of ring-widths of living and preserved larch trees, noting that it has been shown that "the main driver of tree-ring variability at the polar timber-line [where they worked] is temperature (Vaganov *et al.*, 1996; Briffa *et al.*, 1998; Schweingruber and Briffa, 1996)." This work revealed that "the warmest periods over the last two millennia in this region were clearly in the third [Roman Warm Period], tenth to twelfth [Medieval Warm Period] and during the twentieth [Current Warm Period] centuries." With respect to the second of these three periods, they emphasize that "the warmth of the two centuries AD 1058-1157 and 950-1049 attests to the reality of relative mediaeval warmth in this region." Their data also reveal three other important pieces of information: (1) the Roman and Medieval Warm Periods were both warmer than the Current Warm Period has been to date, (2) the beginning of the end of the Little Ice Age was somewhere in the vicinity of 1830, and (3) the Current Warm Period peaked somewhere in the vicinity of 1940.

All of these observations are at odds with what is portrayed in the Northern Hemispheric "hockey stick" temperature history of Mann *et al.* (1998, 1999) and its thousand-year global extension developed by Mann and Jones (2003), wherein (1) the Current

Warm Period is depicted as the warmest such era of the past two millennia, (2) recovery from the Little Ice Age does not begin until after 1910, and (3) the Current Warm Period experiences it highest temperatures in the latter part of the twentieth century's final decade.

Przybylak (2002) conducted a detailed analysis of intraseasonal and interannual variability in maximum, minimum, and average air temperature and diurnal air temperature range for the entire Arctic—as delineated by Treshnikov (1985)—for the period 1951-1990, based on data from 10 stations "representing the majority of the climatic regions in the Arctic." This work indicated that trends in both the intraseasonal and interannual variability of the temperatures studied did not show any significant changes, leading Przybylak to conclude that "this aspect of climate change, as well as trends in average seasonal and annual values of temperature investigated earlier (Przybylak, 1997, 2000), proves that, in the Arctic in the period 1951-90, no tangible manifestations of the greenhouse effect can be identified."

Isaksson et al. (2003) retrieved two ice cores (one from Lomonosovfonna and one from Austfonna) far above the Arctic Circle in Svalbard, Norway, after which the 12 cooperating scientists from Norway, Finland, Sweden, Canada, Japan, Estonia, and the Netherlands used $\delta^{18}O$ data to reconstruct a 600-year temperature history of the region. As would be expected—in light of the earth's transition from the Little Ice Age to the Current Warm Period—the international group of scientists reported that "the $\delta^{18}O$ data from both Lomonosovfonna and Austfonna ice cores suggest that the twentieth century was the warmest during at least the past 600 years." However, the warmest decade of the twentieth century was centered on approximately 1930, while the instrumental temperature record at Longyearbyen also shows the decade of the 1930s to have been the warmest. In addition, the authors remark that, "as on Svalbard, the 1930s were the warmest decade in the Trondheim record." Consequently, there was no net warming over the last seven decades of the twentieth century in the parts of Norway cited in this study.

In the same year, Polyakov et al. (2003) derived a surface air temperature history that stretched from 1875 to 2000, based on measurements carried out at 75 land stations and a number of drifting buoys located poleward of 62°N latitude. From 1875 to about 1917, the team of eight U.S. and Russian scientists found the surface air temperature of the huge northern region rose hardly at all; but then it

climbed 1.7°C in just 20 years to reach a peak in 1937 that was not eclipsed over the remainder of the record. During this 20-year period of rapidly rising air temperature, the atmosphere's CO_2 concentration rose by a mere 8 ppm. But then, over the next six decades, when the air's CO_2 concentration rose by approximately 55 ppm, or nearly seven times more than it did throughout the 20-year period of dramatic warming that preceded it, the surface air temperature of the region poleward of 62°N experienced no net warming and, in fact, may have cooled.

Briffa et al. (2004) reviewed several prior analyses of maximum latewood density data obtained from a widespread network of tree-ring chronologies that spanned three to six centuries and were derived from nearly 400 locations. For the land area of the globe poleward of 20°N latitude, they too found that the warmest period of the past six centuries occurred in the 1930s and early 1940s. Thereafter, the region's temperature dropped dramatically, although it did recover somewhat over the last two decades of the twentieth century. Nevertheless, its final value was still less than the mean value of the entire 1400s and portions of the 1500s.

Averaged across all land area poleward of 50°N latitude, there was a large divergence of reconstructed and instrumental temperatures subsequent to 1960, with measured temperatures rising and reconstructed temperatures falling, such that by the end of the record there was an approximate 1.5°C difference between them. Briffa et al. attempted to relate this large temperature differential to a hypothesized decrease in tree growth that was caused by a hypothesized increase in ultraviolet radiation that they hypothesized to have been caused by declining stratospheric ozone concentrations over this period. The results of their effort, however, proved "equivocal," as they themselves described it, leaving room for a growing urban heat island effect in the instrumental temperature record to be the principal cause of the disconcerting data divergence. The three researchers wrote that these unsettled questions prevented them "from claiming unprecedented hemispheric warming during recent decades on the basis of these tree-ring density data."

About the same time that Briffa et al. were struggling with this perplexing problem, Polyakov et al. (2004) were developing a long-term history of Atlantic Core Water Temperature (ACWT) in the Arctic Ocean using high-latitude hydrographic measurements that were initiated in the late nineteenth century, after which they compared the results of this exercise with the long-term history of Arctic Surface Air Temperature (SAT) developed by Polyakov *et al.* (2003). Their ACWT record, to quote them, revealed the existence of "two distinct warm periods from the late 1920s to 1950s and in the late 1980s-90s and two cold periods, one at the beginning of the record (until the 1920s) and another in the 1960s-70s." The SAT record depicted essentially the same thing, with the peak temperature of the latter warm period being not quite as high as the peak temperature of the former warm period. In the case of the ACWT record, however, this relationship was reversed, with the peak temperature of the latter warm period slightly exceeding the peak temperature of the former warm period. But the most recent temperature peak was very short-lived; and it rapidly declined to hover around a value that was approximately 1°C cooler over the last few years of the record.

In discussing their findings, Polyakov *et al.* say that, like Arctic SATs, Arctic ACWTs are dominated, in their words, "by multidecadal fluctuations with a time scale of 50-80 years." In addition, both records indicate that late twentieth century warmth was basically no different from that experienced in the late 1930s and early 1940s, a time when the air's CO_2 concentration was fully 65 ppm less than it is today.

Knudsen *et al.* (2004) documented climatic changes over the past 1,200 years via high-resolution multi-proxy studies of benthic and planktonic foraminiferal assemblages, stable isotopes, and ice-rafted debris found in three sediment cores retrieved from the North Icelandic shelf. These efforts resulted in their learning that "the time period between 1200 and around 7-800 cal. (years) BP, including the Medieval Warm Period, was characterized by relatively high bottom and surface water temperatures," after which "a general temperature decrease in the area marks the transition to ... the Little Ice Age." They also found that "minimum sea-surface temperatures were reached at around 350 cal. BP, when very cold conditions were indicated by several proxies." Thereafter, they report that "a modern warming of surface waters ... is not registered in the proxy data," and that "there is no clear indication of warming of water masses in the area during the last decades," even in sea surface temperatures measured over the period 1948-2002.

Raspopov *et al.* (2004) presented and analyzed two temperature-related datasets. The first was "a direct and systematic air temperature record for the Kola Peninsula, in the vicinity of Murmansk," which covered the period 1880-2000, while the second was

an "annual tree-ring series generalized for 10 regions (Lovelius, 1997) along the northern timberline, from the Kola Peninsula to Chukotka, for the period 1458-1975 in the longitude range from 30°E to 170°E," which included nearly all of northern Eurasia that borders the Arctic Ocean.

The researchers' primary objectives in this work were to identify any temporal cycles that might be present in the two datasets and to determine what caused them. With respect to this dual goal, they report discovering "climatic cycles with periods of around 90, 22-23 and 11-12 years," which were found to "correlate well with the corresponding solar activity cycles." Of even more interest, however, was what they learned about the temporal development of the Current Warm Period (CWP).

Raspopov *et al.*'s presentation of the mean annual tree-ring series for the northern Eurasia timberline clearly shows that the region's thermal recovery from the coldest temperatures of the Little Ice Age (LIA) may be considered to have commenced as early as 1820 and was in full swing by at least 1840. In addition, it shows that the rising temperature peaked just prior to 1950 and then declined to the end of the record in 1975. Thereafter, however, the Kola-Murmansk instrumental record indicates a significant temperature rise that peaked in the early 1990s at about the same level as the pre-1950 peak; but after that time, the temperature once again declined to the end of the record in 2000.

The latter of these findings (that there has been no net warming of this expansive high-latitude region over the last half of the twentieth century) is in harmony with the findings of the many studies reviewed above, while the former finding (that the thermal recovery of this climatically sensitive region of the planet began in the first half of the nineteenth century) is also supported by a number of other studies (Esper *et al.*, 2002; Moore *et al.*, 2002; Yoo and D'Odorico, 2002; Gonzalez-Rouco *et al.*, 2003; Jomelli and Pech, 2004), all of which demonstrate that the Little Ice Age-to-Current Warm Period transition began somewhere in the neighborhood of 1820 to 1850, well before the date (~1910) that is indicated in the Mann *et al.* (1998, 1999) "hockey stick" temperature history.

One further study from 2004 yields much the same conclusion, but arrives at it by very different means. Benner *et al.* (2004) set the stage for what they did by stating that "thawing of the permafrost which underlies a substantial fraction of the Arctic could accelerate carbon losses from soils (Goulden *et al.*, 1998)." In addition, they report that "freshwater discharge to the Arctic Ocean is expected to increase with increasing temperatures (Peterson *et al.*, 2002), potentially resulting in greater riverine export of terrigenous organic carbon to the ocean." And since the organic carbon in Arctic soils, in their words, "is typically old, with average radiocarbon ages ranging from centuries to millennia (Schell, 1983; Schirrmeister *et al.*, 2002)," they set about to measure the age of dissolved organic carbon (DOC) in Arctic rivers to see if there were any indications of increasing amounts of older carbon being transported to the ocean, which (if there were) would be indicative of enhanced regional warming.

Specifically, they sampled two of the largest Eurasian rivers, the Yenisey and Ob' (which drain vast areas of boreal forest and extensive peat bogs, accounting for about a third of all riverine DOC discharge to the Arctic Ocean), as well as two much smaller rivers on the north slope of Alaska, the Ikpikpuk and Kokolik, whose watersheds are dominated by Arctic tundra. In doing so, they found modern radiocarbon ages for all samples taken from all rivers, which indicates, in their words, that Arctic riverine DOC "is derived primarily from recently fixed plant litter and near-surface soil horizons." Thus, because warming should have caused the average radiocarbon age of the DOC of Arctic rivers to increase, the absence of aging implied by their findings provides strong evidence for the absence of recent large-scale warming there.

Laidre and Heide-Jorgensen (2005) published a most unusual paper, in that it dealt with the danger of oceanic cooling. Using a combination of long-term satellite tracking data, climate data, and remotely sensed sea ice concentrations to detect localized habitat trends of narwhals—a species of whale with a long spear-like tusk—in Baffin Bay between Greenland and Canada, home to the largest narwhal population in the world. They studied the species' vulnerability to recent and possible future climate trends. They found "since 1970, the climate in West Greenland has cooled, reflected in both oceanographic and biological conditions (Hanna and Cappelen, 2003)," with the result that "Baffin Bay and Davis Strait display strong significant increasing trends in ice concentrations and extent, as high as 7.5 percent per decade between 1979 and 1996, with comparable increases detected back to 1953 (Parkinson *et al.*, 1999; Deser *et al.*, 2000; Parkinson, 2000a,b; Parkinson and Cavalieri, 2002; Stern and Heide-Jorgensen, 2003)."

Humlum *et al.* (2005) noted that state-of-the-art climate models were predicting that "the effect of any present and future global climatic change will be amplified in the polar regions as a result of feedbacks in which variations in the extent of glaciers, snow, sea ice and permafrost, as well as atmospheric greenhouse gases, play key roles." However, they also said Polyakov *et al.* (2002a,b) had "presented updated observational trends and variations in Arctic climate and sea-ice cover during the twentieth century, which do not support the modeled polar amplification of surface air-temperature changes observed by surface stations at lower latitudes," and "there is reason, therefore, to evaluate climate dynamics and their respective impacts on high-latitude glaciers." They proceeded to do just that for the Archipelago of Svalbard, focusing on Spitsbergen (the Archipelago's main island) and the Longyearbreen glacier located in its relatively dry central region at 78°13'N latitude.

In reviewing what was already known about the region, Humlum *et al.* report that "a marked warming around 1920 changed the mean annual air temperature (MAAT) at sea level within only 5 years from about -9.5°C to -4.0°C," which change, in their words, "represents the most pronounced increase in MAAT documented anywhere in the world during the instrumental period." Then, they report that "from 1957 to 1968, MAAT dropped about 4°C, followed by a more gradual increase towards the end of the twentieth century."

With respect to the Longyearbreen glacier, their own work revealed that it had "increased in length from about 3 km to its present size of about 5 km during the last c. 1100 years," and they stated that "this example of late-Holocene glacier growth represents a widespread phenomenon in Svalbard and in adjoining Arctic regions," which they describe as a "development towards cooler conditions in the Arctic" that "may explain why the Little Ice Age glacier advance in Svalbard usually represents the Holocene maximum glacier extension."

As for what it all means, climate change in Svalbard over the twentieth century appears to have been a real rollercoaster ride, with temperatures rising more rapidly in the early 1920s than has been documented anywhere else before or since, only to be followed by a nearly equivalent temperature drop four decades later, both of which transitions were totally out of line with what climate models suggest should have occurred. In addition, the current location of the terminus of the Longyearbreen glacier suggests that, even now, Svalbard and "adjoining Arctic regions" are still experiencing some of the lowest temperatures of the entire Holocene, and at a time when atmospheric CO_2 concentrations are higher than they have been for millions of years.

In one final paper from 2005, Soon (2005) explores the question of what was the more dominant driver of twentieth century temperature change in the Arctic: the rising atmospheric CO_2 concentration or variations in solar irradiance. This he did by examining the roles the two variables may have played in forcing decadal, multi-decadal, and longer-term variations in surface air temperature (SAT). He performed a number of statistical analyses on (1) a composite Arctic-wide SAT record constructed by Polyakov et al. (2003), (2) global CO_2 concentrations taken from estimates made by the NASA GISS climate modeling group, and (3) a total solar irradiance (TSI) record developed by Hoyt and Schatten (1993, updated by Hoyt in 2005) over the period 1875-2000.

The results of Soon's analyses indicated a much stronger statistical relationship exists between SAT and TSI than between SAT and atmospheric CO_2 concentration. Solar forcing generally explained well over 75 percent of the variance in decadal-smoothed seasonal and annual Arctic temperatures, while CO_2 forcing explained only between 8 and 22 percent. Wavelet analysis further supported the case for solar forcing of SAT, revealing similar time-frequency characteristics for annual and seasonally averaged temperatures at decadal and multi-decadal time scales. By contrast, wavelet analysis gave little to no indication of a CO_2 forcing of Arctic SSTs. Based on these findings, it would appear that it is the sun, and not atmospheric CO_2, that has been driving temperature change in the Arctic over the twentieth century.

Hanna et al. (2006) developed a 119-year history of Icelandic Sea Surface Temperature (SST) based on measurements made at 10 coastal stations located between latitudes 63°24'N and 66°32'N. This work revealed the existence of past "long-term variations and trends that are broadly similar to Icelandic air temperature records: that is, generally cold conditions during the late nineteenth and early twentieth centuries; strong warming in the 1920s, with peak SSTs typically being attained around 1940; and cooling thereafter until the 1970s, followed once again by warming—but not generally back up to the level of the 1930s/1940s warm period."

Hansen et al. (2006) analyzed meteorological data from Arctic Station (69°15'N, 53°31'W) on Disko Island (West Greenland) for the period 1991-2004, after which their results were correlated, in the words of the researchers, "to the longest record available from Greenland at Ilulissat/Jakobshavn (since 1873)." Once this was done, marked changes were noted over the course of the study period, including "increasing mean annual air temperatures on the order of 0.4°C per year and 50% decrease in sea ice cover." In addition, due to "a high correlation between mean monthly air temperatures at the two stations (1991-2004)," Hansen et al. were able to place the air temperature trend observed at Disko "in a 130 years perspective." This exercise led them to conclude that the climate changes of the past decade were "dramatic," but that "similar changes in air temperatures [had] occurred previous[ly] within the last 130 years." More specifically, they report that the changes they observed over the last decade "are on the same order as changes [that] occurred between 1920 and 1930."

In Iceland, Bradwell et al. (2006) examined the link between late Holocene fluctuations of Lambatungnajokull (an outlet glacier of the Vatnajokull ice cap of southeast Iceland) and variations in climate, using geomorphological evidence to reconstruct patterns of past glacier fluctuations and lichenometry and tephrostratigraphy to date glacial landforms created by the glacier over the past four centuries. This work revealed "there is a particularly close correspondence between summer air temperature and the rate of ice-front recession of Lambatungnajokull during periods of overall retreat," and "between 1930 and 1950 this relationship is striking." They also report that "ice-front recession was greatest during the 1930s and 1940s, when retreat averaged 20 m per year." Thereafter, however, they say the retreat "slowed in the 1960s," and "there has been little overall retreat since the 1980s." The researchers also report that "the 20th-century record of reconstructed glacier-front fluctuations at Lambatungnajokull compares well with those of other similar-sized, non-surging, outlets of southern Vatnajokull," including Skaftafellsjokull, Fjallsjokull, Skalafellsjokull, and Flaajokull. They find "the pattern of glacier fluctuations of Lambatungnajokull over the past 200 years reflects the climatic changes that have occurred in southeast Iceland and the wider region."

Contemporaneously, Drinkwater (2006) decided "to provide a review of the changes to the marine ecosystems of the northern North Atlantic during the 1920s and 1930s and to discuss them in the light of contemporary ideas of regime shifts," where he defined regime shift as "a persistent radical shift in typical levels of abundance or productivity of multiple important components of the marine biological community structure, occurring at multiple trophic levels and on a geographical scale that is at least regional in extent." As a prologue to this effort, he first determined that "in the 1920s and 1930s, there was a dramatic warming of the air and ocean temperatures in the northern North Atlantic and the high Arctic, with the largest changes occurring north of 60°N," which warming "led to reduced ice cover in the Arctic and subarctic regions and higher sea temperatures," as well as northward shifts of multiple marine ecosystems. This change in climate occurred "during the 1920s, and especially after 1925," according to Drinkwater, when he reports that "average air temperatures began to rise rapidly and continued to do so through the 1930s," when "mean annual air temperatures increased by approximately 0.5-1°C and the cumulative sums of anomalies varied from 1.5 to 6°C between 1920 and 1940 with the higher values occurring in West Greenland and Iceland." Thereafter, as he describes it, "through the 1940s and 1950s air temperatures in the northernmost regions varied but generally remained relatively high," declining in the late 1960s in the northwest Atlantic and slightly earlier in the northeast Atlantic, which cooling has only recently begun to be reversed in certain parts of the region.

In the realm of biology, the early twentieth century warming of North Atlantic waters "contributed to higher primary and secondary production," in the words of Drinkwater, and "with the reduced extent of ice-covered waters, more open water allow[ed] for higher production than in the colder periods." As a result, cod "spread approximately 1200 km northward along West Greenland," and "migration of 'warmer water' species also changed with earlier arrivals and later departures." In addition, Drinkwater notes that "new spawning sites were observed farther north for several species or stocks while for others the relative contribution from northern spawning sites increased." Also, he writes that "some southern species of fish that were unknown in northern areas prior to the warming event became occasional, and in some cases, frequent visitors." Consequently, and considering all aspects of the event, Drinkwater states that "the warming in the 1920s and 1930s is considered to constitute the most significant regime shift experienced in the North Atlantic in the 20th century."

Groisman et al. (2006) reported using "a new Global Synoptic Data Network consisting of 2100 stations within the boundaries of the former Soviet Union created jointly by the [U.S.] National Climatic Data Center and Russian Institute for Hydrometeorological Information ... to assess the climatology of snow cover, frozen and unfrozen ground reports, and their temporal variability for the period from 1936 to 2004." They determined that "during the past 69 years (1936-2004 period), an increase in duration of the period with snow on the ground over Russia and the Russian polar region north of the Arctic circle has been documented by 5 days or 3% and 12 days or 5%, respectively," and they note this result "is in agreement with other findings."

In commenting on this development, plus the similar findings of others, the five researchers say "changes in snow cover extent during the 1936-2004 period cannot be linked with 'warming' (particularly with the Arctic warming)." Why? Because, as they continue, "in this particular period the Arctic warming was absent."

A recent essay that appeared in *Ambio: A Journal of the Human Environment*, by Karlén (2005) asks if temperatures in the Arctic are "really rising at an alarming rate," as some have claimed. His answer is a resounding no. Focusing on Svalbard Lufthavn (located at 78°N latitude), which he later shows to be representative of much of the Arctic, Karlén reports that "the Svalbard mean annual temperature increased rapidly from the 1910s to the late 1930s," that "the temperature thereafter became lower, and a minimum was reached around 1970," and that "Svalbard thereafter became warmer, but the mean temperature in the late 1990s was still slightly cooler than it was in the late 1930s," indicative of a cooling trend of 0.11°C per decade over the last seventy years of the twentieth century.

Karlén goes on to say "the observed warming during the 1930s is supported by data from several stations along the Arctic coasts and on islands in the Arctic, e.g. *Nordklim* data from Bjornoya and Jan Mayen in the north Atlantic, Vardo and Tromso in northern Norway, Sodankylaeand Karasjoki in northern Finland, and Stykkisholmur in Iceland," and "there is also [similar] data from other reports; e.g. Godthaab, Jakobshavn, and Egedesmindde in Greenland, Ostrov Dikson on the north coast of Siberia, Salehard in inland Siberia, and Nome in western Alaska." All of these stations, to quote him further, "indicate the same pattern of changes in annual mean temperature: a warm 1930s, a cooling until around 1970, and thereafter a warming, although the temperature remains slightly below the level of the late 1930s." In addition, he says "many stations with records starting later than the 1930s also indicate cooling, e.g. Vize in the Arctic Sea north of the Siberian coast and Frobisher Bay and Clyde on Baffin Island." Finally, Karlén reports that the 250-year temperature record of Stockholm "shows that the fluctuations of the 1900s are not unique," and that "changes of the same magnitude as in the 1900s occurred between 1770 and 1800, and distinct but smaller fluctuations occurred around 1825."

Karlén notes that "during the 50 years in which the atmospheric concentration of CO_2 has increased considerably, the temperature has decreased," which leads him to conclude that "the Arctic temperature data do not support the models predicting that there will be a critical future warming of the climate because of an increased concentration of CO_2 in the atmosphere." And this is especially important, in Karlén's words, because the model-based prediction "is that changes will be strongest and first noticeable in the Arctic."

Chylek *et al.* (2006) provides a more up-to-date report on average summer temperatures recorded at Ammassalik, on Greenland's southeast coast, and Godthab Nuuk on the island's southwestern coast, covering the period 1905 to 2005. They found "the 1955 to 2005 averages of the summer temperatures and the temperatures of the warmest month at both Godthab Nuuk and Ammassalik are significantly lower than the corresponding averages for the previous 50 years (1905-1955). The summers at both the southwestern and the southeastern coast of Greenland were significantly colder within the 1955-2005 period compared to the 1905-1955."

Chylek *et al.* also compared temperatures for the 10-year periods of 1920-1930 and 1995-2005. They found the average summer temperature for 2003 in Ammassalik was a record high since 1895, but "the years 2004 and 2005 were closer to normal being well below temperatures reached in the 1930s and 1940s." Similarly, the record from Godthab Nuuk showed that while temperatures there "were also increasing during the 1995-2005 period,they stayed generally below the values typical for the 1920-1940 period." The authors conclude that "reports of Greenland temperature changes are .... diverse suggesting a long term cooling and shorter warming periods."

Additional information on this topic, including reviews of newer publications as they become available, can be found at http://www.co2science.org/subject/a/arctictemptrends.php.

References

Benner, R., Benitez-Nelson, B., Kaiser, K. and Amon, R.M.W. 2004. Export of young terrigenous dissolved organic carbon from rivers to the Arctic Ocean. *Geophysical Research Letters* **31**: 10.1029/2003GL019251.

Bradwell, T., Dugmore, A.J. and Sugden, D.E. 2006. The Little Ice Age glacier maximum in Iceland and the North Atlantic Oscillation: evidence from Lambatungnajokull, southeast Iceland. *Boreas* **35**: 61-80.

Briffa, K.R., Osborn, T.J. and Schweingruber, F.H. 2004. Large-scale temperature inferences from tree rings: a review. *Global and Planetary Change* **40**: 11-26.

Briffa, K.R., Schweingruber, F.H., Jones, P.D., Osborn, T.J., Shiyatov, S.G. and Vaganov, E.A. 1998. Reduced sensitivity of recent tree-growth to temperature at high northern latitudes. *Nature* **391**: 678-682.

Chylek, P., Dubey, M.K, and Lesins, G. 2006. Greenland warming of 1920-1930 and 1995-2005. *Geophysical Research Letters* **33**: L11707.

Comiso, J.C., Wadhams, P., Pedersen, L.T. and Gersten, R.A. 2001. Seasonal and interannual variability of the Odden ice tongue and a study of environmental effects. *Journal of Geophysical Research* **106**: 9093-9116.

Deser, C., Walsh, J.E. and Timlin, M.S. 2000. Arctic sea ice variability in the context of recent atmospheric circulation trends. *Journal of Climatology* **13**: 617-633.

Drinkwater, K.F. 2006. The regime shift of the 1920s and 1930s in the North Atlantic. *Progress in Oceanography* **68**: 134-151.

Esper, J., Cook, E.R. and Schweingruber, F.H. 2002. Low-frequency signals in long tree-ring chronologies for reconstructing past temperature variability. *Science* **295**: 2250-2253.

Gedalof, Z. and Smith, D.J. 2001. Interdecadal climate variability and regime-scale shifts in Pacific North America. *Geophysical Research Letters* **28**: 1515-1518.

Gonzalez-Rouco, F., von Storch, H. and Zorita, E. 2003. Deep soil temperature as proxy for surface air-temperature in a coupled model simulation of the last thousand years. *Geophysical Research Letters* **30**: 10.1029/2003GL018264.

Goulden, M.L., Wofsy, S.C., Harden, J.W., Trumbore, S.E., Crill, P.M., Gower, S.T., Fries, T., Daube, B.C., Fan,

S., Sutton, D.J., Bazzaz, A. and Munger, J.W. 1998. Sensitivity of boreal forest carbon balance to soil thaw. *Science* **279**: 214-217.

Groisman, P.Ya., Knight, R.W., Razuvaev, V.N., Bulygina, O.N. and Karl, T.R. 2006. State of the ground: Climatology and changes during the past 69 years over northern Eurasia for a rarely used measure of snow cover and frozen land. *Journal of Climate* **19**: 4933-4955.

Hanna, E. and Cappelen, J. 2003. Recent cooling in coastal southern Greenland and relation with the North Atlantic Oscillation. *Geophysical Research Letters* **30**: 10.1029/2002GL015797.

Hanna, E., Jonsson, T., Olafsson, J. and Valdimarsson, H. 2006. Icelandic coastal sea surface temperature records constructed: Putting the pulse on air-sea-climate interactions in the Northern North Atlantic. Part I: Comparison with HadISST1 open-ocean surface temperatures and preliminary analysis of long-term patterns and anomalies of SSTs around Iceland. *Journal of Climate* **19**: 5652-5666.

Hansen, B.U., Elberling, B., Humlum, O. and Nielsen, N. 2006. Meteorological trends (1991-2004) at Arctic Station, Central West Greenland (69°15'N) in a 130 years perspective. *Geografisk Tidsskrift, Danish Journal of Geography* **106**: 45-55.

Hoyt, D.V. and Schatten, K.H. 1993. A discussion of plausible solar irradiance variations, 1700-1992. *Journal of Geophysical Research* **98**: 18,895-18,906.

Humlum, O., Elberling, B., Hormes, A., Fjordheim, K., Hansen, O.H. and Heinemeier, J. 2005. Late-Holocene glacier growth in Svalbard, documented by subglacial relict vegetation and living soil microbes. *The Holocene* **15**: 396-407.

Isaksson, E., Hermanson, M., Hicks, S., Igarashi, M., Kamiyama, K., Moore, J., Motoyama, H., Muir, D., Pohjola, V., Vaikmae, R., van de Wal, R.S.W. and Watanabe, O. 2003. Ice cores from Svalbard—useful archives of past climate and pollution history. *Physics and Chemistry of the Earth* **28**: 1217-1228.

Jomelli, V. and Pech, P. 2004. Effects of the Little Ice Age on avalanche boulder tongues in the French Alps (Massif des Ecrins). *Earth Surface Processes and Landforms* **29**: 553-564.

Karlén, W. 2005. Recent global warming: An artifact of a too-short temperature record? *Ambio* **34**: 263-264.

Kasper, J.N. and Allard, M. 2001. Late-Holocene climatic changes as detected by the growth and decay of ice wedges on the southern shore of Hudson Strait, northern Québec, Canada. *The Holocene* **11**: 563-577.

Knudsen, K.L., Eiriksson, J., Jansen, E., Jiang, H., Rytter, F. and Gudmundsdottir, E.R. 2004. Palaeoceanographic changes off North Iceland through the last 1200 years: foraminifera, stable isotopes, diatoms and ice rafted debris. *Quaternary Science Reviews* **23**: 2231-2246.

Laidre, K.L. and Heide-Jorgensen, M.P. 2005. Arctic sea ice trends and narwhal vulnerability. *Biological Conservation* **121**: 509-517.

Lovelius, N.V. 1997. *Dendroindication of Natural Processes*. World and Family 95. St. Petersburg, Russia.

Mann, M.E., Bradley, R.S. and Hughes, M.K. 1998. Global-scale temperature patterns and climate forcing over the past six centuries. *Nature* **392**: 779-787.

Mann, M.E., Bradley, R.S. and Hughes, M.K. 1999. Northern Hemisphere temperatures during the past millennium: Inferences, uncertainties, and limitations. *Geophysical Research Letters* **26**: 759-762.

Mann, M.E. and Jones, P.D. 2003. Global surface temperatures over the past two millennia. *Geophysical Research Letters* **30**: 10.1029/2003GL017814.

Moore, G.W.K., Holdsworth, G. and Alverson, K. 2002. Climate change in the North Pacific region over the past three centuries. *Nature* **420**: 401-403.

Moore, J.J., Hughen, K.A., Miller, G.H. and Overpeck, J.T. 2001. Little Ice Age recorded in summer temperature reconstruction from varved sediments of Donard Lake, Baffin Island, Canada. *Journal of Paleolimnology* **25**: 503-517.

Naurzbaev, M.M. and Vaganov, E.A. 2000. Variation of early summer and annual temperature in east Taymir and Putoran (Siberia) over the last two millennia inferred from tree rings. *Journal of Geophysical Research* **105**: 7317-7326.

Naurzbaev, M.M., Vaganov, E.A., Sidorova, O.V. and Schweingruber, F.H. 2002. Summer temperatures in eastern Taimyr inferred from a 2427-year late-Holocene tree-ring chronology and earlier floating series. *The Holocene* **12**: 727-736.

Overpeck, J., Hughen, K., Hardy, D., Bradley, R., Case, R., Douglas, M., Finney, B., Gajewski, K., Jacoby, G., Jennings, A., Lamoureux, S., Lasca, A., MacDonald, G., Moore, J., Retelle, M., Smith, S., Wolfe, A. and Zielinski, G. 1997. Arctic environmental change of the last four centuries. *Science* **278**: 1251-1256.

Parkinson, C.L. 2000a. Variability of Arctic sea ice: the view from space, and 18-year record. *Arctic* **53**: 341-358.

Parkinson, C.L. 2000b. Recent trend reversals in Arctic Sea ice extents: possible connections to the North Atlantic oscillation. *Polar Geography* **24**: 1-12.

Parkinson, C.L. and Cavalieri, D.J. 2002. A 21-year record of Arctic sea-ice extents and their regional, seasonal and monthly variability and trends. *Annals of Glaciology* **34**: 441-446.

Parkinson, C., Cavalieri, D., Gloersen, D., Zwally, J. and Comiso, J. 1999. Arctic sea ice extents, areas, and trends, 1978-1996. *Journal of Geophysical Research* **104**: 20,837-20,856.

Peterson, B.J., Holmes, R.M., McClelland, J.W., Vorosmarty, C.J., Lammers, R.B., Shiklomanov, A.I., Shiklomanov, I.A. and Rahmstorf, S. 2002. Increasing river discharge in the Arctic Ocean. *Science* **298**: 2171-2173.

Polyakov, I., Akasofu, S.-I., Bhatt, U., Colony, R., Ikeda, M., Makshtas, A., Swingley, C., Walsh, D. and Walsh, J. 2002a. Trends and variations in Arctic climate system. *EOS: Transactions, American Geophysical Union* **83**: 547-548.

Polyakov, I.V., Alekseev, G.V., Bekryaev, R.V., Bhatt, U., Colony, R.L., Johnson, M.A., Karklin, V.P., Makshtas, A.P., Walsh, D. and Yulin A.V. 2002b. Observationally based assessment of polar amplification of global warming. *Geophysical Research Letters* **29**: 10.1029/2001GL011111.

Polyakov, I.V., Alekseev, G.V., Timokhov, L.A., Bhatt, U.S., Colony, R.L., Simmons, H.L., Walsh, D., Walsh, J.E. and Zakharov, V.F. 2004. Variability of the intermediate Atlantic water of the Arctic Ocean over the last 100 years. *Journal of Climate* **17**: 4485-4497.

Polyakov, I.V., Bekryaev, R.V., Alekseev, G.V., Bhatt, U.S., Colony, R.L., Johnson, M.A., Maskshtas, A.P. and Walsh, D. 2003. Variability and trends of air temperature and pressure in the maritime Arctic, 1875-2000. *Journal of Climate* **16**: 2067-2077.

Przybylak, R. 1997. Spatial and temporal changes in extreme air temperatures in the Arctic over the period 1951-1990. *International Journal of Climatology* **17**: 615-634.

Przybylak, R. 2000. Temporal and spatial variation of surface air temperature over the period of instrumental observations in the Arctic. *International Journal of Climatology* **20**: 587-614.

Przybylak, R. 2002. Changes in seasonal and annual high-frequency air temperature variability in the Arctic from 1951-1990. *International Journal of Climatology* **22**: 1017-1032.

Raspopov, O.M., Dergachev, V.A. and Kolstrom, T. 2004. Periodicity of climate conditions and solar variability derived from dendrochronological and other palaeoclimatic data in high latitudes. *Palaeogeography, Palaeoclimatology, Palaeoecology* **209**: 127-139.

Schell, D.M. 1983. Carbon-13 and carbon-14 abundances in Alaskan aquatic organisms: Delayed production from peat in Arctic food webs. *Science* **219**: 1068-1071.

Schirrmeister, L., Siegert, C., Kuznetsova, T., Kuzmina, S., Andreev, A., Kienast, F., Meyer, H. and Bobrov, A. 2002. Paleoenvironmental and paleoclimatic records from permafrost deposits in the Arctic region of northern Siberia. *Quaternary International* **89**: 97-118.

Schweingruber, F.H. and Briffa, K.R. 1996. Tree-ring density network and climate reconstruction. In: Jones, P.D., Bradley, R.S. and Jouzel, J. (Eds.) *Climatic Variations and Forcing Mechanisms of the Last 2000 Years*, NATO ASI Series 141. Springer-Verlag, Berlin, Germany, pp. 43-66.

Soon, W. W.-H. 2005. Variable solar irradiance as a plausible agent for multidecadal variations in the Arctic-wide surface air temperature record of the past 130 years. *Geophysical Research Letters* **32** L16712, doi:10.1029/2005GL023429.

Stern, H.L. and Heide-Jorgensen, M.P. 2003. Trends and variability of sea ice in Baffin Bay and Davis Strait, 1953-2001. *Polar Research* **22**: 11-18.

Treshnikov, A.F. (Ed.) 1985. *Atlas Arktiki*. Glavnoye Upravlenye Geodeziy i Kartografiy, Moskva.

Vaganov, E.A., Briffa, K.R., Naurzbaev, M.M., Schweingruber, F.H., Shiyatov, S.G. and Shishov, V.V. 2000. Long-term climatic changes in the arctic region of the Northern Hemisphere. *Doklady Earth Sciences* **375**: 1314-1317.

Vaganov, E.A., Shiyatov, S.G. and Mazepa, V.S. 1996. *Dendroclimatic Study in Ural-Siberian Subarctic*. Nauka, Novosibirsk, Russia.

Yoo, J.C. and D'Odorico, P. 2002. Trends and fluctuations in the dates of ice break-up of lakes and rivers in Northern Europe: the effect of the North Atlantic Oscillation. Journal of Hydrology **268**: 100-112.

Zeeberg, J. and Forman, S.L. 2001. Changes in glacier extent on north Novaya Zemlya in the twentieth century. Holocene **11**: 161-175.

3.7. Antarctica

The study of Antarctic temperatures has provided valuable insight and spurred contentious debate on issues pertaining to global climate change. Key among the pertinent findings has been the observation of a large-scale correlation between proxy air temperature and atmospheric CO_2 measurements obtained from ice cores drilled in the interior of the continent. In the mid- to late-1980s, this broad correlation dominated much of the climate change debate. Many jumped on the global warming bandwagon, saying the correlation proved that changes in atmospheric CO_2 concentration caused changes in air temperature, and that future increases in the air's CO_2 content due to anthropogenic CO_2 emissions would therefore intensify global warming.

By the late 1990s and early 2000s, however, ice-coring instrumentation and techniques had improved considerably and newer studies with finer temporal resolution began to reveal that increases (decreases) in air temperature precede increases (decreases) in atmospheric CO_2 content, not vice versa (see Indermuhle et al. (2000), Monnin et al. (2001)). A recent study by Caillon et al. (2003), for example, demonstrated that during Glacial Termination III, "the CO_2 increase lagged Antarctic deglacial warming by 800 ± 200 years." This finding, in the authors' words, "confirms that CO_2 is not the forcing that initially drives the climatic system during a deglaciation."

A second major blow to the CO_2-induced global warming hypothesis comes from the contradiction between observed and model-predicted Antarctic temperature trends of the past several decades. According to nearly all climate models, CO_2-induced global warming should be most evident in earth's polar regions, but analyses of Antarctic near-surface and tropospheric air temperatures contradict this prediction.

Doran et al. (2002) examined temperature trends in the McMurdo Dry Valleys of Antarctica over the period 1986 to 2000, reporting a cooling rate of approximately 0.7°C per decade. This dramatic rate of cooling, they state, "reflects longer term continental Antarctic cooling between 1966 and 2000." In addition, the 14-year temperature decline in the dry valleys occurred in the summer and autumn, just as most of the 35-year cooling over the continent as a whole (which did not include any data from the dry valleys) also occurred in the summer and autumn.

Comiso (2000) assembled and analyzed Antarctic temperature data obtained from 21 surface stations and from infrared satellites operating since 1979. He found that for all of Antarctica, temperatures had declined by 0.08°C and 0.42°C per decade, respectively. Thompson and Solomon (2002) also report a cooling trend for the interior of Antarctica.

In spite of the decades-long cooling that has been observed for the continent as a whole, one region of Antarctica has actually bucked the mean trend and *warmed* over the same time period: the Antarctic Peninsula/Bellingshausen Sea region. But is the temperature increase that has occurred there evidence of CO_2-induced global warming?

According to Vaughan et al. (2001), "rapid regional warming" has led to the loss of seven ice shelves in this region during the past 50 years. However, they note that sediment cores from 6,000 to 1,900 years ago suggest the Prince Gustav Channel Ice Shelf—which collapsed in this region in 1995—"was absent and climate was as warm as it has been recently," when, of course there was much less CO_2 in the air.

Although it is tempting to cite the twentieth century increase in atmospheric CO_2 concentration as the cause of the recent regional warming, "to do so without offering a mechanism," say Vaughan et al., "is superficial." And so it is, as the recent work of Thompson and Solomon (2002) suggests that much of the warming can be explained by "a systematic bias toward the high-index polarity of the SAM," or Southern Hemispheric Annular Mode, such that the ring of westerly winds encircling Antarctica has recently been spending more time in its strong-wind phase.

That is also the conclusion of Kwok and Comiso (2002), who report that over the 17-year period 1982-1998, the SAM index shifted towards more positive values (0.22/decade), noting that a positive polarity of the SAM index "is associated with cold anomalies over most of Antarctica with the center of action over the East Antarctic plateau." At the same time, the SO index shifted in a negative direction, indicating "a drift toward a spatial pattern with warmer temperatures around the Antarctic Peninsula, and cooler temperatures over much of the continent." Together, the authors say the positive trend in the *coupled* mode of variability of these two indices (0.3/decade) represents a "significant bias toward positive polarity" that they describe as "remarkable."

Kwok and Comiso additionally report that "the tropospheric SH annular mode has been shown to be related to changes in the lower stratosphere (Thompson and Wallace, 2000)," noting that "the high index polarity of the SH annular mode is associated with the trend toward a cooling and strengthening of the SH stratospheric polar vortex during the stratosphere's relatively short active season in November," which is pretty much the same theory that has been put forth by Thompson and Solomon (2002).

In another slant on the issue, Yoon et al. (2002) report that "the maritime record on the Antarctic Peninsula shelf suggests close chronological correlation with Holocene glacial events in the Northern Hemisphere, indicating the possibility of coherent climate variability in the Holocene." In the same vein, Khim *et al.* (2002) say that "two of the most significant climatic events during the late Holocene are the Little Ice Age (LIA) and Medieval Warm Period (MWP), both of which occurred globally (Lamb, 1965; Grove, 1988)," noting further that "evidence of the LIA has been found in several studies of Antarctic marine sediments (Leventer and Dunbar, 1988; Leventer *et al.*, 1996; Domack *et al.*, 2000)." To this list of scientific journal articles documenting the existence of the LIA in Antarctica can now be added Khim *et al.*'s own paper, which also demonstrates the presence of the MWP in Antarctica, as well as earlier cold and warm periods of similar intensity and duration.

Further evidence that the Antarctic as a whole is in the midst of a cooling trend comes from Watkins and Simmonds (2000), who analyzed region-wide changes in sea ice. Reporting on trends in a number of Southern Ocean sea ice parameters over the period 1987 to 1996, they found statistically significant increases in sea ice area and total sea ice extent, as well as an increase in sea ice season length since the 1990s. Combining these results with those from a previous study revealed these trends to be consistent back to at least 1978. And in another study of Antarctic sea ice extent, Yuan and Martinson (2000) report that the net trend in the mean Antarctic ice edge over the past 18 years has been an equatorward expansion of 0.011 degree of latitude per year.

The temperature history of Antarctica provides no evidence for the CO_2-induced global warming hypothesis. In fact, it argues strongly against it. But what if the Antarctic *were* to warm as a result of some natural or anthropogenic-induced change in earth's climate? What would the consequences be?

For one thing, it would likely help to increase both the number and diversity of penguin species (Sun *et al.*, 2000; Smith *et al.*, 1999), and it would also tend to increase the size and number of populations of the continent's only two vascular plant species (Xiong *et al.*, 2000). With respect to the continent's great ice sheets, there would not be much of a problem either, as not even a warming event as dramatic as 10°C is predicted to result in a net change in the East Antarctic Ice Sheet (Näslund *et al.*, 2000), which suggests that predictions of catastrophic coastal flooding due to the melting of the world's polar ice sheets are way off the mark.

Additional information on this topic, including reviews of newer publications as they become available, can be found at http://www.co2science.org/subject/a/antarcticatemp.php.

References

Caillon, N., Severinghaus, J.P., Jouzel, J., Barnola, J.-M., Kang, J. and Lipenkov, V.Y. 2003. Timing of atmospheric CO_2 and Antarctic temperature changes across Termination III. *Science* **299**: 1728-1731.

Comiso, J.C. 2000. Variability and trends in Antarctic surface temperatures from *in situ* and satellite infrared measurements. *Journal of Climate* **13**: 1674-1696.

Domack, E.W., Leventer, A., Dunbar, R., Taylor, F., Brachfeld, S. and Sjunneskog, C. 2000. Chronology of the Palmer Deep site, Antarctic Peninsula: A Holocene palaeoenvironmental reference for the circum-Antarctic. *The Holocene* **11**: 1-9.

Doran, P.T., Priscu, J.C., Lyons, W.B., Walsh, J.E., Fountain, A.G., McKnight, D.M., Moorhead, D.L., Virginia, R.A., Wall, D.H., Clow, G.D., Fritsen, C.H., McKay, C.P. and Parsons, A.N. 2002. Antarctic climate cooling and terrestrial ecosystem response. *Nature* advance online publication, 13 January 2002 (DOI 10.1038/nature710).

Grove, J.M. 1988. *The Little Ice Age.* Cambridge University Press, Cambridge, UK.

Indermuhle, A., Monnin, E., Stauffer, B. and Stocker, T.F. 2000. Atmospheric CO_2 concentration from 60 to 20 kyr BP from the Taylor Dome ice core, Antarctica. *Geophysical Research Letters* **27**: 735-738.

Khim, B-K., Yoon, H.I., Kang, C.Y. and Bahk, J.J. 2002. Unstable climate oscillations during the Late Holocene in the Eastern Bransfield Basin, Antarctic Peninsula. *Quaternary Research* **58**: 234-245.

Kwok, R. and Comiso, J.C. 2002. Spatial patterns of variability in Antarctic surface temperature: Connections to the South Hemisphere Annular Mode and the Southern Oscillation. *Geophysical Research Letters* **29**: 10.1029/2002GL015415.

Lamb, H.H. 1965. The early medieval warm epoch and its sequel. *Palaeogeography, Palaeoclimatology, Palaeoecology* **1**: 13-37.

Leventer, A. and Dunbar, R.B. 1988. Recent diatom record of McMurdo Sound, Antarctica: Implications for the history of sea-ice extent. *Paleoceanography* **3**: 373-386.

Leventer, A., Domack, E.W., Ishman, S.E., Brachfeld, S., McClennen, C.E. and Manley, P. 1996. Productivity cycles of 200-300 years in the Antarctic Peninsula region: Understanding linkage among the sun, atmosphere, oceans, sea ice, and biota. *Geological Society of America Bulletin* **108**: 1626-1644.

Monnin, E., Indermühle, A., Dällenbach, A., Flückiger, J., Stauffer, B., Stocker, T.F., Raynaud, D. and Barnola, J.-M. 2001. Atmospheric CO_2 concentrations over the last glacial termination. *Nature* **291**: 112-114.

Näslund, J.O., Fastook, J.L and Holmlund, P. 2000. Numerical modeling of the ice sheet in western Dronning Maud Land, East Antarctica: impacts of present, past and future climates. *Journal of Glaciology* **46**: 54-66.

Smith, R.C., Ainley, D., Baker, K., Domack, E., Emslie, S., Fraser, B., Kennett, J., Leventer, A., Mosley-Thompson, E., Stammerjohn, S. and Vernet M. 1999. Marine ecosystem sensitivity to climate change. *BioScience* **49**: 393-404.

Sun, L., Xie, Z. and Zhao, J. 2000. A 3,000-year record of penguin populations. *Nature* **407**: 858.

Thompson, D.W.J. and Solomon, S. 2002. Interpretation of recent Southern Hemisphere climate change. *Science* **296**: 895-899.

Thompson, D.W.J. and Wallace, J.M. 2000. Annular modes in extratropical circulation, Part II: Trends. *Journal of Climate* **13**: 1018-1036.

Vaughan, D.G., Marshall, G.J., Connolley, W.M., King, J.C. and Mulvaney, R. 2001. Devil in the detail. *Science* **293**: 177-179

Watkins, A.B. and Simmonds, I. 2000. Current trends in Antarctic sea ice: The 1990s impact on a short climatology. *Journal of Climate* **13**: 4441-4451.

Xiong, F.S., Meuller, E.C. and Day, T.A. 2000. Photosynthetic and respiratory acclimation and growth response of Antarctic vascular plants to contrasting temperature regimes. *American Journal of Botany* **87**: 700-710.

Yoon, H.I., Park, B.-K., Kim, Y. and Kang, C.Y. 2002. Glaciomarine sedimentation and its paleoclimatic implications on the Antarctic Peninsula shelf over the last 15,000 years. *Palaeogeography, Palaeoclimatology, Palaeoecology* **185**: 235-254.

Yuan, X. and Martinson, D.G. 2000. Antarctic sea ice extent variability and its global connectivity. *Journal of Climate* **13**: 1697-1717.

4

Observations: Glaciers, Sea Ice, Precipitation, and Sea Level

Introduction

The Intergovernmental Panel on Climate Change (IPCC) alleges that "recent decreases in ice mass are correlated with rising surface air temperatures," and more specifically that "the late 20th-century glacier wastage likely has been a response to post-1970 warming. Strongest mass losses per unit area have been observed in Patagonia, Alaska and northwest USA and southwest Canada. Because of the corresponding large areas, the largest contributions to sea level rise came from Alaska, the Arctic and the Asian high mountains. Taken together, the ice sheets in Greenland and Antarctica have *very likely* been contributing to sea level rise over 1993 to 2003 [italics in the original]" (IPCC, 2007-I, p. 339).

It should be obvious, but apparently is not, that such facts as melting glaciers and disappearing Arctic sea ice, while interesting, are entirely irrelevant to illuminating the causes of warming. Any significant warming, whether anthropogenic or natural, will melt ice—often quite slowly. Therefore, claims that anthropogenic global warming (AGW) is occurring that are backed by such accounts are simply confusing the consequences of warming with the causes—a common logical error. In addition, fluctuations of glacier mass, sea ice, precipitation, and sea level depend on many factors other than temperature and are poor measuring devices for global warming.

This chapter summarizes the extensive scientific literature on glaciers, sea ice, precipitation, and sea level rise that frequently contradicts and rarely reinforces the IPCC's claims quoted above. Glaciers around the world are continuously advancing and retreating, with no evidence of a trend that can be linked to CO_2 concentrations in the air. The same is largely true of sea ice, precipitation patterns, and sea levels: all fluctuate in response to processes that are unrelated to CO_2, and therefore cannot be taken either as signs of anthropogenic global warming or of climate disasters that may be yet to come.

References

IPCC. 2007-I. *Climate Change 2007: The Physical Science Basis. Contribution of Working Group I to the Fourth Assessment Report of the Intergovernmental Panel on Climate Change.* Solomon, S., Qin, D., Manning, M., Chen, Z., Marquis, M., Averyt, K.B., Tignor, M. and Miller, H.L. (Eds.) Cambridge University Press, Cambridge, UK.

4.1. Glaciers

Model studies indicate that CO_2-induced global warming will result in significant melting of earth's glaciers, contributing to a rise in global sea level. In this section, we examine global trends and data from

Africa, Antarctica, the Arctic, Europe, North America, and South America. Additional information on this topic, including reviews of glaciers not discussed here, can be found at http://www. co2science.org/subject/g/subject_g.php under the heading Glaciers.

4.1.1. Global

The full story must begin with a recognition of just how few glacier data exist. Of the 160,000 glaciers presently known to exist, only 67,000 (42 percent) have been inventoried to any degree (Kieffer *et al.*, 2000). Mass balance data (which would be positive for growth, negative for shrinkage) exist for more than a single year for only slightly more than 200 (Braithwaite and Zhang, 2000). When the length of record increases to *five* years, this number drops to 115; and if both winter and summer mass balances are required, the number drops to 79. Furthermore, if *10* years of record is used as a cutoff, only 42 glaciers qualify. This lack of glacial data, in the words of Braithwaite and Zhang, highlights "one of the most important problems for mass-balance glaciology" and demonstrates the "sad fact that many glacierized regions of the world remain unsampled, or only poorly sampled," suggesting we really know very little about the true state of most of the world's glaciers.

During the fifteenth through nineteenth centuries, widespread and major glacier advances occurred during a period of colder global temperature known as the Little Ice Age (Broecker, 2001; Grove, 2001). Many records indicate widespread glacial retreat as temperatures began to rise in the mid- to late-1800s and many glaciers returned to positions characteristic of pre-Little Ice Age times. In many instances the *rate* of glacier retreat has not increased over the past 70 years, during a time when the atmosphere experienced the bulk of the increase in its CO_2 content.

In an analysis of Arctic glacier mass balance, Dowdeswell *et al.* (1997) found that of the 18 glaciers with the longest mass balance histories, just over 80 percent displayed negative mass balances over their periods of record. Yet they additionally report that "almost 80% of the mass balance time series also have a positive trend, toward a less negative mass balance." Although these Arctic glaciers continue to lose mass, as they have probably done since the end of the Little Ice Age, they are losing smaller amounts

each year, which is hardly what one would expect in the face of what some incorrectly call the "unprecedented" warming of the latter part of the twentieth century.

Similar results have been reported by Braithwaite (2002), who reviewed and analyzed mass balance measurements of 246 glaciers from around the world that were made between 1946 and 1995. According to Braithwaite, "there are several regions with highly negative mass balances in agreement with a public perception of 'the glaciers are melting,' but there are also regions with positive balances." Within Europe, for example, he notes that "Alpine glaciers are generally shrinking, Scandinavian glaciers are growing, and glaciers in the Caucasus are close to equilibrium for 1980-95." And when results for the whole world are combined for this most recent period of time, Braithwaite notes that "there is no obvious common or global trend of increasing glacier melt in recent years."

As for the glacier with the longest mass balance record of all, the Storglaciaren in northern Sweden, for the first 15 years of its 50-year record it exhibited a negative mass balance of little trend. Thereafter, however, its mass balance began to trend upward, actually becoming positive over about the last decade (Braithwaite and Zhang, 2000).

Global data on glaciers do not support claims made by the IPCC that most claciers are retreating or melting. Additional information on this topic, including reviews of newer publications as they become available, can be found at http://www.co2science.org/subject/g/glaciers.php.

References

Braithwaite, R.J. 2002. Glacier mass balance: the first 50 years of international monitoring. *Progress in Physical Geography* **26**: 76-95.

Braithwaite, R.J. and Zhang, Y. 2000. Relationships between interannual variability of glacier mass balance and climate. *Journal of Glaciology* **45**: 456-462.

Broecker, W.S. 2001. Glaciers That Speak in Tongues and other tales of global warming. *Natural History* **110** (8): 60-69.

Dowdeswell, J.A., Hagen, J.O., Bjornsson, H., Glazovsky, A.F., Harrison, W.D., Holmlund, P., Jania, J., Koerner, R.M., Lefauconnier, B., Ommanney, C.S.L. and Thomas, R.H. 1997. The mass balance of circum-Arctic glaciers and recent climate change. *Quaternary Research* **48**: 1-14.

Grove, J.M. 2001. The initiation of the "Little Ice Age" in regions round the North Atlantic. *Climatic Change* **48**: 53-82.

Kieffer, H., Kargel, J.S., Barry, R., Bindschadler, R., Bishop, M., MacKinnon, D., Ohmura, A., Raup, B., Antoninetti, M., Bamber, J., Braun, M., Brown, I., Cohen, D., Copland, L., DueHagen, J., Engeset, R.V., Fitzharris, B., Fujita, K., Haeberli, W., Hagen, J.O., Hall, D., Hoelzle, M., Johansson, M., Kaab, A., Koenig, M., Konovalov, V., Maisch, M., Paul, F., Rau, F., Reeh, N., Rignot, E., Rivera, A., Ruyter de Wildt, M., Scambos, T., Schaper, J., Scharfen, G., Shroder, J., Solomina, O., Thompson, D., Van der Veen, K., Wohlleben, T. and Young, N. 2000. New eyes in the sky measure glaciers and ice sheets. *EOS: Transactions, American Geophysical Union* **81**: 265, 270-271.

4.1.2. Africa

On the floor of the U.S. Senate in 2004, Arizona Senator John McCain described his affection for the writings of Ernest Hemingway, especially his famous short story, "The Snows of Kilimanjaro." Then, showing photos of the magnificent landmark taken in 1993 and 2000, he attributed the decline of glacial ice atop the mount during the intervening years to CO_2-induced global warming, calling this attribution a fact "that cannot be refuted by any scientist."

New York Senator Hillary Clinton echoed Senator McCain's sentiments. Displaying a second set of photos taken from the same vantage point in 1970 and 1999—the first depicting "a 20-foot-high glacier" and the second "only a trace of ice"—she said that in those pictures "we have evidence in the most dramatic way possible of the effects of 29 years of global warming." In spite of the absolute certitude with which the two senators expressed their views on the subject, which allowed for no "wiggle room" whatsoever, both of them were wrong.

Modern glacier recession on Kilimanjaro began around 1880, approximately the same time the planet began to recover from the several-hundred-year cold spell of the Little Ice Age. As a result, a number of people, including the aforementioned senators, declared that the ice fields retreated *because* of the rising temperatures, encouraged in this contention by a few reports in the scientific literature (Alverson *et al.*, 2001; Irion, 2001; Thompson *et al.*, 2002). This view of the subject, however, is "highly simplified," in the words of a trio of glaciologists (Molg *et al.*, 2003b), who noted that "glacierization in East Africa is limited to three massifs close to the equator:

Kilimanjaro (Tanzania, Kenya), Mount Kenya (Kenya), and Rwenzori (Zaire, Uganda)." All three sites experienced strong ice field recession over the past century or more. In that part of the world, however, they report "there is no evidence of a sudden change in temperature at the end of the 19th century (Hastenrath, 2001)," and that "East African long-term temperature records of the twentieth century show diverse trends and do not exhibit a uniform warming signal (King'uyu *et al.*, 2000; Hay *et al.*, 2002)." With respect to Kilimanjaro, they say "since February 2000 an automatic weather station has operated on a horizontal glacier surface at the summit's Northern Icefield," and "monthly mean air temperatures only vary slightly around the annual mean of -7.1°C, and air temperatures [measured by ventilated sensors, e.g., Georges and Kaser (2002)] never rise above the freezing point," which makes it pretty difficult to understand how ice could *melt* under such conditions.

So what caused the ice fields of Kilimanjaro to recede so steadily for so many years? Citing "historical accounts of lake levels (Hastenrath, 1984; Nicholson and Yin, 2001), wind and current observations in the Indian Ocean and their relationship to East African rainfall (Hastenrath, 2001), water balance models of lakes (Nicholson and Yin, 2001), and paleolimnological data (Verschuren *et al.*, 2000)," Molg *et al.* say "all data indicate that modern East African climate experienced an abrupt and marked drop in air humidity around 1880," and they add that the resultant "strong reduction in precipitation at the end of the 19th century is the main reason for modern glacier recession in East Africa," as it considerably reduces glacier mass balance accumulation, as has been demonstrated for the region by Kruss (1983) and Hastenrath (1984). In addition, they note that "increased incoming shortwave radiation due to decreases in cloudiness—both effects of the drier climatic conditions—plays a decisive role for glacier retreat by increasing ablation, as demonstrated for Mount Kenya and Rwenzori (Kruss and Hastenrath, 1987; Molg *et al.*, 2003a)."

In further investigating this phenomenon, Molg *et al.* applied a radiation model to an idealized representation of the 1880 ice cap of Kilimanjaro, calculating the spatial extent and geometry of the ice cap for a number of subsequent points in time and finding that "the basic evolution in spatial distribution of ice bodies on the summit is modeled well." The model they used, which specifically addresses the unique configuration of the summit's vertical ice walls, provided "a clear indication that solar radiation is the main climatic parameter governing and maintaining ice retreat on the mountain's summit plateau in the drier climate since ca. 1880." Consequently, Molg et al. concluded that "modern glacier retreat on Kilimanjaro is much more complex than simply attributable to 'global warming only'." Indeed, they say it is "a process driven by a complex combination of changes in several different climatic parameters [e.g., Kruss, 1983; Kruss and Hastenrath, 1987; Hastenrath and Kruss, 1992; Kaser and Georges, 1997; Wagnon et al., 2001; Kaser and Osmaston, 2002; Francou et al., 2003; Molg et al., 2003b], with humidity-related variables dominating this combination."

Kaser et al. (2004) similarly concluded that "changes in air humidity and atmospheric moisture content (e.g. Soden and Schroeder, 2000) seem to play an underestimated key role in tropical high-mountain climate (Broecker, 1997)." Noting that all glaciers in equatorial East Africa exhibited strong recession trends over the past century, they report that "the dominant reasons for this strong recession in modern times are reduced precipitation (Kruss, 1983; Hastenrath, 1984; Kruss and Hastenrath, 1987; Kaser and Noggler, 1996) and increased availability of shortwave radiation due to decreases in cloudiness (Kruss and Hastenrath, 1987; Molg et al., 2003b)," both of which phenomena they relate to a dramatic drying of the regional atmosphere that occurred around 1880 and the ensuing dry climate that subsequently prevailed throughout the twentieth century. Kaser et al. conclude that all relevant "observations and facts" clearly indicate that "climatological processes other than air temperature control the ice recession in a direct manner" on Kilimanjaro, and that "positive air temperatures have not contributed to the recession process on the summit," directly contradicting Irion (2002) and Thompson et al. (2002), who, in their words, see the recession of Kilimanjaro's glaciers as "a direct consequence solely of increased air temperature."

In a subsequent study of the ice fields of Kilimanjaro, Molg and Hardy (2004) derived an energy balance for the horizontal surface of the glacier that comprises the northern ice field of Kibo—the only one of the East African massif's three peaks that is presently glaciated—based on data obtained from an automated weather station. This work revealed, in their words, that "the main energy exchange at the glacier-atmosphere interface results from the terms accounting for net radiation, governed by the variation in net shortwave radiation," which is controlled by surface albedo and, thus, precipitation variability, which determines the reflective characteristics of the glacier's surface. Much less significant, according to the two researchers, is the temperature-driven turbulent exchange of sensible heat, which they say "remains considerably smaller and of little importance."

Molg and Hardy conclude that "modern glacier retreat on Kilimanjaro and in East Africa in general [was] initiated by a drastic reduction in precipitation at the end of the nineteenth century (Hastenrath, 1984, 2001; Kaser et al., 2004)," and that reduced accumulation and increased ablation have "maintained the retreat until the present (Molg et al., 2003b)." Buttressing their findings is the fact, as they report it, that "detailed analyses of glacier retreat in the global tropics uniformly reveal that changes in climate variables related to air humidity prevail in controlling the modern retreat [e.g., Kaser and Georges (1997) for the Peruvian Cordillera Blanca and Francou et al. (2003) for the Bolivian Cordillera Real (both South American Andes); Kruss (1983), Kruss and Hastenrath (1987), and Hastenrath (1995) for Mount Kenya (East Africa); and Molg et al. (2003a) for the Rwenzori massif (East Africa)]." The take-home message of their study is essentially the same as that of Kaser et al. (2004): "Positive air temperatures have not contributed to the recession process on the summit."

Two years later, Cullen et al. (2006) report that "all ice bodies on Kilimanjaro have retreated drastically between 1912-2003," but they add that the highest glacial recession rates on Kilimanjaro "occurred in the first part of the twentieth century, with the most recent retreat rates (1989-2003) smaller than in any other interval." In addition, they say no temperature trends over the period 1948-2005 have been observed at the approximate height of the Kilimanjaro glaciers, but that there has been a small decrease in the region's specific humidity over this period.

In terms of why glacier retreat on Kilimanjaro was so dramatic over the twentieth century, the six researchers note that for the mountain's plateau glaciers, there is no alternative for them "other than to continuously retreat once their vertical margins are exposed to solar radiation," which appears to have happened sometime in the latter part of the nineteenth century. They also say, in this regard, that the "vertical wall retreat that governs the retreat of plateau glaciers is irreversible, and changes in twentieth century climate have not altered their continuous demise." Consequently, the twentieth century retreat of Kilimanjaro's plateau glaciers is a long-term response to what we could call "relict climate change" that likely occurred in the late nineteenth century.

In the case of the mountain's slope glaciers, Cullen *et al.* say that their rapid recession in the first part of the twentieth century shows they "were drastically out of equilibrium," which they take as evidence that the glaciers "were responding to a large prior shift in climate." In addition, they report that "no footprint of multidecadal changes in areal extent of slope glaciers to fluctuations in twentieth century climate is observed, but their ongoing demise does suggest they are still out of equilibrium," and in this regard they add that their continuing but decelerating demise could be helped along by the continuous slow decline in the air's specific humidity. Consequently, and in light of all the facts they present and the analyses they and others have conducted over many years, Cullen *et al.* confidently conclude that the glaciers of Kilimanjaro "are merely remnants of a past climate rather than sensitive indicators of 20th century climate change."

Two more recent studies, Mote and Kaser (2007) and Duane *et al.* (2008) additionally reject the temperature-induced decline hypothesis for Kilimanjaro, with Duane *et al.* concluding that "the reasons for the rapid decline in Kilimanjaro's glaciers are not primarily due to increased air temperatures, but a lack of precipitation," and Mote and Kaser reporting that "warming fails spectacularly to explain the behavior of the glaciers and plateau ice on Africa's Kilimanjaro massif ... and to a lesser extent other tropical glaciers."

Clearly, the misguided rushes to judgment that have elevated Kilimanjaro's predicted demise by CO_2-induced global warming to iconic status should give everyone pause to more carefully evaluate the evidence, or lack thereof, for many similar claims related to the ongoing rise in the air's CO_2 content.

Additional information on this topic, including reviews of newer publications as they become available, can be found at http://www.co2science.org/subject/a/africagla.php.

References

Alverson, K., Bradley, R., Briffa, K., Cole, J., Hughes, M., Larocque, I., Pedersen, T., Thompson, L.G. and Tudhope, S. 2001. A global paleoclimate observing system. *Science* **293**: 47-49.

Broecker, W.S. 1997. Mountain glaciers: records of atmospheric water vapor content? *Global Biogeochemical Cycles* **4**: 589-597.

Cullen, N.J., Molg, T., Kaser, G., Hussein, K., Steffen, K. and Hardy, D.R. 2006. Kilimanjaro glaciers: Recent areal extent from satellite data and new interpretation of observed 20th century retreat rates. *Geophysical Research Letters* **33**: 10.1029/2006GL027084.

Duane, W.J., Pepin, N.C., Losleben, M.L. and Hardy, D.R. 2008. General characteristics of temperature and humidity variability on Kilimanjaro, Tanzania. *Arctic, Antarctic, and Alpine Research* **40**: 323-334.

Francou, B., Vuille, M., Wagnon, P., Mendoza, J. and Sicart, J.E. 2003. Tropical climate change recorded by a glacier in the central Andes during the last decades of the 20th century: Chacaltaya, Bolivia, 16°S. *Journal of Geophysical Research* **108**: 10.1029/2002JD002473.

Georges, C. and Kaser, G. 2002. Ventilated and unventilated air temperature measurements for glacier-climate studies on a tropical high mountain site. *Journal of Geophysical Research* **107**: 10.1029/2002JD002503.

Hastenrath, S. 1984. *The Glaciers of Equatorial East Africa*. D. Reidel, Norwell, MA, USA.

Hastenrath, S. 1995. Glacier recession on Mount Kenya in the context of the global tropics. *Bulletin de l'Institut français d'études andines* **24**: 633-638.

Hastenrath, S. 2001. Variations of East African climate during the past two centuries. *Climatic Change* **50**: 209-217.

Hastenrath, S. and Kruss, P.D. 1992. The dramatic retreat of Mount Kenya's glaciers between 1963 and 1987: Greenhouse forcing. *Annals of Glaciology* **16**: 127-133.

Hay, S.I., Cox, J., Rogers, D.J., Randolph, S.E., Stern, D.I., Shanks, G.D., Myers, M.F. and Snow, R.W. 2002. Climate change and the resurgence of malaria in the East African highlands. *Nature* **415**: 905-909.

Irion, R. 2001. The melting snows of Kilimanjaro. *Science* **291**: 1690-1691.

Kaser, G. and Georges, C. 1997. Changes in the equilibrium line altitude in the tropical Cordillera Blanca (Peru) between 1930 and 1950 and their spatial variations. *Annals of Glaciology* **24**: 344-349.

Kaser, G., Hardy, D.R., Molg, T., Bradley, R.S. and Hyera, T.M. 2004. Modern glacier retreat on Kilimanjaro as evidence of climate change: Observations and facts. *International Journal of Climatology* **24**: 329-339.

Kaser, G. and Noggler, B. 1996. Glacier fluctuations in the Rwenzori Range (East Africa) during the 20th century—a preliminary report. *Zeitschrift fur Gletscherkunde and Glazialgeologie* **32**: 109-117.

Kaser, G. and Osmaston, H. 2002. *Tropical Glaciers*. Cambridge University Press, Cambridge, UK.

King'uyu, S.M., Ogallo, L.A. and Anyamba, E.K. 2000. Recent trends of minimum and maximum surface temperatures over Eastern Africa. *Journal of Climate* **13**: 2876-2886.

Kruss, P.D. 1983. Climate change in East Africa: A numerical simulation from the 100 years of terminus record at Lewis Glacier, Mount Kenya. *Zeitschrift fur Gletscherkunde and Glazialgeologie* **19**: 43-60.

Kruss, P.D. and Hastenrath, S. 1987. The role of radiation geometry in the climate response of Mount Kenya's glaciers, part 1: Horizontal reference surfaces. *International Journal of Climatology* **7**: 493-505.

Molg, T., Georges, C. and Kaser, G. 2003a. The contribution of increased incoming shortwave radiation to the retreat of the Rwenzori Glaciers, East Africa, during the 20th century. *International Journal of Climatology* **23**: 291-303.

Molg, T. and Hardy, D.R. 2004. Ablation and associated energy balance of a horizontal glacier surface on Kilimanjaro. *Journal of Geophysical Research* **109**: 10.1029/2003JD004338.

Molg, T., Hardy, D.R. and Kaser, G. 2003b. Solar-radiation-maintained glacier recession on Kilimanjaro drawn from combined ice-radiation geometry modeling. *Journal of Geophysical Research* **108**: 10.1029/2003JD003546.

Mote, P.W. and Kaser, G. 2007. The shrinking glaciers of Kilimanjaro: Can global warming be blamed? *American Scientist* **95**: 318-325.

Nicholson, S.E. and Yin, X. 2001. Rainfall conditions in Equatorial East Africa during the nineteenth century as inferred from the record of Lake Victoria. *Climatic Change* **48**: 387-398.

Soden, B.J. and Schroeder, S.R. 2000. Decadal variations in tropical water vapor: a comparison of observations and a model simulation. *Journal of Climate* **13**: 3337-3341.

Thompson, L.G., Mosley-Thompson, E., Davis, M.E., Henderson, K.A., Brecher, H.H., Zagorodnov, V.S., Mashiotta, T.A., Lin, P.-N., Mikhalenko, V.N., Hardy, D.R. and Beer, J. 2002. Kilimanjaro ice core records: Evidence of Holocene climate change in tropical Africa. *Science* **298**: 589-593.

Verschuren, D., Laird, K.R. and Cumming, B.F. 2000. Rainfall and drought in equatorial east Africa during the past 1,100 years. *Nature* **403**: 410-414.

Wagnon, P., Ribstein, P., Francou, B. and Sicart, J.E. 2001. Anomalous heat and mass budget of Glaciar Zongo, Bolivia, during the 1997/98 El Niño year. *Journal of Glaciology* **47**: 21-28.

4.1.3. Antarctica

In early November 2001, a large iceberg separated from West Antarctica's Pine Island Glacier. This event was of great interest to scientists because the Pine Island Glacier is currently the fastest-moving glacier in Antarctica and the continent's largest discharger of ice. Some speculate this event could herald the "beginning of the end" of the West Antarctic Ice Sheet. Scientific studies, however, suggest otherwise.

Rignot (1998) employed satellite radar measurements of the grounding line of Pine Island Glacier from 1992 to 1996 to determine whether it was advancing or retreating. The data indicated a retreat rate of 1.2 ± 0.3 kilometers per year over the four-year period of the study. Because the study period was so short, Rignot says the questions the study raises concerning the long-term stability of the West Antarctic Ice Sheet "cannot be answered at present."

In a subsequent study, Stenoien and Bentley (2000) mapped the catchment region of Pine Island Glacier using radar altimetry and synthetic aperture radar interferometry, after which they used the data to develop a velocity map that revealed a system of tributaries that channel ice from the catchment area into the fast-flowing glacier. By combining these velocity data with information on ice thickness and snow accumulation rates, they were able to calculate an approximate mass balance for the glacier within an uncertainty of approximately 30 percent. Their results suggested the mass balance of the catchment region was not significantly different from zero.

Shepherd *et al.* (2001) used satellite altimetry and interferometry to determine the rate of change of thickness of Pine Island Glacier's entire drainage basin between 1992 and 1999, determining that the grounded glacier thinned by up to 1.6 meters per year over this period. They note "the thinning cannot be explained by short-term variability in accumulation and must result from glacier dynamics." And since glacier dynamics are typically driven by phenomena operating on time scales of hundreds to thousands of years, this observation would argue against twentieth century warming being the cause of the thinning. Shepherd *et al.* also say they could "detect no change in the rate of ice thinning across the glacier over [the] 7-year period," which also suggests that a long-term phenomenon of considerable inertia must be at work.

What if the rate of glacier thinning—1.6 meters per year—continues unabated? Shepherd *et al.* state that "if the trunk continues to lose mass at the present rate it will be entirely afloat within 600 years." And if that happens? They say they "estimate the net contribution to eustatic sea level to be 6 mm." This means that for each century of the foreseeable future, we could expect global mean sea level to rise by one millimeter … about the thickness of a common paper clip.

Turning to other glaciers, Hall and Denton (2002) mapped the distribution and elevation of surficial deposits along the southern Scott Coast of Antarctica in the vicinity of the Wilson Piedmont Glacier, which runs parallel to the coast of the western Ross Sea from McMurdo Sound north to Granite Harbor. The chronology of the raised beaches was determined from more than 60 ^{14}C dates of organic materials they had previously collected from hand-dug excavations (Hall and Denton, 1999). They also evaluated more recent changes in snow and ice cover based on aerial photography and observations carried out since the late 1950s.

Near the end of the Medieval Warm Period—"as late as 890 ^{14}C yr BP," as Hall and Denton put it— "the Wilson Piedmont Glacier was still less extensive than it is now." They rightly conclude that the glacier had to have advanced in the past several hundred years, although they note its eastern margin has retreated in the past 50 years. They report a number of similar observations by other investigators. Citing evidence collected by Baroni and Orombelli (1994a), they note there was "an advance of at least one kilometer of the Hell's Gate Ice Shelf ... within the past few hundred years." And they report that Baroni and Orombelli (1994b) "documented post-fourteenth century advance of a glacier near Edmonson's Point." Summarizing these and other findings, they conclude that evidence from the Ross Sea area suggests "late-Holocene climatic deterioration and glacial advance (within the past few hundred years) and twentieth century retreat."

In speaking of the significance of the "recent advance of the Wilson Piedmont Glacier," Hall and Denton report that it "overlaps in time with the readvance phase known in the Alps [of Europe] as the 'Little Ice Age'," which they further note "has been documented in glacial records as far afield as the Southern Alps of New Zealand (Wardle, 1973; Black, 2001), the temperate land mass closest to the Ross Sea region." They further note that "Kreutz *et al.* (1997) interpreted the Siple Dome [Antarctica] glaciochemical record as indicating enhanced atmospheric circulation intensity at AD ~1400, similar to that in Greenland during the 'Little Ice Age' (O'Brien *et al.*, 1995)." In addition, they report that "farther north, glaciers in the South Shetland Islands adjacent to the Antarctic Peninsula underwent a late-Holocene advance, which has been correlated with the 'Little Ice Age' (Birkenmajer, 1981; Clapperton and Sugden, 1988; Martinez de Pison *et al.*, 1996; Björck *et al.*, 1996)."

In summarizing the results of their work, Hall and Denton say "the Wilson Piedmont Glacier appears to have undergone advance at approximately the same time as the main phase of the 'Little Ice Age', followed by twentieth-century retreat at some localities along the Scott Coast." This result and the others they cite make it clear that glacial activity on Antarctica has followed the pattern of millennial-scale variability that is evident elsewhere in the world: recession to positions during the Medieval Warm Period that have not yet been reached in our day, followed by significant advances during the intervening Little Ice Age.

Additional information on this topic, including reviews of newer publications as they become available, can be found at http://www.co2science.org/subject/a/antarcticagla.php

References

Baroni, C. and Orombelli, G. 1994a. Abandoned penguin rookeries as Holocene paleoclimatic indicators in Antarctica. *Geology* **22**: 23-26.

Baroni, C. and Orombelli, G. 1994b. Holocene glacier variations in the Terra Nova Bay area (Victoria Land, Antarctica). *Antarctic Science* **6**: 497-505.

Birkenmajer, K. 1981. Lichenometric dating of raised marine beaches at Admiralty Bay, King George Island (South Shetland Islands, West Antarctica). *Bulletin de l'Academie Polonaise des Sciences* **29**: 119-127.

Björck, S., Olsson, S., Ellis-Evans, C., Hakansson, H., Humlum, O. and de Lirio, J.M. 1996. Late Holocene paleoclimate records from lake sediments on James Ross Island, Antarctica. *Palaeogeography, Palaeoclimatology, Palaeoecology* **121**: 195-220.

Black, J. 2001. Can a Little Ice Age Climate Signal Be Detected in the Southern Alps of New Zealand? MS Thesis, University of Maine.

Clapperton, C.M. and Sugden, D.E. 1988. Holocene glacier fluctuations in South America and Antarctica. *Quaternary Science Reviews* **7**: 195-198.

Hall, B.L. and Denton, G.H. 1999. New relative sea-level curves for the southern Scott Coast, Antarctica: evidence for Holocene deglaciation of the western Ross Sea. *Journal of Quaternary Science* **14**: 641-650.

Hall, B.L. and Denton, G.H. 2002. Holocene history of the Wilson Piedmont Glacier along the southern Scott Coast, Antarctica. *The Holocene* **12**: 619-627.

Kreutz, K.J., Mayewski, P.A., Meeker, L.D., Twickler, M.S., Whitlow, S.I. and Pittalwala, I.I. 1997. Bipolar changes in atmospheric circulation during the Little Ice Age. *Science* **277**: 1294-1296.

Martinez de Pison, E., Serrano, E., Arche, A. and Lopez-Martinez, J. 1996. *Glacial geomorphology. BAS GEOMAP* **5A**: 23-27.

O'Brien, S.R., Mayewski, P.A., Meeker, L.D., Meese, D.A., Twickler, M.S. and Whitlow, S.I. 1995. Complexity of Holocene climate as reconstructed from a Greenland ice core. *Science* **270**: 1962-1964.

Rignot, E.J. 1998. Fast recession of a West Antarctic glacier. *Science* **281**: 549-550.

Shepherd, A., Wingham, D.J., Mansley, J.A.D. and Corr, H.F.J. 2001. Inland thinning of Pine Island Glacier, West Antarctica. *Science* **291**: 862-864.

Stenoien, M.D. and Bentley, C.R. 2000. Pine Island Glacier, Antarctica: A study of the catchment using interferometric synthetic aperture radar measurements and radar altimetry. *Journal of Geophysical Research* **105**: 21,761-21,779.

Wardle, P. 1973. Variations of the glaciers of Westland National Park and the Hooker Range, New Zealand. *New Zealand Journal of Botany* **11**: 349-388.

4.1.4. Arctic

Computer simulations of global climate change have long indicated the world's polar regions should show the first and severest signs of CO_2-induced global warming. If the models are correct, these signs should be especially evident in the second half of the twentieth century, when approximately two-thirds of the modern-era rise in atmospheric CO_2 occurred and earth's temperature supposedly rose to a level unprecedented in the past millennium. In this subsection, we examine historic trends in Arctic glacier behavior to determine the credibility of current climate models with respect to their polar predictions.

In a review of "the most current and comprehensive research of Holocene glaciation," along the northernmost Gulf of Alaska between the Kenai Peninsula and Yakutat Bay, Calkin *et al.* (2001) report there were several periods of glacial advance and retreat over the past 7,000 years. Over the most recent of those seven millennia, there was a general retreat during the Medieval Warm Period that lasted for "at least a few centuries prior to A.D. 1200." Then came three major intervals of Little Ice Age glacial advance: the early fifteenth century, the middle seventeenth century, and the last half of the nineteenth century. During these very cold periods, glacier equilibrium-line altitudes were depressed from 150 to 200 m below present values, as Alaskan glaciers "reached their Holocene maximum extensions."

The mass balance records of the 18 Arctic glaciers with the longest observational histories subsequent to this time, as the planet emerged from the depths of the Little Ice Age, were studied by Dowdeswell *et al.* (1997). Their analysis showed that more than 80 percent of the glaciers displayed negative mass balances over the periods of their observation, as would be expected for glaciers emerging from the coldest part of the past millennium. Nevertheless, the scientists report that "ice-core records from the Canadian High Arctic islands indicate that the generally negative glacier mass balances observed over the past 50 years have probably been typical of Arctic glaciers since the end of the Little Ice Age," when the magnitude of anthropogenic CO_2 emissions was much less than it has been from 1950 onward.

These observations suggest that Arctic glaciers are not experiencing any adverse effects of anthropogenic CO_2 emissions. In fact, Dowdeswell *et al.* say "there is no compelling indication of increasingly negative balance conditions which might, *a priori*, be expected from anthropogenically induced global warming." Quite to the contrary, they report that "almost 80 percent of the mass balance time series also have a positive trend, toward a less negative mass balance." Hence, although most Arctic glaciers continue to lose mass, as they have probably done since the end of the Little Ice Age, they are losing smaller amounts each year.

Additional evidence that the Arctic's glaciers are not responding to human-induced warming comes from the studies of Zeeberg and Forman (2001) and Mackintosh *et al.* (2002), who indicate there has been an *expansion* of glaciers in the European Arctic over the past few decades.

Zeeberg and Forman analyzed twentieth century changes in glacier terminus positions on north Novaya Zemlya—a Russian island located between the Barents and Kara Seas in the Arctic Ocean—providing a quantitative assessment of the effects of temperature and precipitation on glacial mass balance. Their study showed a significant and accelerated post-Little Ice Age glacial retreat in the first and second decades of the twentieth century. By 1952, the region's glaciers had experienced between 75 percent to 100 percent of their net twentieth century retreat; and during the next 50 years, the recession of more than half of the glaciers stopped, while many tidewater glaciers actually began to advance.

These glacial stabilizations and advances were attributed by the authors to observed increases in precipitation and/or decreases in temperature. For the four decades since 1961, weather stations on Novaya Zemlya, for example, show summer temperatures were 0.3 to 0.5°C colder than they were over the prior 40 years, while winter temperatures were 2.3 to 2.8°C colder than they were over that earlier period. These observations, the authors say, are "counter to warming of the Eurasian Arctic predicted for the twenty-first century by climate models, particularly for the winter season."

Other glacier observations that run counter to climate model predictions are discussed by Mackintosh *et al.* (2002), who concentrated on the 300-year history of the Solheimajokull outlet glacier on the southern coast of Iceland. In 1705, this glacier had a length of about 14.8 km; by 1740 it had grown to 15.2 km in length. Thereafter, it began to retreat, reaching a minimum length of 13.2 km in 1783. Rebounding rapidly, however, the glacier returned to its 1705 position by 1794; by 1820 it equaled its 1740 length. This maximum length was maintained for the next half-century, after which the glacier began a slow retreat that continued to about 1932, when its length was approximately 14.75 km. Then it wasted away more rapidly, reaching a second minimum-length value of approximately 13.8 km about 1970, whereupon it began to rapidly expand, growing to 14.3 km by 1995.

The current position of the outlet glacier terminus is by no means unusual. In fact, it is about midway between its maximum and minimum positions of the past three centuries. It is also interesting to note that the glacier has been growing in length since about 1970. Mackintosh *et al.* report that "the recent advance (1970-1995) resulted from a combination of cooling and enhancement of precipitation."

In another study of the Arctic, Humlum *et al.* (2005) evaluated climate dynamics and their respective impacts on high-latitude glaciers for the Archipelago of Svalbard, focusing on Spitsbergen (the Archipelago's main island) and the Longyearbreen glacier located in its relatively dry central region at 78°13'N latitude. In reviewing what was already known about the region, Humlum *et al.* report that "a marked warming around 1920 changed the mean annual air temperature (MAAT) at sea level within only 5 years from about -9.5°C to -4.0°C," which change, in their words, "represents the most pronounced increase in MAAT documented anywhere in the world during the instrumental period." Then, they report that "from 1957 to 1968, MAAT dropped about 4°C, followed by a more gradual increase towards the end of the twentieth century."

With respect to the Longyearbreen glacier, their work reveals it "has increased in length from about 3 km to its present size of about 5 km during the last c. 1100 years," and they say that "the meteorological setting of non-surging Longyearbreen suggest this example of late-Holocene glacier growth represents a widespread phenomenon in Svalbard and in adjoining Arctic regions," which they describe as a "development towards cooler conditions in the Arctic" that "may explain why the Little Ice Age glacier advance in Svalbard usually represents the Holocene maximum glacier extension."

Climate change in Svalbard over the twentieth century was a rollercoaster ride, with temperatures rising more rapidly in the early 1920s than has been documented anywhere else before or since, only to be followed by a nearly equivalent temperature drop four decades later, both of which climatic transitions were totally out of line with what climate models suggest should have occurred. The current location of the terminus of the Longyearbreen glacier suggests that, even now, Svalbard and "adjoining Arctic regions" are experiencing some of the lowest temperatures of the entire Holocene or current interglacial, at a time when atmospheric CO_2 concentrations are higher than they have likely been for millions of years. Both of these observations are at odds with what the IPCC claims about the strong warming power of atmospheric CO_2 enrichment.

Bradwell *et al.* (2006) examined the link between late Holocene fluctuations of Lambatungnajokull (an outlet glacier of the Vatnajokull ice cap of southeast Iceland) and variations in climate, using geomorphological evidence to reconstruct patterns of glacier fluctuations and using lichenometry and tephrostratigraphy to date glacial landforms created by the glacier over the past four centuries. Results indicated that "there is a particularly close correspondence between summer air temperature and the rate of ice-front recession of Lambatungnajokull during periods of overall retreat," and that "between 1930 and 1950 this relationship is striking." They also report that "ice-front recession was greatest during the 1930s and 1940s, when retreat averaged 20 m per year." Thereafter, they say the retreat "slowed in the 1960s," and they report "there has been little overall retreat since the 1980s."

The researchers also report that "the 20th-century record of reconstructed glacier-front fluctuations at Lambatungnajokull compares well with those of other similar-sized, non-surging, outlets of southern Vatnajokull," including Skaftafellsjokull, Fjallsjokull, Skalafellsjokull, and Flaajokull. In fact, they find "the pattern of glacier fluctuations of Lambatungnajokull over the past 200 years reflects the climatic changes that have occurred in southeast Iceland and the wider region."

Bradwell *et al.*'s findings suggest that twentieth century summer air temperature in southeast Iceland and the wider region peaked in the 1930s and 1940s, and was followed by a cooling that persisted through the end of the century. This thermal behavior is about as different as one could imagine from the claim that the warming of the globe over the last two decades of the twentieth century was unprecedented over the past two millennia. Especially is this so for a high-northern-latitude region, where the IPCC claims CO_2-induced global warming should be earliest and most strongly expressed.

Additional information on this topic, including reviews of newer publications as they become available, can be found at http://www.co2science.org/subject/a/arcticgla.php.

References

Bradwell, T., Dugmore, A.J. and Sugden, D.E. 2006. The Little Ice Age glacier maximum in Iceland and the North Atlantic Oscillation: evidence from Lambatungnajokull, southeast Iceland. *Boreas* **35**: 61-80.

Calkin, P.E., Wiles, G.C. and Barclay, D.J. 2001. Holocene coastal glaciation of Alaska. *Quaternary Science Reviews* **20**: 449-461.

Dowdeswell, J.A., Hagen, J.O., Bjornsson, H., Glazovsky, A.F., Harrison, W.D., Holmlund, P., Jania, J., Koerner, R.M., Lefauconnier, B., Ommanney, C.S.L. and Thomas, R.H. 1997. The mass balance of circum-Arctic glaciers and recent climate change. *Quaternary Research* **48**: 1-14.

Humlum, O., Elberling, B., Hormes, A., Fjordheim, K., Hansen, O.H. and Heinemeier, J. 2005. Late-Holocene glacier growth in Svalbard, documented by subglacial relict vegetation and living soil microbes. *The Holocene* **15**: 396-407.

Mackintosh, A.N., Dugmore, A.J. and Hubbard, A.L. 2002. Holocene climatic changes in Iceland: evidence from modeling glacier length fluctuations at Solheimajokull. *Quaternary International* **91**: 39-52.

Zeeberg, J. and Forman, S.L. 2001. Changes in glacier extent on north Novaya Zemlya in the twentieth century. *Holocene* **11**: 161-175.

4.1.5. Europe

Joerin *et al.* (2006) examined glacier recessions in the Swiss Alps over the past ten thousand years based on radiocarbon-derived ages of materials found in proglacial fluvial sediments of subglacial origin, focusing on subfossil remains of wood and peat. Combining their results with earlier data of a similar nature, they then constructed a master chronology of Swiss glacier fluctuations over the course of the Holocene.

Joerin *et al.* first report discovering that "alpine glacier recessions occurred at least 12 times during the Holocene," once again demonstrating that millennial-scale oscillation of climate has reverberated throughout glacial and interglacial periods as far back in time as scientists have searched for the phenomenon. Second, they determined that glacier recessions have been decreasing in frequency since approximately 7,000 years ago, and especially since 3,200 years ago, "culminating in the maximum glacier extent of the 'Little Ice Age'." Third, the last of the major glacier recessions in the Swiss Alps occurred between about 1,400 and 1,200 years ago, according to Joerin *et al.*'s data, but between 1200 and 800 years ago, according to the data of Holzhauser *et al.* (2005) for the Great Aletsch Glacier. Of this discrepancy, Joerin *et al.* say that given the uncertainty of the radiocarbon dates, the two records need not be considered inconsistent with each other. What is more, their presentation of the Great Aletsch Glacier data indicates the glacier's length at about AD 1000—when there was fully 100 ppm *less* CO_2 in the air than there is today—was just slightly less than its length in 2002.

Also in the Swiss Alps, Huss *et al.* (2008) examined various ice and meteorological measurements made between 1865 and 2006 in an effort to compute the yearly mass balances of four glaciers. The results of their computations can be seen in Figure 4.1.5.1.

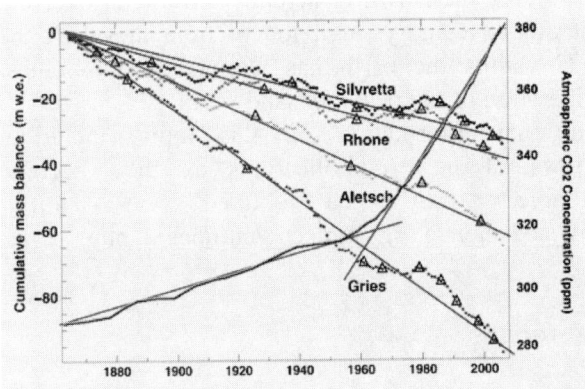

Figure 4.1.5.1. Huss *et al.* (2008) examined various ice and meteorological measurements made between 1865 and 2006 in the Swiss Alps to compute the yearly mass balances of four glaciers.

The most obvious conclusion to be drawn from these data is the fact that each of the four glaciers has decreased in size. But more important is the fact that the rate of shrinkage has not accelerated over time, as evidenced by the long-term trend lines we have fit to the data. There is no compelling evidence that this 14-decade-long glacial decline has had anything to do with the air's CO_2 content.

Consider, for example, the changes in atmospheric CO_2 concentration experienced over the same time period, also shown in the figure. If we compute the mean rate-of-rise of the air's CO_2 content from the start of the record to about 1950, and from about 1970 to 2006, we see that between 1950 and 1970 the rate-of-rise of the atmosphere's CO_2 concentration increased by more than five-fold, yet there were no related increases in the long-term mass balance trends of the four glaciers. It is clear that the ice loss history of the glaciers was not unduly influenced by the increase in the rate-of-rise of the air's CO_2 content that occurred between 1950 and 1970, and that their rate of shrinkage was also not

materially altered by what the IPCC calls the unprecedented warming of the past few decades.

Moving to northern Europe, Linderholm et al. (2007) examined "the world's longest ongoing continuous mass-balance record" of "Storglaciaren in northernmost Sweden," which they report "is generally well correlated to glaciers included in the regional mass balance program (Holmlund and Jansson, 1999), suggesting that it represents northern Swedish glaciers." The results of their work are depicted in Figure 4.1.5.2, where we have also plotted the contemporaneous history of the atmosphere's CO_2 concentration.

In viewing the figure, it should be evident that the historical increase in the air's CO_2 content has had absolutely no discernable impact on the net mass balance history of Sweden's Storglaciaren over the past two-and-a-quarter centuries. Whereas the mean rate-of-rise of the air's CO_2 concentration over the last half-century of Storglaciaren mass balance data is fully 15 times greater than what it was over the first half-century of mass balance data (and some 40 times greater if the first and last quarter-centuries are considered), there has been no sign of any change in the long-term trend of Storglaciaren's net mass balance.

D'Orefice *et al.* (2000) assembled and analyzed a wealth of historical data to derive a history of post-Little Ice Age (LIA) shrinkage of the surface area of the southernmost glacier of Europe, Ghiacciaio del Calderone. From the first available information on the glacier's surface area in 1794, there was a very slow ice wastage that lasted until 1884, whereupon the glacier began to experience a more rapid area reduction that continued, with some irregularities, to 1990, resulting in a loss of just over half the glacier's LIA surface area.

Not all European glaciers, however, have experienced continuous declines since the end of the Little Ice Age. Hormes *et al.* (2001) report that glaciers in the Central Swiss Alps experienced two periods of readvancement, one around 1920 and another as recent as 1980. In addition, Braithwaite (2002) reports that for the period 1980-1995, "Scandinavian glaciers [have been] growing, and glaciers in the Caucasus are close to equilibrium," while "there is no obvious common or global trend of increasing glacier melt."

Fifty years of mass balance data from the storied Storglaciaren of northwestern Sweden also demonstrate a trend reversal in the late twentieth century. According to Braithwaite and Zhang (2000), there has been a significant upward trend in the mass balance of this glacier over the past 30-40 years, and it has been in a state of mass accumulation for at least the past decade.

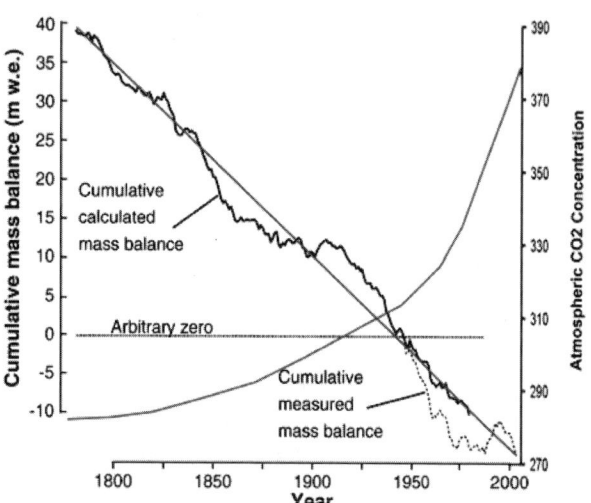

Figure 4.1.5.2. The cumulative reconstructed net mass balance (bN) history of Sweden's Storglaciaren, to which we have added the fit-by-eye descending linear relationship, in blue, and the history of the atmosphere's CO_2 concentration, in red. Adapted from Linderholm et al. (2007).

Additional evidence for post-LIA glacial expansion is provided by the history of the Solheimajokull outlet glacier on the southern coast of Iceland. In a review of its length over the past 300 years, Mackintosh *et al.* (2002) report a post-LIA minimum of 13.8 km in 1970, whereupon the glacier began to expand, growing to a length of about 14.3 km by 1995. The minimum length of 13.8 km observed in 1970 also did not eclipse an earlier minimum in which the glacier had decreased from a 300-year maximum length of 15.2 km in 1740 to a 300-year minimum of 13.2 km in 1783.

More recent glacial advances have been reported in Norway. According to Chin *et al.* (2005), glacial recession in Norway was most strongly expressed in "the middle of the 20th century," ending during the late 1950s to early 1960s." Then, "after some years with more or less stationary glacier front positions, [the glaciers] began to advance, accelerating in the late 1980s." Around 2000, a portion of the glaciers began to slow, while some even ceased moving; but they say that "most of the larger outlets with longer reaction times are continuing to advance." Chin *et al.* report that "the distances regained and the duration of this recent advance episode are both far greater than

any previous readvance since the Little Ice Age maximum, making the recent resurgence a significant event." Mass balance data reveal much the same thing, "especially since 1988" and "at all [western] maritime glaciers in both southern and northern Norway," where "frequent above-average winter balances are a main cause of the positive net balances at the maritime glaciers during the last few decades."

In considering the results of the studies summarized above, it appears there is no correlation between atmospheric CO_2 levels and glacier melting or advancement in Europe. Several European glaciers are holding their own or actually advancing over the past quarter-century, a period of time in which the IPCC claims the earth has warmed to its highest temperature of the past thousand years.

Additional information on this topic, including reviews of newer publications as they become available, can be found at http://www.co2science.org/ subject/e/europegla. php.

References

Braithwaite, R.J. 2002. Glacier mass balance: the first 50 years of international monitoring. *Progress in Physical Geography* **26**: 76-95.

Braithwaite, R.J. and Zhang, Y. 2000. Relationships between interannual variability of glacier mass balance and climate. *Journal of Glaciology* **45**: 456-462.

Chinn, T., Winkler, S., Salinger, M.J. and Haakensen, N. 2005. Recent glacier advances in Norway and New Zealand: A comparison of their glaciological and meteorological causes. *Geografiska Annaler* **87** A: 141-157.

D'Orefice, M., Pecci, M., Smiraglia, C. and Ventura, R. 2000. Retreat of Mediterranean glaciers since the Little Ice Age: Case study of Ghiacciaio del Calderone, central Apennines, Italy. *Arctic, Antarctic, and Alpine Research* **32**: 197-201.

Holmlund, P. and Jansson, P. 1999. The Tarfala mass balance programme. *Geografiska Annaler* **81A**: 621-631.

Holzhauser, H., Magny, M. and Zumbuhl, H.J. 2005. Glacier and lake-level variations in west-central Europe over the last 3500 years. *The Holocene* **15**: 789-801.

Hormes, A., Müller, B.U. and Schlüchter, C. 2001. The Alps with little ice: evidence for eight Holocene phases of reduced glacier extent in the Central Swiss Alps. *The Holocene* **11**: 255-265.

Huss, M., Bauder, A., Funk, M. and Hock, R. 2008. Determination of the seasonal mass balance of four Alpine glaciers since 1865. *Journal of Geophysical Research* **113**: 10.1029/2007JF000803.

Joerin, U.E., Stocker, T.F. and Schlüchter, C. 2006. Multicentury glacier fluctuations in the Swiss Alps during the Holocene. *The Holocene* **16**: 697-704.

Linderholm, H.W., Jansson, P. and Chen, D. 2007. A high-resolution reconstruction of Storglaciaren mass balance back to 1780/81 using tree-ring and circulation indices. *Quaternary Research* **67**: 12-20.

Mackintosh, A.N., Dugmore, A.J. and Hubbard, A.L. 2002. Holocene climatic changes in Iceland: evidence from modeling glacier length fluctuations at Solheimajokull. *Quaternary International* **91**: 39-52.

Petit, J.R., Jouzel, J., Raynaud, D., Barkov, N.I., Barnola, J.-M., Basile, I., Bender, M., Chappellaz, J., Davis, M., Delaygue, G., Delmotte, M., Kotlyakov, V.M., Legrand, M., Lipenkov, V.Y., Lorius, C., Pepin, L., Ritz, C., Saltzman, E. and Stievenard, M. 1999. Climate and atmospheric history of the past 420,000 years from the Vostok ice core, Antarctica. *Nature* **399**: 429-436.

4.1.6. North America

The history of North American glacial activity also fails to support the claim that anthropogenic CO_2 emissions are causing glaciers to melt. Dowdeswell *et al.* (1997) analyzed the mass balance histories of the 18 Arctic glaciers with the longest observational records, finding that just over 80 percent of them displayed negative mass balances over the last half of the twentieth century. However, they note that "ice-core records from the Canadian High Arctic islands indicate that the generally negative glacier mass balances observed over the past 50 years have probably been typical of Arctic glaciers since the end of the Little Ice Age." They say "there is no compelling indication of increasingly negative balance conditions which might, *a priori*, be expected from anthropogenically induced global warming."

Clague *et al.* (2004) documented glacier and vegetation changes at high elevations in the upper Bowser River basin in the northern Coast Mountains of British Columbia, based on studies of the distributions of glacial moraines and trimlines, tree-ring data, cores from two small lakes that were sampled for a variety of analyses (magnetic susceptibility, pollen, diatoms, chironomids, carbon and nitrogen content, ^{210}Pb, ^{137}Cs, ^{14}C), similar analyses of materials obtained from pits and cores from a nearby fen, and by accelerator mass spectrometry radiocarbon dating of plant fossils,

including wood fragments, tree bark, twigs and conifer needles and cones. All this evidence suggested a glacial advance that began about 3,000 years ago and may have lasted for hundreds of years, which would have placed it within the unnamed cold period that preceded the Roman Warm Period. There was also evidence for a second minor phase of activity that began about 1,300 years ago but was of short duration, which would have placed it within the Dark Ages Cold Period. Finally, the third and most extensive Neoglacial interval began shortly after AD 1200, following the Medieval Warm Period, and ended in the late 1800s, which was, of course, the Little Ice Age, during which time Clague *et al.* say "glaciers achieved their greatest extent of the past 3,000 years and probably the last 10,000 years."

These data clearly depict the regular alternation between non-CO_2-forced multi-century cold and warm periods that is the trademark of the millennial-scale oscillation of climate that reverberates throughout glacial and interglacial periods alike. That a significant, but by no means unprecedented, warming followed the most recent cold phase of this cycle is in no way unusual, particularly since the Little Ice Age was likely the coldest period of the last 10,000 years.

Alaska, Calkin *et al.* (2001) reviewed the most current and comprehensive research of Holocene glaciation along the northernmost portion of the Gulf of Alaska between the Kenai Peninsula and Yakutat Bay, where several periods of glacial advance and retreat were noted during the past 7,000 years. Over the latter part of this record, there was a general glacial retreat during the Medieval Warm Period that lasted for a few centuries prior to A.D. 1200, after which there were three major intervals of Little Ice Age glacial advance: the early fifteenth century, the middle seventeenth century, and the last half of the nineteenth century. During these latter time periods, glacier equilibrium line altitudes were depressed from 150 to 200 m below present values as Alaskan glaciers also "reached their Holocene maximum extensions."

Wiles *et al.* (2004) derived a composite Glacier Expansion Index (GEI) for Alaska based on "dendrochronologically derived calendar dates from forests overrun by advancing ice and age estimates of moraines using tree-rings and lichens" for three climatically distinct regions—the Arctic Brooks Range, the southern transitional interior straddled by the Wrangell and St. Elias mountain ranges, and the Kenai, Chugach, and St. Elias coastal ranges—after which they compared this history of glacial activity with "the ^{14}C record preserved in tree rings corrected for marine and terrestrial reservoir effects as a proxy for solar variability" and with the history of the Pacific Decadal Oscillation (PDO) derived by Cook (2002).

As a result of their efforts, Wiles *et al.* discovered that "Alaska shows ice expansions approximately every 200 years, compatible with a solar mode of variability," specifically, the de Vries 208-year solar cycle; and by merging this cycle with the cyclical behavior of the PDO, they obtained a dual-parameter forcing function that was even better correlated with the Alaskan composite GEI, with major glacial advances clearly associated with the Sporer, Maunder, and Dalton solar minima.

Wiles *et al.* said "increased understanding of solar variability and its climatic impacts is critical for separating anthropogenic from natural forcing and for predicting anticipated temperature change for future centuries." They made no mention of possible CO_2-induced global warming in discussing their results, presumably because there was no need to do so. Alaskan glacial activity, which in their words "has been shown to be primarily a record of summer temperature change (Barclay *et al.*, 1999)," appears to be sufficiently well described within the context of centennial (solar) and decadal (PDO) variability superimposed upon the millennial-scale (non-CO_2-forced) variability that produces longer-lasting Medieval Warm Period and Little Ice Age conditions.

Pederson *et al.* (2004) used tree-ring reconstructions of North Pacific surface temperature anomalies and summer drought as proxies for winter glacial accumulation and summer ablation, respectively, to create a 300-year history of regional glacial Mass Balance Potential (MBP), which they compared with historic retreats and advances of Glacier Park's extensively studied Jackson and Agassiz glaciers in northwest Montana..

As they describe it, "the maximum glacial advance of the Little Ice Age coincides with a sustained period of positive MBP that began in the mid-1770s and was interrupted by only one brief ablation phase (~1790s) prior to the 1830s," after which they report "the mid-19th century retreat of the Jackson and Agassiz glaciers then coincides with a period marked by strong negative MBP." From about 1850 onward, they note "Carrara and McGimsey (1981) indicate a modest retreat (~3-14 m/yr) for both glaciers until approximately 1917." At that point, they report that "the MBP shifts to an extreme negative

phase that persists for ~25 yr," during which period the glaciers retreated "at rates of greater than 100 m/yr."

Continuing with their history, Pederson *et al.* report that "from the mid-1940s through the 1970s retreat rates slowed substantially, and several modest advances were documented as the North Pacific transitioned to a cool phase [and] relatively mild summer conditions also prevailed." From the late 1970s through the 1990s, they say, "instrumental records indicate a shift in the PDO back to warmer conditions resulting in continuous, moderate retreat of the Jackson and Agassiz glaciers."

The first illuminating aspect of this glacial history is that the post-Little Ice Age retreat of the Jackson and Agassiz glaciers began just after 1830, in harmony with the findings of a number of other studies from various parts of the world (Vincent and Vallon, 1997; Vincent, 2001, 2002; Moore *et al.*, 2002; Yoo and D'Odorico, 2002; Gonzalez-Rouco *et al.* 2003; Jomelli and Pech, 2004), including the entire Northern Hemisphere (Briffa and Osborn, 2002; Esper *et al.*, 2002). These findings stand in stark contrast to what is suggested by the IPCC-endorsed "hockeystick" temperature history of Mann *et al.* (1998, 1999), which does not portray *any* Northern Hemispheric warming until around 1910.

The second illuminating aspect of the glacial record is that the vast bulk of the glacial retreat in Glacier National Park occurred between 1830 and 1942, over which time the air's CO_2 concentration rose by only 27 ppm, which is less than a third of the total CO_2 increase experienced since the start of glacial recession. Then, from the mid-1940s through the 1970s, when the air's CO_2 concentration rose by another 27 ppm, Pederson *et al.* report that "retreat rates slowed substantially, and several modest advances were documented."

The first 27 ppm increase in atmospheric CO_2 concentration coincided with the great preponderance of glacial retreat experienced since the start of the warming that marked the "beginning of the end" of the Little Ice Age, but the next 27 ppm increase in the air's CO_2 concentration was accompanied by little if any additional glacial retreat, when, of course, there was little if any additional warming.

Something other than the historic rise in the air's CO_2 content was responsible for the disappearing ice fields of Glacier National Park. The historical behavior of North America's glaciers provides no evidence for unprecedented or unnatural CO_2-induced global warming over any part of the twentieth century.

Additional information on this topic, including reviews of newer publications as they become available, can be found at http://www.co2science.org/subject/n/northamgla.php.

References

Barclay, D.J., Wiles, G.C. and Calkin, P.E. 1999. A 1119-year tree-ring-width chronology from western Prince William Sound, southern Alaska. *The Holocene* **9**: 79-84.

Briffa, K.R. and Osborn, T.J. 2002. Blowing hot and cold. *Science* **295**: 2227-2228.

Calkin, P.E., Wiles, G.C. and Barclay, D.J. 2001. Holocene coastal glaciation of Alaska. *Quaternary Science Reviews* **20**: 449-461.

Carrara, P.E. and McGimsey, R.G. 1981. The late neoglacial histories of the Agassiz and Jackson Glaciers, Glacier National Park, Montana. *Arctic and Alpine Research* **13**: 183-196.

Clague, J.J., Wohlfarth, B., Ayotte, J., Eriksson, M., Hutchinson, I., Mathewes, R.W., Walker, I.R. and Walker, L. 2004. Late Holocene environmental change at treeline in the northern Coast Mountains, British Columbia, Canada. *Quaternary Science Reviews* **23**: 2413-2431.

Cook, E.R. 2002. Reconstructions of Pacific decadal variability from long tree-ring records. *EOS: Transactions, American Geophysical Union* **83**: S133.

Dowdeswell, J.A., Hagen, J.O., Bjornsson, H., Glazovsky, A.F., Harrison, W.D., Holmlund, P., Jania, J., Koerner, R.M., Lefauconnier, B., Ommanney, C.S.L. and Thomas, R.H. 1997. The mass balance of circum-Arctic glaciers and recent climate change. *Quaternary Research* **48**: 1-14.

Esper, J., Cook, E.R. and Schweingruber, F.H. 2002. Low-frequency signals in long tree-ring chronologies for reconstructing past temperature variability. *Science* **295**: 2250-2253.

Gonzalez-Rouco, F., von Storch, H. and Zorita, E. 2003. Deep soil temperature as proxy for surface air-temperature in a coupled model simulation of the last thousand years. *Geophysical Research Letters* **30**: 10.1029/2003GL018264.

Jomelli, V. and Pech, P. 2004. Effects of the Little Ice Age on avalanche boulder tongues in the French Alps (Massif des Ecrins). *Earth Surface Processes and Landforms* **29**: 553-564.

Mann, M.E., Bradley, R.S. and Hughes, M.K. 1998. Global-scale temperature patterns and climate forcing over the past six centuries. *Nature* **392**: 779-787.

Mann, M.E., Bradley, R.S. and Hughes, M.K. 1999. Northern Hemisphere temperatures during the past millennium: Inferences, uncertainties, and limitations. *Geophysical Research Letters* **26**: 759-762.

Moore, G.W.K., Holdsworth, G. and Alverson, K. 2002. Climate change in the North Pacific region over the past three centuries. *Nature* **420**: 401-403.

Pederson, G.T., Fagre, D.B., Gray, S.T. and Graumlich, L.J. 2004. Decadal-scale climate drivers for glacial dynamics in Glacier National Park, Montana, USA. *Geophysical Research Letters* **31**: 10.1029/2004GL019770.

Vincent, C. 2001. Fluctuations des bilans de masse des glaciers des Alpes francaises depuis le debut du 20em siecle au regard des variations climatiques. *Colloque SHF variations climatiques et hydrologie*. Paris, France, pp. 49-56.

Vincent, C. 2002. Influence of climate change over the 20th century on four French glacier mass balances. *Journal of Geophysical Research* **107**: 4-12.

Vincent, C. and Vallon, M. 1997. Meteorological controls on glacier mass-balance: empirical relations suggested by Sarennes glaciers measurements (France). *Journal of Glaciology* **43**: 131-137.

Wiles, G.C., D'Arrigo, R.D., Villalba, R., Calkin, P.E. and Barclay, D.J. 2004. Century-scale solar variability and Alaskan temperature change over the past millennium. *Geophysical Research Letters* **31**: 10.1029/2004GL020050.

Yoo, J.C. and D'Odorico, P. 2002. Trends and fluctuations in the dates of ice break-up of lakes and rivers in Northern Europe: the effect of the North Atlantic Oscillation. *Journal of Hydrology* **268**: 100-112.

4.1.7. South America

Harrison and Winchester (2000) used dendrochronology, lichenometry, and aerial photography to date nineteenth and twentieth century fluctuations of the Arco, Colonia, and Arenales glaciers on the eastern side of the Hielo Patagonico Norte in southern Chile. This work revealed that these glaciers, plus four others on the western side of the ice field, began to retreat, in the words of the two researchers, "from their Little Ice Age maximum positions" somewhere between 1850 and 1880, well before the air's CO_2 content began to rise at a significant rate. They also note that the trend continued "through the first half of the 20th century with various still-stands and oscillations between 1925 and 1960 ... with retreat increasing since the

1960s," just as has been observed at many sites in the Northern Hemisphere.

Glasser *et al.* (2004) described a large body of evidence related to glacier fluctuations in the two major ice fields of Patagonia: the Hielo Patagonico Norte and the Hielo Patagonico Sur. This evidence indicates that the most recent glacial advances in Patagonia occurred during the Little Ice Age. Prior to then, their data indicate an interval of higher temperatures known as the Medieval Warm Period, when glaciers decreased in size and extent; this warm interlude was in turn preceded by an era of pronounced glacial activity that is designated the Dark Ages Cold Period, which was also preceded by a period of higher temperatures and retreating glaciers that is denoted the Roman Warm Period.

Glasser *et al.* documented cycles of blacial advances and retreats each lasting hundreds of years going back to sometime between 6,000 and 5,000 ^{14}C years before present (BP). They cited the works of other scientists that reveal a similar pattern of cyclical glacial activity over the preceding millennia in several other locations. Immediately to the east of the Hielo Patagonico Sur in the Rio Guanaco region of the Precordillera, for example, they report that Wenzens (1999) detected five distinct periods of glacial advancement: "4500-4200, 3600-3300, 2300-2000, 1300-1000 ^{14}C years BP and AD 1600-1850." With respect to the glacial advancements that occurred during the cold interval that preceded the Roman Warm Period, they say they constitute "part of a body of evidence for global climatic change around this time (e.g., Grosjean *et al.*, 1998; Wasson and Claussen, 2002) which coincides with an abrupt decrease in solar activity," and they say that this observation was what "led van Geel *et al.* (2000) to suggest that variations in solar irradiance are more important as a driving force in variations in climate than previously believed."

Finally, with respect to the most recent recession of Hielo Patogonico Norte outlet glaciers from their late historic moraine limits at the end of the nineteenth century, Glasser *et al.* say that "a similar pattern can be observed in other parts of southern Chile (e.g., Kuylenstierna *et al.*, 1996; Koch and Kilian, 2001)," to which we would also add the findings of Kaser and Georges (1997) for the Peruvian Cordillera Blanca and Francou *et al.* (2003) for the Bolivian Cordillera Real. Likewise, they note that "in areas peripheral to the North Atlantic and in central Asia the available evidence shows that

glaciers underwent significant recession at this time (cf. Grove, 1988; Savoskul, 1997)."

Georges (2004) constructed a twentieth century history of glacial fluctuations in the Cordillera Blanca of Peru, which is the largest glaciated area within the tropics. This history reveals, in Georges words, that "the beginning of the century was characterized by a glacier recession of unknown extent, followed by a marked readvance in the 1920s that nearly reached the Little Ice Age maximum." Then came the "very strong" 1930s-1940s glacial mass *shrinkage*, after which there was a period of quiescence that was followed by an "intermediate retreat from the mid-1970s until the end of the century."

In comparing the two periods of glacial wasting, Georges says that "the intensity of the 1930s-1940s retreat was more pronounced than that of the one at the end of the century." In fact, his graph of the ice area lost in both time periods suggests that the rate of wastage in the 1930s-1940s was *twice as great* as that of last two decades of the twentieth century.

Georges is quite at ease talking about the Little Ice Age south of the equator in Peru, which is a very long way from the lands that border the North Atlantic Ocean, which is the only region on earth where the IPCC is willing to admit the existence of this chilly era of the planet's climatic history. The glacial extensions of the Cordillera Blanca in the late 1920s were almost equivalent to those experienced there during the depths of the Little Ice Age.

Koch and Kilian (2005) mapped and dated, by dendrochronological means, a number of moraine systems of Glaciar Lengua and neighboring glaciers of Gran Campo Nevado in the southernmost Andes of Chile, after which they compared their results with those of researchers who studied the subject in other parts of South America. According to their findings, in the Patagonian Andes "the culmination of the Little Ice Age glacier advances occurred between AD 1600 and 1700 (e.g., Mercer, 1970; Rothlisberger, 1986; Aniya, 1996)," but "various glaciers at Hielo Patagonico Norte and Hielo Patagonico Sur also formed prominent moraines around 1870 and 1880 (Warren and Sugden, 1993; Winchester *et al.*, 2001; Luckman and Villalba, 2001)." In addition, they note their study "further supports this scenario," and that from their observations at Glaciar Lengua and neighboring glaciers at Gran Campo Nevado, it would appear that "the 'Little Ice Age' advance was possibly the most extensive one during the Holocene for this ice cap."

Working with biogenic silica, magnetic susceptibility, total organic carbon (TOC), total nitrogen (TN), $\delta^{13}CTOC$, $\delta^{15}NTN$, and C/N ratios derived from the sediment records of two Venezuelan watersheds, which they obtained from cores retrieved from Lakes Mucubaji and Blanca, together with ancillary data obtained from other studies that had been conducted in the same general region, Polissar *et al.* (2006) developed continuous decadal-scale histories of glacier activity and moisture balance in that part of the tropical Andes (the Cordillera de Merida) over the past millennium and a half, from which they were able to deduce contemporary histories of regional temperature and precipitation.

The international team of scientists—representing Canada, Spain, the United States, and Venezuela—write that "comparison of the Little Ice Age history of glacier activity with reconstructions of solar and volcanic forcing suggests that solar variability is the primary underlying cause of the glacier fluctuations," because (1) "the peaks and troughs in the susceptibility records match fluctuations of solar irradiance reconstructed from ^{10}Be and $\delta^{14}C$ measurements," (2) "spectral analysis shows significant peaks at 227 and 125 years in both the irradiance and magnetic susceptibility records, closely matching the de Vreis and Gleissberg oscillations identified from solar irradiance reconstructions," and (3) "solar and volcanic forcing are uncorrelated between AD 1520 and 1650, and the magnetic susceptibility record follows the solar-irradiance reconstruction during this interval." In addition, they write that "four glacial advances occurred between AD 1250 and 1810, coincident with solar-activity minima," and that "temperature declines of -3.2 ± 1.4°C and precipitation increases of ~20% are required to produce the observed glacial responses."

In discussing their findings, Polissar *et al.* say their results "suggest considerable sensitivity of tropical climate to small changes in radiative forcing from solar irradiance variability." The six scientists also say their findings imply "even greater probable responses to future anthropogenic forcing," and that "profound climatic impacts can be predicted for tropical montane regions."

With respect to these latter ominous remarks, we note that whereas Polissar *et al.*'s linking of significant climate changes with solar radiation variability is a factual finding of their work, their latter statements with respect to hypothesized CO_2-induced increases in down-welling thermal radiation

are speculations that need not follow from what they learned.

Another point worth noting in this regard is Polissar *et al.*'s acknowledgement that "during most of the past 10,000 years, glaciers were absent from all but the highest peaks in the Cordillera de Merida," which indicates that warmer-than-present temperatures are the norm for this part of the planet, and that any significant warming that might yet occur in this region (as well as most of the rest of the world) would mark only a return to more typical Holocene (or current interglacial) temperatures, which have themselves been significantly lower than those of all four prior interglacials. What is more, atmospheric CO_2 concentrations were much lower during all of those much warmer periods.

Additional information on this topic, including reviews of newer publications as they become available, can be found at http://www.co2science.org/subject/s/southamgla.php.

References

Aniya, M. 1996. Holocene variations of Ameghino Glacier, southern Patagonia. *The Holocene* **6**: 247-252.

Francou, B., Vuille, M., Wagnon, P., Mendoza, J. and Sicart, J.E. 2003. Tropical climate change recorded by a glacier in the central Andes during the last decades of the 20th century: Chacaltaya, Bolivia, 16°S. *Journal of Geophysical Research* **108**: 10.1029/2002JD002473.

Georges, C. 2004. 20th-century glacier fluctuations in the tropical Cordillera Blanca, Peru. *Arctic, Antarctic, and Alpine Research* **35**: 100-107.

Glasser, N.F., Harrison, S., Winchester, V. and Aniya, M. 2004. Late Pleistocene and Holocene palaeoclimate and glacier fluctuations in Patagonia. *Global and Planetary Change* **43**: 79-101.

Grosjean, M., Geyh, M.A., Messerli, B., Schreier, H. and Veit, H. 1998. A late-Holocene (2600 BP) glacial advance in the south-central Andes (29°S), northern Chile. *The Holocene* **8**: 473-479.

Grove, J.M. 1988. *The Little Ice Age*. Routledge, London, UK.

Harrison, S. and Winchester, V. 2000. Nineteenth- and twentieth-century glacier fluctuations and climatic implications in the Arco and Colonia Valleys, Hielo Patagonico Norte, Chile. *Arctic, Antarctic, and Alpine Research* **32**: 55-63.

Kaser, G. and Georges, C. 1997. Changes in the equilibrium line altitude in the tropical Cordillera Blanca

(Peru) between 1930 and 1950 and their spatial variations. *Annals of Glaciology* **24**: 344-349.

Koch, J. and Kilian, R. 2005. "Little Ice Age" glacier fluctuations, Gran Campo Nevado, southernmost Chile. *The Holocene* **15**: 20-28.

Koch, J. and Kilian, R. 2001. Dendroglaciological evidence of Little Ice Age glacier fluctuations at the Gran Campo Nevado, southernmost Chile. In: Kaennel Dobbertin, M. and Braker, O.U. (Eds.) *International Conference on Tree Rings and People*. Davos, Switzerland, p. 12.

Kuylenstierna, J.L., Rosqvist, G.C. and Holmlund, P. 1996. Late-Holocene glacier variations in the Cordillera Darwin, Tierra del Fuego, Chile. *The Holocene* **6**: 353-358.

Luckman, B.H. and Villalba, R. 2001. Assessing the synchroneity of glacier fluctuations in the western Cordillera of the Americas during the last millennium. In: Markgraf, V. (Ed.), *Interhemispheric Climate Linkages*. Academic Press, New York, NY, USA, pp. 119-140.

Mercer, J.H. 1970. Variations of some Patagonian glaciers since the Late-Glacial: II. *American Journal of Science* **269**: 1-25.

Polissar, P.J., Abbott, M.B., Wolfe, A.P., Bezada, M., Rull, V. and Bradley, R.S. 2006. Solar modulation of Little Ice Age climate in the tropical Andes. *Proceedings of the National Academy of Sciences USA* **103**: 8937-8942.

Rothlisberger, F. 1986. *10 000 Jahre Gletschergeschichte der Erde*. Verlag Sauerlander, Aarau.

Savoskul, O.S. 1997. Modern and Little Ice Age glaciers in "humid" and "arid" areas of the Tien Shan, Central Asia: two different patterns of fluctuation. *Annals of Glaciology* **24**: 142-147.

van Geel, B., Heusser, C.J., Renssen, H. and Schuurmans, C.J.E. 2000. Climatic change in Chile at around 2700 B.P. and global evidence for solar forcing: a hypothesis. *The Holocene* **10**: 659-664.

Warren, C.R. and Sugden, D.E. 1993. The Patagonian icefields: a glaciological review. *Arctic and Alpine Research* **25**: 316-331.

Wasson, R.J. and Claussen, M. 2002. Earth systems models: a test using the mid-Holocene in the Southern Hemisphere. *Quaternary Science Reviews* **21**: 819-824.

Wenzens, G. 1999. Fluctuations of outlet and valley glaciers in the southern Andes (Argentina) during the past 13,000 years. *Quaternary Research* **51**: 238-247.

Winchester, V., Harrison, S. and Warren, C.R. 2001. Recent retreat Glacier Nef, Chilean Patagonia, dated by lichenometry and dendrochronology. *Arctic, Antarctic and Alpine Research* **33**: 266-273.

4.2. Sea Ice

A number of claims have been made that CO_2-induced global warming is melting sea ice in the Arctic and Antarctic and that such melting will accelerate as time passes. In this section we analyze Antarctic and Arctic sea ice trends as reported in the scientific literature. We revisit the issue of ice melting in much greater depth in Section 4.5.

4.2.1. Antarctic

Utilizing Special Sensor Microwave Imager (SSM/I) data obtained from the Defense Meteorological Satellite Program (DMSP) for the period December 1987-December 1996, Watkins and Simmonds (2000) analyzed temporal trends in different measures of the sea ice that surrounds Antarctica, noting that "it has been suggested that the Antarctic sea ice may show high sensitivity to any anthropogenic increase in temperature," and that most climate models predict that "any rise in surface temperature would result in a decrease in sea ice coverage."

Contrary to what one would expect on the basis of these predictions, the two scientists observed statistically significant *increases* in both sea ice area and sea ice extent over the period studied; and when they combined their results with results for the preceding period of 1978-1987, both parameters continued to show increases over the sum of the two periods (1978-1996). In addition, they determined that the 1990s also experienced increases in the length of the sea ice season.

Watkins and Simmonds' findings, i.e., that Southern Ocean sea ice has increased in area, extent, and season length since at least 1978, are supported by other studies. Hanna (2001) published an updated analysis of Antarctic sea ice cover based on SSM/I data for the period October 1987-September 1999, finding the serial sea ice data depict "an ongoing slight but significant hemispheric increase of 3.7($\pm$0.3)% in extent and 6.6($\pm$1.5)% in area." Parkinson (2002) utilized satellite passive-microwave data to calculate and map the length of the sea-ice season throughout the Southern Ocean for each year of the period 1979-1999, finding that although there are opposing regional trends, a "much larger area of the Southern Ocean experienced an overall lengthening of the sea-ice season ... than experienced a shortening." Updating the analysis two years later for the period November 1978 through

December 2002, Parkinson (2004) reported a linear increase in 12-month running means of Southern Ocean sea ice extent of 12,380 ± 1,730 km$_2$ per year.

Zwally *et al.* (2002) also utilized passive-microwave satellite data to study Antarctic sea ice trends. Over the 20-year period 1979-1998, they report that the sea ice extent of the entire Southern Ocean increased by 11,181 ± 4,190 square km per year, or by 0.98 ± 0.37 percent per decade, while sea ice area increased by nearly the same amount: 10,860 ± 3,720 square km per year, or by 1.26 ± 0.43 percent per decade. They observed that the variability of monthly sea ice extent declined from 4.0 percent over the first 10 years of the record, to 2.7 percent over the last 10 years.

Yuan and Martinson (2000) analyzed Special SSM/I data together with data derived from brightness temperatures measured by the Nimbus-7 Scanning Multichannel Microwave Radiometer. Among other things, they determined that the mean trend in the latitudinal location of the Antarctic sea ice edge over the prior 18 years was an equatorward expansion of 0.011 degree of latitude per year.

Vyas *et al.* (2003) analyzed data from the multi-channel scanning microwave radiometer carried aboard India's OCEANSAT-1 satellite for the period June 1999-May 2001, which they combined with data for the period 1978-1987 that were derived from space-based passive microwave radiometers carried aboard earlier Nimbus-5, Nimbus-7, and DMSP satellites to study secular trends in sea ice extent about Antarctica over the period 1978-2001. Their work revealed that the mean rate of change of sea ice extent for the entire Antarctic region over this period was an increase of 0.043 M km^2 per year. In fact, they concluded that "the increasing trend in the sea ice extent over the Antarctic region may be slowly accelerating in time, particularly over the last decade," noting that the "continually increasing sea ice extent over the Antarctic Southern Polar Ocean, along with the observed decreasing trends in Antarctic ice surface temperature (Comiso, 2000) over the last two decades, is paradoxical in the global warming scenario resulting from increasing greenhouse gases in the atmosphere."

In a somewhat similar study, Cavalieri *et al.* (2003) extended prior satellite-derived Antarctic sea ice records several years by bridging the gap between Nimbus 7 and earlier Nimbus 5 satellite datasets with National Ice Center digital sea ice data, finding that sea ice extent about the continent increased at a mean rate of 0.10 ± 0.05 x 10^6 km^2 per decade between

1977 and 2002. Likewise, Liu *et al.* (2004) used sea ice concentration data retrieved from the scanning multichannel microwave radiometer on the Nimbus 7 satellite and the spatial sensor microwave/imager on several defense meteorological satellites to develop a quality-controlled history of Antarctic sea ice variability covering the period 1979-2002, which includes different states of the Antarctic Oscillation and several ENSO events, after which they evaluated total sea ice extent and area trends by means of linear least-squares regression. They found that "overall, the total Antarctic sea ice extent (the cumulative area of grid boxes covering at least 15% ice concentrations) has shown an increasing trend (~4,801 km^2/yr)." In addition, they determined that "the total Antarctic sea ice area (the cumulative area of the ocean actually covered by at least 15% ice concentrations) has increased significantly by ~13,295 km^2/yr, exceeding the 95% confidence level," noting that "the upward trends in the total ice extent and area are robust for different cutoffs of 15, 20, and 30% ice concentrations (used to define the ice extent and area)."

Elderfield and Rickaby (2000) concluded that the sea ice cover of the Southern Ocean during glacial periods may have been as much as double the coverage of modern winter ice, suggesting that "by restricting communication between the ocean and atmosphere, sea ice expansion also provides a mechanism for reduced CO_2 release by the Southern Ocean and lower glacial atmospheric CO_2."

Three papers on Antarctic sea ice were published in 2008. Laine (2008) determined 1981-2000 trends of Antarctic sea-ice concentration and extent, based on the Scanning Multichannel Microwave Radiometer (SSMR) and SSM/I for the spring-summer period of November/December/ January. These analyses were carried out for the continent as a whole, as well as five longitudinal sectors emanating from the south pole: 20°E-90°E, 90°E-160°E, 160°E-130°W, 130°W-60°W, and 60°W-20°E. Results indicated that "the sea ice concentration shows slight increasing trends in most sectors, where the sea ice extent trends seem to be near zero." Laine also reports that "the Antarctic region as a whole and all the sectors separately show slightly positive spring-summer albedo trends."

Comiso and Nishio (2008) set out to provide updated and improved estimates of trends in Arctic and Antarctic sea ice cover for the period extending from November 1978 to December 2006, based on data obtained from the Advanced Microwave

Scanning Radiometer (AMSR-E), the SSM/I, and the SMMR, where the data from the last two instruments were adjusted to be consistent with the AMSR-E data. Their findings indicate that sea ice extent and area in the Antarctic grew by +0.9 ± 0.2 and +1.7 ± 0.3 percent per decade, respectively.

A study that "extends the analyses of the sea ice time series reported by Zwally *et al.* (2002) from 20 years (1979-1998) to 28 years (1979-2006)" by Cavalieri and Parkinson (2008) derived new linear trends of Antarctic sea ice extent and area based on satellite-borne passive microwave radiometer data. Results indicate "the total Antarctic sea ice extent trend increased slightly, from 0.96 ± 0.61 percent per decade to 1.0 ± 0.4 percent per decade, from the 20- to 28-year period," noting the latter trend is significant at the 95 percent confidence level. Corresponding numbers for the Antarctic sea ice area trend were 1.2 ± 0.7 percent per decade and 1.2 ± 0.5 percent per decade. Both sets of results indicate a "tightening up" of the two relationships: Over the last eight years of the study period, both the extent and area of Antarctic sea ice have continued to increase, with the former parameter increasing at a more rapid rate than it did over the 1979-1998 period.

Additional information on this topic, including reviews of newer publications as they become available, can be found at http://www.co2science.org/subject/s/seaiceantarctic.php.

References

Cavalieri, D.J. and Parkinson, C.L. 2008. Antarctic sea ice variability and trends, 1979-2006. Journal of Geophysical Research **113**: 10.1029/2007JC004564.

Cavalieri, D.J., Parkinson, C.L. and Vinnikov, K.Y. 2003. 30-Year satellite record reveals contrasting Arctic and Antarctic decadal sea ice variability. *Geophysical Research Letters* **30**: 10.1029/2003GL018031.

Comiso, J.C. 2000. Variability and trends in Antarctic surface temperatures from in situ and satellite infrared measurements. *Journal of Climate* **13**: 1674-1696.

Comiso, J.C. and Nishio, F. 2008. Trends in the sea ice cover using enhanced and compatible AMSR-E, SSM/I, and SMMR data. *Journal of Geophysical Research* **113**: 10.1029/2007JC004257.

Elderfield, H. and Rickaby, R.E.M. 2000. Oceanic Cd/P ratio and nutrient utilization in the glacial Southern Ocean. *Nature* **405**: 305-310.

Hanna, E. 2001. Anomalous peak in Antarctic sea-ice area, winter 1998, coincident with ENSO. *Geophysical Research Letters* **28**: 1595-1598.

Laine, V. 2008. Antarctic ice sheet and sea ice regional albedo and temperature change, 1981-2000, from AVHRR Polar Pathfinder data. *Remote Sensing of Environment* **112**: 646-667.

Liu, J., Curry, J.A. and Martinson, D.G. 2004. Interpretation of recent Antarctic sea ice variability. *Geophysical Research Letters* **31**: 10.1029/2003GL018732.

Parkinson, C.L. 2002. Trends in the length of the Southern Ocean sea-ice season, 1979-99. *Annals of Glaciology* **34**: 435-440.

Parkinson, C.L. 2004. Southern Ocean sea ice and its wider linkages: insights revealed from models and observations. *Antarctic Science* **16**: 387-400.

Vyas, N.K., Dash, M.K., Bhandari, S.M., Khare, N., Mitra, A. and Pandey, P.C. 2003. On the secular trends in sea ice extent over the antarctic region based on OCEANSAT-1 MSMR observations. *International Journal of Remote Sensing* **24**: 2277-2287.

Watkins, A.B. and Simmonds, I. 2000. Current trends in Antarctic sea ice: The 1990s impact on a short climatology. *Journal of Climate* **13**: 4441-4451.

Yuan, X. and Martinson, D.G. 2000. Antarctic sea ice extent variability and its global connectivity. *Journal of Climate* **13**: 1697-1717.

Zwally, H.J., Comiso, J.C., Parkinson, C.L. Cavalieri, D.J. and Gloersen, P. 2002. Variability of Antarctic sea ice 1979-1998. *Journal of Geophysical Research* **107**: 10.1029/2000JC000733.

4.2.2. Arctic

Arctic climate is incredibly complex, varying simultaneously on a number of different timescales for a number of different reasons (Venegas and Mysak, 2000). Against this backdrop of multiple causation and timeframe variability, it is difficult to identify a change in either the extent or thickness of Arctic sea ice that could be attributed to the increase in temperature that has been predicted to result from the burning of fossil fuels. The task is further complicated because many of the records that do exist contain only a few years to a few decades of data, and they yield different trends depending on the period of time studied.

4.2.2.1. Extent

Johannessen *et al.* (1999) analyzed Arctic sea ice extent over the period 1978-1998 and found it to have decreased by about 14 percent. This finding led them to suggest that "the balance of evidence," as small as it then was, indicates "an ice cover in transition," and that "if this apparent transformation continues, it may lead to a markedly different ice regime in the Arctic," as was also suggested by Vinnikov *et al.* (1999).

Reading Johannessen *et al.*'s assessment of the situation, one is left with the impression that a relatively consistent and persistent reduction in the area of Arctic sea ice is in progress. However, and according to their own data, that assessment is highly debatable and possibly false. In viewing their plots of sea ice area, for example, it is readily evident that the decline in this parameter did not occur smoothly over the 20-year period of study. In fact, essentially all of the drop it experienced occurred abruptly over a single period of not more than three years (87/88-90/91) and possibly only one year (89/90-90/91). Furthermore, it could be argued from their data that from 1990/91 onward, sea ice area in the Arctic may have actually *increased*.

Support for this assessment of the data is found in Kwok (2004), who estimated "the time-varying perennial ice zone (PIZ) coverage and construct[s] the annual cycles of multiyear (MY, including second year) ice coverage of the Arctic Ocean using QuikSCAT backscatter, MY fractions from RADARSAT, and the record of ice export from satellite passive microwave observations" for the years 1999-2003. Kwok calculated the coverage of Arctic MY sea ice at the beginning of each year of the study was 3774×10^3 km^2 in 2000, 3896×10^3 km^2 in 2001, 4475×10^3 km^2 in 2002, and 4122×10^3 km^2 in 2003, representing an *increase* in sea ice coverage of 9 percent over a third of a decade.

More questions are raised Parkinson (2000b), who utilized satellite-derived data of sea ice extent to calculate changes in this parameter for the periods 1979-1990 and 1990-1999. He reports that in seven of the nine regions into which he divided the Arctic for his analysis, the "sign of the trend reversed from the 1979-1990 period to the 1990-1999 period," indicative of the ease with which significant decadal trends are often reversed in this part of the world.

In another study, Belchansky *et al.* (2004) report that from 1988 to 2001, total January multiyear ice area declined at a mean rate of 1.4 percent per year. In the autumn of 1996, however, they note that "a large

multiyear ice recruitment of over 10^6 km^2 fully replenished the previous 8-year decline in total area." They add that the replenishment "was followed by an accelerated and compensatory decline during the subsequent 4 years." In addition, they learned that 75 percent of the interannual variation in January multiyear sea area "was explained by linear regression on two atmospheric parameters: the previous winter's Arctic Oscillation index as a proxy to melt duration and the previous year's average sea level pressure gradient across the Fram Strait as a proxy to annual ice export."

Belchansky *et al.* conclude that their 14-year analysis of multiyear ice dynamics is "insufficient to project long-term trends." They also conclude it is insufficient to reveal "whether recent declines in multiyear ice area and thickness are indicators of anthropogenic exacerbations to positive feedbacks that will lead the Arctic to an unprecedented future of reduced ice cover, or whether they are simply ephemeral expressions of natural low frequency oscillations." It should be noted in this regard, however, that low frequency oscillations are what the data actually reveal; and such behavior is not what one would predict from a gradually increasing atmospheric CO_2 concentration.

In another study, Heide-Jorgensen and Laidre (2004) examined changes in the fraction of open-water found within various pack-ice microhabitats of Foxe Basin, Hudson Bay, Hudson Strait, Baffin Bay-Davis Strait, northern Baffin Bay, and Lancaster Sound over a 23-year interval (1979-2001) using remotely sensed microwave measurements of sea-ice extent, after which the trends they documented were "related to the relative importance of each wintering microhabitat for eight marine indicator species and potential impacts on winter success and survival were examined."

Results of the analysis indicate that Foxe Basin, Hudson Bay, and Hudson Strait showed small increasing trends in the fraction of open-water, with the upward trends at all microhabitats studied ranging from 0.2 to 0.7 percent per decade. In Baffin Bay-Davis Straight and northern Baffin Bay, on the other hand, the open-water trend was *downward*, and at a mean rate for all open-water microhabitats studied of fully 1 percent per decade, while the trend in all Lancaster Sound open-water microhabitats was also downward, in this case at a mean rate of 0.6 percent per decade.

With respect to the context of these open-water declines, Heide-Jorgensen and Laidre report that "increasing trends in sea ice coverage in Baffin Bay and Davis Strait (resulting in declining open-water) were as high as 7.5 percent per decade between 1979-1999 (Parkinson *et al.*, 1999; Deser *et al.*, 2000; Parkinson, 2000a,b; Parkinson and Cavalieri, 2002) and comparable significant increases have been detected back to 1953 (Stern and Heide-Jorgensen, 2003)." They additionally note that "similar trends in sea ice have also been detected locally along the West Greenland coast, with slightly lower increases of 2.8 percent per decade (Stern and Heide-Jorgensen, 2003)."

Cavalieri *et al.* (2003) extended prior satellite-derived Arctic sea ice records several years back in time by bridging the gap between Nimbus 7 and earlier Nimbus 5 satellite datasets via comparisons with National Ice Center digital sea ice data. For the newly extended period of 1972-2002, they determined that Arctic sea ice extent had declined at a mean rate of $0.30 \pm 0.03 \times 10^6$ km^2 per decade; while for the shortened period from 1979-2002, they found a mean rate of decline of $0.36 \pm 0.05 \times 10^6$ km^2 per decade, or at a rate that was 20 percent greater than the full-period rate. In addition Serreze *et al.* (2002) determined that the downward trend in Arctic sea ice extent during the passive microwave era culminated with a record minimum value in 2002.

These results could readily be construed to indicate an increasingly greater rate of Arctic sea ice melting during the latter part of the twentieth century. However, the results of these studies are not the end of the story. As Grumet *et al.* (2001) have described the situation, recent trends in Arctic sea ice cover "can be viewed out of context because their brevity does not account for interdecadal variability, nor are the records sufficiently long to clearly establish a climate trend."

In an effort to overcome this "short-sightedness," Grumet *et al.* developed a 1,000-year record of spring sea ice conditions in the Arctic region of Baffin Bay based on sea-salt records from an ice core obtained from the Penny Ice Cap on Baffin Island. In doing so, they determined that after a period of reduced sea ice during the eleventh through fourteenth centuries, enhanced sea ice conditions prevailed during the following 600 years. For the final (twentieth) century of this period, however, they report that "despite warmer temperatures during the turn of the century, sea-ice conditions in the Baffin Bay/Labrador Sea region, at least during the last 50 years, are within 'Little Ice Age' variability," suggesting that sea ice

extent there has not yet emerged from the range of conditions characteristic of the Little Ice Age.

In an adjacent sector of the Arctic, this latter period of time was also studied by Comiso *et al.* (2001), who used satellite imagery to analyze and quantify a number of attributes of the Odden ice tongue—a winter ice-cover phenomenon that occurs in the Greenland Sea with a length of about 1,300 km and an aerial coverage of as much as 330,000 square kilometers—over the period 1979-1998. By utilizing surface air temperature data from Jan Mayen Island, which is located within the region of study, they were able to infer the behavior of this phenomenon over the past 75 years. Trend analyses revealed that the ice tongue has exhibited no statistically significant change in any of the parameters studied over the past 20 years; but the proxy reconstruction of the Odden ice tongue for the past 75 years revealed the ice phenomenon to have been "a relatively smaller feature several decades ago," due to the warmer temperatures that prevailed at that time.

In another study of Arctic climate variability, Omstedt and Chen (2001) obtained a proxy record of the annual maximum extent of sea ice in the region of the Baltic Sea over the period 1720-1997. In analyzing this record, they found that a significant decline in sea ice occurred around 1877. In addition, they reported finding greater variability in sea ice extent in the colder 1720-1877 period than in the warmer 1878-1997 period.

Also at work in the Baltic Sea region, Jevrejeva (2001) reconstructed an even longer record of sea ice duration (and, therefore, extent) by examining historical data for the observed time of ice break-up between 1529 and 1990 in the northern port of Riga, Latvia. The long date-of-ice-break-up time series was best described by a fifth-order polynomial, which identified four distinct periods of climatic transition: (1) 1530-1640, warming with a tendency toward earlier ice break-up of nine days/century, (2) 1640-1770, cooling with a tendency toward later ice break-up of five days/century, (3) 1770-1920, warming with a tendency toward earlier ice break-up of 15 days/century, and (4) 1920-1990, *cooling* with a tendency toward later ice break-up of 12 days/century.

On the other hand, in a study of the Nordic Seas (the Greenland, Iceland, Norwegian, Barents, and Western Kara Seas), Vinje (2001) determined that "the extent of ice in the Nordic Seas measured in April has decreased by 33% over the past 135 years." He notes, however, that "nearly half of this reduction is observed over the period 1860-1900," and we note, in this regard, that the first half of this sea-ice decline occurred over a period of time when the atmosphere's CO_2 concentration rose by only 7 ppm, whereas the second half of the sea-ice decline occurred over a period of time when the air's CO_2 concentration rose by more than 70 ppm. If the historical rise in the air's CO_2 content has been responsible for the historical decrease in sea-ice extent, its impact over the last century has declined to less than a tenth of what its impact was over the preceding four decades. This in turn suggests that the increase in the air's CO_2 content over the past 135 years has likely had nothing to do with the concomitant decline in sea-ice cover.

In a similar study of the Kara, Laptev, East Siberian, and Chuckchi Seas, based on newly available long-term Russian observations, Polyakov *et al.* (2002) found "smaller than expected" trends in sea ice cover that, in their words, "do not support the hypothesized polar amplification of global warming." Likewise, in a study published the following year, Polyakov *et al.* (2003) report that "over the entire Siberian marginal-ice zone the century-long trend is only -0.5% per decade," while "in the Kara, Laptev, East Siberian, and Chukchi Seas the ice extent trends are not large either: -1.1%, -0.4%, +0.3%, and -1.0% per decade, respectively." Moreover, they say "these trends, except for the Chukchi Sea, are not statistically significant."

Divine and Dick (2006) used historical April through August ice observations made in the Nordic Seas—comprised of the Iceland, Greenland, Norwegian, and Barents Seas, extending from 30°W to 70°E—to construct time series of ice-edge position anomalies spanning the period 1750-2002, which they analyzed for evidence of long-term trend and oscillatory behavior. The authors report that "evidence was found of oscillations in ice cover with periods of about 60 to 80 years and 20 to 30 years, superimposed on a continuous negative trend," which observations are indicative of a "persistent ice retreat since the second half of the 19th century" that began well before anthropogenic CO_2 emissions could have had much effect on earth's climate.

Noting that the last cold period observed in the Arctic occurred at the end of the 1960s, the two Norwegian researchers say their results suggest that "the Arctic ice pack is now at the periodical apogee of the low-frequency variability," and that "this could explain the strong negative trend in ice extent during the last decades as a possible superposition of natural low frequency variability and greenhouse gas induced

warming of the last decades." However, as they immediately caution, "a similar shrinkage of ice cover was observed in the 1920s-1930s, during the previous warm phase of the low frequency oscillation, when any anthropogenic influence is believed to have still been negligible." They suggest, therefore, "that during decades to come ... the retreat of ice cover may change to an expansion."

In light of this litany of findings, it is difficult to accept the claim that Northern Hemispheric sea ice is rapidly disintegrating in response to CO_2-induced global warming. Rather, the oscillatory behavior observed in so many of the sea ice studies suggests, in the words of Parkinson (2000b), "the possibility of close connections between the sea ice cover and major oscillatory patterns in the atmosphere and oceans," including connections with: "(1) the North Atlantic Oscillation (e.g., Hurrell and van Loon, 1997; Johannessen *et al.*, 1999; Kwok and Rothrock, 1999; Deser *et al.*, 2000; Kwok, 2000, Vinje, 2001) and the spatially broader Arctic Oscillation (e.g., Deser *et al.*, 2000; Wang and Ikeda, 2000); (2) the Arctic Ocean Oscillation (Polyakov *et al.*, 1999; Proshutinsky *et al.*, 1999); (3) a 'see-saw' in winter temperatures between Greenland and northern Europe (Rogers and van Loon, 1979); and (4) an interdecadal Arctic climate cycle (Mysak *et al.*, 1990; Mysak and Power, 1992)." The likelihood that Arctic sea ice trends are the product of such natural oscillations, Parkinson continues, "provides a strong rationale for considerable caution when extrapolating into the future the widely reported decreases in the Arctic ice cover over the past few decades or when attributing the decreases primarily to global warming," a caution with which we heartily agree.

One final study of note is that of Bamber *et al.* (2004), who used high-accuracy ice-surface elevation measurements (Krabill *et al.*, 2000) of the largest ice cap in the Eurasian Arctic—Austfonna, on the island of Nordaustlandet in northeastern Svalbard—to evaluate ice cap elevation changes between 1996 and 2002. They determined that the central and highest-altitude area of the ice cap, which comprises 15 percent of its total area, "increased in elevation by an average of 50 cm per year between 1996 and 2002," while "to the northeast of this region, thickening of about 10 cm per year was also observed." They further note that the highest of these growth rates represents "as much as a 40% increase in accumulation rate (Pinglot *et al.*, 2001)."

Based on the ancillary sea-ice and meteorological data they analyzed, Bamber *et al.* concluded that the best explanation for the dramatic increase in ice cap growth over the six-year study period was a large increase in precipitation caused by a concomitant reduction in sea-ice cover in this sector of the Arctic. Their way of characterizing this phenomenon is simply to say that it represents the transference of ice from the top of the sea (in this case, the Barents Sea) to the top of the adjacent land (in this case, the Austfonna ice cap). And as what has been observed to date is only the beginning of the phenomenon, which will become even stronger in the absence of nearby sea-ice, "projected changes in Arctic sea-ice cover," as they say in the concluding sentence of their paper, "will have a significant impact on the mass-balance of land ice around the Arctic Basin over at least the next 50 years." Which result, we might add, may be just the opposite of that forecast by the IPCC.

Additional information on this topic, including reviews of newer publications as they become available, can be found at http://www.co2science.org/subject/s/seaicearctic.php.

References

Bamber, J., Krabill, W., Raper, V. and Dowdeswell, J. 2004. Anomalous recent growth of part of a large Arctic ice cap: Austfonna, Svalbard. *Geophysical Research Letters* **31**: 10.1029/2004GL019667.

Belchansky, G.I., Douglas, D.C., Alpatsky, I.V. and Platonov, N.G. 2004. Spatial and temporal multiyear sea ice distributions in the Arctic: A neural network analysis of SSM/I data, 1988-2001. *Journal of Geophysical Research* **109**: 10.1029/2004JC002388.

Cavalieri, D.J., Parkinson, C.L. and Vinnikov, K.Y. 2003. 30-Year satellite record reveals contrasting Arctic and Antarctic decadal sea ice variability. *Geophysical Research Letters* **30**: 10.1029/2003GL018031.

Comiso, J.C., Wadhams, P., Pedersen, L.T. and Gersten, R.A. 2001. Seasonal and interannual variability of the Odden ice tongue and a study of environmental effects. *Journal of Geophysical Research* **106**: 9093-9116.

Deser, C., Walsh, J. and Timlin, M.S. 2000. Arctic sea ice variability in the context of recent atmospheric circulation trends. *Journal of Climate* **13**: 617-633.

Divine, D.V. and Dick, C. 2006. Historical variability of sea ice edge position in the Nordic Seas. *Journal of Geophysical Research* **111**: 10.1029/2004JC002851.

Grumet, N.S., Wake, C.P., Mayewski, P.A., Zielinski, G.A., Whitlow, S.L., Koerner, R.M., Fisher, D.A. and Woollett, J.M. 2001. Variability of sea-ice extent in Baffin Bay over the last millennium. *Climatic Change* **49**: 129-145.

Heide-Jorgensen, M.P. and Laidre, K.L. 2004. Declining extent of open-water refugia for top predators in Baffin Bay and adjacent waters. *Ambio* **33**: 487-494.

Hurrell, J.W. and van Loon, H. 1997. Decadal variations in climate associated with the North Atlantic Oscillation. *Climatic Change* **36**: 301-326.

Jevrejeva, S. 2001. Severity of winter seasons in the northern Baltic Sea between 1529 and 1990: reconstruction and analysis. *Climate Research* **17**: 55-62.

Johannessen, O.M., Shalina, E.V. and Miles, M.W. 1999. Satellite evidence for an Arctic sea ice cover in transformation. *Science* **286**: 1937-1939.

Krabill, W., Abdalati, W., Frederick, E., Manizade, S., Martin, C., Sonntag, J., Swift, R., Thomas, R., Wright, W. and Yungel, J. 2000. Greenland ice sheet: High-elevation balance and peripheral thinning. *Science* **289**: 428-430.

Kwok, R. 2000. Recent changes in Arctic Ocean sea ice motion associated with the North Atlantic Oscillation. *Geophysical Research Letters* **27**: 775-778.

Kwok, R. 2004. Annual cycles of multiyear sea ice coverage of the Arctic Ocean: 1999-2003. *Journal of Geophysical Research* **109**: 10.1029/2003JC002238.

Kwok, R. and Rothrock, D.A. 1999. Variability of Fram Strait ice flux and North Atlantic Oscillation. *Journal of Geophysical Research* **104**: 5177-5189.

Mysak, L.A., Manak, D.K. and Marsden, R.F. 1990. Sea-ice anomalies observed in the Greenland and Labrador Seas during 1901-1984 and their relation to an interdecadal Arctic climate cycle. *Climate Dynamics* **5**: 111-133.

Mysak, L.A. and Power, S.B. 1992. Sea-ice anomalies in the western Arctic and Greenland-Iceland Sea and their relation to an interdecadal climate cycle. *Climatological Bulletin/Bulletin Climatologique* **26**: 147-176.

Omstedt, A. and Chen, D. 2001. Influence of atmospheric circulation on the maximum ice extent in the Baltic Sea. *Journal of Geophysical Research* **106**: 4493-4500.

Parkinson, C.L. 2000a. Variability of Arctic sea ice: the view from space, and 18-year record. *Arctic* **53**: 341-358.

Parkinson, C.L. 2000b. Recent trend reversals in Arctic sea ice extents: possible connections to the North Atlantic Oscillation. *Polar Geography* **24**: 1-12.

Parkinson, C.L. and Cavalieri, D.J. 2002. A 21-year record of Arctic sea-ice extents and their regional, seasonal and monthly variability and trends. *Annals of Glaciology* **34**: 441-446.

Parkinson, C.L., Cavalieri, D.J., Gloersen, P., Zwally, H.J. and Comiso, J.C. 1999. Arctic sea ice extents, areas, and trends, 1978-1996. *Journal of Geophysical Research* **104**: 20,837-20,856.

Pinglot, J.F., Hagen, J.O., Melvold, K., Eiken, T. and Vincent, C. 2001. A mean net accumulation pattern derived from radioactive layers and radar soundings on Austfonna, Nordaustlandet, Svalbard. *Journal of Glaciology* **47**: 555-566.

Polyakov, I.V., Proshutinsky, A.Y. and Johnson, M.A. 1999. Seasonal cycles in two regimes of Arctic climate. *Journal of Geophysical Research* **104**: 25,761-25,788.

Polyakov, I.V., Alekseev, G.V., Bekryaev, R.V., Bhatt, U., Colony, R.L., Johnson, M.A., Karklin, V.P., Makshtas, A.P., Walsh, D. and Yulin, A.V. 2002. Observationally based assessment of polar amplification of global warming. *Geophysical Research Letters* **29**: 10.1029/2001GL011111.

Polyakov, I.V., Alekseev, G.V., Bekryaev, R.V., Bhatt, U.S., Colony, R., Johnson, M.A., Karklin, V.P., Walsh, D. and Yulin, A.V. 2003. Long-term ice variability in Arctic marginal seas. *Journal of Climate* **16**: 2078-2085.

Proshutinsky, A.Y., Polyakov, I.V. and Johnson, M.A. 1999. Climate states and variability of Arctic ice and water dynamics during 1946-1997. *Polar Research* **18**: 135-142.

Rogers, J.C. and van Loon, H. 1979. The seesaw in winter temperatures between Greenland and Northern Europe. Part II: Some oceanic and atmospheric effects in middle and high latitudes. *Monthly Weather Review* **107**: 509-519.

Serreze, M.C., Maslanik, J.A., Scambos, T.A., Fetterer, F., Stroeve, J., Knowles, K., Fowler, C., Drobot, S., Barry, R.G. and Haran, T.M. 2003. A record minimum arctic sea ice extent and area in 2002. *Geophysical Research Letters* **30**: 10.1029/2002GL016406.

Stern, H.L. and Heide-Jorgensen, M.P. 2003. Trends and variability of sea ice in Baffin Bay and Davis Strait. *Polar Research* **22**: 11-18.

Venegas, S.A. and Mysak, L.A. 2000. Is there a dominant timescale of natural climate variability in the Arctic? *Journal of Climate* **13**: 3412-3434.

Vinje, T. 2001. Anomalies and trends of sea ice extent and atmospheric circulation in the Nordic Seas during the period 1864-1998. *Journal of Climate* **14**: 255-267.

Vinnikov, K.Y., Robock, A., Stouffer, R.J., Walsh, J.E., Parkinson, C.L., Cavalieri, D.J., Mitchell, J.F.B., Garrett, D. and Zakharov, V.R. 1999. Global warming and Northern Hemisphere sea ice extent. *Science* **286**: 1934-1937.

Wang, J. and Ikeda, M. 2000. Arctic Oscillation and Arctic Sea-Ice Oscillation. *Geophysical Research Letters* **27**: 1287-1290.

4.2.2.2. Thickness

Based on analyses of submarine sonar data, Rothrock *et al.* (1999) suggested that Arctic sea ice in the mid 1990s had thinned by about 42 percent of the average 1958-1977 thickness. The IPCC reports the Rothrock finding but then reports that other more recent studies found "the reduction in ice thickness was not gradual, but occurred abruptly before 1991," and acknowledges that "ice thickness varies considerably from year to year at a given location and so the rather sparse temporal sampling provided by submarine data makes inferences regarding long term change difficult" (IPCC 2007, p. 353). Johannessen *et al.* (1999), for example, found that essentially all of the drop occurred rather abruptly over a single period of not more than three years (1987/88-1990/91) and possibly only one year (1989/90-1990/91).

Two years after Johannessen *et al.*, Winsor (2001) analyzed a more comprehensive set of Arctic sea-ice data obtained from six submarine cruises conducted between 1991 and 1997 that had covered the central Arctic Basin from 76° N to 90° N, as well as two areas that had been particularly densely sampled, one centered at the North Pole (>87° N) and one in the central part of the Beaufort Sea (centered at approximately 76° N, 145°W). The transect data across the entire Arctic Basin revealed that the mean Arctic sea-ice thickness had remained "almost constant" over the period of study. Data from the North Pole also showed little variability, and a linear regression of the data revealed a "slight increasing trend for the whole period." As for the Beaufort Sea region, annual variability in sea ice thickness was greater than at the North Pole but once again, in Winsor's words, "no significant trend" in mean sea-ice thickness was found. Combining the North Pole results with the results of an earlier study, Winsor concluded that "mean ice thickness has remained on a near-constant level around the North Pole from 1986 to 1997."

The following year, Holloway and Sou (2002) explored "how observations, theory, and modeling work together to clarify perceived changes to Arctic sea ice," incorporating data from "the atmosphere, rivers, and ocean along with dynamics expressed in an ocean-ice-snow model." On the basis of a number of different data-fed model runs, they found that for the last half of the past century, "no linear trend [in Arctic sea ice volume] over 50 years is appropriate," noting their results indicated "increasing volume to the mid-1960s, decadal variability without significant trend from the mid-1960s to the mid-1980s, then a loss of volume from the mid-1980s to the mid-1990s." The net effect of this behavior, in their words, was that "the volume estimated in 2000 is close to the volume estimated in 1950." They suggest that the initial inferred rapid thinning of Arctic sea ice was, as they put it, "unlikely," due to problems arising from under-sampling. They also report that "varying winds that readily redistribute Arctic ice create a recurring pattern whereby ice shifts between the central Arctic and peripheral regions, especially in the Canadian sector," and that the "timing and tracks of the submarine surveys missed this dominant mode of variability."

In the same year, Polyakov *et al.* (2002) employed newly available long-term Russian landfast-ice data obtained from the Kara, Laptev, East Siberian, and Chuckchi Seas to investigate trends and variability in the Arctic environment poleward of 62°N. This study revealed that fast-ice thickness trends in the different seas were "relatively small, positive or negative in sign at different locations, and not statistically significant at the 95% level." A year later, these results were reconfirmed by Polyakov *et al.* (2003), who reported that the available fast-ice records "do not show a significant trend," while noting that "in the Kara and Chukchi Seas trends are positive, and in the Laptev and East Siberian Seas trends are negative," but stating that "these trends are not statistically significant at the 95% confidence level."

Laxon *et al.* (2003) used an eight-year time series (1993-2001) of Arctic sea-ice thickness data derived from measurements of ice freeboard made by radar altimeters carried aboard ERS-1 and 2 satellites to determine the mean thickness and variability of Arctic sea ice between latitudes 65° and 81.5°N, which region covers the entire circumference of the Arctic Ocean, including the Beaufort, Chukchi, East Siberian, Kara, Laptev, Barents, and Greenland Seas. These real-world observations (1) revealed "an interannual variability in ice thickness at higher frequency, and of greater amplitude, than simulated by regional Arctic models," (2) undermined "the conclusion from numerical models that changes in ice thickness occur on much longer timescales than changes in ice extent," and (3) showed that "sea ice mass can change by up to 16% within one year,"

which finding "contrasts with the concept of a slowly dwindling ice pack, produced by greenhouse warming." Laxon *et al.* concluded that "errors are present in current simulations of Arctic sea ice," stating in their closing sentence that "until models properly reproduce the observed high-frequency, and thermodynamically driven, variability in sea ice thickness, simulations of both recent, and future, changes in Arctic ice cover will be open to question."

Pfirman *et al.* (2004) analyzed Arctic sea-ice drift dynamics from 1979-1997, based on monthly fields of ice motion obtained from the International Arctic Buoy Program, using a Lagrangian perspective that "shows the complexities of ice drift response to variations in atmospheric conditions." This analysis indicated that "large amounts of sea ice form over shallow Arctic shelves, are transported across the central basin and are exported primarily through Fram Strait and, to lesser degrees, the Barents Sea and Canadian Archipelago," consistent with the observations of several other investigators. They also determined that within the central Arctic, ice travel times averaged 4.0 years from 1984-85 through 1988-89, but only 3.0 years from 1990-91 through 1996-97. This enhanced rate of export of old ice to Fram Strait from the Beaufort Gyre over the latter period decreased the fraction of thick-ridged ice within the central basin of the Arctic, and was deemed by Pfirman *et al.* to be responsible for some of the sea-ice thinning observed between the 1980s and 1990s. They also note that the rapid change in ice dynamics that occurred between 1988 and 1990 was "in response to a weakening of the Beaufort high pressure system and a strengthening of the European Arctic low (a shift from lower North Atlantic Oscillation/Arctic Oscillation to higher NAO/OA index) [Walsh *et al.*, 1996; Proshutinsky and Johnson, 1997; Kwok, 2000; Zhang *et al.*, 2000; Rigor *et al.*, 2002]."

Lastly, in a paper on landfast ice in Canada's Hudson Bay, Gagnon and Gough (2006) cite nine different studies of sea-ice cover, duration, and thickness in the Northern Hemisphere, noting that the Hudson Bay region "has been omitted from those studies with the exception of Parkinson *et al.* (1999)." For 13 stations located on the shores of Hudson Bay (seven) and surrounding nearby lakes (six), Gagnon and Gough then analyzed long-term weekly measurements of ice thickness and associated weather conditions that began and ended, in the mean, in 1963 and 1993, respectively.

Results of the study revealed that a "statistically significant thickening of the ice cover over time was detected on the western side of Hudson Bay, while a slight thinning lacking statistical significance was observed on the eastern side." This asymmetry, in their words, was "related to the variability of air temperature, snow depth, and the dates of ice freeze-up and break-up," with "increasing maximum ice thickness at a number of stations" being "correlated to earlier freeze-up due to negative temperature trends in autumn," and with high snow accumulation being associated with low ice thickness, "because the snow cover insulates the ice surface, reducing heat conduction and thereby ice growth." Noting that their findings "are in contrast to the projections from general circulation models, and to the reduction in sea-ice extent and thickness observed in other regions of the Arctic," Gagnon and Gough say "this contradiction must be addressed in regional climate change impact assessments."

These observations suggest that much of the reported thinning of Arctic sea ice that occurred in the 1990s—if real, as per Winsor (2001)—was not the result of CO_2-induced global warming. Rather, it was a natural consequence of changes in ice dynamics caused by an atmospheric regime shift, of which there have been several in decades past and will likely be several in decades to come, irrespective of past or future changes in the air's CO_2 content. Whether any portion of possible past sea ice thinning was due to global warming is consequently still impossible to know, for temporal variability in Arctic sea-ice behavior is simply too great to allow such a small and slowly developing signal to be detected yet. In describing an earlier regime shift, for example, Dumas *et al.* (2003) noted that "a sharp decrease in ice thickness of roughly 0.6 m over 4 years (1970-74) [was] followed by an abrupt increase of roughly 0.8 m over 2 years (1974-76)."

It will likely be a number of years before anything definitive can be said about CO_2-induced global warming on the basis of the thickness of Arctic sea-ice, other than that its impact on sea-ice thickness is too small to be detected at the present time.

Additional information on this topic, including reviews of newer publications as they become available, can be found at http://www.co2science.org/subject/s/seaicearcticthick.php.

References

Dumas, J.A., Flato, G.M. and Weaver, A.J. 2003. The impact of varying atmospheric forcing on the thickness of arctic multi-year sea ice. *Geophysical Research Letters* **30**: 10.1029/2003GL017433.

Gagnon, A.S. and Gough, W.A. 2006. East-west asymmetry in long-term trends of landfast ice thickness in the Hudson Bay region, Canada. *Climate Research* **32**: 177-186.

Holloway, G. and Sou, T. 2002. Has Arctic Sea Ice Rapidly Thinned? *Journal of Climate* **15**: 1691-1701.

IPCC. 2007. *Climate Change 2007: The Physical Science Basis. Contribution of Working Group I to the Fourth Assessment Report of the Intergovernmental Panel on Climate Change.* Solomon, S., Qin, D., Manning, M., Chen, Z., Marquis, M., Averyt, K.B., Tignor, M. and Miller, H.L. (Eds.) Cambridge University Press, Cambridge, United Kingdom and New York, NY.

Johannessen, O.M., Shalina, E.V. and Miles, M.W. 1999. Satellite evidence for an Arctic sea ice cover in transformation. *Science* **286**: 1937-1939.

Kwok, R. 2000. Recent changes in Arctic Ocean sea ice motion associated with the North Atlantic Oscillation. *Geophysical Research Letters* **27**: 775-778.

Laxon, S., Peacock, N. and Smith, D. 2003. High interannual variability of sea ice thickness in the Arctic region. *Nature* **425**: 947-950.

Parkinson, C.L., Cavalieri, D.J., Gloersen, P., Zwally, J. and Comiso, J.C. 1999. Arctic sea ice extent, areas, and trends, 1978-1996. *Journal of Geophysical Research* **104**: 20,837-20,856.

Pfirman, S., Haxby, W.F., Colony, R. and Rigor, I. 2004. Variability in Arctic sea ice drift. *Geophysical Research Letters* **31**: 10.1029/2004GL020063.

Polyakov, I.V., Alekseev, G.V., Bekryaev, R.V., Bhatt, U., Colony, R.L., Johnson, M.A., Karklin, V.P., Makshtas, A.P., Walsh, D. and Yulin A.V. 2002. Observationally based assessment of polar amplification of global warming. *Geophysical Research Letters* **29**: 10.1029/2001GL011111.

Polyakov, I.V., Alekseev, G.V., Bekryaev, R.V., Bhatt, U.S., Colony, R., Johnson, M.A., Karklin, V.P., Walsh, D. and Yulin, A.V. 2003. Long-term ice variability in Arctic marginal seas. *Journal of Climate* **16**: 2078-2085.

Proshutinsky, A.Y. and Johnson, M.A. 1997. Two circulation regimes of the wind driven Arctic Ocean. *Journal of Geophysical Research* **102**: 12,493-12,514.

Rigor, I.G., Wallace, J.M. and Colony, R.L. 2002. Response of sea ice to the Arctic oscillation. *Journal of Climate* **15**: 2648-2663.

Rothrock, D.A., Yu, Y. and Maykut, G.A. 1999. Thinning of the Arctic sea ice cover. *Geophysics Research Letters* **26**: 3469-3472.

Walsh, J.E., Chapman, W.L. and Shy, T.L. 1996. Recent decrease of sea level pressure in the central Arctic. *Journal of Climate* **9**: 480-486.

Winsor, P. 2001. Arctic sea ice thickness remained constant during the 1990s. *Geophysical Research Letters* **28**: 1039-1041.

Zhang, J.L., Rothrock, D. and Steele, M. 2000. Recent changes in Arctic sea ice: The interplay between ice dynamics and thermodynamics. *Journal of Climate* **13**: 3099-3114.

4.3. Precipitation Trends

In spite of the fact that global circulation models (GCMs) have failed to accurately reproduce observed patterns and totals of precipitation (Lebel *et al.*, 2000), model predictions of imminent CO_2-induced global warming often suggest that this phenomenon should lead to increases in rainfall amounts and intensities. Rawlins *et al.* (2006) state that "warming is predicted to enhance atmospheric moisture storage resulting in increased net precipitation," citing as the basis for this statement the Arctic Climate Impact Assessment (2005). Peterson *et al.* (2002) have written that "both theoretical arguments and models suggest that net high-latitude precipitation increases in proportion to increases in mean hemispheric temperature," citing the works of Manabe and Stouffer (1994) and Rahmstorf and Ganopolski (1999). Similarly, Kunkel (2003) says "several studies have argued that increasing greenhouse gas concentrations will result in an increase of heavy precipitation (Cubasch *et al.*, 2001; Yonetani and Gordon, 2001; Kharin and Zwiers, 2000; Zwiers and Kharin, 1998; Trenberth, 1998)."

Many scientists are examining historical precipitation records in an effort to determine how temperature changes of the past millennium have impacted these aspects of earth's hydrologic cycle. In this section, we review what some of them have learned about rainfall across the globe, starting with Africa.

Additional information on this subject, including reviews on precipitation topics not discussed here, can be found at http://www.co2science.org/subject/p/subject_p.php.

References

Cubasch, U., Meehl, G.A., Boer, G.J., Stouffer, R.J., Dix, M., Noda, A., Senior, C.A., Raper, S. and Yap, K.S. 2001. Projections of future climate change. In: Houghton, J.T., Ding, Y., Griggs, D.J., Noguer, M., van der Linden, P.J., Dai, X., Maskell, K. and Johnson, C.A. (Eds.) *Climate Change 2001: The Scientific Basis. Contributions of Working Group 1 to the Third Assessment Report of the Intergovernmental Panel on Climate Change.* Cambridge University Press, Cambridge, UK.

Kharin, V.V. and Zwiers, F.W. 2000. Changes in the extremes in an ensemble of transient climate simulations with a coupled atmosphere-ocean GCM. *Journal of Climate* **13**: 3670-3688.

Lebel, T., Delclaux, F., Le Barbé, L. and Polcher, J. 2000. From GCM scales to hydrological scales: rainfall variability in West Africa. *Stochastic Environmental Research and Risk Assessment* **14**: 275-295.

Manabe, S. and Stouffer, R.J. 1994. Multiple-century response of a coupled ocean-atmosphere model to an increase of atmospheric carbon dioxide. *Journal of Climate* **7**: 5-23.

Peterson, B.J., Holmes, R.M., McClelland, J.W., Vorosmarty, C.J., Lammers, R.B., Shiklomanov, A.I., Shiklomanov, I.A. and Rahmstorf, S. 2002. Increasing river discharge to the Arctic Ocean. *Science* **298**: 2171-2173.

Rahmstorf, S. and Ganopolski, A. 1999. Long-term global warming scenarios computed with an efficient coupled climate model. *Climatic Change* **43**: 353-367.

Rawlins, M.A., Willmott, C.J., Shiklomanov, A., Linder, E., Frolking, S., Lammers, R.B. and Vorosmarty, C.J. 2006. Evaluation of trends in derived snowfall and rainfall across Eurasia and linkages with discharge to the Arctic Ocean. *Geophysical Research Letters* **33**: 10.1029/2005GL025231.

Trenberth, K.E. 1998. Atmospheric moisture residence times and cycling: Implications for rainfall rates with climate change. *Climatic Change* **39**: 667-694.

Yonetani, T. and Gordon, H.B. 2001. Simulated changes in the frequency of extremes and regional features of seasonal/annual temperature and precipitation when atmospheric CO_2 is doubled. *Journal of Climate* **14**: 1765-1779.

Zwiers, F.W. and Kharin, V.V. 1998. Changes in the extremes of climate simulated by CCC GCM2 under CO_2-doubling. *Journal of Climate* **11**: 2200-2222.

4.3.1. Global

Huntington (2006) notes there is "a theoretical expectation that climate warming will result in increases in evaporation and precipitation, leading to the hypothesis that one of the major consequences will be an intensification (or acceleration) of the water cycle (DelGenio et al., 1991; Loaciga et al., 1996; Trenberth, 1999; Held and Soden, 2000; Arnell et al., 2001)," and in reviewing the scientific literature on precipitation, he concludes that on a globally averaged basis, "precipitation over land increased by about 2% over the period 1900-1998 (Dai et al., 1997; Hulme et al., 1998)."

New et al. (2001) also reviewed several global precipitation datasets, analyzing the information they contain to obtain a picture of precipitation patterns over the twentieth century. In their case, they determined that precipitation over the land area of the globe was mostly below the century-long mean over the first decade-and-a-half of the record, but that it increased from 1901 to the mid-1950s, whereupon it remained above the century-long mean until the 1970s, after which it declined by about the same amount to 1992 (taking it well below the century-long mean), whereupon it recovered and edged upward towards the century mean. Hence, for the entire century, there was indeed a slight increase in global land area precipitation; but since 1915 there was essentially no net change.

For the oceanic portion of the world between 30°N and 30°S, the record of which begins in 1920, there was an overall decrease of about 0.3 percent per decade. For the world as a whole, which is 70 percent covered by water, there may well have been a slight decrease in precipitation since about 1917 or 1918.

Concentrating on the last half of the twentieth century, Neng et al. (2002) analyzed data from 1948 to 2000 in a quest to determine the effect of warm ENSO years on annual precipitation over the land area of the globe. In doing so, they found some regions experienced more rainfall in warm ENSO years, while others experienced less. However, in the words of the researchers, "in warm event years, the land area where the annual rainfall was reduced is far greater than that where the annual rainfall was increased, and the reduction is more significant than the increase." Consequently, whereas state-of-the-art climate models nearly always predict more precipitation in a warming world, the data of Neng et al.'s study depict just the opposite effect over the land area of the globe.

Most recently—and noting that "the Global Precipitation Climatology Project (GPCP) has produced merged satellite and in situ global precipitation estimates, with a record length now over 26 years beginning 1979 (Huffman et al., 1997; Adler et al., 2003)"—Smith et al. (2006) used empirical orthogonal function (EOF) analysis to study annual GPCP-derived precipitation variations over the period of record. In doing so, they found that the first three EOFs accounted for 52 percent of the observed variance in the precipitation data. Mode 1 was associated with mature ENSO conditions and correlated strongly with the Southern Oscillation Index, while Mode 2 was associated with the strong warm ENSO episodes of 1982/83 and 1997/98. Mode 3 was uncorrelated with ENSO but was associated with tropical trend-like changes that were correlated with interdecadal warming of tropical sea surface temperatures.

Globally, Smith et al. report that "the mode 3 variations average to near zero, so this mode does not represent any net change in the amount of precipitation over the analysis period." Consequently, over the period 1979-2004, when the IPCC claims the world warmed at a rate and to a degree that was unprecedented over the past two millennia, Smith et al. found that most of the precipitation variations in their global dataset were "associated with ENSO and have no trend." As for the variations that were not associated with ENSO and that did exhibit trends, they say that the trends were associated "with increased tropical precipitation over the Pacific and Indian Oceans associated with local warming of the sea." However, they note that this increased precipitation was "balanced by decreased precipitation in other regions," so that "the global average change [was] near zero."

Over the earth as a whole, therefore, it would appear from Smith et al.'s study, as well as from the other studies described above, that one of the major theoretical expectations of the climate modeling community remains unfulfilled, even under the supposedly highly favorable thermal conditions of the last quarter-century.

Additional information on this topic, including reviews of newer publications as they become available, can be found at http://www.co2science.org/subject/p/precipglobal.php

References

Adler, R.F., Susskind, J., Huffman, G.J., Bolvin, D., Nelkin, E., Chang, A., Ferraro, R., Gruber, A., Xie, P.-P., Janowiak, J., Rudolf, B., Schneider, U., Curtis, S. and Arkin, P. 2003. The version-2 global precipitation climatology project (GPCP) monthly precipitation analysis (1979-present). *Journal of Hydrometeorology* **4**: 1147-1167.

Arnell, N.W., Liu, C., Compagnucci, R., da Cunha, L., Hanaki, K., Howe, C., Mailu, G., Shiklomanov, I. and Stakhiv, E. 2001. Hydrology and water resources. In: McCarthy, J.J., Canziani, O.F., Leary, N.A., Dokken, D.J. and White, K.S. (Eds.) *Climate Change 2001: Impacts, Adaptation and Vulnerability, The Third Assessment Report of Working Group II of the Intergovernmental Panel on Climate Change, Cambridge*, University Press, Cambridge, UK, pp. 133-191.

Dai, A., Fung, I.Y. and DelGenio, A.D. 1997. Surface observed global land precipitation variations during 1900-1998. *Journal of Climate* **10**: 2943-2962.

DelGenio, A.D., Lacis, A.A. and Ruedy, R.A. 1991. Simulations of the effect of a warmer climate on atmospheric humidity. *Nature* **351**: 382-385.

Held, I.M. and Soden, B.J. 2000. Water vapor feedback and global warming. *Annual Review of Energy and Environment* **25**: 441-475.

Huffman, G.J., Adler, R.F., Chang, A., Ferraro, R., Gruber, A., McNab, A., Rudolf, B. and Schneider, U. 1997. The Global Precipitation Climatology Project (GPCP) combined data set. *Bulletin of the American Meteorological Society* **78**: 5-20.

Hulme, M., Osborn, T.J. and Johns, T.C. 1998. Precipitation sensitivity to global warming: comparisons of observations with HadCM2 simulations. *Geophysical Research Letters* **25**: 3379-3382.

Huntington, T.G. 2006. Evidence for intensification of the global water cycle: Review and synthesis. *Journal of Hydrology* **319**: 83-95.

Loaciga, H.A., Valdes, J.B., Vogel, R., Garvey, J. and Schwarz, H. 1996. Global warming and the hydrologic cycle. *Journal of Hydrology* **174**: 83-127.

Neng, S., Luwen, C. and Dongdong, X. 2002. A preliminary study on the global land annual precipitation associated with ENSO during 1948-2000. *Advances in Atmospheric Sciences* **19**: 993-1003.

New, M., Todd, M., Hulme, M. and Jones, P. 2001. Precipitation measurements and trends in the twentieth century. *International Journal of Climatology* **21**: 1899-1922.

Smith, T.M., Yin, X. and Gruber, A. 2006. Variations in annual global precipitation (1979-2004), based on the Global Precipitation Climatology Project 2.5° analysis. *Geophysical Research Letters* **33**: 10.1029/2005GL025393.

Trenberth, K.E. 1999. Conceptual framework for changes of extremes of the hydrological cycle with climate change. *Climatic Change* **42**: 327-339.

4.3.2. Africa

Richard *et al.* (2001) analyzed summer (January-March) rainfall totals in southern Africa over the period 1900-1998, finding that interannual variability was higher for the periods 1900-1933 and 1970-1998, but lower for the period 1934-1969. The strongest rainfall anomalies (greater than two standard deviations) were observed at the beginning of the century. However, the authors conclude there were "no significant changes in the January-March rainfall totals," nor any evidence of "abrupt shifts during the 20th century," suggesting that rainfall trends in southern Africa do not appear to have been influenced by CO_2-induced—or any other type of—global warming.

Nicholson and Yin (2001) report there have been "two starkly contrasting climatic episodes" in the equatorial region of East Africa since the late 1700s. The first, which began sometime prior to 1800, was characterized by "drought and desiccation." Extremely low lake levels were the norm, as drought reached its extreme during the 1820s and 1830s. In the mid to latter part of the 1800s, however, the drought began to weaken and floods became "continually high," but by the turn of the century lake levels began to fall as mild drought conditions returned. The drought did not last long, however, and the latter half of the twentieth century has seen an enhanced hydrologic cycle with a return of some lake levels to the high stands of the mid to late 1800s.

Verschuren *et al.* (2000) also examined hydrologic conditions in equatorial East Africa, but over a much longer time scale, i.e., a full thousand years. They report the region was significantly drier than it is today during the Medieval Warm Period from AD 1000 to 1270, while it was relatively wet during the Little Ice Age from AD 1270 to 1850. However, this latter period was interrupted by three episodes of prolonged dryness: 1390-1420, 1560-1625, and 1760-1840. These "episodes of persistent aridity," according to the authors, were "more severe than any recorded drought of the twentieth century."

The dry episode of the late eighteenth/early nineteenth centuries recorded in Eastern Africa has also been identified in Western Africa. In analyzing the climate of the past two centuries, Nicholson (2001) reports that the most significant climatic change that has occurred "has been a long-term reduction in rainfall in the semi-arid regions of West Africa," which has been "on the order of 20 to 40% in parts of the Sahel." There have been, she says, "three decades of protracted aridity," and "nearly all of Africa has been affected ... particularly since the 1980s." However, she goes on to note that "the rainfall conditions over Africa during the last 2 to 3 decades are not unprecedented," and that "a similar dry episode prevailed during most of the first half of the 19th century."

The importance of these findings is best summarized by Nicholson herself, when she states that "the 3 decades of dry conditions evidenced in the Sahel are not in themselves evidence of irreversible global change." Why not? Because an even longer period of similar dry conditions occurred between 1800 and 1850, when the earth was still in the clutches of the Little Ice Age, even in Africa (Lee-Thorp et al., 2001). There is no reason to think that the past two- to three-decade Sahelian drought is unusual or caused by the putative higher temperatures of that period.

Additional information on this topic, including reviews of newer publications as they become available, can be found at http://www.co2science.org/subject/p/precipafrica.php.

References

Lee-Thorp, J.A., Holmgren, K., Lauritzen, S.-E., Linge, H., Moberg, A., Partridge, T.C., Stevenson, C. and Tyson, P.D. 2001. Rapid climate shifts in the southern African interior throughout the mid to late Holocene. *Geophysical Research Letters* **28**: 4507-4510.

Nicholson, S.E. 2001. Climatic and environmental change in Africa during the last two centuries. *Climate Research* **17**: 123-144.

Nicholson, S.E. and Yin, X. 2001. Rainfall conditions in equatorial East Africa during the Nineteenth Century as inferred from the record of Lake Victoria. *Climatic Change* **48**: 387-398.

Richard, Y., Fauchereau, N., Poccard, I., Rouault, M. and Trzaska, S. 2001. 20th century droughts in southern Africa: Spatial and temporal variability, teleconnections with oceanic and atmospheric conditions. *International Journal of Climatology* **21**: 873-885.

Verschuren, D., Laird, K.R. and Cumming, B.F. 2000. Rainfall and drought in equatorial east Africa during the past 1,100 years. *Nature* **403**: 410-414.

4.3.3. Arctic

Curtis *et al.* (1998) examined a number of climatic variables at two first-order Arctic weather stations (Barrow and Barter Island, Alaska) that began in 1949, finding that both the frequency and mean intensity of precipitation at these two locations decreased over the period of record. Contemporaneously, they report that temperatures in the western Arctic increased, but that "the observed mean increase varies strongly from month-to-month making it difficult to explain the annual trend solely on the basis of an anthropogenic effect resulting from the increase in greenhouse gases in the atmosphere." Be that as it may, the four researchers concluded that the theoretical model-based assumption that "increased temperature leads to high precipitation ... is not valid," at least for the part of the western Arctic that was the focus of their analysis.

Lamoureux (2000) analyzed varved lake sediments obtained from Nicolay Lake, Cornwall Island, Nunavut, Canada, which were compared with rainfall events recorded at a nearby weather station over the period 1948-1978 and thereby used to reconstruct a rainfall history for the surrounding region over the 487-year period from 1500 to 1987. The results were suggestive of a small, but statistically insignificant, increase in rainfall over the course of the record. However, *heavy* rainfall was most frequent during the seventeenth and nineteenth centuries, which were the *coldest* periods of the past 400 years in the Canadian High Arctic, as well as the Arctic as a whole. In addition, Lamoureux found that "more frequent extremes and increased variance in yield occurred during the 17th and 19th centuries, likely due to increased occurrences of cool, wet synoptic types during the coldest periods of the Little Ice Age." Here, in a part of the planet predicted to be most impacted by CO_2-induced global warming—the Canadian High Arctic—a warming of the climate is demonstrated to *reduce* weather extremes related to precipitation.

Most recently, Rawlins *et al.* (2006) calculated trends in the spatially averaged water equivalent of annual rainfall and snowfall across the six largest Eurasian drainage basins that feed major rivers that deliver water to the Arctic Ocean for the period 1936-1999. Their results indicated that annual rainfall across the total area of the six basins decreased consistently and significantly over the 64-year period. Annual snowfall, on the other hand, exhibited "a strongly significant increase," but only "until the late 1950s." Thereafter, it exhibited "a moderately significant decrease," so that "no significant change [was] determined in Eurasian-basin snowfall over the entire 64-year period." The researchers' bottom-line finding, therefore, was that annual total precipitation (including both rainfall and snowfall) *decreased* over the period of their study; they note that this finding is "consistent with the reported (Berezovskaya *et al.*, 2004) decline in total precipitation."

In light of the findings reviewed above, either (1) the theoretical arguments and model predictions that suggest that "high-latitude precipitation increases in proportion to increases in mean hemispheric temperature" are not incredibly robust, or (2) late twentieth century temperatures may not have been much warmer than those of the mid-1930s and 40s, or (3) both of the above. Any or all of these choices fail to provide support for a key claim of the IPCC.

Additional information on this topic, including reviews of newer publications as they become available, can be found at http://www.co2science.org/subject/p/preciparctic.php.

References

Arctic Climate Impact Assessment. 2005. *Arctic Climate Impact Assessment—Special Report*. Cambridge University Press, New York, New York, USA.

Berezovskaya, S., Yang, D. and Kane, D.L. 2004. Compatibility analysis of precipitation and runoff trends over the large Siberian watersheds. *Geophysical Research Letters* 31: 10.1029/20004GL021277.

Curtis, J., Wendler, G., Stone, R. and Dutton, E. 1998. Precipitation decrease in the western Arctic, with special emphasis on Barrow and Barter Island, Alaska. *International Journal of Climatology* 18: 1687-1707.

Lamoureux, S. 2000. Five centuries of interannual sediment yield and rainfall-induced erosion in the Canadian High Arctic recorded in lacustrine varves. *Water Resources Research* 36: 309-318.

4.3.4. Asia

Kripalani *et al.* (2003) note that globally averaged temperatures are projected to rise under all scenarios of future energy use, according to the IPCC, leading to "increased variability and strength of the Asian monsoon." To see if there is any sign of such a precipitation response in real-world measurements, they examined Indian monsoon rainfall using observational data for the period 1871-2001 obtained from 306 stations distributed across the country. They discovered "distinct alternate epochs of above and below normal rainfall," which epochs "tend to last for about three decades." In addition, they report "there is no clear evidence to suggest that the strength and variability of the Indian Monsoon Rainfall (IMR) nor the epochal changes are affected by the global warming." They also report that "studies by several authors in India have shown that there is no statistically significant trend in IMR for the country as a whole." They further report that "Singh (2001) investigated the long term trends in the frequency of cyclonic disturbances over the Bay of Bengal and the Arabian Sea using 100-year (1890-1999) data and found significant decreasing trends." As a result, Kripalani *et al.* conclude that "there seem[s] to be no support for the intensification of the monsoon nor any support for the increased hydrological cycle as hypothesized by [the] greenhouse warming scenario in model simulations." In addition, they say that "the analysis of observed data for the 131-year period (1871-2001) suggests no clear role of global warming in the variability of monsoon rainfall over India," much as Kripalani and Kulkarni (2001) had concluded two years earlier.

Kanae *et al.* (2004) note that the number and intensity of heavy precipitation events are projected to increase in a warming world, according to the IPCC. They investigate this climate-model-derived hypothesis with digitalized hourly precipitation data recorded at the Tokyo Observatory of the Japan Meteorological Agency for the period 1890-1999. They report "many hourly heavy precipitation events (above 20 mm/hour) occurred in the 1990s compared with the 1970s and the 1980s," and that against that backdrop, "the 1990s seems to be unprecedented." However, they note that "hourly heavy precipitation around the 1940s is even stronger/more frequent than in the 1990s." In fact, their plots of maximum hourly precipitation and the number of extreme hourly precipitation events rise fairly regularly from the 1890s to peak in the 1940s, after which declines set in that bottom out in the 1970s and then reverse to rise to endpoints in the 1990s that are not yet as high as the peaks of the 1940s.

Taking a longer view of the subject, Pederson *et al.* (2001) used tree-ring chronologies from northeastern Mongolia to reconstruct annual precipitation and streamflow histories for the period 1651-1995. Analyses of both standard deviations and five-year intervals of extreme wet and dry periods of this record revealed that "variations over the recent period of instrumental data are not unusual relative to the prior record." The authors do state, however, that the reconstructions "appear to show more frequent extended wet periods in more recent decades," but they say this observation "does not demonstrate unequivocal evidence of an increase in precipitation as suggested by some climate models." In addition, they report that spectral analysis of the data revealed significant periodicities around 12 and 20-24 years, suggesting, in their words, "possible evidence for solar influences in these reconstructions for northeastern Mongolia."

Going back even further in time, Touchan *et al.* (2003) developed two reconstructions of spring (May-June) precipitation for southwestern Turkey from tree-ring width measurements, one of which extended from 1776 to 1998 and one from 1339 to 1998. These reconstructions, in their words, "show clear evidence of multi-year to decadal variations in spring precipitation," but they report that "dry periods of 1-2 years were well distributed throughout the record" and that the same was true of wet periods of one to two years' duration. With respect to more extreme events, the period that preceded the Industrial Revolution stood out. They say "all of the wettest 5-year periods occurred prior to 1756," while the longest period of reconstructed spring drought was the four-year period 1476-79, and the single driest spring was 1746. Turkey's greatest precipitation extremes, in other words, occurred prior to the Modern Warm Period, which is just the opposite of what the IPCC claims about extreme weather and its response to global warming.

Neff *et al.* (2001) looked much further back in time (from 9,600 to 6,100 years ago), using the relationship between a ^{14}C tree-ring record and a $\delta^{18}O$ proxy record of monsoon rainfall intensity as recorded in calcite $\delta^{18}O$ data obtained from a stalagmite in northern Oman. They found the correlation between the two datasets was "extremely strong," and a spectral analysis of the data revealed statistically significant periodicities centered on 779,

205, 134, and 87 years for the $\delta^{18}O$ record and periodicities of 206, 148, 126, 89, 26, and 10.4 years for the ^{14}C record. Consequently, because variations in ^{14}C tree-ring records are generally attributed to variations in solar activity, and because of the ^{14}C record's strong correlation with the $\delta^{18}O$ record, as well as the closely corresponding results of their spectral analyses, Neff *et al.* conclude there is "solid evidence" that both signals are responding to solar forcing.

In conclusion, evidence from Asia provides no support for the claim that precipitation in a warming world becomes more variable and intense. In fact, in some cases it tends to suggest just the opposite and provides support for the proposition that precipitation responds to cyclical variations in solar activity.

Additional information on this topic, including reviews of newer publications as they become available, can be found at http://www.co2science.org/subject/p/precipasia.php.

References

Kanae, S., Oki, T. and Kashida, A. 2004. Changes in hourly heavy precipitation at Tokyo from 1890 to 1999. *Journal of the Meteorological Society of Japan* **82**: 241-247.

Kripalani, R.H. and Kulkarni, A. 2001. Monsoon rainfall variations and teleconnections over south and east Asia. *International Journal of Climatology* **21**: 603-616.

Kripalani, R.H., Kulkarni, A., Sabade, S.S. and Khandekar, M.L. 2003. Indian monsoon variability in a global warming scenario. *Natural Hazards* **29**: 189-206.

Neff, U., Burns, S.J., Mangini, A., Mudelsee, M., Fleitmann, D. and Matter, A. 2001. Strong coherence between solar variability and the monsoon in Oman between 9 and 6 kyr ago. *Nature* **411**: 290-293.

Pederson, N., Jacoby, G.C., D'Arrigo, R.D., Cook, E.R. and Buckley, B.M. 2001. Hydrometeorological reconstructions for northeastern Mongolia derived from tree rings: 1651-1995. *Journal of Climate* **14**: 872-881.

Singh, O.P. 2001. Long term trends in the frequency of monsoonal cyclonic disturbances over the north Indian ocean. *Mausam* **52**: 655-658.

Touchan, R., Garfin, G.M., Meko, D.M., Funkhouser, G., Erkan, N., Hughes, M.K. and Wallin, B.S. 2003. Preliminary reconstructions of spring precipitation in southwestern Turkey from tree-ring width. *International Journal of Climatology* **23**: 157-171.

4.3.5. Europe

4.3.5.1. Central

Koning and Franses (2005) conducted a detailed analysis of a century of daily precipitation data acquired at the de Bilt meteorological station in the Netherlands. Using what they call "robust nonparametric techniques," they found the cumulative distribution function of annual maximum precipitation levels remained constant throughout the period 1906-2002, leading them to conclude that "precipitation levels are not getting higher." They report that similar analyses they performed for the Netherlands' five other meteorological stations "did not find qualitatively different results."

Wilson *et al.* (2005) developed two versions of a March-August precipitation chronology based on living and historical tree-ring widths obtained from the Bavarian Forest of southeast Germany for the period 1456-2001. The first version, standardized with a fixed 80-year spline function (SPL), was designed to retain decadal and higher frequency variations, while the second version used regional curve standardization (RCS) to retain lower frequency variations. Their efforts revealed significant yearly and decadal variability in the SPL chronology, but there did not appear to be any trend toward either wetter or drier conditions over the 500-year period. The RCS reconstruction, on the other hand, better captured lower frequency variation, suggesting that March-August precipitation was substantially greater than the long-term average during the periods 1730-1810 and 1870-2000 and drier than the long-term average during the periods 1500-1560, 1610-1730, and 1810-1870. Once again, however, there was little evidence of a long-term trend.

Moving still further east in Central Europe, and covering a full millennium and a half, Solomina *et al.* (2005) derived the first tree-ring reconstruction of spring (April-July) precipitation for the Crimean peninsula, located on the northern coast of the Black Sea in the Ukraine, for the period 1620-2002, after which they utilized this chronology to correctly date and correlate with an earlier precipitation reconstruction derived from a sediment core taken in 1931 from nearby Saki Lake, thus ending up with a proxy precipitation record for the region that stretched all the way back to AD 500. In describing their findings, Solomina *et al.* say no trend in precipitation was evident over the period 1896-1988 in an instrumental record obtained at a location adjacent to

the tree-sampling site. Also, the reconstructed precipitation values from the tree-ring series revealed year-to-year and decadal variability, but remained "near-average with relatively few extreme values" from about the middle 1700s to the early 1800s and again since about 1920. The most notable anomaly of the 1500-year reconstruction was an "extremely wet" period that occurred between AD 1050 and 1250, which Solomina *et al.* describe as broadly coinciding with the Medieval Warm Epoch, when humidity was higher than during the instrumental era.

The results of these several analyses demonstrate that over the period of twentieth century global warming, enhanced precipitation was not observed in Central Europe.

Additional information on this topic, including reviews of newer publications as they become available, can be found at http://www.co2science.org/subject/p/precipeurope.php.

References

Koning, A.J. and Franses, P.H. 2005. Are precipitation levels getting higher? Statistical evidence for the Netherlands. *Journal of Climate* **18**: 4701-4714.

Solomina, O., Davi, N., D'Arrigo, R. and Jacoby, G. 2005. Tree-ring reconstruction of Crimean drought and lake chronology correction. *Geophysical Research Letters* **32**: 10.1029/2005GL023335.

Wilson, R.J., Luckman, B.H. and Esper, J. 2005. A 500 year dendroclimatic reconstruction of spring-summer precipitation from the lower Bavarian Forest region, Germany. *International Journal of Climatology* **25**: 611-630.

4.3.5.2. Mediterranean

Starting at the western extreme of the continent, Rodrigo *et al.* (2001) used a variety of documentary data to reconstruct seasonal rainfall in Andalusia (southern Spain) from 1501 to 1997, after which they developed a relationship between seasonal rainfall and the North Atlantic Oscillation (NAO) over the period 1851-1997, which they used to reconstruct a history of the NAO from 1501 to 1997. This work revealed that the NAO influence on climate is stronger in winter than in other seasons of the year in Andalusia, explaining 40 percent of the total variance in precipitation; Rodrigo *et al.* make a point of noting that "the recent positive temperature anomalies over western Europe and recent dry winter conditions over southern Europe and the Mediterranean are strongly related to the persistent and exceptionally strong positive phase of the NAO index since the early 1980s," as opposed to an intensification of global warming.

Also working in the Andalusia region of southern Spain, Sousa and Garcia-Murillo (2003) studied proxy indicators of climatic change in Doñana Natural Park over a period of several hundred years, comparing their results with those of other such studies conducted in neighboring regions. This work revealed that the Little Ice Age (LIA) was by no means uniform in their region of study, as it included both wetter and drier periods. Nevertheless, they cite Rodrigo et al. (2000) as indicating that "the LIA was characterized in the southern Iberian Peninsula by increased rainfall," and they cite Grove (2001) as indicating that "climatic conditions inducing the LIA glacier advances [of Northern Europe] were also responsible for an increase in flooding frequency and sedimentation in Mediterranean Europe." Sousa and Garcia-Murillo's work complements these findings by indicating "an aridization of the climatic conditions after the last peak of the LIA (1830-1870)," which suggests that much of Europe became drier, not wetter, as the earth recovered from the global chill of the Little Ice Age.

Moving eastward into Italy, Crisci et al. (2002) analyzed rainfall data collected from 81 gauges spread throughout the Tuscany region for three different periods: (1) from the beginning of each record through 1994, (2) the shorter 1951-1994 period, and (3) the still-shorter 1970-1994 period. For each of these periods, trends were derived for extreme rainfall durations of 1, 3, 6, 12, and 24 hours. This work revealed that for the period 1970-1994, the majority of all stations exhibited no trends in extreme rainfall at any of the durations tested; four had positive trends at all durations and none had negative trends at all durations. For the longer 1951-1994 period, the majority of all stations exhibited no trends in extreme rainfall at any of the durations tested; none had positive trends at all durations and one had negative trends at all durations. For the still-longer complete period of record, the majority of all stations again continued to exhibit no trends in extreme rainfall at any of the durations tested; none had positive trends at all durations and one had negative trends at all durations, revealing no impact of twentieth century global warming one way or the other.

Working in northern Italy, Tomozeiu et al. (2002) performed a series of statistical tests to investigate the nature and potential causes of trends in winter (Dec-Feb) mean precipitation recorded at 40 stations over the period 1960-1995. This work revealed that nearly all of the stations experienced significant decreases in winter precipitation over the 35-year period of study; and by subjecting the data to a Pettitt test, they detected a significant downward shift at all stations around 1985. An Empirical Orthogonal Function analysis also was performed on the precipitation data, revealing a principal component that represented a common large-scale process that was likely responsible for the phenomenon. Strong correlation between this component and the North Atlantic Oscillation (NAO) suggested, in their words, that the changes in winter precipitation around 1985 "could be due to an intensification of the positive phase of the NAO."

Working in the eastern Basilicata region of southern Italy, where they concentrated on characterizing trends in extreme rainfall events, as well as resultant flood events and landslide events, Clark and Rendell (2006) analyzed 50 years of rainfall records (1951-2000). This work indicated, in their words, that "the frequency of extreme rainfall events in this area declined by more than 50% in the 1990s compared to the 1950s." In addition, they report that "impact frequency also decreased, with landslide-event frequency changing from 1.6/year in the period 1955-1962 to 0.3/year from 1985 to 2005, while flood frequency peaked at 1.0/year in the late 1970s before declining to less than 0.2/year from 1990." They concluded that if the climate-driven changes they observed over the latter part of the twentieth century continue, "the landscape of southern Italy and the west-central Mediterranean will become increasingly stable," or as they say in their concluding paragraph, "increased land-surface stability will be the result."

Alexandrov et al. (2004) analyzed a number of twentieth century datasets from throughout Bulgaria, finding "a decreasing trend in annual and especially summer precipitation from the end of the 1970s" and "variations of annual precipitation in Bulgaria showed an overall decrease." In addition, they report the region stretching from the Mediterranean into European Russia and the Ukraine "has experienced decreases in precipitation by as much as 20% in some areas."

Using analyses of tree-ring data and their connection to large-scale atmospheric circulation,

Touchan *et al.* (2005) developed summer (May-August) precipitation reconstructions for several parts of the eastern Mediterranean region, including Turkey, Syria, Lebanon, Cyprus and Greece, which extend back in time as much as 600 years. Over this period, they found that May-August precipitation varied on multi-annual and decadal timescales, but that on the whole there were no long-term trends. The longest dry period occurred in the late sixteenth century (1591-1595), while there were two extreme wet periods: 1601-1605 and 1751-1755. In addition, both extreme wet and dry precipitation events were found to be more variable over the intervals 1520-1590, 1650-1670, and 1850-1930, indicating that as the globe experienced the supposedly unprecedented warming of the last decades of the twentieth century, May-August precipitation in the eastern Mediterranean region actually became *less* variable than it had been in the earlier part of the century.

In conclusion, these studies of precipitation characteristics of Mediterranean Europe do not find evidence of the rising or more variable precipitation predicted by global climate models.

Additional information on this topic, including reviews of newer publications as they become available, can be found at http://www.co2science.org/subject/p/precipeuropemed.php.

References

Alexandrov, V., Schneider, M., Koleva, E. and Moisselin, J.-M. 2004. Climate variability and change in Bulgaria during the 20th century. *Theoretical and Applied Climatology* **79**: 133-149.

Clarke, M.L. and Rendell, H.M. 2006. Hindcasting extreme events: The occurrence and expression of damaging floods and landslides in southern Italy. *Land Degradation & Development* **17**: 365-380.

Crisci, A., Gozzini, B., Meneguzzo, F., Pagliara, S. and Maracchi, G. 2002. Extreme rainfall in a changing climate: regional analysis and hydrological implications in Tuscany. *Hydrological Processes* **16**: 1261-1274.

Grove, A.T. 2001. The "Little Ice Age" and its geomorphological consequences in Mediterranean Europe. *Climatic Change* **48**: 121-136.

Rodrigo, F.A., Esteban-Parra, M.J., Pozo-Vazquez, D. and Castro-Diez, Y. 2000. Rainfall variability in southern Spain on decadal to centennial time scales. *International Journal of Climatology* **20**: 721-732.

Rodrigo, F.S., Pozo-Vazquez, D., Esteban-Parra, M.J. and Castro-Diez, Y. 2001. A reconstruction of the winter North Atlantic Oscillation index back to A.D. 1501 using documentary data in southern Spain. *Journal of Geophysical Research* **106**: 14,805-14,818.

Sousa, A. and Garcia-Murillo, P. 2003. Changes in the wetlands of Andalusia (Doñana Natural Park, SW Spain) at the end of the Little Ice Age. *Climatic Change* **58**: 193-217.

Tomozeiu, R., Lazzeri, M. and Cacciamani, C. 2002. Precipitation fluctuations during the winter season from 1960 to 1995 over Emilia-Romagna, Italy. *Theoretical and Applied Climatology* **72**: 221-229.

Touchan, R., Xoplaki, E., Funkhouser, G., Luterbacher, J., Hughes, M.K., Erkan, N., Akkemik, U. and Stephan, J. 2005. Reconstructions of spring/summer precipitation for the Eastern Mediterranean from tree-ring widths and its connection to large-scale atmospheric circulation. *Climate Dynamics* **25**: 75-98.

4.3.5.3. Northern

Hanna *et al.* (2004) analyzed variations in several climatic variables in Iceland, including precipitation, over the past century in an effort to determine if there is "possible evidence of recent climatic changes" in that cold island nation. For the period 1923-2002, precipitation appeared to have increased slightly, although they questioned the veracity of the trend, citing several biases that may have corrupted the data base.

Linderholm and Molin (2005) analyzed two independent precipitation proxies, one derived from tree-ring data and one from a farmer's diary, to produce a 250-year record of summer (June-August) precipitation in east central Sweden. This work revealed there had been a high degree of variability in summer precipitation on inter-annual to decadal time scales throughout the record. Over the past century of supposedly unprecedented global warming, however, precipitation was found to have exhibited *less* variability than it did during the 150 years that preceded it.

In a study covering the longest time span of all, Linderholm and Chen (2005) derived a 500-year winter (September-April) precipitation chronology from tree-ring data obtained within the northern boreal forest zone of west-central Scandinavia. They found considerable variability, with the exception of a fairly stable period of above-average precipitation between AD 1730 and 1790. Additionally, above-average winter precipitation was found to have occurred in 1520-1561, 1626-1647, 1670-1695, 1732-1851, 1872-1892, and 1959 to the present, with the highest values reported in the early to mid-1500s; below-average winter precipitation was observed during 1504-1520, 1562-1625, 1648-1669, 1696-1731, 1852-1871, and 1893-1958, with the lowest values occurring at the beginning of the record and the beginning of the seventeenth century.

These findings demonstrate that non-CO_2-forced wetter and drier conditions than those of the present have occurred repeatedly within this region throughout the past five centuries. Similar extreme conditions may therefore be expected to naturally recur in the future.

Additional information on this topic, including reviews of newer publications as they become available, can be found at http://www.co2science.org/subject/p/precipeuropenorth.php.

References

Hanna, H., Jónsson, T. and Box, J.E. 2004. An analysis of Icelandic climate since the nineteenth century. *International Journal of Climatology* **24**: 1193-1210.

Linderholm, H.W. and Chen, D. 2005. Central Scandinavian winter precipitation variability during the past five centuries reconstructed from *Pinus sylvestris* tree rings. *Boreas* **34**: 44-52.

Linderholm, H.W. and Molin, T. 2005. Early nineteenth century drought in east central Sweden inferred from dendrochronological and historical archives. *Climate Research* **29**: 63-72.

4.3.6. United States

Molnar and Ramirez (2001) conducted a detailed watershed-based analysis of precipitation and streamflow trends for the period 1948-97 in the semiarid region of the Rio Puerco Basin of New Mexico. They found "at the annual timescale, a statistically significant increasing trend in precipitation in the basin was detected." This trend was driven primarily by an increase in the number of rainy days in the moderate rainfall intensity range, with essentially no change being observed at the high-intensity end of the spectrum. In the case of streamflow, however, there was no trend at the annual timescale; but monthly totals increased in low-flow months and decreased in high-flow months. Generally speaking, these trends are all positive for plant and animal life.

Cowles *et al.* (2002) analyzed snow water equivalent (SWE) data obtained from four different measuring systems—snow courses, snow telemetry, aerial markers and airborne gamma radiation—at more than 2,000 sites in the eleven westernmost states over the period 1910-1998. This work revealed that the long-term SWE trend of this entire region was negative, but with some significant within-region differences. In the northern Rocky Mountains and Cascades of the Pacific Northwest, for example, the trend was decidedly negative, with SWE decreasing at a rate of 0.1 to 0.2 inches per year. In the intermountain region and southern Rockies, however, there was no change in SWE with time. Cowles *et al.* additionally note that their results "reinforce more tenuous conclusions made by previous authors," citing Changnon *et al.* (1993) and McCabe and Legates (1995), who studied snow course data from 1951-1985 and 1948-1987, respectively, at 275 and 311 sites. They too found a decreasing trend in SWE at most sites in the Pacific Northwest but more ambiguity in the southern Rockies.

These findings are particularly interesting in light of the fact that nearly all climate models suggest the planet's hydrologic cycle will be enhanced in a warming world and that precipitation will increase. This prediction is especially applicable to the Pacific Northwest of the United States, where Kusnierczyk and Ettl (2002) report that climate models predict "increasingly warm and wet winters," as do Leung and Wigmosta (1999). Over the period of Cowles *et al.*'s study, however, when there was well-documented worldwide warming, precipitation that fell and accumulated as snow in the western USA did not respond as predicted. In fact, over the Pacific Northwest, it did just the opposite.

Garbrecht and Rossel (2002) used state divisional monthly precipitation data from the US National Climatic Data Center to investigate the nature of precipitation throughout the US Great Plains from January 1895 through December 1999, finding that regions in the central and southern Great Plains experienced above-average precipitation over the last two decades of the twentieth century. This 20-year span of time was the longest and most intense wet period of the entire 105 years of record, and was primarily the result of a reduction in the number of dry years and an increase in the number of wet years. The number of very wet years, in the words of the authors, "did not increase as much and even showed a decrease for many regions." The northern and northwestern Great Plains also experienced a precipitation increase at the end of this 105-year interval, but it was primarily confined to the final decade of the twentieth century; and again, as Garbrecht and Rossel report, "fewer dry years over the last 10 years, as opposed to an increase in very wet years, were the leading cause of the observed wet conditions."

Looking at the entire conterminous United States from 1895-1999, McCabe and Wolock (2002) evaluated and analyzed (1) values of annual precipitation minus annual potential evapotranspiration, (2) surplus water that eventually becomes streamflow, and (3) the water deficit that must be supplied by irrigation to grow vegetation at an optimum rate. Their work revealed that for the country as a whole, there was a statistically significant increase in the first two of these three parameters, while for the third there was no change. In describing the significance of these findings, McCabe and Wolock say "there is concern that increasing concentrations of atmospheric carbon dioxide and other radiatively active gases may cause global warming and ... adversely affect water resources." The results of their analyses, however, reveal that over the past century of global warming, just the opposite has occurred, at least within the conterminous United States: moisture has become more available, while there has been no change in the amount of water required for optimum plant growth.

Also studying the conterminous United States were Kunkel et al. (2003), who analyzed a new data base of daily precipitation observations for the period 1895-2000. This effort indicated "heavy precipitation frequencies were relatively high during the late 19th/early 20th centuries, decreasing to a minimum in the 1920s and '30s, followed by a general increase into the 1990s." More specifically, they note that "for 1-day duration events, frequencies during 1895-1905 are comparable in magnitude to frequencies in the 1980s and 1990s," while "for 5- and 10-day duration events, frequencies during 1895-1905 are only slightly smaller than late 20th century values."

In commenting on these findings, Kunkel et al. note that since enhanced greenhouse gas forcing of the climate system was very small in the early years of this record, the elevated extreme precipitation frequencies of that time "were most likely a consequence of naturally forced variability," which further suggests, in their words, "the possibility that natural variability could be an important contributor

to the recent increases." This is also the conclusion of Kunkel (2003), who in a review of this and other pertinent studies states that frequencies of extreme precipitation events in the United States in the late 1800s and early 1900s "were about as high as in the 1980s/1990s." Consequently, he too concludes that "natural variability in the frequency of precipitation extremes is quite large on decadal time scales and cannot be discounted as the cause or one of the causes of the recent increases."

Working with proxy data that extend much further back in time, Haston and Michaelsen (1997) developed a 400-year history of precipitation for 29 stations in coastal and near-interior California between San Francisco Bay and the U.S.-Mexican border using tree-ring chronologies. Their research revealed that although region-wide precipitation during the twentieth century was higher than what was experienced during the preceding three centuries, it was also "less variable compared to other periods in the past," both of which characteristics are huge positive developments for both man and nature in this important region of California.

In a similar study, Gray et al. (2003) examined 15 tree ring-width series that had been used in previous reconstructions of drought for evidence of low-frequency variation in precipitation in five regional composite chronologies pertaining to the central and southern Rocky Mountains. They say "strong multidecadal phasing of moisture variation was present in all regions during the late 16th century megadrought," and that "oscillatory modes in the 30-70 year domain persisted until the mid-19th century in two regions, and wet-dry cycles were apparently synchronous at some sites until the 1950s drought." They also note that "severe drought conditions across consecutive seasons and years in the central and southern Rockies may ensue from coupling of the cold phase PDO [Pacific Decadal Oscillation] with the warm phase AMO [Atlantic Multidecadal Oscillation] (Cayan et al., 1998; Barlow et al., 2001; Enfield et al., 2001)," something they envision happening in both the severe drought of the 1950s and the late sixteenth century megadrought.

Going back even further in time, Ni et al. (2002) developed a 1,000-year history of cool-season (November-April) precipitation for each climate division in Arizona and New Mexico from a network of 19 tree-ring chronologies. With respect to drought, they found "sustained dry periods comparable to the 1950s drought" occurred in "the late 1000s, the mid 1100s, 1570-97, 1664-70, the 1740s, the 1770s, and the late 1800s." They also note that the 1950s drought "was large in scale and severity, but it only lasted from approximately 1950 to 1956," whereas the sixteenth century megadrought lasted more than four times longer. With respect to the opposite of drought, Ni *et al.* report that several wet periods comparable to the wet conditions seen in the early 1900s and after 1976 occurred in "1108-20, 1195-1204, 1330-45, the 1610s, and the early 1800s." They also note that "the most persistent and extreme wet interval occurred in the 1330s."

Regarding the causes of the different precipitation extremes, Ni *et al.* say that "the 1950s drought corresponds to La Niña/-PDO [Pacific Decadal Oscillation] and the opposite polarity [+PDO] corresponds to the post-1976 wet period," which leads them to hypothesize that "the prominent shifts seen in the 1,000-year reconstructions in Arizona and New Mexico may also be linked to strong shifts of the coupled ENSO-PDO system." For the particular part of the world covered by their study, therefore, there appears to be nothing unusual about the extremes of both wetness and dryness experienced during the twentieth century.

In another equally long study, but on the opposite side of the country, Cronin *et al.* (2000) measured and analyzed salinity gradients across sediment cores extracted from Chesapeake Bay, the largest estuary in the United Sates, in an effort to examine precipitation variability in the surrounding watershed over the past 1,000 years. They found a high degree of decadal and multidecadal variability between wet and dry conditions throughout the record, where regional precipitation totals fluctuated between 25 percent and 30 percent, often in "extremely rapid [shifts] occurring over about a decade." Precipitation over the last two centuries, however, was on average greater than what it was during the previous eight centuries, with the exception of the Medieval Warm Period (AD 1250-1350), when the climate was judged to have been "extremely wet." In addition, it was determined that this region, like the southwestern United States, had experienced several "mega-droughts," lasting from 60-70 years in length, some of which Cronin *et al.* describe as being "more severe than twentieth century droughts."

Cronin *et al.*'s work, like the study of Ni *et al.*, reveals nothing unusual about precipitation in the U.S. during the twentieth century, the latter two decades of which the IPCC claims comprise the warmest such period of the past two millennia. Cronin *et al.*'s work indicates, for example, that both wetter and drier intervals occurred repeatedly in the past in the Chesapeake Bay watershed. There is reason to believe such intervals will occur in the future ... with or without any further global warming.

Additional information on this topic, including reviews of newer publications as they become available, can be found at http://www.co2science.org/subject/p/precipusa.php.

References

Barlow, M., Nigam, S. and Berberry, E.H. 2001. ENSO, Pacific decadal variability, and U.S. summertime precipitation, drought and streamflow. *Journal of Climate* **14**: 2105-2128.

Cayan, D.R., Dettinger, M.D., Diaz, H.F. and Graham, N.E. 1998. Decadal variability of precipitation over western North America. *Journal of Climate* **11**: 3148-3166.

Changnon, D., McKee, T.B. and Doesken, N.J. 1993. Annual snowpack patterns across the Rockies: Long-term trends and associated 500-mb synoptic patterns. *Monthly Weather Review* **121**: 633-647.

Cowles, M.K., Zimmerman, D.L., Christ, A. and McGinnis, D.L. 2002. Combining snow water equivalent data from multiple sources to estimate spatio-temporal trends and compare measurement systems. *Journal of Agricultural, Biological, and Environmental Statistics* **7**: 536-557.

Cronin, T., Willard, D., Karlsen, A., Ishman, S., Verardo, S., McGeehin, J., Kerhin, R., Holmes, C., Colman, S. and Zimmerman, A. 2000. Climatic variability in the eastern United States over the past millennium from Chesapeake Bay sediments. *Geology* **28**: 3-6.

Enfield, D.B., Mestas-Nuñez, A.M. and Trimble, P.J. 2001. The Atlantic multidecadal oscillation and its relation to rainfall and river flows in the continental U.S. *Geophysical Research Letters* **28**: 277-280.

Garbrecht, J.D. and Rossel, F.E. 2002. Decade-scale precipitation increase in Great Plains at end of 20th century. *Journal of Hydrologic Engineering* **7**: 64-75.

Gray, S.T., Betancourt, J.L., Fastie, C.L. and Jackson, S.T. 2003. Patterns and sources of multidecadal oscillations in drought-sensitive tree-ring records from the central and southern Rocky Mountains. *Geophysical Research Letters* **30**: 10.1029/2002GL016154.

Haston, L. and Michaelsen, J. 1997. Spatial and temporal variability of southern California precipitation over the last 400 yr and relationships to atmospheric circulation patterns. *Journal of Climate* **10**: 1836-1852.

Kunkel, K.E. 2003. North American trends in extreme precipitation. *Natural Hazards* **29**: 291-305.

Kunkel, K.E., Easterling, D.R, Redmond, K. and Hubbard, K. 2003. Temporal variations of extreme precipitation events in the United States: 1895-2000. *Geophysical Research Letters* **30**: 10.1029/2003GL018052.

Kusnierczyk, E.R. and Ettl, G.J. 2002. Growth response of ponderosa pine (*Pinus ponderosa*) to climate in the eastern Cascade Mountain, Washington, U.S.A.: Implications for climatic change. *Ecoscience* **9**: 544-551.

Leung, L.R. and Wigmosta, M.S. 1999. Potential climate change impacts on mountain watersheds in the Pacific Northwest. *Journal of the American Water Resources Association* **35**: 1463-1471.

McCabe, A.J. and Legates, S.R. 1995. Relationships between 700hPa height anomalies and 1 April snowpack accumulations in the western USA. *International Journal of Climatology* **14**: 517-530.

McCabe, G.J. and Wolock, D.M. 2002. Trends and temperature sensitivity of moisture conditions in the conterminous United States. *Climate Research* **20**: 19-29.

Molnar, P. and Ramirez, J.A. 2001. Recent trends in precipitation and streamflow in the Rio Puerco Basin. *Journal of Climate* **14**: 2317-2328.

Ni, F., Cavazos, T., Hughes, M.K., Comrie, A.C. and Funkhouser, G. 2002. Cool-season precipitation in the southwestern USA since AD 1000: Comparison of linear and nonlinear techniques for reconstruction. *International Journal of Climatology* **22**: 1645-1662.

4.3.7. Canada and Mexico

Kunkel (2003) reported that "several studies have argued that increasing greenhouse gas concentrations will result in an increase of heavy precipitation (Cubasch *et al.*, 2001; Yonetani and Gordon, 2001; Kharin and Zwiers, 2000; Zwiers and Kharin, 1998; Trenberth, 1998)." Consequently, Kunkel looked for such a signal in precipitation data from Canada that covered much of the past century. His search, however, was in vain, as the data indicated, in his words, that "there has been no discernible trend in the frequency of the most extreme events in Canada."

Zhang *et al.* (2001) also studied the temporal characteristics of heavy precipitation events across Canada, using what they describe as "the most homogeneous long-term dataset currently available for Canadian daily precipitation." Their efforts revealed that decadal-scale variability was a dominant feature of both the frequency and intensity of the annual number of extreme precipitation events, but they found "no evidence of any significant long-term changes." When the annual data were divided into seasonal data, however, an increasing trend in the number of extreme autumn snowfall events was noted; and an investigation into precipitation totals (extreme plus non-extreme events) revealed a slightly increasing trend that was attributed to increases in the number of non-heavy precipitation events. Zhang *et al.*'s overall conclusion was that "increases in the concentration of atmospheric greenhouse gases during the twentieth century have not been associated with a generalized increase in extreme precipitation over Canada."

Taking a longer view of the subject was Lamoureux (2000), who analyzed varved lake sediments obtained from Nicolay Lake, Cornwall Island, Nunavut, Canada, and compared the results with rainfall events recorded at a nearby weather station over the period 1948-1978, which comparison enabled the reconstruction of a rainfall history for the location over the 487-year period from 1500 to 1987. This history was suggestive of a small, but statistically insignificant, increase in total rainfall over the course of the record. Heavy rainfall was most frequent during the seventeenth and nineteenth centuries, which were the coldest periods of the past 400 years in the Canadian High Arctic, as well as the Arctic as a whole. In addition, Lamoureux says that "more frequent extremes and increased variance in yield occurred during the 17th and 19th centuries, likely due to increased occurrences of cool, wet synoptic types during the coldest periods of the Little Ice Age."

This study, like the others discussed above, contradicts the IPCC's claim that extreme precipitation events become more frequent and more severe with increasing temperature. Here in the Canadian High Arctic, in a part of the planet predicted to be most impacted by CO_2-induced global warming, rising temperatures have been shown to *reduce* precipitation extremes, even in the face of a slight increase in total precipitation.

South of the United States, Diaz *et al.* (2002) created a 346-year history of winter-spring (November-April) precipitation for the Mexican state of Chihuahua, based on earlywood width chronologies of more than 300 Douglas fir trees growing at four locations along the western and southern borders of Chihuahua and at two locations in the United States just above Chihuahua's northeast border. This exercise revealed, in their words, that "three of the 5 worst winter-spring drought years in the past three-and-a-half centuries are estimated to have occurred during the 20th century." Although this fact makes it sound like the twentieth century was highly anomalous in this regard, it was not. Two of those three worst drought years occurred during a decadal period of average to slightly above-average precipitation, so the three years were not representative of long-term droughty conditions.

Diaz *et al.* additionally report that "the longest drought indicated by the smoothed reconstruction lasted 17 years (1948-1964)," which again makes the twentieth century look unusual in this regard. However, for several of the years of that interval, precipitation values were only slightly below normal; and there were four very similar dry periods interspersed throughout the preceding two-and-a-half centuries: one in the late 1850s and early 1860s, one in the late 1790s and early 1800s, one in the late 1720s and early 1730s, and one in the late 1660s and early 1670s.

With respect to the twentieth century alone, there was a long period of high winter-spring precipitation that stretched from 1905 to 1932; and following the major drought of the 1950s, precipitation remained at, or just slightly above, normal for the remainder of the record. Finally, with respect to the entire 346 years, there was no long-term trend in the data, nor was there any evidence of a significant departure from that trend over the course of the twentieth century. Consequently, Chihuahua's precipitation history did not differ in any substantial way during the twentieth century from what it was over the prior quarter of a millennium, suggesting that neither twenteith century anthropogenic CO_2 emissions nor 20th-century warming—whether natural or human-induced—significantly impacted precipitation in that part of North America.

Additional information on this topic, including reviews of newer publications as they become available, can be found at http://www.co2science.org/subject/p/precipnortham.php.

References

Cubasch, U., Meehl, G.A., Boer, G.J., Stouffer, R.J., Dix, M., Noda, A., Senior, C.A., Raper, S. and Yap, K.S. 2001. Projections of future climate change. In: Houghton, J.T., Ding, Y., Griggs, D.J., Noguer, M., van der Linden, P.J., Dai, X., Maskell, K. and Johnson, C.A. (Eds.) *Climate Change 2001: The Scientific Basis. Contributions of Working Group 1 to the Third Assessment Report of the Intergovernmental Panel on Climate Change.* Cambridge University Press, Cambridge, UK.

Diaz, S.C., Therrell, M.D., Stahle, D.W. and Cleaveland, M.K. 2002. Chihuahua (Mexico) winter-spring precipitation reconstructed from tree-rings, 1647-1992. *Climate Research* **22**: 237-244.

Kharin, V.V. and Zwiers, F.W. 2000. Changes in the extremes in an ensemble of transient climate simulations with a coupled atmosphere-ocean GCM. *Journal of Climate* **13**: 3670-3688.

Kunkel, K.E. 2003. North American trends in extreme precipitation. *Natural Hazards* **29**: 291-305.

Lamoureux, S. 2000. Five centuries of interannual sediment yield and rainfall-induced erosion in the Canadian High Arctic recorded in lacustrine varves. *Water Resources Research* **36**: 309-318.

Trenberth, K.E. 1998. Atmospheric moisture residence times and cycling: Implications for rainfall rates with climate change. *Climatic Change* **39**: 667-694.

Yonetani, T. and Gordon, H.B. 2001. Simulated changes in the frequency of extremes and regional features of seasonal/annual temperature and precipitation when atmospheric CO_2 is doubled. *Journal of Climate* **14**: 1765-1779.

Zhang, X., Hogg, W.D. and Mekis, E. 2001. Spatial and temporal characteristics of heavy precipitation events over Canada. *Journal of Climate* **14**: 1923-1936.

Zwiers, F.W. and Kharin, V.V. 1998. Changes in the extremes of climate simulated by CCC GCM2 under CO_2-doubling. *Journal of Climate* **11**: 2200-2222.

4.4. Streamflow

Model projections suggest that CO_2-induced global warming will adversely impact earth's water resources by inducing large changes in global streamflow characteristics. As a result, many scientists are examining proxy streamflow records in an effort to determine how temperature changes of the twentieth century may or may not have impacted this aspect of the planet's hydrologic cycle. This is related to forecasts of droughts, floods, and precipitation variability, issues that are addressed in greater detail in Chapter 6.

A recent global study of this issue is Milliman *et al.* (2008), who computed temporal discharge trends for 137 rivers over the last half of the twentieth century that provide what they call a "reasonable global representation," as their combined drainage basins represent about 55 percent of the land area draining into the global ocean. In the words of the five researchers, "between 1951 and 2000 cumulative discharge for the 137 rivers remained statistically unchanged." In addition, they report that "global on-land precipitation between 1951 and 2000 remained statistically unchanged." Then, in a simple and straightforward conclusion, Milliman *et al.* write that "neither discharge nor precipitation changed significantly over the last half of the 20th century, offering little support to a global intensification of the hydrological cycle," such as is generally claimed to be a consequence of CO_2-induced global warming.

In the rest of this section we review studies for Eurasia and North America, seeking to discover if there have been any twentieth century changes in streamflow regimes that might reasonably have been caused by twentieth century changes in air temperature and atmospheric CO_2 concentration. Additional information on this topic, including reviews on streamflow not discussed here, can be found at http://www.co2science.org/subject/s/subject _s.php under the heading Streamflow.

Reference

Milliman, J.D., Farnsworth, K.L., Jones, P.D., Xu, K.H. and Smith, L.C. 2008. Climatic and anthropogenic factors affecting river discharge to the global ocean, 1951-2000. *Global and Planetary Change* **62**: 187-194.

4.4.1. Eurasia

Pederson *et al.* (2001) used tree-ring chronologies from northeastern Mongolia to develop annual precipitation and streamflow histories for the period 1651-1995. This work revealed, with respect to both standard deviations and five-year intervals of extreme wet and dry periods, that "variations over the recent period of instrumental data are not unusual relative to the prior record," although they say that the reconstructions "appear to show more frequent extended wet periods in more recent decades." Nevertheless, they state that this observation "does not demonstrate unequivocal evidence of an increase in precipitation as suggested by some climate models." Spectral analysis of the data also revealed significant periodicities of 12 and 20-24 years, suggesting, in the researchers' words, "possible

evidence for solar influences in these reconstructions for northeastern Mongolia."

Working in another part of the same region, Davi *et al.* (2006) report that "absolutely dated tree-ring-width chronologies from five sampling sites in west-central Mongolia were used in precipitation models and an individual model was made using the longest of the five tree-ring records (1340-2002)," which effort led to a reconstruction of streamflow that extended from 1637 to 1997. In analyzing these data, the four researchers discovered there was "much wider variation in the long-term tree-ring record than in the limited record of measured precipitation," which for the region they studied covered the period from 1937 to 2003. In addition, they report their streamflow history indicates that "the wettest 5-year period was 1764-68 and the driest period was 1854-58," while "the most extended wet period [was] 1794-1802 and ... extended dry period [was] 1778-83." For this part of Mongolia, therefore—which the researchers say "is representative of the central Asian region"—there is no evidence that the warming of the twentieth century has led to increased variability in precipitation and streamflow.

Pekarova *et al.* (2003) analyzed the annual discharge rates of selected large rivers of the world for recurring cycles of wet and dry periods. For those rivers with sufficiently long and accurate data series, they also derived long-term discharge rate trends. This latter analysis did not show "any significant trend change in long-term discharge series (1810-1990) in representative European rivers," including the Goeta, Rhine, Neman, Loire, Wesaer, Danube, Elbe, Oder, Vistule, Rhone, and Po. These latter observations are most interesting, for they indicate that even over the 180-year time period that saw the demise of the Little Ice Age and the ushering in of the Current Warm Period, there were no long-term trends in the discharge rates of the major rivers of Europe.

In another study, Hisdal *et al.* (2001) performed a series of statistical analyses on more than 600 daily streamflow records from the European Water Archive to examine trends in the severity, duration, and frequency of drought over the following four time periods: 1962-1990, 1962-1995, 1930-1995, and 1911-1995. This protocol indicated that "despite several reports on recent droughts in Europe, there is no clear indication that streamflow drought conditions in Europe have generally become more severe or frequent in the time periods studied." To the contrary, they report discovering that the number of trends pointing towards decreasing streamflow deficits or fewer drought events exceeded the number of trends pointing towards increasing streamflow deficits or more drought events.

Looking back towards Asia, Cluis and Laberge (2001) utilized streamflow records stored in the databank of the Global Runoff Data Center at the Federal Institute of Hydrology in Koblenz (Germany) to see if there were any recent changes in river runoff of the type predicted by IPCC scenarios of global warming, such as increased streamflow and increases in streamflow variability that would lead to more floods and droughts. Spatially, their study encompassed 78 rivers said to be "geographically distributed throughout the whole Asia-Pacific region," while temporally the mean start and end dates of the river flow records were 1936 ± 5 years and 1988 ± 1 year.

As a result of their analyses, the two researchers determined that mean river discharges were unchanged in 67 percent of the cases investigated; where trends did exist 69 percent of them were downward. Likewise, maximum river discharges were unchanged in 77 percent of the cases investigated; where trends did exist 72 percent of them were downward. Minimum river discharges, on the other hand, were unchanged in 53 percent of the cases investigated; where trends did exist, 62 percent of them were upward. All six metrics related to streamflow trends exhibit changes contrary to IPCC-promoted scenarios of climate change.

In another study, MacDonald et al. (2007) used "tree ring records from a network of sites extending across northern Eurasia to provide reconstructions [extending back to AD 1800] of annual discharge for the October to September water year for the major Eurasian rivers entering the Arctic Ocean (S. Dvina, Pechora, Ob', Yenisey, Lena, and Kolyma)." Results indicated that annual discharges of the mid to late twentieth century previously reported are not significantly greater than discharges experienced over the preceding 200 years, and "are thus still within the range of long-term natural variability." In addition, they say their "longer-term discharge records do not indicate a consistent positive significant correlation between discharge [and] Siberian temperature." They report there are actually weak *negative* correlations between discharge and temperature on some of the rivers over the period of their study.

In a contemporaneous study, Smith et al. (2007) present "a first analysis of a new dataset of daily discharge records from 138 small to medium-sized unregulated rivers in northern Eurasia," focusing on providing "a first continental-scale assessment of low-flow trends since the 1930s." Results indicate that "a clear result of this analysis is that, on balance, the monthly minimum values of daily discharge, or 'low flows,' have risen in northern Eurasia during the 20th century," adding that "from 12 unusually complete records from 1935-2002 we see that the minimum flow increases are greatest since ~1985."

Smith et al. reveals that over much of northern Eurasia, predictions of more drought seem rather off the mark, as daily low flows of the majority of northern Eurasian rivers have been *increasing*. Moreover, in the words of the five researchers, they have been increasing "in summer as well as winter and in non-permafrost as well as permafrost terrain," with the greatest increases occurring "since ~1985."

Writing about the Qinghai-Tibet Plateau, where they conducted their streamflow study, Cao et al. (2006) note that "both theoretical arguments and models suggest that net high-latitude precipitation increases in proportion to increases in mean hemispheric temperature (Houghton et al., 2001; Rahmstorf and Ganopolski, 1999; Bruce et al., 2002)," stating that in these scenarios "under global warming, mainly in the middle and west regions of northwest China, precipitation increases significantly," so that "some researchers [have] even advanced the issue of [a] climatic shift from warm-dry to warm-wet in northwest China (Shi, 2003)," with the ultimate expectation that total river discharge within the region would significantly increase in response to global warming.

As a test of these climate-model predictions, Cao et al. analyzed annual discharge data for five large rivers of the Qinghai-Tibet Plateau over the period 1956-2000, using the Mann-Kendall nonparametric trend test; and in doing so, they found that over the period of their study, "river discharges in the Qinghai-Tibet Plateau, in general, have no obvious change with the increase of the Northern Hemisphere surface air temperature." Because they could detect "no increase in the stream discharge in the Qinghai-Tibet Plateau with global warming," Cao et al. concluded that their real-world findings are not "in accordance with the anticipated ideas" that led them to conduct their study. Indeed, the disconnect between streamflow and global warming in this and many other studies argues strongly against the claimed consequences of global warming, the claimed magnitude of global warming, or both of these standard claims.

Worried about the possibility that enhanced freshwater delivery to the Arctic ocean by increased river flow could shut down the ocean's thermohaline circulation, Peterson *et al.* (2002) plotted annual values of the combined discharge of the six largest Eurasian Arctic rivers (Yenisey, Lena, Ob', Pechora, Kolyma, and Severnaya Dvina)—which drain about two-thirds of the Eurasian Arctic landmass—against the globe's mean annual surface air temperature (SAT), after which they ran a simple linear regression through the data and determined that the combined discharge of the six rivers seems to rise by about 212 km³/year in response to a 1°C increase in mean global air temperature. Then, they calculated that for the high-end global warming predicted by the Intergovernmental Panel on Climate Change (IPCC) to occur by AD 2100, i.e., a temperature increase of 5.8°C, the warming-induced increase in freshwater discharge from the six rivers could rise by as much as 1260 km³/year (we calculate 5.8°C x 212 km³/year/°C = 1230 km³/year), which represents a 70 percent increase over the mean discharge rate of the past several years.

The link between this conclusion and the postulated shutting down of the thermohaline circulation of the world's oceans resides in the hypothesis that the delivery of such a large addition of freshwater to the North Atlantic Ocean may slow—or even stop—that location's production of new deep water, which constitutes one of the driving forces of the great oceanic "conveyor belt." Although still discussed, this scenario is currently not as highly regarded as it was when Peterson *et al.* conducted their research, for a number of reasons. One that we have highlighted is the difficulty of accepting the tremendous extrapolation Peterson *et al.* make in extending their Arctic freshwater discharge vs. SAT relationship to the great length that is implied by the IPCC's predicted high-end warming of 5.8°C over the remainder of the current century. Consider, for example, that "over the period of the discharge record, global SAT increased by 0.4°C," according to Peterson *et al.* It is implausible to extend the relationship they derived across that small temperature range fully 14-and-a-half times further, to 5.8°C.

Consider also the Eurasian river discharge anomaly vs. global SAT plot of Peterson *et al.* (their Figure 4), which we have replotted in Figure 4.4.1. Enclosing their data with simple straight-line upper and lower bounds, it can be seen that the upper bound of the data does not change over the entire range of

global SAT variability, suggesting the very real possibility that the upper bound corresponds to a maximum Eurasian river discharge rate that cannot be exceeded in the real world under its current geographic and climatic configuration. The lower bound, on the other hand, rises so rapidly with increasing global SAT that the two bounds intersect less than two-tenths of a degree above the warmest of Peterson et al.'s 63 data points, suggesting that 0.2°C beyond the temperature of their warmest data point may be all the further any relationship derived from their data may validly be extrapolated.

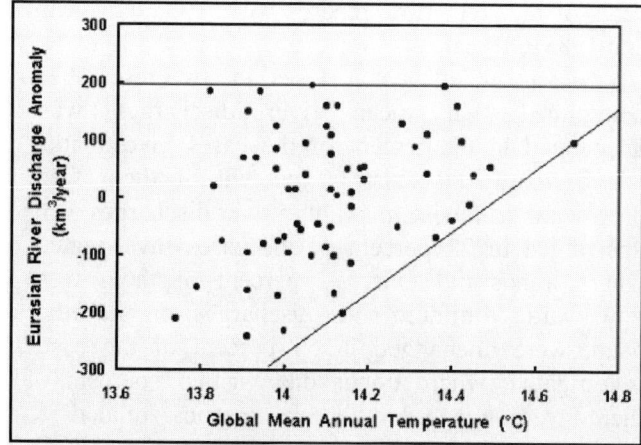

Figure 4.4.1. Annual Eurasian Arctic river discharge anomaly vs. annual global surface air temperature (SAT) over the period 1936 to 1999. Adapted from Peterson et al. (2002).

Clearly, real-world data do not support the hydrologic negativism the IPCC associates with both real-world and simulated global warming in Eurasia.

Additional information on this topic, including reviews of newer publications as they become available, can be found at http://www.co2science.org/subject/s/sfrteurasia.php.

References

Bruce, J.P., Holmes, R.M., McClelland, J.W. *et al.* 2002. Increasing river discharge to the Arctic Ocean. *Science* **298**: 2171-2173.

Cao, J., Qin, D., Kang, E. and Li, Y. 2006. River discharge changes in the Qinghai-Tibet Plateau. *Chinese Science Bulletin* **51**: 594-600.

Cluis, D. and Laberge, C. 2001. Climate change and trend detection in selected rivers within the Asia-Pacific region. *Water International* **26**: 411-424.

Davi, N.K., Jacoby, G.C., Curtis, A.E. and Baatarbileg, N. 2006. Extension of drought records for Central Asia using tree rings: West-Central Mongolia. *Journal of Climate* **19**: 288-299.

Hisdal, H., Stahl, K., Tallaksen, L.M. and Demuth, S. 2001. Have streamflow droughts in Europe become more severe or frequent? *International Journal of Climatology* **21**: 317-333.

Houghton, J.T., Ding, Y., Griggs, D.J. (Eds.) *Climate Change 2001: The Scientific Basis*. Cambridge University Press, Cambridge.

MacDonald, G.M., Kremenetski, K.V., Smith, L.C. and Hidalgo, H.G. 2007. Recent Eurasian river discharge to the Arctic Ocean in the context of longer-term dendrohydrological records. *Journal of Geophysical Research* **112**: 10.1029/2006JG000333.

Pederson, N., Jacoby, G.C., D'Arrigo, R.D., Cook, E.R. and Buckley, B.M. 2001. Hydrometeorological reconstructions for northeastern Mongolia derived from tree rings: 1651-1995. *Journal of Climate* **14**: 872-881.

Pekarova, P., Miklanek, P. and Pekar, J. 2003. Spatial and temporal runoff oscillation analysis of the main rivers of the world during the 19th-20th centuries. *Journal of Hydrology* **274**: 62-79.

Peterson, B.J., Holmes, R.M., McClelland, J.W., Vorosmarty, C.J., Lammers, R.B., Shiklomanov, A.I., Shiklomanov, I.A. and Rahmstorf, S. 2002. Increasing river discharge to the Arctic Ocean. *Science* **298**: 2171-2173.

Rahmstorf, S. and Ganopolski, A. 1999. Long-term global warming scenarios computed with an efficient coupled climate model. *Climatic Change* **43**: 353-367.

Shi, Y. 2003. *An Assessment of the Issues of Climatic Shift from Warm-Dry to Warm-Wet in Northwest China*. Meteorological Press, Beijing.

Smith, L.C., Pavelsky, T.M., MacDonald, G.M., Shiklomanov, A.I. and Lammers, R.B. 2007. Rising minimum daily flows in northern Eurasian rivers: A growing influence of groundwater in the high-latitude hydrologic cycle. *Journal of Geophysical Research* **112**: 10.1029/2006JG000327.

4.4.2. North America

Brown *et al.* (1999) studied siliciclastic sediment grain size, planktonic foraminiferal and pteropod relative frequencies, and the carbon and oxygen isotopic compositions of two species of planktonic foraminifera in cored sequences of hemipelagic muds deposited over the past 5,300 years in the northern Gulf of Mexico for evidence of variations in Mississippi River outflow characteristics over this time period. The results of their research indicated the occurrence of large megafloods—which they describe as having been "almost certainly larger than historical floods in the Mississippi watershed"—at 4,700, 3,500, 3,000, 2,500, 2,000, 1,200, and 300 years before present. These fluvial events, in their estimation, were likely "episodes of multidecadal duration," spawned by an export of extremely moist gulf air to midcontinental North America driven by naturally occurring same-time-scale oscillations in Gulf of Mexico ocean currents. These particular extreme events were in no way related to variations in atmospheric CO_2 concentration, as they occurred over a period of near-constancy in this atmospheric property.

Hidalgo *et al.* (2000) used a form of principal components analysis to reconstruct a history of streamflow in the Upper Colorado River Basin from information obtained from tree-ring data, after which they compared their results with the streamflow reconstruction of Stockton and Jacoby (1976). In doing so, they found their results were similar to those of the earlier 1976 study, but that their newer reconstruction responded with better fidelity to periods of below-average streamflow or regional drought, making it easier for them to see there had been "a near-centennial return period of extreme drought events in this region," going all the way back to the early 1500s. Hidalgo *et al.*'s work provided additional evidence for the existence of past droughts that surpassed the worst of the twentieth century.

Woodhouse *et al.* (2006) generated updated proxy reconstructions of water-year streamflow for four key streamflow gauges in the Upper Colorado River Basin (Green River at Green River, Utah; Colorado near Cisco, Utah; San Juan near Bluff, Utah; and Colorado at Lees Ferry, Arizona), "using an expanded tree-ring network and longer calibration records than in previous efforts." By these means they determined that the major drought of 2000-2004, "as measured by 5-year running means of water-year total flow at Lees Ferry ... is not without precedence in the tree ring record," and that "average reconstructed annual flow for the period 1844-1848 was lower." They also report that "two additional periods, in the early 1500s and early 1600s, have a 25% or greater chance of being as dry as 1999-2004," and that six other periods "have a 10% or greater chance of being drier." Their work revealed that "longer duration droughts have occurred in the past," and "the Lees Ferry reconstruction contains one sequence each of six, eight, and eleven consecutive years with flows below the 1906-1995 average."

"Overall," in the words of the three researchers, "these analyses demonstrate that severe, sustained droughts are a defining feature of Upper Colorado River hydroclimate." They conclude that "droughts more severe than any 20th to 21st century event [have] occurred in the past." This finding is just the opposite of what the IPCC would have us believe.

Woodhouse and Lukas (2006) developed "a network of 14 annual streamflow reconstructions, 300-600 years long, for gages in the Upper Colorado and South Platte River basins in Colorado generated from new and existing tree-ring chronologies." The results indicated that "the 20th century gage record does not fully represent the range of streamflow characteristics seen in the prior two to five centuries." The authors note that "paleoclimatic studies indicate that the natural variability in 20th century [streamflow] gage records is likely only a subset of the full range of natural variability," while citing in support of this statement the studies of Stockton and Jacoby (1976), Smith and Stockton (1981), Meko *et al.* (2001), and Woodhouse (2001). Of greatest significance in this regard was probably the fact that "multi-year drought events more severe than the 1950s drought have occurred," and that "the greatest frequency of extreme low flow events occurred in the 19th century," with a "clustering of extreme event years in the 1840s and 1850s."

Working in an adjacent region of the western United States, Carson and Munroe (2005) used tree-ring data collected by Stockton and Jacoby (1976) from the Uinta Mountains of Utah to reconstruct mean annual discharge in the Ashley Creek watershed for the period 1637 to 1970. Significant persistent departures from the long-term mean were noted throughout the 334-year record of reconstructed streamflow. The periods 1637-1691 and 1741-1897 experienced reduced numbers of extremely large flows and increased numbers of extremely small flows, indicative of persistent drought or near-drought conditions. By contrast, there was an overall

abundance of extremely large flows and relatively few extremely small flows during the periods 1692-1740 and 1898-1945, indicative of wetter conditions.

Lins and Slack (1999) analyzed secular trends in streamflow for 395 climate-sensitive stream gage stations (including data from more than 1,500 individual gages) located throughout the conterminous United States, some of which stations possessed datasets stretching all the way back to 1914. They found many more up-trends than down-trends in streamflow nationally, with slight decreases "only in parts of the Pacific Northwest and the Southeast." These and other of their findings, as they describe them, indicate "the conterminous U.S. is getting wetter, but less extreme," and it is difficult to conceive of a better result. As the world has warmed over the past century, the United States has gotten wetter in the mean, but less variable at the extremes, where floods and droughts occur.

Also studying the conterminous United States were McCabe and Wolock (2002), who for the period 1895-1999 evaluated (1) precipitation minus annual potential evapotranspiration, (2) the surplus water that eventually becomes streamflow, and (3) the water deficit that must be supplied by irrigation to grow vegetation at an optimum rate. This exercise revealed there was a statistically significant increase in the first two of these parameters, while for the third there was no change, indicative of the fact that water has actually become more available within the conterminous United States, and there has been no increase in the amount of water required for optimum plant growth.

Knox (2001) studied how conversion of the U.S. Upper Mississippi River Valley from prairie and forest to crop and pasture land by settlers in the early 1800s influenced subsequent watershed runoff and soil erosion rates. Initially, the conversion of the region's natural landscape to primarily agricultural use boosted surface erosion rates to values three to eight times greater than those characteristic of pre-settlement times. In addition, the land-use conversion increased peak discharges from high-frequency floods by 200 to 400 percent. Since the late 1930s, however, surface runoff has been decreasing. The decrease "is not associated with climatic causes," according to Knox, who reports that "an analysis of temporal variation in storm magnitudes for the same period showed no statistically significant trend."

Other notable findings of Knox's study include the observation that since the 1940s and early 1950s, the magnitudes of the largest daily flows have been decreasing at the same time that the magnitude of the average daily baseflow has been increasing, indicating a trend toward fewer flood and drought conditions. Once again, we have a situation where global warming has coincided with a streamflow trend that is leading to the best of all possible worlds: one of greater water availability, but with fewer and smaller floods and droughts.

Molnar and Ramirez (2001) conducted a detailed watershed-based analysis of precipitation and streamflow trends for the period 1948-97 in a semiarid region of the southwestern United States, the Rio Puerco Basin of New Mexico. "At the annual timescale," as they describe it, "a statistically significant increasing trend in precipitation in the basin was detected." This trend was driven primarily by an increase in the number of rainy days in the moderate rainfall intensity range, with essentially no change at the high-intensity end of the spectrum. In the case of streamflow, there was no trend at the annual timescale; monthly totals increased in *low-flow* months and decreased in high-flow months.

Shifting to a study of snowmelt runoff (SMR), McCabe and Clark (2005) note that most prior studies of this phenomenon in the western United States have depended on trend analyses to identify changes in timing, but they indicate that "trend analyses are unable to determine if a trend is gradual or a step change." This fact is crucial, they say, because when "changes in SMR timing have been identified by linear trends, there is a tendency to attribute these changes to global warming because of large correlations between linear trends in SMR timing and the increasing trend in global temperature." Therefore, using daily streamflow data for 84 stations in the western U.S., each with complete water-year information for the period 1950-2003, they conducted a number of analyses that enabled them to determine each station's mean streamflow trend over the past half century, as well as any stepwise changes that may have occurred in each data series.

As others before them had previously learned, the two researchers found that "the timing of SMR for many rivers in the western United States has shifted to earlier in the snowmelt season." However, they discovered that "the shift to earlier SMR has not been a gradual trend, but appears to have occurred as a step change during the mid-1980s," which shift was "related to a regional step increase in April-July temperatures during the mid-1980s." As a result, and after discussing various other possible reasons for what they had discovered, McCabe and Clark

concluded that "the observed change in the timing of SMR in the western United States is a regional response to natural climatic variability and may not be related to global trends in temperature."

Over in Minnesota, Novotny and Stefan (2006) analyzed streamflow records (extending up to the year 2002, with lengths ranging from 53 to 101 years) obtained from 36 gauging stations in five major river basins of the state, deriving histories of seven annual streamflow statistics: "mean annual flow, 7-day low flow in winter, 7-day low flow in summer, peak flow due to snow melt runoff, peak flow due to rainfall, as well as high and extreme flow days (number of days with flow rates greater than the mean plus one or two standard deviations, respectively)." In doing so, they found significant trends in each of the seven streamflow statistics throughout the state, but that in most cases "the trends are not monotonic but periodic," and they determined, as might have been expected, that "the mean annual stream flow changes are well correlated with total annual precipitation changes."

Most significantly, they found that peak flood flows due to snowmelt runoff "are not changing at a significant rate throughout the state," but that seven-day low flows or base flows are "increasing in the Red River of the North, Minnesota River and Mississippi River basins during both the summer and winter"; that the "low flows are changing at a significant rate in a significant number of stations and at the highest rates in the past 20 years"; and that "this finding matches results of other studies which found low flows increasing in the upper Midwest region including Minnesota (Lins and Slack, 1999; Douglas *et al.*, 2000)."

The two researchers write that "an increase in mean annual streamflow in Minnesota would be welcome," as "it could provide more aquatic habitat, better water quality, and more recreational opportunities, among other benefits." Likewise, they say that "water quality and aquatic ecosystems should benefit from increases in low flows in both the summer and winter, since water quality stresses are usually largest during low flow periods." In addition, they say "other good news is that spring floods (from snowmelt), the largest floods in Minnesota, have not been increasing significantly."

Rood *et al.* (2005) performed an empirical analysis of streamflow trends for rivers fed by relatively pristine watersheds in the central Rocky Mountain Region of North America that extends from Wyoming in the United States through British

Columbia in Canada. They applied both parametric and non-parametric statistical analyses to assess nearly a century of annual discharge (ending about 2002) along 31 river reaches that drain this part of North America. These analyses revealed that river flows in this region *declined* over the past century by an average of 0.22 percent per year, with four of them exhibiting recent decline rates exceeding 0.5 percent per year. This finding, in the words of Rood *et al.*, "contrasts with the many current climate change predictions that [this] region will become warmer and wetter in the near-future."

Working entirely in Canada, where about three-quarters of the country is drained by rivers that discharge their water into the Arctic and North Atlantic Oceans, Déry and Wood (2005) analyzed hydrometric data from 64 northern Canadian rivers that drain more than half of the country's landmass for the period 1964-2003. Then, after assessing both variability and trends, they explored the influence of large-scale teleconnections as possible drivers of the trends they detected. This work indicated there was a statistically significant mean decline of approximately 10 percent in the discharge rates of the 64 rivers over the four decades of their study, which was nearly identical to the decline in precipitation falling over northern Canada between 1964 and 2000. These facts led the two scientists to conclude that the changes in river discharge they observed were driven "primarily by precipitation rather than evapotranspiration." As for the *cause* of the precipitation/river discharge decline, statistically significant links were found between the decline and the Arctic Oscillation, the El Niño/Southern Oscillation, and the Pacific Decadal Oscillation. Consequently, the results of this study indicate there is nothing unusual about the four-decade-long trends in northern Canada river discharge rates, which means there is nothing in these trends that would suggest a global warming impact.

Also in Canada, Campbell (2002) analyzed the grain sizes of sediment cores obtained from Pine Lake, Alberta, to provide a non-vegetation-based high-resolution record of climate variability for this part of North America over the past 4,000 years. This research effort revealed the existence of periods of both increasing and decreasing grain size (a proxy for moisture availability) throughout the 4,000-year record at decadal, centennial, and millennial time scales. The most predominant departures included several-centuries-long epochs that corresponded to the Little Ice Age (about AD 1500-1900), the Medieval Warm Period (about AD 700-1300), the

Dark Ages Cold Period (about BC 100 to AD 700), and the Roman Warm Period (about BC 900-100). In addition, a standardized median grain-size history revealed that the highest rates of stream discharge during the past 4,000 years occurred during the Little Ice Age approximately 300-350 years ago. During this time, grain sizes were about 2.5 standard deviations above the 4,000-year mean. In contrast, the lowest rates of streamflow were observed around AD 1100, when median grain sizes were nearly 2 standard deviations below the 4,000-year mean, while most recently, grain size over the past 150 years has generally remained above average.

The Pine Lake sediment record convincingly demonstrates the reality of the non-CO_2-induced millennial-scale climatic oscillation that has alternately brought several-century-long periods of dryness and wetness to the southern Alberta region of North America during concomitant periods of relative global warmth and coolness, respectively, revealing a relationship that was not evident in the prior streamflow studies reviewed here that did not stretch all the way back in time to the Medieval Warm Period. It also demonstrates there is nothing unusual about the region's current moisture status.

In a final study from Canada, St. George (2007) begins by noting that the study of Burn (1994) suggested that a doubling of the air's CO_2 content could increase the severity and frequency of droughts in the prairie provinces of Canada (Alberta, Saskatchewan, Manitoba), but that results from an ensemble of climate models suggest that runoff in the Winnipeg River region of southern Manitoba, as well as runoff in central and northern Manitoba, could increase 20-30 percent by the middle of the twenty-first century (Milly *et al.*, 2005). To help resolve this dichotomy, St. George obtained daily and monthly streamflow data from nine gauge stations within the Winnipeg River watershed from the Water Survey of Canada's HYDAT data archive, plus precipitation and temperature data from Environment Canada's Adjusted Historical Canadian Climate Data archive, and analyzed them for trends over the period 1924-2003.

This work revealed, in the words of St. George, that "mean annual flows have increased by 58% since 1924 ... with winter streamflow going up by 60-110%," primarily because of "increases in precipitation during summer and autumn." In addition, he notes that similar "changes in annual and winter streamflow are observed in records from both regulated and unregulated portions of the watershed, which point to an underlying cause related to climate." Countering these positive findings, however, St. George says there are "reports of declining flow for many rivers in the adjacent Canadian prairies," citing the studies of Westmacott and Burn (1997), Yulianti and Burn (1998), Dery and Wood (2005), and Rood *et al.* (2005). Consequently, just as there are conflicting predictions about the future water status of portions of the prairie provinces of Canada, especially in Manitoba, so too are there conflicting reports about past streamflow trends in this region. It is anybody's guess as to what will actually occur in the years and decades ahead, although based on the observed trends he discovered, St. George believes "the potential threats to water supply faced by the Canadian Prairie Provinces over the next few decades will not include decreasing streamflow in the Winnipeg River basin."

Thus, we note there appear to be few real-world data that provide any significant support for the contention that CO_2-induced global warming will lead to more frequent and/or more severe increases and decreases in streamflow that result in, or are indicative of, more frequent and/or more severe floods and droughts. In the vast majority of cases, observed trends appear to be just the opposite of what is predicted to occur. Not only are real-world observations nearly all not undesirable, they are positive, and typically extremely so.

Additional information on this topic, including reviews of newer publications as they become available, can be found at http://www.co2science.org/subject/s/sfrtnorthamerica.php.

References

Brown, P., Kennett, J.P. and Ingram B.L. 1999. Marine evidence for episodic Holocene megafloods in North America and the northern Gulf of Mexico. *Paleoceanography* **14**: 498-510.

Burn, D.H. 1994. Hydrologic effects of climate change in western Canada. *Journal of Hydrology* **160**: 53-70.

Campbell, C. 2002. Late Holocene lake sedimentology and climate change in southern Alberta, Canada. *Quaternary Research* **49**: 96-101.

Carson, E.C and Munroe, J.S. 2005. Tree-ring based streamflow reconstruction for Ashley Creek, northeastern Utah: Implications for palaeohydrology of the southern Uinta Mountains. *The Holocene* **15**: 602-611.

Déry, S.J. and Wood, E.F. 2005. Decreasing river discharge in northern Canada. *Geophysical Research Letters* **32**: doi:10.1029/2005GL022845.

Douglas, E.M., Vogel, R.M. and Kroll, C.N. 2000. Trends in floods and low flows in the United States: impact of spatial correlation. *Journal of Hydrology* **240**: 90-105.

Hidalgo, H.G., Piechota, T.C. and Dracup, J.A. 2000. Alternative principal components regression procedures for dendrohydrologic reconstructions. *Water Resources Research* **36**: 3241-3249.

Knox, J.C. 2001. Agricultural influence on landscape sensitivity in the Upper Mississippi River Valley. *Catena* **42**: 193-224.

Lins, H.F. and Slack, J.R. 1999. Streamflow trends in the United States. *Geophysical Research Letters* **26**: 227-230.

McCabe, G.J. and Clark, M.P. 2005. Trends and variability in snowmelt runoff in the western United States. *Journal of Hydrometeorology* **6**: 476-482.

McCabe, G.J. and Wolock, D.M. 2002. Trends and temperature sensitivity of moisture conditions in the conterminous United States. *Climate Research* **20**: 19-29.

Meko, D.M., Therrell, M.D., Baisan, C.H. and Hughes, M.K. 2001. Sacramento River flow reconstructed to A.D. 869 from tree rings. *Journal of the American Water Resources Association* **37**: 1029-1039.

Milly, P.C.D., Dunne, K.A. and Vecchia, A.V. 2005. Global patterns of trends in streamflow and water availability in a changing climate. *Nature* **438**: 347-350.

Molnar, P. and Ramirez, J.A. 2001. Recent trends in precipitation and streamflow in the Rio Puerco Basin. *Journal of Climate* **14**: 2317-2328.

Novotny, E.V. and Stefan, H.G. 2006. Stream flow in Minnesota: Indicator of climate change. *Journal of Hydrology* **334**: 319-333.

Rood, S.B., Samuelson, G.M., Weber, J.K. and Wywrot, K.A. 2005. Twentieth-century decline in streamflow from the hydrographic apex of North America. *Journal of Hydrology* **306**: 215-233.

Smith, L.P. and Stockton, C.W. 1981. Reconstructed stream flow for the Salt and Verde Rivers from tree-ring data. *Water Resources Bulletin* **17**: 939-947.

St. George, S. 2007. Streamflow in the Winnipeg River basin, Canada: Trends, extremes and climate linkages. *Journal of Hydrology* **332**: 396-411.

Stockton, C.W. and Jacoby Jr., G.C. 1976. Long-term surface-water supply and streamflow trends in the Upper Colorado River Basin based on tree-ring analysis. *Lake Powell Research Project Bulletin* **18**, Institute of

Geophysics and Planetary Physics, University of California, Los Angeles.

Westmacott, J.R. and Burn, D.H. 1997. Climate change effects on the hydrologic regime within the Churchill-Nelson River Basin. *Journal of Hydrology* **202**: 263-279.

Woodhouse, C.A. 2001. Tree-ring reconstruction of mean annual streamflow for Middle Boulder Creek, Colorado, USA. *Journal of the American Water Resources Association* **37**: 561-570.

Woodhouse, C.A., Gray, S.T. and Meko, D.M. 2006. Updated streamflow reconstructions for the Upper Colorado River Basin. *Water Resources Research* **42**: 10.1029/2005WR004455.

Woodhouse, C.A. and Lukas, J.J. 2006. Multi-century tree-ring reconstructions of Colorado streamflow for water resource planning. *Climatic Change* **78**: 293-315.

Yulianti, J. and Burn, D.H. 1998. Investigating links between climatic warming and low streamflow in the Prairies region of Canada. *Canadian Water Resources Journal* **23**: 45-60.

4.5. Sea-level Rise

The possibility of large sea-level rises as a result of global warming is featured prominently in presentations of those, such as former U.S. Vice President Al Gore, who call for urgent action to "stop" global warming. In this section we examine historical trends in sea level to see if there is any indication of an increase in the mean rate-of-rise of the global ocean surface in response to the supposedly unprecedented warming of the planet over the course of the twentieth century. We then examine closely the various scenarios proposed whereby melting ice would cause sea levels to rise.

4.5.1 Mean Global Sea Levels

Cazenave *et al.* (2003) studied climate-related processes that cause variations in mean global sea level on interannual to decadal time scales, focusing on thermal expansion of the oceans and continental water mass balance. In doing so, they determined that the rate of thermal-induced sea-level rise over the past 40 years was about 0.5 mm/year. From early 1993 to the end of the twentieth century, however, analyses of TOPEX-Poseidon altimetry data and the global ocean temperature data of Levitus *et al.* (2000) yielded rates-of-rise that were approximately six times greater

than the mean four-decade rate, which suggested to them that "an acceleration took place in the recent past, likely related to warming of the world ocean." However, as they alternatively note, "the recent rise may just correspond to the rising branch of a decadal oscillation." In addition, they say that "satellite altimetry and *in situ* temperature data have their own uncertainties and it is still difficult to affirm with certainty that sea-level rise is indeed accelerating." In fact, they cite the work of Nerem and Mitchum (2001) as indicating that "about 20 years of satellite altimetry data would be necessary to detect, with these data alone, any acceleration in sea-level rise."

Mörner (2004) provided a more expansive setting for his analysis of the subject by noting that "prior to 5000-6000 years before present, all sea-level curves are dominated by a general rise in sea level in true glacial eustatic response to the melting of continental ice caps," but that "sea-level records are now dominated by the irregular redistribution of water masses over the globe ... primarily driven by variations in ocean current intensity and in the atmospheric circulation system and maybe even in some deformation of the gravitational potential surface." With respect to the last 150 years, he reports that "the mean eustatic rise in sea level for the period 1850-1930 was [on] the order of 1.0-1.1 mm/year," but that "after 1930-40, this rise seems to have stopped (Pirazzoli *et al.*, 1989; Mörner, 1973, 2000)." This stasis, in his words, "lasted, at least, up to the mid-60s." Thereafter, with the advent of the TOPEX/Poseidon mission, Mörner notes that "the record can be divided into three parts: (1) 1993-1996 with a clear trend of stability, (2) 1997-1998 with a high-amplitude rise and fall recording the ENSO event of these years and (3) 1998-2000 with an irregular record of no clear tendency." Most important of all, in his words, Mörner states "there is a total absence of any recent 'acceleration in sea-level rise' as often claimed by IPCC and related groups," and, therefore, "there is no fear of any massive future flooding as claimed in most global warming scenarios."

Church *et al.* (2004) used TOPEX/Poseidon satellite altimeter data to estimate global empirical orthogonal functions, which they combined with historical tide gauge data, to estimate monthly distributions of large-scale sea-level variability and change over the period 1950-2000. Their resultant "best estimate" of the rate of globally averaged sea-level rise over the last half of the twentieth century was 1.8 ± 0.3 mm/year. In addition, they noted that "decadal variability in sea level is observed, but to date there is no detectable secular increase in the rate of sea-level rise over the period 1950-2000." What is more, they reported that no increase in the rate of sea-level rise has been detected for the entire twentieth century, citing the work of Woodworth (1990) and Douglas (1992).

Cazenave and Nerem (2004) seemed to dismiss the caveats expressed in Cazenave et al. (2003) when they claimed that "the geocentric rate of global mean sea-level rise over the last decade (1993-2003) is now known to be very accurate, +2.8 ± 0.4 mm/year, as determined from TOPEX/Poseidon and Jason altimeter measurements," and that "this rate is significantly larger than the historical rate of sea-level change measured by tide gauges during the past decades (in the range of 1-2 mm/year)." However, they then admit "the altimetric rate could still be influenced by decadal variations of sea level unrelated to long-term climate change, such as the Pacific Decadal Oscillation, and thus a longer time series is needed to rule this out." They also noted that satellite altimetry had revealed a "non-uniform geographical distribution of sea-level change, with some regions exhibiting trends about 10 times the global mean." In addition, they note that "for the past 50 years, sea-level trends caused by change in ocean heat storage also show high regional variability," which fact "has led to questions about whether the rate of 20th-century sea-level rise, based on poorly distributed historical tide gauges, is really representative of the true global mean." Consequently, and in spite of the many new instruments and techniques that are being used to search for a global warming signal in global sea-level data, Cazenave and Nerem report that "these tools seem to have raised more questions than they have answered."

Noting that global climate models "show an increase in the rate of global average sea-level rise during the twentieth century," but that several prior studies (Douglas, 1991, 1992; Maul and Martin, 1993; Church et al., 2004; Holgate and Woodworth, 2004) had shown the measured rate of global sea-level rise to have been rather stable over the past hundred years, White et al. (2005) compared estimates of coastal and global averaged sea level for 1950 to 2000. Their results confirmed the earlier findings of "no significant increase in the rate of sea-level rise during this 51-year period."

Lombard et al. (2005) investigated the thermosteric or temperature-induced sea-level change of the past 50 years using the global ocean temperature data of Levitus et al. (2000) and Ishii et al. (2003). This work revealed that thermosteric sea-level variations are dominated by decadal oscillations of the planet's chief ocean-atmosphere climatic perturbations (El Niño-Southern Oscillation, Pacific Decadal Oscillation, and North Atlantic Oscillation). In terms of the global mean, as they describe it, thermosteric trends computed over 10-year windows "show large fluctuations in time, with positive values (in the range 1 to 1.5 mm/year for the decade centered on 1970) and negative values (-1 to -1.5 mm/year for the decade centered on 1980)." In the mean, however, and over the full half-century period Lombard et al. investigated, there was a net rise in sea level due to the thermal expansion of sea water, but only because the record began at the bottom of a trough and ended at the top of a peak. In between these two points, there were both higher and lower values, so one cannot be sure what would be implied if earlier data were available or what will be implied as more data are acquired. Noting that sea-level trends derived from TOPEX/Poseidon altimetry over 1993-2003 are "mainly caused by thermal expansion" and are thus "very likely a non-permanent feature," Lombard et al. conclude that "we simply cannot extrapolate sea level into the past or the future using satellite altimetry alone." Even the 50 years of global ocean temperature data we possess are insufficient to tell us much about the degree of global warming that may have occurred over the past half-century, as any long-term increase in global sea level that may have been caused by the temperature increase is dwarfed by decadal-scale variability.

Carton et al. (2005) introduced their study of the subject by noting that "recent altimeter observations indicate an increase in the rate of sea-level rise during the past decade to 3.2 mm/year, well above the centennial estimate of 1.5-2 mm/year," noting further that "this apparent increase could have resulted from enhanced melting of continental ice or from decadal changes in thermosteric and halosteric effects." They explored these opposing options "using the new eddy-permitting Simple Ocean Data Assimilation version 1.2 reanalysis of global temperature, salinity, and sea level spanning the period 1968-2001." They determined that "the effect on global sea-level rise of changing salinity is small except in subpolar regions." However, they found that warming-induced steric effects "are enough to explain much of the observed rate of increase in the rate of sea-level rise in the last decade of the twentieth century without need to invoke acceleration of melting of continental ice."

And as determined by Lombard *et al.*, as described in the preceding paragraph, the high thermosteric-induced rate-of-rise of global sea level over the past decade is likely "a non-permanent feature" of the global ocean's transient thermal behavior. Consequently, and in harmony with the findings of Levitus *et al.* (2005) and Volkov and van Aken (2005), Carton *et al.* found no need to invoke the melting of land-based glacial ice to explain the observed increase in global sea-level rise of the past decade.

Even more revealing was the globally distributed sea-level time series study of Jevrejeva *et al.* (2006), who analyzed information contained in the Permanent Service for Mean Sea Level database using a method based on Monte Carlo Singular Spectrum Analysis and removed 2- to 30-year quasi-periodic oscillations to derive nonlinear long-term trends for 12 large ocean regions, which they combined to produce the mean global sea level (gsl) and gsl rate-of-rise (gsl rate) curves depicted in Figure 4.5.1.1.

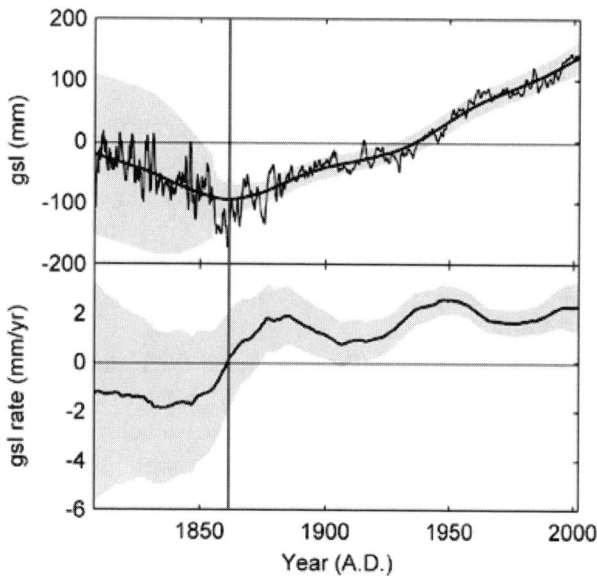

Figure 4.5.1.1. Mean global sea level (top), with shaded 95 percent confidence interval, and mean gsl rate-of-rise (bottom), with shaded standard error interval, adapted from Jevrejeva *et al.* (2006).

The figure clearly shows no acceleration of sea-level rise since the end of the Little Ice Age. Jevrejeva *et al.* say "global sea-level rise is irregular and varies greatly over time," but "it is apparent that rates in the 1920-1945 period are likely to be as large as today's." In addition, they report that their "global sea-level trend estimate of 2.4 ± 1.0 mm/year for the period

from 1993 to 2000 matches the 2.6 ± 0.7 mm/year sea-level rise found from TOPEX/Poseidon altimeter data." With respect to what the four researchers describe as "the discussion on whether sea-level rise is accelerating," their results pretty much answer the question in the negative.

The observations described above make us wonder why late twentieth century global warming—if it were as extreme as the IPCC claims it has been—cannot be detected in global sea-level data. The effects of the warming that led to the demise of the Little Ice Age—which the IPCC contends should have been considerably less dramatic than the warming of the late twentieth century—are readily apparent to the right of the vertical red line in the figure. Likewise, although the atmospheric CO_2 concentration experienced a dramatic increase in its rate-of-rise just after 1950 (shifting from a 1900-1950 mean rate-of-rise of 0.33 ppm/year to a 1950-2000 mean rate-of-rise of 1.17 ppm/year), the mean global sea-level rate-of-rise did not trend upwards after 1950, nor has it subsequently exceeded its 1950 rate-of-rise.

In concluding our examination of the peer-reviewed sea-level science, we report the findings of the most recent study of Holgate (2007). In a previous paper, Holgate and Woodworth (2004) derived a mean global sea-level history from 177 coastal tide gauge records that spanned the period 1955-1998. In an attempt to extend that record back in time another half-century, Holgate chose nine much longer high-quality records from around the world (New York, Key West, San Diego, Balboa, Honolulu, Cascais, Newlyn, Trieste, and Auckland) to see if their combined mean progression over the 1955-1998 period was similar enough to the concomitant mean sea-level history of the 177 stations to employ the mean nine-station record as a reasonable representation of mean global sea-level history for the much longer period stretching from 1904 to 2003.

In comparing the sea-level histories derived from the two datasets, Holgate found their mean rates-of-rise were indeed similar over the second half of the twentieth century; this observation thus implied, in Holgate's words, that "a few high quality records from around the world can be used to examine large spatial-scale decadal variability as well as many gauges from each region are able to [do]."

As a result of this finding, Holgate constructed the nine-station-derived wavering line in Figure 4.5.1.2 as a reasonable best representation of the 1904-2003 mean global sea-level history of the world.

Based on that history, he calculated that the mean rate of global sea-level rise was "larger in the early part of the last century (2.03 ± 0.35 mm/year 1904-1953), in comparison with the latter part (1.45 ± 0.34 mm/year 1954-2003)."

Another way of thinking about the century-long sea-level history portrayed in Figure 4.5.1.2 is suggested by the curve we have fit to it, which indicates that mean global sea level may have been rising, in the mean, ever more slowly with the passage of time throughout the entire last hundred years.

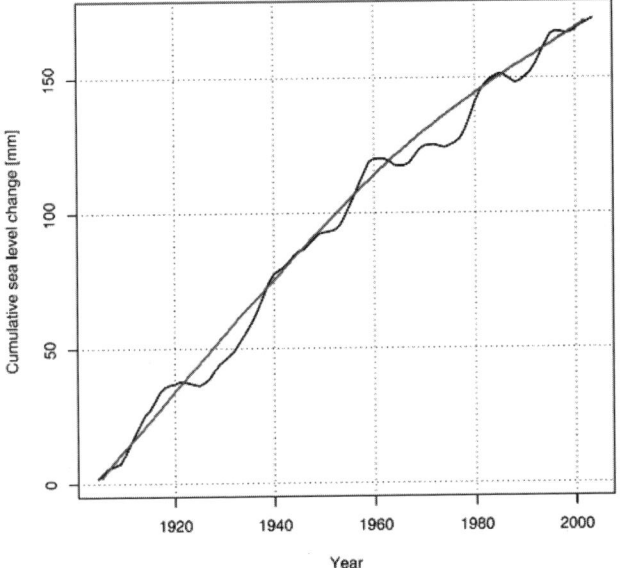

Figure 4.5.1.2. Cumulative increase in mean global sea level (1904-2003) derived from nine high-quality tide gauge records from around the world. Adapted from Holgate (2007).

Whichever way one looks at the findings of Holgate—either as two successive linear trends (representative of the mean rates-of-rise of the first and last halves of the twentieth century) or as one longer continuous curve (such as we have drawn)— the nine select tide gauge records indicate that the mean rate of global sea-level rise has *not* accelerated over the recent past, and has probably fallen.

Additional information on this topic, including reviews on sea level not discussed here, can be found at http://www.co2science.org/subject/s/subject_s.php under the heading Sea Level.

References

Carton, J.A., Giese, B.S. and Grodsky, S.A. 2005. Sea level rise and the warming of the oceans in the Simple Ocean Data Assimilation (SODA) ocean reanalysis. *Journal of Geophysical Research* **110**: 10.1029/2004JC002817.

Cazenave, A., Cabanes, C., Dominh, K., Gennero, M.C. and Le Provost, C. 2003. Present-day sea level change: observations and causes. *Space Science Reviews* **108**: 131-144.

Cazenave, A. and Nerem, R.S. 2004. Present-day sea level change: observations and causes. *Reviews of Geophysics* **42**: 10.1029/2003RG000139.

Church, J.A., White, N.J., Coleman, R., Lambeck, K. and Mitrovica, J.X. 2004. Estimates of the regional distribution of sea level rise over the 1950-2000 period. *Journal of Climate* **17**: 2609-2625.

Douglas, B.C. 1991. Global sea level rise. *Journal of Geophysical Research* **96**: 6981-6992.

Douglas, B.C. 1992. Global sea level acceleration. *Journal of Geophysical Research* **97**: 12,699-12,706.

Holgate, S.J. and Woodworth, P.L. 2004. Evidence for enhanced coastal sea level rise during the 1990s. *Geophysical Research Letters* **31**: 10.1029/2004GL019626.

Holgate, S.J. 2007. On the decadal rates of sea level change during the twentieth century. *Geophysical Research Letters* **34**: 10.1029/2006GL028492.

Holgate, S.J. and Woodworth, P.L. 2004. Evidence for enhanced coastal sea level rise during the 1990s. *Geophysical Research Letters* **31**: 10.1029/2004GL019626.

Ishii, M., Kimoto, M. and Kachi, M. 2003. Historical ocean subsurface temperature analysis with error estimates. *Monthly Weather Review* **131**: 51-73.

Jevrejeva, S., Grinsted, A., Moore, J.C. and Holgate, S. 2006. Nonlinear trends and multiyear cycles in sea level records. *Journal of Geophysical Research* **111**: 10.1029/2005JC003229.

Levitus, S., Antonov, J.I., Boyer, T.P., Garcia, H.E. and Locarnini, R.A. 2005. EOF analysis of upper ocean heat content, 1956-2003. *Geophysical Research Letters* **32**: 10.1029/2005GL023606/.

Levitus, S., Antonov, J.I., Boyer, T.P. and Stephens, C. 2000. Warming of the world ocean. *Science* **287**: 2225-2229.

Lombard, A., Cazenave, A., Le Traon, P.-Y. and Ishii, M. 2005. Contribution of thermal expansion to present-day sea-level change revisited. *Global and Planetary Change* **47**: 1-16.

Maul, G.A. and Martin, D.M. 1993. Sea level rise at Key West, Florida, 1846-1992: America's longest instrument record? *Geophysical Research Letters* **20**: 1955-1958.

Morner, N.-A. 1973. Eustatic changes during the last 300 years. *Palaeogeography, Palaeoclimatology, Palaeoecology* **9**: 153-181.

Morner, N.-A. 2000. Sea level changes along Western Europe. In: *Integrated Coastal Zone Management*, 2nd ed. IPC Publ., London-Hong Kong, pp. 33-37.

Mörner, N.-A. 2004. Estimating future sea level changes from past records. *Global and Planetary Change* **40**: 49-54.

Nerem, R.S. and Mitchum, G.T. 2001. Sea level change. In: Fu, L.L. and Cazenave, A. (Eds.) *Satellite Altimetry and Earth Sciences: A Handbook of Techniques and Applications*. Academic Press, San Diego, CA, pp. 329-349.

Pirazzoli, P.A., Grant, D.R. and Woodworth, P. 1989. Trends of relative sea-level changes: past, present, future. *Quaternary International* **2**: 63-71.

Volkov, D.L. and van Aken, H.M. 2005. Climate-related change of sea level in the extratropical North Atlantic and North Pacific in 1993-2003. *Geophysical Research Letters* **32**: 10.1029/2005GL023097.

White, N.J., Church, J.A. and Gregory, J.M. 2005. Coastal and global averaged sea level rise for 1950 to 2000. *Geophysical Research Letters* **32**: 10.1029/2004GL021391.

Woodworth, P.L. 1990. A search for accelerations in records of European mean sea level. *International Journal of Climatology* **10**: 129-143.

4.5.2. Antarctica Contribution to Sea Level

Vaughn *et al.* (1999) used more than 1,800 published and unpublished *in situ* measurements of the surface mass balance of Antarctica to produce an assessment of yearly ice accumulation over the continent. Their results indicated that the "total net surface mass balance for the conterminous grounded ice sheet is 1811 Gton yr^{-1} (149 kg m^{-2} yr^{-1}) and for the entire ice sheet including ice shelves and embedded ice rises, 2288 Gton yr^{-1} (166 kg m^{-2} yr^{-1})." These values, in their words, "are around 18% and 7% higher than the estimates widely adopted at present [1999]," which were derived about 15 years earlier. They suggest that net icefall on Antarctica may well have increased somewhat over that prior decade and a half. Nevertheless, because of uncertainties in the various numbers, Vaughn *et al.* say "we are still unable to

determine even the sign of the contribution of the Antarctic Ice Sheet to recent sea-level change."

In another review of the subject that was published about the same time, Reeh (1999) found a broad consensus for the conclusion that a 1°C warming would create but little net change in mean global sea level. Greenland's contribution would be a sea-level rise on the order of 0.30 to 0.77 millimeters per year, while Antarctica's contribution would be a fall on the order of 0.20 to 0.70 millimeters per year.

The following year, Wild and Ohmura (2000) studied the mass balance of Antarctica using two general circulation models developed at the Max Planck Institute for Meteorology in Hamburg, Germany: the older ECHAM3 and the new and improved ECHAM4. Under a doubled atmospheric CO_2 scenario, the two models were in close agreement in their mass balance projections, with both of them predicting increases in ice sheet growth, indicative of decreases in sea level.

Two years later, van der Veen (2002) addressed the problem again, noting that "for purposes of formulating policies, some of which could be unpopular or costly, it is imperative that probability density functions be derived for predicted values such as sea-level rise," further stating that with "greater societal relevance comes increased responsibility for geophysical modelers to demonstrate convincingly the veracity of their models to accurately predict future evolution of the earth's natural system or particular components thereof." In stepping forward to perform this task with respect to sea-level change, however, he was forced to conclude that "the validity of the parameterizations used by [various] glaciological modeling studies to estimate changes in surface accumulation and ablation under changing climate conditions has not been convincingly demonstrated." Van der Veen calculated, for example, that uncertainties in model parameters are sufficiently great to yield a 95 percent confidence range of projected meltwater contributions from Greenland and Antarctica that encompass global sea-level lowering as well as rise by 2100 A.D. for low, middle, and high warming scenarios. Hence, even for the worst of the IPCC warming projections, there could well be little to no change in mean global sea level due to the likely rise in the air's CO_2 content that may occur over the rest of this century. As a result, van der Veen concludes that the confidence level that can be placed in current ice sheet mass balance models "is quite low." Paraphrasing an earlier assessment of the subject, he says today's best models "currently reside on the lower rungs of the ladder of excellence" and "considerable improvements are needed before accurate assessments of future sea-level change can be made."

Wadhams and Munk (2004) attempted "an independent estimate of eustatic sea-level rise based on the measured freshening of the global ocean, and with attention to the contribution from melting of sea ice (which affects freshening but not sea level)." Their analysis produced "a eustatic rise of only 0.6 mm/year," and when a steric contribution of 0.5 mm/year is added to the eustatic component, "a total of 1.1 mm/year, somewhat less than IPCC estimates," is the final result. Perhaps the most interesting finding of their analysis, however, is that the continental run-off which is "allowed," after subtracting the effect of sea ice melt, "is considerably lower than current estimates of sub-polar glacial retreat, suggesting a negative contribution from polar ice sheets (Antarctica plus Greenland) or from other non-glacial processes." In this regard, they assert "we do not have good estimates of the mass balance of the Antarctic ice sheet, which could make a much larger positive *or* negative contribution."

The bottom line of Wadhams and Munk's analysis, as well as the other studies we have reviewed, is that there is considerable uncertainty associated with a number of basic parameters related to the water balance of the world's oceans and the meltwater contribution of Antarctica. Until these uncertainties are satisfactorily resolved, we cannot be confident that we know what is happening at the bottom of the world in terms of phenomena related to sea level.

Additional information on this topic, including reviews of newer publications as they become available, can be found at http://www.co2science.org/subject/a/antarcticasealvl.php and http://www.co2science.org/subject/s/sealevelantarctica.php.

References

Reeh, N. 1999. Mass balance of the Greenland ice sheet: Can modern observation methods reduce the uncertainty? *Geografiska Annaler* **81A**: 735-742.

van der Veen, C.J. 2002. Polar ice sheets and global sea level: how well can we predict the future? *Global and Planetary Change* **32**: 165-194.

Vaughn, D.G., Bamber, J.L., Giovinetto, M., Russell, J. and Cooper, A.P.R. 1999. Reassessment of net surface mass balance in Antarctica. *Journal of Climate* **12**: 933-946.

Wadhams, P. and Munk, W. 2004. Ocean freshening, sea level rising, sea ice melting. *Geophysical Research Letters* **31**: 10.1029/2004GL020039.

Wild, M. and Ohmura, A. 2000. Change in mass balance of polar ice sheets and sea level from high-resolution GCM simulations of greenhouse warming. *Annals of Glaciology* **30**: 197-203.

4.5.3. West Antarctic Ice Sheet

4.5.3.1. Collapse and Disintegration

The West Antarctic Ice Sheet (WAIS) is often described as the world's most unstable large ice sheet. As Hillenbrand *et al.* (2002) report, "it was speculated, from observed fast grounding-line retreat and thinning of a glacier in Pine Island Bay (Rignot, 1998; Shepherd *et al.*, 2001), from the timing of late Pleistocene-Holocene deglaciation in the Ross Sea (Bindschadler, 1998; Conway *et al.*, 1999), and from predicted activity of ice-stream drainage in response to presumed future global warming (Oppenheimer, 1998), that the WAIS may disappear in the future, causing the sea-level to rise at a rate of 1 to 10 mm/year (Bindschadler, 1998; Oppenheimer, 1998)."

Cofaigh *et al.* (2001) analyzed five sediment cores from the continental rise west of the Antarctic Peninsula and six from the Weddell and Scotia Seas for their ice rafted debris (IRD) content, in an attempt to see if there are Antarctic analogues of the Heinrich layers of the North Atlantic Ocean, which testify of the repeated collapse of the eastern margin of the Laurentide Ice Sheet and the concomitant massive discharge of icebergs. If such IRD layers exist around Antarctica, the researchers reasoned, they would be evidence of "periodic, widespread catastrophic collapse of basins within the Antarctic Ice Sheet," which could obviously occur again. However, after carefully studying their data, they concluded that "the ice sheet over the Antarctic Peninsula did not undergo widespread catastrophic collapse along its western margin during the late Quaternary," and they say this evidence "argues against pervasive, rapid ice-sheet collapse around the Weddell embayment over the last few glacial cycles." If there was no dramatic break-up of the WAIS over the last few glacial cycles, there's a very good chance there will be none before the current interglacial ends, especially since the data of

Petit *et al.* (1999) indicate that the peak temperatures of each of the previous four interglacials were warmer than the peak temperature of the current interglacial by an average of more than 2°C.

Hillenbrand *et al.* (2002) studied the nature and history of glaciomarine deposits contained in sediment cores recovered from the West Antarctic continental margin in the Amundsen Sea to "test hypotheses of past disintegration of the WAIS." In doing so, they found that all proxies regarded as sensitive to a WAIS collapse changed markedly during the global climatic cycles of the past 1.8 million years, but they "do not confirm a complete disintegration of the WAIS during the Pleistocene" at a place where "dramatic environmental changes linked to such an event should be documented." They say their results "suggest relative stability rather than instability of the WAIS during the Pleistocene climatic cycles," and they note that this conclusion is "consistent with only a minor reduction of the WAIS during the last interglacial period," citing the work of Huybrechts (1990), Cuffey and Marshall (2000) and Huybrechts (2002).

In another paper that addresses the subject of possible WAIS collapse, O'Neill and Oppenheimer (2002) say the ice sheet "may have disintegrated in the past during periods only modestly warmer (~2°C global mean) than today," and they thus claim that setting "a limit of 2°C above the 1990 global average temperature"—above which the mean temperature of the globe should not be allowed to rise—"is justified." In fact, a 2°C warming of the globe would likely have little to no impact on the stability of the WAIS. The average Antarctic peak temperature of all four of the world's prior interglacials was at least 2°C greater than the Antarctic peak temperature of the current interglacial; yet, in the words of the scientists who developed the pertinent temperature record (Petit *et al.*, 1999), the evidence contained in the core "makes it unlikely that the West Antarctic ice sheet collapsed during the past 420,000 years," pretty much the same conclusion that was drawn by Cofaigh *et al.*

In addition, we know from the Vostok ice core record that the peak Antarctic temperature of the most recent prior interglacial was fully 3°C warmer than the peak Antarctic temperature of the interglacial in which we presently live, yet the WAIS still did not disintegrate then. Furthermore, we know that throughout the long central portion of the current interglacial (when the most recent peak Antarctic temperature was reached), it was much warmer than it was in 1990, which is the year from which O'Neill

and Oppenheimer's critical 2°C warming increment is measured; and this fact raises the 3°C temperature elevation of the last interglacial relative to the global temperature of 1990 to something on the order of 4° or 5°C, for which, again, there was no evidence of even a partial WAIS disintegration.

Finally, and in spite of the current interglacial's current relative coolness, the Vostok ice core data indicate that the current interglacial has been by far the longest stable warm period of the entire 420,000-year record, which suggests we are probably long overdue for the next ice age to begin, and that we may not have the "5 to 50 centuries" that O'Neill and Oppenheimer suggest could be needed to bring about the WAIS disintegration subsequent to the attainment of whatever temperature in excess of 4° or 5°C above the current global mean would be needed to initiate the process.

Additional information on this topic, including reviews of newer publications as they become available, can be found at http://www.co2science.org/subject/w/waiscollapse.php.

References

Bindschadler, R. 1998. Future of the West Antarctic Ice Sheet. *Science* **282**: 428-429.

Cofaigh, C.O., Dowdeswell, J.A. and Pudsey, C.J. 2001. Late Quaternary iceberg rafting along the Antarctic Peninsula continental rise in the Weddell and Scotia Seas. *Quaternary Research* **56**: 308-321.

Conway, H., Hall, B.L., Denton, G.H., Gades, A.M. and Waddington, E.D. 1999. Past and future grounding-line retreat of the West Antarctic Ice Sheet. *Science* **286**: 280-283.

Cuffey, K.M. and Marshall, S.J. 2000. Substantial contribution to sea-level rise during the last interglacial from the Greenland ice sheet. *Nature* **404**: 591-594.

Hillenbrand, C-D., Futterer, D.K., Grobe, H. and Frederichs, T. 2002. No evidence for a Pleistocene collapse of the West Antarctic Ice Sheet from continental margin sediments recovered in the Amundsen Sea. *Geo-Marine Letters* **22**: 51-59.

Huybrechts, P. 1990. The Antarctic Ice Sheet during the last glacial-interglacial cycle: a three-dimensional experiment. *Annals of Glaciology* **14**: 115-119.

Huybrechts, P. 2002. Sea-level changes at the LGM from ice-dynamic reconstructions of the Greenland and Antarctic ice sheets during the glacial cycles. *Quaternary Science Reviews* **21**: 203-231.

O'Neill, B.C. and Oppenheimer, M. 2002. Dangerous climate impacts and the Kyoto Protocol. *Science* **296**: 1971-1972.

Oppenheimer, M. 1998. Global warming and the stability of the West Antarctic Ice Sheet. *Nature* **393**: 325-332.

Petit, J.R., Jouzel, J., Raynaud, D., Barkov, N.I., Barnola, J.-M., Basile, I., Bender, M., Chappellaz, J., Davis, M., Delaygue, G., Delmotte, M., Kotlyakov, V.M., Legrand, M., Lipenkov, V.Y., Lorius, C., Pepin, L., Ritz, C., Saltzman, E., and Stievenard, M. 1999. Climate and atmospheric history of the past 420,000 years from the Vostok ice core, Antarctica. *Nature* **399**: 429-436.

Rignot, E.J. 1998. Fast recession of a West Antarctic glacier. *Science* **281**: 549-551.

Shepherd, A., Wingham, D.J., Mansley, J.A.D. and Corr, H.F.J. 2001. Inland thinning of Pine Island Glacier, West Antarctica. *Science* **291**: 862-864.

4.5.3.2. Dynamics

The supposedly imminent demise of the West Antarctic Ice Sheet (WAIS) is what Al Gore apparently had in mind when warned that if "half of Antarctica melted or broke up and slipped into the sea, sea levels worldwide would increase by between 18 and 20 feet" (Gore 2006). A few scientists, as reported by Ackert (2003), believe we are witnessing the CO_2-induced "early stages of rapid ice sheet collapse, with potential near-term impacts on the world's coastlines." However, studies of the dynamics of various components of the WAIS suggest this is highly unlikely.

Bindschadler and Vornberger (1998) utilized satellite imagery taken since 1963 to examine spatial and temporal changes of Ice Stream B, which flows into the Ross Ice Shelf. The data indicated that since that time, the ice stream's width had increased by nearly 4 kilometers, at a rate that was, in their words, an "order of magnitude faster than models have predicted." However, they reported that the flow speed of the ice stream had decreased over this time period by about 50 percent, noting that "such high rates of change in velocity greatly complicate the calculation of mass balance of the ice sheet," and that such changes "do not resolve the overriding question of the stability of the West Antarctic Ice Sheet."

Bindschadler (1998) reviewed what was known about the WAIS and analyzed its historical retreat in terms of its grounding line and ice front. This work revealed that from the time of the Last Glacial Maximum to the present, the retreat of the WAIS's grounding line had been faster than that of its ice front, which resulted in an expanding Ross Ice Shelf. In fact, Bindschadler reported that "the ice front now appears to be nearly stable," although its grounding line appeared to be retreating at a rate that suggested complete dissolution of the WAIS in another 4,000 to 7,000 years. Such a retreat would indeed result in a sustained sea-level rise of 8 to 13 cm per century. However, even the smallest of these sea-level rates-of-rise would require, according to Bindschadler, "a large negative mass balance for all of West Antarctica," and there were no broad-based data to support that scenario.

Oppenheimer (1998) reviewed 122 studies that dealt with the stability of the WAIS and its effects on global sea level, concluding that "human-induced climate change may play a significant role in controlling the long-term stability of the West Antarctic Ice Sheet and in determining its contribution to sea-level change in the near future." Other of his statements, however, detract from this conclusion. He noted, for example, that the Intergovernmental Panel on Climate Change (IPCC) "estimated a zero Antarctic contribution to sea-level rise over the past century, and projected a small negative (about -1 cm) contribution for the twenty-first century." With respect to the state and behavior of the atmosphere and ocean above and around Antarctica, he acknowledged that "measurements are too sparse to enable the observed changes to be attributed to any such [human-induced] global warming." And in the case of sea-ice extent, he admitted "the IPCC assessment is that no trend has yet emerged."

Oppenheimer concluded his review with four scenarios of the future based upon various assumptions. One was that the WAIS will experience a sudden collapse that causes a 4-6 m sea-level rise within the coming century. However, he stated that this scenario "may be put aside for the moment, because no convincing model of it has been presented." A second scenario had the WAIS gradually disintegrating and contributing to a slow sea-level rise over two centuries, followed by a more rapid disintegration over the following 50 to 200 years. Once again, however, he noted that "progress on understanding [the] WAIS over the past two decades has enabled us to lower the relative likelihood of [this] scenario."

In another scenario, the WAIS takes 500-700 years to disappear, as it raises sea-level by 60-120 cm per century. Oppenheimer assesses the relative likelihood of this scenario to be the highest of all, "but with low confidence," as he puts it. Last is what occurs if ice streams slow, as a result of internal ice sheet readjustments, and the discharge of grounded ice decreases, which could well happen even if ice shelves thin and major fast-moving glaciers do not slow. In such a situation, he notes that "the Antarctic contribution to sea-level rise turns increasingly negative," i.e., sea level *falls*. And in commenting upon the suite of scenarios just described, Oppenheimer emphatically states that "it is not possible to place high confidence in any specific prediction about the future of WAIS."

Also writing in *Nature*, Bell *et al.* (1998) used aerogeophysical data to investigate processes that govern fast-moving ice streams on the WAIS. In conjunction with various models, these data suggested a close correlation between the margins of various ice streams and the underlying sedimentary basins, which appeared to act as lubricants for the overlying ice. The seven scientists suggested that the positions of ice-stream margins and their onsets were controlled by features of the underlying sedimentary basins. They concluded that "geological structures beneath the West Antarctic Ice Sheet have the potential to dictate the evolution of the dynamic ice system, modulating the influence of changes in the global climate system," although their work did not indicate what effect, if any, a modest rise in near-surface air temperature might have on this phenomenon.

Rignot (1998) reported on satellite radar measurements of the grounding line of Pine Island Glacier from 1992 to 1996, which were studied to determine whether or not this major ice stream in remote West Antarctica was advancing or retreating. The data indicated that the glacier's grounding line had retreated inland at a rate of 1.2 ± 0.3 kilometers per year over the four-year period of the study; Rignot suggested that this retreat may have been the result of a slight increase in ocean water temperature. Because the study had utilized only four years of data, however, questions concerning the long-term stability of the WAIS, in the words of the researcher, "cannot be answered at present." In addition, although the glacier's grounding line had been found to be retreating, subsequent satellite images suggested that the location of the ice front had remained stable.

Also in the journal *Science*, Conway *et al.* (1999) examined previously reported research, while conducting some of their own, dealing with the retreat of the WAIS since its maximum glacial extent some

20,000 years ago. In doing so, they determined that the ice sheet's grounding line remained near its maximum extent until about 10,000 years ago, whereupon it began to retreat at a rate of about 120 meters per year. This work also indicated that at the end of the twentieth century it was retreating at about the same rate, which suggests that if it continues to behave as it has in the past, complete deglaciation of the WAIS will occur in about 7,000 years. The researchers concluded that the modern-day grounding-line retreat of the WAIS is part of an ongoing recession that has been underway since the early to mid-Holocene, and that "it is not a consequence of anthropogenic warming or recent sea-level rise."

Stenoien and Bentley (2000) mapped the catchment region of Pine Island Glacier using radar altimetry and synthetic aperture radar interferometry, which they used to develop a velocity map that revealed a system of tributaries that channel ice from the catchment area into the fast-flowing glacier. Then, by combining the velocity data with information on ice thickness and snow accumulation rates, they were able to calculate, within an uncertainty of 30 percent, that the mass balance of the catchment region was not significantly different from zero.

One year later, Shepherd *et al.* (2001) used satellite altimetry and interferometry to determine the rate of change of the ice thickness of the entire Pine Island Glacier drainage basin between 1992 and 1999. This work revealed that the grounded glacier thinned by up to 1.6 meters per year between 1992 and 1999. Of this phenomenon, the researchers wrote that "the thinning cannot be explained by short-term variability in accumulation and must result from glacier dynamics," and since glacier dynamics typically respond to phenomena operating on time scales of hundreds to thousands of years, this observation would argue against twentieth century warming being a primary cause of the thinning. Shepherd *et al.* additionally say they could "detect no change in the rate of ice thinning across the glacier over a 7-year period," which also suggests that a long-term phenomenon of considerable inertia must be at work in this particular situation.

But what if the rate of glacier thinning, which sounds pretty dramatic, were to continue unabated? The researchers state that "if the trunk continues to lose mass at the present rate it will be entirely afloat within 600 years." And if that happens, they "estimate the net contribution to eustatic sea level to be 6 mm," which means that over each century, we could expect global sea level to rise by about one millimeter, about the thickness of a paper clip.

Publishing in the same year were Pudsey and Evans (2001), who studied ice-rafted debris obtained from four cores in Prince Gustav Channel, which until 1995 was covered by floating ice shelves. Their efforts indicated that the ice shelves had also retreated in mid-Holocene time, but that, in their words, "colder conditions after about 1.9 ka allowed the ice shelf to reform." Although they concluded that the ice shelves are sensitive indicators of regional climate change, they were careful to state that "we should not view the recent decay as an unequivocal indicator of anthropogenic climate change." Indeed, the disappearance of the ice shelves was not unique; it had happened before without our help, and it could well have happened again on its own. In fact, the breakup of the Prince Gustav Channel ice shelves was likely nothing more than the culmination of the Antarctic Peninsula's natural recovery from the cold conditions of Little Ice Age.

Raymond (2002) presented a brief appraisal of the status of the world's major ice sheets. His primary conclusions relative to the WAIS were that (1) "substantial melting on the upper surface of WAIS would occur only with considerable atmospheric warming," (2) of the three major WAIS drainages, the ice streams that drain northward to the Amundsen Sea have accelerated, widened, and thinned "over substantial distances back into the ice sheet," but that "the eastward drainage toward the Weddell Sea is close to mass balance." And (3) of the westward drainage into the Ross Ice Shelf, "over the last few centuries, margins of active ice streams migrated inward and outward," while the "overall mass balance has changed from loss to gain," as "a currently active ice stream (Whillans) has slowed by about 20% over recent decades."

In a summary statement that takes account of these observations, Raymond says that "the total mass of today's ice sheets is changing only slowly, and even with climate warming increases in snowfall should compensate for additional melting," such as might possibly occur for the WAIS if the planet's temperature continues its post-Little Ice Age rebound.

Stone *et al.* (2003)—working on western Marie Byrd Land—report how they determined cosmogenic [10]Be exposure dates of glacially transported cobbles in elevation transects on seven peaks of the Ford Ranges between the ice sheet's present grounding line and the Clark Mountains some 80 km inland. Based on these ages and the elevations at which the cobbles

were found, they reconstructed a history of ice-sheet thinning over the past 10,000-plus years. This history showed, in their words, that "the exposed rock in the Ford Ranges, up to 700 m above the present ice surface, was deglaciated within the past 11,000 years," and that "several lines of evidence suggest that the maximum ice sheet stood considerably higher than this."

Stone *et al.* additionally report that the consistency of the exposure age versus elevation trends of their data "indicates steady deglaciation since the first of these peaks emerged from the ice sheet some time before 10,400 years ago," and that the mass balance of the region "has been negative throughout the Holocene." The researchers also say their results "add to the evidence that West Antarctic deglaciation continued long after the disappearance of the Northern Hemisphere ice sheets and may still be under way," noting that the ice sheet in Marie Byrd Land "shows the same pattern of steady Holocene deglaciation as the marine ice sheet in the Ross Sea," where ice "has thinned and retreated since 7000 years ago," adding that "there is strong evidence that the limit of grounded ice in both regions—and in Pine Island Bay—is still receding."

The work of Stone *et al.* convincingly demonstrates that the current thinning and retreat of the WAIS are merely manifestations of a slow but steady deglaciation that has been going on ever since the beginning-of-the-end of the last great ice age. Stone *et al.* say "the pattern of recent change is consistent with the idea that thinning of the WAIS over the past few thousand years is continuing," while Ackert (2003) makes the point even plainer when he says "recent ice sheet dynamics appear to be dominated by the ongoing response to deglacial forcing thousands of years ago, rather than by a recent anthropogenic warming or sea-level rise."

Additional information on this topic, including reviews of newer publications as they become available, can be found at http://www.co2science.org/subject/w/waisdynamics.php.

References

Ackert Jr., R.P. 2003. An ice sheet remembers. *Science* **299**: 57-58.

Bell, R.E., Blankenship, D.D., Finn, C.A., Morse, D.L., Scambos, T.A., Brozena, J.M. and Hodge, S.M. 1998. Influence of subglacial geology on the onset of a West

Antarctic ice stream from aerogeophysical observations. *Nature* **394**: 58-62.

Bindschadler, R. 1998. Future of the West Antarctic Ice Sheet. *Science* **282**: 428-429.

Bindschadler, R. and Vornberger, P. 1998. Changes in the West Antarctic Ice Sheet since 1963 from declassified satellite photography. *Science* **279**: 689-692.

Conway, H., Hall, B.L., Denton, G.H., Gades, A.M. and Waddington, E.D. 1999. Past and future grounding-line retreat of the West Antarctic Ice Sheet. *Science* **286**: 280-283.

Gore, A. 2006. An Inconvenient Truth: The Planetary Emergency of Global Warming and What We Can Do About It. Rodale, Emmaus, PA, USA.

Oppenheimer, M. 1998. Global warming and the stability of the West Antarctic Ice Sheet. *Nature* **393**: 325-332.

Pudsey, C.J. and Evans, J. 2001. First survey of Antarctic sub-ice shelf sediments reveals mid-Holocene ice shelf retreat. *Geology* **29**: 787-790.

Raymond, C.F. 2002. Ice sheets on the move. *Science* **298**: 2147-2148.

Rignot, E.J. 1998. Fast recession of a West Antarctic glacier. *Science* **281**: 549-550.

Shepherd, A., Wingham, D.J., Mansley, J.A.D. and Corr, H.F.J. 2001. Inland thinning of Pine Island Glacier, West Antarctica. *Science* **291**: 862-864.

Stenoien, M.D. and Bentley, C.R. 2000. Pine Island Glacier, Antarctica: A study of the catchment using interferometric synthetic aperture radar measurements and radar altimetry. *Journal of Geophysical Research* **105**: 21,761-21,779.

Stone, J.O., Balco, G.A., Sugden, D.E., Caffee, M.W., Sass III, L.C., Cowdery, S.G. and Siddoway, C. 2003. Holocene deglaciation of Marie Byrd Land, West Antarctica. *Science* **299**: 99-102.

4.5.3.2. Mass Balance

Is the West Antarctic Ice Sheet (WAIS) growing or shrinking? In what follows, we briefly review the findings of several researchers who have focused their attention on the mass balance of the WAIS.

Anderson and Andrews (1999) analyzed grain size and foraminiferal contents of radiometrically dated sediment cores collected from the eastern Weddell Sea continental shelf and the western Weddell Sea deep-sea floor in an attempt to better understand the behavior of both the East and West Antarctic ice sheets. In doing so, their data led them to conclude that "significant deglaciation of the Weddell Sea continental shelf took place prior to the last glacial maximum," and that the ice masses that border the Weddell Sea today "are more extensive than they were during the previous glacial minimum." They concluded "that the current interglacial setting is characterized by a more extensive ice margin and larger ice shelves than existed during the last glacial minimum, and that the modern West and East Antarctic ice sheets have not yet shrunk to their minimum." It is thus to be expected—independent of what global air temperature may currently be doing, because of the great inertial forces at work over much longer time scales—that the modern East and West Antarctic Ice Sheets may well continue to shrink and release more icebergs to the Southern Ocean over the coming years, decades and centuries, thereby slowly raising global sea level.

Also studying the combined ice sheets of East and West Antarctica were Wingham *et al.* (1998), who used satellite radar altimeter measurements from 1992 to 1996 to estimate the rate of change of the thickness of nearly two-thirds of the grounded portion of the entire Antarctic Ice Sheet, while using snowfall variability data obtained from ice cores to ultimately calculate the mass balance of the interior of the continental ice sheet over the past century. Their results showed that, at most, the interior of the Antarctic Ice Sheet has been "only a modest source or sink of sea-level mass this century." As a result, Wingham *et al.* concluded that "a large century-scale imbalance for the Antarctic interior is unlikely," noting that this conclusion is in harmony with a body of relative sea-level and geodetic evidence "supporting the notion that the grounded ice has been in balance at the millennial scale." This full set of findings thus suggests that both portions of the Antarctic Ice Sheet may be rather impervious to climate changes of the magnitude characteristic of the Medieval Warm Period and Little Ice Age, which is the type of change most likely to occur—if there is any change at all—in response to the ongoing rise in the air's CO_2 content.

Davis and Ferguson (2004) evaluated elevation changes of the entire Antarctic ice sheet over the five-year period June 1995 to April 2000, based on more than 123 million elevation change measurements made by the European Space Agency's European Remote Sensing 2 satellite radar altimeter. They determined the east Antarctic ice sheet had a five-year trend of 1.0 ± 0.6 cm/year, the west Antarctic ice sheet had a five-year trend of -3.6 ± 1.0 cm/year, and the entire Antarctic continent (north of 81.6°S) had a five-year trend of 0.4 ± 0.4 cm/year. In addition, the Pine Island, Thwaites, DeVicq, and Land glaciers of West Antarctica exhibited five-year trends ranging from -26 to—135 cm/year.

In discussing their findings, Davis and Ferguson noted that the strongly negative trends of the coastal glacier outlets "suggest that the basin results are due to dynamic changes in glacier flow," and that recent observations "indicate strong basal melting, caused by ocean temperature increases, is occurring at the grounding lines of these outlet glaciers." They concluded "there is good evidence that the strongly negative trends at these outlet glaciers, the mass balance of the corresponding drainage basins, and the overall mass balance of the west Antarctic ice sheet may be related to increased basal melting caused by ocean temperature increases." Nevertheless, driven by the significantly positive trend of the much larger east Antarctic ice sheet, the ice volume of the entire continent grew ever larger over the last five years of the twentieth century, the majority of which increase, according to Davis and Ferguson, was due to increased snowfall.

One year later, in an "editorial essay" (i.e., not a peer-reviewed submission) published in the journal *Climatic Change*, Oppenheimer and Alley (2005) discussed "the degree to which warming can affect the rate of ice loss by altering the mass balance between precipitation rates on the one hand, and melting and ice discharge to the ocean through ice streams on the other," with respect to the WAIS and Greenland Ice Sheet (GIS). After a brief overview of the topic, they noted that "the key questions with respect to both WAIS and GIS are: What processes limit ice velocity, and how much can warming affect those processes?" In answer to these questions, they said that "no consensus has emerged about these issues nor, consequently, about the fate of either ice sheet, a state of affairs reflecting the weakness of current models and uncertainty in paleoclimatic reconstructions."

After a cursory review of the science related to these two key questions, Oppenheimer and Alley say their review "leads to a multitude of questions with respect to the basic science of the ice sheets," which we list below. However, instead of listing them in their original question form, we post them in the form of statements that address what we do not know about the various sub-topics mentioned, which is obviously

what prompts the questions in the first place and validates the content of the statements.

(1) We do not know if the apparent response of glaciers and ice streams to surface melting and melting at their termini (e.g., ice shelves) could occur more generally over the ice sheets.

(2) We do not know if dynamical responses are likely to continue for centuries and propagate further inland or if it is more likely that they will be damped over time.

(3) We do not know if surface melting could cause rapid collapse of the Ross or Filchner-Ronne ice shelves, as occurred for the smaller Larsen ice shelf.

(4) We do not know if ice sheets made a significant net contribution to sea-level rise over the past several decades.

(5) We do not know what might be useful paleoclimate analogs for sea level and ice sheet behavior in a warmer world.

(6) We do not know the reliability of Antarctic and Southern Ocean temperatures (and polar amplification) that are projected by current GCMs, nor do we know why they differ so widely among models, nor how these differences might be resolved.

(7) We do not know the prospects for expanding measurements and improving models of ice sheets nor the timescales involved.

(8) We do not know if current uncertainties in future ice sheet behavior can be expressed quantitatively.

(9) We do not know what would be useful early warning signs of impending ice sheet disintegration nor when these might be detectable.

(10) We do not know, given current uncertainties, if our present understanding of the vulnerability of either the WAIS or GIS is potentially useful in defining "dangerous anthropogenic interference" with earth's climate system.

(11) We do not know if the concept of a threshold temperature is useful.

(12) We do not know if either ice sheet seems more vulnerable and thus may provide a more immediate measure of climate "danger" and a more pressing target for research.

(13) We do not know if any of the various temperatures proposed in the literature as demarking danger of disintegration for one or the other ice sheet are useful in contributing to a better understanding of "dangerous anthropogenic interference."

(14) We do not know on what timescale future learning might affect the answers to these questions.

Oppenheimer and Alley describe this list of deficiencies in our knowledge of things related to the WAIS as "gaping holes in our understanding" that "will not be closed unless governments provide adequate resources for research." Nevertheless, they claim that "if emissions of the greenhouse gases are not reduced while uncertainties are being resolved, there is a risk of making ice-sheet disintegration nearly inevitable." Obviously, their own analysis contradicts so dire a warning. Given the degree of deficiency in our knowledge of the matter, it is perhaps as likely as not that a continuation of the planet's recovery from the relative cold of the Little Ice Age will lead to a buildup of polar ice.

The following year also saw the publication of another paper that mixed "gaping holes in our understanding" with warnings of dire-sounding WAIS mass losses. Velicogna and Wahr (2006) used measurements of time-variable gravity from the Gravity Recovery and Climate Experiment (GRACE) satellites to determine mass variations of the Antarctic ice sheet for the 34 months between April 2002 and August 2005. The two researchers concluded that "the ice sheet mass decreased significantly, at a rate of 152 $\pm$ 80 km^3/year of ice, equivalent to 0.4 $\pm$ 0.2 mm/year of global sea-level rise," all of which mass loss came from the WAIS, since they calculated that the East Antarctic Ice Sheet mass balance was 0 $\pm$ 56 km^3/year.

Velicogna and Wahr admit there is "geophysical contamination ... caused by signals outside Antarctica," including "continental hydrology ... and ocean mass variability." The first of these confounding factors, according to them, "is estimated using monthly, global water storage fields from the Global Land Data Assimilation system," while "the ocean contamination is estimated using a JPL version of the Estimating Circulation and Climate of the Ocean (ECCO) general circulation model."

The two researchers note that the GRACE mass solutions "do not reveal whether a gravity variation over Antarctica is caused by a change in snow and ice on the surface, a change in atmospheric mass above Antarctica, or post-glacial rebound (PGR: the viscoelastic response of the solid Earth to glacial unloading over the last several thousand years)."

To adjust for the confounding effect of the variable atmospheric mass above Antarctica, Velicogna and Wahr utilized European Centre for Medium-Range Weather Forecasts (ECMWF) meteorological fields, but they acknowledge that "there are errors in those fields," so they "estimate the secular component of those errors by finding monthly differences between meteorological fields from ECMWF and from the National Centers for Environmental Prediction."

With respect to post-glacial rebound, Velicogna and Wahr say "there are two important sources of error in PGR estimates: the ice history and Earth's viscosity profile." To deal with this problem, they "estimate the PGR contribution and its uncertainties using two ice history models."

All of these estimates and adjustments are convoluted and complex, as well as highly dependent upon various models. Velicogna and Wahr acknowledge that "the PGR contribution is much larger than the uncorrected GRACE trend." In fact, their calculations indicate that the PGR contribution exceeds that of the signal being sought by nearly a factor of five. And they are forced to admit "a significant ice mass trend does not appear until the PGR contribution is removed."

Finally, Velicogna and Wahr's study covered less than a three-year period. Much more likely to be representative of the truth with respect to the WAIS's mass balance are the findings of Zwally *et al.* (2005), who determined Antarctica's contribution to mean global sea level over a recent nine-year period to be only 0.08 mm/year compared to the five-times-greater value of 0.4 mm/year calculated by Velicogna and Wahr.

In a contemporaneous study, van de Berg *et al.* (2006) compared results of model-simulated Antarctic surface mass balance (SMB)—which they derived from a regional atmospheric climate model for the time period 1980 to 2004 that used ERA-40 fields as lateral forcings—with "all available SMB observations from Antarctica (N=1900)" in a recalibration process that ultimately allowed them "to construct a best estimate of contemporary Antarctic SMB," where the many real-world observations employed in this process came from the studies of Vaughan *et al.* (1999), van den Broeke *et al.* (1999), Frezzotti *et al.* (2004), Karlof *et al.* (2000), Kaspari *et al.* (2004), Magand *et al.* (2004), Oerter *et al.* (1999, 2000), Smith *et al.* (2002), and Turner *et al.* (2002). Observations were derived by a number of different measurement techniques, including stake arrays, bomb horizons, and chemical analyses of ice cores that covered time periods ranging from a few years to more than a century.

As a result of this effort, van de Berg *et al.* determined that "the SMB integrated over the grounded ice sheet (171 $\pm$ 3 mm per year) exceeds previous estimates by as much as 15%," with the largest differences between their results and those of others being "up to one meter per year higher in the coastal zones of East and West Antarctica," concluding that "support or falsification of this result can only be found in new SMB observations from poorly covered high accumulation regions in coastal Antarctica."

In the same year, Wingham *et al.* (2006) "analyzed 1.2 x 10^8 European remote sensing satellite altimeter echoes to determine the changes in volume of the Antarctic ice sheet from 1992 to 2003," which survey, in their words, "covers 85% of the East Antarctic ice sheet and 51% of the West Antarctic ice sheet," which together comprise "72% of the grounded ice sheet." In doing so, they found that "overall, the data, corrected for isostatic rebound, show the ice sheet growing at 5 ± 1 mm per year." To calculate the ice sheet's change in mass, however, "requires knowledge of the density at which the volume changes have occurred," and when the researchers' best estimates of regional differences in this parameter were used, they found that "72% of the Antarctic ice sheet is gaining 27 ± 29 Gt per year, a sink of ocean mass sufficient to *lower* [their italics] global sea levels by 0.08 mm per year." This net *extraction* of water from the global ocean, according to Wingham *et al.*, occurs because "mass gains from accumulating snow, particularly on the Antarctic Peninsula and within East Antarctica, exceed the ice dynamic mass loss from West Antarctica."

Ramillien *et al.* (2006) derived new estimates of the mass balances of the East and West Antarctic ice sheets from GRACE data for the period July 2002 to March 2005: a loss of 107 ± 23 km^3/year for West Antarctica and a gain of 67 ± 28 km^3/year for East Antarctica, which results yielded a net ice loss for the entire continent of 40 km^3/year (which translates to a mean sea-level rise of 0.11 mm/year). This is of the same order of magnitude as the 0.08 mm/year Antarctic-induced mean sea-level rise calculated by Zwally *et al.* (2005), which was derived from elevation changes based on nine years of satellite radar altimetry data obtained from the European Remote-sensing Satellites ERS-1 and -2. Even at that, the GRACE approach is still laden with a host of potential errors, as we noted in our discussion of the Velicogna and Wahr paper, and as both they and Ramillien *et al.* readily admit. In addition, as the latter researchers note in their closing paragraph, "the GRACE data time series is still very short and these results must be considered as preliminary since we

cannot exclude that the apparent trends discussed in this study only reflect interannual fluctuations."

Remy and Frezzotti (2006) reviewed "the results given by three different ways of estimating mass balance, first by measuring the difference between mass input and output, second by monitoring the changing geometry of the continent, and third by modeling both the dynamic and climatic evolution of the continent." In describing their findings, the two researchers state that "the East Antarctica ice sheet is nowadays more or less in balance, while the West Antarctica ice sheet exhibits some changes likely to be related to climate change and is in negative balance." In addition, they report that "the current response of the Antarctica ice sheet is dominated by the background trend due to the retreat of the grounding line, leading to a sea-level rise of 0.4 mm/yr over the short-time scale," which they describe in terms of centuries. However, they note that "later, the precipitation increase will counterbalance this residual signal, leading to a thickening of the ice sheet and thus a decrease in sea level."

Van den Broeke *et al.* (2006) employed a regional atmospheric climate model (RACMO2), with snowdrift-related processes calculated offline, to calculate the flux of solid precipitation (Ps), surface sublimation (SU), sublimation from suspended (drifting/saltating) snow particles, horizontal snow drift transport, and surface melt (ME). In doing so, they found that "even without snowdrift-related processes, modeled (Ps-SU-ME) from RACMO2 strongly correlates with 1900 spatially weighted quality-controlled in situ SSMB observations," which result they describe as "remarkable," given that the "model and observations are completely independent." Then, to deal with a remaining systematic elevation bias in the model results, they applied a set of empirical corrections (at 500-m intervals) that "largely eliminated" this final deviation from reality. And after analyzing all of the data-driven results for trends over the period 1980-2004, the four Dutch researchers report that "no trend is found in any of the Antarctic SSMB components, nor in the size of ablation areas."

Krinner *et al.* (2007) used the LMDZ4 atmospheric general circulation model (Hourdin *et al.*, 2006) to simulate Antarctic climate for the periods 1981-2000 (to test the model's ability to adequately simulate present conditions) and 2081-2100 (to see what the future might hold for the mass balance of the Antarctic Ice Sheet and its impact on global sea level). This work revealed, first, that "the simulated present-day surface mass balance is skilful on continental scales," which gave them confidence that their results for the end of the twenty-first century would be reasonably accurate as well. Of that latter period a full century from now, they determined that "the simulated Antarctic surface mass balance increases by 32 mm water equivalent per year," which corresponds "to a sea-level decrease of 1.2 mm per year by the end of the twenty-first century," which would in turn "lead to a cumulated sea-level decrease of about 6 cm." This result, in their words, occurs because the simulated temperature increase "leads to an increased moisture transport towards the interior of the continent because of the higher moisture holding capacity of warmer air," where the extra moisture falls as precipitation, causing the continent's ice sheet to grow.

The results of this study—based on sea surface boundary conditions taken from IPCC Fourth Assessment Report simulations (Dufresne *et al.*, 2005) that were carried out with the IPSL-CM4 coupled atmosphere-ocean general circulation model (Marti *et al.*, 2005), of which the LMDZ4 model is the atmospheric component—argue strongly against predictions of future catastrophic sea-level rise due to mass wastage of the Antarctic Ice Sheet. In fact, they suggest just the opposite, i.e., that CO_2-induced global warming would tend to buffer the world against such an outcome. That seems to be the message of most of the other major studies of the subject.

Additional information on this topic, including reviews of newer publications as they become available, can be found at http://www.co2science.org/subject/w/waisbalance.php.

References

Anderson, J.B. and Andrews, J.T. 1999. Radiocarbon constraints on ice sheet advance and retreat in the Weddell Sea, Antarctica. *Geology* 27: 179-182.

Davis, C.H. and Ferguson, A.C. 2004. Elevation change of the Antarctic ice sheet, 1995-2000, from ERS-2 satellite radar altimetry. *IEEE Transactions on Geoscience and Remote Sensing* 42: 2437-2445.

Dufresne, J.L., Quaas, J., Boucher, O., Denvil, S. and Fairhead, L. 2005. Contrasts in the effects on climate of anthropogenic sulfate aerosols between the 20th and the 21st century. *Geophysical Research Letters* 32: 10.1029/2005GL023619.

Frezzotti, M., Pourchet, M., Flora, O., Gandolfi, S., Gay, M., Urbini, S., Vincent, C., Becagli, S., Gragnani, R.,

Proposito, M., Severi, M., Traversi, R., Udisti, R. and Fily, M. 2004. New estimations of precipitation and surface sublimation in East Antarctica from snow accumulation measurements. *Climate Dynamics* 23: 803-813.

Hourdin, F., Musat, I., Bony, S., Braconnot, P., Codron, F., Dufresne, J.L., Fairhead, L., Filiberti, M.A., Friedlingstein, P., Grandpeix, J.Y., Krinner, G., Le Van, P., Li, Z.X. and Lott, F. 2006. The LMDZ4 general circulation model: climate performance and sensitivity to parameterized physics with emphasis on tropical convection. *Climate Dynamics* 27: 787-813.

Karlof, L., Winther, J.-G., Isaksson, E., Kohler, J., Pinglot, J.F., Wilhelms, F., Hansson, M., Holmlund, P., Nyman, M., Pettersson, R., Stenberg, M., Thomassen, M.P.A., van der Veen, C. and van de Wal, R.S.W. 2000. A 1500-year record of accumulation at Amundsenisen western Dronning Maud Land, Antarctica, derived from electrical and radioactive measurements on a 120-m ice core. *Journal of Geophysical Research* 105: 12,471-12,483.

Kaspari, S., Mayewski, P.A., Dixon, D.A., Spikes, V.B., Sneed, S.B., Handley, M.J. and Hamilton, G.S. 2004. Climate variability in West Antarctica derived from annual accumulation rate records from ITASE firn/ice cores. *Annals of Glaciology* 39: 585-594.

Krinner, G., Magand, O., Simmonds, I., Genthon, C. and Dufresne, J.L. 2007. Simulated Antarctic precipitation and surface mass balance at the end of the twentieth and twenty-first centuries. *Climate Dynamics* 28: 215-230.

Magand, O., Frezzotti, M., Pourchet, M., Stenni, B., Genoni, L. and Fily, M. 2004. Climate variability along latitudinal and longitudinal transects in East Antarctica. *Annals of Glaciology* 39: 351-358.

Marti, O., Braconnot, P., Bellier, J., Benshila, R., Bony, S., Brockmann, P., Cadule, P., Caubel, A., Denvil, S., Dufresne, J.L., Fairhead, L., Filiberti, M.A., Foujols, M.A., Fichefet, T., Friedlingstein, P., Grandpeix, J.Y., Hourdin, F., Krinner, G., Levy, C., Madec, G., Musat, I., de Noblet-Ducoudre, N., Polcher, J. and Talandier, C. 2005. The new IPSL climate system model: IPSL-CM4. Note du Pole de Modelisation n. 26, IPSL, ISSN 1288-1619.

Oerter, H., Graf, W., Wilhelms, F., Minikin, A. and Miller, H. 1999. Accumulation studies on Amundsenisen, Dronning Maud Land, by means of tritium, dielectric profiling and stable-isotope measurements: First results from the 1995-96 and 1996-97 field seasons. *Annals of Glaciology* 29: 1-9.

Oerter, H., Wilhelms, F., Jung-Rothenhausler, F., Goktas, F., Miller, H., Graf, W. and Sommer, S. 2000. Accumulation rates in Dronning Maud Land, Antarctica, as revealed by dielectric-profiling measurements of shallow firn cores. *Annals of Glaciology* 30: 27-34.

Oppenheimer, M. and Alley, R.B. 2005. Ice sheets, global warming, and article 2 of the UNFCCC. *Climatic Change* **68**: 257-267.

Ramillien, G., Lombard, A., Cazenave, A., Ivins, E.R., Llubes, M., Remy, F. and Biancale, R. 2006. Interannual variations of the mass balance of the Antarctica and Greenland ice sheets from GRACE. *Global and Planetary Change* **53**: 198-208.

Remy, F. and Frezzotti, M. 2006. Antarctica ice sheet mass balance. *Comptes Rendus Geoscience* **338**: 1084-1097.

Smith, B.T., van Ommen, T.D. and Morgan, V.I. 2002. Distribution of oxygen isotope ratios and snow accumulation rates in Wilhelm II Land, East Antarctica. *Annals of Glaciology* **35**: 107-110.

Stenoien, M.D. and Bentley, C.R. 2000. Pine Island Glacier, Antarctica: A study of the catchment using interferometric synthetic aperture radar measurements and radar altimetry. *Journal of Geophysical Research* **105**: 21,761-21,779.

Turner, J., Lachlan-Cope, T.A., Marshall, G.J., Morris, E.M., Mulvaney, R. and Winter, W. 2002. Spatial variability of Antarctic Peninsula net surface mass balance. *Journal of Geophysical Research* **107**: 10.1029/JD000755.

Van de Berg, W.J., van den Broeke, M.R., Reijmer, C.H. and van Meijgaard, E. 2006. Reassessment of the Antarctic surface mass balance using calibrated output of a regional atmospheric climate model. *Journal of Geophysical Research* **111**: 10.1029/2005JD006495.

Van den Broeke, M., van de Berg, W.J., van Meijgaard, E. and Reijmer, C. 2006. Identification of Antarctic ablation areas using a regional atmospheric climate model. *Journal of Geophysical Research* **111**: 10.1029/2006JD007127.

Van den Broeke, M.R., Winther, J.-G., Isaksson, E., Pinglot, J.F., Karlof, L., Eiken, T. and Conrads, L. 1999. Climate variables along a traverse line in Dronning Maud Land, East Antarctica. *Journal of Glaciology* **45**: 295-302.

Vaughn, D.G., Bamber, J.L., Giovinetto, M., Russell, J. and Cooper, A.P.R. 1999. Reassessment of net surface mass balance in Antarctica. *Journal of Climate* **12**: 933-946.

Velicogna, I. and Wahr, J. 2006. Measurements of time-variable gravity show mass loss in Antarctica. *Sciencexpress*: 10.1126science.1123785.

Wingham, D.J., Ridout, A.J., Scharroo, R., Arthern, R.J. and Shum, C.K. 1998. Antarctic elevation change from 1992 to 1996. *Science* **282**: 456-458.

Wingham, D.J., Shepherd, A., Muir, A. and Marshall, G.J. 2006. Mass balance of the Antarctic ice sheet.

Philosophical Transactions of the Royal Society A **364**: 1627-1635.

Zwally, H.J., Giovinetto, M.B., Li, J., Cornejo, H.G., Beckley, M.A., Brenner, A.C., Saba, J.L. and Yi, D. 2005. Mass changes of the Greenland and Antarctic ice sheets and shelves and contributions to sea-level rise: 1992-2002. *Journal of Glaciology* **51**: 509-527.

4.5.3.3. *West Antarctic Ice Sheet and Sea Level*

Many of the studies of the West Antarctic Ice Sheet (WAIS) cited in the previous sections of this report address its past and future effects on sea level. In this final section on the WAIS, we bring this body of research together in one place and add other research summaries.

Bindschadler (1998) analyzed the WAIS's historical retreat in terms of its grounding line and ice front. This work revealed that from the time of the Last Glacial Maximum to the present, the retreat of the ice sheet's grounding line had been faster than that of its ice front, which resulted in an expanding Ross Ice Shelf. Although Bindschadler wrote that "the ice front now appears to be nearly stable," there were indications that its grounding line was retreating at a rate that suggested complete dissolution of the WAIS in another 4,000 to 7,000 years. Such a retreat was calculated to result in a sustained sea-level rise of 8-13 cm per century. However, even the smallest of these rates-of-rise would require, in Bindschadler's words, "a large negative mass balance for all of West Antarctica," and there were no broad-based data that supported that scenario.

A year later, Reeh (1999) reviewed what was known about the mass balances of both the Greenland and Antarctic ice sheets, concluding that the future contribution of the Greenland and Antarctic ice sheets to global sea level depends upon their past climate and dynamic histories as much as it does upon future climate. With respect to potential climate change, Reeh determined there was a broad consensus that the effect of a 1°C climatic warming on the Antarctic ice sheet would be a *fall* in global sea level on the order of 0.2 to 0.7 millimeters per year.

The following year, Cuffey and Marshall (2000) reevaluated previous model estimates of the Greenland ice sheet's contribution to sea-level rise during the last interglacial, based on a recalibration of oxygen-isotope-derived temperatures from central Greenland ice cores. Their results suggested that the Greenland ice sheet was much smaller during the last interglacial than previously thought, with melting of the ice sheet contributing somewhere between four and five-and-a-half meters to sea-level rise. According to Hvidberg (2000), this finding suggests that "high sea levels during the last interglacial should not be interpreted as evidence for extensive melting of the West Antarctic Ice Sheet, and so challenges the hypothesis that the West Antarctic is particularly sensitive to climate change."

Oppenheimer and Alley (2005) discussed "the degree to which warming can affect the rate of ice loss by altering the mass balance between precipitation rates on the one hand, and melting and ice discharge to the ocean through ice streams on the other," with respect to both the West Antarctic and Greenland Ice Sheets. Their review of the subject led them to conclude that we simply do not know if these ice sheets had made a significant contribution to sea-level rise over the past several decades.

One year later, however, the world was exposed to a different view of the issue when Velicogna and Wahr (2006) used measurements of time-variable gravity from the Gravity Recovery and Climate Experiment (GRACE) satellites to determine mass variations of the Antarctic ice sheet for the 34 months between April 2002 and August 2005. The two researchers concluded that "the ice sheet mass decreased significantly, at a rate of 152 ± 80 km^3/year of ice, equivalent to 0.4 ± 0.2 mm/year of global sea-level rise," all of which mass loss came from the WAIS, since they calculated that the East Antarctic Ice Sheet mass balance was 0 ± 56 km^3/year.

The many estimates and adjustments used by Velicogna and Wahr to reach this conclusion were described in Section 4.5.3.2. For example, the adjustment for post-glacial rebound alone exceeded the signal being sought by nearly a factor of five. Moreover, the study covers less than a three-year period, which compares poorly with the findings of Zwally *et al.* (2005), who determined Antarctica's contribution to mean global sea level over a recent nine-year period to be only 0.08 mm/year.

Ramillien *et al.* (2006) also used GRACE data to derive estimates of the mass balances of the East and West Antarctic ice sheets for the period July 2002 to March 2005, obtaining a loss of 107 ± 23 km^3/year for West Antarctica and a gain of 67 ± 28 km^3/year for East Antarctica, which results yielded a net ice loss for the entire continent of only 40 km^3/year (which translates to a mean sea-level rise of 0.11 mm/year), as opposed to the 152 km^3/year ice loss calculated by Velicogna and Wahr (which translates to a nearly four times larger mean sea-level rise of

0.40 mm/year). Ramillien *et al.* note in their closing paragraph, "the GRACE data time series is still very short and these results must be considered as preliminary since we cannot exclude that the apparent trends discussed in this study only reflect interannual fluctuations." That caveat also applies to the Velicogna and Wahr analysis.

About the same time, Wingham *et al.* (2006) analyzed European remote sensing satellite altimeter echoes to determine the changes in volume of the Antarctic ice sheet from 1992 to 2003. They found that "72% of the Antarctic ice sheet is gaining 27 ± 29 Gt per year, a sink of ocean mass sufficient to *lower* [their italics] global sea levels by 0.08 mm per year." This net extraction of water from the global ocean, according to Wingham *et al.*, occurs because "mass gains from accumulating snow, particularly on the Antarctic Peninsula and within East Antarctica, exceed the ice dynamic mass loss from West Antarctica."

Remy and Frezzotti (2006) reviewed "the results given by three different ways of estimating mass balance, first by measuring the difference between mass input and output, second by monitoring the changing geometry of the continent, and third by modeling both the dynamic and climatic evolution of the continent." They report that "the current response of the Antarctica ice sheet is dominated by the background trend due to the retreat of the grounding line, leading to a sea-level rise of 0.4 mm/yr over the short-time scale," which they describe in terms of *centuries*. However, they note that "later, the precipitation increase will counterbalance this residual signal, leading to a thickening of the ice sheet and thus a decrease in sea level."

Krinner *et al.* (2007), in a study summarized in Section 5.6.3.3., used the LMDZ4 atmospheric general circulation model of Hourdin *et al.* (2006) to simulate Antarctic climate for the periods 1981-2000 (to test the model's ability to adequately simulate present conditions) and 2081-2100 (to see what the future might hold for the mass balance of the Antarctic Ice Sheet and its impact on global sea level). They determined that "the simulated Antarctic surface mass balance increases by 32 mm water equivalent per year," which corresponds "to a sea-level decrease of 1.2 mm per year by the end of the twenty-first century," which would in turn "lead to a cumulated sea-level decrease of about 6 cm." This result occurs because the simulated temperature increase "leads to an increased moisture transport towards the interior of the continent because of the higher moisture holding capacity of warmer air," where the extra moisture falls as precipitation, causing the continent's ice sheet to grow.

There has been very little change in global sea level due to wastage of the WAIS over the past few decades, and there will probably be little change in both the near and far future. What wastage might occur along the coastal area of the ice sheet over the long term would likely be countered, or more than countered, by greater inland snowfall. In the case of the latter possibility, the entire Antarctic Ice Sheet could well compensate for any long-term wastage of the Greenland Ice Sheet that might occur.

Additional information on this topic, including reviews of newer publications as they become available, can be found at http://www.co2science.org/subject/w/waissealevel.php.

References

Bindschadler, R. 1998. Future of the West Antarctic Ice Sheet. *Science* **282**: 428-429.

Cuffey, K.M. and Marshall, S.J. 2000. Substantial contribution to sea-level rise during the last interglacial from the Greenland ice sheet. *Nature* **404**: 591-594.

Dufresne, J.L., Quaas, J., Boucher, O., Denvil, S. and Fairhead, L. 2005. Contrasts in the effects on climate of anthropogenic sulfate aerosols between the 20th and the 21st century. *Geophysical Research Letters* **32**: 10.1029/2005GL023619.

Hourdin, F., Musat, I., Bony, S., Braconnot, P., Codron, F., Dufresne, J.L., Fairhead, L., Filiberti, M.A., Friedlingstein, P., Grandpeix, J.Y., Krinner, G., Le Van, P., Li, Z.X. and Lott, F. 2006. The LMDZ4 general circulation model: climate performance and sensitivity to parameterized physics with emphasis on tropical convection. *Climate Dynamics* **27**: 787-813.

Hvidberg, C.S. 2000. When Greenland ice melts. *Nature* **404**: 551-552.

Krinner, G., Magand, O., Simmonds, I., Genthon, C. and Dufresne, J.-L. 2007. Simulated Antarctic precipitation and surface mass balance at the end of the twentieth and twenty-first centuries. *Climate Dynamics* **28**: 215-230.

Marti, O., Braconnot, P., Bellier, J., Benshila, R., Bony, S., Brockmann, P., Cadule, P., Caubel, A., Denvil, S., Dufresne, J.L., Fairhead, L., Filiberti, M.A., Foujols, M.A., Fichefet, T., Friedlingstein, P., Grandpeix, J.Y., Hourdin, F., Krinner, G., Levy, C., Madec, G., Musat, I., de Noblet-Ducoudre, N., Polcher, J. and Talandier, C. 2005. The new IPSL climate system model: IPSL-CM4. Note du Pole de Modelisation n. 26, IPSL, ISSN 1288-1619.

Oppenheimer, M. and Alley, R.B. 2005. Ice sheets, global warming, and article 2 of the UNFCCC. *Climatic Change* **68**: 257-267.

Ramillien, G., Lombard, A., Cazenave, A., Ivins, E.R., Llubes, M., Remy, F. and Biancale, R. 2006. Interannual variations of the mass balance of the Antarctica and Greenland ice sheets from GRACE. *Global and Planetary Change* **53**: 198-208.

Reeh, N. 1999. Mass balance of the Greenland ice sheet: Can modern observation methods reduce the uncertainty? *Geografiska Annaler* **81A**: 735-742.

Remy, F. and Frezzotti, M. 2006. Antarctica ice sheet mass balance. *Comptes Rendus Geoscience* **338**: 1084-1097.

Velicogna, I. and Wahr, J. 2006. Measurements of time-variable gravity show mass loss in Antarctica. *Sciencexpress*: 10.1126science.1123785.

Wingham, D.J., Shepherd, A., Muir, A. and Marshall, G.J. 2006. Mass balance of the Antarctic ice sheet. *Philosophical Transactions of the Royal Society A* **364**: 1627-1635.

Zwally, H.J., Giovinetto, M.B., Li, J., Cornejo, H.G., Beckley, M.A., Brenner, A.C., Saba, J.L. and Yi, D. 2005. Mass changes of the Greenland and Antarctic ice sheets and shelves and contributions to sea-level rise: 1992-2002. *Journal of Glaciology* **51**: 509-527.

4.5.4. Greenland Ice Cap

Studies of the growth and decay of polar ice sheets are of great importance because of the relationships of these phenomena to global warming and the impacts they can have on sea level. In this section, we review a number of such studies that pertain to the Greenland Ice Sheet.

In the March 24, 2006 issue of *Science*, several commentaries heralded accelerating discharges of glacial ice from Greenland and Antarctica, while dispensing dire warnings of an imminent large, rapid, and accelerating sea-level rise (Bindschadler, 2006; Joughin, 2006; Kerr, 2006; Kennedy and Hanson, 2006). This distressing news was based largely on three reports published in the same issue (Ekstrom *et al.*, 2006; Otto-Bliesner *et al.*, 2006; Overpeck *et al.*, 2006), wherein the unnerving phenomena were attributed to anthropogenic-induced global warming.

Consider the report of Ekstrom *et al.*, who studied "glacial earthquakes" caused by sudden sliding motions of glaciers on Greenland. Over the period from January 1993 to October 2005, they determined that (1) *all* of the best-recorded quakes were associated with major outlet glaciers on the east and west coasts of Greenland between approximately 65 and 76°N latitude, (2) "a clear increase in the number of events is seen starting in 2002," and (3) "to date in 2005, twice as many events have been detected as in any year before 2002."

With respect to the reason for the recent increase in glacial activity on Greenland, Clayton Sandell of ABC News on March 23, 2006 quoted Ekstrom as saying "I think it is very hard not to associate this with global warming," which sentiment appears to be shared by almost all of the authors of the seven *Science* articles. Unwilling to join that conclusion, however, was Joughin, who in the very same issue presented histories of summer temperature at four coastal Greenland stations located within the same latitude range as the sites of the glacial earthquakes, which histories suggest that it was warmer in this region back in the 1930s than it was over the period of Ekstrom *et al.*'s analysis.

Based on these data, Joughin concluded that the recent warming in Greenland "is too short to determine whether it is an anthropogenic effect or natural variability," a position that is supported by many scientists cited previously in this chapter, and more in the discussion that follows.

A study based on mean monthly temperatures of 37 Arctic and seven sub-Arctic stations and temperature anomalies of 30 grid-boxes from the updated dataset of Jones by Przybylak (2000) found (1) "in the Arctic, the highest temperatures since the beginning of instrumental observation occurred clearly in the 1930s," (2) "even in the 1950s the temperature was higher than in the last 10 years," (3) "since the mid-1970s, the annual temperature shows no clear trend," and (4) "the level of temperature in Greenland in the last 10-20 years is similar to that observed in the 19th century." These findings led him to conclude that the meteorological record "shows that the observed variations in air temperature in the real Arctic are in many aspects not consistent with the projected climatic changes computed by climatic models for the enhanced greenhouse effect," because, in his words, "the temperature predictions produced by numerical climate models significantly differ from those actually observed."

In light of these several other studies of real-world observations, it is clear that the recent upswing in glacial activity on Greenland likely has had nothing to do with anthropogenic-induced global warming, as temperatures there have yet to rise either as fast or as high as they did during the great warming of the

1920s, which was clearly a natural phenomenon. It is also important to recognize the fact that coastal glacial discharge represents only half of the equation relating to sea-level change, the other half being inland ice accumulation derived from precipitation; and when the mass balance of the entire Greenland ice sheet was recently assessed via satellite radar altimetry, quite a different result was obtained than that suggested by the seven *Science* papers.

Zwally *et al.* (2005) found that although "the Greenland ice sheet is thinning at the margins," it is "growing inland with a small overall mass gain." In fact, for the 11-year period 1992-2003, Johannessen *et al.* (2005) found that "below 1500 meters, the elevation-change rate is [a negative] 2.0 ± 0.9 cm/year, in qualitative agreement with reported thinning in the ice-sheet margins," but that "an increase of 6.4 ± 0.2 cm/year is found in the vast interior areas above 1500 meters." Spatially averaged over the bulk of the ice sheet, the net result, according to the latter researchers, was a mean increase of 5.4 ± 0.2 cm/year, "or ~60 cm over 11 years, or ~54 cm when corrected for isostatic uplift." Consequently, the Greenland Ice Sheet would appear to have experienced no net loss of mass over the last decade for which data are available. To the contrary, it was likely host to a net accumulation of ice, which Zwally *et al.* found to be producing a 0.03 ± 0.01 mm/year decline in sea-level.

In an attempt to downplay the significance of these inconvenient findings, Kerr quoted Zwally as saying he believes that "right now" the Greenland Ice Sheet is experiencing a net loss of mass. Why? Kerr says Zwally's belief is "based on his gut feeling about the most recent radar and laser observations." Gut feelings are a poor substitute for comprehensive real-world measurements, and even if Zwally's intestines are ultimately found to be correct, their confirmation would only demonstrate just how rapidly the Greenland environment can change. We would have to wait and see how long the mass losses prevailed in order to assess their significance within the context of the CO_2-induced global warming debate. For the present and immediate future, therefore, we have no choice but to stick with what existent data and analyses suggest; i.e., that cumulatively since the early 1990s and conservatively (since the balance is likely still positive), there has been no net loss of mass from the Greenland Ice Sheet.

The recent study by Eldrett *et al.* (2007) provides further evidence that the IPCC's view of melting sea ice is wrong. The five researchers from the School of

Ocean and Earth Science of the National Oceanography Centre of the University of Southampton in the UK report they "have generated a new stratigraphy for three key Deep Sea Drilling Project/Ocean Drilling Program sites by calibrating dinocyst events to the geomagnetic polarity timescale." In doing so, they say their detailed core observations revealed evidence for "extensive ice-rafted debris, including macroscopic dropstones, in late Eocene to early Oligocene sediments from the Norwegian-Greenland Sea that were deposited between about 38 and 30 million years ago." They further report that their data "indicate sediment rafting by glacial ice, rather than sea ice, and point to East Greenland as the likely source," and they conclude that their data thus suggest "the existence of (at least) isolated glaciers on Greenland about 20 million years earlier than previously documented."

What is particularly interesting about this finding, as Eldrett *et al.* describe it, is that it indicates the presence of glacial ice on Greenland "at a time when temperatures and atmospheric carbon dioxide concentrations were substantially higher." How much higher? According to graphs the researchers present, ocean bottom-water temperatures were 5-8°C warmer, while atmospheric CO_2 concentrations were as much as four times greater than they are today.

The problem these observations provide for those who hold to the view that global warming will melt the Greenland Ice Sheet, to quote Eldrett *et al.*, is that "palaeoclimate model experiments generate substantial ice sheets in the Northern Hemisphere for the Eocene only in runs where carbon dioxide levels are lower (approaching the pre-anthropogenic level) than suggested by proxy records," which records indicate atmospheric CO_2 concentrations fully two to seven times greater than the pre-anthropogenic level during the time of the newly detected ice sheets.

"Regardless," as the researchers say, their data "provide the first stratigraphically extensive evidence for the existence of continental ice in the Northern Hemisphere during the Palaeogene," which "is about 20 million years earlier than previously documented, at a time when global deep water temperatures and, by extension, surface water temperatures at high latitude, were much warmer." Therefore—and also "by extension"—we now have evidence of a much warmer period of time that failed to melt the Greenland Ice Sheet.

Continuing, Krabill *et al.* (2000) used data obtained from aircraft laser-altimeter surveys over northern Greenland in 1994 and 1999, together with

previously reported data from southern Greenland, to evaluate the mass balance of the Greenland Ice Sheet. Above an elevation of 2,000 meters they found areas of both thinning and thickening; and these phenomena nearly balanced each other, so that in the south there was a net thinning of 11 ± 7 mm/year, while in the north there was a net thickening of 14 ± 7 mm/year. Altogether, the entire region exhibited a net thickening of 5 ± 5 mm/year; but in correcting for bedrock uplift, which averaged 4 mm/year in the south and 5 mm/year in the north, the average thickening rate decreased to practically nothing. The word used by Krabill *et al.* to describe the net balance was "zero."

At lower elevations, thinning was found to predominate along approximately 70 percent of the coast. Here, however, flight lines were few and far between; so few and far between, in fact, that the researchers said that "in order to extend our estimates to the edge of the ice sheet in areas not bounded by our surveys, we calculated a hypothetical thinning rate on the basis of the coastal positive degree day anomalies." Then, they interpolated between this calculated coastal thinning rate and the nearest observed elevation changes to obtain their final answer: a total net reduction in ice volume of 51 km^3/year.

Unfortunately, it is difficult to know what estimates derived from interpolations based on calculations of a hypothetical thinning rate mean. We question their significance; and the researchers themselves do the same. They note that they do not have a "satisfactory explanation" for the "widespread thinning at elevations below 2000 m," which suggests that the reason this phenomenon is unexplainable is that it may not be real. The authors further note that even if the thinning was real, it could not be due to global or regional warming, since Greenland temperature records indicate "the 1980s and early 1990s were about half a degree cooler than the 96-year mean."

After discussing some other factors that could be involved, Krabill *et al.* state they are left with changes in ice dynamics as the most likely cause of the hypothetical ice sheet thinning. But they admit in their final sentence that "we have no evidence for such changes, and we cannot explain why they should apply to many glaciers in different parts of Greenland." It would seem logical to admit this study resolves almost nothing about the mass balance of the coastal regions of the Greenland Ice Sheet and nothing about the subject of global warming and its effect or non-effect upon this hypothetical phenomenon.

In a preliminary step required to better understand the relationship of glacier dynamics to climate change in West Greenland, Taurisano *et al.* (2004) described the temperature trends of the Nuuk fjord area during the past century. This analysis of all pertinent regional data led them to conclude that "at all stations in the Nuuk fjord, both the annual mean and the average temperature of the three summer months (June, July and August) exhibit a pattern in agreement with the trends observed at other stations in south and west Greenland (Humlum 1999; Hanna and Cappelen, 2003)." As they describe it, the temperature data "show that a warming trend occurred in the Nuuk fjord during the first 50 years of the 1900s, followed by a cooling over the second part of the century, when the average annual temperatures decreased by approximately 1.5°C." Coincident with this cooling trend there was also what they describe as "a remarkable increase in the number of snowfall days (+59 days)." What is more, they report that "not only did the cooling affect the winter months, as suggested by Hannna and Cappelen (2002), but also the summer mean," noting that "the summer cooling is rather important information for glaciological studies, due to the ablation-temperature relations." Finally, they report there was no significant trend in annual precipitation. In their concluding discussion, Taurisano *et al.* remark that the temperature data they studied "reveal a pattern which is common to most other stations in Greenland."

Rignot and Kanagaratnam (2005) used satellite radar interferometry observations of Greenland to detect what they described as "widespread glacier acceleration." Calculating that this phenomenon had led to a doubling of the ice sheet mass deficit in the past decade and, therefore, a comparable increase in Greenland's contribution to rising sea levels, they went on to claim that "as more glaciers accelerate ... the contribution of Greenland to sea-level rise will continue to increase."

With respect to these contentions, we have no problem with what the two researchers have observed with respect to Greenland's glaciers; but we feel compelled to note that what they have calculated with respect to the mass balance of Greenland's Ice Sheet and what they say it implies about sea level are contradicted by more inclusive real-world data. One reason for this discrepancy is that instead of relying on measurements for this evaluation, Rignot and Kanagaratnam relied on the calculations of Hanna *et* *al.* (2005), who used meteorological models "to retrieve annual accumulation, runoff, and surface mass balance." When actual measurements of the ice sheet via satellite radar altimetry are employed, a decidedly different perspective is obtained, as indicated by the work of Zwally *et al.* (2005) and Johannessen *et al.* (2005), which we cited earlier. Consequently, and contrary to the claim of Rignot and Kanagaratnam, Greenland would appear to have experienced no ice sheet mass deficit in the past decade.

Shepherd and Wingham (2007) reviewed what is known about sea-level contributions arising from wastage of the Antarctic and Greenland Ice Sheets, concentrating on the results of 14 satellite-based estimates of the imbalances of the polar ice sheets that have been derived since 1998. These studies have been of three major types—standard mass budget analyses, altimetry measurements of ice-sheet volume changes, and measurements of the ice sheets' changing gravitational attraction—and they have yielded a diversity of values, ranging from a sea-level-rise-equivalent of 1.0 mm/year to a sea-level-fall-equivalent of 0.15 mm/year. The two researchers conclude that the current "best estimate" of the contribution of polar ice wastage (from both Greenland and Antarctica) to global sea-level change is a rise of 0.35 millimeters per year, which over a century amounts to only 35 millimeters.

Even this unimpressive sea-level increase may be too large an estimate, for although two of Greenland's largest outlet glaciers doubled their rates of mass loss in less than a year in 2004—causing the IPCC to claim the Greenland Ice Sheet was responding much more rapidly to global warming than anyone had ever expected—Howat *et al.* (2007) report that the two glaciers' rates of mass loss "decreased in 2006 to near the previous rates." And these observations, in their words, "suggest that special care must be taken in how mass-balance estimates are evaluated, particularly when extrapolating into the future, because short-term spikes could yield erroneous long-term trends."

In conclusion, the part of the Northern Hemisphere that holds the lion's share of the hemisphere's ice has been cooling for the past half-century, and at a very significant rate, making it unlikely that its frozen water will be released to the world's oceans. In addition, because the annual number of snowfall days over much of Greenland has increased so dramatically over the same time period, it is possible that enhanced accumulation of snow on its huge ice sheet may be compensating for the melting of many of the world's mountain glaciers and keeping global sea level in check for this reason too. Lastly, Greenland's temperature trend of the past half-century has been just the opposite—and strikingly so—of that which is claimed for the Northern Hemisphere and the world by the IPCC.

Additional information on this topic, including reviews of newer publications as they become available, can be found at http://www.co2science.org/subject/s/sealevelgreenland.php.

References

Bindschadler, R. 2006. Hitting the ice sheets where it hurts. *Science* **311**: 1720-1721.

Chylek, P., Box, J.E. and Lesins, G. 2004. Global warming and the Greenland ice sheet. *Climatic Change* **63**: 201-221.

Comiso, J.C., Wadhams, P., Pedersen, L.T. and Gersten, R.A. 2001. Seasonal and interannual variability of the Odden ice tongue and a study of environmental effects. *Journal of Geophysical Research* **106**: 9093-9116.

Cuffey, K.M. and Marshall, S.J. 2000. Substantial contribution to sea-level rise during the last interglacial from the Greenland ice sheet. *Nature* **404**: 591-594.

Ekstrom, G., Nettles, M. and Tsai, V.C. 2006. Seasonality and increasing frequency of Greenland glacial earthquakes. *Science* **311**: 1756-1758.

Eldrett, J.S., Harding, I.C., Wilson, P.A., Butler, E. and Roberts, A.P. 2007. Continental ice in Greenland during the Eocene and Oligocene. *Nature* **446**: 176-179.

Gore, A. 2006. An Inconvenient Truth: The Planetary Emergency of Global Warming and What We Can Do About It. Rodale, Emmaus, PA, USA.

Hanna, E. and Cappelen, J. 2002. Recent climate of Southern Greenland. *Weather* **57**: 320-328.

Hanna, E. and Cappelen, J. 2003. Recent cooling in coastal southern Greenland and relation with the North Atlantic Oscillation. *Geophysical Research Letters* **30**: 1132.

Hanna, E., Huybrechts, P., Janssens, I., Cappelin, J., Steffen, K. and Stephens, A. 2005. *Journal of Geophysical Research* **110**: 10.1029/2004JD005641.

Howat, I.M., Joughin, I. and Scambos, T.A. 2007. Rapid changes in ice discharge from Greenland outlet glaciers. *Science* **315**: 1559-1561.

Humlum, O. 1999. Late-Holocene climate in central West Greenland: meteorological data and rock-glacier isotope evidence. *The Holocene* **9**: 581-594.

Hvidberg, C.S. 2000. When Greenland ice melts. *Nature* **404**: 551-552.

Johannessen, O.M., Khvorostovsky, K., Miles, M.W. and Bobylev, L.P. 2005. Recent ice-sheet growth in the interior of Greenland. *Science* **310**: 1013-1016.

Joughin, I. 2006. Greenland rumbles louder as glaciers accelerate. *Science* **311**: 1719-1720.

Kalnay, E., Kanamitsu, M., Kistler, R., Collins, W., Deaven, D., Gandin, L., Iredell, M., Saha, S., White, G., Woollen, J., Zhu, Y., Chelliah, M., Ebisuzaki, W., Higgins, W., Janowiak, J., Mo, K.C., Ropelewski, C., Wang, J., Leetmaa, A., Reynolds, R., Jenne, R. and Joseph, D. 1996. The NCEP/NCAR 40-year reanalysis project. *Bulletin of the American Meteorological Society* **77**: 437-471.

Kennedy, D. and Hanson, B. 2006. Ice and history. *Science* **311**: 1673.

Kerr, R.A. 2006. A worrying trend of less ice, higher seas. *Science* **311**: 1698-1701.

Krabill, W., Abdalati, W., Frederick, E., Manizade, S., Martin, C., Sonntag, J., Swift, R., Thomas, R., Wright, W. and Yungel, J. 2000. Greenland ice sheet: High-elevation balance and peripheral thinning. *Science* **289**: 428-430.

Otto-Bliesner, B.L., Marshall, S.J., Overpeck, J.T., Miller, G.H., Hu, A., and CAPE Last Interglacial Project members. 2006. Simulating Arctic climate warmth and icefield retreat in the last interglaciation. *Science* **311**: 1751-1753.

Overpeck, J.T., Otto-Bliesner, B.L., Miller, G.H., Muhs, D.R., Alley, R.B. and Kiehl, J.T. 2006. Paleoclimatic evidence for future ice-sheet instability and rapid sea-level rise. *Science* **311**: 1747-1750.

Parker, D.E., Folland, C.K. and Jackson, M. 1995. Marine surface temperature: Observed variations and data requirements. *Climatic Change* **31**: 559-600.

Przybylak, R. 2000. Temporal and spatial variation of surface air temperature over the period of instrumental observations in the Arctic. *International Journal of Climatology* **20**: 587-614.

Rayner, N.A., Horton, E.B., Parker, D.E., Folland, C.K. and Hackett, R.B. 1996. Version 2.2 of the global sea-ice and sea surface temperature data set, 1903-1994. *Climate Research Technical Note 74*, Hadley Centre, U.K. Meteorological Office, Bracknell, Berkshire, UK.

Rignot, E. and Kanagaratnam, P. 2005. Changes in the velocity structure of the Greenland Ice Sheet. *Science* **311**: 986-990.

Shepherd, A. and Wingham, D. 2007. Recent sea-level contributions of the Antarctic and Greenland Ice Sheets. *Science* **315**: 1529-1532.

Steig, E.J., Grootes, P.M. and Stuiver, M. 1994. Seasonal precipitation timing and ice core records. *Science* **266**: 1885-1886.

Stuiver, M., Grootes, P.M. and Braziunas, T.F. 1995. The GISP2 ^{18}O climate record of the past 16,500 years and the role of the sun, ocean and volcanoes. *Quaternary Research* **44**: 341-354.

Taurisano, A., Boggild, C.E. and Karlsen, H.G. 2004. A century of climate variability and climate gradients from coast to ice sheet in West Greenland. *Geografiska Annaler* **86A**: 217-224.

White, J.W.C., Barlow, L.K., Fisher, D., Grootes, P.M., Jouzel, J., Johnsen, S.J., Stuiver, M. and Clausen, H.B. 1997. The climate signal in the stable isotopes of snow from Summit, Greenland: Results of comparisons with modern climate observations. *Journal of Geophysical Research* **102**: 26,425-26,439.

Zwally, H.J., Giovinetto, M.B., Li, J., Cornejo, H.G., Beckley, M.A., Brenner, A.C., Saba, J.L. and Yi, D. 2005. Mass changes of the Greenland and Antarctic ice sheets and shelves and contributions to sea-level rise: 1992-2002. *Journal of Glaciology* **51**: 509-527.

5

Solar Variability and Climate Cycles

Introduction

The Intergovernmental Panel on Climate Change (IPCC) claims "most of the observed increase in global average temperatures since the mid-20th century is *very likely* due to the observed increase in anthropogenic greenhouse gas concentrations [italics in the original]" (IPCC, 2007-I, p. 10). The IPCC's authors even tell us they have decided there is a better-than-90-percent probability that their shared opinion is true. But as we demonstrated in Chapter 1, the general circulation models upon which the IPCC rests its case are notoriously unreliable. In Chapter 2 we documented feedback factors and forcings that the IPCC clearly overlooked. In Chapters 3 and 4 we showed that observations do not confirm the temperatures and weather trends the IPCC said should exist if its theory were true.

In this chapter we set out evidence in favor of an alternative theory of climate change that holds that variations in the sun's output and magnetic field, mediated by cosmic ray fluxes and changes in global cloud cover, play a larger role in regulating the earth's temperature, precipitation, droughts, floods, monsoons, and other climate features than any past or expected human activities, including projected increases in GHG emissions. Unlike the IPCC, we do not invent a measure of our confidence in this theory, nor do we confuse it with a forecast of future weather patterns. Rather, we make the case for this alternative theory to demonstrate how much we *don't* know about earth's climate, and therefore how wrong it is to assume that human activity is responsible for any variability in the climate that we cannot explain by pointing to already known forcings or feedbacks.

According to the IPCC, "changes in solar irradiance since 1750 are estimated to cause a radiative forcing of +0.12 [+0.06 to +0.30] W m^{-2}," which is an order of magnitude smaller than their estimated net anthropogenic forcing of +1.66 W m^{-2} from CO_2 over the same time period (pp. 3,4). However, the studies summarized in this chapter suggest the IPCC has got it backwards, that it is the sun's influence that is responsible for most climate change during the past century and beyond.

In the spirit of genuine scientific inquiry, in contrast to the IPCC's agenda-driven focus on making its case against GHG, we examine some research that is truly on the frontiers of climate research today. We begin with a discussion of cosmic rays, followed by research on irradiance, and then survey the evidence linking solar variability to climate phenomena both ancient and recent.

References

IPCC. 2007-I. *Climate Change 2007: The Physical Science Basis. Contribution of Working Group I to the Fourth Assessment Report of the Intergovernmental Panel on Climate Change.* Solomon, S., Qin, D., Manning, M., Chen, Z., Marquis, M., Averyt, K.B., Tignor, M. and Miller, H.L. (Eds.) Cambridge University Press, Cambridge, UK.

5.1. Cosmic Rays

The study of extraterrestrial climatic forcing factors is primarily a study of phenomena related to the sun. Historically, this field of inquiry began with the work of Milankovitch (1920, 1941), who linked the cyclical glaciations of the past million years to the receipt of solar radiation at the surface of the earth as modulated by variations in earth's orbit and rotational characteristics. Subsequent investigations implicated a number of other solar phenomena that operate on both shorter and longer timescales; in this summary we review the findings of the subset of those studies that involve galactic cosmic rays.

We begin with the review paper of Svensmark (2007), Director of the Center for Sun-Climate Research of the Danish National Space Center, who starts by describing how he and his colleagues experimentally determined that electrons released to the atmosphere by galactic cosmic rays act as catalysts that significantly accelerate the formation of ultra-small clusters of sulfuric acid and water molecules that constitute the building blocks of cloud condensation nuclei. He then discusses how, during periods of greater solar magnetic activity, greater shielding of the earth occurs, resulting in less cosmic rays penetrating to the lower atmosphere, resulting in fewer cloud condensation nuclei being produced, resulting in fewer and less reflective low-level clouds occurring, which leads to more solar radiation being absorbed by the surface of the earth, resulting in increasing near-surface air temperatures and global warming.

Svensmark provides support for key elements of this scenario with graphs illustrating the close correspondence between global low-cloud amount and cosmic-ray counts over the period 1984-2004, as well as by the history of changes in the flux of galactic cosmic rays since 1700, which correlates well with earth's temperature history over the same time period, starting from the latter portion of the Maunder

Minimum (1645-1715), when Svensmark says "sunspots were extremely scarce and the solar magnetic field was exceptionally weak," and continuing on through the twentieth century, over which last hundred-year interval, as noted by Svensmark, "the sun's coronal magnetic field doubled in strength."

Svensmark cites the work of Bond *et al.* (2001), who in studying ice-rafted debris in the North Atlantic Ocean determined, in Svensmark's words, that "over the past 12,000 years, there were many icy intervals like the Little Ice Age" that "alternated with warm phases, of which the most recent were the Medieval Warm Period (roughly AD 900-1300) and the Modern Warm Period (since 1900)." And as Bond's 10-member team clearly indicates, "over the last 12,000 years virtually every centennial time-scale increase in drift ice documented in our North Atlantic records was tied to a solar minimum."

In another expansion of timescale—this one highlighting the work of Shaviv (2002, 2003) and Shaviv and Veizer (2003)—Svensmark presents plots of reconstructed sea surface temperature anomalies and relative cosmic ray flux over the last 550 million years, during which time the solar system experienced four passages through the spiral arms of the Milky Way galaxy, with the climatic data showing "rhythmic cooling of the earth whenever the sun crossed the galactic midplane, where cosmic rays are locally most intense." In addition, he notes that the "Snowball Earth" period of some 2.3 *billion* years ago "coincided with the highest star-formation rate in the Milky Way since the earth was formed, in a mini-starburst 2400-2000 million years ago," when, of course, the cosmic ray flux would have been especially intense. In light of these many diverse observations, Svensmark concludes that "stellar winds and magnetism are crucial factors in the origin and viability of life on wet earth-like planets," as are "ever-changing galactic environments and star-formation rates."

Several studies conducted over the past 10 years have uncovered evidence for several of the linkages described by Svensmark in his overview of what we could call the cosmic ray-climate connection. We start with the work of Lockwood *et al.* (1999), who examined measurements of the near-earth interplanetary magnetic field to determine the total magnetic flux leaving the sun since 1868. In doing so, they were able to show that the total magnetic flux from the sun rose by a factor of 1.41 over the period 1964-1996, while surrogate measurements of the interplanetary magnetic field previous to this time indicate that this parameter had risen by a factor of 2.3 since 1901, which observations led the three researchers to state that the variation in the total solar magnetic flux they found "stresses the importance of understanding the connections between the sun's output and its magnetic field and between terrestrial global cloud cover, cosmic ray fluxes and the heliospheric field." The results of this study lead one to wonder just how much of the 0.8°C global temperature rise of the last century might have been a result of the more than two-fold increase in the total magnetic solar flux over that period.

Next, Parker (1999) noted that the number of sunspots had also doubled over the prior 100 years, and that one consequence of the latter phenomenon would have been "a much more vigorous sun" that was slightly brighter. Parker pointed out that spacecraft measurements suggest that the brightness (B) of the sun varies by an amount $\Delta B/B = 0.15\%$, in step with the 11-year magnetic cycle. He then pointed out that during times of much reduced activity of this sort (such as the Maunder Minimum of 1645-1715) and much increased activity (such as the twelfth century Mediaeval Maximum), brightness variations on the order of $\Delta B/B = 0.5\%$ typically occur, after which he noted the mean temperature (T) of the northern portion of the earth varied by 1 to 2°C in association with these variations in solar activity, stating finally that "we cannot help noting that change in T/T = change in B/B."

Knowing that sea surface temperatures are influenced by the brightness of the sun, and that they had risen since 1900, Parker wrote that "one wonders to what extent the solar brightening [of the past century] has contributed to the increase in atmospheric temperature and CO_2" over that period. It was Parker's "inescapable conclusion" that "we will have to know a lot more about the sun and the terrestrial atmosphere before we can understand the nature of the contemporary changes in climate."

Digging deeper into the subject, Feynman and Ruzmaikin (1999) investigated twentieth century changes in the intensity of cosmic rays incident upon the earth's magnetopause and their transmission through the magnetosphere to the upper troposphere. This work revealed "the intensity of cosmic rays incident on the magnetopause has decreased markedly during this century" and "the pattern of cosmic ray precipitation through the magnetosphere to the upper troposphere has also changed."

With respect to the first and more basic of these changes, they noted that "at 300 MeV the difference between the proton flux incident on the magnetosphere at the beginning of the century and that incident now is estimated to be a factor of 5 decrease between solar minima at the beginning of the century and recent solar minima," and that "at 1 GeV the change is a factor of 2.5." With respect to the second phenomenon, they noted that the part of the troposphere open to cosmic rays of all energies increased by a little over 25 percent and shifted equatorward by about 6.5° of latitude. And with the great decrease in the intensity of cosmic rays incident on earth's magnetosphere over the twentieth century, one would have expected to see a progressive decrease in the presence of low-level clouds and, therefore, an increase in global air temperature, as has indeed been observed.

A number of other pertinent papers also appeared at this time. Black et al. (1999) conducted a high-resolution study of sediments in the southern Caribbean that were deposited over the past 825 years, finding substantial variability of both a decadal and centennial nature, which suggested that such climate regime shifts are a natural aspect of Atlantic variability; and relating these features to other records of climate variability, they concluded that "these shifts may play a role in triggering changes in the frequency and persistence of drought over North America." Another of their findings was a strong correspondence between the changes in North Atlantic climate and similar changes in ^{14}C production; and they concluded that this finding "suggests that small changes in solar output may influence Atlantic variability on centennial time scales."

Van Geel et al. (1999) reviewed what was known at the time about the relationship between variations in the abundances of the cosmogenic isotopes ^{14}C and [10]Be and millennial-scale climate oscillations during the Holocene and portions of the last great ice age. This exercise indicated "there is mounting evidence suggesting that the variation in solar activity is a cause for millennial scale climate change," which is known to operate independently of the glacial-interglacial cycles that are forced by variations in the earth's orbit about the sun. They also reviewed the evidence for various mechanisms by which the postulated solar-climate connection might be implemented, finally concluding that "the climate system is far more sensitive to small variations in solar activity than generally believed" and that "it could mean that the global temperature fluctuations during the last decades are partly, or completely explained by small changes in solar radiation."

Noting that recent research findings in both palaeoecology and solar science "indicate a greater role for solar forcing in Holocene climate change than has previously been recognized," Chambers et al. (1999) reviewed the subject and found much evidence for solar-driven variations in earth-atmosphere processes over a range of timescales stretching from the 11-year solar cycle to century-scale events. They acknowledged, however, that absolute solar flux variations associated with these phenomena are rather small; but they identified a number of "multiplier effects" that can operate on solar rhythms in such a way that, as they describe it, "minor variations in solar activity can be reflected in more significant variations within the earth's atmosphere." They also noted that such nonlinear amplifier responses to solar variability are inadequately represented in the global climate models used by the IPCC to predict future greenhouse gas-induced global warming, even though that organization employs other amplifier effects to model well-known glacial/interglacial cycles of temperature change of the past, as well as the hypothesized CO_2-induced warming of the future.

Noting that "solar magnetic activity exhibits chaotically modulated cycles ... which are responsible for slight variations in solar luminosity and modulation of the solar wind," Tobias and Weiss (2000) attacked the solar forcing of climate problem by means of a model in which the solar dynamo and earth's climate are represented by low-order systems, each of which in isolation supports chaotic oscillations but which when run together sometimes resonate. By this means they determined that "solutions oscillate about either of two fixed points, representing warm and cold states, flipping sporadically between them." They also discovered that a weak nonlinear input from the solar dynamo "has a significant effect when the 'typical frequencies' of each system are in resonance." "It is clear," they say, "that the resonance provides a powerful mechanism for amplifying climate forcing by solar activity."

Contemporaneously, Solanki et al. (2000) developed a model of the long-term evolution of the sun's large-scale magnetic field and compared its predictions against two proxy measures of this parameter. The model proved successful in reproducing the observed century-long doubling of the strength of the part of the sun's magnetic field that

reaches out from the sun's surface into interplanetary space. It also indicated there is a direct connection between the length of the 11-year sunspot cycle and secular variations in solar activity that occur on timescales of centuries, such as the Maunder Minimum of the latter part of the seventeenth century, when sunspots were few in number and earth was in the midst of the Little Ice Age.

In discussing their findings, the solar scientists say their modeled reconstruction of the solar magnetic field "provides the major parameter needed to reconstruct the secular variation of the cosmic ray flux impinging on the terrestrial atmosphere," because, as they continue, a stronger solar magnetic field "more efficiently shields the earth from cosmic rays," and "cosmic rays affect the total cloud cover of the earth and thus drive the terrestrial climate."

One year later, using cosmic ray data recorded by ground-based neutron monitors, global precipitation data from the Climate Predictions Center Merged Analysis of Precipitation project, and estimates of monthly global moisture from the National Centers for Environmental Prediction reanalysis project, Kniveton and Todd (2001) set out to evaluate whether there is empirical evidence to support the hypothesis that solar variability (represented by changes in cosmic ray flux) is linked to climate change (manifested by changes in precipitation and precipitation efficiency) over the period 1979-1999. In doing so, they determined there is "evidence of a statistically strong relationship between cosmic ray flux, precipitation and precipitation efficiency over ocean surfaces at mid to high latitudes," since variations in both precipitation and precipitation efficiency for mid to high latitudes showed a close relationship in both phase and magnitude with variations in cosmic ray flux, varying 7-9 percent during the solar cycle of the 1980s, while other potential forcing factors were ruled out due to poorer statistical relationships.

Also in 2001, Bond et al. (2001) published the results of their study of ice-rafted debris found in three North Atlantic deep-sea sediment cores and cosmogenic nuclides sequestered in the Greenland ice cap (^{10}Be) and Northern Hemispheric tree rings (^{14}C). Based on arduous analyses of deep-sea sediment cores that yielded the variable-with-depth amounts of three proven proxies for the prior presence of overlying drift-ice, the scientists were able to discern and, with the help of an accelerator mass spectrometer, date a number of recurring alternate periods of relative cold and warmth that wended their way through the entire 12,000-year expanse of the Holocene. The mean duration of the several complete climatic cycles thus delineated was 1,340 years, the cold and warm nodes of the latter of which oscillations, in the words of Bond *et al.*, were "broadly correlative with the so called 'Little Ice Age' and 'Medieval Warm Period'."

The signal accomplishment of the scientists' study was the linking of these millennial-scale climate oscillations—and their embedded centennial-scale oscillations—with similar-scale oscillations in cosmogenic nuclide production, which are known to be driven by contemporaneous oscillations in solar activity. In fact, Bond *et al.* were able to report that "over the last 12,000 years virtually every centennial time-scale increase in drift ice documented in our North Atlantic records was tied to a solar minimum." In light of this observation they thus concluded that "a solar influence on climate of the magnitude and consistency implied by our evidence could not have been confined to the North Atlantic," suggesting that the cyclical climatic effects of the variable solar inferno are experienced throughout the world.

At this point in their paper, the international team of scientists cited additional evidence in support of the implications of their work. With respect to the global extent of the climatic impact of the solar radiation variations they detected, they reference studies conducted in Scandinavia, Greenland, the Netherlands, the Faroe Islands, Oman, the Sargasso Sea, coastal West Africa, the Cariaco Basin, equatorial East Africa, and the Yucatan Peninsula, demonstrating thereby that "the footprint of the solar impact on climate we have documented extend[s] from polar to tropical latitudes." They note "the solar-climate links implied by our record are so dominant over the last 12,000 years ... it seems almost certain that the well-documented connection between the Maunder solar minimum and the coldest decades of the Little Ice Age could not have been a coincidence," further noting that their findings supported previous suggestions that both the Little Ice Age and Medieval Warm Period "may have been partly or entirely linked to changes in solar irradiance."

Another point reiterated by Bond *et al.* is that the oscillations in drift-ice they studied "persist across the glacial termination and well into the last glaciation, suggesting that the cycle is a pervasive feature of the climate system." At two of their coring sites, they identified a series of such cyclical variations that extended throughout all of the previous interglacial and were "strikingly similar to those of the

Holocene." Here they could also have cited the work of Oppo *et al.* (1998), who observed similar climatic oscillations in a sediment core that covered the span of time from 340,000 to 500,000 years before present, and that of Raymo *et al.* (1998), who pushed back the time of the cycles' earliest known occurrence to well over one million years ago.

How do the small changes in solar radiation inferred from the cosmogenic nuclide variations bring about such significant and pervasive shifts in earth's global climate? Bond *et al.* described a scenario whereby solar-induced changes high in the stratosphere are propagated downward through the atmosphere to the earth's surface, where they envisioned them provoking changes in North Atlantic deep water formation that alter the thermohaline circulation of the global ocean. In light of the plausibility of this particular scenario, they suggested that "the solar signals thus may have been transmitted through the deep ocean as well as through the atmosphere, further contributing to their amplification and global imprint."

Concluding their landmark paper, the researchers wrote that the results of their study "demonstrate that the earth's climate system is highly sensitive to extremely weak perturbations in the sun's energy output," noting their work "supports the presumption that solar variability will continue to influence climate in the future."

The following year, Carslaw *et al.* (2002) began an essay on "Cosmic Rays, Clouds, and Climate" by noting that the intensity of cosmic rays varies by about 15 percent over a solar cycle, due to changes in the strength of the solar wind, which carries a weak magnetic field into the heliosphere that partially shields the earth from low-energy galactic charged particles. When this shielding is at a minimum, allowing more cosmic rays to impinge upon the planet, more low clouds have been observed to cover the earth, producing a tendency for lower temperatures to occur. When the shielding is maximal, on the other hand, fewer cosmic rays impinge upon the planet and fewer low clouds form, which produces a tendency for the earth to warm.

The three researchers further noted that the total variation in low cloud amount over a solar cycle is about 1.7 percent, which corresponds to a change in the planet's radiation budget of about one watt per square meter (1 Wm^{-2}). This change, they say, "is highly significant when compared ... with the estimated radiative forcing of 1.4 Wm^{-2} from anthropogenic CO_2 emissions." However, because of the short length of a solar cycle (11 years), the large thermal inertia of the world's oceans dampens the much greater global temperature change that would have occurred as a result of this radiative forcing if it had been spread out over a much longer period of time, so that the actual observed warming is something a little less than 0.1°C.

Much of Carslaw et al.'s review focuses on mechanisms by which cosmic rays might induce the synchronous low cloud cover changes that have been observed to accompany their intensity changes. They begin by briefly describing the three principal mechanisms that have been suggested to function as links between solar variability and changes in earth's weather: (1) changes in total solar irradiance that provide variable heat input to the lower atmosphere, (2) changes in solar ultraviolet radiation and its interaction with ozone in the stratosphere that couple dynamically to the lower atmosphere, and (3) changes in cloud processes having significance for condensation nucleus abundances, thunderstorm electrification and thermodynamics, and ice formation in cyclones.

Focusing on the third of these mechanisms, the researchers note that cosmic rays provide the sole source of ions away from terrestrial sources of radioisotopes. Hence, they further refine their focus to concentrate on ways by which cosmic-ray-produced ions may affect cloud droplet number concentrations and ice particles. Here, they concentrate on two specific topics: what they call the ion-aerosol clear-air mechanism and the ion-aerosol near-cloud mechanism. Their review suggests that what we know about these subjects is very much less than what we *could* know about them. Many scientists, as they describe it, believe "it is inconceivable that the lower atmosphere can be globally bombarded by ionizing radiation without producing an effect on the climate system."

Carslaw et al. point out that cosmic ray intensity declined by about 15 percent during the last century "owing to an increase in the solar open magnetic flux by more than a factor of 2." They further report that "this 100-year change in intensity is about the same magnitude as the observed change over the last solar cycle." In addition, *we* note that the cosmic ray intensity was already much lower at the start of the twentieth century than it was just after the start of the nineteenth century, when the Esper et al. (2002) record indicates the planet began its nearly two-century-long recovery from the chilly depths of the Little Ice Age.

These observations strongly suggest that solar-mediated variations in the intensity of cosmic rays bombarding the earth are indeed responsible for the temperature variations of the past three centuries. They provide a much better fit to the temperature data than do atmospheric CO_2 data; and as Carslaw et al. remark, "if the cosmic ray-cloud effect is real, then these long-term changes of cosmic ray intensity could substantially influence climate." It is this possibility, they say, that makes it "all the more important to understand the cause of the cloudiness variations," which is basically the message of their essay; i.e., that we must work hard to deepen our understanding of the cosmic ray-cloud connection, as it may well hold the key to resolving what they call this "fiercely debated geophysical phenomenon."

One year later, and noting that Svensmark and Friis-Christensen (1997), Marsh and Svensmark (2000), and Palle Bago and Butler (2000) had derived positive relationships between global cosmic ray intensity and low-cloud amount from infrared cloud data contained in the International Satellite Cloud Climatology Project (ISCCP) database for the years 1983-1993, Marsden and Lingenfelter (2003) used the ISCCP database for the expanded period 1983-1999 to see if a similar relationship could be detected via cloud amount measurements made in the visible spectrum. This work revealed that there was indeed, in their words, "a positive correlation at low altitudes, which is consistent with the positive correlation between global low clouds and cosmic ray rate seen in the infrared."

That same year, Shaviv and Veizer (2003) suggested that from two-thirds to three-fourths of the variance in earth's temperature (T) over the past 500 million years may be attributable to cosmic ray flux (CRF) variations due to solar system passages through the spiral arms of the Milky Way galaxy. This they did after presenting several half-billion-year histories of T, CRF, and atmospheric CO_2 concentrations derived from various types of proxy data, and after finding that none of the CO_2 curves showed any clear correlation with the T curves, suggesting to them that "CO_2 is not likely to be the principal climate driver." On the other hand, they discovered that the T trends displayed a dominant cyclic component on the order of 135 ± 9 million years, and that "this regular pattern implies that we may be looking at a reflection of celestial phenomena in the climate history of earth."

That such is likely the case is borne out by their identification of a similar CRF cycle of 143 ± 10

million years, together with the fact that the large cold intervals in the T records "appear to coincide with times of high CRF," which correspondence is what would be expected from the likely chain of events: high CRF ==> more low-level clouds ==> greater planetary albedo ==> colder climate, as described by Svensmark and Friis-Christensen (1997), Marsh and Svensmark (2000), Palle Bago and Butler (2000), and Marsden and Lingenfelter (2003).

What do these findings suggest about the role of atmospheric CO_2 variations with respect to global temperature change? Shaviv and Veizer began their analysis of this question by stating that the conservative approach is to assume that the entire residual variance not explained by measurement error is due to CO_2 variations. And when this was done, they found that a doubling of the air's CO_2 concentration could account for only about a 0.5°C increase in T.

This result differs considerably, in their words, "from the predictions of the general circulation models, which typically imply a CO_2 doubling effect of ~1.5-5.5°C," but they say it is "consistent with alternative lower estimates of 0.6-1.6°C (Lindzen, 1997)." We note also, in this regard, that Shaviv and Veizer's result is even more consistent with the results of the eight empirically based "natural experiments" of Idso (1998), which yield an average warming of about 0.4°C for a 300 to 600 ppm doubling of the atmosphere's CO_2 concentration.

In another important test of a critical portion of the cosmic ray-climate connection theory, Usoskin *et al.* (2004b) compared the spatial distributions of low cloud amount (LCA) and cosmic ray-induced ionization (CRII) over the globe for the period 1984-2000. They used observed LCA data obtained from the ISCCP-D2 database limited to infrared radiances, while they employed CRII values calculated by Usoskin *et al.* (2004a) at 3 km altitude, which corresponds roughly to the limiting altitude below which low clouds form. This work revealed that "the LCA time series can be decomposed into a long-term slow trend and inter-annual variations, the latter depicting a clear 11-year cycle in phase with CRII." In addition, they found there was "a one-to-one relation between the relative variations of LCA and CRII over the latitude range 20-55°S and 10-70°N," and that "the amplitude of relative variations in LCA was found to increase polewards, in accordance with the amplitude of CRII variations." These findings of the five-member team of Finnish, Danish, and Russian scientists provide substantial evidence for a

solar-cosmic ray linkage (the 11-year cycle of CRII) and a cosmic ray-cloud linkage (the in-phase cycles of CRII and CLA), making the full solar activity/cosmic ray/low cloud/climate change hypothesis appear to be rather robust.

In a review of the temporal variability of various solar phenomena conducted the following year, Lean (2005) made the following important but disturbing point about climate models and the sun-climate connection: "a major enigma is that general circulation climate models predict an immutable climate in response to decadal solar variability, whereas surface temperatures, cloud cover, drought, rainfall, tropical cyclones, and forest fires show a definite correlation with solar activity (Haigh, 2001, Rind, 2002)."

Lean begins her review by noting that the beginning of the Little Ice Age "coincided with anomalously low solar activity (the so-called Sporer and Maunder minima)," and that "the latter part coincided with both low solar activity (the Dalton minimum) and volcanic eruptions." Then, after discussing the complexities of this potential relationship, she muses about another alternative: "Or might the Little Ice Age be simply the most recent cool episode of millennial climate-oscillation cycles?" Lean cites evidence that reveals the sensitivity of drought and rainfall to solar variability, stating that climate models are unable to reproduce the plethora (her word) of sun-climate connections. She notes that simulations with climate models yield decadal and centennial variability even in the absence of external forcing, stating that "arguably, this very sensitivity of the climate system to unforced oscillation and stochastic noise predisposes it to nonlinear responses to small forcings such as by the sun."

Lean reports that "various high-resolution paleoclimate records in ice cores, tree rings, lake and ocean sediment cores, and corals suggest that changes in the energy output of the sun itself may have contributed to sun-earth system variability," citing the work of Verschuren *et al.* (2000), Hodell *et al.* (2001), and Bond *et al.* (2001). She notes that "many geographically diverse records of past climate are coherent over time, with periods near 2400, 208 and 90 years that are also present in the ^{14}C and ^{10}Be archives," which isotopes (produced at the end of a complex chain of interactions that are initiated by galactic cosmic rays) contain information about various aspects of solar activity (Bard *et al.*, 1997).

Veretenenko *et al.* (2005) examined the potential influence of galactic cosmic rays (GCR) on the long-term variation of North Atlantic sea-level pressure over the period 1874-1995. Their comparisons of long-term variations in cold-season (October-March) sea-level pressure with different solar/geophysical indices revealed that increasing sea-level pressure coincided with a secular rise in solar/geomagnetic activity that was accompanied by a decrease in GCR intensity, whereas long-term decreases in sea-level pressure were observed during periods of decreasing solar activity and rising GCR flux. Spectral analysis further supported a link between sea-level pressure, solar/geomagnetic activity and GCR flux, as similar spectral characteristics (periodicities) were present among all datasets at time scales from approximately 10 to 100 years.

These results support a link between long-term variations in cyclonic activity and trends in solar activity/GCR flux in the extratropical latitudes of the North Atlantic. Concerning how this relationship works, Veretenenko *et al.* hypothesize that GCR-induced changes in cloudiness alter long-term variations in solar and terrestrial radiation receipt in this region, which in turn alters tropospheric temperature gradients and produces conditions more favorable for cyclone formation and development. Although we are still far from possessing a complete understanding of many solar/GCR-induced climatic influences, this study highlights the ever-growing need for such relationships to be explored. As it and others have shown, small changes in solar output can indeed induce significant changes in earth's climate.

Also working in the North Atlantic region, Macklin *et al.* (2005) developed what they call "the first probability-based, long-term record of flooding in Europe, which spans the entire Holocene and uses a large and unique database of ^{14}C-dated British flood deposits," after which they compared their reconstructed flood history "with high-resolution proxy-climate records from the North Atlantic region, northwest Europe and the British Isles to critically test the link between climate change and flooding." By these means they determined that "the majority of the largest and most widespread recorded floods in Great Britain have occurred during cool, moist periods," and that "comparison of the British Holocene palaeoflood series ... with climate reconstructions from tree-ring patterns of subfossil bog oaks in northwest Europe also suggests that a similar relationship between climate and flooding in Great Britain existed during the Holocene, with floods being more frequent and larger during relatively cold, wet periods." In addition, they say that

"an association between flooding episodes in Great Britain and periods of high or increasing cosmogenic ^{14}C production suggests that centennial-scale solar activity may be a key control of non-random changes in the magnitude and recurrence frequencies of floods."

Another intriguing study from this time period, Usoskin *et al.* (2005), noted that "the variation of the cosmic ray flux entering earth's atmosphere is due to a combination of solar modulation and geomagnetic shielding, the latter adding a long-term trend to the varying solar signal," while further noting that "the existence of a geomagnetic signal in the climate data would support a direct effect of cosmic rays on climate." They evaluated this proposition by reproducing 1,000-year reconstructions of two notable solar-heliospheric indices derived from cosmogenic isotope data, i.e., the sunspot number and the cosmic ray flux (Usoskin *et al.*, 2003; Solanki *et al.*, 2004), and by creating a new 1,000-year air temperature history of the Northern Hemisphere by computing annual means of six different thousand-year surface air temperature series: those of Jones *et al.* (1998), Mann *et al.* (1999), Briffa (2000), Crowley (2000), Esper *et al.* (2002), and Mann and Jones (2003).

In comparing these three series (solar activity, cosmic ray, and air temperature), Usoskin *et al.* found that they "indicate higher temperatures during times of more intense solar activity (higher sunspot number, lower cosmic ray flux)." In addition, they report that three different statistical tests "consistently indicate that the long-term trends in the temperature correlate better with cosmic rays than with sunspots," which suggests that something in addition to solar activity must have been influencing the cosmic ray flux, in order to make the cosmic ray flux the better correlate of temperature.

Noting that earth's geomagnetic field strength would be a natural candidate for this "something," Usoskin *et al.* compared their solar activity, cosmic ray, and temperature reconstructions with two long-term reconstructions of geomagnetic dipole moment that they obtained from the work of Hongre *et al.* (1998) and Yang *et al.* (2000). This effort revealed that between AD 1000 and 1700, when there was a substantial downward trend in air temperature associated with a less substantial downward trend in solar activity, there was also a general downward trend in geomagnetic field strength. As a result, Usoskin *et al.* suggested that the substantial upward trend of cosmic ray flux that was needed to sustain the substantial rate of observed cooling (which was more than expected in light of the slow decline in solar activity) was likely due to the positive effect on the cosmic ray flux that was produced by the decreasing geomagnetic field strength.

After 1700, the geomagnetic field strength continued to decline, but air temperature did a dramatic turnabout and began to rise. The reason for this "parting of company" between the two parameters, according to Usoskin *et al.*, was that "the strong upward trend of solar activity during that time overcompensate[d] [for] the geomagnetic effect," leading to a significant warming. In addition, a minor portion of the warming of the last century or so (15-20 percent) may have been caused by the concomitant increase in the atmosphere's CO_2 content, which would have complemented the warming produced by the upward trending solar activity and further decoupled the upward trending temperature from the declining geomagnetic field strength.

In their totality, these several observations tend to strengthen the hypothesis that cosmic ray variability was the major driver of changes in earth's surface air temperature over the past millennium, and that this forcing was primarily driven by variations in solar activity, modulated by the more slowly changing geomagnetic field strength of the planet, which sometimes strengthened the solar forcing and sometimes worked against it. Once again, however, the results leave room for only a small impact of anthropogenic CO_2 emissions on twentieth century global warming.

Publishing in the same year, Versteegh (2005) reviewed what we know about past climatic responses to solar forcing and their geographical coherence based upon proxy records of temperature and the cosmogenic radionuclides ^{10}Be and ^{14}C, which provide a measure of magnetized plasma emissions from the sun that impact earth's exposure to galactic cosmic rays, thereby altering cloud formation and climate. As a result of this exercise, it was concluded that "proxy records provide ample evidence for climate change during the relatively stable and warm Holocene," and that "all frequency components attributed to solar variability re-occur in proxy records of environmental change," emphasizing in this regard "the ~90 years Gleisberg and ~200 years Suess cycles in the ^{10}Be and ^{14}C records," as well as "the ~1500 years Bond cycle which occurs in several proxy records [and] could originate from the interference between centennial-band solar cycles." As a result, Versteegh concludes that "long-term climate change during the preindustrial [era] seems to have been dominated by solar forcing," and that the long-term response to solar forcing "greatly exceeds unforced variability."

Delving further into the subject, Harrison and Stephenson (2005) introduce their contribution by noting that because the net global effect of clouds is cooling (Hartman, 1993), any widespread increase in the amount of overcast days could reduce air temperature globally, while local overcast conditions could do so locally. They compared the ratio of diffuse to total solar radiation (the diffuse fraction, DF)—which had been measured daily at 0900 UT at Whiteknights, Reading (UK) from 1997-2004—with the traditional subjective determination of cloud amount made simultaneously by a human observer, as well as with daily average temperature. They then compared the diffuse fraction measured at Jersey between 1968 and 1994 with corresponding daily mean neutron count rates measured at Climax, Colorado (USA), which provide a globally representative indicator of the galactic cosmic ray flux. The result, as they describe it, was that "across the UK, on days of high cosmic ray flux (which occur 87% of the time on average) compared with low cosmic ray flux, (i) the chance of an overcast day increases by 19% ± 4%, and (ii) the diffuse fraction increases by 2% ± 0.3%." In addition, they found that "during sudden transient reductions in cosmic rays (e.g. Forbush events), simultaneous decreases occur in the diffuse fraction."

As for the implications of their findings, the two researchers note that the latter of these observations indicates that diffuse radiation changes are "unambiguously due to cosmic rays." They also report that "at Reading, the measured sensitivity of daily average temperatures to DF for overcast days is -0.2 K per 0.01 change in DR." Consequently, they suggest that the well-known inverse relationship between galactic cosmic rays and solar activity will lead to cooling at solar minima, and that "this might amplify the effect of the small solar cycle variation in total solar irradiance, believed to be underestimated by climate models (Stott *et al.*, 2003) which neglect a cosmic ray effect." In addition, although the effect they detect is small, they say it is "statistically robust," and that the cosmic ray effect on clouds likely "will emerge on long time scales with less variability than the considerable variability of daily cloudiness."

Next, Shaviv (2005) identified six periods of earth's history (the entire Phanerozoic, the

215

Cretaceous, the Eocene, the Last Glacial Maximum, the twentieth century, and the 11-year solar cycle as manifest over the last three centuries) for which he was able to derive reasonably sound estimates of different time-scale changes in radiative forcing, temperature, and cosmic ray flux. From these sets of data he derived probability distribution functions of whole-earth temperature sensitivity to radiative forcing for each time period and combined them to obtain a mean planetary temperature sensitivity to radiative forcing of 0.28°C per Wm^{-2}. Then, noting that the IPCC (2001) had suggested that the increase in anthropogenic radiative forcing over the twentieth century was about 0.5 Wm^{-2}, Shaviv calculated that the anthropogenic-induced warming of the globe over this period was approximately 0.14°C (0.5 Wm^{-2} x 0.28°C per Wm^{-2}). This result harmonizes perfectly with the temperature increase (0.10°C) that was calculated by Idso (1998) to be due solely to the twentieth century increase in the air's CO_2 concentration (75 ppm), which would have been essentially indistinguishable from Shaviv's result if the warming contributions of the twentieth century concentration increases of all greenhouse gases had been included in the calculation.

Based on information that indicated a solar activity-induced increase in radiative forcing of 1.3 Wm^{-2} over the twentieth century (by way of cosmic ray flux reduction), plus the work of others (Hoyt and Schatten, 1993; Lean *et al.*, 1995; Solanki and Fligge, 1998) that indicated a globally averaged solar luminosity increase of approximately 0.4 Wm^{-2} over the same period, Shaviv calculated an overall and ultimately solar activity-induced warming of 0.47°C (1.7 Wm^{-2} x 0.28°C per Wm^{-2}) over the twentieth century. Added to the 0.14°C of anthropogenic-induced warming, the calculated total warming of the twentieth century thus came to 0.61°C, which was noted by Shaviv to be very close to the 0.57°C temperature increase that was said by the IPCC to have been observed over the past century. Consequently, both Shaviv's and Idso's analyses, which mesh well with real-world data of both the recent and distant past, suggest that only 15-20 percent (0.10°C/0.57°C) of the observed warming of the twentieth century can be attributed to the rise in the air's CO_2 content.

Usoskin *et al.* (2006) note that many solar scientists believe changes in solar activity have been responsible for significant changes in climate, but that to demonstrate that such is truly the case, a record of past variations in solar activity is required. They write

that "long-term solar activity in the past is usually estimated from cosmogenic isotopes, [10]Be or ^{14}C, deposited in terrestrial archives such as ice cores and tree rings," because "the production rate of cosmogenic isotopes in the atmosphere is related to the cosmic ray flux impinging on earth," which "is modulated by the heliospheric magnetic field and is thus a proxy of solar activity." A nagging concern, however, is that the isotope records may suffer from what the five scientists call "uncertainties due to the sensitivity of the data to several *terrestrial* [our italics] processes." Consequently, they devised a plan to attempt to resolve this issue.

Noting that the activity of a cosmogenic isotope in a meteorite represents "the time integrated cosmic ray flux over a period determined by the mean life of the radioisotope," Usoskin *et al.* reasoned that (1) "by measuring abundance of cosmogenic isotopes in meteorites which fell through the ages, one can evaluate the variability of the cosmic ray flux, since the production of cosmogenic isotopes ceases after the fall of the meteorite," and that (2) if they could develop such a meteoritic-based cosmogenic isotope record they could use it "to constrain [other] solar activity reconstructions using cosmogenic [44]Ti activity in meteorites *which is not affected by terrestrial processes* [our italics]."

The researchers' choice of [44]Ti for this purpose was driven by the fact that it has a half-life of about 59 years and is thus "relatively insensitive to variations of the cosmic ray flux on decadal or shorter time scales, but is very sensitive to the level of the cosmic ray flux and its variations on a centennial scale." Hence, they compared the results of different long-term [10]Be- and ^{14}C-based solar activity reconstruction models with measurements of [44]Ti in 19 stony meteorites (chondrites) that fell between 1766 and 2001, as reported by Taricco *et al.* (2006); in doing so, they ultimately determined that "most recent reconstructions of solar activity, in particular those based on [10]Be data in polar ice (Usoskin *et al.*, 2003, 2004c; McCracken *et al.*, 2004) and on ^{14}C in tree rings (Solanki *et al.*, 2004), are consistent with the [44]Ti data." Consequently, the results of this study give ever more credence to the findings of the many studies that have reported strong correlations between various climatic changes and [10]Be- and ^{14}C-based reconstructions of solar activity.

Dergachev *et al.* (2006) reviewed "direct and indirect data on variations in cosmic rays, solar activity, geomagnetic dipole moment, and climate from the present to 10-12 thousand years ago, [as] registered in different natural archives (tree rings, ice layers, etc.)." They found that "galactic cosmic ray levels in the earth's atmosphere are inversely related to the strength of the helio- and geomagnetic fields," and they conclude that "cosmic ray flux variations are apparently the most effective natural factor of climate changes on a large time scale." More specifically, they note that "changes in cloud processes under the action of cosmic rays, which are of importance for abundance of condensation nuclei and for ice formation in cyclones, can act as a connecting link between solar variability and changes in weather and climate," and they cite numerous scientific studies that indicate that "cosmic rays are a substantial factor affecting weather and climate on time scales of hundreds to thousands of years."

Voiculescu *et al.* (2006) observed "there is evidence that solar activity variations can affect the cloud cover at Earth. However, it is still unclear which solar driver plays the most important role in the cloud formation. Here, we use partial correlations to distinguish between the effects of two solar drivers (cosmic rays and the UV irradiance) and the mutual relations between clouds at different altitudes." They found "a strong solar signal in the cloud cover" and that "low clouds are mostly affected by UV irradiance over oceans and dry continental areas and by cosmic rays over some mid-high latitude oceanic areas and moist lands with high aerosol concentration. High clouds respond more strongly to cosmic ray variations, especially over oceans and moist continental areas. These results provide observational constraints on related climate models."

Gallet and Genevey (2007) documented what they call a "good temporal coincidence" between "periods of geomagnetic field intensity increases and cooling events," as measured in western Europe, where cooling events were "marked by glacier advances on land and increases in ice-rafted debris in [North Atlantic] deep-sea sediments." Their analyses revealed "a succession of three cooling periods in western Europe during the first millennium AD," the ages of which were "remarkably coincident with those of the main discontinuities in the history of Maya civilization," confirming the earlier similar work of Gallet *et al.* (2005), who had found a "good temporal coincidence in western Europe between cooling events recovered from successive advances of Swiss glaciers over the past 3000 years and periods of rapid increases in geomagnetic field intensity," the latter of which were "nearly coeval with abrupt changes, or hairpin turns, in magnetic field direction."

Gallet and Genevey concluded that "the most plausible mechanism linking geomagnetic field and climate remains a geomagnetic impact on cloud cover," whereby "variations in morphology of the earth's magnetic field could have modulated the cosmic ray flux interacting with the atmosphere, modifying the nucleation rate of clouds and thus the albedo and earth surface temperatures (Gallet *et al.*, 2005; Courtillot *et al.*, 2007)." These observations clearly suggest a global impact on climate, which is further suggested by the close relationship that has been found to exist between "cooling periods in the North Atlantic and aridity episodes in the Middle East," as well as by the similar relationship demonstrated by Gallet and Genevey to have prevailed between periods of aridity over the Yucatan Peninsula and well-documented times of crisis in Mayan civilization.

In another study that took a look at the *really* big picture, painted by rhythmically interbedded limestone and shale or limestone and chert known as *rhythmites*, Elrick and Hinnov (2007) "(1) review the persistent and widespread occurrence of Palaeozoic rhythmites across North America, (2) demonstrate their primary depositional origin at millennial time scales, (3) summarize the range of paleo-environmental conditions that prevailed during rhythmite accumulation, and (4) briefly discuss the implications primary Palaeozoic rhythmites have on understanding the origin of pervasive late Neogene-Quaternary millennial-scale climate variability." They concluded that "millennial-scale climate changes occurred over a very wide spectrum of paleoceanographic, paleogeographic, paleoclimatic, tectonic, and biologic conditions and over time periods from the Cambrian to the Quaternary," and that given this suite of observations, "it is difficult to invoke models of internally driven thermohaline oceanic oscillations or continental ice sheet instabilities to explain their origin." Consequently, they suggest that "millennial-scale paleoclimate variability is a more permanent feature of the earth's ocean-atmosphere system, which points to an external driver such as solar forcing."

Kirkby (2008) reports that "diverse reconstructions of past climate change have revealed clear associations with cosmic ray variations recorded in cosmogenic isotope archives, providing persuasive evidence for solar or cosmic ray forcing of the climate." He discusses two different classes of microphysical mechanisms that have been proposed to connect cosmic rays with clouds, which interact significantly with fluxes of both solar and thermal radiation and, therefore, climate: "firstly, an influence of cosmic rays on the production of cloud condensation nuclei and, secondly, an influence of cosmic rays on the global electrical circuit in the atmosphere and, in turn, on ice nucleation and other cloud microphysical processes." Kirkby observes that "considerable progress on understanding ion-aerosol-cloud processes has been made in recent years, and the results are suggestive of a physically plausible link between cosmic rays, clouds and climate" and "with new experiments planned or underway, such as the CLOUD facility at CERN, there are good prospects that we will have some firm answers to this question within the next few years."

In one final review paper, Lu (2009) showed that in the period of 1980–2007, two full 11-year cosmic ray cycles clearly correlated with ozone depletion, especially the polar ozone loss (hole) over Antarctica. The temporal correlation is also supported by a strong spatial correlation because the ozone hole is located in the lower polar stratosphere at ~18 km, exactly where the ionization rate of cosmic rays producing electrons is the strongest. The results provide strong evidence that the cosmic ray-driven electron-induced reaction of halogenated molecules plays the dominant role in causing the ozone hole. Changes in ozone then have a global impact on climate.

In conclusion, and as Kirkby (2008) rightly notes, "the question of whether, and to what extent, the climate is influenced by solar and cosmic ray variability remains central to our understanding of the anthropogenic contribution to present climate change." Clearly, carbon dioxide is not the all-important dominating factor in earth's climatic history. Within the context of the Holocene, the only time CO_2 moved in concert with air temperature was over the period of earth's recovery from the global chill of the Little Ice Age (the past century or so), and it does so then only quite imperfectly. The flux of galactic cosmic rays, on the other hand, appears to have influenced ups and downs in both temperature and precipitation over the entire 10-12 thousand years of the Holocene, making it the prime candidate for "prime determinant" of earth's climatic state.

Additional information on this topic, including reviews of newer publications as they become available, can be found at http://www.co2science.org/subject/e/extraterrestrial.php.

References

Bard, E., Raisbeck, G., Yiou, F. and Jouzel, J. 1997. Solar modulation of cosmogenic nuclide production over the last millennium: comparison between ^{14}C and ^{10}Be records. *Earth and Planetary Science Letters* **150**: 453-462.

Black, D.E., Peterson, L.C., Overpeck, J.T., Kaplan, A., Evans, M.N. and Kashgarian, M. 1999. Eight centuries of North Atlantic Ocean atmosphere variability. *Science* **286**: 1709-1713.

Bond, G., Kromer, B., Beer, J., Muscheler, R., Evans, M.N., Showers, W., Hoffmann, S., Lotti-Bond, R., Hajdas, I. and Bonani, G. 2001. Persistent solar influence on North Atlantic climate during the Holocene. *Science* **294**: 2130-2136.

Briffa, K.R. 2000. Annual climate variability in the Holocene: Interpreting the message of ancient trees. *Quaternary Science Review* **19**: 87-105.

Carslaw, K.S., Harrizon, R.G. and Kirkby, J. 2002. Cosmic rays, clouds, and climate. *Science* **298**: 1732-1737.

Chambers, F.M., Ogle, M.I. and Blackford, J.J. 1999. Palaeoenvironmental evidence for solar forcing of Holocene climate: linkages to solar science. *Progress in Physical Geography* **23**: 181-204.

Courtillot, V., Gallet, Y., Le Mouel, J.-L., Fluteau, F. and Genevey, A. 2007. Are there connections between the Earth's magnetic field and climate? *Earth and Planetary Science Letters* **253**: 328-339.

Crowley, T.J. 2000. Causes of climate change over the past 1000 years. *Science* **289**: 270-277.

Dergachev, V.A., Dmitriev, P.B., Raspopov, O.M. and Jungner, H. 2006. Cosmic ray flux variations, modulated by the solar and earth's magnetic fields, and climate changes. 1. Time interval from the present to 10-12 ka ago (the Holocene Epoch). *Geomagnetizm i Aeronomiya* **46**: 123-134.

Elrick M. and Hinnov, L.A. 2007. Millennial-scale paleoclimate cycles recorded in widespread Palaeozoic deeper water rhythmites of North America. *Palaeogeography, Palaeoclimatology, Palaeoecology* **243**: 348-372.

Esper, J., Cook, E.R. and Schweingruber, F.H. 2002. Low-frequency signals in long tree-ring chronologies for reconstructing past temperature variability. *Science* **295**: 2250-2253.

Feynman, J. and Ruzmaikin, A. 1999. Modulation of cosmic ray precipitation related to climate. *Geophysical Research Letters* **26**: 2057-2060.

Gallet, Y. and Genevey A. 2007. The Mayans: Climate determinism or geomagnetic determinism? *EOS: Transactions, American Geophysical Union* **88**: 129-130.

Gallet Y., Genevey, A. and Fluteau, F. 2005. Does Earth's magnetic field secular variation control centennial climate change? *Earth and Planetary Science Letters* **236**: 339-347.

Haigh, J.D. 2001. Climate variability and the influence of the sun. *Science* **294**: 2109-2111.

Harrison, R.G. and Stephenson, D.B. 2005. Empirical evidence for a nonlinear effect of galactic cosmic rays on clouds. *Proceedings of the Royal Society A*: 10.1098/rspa.2005.1628.

Hartman, D.L. 1993. Radiative effects of clouds on earth's climate. In: Hobbs, P.V. (Ed.) *Aerosol-Cloud-Climate Interactions*. Academic Press, New York, NY, USA.

Hodell, D.A., Brenner, M., Curtis, J.H. and Guilderson, T. 2001. Solar forcing of drought frequency in the Maya lowlands. *Science* **292**: 1367-1370.

Hongre, L., Hulot, G. and Khokhlov, A. 1998. An analysis of the geomagnetic field over the past 2000 years. *Physics of the Earth and Planetary Interiors* **106**: 311-335.

Hoyt, D.V. and Schatten, K.H. 1993. A discussion of plausible solar irradiance variations, 1700-1992. *Journal of Geophysical Research* **98**: 18,895-18,906.

Idso, S.B. 1998. Carbon-dioxide-induced global warming: A skeptic's view of potential climate change. *Climate Research* **10**: 69-82.

Intergovernmental Panel on Climate Change (IPCC). 2001. *Climate Change 2001*. Cambridge University Press, New York, NY, USA.

Jones, P.D., Briffa, K.R., Barnett, T.P. and Tett, S.F.B. 1998. High-resolution palaeoclimatic records for the last millennium: interpretation, integration and comparison with general circulation model control-run temperatures. *The Holocene* **8**: 455-471.

Kirkby, J. 2008. Cosmic rays and climate. *Surveys in Geophysics* **28**: 333-375.

Kniveton, D.R. and Todd, M.C. 2001. On the relationship of cosmic ray flux and precipitation. *Geophysical Research Letters* **28**: 1527-1530.

Lean, J. 2005. Living with a variable sun. *Physics Today* **58** (6): 32-38.

Lean, J., Beer, J. and Bradley, R. 1995. Reconstruction of solar irradiance since 1610—Implications for climate change. *Geophysical Research Letters* **22**:3195-3198.

Lindzen, R.S. 1997. Can increasing carbon dioxide cause climate change? *Proceedings of the National Academy of Sciences, USA* **94**: 8335-8342.

Lockwood, M., Stamper, R. and Wild, M.N. 1999. A doubling of the Sun's coronal magnetic field during the past 100 years. *Nature* **399**: 437-439.

Lu, Q.-B. 2009. Correlation between cosmic rays and ozone depletion, *Physical Review Letters,* **102**, 118501.

Macklin, M.G., Johnstone, E. and Lewin, J. 2005. Pervasive and long-term forcing of Holocene river instability and flooding in Great Britain by centennial-scale climate change. *The Holocene* **15**: 937-943.

Mann, M.E., Bradley, R.S. and Hughes, M.K. 1999. Northern Hemisphere temperatures during the past millennium: Inferences, uncertainties, and limitations. *Geophysical Research Letters* **26**: 759-762.

Mann, M.E. and Jones, P.D. 2003. Global surface temperatures over the past two millennia. *Geophysical Research Letters* **30**: 10.1029/2003GL017814.

Marsden, D. and Lingenfelter, R.E. 2003. Solar activity and cloud opacity variations: A modulated cosmic ray ionization model. *Journal of the Atmospheric Sciences* **60**: 626-636.

Marsh, N.D. and Svensmark, H. 2000. Low cloud properties influenced by cosmic rays. *Physical Review Letters* **85**: 5004-5007.

McCracken, K.G., McDonald, F.B., Beer, J., Raisbeck, G. and Yiou, F. 2004. A phenomenological study of the long-term cosmic ray modulation, 850-1958 AD. *Journal of Geophysical Research* **109**: 10.1029/2004JA010685.

Milankovitch, M. 1920. Theorie Mathematique des Phenomenes Produits par la Radiation Solaire. Gauthier-Villars, Paris, France.

Milankovitch, M. 1941. *Canon of Insolation and the Ice-Age Problem*. Royal Serbian Academy, Belgrade, Yugoslavia.

Oppo, D.W., McManus, J.F. and Cullen, J.L. 1998. Abrupt climate events 500,000 to 340,000 years ago: Evidence from subpolar North Atlantic sediments. *Science* **279**: 1335-1338.

Palle Bago, E. and Butler, C.J. 2000. The influence of cosmic rays on terrestrial clouds and global warming. *Astronomy & Geophysics* **41**: 4.18-4.22.

Parker, E.N. 1999. Sunny side of global warming. *Nature* **399**: 416-417.

Raymo, M.E., Ganley, K., Carter, S., Oppo, D.W. and McManus, J. 1998. Millennial-scale climate instability during the early Pleistocene epoch. *Nature* **392**: 699-702.

219

Rind, D. 2002. The sun's role in climate variations. *Science* **296**: 673-677.

Shaviv, N. 2002. Cosmic ray diffusion from the galactic spiral arms, iron meteorites, and a possible climatic connection. *Physics Review Letters* **89**: 051102.

Shaviv, N. 2003. The spiral structure of the Milky Way, cosmic rays, and ice age epochs on Earth. *New Astronomy* **8**: 39-77.

Shaviv, N.J. 2005. On climate response to changes in the cosmic ray flux and radiative budget. *Journal of Geophysical Research* **110**: 10.1029/2004JA010866.

Shaviv, N. and Veizer, J. 2003. Celestial driver of Phanerozoic climate? *GSA Today* **13** (7): 4-10.

Solanki, S.K. and Fligge, M. 1998. Solar irradiance since 1874 revisited. *Geophysical Research Letters* **25**: 341-344.

Solanki, S.K., Schussler, M. and Fligge, M. 2000. Evolution of the sun's large-scale magnetic field since the Maunder minimum. *Nature* **408**: 445-447.

Solanki, S.K., Usoskin, I.G., Kromer, B., Schussler, M. and Beer, J. 2004. Unusual activity of the Sun during recent decades compared to the previous 11,000 years. *Nature* **431**: 1084-1087.

Stott, P.A., Jones, G.S. and Mitchell, J.F.B. 2003. Do models underestimate the solar contribution to recent climate change? *Journal of Climate* **16**: 4079-4093.

Svensmark, H. 2007. Cosmoclimatology: a new theory emerges. *Astronomy & Geophysics* **48**: 1.18-1.24.

Svensmark, H. and Friis-Christensen, E. 1997. Variation of cosmic ray flux and global cloud coverage—A missing link in solar-climate relationships. *Journal of Atmospheric and Solar-Terrestrial Physics* **59**: 1225-1232.

Taricco, C., Bhandari, N., Cane, D., Colombetti, P. and Verma, N. 2006. Galactic cosmic ray flux decline and periodicities in the interplanetary space during the last 3 centuries revealed by Ti-44 in meteorites. *Journal of Geophysical Research* **111**: A08102.

Tobias, S.M. and Weiss, N.O. 2000. Resonant interactions between solar activity and climate. *Journal of Climate* **13**: 3745-3759.

Usoskin, I.G., Gladysheva, O.G. and Kovaltsov, G.A. 2004a. Cosmic ray-induced ionization in the atmosphere: spatial and temporal changes. *Journal of Atmospheric and Solar-Terrestrial Physics* **66**: 1791-1796.

Usoskin, I.G., Marsh, N., Kovaltsov, G.A., Mursula, K. and Gladysheva, O.G. 2004b. Latitudinal dependence of low cloud amount on cosmic ray induced ionization. *Geophysical Research Letters* **31**: 10.1029/2004GL019507.

Usoskin, I.G., Mursula, K., Solanki, S.K., Schussler, M. and Alanko, K. 2004c. Reconstruction of solar activity for the last millennium using Be-10 data. *Astronomy & Astrophysics* **413**: 745-751.

Usoskin, I.G., Schussler, M., Solanki, S.K. and Mursula, K. 2005. Solar activity, cosmic rays, and Earth's temperature: A millennium-scale comparison. *Journal of Geophysical Research* **110**: 10.1029/2004JA010946.

Usoskin, I.G., Solanki, S., Schussler, M., Mursula, K. and Alanko, K. 2003. A millennium scale sunspot number reconstruction: Evidence for an unusually active sun since the 1940s. *Physical Review Letters* **91**: 10.1103/PhysRevLett.91.211101.

Usoskin, I.G., Solanki, S.K., Taricco, C., Bhandari, N. and Kovaltsov, G.A. 2006. Long-term solar activity reconstructions: direct test by cosmogenic [44]Ti in meteorites. *Astronomy & Astrophysics* **457**: 10.1051/0004-6361:20065803.

Van Geel, B., Raspopov, O.M., Renssen, H., van der Plicht, J., Dergachev, V.A. and Meijer, H.A.J. 1999. The role of solar forcing upon climate change. *Quaternary Science Reviews* **18**: 331-338.

Veretenenko, S.V., Dergachev, V.A. and Dmitriyev, P.B. 2005. Long-term variations of the surface pressure in the North Atlantic and possible association with solar activity and galactic cosmic rays. *Advances in Space Research* **35**: 484-490.

Verschuren, D., Laird, K.R. and Cumming, B.F. 2000. Rainfall and drought in equatorial east Africa during the past 1,100 years. *Nature* **403**: 410-414.

Versteegh, G.J.M. 2005. Solar forcing of climate. 2: Evidence from the past. *Space Science Reviews* **120**: 243-286.

Voiculescu, M., Usoskin, I.G. and Mursula, K. 2006. Different response of clouds to solar input. *Geophysical Research Letters* **33**: L21802, doi:10.1029/2006GL027820.

Yang, S., Odah, H. and Shaw, J. 2000. Variations in the geomagnetic dipole moment over the last 12,000 years. *Geophysical Journal International* **140**: 158-162.

5.2. Irradiance

We begin this section of our review of the potential effects of solar activity on earth's climate with the study of Karlén (1998), who examined proxy climate data related to changes in summer temperatures in Scandinavia over the past 10,000 years. This temperature record—derived from analyses of changes in the size of glaciers, changes in the altitude of the alpine tree-limit, and variations in the width of annual tree rings—was compared with contemporaneous solar irradiance data derived from ^{14}C anomalies measured in tree rings. The former record revealed both long- and short-term temperature fluctuations; it was noted by Karlén that during warm periods the temperature was "about 2°C warmer than at present." In addition, the temperature fluctuations were found to be "closely related" to the ^{14}C-derived changes in solar irradiation, leading him to conclude that "the similarity between solar irradiation changes and climate indicate a solar influence on the Scandinavian and Greenland climates." This association led him to further conclude that "the frequency and magnitude of changes in climate during the Holocene [i.e., the current interglacial] do not support the opinion that the climatic change of the last 100 years is unique." He bluntly stated that "there is no evidence of a human influence so far."

Also writing just before the turn of the century, Lockwood et al. (1999) analyzed measurements of the near-earth interplanetary magnetic field to determine the total magnetic flux leaving the sun since 1868. Based on their analysis, they were able to show that the total magnetic flux leaving the sun rose by a factor of 1.41 over the period 1964-1996, while surrogate measurements of the interplanetary magnetic field previous to this time indicated that this parameter had increased by a factor of 2.3 since 1901. These findings and others linking changes in solar magnetic activity with terrestrial climate change led the authors to state that "the variation [in the total solar magnetic flux] found here stresses the importance of understanding the connections between the sun's output and its magnetic field and between terrestrial global cloud cover, cosmic ray fluxes and the heliospheric field."

Parker (1999) noted that the number of sunspots also doubled over the same time period, and that one consequence of this phenomenon is a much more vigorous sun that is slightly brighter. Parker also drew attention to the fact that NASA spacecraft measurements had revealed that the brightness (B) of the sun varies by an amount "change in B/B = 0.15%, in step with the 11-year magnetic cycle." He then pointed out that during times of much reduced activity of this sort (such as the Maunder Minimum of 1645-1715) and much increased activity (such as the twelfth century Mediaeval Maximum), brightness variations on the order of change in B/B = 0.5% typically occur, after which he indicated that the mean temperature (T) of the northern portion of the earth

varied by 1 to 2°C in association with these variations in solar activity, stating finally that "we cannot help noting that change in T/T = change in B/B."

Also in 1999, Chambers et al. noted that recent research findings in both palaeoecology and solar science "indicate a greater role for solar forcing in Holocene climate change than has previously been recognized," which subject they then proceeded to review. In doing so, they found much evidence within the Holocene for solar-driven variations in earth-atmosphere processes over a range of timescales stretching from the 11-year solar cycle to century-scale events. They acknowledge that the absolute solar flux variations associated with these phenomena are rather small; but they identify a number of "multiplier effects" that can operate on solar rhythms in such a way that "minor variations in solar activity can be reflected in more significant variations within the earth's atmosphere."

The three researchers also noted, in this regard, that nonlinear responses to solar variability are inadequately represented (in fact, they are essentially ignored) in the global climate models used by the IPCC to predict future CO_2-induced global warming, while at the same time other amplifier effects are used to model the hypothesized CO_2-induced global warming of the future, where CO_2 is only an initial perturber of the climate system which, according to the IPCC, sets other more powerful forces in motion that produce the bulk of the warming.

At the start of the new millennium, Bard et al. (2000) listed some of the many different types of information that have been used to reconstruct past solar variability, including "the envelope of the SSN [sunspot number] 11-year cycle (Reid, 1991), the length and decay rate of the solar cycle (Hoyt and Schatten, 1993), the structure and decay rate of individual sunspots (Hoyt and Schatten, 1993), the mean level of SSN (Hoyt and Schatten, 1993; Zhang et al., 1994; Reid, 1997), the solar rotation and the solar diameter (Nesme-Ribes et al., 1993), and the geomagnetic aa index (Cliver et al., 1998)." They also noted that "Lean et al. (1995) proposed that the irradiance record could be divided into 2 superimposed components: an 11-year cycle based on the parameterization of sunspot darkening and facular brightening (Lean et al., 1992), and a slowly varying background derived separately from studies of sun-like stars (Baliunas and Jastrow, 1990)," and that Solanki and Fligge (1998) had developed an even more convoluted technique. Bard et al., however, used an entirely different approach.

Rather than directly characterize some aspect of solar variability, they assessed certain consequences of that variability. Specifically, they noted that magnetic fields of the solar wind deflect portions of the primary flux of charged cosmic particles in the vicinity of the earth, leading to reductions in the creation of cosmogenic nuclides in earth's atmosphere. Consequently, they reasoned that histories of the atmospheric concentrations of ^{14}C and ^{10}Be can be used as proxies for solar activity, as noted many years earlier by Lal and Peters (1967).

In employing this approach to the problem, the four researchers first created a 1,200-year history of cosmonuclide production in earth's atmosphere from ^{10}Be measurements of South Pole ice (Raisbeck *et al.*, 1990) and the atmospheric ^{14}C/^{12}C record as measured in tree rings (Bard *et al.*, 1997). This record was then converted to total solar irradiance (TSI) values by "applying a linear scaling using the TSI values published previously for the Maunder Minimum," when cosmonuclide production was 30-50 percent above the modern value.

This approach resulted in an extended TSI record that suggests, in their words, that "solar output was significantly reduced between AD 1450 and 1850, but slightly higher or similar to the present value during a period centered around AD 1200." "It could thus be argued," they say, "that irradiance variations may have contributed to the so-called 'little ice age' and 'medieval warm period'."

In discussing this idea, Bard *et al.* downplay their own suggestion, because, as they report, "some researchers have concluded that the 'little ice age' and/or 'medieval warm period' [were] regional, rather than global events." Noting the TSI variations they developed from their cosmonuclide data "would tend to force global effects," they felt they could not associate this global impetus for climate change with what other people were calling regional climatic anomalies. With respect to these thoughts, we refer the reader to Section 3.2 of this report, where it is demonstrated that the Little Ice Age and Medieval Warm Period were truly global in extent.

Rozelot (2001) conducted a series of analyses designed to determine whether phenomena related to variations in the radius of the sun may have influenced earth's climate over the past four centuries. The results of these analyses revealed, in the words of the researcher, that "at least over the last four centuries, warm periods on the earth correlate well with smaller apparent diameter of the Sun and colder ones with a bigger sun." Although the results

of this study were correlative and did not identify a physical mechanism capable of inducing significant climate change on earth, Rozelot reports that the changes in the sun's radius are "of such magnitude that significant effects on the earth's climate are possible."

Rigozo *et al.* (2001) created a history of sunspot numbers for the last 1,000 years "using a sum of sine waves derived from spectral analysis of the time series of sunspot number R_Z for the period 1700-1999," and from this record they derived the strengths of a number of parameters related to various aspects of solar variability. In describing their results, the researchers say that "the 1000-year reconstructed sunspot number reproduces well the great maximums and minimums in solar activity, identified in cosmonuclides variation records, and, specifically, the epochs of the Oort, Wolf, Sporer, Maunder, and Dalton Minimums, as well [as] the Medieval and Modern Maximums," the latter of which they describe as "starting near 1900." The mean sunspot number for the Wolf, Sporer, and Maunder Minimums was 1.36. For the Oort and Dalton Minimums it was 25.05; for the Medieval Maximum it was 53.00; and for the Modern Maximum it was 57.54. Compared to the average of the Wolf, Sporer, and Maunder Minimums, therefore, the mean sunspot number of the Oort and Dalton Minimums was 18.42 times greater; that of the Medieval Maximum was 38.97 times greater; and that of the Modern Maximum was 42.31 times greater. Similar strength ratios for the solar radio flux were 1.41, 1.89, and 1.97, respectively. For the solar wind velocity the corresponding ratios were 1.05, 1.10, and 1.11; while for the southward component of the interplanetary magnetic field they were 1.70, 2.54, and 2.67. In comparing these numbers, both the Medieval and Modern Maximums in sunspot number and solar variability parameters stand out above all other periods of the past thousand years, with the Modern Maximum slightly besting the Medieval Maximum.

Noting that a number of different spacecraft have monitored total solar irradiance (TSI) for the past 23 years, with at least two of them operating simultaneously at all times, and that TSI measurements made from balloons and rockets supplement the satellite data, Frohlich and Lean (2002) compared the composite TSI record with an empirical model of TSI variations based on known magnetic sources of irradiance variability, such as sunspot darkening and brightening, after which they described how "the TSI record may be extrapolated back to the seventeenth century Maunder Minimum of anomalously lower solar activity, which coincided with the coldest period of the Little Ice Age." This exercise, as they have described it, "enables an assessment of the extent of post-industrial climate change that may be attributable to a varying sun, and how much the sun might influence future climate change."

In reporting their results, Frolich and Lean state that "warming since 1650 due to the solar change is close to 0.4°C, with pre-industrial fluctuations of 0.2°C that are seen also to be present in the temperature reconstructions." From this study, therefore, it would appear that solar variability can explain a significant portion of the warming experienced by the earth in recovering from the global chill of the Little Ice Age, with a modicum of positive feedback accounting for the rest. With respect to the future, however, the two solar scientists say that "solar forcing is unlikely to compensate for the expected forcing due to the increase of anthropogenic greenhouse gases which are projected to be about a factor of 3-6 larger." The magnitude of that anthropogenic forcing, however, has been computed by many different approaches to be much smaller than the value employed by Frohlich and Lean in making this comparison (Idso, 1998).

Contemporaneously, Douglass and Clader (2002) used multiple regression analysis to separate surface and atmospheric temperature responses to solar irradiance variations over the past two-and-a-half solar cycles (1979-2001) from temperature responses produced by variations in ENSO and volcanic activity. Based on the satellite-derived lower tropospheric temperature record, they evaluated the sensitivity (k) of temperature (T) to solar irradiance (I), where temperature sensitivity to solar irradiance is defined as $k = \Delta T/\Delta I$, obtaining the result of $k = 0.11 \pm 0.02°C/(W/m^2)$. Similar analyses based on the radiosonde temperature record of Parker *et al.* (1997) and the surface air temperature records of Jones *et al.* (2001) and Hansen and Lebedeff (1987, with updates) produced k values of 0.13, 0.09, and $0.11°C/(W/m^2)$, respectively, with the identical standard error of $\pm 0.02°C/(W/m^2)$. In addition, they reported that White *et al.* (1997) derived a decadal timescale solar sensitivity of $0.10 \pm 0.02°C/(W/m^2)$ from a study of upper ocean temperatures over the period 1955-1994 and that Lean and Rind (1998) derived a value of $0.12 \pm 0.02°C/(W/m^2)$ from a paleo-reconstructed temperature record spanning the period 1610-1800. They concluded that "the close agreement of these

various independent values with our value of 0.11 ± 0.02 [$°C/(W/m^2)$] suggests that the sensitivity k is the same for both decadal and centennial time scales and for both ocean and lower tropospheric temperatures." They further suggest that if these values of k hold true for centennial time scales, which appears to be the case, their high-end value implies a surface warming of 0.2°C over the last 100 years in response to the 1.5 W/m^2 increase in solar irradiance inferred by Lean (2000) for this period. This warming represents approximately one-third of the total increase in global surface air temperature estimated by Parker *et al.* (1997), 0.55°C, and Hansen *et al.* (1999), 0.65°C, for the same period. It does not, however, include potential indirect effects of more esoteric solar climate-affecting phenomena, such as those discussed in Section 5.1 of this report, that could also have been operative over this period.

Foukal (2002) analyzed the findings of space-borne radiometry and reported that "variations in total solar irradiance, S, measured over the past 22 years, are found to be closely proportional to the difference in projected areas of dark sunspots, AS, and of bright magnetic plage elements, APN, in active regions and in enhanced network," plus the finding that "this difference varies from cycle to cycle and is not simply related to cycle amplitude itself," which facts suggest there is "little reason to expect that S will track any of the familiar indices of solar activity." On the other hand, he notes that "empirical modeling of spectro-radiometric observations indicates that the variability of solar ultraviolet flux, FUV, at wavelengths shorter than approximately 250 nm, is determined mainly by APN alone."

Building upon this conceptual foundation, and using daily data from the Mt. Wilson Observatory that covered the period 1905-1984, plus partially overlapping data from the Sacramento Peak Observatory that extended through 1999, Foukal derived time series of both total solar and UV irradiances between 1915 and 1999, which he then compared with global temperature data for the same time period. This work revealed, in his words, that "correlation of our time series of UV irradiance with global temperature, T, accounts for only 20% of the global temperature variance during the 20th century," but that "correlation of our total irradiance time series with T accounts *statistically* for 80% of the variance in global temperature over that period."

The UV findings of Foukal were not impressive, but the results of his total solar irradiance analysis were, leading him to state that "the possibility of significant driving of twentieth century climate by total irradiance variation cannot be dismissed." Although the magnitude of the total solar effect was determined to be "a factor 3-5 lower than expected to produce a significant global warming contribution based on present-day climate model sensitivities," what Foukal calls the "high correlation between S and T" strongly suggests that changes in S largely determine changes in T, the confirmation of which suggestion likely merely awaits what he refers to as an "improved understanding of possible climate sensitivity to relatively small total irradiance variation."

In the following year, Willson and Mordvinov (2003) analyzed total solar irradiance (TSI) data obtained from different satellite platforms over the period 1978-2002, attempting to resolve various small but important inconsistencies among them. In doing so, they came to the realization that "construction of TSI composite databases will not be without its controversies for the foreseeable future." Nevertheless, their most interesting result, in the estimation of the two researchers, was their confirmation of a +0.05%/decade trend between the minima separating solar cycles 21-22 and 22-23, which they say "appears to be significant."

Willson and Mordvinov say the finding of the 0.05 percent/decade minimum-to-minimum trend "means that TSI variability can be caused by unknown mechanisms other than the solar magnetic activity cycle," which means that "much longer time scales for TSI variations are therefore a possibility," which they say "has obvious implications for solar forcing of climate." Specifically, it means there could be undiscovered long-term variations in total solar irradiance of a magnitude that could possibly explain centennial-scale climate variability, which Bond *et al.* (2001) have already demonstrated to be related to solar activity, as well as the millennial-scale climatic oscillation that pervades both glacial and interglacial periods for essentially as far back in time as paleoclimatologists can see (Oppo *et al.*, 1998; Raymo *et al.*, 1998).

Like Willson and Mordvinov, Foukal (2003) acknowledged that "recent evidence from ocean and ice cores suggests that a significant fraction of the variability in northern hemisphere climate since the last Ice Age correlates with solar activity (Bond *et al.*, 2001)," while additionally noting that "a recent reconstruction of S [total solar irradiance] from archival images of spots and faculae obtained daily from the Mt. Wilson Observatory in California since

1915 shows remarkable agreement with smoothed global temperature in the 20th century," citing his own work of 2002. However, he was forced to acknowledge that the observed variations in S between 1978 and 2002 were not large enough to explain the observed temperature changes on earth within the context of normal radiative forcing. Hence, he proceeded to review the status of research into various subjects that might possibly be able to explain this dichotomy. Specifically, he presented an overview of then-current knowledge relative to the idea that "the solar impact on climate might be driven by other variable solar outputs of ultraviolet radiation or plasmas and fields via more complex mechanisms than direct forcing of tropospheric temperature." As could have been expected, the article contained no grand revelations; when all was said and done, Foukal returned pretty much to where he had started, concluding that "we cannot rule out multi-decadal variations in S sufficiently large to influence climate, yet overlooked so far through limited sensitivity and time span of our present observational techniques."

The following year, Damon and Laut (2004) reported what they described as errors made by Friis-Christensen and Lassen (1991), Svensmark and Friis-Christensen (1997), Svensmark (1998) and Lassen and Friis-Christensen (2000) in their presentation of solar activity data, correlated with terrestrial temperature data. The Danish scientists' error, in the words of Damon and Laut, was "adding to a heavily smoothed ('filtered') curve, four additional points covering the period of global warming, which were only partially filtered or not filtered at all." This in turn led to an apparent dramatic increase in solar activity over the last quarter of the twentieth century that closely matched the equally dramatic rise in temperature manifest by the Northern Hemispheric temperature reconstruction of Mann *et al.* (1998, 1999) over the same period. With the acquisition of additional solar activity data in subsequent years, however, and with what Damon and Laut called the proper handling of the numbers, the late twentieth century dramatic increase in solar activity totally disappears.

This new result, to quote Damon and Laut, means that "the sensational agreement with the recent global warming, which drew worldwide attention, has totally disappeared." In reality, however, it is only the agreement with the last quarter-century of the discredited Mann *et al.* "hockey stick" temperature history that has disappeared. This new disagreement is most welcome, for the Mann *et al.* temperature reconstruction is likely in error over this stretch of time. (See Section 3.2.)

Using a nonlinear non-stationary time series technique called empirical mode decomposition, Coughlin and Tung (2004) analyzed monthly mean geopotential heights and temperatures—obtained from Kalnay *et al.* (1996)—from 1000 hPa to 10 hPa over the period January 1958 to December 2003. This work revealed the existence of five oscillations and a trend in both datasets. The fourth of these oscillations has an average period of 11 years and indicates enhanced warming during times of maximum solar radiation. As the two researchers describe it, "the solar flux is positively correlated with the fourth modes in temperature and geopotential height almost everywhere [and] the overwhelming picture is that of a positive correlation between the solar flux and this mode throughout the troposphere."

Coughlin and Tung concluded that "the atmosphere warms during the solar maximum almost everywhere over the globe." And the unfailing omnipresent impact of this small forcing (a 0.1 percent change in the total energy output of the sun from cycle minimum to maximum) suggests that any longer-period oscillations of the solar inferno could well be causing the even greater centennial- and millennial-scale oscillations of temperature that are observed in paleotemperature data from various places around the world.

Additional light on the subject has been provided by widespread measurements of the flux of solar radiation received at the surface of the earth that have been made since the late 1950s. Nearly all of these measurements reveal a sizeable decline in the surface receipt of solar radiation that was not reversed until the mid-1980s, as noted by Wild *et al.* (2005). During this time, there was also a noticeable dip in earth's surface air temperature, after which temperatures rose at a rate and to a level of warmth that the IPCC claims were both without precedent over the past one to two millennia, which phenomena they attribute to similarly unprecedented increases in greenhouse gas concentrations, the most notable, of course, being CO_2.

This reversal of the decline in the amount of solar radiation incident upon the earth's surface, in the words of Wild *et al.*, "is reconcilable with changes in cloudiness and atmospheric transmission and may substantially affect surface climate." They say, for example, that "whereas the decline in solar energy could have counterbalanced the increase in down-welling longwave energy from the enhanced greenhouse effect before the 1980s, the masking of the greenhouse effect and related impacts may no longer have been effective thereafter, enabling the greenhouse signals to become more evident during the 1990s." Qualitatively, this scenario sounds reasonable; but when the magnitude of the increase in the surface-received flux of solar radiation over the 1990s is considered, the statement is seen to be rather disingenuous.

Over the range of years for which high-quality data were available to them (1992-2002), Wild *et al.* determined that the mean worldwide increase in clear-sky insolation averaged 0.68 Wm^{-2} per year, which increase they found to be "comparable to the increase under all-sky conditions." Consequently, for that specific 10-year period, these real-world data suggest that the total increase in solar radiation received at the surface of the earth should have been something on the order of 6.8 Wm^{-2}, which is not significantly different from what is implied by the satellite and "earthshine" data of Palle *et al.* (2004), although the satellite data of Pinker *et al.* (2005) suggest an increase only about a third as large for this period.

Putting these numbers in perspective, Charlson *et al.* (2005) report that the longwave radiative forcing provided by all greenhouse gas increases since the beginning of the industrial era has amounted to only 2.4 Wm^{-2}, citing the work of Anderson *et al.* (2003), while Palle *et al.* say that "the latest IPCC report argues for a 2.4 Wm^{-2} increase in CO_2 longwave forcing since 1850." Consequently, it can be readily appreciated that the longwave forcing of greenhouse gases over the 1990s would have been but a fraction of a fraction of the observed increase in the contemporary receipt of solar radiation at the surface of the earth. To thus suggest, as Wild *et al.* do, that the increase in insolation experienced at the surface of the earth over the 1990s may have enabled anthropogenic greenhouse gas signals of that period to become more evident, seems incongruous, as their suggestion implies that the bulk of the warming of that period was due to increases in greenhouse gas concentrations, when the solar component of the temperature forcing was clearly much greater. And this incongruity is made all the worse by the fact that methane concentrations rose ever more slowly over this period, apparently actually stabilizing near its end (see Section 2.6. Methane). Consequently, a much more logical conclusion would be that the primary driver of the global warming of the 1990s was the large increase in global surface-level insolation.

A final paper of note from 2005 was that of Soon (2005), who explored the question of which variable was the dominant driver of twentieth century temperature change in the Arctic—rising atmospheric CO_2 concentrations or variations in solar irradiance—by examining what roles the two variables may have played in decadal, multi-decadal, and longer-term variations in surface air temperature (SAT). He performed a number of statistical analyses on (1) a composite Arctic-wide SAT record constructed by Polyakov *et al.* (2003), (2) global CO_2 concentrations taken from estimates given by the NASA GISS climate modeling group, and (3) a total solar irradiance (TSI) record developed by Hoyt and Schatten (1993, updated by Hoyt in 2005) over the period 1875-2000.

The results of these analyses indicated a much stronger statistical relationship between SATs and TSI, as opposed to SATs and CO_2. Solar forcing generally explained well over 75 percent of the variance in decadal-smoothed seasonal and annual Arctic SATs, while CO_2 forcing explained only between 8 and 22 percent of the variance. Wavelet analysis further supported the case for solar forcing of the SAT record, revealing similar time-frequency characteristics for annual and seasonally averaged temperatures at decadal and multi-decadal time scales. By contrast, wavelet analysis gave little to no indication of a CO_2 forcing of Arctic SSTs. Based on these data and analyses, therefore, it would appear that the sun, not atmospheric CO_2, has been the driving force for temperature change in the Arctic.

Lastovicka (2006) summarized recent advancements in the field, saying "new results from various space and ground-based experiments monitoring the radiative and particle emissions of the sun, together with their terrestrial impact, have opened an exciting new era in both solar and atmospheric physics," stating that "these studies clearly show that the variable solar radiative and particle output affects the earth's atmosphere and climate in many fundamental ways." That same year, Bard and Frank (2006) examined "changes on different time scales, from the last million years up to recent decades," and in doing so critically assessed recent claims that "the variability of the sun has had a significant impact on global climate." "Overall," in the judgment of the two researchers, the role of solar activity in causing climate change "remains unproven." However, as they state in the concluding sentence of their abstract, "the weight of evidence suggests that solar changes have contributed to small

climate oscillations occurring on time scales of a few centuries, similar in type to the fluctuations classically described for the last millennium: the so-called Medieval Warm Period (AD 900-1400) followed on by the Little Ice Age (AD 1500-1800)."

In another study from 2006, which also reviewed the scientific literature, Beer *et al.* (2006) explored what we know about solar variability and its possible effects on earth's climate, focusing on two types of variability in the flux of solar radiation incident on the earth. The first type, in their words, "is due to changes in the orbital parameters of the earth's position relative to the sun induced by the other planets," which arises from gravitational perturbations that "induce changes with characteristic time scales in the eccentricity (~100,000 years), the obliquity (angle between the equator and the orbital plane, ~40,000 years) and the precession of the earth's axis (~20,000 years)," while the second type is due to variability within the sun itself.

With respect to the latter category, the three researchers report that direct observations of total solar irradiance above the earth's atmosphere have been made only over the past quarter-century, while observations of sunspots have been made and recorded for approximately four centuries. In between the time scales of these two types of measurements fall neutron count rates and aurora counts. Therefore, [10]Be and other cosmogenic radionuclides (such as ^{14}C) —stored in ice, sediment cores, and tree rings— currently provide our only means of inferring solar irradiance variability on a millennial time scale. These cosmogenic nuclides "clearly reveal that the sun varies significantly on millennial time scales and most likely plays an important role in climate change," especially within this particular time domain. In reference to their [10]Be-based derivation of a 9,000-year record of solar modulation, Beer *et al*. note that its "comparison with paleoclimatic data provides strong evidence for a causal relationship between solar variability and climate change."

We have now reached the work of Nicola Scafetta, a research scientist in the Duke University physics department, and Bruce West, chief scientist in the mathematical and information science directorate of the U.S. Army Research Office in Research Triangle Park, North Carolina. To better follow the arc of their work, we'll temporarily abandon our chronological ordering of this literature review.

Scafetta and West (2006a) developed "two distinct TSI reconstructions made by merging in 1980 the annual mean TSI proxy reconstruction of Lean *et al*. (1995) for the period 1900-1980 and two alternative TSI satellite composites, ACRIM (Willson and Mordvinov, 2003), and PMOD (Frohlich and Lean, 1998), for the period 1980-2000," and then used a climate sensitivity transfer function to create twentieth century temperature histories. The results suggested that the sun contributed some 46 to 49 percent of the 1900-2000 global warming of the earth. Considering that there may have been uncertainties of 20 to 30 percent in their sensitivity parameters, the two researchers suggested the sun may have been responsible for as much as 60 percent of the twentieth century temperature rise.

Scafetta and West say the role of the sun in twentieth century global warming has been significantly underestimated by the climate modeling community, with various energy balance models producing estimates of solar-induced warming over this period that are "two to ten times lower" than what they found. The two researchers say "the models might be inadequate because of the difficulty of modeling climate in general and a lack of knowledge of climate sensitivity to solar variations in particular." They also note that "theoretical models usually acknowledge as solar forcing only the direct TSI forcing," thereby ignoring "possible additional climate effects linked to solar magnetic field, UV radiation, solar flares and cosmic ray intensity modulations." In this regard, we additionally note that some of these phenomena may to some degree be independent of, and thereby add to, the simple TSI forcing Scafetta and West employed, which suggests that the totality of solar activity effects on climate may be even greater than what they calculated.

In a second study published in the same year, Scafetta and West (2006b) begin by noting that nearly all attribution studies begin with pre-determined forcing and feedback mechanisms in the models they employ. "One difficulty with this approach," according to Scafetta and West, "is that the feedback mechanisms and alternative solar effects on climate, since they are only partially known, might be poorly or not modeled at all." Consequently, "to circumvent the lack of knowledge in climate physics," they adopt "an alternative approach that attempts to evaluate the total direct plus indirect effect of solar changes on climate by comparing patterns in the secular temperature and TSI reconstructions," where "a TSI reconstruction is not used as a radiative forcing, but as a proxy [for] the entire solar dynamics." They then proceed on the assumption that "the secular climate sensitivity to solar change can be phenomenologically

estimated by comparing ... solar and temperature records during the pre-industrial era, when, reasonably, only a negligible amount of anthropogenic-added climate forcing was present," and when "the sun was the only realistic force affecting climate on a secular scale."

Scafetta and West used the Northern Hemispheric temperature reconstruction of Moberg *et al*. (2005), three alternative TSI proxy reconstructions developed by Lean *et al*. (1995), Lean (2000), and Wang *et al*. (2005), and a scale-by-scale transfer model of climate sensitivity to solar activity changes created by themselves (Scafetta and West, 2005, 2006a) and found what they called a "good correspondence between global temperature and solar induced temperature curves during the pre-industrial period, such as the cooling periods occurring during the Maunder Minimum (1645-1715) and the Dalton Minimum (1795-1825)." In addition, they note that since the time of the seventeenth century solar minimum, "the sun has induced a warming of $\Delta T \sim 0.7$ K," and that "this warming is of the same magnitude [as] the cooling of $\Delta T \sim 0.7$ K from the medieval maximum to the 17th century minimum," which finding, in their words, "suggests the presence of a millenarian solar cycle, with ... medieval and contemporary maxima, driving the climate of the last millennium," as was first suggested fully three decades ago by Eddy (1976) in his seminal study of the Maunder Minimum.

Scafetta and West say their work provides substantive evidence for the likelihood that "solar change effects are greater than what can be explained by several climate models," citing Stevens and North (1996), the Intergovernmental Panel on Climate Change (2001), Hansen *et al*. (2002), and Foukal *et al*. (2004); and they note that a solar change "might trigger several climate feedbacks and alter the greenhouse gas (H_2O, CO_2, CH_4, etc.) concentrations, as 420,000 years of Antarctic ice core data would also suggest (Petit *et al*., 1999)," once again reiterating that "most of the sun-climate coupling mechanisms are probably still unknown," and that "they might strongly amplify the effects of small solar activity increase." That being said, however, the researchers note that in the twentieth century there was "a clear surplus warming" above and beyond what is suggested by their solar-based temperature reconstruction, such that something in addition to the sun may have been responsible for approximately 50 percent of the total global warming since 1900. This anomalous increase in temperature, it could be argued, was due to anthropogenic greenhouse gas emissions. Scafetta and West say the temperature difference since 1975, where the most noticeable part of the discrepancy occurred, may have been due to "spurious non-climatic contamination of the surface observations such as heat-island and land-use effects (Pielke *et al.*, 2002; Kalnay and Cai, 2003)," which they say is also suggested by "an anomalous warming behavior of the global average land temperature vs. the marine temperature since 1975 (Brohan *et al.*, 2006)."

In their next paper, Scafetta and West (2007) reconstructed a phenomenological solar signature (PSS) of climate for the Northern Hemisphere for the last four centuries that matches relatively well the instrumental temperature record since 1850 and the paleoclimate temperature proxy reconstruction of Moberg (2005). The period from 1950 to 2010 showed excellent agreement between 11- and 22-year PSS cycles when compared to smoothed average global temperature data and the global *cooling* that occurred since 2002. Describing their research in an opinion essay in *Physics Today* (published by the American Institute of Physics), they say "this cooling seems to have been induced by decreased solar activity from the 2001 maximum to the 2007 minimum as depicted in two distinct TSI reconstructions" and "the same patterns are poorly reproduced by present-day GCMs and are dismissively interpreted as internal variability (noise) of climate. The nonequilibrium thermodynamic models we used suggest that the Sun is influencing climate significantly more than the IPCC report claims" (Scafetta and West, 2008).

In 2009, Scafetta and a new coauthor, Richard C. Willson, senior research scientist at Columbia's Center for Climate Systems Research, addressed the issue of whether or not TSI increased from 1980 to 2002. The IPCC assumed there was no increase by adopting the TSI satellite composite produced by the Physikalisch-Meteorologisches Observatorium Davos (PMOD) (see Frohlich, 2006). PMOD assumed that the NIMBUS7 TSI satellite record artificially increased its sensitivity during the ACRIM-gap (1999.5-1991.75), and therefore it reduced the NIMBUS7 record by 0.86 W/m^2 during the ACRIM-gap period, and consequently the TSI results changed little since 1980. However, this PMOD adjustment of NIMBUS7 TSI satellite data was never acknowledged by the experimental teams (Willson and Mordvinov, 2003; supporting material in Scafetta and Willson, 2009).

Scafetta and Willson (2009) proposed to solve the ACRIM-gap calibration controversy by developing a TSI model using a proxy model based on variations of the surface distribution of solar magnetic flux designed by Krivova *et al.* (2007) to bridge the two-year gap between ACRIM1 and ACRIM2. They use this to bridge "mixed" versions of ACRIM and PMOD TSI before and after the ACRIM-gap. Both "mixed" models show, in the authors' words, "a significant TSI increase of 0.033%/decade between the solar activity minima of 1986 and 1996, comparable to the 0.037% found in the TSI satellite ACRIM composite." They conclude that "increasing TSI between 1980 and 2000 could have contributed significantly to global warming during the last three decades. Current climate models have assumed that TSI did not vary significantly during the last 30 years and have, therefore, underestimated the solar contribution and overestimated the anthropogenic contribution to global warming."

Backing up now to 2007, Krivova *et al.* (2007) note there is "strong interest" in the subject of long-term variations of total solar irradiance or TSI "due to its potential influence on global climate," and that "only a reconstruction of solar irradiance for the pre-satellite period with the help of models can aid in gaining further insight into the nature of this influence," which is what they set about to achieve in their paper. They developed a history of TSI "from the end of the Maunder minimum [about AD 1700] to the present based on variations of the surface distribution of the solar magnetic field," which was "calculated from the historical record of the sunspot number using a simple but consistent physical model," i.e., that of Solanki *et al.* (2000, 2002).

Krivova *et al.* report that their model "successfully reproduces three independent datasets: total solar irradiance measurements available since 1978, total photospheric magnetic flux since 1974 and the open magnetic flux since 1868," which was "empirically reconstructed using the geomagnetic *aa*-index." Based on this model, they calculated an increase in TSI since the Maunder minimum somewhere in the range of 0.9-1.5 Wm^{-2}, which encompasses the results of several independent reconstructions that have been derived over the past few years. In the final sentence of their paper, however, they also note that "all the values we obtain are significantly below the ΔTSI values deduced from stellar data and used in older TSI reconstructions," the results of which range from 2 to 16 Wm^{-2}.

Shaviv (2008) has attempted to quantify solar radiative forcing using oceans as a calorimeter. He evaluated three independent measures of net ocean heat flux over five decades, sea level change rate from twentieth century tide gauge records, and sea surface temperature. He found a "very clear correlation between solar activity and sea level" including the 11-year solar periodicity and phase, with a correlation coefficient of r=0.55. He also found "that the total radiative forcing associated with solar cycles variations is about 5 to 7 times larger than those associated with the TSI variations, thus implying the necessary existence of an amplification mechanism, though without pointing to which one." Shaviv claims "the sheer size of the heat flux, and the lack of any phase lag between the flux and the driving force further implies that it cannot be part of an atmospheric feedback and very unlikely to be part of a coupled atmosphere-ocean oscillation mode. It must therefore be the manifestation of real variations in the global radiative forcing." This provides "very strong support for the notion that an amplification mechanism exists. Given that the CRF [Cosmic Ray Flux]/climate links predicts the correct radiation imbalance observed in the cloud cover variations, it is a favorable candidate." These results, Shaviv says, "imply that the climate sensitivity required to explain historic temperature variations is smaller than often concluded."

Pallé *et al.* (2009) re-analyzed the overall reflectance of sunlight from Earth ("earthshine") and re-calibrated the CERES satellite data to obtain consistent results for Earth's solar reflectance. According to the authors, "Earthshine and FD [flux data] analyses show contemporaneous and climatologically significant increases in the Earth's reflectance from the outset of our earthshine measurements beginning in late 1998 roughly until mid-2000.After that and to date, all three show a roughly constant terrestrial albedo, except for the FD data in the most recent years. Using satellite cloud data and Earth reflectance models, we also show that the decadal-scale changes in Earth's reflectance measured by earthshine are reliable and are caused by changes in the properties of clouds rather than any spurious signal, such as changes in the Sun-Earth-Moon geometry."

Ohmura (2009) reviewed surface solar irradiance at 400 sites globally to 2005. They found a brightening phase from the 1920s to 1960s followed by a 20-year dimming phase from 1960 to 1980. Then there is another 15-year brightening phase from 1990

to 2005. Ohmura finds "aerosol direct and indirect effects played about an equal weight in changing global solar radiation. The temperature sensitivity due to radiation change is estimated at 0.05 to 0.06 K/(W m^{-2})."

Long *et al.* (2009) analyzed "all-sky and clear-sky surface downwelling shortwave radiation and bulk cloud properties" from 1995 through 2007. They "show that widespread brightening has occurred over the continental United States ... averaging about 8 W m^{-2}/decade for all-sky shortwave and 5 W m^{-2}/decade for the clear-sky shortwave. This all-sky increase is substantially greater than the (global) 2 W m^{-2}/decade previously reported ..." Their "results show that changes in dry aerosols and/or direct aerosol effects alone cannot explain the observed changes in surface shortwave (SW) radiation, but it is likely that changes in cloudiness play a significant role."

These observations by Shaviv, Pallé, Ohmura, and Long *et al.* point to major variations in earth's radiative budget caused by changes both in aerosols and clouds. Both are affected by natural and anthropogenic causes, including aircraft, power plants, cars, cooking, forest fires, and volcanoes. However, solar activity and cosmic rays also modulate clouds. When GCMs ignore or underestimate causes or modulation by solar cycles, magnetic fields and/or cosmic rays, they overestimate climate sensitivity and anthropogenic impacts.

Although there thus is still significant uncertainty about the true magnitude of the TSI change experienced since the end of the Maunder Minimum, the wide range of possible values suggests that long-term TSI variability cannot be rejected as a plausible cause of the majority of the global warming that has fueled earth's transition from the chilling depths of the Little Ice Age to the much milder weather of the Current Warm Period. Indeed, the results of many of the studies reviewed in this summary argue strongly for this scenario, while others suggest it is the only explanation that fits all the data.

The measured judgment of Bard and Frank (2006) seems to us to be right on the mark. The role of solar activity in causing climate change is so complex that most theories of solar forcing of climate change must be considered to be as yet "unproven." It would also be appropriate for climate scientists to admit the same about the role of rising atmospheric CO_2 concentrations in driving recent global warming. If it is fairly certain that the sun was responsible for creating multi-centennial cold and warm periods, it is clear the sun could easily be responsible for the majority or even the entirety of the global warming of the past century or so.

Additional information on this topic, including reviews of newer publications as they become available, can be found at http://www.co2science.org/subject/s/solarirradiance.php.

References

Anderson, T.L., Charlson, R.J., Schwartz, S.E., Knutti, R., Boucher, O., Rodhe, H. and Heintzenberg, J. 2003. Climate forcing by aerosols—a hazy picture. *Science* **300**: 1103-1104.

Baliunas, S. and Jastrow, R. 1990. Evidence for long-term brightness changes of solar-type stars. *Nature* **348**: 520-522.

Bard, E. and Frank, M. 2006. Climate change and solar variability: What's new under the sun? *Earth and Planetary Science Letters* **248**: 1-14.

Bard, E., Raisbeck, G., Yiou, F. and Jouzel, J. 1997. Solar modulation of cosmogenic nuclide production over the last millennium: comparison between ^{14}C and ^{10}Be records. *Earth and Planetary Science Letters* **150**: 453-462.

Bard, E., Raisbeck, G., Yiou, F. and Jouzel, J. 2000. Solar irradiance during the last 1200 years based on cosmogenic nuclides. *Tellus* **52B**: 985-992.

Beer, J., Vonmoos, M. and Muscheler, R. 2006. Solar variability over the past several millennia. *Space Science Reviews* **125**: 67-79.

Bond, G., Kromer, B., Beer, J., Muscheler, R., Evans, M.N., Showers, W., Hoffmann, S., Lotti-Bond, R., Hajdas, I. and Bonani, G. 2001. Persistent solar influence on North Atlantic climate during the Holocene. *Science* **294**: 2130-2136.

Brohan, P., Kennedy, J.J., Harris, I., Tett, S.F.B. and Jones, P.D. 2006. Uncertainty estimates in regional and global observed temperature changes: A new data set from 1850. *Journal of Geophysical Research* **111**: 10.1029/2005JD006548.

Chambers, F.M., Ogle, M.I. and Blackford, J.J. 1999. Palaeoenvironmental evidence for solar forcing of Holocene climate: linkages to solar science. *Progress in Physical Geography* **23**: 181-204.

Charlson, R.J., Valero, F.P.J. and Seinfeld, J.H. 2005. In search of balance. *Science* **308**: 806-807.

Cliver, E.W., Boriakoff, V. and Feynman, J. 1998. Solar variability and climate change: geomagnetic and aa index and global surface temperature. *Geophysical Research Letters* **25**: 1035-1038.

Coughlin, K. and Tung, K.K. 2004. Eleven-year solar cycle signal throughout the lower atmosphere. *Journal of Geophysical Research* **109**: 10.1029/2004JD004873.

Damon, P.E. and Laut, P. 2004. Pattern of strange errors plagues solar activity and terrestrial climatic data. *EOS: Transactions, American Geophysical Union* **85**: 370, 374.

Douglass, D.H. and Clader, B.D. 2002. Climate sensitivity of the Earth to solar irradiance. *Geophysical Research Letters* **29**: 10.1029/2002GL015345.

Eddy, J.A. 1976. The Maunder Minimum. *Science* **192**: 1189-1202.

Foukal, P. 2002. A comparison of variable solar total and ultraviolet irradiance outputs in the 20th century. *Geophysical Research Letters* **29**: 10.1029/2002GL015474.

Foukal, P. 2003. Can slow variations in solar luminosity provide missing link between the sun and climate? *EOS: Transactions, American Geophysical Union* **84**: 205, 208.

Foukal, P., North, G. and Wigley, T. 2004. A stellar view on solar variations and climate. *Science* **306**: 68-69.

Friis-Christensen, E. and Lassen, K. 1991. Length of the solar cycle: An indicator of solar activity closely associated with climate. *Science* **254**: 698-700.

Frohlich C. 2006. Solar irradiance variability since 1978: revision of the PMOD composite during solar cycle 21. *Space Science Review* **125**: 53–65. doi:10.1007/s11214-006-9046-5.

Frohlich, C. and Lean, J. 1998. The sun's total irradiance: Cycles, trends and related climate change uncertainties since 1976. *Geophysical Research Letters* **25**: 4377-4380.

Frohlich, C. and Lean, J. 2002. Solar irradiance variability and climate. *Astronomische Nachrichten* **323**: 203-212.

Hansen, J. and Lebedeff, S. 1987. Global trends of measured surface air temperature. *Journal of Geophysical Research* **92**: 13,345-13,372.

Hansen, J., Ruedy, R., Glascoe, J. and Sato, M. 1999. GISS analysis of surface temperature change. *Journal of Geophysical Research* **104**: 30,997-31,022.

Hansen, J., Sato, M., Nazarenko, L., Ruedy, R., Lacis, A., Koch, D., Tegen, I., Hall, T., Shindell, D., Santer, B., Stone, P., Novakov, T., Thomason, L., Wang, R., Wang, Y., Jacob, D., Hollandsworth, S., Bishop, L., Logan, J., Thompson, A., Stolarski, R., Lean, J., Willson, R., Levitus, S., Antonov, J., Rayner, N., Parker, D. and Christy, J. 2002. Climate forcings in Goddard Institute for Space Studies S12000 simulations. *Journal of Geophysical Research* **107**: 10.1029/2001JD001143.

Hoyt, D.V. and Schatten, K.H. 1993. A discussion of plausible solar irradiance variations, 1700-1992. *Journal of Geophysical Research* **98**: 18,895-18,906.

Idso, S.B. 1991a. The aerial fertilization effect of CO_2 and its implications for global carbon cycling and maximum greenhouse warming. *Bulletin of the American Meteorological Society* **72**: 962-965.

Idso, S.B. 1991b. Reply to comments of L.D. Danny Harvey, Bert Bolin, and P. Lehmann. *Bulletin of the American Meteorological Society* **72**: 1910-1914.

Idso, S.B. 1998. CO_2-induced global warming: a skeptic's view of potential climate change. *Climate Research* **10**: 69-82.

Intergovernmental Panel on Climate Change (IPCC). 2001. *Climate Change 2001: The Scientific Basis*. Houghton, J.T., Ding, Y., Griggs, D.J., Noguer, M., van der Linden, P.J., Xiaosu, D., Maskell, K. and Johnson, C.A. (Eds.) Cambridge University Press, Cambridge, UK.

Jones, P.D., Parker, D.E., Osborn, T.J. and Briffa, K.R. 2001. Global and hemispheric temperature anomalies—land and marine instrumental records. In: *Trends: A Compendium of Data on Global Change*, Carbon Dioxide Information Analysis Center, Oak Ridge National Laboratory, U.S. Department of Energy, Oak Ridge, TN.

Kalnay, E. and Cai, M. 2003. Impact of urbanization and land-use change on climate. *Nature* **423**: 528-531.

Kalnay, E., Kanamitsu, M., Kistler, R., Collins, W., Deaven, D., Gandin, L., Iredell, M., Saha, S., White, G., Woollen, J., Zhu, Y., Leetmaa, A., Reynolds, R., Chelliah, M., Ebisuzaki, W., Higgins, W., Janowiak, J., Mo, K.C., Ropelewski, C., Wang, J., Jenne, R. and Joseph, D. 1996. The NCEP/NCAR reanalysis 40-year project. *Bulletin of the American Meteorological Society* **77**: 437-471.

Karlén, W. 1998. Climate variations and the enhanced greenhouse effect. *Ambio* **27**: 270-274.

Krivova, N.A., Balmaceda, L. and Solanki, S.K. 2007. Reconstruction of solar total irradiance since 1700 from the surface magnetic flux. *Astronomy & Astrophysics* **467**: 335-346.

Lal, D. and Peters, B. 1967. Cosmic ray produced radio-activity on the Earth. In: *Handbuch der Physik*, XLVI/2. Springer, Berlin, Germany, pp. 551-612.

Lassen, K. and Friis-Christensen, E. 2000. Reply to "Solar cycle lengths and climate: A reference revisited" by P. Laut and J. Gundermann. *Journal of Geophysical Research* **105**: 27,493-27,495.

Lastovicka, J. 2006. Influence of the sun's radiation and particles on the earth's atmosphere and climate—Part 2. *Advances in Space Research* **37**: 1563.

Lean, J. 2000. Evolution of the sun's spectral irradiance since the Maunder Minimum. *Geophysical Research Letters* **27**: 2425-2428.

Lean, J., Beer, J. and Bradley, R. 1995. Reconstruction of solar irradiance since 1610: implications for climate change. *Geophysical Research Letters* **22**: 3195-1398.

Lean, J. and Rind, D. 1998. Climate forcing by changing solar radiation. *Journal of Climate* **11**: 3069-3094.

Lean, J., Skumanich, A. and White, O. 1992. Estimating the sun's radiative output during the maunder minimum. *Geophysical Research Letters* **19**: 1591-1594.

Lockwood, M., Stamper, R. and Wild, M.N. 1999. A doubling of the Sun's coronal magnetic field during the past 100 years. *Nature* **399**: 437-439.

Long, C. N., Dutton, E.G., Augustine, J.A., Wiscombe, W., Wild, M., McFarlane, M.A., and Flynn, C.J. 2009. Significant decadal brightening of downwelling shortwave in the continental United States. *Journal of Geophysical Research* **114**: D00D06, doi:10.1029/2008JD011263.

Mann, M.E., Bradley, R.S. and Hughes, M.K. 1998. Global-scale temperature patterns and climate forcing over the past six centuries. *Nature* **392**: 779-787.

Mann, M.E., Bradley, R.S. and Hughes, M.K. 1999. Northern Hemisphere temperatures during the past millennium: Inferences, uncertainties, and limitations. *Geophysical Research Letters* **26**: 759-762.

Moberg, A., Sonechkin, D.M., Holmgren, K., Datsenko, N.M. and Karlén, W. 2005. Highly variable Northern Hemisphere temperatures reconstructed from low- and high-resolution proxy data. *Nature* **433**: 613-617.

Nesme-Ribes, D., Ferreira, E.N., Sadourny, R., Le Treut, H. and Li, Z.X. 1993. Solar dynamics and its impact on solar irradiance and the terrestrial climate. *Journal of Geophysical Research* **98**: 18,923-18.935.

Ohmura, A. 2009. Observed decadal variations in surface solar radiation and their causes, *Journal of Geophysical Research* **114**: D00D05, doi:10.1029/2008JD011290.

Oppo, D.W., McManus, J.F. and Cullen, J.L. 1998. Abrupt climate events 500,000 to 340,000 years ago: Evidence from subpolar North Atlantic sediments. *Science* **279**: 1335-1338.

Pallé, E., Goode, P.R., Montañés-Rodríguez, P., Koonin, S.E. 2004. Changes in earth's reflectance over the past two decades. *Science* **304**: 1299-1301.

Pallé, E., Goode, P.R., and Montañés-Rodríguez, P. 2009. Interannual variations in Earth's reflectance 1999–2007, *Journal of Geophysical Research* **114**: D00D03, doi:10.1029/2008JD010734.

Parker, D.E., Gordon, M., Cullum, D.P.N., Sexton, D.M.H., Folland, C.K. and Rayner, N. 1997. A new global gridded radiosonde temperature data base and recent temperature trends. *Geophysical Research Letters* **24**: 1499-1502.

Parker, E.N. 1999. Sunny side of global warming. *Nature* **399**: 416-417.

Petit, J.R., Jouzel, J., Raynaud, D., Barkov, N.I., Barnola, J.-M., Basile, I., Bender, M., Chappellaz, J., Davis, M., Delaygue, G., Delmotte, M., Kotlyakov, V.M., Legrand, M., Lipenkov, V.Y., Lorius, C., Pepin, L., Ritz, C., Saltzman, E. and Stievenard, M. 1999. Climate and atmospheric history of the past 420,000 years from the Vostok ice core, Antarctica. *Nature* **399**: 429-436.

Pielke Sr., R.A., Marland, G., Betts, R.A., Chase, T.N., Eastman, J.L., Niles, J.O., Niyogi, D.S. and Running, S.W. 2002. The influence of land-use change and landscape dynamics on the climate system: Relevance to climate-change policy beyond the radiative effects of greenhouse gases. *Philosophical Transactions of the Royal Society of London A* **360**: 1705-1719.

Pinker, R.T., Zhang, B. and Dutton, E.G. 2005. Do satellites detect trends in surface solar radiation? *Science* **308**: 850-854.

Polyakov, I.V., Bekryaev, R.V., Alekseev, G.V., Bhatt, U.S., Colony, R.L., Johnson, M.A., Maskshtas, A.P. and Walsh, D. 2003. Variability and trends of air temperature and pressure in the maritime Arctic, 1875-2000. *Journal of Climate* **16**: 2067-2077.

Raisbeck, G.M., Yiou, F., Jouzel, J. and Petit, J.-R. 1990. ^{10}Be and ^{2}H in polar ice cores as a probe of the solar variability's influence on climate. *Philosophical Transactions of the Royal Society of London* **A300**: 463-470.

Raymo, M.E., Ganley, K., Carter, S., Oppo, D.W. and McManus, J. 1998. Millennial-scale climate instability during the early Pleistocene epoch. *Nature* **392**: 699-702.

Reid, G.C. 1991. Solar total irradiance variations and the global sea surface temperature record. *Journal of Geophysical Research* **96**: 2835-2844.

Reid, G.C. 1997. Solar forcing and global climate change since the mid-17th century. *Climatic Change* **37**: 391-405.

Rigozo, N.R., Echer, E., Vieira, L.E.A. and Nordemann, D.J.R. 2001. Reconstruction of Wolf sunspot numbers on the basis of spectral characteristics and estimates of

associated radio flux and solar wind parameters for the last millennium. *Solar Physics* **203**: 179-191.

Rozelot, J.P. 2001. Possible links between the solar radius variations and the earth's climate evolution over the past four centuries. *Journal of Atmospheric and Solar-Terrestrial Physics* **63**: 375-386.

Scafetta, N. 2008. Comment on "Heat capacity, time constant, and sensitivity of Earth's climate system" by Schwartz. *Journal of Geophysical Research* **113**: D15104 doi:10.1029/2007JD009586.

Scafetta, N. and West, B.J. 2003. Solar flare intermittency and the Earth's temperature anomalies. *Physical Review Letters* **90**: 248701.

Scafetta, N. and West, B.J. 2005. Estimated solar contribution to the global surface warming using the ACRIM TSI satellite composite. *Geophysical Research Letters* **32**: 10.1029/2005GL023849.

Scafetta, N. and West, B.J. 2006a. Phenomenological solar contribution to the 1900-2000 global surface warming. *Geophysical Research Letters* **33**: 10.1029/2005GL025539.

Scafetta, N. and West, B.J. 2006b. Phenomenological solar signature in 400 years of reconstructed Northern Hemisphere temperature record. *Geophysical Research Letters* **33**: 10.1029/2006GL027142.

Scafetta, N. and West, B.J. 2007. Phenomenological reconstructions of the solar signature in the Northern Hemisphere surface temperature records since 1600, *Journal of Geophysical Research* **112**: D24S03, doi:10.1029/2007JD008437.

Scafetta, N. and West, B.J. 2008. Is climate sensitive to solar variability? *Physics Today* **3**: 50-51.

Scafetta, N. and Willson, R.C. 2009. ACRIM-gap and TSI trend issue resolved using a surface magnetic flux TSI proxy model. *Geophysical Research Letters* **36**: L05701, doi:10.1029/2008GL036307.

Shaviv, N.J. 2005. On climate response to changes in the cosmic ray flux and radiative budget. *Journal of Geophysical Research* **110**: 10.1029/2004JA010866.

Shaviv, N.J. 2008. Using the oceans as a calorimeter to quantify the solar radiative forcing, *Journal of Geophysical Research* **113**: A11101, doi:10.1029/2007JA012989.

Solanki, S.K. and Fligge, M. 1998. Solar irradiance since 1874 revisited. *Geophysical Research Letters* **25**: 341-344.

Solanki, S.K., Schussler, M. and Fligge, M. 2000. Evolution of the sun's large-scale magnetic field since the Maunder minimum. *Nature* **408**: 445-447.

Solanki, S.K., Schussler, M. and Fligge, M. 2002. Secular variation of the sun's magnetic flux. *Astronomy & Astrophysics* **383**: 706-712.

Soon, W. W.-H. 2005. Variable solar irradiance as a plausible agent for multidecadal variations in the Arctic-wide surface air temperature record of the past 130 years. *Geophysical Research Letters* **32**:10.1029/2005GL023429.

Stevens, M.J. and North, G.R. 1996. Detection of the climate response to the solar cycle. *Journal of the Atmospheric Sciences* **53**: 2594-2608.

Svensmark, H. 1998. Influence of cosmic rays on Earth's climate. *Physical Review Letters* **22**: 5027-5030.

Svensmark, H. and Friis-Christensen, E. 1997. Variation of cosmic ray flux and global cloud coverage—A missing link in solar-climate relationships. *Journal of Atmospheric and Solar-Terrestrial Physics* **59**: 1225-1232.

Wang, Y.-M., Lean, J.L. and Sheeley Jr., N.R. 2005. Modelling the sun's magnetic field and irradiance since 1713. *Astron. Journal* **625**:522-538.

White, W.B., Lean, J., Cayan, D.R. and Dettinger, M.D. 1997. Response of global upper ocean temperature to changing solar irradiance. *Journal of Geophysical Research* **102**: 3255-3266.

Wild, M., Gilgen, H., Roesch, A., Ohmura, A., Long, C.N., Dutton, E.G., Forgan, B., Kallis, A., Russak, V. and Tsvetkov, A. 2005. From dimming to brightening: Decadal changes in solar radiation at earth's surface. *Science* **308**: 847-850.

Willson, R.C. and Mordvinov, A.V. 2003. Secular total solar irradiance trend during solar cycles 21-23. *Geophysical Research Letters* **30**: 10.1029/2002GL 016038.

Zhang, Q., Soon, W.H., Baliunas, S.L., Lockwood, G.W., Skiff, B.A. and Radick, R.R. 1994. A method of determining possible brightness variations of the sun in past centuries from observations of solar-type stars. *Astrophysics Journal* **427**: L111-L114.

5.3. Temperature

5.3.1. Global

The IPCC's claim that anthropogenic greenhouse gas emissions have been responsible for the warming detected in the twentieth century is based on what Loehle (2004) calls "the standard assumption in climate research, including the IPCC reports," that "over a century time interval there is not likely to be any recognizable trend to global temperatures (Risbey *et al.*, 2000), and thus the null model for climate signal detection is a flat temperature trend with some autocorrelated noise," so that "any warming trends in excess of that expected from normal climatic variability are then assumed to be due to anthropogenic effects." If, however, there are significant underlying climate trends or cycles—or both—either known or unknown, that assumption is clearly invalid.

Loehle used a pair of 3,000-year proxy climate records with minimal dating errors to characterize the pattern of climate change over the past three millennia simply as a function of time, with no attempt to make the models functions of solar activity or any other physical variable. The first of the two temperature series is the sea surface temperature (SST) record of the Sargasso Sea, derived by Keigwin (1996) from a study of the oxygen isotope ratios of foraminifera and other organisms contained in a sediment core retrieved from a deep-ocean drilling site on the Bermuda Rise. This record provides SST data for about every 67th year from 1125 BC to 1975 AD. The second temperature series is the ground surface temperature record derived by Holmgren *et al.* (1999, 2001) from studies of color variations of stalagmites found in a cave in South Africa, which variations are caused by changes in the concentrations of humic materials entering the region's ground water that have been reliably correlated with regional near-surface air temperature.

Why does Loehle use these two specific records? He says "most other long-term records have large dating errors, are based on tree rings, which are not reliable for this purpose (Broecker, 2001), or are too short for estimating long-term cyclic components of climate." Also, in a repudiation of the approach employed by Mann *et al.* (1998, 1999) and Mann and Jones (2003), he reports that "synthetic series consisting of hemispheric or global mean temperatures are not suitable for such an analysis because of the inconsistent timescales in the various data sets," noting further, as a result of his own testing, that "when dating errors are present in a series, and several series are combined, the result is a smearing of the signal." But can only two temperature series reveal the pattern of global temperature change? According to Loehle, "a comparison of the Sargasso and South Africa series shows some remarkable similarities of pattern, especially considering the distance separating the two locations," and he says that this fact "suggests that the climate signal reflects some global pattern rather than being a regional signal only." He also notes that a comparison of the mean record with the South Africa and Sargasso series from which it was derived "shows excellent agreement," and that "the patterns match closely," concluding that "this would not be the case if the two series were independent or random."

Loehle fit seven different time-series models to the two temperature series and to the average of the two series, using no data from the twentieth century. In all seven cases, he reports that good to excellent fits were obtained. As an example, the three-cycle model he fit to the averaged temperature series had a simple correlation of 0.58 and an 83 percent correspondence of peaks when evaluated by a moving window count.

Comparing the forward projections of the seven models through the twentieth century leads directly to the most important conclusions of Loehle's paper. He notes, first of all, that six of the models "show a warming trend over the 20th century similar in timing and magnitude to the Northern Hemisphere instrumental series," and that "one of the models passes right through the 20th century data." These results suggest, in his words, "that 20th century warming trends are plausibly a continuation of past climate patterns" and, therefore, that "anywhere from a major portion to all of the warming of the 20th century could plausibly result from natural causes."

As dramatic and important as these observations are, they are not the entire story of Loehle's insightful paper. His analyses also reveal a long-term linear cooling trend of 0.25°C per thousand years since the peak of the interglacial warm period that occurred some 7,000 years ago, which result is essentially identical to the mean value of this trend that was derived from seven prior assessments of its magnitude and five prior climate reconstructions. In addition, Loehle's analyses reveal the existence of the Medieval Warm Period of 800-1200 AD, which is shown to have been significantly warmer than the portion of the Current Warm Period we have so far experienced, as well as the existence of the Little Ice Age of 1500-1850 AD, which is shown to have been the coldest period of the entire 3,000-year record.

As corroborating evidence for the global nature of these major warm and cold intervals, Loehle cites 16 peer-reviewed scientific journal articles that document the existence of the Medieval Warm Period in all parts of the world, as well as 18 other articles that document the worldwide occurrence of the Little Ice Age. And in one of the more intriguing aspects of his study—of which Loehle makes no mention, however—both the Sargasso Sea and South African temperature records reveal the existence of a major temperature spike that began sometime in the early 1400s. This abrupt warming pushed temperatures considerably above the peak warmth of the twentieth century before falling back to pre-spike levels in the mid 1500s, providing support for the similar finding of higher-than-current temperatures in that time interval by McIntyre and McKitrick (2003) in their reanalysis of the data employed by Mann et al. to create their controversial "hockey stick" temperature history, which gives no indication of the occurrence of this high-temperature regime.

In another accomplishment of note, the models developed by Loehle reveal the existence of three climate cycles previously identified by others. In his culminating seventh model, for example, there is a 2,388-year cycle that he describes as comparing "quite favorably to a cycle variously estimated as 2200, 2300, and 2500 years (Denton and Karlén, 1973; Karlén and Kuylenstierna, 1996; Magny, 1993; Mayewski et al., 1997)." There is also a 490-year cycle that likely "corresponds to a 500-year cycle found previously (e.g. Li et al., 1997; Magny, 1993; Mayewski et al., 1997)" and a 228-year cycle that "approximates the 210-year cycle found by Damon and Jirikowic (1992)."

The compatibility of these findings with those of several studies that have identified similar solar forcing signals caused Loehle to conclude that "solar forcing (and/or other natural cycles) is plausibly responsible for some portion of 20th century warming" or, as he indicates in his abstract, maybe even all of it.

In spite of potential smearing and dating errors, other globally represented datasets have provided additional evidence of a solar influence on temperature. The 16 authors of Mayewski et al. (2004) examined some 50 globally distributed paleoclimate records in search of evidence for what they call rapid climate change (RCC) over the Holocene. This terminology is not to be confused with the rapid climate changes typical of glacial periods, but is used in the place of what the authors call the "more geographically or temporally restrictive terminology such as 'Little Ice Age' and 'Medieval Warm Period'." RCC events, as they also call them, are multi-century periods of time characterized by extremes of thermal and/or hydrological properties, rather than the much shorter periods of time during which the changes that led to these situations took place.

Mayewski *et al.* identify six RCCs during the Holocene: 9,000-8,000, 6,000-5,000, 4,200-3,800, 3,500-2,500, 1,200-1,000, and 600-150 cal yr BP, the last two of which intervals are, in fact, the "globally distributed" Medieval Warm Period and Little Ice Age, respectively. In speaking further of these two periods, they say that "the short-lived 1200-1000 cal yr BP RCC event coincided with the drought-related collapse of Maya civilization and was accompanied by a loss of several million lives (Hodell *et al.*, 2001; Gill, 2000), while the collapse of Greenland's Norse colonies at ~600 cal yr BP (Buckland *et al.*, 1995) coincides with a period of polar cooling."

With respect to the causes of these and other Holocene RCCs, the international team of scientists says that "of all the potential climate forcing mechanisms, solar variability superimposed on long-term changes in insolation (Bond *et al.*, 2001; Denton and Karlén, 1973; Mayewski *et al.*, 1997; O'Brien *et al.*, 1995) seems to be the most likely important forcing mechanism." In addition, they note that "negligible forcing roles are played by CH_4 and CO_2," and that "changes in the concentrations of CO_2 and CH_4 appear to have been more the result than the cause of the RCCs."

In another study with global implications, eight researchers hailing from China, Finland, Russia, and Switzerland published a paper wherein they describe evidence that makes the case for a causative link, or set of links, between solar forcing and climate change. Working with tree-ring width data obtained from two types of juniper found in Central Asia— *Juniperus turkestanica* (related to variations in summer temperature in the Tien Shan Mountains) and *Sabina przewalskii* (related to variations in precipitation on the Qinghai-Tibetan Plateau)— Raspopov *et al.* (2008) employed band-pass filtering in the 180- to 230-year period range, wavelet transformation (Morlet basis) for the range of periods between 100 and 300 years, as well as spectral analysis, in order to compare the variability in the two tree-ring records with independent $\Delta^{14}C$ variations representative of the approximate 210-year de Vries solar cycle over the past millennium. These analyses indicated that the approximate 200-year cyclical variations present in the palaeoclimatic reconstructions were well correlated ($R^2 = 0.58$-0.94) with similar variations in the $\Delta^{14}C$ data, which obviously suggests the existence of a solar-climate connection. In addition, they say "the de Vries cycle has been found to occur not only during the last

millennia but also in earlier epochs, up to hundreds of millions [of] years ago."

After reviewing additional sets of published palaeoclimatic data from various parts of the world, the eight researchers satisfied themselves that the same periodicity is evident in Europe, North and South America, Asia, Tasmania, Antarctica, and the Arctic, as well as "sediments in the seas and oceans," citing 20 independent research papers in support of this statement. This fact led them to conclude there is "a pronounced influence of solar activity on global climatic processes" related to "temperature, precipitation and atmospheric and oceanic circulation."

Complicating the matter, however, Raspopov *et al.* report there can sometimes be "an appreciable delay in the climate response to the solar signal," which can be as long as 150 years, and they note that regional climate responses to the de Vries cycle "can markedly differ in phase," even at distances of only hundreds of kilometers, due to "the nonlinear character of the atmosphere-ocean system response to solar forcing." Nevertheless, the many results they culled from the scientific literature, as well as their own findings, all testify to the validity of their primary conclusion, that throughout the past millennium, and stretching back in time as much as 250 million years, the de Vries cycle has been "one of the most intense solar activity periodicities that affected climatic processes."

As for the more recent historical significance of the de Vries cycle, Raspopov *et al.* write that "the temporal synchrony between the Maunder, Sporer, and Wolf minima and the expansion of Alpine glaciers (Haeberlie and Holzhauser, 2003) further points to a climate response to the deep solar minima." In this regard, we again add that Earth's recent recovery from those deep solar minima could well have played a major role in the planet's emergence from the Little Ice Age, and, therefore, could well have accounted for much of twentieth century global warming, as suggested fully 20 years ago by Idso (1988).

Clearly, there is much to recommend the overriding concept that is suggested by the data of these several papers, i.e., that the sun rules the earth when it comes to orchestrating major changes in the planet's climate. It is becoming ever more clear that the millennial-scale oscillation of climate that has reverberated throughout the Holocene is indeed the result of similar-scale oscillations in some aspect of solar activity. Consequently, Mayewski *et al.* suggest that "significantly more research into the potential role of solar variability is warranted, involving new assessments of potential transmission mechanisms to induce climate change and potential enhancement of natural feedbacks that may amplify the relatively weak forcing related to fluctuations in solar output."

Additional information on this topic, including reviews of newer publications as they become available, can be found at http://www.co2science.org/subject/s/solartempglobal.php.

References

Bond, G., Kromer, B., Beer, J., Muscheler, R., Evans, M.N., Showers, W., Hoffmann, S., Lotti-Bond, R., Hajdas, I. and Bonani, G. 2001. Persistent solar influence on North Atlantic climate during the Holocene. *Science* **294**: 2130-2136.

Broecker, W.S. 2001. Was the Medieval Warm Period global? *Science* **291**: 1497-1499.

Buckland, P.C., Amorosi, T., Barlow, L.K., Dugmore, A.J., Mayewski, P.A., McGovern, T.H., Ogilvie, A.E.J., Sadler, J.P. and Skidmore, P. 1995. Bioarchaeological evidence and climatological evidence for the fate of Norse farmers in medieval Greenland. *Antiquity* **70**: 88-96.

Damon, P.E. and Jirikowic, J.L. 1992. Solar forcing of global climate change? In: Taylor, R.E., Long, A. and Kra, R.S. (Eds.) *Radiocarbon After Four Decades.* Springer-Verlag, Berlin, Germany, pp. 117-129.

Denton, G.H. and Karlén, W. 1973. Holocene climate variations—their pattern and possible cause. *Quaternary Research* **3**: 155-205.

Gill, R.B. 2000. *The Great Maya Droughts: Water, Life, and Death*. University of New Mexico Press, Albuquerque, New Mexico, USA.

Haeberli, W. and Holzhauser, H. 2003. Alpine glacier mass changes during the past two millennia. *PAGES News* **1** (1): 13-15.

Hodell, D.A., Brenner, M., Curtis, J.H. and Guilderson, T. 2001. Solar forcing of drought frequency in the Maya lowlands. *Science* **292**: 1367-1369.

Holmgren, K., Karlén, W., Lauritzen, S.E., Lee-Thorp, J.A., Partridge, T.C., Piketh, S., Repinski, P., Stevenson, C., Svanered, O. and Tyson, P.D. 1999. A 3000-year high-resolution stalagmite-based record of paleoclimate for northeastern South Africa. *The Holocene* **9**: 295-309.

Holmgren, K., Tyson, P.D., Moberg, A. and Svanered, O. 2001. A preliminary 3000-year regional temperature reconstruction for South Africa. *South African Journal of Science* **99**: 49-51.

Idso, S.B. 1988. Greenhouse warming or Little Ice Age demise: A critical problem for climatology. *Theoretical and Applied Climatology* **39**: 54-56.

Karlén, W. and Kuylenstierna, J. 1996. On solar forcing of Holocene climate: evidence from Scandinavia. *The Holocene* **6**: 359-365.

Keigwin, L.D. 1996. The Little Ice Age and Medieval Warm Period in the Sargasso Sea. *Science* **274**: 1504-1508.

Li, H., Ku, T.-L., Wenji, C. and Tungsheng, L. 1997. Isotope studies of Shihua Cave; Part 3, Reconstruction of paleoclimate and paleoenvironment of Beijing during the last 3000 years from delta and ^{13}C records in stalagmite. *Dizhen Dizhi* **19**: 77-86.

Loehle, C. 2004. Climate change: detection and attribution of trends from long-term geologic data. *Ecological Modelling* **171**: 433-450.

Magny, M. 1993. Solar influences on Holocene climatic changes illustrated by correlations between past lake-level fluctuations and the atmospheric ^{14}C record. *Quaternary Research* **40**: 1-9.

Mann, M.E., Bradley, R.S. and Hughes, M.K. 1998. Global-scale temperature patterns and climate forcing over the past six centuries. *Nature* **392**: 779-787.

Mann, M.E., Bradley, R.S. and Hughes, M.K. 1999. Northern Hemisphere temperatures during the past millennium: Inferences, uncertainties, and limitations. *Geophysical Research Letters* **26**: 759-762.

Mann, M.E. and Jones, P.D. 2003. Global surface temperatures over the past two millennia. *Geophysical Research Letters* **30**: 10.1029/2003GL017814.

Mayewski, P.A., Meeker, L.D., Twickler, M.S., Whitlow, S., Yang, Q., Lyons, W.B. and Prentice, M. 1997. Major features and forcing of high-latitude northern hemisphere atmospheric circulation using a 110,000-year-long glaciochemical series. *Journal of Geophysical Research* **102**: 26,345-26,366.

Mayewski, P.A., Rohling, E.E., Stager, J.C., Karlén, W., Maasch, K.A., Meeker, L.D., Meyerson, E.A., Gasse, F., van Kreveld, S., Holmgren, K., Lee-Thorp, J., Rosqvist, G., Rack, F., Staubwasser, M., Schneider, R.R. and Steig, E.J. 2004. Holocene climate variability. *Quaternary Research* **62**: 243-255.

McIntyre, S. and McKitrick, R. 2003. Corrections to the Mann *et al.* (1998) proxy data base and Northern Hemispheric average temperature series. *Energy and Environment* **14**: 751-771.

O'Brien, S.R., Mayewski, P.A., Meeker, L.D., Meese, D.A., Twickler, M.S. and Whitlow, S.E. 1995. Complexity of Holocene climate as reconstructed from a Greenland ice core. *Science* **270**: 1962-1964.

Raspopov, O.M., Dergachev, V.A., Esper, J., Kozyreva, O.V., Frank, D., Ogurtsov, M., Kolstrom, T. and Shao, X. 2008. The influence of the de Vries (~200-year) solar cycle on climate variations: Results from the Central Asian Mountains and their global link. *Palaeogeography, Palaeoclimatology, Palaeoecology* **259**: 6-16.

Risbey, J.S., Kandlikar, M. and Karoly, D.J. 2000. A protocol to articulate and quantify uncertainties in climate change detection and attribution. *Climate Research* **16**: 61-78.

5.3.2. Northern Hemisphere

Evidence of the influence of the sun on Northern Hemisphere temperatures can be found in the seminar research of Bond *et al.* (2001), who examined ice-rafted debris found in three North Atlantic deep-sea sediment cores and cosmogenic nuclides (^{10}Be and ^{14}C) sequestered in the Greenland ice cap (^{10}Be) and Northern Hemispheric tree rings (^{14}C). This study is described in depth in Section 5.1.

Bond *et al.* found that "over the last 12,000 years virtually every centennial time-scale increase in drift ice documented in our North Atlantic records was tied to a solar minimum," and "a solar influence on climate of the magnitude and consistency implied by our evidence could not have been confined to the North Atlantic," suggesting that the cyclical climatic effects of the variable solar inferno are experienced throughout the world. Bond *et al.* also observed that the oscillations in drift-ice they studied "persist across the glacial termination and well into the last glaciation, suggesting that the cycle is a pervasive feature of the climate system."

Björck *et al.* (2001) assembled a wide range of lacustrine, tree-ring, ice-core, and marine records that reveal a Northern Hemispheric, and possibly global, cooling event of less than 200 years' duration with a 50-year cooling-peak centered at approximately 10,300 years BP. According to the authors, the onset of the cooling event broadly coincided with rising ^{10}Be fluxes, which are indicative of either decreased solar or geomagnetic forcing; and since the authors note that "no large magnetic field variation that could have caused this event has been found," they postulate that "the ^{10}Be maximum was caused by distinctly reduced solar forcing." They also note that

the onset of the Younger Dryas is coeval with a rise in ^{10}Be flux, as is the Preboreal climatic oscillation.

Pang and Yau (2002) assembled and analyzed a vast amount of data pertaining to phenomena that have been reliably linked to variations in solar activity, including frequencies of sunspot and aurora sightings, the abundance of carbon-14 in the rings of long-lived trees, and the amount of beryllium-10 in the annual ice layers of polar ice cores. In the case of sunspot sightings, the authors used a catalogue of 235 Chinese, Korean, and Japanese records compiled by Yau (1988), a catalogue of 270 Chinese records compiled by Zhuang and Wang (1988), and a time chart of 139 records developed by Clark and Stephenson (1979), as well as a number of later catalogues that made the overall record more complete.

Over the past 1,800 years, the authors identified "some nine cycles of solar brightness change," which include the well-known Oort, Wolf, Sporer, Maunder, and Dalton Minima. With respect to the Maunder Minimum—which occurred between 1645 and 1715 and is widely acknowledged to have been responsible for some of the coldest weather of the Little Ice Age—they report that the temperatures of that period "were about one-half of a degree Celsius lower than the mean for the 1970s, consistent with the decrease in the decadal average solar irradiance." Then, from 1795 to 1825 came the Dalton Minimum, along with another dip in Northern Hemispheric temperatures. Since that time, however, the authors say "the sun has gradually brightened" and "we are now in the Modern Maximum," which is likely responsible for the warmth of the Current Warm Period.

The authors say that although the long-term variations in solar brightness they identified "account for less than 1% of the total irradiance, there is clear evidence that they affect the Earth's climate." Pang and Yau's dual plot of total solar irradiance and Northern Hemispheric temperature from 1620 to the present (their Fig. 1c) indicates that the former parameter (when appropriately scaled, but without reference to any specific climate-change mechanism) can account for essentially all of the net change experienced by the latter parameter up to about 1980. After that time, however, the IPCC surface air temperature record rises dramatically, although radiosonde and satellite temperature histories largely match what would be predicted from the solar irradiance record. These facts could be interpreted as new evidence of the corruptness of the IPCC temperature history.

In a separate study, Rohling *et al.* (2003) "narrow down" temporal constraints on the millennial-scale variability of climate evident in ice-core $\delta^{18}O$ records by "determining statistically significant anomalies in the major ion series of the GISP2 ice core," after which they conduct "a process-oriented synthesis of proxy records from the Northern Hemisphere." With respect to the temporal relationships among various millennial-scale oscillations in Northern Hemispheric proxy climate records, the authors conclude that a "compelling case" can be made for their being virtually in-phase, based on (1) "the high degree of similarity in event sequences and structures over a very wide spatial domain," and (2) "the fact that our process-oriented synthesis highlights a consistent common theme of relative dominance shifts between winter-type and summer-type conditions, ranging all the way across the Northern Hemisphere from polar into monsoonal latitudes." These findings, they additionally note, "corroborate the in-phase relationship between climate variabilities in the high northern latitudes and the tropics suggested in Blunier *et al.* (1998) and Brook *et al.* (1999)."

Rohling *et al.* further report that although individual cycles of the persistent climatic oscillation "appear to have different intensities and durations, a mean periodicity appears around ~1500 years (Mayewski *et al.*, 1997; Van Kreveld *et al.*, 2000; Alley *et al.*, 2001)." They further report that "this cycle seems independent from the global glaciation state (Mayewski *et al.*, 1997; Bond *et al.*, 1999)," and that "^{10}Be and delta ^{14}C records may imply a link with solar variability (Mayewski *et al.*, 1997; Bond *et al.*, 2001)."

Lastly, we come to the study of Usoskin *et al.* (2003), who note that "sunspots lie at the heart of solar active regions and trace the emergence of large-scale magnetic flux, which is responsible for the various phenomena of solar activity" that may influence earth's climate. They say "the sunspot number (SN) series represents the longest running direct record of solar activity, with reliable observations starting in 1610, soon after the invention of the telescope." To compare SN data with the millennial-scale temperature reconstruction of Mann *et al.* (1999), the directly measured SN record must be extended back in time at least another 600 years, which Usoskin *et al.* did using records of ^{10}Be cosmonuclide concentration derived from polar ice cores dating back to AD 850. In accomplishing this task, they employed detailed physical models that they say were "developed for each individual link in

the chain connecting the SN with the cosmogenic isotopes," and they combined these models in such a way that "the output of one model [became] the input for the next step."

The reconstructed SN history of the past millennium looks very much like the infamous "hockey stick" temperature history of Mann *et al.* (1999). It slowly declines over the entire time period—with numerous modest oscillations associated with well-known solar maxima and minima—until the end of the Little Ice Age, whereupon it rises dramatically. Usoskin *et al.* report, for example, that "while the average value of the reconstructed SN between 850 and 1900 is about 30, it reaches values of 60 since 1900 and 76 since 1944." In addition, they report that "the largest 100-year average of the reconstructed SN prior to 1900 is 44, which occurs in 1140-1240, i.e., during the medieval maximum," but they note that "even this is significantly less than the level reached in the last century." Hence, they readily and correctly conclude, on the basis of their work, that "the high level of solar activity since the 1940s is unique since the year 850."

The studies reported in this section show that the temperature record of the Northern Hemisphere supports the theory that solar cycles strongly influence temperatures. Additional information on this topic, including reviews of newer publications as they become availables, can be found at http://www.co2science.org/subject/s/solartempnhemis.php.

References

Alley, R.B., Anandakrishnan, S. and Jung, P. 2001. Stochastic resonance in the North Atlantic. *Paleoceanography* **16**: 190-198.

Björck, S., Muscheler, R., Kromer, B., Andresen, C.S., Heinemeier, J., Johnsen, S.J., Conley, D., Koc, N., Spurk, M. and Veski, S. 2001. High-resolution analyses of an early Holocene climate event may imply decreased solar forcing as an important climate trigger. *Geology* **29**: 1107-1110.

Blunier, T., Chapellaz, J., Schwander, J., Dallenbach, A., Stauffer, B., Stocker, T.F., Raynaud, D., Jouzel, J., Clausen, H.B., Hammer, C.U. and Johnsen, S.J. 1998. Asynchrony of Antarctic and Greenland climate change during the last glacial period. *Nature* **394**: 739-743.

Bond, G., Kromer, B., Beer, J., Muscheler, R., Evans, M.N., Showers, W., Hoffmann, S., Lotti-Bond, R., Hajdas, I. and Bonani, G. 2001. Persistent solar influence on North

Atlantic climate during the Holocene. *Science* **294**: 2130-2136.

Bond, G.C., Showers, W., Elliot, M., Evans, M., Lotti, R., Hajdas, I., Bonani, G. and Johnson, S. 1999. The North Atlantic's 1-2kyr climate rhythm: relation to Heinrich events, Dansgaard/Oeschger cycles and the little ice age. In: Clark, P.U., Webb, R.S. and Keigwin, L.D. (Eds.) *Mechanisms of Global Climate Change at Millennial Time Scales*. American Geophysical Union *Geophysical Monographs* **112**: 35-58.

Brook, E.J., Harder, S., Severinghaus, J. and Bender, M. 1999. Atmospheric methane and millennial-scale climate change. In: Clark, P.U., Webb, R.S. and Keigwin, L.D. (Eds.), *Mechanisms of Global Climate Change at Millennial Time Scales*. American Geophysical Union *Geophysical Monographs* **112**: 165-175.

Clark, D.H. and Stephenson, F.R. 1979. A new revolution in solar physics. *Astronomy* **7**(2): 50-54.

Mann, M.E., Bradley, R.S. and Hughes, M.K. 1999. Northern Hemisphere temperatures during the past millennium: Inferences, uncertainties, and limitations. *Geophysical Research Letters* **26**: 759-762.

Mayewski, P.A., Meeker, L.D., Twickler, M.S., Whitlow, S., Yang, Q., Lyons, W.B. and Prentice, M. 1997. Major features and forcing of high-latitude northern hemisphere atmospheric circulation using a 110,000-year-long glaciochemical series. *Journal of Geophysical Research* **102**: 26,345-26,366.

Oppo, D.W., McManus, J.F. and Cullen, J.L. 1998. Abrupt climate events 500,000 to 340,000 years ago: Evidence from subpolar North Atlantic sediments. *Science* **279**: 1335-1338.

Pang, K.D. and Yau, K.K. 2002. Ancient observations link changes in sun's brightness and earth's climate. *EOS: Transactions, American Geophysical Union* **83**: 481, 489-490.

Raymo, M.E., Ganley, K., Carter, S., Oppo, D.W. and McManus, J. 1998. Millennial-scale climate instability during the early Pleistocene epoch. *Nature* **392**: 699-702.

Rohling, E.J., Mayewski, P.A. and Challenor, P. 2003. On the timing and mechanism of millennial-scale climate variability during the last glacial cycle. *Climate Dynamics* **20**: 257-267.

Usoskin, I.G., Solanki, S.K., Schussler, M., Mursula, K. and Alanko, K. 2003. Millennium-scale sunspot number reconstruction: Evidence for an unusually active sun since the 1940s. *Physical Review Letters* **91**: 10.1103/PhysRevLett.91.211101.

Van Kreveld, S., Sarnthein, M., Erlenkeuser, H., Grootes, P., Jung, S., Nadeau, M.J., Pflaumann, U. and Voelker, A.

2000. Potential links between surging ice sheets, circulation changes, and the Dansgaard-Oeschger cycles in the Irminger Sea, 60-18 kyr. *Paleoceanography* **15**: 425-442.

Yau, K.K.C. 1988. A revised catalogue of Far Eastern observations of sunspots (165 B.C. to A.D. 1918). *Quarterly Journal of the Royal Astronomical Society* **29**: 175-197.

Zhuang, W.F. and Wang, L.Z. 1988. *Union Compilation of Ancient Chinese Records of Celestial Phenomena*. Jiangsu Science and Technology Press, Jiangsu Province, China.

5.3.3. North America

We begin our review of the influence of the sun on North American temperatures with the study of Wiles *et al.* (2004), who derived a composite Glacier Expansion Index (GEI) for Alaska based on "dendrochronologically derived calendar dates from forests overrun by advancing ice and age estimates of moraines using tree-rings and lichens," after which they compared this history of glacial activity with "the ^{14}C record preserved in tree rings corrected for marine and terrestrial reservoir effects as a proxy for solar variability" and with the history of the Pacific Decadal Oscillation (PDO) derived by Cook (2002).

Results of the study showed Alaska ice expansions "approximately every 200 years, compatible with a solar mode of variability," specifically, the de Vries 208-year solar cycle; and by merging this cycle with the cyclical behavior of the PDO, Wiles *et al.* obtained a dual-parameter forcing function that was even better correlated with the Alaskan composite GEI, with major glacial advances clearly associated with the Sporer, Maunder, and Dalton solar minima.

In introducing the rational for their study, Wiles *et al.* say that "increased understanding of solar variability and its climatic impacts is critical for separating anthropogenic from natural forcing and for predicting anticipated temperature change for future centuries." In this regard, it is most interesting that they make no mention of possible CO_2-induced global warming in discussing their results, presumably because there is no need to do so. Alaskan glacial activity, which, in their words, "has been shown to be primarily a record of summer temperature change (Barclay *et al.*, 1999)," appears to be sufficiently well described within the context of solar and PDO variability alone. Four years later, Wiles *et al.* (2008)

reconfirmed this Alaska solar-climate link in a separate study.

Nearby in the Columbia Icefield area of the Canadian Rockies, Luckman and Wilson (2005) used new tree-ring data to present a significant update to a millennial temperature reconstruction published for this region in 1997. The new update employed different standardization techniques, such as the regional curve standardization method, in an effort to capture a greater degree of low frequency variability (centennial to millennial scale) than reported in the initial study. In addition, the new dataset added more than one hundred years to the chronology and now covers the period AD 950-1994.

The updated proxy indicator of temperature showed considerable decadal- and centennial-scale variability, where generally warmer conditions prevailed during the eleventh and twelfth centuries, between about AD 1350-1450 and from about 1875 through the end of the record, while persistent cold conditions prevailed between 1200-1350, 1450-1550, and 1650-1850, with the 1690s being exceptionally cold (more than 0.4°C colder than the other intervals).

The revised Columbia Icefield temperature reconstruction provides further evidence for natural climate fluctuations on centennial-to-millennial timescales and demonstrates, once again, that temperatures during the Current Warm Period are no different from those observed during the Medieval Warm Period (eleventh—twelfth centuries) or the Little Medieval Warm Period (1350-1450). And since we know that atmospheric CO_2 concentrations had nothing to do with the warm temperatures of those earlier periods, we cannot rule out the possibility that they also have nothing to do with the warm temperatures of the modern era.

But if not CO_2, then what? According to Luckman and Wilson, the Columbia Icefield reconstruction "appears to indicate a reasonable response of local trees to large-scale forcing of climates, with reconstructed cool conditions comparing well with periods of known low solar activity," which is a nice way of suggesting that the *sun* is the main driver of these low frequency temperature trends.

Heading south to the warmer regions of North America, Barron and Bukry (2007) extracted sediment cores from three sites on the eastern slope of the Gulf of California. By examining these high-resolution records of diatoms and silicoflagellate assemblages, they were able to reconstruct sea surface temperatures there over the past 2,000 years. In all

three of the sediment cores, the relative abundance of *Azpeitia nodulifera* (a tropical diatom whose presence suggests the occurrence of higher sea surface temperatures), was found to be greater during the Medieval Warm Period than at any other time over the 2,000-year period studied, while during the Current Warm Period its relative abundance was actually lower than the 2,000-year mean, also in all three of the sediment cores. In addition, the first of the cores exhibited elevated *A. nodulifera* abundances from the start of the record to about AD 350, during the latter part of the Roman Warm Period, as well as between AD 1520 and 1560, during what we have denominated the Little Medieval Warm Period. By analyzing radiocarbon production data, Barron and Bukry determined that "intervals of increased radiocarbon production (sunspot minima) correlate with intervals of enhanced biosilica productivity," leading the two authors to conclude that "solar forcing played a major role in determining surface water conditions in the Gulf of California during the past 2000 yr." As for how this was accomplished, Barron and Bukry say that "reduced solar irradiance (sunspot minima) causes cooling of winter atmospheric temperatures above the southwest US," and that "this strengthens the atmospheric low and leads to intensification of northwest winds blowing down the Gulf, resulting in increased overturn of surface waters, increased productivity, and cooler SST."

Richey *et al.* (2007) constructed "a continuous decadal-scale resolution record of climate variability over the past 1400 years in the northern Gulf of Mexico" from a box core recovered in the Pigmy Basin, northern Gulf of Mexico [27°11.61'N, 91°24.54'W]," based on "paired analyses of Mg/Ca and $\delta^{18}O$ in the white variety of the planktic foraminifer *Globigerinoides ruber* and relative abundance variations of *G. sacculifer* in the foraminifer assemblages."

Results revealed that "two multi-decadal intervals of sustained high Mg/Ca indicate that Gulf of Mexico sea surface temperatures (SSTs) were as warm or warmer than near-modern conditions between 1000 and 1400 yr B.P.," while "foraminiferal Mg/Ca during the coolest interval of the Little Ice Age (ca. 250 yr B.P.) indicate that SST was 2-2.5°C below modern SST." In addition, they found that "four minima in the Mg/Ca record between 900 and 250 yr. B.P. correspond with the Maunder, Sporer, Wolf, and Oort sunspot minima," providing additional evidence that the historic warmth of earth's past was likely solar-induced.

Also in the Gulf of Mexico, Poore *et al.* (2003) developed a 14,000-year record of Holocene climate based primarily on the relative abundance of the planktic foraminifer *Globigerinoides sacculifer* found in two sediment cores. In reference to North Atlantic millennial-scale cool events 1-7 identified by Bond *et al.* (2001) as belonging to a pervasive climatic oscillation with a period of approximately 1,500 years, Poore *et al.* say of their own study that distinct excursions to lower abundances of *G. sacculifer* "match within 200 years the ages of Bond events 1-6," noting that "major cooling events detected in the subpolar North Atlantic can be recognized in the GOM record." They additionally note that "the GOM record includes more cycles than can be explained by a quasiperiodic 1500-year cycle," but that such centennial-scale cycles with periods ranging from 200 to 500 years are also observed in the study of Bond *et al.*, noting further that their results "are in agreement with a number of studies indicating the presence of substantial century-scale variability in Holocene climate records from different areas," specifically citing the reports of Campbell *et al.* (1998), Peterson *et al.* (1991), and Hodell *et al.* (2001). Last, they discuss evidence that leads them to conclude that "some of the high-frequency variation (century scale) in *G. sacculifer* abundance in our GOM records is forced by solar variability."

In still another example of a solar-temperature connection, Lund and Curry (2004) analyzed a planktonic foraminiferal $\delta^{18}O$ time series obtained from three well-dated sediment cores retrieved from the seabed near the Florida Keys (24.4°N, 83.3°W) that covered the past 5,200 years. As they describe it, isotopic data from the three cores "indicate the surface Florida Current was denser (colder, saltier or both) during the Little Ice Age than either the Medieval Warm Period or today," and that "when considered with other published results (Keigwin, 1996; deMenocal *et al.*, 2000), it is possible that the entire subtropical gyre of the North Atlantic cooled during the Little Ice Age ... perhaps consistent with the simulated effects of reduced solar irradiance (Rind and Overpeck, 1993; Shindell *et al.*, 2001)." In addition, they report that "the coherence and phasing of atmospheric ^{14}C production and Florida Current $\delta^{18}O$ during the Late Holocene implies that solar variability may influence Florida Current surface density at frequencies between 1/300 and 1/100 years," demonstrating once again a situation where

both centennial- and millennial-scale climatic variability is explained by similar-scale variability in solar activity.

We conclude with the study of Li *et al.* (2006), who "recovered a 14,000-year mineral-magnetic record from White Lake (~41°N, 75°W), a hardwater lake containing organic-rich sediments in northwestern New Jersey, USA." According to these researchers, a comparison of the White Lake data with climate records from the North Atlantic sediments "shows that low lake levels at ~1.3, 3.0, 4.4, and 6.1 ka [1000 years before present] in White Lake occurred almost concurrently with the cold events at ~1.5, 3.0, 4.5, and 6.0 ka in the North Atlantic Ocean (Bond *et al.*, 2001)," and that "these cold events are associated with the 1500-year warm/cold cycles in the North Atlantic during the Holocene" that have "been interpreted to result from solar forcing (Bond *et al.*, 2001)."

It is clear that broad-scale periods of warmth in North America have occurred over and over again throughout the Holocene—and beyond (Oppo *et al.*, 1998; Raymo *et al.*, 1998)—forced by variable solar activity. This suggests that the Current Warm Period was also instigated by this recurring phenomenon, not the CO_2 output of the Industrial Revolution.

Additional information on this topic, including reviews of newer publications as they become available, can be found at http://www.co2science.org/subject/s/solartempnamer.php.

References

Barclay, D.J., Wiles, G.C. and Calkin, P.E. 1999. A 1119-year tree-ring-width chronology from western Prince William Sound, southern Alaska. *The Holocene* **9**: 79-84.

Barron, J.A. and Bukry, D. 2007. Solar forcing of Gulf of California climate during the past 2000 yr suggested by diatoms and silicoflagellates. *Marine Micropaleontology* **62**: 115-139.

Bond, G., Kromer, B., Beer, J., Muscheler, R., Evans, M.N., Showers, W., Hoffmann, S., Lotti-Bond, R., Hajdas, I. and Bonani, G. 2001. Persistent solar influence on North Atlantic climate during the Holocene. *Science* **294**: 2130-2136.

Campbell, I.D., Campbell, C., Apps, M.J., Rutter, N.W. and Bush, A.B.G. 1998. Late Holocene ca.1500 yr climatic periodicities and their implications. *Geology* **26**: 471-473.

Cook, E.R. 2002. Reconstructions of Pacific decadal variability from long tree-ring records. *EOS: Transactions, American Geophysical Union* **83**: S133.

deMenocal, P., Ortiz, J., Guilderson, T. and Sarnthein, M. 2000. Coherent high- and low-latitude variability during the Holocene warm period. *Science* **288**: 2198-2202.

Hodell, D.A., Brenner, M., Curtis, J.H. and Guilderson, T. 2001. Solar forcing of drought frequency in the Maya lowlands. *Science* **292**: 1367-1370.

Keigwin, L. 1996. The Little Ice Age and Medieval Warm Period in the Sargasso Sea. *Science* **274**: 1504-1508.

Li, Y.-X., Yu, Z., Kodama, K.P. and Moeller, R.E. 2006. A 14,000-year environmental change history revealed by mineral magnetic data from White Lake, New Jersey, USA. *Earth and Planetary Science Letters* **246**: 27-40.

Luckman, B.H. and Wilson, R.J.S. 2005. Summer temperatures in the Canadian Rockies during the last millennium: a revised record. *Climate Dynamics* **24**: 131-144.

Lund, D.C. and Curry, W.B. 2004. Late Holocene variability in Florida Current surface density: Patterns and possible causes. *Paleoceanography* **19**: 10.1029/2004 PA001008.

Oppo, D.W., McManus, J.F. and Cullen, J.L. 1998. Abrupt climate events 500,000 to 340,000 years ago: Evidence from subpolar North Atlantic sediments. *Science* **279**: 1335-1338.

Peterson, L.C., Overpeck, J.T., Kipp, N.G. and Imbrie, J. 1991. A high-resolution Late Quaternary upwelling record from the anoxic Cariaco Basin, Venezuela. *Paleoceanography* **6**: 99-119.

Poore, R.Z., Dowsett, H.J., Verardo, S. and Quinn, T.M. 2003. Millennial- to century-scale variability in Gulf of Mexico Holocene climate records. *Paleoceanography* **18**: 10.1029/2002PA000868.

Raymo, M.E., Ganley, K., Carter, S., Oppo, D.W. and McManus, J. 1998. Millennial-scale climate instability during the early Pleistocene epoch. *Nature* **392**: 699-702.

Richey, J.N., Poore, R.Z., Flower, B.P. and Quinn, T.M. 2007. 1400 yr multiproxy record of climate variability from the northern Gulf of Mexico. *Geology* **35**: 423-426.

Rind, D. and Overpeck, J. 1993. Hypothesized causes of decade- to century-scale climate variability: Climate model results. *Quaternary Science Reviews* **12**: 357-374.

Shindell, D.T., Schmidt, G.A., Mann, M.E., Rind, D. and Waple, A. 2001. Solar forcing of regional climate during the Maunder Minimum. *Science* **294**: 2149-2152.

Wiles, G.C., Barclay, D.J., Calkin, P.E. and Lowell, T.V. 2008. Century to millennial-scale temperature variations for the last two thousand years indicated from glacial geologic records of Southern Alaska. *Global and Planetary Change* **60**: 115-125.

Wiles, G.C., D'Arrigo, R.D., Villalba, R., Calkin, P.E. and Barclay, D.J. 2004. Century-scale solar variability and Alaskan temperature change over the past millennium. Geophysical Research Letters 31: 10.1029/2004GL020050.

5.3.4. South America

Nordemann *et al.* (2005) examined tree rings from species sensitive to fluctuations in temperature and precipitation throughout the southern region of Brazil and Chile, along with sunspot data, via harmonic spectral and wavelet analysis in an effort to obtain a greater understanding of the effects of solar activity, climate, and geophysical phenomena on the continent of South America, where the time interval covered by the tree-ring samples from Brazil was 200 years and that from Chile was 2,500 years. Results of the spectral analysis revealed periodicities in the tree rings that corresponded well with the DeVries-Suess (~200 yr), Gleissberg (~80 yr), Hale (~22 yr), and Schwabe (~11 yr) solar activity cycles, while wavelet cross-spectrum analysis of sunspot number and tree-ring growth revealed a clear relation between the tree-ring and solar series.

Next, utilizing a lichenometric method for dating glacial moraines, the Bolivian and French research team of Rabatel *et al.* (2005) developed what they call "the first detailed chronology of glacier fluctuations in a tropical area during the Little Ice Age," focusing on fluctuations of the Charquini glaciers of the Cordillera Real in Bolivia, where they studied a set of 10 moraines that extend below the present glacier termini. Based on the chronology, the researchers determined that the maximum glacier extension in Bolivia "occurred in the second half of the 17th century, as observed in many mountain areas of the Andes and the Northern Hemisphere." In addition, they found that "this expansion has been of a comparable magnitude to that observed in the Northern Hemisphere, with the equilibrium line altitude depressed by 100-200 m during the glacier maximum." They say "the synchronization of glacier expansion with the Maunder and Dalton minima supports the idea that solar activity could have cooled enough the tropical atmosphere to provoke this evolution."

As for the magnitude and source of the cooling in the Bolivian Andes during the Little Ice Age, three years later Rabatal *et al.* (2008) estimated it to have been 1.1 to 1.2°C below that of the present, while once again noting that at that time there was a "striking coincidence between the glacier expansion in this region of the tropics and the decrease in solar irradiance: the so-called 'Maunder minimum' (AD 1645-1715) during which irradiance might have decreased by around 0.24% (Lean and Rind, 1998) and could have resulted in an atmospheric cooling of 1°C worldwide (Rind *et al.*, 2004)."

Further south, Glasser *et al.* (2004) analyzed a large body of evidence related to glacier fluctuations in the two major ice fields of Patagonia: the Hielo Patagonico Norte (47°00'S, 73°39'W) and the Hielo Patagonico Sur (between 48°50'S and 51°30'S). With respect to the glacial advancements that occurred during the cold interval that preceded the Roman Warm Period, they say they are "part of a body of evidence for global climatic change around this time (e.g., Grosjean *et al.*, 1998; Wasson and Claussen, 2002), which coincides with an abrupt decrease in solar activity," adding that this observation "led van Geel *et al.* (2000) to suggest that variations in solar irradiance are more important as a driving force in variations in climate than previously believed."

With respect to the most recent recession of Hielo Patagonico Norte outlet glaciers from their late historic moraine limits at the end of the nineteenth century, Glasser *et al.* say that "a similar pattern can be observed in other parts of southern Chile (e.g., Kuylenstierna *et al.*, 1996; Koch and Kilian, 2001)." Likewise, they note that "in areas peripheral to the North Atlantic and in central Asia the available evidence shows that glaciers underwent significant recession at this time (cf. Grove, 1988; Savoskul, 1997)," which again suggests the operation of a globally distributed forcing factor such as cyclically variable solar activity.

Working on a bog, as opposed to a glacier, Chambers *et al.* (2007) presented new proxy climate data they obtained from the Valle de Andorra northeast of Ushuaia, Tierra del Fuego, Argentina, which data, they emphasize, are "directly comparable" with similar proxy climate data obtained in numerous studies conducted in European bogs, "as they were produced using identical laboratory methods." This latter point is very important because Chambers *et al.* say their new South American data show there was "a major climate perturbation at the same time as in northwest Europe," which they describe as "an abrupt climate cooling" that occurred approximately 2,800 years ago, and that "its timing, nature and apparent global synchronicity lend support to the notion of solar forcing of past climate change, amplified by oceanic circulation."

The five European researchers further state their finding that "rapid, high-magnitude climate changes might be produced within the Holocene by an inferred *decline* in solar activity (van Geel *et al.*, 1998, 2000, 2003; Bond *et al.*, 2001; Blaauw *et al.*, 2004; Renssen *et al.*, 2006) has implications for rapid, high-magnitude climate changes of the opposite direction—climatic warmings, possibly related to *increases* in solar activity." In this regard, they further note that "for the past 100 years any solar influence would for the most part have been in the opposite direction (i.e., to help generate a global climate warming) to that inferred for c. 2800-2710 cal. BP." And they conclude that this observation "has implications for interpreting the relative contribution of climate drivers of recent 'global warming'," implying that a solar-induced, rather than a CO_2-induced, climate driver may have been the primary cause of twentieth century global warming.

Polissar *et al.* (2006) worked with data derived from sediment records of two Venezuelan watersheds along with ancillary data obtained from other studies that had been conducted in the same general region. They developed continuous decadal-scale histories of glacier activity and moisture balance in a part of the tropical Andes (the Cordillera de Merida) over the past millennium and a half, from which they were able to deduce contemporary histories of regional temperature and precipitation. The international (Canada, Spain, United States, Venezuela) team of scientists write that "comparison of the Little Ice Age history of glacier activity with reconstructions of solar and volcanic forcing suggest that solar variability is the primary underlying cause of the glacier fluctuations," because (1) "the peaks and troughs in the susceptibility records match fluctuations of solar irradiance reconstructed from ^{10}Be and $\delta^{14}C$ measurements," (2) "spectral analysis shows significant peaks at 227 and 125 years in both the irradiance and magnetic susceptibility records, closely matching the de Vreis and Gleissberg oscillations identified from solar irradiance reconstructions," and (3) "solar and volcanic forcing are uncorrelated between AD 1520 and 1650, and the magnetic susceptibility record follows the solar-irradiance reconstruction during this interval." In addition, they write that "four glacial advances occurred between AD 1250 and 1810, coincident with solar-activity minima," and that "temperature declines of -3.2 ± 1.4°C and precipitation increases of ~20% are required to produce the observed glacial responses."

In discussing their findings, Polissar *et al.* say their results "suggest considerable sensitivity of tropical climate to small changes in radiative forcing from solar irradiance variability." This research from South American strongly suggests that the IPCC is failing to take into account the effect of solar cycles on temperatures.

Additional information on this topic, including reviews of newer publications as they become available, can be found at http://www.co2science.org/subject/s/solartempsamer.php.

References

Blaauw, M., van Geel, B. and van der Plicht, J. 2004. Solar forcing of climate change during the mid-Holocene: indications from raised bogs in The Netherlands. *The Holocene* **14**: 35-44.

Bond, G., Kromer, B., Beer, J., Muscheler, R., Evans, M.N., Showers, W., Hoffmann, S., Lotti-Bond, R., Hajdas, I. and Bonani, G. 2001. Persistent solar influence on North Atlantic climate during the Holocene. *Science* **294**: 2130-2136.

Chambers, F.M., Mauquoy, D., Brain, S.A., Blaauw, M. and Daniell, J.R.G. 2007. Globally synchronous climate change 2800 years ago: Proxy data from peat in South America. *Earth and Planetary Science Letters* **253**: 439-444.

Glasser, N.F., Harrison, S., Winchester, V. and Aniya, M. 2004. Late Pleistocene and Holocene palaeoclimate and glacier fluctuations in Patagonia. *Global and Planetary Change* **43**: 79-101.

Grosjean, M., Geyh, M.A., Messerli, B., Schreier, H. and Veit, H. 1998. A late-Holocene (?2600 BP) glacial advance in the south-central Andes (29°S), northern Chile. *The Holocene* **8**: 473-479.

Grove, J.M. 1988. *The Little Ice Age*. Routledge, London, UK.

Koch, J. and Kilian, R. 2001. Dendroglaciological evidence of Little Ice Age glacier fluctuations at the Gran Campo Nevado, southernmost Chile. In: Kaennel Dobbertin, M. and Braker, O.U. (Eds.) *International Conference on Tree Rings and People*. Davos, Switzerland, p. 12.

Kuylenstierna, J.L., Rosqvist, G.C. and Holmlund, P. 1996. Late-Holocene glacier variations in the Cordillera Darwin, Tierra del Fuego, Chile. *The Holocene* **6**: 353-358.

Lean, J. and Rind, D. 1998. Climate forcing by changing solar radiation. *Journal of Climate* **11**: 3069-3094.

Nordemann, D.J.R., Rigozo, N.R. and de Faria, H.H. 2005. Solar activity and El-Niño signals observed in Brazil and Chile tree ring records. *Advances in Space Research* **35**: 891-896.

Polissar, P.J., Abbott, M.B., Wolfe, A.P., Bezada, M., Rull, V. and Bradley, R.S. 2006. Solar modulation of Little Ice Age climate in the tropical Andes. *Proceedings of the National Academy of Sciences USA* **103**: 8937-8942.

Rabatel, A., Francou, B., Jomelli, V., Naveau, P. and Grancher, D. 2008. A chronology of the Little Ice Age in the tropical Andes of Bolivia (16°S) and its implications for climate reconstruction. *Quaternary Research* **70**: 198-212.

Rabatel, A., Jomelli, V., Naveau, P., Francou, B. and Grancher, D. 2005. Dating of Little Ice Age glacier fluctuations in the tropical Andes: Charquini glaciers, Bolivia, 16°S. *Comptes Rendus Geoscience* **337**: 1311-1322.

Renssen, H., Goosse, H. and Muscheler, R. 2006. Coupled climate model simulation of Holocene cooling events: solar forcing triggers oceanic feedback. *Climate Past Discuss.* **2**: 209-232.

Rind, D., Shindell, D., Perlwitz, J., Lerner, J., Lonergan, P., Lean, J. and McLinden, C. 2004. The relative importance of solar and anthropogenic forcing of climate change between the Maunder minimum and the present. *Journal of Climate* **17**: 906-929.

Savoskul, O.S. 1997. Modern and Little Ice Age glaciers in "humid" and "arid" areas of the Tien Shan, Central Asia: two different patterns of fluctuation. *Annals of Glaciology* **24**: 142-147.

Van Geel, B., Heusser, C.J., Renssen, H. and Schuurmans, C.J.E. 2000. Climatic change in Chile at around 2700 BP and global evidence for solar forcing: a hypothesis. *The Holocene* **10**: 659-664.

Van Geel, B., van der Plicht, J., Kilian, M.R., Klaver, E.R., Kouwenberg, J.H.M., Renssen, H., Reynaud-Farrera, I. and Waterbolk, H.T. 1998. The sharp rise of $\delta^{14}C$ ca. 800 cal BC: possible causes, related climatic teleconnections and the impact on human environments. *Radiocarbon* **40**: 535-550.

Van Geel, B., van der Plicht, J. and Renssen, H. 2003. Major $\delta^{14}C$ excursions during the Late Glacial and early Holocene: changes in ocean ventilation or solar forcing of climate change? *Quaternary International* **105**: 71-76.

Wasson, R.J. and Claussen, M. 2002. Earth systems models: a test using the mid-Holocene in the Southern Hemisphere. *Quaternary Science Reviews* **21**: 819-824.

5.3.5. Asia

We begin our study of Asia with a 2003 paper published in the Russian journal *Geomagnetizm i Aeronomiya*, where two scientists from the Institute of Solar-Terrestrial Physics of the Siberian Division of the Russian Academy of Sciences, Bashkirtsev and Mashnich (2003), say "a number of publications report that the anthropogenic impact on the Earth's climate is an obvious and proven fact," when in their opinion "none of the investigations dealing with the anthropogenic impact on climate convincingly argues for such an impact."

In the way of contrary evidence, they begin by citing the work of Friis-Christensen and Lassen (1991), who first noted the close relationship (r = -0.95) between the length of the sunspot cycle and the surface air temperature of the Northern Hemisphere over the period 1861-1989, where "warming and cooling corresponded to short (~10 yr) and prolonged (~11.5 yr) solar cycles, respectively." They then cite the work of Zherebtsov and Kovalenko (2000), who they say established a high correlation (r = 0.97) between "the average power of the solar activity cycle and the surface air temperature in the Baikal region averaged over the solar cycle." These two findings, they contend, "leave little room for the anthropogenic impact on the Earth's climate." In addition, they note that "solar variations naturally explain global cooling observed in 1950-1970, which cannot be understood from the standpoint of the greenhouse effect, since CO_2 was intensely released into the atmosphere in this period," citing in support of this statement the work of Dergachev and Raspopov (2000).

Bashkirtsev and Mashnich conducted their own wavelet-spectra and correlation analyses of Irkutsk and world air temperatures and Wolf number data for the period 1882-2000, finding periodicities of 22 (Hale cycle) and 52 (Fritz cycle) years and reporting that "the temperature response of the air lags behind the sunspot cycles by approximately 3 years in Irkutsk and by 2 years over the entire globe."

Noting that one could thus expect the upper envelope of sunspot cycles to reproduce the global temperature trend, they created such a plot and found that such is indeed the case. As they describe their results, "the lowest temperatures in the early 1900s correspond to the lowest solar activity (weak cycle 14), the further temperature rise follows the increase in solar activity; the decrease in solar activity in cycle 20 is accompanied by the temperature fall [from 1950-1970], and the subsequent growth of solar activity in cycles 21 and 22 entails the temperature rise [of the last quarter century]."

Bashkirtsev and Mashnich say "it has become clear that the current sunspot cycle (cycle 23) is weaker than the preceding cycles (21 and 22)," and that "solar activity during the subsequent cycles (24 and 25) will be, as expected, even lower," noting that "according to Chistyakov (1996, 2000), the minimum of the secular cycle of solar activity will fall on cycle 25 (2021-2026), which will result in the minimum global temperature of the surface air (according to our prediction)." Only time will tell if such predictions will prove correct.

Turning our attention back toward the past, but staying in the Asian subarctic, Vaganov *et al.* (2000) utilized tree-ring width as a proxy for temperature to examine temperature variations in this region over the past 600 years. According to a graph of the authors' data, temperatures in the Asian subarctic exhibited a small positive trend from the start of the record until about 1750. Thereafter, a severe cooling trend ensued, followed by a 130-year warming trend from about 1820 through 1950, after which temperatures fell once again. In considering the entire record, the authors state that the amplitude of twentieth century warming "does not go beyond the limits of reconstructed natural temperature fluctuations in the Holocene subarctic zone."

In attempting to determine the cause or causes of the temperature fluctuations, the authors report finding a significant correlation with solar radiation and volcanic activity over the entire 600-year period (R = 0.32 for solar radiation, R = -0.41 for volcanic activity), which correlation *improved* over the shorter interval of the industrial period—1800 to 1990—(R = 0.68 for solar radiation, R = -0.59 for volcanic activity).

It is interesting to note that in this region of the world, where climate models predict large increases in temperature as a result of the historical rise in the air's CO_2 concentration, real-world data show a *cooling* trend since around 1940, when the greenhouse effect of CO_2 should have been most prevalent. And, where warming does exist in the record (between about 1820 and 1940), much of it correlates with changes in solar irradiance and volcanic activity—two factors free of anthropogenic influence.

In two additional paleoclimate studies from the continental interior of Russia's Siberia, Kalugin *et al.* (2005) and Kalugin *et al.* (2007) analyzed sediment cores from Lake Teletskoye in the Altai Mountains

(51°42.90'N, 87°39.50'E) to produce multi-proxy climate records spanning the past 800 years. Analyses of the multi-proxy records revealed several distinct climatic periods over the past eight centuries. With respect to temperature, the regional climate was relatively warm with high terrestrial productivity from AD 1210 to 1380. Thereafter, temperatures cooled, reaching peak deterioration between 1660 and 1700, which time period, in the words of Kalugin *et al.* (2005), "corresponds to the age range of the well-known Maunder Minimum (1645-1715)" of solar sunspot activity.

Moving to Japan, an uninterrupted 1,100-year history of March mean temperature at Kyoto was developed by Aono and Kazui (2008), who used phenological data on the times of full-flowering of cherry trees (*Prunus jamasakura*) acquired from old diaries and chronicles written at Kyoto. Upon calibration with instrumental temperature measurements obtained over the period 1881-2005, the results were compared with the sunspot number history developed by Solanki *et al.* (2004).

The results of the study suggest "the existence of four cold periods, 1330-1350, 1520-1550, 1670-1700, and 1825-1830, during which periods the estimated March mean temperature was 4-5°C, about 3-4°C lower than the present normal temperature," and that "these cold periods coincided with the less extreme periods [of solar activity], known as the Wolf, Spoerer, Maunder, and Dalton minima, in the long-term solar variation cycle, which has a periodicity of 150-250 years." In addition, they report that "a time lag of about 15 years was detected in the climatic temperature response to short-term solar variation."

Also in Japan, Kitagawa and Matsumoto (1995) analyzed $\delta^{13}C$ variations of Japanese cedars growing on Yakushima Island (30°20'N, 130°30'E), in an effort to reconstruct a high-resolution proxy temperature record over the past two thousand years. In addition, they applied spectral analysis to the $\delta^{13}C$ time series in an effort to learn if any significant periodicities were present in the record.

Results indicated significant decadal to centennial-scale variability throughout the record, with temperatures fluctuating by about 5°C across the series. Most notable among the fluctuations were multi-century warm and cold epochs. Between AD 700-1200, for example, there was about a 1°C rise in average temperature (pre-1850 average), which the authors state "appears to be related to the 'Medieval Warm Period'." In contrast, temperatures were about 2°C below the long-term pre-1850 average during the

multi-century Little Ice Age that occurred between AD 1580 and 1700. Kitagawa and Matsumoto also report finding significant temperature periodicities of 187, 89, 70, 55, and 44 years. Noting that the 187-year cycle closely corresponds to the well-known Suess cycle of solar activity and that the 89-year cycle compares well with the Gleissberg solar cycle, they conclude that their findings provide further support for a sun-climate relationship.

Ten years later, Cini Castagnoli *et al.* (2005) re-examined the Kitagawa and Matsumoto dataset for evidence of recurring cycles using Singular Spectrum Analysis and Wavelet Transform, after which it was compared with a 300-year record of sunspots. Results of the newer analyses showed a common 11-year oscillation in phase with the Schwabe cycle of solar activity, plus a second multi-decadal oscillation (of about 87 years for the tree-ring series) in phase with the amplitude modulation of the sunspot number series over the plast 300 years, which led this second group of authors to conclude that the overall phase agreement between the climate reconstruction and variation in the sunspot number series "favors the hypothesis that the [multi-decadal] oscillation" revealed in the record "is connected to the solar activity."

Turning to China, there have been several studies documenting a solar influence on temperature from several proxy temperature indicators. Beginning with stalagmite-derived proxies, Paulsen *et al.* (2003) utilized high-resolution records of $\delta^{13}C$ and $\delta^{18}O$ from a stalagmite in Buddha Cave, central China [33°40'N, 109°05'E], to infer changes in climate there over the past 1,270 years. Among the climatic episodes evident in the authors' data were "those corresponding to the Medieval Warm Period, Little Ice Age and 20th-century warming, lending support to the global extent of these events." The authors' data also revealed a number of other cycles superimposed on these major millennial-scale temperature cycles, which they attributed to cyclical solar and lunar phenomena.

In a separate study, Tan *et al.* (2004) established an annual layer thickness chronology for a stalagmite from Beijing Shihua Cave and reconstructed a 2,650-year (BC 665-AD 1985) warm season (MJJA: May, June, July, August) temperature record for Beijing by calibrating the thickness chronology with the observed MJJA temperature record (Tan *et al.*, 2003). Results of the analysis showed that the warm season temperature record was "consistent with oscillations in total solar irradiance inferred from cosmogenic

[10]Be and ^{14}C," and that it also "is remarkably consistent with Northern Atlantic drift ice cycles that were identified to be controlled by the sun through the entire Holocene [Bond *et al.*, 2001]." Going backwards in time, both records clearly depict the start of the Current Warm Period, the prior Little Ice Age, the Medieval Warm Period, the Dark Ages Cold Period, the Roman Warm Period, and the cold climate at the start of both records.

The authors conclude that "the synchronism between the two independent sun-linked climate records therefore suggests that the sun may directly couple hemispherical climate changes on centennial to millennial scales." It stands to reason that the cyclical nature of the millennial-scale oscillation of climate evident in both climate records suggests there is no need to invoke rising atmospheric CO_2 concentrations as a cause of the Current Warm Period.

Working with a stalagmite found in another China cave, Wanxiang Cave (33°19'N, 105°00'E), Zhang *et al.* (2008) developed a $\delta^{18}O$ record with an average resolution of 2.5 years covering the period AD 190 to 2003. According to the 17 authors of this study, the $\delta^{18}O$ record "exhibits a series of centennial to multi-centennial fluctuations broadly similar to those documented in Northern Hemisphere temperature reconstructions, including the Current Warm Period, Little Ice Age, Medieval Warm Period and Dark Age Cold Period."

In addition, Zhang *et al.* state that it "correlates with solar variability, Northern Hemisphere and Chinese temperature, Alpine glacial retreat, and Chinese cultural changes." And since none of the last four phenomena can influence the first one, solar variability appears to have driven the variations in the other factors mentioned. In a commentary that accompanied Zhang *et al.*'s article, Kerr (2008) quotes other researchers calling the Zhang *et al.* record "amazing," "fabulous," and "phenomenal," and it "provides the strongest evidence yet for a link among sun, climate, and culture."

Still in China, we turn next to the study of Hong *et al.* (2000), who developed a 6,000-year high-resolution $\delta^{18}O$ record from plant cellulose deposited in a peat bog in the Jilin Province of China (42° 20' N, 126° 22' E), from which they inferred the temperature history of that location over the past six millennia. They then compared this record with a previously derived $\delta^{14}C$ tree-ring record that is representative of the intensity of solar activity over this period.

Results indicated the study area was relatively cold between 4000 and 2600 BC. Then it warmed fairly continuously until it reached the maximum warmth of the record about 1600 BC, after which it fluctuated about this warm mean for approximately 2,000 years. Starting about AD 350, however, the climate began to cool, with the most dramatic cold associated with three temperature minima centered at about AD 1550, 1650, and 1750, corresponding to the most severe cold of the Little Ice Age.

Of particular note is the authors' finding of "an obvious warm period represented by the high $\delta^{18}O$ from around AD 1100 to 1200 which may correspond to the Medieval Warm Epoch of Europe." They also report that "at that time, the northern boundary of the cultivation of citrus tree (*Citrus reticulata Blanco*) and *Boehmeria nivea* (a perennial herb), both subtropical and thermophilous plants, moved gradually into the northern part of China, and it has been estimated that the annual mean temperature was 0.9-1.0°C higher than at present."

Hong *et al.* also note "there is a remarkable, nearly one to one, correspondence between the changes of atmospheric $\delta^{14}C$ and the variation in $\delta^{18}O$ of the peat cellulose," which led them to conclude that the temperature history of the past 6,000 years at the site of their study has been "forced mainly by solar variability."

In another study, 18 radiocarbon-dated aeolian and paleosol profiles within a 1,500-km-long belt along the arid to semi-arid transition zone of north-central China were analyzed by Porter and Weijian (2006) to determine variations in the extent and strength of the East Asian summer monsoon throughout the Holocene.

The dated paleosols and peat layers, in the words of Porter and Weijian, "represent intervals when the zone was dominated by a mild, moist summer monsoon climate that favored pedogenesis and peat accumulation," while "brief intervals of enhanced aeolian activity that resulted in the deposition of loess and aeolian sand were times when strengthened winter monsoon conditions produced a colder, drier climate." They also report that the climatic variations they discovered "correlate closely with variations in North Atlantic drift-ice tracers that represent episodic advection of drift ice and cold polar surface water southward and eastward into warmer subpolar water."

The researchers state that "the correspondence of these records over the full span of Holocene time implies a close relationship between North Atlantic climate and the monsoon climate of central China."

They also state that the most recent of the episodic cold periods, which they identify as the Little Ice Age, began about AD 1370, while the preceding cold period ended somewhere in the vicinity of AD 810. Consequently, their work implies the existence of a medieval warm period that began some time after AD 810 and ended some time before AD 1370. In addition, their relating of this millennial-scale climate cycle to the similar-scale drift-ice cycle of Bond *et al.* (2001) implies they accept solar forcing as the most likely cause of the alternating multi-century mild/moist and cold/dry periods of North-Central China. As a result, Porter and Weijian's work helps to establish the global extent of the Medieval Warm Period, as well as its likely solar origin.

Much more evidence of a solar-climate link has been obtained from the Tibetan Plateau in China. Wang *et al.* (2002), for example, studied changes in $\delta^{18}O$ and NO_3^- in an ice core retrieved from the Guliya Ice Cap (35°17'N, 81°29'E) there, comparing the results they obtained with ancillary data from Greenland and Antarctica. Two cold events—a weak one around 9.6-9.2 thousand years ago (ka) and a strong one universally referred to as the "8.2 ka cold event"—were identified in the Guliya ice core record. The authors report that these events occurred "nearly simultaneously with two ice-rafted episodes in the North Atlantic Ocean." They additionally report that both events occurred during periods of weakened solar activity.

Remarking that evidence for the 8.2 ka cold event "occurs in glacial and lacustrine deposits from different areas," the authors say this evidence "suggests that the influence of this cold event may have been global." They also say that "comprehensive analyses indicate that the weakening of solar insolation might have been the external cause of the '8.2 ka cold event'," and that "the cause of the cold event around 9.6-9.2 ka was also possibly related to the weaker solar activity." The authors thus conclude that all of these things considered together imply that "millennial-scale climatic cyclicity might exist in the Tibetan Plateau as well as in the North Atlantic."

In a contemporaneous paper enlarging this thesis, Xu *et al.* (2002) studied plant cellulose $\delta^{18}O$ variations in cores retrieved from peat deposits west of Hongyuan County at the northeastern edge of the Qinghai-Tibetan Plateau (32° 46'N, 102° 30'E). Based on their analysis, the authors report finding the existence of three consistently cold events that were centered at approximately 500, 700, and 900 AD, during what is sometimes referred to as the Dark

Ages Cold Period. Then, from 1100-1300 AD, they report "the $\delta^{18}O$ of Hongyuan peat cellulose increased, consistent with that of Jinchuan peat cellulose and corresponding to the 'Medieval Warm Period'." Finally, they note that "the periods 1370-1400 AD, 1550-1610 AD, [and] 1780-1880 AD recorded three cold events, corresponding to the 'Little Ice Age'."

Regarding the origins of these climatic fluctuations, power spectrum analyses of their data revealed periodicities of 79, 88, and 123-127 years, "suggesting," in the words of the authors, "that the main driving force of Hongyuan climate change is from solar activities." In a subsequent paper by the same authors, Xu *et al.* (2006) compared the Hongyuan temperature variations with solar activity inferred from atmospheric ^{14}C and ^{10}Be concentrations measured in a South Pole ice core, after which they performed cross-spectral analyses to determine the relationship between temperature and solar variability, comparing their results with similar results obtained other researchers around the world. What did they learn this time?

Xu *et al.* (2006) report that "during the past 6000 years, temperature variations in China exhibit high synchrony among different regions, and importantly, are in-phase with those discovered in other regions in the northern hemisphere." They also say that their "comparisons between temperature variations and solar activities indicate that both temperature trends on centennial/millennial timescales and climatic events are related to solar variability."

The researchers' final conclusion was that "quasi-100-year fluctuations of solar activity may be the primary driving force of temperature during the past 6000 years in China." And since their data indicate that peak Medieval Warm Period temperatures were higher than those of the recent past, it is not unreasonable to assume that the planet's recent warmth may have been solar-induced as well.

Still in the northeast edge of the Tibetan Plateau, two years later Tan *et al.* (2008) developed a precipitation history of the Longxi area of the plateau's northeast margin since AD 960 based on an analysis of Chinese historical records, after which they compared the result with the same-period Northern Hemisphere temperature record and contemporaneous atmospheric ^{14}C and ^{10}Be histories.

In their words, Tan *et al.* discovered that "high precipitation of Longxi corresponds to high temperature of the Northern Hemisphere, and low precipitation of Longxi corresponds to low temperature of the Northern Hemisphere." Consequently, their precipitation record may be used to infer a Medieval Warm Period that stretched from approximately AD 960 to 1230, with temperature peaks in the vicinity of AD 1000 and 1215 that clearly exceeded the twentieth century peak temperature of the Current Warm Period. They also found "good coherences among the precipitation variations of Longxi and variations of atmospheric ^{14}C concentration, the averaged ^{10}Be record and the reconstructed solar modulation record," which findings harmonize, in their words, with "numerous studies [that] show that solar activity is the main force that drives regional climate changes in the Holocene," in support of which statement they attach 22 other scientific references.

The researchers ultimately concluded that the "synchronous variations between Longxi precipitation and Northern Hemisphere temperature may be ascribed to solar activity," which apparently produced a Medieval Warm Period that was both longer and stronger than what has been experienced to date during the Current Warm Period in the northeast margin of the Tibetan Plateau.

Lastly, Xu *et al.* (2008) studied decadal-scale temperature variations of the past six centuries derived from four high-resolution temperature indicators—the δ^{18}O and δ^{13}C of bulk carbonate, total carbonate content, and the detrended δ^{15}N of organic matter—which they extracted from Lake Qinghai (36°32'-37°15'N, 99°36'-100°47'E) on the northeast Qinghai-Tibet plateau, comparing the resultant variations with proxy temperature indices derived from nearby tree rings and reconstructed solar activity. Results of the analysis showed that "there are four obvious cold intervals during the past 600 years at Lake Qinghai, namely 1430-1470, 1650-1715, 1770-1820 and 1920-1940," and that "these obvious cold intervals are also synchronous with the minimums of the sunspot numbers during the past 600 years," namely, "the Sporer, the Maunder, and the Dalton minimums," which facts strongly suggest, in their words, "that solar activities may dominate temperature variations on decadal scales at the northeastern Qinghai-Tibet plateau."

If the development of the significant cold of the worldwide Little Ice Age was driven by a concomitant change in some type of solar activity, which seems fairly well proven by a wealth of real-world data, it logically follows that the global warming of the twentieth century was driven primarily by the reversal of that change in solar

activity, and not by the historical rise in the air's CO_2 content. However, as also noted by Xu *et al.*, how small perturbations of solar activity have led "to the observed global warming, what is the mechanism behind it, etc., are still open questions."

Additional information on this topic, including reviews of newer publications as they become available, can be found at http://www.co2science.org/subject/s/solartempasia.php.

References

Aono, Y. and Kazui, K. 2008. Phenological data series of cherry tree flowering in Kyoto, Japan, and its application to reconstruction of springtime temperatures since the 9th century. *International Journal of Climatology* **28**: 905-914.

Bashkirtsev, V.S. and Mashnich, G.P. 2003. Will we face global warming in the nearest future? *Geomagnetiz i Aeronomija* **43**: 132-135.

Bond, G., Kromer, B., Beer, J., Muscheler, R., Evans, M.N., Showers, W., Hoffmann, S., Lotti-Bond, R., Hajdas, I. and Bonani, G. 2001. Persistent solar influence on North Atlantic climate during the Holocene. *Science* **294**: 2130-2136.

Chistyakov, V.F. 1996. On the structure of the secular cycles of solar activity. In: *Solar Activity and Its Effect on the Earth* (Chistyakov, V.F., Asst. Ed.), Dal'nauka, Vladivostok, Russia, pp. 98-105.

Chistyakov, V.F. 2000. On the sun's radius oscillations during the Maunder and Dalton Minimums. In: *Solar Activity and Its Effect on the Earth* (Chistyakov, V.F., Asst. Ed.), Dal'nauka, Vladivostok, Russia, pp. 84-107.

Cini Castagnoli, G., Taricco, C. and Alessio, S. 2005. Isotopic record in a marine shallow-water core: Imprint of solar centennial cycles in the past 2 millennia. *Advances in Space Research* **35**: 504-508.

Dergachev, V.A. and Raspopov, O.M. 2000. Long-term processes on the sun controlling trends in the solar irradiance and the earth's surface temperature. *Geomagnetism and Aeronomy* **40**: 9-14.

Friis-Christensen, E. and Lassen, K. 1991. Length of the solar cycle: An indicator of solar activity closely associated with climate. *Science* **254**: 698-700.

Hong, Y.T., Jiang, H.B., Liu, T.S., Zhou, L.P., Beer, J., Li, H.D., Leng, X.T., Hong, B. and Qin, X.G. 2000. Response of climate to solar forcing recorded in a 6000-year δ^{18}O time-series of Chinese peat cellulose. *The Holocene* **10**: 1-7.

Kalugin, I., Daryin, A., Smolyaninova, L., Andreev, A., Diekmann, B. and Khlystov, O. 2007. 800-yr-long records of annual air temperature and precipitation over southern Siberia inferred from Teletskoye Lake sediments. *Quaternary Research* **67**: 400-410.

Kalugin, I., Selegei, V., Goldberg, E. and Seret, G. 2005. Rhythmic fine-grained sediment deposition in Lake Teletskoye, Altai, Siberia, in relation to regional climate change. *Quaternary International* **136**: 5-13.

Kerr, R.A. 2008. Chinese cave speaks of a fickle sun bringing down ancient dynasties. *Science* **322**: 837-838.

Kitagawa, H. and Matsumoto, E. 1995. Climatic implications of $\delta^{13}C$ variations in a Japanese cedar (*Cryptomeria japonica*) during the last two millennia. *Geophysical Research Letters* **22**: 2155-2158.

Paulsen, D.E., Li, H.-C. and Ku, T.-L. 2003. Climate variability in central China over the last 1270 years revealed by high-resolution stalagmite records. *Quaternary Science Reviews* **22**: 691-701.

Porter, S.C. and Weijian, Z. 2006. Synchronism of Holocene East Asian monsoon variations and North Atlantic drift-ice tracers. *Quaternary Research* **65**: 443-449.

Solanki, S.K., Usoskin, I.G., Kromer, B., Schussler, M. and Beer, J. 2004. Unusual activity of the Sun during recent decades compared to the previous 11,000 years. *Nature* **431**: 1084-1087.

Tan, L., Cai, Y., An, Z. and Ai, L. 2008. Precipitation variations of Longxi, northeast margin of Tibetan Plateau since AD 960 and their relationship with solar activity. *Climate of the Past* **4**: 19-28.

Tan, M., Hou, J. and Liu, T. 2004. Sun-coupled climate connection between eastern Asia and northern Atlantic. *Geophysical Research Letters* **31**: 10.1029/2003GL019085.

Tan, M., Liu, T.S., Hou, J., Qin, X., Zhang, H. and Li, T. 2003. Cyclic rapid warming on centennial-scale revealed by a 2650-year stalagmite record of warm season temperature. *Geophysical Research Letters* **30**: 10.1029/2003GL017352.

Vaganov, E.A., Briffa, K.R., Naurzbaev, M.M., Schweingruber, F.H., Shiyatov, S.G. and Shishov, V.V. 2000. Long-term climatic changes in the arctic region of the Northern Hemisphere. *Doklady Earth Sciences* **375**: 1314-1317.

Wang, N., Yao, T., Thompson, L.G., Henderson, K.A. and Davis, M.E. 2002. Evidence for cold events in the early Holocene from the Guliya ice core, Tibetean Plateau, China. *Chinese Science Bulletin* **47**: 1422-1427.

Xu, H., Hong, Y.T., Lin, Q.H., Hong, B., Jiang, H.B. and Zhu, Y.X. 2002. Temperatures in the past 6000 years inferred from $\delta^{18}O$ of peat cellulose from Hongyuan, China. *Chinese Science Bulletin* **47**: 1578-1584.

Xu, H., Hong, Y., Lin, Q., Zhu, Y., Hong, B. and Jiang, H. 2006. Temperature responses to quasi-100-yr solar variability during the past 6000 years based on $\delta^{18}O$ of peat cellulose in Hongyuan, eastern Qinghai-Tibet plateau, China. *Palaeogeography, Palaeoclimatology, Palaeoecology* **230**: 155-164.

Xu, H., Liu, X. and Hou, Z. 2008. Temperature variations at Lake Qinghai on decadal scales and the possible relation to solar activities. *Journal of Atmospheric and Solar-Terrestrial Physics* **70**: 138-144.

Zhang, P., Cheng, H., Edwards, R.L., Chen, F., Wang, Y., Yang, X., Liu, J., Tan, M., Wang, X., Liu, J., An, C., Dai, Z., Zhou, J., Zhang, D., Jia, J., Jin, L. and Johnson, K.R. 2008. A test of climate, sun, and culture relationships from an 1810-year Chinese cave record. *Science* **322**: 940-942.

Zherebtsov, G.A. and Kovalenko, V.A. 2000. Effect of solar activity on hydrometeorological characteristics in the Baikal region. *Proceedings of the International Conference "Solar Activity and Its Terrestrial Manifestations,"* Irkutsk, Russia, p. 54.

5.3.6. Europe

We begin our review of the sun's influence on Europe's temperatures with the study of Holzhauser *et al.* (2005), who presented high-resolution records of variations in glacier size in the Swiss Alps together with lake-level fluctuations in the Jura mountains, the northern French Pre-Alps, and the Swiss Plateau in developing a 3,500-year climate history of west-central Europe, beginning with an in-depth analysis of the Great Aletsch glacier, which is the largest of all glaciers located in the European Alps.

Near the beginning of the time period studied, the three researchers report that "during the late Bronze Age Optimum from 1350 to 1250 BC, the Great Aletsch glacier was approximately 1000 m shorter than it is today," noting that "the period from 1450 to 1250 BC has been recognized as a warm-dry phase in other Alpine and Northern Hemisphere proxies (Tinner *et al.*, 2003)." Then, after an intervening unnamed cold-wet phase, when the glacier grew in both mass and length, they say that "during the Iron/Roman Age Optimum between c. 200 BC and AD 50," which is perhaps better known as the Roman Warm Period, the glacier again retreated and "reached today's extent or was even somewhat shorter than today." Next came the Dark Ages Cold Period, which they say was followed by "the Medieval Warm Period, from around AD 800 to the onset of the Little Ice Age around AD 1300," which latter cold-wet phase was "characterized by three successive [glacier length] peaks: a first maximum after 1369 (in the late 1370s), a second between 1670 and 1680, and a third at 1859/60," following which the glacier began its latest and still-ongoing recession in 1865. In addition, they state that written documents from the fifteenth century AD indicate that at some time during that hundred-year interval "the glacier was of a size similar to that of the 1930s," which latter period in many parts of the world was as warm as, or even warmer than, it is today. Data pertaining to the Gorner glacier (the second largest of the Swiss Alps) and the Lower Grindelwald glacier of the Bernese Alps tell much the same story, as Holzhauser et al. report that these glaciers and the Great Aletsch glacier "experienced nearly synchronous advances" throughout the study period.

With respect to what was responsible for the millennial-scale climatic oscillation that produced the alternating periods of cold-wet and warm-dry conditions that fostered the similarly paced cycle of glacier growth and retreat, the Swiss and French scientists report that "glacier maximums coincided with radiocarbon peaks, i.e., periods of weaker solar activity," which in their estimation "suggests a possible solar origin of the climate oscillations punctuating the last 3500 years in west-central Europe, in agreement with previous studies (Denton and Karlén, 1973; Magny, 1993; van Geel et al., 1996; Bond et al., 2001)." And to underscore that point, they conclude their paper by stating that "a comparison between the fluctuations of the Great Aletsch glacier and the variations in the atmospheric residual ^{14}C records supports the hypothesis that variations in solar activity were a major forcing factor of climate oscillations in west-central Europe during the late Holocene."

In another study of paleoclimate in western Europe, Mauquoy et al. (2002a) extracted peat monoliths from ombrotrophic mires at Lille Vildmose, Denmark (56°50'N, 10°15'E) and Walton Moss, UK (54°59'N, 02°46'W), which sites, being separated by about 800 km, "offer the possibility of detecting supraregional changes in climate." From these monoliths, vegetative macrofossils were extracted at contiguous 1-cm intervals and examined using light microscopy. Where increases in the abundances of Sphagnum tenellum and Sphagnum

cuspidatum were found, a closely spaced series of ^{14}C AMS-dated samples immediately preceding and following each increase was used to "wiggle-match" date them (van Geel and Mook, 1989), thereby enabling comparison of the climate-induced shifts with the history of ^{14}C production during the Holocene.

Results indicated the existence of a climatic deterioration that marked the beginning of a period of inferred cool, wet conditions that correspond fairly closely in time with the Wolf, Sporer, and Maunder Minima of solar activity, as manifest in contemporary $\delta^{14}C$ data. The authors report "these time intervals correspond to periods of peak cooling in 1000-year Northern Hemisphere climate records," adding to the "increasing body of evidence" that "variations in solar activity may well have been an important factor driving Holocene climate change."

Two years later, Mauquoy et al. (2004) reviewed the principles of ^{14}C wiggle-match dating, its limitations, and the insights it has provided about the timing and possible causes of climate change during the Holocene. Based upon their review, the authors stated that "analyses of microfossils and macrofossils from raised peat bogs by Kilian et al. (1995), van Geel et al. (1996), Speranza et al. (2000), Speranza (2000) and Mauquoy et al. (2002a, 2002b) have shown that climatic deteriorations [to cooler and wetter conditions] occurred during periods of transition from low to high delta ^{14}C (the relative deviation of the measured ^{14}C activity from the standard after correction for isotope fractionation and radioactive decay; Stuiver and Polach, 1977)." This close correspondence, in the words of the authors, again suggests that "changes in solar activity may well have driven these changes during the Bronze Age/Iron Age transition around c. 850 cal. BC (discussed in detail by van Geel et al., 1996, 1998, 1999, 2000) and the 'Little Ice Age' series of palaeoclimatic changes."

Working with a marine sediment core retrieved from the southern Norwegian continental margin, Berstad et al. (2003) reconstructed sea surface temperatures (SSTs) from $\delta^{18}O$ data derived from the remains of the planktonic foraminifera species Neogloboquadrina pachyderma (summer temperatures) and Globigerina bulloides (spring temperatures). Among other things, the authors' work depicted a clear connection between the cold temperatures of the Little Ice Age and the reduced solar activity of the concomitant Maunder and Sporer solar minima, as well as between the warm

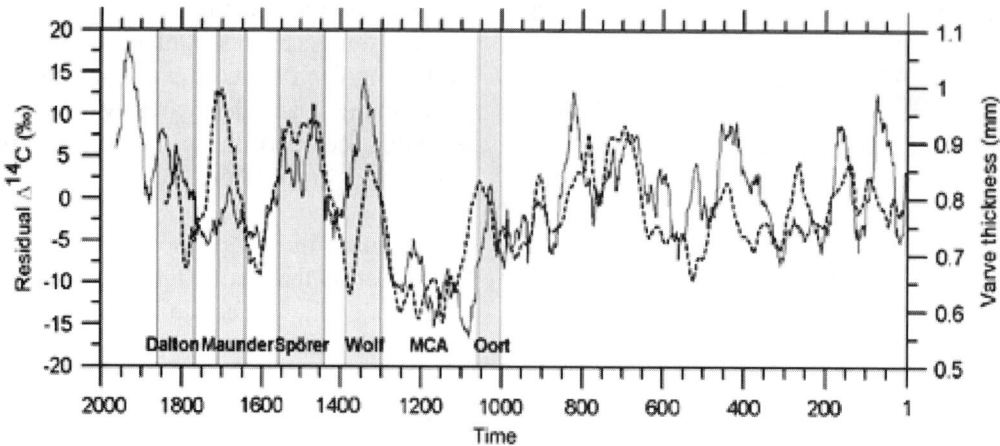

Figure 5.3.6. Residual $\Delta^{14}C$ data (dashed line) and varve thickness (smooth line) vs. time, specifically highlighting the Oort, Wolf, Sporer, Maunder and Dalton solar activity minima, as well as the "Medieval Climate Anomaly (also referred to as Medieval Warm Period)," during the contemporaneous "solar activity maxima in the Middle Ages." Adapted from Haltia-Hovi *et al.* (2007).

temperatures of the most recent 70 years and the enhanced solar activity of the concomitant Modern solar maximum, which they clearly implied in their paper is a causative connection, as is also implied by the recent sunspot number reconstruction of Usoskin *et al.* (2003).

Nearby in Finland, Haltia-Hovi *et al.* (2007) extracted sediment cores from beneath the 0.7-m-thick ice platform on Lake Lehmilampi (63°37'N, 29°06'E) in North Karelia, eastern Finland, after which they identified and counted the approximately 2,000 annual varves contained in the cores and measured their individual thicknesses and mineral and organic matter contents. These climate-related data were then compared with residual $\Delta^{14}C$ data derived from tree rings, which serve as a proxy for solar activity.

According to Haltia-Hovi *et al.*, their "comparison of varve parameters (varve thickness, mineral and organic matter accumulation) and the activity of the sun, as reflected in residual $\Delta^{14}C$ [data] appears to coincide remarkably well in Lake Lehmilampi during the last 2000 years, suggesting solar forcing of the climate," as depicted in Figure 5.3.6 for the case of varve thickness. What is more, the low deposition rate of mineral matter in Lake Lehmilampi in AD 1060-1280 "possibly implies mild winters with a short ice cover period during that time with minor snow accumulation interrupted by thawing periods." Likewise, they say that the low accumulation of organic matter during this period "suggests a long open water season and a high decomposition rate of organic matter." Consequently,

since the AD 1060-1280 period shows the lowest levels of both mineral and organic matter content, and since "the thinnest varves of the last 2000 years were deposited during [the] solar activity maxima in the Middle Ages," it is difficult not to conclude that that period was likely the warmest of the past two millennia in the part of the world studied by the three scientists.

Hanna *et al.* (2004) analyzed several climatic variables over the past century in Iceland in an effort to determine if there is "possible evidence of recent climatic changes" in that cold island nation. Results indicated that for the period 1923-2002, no trend was found in either annual or monthly sunshine data. Similar results were reported for annual and monthly pressure data, which exhibited semi-decadal oscillations throughout the 1820-2002 period but no significant upward or downward trend. Precipitation, on the other hand, appears to have increased slightly, although the authors question the veracity of the trend, citing a number of biases that have potentially corrupted the database.

With respect to temperature, however, the authors indicate that of the handful of locations they examined for this variable, all stations experienced a net warming since the mid-1800s. The warming, however, was not linear over the entire time period. Rather, temperatures rose from their coldest levels in the mid-1800s to their warmest levels in the 1930s, whereupon they remained fairly constant for approximately three decades. Then came a period of rapid cooling, which ultimately gave way to the warming of the 1980s and 1990s. However, it is important to note that the warming of the past two decades has not resulted in temperatures rising above those observed in the 1930s. In this point the authors are particularly clear, stating emphatically that "the 1990s was definitely *not* the warmest decade of the 20th century in Iceland, in contrast to the Northern Hemisphere land average." In fact, a linear trend fit to the post-1930 data would indicate an overall temperature decrease since that time.

As for what may be responsible for the various trends evident in the data, Hanna *et al.* note the likely influence of the sun on temperature and pressure values in consequence of their finding a significant correlation between 11-year running temperature means and sunspot numbers, plus the presence of a 12-year peak in their spectral analysis of the pressure data, which they say is "suggestive of solar activity."

In another study, Mangini *et al.* (2005) develop a highly resolved 2,000-year $\delta^{18}O$ proxy record of temperature obtained from a stalagmite recovered from Spannagel Cave in the Central Alps of Austria. Results indicated that the lowest temperatures of the past two millennia occurred during the Little Ice Age (AD 1400-1850), while the highest temperatures were found in the Medieval Warm Period (MWP: AD 800-1300). Furthermore, Mangini *et al.* say that the highest temperatures of the MWP were "slightly higher than those of the top section of the stalagmite (1950 AD) and higher than the present-day temperature." At three different points during the MWP, their data indicate temperature spikes in excess of 1°C above present (1995-1998) temperatures.

Mangini *et al.* additionally report that their temperature reconstruction compares well with reconstructions developed from Greenland ice cores (Muller and Gordon, 2000), Bermuda Rise ocean-bottom sediments (Keigwin, 1996), and glacier tongue advances and retreats in the Alps (Holzhauser, 1997; Wanner *et al.*, 2000), as well as with the Northern Hemispheric temperature reconstruction of Moberg *et al.* (2005). Considered together, they say these several datasets "indicate that the MWP was a climatically distinct period in the Northern Hemisphere," emphasizing that "this conclusion is in strong contradiction to the temperature reconstruction by the IPCC, which only sees the last 100 years as a period of increased temperature during the last 2000 years."

In a second severe blow to the theory of CO_2-induced global warming, Mangini *et al.* found "a high correlation between $\delta^{18}O$ and $\delta^{14}C$, that reflects the amount of radiocarbon in the upper atmosphere," and they note that this correlation "suggests that solar variability was a major driver of climate in Central Europe during the past 2 millennia." In this regard, they report that "the maxima of $\delta^{18}O$ coincide with solar minima (Dalton, Maunder, Sporer, Wolf, as well as with minima at around AD 700, 500 and 300)," and that "the coldest period between 1688 and 1698 coincided with the Maunder Minimum." Also, in a linear-model analysis of the percent of variance of their full temperature reconstruction that is individually explained by solar and CO_2 forcing, they found that the impact of the sun was fully 279 times greater than that of the air's CO_2 concentration, noting that "the flat evolution of CO_2 during the first 19 centuries yields almost vanishing correlation coefficients with the temperature reconstructions."

Two years later, Mangini *et al.* (2007) updated the 2005 study with additional data after which they compared it with the Hematite-Stained-Grain (HSG) history of ice-rafted debris in North Atlantic Ocean sediments developed by Bond *et al.* (2001), finding an undeniably good correspondence between the peaks and valleys of their $\delta^{18}O$ curve and the HSG curve. The significance of such correspondence is evidenced by the fact that Bond *et al.* reported that "over the last 12,000 years virtually every centennial time-scale increase in drift ice documented in our North Atlantic records was tied to a solar minimum."

Other researchers have found similar periodicities in their climate proxies. Turner *et al.* (2008), for example, found an ~1500 year cycle in a climate history reconstructed from sediment cores extracted from two crater lake basins in central Turkey, which they indicate "may be linked with large-scale climate forcing" such as that found in the North Atlantic by Bond *et al.* (1997, 2001). McDermott *et al.* (2001) found evidence of millennial-scale climate cycles in a $\delta^{18}O$ record from a stalagmite in southwestern Ireland, as did Sbaffi *et al.* (2004) from two deep-sea sediment cores recovered from the Tyrrhenian Sea, which latter proxy corresponded well with the North Atlantic solar-driven cycles of Bond *et al.* (1997).

Nearby in the Mediterranean Sea, Cini Castagnoli *et al.* (2002) searched for possible solar-induced variations in the $\delta^{13}C$ record of the foraminifera *Globigerinoides rubber* obtained from a sea core located in the Gallipoli terrace of the Gulf of Taranto (39°45'53"N, 17°53'33"E, depth of 178 m) over the past 1,400 years. Starting at the beginning of the 1,400-year record, the $\delta^{13}C$ values increased from about 0.4 per mil around 600 A.D. to a value of 0.8 per mil by 900 A.D. Thereafter, the $\delta^{13}C$ record remained relatively constant until about 1800, when it rose another 0.2 per mil to its present-day value of around 1.0 per mil.

Using statistical procedures, the authors were able to identify three important cyclical components in their record, with periods of approximately 11.3, 100, and 200 years. Comparison of both the raw $\delta^{13}C$ and component data with the historical aurorae and sunspot time series, respectively, revealed that the records are "associable in phase" and "disclose a statistically significant imprint of the solar activity in a climate record." Three years later, Cini Castagnoli *et al.* (2005) extended the $\delta^{13}C$ temperature proxy from the Gulf of Taranto an additional 600 years, reporting an overall phase agreement between the climate reconstruction and variations in the sunspot number series that "favors the hypothesis that the [multi-decadal] oscillation revealed in $\delta^{13}C$ is connected to the solar activity."

Finally, we report on the study of Desprat *et al.* (2003), who studied the climatic variability of the last three millennia in northwest Iberia via a high-resolution pollen analysis of a sediment core retrieved from the central axis of the Ria de Vigo in the south of Galicia (42°14.07'N, 8°47.37'W). According to the authors, over the past 3,000 years there was "an alternation of three relatively cold periods with three relatively warm episodes." In order of their occurrence, these periods are described by the authors as the "first cold phase of the Subatlantic period (975-250 BC)," which was "followed by the Roman Warm Period (250 BC-450 AD)," which was followed by "a successive cold period (450-950 AD), the Dark Ages," which "was terminated by the onset of the Medieval Warm Period (950-1400 AD)," which was followed by "the Little Ice Age (1400-1850 AD), including the Maunder Minimum (at around 1700 AD)," which "was succeeded by the recent warming (1850 AD to the present)." Based upon this "millennial-scale climatic cyclicity over the last 3000 years," which parallels "global climatic changes recorded in North Atlantic marine records (Bond *et al.*, 1997; Bianchi and McCave, 1999; Chapman and Shackelton, 2000)," Desprat *et al.* conclude that "solar radiative budget and oceanic circulation seem to be the main mechanisms forcing this cyclicity in NW Iberia."

In conclusion, paleoclimatic studies from Europe provide more evidence is for the global reality of the solar-induced millennial-scale oscillation of temperatures pervading both glacial and interglacial periods. The Current Warm Period can consequently be viewed as the most recent manifestation of this recurring phenomenon and unrelated to the concurrent historical increase in the air's CO_2 content.

Additional information on this topic, including reviews of newer publications as they become available, can be found at http://www.co2science.org/subject/s/solartempeurope.php.

References

Berstad, I.M., Sejrup, H.P., Klitgaard-Kristensen, D. and Haflidason, H. 2003. Variability in temperature and geometry of the Norwegian Current over the past 600 yr; stable isotope and grain size evidence from the Norwegian margin. *Journal of Quaternary Science* **18**: 591-602.

Bianchi, G.G. and McCave, I.N. 1999. Holocene periodicity in North Atlantic climate and deep-ocean flow south of Iceland. *Nature* **397**: 515-517.

Bond, G., Showers, W., Cheseby, M., Lotti, R., Almasi, P., deMenocal, P., Priore, P., Cullen, H., Hajdas, L. and Bonani, G. 1997. A pervasive millennial-scale cycle in North Atlantic Holocene and Glacial climates. *Science* **278**: 1257-1266.

Bond, G., Kromer, B., Beer, J., Muscheler, R., Evans, M.N., Showers, W., Hoffmann, S., Lotti-Bond, R., Hajdas, I. and Bonani, G. 2001. Persistent solar influence on North Atlantic climate during the Holocene. *Science* **294**: 2130-2136.

Chapman, M.R. and Shackelton, N.L. 2000. Evidence of 550-year and 1000-year cyclicities in North Atlantic circulation patterns during the Holocene. *The Holocene* **10**: 287-291.

Cini Castagnoli, G.C., Bonino, G., Taricco, C. and Bernasconi, S.M. 2002. Solar radiation variability in the last 1400 years recorded in the carbon isotope ratio of a Mediterranean sea core. *Advances in Space Research* **29**: 1989-1994.

Cini Castagnoli, G., Taricco, C. and Alessio, S. 2005. Isotopic record in a marine shallow-water core: Imprint of solar centennial cycles in the past 2 millennia. *Advances in Space Research* **35**: 504-508.

Denton, G.H. and Karlén, W. 1973. Holocene climate variations—their pattern and possible cause. *Quaternary Research* **3**: 155-205.

Desprat, S., Goñi, M.F.S. and Loutre, M.-F. 2003. Revealing climatic variability of the last three millennia in northwestern Iberia using pollen influx data. *Earth and Planetary Science Letters* **213**: 63-78.

Haltia-Hovi, E., Saarinen, T. and Kukkonen, M. 2007. A 2000-year record of solar forcing on varved lake sediment in eastern Finland. *Quaternary Science Reviews* **26**: 678-689.

Hanna, H., Jónsson, T. and Box, J.E. 2004. An analysis of Icelandic climate since the nineteenth century. *International Journal of Climatology* **24**: 1193-1210.

Holzhauser, H. 1997. Fluctuations of the Grosser Aletsch Glacier and the Gorner Glacier during the last 3200 years: new results. In: Frenzel, B. (Ed.) *Glacier Fluctuations During the Holocene*. Fischer, Stuttgart, Germany, pp. 35-58.

Holzhauser, H., Magny, M. and Zumbuhl, H.J. 2005. Glacier and lake-level variations in west-central Europe over the last 3500 years. *The Holocene* **15**: 789-801.

Keigwin, L.D. 1996. The Little Ice Age and Medieval Warm Period in the Sargasso Sea. *Science* **274**: 1503-1508.

Kilian, M.R., van der Plicht, J. and van Geel, B. 1995. Dating raised bogs: new aspects of AMS ^{14}C wiggle matching, a reservoir effect and climatic change. *Quaternary Science Reviews* **14**: 959-966.

Magny, M. 1993. Solar influences on Holocene climatic changes illustrated by correlations between past lake-level fluctuations and the atmospheric ^{14}C record. *Quaternary Research* **40**: 1-9.

Mangini, A., Spotl, C. and Verdes, P. 2005. Reconstruction of temperature in the Central Alps during the past 2000 yr from a δ^{18}O stalagmite record. *Earth and Planetary Science Letters* **235**: 741-751.

Mangini, A., Verdes, P., Spotl, C., Scholz, D., Vollweiler, N. and Kromer, B. 2007. Persistent influence of the North Atlantic hydrography on central European winter temperature during the last 9000 years. *Geophysical Research Letters* **34**: 10.1029/2006GL028600.

Mauquoy, D., Engelkes, T., Groot, M.H.M., Markesteijn, F., Oudejans, M.G., van der Plicht, J. and van Geel, B. 2002b. High resolution records of late Holocene climate change and carbon accumulation in two north-west European ombrotrophic peat bogs. *Palaeogeography, Palaeoclimatology, Palaeoecology* **186**: 275-310.

Mauquoy, D., van Geel, B., Blaauw, M., Speranza, A. and van der Plicht, J. 2004. Changes in solar activity and Holocene climatic shifts derived from ^{14}C wiggle-match dated peat deposits. *The Holocene* **14**: 45-52.

Mauquoy, D., van Geel, B., Blaauw, M. and van der Plicht, J. 2002a. Evidence from North-West European bogs shows 'Little Ice Age' climatic changes driven by changes in solar activity. *The Holocene* **12**: 1-6.

McDermott, F., Mattey, D.P. and Hawkesworth, C. 2001. Centennial-scale Holocene climate variability revealed by a high-resolution speleothem δ^{18}O record from SW Ireland. *Science* **294**: 1328-1331.

Moberg, A., Sonechkin, D.M., Holmgren, K., Datsenko, N.M. and Karlén, W. 2005. Highly variable Northern Hemisphere temperatures reconstructed from low- and high-resolution proxy data. *Nature* **433**: 613-617.

Muller, R.A. and Gordon, J.M. 2000. *Ice Ages and Astronomical Causes*. Springer-Verlag, Berlin, Germany.

Sbaffi, L., Wezel, F.C., Curzi, G. and Zoppi, U. 2004. Millennial- to centennial-scale palaeoclimatic variations during Termination I and the Holocene in the central Mediterranean Sea. *Global and Planetary Change* **40**: 201-217.

Speranza, A. 2000. Solar and Anthropogenic Forcing of Late-Holocene Vegetation Changes in the Czech Giant Mountains. PhD thesis. University of Amsterdam, Amsterdam, The Netherlands.

Speranza, A.O.M., van der Plicht, J. and van Geel, B. 2000. Improving the time control of the Subboreal/Subatlantic transition in a Czech peat sequence by ^{14}C wiggle-matching. *Quaternary Science Reviews* **19**: 1589-1604.

Stuiver, M. and Polach, H.A. 1977. Discussion: reporting ^{14}C data. *Radiocarbon* **19**: 355-363.

Tinner, W., Lotter, A.F., Ammann, B., Condera, M., Hubschmied, P., van Leeuwan, J.F.N. and Wehrli, M. 2003. Climatic change and contemporaneous land-use phases north and south of the Alps 2300 BC to AD 800. *Quaternary Science Reviews* **22**: 1447-1460.

Turner, R., Roberts, N. and Jones, M.D. 2008. Climatic pacing of Mediterranean fire histories from lake sedimentary microcharcoal. *Global and Planetary Change* **63**: 317-324.

Usoskin, I.G., Solanki, S.K., Schussler, M., Mursula, K. and Alanko, K. 2003. Millennium-scale sunspot number reconstruction: Evidence for an unusually active sun since the 1940s. *Physical Review Letters* **91**: 10.1103/PhysRevLett.91.211101.

van Geel, B. and Mook, W.G. 1989. High resolution ^{14}C dating of organic deposits using natural atmospheric ^{14}C variations. *Radiocarbon* **31**: 151-155.

van Geel, B., Buurman, J. and Waterbolk, H.T. 1996. Archaeological and palaeoecological indications of an abrupt climate change in the Netherlands and evidence for climatological teleconnections around 2650 BP. *Journal of Quaternary Science* **11**: 451-460.

van Geel, B., Heusser, C.J., Renssen, H. and Schuurmans, C.J.E. 2000. Climatic change in Chile at around 2700 BP and global evidence for solar forcing: a hypothesis. *The Holocene* **10**: 659-664.

van Geel, B., Raspopov, O.M., Renssen, H., van der Plicht, J., Dergachev, V.A. and Meijer, H.A.J. 1999. The role of solar forcing upon climate change. *Quaternary Science Reviews* **18**: 331-338.

van Geel, B., van der Plicht, J., Kilian, M.R., Klaver, E.R., Kouwenberg, J.H.M., Renssen, H., Reynaud-Farrera, I. and Waterbolk, H.T. 1998. The sharp rise of delta ^{14}C c. 800 cal BC: possible causes, related climatic teleconnections and the impact on human environments. *Radiocarbon* **40**: 535-550.

Wanner, H., Dimitrios, G., Luterbacher, J., Rickli, R., Salvisberg, E. and Schmutz, C. 2000. *Klimawandel im Schweizer Alpenraum*. VDF Hochschulverlag, Zurich, Switzerland.

5.3.7. Other

Rounding out our examination of the influence of the sun on earth's temperatures, we begin with the review study of Van Geel *et al.* (1999), who examined what is known about the relationship between variations in the abundances of the cosmogenic isotopes ^{14}C and ^{10}Be and millennial-scale climate oscillations during the Holocene and portions of the last great ice age. As they describe it, "there is mounting evidence suggesting that the variation in solar activity is a cause for millennial-scale climate change," which is known to operate independently of the glacial-interglacial cycles that are forced by variations in the earth's orbit about the sun. Continuing, they add that "accepting the idea of solar forcing of Holocene and Glacial climatic shifts has major implications for our view of present and future climate," for it implies, as they note, that "the climate system is far more sensitive to small variations in solar activity than generally believed" and that "it could mean that the global temperature fluctuations during the last decades are partly, or completely explained by small changes in solar radiation." These observations, of course, call into question the conventional wisdom of attributing the global warming of the past century or so to the ongoing rise in the air's CO_2 content.

In a study published the following year, Tyson *et al.* (2000) obtained a quasi-decadal-resolution record of oxygen and carbon-stable isotope data from a well-dated stalagmite recovered from Cold Air Cave in the Makapansgat Valley, 30 km southwest of Pietersburg, South Africa, which they augmented with temperature data reconstructed from color variations in banded growth-layer laminations of the stalagmite that were derived from a relationship calibrated against actual air temperatures obtained from a surrounding 49-station climatological network over the period 1981-1995, which had a correlation of +0.78 that was significant at the 99 percent confidence level.

According to the authors, both the Little Ice Age (prevailing from about AD 1300 to 1800) and the Medieval Warm Period (prevailing from before AD 1000 to around 1300) were found to be distinctive features of the climate of the last millennium. Relative to the period 1961-1990, in fact, the Little Ice Age, which "was a widespread event in South Africa specifically and southern Africa generally," was characterized by a mean annual temperature depression of about 1°C at its coolest point. The Medieval Warm Period, on the other hand, was as much as 3-4°C warmer at its warmest point. The researchers also note that the coolest point of the Little Ice Age corresponded in time with the Maunder Minimum of sunspot activity and that the Medieval Warm Period corresponded with the Medieval Maximum in solar activity.

In a study demonstrating a solar-climate link on shorter decadal to centennial time scales, Domack *et al.* (2001) examined ocean sediment cores obtained from the Palmer Deep on the inner continental shelf of the western Antarctic Peninsula (64° 51.71' S, 64° 12.47' W) to produce a high-resolution proxy temperature history of that area spanning the past 13,000 years. Results indicated the presence of five prominent palaeoenvironmental intervals over the past 14,000 years: (1) a "Neoglacial" cool period beginning 3,360 years ago and continuing to the present, (2) a mid-Holocene climatic optimum from 9,070 to 3,360 years ago, (3) a cool period beginning 11,460 years ago and ending at 9,070 years ago, (4) a warm period from 13,180 to 11,460 years ago, and (5) cold glacial conditions prior to 13,180 years ago. Spectral analyses of the data revealed that superimposed upon these broad climatic intervals were decadal and centennial-scale temperature cycles. Throughout the current Neoglacial period, they report finding "very significant" (above the 99 percent confidence level) peaks, or oscillations, that occurred at intervals of 400, 190, 122, 85, and 70 years, which they suggest are perhaps driven by solar variability.

Moving upward to the warmer ocean waters off the Cook Islands, South Pacific Ocean, Dima *et al.* (2005) performed Singular Spectrum Analysis on a Rarotonga coral-based sea surface temperature (SST) reconstruction in an effort to determine the dominant periods of multi-decadal variability in the series over the period 1727-1996. Results of the analysis revealed two dominant multi-decadal cycles, with periods of about 25 and 80 years. These modes of variability were determined to be similar to multi-decadal modes found in the global SST field of Kaplan *et al.* (1998) for the period 1856-1996. The ~25-year cycle was found to be associated with the well-known Pacific Decadal Oscillation, whereas the ~80-year cycle was determined to be "almost identical" to a pattern of solar forcing found by Lohmann *et al.* (2004), which, according to Dima *et al.*, "points to a possible solar origin" of this mode of SST variability.

We conclude this brief review with the study of Bard and Frank (2006), who reviewed what is known, and unknown, about solar variability and its effects on earth's climate, focusing on the past few decades, the past few centuries, the entire Holocene, and orbital timescales. Of greatest interest to the present discussion are Bard and Frank's conclusions about sub-orbital time scales, i.e., the first three of their four major focal points. Within this context, as they say in the concluding section of their review, "it appears that solar fluctuations were involved in causing widespread but limited climatic changes, such as the Little Ice Age (AD 1500-1800) that followed the Medieval Warm Period (AD 900-1400)." Or as they say in the concluding sentence of their abstract, "the weight of evidence suggests that solar changes have contributed to small climate oscillations occurring on time scales of a few centuries, similar in type to the fluctuations classically described for the last millennium: The so-called Medieval Warm Period (AD 900-1400) followed on by the Little Ice Age (AD 1500-1800)."

In the words of Bard and Frank, "Bond *et al.* (1997, 2001) followed by Hu *et al.* (2003) proposed that variations of solar activity are responsible for quasi-periodic climatic and oceanographic fluctuations that follow cycles of about one to two millennia." As a result, they say that "the succession from the Medieval Warm Period to the Little Ice Age would thus represent the last [such] cycle," leading to the conclusion that "our present climate is in an ascending phase on its way to attaining a new warm optimum," due to some form of solar variability. In addition, they note that "a recent modeling study suggests that an apparent 1500-year cycle could arise from the superimposed influence of the 90 and 210 year solar cycles on the climate system, which is characterized by both nonlinear dynamics and long time scale memory effects (Braun *et al.* 2005)."

These studies demonstrate that the warming of the earth since the termination of the Little Ice Age is not unusual or different from other climate changes of the past millennium, when atmospheric CO_2 concentrations were stable, lower than at present, and obviously not responsible for the observed variations in temperature. This further suggests that the warming of the past century was not due to the contemporaneous historical increase in the air's CO_2 content.

Additional information on this topic, including reviews of newer publications as they become available, can be found at http://www.co2science.org/subject/s/solartempmisc.php.

References

Bard, E. and Frank, M. 2006. Climate change and solar variability: What's new under the sun? *Earth and Planetary Science Letters* **248**: 1-14.

Bond, G., Kromer, B., Beer, J., Muscheler, R., Evans, M.N., Showers, W., Hoffmann, S., Lotti-Bond, R., Hajdas, I. and Bonani, G. 2001. Persistent solar influence on North Atlantic climate during the Holocene. *Science* **294**: 2130-2136.

Bond, G., Showers, W., Cheseby, M., Lotti, R., Almasi, P., deMenocal, P., Priore, P., Cullen, H., Hajdas, I. and Bonani, G. 1997. A pervasive millennial-scale cycle in North Atlantic Holocene and Glacial climate. *Science* **278**: 1257-1266.

Braun, H., Christl, M., Rahmstorf, S., Ganopolski, A., Mangini, A., Kubatzki, C., Roth, K. and Kromer, B. 2005. Possible solar origin of the 1470-year glacial climate cycle demonstrated in a coupled model. *Nature* **438**: 208-211.

Dima, M., Felis, T., Lohmann, G. and Rimbu, N. 2005. Distinct modes of bidecadal and multidecadal variability in a climate reconstruction of the last centuries from a South Pacific coral. *Climate Dynamics* **25**: 329-336.

Domack, E., Leventer, A., Dunbar, R., Taylor, F., Brachfeld, S., Sjunneskog, C. and ODP Leg 178 Scientific Party. 2001. Chronology of the Palmer Deep site, Antarctic Peninsula: A Holocene palaeoenvironmental reference for the circum-Antarctic. *The Holocene* **11**: 1-9.

Hu, F.S., Kaufman, D., Yoneji, S., Nelson, D., Shemesh, A., Huang, Y., Tian, J., Bond, G., Clegg, B. and Brown, T. 2003. Cyclic variation and solar forcing of Holocene climate in the Alaskan subarctic. *Science* **301**: 1890-1893.

Kaplan, A., Cane, M.A., Kushnir, Y., Clement, A.C., Blumenthal, M.B. and Rajagopalan, B. 1998. Analyses of global sea surface temperature 1856-1991. *Journal of Geophysical Research* **103**: 18,567-18,589.

Lohmann, G., Rimbu, N. and Dima, M. 2004. Climate signature of solar irradiance variations: analysis of long-term instrumental, historical, and proxy data. *International Journal of Climatology* **24**: 1045-1056.

Tyson, P.D., Karlén, W., Holmgren, K. and Heiss, G.A. 2000. The Little Ice Age and medieval warming in South Africa. *South African Journal of Science* **96**: 121-126.

Van Geel, B., Raspopov, O.M., Renssen, H., van der Plicht, J., Dergachev, V.A. and Meijer, H.A.J. 1999. The role of solar forcing upon climate change. *Quaternary Science Reviews* **18**: 331-338.

5.4. Precipitation

The IPCC claims to have found a link between CO_2 concentrations in the air and precipitation trends. In this section of our report, we show that solar variability offers a superior explanation of past trends in precipitation.

5.4.1. North America

We begin our review of the influence of the sun on North American precipitation with a study that examines the relationship between the sun and low-level clouds, considering the presence of low-level clouds to be correlated with precipitation. According to Kristjansson *et al.* (2002), solar irradiance "varies by about 0.1% over the 11-year solar cycle, which would appear to be too small to have an impact on climate." Nevertheless, they report that "persistent claims have been made of 11-year signals in various meteorological time series, e.g., sea surface temperature (White *et al.*, 1997) and cloudiness over North America (Udelhofen and Cess, 2001)." Kristjansson *et al.* purposed to "re-evaluate the statistical relationship between low cloud cover and solar activity adding 6 years of ISCCP [International Satellite Cloud Climatology Project] data that were recently released."

For the period 1983-1999, the authors compared temporal trends of solar irradiance at the top of the atmosphere with low cloud cover derived from different sets of satellite-borne instruments that provided two measures of the latter parameter: full temporal coverage and daytime-only coverage. Results indicated that "solar irradiance correlates well with low cloud cover," with the significance level of the correlation being 98 percent for the case of full temporal coverage and 90 percent for the case of daytime-only coverage. As would be expected if the variations in cloud cover were driven by variations in solar irradiance, they also report that lagged correlations between the two parameters reveal a maximum correlation between solar irradiance and low cloud cover when the former leads the latter by one month for the full temporal coverage case and by four months for the daytime-only situation.

The authors' observation that "low clouds appear to be significantly inversely correlated with solar irradiance" compelled them to suggest a possible physical mechanism that could explain this phenomenon. Very briefly, this mechanism, in their words, "acts through UV [ultraviolet radiation] in the stratosphere affecting tropospheric planetary waves and hence the subtropical highs, modulated by an interaction between sea surface temperature [SST] and lower tropospheric static stability," which "relies on a positive feedback between changes in SST and low cloud cover changes of opposite sign, in the subtropics." Based on experimentally determined values of factors that enter into this scenario, they obtain a value for the amplitude of the variation in low cloud cover over a solar cycle that "is very close to the observed amplitude."

In pursuing other indirect means of ferreting out a solar influence on precipitation, several authors have examined lake level fluctuations, which are generally highly dependent on precipitation levels. Cumming *et al.* (2002), for example, studied a sediment core retrieved from Big Lake (51°40'N, 121°27'W) on the Cariboo Plateau of British Columbia, Canada, carefully dating it and deriving estimates of changes in precipitation-sensitive limnological variables (salinity and lake depth) from transfer functions based on modern distributions of diatom assemblages in 219 lakes from western Canada.

On the basis of observed changes in patterns of the floristic composition of diatoms over the past 5,500 years, the authors report that "alternating millennial-scale periods of high and low moisture availability were inferred, with *abrupt* [our italics] transitions in diatom communities occurring 4960, 3770, 2300 and 1140 cal. yrs. BP." They also indicate that "periods of inferred lower lake depth correspond closely to the timing of worldwide Holocene glacier expansions," and that the mean length of "the relatively stable intervals between the abrupt transitions ... is similar to the mean Holocene pacing of IRD [ice rafted debris] events ... in the North Atlantic," which have been described by Bond *et al.* (1997) and attributed to "solar variability amplified through oceanic and atmospheric dynamics," as detailed by Bond *et al.* (2001).

Li *et al.* (2006, 2007) also developed a precipitation proxy from a lake-level record, based on lithologic and mineral magnetic data from the Holocene sediments of White Lake, New Jersey, northeastern USA (41°N, 74.8°W), the characteristics of which they compared with a host of other paleoclimatic reconstructions from this region and beyond.

According to the authors of these two papers, the lake-level history revealed low lake levels at ~1.3, 3.0, 4.4, and 6.1 thousand years before present; comparison of the results with drift-ice records from the North Atlantic Ocean according to Li *et al.* (2007) "indicates a striking correspondence," as they "correlate well with cold events 1, 2, 3 and 4 of Bond *et al.* (2001)." They also report that a comparison of their results with those of other land-based studies suggests "a temporally coherent pattern of climate variations at a quasi-1500-year periodicity at least in the Mid-Atlantic region, if not the entire northeastern USA." In addition, and with respect to the other node of the climatic cycle, they note that "the Mid-Atlantic region was dominated by wet conditions, while most parts of the conterminous USA experienced droughts, when the North Atlantic Ocean was warm."

In discussing their findings, the three researchers say the dry-cold correlation they found "resembles the modern observed relationship between moisture conditions in eastern North America and the North Atlantic Oscillation (NAO), but operates at millennial timescales, possibly through modulation of atmospheric dynamics by solar forcing," and in this regard they write that the sun-climate link on millennial timescales has "been demonstrated in several records (e.g., Bond *et al.*, 2001; Hu *et al.*, 2003; Niggemann *et al.*, 2003), supporting solar forcing as a plausible mechanism for modulating the AO [Arctic Oscillation]/NAO at millennial timescales."

Along the same vein of research, Dean *et al.* (2002) analyzed the varve thickness and continuous gray-scale density of sediment cores taken from Elk Lake, MN (47°12'N, 95°15'W) for the past 1,500 years. Results indicated the presence of significant periodicities throughout the record, including multidecadal periodicities of approximately 10, 29, 32, 42, and 96 years, and a strong multicentennial

periodicity of about 400 years, leading the authors to wonder whether the observed periodicities are manifestations of solar-induced climate signals, upon which they present strong correlative evidence that they are. The 10-year oscillation was found to be strongest in the time series between the fourteenth and nineteenth centuries, during the Little Ice Age, and may well have been driven by the 11-year sunspot cycle.

Further south in Mexico, Lozano-Garcia *et al.* (2007) conducted a high-resolution multi-proxy analysis of pollen, charcoal particles, and diatoms found in the sediments of Lago Verde (18°36'46" N, 95°20'52"W)—a small closed-basin lake on the outskirts of the Sierra de Los Tuxtlas (a volcanic field on the coast of the Gulf of Mexico)—which covered the past 2,000 years. The five Mexican researchers who conducted the study say their data "provide evidence that the densest tropical forest cover and the deepest lake of the last two millennia were coeval with the Little Ice Age, with two deep lake phases that follow the Sporer and Maunder minima in solar activity." In addition, they suggest that "the high tropical pollen accumulation rates limit the Little Ice Age's winter cooling to a maximum of 2°C," and they conclude that the "tropical vegetation expansion during the Little Ice Age is best explained by a reduction in the extent of the dry season as a consequence of increased meridional flow leading to higher winter precipitation." Concluding their study, Lozano-Garcia *et al.* state "the data from Lago Verde strongly suggest that during the Little Ice Age lake levels and vegetation at Los Tuxtlas were responding to solar forcing and provide further evidence that solar activity is an important element controlling decadal to centennial scale climatic variability in the tropics (Polissar *et al.*, 2006) and in general over the North Atlantic region (Bond *et al.*, 2001; Dahl-Jensen *et al.*, 1998)."

Across the Gulf of Mexico in a study from the sea, as opposed to from a lake, Lund and Curry (2004) analyzed planktonic foraminiferal $\delta^{18}O$ time series obtained from three well-dated sediment cores retrieved from the seabed near the Florida Keys (24.4°N, 83.3°W) in an effort to examine centennial to millennial timescale changes in the Florida Current-Gulf Stream system over the past 5,200 years. Based upon the analysis, isotopic data from the three cores indicated that "the surface Florida Current was denser (colder, saltier or both) during the Little Ice Age than either the Medieval Warm Period or today," and that "when considered with other published results (Keigwin, 1996; deMenocal *et al.*, 2000), it is possible that the entire subtropical gyre of the North Atlantic cooled during the Little Ice Age ... perhaps consistent with the simulated effects of reduced solar irradiance (Rind and Overpeck, 1993; Shindell *et al.*, 2001)." In addition, they report that "the coherence and phasing of atmospheric ^{14}C production and Florida Current δ^{18}O during the Late Holocene implies that solar variability may influence Florida Current surface density at frequencies between 1/300 and 1/100 years." Hence, we once again have a situation where both centennial- and millennial-scale climatic variability is explained by similar-scale variability in solar activity.

Moving up the Atlantic coast to Newfoundland, Hughes *et al.* (2006) derived a multi-proxy palaeoclimate record from Nordan's Pond Bog, a coastal plateau bog in Newfoundland, based upon "analyses of plant macrofossils, testate amoebae and the degree of peat humification," which enabled them to create "a single composite reconstruction of bog surface wetness (BSW)" that they compare with "records of cosmogenic isotope flux."

Results of the analysis indicated that "at least 14 distinctive phases of increased BSW may be inferred from the Nordan's Pond Bog record," commencing at 8,270 cal. years BP, and that "comparisons of the BSW reconstruction with records of cosmogenic isotope flux ... suggest a persistent link between reduced solar irradiance and increased BSW during the Holocene." Furthermore, Hughes *et al.* conclude that the "strong correlation between increased ^{14}C production [which accompanies reduced solar activity] and phases of maximum BSW supports the role of solar forcing as a persistent driver of changes to the atmospheric moisture balance throughout the Holocene," which finding further suggests that the sun likely orchestrated the Little Ice Age to Current Warm Period transition, which altered precipitation regimes around the globe. Consequently, the authors state that "evidence suggesting a link between solar irradiance and sub-Milankovitch-scale palaeoclimatic change has mounted," and the "solar hypothesis, as an explanation for Holocene climate change, is now gaining wider acceptance."

Asmerom *et al.* (2007) developed a high-resolution Holocene climate proxy for the southwest United States from δ^{18}O variations in a stalagmite obtained in Pink Panther Cave in the Guadalupe Mountains of New Mexico. Spectral analysis performed on the raw δ^{18}O data revealed significant peaks that the researchers say "closely match previously reported periodicities in the ^{14}C content of the atmosphere, which have been attributed to periodicities in the solar cycle (Stuiver and Braziunas, 1993)." More specifically, they say that cross-spectral analysis of the Δ^{14}C and δ^{18}O data confirms that the two records have matching periodicities at 1,533 years (the Bond cycle), 444 years, 170 years, 146 years, and 88 years (the Gleissberg cycle). In addition, they report that periods of increased solar radiation correlate with periods of decreased rainfall in the southwestern United States (via changes in the North American monsoon), and that this behavior is just the opposite of what is observed with the Asian monsoon. These observations lead them to suggest that the proposed solar link to Holocene climate operates "through changes in the Walker circulation and the Pacific Decadal Oscillation and El Niño-Southern Oscillation systems of the tropical Pacific Ocean."

Since the warming of the twentieth century appears to represent the most recent rising phase of the Bond cycle, which in its previous rising phase produced the Medieval Warm Period (see Bond *et al.*, 2001), and since we could still be embedded in that rising temperature phase, which could well continue for some time to come, there is a reasonable probability that the desert southwest of the United States could experience an intensification of aridity in the not-too-distant future, and that wetter conditions could be expected in the monsoon regions of Asia, without atmospheric greenhouse gases playing any role.

Additional information on this topic, including reviews of newer publications as they become available, can be found at http://www.co2science.org/subject/s/solarpcpnamer.php.

References

Asmerom, Y., Polyak, V., Burns, S. and Rassmussen, J. 2007. Solar forcing of Holocene climate: New insights from a speleothem record, southwestern United States. *Geology* **35**: 1-4.

Bond, G., Kromer, B., Beer, J., Muscheler, R., Evans, M.N., Showers, W., Hoffmann, S., Lotti-Bond, R., Hajdas, I. and Bonani, G. 2001. Persistent solar influence on North Atlantic climate during the Holocene. *Science* **294**: 2130-2136.

Bond, G., Showers, W., Chezebiet, M., Lotti, R., Almasi, P., deMenocal, P., Priore, P., Cullen, H., Hajdas, I. and Bonani, G. 1997. A pervasive millennial scale cycle in

North-Atlantic Holocene and glacial climates. *Science* **278**: 1257-1266.

Cumming, B.F., Laird, K.R., Bennett, J.R., Smol, J.P. and Salomon, A.K. 2002. Persistent millennial-scale shifts in moisture regimes in western Canada during the past six millennia. *Proceedings of the National Academy of Sciences, USA* **99**: 16,117-16,121.

Dahl-Jensen, D., Mosegaard, K., Gundestrup, N., Clow, G.D., Johnsen, S.J., Hansen, A.W. and Balling, N. 1998. Past temperatures directly from the Greenland Ice Sheet. *Science* **282**: 268-271.

Dean, W., Anderson, R., Bradbury, J.P. and Anderson, D. 2002. A 1500-year record of climatic and environmental change in Elk Lake, Minnesota I: Varve thickness and gray-scale density. *Journal of Paleolimnology* **27**: 287-299.

deMenocal, P., Ortiz, J., Guilderson, T. and Sarnthein, M. 2000. Coherent high- and low-latitude variability during the Holocene warm period. *Science* **288**: 2198-2202.

Hu, F.S., Kaufman, D., Yoneji, S., Nelson, D., Shemesh, A., Huang, Y., Tian, J., Bond, G., Clegg, B. and Brown, T. 2003. Cyclic variation and solar forcing of Holocene climate in the Alaskan subarctic. *Science* **301**: 1890-1893.

Hughes, P.D.M., Blundell, A., Charman, D.J., Bartlett, S., Daniell, J.R.G., Wojatschke, A. and Chambers, F.M. 2006. An 8500 cal. year multi-proxy climate record from a bog in eastern Newfoundland: contributions of meltwater discharge and solar forcing. *Quaternary Science Reviews* **25**: 1208-1227.

Keigwin, L. 1996. The Little Ice Age and Medieval Warm Period in the Sargasso Sea. *Science* **274**: 1504-1508.

Kristjansson, J.E., Staple, A. and Kristiansen, J. 2002. A new look at possible connections between solar activity, clouds and climate. *Geophysical Research Letters* **29**: 10.1029/2002GL015646.

Li, Y.-X., Yu, Z. and Kodama, K.P. 2007. Sensitive moisture response to Holocene millennial-scale climate variations in the Mid-Atlantic region, USA. *The Holocene* **17**: 3-8.

Li, Y.-X., Yu, Z., Kodama, K.P. and Moeller, R.E. 2006. A 14,000-year environmental change history revealed by mineral magnetic data from White Lake, New Jersey, USA. *Earth and Planetary Science Letters* **246**: 27-40.

Lozano-Garcia, Ma. del S., Caballero, M., Ortega, B., Rodriguez, A. and Sosa, S. 2007. Tracing the effects of the Little Ice Age in the tropical lowlands of eastern Mesoamerica. *Proceedings of the National Academy of Sciences, USA* **104**: 16,200-16,203.

Lund, D.C. and Curry, W.B. 2004. Late Holocene variability in Florida Current surface density: Patterns and possible causes. *Paleoceanography* **19**: 10.1029/2004PA001008.

Niggemann, S., Mangini, A., Mudelsee, M., Richter, D.K., and Wurth, G. 2003. Sub-Milankovitch climatic cycles in Holocene stalagmites from Sauerland, Germany. *Earth and Planetary Science Letters* **216**: 539-547.

Polissar, P.J., Abbott, M.B., Wolfe, A.P., Bezada, M., Rull, V. and Bradley, R.S. 2006. Solar modulation of Little Ice Age climate in the tropical Andes. *Proceedings of the National Academy of Sciences USA* **103**: 8937-8942.

Rind, D. and Overpeck, J. 1993. Hypothesized causes of decade- to century-scale climate variability: Climate model results. *Quaternary Science Reviews* **12**: 357-374.

Shindell, D.T., Schmidt, G.A., Mann, M.E., Rind, D. and Waple, A. 2001. Solar forcing of regional climate during the Maunder Minimum. *Science* **294**: 2149-2152.

Stuiver, M. and Braziunas, T.F. 1993. Sun, ocean climate and atmospheric $^{14}CO_2$: An evaluation of causal and spectral relationships. *The Holocene* **3**: 289-305.

Udelhofen, P.M. and Cess, R.D. 2001. Cloud cover variations over the United States: An influence of cosmic rays or solar variability? *Geophysical Research Letters* **28**: 2617-2620.

White, W.B., Lean, J., Cayan, D.R. and Dettinger, M.D. 1997. Response of global upper ocean temperature to changing solar irradiance. *Journal of Geophysical Research* **102**: 3255-3266.

5.4.2. South America

We begin our investigation of the influence of the sun on precipitation in South America with the study of Nordemann *et al.* (2005), who analyzed tree rings from species sensitive to fluctuations in precipitation from the southern region of Brazil and Chile along with sunspot data via harmonic spectral and wavelet analysis in an effort to obtain a greater understanding of the effects of solar activity, climate, and geophysical phenomena on the continent of South America, where the time interval covered by the tree-ring samples from Brazil was 200 years and that from Chile was 2,500 years. Results from the spectral analysis revealed periodicities in the tree rings that corresponded well with the DeVries-Suess (~200 yr), Gleissberg (~80 yr), Hale (~22 yr), and Schwabe (~11 yr) solar activity cycles; while wavelet cross-spectrum analysis of sunspot number and tree-ring growth revealed a clear relation between the tree-ring and solar series.

Working with a sediment core retrieved from the main basin of Lake Titicaca (16°S, 69°W) on the Altiplano of Bolivia and Peru, Baker *et al.* (2005) reconstructed the lake-level history of the famous South American water body over the past 13,000 years at decadal to multi-decadal resolution based on $\delta^{13}C$ measurements of sediment bulk organic matter.

Baker *et al.* report that "the pattern and timing of lake-level change in Lake Titicaca is similar to the ice-rafted debris record of Holocene Bond events, demonstrating a possible coupling between precipitation variation on the Altiplano and North Atlantic sea-surface temperatures." Noting that "cold periods of the Holocene Bond events correspond with periods of increased precipitation on the Altiplano," they further conclude that "Holocene precipitation variability on the Altiplano is anti-phased with respect to precipitation in the Northern Hemisphere monsoon region." In further support of these findings, they add that "the relationship between lake-level variation at Lake Titicaca and Holocene Bond events also is supported by the more coarsely resolved (but very well documented) record of water-level fluctuations over the past 4000 years based on the sedimentology of cores from the shallow basin of the lake (Abbott *et al.*, 1997)."

Also examining a time interval spanning the length of the Holocene, Haug *et al.* (2001) utilized the titanium and iron concentrations of an ocean sediment core taken from a depth of 893 meters in the Cariaco Basin on the Northern Shelf of Venezuela (10°42.73'N, 65°10.18'W) to infer variations in the hydrologic cycle over northern South America over the past 14,000 years. Results indicated that titanium and iron concentrations were lower during the Younger Dryas cold period between 12.6 and 11.5 thousand years ago, corresponding to a weakened hydrologic cycle with less precipitation and runoff. During the Holocene Optimum (10.5 to 5.4 thousand years ago), however, concentrations of these metals remained at or near their highest values, suggesting wet conditions and an enhanced hydrologic cycle for more than five thousand years. Closer to the present, the largest century-scale variations in precipitation are inferred in the record between approximately 3.8 and 2.8 thousand years ago, as the amounts of these metals in the sediment record varied widely over short time intervals. Higher precipitation was noted during the Medieval Warm Period from 1.05 to 0.7 thousand years ago, followed by drier conditions associated with the Little Ice Age (between 550 and 200 years ago).

With respect to what factor(s) might best explain the regional changes in precipitation inferred from the Cariaco metals' records of the past 14,000 years, the authors state that the regional changes in precipitation "are best explained by shifts in the mean latitude of the Atlantic Intertropical Convergence Zone," which, in turn, "can be explained by the Holocene history of insolation, both directly and through its effect on tropical Pacific sea surface conditions."

The results of these studies in South America demonstrate the important influence of solar variations on precipitation, necessarily implying a smaller possible role or no role at all for greenhouse gases.

Additional information on this topic, including reviews of newer publications as they become available, can be found at http://www.co2science.org/subject/s/solarpcpsamer.php.

References

Abbott, M., Binford, M.B., Brenner, M.W. and Kelts, K.R. 1997. A 3500 ^{14}C yr high resolution record of lake level changes in Lake Titicaca, South America. *Quaternary Research* **47**: 169-180.

Baker, P.A., Fritz, S.C., Garland, J. and Ekdahl, E. 2005. Holocene hydrologic variation at Lake Titicaca, Bolivia/Peru, and its relationship to North Atlantic climate variation. *Journal of Quaternary Science* **207**: 655-662.

Bond, G., Kromer, B., Beer, J., Muscheler, R., Evans, M.N., Showers, W., Hoffmann, S., Lotti- Bond, R., Hajdas, I. and Bonani, G. 2001. Persistent solar influence on North Atlantic climate during the Holocene. *Science* **294**: 2130-2136.

Haug, G.H., Hughen, K.A., Sigman, D.M., Peterson, L.C. and Rohl, U. 2001. Southward migration of the intertropical convergence zone through the Holocene. *Science* **293**: 1304-1308.

Nordemann, D.J.R., Rigozo, N.R. and de Faria, H.H. 2005. Solar activity and El-Niño signals observed in Brazil and Chile tree ring records. *Advances in Space Research* **35**: 891-896.

5.4.3. Africa

We begin our examination of Africa with a study covering the period 9,600-6,100 years before present, where Neff *et al.* (2001) investigated the relationship between a ^{14}C tree-ring record and a delta^{18}O proxy record of monsoon rainfall intensity as recorded in calcite delta[18]O data obtained from a stalagmite in northern Oman. The correlation between the two datasets was reported to be "extremely strong," and a spectral analysis of the data revealed statistically significant periodicities centered on 779, 205, 134, and 87 years for the delta[18]O record and periodicities of 206, 148, 126, 89, 26, and 10.4 years for the ^{14}C record. Because variations in ^{14}C tree-ring records are generally attributed to variations in solar activity and intensity, and because of this particular ^{14}C record's strong correlation with the delta[18]O record, as well as the closely corresponding results of the spectral analyses, Neff *et al.* conclude there is "solid evidence" that both signals (the ^{14}C and delta[18]O records) are responding to solar forcing.

In another study from eastern Africa, Stager *et al.* (2003) studied changes in diatom assemblages preserved in a sediment core extracted from Pilkington Bay, Lake Victoria, together with diatom and pollen data acquired from two nearby sites. According to the authors, the three coherent datasets revealed a "roughly 1400- to 1500-year spacing of century-scale P:E [precipitation:evaporation] fluctuations at Lake Victoria," which they say "may be related to a ca. 1470-year periodicity in northern marine and ice core records that has been linked to solar variability (Bond *et al.*, 1997; Mayewski *et al.*, 1997)."

Further support of Stager *et al.*'s thesis comes from Verschuren *et al.* (2000), who developed a decadal-scale history of rainfall and drought in equatorial east Africa for the past thousand years based on lake-level and salinity fluctuations of a small crater-lake basin in Kenya, as reconstructed from sediment stratigraphy and the species compositions of fossil diatom and midge assemblages. They compared this history with an equally long record of atmospheric [14]CO_2 production, which is a proxy for solar radiation variations.

They found equatorial east Africa was significantly drier than today during the Medieval Warm Period from AD 1000 to 1270, while it was relatively wet during the Little Ice Age from AD 1270 to 1850. However, this latter period was interrupted by three periods of prolonged dryness: 1390-1420, 1560-1625, and 1760-1840. These "episodes of persistent aridity," in the words of the authors, were "more severe than any recorded drought of the twentieth century." In addition, they discovered that "all three severe drought events of the past 700 years were broadly coeval with phases of high solar radiation, and the intervening periods of increased moisture were coeval with phases of low solar radiation."

Verschuren *et al.* note that their results "corroborate findings from north-temperate dryland regions that instrumental climate records are inadequate to appreciate the full range of natural variation in drought intensity at timescales relevant to socio-economic activity." This point is important, for with almost every new storm of significant size, with every new flood, or with every new hint of drought almost anywhere in the world, there are claims that the weather is becoming more extreme than ever before as a consequence of global warming. What we can learn from this study, however, is that there were more intense droughts in the centuries preceding the recent rise in the air's CO_2 content.

Verschuren *et al.* state that variations in solar radiative output "may have contributed to decade-scale rainfall variability in equatorial east Africa." This conclusion is hailed as robust by Oldfield (2000), who states that the thinking of the authors on this point is "not inconsistent with current views." Indeed, he too suggests that their results "provide strong evidence for a link between solar and climate variability."

Additional information on this topic, including reviews of newer publications as they become available, can be found at http://www.co2science.org/subject/s/solarpcpafrica.php.

References

Bond, G., Showers, W., Chezebiet, M., Lotti, R., Almasi, P., deMenocal, P., Priore, P., Cullen, H., Hajdas, I. and Bonani, G. 1997. A pervasive millennial scale cycle in North-Atlantic Holocene and glacial climates. *Science* **278**: 1257-1266.

Mayewski, P.A., Meeker, L.D., Twickler, M.S., Whitlow, S., Yang, Q., Lyons, W.B. and Prentice, M. 1997. Major features and forcing of high-latitude northern hemisphere atmospheric circulation using a 110,000-year-long glaciochemical series. *Journal of Geophysical Research* **102**: 26,345-26,366.

Neff, U., Burns, S.J., Mangini, A., Mudelsee, M., Fleitmann, D. and Matter, A. 2001. Strong coherence between solar variability and the monsoon in Oman between 9 and 6 kyr ago. *Nature* **411**: 290-293.

Oldfield, F. 2000. Out of Africa. *Nature* **403**: 370-371.

Stager, J.C., Cumming, B.F. and Meeker, L.D. 2003. A 10,000-year high-resolution diatom record from Pilkington

Bay, Lake Victoria, East Africa. *Quaternary Research* **59**: 172-181.

Verschuren, D., Laird, K.R. and Cumming, B.F. 2000. Rainfall and drought in equatorial east Africa during the past 1,100 years. *Nature* **403**: 410-414.

5.4.4. Asia

Our review of the influence of the sun on Asian precipitation trends begins with the study of Pederson *et al.* (2001), who utilized tree-ring chronologies from northeastern Mongolia to reconstruct annual precipitation and streamflow histories for this region over the period 1651-1995.

Analyses of both standard deviations and five-year intervals of extreme wet and dry periods revealed that "variations over the recent period of instrumental data are not unusual relative to the prior record." (See Figures 5.4.4.1 and 5.4.4.2.) The authors state, however, that the reconstructions "appear to show more frequent extended wet periods in more recent decades," but they note that this observation "does not demonstrate unequivocal evidence of an increase in precipitation as suggested by some climate models." More important to the present discussion, however, is the observation that spectral analysis of the data revealed significant periodicities of around 12 and 20-24 years, suggesting "possible evidence for solar influences in these reconstructions for northeastern Mongolia."

Nearby, Tan *et al.* (2008) developed a precipitation history of the Longxi area of the Tibetan Plateau's northeast margin since AD 960 based on an analysis of Chinese historical records, after which they compared the result with the same-period Northern Hemisphere temperature record and contemporaneous atmospheric ^{14}C and ^{10}Be histories. Results indicated that "high precipitation of Longxi corresponds to high temperature of the Northern Hemisphere, and low precipitation of Longxi corresponds to low temperature of the Northern Hemisphere." They also found "good coherences among the precipitation variations of Longxi and variations of atmospheric ^{14}C concentration, the averaged ^{10}Be record and the reconstructed solar modulation record," which findings harmonize, in their words, with "numerous studies [that] show that solar activity is the main force that drives regional climate changes in the Holocene," in support of which statement they attach 22 other scientific references. As a result, the researchers ultimately concluded that the "synchronous variations between Longxi

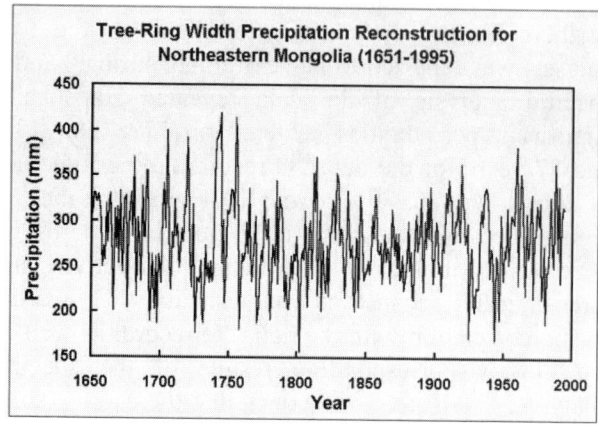

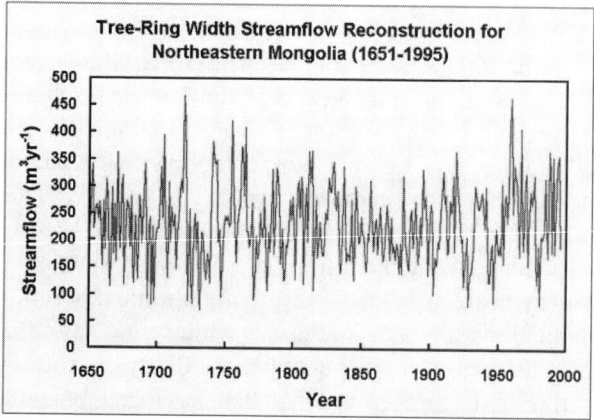

Figure 5.4.4.1. Pederson *et al.* (2001) reconstruction of annual precipitation in mm in northeastern Mongolia over the period 1651-1995.

Figure 5.4.4.2. Pederson *et al.* (2001) reconstruction of annual variation in streamflow (m3yr-1) in northeastern Mongolia over the period 1651-1995.

precipitation and Northern Hemisphere temperature may be ascribed to solar activity."

Paulsen *et al.* (2003) utilized "high-resolution records of δ^{13}C and δ^{18}O in stalagmite SF-1 from Buddha Cave [33°40'N, 109°05'E] ... to infer changes in climate in central China for the last 1270 years in terms of warmer, colder, wetter and drier conditions." Results indicated the presence of several major climatic episodes, including the Dark Ages Cold Period, Medieval Warm Period, Little Ice Age, and 20th-century warming, "lending support to the global extent of these events." With respect to hydrologic balance, the last part of the Dark Ages Cold Period was characterized as wet, followed by a dry, a wet, and another dry interval in the Medieval Warm Period, which was followed by a wet and a dry interval in the Little Ice Age, and finally a mostly wet but highly moisture-variable Current Warm Period. Some of this latter enhanced variability is undoubtedly due to the much finer one-year time resolution of the last 150 years of the record as compared to the three- to four-year resolution of the prior 1,120 years. This most recent improved resolution thus led to the major droughts centered on AD 1835, 1878, and 1955 being very well delineated.

The authors' data also revealed a number of other cycles superimposed on the major millennial-scale cycle of temperature and the centennial-scale cycle of moisture. They attributed most of these higher-frequency cycles to cyclical solar and lunar phenomena, concluding that the summer monsoon over eastern China, which brings the region much of its precipitation, may thus "be related to solar irradiance."

In conclusion, as we have consistently seen throughout this section, precipitation in Asia is determined by a conglomerate of cycles, many of which are solar-driven, but nearly all of which are independent of the air's CO_2 concentration.

Additional information on this topic, including reviews of newer publications as they become available, can be found at http://www.co2science.org/subject/s/solarpcpasia.php.

References

Paulsen, D.E., Li, H.-C. and Ku, T.-L. 2003. Climate variability in central China over the last 1270 years revealed by high-resolution stalagmite records. *Quaternary Science Reviews* **22**: 691-701.

Pederson, N., Jacoby, G.C., D'Arrigo, R.D., Cook, E.R. and Buckley, B.M. 2001. Hydrometeorological reconstructions for northeastern Mongolia derived from tree rings: 1651-1995. *Journal of Climate* **14**: 872-881.

Tan, L., Cai, Y., An, Z. and Ai, L. 2008. Precipitation variations of Longxi, northeast margin of Tibetan Plateau since AD 960 and their relationship with solar activity. *Climate of the Past* **4**: 19-28.

5.4.5. Europe

In this section, we investigate the natural influence of the sun on precipitation in Europe.

"Raised bogs," in the words of Blaauw *et al.* (2004), "are dependent on precipitation alone for water and nutrients." In addition, the various species of plants that are found in them each have their own requirements with respect to depth of water table. As a result, the vertical distribution of macro- and micro-fossils in raised bogs reveals much about past changes in local moisture conditions, especially, as the authors note, "about changes in effective precipitation (precipitation minus evapotranspiration)."

At the same time, changes in the carbon-14 content of bog deposits reveal something about solar activity, because, as Blaauw *et al.* describe the connection, "a decreased solar activity leads to less solar wind, reduced shielding against cosmic rays, and thus to increased production of cosmogenic isotopes [such as ^{14}C]." Consequently, it is possible to compare the histories of the two records (effective precipitation and δ^{14}C) and see if inferred changes in climate bear any relationship to inferred changes in solar activity, which is exactly what Blaauw *et al.* did.

Two cores of mid-Holocene raised-bog deposits in the Netherlands were ^{14}C "wiggle-match" dated by the authors at high precision, as per the technique described by Kilian *et al.* (1995, 2000) and Blaauw *et al.* (2003), while changes in local moisture conditions were inferred from the changing species composition of consecutive series of macrofossil samples. Results indicated that nine of 11 major mid-Holocene δ^{14}C increases (for which they provide evidence that the rises were "probably caused by declines in solar activity") were coeval with major wet-shifts (for which they provide evidence that the shifts were "probably caused by climate getting cooler and/or wetter"). In the case of the significant wet-shift at the major δ^{14}C rise in the vicinity of 850 BC, they additionally note that this prominent climatic cooling has been independently documented in many parts of the world, including "the North Atlantic Ocean (Bond *et al.*, 2001), the Norwegian Sea (Calvo *et al.*, 2002) [see also Andersson *et al.* (2003)], Northern Norway (Vorren, 2001), England (Waller *et al.*, 1999), the Czech Republic (Speranza *et al.*, 2000, 2002), central southern Europe (Magny, 2004), Chile (van Geel *et al.*, 2000), New Mexico (Armour *et al.*, 2002) and across the continent of North America (Viau *et al.*, 2002)." This close correspondence between major shifts in δ^{14}C and precipitation has additionally led Mauquoy *et al.* (2004) to conclude that "changes in solar activity may well have driven these changes during the Bronze Age/Iron Age transition around c. 850 cal. BC (discussed in detail by van Geel *et al.*, 1996, 1998, 1999, 2000) and the 'Little Ice Age' series of palaeoclimatic changes."

Based upon these considerations, Blaauw *et al.* (2003) state that their findings "add to the accumulating evidence that solar variability has played an important role in forcing climatic change during the Holocene." The broad geographic extent of the observations of the 850 BC phenomenon also adds to the accumulating evidence that many of the climate changes that have occurred in the vicinity of the North Atlantic Ocean have been truly global in scope, providing even more evidence for their having been forced by variations in solar activity, which would be expected to have considerably more than a regional climatic impact.

In another bog study, cores of peat taken from two raised bogs in the near-coastal part of Halland, Southwest Sweden (Boarps Mosse and Hyltemossen) by Björck and Clemmensen (2004) were examined for their content of wind-transported clastic material via a systematic count of quartz grains of diameter 0.2-0.35 mm and larger than 0.35 mm to determine temporal variations in Aeolian Sand Influx (ASI), which is correlated with winter storminess in that part of the world.

According to the authors, "the ASI records of the last 2500 years (both sites) indicate two timescales of winter storminess variation in southern Scandinavia." Specifically, they note that "decadal-scale variation (individual peaks) seems to coincide with short-term variation in sea-ice cover in the North Atlantic and is thus related to variations in the position of the North Atlantic winter season storm tracks," while "centennial-scale changes—peak families, like high peaks 1, 2 and 3 during the Little Ice Age, and low peaks 4 and 5 during the Medieval Warm Period— seem to record longer-scale climatic variation in the frequency and severity of cold and stormy winters."

Björck and Clemmensen also found a striking association between the strongest of these winter storminess peaks and periods of reduced solar activity. They specifically note, for example, that the solar minimum between AD 1880 and 1900 "is almost exactly coeval with the period of increased storminess at the end of the nineteenth century, and the Dalton Minimum between AD 1800 and 1820 is almost coeval with the period of peak storminess reported here." In addition, they say that an event of increased storminess they dated to AD 1650 "falls at the beginning of the Maunder solar minimum (AD 1645-1715)," while further back in time they report high ASI values between AD 1450 and 1550 with "a very distinct peak at AD 1475," noting that this period coincides with the Sporer Minimum of AD 1420-1530. In addition, they call our attention to the fact that the latter three peaks in winter storminess all

occurred during the Little Ice Age and "are among the most prominent in the complete record."

Shifting gears, several researchers have studied the precipitation histories of regions along the Danube River in western Europe and the effects they have on river discharge, with some of them suggesting that an anthropogenic signal is present in the latter decades of the twentieth century and is responsible for that period's drier conditions. Ducic (2005) examined such claims by analyzing observed and reconstructed river discharge rates near Orsova, Serbia over the period 1731-1990. He notes that the lowest five-year discharge value in the pre-instrumental era (period of occurrence: 1831-1835) was practically equal to the lowest five-year discharge value in the instrumental era (period of occurrence: 1946-1950), and that the driest decade of the entire 260-year period was 1831-1840. Similarly, the highest five-year discharge value for the pre-instrumental era (period of occurrence: 1736-1740) was nearly equal to the five-year maximum discharge value for the instrumental era (period of occurrence: 1876-1880), differing by only 0.7 percent. What is more, the discharge rate for the last decade of the record (1981-1990), which prior researchers had claimed was anthropogenically influenced, was found to be "completely inside the limits of the whole series," in Ducic's words, and only slightly (38 m^3s^{-1} or 0.7 percent) less than the 260-year mean of 5356 m^3s^{-1}. Thus, Ducic concludes that "modern discharge fluctuations do not point to [a] dominant anthropogenic influence." Ducic's correlative analysis suggests that the detected cyclicity in the record could "point to the domination of the influence of solar activity."

Next, we turn to the study of Holzhauser et al. (2005), who presented high-resolution records of variations in glacier size in the Swiss Alps together with lake-level fluctuations in the Jura mountains, the northern French Pre-Alps, and the Swiss Plateau in developing a 3,500-year temperature and precipitation climate history of west-central Europe, beginning with an in-depth analysis of the Great Aletsch glacier, which is the largest of all glaciers located in the European Alps.

Near the beginning of the time period studied, the three researchers report that "during the late Bronze Age Optimum from 1350 to 1250 BC, the Great Aletsch glacier was approximately 1000 m shorter than it is today," noting that "the period from 1450 to 1250 BC has been recognized as a warm-dry phase in other Alpine and Northern Hemisphere proxies (Tinner et al., 2003)." Then, after an intervening unnamed cold-wet phase, when the glacier grew in both mass and length, they say that "during the Iron/Roman Age Optimum between c. 200 BC and AD 50," which is perhaps better known as the Roman Warm Period, the glacier again retreated and "reached today's extent or was even somewhat shorter than today." Next came the Dark Ages Cold Period, which they say was followed by "the Medieval Warm Period, from around AD 800 to the onset of the Little Ice Age around AD 1300," which latter cold-wet phase was "characterized by three successive [glacier length] peaks: a first maximum after 1369 (in the late 1370s), a second between 1670 and 1680, and a third at 1859/60," following which the glacier began its latest and still-ongoing recession in 1865.

Data pertaining to the Gorner glacier (the second largest of the Swiss Alps) and the Lower Grindelwald glacier of the Bernese Alps tell much the same story, as Holzhauser et al. report that these glaciers and the Great Aletsch glacier "experienced nearly synchronous advances" throughout the study period. With respect to what was responsible for the millennial-scale climatic oscillation that produced the alternating periods of cold-wet and warm-dry conditions that fostered the similarly paced cycle of glacier growth and retreat, the Swiss and French scientists report that "glacier maximums coincided with radiocarbon peaks, i.e., periods of weaker solar activity," which in their estimation "suggests a possible solar origin of the climate oscillations punctuating the last 3500 years in west-central Europe, in agreement with previous studies (Denton and Karlén, 1973; Magny, 1993; van Geel et al., 1996; Bond et al., 2001)." And to underscore that point, they conclude their paper by stating that "a comparison between the fluctuations of the Great Aletsch glacier and the variations in the atmospheric residual ^{14}C records supports the hypothesis that variations in solar activity were a major forcing factor of climate oscillations in west-central Europe during the late Holocene."

In one final study, Lamy et al. (2006), utilized paleoenvironmental proxy data for ocean properties, eolian sediment input, and continental rainfall based on high-resolution analyses of sediment cores from the southwestern Black Sea and the northernmost Gulf of Aqaba to infer hydroclimatic changes in northern Anatolia and the northern Red Sea region during the last ~7500 years, after which the reconstructed hydroclimatic history was compared with Δ^{14}C periodicities evident in the tree-ring data of Stuiver et al. (1998).

Analysis of the data showed "pronounced and coherent" multi-centennial variations that the authors say "strongly resemble modern temperature and rainfall anomalies related to the Arctic Oscillation/North Atlantic Oscillation (AO/NAO)." In addition, they say that "the multicentennial variability appears to be similar to changes observed in proxy records for solar output changes," although "the exact physical mechanism that transfers small solar irradiance changes either to symmetric responses in the North Atlantic circulation or to atmospheric circulation changes involving an AO/NAO-like pattern, remains unclear."

In conclusion, each of the studies above indicates cyclical solar activity induces similar cyclical precipitation activity in Europe. They indicate, in Lamy et al.'s words, that "the impact of (natural) centennial-scale climate variability on future climate projections could be more substantial than previously thought."

Additional information on this topic, including reviews of newer publications as they become available, can be found at http://www.co2science.org/subject/s/solarpceurope.php.

References

Andersson, C., Risebrobakken, B., Jansen, E. and Dahl, S.O. 2003. Late Holocene surface ocean conditions of the Norwegian Sea (Voring Plateau). *Paleoceanography* **18**: 10.1029/2001PA000654.

Armour, J., Fawcett, P.J. and Geissman, J.W. 2002. 15 k.y. palaeoclimatic and glacial record from northern New Mexico. *Geology* **30**: 723-726.

Björck, S. and Clemmensen, L.B. 2004. Aeolian sediment in raised bog deposits, Halland, SW Sweden: a new proxy record of Holocene winter storminess variation in southern Scandinavia? *The Holocene* **14**: 677-688.

Blaauw, M., Heuvelink, G.B.M., Mauquoy, D., van der Plicht, J. and van Geel, B. 2003. A numerical approach to ^{14}C wiggle-match dating of organic deposits: best fits and confidence intervals. *Quaternary Science Reviews* **22**: 1485-1500.

Blaauw, M., van Geel, B. and van der Plicht, J. 2004. Solar forcing of climatic change during the mid-Holocene: indications from raised bogs in The Netherlands. *The Holocene* **14**: 35-44.

Bond, G., Kromer, B., Beer, J., Muscheler, R., Evans, M.N., Showers, W., Hoffmann, S., Lotti-Bond, R., Hajdas, I. and Bonani, G. 2001. Persistent solar influence on North

Atlantic climate during the Holocene. *Science* **294**: 2130-2136.

Calvo, E., Grimalt, J. and Jansen, E. 2002. High resolution U_{37}^{K} sea temperature reconstruction in the Norwegian Sea during the Holocene. *Quaternary Science Reviews* **21**: 1385-1394.

Denton, G.H. and Karlén, W. 1973. Holocene climate variations—their pattern and possible cause. *Quaternary Research* **3**: 155-205.

Ducic, V. 2005. Reconstruction of the Danube discharge on hydrological station Orsova in pre-instrumental period: Possible causes of fluctuations. *Edition Physical Geography of Serbia* **2**: 79-100.

Holzhauser, H., Magny, M. and Zumbuhl, H.J. 2005. Glacier and lake-level variations in west-central Europe over the last 3500 years. *The Holocene* **15**: 789-801.

Kilian, M.R., van Geel, B. and van der Plicht, J. 2000. [14]C AMS wiggle matching of raised bog deposits and models of peat accumulation. *Quaternary Science Reviews* **19**: 1011-1033.

Kilian, M.R., van der Plicht, J. and van Geel, B. 1995. Dating raised bogs: new aspects of AMS ^{14}C wiggle matching, a reservoir effect and climatic change. *Quaternary Science Reviews* **14**: 959-966.

Lamy, F., Arz, H.W., Bond, G.C., Bahr, A. and Patzold, J. 2006. Multicentennial-scale hydrological changes in the Black Sea and northern Red Sea during the Holocene and the Arctic/North Atlantic Oscillation. *Paleoceanography* **21**: 10.1029/2005PA001184l.

Magny, M. 1993. Solar influences on Holocene climatic changes illustrated by correlations between past lake-level fluctuations and the atmospheric ^{14}C record. *Quaternary Research* **40**: 1-9.

Magny, M. 2004. Holocene climate variability as reflected by mid-European lake-level fluctuations, and its probable impact on prehistoric human settlements. *Quaternary International* **113**: 65-79.

Mauquoy, D., van Geel, B., Blaauw, M., Speranza, A. and van der Plicht, J. 2004. Changes in solar activity and Holocene climatic shifts derived from ^{14}C wiggle-match dated peat deposits. *The Holocene* **14**: 45-52.

Speranza, A.O.M., van der Plicht, J. and van Geel, B. 2000. Improving the time control of the Subboreal/Subatlantic transition in a Czech peat sequence by ^{14}C wiggle-matching. *Quaternary Science Reviews* **19**: 1589-1604.

Speranza, A.O.M., van Geel, B. and van der Plicht, J. 2002. Evidence for solar forcing of climate change at ca. 850 cal. BC from a Czech peat sequence. *Global and Planetary Change* **35**: 51-65.

Stuiver, M., Reimer, P.J., Bard, E., Beck, J.W., Burr, G.S., Hughen, K.A., Kromer, B., McCormac, G., van der Plicht, J. and Spurk, M. 1998. INTCAL98 radiocarbon age calibration, 24,000-0 cal PB. *Radiocarbon* **40**: 1041-1083.

Tinner, W., Lotter, A.F., Ammann, B., Condera, M., Hubschmied, P., van Leeuwan, J.F.N. and Wehrli, M. 2003. Climatic change and contemporaneous land-use phases north and south of the Alps 2300 BC to AD 800. *Quaternary Science Reviews* **22**: 1447-1460.

van Geel, B., Buurman, J. and Waterbolk, H.T. 1996. Archaeological and palaeoecological indications of an abrupt climate change in the Netherlands and evidence for climatological teleconnections around 2650 BP. *Journal of Quaternary Science* **11**: 451-460.

van Geel, B., Heusser, C.J., Renssen, H. and Schuurmans, C.J.E. 2000. Climatic change in Chile at around 2700 BP and global evidence for solar forcing: a hypothesis. *The Holocene* **10**: 659-664.

van Geel, B., Raspopov, O.M., Renssen, H., van der Plicht, J., Dergachev, V.A. and Meijer, H.A.J. 1999. The role of solar forcing upon climate change. *Quaternary Science Reviews* **18**: 331-338.

van Geel, B., van der Plicht, J., Kilian, M.R., Klaver, E.R., Kouwenberg, J.H.M., Renssen, H., Reynaud-Farrera, I. and Waterbolk, H.T. 1998. The sharp rise of delta [14]C c. 800 cal BC: possible causes, related climatic teleconnections and the impact on human environments. *Radiocarbon* **40**: 535-550.

Viau, A.E., Gajewski, K., Fines, P., Atkinson, D.E. and Sawada, M.C. 2002. Widespread evidence of 1500 yr climatic variability in North America during the past 14000 yr. *Geology* **30**: 455-458.

Vorren, K.-D. 2001. Development of bogs in a coast-inland transect in northern Norway. *Acta Palaeobotanica* **41**: 43-67.

Waller, M.P., Long, A.J., Long, D. and Innes, J.B. 1999. Patterns and processes in the development of coastal mire vegetation multi-site investigations from Walland Marsh, Southeast England. *Quaternary Science Reviews* **18**: 1419-1444.

5.5. Droughts

The IPCC claims earth's climate is becoming more variable and extreme as a result of CO_2-induced global warming, and it forecasts increasing length and severity of drought as one of the consequences. In the next chapter of this report, we report evidence that modern drought frequency and severity fall well within the range of natural variability. In the present section, however, we examine the issue of attribution, specifically investigating the natural influence of the sun on drought. We begin with a review of the literature on droughts in the United States.

According to Cook *et al.* (2007), recent advances in the reconstruction of past drought over North America "have revealed the occurrence of a number of unprecedented megadroughts over the past millennium that clearly exceed any found in the instrumental records." Indeed, they state that "these past megadroughts dwarf the famous droughts of the twentieth century, such as the Dust Bowl drought of the 1930s, the southern Great Plains drought of the 1950s, and the current one in the West that began in 1999," all of which dramatic droughts pale when compared to "an epoch of significantly elevated aridity that persisted for almost 400 years over the AD 900-1300 period."

Of central importance to North American drought formation, in the words of the four researchers, "is the development of cool 'La Niña-like' SSTs in the eastern tropical Pacific." Paradoxically, as they describe the situation, "warmer conditions over the tropical Pacific region lead to the development of cool La Niña-like SSTs there, which is drought inducing over North America." In further explaining the mechanics of this phenomenon, on which both "model and data agree," Cook *et al.* state that "if there is a heating over the entire tropics then the Pacific will warm more in the west than in the east because the strong upwelling and surface divergence in the east moves some of the heat poleward," with the result that "the east-west temperature gradient will strengthen, so the winds will also strengthen, so the temperature gradient will increase further ... leading to a more La Niña-like state." They add that "La Niña-like conditions were apparently the norm during much of the Medieval period when the West was in a protracted period of elevated aridity and solar irradiance was unusually high."

Shedding some more light on the subject, Yu and Ito (1999) studied a sediment core from a closed-basin lake in the northern Great Plains of North America, producing a 2,100-year record that revealed four dominant periodicities of drought that matched "in surprising detail" similar periodicities of various solar indices. The correspondence was so close, in fact, that they say "this spectral similarity forces us to consider solar variability as the major cause of century-scale drought frequency in the northern Great Plains."

One year later, Dean and Schwalb (2000) derived a similar-length record of drought from sediment cores extracted from Pickerel Lake, South Dakota, which also exhibited recurring incidences of major drought on the northern Great Plains. They too reported that the cyclical behavior appeared to be in synchrony with similar variations in solar irradiance. After making a case for "a direct connection between solar irradiance and weather and climate," they thus concluded that "it seems reasonable that the cycles in aridity and eolian activity over the past several thousand years recorded in the sediments of lakes in the northern Great Plains might also have a solar connection."

Moving to east-central North America, Springer *et al.* (2008) derived a multi-decadal-scale record of Holocene drought based on Sr/Ca ratios and $\delta^{13}C$ data obtained from stalagmite BCC-002 from Buckeye Creek Cave (BCC), West Virginia (USA) that "grew continuously from ~7000 years B.P. to ~800 years B.P." and then again "from ~800 years B.P. until its collection in 2002."

Results of their study indicated the presence of seven significant Mid- to Late-Holocene droughts, six of which "correlate with cooling of the Atlantic and Pacific Oceans as part of the North Atlantic Ocean ice-rafted debris [IRD] cycle, which has been linked to the solar irradiance cycle," as per Bond *et al.* (2001). In addition, they determined that the Sr/Ca and $\delta^{13}C$ time series "display periodicities of ~200 and ~500 years and are coherent in those frequency bands." They also say "the ~200-year periodicity is consistent with the de Vries (Suess) solar irradiance cycle," and they "interpret the ~500-year periodicity to be a harmonic of the IRD oscillations." Noting further that "cross-spectral analysis of the Sr/Ca and IRD time series yields statistically significant coherencies at periodicities of 455 and 715 years," they go on to note that "these latter values are very similar to the second (725-years) and third (480-years) harmonics of the 1450 ± 500-years IRD periodicity." As a result of these observations, the five researchers conclude their report by saying their findings "corroborate works indicating that millennial-scale solar-forcing is responsible for droughts and ecosystem changes in central and eastern North America (Viau *et al.*, 2002; Willard *et al.*, 2005; Denniston *et al.*, 2007)," adding that their high-resolution time series now provide even stronger evidence "in favor of solar-forcing of North American drought by yielding unambiguous spectral analysis results."

In Nevada, Mensing *et al.* (2004) analyzed a set of sediment cores extracted from Pyramid Lake for pollen and algal microfossils deposited there over the past 7,630 years that allowed them to infer the hydrological history of the area over that time period. According to the authors, "sometime after 3430 but before 2750 cal yr B.P., climate became cool and wet," but "the past 2500 yr have been marked by recurring persistent droughts." The longest of these droughts, according to them, "occurred between 2500 and 2000 cal yr B.P.," while others occurred "between 1500 and 1250, 800 and 725, and 600 and 450 cal yr B.P." They also note that "the timing and magnitude of droughts identified in the pollen record compares favorably with previously published $\delta^{18}O$ data from Pyramid Lake" and with "the ages of submerged rooted stumps in the Eastern Sierra Nevada and woodrat midden data from central Nevada." When they compared the pollen record of droughts from Pyramid Lake with the stacked petrologic record of North Atlantic drift ice of Bond *et al.* (2001), like other researchers they too found "nearly every occurrence of a shift from ice maxima (reduced solar output) to ice minima (increased solar output) corresponded with a period of prolonged drought in the Pyramid Lake record." Mensing *et al.* conclude that "changes in solar irradiance may be a possible mechanism influencing century-scale drought in the western Great Basin [of the United States]." Indeed, it would appear that variable solar activity is the major factor in determining the hydrological state of the region and all of North America.

Moving slightly south geographically, Asmerom *et al.* (2007) developed a high-resolution climate proxy for the southwest United States from $\delta^{18}O$ variations in a stalagmite found in Pink Panther Cave in the Guadalupe Mountains of New Mexico. Spectral analysis performed on the raw $\delta^{18}O$ data revealed significant peaks that the researchers say "closely match previously reported periodicities in the ^{14}C content of the atmosphere, which have been attributed to periodicities in the solar cycle (Stuiver and Braziunas, 1993)." More specifically, they say that cross-spectral analysis of the $\Delta^{14}C$ and $\delta^{18}O$ data confirms that the two records have matching periodicities at 1,533 years (the Bond cycle), 444 years, 170 years, 146 years, and 88 years (the Gleissberg cycle). In addition, they report that periods of increased solar radiation correlate with periods of decreased rainfall in the southwestern United States (via changes in the North American monsoon), and

that this behavior is just the opposite of what is observed with the Asian monsoon. These observations thus lead them to suggest that the proposed solar link to Holocene climate operates "through changes in the Walker circulation and the Pacific Decadal Oscillation and El Niño-Southern Oscillation systems of the tropical Pacific Ocean."

Making our way to Mexico, Hodell *et al.* (2001) analyzed sediment cores obtained from Lake Chichancanab on the Yucatan Peninsula, reconstructing the climatic history of this region over the past 2,600 years. Long episodes of drought were noted throughout the entire record, and spectral analysis revealed a significant periodicity that matched well with a cosmic ray-produced ^{14}C record preserved in tree rings that is believed to reflect variations in solar activity. Hence, they too concluded that "a significant component of century-scale variability in Yucatan droughts is explained by solar forcing."

Expanding the geographical scope of such studies, Black *et al.* (1999) found evidence of substantial decadal and centennial climate variability in a study of ocean sediments in the southern Caribbean that were deposited over the past 825 years. Their data suggested that climate regime shifts are a natural aspect of Atlantic variability; in relating these features to records of terrestrial climate, they concluded that "these shifts may play a role in triggering changes in the frequency and persistence of drought over North America." In addition, because there was a strong correspondence between these phenomena and similar changes in ^{14}C production rate, they further concluded that "small changes in solar output may influence Atlantic variability on centennial time scales."

Moving to Africa, Verschuren *et al.* (2000) conducted a similar study in a small lake in Kenya, documenting the existence of three periods of prolonged dryness during the Little Ice Age that were, in their words, "more severe than any recorded drought of the twentieth century." In addition, they discovered that all three of these severe drought events "were broadly coeval with phases of high solar radiation"—as inferred from ^{14}C production data—"and the intervening periods of increased moisture were coeval with phases of low solar radiation." They thus concluded that variations in solar activity "may have contributed to decade-scale rainfall variability in equatorial east Africa."

Also in Africa, working with three sediment cores extracted from Lake Edward (0°N, 30°E), Russell and

Johnson (2005) developed a continuous 5,400-year record of Mg concentration and isotopic composition of authigenic inorganic calcite as proxies for the lake's water balance, which is itself a proxy for regional drought conditions in equatorial Africa. They found "the geochemical record from Lake Edward demonstrates a consistent pattern of equatorial drought during both cold and warm phases of the North Atlantic's '1500-year cycle' during the late Holocene," noting that similar "725-year climate cycles" are found in several records from the Indian and western Pacific Oceans and the South China Sea, citing as authority for the latter statement the studies of von Rad *et al.* (1999), Wang *et al.* (1999), Russell *et al.* (2003) and Staubwasser *et al.* (2003). In light of these findings, the two scientists say their results "show that millennial-scale high-latitude climate events are linked to changes in equatorial terrestrial climate ... during the late Holocene," or as they phrase it in another place, that their results "suggest a spatial footprint in the tropics for the '1500-year cycle' that may help to provide clues to discern the cycle's origin," noting there is already reason to believe that it may be solar-induced.

Lastly, Garcin *et al.* (2007) explored hydrologic change using late-Holocene paleoenvironmental data derived from several undisturbed sediment cores retrieved from the deepest central part of Lake Masoko (9°20.0'S, 33°45.3'E), which occupies a maar crater in the Rungwe volcanic highlands of the western branch of Africa's Rift Valley, where it is situated approximately 35 km north of Lake Malawi.

According to the 10 researchers who conducted the work, "magnetic, organic carbon, geochemical proxies and pollen assemblages indicate a dry climate during the 'Little Ice Age' (AD 1550-1850), confirming that the LIA in eastern Africa resulted in marked and synchronous hydrological changes," although "the direction of response varies between different African lakes." In this regard, for example, they report that "to the south (9.5-14.5°S), sediment cores from Lake Malawi have revealed similar climatic conditions (Owen *et al.*, 1990; Johnson *et al.*, 2001; Brown and Johnson, 2005)" that are "correlated with the dry period of Lakes Chilwa and Chiuta (Owen and Crossley, 1990)," and they say that "lowstands have been also observed during the LIA at Lake Tanganyika ... from AD 1500 until AD 1580, and from ca. AD 1650 until the end of the 17th century, where the lowest lake-levels are inferred (Cohen *et al.*, 1997; Alin and Cohen, 2003)." By contrast, however, they report that "further north,

evidence from Lakes Naivasha (0.7°S) and Victoria (2.5°S-0.5°N) indicates relatively wet conditions with high lake-levels during the LIA, interrupted by short drought periods (Verschuren *et al.*, 2000; Verschuren, 2004; Stager *et al.*, 2005)." Lastly, Garcin *et al.* state that "inferred changes of the Masoko hydrology are positively correlated with the solar activity proxies."

In discussing their findings, the African and French scientists note the Little Ice Age in Africa appears to have had a greater thermal amplitude than it did in the Northern Hemisphere, citing in support of this statement the paleoclimate studies of Bonnefille and Mohammed (1994), Karlén *et al.* (1999), Holmgren *et al.* (2001), and Thompson *et al.* (2002). Nevertheless, the more common defining parameter of the Little Ice Age in Africa was the moisture status of the continent, which appears to have manifested opposite directional trends in different latitudinal bands. In addition, the group of scientists emphasizes that the positive correlation of Lake Masoko hydrology with various solar activity proxies "implies a forcing of solar activity on the atmospheric circulation and thus on the regional climate of this part of East Africa."

There seems to be little question but what variations in solar activity have been responsible for much of the drought variability of the Holocene in many parts of the world.

Additional information on this topic, including reviews of newer publications as they become available, can be found at http://www.co2science.org/subject/d/droughtsolar.php.

References

Alin, S.R. and Cohen, A.S. 2003. Lake-level history of Lake Tanganyika, East Africa, for the past 2500 years based on ostracode-inferred water-depth reconstruction. *Palaeogeography, Palaeoclimatology, Palaeoecology* **1999**: 31-49.

Asmerom, Y., Polyak, V., Burns, S. and Rassmussen, J. 2007. Solar forcing of Holocene climate: New insights from a speleothem record, southwestern United States. *Geology* **35**: 1-4.

Black, D.E., Peterson, L.C., Overpeck, J.T., Kaplan, A., Evans, M.N. and Kashgarian, M. 1999. Eight centuries of North Atlantic Ocean atmosphere variability. *Science* **286**: 1709-1713.

Bond, G., Kromer, B., Beer, J., Muscheler, R., Evans, M.N., Showers, W., Hoffmann, S., Lotti-Bond, R., Hajdas,

I. and Bonani, G. 2001. Persistent solar influence on North Atlantic climate during the Holocene. *Science* **294**: 2130-2136.

Bonnefille, R. and Mohammed, U. 1994. Pollen-inferred climatic fluctuations in Ethiopia during the last 2000 years. *Palaeogeography, Palaeoclimatology, Palaeoecology* **109**: 331-343.

Brown, E.T. and Johnson, T.C. 2005. Coherence between tropical East African and South American records of the Little Ice Age. *Geochemistry, Geophysics, Geosystems* **6**: 10.1029/2005GC000959.

Cohen, A.S., Talbot, M.R., Awramik, S.M., Dettmen, D.L. and Abell, P. 1997. Lake level and paleoenvironmental history of Lake Tanganyika, Africa, as inferred from late Holocene and modern stromatolites. *Geological Society of American Bulletin* **109**: 444-460.

Cook, E.R., Seager, R., Cane, M.A. and Stahle, D.W. 2007. North American drought: Reconstructions, causes, and consequences. *Earth-Science Reviews* **81**: 93-134.

Dean, W.E. and Schwalb, A. 2000. Holocene environmental and climatic change in the Northern Great Plains as recorded in the geochemistry of sediments in Pickerel Lake, South Dakota. *Quaternary International* **67**: 5-20.

Denniston, R.F., DuPree, M., Dorale, J.A., Asmerom, Y., Polyak, V.J. and Carpenter, S.J. 2007. Episodes of late Holocene aridity recorded by stalagmites from Devil's Icebox Cave, central Missouri, USA. *Quaternary Research* **68**: 45-52.

Garcin, Y., Williamson, D., Bergonzini, L., Radakovitch, O., Vincens, A., Buchet, G., Guiot, J., Brewer, S., Mathe, P.-E. and Majule, A. 2007. Solar and anthropogenic imprints on Lake Masoko (southern Tanzania) during the last 500 years. *Journal of Paleolimnology* **37**: 475-490.

Hodell, D.A., Brenner, M., Curtis, J.H. and Guilderson, T. 2001. Solar forcing of drought frequency in the Maya lowlands. *Science* **292**: 1367-1370.

Holmgren, K., Moberg, A., Svanered, O. and Tyson, P.D. 2001. A preliminary 3000-year regional temperature reconstruction for South Africa. *South African Journal of Science* **97**: 49-51.

Johnson, T.C., Barry, S.L., Chan, Y. and Wilkinson, P. 2001. Decadal record of climate variability spanning the past 700 years in the Southern Tropics of East Africa. *Geology* **29**: 83-86.

Karlén, W., Fastook, J.L., Holmgren, K., Malmstrom, M., Matthews, J.A., Odada, E., Risberg, J., Rosqvist, G., Sandgren, P., Shemesh, A. and Westerberg, L.O. 1999. Glacier Fluctuations on Mount Kenya since ~6000 cal.

years BP: implications for Holocene climatic change in Africa. *Ambio* **28**: 409-418.

Mensing, S.A., Benson, L.V., Kashgarian, M. and Lund, S. 2004. A Holocene pollen record of persistent droughts from Pyramid Lake, Nevada, USA. *Quaternary Research* **62**: 29-38.

Owen, R.B. and Crossley, R. 1990. Recent sedimentation in Lakes Chilwa and Chiuata, Malawi. *Palaeoecology of Africa* **20**: 109-117.

Owen, R.B., Crossley, R., Johnson, T.C., Tweddle, D., Kornfield, I., Davison, S., Eccles, D.H. and Engstrom, D.E. 1990. Major low levels of Lake Malawi and their implications for speciation rates in Cichlid fishes. *Proceedings of the Royal Society of London Series B* **240**: 519-553.

Russell, J.M and Johnson, T.C. 2005. Late Holocene climate change in the North Atlantic and equatorial Africa: Millennial-scale ITCZ migration. *Geophysical Research Letters* **32**: 10.1029/2005GL023295.

Russell, J.M., Johnson, T.C. and Talbot, M.R. 2003. A 725-year cycle in the Indian and African monsoons. *Geology* **31**: 677-680.

Springer, G.S., Rowe, H.D., Hardt, B., Edwards, R.L. and Cheng, H. 2008. Solar forcing of Holocene droughts in a stalagmite record from West Virginia in east-central North America. *Geophysical Research Letters* **35**: 10.1029/2008GL034971.

Stager, J.C., Ryves, D., Cumming, B.F., Meeker, L.D. and Beer, J. 2005. Solar variability and the levels of Lake Victoria, East Africa, during the last millennium. *Journal of Paleolimnology* **33**: 243-251.

Staubwasser, M., Sirocko, F., Grootes, P. and Segl, M. 2003. Climate change at the 4.2 ka BP termination of the Indus valley civilization and Holocene south Asian monsoon variability. *Geophysical Research Letters* **30**: 10.1029/2002GL016822.

Stuiver, M. and Braziunas, T.F. 1993. Sun, ocean climate and atmospheric $^{14}CO_2$: An evaluation of causal and spectral relationships. *The Holocene* **3**: 289-305.

Thompson, L.G., Mosley-Thompson, E., Davis, M.E., Henderson, K.A., Brecher, H.H., Zagorodnov, V.S., Mashiotta, T.A., Lin, P.N., Mikhalenko, V.N., Hardy, D.R. and Beer, J. 2002. Kilimanjaro ice core records: evidence of Holocene climate change in tropical Africa. *Science* **298**: 589-593.

Verschuren, D. 2004. Decadal and century-scale climate variability in tropical Africa during the past 2000 years. In: Battarbee, R.W., Gasse, F. and Stickley, C.E. (Eds.) *Past Climate Variability Through Europe and Africa*. Springer, Dordrecht, The Netherlands, pp.139-158.

Verschuren, D., Laird, K.R. and Cumming, B.F. 2000. Rainfall and drought in equatorial east Africa during the past 1,100 years. *Nature* **403**: 410-414.

Viau, A.E., Gajewski, K., Fines, P., Atkinson, D.E. and Sawada, M.C. 2002. Widespread evidence of 1500 yr climate variability in North America during the past 14,000 yr. *Geology* **30**: 455-458.

von Rad, U., *et al.* 1999. A 5000-yr record of climate changes in varved sediments from the oxygen minimum zone off Pakistan, northeastern Arabian Sea. *Quaternary Research* **51**: 39-53.

Wang, L., *et al.* 1999. East Asian monsoon climate during the late Pleistocene: High resolution sediment records from the South China Sea. *Marine Geology* **156**: 245-284.

Willard, D.A., Bernhardt, C.E., Korejwo, D.A. and Meyers, S.R. 2005. Impact of millennial-scale Holocene climate variability on eastern North American terrestrial ecosystems: Pollen-based climatic reconstruction. *Global and Planetary Change* **47**: 17-35.

Yu, Z. and Ito, E. 1999. Possible solar forcing of century-scale drought frequency in the northern Great Plains. *Geology* **27**: 263-266.

5.6. Floods

The IPCC claims that floods will become more variable and extreme as a result of CO_2-induced global warming. In the next chapter of this report, we show modern flood frequency and severity fall well within the range of natural variability. In the present section, we limit our examination once again to the issue of attribution, specifically investigating the influence of the sun on floods.

We begin by reviewing what is known about the relationship of extreme weather events to climate in Europe during the Holocene. According to Starkel (2002), in general, more extreme fluvial activity, of both the erosional and depositional type, is associated with cooler climates. "Continuous rains and high-intensity downpours," writes Starkel, were most common during the Little Ice Age. Such "flood phases," the researcher reports, "were periods of very unstable weather and frequent extremes of various kinds." More related to the present discussion, Starkel also notes that "most of the phases of high frequency of extreme events during the Holocene coincide with the periods of declined solar activity."

Noren *et al.* (2002) extracted sediment cores from 13 small lakes distributed across a 20,000-km^2 region in Vermont and eastern New York, after which several techniques were used to identify and date terrigenous in-wash layers that depict the frequency of storm-related floods. Results of the analysis showed that "the frequency of storm-related floods in the northeastern United States has varied in regular cycles during the past 13,000 years (13 kyr), with a characteristic period of about 3 kyr." There were four major storminess peaks during this period; they occurred approximately 2.6, 5.8, 9.1, and 11.9 kyr ago, with the most recent upswing in storminess beginning "at about 600 yr BP [before present], coincident with the beginning of the Little Ice Age." With respect to the causative factor(s) behind the cyclic behavior, Noren *et al.* state that the pattern they observed "is consistent with long-term changes in the average sign of the Arctic Oscillation [AO]," further suggesting that "changes in the AO, perhaps modulated by solar forcing, may explain a significant portion of Holocene climate variability in the North Atlantic region."

Also working in the United States, Schimmelmann *et al.* (2003) identified conspicuous gray clay-rich flood deposits in the predominantly olive varved sediments of the Santa Barbara Basin off the coast of California, which they accurately dated by varve-counting. Analysis of the record revealed six prominent flood events that occurred at approximately AD 212, 440, 603, 1029, 1418, and 1605, "suggesting," in their words, "a quasi-periodicity of ~200 years," with "skipped" flooding just after AD 800, 1200, and 1800. They further note that "the floods of ~AD 1029 and 1605 seem to have been associated with brief cold spells," that "the flood of ~AD 440 dates to the onset of the most unstable marine climatic interval of the Holocene (Kennett and Kennett, 2000)," and that "the flood of ~AD 1418 occurred at a time when the global atmospheric circulation pattern underwent fundamental reorganization at the beginning of the 'Little Ice Age' (Kreutz *et al.*, 1997; Meeker and Mayewski, 2002)." Lastly, they report that "the quasi-periodicity of ~200 years for southern California floods matches the ~200-year periodicities found in a variety of high-resolution palaeoclimate archives and, more importantly, a c.208-year cycle of solar activity and inferred changes in atmospheric circulation."

As a result of these findings, Schimmelmann *et al.* "hypothesize that solar-modulated climatic background conditions are opening a ~40-year window of opportunity for flooding every ~200 years," and that "during each window, the danger of flooding is exacerbated by additional climatic and environmental cofactors." They also note that "extrapolation of the ~200-year spacing of floods into the future raises the uncomfortable possibility for historically unprecedented flooding in southern California during the first half of this century." When such flooding occurs, there will be no need to suppose it came as a consequence of what the IPCC calls the unprecedented warming of the past century.

Additional information on this topic, including reviews of newer publications as they become available, can be found at http://www.co2science.org/subject/s/solarflood.php.

References

Kennett, D.J. and Kennett, J.P. 2000. Competitive and cooperative responses to climatic instability in coastal southern California. *American Antiquity* **65**: 379-395.

Kreutz, K.J., Mayewski, P.A., Meeker, L.D., Twickler, M.S., Whitlow, S.I. and Pittalwala, I.I. 1997. Bipolar changes in atmospheric circulation during the Little Ice Age. *Science* **277**: 1294-1296.

Meeker, L.D. and Mayewski, P.A. 2002. A 1400-year high-resolution record of atmospheric circulation over the North Atlantic and Asia. *The Holocene* **12**: 257-266.

Noren, A.J., Bierman, P.R., Steig, E.J., Lini, A. and Southon, J. 2002. Millennial-scale storminess variability in the northeastern Unites States during the Holocene epoch. *Nature* **419**: 821-824.

Schimmelmann, A., Lange, C.B. and Meggers, B.J. 2003. Palaeoclimatic and archaeological evidence for a 200-yr recurrence of floods and droughts linking California, Mesoamerica and South America over the past 2000 years. *The Holocene* **13**: 763-778.

Starkel, L. 2002. Change in the frequency of extreme events as the indicator of climatic change in the Holocene (in fluvial systems). *Quaternary International* **91**: 25-32.

5.7. Monsoons

The IPCC's computer models fail to predict variability in monsoon weather. One reason is because they underestimate the sun's role.

For the period 9,600-6,100 years before present, Neff *et al.* (2001) investigated the relationship between a ^{14}C tree-ring record and a $\delta^{18}O$ proxy record of monsoon rainfall intensity as recorded in calcite $\delta^{18}O$ data obtained from a stalagmite in northern Oman. According to the authors, the correlation between the two datasets was reported to be "extremely strong," and a spectral analysis of the data revealed statistically significant periodicities centered on 779, 205, 134, and 87 years for the $\delta^{18}O$ record and periodicities of 206, 148, 126, 89, 26, and 10.4 years for the ^{14}C record.

Because variations in ^{14}C tree-ring records are generally attributed to variations in solar activity and intensity, and because of this particular ^{14}C record's strong correlation with the $\delta^{18}O$ record, as well as the closely corresponding results of the spectral analyses, the authors conclude there is "solid evidence" that both signals (the ^{14}C and $\delta^{18}O$ records) are responding to solar forcing.

Similar findings were reported by Lim *et al.* (2005), who examined the eolian quartz content (EQC) of a high-resolution sedimentary core taken from Cheju Island, Korea, creating a 6,500-year proxy record of major Asian dust events produced by northwesterly winter monsoonal winds that carry dust from the inner part of China all the way to Korea and the East China Sea. The Asian dust time series was found to contain both millennial- and centennial-scale periodicities; cross-spectral analysis between the EQC and a solar activity record showed significant coherent cycles at 700, 280, 210, and 137 years with nearly the same phase changes, leading the researchers to conclude that centennial-scale periodicities in the EQC could be ascribed primarily to fluctuations in solar activity.

In another study, Ji *et al.* (2005) used reflectance spectroscopy on a sediment core taken from Qinghai Lake in the northeastern part of the Qinghai-Tibet Plateau to construct a continuous high-resolution proxy record of the Asian monsoon over the past 18,000 years. As a result of this effort, monsoonal moisture since the late glacial period was shown to be subject to "continual and cyclic variations," including the well-known centennial-scale cold and dry spells of the Dark Ages Cold Period (DACP) and Little Ice Age, which lasted from 2,100 to 1,800 yr BP and 780 to 400 yr BP, respectively. Sandwiched between them was the warmer and wetter Medieval Warm Period, while preceding the DACP was the Roman Warm Period. Also, time series analysis of the sediment record revealed statistically significant periodicities (above the 95 percent level) of 123, 163, 200, and 293 years. The third of these periodicities corresponds well with the de Vries or Suess solar cycle, which suggests cyclical changes in solar activity are important triggers for some of the cyclical changes in monsoon moisture at Qinghai Lake.

Citing studies that suggest the Indian summer monsoon may be sensitive to changes in solar forcing of as little as 0.25 percent (Overpeck et al., 1996; Neff et al., 2001; Fleitmann et al., 2003), Gupta et al. (2005) set out to test this hypothesis by comparing trends in the Indian summer monsoon with trends in solar activity across the Holocene. In this endeavor, temporal trends in the Indian summer monsoon were inferred from relative abundances of fossil shells of the planktic foraminifer Globigerina bulloides in sediments of the Oman margin, while temporal trends in solar variability were inferred from relative abundances of ^{14}C, ^{10}Be and haematite-stained grains.

Spectral analyses of the various datasets revealed statistically significant periodicities in the G. bulloides time series centered at 1550, 152, 137, 114, 101, 89, 83, and 79 years, all but the first of which periodicities closely matched periodicities of sunspot numbers centered at 150, 132, 117, 104, 87, 82, and 75 years. This close correspondence, in the words of Gupta et al., provides strong evidence for a "century-scale relation between solar and summer monsoon variability." In addition, they report that intervals of monsoon minima correspond to intervals of low sunspot numbers, increased production rates of the cosmogenic nuclides ^{14}C and ^{10}Be, and increased advection of drift ice in the North Atlantic, such that over the past 11,100 years "almost every multi-decadal to centennial scale decrease in summer monsoon strength is tied to a distinct interval of reduced solar output," and nearly every increase "coincides with elevated solar output," including a stronger monsoon (high solar activity) during the Medieval Warm Period and a weaker monsoon (low solar activity) during the Little Ice Age.

As for the presence of the 1,550-year cycle in the Indian monsoon data, Gupta et al. consider it to be "remarkable," since this cycle has been identified in numerous climate records of both the Holocene and the last glacial epoch (including Dansgaard/Oeschger cycles in the North Atlantic), strengthening the case for a sun-monsoon-North Atlantic link. Given the remarkable findings of this study, it is no wonder the researchers who conducted it say they are "convinced" there is a direct solar influence on the Indian summer monsoon in which small changes in solar output bring about pronounced changes in tropical climate.

In still another study, Khare and Nigam (2006) examined variations in angular-asymmetrical forms of benthic foraminifera and planktonic foraminiferal populations in a shallow-water sediment core

obtained just off Kawar (14°49'43"N, 73°59'37"E) on the central west coast of India, which receives heavy river discharge during the southwest monsoon season (June to September) from the Kali and Gangavali rivers.

Down-core plots of the data showed three major troughs separated by intervening peaks; "since angular-asymmetrical forms and planktonic foraminiferal population are directly proportional to salinity fluctuations," according to Khare and Nigam, "the troughs ... suggest low salinity (increased river discharge and thus more rainfall)," and that "these wet phases are alternated by dry conditions." They further report that the dry episodes of higher salinity occurred from AD 1320-1355, 1445-1535, and 1625-1660, and that the wet phases were centered at approximately AD 1410, 1590, and 1750, close to the ending of the sunspot minima of the Wolf Minima (AD 1280-1340), the Sporer Minima (AD 1420-1540), and the Maunder Minima (AD 1650-1710), respectively.

Although Khare and Nigam say that "providing a causal mechanism is beyond the scope of the present study," they note that "the occurrence of periods of enhanced monsoonal precipitation slightly after the termination of the Wolf, Sporer and Maunder minima periods (less sun activity) and concomitant temperature changes could be a matter of further intense research." The correspondences seem to be more than merely coincidental, especially when the inferences of the two researchers are said by them to be "in agreement with the findings of earlier workers, who reported high lake levels from Mono Lake and Chad Lake in the vicinity of solar minima," as well as the Nile river in Africa, which "witnessed high level at around AD 1750 and AD 1575."

Nearby in the Arabian Sea, Tiwari et al. (2005) conducted a high-resolution (~50 years) oxygen isotope analysis of three species of planktonic foraminifera (Globigerinoides ruber, Gs. sacculifer and Globarotalia menardii) contained in a sediment core extracted from the eastern continental margin (12.6°N, 74.3°E) that covered the past 13,000 years. Data for the final 1,200 years of this period were compared with the reconstructed total solar irradiance (TSI) record developed by Bard et al. (2000), which is based on fluctuations of ^{14}C and ^{10}Be production rates obtained from tree rings and polar ice sheets.

Results of the analysis showed that the Asian SouthWest Monsoon (SWM) "follows a dominant quasi periodicity of ~200 years, which is similar to that of the 200-year Suess solar cycle (Usokin et al.,

2003)." This finding indicates, in their words, "that SWM intensity on a centennial scale is governed by variation in TSI," which "reinforces the earlier findings of Agnihotri *et al.* (2002) from elsewhere in the Arabian Sea." However, in considering the SWM/TSI relationship, the five researchers note that "variations in TSI (~0.2%) seem to be too small to perturb the SWM, unless assisted by some internal amplification mechanism with positive feedback." In this regard, they discuss two possible mechanisms. The first, in their words, "involves heating of the earth's stratosphere by increased absorption of solar ultraviolet (UV) radiation by ozone during periods of enhanced solar activity (Schneider, 2005)." According to this scenario, more UV reception leads to more ozone production in the stratosphere, which leads to more heat being transferred to the troposphere, which leads to enhanced evaporation from the oceans, which finally enhances monsoon winds and precipitation. The second mechanism, as they describe it, is that "during periods of higher solar activity, the flux of galactic cosmic rays to the earth is reduced, providing less cloud condensation nuclei, resulting in less cloudiness (Schneider, 2005; Friis-Christensen and Svensmark, 1997)," which then allows for "extra heating of the troposphere" that "increases the evaporation from the oceans."

In another study, Dykoski *et al.* (2005) obtained high-resolution records of stable oxygen and carbon isotope ratios from a stalagmite recovered from Dongge Cave in southern China and utilized them to develop a proxy history of Asian monsoon variability over the last 16,000 years. In doing so, they discovered numerous centennial- and multi-decadal-scale oscillations in the record that were up to half the amplitude of interstadial events of the last glacial age, indicating that "significant climate variability characterizes the Holocene." As to what causes this variability, spectral analysis of $\delta^{14}C$ data revealed significant peaks at solar periodicities of 208, 86, and 11 years, which they say is "clear evidence that some of the variability in the monsoon can be explained by solar variability."

Building upon this work, as well as that of Yuan *et al.* (2004), Wang *et al.* (2005) developed a shorter (9,000-year) but higher-resolution (4.5-year) absolute-dated $\delta^{18}O$ monsoon record for the same location, which they compared with atmospheric ^{14}C data and climate records from lands surrounding the North Atlantic Ocean. This work indicated their monsoon record broadly followed summer insolation but was punctuated by eight significantly weaker monsoon periods lasting from one to five centuries, most of which coincided with North Atlantic ice-rafting events. In addition, they found that "cross-correlation of the decadal- to centennial-scale monsoon record with the atmospheric ^{14}C record shows that some, but not all, of the monsoon variability at these frequencies results from changes in solar output," similar to "the relation observed in the record from a southern Oman stalagmite (Fleitmann *et al.*, 2003)."

In a news item by Kerr (2005) accompanying the report of Wang *et al.*, one of the report's authors (Hai Cheng of the University of Minnesota) was quoted as saying their study suggests that "the intensity of the summer [East Asian] monsoon is affected by solar activity." Dominik Fleitman, who worked with the Oman stalagmite, also said that "the correlation is very strong," stating that it is probably the best monsoon record he had seen, calling it "even better than ours." Lastly, Gerald North of Texas A & M University, who Kerr described as a "longtime doubter," admitted that he found the monsoon's solar connection "very hard to refute," although he stated that "the big mystery is that the solar signal should be too small to trigger anything."

Next, Porter and Weijian (2006) used 18 radiocarbon-dated aeolian and paleosol profiles (some obtained by the authors and some by others) within a 1,500-km-long belt along the arid to semi-arid transition zone of north-central China to determine variations in the extent and strength of the East Asian summer monsoon throughout the Holocene.

In the words of the authors, the dated paleosols and peat layers "represent intervals when the zone was dominated by a mild, moist summer monsoon climate that favored pedogenesis and peat accumulation," while "brief intervals of enhanced aeolian activity that resulted in the deposition of loess and aeolian sand were times when strengthened winter monsoon conditions produced a colder, drier climate." The most recent of the episodic cold periods, which they identify as the Little Ice Age, began about AD 1370, while the preceding cold period ended somewhere in the vicinity of AD 810. Consequently, their work implies the existence of a Medieval Warm Period that began some time after AD 810 and ended some time before AD 1370. They also report that the climatic variations they discovered "correlate closely with variations in North Atlantic drift-ice tracers that represent episodic advection of drift ice and cold polar surface water southward and eastward into warmer subpolar water," which correlation implies solar forcing (see Bond *et al.*, 2001) as the most likely cause of the alternating multi-century mild/moist and cold/dry periods of North-Central China. As a result, Porter and Weijian's work helps to establish the global extent of the Medieval Warm Period, as well as its likely solar origin.

We end with a study of the North American monsoon by Asmerom *et al.* (2007), who developed a high-resolution climate proxy for the southwest United States in the form of $\delta^{18}O$ variations in a stalagmite found in Pink Panther Cave in the Guadalupe Mountains of New Mexico.

Spectral analysis performed on the raw $\delta^{18}O$ data revealed significant peaks that the researchers say "closely match previously reported periodicities in the ^{14}C content of the atmosphere, which have been attributed to periodicities in the solar cycle (Stuiver and Braziunas, 1993)." More specifically, they say that cross-spectral analysis of the $\Delta^{14}C$ and $\delta^{18}O$ data confirms that the two records have matching periodicities at 1,533 years (the Bond cycle), 444 years, 170 years, 146 years, and 88 years (the Gleissberg cycle). In addition, they report that periods of increased solar radiation correlate with periods of decreased rainfall in the southwestern United States (via changes in the North American monsoon), and that this behavior is just the opposite of what is observed with the Asian monsoon. These observations thus lead them to suggest that the proposed solar link to Holocene climate operates "through changes in the Walker circulation and the Pacific Decadal Oscillation and El Niño-Southern Oscillation systems of the tropical Pacific Ocean."

In conclusion, research conducted in countries around the world has found a significant effect of solar variability on monsoons. This necessarily implies a small role, or no role, for anthropogenic causes.

Additional information on this topic, including reviews of newer publications as they become available, can be found at http://www.co2science.org/subject/m/monsoonsolar.php.

References

Agnihotri, R., Dutta, K., Bhushan, R. and Somayajulu, B.L.K. 2002. Evidence for solar forcing on the Indian monsoon during the last millennium. *Earth and Planetary Science Letters* **198**: 521-527.

Asmerom, Y., Polyak, V., Burns, S. and Rassmussen, J. 2007. Solar forcing of Holocene climate: New insights from a speleothem record, southwestern United States. *Geology* **35**: 1-4.

Bard, E., Raisbeck, G., Yiou, F. and Jouzel, J. 2000. Solar irradiance during the last 1200 years based on cosmogenic nuclides. *Tellus B* **52**: 985-992.

Bond, G., Kromer, B., Beer, J., Muscheler, R., Evans, M.N., Showers, W., Hoffmann, S., Lotti-Bond, R., Hajdas, I. and Bonani, G. 2001. Persistent solar influence on North Atlantic climate during the Holocene. *Science* **294**: 2130-2136.

Dykoski, C.A., Edwards, R.L., Cheng, H., Yuan, D., Cai, Y., Zhang, M., Lin, Y., Qing, J., An, Z. and Revenaugh, J. 2005. A high-resolution, absolute-dated Holocene and deglacial Asian monsoon record from Dongge Cave, China. *Earth and Planetary Science Letters* **233**: 71-86.

Fleitmann, D., Burns, S.J., Mudelsee, M., Neff, U., Kramers, J., Mangini, A. and Matter, A. 2003. Holocene forcing of the Indian monsoon recorded in a stalagmite from southern Oman. *Science* **300**: 1737-1739.

Friis-Christensen, E. and Svensmark, H. 1997. What do we really know about the sun-climate connection? *Advances in Space Research* **20**: 913-921.

Gupta, A.K., Das, M. and Anderson, D.M. 2005. Solar influence on the Indian summer monsoon during the Holocene. *Geophysical Research Letters* **32**: 10.1029/2005GL022685.

Ji, J., Shen, J., Balsam, W., Chen, J., Liu, L. and Liu, X. 2005. Asian monsoon oscillations in the northeastern Qinghai-Tibet Plateau since the late glacial as interpreted from visible reflectance of Qinghai Lake sediments. *Earth and Planetary Science Letters* **233**: 61-70.

Kerr, R.A. 2005. Changes in the sun may sway the tropical monsoon. *Science* **308**: 787.

Khare, N. and Nigam, R. 2006. Can the possibility of some linkage of monsoonal precipitation with solar variability be ignored? Indications from foraminiferal proxy records. *Current Science* **90**: 1685-1688.

Lim, J., Matsumoto, E. and Kitagawa, H. 2005. Eolian quartz flux variations in Cheju Island, Korea, during the last 6500 yr and a possible Sun-monsoon linkage. *Quaternary Research* **64**: 12-20.

Neff, U., Burns, S.J., Mangini, A., Mudelsee, M., Fleitmann, D. and Matter, A. 2001. Strong coherence between solar variability and the monsoon in Oman between 9 and 6 kyr ago. *Nature* **411**: 290-293.

Overpeck, J.T., Anderson, D.M., Trumbore, S. and Prell, W.L. 1996. The southwest monsoon over the last 18,000 years. *Climate Dynamics* **12**: 213-225.

Porter, S.C. and Weijian, Z. 2006. Synchronism of Holocene East Asian monsoon variations and North Atlantic drift-ice tracers. *Quaternary Research* **65**: 443-449.

Schneider, D. 2005. Living in sunny times. *American Scientist* **93**: 22-24.

Stuiver, M. and Braziunas, T.F. 1993. Sun, ocean climate and atmospheric $^{14}CO_2$: An evaluation of causal and spectral relationships. *The Holocene* **3**: 289-305.

Tiwari, M., Ramesh, R., Somayajulu, B.L.K., Jull, A.J.T. and Burr, G.S. 2005. Solar control of southwest monsoon on centennial timescales. *Current Science* **89**: 1583-1588.

Usoskin, I.G. and Mursula, K. 2003. Long-term solar cycle evolution: Review of recent developments. *Solar Physics* **218**: 319-343.

Wang, Y., Cheng, H., Edwards, R.L., He, Y., Kong, X., An, Z., Wu, J., Kelly, M.J., Dykoski, C.A. and Li, X. 2005. The Holocene Asian monsoon: Links to solar changes and North Atlantic climate. *Science* **308**: 854-857.

Yuan, D., Cheng, H., Edwards, R.L., Dykoski, C.A., Kelly, M.J., Zhang, M., Qing, J., Lin, Y., Wang, Y., Wu, J., Dorale, J.A., An, Z. and Cai, Y. 2004. Timing, duration, and transitions of the last interglacial Asian monsoon. *Science* **304**: 575-578.

5.8. Streamflow

In this section we highlight some of the scientific literature that demonstrates the influence of solar viability on streamflow, a climate variable related to precipitation, droughts, and floods. If a significant influence exists, it follows that greenhouse gases or global warming (regardless of its cause) played a smaller or even non-existent role in streamflow trends during the twentieth century.

Starting in northeastern Mongolia, Pederson *et al.* (2001) used tree-ring chronologies to reconstruct annual precipitation and streamflow histories in this region over the period 1651-1995. Analyses of both standard deviations and five-year intervals of extreme wet and dry periods revealed that "variations over the recent period of instrumental data are not unusual relative to the prior record." The authors state, however, that the reconstructions "appear to show more frequent extended wet periods in more recent decades," but they note this observation "does not

demonstrate unequivocal evidence of an increase in precipitation as suggested by some climate models." More relevant to the present discussion is the researchers' observation that spectral analysis of the data revealed significant periodicities around 12 and 20-24 years, suggesting "possible evidence for solar influences in these reconstructions for northeastern Mongolia."

In Western Europe, several researchers have studied precipitation histories of regions along the Danube River and the effects they have on river discharge, with some of them suggesting that an anthropogenic signal is present in the latter decades of the twentieth century and is responsible for that period's drier conditions. In an effort to validate such claims, Ducic (2005) analyzed observed and reconstructed river discharge rates near Orsova, Serbia over the period 1731-1990.

Results of the study indicated that the lowest five-year discharge value in the pre-instrumental era (period of occurrence: 1831-1835) was practically equal to the lowest five-year discharge value in the instrumental era (period of occurrence: 1946-1950), and that the driest decade of the entire 260-year period was 1831-1840. Similarly, the highest five-year discharge value for the pre-instrumental era (period of occurrence: 1736-1740) was nearly equal to the five-year maximum discharge value for the instrumental era (period of occurrence: 1876-1880), differing by only 0.7 percent. What is more, the discharge rate for the last decade of the record (1981-1990), which prior researchers had claimed was anthropogenically influenced, was found to be "completely inside the limits of the whole series," in Ducic's words, and only slightly (38 m^3s^{-1} or 0.7 percent) less than the 260-year mean of 5356 m^3s^{-1}. In conclusion, Ducic states that "modern discharge fluctuations do not point to [a] dominant anthropogenic influence." Ducic's correlative analysis suggests that the detected cyclicity in the record may "point to the domination of the influence of solar activity."

Solar-related streamflow oscillations have also been reported for the Nile. Kondrashov *et al.* (2005), for example, applied advanced spectral methods to fill in data gaps and locate interannual and interdecadal periodicities in historical records of annual low- and high-water levels on the Nile over the 1,300-year period AD 622 to 1922. In doing so, they found several statistically significant periodicities, including cycles of 256, 64, 19, 12, 7, 4.2, and 2.2 years. With respect to the causes of these cycles, Kondrashov *et al.* say the 4.2- and 2.2-year oscillations are likely the product of El Niño-Southern Oscillation variations. They find the 7-year cycle may be related to North Atlantic influences, according to them, while the longer-period oscillations may be due to solar-related forcings.

Mauas *et al.* (2008) write that river streamflows "are excellent climatic indicators," especially in the case of rivers "with continental scale basins" that "smooth out local variations" and can thus "be particularly useful to study global forcing mechanisms." Focusing on South America's Parana River—the world's fifth largest in terms of drainage area and its fourth largest with respect to streamflow—Mauas *et al.* analyzed streamflow data collected continuously on a daily basis since 1904. With respect to periodicities, they report that the detrended time series of the streamflow data were correlated with the detrended times series of both sunspot number and total solar irradiance, yielding Pearson's correlation coefficients between streamflow and the two solar parameters of 0.78 and 0.69, respectively, at "a significance level higher than 99.99% in both cases." This is strong evidence indeed that solar variability, and nont man-made greenhouse gas emissions, were responsible for variation in streamflow during the modern industrial era.

Additional information on this topic, including reviews of newer publications as they become available, can be found at http://www.co2science.org/subject/s/streamflowsolarin.php.

References

Ducic, V. 2005. Reconstruction of the Danube discharge on hydrological station Orsova in pre-instrumental period: Possible causes of fluctuations. *Edition Physical Geography of Serbia* **2**: 79-100.

Kondrashov, D., Feliks, Y. and Ghil, M. 2005. Oscillatory modes of extended Nile River records (A.D. 622-1922). *Geophysical Research Letters* **32**: doi:10.1029/2004 GL022156.

Mauas, P.J.D., Flamenco, E. and Buccino, A.P. 2008. Solar forcing of the stream flow of a continental scale South American river. *Physical Review Letters* **101**: 168501.

Pederson, N., Jacoby, G.C., D'Arrigo, R.D., Cook, E.R. and Buckley, B.M. 2001. Hydrometeorological reconstructions for northeastern Mongolia derived from tree rings: 1651-1995. *Journal of Climate* **14**: 872-881.

[this page intentionally blank]

6

OBSERVATIONS: EXTREME WEATHER

Introduction

The Intergovernmental Panel on Climate Change (IPCC) claims, in Section 3.8 of the report of Working Group I to the Fourth Assessment Report, that global warming will cause (or already is causing) more extreme weather: droughts, floods, tropical cyclones, storms, and more (IPCC, 2007-I). Chapter 5 of the present report presented extensive evidence that solar variability, not CO_2 concentrations in the air or rising global temperatures (regardless of their cause) is responsible for trends in many of these weather variables. In this chapter we ask if there is evidence that the twentieth century, which the IPCC claims was the warmest century in a millennium, experienced more severe weather than was experienced in previous, cooler periods. We find no support for the IPCC's predictions. In fact, we find more evidence to support the opposite prediction: that weather would be *less* extreme in a warmer world.

References

IPCC 2007-I. *Climate Change 2007: The Physical Science Basis. Contribution of Working Group I to the Fourth Assessment Report of the Intergovernmental Panel on Climate Change.* Solomon, S., Qin, D., Manning, M.,

Chen, Z., Marquis, M., Averyt, K.B., Tignor, M. and Miller, H.L. (Eds.) Cambridge University Press, Cambridge, UK.

6.1. Droughts

One of the many dangers of global warming, according to the IPCC, is more frequent, more severe, and longer-lasting droughts. In this section, we discuss the findings of scientific papers that compared droughts in the twentieth century with those longer ago, beginning with Africa and then Asia, Europe, and finally North America.

Additional information on this topic, including reviews on drought not discussed here, can be found at http://www.co2science.org/subject/d/subject_d .php under the heading Drought.

6.1.1. Africa

Lau *et al.*, 2006 explored "the roles of sea surface temperature coupling and land surface processes in producing the Sahel drought" in the computer models used by the IPCC for its Fourth Assessment Report. These 19 computer models were "driven by combinations of realistic prescribed external forcing, including anthropogenic increase in greenhouse gases and sulfate aerosols, long-term variation in solar radiation, and volcanic eruptions." This work revealed that "only eight models produce a reasonable Sahel drought signal, seven models produce excessive rainfall over [the] Sahel during the observed drought period, and four models show no significant deviation from normal." In addition, they report that "even the model with the highest skill for the Sahel drought could only simulate the increasing trend of severe drought events but not the magnitude, nor the beginning time and duration of the events."

All 19 of the models used in preparing the IPCC's Fourth Assessment Report were unable to adequately simulate the basic characteristics of what Lau *et al.* call one of the past century's "most pronounced signals of climate change." This failure of what the authors call an "ideal test" for evaluating the models' abilities to accurately simulate "long-term drought" and "coupled atmosphere-ocean-land processes and their interactions" vividly illustrates the fallibility of computer climate models.

In a review of information pertaining to the past two centuries, Nicholson (2001) reports there has been "a long-term reduction in rainfall in the semi-arid regions of West Africa" that has been "on the order of 20 to 40% in parts of the Sahel." Describing the phenomenon as "three decades of protracted aridity," she reports that "nearly all of Africa has been affected ... particularly since the 1980s." Nevertheless, Nicholson says that "rainfall conditions over Africa during the last 2 to 3 decades are not unprecedented," and that "a similar dry episode prevailed during most of the first half of the 19th century," when much of the planet was still experiencing Little Ice Age conditions.

Therrell *et al.* (2006) developed what they describe as "the first tree-ring reconstruction of rainfall in tropical Africa using a 200-year regional chronology based on samples of *Pterocarpus angolensis* [a deciduous tropical hardwood known locally as Mukwa] from Zimbabwe." This project revealed that "a decadal-scale drought reconstructed from 1882 to 1896 matches the most severe sustained drought during the instrumental period (1989-1995)," and that "an even more severe drought is indicated from 1859 to 1868 in both the tree-ring and documentary data." They report, for example, that the year 1860 (which was the most droughty year of the entire period), was described in a contemporary account from Botswana (where part of their tree-ring chronology originated) as "a season of 'severe and

universal drought' with 'food of every description' being 'exceedingly scarce' and the losses of cattle being 'very severe' (Nash and Endfield, 2002)." At the other end of the moisture spectrum, Therrel *et al.* report that "a 6-year wet period at the turn of the nineteenth century (1897-1902) exceeds any wet episode during the instrumental era." Consequently, for a large part of central southern Africa, it is clear that the supposedly unprecedented global warming of the twentieth century did not result in an intensification of either extreme dry or wet periods.

Looking further back in time, Verschuren *et al.* (2000) developed a decadal-scale history of rainfall and drought in equatorial east Africa for the past thousand years, based on level and salinity fluctuations of a small crater-lake in Kenya that were derived from diatom and midge assemblages retrieved from the lake's sediments. Once again, they found that the Little Ice Age was generally wetter than the Current Warm Period; but they identified three intervals of prolonged dryness within the Little Ice Age (1390-1420, 1560-1625, and 1760-1840), and of these "episodes of persistent aridity," as they refer to them, *all* were determined to have been "more severe than any recorded drought of the twentieth century."

Probing some 1,500 years into the past was the study of Holmes *et al.* (1997), who wrote that since the late 1960s, the African Sahel had experienced "one of the most persistent droughts recorded by the entire global meteorological record." However, in a high-resolution study of a sediment sequence extracted from an oasis in the Manga Grasslands of northeast Nigeria, they too determined that "the present drought is not unique and that drought has recurred on a centennial to interdecadal timescale during the last 1500 years."

Last, and going back in time almost 5,500 years, Russell and Johnson (2005) analyzed sediment cores that had been retrieved from Lake Edward—the smallest of the great rift lakes of East Africa, located on the border that separates Uganda and the Democratic Republic of the Congo—to derive a detailed precipitation history for that region. In doing so, they discovered that from the start of the record until about 1,800 years ago, there was a long-term trend toward progressively more arid conditions, after which there followed what they term a "slight trend" toward wetter conditions that has persisted to the present. In addition, superimposed on these long-term trends were major droughts of "at least century-scale duration," centered at approximately 850, 1,500, 2,000, and 4,100 years ago. Consequently, it would not be unnatural for another such drought to grip the region in the not-too-distant future.

In summation, real-world evidence from Africa suggests that the global warming of the past century or so has not led to a greater frequency or greater severity of drought in that part of the world. Indeed, even the continent's worst drought in recorded meteorological history was much milder than droughts that occurred periodically during much colder times.

Additional information on this topic, including reviews of newer publications as they become available, can be found at http://www.co2science.org/subject/d/droughtafrica.php.

References

Holmes, J.A., Street-Perrott, F.A., Allen, M.J., Fothergill, P.A., Harkness, D.D., Droon, D. and Perrott, R.A. 1997. Holocene palaeolimnology of Kajemarum Oasis, Northern Nigeria: An isotopic study of ostracodes, bulk carbonate and organic carbon. *Journal of the Geological Society, London* **154**: 311-319.

Lau, K.M., Shen, S.S.P., Kim, K.-M. and Wang, H. 2006. A multimodel study of the twentieth-century simulations of Sahel drought from the 1970s to 1990s. *Journal of Geophysical Research* **111**: 10.1029/2005JD006281.

Nash, D.J. and Endfield, G.H. 2002. A 19th-century climate chronology for the Kalahari region of central southern Africa derived from missionary correspondence. *International Journal of Climatology* **22**: 821-841.

Nicholson, S.E. 2001. Climatic and environmental change in Africa during the last two centuries. *Climate Research* **17**: 123-144.

Russell, J.M. and Johnson, T.C. 2005. A high-resolution geochemical record from Lake Edward, Uganda Congo and the timing and causes of tropical African drought during the late Holocene. *Quaternary Science Reviews* **24**: 1375-1389.

Therrell, M.D., Stahle, D.W., Ries, L.P. and Shugart, H.H. 2006. Tree-ring reconstructed rainfall variability in Zimbabwe. *Climate Dynamics* **26**: 677-685.

Verschuren, D., Laird, K.R. and Cumming, B.F. 2000. Rainfall and drought in equatorial east Africa during the past 1,100 years. *Nature* **403**: 410-414.

6.1.2. Asia

Paulsen *et al.* (2003) employed high-resolution stalagmite records of $\delta^{13}C$ and $\delta^{18}O$ from Buddha Cave "to infer changes in climate in central China for the last 1270 years in terms of warmer, colder, wetter and drier conditions." Among the climatic episodes evident in their data were "those corresponding to the Medieval Warm Period, Little Ice Age and twentieth century warming, lending support to the global extent of these events." More specifically, their record begins in the depths of the Dark Ages Cold Period, which ends about AD 965 with the commencement of the Medieval Warm Period. The warming trend continues until approximately AD 1475, whereupon the Little Ice Age sets in. That cold period holds sway until about AD 1825, after which the warming responsible for the Current Warm Period begins.

With respect to hydrologic balance, the last part of the Dark Ages Cold Period was characterized as wet. It, in turn, was followed by a dry, a wet, and another dry interval in the Medieval Warm Period, which was followed by a wet and a dry interval in the Little Ice Age, and finally a mostly wet but highly moisture-variable Current Warm Period. Paulsen *et al.*'s data also reveal a number of other cycles superimposed on the major millennial-scale cycle of temperature and the centennial-scale cycle of moisture. They attribute most of these higher-frequency cycles to solar phenomena and not CO_2 concentrations in the air. Paulsen *et al.* conclude that the summer monsoon over eastern China, which brings the region much of its precipitation, may "be related to solar irradiance."

Kalugin *et al.* (2005) worked with sediment cores from Lake Teletskoye in the Altai Mountains of Southern Siberia to produce a multi-proxy climate record spanning the past 800 years. This record revealed that the regional climate was relatively warm with high terrestrial productivity from AD 1210 to 1380. Thereafter, however, temperatures cooled and productivity dropped, reaching a broad minimum between 1660 and 1700, which interval, in their words, "corresponds to the age range of the well-known Maunder Minimum (1645-1715)" and is "in agreement with the timing of the Little Ice Age in Europe (1560-1850)."

With respect to moisture and precipitation, Kalugin *et al.* state that the period between 1210 and 1480 was more humid than that of today, while the period between 1480 and 1840 was more arid. In addition, they report three episodes of multi-year drought (1580-1600, 1665-1690, and 1785-1810), which findings are in agreement with other historical data and tree-ring records from the Mongolia-Altai region (Butvilovskii, 1993; Jacoby *et al.*, 1996; Panyushkina *et al.*, 2000). It is problematic for the IPCC to claim that global warming will lead to more frequent and more severe droughts, as *all* of the major multi-year droughts detected in this study occurred during the cool phase of the 800-year record.

Touchan *et al.* (2003) developed two reconstructions of spring precipitation for southwestern Turkey from tree-ring width measurements, one of them (1776-1998) based on nine chronologies of *Cedrus libani, Juniperus excelsa, Pinus brutia,* and *Pinus nigra*, and the other one (1339-1998) based on three chronologies of *Juniperus excelsa*. These records, according to them, "show clear evidence of multi-year to decadal variations in spring precipitation." Nevertheless, they report that "dry periods of 1-2 years were well distributed throughout the record" and that the same was largely true of similar wet periods. With respect to more extreme events, the period preceding the Industrial Revolution stood out. They note, for example, that "all of the wettest 5-year periods occurred prior to 1756." Likewise, the longest period of reconstructed spring drought was the four-year period 1476-79, while the single driest spring was 1746. We see no evidence that the past century produced distinctive changes in the nature of drought in this part of Asia.

Cluis and Laberge (2001) analyzed streamflow records stored in the databank of the Global Runoff Data Center at the Federal Institute of Hydrology in Koblenz (Germany) to see if there were any changes in Asian river runoff of the type predicted by the IPCC to lead to more frequent and more severe drought. This study was based on the streamflow histories of 78 rivers said to be "geographically distributed throughout the whole Asia-Pacific region." The mean start and end dates of these series were 1936 ± 5 years and 1988 ± 1 year, respectively, representing an approximate half-century time span. In the case of the annual minimum discharges of these rivers, which are the ones associated with drought, 53 percent of them were unchanged over the period of the study; where there were trends, 62 percent of them were upward, indicative of a growing likelihood of both less frequent and less severe drought.

Ducic (2005) analyzed observed and reconstructed discharge rates of the Danube River near Orsova, Serbia, over the period 1731-1990,

finding that the lowest five-year discharge value in the pre-instrumental era (period of occurrence: 1831-1835) was practically equal to the lowest five-year discharge value in the instrumental era (period of occurrence: 1946-1950), and that the driest decade of the entire 260-year period was 1831-1840. The discharge rate for the last decade of the record (1981-1990), was "completely inside the limits of the whole series," in Ducic's words, and only slightly (0.7 percent) less than the 260-year mean. As a result, Ducic concluded that "modern discharge fluctuations do not point to [a] dominant anthropogenic influence." Ducic's correlative analysis suggests that the detected cyclicity in the record could "point to the domination of the influence of solar activity."

Jiang *et al.* (2005) analyzed historical documents to produce a time series of flood and drought occurrences in eastern China's Yangtze Delta since AD 1000. Their work revealed that alternating wet and dry episodes occurred throughout this period; the data demonstrate that droughts and floods usually occurred in the spring and autumn seasons of the same year, with the most rapid and strongest of these fluctuations occurring during the Little Ice Age (1500-1850), as opposed to the preceding Medieval Warm Period and the following Current Warm Period.

Davi *et al.* (2006) employed absolutely dated tree-ring-width chronologies from five sampling sites in west-central Mongolia—all of them "in or near the Selenge River basin, the largest river in Mongolia"— to develop a reconstruction of streamflow that extends from 1637 to 1997. Of the 10 driest five-year periods of the 360-year record, only one occurred during the twentieth century (and that just barely: 1901-1905, sixth driest of the 10 extreme periods), while of the 10 wettest five-year periods, only two occurred during the twentieth century (1990-1994 and 1917-1921, the second and eighth wettest of the 10 extreme periods, respectively). Consequently, as Davi *et al.* describe the situation, "there is much wider variation in the long-term tree-ring record than in the limited record of measured precipitation," such that over the course of the twentieth century, extremes of both dryness and wetness were less frequent and less severe.

Sinha *et al.* (2007) derived a nearly annually resolved record of Indian summer monsoon (ISM) rainfall variations for the core monsoon region of India that stretches from AD 600 to 1500 based on a ^{230}Th-dated stalagmite oxygen isotope record from Dandak Cave, which is located at 19°00'N, 82°00'E. This work revealed that "the short instrumental record of ISM underestimates the magnitude of monsoon rainfall variability," and they state that "nearly every major famine in India [over the period of their study] coincided with a period of reduced monsoon rainfall as reflected in the Dandak $\delta^{18}O$ record," noting two particularly devastating famines that "occurred at the beginning of the Little Ice Age during the longest duration and most severe ISM weakening of [their] reconstruction."

Sinha *et al.* state that "ISM reconstructions from Arabian Sea marine sediments (Agnihotri *et al.*, 2002; Gupta *et al.*, 2003; von Rad *et al.*, 1999), stalagmite $\delta^{18}O$ records from Oman and Yemen (Burns *et al.*, 2002; Fleitmann *et al.*, 2007), and a pollen record from the western Himalaya (Phadtare and Pant, 2006) also indicate a weaker monsoon during the Little Ice Age and a relatively stronger monsoon during the Medieval Warm Period." As a result, the eight researchers note that "since the end of the Little Ice Age, ca 1850 AD, the human population in the Indian monsoon region has increased from about 200 million to over 1 billion," and that "a recurrence of weaker intervals of ISM comparable to those inferred in our record would have serious implications to human health and economic sustainability in the region." Thus the Current Warm Period is beneficial to the population of India.

Zhang *et al.* (2007) developed flood and drought histories of the past thousand years in China's Yangtze Delta, based on "local chronicles, old and very comprehensive encyclopaedia, historic agricultural registers, and official weather reports," after which "continuous wavelet transform was applied to detect the periodicity and variability of the flood/drought series"—which they describe as "a powerful way to characterize the frequency, the intensity, the time position, and the duration of variations in a climate data series"—and, finally, the results of the entire set of operations were compared with two one-thousand-year temperature histories of the Tibetan Plateau: northeastern Tibet and southern Tibet.

As a result of this effort, Zhang *et al.* report that "during AD 1400-1700 [the coldest portion of their record, corresponding to much of the Little Ice Age], the proxy indicators showing the annual temperature experienced larger variability (larger standard deviation), and this time interval exactly corresponds to the time when the higher and significant wavelet variance occurred." By contrast, they report that "during AD 1000-1400 [the warmest portion of their record, corresponding to much of the Medieval Warm

Period], relatively stable changes of climatic changes reconstructed from proxy indicators in Tibet correspond to lower wavelet variance of flood/drought series in the Yangtze Delta region."

The research summarized in this section shows the frequency of drought in Asia varies according to millennial, centennial, and decadal cycles. Since those cycles predate any possible human influence on climate, they serve to refute the claim that today's relatively dry climate is the result of human activity.

Additional information on this topic, including reviews of newer publications as they become available, can be found at http://www.co2science.org/subject/d/droughtasia.php.

References

Agnihotri, R., Dutta, K., Bhushan, R. and Somayajulu, B.L.K. 2002. Evidence for solar forcing on the Indian monsoon during the last millennium. *Earth and Planetary Science Letters* **198**: 521-527.

Burns, S.J., Fleitmann, D., Mudelsee, M., Neff, U., Matter, A. and Mangini, A. 2002. A 780-year annually resolved record of Indian Ocean monsoon precipitation from a speleothem from south Oman. *Journal of Geophysical Research* **107**: 10.1029/2001JD001281.

Butvilovskii, V.V. 1993. Paleogeography of the Late Glacial and Holocene on Altai. Tomsk University, Tomsk.

Cluis, D. and Laberge, C. 2001. Climate change and trend detection in selected rivers within the Asia-Pacific region. *Water International* **26**: 411-424.

Davi, N.K., Jacoby, G.C., Curtis, A.E. and Baatarbileg, N. 2006. Extension of drought records for central Asia using tree rings: West-Central Mongolia. *Journal of Climate* **19**: 288-299.

Ducic, V. 2005. Reconstruction of the Danube discharge on hydrological station Orsova in pre-instrumental period: Possible causes of fluctuations. *Edition Physical Geography of Serbia* **2**: 79-100.

Fleitmann, D., Burns, S.J., Mangini, A., Mudelsee, M., Kramers, J., Neff, U., Al-Subbary, A.A., Buettner, A., Hippler, D. and Matter, A. 2007. Holocene ITCZ and Indian monsoon dynamics recorded in stalagmites from Oman and Yemen (Socotra). *Quaternary Science Reviews* **26**: 170-188.

Gupta, A.K., Anderson, D.M. and Overpeck, J.T. 2003. Abrupt changes in the Asian southwest monsoon during the Holocene and their links to the North Atlantic Ocean. *Nature* **421**: 354-356.

Jacoby, G.C., D'Arrigo, R.D. and Davaajatms, T. 1996. Mongolian tree rings and 20th century warming. *Science* **273**: 771-773.

Jiang, T., Zhang, Q., Blender, R. and Fraedrich, K. 2005. Yangtze Delta floods and droughts of the last millennium: Abrupt changes and long term memory. *Theoretical and Applied Climatology* **82**: 131-141.

Kalugin, I., Selegei, V., Goldberg, E. and Seret, G. 2005. Rhythmic fine-grained sediment deposition in Lake Teletskoye, Altai, Siberia, in relation to regional climate change. *Quaternary International* **136**: 5-13.

Panyushkina, I.P., Adamenko, M.F., Ovchinnikov, D.V. 2000. Dendroclimatic net over Altai Mountains as a base for numerical paleogeographic reconstruction of climate with high time resolution. In: *Problems of Climatic Reconstructions in Pliestocene and Holocene 2*. Institute of Archaeology and Ethnography, Novosibirsk, pp. 413-419.

Paulsen, D.E., Li, H.-C. and Ku, T.-L. 2003. Climate variability in central China over the last 1270 years revealed by high-resolution stalagmite records. *Quaternary Science Reviews* **22**: 691-701.

Phadtare, N.R. and Pant, R.K. 2006. A century-scale pollen record of vegetation and climate history during the past 3500 years in the Pinder Valley, Kumaon Higher Himalaya, India. *Journal of the Geological Society of India* **68**: 495-506.

Sinha, A., Cannariato, K.G., Stott, L.D., Cheng, H., Edwards, R.L., Yadava, M.G., Ramesh, R. and Singh, I.B. 2007. A 900-year (600 to 1500 A.D.) record of the Indian summer monsoon precipitation from the core monsoon zone of India. *Geophysical Research Letters* **34**: 10.1029/2007GL030431.

Touchan, R., Garfin, G.M., Meko, D.M., Funkhouser, G., Erkan, N., Hughes, M.K. and Wallin, B.S. 2003. Preliminary reconstructions of spring precipitation in southwestern Turkey from tree-ring width. *International Journal of Climatology* **23**: 157-171.

von Rad, U., Michels, K.H., Schulz, H., Berger, W.H. and Sirocko, F. 1999. A 5000-yr record of climate change in varved sediments from the oxygen minimum zone off Pakistan, northeastern Arabian Sea. *Quaternary Research* **51**: 39-53.

Zhang, Q., Chen, J. and Becker, S. 2007. Flood/drought change of last millennium in the Yangtze Delta and its possible connections with Tibetan climatic changes. *Global and Planetary Change* **57**: 213-221.

6.1.3. Europe

Linderholm and Chen (2005) derived a 500-year history of winter (September-April) precipitation from tree-ring data obtained within the Northern Boreal zone of Central Scandinavia. This chronology indicated that below-average precipitation was observed during the periods 1504-1520, 1562-1625, 1648-1669, 1696-1731, 1852-1871, and 1893-1958, with the lowest values occurring at the beginning of the record and at the beginning of the seventeenth century. These results demonstrate that for this portion of the European continent, twentieth century global warming did not result in more frequent or more severe droughts.

Another five-century perspective on the issue was provided by Wilson *et al.* (2005), who used a regional curve standardization technique to develop a summer (March-August) precipitation chronology from living and historical ring-widths of trees in the Bavarian Forest region of southeast Germany for the period 1456-2001. This technique captured low frequency variations that indicated the region was substantially drier than the long-term average during the periods 1500-1560, 1610-1730, and 1810-1870, all of which intervals were much colder than the bulk of the twentieth century.

A third study of interest concerns the Danube River in western Europe, where some researchers had previously suggested that an anthropogenic signal was present in the latter decades of the twentieth century, and that it was responsible for that period's supposedly drier conditions. Ducic (2005) tested these claims by analyzing observed and reconstructed discharge rates of the river near Orsova, Serbia over the period 1731-1990. This work revealed that the lowest five-year discharge value in the pre-instrumental era (1831-1835) was practically equal to the lowest five-year discharge value in the instrumental era (1946-1950), and that the driest decade of the entire 260-year period was 1831-1840. What is more, the discharge rate for the last decade of the record (1981-1990), which prior researchers had claimed was anthropogenically influenced, was found to be "completely inside the limits of the whole series," in Ducic's words, and only 0.7 percent less than the 260-year mean, leading to the conclusion that "modern discharge fluctuations do not point to dominant anthropogenic influence." In fact, Ducic's correlative analysis suggests that the detected cyclicity in the record could "point to the domination of the influence of solar activity."

In much the same vein and noting that "the media often reflect the view that recent severe drought events are signs that the climate has in fact already changed owing to human impacts," Hisdal *et al.* (2001) examined pertinent data from many places in Europe. They performed a series of statistical analyses on more than 600 daily streamflow records from the European Water Archive to examine trends in the severity, duration, and frequency of drought over the following four time periods: 1962-1990, 1962-1995, 1930-1995, and 1911-1995. This work revealed, in their words, that "despite several reports on recent droughts in Europe, there is no clear indication that streamflow drought conditions in Europe have generally become more severe or frequent in the time periods studied." To the contrary, they report that "overall, the number of negative significant trends pointing towards decreasing drought deficit volumes or fewer drought events exceeded the number of positive significant trends (increasing drought deficit volumes or more drought events)."

In conclusion, there is no evidence that droughts in Europe became more frequent or more severe due to global warming in the twentieth century. Additional information on this topic, including reviews of newer publications as they become available, can be found at http://www.co2science.org/subject/d/droughteurope.php.

References

Ducic, V. 2005. Reconstruction of the Danube discharge on hydrological station Orsova in pre-instrumental period: Possible causes of fluctuations. *Edition Physical Geography of Serbia* **2**: 79-100.

Hisdal, H., Stahl, K., Tallaksen, L.M. and Demuth, S. 2001. Have streamflow droughts in Europe become more severe or frequent? *International Journal of Climatology* **21**: 317-333.

Linderholm, H.W. and Chen, D. 2005. Central Scandinavian winter precipitation variability during the past five centuries reconstructed from Pinus sylvestris tree rings. *Boreas* **34**: 44-52.

Wilson, R.J., Luckman, B.H. and Esper, J. 2005. A 500 year dendroclimatic reconstruction of spring-summer precipitation from the lower Bavarian Forest region, Germany. *International Journal of Climatology* **25**: 611-630.

6.1.4. North America

6.1.4.1. Canada

Gan (1998) performed several statistical tests on datasets pertaining to temperature, precipitation, spring snowmelt dates, streamflow, potential and actual evapotranspiration, and the duration, magnitude, and severity of drought throughout the Canadian Prairie Provinces of Alberta, Saskatchewan, and Manitoba. The results of these several tests suggest that the Prairies have become somewhat warmer and drier over the past four to five decades, although there are regional exceptions to this generality. After weighing all of the pertinent facts, however, Gan reports "there is no solid evidence to conclude that climatic warming, if it occurred, has caused the Prairie drought to become more severe," further noting, "the evidence is insufficient to conclude that warmer climate will lead to more severe droughts in the Prairies."

Working in the same general area, Quiring and Papakyriakou (2005) used an agricultural drought index (Palmer's Z-index) to characterize the frequency, severity, and spatial extent of June-July moisture anomalies for 43 crop districts from the agricultural region of the Canadian prairies over the period 1920-99. This work revealed that for the 80-year period of their study, the single most severe June-July drought on the Canadian prairies occurred in 1961, and that the next most severe droughts, in descending order of severity, occurred in 1988, 1936, 1929, and 1937, for little net overall trend. Simultaneously, however, they say there was an upward trend in mean June-July moisture conditions. In addition, they note that "reconstructed July moisture conditions for the Canadian prairies demonstrate that droughts during the 18th and 19th centuries were more persistent than those of the 20th century (Sauchyn and Skinner, 2001)."

In a subsequent study that covered an even longer span of time, St. George and Nielsen (2002) used "a ringwidth chronology developed from living, historical and subfossil bur oak in the Red River basin to reconstruct annual precipitation in southern Manitoba since AD 1409." They say that "prior to the 20th century, southern Manitoba's climate was more extreme and variable, with prolonged intervals that were wetter and drier than any time following permanent Euro-Canadian settlement." In other words, the twentieth century had more stable climatic conditions with fewer hydrologic extremes (floods and droughts) than was typical of prior conditions. St. George and Nielsen conclude that "climatic case studies in regional drought and flood planning based exclusively on experience during the 20th century may dramatically underestimate true worst-case scenarios." They also indicate that "multidecadal fluctuations in regional hydroclimate have been remarkably coherent across the northeastern Great Plains during the last 600 years," and that "individual dry years in the Red River basin were usually associated with larger scale drought across much of the North American interior," which suggests that their results for the Red River basin are likely representative of this entire larger region.

Taking an even longer look back in time, Campbell (2002) analyzed the grain sizes of sediment cores obtained from Pine Lake, Alberta, Canada to derive a non-vegetation-based high-resolution record of climate variability over the past 4,000 years. Throughout this record, periods of both increasing and decreasing moisture availability, as determined from grain size, were evident at decadal, centennial, and millennial time scales, as was also found by Laird et al. (2003) in a study of diatom assemblages in sediment cores taken from three additional Canadian lakes. Over the most recent 150 years, however, the grain size of the Pine Lake study generally remained above the 4,000-year average, indicative of relatively stable and less droughty conditions than the mean of the past four millennia.

Also working in eastern Canada, Girardin et al. (2004) developed a 380-year reconstruction of the Canadian Drought Code (CDC, a daily numerical rating of the average moisture content of deep soil organic layers in boreal conifer stands that is used to monitor forest fire danger) for the month of July, based on 16 well-replicated tree-ring chronologies from the Abitibi Plains of eastern Canada just below James Bay. Cross-continuous wavelet transformation analyses of these data, in their words, "indicated coherency in the 8-16 and 17-32-year per cycle oscillation bands between the CDC reconstruction and the Pacific Decadal Oscillation prior to 1850," while "following 1850, the coherency shifted toward the North Atlantic Oscillation." These results led them to suggest that "the end of [the] 'Little Ice Age' over the Abitibi Plains sector corresponded to a decrease in the North Pacific decadal forcing around the 1850s," and that "this event could have been followed by an inhibition of the Arctic air outflow and an incursion of more humid air masses from the subtropical Atlantic climate sector," which may have

helped reduce fire frequency and drought severity. In this regard, they note that several other paleo-climate and ecological studies have suggested that "climate in eastern Canada started to change with the end of the 'Little Ice Age' (~1850)," citing the works of Tardif and Bergeron (1997, 1999), Bergeron (1998, 2000) and Bergeron et al. (2001), while further noting that Bergeron and Archambault (1993) and Hofgaard et al. (1999) have "speculated that the poleward retreat of the Arctic air mass starting at the end of the 'Little Ice Age' contributed to the incursion of moister air masses in eastern Canada."

Moving back towards the west, Wolfe et al. (2005) conducted a multi-proxy hydro-ecological analysis of Spruce Island Lake in the northern Peace sector of the Peace-Athabasca Delta in northern Alberta. Their research revealed that hydro-ecological conditions in that region varied substantially over the past 300 years, especially in terms of multi-decadal dry and wet periods. More specifically, they found that (1) recent drying in the region was not the product of Peace River flow regulation that began in 1968, but rather the product of an extended drying period that was initiated in the early to mid-1900s, (2) the multi-proxy hydro-ecological variables they analyzed were well correlated with other reconstructed records of natural climate variability, and (3) hydro-ecological conditions after 1968 have remained well within the broad range of natural variability observed over the past 300 years, with the earlier portion of the record actually depicting "markedly wetter and drier conditions compared to recent decades."

Moving to the Pacific coast of North America (Heal Lake near the city of Victoria on Canada's Vancouver Island), Zhang and Hebda (2005) conducted dendroclimatological analyses of 121 well-preserved subfossil logs discovered at the bottom of the lake plus 29 Douglas-fir trees growing nearby that led to the development of an ~ 4,000-year chronology exhibiting sensitivity to spring precipitation. In doing so, they found that "the magnitude and duration of climatic variability during the past 4000 years are not well represented by the variation in the brief modern period." As an example of this fact, they note that spring droughts represented by ring-width departures exceeding two standard deviations below the mean in at least five consecutive years occurred in the late AD 1840s and mid 1460s, as well as the mid 1860s BC, and were more severe than any drought of the twentieth century. In addition, the most persistent drought occurred during the 120-year period between about AD 1440 and 1560. Other severe droughts of multi-decadal duration occurred in the mid AD 760s-800s, the 540s-560s, the 150s-late 190s, and around 800 BC. Wavelet analyses of the tree-ring chronology also revealed a host of natural oscillations on timescales of years to centuries, demonstrating that the twentieth century was in no way unusual in this regard, as there were many times throughout the prior 4,000 years when it was both wetter and drier than it was during the last century of the past millennium.

Additional information on this topic, including reviews of newer publications as they become available, can be found at http://www.co2science.org/subject/d/droughtcanada.php.

References

Bergeron, Y. 1998. Les consequences des changements climatiques sur la frequence des feux et la composition forestiere au sud-ouest de la foret boreale quebecoise. *Geogr. Phy. Quaternary* **52**: 167-173.

Bergeron, Y. 2000. Species and stand dynamics in the mixed woods of Quebec's boreal forest. *Ecology* **81**: 1500-1516.

Bergeron, Y. and Archambault, S. 1993. Decreasing frequency of forest fires in the southern boreal zone of Quebec and its relation to global warming since the end of the 'Little Ice Age'. *The Holocene* **3**: 255-259.

Bergeron, Y., Gauthier, S., Kafka, V., Lefort, P. and Lesieur, D. 2001. Natural fire frequency for the eastern Canadian boreal forest: consequences for sustainable forestry. *Canadian Journal of Forest Research* **31**: 384-391.

Campbell, C. 2002. Late Holocene lake sedimentology and climate change in southern Alberta, Canada. *Quaternary Research* **49**: 96-101.

Gan, T.Y. 1998. Hydroclimatic trends and possible climatic warming in the Canadian Prairies. *Water Resources Research* **34**: 3009-3015.

Girardin, M-P., Tardif, J., Flannigan, M.D. and Bergeron, Y. 2004. Multicentury reconstruction of the Canadian Drought Code from eastern Canada and its relationship with paleoclimatic indices of atmospheric circulation. *Climate Dynamics* **23**: 99-115.

Hofgaard, A., Tardif, J. and Bergeron, Y. 1999. Dendroclimatic response of *Picea mariana* and *Pinus banksiana* along a latitudinal gradient in the eastern Canadian boreal forest. *Canadian Journal of Forest Research* **29**: 1333-1346.

Laird, K.R., Cumming, B.F., Wunsam, S., Rusak, J.A., Oglesby, R.J., Fritz, S.C. and Leavitt, P.R. 2003. Lake sediments record large-scale shifts in moisture regimes across the northern prairies of North America during the past two millennia. *Proceedings of the National Academy of Sciences USA* **100**: 2483-2488.

Quiring, S.M. and Papakyriakou, T.N. 2005. Characterizing the spatial and temporal variability of June-July moisture conditions in the Canadian prairies. *International Journal of Climatology* **25**: 117-138.

Sauchyn, D.J. and Skinner, W.R. 2001. A proxy record of drought severity for the southwestern Canadian plains. *Canadian Water Resources Journal* **26**: 253-272.

St. George, S. and Nielsen, E. 2002. Hydroclimatic change in southern Manitoba since A.D. 1409 inferred from tree rings. *Quaternary Research* **58**: 103-111.

Tardif, J. and Bergeron, Y. 1997. Ice-flood history reconstructed with tree-rings from the southern boreal forest limit, western Quebec. *The Holocene* **7**: 291-300.

Tardif, J. and Bergeron, Y. 1999. Population dynamics of *Fraxinus nigra* in response to flood-level variations, in northwestern Quebec. *Ecological Monographs* **69**: 107-125.

Wolfe, B.B., Karst-Riddoch, T.L., Vardy, S.R., Falcone, M.D., Hall, R.I. and Edwards, T.W.D. 2005. Impacts of climate and river flooding on the hydro-ecology of a floodplain basin, Peace-Athabasca Delta, Canada since A.D. 1700. *Quaternary Research* **64**: 147-162.

Zhang, Q.-B. and Hebda, R.J. 2005. Abrupt climate change and variability in the past four millennia of the southern Vancouver Island, Canada. *Geophysical Research Letters* **32** L16708, doi:10.1029/2005GL022913.

6.1.4.2. Mexico

Stahle *et al.* (2000) developed a long-term history of drought over much of North America from reconstructions of the Palmer Drought Severity Index, based on analyses of many lengthy tree-ring records. This history reveals the occurrence of a sixteenth century drought in Mexico that persisted from the 1540s to the 1580s. Writing of this anomalous period of much reduced precipitation, they say that "the 'megadrought' of the sixteenth century far exceeded any drought of the 20th century."

Diaz *et al.* (2002) constructed a history of winter-spring (November-April) precipitation—which accounts for one-third of the yearly total—for the Mexican state of Chihuahua for the period 1647-1992, based on earlywood width chronologies of more than 300 Douglas fir trees growing at four locations along the western and southern borders of Chihuahua and at two locations in the United States just above Chihuahua's northeast border. On the basis of these reconstructions, they note that "three of the 5 worst winter-spring drought years in the past three-and-a-half centuries are estimated to have occurred during the 20th century." Although this observation tends to make the twentieth century look highly anomalous in this regard, it is not, for two of those three worst drought years occurred during a period of average to slightly above-average precipitation.

Diaz *et al.* also note that "the longest drought indicated by the smoothed reconstruction lasted 17 years (1948-1964)," which is again correct and seemingly indicative of abnormally dry conditions during the twentieth century. However, for several of the 17 years of that below-normal-precipitation interval, precipitation values were only slightly below normal. For all practical purposes, there were four very similar dry periods interspersed throughout the preceding two-and-a-half centuries: one in the late 1850s and early 1860s, one in the late 1790s and early 1800s, one in the late 1720s and early 1730s, and one in the late 1660s and early 1670s.

With respect to the twentieth century alone, there was also a long period of high winter-spring precipitation that stretched from 1905 to 1932; following the major drought of the 1950s, precipitation remained at or just slightly above normal for the remainder of the record. Finally, with respect to the entire 346 years, there is no long-term trend in the data, nor is there any evidence of any sustained departure from that trend over the course of the twentieth century.

Cleaveland *et al.* (2003) constructed a winter-spring (November-March) precipitation history for the period 1386-1993 for Durango, Mexico, based on earlywood width chronologies of Douglas-fir tree rings collected at two sites in the Sierra Madre Occidental. They report that this record "shows droughts of greater magnitude and longer duration than the worst historical drought that occurred in the 1950s and 1960s." These earlier dramatic droughts include the long dry spell of the 1850s-1860s and what they call the megadrought of the mid- to late-sixteenth century. Their work clearly demonstrates that the worst droughts of the past 600 years did *not* occur during the period of greatest warmth. Instead, they occurred during the Little Ice Age, which was perhaps the coldest period of the current interglacial.

Investigating the same approximate time period, Hodell *et al.* (2005b) analyzed a 5.1-m sediment core they retrieved from Aguada X'caamal, a small sinkhole lake in northwest Yucatan, Mexico, finding that an important hydrologic change occurred there during the fifteenth century AD, as documented by the appearance of *A. beccarii* in the sediment profile, a decline in the abundance of charophytes, and an increase in the $\delta^{18}O$ of gastropods and ostracods. In addition, they report that "the salinity and ^{18}O content of the lake water increased as a result of reduced precipitation and/or increased evaporation in the mid- to late 1500s." These several changes, as well as many others they cite, were, as they describe it, "part of a larger pattern of oceanic and atmospheric change associated with the Little Ice Age that included cooling throughout the subtropical gyre (Lund and Curry, 2004)." Their assessment of the situation was that the "climate became drier on the Yucatan Peninsula in the 15th century AD near the onset of the Little Ice Age," as is also suggested by Maya and Aztec chronicles that "contain references to cold, drought and famine in the period AD 1441-1460."

Going back even further in time, Hodell *et al.* (1995) had provided evidence for a protracted drought during the Terminal Classic Period of Mayan civilization (AD 800-1000), based on their analysis of a single sediment core retrieved in 1993 from Lake Chichanacanab in the center of the northern Yucatan Peninsula of Mexico. Subsequently, based on two additional sediment cores retrieved from the same location in 2000, Hodell *et al.* (2001) determined that the massive drought likely occurred in two distinct phases (750-875 and 1000-1075). Reconstructing the climatic history of the region over the past 2,600 years and applying spectral analysis to the data also revealed a significant recurrent drought periodicity of 208 years that matched well with a cosmic ray-produced ^{14}C record preserved in tree rings, which is believed to reflect variations in solar activity. Because of the good correspondence between the two datasets, they concluded that "a significant component of century-scale variability in Yucatan droughts is explained by solar forcing."

Hodell *et al.* (2005a) returned to Lake Chichanacanab in March 2004 and retrieved a number of additional sediment cores in some of the deeper parts in the lake, with multiple cores being taken from its deepest point, from which depth profiles of bulk density were obtained by means of gamma-ray attenuation, as were profiles of reflected red, green, and blue light via a digital color line-scan camera. As they describe their findings, "the data reveal in great detail the climatic events that comprised the Terminal Classic Drought and coincided with the demise of Classic Maya civilization." In this regard, they again report that "the Terminal Classic Drought was not a single, two-century-long megadrought, but rather consisted of a series of dry events separated by intervening periods of relatively moister conditions," and that it "included an early phase (ca 770-870) and late phase (ca 920-1100)." Last of all, they say that "the bipartite drought history inferred from Chichancanab is supported by oxygen isotope records from nearby Punta Laguna," and that "the general pattern is also consistent with findings from the Cariaco Basin off northern Venezuela (Haug *et al.*, 2003), suggesting that the Terminal Classic Drought was a widespread phenomenon and not limited to north-central Yucatan."

Concurrent with the study of Hodell *et al.* (2005a), Almeida-Lenero *et al.* (2005) analyzed pollen profiles derived from sediment cores retrieved from Lake Zempoala and nearby Lake Quila in the central Mexican highlands about 65 km southwest of Mexico City, determining that it was generally more humid than at present in the central Mexican highlands during the mid-Holocene. Thereafter, however, there was a gradual drying of the climate; their data from Lake Zempoala indicate that "the interval from 1300 to 1100 cal yr BP was driest and represents an extreme since the mid-Holocene," noting further that this interval of 200 years "coincides with the collapse of the Maya civilization." Likewise, they report that their data from Lake Quila are also "indicative of the most arid period reported during the middle to late Holocene from c. 1300 to 1100 cal yr BP." In addition, they note that "climatic aridity during this time was also noted by Metcalfe *et al.* (1991) for the Lerma Basin [central Mexico]," that "dry climatic conditions were also reported from Lake Patzcuaro, central Mexico by Watts and Bradbury (1982)," and that "dry conditions were also reported for [Mexico's] Zacapu Basin (Metcalfe, 1995) and for [Mexico's] Yucatan Peninsula (Curtis *et al.*, 1996, 1998; Hodell *et al.*, 1995, 2001)."

Based on the many results described above, it is evident that throughout much of Mexico some of the driest conditions and worst droughts of the Late Holocene occurred during the Little Ice Age and the latter part of the Dark Ages Cold Period. These observations do much to discredit the model-based claim that droughts will get worse as air temperatures rise. All of the Mexican droughts of the twentieth century (when the IPCC claims the planet warmed at a rate and to a level that were unprecedented over the past two millennia) were much milder than many of the droughts that occurred during much colder centuries.

Additional information on this topic, including reviews of newer publications as they become available, can be found at http://www.co2science.org/subject/d/droughtmexico.php.

References

Almeida-Lenero, L., Hooghiemstra, H., Cleef, A.M. and van Geel, B. 2005. Holocene climatic and environmental change from pollen records of Lakes Zempoala and Quila, central Mexican highlands. *Review of Palaeobotany and Palynology* **136**: 63-92.

Cleaveland, M.K., Stahle, D.W., Therrell, M.D., Villanueva-Diaz, J. and Burns, B.T. 2003. Tree-ring reconstructed winter precipitation and tropical teleconnections in Durango, Mexico. *Climatic Change* **59**: 369-388.

Curtis, J., Hodell, D. and Brenner, M. 1996. Climate variability on the Yucatan Peninsula (Mexico) during the past 3500 years, and implications for Maya cultural evolution. *Quaternary Research* **46**: 37-47.

Curtis, J., Brenner, M., Hodell, D., Balser, R., Islebe, G.A. and Hooghiemstra, H. 1998. A multi-proxy study of Holocene environmental change in the Maya Lowlands of Peten Guatemala. *Journal of Paleolimnology* **19**: 139-159.

Diaz, S.C., Therrell, M.D., Stahle, D.W. and Cleaveland, M.K. 2002. Chihuahua (Mexico) winter-spring precipitation reconstructed from tree-rings, 1647-1992. *Climate Research* **22**: 237-244.

Haug, G.H., Gunther, D., Peterson, L.C., Sigman, D.M., Hughen, K.A. and Aeschlimann, B. 2003. Climate and the collapse of Maya civilization. *Science* **299**: 1731-1735.

Hodell, D.A., Brenner, M. and Curtis, J.H. 2005a. Terminal Classic drought in the northern Maya lowlands inferred from multiple sediment cores in Lake Chichancanab (Mexico). *Quaternary Science Reviews* **24**: 1413-1427.

Hodell, D.A., Brenner, M., Curtis, J.H. and Guilderson, T. 2001. Solar forcing of drought frequency in the Maya lowlands. *Science* **292**: 1367-1370.

Hodell, D.A., Brenner, M., Curtis, J.H., Medina-Gonzalez, R., Can, E.I.-C., Albornaz-Pat, A. and Guilderson, T.P. 2005b. Climate change on the Yucatan Peninsula during the Little Ice Age. *Quaternary Research* **63**: 109-121.

Hodell, D.A., Curtis, J.H. and Brenner, M. 1995. Possible role of climate in the collapse of Classic Maya civilization. *Nature* **375**: 391-394.

Lund, D.C. and Curry, W.B. 2004. Late Holocene variability in Florida Current surface density: patterns and possible causes. *Paleoceanography* **19**: 10.1029/2004PA001008.

Metcalfe, S.E. 1995. Holocene environmental change in the Zacapu Basin, Mexico: a diatom based record. *The Holocene* **5**: 196-208.

Metcalfe, S.E., Street-Perrott, F.A., Perrott, R.A. and Harkness, D.D. 1991. Palaeolimnology of the Upper Lerma Basin, central Mexico: a record of climatic change and anthropogenic disturbance since 11,600 yr B.P. *Journal of Paleolimnology* **5**: 197-218.

Stahle, D.W., Cook, E.R., Cleaveland, M.K, Therrell, M.D., Meko, D.M., Grissino-Mayer, H.D., Watson, E. and Luckman, B.H. 2000. Tree-ring data document 16th century megadrought over North America. *EOS: Transactions, American Geophysical Union* **81**: 121, 125.

Watts, W.A. and Bradbury, J.P. 1982. Paleoecological studies at Lake Patzcuaro on the West Central Mexican plateau and at Chalco in the Basin of Mexico. *Quaternary Research* **17**: 56-70.

6.1.4.3. United States

6.1.4.3.1. Central United States

Starting at the U.S.-Canadian border and working our way south, we begin with the study of Fritz *et al.* (2000), who utilized data derived from sediment cores retrieved from three North Dakota lakes to reconstruct a 2,000-year history of drought in this portion of the Northern Great Plains. This work suggested, in their words, "that droughts equal or greater in magnitude to those of the Dust Bowl period were a common occurrence during the last 2000 years."

Also working in the Northern Great Plains, but extending down into South Dakota, Shapley *et al.* (2005) developed a 1,000-year hydroclimate reconstruction from local bur oak tree-ring records and various lake sediment cores. Based on this record, they determined that prior to 1800, "droughts tended towards greater persistence than during the past two centuries," suggesting that droughts of the region became shorter-lived as opposed to longer-lasting as the earth gradually recovered from the cold temperatures of the Little Ice Age.

The above observations are significant because the United States' Northern Great Plains is an important agricultural region, providing a significant source of grain for both local and international consumption. However, the region is susceptible to periodic extreme droughts that tend to persist longer than those in any other part of the country (Karl *et al.*, 1987; Soule, 1992). Because of this fact, Laird *et al.* (1998) examined the region's historical record of drought in an attempt to establish a baseline of natural drought variability that could help in attempts to determine if current and future droughts might be anthropogenically influenced.

Working with a high-resolution sediment core obtained from Moon Lake, North Dakota, which provided a sub-decadal record of salinity (drought) over the past 2,300 years, Laird *et al.* discovered that the U.S. Northern Great Plains were relatively wet during the final 750 years of this period. Throughout the 1,550 prior years, they determined that "recurring severe droughts were more the norm," and that they were "of much greater intensity and duration than any in the 20th century," including the great Dust Bowl event of the 1930s. There were, as they put it, "no modern equivalents" to Northern Great Plains droughts experienced prior to AD 1200, which means the human presence has not led to unusual drought conditions in this part of the world.

Continuing our southward trek, we encounter the work of Forman *et al.* (2005), who note that "periods of dune reactivation reflect sustained moisture deficits for years to decades and reflect broader environmental change with diminished surface- and ground-water resources." This observation prompted them to focus on "the largest dune system in North America, the Nebraska Sand Hills," where they utilized "recent advances in optically stimulated luminescence dating (Murray and Wintle, 2000) to improve chronologic control on the timing of dune reactivation." They also linked landscape response to drought over the past 1,500 years to tree-ring records of aridity.

Forman *et al.* identified six major aeolian depositional events in the past 1,500 years, all but one of which (the 1930s "Dust Bowl" drought) occurred prior to the twentieth century. Moving backwards in time from the Dust Bowl, the next three major events occurred during the depths of the Little Ice Age, the next one near the Little Ice Age's inception, and the earliest one near the end of the Dark Ages Cold Period. As for how the earlier droughts compare with those of the past century, the researchers say the

1930s drought (the twentieth century's worst depositional event) was less severe than the others, especially the one that has come to be known as the sixteenth century megadrought. Forman *et al.* thus conclude that the aeolian landforms they studied "are clear indicators of climate variability beyond twentieth century norms, and signify droughts of greater severity and persistence than thus far instrumentally recorded."

In a study that covered the entirety of the U.S. Great Plains, Daniels and Knox (2005) analyzed the alluvial stratigraphic evidence for an episode of major channel incision in tributaries of the upper Republican River that occurred between 1,100 and 800 years ago, after which they compared their findings with proxy drought records from 28 other locations throughout the Great Plains and surrounding regions. This work revealed that channel incision in the Republican River between about AD 900 and 1200 was well correlated with a multi-centennial episode of widespread drought, which in the words of Daniels and Knox, "coincides with the globally recognized Medieval Warm Period." Of great interest, however, is the fact that modern twentieth century warming has *not* led to a repeat of those widespread drought conditions.

Working in pretty much the same area some seven years earlier, Woodhouse and Overpeck (1998) reviewed what we know about the frequency and severity of drought in the central United States over the past two thousand years based upon empirical evidence of drought from various proxy indicators. Their study indicated the presence of numerous "multidecadal- to century-scale droughts," leading them to conclude that "twentieth-century droughts are not representative of the full range of drought variability that has occurred over the last 2000 years." In addition, they noted that the twentieth century was characterized by droughts of "moderate severity and comparatively short duration, relative to the full range of past drought variability."

With respect to the causes of drought, Woodhouse and Overpeck suggest a number of different possibilities that either directly or indirectly induce changes in atmospheric circulation and moisture transport. However, they caution that "the causes of droughts with durations of years (i.e., the 1930s) to decades or centuries (i.e., paleodroughts) are not well understood." They conclude that "the full range of past natural drought variability, deduced from a comprehensive review of the paleoclimatic literature, suggests that droughts more severe than those of the 1930s and 1950s are likely to occur in the

future," whatever the air's CO_2 concentration or temperature might.

Mauget (2004) looked for what he called "initial clues" to the commencement of the great drying of the U.S. Heartland that had been predicted to occur in response to CO_2-induced global warming by Manabe and Wetherald (1987), Rind *et al.* (1990), Rosenzweig and Hillel (1993), and Manabe *et al.* (2004), which Mauget reasoned would be apparent in the observational streamflow record of the region. In this endeavor, he employed data obtained from the archives of the U.S. Geological Survey's Hydro-Climatic Data Network, which come from 42 stations covering the central third of the United States that stretch from the Canadian border on the north to the Gulf of Mexico on the south, with the most dense coverage being found within the U.S. Corn Belt.

Mauget reports finding "an overall pattern of low flow periods before 1972, and high flow periods occurring over time windows beginning after 1969." Of the 42 stations' high flow periods, he says that "34 occur during 1969-1998, with 25 of those periods ending in either 1997 or 1998," and that "of those 25 stations 21 are situated in the key agricultural region known as the Corn Belt." He also reports that "among most of the stations in the western portions of the Corn Belt during the 1980s and 1990s there is an unprecedented tendency toward extended periods of daily high flow conditions, which lead to marked increases in the mean annual frequency of hydrological surplus conditions relative to previous years." What is more, he notes that "in 15 of the 18 Corn Belt gage stations considered here at daily resolution, a more than 50 percent reduction in the mean annual incidence of hydrological drought conditions is evident during those periods." Last, Mauget reports that "the gage station associated with the largest watershed area—the Mississippi at Vicksburg—shows more than a doubling of the mean annual frequency of hydrological surplus days during its 1973-1998 high flow period relative to previous years, and more than a 50% reduction in the mean annual incidence of hydrological drought condition."

In summarizing his findings, Mauget states that the overall pattern of climate variation "is that of a reduced tendency to hydrological drought and an increased incidence of hydrological surplus over the Corn Belt and most of the Mississippi River basin during the closing decades of the 20th century," noting further that "some of the most striking evidence of a transition to wetter conditions in the streamflow analyses is found among streams and rivers situated along the Corn Belt's climatologically drier western edge."

Mauget states that the streamflow data "suggest a fundamental climate shift, as the most significant incidence of high ranked annual flow was found over relatively long time scales at the end of the data record." In other words, the shift is *away from* the droughty conditions predicted by the IPCC to result from CO_2-induced global warming in this important agricultural region of the United States.

Additional information on this topic, including reviews of newer publications as they become available, can be found at http://www.co2science.org/subject/d/droughtusacentral.php.

References

Daniels, J.M. and Knox, J.C. 2005. Alluvial stratigraphic evidence for channel incision during the Mediaeval Warm Period on the central Great plains, USA. *The Holocene* **15**: 736-747.

Forman, S.L., Marin, L., Pierson, J., Gomez, J., Miller, G.H. and Webb, R.S. 2005. Aeolian sand depositional records from western Nebraska: landscape response to droughts in the past 1500 years. *The Holocene* **15**: 973-981.

Fritz, S.C., Ito, E., Yu, Z., Laird, K.R. and Engstrom, D.R. 2000. Hydrologic variation in the Northern Great Plains during the last two millennia. *Quaternary Research* **53**: 175-184.

Karl, T., Quinlan, F. and Ezell, D.S. 1987. Drought termination and amelioration: its climatological probability. *Journal of Climate and Applied Meteorology* **26**: 1198-1209.

Laird, K.R., Fritz, S.C. and Cumming, B.F. 1998. A diatom-based reconstruction of drought intensity, duration, and frequency from Moon Lake, North Dakota: a sub-decadal record of the last 2300 years. *Journal of Paleolimnology* **19**: 161-179.

Manabe, S., Milly, P.C.D. and Wetherald, R. 2004. Simulated long-term changes in river discharge and soil moisture due to global warming. *Hydrological Sciences Journal* **49**: 625-642.

Manabe, S. and Wetherald, R.T. 1987. Large-scale changes of soil wetness induced by an increase in atmospheric carbon dioxide. *Journal of the Atmospheric Sciences* **44**: 1211-1235.

Mauget, S.A. 2004. Low frequency streamflow regimes over the central United States: 1939-1998. *Climatic Change* **63**: 121-144.

Murray, A.S. and Wintle, A.G. 2000. Luminescence dating of quartz using an improved single-aliquot regenerative-dose protocol. *Radiation Measurements* **32**: 57-73.

Rind, D., Goldberg, R., Hansen, J., Rosenzweig, C. and Ruedy, R. 1990. Potential evapotranspiration and the likelihood of future drought. *Journal of Geophysical Research* **95**: 9983-10004.

Rosenzweig, C. and Hillel, D. 1993. The Dust Bowl of the 1930s: Analog of greenhouse effect in the Great Plains? *Journal of Environmental Quality* **22**: 9-22.

Shapley, M.D., Johnson, W.C., Engstrom, D.R. and Osterkamp, W.R. 2005. Late-Holocene flooding and drought in the Northern Great Plains, USA, reconstructed from tree rings, lake sediments and ancient shorelines. *The Holocene* **15**: 29-41.

Soule, P.T. 1992. Spatial patterns of drought frequency and duration in the contiguous USA based on multiple drought event definitions. *International Journal of Climatology* **12**: 11-24.

Woodhouse, C.A. and Overpeck, J.T. 1998. 2000 years of drought variability in the central United States. *Bulletin of the American Meteorological Society* **79**: 2693-2714.

6.1.4.3.2. Eastern United States

Cronin *et al.* (2000) studied the salinity gradient across sediment cores from Chesapeake Bay, the largest estuary in the United Sates, in an effort to examine precipitation variability in the surrounding watershed over the past millennium. Their work revealed the existence of a high degree of decadal and multidecadal variability between wet and dry conditions throughout the 1,000-year record, where regional precipitation totals fluctuated by 25 to 30 percent, often in "extremely rapid [shifts] occurring over about a decade." In addition, precipitation over the past two centuries of the record was found to be generally greater than it was during the previous eight centuries, with the exception of the Medieval Warm Period (AD 1250-1350) when the [local] climate was found to be "extremely wet." Equally significant was the 10 researchers' finding that the region had experienced several "mega-droughts" that had lasted for 60 to 70 years, several of which they judged to have been "more severe than twentieth century droughts."

Building upon the work of Cronin *et al.* were Willard *et al.* (2003), who examined the last 2,300 years of the Holocene record of Chesapeake Bay and the adjacent terrestrial ecosystem "through the study of fossil dinoflagellate cysts and pollen from sediment cores." In doing so, they found that "several dry periods ranging from decades to centuries in duration are evident in Chesapeake Bay records." The first of these periods of lower-than-average precipitation (200 BC-AD 300) occurred during the latter part of the Roman Warm Period, while the next such period (~AD 800-1200), according to Willard *et al.*, "corresponds to the 'Medieval Warm Period'." In addition, they identified several decadal-scale dry intervals that spanned the years AD 1320-1400 and 1525-1650.

In discussing their findings, Willard *et al.* note that "mid-Atlantic dry periods generally correspond to central and southwestern USA 'megadroughts', which are described by Woodhouse and Overpeck (1998) as major droughts of decadal or more duration that probably exceeded twentieth-century droughts in severity." Emphasizing this important point, they additionally indicate that "droughts in the late sixteenth century that lasted several decades, and those in the 'Medieval Warm Period' and between ~AD 50 and AD 350 spanning a century or more have been indicated by Great Plains tree-ring (Stahle *et al.*, 1985; Stahle and Cleaveland, 1994), lacustrine diatom and ostracode (Fritz *et al.*, 2000; Laird *et al.*, 1996a, 1996b) and detrital clastic records (Dean, 1997)." Their work in the eastern United States, together with the work of other researchers in still other parts of the country, demonstrates that twentieth century global warming has not led to the occurrence of unusually strong wet or dry periods.

Quiring (2004) introduced his study of the subject by describing the drought of 2001-2002, which had produced anomalously dry conditions along most of the east coast of the U.S., including severe drought conditions from New Jersey to northern Florida that forced 13 states to ration water. Shortly after the drought began to subside in October 2002, however, moist conditions returned and persisted for about a year, producing the wettest growing-season of the instrumental record. These observations, in Quiring's words, "raise some interesting questions," including the one we are considering here. As he phrased the call to inquiry, "are moisture conditions in this region becoming more variable?"

Using an 800-year tree-ring-based reconstruction of the Palmer Hydrological Drought Index to address this question, Quiring documented the frequency, severity, and duration of growing-season moisture anomalies in the southern mid-Atlantic region of the United States. Among other things, this work

revealed, in Quiring's words, that "conditions during the 18th century were much wetter than they are today, and the droughts that occurred during the sixteenth century tended to be both longer and more severe." He concluded that "the recent growing-season moisture anomalies that occurred during 2002 and 2003 can only be considered rare events if they are evaluated with respect to the relatively short instrumental record (1895-2003)," for when compared to the 800-year reconstructed record, he notes that "neither of these events is particularly unusual." In addition, Quiring reports that "although climate models predict decreases in summer precipitation and significant increases in the frequency and duration of extreme droughts, the data indicate that growing-season moisture conditions during the 20th century (and even the last 19 years) appear to be near normal (well within the range of natural climate variability) when compared to the 800-year record."

Additional information on this topic, including reviews of newer publications as they become available, can be found at http://www.co2science.org/subject/d/droughtusaeast.php.

References

Cronin, T., Willard, D., Karlsen, A., Ishman, S., Verardo, S., McGeehin, J., Kerhin, R., Holmes, C., Colman, S. and Zimmerman, A. 2000. Climatic variability in the eastern United States over the past millennium from Chesapeake Bay sediments. *Geology* **28**: 3-6.

Dean, W.E. 1997. Rates, timing, and cyclicity of Holocene aeolian activity in north-central United States: evidence from varved lake sediments. *Geology* **25**: 331-334.

Fritz, S.C., Ito, E., Yu, Z., Laird, K.R. and Engstrom, D.R. 2000. Hydrologic variation in the northern Great Plains during the last two millennia. *Quaternary Research* **53**: 175-184.

Laird, K.R., Fritz, S.C., Grimm, E.C. and Mueller, P.G. 1996a. Century-scale paleoclimatic reconstruction from Moon Lake, a closed-basin lake in the northern Great Plains. *Limnology and Oceanography* **41**: 890-902.

Laird, K.R., Fritz, S.C., Maasch, K.A. and Cumming, B.F. 1996b. Greater drought intensity and frequency before AD 1200 in the Northern Great Plains, USA. *Nature* **384**: 552-554.

Quiring, S.M. 2004. Growing-season moisture variability in the eastern USA during the last 800 years. *Climate Research* **27**: 9-17.

Stahle, D.W. and Cleaveland, M.K. 1994. Tree-ring reconstructed rainfall over the southeastern U.S.A. during the Medieval Warm Period and Little Ice Age. *Climatic Change* **26**: 199-212.

Stahle, D.W., Cleaveland, M.K. and Hehr, J.G. 1985. A 450-year drought reconstruction for Arkansas, United States. *Nature* **316**: 530-532.

Willard, D.A., Cronin, T.M. and Verardo, S. 2003. Late-Holocene climate and ecosystem history from Chesapeake Bay sediment cores, USA. *The Holocene* **13**: 201-214.

Woodhouse, C.A. and Overpeck, J.T. 1998. 2000 years of drought variability in the Central United States. *Bulletin of the American Meteorological Society* **79**: 2693-2714.

6.1.4.3.3. *Western United States*

Is there evidence of more severe and longer-lasting droughts in the western United States? We begin our journey of inquiry just below Canada, in the U.S. Pacific Northwest, from whence we gradually wend our way to the U.S./Mexico border.

Knapp *et al.* (2002) created a 500-year history of severe single-year Pacific Northwest droughts from a study of 18 western juniper tree-ring chronologies that they used to identify what they call extreme Climatic Pointer Years or CPYs, which are indicative of severe single-year droughts. As they describe it, this procedure revealed that "widespread and extreme CPYs were concentrated in the 16th and early part of the 17th centuries," while "both the 18th and 19th centuries were largely characterized by a paucity of drought events that were severe and widespread." Thereafter, however, they say that "CPYs became more numerous during the 20th century," although the number of twentieth century extreme CPYs (26) was still substantially less than the mean of the number of sixteenth and seventeenth century extreme CPYs (38), when the planet was colder. The data of this study fail to support the IPCC's claim that global warming increases the frequency of severe droughts.

Gedalof *et al.* (2004) used a network of 32 drought-sensitive tree-ring chronologies to reconstruct mean water-year flow on the Columbia River at The Dales in Oregon since 1750. This study of the second-largest drainage basin in the United States is stated by them to have been done "for the purpose of assessing the representativeness of recent observations, especially with respect to low frequency changes and extreme events." When finished, it revealed, in their words, that "persistent low flows during the 1840s were probably the most severe of the

past 250 years," and that "the drought of the 1930s is probably the second most severe."

More recent droughts, in the words of the researchers, "have led to conflicts among uses (e.g., hydroelectric production versus protecting salmon runs), increased costs to end users (notably municipal power users), and in some cases the total loss of access to water (in particular junior water rights holders in the agricultural sector)." Nevertheless, they say that "these recent droughts were not exceptional in the context of the last 250 years and were of shorter duration than many past events." In fact, they say, "the period from 1950 to 1987 is anomalous in the context of this record for having no notable multiyear drought events."

Working in the Bighorn Basin of north-central Wyoming and south-central Montana, Gray *et al.* (2004a) used cores and cross sections from 79 Douglas fir and limber pine trees at four different sites to develop a proxy for annual precipitation spanning the period AD 1260-1998. This reconstruction, in their words, "exhibits considerable nonstationarity, and the instrumental era (post-1900) in particular fails to capture the full range of precipitation variability experienced in the past ~750 years." More specifically, they say that "both single-year and decadal-scale dry events were more severe before 1900," and that "dry spells in the late thirteenth and sixteenth centuries surpass both [the] magnitude and duration of any droughts in the Bighorn Basin after 1900." They say that "single- and multi-year droughts regularly surpassed the severity and magnitude of the 'worst-case scenarios' presented by the 1930s and 1950s droughts." If twentieth century global warming had any effect at all on Bighorn Basin precipitation, it was to make it less extreme rather than more extreme.

Moving further south, Benson *et al.* (2002) developed continuous high-resolution $\delta^{18}O$ records from cored sediments of Pyramid Lake, Nevada, which they used to help construct a 7,600-year history of droughts throughout the surrounding region. Oscillations in the hydrologic balance that were evident in this record occurred, on average, about every 150 years, but with significant variability. Over the most recent 2,740 years, for example, intervals between droughts ranged from 80 to 230 years; while drought durations ranged from 20 to 100 years, with some of the larger ones forcing mass migrations of indigenous peoples from lands that could no longer support them. In contrast, historical droughts typically have lasted less than a decade.

In another study based on sediment cores extracted from Pyramid Lake, Nevada, Mensing *et al.* (2004) analyzed pollen and algal microfossils deposited there over the prior 7,630 years that allowed them to infer the hydrologic history of the area over that time period. Their results indicated that "sometime after 3430 but before 2750 cal yr B.P., climate became cool and wet," but, paradoxically, that "the past 2500 yr have been marked by recurring persistent droughts." The longest of these droughts, according to them, "occurred between 2500 and 2000 cal yr B.P.," while others occurred "between 1500 and 1250, 800 and 725, and 600 and 450 cal yr B.P," with none recorded in more recent warmer times.

The researchers also note that "the timing and magnitude of droughts identified in the pollen record compares favorably with previously published $\delta^{18}O$ data from Pyramid Lake" and with "the ages of submerged rooted stumps in the Eastern Sierra Nevada and woodrat midden data from central Nevada." Noting that Bond *et al.* (2001) "found that over the past 12,000 yr, decreases in [North Atlantic] drift ice abundance corresponded to increased solar output," they report that when they "compared the pollen record of droughts from Pyramid Lake with the stacked petrologic record of North Atlantic drift ice ... nearly every occurrence of a shift from ice maxima (reduced solar output) to ice minima (increased solar output) corresponded with a period of prolonged drought in the Pyramid Lake record." As a result, Mensing *et al.* concluded that "changes in solar irradiance may be a possible mechanism influencing century-scale drought in the western Great Basin [of the United States]."

Only a state away, Gray *et al.* (2004b) used samples from 107 piñon pines at four different sites to develop a proxy record of annual precipitation spanning the AD 1226- 2001 interval for the Uinta Basin watershed of northeastern Utah. This effort revealed, in their words, that "single-year dry events before the instrumental period tended to be more severe than those after 1900," and that decadal-scale dry events were longer and more severe prior to 1900 as well. In particular, they found that "dry events in the late 13th, 16th, and 18th Centuries surpass the magnitude and duration of droughts seen in the Uinta Basin after 1900."

At the other end of the moisture spectrum, Gray *et al.* report that the twentieth century was host to two of the strongest wet intervals (1938-1952 and 1965-1987), although these two periods were only the seventh and second most intense wet regimes,

respectively, of the entire record. Hence, it would appear that in conjunction with twentieth century global warming, precipitation extremes (both high and low) within northeastern Utah's Uinta Basin have become more attenuated as opposed to more amplified.

Working in the central and southern Rocky Mountains, Gray *et al.* (2003) examined 15 tree ring-width chronologies that had been used in previous reconstructions of drought for evidence of low-frequency variations in five regional composite precipitation histories. In doing so, they found that "strong multidecadal phasing of moisture variation was present in all regions during the late 16th-century megadrought," and that "oscillatory modes in the 30-70 year domain persisted until the mid-19th century in two regions, and wet-dry cycles were apparently synchronous at some sites until the 1950s drought." They thus speculate that "severe drought conditions across consecutive seasons and years in the central and southern Rockies may ensue from coupling of the cold phase Pacific Decadal Oscillation with the warm phase Atlantic Multidecadal Oscillation," which is something they envision as having happened in both the severe 1950s drought and the late sixteenth century megadrought. Hence, there is reason to believe that episodes of extreme dryness in this part of the country may be driven in part by naturally recurring climate "regime shifts" in the Pacific and Atlantic Oceans.

Hidalgo *et al.* (2000) used a new form of principal components analysis to reconstruct a history of streamflow in the Upper Colorado River Basin based on information obtained from tree-ring data, after which they compared their results to those of Stockton and Jacoby (1976). In doing so, they found the two approaches to yield similar results, except that Hidalgo *et al.*'s approach responded with more intensity to periods of below-average streamflow or regional drought. Hence, it was easier for them to determine there has been "a near-centennial return period of extreme drought events in this region," going all the way back to the early 1500s. It is reasonable to assume that if such an extreme drought were to commence today, it would not be related to either the air's CO_2 content or its temperature.

Woodhouse *et al.* (2006) also generated updated proxy reconstructions of water-year streamflow for the Upper Colorado River Basin, based on four key gauges (Green River at Green River, Utah; Colorado near Cisco, Utah; San Juan near Bluff, Utah; and Colorado at Lees Ferry, Arizona) and using an expanded tree-ring network and longer calibration records than in previous efforts. The results of this program indicated that the major drought of 2000-2004, "as measured by 5-year running means of water-year total flow at Lees Ferry ... is not without precedence in the tree ring record," and that "average reconstructed annual flow for the period 1844-1848 was lower." They also report that "two additional periods, in the early 1500s and early 1600s, have a 25% or greater chance of being as dry as 1999-2004," and that six other periods "have a 10% or greater chance of being drier." In addition, their work revealed that "longer duration droughts have occurred in the past," and that "the Lees Ferry reconstruction contains one sequence each of six, eight, and eleven consecutive years with flows below the 1906-1995 average."

"Overall," in the words of the three researchers, "these analyses demonstrate that severe, sustained droughts are a defining feature of Upper Colorado River hydroclimate." In fact, they conclude from their work that "droughts more severe than any 20th to 21st century event occurred in the past," meaning the preceding few centuries.

Moving closer still to the U.S. border with Mexico, Ni *et al.* (2002) developed a 1,000-year history of cool-season (November-April) precipitation for each climate division of Arizona and New Mexico from a network of 19 tree-ring chronologies. They determined that "sustained dry periods comparable to the 1950s drought" occurred in "the late 1000s, the mid 1100s, 1570-97, 1664-70, the 1740s, the 1770s, and the late 1800s." They also note that although the 1950s drought was large in both scale and severity, "it only lasted from approximately 1950 to 1956," whereas the sixteenth century mega-drought lasted more than four times longer.

With respect to the opposite of drought, Ni *et al.* report that "several wet periods comparable to the wet conditions seen in the early 1900s and after 1976" occurred in "1108-20, 1195-1204, 1330-45, the 1610s, and the early 1800s," and they add that "the most persistent and extreme wet interval occurred in the 1330s." Consequently, for the particular part of the world covered by Ni *et al.*'s study, there appears to be nothing unusual about the extremes of both wetness and dryness experienced during the twentieth century.

Also working in New Mexico, Rasmussen *et al.* (2006) derived a record of regional relative moisture from variations in the annual band thickness and mineralogy of two columnar stalagmites collected from Carlsbad Cavern and Hidden Cave in the Guadalupe Mountains near the New Mexico/Texas border. From this work they discovered that both records "suggest periods of dramatic precipitation variability over the last 3000 years, exhibiting large shifts unlike anything seen in the modern record."

We come now to two papers that deal with the western United States as a whole. In the first, Cook *et al.* (2004) developed a 1,200-year history of drought for the western half of the country and adjacent parts of Canada and Mexico (hereafter the "West"), based on annually resolved tree-ring records of summer-season Palmer Drought Severity Index that were derived for 103 points on a 2.5° x 2.5° grid, 68 of which grid points (66 percent of them) possessed data that extended back to AD 800. This reconstruction, in the words of Cook *et al.*, revealed "some remarkable earlier increases in aridity that *dwarf* [our italics] the comparatively short-duration current drought in the 'West'." Interestingly, they report that "the four driest epochs, centered on AD 936, 1034, 1150 and 1253, all occur during a ~400 year interval of overall elevated aridity from AD 900 to 1300," which they say is "broadly consistent with the Medieval Warm Period."

Commenting on their findings, the five scientists say "the overall coincidence between our megadrought epoch and the Medieval Warm Period suggests that anomalously warm climate conditions during that time may have contributed to the development of more frequent and persistent droughts in the 'West'," as well as the megadrought that was discovered by Rein *et al.* (2004) to have occurred in Peru at about the same time (AD 800-1250); and after citing nine other studies that provide independent evidence of drought during this time period for various sub-regions of the West, they warn that "any trend toward warmer temperatures in the future could lead to a serious long-term increase in aridity over western North America," noting that "future droughts in the 'West' of similar duration to those seen prior to AD 1300 would be disastrous."

While we agree with Cook *et al.*'s conclusion, we cannot help but note that the droughts that occurred during the Medieval Warm Period were obviously not CO_2-induced. If the association between global warmth and drought in the western United States is robust, it suggests that current world temperatures are still well below those experienced during large segments of the Medieval Warm Period.

The last of the two papers to cover the western United States as a whole is that of Woodhouse

(2004), who reports what is known about natural hydroclimatic variability throughout the region via descriptions of several major droughts that occurred there over the past three millennia, all but the last century of which had atmospheric CO_2 concentrations that never varied by more than about 10 ppm from a mean value of 280 ppm.

For comparative purposes, Woodhouse begins by noting that "the most extensive U.S. droughts in the 20th century were the 1930s Dust Bowl and the 1950s droughts." The first of these lasted "most of the decade of the 1930s" and "occurred in several waves," while the latter "also occurred in several waves over the years 1951-1956." More severe than either of these two droughts was what has come to be known as the Sixteenth Century Megadrought, which lasted from 1580 to 1600 and included northwestern Mexico in addition to the southwestern United States and the western Great Plains. Then there was what is simply called The Great Drought, which spanned the last quarter of the thirteenth century and was actually the last in a series of three thirteenth century droughts, the first of which may have been even more severe than the last. In addition, Woodhouse notes there was a period of remarkably sustained drought in the second half of the twelfth century.

It is evident from these observations, according to Woodhouse, that "the 20th century climate record contains only a subset of the range of natural climate variability in centuries-long and longer paleoclimatic records." This subset, as it pertains to water shortage, does not approach the level of drought severity and duration experienced in prior centuries and millennia. A drought much more extreme than the most extreme droughts of the twentieth century would be required to propel the western United States and adjacent portions of Canada and Mexico into a truly unprecedented state of dryness.

Additional information on this topic, including reviews of newer publications as they become available, can be found at http://www.co2science.org/subject/d/droughtusawest.php.

References

Benson, L., Kashgarian, M., Rye, R., Lund, S., Paillet, F., Smoot, J., Kester, C., Mensing, S., Meko, D. and Lindstrom, S. 2002. Holocene multidecadal and multicentennial droughts affecting Northern California and Nevada. *Quaternary Science Reviews* **21**: 659-682.

Bond, G., Kromer, B., Beer, J., Muscheler, R., Evans, M.N., Showers, W., Hoffmann, S., Lotti-Bond, R., Hajdas, I. and Bonani, G. 2001. Persistent solar influence on North Atlantic climate during the Holocene. *Science* **294**: 2130-2136.

Cook, E.R., Woodhouse, C., Eakin, C.M., Meko, D.M. and Stahle, D.W. 2004. Long-term aridity changes in the western United States. *Sciencexpress.org* / 7 October 2004.

Gedalof, Z., Peterson, D.L. and Mantua, N.J. 2004. Columbia River flow and drought since 1750. *Journal of the American Water Resources Association* **40**: 1579-1592.

Gray, S.T., Betancourt, J.L., Fastie, C.L. and Jackson, S.T. 2003. Patterns and sources of multidecadal oscillations in drought-sensitive tree-ring records from the central and southern Rocky Mountains. *Geophysical Research Letters* **30**: 10.1029/2002GL016154.

Gray, S.T., Fastie, C.L., Jackson, S.T. and Betancourt, J.L. 2004a. Tree-ring-based reconstruction of precipitation in the Bighorn Basin, Wyoming, since 1260 A.D. *Journal of Climate* **17**: 3855-3865.

Gray, S.T., Jackson, S.T. and Betancourt, J.L. 2004b. Tree-ring based reconstructions of interannual to decadal scale precipitation variability for northeastern Utah since 1226 A.D. *Journal of the American Water Resources Association* **40**: 947-960.

Hidalgo, H.G., Piechota, T.C. and Dracup, J.A. 2000. Alternative principal components regression procedures for dendrohydrologic reconstructions. *Water Resources Research* **36**: 3241-3249.

Knapp, P.A., Grissino-Mayer, H.D. and Soule, P.T. 2002. Climatic regionalization and the spatio-temporal occurrence of extreme single-year drought events (1500-1998) in the interior Pacific Northwest, USA. *Quaternary Research* **58**: 226-233.

Mensing, S.A., Benson, L.V., Kashgarian, M. and Lund, S. 2004. A Holocene pollen record of persistent droughts from Pyramid Lake, Nevada, USA. *Quaternary Research* **62**: 29-38.

Ni, F., Cavazos, T., Hughes, M.K., Comrie, A.C. and Funkhouser, G. 2002. Cool-season precipitation in the southwestern USA since AD 1000: Comparison of linear and nonlinear techniques for reconstruction. *International Journal of Climatology* **22**: 1645-1662.

Rasmussen, J.B.T., Polyak, V.J. and Asmerom, Y. 2006. Evidence for Pacific-modulated precipitation variability during the late Holocene from the southwestern USA. *Geophysical Research Letters* **33**: 10.1029/2006GL025714.

Rein, B., Luckge, A. and Sirocko, F. 2004. A major Holocene ENSO anomaly during the Medieval period. *Geophysical Research Letters* **31**: 10.1029/2004GL020161.

Stockton, C.W. and Jacoby Jr., G.C. 1976. Long-term surface-water supply and streamflow trends in the Upper Colorado River Basin based on tree-ring analysis. *Lake Powell Research Project Bulletin* 18, Institute of Geophysics and Planetary Physics, University of California, Los Angeles.

Woodhouse, C.A. 2004. A paleo perspective on hydroclimatic variability in the western United States. *Aquatic Sciences* **66**: 346-356.

Woodhouse, C.A., Gray, S.T. and Meko, D.M. 2006. Updated streamflow reconstructions for the Upper Colorado River Basin. *Water Resources Research* **42**: 10.1029/2005WR004455.

6.1.4.3.4. Entire United States

Andreadis and Lettenmaier (2006) examined twentieth century trends in soil moisture, runoff, and drought over the conterminous United States with a hydro-climatological model forced by real-world measurements of precipitation, air temperature, and wind speed over the period 1915-2003. This work revealed, in their words, that "droughts have, for the most part, become shorter, less frequent, less severe, and cover a smaller portion of the country over the last century."

Using the self-calibrating Palmer (1965) drought severity index (SCPDSI), as described by Wells *et al.* (2004), Van der Schrier *et al.* (2006) constructed maps of summer moisture availability across a large portion of North America (20-50°N, 130-60°W) for the period 1901-2002 with a spatial latitude/longitude resolution of 0.5° x 0.5°. This operation revealed, in their words, that over the area as a whole, "the 1930s and 1950s stand out as times of persistent and exceptionally dry conditions, whereas the 1970s and the 1990s were generally wet." However, they say that "no statistically significant trend was found in the mean summer SCPDSI over the 1901-2002 period, nor in the area percentage with moderate or severe moisture excess or deficit." In fact, they could not find a single coherent area within the SCPDSI maps that "showed a statistically significant trend over the 1901-2002 period."

Going back considerably further in time, Fye *et al.* (2003) developed gridded reconstructions of the summer (June-August) basic Palmer Drought Severity Index over the continental United States, based on "annual proxies of drought and wetness provided by 426 climatically sensitive tree-ring chronologies." This work revealed that the greatest twentieth century

moisture anomalies across the United States were the 13-year pluvial over the West in the early part of the century, and the epic droughts of the 1930s (the Dust Bowl years) and 1950s, which lasted 12 and 11 years, respectively.

The researchers found the 13-year pluvial from 1905 to 1917 had three earlier analogs: an extended 16-year pluvial from 1825 to 1840, a prolonged 21-year wet period from 1602 to 1622, and a 10-year pluvial from 1549 to 1558. The 11-year drought from 1946 to 1956, on the other hand, had at least 12 earlier analogs in terms of location, intensity, and duration; but the Dust Bowl drought was greater than all of them, except for a sixteenth century "megadrought" which lasted some 18 years and was, in the words of Fye *et al.*, "the most severe sustained drought to impact North America in the past 500 to perhaps 1000 years."

In another long-term study, Stahle *et al.* (2000) developed a long-term history of North American drought from reconstructions of the Palmer Drought Severity Index based on analyses of many lengthy tree-ring records. This history also revealed that the 1930s Dust Bowl drought in the United States was eclipsed in all three of these categories by the sixteenth century megadrought. This incredible period of dryness, as they describe it, persisted "from the 1540s to 1580s in Mexico, from the 1550s to 1590s over the [U.S.] Southwest, and from the 1570s to 1600s over Wyoming and Montana." In addition, it "extended across most of the continental United States during the 1560s," and it recurred with greater intensity over the Southeast during the 1580s to 1590s. So horrendous were its myriad impacts, Stahle *et al.* unequivocally state that "the 'megadrought' of the sixteenth century far exceeded any drought of the 20th century." They state that a "precipitation reconstruction for western New Mexico suggests that the sixteenth century drought was the most extreme prolonged drought in the past 2000 years."

Last, we come to the intriguing study of Herweijer *et al.* (2006), who begin the report of their work by noting that "drought is a recurring major natural hazard that has dogged civilizations through time and remains the 'world's costliest natural disaster'." With respect to the twentieth century, they report that the "major long-lasting droughts of the 1930s and 1950s covered large areas of the interior and southern states and have long served as paradigms for the social and economic cost of sustained drought in the USA." However, they add that "these events are not unique to the twentieth century," and they go on to describe three periods of widespread and persistent drought in the latter half of the nineteenth century—1856-1865 (the "Civil War" drought), 1870-1877 and 1890-1896—based on evidence obtained from proxy, historical, and instrumental data.

With respect to the first of these impressive mid- to late-nineteenth century droughts, Herweijer *et al.* say it "is likely to have had a profound ecological and cultural impact on the interior USA, with the persistence and severity of drought conditions in the Plains surpassing those of the infamous 1930s Dust Bowl drought." In addition, they report that "drought conditions during the Civil War, 1870s and 1890s droughts were not restricted to the summer months, but existed year round, with a large signal in the winter and spring months."

Taking a still longer look back in time, the three researchers cite the work of Cook and Krusic (2004), who constructed a North American Drought Atlas using hundreds of tree-ring records. This atlas reveals what Herweijer *et al.* describe as "a 'Mediaeval Megadrought' that occurred from AD 900 to AD 1300," along with "an abrupt shift to wetter conditions after AD 1300, coinciding with the 'Little Ice Age', a time of globally cooler temperatures" that ultimately gave way to "a return to more drought-prone conditions beginning in the nineteenth century."

The broad picture that emerges from these observations is one where the most severe North American droughts of the past millennium were associated with the globally warmer temperatures of the Medieval Warm Period plus the initial stage of the globally warmer Current Warm Period. Superimposed upon this low-frequency behavior, however, Herweijer *et al.* find evidence for a "linkage between a colder eastern equatorial Pacific and persistent North American drought over the last 1000 years," noting further that "Rosby wave propagation from the cooler equatorial Pacific amplifies dry conditions over the USA." In addition, they report that after using "published coral data for the last millennium to reconstruct a NINO 3.4 history," they applied "the modern-day relationship between NINO 3.4 and North American drought ... to recreate two of the severest Mediaeval 'drought epochs' in the western USA."

But how is it that simultaneous global-scale warmth and regional-scale cold combine to produce the most severe North American droughts? One possible answer is variable solar activity. When solar activity is in an ascending mode, the globe as a whole warms; but at the same time, to quote from Herweijer *et al.*'s concluding sentence, increased irradiance typically "corresponds to a colder eastern equatorial Pacific and, by extension, increased drought occurrence in North America and other mid-latitude continental regions."

An important implication of these observations is that the most severe North American droughts should occur during major multi-centennial global warm periods, as has in fact been observed to be the case. Since the greatest such droughts of the Current Warm Period have not yet approached the severity of those that occurred during the Medieval Warm Period, it seems a good bet that the global temperature of the Current Warm Period is not yet as high as the global temperature that prevailed throughout the Medieval Warm Period.

Additional information on this topic, including reviews of newer publications as they become available, can be found at http://www.co2science.org/subject/d/droughtusa.php.

References

Andreadis, K.M. and Lettenmaier, D.P. 2006. Trends in 20th century drought over the continental United States. *Geophysical Research Letters* **33**: 10.1029/2006GL025711.

Cook, E.R. and Krusic, P.J. 2004. *North American Summer PDSI Reconstructions*. IGBP PAGES/World Data Center for Paleoclimatology Data Contribution Series # 2004-045. NOAA/NGDC Paleoclimatology Program.

Fye, F.K., Stahle, D.W. and Cook, E.R. 2003. Paleoclimatic analogs to twentieth-century moisture regimes across the United States. *Bulletin of the American Meteorological Society* **84**: 901-909.

Herweijer, C., Seager, R. and Cook, E.R. 2006. North American droughts of the mid to late nineteenth century: a history, simulation and implication for Mediaeval drought. *The Holocene* **16**: 159-171.

Palmer, W.C. 1965. *Meteorological Drought*. Office of Climatology Research Paper 45. U.S. Weather Bureau, Washington, DC, USA.

Stahle, D.W., Cook, E.R., Cleaveland, M.K, Therrell, M.D., Meko, D.M., Grissino-Mayer, H.D., Watson, E. and Luckman, B.H. 2000. Tree-ring data document 16th century megadrought over North America. *EOS: Transactions, American Geophysical Union* **81**: 121, 125.

Van der Schrier, G., Briffa, K.R., Osborn, T.J. and Cook, E.R. 2006. Summer moisture availability across North

America. *Journal of Geophysical Research* **111**: 10.1029/2005JD006745.

Wells, N., Goddard, S. and Hayes, M.J. 2004. A self-calibrating Palmer drought severity index. *Journal of Climate* 17: 2335-2351.

6.2. Floods

In the midst of 2002's massive flooding in Europe, Gallus Cadonau, the managing director of the Swiss Greina Foundation, called for a punitive tariff on U.S. imports to force cooperation on reducing greenhouse gas emissions, claiming that the flooding "definitely has to do with global warming" and stating that "we must change something now" (Hooper, 2002). Cadonau was joined in this sentiment by Germany's environment minister, Jurgen Trittin, who implied much the same thing when he said "if we don't want this development to get worse, then we must continue with the consistent reduction of environmentally harmful greenhouse gasses" (Ibid.).

The IPCC seems to agree with Cadonau and Trittin. Its authors report "a catastrophic flood occurred along several central European rivers in August 2002. The floods resulting from extraordinarily high precipitation were enhanced by the fact that the soils were completely saturated and the river water levels were already high because of previous rain. Hence, it was part of a pattern of weather over an extended period" (IPCC, 2007-I, p. 311). While admitting "there is no significant trend in flood occurrences of the Elbe within the last 500 years," the IPCC nevertheless says the "observed increase in precipitation variability at a majority of German precipitation stations during the last century is indicative of an enhancement of the probability of both floods and droughts" (Ibid.)

In evaluating these claims it is instructive to see how flood activity has responded to the global warming of the past century. In the sections below we review studies of the subject that have been conducted in Asia, Europe, and North America.

Additional information on this topic, including reviews on floods not discussed here, can be found at http://www.co2science.org/subject/f/subject_f.php under the heading Floods.

6.2.1. Asia

In a study that covered the entire continent, Cluis and Laberge (2001) analyzed the flow records of 78 rivers distributed throughout the Asia-Pacific region to see if there had been any enhancement of earth's hydrologic cycle coupled with an increase in variability that might have led to more floods between the mean beginning and end dates of the flow records: 1936 ± 5 years and 1988 ± 1 year, respectively. Over this period, the two scientists determined that mean river discharges were unchanged in 67 percent of the cases investigated; where there were trends, 69 percent of them were downward. In addition, maximum river discharges were unchanged in 77 percent of the cases investigated; where there were trends, 72 percent of them were downward. Consequently, the two researchers observed no changes in both of these flood characteristics in the majority of the rivers they studied; where there were changes, more of them were of the type that typically leads to less flooding and less severe floods.

Two years later, Kale *et al.* (2003) conducted geomorphic studies of slackwater deposits in the bedrock gorges of the Tapi and Narmada Rivers of central India, which allowed them to assemble long chronologies of large floods of these rivers. In doing so, they found that "since 1727 at least 33 large floods have occurred on the Tapi River and the largest on the river occurred in 1837." With respect to large floods on the Narmada River, they reported at least nine or 10 floods between the beginning of the Christian era and AD 400; between AD 400 and 1000 they documented six to seven floods; between AD 1000 and 1400 eight or nine floods; and after 1950 three more such floods. In addition, on the basis of texture, elevation, and thickness of the flood units, they concluded that "the periods AD 400-1000 and post-1950 represent periods of extreme floods."

What do these findings imply about the effects of global warming on central India flood events? The post-1950 period would likely be claimed by the IPCC to have been the warmest of the past millennium; it has indeed experienced some extreme floods. However, the flood characteristics of the AD 400-1000 period are described in equivalent terms, and this was a rather cold climatic interval known as the Dark Ages Cold Period. See, for example, McDermott *et al.* (2001) and Andersson *et al.* (2003). In addition, the most extreme flood in the much shorter record of the Tapi River occurred in 1837, near the beginning of one of the colder periods of the

Little Ice Age. There appears to be little correlation between the flood characteristics of the Tapi and Narmada Rivers of central India and the thermal state of the global climate.

Focusing on the much smaller area of southwestern Turkey, Touchan *et al.* (2003) developed two reconstructions of spring (May-June) precipitation from tree-ring width measurements, one of them (1776-1998) based on nine chronologies of *Cedrus libani, Juniperus excelsa, Pinus brutia* and *Pinus nigra*, and the other one (1339-1998) based on three chronologies of *Juniperus excelsa*. These reconstructions, in their words, "show clear evidence of multi-year to decadal variations in spring precipitation," with both wet and dry periods of 1-2 years duration being well distributed throughout the record. However, in the case of more extreme hydrologic events, they found that all of the wettest five-year periods preceded the Industrial Revolution, manifesting themselves at times when the air's carbon dioxide content was largely unaffected by anthropogenic CO_2 emissions.

Two years later, Jiang *et al.* (2005) analyzed pertinent historical documents to produce a 1,000-year time series of flood and drought occurrence in the Yangtze Delta of Eastern China (30 to 33°N, 119 to 122°E), which with a nearly level plain that averages only two to seven meters above sea level across 75 percent of its area is vulnerable to flooding and maritime tidal hazards. This work demonstrated that alternating wet and dry episodes occurred throughout the 1,000-year period, with the most rapid and strongest of these fluctuations occurring during the Little Ice Age (1500-1850).

The following year, Davi *et al.* (2006) developed a reconstruction of streamflow that extended from 1637 to 1997, based on absolutely dated tree-ring-width chronologies from five sampling sites in west-central Mongolia, all of which sites were in or near the Selenge River basin, the largest river in Mongolia. Of the 10 wettest five-year periods, only two occurred during the twentieth century (1990-1994 and 1917-1921, the second and eighth wettest of the 10 extreme periods, respectively), once again indicative of a propensity for less flooding during the warmest portion of the 360-year period.

The year 2007 produced a second study of the Yangtze Delta of Eastern China, when Zhang *et al.* (2007) developed flood and drought histories of the past thousand years "from local chronicles, old and very comprehensive encyclopaedia, historic agricultural registers, and official weather reports,"

after which "continuous wavelet transform was applied to detect the periodicity and variability of the flood/drought series" and, finally, the results of the entire set of operations were compared with 1,000-year temperature histories of northeastern Tibet and southern Tibet. This work revealed, in the words of the researchers, that "colder mean temperature in the Tibetan Plateau usually resulted in higher probability of flood events in the Yangtze Delta region."

Contemporaneously, Huang *et al.* (2007) constructed a complete catalog of Holocene overbank flooding events at a watershed scale in the headwater region of the Sushui River within the Yuncheng Basin in the southeast part of the middle reaches of China's Yellow River, based on pedo-sedimentary records of the region's semiarid piedmont alluvial plains, including the color, texture, and structure of the sediment profiles, along with determinations of particle-size distributions, magnetic susceptibilities, and elemental concentrations. This work revealed there were six major episodes of overbank flooding. The first occurred at the onset of the Holocene, the second immediately before the mid-Holocene Climatic Optimum, and the third in the late stage of the mid-Holocene Climatic Optimum, while the last three episodes coincided with "the cold-dry stages during the late Holocene," according to the six scientists. Speaking of the last of the overbank flooding episodes, they note that it "corresponds with the well documented 'Little Ice Age,' when "climate departed from its long-term average conditions and was unstable, irregular, and disastrous," which is pretty much like the Little Ice Age has been described in many other parts of the world as well.

The history of floods in Asia provides no evidence of increased frequency or severity during the Current Warm Period. Additional information on this topic, including reviews of newer publications as they become available, can be found at http://www.co2science.org/subject/f/floodsasia.php.

References

Andersson, C., Risebrobakken, B., Jansen, E. and Dahl, S.O. 2003. Late Holocene surface ocean conditions of the Norwegian Sea (Voring Plateau). *Paleoceanography* **18**: 10.1029/2001PA000654.

Cluis, D. and Laberge, C. 2001. Climate change and trend detection in selected rivers within the Asia-Pacific region. *Water International* **26**: 411-424.

Davi, N.K., Jacoby, G.C., Curtis, A.E. and Baatarbileg, N. 2006. Extension of drought records for central Asia using tree rings: West-Central Mongolia. *Journal of Climate* **19**: 288-299.

Huang, C.C., Pang, J., Zha, X., Su, H., Jia, Y. and Zhu, Y. 2007. Impact of monsoonal climatic change on Holocene overbank flooding along Sushui River, middle reach of the Yellow River, China. *Quaternary Science Reviews* **26**: 2247-2264.

Jiang, T., Zhang, Q., Blender, R. and Fraedrich, K. 2005. Yangtze Delta floods and droughts of the last millennium: Abrupt changes and long term memory. *Theoretical and Applied Climatology* **82**: 131-141.

Kale, V.S., Mishra, S. and Baker, V.R. 2003. Sedimentary records of palaeofloods in the bedrock gorges of the Tapi and Narmada rivers, central India. *Current Science* **84**: 1072-1079.

McDermott, F., Mattey, D.P. and Hawkesworth, C. 2001. Centennial-scale Holocene climate variability revealed by a high-resolution speleothem $\delta^{18}O$ record from SW Ireland. *Science* **294**: 1328-1331.

Touchan, R., Garfin, G.M., Meko, D.M., Funkhouser, G., Erkan, N., Hughes, M.K. and Wallin, B.S. 2003. Preliminary reconstructions of spring precipitation in southwestern Turkey from tree-ring width. *International Journal of Climatology* **23**: 157-171.

Zhang, Q., Chen, J. and Becker, S. 2007. Flood/drought change of last millennium in the Yangtze Delta and its possible connections with Tibetan climatic changes. *Global and Planetary Change* **57**: 213-221.

6.2.2. Europe

Nesje *et al.* (2001) analyzed a sediment core from a lake in southern Norway in an attempt to determine the frequency and magnitude of floods in that region. The last thousand years of the record revealed "a period of little flood activity around the Medieval period (AD 1000-1400)," which was followed by a period of extensive flood activity that was associated with the "post-Medieval climate deterioration characterized by lower air temperature, thicker and more long-lasting snow cover, and more frequent storms associated with the 'Little Ice Age'." This particular study suggests that the post-Little Ice Age warming the earth has experienced for the past century or two—and which could well continue for some time to come—should be leading this portion of the planet into a period of less-extensive floods.

Pirazzoli (2000) analyzed tide-gauge and meteorological data over the period 1951-1997 for the northern portion of the Atlantic coast of France, discovering that the number of atmospheric depressions and strong surge winds in this region "are becoming less frequent." The data also revealed that "ongoing trends of climate variability show a decrease in the frequency and hence the gravity of coastal flooding," which is what would be expected in view of the findings of Nesje *et al.*

Reynard *et al.* (2001) used a continuous flow simulation model to assess the impacts of potential climate and land use changes on flood regimes of the UK's Thames and Severn Rivers; and, as might have been expected of a model study, it predicted modest increases in the magnitudes of 50-year floods on these rivers when the climate was forced to change as predicted for various global warming scenarios. However, when the modelers allowed forest cover to rise concomitantly, they found that this land use change "acts in the opposite direction to the climate changes and under some scenarios is large enough to fully compensate for the shifts due to climate." As the air's CO_2 content continues to rise, there will be a natural impetus for forests to expand their ranges and grow in areas where grasses now dominate the landscape. If public policies cooperate, forests will indeed expand their presence on the river catchments in question and neutralize any predicted increases in flood activity in a future high-CO_2 world.

Starkel (2002) reviewed what is known about the relationship between extreme weather events and the thermal climate of Europe during the Holocene. This review demonstrated that more extreme fluvial activity was typically associated with cooler time intervals. In recovering from one such period (the Younger Dryas), for example, temperatures in Germany and Switzerland rose by 3-5°C over several decades; "this fast shift," in Starkel's words, "caused a rapid expansion of forest communities, [a] rise in the upper treeline and higher density of vegetation cover," which led to a "drastic" reduction in sediment delivery from slopes to river channels.

Mudelsee *et al.* (2003) analyzed historical documents from the eleventh century to 1850, plus subsequent water stage and daily runoff records from then until 2002, for two of the largest rivers in central Europe: the Elbe and Oder Rivers. The team of German scientists reported that "for the past 80 to 150 years"—which the IPCC claims was a period of unprecedented global warming—"we find a decrease in winter flood occurrence in both rivers, while summer floods show no trend, consistent with trends in extreme precipitation occurrence." As the world has recovered from the global chill of the Little Ice Age, flooding of the Elbe and Oder rivers has not materially changed in summer and has actually decreased in winter. Blaming anthropogenic CO_2 emissions for the European flooding of 2002, then, is not a reasoned deduction based on scientific evidence.

On September 8 and 9, 2002, extreme flooding of the Gardon River in southern France occurred as a result of half-a-year's rainfall being received in approximately 20 hours. Floods claimed the lives of a number of people and caused much damage to towns and villages situated adjacent to its channel. The event elicited much coverage in the press; in the words of Sheffer et al. (2003), "this flood is now considered by the media and professionals to be 'the largest flood on record'," which record extends all the way back to 1890. Coincidently, Sheffer et al. were in the midst of a study of prior floods of the Gardon River, so they had data spanning a much longer time period. They report that "the extraordinary flood of September 2002 was not the largest by any means," noting that "similar, and even larger floods have occurred several times in the recent past," with three of the five greatest floods they had identified to that point in time occurring over the period AD 1400-1800 during the Little Ice Age. Commenting on these facts, Sheffer et al. stated that "using a longer time scale than human collective memory, paleoflood studies can put in perspective the occurrences of the extreme floods that hit Europe and other parts of the world during the summer of 2002."

Lindstrom and Bergstrom (2004) analyzed runoff and flood data from more than 60 discharge stations scattered throughout Sweden, some of which provide information stretching to the early- to mid-1800s, when Sweden and the world were still experiencing the cold of the Little Ice Age. This analysis led them to discover that the last 20 years of the past century were indeed unusually wet, with a runoff anomaly of +8 percent compared with the century average. But they also found that "the runoff in the 1920s was comparable to that of the two latest decades," and that "the few observation series available from the 1800s show that the runoff was even higher than recently." What is more, they note that "flood peaks in old data are probably underestimated," which "makes it difficult to conclude that there has really been a significant increase in average flood levels." In addition, they say "no increased frequency of floods

with a return period of 10 years or more, could be determined."

With respect to the generality of their findings, Lindstrom and Bergstrom say that conditions in Sweden "are consistent with results reported from nearby countries: e.g. Forland et al. (2000), Bering Ovesen et al. (2000), Klavins et al. (2002) and Hyvarinen (2003)," and that, "in general, it has been difficult to show any convincing evidence of an increasing magnitude of floods (e.g. Roald, 1999) in the near region, as is the case in other parts of the world (e.g. Robson et al., 1998; Lins and Slack, 1999; Douglas et al., 2000; McCabe and Wolock, 2002; Zhang et al., 2001)."

It is clear that for most of Europe, there are no compelling real-world data to support the claim that the global warming of the past two centuries led to more frequent or severe flooding. Additional information on this topic, including reviews of newer publications as they become available, can be found at http://www.co2science.org/subject/f/floodseuro.php.

References

Bering Ovesen, N., Legard Iversen, H., Larsen, S., Muller-Wohlfeil, D.I. and Svendsen, L. 2000. *Afstromningsforhold i danske vandlob.* Faglig rapport fra DMU, no. 340. Miljo- og Energiministeriet. Danmarks Miljoundersogelser, Silkeborg, Denmark.

Douglas, E.M., Vogel, R.M. and Kroll, C.N. 2000. Trends in floods and low flows in the United States: impact of spatial correlation. *Journal of Hydrology* **240**: 90-105.

Forland, E., Roald, L.A., Tveito, O.E. and Hanssen-Bauer, I. 2000. *Past and future variations in climate and runoff in Norway.* DNMI Report no. 1900/00 KLIMA, Oslo, Norway.

Hyvarinen, V. 2003. Trends and characteristics of hydrological time series in Finland. *Nordic Hydrology* **34**: 71-90.

Klavins, M., Briede, A., Rodinov, V., Kokorite, I. and Frisk, T. 2002. Long-term changes of the river runoff in Latvia. *Boreal Environmental Research* **7**: 447-456.

Lindstrom, G. and Bergstrom, S. 2004. Runoff trends in Sweden 1807-2002. *Hydrological Sciences Journal* **49**: 69-83.

Lins, H.F. and Slack, J.R. 1999. Streamflow trends in the United States. *Geophysical Research Letters* **26**: 227-230.

McCabe, G.J. and Wolock, D.M. 2002. A step increase in streamflow in the conterminous United States. *Geophysical Research Letters* **29**: 2185-2188.

Mudelsee, M., Borngen, M., Tetzlaff, G. and Grunewald, U. 2003. No upward trends in the occurrence of extreme floods in central Europe. *Nature* **425**: 166-169.

Nesje, A., Dahl, S.O., Matthews, J.A. and Berrisford, M.S. 2001. A ~4500-yr record of river floods obtained from a sediment core in Lake Atnsjoen, eastern Norway. *Journal of Paleolimnology* **25**: 329-342.

Pirazzoli, P.A. 2000. Surges, atmospheric pressure and wind change and flooding probability on the Atlantic coast of France. *Oceanologica Acta* **23**: 643-661.

Reynard, N.S., Prudhomme, C. and Crooks, S.M. 2001. The flood characteristics of large UK rivers: Potential effects of changing climate and land use. *Climatic Change* **48**: 343-359.

Roald, L.A. 1999. *Analyse av lange flomserier*. HYDRA-rapport no. F01, NVE, Oslo, Norway.

Robson, A.J., Jones, T.K., Reed, D.W. and Bayliss, A.C. 1998. A study of national trends and variation in UK floods. *International Journal of Climatology* **18**: 165-182.

Sheffer, N.A., Enzel, Y., Waldmann, N., Grodek, T. and Benito, G. 2003. Claim of largest flood on record proves false. *EOS: Transactions, American Geophysical Union* **84**: 109.

Starkel, L. 2002. Change in the frequency of extreme events as the indicator of climatic change in the Holocene (in fluvial systems). *Quaternary International* **91**: 25-32.

Zhang, X., Harvey, K.D., Hogg, W.D. and Yuzyk, T.R. 2001. Trends in Canadian streamflow. *Water Resources Research* **37**: 987-998.

6.2.3. North America

Lins and Slack (1999) analyzed secular streamflow trends in 395 different parts of the United States that were derived from more than 1,500 individual streamgauges, some of which had continuous data stretching to 1914. In the mean, they found that "the conterminous U.S. is getting wetter, but less extreme." That is to say, as the near-surface air temperature of the planet gradually rose throughout the course of the twentieth century, the United States became wetter in the mean but less variable at the extremes, which is where floods and droughts occur, leading to what could well be called the best of both worlds, i.e., more water with fewer floods and droughts.

In a similar but more regionally focused study, Molnar and Ramirez (2001) conducted a detailed analysis of precipitation and streamflow trends for the period 1948-1997 in the semiarid Rio Puerco Basin of New Mexico. At the annual timescale, they reported finding "a statistically significant increasing trend in precipitation," which was driven primarily by an increase in the number of rainy days in the moderate rainfall intensity range, with essentially no change at the high-intensity end of the spectrum. In the case of streamflow, there was no trend at the annual timescale, but monthly totals increased in low-flow months and decreased in high-flow months, once again reducing the likelihood of both floods and droughts.

Knox (2001) identified an analogous phenomenon in the more mesic Upper Mississippi River Valley, but with a slight twist. Since the 1940s and early 1950s, the magnitudes of the largest daily flows in this much wetter region have been decreasing at the same time that the magnitude of the average daily baseflow has been increasing, once again manifesting simultaneous trends towards lessened flood and drought conditions.

Much the same story is told by the research of Garbrecht and Rossel (2002), who studied the nature of precipitation throughout the U.S. Great Plains over the period 1895-1999. For the central and southern Great Plains, the last two decades of this period were found to be the longest and wettest of the entire 105 years of record, due primarily to a reduction in the number of dry years and an increase in the number of wet years. Once again, however, the number of very wet years—which would be expected to produce flooding—"did not increase as much and even showed a decrease for many regions."

The northern and northwestern Great Plains also experienced a precipitation increase near the end of Garbrecht and Rossel's 105-year record, but it was primarily confined to the final decade of the twentieth century. And again, as they report, "fewer dry years over the last 10 years, as opposed to an increase in very wet years, were the leading cause of the observed wet conditions."

In spite of the general tendencies described in these several papers, there still were some significant floods during the last decade of the past century, such as the 1997 flooding of the Red River of the North, which devastated Grand Forks, North Dakota, as well as parts of Canada. However, as Haque (2000) reports, although this particular flood was indeed the largest experienced by the Red River over the past century, it was not the largest to occur in historic times. In 1852 there was a slightly larger Red River flood, and in 1826 there was a flood that was nearly 40 percent greater than the flood of 1997. The temperature of the globe was colder at the times of these earlier catastrophic floods than it was in 1997, indicating that one cannot attribute the strength of the 1997 flood to higher temperatures that year or the warming of the preceding decades. We also note that Red River flooding is also linked to snow melt and ice jams because itflows northward into frozen areas.

Olsen *et al.* (1999) report that some upward trends in flood-flows have been found in certain places along the Mississippi and Missouri Rivers, which is not at all surprising, as there will always be exceptions to the general rule. They note that many of the observed upward trends were highly dependent upon the length of the data record and when the trends began and ended. They say of these trends that they "were not necessarily there in the past and they may not be there tomorrow."

Expanding the scope of our survey to much longer intervals of time is Fye *et al.* (2003), who developed multi-century reconstructions of summer (June-August) Palmer Drought Severity Index over the continental United States from annual proxies of moisture status provided by 426 climatically sensitive tree-ring chronologies. This exercise indicated that the greatest twentieth century wetness anomaly across the United States was a 13-year period in the early part of the century when it was colder than it is now. Fye *et al.*'s analysis also revealed the existence of a 16-year pluvial from 1825 to 1840 and a prolonged 21-year wet period from 1602 to 1622, both of which anomalies occurred during the Little Ice Age, when, of course, it was colder still.

St. George and Nielsen (2002) used "a ringwidth chronology developed from living, historical and subfossil bur oak (*Quercus macrocarpa* (Michx.)) in the Red River basin to reconstruct annual precipitation in southern Manitoba since A.D. 1409." Their analysis indicated, in their words, that "prior to the 20th century, southern Manitoba's climate was more extreme and variable, with prolonged intervals that were wetter and drier than any time following permanent Euro-Canadian settlement."

Also working with tree-ring chronologies, Ni *et al.* (2002) developed a 1,000-year history of cool-season (November-April) precipitation for each climate division in Arizona and New Mexico, USA. In doing so, they found that several wet periods comparable to the wet conditions seen in the early

1900s and post-1976 occurred in 1108-20, 1195-1204, 1330-45 (which they denominate "the most persistent and extreme wet interval"), the 1610s, and the early 1800s, all of which wet periods are embedded in the long cold expanse of the Little Ice Age, which is clearly revealed in the work of Esper *et al.* (2002).

Doubling the temporal extent of Ni *et al.*'s investigation, Schimmelmann *et al.* (2003) analyzed gray clay-rich flood deposits in the predominantly olive varved sediments of the Santa Barbara Basin off the coast of California, USA, which they accurately dated by varve-counting. Their analysis indicated that six prominent flood events occurred at approximately AD 212, 440, 603, 1029, 1418, and 1605, "suggesting," in their words, "a quasi-periodicity of ~200 years," with "skipped" flooding just after AD 800, 1200, and 1800. They further note that "the floods of ~AD 1029 and 1605 seem to have been associated with brief cold spells," that "the flood of ~AD 440 dates to the onset of the most unstable marine climatic interval of the Holocene (Kennett and Kennett, 2000)," and that "the flood of ~AD 1418 occurred at a time when the global atmospheric circulation pattern underwent fundamental reorganization at the beginning of the 'Little Ice Age' (Kreutz *et al.*, 1997; Meeker and Mayewski, 2002)." As a result, they hypothesize that "solar-modulated climatic background conditions are opening a ~40-year window of opportunity for flooding every ~200 years," and that "during each window, the danger of flooding is exacerbated by additional climatic and environmental cofactors." They also note that "extrapolation of the ~200-year spacing of floods into the future raises the uncomfortable possibility for historically unprecedented flooding in southern California during the first half of this century." Consequently, if such flooding does occur in the near future, there will be no need to suppose it came as a consequence of what the IPCC calls the unprecedented warming of the past century.

Once again doubling the length of time investigated, Campbell (2002) analyzed the grain sizes of sediment cores obtained from Pine Lake, Alberta, Canada, to provide a non-vegetation-based high-resolution record of streamflow variability for this part of North America over the past 4,000 years. This work revealed that the highest rates of stream discharge during this period occurred during the Little Ice Age, approximately 300-350 years ago, at which time grain sizes were about 2.5 standard deviations above the 4,000-year mean. In contrast, the lowest rates of streamflow were observed around AD 1100, during the Medieval Warm Period, when median grain sizes were nearly 2.0 standard deviations below the 4,000-year mean.

Further extending the temporal scope of our review, Brown *et al.* (1999) analyzed various properties of cored sequences of hemipelagic muds deposited in the northern Gulf of Mexico for evidence of variations in Mississippi River outflow over the past 5,300 years. This group of researchers found evidence of seven large megafloods, which they describe as "almost certainly larger than historical floods in the Mississippi watershed." In fact, they say these fluvial events were likely "episodes of multidecadal duration," five of which occurred during cold periods similar to the Little Ice Age.

Last, in a study that covered essentially the entire Holocene, Noren *et al.* (2002) employed several techniques to identify and date terrigenous in-wash layers found in sediment cores extracted from 13 small lakes distributed across a 20,000-km^2 region in Vermont and eastern New York that depict the frequency of storm-related floods. They found that "the frequency of storm-related floods in the northeastern United States has varied in regular cycles during the past 13,000 years (13 kyr), with a characteristic period of about 3 kyr." Specifically, they found there were four major peaks in the data during this period, with the most recent upswing in storm-related floods beginning "at about 600 yr BP [Before Present], coincident with the beginning of the Little Ice Age." In addition, they note that several "independent records of storminess and flooding from around the North Atlantic show maxima that correspond to those that characterize our lake records [Brown *et al.*, 1999; Knox, 1999; Lamb, 1979; Liu and Fearn, 2000; Zong and Tooley, 1999]."

Taken together, the research described in this section suggests that North American flooding tends to become both less frequent and less severe when the planet warms, although there have been some exceptions to this general rule. We would expect that any further warming of the globe would tend to further reduce both the frequency and severity of flooding in North America.

Additional information on this topic, including reviews of newer publications as they become available, can be found at http://www.co2science.org/subject/f/floodsnortham.php.

References

Brown, P., Kennett, J.P. and Ingram, B.L. 1999. Marine evidence for episodic Holocene megafloods in North America and the northern Gulf of Mexico. *Paleoceanography* **14**: 498-510.

Campbell, C. 2002. Late Holocene lake sedimentology and climate change in southern Alberta, Canada. *Quaternary Research* **49**: 96-101.

Esper, J., Cook, E.R. and Schweingruber, F.H. 2002. Low-frequency signals in long tree-ring chronologies for reconstructing past temperature variability. *Science* **295**: 2250-2253.

Fye, F.K., Stahle, D.W. and Cook, E.R. 2003. Paleoclimatic analogs to twentieth-century moisture regimes across the United States. *Bulletin of the American Meteorological Society* **84**: 901-909.

Garbrecht, J.D. and Rossel, F.E. 2002. Decade-scale precipitation increase in Great Plains at end of 20th century. *Journal of Hydrologic Engineering* **7**: 64-75.

Haque, C.E. 2000. Risk assessment, emergency preparedness and response to hazards: The case of the 1997 Red River Valley flood, Canada. *Natural Hazards* **21**: 225-245.

Kennett, D.J. and Kennett, J.P. 2000. Competitive and cooperative responses to climatic instability in coastal southern California. *American Antiquity* **65**: 379-395.

Knox, J.C. 1999. Sensitivity of modern and Holocene floods to climate change. *Quaternary Science Reviews* **19**: 439-457.

Knox, J.C. 2001. Agricultural influence on landscape sensitivity in the Upper Mississippi River Valley. *Catena* **42**: 193-224.

Kreutz, K.J., Mayewski, P.A., Meeker, L.D., Twickler, M.S., Whitlow, S.I. and Pittalwala, I.I. 1997. Bipolar changes in atmospheric circulation during the Little Ice Age. *Science* **277**: 1294-1296.

Lamb, H.H. 1979. Variation and changes in the wind and ocean circulation: the Little Ice Age in the northeast Atlantic. *Quaternary Research* **11**: 1-20.

Lins, H.F. and Slack, J.R. 1999. Streamflow trends in the United States. *Geophysical Research Letters* **26**: 227-230.

Liu, K.B. and Fearn, M.L. 2000. Reconstruction of prehistoric landfall frequencies of catastrophic hurricanes in northwestern Florida from lake sediment records. *Quaternary Research* **54**: 238-245.

Meeker, L.D. and Mayewski, P.A. 2002. A 1400-year high-resolution record of atmospheric circulation over the North Atlantic and Asia. *The Holocene* **12**: 257-266.

Molnar, P. and Ramirez, J.A. 2001. Recent trends in precipitation and streamflow in the Rio Puerco Basin. *Journal of Climate* **14**: 2317-2328.

Ni, F., Cavazos, T., Hughes, M.K., Comrie, A.C. and Funkhouser, G. 2002. Cool-season precipitation in the southwestern USA since AD 1000: Comparison of linear and nonlinear techniques for reconstruction. *International Journal of Climatology* **22**: 1645-1662.

Noren, A.J., Bierman, P.R., Steig, E.J., Lini, A. and Southon, J. 2002. Millennial-scale storminess variability in the northeastern Unites States during the Holocene epoch. *Nature* **419**: 821-824.

Olsen, J.R., Stedinger, J.R., Matalas, N.C. and Stakhiv, E.Z. 1999. Climate variability and flood frequency estimation for the Upper Mississippi and Lower Missouri Rivers. *Journal of the American Water Resources Association* **35**: 1509-1523.

Schimmelmann, A., Lange, C.B. and Meggers, B.J. 2003. Palaeoclimatic and archaeological evidence for a 200-yr recurrence of floods and droughts linking California, Mesoamerica and South America over the past 2000 years. *The Holocene* **13**: 763-778.

St. George, S. and Nielsen, E. 2002. Hydroclimatic change in southern Manitoba since A.D. 1409 inferred from tree rings. *Quaternary Research* **58**: 103-111.

Zong, Y. and Tooley, M.J. 1999. Evidence of mid-Holocene storm-surge deposits from Morecambe Bay, northwest England: A biostratigraphical approach. *Quaternary International* **55**: 43-50.

6.3. Tropical Cyclones

The IPCC contends that global warming is likely to increase the frequency and intensity of hurricanes. For example, it states "it is *likely* that future tropical cyclones (typhoons and hurricanes) will become more intense, with larger peak wind speeds and more heavy precipitation associated with ongoing increases of tropical sea surface temperatures [italics in the original]" (IPCC, 2007-I, p. 15). However, numerous peer-reviewed studies suggest otherwise. In the following sections we examine such claims as they pertain to hurricane activity in the Atlantic, Pacific, and Indian Ocean basins, and the globe as a whole.

Reference

IPCC. 2007-I. *Climate Change 2007: The Physical Science Basis. Contribution of Working Group I to the Fourth Assessment Report of the Intergovernmental Panel on Climate Change.* Solomon, S., Qin, D., Manning, M., Chen, Z., Marquis, M., Averyt, K.B., Tignor, M. and Miller, H.L. (Eds.) Cambridge University Press, Cambridge, UK.

6.3.1. Atlantic Ocean

6.3.1.1. Intensity

Free *et al.* (2004) write that "increases in hurricane intensity are expected to result from increases in sea surface temperature and decreases in tropopause-level temperature accompanying greenhouse warming (Emanuel, 1987; Henderson-Sellers *et al.*, 1998; Knutson *et al.*, 1998)," but that "because the predicted increase in intensity for doubled CO_2 is only 5%-20%, changes over the past 50 years would likely be less than 2%—too small to be detected easily." They report that "studies of observed frequencies and maximum intensities of tropical cyclones show no consistent upward trend (Landsea *et al.*, 1996; Henderson-Sellers *et al.*, 1998; Solow and Moore, 2002)," and set out to find increases in what they call "potential" hurricane intensity, because, as they describe it, "changes in potential intensity (PI) can be estimated from thermodynamic principles as shown in Emanuel (1986, 1995) given a record of SSTs [sea surface temperatures] and profiles of atmospheric temperature and humidity." Using radiosonde and SST data from 14 island radiosonde stations in the tropical Atlantic and Pacific Oceans, they compare their results with those of Bister and Emanuel (2002) at grid points near the selected stations. They report that their results "show no significant trend in potential intensity from 1980 to 1995 and no consistent trend from 1975 to 1995." What is more, they report that between 1975 and 1980, "while SSTs rose, PI decreased, illustrating the hazards of predicting changes in hurricane intensity from projected SST changes alone."

In the following year, some important new studies once again promoted the IPCC's claim that warming would enhance tropical cyclone intensity (Emanuel, 2005; Webster *et al.*, 2005), but a new review of the subject once again cast doubt on this contention. Pielke *et al.* (2005) began their discussion by noting that "globally there has been no increase in tropical cyclone frequency over at least the past several decades," citing the studies of Lander and Guard (1998), Elsner and Kocher (2000) and Webster *et al.* (2005). They noted that research on possible future changes in hurricane frequency due to global warming has produced studies that "give such contradictory results as to suggest that the state of understanding of tropical cyclogenesis provides too poor a foundation to base any projections about the future."

With respect to hurricane intensity, Pielke *et al.* noted that Emanuel (2005) claimed to have found "a very substantial upward trend in power dissipation (i.e., the sum over the life-time of the storm of the maximum wind speed cubed) in the North Atlantic and western North Pacific." However, they report that "other studies that have addressed tropical cyclone intensity variations (Landsea *et al.*, 1999; Chan and Liu, 2004) show no significant secular trends during the decades of reliable records." In addition, they indicate that although early theoretical work by Emanuel (1987) "suggested an increase of about 10% in wind speed for a 2°C increase in tropical sea surface temperature," more recent work by Knutson and Tuleya (2004) points to only a 5 percent increase in hurricane windspeeds by 2080, and that Michaels *et al.* (2005) conclude that even this projection is likely twice as great as it should be.

By 2050, Pielke *et al.* report that "for every additional dollar in damage that the Intergovernmental Panel on Climate Change expects to result from the effects of global warming on tropical cyclones, we should expect between $22 and $60 of increase in damage due to population growth and wealth," citing the findings of Pielke *et al.* (2000) in this regard. Based on this evidence, they state without equivocation that "the primary factors that govern the magnitude and patterns of future damages and casualties are how society develops and prepares for storms rather than any presently conceivable future changes in the frequency and intensity of the storms."

In concluding their review, Pielke *et al.* note that massive reductions of anthropogenic CO_2 emissions "simply will not be effective with respect to addressing future hurricane impacts," and that "there are much, much better ways to deal with the threat of hurricanes than with energy policies (e.g., Pielke and Pielke, 1997)."

Michaels *et al.* (2006) subsequently analyzed Emanuel's (2005) and Webster *et al.*'s (2005) claims

that "rising sea surface temperatures (SSTs) in the North Atlantic hurricane formation region are linked to recent increases in hurricane intensity, and that the trend of rising SSTs during the past 3 to 4 decades bears a strong resemblance to that projected to occur from increasing greenhouse gas concentrations." The researchers used weekly averaged 1° latitude by 1° longitude SST data together with hurricane track data of the National Hurricane Center that provide hurricane-center locations (latitude and longitude in tenths of a degree) and maximum 1-minute surface wind speeds (both at six-hour intervals) for all tropical storms and hurricanes in the Atlantic basin that occurred between 1982 (when the SST dataset begins) through 2005. Plotting maximum cyclone wind speed against the maximum SST that occurred prior to (or concurrent with) the maximum wind speed of each of the 270 Atlantic tropical cyclones of their study period, they found that for each 1°C increase in SST between 21.5°C and 28.25°C, the maximum wind speed attained by Atlantic basin cyclones rises, in the mean, by 2.8 m/s, and that thereafter, as SSTs rise still further, the first category-3-or-greater storms begin to appear. However, they report "there is no significant relationship between SST and maximum winds at SST exceeding 28.25°C."

From these observations, Michaels *et al.* conclude that "while crossing the 28.25°C threshold is a virtual necessity for attaining category 3 or higher winds, SST greater than 28.25°C does not act to further increase the intensity of tropical cyclones." The comparison of SSTs actually encountered by individual storms performed by Michaels *et al.*—as opposed to the comparisons of Emanuel (2005) and Webster *et al.* (2005), which utilized basin-wide averaged monthly or seasonal SSTs—refutes the idea that anthropogenic activity has detectably influenced the severity of Atlantic basin hurricanes over the past quarter-century.

Simultaneously, Balling and Cerveny (2006) examined temporal patterns in the frequency of intense tropical cyclones (TCs), the rates of rapid intensification of TCs, and the average rate of intensification of hurricanes in the North Atlantic Basin, including the tropical and subtropical North Atlantic, Caribbean Sea, and Gulf of Mexico, where they say there was "a highly statistically significant warming of 0.12°C decade[-1] over the period 1970-2003 ... based on linear regression analysis and confirmed by a variety of other popular trend identification techniques." In doing so, they found "no increase in a variety of TC intensification indices," and that "TC intensification and/or hurricane intensification rates ... are not explained by current month or antecedent sea surface temperatures (despite observed surface warming over the study period)." They concluded that "while some researchers have hypothesized that increases in long-term sea surface temperature may lead to marked increases in TC storm intensity, our findings demonstrate that various indicators of TC intensification show no significant trend over the recent three decades."

Klotzbach and Gray (2006) note that still other papers question the validity of the findings of Emanuel (2005) and Webster *et al.* (2005) "due to potential bias-correction errors in the earlier part of the data record for the Atlantic basin (Landsea, 2005)," and that "while major hurricane activity in the Atlantic has shown a large increase since 1995, global tropical-cyclone activity, as measured by the accumulated cyclone energy index, has decreased slightly during the past 16 years (Klotzbach, 2006)." And as a result of these and other data and reasoning described in their paper, they "attribute the heightened Atlantic major hurricane activity of the 2004 season as well as the increased Atlantic major hurricane activity of the previous nine years to be a consequence of multidecadal fluctuations in the strength of the Atlantic multidecadal mode and strength of the Atlantic Ocean thermohaline circulation." In this regard, they say "historical records indicate that positive and negative phases of the Atlantic multidecadal mode and thermohaline circulation last about 25-30 years (typical period ~50-60 years; Gray *et al.*, 1997; Latif *et al.*, 2004)," and "since we have been in this new active thermohaline circulation period for about 11 years, we can likely expect that most of the next 15-20 hurricane seasons will also be active, particularly with regard to increased major hurricane activity."

Vecchi and Soden (2007a) explored twenty first century projected changes in vertical wind shear (VS) over the tropical Atlantic and its ties to the Pacific Walker circulation via a suite of coupled ocean-atmosphere models forced by emissions scenario A1B (atmospheric CO_2 stabilization at 720 ppm by 2100) of the Intergovernmental Panel on Climate Change's Fourth Assessment Report, where VS is defined as the magnitude of the vector difference between monthly mean winds at 850 and 200 hPa, and where changes are computed between the two 20-year periods 2001-2020 and 2081-2100. The 18-model mean result indicated a prominent increase in VS over

the topical Atlantic and East Pacific (10°N-25°N). Noting that "the relative amplitude of the shear increase in these models is comparable to or larger than model-projected changes in other large-scale parameters related to tropical cyclone activity," the two researchers went on to state that the projected changes "would not suggest a strong anthropogenic increase in tropical Atlantic or Pacific hurricane activity during the 21st Century," and that "in addition to impacting cyclogenesis, the increase in SER [shear enhancement region] shear could act to inhibit the intensification of tropical cyclones as they traverse from the MDR [main development region] to the Caribbean and North America." Consequently, and in addition to the growing body of empirical evidence that indicates global warming has little to no impact on the intensity of hurricanes (Donnelly and Woodruff, 2007; Nyberg *et al.*, 2007), there is now considerable up-to-date model-based evidence for the same conclusion.

In a closely related paper, Vecchi and Soden (2007b) used both climate models and observational reconstructions "to explore the relationship between changes in sea surface temperature and tropical cyclone 'potential intensity'—a measure that provides an upper bound on cyclone intensity and can also reflect the likelihood of cyclone development." They found "changes in local sea surface temperature are inadequate for characterizing *even the sign* [our italics] of changes in potential intensity." Instead, they report that "long-term changes in potential intensity are closely related to the regional structure of warming," such that "regions that warm more than the tropical average are characterized by increased potential intensity, and vice versa." Using this relationship to reconstruct changes in potential intensity over the twentieth century, based on observational reconstructions of sea surface temperature, they further found that "even though tropical Atlantic sea surface temperatures are currently at a historical high, Atlantic potential intensity probably peaked in the 1930s and 1950s," noting that "recent values are near the historical average." The two scientists' conclusion was that the response of tropical cyclone activity to natural climate variations "may be larger than the response to the more uniform patterns of greenhouse-gas-induced warming."

Also in the year 2007, and at the same time Vecchi and Soden were conducting their studies of the subject, Latif *et al.* (2007) were analyzing the 1851-2005 history of Accumulated Cyclone Energy

(ACE) Index for the Atlantic basin, which parameter, in their words, "takes into account the number, strength and duration of all tropical storms in a season," after which they "analyzed the results of an atmospheric general circulation model forced by the history of observed global monthly sea surface temperatures for the period 1870-2003."

With respect to the first part of their study, they report that "the ACE Index shows pronounced multidecadal variability, with enhanced tropical storm activity during the 1890s, 1950s and at present, and mostly reduced activity in between, but no sustained long-term trend," while with respect to the second part of their study, they report that "a clear warming trend is seen in the tropical North Atlantic sea surface temperature," but that this warming trend "does not seem to influence the tropical storm activity."

This state of affairs seemed puzzling at first, because a warming of the tropical North Atlantic is known to reduce vertical wind shear there and thus promote the development of tropical storms. However, Latif *et al.*'s modeling work revealed that a warming of the tropical Pacific enhances the vertical wind shear over the Atlantic, as does a warming of the tropical Indian Ocean. Consequently, they learned, as they describe it, that "the response of the vertical wind shear over the tropical Atlantic to a warming of all three tropical oceans, as observed during the last decades, will depend on the warming of the Indo-Pacific relative to that of the tropical North Atlantic," and "apparently," as they continue, "the warming trends of the three tropical oceans cancel with respect to their effects on the vertical wind shear over the tropical North Atlantic, so that the tropical cyclone activity [has] remained rather stable and mostly within the range of the natural multidecadal variability."

Nevertheless, a striking exception to this general state of affairs occurred in 2005, when the researchers report that "the tropical North Atlantic warmed more rapidly than the Indo-Pacific," which reduced vertical wind shear over the North Atlantic, producing the most intense Atlantic hurricane season of the historical record. By contrast, they say that the summer and fall of 2006 were "characterized by El Niño conditions in the Indo-Pacific, leading to a rather small temperature difference between the tropical North Atlantic and the tropical Indian and Pacific Oceans," and they say that "this explains the weak tropical storm activity [of that year]."

Latif *et al.* say "the future evolution of Atlantic tropical storm activity will critically depend on the warming of the tropical North Atlantic relative to that in the Indo-Pacific region," and "changes in the meridianal overturning circulation and their effect on tropical Atlantic sea surface temperatures have to be considered," and that "changes in ENSO statistics in the tropical Pacific may become important." Consequently, it is anyone's guess as to what would actually occur in the real world if the earth were to experience additional substantial warming. However, since the global temperature rise of the twentieth century—which the IPCC contends was unprecedented over the past two millennia—did not lead to a sustained long-term increase in hurricane intensity, there is little reason to believe any further warming would do so.

In one final concurrent study, Scileppi and Donnelly (2007) note that "when a hurricane makes landfall, waves and storm surge can overtop coastal barriers, depositing sandy overwash fans on backbarrier salt marshes and tidal flats," and that long-term records of hurricane activity are thus formed "as organic-rich sediments accumulate over storm-induced deposits, preserving coarse overwash layers." Based on this knowledge, they refined and lengthened the hurricane record of the New York City area by first calibrating the sedimentary record of surrounding backbarrier environments to documented hurricanes—including those of 1893, 1821, 1788, and 1693—and then extracting several thousand additional years of hurricane history from this important sedimentary archive.

As a result of these efforts, the two researchers determined that "alternating periods of quiescent conditions and frequent hurricane landfall are recorded in the sedimentary record and likely indicate that climate conditions may have modulated hurricane activity on millennial timescales." Of special interest in this regard, as they describe it, is the fact that "several major hurricanes occur in the western Long Island record during the latter part of the Little Ice Age (~1550-1850 AD) when sea surface temperatures were generally colder than present," but that "no major hurricanes have impacted this area since 1893," when the earth experienced the warming that took it from the Little Ice Age to the Current Warm Period.

Noting that Emanuel (2005) and Webster *et al.* (2005) had produced analyses that suggest that "cooler climate conditions in the past may have resulted in fewer strong hurricanes," but that their own findings suggest just the opposite, Scileppe and Donnelly concluded that "other climate phenomena, such as atmospheric circulation, may have been favorable for intense hurricane development despite lower sea surface temperatures" prior to the development of the Current Warm Period.

Last, Briggs (2008) developed Bayesian statistical models for the number of tropical cyclones, the rate at which these cyclones became hurricanes, and the rate at which the hurricanes became category 4+ storms in the North Atlantic, based on data from 1966 to 2006; this work led him to conclude that there is "no evidence that the distributional mean of individual storm intensity, measured by storm days, track length, or individual storm power dissipation index, has changed (increased or decreased) through time."

In light of the many real-world observations (as well as certain modeling work) discussed above, it would appear that even the supposedly unprecedented global warming of the past century or more has not led to an increase in the intensity of Atlantic hurricanes.

Additional information on this topic, including reviews of newer publications as they become available, can be found at http://www.co2science.org/subject/h/hurratlanintensity.php.

References

Balling Jr., R.C. and Cerveny, R.S. 2006. Analysis of tropical cyclone intensification trends and variability in the North Atlantic Basin over the period 1970-2003. *Meteorological and Atmospheric Physics* **93**: 45-51.

Bister, M. and Emanuel, K. 2002. Low frequency variability of tropical cyclone potential intensity. 1. Interannual to interdecadal variability. *Journal of Geophysical Research* **107**: 10.1029/2001JD000776.

Briggs, W.M. 2008. On the changes in the number and intensity of North Atlantic tropical cyclones. *Journal of Climate* **21**: 1387-1402.

Chan, J.C.L. and Liu, S.L. 2004. Global warming and western North Pacific typhoon activity from an observational perspective. *Journal of Climate* **17**: 4590-4602.

Donnelly, J.P. and Woodruff, J.D. 2007. Intense hurricane activity over the past 5,000 years controlled by El Niño and the West African Monsoon. *Nature* **447**: 465-468.

Elsner, J.B. and Kocher, B. 2000. Global tropical cyclone activity: A link to the North Atlantic Oscillation. *Geophysical Research Letters* **27**: 129-132.

Emanuel, K.A. 1986. An air-sea interaction theory for tropical cyclones. Part I: Steady-state maintenance. *Journal of the Atmospheric Sciences* **43**: 585-604.

Emanuel, K.A. 1987. The dependence of hurricane intensity on climate. *Nature* **326**: 483-485.

Emanuel, K.A. 1995. Sensitivity of tropical cyclones to surface exchange coefficients and a revised steady-state model incorporating eye dynamics. *Journal of the Atmospheric Sciences* **52**: 3969-3976.

Emanuel, K. 2005. Increasing destructiveness of tropical cyclones over the past 30 years. *Nature* **436**: 686-688.

Free, M., Bister, M. and Emanuel, K. 2004. Potential intensity of tropical cyclones: Comparison of results from radiosonde and reanalysis data. *Journal of Climate* **17**: 1722-1727.

Gray, W.M., Sheaffer, J.D. and Landsea, C.W. 1997. Climate trends associated with multi-decadal variability of Atlantic hurricane activity. In: Diaz, H.F. and Pulwarty, R.S. (Eds.) *Hurricanes: Climate and Socioeconomic Impacts*, Springer-Verlag, pp. 15-52.

Henderson-Sellers, A., Zhang, H., Berz, G., Emanuel, K., Gray, W., Landsea, C., Holland, G., Lighthill, J., Shieh, S.-L., Webster, P. and McGuffie, K. 1998. Tropical cyclones and global climate change: A post-IPCC assessment. *Bulletin of the American Meteorological Society* **79**: 19-38.

Klotzbach, P.J. 2006. Trends in global tropical cyclone activity over the past 20 years (1986-2005). *Geophysical Research Letters* **33**: 10.1029/2006GL025881.

Klotzbach, P.J. and Gray, W.M. 2006. Causes of the unusually destructive 2004 Atlantic basin hurricane season. *Bulletin of the American Meteorological Society* **87**: 1325-1333.

Knutson, T.R. and Tuleya, R.E. 2004. Impact of CO_2-induced warming on simulated hurricane intensity and precipitation: Sensitivity to the choice of climate model and convective parameterization. *Journal of Climate* **17**: 3477-3495.

Knutson, T., Tuleya, R. and Kurihara, Y. 1998. Simulated increase of hurricane intensities in a CO_2-warmed climate. *Science* **279**: 1018-1020.

Lander, M.A. and Guard, C.P. 1998. A look at global tropical cyclone activity during 1995: Contrasting high Atlantic activity with low activity in other basins. *Monthly Weather Review* **126**: 1163-1173.

Landsea, C.W. 2005. Hurricanes and global warming. *Nature* **438**: E11-13, doi:10.1038/nature04477.

Landsea, C., Nicholls, N., Gray, W. and Avila, L. 1996. Downward trends in the frequency of intense Atlantic hurricanes during the past five decades. *Geophysical Research Letters* **23**: 1697-1700.

Landsea, C.W., Pielke Jr., R.A., Mestas-Nunez, A.M. and Knaff, J.A. 1999. Atlantic basin hurricanes: Indices of climatic changes. *Climatic Change* **42**: 89-129.

Latif, M., Keenlyside, N. and Bader, J. 2007. Tropical sea surface temperature, vertical wind shear, and hurricane development. *Geophysical Research Letters* **34**: 10.1029/2006GL027969.

Latif, M., Roeckner, E., Botzet, M., Esch, M., Haak, H., Hagemann, S., Jungclaus, J., Legutke, S., Marsland, S., Mikolajewicz, U. and Mitchell, J. 2004. Reconstructing, monitoring, and predicting multidecadal-scale changes in the North Atlantic thermohaline circulation with sea surface temperature. *Journal of Climate* **17**: 1605-1614.

Michaels, P.J., Knappenberger, P.C. and Davis, R.E. 2006. Sea-surface temperatures and tropical cyclones in the Atlantic basin. *Geophysical Research Letters* **33**: 10.1029/2006GL025757.

Michaels, P.J., Knappenberger, P.C. and Landsea, C.W. 2005. Comments on "Impacts of CO2-induced warming on simulated hurricane intensity and precipitation: Sensitivity to the choice of climate model and convective scheme." *Journal of Climate* **18**: 5179-5182

Nyberg, J., Malmgren, B.A., Winter, A., Jury, M.R., Kilbourne, K.H. and Quinn, T.M. 2007. Low Atlantic hurricane activity in the 1970s and 1980s compared to the past 270 years. *Nature* **447**: 698-701.

Pielke Jr., R.A., Landsea, C., Mayfield, M., Laver, J. and Pasch, R. 2005. Hurricanes and global warming. *Bulletin of the American Meteorological Society* **86**: 1571-1575.

Pielke Jr., R.A. and Pielke Sr., R.A. 1997. *Hurricanes: Their Nature and Impacts on Society*. John Wiley and Sons.

Pielke Jr., R.A., Pielke, Sr., R.A., Klein, R. and Sarewitz, D. 2000. Turning the big knob: Energy policy as a means to reduce weather impacts. *Energy and Environment* **11**: 255-276.

Scileppi, E. and Donnelly, J.P. 2007. Sedimentary evidence of hurricane strikes in western Long Island, New York. *Geochemistry, Geophysics, Geosystems* **8**: 10.1029/2006GC001463.

Solow, A.R. and Moore, L.J. 2002. Testing for trend in North Atlantic hurricane activity, 1900-98. *Journal of Climate* **15**: 3111-3114.

Vecchi, G.A. and Soden, B.J. 2007a. Increased tropical Atlantic wind shear in model projections of global warming. *Geophysical Research Letters* **34**: 10.1029/2006GL028905.

Vecchi, G.A. and Soden, B.J. 2007b. Effect of remote sea surface temperature change on tropical cyclone potential intensity. *Nature* **450**: 1066-1070.

Webster, P.J., Holland, G.J., Curry, J.A. and Chang, H.-R. 2005. Changes in tropical cyclone number, duration and intensity in a warming environment. *Science* **309**: 1844-1846.

6.3.1.2. Frequency

6.3.1.2.1. The Past Few Millennia

Has the warming of the past century increased the yearly number of intense Atlantic Basin hurricanes? We offer a brief review of some studies that have explored this question via thousand-year reconstructions of the region's intense hurricane activity.

Liu and Fearn (1993) analyzed sediment cores retrieved from the center of Lake Shelby in Alabama (USA) to determine the history of intense (category 4 and 5) hurricane activity there over the past 3,500 years. This work revealed that over the period of their study, "major hurricanes of category 4 or 5 intensity directly struck the Alabama coast ... with an average recurrence interval of ~600 years." They also report that the last of these hurricane strikes occurred about 700 years ago. Hence, it would appear that twentieth century global warming has not accelerated the occurrence of such severe storm activity.

Seven years later, Liu and Fearn (2000) conducted a similar study based on 16 sediment cores retrieved from Western Lake, Florida (USA), which they used to produce a proxy record of intense hurricane strikes for this region of the Gulf of Mexico that covered the past 7,000 years. In this study, 12 major hurricanes of category 4 or 5 intensity were found to have struck the Western Lake region. Nearly all of these events were centered on a 2,400-year period between 1,000 and 3,400 years ago, when 11 of the 12 events were recorded. In contrast, between 0 to 1,000 and 3,400 to 7,000 years ago, only one and zero major hurricane strikes were recorded, respectively. According to the two researchers, a probable explanation for the "remarkable increase in hurricane frequency and intensity" that affected the Florida Panhandle and the Gulf Coast after 1400 BC would have been a continental-scale shift in circulation patterns that caused the jet stream to shift south and the Bermuda High southwest of their earlier Holocene positions, such as would be expected with global cooling, giving strength to their contention that "paleohurricane records from the past century or even the past millennium are not long enough to capture the full range of variability of catastrophic hurricane activities inherent in the Holocene climatic regime."

Last, we have the study of Donnelly and Woodruff (2007), who state that "it has been proposed that an increase in sea surface temperatures caused by anthropogenic climate change has led to an increase in the frequency of intense tropical cyclones," citing the studies of Emanuel (2005) and Webster *et al.* (2005). Donnelly and Woodruff developed "a record of intense [category 4 and greater] hurricane activity in the western North Atlantic Ocean over the past 5,000 years based on sediment cores from a Caribbean lagoon [Laguna Playa Grande on the island of Vieques, Puerto Rico] that contains coarse-grained deposits associated with intense hurricane landfalls."

Based on this work, the two researchers from the Woods Hole Oceanographic Institution detected three major intervals of intense hurricane strikes: one between 5,400 and 3,600 calendar years before present (yr BP, where "present" is AD 1950), one between 2,500 and 1,000 yr BP, and one after 250 yr BP. They also report that coral-based sea surface temperature (SST) data from Puerto Rico "indicate that mean annual Little Ice Age (250-135 yr BP or AD 1700-1815) SSTs were 2-3°C cooler than they are now," and they say that "an analysis of Caribbean hurricanes documented in Spanish archives indicates that 1766-1780 was one of the most active intervals in the period between 1500 and 1800 (Garcia-Herrera *et al.*, 2005), when tree-ring-based reconstructions indicate a negative (cooler) phase of the Atlantic Multidecadal Oscillation (Gray *et al.*, 2004)."

In light of these findings, Donnelly and Woodruff concluded that "the information available suggests that tropical Atlantic SSTs were probably not the principal driver of intense hurricane activity over the past several millennia." Indeed, there is no compelling reason to believe that the current level of intense hurricane activity is in any way unprecedented or that it has been caused by global warming. Quite to the contrary, the two researchers write that "studies relying on recent climatology indicate that North Atlantic hurricane activity is greater during [cooler] La Niña years and suppressed during [warmer] El Niño years (Gray, 1984; Bove *et al.*, 1998), due primarily to increased vertical wind shear in strong El Niño years hindering hurricane development."

In summary, millennial-scale reconstructions of intense hurricane activity within the Atlantic Basin provide no support for the claim that global warming will lead to the creation of more intense Atlantic hurricanes that will batter the east, southeast, and southern coasts of the United States. In fact, they suggest just the opposite.

Additional information on this topic, including reviews of newer publications as they become available, can be found at http://www.co2science.org/subject/h/hurratlanmill.php.

References

Bove, M.C., Elsner, J.B., Landsea, C.W., Niu, X.F. and O'Brien, J.J. 1998. Effect of El Niño on US landfalling hurricanes, revisited. *Bulletin of the American Meteorological Society* **79**: 2477-2482.

Donnelly, J.P. and Woodruff, J.D. 2007. Intense hurricane activity over the past 5,000 years controlled by El Niño and the West African Monsoon. *Nature* **447**: 465-468.

Emanuel, K. 2005. Increasing destructiveness of tropical cyclones over the past 30 years. *Nature* **436**: 686-688.

Garcia-Herrera, R., Gimeno, L., Ribera, P. and Hernandez, E. 2005. New records of Atlantic hurricanes from Spanish documentary sources. *Journal of Geophysical Research* **110**: 1-7.

Gray, S.T., Graumlich, L.J., Betancourt, J.L. and Pederson, G.T. 2004. A tree-ring-based reconstruction of the Atlantic Multidecadal Oscillation since 1567 A.D. *Geophysical Research Letters* **31**: 1-4.

Gray, W.M. 1984. Atlantic seasonal hurricane frequency. Part I: El Niño and 30 mb quasi-biennial oscillation influences. *Monthly Weather Review* **112**: 1649-1668.

Liu, K.-B. and Fearn, M.L. 1993. Lake-sediment record of late Holocene hurricane activities from coastal Alabama. *Geology* **21**: 793-796.

Liu, K.-B. and Fearn, M.L. 2000. Reconstruction of prehistoric landfall frequencies of catastrophic hurricanes in northwestern Florida from lake sediment records. *Quaternary Research* **54**: 238-245.

Webster, P.J., Holland, G.J., Curry, J.A. and Chang, H.-R. 2005. Changes in tropical cyclone number, duration, and intensity in a warming environment. *Science* **309**: 1844-1846.

6.3.1.2.2. The Past Few Centuries

Has the warming of the past century, which rescued the world from the extreme cold of the Little Ice Age, led to the formation of more numerous Atlantic Basin tropical storms and hurricanes? We review several studies that have broached this question with sufficiently long databases to provide reliable answers.

Elsner *et al.* (2000) provided a statistical and physical basis for understanding regional variations in major hurricane activity along the U.S. coastline on long timescales; in doing so, they presented data on major hurricane occurrences in 50-year intervals for Bermuda, Jamaica, and Puerto Rico. These data revealed that hurricanes occurred at lower frequencies in the last half of the twentieth century than they did in the preceding five 50-year periods, at all three of the locations studied. From 1701 to 1850, for example, when the earth was locked in the icy grip of the Little Ice Age, major hurricane frequency was 2.77 times greater at Bermuda, Jamaica, and Puerto Rico than it was from 1951 to 1998; from 1851 to 1950, when the planet was in transition from Little Ice Age to current conditions, the three locations experienced a mean hurricane frequency that was 2.15 times greater than what they experienced from 1951 to 1998.

Boose *et al.* (2001) used historical records to reconstruct hurricane damage regimes for an area composed of the six New England states plus adjoining New York City and Long Island for the period 1620-1997. In describing their findings, they wrote that "there was no clear century-scale trend in the number of major hurricanes." At lower damage levels, however, fewer hurricanes were recorded in the seventeenth and eighteenth centuries than in the nineteenth and twentieth centuries; but the three researchers concluded that "this difference is probably the result of improvements in meteorological observations and records since the early 19th century." Confining ourselves to the better records of the past 200 years, we note that the cooler nineteenth century had five of the highest-damage storms, while the warmer twentieth century had only one such storm.

Nyberg *et al.* (2007) developed a history of major (category 3-5) Atlantic hurricanes over the past 270 years based on proxy records of vertical wind shear and sea surface temperature that they derived from corals and a marine sediment core. These parameters are the primary controlling forces that set the stage for the formation of major hurricanes in the main development region westward of Africa across the tropical Atlantic and Caribbean Sea between latitudes 10 and 20°N, where 85 percent of all major hurricanes and 60 percent of all non-major hurricanes and tropical storms of the Atlantic are formed. This effort resulted in their discovering that the average frequency of major Atlantic hurricanes "decreased gradually from the 1760s until the early 1990s, reaching anomalously low values during the 1970s and 1980s." More specifically, they note that "a gradual downward trend is evident from an average of ~4.1 (1775-1785) to ~1.5 major hurricanes [per year] during the late 1960s to early 1990s," and that "the current active phase (1995-2005) is unexceptional compared to the other high-activity periods of ~1756-1774, 1780-1785, 1801-1812, 1840-1850, 1873-1890 and 1928-1933." They conclude that the recent ratcheting up of Atlantic major hurricane activity appears to be simply "a recovery to normal hurricane activity." In a commentary on Nyberg *et al.*'s paper, Elsner (2007) states that "the assumption that hurricanes are simply passive responders to climate change should be challenged."

Also noting that "global warming is postulated by some researchers to increase hurricane intensity in the north basin of the Atlantic Ocean," with the implication that "a warming ocean may increase the frequency, intensity, or timing of storms of tropical origin that reach New York State," Vermette (2007) employed the Historical Hurricane Tracks tool of the National Oceanic and Atmospheric Administration's Coastal Service Center to document all Atlantic Basin tropical cyclones that reached New York State between 1851 and 2005, in order to assess the degree of likelihood that twentieth century global warming might be influencing these storms, particularly for hurricanes but also for tropical storms, tropical depressions and extratropical storms.

This work revealed, in Vermette's words, that "a total of 76 storms of tropical origin passed over New York State between 1851 and 2005," and that of these storms, 14 were hurricanes, 27 were tropical storms, seven were tropical depressions and 28 were extratropical storms. For Long Island, he further reports that "the average frequency of hurricanes and storms of tropical origin (all types) is one in every 11 years and one in every 2 years, respectively." Also of note is his finding that storm activity was greatest in both the late nineteenth century and the late twentieth century, and the fact that "the frequency and intensity of storms in the late 20th century are similar to those of the late 19th century." As a result, Vermette concludes that "rather than a linear change, that may be associated with a global warming, the changes in recent time are following a multidecadal cycle and returning to conditions of the latter half of the 19th century." He also concludes that "yet unanswered is whether a warmer global climate of the future will take hurricane activity beyond what has been experienced in the observed record."

In a similar study, Mock (2008) developed a "unique documentary reconstruction of tropical cyclones for Louisiana, U.S.A. that extends continuously back to 1799 for tropical cyclones, and to 1779 for hurricanes." This record—which was derived from daily newspaper accounts, private diaries, plantation diaries, journals, letters, and ship records, and which was augmented "with the North Atlantic hurricane database as it pertains to all Louisiana tropical cyclones up through 2007"—is, in Mock's words, "the longest continuous tropical cyclone reconstruction conducted to date for the United States Gulf Coast." And this record reveals that "the 1820s/early 1830s and the early 1860s are the most active periods for the entire record."

In discussing his findings, the University of South Carolina researcher says that "the modern records which cover just a little over a hundred years is too short to provide a full spectrum of tropical cyclone variability, both in terms of frequency and magnitude." In addition, he states that "if a higher frequency of major hurricanes occurred in the near future in a similar manner as the early 1800s or in single years such as in 1812, 1831, and 1860, [they] would have devastating consequences for New Orleans, perhaps equaling or exceeding the impacts such as in hurricane Katrina in 2005." We also observe that the new record clearly indicates that the planet's current high levels of air temperature and CO_2 concentration cannot be blamed for the 2005 Katrina catastrophe, as both parameters were much lower when tropical cyclone and hurricane activity in that region were much higher in the early- to mid-1800s.

Around the same time, Wang and Lee (2008) used the "improved extended reconstructed" sea surface temperature (SST) data described by Smith and Reynolds (2004) for the period 1854-2006 to examine historical temperature changes over the global ocean, after which they regressed vertical wind shear—"calculated as the magnitude of the vector difference between winds at 200 mb and 850 mb during the Atlantic hurricane season (June to November), using NCEP-NCAR reanalysis data"—onto a temporal variation of global warming defined by the SST data. This work led to their discovery that warming of the surface of the global ocean is typically associated with a secular increase of tropospheric vertical wind shear in the main development region (MDR) for Atlantic hurricanes, and that the long-term increased wind shear of that region has coincided with a weak but robust downward trend in U.S. landfalling hurricanes. However, this relationship has a pattern to it, whereby local ocean warming in the Atlantic MDR actually reduces the vertical wind shear there, while "warmings in the tropical Pacific and Indian Oceans produce an opposite effect, i.e., they increase the vertical wind shear in the MDR for Atlantic hurricanes."

In light of these findings, the two researchers conclude that "the tropical oceans compete with one another for their impacts on the vertical wind shear over the MDR for Atlantic hurricanes," and they say that to this point in time, "warmings in the tropical Pacific and Indian Oceans win the competition and produce increased wind shear which reduces U.S. landfalling hurricanes." As for the years and decades ahead, they write that "whether future global warming increases the vertical wind shear in the MDR for Atlantic hurricanes will depend on the relative role induced by secular warmings over the tropical oceans."

Vecchi and Knutson (2008) write in the introduction to their study of the subject that "there is currently disagreement within the hurricane/climate community on whether anthropogenic forcing (greenhouse gases, aerosols, ozone depletion, etc.) has caused an increase in Atlantic tropical storm or hurricane frequency." In further exploring this question, they derived an estimate of the expected number of North Atlantic tropical cyclones (TCs) that were missed by the observing system in the pre-satellite era (1878-1965), after which they analyzed trends of both reconstructed TC numbers and duration over various time periods and looked at how they may or may not have been related to trends in sea surface temperature over the main development region of North Atlantic TCs. This work revealed, in their words, that "the estimated trend for 1900-2006 is highly significant (+~4.2 storms century^{-1})," but they say that the trend "is strongly influenced by a minimum in 1910-30, perhaps artificially enhancing significance." When using their base case adjustment for missed TCs and considering the entire 1878-2006 record, they find that the trend in the number of TCs is only "weakly positive" and "not statistically significant," while they note that the trend in average TC duration over the 1878-2006 period "is negative and highly significant."

Elsner (2008), in his summary of the *International Summit on Hurricanes and Climate Change* held in May 2007 on the Greek island of Crete, said the presence of more hurricanes in the northeastern Caribbean Sea "during the second half of the Little Ice Age when sea temperatures near Puerto Rico were a few degrees (Celsius) cooler than today" provides evidence that "today's warmth is not needed for increased storminess."

In conclusion, the bulk of the evidence that has been accumulated to date over multi-century timescales indicates that late twentieth century yearly hurricane numbers were considerably lower than those observed in colder prior centuries. It is by no means clear that further global warming, due to any cause, would lead to an increase or decrease in U.S. landfalling hurricanes. All we can say is that up to this point in time, global warming appears to have had a weak negative impact on their numbers.

Additional information on this topic, including reviews of newer publications as they become available, can be found at http://www.co2science.org/subject/h/hurratlancent.php.

References

Boose, E.R., Chamberlin, K.E. and Foster, D.R. 2001. Landscape and regional impacts of hurricanes in New England. *Ecological Monographs* **71**: 27-48.

Elsner, J.B. 2007. Tempests in time. *Nature* **447**: 647-649.

Elsner, J.B. 2008. Hurricanes and climate change. *Bulletin of the American Meteorological Society* **89**: 677-679.

Elsner, J.B., Liu, K.-B. and Kocher, B. 2000. Spatial variations in major U.S. hurricane activity: Statistics and a physical mechanism. *Journal of Climate* **13**: 2293-2305.

Mock, C.J. 2008. Tropical cyclone variations in Louisiana, U.S.A., since the late eighteenth century. *Geochemistry, Geophysics, Geosystems* **9**: 10.1029/2007GC001846.

Nyberg, J., Malmgren, B.A., Winter, A., Jury, M.R., Kilbourne, K.H. and Quinn, T.M. 2007. Low Atlantic hurricane activity in the 1970s and 1980s compared to the past 270 years. *Nature* **447**: 698-701.

Smith, T.M. and Reynolds, R.W. 2004. Improved extended reconstruction of SST (1854-1997). *Journal of Climate* **17**: 2466-2477.

Vecchi, G.A. and Knutson, T.R. 2008. On estimates of historical North Atlantic tropical cyclone activity. *Journal of Climate* **21**: 3580-3600.

Vermette, S. 2007. Storms of tropical origin: a climatology for New York State, USA (1851-2005). *Natural Hazards* **42**: 91-103.

Wang, C. and Lee, S.-K. 2008. Global warming and United States landfalling hurricanes. *Geophysical Research Letters* **35**: 10.1029/2007GL032396.

6.3.1.2.3. The Past Century

Have tropical storms and hurricanes of the Atlantic Ocean become more numerous over the past century, in response to what the IPCC describes as unprecedented global warming? This became a matter of intense speculation following a spike of storm occurrences in 2004-2005, but once again it is instructive to approach the question by starting with the findings of earlier research.

Bove *et al.* (1998) examined the characteristics of all recorded landfalling U.S. Gulf Coast hurricanes—defined as those whose eyes made landfall between Cape Sable, Florida and Brownsville, Texas—from 1896 to 1995. They found that the first half of this period saw considerably more hurricanes than the last half: 11.8 per decade vs. 9.4 per decade. The same was true for intense hurricanes of category 3 or more on the Saffir-Simpson storm scale: 4.8 vs. 3.6. The numbers of all hurricanes and the numbers of intense hurricanes both tended downward from 1966 to the end of the period investigated, with the decade 1986-1995 exhibiting the fewest intense hurricanes of the entire century. The three researchers concluded that "fears of increased hurricane activity in the Gulf of Mexico are premature."

Noting that the 1995 Atlantic hurricane season was one of near-record tropical storm and hurricane activity, but that during the preceding four years (1991-94) such activity over the Atlantic basin was the lowest since the keeping of reliable records began in the mid-1940s, Landsea *et al.* (1998) studied the meteorological characteristics of the two periods to determine what might have caused the remarkable upswing in storm activity in 1995. In doing so, they found that "perhaps the primary factor for the increased hurricane activity during 1995 can be attributed to a favorable large-scale pattern of extremely low vertical wind shear throughout the main development region." They also noted that "in addition to changes in the large-scale flow fields, the enhanced Atlantic hurricane activity has also been linked to below-normal sea-level pressure, abnormally warm ocean waters, and very humid values of total precipitable water."

An additional factor that may have contributed to the enhanced activity of the 1995 Atlantic hurricane season was the westerly phase of the stratospheric quasi-biennial oscillation, which is known to enhance Atlantic basin storm activity. Possibly the most important factor of all, however, was what Landsea *et al.* called the "dramatic transition from the prolonged late 1991-early 1995 warm episode (El Niño) to cold episode (La Niña) conditions," which contributed to what they described as "the dramatic reversal" of weather characteristics "which dominated during the [prior] four hurricane seasons."

"Some have asked," in the words of the four researchers, "whether the increase in hurricanes during 1995 is related to the global surface temperature increases that have been observed over the last century, some contribution of which is often ascribed to increases in anthropogenic 'greenhouse' gases." In reply, they stated that "such an interpretation is not warranted," because the various factors noted above seem sufficient to explain the observations. "Additionally," as they further wrote, "Atlantic hurricane activity has actually decreased significantly in both frequency of intense hurricanes and mean intensity of all named storms over the past few decades," and "this holds true even with the inclusion of 1995's Atlantic hurricane season."

In a major synthesis of Atlantic basin hurricane indices published the following year, Landsea *et al.* (1999) reported long-term variations in tropical cyclone activity for this region (North Atlantic Ocean, Gulf of Mexico, and Caribbean Sea). Over the period 1944-1996, decreasing trends were found for (1) the total number of hurricanes, (2) the number of intense hurricanes, (3) the annual number of hurricane days, (4) the maximum attained wind speed of all hurricane storms averaged over the course of a year, and (5) the highest wind speed associated with the strongest hurricane recorded in each year. In addition, they reported that the total number of Atlantic hurricanes making landfall in the United States had decreased over the 1899-1996 time period, and that normalized trends in hurricane damage in the United States between 1925 and 1996 revealed such damage to be decreasing at a rate of $728 million per decade.

In a similar study that included a slightly longer period of record (1935-1998), Parisi and Lund (2000) conducted a number of statistical tests on all Atlantic Basin hurricanes that made landfall in the contiguous United States, finding that "a simple linear regression of the yearly number of landfalling hurricanes on the years of study produces a trend slope estimate of -0.011 ± 0.0086 storms per year." To drive home the significance of that result, they expressly called attention to the fact that "the estimated trend slope is negative," which means, of course, that the yearly number of such storms is decreasing, which is just the opposite of what they described as the "frequent hypothesis ... that global warming is causing increased storm activity." Their statistical analysis indicates that "the trend slope is not significantly different from zero."

Contemporaneously, Easterling *et al.* (2000) noted that the mean temperature of the globe rose by about 0.6°C over the past century, and they thus looked for possible impacts of this phenomenon on extreme weather events, which if found to be increasing, as they describe it, "would add to the body of evidence that there is a discernable human affect on the climate." Their search, however, revealed few changes of significance, although they did determine that "the number of intense and landfalling Atlantic hurricanes has declined."

Lupo and Johnston (2000) found "there has been relatively little trend in the overall occurrence of hurricanes within the Atlantic Ocean Basin (62 year period)," reflecting an upward trend in category 1 hurricanes which is countered by downward or weak trends in the occurrence of category 2-5 hurricanes. Stratifying by hurricane genesis region indicated the tendency for more hurricanes to form in La Niña years during PDO1 (1977-1999) was strongly influenced by more storms being generated in the Caribbean and Eastern Atlantic. Only two storms formed in these regions during El Niño years. During PDO2 (1947-1976) there was a weak tendency for more (fewer) storms forming in the Gulf and Caribbean (West and East Atlantic) sub-regions during La Niña years, while the reverse occurred for El Niño years.

Three years later, Balling and Cerveny (2003) wrote that "many numerical modeling papers have appeared showing that a warmer world with higher sea surface temperatures and elevated atmospheric moisture levels could increase the frequency, intensity, or duration of future tropical cyclones," but that empirical studies had failed to reveal any such relationships. They also noted that "some scientists have suggested that the buildup of greenhouse gases can ultimately alter other characteristics of tropical cyclones, ranging from timing of the active season to the location of the events," and that these relationships have not been thoroughly studied with historical real-world data. They proceeded to fill this void by conducting such a study for tropical storms in the Caribbean Sea, the Gulf of Mexico, and the western North Atlantic Ocean.

More specifically, the two Arizona State University climatologists constructed a daily database of tropical storms that occurred within their study area over the period 1950-2002, generating "a variety of parameters dealing with duration, timing, and location of storm season," after which they tested for trends in these characteristics, attempting to explain the observed variances in the variables using regional, hemispheric, and global temperatures. In doing so, they "found no trends related to timing and duration of the hurricane season and geographic position of storms in the Caribbean Sea, Gulf of Mexico and tropical sector of the western North Atlantic Ocean." Likewise, they said they "could find no significant trends in these variables and generally no association with them and the local ocean, hemispheric, and global temperatures."

Elsner *et al.* (2004) conducted a changepoint analysis of time series of annual major North Atlantic hurricane counts and annual major U.S. hurricane counts for the twentieth century, which technique, in their words, "quantitatively identifies temporal shifts in the mean value of the observations." This work revealed that "major North Atlantic hurricanes have become more frequent since 1995," but at "a level reminiscent of the 1940s and 1950s." In actuality, however, they had not quite reached that level, nor had they maintained it for as long a time. Their data indicate that the mean annual hurricane count for the seven-year period 1995-2001 was 3.86, while the mean count for the 14-year period 1948-1961 was 4.14. They also reported that, "in general, twentieth-century U.S. hurricane activity shows no abrupt shifts," noting, however, that there was an exception over Florida, "where activity decreased during the early 1950s and again during the late 1960s." Last, they found that "El Niño events tend to suppress hurricane activity along the entire coast with the most pronounced effects over Florida."

In contradiction of the IPCC's claim that global warming leads to more intense hurricane activity, the results of Elsner *et al.*'s study found that not only did

North Atlantic hurricane activity not increase over the entire twentieth century, hurricane activity also did not increase in response to the more sporadic warming associated with periodic El Niño conditions.

Two years later, things got a bit more interesting. "The 2005 hurricane season," in the words of Virmani and Weisberg (2006), "saw an unprecedented number of named tropical storms since records began in 1851." Moreover, they said it followed "on the heels of the unusual 2004 hurricane season when, in addition to the first South Atlantic hurricane, a record-breaking number of major hurricanes made landfall in the United States, also causing destruction on the Caribbean islands in their path." The question they thus posed was whether these things occurred in response to recent global warming or if they bore sufficient similarities with hurricane seasons of years past to preclude such an attribution.

The two researchers determined that "latent heat loss from the tropical Atlantic and Caribbean was less in late spring and early summer 2005 than preceding years due to anomalously weak trade winds associated with weaker sea-level pressure," which phenomenon "resulted in anomalously high sea surface temperatures" that "contributed to earlier and more intense hurricanes in 2005." However, they went on to note that "these conditions in the Atlantic and Caribbean during 2004 and 2005 were not unprecedented and were equally favorable during the active hurricane seasons of 1958, 1969, 1980, 1995 and 1998." In addition, they said there was "not a clear link between the Atlantic Multidecadal Oscillation or the long term trend [of temperature] and individual active hurricane years, confirming the importance of other factors in hurricane formation."

The following year, Mann and Emanuel (2006) used quantitative records stretching back to the mid-nineteenth century to develop a positive correlation between sea surface temperatures and Atlantic basin tropical cyclone frequency for the period 1871-2005, while Holland and Webster (2007) had analyzed Atlantic tropical cyclone frequency back to 1855 and found a doubling of the number of tropical cyclones over the past 100 years. Both of these papers linked these changes to anthropogenic greenhouse warming. In a compelling rebuttal of those conclusions, however, Landsea (2007) cited a number of possible biases that may exist in the cyclone frequency trends derived in the two studies, concluding that "improved monitoring in recent years is responsible for most, if not all, of the observed trend in increasing frequency of tropical cyclones."

Parisi and Lund (2008) calculated return periods of Atlantic-basin U.S. landfalling hurricanes based on "historical data from the 1900 to 2006 period via extreme value methods and Poisson regression techniques" for each of the categories (1-5) of the Saffir-Simpson Hurricane Scale. This work revealed that return periods (in years) for these hurricanes were, in ascending Saffir-Simpson Scale category order: (1) 0.9, (2) 1.3, (3) 2.0, (4) 4.7, and (5) 23.1. In addition, the two researchers reported that corresponding non-encounter probabilities in any one hurricane season were calculated to be (1) 0.17, (2) 0.37, (3) 0.55, (4) 0.78, and (5) 0.95. They stated that the hypothesis that U.S. hurricane strike frequencies are "increasing in time" is "statistically rejected."

Lupo et al. (2008) added data for seven more years to the data originally analyzed by Lupo and Johnston (2000) and found it "did not change the major findings." The authors hypothesized that the Atlantic hurricane season of 2005 was so active, not only because of the recent increase in hurricane activity which may be associated with the PDO, but also possibly due to decreased upper tropospheric shear over the Atlantic which may have been associated with a stronger easterly phase of the quasi-biennial oscillation along with warmer-than-normal SSTs.

In light of the long history of multi-decadal to century-scale analyses that have come to the same conclusion, we must reject the oft-heard claim that Atlantic hurricanes have increased in frequency in response to twentieth century global warming.

Additional information on this topic, including reviews of newer publications as they become available, can be found at http://www.co2science.org/subject/h/hurratlangwe.php.

References

Balling Jr., R.C. and Cerveny, R.S. 2003. Analysis of the duration, seasonal timing, and location of North Atlantic tropical cyclones: 1950-2002. *Geophysical Research Letters* **30**: 10.1029/2003GL018404.

Bove, M.C., Zierden, D.F. and O'Brien, J.J. 1998. Are gulf landfalling hurricanes getting stronger? *Bulletin of the American Meteorological Society* **79**: 1327-1328.

Easterling, D.R., Evans, J.L., Groisman, P. Ya., Karl, T.R., Kunkel, K.E. and Ambenje, P. 2000. Observed variability and trends in extreme climate events: A brief review. *Bulletin of the American Meteorological Society* **81**: 417-425.

Elsner, J.B., Niu, X. and Jagger, T.H. 2004. Detecting shifts in hurricane rates using a Markov Chain Monte Carlo approach. *Journal of Climate* **17**: 2652-2666.

Holland, G.J. and Webster, P.J. 2007. Heightened tropical cyclone activity in the North Atlantic: Natural variability or climate trend? *Philosophical Transactions of the Royal Society of London, Series A* **365**: 10.1098/rsta.2007.2083.

Landsea, C.W. 2007. Counting Atlantic tropical cyclones back to 1900. *EOS: Transactions, American Geophysical Union* **88**: 197, 202.

Landsea, C.W, Bell, G.D., Gray, W.M. and Goldenberg, S.B. 1998. The extremely active 1995 Atlantic hurricane season: environmental conditions and verification of seasonal forecasts. *Monthly Weather Review* **126**: 1174-1193.

Landsea, C.N., Pielke Jr., R.A., Mestas-Nuñez, A.M. and Knaff, J.A. 1999. Atlantic basin hurricanes: Indices of climatic changes. *Climatic Change* **42**: 89-129.

Lupo, A.R., and Johnston, G. 2000. The interannual variability of Atlantic ocean basin hurricane occurrence and intensity. *National Weather Digest* **24** (1): 1-11.

Lupo, A.R., Latham, T.K., Magill, T., Clark, J.V., Melick, C.J., and Market, P.S. 2008. The interannual variability of hurricane activity in the Atlantic and east Pacific regions. *National Weather Digest* **32** (2): 119-135.

Mann, M. and Emanuel, K. 2006. Atlantic hurricane trends linked to climate change. *EOS: Transactions, American Geophysical Union* **87**: 233, 238, 241.

Parisi, F. and Lund, R. 2000. Seasonality and return periods of landfalling Atlantic basin hurricanes. *Australian & New Zealand Journal of Statistics* **42**: 271-282.

Parisi, F. and Lund, R. 2008. Return periods of continental U.S. hurricanes. *Journal of Climate* **21**: 403-410.

Virmani, J.I. and Weisberg, R.H. 2006. The 2005 hurricane season: An echo of the past or a harbinger of the future? *Geophysical Research Letters* **33**: 10.1029/2005GL025517.

6.3.1.2.4. The El Niño Effect

How does the frequency of Atlantic basin hurricanes respond to increases in ocean temperature? In exploring this important question one has to look not only at Atlantic Ocean temperatures, but also those in the eastern tropical Pacific, in particular during La Niña and El Niño conditions. Wilson (1999) utilized data from the last half of the twentieth century to determine that the probability of having three or more intense Atlantic hurricanes was only 14 percent during an El Niño year (warm temperatures in the eastern tropical Pacific), but fully 53 percent during a La Niña year (cold ocean temperatures in the eastern tropical Pacific). When ocean temperatures warm in the eastern tropical Pacific, they cause stronger upper level winds in the tropical Atlantic and a greater likelihood that storms would become sheared, and hence weaker. The opposite (weaker upper level winds) occurs during La Niña years.

Muller and Stone (2001) conducted a similar study of tropical storm and hurricane strikes along the southeast U.S. coast from South Padre Island (Texas) to Cape Hatteras (North Carolina), using data from the entire past century. For tropical storms and hurricanes together, they found an average of 3.3 strikes per La Niña season, 2.6 strikes per neutral season, and 1.7 strikes per El Niño season. For hurricanes alone, the average rate of strike occurrence ranged from 1.7 per La Niña season to 0.5 per El Niño season, which represents a frequency-of-occurrence decline of fully 70 percent in going from cooler La Niña conditions to warmer El Niño conditions. Likewise, Elsner *et al.* (2001)—who also worked with data from the entire past century—found that when there are below normal sea surface temperatures in the equatorial Pacific, "the probability of a U.S. hurricane increases."

Lyons (2004) also conducted a number of analyses of U.S. landfalling tropical storms and hurricanes, dividing them into three different groupings: the 10 highest storm and hurricane landfall years, the nine lowest such years, and all other years. These groupings revealed, in Lyons' words, that "La Niña conditions occurred 19% more often during high U.S. landfall years than during remaining years," and that "El Niño conditions occurred 10% more often during low U.S. landfall years than during remaining years." In addition, it was determined that "La Niña (El Niño) conditions were 18% (25%) more frequent during high (low) U.S. landfall years than during low (high) U.S. landfall years."

An analogous approach was used by Pielke and Landsea (1999) to study the effect of warming on the intensity of Atlantic basin hurricanes, using data from the period 1925 to 1997. In their analysis, they first determined that 22 years of this period were El Niño years, 22 were La Niña years, and 29 were neither El Niño nor La Niña years. Then, they compared the average hurricane wind speed of the cooler La Niña years with that of the warmer El Niño years, finding that in going from the cooler climatic state to the warmer climatic state, average hurricane wind speed dropped by about 6 meters per second.

Independent confirmation of these findings was provided by Pielke and Landsea's assessment of concurrent hurricane damage in the United States: El Niño years experienced only half the damage of La Niña years. And in a 10-year study of a Mediterranean waterbird (Cory's Shearwater) carried out on the other side of the Atlantic, Brichetti *et al.* (2000) determined—contrary to their own expectation—that survival rates during warmer El Niño years were greater than during cooler La Niña years.

In another pertinent study, Landsea *et al.* (1998) analyzed the meteorological circumstances associated with the development of the 1995 Atlantic hurricane season, which was characterized by near-record tropical storm and hurricane activity after four years (1991-94) that had exhibited the lowest such activity since the keeping of reliable records began. They determined that the most important factor behind this dramatic transition from extreme low to extreme high tropical storm and hurricane activity was what they called the "dramatic transition from the prolonged late 1991-early 1995 warm episode (El Niño) to cold episode (La Niña) conditions."

Last, in a twentieth century changepoint analysis of time series of major North Atlantic and U.S. annual hurricane counts, which in the words of its authors, "quantitatively identifies temporal shifts in the mean value of the observations," Elsner *et al.* (2004) found that "El Niño events tend to suppress hurricane activity along the entire coast with the most pronounced effects over Florida."

Additional information on this topic, including reviews of newer publications as they become available, can be found at http://www.co2science.org/subject/h/hurratlanelnino.php.

References

Brichetti, P., Foschi, U.F. and Boano, G. 2000. Does El Niño affect survival rate of Mediterranean populations of Cory's Shearwater? *Waterbirds* **23**: 147-154.

Elsner, J.B., Bossak, B.H. and Niu, X.F. 2001. Secular changes to the ENSO-U.S. hurricane relationship. *Geophysical Research Letters* **28**: 4123-4126.

Elsner, J.B., Niu, X. and Jagger, T.H. 2004. Detecting shifts in hurricane rates using a Markov Chain Monte Carlo approach. *Journal of Climate* **17**: 2652-2666.

Landsea, C.W, Bell, G.D., Gray, W.M. and Goldenberg, S.B. 1998. The extremely active 1995 Atlantic hurricane season: environmental conditions and verification of seasonal forecasts. *Monthly Weather Review* **126**: 1174-1193.

Lyons, S.W. 2004. U.S. tropical cyclone landfall variability: 1950-2002. *Weather and Forecasting* **19**: 473-480.

Muller, R.A. and Stone, G.W. 2001. A climatology of tropical storm and hurricane strikes to enhance vulnerability prediction for the southeast U.S. coast. *Journal of Coastal Research* **17**: 949-956.

Pielke Jr., R.A. and Landsea, C.N. 1999. La Niña, El Niño, and Atlantic hurricane damages in the United States. *Bulletin of the American Meteorological Society* **80**: 2027-2033.

Wilson, R.M. 1999. Statistical aspects of major (intense) hurricanes in the Atlantic basin during the past 49 hurricane seasons (1950-1998): Implications for the current season. *Geophysical Research Letters* **26**: 2957-2960.

6.3.2. Indian Ocean

Singh *et al.* (2000, 2001) analyzed 122 years of tropical cyclone data from the North Indian Ocean over the period 1877-1998. Since this was the period of time during which the planet recovered from the global chill of the Little Ice Age, it is logical to assume that their findings would be indicative of changes in hurricane characteristics we might expect if the earth were to warm by that amount again, which is what the IPCC is projecting.

Singh *et al.* found that on an annual basis, there was a slight decrease in tropical cyclone frequency, such that the North Indian Ocean, on average, experienced about one less hurricane per year at the end of the 122-year record in 1998 than it did at its start in 1877. In addition, based on data from the Bay of Bengal, they found that tropical cyclone numbers dropped during the months of most severe cyclone formation (November and May), when the El Niño-Southern Oscillation was in a warm phase. In light of these real-world observations, it would thus appear that if tropical cyclones of the North Indian Ocean were to change at all in response to global warming, their overall frequency and the frequency of the most intense such storms would likely decrease.

Hall (2004) analyzed characteristics of cyclones occurring south of the equator from longitude 90°E to 120°W in the South Pacific and southeast Indian Oceans, concentrating on the 2001-2002 cyclone season and comparing the results with those of the preceding four years and the 36 years before that. This work revealed that "the 2001-2002 tropical cyclone season in the South Pacific and southeast Indian Ocean was one of the quietest on record, in terms of both the number of cyclones that formed, and the impact of those systems on human affairs." In the southeast Indian Ocean, for example, Hall determined that "the overall number of depressions and tropical cyclones was below the long-term mean." Further east, he found that broad-scale convection was near or slightly above normal, but that "the proportion of tropical depressions and weak cyclones developing into severe cyclones was well below average," which result represented "a continuation of the trend of the previous few seasons." What is more, Hall writes that "in the eastern Australian region, the four-year period up to 2001-2002 was by far the quietest recorded in the past 41 years."

Raghavan and Rajesh (2003) reviewed the general state of scientific knowledge relative to trends in the frequency and intensity of tropical cyclones throughout the world, giving special attention to the Indian state of Andhra Pradesh, which borders on the Bay of Bengal. For the North Indian Ocean (NIO), comprising both the Bay of Bengal and the Arabian Sea, they report that for the period 1891-1997 there was a significant decreasing trend (at the 99 percent confidence level) in the frequency of cyclones with the designation of "cyclonic storm" and above, and that "the maximum decrease was in the last four decades," citing the work of Srivastava *et al.* (2000). In addition, they note that Singh and Khan (1999), who studied 122 years of data, also found the annual frequency of NIO-basin tropical cyclones to be decreasing.

As in other parts of the world, they found increasing impacts of tropical cyclones; but their economic analysis led them to conclude that "increasing damage due to tropical cyclones over Andhra Pradesh, India, is attributable mainly to economic and demographic factors and not to any increase in frequency or intensity of cyclones." With no equivocation, they state that "inflation, growth in population, and the increased wealth of people in the coastal areas (and not global warming) are the factors contributing to the increased impact."

Commenting on their findings, the researchers say "there is a common perception in the media, and even government and management circles, that [increased property damage from tropical cyclones] is due to an increase in tropical cyclone frequency and perhaps in intensity, probably as a result of global climate change." However, as they continue, "studies all over the world show that though there are decadal variations, there is no definite long-term trend in the frequency or intensity of tropical cyclones." They confidently state that "the specter of tropical cyclones increasing alarmingly due to global climate change, portrayed in the popular media and even in some more serious publications, does not therefore have a sound scientific basis."

Additional information on this topic, including reviews of newer publications as they become available, can be found at http://www.co2science.org/subject/h/hurricaneindian.php.

References

Hall, J.D. 2004. The South Pacific and southeast Indian Ocean tropical cyclone season 2001-02. *Australian Meteorological Magazine* **53**: 285-304.

Raghavan, S. and Rajesh, S. 2003. Trends in tropical cyclone impact: A study in Andhra Pradesh, India. *Bulletin of the American Meteorological Society* **84**: 635-644.

Singh, O.P. and Ali Khan, T.M. 1999. *Changes in the frequencies of cyclonic storms and depressions over the Bay of Bengal and the Arabian Sea*. SMRC Report 2. South Asian Association for Regional Cooperation, Meteorological Research Centre, Agargaon, Dhaka, Bangladesh.

Singh, O.P., Ali Khan, T.M. and Rahman, S. 2000. Changes in the frequency of tropical cyclones over the North Indian Ocean. *Meteorology and Atmospheric Physics* **75**: 11-20.

Singh, O.P., Ali Kahn, T.M. and Rahman, S. 2001. Has the frequency of intense tropical cyclones increased in the North Indian Ocean? *Current Science* **80**: 575-580.

Srivastava, A.K., Sinha Ray, K.C. and De, U.S. 2000. Trends in the frequency of cyclonic disturbances and their intensification over Indian seas. *Mausam* **51**: 113-118.

6.3.3. Pacific Ocean

Chu and Clark (1999) analyzed the frequency and intensity of tropical cyclones that either originated in or entered the central North Pacific (0-70°N, 140-180°W) over the 32-year period 1966-1997. They determined that "tropical cyclone activity (tropical depressions, tropical storms, and hurricanes combined) in the central North Pacific [was] on the rise." This increase, however, appears to have been due to a step-change that led to the creation of "fewer cyclones during the first half of the record (1966-81) and more during the second half of the record (1982-1997)," and accompanying the abrupt rise in tropical cyclone numbers was a similar abrupt increase in maximum hurricane intensity. Chu and Clark say the observed increase in tropical cyclone activity cannot be due to CO_2-induced global warming, because, in their words, "global warming is a gradual process" and "it cannot explain why there is a steplike change in the tropical cyclone incidences in the early 1980s."

Clearly, a much longer record of tropical cyclone activity is needed to better understand the nature of the variations documented by Chu and Clark, as well as their relationship to mean global air temperature. The beginnings of such a history were presented by Liu *et al.* (2001), who meticulously waded through a wealth of weather records from Guangdong Province in southern China, extracting data pertaining to the landfall of typhoons there since AD 975. Calibrating the historical data against instrumental observations over the period 1884-1909, they found the trends of the two datasets to be significantly correlated (r = 0.71). This observation led them to conclude that "the time series reconstructed from historical documentary evidence contains a reliable record of variability in typhoon landfalls." They proceeded to conduct a spectral analysis of the Guangdong time series and discovered an approximate 50-year cycle in the frequency of typhoon landfall that "suggests an external forcing mechanism, which remains to be identified." They also found that "the two periods of most frequent typhoon strikes in Guangdong (AD 1660-1680, 1850-1880) coincide with two of the coldest and driest periods in northern and central China during the Little Ice Age."

Looking even further back in time into the Southern Hemisphere, Hayne and Chappell (2001) studied a series of storm ridges at Curacoa Island, which were deposited over the past 5,000 years on the central Queensland shelf (18°40'S; 146°33'E), in an attempt to create a long-term history of major cyclonic events that have impacted that area. One of their stated reasons for doing so was to test the climate-model-based hypothesis that "global warming leads to an increase of cyclone frequency or intensity." They found that "cyclone frequency was statistically constant over the last 5,000 years." In addition, they could find "no indication that cyclones have changed in intensity," a finding that is inconsistent with the climate-model-based hypothesis.

In a similar study, Nott and Hayne (2001) produced a 5,000-year record of tropical cyclone frequency and intensity along a 1,500-km stretch of coastline in northeast Australia located between latitudes 13 and 24°S by geologically dating and tropographically surveying landform features left by historic hurricanes, and running numerical models to estimate storm surge and wave heights necessary to reach the landform locations. These efforts revealed that several "super-cyclones" with central pressures less than 920 hPa and wind speeds in excess of 182 kilometers per hour had occurred over the past 5,000 years at intervals of roughly 200 to 300 years in all parts of the region of their study. They also report that the Great Barrier Reef "experienced at least five such storms over the past 200 years, with the area now occupied by Cairns experiencing two super-cyclones between 1800 and 1870." The twentieth century, however, was *totally devoid* of such storms, "with only one such event (1899) since European settlement in the mid-nineteenth century."

Also noting that "many researchers have suggested that the buildup of greenhouse gases (Watson *et al.*, 2001) will likely result in a rise in sea surface temperature (SST), subsequently increasing both the number and maximum intensity of tropical cyclones (TCs)," Chan and Liu (2004) explored the validity of this assertion via an examination of pertinent real-world data. As they put it, "if the frequency of TC occurrence were to increase with increasing global air temperature, one would expect to see an increase in the number of TCs during the past few decades." Their efforts, which focused on the last four decades of the twentieth century, resulted in their finding that a number of parameters related to SST and TC activity in the Western North Pacific (WNP) "have gone through large interannual as well as interdecadal variations," and that "they also show a slight decreasing trend." In addition, they say that "no significant correlation was found between the typhoon activity parameters and local SST," and "an increase in local SST does not lead to a significant change of the number of intense TCs in the WNP, which is contrary to the results produced by many of the numerical climate models." Instead, they found that "the interannual variation of annual typhoon activity is mainly constrained by the ENSO phenomenon through the alteration of the large-scale circulation induced by the ENSO event."

In discussing their results, Chan and Liu write that the reason for the discrepancies between their real-world results and those of many of the numerical climate models likely lies in the fact that the models assume TCs are generated primarily from energy from the oceans and that a higher SST therefore would lead to more energy being transferred from the ocean to the atmosphere. "In other words," as they say, "the typhoon activity predicted in these models is almost solely determined by thermodynamic processes, as advocated by Emanuel (1999)," whereas "in the real atmosphere, dynamic factors, such as the vertical variation of the atmospheric flow (vertical wind shear) and the juxtaposition of various flow patterns that lead to different angular momentum transports, often outweigh the thermodynamic control in limiting the intensification process." Their final conclusion is that "at least for the western North Pacific, observational evidence does not support the notion that increased typhoon activity will occur with higher local SSTs."

Much the same thing was found by Free *et al.* (2004), who looked for increases in potential hurricane intensity, as they put it, "estimated from thermodynamic principles as shown in Emanuel (1986, 1995) given a record of SSTs and profiles of atmospheric temperature and humidity." This they did using radiosonde and SST data from 14 island radiosonde stations in both the tropical Pacific and Atlantic Oceans, after which they compared their results with those of Bister and Emanuel (2002) at grid points near the selected stations. They found "no significant trend in potential intensity from 1980 to 1995 and no consistent trend from 1975 to 1995." What is more, they report that between 1975 and 1980, "while SSTs rose, PI decreased, illustrating the hazards of predicting changes in hurricane intensity from projected SST changes alone."

Hall (2004) reviewed the characteristics of cyclones occurring south of the equator and eastward from longitude 90°E to 120°W in the South Pacific and southeast Indian Oceans, concentrating on the 2001-2002 cyclone season and comparing the results with those of the preceding four years and the 36 years before that. This analysis indicated that "the 2001-2002 tropical cyclone season in the South Pacific and southeast Indian Ocean was one of the quietest on record, in terms of both the number of cyclones that formed, and the impact of those systems on human affairs." In the southeast Indian Ocean, for example, he writes that "the overall number of depressions and tropical cyclones was below the long-term mean," while further east he found that broad-scale convection was near or slightly above normal, but that "the proportion of tropical depressions and weak cyclones developing into severe cyclones was well below average," which result represented "a continuation of the trend of the previous few seasons." Hall writes that "in the eastern Australian region, the four-year period up to 2001-2002 was by far the quietest recorded in the past 41 years."

Noting that "according to Walsh and Ryan (2000), future global climate trends may result in an increased incidence of cyclones," and realizing that "understanding the behavior and frequency of severe storms in the past is crucial for the prediction of future events," Yu *et al.* (2004) devised a way to decipher the history of severe storms in the southern South China Sea. Working at Youngshu Reef (9°32'-9°42'N, 112°52 -113°04'E), they used standard radiocarbon dating together with TIMS U-series dating to determine the times of occurrence of storms that were strong enough to relocate large *Porites* coral blocks that are widespread on the reef flats there. This program revealed that "during the past 1000 years, at least six exceptionally strong storms occurred," which they dated to approximately AD 1064 ± 30, 1218 ± 5, 1336 ± 9, 1443 ± 9, 1682 ± 7, and 1872 ± 15, yielding an average recurrence time of 160 years. Interestingly, none of these six severe storms occurred during the past millennium's last century, which the IPCC claims was the warmest such period of that thousand-year interval.

Noting that Emanuel (2005) and Webster *et al.* (2005) have claimed that "tropical cyclone intensity has increased markedly in recent decades," and saying that because they specifically argued that "tropical cyclone activity over the western North Pacific has been changed in response to the ongoing global warming," Ren *et al.* (2006) decided to see if any increases in tropical cyclone activity had occurred over China between 1957 and 2004. This they did by analyzing tropical cyclone (TC) precipitation (P) data from 677 Chinese weather stations for the period 1957 to 2004, searching for evidence of long-term changes in TCP and TC-induced torrential precipitation events. This search indicated, in their words, that "significant downward trends are found in the TCP volume, the annual frequency of torrential TCP events, and the contribution of TCP to the annual precipitation over the past 48 years." Also, they say that the downward trends were accompanied by "decreases in the numbers of TCs and typhoons that affected China during the period 1957-2004." In

a conclusion that consequently differs dramatically from the claims of Emanuel (2005) and Webster *et al.* (2005) relative to inferred increases in tropical cyclone activity over the western North Pacific in recent decades, Ren *et al.* say their findings "strongly suggest that China has experienced decreasing TC influence over the past 48 years, especially in terms of the TCP."

Nott *et al.* (2007) developed a 777-year-long annually resolved record of landfalling tropical cyclones in northeast Australia based on analyses of isotope records of tropical cyclone rainfall in an annually layered carbonate stalagmite from Chillagoe (17.2°S, 144.6°E) in northeast Queensland. Perhaps their most important discovery in doing so was their finding that "the period between AD 1600 to 1800"—when the Little Ice Age held sway throughout the world—"had many more intense or hazardous cyclones impacting the site than the post AD 1800 period," when the planet gradually began to warm.

Li *et al.* (2007) analyzed real-world tropical cyclone data pertaining to the western North Pacific basin archived in the *Yearbook of Typhoon* published by the China Meteorological Administration for the period 1949-2003, together with contemporaneous atmospheric information obtained from the National Center for Environmental Protection reanalysis dataset for the period 1951-2003. Following this endeavor, they used their empirical findings to infer future tropical cyclone activity in the region based upon climate-model simulations of the state of the general circulation of the atmosphere over the next half-century. This protocol revealed, first, that there were "more tropical cyclones generated over the western North Pacific from the early 1950s to the early 1970s in the 20th century and less tropical cyclones from the mid-1970s to the present." They further found that "the decadal changes of tropical cyclone activities are closely related to the decadal changes of atmospheric general circulation in the troposphere, which provide favorable or unfavorable conditions for the formation of tropical cyclones." Based on simulations of future occurrences of these favorable and unfavorable conditions derived from "a coupled climate model under the [A2 and B2] schemes of the Intergovernmental Panel on Climate Change special report on emission scenarios," they then determined that "the general circulation of the atmosphere would become unfavorable for the formation of tropical cyclones as a whole and the frequency of tropical cyclone formation would likely decrease by 5% within the next half century, although more tropical cyclones would appear during a short period of it."

Last, an analysis by Lupo *et al.* (2008) of 69 years of East Pacific tropical cyclone activity (1970 – 2007) found that there were 16.3 storms per year (9.0 hurricanes and 7.3 tropical storms), which was a greater amount of activity than found in the Atlantic Ocean basin. The long-term trend showed a slight decrease (not statistically significant) in East Pacific tropical cyclone activity. An examination of the interannual variability demonstrated that there were more East Pacific tropical cyclones during El Niño years, and that this was mainly accounted for by more storms becoming intense hurricanes than during La Niña years. The tropical cyclone season was one or two months longer in El Niño years, while more storms formed in the southeast and southwest part of the East Pacific Ocean Basin. This is likely due to the fact that ENSO years bring warmer waters to the East Pacific region. When breaking down the ENSO years by phase of the PDO, the ENSO-related differences in occurrence and intensity and geographic formation region are accentuated in PDO1 years (1977-1999), but were blurred in PDO2 (1947-1976) years. This ENSO and PDO related variability is similar to that occurring in the Atlantic (LJ00), except that in the Atlantic more storms occurred in La Niña years and they were more intense.

Additional information on this topic, including reviews of newer publications as they become available, can be found at http://www.co2science.org/subject/h/hurricanepacific.php.

References

Bister, M. and Emanuel, K. 2002. Low frequency variability of tropical cyclone potential intensity. 1. Interannual to interdecadal variability. *Journal of Geophysical Research* **107**: 10.1029/2001JD000776.

Chan, J.C.L. and Liu, K.S. 2004. Global warming and western North Pacific typhoon activity from an observational perspective. *Journal of Climate* **17**: 4590-4602.

Chu, P.-S. and Clark, J.D. 1999. Decadal variations of tropical cyclone activity over the central North Pacific. *Bulletin of the American Meteorological Society* **80**: 1875-1881.

Emanuel, K.A. 1986. An air-sea interaction theory for tropical cyclones. Part I: Steady-state maintenance. *Journal of the Atmospheric Sciences* **43**: 585-604.

Emanuel, K.A. 1995. Sensitivity of tropical cyclones to surface exchange coefficients and a revised steady-state model incorporating eye dynamics. *Journal of the Atmospheric Sciences* **52**: 3969-3976.

Emanuel, K.A. 1999. Thermodynamic control of hurricane intensity. *Nature* **401**: 665-669.

Emanuel, K.A. 2005. Increasing destructiveness of tropical cyclones over the past 30 years. *Nature* **436**: 686-688.

Free, M., Bister, M. and Emanuel, K. 2004. Potential intensity of tropical cyclones: Comparison of results from radiosonde and reanalysis data. *Journal of Climate* **17**: 1722-1727.

Hall, J.D. 2004. The South Pacific and southeast Indian Ocean tropical cyclone season 2001-02. *Australian Meteorological Magazine* **53**: 285-304.

Hayne, M. and Chappell, J. 2001. Cyclone frequency during the last 5000 years at Curacoa Island, north Queensland, Australia. *Palaeogeography, Palaeoclimatology, Palaeoecology* **168**: 207-219.

Li, Y., Wang, X., Yu, R. and Qin, Z. 2007. Analysis and prognosis of tropical cyclone genesis over the western North Pacific on the background of global warming. *Acta Oceanologica Sinica* **26**: 23-34.

Liu, K.-B., Shen, C. and Louie, K.-S. 2001. A 1,000-year history of typhoon landfalls in Guangdong, southern China, reconstructed from Chinese historical documentary records. *Annals of the Association of American Geographers* **91**: 453-464.

Lupo, A.R., Latham, T.K., Magill, T., Clark, J.V., Melick, C.J., and Market, P.S. 2008. The interannual variability of hurricane activity in the Atlantic and east Pacific regions. *National Weather Digest* **32** (2): 119-135.

Nott, J., Haig, J., Neil, H. and Gillieson, D. 2007. Greater frequency variability of landfalling tropical cyclones at centennial compared to seasonal and decadal scales. *Earth and Planetary Science Letters* **255**: 367-372.

Nott, J. and Hayne, M. 2001. High frequency of 'super-cyclones' along the Great Barrier Reef over the past 5,000 years. *Nature* **413**: 508-512.

Ren, F., Wu, G., Dong, W., Wang, X., Wang, Y., Ai, W. and Li, W. 2006. Changes in tropical cyclone precipitation over China. *Geophysical Research Letters* **33**: 10.1029/2006GL027951.

Walsh, K.J.E. and Ryan, B.F. 2000. Tropical cyclone intensity increase near Australia as a result of climate change. *Journal of Climate* **13**: 3029-3036.

Watson, R.T. and the Core Writing Team (Eds.) 2001. *Climate Change 2001: Synthesis Report.* Cambridge University Press, Cambridge, UK.

Webster, P.J., Holland, G.J, Curry, J.A. and Chang, H.-R. 2005. Changes in tropical cyclone number, duration, and intensity in a warming environment. *Science* **309**: 1844-1846.

Yu, K.-F., Zhao, J.-X., Collerson, K.D., Shi, Q., Chen, T.-G., Wang, P.-X. and Liu, T.-S. 2004. Storm cycles in the last millennium recorded in Yongshu Reef, southern South China Sea. *Palaeogeography, Palaeoclimatology, Palaeoecology* **210**: 89-100.

6.3.4. Global

Although some climate models suggest the intensity and frequency of tropical cyclones on a global scale may be significantly reduced in response to global warming (Bengtsson *et al.*, 1996), thus implying a "decrease in the global total number of tropical cyclones on doubling CO_2," as noted by Sugi *et al.* (2002), most of them suggest otherwise. Free *et al.* (2004) state that "increases in hurricane intensity are expected to result from increases in sea surface temperature and decreases in tropopause-level temperature accompanying greenhouse warming (Emanuel, 1987; Henderson-Sellers *et al.*, 1998; Knutson *et al.*, 1998)."

In an early review of empirical evidence related to the subject, Walsh and Pittock (1998) concluded that "the effect of global warming on the number of tropical cyclones is presently unknown," and "there is little relationship between SST (sea surface temperature) and tropical cyclone numbers in several regions of the globe." They opined there was "little evidence that changes in SSTs, by themselves, could cause change in tropical cyclone numbers."

In a second early analysis of the topic, Henderson-Sellers *et al.* (1998) determined that (1) "there are no discernible global trends in tropical cyclone number, intensity, or location from historical data analyses," (2) "global and mesoscale-model-based predictions for tropical cyclones in greenhouse conditions have not yet demonstrated prediction skill," and (3) "the popular belief that the region of cyclogenesis will expand with the 26°C SST isotherm is a fallacy."

Six years later, Free *et al.* (2004) looked for increases in "potential" hurricane intensity and found "no significant trend in potential intensity from 1980 to 1995 and no consistent trend from 1975 to 1995."

What is more, they report that between 1975 and 1980, "while SSTs rose, PI decreased, illustrating the hazards of predicting changes in hurricane intensity from projected SST changes alone."

In another review of what real-world data have to say about the subject, Walsh (2004) was once again forced to report "there is as yet no convincing evidence in the observed record of changes in tropical cyclone behavior that can be ascribed to global warming." Nevertheless, Walsh continued to believe that (1) "there is likely to be some increase in maximum tropical cyclone intensities in a warmer world," (2) "it is probable that this would be accompanied by increases in mean tropical cyclone intensities," and (3) "these increases in intensities are likely to be accompanied by increases in peak precipitation rates of about 25%," putting the date of possible detection of these increases "some time after 2050," little knowing that two such claims would actually be made the very next year.

Emanuel (2005) claimed to have found that a hurricane power dissipation index had increased by approximately 50 percent for both the Atlantic basin and the Northwest Pacific basin since the mid 1970s, and Webster *et al.* (2005) contended the numbers of Category 4 and 5 hurricanes for all tropical cyclone basins had nearly doubled between an earlier (1975-1989) and a more recent (1990-2004) 15-year period. However, in a challenge to both of these claims, Klotzbach (2006) wrote that "many questions have been raised regarding the data quality in the earlier part of their analysis periods," and he thus proceeded to perform a new analysis based on a "near-homogeneous" global dataset for the period 1986-2005.

Klotzbach first tabulated global tropical cyclone (TC) activity using best track data—which he describes as "the best estimates of the locations and intensities of TCs at six-hour intervals produced by the international warning centers"—for all TC basins (North Atlantic, Northeast Pacific, Northwest Pacific, North Indian, South Indian, and South Pacific), after which he determined trends of worldwide TC frequency and intensity over the period 1986-2005, during which time global SSTs are purported to have risen by about 0.2-0.4°C. This work did indeed indicate, in his words, "a large increasing trend in tropical cyclone intensity and longevity for the North Atlantic basin," but it also indicated "a considerable decreasing trend for the Northeast Pacific." Combining these observations with the fact that "all other basins showed small trends," he determined there had been "no significant change in global net tropical cyclone activity" over the past two decades. With respect to Category 4 and 5 hurricanes, however, he found there had been a "small increase" in their numbers from the first half of the study period (1986-1995) to the last half (1996-2005); but he noted that "most of this increase is likely due to improved observational technology." Klotzbach said his findings were "contradictory to the conclusions drawn by Emanuel (2005) and Webster *et al.* (2005)," in that the global TC data did "not support the argument that global TC frequency, intensity and longevity have undergone increases in recent years."

Following close on the heels of Klotzbach's study came the paper of Kossin *et al.* (2007), who wrote that "the variability of the available data combined with long time-scale changes in the availability and quality of observing systems, reporting policies, and the methods utilized to analyze the data make the best track records inhomogeneous," and stated that this "known lack of homogeneity in both the data and techniques applied in the post-analyses has resulted in skepticism regarding the consistency of the best track intensity estimates." Consequently, as an important first step in resolving this problem, Kossin *et al.* "constructed a more homogeneous data record of hurricane intensity by first creating a new consistently analyzed global satellite data archive from 1983 to 2005 and then applying a new objective algorithm to the satellite data to form hurricane intensity estimates," after which they analyzed the resultant homogenized data for temporal trends over the period 1984-2004 for all major ocean basins and the global ocean as a whole.

The five scientists who conducted the work said that "using a homogeneous record, we were not able to corroborate the presence of upward trends in hurricane intensity over the past two decades in any basin other than the Atlantic." Therefore, noting that "the Atlantic basin accounts for less than 15% of global hurricane activity," they concluded that "this result poses a challenge to hypotheses that directly relate globally increasing tropical sea surface temperatures to increases in long-term mean global hurricane intensity." They deliver another major blow to the contentions of Emanuel (2005) and Webster *et al.* (2005) when they say "the question of whether hurricane intensity is globally trending upwards in a warming climate will likely remain a point of debate in the foreseeable future."

As a result of the many investigations of the subject that have been conducted over the past several years, there currently appears to be no factual basis for claiming that planet-wide hurricane frequency and/or intensity will rise in response to potential future global warming. Nevertheless, parties pushing for restrictions on anthropogenic CO_2 emissions continue to do so, citing the now-rebutted claims of Emanuel (2005) and Webster *et al.* (2005).

Additional information on this topic, including reviews of newer publications as they become available, can be found at http://www.co2science.org/subject/h/hurricaneglobal.php.

References

Bengtsson, L., Botzet, M. and Esch, M. 1996. Will greenhouse gas-induced warming over the next 50 years lead to higher frequency and greater intensity of hurricanes? *Tellus* **48A**: 57-73.

Bister, M. and Emanuel, K. 2002. Low frequency variability of tropical cyclone potential intensity. 1. Interannual to interdecadal variability. *Journal of Geophysical Research* **107**: 10.1029/2001JD000776.

Emanuel, K.A. 1986. An air-sea interaction theory for tropical cyclones. Part I: Steady-state maintenance. *Journal of the Atmospheric Sciences* **43**: 585-604.

Emanuel, K.A. 1987. The dependence of hurricane intensity on climate. *Nature* **326**: 483-485.

Emanuel, K.A. 1995. Sensitivity of tropical cyclones to surface exchange coefficients and a revised steady-state model incorporating eye dynamics. *Journal of the Atmospheric Sciences* **52**: 3969-3976.

Emanuel, K. 2005. Increasing destructiveness of tropical cyclones over the past 30 years. *Nature* **436**: 686-688.

Free, M., Bister, M. and Emanuel, K. 2004. Potential intensity of tropical cyclones: Comparison of results from radiosonde and reanalysis data. *Journal of Climate* **17**: 1722-1727.

Henderson-Sellers, A., Zhang, H., Berz, G., Emanuel, K., Gray, W., Landsea, C., Holland, G., Lighthill, J., Shieh, S.-L., Webster, P. and McGuffie, K. 1998. Tropical cyclones and global climate change: A post-IPCC assessment. *Bulletin of the American Meteorological Society* **79**: 19-38.

Klotzbach, P.J. 2006. Trends in global tropical cyclone activity over the past twenty years (1986-2005). *Geophysical Research Letters* **33**: 10.1029/2006GL025881.

Knutson, T., Tuleya, R. and Kurihara, Y. 1998. Simulated increase of hurricane intensities in a CO_2-warmed climate. *Science* **279**: 1018-1020.

Kossin, J.P., Knapp, K.R., Vimont, D.J., Murnane, R.J. and Harper, B.A. 2007. A globally consistent reanalysis of hurricane variability and trends. *Geophysical Research Letters* **34**: 10.1029/2006GL028836.

Sugi, M., Noda, A. and Sato, N. 2002. Influence of the global warming on tropical cyclone climatology: an experiment with the JMA global model. *Journal of the Meteorological Society of Japan* **80**: 249-272.

Walsh, K. 2004. Tropical cyclones and climate change: unresolved issues. *Climate Research* **27**: 77-83.

Walsh, K. and Pittock, A.B. 1998. Potential changes in tropical storms, hurricanes, and extreme rainfall events as a result of climate change. *Climatic Change* **39**: 199-213.

Webster, P.J., Holland, G.J., Curry, J.A. and Chang, H.-R. 2005. Changes in tropical cyclone number, duration, and intensity in a warming environment. *Science* **309**: 1844-1846.

6.4. ENSO

Computer model simulations have given rise to three claims regarding the influence of global warming on El Niño/Southern Oscillation (ENSO) events: (1) global warming will increase the frequency of ENSO events, (2) global warming will increase the intensity of ENSO events, and (3) weather-related disasters will be exacerbated under El Niño conditions. Here, we test the validity of these assertions, demonstrating they are in conflict with the observational record. We begin by highlighting studies that suggest the virtual world of ENSO, as simulated by state-of-the-art climate models, is at variance with reality.

Additional information on this topic, including reviews on ENSO not discussed here, can be found at http://www.co2science.org/subject/e/subject_e.php under the heading ENSO.

6.4.1. Model Inadequacies

In a comparison of 24 coupled ocean-atmosphere climate models, Latif *et al.* (2001) report that "almost all models (even those employing flux corrections) still have problems in simulating the SST [sea surface temperature] climatology." They also note that "only a few of the coupled models simulate the El Niño/Southern Oscillation (ENSO) in terms of gross equatorial SST anomalies realistically." And they state that "no model has been found that simulates realistically all aspects of the interannual SST variability." Because "changes in sea surface temperature are both the cause and consequence of wind fluctuations," and because these phenomena figure prominently in the El Niño-La Niña oscillation, it is not surprising that Fedorov and Philander (2000) conclude that current climate models do not do a good job of determining the potential effects of global warming on ENSO.

Human ignorance likely also plays a role in the models' failure to simulate ENSO. According to Overpeck and Webb (2000), there is evidence that "ENSO may change in ways that we do not yet understand," which "ways" have clearly not yet been modeled. White *et al.* (2001), for example, found that "global warming and cooling during earth's internal mode of interannual climate variability [the ENSO cycle] arise from fluctuations in the global hydrological balance, not the global radiation balance," and that these fluctuations are the result of no known forcing of either anthropogenic or extraterrestrial origin, although Cerveny and Shaffer (2001) make a case for a lunar forcing of ENSO activity, which also is not included in any climate model.

Another example of the inability of today's most sophisticated climate models to properly describe El Niño events is provided by Landsea and Knaff (2000), who employed a simple statistical tool to evaluate the skill of 12 state-of-the-art climate models in real-time predictions of the development of the 1997-98 El Niño. They found that the models exhibited essentially no skill in forecasting this very strong event at lead times ranging from 0 to eight months. They also determined that no models were able to anticipate even one-half of the actual amplitude of the El Niño's peak at a medium range lead-time of six to 11 months. They state that "since no models were able to provide useful predictions at the medium and long ranges, *there were no models that provided both useful and skillful forecasts for the entirety of the 1997-98 El Niño*" [italics in the original].

Given the inadequacies listed above, it is little wonder several scientists have criticized model simulations of current ENSO behavior, including Walsh and Pittock (1998), who say "there is insufficient confidence in the predictions of current models regarding any changes in ENSO," and Fedorov and Philander (2000), who say "at this time, it is impossible to decide which, if any, are correct." As a result, there is also little reason to believe that current climate models can correctly predict ENSO behavior under future conditions of changed climate.

Additional information on this topic, including reviews of newer publications as they become available, can be found at http://www.co2science.org/subject/e/ensomo.php.

References

Cerveny, R.S. and Shaffer, J.A. 2001. The moon and El Niño. *Geophysical Research Letters* **28**: 25-28.

Fedorov, A.V. and Philander, S.G. 2000. Is El Niño changing? *Science* **288**: 1997-2002.

Landsea, C.W. and Knaff, J.A. 2000. How much skill was there in forecasting the very strong 1997-98 El Niño? *Bulletin of the American Meteorological Society* **81**: 2107-2119.

Latif, M., Sperber, K., Arblaster, J., Braconnot, P., Chen, D., Colman, A., Cubasch, U., Cooper, C., Delecluse, P., DeWitt, D., Fairhead, L., Flato, G., Hogan, T., Ji, M., Kimoto, M., Kitoh, A., Knutson, T., Le Treut, H., Li, T., Manabe, S., Marti, O., Mechoso, C., Meehl, G., Power, S., Roeckner, E., Sirven, J., Terray, L., Vintzileos, A., Voss, R., Wang, B., Washington, W., Yoshikawa, I., Yu, J. and Zebiak, S. 2001. ENSIP: the El Niño simulation intercomparison project. *Climate Dynamics* **18**: 255-276.

Overpeck, J. and Webb, R. 2000. Nonglacial rapid climate events: Past and future. *Proceedings of the National Academy of Sciences USA* **97**: 1335-1338.

Walsh, K. and Pittock, A.B. 1998. Potential changes in tropical storms, hurricanes, and extreme rainfall events as a result of climate change. *Climatic Change* **39**: 199-213.

White, W.B., Cayan, D.R., Dettinger, M.D. and Auad, G. 2001. Sources of global warming in upper ocean temperature during El Niño. *Journal of Geophysical Research* **106**: 4349-4367.

6.4.2. Relationship to Extreme Weather

Changnon (1999) determined that adverse weather events attributed to the El Niño of 1997-98 negatively affected the United States economy to the tune of $4.5 billion and contributed to the loss of 189 lives, which is serious indeed. On the other hand, he determined that El Niño-related benefits amounted to approximately $19.5 billion—resulting primarily from reduced energy costs, increased industry sales, and the lack of normal hurricane damage—and that a total of 850 lives were *saved* due to the reduced

amount of bad winter weather. Thus, the net effect of the 1997-98 El Niño on the United States, according to Changnon, was "surprisingly positive," in stark contrast to what was often reported in the media and by some commentators who tend, in his words, "to focus only on the negative outcomes."

Another of the "surprisingly positive" consequences of El Niños is their tendency to moderate Atlantic hurricane frequencies. Working with data from 1950 to 1998, Wilson (1999) determined that the probability of having three or more intense hurricanes during a warmer El Niño year was approximately 14 percent, while during a cooler non-El Niño year the probability jumped to 53 percent. Similarly, in a study of tropical storm and hurricane strikes along the southeast coast of the United States over the entire last century, Muller and Stone (2001) determined that "more tropical storm and hurricane events can be anticipated during La Niña seasons [3.3 per season] and fewer during El Niño seasons [1.7 per season]." And in yet another study of Atlantic basin hurricanes, this one over the period 1925 to 1997, Pielke and Landsea (1999) reported that average hurricane wind speeds during warmer El Niño years were about six meters per second lower than during cooler La Niña years. In addition, they reported that hurricane damage during cooler La Niña years was twice as great as during warmer El Niño years. These year-to-year variations thus indicate that, if anything, hurricane frequency and intensity—as well as damage—tend to decrease under warmer El Niño conditions.

Much the same story is being said of other parts of the world. In the North Indian Ocean, Singh *et al*. (2000) studied tropical cyclone data pertaining to the period 1877-1998, finding that tropical cyclone frequency there declined during the months of most severe cyclone formation—November and May—when ENSO was in a warm phase. In New Zealand, De Lange and Gibb (2000) studied storm surge events recorded by several tide gauges in Tauranga Harbor over the period 1960-1998, finding a considerable decline in both the annual number of such events and their magnitude in the latter (warmer) half of the nearly four-decade-long record, additionally noting that La Niña seasons typically experienced more storm surge days than El Niño seasons. And in Australia, Kuhnel and Coates (2000) found that over the period 1876-1991, yearly fatality event-days due to floods, bushfires, and heatwaves were greater in cooler La Niña years than in warmer El Niño years.

Zuki and Lupo (2008), when examining Southern South China Sea (SSCS) data on tropical storms and cyclones for interannual variability, found La Niña years were more active and El Niño years were less active than other years, and this result was significant at the 95 percent confidence level when examining the total sample. The variability of tropical storms and tropical cyclones of local origin was similar to that of the total sample. There was no apparent climatic variability (statistically significant) in the SSCS that could be attributed to interdecadal variability such as the PDO. A spectral analysis of the filtered climatological background variables such as SST, SLP, 200–850 hPa wind shear, 850 hPa divergence and 850 hPa vorticity showed that there was significant variability found in the 3–7 year period, which is consistent with that of the ENSO period.

Zuki and Lupo then examined a subset of the most active years (all La Niña and "cold" neutral years) versus those years with no tropical cyclone activity for the five years of warmest SSTs (predominantly El Niño years) and coolest SSTs. They found that during warm non-active SST years, tropical cyclone activity was likely suppressed as the low-level relative vorticity was considerably more anticyclonic, even though SSTs were about one standard deviation warmer and wind shears were similar to those of active years. The SSCS atmospheric environment for warm SST non-active years was drier than that of the active years, and did not exhibit a surface–500 hPa structure that would be as supportive of warm-core tropical cyclones. Most of these years were also ENSO years, and two thirds of all years with no activity were El Nino or warm neutral.

Apparently, even birds seem to know the dangers of La Niña vs. El Niño. In a study of breeding populations of Cory's Shearwaters on the Tremiti Islands of Italy, for example, Brichetti *et al.* (2000) found that, contrary to even their hypothesis, survival rates during El Niño years were greater than during La Niña years.

Additional information on this topic, including reviews of newer publications as they become available, can be found at http://www.co2science.org/subject/e/ensoew.php.

References

Brichetti, P., Foschi, U.F. and Boano, G. 2000. Does El Niño affect survival rate of Mediterranean populations of Cory's Shearwater? *Waterbirds* **23**: 147-154.

Changnon, S.A. 1999. Impacts of 1997-98 El Niño-generated weather in the United States. *Bulletin of the American Meteorological Society* **80**: 1819-1827.

De Lange, W.P. and Gibb, J.G. 2000. Seasonal, interannual, and decadal variability of storm surges at Tauranga, New Zealand. *New Zealand Journal of Marine and Freshwater Research* **34**: 419-434.

Kuhnel, I. and Coates, L. 2000. El Niño-Southern Oscillation: Related probabilities of fatalities from natural perils in Australia. *Natural Hazards* **22**: 117-138.

Muller, R.A. and Stone, G.W. 2001. A climatology of tropical storm and hurricane strikes to enhance vulnerability prediction for the southeast U.S. coast. *Journal of Coastal Research* **17**: 949-956.

Pielke Jr., R.A. and Landsea, C.N. 1999. La Niña, El Niño, and Atlantic hurricane damages in the United States. *Bulletin of the American Meteorological Society* **80**: 2027-2033.

Singh, O.P., Ali Khan, T.M. and Rahman, M.S. 2000. Changes in the frequency of tropical cyclones over the North Indian Ocean. *Meteorology and Atmospheric Physics* **75**: 11-20.

Wilson, R.M. 1999. Statistical aspects of major (intense) hurricanes in the Atlantic basin during the past 49 hurricane seasons (1950-1998): Implications for the current season. *Geophysical Research Letters* **26**: 2957-2960.

Zuki, Z. and Lupo, A.R. 2008. The interannual variability of tropical cyclone activity in the southern South China Sea. *Journal of Geophysical Research* **113**: D06106, doi:10.1029/2007JD009218.

6.4.3. Relationship to Global Warming

All of the claims regarding the influence of global warming on ENSO events are derived from climate model simulations. Timmermann *et al.* (1999), for example, developed a global climate model which, according to them, operates with sufficient resolution to address the issue of whether "human-induced 'greenhouse' warming affects, or will affect, ENSO." When running this model with increasing greenhouse-gas concentrations, more frequent El-Niño-like conditions do indeed occur. However, this is not what observational data reveal to be the case. The frequent and strong El Niño activity of the recent past is no different from that of a number of other such episodes of prior centuries, when it was colder than it is today, as described in several of the papers highlighted below. And in many instances, the El Niño activity of the recent past is shown to be inferior to that of colder times.

Evans *et al.* (2002) reconstructed gridded Pacific Ocean sea surface temperatures from coral stable isotope ($\delta^{18}O$) data, from which they assessed ENSO activity over the period 1607-1990. The results of their analysis showed that a period of relatively vigorous ENSO activity over the colder-than-present period of 1820-1860 was "similar to [that] observed in the past two decades." Likewise, in a study that was partly based upon the instrumental temperature record for the period 1876-1996, Allan and D'Arrigo (1999) found four persistent El Niño sequences similar to that of the 1990s; using tree-ring proxy data covering the period 1706 to 1977, they found several other ENSO events of prolonged duration. There were four or five persistent El Niño sequences in each of the eighteenth and nineteenth centuries, which were both significantly colder than the final two decades of the twentieth century, leading them to conclude there is "no evidence for an enhanced greenhouse influence in the frequency or duration of 'persistent' ENSO event sequences."

Brook *et al.* (1999) analyzed the layering of couplets of inclusion-rich calcite over inclusion-free calcite, and darker aragonite over clear aragonite, in two stalagmites from Anjohibe Cave in Madagascar, comparing their results with historical records of El Niño events and proxy records of El Niño events and sea surface temperatures derived from ice core and coral data. This exercise revealed that the cave-derived record of El Niño events compared well with the historical and proxy ice core and coral records; these data indicated, in Brook *et al.*'s words, that "the period 1700-50 possibly witnessed the highest frequency of El Niño events in the last four and a half centuries while the period 1780-1930 was the longest period of consistently high El Niño occurrences," both of which periods were cooler than the 1980s and 1990s.

In another multi-century study, Meyerson *et al.* (2003) analyzed an annually dated ice core from the South Pole that covered the period 1487-1992, specifically focusing on the marine biogenic sulfur species methanesulfonate (MS), after which they used orthogonal function analysis to calibrate the high-resolution MS series with associated environmental

series for the period of overlap (1973-92). This procedure allowed them to derive a five-century history of ENSO activity and southeastern Pacific sea-ice extent, the latter of which parameters they say "is indicative of regional temperatures within the Little Ice Age period in the southeastern Pacific sea-ice sector."

In analyzing these records, Meyerson *et al.* noted a shift at about 1800 towards generally cooler conditions. This shift was concurrent with an increase in the frequency of El Niño events in the ice core proxy record, which is contrary to what is generally predicted by climate models. On the other hand, their findings were harmonious with the historical El Niño chronologies of both South America (Quinn and Neal, 1992) and the Nile region (Quinn, 1992; Diaz and Pulwarty, 1994), which depict, in their words, "increased El Niño activity during the period of the Little Ice Age (nominally 1400-1900) and decreased El Niño activity during the Medieval Warm Period (nominally 950-1250)," as per Anderson (1992) and de Putter *et al.*, 1998).

Taking a little longer look back in time were Cobb *et al.* (2003), who generated multi-century monthly resolved records of tropical Pacific climate variability over the last millennium by splicing together overlapping fossil-coral records from the central tropical Pacific, which exercise allowed them "to characterize the range of natural variability in the tropical Pacific climate system with unprecedented fidelity and detail." In doing so, they discovered that "ENSO activity in the seventeenth-century sequence [was] not only stronger, but more frequent than ENSO activity in the late twentieth century." They also found "there [were] 30-yr intervals during both the twelfth and fourteenth centuries when ENSO activity [was] greatly reduced relative to twentieth-century observations." Once again, we have evidence of a situation where ENSO activity was much greater and more intense during the cold of the Little Ice Age than the warmth of the late twentieth century.

Inching still further back in time, Eltahir and Wang (1999) used water-level records of the Nile River as a proxy for El Niño episodes over the past 14 centuries. This approach indicated that although the frequency of El Niño events over the 1980s and 1990s was high, it was not without precedent, being similar to values observed near the start of the twentieth century and much the same as those "experienced during the last three centuries of the first millennium," which latter period, according to Esper *et al.* (2002), was also cooler than the latter part of the twentieth century.

Woodroffe *et al.* (2003) found pretty much the same thing, but over an even longer period of time. Using oxygen isotope ratios obtained from *Porites* microatolls at Christmas Island in the central Pacific to provide high-resolution proxy records of ENSO variability since 3.8 thousand years ago (ka), they found, in their words, that "individual ENSO events in the late Holocene [3.8-2.8 ka] appear at least as intense as those experienced in the past two decades." In addition, they note that "geoarcheological evidence from South America (Sandweiss *et al.*, 1996), Ecuadorian varved lake sediments (Rodbell *et al.*, 1999), and corals from Papua New Guinea (Tudhope *et al.*, 2001) indicate that ENSO events were considerably weaker or absent between 8.8 and 5.8 ka," which was the warmest part of the Holocene. In fact, they report that "faunal remains from archeological sites in Peru (Sandweiss *et al.*, 2001) indicate that the onset of modern, rapid ENSO recurrence intervals was achieved only after ~4-3 ka," or during the long cold interlude that preceded the Roman Warm Period (McDermott *et al.*, 2001).

Also concentrating on the mid to late Holocene were McGregor and Gagan (2004), who used several annually resolved fossil *Porites* coral $\delta^{18}O$ records to investigate the characteristics of ENSO events over a period of time in which the earth cooled substantially. For comparison, study of a modern coral core provided evidence of ENSO events for the period 1950-1997, the results of which analysis suggest they occurred at a rate of 19 events/century. The mid-Holocene coral $\delta^{18}O$ records, on the other hand, showed reduced rates of ENSO occurrence: 12 events/century for the period 7.6-7.1 ka, eight events/century for the period 6.1-5.4 ka, and six events/century at 6.5 ka. For the period 2.5-1.7 ka, however, the results were quite different, with all of the coral records revealing, in the words of McGregor and Gagan, "large and protracted $\delta^{18}O$ anomalies indicative of particularly severe El Niño events." They note specifically that "the 2.5 ka Madang PNG coral records a protracted 4-year El Niño, like the 1991-1994 event, but almost twice the amplitude of [the] 1997-1998 event (Tudhope *et al.*, 2001)." In addition, they say that "the 2 ka Muschu Island coral $\delta^{18}O$ record shows a severe 7-year El Niño, longer than any recorded Holocene or modern event." And they add that "the 1.7 ka *Porites* microatoll of Woodroffe *et al.* (2003) also records an extreme El Niño that was twice the amplitude of the 1997-1998

event." Taken together, these several sets of results portray what McGregor and Gagan describe as a "mid-Holocene El Niño suppression and late Holocene amplification."

That there tend to be fewer and weaker ENSO events during warm periods has further been documented by Riedinger *et al.* (2002). In a 7,000-year study of ENSO activity in the vicinity of the Galapagos Islands, they determined that "mid-Holocene [7,130 to 4,600 yr BP] El Niño activity was infrequent," when, of course, global air temperature was significantly warmer than it is now, but that both the "frequency and intensity of events increased at about 3100 yr BP," when it finally cooled below current temperatures. Throughout the former 2,530-year warm period, their data revealed the existence of 23 strong to very strong El Niños and 56 moderate events; while throughout the most recent (and significantly colder) 3,100-year period, they identified 80 strong to very strong El Niños and 186 moderate events. These numbers correspond to rates of 0.9 strong and 2.2 moderate occurrences per century in the earlier warm period and 2.7 strong and 6.0 moderate occurrences per century in the latter cool period, suggestive of an approximate tripling of the rate of occurrence of both strong and moderate El Niños in going from the warmth of the Holocene "Climatic Optimum" to the colder conditions of the past three millennia.

Similar results have been reported by Andrus *et al.* (2002) and Moy *et al.* (2002). According to Andrus *et al.*, sea surface temperatures off the coast of Peru some 6,000 years ago were 3° to 4°C warmer than what they were over the decade of the 1990s and provided little evidence of any El Niño activity. Nearby, Moy *et al.* analyzed a sediment core from lake Laguna Pallcacocha in the southern Ecuadorian Andes, producing a proxy measure of ENSO over the past 12,000 years. For the moderate and strong ENSO events detected by their analytical techniques (weaker events are not registered), these researchers state that "the overall trend exhibited in the Pallcacocha record includes a low concentration of events in the early Holocene, followed by increasing occurrence after 7,000 cal. yr BP, with peak event frequency occurring at ~1,200 cal. yr BP," after which the frequency of events declines dramatically to the present.

With respect to the last 1,200 years of this record, the decline in the frequency of ENSO events is anything but smooth. In coming out of the Dark Ages Cold Period, which was one of the coldest intervals of the Holocene (McDermott *et al.*, 2001), the number of

ENSO events experienced by the earth drops by an order of magnitude, from a high of approximately 33 events per 100 yr to a low of about three events per 100 yr, centered approximately on the year AD 1000, which is right in the middle of the Medieval Warm Period, as delineated by the work of Esper *et al.* (2002). Then, at approximately AD 1250, the frequency of ENSO events exhibits a new peak of approximately 27 events per 100 yr in the midst of the longest sustained cold period of the Little Ice Age, again as delineated by the work of Esper *et al.* Finally, ENSO event frequency declines in zigzag fashion to a low on the order of four to five events per 100 yr at the start of the Current Warm Period, which according to the temperature history of Esper *et al.* begins at about 1940.

Going even further back in time, in a study of a recently revised New England varve chronology derived from proglacial lakes formed during the recession of the Laurentide ice sheet some 17,500 to 13,500 years ago, Rittenour *et al.* (2000) determined that "the chronology shows a distinct interannual band of enhanced variability suggestive of El Niño-Southern Oscillation (ENSO) teleconnections into North America during the late Pleistocene, when the Laurentide ice sheet was near its maximum extent ... during near-peak glacial conditions." But during the middle of the Holocene, when it was considerably warmer, even than it is today, Overpeck and Webb (2000) report that data from corals suggest that "interannual ENSO variability, as we now know it, was substantially reduced, or perhaps even absent."

In summing up the available evidence pertaining to the effect of temperature on the frequency of occurrence and strength of ENSO events, we have one of the most sophisticated climate models ever developed to deal with the ENSO phenomenon implying that global warming will promote more frequent El Niño-like conditions. But we also have real-world observations demonstrating that El Niño-like conditions during the latter part of the twentieth century (claimed by the IPCC to be the warmest period of the past 1,300 years) are not much different from those that occurred during much colder times. In addition, we have a number of long-term records that suggest that when the earth was significantly warmer than it is currently, ENSO events were substantially reduced or perhaps even absent.

Additional information on this topic, including reviews of newer publications as they become available, can be found at http://www.co2science.org/subject/e/ensogw.php.

References

Allan, R.J. and D'Arrigo, R.D. 1999. "Persistent" ENSO sequences: How unusual was the 1990-1995 El Niño? *The Holocene* **9**: 101-118.

Anderson, R.Y. 1992. Long-term changes in the frequency of occurrence of El Niño events. In: Diaz, H.F. and Markgraf, V. (Eds.) *El Niño. Historical and Paleoclimatic Aspects of the Southern Oscillation.* Cambridge University Press, Cambridge, UK, pp. 193-200.

Andrus, C.F.T., Crowe, D.E., Sandweiss, D.H., Reitz, E.J. and Romanek, C.S. 2002. Otolith $\delta^{18}O$ record of mid-Holocene sea surface temperatures in Peru. *Science* **295**: 1508-1511.

Brook, G.A., Rafter, M.A., Railsback, L.B., Sheen, S.-W. and Lundberg, J. 1999. A high-resolution proxy record of rainfall and ENSO since AD 1550 from layering in stalagmites from Anjohibe Cave, Madagascar. *The Holocene* **9**: 695-705.

Cobb, K.M., Charles, C.D., Cheng, H. and Edwards, R.L. 2003. El Niño/Southern Oscillation and tropical Pacific climate during the last millennium. *Nature* **424**: 271-276.

de Putter, T., Loutre, M.-F. and Wansard, G. 1998. Decadal periodicities of Nile River historical discharge (A.D. 622-1470) and climatic implications. *Geophysical Research Letters* **25**: 3195-3197.

Diaz, H.F. and Pulwarty, R.S. 1994. An analysis of the time scales of variability in centuries-long ENSO-sensitive records of the last 1000 years. *Climatic Change* **26**: 317-342.

Eltahir, E.A.B. and Wang, G. 1999. Nilometers, El Niño, and climate variability. *Geophysical Research Letters* **26**: 489-492.

Esper, J., Cook, E.R. and Schweingruber, F.H. 2002. Low-frequency signals in long tree-ring chronologies for reconstructing past temperature variability. *Science* **295**: 2250-2253.

Evans, M.N., Kaplan, A. and Cane, M.A. 2002. Pacific sea surface temperature field reconstruction from coral $\delta^{18}O$ data using reduced space objective analysis. *Paleoceanography* **17**: U71-U83.

McDermott, F., Mattey, D.P. and Hawkesworth, C. 2001. Centennial-scale Holocene climate variability revealed by a high-resolution speleothem delta^{18}O record from SW Ireland. *Science* **294**: 1328-1331.

McGregor, H.V. and Gagan, M.K. 2004. Western Pacific coral $\delta^{18}O$ records of anomalous Holocene variability in the El Niño-Southern Oscillation. *Geophysical Research Letters* **31**: 10.1029/2004GL019972.

Meyerson, E.A., Mayewski, P.A., Kreutz, K.J., Meeker, D., Whitlow, S.I. and Twickler, M.S. 2003. The polar expression of ENSO and sea-ice variability as recorded in a South Pole ice core. *Annals of Glaciology* **35**: 430-436.

Moy, C.M., Seltzer, G.O., Rodbell, D.T. and Anderson D.M. 2002. Variability of El Niño/Southern Oscillation activity at millennial timescales during the Holocene epoch. *Nature* **420**: 162-165.

Overpeck, J. and Webb, R. 2000. Nonglacial rapid climate events: Past and future. *Proceedings of the National Academy of Sciences USA* **97**: 1335-1338.

Quinn, W.H. 1992. A study of Southern Oscillation-related climatic activity for A.D. 622-1990 incorporating Nile River flood data. In: Diaz, H.F. and Markgraf, V. (Eds.) *El Niño. Historical and Paleoclimatic Aspects of the Southern Oscillation.* Cambridge University Press, Cambridge, UK, pp. 119-149.

Quinn, W.H. and Neal, V.T. 1992. The historical record of El Niño events. In: Bradley, R.S. and Jones, P.D. (Eds.) *Climate Since A.D. 1500.* Routledge, London, UK, pp. 623-648.

Riedinger, M.A., Steinitz-Kannan, M., Last, W.M. and Brenner, M. 2002. A ~6100 ^{14}C yr record of El Niño activity from the Galapagos Islands. *Journal of Paleolimnology* **27**: 1-7.

Rittenour, T.M., Brigham-Grette, J. and Mann, M.E. 2000. El Niño-like climate teleconnections in New England during the late Pleistocene. *Science* **288**: 1039-1042.

Rodbell, D.T., Seltzer, G.O., Abbott, M.B., Enfield, D.B. and Newman, J.H. 1999. A 15,000-year record of El Niño-driven alluviation in southwestern Ecuador. *Science* **283**: 515-520.

Sandweiss, D.H., Richardson III, J.B., Reitz, E.J., Rollins, H.B. and Maasch, K.A. 1996. Geoarchaeological evidence from Peru for a 5000 years BP onset of El Niño. *Science* **273**: 1531-1533.

Sandweiss, D.H., Maasch, K.A., Burger, R.L., Richardson III, J.B., Rollins, H.B. and Clement, A. 2001. Variation in Holocene El Niño frequencies: Climate records and cultural consequences in ancient Peru. *Geology* **29**: 603-606.

Timmermann, A., Oberhuber, J., Bacher, A., Esch, M., Latif, M. and Roeckner, E. 1999. Increased El Niño frequency in a climate model forced by future greenhouse warming. *Nature* **398**: 694-696.

Tudhope, A.W., Chilcott, C.P., McCuloch, M.T., Cook, E.R., Chappell, J., Ellam, R.M., Lea, D.W., Lough, J.M. and Shimmield, G.B. 2001. Variability in the El Niño-Southern Oscillation through a glacial-interglacial cycle. *Science* **291**: 1511-1517.

Woodroffe, C.D., Beech, M.R. and Gagan, M.K. 2003. Mid-late Holocene El Niño variability in the equatorial Pacific from coral microatolls. *Geophysical Research Letters* **30**: 10.1029/2002GL 015868.

6.5. Precipitation Variability

The IPCC contends that global warming is responsible for causing greater variability in precipitation, leading to more droughts and floods. In this section we review empirical research on precipitation patterns in Africa, Asia, and North America.

Additional information on this topic, including reviews on precipitation not discussed here, can be found at http://www.co2science.org/subject/p/subject _p.php under the heading Precipitation.

6.5.1. Africa

Nicholson and Yin (2001) described climatic and hydrologic conditions in equatorial East Africa from the late 1700s to close to the present, based on histories of the levels of 10 major African lakes. They also used a water balance model to infer changes in rainfall associated with the different conditions, concentrating most heavily on Lake Victoria. This work revealed "two starkly contrasting climatic episodes." The first, which began sometime prior to 1800 and was characteristic of Little Ice Age conditions, was one of "drought and desiccation throughout Africa." This arid episode, which was most extreme during the 1820s and 1830s, was accompanied by extremely low lake levels. As the two researchers describe it, "Lake Naivash was reduced to a puddle ... Lake Chad was desiccated ... Lake Malawi was so low that local inhabitants traversed dry land where a deep lake now resides ... Lake Rukwa [was] completely desiccated ... Lake Chilwa, at its southern end, was very low and nearby Lake Chiuta almost dried up." Throughout this period, they report that "intense droughts were ubiquitous." Some were "long and severe enough to force the migration of peoples and create warfare among various tribes."

As the Little Ice Age's grip on the world began to loosen in the mid to latter part of the 1800s, however, things began to improve for most of the continent. Nicholson and Yin report that "semi-arid regions of

Mauritania and Mali experienced agricultural prosperity and abundant harvests ... the Niger and Senegal Rivers were continually high; and wheat was grown in and exported from the Niger Bend region." Across the length of the northern Sahel, maps and geographical reports described the presence of "forests." As the nineteenth century came to an end and the twentieth century began, there was a slight lowering of lake levels, but nothing like what had occurred a century earlier (i.e., variability was much reduced). And then, in the latter half of the twentieth century, things once again began to pick up for the Africans, with the levels of some of the lakes rivaling the high-stands characteristic of the years of transition to the Current Warm Period.

Concentrating on the more recent past, Nicholson (2001) says the most significant climatic change has been "a long-term reduction in rainfall in the semi-arid regions of West Africa," which has been "on the order of 20 to 40% in parts of the Sahel." There have been, she says, "three decades of protracted aridity," and "nearly all of Africa has been affected ... particularly since the 1980s." However, she goes on to note that "the rainfall conditions over Africa during the last 2 to 3 decades are not unprecedented," and that "a similar dry episode prevailed during most of the first half of the 19th century."

Describing the situation in more detail, Nicholson says "the 3 decades of dry conditions evidenced in the Sahel are not in themselves evidence of irreversible global change," because a longer historical perspective indicates an even longer period of similar dry conditions occurred between 1800 and 1850. This remarkable dry period occurred when the earth was still in the clutches of the Little Ice Age, a period of cold that is without precedent in at least the past 6,500 years, even in Africa (Lee-Thorp et al., 2001). There is no reason to think that the most recent two- to three-decade Sahelian drought was unusual or that it was caused by the higher temperatures of that period.

Also taking a longer view of the subject were Nguetsop et al. (2004), who developed a high-resolution proxy record of West African precipitation based on analyses of diatoms recovered from a sediment core retrieved from Lake Ossa, West Cameroon, which they describe as "the first paleohydrological record for the last 5500 years in the equatorial near-coastal area, east of the Guinean Gulf." They reported that this record provides evidence for alternating periods of increasing and decreasing precipitation "at a millennial time scale for the last 5500 years," which oscillatory behavior they interpret as being "a result of south/northward shifts of the Intertropical Convergence Zone," specifically noting that "a southward shift of the ITCZ, combined with strengthened northern trade winds, was marked by low and high precipitation at the northern subtropics and the subequatorial zone, respectively," and that "these events occurred in coincidence with cold spells in the northern Atlantic."

Most recently, Therrell et al. (2006) developed "the first tree-ring reconstruction of rainfall in tropical Africa using a 200-year regional chronology based on samples of Pterocarpus angolensis [a deciduous tropical hardwood known locally as Mukwa] from Zimbabwe." This record revealed that "a decadal-scale drought reconstructed from 1882 to 1896 matches the most severe sustained drought during the instrumental period (1989-1995)," and that "an even more severe drought is indicated from 1859 to 1868 in both the tree-ring and documentary data." They report, for example, that the year 1860, which exhibited the lowest reconstructed rainfall value during this period, was described in a contemporary account from Botswana (where part of their tree-ring chronology originated) as "a season of 'severe and universal drought' with 'food of every description' being 'exceedingly scarce' and the losses of cattle being 'very severe' (Nash and Endfield, 2002)." At the other end of the moisture spectrum, they report that "a 6-year wet period at the turn of the nineteenth century (1897-1902) exceeds any wet episode during the instrumental era."

Additional information on this topic, including reviews of newer publications as they become available, can be found at http://www.co2science.org/subject/p/variabilafrica.php.

References

Lee-Thorp, J.A., Holmgren, K., Lauritzen, S.-E., Linge, H., Moberg, A., Partridge, T.C., Stevenson, C. and Tyson, P.D. 2001. Rapid climate shifts in the southern African interior throughout the mid to late Holocene. *Geophysical Research Letters* **28**: 4507-4510.

Nash, D.J. and Endfield, G.H. 2002. A 19th-century climate chronology for the Kalahari region of central southern Africa derived from missionary correspondence. *International Journal of Climatology* **22**: 821-841.

Nguetsop, V.F., Servant-Vildary, S. and Servant, M. 2004. Late Holocene climatic changes in west Africa, a high resolution diatom record from equatorial Cameroon. *Quaternary Science Reviews* **23**: 591-609.

Nicholson, S.E. 2001. Climatic and environmental change in Africa during the last two centuries. *Climate Research* **17**: 123-144.

Nicholson, S.E. and Yin, X. 2001. Rainfall conditions in equatorial East Africa during the Nineteenth Century as inferred from the record of Lake Victoria. *Climatic Change* **48**: 387-398.

Therrell, M.D., Stahle, D.W., Ries, L.P. and Shugart, H.H. 2006. Tree-ring reconstructed rainfall variability in Zimbabwe. *Climate Dynamics* **26**: 677-685.

6.5.2. Asia

Pederson *et al.* (2001) developed tree-ring chronologies for northeastern Mongolia and used them to reconstruct annual precipitation and streamflow histories for the period 1651-1995. Working with both standard deviations and five-year intervals of extreme wet and dry periods, they found that "variations over the recent period of instrumental data are not unusual relative to the prior record." They note, however, that their reconstructions "appear to show more frequent extended wet periods in more recent decades," but they say that this observation "does not demonstrate unequivocal evidence of an increase in precipitation as suggested by some climate models." Spectral analysis of the data also revealed significant periodicities around 12 and 20-24 years, which they suggested may constitute "possible evidence for solar influences in these reconstructions for northeastern Mongolia."

Kripalani and Kulkarni (2001) studied seasonal summer monsoon (June-September) rainfall data from 120 east Asia stations for the period 1881-1998. A series of statistical tests they applied to these data revealed the presence of short-term variability in rainfall amounts on decadal and longer time scales, the longer "epochs" of which were found to last for about three decades over China and India and for approximately five decades over Japan. With respect to long-term trends, however, none was detected. Consequently, the history of summer rainfall trends in east Asia does not support claims of intensified monsoonal conditions in this region as a result of CO_2-induced global warming. As for the decadal variability inherent in the record, the two researchers say it "appears to be just a part of natural climate variations."

Taking a much longer look at the Asian monsoon were Ji *et al.* (2005), who used reflectance spectroscopy on a sediment core taken from a lake in the northeastern part of the Qinghai-Tibetan Plateau to obtain a continuous high-resolution proxy record of the Asian monsoon over the past 18,000 years. This project indicated that monsoonal moisture since the late glacial period had been subject to "continual and cyclic variations," among which was a "very abrupt onset and termination" of a 2,000-year dry spell that started about 4,200 years ago (yr BP) and ended around 2,300 yr BP. Other variations included the well-known centennial-scale cold and dry spells of the Dark Ages Cold Period (DACP) and Little Ice Age (LIA), which lasted from 2,100 yr BP to 1,800 yr BP and 780 yr BP to 400 yr BP, respectively, while sandwiched between them was the warmer and wetter Medieval Warm Period, and preceding the DACP was the Roman Warm Period. Time series analyses of the sediment record also revealed several statistically significant periodicities (123, 163, 200, and 293 years, all above the 95 percent level), with the 200-year cycle matching the de Vries or Suess solar cycle, implying that changes in solar activity are important triggers for some of the recurring precipitation changes in that part of Asia. It is clear that large and abrupt fluctuations in the Asian monsoon have occurred numerous times and with great regularity throughout the Holocene, and that the sun played an important role in orchestrating them.

Also working on the Tibetan Plateau were Shao *et al.* (2005), who used seven Qilian juniper ring-width chronologies from the northeastern part of the Qaidam Basin to reconstruct a thousand-year history of annual precipitation there. In doing so, they discovered that annual precipitation had fluctuated at various intervals and to various degrees throughout the entire past millennium. Wetter periods occurred between 1520 and 1633, as well as between 1933 and 2001, although precipitation has declined somewhat since the 1990s. Drier periods, on the other hand, occurred between 1429 and 1519 and between 1634 and 1741. With respect to variability, the scientists report that the magnitude of variation in annual precipitation was about 15 mm before 1430, increased to 30 mm between 1430 and 1850, and declined thereafter to the present.

Based on analyses of tree-ring width data and their connection to large-scale atmospheric circulation, Touchan *et al.* (2005) developed summer (May-August) precipitation reconstructions for several parts of the eastern Mediterranean region (Turkey, Syria, Lebanon, Cyprus, and Greece) that extend back in time anywhere from 115 to 600 years. Over the latter length of time, they found that May-

August precipitation varied on multiannual and decadal timescales, but that on the whole there were no long-term trends. The longest dry period occurred in the late sixteenth century (1591-1595), while there were two extreme wet periods: 1601-1605 and 1751-1755. In addition, both extremely strong and weak precipitation events were found to be more variable over the intervals 1520-1590, 1650-1670, and 1850-1930. The results of this study demonstrate there was nothing unusual or unprecedented about late twentieth century precipitation events in the eastern Mediterranean part of Asia that would suggest a CO_2 influence.

Last, Davi *et al.* (2006) used absolutely dated tree-ring-width chronologies obtained from five sampling sites in west-central Mongolia to derive individual precipitation models, the longest of which stretches from 1340 to 2002, additionally developing a reconstruction of streamflow that extends from 1637 to 1997. In the process of doing so, they discovered there was "much wider variation in the long-term tree-ring record than in the limited record of measured precipitation," which for the region they studied covers the period from 1937 to 2003. In addition, they say their streamflow history indicates that "the wettest 5-year period was 1764-68 and the driest period was 1854-58," while "the most extended wet period [was] 1794-1802 and ... extended dry period [was] 1778-83." For this part of Mongolia, therefore, which the researchers say is "representative of the central Asian region," there is no support to be found for the contention that the "unprecedented warming" of the twentieth century has led to increased variability in precipitation and streamflow.

These several findings suggest that either there is nothing unusual about Asia's current degree of warmth, i.e., it is not unprecedented relative to that of the early part of the past millennium, or unprecedented warming need not lead to unprecedented precipitation or unprecedented precipitation variability ... or both of the above. We conclude that the findings of this study and of others reviewed in this section provide no support for the contention that global warming leads to greater and more frequent precipitation extremes in Asia.

Additional information on this topic, including reviews of newer publications as they become available, can be found at http://www.co2science.org/subject/p/variabilasia.php.

References

Davi, N.K., Jacoby, G.C., Curtis, A.E. and Baatarbileg, N. 2006. Extension of drought records for Central Asia using tree rings: West-Central Mongolia. *Journal of Climate* **19**: 288-299.

Ji, J., Shen, J., Balsam, W., Chen, J., Liu, L. and Liu, X. 2005. Asian monsoon oscillations in the northeastern Qinghai-Tibet Plateau since the late glacial as interpreted from visible reflectance of Qinghai Lake sediments. *Earth and Planetary Science Letters* **233**: 61-70.

Kripalani, R.H. and Kulkarni, A. 2001. Monsoon rainfall variations and teleconnections over south and east Asia. *International Journal of Climatology* **21**: 603-616.

Pederson, N., Jacoby, G.C., D'Arrigo, R.D., Cook, E.R. and Buckley, B.M. 2001. Hydrometeorological reconstructions for northeastern Mongolia derived from tree rings: 1651-1995. *Journal of Climate* **14**: 872-881.

Shao, X., Huang, L., Liu, H., Liang, E., Fang, X. and Wang, L. 2005. Reconstruction of precipitation variation from tree rings in recent 1000 years in Delingha, Qinghai. *Science in China Series D: Earth Sciences* **48**: 939-949.

Touchan, R., Xoplaki, E., Funkhouser, G., Luterbacher, J., Hughes, M.K., Erkan, N., Akkemik, U. and Stephan, J. 2005. Reconstructions of spring/summer precipitation for the Eastern Mediterranean from tree-ring widths and its connection to large-scale atmospheric circulation. *Climate Dynamics* **25**: 75-98.

6.5.3. North America

Cronin *et al.* (2000) studied salinity gradients across sediment cores extracted from Chesapeake Bay, the largest estuary in the United Sates, in an effort to determine precipitation variability in the surrounding watershed over the prior millennium. They discovered there was a high degree of decadal and multidecadal variability in moisture conditions over the 1,000-year period, with regional precipitation totals fluctuating by between 25 percent and 30 percent, often in extremely rapid shifts occurring over about a decade. They also determined that precipitation was generally greater over the past two centuries than it was over the eight previous centuries, with the exception of a portion of the Medieval Warm Period (AD 1250-1350), when the climate was extremely wet. In addition, they found that the region surrounding Chesapeake Bay had experienced several "mega-droughts" lasting from 60-70 years, some of which the researchers say "were more severe than twentieth century droughts." Likewise, across the continent, Haston and Michaelsen (1997) developed a 400-year history of precipitation for 29 stations in coastal and near-interior California between San Francisco Bay and the U.S.-Mexican border using tree-ring chronologies; their work also revealed that "region-wide precipitation during the last 100 years has been unusually high and less variable compared to other periods in the past."

Crossing the continent yet again, and dropping down to the Caribbean Sea, Watanabe *et al.* (2001) analyzed delta $^{18}O/^{16}O$ and Mg/Ca ratios in cores obtained from a coral in an effort designed to examine seasonal variability in sea surface temperature and salinity there during the Little Ice Age. In doing so, they found that sea surface temperatures during this period were about 2°C colder than they are currently, while sea surface salinity exhibited greater variability than it does now, indicating that during the Little Ice Age "wet and dry seasons were more pronounced."

In Canada, Zhang *et al.* (2001) analyzed the spatial and temporal characteristics of extreme precipitation events for the period 1900-1998, using what they describe as "the most homogeneous long-term dataset currently available for Canadian daily precipitation." This exercise indicated that decadal-scale variability was a dominant feature of both the frequency and intensity of extreme precipitation events, but it provided "no evidence of any significant long-term changes" in these indices during the twentieth century. Their analysis of precipitation totals (extreme and non-extreme) did reveal a slightly increasing trend across Canada during the period of study, but it was found to be due to increases in the number of non-heavy precipitation events. Consequently, the researchers concluded that "increases in the concentration of atmospheric greenhouse gases during the twentieth century have not been associated with a generalized increase in extreme precipitation over Canada."

Dropping down into the Uinta Basin Watershed of northeastern Utah, Gray *et al.* (2004) used cores extracted from 107 piñon pines at four different sites to develop a proxy record of annual (June to June) precipitation spanning the period AD 1226-2001. They report that "single-year dry events before the instrumental period tended to be more severe than those after 1900," and that decadal-scale dry events were longer and more severe prior to 1900 as well. In particular, they found that "dry events in the late 13th, 16th, and 18th centuries surpass the magnitude and duration of droughts seen in the Uinta Basin after

1900." At the other end of the spectrum, they report that the twentieth century contained two of the strongest wet intervals (1938-1952 and 1965-1987), although the two periods were only the seventh and second most intense wet regimes, respectively, of the entire record. Hence, it would appear that in conjunction with twentieth century global warming, precipitation extremes (both high and low) within the Uinta Basin of northeastern Utah have become attenuated as opposed to amplified.

Last, we come to the study of Rasmussen *et al.* (2006), who had previously demonstrated that "speleothems from the Guadalupe Mountains in southeastern New Mexico are annually banded, and variations in band thickness and mineralogy can be used as a record of regional relative moisture (Asmerom and Polyak, 2004)." In their new study, they continued this tack, concentrating on "two columnar stalagmites collected from Carlsbad Cavern (BC2) and Hidden Cave (HC1) in the Guadalupe Mountains."

The three researchers report that "both records, BC2 and HC1, suggest periods of dramatic precipitation variability over the last 3000 years, exhibiting large shifts *unlike anything seen in the modern record* [our italics]." Second, they report that the time interval from AD 900-1300 coincides with the well-known Medieval Warm Period and "shows dampened precipitation variability and overall drier conditions" that are "consistent with the idea of more frequent La Niña events and/or negative PDO phases causing elevated aridity in the region during this time." Third, they indicate that the preceding and following colder centuries "show increased precipitation variability ... coinciding with increased El Niño flooding events."

Clearly, moisture extremes in North America much greater than those observed in the modern era are neither unusual nor manmade; they are simply a normal part of earth's natural climatic variability. In this regard, North America is like Africa and Asia: Precipitation variability in the Current Warm Period is no greater than what was experienced in earlier times.

Additional information on this topic, including reviews of newer publications as they become available, can be found at http://www.co2science.org/subject/p/variabilnortham.php.

References

Asmerom, Y. and Polyak, V.J. 2004. Comment on "A test of annual resolution in stalagmites using tree rings." *Quaternary Research* **61**: 119-121.

Cronin, T., Willard, D., Karlsen, A., Ishman, S., Verardo, S., McGeehin, J., Kerhin, R., Holmes, C., Colman, S. and Zimmerman, A. 2000. Climatic variability in the eastern United States over the past millennium from Chesapeake Bay sediments. *Geology* **28**: 3-6.

Gray, S.T., Jackson, S.T. and Betancourt, J.L. 2004. Tree-ring based reconstructions of interannual to decadal scale precipitation variability for northeastern Utah since 1226 A.D. *Journal of the American Water Resources Association* **40**: 947-960.

Haston, L. and Michaelsen, J. 1997. Spatial and temporal variability of southern California precipitation over the last 400 yr and relationships to atmospheric circulation patterns. *Journal of Climate* **10**: 1836-1852.

Rasmussen, J.B.T., Polyak, V.J. and Asmerom, Y. 2006. Evidence for Pacific-modulated precipitation variability during the late Holocene from the southwestern USA. *Geophysical Research Letters* **33**: 10.1029/2006GL025714.

Watanabe, T., Winter, A. and Oba, T. 2001. Seasonal changes in sea surface temperature and salinity during the Little Ice Age in the Caribbean Sea deduced from Mg/Ca and $^{18}O/^{16}O$ ratios in corals. *Marine Geology* **173**: 21-35.

Zhang, X., Hogg, W.D. and Mekis, E. 2001. Spatial and temporal characteristics of heavy precipitation events over Canada. *Journal of Climate* **14**: 1923-1936.

6.6. Storms

Among the highly publicized changes in weather phenomena that are predicted to attend the ongoing rise in the air's CO_2 content are increases in the frequency and severity of all types of storms. Many researchers have examined historical and proxy records in an attempt to determine how temperature changes over the past millennium or two have affected this aspect of earth's climate. This section reviews what some of them have learned about storm trends, focusing on Europe and North America.

A number of studies have reported increases in North Atlantic storminess over the last two decades of the twentieth century (Jones *et al.*, 1997; Gunther *et al.*, 1998; Dickson *et al.*, 2000). Since the IPCC claims this period was the warmest of the past millennium, this observation might appear to vindicate their view of the subject. When much longer time periods are considered, however, the storminess of the twentieth century is found to be not uncommon and even mild compared to times when temperatures and CO_2 levels were lower.

Dawson *et al.* (2002) searched daily meteorological records from Stornoway (Outer Hebrides), Lerwick (Shetland Islands), Wick (Caithness), and Fair Isle (west of the Shetland Islands) for all data pertaining to gale-force winds over the period 1876-1996, which they used to construct a history of storminess for that period for northern and northwestern Scotland. This history indicated that although North Atlantic storminess and associated wave heights had indeed increased over the prior two decades, storminess in the North Atlantic region "was considerably more severe during parts of the nineteenth century than in recent decades." In addition, whereas the modern increase in storminess appeared to be associated with a spate of substantial positive values of the North Atlantic Oscillation (NAO) index, they say "this was not the case during the period of exceptional storminess at the close of the nineteenth century." During that earlier period, the conditions that fostered modern storminess were apparently overpowered by something even more potent, i.e., cold temperatures, which in the view of Dawson *et al.* led to an expansion of sea ice in the Greenland Sea that expanded and intensified the Greenland anticyclone, which in turn led to the North Atlantic cyclone track being displaced farther south. Additional support for this view is provided by the hypothesis propounded by Clarke *et al.* (2002), who postulated that a southward spread of sea ice and polar water results in an increased thermal gradient between 50°N and 65°N that intensifies storm activity in the North Atlantic and supports dune formation in the Aquitaine region of southwest France.

The results of these two studies suggest that the increased storminess and wave heights observed in the European sector of the North Atlantic Ocean over the past two decades are not the result of global warming. Rather, they are associated with the most recent periodic increase in the NAO index. Furthermore, a longer historical perspective reveals that North Atlantic storminess was even more severe than it is now during the latter part of the nineteenth century, when it was significantly colder than it is now. In fact, the storminess of that much colder period was so great that it was actually decoupled from the NAO index. Hence, the long view of history suggests that the global warming of the past century or so has actually led to an overall *decrease* in North Atlantic storminess.

Additional evidence for the recent century-long decrease in storminess in and around Europe comes from the study of Bijl *et al.* (1999), who analyzed long-term sea-level records from several coastal stations in northwest Europe. According to these researchers, "although results show considerable natural variability on relatively short (decadal) time scales," there is "no sign of a significant increase in storminess ... over the complete time period of the data sets." In the southern portion of the North Sea, however, where natural variability was more moderate, they did find a trend, but it was "a tendency towards a weakening of the storm activity over the past 100 years."

Much the same results were obtained by Pirazzoli (2000), who analyzed tide-gauge, wind, and atmospheric pressure data over the period 1951-1997 for the northern portion of the Atlantic coast of France. In that study, the number of atmospheric depressions (storms) and strong surge winds were found to be decreasing in frequency. In addition, it was reported that "ongoing trends of climate variability show a decrease in the frequency and hence the gravity of coastal flooding."

Tide-gauge data also have been utilized as proxies for storm activity in England. Based on high-water measurements made at the Liverpool waterfront over the period 1768-1999, Woodworth and Blackman (2002) found that the annual maximum surge-at-high-water declined at a rate of 0.11 ± 0.04 meters per century, suggesting that the winds responsible for producing high storm surges were much stronger and/or more common during the early part of the record (colder Little Ice Age) than the latter part (Current Warm Period).

On a somewhat different front, and quite a ways inland, Bielec (2001) analyzed thunderstorm data from Cracow, Poland for the period 1896-1995, finding an average of 25 days of such activity per year, with a non-significant linear-regression-derived increase of 1.6 storm days from the beginning to the end of the record. From 1930 onward, however, the trend was negative, revealing a similarly derived decrease of 1.1 storm days. In addition, there was a decrease in the annual number of thunderstorms with hail over the entire period and a decrease in the frequency of storms producing precipitation in excess of 20 mm.

In introducing a study they conducted in Switzerland, Stoffel *et al.* (2005) noted that debris

flows are a type of mass movement that frequently causes major destruction in alpine areas; and they reported that since 1987 there had been an apparent above-average occurrence of such events in the Valais region of the Swiss Alps, which had prompted some researchers to suggest that the increase was the result of global warming (Rebetez *et al.*, 1997). Consequently, Stoffel *et al.* used dendrochronological methods to determine if the recent increase in debris-flow events was indeed unusual, and if it appeared that it was, to see if it made sense to attribute it to CO_2-induced global warming.

In extending the history of debris-flow events (1922-2002) back to the year 1605, they found that "phases with accentuated activity and shorter recurrence intervals than today existed in the past, namely after 1827 and until the late nineteenth century." What is more, the nineteenth century period of high-frequency debris flow was shown to coincide with a period of higher flood activity in major Swiss rivers, while less frequent debris flow activity after 1922 corresponded with lower flooding frequencies. In addition, debris flows from extremely large mass movement events, similar to what occurred in 1993, were found to have "repeatedly occurred" in the historical past, and to have been of such substantial magnitude that, in the opinion of Stoffel *et al.*, the "importance of the 1993 debris-flow surges has to be thoroughly revised." They reported that debris flows occurred "ever more frequently in the nineteenth century than they do today" and concluded that "correlations between global warming and modifications in the number or the size of debris-flow events, as hypothesized by, e.g., Haeberli and Beniston (1998), cannot, so far, be confirmed in the study area."

Noting that "a great amount of evidence for changing storminess over northwestern Europe is based on indirect data and reanalysis data rather than on station wind data," Smits *et al.* (2005) investigated trends in storminess over the Netherlands based on hourly records of 10-m wind speed observations made at 13 meteorological stations scattered across the country that have uninterrupted records for the time period 1962-2002. This effort led to their discovery that "results for moderate wind events (that occur on average 10 times per year) and strong wind events (that occur on average twice a year) indicate a decrease in storminess over the Netherlands [of] between 5 and 10% per decade."

Moving cross-continent to the south and west, Raicich (2003) analyzed 62 years of sea-level data for the period 1 July 1939 to 30 June 2001 at Trieste, in the Northern Adriatic, to determine historical trends of surges and anomalies. This work revealed that weak and moderate positive surges did not exhibit any definite trends, while strong positive surges clearly became less frequent, even in the face of a gradually rising sea level, "presumably," in the words of Raicich, "as a consequence of a general weakening of the atmospheric activity," which was also found to have been the case for Brittany by Pirazzoli (2000).

Further north, Björck and Clemmensen (2004) extracted cores of peat from two raised bogs in the near-coastal part of southwest Sweden, from which they derived histories of wind-transported clastic material via systematic counts of quartz grains of various size classes that enabled them to calculate temporal variations in Aeolian Sand Influx (ASI), which has been shown to be correlated with winter wind climate in that part of the world. In doing so, they found that "the ASI records of the last 2500 years (both sites) indicate two timescales of winter storminess variation in southern Scandinavia." Specifically, they note that "decadal-scale variation (individual peaks) seems to coincide with short-term variation in sea-ice cover in the North Atlantic and is thus related to variations in the position of the North Atlantic winter season storm tracks," while "centennial-scale changes—peak families, like high peaks 1, 2 and 3 during the Little Ice Age, and low peaks 4 and 5 during the Medieval Warm Period— seem to record longer-scale climatic variation in the frequency and severity of cold and stormy winters."

Björck and Clemmensen also found a striking association between the strongest of these winter storminess peaks and periods of reduced solar activity. They specifically note, for example, that the solar minimum between AD 1880 and 1900 "is almost exactly coeval with the period of increased storminess at the end of the nineteenth century, and the Dalton Minimum between AD 1800 and 1820 is almost coeval with the period of peak storminess reported here." In addition, they say that an event of increased storminess they dated to AD 1650 "falls at the beginning of the Maunder solar minimum (AD 1645-1715)," while further back in time they report high ASI values between AD 1450 and 1550 with "a very distinct peak at AD 1475," noting that this period coincides with the Sporer Minimum of AD 1420-1530. In addition, they call attention to the fact that the latter three peaks in winter storminess all occurred during the Little Ice Age and "are among the most prominent in the complete record."

Last, the two researchers report that degree of humification (DOH) intervals "correlate well with the classic late-Holocene climatic intervals," which they specifically state to include the Modern Climate Optimum (100-0 cal. yr BP), the Little Ice Age (600-100 cal. yr BP), the Medieval Warm Period (1,250-600 cal. yr BP), the Dark Ages Cold Period (1,550-1,250 cal. yr BP) and the Roman Climate Optimum (2,250-1,550 cal. yr BP). There would thus appear to be little doubt that winter storms throughout southern Scandinavia were more frequent and intense during the multi-century Dark Ages Cold Period and Little Ice Age than they were during the Roman Warm Period, the Medieval Warm Period, and the Current Warm Period, providing strong evidence to refute the IPCC's contention that storminess tends to increase during periods of greater warmth. In the real world, just the opposite would appear to be the case.

Also working in Sweden were Barring and von Storch (2004), who say the occurrence of extreme weather events may "create the perception that ... the storms lately have become more violent, a trend that may continue into the future." These two researchers analyzed long time series of pressure readings for Lund (since 1780) and Stockholm (since 1823), analyzing (1) the annual number of pressure observations below 980 hPa, (2) the annual number of absolute pressure tendencies exceeding 16 hPa/12h, and (3) intra-annual 95th and 99th percentiles of the absolute pressure differences between two consecutive observations. They determined that the storminess time series they developed "are remarkably stationary in their mean, with little variations on time scales of more than one or two decades." They note that "the 1860s-70s was a period when the storminess indices showed general higher values," as was the 1980s-90s, but that, subsequently, "the indices have returned to close to their long-term mean."

Barring and von Storch conclude that their storminess proxies "show no indication of a long-term robust change towards a more vigorous storm climate." In fact, during "the entire historical period," in their words, storminess was "remarkably stable, with no systematic change and little transient variability." We can conclude that for much of Sweden, at least, there was no warming-induced increase in windstorms over the entire transitional period between the Little Ice Age and the Current Warm Period.

Dawson et al. (2004b) examined the sedimentary characteristics of a series of Late Holocene coastal windstorm deposits found on the Scottish Outer Hebrides, an island chain that extends across the latitudinal range 56-58°N. These deposits form part of the landward edges of coastal sand accumulations that are intercalated with peat, the radiocarbon dating of which was used to construct a local chronology of the windstorms. This work revealed that "the majority of the sand units were produced during episodes of climate deterioration both prior to and after the well-known period of Medieval warmth." The researchers also say "the episodes of sand blow indicated by the deposits may reflect periods of increased cyclogenesis in the Atlantic associated with increased sea ice cover and an increase in the thermal gradient across the North Atlantic region." In addition, they report that "dated inferred sand drift episodes across Europe show synchroneity with increased sand mobilization in SW France, NE England, SW Ireland and the Outer Hebrides, implying a regional response to storminess with increased sand invasion during the cool periods of the Little Ice Age," citing the corroborative studies of Lamb (1995), Wintle *et al.* (1998), Gilbertson *et al.* (1999), and Wilson *et al.* (2001). Throughout a vast portion of the North Atlantic Ocean and adjacent Europe, therefore, storminess and wind strength appear to have been inversely related to mean global air temperature over most of the past two millennia, with the most frequent and intense events occurring both prior to and following the Medieval Warm Period.

Dawson *et al.* (2004a) examined 120- to 225-year records of gale-days per year from five locations scattered across Scotland, northwest Ireland, and Iceland, which they compared with a much longer 2,000-year record for the same general region. In doing so, they found that four of the five century-scale records showed a greater frequency of storminess in the cooler 1800s and early 1900s than throughout the remainder of the warmer twentieth century. In addition, they report that "considered over the last ca. 2000 years, it would appear that winter storminess and climate-driven coastal erosion was at a minimum during the Medieval Warm Period," which again is just the opposite of what the IPCC forecasts, i.e., more storminess with warmer temperatures.

Moving over to the United States, Zhang *et al.* (2000) used 10 long-term records of storm surges derived from hourly tide gauge measurements to calculate annual values of the number, duration, and integrated intensity of storms in eastern North America. Their analysis did not reveal any trends in storm activity during the twentieth century, which

they say is suggestive of "a lack of response of storminess to minor global warming along the U.S. Atlantic coast during the last 100 yr."

Similar results were found by Boose *et al.* (2001). After scouring historical records to reconstruct hurricane damage regimes for an area composed of the six New England states plus adjoining New York City and Long Island for the period 1620-1997, they could discern "no clear century-scale trend in the number of major hurricanes." For the most recent and reliable 200-year portion of the record, however, the cooler nineteen century had five of the highest-damage category 3 storms, while the warmer twentieth century had only one such storm. Hence, as the earth experienced the warming associated with its recovery from the cold temperatures of the Little Ice Age, it would appear this part of the planet (New England, USA) experienced a decline in the intensity of severe hurricanes.

Going back further in time, Noren *et al.* (2002) extracted sediment cores from 13 small lakes distributed across a 20,000-km^2 region of Vermont and eastern New York, finding that "the frequency of storm-related floods in the northeastern United States has varied in regular cycles during the past 13,000 years (13 kyr), with a characteristic period of about 3 kyr." The most recent upswing in storminess did not begin with what the supposedly unprecedented warming of the twentieth century, but "at about 600 yr BP [Before Present], coincident with the beginning of the Little Ice Age." According to the authors, the increase in storminess was likely a product of natural changes in the Arctic Oscillation.

Moving to southern North America, land-falling hurricanes whose eyes crossed the coast between Cape Sable, Florida and Brownsville, Texas between 1896 and 1995 were the subject of investigation by Bove *et al.* (1998). The authors note that the first half of the twentieth century saw considerably more hurricanes than the last half: 11.8 per decade vs. 9.4 per decade. The same holds true for intense hurricanes of category 3 on the Saffir-Simpson storm scale: 4.8 vs. 3.6. The numbers of all hurricanes and the numbers of intense hurricanes have both been trending downward since 1966, with the decade starting in 1986 exhibiting the fewest intense hurricanes of the century.

Liu and Fearn (1993) also studied major storms along the U.S. Gulf Coast, but over a much longer time period: the past 3,500 years. Using sediment cores taken from the center of Lake Shelby in Alabama they determined that "major hurricanes of category 4 or 5 intensity directly struck the Alabama coast ... with an average recurrence interval of ~600 years," the last of which super-storms occurred around 700 years ago. They further note that "climate modeling results based on scenarios of greenhouse warming predict a 40%—50% increase in hurricane intensities in response to warmer tropical oceans." If one of these severe storms (which is now about a century overdue) were to hit the Alabama coast again, some commentators would probably cite its occurrence as vindication of the IPCC's doomsday predictions. In reality, it would be the result of natural (not man-made) causes.

Muller and Stone (2001) examined historical data relating to tropical storm and hurricane strikes along the southeast U.S. coast from South Padre Island, Texas to Cape Hatteras, North Carolina for the 100-year period 1901-2000. Their analysis revealed that the temporal variability of tropical storm and hurricane strikes was "great and significant," with most coastal sites experiencing "pronounced clusters of strikes separated by tens of years with very few strikes." With respect to the claim of a tendency for increased storminess during warmer El Niño years, the data didn't cooperate. For tropical storms and hurricanes together, the authors found an average of 1.7 storms per El Niño season, 2.6 per neutral season, and 3.3 per La Niña season. For hurricanes only, the average rate of occurrence ranged from 0.5 per El Niño season to 1.7 per La Niña season.

In the interior of the United States, Changnon and Changnon (2000) examined hail-day and thunder-day occurrences over the 100-year period 1896-1995 in terms of 20-year averages obtained from records of 66 first-order weather stations distributed across the country. They found that the frequency of thunder-days peaked in the second of the five 20-year intervals, while hail-day frequency peaked in the third or middle interval. Thereafter, both parameters declined to their lowest values of the century in the final 20-year period. Hail-day occurrence, in fact, decreased to only 65 percent of what it was at mid-century, accompanied by a drop in national hail insurance losses over the same period.

After completing this large regional study, S.A. Changnon (2001) turned his attention to an urban and a more rural site in Chicago in an attempt to determine if there might be an urban heat island influence on thunderstorm activity. Over the 40-year period investigated (1959-1998), he found the urban station experienced an average of 4.5 (12 percent) more thunderstorm days per year than the more rural

station; statistical tests revealed this difference to be significant at the 99 percent level in all four seasons of the year. This suggests that the actual decreases in hail- and thunder-days Changnon and Changnon found for the interior of the United States over the last half of the twentieth century may well have been even greater than what they reported in 2000.

Also in the interior of North America, Schwartz and Schmidlin (2002) compiled a blizzard database for the years 1959-2000 for the conterminous United States. A total of 438 blizzards were identified in the 41-year record, yielding an average of 10.7 blizzards per year. Year-to-year variability was significant, with the number of annual blizzards ranging from a low of 1 in the winter of 1980/81 to a high of 27 during the winter of 1996/97. Linear regression analysis revealed a statistically significant increase in the annual number of blizzards during the 41-year period; but the total area affected by blizzards each winter remained relatively constant and showed no trend. If these observations are both correct, then average blizzard size is much smaller now than it was four decades ago. As the authors note, however, "it may also be that the NWS is recording smaller, weaker blizzards in recent years that went unrecorded earlier in the period, as occurred also in the official record of tornadoes in the United States."

The results of this study thus suggest that—with respect to U.S. blizzards—frequency may have increased, but if it did, intensity likely decreased. On the other hand, the study's authors suggest that the reported increase in blizzard frequency may well be due to an observational bias that developed over the years, for which there is a known analogue in the historical observation of tornados. For example, Gulev et al. (2001) analyzed trends in Northern Hemispheric winter cyclones over essentially the same time period (1958-1999) and found a statistically significant decline of 1.2 cyclones per year using NCEP/NCAR reanalysis pressure data.

Further evidence that the blizzard frequency data are observationally biased can be deduced from the study of Hayden (1999), who investigated storm frequencies over North America between 25° and 55°N latitude and 60° and 125°W longitude from 1885 to 1996. Over this 112-year period, Hayden reports that large regional changes in storm occurrences were observed; but when integrated over the entire geographic area, no net change in storminess was evident.

To the north, Mason and Jordan (2002) studied numerous depositional environments along the tectonically stable, unglaciated eastern Chuckchi Sea coast that stretches across northwest Alaska, deriving a 6,000-year record of sea-level change while simultaneously learning some interesting things about the correlation between storminess and climate in that part of the world. With respect to storminess, they learned that "in the Chukchi Sea, storm frequency is correlated with colder rather than warmer climatic conditions." Consequently, they say that their data "do not therefore support predictions of more frequent or intense coastal storms associated with atmospheric warming for this region."

Hudak and Young (2002) examined the number of fall (June-November) storms in the southern Beaufort Sea region based on criteria of surface wind speed for the relatively short period of 1970-1995. Although there was considerable year-to-year variability in the number of storms, there was no discernible trend over the 26-year period in this region of the globe, where climate models predict the effects of CO_2-induced global warming would be most evident.

In conclusion, as the earth has warmed over the past 150 years during its recovery from the global chill of the Little Ice Age there has been no significant increase in either the frequency or intensity of stormy weather in Europe and North America. In fact, most studies suggest just the opposite has probably occurred.

Additional information on this topic, including reviews of storms not discussed here, can be found at http://www.co2science.org/subject/s/subject_s.php under the heading Storms.

References

Barring, L. and von Storch, H. 2004. Scandinavian storminess since about 1800. *Geophysical Research Letters* **31**: 10.1029/2004GL020441.

Bielec, Z. 2001. Long-term variability of thunderstorms and thunderstorm precipitation occurrence in Cracow, Poland, in the period 1896-1995. *Atmospheric Research* **56**: 161-170.

Bijl, W., Flather, R., de Ronde, J.G. and Schmith, T. 1999. Changing storminess? An analysis of long-term sea level data sets. *Climate Research* **11**: 161-172.

Björck, S. and Clemmensen, L.B. 2004. Aeolian sediment in raised bog deposits, Halland, SW Sweden: a new proxy record of Holocene winter storminess variation in southern Scandinavia? *The Holocene* **14**: 677-688.

Boose, E.R., Chamberlin, K.E. and Foster, D.R. 2001. Landscape and regional impacts of hurricanes in New England. *Ecological Monographs* **71**: 27-48.

Bove, M.C., Zierden, D.F. and O'Brien, J.J. 1998. Are gulf landfalling hurricanes getting stronger? *Bulletin of the American Meteorological Society* **79**: 1327-1328.

Changnon, S.A. 2001. Assessment of historical thunderstorm data for urban effects: the Chicago case. *Climatic Change* **49**: 161-169.

Changnon, S.A. and Changnon, D. 2000. Long-term fluctuations in hail incidences in the United States. *Journal of Climate* **13**: 658-664.

Clarke, M., Rendell, H., Tastet, J-P., Clave, B. and Masse, L. 2002. Late-Holocene sand invasion and North Atlantic storminess along the Aquitaine Coast, southwest France. *The Holocene* **12**: 231-238.

Dawson, A., Elliott, L., Noone, S., Hickey, K., Holt, T., Wadhams, P. and Foster, I. 2004a. Historical storminess and climate 'see-saws' in the North Atlantic region. *Marine Geology* **210**: 247-259.

Dawson, A.G., Hickey, K., Holt, T., Elliott, L., Dawson, S., Foster, I.D.L., Wadhams, P., Jonsdottir, I., Wilkinson, J., McKenna, J., Davis, N.R. and Smith, D.E. 2002. Complex North Atlantic Oscillation (NAO) Index signal of historic North Atlantic storm-track changes. *The Holocene* **12**: 363-369.

Dawson, S., Smith, D.E., Jordan, J. and Dawson, A.G. 2004b. Late Holocene coastal sand movements in the Outer Hebrides, N.W. Scotland. *Marine Geology* **210**: 281-306.

Dickson, R.R., Osborn, T.J., Hurrell, J.W., Meincke, J., Blindheim, J., Adlandsvik, B., Vinje, T., Alekseev, G. and Maslowski, W. 2000. The Arctic Ocean response to the North Atlantic Oscillation. *Journal of Climate* **13**: 2671-2696.

Gilbertson, D.D., Schwenninger, J.L., Kemp, R.A. and Rhodes, E.J. 1999. Sand-drift and soil formation along an exposed North Atlantic coastline: 14,000 years of diverse geomorphological, climatic and human impacts. *Journal of Archaeological Science* **26**: 439-469.

Gulev, S.K., Zolina, O. and Grigoriev, S. 2001. Extratropical cyclone variability in the Northern Hemisphere winter from the NCEP/NCAR reanalysis data. *Climate Dynamics* **17**: 795-809.

Gunther, H., Rosenthal, W., Stawarz, M., Carretero, J.C., Gomez, M., Lozano, I., Serrano, O. and Reistad, M. 1998. The wave climate of the northeast Atlantic over the period 1955-1994: the WASA wave hindcast. *The Global Atmosphere and Ocean System* **6**: 121-163.

Haeberli, W. and Beniston, M. 1998. Climate change and its impacts on glaciers and permafrost in the Alps. *Ambio* **27**: 258-265.

Hayden, B.P. 1999. Climate change and extratropical storminess in the United States: An assessment. *Journal of the American Water Resources Association* **35**: 1387-1397.

Hudak, D.R. and Young, J.M.C. 2002. Storm climatology of the southern Beaufort Sea. *Atmosphere-Ocean* **40**: 145-158.

Jones, P.D., Jonsson, T. and Wheeler, D. 1997. Extension to the North Atlantic Oscillation using early instrumental pressure observations from Gibraltar and South-West Iceland. *International Journal of Climatology* **17**: 1433-1450.

Lamb, H.H. 1995. *Climate, History and the Modern World*. Routledge, London, UK.

Liu, K.-B. and Fearn, M.L. 1993. Lake-sediment record of late Holocene hurricane activities from coastal Alabama. *Geology* **21**: 793-796.

Mason, O.W. and Jordan, J.W. 2002. Minimal late Holocene sea level rise in the Chukchi Sea: Arctic insensitivity to global change? *Global and Planetary Changes* **32**: 13-23.

Muller, R.A. and Stone, G.W. 2001. A climatology of tropical storm and hurricane strikes to enhance vulnerability prediction for the southeast U.S. coast. *Journal of Coastal Research* **17**: 949-956.

Noren, A.J., Bierman, P.R., Steig, E.J., Lini, A. and Southon, J. 2002. Millennial-scale storminess variability in the northeastern Unites States during the Holocene epoch. *Nature* **419**: 821-824.

Pirazzoli, P.A. 2000. Surges, atmospheric pressure and wind change and flooding probability on the Atlantic coast of France. *Oceanologica Acta* **23**: 643-661.

Raicich, F. 2003. Recent evolution of sea-level extremes at Trieste (Northern Adriatic). *Continental Shelf Research* **23**: 225-235.

Rebetez, M., Lugon, R. and Baeriswyl, P.-A. 1997. Climatic change and debris flows in high mountain regions: the case study of the Ritigraben torrent (Swiss Alps). *Climatic Change* **36**: 371-389.

Schwartz, R.M. and Schmidlin, T.W. 2002. Climatology of blizzards in the conterminous United States, 1959-2000. *Journal of Climate* **15**: 1765-1772.

Smits, A., Klein Tank, A.M.G. and Konnen, G.P. 2005. Trends in storminess over the Netherlands, 1962-2002. *International Journal of Climatology* **25**: 1331-1344.

Stoffel, M., Lièvre, I., Conus, D., Grichting, M.A., Raetzo, H., Gärtner, H.W. and Monbaron, M. 2005. 400 years of debris-flow activity and triggering weather conditions: Ritigraben, Valais, Switzerland. *Arctic, Antarctic, and Alpine Research* **37**: 387-395.

Wilson, P., Orford, J.D., Knight, J., Bradley, S.M. and Wintle, A.G. 2001. Late Holocene (post-4000 yrs BP) coastal development in Northumberland, northeast England. *The Holocene* **11**: 215-229.

Wintle, A.G., Clarke, M.L., Musson, F.M., Orford, J.D. and Devoy, R.J.N. 1998. Luminescence dating of recent dune formation on Inch Spit, Dingle Bay, southwest Ireland. *The Holocene* **8**: 331-339.

Woodworth, P.L. and Blackman, D.L. 2002. Changes in extreme high waters at Liverpool since 1768. *International Journal of Climatology* **22**: 697-714.

Zhang, K., Douglas, B.C. and Leatherman, S.P. 2000. Twentieth-Century storm activity along the U.S. East Coast. *Journal of Climate* **13**: 1748-1761.

6.7. Snow

The IPCC claims "snow cover has decreased in most regions, especially in spring," and "decreases in snowpack have been documented in several regions worldwide based upon annual time series of mountain snow water equivalent and snow death. (IPCC, 2007-I, p. 43). Later in the report, the authors claim "observations show a global-scale decline of snow and ice over many years, especially since 1980 and increasing during the past decade, despite growth in some places and little change in others" (p. 376). Has global warming really caused there to be less snow? We addressed this question regarding polar regions in Chapters 3 and 4 of this report. In this section, we focus (as the IPCC does) on studies conducted in North America.

Brown (2000) employed data from Canada and the United States to reconstruct monthly snow cover properties over mid-latitude (40-60°N) regions of North America back to the early 1900s, finding evidence of what he described as "a general twentieth century increase in North American snow cover extent, with significant increases in winter (December-February) snow water equivalent averaging 3.9% per decade." This finding is consistent with climate model simulations that indicate increased precipitation in response to global warming, but it covers too little time to tell us much about the *cause* of increased snow cover.

Moore *et al.* (2002) studied a longer period of time in their analysis of a 103-meter ice core retrieved from a high elevation site on Mount Logan—Canada's highest mountain—which is located in the heavily glaciated Saint Elias region of the Yukon. From this deep core, as well as from some shallow coring and snow-pit sampling, they derived a snow accumulation record that extended over three centuries (from 1693 to 2000), which indicated that heavier snow accumulation at their study site was generally associated with warmer tropospheric temperatures over northwestern North America.

Over the first half of their record, there is no significant trend in the snow accumulation data. From 1850 onward, however, there is a positive trend that is significant at the 95 percent confidence level, which indicates that recovery from the cold conditions of the Little Ice Age began in the mid-1800s, well before there was a large increase in the air's CO_2 concentration. This finding is further strengthened by the temperature reconstruction of Esper *et al.* (2002), which places the start of modern warming at about the same time as that suggested by Moore *et al.*'s snow data, contradicting the temperature record of Mann *et al.* (1998, 1999), which puts the beginning of the modern warming trend at about 1910.

Cowles *et al.* (2002) analyzed snow water equivalent (SWE) data obtained from four different measuring systems—snow courses, snow telemetry, aerial markers, and airborne gamma radiation—at more than 2,000 sites in the 11 westernmost states of the conterminous USA over the period 1910-1998, finding that the long-term SWE trend of the region was negative, indicative of declining winter precipitation. In addition, they report that their results "reinforce more tenuous conclusions made by previous authors," citing Changnon *et al.* (1993) and McCabe and Legates (1995), who studied snow course data from 1951-1985 and 1948-1987, respectively, at 275 and 311 sites, and who also found a decreasing trend in SWE at most sites in the Pacific Northwest.

Schwartz and Schmidlin (2002) examined past issues of *Storm Data*—a publication of the U.S. National Weather Service (NWS)—to compile a blizzard database for the years 1959-2000 for the conterminous United States. This effort resulted in a total of 438 blizzards being identified in the 41-year record, yielding an average of 10.7 blizzards per year; linear regression analysis revealed a statistically significant increase in the annual number of blizzards during the 41-year period. However, the total *area*

affected by blizzards each winter remained relatively constant. If these observations are both correct, average blizzard size must have decreased over the four-decade period. On the other hand, as the researchers note, "it may also be that the NWS is recording smaller, weaker blizzards in recent years that went unrecorded earlier in the period, as occurred also in the official record of tornadoes in the United States."

In a study of winter weather variability, Woodhouse (2003) generated a tree-ring based reconstruction of SWE for the Gunnison River basin of western Colorado that spans the period 1569-1999. This work revealed, in her words, that "the twentieth century is notable for several periods that lack extreme years." She reports that "the twentieth century is notable for several periods that contain few or no extreme years, for both low and high SWE extremes."

Lawson (2003) examined meteorological records for information pertaining to the occurrence and severity of blizzards within the Prairie Ecozone of western Canada. Over the period 1953-1997, no significant trends were found in central and eastern locations. However, there was a significant *downward* trend in blizzard frequency in the western prairies; Lawson remarks that "this trend is consistent with results found by others that indicate a decrease in cyclone frequency over western Canada." He also notes that the blizzards that do occur there "exhibit no trend in the severity of their individual weather elements." These findings, in his words, "serve to illustrate that the changes in extreme weather events anticipated under Climate Change may not always be for the worse."

Berger *et al.* (2003) collected a 50-year record (1949/1950 to 1998/1999) of snowfall occurrences using data from a dense network of cooperative station observations covering northwest and central Missouri provided by the Missouri Climate Center. The study looked at long-term trends and interannual variability in snowfall occurrence as related to sea surface temperature variations in the Pacific Ocean basin associated with the El Niño and Southern Oscillation (ENSO) and the North Pacific Oscillation (NPO). These trends and variations were then related to four synoptic-scale flow regimes that produce these snowfalls in the Midwest. The authors found no significant long-term trend in overall snowfall occurrence and a decrease in the number of extreme events (≥10 in, >25 cm) was noted. Two years later, Lupo *et al.* (2005) assembled a similar 54-year database (1948/1949 to 2002/2003) of snowfall occurrences for southwest Missouri and found "no variability or trends with respect to longer-term climatic variability and/or climate change."

Bartlett *et al.* (2005) set out to determine what changes might have occurred in the mean onset date of snow and its yearly duration in North America over the period 1950-2002, based on data for the contiguous United States that come from the 1,062 stations of the U.S. Historical Climatology Network, data for Canada that come from the 3,785 stations of the Canadian Daily Climate Dataset, and data for Alaska that come from the 543 stations of the National Weather Service cooperative network in that state. As a result of their efforts, the three researchers found that "for the period 1961-1990 the mean snow onset date in North America [was] 15 December, with mean snow cover duration of 81 days." They report there were "no significant trends in either onset or duration from 1950 to 2002." However, interannual variations of as much as 18 and 15 days in onset and duration, respectively, were present in the data; for both parameters they report that "no net trend was observed."

We find it significant that from 1950 to 2002, during which time the air's CO_2 concentration rose by 20 percent (from approximately 311 to 373 ppm), there was no net change in either the mean onset or duration of snow cover for the entire continent of North America. To provide some context for this 62-ppm increase in atmospheric CO_2 concentration, we note that it is essentially identical to the mean difference between the highs and lows of the three interglacials and glacials reported by Siegenthaler *et al.* (2005) for the period prior to 430,000 years ago. Surely, one would expect that such a change should have had some effect on North American snow cover—unless, of course, atmospheric CO_2 enrichment has very little or no impact on climate.

Julander and Bricco (2006) reported that snowpack data were being consistently used as indicators of global warming, and that individuals doing so should quantify, as best they could, all other influences embedded in their data. That meeting this requirement is no trivial undertaking is indicated by their statement that "snow data may be impacted by site physical changes, vegetation changes, weather modification, pollution, sensor changes, changes in transportation or sampling date, comparisons of snow course to SNOTEL data, changes in measurement personnel or recreational and other factors," including sensors that "do not come back to zero at the end of

the snow season." In an analysis of 134 sites (some having pertinent data stretching back to at least 1912), they thus selected 15 long-term Utah snow courses representing complete elevational and geographic coverage of the dominant snowpacks within the state and adjusted them for the major known site conditions affecting the data, after which the adjusted data for the period 1990-2005 were "compared to earlier portions of the historic record to determine if there were statistically significant differences in snowpack characteristics, particularly those that could indicate the impacts of global warming."

Of the 15 sites studied in greatest detail, the two researchers found that seven of them exhibited increased snowpack in recent years, while eight exhibited decreased snow accumulation. They also report that "six of the seven sites with increases have significant vegetative or physical conditions leading us to believe that the impacts associated with this analysis are overstated." The ultimate conclusion of Julander and Bricco, therefore, was that "any signature of global warming currently present in the snowpack data of Utah is not yet at a level of statistical significance ... and will likely be very difficult to isolate from other causes of snowpack decline."

Changnon and Changnon (2006) analyzed the spatial and temporal distributions of damaging snowstorms and their economic losses by means of property-casualty insurance data pertaining to "highly damaging storm events, classed as catastrophes by the insurance industry, during the 1949-2000 period." This work indicated, as they describe it, that "the incidence of storms peaked in the 1976-1985 period," but that snowstorm incidence "exhibited no up or down trend during 1949-2000." The two researchers concluded their paper by stating that "the temporal frequency of damaging snowstorms during 1949-2000 in the United States does not display any increase over time, indicating that either no climate change effect on cyclonic activity has begun, or if it has begun, altered conditions have not influenced the incidence of snowstorms."

Evidence supporting Changnon and Changnon's conclusion can be found in the work of Paul Kocin of The Weather Channel and Louis Uccellini of the National Weather Service (Kocin and Uccellini, 2004; Squires and Lawrimore, 2006). The authors created a scale to classify snowstorms, called the Northeast Snowfall Impact Scale (NESIS), that characterizes and ranks high-impact Northeast snowstorms. These storms typically cover large areas with snowfall accumulations of 10 inches or more. NESIS uses population information in addition to meteorological measurements to help communicate the social and economic impact of snowstorms. Storms are put into five categories with 1 being the smallest ("notable") and 5 being the largest ("extreme").

Using the NESIS scale, the National Oceanic and Atmospheric Administration's National Climatic Data Center created a list of 36 "high-impact snowstorms that affected the Northeast urban corridor," with the earliest storm occurring in 1956 and the most recent on March 1-3, 2009 (NOAA, 2009). Since population has increased in the Northeast over time, more recent storms rank higher on the NESIS scale even if they are no more severe than storms of the past. Nevertheless, fully half (6) of the highest rated (most severe) snowstorms on this record occurred before 1970, as did 14 of all 36 storms in the record. The three most severe storms occurred in 1993, 1996, and 2003, but the next three worst happened in 1960, 1961, and 1964.

Similarly, the National Weather Service tracks the "biggest snowstorms on record" for several cities, with tables showing the dates of the storms and number of inches of snowfall for each posted on its Web site (NWS, 2009). The table for Washington DC shows 15 snowstorms, with five storms having occurred since 1970, four between 1930 and 1970, and six prior to 1930. The biggest snowstorm ever to hit Washington DC arrived in January 1772, when 36 inches fell in the Washington-Baltimore area. It has been called the Washington-Jefferson snowstorm because it was recorded in both of their diaries. It is unlikely that human activity could have contributed to the severity of that storm, or to any other storm prior to the start of significant anthropogenic greenhouse gas emissions in the 1940s.

The research summarized in this section reveals that there has been no trend toward less snowfall or snow accumulation, or toward more snowstorms, in North America during the second half of the twentieth century. This record contradicts either the claim that warmer temperatures will lead to more snowfall and winter storms, or the claim that North America experienced warmer winter temperatures during the past half-century. In either case, the IPCC's claims in this regard must be erroneous.

Additional information on this topic, including reviews on snow not discussed here, can be found at http://www.co2science.org/subject/s/subject_s.php under the heading Snow.

References

Bartlett, M.G., Chapman, D.S. and Harris, R.N. 2005. Snow effect on North American ground temperatures, 1950-2002. *Journal of Geophysical Research* **110**: F03008, 10.1029/2005JF000293.

Brown, R.D. 2000. Northern hemisphere snow cover variability and change, 1915-97. *Journal of Climate* **13**: 2339-2355.

Changnon, D., McKee, T.B. and Doesken, N.J. 1993. Annual snowpack patterns across the Rockies: Long-term trends and associated 500-mb synoptic patterns. *Monthly Weather Review* **121**: 633-647.

Changnon, S.A. and Changnon, D. 2006. A spatial and temporal analysis of damaging snowstorms in the United States. *Natural Hazards* **37**: 373-389.

Cowles, M.K., Zimmerman, D.L., Christ, A. and McGinnis, D.L. 2002. Combining snow water equivalent data from multiple sources to estimate spatio-temporal trends and compare measurement systems. *Journal of Agricultural, Biological, and Environmental Statistics* **7**: 536-557.

Esper, J., Cook, E.R. and Schweingruber, F.H. 2002. Low-frequency signals in long tree-ring chronologies for reconstructing past temperature variability. *Science* **295**: 2250-2253.

IPCC. 2007-I. *Climate Change 2007: The Physical Science Basis. Contribution of Working Group I to the Fourth Assessment Report of the Intergovernmental Panel on Climate Change.* Solomon, S., Qin, D., Manning, M., Chen, Z., Marquis, M., Averyt, K.B., Tignor, M. and Miller, H.L. (Eds.) Cambridge University Press, Cambridge, UK.

Julander, R.P. and Bricco, M. 2006. An examination of external influences imbedded in the historical snow data of Utah. In: *Proceedings of the Western Snow Conference 2006*, pp. 61-72.

Kocin, P. J. and Uccellini, L.W. 2004. A snowfall impact scale derived from Northeast storm snowfall distributions. *Bulletin of the American Meteorological Society* **85**: 177-194.

Lawson, B.D. 2003. Trends in blizzards at selected locations on the Canadian prairies. *Natural Hazards* **29**: 123-138.

Lupo, A.R., Albert, D., Hearst, R., Market, P.S., Adnan Akyuz, F., and Almeyer, C.L. 2005. Interannual variability of snowfall events and snowfall-to-liquid water equivalents in Southwest Missouri. *National Weather Digest* **29**: 13 – 24.

Mann, M.E., Bradley, R.S. and Hughes, M.K. 1998. Global-scale temperature patterns and climate forcing over the past six centuries. *Nature* **392**: 779-787.

Mann, M.E., Bradley, R.S. and Hughes, M.K. 1999. Northern Hemisphere temperatures during the past millennium: Inferences, uncertainties, and limitations. *Geophysical Research Letters* **26**: 759-762.

McCabe, A.J. and Legates, S.R. 1995. Relationships between 700hPa height anomalies and 1 April snowpack accumulations in the western USA. *International Journal of Climatology* **14**: 517-530.

Moore, G.W.K., Holdsworth, G. and Alverson, K. 2002. Climate change in the North Pacific region over the past three centuries. *Nature* **420**: 401-403.

NOA. 2009. The Northeast snowfall impact scale (NESIS). U.S. National Oceanic and Atmospheric Administration, National Climatic Data Center. http://www.ncdc.noaa.gov/snow-and-ice/nesis.php. Last accessed May 6, 2009.

NWS. 2009. Biggest snowstorms on record, Baltimore/Washington. U.S. National Weather Service. http://www.erh.noaa.gov/lwx/Historic_Events/snohist.htm. Last accessed May 6, 2009.

Schwartz, R.M. and Schmidlin, T.W. 2002. Climatology of blizzards in the conterminous United States, 1959-2000. *Journal of Climate* **15**: 1765-1772.

Siegenthaler, U., Stocker, T.F., Monnin, E., Luthi, D., Schwander, J., Stauffer, B., Raynaud, D., Barnola, J.-M., Fischer, H., Masson-Delmotte, V. and Jouzel, J. 2005. Stable carbon cycle-climate relationship during the late Pleistocene. *Science* **310**: 1313-1317.

Squires, M. F. and Lawrimore, J. H. 2006. Development of an operational snowfall impact scale. Presentation at 22nd International Conference on Interactive Information Processing Systems for Meteorology, Oceanography, and Hydrology. Atlanta, GA.

Woodhouse, C.A. 2003. A 431-yr reconstruction of western Colorado snowpack from tree rings. *Journal of Climate* **16**: 1551-1561.

6.8. Storm Surges

One of the many catastrophes said to be caused by global warming is the heaving of the world's seas beyond their normal bounds in more frequent and increasingly violent storm surges. The following section summarizes the findings of a number of pertinent papers.

De Lange and Gibb (2000) analyzed trends in sea-level data obtained from several tide gauges located within Tauranga Harbor, New Zealand, over the period 1960-1998. In studying seasonal, interannual, and decadal distributions of storm surge data, they discovered a considerable decline in the annual number of storm surge events in the latter half of the nearly four-decade-long record. A similar trend was noted in the magnitude of storm surges. In addition, maximum water levels, including tides, also declined over the past two decades.

Decadal variations in the data were linked to the Inter-decadal Pacific Oscillation (IPO) and the El Niño-Southern Oscillation (ENSO), with La Niña events producing more storm surge days than El Niño events. In addition, wavelet analyses of annual storm surge frequency data indicated that before 1978 the frequency "was enhanced by the IPO, and subsequently it has been attenuated."

Pirazzoli (2000) analyzed tide-gauge and meteorological (wind and atmospheric pressure) data for the slightly longer period of 1951-1997 along the northern portion of the Atlantic coast of France. This effort revealed that the number of atmospheric depressions (storms) and strong surge winds in this region "are becoming less frequent" and that "ongoing trends of climate variability show a decrease in the frequency and hence the gravity of coastal flooding." Pirazzoli suggests these findings should be "reassuring," especially for those concerned about coastal flooding.

Raicich (2003) analyzed 62 years of sea-level data for the period 1 July 1939 to 30 June 2001 at Trieste, in the Northern Adriatic, in an attempt to determine historical trends of positive and negative surge anomalies. This work led to the discovery that weak and moderate positive surges did not exhibit any definite trends, while strong positive surges clearly became less frequent over the period of study, even in the face of a gradually rising sea level, "presumably," in Raicich's words, "as a consequence of a general weakening of the atmospheric activity," which was likewise found by Pirazzoli to be the case for Brittany.

Based on data for the somewhat longer period of 1901-1990, Wroblewski (2001) determined there was a linear increase in mean annual sea level at the southern Baltic seaport of Kolobrzeg of 12 ± 2 cm per century. Over this same period, however, there was no trend in annual sea-level maxima. Two high values stood out above the rest in the 1980s, but two similar spikes occurred in the 1940s; there were half-a-dozen comparable high values in the first two decades of the record. In light of the slow upward trend in mean sea level, therefore, it is extremely surprising that annual maximum sea levels due to storm surges did not likewise rise over the past century. One can only conclude these events must have become less intense over the same time interval.

Utilizing a full century of data, Zhang *et al.* (2000) analyzed 10 very long records of storm surges derived from hourly tide gauge measurements made along the east coast of the United States, in order to calculate indexes of count, duration and integrated intensity of surge-producing storms that provide objective, quantitative, and comprehensive measures of historical storm activities in this region. The end result of their comprehensive undertaking was a demonstrable lack of "any discernible long-term secular trend in storm activity during the twentieth century," which finding, in their words, "suggests a lack of response of storminess to minor global warming along the U.S. Atlantic coast during the last 100 years."

Looking considerably further back in time, Woodworth and Blackman (2002) analyzed four discontinuous sets of high-water data from the UK's Liverpool waterfront that span the period 1768-1999, looking for changes in annual maximum high water (tide plus surge), surge at annual maximum high water (surge component of annual maximum high water), and annual maximum surge-at-high-water. They could detect no significant trends in the first two parameters over the period of study; but they found that the annual maximum surge-at-high-water declined at a rate of 0.11 ± 0.04 meters per century. This finding suggests the winds responsible for producing high storm surges were much stronger and/or more common during the early part of the record (Little Ice Age) than during the latter part (Current Warm Period).

Last, in what is the longest look back in time of the papers treating this subject, Nott and Hayne (2001) produced a 5,000-year record of tropical cyclone frequency and intensity along a 1,500-km stretch of coastline in northeastern Australia located between latitudes 13 and 24°S by (1) geologically dating and topographically surveying landform features left by surges produced by historic hurricanes and (2) running numerical models to estimate storm surge and wave heights necessary to reach the landform locations. This work revealed that several "super-cyclones" with central pressures less than 920 hPa and wind speeds in excess of 182 kilometers per hour had occurred over the past 5,000 years at intervals of roughly 200 to 300 years in all parts of

the region of study. The two researchers also report that the Great Barrier Reef "experienced at least five such storms over the past 200 years, with the area now occupied by Cairns experiencing two super-cyclones between 1800 and 1870." The twentieth century, however, was totally devoid of such storms, "with only one such event (1899) since European settlement in the mid-nineteenth century."

It seems safe to conclude that storm surges around the world have not responded to rising temperatures during the twentieth century. In the majority of cases investigated, they have decreased.

Additional information on this topic, including reviews of newer publications as they become available, can be found at http://www.co2science.org/subject/o/oceanstormsurge.php.

References

De Lange, W.P. and Gibb, J.G. 2000. Seasonal, interannual, and decadal variability of storm surges at Tauranga, New Zealand. *New Zealand Journal of Marine and Freshwater Research* **34**: 419-434.

Nott, J. and Hayne, M. 2001. High frequency of 'super-cyclones' along the Great Barrier Reef over the past 5,000 years. *Nature* **413**: 508-512.

Pirazzoli, P.A. 2000. Surges, atmospheric pressure and wind change and flooding probability on the Atlantic coast of France. *Oceanologica Acta* **23**: 643-661.

Raicich, F. 2003. Recent evolution of sea-level extremes at Trieste (Northern Adriatic). *Continental Shelf Research* **23**: 225-235.

Woodworth, P.L. and Blackman, D.L. 2002. Changes in extreme high waters at Liverpool since 1768. *International Journal of Climatology* **22**: 697-714.

Wroblewski, A. 2001. A probabilistic approach to sea level rise up to the year 2100 at Kolobrzeg, Poland. *Climate Research* **18**: 25-30.

Zhang, K., Douglas, B.C. and Leatherman, S.P. 2000. Twentieth-Century storm activity along the U.S. East Coast. *Journal of Climate* **13**: 1748-1761.

6.9. Temperature Variability

One more measure of climatic change is temperature variability. Has the earth experienced more record highs or lows of temperature during the Current Warm Period?

Oppo *et al.* (1998) studied sediments from Ocean Drilling Project site 980 on the Feni Drift (55.5°N, 14.7°W) in the North Atlantic. Working with a core pertaining to the period from 500,000 to 340,000 years ago, they analyzed $\delta^{18}O$ and $\delta^{13}C$ obtained from benthic foraminifera and $\delta^{18}O$ obtained from planktonic foraminifera to develop histories of deep water circulation and sea surface temperature (SST), respectively. In doing so, they discovered a number of persistent climatic oscillations with periods of 6,000, 2,600, 1,800 and 1,400 years that traversed the entire length of the sediment core record, extending through glacial and interglacial epochs alike. These SST variations, which were found to be in phase with deep-ocean circulation changes, were on the order of 3°C during cold glacial maxima but only 0.5 to 1°C during warm interglacials.

Similar results were obtained by McManus *et al.* (1999), who also examined a half-million-year-old deep-sea sediment core from the eastern North Atlantic. Significant SST oscillations were again noted throughout the record, and they too were of much greater amplitude during glacial periods (4 to 6°C) than during interglacials (1 to 2°C). Likewise, in another study of a half-million-year-long sediment core from the same region, Helmke *et al.* (2002) found that the most stable of all climates held sway during what they called "peak interglaciations" or periods of greatest warmth. In this regard, we note that the temperatures of all four of the interglacials that preceded the one in which we currently live were warmer than the present one, and by an average temperature in excess of 2°C, as determined by Petit *et al.* (1999). Hence, even if the earth were to continue its recent (and possibly ongoing) recovery from the global chill of the Little Ice Age, that warming likely would lead to a state of reduced temperature variability, as evidenced by real-world data pertaining to the past half-million years.

Shifting our focus to the past millennium, Cook *et al.* (2002) report the results of a tree-ring study of long-lived silver pines on the West Coast of New Zealand's South Island. The chronology they derived provides a reliable history of Austral summer temperatures from AD 1200 to 1957, after which measured temperatures were used to extend the history to 1999. Cook *et al.* say their reconstruction indicates "there have been several periods of above and below average temperature that have not been experienced in the 20th century." This finding indicates that New Zealand temperatures grew less variable over the twentieth century.

Manrique and Fernandez-Cancio (2000) employed a network of approximately 1,000 samples of tree-ring series representative of a significant part of Spain to reconstruct thousand-year chronologies of temperature and precipitation, after which they used this database to identify anomalies in these parameters that varied from their means by more than four standard deviations. In doing so, they found that the greatest concentration of extreme climatic excursions, which they describe as "the outstanding oscillations of the Little Ice Age," occurred between AD 1400 and 1600, during a period when extreme low temperatures reached their maximum frequency.

In yet another part of the world, many long tree-ring series obtained from widely spaced Himalayan cedar trees were used by Yadav *et al.* (2004) to develop a temperature history of the western Himalayas for the period AD 1226-2000. "Since the 16th century," to use their words, "the reconstructed temperature shows higher variability as compared to the earlier part of the series (AD 1226-1500), reflecting unstable climate during the Little Ice Age (LIA)." With respect to this greater variability of climate during colder conditions, they note that similar results have been obtained from juniper tree-ring chronologies from central Tibet (Braeuning, 2001), and that "historical records on the frequency of droughts, dust storms and floods in China also show that the climate during the LIA was highly unstable (Zhang and Crowley, 1989)." Likewise, in a study of the winter half-year temperatures of a large part of China, Ge *et al.* (2003) identified greater temperature anomalies during the 1600s than in the 1980s and 1990s.

Focusing on just the past century, Rebetez (2001) analyzed day-to-day variability in two temperature series from Switzerland over the period 1901-1999, during which time the two sites experienced temperatures increases of 1.2 and 1.5°C. Their work revealed that warmer temperatures led to a reduction in temperature variability at both locations. As they describe it, "warmer temperatures are accompanied by a general reduction of variability, both in daily temperature range and in the monthly day-to-day variability." We see that even on this much finer time scale, it is cooling, not warming, that brings an increase in temperature variability.

In a study based on daily maximum (max), minimum (min), and mean air temperatures (T) from 1,062 stations of the U.S. Historical Climatology Network, Robeson (2002) computed the slopes of the relationships defined by plots of daily air temperature standard deviation vs. daily mean air temperature for each month of the year for the period 1948-1997. This protocol revealed, in Robeson's words, that "for most of the contiguous USA, the slope of the relationship between the monthly mean and monthly standard deviation of daily Tmax and Tmin—the variance response—is either negative or near-zero," which means, he describes it, that "for most of the contiguous USA, a warming climate should produce either reduced air-temperature variability or no change in air-temperature variability." He also reports that the negative relationships are "fairly strong, with typical reductions in standard deviation ranging from 0.2 to 0.5°C for every 1°C increase in mean temperature."

In Canada, according to Shabbar and Bonsal (2003), "extreme temperature events, especially those during winter, can have many adverse environmental and economic impacts." They chose to examine trends and variability in the frequency, duration, and intensity of winter (January-March) cold and warm spells during the second half of the twentieth century. From 1950-1998, they found that western Canada experienced decreases in the frequency, duration, and intensity of winter cold spells. In the east, however, distinct increases in the frequency and duration of winter cold spells occurred. With respect to winter warm spells, significant increases in both their frequency and duration were observed across most of Canada, with the exception of the extreme northeastern part of the country, where warm spells appear to be becoming shorter and less frequent. In the mean, therefore, there appear to be close-to-compensating trends in the frequency and intensity of winter cold spells in different parts of Canada, while winter warm spells appear to be increasing somewhat. As a result, Canada appears to have experienced a slight amelioration of extreme winter weather over the past half-century.

In another study that suffers from the difficulty of having but a few short decades of data to analyze, Iskenderian and Rosen (2000) studied two mid-tropospheric temperature datasets spanning the past 40 years, calculating day-to-day variability within each month, season, and year. Averaged over the entire Northern Hemisphere, they found that mid-tropospheric temperature variability exhibited a slight upward trend since the late 1950s in one of the datasets; but, as they note, "this trend is significant in the spring season only." They also admit that "the robustness of this springtime trend is in doubt," because the trend obtained from the other dataset was

negative. For the conterminous United States, however, the two datasets both showed "mostly small positive trends in most seasons." But, again, none of these trends was statistically significant. Therefore, Iskenderian and Rosen acknowledge they "cannot state with confidence that there has been a change in synoptic-scale temperature variance in the mid-troposphere over the United States since 1958."

In an attempt to determine the role that might have been played by the planet's mean temperature in influencing temperature variability over the latter half of the twentieth century, Higgins *et al.* (2002) examined the influence of two important sources of Northern Hemispheric climate variability—the El Niño/Southern Oscillation (ENSO) and the Arctic Oscillation—on winter (Jan-Mar) daily temperature extremes over the conterminous United States from 1950 to 1999. With respect to the Arctic Oscillation, there was basically no difference in the number of extreme temperature days between its positive and negative phases. With respect to the ENSO phenomenon, however, Higgins *et al.* found that during El Niño years, the total number of extreme temperature days was found to decrease by around 10 percent, while during La Niña years they increased by around 5 percent. With respect to winter temperatures across the conterminous United States, therefore, the contention that warmer global temperatures—such as are typically experienced during El Niño years—would produce more extreme weather conditions is found to be false.

Over the same time period, Zhai and Pan (2003) derived trends in the frequencies of warm days and nights, cool days and nights, and hot days and frost days for the whole of China, based on daily surface air temperature data obtained from approximately 200 weather observation stations scattered across the country. Over the period of record, and especially throughout the 1980s and 1990s, there were increases in the numbers of warm days and nights, while there were decreases in the numbers of cool days and nights, consistent with an overall increase in mean daily temperature. At the extreme hot end of the temperature spectrum, however, the authors report that "the number of days with daily maximum temperature above 35°C showed a slightly decreasing trend for China as a whole," while at the extreme cold end of the spectrum, the number of frost days with daily minimum temperature below 0°C declined at the remarkable rate of 2.4 days per decade.

In considering this entire body of research, it is evident that air temperature variability almost always decreases when mean air temperature rises.

Additional information on this topic, including reviews of newer publications as they become available, can be found at http://www.co2science.org/subject/e/extremetemp.php.

References

Braeuning, A. 2001. Climate history of Tibetan Plateau during the last 1000 years derived from a network of juniper chronologies. *Dendrochronologia* **19**: 127-137.

Cook, E.R., Palmer, J.G., Cook, B.I., Hogg, A. and D'Arrigo, R.D. 2002. A multi-millennial palaeoclimatic resource from Lagarostrobos colensoi tree-rings at Oroko Swamp, New Zealand. *Global and Planetary Change* **33**: 209-220.

Ge, Q., Fang, X. and Zheng, J. 2003. Quasi-periodicity of temperature changes on the millennial scale. *Progress in Natural Science* **13**: 601-606.

Helmke, J.P., Schulz, M. and Bauch, H.A. 2002. Sediment-color record from the northeast Atlantic reveals patterns of millennial-scale climate variability during the past 500,000 years. *Quaternary Research* **57**: 49-57.

Higgins, R.W., Leetmaa, A. and Kousky, V.E. 2002. Relationships between climate variability and winter temperature extremes in the United States. *Journal of Climate* **15**: 1555-1572.

Iskenderian, H. and Rosen, R.D. 2000. Low-frequency signals in midtropospheric submonthly temperature variance. *Journal of Climate* **13**: 2323-2333.

Manrique, E. and Fernandez-Cancio, A. 2000. Extreme climatic events in dendroclimatic reconstructions from Spain. *Climatic Change* **44**: 123-138.

McManus, J.F., Oppo, D.W. and Cullen, J.L. 1999. A 0.5-million-year record of millennial-scale climate variability in the North Atlantic. *Science* **283**: 971-974.

Oppo, D.W., McManus, J.F. and Cullen, J.L. 1998. Abrupt climate events 500,000 to 340,000 years ago: Evidence from subpolar North Atlantic sediments. *Science* **279**: 1335-1338.

Petit, J.R., Jouzel, J., Raynaud, D., Barkov, N.I., Barnola, J.-M., Basile, I., Bender, M., Chappellaz, J., Davis, M., Delaygue, G., Delmotte, M., Kotlyakov, V.M., Legrand, M., Lipenkov, V.Y., Lorius, C., Pepin, L., Ritz, C., Saltzman, E., and Stievenard, M. 1999. Climate and atmospheric history of the past 420,000 years from the Vostok ice core, Antarctica. *Nature* **399**: 429-436.

Rebetez, M. 2001. Changes in daily and nightly day-to-day temperature variability during the twentieth century for two stations in Switzerland. *Theoretical and Applied Climatology* **69**: 13-21.

Robeson, S.M. 2002. Relationships between mean and standard deviation of air temperature: implications for global warming. *Climate Research* **22**: 205-213.

Shabbar, A. and Bonsal, B. 2003. An assessment of changes in winter cold and warm spells over Canada. *Natural Hazards* **29**: 173-188.

Yadav, R.R., Park, W.K., Singh, J. and Dubey, B. 2004. Do the western Himalayas defy global warming? *Geophysical Research Letters* **31**: 10.1029/2004GL020201.

Zhai, P. and Pan, X. 2003. Trends in temperature extremes during 1951-1999 in China. *Geophysical Research Letters* **30**: 10.1029/2003GL018004.

Zhang, J. and Crowley, T.J. 1989. Historical climate records in China and reconstruction of past climates (1470-1970). *Journal of Climate* **2**: 833-849.

6.10. Wildfires

As stated by the IPCC, "an intensification and expansion of wildfires is likely globally, as temperatures increase and dry spells become more frequent and more persistent" (IPCC, 2007-II). Below, we test this claim by reviewing the results of studies conducted in various parts of the world, ending with a recent satellite study that evaluated the globe as a whole.

Carcaillet *et al.* (2001) developed high-resolution charcoal records from laminated sediment cores extracted from three small kettle lakes located within the mixed-boreal and coniferous-boreal forest region of eastern Canada, after which they determined whether vegetation change or climate change was the primary determinant of changes in fire frequency, comparing their fire history with hydro-climatic reconstructions derived from $\delta^{18}O$ and lake-level data. Throughout the Climatic Optimum of the mid-Holocene, between about 7,000 and 3,000 years ago, when it was significantly warmer than it is today, they report that "fire intervals were double those in the last 2000 years," meaning fires were only half as frequent throughout the earlier warmer period as they were during the subsequent cooler period. They also determined that "vegetation does not control the long-term fire regime in the boreal forest," but that "climate appears to be the main process triggering fire." In addition, they report that "dendroecological

355

studies show that both frequency and size of fire decreased during the 20th century in both west (e.g. Van Wagner, 1978; Johnson *et al.*, 1990; Larsen, 1997; Weir *et al.*, 2000) and east Canadian coniferous forests (e.g. Cwynar, 1997; Foster, 1983; Bergeron, 1991; Bergeron *et al.*, 2001), possibly due to a drop in drought frequency and an increase in long-term annual precipitation (Bergeron and Archambault, 1993)." These several findings thus led them to conclude that a "future warmer climate is likely to be less favorable for fire ignition and spread in the east Canadian boreal forest than over the last 2 millennia."

In another Canadian study that sheds important new light on this subject, four forest scientists investigated "regional fire activity as measured by the decadal proportion of area burned and the frequency of fire years vs. non-fire years in the Waswanipi area of northeastern Canada [49.5-50.5°N, 75-76.5°W], and the long-term relationship with large-scale climate variations ... using dendroecological sampling along with forest inventories, aerial photographs, and ecoforest maps." The results of their investigation revealed that instead of the interval of time between wildfires shortening as time progressed and the climate warmed, there was "a major lengthening of the fire cycle." In addition, the four researchers note that "in the context of the past 300 years, many regional fire regimes of the Canadian boreal forest, as reconstructed from dendroecological analysis, experienced a decrease in fire frequency after 1850 [or the "end of the Little Ice Age," as they describe it] (Bergeron and Archambault, 1993; Larsen, 1996) and a further decrease after 1940 (Bergeron *et al.*, 2001, 2004a,b, 2006)."

In further study of this subject, Lauzon *et al.* (2007) investigated the fire history of a 6,480-km^2 area located in the Baie-Des-Chaleurs region of Gaspesie at the southeastern edge of Quebec, "using Quebec Ministry of Natural Resource archival data and aerial photographs combined with dendrochronological data." Results indicated that coincident with the 150-year warming that led to the demise of the Little Ice Age and the establishment of the Current Warm Period, there was "an increase in the fire cycle from the pre-1850 period (89 years) to the post-1850 period (176 years)," and that "both maximum and mean values of the Fire Weather Index decreased statistically between 1920 and 2003," during which period "extreme values dropped from the very high to high categories, while mean values changed from moderate to low categories." In this particular part of the world, therefore, twentieth

century global warming has led to a significant decrease in the frequency of forest fires, as weather conditions conducive to their occurrence have gradually become less prevalent and extreme.

Pitkanen *et al.* (2003) constructed a Holocene fire history of dry heath forests in eastern Finland on the basis of charcoal layer data obtained from two small mire basins and fire scars on living and dead pine trees. This work revealed a "decrease in fires during climatic warming in the Atlantic chronozone (about 9000-6000 cal. yr. BP)," prompting them to conclude that "the very low fire frequency during the Atlantic chronozone despite climatic warming with higher summer temperatures, is contrary to assumptions about possible implications of the present climatic warming due to greenhouse gases."

Thereafter, the researchers observed an increase in fire frequency at the transition between the Atlantic and Subboreal chronozones around 6,000 cal. yr. BP, noting that "the climatic change that triggered the increase in fire frequency was cooling and a shift to a more continental climate." In addition, they report that the data of Bergeron and Archambault (1993) and Carcaillet *et al.* (2001) from Canada suggest much the same thing; i.e., fewer boreal forest fires during periods of greater warmth. Consequently, "as regards the concern that fire frequency will increase in [the] near future owing to global warming," the researchers say their data "suggest that fires from 'natural' causes (lightning) are not likely to increase significantly in eastern Finland and in geographically and climatically related areas."

Back in Canada, Girardin *et al.* (2006) introduced their study of the subject by citing a number of predictions that "human-induced climate change could lead to an increase in forest fire activity in Ontario, owing to the increased frequency and severity of drought years, increased climatic variability and incidence of extreme climatic events, and increased spring and fall temperatures," noting that "climate change therefore could cause longer fire seasons (Wotton and Flannigan, 1993), with greater fire activity and greater incidence of extreme fire activity years (Colombo *et al.*, 1998; Parker *et al.*, 2000)." To see if any of these predictions might have recently come to pass, they reconstructed a history of area burned within the province of Ontario for the period AD 1781-1982 from 25 tree-ring width chronologies obtained from various sites throughout the province. An increase in area burned within Ontario is known to have occurred from 1970 through 1981 (Podur *et al.*, 2002).

The three researchers report that "while in recent decades area burned has increased, it remained below the level recorded prior to 1850 and particularly below levels recorded in the 1910s and 1920s," noting further that "the most recent increase in area burned in the province of Ontario (circa 1970-1981) [Podur *et al.*, 2002] was preceded by the period of lowest fire activity ever estimated for the past 200 years (1940s-1960s)."

Schoennagel *et al.* (2007) investigated "climatic mechanisms influencing subalpine forest fire occurrence in western Colorado, which provide a key to the intuitive link between drought and large, high-severity fires that are keystone disturbance processes in many high-elevation forests in the western United States," focusing on three major climatic oscillations: the El Niño Southern Oscillation (ENSO), the Pacific Decadal Oscillation (PDO), and the Atlantic Multidecadal Oscillation (AMO).

Results of this analysis revealed that "fires occurred during short-term periods of significant drought and extreme cool (negative) phases of ENSO and PDO and during positive departures from [the] mean AMO index," while "at longer time scales, fires exhibited 20-year periods of synchrony with the cool phase of the PDO, and 80-year periods of synchrony with extreme warm (positive) phases of the AMO." In addition, they say that "years of combined positive AMO and negative ENSO and PDO phases represent 'triple whammies' that significantly increased the occurrence of drought-induced fires." On the other hand, they write that "drought and wildfire are associated with warm phases of ENSO and PDO in the Pacific Northwest and northern Rockies while the opposite occurs in the Southwest and southern Rockies," citing the findings of Westerling and Swetnam (2003), McCabe *et al.* (2004), and Schoennagel *et al.* (2005). Schoennagel *et al.* conclude that "there remains considerable uncertainty regarding the effects of CO_2-induced warming at regional scales." Nevertheless, they report "there is mounting evidence that the recent shift to the positive phase of the AMO will promote higher fire frequencies" in the region of their study, i.e., high-elevation western U.S. forests. The body of their work clearly suggests that such a consequence should not be viewed as a response to CO_2-induced global warming.

A contrary example, where warming does appear to enhance fire occurrence, is provided by Pierce *et al.* (2004), who dated fire-related sediment deposits in alluvial fans in central Idaho, USA, in a research program designed to reconstruct Holocene fire history in xeric ponderosa pine forests and to look for links to past climate change. This endeavor focused on tributary alluvial fans of the South Fork Payette (SFP) River area, where fans receive sediment from small but steep basins in weathered batholith granitic rocks that are conducive to post-fire erosion. They obtained 133 AMS ^{14}C-derived dates from 33 stratigraphic sites in 32 different alluvial fans. In addition, they compared their findings with those of Meyer *et al.* (1995), who had earlier reconstructed a similar fire history for nearby Yellowstone National Park in Wyoming, USA.

Pierce *et al.*'s work revealed, in their words, that "intervals of stand-replacing fires and large debris-flow events are largely coincident in SFP ponderosa pine forests and Yellowstone, most notably during the 'Medieval Climatic Anomaly' (MCA), ~1,050-650 cal. yr BP." What is more, they note that "in the western USA, the MCA included widespread, severe miltidecadal droughts (Stine, 1998; Woodhouse and Overpeck, 1998), with increased fire activity across diverse northwestern conifer forests (Meyer *et al.*, 1995; Rollins *et al.*, 2002)."

Following the Medieval Warm Period and its frequent large-event fires was the Little Ice Age, when, as Pierce *et al.* describe it, "colder conditions maintained high canopy moisture, inhibiting stand-replacing fires in both Yellowstone lodgepole pine forests and SFP ponderosa pine forests (Meyer *et al.*, 1995; Rollins *et al.*, 2002; Whitlock *et al.*, 2003)." Subsequently, however, they report that "over the twentieth century, fire size and severity have increased in most ponderosa pine forests," which they suggest may be largely due to "the rapidity and magnitude of twentieth-century global climate change."

With respect to their central thesis, which appears to be well supported by both the SFP and Yellowstone data, we agree with Pierce *et al.* that both the size and severity of large-event stand-replacing fires tend to increase with increasing temperature in the part of the world and for the specific forests they studied. We note that the Yellowstone data also depict a sharp drop in large-event fire frequency and severity during the earlier Dark Ages Cold Period, which followed on the heels of the preceding peak in such fires that was concomitant with the still earlier Roman Warm Period.

Also working in the United States, and coming to much the same general conclusion, were Westerling *et al.* (2006), who compiled a comprehensive database of large wildfires in western United States forests since 1970 and compared it to hydro-climatic and land-surface data. Their findings are succinctly summarized by Running (2006) in an accompanying Perspective, wherein he writes that "since 1986, longer warmer summers have resulted in a fourfold increase of major wildfires and a sixfold increase in the area of forest burned, compared to the period from 1970 to 1986," noting also that "the length of the active wildfire season in the western United States has increased by 78 days, and that the average burn duration of large fires has increased from 7.5 to 37.1 days." In addition, he notes that "four critical factors—earlier snowmelt [by one to four weeks], higher summer temperatures [by about 0.9°C], longer fire season, and expanded vulnerable area of high-elevation forests—are combining to produce the observed increase in wildfire activity."

So what is the case for the world as a whole—i.e., what is the net result of the often-opposite wildfire responses to warming that are typical of different parts of the planet?

This question was recently explored by Riano *et al.* (2007), who conducted "an analysis of the spatial and temporal patterns of global burned area with the Daily Tile US National Oceanic and Atmospheric Administration-Advanced Very High-Resolution Radiometer Pathfinder 8 km Land dataset between 1981 and 2000." For several areas of the world, this effort revealed there were indeed significant upward trends in land area burned. Some parts of Eurasia and western North America, for example, had annual upward trends as high as 24.2 pixels per year, where a pixel represents an area of 64 km^2. These increases in burned area, however, were offset by equivalent decreases in burned area in tropical southeast Asia and Central America. Consequently, in the words of Riano *et al.*, "there was no significant global annual upward or downward trend in burned area." In fact, they say "there was also no significant upward or downward global trend in the burned area for any individual month." In addition, they say that "latitude was not determinative, as divergent fire patterns were encountered for various land cover areas at the same latitude."

Although one can readily identify specific parts of the planet that have experienced both significant increases and decreases in land area burned over the last two or three decades of the twentieth century, as we have done in the materials reviewed above, for the globe as a whole there was no relationship between global warming and total area burned over this latter period, during which time it is claimed the world warmed at a rate and to a degree that were both unprecedented over the past millennium.

Additional information on this topic, including reviews of newer publications as they become available, can be found at http://www.co2science.org/subject/f/firegw.php.

References

Bergeron, Y. 1991. The influence of island and mainland lakeshore landscape on boreal forest fire regime. *Ecology* **72**: 1980-1992.

Bergeron, Y. and Archambault, S. 1993. Decreasing frequency of forest fires in the southern boreal zone of Quebec and its relation to global warming since the end of the "Little Ice Age." *The Holocene* **3**: 255-259.

Bergeron, Y., Gauthier, S., Kafka, V., Lefort, P. and Lesieur, D. 2001. Natural fire frequency for the eastern Canadian boreal forest: consequences for sustainable forestry. *Canadian Journal of Forest Research* **31**: 384-391.

Bergeron, Y., Flannigan, M., Gauthier, S., Leduc, A. and Lefort, P. 2004a. Past, current and future fire frequency in the Canadian boreal forest: Implications for sustainable forest management. *Ambio* **33**: 356-360.

Bergeron, Y., Gauthier, S., Flannigan, M. and Kafka, V. 2004b. Fire regimes at the transition between mixedwood and coniferous boreal forest in northwestern Quebec. *Ecology* **85**: 1916-1932.

Bergeron, Y., Cyr, D., Drever, C.R., Flannigan, M., Gauthier, S., Kneeshaw, D., Lauzon, E., Leduc, A., Le Goff, H., Lesieur, D. and Logan, K. 2006. Past, current, and future fire frequencies in Quebec's commercial forests: implications for the cumulative effects of harvesting and fire on age-class structure and natural disturbance-based management. *Canadian Journal of Forest Research* **36**: 2737-2744.

Carcaillet, C., Bergeron, Y., Richard, P.J.H., Frechette, B., Gauthier, S. and Prairie, Y. 2001. Change of fire frequency in the eastern Canadian boreal forests during the Holocene: Does vegetation composition or climate trigger the fire regime? *Journal of Ecology* **89**: 930-946.

Colombo, S.J., Cherry, M.L., Graham, C., Greifenhagen, S., McAlpine, R.S., Papadopol, C.S., Parker, W.C., Scarr, T., Ter-Mikaelien, M.T. and Flannigan, M.D. 1998. *The Impacts of Climate Change on Ontario's Forests*. Forest Research Information Paper 143, Ontario Forest Research Institute, Ontario Ministry of Natural Resources, Sault Ste. Marie, Ontario, Canada.

Cwynar, L.C. 1977. Recent history of fire of Barrow Township, Algonquin Park. *Canadian Journal of Botany* **55**: 10-21.

Foster, D.R. 1983. The history and pattern of fire in the boreal forest of southeastern Labrador. *Canadian Journal of Botany* **61**: 2459-2471.

Girardin, M.P., Tardif, J. and Flannigan, M.D. 2006. Temporal variability in area burned for the province of Ontario, Canada, during the past 2000 years inferred from tree rings. *Journal of Geophysical Research* **111**: 10.1029/2005JD006815.

IPCC 2007-II. *Climate Change 2007: Impacts, Adaptation and Vulnerability. Contribution of Working Group II to the Fourth Assessment Report of the Intergovernmental Panel on Climate Change.* Parry, M.L., Canziani, O.F., Palutikof, J.P., van der Linden, P.J. and Hanson, C.E. (Eds.) Cambridge University Press, Cambridge, UK.

Johnson, E.A., Fryer, G.I. and Heathcott, J.M. 1990. The influence of Man and climate on frequency of fire in the interior wet belt forest, British Columbia. *Journal of Ecology* **78**: 403-412.

Larsen, C.P.S. 1996. Fire and climate dynamics in the boreal forest of northern Alberta, Canada, from AD 1850 to 1985. *The Holocene* **6**: 449-456.

Larsen, C.P.S. 1997. Spatial and temporal variations in boreal forest fire frequency in northern Alberta. *Journal of Biogeography* **24**: 663-673.

Lauzon, E., Kneeshaw, D. and Bergeron, Y. 2007. Reconstruction of fire history (1680-2003) in Gaspesian mixedwood boreal forests of eastern Canada. *Forest Ecology and Management* **244**: 41-49.

Le Goff, H., Flannigan, M.D., Bergeron, Y. and Girardin, M.P. 2007. Historical fire regime shifts related to climate teleconnections in the Waswanipi area, central Quebec, Canada. *International Journal of Wildland Fire* **16**: 607-618.

McCabe, G.J., Palecki, M.A. and Betancourt, J.L. 2004. Pacific and Atlantic Ocean influences on multidecadal drought frequency in the United States. *Proceedings of the National Academy of Sciences (USA)* **101**: 4136-4141.

Meyer, G.A., Wells, S.G. and Jull, A.J.T. 1995. Fire and alluvial chronology in Yellowstone National Park: Climatic and intrinsic controls on Holocene geomorphic processes. *Geological Society of America Bulletin* **107**: 1211-1230.

Parker, W.C., Colombo, S.J., Cherry, M.L., Flannigan, M.D., Greifenhagen, S., McAlpine, R.S., Papadopol, C. and Scarr, T. 2000. Third millennium forestry: What climate change might mean to forests and forest management in Ontario. *Forest Chronicles* **76**: 445-463.

Pierce, J.L., Meyer, G.A. and Jull, A.J.T. 2004. Fire-induced erosion and millennial-scale climate change in northern ponderosa pine forests. *Nature* **432**: 87-90.

Pitkanen, A., Huttunen, P., Jungner, H., Merilainen, J. and Tolonen, K. 2003. Holocene fire history of middle boreal pine forest sites in eastern Finland. *Annales Botanici Fennici* **40**: 15-33.

Podur, J., Martell, D.L. and Knight, K. 2002. Statistical quality control analysis of forest fire activity in Canada. *Canadian Journal of Forest Research* **32**: 195-205.

Riano, D., Moreno Ruiz, J.A., Isidoro, D. and Ustin, S.L. 2007. Global spatial patterns and temporal trends of burned area between 1981 and 2000 using NOAA-NASA Pathfinder. *Global Change Biology* **13**: 40-50.

Rollins, M.G., Morgan, P. and Swetnam, T. 2002. Landscape-scale controls over 20th century fire occurrence in two large Rocky Mountain (USA) wilderness areas. *Landscape Ecology* **17**: 539-557.

Running, S.W. 2006. Is global warming causing more, larger wildfires? Scienc*express* 6 July 2006 10.1126/science.1130370.

Schoennagel, T., Veblen, T.T., Kulakowski, D. and Holz, A. 2007. Multidecadal climate variability and climate interactions affect subalpine fire occurrence, western Colorado (USA). *Ecology* **88**: 2891-2902.

Schoennagel, T., Veblen, T.T., Romme, W.H., Sibold, J.S. and Cook, E.R. 2005. ENSO and PDO variability affect drought-induced fire occurrence in Rocky Mountain subalpine forests. *Ecological Applications* **15**: 2000-2014.

Stine, S. 1998. In: Issar, A.S. and Brown, N. (Eds.) *Water, Environment and Society in Times of Climatic Change.* Kluwer, Dordrecth, The Netherlands, pp. 43-67.

Van Wagner, C.E. 1978. Age-class distribution and the forest fire cycle. *Canadian Journal of Forest Research* **8**: 220-227.

Weir, J.M.H., Johnson, E.A. and Miyanishi, K. 2000. Fire frequency and the spatial age mosaic of the mixed-wood boreal forest in western Canada. *Ecological Applications* **10**: 1162-1177.

Westerling, A.L., Hidalgo, H.G., Cayan, D.R. and Swetnam, T.W. 2006. Warming and earlier spring increases western U.S. Forest wildfire activity. Scienc*express* 6 July 2006 10.1126/science.1128834.

Westerling, A.L. and Swetnam, T.W. 2003. Interannual to decadal drought and wildfire in the western United States. *EOS: Transactions, American Geophysical Union* **84**: 545-560.

Whitlock, C., Shafer, S.L. and Marlon, J. 2003. The role of climate and vegetation change in shaping past and future

359

fire regimes in the northwestern US and the implications for ecosystem management. *Forest Ecology and Management* **178**: 163-181.

Woodhouse, C.A. and Overpeck, J.T. 1998. 2000 years of drought variability in the central United States. *Bulletin of the American Meteorological Society* **79**: 2693-2714.

Wotton, B.M. and Flanigan, M.D. 1993. Length of the fire season in a changing climate. *Forest Chronicles* **69**: 187-192.

7

Biological Effects of Carbon Dioxide Enrichment

Introduction

The Contribution of Group 1 to the Fourth Assessment Report of the Intergovernmental Panel on Climate Change (IPCC) hardly mentions the beneficial effects of earth's rising atmospheric carbon dioxide (CO_2) concentration on the biosphere. In a chapter titled "Changes in Atmospheric Constituents and in Radiative Forcing" the authors say the following (IPCC, 2007–I, p. 186):

> Increased CO_2 concentrations can also "fertilize" plants by stimulating photosynthesis, which models suggest has contributed to increased vegetation cover and leaf area over the 20th century (Cramer *et al.*, 2001). Increases in the Normalized Difference Vegetation Index, a remote sensing product indicative of leaf area, biomass and potential photosynthesis, have been observed (Zhou *et al.*, 2001), although other causes including climate change itself are also likely to have contributed. Increased vegetative cover and leaf area would decrease surface albedo, which would act to oppose the increase in albedo due to deforestation. The RF due to this process has not been evaluated and there is a very low scientific understanding of these effects.

Later in that report, in a chapter titled "Couplings Between Changes in the Climate System and Biogeochemistry," a single paragraph is devoted to the "effects of elevated carbon dioxide" on plants. The paragraph concludes, "it is not yet clear how strong the CO_2 fertilization effect actually is" (p. 527).

Since CO_2 fertilization could affect crop yields and how efficiently plants use mineral nutrients and water, one would expect the subject to be addressed in the Contribution of Group 2, on "Impacts, Adaption and Vulnerability," and indeed it is, in a chapter titled "Food, Fibre and Forest Products" (IPCC, 2007-II). But that chapter belittles and largely ignores research on the benefits of enhanced CO_2 while exaggerating the possible negative effects of rapidly rising temperatures and extreme weather events predicted by computer models. The subject is not even mentioned in Chapter 8 of that report, on "Human Health," although even a modest effect on crops would have some effect on human health. (See Chapter 9 of the present report for our own, more complete, discussion of the health effects of climate change.)

The IPCC's failure to report the beneficial effects of rising CO_2 concentrations is surprising when literally thousands of peer-reviewed journal articles exist on the subject. It is also a major defect of the IPCC report and one reason why it is not a reliable summary of the science of climate change. In this chapter, we seek to provide the balance that eluded the IPCC.

The chapter begins with a survey of the scientific literature on the productivity responses of plants to higher CO_2 concentrations, and then reviews research on the effect of enhanced CO_2 on plant water-use efficiency, responsiveness to environmental stress, acclimation, competition among species (e.g., crops versus weeds), respiration, and other effects. We end with a survey of literature showing the general "greening of the Earth" that has occurred during the Current Warm Period.

References

Cramer, W., A. Bondeau, F. I. Woodward, I. C. Prentice, R. A. Betts, V. Brovkin, P. M. Cox, V. Fisher, J. Foley, A. D. Friend, C. Kucharik, M. R. Lomas, N. Ramankutty, S. Sitch, B. Smith, A. White, and C. Young-Molling. 2001. Global response of terrestrial ecosystem structure and function to CO_2 and climate change: Results from six dynamic global vegetation models. *Global Change Biol.* **7**: 357-373.

IPCC. 2007-I. *Climate Change 2007: The Physical Science Basis. Contribution of Working Group I to the Fourth Assessment Report of the Intergovernmental Panel on Climate Change.* Solomon, S., D. Qin, M. Manning, Z. Chen, M. Marquis, K.B. Averyt, M. Tignor and H.L. Miller. (Eds.) Cambridge University Press, Cambridge, UK.

IPCC. 2007-II. *Climate Change 2007: Impacts, Adaptation and Vulnerability. Contribution of Working Group II to the Fourth Assessment Report of the Intergovernmental Panel on Climate Change.* M.L. Parry, O.F. Canziani, J.P. Palutikof, P.J. van der Linden and Hanson, C.D. (Eds.) Cambridge University Press, Cambridge, UK.

Zhou, L.M., Tucker, C.J., Kaufmann, R.K., Slayback, D., Shabanov, N.V., and Myneni, R.B. 2001. Variations in northern vegetation activity inferred from satellite data of vegetation index during 1981 to 1999. *Journal of Geophysical Research* **106** (D17): 20069-20083.

7.1. Plant Productivity Responses

Perhaps the best-known consequence of the rise in atmospheric CO_2 is the stimulation of plant productivity. This growth enhancement occurs because carbon dioxide is the primary raw material utilized by plants to produce the organic matter out of which they construct their tissues. Consequently, the more CO_2 there is in the air, the better plants grow. Over the past decade, the Center for the Study of Carbon Dioxide and Global Change, has archived thousands of results from hundreds of peer-reviewed research studies conducted by hundreds of researchers demonstrating this fact. The archive is available free of charge at http://www.co2science.org/data/plantgrowth/plantgrowth.php.

The Center's Web site lists the photosynthetic and dry weight responses of plants growing in CO_2-enriched air, arranged by scientific or common plant name. It also provides the full peer-reviewed journal references and experimental conditions in which each study was conducted for each record. We have summarized those results in two tables appearing in Appendix 2 and Appendix 3 to the current report. The first table, Table 7.1.1, indicates the mean biomass response of nearly 1,000 plants to a 300-ppm increase in atmospheric CO_2 concentration. The second table, Table 7.1.2, indicates the photosynthetic response to the same CO_2 enrichment for a largely similar list of plants.

In the rest of this section of Chapter 7, we provide a review of research on a representative sample of herbaceous and woody plants, chosen with an eye toward their importance to agriculture and the forestry and papermaking industries, followed by several acquatic plants.

7.1.1. Herbaceous Plants

A 300 ppm increase in the air's CO_2 content typically raises the productivity of most herbaceous plants by about one-third (Cure and Acock, 1986; Mortensen, 1987). This positive response occurs in plants that utilize all three of the major biochemical pathways (C_3, C_4, and crassulacean acid metabolism (CAM)) of photosynthesis (Poorter, 1993). Thus, with more CO_2 in the air, the productivity of nearly all crops rises, as they produce more branches and tillers, more and thicker leaves, more extensive root systems, and more flowers and fruit (Idso, 1989).

On average, a 300 ppm increase in atmospheric CO_2 enrichment leads to yield increases of 15 percent for CAM crops, 49 percent for C_3 cereals, 20 percent for C_4 cereals, 24 percent for fruits and melons, 44 percent for legumes, 48 percent for roots and tubers, and 37 percent for vegetables (Idso and Idso, 2000). It should come as no surprise, therefore, that the father of modern research in this area—Sylvan H. Wittwer—has said "it should be considered good fortune that we are living in a world of gradually increasing levels of atmospheric CO_2," and "the rising level of atmospheric CO_2 is a universally free premium, gaining in magnitude with time, on which we can all reckon for the future."

Additional information on this topic, including reviews of herbaceous plants not discussed here, can be found at http://www.co2science.org/subject/a/subject_a.php under the heading Agriculture.

References

Cure, J.D., and Acock, B. (1986). Crop Responses to Carbon Dioxide Doubling: A Literature Survey. *Agric. For. Meteorol.* **38**, 127-145.

Idso, C.D. and Idso, K.E. (2000) Forecasting world food supplies: The impact of rising atmospheric CO_2 concentration. Technology **7** (suppl): 33-56.

Idso, S.B. (1989) *Carbon Dioxide: Friend or Foe?* IBR Press, Tempe, AZ.

Mortensen, L.M. (1987). Review: CO_2 Enrichment in Greenhouses. Crop Responses. *Sci. Hort.* **33**, 1-25.

Poorter, H. (1993). Interspecific Variation in the Growth Response of Plants to an Elevated Ambient CO_2 Concentration. *Vegetatio* **104/105** 77-97.

7.1.1.1. Alfalfa

Morgan *et al.* (2001) grew the C_3 legume alfalfa (*Medicago sativa* L.) for 20 days post-defoliation in growth chambers maintained at atmospheric CO_2 concentrations of 355 and 700 ppm and low or high levels of soil nitrogen to see how these factors affected plant regrowth. They determined that the plants in the elevated CO_2 treatment attained total dry weights over the 20-day regrowth period that were 62 percent greater than those reached by the plants grown in ambient air, irrespective of soil nitrogen concentration.

De Luis *et al.* (1999) grew alfalfa plants in controlled environment chambers in air of 400 and 700 ppm CO_2 for two weeks before imposing a two-week water treatment on them, wherein the soil in which half of the plants grew was maintained at a moisture content approaching field capacity while the soil in which the other half grew was maintained at a moisture content that was only 30 percent of field capacity. Under these conditions, the CO_2-enriched water-stressed plants displayed an average water-use

efficiency that was 2.6 and 4.1 times greater than that of the water-stressed and well-watered plants, respectively, growing in ambient 400-ppm-CO_2 air. In addition, under ambient CO_2 conditions, the water stress treatment increased the mean plant root:shoot ratio by 108 percent, while in the elevated CO_2 treatment it increased it by 269 percent. As a result, the nodule biomass on the roots of the CO_2-enriched water-stressed plants was 40 and 100 percent greater than the nodule biomass on the roots of the well-watered and water-stressed plants, respectively, growing in ambient air. Hence, the CO_2-enriched water-stressed plants acquired 31 and 97 percent more total plant nitrogen than the well-watered and water-stressed plants, respectively, growing in ambient air. The CO_2-enriched water-stressed plants attained 2.6 and 2.3 times more total biomass than the water-stressed and well-watered plants, respectively, grown at 400 ppm CO_2.

Luscher *et al.* (2000) grew effectively and ineffectively nodulating (good nitrogen-fixing vs. poor nitrogen-fixing) alfalfa plants in large free-air CO_2 enrichment (FACE) plots for multiple growing seasons at atmospheric CO_2 concentrations of 350 and 600 ppm, while half of the plants in each treatment received a high supply of soil nitrogen and the other half received only minimal amounts of this essential nutrient. The extra CO_2 increased the yield of effectively nodulating plants by about 50 percent, regardless of soil nitrogen supply; caused a 25 percent yield reduction in ineffectively nodulating plants subjected to low soil nitrogen; and produced an intermediate yield stimulation of 11 percent for the same plants under conditions of high soil nitrogen, which suggests that the ability to symbiotically fix nitrogen is an important factor in eliciting strong positive growth responses to elevated CO_2 under conditions of low soil nitrogen supply.

Sgherri *et al.* (1998) grew alfalfa in open-top chambers at ambient (340 ppm) and enriched (600 ppm) CO_2 concentrations for 25 five days, after which water was withheld for five additional days so they could investigate the interactive effects of elevated CO_2 and water stress on plant water status, leaf soluble protein and carbohydrate content, and chloroplast thylakoid membrane composition. They found that the plants grown in elevated CO_2 exhibited the best water status during the moisture deficit part of the study, as indicated by leaf water potentials that were approximately 30 percent higher (less negative) than those observed in plants grown in ambient CO_2. This beneficial adjustment was achieved by partial closure of leaf stomata and by greater production of nonstructural carbohydrates (a CO_2-induced enhancement of 50 percent was observed), both of which phenomena can lead to decreases in transpirational water loss, the former by guard cells physically regulating stomatal apertures to directly control the exodus of water from leaves, and the latter by nonstructural carbohydrates influencing the amount of water available for transpiration. This latter phenomenon occurs because many nonstructural carbohydrates are osmotically active solutes that chemically associate with water through the formation of hydrogen bonds, thereby effectively reducing the amount of unbound water available for bulk flow during transpiration. Under water-stressed conditions, however, the CO_2-induced difference in total leaf nonstructural carbohydrates disappeared. This may have resulted from an increased mobilization of nonstructural carbohydrates to roots in the elevated CO_2 treatment, which would decrease the osmotic potential in that part of the plant, thereby causing an increased influx of soil moisture into the roots. If this did indeed occur, it would also contribute to a better overall water status of CO_2-enriched plants during drought conditions.

The plants grown at elevated CO_2 also maintained greater leaf chlorophyll contents and lipid to protein ratios, especially under conditions of water stress. Leaf chlorophyll content, for example, decreased by a mere 6 percent at 600 ppm CO_2, while it plummeted by approximately 30 percent at 340 ppm, when water was withheld. Moreover, leaf lipid contents in plants grown with atmospheric CO_2 enrichment were about 22 and 83 percent higher than those measured in plants grown at ambient CO_2 during periods of ample and insufficient soil moisture supply, respectively. Furthermore, at elevated CO_2 the average amounts of unsaturation for two of the most important lipids involved in thylakoid membrane composition were approximately 20 and 37 percent greater than what was measured in plants grown at 340 ppm during times of adequate and inadequate soil moisture, respectively. The greater lipid contents observed at elevated CO_2, and their increased amounts of unsaturation, may allow thylakoid membranes to maintain a more fluid and stable environment, which is critical during periods of water stress in enabling plants to continue photosynthetic carbon uptake.

Additional information on this topic, including reviews of newer publications as they become available, can be found at http://www.co2science.org/subject/a/agriculturealfalfa.php.

References

De Luis, J., Irigoyen, J.J. and Sanchez-Diaz, M. 1999. Elevated CO_2 enhances plant growth in droughted N_2-fixing alfalfa without improving water stress. *Physiologia Plantarum* **107**: 84-89.

Luscher, A., Hartwig, U.A., Suter, D. and Nosberger, J. 2000. Direct evidence that symbiotic N_2 fixation in fertile grassland is an important trait for a strong response of plants to elevated atmospheric CO_2. *Global Change Biology* **6**: 655-662.

Morgan, J.A., Skinner, R.H. and Hanson, J.D. 2001. Nitrogen and CO_2 affect regrowth and biomass partitioning differently in forages of three functional groups. *Crop Science* **41**: 78-86.

Sgherri, C.L.M., Quartacci, M.F., Menconi, M., Raschi, A. and Navari-Izzo, F. 1998. Interactions between drought and elevated CO_2 on alfalfa plants. *Journal of Plant Physiology* **152**: 118-124.

7.1.1.2. Cotton

As the CO_2 content of the air increases, cotton (*Gossypium hirsutum* L.) plants typically display enhanced rates of photosynthetic carbon uptake, as noted by Reddy *et al.* (1999), who reported that twice-ambient atmospheric CO_2 concentrations boosted photosynthetic rates of cotton by 137 to 190 percent at growth temperatures ranging from 2°C below ambient to 7°C above ambient.

Elevated CO_2 also enhances total plant biomass and harvestable yield. Reddy *et al.* (1998), for example, reported that plant biomass at 700 ppm CO_2 was enhanced by 31 to 78 percent at growth temperatures ranging from 20 to 40°C, while boll production was increased by 40 percent. Similarly, Tischler *et al.* (2000) found that a doubling of the atmospheric CO_2 concentration increased seedling biomass by at least 56 percent.

These results indicate that elevated CO_2 concentrations tend to ameliorate the negative effects of heat stress on productivity and growth in cotton. In addition, Booker (2000) discovered that elevated CO_2 reduced the deleterious effects of elevated ozone on leaf biomass and starch production.

Atmospheric CO_2 enrichment also can induce changes in cotton leaf chemistry that tend to increase carbon sequestration in plant biolitter and soils. Booker *et al.* (2000), for example, observed that biolitter produced from cotton plants grown at 720 ppm CO_2 decomposed at rates that were 10 to 14 percent slower than those displayed by ambiently grown plants; a after three years of exposure to air containing 550 ppm CO_2, Leavitt *et al.* (1994) reported that 10 percent of the organic carbon present in soils beneath CO_2-enriched FACE plots resulted from the extra CO_2 supplied to them.

In summary, as the CO_2 content of the air increases, cotton plants will display greater rates of photosynthesis and biomass production, which should lead to greater boll production in this important fiber crop, even under conditions of elevated air temperature and ozone concentration. In addition, carbon sequestration in fields planted to cotton should also increase with future increases in the air's CO_2 content.

Additional information on this topic, including reviews of newer publications as they become available, can be found at http://www.co2science.org/subject/a/agriculturecotton.php.

References

Booker, F.L. 2000. Influence of carbon dioxide enrichment, ozone and nitrogen fertilization on cotton (*Gossypium hirsutum* L.) leaf and root composition. *Plant, Cell and Environment* **23**: 573-583.

Booker, F.L., Shafer, S.R., Wei, C.-M. and Horton, S.J. 2000. Carbon dioxide enrichment and nitrogen fertilization effects on cotton (*Gossypium hirsutum* L.) plant residue chemistry and decomposition. *Plant and Soil* **220**: 89-98.

Leavitt, S.W., Paul, E.A., Kimball, B.A., Hendrey, G.R., Mauney, J.R., Rauschkolb, R., Rogers, H., Lewin, K.F., Nagy, J., Pinter Jr., P.J. and Johnson, H.B. 1994. Carbon isotope dynamics of free-air CO_2-enriched cotton and soils. *Agricultural and Forest Meteorology* **70**: 87-101.

Reddy, K.K., Davidonis, G.H., Johnson, A.S. and Vinyard, B.T. 1999. Temperature regime and carbon dioxide enrichment alter cotton boll development and fiber properties. *Agronomy Journal* **91**: 851-858.

Reddy, K.R., Robana, R.R., Hodges, H.F., Liu, X.J. and McKinion, J.M. 1998. Interactions of CO_2 enrichment and temperature on cotton growth and leaf characteristics. *Environmental and Experimental Botany* **39**: 117-129.

Tischler, C.R., Polley, H.W., Johnson, H.B. and Pennington, R.E. 2000. Seedling response to elevated CO_2 in five epigeal species. *International Journal of Plant Science* **161**: 779-783.

7.1.1.3. Maize

Maroco *et al.* (1999) grew corn (*Zea mays* L.), or maize as it is often called, for 30 days in plexiglass chambers maintained at either ambient or triple-ambient concentrations of atmospheric CO_2 to determine the effects of elevated CO_2 on the growth of this important agricultural C_4 species. This exercise revealed that elevated CO_2 (1,100 ppm) increased maize photosynthetic rates by about 15 percent relative to those measured in plants grown at 350 ppm CO_2, in spite of the fact that both rubisco and PEP-carboxylase were down-regulated. This increase in carbon fixation likely contributed to the 20 percent greater biomass accumulation observed in the CO_2-enriched plants. In addition, leaves of CO_2-enriched plants contained approximately 10 percent fewer stomates per unit leaf area than leaves of control plants, and atmospheric CO_2 enrichment reduced stomatal conductance by as much as 71 percent in elevated-CO_2-grown plants. As a result of these several different phenomena, the higher atmospheric CO_2 concentration greatly increased the intrinsic water-use efficiency of the CO_2-enriched plants.

In a study designed to examine the effects of elevated CO_2 under real-world field conditions, Leakey *et al.* (2004) grew maize out-of-doors at the SoyFACE facility in the heart of the United States Corn Belt, while exposing different sections of the field to atmospheric CO_2 concentrations of either 354 or 549 ppm. The crop was grown, in the words of the researchers, using cultural practices deemed "typical for this region of Illinois," during a year that turned out to have experienced summer rainfall that was "very close to the 50-year average for this site, indicating that the year was not atypical or a drought year." Then, on five different days during the growing season (11 and 22 July, 9 and 21 August, and 5 September), they measured diurnal patterns of photosynthesis, stomatal conductance, and microclimatic conditions.

Contrary to what many people had long assumed would be the case for a C_4 crop such as corn growing under even the best of natural conditions, Leakey *et al.* found that "growth at elevated CO_2 significantly increased leaf photosynthetic CO_2 uptake rate by up to 41 percent." The greatest whole-day increase was 21 percent (11 July) followed by 11 percent (22 July), during a period of low rainfall. Thereafter, however, during a period of greater rainfall, there were no significant differences between the photosynthetic rates of the plants in the two CO_2 treatments, so that over the entire growing season, the CO_2-induced increase in leaf photosynthetic rate averaged 10 percent.

Additionally, on all but the first day of measurements, stomatal conductance was significantly lower (-23 percent on average) under elevated CO_2 compared to ambient CO_2, which led to reduced transpiration rates in the CO_2-enriched plants on those days as well. Since "low soil water availability and high evaporative demand can both generate water stress and inhibit leaf net CO_2 assimilation in C_4 plants," they state that the lower stomatal conductance and transpiration rate they observed under elevated CO_2 "may have counteracted the development of water stress under elevated CO_2 and prevented the inhibition of leaf net CO_2 assimilation observed under ambient CO_2."

The implication of their research, in the words of Leakey *et al.*, was that "contrary to expectations, this US Corn Belt summer climate appeared to cause sufficient water stress under ambient CO_2 to allow the ameliorating effects of elevated CO_2 to significantly enhance leaf net CO_2 assimilation." They concluded that "this response of *Z. mays* to elevated CO_2 indicates the potential for greater future crop biomass and harvestable yield across the US Corn Belt."

Also germane to this subject and supportive of the above conclusion are the effects of elevated CO_2 on weeds associated with corn. Conway and Toenniessen (2003), for example, speak of maize in Africa being attacked by the parasitic weed *Striga hermonthica*, which sucks vital nutrients from its roots, as well as from the roots of many other C_4 crops of the semi-arid tropics, including sorghum, sugar cane, and millet, plus the C_3 crop rice, particularly throughout much of Africa, where *Striga* is one of the region's most economically important parasitic weeds. Research shows how atmospheric CO_2 enrichment greatly reduces the damage done by this devastating weed (Watling and Press, 1997; Watling and Press, 2000).

Baczek-Kwinta and Koscielniak (2003) studied another phenomenon that is impacted by atmospheric CO_2 enrichment and that can affect the productivity of maize. Noting the tropical origin of maize and that the crop "is extremely sensitive to chill (temperatures 0-15°C)," they report that it is nevertheless often grown in cooler temperate zones. In such circumstances, however, maize can experience a variety of maladies associated with exposure to periods of low air temperature. To see if elevated CO_2 either exacerbates or ameliorates this problem, they grew two hybrid genotypes—KOC 9431 (chill-resistant) and

K103xK85 (chill-sensitive)—from seed in air of either ambient (350 ppm) or elevated (700 ppm) CO_2 concentration (AC or EC, respectively), after which they exposed the plants to air of 7°C for eleven days, whereupon they let them recover for one day in ambient air of 20°C, all the while measuring several physiological and biochemical parameters pertaining to the plants' third fully expanded leaves.

The two researchers' protocol revealed that "EC inhibited chill-induced depression of net photosynthetic rate (PN), especially in leaves of chill-resistant genotype KOC 9431," which phenomenon "was distinct not only during chilling, but also during the recovery of plants at 20°C." In fact, they found that "seedlings subjected to EC showed 4-fold higher PN when compared to AC plants." They also determined that "EC diminished the rate of superoxide radical formation in leaves in comparison to the AC control." In addition, they found that leaf membrane injury "was significantly lower in samples of plants subjected to EC than AC." Last, they report that enrichment of the air with CO_2 successfully inhibited the decrease in the maximal quantum efficiency of photosystem 2, both after chilling and during the one-day recovery period. And in light of all of these positive effects of elevated CO_2, they concluded that "the increase in atmospheric CO_2 concentration seems to be one of the protective factors for maize grown in cold temperate regions."

But what about the effects of climate change, both past and possibly future, on corn production? For nine areas of contrasting environment within the Pampas region of Argentina, Magrin *et al.* (2005) evaluated changes in climate over the twentieth century along with changes in the yields of the region's chief crops. Then, after determining upward low-frequency trends in yield due to technological improvements in crop genetics and management techniques, plus the aerial fertilization effect of the historical increase in the air's CO_2 concentration, annual yield anomalies and concomitant climatic anomalies were calculated and used to develop relations describing the effects of changes in precipitation, temperature and solar radiation on crop yields, so that the effects of long-term changes in these climatic parameters on Argentina agriculture could be determined.

Noting that "technological improvements account for most of the observed changes in crop yields during the second part of the twentieth century," which totaled 110 percent for maize, Magrin *et al.* report that due to changes in climate between the periods 1950-70 and 1970-99, maize yields increased by 18 percent.

Much the same has been found to be true in Alberta, Canada, where Shen *et al.* (2005) derived and analyzed long-term (1901-2002) temporal trends in the agroclimate of the region. They report, for example, that "an earlier last spring frost, a later first fall frost, and a longer frost-free period are obvious all over the province." They also found that May-August precipitation in Alberta increased 14 percent from 1901 to 2002, and that annual precipitation exhibited a similar increasing trend, with most of the increase coming in the form of low-intensity events. In addition, the researchers note that "the area with sufficient corn heat units for corn production, calculated according to the 1973-2002 normal, has extended to the north by about 200-300 km, when compared with the 1913-32 normal, and by about 50-100 km, when compared with the 1943-72 normal."

In light of these findings, Shen *et al.* conclude that "the changes of the agroclimatic parameters imply that Alberta agriculture has benefited from the last century's climate change," emphasizing that "the potential exists to grow crops and raise livestock in more regions of Alberta than was possible in the past." They also note that the increase in the length of the frost-free period "can greatly reduce the frost risks to crops and bring economic benefits to Alberta agricultural producers," and that the northward extension of the corn heat unit boundary that is sufficient for corn production "implies that Alberta farmers now have a larger variety of crops to choose from than were available previously." Hence, they say "there is no hesitation for us to conclude that the warming climate and increased precipitation benefit agriculture in Alberta."

With respect to the future, Bootsma *et al.* (2005) derived relationships between agroclimatic indices and average yields of major grain crops, including corn, from field trials conducted in eastern Canada, after which they used them to estimate potential impacts of projected climate change scenarios on anticipated average yields for the period 2040 to 2069. Based on a range of available heat units projected by multiple General Circulation Model (GCM) experiments, they determined that average yields achievable in field trials could increase by 40 to 115 percent for corn, "not including the direct effect of increased atmospheric CO_2 concentrations." Adding expected CO_2 increases to the mix, along with gains in yield anticipated to be achieved through breeding and improved technology, these numbers rose to 114 to 186 percent.

In light of their findings, Bootsma *et al.* predict there will be a "switch to high-energy and high-protein-content crops (corn and soybeans) that are better adapted to the warmer climate."

In summary, as the air's CO_2 content continues to rise, and even if the climate of the world changes in the ways suggested by GCM and IPCC calculations, maize plants will likely display greater rates of photosynthesis and biomass production, as well as reduced transpirational water losses and increased water-use efficiencies, and more areas of the world will likely become suitable for growing this important crop.

Additional information on this topic, including reviews of newer publications as they become available, can be found at http://www.co2science.org/subject/a/agriculturemaize.php.

References

Baczek-Kwinta, R. and Koscielniak, J. 2003. Anti-oxidative effect of elevated CO_2 concentration in the air on maize hybrids subjected to severe chill. *Photosynthetica* **41**: 161-165.

Bootsma, A., Gameda, S. and McKenney, D.W. 2005. Potential impacts of climate change on corn, soybeans and barley yields in Atlantic Canada. *Canadian Journal of Plant Science* **85**: 345-357.

Conway, G. and Toenniessen, G. 2003. Science for African food security. *Science* **299**: 1187-1188.

Leakey, A.D.B., Bernacchi, C.J., Dohleman, F.G., Ort, D.R. and Long, S.P. 2004. Will photosynthesis of maize (*Zea mays*) in the US Corn Belt increase in future [CO_2] rich atmospheres? An analysis of diurnal courses of CO_2 uptake under free-air concentration enrichment (FACE). *Global Change Biology* **10**: 951-962.

Magrin, G.O., Travasso, M.I. and Rodriguez, G.R. 2005. Changes in climate and crop production during the twentieth century in Argentina. *Climatic Change* **72**: 229-249.

Maroco, J.P., Edwards, G.E. and Ku, M.S.B. 1999. Photosynthetic acclimation of maize to growth under elevated levels of carbon dioxide. *Planta* **210**: 115-125.

Shen, S.S.P., Yin, H., Cannon, K., Howard, A., Chetner, S. and Karl, T.R. 2005. Temporal and spatial changes of the agroclimate in Alberta, Canada, from 1901 to 2002. *Journal of Applied Meteorology* **44**: 1090-1105.

Watling, J.R. and Press, M.C. 1997. How is the relationship between the C$_4$ cereal *Sorghum bicolor* and the C$_3$ root hemi-parasites *Striga hermonthica* and *Striga asiatica* affected by elevated CO$_2$? *Plant, Cell and Environment* **20**: 1292-1300.

Watling, J.R. and Press, M.C. 2000. Infection with the parasitic angiosperm Striga hermonthica influences the response of the C$_3$ cereal *Oryza sativa* to elevated CO$_2$. *Global Change Biology* **6**: 919-930.

7.1.1.4. Peanut

Stanciel *et al.* (2000) grew peanuts (*Arachis hypogaea* L.) hydroponically for 110 days in controlled environment chambers maintained at atmospheric CO$_2$ concentrations of 400, 800 and 1200 ppm, finding that the net photosynthetic rates of plants grown at 800 ppm CO$_2$ were 29 percent greater than those of plants grown at 400 ppm CO$_2$, but that plants grown at 1200 ppm CO$_2$ displayed photosynthetic rates that were 24 percent *lower* than those exhibited by plants grown in 400-ppm CO$_2$ air. Nevertheless, the number of pods, pod weight and seed dry weight per unit area were all greater at 1200 ppm than at 400 ppm CO$_2$. Also, harvest index, which is the ratio of seed dry weight to pod dry weight, was 19 and 31 percent greater at 800 and 1200 ppm CO$_2$, respectively, than it was at 400 ppm CO$_2$. In addition, as the atmospheric CO$_2$ concentration increased, stomatal conductance decreased, becoming 44 and 50 percent lower at 800 and 1200 ppm than it was at 400 ppm CO$_2$. Thus, atmospheric CO$_2$ enrichment also reduced transpirational water loss, leading to a significant increase in plant water use efficiency.

Prasad *et al.* (2003) grew Virginia Runner (Georgia Green) peanuts from seed to maturity in sunlit growth chambers maintained at atmospheric CO$_2$ concentrations of 350 and 700 ppm and daytime-maximum/nighttime-minimum air temperatures of 32/22, 36/26, 40/30 and 44/34°C, while they assessed various aspects of vegetative and reproductive growth. In doing so, they found that leaf photosynthetic rates were unaffected by air temperature over the range investigated, but they rose by 27 percent in response to the experimental doubling of the air's CO$_2$ content. Vegetative biomass, on the other hand, increased by 51 percent and 54 percent in the ambient and CO$_2$-enriched air, respectively, as air temperature rose from 32/22 to 40/30°C. A further air temperature increase to 44/34°C, however, caused moderate to slight decreases in vegetative biomass in both the ambient

and CO$_2$-enriched air, so that the final biomass increase over the entire temperature range investigated was 27 percent in ambient air and 53 percent in CO$_2$-enriched air. When going from the lowest temperature ambient CO$_2$ treatment to the highest temperature elevated CO$_2$ treatment, however, there was a whopping 106 percent increase in vegetative biomass.

By contrast, seed yields in both the ambient and CO$_2$-enriched air dropped dramatically with each of the three temperature increases studied, declining at the highest temperature regime to but a small percentage of what they were at the lowest temperature regime. Nevertheless, Prasad *et al.* report that "seed yields at 36.4/26.4°C under elevated CO$_2$ were similar to those obtained at 32/22°C under ambient CO$_2$," the latter pair of which temperatures they describe as "present-day seasonal temperatures."

It would appear that a warming of 4.4°C above present-day seasonal temperatures for peanut production would have essentially no effect on peanut seed yields, as long as the atmosphere's CO$_2$ concentration rose concurrently, by something on the order of 350 ppm. It is also important to note, according to Prasad *et al.*, that "maximum/minimum air temperatures of 32/22°C and higher are common in many peanut-producing countries across the globe." In fact, they note that "the Anantapur district in Andhra Pradesh, which is one of the largest peanut-producing regions in India, experiences season-long temperatures considerably greater than 32/22°C from planting to maturity."

In light of these real-world observations, i.e., that some of the best peanut-producing regions in the world currently experience air temperatures considerably greater than what Prasad *et al.* suggest is optimum for peanuts (something *less* than 32/22°C), it would appear that real-world declines in peanut seed yields in response to a degree or two of warming, even in air of ambient CO$_2$ concentration, must be very slight or even non-existent, for how else could the places that commonly experience these considerably higher temperatures remain some of the best peanut-producing areas in the world? This in turn suggests that for more realistic values of CO$_2$-induced global warming, i.e., temperature increases on the order of 0.4°C for a doubling of the air's CO$_2$ content (Idso, 1998), there would likely be a significant increase in real-world peanut production.

In another pertinent study, Vu (2005) grew peanut plants from seed to maturity in greenhouses maintained at atmospheric CO$_2$ concentrations of 360 and 720 ppm and at air temperatures that were 1.5 and 6.0°C above outdoor air temperatures, while he measured a number of parameters related to the plants' photosynthetic performance. His work revealed that although Rubisco protein content and activity were down-regulated by elevated CO_2, the Rubisco photosynthetic efficiency (the ratio of midday light-saturated carbon exchange rate to Rubisco initial or total activity) of the elevated-CO_2 plants "was 1.3- to 1.9-fold greater than that of the ambient-CO_2 plants at both growth temperatures." He also determined that "leaf soluble sugars and starch of plants grown at elevated CO_2 were 1.3- and 2-fold higher, respectively, than those of plants grown at ambient CO_2." In addition, he discovered that the leaf transpiration of the elevated-CO_2 plants relative to that of the ambient-CO_2 plants was 12 percent less at near-ambient temperatures and 17 percent less in the higher temperature regime, while the water use efficiency of the elevated-CO_2 plants relative to the ambient-CO_2 plants was 56 percent greater at near-ambient temperatures and 41 percent greater in the higher temperature environment.

In commenting on his findings, Vu notes that because less Rubisco protein was required by the elevated-CO_2 plants, the subsequent redistribution of excess leaf nitrogen "would increase the efficiency of nitrogen use for peanut under elevated CO_2," just as the optimization of inorganic carbon acquisition and greater accumulation of the primary photosynthetic products in the CO_2-enriched plants "would be beneficial for peanut growth at elevated CO_2." Indeed, in the absence of other stresses, Vu's conclusion was that "peanut photosynthesis would perform well under rising atmospheric CO_2 and temperature predicted for this century."

In a somewhat different type of study, Alexandrov and Hoogenboom (2000) studied how year-to-year changes in temperature, precipitation and solar radiation had influenced the yields of peanuts over a 30-year period in the southeastern United States, after which they used the results to predict future crop yields based on climate-change output from various global circulation models (GCMs) of the atmosphere. At ambient CO_2 concentrations, the GCM scenarios suggested a decrease in peanut yields by the year 2020, due in part to predicted changes in temperature and precipitation. However, when the yield-enhancing effects of a doubling of the atmospheric CO_2 concentration were included, a totally different result was obtained: a yield *increase*.

Although we have little faith in GCM scenarios, it is interesting to note that their climate change predictions often result in positive outcomes for agricultural productivity when the direct effects of elevated CO_2 on plant growth and development are included in the analyses. These results support the research reported later in this chapter describing the stress-ameliorating effects of atmospheric CO_2 enrichment on plant growth and development under unfavorable growing conditions characterized by high air temperatures and inadequate soil moisture.

In conclusion, it would appear that even if the climate changes that are typically predicted to result from anticipated increases in the air's CO_2 content were to materialize, the concurrent rise in the air's CO_2 concentration should more than compensate for any deleterious effects those changes in climate might otherwise have had on the growth and yield of peanuts.

Additional information on this topic, including reviews of newer publications as they become available, can be found at http://www.co2science.org/subject/a/peanut.php.

References

Alexandrov, V.A. and Hoogenboom, G. 2000. Vulnerability and adaptation assessments of agricultural crops under climate change in the Southeastern USA. *Theoretical and Applied Climatology* **67**: 45-63.

Idso, S.B. 1998. CO_2-induced global warming: a skeptic's view of potential climate change. *Climate Research* **10**: 69-82.

Prasad, P.V.V., Boote, K.J., Allen Jr., L.H. and Thomas, J.M.G. 2003. Super-optimal temperatures are detrimental to peanut (*Arachis hypogaea* L.) reproductive processes and yield at both ambient and elevated carbon dioxide. *Global Change Biology* **9**: 1775-1787.

Stanciel, K., Mortley, D.G., Hileman, D.R., Loretan, P.A., Bonsi, C.K. and Hill, W.A. 2000. Growth, pod and seed yield, and gas exchange of hydroponically grown peanut in response to CO_2 enrichment. *HortScience* **35**: 49-52.

Vu, J.C.V. 2005. Acclimation of peanut (*Arachis hypogaea* L.) leaf photosynthesis to elevated growth CO_2 and temperature. *Environmental and Experimental Botany* **53**: 85-95.

7.1.1.5. Potato

In the study of Sicher and Bunce (1999), exposure to twice-ambient atmospheric CO_2 concentrations enhanced rates of net photosynthesis in potato plants (*Solanum tuberosum* L.) by 49 percent; while in the study of Schapendonk *et al.* (2000), a doubling of the air's CO_2 content led to an 80 percent increase in net photosynthesis. In a study that additionally considered the role of the air's vapor pressure deficit (VPD), Bunce (2003) found that exposure to twice-ambient atmospheric CO_2 concentrations boosted net photosynthesis by 36 percent at a VPD of 0.5 kPa (moist air) but by 70 percent at a VPD of 3.5 kPa (dry air). Yet another complexity was investigated by Olivo *et al.* (2002), who assessed the effect of a doubling of the air's CO_2 content on the net photosynthetic rates of high-altitude (*Solanum curtilobum*) and low-altitude (*S. tuberosum*) and found the rate of the former to be enhanced by 56 percent and that of the latter by 53 percent. In addition, although they did not directly report photosynthetic rates, Louche-Tessandier *et al.* (1999) noted that photosynthetic acclimation was reduced in CO_2-enriched plants that were inoculated with a fungal symbiont, which consequently allowed them to produce greater amounts of biomass than non-inoculated control plants grown in ambient air.

Because elevated CO_2 concentrations stimulate photosynthesis in potatoes, it is to be expected they would also increase potato biomass production. Miglietta *et al.* (1998), for example, reported that potatoes grown at 660 ppm CO_2 produced 40 percent more tuber biomass than control plants grown in ambient air. Such reports are common, in fact, with twice-ambient atmospheric CO_2 concentrations having been reported to produce yield increases of 25 percent (Lawson *et al.*, 2001), 36 percent (Chen and Setter, 2003), 37 percent (Schapendonk *et al.*, 2000), 40 percent (Olivo *et al.*, 2002), 44 percent (Sicher and Bunce, 1999), 85 percent (Olivo *et al.*, 2002) and 100 percent (Ludewig *et al.*, 1998).

A few studies have been conducted at even higher atmospheric CO_2 concentrations. Kauder *et al.* (2000), for example, grew plants for up to seven weeks in controlled environments receiving an extra 600 ppm CO_2, obtaining final tuber yields that were 30 percent greater than those of ambiently grown plants. Also, in a study of potato microcuttings grown for four weeks in environmental chambers maintained at ambient air and air enriched with an extra 1200 ppm CO_2, Pruski *et al.* (2002) found that the average number of nodes per stem was increased by 64 percent, the average stem dry weight by 92 percent, and the average shoot length by 131 percent.

Atmospheric CO_2 enrichment also leads to reductions in transpirational water loss by potato plants. Magliulo *et al.* (2003), for example, grew potatoes in the field within FACE rings maintained at either ambient (370 ppm) or enriched (550 ppm) atmospheric CO_2 concentrations for two consecutive years, finding that the CO_2-enriched plants used 12 percent less water than the ambient-treatment plants, while they produced 47 percent more tuber biomass. Hence, the CO_2-enriched plants experienced a 68 percent increase in water use efficiency, or the amount of biomass produced per unit of water used in producing it. Likewise, Olivo *et al.* (2002) found that a doubling of the air's CO_2 content increased the instantaneous water-use efficiencies of high-altitude and low-altitude potato species by 90 percent and 80 percent, respectively.

In the final phenomenon considered here, we review the findings of three studies that evaluated the ability of atmospheric CO_2 enrichment to mitigate the deleterious effects of ozone pollution on potato growth. Fangmeier and Bender (2002) determined the mean tuber yield of potato as a function of atmospheric CO_2 concentration for conditions of ambient and high atmospheric O_3 concentrations, as derived from a trans-European set of experiments. At the mean ambient CO_2 concentration of 380 ppm, the high O_3 stress reduced mean tuber yield by approximately 9 percent. At CO_2 concentrations of 540 and 680 ppm, however, the high O_3 stress had no significant effect on tuber yield.

Much the same results were obtained by Wolf and van Oijen (2002, 2003), who used the validated potato model LPOTCO to project future European tuber yields. Under two climate change scenarios that incorporated the effects of increased greenhouse gases on climate (i.e., increased air temperature and reduced precipitation), the model generated increases in irrigated tuber production ranging from 2,000 to 4,000 kg of dry matter per hectare across Europe, with significant reductions in the negative effects of O_3 pollution.

Additional information on this topic, including reviews of newer publications as they become available, can be found at http://www.co2science.org/subject/a/agriculturepotato.php.

References

Bunce, J.A. 2003. Effects of water vapor pressure difference on leaf gas exchange in potato and sorghum at ambient and elevated carbon dioxide under field conditions. *Field Crops Research* **82**: 37-47.

Chen, C.-T. and Setter, T.L. 2003. Response of potato tuber cell division and growth to shade and elevated CO_2. *Annals of Botany* **91**: 373-381.

Fangmeier, A. and Bender, J. 2002. Air pollutant combinations—Significance for future impact assessments on vegetation. *Phyton* **42**: 65-71.

Kauder, F., Ludewig, F. and Heineke, D. 2000. Ontogenetic changes of potato plants during acclimation to elevated carbon dioxide. *Journal of Experimental Botany* **51**: 429-437.

Lawson, T., Craigon, J., Black, C.R., Colls, J.J., Tulloch, A.-M. and Landon, G. 2001. Effects of elevated carbon dioxide and ozone on the growth and yield of potatoes (*Solanum tuberosum*) grown in open-top chambers. *Environmental Pollution* **111**: 479-491.

Louche-Tessandier, D., Samson, G., Hernandez-Sebastia, C., Chagvardieff, P. and Desjardins, Y. 1999. Importance of light and CO_2 on the effects of endomycorrhizal colonization on growth and photosynthesis of potato plantlets (*Solanum tuberosum*) in an *in vitro* tripartite system. *New Phytologist* **142**: 539-550.

Ludewig, F., Sonnewald, U., Kauder, F., Heineke, D., Geiger, M., Stitt, M., Muller-Rober, B.T., Gillissen, B., Kuhn, C. and Frommer, W.B. 1998. The role of transient starch in acclimation to elevated atmospheric CO_2. *FEBS Letters* **429**: 147-151.

Magliulo, V., Bindi, M. and Rana, G. 2003. Water use of irrigated potato (*Solanum tuberosum* L.) grown under free air carbon dioxide enrichment in central Italy. *Agriculture, Ecosystems and Environment* **97**: 65-80.

Miglietta, F., Magliulo, V., Bindi, M., Cerio, L., Vaccari, F.P., Loduca, V. and Peressotti, A. 1998. Free Air CO_2 Enrichment of potato (*Solanum tuberosum* L.): development, growth and yield. *Global Change Biology* **4**: 163-172.

Olivo, N., Martinez, C.A. and Oliva, M.A. 2002. The photosynthetic response to elevated CO_2 in high altitude potato species (*Solanum curtilobum*). *Photosynthetica* **40**: 309-313.

Pruski, K., Astatkie, T., Mirza, M. and Nowak, J. 2002. Photoautotrophic micropropagation of Russet Burbank potato. *Plant, Cell and Environment* **69**: 197-200.

Schapendonk, A.H.C.M., van Oijen, M., Dijkstra, P., Pot, C.S., Jordi, W.J.R.M. and Stoopen, G.M. 2000. Effects of elevated CO_2 concentration on photosynthetic acclimation and productivity of two potato cultivars grown in open-top chambers. *Australian Journal of Plant Physiology* **27**: 1119-1130.

Sicher, R.C. and Bunce, J.A. 1999. Photosynthetic enhancement and conductance to water vapor of field-grown *Solanum tuberosum* (L.) in response to CO_2 enrichment. *Photosynthesis Research* **62**: 155-163.

Wolf, J. and van Oijen, M. 2002. Modelling the dependence of European potato yields on changes in climate and CO_2. *Agricultural and Forest Meteorology* **112**: 217-231.

Wolf, J. and van Oijen, M. 2003. Model simulation of effects of changes in climate and atmospheric CO_2 and O_3 on tuber yield potential of potato (cv. Bintje) in the European Union. *Agriculture, Ecosystems and Environment* **94**: 141-157.

7.1.1.6. Rice

DeCosta *et al.* (2003a) grew two crops of rice (*Oryza sativa* L.) at the Rice Research and Development Institute of Sri Lanka from January to March (the maha season) and from May to August (the yala season) in open-top chambers in air of either ambient or ambient plus 200 pppm CO_2, determining that leaf net photosynthetic rates were significantly higher in the CO_2-enriched chambers than in the ambient-air chambers: 51-75 percent greater in the maha season and 22-33 percent greater in the yala season. Likewise, in the study of Gesch *et al.* (2002), where one-month-old plants were maintained at either 350 ppm CO_2 or switched to a concentration of 700 ppm for 10 additional days, the plants switched to CO_2-enriched air immediately displayed large increases in their photosynthetic rates that at the end of the experiment were still 31 percent greater than those exhibited by unswitched control plants.

With respect to the opposite of photosynthesis, Baker *et al.* (2000) reported that rates of carbon loss via dark respiration in rice plants decreased with increasing nocturnal CO_2 concentrations. As a result, it is not surprising that in the study of Weerakoon *et al.* (2000), rice plants exposed to an extra 300 ppm of atmospheric CO_2 exhibited a 35 percent increase in mean season-long radiation-use efficiency, defined as the amount of biomass produced per unit of solar radiation intercepted. In light of these several observations, therefore, one would logically expect rice plants to routinely produce more biomass at elevated levels of atmospheric CO_2.

In conjunction with the study of DeCosta et al. (2003a), DeCosta et al. (2003b) found that CO_2-enriched rice plants produced more leaves per hill, more tillers per hill, more total plant biomass, greater root dry weight, and more panicles per plant and had harvest indices that were increased by 4 percent and 2 percent, respectively, in the maha and yala seasons, which suite of benefits led to grain yield increases of 24 percent and 39 percent in those two periods. In another study, Kim et al. (2003) grew rice crops from the seedling stage to maturity at atmospheric CO_2 concentrations of ambient and ambient plus 200 ppm using FACE technology and three levels of applied nitrogen—low (LN, 4 g N m^{-2}), medium (MN, 8 and 9 g N m^{-2}), and high (HN, 15 g N m^{-2})—for three cropping seasons (1998-2000). They found that "the yield response to elevated CO_2 in crops supplied with MN (+14.6 percent) or HN (+15.2 percent) was about twice that of crops supplied with LN (+7.4 percent)," confirming the importance of N availability to the response of rice to atmospheric CO_2 enrichment that had previously been determined by Kim et al. (2001) and Kobaysahi et al. (2001).

Various environmental stresses can significantly alter the effect of elevated CO_2 on rice. In the study of Tako et al. (2001), rice plants grown at twice-ambient CO_2 concentrations and ambient temperatures displayed no significant increases in biomass production; but when air temperatures were raised by 2°C, the CO_2-enriched plants produced 22 percent more biomass than the plants grown in non-CO_2-enriched air. By contrast, Ziska et al. (1997) reported that CO_2-enriched rice plants grown at elevated air temperatures displayed no significant increases in biomass; but when the plants were grown at ambient air temperatures, the additional 300 ppm of CO_2 boosted their rate of biomass production by 40 percent. In light of these observations, rice growers should select cultivars that are most responsive to elevated CO_2 concentrations at the air temperatures likely to prevail in their locality in order to maximize their yield production in a future high-CO_2 world.

Water stress can also severely reduce rice production. As an example, Widodo et al. (2003) grew rice plants in eight outdoor, sunlit, controlled-environment chambers at daytime atmospheric CO_2 concentrations of 350 and 700 ppm for an entire season. In one set of chambers, the plants were continuously flooded. In another set, drought stress was imposed during panicle initiation. In another, it was imposed during anthesis; and in the last set, drought stress was imposed at both stages. The resultant drought-induced effects, according to the scientists, "were more severe for plants grown at ambient than at elevated CO_2." They report, for example, that "plants grown under elevated CO_2 were able to maintain midday leaf photosynthesis, and to some extent other photosynthetic-related parameters, longer into the drought period than plants grown at ambient CO_2."

Recovery from the drought-induced water stress was also more rapid in the elevated CO_2 treatment. At panicle initiation, for example, Widodo et al. observed that "as water was added back following a drought induction, it took more than 24 days for the ambient CO_2 [water]-stressed plants to recuperate in midday leaf CER, compared with only 6-8 days for the elevated CO_2 [water]-stressed plants." Similarly, they report that "for the drought imposed during anthesis, midday leaf CER of the elevated CO_2 [water]-stressed plants were fully recovered after 16 days of re-watering, whereas those of the ambient CO_2 [water]-stressed plants were still 21 percent lagging behind their unstressed controls at that date." Hence, they concluded that "rice grown under future rising atmospheric CO_2 should be better able to tolerate drought situations."

In a somewhat different type of study, Watling and Press (2000) found that rice plants growing in ambient air and infected with a root hemiparasitic angiosperm obtained final biomass values that were only 35 percent of those obtained by uninfected plants. In air of 700 ppm CO_2, however, the infected plants obtained biomass values that were 73 percent of those obtained by uninfected plants. Thus, atmospheric CO_2 enrichment significantly reduced the negative impact of this parasite on biomass production in rice.

In summary, as the CO_2 concentration of the air continues to rise, rice plants will likely experience greater photosynthetic rates, produce more biomass, be less affected by root parasites, and better deal with environmental stresses, all of which effects should lead to greater grain yields.

Additional information on this topic, including reviews of newer publications as they become available, can be found at http://www.co2science.org/subject/a/agriculturerice.php.

References

Baker, J.T., Allen Jr., L.H., Boote, K.J. and Pickering, N.B. 2000. Direct effects of atmospheric carbon dioxide concentration on whole canopy dark respiration of rice. *Global Change Biology* **6**: 275-286.

De Costa, W.A.J.M., Weerakoon, W.M.W., Abeywardena, R.M.I. and Herath, H.M.L.K. 2003a. Response of photosynthesis and water relations of rice (*Oryza sativa*) to elevated atmospheric carbon dioxide in the subhumid zone of Sri Lanka. *Journal of Agronomy and Crop Science* **189**: 71-82.

De Costa, W.A.J.M., Weerakoon, W.M.W., Herath, H.M.L.K. and Abeywardena, R.M.I. 2003b. Response of growth and yield of rice (*Oryza sativa*) to elevated atmospheric carbon dioxide in the subhumid zone of Sri Lanka. *Journal of Agronomy and Crop Science* **189**: 83-95.

Gesch, R.W., Vu, J.C., Boote, K.J., Allen Jr., L.H. and Bowes, G. 2002. Sucrose-phosphate synthase activity in mature rice leaves following changes in growth CO_2 is unrelated to sucrose pool size. *New Phytologist* **154**: 77-84.

Kim, H.-Y., Lieffering, M., Kobayashi, K., Okada, M., Mitchell, M.W. and Gumpertz, M. 2003. Effects of free-air CO_2 enrichment and nitrogen supply on the yield of temperate paddy rice crops. *Field Crops Research* **83**: 261-270.

Kim, H.-Y., Lieffering, M., Miura, S., Kobayashi, K. and Okada, M. 2001. Growth and nitrogen uptake of CO_2-enriched rice under field conditions. *New Phytologist* **150**: 223-229.

Kobayashi, K., Lieffering, M. and Kim, H.-Y. 2001. Growth and yield of paddy rice under free-air CO_2 enrichment. In: Shiyomi, M. and Koizumi, H. (Eds.) *Structure and Function in Agroecosystem Design and Management.* CRC Press, Boca Raton, FL, USA, pp. 371-395.

Tako, Y., Arai, R., Otsubo, K. and Nitta, K. 2001. Application of crop gas exchange and transpiration data obtained with CEEF to global change problem. *Advances in Space Research* **27**: 1541-1545.

Watling, J.R. and Press, M.C. 2000. Infection with the parasitic angiosperm *Striga hermonthica* influences the response of the C_3 cereal *Oryza sativa* to elevated CO_2. *Global Change Biology* **6**: 919-930.

Weerakoon, W.M.W., Ingram, K.T. and Moss, D.D. 2000. Atmospheric carbon dioxide and fertilizer nitrogen effects on radiation interception by rice. *Plant and Soil* **220**: 99-106.

Widodo, W., Vu, J.C.V., Boote, K.J., Baker, J.T. and Allen Jr., L.H. 2003. Elevated growth CO_2 delays drought stress and accelerates recovery of rice leaf photosynthesis. *Environmental and Experimental Botany* **49**: 259-272.

Ziska, L.H., Namuco, O., Moya, T. and Quilang, J. 1997. Growth and yield response of field-grown tropical rice to

increasing carbon dioxide and air temperature. *Agronomy Journal* **89**: 45-53.

7.1.1.7. Sorghum

Many laboratory and field experiments have demonstrated a significant positive impact of elevated levels of atmospheric CO_2 on total biomass and grain production in the C_4 crop sorghum (*Sorghum bicolor* (L.) Moench).

Ottman *et al.* (2001) grew sorghum plants in a FACE experiment conducted near Phoenix, Arizona, USA, where plants were fumigated with air containing either 360 or 560 ppm CO_2 and where they were further subjected to irrigation regimes resulting in both adequate and inadequate levels of soil moisture. Averaged over the two years of their study, the extra CO_2 increased grain yield by only 4 percent in the plots receiving adequate levels of soil moisture but by 16 percent in the dry soil moisture plots.

Prior *et al.* (2005) grew sorghum in two different years in 7-meter-wide x 76-meter-long x 2-m-deep bins filled with a silt loam soil, upon which they constructed a number of clear-plastic-wall open-top chambers they maintained at ambient CO_2 concentrations and ambient concentrations plus 300 ppm. In the first of the two years, the extra CO_2 increased sorghum residue production by 14 percent, while in the second year it increased crop residue production by 24 percent and grain production by 22 percent. For a CO_2 increase of 200 ppm comparable to that employed in the study of Ottman *et al.*, these figures translate to crop residue increases of 9 percent and 16 percent and a grain increase of 15 percent.

In a review of primary research papers describing results obtained from large-scale FACE experiments conducted over the prior 15 years, Ainsworth and Long (2005) determined that, in the mean, sorghum grain yield was increased by approximately 7 percent in response to a 200-ppm increase in the atmosphere's CO_2 concentration.

An experiment with a bit more complexity was carried out several years earlier by Watling and Press (1997), who grew sorghum with and without infection by the parasitic C_3 weeds *Striga hermonthica* and *S. asiatica*. The study lasted for about two months and was conducted in controlled environment cabinets fumigated with air of either 350 or 700 ppm CO_2. In the absence of parasite infection, the extra 350 ppm of CO_2 boosted plant biomass production by 35 percent, which adjusted downward to make it compatible with the 200-ppm increase employed in most FACE studies corresponds to an increase of just under 21 percent. When infected with *S. asiatica*, the biomass stimulation provided by the extra CO_2 was about the same; but when infected with *S. hermonthica*, it was almost 80 percent, which corresponds to a similarly downward adjusted biomass increase of 45 percent.

In light of these several observations, it would appear that although the CO_2-induced increase in total biomass and grain yield of sorghum is rather modest, ranging from 4 to 16 percent under well-watered conditions, it can be on the high end of this range when the plants are stressed by a shortage of water (16 percent has been observed) and by parasitic infection (45 percent has been observed). Consequently, elevated levels of atmospheric CO_2 seem to help sorghum most when help is most needed.

Additional information on this topic, including reviews on sorghum not discussed here, can be found at http://www.co2science.org/subject/a/subject_a.php under the main heading Agriculture, sub heading Sorghum.

References

Ainsworth, E.A. and Long, S.P. 2005. What have we learned from 15 years of free-air CO_2 enrichment (FACE)? A meta-analytic review of the responses of photosynthesis, canopy properties and plant production to rising CO_2. *New Phytologist* **165**: 351-372.

Ottman, M.J., Kimball, B.A., Pinter Jr., P.J., Wall, G.W., Vanderlip, R.L., Leavitt, S.W., LaMorte, R.L., Matthias, A.D. and Brooks, T.J. 2001. Elevated CO_2 increases sorghum biomass under drought conditions. *New Phytologist* **150**: 261-273.

Prior, S.A., Runion, G.B., Rogers, H.H., Torbert, H.A. and Reeves, D.W. 2005. Elevated atmospheric CO_2 effects on biomass production and soil carbon in conventional and conservation cropping systems. *Global Change Biology* **11**: 657-665.

Watling, J.R. and Press, M.C. 1997. How is the relationship between the C_4 cereal *Sorghum bicolor* and the C_3 root hemi-parasites *Striga hermonthica* and *Striga asiatica* affected by elevated CO_2? *Plant, Cell and Environment* **20**: 1292-1300.

7.1.1.8. Soybean

Wittwer (1995) reports that the common soybean (*Glycine max* L.) "provides about two-thirds of the world's protein concentrate for livestock feeding, and is a valuable ingredient in formulated feeds for poultry and fish." Bernacchi *et al.* (2005) characterize the soybean as "the world's most important seed legume." Consequently, it is important to determine how soybeans will likely respond to rising atmospheric CO_2 concentrations with and without concomitant increases in air temperature and under both well-watered and water-stressed conditions.

Rogers *et al.* (2004) grew soybeans from emergence to grain maturity in ambient and CO_2-enriched air (372 and 552 ppm CO_2, respectively) at the SoyFACE facility of the University of Illinois at Urbana-Champaign, Illinois, USA, while CO_2 uptake and transpiration measurements were made from pre-dawn to post-sunset on seven days representative of different developmental stages of the crop. Across the growing season, they found that the mean daily integral of leaf net photosynthesis rose by 24.6 percent in the elevated CO_2 treatment, while mid-day stomatal conductance dropped by 21.9 percent, in response to the 48 percent increase in atmospheric CO_2 employed in their study. With respect to photosynthesis, they additionally report "there was no evidence of any loss of stimulation toward the end of the growing season," noting that the largest stimulation actually occurred during late seed filling. Nevertheless, they say that the photosynthetic stimulation they observed was only "about half the 44.5 percent theoretical maximum increase calculated from Rubisco kinetics." Thus, there is an opportunity for soybeans to perhaps become even more responsive to atmospheric CO_2 enrichment than they are currently, which potential could well be realized via future developments in the field of genetic engineering.

Bunce (2005) grew soybeans in the field in open-top chambers maintained at atmospheric CO_2 concentrations of ambient and ambient +350 ppm at the Beltsville Agricultural Research Center in Maryland, USA, where net CO_2 exchange rate measurements were performed on a total of 16 days between 18 July and 11 September of 2000 and 2003, during flowering to early pod-filling. Over the course of this study, daytime net photosynthesis per unit leaf area was 48 percent greater in the plants growing in the CO_2-enriched air, while nighttime respiration per unit leaf area was not affected by elevated CO_2.

However, because the elevated CO_2 increased leaf dry mass per unit area by an average of 23 percent, respiration per unit mass was significantly lower for the leaves of the soybeans growing in the CO_2-enriched air, producing a sure recipe for accelerated growth and higher soybean seed yields.

Working in Australia, Japan, and the United States, Ziska *et al.* (2001b) observed a recurrent diurnal pattern of atmospheric CO_2 concentration, whereby maximum values of 440-540 ppm occurred during a three-hour pre-dawn period that was followed by a decrease to values of 350-400 ppm by mid-morning, after which there was a slow but steady increase in the late afternoon and early evening that brought the air's CO_2 concentration back to its pre-dawn maximum value. In an attempt to see if the pre-dawn CO_2 spikes they observed affected plant growth, they grew soybeans for one month in controlled-environment chambers under three different sets of conditions: a constant 24-hour exposure to 370 ppm CO_2, a constant 370 ppm CO_2 exposure during the day followed by a constant 500 ppm CO_2 exposure at night, and a CO_2 exposure of 500 ppm from 2200 to 0900 followed by a decrease to 370 ppm by 1000, which was maintained until 2200, somewhat mimicking the CO_2 cycle they observed in nature. This program revealed that the 24-hour exposure to 370 ppm CO_2 and the 370-ppm-day/500-ppm-night treatments produced essentially the same results in terms of biomass production after 29 days. However, the CO_2 treatment that mimicked the observed atmospheric CO_2 pattern resulted in a plant biomass increase of 20 percent.

In a study that evaluated a whole range of atmospheric CO_2 concentrations, from far below ambient levels to high above them, Allen *et al.* (1998) grew soybeans for an entire season in growth chambers maintained at atmospheric CO_2 concentrations of 160, 220, 280, 330, 660, and 990 ppm. In doing so, they observed a consistent increase in total nonstructural carbohydrates in all vegetative components including roots, stems, petioles, and especially the leaves, as CO_2 concentrations rose. There was, however, no overall significant effect of treatment CO_2 concentration on nonstructural carbohydrate accumulation in soybean reproductive components, including podwalls and seeds, which observations indicate that the higher yields reported in the literature for soybeans exposed to elevated CO_2 most likely result from increases in the number of pods produced per plant, and not from the production of larger individual pods or seeds.

The increasing amounts of total nonstructural carbohydrates that were produced with each additional increment of CO_2 provided the raw materials to support greater biomass production at each CO_2 level. Although final biomass and yield data were not reported in this paper, the authors did present biomass data obtained at 66 days into the experiment. Relative to above-ground biomass measured at 330 ppm CO_2, the plants that were grown in sub-ambient CO_2 concentrations of 280, 220, and 160 ppm exhibited 12, 33, and 60 percent *less* biomass, respectively, while plants grown in atmospheric CO_2 concentrations of 660 and 990 ppm displayed 46 and 66 percent more biomass.

In a study of two contrasting soybean cultivars, Ziska and Bunce (2000) grew *Ripley*, which is semi-dwarf and determinate in growth, and *Spencer*, which is standard-size and indeterminate in growth, for two growing seasons in open-top chambers maintained at atmospheric CO_2 concentrations of ambient and ambient plus 300 ppm. Averaged over both years, the elevated CO_2 treatment increased photosynthetic rates in the Ripley and Spencer varieties by 76 and 60 percent, respectively. However, Spencer showed a greater CO_2-induced increase in vegetative biomass than Ripley (132 vs. 65 percent). Likewise, elevated CO_2 enhanced seed yield in Spencer by 60 percent but by only 35 percent in Ripley, suggesting that cultivar selection for favorable yield responses to atmospheric CO_2 enrichment could have a big impact on future farm productivity.

In another study of contrasting types of soybeans, Nakamura *et al.* (1999) grew nodulated and non-nodulated plants in pots within controlled-environmental cabinets maintained at atmospheric CO_2 concentrations of 360 and 700 ppm in combination with low and high soil nitrogen supply for three weeks. They found that at low nitrogen, elevated CO_2 increased total plant dry mass by approximately 40 and 80 percent in nodulated soybeans grown at low and high nitrogen supply, respectively, while non-nodulated plants exhibited no CO_2-induced growth response at low nitrogen but an approximate 60 percent growth enhancement at high nitrogen supply. Hence, it would appear that as the air's CO_2 content continues to rise, non-nodulated soybeans will display increases in biomass only if they are grown in nitrogen-rich soils. Nodulated soybeans, however, should display increased growth in both nitrogen-rich and nitrogen-poor soils, with their responses being about twice as large in high as in low soil nitrogen conditions.

In yet another study of soybeans with different genetic characteristics, Ziska *et al.* (2001a) grew one modern and eight ancestral soybean genotypes in glasshouses maintained at atmospheric CO_2 concentrations of 400 and 710 ppm, finding that the elevated CO_2 increased photosynthetic rates in all cultivars by an average of 75 percent. This photosynthetic enhancement led to CO_2-induced increases in seed yield that averaged 40 percent, except for one of the ancestral varieties that exhibited an 80 percent increase in seed yield.

To get a glimpse of what might happen if future temperatures also continue to rise, Ziska (1998) grew soybeans for 21 days in controlled environments having atmospheric CO_2 concentrations of approximately 360 (ambient) or 720 ppm and soil temperatures of 25° (ambient) or 30°C. He found that elevated CO_2 significantly increased whole plant net photosynthesis at both temperatures, with the greatest effect occurring at 30°C. As time progressed, however, this photosynthetic stimulation dropped from 50 percent at 13 days into the experiment to 30 percent at its conclusion eight days later; in spite of this partial acclimation, which was far from complete, atmospheric CO_2 enrichment significantly enhanced total plant dry weight at final harvest by 36 and 42 percent at 25° and 30°C, respectively.

Studying the complicating effects of water stress were Serraj *et al.* (1999), who grew soybeans from seed in pots within a glasshouse until they were four weeks old, after which half of the plants were subjected to an atmospheric CO_2 concentration of 360 ppm, while the other half were exposed to an elevated concentration of 700 ppm. In addition, half of the plants at each CO_2 concentration were well-watered and half of them were allowed to experience water stress for a period of 18 days. This protocol revealed that short-term (18-day) exposure of soybeans to elevated CO_2 significantly decreased daily and cumulative transpirational water losses compared to plants grown at 360 ppm CO_2, regardless of water treatment. In fact, elevated CO_2 reduced total water loss by 25 and 10 percent in well-watered and water-stressed plants, respectively. Also, drought stress significantly reduced rates of net photosynthesis among plants of both CO_2 treatments. However, plants grown in elevated CO_2 consistently exhibited higher photosynthetic rates than plants grown at ambient CO_2, regardless of soil water status.

At final harvest, the elevated CO_2 treatment had little effect on the total dry weight of plants grown at optimal soil moisture, but it increased the total dry weight of water-stressed plants by about 33 percent. Also, while root dry weight declined for plants grown under conditions of water stress and ambient CO_2 concentration, no such decline was exhibited by plants subjected to atmospheric CO_2 enrichment and water stress.

Studying *both* water and high-temperature stress were Ferris *et al.* (1999), who grew soybeans in glasshouses maintained at atmospheric CO_2 concentrations of 360 and 700 ppm for 52 days, before having various environmental stresses imposed on them for eight days during early seed filling. For the eight-day stress period, some plants were subjected to air temperatures that were 15°C higher than those to which the control plants were exposed, while some were subjected to a water stress treatment in which their soil moisture contents were maintained at 40 percent of that experienced by the control plants. Averaged across all stress treatments and harvests, this protocol revealed that the high CO_2 treatment increased total plant biomass by 41 percent. Both high-temperature and water-deficit treatments, singly or in combination, reduced overall biomass by approximately the same degree, regardless of CO_2 treatment. Thus, even when the greatest biomass reductions of 17 percent occurred in the CO_2-enriched and ambiently grown plants, in response to the combined stresses of high temperature and low soil moisture, plants grown in elevated CO_2 still exhibited an average biomass that was 24 percent greater than that displayed by plants grown in ambient CO_2.

Averaged across all stress treatments and harvests, elevated CO_2 increased seed yield by 32 percent. In addition, it tended to ameliorate the negative effects of environmental stresses. CO_2-enriched plants that were water stressed, for example, had an average seed yield that was 34 percent greater than that displayed by water-stressed controls grown at ambient CO_2, while CO_2-enriched plants exposed to high temperatures produced 38 percent more seed than their respectively stressed counterparts. In fact, the greatest relative impact of elevated CO_2 on seed yield occurred in response to the combined stresses of high temperature and low soil moisture, with CO_2-enriched plants exhibiting a seed yield that was 50 percent larger than that of similarly stressed plants grown in ambient CO_2.

In a predictive application of this type of knowledge, but based on a different means of obtaining it, Alexandrov and Hoogenboom (2000) studied how temperature, precipitation, and solar radiation influenced soybean yields over a 30-year period in the southeastern United States, after which they used the results they obtained to predict future crop yields based on climate output from various global circulation models of the atmosphere. At ambient CO_2 concentrations, the model-derived scenarios pointed to a decrease in soybean yields by the year 2020, due in part to predicted changes in temperature and precipitation. However, when the yield-enhancing effects of a doubling of the air's CO_2 concentration were included in the simulations, a completely different projection was obtained: a yield *increase*.

Shifting to the subject of soybean seed *quality*, Caldwell *et al.* (2005) write that "the beneficial effects of isoflavone-rich foods have been the subject of numerous studies (Birt *et al.*, 2001; Messina, 1999)," and that "foods derived from soybeans are generally considered to provide both specific and general health benefits," presumably via these substances. Hence, it is only natural they would wonder how the isoflavone content of soybean seeds may be affected by the ongoing rise in the air's CO_2 content, and that they would conduct a set of experiments to find the answer.

The scientists grew well-watered and fertilized soybean plants from seed to maturity in pots within two controlled-environment chambers, one maintained at an atmospheric CO_2 concentration of 400 ppm and one at 700 ppm. The chambers were initially kept at a constant air temperature of 25°C. At the onset of seed fill, however, air temperature was reduced to 18°C until seed development was complete, in order to simulate average outdoor temperatures at this stage of plant development. In a second experiment, this protocol was repeated, except the temperature during seed fill was maintained at 23°C, with and without drought (a third treatment), while in a third experiment, seed-fill temperature was maintained at 28°C, with or without drought.

In the first experiment, where air temperature during seed fill was 18°C, the elevated CO_2 treatment increased the total isoflavone content of the soybean seeds by 8 percent. In the second experiment, where air temperature during seed fill was 23°C, the extra CO_2 increased total seed isoflavone content by 104 percent, while in the third experiment, where air temperature during seed fill was 28°C, the CO_2-induced isoflavone increase was 101 percent. Finally, when drought-stress was added as a third environmental variable, the extra CO_2 boosted total seed isoflavone content by 186 percent when seed-fill air temperature was 23°C, while at a seed-fill

temperature of 28°C, it increased isoflavone content by 38 percent.

Under all environmental circumstances studied, enriching the air with an extra 300 ppm of CO_2 increased the total isoflavone content of soybean seeds. In addition, the percent increases measured under the stress situations investigated were always greater than the percent increase measured under optimal growing conditions.

Also writing on the subject of soybean seed quality, Thomas *et al.* (2003) say "the unique chemical composition of soybean has made it one of the most valuable agronomic crops worldwide," noting that "oil and protein comprise ~20 and 40%, respectively, of the dry weight of soybean seed." Consequently, they explored the effects of elevated CO_2 plus temperature on soybeans that were grown to maturity in sunlit controlled-environment chambers with sinusoidally varying day/night max/min temperatures of 28/18°, 32/22°, 36/26°, 40/30°, and 44/34°C and atmospheric CO_2 concentrations of 350 and 700 ppm. This work revealed that the effect of temperature on seed composition and gene expression was "pronounced," but that "there was no effect of CO_2." In this regard, however, they note that "Heagle *et al.* (1998) observed a positive significant effect of CO_2 enrichment on soybean seed oil and oleic acid concentration," the latter of which parameters their own study found to rise with increasing temperature all the way from 28/18° to 44/34°C. In addition, they determined that "32/22°C is optimum for producing the highest oil concentration in soybean seed," that "the degree of fatty acid saturation in soybean oil was significantly increased by increasing temperature," and that crude protein concentration increased with temperature to 40/30°C.

In commenting on these observations, Thomas *et al.* note that "the intrinsic value of soybean seed is in its supply of essential fatty acids and amino acids in the oil and protein, respectively." Hence, we conclude that the temperature-driven changes they identified in these parameters, as well as the CO_2 effect observed by Heagle *et al.*, bode well for the future production of this important crop and its value to society in a CO_2-enriched and warming world. Thomas *et al.* note, however, that "temperatures during the soybean-growing season in the southern USA are at, or slightly higher than, 32/22°C," and that warming could negatively impact the soybean oil industry in this region. For the world as a whole, however, warming would be a positive development for soybean production; while in the southern United States, shifts in planting zones could readily accommodate changing weather patterns associated with this phenomenon.

In conclusion, as the air's CO_2 content continues to rise, soybeans will likely respond by displaying significant increases in growth and yield, with possible improvements in seed quality; these beneficial effects will likely persist even if temperatures rise or soil moisture levels decline, regardless of their cause.

Additional information on this topic, including reviews on sorghum not discussed here, can be found at http://www.co2science.org/subject/a/subject_a.php under the main heading Agriculture, sub heading Soybean.

References

Alexandrov, V.A. and Hoogenboom, G. 2000. Vulnerability and adaptation assessments of agricultural crops under climate change in the Southeastern USA. *Theoretical and Applied Climatology* **67**: 45-63.

Allen Jr., L.H., Bisbal, E.C. and Boote, K.J. 1998. Nonstructural carbohydrates of soybean plants grown in subambient and superambient levels of CO_2. *Photosynthesis Research* **56**: 143-155.

Bernacchi, C.J., Morgan, P.B., Ort, D.R. and Long, S.P. 2005. The growth of soybean under free air [CO_2] enrichment (FACE) stimulates photosynthesis while decreasing in vivo Rubisco capacity. *Planta* **220**: 434-446.

Birt, D.F., Hendrich, W. and Wang, W. 2001. Dietary agents in cancer prevention: flavonoids and isoflavonoids. *Pharmacology & Therapeutics* **90**: 157-177.

Bunce, J.A. 2005. Response of respiration of soybean leaves grown at ambient and elevated carbon dioxide concentrations to day-to-day variation in light and temperature under field conditions. *Annals of Botany* **95**: 1059-1066.

Caldwell, C.R., Britz, S.J. and Mirecki, R.M. 2005. Effect of temperature, elevated carbon dioxide, and drought during seed development on the isoflavone content of dwarf soybean [*Glycine max* (L.) Merrill] grown in controlled environments. *Journal of Agricultural and Food Chemistry* **53**: 1125-1129.

Ferris, R., Wheeler, T.R., Ellis, R.H. and Hadley, P. 1999. Seed yield after environmental stress in soybean grown under elevated CO_2. *Crop Science* **39**: 710-718.

Heagle, A.S., Miller, J.E. and Pursley, W.A. 1998. Influence of ozone stress on soybean response to carbon dioxide enrichment: III. Yield and seed quality. *Crop Science* **38**: 128-134.

Messina, M.J. 1999. Legumes and soybeans: overview of their nutritional profiles and health effects. *American Journal of Clinical Nutrition* **70**(S): 439s-450s.

Nakamura, T., Koike, T., Lei, T., Ohashi, K., Shinano, T. and Tadano, T. 1999. The effect of CO_2 enrichment on the growth of nodulated and non-nodulated isogenic types of soybean raised under two nitrogen concentrations. *Photosynthetica* **37**: 61-70.

Rogers, A., Allen, D.J., Davey, P.A., Morgan, P.B., Ainsworth, E.A., Bernacchi, C.J., Cornic, G., Dermody, O., Dohleman, F.G., Heaton, E.A., Mahoney, J., Zhu, X.-G., DeLucia, E.H., Ort, D.R. and Long, S.P. 2004. Leaf photosynthesis and carbohydrate dynamics of soybeans grown throughout their life-cycle under Free-Air Carbon dioxide Enrichment. *Plant, Cell and Environment* **27**: 449-458.

Serraj, R., Allen Jr., L.H., Sinclair, T.R. 1999. Soybean leaf growth and gas exchange response to drought under carbon dioxide enrichment. *Global Change Biology* **5**: 283-291.

Thomas, J.M.G., Boote, K.J., Allen Jr., L.H., Gallo-Meagher, M. and Davis, J.M. 2003. Elevated temperature and carbon dioxide effects on soybean seed composition and transcript abundance. *Crop Science* **43**: 1548-1557.

Wittwer, S.H. 1995. Food, Climate, and Carbon Dioxide: The Global Environment and World Food Production. CRC Press, Boca Raton, FL.

Ziska, L.H. 1998. The influence of root zone temperature on photosynthetic acclimation to elevated carbon dioxide concentrations. *Annals of Botany* **81**: 717-721.

Ziska, L.W. and Bunce, J.A. 2000. Sensitivity of field-grown soybean to future atmospheric CO_2: selection for improved productivity in the 21st century. *Australian Journal of Plant Physiology* **27**: 979-984.

Ziska, L.H., Bunce, J.A. and Caulfield, F.A. 2001a. Rising atmospheric carbon dioxide and seed yields of soybean genotypes. *Crop Science* **41**: 385-391.

Ziska, L.H., Ghannoum, O., Baker, J.T., Conroy, J., Bunce, J.A., Kobayashi, K. and Okada, M. 2001b. A global perspective of ground level, 'ambient' carbon dioxide for assessing the response of plants to atmospheric CO_2. *Global Change Biology* **7**: 789-796.

7.1.1.9. Strawberry

In the open-top chamber study of Bunce (2001), strawberry plants (*Fragaria* x *ananassa*) exposed to air containing an extra 300 and 600 ppm CO_2

displayed photosynthetic rates that were 77 and 106 percent greater, respectively, than rates displayed by plants grown in ambient air containing 350 ppm CO_2. Similarly, Bushway and Pritts (2002) reported that strawberry plants grown at atmospheric CO_2 concentrations between 700 and 1,000 ppm exhibited photosynthetic rates that were consistently more than 50 percent greater than rates displayed by control plants.

Because elevated CO_2 stimulates rates of photosynthesis in strawberry plants, it is expected that it would also increase biomass production in this important agricultural species. After growing plants in air containing an additional 170 ppm CO_2 above ambient concentrations, Deng and Woodward (1998) reported that total fresh fruit weights were 42 and 17 percent greater than weights displayed by control plants receiving high and low soil nitrogen inputs, respectively. In addition, Bushway and Pritts (2002) reported that a two- to three-fold increase in the air's CO_2 content boosted strawberry fruit yield by 62 percent.

As the air's CO_2 content continues to rise, strawberry plants will likely exhibit enhanced rates of photosynthesis and biomass production, which should lead to greater fruit yields.

Additional information on this topic, including reviews of newer publications as they become available, can be found at http://www.co2science.org/subject/a/agriculturestraw.php.

References

Bunce, J.A. 2001. Seasonal patterns of photosynthetic response and acclimation to elevated carbon dioxide in field-grown strawberry. *Photosynthesis Research* **68**: 237-245.

Bushway, L.J. and Pritts, M.P. 2002. Enhancing early spring microclimate to increase carbon resources and productivity in June-bearing strawberry. *Journal of the American Society for Horticultural Science* **127**: 415-422.

Deng, X. and Woodward, F.I. 1998. The growth and yield responses of *Fragaria ananassa* to elevated CO_2 and N supply. *Annals of Botany* **81**: 67-71.

7.1.1.10. Sunflower

As the CO_2 content of the air increases, sunflower plants (*Helianthus annus* L.) will likely display enhanced rates of photosynthetic carbon uptake. In the study of Sims *et al.* (1999), exposure to twice-ambient atmospheric CO_2 concentrations enhanced rates of net photosynthesis in individual upper-canopy sunflower leaves by approximately 50 percent. Similarly, Luo *et al.* (2000) reported that sunflowers grown at 750 ppm CO_2 displayed canopy carbon uptake rates that were fully 53 percent greater than those exhibited by plants grown at 400 ppm CO_2.

The study of Zerihun *et al.* (2000) reported that twice-ambient CO_2 concentrations increased whole plant biomass in sunflowers by 44, 13, and 115 percent when the plants were simultaneously exposed to low, medium, and high levels of soil nitrogen, respectively.

Additional information on this topic, including reviews of newer publications as they become available, can be found at http://www.co2science.org/subject/a/agriculturesun.php.

References

Luo, Y., Hui, D., Cheng, W., Coleman, J.S., Johnson, D.W. and Sims, D.A. 2000. Canopy quantum yield in a mesocosm study. *Agricultural and Forest Meteorology* **100**: 35-48.

Sims, D.A., Cheng, W., Luo, Y. and Seeman, J.R. 1999. Photosynthetic acclimation to elevated CO_2 in a sunflower canopy. *Journal of Experimental Botany* **50**: 645-653.

Zerihun, A., Gutschick, V.P. and BassiriRad, H. 2000. Compensatory roles of nitrogen uptake and photosynthetic N-use efficiency in determining plant growth response to elevated CO_2: Evaluation using a functional balance model. *Annals of Botany* **86**: 723-730.

7.1.1.11. Tomato

In the study of Ziska *et al.* (2001), tomato plants (*Lycopersicon esculentum* Mill.) grown at a nocturnal atmospheric CO_2 concentration of 500 ppm displayed total plant biomass values that were 10 percent greater than those exhibited by control plants growing in air containing 370 ppm CO_2. This result was likely the consequence of the elevated CO_2 reducing the rate of nocturnal respiration in the plants, which would have allowed them to utilize the retained carbon to produce more biomass.

This CO_2-induced benefit, as well as a host of other positive effects of atmospheric CO_2 enrichment, are also manifest under unfavorable growing conditions. Jwa and Walling (2001), for example, reported that fungal infection reduced plant biomass in tomatoes growing in normal air by about 30 percent. However, in fungal-infected plants grown at twice-ambient atmospheric CO_2 concentrations, the elevated CO_2 completely ameliorated the growth-reducing effects of the pathogen.

In another stressful situation, Maggio *et al.* (2002) reported that a 500-ppm increase in the air's CO_2 concentration increased the average value of the root-zone salinity threshold in tomato plants by about 60 percent. In addition, they reported that the water-use efficiency of the CO_2-enriched plants was about twice that of the ambiently grown plants.

As the CO_2 content of the air increases, tomato plants will likely display greater rates of photosynthesis and biomass production, which should consequently lead to greater fruit yields, even under stressful conditions of fungal infection and high soil salinity.

Additional information on this topic, including reviews of newer publications as they become available, can be found at http://www.co2science.org/subject/a/agriculturetomato.php.

References

Jwa, N.-S. and Walling, L.L. 2001. Influence of elevated CO_2 concentration on disease development in tomato. *New Phytologist* **149**: 509-518.

Maggio, A., Dalton, F.N. and Piccinni, G. 2002. The effects of elevated carbon dioxide on static and dynamic indices for tomato salt tolerance. *European Journal of Agronomy* **16**: 197-206.

Ziska, L.H., Ghannoum, O., Baker, J.T., Conroy, J., Bunce, J.A., Kobayashi, K. and Okada, M. 2001. A global perspective of ground level, 'ambient' carbon dioxide for assessing the response of plants to atmospheric CO_2. *Global Change Biology* **7**: 789-796.

7.1.1.12. Wheat

In one study, Dijkstra *et al.* (1999) grew winter wheat (*Triticum aestivum* L.) in open-top chambers and field-tracking sun-lit climatized enclosures maintained at atmospheric CO_2 concentrations of ambient and ambient plus 350 ppm CO_2 for two years, determining that the elevated CO_2 increased both final grain yield and total above-ground biomass by 19 percent. In another study, Masle (2000) grew two varieties of wheat for close to a month in greenhouses maintained at atmospheric CO_2 concentrations of 350 and 900 ppm, finding that the CO_2-enriched plants exhibited biomass increases of 52 to 93 percent, depending upon variety and vernalization treatment.

Based on a plethora of experimental observations of this nature, many scientists have developed yield prediction models for wheat. Using the output of several such models, Alexandrov and Hoogenboom (2000) estimated the impact of typically predicted climate changes on wheat production in Bulgaria in the twenty-first century, finding that a doubling of the air's CO_2 concentration would likely enhance wheat yields there between 12 and 49 percent in spite of a predicted 2.9° to 4.1°C increase in air temperature. Likewise, Eitzinger *et al.* (2001) employed the WOFOST crop model to estimate wheat production in northeastern Austria in the year 2080. For a doubled atmospheric CO_2 concentration with concomitant climate changes derived from five different general circulation models of the atmosphere, they obtained simulated yield increases of 30 to 55 percent, even in the face of predicted changes in both temperature and precipitation.

Southworth *et al.* (2002) used the CERES-Wheat growth model to calculate winter wheat production during the period 2050-2059 for 10 representative farm locations in Indiana, Illinois, Ohio, Michigan, and Wisconsin, USA, for six future climate scenarios. They report that some of the southern portions of this group of states would have exhibited climate-induced yield decreases had the aerial fertilization effect of the CO_2 increase that drove the predicted changes in climate not been included in the model. When they did include the increase in the air's CO_2 concentration (to a value of 555 ppm), however, they note that "wheat yields increased 60 to 100% above current yields across the central and northern areas of the study region," while in the southern areas "small increases and small decreases were found." The few minor decreases, however, were associated with the more extreme Hadley Center greenhouse run that presumed a 1 percent increase in greenhouse gases per year and a doubled climate variability; hence, they would have to be considered highly unlikely.

In discussing their findings, Southworth *et al.* note that other modeling studies have obtained similar results for other areas. They report, for example, that Brown and Rosenberg (1999) found winter wheat yields across other parts of the United States to increase "under all climate change scenarios modeled (1, 2.5, and 5°C temperature increases)," and that

Cuculeanu *et al.* (1999) found modeled yields of winter wheat in southern Romania to increase by 15 to 21 percent across five sites. Also, they note that Harrison and Butterfield (1996) "found increased yields of winter wheat across Europe under all the climate change scenarios they modeled."

Van Ittersum *et al.* (2003) performed a number of simulation experiments with the Agricultural Production Systems Simulator (APSIM)-Nwheat model in which they explored the implications of possible increases in atmospheric CO_2 concentration and near-surface air temperature for wheat production and deep drainage at three sites in Western Australia differing in precipitation, soil characteristics, nitrogenous fertilizer application rates, and wheat cultivars. They first assessed the impact of the ongoing rise in the air's CO_2 content, finding that wheat grain yield increased linearly at a rate of 10-16 percent for each 100-ppm increase in atmospheric CO_2 concentration, with only a slight concomitant increase in deep drainage (a big win, small loss outcome). For a likely future CO_2 increase of 200 ppm, increases in grain yield varied between 3 and 17 percent for low nitrogen fertilizer application rates and between 21 and 34 percent for high rates of nitrogen application, with the greatest relative yield response being found for the driest site studied.

When potential warming was factored into the picture, the results proved even better. The positive effects of the CO_2 increase on wheat grain yield were enhanced an extra 3-8 percent when temperatures were increased by 3°C in the model simulations. These yield increases were determined to result in an increased financial return to the typical Western Australian wheat farmer of 15-35 percent. In addition, the imposition of the simultaneous temperature increase led to a significant decline in deep drainage, producing a truly win-win situation that enhanced the average farmer's net income by an additional 10-20 percent. Consequently, it was determined that the CO_2-induced increase in temperature predicted by the IPCC could well increase the net profitability of Western Australian wheat farmers by anywhere from 25-55 percent, while at the same time mitigating what van Ittersum *et al.* refer to as "one of Australia's most severe land degradation problems."

In a wide variety of circumstances, atmospheric CO_2 enrichment significantly increases the biomass production and yield of wheat plants, thereby benefiting both wheat producers and consumers alike.

Additional information on this topic, including reviews on sorghum not discussed here, can be found at http://www.co2science.org/subject/a/subject_a.php under the main heading Agriculture, sub heading Wheat.

References

Alexandrov, V.A. and Hoogenboom, G. 2000. The impact of climate variability and change on crop yield in Bulgaria. *Agricultural and Forest Meteorology* **104**: 315-327.

Brown, R.A. and Rosenberg, N.J. 1999. Climate change impacts on the potential productivity of corn and winter wheat in their primary United States growing regions. *Climatic Change* **41**: 73-107.

Cuculeanu, V., Marcia, A. and Simota, C. 1999. Climate change impact on agricultural crops and adaptation options in Romania. *Climate Research* **12**: 153-160.

Dijkstra, P., Schapendonk, A.H.M.C., Groenwold, K., Jansen, M. and Van de Geijn, S.C. 1999. Seasonal changes in the response of winter wheat to elevated atmospheric CO_2 concentration grown in open-top chambers and field tracking enclosures. *Global Change Biology* **5**: 563-576.

Eitzinger, J., Zalud, Z., Alexandrov, V., van Diepen, C.A., Trnka, M., Dubrovsky, M., Semeradova, D. and Oberforster, M. 2001. A local simulation study on the impact of climate change on winter wheat production in north-east Austria. *Ecology and Economics* **52**: 199-212.

Harrison, P.A. and Butterfield, R.E. 1996. Effects of climate change on Europe-wide winter wheat and sunflower productivity. *Climate Research* **7**: 225-241.

Masle, J. 2000. The effects of elevated CO_2 concentrations on cell division rates, growth patterns, and blade anatomy in young wheat plants are modulated by factors related to leaf position, vernalization, and genotype. *Plant Physiology* **122**: 1399-1415.

Southworth, J., Pfeifer, R.A., Habeck, M., Randolph, J.C., Doering, O.C. and Rao, D.G. 2002. Sensitivity of winter wheat yields in the Midwestern United States to future changes in climate, climate variability, and CO_2 fertilization. *Climate Research* **22**: 73-86.

van Ittersum, M.K., Howden, S.M. and Asseng, S. 2003. Sensitivity of productivity and deep drainage of wheat cropping systems in a Mediterranean environment to changes in CO_2, temperature and precipitation. *Agriculture, Ecosystems and Environment* **97**: 255-273.

7.1.2. Woody Plants

The growth response of woody plants to atmospheric CO_2 enrichment has also been extensively studied. Ceulemans and Mousseau (1994), for example, tabulated the results of 95 separate experimental investigations related to this topic. The review of Poorter (1993) includes 41 additional sets of pertinent results, and the two reviews of Wullschleger et al. (1995, 1997) contain 40 other sets of applicable data. When averaged together, these 176 individual woody plant experiments reveal a mean growth enhancement on the order of 50 percent for an approximate doubling of the air's CO_2 content, which is about one-and-a-half times as much as the response of non-woody herbaceous plants.

It is possible, however, that this larger result is still an underestimate of the capacity of trees and shrubs to respond to atmospheric CO_2 enrichment; for the mean duration of the 176 woody plant experiments described above was only five months, which may not have been sufficient for the long-term equilibrium effects of the CO_2 enrichment of the air to be manifest. In the world's longest such experiment, for example, Kimball et al. (2007) observed a 70 percent sustained increase in biomass production over the entire last decade of a 17-year study in response to a 75 percent increase in the air's CO_2 content employed throughout the experiment. Likewise, studies of Eldarica pine trees conducted at the same location have revealed a similarly increasing growth response over the same length of time (Idso and Kimball, 1994).

In the subsections that follow, we highlight the results of studies that have examined the growth response of several woody plants to atmospheric CO_2 enrichment. We end the section with discussions of the effect of CO_2 enhancement on wood density and forest productivity and carbon sequestration. For more information on this topic, see http://www.co2science.org/data/plant_growth/plantgrowth.php.

References

Ceulemans, R. and Mousseau, M. (1994). Effects of elevated atmospheric CO_2 on woody plants. *New Phytologist* **127**: 425-446.

Idso, S.B. and Kimball, B.A. (1994). Effects of atmospheric CO_2 enrichment on biomass accumulation and distribution in Eldarica pine trees. *Journal of Experimental Botany* **45**: 1669-1672.

Kimball, B.A., Idso, S.B., Johnson, S. and Rillig, M.C. 2007. Seventeen years of carbon dioxide enrichment of sour orange trees: final results. *Global Change Biology* **13**: 2171-2183.

Poorter, H. (1993). Interspecific Variation in the Growth Response of Plants to an Elevated Ambient CO_2 Concentration. *Vegetatio* **104/105** 77-97.

Wullschleger, S.D., Post, W.M. and King, A.W. (1995). On the potential for a CO_2 fertilization effect in forests: Estimates of the biotic growth factor based on 58 controlled-exposure studies. In: Woodwell, G.M. and Mackenzie, F.T. (Eds.) *Biotic Feedbacks in the Global Climate System*. Oxford University Press, Oxford, pp. 85-107.

Wullschleger, S.D., Norby, R.J. and Gunderson, C.A. (1997). Forest trees and their response to atmospheric CO_2 enrichment: A compilation of results. In: Allen Jr., L.H., Kirkham, M.B., Olszyk, D.M. and Whitman, C.E. (Eds.) *Advances in Carbon Dioxide Effects Research*. American Society of Agronomy, Madison, WI, pp. 79-100.

7.1.2.1. Aspen-Poplar

Several studies have documented the effects of elevated levels of atmospheric CO_2 on photosynthesis in various aspen clones (*Populus tremuloides*). In the short-term study of Kruger et al. (1998), aspen seedlings grown for 70 days at atmospheric CO_2 concentrations of 650 ppm exhibited photosynthetic rates that were approximately 10 percent greater than those displayed by seedlings maintained at ambient CO_2 concentrations. In the longer five-month study of Kubiske et al. (1998), atmospheric CO_2 enrichment significantly increased photosynthetic rates in four aspen genotypes, regardless of soil nitrogen status.

In an even longer 2.5-year study, Wang and Curtis (2001) also observed significant CO_2-induced photosynthetic increases in two male and two female aspen clones; when six aspen genotypes were grown in open-top chambers for 2.5 years at atmospheric CO_2 concentrations of 350 and 700 ppm, Curtis et al. (2000) reported that the elevated CO_2 concentrations increased rates of net photosynthesis by 128 and 31 percent at high and low soil nitrogen contents, respectively. In addition, in a study that looked only at air temperature effects that was conducted at ambient CO_2 concentrations, King et al. (1999) determined that increasing the air temperature from 13° to 29°C enhanced photosynthetic rates in four different aspen clones by an average of 35 percent.

In a FACE study, where O_3-sensitive and O_3-tolerant clones were grown for six months in field plots receiving 360 and 560 ppm CO_2 in combination with ambient and enriched (1.5 times ambient) O_3 levels, Noormets et al. (2001) reported that CO_2-induced increases in photosynthetic rates were at least maintained, and sometimes even increased, when clones were simultaneously exposed to elevated O_3. After an entire year of treatment exposure, in fact, Karnosky et al. (1999) noted that the powerful ameliorating effect of elevated CO_2 on ozone-induced damage was still operating strongly in this system. O_3-induced foliar damages in O_3-sensitive and O_3-tolerant clones were reduced from 55 and 17 percent, respectively, at ambient CO_2, to 38 and 3 percent, respectively, at elevated CO_2.

With respect to biomass production, Pregitzer et al. (2000) reported that 2.5 years of exposure to twice-ambient concentrations of atmospheric CO_2 increased fine-root biomass in six aspen genotypes by an average of 65 and 17 percent on nitrogen-rich and nitrogen-poor soils, respectively. Using this same experimental system, Zak et al. (2000) determined that elevated CO_2 enhanced total seedling biomass by 38 percent at high soil nitrogen and by 16 percent at low soil nitrogen. Similar results were reported in the two-year open-top chamber study of Mikan et al. (2000), who observed 50 and 25 percent CO_2-induced increases in total seedling biomass at high and low soil nitrogen levels, respectively.

As the air's CO_2 content continues to increase, aspen seedlings will likely display enhanced rates of photosynthesis and biomass production, regardless of genotype, gender, O_3-sensitvity, and soil nitrogen status. Consequently, greater amounts of carbon will likely be sequestered in the tissues of this most abundant of North American tree species and in the soils in which they are rooted in the years and decades ahead.

Additional information on this topic, including reviews of newer publications as they become available, can be found at http://www.co2science.org/subject/t/treesaspen.php.

References

Curtis, P.S., Vogel, C.S., Wang, X.Z., Pregitzer, K.S., Zak, D.R., Lussenhop, J., Kubiske, M. and Teeri, J.A. 2000. Gas exchange, leaf nitrogen, and growth efficiency of *Populus tremuloides* in a CO_2-enriched atmosphere. *Ecological Applications* 10: 3-17.

Karnosky, D.F., Mankovska, B., Percy, K., Dickson, R.E., Podila, G.K., Sober, J., Noormets, A., Hendrey, G., Coleman, M.D., Kubiske, M., Pregitzer, K.S. and Isebrands, J.G. 1999. Effects of tropospheric O_3 on trembling aspen and interaction with CO_2: results from an O_3-gradient and a FACE experiment. *Water, Air, and Soil Pollution* 116: 311-322.

King, J.S., Pregitzer, K.S. and Zak, D.R. 1999. Clonal variation in above- and below-ground responses of *Populus tremuloides* Michaux: Influence of soil warming and nutrient availability. *Plant and Soil* 217: 119-130.

Kruger, E.L., Volin, J.C. and Lindroth, R.L. 1998. Influences of atmospheric CO_2 enrichment on the responses of sugar maple and trembling aspen to defoliation. *New Phytologist* 140: 85-94.

Kubiske, M.E., Pregitzer, K.S., Zak, D.R. and Mikan, C.J. 1998. Growth and C allocation of *Populus tremuloides* genotypes in response to atmospheric CO_2 and soil N availability. *New Phytologist* 140: 251-260.

Mikan, C.J., Zak, D.R., Kubiske, M.E. and Pregitzer, K.S. 2000. Combined effects of atmospheric CO_2 and N availability on the belowground carbon and nitrogen dynamics of aspen mesocosms. *Oecologia* 124: 432-445.

Noormets, A., Sober, A., Pell, E.J., Dickson, R.E., Podila, G.K., Sober, J., Isebrands, J.G. and Karnosky, D.F. 2001. Stomatal and non-stomatal limitation to photosynthesis in two trembling aspen (*Populus tremuloides* Michx.) clones exposed to elevated CO_2 and O_3. *Plant, Cell and Environment* 24: 327-336.

Pregitzer, K.S., Zak, D.R., Maziasz, J., DeForest, J., Curtis, P.S. and Lussenhop, J. 2000. Interactive effects of atmospheric CO_2 and soil-N availability on fine roots of *Populus tremuloides*. *Ecological Applications* 10: 18-33.

Wang, X. and Curtis, P.S. 2001. Gender-specific responses of *Populus tremuloides* to atmospheric CO_2 enrichment. *New Phytologist* 150: 675-684.

Zak, D.R., Pregitzer, K.S., Curtis, P.S., Vogel, C.S., Holmes, W.E. and Lussenhop, J. 2000. Atmospheric CO_2, soil-N availability, and allocation of biomass and nitrogen by *Populus tremuloides*. *Ecological Applications* 10: 34-46.

7.1.2.2. Beech

Egli and Korner (1997) rooted eight beech saplings (genus *Fagus*) directly into calcareous or acidic soils in open-top chambers and exposed them to atmospheric CO_2 concentrations of either 370 or 570 ppm. Over the first year of their study, the saplings growing on calcareous soil in CO_2-enriched air exhibited a 9 percent increase in stem diameter; they speculated that this initial small difference may "cumulate to higher 'final' tree biomass through compounding interest." At the end of three years of differential CO_2 exposure, the trees in the CO_2-enriched chambers were experiencing net ecosystem carbon exchange rates that were 58 percent greater than the rates of the trees in the ambient CO_2 chambers, regardless of soil type; the stem dry mass of the CO_2-enriched trees was increased by about 13 percent over that observed in the ambient-air chambers (Maurer *et al.*, 1999).

In a similar but much shorter experiment, Dyckmans *et al.* (2000) grew three-year-old seedlings of beech for six weeks in controlled environment chambers maintained at atmospheric CO_2 concentrations of 350 and 700 ppm, finding that the doubling of the air's CO_2 content increased seedling carbon uptake by 63 percent. They also noted that the majority of the assimilated carbon was allocated to the early development of leaves, which would be expected to subsequently lead to greater absolute amounts of photosynthetic carbon fixation.

In the two-year study of Grams *et al.* (1999), beech seedlings grown at ambient CO_2 concentrations displayed large reductions in photosynthetic rates when simultaneously exposed to twice-ambient levels of ozone. However, at twice-ambient CO_2 concentrations, twice-ambient ozone concentrations had no negative effects on the trees' photosynthetic rates. Thus, atmospheric CO_2 enrichment completely ameliorated the negative effects of ozone on photosynthesis in this species.

Similarly, Polle *et al.* (1997) reported that beech seedlings grown at 700 ppm CO_2 for two years displayed significantly reduced activities of catalase and superoxide dismutase, which are antioxidative enzymes responsible for detoxifying highly reactive oxygenated compounds within cells. Their data imply that CO_2-enriched atmospheres are conducive to less oxidative stress and, therefore, less production of harmful oxygenated compounds than typically occurs in ambient air. Consequently, the seedlings growing in the CO_2-enriched air were likely able to remobilize a portion of some of their valuable raw materials away from the production of detoxifying enzymes and reinvest them into other processes required for facilitating optimal plant development and growth.

With respect to this concept of resource optimization, Duquesnay *et al.* (1998) studied the relative amounts of ^{12}C and ^{13}C in tree rings of beech growing for the past century in northeastern France

and determined that the intrinsic water-use efficiency of the trees had increased by approximately 33 percent over that time period, no doubt in response to the concomitant rise in the air's CO_2 concentration over the past 100 years.

In conclusion, as the CO_2 content of the air increases, beech trees will likely display enhanced rates of photosynthesis and decreased damage resulting from oxidative stress. Together, these phenomena should allow greater optimization of raw materials within beech, allowing them to produce greater amounts of biomass ever more efficiently as the atmospheric CO_2 concentration increases.

Additional information on this topic, including reviews of newer publications as they become available, can be found at http://www.co2science.org/subject/t/treesbeech.php.

References

Duquesnay, A., Breda, N., Stievenard, M. and Dupouey, J.L. 1998. Changes of tree-ring $\delta^{13}C$ and water-use efficiency of beech (*Fagus sylvatica* L.) in north-eastern France during the past century. *Plant, Cell and Environment* **21**: 565-572.

Dyckmans, J., Flessa, H., Polle, A. and Beese, F. 2000. The effect of elevated [CO_2] on uptake and allocation of ^{13}C and ^{15}N in beech (*Fagus sylvatica* L.) during leafing. *Plant Biology* **2**: 113-120.

Egli, P. and Korner, C. 1997. Growth responses to elevated CO_2 and soil quality in beech-spruce model ecosystems. *Acta Oecologica* **18**: 343-349.

Grams, T.E.E., Anegg, S., Haberle, K.-H., Langebartels, C. and Matyssek, R. 1999. Interactions of chronic exposure to elevated CO_2 and O_3 levels in the photosynthetic light and dark reactions of European beech (*Fagus sylvatica*). *New Phytologist* **144**: 95-107.

Maurer, S., Egli, P., Spinnler, D. and Korner, C. 1999. Carbon and water fluxes in beech-spruce model ecosystems in response to long-term exposure to atmospheric CO_2 enrichment and increased nitrogen deposition. *Functional Ecology* **13**: 748-755.

Polle, A., Eiblmeier, M., Sheppard, L. and Murray, M. 1997. Responses of antioxidative enzymes to elevated CO_2 in leaves of beech (*Fagus sylvatica* L.) seedlings grown under range of nutrient regimes. *Plant, Cell and Environment* **20**: 1317-1321.

7.1.2.3. Birch

In the relatively short-term study of Wayne *et al.* (1998), yellow birch seedlings (*Betula pendula*) grown for two months at atmospheric CO_2 concentrations of 800 ppm exhibited photosynthetic rates that were about 50 percent greater than those displayed by control seedlings fumigated with air containing 400 ppm CO_2. Similarly, in the three-month study of Tjoelker *et al.* (1998a), paper birch seedlings grown at 580 ppm CO_2 displayed photosynthetic rates that were approximately 30 percent greater than those exhibited by seedlings exposed to 370 ppm CO_2. Likewise, Kellomaki and Wang (2001) reported that birch seedlings exposed to an atmospheric CO_2 concentration of 700 ppm for five months displayed photosynthetic rates that were about 25 percent greater than seedlings grown at 350 ppm CO_2. Finally, in the much longer four-year study conducted by Wang *et al.* (1998), silver birch seedlings grown in open-top chambers receiving twice-ambient concentrations of atmospheric CO_2 displayed photosynthetic rates that were fully 110 percent greater than rates displayed by their ambiently grown counterparts. Thus, short-term photosynthetic enhancements resulting from atmospheric CO_2 enrichment appear to persist for several years or longer.

Because elevated CO_2 enhances photosynthetic rates in birch trees, it likely will also lead to increased biomass production in these important deciduous trees, as it has in several experiments. In the three-month study of Tjoelker *et al.* (1998b), for example, a 57 percent increase in the air's CO_2 content increased the biomass of paper birch seedlings by 50 percent. When similar seedlings were grown at 700 ppm CO_2 for four months, Catovsky and Bazzaz (1999) reported that elevated CO_2 increased total seedling biomass by 27 and 130 percent under wet and dry soil moisture regimes, respectively. In the interesting study of Godbold *et al.* (1997), paper birch seedlings grown at 700 ppm for six months not only increased their total biomass, but also increased the number of root tips per plant by more than 50 percent. In the longer two-year study of Berntson and Bazzaz (1998), twice-ambient levels of CO_2 increased the biomass of a mixed yellow and white birch mesocosm by 31 percent; and in another two-year study, Wayne *et al.* (1998) reported that yellow birch seedlings grown at 800 ppm CO_2 produced 60 and 227 percent more biomass than seedlings grown at 400 ppm CO_2 at ambient and elevated air temperatures, respectively.

Finally, after exposing silver birch seedlings to twice-ambient CO_2 concentrations for four years, Wang *et al.* (1998) noted that CO_2-enriched seedlings produced 60 percent more biomass than ambiently grown seedlings. Hence, atmospheric CO_2 enrichment clearly enhances birch biomass in both short- and medium-term experiments.

In some studies, elevated CO_2 also reduced stomatal conductances in birch trees, thereby boosting their water-use efficiencies. Tjoelker *et al.* (1998a), for example, reported that paper birch seedlings grown at 580 ppm CO_2 for three months experienced 10-25 percent reductions in stomatal conductance, which contributed to 40-80 percent increases in water-use efficiency. Similar CO_2-induced reductions in stomatal conductance (21 percent) were reported in silver birch seedlings grown for four years at 700 ppm CO_2 by Rey and Jarvis (1998).

The results of these several studies suggest that the ongoing rise in the air's CO_2 content will likely increase rates of photosynthesis and biomass production in birch trees, as well as improve their water use efficiencies, irrespective of any concomitant changes in air temperature and/or soil moisture status that might occur. Consequently, rates of carbon sequestration by this abundant temperate forest species should also increase in the years and decades ahead.

Additional information on this topic, including reviews of newer publications as they become available, can be found at http://www.co2science.org/subject/t/treesbirch.php.

References

Berntson, G.M. and Bazzaz, F.A. 1998. Regenerating temperate forest mesocosms in elevated CO_2: belowground growth and nitrogen cycling. *Oecologia* **113**: 115-125.

Catovsky, S. and Bazzaz, F.A. 1999. Elevated CO_2 influences the responses of two birch species to soil moisture: implications for forest community structure. *Global Change Biology* **5**: 507-518.

Godbold, D.L., Berntson, G.M. and Bazzaz, F.A. 1997. Growth and mycorrhizal colonization of three North American tree species under elevated atmospheric CO_2. *New Phytologist* **137**: 433-440.

Kellomaki, S. and Wang, K.-Y. 2001. Growth and resource use of birch seedlings under elevated carbon dioxide and temperature. *Annals of Botany* **87**: 669-682.

Rey, A. and Jarvis, P.G. 1998. Long-Term photosynthetic acclimation to increased atmospheric CO_2 concentration in young birch (*Betula pendula*) trees. *Tree Physiology* **18**: 441-450.

Tjoelker, M.G., Oleksyn, J. and Reich, P.B. 1998a. Seedlings of five boreal tree species differ in acclimation of net photosynthesis to elevated CO_2 and temperature. *Tree Physiology* **18**: 715-726.

Tjoelker, M.G., Oleksyn, J. and Reich, P.B. 1998b. Temperature and ontogeny mediate growth response to elevated CO_2 in seedlings of five boreal tree species. *New Phytologist* **140**: 197-210.

Wang, Y.-P., Rey, A. and Jarvis, P.G. 1998. Carbon balance of young birch trees grown in ambient and elevated atmospheric CO_2 concentrations. *Global Change Biology* **4**: 797-807.

Wayne, P.M., Reekie, E.G. and Bazzaz, F.A. 1998. Elevated CO_2 ameliorates birch response to high temperature and frost stress: implications for modeling climate-induced geographic range shifts. *Oecologia* **114**: 335-342.

7.1.2.4. Citrus Trees

How does atmospheric CO_2 enrichment affect the growth and development of citrus trees and the fruit they produce?

In the study of Keutgen and Chen (2001), cuttings of *Citrus madurensis* grown for three months at 600 ppm CO_2 displayed rates of photosynthesis that were more than 300 percent greater than those measured on control cuttings grown at 300 ppm CO_2. In addition, elevated CO_2 concentrations have been shown to increase photosynthetic rates in mango (Schaffer *et al.*, 1997), mangosteen (Schaffer *et al.*, 1999), and sweet orange (Jifon *et al.*, 2002). In the study of Jifon *et al.*, it was further reported that twice-ambient CO_2 concentrations increased photosynthetic rates in mycorrhizal- and non-mycorrhizal-treated sour orange seedlings by 118 and 18 percent, respectively.

Such CO_2-induced increases in photosynthesis should lead to enhanced biomass production; and so they do. Idso and Kimball (2001), for example, have documented how a 75 percent increase in the air's CO_2 content has boosted the long-term production of above-ground wood and fruit biomass in sour orange trees by 80 percent in a study that has been ongoing since November 1987. Furthermore, Idso *et al.* (2002) have additionally demonstrated that the 300-ppm increase in the air's CO_2 content has increased the fresh weight of individual oranges by an average of 4

percent and the vitamin C content of their juice by an average of 5 percent.

In summary, these peer-reviewed studies suggest that as the air's CO_2 content slowly but steadily rises, citrus trees will respond by increasing their rates of photosynthesis and biomass production. In addition, they may also increase the vitamin C content of their fruit, which may help to prevent an array of human health problems brought about by insufficient intake of vitamin C.

Additional information on this topic, including reviews of newer publications as they become available, can be found at http://www.co2science.org/subject/t/treescitrus.php.

References

Idso, S.B. and Kimball, B.A. 2001. CO_2 enrichment of sour orange trees: 13 years and counting. *Environmental and Experimental Botany* **46**: 147-153.

Idso, S.B., Kimball, B.A., Shaw, P.E., Widmer, W., Vanderslice, J.T., Higgs, D.J., Montanari, A. and Clark, W.D. 2002. The effect of elevated atmospheric CO_2 on the vitamin C concentration of (sour) orange juice. *Agriculture, Ecosystems and Environment* **90**: 1-7.

Jifon, J.L., Graham, J.H., Drouillard, D.L. and Syvertsen, J.P. 2002. Growth depression of mycorrhizal Citrus seedlings grown at high phosphorus supply is mitigated by elevated CO_2. *New Phytologist* **153**: 133-142.

Keutgen, N. and Chen, K. 2001. Responses of citrus leaf photosynthesis, chlorophyll fluorescence, macronutrient and carbohydrate contents to elevated CO_2. *Journal of Plant Physiology* **158**: 1307-1316.

Schaffer, B., Whiley, A.W. and Searle, C. 1999. Atmospheric CO_2 enrichment, root restriction, photosynthesis, and dry-matter partitioning in subtropical and tropical fruit crops. *HortScience* **34**: 1033-1037.

Schaffer, B., Whiley, A.W., Searle, C. and Nissen, R.J. 1997. Leaf gas exchange, dry matter partitioning, and mineral element concentrations in mango as influenced by elevated atmospheric carbon dioxide and root restriction. *Journal of the American Society of Horticultural Science* **122**: 849-855.

7.1.2.5. Eucalyptus

In the eight-month study of Roden *et al.* (1999), *Eucalyptus pauciflora* seedlings growing at 700 ppm CO_2 displayed seasonal rates of net photosynthesis that were approximately 30 percent greater than those exhibited by their ambiently grown counterparts. In another eight-month study, Palanisamy (1999) reported that well-watered *Eucalyptus cladocalyx* seedlings exposed to 800 ppm CO_2 exhibited photosynthetic rates that were 120 percent higher than those observed in control plants growing at 380 ppm CO_2. Moreover, after a one-month period of water stress, photosynthetic rates of CO_2-enriched seedlings were still 12 percent greater than rates displayed by ambiently grown water-stressed seedlings.

Because elevated CO_2 enhances photosynthetic rates in eucalyptus species, this phenomenon should lead to increased biomass production in these rapidly growing trees. And so it does. In the eight-month experiment of Gleadow *et al.* (1998), for example, *Eucalyptus cladocalyx* seedlings growing at 800 ppm CO_2 displayed 134 and 98 percent more biomass than seedlings growing at 400 ppm CO_2 at low and high soil nitrogen concentrations, respectively. Similarly, *Eucalyptus pauciflora* seedlings growing at twice-ambient CO_2 concentrations for eight months produced 53 percent more biomass than control seedlings (Roden *et al.*, 1999).

In summary, as the CO_2 content of the air increases, eucalyptus seedlings will likely display enhanced rates of photosynthesis and biomass production, regardless of soil moisture and nutrient status. Consequently, greater amounts of carbon will likely be sequestered by this rapidly growing tree species.

Additional information on this topic, including reviews of newer publications as they become available, can be found at http://www.co2science.org/subject/t/treeseuc.php.

References

Gleadow, R.M., Foley, W.J. and Woodrow, I.E. 1998. Enhanced CO_2 alters the relationship between photosynthesis and defense in cyanogenic *Eucalyptus cladocalyx* F. Muell. *Plant, Cell and Environment* **21**: 12-22.

Palanisamy, K. 1999. Interactions of elevated CO_2 concentration and drought stress on photosynthesis in *Eucalyptus cladocalyx* F. Muell. *Photosynthetica* **36**: 635-638.

Roden, J.S., Egerton, J.J.G. and Ball, M.C. 1999. Effect of elevated [CO_2] on photosynthesis and growth of snow gum (*Eucalyptus pauciflora*) seedlings during winter and spring. *Australian Journal of Plant Physiology* **26**: 37-46.

7.1.2.6. Fruit-Bearing

Several studies have recently documented the effects of elevated atmospheric CO_2 concentrations on photosynthesis in various fruiting trees. In an eight-day experiment, Pan *et al.* (1998) found that twice-ambient CO_2 concentrations increased rates of net photosynthesis in one-year-old apple seedlings by 90 percent. In a longer three-month study, Keutgen and Chen (2001) noted that cuttings of *Citrus madurensis* exposed to 600 ppm CO_2 displayed rates of photosynthesis that were more than 300 percent greater than rates observed in control cuttings exposed to 300 ppm CO_2. Likewise, in the review paper of Schaffer *et al.* (1997), it was noted that atmospheric CO_2 enrichment had previously been shown to enhance rates of net photosynthesis in various tropical and sub-tropical fruit trees, including avocado, banana, citrus, mango, and mangosteen. Finally in the two-year study of Centritto *et al.* (1999a), cherry seedlings grown at 700 ppm CO_2 exhibited photosynthetic rates that were 44 percent greater than those displayed by seedlings grown in ambient air, independent of a concomitant soil moisture treatment.

Because elevated CO_2 enhances the photosynthetic rates of fruiting trees, it should also lead to increased biomass production in them. In the two-year study of Centritto *et al.* (1999b), for example, well-watered and water-stressed seedlings growing at twice-ambient CO_2 concentrations displayed basal trunk areas that were 47 and 51 percent larger than their respective ambient controls. Similarly, in a study spanning more than 13 years, Idso and Kimball (2001) demonstrated that the above-ground wood biomass of mature sour orange trees growing in air enriched with an additional 300 ppm of CO_2 was 80 percent greater than that attained by control trees growing in ambient air.

As the CO_2 content of the air increases, fruit trees will likely display enhanced rates of photosynthesis and biomass production, regardless of soil moisture conditions. Consequently, greater amounts of carbon will likely be sequestered in the woody trunks and branches of such species. Moreover, fruit yields may increase as well. In the study of Idso and Kimball, for example, fruit yields were stimulated to essentially the same degree as above-ground wood biomass; i.e., by 80 percent in response to a 75 percent increase in the air's CO_2 content.

Additional information on this topic, including reviews of newer publications as they become available, can be found at http://www.co2science.org/subject/t/treesfruit.php.

References

Centritto, M., Magnani, F., Lee, H.S.J. and Jarvis, P.G. 1999a. Interactive effects of elevated [CO_2] and drought on cherry (*Prunus avium*) seedlings. II. Photosynthetic capacity and water relations. *New Phytologist* **141**: 141-153.

Centritto, M., Lee, H.S.J. and Jarvis, P.G. 1999b. Interactive effects of elevated [CO_2] and drought on cherry (*Prunus avium*) seedlings. I. Growth, whole-plant water use efficiency and water loss. *New Phytologist* **141**: 129-140.

Idso, S.B. and Kimball, B.A. 2001. CO_2 enrichment of sour orange trees: 13 years and counting. *Environmental and Experimental Botany* **46**: 147-153.

Keutgen, N. and Chen, K. 2001. Responses of citrus leaf photosynthesis, chlorophyll fluorescence, macronutrient and carbohydrate contents to elevated CO_2. *Journal of Plant Physiology* **158**: 1307-1316.

Pan, Q., Wang, Z. and Quebedeaux, B. 1998. Responses of the apple plant to CO_2 enrichment: changes in photosynthesis, sorbitol, other soluble sugars, and starch. *Australian Journal of Plant Physiology* **25**: 293-297.

Schaffer, B., Whiley, A.W., Searle, C. and Nissen, R.J. 1997. Leaf gas exchange, dry matter partitioning, and mineral element concentrations in mango as influenced by elevated atmospheric carbon dioxide and root restriction. *Journal of the American Society of Horticultural Science* **122**: 849-855.

7.1.2.7. Nitrogen-Fixing Trees

In the six-week study of Schortemeyer *et al.* (1999), seedlings of Australian blackwood (*Acacia melanoxylon*) grown at twice-ambient atmospheric CO_2 concentrations displayed photosynthetic rates that were 22 percent greater than those of ambiently grown seedlings. In addition, the CO_2-enriched seedlings exhibited biomass values that were twice as large as those displayed by control seedlings grown in air of 350 ppm CO_2. Likewise, Polley *et al.* (1999) reported that a doubling of the atmospheric CO_2 concentration for three months increased honey mesquite (*Prosopis glandulosa*) seedling root and shoot biomass by 37 and 46 percent, respectively.

Several studies have investigated the effects of elevated CO_2 on black locust (*Robinia pseudoacacia*) seedlings. Uselman *et al.* (2000), grew seedlings for three months at 700 ppm CO_2 and reported that this treatment increased the root exudation of organic carbon compounds by 20 percent, while Uselman *et al.* (1999) reported no CO_2-induced increases in the root exudation of organic nitrogen compounds. Nonetheless, elevated CO_2 enhanced total seedling biomass by 14 percent (Uselman *et al.*, 2000).

In the study of Olesniewicz and Thomas (1999), black locust seedlings grown at twice-ambient CO_2 concentrations for two months exhibited a 69 percent increase in their average rate of nitrogen-fixation when they were not inoculated with an arbuscular mycorrhizal fungal species. It was further determined that the amount of seedling nitrogen derived from nitrogen-fixation increased in CO_2-enriched plants by 212 and 90 percent in non-inoculated and inoculated seedlings, respectively. Elevated CO_2 enhanced total plant biomass by 180 and 51 percent in non-inoculated and inoculated seedlings, respectively.

As the CO_2 content of the air increases, nitrogen-fixing trees respond by exhibiting enhanced rates of photosynthesis and biomass production, as well as enhanced rates of nitrogen fixation.

Additional information on this topic, including reviews of newer publications as they become available, can be found at http://www.co2science.org/subject/t/treesnitrofix.php.

References

Olesniewicz, K.S. and Thomas, R.B. 1999. Effects of mycorrhizal colonization on biomass production and nitrogen fixation of black locust (*Robinia pseudoacacia*) seedlings grown under elevated atmospheric carbon dioxide. *New Phytologist* **142**: 133-140.

Polley, H.W., Tischler, C.R., Johnson, H.B. and Pennington, R.E. 1999. Growth, water relations, and survival of drought-exposed seedlings from six maternal families of honey mesquite (*Prosopis glandulosa*): responses to CO_2 enrichment. *Tree Physiology* **19**: 359-366.

Schortemeyer, M., Atkin, O.K., McFarlane, N. and Evans, J.R. 1999. The impact of elevated atmospheric CO_2 and nitrate supply on growth, biomass allocation, nitrogen partitioning and N_2 fixation of *Acacia melanoxylon*. *Australian Journal of Plant Physiology* **26**: 737-774.

Uselman, S.M., Qualls, R.G. and Thomas, R.B. 1999. A test of a potential short cut in the nitrogen cycle: The role of exudation of symbiotically fixed nitrogen from the roots of a N-fixing tree and the effects of increased atmospheric CO_2 and temperature. *Plant and Soil* **210**: 21-32.

Uselman, S.M., Qualls, R.G. and Thomas, R.B. 2000. Effects of increased atmospheric CO_2, temperature, and soil N availability on root exudation of dissolved organic carbon by a N-fixing tree (*Robinia pseudoacacia* L.). *Plant and Soil* **222**: 191-202.

7.1.2.8. Oak

How do oak (genus *Quercus*) trees respond to atmospheric CO_2 enrichment? In the two-month study of Anderson and Tomlinson (1998), northern red oak seedlings exposed to 700 ppm CO_2 displayed photosynthetic rates that were 34 and 69 percent greater than those displayed by control plants growing under well-watered and water-stressed conditions, respectively. Similarly, in the four-month study of Li *et al.* (2000), *Quercus myrtifolia* seedlings growing at twice-ambient CO_2 concentrations exhibited rates of photosynthesis at the onset of senescence that were 97 percent greater than those displayed by ambiently growing seedlings.

In the year-long study of Staudt *et al.* (2001), *Quercus ilex* seedlings grown at 700 ppm CO_2 displayed trunk and branch biomasses that were 90 percent greater than those measured on seedlings growing at 350 ppm CO_2. Also, in the eight-month inter-generational study performed by Polle *et al.* (2001), seedlings produced from acorns collected from ambient and CO_2-enriched mother trees and germinated in air of either ambient or twice-ambient atmospheric CO_2 concentration displayed whole-plant biomass values that were 158 and 246 percent greater, respectively, than those exhibited by their respective control seedlings growing in ambient air.

In another study, Schulte *et al.* (1998) grew oak seedlings for 15 weeks at twice-ambient CO_2 concentrations, finding that elevated CO_2 enhanced seedling biomass by 92 and 128 percent under well-watered and water-stressed conditions, respectively. In a similar study conducted by Tomlinson and Anderson (1998), water-stressed seedlings growing at 700 ppm CO_2 displayed biomass values that were similar to those exhibited by well-watered plants growing in ambient air. Thus, atmospheric CO_2 enrichment continues to benefit oak trees even under water-stressed conditions.

Additional studies have demonstrated that oak seedlings also respond positively to atmospheric CO_2 enrichment when they are faced with other environmental stresses and resource limitations. When pedunculate oak seedlings were subjected to two different soil nutrient regimes, for example,

Maillard *et al.* (2001) reported that a doubling of the atmospheric CO_2 concentration enhanced seedling biomass by 140 and 30 percent under high and low soil nitrogen conditions, respectively. And in the study of Usami *et al.* (2001), saplings of *Quercus myrsinaefolia* that were grown at 700 ppm CO_2 displayed biomass increases that were 110 and 140 percent greater than their ambiently grown counterparts when they were simultaneously subjected to air temperatures that were 3° and 5°C greater than ambient temperature, respectively. Thus, elevated CO_2 concentrations tend to ameliorate some of the negative effects caused by growth-reducing stresses in oaks. In fact, when Schwanz and Polle (1998) reported that elevated CO_2 exposure caused reductions in the amounts of several foliar antioxidative enzymes in mature oak trees, they suggested that this phenomenon was the result of atmospheric CO_2 enrichment causing the trees to experience less oxidative stress and, therefore, they had less need for antioxidative enzymes.

In some studies, elevated CO_2 has been shown to reduce stomatal conductances in oak trees, thus contributing to greater tree water-use efficiencies. Tognetti *et al.* (1998a), for example, reported that oak seedlings growing near a natural CO_2-emitting spring exhibited less water loss and more favorable turgor pressures than trees growing further away from the spring. The resulting improvement in water-use efficiency was so significant that Tognetti *et al.* (1998b) stated, "such marked increases in water-use efficiency under elevated CO_2 might be of great importance in Mediterranean environments in the perspective of global climate change."

In summary, it is clear that as the CO_2 content of the air increases, oak seedlings will likely display enhanced rates of photosynthesis and biomass production, regardless of air temperature, soil moisture, and soil nutrient status. Consequently, greater amounts of carbon will likely be removed from the atmosphere by the trees of this abundant genus and stored in their tissues and the soils in which they are rooted.

Additional information on this topic, including reviews of newer publications as they become available, can be found at http://www.co2science.org/subject/t/treesoak.php.

References

Anderson, P.D. and Tomlinson, P.T. 1998. Ontogeny affects response of northern red oak seedlings to elevated CO_2 and water stress. I. Carbon assimilation and biomass production. *New Phytologist* **140**: 477-491.

Li, J.-H., Dijkstra, P., Hymus, G.J., Wheeler, R.M., Piastuchi, W.C., Hinkle, C.R. and Drake, B.G. 2000. Leaf senescence of *Quercus myrtifolia* as affected by long-term CO_2 enrichment in its native environment. *Global Change Biology* **6**: 727-733.

Maillard, P., Guehl, J.-M., Muller, J.-F. and Gross, P. 2001. Interactive effects of elevated CO_2 concentration and nitrogen supply on partitioning of newly fixed ^{13}C and ^{15}N between shoot and roots of pedunculate oak seedlings (*Quercus robur* L.). *Tree Physiology* **21**: 163-172.

Polle, A., McKee, I. and Blaschke, L. 2001. Altered physiological and growth responses to elevated [CO_2] in offspring from holm oak (*Quercus ilex* L.) mother trees with lifetime exposure to naturally elevated [CO_2]. *Plant, Cell and Environment* **24**: 1075-1083.

Schwanz, P. and Polle, A. 1998. Antioxidative systems, pigment and protein contents in leaves of adult mediterranean oak species (*Quercus pubescens* and *Q. ilex*) with lifetime exposure to elevated CO_2. *New Phytologist* **140**: 411-423.

Schulte, M., Herschbach, C. and Rennenberg, H. 1998. Interactive effects of elevated atmospheric CO_2, mycorrhization and drought on long-distance transport of reduced sulphur in young pedunculate oak trees (*Quercus robur* L.). *Plant, Cell and Environment* **21**: 917-926.

Staudt, M., Joffre, R., Rambal, S. and Kesselmeier, J. 2001. Effect of elevated CO_2 on monoterpene emission of young *Quercus ilex* trees and its relation to structural and ecophysiological parameters. *Tree Physiology* **21**: 437-445.

Tognetti, R., Longobucco, A., Miglietta, F. and Raschi, A. 1998a. Transpiration and stomatal behaviour of *Quercus ilex* plants during the summer in a Mediterranean carbon dioxide spring. *Plant, Cell and Environment* **21**: 613-622.

Tognetti, R., Johnson, J.D., Michelozzi, M. and Raschi, A. 1998b. Response of foliar metabolism in mature trees of *Quercus pubescens* and *Quercus ilex* to long-term elevated CO_2. *Environmental and Experimental Botany* **39**: 233-245.

Tomlinson, P.T. and Anderson, P.D. 1998. Ontogeny affects response of northern red oak seedlings to elevated CO_2 and water stress. II. Recent photosynthate distribution and growth. *New Phytologist* **140**: 493-504.

Usami, T., Lee, J. and Oikawa, T. 2001. Interactive effects of increased temperature and CO_2 on the growth of *Quercus myrsinaefolia* saplings. *Plant, Cell and Environment* **24**: 1007-1019.

7.1.2.9. Pine

7.1.2.9.1. Loblolly

Tissue *et al.* (1997) grew seedlings of loblolly pine trees (*Pinus taeda* L.) for a period of four years in open-top chambers maintained at atmospheric CO_2 concentrations of either 350 or 650 ppm in a study of the long-term effects of elevated CO_2 on the growth of this abundant pine species. This experiment indicated there was a mean biomass accumulation in the seedlings grown in CO_2-enriched air that was 90 percent greater than that attained by the seedlings grown in ambient air.

Johnson *et al.* (1998) reviewed 11 of their previously published papers, describing the results of a series of greenhouse and open-top chamber studies of the growth responses of loblolly pine seedlings to a range of atmospheric CO_2 and soil nitrogen concentrations. This work indicated that when soil nitrogen levels were so low as to be extremely deficient, or so high as to be toxic, growth responses to atmospheric CO_2 enrichment were negligible. For moderate soil nitrogen deficiencies, however, a doubling of the air's CO_2 content sometimes boosted growth by as much as 1,000 percent. Consequently, since the nitrogen status of most of earth's ecosystems falls somewhere between extreme deficiency and toxicity, these results suggest that loblolly pine trees may experience large increases in growth as the air's CO_2 content continues to climb.

Naidu and DeLucia (1999) described the results of working one full year in 30-meter-diameter circular FACE plots maintained at atmospheric CO_2 concentrations of either 350 or 560 ppm in an originally 13-year-old loblolly pine plantation in North Carolina, USA, where they determined the effects of the elevated CO_2 treatment on the productivity of the trees, which were growing in soil that was characteristically low in nitrogen and phosphorus. After the first year of atmospheric CO_2 enrichment in this Duke Forest Face Study, the growth rate of the CO_2-enriched trees was about 24 percent greater than that of the trees exposed to ambient CO_2, in spite of the likelihood of soil nutrient limitations and a severe summer drought (rainfall in August 1997 was about 90 percent below the 50-year average).

After four years of work at the Duke Forest Face Site, Finzi *et al.* (2002) reported that the extra 200 or so ppm of CO_2 had increased the average yearly dry matter production of the CO_2-enriched trees by 32 percent, while at the eight-year point of the experiment Moore *et al.* (2006) reported there had been a sustained increase in trunk basal area increment that varied between 13 and 27 percent with variations in weather and the timing of growth. What is more, they say "there was no evidence of a decline in the relative enhancement of tree growth by elevated CO_2 as might be expected if soil nutrients were becoming progressively more limiting," which many people had expected would occur in light of the site's low soil nitrogen and phosphorus content. In addition, at the six-year point of the study Pritchard *et al.* (2008) determined that the extra CO_2 had increased the average standing crop of fine roots by 23 percent.

Gavazzi *et al.* (2000) grew one-year-old loblolly pine seedlings for about four months in pots placed within growth chambers maintained at atmospheric CO_2 concentrations of either 360 or 660 ppm and adequate or inadequate levels of soil moisture, while the pots were seeded with a variety of C_3 and C_4 weeds. In the course of this experiment, they found that total seedling biomass was always greater under well-watered as opposed to water-stressed conditions, and that elevated CO_2 increased total seedling biomass by 22 percent in both water treatments. In the elevated CO_2 and water-stressed treatment, however, they also found that seedling root-to-shoot ratios were about 80 percent greater than they were in the elevated CO_2 and well-watered treatment, due to a 63 percent increase in root biomass. In the case of the weeds, total biomass was also always greater under well-watered compared to water-stressed conditions. However, the elevated CO_2 did *not* increase weed biomass; in fact, it reduced it by approximately 22 percent. Consequently, in assessing the effects of elevated CO_2 on competition between loblolly pine seedlings and weeds, the seedlings were definitely the winners, with the researchers concluding that the CO_2-induced increase in root-to-shoot ratio under water-stressed conditions may "contribute to an improved ability of loblolly pine to compete against weeds on dry sites."

Working with data obtained from stands of loblolly pine plantations at 94 locations scattered throughout the southeastern United States, Westfall and Amateis (2003) employed mean height measurements made at three-year intervals over a period of 15 years to calculate a site index related to mean growth rate for each of the five three-year periods, which index would be expected to increase monotonically if growth rates were being enhanced above "normal" by some monotonically increasing growth-promoting factor. This protocol indicated, in their words, that "mean site index over the 94 plots consistently increased at each remeasurement period," which would suggest, as they phrase it, that "loblolly pine plantations are realizing greater than expected growth rates," and, we would add, that the growth rate increases are growing larger and larger with each succeeding three-year period.

As for what might be causing the monotonically increasing growth rates of loblolly pine trees over the entire southeastern region of the United States, the two researchers say that in addition to rising atmospheric CO_2 concentrations, "two other likely factors that could affect growth are temperature and precipitation." However, they report that a review of annual precipitation amounts and mean ground surface temperatures showed no trends in these factors over the period of their study. They also suggest that if increased nitrogen deposition were the cause, "such a factor would have to be acting on a regional scale to produce growth increases over the range of study plots." Hence, they are partial to the aerial fertilization effect of atmospheric CO_2 enrichment as the explanation. What is more, they note that "similar results were reported by Boyer (2001) for natural stands of longleaf pine, where increases in dominant stand height are occurring over generations on the same site."

The studies reported here indicate that as the CO_2 content of the air continues to rise, loblolly pine trees will likely experience significant increases in biomass production, even on nutrient-poor soils, during times of drought, and in competition with weeds.

Additional information on this topic, including reviews loblolly pine trees not discussed here, can be found at http://www.co2science.org/subject/t/subject_t.php, under the heading Trees, Types, Pine, Loblolly.

References

Boyer, W.D. 2001. A generational change in site index for naturally established longleaf pine on a south Alabama coastal plain site. *Southern Journal of Applied Forestry* **25**: 88-92.

Finzi, A.C., DeLucia, E.H., Hamilton, J.G., Richter, D.D. and Schlesinger, W.H. 2002. The nitrogen budget of a pine forest under free air CO_2 enrichment. *Oecologia* **132**: 567-578.

Gavazzi, M., Seiler, J., Aust, W. and Zedaker, S. 2000. The influence of elevated carbon dioxide and water availability on herbaceous weed development and growth of transplanted loblolly pine (*Pinus taeda*). *Environmental and Experimental Botany* **44**: 185-194.

Johnson, D.W., Thomas, R.B., Griffin, K.L., Tissue, D.T., Ball, J.T., Strain, B.R. and Walker, R.F. 1998. Effects of carbon dioxide and nitrogen on growth and nitrogen uptake in ponderosa and loblolly pine. *Journal of Environmental Quality* **27**: 414-425.

Moore, D.J.P., Aref, S., Ho, R.M., Pippen, J.S., Hamilton, J.G. and DeLucia, E.H. 2006. Annual basal area increment and growth duration of *Pinus taeda* in response to eight years of free-air carbon dioxide enrichment. *Global Change Biology* **12**: 1367-1377.

Naidu, S.L. and DeLucia, E.H. 1999. First-year growth response of trees in an intact forest exposed to elevated CO_2. *Global Change Biology* **5**: 609-613.

Pritchard, S.G., Strand, A.E., McCormack, M.L., Davis, M.A., Finzi, A.C., Jackson, R.B., Matamala, R., Rogers, H.H. and Oren, R. 2008. Fine root dynamics in a loblolly pine forest are influenced by free-air-CO_2-enrichment: a six-year-minirhizotron study. *Global Change Biology* **14**: 588-602.

Tissue, D.T., Thomas, R.B. and Strain, B.R. 1997. Atmospheric CO_2 enrichment increases growth and photosynthesis of *Pinus taeda*: a 4-year experiment in the field. *Plant, Cell and Environment* **20**: 1123-1134.

Westfall, J.A. and Amateis, R.L. 2003. A model to account for potential correlations between growth of loblolly pine and changing ambient carbon dioxide concentrations. *Southern Journal of Applied Forestry* **27**: 279-284.

7.1.2.9.2. Ponderosa

Walker *et al.* (1998b) grew seedlings of Ponderosa pine (*Pinus ponderosa* Dougl. ex P. Laws & C. Laws) for an entire year in controlled environment chambers with atmospheric CO_2 concentrations of either 350 (ambient), 525, or 700 ppm. In addition, low or high levels of nitrogen and phosphorus were supplied to determine the main and interactive effects of atmospheric CO_2 enrichment and soil nutrition on seedling growth and fungal colonization of the seedlings' roots. After 12 months, they found that phosphorus supply had little impact on overall seedling growth, while high nitrogen increased nearly every parameter measured, including root, shoot, and total biomass, as did atmospheric CO_2 enrichment. Averaged over all nitrogen and phosphate treatments, total root dry weights at 525 and 700 ppm CO_2 were 92 and 49 percent greater, respectively, than those observed at ambient CO_2, while shoot dry weights were 83 and 26 percent greater. Consequently, seedlings grown at 525 and 700 ppm CO_2 had total dry weights that were 86 and 35 percent greater, respectively, than those measured at ambient CO_2. In addition, elevated CO_2 increased the total number of ectomycorrhizal fungi on roots by 170 percent at 525 ppm CO_2 and 85 percent at 700 ppm CO_2 relative to the number observed at ambient CO_2.

Walker *et al.* (1998a) grew Ponderosa pine seedlings for two growing seasons out-of-doors in open-top chambers having atmospheric CO_2 concentrations of 350, 525, and 700 ppm on soils of low, medium, and high nitrogen content to determine the interactive effects of these variables on juvenile tree growth. The elevated CO_2 concentrations had little effect on most growth parameters after the first growing season, with the one exception of below-ground biomass, which increased with both CO_2 and soil nitrogen. After two growing seasons, however, elevated CO_2 significantly increased all growth parameters, including tree height, stem diameter, shoot weight, stem volume, and root volume, with the greatest responses typically occurring at the highest CO_2 concentration in the highest soil nitrogen treatment. Root volume at 700 ppm CO_2 and high soil nitrogen, for example, exceeded that of all other treatments by at least 45 percent, as did shoot volume by 42 percent. Similarly, at high CO_2 and soil nitrogen coarse root and shoot weights exceeded those at ambient CO_2 and high nitrogen by 80 and 88 percent, respectively.

Johnson *et al.* (1998) reviewed 11 of their previously published papers (including the two discussed above) in which they describe the results of a series of greenhouse and open-top chamber studies of the growth responses of Ponderosa pine seedlings to a range of atmospheric CO_2 and soil nitrogen concentrations. These studies indicated that when soil nitrogen levels were so low as to be extremely deficient, or so high as to be toxic, growth responses to atmospheric CO_2 enrichment were negligible. For moderate soil nitrogen deficiencies, however, a doubling of the air's CO_2 content sometimes boosted growth by as much as 1,000 percent. In addition, atmospheric CO_2 enrichment mitigated the negative growth response of ponderosa pine to extremely high soil nitrogen in two separate studies.

Maherali and DeLucia (2000) grew Ponderosa pine seedlings for six months in controlled environment chambers maintained at atmospheric CO_2 concentrations ranging from 350 to 1,100 ppm, while they were subjected to either low (15/25°C night/day) or high (20/30°C night/day) temperatures. This study revealed that although elevated CO_2 had no significant effect on stomatal conductance, seedlings grown in the high temperature treatment exhibited a 15 percent increase in this parameter relative to seedlings grown in the low temperature treatment. Similarly, specific hydraulic conductivity, which is a measure of the amount of water moving through a plant relative to its leaf or needle area, also increased in the seedlings exposed to the high temperature treatment. In addition, biomass production rose by 42 percent in the low temperature treatment and 62 percent in the high temperature treatment when the atmospheric CO_2 concentration was raised from 350 to 1,100 ppm.

Tingey *et al.* (2005) studied the effects of atmospheric CO_2 enrichment (to approximately 350 ppm above ambient) on the fine-root architecture of Ponderosa pine seedlings growing in open-top chambers via minirhizotron tubes over a period of four years. This experiment showed that "elevated CO_2 increased both fine root extensity (degree of soil exploration) and intensity (extent that roots use explored areas) but had no effect on mycorrhizae," the latter of which observations was presumed to be due to the fact that soil nitrogen was not limiting to growth in this study. More specifically, they report that "extensity increased 1.5- to 2-fold in elevated CO_2 while intensity increased only 20 percent or less," noting that similar extensity results had been obtained over shorter periods of four months to two years by Arnone (1997), Berntson and Bazzaz (1998), DeLucia *et al.* (1997) and Runion *et al.* (1997), while similar intensity results had been obtained by Berntson (1994).

Last, Soule and Knapp (2006) studied Ponderosa pine trees growing naturally at eight sites within the Pacific Northwest of the United States, in order to see how they may have responded to the increase in the atmosphere's CO_2 concentration that occurred after 1950. In selecting these sites, they chose locations that "fit several criteria designed to limit potential confounding influences associated with anthropogenic disturbance." They also say they selected locations with "a variety of climatic and topoedaphic conditions, ranging from extremely water-limiting environments ... to areas where soil moisture should be a limiting factor for growth only during extreme drought years," additionally noting that all sites were located in areas "where ozone concentrations and nitrogen deposition are typically low."

At each of the eight sites that met all of these criteria, Soule and Knapp obtained core samples from about 40 mature trees that included "the potentially oldest trees on each site," so that their results would indicate, as they put it, "the response of mature, naturally occurring ponderosa pine trees that germinated before anthropogenically elevated CO_2 levels, but where growth, particularly post-1950, has occurred under increasing and substantially higher atmospheric CO_2 concentrations." Utilizing meteorological evaluations of the Palmer Drought Severity Index, they thus compared ponderosa pine radial growth rates during matched wet and dry years pre- and post-1950.

So what did they find? Overall, the two researchers report finding a post-1950 radial growth enhancement that was "more pronounced during drought years compared with wet years, and the greatest response occurred at the most stressed site." As for the magnitude of the response, they determined that "the relative change in growth [was] upward at seven of our [eight] sites, ranging from 11 to 133%."

With respect to the meaning and significance of their observations, Soule and Knapp say their results "showing that radial growth has increased in the post-1950s period ... while climatic conditions have generally been unchanged, suggest that nonclimatic driving forces are operative." In addition, they say that "these radial growth responses are generally consistent with what has been shown in long-term open-top chamber (Idso and Kimball, 2001) and FACE studies (Ainsworth and Long, 2005)." Hence, they say their findings suggest that "elevated levels of atmospheric CO_2 are acting as a driving force for increased radial growth of ponderosa pine, but that the overall influence of this effect may be enhanced, reduced or obviated by site-specific conditions."

Summarizing their findings Soule and Knapp recount how they had "hypothesized that ponderosa pine ... would respond to gradual increases in atmospheric CO_2 over the past 50 years, and that these effects would be most apparent during drought stress and on environmentally harsh sites," and they state in their very next sentence that their results "support these hypotheses." Hence, they conclude their paper by stating it is likely that "an atmospheric CO_2-driven growth-enhancement effect exists for ponderosa pine growing under specific natural conditions within the [USA's] interior Pacific Northwest."

Additional information on this topic, including reviews of newer publications as they become available, can be found at http://www.co2science.org/subject/t/treesponderosa.php.

References

Ainsworth, E.A. and Long, S.P. 2005. What have we learned from 15 years of free-air CO_2 enrichment (FACE)? A meta-analytic review of the responses of photosynthesis, canopy properties and plant production to rising CO_2. *New Phytologist* **165**: 351-372.

Arnone, J.A. 1997. Temporal responses of community fine root populations to long-term elevated atmospheric CO_2 and soil nutrient patches in model tropical ecosystems. *Acta Oecologia* **18**: 367-376.

Berntson, G.M. 1994. Modeling root architecture: are there tradeoffs between efficiency and potential of resource acquisition? *New Phytologist* **127**: 483-493.

Berntson, G.M. and Bazzaz, F.A. 1998. Regenerating temperate forest mesocosms in elevated CO_2: belowground growth and nitrogen cycling. *Oecologia* **113**: 115-125.

DeLucia, E.H., Callaway, R.M., Thomas, E.M. and Schlesinger, W.H. 1997. Mechanisms of phosphorus acquisition for ponderosa pine seedlings under high CO_2 and temperature. *Annals of Botany* **79**: 111-120.

Idso, S.B. and Kimball, B.A. 2001. CO_2 enrichment of sour orange trees: 13 years and counting. *Environmental and Experimental Botany* **46**: 147-153.

Johnson, D.W., Thomas, R.B., Griffin, K.L., Tissue, D.T., Ball, J.T., Strain, B.R. and Walker, R.F. 1998. Effects of carbon dioxide and nitrogen on growth and nitrogen uptake in ponderosa and loblolly pine. *Journal of Environmental Quality* **27**: 414-425.

Maherali, H. and DeLucia, E.H. 2000. Interactive effects of elevated CO_2 and temperature on water transport in ponderosa pine. *American Journal of Botany* **87**: 243-249.

Runion, G.B., Mitchell, R.J., Rogers, H.H., Prior, S.A. and Counts, T.K. 1997. Effects of nitrogen and water limitation and elevated atmospheric CO_2 on ectomycorrhiza of longleaf pine. *New Phytologist* **137**: 681-689.

Soule, P.T. and Knapp, P.A. 2006. Radial growth rate increases in naturally occurring ponderosa pine trees: a late-twentieth century CO_2 fertilization effect? *New Phytologist*: 10.1111/j.1469-8137.2006.01746.x.

Tingey, D.T., Johnson, M.G. and Phillips, D.L. 2005. Independent and contrasting effects of elevated CO_2 and N-fertilization on root architecture in *Pinus ponderosa*. *Trees* **19**: 43-50.

Walker, R.F., Geisinger, D.R., Johnson, D.W. and Ball, J.T. 1998a. Atmospheric CO_2 enrichment and soil N fertility effects on juvenile ponderosa pine: Growth, ectomycorrhizal development, and xylem water potential. *Forest Ecology and Management* **102**: 33-44.

Walker, R.F., Johnson, D.W., Geisinger, D.R. and Ball, J.T. 1998b. Growth and ectomycorrhizal colonization of ponderosa pine seedlings supplied different levels of atmospheric CO_2 and soil N and P. *Forest Ecology and Management* **109**: 9-20.

7.1.2.9.3. Scots

Rouhier and Read (1998) grew seedlings of Scots pine (*Pinus sylvestris* L.) for four months in growth cabinets maintained at atmospheric CO_2 concentrations of either 350 or 700 ppm. In addition, one-third of the seedlings were inoculated with one species of mycorrhizal fungi, one-third were inoculated with another species, and one-third were not inoculated at all, in order to determine the effects of elevated CO_2 on mycorrhizal fungi and their interactive effects on seedling growth. These procedures resulted in the doubled atmospheric CO_2 content increasing seedling dry mass by an average of 45 percent regardless of fungal inoculation. In addition, the extra CO_2 increased the number of hyphal tips associated with seedling roots by about 62 percent for both fungal species. Hyphal growth was also accelerated by elevated CO_2; after 55 days of treatment, the mycorrhizal network produced by one of the fungal symbionts occupied 444 percent more area than its counterpart exposed to ambient CO_2.

These results suggest that as the air's CO_2 content continues to rise, fungal symbionts of Scots pine will likely receive greater allocations of carbon from their host. This carbon can be used to increase their mycorrhizal networks, which would enable the fungi to explore greater volumes of soil in search of minerals and nutrients to benefit the growth of its host. In addition, by receiving greater allocations of carbon, fungal symbionts may keep photosynthetic down regulation from occurring, as they provide an additional sink for leaf-produced carbohydrates.

Janssens *et al.* (1998) grew three-year-old Scots pine seedlings in open-top chambers kept at ambient and 700 ppm atmospheric CO_2 concentrations for six months while they studied the effects of elevated CO_2 on root growth and respiration. In doing so, they learned that the elevated CO_2 treatment significantly increased total root length by 122 percent and dry mass by 135 percent relative to the roots of seedlings grown in ambient-CO_2 air. In addition, although starch accumulation in the CO_2-enriched roots was nearly 90 percent greater than that observed in the roots produced in the ambient-CO_2 treatment, the carbon-to-nitrogen ratio of the CO_2-enriched roots was significantly *lower* than that of the control-plant roots, indicative of the fact that they contained an even greater relative abundance of nitrogen. The most important implication of this study, therefore, was that Scots pine seedlings will likely be able to find the nitrogen they need to sustain large growth responses to atmospheric CO_2 enrichment with the huge root systems they typically produce in CO_2-enriched air.

Kainulainen *et al.* (1998) constructed open-top chambers around Scots pine trees that were about 20 years old and fumigated them with combinations of ambient or CO_2-enriched air (645 ppm) and ambient or twice-ambient (20 to 40 ppb) ozone-enriched air for three growing seasons to study the interactive effects of these gases on starch and secondary metabolite production. In doing so, they determined that elevated CO_2 and O_3 (ozone) had no significant impact on starch production in Scots pine, even after two years of treatment exposure. However, near the end of the third year, the elevated CO_2 alone significantly enhanced starch production in current-year needles, although neither extra CO_2, extra O_3, nor combinations thereof had any significant effects on the concentrations of secondary metabolites they investigated.

Kellomaki and Wang (1998) constructed closed-top chambers around 30-year-old Scots pine trees, which they fumigated with air containing either 350 or 700 ppm CO_2 at ambient and elevated (ambient plus 4°C) air temperatures for one full year, after which they assessed tree water-use by measuring cumulative sap flow for 32 additional days. This protocol revealed that the CO_2-enriched air reduced cumulative sap flow by 14 percent at ambient air temperatures, but that sap flow was unaffected by atmospheric CO_2 concentration in the trees growing at the elevated air temperatures. These findings suggest that cumulative water-use by Scotts pine trees in a CO_2-enriched world of the future will likely be less than or equal to—but no more than—what it is today.

Seven years later, Wang *et al.* (2005) published a report of a study in which they measured sap flow, crown structure, and microclimatic parameters in order to calculate the transpiration rates of individual 30-year-old Scots pine trees that were maintained for a period of *three* years in ambient air and air enriched with an extra 350 ppm of CO_2 and/or warmed by 2° to 6°C in closed-top chambers constructed within a naturally seeded stand of the trees. As they describe it, the results of this experiment indicated that "(i) elevated CO_2 significantly enhanced whole-tree transpiration rate during the first measuring year [by 14%] due to a large increase in whole-tree foliage area, 1998, but reduced it in the subsequent years of 1999 and 2000 [by 13% and 16%, respectively] as a consequence of a greater decrease in crown conductance which off-set the increase in foliage area per tree; (ii) trees growing in elevated temperature always had higher sap flow rates throughout three measuring years [by 54%, 45% and 57%, respectively]; and (iii) the response of sap flow to the combination of elevated temperature and CO_2 was similar to that of elevated temperature alone, indicating a dominant role for temperature and a lack of interaction between elevated CO_2 and temperature." These observations suggest that as the air's CO_2 content continues to rise, we probably can expect to see a decrease in evaporative water loss rates from naturally occurring stands of Scots pine trees ... unless there is a large concurrent increase in air temperature.

Also working with closed-top chambers that were constructed around 20-year-old Scots pines and fumigated with air containing 350 and 700 ppm CO_2 at ambient and elevated (ambient plus 4°C) air temperatures for a period of three years were Peltola *et al.* (2002), who studied the effects of elevated CO_2 and air temperature on stem growth in this coniferous species when it was growing on a soil low in nitrogen. After three years of treatment, they found that cumulative stem diameter growth in the CO_2-enriched trees growing at ambient air temperatures was 57 percent greater than that displayed by control trees growing at ambient CO_2 and ambient air temperatures, while the trees exposed to elevated CO_2 *and* elevated air temperature exhibited cumulative stem-diameter growth that was 67 percent greater than that displayed by trees exposed to ambient-CO_2 air and ambient air temperatures. Consequently, as the air's CO_2 content continues to rise, Scots pine trees will likely respond by increasing stem-diameter growth, even if growing on soils low in nitrogen, and even if air temperatures rise by as much as 4°C.

In a somewhat different type of study, Kainulainen *et al.* (2003) collected needle litter beneath 22-year-old Scots pines that had been growing for the prior three years in open-top chambers that had been maintained at atmospheric CO_2 concentrations of 350 and 600 ppm in combination with ambient and elevated (approximately 1.4 x ambient) ozone concentrations to determine the impacts of these variables on the subsequent decomposition of senesced needles. This they did by enclosing the needles in litterbags and placing the bags within a native litter layer in a Scots pine forest, where decomposition rates were assessed by measuring accumulated litterbag mass loss over a period of 19 months. Interestingly, the three researchers found that exposure to elevated CO_2 during growth did *not* affect subsequent rates of needle decomposition, nor did elevated O_3 exposure affect decomposition, nor did exposure to elevated concentrations of the two gases together affect it.

Finally, Bergh *et al.* (2003) used a boreal version of the process-based BIOMASS simulation model to quantify the individual and combined effects of elevated air temperature (2° and 4°C above ambient) and CO_2 concentration (350 ppm above ambient) on the net primary production (NPP) of Scots pine forests growing in Denmark, Finland, Iceland, Norway, and Sweden. This work revealed that air temperature increases of 2° and 4°C led to mean NPP increases of 11 and 20 percent, respectively. However, when the air's CO_2 concentration was simultaneously increased from 350 to 700 ppm, the corresponding mean NPP increases rose to 41 and 55 percent. Last, when the air's CO_2 content was doubled at the prevailing ambient temperature, the mean value of the NPP rose by 27 percent. Consequently, as the air's CO_2 content continues to rise, Ponderosa pines of Denmark, Finland, Iceland, Norway, and Sweden should grow ever more productively; and if air temperature also rises, they will likely grow better still.

Given the above results, as the air's CO_2 content continues to rise, we can expect to see the root systems of Scots pines significantly enhanced, together with the mycorrhizal fungal networks that live in close association with them and help secure the nutrients the trees need to sustain large CO_2-induced increases in biomass production. Concurrently, we can expect to see much smaller changes in total evaporative water loss, which means that whole-tree water use efficiency should also be significantly enhanced.

Additional information on this topic, including reviews of newer publications as they become available, can be found at http://www.co2science.org/subject/t/treesscots.php.

References

Bergh, J., Freeman, M., Sigurdsson, B., Kellomaki, S., Laitinen, K., Niinisto, S., Peltola, H. and Linder, S. 2003. Modelling the short-term effects of climate change on the productivity of selected tree species in Nordic countries. *Forest Ecology and Management* **183**: 327-340.

Janssens, I.A., Crookshanks, M., Taylor, G. and Ceulemans, R. 1998. Elevated atmospheric CO_2 increases fine root production, respiration, rhizosphere respiration and soil CO_2 efflux in Scots pine seedlings. *Global Change Biology* **4**: 871-878.

Kainulainen, P., Holopainen, J.K. and Holopainen, T. 1998. The influence of elevated CO_2 and O_3 concentrations on Scots pine needles: Changes in starch and secondary metabolites over three exposure years. *Oecologia* **114**: 455-460.

Kainulainen, P., Holopainen, T. and Holopainen, J.K. 2003. Decomposition of secondary compounds from needle litter of Scots pine grown under elevated CO_2 and O_3. *Global Change Biology* **9**: 295-304.

Kellomaki, S. and Wang, K.-Y. 1998. Sap flow in Scots pines growing under conditions of year-round carbon dioxide enrichment and temperature elevation. *Plant, Cell and Environment* **21**: 969-981.

Peltola, H., Kilpelainen, A. and Kellomaki, S. 2002. Diameter growth of Scots pine (*Pinus sylvestris*) trees grown at elevated temperature and carbon dioxide concentration under boreal conditions. *Tree Physiology* **22**: 963-972.

Rouhier, H. and Read, D.J. 1998. Plant and fungal responses to elevated atmospheric carbon dioxide in mycorrhizal seedlings of *Pinus sylvestris*. *Environmental and Experimental Botany* **40**: 237-246.

Wang, K.-Y., Kellomaki, S., Zha, T. and Peltola, H. 2005. Annual and seasonal variation of sap flow and conductance of pine trees grown in elevated carbon dioxide and temperature. *Journal of Experimental Botany* **56**: 155-165.

7.1.2.10. Spruce

Several studies have recently documented the effects of elevated CO_2 on photosynthesis in various varieties of spruce (genus *Picea*). In the relatively short-term study of Tjoelker *et al.* (1998a), black spruce (*Picea mariana*) seedlings grown for three months at atmospheric CO_2 concentrations of 580 ppm exhibited photosynthetic rates that were about 28 percent greater than those displayed by control seedlings fumigated with air containing 370 ppm CO_2. Similarly, Egli *et al.* (1998) reported that Norway spruce (*Picea abies*) seedlings grown at 570 ppm CO_2 displayed photosynthetic rates that were 35 percent greater than those exhibited by seedlings grown at 370 ppm. In two branch bag studies conducted on mature trees, it was demonstrated that twice-ambient levels of atmospheric CO_2 enhanced rates of photosynthesis in current-year needles by 50 percent in Norway spruce (Roberntz and Stockfors, 1998) and 100 percent in Sitka spruce (*Picea sitchensis*) (Barton and Jarvis, 1999). Finally, in the four-year open-top chamber study of Murray *et al.* (2000), the authors reported that Sitka spruce seedlings growing at 700 ppm CO_2 exhibited photosynthetic rates that were 19 and 33 percent greater than those observed in control trees growing in ambient air and receiving low and high amounts of nitrogen fertilization, respectively.

Because elevated CO_2 enhances photosynthetic rates in spruce species, this phenomenon should lead to increased biomass production in these important coniferous trees. In the short-term three-month study of Tjoelker *et al.* (1998b), for example, black spruce seedlings receiving an extra 210 ppm CO_2 displayed final dry weights that were about 20 percent greater than those of seedlings growing at ambient CO_2. Similarly, after growing Sitka spruce for three years in open-top chambers, Centritto *et al.* (1999) reported that a doubling of the atmospheric CO_2 concentration enhanced sapling dry mass by 42 percent.

In summary, it is clear that as the CO_2 content of the air increases, spruce trees will likely display enhanced rates of photosynthesis and biomass production, regardless of soil nutrient status. Consequently, rates of carbon sequestration by this abundant coniferous forest species will likely be enhanced.

Additional information on this topic, including reviews of newer publications as they become available, can be found at http://www.co2science.org/subject/t/treespruce.php.

References

Barton, C.V.M. and Jarvis, P.G. 1999. Growth response of branches of *Picea sitchensis* to four years exposure to elevated atmospheric carbon dioxide concentration. *New Phytologist* **144**: 233-243.

Centritto, M., Lee, H.S.J. and Jarvis, P.G. 1999. Long-term effects of elevated carbon dioxide concentration and provenance on four clones of Sitka spruce (*Picea sitchensis*). I. Plant growth, allocation and ontogeny. *Tree Physiology* **19**: 799-806.

Egli, P., Maurer, S., Gunthardt-Goerg, M.S. and Korner, C. 1998. Effects of elevated CO_2 and soil quality on leaf gas exchange and aboveground growth in beech-spruce model ecosystems. *New Phytologist* **140**: 185-196.

Murray, M.B., Smith, R.I., Friend, A. and Jarvis, P.G. 2000. Effect of elevated [CO_2] and varying nutrient application rates on physiology and biomass accumulation of Sitka spruce (*Picea sitchensis*). *Tree Physiology* **20**: 421-434.

Roberntz, P. and Stockfors, J. 1998. Effects of elevated CO_2 concentration and nutrition on net photosynthesis, stomatal conductance and needle respiration of field-grown Norway spruce trees. *Tree Physiology* **18**: 233-241.

Tjoelker, M.G., Oleksyn, J. and Reich, P.B. 1998a. Seedlings of five boreal tree species differ in acclimation of net photosynthesis to elevated CO_2 and temperature. *Tree Physiology* **18**: 715-726.

Tjoelker, M.G., Oleksyn, J. and Reich, P.B. 1998b. Temperature and ontogeny mediate growth response to elevated CO_2 in seedlings of five boreal tree species. *New Phytologist* **140**: 197-210.

7.1.2.11. Tropical

Several studies have recently documented the effects of elevated atmospheric CO_2 concentrations on photosynthesis in various tropical and sub-tropical trees. In the relatively short-term study of Lovelock *et al.* (1999a), for example, seedlings of the tropical tree *Copaifera aromatica* that were grown for two months at an atmospheric CO_2 concentration of 860 ppm exhibited photosynthetic rates that were consistently 50-100 percent greater than those displayed by control seedlings fumigated with air containing 390 ppm CO_2. Similarly, Lovelock *et al.* (1999b) reported that a 10-month 390-ppm increase in the air's CO_2 content boosted rates of net photosynthesis in 30-m tall *Luehea seemannii* trees by 30 percent. Likewise, in the review paper of Schaffer *et al.* (1999), it was noted that atmospheric CO_2 enrichment had previously been shown to enhance rates of net photosynthesis in a number of tropical and sub-tropical fruit trees, including avocado, banana, citrus, mango, and mangosteen. Even at the ecosystem level, Lin *et al.* (1998) found that a 1,700-m^2 synthetic rainforest mesocosm displayed a 79 percent enhancement in net ecosystem carbon exchange rate in response to a 72 percent increase in the air's CO_2 content.

Because elevated CO_2 enhances photosynthetic rates in tropical and sub-tropical trees, it should also lead to increased carbohydrate and biomass production in these species. At a tropical forest research site in Panama, twice-ambient CO_2 concentrations enhanced foliar sugar concentrations by up to 30 percent (Wurth *et al.*, 1998), while doubling the foliar concentrations of starch (Lovelock *et al.*, 1998) in a number of tree species. Also, in the study of Hoffmann *et al.* (2000), elevated CO_2 (700 ppm) enhanced dry weights of an "uncut" Brazilian savannah tree species (*Keilmeyera coriacea*) by about 50 percent, while it enhanced the dry weight of the same "cut" species by nearly 300 percent. Although not specifically quantified, Schaffer *et al.* (1997) noted that twice-ambient CO_2 exposure for one year obviously enhanced dry mass production in two mango ecotypes. Finally, in the six-month study of Sheu *et al.* (1999), a doubling of the atmospheric CO_2 concentration increased seedling dry weight in *Schima superba* by 14 and 49 percent when grown at ambient and elevated (5°C above ambient) air temperatures, respectively.

It is clear that as the air's CO_2 content rises, tropical and sub-tropical trees will likely display enhanced rates of photosynthesis and biomass production, even under conditions of herbivory and elevated air temperature. Consequently, greater carbon sequestration will also likely occur within earth's tropical and sub-tropical forests as ever more CO_2 accumulates in the atmosphere.

Additional information on this topic, including reviews of newer publications as they become available, can be found at http://www.co2science.org/subject/t/treestropical.php.

References

Hoffmann, W.A., Bazzaz, F.A., Chatterton, N.J., Harrison, P.A. and Jackson, R.B. 2000. Elevated CO_2 enhances resprouting of a tropical savanna tree. *Oecologia* **123**: 312-317.

Lin, G., Marino, B.D.V., Wei, Y., Adams, J., Tubiello, F. and Berry, J.A. 1998. An experimental and modeling study of responses in ecosystems carbon exchanges to increasing CO_2 concentrations using a tropical rainforest mesocosm. *Australian Journal of Plant Physiology* **25**: 547-556.

Lovelock, C.E., Posada, J. and Winter, K. 1999a. Effects of elevated CO_2 and defoliation on compensatory growth and photosynthesis of seedlings in a tropical tree, *Copaifera aromatica*. *Biotropica* **31**: 279-287.

Lovelock, C.E., Virgo, A., Popp, M. and Winter, K. 1999b. Effects of elevated CO_2 concentrations on photosynthesis, growth and reproduction of branches of the tropical canopy trees species, *Luehea seemannii* Tr. & Planch. *Plant, Cell and Environment* **22**: 49-59.

Lovelock, C.E., Winter, K., Mersits, R. and Popp, M. 1998. Responses of communities of tropical tree species to elevated CO_2 in a forest clearing. *Oecologia* **116**: 207-218.

Schaffer, B., Whiley, A.W. and Searle, C. 1999. Atmospheric CO_2 enrichment, root restriction, photosynthesis, and dry-matter partitioning in subtropical and tropical fruit crops. *HortScience* **34**: 1033-1037.

Schaffer, B., Whiley, A.W., Searle, C. and Nissen, R.J. 1997. Leaf gas exchange, dry matter partitioning, and mineral element concentrations in mango as influenced by elevated atmospheric carbon dioxide and root restriction. *Journal of the American Society of Horticultural Science* **122**: 849-855.

Sheu, B.-H. and Lin, C.-K. 1999. Photosynthetic response of seedlings of the sub-tropical tree *Schima superba* with exposure to elevated carbon dioxide and temperature. *Environmental and Experimental Botany* **41**: 57-65.

Wurth, M.K.R., Winter, K. and Korner, C. 1998. Leaf carbohydrate responses to CO_2 enrichment at the top of a tropical forest. *Oecologia* **116**: 18-25.

7.1.2.12. Wood Density

Numerous experiments have demonstrated that trees grown in air enriched with CO_2 nearly always sequester more biomass in their trunks and branches than do trees grown in ambient air. Several studies also have looked at the effects of elevated CO_2 on the *density* of that sequestered biomass.

Rogers *et al.* (1983) observed no difference in the wood density of loblolly pine (*Pinus taeda*) trees grown at 340 and 718 ppm CO_2 for 10 weeks; but they found a 33 percent CO_2-induced increase in the wood density of sweetgum (*Liquidambar styraciflua*) trees that were grown at these concentrations for only eight weeks. Doyle (1987) and Telewski and Strain

(1987) studied the same two tree species over three growing seasons in air of 350 and 650 ppm CO_2, finding no effect of atmospheric CO_2 enrichment on the stem density of sweetgum, but a mean increase of 9 percent in the stem density of loblolly pine.

Conroy *et al.* (1990) grew seedlings of two *Pinus radiata* families at 340 and 660 ppm CO_2 for 114 weeks, finding CO_2-induced trunk density increases for the two families of 5.4 and 5.6 percent when soil phosphorus was less than adequate and increases of 5.6 and 1.2 percent when it was non-limiting. In a similar study, Hattenschwiler *et al.* (1996) grew six genotypes of clonally propagated four-year-old Norway spruce (*Picea abies*) for three years at CO_2 concentrations of 280, 420, and 560 ppm at three different rates of wet nitrogen deposition. On average, they found that wood density was 12 percent greater in the trees grown at the two higher CO_2 concentrations than it was in the trees grown at 280 ppm.

Norby *et al.* (1996) grew yellow poplar or "tulip" trees (*Liriodendron tulipifera*) at ambient and ambient plus 300 ppm CO_2 for three years, during which time the wood density of the trees increased by approximately 7 percent. Tognetti *et al.* (1998) studied two species of oak tree—one deciduous (*Quercus pubescens*) and one evergreen (*Quercus ilex*)—growing in the vicinity of CO_2 springs in central Italy that raised the CO_2 concentration of the surrounding air by approximately 385 ppm. This increase in the air's CO_2 content increased the wood density of the deciduous oaks by 4.2 percent and that of the evergreen oaks by 6.4 percent.

Telewski *et al.* (1999) grew loblolly pine trees for four years at ambient and ambient plus 300 ppm CO_2. In their study, wood density determined directly from mass and volume measurements was increased by 15 percent by the extra CO_2; average ring density determined by X-ray densitometry was increased by 4.5 percent.

Beismann *et al.* (2002) grew different genotypes of spruce and beech (*Fagus sylvatica*) seedlings for four years in open-top chambers maintained at atmospheric CO_2 concentrations of 370 and 590 ppm in combination with low and high levels of wet nitrogen application on both rich calcareous and poor acidic soils to study the effects of these factors on seedling toughness (fracture characteristics) and rigidity (bending characteristics such as modulus of elasticity). They found that some genotypes of each species were sensitive to elevated CO_2, while others were not. Similarly, some were responsive to elevated

nitrogen deposition, while others were not. Moreover, such responses were often dependent upon soil type. Averaged across all tested genotypes, however, atmospheric CO_2 enrichment increased wood toughness in spruce seedlings grown on acidic soils by 12 and 18 percent at low and high levels of nitrogen deposition, respectively. In addition, atmospheric CO_2 enrichment increased this same wood property in spruce seedlings grown on calcareous soils by about 17 and 14 percent with low and high levels of nitrogen deposition, respectively. By contrast, elevated CO2 had no significant effects on the mechanical wood properties of beech seedlings, regardless of soil type.

Finally, Kilpelainen *et al.* (2003) erected 16 open-top chambers within a 15-year-old stand of Scots pines growing on a nutrient-poor sandy soil of low nitrogen content near the Mekrijarvi Research Station of the University of Joensuu, Finland. Over the next three years they maintained the trees within these chambers in a well-watered condition, while they enriched the air in half of the chambers to a mean daytime CO_2 concentration of approximately 580 ppm and maintained the air in half of each of the two CO_2 treatments at 2°C above ambient. In the ambient temperature treatment the 60 percent increase in the air's CO_2 concentration increased latewood density by 27 percent and maximum wood density by 11 percent, while in the elevated-temperature treatment it increased latewood density by 25 percent and maximum wood density by 15 percent. These changes led to mean overall CO_2-induced wood density increases of 2.8 percent in the ambient-temperature treatment and 5.6 percent in the elevated-temperature treatment.

In light of these several observations, it is clear that different species of trees respond differently to atmospheric CO_2 enrichment, and that they respond with still greater variety under different sets of environmental conditions. In general, however, atmospheric CO_2 enrichment tends to increase wood density in both seedlings and mature trees more often than not, thereby also increasing a number of strength properties of their branches and trunks.

Additional information on this topic, including reviews of newer publications as they become available, can be found at http://www.co2science.org/subject/w/wooddensity.php.

References

Beismann, H., Schweingruber, F., Speck, T. and Korner, C. 2002. Mechanical properties of spruce and beech wood grown in elevated CO_2. *Trees* **16**: 511-518.

Conroy, J.P., Milham, P.J., Mazur, M., Barlow, E.W.R. 1990. Growth, dry weight partitioning and wood properties of *Pinus radiata* D. Don after 2 years of CO_2 enrichment. *Plant, Cell and Environment* **13**: 329-337.

Doyle, T.W. 1987. Seedling response to CO_2 enrichment under stressed and non-stressed conditions. In: Jacoby Jr., G.C. and Hornbeck, J.W. (Eds.) *Proceedings of the International Symposium on Ecological Aspects of Tree-Ring Analysis*. National Technical Information Service, Springfield, VA, pp. 501-510.

Hattenschwiler, S., Schweingruber, F.H., Korner, C. 1996. Tree ring responses to elevated CO_2 and increased N deposition in *Picea abies*. *Plant, Cell and Environment* **19**: 1369-1378.

Kilpelainen A., Peltola, H., Ryyppo, A., Sauvala, K., Laitinen, K. and Kellomaki, S. 2003. Wood properties of Scots pines (*Pinus sylvestris*) grown at elevated temperature and carbon dioxide concentration. *Tree Physiology* **23**: 889-897.

Norby, R.J., Wullschleger, S.D., Gunderson, C.A. 1996. Tree responses to elevated CO_2 and implications for forests. In: Koch, G.W. and Mooney, H.A. (Eds.) *Carbon Dioxide and Terrestrial Ecosystems*. Academic Press, New York, NY, pp. 1-21.

Rogers, H.H., Bingham, G.E., Cure, J.D., Smith, J.M. and Surano, K.A. 1983. Responses of selected plant species to elevated carbon dioxide in the field. *Journal of Environmental Quality* **12**: 569-574.

Telewski, F.W. and Strain, B.R. 1987. Densitometric and ring width analysis of 3-year-old *Pinus taeda* L. and *Liquidambar styraciflua* L. grown under three levels of CO_2 and two water regimes. In: Jacoby Jr., G.C. and Hornbeck, J.W. (Eds.) *Proceedings of the International Symposium on Ecological Aspects of Tree-Ring Analysis*. National Technical Information Service, Springfield, VA, pp. 494-500.

Telewski, F.W., Swanson, R.T., Strain, B.R. and Burns, J.M. 1999. Wood properties and ring width responses to long-term atmospheric CO_2 enrichment in field-grown loblolly pine (*Pinus taeda* L.). *Plant, Cell and Environment* **22**: 213-219.

Tognetti, R., Johnson, J.D., Michelozzi, M. and Raschi, A. 1998. Response of foliar metabolism in mature trees of *Quercus pubescens* and *Quercus ilex* to long-term elevated CO_2. *Environmental and Experimental Botany* **39**: 233-245.

7.1.2.13. Forests

Forests contain perennial trees that remove CO_2 from the atmosphere during photosynthesis and store its carbon within their woody tissues for decades to periods sometimes in excess of a thousand years. Thus, it is important to understand how increases in the air's CO_2 content affect forest productivity and carbon sequestration, which has a great effect on the rate of rise of the air's CO_2 concentration. In this summary, we review several recent scientific publications pertaining to these subjects.

By examining various properties of tree rings, researchers can deduce how historical increases in the air's CO_2 concentration have already affected tree productivity and water use efficiency. Duquesnay *et al.* (1998), for example, analyzed the relative amounts of ^{12}C and ^{13}C present in yearly growth rings of beech trees raised in silviculture regimes in northeastern France, determining that their intrinsic water use efficiencies rose by approximately 33 percent during the past century, as the air's CO_2 concentration rose from approximately 280 to 360 ppm. In another case, Rathgeber *et al.* (2000) used tree-ring density data to create a historical productivity baseline for forest stands of *Pinus halepensis* in southeastern France, from which they determined that the net productivity of such forests would increase by 8 to 55 percent with a doubling of the air's CO_2 content. Finally, when running a forest growth model based on empirical observations reported in the literature, Lloyd (1999) determined that the rise in the atmospheric CO_2 concentration since the onset of the Industrial Revolution likely increased the net primary productivity of mature temperate deciduous forests by about 7 percent. In addition, he determined that a proportional increase in anthropogenic nitrogen deposition likely increased forest net primary productivity by 25 percent. And when he combined the two effects, the net primary productivity stimulation rose to 40 percent, which is more than the sum of the individual growth enhancements resulting from the increases in CO_2 and nitrogen.

The results of these studies demonstrate that historic increases in the air's CO_2 content have already conferred great benefits upon earth's forests. But will future increases in the air's CO_2 concentration continue to do so? Several research teams have embarked on long-term studies of various forest communities in an attempt to address this important question. What follows are some important observations that have been made from their mostly ongoing CO_2-enrichment studies.

In 1996, circular FACE plots (30-m diameter) receiving atmospheric CO_2 concentrations of 360 and 560 ppm were established in a 15-year-old loblolly pine (*Pinus taeda*) plantation in North Carolina, USA, to study the effects of elevated CO_2 on the growth and productivity of this particular forest community, which also had several hardwood species present in the understory beneath the primary coniferous canopy. Using this experimental set-up as a platform for several experiments, Hymus *et al.* (1999) reported that net photosynthetic rates of CO_2-enriched loblolly pines trees were 65 percent greater than rates observed in control trees exposed to ambient air. These greater rates of carbon fixation contributed to the 24 percent greater growth rates observed in the CO_2-enriched pine trees in the first year of this long-term study (Naidu and DeLucia 1999). In addition, DeLucia and Thomas (2000) reported that the elevated CO_2 increased rates of net photosynthesis by 50 to 160 percent in four subdominant hardwood species present in the forest understory. Moreover, for one species—sweetgum (*Liquidambar styraciflua*)—the extra CO_2 enhanced rates of net photosynthesis in sun and shade leaves by 166 and 68 percent, respectively, even when the trees were naturally subjected to summer seasonal stresses imposed by high temperature and low soil water availability. Consequently, after two years of atmospheric CO_2 enrichment, total ecosystem net primary productivity in the CO_2-enriched plots was 25 percent greater than that measured in control plots fumigated with ambient air.

In a similar large-scale study, circular (25-m diameter) FACE plots receiving atmospheric CO_2 concentrations of 400 and 530 ppm were constructed within a 10-year-old sweetgum plantation in Tennessee, USA, to study the effects of elevated CO_2 on the growth and productivity of this forest community. After two years of treatment, Norby *et al.* (2001) reported that the modest 35 percent increase in the air's CO_2 content boosted tree biomass production by an average of 24 percent. In addition, Wullschleger and Norby (2001) noted that CO_2-enriched trees displayed rates of transpirational water loss that were approximately 10 percent lower than those exhibited by control trees grown in ambient air. Consequently, elevated CO_2 enhanced seasonal water use efficiencies of these mature sweetgum trees by 28 to 35 percent.

On a smaller scale, Pritchard *et al.* (2001) constructed idealized ecosystems (containing five different species) representative of regenerating longleaf pine (*Pinus palustris* Mill.) communities of the southeastern USA, fumigating them for 18 months with air containing 365 and 720 ppm CO_2 to study the effects of elevated CO_2 on this forest community. They reported that elevated CO_2 increased the above- and below-ground biomass of the dominant longleaf pine individuals by 20 and 62 percent, respectively. At the ecosystem level, elevated CO_2 stimulated total above-ground biomass production by an average of 35 percent. Similar results for regenerating temperate forest communities have been reported by Berntson and Bazzaz (1998), who documented a 31 percent increase in Transition Hardwood-White Pine-Hemlock forest mesocosm biomass in response to two years of fumigation with twice-ambient concentrations of atmospheric CO_2.

It is clear that as the air's CO_2 concentration continues to rise, forests will likely respond by exhibiting significant increases in total primary productivity and biomass production. Consequently, forests will likely grow much more robustly and significantly expand their ranges, as has already been documented in many parts of the world, including gallery forest in Kansas, USA (Knight *et al.*, 1994) and the Budal and Sjodal valleys in Norway (Olsson *et al.*, 2000). Such CO_2-induced increases in growth and range expansion should result in large increases in global carbon sequestration within forests.

Additional information on this topic, including reviews on forests not discussed here, can be found at http://www.co2science.org/subject/f/subject_f.php under the heading Forests.

References

Berntson, G.M. and Bazzaz, F.A. 1998. Regenerating temperate forest mesocosms in elevated CO_2: belowground growth and nitrogen cycling. *Oecologia* **113**: 115-125.

DeLucia, E.H. and Thomas, R.B. 2000. Photosynthetic responses to CO_2 enrichment of four hardwood species in a forest understory. *Oecologia* **122**: 11-19.

Duquesnay, A., Breda, N., Stievenard, M. and Dupouey, J.L. 1998. Changes of tree-ring $\delta^{13}C$ and water-use efficiency of beech (*Fagus sylvatica* L.) in north-eastern France during the past century. *Plant, Cell and Environment* **21**: 565-572.

Hymus, G.J., Ellsworth, D.S., Baker, N.R. and Long, S.P. 1999. Does free-air carbon dioxide enrichment affect photochemical energy use by evergreen trees in different seasons? A chlorophyll fluorescence study of mature loblolly pine. *Plant Physiology* **120**: 1183-1191.

Knight, C.L., Briggs, J.M. and Nellis, M.D. 1994. Expansion of gallery forest on Konza Prairie Research Natural Area, Kansas, USA. *Landscape Ecology* **9**: 117-125.

Lloyd, J. 1999. The CO_2 dependence of photosynthesis, plant growth responses to elevated CO_2 concentrations and their interaction with soil nutrient status, II. Temperate and boreal forest productivity and the combined effects of increasing CO_2 concentrations and increased nitrogen deposition at a global scale. *Functional Ecology* **13**: 439-459.

Naidu, S.L. and DeLucia, E.H. 1999. First-year growth response of trees in an intact forest exposed to elevated CO_2. *Global Change Biology* **5**: 609-613.

Norby, R.J., Todd, D.E., Fults, J. and Johnson, D.W. 2001. Allometric determination of tree growth in a CO_2-enriched sweetgum stand. *New Phytologist* **150**: 477-487.

Olsson, E.G.A., Austrheim, G. and Grenne, S.N. 2000. Landscape change patterns in mountains, land use and environmental diversity, Mid-Norway 1960-1993. *Landscape Ecology* **15**: 155-170.

Pritchard, S.G., Davis, M.A., Mitchell, R.J., Prior, A.S., Boykin, D.L., Rogers, H.H. and Runion, G.B. 2001. Root dynamics in an artificially constructed regenerating longleaf pine ecosystem are affected by atmospheric CO_2 enrichment. *Environmental and Experimental Botany* **46**: 35-69.

Rathgeber, C., Nicault, A., Guiot, J., Keller, T., Guibal, F. and Roche, P. 2000. Simulated responses of *Pinus halepensis* forest productivity to climatic change and CO_2 increase using a statistical model. *Global and Planetary Change* **26**: 405-421.

Wullschleger, S.D. and Norby, R.J. 2001. Sap velocity and canopy transpiration in a sweetgum stand exposed to free-air CO_2 enrichment (FACE). *New Phytologist* **150**: 489-498.

7.1.3. Aquatic Plants

We have shown how atmospheric CO_2 enrichment typically enhances the growth and productivity of nearly all terrestrial plants. But what about *aquatic* plants? In this section we seek to answer that question.

7.1.3.1. Freshwater Algae

How do freshwater algae respond to increases in the air's CO_2 content? The subject has not been thoroughly researched, but the results of the studies discussed below provide a glimpse of what the future may hold as the atmosphere's CO_2 concentration continues its upward course.

Working with cells of the freshwater alga *Chlorella pyrenoidosa*, Xia and Gao (2003) cultured them in Bristol's solution within controlled environment chambers maintained at low and high light levels (50 and 200 $\mu mol/m^2/s$) during 12-hour light periods that were followed by 12-hour dark periods for a total of 13 days, while the solutions in which the cells grew were continuously aerated with air of either 350 or 700 ppm CO_2. When the cells were harvested (in the exponential growth phase) at the conclusion of this period, the biomass (cell density) of the twice-ambient CO_2 treatment was found to be 10.9 percent and 8.3 percent greater than that of the ambient-air treatment in the low- and high-light regimes, respectively, although only the high-light result was statistically significant. The two scientists concluded from these observations that a "doubled atmospheric CO_2 concentration would affect the growth of *C. pyrenoidosa* when it grows under bright solar radiation, and such an effect would increase by a great extent when the cell density becomes high." Their data also suggest the same may well be true when the alga grows under *not*-so-bright conditions.

Working on a much larger scale "in the field" with six 1.5-m-diameter flexible plastic cylinders placed in the littoral zone of Lake Hampen in central Jutland, Denmark (three maintained at the ambient CO_2 concentration of the air and three enriched to 10 times the ambient CO_2 concentration), Andersen and Andersen (2006) measured the CO_2-induced growth response of a mixture of several species of filamentous freshwater algae dominated by *Zygnema* species, but containing some *Mougeotia* and *Spirogyra*. After one full growing season (May to November), they determined that the biomass of the microalgal mixture in the CO_2-enriched cylinders was increased by 220 percent in early July, by 90 percent in mid-August, and by a whopping 3,750 percent in mid-November.

In another study of the subject, Schippers *et al.* (2004a) say "it is usually thought that unlike terrestrial plants, phytoplankton will not show a significant response to an increase of atmospheric

CO_2," but they note, in this regard, that "most analyses have not examined the full dynamic interaction between phytoplankton production and assimilation, carbon-chemistry and the air-water flux of CO_2," and that "the effect of photosynthesis on pH and the dissociation of carbon (C) species have been neglected in most studies."

In an attempt to rectify this situation, Schippers *et al.* developed "an integrated model of phytoplankton growth, air-water exchange and C chemistry to analyze the potential increase of phytoplankton productivity due to an atmospheric CO_2 elevation," and as a test of their model, they let the freshwater alga *Chlamydomonas reinhardtii* grow in 300-ml bottles filled with 150 ml of a nutrient-rich medium at enclosed atmospheric CO_2 concentrations of 350 and 700 ppm that they maintained at two air-water exchange rates characterized by CO_2 exchange coefficients of 2.1 and 5.1 m day^{-1}, as described by Shippers *et al.* (2004b), while periodically measuring the biovolume of the solutions by means of an electronic particle counter. The results of this effort, as they describe it, "confirm the theoretical prediction that if algal effects on C chemistry are strong, increased phytoplankton productivity because of atmospheric CO_2 elevation should become proportional to the increased atmospheric CO_2," which suggests that algal productivity "would double at the predicted increase of atmospheric CO_2 to 700 ppm." Although they note that "strong algal effects (resulting in high pH levels) at which this occurs are rare under natural conditions," they still predict that effects on algal production in freshwater systems could be such that a "doubling of atmospheric CO_2 may result in an increase of the productivity of more than 50%."

In the last of the few papers we have reviewed in this area, Logothetis *et al.* (2004) note that "the function and structure of the photosynthetic apparatus of many algal species resembles that of higher plants (Plumley and Smidt, 1984; Brown, 1988; Plumley *et al.*, 1993)," and that "unicellular green algae demonstrate responses to increased CO_2 similar to those of higher plants in terms of biomass increases (Muller *et al.*, 1993)." However, they also note that "little is known about the changes to their photosynthetic apparatus during exposure to high CO_2," which deficiency they began to correct via a new experiment, wherein batches of the unicellular green alga *Scenedesmus obliquus* (wild type strain D3) were grown autotrophically in liquid culture medium for several days in a temperature-controlled

water bath of 30°C at low (55 µmol m^{-2} s^{-1}) and high (235 µmol m^{-2} s^{-1}) light intensity while they were continuously aerated with air of either 300 or 100,000 ppm CO_2. This protocol revealed that exposure to the latter high CO_2 concentration produces, in their words, a "reorganization of the photosynthetic apparatus" that "leads to enhanced photosynthetic rates, which ... leads to an immense increase of biomass." After five days under low light conditions, for example, the CO_2-induced increase in biomass was approximately 300 percent, while under high light conditions it was approximately 600 percent.

Based on these few observations, it is not possible to draw any sweeping conclusions about the subject. However, they do indicate there may be a real potential for the ongoing rise in the air's CO_2 content to significantly stimulate the productivity of this freshwater contingent of earth's plants.

Additional information on this topic, including reviews of newer publications as they become available, can be found at http://www.co2science.org/subject/a/aquaticplants.php.

References

Andersen, T. and Andersen, F.O. 2006. Effects of CO_2 concentration on growth of filamentous algae and *Littorella uniflora* in a Danish softwater lake. *Aquatic Botany* **84**: 267-271.

Brown, J.S. 1988. Photosynthetic pigment organization in diatoms (Bacillariophyceae). *Journal of Phycology* **24**: 96-102.

Logothetis, K., Dakanali, S., Ioannidis, N. and Kotzabasis, K. 2004. The impact of high CO_2 concentrations on the structure and function of the photosynthetic apparatus and the role of polyamines. *Journal of Plant Physiology* **161**: 715-724.

Muller, C., Reuter, W. and Wehrmeyer, W. 1993. Adaptation of the photosynthetic apparatus of *Anacystis nidulans* to irradiance and CO_2-concentration. *Botanica Acta* **106**: 480-487.

Plumley, F.G., Marinson, T.A., Herrin, D.L., Ideuchi, M. and Schmidt, G.W. 1993. Structural relationships of the photosystem I and photosystem II chlorophyll a/b and a/c light-harvesting apoproteins of plants and algae. *Photochemistry and Photobiology* **57**: 143-151.

Plumley, F.G. and Smidt, G.W. 1984. Immunochemical characterization of families of light-harvesting pigment-protein complexes in several groups of algae. *Journal of Phycology* **20**: 10.

Schippers, P., Lurling, M. and Scheffer, M. 2004a. Increase of atmospheric CO_2 promotes phytoplankton productivity. *Ecology Letters* **7**: 446-451.

Schippers, P., Vermaat, J.E., de Klein, J. and Mooij, W.M. 2004b. The effect of atmospheric carbon dioxide elevation on plant growth in freshwater ecosystems. *Ecosystems* **7**: 63-74.

Xia, J. and Gao, K. 2003. Effects of doubled atmospheric CO_2 concentration on the photosynthesis and growth of *Chlorella pyrenoidosa* cultured at varied levels of light. *Fisheries Science* **69**: 767-771.

7.1.3.2. Freshwater Macrophytes

In this section we discuss the findings of papers that investigate the influence of CO_2 concentrations as it applies to submersed, floating, and emergent freshwater macrophytes, beginning with studies of aquatic plants that live their lives totally submersed in freshwater environments.

For several multi-week periods, Idso (1997) grew specimens of corkscrew vallisneria (*Vallisneria tortifolia*) in several 10- and 29-gallon glass tanks (containing 10-cm bottom-layers of common aquarium gravel) that were filled with tap water maintained within 0.5°C of either 18.2°C or 24.5°C, while the semi-sealed air spaces above these "Poor Man's Biospheres," as he christened them, were maintained at a number of different CO_2 concentrations. With the harvesting of plants at the end of the study, this protocol revealed that the CO_2-induced growth enhancement of the plants was *linear* (in contrast to the gradually declining CO_2-induced growth enhancements typically exhibited by most terrestrial plants as the air's CO_2 content climbs ever higher), and that the linear relationship extended to the highest atmospheric CO_2 concentration studied: 2,100 ppm. In addition, he found that the CO_2-induced growth increase experienced by the plants in the higher of the two water temperature treatments (a 128 percent increase in going from an atmospheric CO_2 concentration of 365 ppm to one of 2,100 ppm) was 3.5 times greater than that of the plants in the lower water temperature treatment. Although this response may seem rather dramatic, it is not unique; Idso reports that Titus *et al.* (1990), who studied the closely related *Vallisneria americana*, "observed that the biomass of their experimental plants also rose linearly with the CO_2 content of the air above the water within which they grew, and that [it] did so

from the value of the [then] current global mean (365 ppm) to a concentration fully ten times larger."

In another study of a closely allied species, Yan *et al.* (2006) collected turions of *Vallisneria spinulosa* from Liangzi Lake, Hubei Province, China, and planted them in tanks containing 15-cm-deep layers of fertile lake sediments, topped with 40 cm of lake water, that were placed in two glasshouses—one maintained at the ambient atmospheric CO_2 concentration of 390 ppm and the other maintained at an elevated concentration of 1,000 ppm—where the plants grew for a period of 120 days, after which they were harvested and the dry weights of their various organs determined. As they describe it, this work indicated that the "total biomass accumulation of plants grown in the elevated CO_2 was 2.3 times that of plants grown in ambient CO_2, with biomass of leaves, roots and rhizomes increasing by 106%, 183% and 67%, respectively." Most spectacularly of all, they report that "turion biomass increased 4.5-fold," because "the mean turion numbers per ramet and mean biomass per turion in elevated CO_2 were 1.7-4.3 and 1.9-3.4 times those in ambient CO_2."

In Denmark, in a study of small slow-growing evergreen perennials called *isoetids* that live submersed along the shores of numerous freshwater lakes, Andersen *et al.* (2006) grew specimens of *Littorella uniflora* in sediment cores removed from Lake Hampen in 75-liter tanks with 10-cm overburdens of filtered lake water for a period of 53 days, while measuring various plant, water, and sediment properties, after which they destructively harvested the plants and measured their biomass. Throughout this period, half of the tanks had ambient air bubbled through their waters, while the other half were similarly exposed to a mixture of ambient air and pure CO_2 that produced a 10-fold increase in the air's CO_2 concentration. This ultra-CO_2-enrichment led to a 30 percent increase in plant biomass, as well as "higher O_2 release to the sediment which is important for the cycling and retention of nutrients in sediments of oligotrophic softwater lakes." And when the ultra-CO_2-enrichment was maintained for an entire growing season (May-November), Andersen and Andersen (2006) report that the 10-fold increase in aquatic CO_2 concentration enhanced the biomass production of *Littorella uniflora* by a much larger 78 percent.

In a study of an "in-between" type of plant that has submersed roots and rhizomes that are anchored in water-body sediments, but which has floating leaves on the surface of the water and emergent flowers that protrude above the water surface, Idso *et al.* (1990) grew water lilies (*Nymphaea marliac*) for two consecutive years in sunken metal stock tanks located out-of-doors at Phoenix, Arizona (USA) and enclosed within clear-plastic-wall open-top chambers through which air of either 350 or 650 ppm CO_2 was continuously circulated. This work revealed that in addition to the leaves of the plants being larger in the CO_2-enriched treatment, there were 75 percent more of them than there were in the ambient-air tanks at the conclusion of the initial five-month-long growing season. Each of the plants in the high-CO_2 tanks also produced twice as many flowers as the plants growing in ambient air; and the flowers that blossomed in the CO_2-enriched air were more substantial than those that bloomed in the air of ambient CO_2 concentration: they had more petals, the petals were longer, and they had a greater percent dry matter content, such that each flower consequently weighed about 50 percent more than each flower in the ambient-air treatment. In addition, the stems that supported the flowers were slightly longer in the CO_2-enriched tanks, and the percent dry matter contents of both the flower and leaf stems were greater, so that the total dry matter in the flower and leaf stems in the CO_2-enriched tanks exceeded that of the flower and leaf stems in the ambient-air tanks by approximately 60 percent.

Just above the surface of the soil that covered the bottoms of the tanks, there were also noticeable differences. Plants in the CO_2-enriched tanks had more and bigger basal rosette leaves, which were attached to longer stems of greater percent dry matter content, which led to the total biomass of these portions of the plants being 2.9 times greater than the total biomass of the corresponding portions of the plants in the ambient-air tanks. In addition, plants in the CO_2-enriched tanks had more than twice as many unopened basal rosette leaves.

The greatest differences of all, however, were hidden within the soil that covered the bottoms of the stock tanks. When half of the plants were harvested at the conclusion of the first growing season, for example, the number of new rhizomes produced over that period was discovered to be 2.4 times greater in the CO_2-enriched tanks than it was in the ambient-air tanks; the number of major roots produced there was found to be 3.2 times greater. As with all other plant parts, the percent dry matter contents of the new roots and rhizomes were also greater in the CO_2-enriched tanks. Overall, therefore, the total dry matter production within the submerged soils of the water lily ecosystems was 4.3 times greater in the CO_2-enriched tanks than it was in the ambient-air tanks; the total dry matter production of all plant parts—those in the submerged soil, those in the free water, and those in the air above—was 3.7 times greater in the high-CO_2 enclosures.

Over the second growing season, the growth enhancement in the high-CO_2 tanks was somewhat less; but the plants in those tanks were so far ahead of the plants in the ambient-air tanks that in their first five months of growth, they produced what it took the plants in the ambient-air tanks fully 21 months to produce.

Moving on to plants that are exclusively floating freshwater macrophytes, Idso (1997) grew many batches of the common water fern (*Azolla pinnata*) over a wide range of atmospheric CO_2 concentrations at two different water temperatures (18.2°C and 24.5°C) in Poor Man's Biospheres for periods of several weeks. This work revealed that a 900-ppm increase in the CO_2 concentration of the air above the tanks led to only a 19 percent increase in the biomass production of the plants floating in the cooler water, but that it led to a 66 percent biomass increase in the plants floating in the warmer water.

In an earlier study of *Azolla pinnata*, Idso *et al.* (1989) conducted three separate two- to three-month experiments wherein they grew batches of the floating fern out-of-doors in adequately fertilized water contained in sunken metal stock tanks located within clear-plastic-wall open-top chambers that were continuously maintained at atmospheric CO_2 concentrations of either 340 or 640 ppm, during which time the plants were briefly removed from the water and weighed at weekly intervals, while their photosynthetic rates were measured at hourly intervals from dawn to dusk on selected cloudless days. As a result of this protocol, they found the photosynthetic and growth rates of the plants growing in ambient air "first decreased, then stagnated, and finally became negative when mean air temperature rose above 30°C." In the high CO_2 treatment, on the other hand, they found that "the debilitating effects of high temperatures were reduced: in one case to a much less severe negative growth rate, in another case to merely a short period of zero growth rate, and in a third case to no discernible ill effects whatsoever—in spite of the fact that the ambient treatment plants in this instance all died."

Last, in a study of an emergent freshwater macrophyte, Ojala *et al.* (2002) grew water horsetail (*Equisetum fluviatile*) plants at ambient and double-ambient atmospheric CO_2 concentrations and ambient and ambient + 3°C air temperatures for three years, although the plants were subjected to the double-ambient CO_2 condition for only approximately five months of each year. This work revealed that the increase in air temperature boosted maximum shoot biomass by 60 percent, but the elevated CO_2 had no effect on this aspect of plant growth. However, elevated CO_2 and temperature—both singly and in combination—positively affected *root* growth, which was enhanced by 10, 15, and 25 percent by elevated air temperature, CO_2, and the two factors together, respectively.

In light of the several experimental findings discussed above, we conclude that the ongoing rise in the air's CO_2 content will likely have significant positive impacts on most freshwater macrophytes, including submersed, floating, and emergent species.

Additional information on this topic, including reviews of newer publications as they become available, can be found at http://www.co2science.org/subject/a/aquaticmacrophytes.php.

References

Andersen, T. and Andersen, F.O. 2006. Effects of CO_2 concentration on growth of filamentous algae and *Littorella uniflora* in a Danish softwater lake. *Aquatic Botany* **84**: 267-271.

Andersen, T., Andersen, F.O. and Pedersen, O. 2006. Increased CO_2 in the water around *Littorella uniflora* raises the sediment O_2 concentration. *Aquatic Botany* **84**: 294-300.

Idso, S.B. 1997. The Poor Man's Biosphere, including simple techniques for conducting CO_2 enrichment and depletion experiments on aquatic and terrestrial plants. *Environmental and Experimental Botany* **38**: 15-38.

Idso, S.B., Allen, S.G., Anderson, M.G. and Kimball, B.A. 1989. Atmospheric CO_2 enrichment enhances survival of Azolla at high temperatures. *Environmental and Experimental Botany* **29**: 337-341.

Idso, S.B., Allen, S.G. and Kimball, B.A. 1990. Growth response of water lily to atmospheric CO_2 enrichment. *Aquatic Botany* **37**: 87-92.

Ojala, A., Kankaala, P. and Tulonen, T. 2002. Growth response of *Equisetum fluviatile* to elevated CO_2 and temperature. *Environmental and Experimental Botany* **47**: 157-171.

Titus, J.E., Feldman, R.S. and Grise, D. 1990. Submersed macrophyte growth at low pH. I. CO_2 enrichment effects with fertile sediment. *Oecologia* **84**: 307-313.

Yan, X., Yu, D. and Li, Y.-K. 2006. The effects of elevated CO_2 on clonal growth and nutrient content of submerged plant *Vallisneria spinulosa*. *Chemosphere* **62**: 595-601.

7.1.3.3. Marine Macroaglae

How do marine macroalgae respond to increases in the air's CO_2 content? The results of the studies discussed below provide a glimpse of what the future may hold in this regard, as the atmosphere's CO_2 concentration continues its upward climb.

Gao *et al.* (1993) grew cultures of the red macroalgae *Gracilaria* sp. and *G. chilensis* in vessels enriched with nitrogen and phosphorus that were continuously aerated with normal air containing 350 ppm CO_2, air enriched with an extra 650 ppm CO_2, and air enriched with an extra 1,250 ppm CO_2 for a period of 19 days. Compared to the control treatments, the relative growth enhancements in the + 650-ppm and +1250-ppm CO_2 treatments were 20 percent and 60 percent, respectively, for *G. chilensis*, and 130 percent and 190 percent, respectively, for the *Gracilaria* sp.

With respect to these findings, the researchers comment that "in their natural habitats, photosynthesis and growth of *Gracilaria* species are likely to be CO_2-limited, especially when the population density is high and water movement is slow." Hence, as the air's CO_2 content continues to rise, these marine macroalgae should grow ever better in the years ahead. Such should also be the case with many other macroalgae, for Gao *et al.* note that "photosynthesis by most macroalgae is probably limited by inorganic carbon sources in natural seawater," citing the studies of Surif and Raven (1989), Maberly (1990), Gao *et al.* (1991), and Levavasseur *et al.* (1991) as evidence for this statement.

In a subsequent study, Kubler *et al.* (1999) grew *Lomentaria articulata*, a red seaweed common to the Northeast Atlantic intertidal zone, for three weeks in hydroponic cultures subjected to various atmospheric CO_2 and O_2 concentrations. In doing so, they found that oxygen concentrations ranging from 10 to 200 percent of ambient had no significant effect on either the seaweed's daily net carbon gain or its total wet biomass production rate. By contrast, CO_2 concentrations ranging from 67 to 500 percent of ambient had highly significant effects on these parameters. At twice the ambient CO_2 concentration, for example, daily net carbon gain and total wet biomass production rates were 52 and 314 percent greater than they were at ambient CO_2.

More recently, Zou (2005) collected specimens of the brown seaweed *Hizikia fusiforme* from intertidal rocks along the coast of Nanao Island, Shantou, China, and maintained them in glass aquariums that contained filtered seawater enriched with 60 µM $NaNO_3$ and 6.0 µM NaH_2PO_4, while continuously aerating the aquariums with air of either 360 or 700 ppm CO_2 and periodically measuring seaweed growth and nitrogen assimilation rates, as well as nitrate reductase activities. By these means they determined that the slightly less than a doubling of the air's CO_2 concentration increased the seaweed's mean relative growth rate by about 50 percent, its mean rate of nitrate uptake during the study's 12-hour light periods by some 200 percent, and its nitrate reductase activity by approximately 20 percent over a wide range of substrate nitrate concentrations.

As a subsidiary aspect of the study, Zou notes that "the extract of *H. fusiforme* has an immunomodulating activity on humans and this ability might be used for clinical application to treat several diseases such as tumors (Suetsuna, 1998; Shan *et al.*, 1999)." He also reports that the alga "has been used as a food delicacy and an herbal ingredient in China, Japan and Korea." In fact, he says that it "is now becoming one of the most important species for seaweed mariculture in China, owing to its high commercial value and increasing market demand." The ongoing rise in the air's CO_2 content bodes well for all of these applications. In addition, Zou notes that "the intensive cultivation of *H. fusiforme* would remove nutrients more efficiently with the future elevation of CO_2 levels in seawater, which could be a possible solution to the problem of ongoing coastal eutrophication," suggesting that rising CO_2 levels may also assist in the amelioration of this environmental problem.

In light of these several observations, the ongoing rise in the air's CO_2 content should help marine macroalgae to become more productive with the passage of time.

Additional information on this topic, including reviews of newer publications as they become available, can be found at http://www.co2science.org/subject/a/aquaticmacroalgae.php.

References

Gao, K., Aruga, Y., Asada, K., Ishihara, T., Akano, T. and Kiyohara, M. 1991. Enhanced growth of the red alga *Porphyra yezoensis* Ueda in high CO_2 concentrations. *Journal of Applied Phycology* **3**: 355-362.

Gao, K., Aruga, Y., Asada, K. and Kiyohara, M. 1993. Influence of enhanced CO_2 on growth and photosynthesis of the red algae *Gracilaria* sp. and *G. chilensis*. *Journal of Applied Phycology* **5**: 563-571.

Kubler, J.E., Johnston, A.M. and Raven, J.A. 1999. The effects of reduced and elevated CO_2 and O_2 on the seaweed *Lomentaria articulata*. *Plant, Cell and Environment* **22**: 1303-1310.

Levavasseur, G., Edwards, G.E., Osmond, C.B. and Ramus, J. 1991. Inorganic carbon limitation of photosynthesis in *Ulva rotundata* (Chlorophyta). *Journal of Phycology* **27**: 667-672.

Maberly, S.C. 1990. Exogenous sources of inorganic carbon for photosynthesis by marine macroalgae. *Journal of Phycology* **26**: 439-449.

Shan, B.E., Yoshida, Y., Kuroda, E. and Yamashita, U. 1999. Immunomodulating activity of seaweed extract on human lymphocytes *in vitro*. *International Journal of Immunopharmacology* **21**: 59-70.

Suetsuna, K. 1998. Separation and identification of angiotensin I-converting enzyme inhibitory peptides from peptic digest of *Hizikia fusiformis* protein. *Nippon Suisan Gakkaishi* **64**: 862-866.

Surif, M.B. and Raven, J.A. 1989. Exogenous inorganic carbon sources for photosynthesis in seawater by members of the Fucales and the Laminariales (Phaeophyta): ecological and taxonomic implications. *Oecologia* **78**: 97-103.

Zou, D. 2005. Effects of elevated atmospheric CO_2 on growth, photosynthesis and nitrogen metabolism in the economic brown seaweed, *Hizikia fusiforme* (Sargassaceae, Phaeophyta). *Aquaculture* **250**: 726-735.

7.1.3.4. Marine Microalgae

How do marine microalgae respond to increases in the air's CO_2 content? Based on the late twentieth century work of Riebesell *et al.* (1993), Hein and Sand-Jensen (1997), and Wolf-Gladrow *et al.*, (1999), it would appear that the productivity of earth's marine microalgae may be significantly enhanced by elevated concentrations of atmospheric CO_2. More recent work by other researchers suggests the same.

In a study of the unicellular marine diatom *Skeletonema costatus*, which is widely distributed in coastal waters throughout the world and is a major component of most natural assemblages of marine phytoplankton, Chen and Gao (2004) grew cell cultures of the species in filtered nutrient-enriched seawater maintained at 20°C under a light/dark cycle of 12/12 hours at a light intensity of 200 $\mu mol\ m^{-2}\ s^{-1}$, while continuously aerating the culture solutions with air of either 350 or 1,000 ppm CO_2 and measuring a number of physiological parameters related to the diatom's photosynthetic activity. They report that cell numbers of the alga "increased steadily throughout the light period and they were 1.6 and 2.1 times higher after the 12 h light period for the alga grown at 350 and 1000 ppm CO_2, respectively." They also say that chlorophyll *a* concentrations "increased 4.4- and 5.4-fold during the middle 8 h of the light period for the alga grown at 350 and 1000 ppm CO_2, respectively," and that "the contents of cellular chlorophyll a were higher for the alga grown at 1000 ppm CO_2 than that at 350 ppm CO_2." In addition, they note that the initial slope of the light saturation curve of photosynthesis and the photochemical efficiency of photosystem II "increased with increasing CO_2, indicating that the efficiency of light-harvesting and energy conversion in photosynthesis were increased." The end result of these several responses, in the words of Chen and Gao, was that "*S. costatum* benefited from CO_2 enrichment."

In another report of a study of marine microalgae that would appear to have enormous implications, Gordillo *et al.* (2003) begin by noting that "one of the main queries for depicting future scenarios of evolution of atmospheric composition and temperature is whether an atmospheric CO_2 increase stimulates primary production, especially in aquatic plants." Why do they say that? They say it because, as they put it, "aquatic primary producers account for about 50 percent of the total carbon fixation in the biosphere (Falkowski and Raven, 1997)."

Although the question addressed by Gordillo *et al.* sounds simple enough, its answer is not straightforward. In many phytoplankton, both freshwater and marine, photosynthesis appears to be saturated under current environmental conditions. Raven (1991), however, has suggested that those very same species, many of which employ carbon-concentrating mechanisms, could well decrease the amount of energy they expend in this latter activity in a CO_2-enriched world, which metabolic readjustment would leave a larger proportion of their captured energy available for fueling enhanced growth.

To explore this possibility, the four researchers studied various aspects of the growth response of the microalgal chlorophyte *Dunaliella viridis* (which possesses a carbon concentrating mechanism and has been used as a model species for the study of inorganic carbon uptake) to atmospheric CO_2 enrichment. Specifically, they batch-cultured the chlorophyte, which is one of the most ubiquitous eukaryotic organisms in hypersaline environments, in 250-ml Perspex cylinders under controlled laboratory conditions at high (5 m*M*) and low (0.5 m*M*) nitrate concentrations, while continuously aerating half of the cultures with ambient air of approximately 350 ppm CO_2 and the other half with air of approximately 10,000 ppm CO_2. In doing so, they discovered that atmospheric CO_2 enrichment had little effect on dark respiration in both N treatments. Likewise, it had little effect on photosynthesis in the low-N treatment. In the high-N treatment, the extra CO_2 increased photosynthesis by 114 percent. In the case of biomass production, the results were even more extreme: in the low-N treatment elevated CO_2 had no effect at all, while in the high-N treatment it nearly tripled the cell density of the culture solution.

In discussing their findings, Gordillo *et al.* note that "it has long been debated whether phytoplankton species are growth-limited by current levels of CO_2 in aquatic systems, i.e. whether an increase in atmospheric CO_2 could stimulate growth (Riebesell *et al.*, 1993)." Their results clearly indicate that it can, as long as sufficient nitrogen is available. But that was not all that Gordillo *et al.* learned. In the high-N treatment, where elevated CO_2 greatly stimulated photosynthesis and biomass production, once the logarithmic growth phase had run its course and equilibrium growth was attained, approximately 70 percent of the carbon assimilated by the chlorophyte was released to the water, while in the low-CO_2 treatment only 35 percent was released.

With respect to this suite of observations, Gordillo *et al.* say "the release of organic carbon to the external medium has been proposed as a mechanism for maintaining the metabolic integrity of the cell (Ormerod, 1983)," and that "according to Wood and Van Valen (1990), organic carbon release would be a sink mechanism protecting the photosynthetic apparatus from an overload of products that cannot be invested in growth or stored." They additionally state that stores of photosynthetic products "are reduced to avoid overload and produce a high demand for photosynthates." Under these conditions, they conclude that "the process would then divert assimilated C to either the production of new biomass, or the release to the external medium once the culture conditions do not allow further exponential growth."

A second consequence of enhanced organic carbon release in the face of atmospheric CO_2 enrichment and sufficient N availability is that the internal C:N balance of the phytoplankton is maintained within a rather tight range. This phenomenon has also been observed in the green seaweed *Ulva rigida* (Gordillo *et al.*, 2001) and the cyanobacterium *Spirulina platensis* (Gordillo *et al.*, 1999). Hence, what the study of Gordillo *et al.* implies about the response of *Dunaliella viridis* to atmospheric CO_2 enrichment may well be widely applicable to many, if not most, aquatic plants, not the least of which may be the zooxanthellae that by this means (enhanced organic carbon release) could provide their coral hosts with the source of extra energy they need to continue building their skeletons at a non-reduced rate in the face of the negative calcification pressure produced by the changes in seawater chemistry that have been predicted to result from the ongoing rise in the air's CO_2 concentration.

In light of these several observations, there would appear to be ample reason to be optimistic about the response of earth's marine macroalgae to the ongoing rise in the air's CO_2 content.

Additional information on this topic, including reviews of newer publications as they become available, can be found at http://www.co2science.org/subject/a/aquaticmicroalgae.php.

References

Chen, X. and Gao, K. 2004. Characterization of diurnal photosynthetic rhythms in the marine diatom *Skeletonema costatum* grown in synchronous culture under ambient and elevated CO_2. *Functional Plant Biology* **31**: 399-404.

Falkowski, P.G. and Raven, J.A. 1997. *Aquatic Photosynthesis*. Blackwell Science, Massachusetts, USA.

Gordillo, F.J.L., Jimenez, C., Figueroa, F.L. and Niell, F.X. 1999. Effects of increased atmospheric CO_2 and N supply on photosynthesis, growth and cell composition of the cyanobacterium *Spirulina platensis* (Arthrospira). *Journal of Applied Phycology* **10**: 461-469.

Gordillo, F.J.L., Jimenez, C., Figueroa, F.L. and Niell, F.X. 2003. Influence of elevated CO_2 and nitrogen supply on the carbon assimilation performance and cell composition of the unicellular alga *Dunaliella viridis*. *Physiologia Plantarum* **119**: 513-518.

Gordillo, F.J.L., Niell, F.X. and Figueroa, F.L. 2001. Non-photosynthetic enhancement of growth by high CO_2 level in the nitrophilic seaweed *Ulva rigida* C. Agardh (Chlorophyta). *Planta* **213**: 64-70.

Hein, M. and Sand-Jensen, K. 1997. CO_2 increases oceanic primary production. *Nature* **388**: 988-990.

Ormerod, J.G. 1983. The carbon cycle in aquatic ecosystems. In: Slater, J.H., Whittenbury, R. and Wimpeny, J.W.T. (Eds.) *Microbes in Their Natural Environment*. Cambridge University Press, Cambridge, UK, pp. 463-482.

Raven, J.A. 1991. Physiology of inorganic carbon acquisition and implications for resource use efficiency by marine phytoplankton: Relation to increased CO_2 and temperature. *Plant, Cell and Environment* **14**: 774-794.

Riebesell, U., Wolf-Gladrow, D.A. and Smetacek, V. 1993. Carbon dioxide limitation of marine phytoplankton growth rates. *Nature* **361**: 249-251.

Wood, A.M. and Van Valen, L.M. 1990. Paradox lost? On the release of energy rich compounds by phytoplankton. *Marine Microbial Food Webs* **4**: 103-116.

Wolf-Gladrow, D.A., Riebesell, U., Burkhardt, S. and Bijma, J. 1999. Direct effects of CO_2 concentration on growth and isotopic composition of marine plankton. *Tellus* **51B**: 461-476.

7.2. Water Use Efficiency

Another major consequence of atmospheric CO_2 enrichment is that plants exposed to elevated levels of atmospheric CO_2 generally do not open their leaf stomatal pores—through which they take in carbon dioxide and give off water vapor—as wide as they do at lower CO_2 concentrations and tend to produce fewer of these pores per unit area of leaf surface. Both changes tend to reduce most plants' rates of water loss by transpiration. The amount of carbon they gain per unit of water lost—or water-use efficiency— therefore typically rises, increasing their ability to withstand drought. In this section, we explore the phenomena of water use efficiency as it pertains to agricultural, grassland, and woody species.

Additional information on this topic, including reviews water use efficiency not discussed here, can be found at http://www.co2science.org/subject/w/subject_w.php under the heading Water Use Efficiency.

7.2.1. Agricultural Species

In the study of Serraj *et al.* (1999), soybeans grown at 700 ppm CO_2 displayed 10 to 25 percent reductions in total water loss while simultaneously exhibiting increases in dry weight of as much as 33 percent. Thus, elevated CO_2 significantly increased the water-use efficiencies of the studied plants. Likewise, Garcia *et al.* (1998) determined that spring wheat grown at 550 ppm CO_2 exhibited a water-use efficiency that was about one-third greater than that exhibited by plants grown at 370 ppm CO_2. Similarly, Hakala *et al.* (1999) reported that twice-ambient CO_2 concentrations increased the water-use efficiency of spring wheat by 70 to 100 percent, depending on experimental air temperature. In addition, Hunsaker *et al.* (2000) reported CO_2-induced increases in water-use efficiency for field-grown wheat that were 20 and 10 percent higher than those displayed by ambiently grown wheat subjected to high and low soil nitrogen regimes, respectively. Also, pea plants grown for two months in growth chambers receiving atmospheric CO_2 concentrations of 700 ppm displayed an average water-use efficiency that was 27 percent greater than that exhibited by ambiently grown control plants (Gavito *et al.*, 2000).

In some cases, the water-use efficiency increases caused by atmospheric CO_2 enrichment are spectacularly high. De Luis *et al.* (1999), for example, demonstrated that alfalfa plants subjected to atmospheric CO_2 concentrations of 700 ppm had water-use efficiencies that were 2.6 and 4.1 times greater than those displayed by control plants growing at 400 ppm CO_2 under water-stressed and well-watered conditions, respectively. Also, when grown at an atmospheric CO_2 concentration of 700 ppm, a 2.7-fold increase in water-use efficiency was reported by Malmstrom and Field (1997) for oats infected with the barley yellow dwarf virus.

In addition to enhancing the water-use efficiencies of agricultural C_3 crops, as reported in the preceding paragraphs, elevated CO_2 also enhances the water-use efficiencies of crops possessing alternate carbon fixation pathways. Maroco *et al.* (1999), for example, demonstrated that maize—a C_4 crop—grown for 30 days at an atmospheric CO_2 concentration of 1,100 ppm exhibited an intrinsic water-use efficiency that was 225 percent higher than that of plants grown at 350 ppm CO_2. In addition, Conley *et al.* (2001) reported that a 200-ppm increase in the air's CO_2 content boosted the water-use efficiency of field-grown sorghum by 9 and 19 percent under well-watered and water-stressed conditions, respectively. Also, Zhu *et al.* (1999) reported that pineapple—a CAM plant—grown at 700 ppm CO_2 exhibited water-use efficiencies that were always significantly greater than those displayed by control plants grown at 350 ppm CO_2 over a range of growth temperatures.

It is clear from the studies above that as the CO_2 content of the air continues to rise, earth's agricultural species will respond favorably by exhibiting increases in water-use efficiency. It is likely that food and fiber production will increase on a worldwide basis, even in areas where productivity is severely restricted due to limited availability of soil moisture.

Additional information on this topic, including reviews of newer publications as they become available, can be found at http://www.co2science.org/subject/w/wateruseag.php.

References

Conley, M.M., Kimball, B.A., Brooks, T.J., Pinter Jr., P.J., Hunsaker, D.J., Wall, G.W., Adams, N.R., LaMorte, R.L., Matthias, A.D., Thompson, T.L., Leavitt, S.W., Ottman, M.J., Cousins, A.B. and Triggs, J.M. 2001. CO_2 enrichment increases water-use efficiency in sorghum. *New Phytologist* **151**: 407-412.

De Luis, J., Irigoyen, J.J. and Sanchez-Diaz, M. 1999. Elevated CO_2 enhances plant growth in droughted N_2-fixing alfalfa without improving water stress. *Physiologia Plantarum* **107**: 84-89.

Garcia, R.L., Long, S.P., Wall, G.W., Osborne, C.P., Kimball, B.A., Nie, G.Y., Pinter Jr., P.J., LaMorte, R.L. and Wechsung, F. 1998. Photosynthesis and conductance of spring-wheat leaves: field response to continuous free-air atmospheric CO_2 enrichment. *Plant, Cell and Environment* **21**: 659-669.

Gavito, M.E., Curtis, P.S., Mikkelsen, T.N. and Jakobsen, I. 2000. Atmospheric CO_2 and mycorrhiza effects on biomass allocation and nutrient uptake of nodulated pea (*Pisum sativum* L.) plants. *Journal of Experimental Botany* **52**: 1931-1938.

Hakala, K., Helio, R., Tuhkanen, E. and Kaukoranta, T. 1999. Photosynthesis and Rubisco kinetics in spring wheat and meadow fescue under conditions of simulated climate change with elevated CO_2 and increased temperatures. *Agricultural and Food Science in Finland* **8**: 441-457.

Hunsaker, D.J., Kimball. B.A., Pinter Jr., P.J., Wall, G.W., LaMorte, R.L., Adamsen, F.J., Leavitt, S.W., Thompson, T.L., Matthias, A.D. and Brooks, T.J. 2000. CO_2 enrichment and soil nitrogen effects on wheat evapotranspiration and water use efficiency. *Agricultural and Forest Meteorology* **104**: 85-105.

Malmstrom, C.M. and Field, C.B. 1997. Virus-induced differences in the response of oat plants to elevated carbon dioxide. *Plant, Cell and Environment* **20**: 178-188.

Maroco, J.P., Edwards, G.E. and Ku, M.S.B. 1999. Photosynthetic acclimation of maize to growth under elevated levels of carbon dioxide. *Planta* **210**: 115-125.

Serraj, R., Allen Jr., L.H. and Sinclair, T.R. 1999. Soybean leaf growth and gas exchange response to drought under carbon dioxide enrichment. *Global Change Biology* **5**: 283-291.

Zhu, J., Goldstein, G. and Bartholomew, D.P. 1999. Gas exchange and carbon isotope composition of *Ananas comosus* in response to elevated CO_2 and temperature. *Plant, Cell and Environment* **22**: 999-1007.

7.2.2. Grassland Species

In the study of Grunzweig and Korner (2001), model grasslands representative of the semi-arid Negev of Israel, which were grown for five months at atmospheric CO_2 concentrations of 440 and 600 ppm, exhibited cumulative water-use efficiencies that were 17 and 28 percent greater, respectively, than control communities grown at 280 ppm CO_2. Similarly, Szente *et al.* (1998) reported a doubling of the atmospheric CO_2 concentration increased the water-use efficiency of two C_3 grasses and two broad-leaved species common to the loess grasslands of Budapest by 72 and 266 percent, respectively. In addition, Leymarie *et al.* (1999) calculated that twice-ambient CO_2 concentrations increased the water-use efficiency of the herbaceous weedy species *Arabidopsis thaliana* by 41 and 120 percent under well-watered and water-stressed conditions, respectively. Other CO_2-induced increases in C_3 plant water-use efficiency have been documented by Clark *et al.* (1999) for several New Zealand pasture species and Roumet *et al.* (2000) for various Mediterranean herbs.

Elevated CO_2 also has been shown to substantially increase the water-use efficiency of C_4 grassland species. Adams *et al.* (2000), for example, reported that twice-ambient CO_2 concentrations enhanced the daily water-use efficiency of a C_4 tallgrass prairie in Kansas, USA, dominated by *Andropogon gerardii*. LeCain and Morgan (1998) also documented enhanced water-use efficiencies for six different C_4 grasses grown with twice-ambient CO_2 concentrations. Likewise, Seneweera *et al.*

(1998) reported that a 650-ppm increase in the air's CO_2 content dramatically increased the water-use efficiency of the perennial C_4 grass *Panicum coloratum*.

As the air's CO_2 content continues to rise, nearly all of earth's grassland species—including both C_3 and C_4 plants—will likely experience increases in water-use efficiency. Concomitantly, the productivity of the world's grasslands should increase, even if available moisture decreases in certain areas. Moreover, such CO_2-induced increases in water-use efficiency will likely allow grassland species to expand their ranges into desert areas where they previously could not survive due to lack of sufficient moisture.

Additional information on this topic, including reviews of newer publications as they become available, can be found at http://www.co2science.org/subject/w/waterusegrass.php.

References

Adams, N.R., Owensby, C.E. and Ham, J.M. 2000. The effect of CO_2 enrichment on leaf photosynthetic rates and instantaneous water use efficiency of *Andropogon gerardii* in the tallgrass prairie. *Photosynthesis Research* **65**: 121-129.

Clark, H., Newton, P.C.D. and Barker, D.J. 1999. Physiological and morphological responses to elevated CO_2 and a soil moisture deficit of temperate pasture species growing in an established plant community. *Journal of Experimental Botany* **50**: 233-242.

Grunzweig, J.M. and Korner, C. 2001. Growth, water and nitrogen relations in grassland model ecosystems of the semi-arid Negev of Israel exposed to elevated CO_2. *Oecologia* **128**: 251-262.

LeCain, D.R. and Morgan, J.A. 1998. Growth, gas exchange, leaf nitrogen and carbohydrate concentrations in NAD-ME and NADP-ME C_4 grasses grown in elevated CO_2. *Physiologia Plantarum* **102**: 297-306.

Leymarie, J., Lasceve, G. and Vavasseur, A. 1999. Elevated CO_2 enhances stomatal responses to osmotic stress and abscisic acid in *Arabidopsis thaliana*. *Plant, Cell and Environment* **22**: 301-308.

Roumet, C., Garnier, E., Suzor, H., Salager, J.-L. and Roy, J. 2000. Short and long-term responses of whole-plant gas exchange to elevated CO_2 in four herbaceous species. *Environmental and Experimental Botany* **43**: 155-169.

Seneweera, S.P., Ghannoum, O. and Conroy, J. 1998. High vapor pressure deficit and low soil water availability enhance shoot growth responses of a C_4 grass (*Panicum coloratum* cv. Bambatsi) to CO_2 enrichment. *Australian Journal of Plant Physiology* **25**: 287-292.

Szente, K., Nagy, Z. and Tuba, Z. 1998. Enhanced water use efficiency in dry loess grassland species grown at elevated air CO_2 concentration. *Photosynthetica* **35**: 637-640.

7.2.3. Woody Species

The effect of elevated atmospheric CO_2 concentrations on the water-use efficiencies of trees is clearly positive, having been documented in a number of different single-species studies of longleaf pine (Runion *et al.*, 1999), red oak (Anderson and Tomlinson, 1998), scrub oak (Lodge *et al.*, 2001), silver birch (Rey and Jarvis, 1998), beech (Bucher-Wallin *et al.*, 2000; Egli *et al.*, 1998), sweetgum (Gunderson *et al.*, 2002; Wullschleger and Norby, 2001), and spruce (Roberntz and Stockfors, 1998). Likewise, in a multi-species study performed by Tjoelker *et al.* (1998), seedlings of quaking aspen, paper birch, tamarack, black spruce, and jack pine, which were grown at 580 ppm CO_2 for three months, displayed water-use efficiencies that were 40 to 80 percent larger than those exhibited by their respective controls grown at 370 ppm CO_2.

Similar results are also obtained when trees are exposed to different environmental stresses. In a study conducted by Centritto *et al.* (1999), for example, cherry seedlings grown at twice-ambient levels of atmospheric CO_2 displayed water-use efficiencies that were 50 percent greater than their ambient controls, regardless of soil moisture status. And in the study of Wayne *et al.* (1998), yellow birch seedlings grown at 800 ppm CO_2 had water-use efficiencies that were 52 and 94 percent greater than their respective controls, while simultaneously subjected to uncharacteristically low and high air temperature regimes.

In some parts of the world, perennial woody species have been exposed to elevated atmospheric CO_2 concentrations for decades, due to their proximity to CO_2-emitting springs and vents in the earth, allowing scientists to assess the long-term effects of this phenomenon. In Venezuela, for example, the water-use efficiency of a common tree exposed to a lifetime atmospheric CO_2 concentration of approximately 1,000 ppm rose 2-fold and 19-fold during the local wet and dry seasons, respectively (Fernandez *et al.*, 1998). Similarly, Bartak *et al.*

(1999) reported that 30-year-old *Arbutus unedo* trees growing in central Italy at a lifetime atmospheric CO_2 concentration around 465 ppm exhibited water-use efficiencies that were 100 percent greater than control trees growing at a lifetime CO_2 concentration of 355 ppm. In addition, two species of oaks in central Italy that had been growing for 15 to 25 years at an atmospheric CO_2 concentration ranging from 500 to 1,000 ppm displayed "such marked increases in water-use efficiency under elevated CO_2," in the words of the scientists who studied them, that this phenomenon "might be of great importance in Mediterranean environments in the perspective of global climate change" (Blaschke *et al.*, 2001; Tognetti *et al.*, 1998). Thus, the long-term effects of elevated CO_2 concentrations on water-use efficiency are likely to persist and increase with increasing atmospheric CO_2 concentrations.

In some cases, scientists have looked to the past and determined the positive impact the historic rise in the air's CO_2 content has already had on plant water-use efficiency. Duquesnay *et al.* (1998), for example, used tree-ring data derived from beech trees to determine that over the past century the water-use efficiency of such trees in north-eastern France increased by approximately 33 percent. Similarly, Feng (1999) used tree-ring chronologies derived from a number of trees in western North America to calculate a 10 to 25 percent increase in tree water-use efficiency from 1750 to 1970, during which time the atmospheric CO_2 concentration rose by approximately 16 percent. In another study, Knapp *et al.* (2001) developed tree-ring chronologies from western juniper stands located in Oregon, USA, for the past century, determining that growth recovery from drought was much greater in the latter third of their chronologies (1964-1998) than it was in the first third (1896-1930). In this case, the authors suggested that the greater atmospheric CO_2 concentrations of the latter period allowed the trees to more quickly recover from water stress. Finally, Beerling *et al.* (1998) grew *Gingko* saplings at 350 and 650 ppm CO_2 for three years, finding that elevated atmospheric CO_2 concentrations reduced leaf stomatal densities to values comparable to those measured on fossilized *Gingko* leaves dating back to the Triassic and Jurassic periods, implying greater water-use efficiencies for those times too.

On another note, Prince *et al.* (1998) demonstrated that rain-use efficiency, which is similar to water-use efficiency, slowly increased in the African Sahel from 1982 to 1990, while Nicholson *et al.* (1998) observed neither an increase nor a decrease in this parameter from 1980 to 1995 for the central and western Sahel.

In summary, it is clear that as the CO_2 content of the air continues to rise, nearly all of earth's trees will respond favorably by exhibiting increases in water-use efficiency. It is thus likely that as time progresses, earth's woody species will expand into areas where they previously could not exist due to limiting amounts of available moisture. Therefore, one can expect the earth to become a greener biospheric body with greater carbon sequestering capacity as the atmospheric CO_2 concentration continues to rise.

Additional information on this topic, including reviews of newer publications as they become available, can be found at http://www.co2science.org/subject/w/waterusetrees.php.

References

Anderson, P.D. and Tomlinson, P.T. 1998. Ontogeny affects response of northern red oak seedlings to elevated CO_2 and water stress. I. Carbon assimilation and biomass production. *New Phytologist* **140**: 477-491.

Bartak, M., Raschi, A. and Tognetti, R. 1999. Photosynthetic characteristics of sun and shade leaves in the canopy of *Arbutus unedo* L. trees exposed to *in situ* long-term elevated CO_2. *Photosynthetica* **37**: 1-16.

Beerling, D.J., McElwain, J.C. and Osborne, C.P. 1998. Stomatal responses of the 'living fossil' *Ginkgo biloba* L. to changes in atmospheric CO_2 concentrations. *Journal of Experimental Botany* **49**: 1603-1607.

Blaschke, L., Schulte, M., Raschi, A., Slee, N., Rennenberg, H. and Polle, A. 2001. Photosynthesis, soluble and structural carbon compounds in two Mediterranean oak species (*Quercus pubescens* and *Q. ilex*) after lifetime growth at naturally elevated CO_2 concentrations. *Plant Biology* **3**: 288-298.

Bucher-Wallin, I.K., Sonnleitner, M.A., Egli, P., Gunthardt-Goerg, M.S., Tarjan, D., Schulin, R. and Bucher, J.B. 2000. Effects of elevated CO_2, increased nitrogen deposition and soil on evapotranspiration and water use efficiency of spruce-beech model ecosystems. *Phyton* **40**: 49-60.

Centritto, M., Lee, H.S.J. and Jarvis, P.G. 1999. Interactive effects of elevated [CO_2] and drought on cherry (*Prunus avium*) seedlings. I. Growth, whole-plant water use efficiency and water loss. *New Phytologist* **141**: 129-140.

Duquesnay, A., Breda, N., Stievenard, M. and Dupouey, J.L. 1998. Changes of tree-ring $d^{13}C$ and water-use efficiency of beech (*Fagus sylvatica* L.) in north-eastern France during the past century. *Plant, Cell and Environment* **21**: 565-572.

Egli, P., Maurer, S., Gunthardt-Goerg, M.S. and Korner, C. 1998. Effects of elevated CO_2 and soil quality on leaf gas exchange and aboveground growth in beech-spruce model ecosystems. *New Phytologist* **140**: 185-196.

Feng, X. 1999. Trends in intrinsic water-use efficiency of natural trees for the past 100-200 years: A response to atmospheric CO_2 concentration. *Geochimica et Cosmochimica Acta* **63**: 1891-1903.

Fernandez, M.D., Pieters, A., Donoso, C., Tezara, W., Azuke, M., Herrera, C., Rengifo, E. and Herrera, A. 1998. Effects of a natural source of very high CO_2 concentration on the leaf gas exchange, xylem water potential and stomatal characteristics of plants of *Spatiphylum cannifolium* and *Bauhinia multinervia*. *New Phytologist* **138**: 689-697.

Gunderson, C.A., Sholtis, J.D., Wullschleger, S.D., Tissue, D.T., Hanson, P.J. and Norby, R.J. 2002. Environmental and stomatal control of photosynthetic enhancement in the canopy of a sweetgum (*Liquidambar styraciflua* L.) plantation during 3 years of CO_2 enrichment. *Plant, Cell and Environment* **25**: 379-393.

Knapp, P.A., Soule, P.T. and Grissino-Mayer, H.D. 2001. Post-drought growth responses of western juniper (*Junipers occidentalis* var. *occidentalis*) in central Oregon. *Geophysical Research Letters* **28**: 2657-2660.

Lodge, R.J., Dijkstra, P., Drake, B.G. and Morison, J.I.L. 2001. Stomatal acclimation to increased CO_2 concentration in a Florida scrub oak species *Quercus myrtifolia* Willd. *Plant, Cell and Environment* **24**: 77-88.

Nicholson, S.E., Tucker, C.J. and Ba, M.B. 1998. Desertification, drought, and surface vegetation: An example from the West African Sahel. *Bulletin of the American Meteorological Society* **79**: 815-829.

Prince, S.D., Brown De Colstoun, E. and Kravitz, L.L. 1998. Evidence from rain-use efficiencies does not indicate extensive Sahelian desertification. *Global Change Biology* **4**: 359-374.

Rey, A. and Jarvis, P.G. 1998. Long-term photosynthetic acclimation to increased atmospheric CO_2 concentration in young birch (*Betula pendula*) trees. *Tree Physiology* **18**: 441-450.

Roberntz, P. and Stockfors, J. 1998. Effects of elevated CO_2 concentration and nutrition on net photosynthesis, stomatal conductance and needle respiration of field-grown Norway spruce trees. *Tree Physiology* **18**: 233-241.

Runion, G.B., Mitchell, R.J., Green, T.H., Prior, S.A., Rogers, H.H. and Gjerstad, D.H. 1999. Longleaf pine photosynthetic response to soil resource availability and elevated atmospheric carbon dioxide. *Journal of Environmental Quality* **28**: 880-887.

Tjoelker, M.G., Oleksyn, J. and Reich, P.B. 1998. Seedlings of five boreal tree species differ in acclimation of net photosynthesis to elevated CO_2 and temperature. *Tree Physiology* **18**: 715-726.

Tognetti, R., Johnson, J.D., Michelozzi, M. and Raschi, A. 1998. Response of foliar metabolism in mature trees of *Quercus pubescens* and *Quercus ilex* to long-term elevated CO_2. *Environmental and Experimental Botany* **39**: 233-245.

Wayne, P.M., Reekie, E.G. and Bazzaz, F.A. 1998. Elevated CO_2 ameliorates birch response to high temperature and frost stress: implications for modeling climate-induced geographic range shifts. *Oecologia* **114**: 335-342.

Wullschleger, S.D. and Norby, R.J. 2001. Sap velocity and canopy transpiration in a sweetgum stand exposed to free-air CO_2 enrichment (FACE). *New Phytologist* **150**: 489-498.

7.3. Amelioration of Environmental Stresses

Atmospheric CO_2 enrichment has been shown to help ameliorate the detrimental effects of several environmental stresses on plant growth and development, including disease, herbivory (predation by insects), shade (caused by increased cloudiness), ozone (a common air pollutant), low temperatures, nitrogen deficiency, UV-B radiation, and water stress. In this section we survey research on each of these types of stress.

Additional information on this topic, including reviews on stresses not discussed here, can be found at http://www.co2science.org/subject/g/subject_g.php under the heading Growth Response to CO_2 with Other Variables.

7.3.1. Disease

According to the IPCC, CO_2-induced global warming will increase the risk of plant disease outbreaks, resulting in negative consequences for food, fiber, and forestry across all world regions (IPCC, 2007-II). But it appears the IPCC has omitted the results of real-world observations that contradict this forecast.

Chakraborty and Datta (2003) note there are a number of CO_2-induced changes in plant physiology, anatomy and morphology that have been implicated in increased plant resistance to disease and that "can potentially enhance host resistance at elevated CO_2." Among these phenomena they list "increased net photosynthesis allowing mobilization of resources into host resistance (Hibberd *et al.*, 1996a.); reduced stomatal density and conductance (Hibberd *et al.*, 1996b); greater accumulation of carbohydrates in leaves; more waxes, extra layers of epidermal cells and increased fibre content (Owensby, 1994); production of papillae and accumulation of silicon at penetration sites (Hibberd *et al.*, 1996a); greater number of mesophyll cells (Bowes, 1993); and increased biosynthesis of phenolics (Hartley *et al.*, 2000), among others."

Malmstrom and Field (1997) grew individual oat plants for two months in pots placed within phytocells maintained at atmospheric CO_2 concentrations of 350 and 700 ppm, while they infected one-third of the plants with the barley yellow dwarf virus (BYDV), which plagues more than 150 plant species worldwide, including all major cereal crops. Over the course of their study, they found that elevated CO_2 stimulated rates of net photosynthesis in all plants, regardless of pathogen infection. However, the greatest percentage increase occurred in diseased individuals (48 percent vs. 34 percent). Moreover, atmospheric CO_2 enrichment decreased stomatal conductance by 50 percent in infected plants but by only 34 percent in healthy ones, which led to a CO_2-induced doubling of the instantaneous water-use efficiency of the healthy plants, but an increase of fully 2.7-fold in the diseased plants. Last, after 60 days of growth under these conditions, they determined that the extra CO_2 increased total plant biomass by 36 percent in infected plants, but by only 12 percent in healthy plants. In addition, while elevated CO_2 had little effect on root growth in the healthy plants, it increased root biomass in the infected plants by up to 60 percent. Consequently, it can be appreciated that as the CO_2 content of the air continues to rise, its many positive effects will likely offset some, if not most, of the negative effects of the destructive BYDV. Quoting Malmstrom and Field with respect to two specific examples, they say in their concluding remarks that CO_2 enrichment "may reduce losses of infected plants to drought" and "may enable diseased plants to compete better with healthy neighbors."

Tiedemann and Firsching (2000) grew spring wheat plants from germination to maturity in controlled-environment chambers maintained at ambient (377 ppm) and elevated (612 ppm) concentrations of atmospheric CO_2 and at ambient (20 ppb) and elevated (61 ppb) concentrations of ozone (and combinations thereof), the latter of which gases is typically toxic to most plants. In addition, half of the plants in each treatment were inoculated with a leaf rust-causing fungus. Under these conditions, the elevated CO_2 increased the photosynthetic rates of the diseased plants by 20 and 42 percent at the ambient and elevated ozone concentrations, respectively. It also enhanced the yield of the infected plants, increasing it by 57 percent, even in the presence of high ozone concentrations.

Jwa and Walling (2001) grew tomato plants hydroponically for eight weeks in controlled-environment chambers maintained at atmospheric CO_2 concentrations of 350 and 700 ppm. In addition, at week five of the study, half of all plants growing in each CO_2 concentration were infected with a fungal pathogen that attacks plant roots and induces a water stress that decreases growth and yield. At the end of the study, they found that the pathogenic infection had reduced total plant biomass by nearly 30 percent at both atmospheric CO_2 concentrations. However, the elevated CO_2 had increased the total biomass of the healthy and diseased plants by the same amount (+30 percent), with the result that the infected tomato plants grown at 700 ppm CO_2 had biomass values that were essentially identical to those of the healthy tomato plants grown at 350 ppm CO_2. Thus, the extra CO_2 completely counterbalanced the negative effect of the pathogenic infection on overall plant productivity.

Chakraborty and Datta (2003) studied the aggressiveness of the fungal anthracnose pathogen *Colletotrichum gloeosporioides* by inoculating two isolates of the pathogen onto two cultivars of the tropical pasture legume *Stylosanthes scabra* (Fitzroy, which is susceptible to the fungal pathogen, and Seca, which is more resistant) over 25 sequential infection cycles in controlled-environment chambers filled with air of either 350 or 700 ppm CO_2. By these means they determined that the aggressiveness of the pathogen was reduced at the twice-ambient level of atmospheric CO_2, where aggressiveness is defined as "a property of the pathogen reflecting the relative amount of damage caused to the host without regard to resistance genes (Shaner *et al.*, 1992)." As they describe it, "at twice-ambient CO_2 the overall level of

aggressiveness of the two [pathogen] isolates was significantly reduced on both cultivars."

Simultaneously, however, pathogen fecundity was found to increase at twice-ambient CO_2. Of this finding, Chakraborty and Datta report that their results "concur with the handful of studies that have demonstrated increased pathogen fecundity at elevated CO_2 (Hibberd *et al.*, 1996a; Klironomos *et al.*, 1997; Chakraborty *et al.*, 2000)." How this happened in the situation they investigated, according to Chakraborty and Datta, is that the overall increase in fecundity at high CO_2 "is a reflection of the altered canopy environment," wherein "the 30% larger *S. scabra* plants at high CO_2 (Chakraborty *et al.*, 2000) makes the canopy microclimate more conducive to anthracnose development."

In view of these opposing changes in pathogen behavior at elevated levels of atmospheric CO_2, it is difficult to know the outcome of atmospheric CO_2 enrichment for this specific pathogen-host relationship. More research, especially under realistic field conditions, will be needed to clarify the situation; and, of course, different results are likely to be observed for different pathogen-host associations. What is more, results could also differ under different climatic conditions. Nevertheless, the large number of ways in which elevated CO_2 has been demonstrated to increase plant resistance to pathogen attack gives us reason to believe that plants will gain the advantage as the air's CO_2 content continues to climb in the years ahead.

Another study that fuels this optimism was conducted by Parsons *et al.* (2003), who grew two-year-old saplings of paper birch and three-year-old saplings of sugar maple in well-watered and fertilized pots from early May until late August in glasshouse rooms maintained at either 400 or 700 ppm CO_2. In these circumstances, the whole-plant biomass of paper birch was increased by 55 percent in the CO_2-enriched portions of the glasshouse, while that of sugar maple was increased by 30 percent. Also, concentrations of condensed tannins were increased by 27 percent in the paper birch (but not the sugar maple) saplings grown in the CO_2-enriched air; in light of this finding, Parsons *et al.* conclude that "the higher condensed tannin concentrations that were present in the birch fine roots may offer these tissues greater protection against soil-borne pathogens and herbivores."

Within this context, it is interesting to note that Parsons *et al.* report that CO_2-induced increases in fine root concentrations of total phenolics and condensed tannins have also been observed in warm temperate conifers by King *et al.* (1997), Entry *et al.* (1998), Gebauer *et al.* (1998), and Runion *et al.* (1999), as well as in cotton by Booker (2000).

In another intriguing study, Gamper *et al.* (2004) begin by noting that arbuscular mycorrhizal fungi (AMF) are expected to modulate plant responses to elevated CO_2 by "increasing resistance/tolerance of plants against an array of environmental stressors (Smith and Read, 1997)." In investigating this subject in a set of experiments conducted over a seven-year period of free-air CO_2-enrichment on two of the world's most extensively grown cool-season forage crops (*Lolium perenne* and *Trifolium repens*) at the Swiss free-air CO_2 enrichment (FACE) facility near Zurich, they determined that "at elevated CO_2 and under [two] N treatments, AMF root colonization of both host plant species was increased," and that "colonization levels of all three measured intraradical AMF structures (hyphae, arbuscules and vesicles) tended to be higher." Hence, they concluded that these CO_2-induced benefits may lead to "increased protection against pathogens and/or herbivores."

Pangga *et al.* (2004) grew well-watered and fertilized seedlings of a cultivar (Fitzroy) of the pencilflower (*Stylosanthes scabra*)—an important legume crop that is susceptible to anthracnose disease caused by *Colletotrichum gloeosporioides* (Penz.) Penz. & Sacc.—within a controlled-environment facility maintained at atmospheric CO_2 concentrations of either 350 or 700 ppm, where they inoculated six-, nine- and 12-week-old plants with conidia of *C. gloeosporioides*. Then, 10 days after inoculation, they counted the anthracnose lesions on the plants and classified them as either resistant or susceptible.

Adherence to this protocol revealed, in their words, that "the mean number of susceptible, resistant, and total lesions per leaf averaged over the three plant ages was significantly (P<0.05) greater at 350 ppm than at 700 ppm CO_2, reflecting the development of a level of resistance in susceptible cv. Fitzroy at high CO_2." In fact, with respect to the plants inoculated at 12 weeks of age, they say that those grown "at 350 ppm had 60 and 75% more susceptible and resistant lesions per leaf, respectively, than those [grown] at 700 ppm CO_2."

In terms of infection efficiency (IE), the Australian scientists say their work "clearly shows that at 350 ppm overall susceptibility of the canopy increases with increasing age because more young leaves are produced on secondary and tertiary branches of the more advanced plants." However,

they report that "at 700 ppm CO_2, IE did not increase with increasing plant age despite the presence of many more young leaves in the enlarged canopy," which finding, in their words, "points to reduced pathogen efficiency or an induced partial resistance to anthracnose in Fitzroy at 700 ppm CO_2." Consequently, as the air's CO_2 content continues to rise, it would appear that the Fitzroy cultivar of the pasture legume *Stylosanthes scabra* will acquire a greater intrinsic resistance to the devastating anthracnose disease.

McElrone *et al.* (2005) "assessed how elevated CO_2 affects a foliar fungal pathogen, *Phyllosticta minima*, of *Acer rubrum* [red maple] growing in the understory at the Duke Forest free-air CO_2 enrichment experiment in Durham, North Carolina, USA ... in the 6th, 7th, and 8th years of the CO_2 exposure." Surveys conducted in those years, in their words, "revealed that elevated CO_2 [to 200 ppm above ambient] significantly reduced disease incidence, with 22%, 27% and 8% fewer saplings and 14%, 4%, and 5% fewer leaves infected per plant in the three consecutive years, respectively." In addition, they report that the elevated CO_2 "also significantly reduced disease severity in infected plants in all years (e.g. mean lesion area reduced 35%, 50%, and 10% in 2002, 2003, and 2004, respectively)."

What underlying mechanism or mechanisms produced these beneficent consequences? Thinking it could have been a direct deleterious effect of elevated CO_2 on the fungal pathogen, McElrone *et al.* performed some side experiments in controlled-environment chambers. However, they found that the elevated CO_2 benefited the fungal pathogen as well as the red maple saplings, observing that "exponential growth rates of *P. minima* were 17% greater under elevated CO_2." And they obtained similar results when they repeated the *in vitro* growth analysis two additional times in different growth chambers.

Taking another tack when "scanning electron micrographs verified that conidia germ tubes of *P. minima* infect *A. rubrum* leaves by entering through the stomata," the researchers turned their attention to the pathogen's mode of entry into the saplings' foliage. In this investigation they found that both stomatal size and density were unaffected by atmospheric CO_2 enrichment, but that "stomatal conductance was reduced by 21-36% under elevated CO_2, providing smaller openings for infecting germ tubes." In addition, they concluded that reduced disease severity under elevated CO_2 was also likely due to altered leaf chemistry, as elevated CO_2

increased total leaf phenolic concentrations by 15 percent and tannin concentrations by 14 percent.

Because the phenomena they found to be important in reducing the amount and severity of fungal pathogen infection (leaf spot disease) of red maple have been demonstrated to be operative in most other plants as well, McElrone *et al.* say these CO_2-enhanced leaf defensive mechanisms "may be prevalent in many plant pathosystems where the pathogen targets the stomata." Indeed, they state that their results "provide concrete evidence for a potentially generalizable mechanism to predict disease outcomes in other pathosystems under future climatic conditions."

Matros *et al.* (2006) grew tobacco plants (*Nicotiana tabacum* L.) in 16-cm-diameter pots filled with quartz sand in controlled-climate chambers maintained at either 350 or 1,000 ppm CO_2 for a period of eight weeks, where they were irrigated daily with a complete nutrient solution containing either 5 or 8 mM NH_4NO_3. In addition, some of the plants in each treatment were mechanically infected with the *potato virus Y* (PVY) when they were six weeks old. Then, at the end of the study, the plants were harvested and a number of their chemical constitutes identified and quantified.

This work revealed, in the researchers words, that "plants grown at elevated CO_2 and 5 mM NH_4NO_3 showed a marked and significant decrease in content of nicotine in leaves as well as in roots," while at 8 mM NH_4NO_3 the same was found to be true of upper leaves but not of lower leaves and roots. With respect to the PVY part of the study, they further report that the "plants grown at high CO_2 showed a markedly decreased spread of virus." Both of these findings would likely be considered beneficial by most people, as *potato virus Y* is an economically important virus that infects many crops and ornamental plants throughout the world, while nicotine is nearly universally acknowledged to have significant negative impacts on human health (Topliss *et al.*, 2002).

Braga *et al.* (2006) conducted three independent experiments where they grew well-watered soybean (*Glycine max* (L.) Merr) plants of two cultivars (IAC-14, susceptible to stem canker disease, and IAC-18, resistant to stem canker disease) from seed through the cotyledon stage in five-liter pots placed within open-top chambers maintained at atmospheric CO_2 concentrations of either 360 or 720 ppm in a glasshouse, while they measured various plant properties and processes, concentrating on the production of *glyceollins* (the major *phytoalexins*, or

anti-microbial compounds, produced in soybeans) in response to the application of ß-glucan elicitor (derived from mycelial walls of *Phythophthora sojae*) to carefully created and replicated wounds in the surfaces of several soybean cotyledons. In doing so, they found that the IAC-14 cultivar did *not* exhibit a CO_2-induced change in glyceollin production in response to elicitation—as Braga *et al.* had hypothesized would be the case, since this cultivar is susceptible to stem canker disease—but they found that the IAC-18 cultivar (which has the potential to resist the disease to varying degrees) experienced a 100 percent CO_2-induced increase in the amount of glyceollins produced after elicitation, a response the researchers described as *remarkable*. As for its significance, Braga *et al.* say the CO_2-induced response they observed "may increase the potential of the soybean defense since infection at early stages of plant development, followed by a long incubation period before symptoms appear, is characteristic of the stem canker disease cycle caused by Dpm [*Diaporthe phaseolorum* (Cooke & Ellis) Sacc. f. sp. *meridionalis* Morgan-Jones]." They say the response they observed "indicates that raised CO_2 levels forecasted for next decades may have a real impact on the defensive chemistry of the cultivars."

Last, in a study conducted within the BioCON (Biodiversity, Carbon dioxide, and Nitrogen effects on ecosystem functioning) FACE facility located at the Cedar Creek Natural History Area in east-central Minnesota, USA, Strengbom and Reich (2006) evaluated the effects of an approximate 190-ppm increase in the air's daytime CO_2 concentration on leaf photosynthetic rates of stiff goldenrod (*Solidago rigida*) growing in monoculture for two full seasons, together with its concomitant effects on the incidence and severity of leaf spot disease. Although they found that elevated CO_2 had no significant effect on plant photosynthetic rate in their study, they report that "both disease incidence and severity were lower on plants grown under elevated CO_2." More specifically, they found that "disease incidence was on average *more than twice as high* [our italics] under ambient as under elevated CO_2," and that "disease severity (proportion of leaf area with lesions) was on average 67% lower under elevated CO_2 compared to ambient conditions."

In discussing their results, Strengbom and Reich say the "indirect effects from elevated CO_2, i.e., lower disease incidence, had a stronger effect on realized photosynthetic rate than the direct effect of higher CO_2," which as noted above was negligible in their study. They conclude "it may be necessary to consider potential changes in susceptibility to foliar diseases to correctly estimate the effects on plant photosynthetic rates of elevated CO_2." In addition, they note that the plants grown in CO_2-enriched air had lower leaf nitrogen concentrations than the plants grown in ambient air, as is often observed in studies of this type; and they say that their results "are, thus, also in accordance with other studies that have found reduced pathogen performance following reduced nitrogen concentration in plants grown under elevated CO_2 (Thompson and Drake, 1994)." What is more, they conclude that their results are "also in accordance with studies that have found increased [disease] susceptibility following increased nitrogen concentration of host plants (Huber and Watson, 1974; Nordin et al., 1998; Strengbom et al., 2002)." It is possible, therefore, that the ongoing rise in the air's CO_2 content may help many plants of the future reduce the deleterious impacts of various pathogenic fungal diseases that currently beset them, thereby enabling them to increase their productivities above and beyond what is typically provided by the more direct growth stimulation resulting from the aerial fertilization effect of elevated atmospheric CO_2 concentrations.

In summation, the bulk of the available data shows atmospheric CO_2 enrichment asserts its greatest positive influence on infected as opposed to healthy plants. Moreover, it would appear that elevated CO_2 has the ability to significantly ameliorate the deleterious effects of various stresses imposed upon plants by numerous pathogenic invaders. Consequently, as the atmosphere's CO_2 concentration continues its upward climb, earth's vegetation should be increasingly better equipped to successfully deal with pathogenic organisms and the damage they have traditionally done to mankind's crops, as well as to the plants that sustain the rest of the planet's animal life.

Additional information on this topic, including reviews of newer publications as they become available, can be found at http://www.co2science.org/subject/g/disease.php

References

Booker, F.L. 2000. Influence of carbon dioxide enrichment, ozone and nitrogen fertilization on cotton (Gossypium hirsutum L.) leaf and root composition. Plant, Cell and Environment 23: 573-583.

Bowes, G. 1993. Facing the inevitable: Plants and increasing atmospheric CO_2. Annual Review of Plant Physiology and Plant Molecular Biology 44: 309-332.

Braga, M.R., Aidar, M.P.M., Marabesi, M.A. and de Godoy, J.R.L. 2006. Effects of elevated CO_2 on the phytoalexin production of two soybean cultivars differing in the resistance to stem canker disease. Environmental and Experimental Botany 58: 85-92.

Chakraborty, S. and Datta, S. 2003. How will plant pathogens adapt to host plant resistance at elevated CO_2 under a changing climate? New Phytologist 159: 733-742.

Chakraborty, S., Pangga, I.B., Lupton, J., Hart, L., Room, P.M. and Yates, D. 2000. Production and dispersal of Colletotrichum gloeosporioides spores on Stylosanthes scabra under elevated CO_2. Environmental Pollution 108: 381-387.

Entry, J.A., Runion, G.B., Prior, S.A., Mitchell, R.J. and Rogers, H.H. 1998. Influence of CO_2 enrichment and nitrogen fertilization on tissue chemistry and carbon allocation in longleaf pine seedlings. Plant and Soil 200: 3-11.

Gamper, H., Peter, M., Jansa, J., Luscher, A., Hartwig, U.A. and Leuchtmann, A. 2004. Arbuscular mycorrhizal fungi benefit from 7 years of free air CO_2 enrichment in well-fertilized grass and legume monocultures. Global Change Biology 10: 189-199.

Gebauer, R.L., Strain, B.R. and Reynolds, J.F. 1998. The effect of elevated CO_2 and N availability on tissue concentrations and whole plant pools of carbon-based secondary compounds in loblolly pine. Oecologia 113: 29-36.

Hartley, S.E., Jones, C.G. and Couper, G.C. 2000. Biosynthesis of plant phenolic compounds in elevated atmospheric CO_2. Global Change Biology 6: 497-506.

Hibberd, J.M., Whitbread, R. and Farrar, J.F. 1996a. Effect of elevated concentrations of CO_2 on infection of barley by Erysiphe graminis. Physiological and Molecular Plant Pathology 48: 37-53.

Hibberd, J.M., Whitbread, R. and Farrar, J.F. 1996b. Effect of 700 μmol per mol CO_2 and infection of powdery mildew on the growth and partitioning of barley. New Phytologist 134: 309-345.

Huber, D.M. and Watson, R.D. 1974. Nitrogen form and plant disease. Annual Reviews of Phytopathology 12: 139-155.

IPCC. 2007-II. Climate Change 2007: Impacts, Adaptation and Vulnerability. Contribution of Working Group II to the Fourth Assessment Report of the Intergovernmental Panel on Climate Change, M.L. Parry, O.F. Canziani, J.P.

Palutikof, P.J. van der Linden and C.E. Hanson (Eds.) Cambridge University Press, Cambridge, UK.

Jwa, N.-S. and Walling, L.L. 2001. Influence of elevated CO_2 concentration on disease development in tomato. *New Phytologist* **149**: 509-518.

King, J.S., Thomas, R.B. and Strain, B.R. 1997. Morphology and tissue quality of seedling root systems of *Pinus taeda* and *Pinus ponderosa* as affected by varying CO_2, temperature, and nitrogen. *Plant and Soil* **195**: 107-119.

Klironomos, J.N., Rillig, M.C., Allen, M.F., Zak, D.R., Kubiske, M. and Pregitzer, K.S. 1997. Soil fungal-arthropod responses to *Populus tremuloides* grown under enriched atmospheric CO_2 under field conditions. *Global Change Biology* **3**: 473-478.

Malmstrom, C.M. and Field, C.B. 1997. Virus-induced differences in the response of oat plants to elevated carbon dioxide. *Plant, Cell and Environment* **20**: 178-188.

Matros, A., Amme, S., Kettig, B., Buck-Sorlin, G.H., Sonnewald, U. and Mock, H.-P. 2006. Growth at elevated CO_2 concentrations leads to modified profiles of secondary metabolites in tobacco cv. SamsunNN and to increased resistance against infection with *potato virus Y*. *Plant, Cell and Environment* **29**: 126-137.

McElrone, A.J., Reid, C.D., Hoye, K.A., Hart, E. and Jackson, R.B. 2005. Elevated CO_2 reduces disease incidence and severity of a red maple fungal pathogen via changes in host physiology and leaf chemistry. *Global Change Biology* **11**: 1828-1836.

Nordin, A., Nasholm, T. and Ericson, L. 1998. Effects of simulated N deposition on understorey vegetation of a boreal coniferous forest. *Functional Ecology* **12**: 691-699.

Owensby, C.E. 1994. Climate change and grasslands: ecosystem-level responses to elevated carbon dioxide. *Proceedings of the XVII International Grassland Congress.* Palmerston North, New Zealand: New Zealand Grassland Association, pp. 1119-1124.

Pangga, I.B., Chakraborty, S. and Yates, D. 2004. Canopy size and induced resistance in *Stylosanthes scabra* determine anthracnose severity at high CO_2. *Phytopathology* **94**: 221-227.

Parsons, W.F.J., Kopper, B.J. and Lindroth, R.L. 2003. Altered growth and fine root chemistry of *Betula papyrifera* and *Acer saccharum* under elevated CO_2. *Canadian Journal of Forest Research* **33**: 842-846.

Runion, G.B., Entry, J.A., Prior, S.A., Mitchell, R.J. and Rogers, H.H. 1999. Tissue chemistry and carbon allocation in seedlings of *Pinus palustris* subjected to elevated atmospheric CO_2 and water stress. *Tree Physiology* **19**: 329-335.

Shaner, G., Stromberg, E.L., Lacy, G.H., Barker, K.R. and Pirone, T.P. 1992. Nomenclature and concepts of aggressiveness and virulence. *Annual Review of Phytopathology* **30**: 47-66.

Smith, S.E. and Read, D.J. 1997. *Mycorrhizal Symbioses.* Academic Press, London, UK.

Strengbom, J. and Reich, P.B. 2006. Elevated [CO_2] and increased N supply reduce leaf disease and related photosynthetic impacts on *Solidago rigida*. *Oecologia* **149**: 519-525.

Strengbom, J., Nordin, A., Nasholm, T. and Ericson, L. 2002. Parasitic fungus mediates change in nitrogen-exposed boreal forest vegetation. *Journal of Ecology* **90**: 61-67.

Thompson, G.B. and Drake, B.G. 1994. Insect and fungi on a C3 sedge and a C4 grass exposed to elevated atmospheric CO_2 concentrations in open-top chambers in the field. *Plant, Cell and Environment* **17**: 1161-1167.

Tiedemann, A.V. and Firsching, K.H. 2000. Interactive effects of elevated ozone and carbon dioxide on growth and yield of leaf rust-infected versus non-infected wheat. *Environmental Pollution* **108**: 357-363.

Topliss, J.G., Clark, A.M., Ernst, E. *et al.* 2002. Natural and synthetic substances related to human health. *Pure and Applied Chemistry* **74**: 1957-1985.

7.3.2. Herbivory

Insect pests have greatly vexed earth's plants in the past and will likely continue to do so in the future. It is possible, however, that the ongoing rise in the atmosphere's CO_2 content may affect this phenomenon, for better or for worse. In this section we review the results of several studies that have addressed this subject as it applies to herbaceous and woody plants.

Additional information on this topic, including reviews on herbivory not discussed here, can be found at http://www.co2science.org/subject/h/subject_h.php under the heading Herbivory.

7.3.2.1. Herbaceous Plants

Kerslake *et al.* (1998) grew five-year-old heather (*Calluna vulgaris*) plants collected from a Scottish moor in open-top chambers maintained at atmospheric CO_2 concentrations of 350 and 600 ppm. At two different times during the study, larvae of the destructive winter moth *Operophtera brumata*—whose outbreaks periodically cause extensive damage to heather moorland—were allowed to feed upon current-year shoots. Feeding upon the high-CO_2-grown foliage did not affect larval growth rates, development, or final pupal weights; neither was moth survivorship significantly altered. The authors concluded that their study "provides no evidence that increasing atmospheric CO_2 concentrations will affect the potential for outbreak of *Operophtera brumata* on this host." What it did show, however, was a significant CO_2-induced increase in heather water use efficiency.

Newman *et al.* (1999) inoculated tall fescue (*Festuca arundinacea*) plants growing in open-top chambers maintained at atmospheric CO_2 concentrations of 350 and 700 ppm with bird cherry-oat aphids (*Rhopalosiphum padi*). After nine weeks, the plants growing in the CO_2-enriched air had experienced a 37 percent increase in productivity and were covered with many fewer aphids than the plants growing in ambient air.

Goverde *et al.* (1999) collected four genotypes of *Lotus corniculatus* near Paris and grew them in controlled environment chambers kept at atmospheric CO_2 concentrations of 350 and 700 ppm. Larvae of the Common Blue Butterfly (*Polyommatus icarus*) that were allowed to feed upon the foliage produced in the CO_2-enriched air ate more, grew larger, and experienced shorter development times than larvae feeding on the foliage produced in the ambient-air treatment, suggesting that this butterfly species will likely become more robust and plentiful as the air's CO_2 content continues to rise.

Brooks and Whittaker (1999) removed grassland monoliths containing eggs of the xylem-feeding spittlebug *Neophilaenus lineatus* from the UK's Great Dun Fell in Cumbria and placed them in glasshouses maintained at atmospheric CO_2 concentrations of 350 and 600 ppm for two years. Survival of the spittlebug's nymphal states was reduced by 24 percent in both of the generations produced in their experiment, suggesting that this particular insect will likely cause less tissue damage to the plants of this species-poor grassland in a CO_2-enriched world of the future.

Joutei *et al.* (2000) grew bean (*Phaseolus vulgaris*) plants in controlled environments kept at atmospheric CO_2 concentrations of 350 and 700 ppm, to which they introduced the destructive agricultural mite *Tetranychus urticae*, observing that female mites produced 34 percent and 49 percent less offspring in

the CO_2-enriched chambers in their first and second generations, respectively. This CO_2-induced reduction in the reproductive success of this invasive insect, which negatively affects more than 150 crop species worldwide, bodes well for mankind's ability to grow the food we will need to feed our growing numbers in the years ahead.

Peters *et al.* (2000) fed foliage derived from FACE plots of calcareous grasslands of Switzerland (maintained at 350 and 650 ppm CO_2) to terrestrial slugs, finding they exhibited no preference with respect to the CO_2 treatment from which the foliage was derived. Also, in a study that targeted no specific insect pest, Castells *et al.* (2002) found that a doubling of the air's CO_2 content enhanced the total phenolic concentrations of two Mediterranean perennial grasses (*Dactylis glomerata* and *Bromus erectus*) by 15 percent and 87 percent, respectively, which compounds tend to enhance plant defensive and resistance mechanisms to attacks by both herbivores and pathogens.

Coviella and Trumbel (2000) determined that toxins produced by *Bacillus thuringiensis* (Bt), which are applied to crop plants by spraying as a means of combating various crop pests, were "more efficacious" in cotton grown in an elevated CO_2 environment than in ambient air, which is a big plus for modern agriculture. In addition, Coviella *et al.* (2000) determined that "elevated CO_2 appears to eliminate differences between transgenic [Bt-containing] and nontransgenic plants for some key insect developmental/fitness variables including length of the larval stage and pupal weight."

In summary, the majority of evidence that has been accumulated to date suggests that rising atmospheric CO_2 concentrations may reduce the frequency and severity of pest outbreaks that are detrimental to agriculture, while not seriously impacting herbivorous organisms found in natural ecosystems that are normally viewed in a more favorable light.

Additional information on this topic, including reviews of newer publications as they become available, can be found at http://www.co2science.org/subject/h/herbivoresherbplants.php.

References

Brooks, G.L. and Whittaker, J.B. 1999. Responses of three generations of a xylem-feeding insect, *Neophilaenus lineatus* (Homoptera), to elevated CO_2. *Global Change Biology* **5**: 395-401.

Castells, E., Roumet, C., Penuelas, J. and Roy, J. 2002. Intraspecific variability of phenolic concentrations and their responses to elevated CO_2 in two mediterranean perennial grasses. *Environmental and Experimental Botany* **47**: 205-216.

Coviella, C.E. and Trumble, J.T. 2000. Effect of elevated atmospheric carbon dioxide on the use of foliar application of *Bacillus thuringiensis*. *BioControl* **45**: 325-336.

Coviella, C.E., Morgan, D.J.W. and Trumble, J.T. 2000. Interactions of elevated CO_2 and nitrogen fertilization: Effects on production of *Bacillus thuringiensis* toxins in transgenic plants. *Environmental Entomology* **29**: 781-787.

Goverde, M., Bazin, A., Shykoff, J.A. and Erhardt, A. 1999. Influence of leaf chemistry of *Lotus corniculatus* (Fabaceae) on larval development of *Polyommatus icarus* (Lepidoptera, Lycaenidae): effects of elevated CO_2 and plant genotype. *Functional Ecology* **13**: 801-810.

Joutei, A.B., Roy, J., Van Impe, G. and Lebrun, P. 2000. Effect of elevated CO_2 on the demography of a leaf-sucking mite feeding on bean. *Oecologia* **123**: 75-81.

Kerslake, J.E., Woodin, S.J. and Hartley, S.E. 1998. Effects of carbon dioxide and nitrogen enrichment on a plant-insect interaction: the quality of *Calluna vulgaris* as a host for *Operophtera brumata*. *New Phytologist* **140**: 43-53.

Newman, J.A., Gibson, D.J., Hickam, E., Lorenz, M., Adams, E., Bybee, L. and Thompson, R. 1999. Elevated carbon dioxide results in smaller populations of the bird cherry-oat aphid *Rhopalosiphum padi*. *Ecological Entomology* **24**: 486-489.

Peters, H.A., Baur, B., Bazzaz, F. and Korner, C. 2000. Consumption rates and food preferences of slugs in a calcareous grassland under current and future CO_2 conditions. *Oecologia* **125**: 72-81.

7.3.2.2. Woody Plants

7.3.2.2.1. Maple

Working with *Acer rubrum* saplings beginning their fourth year of growth in open-top chambers maintained at four different atmospheric CO_2/temperature conditions—(1) ambient temperature, ambient CO_2, (2) ambient temperature, elevated CO_2 (ambient + 300 ppm), (3) elevated temperature (ambient + 3.5°C), ambient CO_2, and (4) elevated temperature, elevated CO_2—Williams *et al.* (2003) bagged first instar gypsy moth larvae on various branches of the trees and observed their behavior. The data they obtained demonstrated, in

their words, "that larvae feeding on CO_2-enriched foliage ate a comparably poorer food source than those feeding on ambient CO_2-grown plants, irrespective of temperature," and that there was a minor reduction in leaf water content due to CO_2 enrichment. Nevertheless, they found the "CO_2-induced reductions in foliage quality (e.g. nitrogen and water) were unrelated [our italics] to insect mortality, development rate and pupal weight," and that these and any other phytochemical changes that may have occurred "resulted in no negative effects on gypsy moth performance." They also found that "irrespective of CO_2 concentration, on average, male larvae pupated 7.5 days earlier and female larvae 8 days earlier at elevated temperature," and noting that anything that prolongs the various development stages of insects potentially exposes them to greater predation and parasitism risk, they concluded that the observed temperature-induced hastening of the insects' development would likely expose them to *less* predation and parasitism risk.

One year later, Hamilton *et al.* (2004) began the report of their study of this important subject by noting that many single-species investigations have suggested that increases in atmospheric CO_2 will increase herbivory (Bezemer and Jones, 1998; Cannon, 1998; Coviella and Trumble, 1999; Hunter, 2001; Lincoln *et al.*, 1993; Whittaker, 1999). However, because there are so many feedbacks and complex interactions among the numerous components of real-world ecosystems, they warned that one ought not put too much faith in these predictions until relevant real-world ecosystem-level experiments have been completed.

In one such study they conducted at the Duke Forest FACE facility near Chapel Hill, North Carolina, USA, Hamilton *et al.* "measured the amount of leaf tissue damaged by insects and other herbivorous arthropods during two growing seasons in a deciduous forest understory continuously exposed to ambient (360 ppm) and elevated (560 ppm) CO_2 conditions." This forest is dominated by loblolly pine trees that account for fully 92 percent of the ecosystem's total woody biomass. In addition, it contains 48 species of other woody plants (trees, shrubs, and vines) that have naturally established themselves in the forest's understory. In their study of this ecosystem, Hamilton *et al.* quantified the loss of foliage due to herbivory that was experienced by three deciduous tree species, one of which was *Acer rubrum*.

421

As Hamilton *et al.* describe it, "we found that elevated CO_2 led to a trend toward *reduced herbivory* [our italics] in [the] deciduous understory in a situation that included the full complement of naturally occurring plant and insect species." In 1999, for example, they determined that "elevated CO_2 reduced overall herbivory by more than 40 percent," while in 2000 they say they observed "the same pattern and magnitude of reduction."

With respect to changes in foliage properties that might have been expected to lead to increases in herbivory, Hamilton *et al.* report they "found no evidence for significant changes in leaf nitrogen, C/N ratio, sugar, starch or total leaf phenolics in either year of [the] study," noting that these findings agree with those of "another study performed at the Duke Forest FACE site that also found no effect of elevated CO_2 on the chemical composition of leaves of understory trees (Finzi and Schlesinger, 2002)."

Hamilton *et al.* thus concluded their landmark paper by emphasizing that "despite the large number of studies that predict increased herbivory, particularly from leaf chewers, under elevated CO_2, our study found a trend toward reduced herbivory two years in a row." In addition, they note that their real-world results "agree with the only other large-scale field experiment that quantified herbivory for a community exposed to elevated CO_2 (Stiling *et al.*, 2003)."

Consequently, and in spite of the predictions of increased destruction of natural ecosystems by insects and other herbivorous arthropods in a CO_2-enriched world of the future, just the opposite would appear to be the more likely outcome; i.e., greater plant productivity plus less foliage consumption by herbivores, "whether expressed on an absolute or a percent basis," as Hamilton *et al.* found to be the case in their study.

In another study conducted at the same site, Knepp *et al.* (2005) quantified leaf damage by chewing insects on saplings of seven species (including *Acer rubrum*) in 2001, 2002, and 2003, while five additional species (including *Acer barbatum*) were included in 2001 and 2003. This work revealed, in their words, that "across the seven species that were measured in each of the three years, elevated CO_2 caused a reduction in the percentage of leaf area removed by chewing insects," which was such that "the percentage of leaf tissue damaged by insect herbivores was 3.8 percent per leaf under ambient CO_2 and 3.3 percent per leaf under elevated CO_2." Greatest effects were observed in 2001, when

they report that "across 12 species the average damage per leaf under ambient CO_2 was 3.1 percent compared with 1.7 percent for plants under elevated CO_2," which was "indicative of a 46 percent decrease in the total area and total mass of leaf tissue damaged by chewing insects in the elevated CO_2 plots."

What was responsible for these welcome results? Knepp *et al.* say that "given the consistent reduction in herbivory under high CO_2 across species in 2001, it appears that some universal feature of chemistry or structure that affected leaf suitability was altered by the treatment." Another possibility they discuss is that "forest herbivory may decrease under elevated CO_2 because of a decline in the abundance of chewing insects," citing the observations of Stiling *et al.* (2002) to this effect and noting that "slower rates of development under elevated CO_2 prolongs the time that insect herbivores are susceptible to natural enemies, which may be abundant in open-top chambers and FACE experiments but absent from greenhouse experiments." In addition, they suggest that "decreased foliar quality and increased per capita consumption under elevated CO_2 may increase exposure to toxins and insect mortality," also noting that "CO_2-induced changes in host plant quality directly decrease insect fecundity," citing the work of Coviella and Trumble (1999) and Awmack and Leather (2002).

So what's the bottom line with respect to the outlook for earth's forests, and especially its maple trees, in a high-CO_2 world of the future? In their concluding paragraph, Knepp *et al.* say that "By contrast to the view that herbivore damage will increase under elevated CO_2 as a result of compensatory feeding on lower quality foliage, our results and those of Stiling *et al.* (2002) and Hamilton *et al.* (2004) in open experimental systems suggest that damage to trees may decrease."

But what if herbivore-induced damage in fact increases in a future CO_2-enriched world? The likely answer is provided by the work of Kruger *et al.* (1998), who grew well-watered and fertilized one-year-old *Acer saccharum* saplings in glasshouses maintained at atmospheric CO_2 concentrations of either 356 or 645 ppm for 70 days to determine the effects of elevated CO_2 on photosynthesis and growth. In addition, on the 49th day of differential CO_2 exposure, 50 percent of the saplings' leaf area was removed from half of the trees in order to study the impact of simulated herbivory. This protocol revealed that the 70-day CO_2 enrichment treatment increased the total dry weight of the non-defoliated seedlings by about 10 percent. When the trees were stressed by simulated herbivory, however, the CO_2-enriched maples produced 28 percent more dry weight over the final phase of the study than the maples in the ambient-air treatment did. This result thus led Kruger *et al.* to conclude that in a high-CO_2 world of the future "sugar maple might be more capable of tolerating severe defoliation events which in the past have been implicated in widespread maple declines."

It appears that earth's maple trees—and probably many, if not most, other trees—will fare much better in the future with respect to the periodic assaults of leaf-damaging herbivores, as the air's CO_2 content continues its upward climb.

Additional information on this topic, including reviews of newer publications as they become available, can be found at http://www.co2science.org/subject/h/herbivoresmaple.php.

References

Awmack, C.S. and Leather, S.R. 2002. Host plant quality and fecundity in herbivorous insects. *Annual Review of Entomology* **47**: 817-844.

Bezemer, T.M. and Jones, T.H. 1998. Plant-insect herbivore interactions in elevated atmospheric CO_2: quantitative analyses and guild effects. *Oikos* **82**: 212-222.

Cannon, R.J. 1998. The implications of predicted climate change for insect pests in the UK, with emphasis on non-indigenous species. *Global Change Biology* **4**: 785-796.

Coviella, C.E. and Trumble, J.T. 1999. Effects of elevated atmospheric carbon dioxide on insect-plant interactions. *Conservation Biology* **13**: 700-712.

Finzi, A.C. and Schlesinger, W.H. 2002. Species control variation in litter decomposition in a pine forest exposed to elevated CO_2. *Global Change Biology* **8**: 1217-1229.

Hamilton, J.G., Zangerl, A.R., Berenbaum, M.R., Pippen, J., Aldea, M. and DeLucia, E.H. 2004. Insect herbivory in an intact forest understory under experimental CO_2 enrichment. *Oecologia* **138**: 10.1007/s00442-003-1463-5.

Hunter, M.D. 2001. Effects of elevated atmospheric carbon dioxide on insect-plant interactions. *Agricultural and Forest Entomology* **3**: 153-159.

Knepp, R.G., Hamilton, J.G., Mohan, J.E., Zangerl, A.R., Berenbaum, M.R. and DeLucia, E.H. 2005. Elevated CO_2 reduces leaf damage by insect herbivores in a forest community. *New Phytologist* **167**: 207-218.

Kruger, E.L., Volin, J.C. and Lindroth, R.L. 1998. Influences of atmospheric CO_2 enrichment on the responses of sugar maple and trembling aspen to defoliation. *New Phytologist* **140**: 85-94.

Lincoln, D.E., Fajer, E.D. and Johnson, R.H. 1993. Plant-insect herbivore interactions in elevated CO_2 environments. *Trends in Ecology and Evolution* **8**: 64-68.

Stiling, P., Cattell, M., Moon, D.C., Rossu, A., Hungate, B.A., Hymuss, G. and Drake, B. 2002. Elevated atmospheric CO_2 lowers herbivore abundance, but increases leaf abscission rates. *Global Change Biology* **8**: 658-667.

Stiling, P., Moon, D.C., Hunter, M.D., Colson, J., Rossi, A.M., Hymus, G.J. and Drake, B.G. 2003. Elevated CO_2 lowers relative and absolute herbivore density across all species of a scrub-oak forest. *Oecologia* **134**: 82-87.

Whittaker, J.B. 1999. Impacts and responses at population level of herbivorous insects to elevated CO_2. *European Journal of Entomology* **96**: 149-156.

Williams, R.S., Lincoln, D.E. and Norby, R.J. 2003. Development of gypsy moth larvae feeding on red maple saplings at elevated CO_2 and temperature. *Oecologia* **137**: 114-122.

7.3.2.2.2. Oak

Dury *et al.* (1998) grew four-year-old *Quercus robur* seedlings in pots in greenhouses maintained at ambient and twice-ambient atmospheric CO_2 concentrations in combination with ambient and elevated (ambient plus 3°C) air temperatures for approximately one year to study the interactive effects of elevated CO_2 and temperature on leaf nutritional quality. In doing so, they found that the elevated air temperature treatment significantly reduced leaf palatability, and that leaf toughness increased as a consequence of temperature-induced increases in condensed tannin concentrations. In addition, the higher temperatures significantly reduced leaf nitrogen content, while elevated CO_2 caused a temporary increase in leaf phenolic concentrations and a decrease in leaf nitrogen content.

In one of the first attempts to move outside the laboratory/greenhouse and study the effects of atmospheric CO_2 enrichment on trophic food webs in a natural ecosystem, Stiling *et al.* (1999) enclosed portions of a native scrub-oak community in Florida (USA) within 3.6-m-diameter open-top chambers and fumigated them with air having CO_2 concentrations of either 350 or 700 ppm for approximately one year, in order to see if elevated CO_2 would impact leaf miner densities, feeding rates, and mortality in this nutrient-poor ecosystem.

Adherence to this protocol led to the finding that total leaf miner densities were 38 percent less on the foliage of trees growing in CO_2-enriched air than on the foliage of trees growing in ambient air. In addition, atmospheric CO_2 enrichment consistently reduced the absolute numbers of the study's six leaf miner species. At the same time, however, the elevated CO_2 treatment increased the leaf area consumed by the less abundant herbivore miners by approximately 40 percent relative to the areas mined by the more abundant herbivores present on the foliage exposed to ambient air; but in spite of this increase in feeding, the leaf miners in the CO_2-enriched chambers experienced significantly greater mortality than those in the ambient-air chambers. Although CO_2-induced reductions in leaf nitrogen content played a role in this phenomenon, the greatest factor contributing to increased herbivore mortality was a four-fold increase in parasitization by various wasps, which could more readily detect the more-exposed leaf miners on the CO_2-enriched foliage.

If extended to agricultural ecosystems, these findings suggest that crops may experience less damage from such herbivores in a high-CO_2 world of the future, thus increasing potential harvest and economic gains. In addition, with reduced numbers of leaf miners in CO_2-enriched air, farmers could reduce their dependency upon chemical pesticides to control them.

In another study conducted on five scrub-oak forest species at the same experimental facility, Stiling *et al.* (2003) investigated the effects of an approximate doubling of the air's CO_2 concentration on a number of characteristics of several insect herbivores. As before, they found that the "relative levels of damage by the two most common herbivore guilds, leaf-mining moths and leaf-chewers (primarily larval lepidopterans and grasshoppers), were significantly lower in elevated CO_2 than in ambient CO_2, for all five plant species," *and* they found that "the response to elevated CO_2 was the same across all plant species." In addition, they report that "more host-plant induced mortality was found for all miners on all plants in elevated CO_2 than in ambient CO_2." These effects were so powerful that in addition to the relative densities of insect herbivores being reduced in the CO_2-enriched chambers, and "even though there were more leaves of most plant species in the elevated CO_2 chambers," the total densities of leaf miners in the high-CO_2 chambers were also lower for all plant species. Consequently, it would appear that in a high-CO_2 world of the future, many of earth's plants may be able to better withstand the onslaughts of various insect pests that have plagued them in the past. Another intriguing implication of this finding, as Stiling *et al.* note, is that "reductions in herbivore loads in elevated CO_2 could boost plant growth beyond what might be expected based on pure plant responses to elevated CO_2."

Continuing to investigate the same ecosystem, which is dominated by two species of scrub oak (*Quercus geminata* and *Q. myrtifolia*) that account for more than 90 percent of the ecosystem's biomass, and focusing on the abundance of a guild of lepidopteran leafminers that attack the leaves of *Q. myrtifolia*, as well as various leaf chewers that also like to munch on this species, Rossi *et al.* (2004) followed 100 marked leaves in each of 16 open-top chambers (half exposed to ambient air and half exposed to air containing an extra 350 ppm of CO_2) for a total of nine months, after which, in their words, "differences in mean percent of leaves with leafminers and chewed leaves on trees from ambient and elevated chambers were assessed using paired *t*-tests."

In reporting their findings the researchers wrote that "both the abundance of the guild of leafmining lepidopterans and damage caused by leaf chewing insects attacking myrtle oak were depressed in elevated CO_2." Specifically, they found that leafminer abundance was 44 percent lower ($P = 0.096$) in the CO_2-enriched chambers compared to the ambient-air chambers, and that the abundance of leaves suffering chewing damage was 37 percent lower ($P = 0.072$) in the CO_2-enriched air. The implications of these findings are obvious: Myrtle oak trees growing in their natural habitat will likely suffer less damage from both leaf miners and leaf chewers as the air's CO_2 concentration continues to rise in the years and decades ahead.

Still concentrating on the same ecosystem, where atmospheric enrichment with an extra 350 ppm of CO_2 was begun in May 1996, Hall *et al.* (2005b) studied the four species that dominate the community and are present in every experimental chamber: the three oaks (*Quercus myrtifolia*, *Q. chapmanii* and *Q. geminata*) plus the nitrogen-fixing legume *Galactia elliottii*. At three-month intervals from May 2001 to May 2003, undamaged leaves were removed from each of these species in all chambers and analyzed for various chemical constituents, while 200 randomly selected leaves of each species in each chamber were scored for the presence of six types of herbivore damage.

Throughout the study there were no significant differences between the CO_2-enriched and ambient-treatment leaves of any single species in terms of either condensed tannins, hydrolyzable tannins, total phenolics, or lignin. However, in all four species together there were always greater concentrations of all four leaf constituents in the CO_2-enriched leaves, with accross-species mean increases of 6.8 percent for condensed tannins, 6.1 percent for hydrolyzable tannins, 5.1 percent for total phenolics, and 4.3 percent for lignin. In addition, there were large and often significant CO_2-induced decreases in all leaf damage categories among all species: chewing (-48 percent, $P < 0.001$), mines (-37 percent, $P = 0.001$), eye spot gall (-45 percent, $P < 0.001$), leaf tier (-52 percent, $P = 0.012$), leaf mite (-23 percent, $P = 0.477$), and leaf gall (-16 percent, $P = 0.480$). Hall *et al.* thus concluded that the changes they observed in leaf chemical constituents and herbivore damage "suggest that damage to plants may decline as atmospheric CO_2 levels continue to rise."

In one additional study to come out of the Florida scrub-oak ecosystem, Hall *et al.* (2005a) studied the effects of an extra 350 ppm of CO_2 on litter quality, herbivore activity and their interactions. Over the three years of this experiment (2000, 2001, 2002), they determined that "changes in litter chemistry from year to year were far larger than effects of CO_2 or insect damage, suggesting that these may have only minor effects on litter decomposition." The one exception to this finding, in their words, was that "condensed tannin concentrations increased under elevated CO_2 regardless of species, herbivore damage, or growing season," rising by 11 percent in 2000, 18 percent in 2001, and 41 percent in 2002 as a result of atmospheric CO_2 enrichment, as best we can determine from their bar graphs. Also, the five researchers report that "lepidopteran larvae can exhibit slower growth rates when feeding on elevated CO_2 plants (Fajer *et al.*, 1991) and become more susceptible to pathogens, parasitoids, and predators (Lindroth, 1996; Stiling *et al.*, 1999)," noting further that at their field site, "which hosts the longest continuous study of the effects of elevated CO_2 on insects, herbivore populations decline[d] markedly under elevated CO_2 (Stiling *et al.*, 1999, 2002, 2003; Hall *et al.*, 2005b)."

In conclusion, from the evidence accumulated to date with respect to herbivory in oak trees, it would appear that ever less damage will be done to such trees by various insect pests as the air's CO_2 concentration continues to climb ever higher.

Additional information on this topic, including reviews of newer publications as they become available, can be found at http://www.co2science.org/subject/h/herbivoreswoodoak.php.

References

Dury, S.J., Good, J.E.G., Perrins, C.M., Buse, A. and Kaye, T. 1998. The effects of increasing CO_2 and temperature on oak leaf palatability and the implications for herbivorous insects. *Global Change Biology* **4**: 55-61.

Fajer, E.D., Bowers, M.D. and Bazzaz, F.A. 1991. The effects of enriched CO_2 atmospheres on the buckeye butterfly, *Junonia coenia*. *Ecology* **72**: 751-754.

Hall, M.C., Stiling, P., Hungate, B.A., Drake, B.G. and Hunter, M.D. 2005a. Effects of elevated CO_2 and herbivore damage on litter quality in a scrub oak ecosystem. *Journal of Chemical Ecology* **31**: 2343-2356.

Hall, M.C., Stiling, P., Moon, D.C., Drake, B.G. and Hunter, M.D. 2005b. Effects of elevated CO_2 on foliar quality and herbivore damage in a scrub oak ecosystem. *Journal of Chemical Ecology* **31**: 267-285.

Lindroth, R.L. 1996. CO_2-mediated changes in tree chemistry and tree-Lepidoptera interactions. In: Koch, G.W. and Mooney, H.A. (Eds.) *Carbon Dioxide and Terrestrial Ecosystems*. Academic Press, San Diego, California, USA, pp. 105-120.

Rossi, A.M., Stiling, P., Moon, D.C., Cattell, M.V. and Drake, B.G. 2004. Induced defensive response of myrtle oak to foliar insect herbivory in ambient and elevated CO_2. *Journal of Chemical Ecology* **30**: 1143-1152.

Stiling, P., Cattell, M., Moon, D.C., Rossi, A., Hungate, B.A., Hymus, G. and Drake, B.G. 2002. Elevated atmospheric CO_2 lowers herbivore abundance, but increases leaf abscission rates. *Global Change Biology* **8**: 658-667.

Stiling, P., Moon, D.C., Hunter, M.D., Colson, J., Rossi, A.M., Hymus, G.J. and Drake, B.G. 2003. Elevated CO_2 lowers relative and absolute herbivore density across all species of a scrub-oak forest. *Oecologia* **134**: 82-87.

Stiling, P., Rossi, A.M., Hungate, B., Dijkstra, P., Hinkle, C.R., Knot III, W.M., and Drake, B. 1999. Decreased leaf-miner abundance in elevated CO_2: Reduced leaf quality and increased parasitoid attack. *Ecological Applications* **9**: 240-244.

7.3.2.2.3. Other

Stiling *et al.* (1999) enclosed portions of a Florida scrub-oak community in open-top chambers and maintained them at atmospheric CO_2 concentrations of 350 and 700 ppm for approximately one year, while they studied the effects of this treatment on destructive leaf miners. Among their many findings, the researchers noted that the individual areas consumed by leaf miners munching on leaves in the CO_2-enriched chambers were larger than those created by leaf miners dining on leaves in the ambient-air chambers. As a result, there was a *four-fold increase* in parasitization by various wasps that could more readily detect the more-exposed leaf miners on the CO_2-enriched foliage. Consequently, leaf miners in the elevated CO_2 chambers suffered significantly greater mortality than those in the control chambers.

In a subsequent and much expanded study of the same ecosystem, Stiling *et al.* (2002) investigated several characteristics of a number of insect herbivores found on the five species of plants that accounted for more than 98 percent of the total plant biomass within the chambers. As they describe their results, the "relative levels of damage by the two most common herbivore guilds, leaf-mining moths and leaf-chewers (primarily larval lepidopterans and grasshoppers), were significantly lower in elevated CO_2 than in ambient CO_2, for all five plant species."

In another study that did not involve herbivores, Gleadow *et al.* (1998) grew eucalyptus seedlings in glasshouses maintained at 400 and 800 ppm CO_2 for a period of six months, observing biomass increases of 98 percent and 134 percent in high and low nitrogen treatments, respectively. They also studied a sugar-based compound called *prunasin*, which produces cyanide in response to tissue damage caused by foraging herbivores. Although elevated CO_2 caused no significant change in leaf prunasin content, it was determined that the proportion of nitrogen allocated to prunasin increased by approximately 20 percent in the CO_2-enriched saplings, suggestive of a *potential* for increased prunasin production had the saplings been under attack by herbivores.

Lovelock *et al.* (1999) grew seedlings of the tropical tree *Copaifera aromatica* for 50 days in pots placed within open-top chambers maintained at atmospheric CO_2 concentrations of 390 and 860 ppm. At the 14-day point of the experiment, half of the seedlings in each treatment had about 40 percent of their total leaf area removed. In this case, none of the

defoliated trees of either CO_2 treatment fully recovered from this manipulation, but at the end of the experiment, the total plant biomass of the defoliated trees in the CO_2-enriched treatment was 15 percent greater than that of the defoliated trees in the ambient-CO_2 treatment, again attesting to the benefits of atmospheric CO_2 enrichment in helping trees to deal with herbivory.

Docherty *et al.* (1997) grew beech and sycamore saplings in glasshouses maintained at atmospheric CO_2 concentrations of 350 and 600 ppm, while groups of three sap-feeding aphid species and two sap-feeding leafhopper species were allowed to feed on them. Overall, elevated CO_2 had few significant effects on the performance of the insects, although there was a non-significant tendency for elevated CO_2 to reduce the individual weights and population sizes of the aphids.

Finally, Hattenschwiler and Schafellner (1999) grew seven-year-old spruce (*Picea abies*) trees at atmospheric CO_2 concentrations of 280, 420, and 560 ppm and various nitrogen deposition treatments for three years, allowing nun moth larvae to feed on current-year needles for a period of 12 days. Larvae placed upon the CO_2-enriched foliage consumed less needle biomass than larvae placed upon the ambiently grown foliage, regardless of nitrogen treatment. This effect was so pronounced that the larvae feeding on needles produced by the CO_2-enriched trees attained an average final biomass that was only two-thirds of that attained by the larvae that fed on needles produced at 280 ppm CO_2. Since the nun moth is a deadly defoliator that resides in most parts of Europe and East Asia between 40° and 60°N latitude and is commonly regarded as the coniferous counterpart of its close relative the gypsy moth, which feeds primarily on deciduous trees, the results of this study suggest that the ongoing rise in the air's CO_2 content will likely lead to significant reductions in damage to spruce and other coniferous trees by this voracious insect pest in the years and decades ahead.

In light of these several observations, the balance of evidence seems to suggest that earth's woody plants will be better able to deal with the challenges provided by herbivorus insects as the air's CO_2 content continues to rise.

Additional information on this topic, including reviews of newer publications as they become available, can be found at http://www.co2science.org/subject/h/herbivoreswoodyplants.php.

References

Docherty, M., Wade, F.A., Hurst, D.K., Whittaker, J.B. and Lea, P.J. 1997. Responses of tree sap-feeding herbivores to elevated CO_2. *Global Change Biology* **3**: 51-59.

Gleadow, R.M., Foley, W.J. and Woodrow, I.E. 1998. Enhanced CO_2 alters the relationship between photosynthesis and defense in cyanogenic *Eucalyptus cladocalyx* F. Muell. *Plant, Cell and Environment* **21**: 12-22.

Hattenschwiler, S. and Schafellner, C. 1999. Opposing effects of elevated CO_2 and N deposition on *Lymantria monacha* larvae feeding on spruce trees. *Oecologia* **118**: 210-217.

Lovelock, C.E., Posada, J. and Winter, K. 1999. Effects of elevated CO_2 and defoliation on compensatory growth and photosynthesis of seedlings in a tropical tree, *Copaifera aromatica*. *Biotropica* **31**: 279-287.

Stiling, P., Moon, D.C., Hunter, M.D., Colson, J., Rossi, A.M., Hymus, G.J. and Drake, B.G. 2002. Elevated CO_2 lowers relative and absolute herbivore density across all species of a scrub-oak forest. *Oecologia* **10.1007**/s00442-002-1075-5.

Stiling, P., Rossi, A.M., Hungate, B., Dijkstra, P., Hinkle, C.R., Knot III, W.M., and Drake, B. 1999. Decreased leaf-miner abundance in elevated CO_2: Reduced leaf quality and increased parasitoid attack. *Ecological Applications* **9**: 240-244.

7.3.3. Insects

As the atmosphere's CO_2 concentration climbs ever higher, it is important to determine how this phenomenon will affect the delicate balance that exists between earth's plants and the insects that feed on them. In this section we thus review what has been learned about this subject with respect to aphids, moths, and other insects.

Additional information on this topic, including reviews on insects not discussed here, can be found at http://www.co2science.org/subject/i/subject_i.php under the heading Insects.

7.3.3.1. Aphids

Docherty *et al.* (1997) grew beech and sycamore saplings in glasshouses maintained at atmospheric CO_2 concentrations of 350 and 600 ppm, while groups of three sap-feeding aphid species were allowed to feed on the saplings. Overall, the elevated CO_2 had few significant effects on aphid feeding and performance. There was, however, a non-significant tendency for elevated CO_2 to reduce the individual weights and population sizes of the aphids, suggesting that future increases in the air's CO_2 content *might* reduce aphid feeding pressures on beech and sycamore saplings, and possibly other plants as well.

Whittaker (1999) reviewed the scientific literature dealing with population responses of herbivorous insects to atmospheric CO_2 enrichment, concentrating on papers resulting from relatively long-term studies. Based on all pertinent research reports available at that time, the only herbivorous insects that exhibited population increases in response to elevated CO_2 exposure were those classified as phloem feeders, specifically, aphids. Although this finding appeared to favor aphids over plants, additional studies would complicate the issue and swing the pendulum back the other way.

Newman *et al.* (1999) grew tall fescue plants for two weeks in open-top chambers maintained at atmospheric CO_2 concentrations of 350 and 700 ppm before inoculating them with aphids (*Rhopalosiphum padi*). After nine additional weeks of differential CO_2 exposure, the plants were harvested and their associated aphids counted. Although elevated CO_2 increased plant dry matter production by 37 percent, the plants grown in air of elevated CO_2 concentration contained fewer aphids than the plants grown in ambient air.

Percy *et al.* (2002) grew the most widely distributed tree species in all of North America—trembling aspen—in twelve 30-m-diameter FACE rings in air maintained at (1) ambient CO_2 and O_3 concentrations, (2) ambient O_3 and elevated CO_2 (560 ppm during daylight hours), (3) ambient CO_2 and elevated O_3 (46.4-55.5 ppb during daylight hours), and (4) elevated CO_2 and O_3 over each growing season from 1998 through 2001. Throughout their experiment they assessed a number of the young trees' growth characteristics, as well as the responses of the sap-feeding aphid *Chaitophorus stevensis*, which they say "infests aspen throughout its range." This experiment revealed that, by itself, elevated CO_2 did not affect aphid abundance, but it increased the densities of natural enemies of the aphids, which over the long term would tend to reduce aphid numbers. Also, by itself, elevated O_3 did not affect aphid abundance, but it had a strong negative effect on natural enemies of aphids, which over the long term would tend to increase aphid numbers. When both

427

trace gases were applied together, elevated CO_2 *completely counteracted* the reduction in the abundance of natural enemies of aphids caused by elevated O_3. Hence, elevated CO_2 tended to reduce the negative impact of aphids on trembling aspen in this comprehensive study.

At about the same time, Holopainen (2002) reviewed the scientific literature dealing with the joint effects of elevated concentrations of atmospheric O_3 and CO_2 on aphid-plant interactions. After compiling the results of 26 pertinent studies, it was found that atmospheric CO_2 enrichment increased aphid performance in six studies, decreased it in six studies, and had no significant impact on it in the remaining 14 studies. Similar results were found for aphid-plant interactions in the presence of elevated O_3 concentrations.

Newman (2003) reviewed what was known and not known about aphid responses to concurrent increases in atmospheric CO_2 and air temperature, while also investigating the subject via the aphid population model of Newman *et al.* (2003). This literature review and model analysis led him to conclude that when the air's CO_2 concentration and temperature are both elevated, "aphid population dynamics will be more similar to current ambient conditions than expected from the results of experiments studying either factor alone." We can draw only the general conclusion, according to Newman, that "insect responses to CO_2 are *unlikely* to all be in the same direction." Nevertheless, he says that "the lack of a simple common phenomenon does not deny that there is some overriding generality in the responses by the system." It's just that we did not at that time know what that overriding generality was, which is why experimental work on the subject has continued apace.

Concentrating on thermal effects alone, Ma *et al.* (2004) conducted detailed experiments on the effects of high temperature, period of exposure, and developmental stage on the survival of the aphid *Metopolophium dirhodum*, which they say "is the most abundant of the three cereal aphid species in Germany and central European countries." This protocol revealed, in their words, that "temperatures over 29°C for 8 hours significantly reduced survival, which decreased generally as the temperature increased." They also determined that "exposing aphids to 32.5°C for 4 hours or longer significantly reduced survival," and that "mature aphids had a lower tolerance of high temperatures than nymphs." In light of what they observed, therefore, as well as

what a number of other scientists had observed, Ma *et al.* concluded that "global warming may play a role in the long-term changes in the population abundance of *M. dirhodum*." Specifically, they say that "an increase in TX [daily average temperature] of 1°C and MaxT [maximum daily temperature] of 1.3°C during the main period of the aphid population increase would result in a 33 percent reduction in peak population size," while "an increase in TX of 2°C and MaxT of 2.6°C would result in an early population collapse (74 percent reduction of population size)." It would appear that a little global warming could greatly decrease aphid infestations of cereal crops grown throughout Germany and Central Europe.

Returning to the subject of joint CO_2 and O_3 effects on aphids, Awmack *et al.* (2004) conducted a two-year study at the Aspen FACE site near Rhinelander, Wisconsin, USA, of the individual and combined effects of elevated CO_2 (+200 ppm) and O_3 (1.5 x ambient) on the performance of *Cepegillettea betulaefoliae* aphids feeding on paper birch trees in what they call "the first investigation of the long-term effects of elevated CO_2 and O_3 atmospheres on natural insect herbivore populations." At the individual scale, they report that "elevated CO_2 and O_3 did not significantly affect [aphid] growth rates, potential fecundity (embryo number) or offspring quality." At the population scale, on the other hand, they found that "elevated O_3 had a strong positive effect," but that "elevated CO_2 did not significantly affect aphid populations."

In comparing their results with those of prior related studies, the three scientists report that "the responses of other aphid species to elevated CO_2 or O_3 are also complex." In particular, they note that "tree-feeding aphids show few significant responses to elevated CO_2 (Docherty *et al.*, 1997), while crop-feeding species may respond positively (Awmack *et al.*, 1997; Bezemer *et al.*, 1998; Hughes and Bazzaz, 2001; Zhang *et al.*, 2001; Stacey and Fellowes, 2002), negatively (Newman *et al.*, 1999), or not at all (Hughes and Bazzaz, 2001), and the same species may show different responses on different host plant species (Awmack *et al.*, 1997; Bezemer *et al.*, 1999)." In summarizing their observations, they stated that "aphid individual performance did not predict population responses to CO_2 and O_3," and they concluded that "elevated CO_2 and O_3 atmospheres are unlikely to affect *C. betulaefoliae* populations in the presence of natural enemy communities."

In a study of a different aphid (*Chaitophorus stevensis*) conducted at the same FACE site, Mondor *et al.* (2004) focused on the subject of pheromones, which they say "are utilized by insects for several purposes, including alarm signaling," and which in the case of phloem-feeding aphids induces high-density groups of them on exposed leaves of trembling aspen trees to disperse and move to areas of lower predation risk. In this experiment the four treatments were: control (367 ppm CO_2, 38 ppb O_3), elevated CO_2 (537 ppm), elevated O_3 (51 ppb), and elevated CO_2 and O_3 (537 ppm CO_2, 51 ppb O_3). Within each treatment, several aspen leaves containing a single aphid colony of 25 ± 2 individuals were treated in one of two different ways: (1) an aphid was prodded lightly on the thorax so as to *not* produce a visible pheromone droplet, or (2) an aphid was prodded more heavily on the thorax and induced to emit a visible pheromone droplet, after which, in the words of the scientists, "aphids exhibiting any dispersal reactions in response to pheromone emission as well as those exhibiting the most extreme dispersal response, walking down the petiole and off the leaf, were recorded over 5 min."

Mondor *et al.*'s observations were striking. They found that the aphids they studied "have diminished escape responses in enriched carbon dioxide environments, while those in enriched ozone have augmented escape responses, to alarm pheromone." In fact, they report that "0 percent of adults dispersed from the leaf under elevated CO_2, while 100 percent dispersed under elevated O_3," indicating that the effects of elevated CO_2 and elevated O_3 on aphid response to pheromone alarm signaling are diametrically opposed to each other, with elevated O3 (which is detrimental to vegetation) helping aphids to escape predation and therefore live to do further harm to the leaves they infest, but with elevated CO_2 (which is beneficial to vegetation) making it more difficult for aphids to escape predation and thereby providing yet an additional benefit to plant foliage. Within this context, therefore, ozone may be seen to be doubly bad for plants, while carbon dioxide may be seen to be doubly good. In addition, Mondor *et al.* state that this phenomenon may be of broader scope than what is revealed by their specific study, noting that other reports suggest that "parasitoids and predators are more abundant and/or efficacious under elevated CO_2 levels (Stiling *et al.*, 1999; Percy *et al.*, 2002), but are negatively affected by elevated O_3 (Gate *et al.*, 1995; Percy *et al.*, 2002)."

In another intriguing study, Chen *et al.* (2004) grew spring wheat from seed to maturity in high-fertility well-watered pots out-of-doors in open-top chambers (OTCs) maintained at atmospheric CO_2 concentrations of 370, 550, and 750 ppm. Approximately two months after seeding, 20 apterous adult aphids (*Sitobion avenae*) from an adjacent field were placed upon the wheat plants of each of 25 pots in each OTC, while 15 pots were left as controls; and at subsequent 5-day intervals, both apterous and alate aphids were counted. Then, about one month later, 10 alate morph fourth instar nymphs were introduced onto the plants of each of nine control pots; for the next two weeks the number of offspring laid on those plants were recorded and removed daily to measure reproductive activity. At the end of the study, the wheat plants were harvested and their various growth responses determined.

Adherence to these protocols revealed that the introduced aphid populations increased after infestation, peaked during the grain-filling stage, and declined a bit as the wheat matured. On the final day of measurement, aphids in the 550-ppm CO_2 treatment were 32 percent more numerous than those in ambient air, while aphids in the 750-ppm treatment were 50 percent more numerous. Alate aphids also produced more offspring on host plants grown in elevated CO_2: 13 percent more in the 550-ppm treatment and 19 percent more in the 750-ppm treatment. As for the wheat plants, Chen *et al.* report that "elevated CO_2 generally enhanced plant height, aboveground biomass, ear length, and number of and dry weight of grains per ear, consistent with most other studies." With respect to above-ground biomass, for example, the 550-ppm treatment displayed an increase of 36 percent, while the 750-ppm treatment displayed an increase of 50 percent, in the case of both aphid-infested and non-infested plants.

In commenting on their findings, Chen *et al.* report that "aphid infestation caused negative effects on all the plant traits measured ... but the negative effects were smaller than the positive effects of elevated CO_2 on the plant traits." Hence, they concluded that "the increased productivity occurring in plants exposed to higher levels of CO_2 more than compensate for the increased capacity of the aphids to cause damage." In this experiment, therefore, we have a situation where both the plant and the insect that feeds on it were simultaneously benefited by the applied increases in atmospheric CO_2 concentration.

Last, in a study that investigated a number of plant-aphid-predator relationships, Chen *et al.* (2005) grew transgenic cotton plants for 30 days in well watered and fertilized sand/vermiculite mixtures in pots set in controlled-environment chambers maintained at atmospheric CO_2 concentrations of 370, 700, and 1050 ppm. A subset of aphid-infected plants was additionally supplied with predatory ladybugs, while three generations of cotton aphids (*Aphis gossypii*) were subsequently allowed to feed on some of the plants. Based on measurements made throughout this complex set of operations, Chen *et al.* found that (1) "plant height, biomass, leaf area, and carbon:nitrogen ratios were significantly higher in plants exposed to elevated CO_2 levels," (2) "more dry matter and fat content and less soluble protein were found in *A. gossypii* in elevated CO_2," (3) "cotton aphid fecundity significantly increased ... through successive generations reared on plants grown under elevated CO_2," (4) "significantly higher mean relative growth rates were observed in lady beetle larvae under elevated CO_2," and (5) "the larval and pupal durations of the lady beetle were significantly shorter and [their] consumption rates increased when fed *A. gossypii* from elevated CO_2 treatments." In commenting on the significance of their findings, Chen *et al.* say their study "provides the first empirical evidence that changes in prey quality mediated by elevated CO_2 can alter the prey preference of their natural enemies," and in this particular case, they found that this phenomenon could "enhance the biological control of aphids by lady beetle."

In considering the totality of these many experimental findings, it would appear that the ongoing rise in the air's CO_2 content will likely not have a major impact, one way or the other, on aphid-plant interactions, although the scales do appear to be slightly tipped in favor of plants over aphids. Yet a third possibility is that both plants and aphids will be benefited by atmospheric CO_2 enrichment, but with plants benefiting more.

Additional information on this topic, including reviews of newer publications as they become available, can be found at http://www.co2science.org/subject/i/insectsaphids.php.

References

Awmack, C.S., Harrington, R. and Leather, S.R. 1997. Host plant effects on the performance of the aphid *Aulacorthum solani* (Homoptera: Aphididae) at ambient and elevated CO_2. *Global Change Biology* 3: 545-549.

Awmack, C.S., Harrington, R. and Lindroth, R.L. 2004. Aphid individual performance may not predict population

responses to elevated CO_2 or O_3. *Global Change Biology* 10: 1414-1423.

Bezemer, T.M., Jones, T.H. and Knight, K.J. 1998. Long-term effects of elevated CO_2 and temperature on populations of the peach potato aphid *Myzus persicae* and its parasitoid *Aphidius matricariae*. *Oecologia* 116: 128-135.

Bezemer, T.M., Knight, K.J., Newington, J.E. *et al.* 1999. How general are aphid responses to elevated atmospheric CO_2? *Annals of the Entomological Society of America* 92: 724-730.

Chen, F., Ge, F., and Parajulee, M.N. 2005. Impact of elevated CO_2 on tri-trophic interaction of *Gossypium hirsutum*, *Aphis gossypii*, and *Leis axyridis*. *Environmental Entomology* 34: 37-46.

Chen, F.J., Wu, G. and Ge, F. 2004. Impacts of elevated CO_2 on the population abundance and reproductive activity of aphid *Sitobion avenae* Fabricius feeding on spring wheat. *JEN* 128: 723-730.

Docherty, M., Wade, F.A., Hurst, D.K., Whittaker, J.B. and Lea, P.J. 1997. Responses of tree sap-feeding herbivores to elevated CO_2. *Global Change Biology* 3: 51-59.

Gate, I.M., McNeill, S. and Ashmore, M.R. 1995. Effects of air pollution on the searching behaviour of an insect parasitoid. *Water, Air and Soil Pollution* 85: 1425-1430.

Holopainen, J.K. 2002. Aphid response to elevated ozone and CO_2. *Entomologia Experimentalis et Applicata* 104: 137-142.

Hughes, L. and Bazzaz, F.A. 2001. Effects of elevated CO_2 on five plant-aphid interactions. *Entomologia Experimentalis et Applicata* 99: 87-96.

Ma, C.S., Hau, B. and Poehling, M.-M. 2004. The effect of heat stress on the survival of the rose grain aphid, *Metopolophium dirhodum* (Hemiptera: Aphididae). *European Journal of Entomology* 101: 327-331.

Mondor, E.B., Tremblay, M.N., Awmack, C.S. and Lindroth, R.L. 2004. Divergent pheromone-mediated insect behavior under global atmospheric change. *Global Change Biology* 10: 1820-1824.

Newman, J.A. 2003. Climate change and cereal aphids: the relative effects of increasing CO_2 and temperature on aphid population dynamics. *Global Change Biology* 10: 5-15.

Newman, J.A., Gibson, D.J., Hickam, E., Lorenz, M., Adams, E., Bybee, L. and Thompson, R. 1999. Elevated carbon dioxide results in smaller populations of the bird cherry-oat aphid *Rhopalosiphum padi*. *Ecological Entomology* 24: 486-489.

Newman, J.A., Gibson, D.J., Parsons, A.J. and Thornley, J.H.M. 2003. How predictable are aphid population responses to elevated CO_2? *Journal of Animal Ecology* **72**: 556-566.

Percy, K.E., Awmack, C.S., Lindroth, R.L., Kubiske, M.E., Kopper, B.J., Isebrands, J.G., Pregitzer, K.S., Hendrey, G.R., Dickson, R.E., Zak, D.R., Oksanen, E., Sober, J., Harrington, R. and Karnosky, D.F. 2002. Altered performance of forest pests under atmospheres enriched by CO_2 and O_3. *Nature* **420**: 403-407.

Stacey, D. and Fellowes, M. 2002. Influence of elevated CO_2 on interspecific interactions at higher trophic levels. *Global Change Biology* **8**: 668-678.

Stiling, P., Rossi, A.M., Hungate, B. *et al.* 1999. Decreased leaf-miner abundance in elevated CO_2: reduced leaf quality and increased parasitoid attack. *Ecological Applications* **9**: 240-244.

Whittaker, J.B. 1999. Impacts and responses at population level of herbivorous insects to elevated CO_2. *European Journal of Entomology* **96**: 149-156.

Zhang, J., Liu, J., Wang, G. *et al.* 2001. Effect of elevated atmospheric CO_2 concentration on *Rhopalsiphum padi* population under different soil water levels. *Yingyong Shengtai Xuebao* **12**: 253-256.

7.3.3.2. Butterflies

How will earth's butterflies respond to atmospheric CO_2 enrichment and global warming? We here explore what has been learned about the question over the past few years, beginning with a review of studies that focus on carbon dioxide and concluding with studies that focus on temperature.

In a study of *Lotus corniculatus* (a cyanogenic plant that produces foliar cyanoglycosides to deter against herbivory by insects) and the Common Blue Butterfly (*Polyommatus icarus*, which regularly feeds upon *L. corniculatus* because it possesses an enzyme that detoxifies cyanide-containing defensive compounds), Goverde *et al.* (1999) collected four genotypes of *L. corniculatus* differing in their concentrations of cyanoglycosides and tannins (another group of defensive compounds) near Paris, France. They then grew them in controlled-environment chambers maintained at atmospheric CO_2 concentrations of 350 and 700 ppm, after which they determined the effects of the doubled CO_2 concentration on leaf quality and allowed the larvae of the Common Blue Butterfly to feed upon the plants' leaves. This work revealed that elevated CO_2 significantly increased leaf tannin and starch contents in a genotypically dependent and -independent

manner, respectively, while decreasing leaf cyanoglycoside contents independent of genotype. These CO_2-induced changes in leaf chemistry increased leaf palatability, as indicated by greater dry weight consumption of CO_2-enriched leaves by butterfly larvae. In addition, the increased consumption of CO_2-enriched leaves led to greater larval biomass and shorter larval development times, positively influencing the larvae of the Common Blue Butterfly. Hence, it is not surprising that larval mortality was lower when feeding upon CO_2-enriched as opposed to ambiently grown leaves.

Goverde *et al.* (2004) grew *L. corniculatus* plants once again, this time from seed in tubes recessed into the ground under natural conditions in a nutrient-poor calcareous grassland, where an extra 232 ppm of CO_2 was supplied to them via a Screen-Aided CO_2 Control (SACC) system (Leadley *et al.*, 1997, 1999), and where insect larvae were allowed to feed on the plants (half of which received extra phosphorus fertilizer) for the final month of the experiment. The atmospheric CO_2 enrichment employed in this experiment increased the total dry weight of plants growing on the unfertilized soil by 21.5 percent and that of plants growing on the phosphorus-enriched soil by 36.3 percent. However, the elevated CO_2 treatment had no effect on pupal and adult insect mass, although Goverde *et al.* report there were "genotype-specific responses in the development time of *P. icarus* to elevated CO_2 conditions," with larvae originating from different mothers developing better under either elevated CO_2 or ambient CO_2, while for still others the air's CO_2 concentration had no effect on development.

In another study by some of the same researchers, Goverde *et al.* (2002) raised larvae of the satyrid butterfly (*Coenonympha pamphilus*) in semi-natural undisturbed calcareous grassland plots exposed to atmospheric CO_2 concentrations of 370 and 600 ppm for five growing seasons. In doing so, they found that the elevated CO_2 concentration increased foliar concentrations of total nonstructural carbohydrates and condensed tannins in the grassland plants; but in what is often considered a negative impact, they found that it also decreased foliar nitrogen concentrations. Nevertheless, this phenomenon had no discernible effect on butterfly growth and performance. Larval development time, for example, was not affected by elevated CO_2, nor was adult dry mass. In fact, the elevated CO_2 increased lipid concentrations in adult male butterflies by nearly 14 percent, while it marginally increased the number of eggs in female butterflies. The former of these responses is especially important, because lipids are used as energy resources in these and other butterflies, while increased egg numbers in females also suggests an increase in fitness.

Turning to the study of temperature effects on butterflies, Parmesan *et al.* (1999) analyzed distributional changes over the past century of non-migratory species whose northern boundaries were in northern Europe (52 species) and whose southern boundaries were in southern Europe or northern Africa (40 species). This work revealed that the northern boundaries of the first group shifted northward for 65 percent of them, remained stable for 34 percent, and shifted southward for 2 percent, while the southern boundaries of the second group shifted northward for 22 percent of them, remained stable for 72 percent, and shifted southward for 5 percent, such that "nearly all northward shifts involved extensions at the northern boundary with the southern boundary remaining stable."

This behavior is precisely what we would expect to see if the butterflies were responding to shifts in the ranges of the plants upon which they depend for their sustenance, because increases in atmospheric CO_2 concentration tend to ameliorate the effects of heat stress in plants and induce an upward shift in the temperature at which they function optimally. These phenomena tend to cancel the impetus for poleward migration at the warm edge of a plant's territorial range, yet they continue to provide the opportunity for poleward expansion at the cold edge of its range. Hence, it is possible that the observed changes in butterfly ranges over the past century of concomitant warming and rising atmospheric CO_2 concentration are related to matching changes in the ranges of the plants upon which they feed. Or, this similarity could be due to some more complex phenomenon, possibly even some direct physiological effect of temperature and atmospheric CO_2 concentration on the butterflies themselves.

In any event, and in the face of the 0.8°C of global warming that occurred in Europe over the twentieth century, the consequences for European butterflies were primarily beneficial because, as Parmesan *et al.* describe the situation, "most species effectively expanded the size of their range when shifting northwards," since "nearly all northward shifts involved extensions at the northern boundary with the southern boundary remaining stable."

Across the Atlantic in America, Fleishman *et al.* (2001) used comprehensive data on butterfly distributions from six mountain ranges in the U.S. Great Basin to study how butterfly assemblages of that region may respond to IPCC-projected climate change. Whereas prior, more-simplistic analyses have routinely predicted the extirpation of great percentages of the butterfly species in this region in response to model-predicted increases in air temperature, Fleishman *et al.*'s study revealed that "few if any species of montane butterflies are likely to be extirpated from the entire Great Basin (i.e., lost from the region as a whole)."

In further discussing their results, the three researchers note that "during the Middle Holocene, approximately 8000-5000 years ago, temperatures in the Great Basin were several degrees warmer than today." Thus, they go on say that "we might expect that most of the montane species—including butterflies—that currently inhabit the Great Basin would be able to tolerate the magnitude of climatic warming forecast over the next several centuries." Consequently, it would appear that even if the global warming projections of the IPCC were true, the predictions of butterfly extinctions associated with those projections are almost certainly false.

Returning to the British Isles, Thomas *et al.* (2001) documented an unusually rapid expansion of the ranges of two butterfly species (the silver-spotted skipper butterfly and the brown argus butterfly) along with two cricket species (the long-winged cone-head and Roesel's bush cricket). They write that the warming-induced "increased habitat breadth and dispersal tendencies have resulted in about 3- to 15-fold increases in expansion rates."

In commenting on these findings, Pimm (2001) truly states the obvious when he says the geographical ranges of these insects are "expanding faster than expected," and that the synergies involved in the many intricacies of the range expansion processes are also "unexpected."

Crozier (2004) writes that "*Atalopedes campestris*, the sachem skipper butterfly, expanded its range from northern California into western Oregon in 1967, and into southwestern Washington in 1990," where she reports that temperatures rose by 2-4°C over the prior half-century. Thus intrigued, and in an attempt to assess the importance of this regional warming for the persistence of *A. campestris* in the recently colonized areas, Crozier "compared population dynamics at two locations (the butterfly's current range edge and just inside the range) that differ by 2-3°C." Then, to determine the role of over-winter larval survivorship, she "transplanted larvae over winter to both sites."

This work revealed, in her words, that "combined results from population and larval transplant analyses indicate that winter temperatures directly affect the persistence of *A. campestris* at its northern range edge, and that winter warming was a prerequisite for this butterfly's range expansion." Noting that "populations are more likely to go extinct in colder climates," Crozier says "the good news about rapid climate change [of the warming type] is that new areas may be available for the introduction of endangered species." Her work also demonstrates that the species she studied has responded to regional warming by extending its northern range boundary and thereby expanding its range.

Two years later, Davies *et al.* (2006) introduced their study of the silver-spotted skipper butterfly (*Hesperia comma* L.) by noting that during the twentieth century it "became increasingly rare in Britain [as] a result of the widespread reduction of sparse, short-turfed calcareous grassland containing the species' sole larval host plant, sheep's fescue grass [*Festuca ovina* L.]." As a result, they describe the "refuge" colonies of 1982 as but a "remnant" of what once had been. The four researchers analyzed population density data together with estimates of the percentage bare ground and the percentage of sheep's fescue available to the butterflies, based on surveys conducted in Surrey in the chalk hills of the North Downs, south of London, in *1982* (Thomas *et al.*, 1986), *1991* (Thomas and Jones, 1993), *2000* (Thomas *et al.*, 2001; Davies *et al.*, 2005), and *2001* (R.J. Wilson, unpublished data). In addition, they assessed egg-laying rates in different microhabitats, as well as the effects of ambient and oviposition site temperatures on egg laying, and the effects of sward composition on egg location. This work revealed, in their words, that "in 1982, 45 habitat patches were occupied by *H. comma* [but] in the subsequent 18-year period, the species expanded and, by 2000, a further 29 patches were colonized within the habitat network." In addition, they found that "the mean egg-laying rate of *H. comma* females increased with rising ambient temperatures," and that "a wider range of conditions have become available for egg-laying."

In discussing their findings, Davies *et al.* state that "climate warming has been an important driving force in the recovery of *H. comma* in Britain [as] the rise in ambient temperature experienced by the butterfly will have aided the metapopulation re-expansion in a number of ways." First, they suggest that "greater temperatures should increase the potential fecundity of *H. comma* females," and that "if this results in larger populations, for which there is some evidence (e.g. 32 of the 45 habitat patches occupied in the Surrey network experienced site-level increases in population density between 1982 and 2000), they will be less prone to extinction," with "larger numbers of dispersing migrant individuals being available to colonize unoccupied habitat patches and establish new populations." Second, they state that "the wider range of thermal and physical microhabitats used for egg-laying increased the potential resource density within each grassland habitat fragment," and that "this may increase local population sizes." Third, they argue that "colonization rates are likely to be greater as a result of the broadening of the species realized niche, [because] as a larger proportion of the calcareous grassland within the species' distribution becomes thermally suitable, the relative size and connectivity of habitat patches within the landscape increases." Fourth, they note that "higher temperatures may directly increase flight (dispersal) capacity, and the greater fecundity of immigrants may improve the likelihood of successful population establishment." Consequently, Davies *et al.* conclude that "the warmer summers predicted as a consequence of climate warming are likely to be beneficial to *H. comma* within Britain," and they suggest that "warmer winter temperatures could also allow survival in a wider range of microhabitats."

In a concurrent study, Menendez *et al.* (2006) provided what they call "the first assessment, at a geographical scale, of how species richness has changed in response to climate change," concentrating on British butterflies. This they did by testing "whether average species richness of resident British butterfly species has increased in recent decades, whether these changes are as great as would be expected given the amount of warming that has taken place, and whether the composition of butterfly communities is changing towards a dominance by generalist species." By these means they determined that "average species richness of the British butterfly fauna at 20 x 20 km grid resolution has increased since 1970-82, during a period when climate warming would lead us to expect increases." They also found, as expected, that "southerly habitat generalists increased more than specialists," which require a specific type of habitat that is sometimes difficult for them to find, especially in the modern world where habitat destruction is commonplace. In addition, they were able to determine that observed species richness increases lagged behind those expected on the basis of climate change.

These results "confirm," according to the nine UK researchers, "that the average species richness of British butterflies has increased since 1970-82." However, some of the range shifts responsible for the increase in species richness take more time to occur than those of other species; they say their results imply "it may be decades or centuries before the species richness and composition of biological communities adjusts to the current climate."

Also working in Britain, Hughes *et al.* (2007) examined evolutionary changes in adult flight morphology in six populations of the speckled wood butterfly—*Pararge aegeria* L. (Satyrinae)—along a transect from its distribution core to its warming-induced northward expanding range margin. The results of this exercise were then compared with the output of an individual-based spatially explicit model that was developed "to investigate impacts of habitat availability on the evolution of dispersal in expanding populations." This work indicated that the empirical data the researchers gathered "were in agreement with model output," and that they "showed increased dispersal ability with increasing distance from the distribution core," which included favorable changes in thorax shape, abdomen mass, and wing aspect ratio for both males and females, as well as thorax mass and wing loading for females. In addition, they say that "increased dispersal ability was evident in populations from areas colonized >30 years previously."

In discussing their findings, Hughes *et al.* suggest that "evolutionary increases in dispersal ability in expanding populations may help species track future climate changes and counteract impacts of habitat fragmentation by promoting colonization." However, they report that in the specific situation they investigated, "at the highest levels of habitat loss, increased dispersal was less evident during expansion and reduced dispersal was observed at equilibrium, indicating that for many species, continued habitat fragmentation is likely to outweigh any benefits from dispersal." Put another way, it would appear that global warming is proving not to be an insurmountable problem for the speckled wood butterfly, which is evolving physical characteristics that allow it to better keep up with the poleward migration of its current environmental niche, but that the direct destructive assaults of humanity upon its natural habitat could still end up driving it to extinction.

Analyzing data pertaining to the general abundance of Lepidoptera in Britain over the period 1864-1952, based on information assembled by Beirne (1955) via his examination of "several thousand papers in entomological journals describing annual abundances of moths and butterflies," were Dennis and Sparks (2007), who report that "abundances of British Lepidoptera were significantly positively correlated with Central England temperatures in the current year for each month from May to September and November," and that "increased overall abundance in Lepidoptera coincided significantly with increased numbers of migrants," which latter data were derived from the work of Williams (1965). In addition, they report that Pollard (1988) subsequently found much the same thing for 31 butterfly species over the period 1976-1986, and that Roy *et al.* (2001) extended the latter investigation to 1997, finding "strong associations between weather and population fluctuations and trends in 28 of 31 species which confirmed Pollard's (1988) findings," all of which observations indicate that the warming-driven increase in Lepidopteran species and numbers in Britain has been an ongoing phenomenon ever since the end of the Little Ice Age.

Returning to North America for one final study, White and Kerr (2006), as they describe it, "report butterfly species' range shifts across Canada between 1900 and 1990 and develop spatially explicit tests of the degree to which observed shifts result from climate or human population density," the latter of which factors they describe as "a reasonable proxy for land use change," within which broad category they include such things as "habitat loss, pesticide use, and habitat fragmentation," all of which anthropogenic-driven factors have been tied to declines of various butterfly species. In addition, they say that to their knowledge, "this is the broadest scale, longest term dataset yet assembled to quantify global change impacts on patterns of species richness."

This exercise led White and Kerr to discover that butterfly species richness "generally increased over the study period, a result of range expansion among the study species," and they further found that this increase "from the early to late part of the twentieth century was positively correlated with temperature change," which had to have been the *cause* of the change, for they also found that species richness was "negatively correlated with human population density change."

Contrary to the doom-and-gloom prognostications of some experts, the supposedly unprecedented global warming of the twentieth century has been beneficial for the butterfly species that inhabit Canada, Britain, and the United States, as their ranges have expanded and greater numbers of species are now being encountered in each country.

Additional information on this topic, including reviews of newer publications as they become available, can be found at http://www.co2science.org/subject/i/summaries/butterflies.php.

References

Beirne, B.P. 1955. Natural fluctuations in abundance of British Lepidoptera. *Entomologist's Gazette* **6**: 21-52.

Crozier, L. 2004. Warmer winters drive butterfly range expansion by increasing survivorship. *Ecology* **85**: 231-241.

Davies, Z.G., Wilson, R.J., Brereton, T.M. and Thomas, C.D. 2005. The re-expansion and improving status of the silver-spotted skipper butterfly (*Hesperia comma*) in Britain: a metapopulation success story. *Biological Conservation* **124**: 189-198.

Davies, Z.G., Wilson, R.J., Coles, S. and Thomas, C.D. 2006. Changing habitat associations of a thermally constrained species, the silver-spotted skipper butterfly, in response to climate warming. *Journal of Animal Ecology* **75**: 247-256.

Dennis, R.L.H. and Sparks, T.H. 2007. Climate signals are reflected in an 89 year series of British Lepidoptera records. *European Journal of Entomology* **104**: 763-767.

Fleishman, E., Austin, G.T. and Murphy, D.D. 2001. Biogeography of Great Basin butterflies: revisiting patterns, paradigms, and climate change scenarios. *Biological Journal of the Linnean Society* **74**: 501-515.

Goverde, M., Bazin, A., Shykoff, J.A. and Erhardt, A. 1999. Influence of leaf chemistry of *Lotus corniculatus* (Fabaceae) on larval development of *Polyommatus icarus* (Lepidoptera, Lycaenidae): effects of elevated CO_2 and plant genotype. *Functional Ecology* **13**: 801-810.

Goverde, M., Erhardt, A. and Niklaus, P.A. 2002. In situ development of a satyrid butterfly on calcareous grassland exposed to elevated carbon dioxide. *Ecology* **83**: 1399-1411.

Goverde, M., Erhardt, A. and Stocklin, J. 2004. Genotype-specific response of a lycaenid herbivore to elevated carbon dioxide and phosphorus availability in calcareous grassland. *Oecologia* **139**: 383-391.

Hughes, C.L., Dytham, C. and Hill, J.K. 2007. Modelling and analyzing evolution of dispersal in populations at

expanding range boundaries. *Ecological Entomology* **32**: 437-445.

Leadley, P.W., Niklaus, P., Stocker, R. and Korner, C. 1997. Screen-aided CO_2 control (SACC): a middle-ground between FACE and open-top chamber. *Acta Oecologia* **18**: 207-219.

Leadley, P.W., Niklaus, P.A., Stocker, R. and Korner, C. 1999. A field study of the effects of elevated CO2 on plant biomass and community structure in a calcareous grassland. *Oecologia* **118**: 39-49.

Menendez, R., Gonzalez-Megias, A., Hill, J.K., Braschler, B., Willis, S.G., Collingham, Y., Fox, R., Roy, D.B. and Thomas, C.D. 2006. Species richness changes lag behind climate change. *Proceedings of the Royal Society B* **273**: 1465-1470.

Parmesan, C., Ryrholm, N., Stefanescu, C., Hill, J.K., Thomas, C.D., Descimon, H., Huntley, B., Kaila, L., Kullberg, J., Tammaru, T., Tennent, W.J., Thomas, J.A. and Warren, M. 1999. Poleward shifts in geographical ranges of butterfly species associated with regional warming. *Nature* **399**: 579-583.

Pimm, S.L. 2001. Entrepreneurial insects. *Nature* **411**: 531-532.

Pollard, E. 1988. Temperature, rainfall and butterfly numbers. *Journal of Applied Ecology* **25**: 819-828.

Roy, D.B., Rothery, P., Moss, D., Pollard, E. and Thomas, J.A. 2001. Butterfly numbers and weather: predicting historical trends in abundance and the future effects of climate change. *Journal of Animal Ecology* **70**: 201-217.

Thomas, C.D., Bodsworth, E.J., Wilson, R.J., Simmons, A.D., Davies, Z.G., Musche, M. and Conradt, L. 2001. Ecological and evolutionary processes at expanding range margins. *Nature* **411**: 577-581.

Thomas, C.D. and Jones, T.M. 1993. Partial recovery of a skipper butterfly (*Hesperia comma*) from population refuges: lessons for conservation in a fragmented landscape. *Journal of Animal Ecology* **62**: 472-481.

Thomas, C.D. and Lennon, J.J. 1999. Birds extend their ranges northwards. *Nature* **399**: 213.

Thomas, J.A., Thomas, C.D., Simcox, D.J. and Clarke, R.T. 1986. Ecology and declining status of the silver-spotted skipper butterfly (*Hesperia comma*) in Britain. *Journal of Applied Ecology* **23**: 365-380.

White, P. and Kerr, J.T. 2006. Contrasting spatial and temporal global change impacts on butterfly species richness during the twentieth century. *Ecography* **29**: 908-918.

Williams, C.B. 1965. *Insect Migration*. Collins, London, UK, 237 pp.

7.3.3.3. Moths

Kerslake *et al.* (1998) collected five-year-old heather plants from a Scottish moor and grew them in open-top chambers maintained at atmospheric CO_2 concentrations of 350 and 600 ppm for 20 months, with and without soil nitrogen fertilization. At two different times during the study, larvae of *Operophtera brumata*, a voracious winter moth whose outbreaks have caused extensive damage to heather moorland in recent years, were allowed to feed upon current-year shoots for up to one month. The results obtained from this experiment revealed that the survivorship of larvae placed on CO_2-enriched foliage was not significantly different from that of larvae placed on foliage produced in ambient air, regardless of nitrogen treatment. In addition, feeding upon CO_2-enriched foliage did not affect larval growth rate, development, or final pupal weight. Consequently, Kerslake *et al.* concluded that their study "provides no evidence that increasing atmospheric CO_2 concentrations will affect the potential for outbreak of *Operophtera brumata* on this host."

Hattenschwiler and Schafellner (1999) grew seven-year-old spruce trees at atmospheric CO_2 concentrations of 280, 420, and 560 ppm in various nitrogen deposition treatments for three years, after which they performed needle quality assessments and allowed nun moth (*Lymantria monacha*) larvae to feed upon current-year needles for 12 days. This moth is an especially voracious defoliator that resides in most parts of Europe and East Asia between 40 and 60° N latitude, and it is commonly regarded as the "coniferous counterpart" of its close relative the gypsy moth, which feeds primarily upon deciduous trees.

The two scientists determined from their observations that elevated CO_2 significantly enhanced needle starch, tannin, and phenolic concentrations, while significantly decreasing needle water and nitrogen contents. Thus, atmospheric CO_2 enrichment reduced overall needle quality from the perspective of this foliage-consuming moth, as nitrogen content is the primary factor associated with leaf quality. Increasing nitrogen deposition, on the other hand, tended to enhance needle quality, for it lowered starch, tannin, and phenolic concentrations while boosting needle nitrogen content. Nevertheless, the positive influence of nitrogen deposition on needle quality was not large enough to completely offset the quality reduction caused by elevated CO_2.

In light of these observations, it was no surprise that larvae placed on CO_2-enriched foliage consumed less needle biomass than larvae placed on low-CO_2-grown foliage, regardless of nitrogen treatment, and that the larvae feeding on CO_2-enriched foliage exhibited reduced relative growth rates and attained an average biomass that was only two-thirds of that attained by larvae consuming foliage produced at 280 ppm CO_2. As a result, Hattenschwiler and Schafellner concluded that "altered needle quality in response to elevated CO_2 will impair the growth and development of *Lymantria monacha* larvae," which should lead to reductions in the degree of spruce tree destruction caused by this voracious defoliator.

Stiling *et al.* (2002) studied the effects of an approximate doubling of the air's CO_2 concentration on a number of characteristics of several insect herbivores feeding on plants native to a scrub-oak forest ecosystem at the Kennedy Space Center, Florida, USA, in eight ambient and eight CO_2-enriched open-top chambers. They say that the "relative levels of damage by the two most common herbivore guilds, leaf-mining moths and leaf-chewers (primarily larval lepidopterans and grasshoppers), were significantly lower in elevated CO_2 than in ambient CO_2," and that "the response to elevated CO_2 was the same across all plant species."

In a follow-up study to that of Stiling *et al.*, which was conducted at the same facilities, Rossi *et al.* (2004), focused on the abundance of a guild of lepidopteran leafminers that attack the leaves of myrtle oak, as well as various leaf chewers that also like to munch on this species. Specifically, they periodically examined 100 marked leaves in each of the 16 open-top chambers for a total of nine months, after which, in their words, "differences in mean percent of leaves with leafminers and chewed leaves on trees from ambient and elevated chambers were assessed using paired *t*-tests." This protocol revealed, in their words, that "both the abundance of the guild of leafmining lepidopterans and damage caused by leaf chewing insects attacking myrtle oak were depressed in elevated CO_2." Leafminer abundance was 44 percent lower ($P = 0.096$) in the CO_2-enriched chambers compared to the ambient-air chambers, while the abundance of leaves suffering chewing damage was 37 percent lower ($P = 0.072$) in the CO_2-enriched air.

Working with red maple saplings, Williams *et al.* (2003) bagged first instar gypsy moth larvae on branches of trees that were entering their fourth year of growth within open-top chambers maintained at four sets of CO_2/temperature conditions: (1) ambient temperature, ambient CO_2, (2) ambient temperature, elevated CO_2 (ambient + 300 ppm), (3) elevated temperature (ambient + 3.5°C), ambient CO_2, and (4) elevated temperature, elevated CO_2. For these conditions they measured several parameters that were required to test their hypothesis that a CO_2-enriched atmosphere would lead to reductions in foliar nitrogen concentrations and increases in defensive phenolics that would in turn lead to increases in insect mortality. The results they obtained indicated, in their words, "that larvae feeding on CO_2-enriched foliage ate a comparably poorer food source than those feeding on ambient CO_2-grown plants, irrespective of temperature." Nevertheless, they determined that "CO_2-induced reductions in foliage quality were unrelated to insect mortality, development rate and pupal weight." As a result, they were forced to conclude that "phytochemical changes resulted in no negative effects on gypsy moth performance," but neither did they help them.

Noting that increases in the atmosphere's CO_2 concentration typically lead to greater decreases in the concentrations of nitrogen in the foliage of C_3 as opposed to C_4 grasses, Barbehenn et al. (2004) say "it has been predicted that insect herbivores will increase their feeding damage on C_3 plants to a greater extent than on C_4 plants (Lincoln et al., 1984, 1986; Lambers, 1993). To test this hypothesis, they grew *Lolium multiflorum* (Italian ryegrass, a common C_3 pasture grass) and *Bouteloua curtipendula* (sideoats gramma, a native C_4 rangeland grass) in chambers maintained at either the ambient atmospheric CO_2 concentration of 370 ppm or the doubled CO_2 concentration of 740 ppm for two months, after which newly molted sixth-instar larvae of *Pseudaletia unipuncta* (a grass-specialist noctuid) and *Spodoptera frugiperda* (a generalist noctuid) were allowed to feed upon the two grasses.

As expected, Barbehenn et al. found that foliage protein concentration decreased by 20 percent in the C_3 grass, but by only 1 percent in the C_4 grass, when they were grown in CO_2-enriched air; and they say that "to the extent that protein is the most limiting of the macronutrients examined, these changes represent a decline in the nutritional quality of the C_3 grass." However, and "contrary to our expectations," in the words of Barbehenn et al., "neither caterpillar species significantly increased its consumption rate to compensate for the lower concentration of protein in [the] C_3 grass," and they note that "this result does not

support the hypothesis that C_3 plants will be subject to greater rates of herbivory relative to C_4 plants in future [high-CO_2] atmospheric conditions (Lincoln et al., 1984)." In addition, and "despite significant changes in the nutritional quality of *L. multiflorum* under elevated CO_2," they note that "no effect on the relative growth rate of either caterpillar species on either grass species resulted," and that there were "no significant differences in insect performance between CO_2 levels." By way of explanation of these results, they suggest that "post-ingestive mechanisms could provide a sufficient means of compensation for the lower nutritional quality of C_3 plants grown under elevated CO_2."

In light of these observations, Barbehenn et al. suggest "there will not be a single pattern that characterizes all grass feeders" with respect to their feeding preferences and developmental responses in a world where certain C_3 plants may experience foliar protein concentrations that are lower than those they exhibit today, nor will the various changes that may occur necessarily be detrimental to herbivore development or to the health and vigor of their host plants. Nevertheless, subsequent studies continue to suggest that various moth species will likely be negatively affected by the ongoing rise in the air's CO_2 content.

A case in point is the study of Chen et al. (2005), who grew well watered and fertilized cotton plants of two varieties (one expressing *Bacillus thurigiensis* toxin genes and one a non-transgenic cultivar from the same recurrent parent) in pots placed within open-top chambers maintained at either 376 or 754 ppm CO_2 in Sanhe County, Hebei Province, China, from planting in mid-May to harvest in October, while immature bolls were periodically collected and analyzed for various chemical characteristics and others were stored under refrigerated conditions for later feeding to larvae of the cotton bollworm. By these means they found that the elevated CO_2 treatment increased immature boll concentrations of condensed tannins by approximately 22 percent and 26 percent in transgenic and non-transgenetic cotton, respectively, and that it slightly decreased the body biomass of the cotton bollworm and reduced moth fecundity. The Bt treatment was even more effective in this regard; and in the combined Bt-high-CO_2 treatment, the negative cotton bollworm responses were expressed most strongly of all.

Bidart-Bouzat et al. (2005) grew three genotypes of mouse-ear cress (*Arabidopsis thaliana*) from seed in pots within controlled-environment chambers maintained at either ambient CO_2 (360 ppm) or elevated CO_2 (720 ppm). On each of half of the plants (the herbivory treatment) in each of these CO_2 treatments, they placed two second-instar larvae of the diamondback moth (*Plutella xylostella*) at bolting time and removed them at pupation, which resulted in an average of 20 percent of each plant's total leaf area in the herbivory treatment being removed. Then, each pupa was placed in a gelatin capsule until adult emergence and ultimate death, after which insect gender was determined and the pupa's weight recorded. At the end of this herbivory trial, the leaves of the control and larvae-infested plants were analyzed for concentrations of individual glucosinolates—a group of plant-derived chemicals that can act as herbivore deterrents (Maruicio and Rausher, 1997)—while total glucosinolate production was determined by summation of the individual glucosinolate assays. Last, influences of elevated CO_2 on moth performance and its association with plant defense-related traits were evaluated.

Overall, it was determined by these means that herbivory by larvae of the diamondback moth did not induce any increase in the production of glucosinolates in the mouse-ear cress in the ambient CO_2 treatment. However, the three scientists report that "herbivory-induced increases in glucosinolate contents, ranging from 28% to 62% above basal levels, were found under elevated CO_2 in two out of the three genotypes studied." In addition, they determined that "elevated CO_2 decreased the overall performance of diamondback moths." And because "induced defenses can increase plant fitness by reducing subsequent herbivore attacks (Agrawal, 1999; Kessler and Baldwin, 2004)," according to Bidart-Bouzat *et al.*, they suggest that "the pronounced increase in glucosinolate levels under CO_2 enrichment may pose a threat not only for insect generalists that are likely to be more influenced by rapid changes in the concentration of these chemicals, but also for other insect specialists more susceptible than diamondback moths to high glucosinolate levels (Stowe, 1998; Kliebenstein *et al.*, 2002)."

In a study of a major crop species, Wu *et al.* (2006) grew spring wheat (*Triticum aestivum* L.) from seed to maturity in pots placed within open-top chambers maintained at either 370 or 750 ppm CO_2 in Sanhe County, Hebei Province, China, after which they reared three generations of cotton bollworms (*Helicoverpa armigera* Hubner) on the milky grains of the wheat, while monitoring a number of different bollworm developmental characteristics. In doing so,

as they describe it, "significantly lower pupal weights were observed in the first, second and third generations," and "the fecundity of *H. armigera* decreased by 10% in the first generation, 13% in the second generation and 21% in the third generation," resulting in a "potential population decrease in cotton bollworm by 9% in the second generation and 24% in the third generation." In addition, they say that "population consumption was significantly reduced by 14% in the second generation and 24% in the third generation," and that the efficiency of conversion of ingested food was reduced "by 18% in the first generation, 23% in the second generation and 30% in the third generation." As a result, they concluded that the "net damage of cotton bollworm on wheat will be less under elevated atmospheric CO_2," while noting that "at the same time, gross wheat production is expected to increase by 63% under elevated CO_2."

In another report of their work, Wu *et al.* (2007) write that "significant decreases in the protein, total amino acid, water and nitrogen content by 15.8%, 17.7%, 9.1% and 20.6% and increases in free fatty acid by 16.1% were observed in cotton bolls grown under elevated CO_2." And when fed with these cotton bolls, they say that the larval survival rate of *H. armigera* "decreased by 7.35% in the first generation, 9.52% in the second generation and 11.48% in the third generation under elevated CO_2 compared with ambient CO_2." In addition, they observed that "the fecundity of *H. armigera* decreased by 7.74% in the first generation, 14.23% in the second generation and 16.85% in the third generation," while noting that "fecundity capacity is likely to be reduced even further in the next generation."

The synergistic effects of these several phenomena, in the words of Wu *et al.*, "resulted in a potential population decrease in cotton bollworm by 18.1% in the second generation and 52.2% in the third generation under elevated CO_2," with the result that "the potential population consumption of cotton bollworm decreased by 18.0% in the second generation and 55.6% in the third generation ... under elevated CO_2 compared with ambient CO_2." And in light of these several findings, they concluded that "the potential population dynamics and potential population consumption of cotton bollworm will alleviate the harm to [cotton] plants in the future rising-CO_2 atmosphere."

In a different type of study, Esper *et al.* (2007) reconstructed an annually resolved history of population cycles of a foliage-feeding Lepidopteran commonly known as the larch budmoth (*Zeiraphera diniana* Gn.)—or LBM for short—within the European Alps in the southern part of Switzerland. As is typical of many such insect pests, they note that "during peak activity, populations may reach very high densities over large areas," resulting in "episodes of massive defoliation and/or tree mortality" that could be of great ecological and economic significance.

The first thing the team of Swiss and US researchers thus did in this regard was develop a history of LBM outbreaks over the 1,173-year period AD 832-2004, which they describe as "the longest continuous time period over which any population cycle has ever been documented."

They accomplished this feat using radiodensitometric techniques to characterize the tree-ring density profiles of 180 larch (*Larix deciduas* Mill.) samples, where "LBM outbreaks were identified based upon characteristic maximum latewood density (MXD) patterns in wood samples, and verified using more traditional techniques of comparison with tree-ring chronologies from non-host species," i.e., fir and spruce. Then, they developed a matching temperature history for the same area, which was accomplished by combining "a tree-ring width-based reconstruction from AD 951 to 2002 integrating 1527 pine and larch samples (Buntgen *et al.*, 2005) and a MXD-based reconstruction from AD 755 to 2004 based upon the same 180 larch samples used in the current study for LBM signal detection (Buntgen *et al.*, 2006)."

Over almost the entire period studied, from its start in AD 832 to 1981, there were a total of 123 LBM outbreaks with a mean reoccurrence time of 9.3 years. In addition, the researchers say "there was never a gap that lasted longer than two decades." From 1981 to the end of their study in 2004, however, there were *no* LBM outbreaks; since there had never before (within their record) been such a long outbreak hiatus, they concluded that "the absence of mass outbreaks since the 1980s is truly exceptional."

To what do Esper *et al.* attribute this unprecedented recent development? Noting that "conditions during the late twentieth century represent the warmest period of the past millennium"—as per their temperature reconstruction for the region of the Swiss Alps within which they worked—they point to "the role of extraordinary climatic conditions as the cause of outbreak failure," and they discuss what they refer to as the "probable hypothesis" of Baltensweiler (1993), who described a

scenario by which local warmth may lead to reduced LBM populations.

Such may well be the case, but we hasten to add that atmospheric CO_2 concentrations since 1980 have also been unprecedented over the 1,173-year period of Esper *et al.*'s study. Hence, the suppression of LBM outbreaks over the past quarter-century may have been the result of some *synergistic* consequence of the two factors (temperature and CO_2) acting in unison, while a third possibility may involve only the increase in the air's CO_2 content.

Esper *et al.* say their findings highlight the "vulnerability of an otherwise stable ecological system in a warming environment," in what would appear to be an attempt to attach an undesirable connotation to the observed outcome. This wording seems strange indeed, for it is clear that the "recent disruption of a major disturbance regime," as Esper *et al.* refer to the suppression of LBM outbreaks elsewhere in their paper, would be considered by most people to be a positive outcome, and something to actually be welcomed.

Working with *Antheraea polyphemus*—a leaf-chewing generalist lepidopteran herbivore that represents the most abundant feeding guild in the hardwood trees that grow beneath the canopy of the unmanaged loblolly pine plantation that hosts the Forest Atmosphere Carbon Transfer and Storage (FACTS-1) research site in the Piedmont region of North Carolina, USA, where the leaf-chewer can consume 2-15 percent of the forest's net primary production in any given year—Knepp *et al.* (2007) focused their attention on two species of oak tree—*Quercus alba* L. (white oak) and *Quercus velutina* Lam. (black oak)—examining host plant preference and larval performance of *A. polyphemus* when fed foliage of the two tree species that had been grown in either ambient or CO_2-enriched air (to 200 ppm above ambient) in this long-running FACE experiment. In doing so, they determined that "growth under elevated CO_2 reduced the food quality of oak leaves for caterpillars," while "consuming leaves of either oak species grown under elevated CO_2 slowed the rate of development of *A. polyphemus* larvae." In addition, they found that feeding on foliage of *Q. velutina* that had been grown under elevated CO_2 led to reduced consumption by the larvae and greater mortality. As a result, they concluded that "reduced consumption, slower growth rates, and increased mortality of insect larvae may explain [the] lower total leaf damage observed previously in plots of this forest exposed to elevated CO_2," as documented by Hamilton *et al.*

(2004) and Knepp *et al.* (2005), which finding bodes well indeed for the growth and vitality of such forests in the years and decades ahead, as the air's CO_2 content continues to rise.

Kampichler *et al.* (2008) also worked with oak trees. Noting, however, that "systems studied so far have not included mature trees," they attempted to remedy this situation by determining "the abundance of dominant leaf-galls (spangle-galls induced by the cynipid wasps *Neuroterus quercusbaccarum* and *N. numismalis*) and leaf-mines (caused by the larvae of the moth *Tischeria ekebladella*) on freely colonized large oaks in a mixed forest in Switzerland, which received CO_2 enrichment [540 ppm vs. 375 ppm during daylight hours] from 2001 to 2004" via "the Swiss Canopy Crane (SCC) and a new CO_2 enrichment technique (web-FACE)" in a forest that they say "is 80-120 years old with a canopy height of 32-38 m, consisting of seven deciduous and four coniferous species." This work allowed the German, Mexican, and Swiss researchers to discover that although elevated CO_2 reduced various leaf parameters (water content, proteins, non-structural carbohydrates, tannins, etc.) at the SCC site, "on the long term, their load with cynipid spangle-galls and leaf-mines of *T. ekebladella* was not distinguishable from that in oaks exposed to ambient CO_2 after 4 years of treatment." Kampichler *et al.* concluded that in the situation they investigated, "CO_2 enrichment had no lasting effect in all three [animal] taxa, despite the substantial and consistent change in leaf chemistry of oak due to growth in elevated CO_2."

In conclusion, therefore, and considering the results of all of the studies reviewed in this section, it would appear that the ongoing rise in the air's CO_2 content will *not* result in greater damage to earth's vegetation by the larvae of the many moths that inhabit the planet, and could reduce the damage they cause.

Additional information on this topic, including reviews of newer publications as they become available, can be found at http://www.co2science.org/subject/i/summaries/moths.php.

References

Agrawal, A.A. 1999. Induced-responses to herbivory in wild radish: effects on several herbivores and plant fitness. *Ecology* **80**: 1713-1723.

Baltensweiler, W. 1993. Why the larch bud moth cycle collapsed in the subalpine larch-cembran pine forests in the year 1990 for the first time since 1850. *Oecologia* **94**: 62-66.

Barbehenn, R.V., Karowe, D.N. and Spickard, A. 2004. Effects of elevated atmospheric CO_2 on the nutritional ecology of C_3 and C_4 grass-feeding caterpillars. *Oecologia* **140**: 86-95.

Bidart-Bouzat, M.G., Mithen, R. and Berenbaum, M.R. 2005. Elevated CO_2 influences herbivory-induced defense responses of *Arabidopsis thaliana*. *Oecologia* **145**: 415-424.

Buntgen, U., Esper, J., Frank, D.C., Nicolussi, K. and Schmidhalter, M. 2005. A 1052-year tree-ring proxy for alpine summer temperatures. *Climate Dynamics* **25**: 141-153.

Buntgen, U., Frank, D.C., Nievergelt, D. and Esper, J. 2006. Alpine summer temperature variations, AD 755-2004. *Journal of Climate* **19**: 5606-5623.

Chen, F., Wu, G., Ge, F., Parajulee, M.N. and Shrestha, R.B. 2005. Effects of elevated CO_2 and transgenic Bt cotton on plant chemistry, performance, and feeding of an insect herbivore, the cotton bollworm. *Entomologia Experimentalis et Applicata* **115**: 341-350.

Esper, J., Buntgen, U., Frank, D.C., Nievergelt, D. and Liebhold, A. 2007. 1200 years of regular outbreaks in alpine insects. *Proceedings of the Royal Society B* **274**: 671-679.

Hamilton, J.G., Zangerl, A.R., Berenbaum, M.R., Pippen, J.S., Aldea, M. and DeLucia, E.H. 2004. Insect herbivory in an intact forest understory under experimental CO_2 enrichment. *Oecologia* **138**: 566-573.

Hattenschwiler, S. and Schafellner, C. 1999. Opposing effects of elevated CO_2 and N deposition on *Lymantria monacha* larvae feeding on spruce trees. *Oecologia* **118**: 210-217.

Kampichler, C., Teschner, M., Klein, S. and Korner, C. 2008. Effects of 4 years of CO_2 enrichment on the abundance of leaf-galls and leaf-mines in mature oaks. *Acta Oecologica* **34**: 139-146.

Kerslake, J.E., Woodin, S.J. and Hartley, S.E. 1998. Effects of carbon dioxide and nitrogen enrichment on a plant-insect interaction: the quality of *Calluna vulgaris* as a host for *Operophtera brumata*. *New Phytologist* **140**: 43-53.

Kessler, A. and Baldwin, I.T. 2004. Herbivore-induced plant vaccination. Part I. The orchestration of plant defenses in nature and their fitness consequences in the wild tobacco, *Nicotiana attenuata*. *Plant Journal* **38**: 639-649.

Kliebenstein, D., Pedersen, D., Barker, B. and Mitchell-Olds, T. 2002. Comparative analysis of quantitative trait loci controlling glucosinolates, myrosinase and insect resistance in *Arabidopsis thaliana*. *Genetics* **161**: 325-332.

Knepp, R.G., Hamilton, J.G., Mohan, J.E., Zangerl, A.R., Berenbaum, M.R. and DeLucia, E.H. 2005. Elevated CO_2 reduces leaf damage by insect herbivores in a forest community. *New Phytologist* **167**: 207-218.

Knepp, R.G., Hamilton, J.G., Zangerl, A.R., Berenbaum, M.R. and DeLucia, E.H. 2007. Foliage of oaks grown under elevated CO_2 reduces performance of *Antheraea polyphemus* (Lepidoptera: Saturniidae). *Environmental Entomology* **36**: 609-617.

Lambers, H. 1993. Rising CO_2, secondary plant metabolism, plant-herbivore interactions and litter decomposition. Theoretical considerations. *Vegetatio* **104/105**: 263-271.

Lincoln, D.E., Couvet, D. and Sionit, N. 1986. Responses of an insect herbivore to host plants grown in carbon dioxide enriched atmospheres. *Oecologia* **69**: 556-560.

Lincoln, D.E., Sionit, N. and Strain, B.R. 1984. Growth and feeding response of *Pseudoplusia includens* (Lepidoptera: Noctuidae) to host plants grown in controlled carbon dioxide atmospheres. *Environmental Entomology* **13**: 1527-1530.

Mauricio, R. and Rausher, M.D. 1997. Experimental manipulation of putative selective agents provides evidence for the role of natural enemies in the evolution of plant defense. *Evolution* **51**: 1435-1444.

Rossi, A.M., Stiling, P., Moon, D.C., Cattell, M.V. and Drake, B.G. 2004. Induced defensive response of myrtle oak to foliar insect herbivory in ambient and elevated CO_2. *Journal of Chemical Ecology* **30**: 1143-1152.

Stiling, P., Moon, D.C., Hunter, M.D., Colson, J., Rossi, A.M., Hymus, G.J. and Drake, B.G. 2002. Elevated CO_2 lowers relative and absolute herbivore density across all species of a scrub-oak forest. *Oecologia* DOI 10.1007/s00442-002-1075-5.

Stowe, K.A. 1998. Realized defense of artificially selected lines of *Brassica rapa*: effects of quantitative genetic variation in foliar glucosinolate concentration. *Environmental Entomology* **27**: 1166-1174.

Williams, R.S., Lincoln, D.E. and Norby, R.J. 2003. Development of gypsy moth larvae feeding on red maple saplings at elevated CO_2 and temperature. *Oecologia* **137**: 114-122.

Wu, G., Chen, F.-J. and Ge, F. 2006. Response of multiple generations of cotton bollworm *Helicoverpa armigera* Hubner, feeding on spring wheat, to elevated CO_2. *Journal of Applied Entomology* **130**: 2-9.

Wu, G., Chen, F.-J., Sun, Y.-C. and Ge, F. 2007. Response of successive three generations of cotton bollworm, *Helicoverpa armigera* (Hubner), fed on cotton bolls under elevated CO_2. *Journal of Environmental Sciences* **19**: 1318-1325.

7.3.3.4. Other Insects

Docherty *et al.* (1997), in addition to studying aphids, studied two sap-feeding leafhopper species that were allowed to feed on saplings of beech and sycamore that were grown in glasshouses maintained at atmospheric CO_2 concentrations of 350 and 600 ppm. As far as they could determine, there were no significant effects of the extra CO_2 on either the feeding or performance characteristics of either leafhopper species.

In a literature review of more than 30 studies published two years later, Whittaker (1999) found that chewing insects (leaf chewers and leaf miners) showed either no change or reductions in abundance in response to atmospheric CO_2 enrichment, noting, however, that population reductions in this feeding guild were often accompanied by increased herbivory in response to CO_2-induced reductions in leaf nitrogen content.

In an experiment conducted on a natural ecosystem in Wisconsin, USA—comprised predominantly of trembling aspen (*Populus tremuloides* Michx.)—Percy *et al.* (2002) studied the effects of increases in CO_2 alone (to 560 ppm during daylight hours), O_3 alone (to 46.4-55.5 ppb during daylight hours), and CO_2 and O_3 together on the forest tent caterpillar (*Malacosoma disstria*), a common leaf-chewing lepidopteran found in North American hardwood forests. By itself, elevated CO_2 reduced caterpillar performance by reducing female pupal mass; while elevated O_3 alone improved caterpillar performance by increasing female pupal mass. When both gases were applied together, however, the elevated CO_2 completely counteracted the enhancement of female pupal mass caused by elevated O_3. Hence, either alone or in combination with undesirable increases in the air's O_3 concentration, elevated CO_2 tended to reduce the performance of the forest tent caterpillar. This finding is particularly satisfying because, in the words of Percy *et al.*, "historically, the forest tent catepillar has defoliated more deciduous forest than any other insect in North America," and because "outbreaks can reduce timber yield up to 90% in one year, and increase tree vulnerability to disease and environmental stress."

In a study of yet another type of insect herbivore, Brooks and Whittaker (1999) removed grassland monoliths from the Great Dun Fell of Cumbria, UK—which contained eggs of a destructive xylem-feeding spittlebug (*Neophilaenus lineatus*)—and grew them in glasshouses maintained for two years at atmospheric CO_2 concentrations of 350 and 600 ppm. During the course of their experiment, two generations of the xylem-feeding insect were produced; in each case, elevated CO_2 reduced the survival of nymphal stages by an average of 24 percent. Brooks and Whittaker suggest that this reduction in survival rate may have been caused by CO_2-induced reductions in stomatal conductance and transpirational water loss, which may have reduced xylem nutrient-water availability. Whatever the mechanism, the results of this study bode well for the future survival of these species-poor grasslands as the air's CO_2 content continues to rise.

In summing up the implications of the various phenomena described in this section, it would appear that both CO_2-induced and warming-induced changes in the physical characteristics and behavioral patterns of a diverse assemblage of insect types portend good things for the biosphere in the years and decades to come.

Additional information on this topic, including reviews of newer publications as they become available, can be found at http://www.co2science.org/subject/i/insectsother.php.

References

Brooks, G.L. and Whittaker, J.B. 1999. Responses of three generations of a xylem-feeding insect, *Neophilaenus lineatus* (Homoptera), to elevated CO_2. *Global Change Biology* **5**: 395-401.

Docherty, M., Wade, F.A., Hurst, D.K., Whittaker, J.B. and Lea, P.J. 1997. Responses of tree sap-feeding herbivores to elevated CO_2. *Global Change Biology* **3**: 51-59.

Percy, K.E., Awmack, C.S., Lindroth, R.L., Kubiske, M.E., Kopper, B.J., Isebrands, J.G., Pregitzer, K.S., Hendrey, G.R., Dickson, R.E., Zak, D.R., Oksanen, E., Sober, J., Harrington, R. and Karnosky, D.F. 2002. Altered performance of forest pests under atmospheres enriched by CO_2 and O_3. *Nature* **420**: 403-407.

Stiling, P., Moon, D.C., Hunter, M.D., Colson, J., Rossi, A.M., Hymus, G.J. and Drake, B.G. 2002. Elevated CO_2 lowers relative and absolute herbivore density across all species of a scrub-oak forest. *Oecologia* DOI 10.1007/s00442-002-1075-5.

Stiling, P., Rossi, A.M., Hungate, B., Dijkstra, P., Hinkle, C.R., Knot III, W.M., and Drake, B. 1999. Decreased leaf-miner abundance in elevated CO_2: Reduced leaf quality and increased parasitoid attack. *Ecological Applications* **9**: 240-244.

Thomas, C.D., Bodsworth, E.J., Wilson, R.J., Simmons, A.D., Davies, Z.G., Musche, M. and Conradt, L. 2001. Ecological and evolutionary processes at expanding range margins. *Nature* **411**: 577-581.

Whittaker, J.B. 1999. Impacts and responses at population level of herbivorous insects to elevated CO_2. *European Journal of Entomology* **96**: 149-156.

7.3.4. Shade

Is the growth-enhancing effect of atmospheric CO_2 enrichment reduced when light intensities are less than optimal? The question may be important if a warmer world is also a cloudier world, as some climate models predict.

In a review of the scientific literature designed to answer this question, Kerstiens (1998) analyzed the results of 15 previously published studies of trees having differing degrees of shade tolerance, finding that elevated CO_2 caused greater relative biomass increases in shade-tolerant species than in shade-intolerant or sun-loving species. In more than half of the studies analyzed, shade-tolerant species experienced CO_2-induced relative growth increases that were two to three times greater than those of less shade-tolerant species.

In an extended follow-up review analyzing 74 observations from 24 studies, Kerstiens (2001) reported that twice-ambient CO_2 concentrations increased the relative growth response of shade-tolerant and shade-intolerant woody species by an average of 51 and 18 percent, respectively. Similar results were reported by Poorter and Perez-Soba (2001), who performed a detailed meta-analysis of research results pertaining to this topic, and more recently by Kubiske *et al.* (2002), who measured photosynthetic acclimation in aspen and sugar maple trees. On the other hand, a 200-ppm increase in the air's CO_2 concentration was found to enhance the photosynthetic rates of sunlit and shaded leaves of sweetgum trees by 92 and 54 percent, respectively, at one time of year, and by 166 and 68 percent at another time (Herrick and Thomas, 1999). Likewise,

Naumburg and Ellsworth (2000) reported that a 200-ppm increase in the air's CO_2 content boosted steady-state photosynthetic rates in leaves of four hardwood understory species by an average of 60 and 40 percent under high and low light intensities, respectively. Even though these photosynthetic responses were significantly less in shaded leaves, they were still substantial, with mean increases ranging from 40 to 68 percent for a 60 percent increase in atmospheric CO_2 concentration.

Under extremely low light intensities, the benefits arising from atmospheric CO_2 enrichment may be small, but oftentimes they are very important in terms of plant carbon budgeting. In the study of Hattenschwiler (2001), for example, seedlings of five temperate forest species subjected to an additional 200-ppm CO_2 under light intensities that were only 3.4 and 1.3 percent of full sunlight exhibited CO_2-induced biomass increases that ranged from 17 to 74 percent. Similarly, in the study of Naumburg *et al.* (2001), a 200-ppm increase in the air's CO_2 content enhanced photosynthetic carbon uptake in three of four hardwood understory species by more than two-fold in three of the four species under light irradiances that were as low as 3 percent of full sunlight.

In a final study, in which potato plantlets inoculated with an arbuscular mycorrhizal fungus were grown at various light intensities and *super* CO_2 enrichment of approximately 10,000 ppm, Louche-Tessandier *et al.* (1999) found that the unusually high CO_2 concentration produced an unusually high degree of root colonization by the beneficial mycorrhizal fungus, which typically helps supply water and nutrients to plants. And it did so irrespective of the degree of light intensity to which the potato plantlets were exposed.

So, whether light intensity is high or low, or leaves are shaded or sunlit, when the CO_2 content of the air is increased, so too are the various biological processes that lead to plant robustness. Less than optimal light intensities do not negate the beneficial effects of atmospheric CO_2 enrichment.

Additional information on this topic, including reviews of newer publications as they become available, can be found at http://www.co2science.org/subject/g/lightinteraction.php.

References

Hattenschwiler, S. 2001. Tree seedling growth in natural deep shade: functional traits related to interspecific variation in response to elevated CO_2. *Oecologia* **129**: 31-42.

Herrick, J.D. and Thomas, R.B. 1999. Effects of CO_2 enrichment on the photosynthetic light response of sun and shade leaves of canopy sweetgum trees (*Liquidambar styraciflua*) in a forest ecosystem. *Tree Physiology* **19**: 779-786.

Kerstiens, G. 1998. Shade-tolerance as a predictor of responses to elevated CO_2 in trees. *Physiologia Plantarum* **102**: 472-480.

Kerstiens, G. 2001. Meta-analysis of the interaction between shade-tolerance, light environment and growth response of woody species to elevated CO_2. *Acta Oecologica* **22**: 61-69.

Kubiske, M.E., Zak, D.R., Pregitzer, K.S. and Takeuchi, Y. 2002. Photosynthetic acclimation of overstory *Populus tremuloides* and understory *Acer saccharum* to elevated atmospheric CO_2 concentration: interactions with shade and soil nitrogen. *Tree Physiology* **22**: 321-329.

Louche-Tessandier, D., Samson, G., Hernandez-Sebastia, C., Chagvardieff, P. and Desjardins, Y. 1999. Importance of light and CO_2 on the effects of endomycorrhizal colonization on growth and photosynthesis of potato plantlets (*Solanum tuberosum*) in an in vitro tripartite system. *New Phytologist* **142**: 539-550.

Naumburg, E. and Ellsworth, D.S. 2000. Photosynthetic sunfleck utilization potential of understory saplings growing under elevated CO_2 in FACE. *Oecologia* **122**: 163-174.

Naumburg, E., Ellsworth, D.S. and Katul, G.G. 2001. Modeling dynamic understory photosynthesis of contrasting species in ambient and elevated carbon dioxide. *Oecologia* **126**: 487-499.

Poorter, H. and Perez-Soba, M. 2001. The growth response of plants to elevated CO_2 under non-optimal environmental conditions. *Oecologia* **129**: 1-20.

7.3.5. Ozone

Plants grown in CO_2-enriched air nearly always exhibit increased photosynthetic rates and biomass production relative to plants grown at the current ambient CO_2 concentration. By contrast, plants exposed to elevated ozone concentrations typically display reductions in photosynthesis and growth in comparison with plants grown at the current ambient ozone concentration.

In discussing the problem of elevated tropospheric ozone (O_3) concentrations, Liu *et al.*

(2004) wrote that "ozone is considered to be one of the air pollutants most detrimental to plant growth and development in both urban and rural environments (Lefohn, 1992; Skarby *et al.*, 1998; Matyssek and Innes, 1999)," because it "reduces the growth and yield of numerous agronomic crops as well as fruit and forest trees (Retzlaff *et al.*, 1997; Fumagalli *et al.*, 2001; Matyssek and Sandermann, 2003)." In addition, they say that ozone concentrations are "currently two to three times higher than in the early 1900s (Galloway, 1998; Fowler *et al.*, 1999)," and that they likely "will remain high in the future (Elvingson, 2001)."

It is important to determine how major plants respond to concomitant increases in the abundances of these two trace gases of the atmosphere, as their concentrations will likely continue to increase for many years to come. We begin with a review of the literature with respect to various agriculture species, followed by a discussion on trees.

Additional information on this topic, including reviews on the interaction of CO_2 and O_3 not discussed here, can be found at http://www.co2 science.org/subject/o/subject_o.php under the heading Ozone.

References

Elvingson, P. 2001. For the most parts steadily down. *Acid News* **3**: 20-21.

Fowler, D., Cape, J.N., Coyle, M., Flechard, C., Kuylenstrierna, J., Hicks, K., Derwent, D., Johnson, C. and Stevenson, D. 1999. The global exposure of forests to air pollutants. In: Sheppard, L.J. and Cape, J.N. (Eds.) *Forest Growth Responses to the Pollution Climate of the 21st Century*. Kluwer Academic Publisher, Dordrecht, The Netherlands, p. 5032.

Fumagalli, I., Gimeno, B.S., Velissariou, D., De Temmerman, L. and Mills, G. 2001. Evidence of ozone-induced adverse effects on crops in the Mediterranean region. *Atmospheric Environment* **35**: 2583-2587.

Galloway, J.N. 1998. The global nitrogen cycle: changes and consequences. *Environmental Pollution* **102**: 15-24.

Lefohn, A.S. 1992. Surface Level Ozone Exposure and Their Effects on Vegetation. Lewis Publishers, Chelsea, UK.

Liu, X.-P., Grams, T.E.E., Matyssek, R. and Rennenberg, H. 2005. Effects of elevated pCO_2 and/or pO_3 on C-, N-, and S-metabolites in the leaves of juvenile beech and spruce differ between trees grown in monoculture and mixed culture. *Plant Physiology and Biochemistry* **43**: 147-154.

Liu, X., Kozovits, A.R., Grams, T.E.E., Blaschke, H., Rennenberg, H. and Matyssek, R. 2004. Competition modifies effects of ozone/carbon dioxide concentrations on carbohydrate and biomass accumulation in juvenile Norway spruce and European beech. *Tree Physiology* **24**: 1045-1055.

Matyssek, R. and Innes, J.L. 1999. Ozone—a risk factor for trees and forests in Europe? *Water, Air and Soil Pollution* **116**: 199-226.

Matyssek, R. and Sandermann, H. 2003. Impact of ozone on trees: an ecophysiological perspective. *Progress in Botany* **64**: 349-404.

Retzlaff, W.A., Williams, L.E. and DeJong, T.M. 1997. Growth and yield response of commercial bearing-age "Casselman" plum trees to various ozone partial pressures. *Journal of Environmental Quality* **26**: 858-865.

Skarby, L., Ro-Poulsen, H., Wellburn, F.A.M. and Sheppard, L.J. 1998. Impacts of ozone on forests: a European perspective. *New Phytologist* **139**: 109-122.

7.3.5.1. Agricultural Species

Several studies have used soybean as a model plant to study the effects of elevated CO_2 and ozone on photosynthesis and growth. Reid *et al.* (1998), for example, grew soybeans for an entire season at different combinations of atmospheric CO_2 and ozone, reporting that elevated CO_2 enhanced rates of photosynthesis in the presence or absence of ozone and that it typically ameliorated the negative effects of elevated ozone on carbon assimilation. At the cellular level, Heagle *et al.* (1998a) reported that at twice the current ambient ozone concentration, soybeans simultaneously exposed to twice the current ambient atmospheric CO_2 concentration exhibited less foliar injury while maintaining significantly greater leaf chlorophyll contents than control plants exposed to elevated ozone and ambient CO_2 concentrations. By harvest time, the plants grown in the elevated ozone/elevated CO_2 treatment combination had produced 53 percent more total biomass than their counterparts did at elevated ozone and ambient CO_2 concentrations (Miller *et al.*, 1998). Finally, in analyzing seed yield, it was determined that atmospheric CO_2 enrichment enhanced this parameter by 20 percent at ambient ozone, while it increased it by 74 percent at twice the ambient ozone concentration (Heagle *et al.*, 1998b). Thus, elevated CO_2 completely ameliorated the negative effects of elevated ozone concentration on photosynthetic rate and yield production in soybean.

The ameliorating responses of elevated CO_2 to ozone pollution also have been reported for various cultivars of spring and winter wheat. In the study of Tiedemann and Firsching (2000), for example, atmospheric CO_2 enrichment not only overcame the detrimental effects of elevated ozone on photosynthesis and growth, it overcame the deleterious consequences resulting from inoculation with a biotic pathogen as well. Although infected plants displayed less absolute yield than non-infected plants at elevated ozone concentrations, atmospheric CO_2 enrichment caused the greatest relative yield increase in infected plants (57 percent vs. 38 percent).

McKee *et al.* (2000) reported that O_3-induced reductions in leaf rubisco contents in spring wheat were reversed when plants were simultaneously exposed to twice-ambient concentrations of atmospheric CO_2. In the study of Vilhena-Cardoso and Barnes (2001), elevated ozone concentrations reduced photosynthetic rates in spring wheat grown at three different soil nitrogen levels. However, when concomitantly exposed to twice-ambient atmospheric CO_2 concentrations, elevated ozone had no effect on rates of photosynthesis, regardless of soil nitrogen. Going a step further, Pleijel *et al.* (2000) observed that ozone-induced reductions in spring wheat yield were partially offset by concomitant exposure to elevated CO_2 concentrations. Similar results have been reported in spring wheat by Hudak *et al.* (1999) and in winter wheat by Heagle *et al.* (2000).

Cotton plants grown at elevated ozone concentrations exhibited 25 and 48 percent reductions in leaf mass per unit area and foliar starch concentration, respectively, relative to control plants grown in ambient air. When simultaneously exposed to twice-ambient CO_2 concentrations, however, the reductions in these parameters were only 5 and 7 percent, respectively (Booker, 2000). With respect to potato, Wolf and van Oijen (2002) used a validated potato model to predict increases in European tuber production ranging from 1,000 to 3,000 kg of dry matter per hectare in spite of concomitant increases in ozone concentrations and air temperatures.

It is clear from these studies that elevated CO_2 reduces, and nearly always completely overrides, the negative effects of ozone pollution on plant photosynthesis, growth, and yield. When explaining the mechanisms behind such responses, most authors suggest that atmospheric CO_2 enrichment tends to reduce stomatal conductance, which causes less indiscriminate uptake of ozone into internal plant air spaces and reduces subsequent conveyance to tissues where damage often results to photosynthetic pigments and proteins, reducing plant growth and biomass production.

Additional information on this topic, including reviews of newer publications as they become available, can be found at http://www.co2science.org/subject/o/ozoneplantsag.php.

References

Booker, F.L. 2000. Influence of carbon dioxide enrichment, ozone and nitrogen fertilization on cotton (*Gossypium hirsutum* L.) leaf and root composition. *Plant, Cell and Environment* **23**: 573-583.

Heagle, A.S., Miller, J.E. and Booker, F.L. 1998a. Influence of ozone stress on soybean response to carbon dioxide enrichment: I. Foliar properties. *Crop Science* **38**: 113-121.

Heagle, A.S., Miller, J.E. and Pursley, W.A. 1998b. Influence of ozone stress on soybean response to carbon dioxide enrichment: III. Yield and seed quality. *Crop Science* **38**: 128-134.

Heagle, A.S., Miller, J.E. and Pursley, W.A. 2000. Growth and yield responses of winter wheat to mixtures of ozone and carbon dioxide. *Crop Science* **40**: 1656-1664.

Hudak, C., Bender, J., Weigel, H.-J. and Miller, J. 1999. Interactive effects of elevated CO_2, O_3, and soil water deficit on spring wheat (*Triticum aestivum* L. cv. Nandu). *Agronomie* **19**: 677-687.

McKee, I.F., Mulholland, B.J., Craigon, J., Black, C.R. and Long, S.P. 2000. Elevated concentrations of atmospheric CO_2 protect against and compensate for O_3 damage to photosynthetic tissues of field-grown wheat. *New Phytologist* **146**: 427-435.

Miller, J.E., Heagle, A.S. and Pursley, W.A. 1998. Influence of ozone stress on soybean response to carbon dioxide enrichment: II. Biomass and development. *Crop Science* **38**: 122-128.

Pleijel, H., Gelang, J., Sild, E., Danielsson, H., Younis, S., Karlsson, P.-E., Wallin, G., Skarby, L. and Sellden, G. 2000. Effects of elevated carbon dioxide, ozone and water availability on spring wheat growth and yield. *Physiologia Plantarum* **108**: 61-70.

Reid, C.D. and Fiscus, E.L. 1998. Effects of elevated [CO_2] and/or ozone on limitations to CO_2 assimilation in soybean

(*Glycine max*). *Journal of Experimental Botany* **18**: 885-895.

Tiedemann, A.V. and Firsching, K.H. 2000. Interactive effects of elevated ozone and carbon dioxide on growth and yield of leaf rust-infected versus non-infected wheat. *Environmental Pollution* **108**: 357-363.

Vilhena-Cardoso, J. and Barnes, J. 2001. Does nitrogen supply affect the response of wheat (*Triticum aestivum* cv. Hanno) to the combination of elevated CO_2 and O_3? *Journal of Experimental Botany* **52**: 1901-1911.

Wolf, J. and van Oijen, M. 2002. Modelling the dependence of European potato yields on changes in climate and CO_2. *Agricultural and Forest Meteorology* **112**: 217-231.

7.3.5.2. Woody Species

7.3.5.2.1. Aspen

Karnosky *et al.* (1999) grew O_3-sensitive and O_3-tolerant aspen clones in 30-m diameter plots at the Aspen FACE site near Rhinelander, Wisconsin, USA, which were maintained at atmospheric CO_2 concentrations of 360 and 560 ppm with and without exposure to elevated O_3 (1.5 times ambient ozone concentration). After one year of growth at ambient CO_2, elevated O_3 had caused visible injury to leaves of both types of aspen, with the average percent damage in O_3-sensitive clones being more than three times as great as that observed in O_3-tolerant clones (55 percent vs. 17 percent, respectively). In combination with elevated CO_2, however, O_3-induced damage to leaves of these same clones was only 38 percent and 3 percent, respectively. Thus, elevated CO_2 ameliorated much of the foliar damage induced by high O_3 concentrations.

King *et al.* (2001) studied the same plants for a period of two years, concentrating on below-ground growth, where elevated O_3 alone had no effect on fine-root biomass. When the two aspen clones were simultaneously exposed to elevated CO_2 and O_3, however, there was an approximate 66 percent increase in the fine-root biomass of both of them.

Also in the same experiment, Noormets *et al.* (2001) studied the interactive effects of O_3 and CO_2 on photosynthesis, finding that elevated CO_2 increased rates of photosynthesis in both clones at all leaf positions. Maximum rates of photosynthesis were increased in the O_3-tolerant clone by averages of 33 and 49 percent due to elevated CO_2 alone and in

combination with elevated O_3, respectively, while in the O_3-sensitive clone they were increased by 38 percent in both situations. Hence, CO_2-induced increases in maximal rates of net photosynthesis were typically maintained, and sometimes increased, during simultaneous exposure to elevated O_3.

Yet again in the same experiment, Oksanen *et al.* (2001) reported that after three years of treatment, ozone exposure caused significant structural injuries to thylakoid membranes and the stromal compartment within chloroplasts, but that these injuries were largely ameliorated by atmospheric CO_2 enrichment. Likewise, leaf thickness, mesophyll tissue thickness, the amount of chloroplasts per unit cell area, and the amount of starch in leaf chloroplasts were all decreased in the high ozone treatment, but simultaneous exposure of the ozone-stressed trees to elevated CO_2 more than compensated for the ozone-induced reductions.

After four years of growing five aspen clones with varying degrees of tolerance to ozone under the same experimental conditions, McDonald *et al.* (2002) developed what they termed a "competitive stress index," based on the heights of the four nearest neighbors of each tree, to study the influence of competition on the CO_2 growth response of the various clones as modified by ozone. In general, elevated O_3 reduced aspen growth independent of competitive status, while the authors noted an "apparent convergence of competitive performance responses in $+CO_2$ $+O_3$ conditions," which they say suggests that "stand diversity may be maintained at a higher level" in such circumstances.

Percy *et al.* (2002) utilized the same experimental setting to assess a number of the trees' growth characteristics, as well as the responses of one plant pathogen and two insects with different feeding strategies that typically attack the trees. Of the plant pathogen studied, they say that "the poplar leaf rust, *Melampsora medusae*, is common on aspen and belongs to the most widely occurring group of foliage diseases." As for the two insects, they report that "the forest tent caterpillar, *Malacosoma disstria*, is a common leaf-chewing lepidopteran in North American hardwood forests" and that "the sap-feeding aphid, *Chaitophorus stevensis*, infests aspen throughout its range." Hence, the rust and the two insect pests the scientists studied are widespread and have significant deleterious impacts on trembling aspen and other tree species. As but one example of this fact, the authors note that, "historically, the forest tent caterpillar has defoliated more deciduous forest than any other insect in North America" and that "outbreaks can reduce timber yield up to 90% in one year, and increase tree vulnerability to disease and environmental stress."

Percy *et al.* found that by itself, elevated O_3 decreased tree height and trunk diameter, increased rust occurrence by nearly fourfold, improved tent caterpillar performance by increasing female pupal mass by 31 percent, and had a strong negative effect on the natural enemies of aphids. The addition of the extra CO_2, however, completely ameliorated the negative effects of elevated O_3 on tree height and trunk diameter, reduced the O_3-induced enhancement of rust development from nearly fourfold to just over twofold, completely ameliorated the enhancement of female tent caterpillar pupal mass caused by elevated O_3, and completely ameliorated the reduction in the abundance of natural enemies of aphids caused by elevated O_3.

In a final study from the Aspen FACE site, Holton *et al.* (2003) raised parasitized and non-parasitized forest tent caterpillars on two quaking aspen genotypes (O_3-sensitive and O_3-tolerant) alone and in combination for one full growing season; they, too, found that elevated O_3 improved tent caterpillar performance under ambient CO_2 conditions, but not in CO_2-enriched air.

In summary, it is clear that elevated ozone concentrations have a number of significant negative impacts on the well-being of North America's most widely distributed tree species, while elevated carbon dioxide concentrations have a number of significant positive impacts. In addition, elevated CO_2 often completely eliminates the negative impacts of elevated O_3. If the tropospheric O_3 concentration continues to rise as expected (Percy *et al.* note that "damaging O_3 concentrations currently occur over 29% of the world's temperate and subpolar forests but are predicted to affect fully 60% by 2100"), we might hope the air's CO_2 content continues to rise as well.

Additional information on this topic, including reviews of newer publications as they become available, can be found at http://www.co2science.org/subject/o/ozoneaspen.php.

References

Holton, M.K., Lindroth, R.L. and Nordheim, E.V. 2003. Foliar quality influences tree-herbivore-parasitoid interactions: effects of elevated CO_2, O_3, and plant genotype. *Oecologia* **137**: 233-244.

Karnosky, D.F., Mankovska, B., Percy, K., Dickson, R.E., Podila, G.K., Sober, J., Noormets, A., Hendrey, G., Coleman, M.D., Kubiske, M., Pregitzer, K.S. and Isebrands, J.G. 1999. Effects of tropospheric O_3 on trembling aspen and interaction with CO_2: results from an O_3-gradient and a FACE experiment. *Water, Air, and Soil Pollution* **116**: 311-322.

King, J.S., Pregitzer, K.S., Zak, D.R., Sober, J., Isebrands, J.G., Dickson, R.E., Hendrey, G.R. and Karnosky, D.F. 2001. Fine-root biomass and fluxes of soil carbon in young stands of paper birch and trembling aspen as affected by elevated atmospheric CO_2 and tropospheric O_3. *Oecologia* **128**: 237-250.

McDonald, E.P., Kruger, E.L., Riemenschneider, D.E. and Isebrands, J.G. 2002. Competitive status influences tree-growth responses to elevated CO_2 and O_3 in aggrading aspen stands. *Functional Ecology* **16**: 792-801.

Noormets, A., Sober, A., Pell, E.J., Dickson, R.E., Podila, G.K., Sober, J., Isebrands, J.G. and Karnosky, D.F. 2001. Stomatal and non-stomatal limitation to photosynthesis in two trembling aspen (*Populus tremuloides* Michx.) clones exposed to elevated CO_2 and O_3. *Plant, Cell and Environment* **24**: 327-336.

Oksanen, E., Sober, J. and Karnosky, D.F. 2001. Impacts of elevated CO_2 and/or O_3 on leaf ultrastructure of aspen (*Populus tremuloides*) and birch (*Betula papyrifera*) in the Aspen FACE experiment. *Environmental Pollution* **115**: 437-446.

Percy, K.E., Awmack, C.S., Lindroth, R.L., Kubiske, M.E., Kopper, B.J., Isebrands, J.G., Pregitzer, K.S., Hendrey, G.R., Dickson, R.E., Zak, D.R., Oksanen, E., Sober, J., Harrington, R. and Karnosky, D.F. 2002. Altered performance of forest pests under atmospheres enriched by CO_2 and O_3. *Nature* **420**: 403-407.

7.3.5.2.2. Beech

Liu *et al.* (2005) grew three- and four-year-old seedlings of European beech (*Fagus sylvatica* L.) for five months in well watered and fertilized soil in containers located within walk-in phytotrons maintained at either ambient or ambient + 300 ppm CO_2 (each subdivided into ambient and double-ambient O_3 concentration treatments, with maximum ozone levels restricted to <150 ppb), in both monoculture and in competition with Norway spruce, after which they examined the effects of each treatment on leaf non-structural carbohydrate levels (soluble sugars and starch). They found that the effects of elevated O_3 alone on non-structural carbohydrate levels were small when the beech seedlings were grown in monoculture. When they were grown in mixed culture, however, the elevated O_3 slightly enhanced leaf sugar levels, but reduced starch levels by 50 percent.

With respect to elevated CO_2 alone, for the beech seedlings grown in both monoculture and mixed culture, levels of sugar and starch were significantly enhanced. Hence, when elevated O_3 and CO_2 significantly affected non-structural carbohydrate levels, elevated CO_2 tended to enhance them, whereas elevated O_3 tended to reduce them. In addition, the combined effects of elevated CO_2 and O_3 acting together were such as to produce a significant increase in leaf non-structural carbohydrates in both mixed and monoculture conditions. As a result, the researchers concluded that "since the responses to the combined exposure were more similar to elevated pCO_2 than to elevated pO_3, apparently elevated pCO_2 overruled the effects of elevated pO_3 on non-structural carbohydrates."

In a slightly longer study, Grams et al. (1999) grew European beech seedlings in glasshouses maintained at average atmospheric CO_2 concentrations of either 367 or 667 ppm for a period of one year. Then, throughout the following year, in addition to being exposed to the same set of CO_2 concentrations the seedlings were exposed to either ambient or twice-ambient levels of O_3. This protocol revealed that elevated O_3 significantly reduced photosynthesis in beech seedlings grown at ambient CO_2 concentrations by a factor of approximately three. By contrast, in the CO_2-enriched air the seedlings did not exhibit any photosynthetic reduction due to the doubled O_3 concentrations. In fact, the photosynthetic rates of the CO_2-enriched seedlings actually rose by 8 percent when simultaneously fumigated with elevated O_3, leading the researchers to conclude that "long-term acclimation to elevated CO_2 supply does counteract the O_3-induced decline of photosynthetic light and dark reactions."

In a still longer study, Liu et al. (2004) grew three- and four-year-old beech seedlings for two growing seasons under the same experimental conditions as Liu et al. (2005) after the seedlings had been pre-acclimated for one year to either the ambient or elevated CO_2 treatment. At the end of the study, the plants were harvested and fresh weights and dry biomass values were determined for leaves, shoot axes, coarse roots, and fine roots, as were carbohydrate (starch and soluble sugar) contents and concentrations for the same plant parts. This work falsified the hypothesis that "prolonged exposure to

elevated CO_2 does not compensate for the adverse ozone effects on European beech," as it revealed that all "adverse effects of ozone on carbohydrate concentrations and contents were counteracted when trees were grown in elevated CO_2."

Additional information on this topic, including reviews of newer publications as they become available, can be found at http://www.co2science.org/subject/o/ozonebeech.php.

References

Grams, T.E.E., Anegg, S., Haberle, K.-H., Langebartels, C. and Matyssek, R. 1999. Interactions of chronic exposure to elevated CO_2 and O_3 levels in the photosynthetic light and dark reactions of European beech (*Fagus sylvatica*). *New Phytologist* **144**: 95-107.

Liu, X.-P., Grams, T.E.E., Matyssek, R. and Rennenberg, H. 2005. Effects of elevated pCO_2 and/or pO_3 on C-, N-, and S-metabolites in the leaves of juvenile beech and spruce differ between trees grown in monoculture and mixed culture. *Plant Physiology and Biochemistry* **43**: 147-154.

Liu, X., Kozovits, A.R., Grams, T.E.E., Blaschke, H., Rennenberg, H. and Matyssek, R. 2004. Competition modifies effects of ozone/carbon dioxide concentrations on carbohydrate and biomass accumulation in juvenile Norway spruce and European beech. *Tree Physiology* **24**: 1045-1055.

7.3.5.2.3. Birch

At the FACE facility near Rhinelander, Wisconsin, USA, King et al. (2001) grew a mix of paper birch and quaking aspen trees in 30-m diameter plots that were maintained at atmospheric CO_2 concentrations of 360 and 560 ppm with and without exposure to elevated O_3 (1.5 times the ambient O_3 concentration) for a period of two years. In their study of the below-ground environment of the trees, they found that the extra O_3 had no effect on the growth of fine roots over that time period, but that elevated O_3 and CO_2 together increased the fine-root biomass of the mixed stand by 83 percent.

One year later at the same FACE facility, Oksanen et al. (2001) observed O_3-induced injuries in the thylokoid membranes of the chloroplasts of the birch trees' leaves; the injuries were partially ameliorated in the elevated CO_2 treatment. And in a study conducted two years later, Oksanen et al.

(2003) say they "were able to visualize and locate ozone-induced H_2O_2 accumulation within leaf mesophyll cells, and relate oxidative stress with structural injuries." However, they report that "H_2O_2 accumulation was found only in ozone-exposed leaves and not in the presence of elevated CO_2," adding that "CO_2 enrichment appears to alleviate chloroplastic oxidative stress."

Across the Atlantic in Finland, Kull et al. (2003) constructed open-top chambers around two clones (V5952 and K1659) of silver birch saplings that were rooted in the ground and had been growing there for the past seven years. These chambers were fumigated with air containing 360 and 720 ppm CO_2 in combination with 30 and 50 ppb O_3 for two growing seasons, after which it was noted that the extra O_3 had significantly decreased branching in the trees' crowns. This malady, however, was almost completely ameliorated by a doubling of the air's CO_2 concentration. In addition, after one more year of study, Eichelmann et al. (2004) reported that, by itself, the increase in the air's CO_2 content increased the average net photosynthetic rates of both clones by approximately 16 percent, while the increased O_3 by itself caused a 10 percent decline in the average photosynthetic rate of clone V5952, but not of clone K1659. When both trace gases were simultaneously increased, however, the photosynthetic rate of clone V5952 once again experienced a 16 percent increase in net photosynthesis, as if the extra O_3 had had no effect when applied in the presence of the extra CO_2.

After working with the same trees for one additional year, Riikonen et al. (2004) harvested them and reported finding that "the negative effects of elevated O_3 were found mainly in ambient CO_2, not in elevated CO_2." In fact, whereas doubling the air's O_3 concentration decreased total biomass production by 13 percent across both clones, simultaneously doubling the air's CO_2 concentration increased total biomass production by 30 percent, thereby more than compensating for the deleterious consequences of doubling the atmospheric ozone concentration.

In commenting on this ameliorating effect of elevated CO_2, the team of Finnish scientists said it "may be associated with either increased detoxification capacity as a consequence of higher carbohydrate concentrations in leaves grown in elevated CO_2, or decreased stomatal conductance and thus decreasing O_3 uptake in elevated CO_2 conditions (e.g., Rao et al., 1995)." They also noted that "the ameliorating effect of elevated CO_2 is in accordance with the results of single-season open-top chamber and growth chamber studies on small saplings of various deciduous tree species (Mortensen 1995; Dickson et al., 1998; Loats and Rebbeck, 1999) and long-term open-field and OTC studies with aspen and yellow-poplar (Percy et al., 2002; Rebbeck and Scherzer, 2002)."

In another paper to come out of the Finnish silver birch study, Peltonen et al. (2005) evaluated the impacts of doubled atmospheric CO_2 and O_3 concentrations on the accumulation of 27 phenolic compounds in the leaves of the trees, finding that elevated CO_2 increased the concentration of phenolic acids (+25 percent), myricetin glycosides (+18 percent), catechin derivatives (+13 percent), and soluble condensed tannins (+19 percent). Elevated O_3, on the other hand, increased the concentration of one glucoside by 22 percent, chlorogenic acid by 19 percent, and flavone aglycons by 4 percent. However, Peltonen et al. say this latter O_3-induced production of antioxidant phenolic compounds "did not seem to protect the birch leaves from detrimental O_3 effects on leaf weight and area, but may have even exacerbated them." Last, in the combined elevated CO_2 and O_3 treatment, they found that "elevated CO_2 did seem to protect the leaves from elevated O_3 because all the O_3-derived effects on the leaf phenolics and traits were prevented by elevated CO_2."

Meanwhile, back at the FACE facility near Rhinelander, Wisconsin, USA, Agrell et al. (2005) examined the effects of ambient and elevated concentrations of atmospheric CO_2 and O_3 on the foliar chemistry of birch and aspen trees, plus the consequences of these effects for host plant preferences of forest tent caterpillar larvae. In doing so, they found that "the only chemical component showing a somewhat consistent co-variation with larval preferences was condensed tannins," and they discovered that "the tree becoming relatively less preferred as a result of CO_2 or O_3 treatment was in general also the one for which average levels of condensed tannins were most positively (or least negatively) affected by that treatment." The mean condensed tannin concentration of birch leaves was 18 percent higher in the elevated CO_2 and O_3 treatment. Consequently, as atmospheric concentrations of CO_2 and O_3 continue to rise, the increases in condensed tannin concentrations likely to occur in the foliage of birch trees should lead to their leaves becoming less preferred for consumption by the forest tent caterpillar.

Concurrent with the work of Agrell et al., King et al. (2005) evaluated the effect of CO_2 enrichment alone, O_3 enrichment alone, and the net effect of both CO_2 and O_3 enrichment together on the growth of the Rhinelander birch trees, finding that relative to the ambient-air control treatment, elevated CO_2 increased total biomass by 45 percent in the aspen-birch community, while elevated O_3 caused a 13 percent reduction in total biomass relative to the control. Of most interest, the combination of elevated CO_2 and O_3 resulted in a total biomass increase of 8.4 percent relative to the control aspen-birch community. King *et al.* thus concluded that "exposure to even moderate levels of O_3 significantly reduces the capacity of net primary productivity to respond to elevated CO_2 in some forests." Consequently, they suggested it makes sense to move forward with technologies that reduce anthropogenic precursors to photochemical O_3 formation, because the implementation of such a policy would decrease an important constraint on the degree to which forest ecosystems can positively respond to the ongoing rise in the air's CO_2 concentration.

Another paper to come out of the Finnish silver birch study was that of Kostiainen *et al.* (2006), who studied the effects of elevated CO_2 and O_3 on various wood properties. Their work revealed that the elevated CO_2 treatment had no effect on wood structure, but that it increased annual ring width by 21 percent, woody biomass by 23 percent, and trunk starch concentration by 7 percent. Elevated O_3, on the other hand, decreased stem vessel percentage in one of the clones by 10 percent; it had no effect on vessel percentage in the presence of elevated CO_2.

In discussing their results, Kostiainen *et al.* note that "in the xylem of angiosperms, water movement occurs principally in vessels (Kozlowski and Pallardy, 1997)," and that "the observed decrease in vessel percentage by elevated O_3 may affect water transport," lowering it. However, as they continue, "elevated CO_2 ameliorated the O_3-induced decrease in vessel percentage." In addition, they note that "the concentration of nonstructural carbohydrates (starch and soluble sugars) in tree tissues is considered a measure of carbon shortage or surplus for growth (Korner, 2003)." They conclude that "starch accumulation observed under elevated CO_2 in this study indicates a surplus of carbohydrates produced by enhanced photosynthesis of the same trees (Riikonen *et al.*, 2004)." In addition, they report that "during winter, starch reserves in the stem are gradually transformed to soluble carbohydrates involved in freezing tolerance (Bertrand *et al.*, 1999; Piispanen and Saranpaa, 2001), so the increase in

starch concentration may improve acclimation in winter."

Rounding out the suite of Rhinelander FACE studies of paper birch is the report of Darbah *et al.* (2007), who found that the total number of trees that flowered increased by 139 percent under elevated CO_2 but only 40 percent under elevated O_3. Likewise, with respect to the quantity of flowers produced, they found that elevated CO_2 led to a 262 percent increase, while elevated O_3 led to only a 75 percent increase. They also determined that elevated CO_2 had significant positive effects on birch catkin size, weight, and germination success rate, with elevated CO_2 increasing the germination rate of birch by 110 percent, decreasing seedling mortality by 73 percent, increasing seed weight by 17 percent, and increasing new seedling root length by 59 percent. They found just the *opposite* was true of elevated O_3, as it decreased the germination rate of birch by 62 percent, decreased seed weight by 25 percent, and increased new seedling root length by only 15 percent.

In discussing their findings, Darbah *et al.* additionally report that "the seeds produced under elevated O_3 had much less stored carbohydrate, lipids, and proteins for the newly developing seedlings to depend on and, hence, the slow growth rate." As a result, they conclude that "seedling recruitment will be enhanced under elevated CO_2 but reduced under elevated O_3."

In summary, from their crowns to their roots, birch trees are generally negatively affected by rising ozone concentrations. When the air's CO_2 content is also rising, however, these negative consequences may often be totally eliminated and replaced by positive responses.

Additional information on this topic, including reviews of newer publications as they become available, can be found at http://www.co2science.org/subject/o/ozonebirch.php.

References

Agrell, J., Kopper, B., McDonald, E.P. and Lindroth, R.L. 2005. CO_2 and O_3 effects on host plant preferences of the forest tent caterpillar (*Malacosoma disstria*). *Global Change Biology* **11**: 588-599.

Bertrand, A., Robitaille, G., Nadeau, P. and Castonguay, Y. 1999. Influence of ozone on cold acclimation in sugar maple seedlings. *Tree Physiology* **19**: 527-534.

Darbah, J.N.T., Kubiske, M.E., Nelson, N., Oksanen, E., Vaapavuori, E. and Karnosky, D.F. 2007. Impacts of elevated atmospheric CO_2 and O_3 on paper birch (*Betula papyrifera*): Reproductive fitness. *The Scientific World Journal* 7(S1): 240-246.

Dickson, R.E., Coleman, M.D., Riemenschneider, D.E., Isebrands, J.G., Hogan, G.D. and Karnosky, D.F. 1998. Growth of five hybrid poplar genotypes exposed to interacting elevated [CO_2] and [O_3]. *Canadian Journal of Forest Research* 28: 1706-1716.

Eichelmann, H., Oja, V., Rasulov, B., Padu, E., Bichele, I., Pettai, H., Mols, T., Kasparova, I., Vapaavuori, E. and Laisk, A. 2004. Photosynthetic parameters of birch (*Betula pendula* Roth) leaves growing in normal and in CO_2- and O_3-enriched atmospheres. *Plant, Cell and Environment* 27: 479-495.

King, J.S., Kubiske, M.E., Pregitzer, K.S., Hendrey, G.R., McDonald, E.P., Giardina, C.P., Quinn, V.S. and Karnosky, D.F. 2005. Tropospheric O_3 compromises net primary production in young stands of trembling aspen, paper birch and sugar maple in response to elevated atmospheric CO_2. *New Phytologist* 168: 623-636.

King, J.S., Pregitzer, K.S., Zak, D.R., Sober, J., Isebrands, J.G., Dickson, R.E., Hendrey, G.R. and Karnosky, D.F. 2001. Fine-root biomass and fluxes of soil carbon in young stands of paper birch and trembling aspen as affected by elevated atmospheric CO_2 and tropospheric O_3. *Oecologia* 128: 237-250.

Korner, C. 2003. Carbon limitation in trees. *Journal of Ecology* 91: 4-17.

Kostiainen, K., Jalkanen, H., Kaakinen, S., Saranpaa, P. and Vapaavuori, E. 2006. Wood properties of two silver birch clones exposed to elevated CO_2 and O_3. *Global Change Biology* 12: 1230-1240.

Kozlowski, T.T. and Pallardy, S.G. 1997. *Physiology of Woody Plants*. Academic Press, San Diego, CA, USA.

Kull, O., Tulva, I. and Vapaavuori, E. 2003. Influence of elevated CO_2 and O_3 on *Betula pendula* Roth crown structure. *Annals of Botany* 91: 559-569.

Loats, K.V. and Rebbeck, J. 1999. Interactive effects of ozone and elevated carbon dioxide on the growth and physiology of black cherry, green ash, and yellow-poplar seedlings. *Environmental Pollution* 106: 237-248.

Mortensen, L.M. 1995. Effects of carbon dioxide concentration on biomass production and partitioning in *Betula pubescens* Ehrh. seedlings at different ozone and temperature regimes. *Environmental Pollution* 87: 337-343.

Oksanen, E., Haikio, E., Sober, J. and Karnosky, D.F. 2003. Ozone-induced H_2O_2 accumulation in field-grown aspen and birch is linked to foliar ultrastructure and peroxisomal activity. *New Phytologist* 161: 791-799.

Oksanen, E., Sober, J. and Karnosky, D.F. 2001. Impacts of elevated CO_2 and/or O_3 on leaf ultrastructure of aspen (*Populus tremuloides*) and birch (*Betula papyrifera*) in the Aspen FACE experiment. *Environmental Pollution* 115: 437-446.

Peltonen, P.A., Vapaavuori, E. and Julkunen-Tiitto, R. 2005. Accumulation of phenolic compounds in birch leaves is changed by elevated carbon dioxide and ozone. *Global Change Biology* 11: 1305-1324.

Percy, K.E., Awmack, C.S., Lindroth, R.L., Kubiske, M.E., Kopper, B.J., Isebrands, J.G., Pregitzer, K.S., Hendrey, G.R., Dickson, R.E., Zak, D.R., Oksanen, E., Sober, J., Harrington, R. and Karnosky, D.F. 2002. Altered performance of forest pests enriched by CO_2 and O_3. *Nature* 420: 403-407.

Piispanen, R. and Saranpaa, P. 2001. Variation of non-structural carbohydrates in silver birch (*Betula pendula* Roth) wood. *Trees* 15: 444-451.

Rao, M.V., Hale, B.A. and Ormrod, D.P. 1995. Amelioration of ozone-induced oxidative damage in wheat plants grown under high carbon dioxide. *Plant Physiology* 109: 421-432.

Rebbeck, J. and Scherzer, A.J. 2002. Growth responses of yellow-poplar (*Liriodendron tulipifera* L.) exposed to 5 years of [O_3] alone or combined with elevated [CO_2]. *Plant, Cell and Environment* 25: 1527-1537.

Riikonen, J., Lindsberg, M.-M., Holopainen, T., Oksanen, E., Lappi, J., Peltonen, P. and Vapaavuori, E. 2004. Silver birch and climate change: variable growth and carbon allocation responses to elevated concentrations of carbon dioxide and ozone. *Tree Physiology* 24: 1227-1237.

7.3.5.2.4. Yellow-Poplar

Scherzel *et al.* (1998) grew yellow-poplar seedlings in open-top chambers for four years at three different combinations of atmospheric O_3 and CO_2—(1) ambient O_3 and ambient CO_2, (2) doubled O_3 and ambient CO_2, and (3) doubled O_3 and doubled CO_2—to study the interactive effects of these gases on leaf-litter decomposition. This experiment revealed that the decomposition rates of yellow-poplar leaves were similar for all three treatments for nearly five months, after which time litter produced in the elevated O_3 and elevated CO_2 air decomposed at a significantly slower rate, such that even after two years of decomposition, litter from the elevated O_3 and elevated CO_2 treatment still contained about 12 percent more biomass than litter produced in the other two treatments. This reduced rate of decomposition under elevated O_3 and CO_2 conditions will likely result in greater carbon sequestration in soils supporting yellow-poplar trees over the next century or more.

Loats and Rebbeck (1999) grew yellow-poplar seedlings for ten weeks in pots they placed within growth chambers filled with ambient air, air with twice the ambient CO_2 concentration, air with twice the ambient O_3 concentration, and air with twice the ambient CO_2 and O_3 concentrations to determine the effects of elevated CO_2 and O_3 on photosynthesis and growth in this deciduous tree species. In doing so, they found that doubling the air's CO_2 concentration increased the rate of net photosynthesis by 55 percent in ambient O_3 air, and that at twice the ambient level of O_3 it stimulated net photosynthesis by an average of 50 percent. Similarly, the doubled CO_2 concentration significantly increased total biomass by 29 percent, while the doubled O_3 concentration had little impact on growth.

Last, Rebbeck et al. (2004) grew yellow poplar seedlings for five years within open-top chambers in a field plantation at Delaware, Ohio, USA, exposing them continuously from mid-May through mid-October of each year to either (1) charcoal-filtered air to remove ambient O_3, (2) ambient O_3, (3) 1.5 times ambient O_3, and (4) 1.5 times ambient O_3 plus 350 ppm CO_2 above ambient CO_2, while they periodically measured a number of plant parameters and processes. Throughout the study, the trees were never fertilized, and they received no supplemental water beyond some given in the first season.

Averaged over the experiment's five growing seasons, the midseason net photosynthetic rate of upper canopy foliage at saturating light intensities declined by 10 percent when the trees were grown in ambient O_3-air and by 14 percent when they were grown in elevated O_3-air, when compared to the trees that were grown in the charcoal-filtered air, while seasonal net photosynthesis of foliage grown in the combination of elevated O_3 and elevated CO_2 was 57-80 percent higher than it was in the trees exposed to elevated O_3 alone. There was also no evidence of any photosynthetic down regulation in the trees exposed to the elevated O_3 and CO_2 air, with some of the highest rates being observed during the final growing season. Consequently, Rebbeck et al. concluded that "elevated CO_2 may ameliorate the negative effects of increased tropospheric O_3 on yellow-poplar." In fact, their results suggest that a nominally doubled atmospheric CO_2 concentration more than compensates for the deleterious effects of a 50 percent increase in ambient O_3 levels.

As the air's CO_2 content continues to rise, earth's yellow-poplar trees will likely display substantial increases in photosynthetic rate and biomass production, even under conditions of elevated O_3 concentrations; and the soils in which the trees grow should sequester ever greater quantities of carbon.

Additional information on this topic, including reviews of newer publications as they become available, can be found at http://www.co2science.org/subject/o/ozoneyellowpoplar.php.

References

Loats, K.V. and Rebbeck, J. 1999. Interactive effects of ozone and elevated carbon dioxide on the growth and physiology of black cherry, green ash, and yellow-poplar seedlings. *Environmental Pollution* **106**: 237-248.

Rebbeck, J., Scherzer, A.J. and Loats, K.V. 2004. Foliar physiology of yellow-poplar (*Liriodendron tulipifera* L.) exposed to O_3 and elevated CO_2 over five seasons. *Trees* **18**: 253-263.

Scherzel, A.J., Rebbeck, J. and Boerner, R.E.J. 1998. Foliar nitrogen dynamics and decomposition of yellow-poplar and eastern white pine during four seasons of exposure to elevated ozone and carbon dioxide. *Forest Ecology and Management* **109**: 355-366.

7.3.6. Low Temperatures

Only a handful of studies have attempted to determine what relationship, if any, exists between atmospheric CO_2 enrichment and the ability of plants to withstand the rigors of low temperatures.

Loik et al. (2000) grew three *Yucca* species (*brevifolia*, *schidigera*, and *whipplei*) in pots placed within glasshouses maintained at atmospheric CO_2 concentrations of 360 and 700 ppm and day/night air temperatures of 40/24°C for seven months, after which some of the plants were subjected to a two-week day/night air temperature treatment of 20/5°C. In addition, leaves from each *Yucca* species were removed and placed in a freezer that was cooled at a rate of 3°C per hour until a minimum temperature of -15°C was reached. These manipulations indicated that elevated CO_2 lowered the air temperature at which 50 percent low-temperature-induced cell mortality occurred by 1.6, 1.4 and 0.8°C in *brevifolia*, *schidigera* and *whipplei*, respectively. On the basis of the result obtained for *Y. brevifolia*, Dole et al. (2003) estimated that "the increase in freezing tolerance caused by doubled CO_2 would increase the potential habitat of this species by 14%."

By contrast, Obrist *et al.* (2001) observed the opposite response. In an open-top chamber study of a temperate grass ecosystem growing on a nutrient-poor calcareous soil in northwest Switzerland, portions of which had been exposed to atmospheric CO_2 concentrations of 360 and 600 ppm for a period of six years, they determined that the average temperature at which 50 percent low-temperature-induced leaf mortality occurred in five prominent species actually *rose* by an average of 0.7°C in response to the extra 240 ppm of CO_2 employed in their experiment.

Most relevant investigations, however, have produced evidence of positive CO_2 effects on plant low temperature tolerance. Sigurdsson (2001), for example, grew black cottonwood seedlings near Gunnarsholt, Iceland within closed-top chambers maintained at ambient and twice-ambient atmospheric CO_2 concentrations for a period of three years, finding that elevated CO_2 tended to hasten the end of the growing season. This effect was interpreted as enabling the seedlings to better avoid the severe cold-induced dieback of newly produced tissues that often occurs with the approach of winter in this region. Likewise, Wayne *et al.* (1998) found that yellow birch seedlings grown at an atmospheric CO_2 concentration of 800 ppm exhibited greater dormant bud survivorship at low air temperatures than did seedlings grown at 400 ppm CO_2.

Schwanz and Polle (2001) investigated the effects of elevated CO_2 on chilling stress in micropropagated hybrid poplar clones that were subsequently potted and transferred to growth chambers maintained at either ambient (360 ppm) or elevated (700 ppm) CO_2 for a period of three months. They determined that "photosynthesis was less diminished and electrolyte leakage was lower in stressed leaves from poplar trees grown under elevated CO_2 as compared with those from ambient CO_2." Although severe chilling did cause pigment and protein degradation in all stressed leaves, the damage was expressed to a lower extent in leaves from the elevated CO_2 treatment. This CO_2-induced chilling protection was determined to be accompanied by a rapid induction of superoxide dismutase activity, as well as by slightly higher stabilities of other antioxidative enzymes.

Another means by which chilling-induced injury may be reduced in CO_2-enriched air is suggested by the study of Sgherri *et al.* (1998), who reported that raising the air's CO_2 concentration from 340 to 600 ppm increased lipid concentrations in alfalfa

thylakoid membranes while simultaneously inducing a higher degree of unsaturation in the most prominent of those lipids. Under well-watered conditions, for example, the 76 percent increase in atmospheric CO_2 enhanced overall thylakoid lipid concentration by about 25 percent, while it increased the degree of unsaturation of the two main lipids by approximately 17 percent and 24 percent. Under conditions of water stress, these responses were found to be even greater, as thylakoid lipid concentration rose by approximately 92 percent, while the degree of unsaturation of the two main lipids rose by about 22 percent and 53 percent.

Several studies conducted over the past decade explain what these observations have to do with a plant's susceptibility to chilling injury. Working with wild-type *Arabidopsis thaliana* and two mutants deficient in thylakoid lipid unsaturation, Hugly and Somerville (1992) found that "chloroplast membrane lipid polyunsaturation contributes to the low-temperature fitness of the organism," and that it "is required for some aspect of chloroplast biogenesis." When lipid polyunsaturation was low, they observed "dramatic reductions in chloroplast size, membrane content, and organization in developing leaves." There was a positive correlation "between the severity of chlorosis in the two mutants at low temperatures and the degree of reduction in polyunsaturated chloroplast lipid composition."

Working with tobacco, Kodama *et al.* (1994) demonstrated that the low-temperature-induced suppression of leaf growth and concomitant induction of chlorosis observed in wild-type plants was much less evident in transgenic plants containing a gene that allowed for greater expression of unsaturation in the fatty acids of leaf lipids. This observation and others led them to conclude that substantially unsaturated fatty acids "are undoubtedly an important factor contributing to cold tolerance."

In a closely related study, Moon *et al.* (1995) found that heightened unsaturation of the membrane lipids of chloroplasts stabilized the photosynthetic machinery of transgenic tobacco plants against low-temperature photoinibition "by accelerating the recovery of the photosystem II protein complex." Likewise, Kodama *et al.* (1995), also working with transgenic tobacco plants, showed that increased fatty acid desaturation is one of the prerequisites for normal leaf development at low, nonfreezing temperatures; and Ishizaki-Nishizawa *et al.* (1996) demonstrated that transgenic tobacco plants with a reduced level of saturated fatty acids in most membrane lipids "exhibited a significant increase in chilling resistance."

Many economically important crops, such as rice, maize and soybeans, are classified as chilling-sensitive and experience injury or death at temperatures between 0 and 15°C (Lyons, 1973). If atmospheric CO_2 enrichment enhances their production and degree-of-unsaturation of thylakoid lipids, as it does in alfalfa, a continuation of the ongoing rise in the air's CO_2 content could increase the abilities of these important agricultural species to withstand periodic exposure to debilitating low temperatures. This phenomenon could provide the extra boost in food production that will be needed to sustain an increasing population in the years and decades ahead.

Earth's natural ecosystems would also benefit from a CO_2-induced increase in thylakoid lipids containing more-highly unsaturated fatty acids. Many plants of tropical origin, for example, suffer cold damage when temperatures fall below 20°C (Graham and Patterson, 1982). With improved lipid characteristics provided by the ongoing rise in the air's CO_2 content, such plants would be able to expand their ranges both poleward and upward in a higher-CO_2 world.

More research remains to be done before we can accurately assess the extent of these potential biological benefits. In particular, we must conduct more studies of the effects of atmospheric CO_2 enrichment on the properties of thylakoid lipids in a greater variety of plants. In the same experiments, we must assess the efficacy of these lipid property changes in enhancing plant tolerance of low temperatures. Such studies should rank high on the to-do list of relevant funding agencies.

Additional information on this topic, including reviews of newer publications as they become available, can be found at http://www.co2science.org/subject/f/frosthardiness.php.

References

Dole, K.P., Loik, M.E. and Sloan, L.C. 2003. The relative importance of climate change and the physiological effects of CO_2 on freezing tolerance for the future distribution of *Yucca brevifolia*. *Global and Planetary Change* **36**: 137-146.

Graham, D. and Patterson, B.D. 1982. Responses of plants to low, non-freezing temperatures: proteins, metabolism,

and acclimation. *Annual Review of Plant Physiology* **33**: 347-372.

Hugly, S. and Somerville, C. 1992. A role for membrane lipid polyunsaturation in chloroplast biogenesis at low temperature. *Plant Physiology* **99**: 197-202.

Ishizaki-Nishizawa, O., Fujii, T., Azuma, M., Sekiguchi, K., Murata, N., Ohtani, T. and Toguri T. 1996. Low-temperature resistance of higher plants is significantly enhanced by a nonspecific cyanobacterial desaturase. *Nature Biotechnology* **14**: 1003-1006.

Kodama, H., Hamada, T., Horiguchi, G., Nishimura, M. and Iba, K. 1994. Genetic enhancement of cold tolerance by expression of a gene for chloroplast w-3 fatty acid desaturase in transgenic tobacco. *Plant Physiology* **105**: 601-605.

Kodama, H., Horiguchi, G., Nishiuchi, T., Nishimura, M. and Iba, K. 1995. Fatty acid desaturation during chilling acclimation is one of the factors involved in conferring low-temperature tolerance to young tobacco leaves. *Plant Physiology* **107**: 1177-1185.

Loik, M.E., Huxman, T.E., Hamerlynck, E.P. and Smith, S.D. 2000. Low temperature tolerance and cold acclimation for seedlings of three Mojave Desert *Yucca* species exposed to elevated CO_2. *Journal of Arid Environments* **46**: 43-56.

Lyons, J.M. 1973. Chilling injury in plants. *Annual Review of Plant Physiology* **24**: 445-466.

Moon, B.Y., Higashi, S.-I., Gombos, Z. and Murata, N. 1995. Unsaturation of the membrane lipids of chloroplasts stabilizes the photosynthetic machinery against low-temperature photoinhibition in transgenic tobacco plants. *Proceedings of the National Academy of Sciences, USA* **92**: 6219-6223.

Obrist, D., Arnone III, J.A. and Korner, C. 2001. *In situ* effects of elevated atmospheric CO_2 on leaf freezing resistance and carbohydrates in a native temperate grassland. *Annals of Botany* **87**: 839-844.

Schwanz, P. and Polle, A. 2001. Growth under elevated CO_2 ameliorates defenses against photo-oxidative stress in poplar (*Populus alba x tremula*). *Environmental and Experimental Botany* **45**: 43-53.

Sgherri, C.L.M., Quartacci, M.F., Menconi, M., Raschi, A. and Navari-Izzo, F. 1998. Interactions between drought and elevated CO_2 on alfalfa plants. *Journal of Plant Physiology* **152**: 118-124.

Sigurdsson, B.D. 2001. Elevated [CO_2] and nutrient status modified leaf phenology and growth rhythm of young *Populus trichocarpa* trees in a 3-year field study. *Trees* **15**: 403-413.

Wayne, P.M., Reekie, E.G. and Bazzaz, F.A. 1998. Elevated CO_2 ameliorates birch response to high temperature and frost stress: implications for modeling climate-induced geographic range shifts. *Oecologia* **114**: 335-342.

7.3.7. Nitrogen Deficiency

Numerous studies have investigated the effects of different soil nitrogen (N) concentrations on plant responses to increases in the air's CO_2 content, as it has been claimed that a deficiency of soil nitrogen lessens the relative growth stimulation in plants that is typically provided by elevated concentrations of atmospheric CO_2. In this section, we evaluate the credibility of that claim for various crops, fungi, grasses and trees.

The results of these experiments indicate that some plants sometimes will not respond at all to atmospheric CO_2 enrichment at low levels of soil N, while some will. Some plants respond equally well to increases in the air's CO_2 content when growing in soils exhibiting a whole range of N concentrations. Most common, however, is the observation that plants respond ever better to rising atmospheric CO_2 concentrations as soil N concentrations rise. Interestingly, the current state of earth's atmosphere and land surface is one of jointly increasing CO_2 and N concentrations. Hence, the outlook is good for continually increasing terrestrial vegetative productivity in the years and decades ahead, as these trends continue.

Additional information on this topic, including reviews on nitrogen not discussed here, can be found at http://www.co2science.org/subject/g/subject_g.php under the heading Growth Response to CO_2 With Other Variables: Nutrients: Nitrogen, as well as at http://www.co2science.org/subject/n/subject_n.php under the headings Nitrogen, Nitrogen Fixation and Nitrogen Use Efficiency.

7.3.7.1. Crops

7.3.7.1.1. Rice

Does a deficiency of soil nitrogen lessen the relative growth and yield stimulation of rice that is typically provided by elevated levels of atmospheric CO_2? In exploring this question, Weerakoon *et al.* (1999) grew seedlings of two rice cultivars for 28 days in glasshouses maintained at atmospheric CO_2 concentrations of 373, 545, 723 and 895 ppm under conditions of low, medium and high soil nitrogen content. After four weeks of treatment, photosynthesis was found to significantly increase with increasing nitrogen availability and atmospheric CO_2 concentration. Averaged across all nitrogen regimes, plants grown at 895 ppm CO_2 exhibited photosynthetic rates that were 50 percent greater than those observed in plants grown at ambient CO_2. Total plant dry weight also increased with increasing atmospheric CO_2. In addition, the percentage growth enhancement resulting from CO_2 enrichment increased with increasing soil nitrogen; from 21 percent at the lowest soil nitrogen concentration to 60 percent at the highest concentration.

Using a different CO_2 enrichment technique, Weerakoon *et al.* (2000) grew rice in open-top chambers maintained at atmospheric CO_2 concentrations of approximately 350 and 650 ppm during a wet and dry growing season and under a range of soil nitrogen contents. Early in both growing seasons, plants exposed to elevated atmospheric CO_2 concentrations intercepted significantly more sunlight than plants fumigated with ambient air, due to CO_2-induced increases in leaf area index. This phenomenon occurred regardless of soil nitrogen content, but disappeared shortly after canopy closure in all treatments. Later, mature canopies achieved similar leaf area indexes at identical levels of soil nitrogen supply; but mean season-long radiation use efficiency, which is the amount of biomass produced per unit of solar radiation intercepted, was 35 percent greater in CO_2-enriched vs. ambiently grown plants and tended to increase with increasing soil nitrogen content.

Utilizing a third approach to enriching the air about a crop with elevated levels of atmospheric CO_2, Kim *et al.* (2003) grew rice crops from the seedling stage to maturity at atmospheric CO_2 concentrations of ambient and ambient plus 200 ppm using FACE technology and three levels of applied nitrogen—low (LN, 4 g N m^{-2}), medium (MN, 8 and 9 g N m^{-2}) and high (HN, 15 g N m^{-2})—for three cropping seasons (1998-2000). They report that "the yield response to elevated CO_2 in crops supplied with MN (+14.6%) or HN (+15.2%) was about twice that of crops supplied with LN (+7.4%)," confirming the importance of nitrogen availability to the response of rice to atmospheric CO_2 enrichment previously determined by Kim *et al.* (2001) and Kobaysahi *et al.* (2001).

In light of these observations, it would appear that the maximum benefits of elevated levels of

455

atmospheric CO_2 for the growth and grain production of rice cannot be realized in soils that are highly deficient in nitrogen. Increasing nitrogen concentrations above what is considered adequate may not result in proportional gains in CO_2-induced growth and yield enhancement.

Additional information on this topic, including reviews of newer publications as they become available, can be found at http://www.co2science.org/subject/n/nitrogenrice.php.

References

Kim, H.-Y., Lieffering, M., Kobayashi, K., Okada, M., Mitchell, M.W. and Gumpertz, M. 2003. Effects of free-air CO_2 enrichment and nitrogen supply on the yield of temperate paddy rice crops. *Field Crops Research* **83**: 261-270.

Kim, H.-Y., Lieffering, M., Miura, S., Kobayashi, K. and Okada, M. 2001. Growth and nitrogen uptake of CO_2-enriched rice under field conditions. *New Phytologist* **150**: 223-229.

Kobayashi, K., Lieffering, M. and Kim, H.-Y. 2001. Growth and yield of paddy rice under free-air CO_2 enrichment. In: Shiyomi, M. and Koizumi, H. (Eds.), *Structure and Function in Agroecosystem Design and Management*. CRC Press, Boca Raton, FL, USA, pp. 371-395.

Weerakoon, W.M.W., Ingram, K.T. and Moss, D.D. 2000. Atmospheric carbon dioxide and fertilizer nitrogen effects on radiation interception by rice. *Plant and Soil* **220**: 99-106.

Weerakoon, W.M., Olszyk, D.M. and Moss, D.N. 1999. Effects of nitrogen nutrition on responses of rice seedlings to carbon dioxide. *Agriculture, Ecosystems and Environment* **72**: 1-8.

7.3.7.1.2. Wheat

Smart *et al.* (1998) grew wheat from seed for 23 days in controlled environment chambers maintained at atmospheric CO_2 concentrations of 360 and 1000 ppm and two concentrations of soil nitrate, finding that the extra CO_2 increased average plant biomass by approximately 15 percent, irrespective of soil nitrogen content. In a more realistic FACE experiment, however, Brooks *et al.* (2000) grew spring wheat for two seasons at atmospheric CO_2 concentrations of 370 and 570 ppm at both high and low levels of nitrogen fertility; and they obtained *twice* the yield

enhancement (16 percent vs. 8 percent) in the high nitrogen treatment.

In an experiment with one additional variable, Vilhena-Cardoso and Barnes (2001) grew spring wheat for two months in environmental chambers fumigated with air containing atmospheric CO_2 concentrations of either 350 or 700 ppm at ambient and elevated (75 ppb) ozone concentrations, while the plants were simultaneously subjected to either low, medium or high levels of soil nitrogen. With respect to biomass production, the elevated CO_2 treatment increased total plant dry weight by 44, 29 and 12 percent at the high, medium and low soil nitrogen levels, respectively. In addition, although elevated ozone alone reduced plant biomass, the simultaneous application of elevated CO_2 completely ameliorated its detrimental effects on biomass production, irrespective of soil nitrogen supply.

Why do the plants of some studies experience a major reduction in the relative growth stimulation provided by atmospheric CO_2 enrichment under low soil nitrogen conditions, while other studies find the aerial fertilization effect of elevated CO_2 to be independent of root-zone nitrogen concentration? Based on studies of both potted and hydroponically grown plants, Farage *et al.* (1998) determined that low root-zone nitrogen concentrations need not lead to photosynthetic acclimation (less than maximum potential rates of photosynthesis) in elevated CO_2, as long as root-zone nitrogen *supply* is adequate to meet plant nitrogen *needs* to maintain the enhanced relative growth rate that is made possible by atmospheric CO_2 enrichment. When supply cannot meet this need, as is often the case in soils with limited nitrogen reserves, the aerial fertilization effect of atmospheric CO_2 enrichment begins to be reduced and less-than-potential CO_2-induced growth stimulation is observed. Nevertheless, the acclimation process is the plant's "first line of defense" to keep its productivity from falling even further than it otherwise would, as it typically mobilizes nitrogen from "excess" rubisco and sends it to more needy plant sink tissues to allow for their continued growth and development (Theobald *et al.*, 1998).

In conclusion, although atmospheric CO_2 enrichment tends to increase the growth and yield of wheat under a wide range of soil nitrogen concentrations, including some that are very low, considerably greater CO_2-induced enhancements are possible when more soil nitrogen is available, although the response can saturate at high soil nitrogen levels, with excess nitrogen providing little to no extra yield.

Additional information on this topic, including reviews of newer publications as they become available, can be found at http://www.co2science.org/subject/n/nitrogenwheat.php.

References

Brooks, T.J., Wall, G.W., Pinter Jr., P.J., Kimball, B.A., LaMorte, R.L., Leavitt, S.W., Matthias, A.D., Adamsen, F.J., Hunsaker, D.J. and Webber, A.N. 2000. Acclimation response of spring wheat in a free-air CO_2 enrichment (FACE) atmosphere with variable soil nitrogen regimes. 3. Canopy architecture and gas exchange. *Photosynthesis Research* **66**: 97-108.

Farage, P.K., McKee, I.F. and Long, S.P. 1998. Does a low nitrogen supply necessarily lead to acclimation of photosynthesis to elevated CO_2? *Plant Physiology* **118**: 573-580.

Smart, D.R., Ritchie, K., Bloom, A.J. and Bugbee, B.B. 1998. Nitrogen balance for wheat canopies (*Triticum aestivum* cv. Veery 10) grown under elevated and ambient CO_2 concentrations. *Plant, Cell and Environment* **21**: 753-763.

Theobald, J.C., Mitchell, R.A.C., Parry, M.A.J. and Lawlor, D.W. 1998. Estimating the excess investment in ribulose-1,5-bisphosphate carboxylase/oxygenase in leaves of spring wheat grown under elevated CO_2. *Plant Physiology* **118**: 945-955.

Vilhena-Cardoso, J. and Barnes, J. 2001. Does nitrogen supply affect the response of wheat (*Triticum aestivum* cv. Hanno) to the combination of elevated CO_2 and O_3? *Journal of Experimental Botany* **52**: 1901-1911.

7.3.7.1.3. Other

Zerihun *et al.* (2000) grew sunflowers for one month in pots of three different soil nitrogen concentrations that were placed within open-top chambers maintained at atmospheric CO_2 concentrations of 360 and 700 ppm. The extra CO_2 of the CO_2-enriched chambers reduced average rates of root nitrogen uptake by about 25 percent, which reduction, by itself, would normally tend to reduce tissue nitrogen contents and the relative growth rates of the seedlings. However, the elevated CO_2 also increased photosynthetic nitrogen-use efficiency by an average of 50 percent, which increase normally tends to increase the relative growth rates of seedlings. Of

these two competing effects, the latter was the more powerful, leading to an increase in whole plant biomass. After the one month of the study, for example, the CO_2-enriched plants exhibited whole plant biomass values that were 44, 13 and 115 percent greater than those of the plants growing in ambient air at low, medium and high levels of soil nitrogen, respectively, thus demonstrating that low tissue nitrogen contents do not necessarily preclude a growth response to atmospheric CO_2 enrichment, particularly if photosynthetic nitrogen-use efficiency is enhanced, which is typically the case, as it was in this study. Nevertheless, the greatest CO_2-induced growth increase of Zerihun *et al.*'s study was exhibited by the plants growing in the high soil nitrogen treatment.

Deng and Woodward (1998) grew strawberries in environment-controlled glasshouses maintained at atmospheric CO_2 concentrations of 390 and 560 ppm for nearly three months. In addition, the strawberries were supplied with fertilizers containing three levels of nitrogen. The extra CO_2 increased rates of net photosynthesis and total plant dry weight at all three nitrogen levels, but the increases were not significant. Nevertheless, they provided the CO_2-enriched plants with enough additional sugar and physical mass to support significantly greater numbers of flowers and fruits than the plants grown at 390 ppm CO_2. This effect consequently led to total fresh fruit weights that were 42 and 17 percent greater in the CO_2-enriched plants that received the highest and lowest levels of nitrogen fertilization, respectively, once again indicating a greater growth response at higher nitrogen levels.

Newman *et al.* (2003) investigated the effects of two levels of nitrogen fertilization and an approximate doubling of the air's CO_2 concentration on the growth of tall fescue, which is an important forage crop. The plants with which they worked were initially grown from seed in greenhouse flats, but after sixteen weeks they were transplanted into 19-liter pots filled with potting media that received periodic applications of a slow-release fertilizer. Then, over the next two years of outdoor growth, they were periodically clipped, divided and repotted to ensure they did not become root-bound; and at the end of that time, they were placed within twenty 1.3-m-diameter open-top chambers, half of which were maintained at the ambient atmospheric CO_2 concentration and half of which were maintained at an approximately doubled CO_2 concentration of 700 ppm. In addition, half of the pots in each CO_2

treatment received 0.0673 kg N m^{-2} applied over a period of three consecutive days, while half of them received only one-tenth that amount, with the entire procedure being repeated three times during the course of the 12-week experiment. Newman *et al.* report that the plants grown in the high-CO_2 air photosynthesized 15 percent more and produced 53 percent more dry matter (DM) under low N conditions and 61 percent more DM under high N conditions. In addition, they report that the percent organic matter (OM) was little changed, except under elevated CO_2 and high N, when %OM (as %DM) increased by 3 percent. In this study too, therefore, the greatest relative increase in productivity occurred under high, as opposed to low, soil N availability.

Demmers-Derks *et al.* (1998) grew sugar beets as an annual crop in controlled-environment chambers at atmospheric CO_2 concentrations of 360 and 700 ppm and air temperatures of ambient and ambient plus 3°C for three consecutive years. In addition to being exposed to these CO_2 and temperature combinations, the sugar beets were supplied with solutions of low and high nitrogen content. Averaged across all three years and both temperature regimes, the extra CO_2 of this study enhanced total plant biomass by 13 and 25 percent in the low and high nitrogen treatments, respectively. In addition, it increased root biomass by 12 and 26 percent for the same situations. As was the case with sunflowers, strawberries and tall fescue, elevated CO_2 elicited the largest growth responses in the sugar beets that received a high, as opposed to a low, supply of nitrogen.

Also working with sugar beets were Romanova *et al.* (2002), who grew them from seed for one month in controlled environment chambers maintained at atmospheric CO_2 concentrations of 350 and 700 ppm, while fertilizing them with three different levels of nitrate-nitrogen. In this study, the plants grown in CO_2-enriched air exhibited rates of net photosynthesis that were approximately 50 percent greater than those displayed by the plants grown in ambient air, regardless of soil nitrate availability. These CO_2-induced increases in photosynthetic carbon uptake contributed to 60, 40 and 30 percent above-ground organ dry weight increases in plants receiving one-half, standard, and three-fold levels of soil nitrate, respectively. Root weights, however, were less responsive to atmospheric CO_2 enrichment, displaying 10 and 30 percent increases in dry weight at one-half and standard nitrate levels, but no increase at the high soil nitrate concentration. In this study, therefore, the role of soil nitrogen fertility was clearly opposite to that observed in the four prior studies in the case of above-ground biomass production, but was mixed in the case of belowground biomass production.

Switching to barley, Fangmeier *et al.* (2000) grew plants in containers placed within open-top chambers maintained at atmospheric CO_2 concentrations of either 360 or 650 ppm and either a high or low nitrogen fertilization regime. As in the case of the above-ground biomass response of the sugar beets of Romanova *et al.*, the elevated CO_2 had the greatest relative impact on yield when the plants were grown under the *less*-than-optimum *low*-nitrogen regime, i.e., a 48 percent increase vs. 31 percent under high-nitrogen conditions.

Last, we report the pertinent results of the review and analysis of Kimball *et al.* (2002), who summarized the findings of most FACE studies conducted on agricultural crops since the introduction of that technology back in the late 1980s. In response to a 300-ppm increase in the air's CO_2 concentration, rates of net photosynthesis in several C_3 grasses were enhanced by an average of 46 percent under conditions of ample soil nitrogen supply and by 44 percent when nitrogen was limiting to growth. With respect to above-ground biomass production, the differential was much larger, with the C_3 grasses wheat, rice and ryegrass experiencing an average increase of 18 percent at ample nitrogen but only 4 percent at low nitrogen; while with respect to belowground biomass production, they experienced an average increase of 70 percent at ample nitrogen and 58 percent at low nitrogen. Similarly, clover experienced a 38 percent increase in belowground biomass production at ample soil nitrogen, and a 32 percent increase at low soil nitrogen. Finally, with respect to agricultural yield, which is the bottom line in terms of food and fiber production, wheat and ryegrass experienced an average increase of 18 percent at ample nitrogen, while wheat experienced only a 10 percent increase at low nitrogen.

In light of these several results, it can be safely concluded that although there are some significant exceptions to the rule, most agricultural crops generally experience somewhat greater CO_2-induced relative (percentage) increases in net photosynthesis and biomass production even when soil nitrogen concentrations are a limiting factor.

Additional information on this topic, including reviews of newer publications as they become available, can be found at http://www.co2science.org/subject/n/nitrogenagriculture.php.

References

Demmers-Derks, H., Mitchell, R.A.G., Mitchell, V.J. and Lawlor, D.W. 1998. Response of sugar beet (*Beta vulgaris* L.) yield and biochemical composition to elevated CO_2 and temperature at two nitrogen applications. *Plant, Cell and Environment* **21**: 829-836.

Deng, X. and Woodward, F.I. 1998. The growth and yield responses of *Fragaria ananassa* to elevated CO_2 and N supply. *Annals of Botany* **81**: 67-71.

Fangmeier, A., Chrost, B., Hogy, P. and Krupinska, K. 2000. CO_2 enrichment enhances flag leaf senescence in barley due to greater grain nitrogen sink capacity. *Environmental and Experimental Botany* **44**: 151-164.

Kimball, B.A., Kobayashi, K. and Bindi, M. 2002. Responses of agricultural crops to free-air CO_2 enrichment. *Advances in Agronomy* **77**: 293-368.

Newman, J.A., Abner, M.L., Dado, R.G., Gibson, D.J., Brookings, A. and Parsons, A.J. 2003. Effects of elevated CO_2, nitrogen and fungal endophyte-infection on tall fescue: growth, photosynthesis, chemical composition and digestibility. *Global Change Biology* **9**: 425-437.

Romanova, A.K., Mudrik, V.A., Novichkova, N.S., Demidova, R.N. and Polyakova, V.A. 2002. Physiological and biochemical characteristics of sugar beet plants grown at an increased carbon dioxide concentration and at various nitrate doses. *Russian Journal of Plant Physiology* **49**: 204-210.

Zerihun, A., Gutschick, V.P. and BassiriRad, H. 2000. Compensatory roles of nitrogen uptake and photosynthetic N-use efficiency in determining plant growth response to elevated CO_2: Evaluation using a functional balance model. *Annals of Botany* **86**: 723-730.

7.3.7.2. Fungi

Nearly all of earth's plants become involved in intimate relationships with different fungal species at one point or another in their life cycles. Among other things, the fungi commonly aid plants in the acquisition of water and important soil nutrients. In addition, fungal-plant interactions are often impacted by variations in both atmospheric CO_2 and soil nitrogen concentrations. In this subsection, we review how various aspects of fungal-plant interactions are influenced by elevated CO_2 under varying soil nitrogen regimes.

In a one-year study conducted by Walker *et al.* (1998), ponderosa pine seedlings exposed to atmospheric CO_2 concentrations of 525 and 700 ppm displayed total numbers of ectomycorrhizal fungi on their roots that were 170 and 85 percent greater, respectively, than those observed on roots of ambiently grown seedlings.

In the study of Rillig *et al.* (1998), three grasses and two herbs fumigated with ambient air and air containing an extra 350 ppm CO_2 for four months displayed various root infection responses by arbuscular mycorrhizal fungi, which varied with soil nitrogen supply. At low soil nitrogen contents, elevated CO_2 increased the percent root infection by this type of fungi in all five annual grassland species. However, at high soil nitrogen contents, this trend was reversed in four of the five species.

Finally, in the study of Rillig and Allen (1998), several important observations were made with respect to the effects of elevated CO_2 and soil nitrogen status on fungal-plant interactions. First, after growing three-year-old shrubs at an atmospheric CO_2 concentration of 750 ppm for four months, they reported insignificant 19 and 9 percent increases in percent root infected by arbuscular mycorrhizal fungi at low and high soil nitrogen concentrations, respectively. However, elevated CO_2 significantly increased the percent root infection by arbuscules, which are the main structures involved in the symbiotic exchange of carbon and nutrients between a host plant and its associated fungi, by more than 14-fold at low soil nitrogen concentrations. In addition, the length of fungal hyphae more than doubled with atmospheric CO_2 enrichment in the low soil nitrogen regime. In the high soil nitrogen treatment, elevated CO_2 increased the percent root infection by vesicles, which are organs used by arbuscular mycorrhizal fungi for carbon storage, by approximately 2.5-fold.

In conclusion, these observations suggest that elevated CO_2 will indeed affect fungal-plant interactions in positive ways that often depend upon soil nitrogen status. Typically, it appears that CO_2-induced stimulations of percent root infection by various fungal components is greater under lower, rather than higher, soil nitrogen concentrations. This tendency implies that elevated CO_2 will enhance fungal-plant interactions to a greater extent when soil nutrition is less-than-optimal for plant growth, which is the common case for most of earth's ecosystems that are not subjected to cultural fertilization practices typical of intensive agricultural production.

Additional information on this topic, including reviews of newer publications as they become available, can be found at http://www.co2science.org/subject/n/nitrogenfungi.php.

References

Rillig, M.C. and Allen, M.F. 1998. Arbuscular mycorrhizae of *Gutierrezia sarothrae* and elevated carbon dioxide: evidence for shifts in C allocation to and within the mycobiont. *Soil Biology and Biochemistry* **30**: 2001-2008.

Rillig, M.C., Allen, M.F., Klironomous, J.N., Chiariello, N.R. and Field, C.B. 1998. Plant species-specific changes in root-inhabiting fungi in a California annual grassland: responses to elevated CO₂ and nutrients. *Oecologia* **113**: 252-259.

Walker, R.F., Johnson, D.W., Geisinger, D.R. and Ball, J.T. 1998. Growth and ectomycorrhizal colonization of ponderosa pine seedlings supplied different levels of atmospheric CO₂ and soil N and P. *Forest Ecology and Management* **109**: 9-20.

7.3.7.3. Grasses

Perennial ryegrass (*Lolium perenne* L.) has been used as a model species in many experiments to help elucidate grassland responses to atmospheric CO_2 enrichment and soil nitrogen availability. In the FACE study of Rogers *et al.* (1998), for example, plants exposed to 600 ppm CO_2 exhibited a 35 percent increase in their photosynthetic rates without regard to soil nitrogen availability. However, when ryegrass was grown in plastic ventilated tunnels at twice-ambient concentrations of atmospheric CO_2, the CO_2-induced photosynthetic response was about 3-fold greater in a higher, as opposed to a lower, soil nitrogen regime (Casella and Soussana, 1997). Similarly, in an open-top chamber study conducted by Davey *et al.* (1999), it was reported that an atmospheric CO_2 concentration of 700 ppm stimulated photosynthesis by 30 percent in this species when it was grown with moderate, but not low, soil nitrogen availability. Thus, CO_2-induced photosynthetic stimulations in perennial ryegrass can be influenced by soil nitrogen content, with greater positive responses typically occurring under higher, as opposed to lower, soil nitrogen availability.

With respect to biomass production, van Ginkel and Gorissen (1998) reported that a doubling of the atmospheric CO_2 concentration increased shoot biomass of perennial ryegrass by 28 percent, regardless of soil nitrogen concentration. In the more revealing six-year FACE study of Daepp *et al.* (2000), plants grown at 600 ppm CO_2 and high soil nitrogen availability continually increased their dry matter production over that observed in ambient-treatment plots, from 8 percent more in the first year to 25 percent more at the close of year six. When grown at a low soil nitrogen availability, however, CO_2-enriched plants exhibited an initial 5 percent increase in dry matter production, which dropped to a negative 11 percent in year two; but this negative trend was thereafter turned around, and it continually rose to reach a 9 percent stimulation at the end of the study. Thus, these data demonstrate that elevated CO_2 increases perennial ryegrass biomass, even under conditions of low soil nitrogen availability, especially under conditions of long-term atmospheric CO_2 enrichment.

Lutze *et al.* (1998) reported that microcosms of the C_3 grass *Danthonia richardsonii* grown for four years in glasshouses fumigated with air containing 720 ppm CO_2 displayed total photosynthetic carbon gains that were 15-34 percent higher than those of ambiently grown microcosms, depending on the soil nitrogen concentration. In a clearer depiction of photosynthetic responses to soil nitrogen, Davey *et al.* (1999) noted that photosynthetic rates of *Agrostis capillaries* subjected to twice-ambient levels of atmospheric CO_2 for two years were 12 and 38 percent greater than rates measured in control plants grown at 350 ppm CO_2 under high and low soil nitrogen regimes, respectively. They also reported CO_2-induced photosynthetic stimulations of 25 and 74 percent for *Trifolium repens* subjected to high and low soil nitrogen regimes, respectively. Thus, we see that the greatest CO_2-induced percentage increase in photosynthesis can sometimes occur under the *least* favorable soil nitrogen conditions.

With respect to biomass production, Navas *et al.* (1999) reported that 60 days' exposure to 712 ppm CO_2 increased biomass production of *Danthonia richardsonii*, *Phalaris aquatica*, *Lotus pedunculatus*, and *Trifolium repens* across a large soil nitrogen gradient. With slightly more detail, Cotrufo and Gorissen (1997) reported average CO_2-induced increases in whole-plant dry weights of *Agrostis capillaries* and *Festuca ovina* that were 20 percent greater than those of their respective controls, regardless of soil nitrogen availability. In the study of Ghannoum and Conroy (1998), three *Panicum* grasses grown for two months at twice-ambient levels of atmospheric CO_2 and high soil nitrogen availability displayed similar increases in total plant dry mass that were about 28 percent greater than those of their respective ambiently grown controls. At low nitrogen, however, elevated CO_2 had no significant effect on the dry mass of two of the species, while it actually decreased that of the third species.

In summary, it is clear that atmospheric CO_2 enrichment stimulates photosynthesis and biomass production in grasses and grassland species when soil nitrogen availability is high and/or moderate. Under lower soil nitrogen conditions, it is also clear that atmospheric CO_2 enrichment can have the same positive effect on these parameters, but that it can also have a reduced positive effect, no effect, or (in one case) a negative effect. In light of the one long-term study that lasted six years, however, it is likely that—given enough time—grasslands have the ability to overcome soil nitrogen limitations and produce positive CO_2-induced growth responses. Thus, because the rising CO_2 content of the air is likely to continue for a long time to come, occasional nitrogen limitations on the aerial fertilization effect of atmospheric CO_2 enrichment of grasslands will likely become less and less restrictive as time progresses.

Additional information on this topic, including reviews of newer publications as they become available, can be found at http://www.co2science.org/subject/n/nitrogengrass.php.

References

Casella, E. and Soussana, J-F. 1997. Long-term effects of CO_2 enrichment and temperature increase on the carbon balance of a temperate grass sward. *Journal of Experimental Botany* **48**: 1309-1321.

Cotrufo, M.F. and Gorissen, A. 1997. Elevated CO_2 enhances below-ground C allocation in three perennial grass species at different levels of N availability. *New Phytologist* **137**: 421-431.

Daepp, M., Suter, D., Almeida, J.P.F., Isopp, H., Hartwig, U.A., Frehner, M., Blum, H., Nosberger, J. and Luscher, A. 2000. Yield response of *Lolium perenne* swards to free air CO_2 enrichment increased over six years in a high N input system on fertile soil. *Global Change Biology* **6**: 805-816.

Davey, P.A., Parsons, A.J., Atkinson, L., Wadge, K. and Long, S.P. 1999. Does photosynthetic acclimation to elevated CO_2 increase photosynthetic nitrogen-use efficiency? A study of three native UK grassland species in open-top chambers. *Functional Ecology* **13**: 21-28.

Ghannoum, O. and Conroy, J.P. 1998. Nitrogen deficiency precludes a growth response to CO_2 enrichment in C_3 and C_4 *Panicum* grasses. *Australian Journal of Plant Physiology* **25**: 627-636.

Lutze, J.L. and Gifford, R.M. 1998. Carbon accumulation, distribution and water use of *Danthonia richardsonii* swards in response to CO_2 and nitrogen supply over four years of growth. *Global Change Biology* **4**: 851-861.

Navas, M.-L., Garnier, E., Austin, M.P. and Gifford, R.M. 1999. Effect of competition on the responses of grasses and legumes to elevated atmospheric CO_2 along a nitrogen gradient: differences between isolated plants, monocultures and multi-species mixtures. *New Phytologist* **143**: 323-331.

Rogers, A., Fischer, B.U., Bryant, J., Frehner, M., Blum, H., Raines, C.A. and Long, S.P. 1998. Acclimation of photosynthesis to elevated CO_2 under low-nitrogen nutrition is affected by the capacity for assimilate utilization. Perennial ryegrass under free-air CO_2 enrichment. *Plant Physiology* **118**: 683-689.

Van Ginkel, J.H. and Gorissen, A. 1998. In situ decomposition of grass roots as affected by elevated atmospheric carbon dioxide. *Soil Science Society of America Journal* **62**: 951-958.

7.3.7.4. Trees

7.3.7.4.1. Aspen

Does a deficiency of soil nitrogen lessen the relative growth stimulation of quaking aspen (*Populus tremuloides* Michx) that is typically provided by elevated concentrations of atmospheric CO_2?

In exploring this question, Kubiske *et al.* (1998) grew cuttings of four quaking aspen genotypes for five months at CO_2 concentrations of 380 or 720 ppm and low or high soil nitrogen in open-top chambers in the field in Michigan, USA. They found that the elevated CO_2 treatment significantly increased net photosynthesis, regardless of soil nitrogen content, although there were no discernible increases in above-ground growth within the five-month study period. Belowground, however, elevated CO_2 significantly increased fine root production, but only in the high soil nitrogen treatment.

Working at the same site, Zak *et al.* (2000) and Curtis *et al.* (2000) grew six aspen genotypes from cuttings in open-top chambers for 2.5 growing seasons at atmospheric CO_2 concentrations of 350 and 700 ppm on soils containing either adequate or inadequate supplies of nitrogen. Curtis *et al.* report that at the end of this period the trees growing in the doubled-CO_2 treatment exhibited rates of net photosynthesis that were 128 percent and 31 percent greater than those of the trees growing in the ambient-air treatment on the high- and low-nitrogen soils, respectively, while Zak *et al.* determined the CO_2-induced biomass increases of the trees in the high- and low-nitrogen soils to be 38 percent and 16 percent, respectively.

In yet another study from the Michigan site, Mikan *et al.* (2000) grew aspen cuttings for two years in open-top chambers receiving atmospheric CO_2 concentrations of 367 and 715 ppm in soils of low and high soil nitrogen concentrations. They report finding that elevated CO_2 increased the total biomass of the aspen cuttings by 50 percent and 26 percent in the high and low soil nitrogen treatments, respectively, and that it increased coarse root biomass by 78 percent and 24 percent in the same respective treatments.

Last, but again at the same site, Wang and Curtis (2001) grew cuttings of two male and two female aspen trees for about five months in open-top chambers maintained at atmospheric CO_2 concentrations of 380 and 765 ppm on soils of high and low nitrogen content. In the male cuttings, there was a modest difference in the CO_2-induced increase in total biomass (58 percent and 66 percent in the high- and low-nitrogen soils, respectively), while in the female cuttings the difference was much greater (82 percent and 22 percent in the same respective treatments).

Considering the totality of these several observations, it would appear that the degree of soil nitrogen availability does indeed impact the aerial fertilization effect of atmospheric CO_2 enrichment on the growth of aspen trees by promoting a greater CO_2-induced growth enhancement in soils of adequate, as opposed to insufficient, nitrogen content.

Additional information on this topic, including reviews of newer publications as they become available, can be found at http://www.co2science.org/subject/n/nitrogenaspen.php.

References

Curtis, P.S., Vogel, C.S., Wang, X.Z., Pregitzer, K.S., Zak, D.R., Lussenhop, J., Kubiske, M. and Teeri, J.A. 2000. Gas exchange, leaf nitrogen, and growth efficiency of *Populus tremuloides* in a CO_2-enriched atmosphere. *Ecological Applications* **10**: 3-17.

Kubiske, M.E., Pregitzer, K.S., Zak, D.R. and Mikan, C.J. 1998. Growth and C allocation of *Populus tremuloides* genotypes in response to atmospheric CO_2 and soil N availability. *New Phytologist* **140**: 251-260.

Mikan, C.J., Zak, D.R., Kubiske, M.E. and Pregitzer, K.S. 2000. Combined effects of atmospheric CO_2 and N availability on the belowground carbon and nitrogen dynamics of aspen mesocosms. *Oecologia* **124**: 432-445.

Wang, X. and Curtis, P.S. 2001. Gender-specific responses of *Populus tremuloides* to atmospheric CO_2 enrichment. *New Phytologist* **150**: 675-684.

Zak, D.R., Pregitzer, K.S., Curtis, P.S., Vogel, C.S., Holmes, W.E. and Lussenhop, J. 2000. Atmospheric CO_2, soil-N availability, and allocation of biomass and nitrogen by *Populus tremuloides*. *Ecological Applications* **10**: 34-46.

7.3.7.4.2. Pine

In a review of eleven of their previously published papers dealing with both loblolly pine (*Pinus taeda* L.) and ponderosa pine (*Pinus ponderosa* Dougl.), Johnson *et al.* (1998) report that when soil nitrogen levels were so low as to be extremely deficient, or so high as to be toxic, growth responses to atmospheric CO_2 enrichment in both species were negligible. For moderate soil nitrogen deficiencies, however, a doubling of the air's CO_2 content sometimes boosted growth by as much as 1,000 percent. In addition, atmospheric CO_2 enrichment mitigated the negative growth response of ponderosa pine to extremely high soil nitrogen concentrations.

In a second paper published by some of the same scientists in the same year, Walker *et al.* (1998) describe how they raised ponderosa pine tree seedlings for two growing seasons in open-top chambers having CO_2 concentrations of 350, 525 and 700 ppm on soils of low, medium and high nitrogen content. They report that elevated CO_2 had little effect on most growth parameters after the first growing season, the one exception being belowground biomass, which increased with both CO_2 and soil nitrogen. After two growing seasons, however, elevated CO_2 significantly increased all growth parameters, including tree height, stem diameter, shoot weight, stem volume and root volume, with the greatest responses typically occurring at the highest CO_2 concentration in the highest soil nitrogen treatment. Root volume at 700 ppm CO_2 and high soil nitrogen, for example, exceeded that of all other treatments by at least 45 percent, as did shoot volume by 42 percent. Similarly, at high CO_2 and soil nitrogen, coarse root and shoot weights exceeded those at ambient CO_2 and high nitrogen by 80 and 88 percent, respectively.

Walker *et al.* (2000) published another paper on the same trees and treatments after five years of growth. At this time, the trees exposed to the twice-ambient levels of atmospheric CO_2 had heights that were 43, 64 and 25 percent greater than those of the trees exposed to ambient air and conditions of high, medium and low levels of soil nitrogen, respectively. Similarly, the trunk diameters of the 700-ppm-trees were 24, 73 and 20 percent greater than the trunk diameters of the ambiently grown trees exposed to high, medium and low levels of soil nitrogen.

Switching to a different species, Entry *et al.* (1998) grew one-year-old longleaf pine seedlings for 20 months in pots of high and low soil nitrogen content within open-top chambers maintained at atmospheric CO_2 concentrations of 365 or 720 ppm, finding that the elevated CO_2 caused no overall change in whole-plant biomass at low soil nitrogen, but that at high soil nitrogen, it increased it by 42 percent. After two years of these treatments, Runion *et al.* (1999) also reported that rates of net photosynthesis were about 50 percent greater in the high CO_2 treatment, irrespective of soil nitrogen content ... and water content too.

Last, Finzi and Schlesinger (2003) measured and analyzed the pool sizes and fluxes of inorganic and organic nitrogen (N) in the floor and top 30 cm of mineral soil of the Duke Forest at the five-year point of a long-term FACE study, where half of the experimental plots are enriched with an extra 200 ppm of CO_2. In commencing this study, they had originally hypothesized that "the increase in carbon fluxes to the microbial community under elevated CO_2 would increase the rate of N immobilization over mineralization," leading to a decline in the significant CO_2-induced stimulation of forest net primary production that developed over the first two years of the experiment (DeLucia *et al.*, 1999; Hamilton *et al.*, 2002). Quite to the contrary, however, they discovered "there was no statistically significant change in the cycling rate of N derived from soil organic matter under elevated CO_2." Neither was the rate of net N mineralization significantly altered by elevated CO_2, nor was there any statistically significant difference in the concentration or net flux of organic and inorganic N in the forest floor and top 30-cm of mineral soil after 5 years of CO_2 fumigation. Hence, at this stage of the study, they could find no support for their original hypothesis, which suggests that the growth stimulation provided by elevated levels of atmospheric CO_2 would gradually dwindle away to something rather insignificant before the

stand reached its equilibrium biomass, although they continue to cling to this unsubstantiated belief.

Considering the totality of these several observations, it would appear that the degree of soil nitrogen availability impacts the effect of atmospheric CO_2 enrichment on the growth of pine trees, with greater CO_2-induced growth enhancement occurring in soils of adequate, as opposed to insufficient, nitrogen content. As in the case of aspen, however, there is evidence to suggest that at some point the response to increasing soil nitrogen saturates, and beyond that point, higher N concentrations may sometimes even reduce the forest growth response to elevated CO_2.

Additional information on this topic, including reviews of newer publications as they become available, can be found at http://www.co2science.org/subject/n/nitrogenpine.php.

References

DeLucia, E.H., Hamilton, J.G., Naidu, S.L., Thomas, R.B., Andrews, J.A., Finzi, A., Lavine, M., Matamala, R., Mohan, J.E., Hendrey, G.R. and Schlesinger, W.H. 1999. Net primary production of a forest ecosystem with experimental CO_2 enrichment. *Science* **284**: 1177-1179.

Entry, J.A., Runion, G.B., Prior, S.A., Mitchell, R.J. and Rogers, H.H. 1998. Influence of CO_2 enrichment and nitrogen fertilization on tissue chemistry and carbon allocation in longleaf pine seedlings. *Plant and Soil* **200**: 3-11.

Finzi, A.C. and Schlesinger, W.H. 2003. Soil-nitrogen cycling in a pine forest exposed to 5 years of elevated carbon dioxide. *Ecosystems* **6**: 444-456.

Hamilton, J.G., DeLucia, E.H., George, K., Naidu, S.L., Finzi, A.C. and Schlesinger, W.H. 2002. Forest carbon balance under elevated CO_2. *Oecologia* **131**: 250-260.

Johnson, D.W., Thomas, R.B., Griffin, K.L., Tissue, D.T., Ball, J.T., Strain, B.R. and Walker, R.F. 1998. Effects of carbon dioxide and nitrogen on growth and nitrogen uptake in ponderosa and loblolly pine. *Journal of Environmental Quality* **27**: 414-425.

Runion, G.B., Mitchell, R.J., Green, T.H., Prior, S.A., Rogers, H.H. and Gjerstad, D.H. 1999. Longleaf pine photosynthetic response to soil resource availability and elevated atmospheric carbon dioxide. *Journal of Environmental Quality* **28**: 880-887.

Walker, R.F., Geisinger, D.R., Johnson, D.W. and Ball, J.T. 1998. Atmospheric CO_2 enrichment and soil N fertility effects on juvenile ponderosa pine: Growth, ectomycorrhizal development, and xylem water potential. *Forest Ecology and Management* **102**: 33-44.

Walker, R.F., Johnson, D.W., Geisinger, D.R. and Ball, J.T. 2000. Growth, nutrition, and water relations of ponderosa pine in a field soil as influenced by long-term exposure to elevated atmospheric CO_2. *Forest Ecology and Management* **137**: 1-11.

7.3.7.4.3. Spruce

Egli *et al.* (1998) rooted saplings of different genotypes of Norway spruce (*Picea abies* L. Karst.) directly into calcareous or acidic soils in open-top chambers and exposed them to atmospheric CO_2 concentrations of 370 or 570 ppm and low or high soil nitrogen contents. They found that elevated CO_2 generally stimulated light-saturated rates of photosynthesis under all conditions by as much as 35 percent, regardless of genotype, which consistently led to increased above-ground biomass production, also regardless of genotype, as well as without respect to soil type or nitrogen content.

Murray *et al.* (2000) grew Sitka spruce (*Picea sitchensis* (Bong.) Carr.) seedlings for two years in pots within open-top chambers maintained at atmospheric CO_2 concentrations of 355 and 700 ppm. In the last year of the study, half of the seedlings received one-tenth of the optimal soil nitrogen supply recommended for this species, while the other half received twice the optimal amount. Under this protocol, the extra CO_2 increased the seedlings' light-saturated rates of net photosynthesis by 19 percent and 33 percent in the low- and high-nitrogen treatments, respectively, while it increased their total biomass by 0 percent and 37 percent in these same treatments. Nevertheless, Murray *et al.* note there was a reallocation of biomass from above-ground organs (leaves and stems) into roots in the low-nitrogen treatment; and they remark that this phenomenon "may provide a long-term mechanism by which Sitka spruce could utilize limited resources both more efficiently and effectively," which suggests that although low soil nitrogen precluded a short-term CO_2-induced growth response in this tree species, it is possible that the negative impact of nitrogen deficiency could be overcome in the course of much longer-term atmospheric CO_2 enrichment.

In a related experiment, Liu *et al.* (2002) grew Sitka spruce seedlings in well-watered and fertilized pots within open-top chambers that were maintained for three years at atmospheric CO_2 concentrations of either 350 or 700 ppm, after which the seedlings were planted directly into native nutrient-deficient forest soil and maintained at the same atmospheric CO_2 concentrations for two more years in larger open-top chambers either with or without extra nitrogen being supplied to the soil. After the first three years of the study, they determined that the CO_2-enriched trees possessed 11.6 percent more total biomass than the ambient-treatment trees. At the end of the next two years, however, the CO_2-enriched trees supplied with extra nitrogen had 15.6 percent more total biomass than their similarly treated ambient-air counterparts, while the CO_2-enriched trees receiving no extra nitrogen had 20.5 percent more biomass than their ambient-treatment counterparts.

In light of these several observations, it would appear that the degree of soil nitrogen availability affects the growth of spruce trees by promoting a greater CO_2-induced growth enhancement in soils of adequate, as opposed to insufficient, nitrogen content. As in the cases of aspen and pine, however, at some point the response of spruce trees to increasing soil nitrogen saturates, and even higher nitrogen concentrations may reduce the growth response to elevated CO_2 below that observed at optimal or low soil nitrogen concentrations.

Additional information on this topic, including reviews of newer publications as they become available, can be found at http://www.co2science.org/subject/n/nitrogenspruce.php.

References

Egli, P., Maurer, S., Gunthardt-Goerg, M.S. and Korner, C. 1998. Effects of elevated CO_2 and soil quality on leaf gas exchange and aboveground growth in beech-spruce model ecosystems. *New Phytologist* **140**: 185-196.

Liu, S.R., Barton, C., Lee, H., Jarvis, P.G. and Durrant, D. 2002. Long-term response of Sitka spruce (*Picea sitchensis* (Bong.) Carr.) to CO_2 enrichment and nitrogen supply. I. Growth, biomass allocation and physiology. *Plant Biosystems* **136**: 189-198.

Murray, M.B., Smith, R.I., Friend, A. and Jarvis, P.G. 2000. Effect of elevated [CO_2] and varying nutrient application rates on physiology and biomass accumulation of Sitka spruce (*Picea sitchensis*). *Tree Physiology* **20**: 421-434.

7.3.7.4.4. Other

Maillard *et al.* (2001) grew pedunculate oak seedlings for three to four months in greenhouses maintained at atmospheric CO_2 concentrations of either 350 or 700 ppm under conditions of either low or high soil nitrogen concentration. The elevated CO_2 of their study stimulated belowground growth in the seedlings growing in the nitrogen-poor soil, significantly increasing their root-to-shoot ratios. However, it increased both the below- and above-ground biomass of seedlings growing in nitrogen-rich soil. In fact, the CO_2-enriched seedlings growing in the nitrogen-rich soil produced 217 and 533 percent more stem and coarse-root biomass, respectively, than their ambient-air counterparts growing in the same fertility treatment. Overall, the doubled CO_2 concentration of the air in their study enhanced total seedling biomass by approximately 30 and 140 percent under nitrogen-poor and nitrogen-rich soil conditions, respectively.

Schortemeyer *et al.* (1999) grew seedlings of *Acacia melanoxylon* (a leguminous nitrogen-fixing tree native to south-eastern Australia) in hydroponic culture for six weeks in growth cabinets, where the air was maintained at CO_2 concentrations of either 350 or 700 ppm and the seedlings were supplied with water containing nitrogen in a number of discrete concentrations ranging from 3 to 6,400 mmol m^{-3}. In the two lowest of these nitrogen concentration treatments, final biomass was unaffected by atmospheric CO_2 enrichment; but, as in the study of Maillard *et al.*, it was increased by 5- to 10-fold at the highest nitrogen concentration.

Temperton *et al.* (2003) measured total biomass production in another N_2-fixing tree—*Alnus glutinosa* (the common alder)—seedlings of which had been grown for three years in open-top chambers in either ambient or elevated (ambient + 350 ppm) concentrations of atmospheric CO_2 and one of two soil nitrogen regimes (full nutrient solution or no fertilizer). In their study, by contrast, they found that the trees growing under low soil nutrient conditions exhibited essentially the *same* growth enhancement as that of the well-fertilized trees.

Rounding out the full gamut of growth responses, Gleadow *et al.* (1998) grew eucalyptus seedlings for six months in glasshouses maintained at atmospheric CO_2 concentrations of either 400 or 800 ppm, fertilizing them twice daily with low or high nitrogen solutions. They found that their doubling of the air's CO_2 concentration increased total seedling biomass by 134 percent in the low nitrogen treatment but by a

smaller 98 percent in the high nitrogen treatment. In addition, the elevated CO_2 led to greater root growth in the low nitrogen treatment, as indicated by a 33 percent higher root:shoot ratio.

In conclusion, different species of trees respond differently to atmospheric CO_2 enrichment under conditions of low vs. high soil nitrogen fertility. The most common response is for the growth-promoting effects of atmospheric CO_2 enrichment to be expressed to a greater degree when soil nitrogen fertility is optimal as opposed to less-than-optimal.

Additional information on this topic, including reviews of newer publications as they become available, can be found at http://www.co2science.org/subject/o/ozonetreemisc.php.

References

Gleadow, R.M., Foley, W.J. and Woodrow, I.E. 1998. Enhanced CO_2 alters the relationship between photosynthesis and defense in cyanogenic *Eucalyptus cladocalyx* F. Muell. *Plant, Cell and Environment* **21**: 12-22.

Maillard, P., Guehl, J.-M., Muller, J.-F. and Gross, P. 2001. Interactive effects of elevated CO_2 concentration and nitrogen supply on partitioning of newly fixed ^{13}C and ^{15}N between shoot and roots of pedunculate oak seedlings (*Quercus robur* L.). *Tree Physiology* **21**: 163-172.

Schortemeyer, M., Atkin, O.K., McFarlane, N. and Evans, J.R. 1999. The impact of elevated atmospheric CO_2 and nitrate supply on growth, biomass allocation, nitrogen partitioning and N2 fixation of *Acacia melanoxylon*. *Australian Journal of Plant Physiology* **26**: 737-774.

Temperton, V.M., Grayston, S.J., Jackson, G., Barton, C.V.M. , Millard, P. and Jarvis, P.G. 2003. Effects of elevated carbon dioxide concentration on growth and nitrogen fixation in *Alnus glutinosa* in a long-term field experiment. *Tree Physiology* **23**: 1051-1059.

7.3.8. High Salinity

In managed agricultural ecosystems, the buildup of soil salinity from repeated irrigations can sometimes reduce crop yields. Similarly, in natural ecosystems where exposure to brackish or salty water is commonplace, saline soils can induce growth stress in plants not normally adapted to coping with this problem. Thus, it is important to understand how rising atmospheric CO_2 concentrations may interact with soil salinity to affect plant growth.

In the study of Ball *et al.* (1997), it was found that two Australian mangrove species with differing tolerance to salinity exhibited increased rates of net photosynthesis in response to a doubling of the atmospheric CO_2 concentration, but only when exposed to salinity levels that were 25 percent, but not 75 percent, of full-strength seawater.

Mavrogianopoulos *et al.* (1999) reported that atmospheric CO_2 concentrations of 800 and 1200 ppm stimulated photosynthesis in parnon melons by 75 and 120 percent, respectively, regardless of soil salinity, which ranged from 0 to 50 mM NaCl. Moreover, the authors noted that atmospheric CO_2 enrichment partially alleviated the negative effects of salinity on melon yield, which increased with elevated CO_2 at all salinity levels.

Maggio *et al.* (2002) grew tomatos at 400 and 900 ppm in combination with varying degrees of soil salinity and noted that plants grown in elevated CO_2 tolerated an average root-zone salinity threshold value that was about 60 percent greater than that exhibited by plants grown at 400 ppm CO_2 (51 vs. 32 mmol dm^{-3} Cl).

The review of Poorter and Perez-Soba (2001) found no changes in the effect of elevated CO_2 on the growth responses of most plants over a wide range of soil salinities, in harmony with the earlier findings of Idso and Idso (1994).

These various studies suggest that elevated CO_2 concentrations have either positive or no effects on plan growth where mild to moderate stresses may be present due to high soil salinity levels. Additional information on this topic, including reviews of newer publications as they become available, can be found at http://www.co2science.org/subject/s/salinity stress. php.

References

Ball, M.C., Cochrane, M.J. and Rawson, H.M. 1997. Growth and water use of the mangroves *Rhizophora apiculata* and *R. stylosa* in response to salinity and humidity under ambient and elevated concentrations of atmospheric CO_2. *Plant, Cell and Environment* **20**: 1158-1166.

Idso, K.E. and Idso, S.B. 1994. Plant responses to atmospheric CO_2 enrichment in the face of environmental constraints: A review of the past 10 years' research. *Agricultural and Forest Meteorology* **69**: 153-203.

Maggio, A., Dalton, F.N. and Piccinni, G. 2002. The effects of elevated carbon dioxide on static and dynamic

indices for tomato salt tolerance. *European Journal of Agronomy* **16**: 197-206.

Mavrogianopoulos, G.N., Spanakis, J. and Tsikalas, P. 1999. Effect of carbon dioxide enrichment and salinity on photosynthesis and yield in melon. *Scientia Horticulturae* **79**: 51-63.

Poorter, H. and Perez-Soba, M. 2001. The growth response of plants to elevated CO_2 under non-optimal environmental conditions. *Oecologia* **129**: 1-20.

7.3.9. Elevated Temperature

Will plants continue to exhibit CO_2-induced growth increases under conditions of elevated air temperature? In this section, we review the photosynthetic and growth responses of agricultural crops, grasslands and woody species to answer this question.

7.3.9.1. Agricultural Crops

The optimum growth temperature for several plants has been shown to rise substantially with increasing levels of atmospheric CO_2 (McMurtrie and Wang, 1993; McMurtrie *et al.*, 1992; Stuhlfauth and Fock, 1990; Berry and Bjorkman, 1980). This phenomenon was predicted by Long (1991), who calculated from well-established plant physiological principles that most C_3 plants should increase their optimum growth temperature by approximately 5°C for a 300 ppm increase in the air's CO_2 content. One would thus also expect plant photosynthetic rates to rise with concomitant increases in the air's CO_2 concentration and temperature, as has indeed been previously shown to be true by Idso and Idso (1994). We here proceed to see if these positive CO_2 and temperature interactions are still being supported in the recent scientific literature.

In the study of Zhu *et al.* (1999), pineapples grown at 700 ppm CO_2 assimilated 15, 97 and 84 percent more total carbon than pineapples grown at the current ambient CO_2 concentration in day/night air temperature regimes of 30/20 (which is optimal for pineapple growth at ambient CO_2), 30/25, and 35/25 °C, respectively. Similarly, Taub *et al.* (2000) demonstrated that net photosynthetic rates of cucumbers grown at twice-ambient levels of atmospheric CO_2 and air temperatures of 40°C were 3.2 times greater than those displayed by control plants grown at ambient CO_2 and this same elevated air temperature. Thus, at air temperatures normally considered to be deleterious to plant growth, rates of photosynthesis are typically considerably greater for CO_2 enriched vs. ambiently grown plants.

Reddy et al. (1999) grew cotton plants at air temperatures ranging from 2°C below to 7°C above ambient air temperatures and reported that plants simultaneously exposed to 720 ppm CO_2 displayed photosynthetic rates that were 137 to 190 percent greater than those displayed by plants exposed to ambient CO_2 concentrations across this temperature spectrum. Similarly, Cowling and Sage (1998) reported that a 200-ppm increase in the air's CO_2 concentration boosted photosynthetic rates of young bean plants by 58 and 73 percent at growth temperatures of 25 and 36°C, respectively. In addition, Bunce (1998) grew wheat and barley at 350 and 700 ppm CO_2 across a wide range of temperatures and reported that elevated CO_2 stimulated photosynthesis in these species by 63 and 74 percent, respectively, at an air temperature of 10°C and by 115 and 125 percent at 30°C. Thus, the percentage increase in photosynthetic rate resulting from atmospheric CO_2 enrichment often increases substantially with increasing air temperature.

Elevated CO_2 often aids in the recovery of plants from high temperature-induced reductions in photosynthetic capacity, as noted by Ferris et al. (1998), who grew soybeans for 52 days under normal air temperature and soil water conditions at atmospheric CO_2 concentrations of 360 and 700 ppm, but then subjected them to an 8-day period of high temperature and water stress. After normal air temperature and soil water conditions were restored, the CO_2-enriched plants attained photosynthetic rates that were 72 percent of their unstressed controls, while stressed plants grown at ambient CO_2 attained photosynthetic rates that were only 52 percent of their respective controls.

CO_2-induced increases in plant growth under high air temperatures have also been observed in a number of other agricultural species. In the previously mentioned study of Cowling and Sage (1998), for example, the 200-ppm increase in the air's CO_2 content boosted total plant biomass for wheat and barley by a combined average of 59 and 200 percent at air temperatures of 25 and 36°C. Similarly, Ziska (1998) reported that a doubling of the atmospheric CO_2 concentration increased the total dry weight of soybeans by 36 and 42 percent at root zone temperatures of 25 and 30°C, respectively. Likewise,

Hakala (1998) noted that spring wheat grown at 700 ppm CO_2 attained total biomass values that were 17 and 23 percent greater than those attained by ambiently grown plants exposed to ambient and elevated (ambient plus 3°C) air temperatures. In addition, after inputting various observed CO_2-induced growth responses of winter wheat into plant growth models, Alexandrov and Hoogenboom (2000) predicted 12 to 49 percent increases in wheat yield in Bulgaria even if air temperatures rise by as much as 4°C. Finally, in the study of Reddy et al. (1998), it was shown that elevated CO_2 (700 ppm) increased total cotton biomass by 31 to 78 percent across an air temperature range from 20 to 40°C. Thus, the beneficial effects of elevated CO_2 on agricultural crop yield is often enhanced by elevated air temperature.

In some cases, however, elevated CO_2 does not interact with air temperature to further increase the growth-promoting effects of atmospheric CO_2 enrichment, but simply allows the maintenance of the status quo. In the study of Demmers-Derks et al. (1998), for example, sugar beets grown at 700 ppm CO_2 attained 25 percent more biomass than ambiently grown plants, regardless of air temperature, which was increased by 3°C. Similarly, in the study of Fritschi et al. (1999), elevated CO_2 concentrations did not significantly interact with air temperature (4.5°C above ambient) to impact the growth of rhizoma peanut. Nonetheless, the 300-ppm increase in the air's CO_2 content increased total biomass by 52 percent, regardless of air temperature.

Finally, even if the air's CO_2 content were to cease rising or have no effect on plants, it is possible that temperature increases alone would promote plant growth and development. This was the case in the study of Wurr et al. (2000), where elevated CO_2 had essentially no effect on the yield of French bean. However, a 4°C increase in air temperature increased yield by approximately 50 percent.

In conclusion, the recent scientific literature continues to indicate that as the air's CO_2 content rises, agricultural crops will likely exhibit enhanced rates of photosynthesis and biomass production that will not be diminished by any global warming that might occur concurrently. In fact, if the ambient air temperature rises, the growth-promoting effects of atmospheric CO_2 enrichment will likely rise along with it.

Additional information on this topic, including reviews of newer publications as they become available, can be found at http://www.co2science.org/subject/g/tempco2ag.php.

References

Alexandrov, V.A. and Hoogenboom, G. 2000. The impact of climate variability and change on crop yield in Bulgaria. *Agricultural and Forest Meteorology* **104**: 315-327.

Berry, J. and Bjorkman, O. 1980. Photosynthetic response and adaptation to temperature in higher plants. *Annual Review of Plant Physiology* **31**: 491-543.

Bunce, J.A. 1998. The temperature dependence of the stimulation of photosynthesis by elevated carbon dioxide in wheat and barley. *Journal of Experimental Botany* **49**: 1555-1561.

Cowling, S.A. and Sage, R.F. 1998. Interactive effects of low atmospheric CO_2 and elevated temperature on growth, photosynthesis and respiration in *Phaseolus vulgaris*. *Plant, Cell and Environment* **21**: 427-435.

Demmers-Derks, H., Mitchell, R.A.G., Mitchell, V.J. and Lawlor, D.W. 1998. Response of sugar beet (*Beta vulgaris* L.) yield and biochemical composition to elevated CO_2 and temperature at two nitrogen applications. *Plant, Cell and Environment* **21**: 829-836.

Ferris, R., Wheeler, T.R., Hadley, P. and Ellis, R.H. 1998. Recovery of photosynthesis after environmental stress in soybean grown under elevated CO_2. *Crop Science* **38**: 948-955.

Fritschi, F.B., Boote, K.J., Sollenberger, L.E., Allen, Jr. L.H. and Sinclair, T.R. 1999. Carbon dioxide and temperature effects on forage establishment: photosynthesis and biomass production. *Global Change Biology* **5**: 441-453.

Hakala, K. 1998. Growth and yield potential of spring wheat in a simulated changed climate with increased CO_2 and higher temperature. *European Journal of Agronomy* **9**: 41-52.

Idso, K.E. and Idso, S.B. 1994. Plant responses to atmospheric CO_2 enrichment in the face of environmental constraints: A review of the past 10 years' research. *Agricultural and Forest Meteorology* **69**: 153-203.

Long, S.P. 1991. Modification of the response of photosynthetic productivity to rising temperature by atmospheric CO_2 concentrations: Has its importance been underestimated? *Plant, Cell and Environment* **14**: 729-739.

McMurtrie, R.E. and Wang, Y.-P. 1993. Mathematical models of the photosynthetic response of tree stands to rising CO_2 concentrations and temperatures. *Plant, Cell and Environment* **16**: 1-13.

McMurtrie, R.E., Comins, H.N., Kirschbaum, M.U.F. and Wang, Y.-P. 1992. Modifying existing forest growth models to take account of effects of elevated CO_2. *Australian Journal of Botany* **40**: 657-677.

Reddy, K.K., Davidonis, G.H., Johnson, A.S. and Vinyard, B.T. 1999. Temperature regime and carbon dioxide enrichment alter cotton boll development and fiber properties. *Agronomy Journal* **91**: 851-858.

Reddy, K.R., Robana, R.R., Hodges, H.F., Liu, X.J. and McKinion, J.M. 1998. Interactions of CO_2 enrichment and temperature on cotton growth and leaf characteristics. *Environmental and Experimental Botany* **39**: 117-129.

Stuhlfauth, T. and Fock, H.P. 1990. Effect of whole season CO_2 enrichment on the cultivation of a medicinal plant, *Digitalis lanata*. *Journal of Agronomy and Crop Science* **164**: 168-173.

Taub, D.R., Seeman, J.R. and Coleman, J.S. 2000. Growth in elevated CO_2 protects photosynthesis against high-temperature damage. *Plant, Cell and Environment* **23**: 649-656.

Wurr, D.C.E., Edmondson, R.N. and Fellows, J.R. 2000. Climate change: a response surface study of the effects of CO_2 and temperature on the growth of French beans. *Journal of Agricultural Science* **135**: 379-387.

Zhu, J., Goldstein, G. and Bartholomew, D.P. 1999. Gas exchange and carbon isotope composition of *Ananas comosus* in response to elevated CO_2 and temperature. *Plant, Cell and Environment* **22**: 999-1007.

Ziska, L.H. 1998. The influence of root zone temperature on photosynthetic acclimation to elevated carbon dioxide concentrations. *Annals of Botany* **81**: 717-721.

7.3.9.2. Grassland Species

In the study of Lilley *et al.* (2001), swards of *Trifolium subterraneum* were grown at 380 and 690 ppm CO_2 in combination with simultaneous exposure to ambient and elevated (ambient plus 3.4°C) air temperature. After one year of treatment, they reported that elevated CO_2 increased foliage growth by 19 percent at ambient air temperature. At elevated air temperature, however, plants grown at ambient CO_2 exhibited a 28 percent reduction in foliage growth, while CO_2-enriched plants still displayed a growth enhancement of 8 percent. Similarly, Morgan *et al.* (2001) reported that twice-ambient levels of atmospheric CO_2 increased above-ground biomass in native shortgrass steppe ecosystems by an average of 38 percent, in spite of an average air temperature increase of 2.6°C. Likewise, when bahiagrass was grown across a temperature gradient of 4.5°C, Fritschi *et al.* (1999) reported that a 275 ppm increase in the air's CO_2 content boosted photosynthesis and above-ground biomass by 22 and 17 percent, respectively, independent of air temperature. Thus, at elevated air temperature, CO_2-induced increases in rates of photosynthesis and biomass production are typically equal to or greater than what they are at ambient air temperature.

Other studies report similar results. Greer *et al.* (2000), for example, grew five pasture species at 18 and 28°C and reported that plants concomitantly exposed to 700 ppm CO_2 displayed average photosynthetic rates that were 36 and 70 percent greater, respectively, than average rates exhibited by control plants subjected to ambient CO_2 concentrations. Moreover, the average CO_2-induced biomass increase for these five species rose dramatically with increasing air temperature: from only 8 percent at 18°C to 95 percent at 28°C. Thus, the beneficial effects of elevated CO_2 on grassland productivity is often significantly enhanced by elevated air temperature.

Finally, temperature increases alone can promote grass growth and development. Norton *et al.* (1999) found elevated CO_2 had essentially no effect on the growth of the perennial grass *Agrostis curtisii* after two years of fumigation; however, a 3°C increase in air temperature increased the growth of this species considerably.

In conclusion, grassland plants will likely exhibit enhanced rates of photosynthesis and biomass production as the air's CO_2 content rises that will not be diminished by any global warming that might occur concurrently. If the ambient air temperature rises, the growth-promoting effects of atmospheric CO_2 enrichment will likely rise along with it.

Additional information on this topic, including reviews of newer publications as they become available, can be found at http://www.co2science.org/subject/g/tempco2grass.php.

References

Fritschi, F.B., Boote, K.J., Sollenberger, L.E., Allen, Jr. L.H. and Sinclair, T.R. 1999. Carbon dioxide and temperature effects on forage establishment: photosynthesis and biomass production. *Global Change Biology* **5**: 441-453.

Greer, D.H., Laing, W.A., Campbell, B.D. and Halligan, E.A. 2000. The effect of perturbations in temperature and photon flux density on the growth and photosynthetic responses of five pasture species. *Australian Journal of Plant Physiology* **27**: 301-310.

Idso, K.E. and Idso, S.B. 1994. Plant responses to atmospheric CO_2 enrichment in the face of environmental constraints: A review of the past 10 years' research. *Agricultural and Forest Meteorology* **69**: 153-203.

Lilley, J.M., Bolger, T.P. and Gifford, R.M. 2001. Productivity of *Trifolium subterraneum* and *Phalaris aquatica* under warmer, higher CO_2 conditions. *New Phytologist* **150**: 371-383.

Morgan, J.A., LeCain, D.R., Mosier, A.R. and Milchunas, D.G. 2001. Elevated CO_2 enhances water relations and productivity and affects gas exchange in C_3 and C_4 grasses of the Colorado shortgrass steppe. *Global Change Biology* **7**: 451-466.

Norton, L.R., Firbank, L.G., Gray, A.J. and Watkinson, A.R. 1999. Responses to elevated temperature and CO_2 in the perennial grass *Agrostis curtisii* in relation to population origin. *Functional Ecology* **13**: 29-37.

7.3.9.3. Trees

In the study of Kellomaki and Wang (2001), birch seedlings were grown at atmospheric CO_2 concentrations of 350 and 700 ppm in combination with ambient and elevated (ambient plus 3°C) air temperatures. After five months of treatment, the authors reported that photosynthetic rates of CO_2-enriched seedlings were 21 and 28 percent greater than those displayed by their ambiently grown counterparts at ambient and elevated air temperatures, respectfully. In another study, Carter *et al.* (2000) observed that a 300 ppm increase in the air's CO_2 content allowed leaves of sugar maple seedlings to remain green and non-chlorotic when exposed to air temperatures 3°C above ambient air temperature. On the other hand, seedlings fumigated with ambient air exhibited severe foliar chlorosis when exposed to the same elevated air temperatures. These results thus indicate that at elevated air temperatures, rates of photosynthesis are greater and foliar health is typically better in birch and sugar maples trees in CO_2-enriched as opposed to ambient air.

Other studies report similar results. Sheu *et al.* (1999) grew a sub-tropical tree at day/night temperatures of 25/20°C (ambient) and 30/25°C (elevated) for six months and reported that seedlings exposed to 720 ppm CO_2 displayed photosynthetic rates that were 20 and 40 percent higher, respectively, than that of their ambiently grown controls. In addition, the CO_2-induced increases in total dry weight for this species were 14 and 49 percent, respectively, at ambient and elevated air temperatures. Likewise, Maherali *et al.* (2000) observed that a 5°C increase in ambient air temperature increased the CO_2-induced biomass enhancement resulting from a 750 ppm CO_2 enrichment of ponderosa pine seedlings from 42 to 62 percent. Wayne *et al.* (1998) reported that a 5°C increase in the optimal growth temperature of yellow birch seedlings fumigated with an extra 400 ppm CO_2 increased the CO_2-induced increase in biomass from 60 to 227 percent. The beneficial effects of elevated CO_2 on tree species photosynthesis and growth can also be assessed during natural seasonal temperature changes, as documented by Hymus *et al.* (1999) for loblolly pine and Roden *et al.* (1999) for snow gum seedlings.

In some cases, however, there appear to be little interactive effects between elevated CO_2 and temperature on photosynthesis and growth in tree species. When Tjoelker *et al.* (1998a), for example, grew seedlings of quaking aspen, paper birch, tamarack, black spruce and jack pine at atmospheric CO_2 concentrations of 580 ppm, they reported average increases in photosynthetic rates of 28 percent, regardless of temperature, which varied from 18 to 30°C. After analyzing the CO_2-induced increases in dry mass for these seedlings, Tjoelker *et al.* (1998b) further reported that dry mass values were about 50 and 20 percent greater for the deciduous and coniferous species, respectively, regardless of air temperature.

Additional information on this topic, including reviews of newer publications as they become available, can be found at http://www.co2science.org/subject/g/tempco2trees.php.

References

Carter, G.A., Bahadur, R. and Norby, R.J. 2000. Effects of elevated atmospheric CO_2 and temperature on leaf optical properties in *Acer saccharum*. *Environmental and Experimental Botany* **43**: 267-273.

Hymus, G.J., Ellsworth, D.S., Baker, N.R. and Long, S.P. 1999. Does free-air carbon dioxide enrichment affect photochemical energy use by evergreen trees in different seasons? A chlorophyll fluorescence study of mature loblolly pine. *Plant Physiology* **120**: 1183-1191.

Idso, K.E. and Idso, S.B. 1994. Plant responses to atmospheric CO_2 enrichment in the face of environmental constraints: A review of the past 10 years' research. *Agricultural and Forest Meteorology* **69**: 153-203.

Kellomaki, S. and Wang, K.-Y. 2001. Growth and resource use of birch seedlings under elevated carbon dioxide and temperature. *Annals of Botany* **87**: 669-682.

Maherali, H. and DeLucia, E.H. 2000. Interactive effects of elevated CO_2 and temperature on water transport in ponderosa pine. *American Journal of Botany* **87**: 243-249.

Roden, J.S., Egerton, J.J.G. and Ball, M.C. 1999. Effect of elevated [CO_2] on photosynthesis and growth of snow gum (*Eucalyptus pauciflora*) seedlings during winter and spring. *Australian Journal of Plant Physiology* **26**: 37-46.

Sheu, B.-H. and Lin, C.-K. 1999. Photosynthetic response of seedlings of the sub-tropical tree *Schima superba* with exposure to elevated carbon dioxide and temperature. *Environmental and Experimental Botany* **41**: 57-65.

Tjoelker, M.G., Oleksyn, J. and Reich, P.B. 1998a. Seedlings of five boreal tree species differ in acclimation of net photosynthesis to elevated CO_2 and temperature. *Tree Physiology* **18**: 715-726.

Tjoelker, M.G., Oleksyn, J. and Reich, P.B. 1998b. Temperature and ontogeny mediate growth response to elevated CO_2 in seedlings of five boreal tree species. *New Phytologist* **140**: 197-210.

Wayne, P.M., Reekie, E.G. and Bazzaz, F.A. 1998. Elevated CO_2 ameliorates birch response to high temperature and frost stress: implications for modeling climate-induced geographic range shifts. *Oecologia* **114**: 335-342.

7.3.9.4. Other

In a mechanistic model study of Mediterranean shrub vegetation, Osborne *et al.* (2000) reported that increased warming and reduced precipitation would likely decrease net primary production. However, when the same model was run at twice the ambient atmospheric CO_2 concentration, it predicted a 25 percent increase in vegetative productivity, in spite of the increased warming and reduced precipitation. Although we tend to not review studies based on mechanistic models, it is also interesting to note that Bunce (2000) demonstrated that field-grown *Taraxacum officinale* plants exposed to 525 ppm CO_2 and low air temperatures (between 15 and 25°C) displayed photosynthetic rates that were 10 to 30 percent greater than what was predicted by state-of-the-art biochemical models of photosynthesis for this range of temperatures. Thus, at both high and low air temperatures, elevated CO_2 appears to be capable of significantly increasing the photosynthetic prowess of some plants.

In the real world, Stirling *et al.* (1998) grew five fast-growing native species at various atmospheric CO_2 concentrations and air temperatures, finding that twice-ambient levels of atmospheric CO_2 increased photosynthetic rates by 18-36 percent for all species regardless of air temperature, which was up to 3°C higher than ambient air temperature. In addition, atmospheric CO_2 enrichment increased average plant biomass by 25 percent, also regardless of air temperature. Likewise, in a study of vascular plants from Antarctica, Xiong *et al.* (2000) reported that a 13°C rise in air temperature increased plant biomass by 2- to 3-fold. We can only imagine what the added benefit of atmospheric CO_2 enrichment would do for these species.

Hamerlynck *et al.* (2000) demonstrated that the desert perennial shrub *Larrea tridentata* maintained more favorable midday leaf water potentials during a nine-day high-temperature treatment when fumigated with 700 ppm CO_2, as compared to 350 ppm.

Additional information on this topic, including reviews of newer publications as they become available, can be found at http://www.co2science.org/subject/g/tempco2growthres.php.

References

Bunce, J.A. 2000. Acclimation to temperature of the response of photosynthesis to increased carbon dioxide concentration in *Taraxacum officinale*. *Photosynthesis Research* **64**: 89-94.

Hamerlynck, E.P., Huxman, T.E., Loik, M.E. and Smith, S.D. 2000. Effects of extreme high temperature, drought and elevated CO_2 on photosynthesis of the Mojave Desert evergreen shrub, *Larrea tridentata*. *Plant Ecology* **148**: 183-193.

Idso, K.E. and Idso, S.B. 1994. Plant responses to atmospheric CO_2 enrichment in the face of environmental constraints: A review of the past 10 years' research. *Agricultural and Forest Meteorology* **69**: 153-203.

Osborne, C.P., Mitchell, P.L., Sheehy, J.E. and Woodward, F.I. 2000. Modeling the recent historical impacts of atmospheric CO_2 and climate change on Mediterranean vegetation. *Global Change Biology* **6**: 445-458.

Stirling, C.M., Heddell-Cowie, M., Jones, M.L., Ashenden, T.W. and Sparks, T.H. 1998. Effects of elevated CO_2 and temperature on growth and allometry of five native fast-growing annual species. *New Phytologist* **140**: 343-354.

Xiong, F.S., Meuller, E.C. and Day, T.A. 2000. Photosynthetic and respiratory acclimation and growth response of Antarctic vascular plants to contrasting temperature regimes. *American Journal of Botany* **87**: 700-710.

7.3.10. UV-B Radiation

Zhao *et al.* (2004) report that "as a result of stratospheric ozone depletion, UV-B radiation (280-320 nm) levels are still high at the Earth's surface and are projected to increase in the near future (Madronich *et al.*, 1998; McKenzie *et al.*, 2003)." In reference to this potential development, they note that "increased levels of UV-B radiation are known to affect plant growth, development and physiological processes (Dai *et al.*, 1992; Nogués *et al.*, 1999)," stating that high UV-B levels often result in "inhibition of photosynthesis, degradation of protein and DNA, and increased oxidative stress (Jordan *et al.*, 1992; Stapleton, 1992)." In light of the above observations, it is important to clarify how the ongoing rise in the air's CO_2 content might affect the deleterious effects of UV-B radiation on earth's vegetation.

To investigate this question, Zhao *et al.* grew well watered and fertilized cotton plants in sunlit controlled environment chambers maintained at atmospheric CO_2 concentrations of 360 or 720 ppm from emergence until three weeks past first-flower stage under three levels of UV-B radiation (0, 8 and 16 kJ m^{-2} d^{-1}). On five dates between 21 and 62 days after emergence, they measured a number of plant physiological processes and parameters. Over the course of the experiment, the mean net photosynthetic rate of the upper-canopy leaves in the CO_2-enriched chambers was increased—relative to that in the ambient-air chambers—by 38.3 percent in the low UV-B treatment (from 30.3 to 41.9 m m^{-2} s^{-1}), 41.1 percent in the medium UV-B treatment (from 28.7 to 40.5 m m^{-2} s^{-1}), and 51.5 percent in the high UV-B treatment (from 17.1 to 25.9 m m^{-2} s^{-1}). In the medium UV-B treatment, the growth stimulation from the elevated CO_2 was sufficient to raise net photosynthesis rates 33.7 percent above the rates experienced in the ambient air and no UV-B treatment (from 30.3 to 40.5 m m^{-2} s^{-1}); but in the high UV-B treatment the radiation damage was so great that even with the help of the 51.5 percent increase in net photosynthesis provided by the doubled-CO_2 air, the mean net photosynthesis rate of the cotton leaves was 14.5 percent less than that experienced in the ambient air and no UV-B treatment (dropping from 30.3 to 25.9 m m^{-2} s^{-1}).

It should be noted that the medium UV-B treatment of this study was chosen to represent the intensity of UV-B radiation presently received on a clear summer day in the major cotton production region of Mississippi, USA, under current stratospheric ozone conditions, while the high UV-B treatment was chosen to represent what might be expected there following a 30 percent depletion of the ozone layer, which has been predicted to double the region's reception of UV-B radiation from 8 to 16 kJ m^{-2} d^{-1}. Consequently, a doubling of the current CO_2 concentration and the current UV-B radiation level would reduce the net photosynthetic rate of cotton leaves by just under 10 percent (from 28.7 to 25.9 m m^{-2} s^{-1}), whereas in the absence of a doubling of the air's CO_2 content, a doubling of the UV-B radiation level would reduce cotton net photosynthesis by just over 40 percent (from 28.7 to 17.1 m m^{-2} s^{-1}).

Viewed in this light, it can be seen that a doubling the current atmospheric CO_2 concentration would compensate for over three-fourths of the loss of cotton photosynthetic capacity caused by a doubling of the current UV-B radiation intensity. It may do better than that, for in the study of Zhao *et al.* (2003), it was reported that both Adamse and Britz (1992) and Rozema *et al.* (1997) found that doubled CO_2 totally compensated for the negative effects of equally high UV-B radiation.

In another study (Qaderi and Reid, 2005), well watered and fertilized canola (*Brassica napus* L.) plants were grown from seed to maturity in pots within controlled environment chambers maintained at either 370 or 740 ppm CO_2 with and without a daily dose of UV-B radiation in the amount of 4.2 kJ m^{-2}, while a number of plant parameters were measured at various times throughout the growing season. With respect to the bottom-line result of final seed yield, this parameter was determined to be 0.98 g/plant in the control treatment (ambient CO_2, with UV-B). Doubling the CO_2 concentration increased yield by 25.5 percent to 1.23 g/plant. Alternatively, removing the UV-B radiation flux increased yield by 91.8 percent to 1.88 g/plant. Doing both (doubling the CO_2 concentration while simultaneously removing the UV-B flux) increased final seed yield most of all, by 175.5 percent to 2.7 g/plant. Viewed from a different perspective, doubling the air's CO_2 concentration in the *presence* of the UV-B radiation flux enhanced final seed yield by 25.5 percent, while doubling CO_2 in the *absence* of the UV-B radiation flux increased

seed yield by 43.6 percent. In concluding their paper, the authors note that "previous studies have shown that elevated CO_2 increases biomass and seed yield, whereas UV-B decreases them (Sullivan, 1997; Teramura *et al.*, 1990)." Finding much the same thing in their study, they thus reckoned that "elevated CO_2 may have a positive effect on plants by mitigating the detrimental effects caused by UV-B radiation."

Two years later in a similar study of the same plant, Qaderi *et al.* (2007) grew well watered and fertilized canola plants from the 30-day-old stage until 25 days after anthesis in 1-L pots within controlled environment chambers exposed to either 4.2 kJ m^{-2} d^{-1} of UV-B radiation or no such radiation in air of either 370 or 740 ppm CO_2, in order to determine the effects of these two parameters on the photosynthetic rates and water use efficiency of the maturing husks or *siliquas* that surround the plants' seeds. Results indicated that for the plants exposed to 4.2 kJ m^{-2} d^{-1} of UV-B radiation, the experimental doubling of the air's CO_2 concentration led to a 29 percent increase in siliqua net photosynthesis, an 18 percent decrease in siliqua transpiration, and a 58 percent increase in siliqua water use efficiency; while for the plants exposed to no UV-B radiation, siliqua net photosynthesis was increased by a larger 38 percent, transpiration was decreased by a larger 22 percent and water use efficiency was increased by a larger 87 percent in the CO_2-enriched air.

In another noteworthy study, Deckmyn *et al.* (2001) grew white clover plants for four months in four small greenhouses, two of which allowed 88 percent of the incoming UV-B radiation to pass through their roofs and walls and two of which allowed 82 percent to pass through, while one of the two greenhouses in each of the UV-B treatments was maintained at ambient CO_2 (371 ppm) and the other at elevated CO_2 (521 ppm). At the mid-season point of their study, they found that the 40 percent increase in atmospheric CO_2 concentration stimulated the production of flowers in the low UV-B treatment by 22 percent and in the slightly higher UV-B treatment by 43 percent; while at the end of the season, the extra CO_2 was determined to have provided no stimulation of biomass production in the low UV-B treatment, but it significantly stimulated biomass production by 16 percent in the high UV-B treatment.

The results of this study indicate that the positive effects of atmospheric CO_2 enrichment on flower and biomass production in white clover are greater at more realistic or natural values of UV-B radiation than those found in many greenhouses. As a result,

Deckmyn *et al.* say their results "clearly indicate the importance of using UV-B transmittant greenhouses or open-top chambers when conducting CO_2 studies," for if this is not done, their work suggests that the results obtained could significantly underestimate the magnitude of the benefits that are being continuously accrued by earth's vegetation as a result of the ongoing rise in the air's CO_2 content.

In 2007, Koti *et al.* (2007) used Soil-Plant-Atmosphere-Research (SPAR) chambers at Mississippi State University (USA) to investigate the effects of doubled atmospheric CO_2 concentration (720 vs. 360 ppm) on the growth and development of six well watered and fertilized soybean (*Glycine max* L.) genotypes grown from seed in pots filled with fine sand and exposed to the dual stresses of high day/night temperatures (38/30°C vs. 30/22°C) and high UV-B radiation levels (10 vs. 0 kJ/m^2/day). Results led this group of authors to report that "elevated CO_2 partially compensated [for] the damaging effects on vegetative growth and physiology caused by negative stressors such as high temperatures and enhanced UV-B radiation levels in soybean," specifically noting, in this regard, CO_2's positive influence on the physiological parameters of plant height, leaf area, total biomass, net photosynthesis, total chlorophyll content, phenolic content and wax content, as well as relative plant injury.

In a study that did not include UV-B radiation as an experimental parameter, Estiarte *et al.* (1999) grew spring wheat in FACE plots in Arizona, USA, at atmospheric CO_2 concentrations of 370 and 550 ppm and two levels of soil moisture (50 and 100 percent of potential evapotranspiration). They found that leaves of plants grown in elevated CO_2 had 14 percent higher total flavonoid concentrations than those of plants grown in ambient air, and that soil water content did not affect the relationship. An important aspect of this finding is that one of the functions of flavonoids in plant leaves is to protect them against UV-B radiation. More studies of this nature should be conducted to see how general this beneficial response may be throughout the plant world.

In a study of UV-B and CO_2 effects on a natural ecosystem, which was conducted at the Abisko Scientific Research Station in Swedish Lapland, Johnson *et al.* (2002) studied plots of subarctic heath composed of open canopies of downy birch and dense dwarf-shrub layers containing scattered herbs and grasses. For a period of five years, they exposed the plots to factorial combinations of UV-B radiation—ambient and that expected to result from a 15 percent stratospheric ozone depletion—and atmospheric CO_2 concentration—ambient (around 365 ppm) and enriched (around 600 ppm)—after which they determined the amounts of microbial carbon (C_{mic}) and nitrogen (N_{mic}) in the soils of the plots.

When the plots were exposed to the enhanced UV-B radiation, the amount of C_{mic} in the soil was reduced to only 37 percent of what it was at the ambient UV-B level when the air's CO_2 content was maintained at the ambient concentration. When the UV-B increase was accompanied by the CO_2 increase, however, not only was there not a decrease in C_{mic}, there was an actual increase of 37 percent. The amount of N_{mic} in the soil experienced a 69 percent increase when the air's CO_2 content was maintained at the ambient concentration; and when the UV-B increase was accompanied by the CO_2 increase, N_{mic} rose even more, by 138 percent.

These findings, in the words of Johnson *et al.*, "may have far-reaching implications ... because the productivity of many semi-natural ecosystems is limited by N (Ellenberg, 1988)." The 138 percent increase in soil microbial N observed in this study to accompany a 15 percent reduction in stratospheric ozone and a 64 percent increase in atmospheric CO_2 concentration (experienced in going from 365 ppm to 600 ppm) should significantly enhance the input of plant litter to the soils of these ecosystems, which phenomenon represents the first half of the carbon sequestration process, i.e., the carbon input stage. With respect to the second stage of *keeping* as much of that carbon as possible in the soil, Johnson *et al.* note that "the capacity for subarctic semi-natural heaths to act as major sinks for fossil fuel-derived carbon dioxide is [also] likely to be critically dependent on the supply of N," as is indeed indicated to be the case in the literature review of Berg and Matzner (1997), who report that with more nitrogen in the soil, the long-term storage of carbon is significantly enhanced, as more litter is chemically transformed into humic substances when nitrogen is more readily available, and these more recalcitrant carbon compounds can be successfully stored in the soil for many millennia.

In light of these several findings, we conclude that the ongoing rise in the air's CO_2 content is a powerful antidote for the deleterious biological impacts that might possibly be caused by an increase in the flux of UV-B radiation at the surface of the earth due to any further depletion of the planet's stratospheric ozone layer.

473

Additional information on this topic, including reviews of newer publications as they become available, can be found at http://www.co2science.org/subject/u/uvbradiation.php.

References

Adamse, P. and Britz, S.J. 1992. Amelioration of UV-B damage under high irradiance. I. Role of photosynthesis. *Photochemistry and Photobiology* **56**: 645-650.

Berg, B. and Matzner, E. 1997. Effect of N deposition on decomposition of plant litter and soil organic matter in forest ecosystems. *Environmental Reviews* **5**: 1-25.

Dai, Q., Coronal, V.P., Vergara, B.S., Barnes, P.W. and Quintos, A.T. 1992. Ultraviolet-B radiation effects on growth and physiology of four rice cultivars. *Crop Science* **32**: 1269-1274.

Deckmyn, G., Caeyenberghs, E. and Ceulemans, R. 2001. Reduced UV-B in greenhouses decreases white clover response to enhanced CO₂. *Environmental and Experimental Botany* **46**: 109-117.

Ellenberg, H. 1988. *Vegetation Ecology of Central Europe.* Cambridge University Press, Cambridge, UK.

Estiarte, M., Penuelas, J., Kimball, B.A., Hendrix, D.L., Pinter Jr., P.J., Wall, G.W., LaMorte, R.L. and Hunsaker, D.J. 1999. Free-air CO₂ enrichment of wheat: leaf flavonoid concentration throughout the growth cycle. *Physiologia Plantarum* **105**: 423-433.

Johnson, D., Campbell, C.D., Lee, J.A., Callaghan, T.V. and Gwynn-Jones, D. 2002. Arctic microorganisms respond more to elevated UV-B radiation than CO₂. *Nature* **416**: 82-83.

Jordan, B.R., Chow, W.S. and Anderson, J.M. 1992. Changes in mRNA levels and polypeptide subunits of ribulose 1,5-bisphosphate carboxylase in response to supplementary ultraviolet-B radiation. *Plant, Cell and Environment* **15**: 91-98.

Koti, S., Reddy, K.R., Kakani, V.G., Zhao, D. and Gao, W. 2007. Effects of carbon dioxide, temperature and ultraviolet-B radiation and their interactions on soybean (*Glycine max* L.) growth and development. *Environmental and Experimental Botany* **60**: 1-10.

Madronich, S., McKenzie, R.L., Bjorn, L.O. and Caldwell, M.M. 1998. Changes in biologically active ultraviolet radiation reaching the Earth's surface. *Journal of Photochemistry and Photobiology B* **46**: 5-19.

McKenzie, R.L., Bjorn, L.O., Bais, A. and Ilyasd, M. 2003. Changes in biologically active ultraviolet radiation

reaching the earth's surface. *Photochemical and Photobiological Sciences* **2**: 5-15.

Nogués, S., Allen, D.J., Morison, J.I.L. and Baker, N.R. 1999. Characterization of stomatal closure caused by ultraviolet-B radiation. *Plant Physiology* **121**: 489-496.

Qaderi, M.M. and Reid, D.M. 2005. Growth and physiological responses of canola (*Brassica napus*) to UV-B and CO₂ under controlled environment conditions. *Physiologia Plantarum* **125**: 247-259.

Qaderi, M.M., Reid, D.M. and Yeung, E.C. 2007. Morphological and physiological responses of canola (*Brassica napus*) siliquas and seeds to UVB and CO₂ under controlled environment conditions. *Environmental and Experimental Botany* **60**: 428-437.

Rozema, J., Lenssen, G.M., Staaij, J.W.M., Tosserams, M., Visser, A.J. and Brockman, R.A. 1997. Effects of UV-B radiation on terrestrial plants and ecosystems: interaction with CO₂ enrichment. *Plant Ecology* **128**: 182-191.

Stapleton, A.E. 1992. Ultraviolet radiation and plants: Burning questions. *The Plant Cell* **105**: 881-889.

Sullivan, J.H. 1997. Effects of increasing UV-B radiation and atmospheric CO₂ on photosynthesis and growth: implications for terrestrial ecosystems. *Plant Ecology* **128**: 195-206.

Teramura, A.H., Sullivan, J.H. and Ziska, L.H. 1990. Interaction of elevated UV-B radiation and CO₂ on productivity and photosynthetic characteristics in wheat, rice, and soybean. *Plant Physiology* **94**: 470-475.

Zhao, D., Reddy, K.R., Kakani, V.G., Mohammed, A.R., Read, J.J. and Gao, W. 2004. Leaf and canopy photosynthetic characteristics of cotton (*Gossypiuym hirsutum*) under elevated CO₂ concentration and UV-B radiation. *Journal of Plant Physiology* **161**: 581-590.

Zhao, D., Reddy, K.R., Kakani, V.G., Read, J.J. and Sullivan, J.H. 2003. Growth and physiological responses of cotton (*Gossypium hirsutum* L.) to elevated carbon dioxide and ultraviolet-B radiation under controlled environmental conditions. *Plant, Cell and Environment* **26**: 771-782.

7.3.11. Water Stress

As the CO₂ content of the air continues to rise, nearly all of earth's plants will exhibit increases in photosynthesis and biomass production. However, some experts claim that water stress will negate these benefits. In reviewing the scientific literature of the ten-year period 1983-1994, however, Idso and Idso (1994) found that water stress typically will *not* negate the CO_2-induced stimulation of plant productivity. They found that the CO_2-induced percentage increase in plant biomass production was often greater under water-stressed conditions than it was when plants were well-watered. We here review some more recent scientific literature in this area for agricultural, grassland and woody plant species.

7.3.11.1. Agricultural Species

During times of water stress, atmospheric CO_2 enrichment often stimulates plants to develop larger-than-usual and more robust root systems to probe greater volumes of soil for scarce and much-needed moisture. Wechsung *et al.* (1999), for example, observed a 70 percent increase in lateral root dry weights of water-stressed wheat grown at 550 ppm CO_2, while De Luis *et al.* (1999) reported a 269 percent increase in root-to-shoot ratio of water-stressed alfalfa growing at 700 ppm CO_2. Thus, elevated CO_2 elicits stronger-than-usual positive root responses in agricultural species under conditions of water stress.

Elevated levels of atmospheric CO_2 also tend to reduce the openness of stomatal pores on leaves, thus decreasing plant stomatal conductance. This phenomenon, in turn, reduces the amount of water lost to the atmosphere by transpiration and, consequently, lowers overall plant water use. Serraj *et al.* (1999) report that water-stressed soybeans grown at 700 ppm CO_2 reduced their total seasonal water loss by 10 percent relative to that of water-stressed control plants grown at 360 ppm CO_2. In addition, Conley *et al.* (2001) noted that a 200-ppm increase in the air's CO_2 concentration reduced cumulative evapotranspiration in water-stressed sorghum by approximately 4 percent. Atmospheric CO_2 enrichment thus increases plant water acquisition, by stimulating root growth, while it reduces plant water loss, by constricting stomatal apertures; and these dual effects typically enhance plant water-use efficiency, even under conditions of less-than-optimal soil water content. But these phenomena have other implications as well.

CO_2-induced increases in root development together with CO_2-induced reductions in stomatal conductance often contribute to the maintenance of a more favorable plant water status during times of drought. Sgherri *et al.* (1998) reported that leaf water potential, which is a good indicator of overall plant water status, was 30 percent higher (less negative and therefore more favorable) in water-stressed alfalfa grown at an atmospheric CO_2 concentration of 600 ppm CO_2 versus 340 ppm CO_2. Wall (2001) reports that leaf water potentials were similar in CO_2-enriched water-stressed plants and ambiently grown well-watered control plants, which implies a complete CO_2-induced amelioration of water stress in the CO_2-enriched plants. Similarly, Lin and Wang (2002) demonstrated that elevated CO_2 caused a several-day delay in the onset of the water stress-induced production of the highly reactive oxygenated compound H_2O_2 in spring wheat.

If atmospheric CO_2 enrichment thus allows plants to maintain a better water status during times of water stress, it is only logical to expect that such plants should exhibit greater rates of photosynthesis than ambiently grown plants. And so they do. With the onset of water stress in *Brassica juncea*, for example, photosynthetic rates dropped by 40 percent in plants growing in ambient air, while plants growing in air containing 600 ppm CO_2 only experienced a 30 percent reduction in net photosynthesis (Rabha and Uprety, 1998). Ferris *et al.* (1998) reported that after imposing water-stress conditions on soybeans and allowing them to recover following complete rewetting of the soil, plants grown in air containing 700 ppm CO_2 reached pre-stressed rates of photosynthesis after six days, while plants grown in ambient air never recovered to pre-stressed rates.

Reasoning analogously, it is also to be expected that plant biomass production would be enhanced by elevated CO_2 concentrations under drought conditions. In exploring this idea, Ferris *et al.* (1999) reported that water-stressed soybeans grown at 700 ppm CO_2 attained seed yields that were 24 percent greater than those of similarly water-stressed plants grown at ambient CO_2 concentrations, while Hudak *et al.* (1999) reported that water-stress had no effect on yield in CO_2-enriched spring wheat.

In some cases, the CO_2-induced percentage biomass increase is actually greater for water-stressed plants than it is for well-watered plants. Li *et al.* (2000), for example, reported that a 180-ppm increase in the air's CO_2 content increased lower stem grain weights in water-stressed and well-watered spring wheat by 24 and 14 percent, respectively. Similarly, spring wheat grown in air containing an additional 280 ppm CO_2 exhibited 57 and 40 percent increases in grain yield under water-stressed and well-watered conditions, respectively (Schutz and Fangmeier, 2001). Likewise, Ottman *et al.* (2001) noted that elevated CO_2 increased plant biomass in water-stressed sorghum by 15 percent, while no biomass increase occurred in well-watered sorghum.

In summary, the conclusions of Idso and Idso (1994) are well supported by the recent peer-reviewed scientific literature, which indicates that the ongoing rise in the air's CO_2 content will likely lead to substantial increases in plant photosynthetic rates and biomass production, even in the face of stressful conditions imposed by less-than-optimum soil moisture conditions.

Additional information on this topic, including reviews of newer publications as they become available, can be found at http://www.co2science.org/subject/g/growthwaterag.php.

References

Conley, M.M., Kimball, B.A., Brooks, T.J., Pinter Jr., P.J., Hunsaker, D.J., Wall, G.W., Adams, N.R., LaMorte, R.L., Matthias, A.D., Thompson, T.L., Leavitt, S.W., Ottman, M.J., Cousins, A.B. and Triggs, J.M. 2001. CO_2 enrichment increases water-use efficiency in sorghum. *New Phytologist* **151**: 407-412.

De Luis, J., Irigoyen, J.J. and Sanchez-Diaz, M. 1999. Elevated CO_2 enhances plant growth in droughted N_2-fixing alfalfa without improving water stress. *Physiologia Plantarum* **107**: 84-89.

Ferris, R., Wheeler, T.R., Hadley, P. and Ellis, R.H. 1998. Recovery of photosynthesis after environmental stress in soybean grown under elevated CO_2. *Crop Science* **38**: 948-955.

Ferris, R., Wheeler, T.R., Ellis, R.H. and Hadley, P. 1999. Seed yield after environmental stress in soybean grown under elevated CO_2. *Crop Science* **39**: 710-718.

Hudak, C., Bender, J., Weigel, H.-J. and Miller, J. 1999. Interactive effects of elevated CO_2, O_3, and soil water deficit on spring wheat (*Triticum aestivum* L. cv. Nandu). *Agronomie* **19**: 677-687.

Idso, K.E. and Idso, S.B. 1994. Plant responses to atmospheric CO_2 enrichment in the face of environmental constraints: A review of the past 10 years' research. *Agricultural and Forest Meteorology* **69**: 153-203.

Li, A.-G., Hou, Y.-S., Wall, G.W., Trent, A., Kimball, B.A. and Pinter Jr., P.J. 2000. Free-air CO_2 enrichment and drought stress effects on grain filling rate and duration in spring wheat. *Crop Science* **40**: 1263-1270.

Lin, J.-S and Wang, G.-X. 2002. Doubled CO_2 could improve the drought tolerance better in sensitive cultivars than in tolerant cultivars in spring wheat. *Plant Science* **163**: 627-637.

Ottman, M.J., Kimball, B.A., Pinter Jr., P.J., Wall, G.W., Vanderlip, R.L., Leavitt, S.W., LaMorte, R.L., Matthias, A.D. and Brooks, T.J. 2001. Elevated CO_2 increases sorghum biomass under drought conditions. *New Phytologist* **150**: 261-273.

Rabha, B.K. and Uprety, D.C. 1998. Effects of elevated CO_2 and moisture stress on *Brassica juncea*. *Photosynthetica* **35**: 597-602.

Serraj, R., Allen, L.H., Jr., Sinclair, T.R. 1999. Soybean leaf growth and gas exchange response to drought under carbon dioxide enrichment. *Global Change Biology* **5**: 283-291.

Sgherri, C.L.M., Quartacci, M.F., Menconi, M., Raschi, A. and Navari-Izzo, F. 1998. Interactions between drought and elevated CO_2 on alfalfa plants. *Journal of Plant Physiology* **152**: 118-124.

Schutz, M. and Fangmeier, A. 2001. Growth and yield responses of spring wheat (*Triticum aestivum* L. cv. Minaret) to elevated CO_2 and water limitation. *Environmental Pollution* **114**: 187-194.

Wall, G.W. 2001. Elevated atmospheric CO_2 alleviates drought stress in wheat. *Agriculture, Ecosystems and Environment* **87**: 261-271.

Wechsung, G., Wechsung, F., Wall, G.W., Adamsen, F.J., Kimball, B.A., Pinter Jr., P.J., LaMorte, R.L., Garcia, R.L. and Kartschall, T. 1999. The effects of free-air CO_2 enrichment and soil water availability on spatial and seasonal patterns of wheat root growth. *Global Change Biology* **5**: 519-529.

7.3.11.2. Grassland Species

In the study of Leymarie *et al.* (1999), twice-ambient levels of atmospheric CO_2 caused significant reductions in the stomatal conductance of water-stressed *Arabidopsis thaliana*. Similarly, Volk *et al.* (2000) reported that calcareous grassland species exposed to elevated CO_2 concentrations (600 ppm) consistently exhibited reduced stomatal conductance, regardless of soil moisture availability. Thus, atmospheric CO_2 enrichment clearly reduces stomatal conductance and plant transpiration and soil water depletion in grassland ecosystems.

In the case of four grassland species comprising a pasture characteristic of New Zealand, Clark *et al.* (1999) found that leaf water potential, which is a good indicator of plant water status, was consistently higher (less negative and, therefore, less stressful) under elevated atmospheric CO_2 concentrations. Similarly, leaf water potentials of the water-stressed

C_4 grass *Panicum coloratum* grown at 1000 ppm CO_2 were always higher than those of their water-stressed counterparts growing in ambient air (Seneweera *et al.*, 2001). Indeed, Seneweera *et al.* (1998) reported that leaf water potentials observed in CO_2-enriched water-stressed plants were an amazing three-and-a-half times greater than those observed in control plants grown at 350 ppm during drought conditions (Seneweera *et al.*, 1998).

If atmospheric CO_2 enrichment thus allows plants to maintain improved water status during times of water stress, it is only logical to expect that such plants will exhibit greater photosynthetic rates than similar plants growing in ambient air. In a severe test of this concept, Ward *et al.* (1999) found that extreme water stress caused 93 and 85 percent reductions in the photosynthetic rates of two CO_2-enriched grassland species; yet their rates of carbon fixation were still greater than those observed under ambient CO_2 conditions.

In view of the fact that elevated CO_2 enhances photosynthetic rates during times of water stress, one would expect that plant biomass production would also be enhanced by elevated CO_2 concentrations under drought conditions. And so it is. On the American prairie, for example, Owensby *et al.* (1999) reported that tallgrass ecosystems exposed to twice-ambient concentrations of atmospheric CO_2 for eight years only exhibited significant increases in above- and below-ground biomass during years of less-than-average rainfall. Also, in the study of Derner *et al.* (2001), the authors reported that a 150-ppm increase in the CO_2 content of the air increased shoot biomass in two C_4 grasses by 57 percent, regardless of soil water content. Seneweera *et al.* (2001) reported that a 640-ppm increase in the air's CO_2 content increased shoot dry mass in a C_4 grass by 44 and 70 percent under well-watered and water-stressed conditions, respectively. Likewise, Volk *et al.* (2000) grew calcareous grassland assemblages at 360 and 600 ppm CO_2 and documented 18 and 40 percent CO_2-induced increases in whole-community biomass under well-watered and water-stressed conditions, respectively.

In summary, the peer-reviewed scientific literature indicates that the ongoing rise in the air's CO_2 content will likely lead to substantial increases in plant photosynthetic rates and biomass production for grassland species even in the face of stressful environmental conditions imposed by less-than-optimum soil moisture contents.

Additional information on this topic, including reviews of newer publications as they become available, can be found at http://www.co2science.org/subject/g/growthwatergrass.php.

References

Clark, H., Newton, P.C.D. and Barker, D.J. 1999. Physiological and morphological responses to elevated CO_2 and a soil moisture deficit of temperate pasture species growing in an established plant community. *Journal of Experimental Botany* **50**: 233-242.

Derner, J.D., Polley, H.W., Johnson, H.B. and Tischler, C.R. 2001. Root system response of C_4 grass seedlings to CO_2 and soil water. *Plant and Soil* **231**: 97-104.

Leymarie, J., Lasceve, G. and Vavasseur, A. 1999. Elevated CO_2 enhances stomatal responses to osmotic stress and abscisic acid in *Arabidopsis thaliana*. *Plant, Cell and Environment* **22**: 301-308.

Owensby, C.E., Ham, J.M., Knapp, A.K. and Auen, L.M. 1999. Biomass production and species composition change in a tallgrass prairie ecosystem after long-term exposure to elevated atmospheric CO_2. *Global Change Biology* **5**: 497-506.

Seneweera, S.P., Ghannoum, O. and Conroy, J. 1998. High vapor pressure deficit and low soil water availability enhance shoot growth responses of a C_4 grass (*Panicum coloratum* cv. Bambatsi) to CO_2 enrichment. *Australian Journal of Plant Physiology* **25**: 287-292.

Seneweera, S., Ghannoum, O. and Conroy, J.P. 2001. Root and shoot factors contribute to the effect of drought on photosynthesis and growth of the C_4 grass *Panicum coloratum* at elevated CO_2 partial pressures. *Australian Journal of Plant Physiology* **28**: 451-460.

Volk, M., Niklaus, P.A. and Korner, C. 2000. Soil moisture effects determine CO_2 responses of grassland species. *Oecologia* **125**: 380-388.

Ward, J.K., Tissue, D.T., Thomas, R.B. and Strain, B.R. 1999. Comparative responses of model C_3 and C_4 plants to drought in low and elevated CO_2. *Global Change Biology* **5**: 857-867.

7.3.11.3. Woody Species

During times of water stress, atmospheric CO_2 enrichment often stimulates the development of larger-than-usual and more robust root systems in woody perennial species, which allows them to probe greater volumes of soil for scarce and much-needed moisture. Tomlinson and Anderson (1998), for example, report that greater root development in water-stressed red oak seedlings grown at 700 ppm CO_2 helped them effectively deal with the reduced availability of moisture. These trees eventually produced just as much biomass as well-watered controls exposed to ambient air containing 400 ppm CO_2. In addition, Polley *et al.* (1999) note that water-stressed honey mesquite trees subjected to an atmospheric CO_2 concentration of 700 ppm produced 37 percent more root biomass than water-stressed control seedlings growing at 370 ppm.

Elevated levels of atmospheric CO_2 also tend to reduce the area of open stomatal pore space on leaf surfaces, thus reducing plant stomatal conductance. This phenomenon, in turn, reduces the amount of water lost to the atmosphere via transpiration. Tognetti *et al.* (1998), for example, determined that stomatal conductances of mature oak trees growing near natural CO_2 springs in central Italy were significantly lower than those of similar trees growing further away from the springs during periods of severe summer drought.

CO_2-induced increases in root development together with CO_2-induced reductions in stomatal conductance often contribute to the maintenance of a more favorable plant water status during times of drought. In the case of three Mediterranean shrubs, Tognetti *et al.* (2002) found that leaf water potential, which is a good indicator of plant water status, was consistently higher (less negative and, hence, less stressful) under twice-ambient CO_2 concentrations. Similarly, leaf water potentials of water-stressed mesquite seedlings grown at 700 ppm CO_2 were 40 percent higher than those of their water-stressed counterparts growing in ambient air (Polley *et al.*, 1999), which is comparable to values of -5.9 and -3.4 MPa observed in water-stressed evergreen shrubs (*Larrea tridentata*) exposed to 360 and 700 ppm CO_2, respectively (Hamerlynck *et al.*, 2000).

Palanisamy (1999) observed water-stressed *Eucalyptus* seedlings grown at 800 ppm CO_2 display greater net photosynthetic rates than their ambiently grown and water-stressed counterparts. Runion *et al.* (1999) observed the CO_2-induced photosynthetic stimulation of water-stressed pine seedlings grown at 730 ppm CO_2 to be nearly 50 percent greater than that of similar water-stressed pine seedlings grown at 365 ppm CO_2. Similarly, Centritto *et al.* (1999a) found that water-stressed cherry trees grown at 700 ppm CO_2 displayed net photosynthetic rates that were 44 percent greater than those of water-stressed trees grown at 350 ppm CO_2. And Anderson and

Tomlinson (1998) found that a 300-ppm increase in the air's CO_2 concentration boosted photosynthetic rates in well-watered and water-stressed red oak seedlings by 34 and 69 percent, respectively, demonstrating that the CO_2-induced percentage enhancement in net photosynthesis in this species was essentially twice as great in water-stressed seedlings as in well-watered ones.

Sometimes, plants suffer drastically when subjected to extreme water stress. However, the addition of CO_2 to the atmosphere often gives them an edge over ambiently growing plants. Tuba *et al.* (1998), for example, reported that leaves of a water-stressed woody shrub exposed to an atmospheric CO_2 concentration of 700 ppm continued to maintain positive rates of net carbon fixation for a period that lasted three times longer than that observed for leaves of equally water-stressed control plants growing in ambient air. Similarly, Fernandez *et al.* (1998) discovered that herb and tree species growing near natural CO_2 vents in Venezuela continued to maintain positive rates of net photosynthesis during that location's dry season, while the same species growing some distance away from the CO_2 source displayed net losses of carbon during this stressful time. Likewise, Fernandez *et al.* (1999) noted that after four weeks of drought, the deciduous Venezuelan shrub *Ipomoea carnea* continued to exhibit positive carbon gains under elevated CO_2 conditions, whereas ambiently growing plants displayed net carbon losses. Polley *et al.* (2002) reported that seedlings of five woody species grown at twice-ambient CO_2 concentrations survived 11 days longer (on average) than control seedlings when subjected to maximum drought conditions. Thus, in some cases of water stress, enriching the air with CO_2 can mean the difference between life or death.

Arp *et al.* (1998) reported that six perennial plants common to the Netherlands increased their biomass under CO_2-enriched conditions even when suffering from lack of water. In other cases, the CO_2-induced percentage biomass increase is sometimes even greater for water-stressed plants than it is for well-watered plants. Catovsky and Bazzaz (1999), for example, reported that the CO_2-induced biomass increase for paper birch was 27 percent and 130 percent for well-watered and water-stressed seedlings, respectively. Similarly, Schulte *et al.* (1998) noted that the CO_2-induced biomass increase of oak seedlings was greater under water-limiting conditions than under well-watered conditions (128 percent vs. 92 percent), as did Centritto *et al.* (1999b) for basal trunk area in cherry seedlings (69 percent vs. 22 percent).

Finally, Knapp *et al.* (2001) developed tree-ring index chronologies from western juniper stands in Oregon, USA, finding that the trees recovered better from the effects of drought in the 1990's, when the air's CO_2 concentration was around 340 ppm, than they did from 1900-1930, when the atmospheric CO_2 concentration was around 300 ppm. In a loosely related study, Osborne *et al.* (2002) looked at the warming and reduced precipitation experienced in Mediterranean shrublands over the last century and concluded that primary productivity should have been negatively impacted in those areas. However, when the concurrent increase in atmospheric CO_2 concentration was factored into their mechanistic model, a 25 percent increase in primary productivity was projected.

In summary, the peer-reviewed scientific literature indicates that the ongoing rise in the air's CO_2 content will likely lead to substantial increases in photosynthetic rates and biomass production in earth's woody species in the years and decades ahead, even in the face of stressful conditions imposed by less-than-optimal availability of soil moisture.

Additional information on this topic, including reviews of newer publications as they become available, can be found at http://www.co2science.org/subject/g/growthwaterwood.php.

References

Anderson, P.D. and Tomlinson, P.T. 1998. Ontogeny affects response of northern red oak seedlings to elevated CO_2 and water stress. I. Carbon assimilation and biomass production. *New Phytologist* **140**: 477-491.

Arp, W.J., Van Mierlo, J.E.M., Berendse, F. and Snijders, W. 1998. Interactions between elevated CO_2 concentration, nitrogen and water: effects on growth and water use of six perennial plant species. *Plant, Cell and Environment* **21**: 1-11.

Catovsky, S. and Bazzaz, F.A. 1999. Elevated CO_2 influences the responses of two birch species to soil moisture: implications for forest community structure. *Global Change Biology* **5**: 507-518.

Centritto, M., Magnani, F., Lee, H.S.J. and Jarvis, P.G. 1999a. Interactive effects of elevated [CO_2] and drought on cherry (*Prunus avium*) seedlings. II. Photosynthetic capacity and water relations. *New Phytologist* **141**: 141-153.

Centritto, M., Lee, H.S.J. and Jarvis, P.G. 1999b. Interactive effects of elevated [CO_2] and drought on cherry (*Prunus avium*) seedlings. I. Growth, whole-plant water use efficiency and water loss. *New Phytologist* **141**: 129-140.

Fernandez, M.D., Pieters, A., Azuke, M., Rengifo, E., Tezara, W., Woodward, F.I. and Herrera, A. 1999. Photosynthesis in plants of four tropical species growing under elevated CO_2. *Photosynthetica* **37**: 587-599.

Fernandez, M.D., Pieters, A., Donoso, C., Tezara, W., Azuke, M., Herrera, C., Rengifo, E. and Herrera, A. 1998. Effects of a natural source of very high CO_2 concentration on the leaf gas exchange, xylem water potential and stomatal characteristics of plants of *Spatiphylum cannifolium* and *Bauhinia multinervia*. *New Phytologist* **138**: 689-697.

Hamerlynck, E.P., Huxman, T.E., Loik, M.E. and Smith, S.D. 2000. Effects of extreme high temperature, drought and elevated CO_2 on photosynthesis of the Mojave Desert evergreen shrub, *Larrea tridentata*. *Plant Ecology* **148**: 183-193.

Knapp, P.A., Soule, P.T. and Grissino-Mayer, H.D. 2001. Post-drought growth responses of western juniper (*Juniperus occidentalis* var. *occidentalis*) in central Oregon. *Geophysical Research Letters* **28**: 2657-2660.

Osborne, C.P., Mitchell, P.L., Sheehy, J.E. and Woodward, F.I. 2000. Modellng the recent historical impacts of atmospheric CO_2 and climate change on Mediterranean vegetation. *Global Change Biology* **6**: 445-458.

Palanisamy, K. 1999. Interactions of elevated CO_2 concentration and drought stress on photosynthesis in *Eucalyptus cladocalyx* F. Muell. *Photosynthetica* **36**: 635-638.

Polley, H.W., Tischler, C.R., Johnson, H.B. and Derner, J.D. 2002. Growth rate and survivorship of drought: CO_2 effects on the presumed tradeoff in seedlings of five woody legumes. *Tree Physiology* **22**: 383-391.

Polley, H.W., Tischler, C.R., Johnson, H.B. and Pennington, R.E. 1999. Growth, water relations, and survival of drought-exposed seedlings from six maternal families of honey mesquite (*Prosopis glandulosa*): responses to CO_2 enrichment. *Tree Physiology* **19**: 359-366.

Runion, G.B., Mitchell, R.J., Green, T.H., Prior, S.A., Rogers, H.H. and Gjerstad, D.H. 1999. Longleaf pine photosynthetic response to soil resource availability and elevated atmospheric carbon dioxide. *Journal of Environmental Quality* **28**: 880-887.

Schulte, M., Herschbach, C. and Rennenberg, H. 1998. Interactive effects of elevated atmospheric CO_2, mycorrhization and drought on long-distance transport of reduced sulphur in young pedunculate oak trees (*Quercus robur* L.). *Plant, Cell and Environment* **21**: 917-926.

Tognetti, R., Longobucco, A., Miglietta, F. and Raschi, A. 1998. Transpiration and stomatal behaviour of *Quercus ilex* plants during the summer in a Mediterranean carbon dioxide spring. *Plant, Cell and Environment* **21**: 613-622.

Tognetti, R., Raschi, A. and Jones M.B. 2002. Seasonal changes in tissue elasticity and water transport efficiency in three co-occurring Mediterranean shrubs under natural long-term CO_2 enrichment. *Functional Plant Biology* **29**: 1097-1106.

Tomlinson, P.T. and Anderson, P.D. 1998. Ontogeny affects response of northern red oak seedlings to elevated CO_2 and water stress. II. Recent photosynthate distribution and growth. *New Phytologist* **140**: 493-504.

Tuba, Z., Csintalan, Z., Szente, K., Nagy, Z. and Grace, J. 1998. Carbon gains by desiccation-tolerant plants at elevated CO_2. *Functional Ecology* **12**: 39-44.

7.4. Acclimation

Plants grown in elevated CO_2 environments often exhibit some degree of photosynthetic acclimation or down regulation, which is typically characterized by long-term rates of photosynthesis that are somewhat lower than what would be expected on the basis of measurements made during short-term exposure to CO_2-enriched air. These downward adjustments result from modest long-term decreases in the activities and/or amounts of the primary plant carboxylating enzyme rubisco. Acclimation is said to be present when the photosynthetic rates of long-term CO_2-enriched plants are found to be lower than those of long-term *non*-CO_2-enriched plants when the normally CO_2-enriched plants are measured during brief exposures to ambient CO_2 concentrations. In this section, we review research that has been published on acclimation in agricultural, desert, grassland and woody species.

Additional information on this topic, including reviews on acclimation not discussed here, can be found at http://www.co2science.org/subject/a/subject_a.php under the heading Acclimation.

7.4.1. Agricultural Species

Several studies have examined the effects of elevated CO_2 on acclimation in agricultural crops. Ziska

(1998), for example, reported that soybeans grown at an atmospheric CO_2 concentration of 720 ppm initially exhibited photosynthetic rates that were 50 percent greater than those observed in control plants grown at 360 ppm. However, after the onset of photosynthetic acclimation, CO_2-enriched plants displayed subsequent photosynthetic rates that were only 30 percent greater than their ambiently grown counterparts. In another study, Theobald *et al.* (1998) grew spring wheat at twice-ambient atmospheric CO_2 concentrations and determined that elevated CO_2 reduced the amount of rubisco required to sustain enhanced rates of photosynthesis, which led to a significant increase in plant photosynthetic nitrogen-use efficiency. CO_2-induced increases in photosynthetic nitrogen-use efficiency have also been reported in spring wheat by Osborne *et al.* (1998).

In an interesting study incorporating both hydroponically and pot-grown wheat plants, Farage *et al.* (1998) demonstrated that low nitrogen fertilization does not lead to photosynthetic acclimation in elevated CO_2 environments, as long as the nitrogen supply keeps pace with the relative growth rate of the plants. Indeed, when spring wheat was grown at an atmospheric CO_2 concentration of 550 ppm in a free-air CO_2 enrichment (FACE) experiment with optimal soil nutrition and unlimited rooting volume, Garcia *et al.* (1998) could find no evidence of photosynthetic acclimation.

CO_2-induced photosynthetic acclimation often results from insufficient plant sink strength, which can lead to carbohydrate accumulation in source leaves and the triggering of photosynthetic end product feedback inhibition, which reduces net photosynthetic rates. Indeed, Gesch *et al.* (1998) reported that rice plants—which have relatively limited potential for developing additional carbon sinks—grown at an atmospheric CO_2 concentration of 700 ppm exhibited increased leaf carbohydrate contents, which likely reduced *rbcS* mRNA levels and rubisco protein content. Similarly, Sims *et al.* (1998) reported that photosynthetic acclimation was induced in CO_2-enriched soybean plants from the significant accumulation of nonstructural carbohydrates in their leaves. However, in growing several different *Brassica* species at 1,000 ppm CO_2, Reekie *et al.* (1998) demonstrated that CO_2-induced acclimation was avoided in species having well-developed carbon sinks (broccoli and cauliflower) and only appeared in those lacking significant sink strength (rape and mustard). Thus, acclimation does not appear to be a direct consequence of atmospheric CO_2 enrichment but rather an indirect effect of low sink strength, which results in leaf carbohydrate accumulation that can trigger acclimation.

In some cases, plants can effectively increase their sink strength, and thus reduce the magnitude of CO_2-induced acclimation, by forming symbiotic relationships with certain species of soil fungi. Under such conditions, photosynthetic down regulation is not triggered as rapidly, or as frequently, by end product feedback inhibition, as excess carbohydrates are mobilized out of source leaves and sent belowground to symbiotic fungi. Indeed, Louche-Tessandier et al. (1999) report that photosynthetic acclimation in CO_2-enriched potatoes was less apparent when plants were simultaneously colonized by a mycorrhizal fungus. Thus, CO_2-induced acclimation appears to be closely related to the source:sink balance that exists within plants, being triggered when sink strength falls below, and source strength rises above, critical thresholds in a species-dependent manner.

Acclimation is generally regarded as a process that reduces the amount of rubisco and/or other photosynthetic proteins, which effectively increases the amount of nitrogen available for enhancing sink development or stimulating other nutrient-limited processes. In the study of Watling et al. (2000), for example, the authors reported a 50 percent CO_2-induced reduction in the concentration of PEP-carboxylase, the primary carboxylating enzyme in C_4 plants, within sorghum leaves. Similarly, Maroco et al. (1999) documented CO_2-induced decreases in both PEP-carboxylase and rubisco in leaves of the C_4 crop maize.

In some cases, however, acclimation to elevated CO_2 is manifested by an "up-regulation" of certain enzymes. When Gesch et al. (2002) took rice plants from ambient air and placed them in air containing 700 ppm, for example, they noticed a significant increase in the activity of sucrose-phosphate synthase (SPS), which is a key enzyme involved in the production of sucrose. Similarly, Hussain et al. (1999) reported that rice plants grown at an atmospheric CO_2 concentration of 660 ppm displayed 20 percent more SPS activity during the growing season than did ambiently grown rice plants. Such increases in the activity of this enzyme could allow CO_2-enriched plants to avoid the onset of photosynthetic acclimation by synthesizing and subsequently exporting sucrose from source leaves into sink tissues before they accumulate and trigger end product feedback inhibition.

In an interesting experiment, Gesch et al. (2000) took ambiently growing rice plants and placed them in an atmospheric CO_2 concentration of 175 ppm, which reduced photosynthetic rates by 45 percent. However, after five days exposure to this sub-ambient CO_2 concentration, the plants manifested an up-regulation of rubisco, which stimulated photosynthetic rates by 35 percent. Thus, plant acclimation responses can involve both an increase or decrease in specific enzymes, depending on the atmospheric CO_2 concentration.

In summary, many peer-reviewed studies suggest that as the CO_2 content of the air slowly but steadily rises, agricultural species may not necessarily exhibit photosynthetic acclimation, even under conditions of low soil nitrogen. If a plant can maintain a balance between its sources and sinks for carbohydrates at the whole-plant level, acclimation should not be necessary. Because earth's atmospheric CO_2 content is rising by an average of only 1.5 ppm per year, most plants should be able to either (1) adjust their relative growth rates by the small amount that would be needed to prevent low nitrogen-induced acclimation from ever occurring, or (2) expand their root systems by the small amount that would be needed to supply the extra nitrogen required to take full advantage of the CO_2-induced increase in leaf carbohydrate production. In the event a plant cannot initially balance its sources and sinks for carbohydrates at the whole-plant level, CO_2-induced acclimation represents a beneficial secondary mechanism for achieving that balance through redistributing limiting resources away from the plant's photosynthetic machinery to strengthen sink development or enhance other nutrient-limiting processes.

Additional information on this topic, including reviews of newer publications as they become available, can be found at http://www.co2science.org/subject/a/acclimationag.php.

References

Farage, P.K., McKee, I.F. and Long, S.P. 1998. Does a low nitrogen supply necessarily lead to acclimation of photosynthesis to elevated CO_2? *Plant Physiology* **118**: 573-580.

Garcia, R.L., Long, S.P., Wall, G.W., Osborne, C.P., Kimball, B.A., Nie, G.Y., Pinter Jr., P.J., LaMorte, R.L. and Wechsung, F. 1998. Photosynthesis and conductance of spring-wheat leaves: field response to continuous free-air atmospheric CO_2 enrichment. *Plant, Cell and Environment* **21**: 659-669.

Gesch, R.W., Boote, K.J., Vu, J.C.V., Allen Jr., L.H. and Bowes, G. 1998. Changes in growth CO_2 result in rapid adjustments of ribulose-1,5-bisphosphate carboxylase/oxygenase small subunit gene expression in expanding and mature leaves of rice. *Plant Physiology* **118**: 521-529.

Gesch, R.W., Vu, J.C.V., Boote, K.J., Allen Jr., L.H. and Bowes, G. 2000. Subambient growth CO_2 leads to increased Rubisco small subunit gene expression in developing rice leaves. *Journal of Plant Physiology* **157**: 235-238.

Gesch, R.W., Vu, J.C.V., Boote, K.J., Allen Jr., L.H. and Bowes, G. 2002. Sucrose-phosphate synthase activity in mature rice leaves following changes in growth CO_2 is unrelated to sucrose pool size. *New Phytologist* **154**: 77-84.

Hussain, M.W., Allen, L.H., Jr. and Bowes, G. 1999. Up-regulation of sucrose phosphate synthase in rice grown under elevated CO_2 and temperature. *Photosynthesis Research* **60**: 199-208.

Louche-Tessandier, D., Samson, G., Hernandez-Sebastia, C., Chagvardieff, P. and Desjardins, Y. 1999. Importance of light and CO_2 on the effects of endomycorrhizal colonization on growth and photosynthesis of potato plantlets (*Solanum tuberosum*) in an *in vitro* tripartite system. *New Phytologist* **142**: 539-550.

Maroco, J.P., Edwards, G.E. and Ku, M.S.B. 1999. Photosynthetic acclimation of maize to growth under elevated levels of carbon dioxide. *Planta* **210**: 115-125.

Osborne, C.P., LaRoche, J., Garcia, R.L., Kimball, B.A., Wall, G.W., Pinter, P.J., Jr., LaMorte, R.L., Hendrey, G.R. and Long, S.P. 1998. Does leaf position within a canopy affect acclimation of photosynthesis to elevated CO_2? *Plant Physiology* **117**: 1037-1045.

Reekie, E.G., MacDougall, G., Wong, I. and Hicklenton, P.R. 1998. Effect of sink size on growth response to elevated atmospheric CO_2 within the genus *Brassica*. *Canadian Journal of Botany* **76**: 829-835.

Sims, D.A., Luo, Y. and Seeman, J.R. 1998. Comparison of photosynthetic acclimation to elevated CO_2 and limited nitrogen supply in soybean. *Plant, Cell and Environment* **21**: 945-952.

Theobald, J.C., Mitchell, R.A.C., Parry, M.A.J. and Lawlor, D.W. 1998. Estimating the excess investment in ribulose-1,5-bisphosphate carboxylase/oxygenase in leaves of spring wheat grown under elevated CO_2. *Plant Physiology* **118**: 945-955.

Watling, J.R., Press, M.C. and Quick, W.P. 2000. Elevated CO_2 induces biochemical and ultrastructural changes in leaves of the C4 cereal sorghum. *Plant Physiology* **123**: 1143-1152.

Ziska, L.H. 1998. The influence of root zone temperature on photosynthetic acclimation to elevated carbon dioxide concentrations. *Annals of Botany* **81**: 717-721.

7.4.2. Chaparral and Desert Species

Roberts *et al.* (1998) conducted a FACE experiment in southern California, USA, exposing *Adenostoma fassciculatum* shrubs to atmospheric CO_2 concentrations of 360 and 550 ppm while they studied the nature of gas-exchange in this chaparral species. After six months of CO_2 fumigation, photosynthetic acclimation occurred. However, because of reductions in stomatal conductance and transpirational water loss, the CO_2-enriched shrubs exhibited leaf water potentials that were less negative (and, hence, less stressful) than those of control plants. This CO_2-induced water conservation phenomenon should enable this woody perennial to better withstand the periods of drought that commonly occur in this southern California region, while the photosynthetic down regulation it exhibits should allow it to more equitably distribute the limiting resources it possesses among different essential plant physiological processes.

Huxman and Smith (2001) measured seasonal gas exchange during an unusually wet El Niño year in an annual grass (*Bromus madritensis* ssp. *rubens*) and a perennial forb (*Eriogonum inflatum*) growing within FACE plots established in the Mojave Desert, USA, which they maintained at atmospheric CO_2 concentrations of 350 and 550 ppm. The elevated CO_2 consistently increased net photosynthetic rates in the annual grass without inducing photosynthetic acclimation. In fact, even as seasonal photosynthetic rates declined post-flowering, the reduction was much less in the CO_2-enriched plants. However, elevated CO_2 had no consistent effect on stomatal conductance in this species. By contrast, *Eriogonum* plants growing at 550 ppm CO_2 exhibited significant photosynthetic acclimation, especially late in the season, which led to similar rates of net photosynthesis in these plants in both CO_2 treatments. But in this species, elevated CO_2 reduced stomatal conductance over most of the growing season. Although the two desert plants exhibited different stomatal and photosynthetic responses to elevated CO_2, both experienced significant CO_2-induced increases in water use efficiency and biomass production, thus highlighting the existence of different, but equally effective, species-specific mechanisms for responding positively to atmospheric CO_2 enrichment in a desert environment.

In another study conducted at the Mojave Desert FACE site, Hamerlynck et al. (2002) determined that plants of the deciduous shrub *Lycium andersonii* grown in elevated CO_2 displayed photosynthetic acclimation, as maximum rubisco activity in the plants growing in the CO_2-enriched air was 19 percent lower than in the plants growing in ambient air. Also, the elevated CO_2 did not significantly impact rates of photosynthesis. Leaf stomatal conductance, on the other hand, was consistently about 27 percent lower in the plants grown in the CO_2-enriched air; and during the last month of the spring growing season, the plants in the elevated CO_2 plots displayed leaf water potentials that were less negative than those exhibited by the control plants growing in ambient air. Hence, as the CO_2 content of the air increases, *Lycium andersonii* will likely respond by exhibiting significantly enhanced water use efficiency, which should greatly increase its ability to cope with the highly variable precipitation and temperature regimes of the Mojave Desert. The acclimation observed within the shrub's photosynthetic apparatus should allow it to reallocate more resources to producing and sustaining greater amounts of biomass. Thus, it is likely that future increases in the air's CO_2 content will favor a "greening" of the American Mojave Desert.

In summary, the few studies of the acclimation phenomenon that have been conducted on chaparral and desert plants indicate that although it can sometimes be complete, other physiological changes, such as the reductions in stomatal conductance that typically produce large increases in water use efficiency, often more than compensate for the sometimes small to negligible increases in photosynthesis.

Additional information on this topic, including reviews of newer publications as they become available, can be found at http://www.co2science.org/subject/a/acclimationdesert.php.

References

Hamerlynck, E.P., Huxman, T.E., Charlet, T.N. and Smith, S.D. 2002. Effects of elevated CO_2 (FACE) on the functional ecology of the drought-deciduous Mojave Desert shrub, *Lycium andersonii*. *Environmental and Experimental Botany* **48**: 93-106.

Huxman, T.E. and Smith, S.D. 2001. Photosynthesis in an invasive grass and native forb at elevated CO_2 during an El Niño year in the Mojave Desert. *Oecologia* **128**: 193-201.

Roberts, S.W., Oechel, W.C., Bryant, P.J., Hastings, S.J., Major, J. and Nosov, V. 1998. A field fumigation system for elevated carbon dioxide exposure in chaparral shrubs. *Functional Ecology* **12**: 708-719.

7.4.3. Grassland Species

In nearly every reported case of photosynthetic acclimation in CO_2-enriched plants, rates of photosynthesis displayed by grassland species grown and measured at elevated CO_2 concentrations are typically greater than those exhibited by control plants grown and measured at ambient CO_2 concentrations (Davey et al., 1999; Bryant et al., 1998).

As mentioned in prior sections, CO_2-induced photosynthetic acclimation often results from insufficient plant sink strength, which can lead to carbohydrate accumulation in source leaves and the triggering of photosynthetic end-product feedback inhibition, which reduces rubisco activity and rates of net photosynthesis (Roumet et al., 2000). As one example of this phenomenon, Rogers et al. (1998) reported that perennial ryegrass grown at an atmospheric CO_2 concentration of 600 ppm and low soil nitrogen exhibited leaf carbohydrate contents and rubisco activities that were 100 percent greater and 25 percent less, respectively, than those observed in control plants grown at 360 ppm CO_2, prior to a cutting event. Following the cutting, which effectively reduced the source:sink ratio of the plants, leaf carbohydrate contents in CO_2-enriched plants decreased and rubisco activities increased, completely ameliorating the photosynthetic acclimation in this species. However, at high soil nitrogen, photosynthetic acclimation to elevated CO_2 did not occur. Thus, photosynthetic acclimation appears to result from the inability of plants to develop adequate sinks at low soil nitrogen, and is not necessarily induced directly by atmospheric CO_2 enrichment.

In some cases, plants can effectively increase their sink strength and thus reduce the magnitude of CO_2-induced acclimation by forming symbiotic relationships with certain species of soil fungi. Under such conditions, photosynthetic down regulation is not triggered as rapidly, or as frequently, by end-product feedback inhibition, as excess carbohydrates are mobilized out of source leaves and sent belowground to symbiotic fungi. Staddon *et al.* (1999) reported that photosynthetic acclimation was not induced in CO_2-enriched *Plantago lanceolata* plants that were inoculated with a mycorrhizal fungus, while it was induced in control plants that were not inoculated with the fungus. Thus, CO_2-induced acclimation appears to be closely related to the source:sink balance that exists within plants, and is triggered when sink strength falls below, and source strength rises above, certain critical thresholds in a species-dependent manner.

As the CO_2 content of the air slowly but steadily rises, these peer-reviewed studies suggest that grassland species may not exhibit photosynthetic acclimation if they can maintain a balance between their sources and sinks for carbohydrates at the whole-plant level. But in the event this balancing act is not initially possible, acclimation represents a beneficial secondary mechanism for achieving that balance by redistributing limiting resources away from the plant's photosynthetic machinery to strengthen its sink development and/or nutrient-gathering activities.

Additional information on this topic, including reviews of newer publications as they become available, can be found at http://www.co2science.org/subject/a/acclimationgrass.php.

References

Bryant, J., Taylor, G. and Frehner, M. 1998. Photosynthetic acclimation to elevated CO_2 is modified by source:sink balance in three component species of chalk grassland swards grown in a free air carbon dioxide enrichment (FACE) experiment. *Plant, Cell and Environment* **21**: 159-168.

Davey, P.A., Parsons, A.J., Atkinson, L., Wadge, K. and Long, S.P. 1999. Does photosynthetic acclimation to elevated CO_2 increase photosynthetic nitrogen-use efficiency? A study of three native UK grassland species in open-top chambers. *Functional Ecology* **13**: 21-28.

Rogers, A., Fischer, B.U., Bryant, J., Frehner, M., Blum, H., Raines, C.A. and Long, S.P. 1998. Acclimation of photosynthesis to elevated CO_2 under low-nitrogen nutrition is affected by the capacity for assimilate utilization. Perennial ryegrass under free-air CO_2 enrichment. *Plant Physiology* **118**: 683-689.

Roumet, C., Garnier, E., Suzor, H., Salager, J.-L. and Roy, J. 2000. Short and long-term responses of whole-plant gas exchange to elevated CO_2 in four herbaceous species. *Environmental and Experimental Botany* **43**: 155-169.

Staddon, P.L., Fitter, A.H. and Robinson, D. 1999. Effects of mycorrhizal colonization and elevated atmospheric carbon dioxide on carbon fixation and below-ground carbon partitioning in *Plantago lanceolata*. *Journal of Experimental Botany* **50**: 853-860.

7.4.4. Tree Species

Trees grown for long periods of time in elevated CO_2 environments often, but not always (Marek *et al.*, 2001; Stylinski *et al.*, 2000; Bartak *et al.*, 1999; Schortemeyer *et al.*, 1999), exhibit some degree of photosynthetic acclimation or down regulation, which is typically characterized by modestly reduced rates of photosynthesis (compared to what might be expected on the basis of short-term exposure to CO_2-enriched air) that result from a long-term decrease in the activity and/or amount of the primary plant carboxylating enzyme rubisco (Kubiske *et al.*, 2002; Egli *et al.*, 1998). This acclimation response in plants accustomed to growing in CO_2-enriched air is characterized by short-term reductions in their rates of net photosynthesis when measured during short-term exposure to ambient air relative to net photosynthesis rates of comparable plants that have always been grown in ambient air.

Jach and Ceulemans (2000) grew one-year-old Scots pine seedlings for two additional years at twice-ambient atmospheric CO_2 concentrations and reported that the elevated CO_2 increased the trees' mean rate of net photosynthesis by 64 percent. However, when measured during a brief return to ambient CO_2 concentrations, the normally CO_2-enriched seedlings exhibited an approximate 21 percent reduction in average net photosynthesis rate relative to that of seedlings that had always been exposed to ambient air. Similarly, Spunda *et al.* (1998) noted that a 350-ppm increase in the air's CO_2 concentration boosted rates of net photosynthesis in fifteen-year-old Norway spruce trees by 78 percent; but when net photosynthesis in the normally CO_2-enriched trees was measured at a temporary atmospheric CO_2 concentration of 350 ppm, an 18 percent reduction was observed relative to what was observed in comparable trees that had always been grown in ambient air. After reviewing the results of 15 different atmospheric CO_2 enrichment studies of European forest species growing in field environments maintained at twice-ambient CO_2 concentrations, Medlyn *et al.* (1999) found that the mean photosynthetic acclimation effect in the CO_2-enriched trees was characterized by an average reduction of 19 percent in their rates of net photosynthesis when measured at temporary ambient CO_2 concentrations relative to the mean rate of net photosynthesis exhibited by those trees that had always been exposed to ambient air. Nonetheless, in nearly every reported case of CO_2-induced photosynthetic acclimation in trees, the photosynthetic rates of trees growing and measured in CO_2-enriched air have still been much greater than those exhibited by trees growing and measured in ambient air.

CO_2-induced photosynthetic acclimation, when it occurs, often results from insufficient plant sink strength, which can lead to carbohydrate accumulation in source leaves and the triggering of photosynthetic end-product feedback inhibition, which results in reduced rates of net photosynthesis. Pan et al. (1998), for example, reported that apple seedlings grown at an atmospheric CO_2 concentration of 1600 ppm had foliar starch concentrations 17-fold greater than those observed in leaves of seedlings grown at 360 ppm CO_2, suggesting that this phenomenon likely triggered the reductions in leaf net photosynthesis rates they observed. Similarly, Rey and Jarvis (1998) reported that the accumulation of starch within leaves of CO_2-enriched silver birch seedlings (100 percent above ambient) may have induced photosynthetic acclimation in that species. Also, in the study of Wiemken and Ineichen (2000), a 300-ppm increase in the air's CO_2 concentration induced seasonal acclimation in young spruce trees, where late-summer, fall and winter rates of net photosynthesis declined in conjunction with 40 to 50 percent increases in foliar glucose levels.

In some cases, trees can effectively increase their sink strength, and thus reduce the magnitude of CO_2-induced photosynthetic acclimation, by forming symbiotic relationships with certain species of soil fungi. Under such conditions, photosynthetic down regulation is not triggered as rapidly, or as frequently, by end-product feedback inhibition, as excess carbohydrates are mobilized out of the trees' source leaves and sent belowground to support the growth of symbiotic fungi. Jifon et al. (2002), for example, reported that the degree of CO_2-induced photosynthetic acclimation in sour orange tree seedlings was significantly reduced by the presence of mycorrhizal fungi, which served as sinks for excess carbohydrates synthesized by the CO_2-enriched seedlings.

During acclimation to elevated CO_2, the amounts and activities of rubisco and/or other photosynthetic proteins are often reduced, which effectively increases the amount of nitrogen available for enhancing sink development or stimulating other nutrient-limited processes. As an example of this phenomenon operating in trees, in the study of Blaschke et al. (2001), the authors reported that a doubling of the atmospheric CO_2 concentration reduced foliar rubisco concentrations by 15 and 30 percent in two mature oak species growing near CO_2-emitting springs. Similarly, a 200-ppm increase in the air's CO_2 content reduced foliar rubisco concentrations in young aspen and birch seedlings by 39 percent (Takeuchi et al., 2001) and 24 percent (Tjoelker et al., 1998), respectively. Also, in two Pinus radiata studies, seedlings fumigated with air containing 650 ppm CO_2 displayed 30 to 40 percent reductions in rubisco concentration and rubisco activity relative to measurements made on seedlings grown in ambient air (Griffin et al., 2000; Turnbull et al., 1998). Other studies of CO_2-enriched Norway spruce and Scots pines have documented CO_2-induced reductions in foliar chlorophyll contents of 17 percent (Spunda et al., 1998) and 26 percent (Gielen et al., 2000), respectively. And in the study of Gleadow et al. (1998), elevated CO_2 led to acclimation in eucalyptus seedlings, which mobilized nitrogen away from rubisco and into prunasin, a sugar-based defense compound that deters herbivory.

In another interesting experiment, Polle et al. (2001) germinated acorns from oak trees exposed to ambient and CO_2-enriched air and subsequently grew them at ambient and twice-ambient atmospheric CO_2 concentrations. They discovered that seedlings derived from acorns produced on CO_2-enriched trees exhibited less-pronounced photosynthetic acclimation to elevated CO_2 than did seedlings derived from acorns produced on ambiently grown trees, suggesting the possibility of generational adaptation to higher atmospheric CO_2 concentrations over even longer periods of time.

In summary, these many peer-reviewed scientific studies suggest that as the air's CO_2 content slowly but steadily rises, trees may be able to avoid photosynthetic acclimation if they maintain a proper balance between carbohydrate sources and sinks at the whole-tree level, which they may well be able to do in response to the current rate of rise of the air's CO_2 concentration (a mere 1.5 ppm per year). If a tree cannot *initially* balance its sources and sinks of carbohydrates, however, acclimation is an important and effective means of achieving that balance through redistributing essential resources away from the tree's photosynthetic machinery in an effort to strengthen sink development, enhance various nutrient-limited processes, and increase nutrient acquisition by, for example, stimulating the development of roots and their symbiotic fungal partners. And if those adjustments are not entirely successful, it is still the case that the acclimation process is hardly ever 100 percent complete (in the studies reviewed here it was on the order of 20 percent), so that tree growth is almost always significantly enhanced in CO_2-enriched air in nearly every real-world setting.

Additional information on this topic, including reviews of newer publications as they become available, can be found at http://www.co2science.org/subject/a/acclimationtree.php.

References

Bartak, M., Raschi, A. and Tognetti, R. 1999. Photosynthetic characteristics of sun and shade leaves in the canopy of *Arbutus unedo* L. trees exposed to *in situ* long-term elevated CO_2. *Photosynthetica* **37**: 1-16.

Blaschke, L., Schulte, M., Raschi, A., Slee, N., Rennenberg, H. and Polle, A. 2001. Photosynthesis, soluble and structural carbon compounds in two Mediterranean oak species (*Quercus pubescens* and *Q. ilex*) after lifetime growth at naturally elevated CO_2 concentrations. *Plant Biology* **3**: 288-297.

Egli, P., Maurer, S., Gunthardt-Goerg, M.S. and Korner, C. 1998. Effects of elevated CO_2 and soil quality on leaf gas exchange and aboveground growth in beech-spruce model ecosystems. *New Phytologist* **140**: 185-196.

Gielen, B., Jach, M.E. and Ceulemans, R. 2000. Effects of season, needle age and elevated atmospheric CO_2 on chlorophyll fluorescence parameters and needle nitrogen concentration in (*Pinus sylvestris* L.). *Photosynthetica* **38**: 13-21.

Gleadow, R.M., Foley, W.J. and Woodrow, I.E. 1998. Enhanced CO_2 alters the relationship between photosynthesis and defense in cyanogenic *Eucalyptus cladocalyx* F. Muell. *Plant, Cell and Environment* **21**: 12-22.

Griffin, K.L., Tissue, D.T., Turnbull, M.H. and Whitehead, D. 2000. The onset of photosynthetic acclimation to elevated CO_2 partial pressure in field-grown *Pinus radiata* D. Don. after 4 years. *Plant, Cell and Environment* **23**: 1089-1098.

Jach, M.E. and Ceulemans, R. 2000. Effects of season, needle age and elevated atmospheric CO_2 on photosynthesis in Scots pine (*Pinus sylvestris* L.). *Tree Physiology* **20**: 145-157.

Jifon, J.L., Graham, J.H., Drouillard, D.L. and Syvertsen, J.P. 2002. Growth depression of mycorrhizal Citrus seedlings grown at high phosphorus supply is mitigated by elevated CO_2. *New Phytologist* **153**: 133-142.

Kubiske, M.E., Zak, D.R., Pregitzer, K.S. and Takeuchi, Y. 2002. Photosynthetic acclimation of overstory *Populus tremuloides* and understory *Acer saccharum* to elevated atmospheric CO_2 concentration: interactions with shade and soil nitrogen. *Tree Physiology* **22**: 321-329.

Marek, M.V., Sprtova, M., De Angelis, P. and Scarascia-Mugnozza, G. 2001. Spatial distribution of photosynthetic response to long-term influence of elevated CO_2 in a Mediterranean *macchia* mini-ecosystem. *Plant Science* **160**: 1125-1136.

Medlyn, B.E., Badeck. F.-W., De Pury, D.G.G., Barton, C.V.M., Broadmeadow, M., Ceulemans, R., De Angelis, P., Forstreuter, M., Jach, M.E., Kellomaki, S., Laitat, E., Marek, M., Philippot, S., Rey, A., Strassemeyer, J., Laitinen, K., Liozon, R., Portier, B., Roberntz, P., Wang, K. and Jarvis, P.G. 1999. Effects of elevated [CO_2] on photosynthesis in European forest species: a meta-analysis of model parameters. *Plant, Cell and Environment* **22**: 1475-1495.

Pan, Q., Wang, Z. and Quebedeaux, B. 1998. Responses of the apple plant to CO_2 enrichment: changes in photosynthesis, sorbitol, other soluble sugars, and starch. *Australian Journal of Plant Physiology* **25**: 293-297.

Polle, A., McKee, I. and Blaschke, L. 2001. Altered physiological and growth responses to elevated [CO_2] in offspring from holm oak (*Quercus ilex* L.) mother trees with lifetime exposure to naturally elevated [CO_2]. *Plant, Cell and Environment* **24**: 1075-1083.

Rey, A. and Jarvis, P.G. 1998. Long-Term photosynthetic acclimation to increased atmospheric CO_2 concentration in young birch (*Betula pendula*) trees. *Tree Physiology* **18**: 441-450.

Schortemeyer, M., Atkin, O.K., McFarlane, N. and Evans, J.R. 1999. The impact of elevated atmospheric CO_2 and nitrate supply on growth, biomass allocation, nitrogen partitioning and N2 fixation of *Acacia melanoxylon*. *Australian Journal of Plant Physiology* **26**: 737-774.

Spunda, V., Kalina, J., Cajanek, M., Pavlickova, H. and Marek, M.V. 1998. Long-term exposure of Norway spruce to elevated CO_2 concentration induces changes in photosystem II mimicking an adaptation to increased irradiance. *Journal of Plant Physiology* **152**: 413-419.

Stylinski, C.D., Oechel, W.C., Gamon, J.A., Tissue, D.T., Miglietta, F. and Raschi, A. 2000. Effects of lifelong [CO_2] enrichment on carboxylation and light utilization of *Quercus pubescens* Willd. examined with gas exchange, biochemistry and optical techniques. *Plant, Cell and Environment* **23**: 1353-1362.

Takeuchi, Y., Kubiske, M.E., Isebrands, J.G., Pregitzer, K.S., Hendrey, G. and Karnosky, D.F. 2001. Photosynthesis, light and nitrogen relationships in a young deciduous forest canopy under open-air CO_2 enrichment. *Plant, Cell and Environment* **24**: 1257-1268.

Tjoelker, M.G., Oleksyn, J. and Reich, P.B. 1998. Seedlings of five boreal tree species differ in acclimation of net photosynthesis to elevated CO_2 and temperature. *Tree Physiology* **18**: 715-726.

Turnbull, M.H., Tissue, D.T., Griffin, K.L., Rogers, G.N.D. and Whitehead, D. 1998. Photosynthetic acclimation to long-term exposure to elevated CO_2 concentration in *Pinus radiata* D. Don. is related to age of needles. *Plant, Cell and Environment* **21**: 1019-1028.

Wiemken, V. and Ineichen, K. 2000. Seasonal fluctuations of the levels of soluble carbohydrates in spruce needles exposed to elevated CO_2 and nitrogen fertilization and glucose as a potential mediator of acclimation to elevated CO_2. *Journal of Plant Physiology* **156**: 746-750.

7.5. Competition

Do higher levels of atmospheric CO_2 favor some plants over others? Could this result in ecological changes that could be judged "bad" because of their effects on wildlife or plants that are beneficial to mankind? This section seeks to answer these questions by surveying research on the effects of CO_2 enhancement on C_3 versus C_4 plants, and weeds versus crops.

Additional information on this topic, including reviews on competition not discussed here, can be found at http://www.co2science.org/ subject/c/subject _c.php under the heading Competition.

7.5.1. C₃ vs C₄ Plants

C_3 plants typically respond better to atmospheric CO_2 enrichment than do C_4 plants in terms of increasing their rates of photosynthesis and biomass production. Hence, it has periodically been suggested that in a world of rising atmospheric CO_2 concentration, C_3 plants may out-compete C_4 plants and displace them, thereby decreasing the biodiversity of certain ecosystems. However, the story is much more complex that what is suggested by this simple scenario.

Wilson *et al.* (1998) grew 36 species of perennial grass common to tallgrass prairie ecosystems with and without arbuscular mycorrhizal fungi, finding that the dry matter production of the C_3 species that were colonized by the fungi was the same as that of the non-inoculated C_3 species, but that the fungal-colonized C_4 species produced, on average, 85 percent *more* dry matter than the non-inoculated C_4 species. This finding is of pertinence to the relative responsiveness of C_3 and C_4 plants to atmospheric CO_2 enrichment; for elevated levels of atmospheric CO_2 tend to enhance the mycorrhizal colonization of plant roots, which is known to make soil minerals and water more available for plant growth. Hence, this CO_2-induced fungal-mediated growth advantage, which from this study appears to be more readily available to C_4 plants, could well counter the inherently greater biomass response of C_3 plants relative to that of C_4 plants, leveling the playing field relative to their competition for space in any given ecosystem.

Another advantage that may come to C_4 plants as a consequence of the ongoing rise in the air's CO_2 content was elucidated by BassiriRad *et al.* (1998), who found that elevated CO_2 enhanced the ability of the perennial C_4 grass *Bouteloua eriopoda* to increase its uptake of NO_3^- and PO_4^{3-} considerably more than the perennial C_3 shrubs *Larrea tridentata* and *Prosopis glandulosa*. In an eight-year study of the effects of twice-ambient atmospheric CO_2 concentrations on a pristine tallgrass prairie in Kansas, Owensby *et al.* (1999) found that the elevated CO_2 did not affect the basal coverage of its C_4 species or their relative contribution to the composition of the ecosystem.

The antitranspirant effect of atmospheric CO_2 enrichment (Pospisilova and Catsky, 1999) is often more strongly expressed in C_4 plants than in C_3 plants and typically allows C_4 plants to better cope with water stress. In a study of the C_3 dicot *Abutilon theophrasti* and the C_4 dicot *Amaranthus retroflexus*, for example, Ward *et al.* (1999) found that *Amaranthus retroflexus* exhibited a greater relative recovery from drought than did the C_3 species, which suggests, in their words, that "the C_4 species would continue to be more competitive than the C_3 species in regions receiving more frequent and severe droughts," which basically characterizes regions where C_4 plants currently exist.

Two years later, Morgan *et al.* (2001) published the results of an open-top chamber study of a native shortgrass steppe ecosystem in Colorado, USA, where

487

they had exposed the enclosed ecosystems to atmospheric CO_2 concentrations of 360 and 720 ppm for two six-month growing seasons. In spite of an average air temperature increase of 2.6°C, which was caused by the presence of the open-top chambers, the elevated CO_2 increased above-ground biomass production by an average of 38 percent in both years of the study; and when 50 percent of the standing green plant biomass was defoliated to simulate grazing halfway through the growing season, atmospheric CO_2 enrichment still increased above-ground biomass by 36 percent. It was also found that the communities enriched with CO_2 tended to have greater amounts of moisture in their soils than communities exposed to ambient air; and this phenomenon likely contributed to the less negative and, therefore, less stressful plant water potentials that were measured in the CO_2-enriched plants. Last, the elevated CO_2 did not preferentially stimulate the growth of C_3 species over that of C_4 species in these communities. Elevated CO_2 did not significantly affect the percentage composition of C_3 and C_4 species in these grasslands; they maintained their original level of vegetative biodiversity.

This would also appear to be the conclusion of the study of Wand *et al.* (1999), who in a massive review of the scientific literature published between 1980 and 1997 analyzed nearly 120 individual responses of C_3 and C_4 grasses to elevated CO_2. On average, they found photosynthetic enhancements of 33 and 25 percent, respectively, for C_3 and C_4 plants, along with biomass enhancements of 44 and 33 percent, respectively, for a doubling of the air's CO_2 concentration. These larger-than-expected growth responses in the C_4 species led them to conclude that "it may be premature to predict that C_4 grass species will lose their competitive advantage over C_3 grass species in elevated CO_2."

Further support for this conclusion comes from the study of Campbell *et al.* (2000), who reviewed research work done between 1994 and 1999 by a worldwide network of 83 scientists associated with the Global Change and Terrestrial Ecosystems (GCTE) Pastures and Rangelands Core Research Project 1, which resulted in the publication of over 165 peer-reviewed scientific journal articles. After analyzing this body of research, they concluded that the "growth of C_4 species is about as responsive to CO_2 concentration as [is that of] C_3 species when water supply restricts growth, as is usual in grasslands containing C_4 species." The work of this group of scientists also provides no evidence for the suggestion

that C_3 plants may out-compete C_4 plants and thereby replace them in a high-CO_2 world of the future.

Additional information on this topic, including reviews of newer publications as they become available, can be found at http://www.co2science.org/subject/b/biodivc3vsc4.php.

References

BassiriRad, H., Reynolds, J.F., Virginia, R.A. and Brunelle, M.H. 1998. Growth and root NO_3^- and PO_4^{3-} uptake capacity of three desert species in response to atmospheric CO_2 enrichment. *Australian Journal of Plant Physiology* **24**: 353-358.

Campbell, B.D., Stafford Smith, D.M., Ash, A.J., Fuhrer, J., Gifford, R.M., Hiernaux, P., Howden, S.M., Jones, M.B., Ludwig, J.A., Manderscheid, R., Morgan, J.A., Newton, P.C.D., Nosberger, J., Owensby, C.E., Soussana, J.F., Tuba, Z. and ZuoZhong, C. 2000. A synthesis of recent global change research on pasture and rangeland production: reduced uncertainties and their management implications. *Agriculture, Ecosystems and Environment* **82**: 39-55.

Morgan, J.A., Lecain, D.R., Mosier, A.R. and Milchunas, D.G. 2001. Elevated CO_2 enhances water relations and productivity and affects gas exchange in C_3 and C_4 grasses of the Colorado shortgrass steppe. *Global Change Biology* **7**: 451-466.

Owensby, C.E., Ham, J.M., Knapp, A.K. and Auen, L.M. 1999. Biomass production and species composition change in a tallgrass prairie ecosystem after long-term exposure to elevated atmospheric CO_2. *Global Change Biology* **5**: 497-506.

Pospisilova, J. and Catsky, J. 1999. Development of water stress under increased atmospheric CO_2 concentration. *Biologia Plantarum* **42**: 1-24.

Wand, S.J.E., Midgley, G.F., Jones, M.H. and Curtis, P.S. 1999. Responses of wild C_4 and C_3 grass (Poaceae) species to elevated atmospheric CO_2 concentration: a meta-analytic test of current theories and perceptions. *Global Change Biology* **5**: 723-741.

Ward, J.K., Tissue, D.T., Thomas, R.B. and Strain, B.R. 1999. Comparative responses of model C_3 and C_4 plants to drought in low and elevated CO_2. *Global Change Biology* **5**: 857-867.

Wilson, G.W.T. and Hartnett, D.C. 1998. Interspecific variation in plant responses to mycorrhizal colonization in tallgrass prairie. *American Journal of Botany* **85**: 1732-1738.

7.5.2. N-Fixers vs. Non-N-Fixers

Will nitrogen-fixing (N-fixing) plants benefit more from atmospheric CO_2 enrichment than non-N-fixers and thus obtain a competitive advantage over them that could lead to some non-N-fixers being excluded from certain plant communities, thereby decreasing the biodiversity of those ecosystems?

In a two-year glasshouse study of simulated low-fertility ecosystems composed of grassland species common to Switzerland, Stocklin and Korner (1999) found that atmospheric CO_2 enrichment gave nitrogen-fixing legumes an initial competitive advantage over non-N-fixers. However, it would be expected that, over time, a portion of the extra nitrogen fixed by these legumes would become available to neighboring non-N-fixing species, which would then be able to use it to their own advantage, thereby preserving the species richness of the ecosystem over the long haul. Indeed, in a four-year study of an established (non-simulated) high grassland ecosystem located in the Swiss Alps, Arnone (1999) found no difference between the minimal to non-existent growth responses of N-fixing and non-N-fixing species to elevated levels of atmospheric CO_2.

In a study of mixed plantings of the grass *Lolium perenne* and the legume *Medicago sativa*, Matthies and Egli (1999) found that elevated CO_2 did not influence the competition between the two plants, either directly or indirectly via its effects upon the root hemiparasite *Rhinanthus alectorolophus*. In a study of mixed plantings of two grasses and two legumes, Navas *et al.* (1999) observed that plant responses to atmospheric CO_2 enrichment are more dependent upon neighboring plant density than they are upon neighboring plant identity.

In the few studies of this question that have been conducted to date, therefore, it would appear that there is little evidence to suggest that N-fixing legumes will out-compete non-N-fixing plants.

Additional information on this topic, including reviews of newer publications as they become available, can be found at http://www.co2science.org/subject/b/biodivnfixers.php.

References

Arnone III, J.A. 1999. Symbiotic N_2 fixation in a high Alpine grassland: effects of four growing seasons of elevated CO_2. *Functional Ecology* **13**: 383-387.

Matthies, D. and Egli, P. 1999. Response of a root hemiparasite to elevated CO_2 depends on host type and soil nutrients. *Oecologia* **120**: 156-161.

Navas, M.-L., Garnier, E., Austin, M.P. and Gifford, R.M. 1999. Effect of competition on the responses of grasses and legumes to elevated atmospheric CO_2 along a nitrogen gradient: differences between isolated plants, monocultures and multi-species mixtures. *New Phytologist* **143**: 323-331.

Stocklin, J. and Korner, C. 1999. Interactive effects of elevated CO_2, P availability and legume presence on calcareous grassland: results of a glasshouse experiment. *Functional Ecology* **13**: 200-209.

7.5.3. Weeds vs. Non-Weeds

Elevated CO_2 typically stimulates the growth of nearly all plant species in monoculture, including those deemed undesirable by humans, i.e., weeds. Consequently, it is important to determine how future increases in the air's CO_2 content may influence relationships between weeds and non-weeds when they grow competitively in mixed-species stands.

Dukes (2002) grew model serpentine grasslands common to California, USA, in competition with the invasive forb *Centaurea solstitialis* at atmospheric CO_2 concentrations of 350 and 700 ppm for one year, determining that elevated CO_2 increased the biomass proportion of this weedy species in the community by a mere 1.2 percent, while total community biomass increased by 28 percent. Similarly, Gavazzi *et al.* (2000) grew loblolly pine seedlings for four months in competition with both C_3 and C_4 weeds at atmospheric CO_2 concentrations of 260 and 660 ppm, reporting that elevated CO_2 increased pine biomass by 22 percent while eliciting no response from either type of weed.

In a study of pasture ecosystems near Montreal, Canada, Taylor and Potvin (1997) found that elevated CO_2 concentrations did not influence the number of native species returning after their removal (to simulate disturbance), even in the face of the introduced presence of the C_3 weed *Chenopodium album*, which normally competes quite effectively with several slower-growing crops in ambient air. In fact, atmospheric CO_2 enrichment did not impact the growth of this weed in any measurable way.

Ziska *et al.* (1999) also studied the C_3 weed *C. album*, along with the C_4 weed *Amaranthus retroflexus*, in glasshouses maintained at atmospheric CO_2 concentrations of 360 and 720 ppm. They determined that elevated CO_2 significantly increased the photosynthetic rate and total dry weight of the C_3 weed, but that it had no effect on the C_4 weed. Also, they found that the growth response of the C_3 weed to a doubling of the air's CO_2 content was approximately 51 percent, which is about the same as the average 52 percent growth response tabulated by Idso (1992), and that obtained by Poorter (1993) for rapidly growing wild C_3 species (54 percent), which finding suggests there is no enhanced dominance of the C_3 weed over other C_3 plants in a CO_2-enriched environment.

Wayne *et al.* (1999) studied another agricultural weed, field mustard (*Brassica kaber*), which was sown in pots at six densities, placed in atmospheric CO_2 concentrations of 350 and 700 ppm, and sequentially harvested during the growing season. Early in stand development, elevated CO_2 increased above-ground weed biomass in a density-dependent manner, with the greatest stimulation of 141 percent occurring at the lowest density (corresponding to 20 plants per square meter) and the smallest stimulation of 59 percent occurring at the highest density (corresponding to 652 plants per square meter). However, as stands matured, the density-dependence of the CO_2-induced growth response disappeared, and CO_2-enriched plants exhibited an average above-ground biomass that was 34 percent greater than that of ambiently grown plants across a broad range of plant densities. This final growth stimulation was similar to that of most other herbaceous plants exposed to atmospheric CO_2 enrichment (30 to 50 percent biomass increases for a doubling of the air's CO_2 content), evidence once again that atmospheric CO_2 enrichment confers no undue advantage upon weeds at the expense of other plants.

In a study of a weed that affects both plants and animals, Caporn *et al.* (1999) examined bracken (*Pteridium aquilinum*), which poses a serious weed problem and potential threat to human health in the United Kingdom and other regions, growing specimens for 19 months in controlled environment chambers maintained at atmospheric CO_2 concentrations of 370 and 570 ppm and normal or high levels of soil fertility. They found that the high CO_2 treatment consistently increased rates of net photosynthesis by 30 to 70 percent, depending on soil fertility and time of year. However, elevated CO_2 did not increase total plant dry mass or the dry mass of any plant organ, including rhizomes, roots and fronds. In fact, the only significant effect of elevated CO_2 on

bracken growth was observed in the normal nutrient regime, where elevated CO_2 reduced mean frond area.

Finally, in a study involving two parasitic species (*Striga hermonthica* and *Striga asiatica*), Watling and Press (1997) reported that total parasitic biomass per host plant at an atmospheric CO_2 concentration of 700 ppm was 65 percent less than it was in ambient air. And in a related study, Dale and Press (1999) observed that the presence of a parasitic plant (*Orobanche minor*) reduced its host's biomass by 47 percent in ambient air of 360 ppm CO_2, but by only 20 percent in air of 550 ppm CO_2.

These several studies suggest that, contrary to what is claimed by the IPCC, the ongoing rise in the air's CO_2 content will not favor the growth of weedy species over that of crops and native plants. In fact, it may provide non-weeds greater protection against weed-induced decreases in their productivity and growth. Future increases in the air's CO_2 content may actually increase the competitiveness of non-weeds over weeds.

Additional information on this topic, including reviews of newer publications as they become available, can be found at http://www.co2science.org/subject/b/weedsvsnonw.php.

References

Caporn, S.J.M., Brooks, A.L., Press, M.C. and Lee, J.A. 1999. Effects of long-term exposure to elevated CO_2 and increased nutrient supply on bracken (*Pteridium aquilinum*). *Functional Ecology* **13**: 107-115.

Dale, H. and Press, M.C. 1999. Elevated atmospheric CO_2 influences the interaction between the parasitic angiosperm *Orobanche minor* and its host *Trifolium repens*. *New Phytologist* **140**: 65-73.

Dukes, J.S. 2002. Comparison of the effect of elevated CO_2 on an invasive species (*Centaurea solstitialis*) in monoculture and community settings. *Plant Ecology* **160**: 225-234.

Gavazzi, M., Seiler, J., Aust, W. and Zedaker, S. 2000. The influence of elevated carbon dioxide and water availability on herbaceous weed development and growth of transplanted loblolly pine (*Pinus taeda*). *Environmental and Experimental Botany* **44**: 185-194.

Idso, K.E. 1992. *Plant Responses to Rising Levels of Carbon Dioxide: A Compilation and Analysis of the Results of a Decade of International Research into the Direct Biological Effects of Atmospheric CO_2 Enrichment.* Climatological Publications Scientific Paper #23, Office of Climatology, Arizona State University, Tempe, AZ.

Poorter, H. 1993. Interspecific variation in the growth response of plants to an elevated and ambient CO_2 concentration. *Vegetatio* **104/105**: 77-97.

Taylor, K. and Potvin, C. 1997. Understanding the long-term effect of CO_2 enrichment on a pasture: the importance of disturbance. *Canadian Journal of Botany* **75**: 1621-1627.

Watling, J.R. and Press, M.C. 1997. How is the relationship between the C_4 cereal *Sorghum bicolor* and the C_3 root hemi-parasites *Striga hermonthica* and *Striga asiatica* affected by elevated CO_2? *Plant, Cell and Environment* **20**: 1292-1300.

Wayne, P.M., Carnelli, A.L., Connolly, J. and Bazzaz, F.A. 1999. The density dependence of plant responses to elevated CO_2. *Journal of Ecology* **87**: 183-192.

Ziska, L.H., Teasdale, J.R. and Bunce, J.A. 1999. Future atmospheric carbon dioxide may increase tolerance to glyphosate. *Weed Science* **47**: 608-615.

7.6. Respiration

Nearly all of earth's plants respond favorably to increases in the air's CO_2 concentration by exhibiting enhanced rates of net photosynthesis and biomass production during the light part of each day. In many cases, observed increases in these parameters (especially biomass production) are believed to be due, in part, to CO_2-induced reductions in carbon losses via respiration during the day and especially at night (called "dark respiration"). In this summary, we examine what has been learned about this subject from experiments conducted on various herbaceous and woody plants.

Additional information on this topic, including reviews on respiration not discussed here, can be found at http://www.co2science.org/subject/r/subject_r.php under the heading Respiration.

7.6.1. Herbaceous Plants

7.6.1.1. Crops

Baker *et al.* (2000) grew rice in Soil-Plant-Atmosphere Research (SPAR) units at atmospheric CO_2 concentrations of 350 and 700 ppm during daylight hours. Under these conditions, rates of dark respiration decreased in both CO_2 treatments with short-term increases in the air's CO_2 concentration at night. However, when dark respiration rates were measured at the CO_2 growth concentrations of the plants, they were not significantly different from each other.

Cousins *et al.* (2001) grew sorghum at atmospheric CO_2 concentrations of 370 and 570 ppm within a free-air CO_2 enrichment (FACE) facility near Phoenix, Arizona, USA. Within six days of planting, the photosynthetic rates of the second leaves of the CO_2-enriched plants were 37 percent greater than those of the second leaves of the ambiently grown plants. However, this CO_2-induced photosynthetic enhancement slowly declined with time, stabilizing at approximately 15 percent between 23 and 60 days after planting. In addition, when measuring photosynthetic rates at a reduced oxygen concentration of 2 percent, they observed 16 and 9 percent increases in photosynthesis for the ambient and CO_2-enriched plants, respectively. These observations suggest that the extra 200 ppm of CO_2 was reducing photorespiratory carbon losses, although this phenomenon did not account for all of the CO_2-induced stimulation of photosynthesis.

Das *et al.* (2002) grew tropical nitrogen-fixing mungbean plants in open-top chambers maintained at atmospheric CO_2 concentrations of either 350 or 600 ppm for two growing seasons, with the extra CO_2 being provided between either days 0 and 20 or days 21 and 40 after germination. This work revealed that the elevated CO_2 decreased rates of respiration by 54-62 percent, with the greatest declines occurring during the first 20 days after germination.

Wang *et al.* (2004) grew well-watered and fertilized South American tobacco plants from seed in 8.4-liter pots (one plant per pot) filled with sand and housed in controlled-environment growth chambers maintained at atmospheric CO_2 concentrations of either 365 or 730 ppm for a total of nine weeks. Over this period they found that the ratio of net photosynthesis per unit leaf area (A) to dark respiration per unit leaf area (Rd) "changed dramatically." Whereas A/Rd was the same in both treatments at the beginning of the measurement period, a month later it had doubled in the CO_2-enriched environment but had risen by only 58 percent in the ambient treatment. Speaking of this finding, the three researchers say that "if the dynamic relationship between A and Rd observed in *N. sylvestris* is applicable to other species, it will have important implications for carbon cycling in terrestrial ecosystems, since plants will assimilate CO_2 more efficiently as they mature."

Bunce (2005) grew soybeans in the field in open-top chambers maintained at atmospheric CO_2 concentrations of ambient and ambient +350 ppm at the Beltsville Agricultural Research Center in Maryland (USA), where net carbon dioxide exchange rate measurements were performed on a total of 16 days between 18 July and 11 September of 2000 and 2003, during the flowering to early pod-filling stages of the growing season. Averaged over the course of the study, he found that daytime net photosynthesis per unit leaf area was 48 percent greater in the plants growing in the CO_2-enriched air, while nighttime respiration per unit leaf area was unaffected by elevated CO_2. However, because the extra 350 ppm of CO_2 increased leaf dry mass per unit area by an average of 23 percent, respiration per unit of mass was significantly lower for the leaves of the soybeans growing in the CO_2-enriched air.

Wang and Curtis (2002) conducted a meta-analysis of the results of 45 area-based dark respiration (Rda) and 44 mass-based dark respiration (Rdm) assessments of the effects of a doubling of the air's CO_2 concentration on 33 species of plants derived from 37 scientific studies. This work revealed that the mean leaf Rda of the suite of herbaceous plants studied was significantly higher (+29 percent, $P < 0.01$) at elevated CO_2 than at ambient CO_2. However, when the herbaceous plants were separated into groups that had experienced durations of CO_2 enrichment that were either shorter or longer than 60 days, it was found that the short-term studies exhibited a mean Rda increase of 51 percent ($P < 0.05$), while the long-term studies exhibited no effect. Hence, for conditions of continuous atmospheric CO_2 enrichment, herbaceous plants would likely experience no change in leaf Rda. In addition, the two researchers found that plants exposed to elevated CO_2 for < 100 days "showed significantly less of a reduction in leaf Rdm due to CO_2 enrichment (-12%) than did plants exposed for longer periods (-35%, $P < 0.01$)." Hence, for long-term conditions of continuous atmospheric CO_2 enrichment, herbaceous crops would likely experience an approximate 35 percent decrease in leaf Rdm.

Bunce (2004) grew six different 16-plant batches of soybeans within a single controlled-environment chamber, one to a pot filled with 1.8 liters of vermiculite that was flushed daily with a complete nutrient solution. In three experiments conducted at day/night atmospheric CO_2 concentrations of 370/390 ppm, air temperatures were either 20, 25 or 30°C, while in three other experiments conducted at an air temperature of 25°C, atmospheric CO_2 concentrations were either 40, 370 or 1400 ppm. At the end of the normal 16 hours of light on the 17th day after planting, half of the plants were harvested and used for the measurement of a number of physical parameters, while measurements of the plant physiological processes of respiration, translocation and nitrate reduction were made on the other half of the plants over the following 8-hour dark period.

Plotting translocation and nitrate reduction as functions of respiration, Bunce found that "a given change in the rate of respiration was accompanied by the same change in the rate of translocation or nitrate reduction, regardless of whether the altered respiration was caused by a change in temperature or by a change in atmospheric CO_2 concentration." As a result, and irrespective of whatever mechanisms may have been involved in eliciting the responses observed, Bunce concluded that "the parallel responses of translocation and nitrate reduction for both the temperature and CO_2 treatments make it unlikely that the response of respiration to one variable [CO_2] was an artifact while the response to the other [temperature] was real." Hence, there is reason to believe that the oft-observed decreases in dark respiration experienced by plants exposed to elevated levels of atmospheric CO_2, as per the review and analysis studies of Drake et al. (1999) and Wang and Curtis (2002), are indeed real and not the result of measurement system defects.

In light of these several findings, it can be concluded that the balance of evidence suggests that the growth of herbaceous crops is generally enhanced by CO_2-induced decreases in respiration during the dark period.

Additional information on this topic, including reviews of newer publications as they become available, can be found at http://www.co2science.org/subject/r/respirationcrops.php.

References

Baker, J.T., Allen, L.H., Jr., Boote, K.J. and Pickering, N.B. 2000. Direct effects of atmospheric carbon dioxide concentration on whole canopy dark respiration of rice. *Global Change Biology* **6**: 275-286.

Bunce, J.A. 2004. A comparison of the effects of carbon dioxide concentration and temperature on respiration, translocation and nitrate reduction in darkened soybean leaves. *Annals of Botany* **93**: 665-669.

Bunce, J.A. 2005. Response of respiration of soybean leaves grown at ambient and elevated carbon dioxide concentrations to day-to-day variation in light and temperature under field conditions. *Annals of Botany* **95**: 1059-1066.

Cousins, A.B., Adam, N.R., Wall, G.W., Kimball, B.A., Pinter Jr., P.J., Leavitt, S.W., LaMorte, R.L., Matthias, A.D., Ottman, M.J., Thompson, T.L. and Webber, A.N. 2001. Reduced photorespiration and increased energy-use efficiency in young CO_2-enriched sorghum leaves. *New Phytologist* **150**: 275-284.

Das, M., Zaidi, P.H., Pal, M. and Sengupta, U.K. 2002. Stage sensitivity of mungbean (*Vigna radiata* L. Wilczek) to an elevated level of carbon dioxide. *Journal of Agronomy and Crop Science* **188**: 219-224.

Drake, B.G., Azcon-Bieto, J., Berry, J., Bunce, J., Dijkstra, P., Farrar, J., Gifford, R.M., Gonzalez-Meler, M.A., Koch, G., Lambers, H., Siedow, J. and Wullschleger, S. 1999. Does elevated atmospheric CO_2 inhibit mitochondrial respiration in green plants? *Plant, Cell and Environment* **22**: 649-657.

Wang, X., Anderson, O.R. and Griffin, K.L. 2004. Chloroplast numbers, mitochondrion numbers and carbon assimilation physiology of *Nicotiana sylvestris* as affected by CO_2 concentration. *Environmental and Experimental Botany* **51**: 21-31.

Wang, X. and Curtis, P. 2002. A meta-analytical test of elevated CO_2 effects on plant respiration. *Plant Ecology* **161**: 251-261.

7.6.1.2. Other Herbaceous Plants

In this section we review the results of studies of non-crop herbaceous plants to determine if atmospheric CO_2 enrichment tends to increase or decrease (or leave unaltered) their respiration rates.

Rabha and Uprety (1998) grew India mustard plants for an entire season in open-top chambers with either ambient or enriched (600 ppm) atmospheric CO_2 concentrations and adequate or inadequate soil moisture levels. Their work revealed that the elevated CO_2 concentration reduced leaf dark respiration rates by about 25 percent in both soil moisture treatments, which suggests that a greater proportion of the increased carbohydrate pool in the CO_2-enriched plants remained within them to facilitate increases in growth and development.

Ziska and Bunce (1999) grew four C4 plants in controlled environment chambers maintained at either full-day (24-hour) atmospheric CO_2 concentrations of 350 and 700 ppm or a nocturnal-only CO_2 concentration of 700 ppm (with 350 ppm CO_2 during the day) for about three weeks. In this particular study, 24-hour CO_2 enrichment caused a significant increase in the photosynthesis (+13 percent) and total dry mass (+21 percent) of only one of the four C4 species (*Amaranthus retroflexus*). However, there was no significant effect of nocturnal-only CO_2 enrichment on this species, indicating that the observed increase in biomass, resulting from 24-hour atmospheric CO_2 enrichment, was not facilitated by greater carbon conservation stemming from a CO_2-induced reduction in dark respiration.

In an experiment that produced essentially the same result, Grunzweig and Korner (2001) constructed model grasslands representative of the Negev of Israel and placed them in growth chambers maintained at atmospheric CO_2 concentrations of 280, 440 and 600 ppm for five months. This study also revealed that atmospheric CO_2 enrichment had no effect on nighttime respiratory carbon losses.

Moving to the other end of the moisture spectrum, Van der Heijden *et al.* (2000) grew peat moss hydroponically within controlled environment chambers maintained at atmospheric CO_2 concentrations of 350 and 700 ppm for up to six months, while simultaneously subjecting the peat moss to three different levels of nitrogen deposition. In all cases, they found that the elevated CO_2 reduced rates of dark respiration consistently throughout the study by 40 to 60 percent.

In a final multi-species study, Gonzalez-Meler *et al.* (2004) reviewed the scientific literature pertaining to the effects of atmospheric CO_2 enrichment on plant respiration from the cellular level to the level of entire ecosystems. They report finding that "contrary to what was previously thought, specific respiration rates are generally not reduced when plants are grown at elevated CO_2." Nevertheless, they note that "whole ecosystem studies show that canopy respiration does not increase proportionally to increases in biomass in response to elevated CO_2," which suggests that respiration per unit biomass is likely somewhat reduced by atmospheric CO_2 enrichment. However, they also find that "a larger proportion of respiration takes place in the root system [when plants are grown in CO_2-enriched air]," which once again obfuscates the issue.

The three researchers say "fundamental information is still lacking on how respiration and the processes supported by it are physiologically controlled, thereby preventing sound interpretations of what seem to be species-specific responses of respiration to elevated CO_2." They conclude that "the role of plant respiration in augmenting the sink capacity of terrestrial ecosystems is still uncertain."

Additional information on this topic, including reviews of newer publications as they become available, can be found at http://www.co2science.org/subject/r/respirationherbaceous.php.

References

Gonzalez-Meler, M.A., Taneva, L. and Trueman, R.J. 2004. Plant respiration and elevated atmospheric CO_2 concentration: Cellular responses and global significance. *Annals of Botany* **94**: 647-656.

Grunzweig, J.M. and Korner, C. 2001. Growth, water and nitrogen relations in grassland model ecosystems of the semi-arid Negev of Israel exposed to elevated CO_2. *Oecologia* **128**: 251-262.

Rabha, B.K. and Uprety, D.C. 1998. Effects of elevated CO_2 and moisture stress on Brassica juncea. *Photosynthetica* **35**: 597-602.

Van der Heijden, E., Verbeek, S.K. and Kuiper, P.J.C. 2000. Elevated atmospheric CO_2 and increased nitrogen deposition: effects on C and N metabolism and growth of the peat moss *Sphagnum recurvum* P. Beauv. Var. *mucronatum* (Russ.) Warnst. *Global Change Biology* **6**: 201-212.

Ziska, L.H. and Bunce, J.A. 1999. Effect of elevated carbon dioxide concentration at night on the growth and gas exchange of selected C4 species. *Australian Journal of Plant Physiology* **26**: 71-77.

7.6.2. Woody Plants

7.6.2.1. Coniferous Trees

Jach and Ceulemans (2000) grew three-year old Scots pine seedlings out-of-doors and rooted in the ground in open-top chambers maintained at atmospheric CO_2 concentrations of either 350 or 750 ppm for two years. To make the experiment more representative of the natural world, they applied no nutrients or irrigation water to the soils in which the trees grew for the duration of the study. After two years of growth under these conditions, dark respiration on a needle mass basis in the CO_2-enriched seedlings was 27 percent and 33 percent lower in current-year and one-year-old needles, respectively, with the greater reduction in the older needles being thought to arise

from the greater duration of elevated CO_2 exposure experienced by those needles.

Hamilton *et al.* (2001) studied the short- and long-term respiratory responses of loblolly pines in a free-air CO_2-enrichment (FACE) study that was established in 1996 on 13-year-old trees in a North Carolina (USA) plantation, where the CO_2-enriched trees were exposed to an extra 200 ppm of CO_2. This modest increase in the atmosphere's CO_2 concentration produced no significant short-term suppression of dark respiration rates in the trees' needles. Neither did long-term exposure to elevated CO_2 alter maintenance respiration, which is the amount of CO_2 respired to maintain existing plant tissues. However, growth respiration, which is the amount of CO_2 respired when constructing new tissues, was reduced by 21 percent.

McDowell *et al.* (1999) grew five-month-old seedlings of western hemlock in root boxes subjected to various root-space CO_2 concentrations (ranging from 90 to 7000 ppm) for periods of several hours to determine the effects of soil CO_2 concentration on growth, maintenance and total root respiration. In doing so, they found that although elevated CO_2 had no effect on growth respiration, it significantly impacted maintenance and total respiration. At a soil CO_2 concentration of 1585 ppm, for example, total and maintenance respiration rates of roots were 55 percent and 60 percent lower, respectively, than they were at 395 ppm. The impact of elevated CO_2 on maintenance respiration was so strong that it exhibited an exponential decline of about 37 percent for every doubling of soil CO_2 concentration. The implications of this observation are especially important because maintenance respiration comprised 85 percent of total root respiration in this study.

The results of these experiments suggest that both above and below the soil surface, coniferous trees may exhibit reductions in total respiration in a high-CO_2 world of the future. Three studies of three species, however, is insufficient evidence to reach a firm conclusion.

Additional information on this topic, including reviews of newer publications as they become available, can be found at http://www.co2science.org/subject/r/respirationconifers.php.

References

Hamilton, J.G., Thomas, R.B. and DeLucia, E.H. 2001. Direct and indirect effects of elevated CO_2 on leaf respiration in a forest ecosystem. *Plant, Cell and Environment* **24**: 975-982.

Jach, M.E. and Ceulemans, R. 2000. Short- *versus* long-term effects of elevated CO_2 on night-time respiration of needles of Scots pine (*Pinus sylvestris* L.). *Photosynthetica* **38**: 57-67.

McDowell, N.G., Marshall, J.D., Qi, J. and Mattson, K. 1999. Direct inhibition of maintenance respiration in western hemlock roots exposed to ambient soil carbon dioxide concentrations. *Tree Physiology* **19**: 599-605.

7.6.2.2. Deciduous Trees

Wang and Curtis (2001) grew cuttings of two male and two female trembling aspen trees for about five months on soils containing low and high nitrogen contents in open-top chambers maintained at atmospheric CO_2 concentrations of 380 and 765 ppm, finding that gender had little effect on dark respiration rates, but that elevated CO_2 increased them, by 6 percent and 32 percent in the low and high soil nitrogen treatments, respectively. On the other hand, Karnosky *et al.* (1999) grew both O_3-sensitive and O_3-tolerant aspen clones for one full year in free-air CO_2-enrichment (FACE) plots maintained at atmospheric CO_2 concentrations of 360 and 560 ppm, finding that the extra CO_2 decreased dark respiration rates by 24 percent.

Gielen *et al.* (2003) measured stem respiration rates of white, black and robusta poplar trees in a high-density forest plantation in the third year of a FACE experiment in which the CO_2 concentration of the air surrounding the trees was increased to a value of approximately 550 ppm. This study revealed, in their words, that "stem respiration rates were not affected by the FACE treatment," and that "FACE did not influence the relationships between respiration rate and both stem temperature and relative growth rate." In addition, they say they could find "no effect of the FACE treatment on Rm [maintenance respiration, which is related to the sustaining of existing cells] and Rg [growth respiration, which is related to the synthesis of new tissues]."

Hamilton *et al.* (2001) studied respiratory responses of sweetgum trees growing in the understory of a loblolly pine plantation (but occasionally reaching the top of the canopy) to an extra 200 ppm of CO_2 in a FACE study conducted in North Carolina, USA. As a result of their measurement program, they determined that the modest increase in atmospheric CO_2 concentration did

not appear to alter maintenance respiration to any significant degree, but that it reduced dark respiration by an average of 10 percent and growth respiration of leaves at the top of the canopy by nearly 40 percent.

In reviewing the results of these several deciduous tree studies, we see cases of both increases and decreases in respiration rates in response to atmospheric CO_2 enrichment, as well as cases of no change in respiration. More data are needed before any general conclusions may safely be drawn.

Additional information on this topic, including reviews of newer publications as they become available, can be found at http://www.co2science.org/subject/r/respirationdeciduous.php.

References

Gielen, B., Scarascia-Mugnozza, G. and Ceulemans, R. 2003. Stem respiration of *Populus* species in the third year of free-air CO_2 enrichment. *Physiologia Plantarum* **117**: 500-507.

Hamilton, J.G., Thomas, R.B. and DeLucia, E.H. 2001. Direct and indirect effects of elevated CO_2 on leaf respiration in a forest ecosystem. *Plant, Cell and Environment* **24**: 975-982.

Karnosky, D.F., Mankovska, B., Percy, K., Dickson, R.E., Podila, G.K., Sober, J., Noormets, A., Hendrey, G., Coleman, M.D., Kubiske, M., Pregitzer, K.S. and Isebrands, J.G. 1999. Effects of tropospheric O_3 on trembling aspen and interaction with CO_2: results from an O_3-gradient and a FACE experiment. *Water, Air, and Soil Pollution* **116**: 311-322.

Wang, X. and Curtis, P.S. 2001. Gender-specific responses of *Populus tremuloides* to atmospheric CO_2 enrichment. *New Phytologist* **150**: 675-684.

7.6.2.3. Multiple Tree Studies

Amthor (2000) measured dark respiration rates of intact leaves of nine different tree species growing naturally in an American deciduous forest. Within a specially designed leaf chamber, the CO_2 concentration surrounding individual leaves was stabilized at 400 ppm for 15 minutes, whereupon their respiration rates were measured for 30 minutes, after which the CO_2 concentration in the leaf chamber was raised to 800 ppm for 15 minutes and respiration data were again recorded for the same leaves. This protocol revealed that elevated CO_2 had little effect on leaf dark respiration rates. The extra 400 ppm of

CO_2 within the measurement cuvette decreased the median respiration rate by only 1.5 percent across the nine tree species. This observation led Amthor to state that the "rising atmospheric CO_2 concentration has only a small direct effect on tree leaf respiration in deciduous forests;" and he calculated that it can be "more than eliminated by a 0.22°C temperature increase." Upon this premise, he concluded that "future direct effects of increasing CO_2 in combination with warming could stimulate tree leaf respiration in their sum," and that this consequence "would translate into only slight, if any, effects on the carbon balance of temperate deciduous forests in a future atmosphere containing as much as [800 ppm] CO_2."

Amthor's conclusion is debatable, for it is based upon the extrapolation of the short-term respiratory responses of individual leaves, exposed to elevated CO_2 for only an hour or two, to that of entire trees, many of which will experience rising CO_2 levels for a century or more during their lifetimes. Trees are long-lived organisms that should not be expected to reveal the nature of their long-term responses to elevated atmospheric CO_2 concentrations on as short a time scale as 15 minutes. Indeed, their respiratory responses may change significantly with the passage of time as they acclimate and optimize their physiology and growth patterns to the gradually rising CO_2 content of earth's atmosphere, as evidenced by the findings of the following two studies.

Wang and Curtis (2002) conducted a meta-analysis of the results of 45 area-based dark respiration (Rda) and 44 mass-based dark respiration (Rdm) assessments of the effects of an approximate doubling of the air's CO_2 concentration on 33 species of plants (both herbaceous and woody) derived from 37 scientific publications. This effort revealed that the mean leaf Rda of the woody plants they analyzed was unaffected by elevated CO_2. However, there was an effect on mean leaf Rdm, and it was determined to be time-dependent. The woody plants exposed to elevated CO_2 for < 100 days, in the reviewing scientists' words, "showed significantly less of a reduction in leaf Rdm due to CO_2 enrichment (-12%) than did plants exposed for longer periods (-35%, $P < 0.01$)." Hence, for conditions of continuous long-term atmospheric CO_2 enrichment, the results of Wang and Curtis' analysis suggest woody plants may experience an approximate 35 percent decrease in leaf Rdm.

Drake *et al.* (1999) also conducted a comprehensive analysis of the peer-reviewed scientific literature to determine the effects of elevated atmospheric CO_2 concentrations on plant respiration rates. They found that atmospheric CO_2 enrichment typically decreased respiration rates in mature foliage, stems, and roots of CO_2-enriched plants relative to rates measured in plants grown in ambient air; and when normalized on a biomass basis, they determined that a doubling of the atmosphere's CO_2 concentration would likely reduce plant respiration rates by an average of 18 percent. To determine the potential effects of this phenomenon on annual global carbon cycling, which the twelve researchers say "will enhance the quantity of carbon stored by forests," they input a 15 percent CO_2-induced respiration reduction into a carbon sequestration model, finding that an additional 6 to 7 Gt of carbon would remain sequestered within the terrestrial biosphere each year, thus substantially strengthening the terrestrial carbon sink.

Davey *et al.* (2004) reached a different conclusion. "Averaged across many previous investigations, doubling the CO_2 concentration has frequently been reported to cause an instantaneous reduction of leaf dark respiration measured as CO_2 efflux." However, as they continue, "no known mechanism accounts for this effect, and four recent studies [Amthor (2000); Anthor *et al.* (2001); Jahnke (2001); Jahnke and Krewitt (2002)] have shown that the measurement of respiratory CO_2 efflux is prone to experimental artifacts that could account for the reported response."

Using a technique that avoids the potential artifacts of prior attempts to resolve the issue, Davey *et al.* employed a high-resolution dual channel oxygen analyzer in an open gas exchange system to measure the respiratory O_2 uptake of nine different species of plants in response to a short-term increase in atmospheric CO_2 concentration, as well as the response of seven species to long-term elevation of the air's CO_2 content in four different field experiments. In doing so, they found that "over six hundred separate measurements of respiration failed to reveal any decrease in respiratory O_2 uptake with an instantaneous increase in CO_2." Neither could they detect any response to a five-fold increase in the air's CO_2 concentration nor to the total removal of CO_2 from the air. They also note that "this lack of response of respiration to elevated CO_2 was independent of treatment method, developmental stage, beginning or end of night, and the CO_2 concentration at which the plants had been grown." In the long-term field studies, however, there was a respiratory response;

but it was small (7 percent on a leaf mass basis), and it was positive, not negative.

In light of these contradicgtory results, the most reasonable conclusion is that atmospheric CO_2 enrichment may either increase or decrease woody-plant respiration, but not to any great degree, and that in the mean, the net result for the conglomerate of earth's trees would likely be something of little impact.

Additional information on this topic, including reviews of newer publications as they become available, can be found at http://www.co2science.org/subject/r/respirationtrees.php.

References

Amthor, J.S. 2000. Direct effect of elevated CO_2 on nocturnal in situ leaf respiration in nine temperate deciduous tree species is small. *Tree Physiology* **20**: 139-144.

Amthor, J.S., Koch, G.W., Willms, J.R. and Layzell, D.B. 2001. Leaf O_2 uptake in the dark is independent of coincident CO_2 partial pressure. *Journal of Experimental Botany* **52**: 2235-2238.

Davey, P.A., Hunt, S., Hymus, G.J., DeLucia, E.H., Drake, B.G., Karnosky, D.F. and Long, S.P. 2004. Respiratory oxygen uptake is not decreased by an instantaneous elevation of [CO_2], but is increased with long-term growth in the field at elevated [CO_2]. *Plant Physiology* **134**: 520-527.

Drake, B.G., Azcon-Bieto, J., Berry, J., Bunce, J., Dijkstra, P., Farrar, J., Gifford, R.M., Gonzalez-Meler, M.A., Koch, G., Lambers, H., Siedow, J. and Wullschleger, S. 1999. Does elevated atmospheric CO_2 inhibit mitochondrial respiration in green plants? *Plant, Cell and Environment* **22**: 649-657.

Jahnke, S. 2001. Atmospheric CO_2 concentration does not directly affect leaf respiration in bean or poplar. *Plant, Cell and Environment* **24**: 1139-1151.

Jahnke, S. and Krewitt, M. 2002. Atmospheric CO_2 concentration may directly affect leaf respiration measurement in tobacco, but not respiration itself. *Plant, Cell and Environment* **25**: 641-651.

Wang, X. and Curtis, P. 2002. A meta-analytical test of elevated CO_2 effects on plant respiration. *Plant Ecology* **161**: 251-261.

7.7. Carbon Sequestration

As the CO_2 content of the air continues to rise, nearly all of earth's plants respond by increasing their photosynthetic rates and producing more biomass. This results in more carbon being captured and stored, or sequestered, in plant fibers and soil, which counterbalances some of the CO_2 emissions produced by mankind's use of fossil fuels.

In this section we begin with a research review of what is known about forest and forest-species responses to atmospheric CO_2 enrichment and subsequent carbon sequestration. Then we survey research on how temperatures affect sequestration and whether CO_2 enhancement offsets the rate at which carbon is re-released from soil (decomposition).

7.7.1. CO$_2$ Enhancement and Carbon Sequestration

7.7.1.1 Forests

The planting and preservation of forests has long been acknowledged to be an effective and environmentally friendly (indeed, *enhancing*) means for slowing climate-model-predicted CO_2-induced global warming. This prescription for moderating potential climate change is based on two well-established and very straightforward facts: (1) the carbon trees use to construct their tissues comes from the air, and (2) its extraction from the atmosphere slows the rate of rise of the air's CO_2 content.

In an open-top chamber experiment conducted in Switzerland, Nitschelm *et al.* (1997) reported that a 71 percent increase in the atmospheric CO_2 concentration above white clover monocultures led to a 50 percent increase in soil organic carbon content. Related studies on wheat and soybean agroecosystems (Islam *et al.*, 1999) provided similar results, as did a free-air CO_2 enrichment (FACE) experiment on cotton, which documented a 10 percent increase in soil organic carbon content in plots receiving 550 ppm CO_2 relative to those receiving 370 ppm (Leavitt *et al.*, 1994). These phenomena will allow long-lived perennial species characteristic of forest ecosystems to sequester large amounts of carbon within their wood for extended periods of time (Chambers *et al.*, 1998).

In reviewing studies that have been conducted on individual trees, it is clear that elevated levels of atmospheric CO_2 increase photosynthesis and growth in both broad-leaved and coniferous species. When broad-leaved trembling aspen (*Populus tremuloides*) were exposed to twice-ambient levels of atmospheric CO_2 for 2.5 years, for example, Pregitzer *et al.* (2000) reported 17 and 65 percent increases in fine root biomass at low and high levels of soil nitrogen, respectively; while Zak *et al.* (2000) observed 16 and 38 percent CO_2-induced increases in total tree biomass when subjected to the same respective levels of soil nitrogen.

Similar results have been reported for coniferous trees. When branches of Sitka spruce (*Picea sitchensis*) were fumigated with air of 700 ppm CO_2 for four years, rates of net photosynthesis in current and second-year needles were 100 and 43 percent higher, respectively, than photosynthetic rates of needles exposed to ambient air (Barton and Jarvis, 1999). In addition, ponderosa pine (*Pinus ponderosa*) grown at 700 ppm CO_2 for close to 2.5 years exhibited rates of net photosynthesis in current-year needles that were 49 percent greater than those of needles exposed to air containing 350 ppm CO_2 (Houpis *et al.*, 1999).

Elevated CO_2 also enhances carbon sequestration by reducing carbon losses arising from plant respiration. Karnosky *et al.* (1999) reported that aspen seedlings grown for one year at 560 ppm CO_2 displayed dark respiration rates that were 24 percent lower than rates exhibited by trembling aspen grown at 360 ppm CO_2. Also, elevated CO_2 has been shown to decrease maintenance respiration, which it did by 60 percent in western hemlock seedlings exposed to an atmospheric CO_2 concentration of nearly 1600 ppm (McDowell *et al.*, 1999).

In a thorough review of these topics, Drake *et al.* (1999) concluded that, on average, a doubling of the atmospheric CO_2 concentration reduces plant respiration rates by approximately 17 percent. This finding contrasts strikingly with the much smaller effects reported by Amthor (2000), who found an average reduction in dark respiration of only 1.5 percent for nine deciduous tree species exposed to 800 ppm CO_2. The period of CO_2 exposure in his much shorter experiments, however, was a mere 15 minutes. If the air's CO_2 content doubles, plants will likely sequester something on the order of 17 percent more carbon than ambiently grown plants, solely as a consequence of CO_2-induced reductions in respiration. And it is good to remember that this stored carbon is in addition to that sequestered as a result of CO_2-induced increases in plant photosynthetic rates.

Based upon several different types of empirical data, a number of researchers have concluded that current rates of carbon sequestration are robust and that future rates will increase with increasing atmospheric CO_2 concentrations. In the study of Fan *et al.* (1998) based on atmospheric measurements, for example, the broad-leaved forested region of North America between 15 and 51°N latitude was calculated to possess a current carbon sink that can annually remove all the CO_2 emitted into the air from fossil fuel combustion in both Canada and the United States. On another large scale, Phillips *et al.* (1998) used data derived from tree basal area to show that average forest biomass in the tropics has increased substantially over the past 40 years and that growth in the Neotropics alone (south and central South America, the Mexican lowlands, the Carribean islands, and southern Florida) can account for 40 percent of the missing carbon of the entire globe. And in looking to the future, White *et al.* (2000) have calculated that coniferous and mixed forests north of 50°N latitude will likely expand their northern and southern boundaries by about 50 percent due to the combined effects of increasing atmospheric CO_2, rising temperature, and nitrogen deposition.

The latter of these factors, nitrogen deposition, is an important variable. As indicated in the study of White *et al.*, it can play an interactive role with increasing atmospheric CO_2 to increase plant growth and carbon sequestration. However, the magnitude of that role is still being debated. Nadelhoffer *et al.* (1999), for example, concluded that nitrogen deposition from human activities is "unlikely to be a major contributor" to the large CO_2 sink that exists in northern temperate forests. Houghton *et al.* (1998), however, feel that nitrogen deposition holds equal weight with CO_2 fertilization in the production of terrestrial carbon sinks; and Lloyd (1999) demonstrated that when CO_2 and nitrogen increase together, modeled forest productivity is greater than that predicted by the sum of the individual contributions of these two variables. Thus, anthropogenic nitrogen deposition can have anywhere from small to large positive effects on carbon sequestration, as well as everything in between.

In conclusion, as the air's CO_2 content continues to rise, the ability of earth's forests to sequester carbon should also rise. With more CO_2 in the atmosphere, trees will likely exhibit greater rates of photosynthesis and reduced rates of respiration. Together, these observations suggest that biologically fixed carbon will experience greater residency times within plant tissues. And if this carbon is directed into wood production, which increases substantially with atmospheric CO_2 enrichment, some of it can be kept out of circulation for a *very* long time, possibly even a millennium or more.

Additional information on this topic, including reviews of newer publications as they become available, can be found at http://www.co2science.org/subject/c/carbonforests.php.

References

Amthor, J.S. 2000. Direct effect of elevated CO_2 on nocturnal in situ leaf respiration in nine temperate deciduous tree species is small. *Tree Physiology* **20**: 139-144.

Barton, C.V.M. and Jarvis, P.G. 1999. Growth response of branches of *Picea sitchensis* to four years exposure to elevated atmospheric carbon dioxide concentration. *New Phytologist* **144**: 233-243.

Chambers, J.Q., Higuchi, N. and Schimel, J.P. 1998. Ancient trees in Amazonia. *Nature* **391**: 135-136.

De Angelis, P., Chigwerewe, K.S. and Mugnozza, G.E.S. 2000. Litter quality and decomposition in a CO_2-enriched Mediterranean forest ecosystem. *Plant and Soil* **224**: 31-41.

Drake, B.G., Azcon-Bieto, J., Berry, J., Bunce, J., Dijkstra, P., Farrar, J., Gifford, R.M., Gonzalez-Meler, M.A., Koch, G., Lambers, H., Siedow, J. and Wullschleger, S. 1999. Does elevated atmospheric CO_2 inhibit mitochondrial respiration in green plants? *Plant, Cell and Environment* **22**: 649-657.

Fan, S., Gloor, M., Mahlman, J., Pacala, S., Sarmiento, J., Takahashi, T. and Tans, P. 1998. A large terrestrial carbon sink in North America implied by atmospheric and oceanic carbon dioxide data and models. *Science* **282**: 442-446.

Hirschel, G., Korner, C. and Arnone III, J.A. 1997. Will rising atmospheric CO_2 affect leaf litter quality and in situ decomposition rates in native plant communities? *Oecologia* **110**: 387-392.

Houghton, R.A., Davidson, E.A. and Woodwell, G.M. 1998. Missing sinks, feedbacks, and understanding the role of terrestrial ecosystems in the global carbon balance. *Global Biogeochemical Cycles* **12**: 25-34.

Houpis, J.L.J., Anderson, P.D., Pushnik, J.C. and Anschel, D.J. 1999. Among-provenance variability of gas exchange and growth in response to long-term elevated CO_2 exposure. *Water, Air, and Soil Pollution* **116**: 403-412.

Islam, K.R., Mulchi, C.L. and Ali, A.A. 1999. Tropospheric carbon dioxide or ozone enrichments and moisture effects on soil organic carbon quality. *Journal of Environmental Quality* **28**: 1629-1636.

Karnosky, D.F., Mankovska, B., Percy, K., Dickson, R.E., Podila, G.K., Sober, J., Noormets, A., Hendrey, G., Coleman, M.D., Kubiske, M., Pregitzer, K.S. and Isebrands, J.G. 1999. Effects of tropospheric O_3 on trembling aspen and interaction with CO_2: results from an O_3-gradient and a FACE experiment. *Water, Air, and Soil Pollution* **116**: 311-322.

King, J.S., Pregitzer, K.S. and Zak, D.R. 1999. Clonal variation in above- and below-ground responses of *Populus tremuloides* Michaux: Influence of soil warming and nutrient availability. *Plant and Soil* **217**: 119-130.

Leavitt, S.W., Paul, E.A., Kimball, B.A., Hendrey, G.R., Mauney, J.R., Rauschkolb, R., Rogers, H., Nagy, J., Pinter Jr., P.J. and Johnson, H.B. 1994. Carbon isotope dynamics of free-air CO_2-enriched cotton and soils. *Agricultural and Forest Meteorology* **70**: 87-101.

Lloyd, J. 1999. The CO_2 dependence of photosynthesis, plant growth responses to elevated CO_2 concentrations and their interaction with soil nutrient status, II. Temperate and boreal forest productivity and the combined effects of increasing CO_2 concentrations and increased nitrogen deposition at a global scale. *Functional Ecology* **13**: 439-459.

McDowell, N.G., Marshall, J.D., Qi, J. and Mattson, K. 1999. Direct inhibition of maintenance respiration in western hemlock roots exposed to ambient soil carbon dioxide concentrations. *Tree Physiology* **19**: 599-605.

Nadelhoffer, K.J., Emmett, B.A., Gundersen, P., Kjonaas, O.J., Koopmans, C.J., Schleppi, P., Tietema, A. and Wright, R.F. 1999. Nitrogen deposition makes a minor contribution to carbon sequestration in temperate forests. *Nature* **398**: 145-148.

Nitschelm, J.J., Luscher, A., Hartwig, U.A. and van Kessel, C. 1997. Using stable isotopes to determine soil carbon input differences under ambient and elevated atmospheric CO_2 conditions. *Global Change Biology* **3**: 411-416.

Phillips, O.L., Malhi, Y., Higuchi, N., Laurance, W.F., Nunez, P.V., Vasquez, R.M., Laurance, S.G., Ferreira, L.V., Stern, M., Brown, S. and Grace, J. 1998. Changes in the carbon balance of tropical forests: Evidence from long-term plots. *Science* **282**: 439-442.

Pregitzer, K.S., Zak, D.R., Maziaasz, J., DeForest, J., Curtis, P.S. and Lussenhop, J. 2000. Interactive effects of atmospheric CO_2 and soil-N availability on fine roots of *Populus tremuloides*. *Ecological Applications* **10**: 18-33.

Scherzel, A.J., Rebbeck, J. and Boerner, R.E.J. 1998. Foliar nitrogen dynamics and decomposition of yellow-poplar and eastern white pine during four seasons of exposure to elevated ozone and carbon dioxide. *Forest Ecology and Management* **109**: 355-366.

White, A., Cannell, M.G.R. and Friend, A.D. 2000. The high-latitude terrestrial carbon sink: a model analysis. *Global Change Biology* **6**: 227-245.

White, M.A., Running, S.W. and Thornton, P.E. 1999. The impact of growing-season length variability on carbon assimilation and evapotranspiration over 88 years in the eastern US deciduous forest. *International Journal of Biometeorology* **42**: 139-145.

Zak, D.R., Pregitzer, K.S., Curtis, P.S., Vogel, C.S., Holmes, W.E. and Lussenhop, J. 2000. Atmospheric CO_2, soil-N availability, and allocation of biomass and nitrogen by *Populus tremuloides*. *Ecological Applications* **10**: 34-46.

7.7.1.2. Old Forests

In addition to enhancing the growth and production of young forests, available research indicates that rising atmospheric CO_2 concentrations will also increase the productivity and growth of older forests. For most of the past century it was believed that old-growth forests, such as those of Amazonia, should be close to dynamic equilibrium. Just the opposite, however, has been repeatedly observed over the past two decades.

In one of the first studies to illuminate this reality, Phillips and Gentry (1994) analyzed the turnover rates—which are close correlates of net productivity (Weaver and Murphy, 1990)—of 40 tropical forests from all around the world. They found that the growth rates of these already highly productive forests had been rising ever higher since at least 1960, and that they had experienced an apparent acceleration in growth rate sometime after 1980. Commenting on these findings, Pimm and Sugden (1994) reported that the consistency and simultaneity of the forest growth trends that Phillips and Gentry had documented on several continents led them to conclude that "enhanced productivity induced by increased CO_2 is the most plausible candidate for the cause of the increased turnover."

A few years later, Phillips *et al.* (1998) analyzed forest growth rate data for the period 1958 to 1996 for several hundred plots of mature tropical trees scattered around the world, finding that tropical forest biomass, as a whole, increased substantially over the period of record. In fact, the increase in the Neotropics was equivalent to approximately 40 percent of the missing terrestrial carbon sink of the entire globe. Consequently, they concluded that tropical forests "may be helping to buffer the rate of increase in atmospheric CO_2, thereby reducing the impacts of global climate change." And, again, they identified the aerial fertilization effect of the ongoing rise in the air's CO_2 content as one of the primary factors likely to be responsible for this phenomenon.

More recently, Laurance *et al.* (2004a) reported accelerated growth in the 1990s relative to the 1980s for the large majority (87 percent) of tree genera in 18 one-hectare plots spanning an area of about 300 km^2 in central Amazonia, while Laurance *et al.* (2004b) observed similarly accelerated tree community dynamics in the 1990s relative to the 1980s. In addition, Baker *et al.* (2004) reported there has been a net increase in biomass in old-growth Amazonian forests in recent decades at a rate of 1.22 ± 0.42 Mg ha^{-1} yr^{-1}, slightly greater than that originally estimated by Phillips *et al.* And once again, it was suggested, in the words of Laurance *et al.* (2005), that these "pervasive changes in central Amazonian tree communities were most likely caused by global- or regional-scale drivers, such as increasing atmospheric CO_2 concentrations (Laurance *et al.*, 2004a,b)."

Expanding upon this theme, Laurance *et al.* (2005) say they "interpreted these changes as being consistent with an ecological 'signature' expected from increasing forest productivity (cf., Phillips and Gentry, 1994; Lewis *et al.* 2004a,b; Phillips *et al.*, 2004)." They note, however, that they have been challenged in this conclusion by Nelson (2005), and they thus go on to consider his arguments in some detail, methodically dismantling them one by one.

Evidence of increasing dynamism and productivity of intact tropical forests continued, with Lewis *et al.*, (2004a) reporting that among 50 old-growth plots scattered across tropical South America, "stem recruitment, stem mortality, and biomass growth, and loss, *all* increased significantly." In summarizing these and other findings, Lewis (2006) reports that over the past two decades, "these forests have shown concerted changes in their ecology, becoming, on average, faster growing—more productive—and more dynamic, and showing a net increase in above-ground biomass," all of which rates of increase are greater than the previously documented increases in the rates of these phenomena. What is more, Lewis says that "preliminary analyses also suggest the African and Australian forests are showing structural changes similar to South American forests."

So why should we care about growth trends of old forests? People who seek to address the issue solely on the basis of forced reductions in anthropogenic CO_2 emissions claim that carbon sequestration by forests is viable only when forests are young and growing vigorously. (Pearce, 1999) As forests age, as claimed by the IPCC, they gradually lose their carbon-sequestering prowess, such that forests more than one hundred years old become essentially useless for removing CO_2 from the air, as they claim such ancient and decrepit stands yearly lose as much CO_2 via respiration as they take in via photosynthesis.

Although demonstrably erroneous, with repeated telling the twisted tale actually begins to sound reasonable. After all, doesn't the metabolism of every living thing slow down as it gets older? We grudgingly admit that it does—even with trees—but some trees live a remarkably long time. In Panama (Condit *et al.*, 1995), Brazil (Chambers *et al.*, 1998; Laurance *et al.*, 2004; Chambers *et al.*, 2001), and many parts of the southwestern United States (Graybill and Idso, 1993), for example, individuals of a number of different species have been shown to live for nearly one and a half millennia. At a hundred years of age, these super-slurpers of CO_2 are mere youngsters. And in their really old age, their appetite for the vital gas, though diminished, is not lost. In fact, Chambers *et al.* (1998) indicate that the long-lived trees of Brazil continue to experience protracted slow growth even at 1,400 years of age. And protracted slow growth (evident in yearly increasing trunk diameters) of very old and *large* trees can absorb a huge amount of CO_2 out of the air each year, especially when, as noted by Chanbers *et al.* (1998) with respect to the Brazilian forests in the central Amazon, about 50 percent of their above-ground biomass is contained in less than the largest 10 percent of their trees. Consequently, since the life span of these massive long-lived trees is considerably greater than the projected life span of the entire "Age of Fossil Fuels," their cultivation and preservation represents an essentially permanent, though only partial, solution to the perceived problem of the anthropogenic global warming.

Another important fact about forests and their ability to sequester carbon over long periods of time is that the forest itself is the unit of primary importance when it comes to determining the amount of carbon that can be sequestered on a unit area of land. Cary *et al.* (2001) note most models of forest carbon sequestration wrongly assume that "age-related growth trends of individual trees and even-aged, monospecific stands can be extended to natural forests." When they compared the predictions of such models against real-world data they gathered from northern Rocky Mountain subalpine forests that ranged in age from 67 to 458 years, for example, they found that above-ground net primary productivity in 200-year-old natural stands was almost twice as great as that of modeled stands, and that the difference between the two increased linearly throughout the entire sampled age range.

So what's the explanation for the huge discrepancy? Cary *et al.* suggest that long-term recruitment and the periodic appearance of additional late-successional species (increasing biodiversity) may have significant effects on stand productivity, infusing the primary unit of concern, i.e., the ever-evolving forest super-organism, with greater vitality than would have been projected on the basis of characteristics possessed by the unit earlier in its life. They also note that by not including effects of size- or age-dependent decreases in stem and branch respiration per unit of sapwood volume in models of forest growth, respiration in older stands can be over-estimated by a factor of two to five.

How serious are these model shortcomings? For the real-world forests studied by Cary *et al.*, they produce predictions of carbon sequestration that are only a little over half as large as what is observed in nature for 200-year-old forests, while for 400-year-old forests they produce results that are only about a third as large as what is characteristic of the real world. And as the forests grow older still, the difference between reality and model projections grows with them.

Another study relevant to the ability of forests to act as long-term carbon sinks was conducted by Luo *et al.* (2003), who analyzed data obtained from the Duke Forest FACE experiment, in which three 30-meter-diameter plots within a 13-year-old forest (composed primarily of loblolly pines with sweetgum and yellow poplar trees as sub-dominants, together with numerous other trees, shrubs, and vines that occupy still smaller niches) began to be enriched with an extra 200 ppm of CO_2 in August 1996, while three similar plots were maintained at the ambient atmospheric CO_2 concentration. A number of papers describing different facets of this still-ongoing long-term study have been published; and as recounted by Luo *et al.*, they have revealed the existence of a CO_2-induced "sustained photosynthetic stimulation at leaf and canopy levels [Myers *et al.*, 1999; Ellsworth, 2000; Luo *et al.*, 2001; Lai *et al.*, 2002], which resulted in sustained stimulation of wood biomass increment [Hamilton *et al.*, 2002] and a larger carbon accumulation in the forest floor at elevated CO_2 than at ambient CO_2 [Schlesinger and Lichter, 2001]."

Based upon these findings and what they imply about rates of carbon removal from the atmosphere and its different residence times in plant, litter, and soil carbon pools, Luo *et al.* developed a model for studying the sustainability of forest carbon sequestration. Applying this model to a situation where the atmospheric CO_2 concentration gradually rises from a value of 378 ppm in 2000 to a value of 710 ppm in 2100, they calculated that the carbon sequestration rate of the Duke Forest would rise from an initial value of 69 g m^{-2} yr^{-1} to a final value of 201 g m^{-2} yr^{-1}, which is a far cry from the scenario promulgated by those who claim earth's forests will have released much of the carbon they had previously absorbed as early as the year 2050 (Pearce, 1999).

Another study that supports the long-term viability of carbon sequestration by forests was conducted by Paw U *et al.* (2004), who also note that old-growth forests have generally been considered to "represent carbon sources or are neutral (Odum, 1963, 1965)," stating that "it is generally assumed that forests reach maximum productivity at an intermediate age and productivity declines in mature and old-growth stands (Franklin, 1988), presumably as dead woody debris and other respiratory demands increase." More particularly, they report that a number of articles have suggested that "old-growth conifer forests are at equilibrium with respect to net ecosystem productivity or net ecosystem exchange (DeBell and Franklin, 1987; Franklin and DeBell, 1988; Schulze *et al.*, 1999), as an age-class end point of ecosystem development."

To see if these claims had any merit, Paw U *et al.* used an eddy covariance technique to estimate the CO_2 exchange rate of the oldest forest ecosystem (500 years old) in the AmeriFlux network of carbon-flux measurement stations—the Wind River old-growth forest in southwestern Washington, USA, which is composed mainly of Douglas fir and western hemlock—over a period of 16 months, from May 1998 to August 1999. Throughout this period, the 14 scientists report "there were no monthly averages with net release of CO_2," and that the cumulative net ecosystem exchange showed "remarkable sequestration of carbon, comparable to many younger forests." They concluded that "in contrast to frequently stated opinions, old-growth forests can be significant carbon sinks," noting that "the old-growth forests of the Pacific Northwest can contribute to

optimizing carbon sequestration strategies while continuing to provide ecosystem services essential to supporting biodiversity."

Binkley *et al.* (2004) revisited an aging aspen forest in the Tesuque watershed of northern New Mexico, USA—which between 1971 and 1976 (when it was between 90 and 96 years old) was thought to have had a *negative* net ecosystem production rate of -2.0 Mg ha^{-1} yr^{-1}—and measured the basal diameters of all trees in the central 0.01 ha of each of 27 plots arrayed across the watershed, after which they used the same regression equations employed in the earlier study to calculate live tree biomass as of 2003.

"Contrary to expectation," as they describe it, Binkley *et al.* report that "live tree mass in 2003 [186 Mg ha^{-1}] was significantly greater than in 1976 [149 Mg ha^{-1}] (P = 0.02), refuting the hypothesis that live tree mass declined." In fact, they found that the annual net increment of live tree mass was about 1.37 Mg ha^{-1} yr^{-1} from age 96 to age 123 years, which is only 12 percent less than the mean annual increment of live tree mass experienced over the forest's initial 96 years of existence (149 Mg ha^{-1} / 96 yr = 1.55 Mg ha^{-1} yr^{-1}). Consequently, in response to the question they posed when embarking on their study—"Do old forests gain or lose carbon?"—Binkley *et al.* concluded that "old aspen forests continue to accrue live stem mass well into their second century, despite declining current annual increments," which, we might add, are not all that much smaller than those the forests exhibited in their younger years.

Similar results have been obtained by Hollinger *et al.* (1994) for a 300-year-old *Nothofagus* site in New Zealand, by Law *et al.* (2001) for a 250-year-old ponderosa pine site in the northwestern United States, by Falk *et al.* (2002) for a 450-year-old Douglas fir/western hemlock site in the same general area, and by Knohl *et al.* (2003) for a 250-year-old deciduous forest in Germany. In commenting on these findings, the latter investigators say they found "unexpectedly high carbon uptake rates during 2 years for an unmanaged 'advanced' beech forest, which is in contrast to the widely spread hypothesis that 'advanced' forests are insignificant as carbon sinks." For the forest they studied, as they describe it, "assimilation is clearly not balanced by respiration, although this site shows typical characteristics of an 'advanced' forest at a comparatively late stage of development."

These observations about forests are remarkably similar to recent findings regarding humans; i.e., that nongenetic interventions, even late in life, can put one on a healthier trajectory that extends productive lifespan. So what is the global "intervention" that has put the planet's trees on the healthier trajectory of being able to sequester significant amounts of carbon in their old age, when past theory (which was obviously based on past observations) decreed they should be in a state of no-net-growth or even negative growth? The answer is probably CO_2 enhancement.

For any tree of age 250 years or more, the greater portion of its life (at least two-thirds of it) has been spent in an atmosphere of reduced CO_2 content. Up until 1920, for example, the air's CO_2 concentration had never been above 300 ppm throughout the lives of such trees, whereas it is currently 375 ppm or 25 percent higher. And for older trees, even greater portions of their lives have been spent in air of even lower CO_2 concentration. Hence, the "intervention" that has given new life to old trees would appear to be the aerial fertilization effect produced by the CO_2 that resulted from the Industrial Revolution and is being maintained by its ever-expanding aftermath (Idso, 1995).

Based on these many observations, as well as the results of the study of Greenep *et al.* (2003), which strongly suggest, in their words, that "the capacity for enhanced photosynthesis in trees growing in elevated CO_2 is unlikely to be lost in subsequent generations," it would appear that earth's forests will remain strong sinks for atmospheric carbon well into the distant future. A wealth of scientific data confirms the reality of the ever-increasing productivity of earth's older forests, especially those of Amazonia, concomitant with the rise in the air's CO_2 content. An even greater wealth of laboratory and field data demonstrates that rising forest productivity is what one would expect to observe in response to the stimulus provided by the ongoing rise in the atmosphere's CO_2 concentration.

Additional information on this topic, including reviews of newer publications as they become available, can be found at http://www.co2science.org/subject/f/forestold.php.

References

Baker, T.R., Phillips, O.L., Malhi, Y., Almeida, S., Arroyo, L., Di Fiore, A., Erwin, T., Higuchi, N., Killeen, T.J., Laurance, S.G., Laurance, W.F., Lewis, S.L., Monteagudo, A., Neill, D.A., Núñez Vargas, P., Pitman, N.C.A., Silva, J.N.M. and Vásquez Martínez, R. 2004. Increasing biomass in Amazonian forest plots. *Philosophical Transactions of the Royal Society of London Series B - Biological Sciences* **359**: 353-365.

Binkley, D., White, C.S. and Gosz, J.R. 2004. Tree biomass and net increment in an old aspen forest in New Mexico. *Forest Ecology and Management* **203**: 407-410.

Carey, E.V., Sala, A., Keane, R. and Callaway, R.M. 2001. Are old forests underestimated as global carbon sinks? *Global Change Biology* **7**: 339-344.

Chambers, J.Q., Higuchi, N. and Schimel, J.P. 1998. Ancient trees in Amazonia. *Nature* **391**: 135-136.

Chambers, J.Q., Van Eldik, T., Southon, J., Higuchi, N. 2001. Tree age structure in tropical forests of central Amazonia. In: Bierregaard, R.O., Gascon, C., Lovejoy, T., and Mesquita, R. (Eds.). *Lessons from Amazonia: Ecology and Conservation of a Fragmented Forest*. Yale University Press, New Haven, CT, USA, pp. 68-78.

Condit, R., Hubbell, S.P. and Foster, R.B. 1995. Mortality-rates of 205 neotropical tree and shrub species and the impact of a severe drought. *Ecological Monographs* **65**: 419-439.

DeBell, D.S. and Franklin, J.S. 1987. Old-growth Douglas-fir and western hemlock: a 36-year record of growth and mortality. *Western Journal of Applied Forestry* **2**: 111-114.

Ellsworth, D.S. 2000. Seasonal CO_2 assimilation and stomatal limitations in a *Pinus taeda* canopy with varying climate. *Tree Physiology* **20**: 435-444.

Falk, M., Paw, U.K.T., Schroeder, M. 2002. Interannual variability of carbon and energy fluxes for an old-growth rainforest. In: *Proceedings of the 25th Conference on Agricultural and Forest Meteorology*. American Meteorological Society, Boston, Massachusetts, USA.

Franklin, J.F. 1988. Pacific Northwest Forests. In: Barbour, M.G. and Billings, W.D. (Eds.) *North American Terrestrial Vegetation*. Cambridge University Press, New York, New York, USA, pp. 104-131.

Franklin, J.F. and DeBell, D.S. 1988. Thirty-six years of tree population change in an old-growth Pseudotsuga-Tsuga forest. *Canadian Journal of Forest Research* **18**: 633-639.

Graybill, D.A. and Idso, S.B. 1993. Detecting the aerial fertilization effect of atmospheric CO_2 enrichment in tree-ring chronologies. *Global Biogeochemical Cycles* **7**: 81-95.

Greenep, H., Turnbull, M.H. and Whitehead, D. 2003. Response of photosynthesis in second-generation *Pinus radiata* trees to long-term exposure to elevated carbon dioxide partial pressure. *Tree Physiology* **23**: 569-576.

Hamilton, J.G., DeLucia, E.H., George, K., Naidu, S.L., Finzi, A.C. and Schlesinger, W.H. 2002. Forest carbon balance under elevated CO_2. *Oecologia* **10.1007**/s00442-002-0884-x.

Hollinger, D.Y., Kelliher, F.M., Byers, J.N., Hunt, J.E., McSeveny, T.M. and Weir, P.L. 1994. Carbon dioxide exchange between an undisturbed old-growth temperate forest and the atmosphere. *Ecology* 75: 143-150.

Idso, S.B. 1995. *CO_2 and the Biosphere: The Incredible Legacy of the Industrial Revolution*. Department of Soil, Water and Climate, University of Minnesota, St. Paul, Minnesota, USA.

Knohl, A., Schulze, E.-D., Kolle, O. and Buchmann, N. 2003. Large carbon uptake by an unmanaged 250-year-old deciduous forest in Central Germany. *Agricultural and Forest Meteorology* 118: 151-167.

Lai, C.T., Katul, G., Butnor, J., Ellsworth, D. and Oren, R. 2002. Modeling nighttime ecosystem respiration by a constrained source optimization method. *Global Change Biology* 8: 124-141.

Laurance, W.F., Nascimento, H.E.M., Laurance, S.G., Condit, R., D'Angelo, S. and Andrade, A. 2004. Inferred longevity of Amazonian rainforest trees based on a long-term demographic study. *Forest Ecology and Management* 190: 131-143.

Laurance, W.F., Oliveira, A.A., Laurance, S.G., Condit, R., Dick, C.W., Andrade, A., Nascimento, H.E.M., Lovejoy, T.E. and Ribeiro, J.E.L.S. 2005. Altered tree communities in undisturbed Amazonian forests: A consequence of global change? *Biotropica* 37: 160-162.

Laurance, W.F., Oliveira, A.A., Laurance, S.G., Condit, R., Nascimento, H.E.M., Sanchez-Thorin, A.C., Lovejoy, T.E., Andrade, A., D'Angelo, S. and Dick, C. 2004a. Pervasive alteration of tree communities in undisturbed Amazonian forests. *Nature* 428: 171-175.

Law, B.E., Goldstein, A.H., Anthoni, P.M., Unsworth, M.H., Panek, J.A., Bauer, M.R., Fracheboud, J.M. and Hultman, N. 2001. Carbon dioxide and water vapor exchange by young and old ponderosa pine ecosystems during a dry summer. *Tree Physiology* 21: 299-308.

Lewis, S.L. 2006. Tropical forests and the changing earth system. *Philosophical Transactions of the Royal Society Series B - Biological Science* 361: 195-210.

Lewis, S.L., Malhi, Y. and Phillips, O.L. 2004a. Fingerprinting the impacts of global change on tropical forests. *Philosophical Transactions of the Royal Society of London Series B - Biological Sciences* 359: 437-462.

Lewis, S.L., Phillips, O.L., Baker, T.R., Lloyd, J., Malhi, Y., Almeida, S., Higuchi, N., Laurance, W.F., Neill, D.A., Silva, J.N.M., Terborgh, J., Lezama, A.T., Vásquez Martinez, R., Brown, S., Chave, J., Kuebler, C., Núñez Vargas, P. and Vinceti, B. 2004b. Concerted changes in tropical forest structure and dynamics: evidence from 50 South American long-term plots. *Philosophical*

Transactions of the Royal Society of London Series B - Biological Sciences 359: 421-436.

Luo, Y., Medlyn, B., Hui, D., Ellsworth, D., Reynolds, J. and Katul, G. 2001. Gross primary productivity in the Duke Forest: Modeling synthesis of the free-air CO_2 enrichment experiment and eddy-covariance measurements. *Ecological Applications* 11: 239-252.

Luo, Y., White, L.W., Canadell, J.G., DeLucia, E.H., Ellsworth, D.S., Finzi, A., Lichter, J. and Schlesinger, W.H. 2003. Sustainability of terrestrial carbon sequestration: A case study in Duke Forest with inversion approach. *Global Biogeochemical Cycles* 17: 10.1029/2002GB001923.

Malhi, Y. and Phillips, O.L. 2004. Tropical forests and global atmospheric change: a synthesis. *Philosophical Transactions of the Royal Society of London Series B - Biological Sciences* 359: 549-555.

Myers, D.A., Thomas, R.B. and DeLucia, E.H. 1999. Photosynthetic capacity of loblolly pine (*Pinus taeda* L.) trees during the first year of carbon dioxide enrichment in a forest ecosystem. *Plant, Cell and Environment* 22: 473-481.

Nelson, B.W. 2005. Pervasive alteration of tree communities in undisturbed Amazonian forests. *Biotropica* 37: 158-159.

Odum, E.P. 1963. *Ecology*. Holt, Rinehart and Winston, New York, New York, USA.

Odum E.P. 1965. *Fundamentals of Ecology*. Saunders, Philadelphia, Pennsylvania, USA.

Paw U, K.T., Falk, M., Suchanek, T.H., Ustin, S.L., Chen, J., Park, Y.-S., Winner, W.E., Thomas, S.C., Hsiao, T.C., Shaw, R.H., King, T.S., Pyles, R.D., Schroeder, M. and Matista, A.A. 2004. Carbon dioxide exchange between an old-growth forest and the atmosphere. *Ecosystems* 7: 513-524.

Pearce, F. 1999. That sinking feeling. *New Scientist* 164 (2209): 20-21.

Phillips, O.L., Baker, T.R., Arroyo, L., Higuchi, N., Killeen, T.J., Laurance, W.F., Lewis, S.L., Lloyd, J., Malhi, Y., Monteagudo, A., Neill, D.A., Núñez Vargas, P., Silva, J.N.M., Terborgh, J., Vásquez Martínez, R., Alexiades, M., Almeida, S., Brown, S., Chave, J., Comiskey, J.A., Czimczik, C.I., Di Fiore, A., Erwin, T., Kuebler, C., Laurance, S.G., Nascimento, H.E.M., Olivier, J., Palacios, W., Patiño, S., Pitman, N.C.A., Quesada, C.A., Saldias, M., Torres Lezama, A., B. and Vinceti, B. 2004. Pattern and process in Amazon tree turnover: 1976-2001. *Philosophical Transactions of the Royal Society of London Series B - Biological Sciences* 359: 381-407.

Phillips, O.L. and Gentry, A.H. 1994. Increasing turnover through time in tropical forests. *Science* **263**: 954-958.

Phillips, O.L., Malhi, Y., Higuchi, N., Laurance, W.F., Nunez, P.V., Vasquez, R.M., Laurance, S.G., Ferreira, L.V., Stern, M., Brown, S. and Grace, J. 1998. Changes in the carbon balance of tropical forests: Evidence from long-term plots. *Science* **282**: 439-442.

Pimm, S.L. and Sugden, A.M. 1994. Tropical diversity and global change. *Science* **263**: 933-934.

Schlesinger, W.H. and Lichter, J. 2001. Limited carbon storage in soil and litter of experimental forest plots under increased atmospheric CO_2. *Nature* **411**: 466-469.

Schulze, E.-D., Lloyd, J., Kelliher, F.M., Wirth, C., Rebmann, C., Luhker, B., Mund, M., Knohl, A., Milyuokova, I.M. and Schulze, W. 1999. Productivity of forests in the Eurosiberian boreal region and their potential to act as a carbon sink: a synthesis. *Global Change Biology* **5**: 703-722.

Weaver, P.L. and Murphy, P.G. 1990. Forest structure and productivity in Puerto Rico's Luquillo Mountains. *Biotropica* **22**: 69-82.

7.7.2. Decomposition

What is the fate of the extra carbon that is stored within plant tissues as a consequence of atmospheric CO_2 enrichment? Is it rapidly returned to the atmosphere following tissue senescence and decomposition? Or is it locked away for long periods of time? Experiments and real-world observations reveal that atmospheric CO_2 enrichment typically reduces, or has no effect upon, decomposition rates of senesced plant material.

7.7.2.1. Processes and Properties

Atmospheric CO_2 enrichment stimulates photosynthesis and growth in nearly all plants, typically producing more non-structural carbohydrates, which can be used to manufacture more carbon-based secondary compounds (CBSCs) or phenolics. This observation is important because phenolics tend to inhibit the decomposition of the organic matter in which they are found (Freeman *et al.*, 2001). If elevated levels of atmospheric CO_2 lead to the production of more of these decay-resistant substances, one would expect the ongoing rise in the air's CO_2 content to lead to the enhanced

sequestration of plant-litter-derived carbon in the world's soils, producing a negative feedback phenomenon that would tend to slow the rate of rise of the air's CO_2 content and thereby moderate CO_2-induced global warming.

For a long time, research on this matter was rather muddled. Many studies reported the expected increases in CBSC concentrations with experimentally created increases in the air's CO_2 content. Others, however, could find no significant plant phenolic content changes; a few even detected CO_2-induced decreases in CBSC concentrations. Penuelas *et al.* (1997) finally brought order to the issue when they identified the key role played by soil nitrogen.

In analyzing the results of several different studies, Penuelas *et al.* noticed that when soil nitrogen supply was less than adequate, some of the CBSC responses to a doubling of the air's CO_2 content were negative, i.e., a portion of the studies indicated that plant CBSC concentrations declined as the air's CO_2 content rose. When soil nutrient supply was more than adequate, however, the responses were almost all positive, with plant CBSC concentrations rising in response to a doubling of the air's CO_2 concentration. In addition, when the CO_2 content of the air was tripled, *all* CBSC responses, under both low and high soil nitrogen conditions, were positive.

The solution to the puzzle was thus fairly simple. With a tripling of the air's CO_2 content, nearly all plants exhibited increases in CBSC production; but with only a doubling of the atmospheric CO_2 concentration, adequate nitrogen is needed to ensure a positive CBSC response.

What makes these observations exciting is that atmospheric CO_2 enrichment, in addition to enhancing plant growth, typically stimulates nitrogen fixation in both woody (Olesniewicz and Thomas, 1999) and non-woody (Niklaus *et al.*, 1998; Dakora and Drake, 2000) legumes. As the air's CO_2 content continues to rise, earth's nitrogen-fixing plants should become ever more proficient in this important enterprise. In addition, some of the extra nitrogen thus introduced into earth's ecosystems will likely be shared with non-nitrogen-fixing plants. Also, since the microorganisms responsible for nitrogen fixation are found in nearly all natural ecosystems (Gifford, 1992), and since atmospheric CO_2 enrichment can directly stimulate the nitrogen-fixing activities of these microbes (Lowe and Evans, 1962), it can be appreciated that the ongoing rise in the air's CO_2 content will likely provide more nitrogen for the production of more CBSCs in all of earth's plants. And with ever-increasing concentrations of decay-resistant materials being found throughout plant tissues, the plant-derived organic matter that is incorporated into soils should remain there for ever longer periods of time.

On the other hand, in a meta-analysis of the effects of atmospheric CO_2 enrichment on leaf-litter chemistry and decomposition rate that was based on a total of 67 experimental observations, Norby *et al.* (2001) found that elevated atmospheric CO_2 concentrations—mostly between 600 and 700 ppm—reduced leaf-litter nitrogen concentration by about 7 percent. But in experiments where plants were grown under as close to natural conditions as possible, such as in open-top chambers, free-air CO_2 enrichment (FACE) plots, or in the proximity of CO_2-emitting springs, there were no significant effects of elevated CO_2 on leaf-litter nitrogen content.

In addition, based on a total of 46 experimental observations, Norby *et al.* determined that elevated atmospheric CO_2 concentrations increased leaf-litter lignin concentrations by an average of 6.5 percent. However, these increases in lignin content occurred in woody but not in herbaceous species. And again, the lignin concentrations of leaf litter were not affected by elevated CO_2 when plants were grown in open-top chambers, FACE plots, or in the proximity of CO_2-emitting springs.

In an analysis of a total of 101 observations, Norby *et al.* found elevated CO_2 had no consistent effect on leaf-litter decomposition rate in any type of experimental setting. As the air's CO_2 content continues to rise, it will likely have little to no impact on leaf-litter chemistry and rates of leaf-litter decomposition. Since there will be more leaf litter produced in a high-CO_2 world of the future, however, that fact alone will ensure that more carbon is sequestered in the world's soils for longer periods of time.

Additional information on this topic, including reviews of newer publications as they become available, can be found at http://www.co2science.org/subject/d/decompprocesses.php.

References

Agren, G.I. and Bosatta, E. 2002. Reconciling differences in predictions of temperature response of soil organic matter. *Soil Biology & Biochemistry* **34**: 129-132.

Dakora, F.D. and Drake, B.G. 2000. Elevated CO_2 stimulates associative N_2 fixation in a C_3 plant of the Chesapeake Bay wetland. *Plant, Cell and Environment* **23**: 943-953.

Fitter, A.H., Self, G.K., Brown, T.K., Bogie, D.S., Graves, J.D., Benham, D. and Ineson, P. 1999. Root production and turnover in an upland grassland subjected to artificial soil warming respond to radiation flux and nutrients, not temperature. *Oecologia* **120**: 575-581.

Freeman, C., Ostle, N. and Kang, H. 2001. An enzymic 'latch' on a global carbon store. *Nature* **409**: 149.

Giardina, C.P. and Ryan, M.G. 2000. Evidence that decomposition rates of organic carbon in mineral soil do not vary with temperature. *Nature* **404**: 858-861.

Gifford, R.M. 1992. Interaction of carbon dioxide with growth-limiting environmental factors in vegetative productivity: Implications for the global carbon cycle. *Advances in Bioclimatology* **1**: 24-58.

Grisi, B., Grace, C., Brookes, P.C., Benedetti, A. and Dell'abate, M.T. 1998. Temperature effects on organic matter and microbial biomass dynamics in temperate and tropical soils. *Soil Biology & Biochemistry* **30**: 1309-1315.

Johnson, L.C., Shaver, G.R., Cades, D.H., Rastetter, E., Nadelhoffer, K., Giblin, A., Laundre, J. and Stanley, A. 2000. Plant carbon-nutrient interactions control CO_2 exchange in Alaskan wet sedge tundra ecosystems. *Ecology* **81**: 453-469.

Liski, J., Ilvesniemi, H., Makela, A. and Westman, C.J. 1999. CO_2 emissions from soil in response to climatic warming are overestimated—The decomposition of old soil organic matter is tolerant of temperature. *Ambio* **28**: 171-174.

Lowe, R.H. and Evans, H.J. 1962. Carbon dioxide requirement for growth of legume nodule bacteria. *Soil Science* **94**: 351-356.

Niklaus, P.A., Leadley, P.W., Stocklin, J. and Korner, C. 1998. Nutrient relations in calcareous grassland under elevated CO_2. *Oecologia* **116**: 67-75.

Norby, R.J., Cotrufo, M.F., Ineson, P., O'Neill, E.G. and Canadell, J.G. 2001. Elevated CO_2, litter chemistry, and decomposition: a synthesis. *Oecologia* **127**: 153-165.

Olesniewicz, K.S. and Thomas, R.B. 1999. Effects of mycorrhizal colonization on biomass production and nitrogen fixation of black locust (*Robinia pseudoacacia*) seedlings grown under elevated atmospheric carbon dioxide. *New Phytologist* **142**: 133-140.

Peñuelas, J., Estiarte, M. and Llusia, J. 1997. Carbon-based secondary compounds at elevated CO_2. *Photosynthetica* **33**: 313-316.

Thornley, J.H.M. and Cannell, M.G.R. 2001. Soil carbon storage response to temperature: an hypothesis. *Annals of Botany* **87**: 591-598.

7.7.2.2. Agricultural Crops

In the study by Booker *et al.* (2000), leaves from defoliated cotton plants grown at an atmospheric CO_2 concentration of 720 ppm displayed significantly greater amounts of starch and soluble sugars and significantly lower concentrations of nitrogen than the leaves of plants grown in ambient air. These changes in the quality of the leaf litter produced under high CO_2 likely affected its subsequent decomposition rate, which was 10 to 14 percent slower than that observed for leaf litter collected from plants grown in air of normal CO_2 concentration. Likewise, when crop residues from soybean and sorghum plants that were raised in twice-ambient CO_2 environments were mixed with soils to study their decomposition rates, Torbert *et al.* (1998) noted they lost significantly less carbon – up to 40 percent less – than similarly treated crop residues from ambiently grown crops.

In contrast to the aforementioned studies, neither Van Vuuren *et al.* (2000), for spring wheat, nor Henning *et al.* (1996), for soybean and sorghum, found any significant differences in the decomposition rates of the residues of crops grown under conditions of high or normal atmospheric CO_2 concentration.

As the air's CO_2 content continues to rise, therefore, and agricultural crops grow more robustly and return greater amounts of litter to the soil, it is likely that greater amounts of carbon will be sequestered in the soil in which they grew, as crop residue decomposition rates are significantly decreased or remain unchanged.

Additional information on this topic, including reviews of newer publications as they become available, can be found at http://www.co2science.org/subject/d/decompositionagri.php.

References

Booker, F.L., Shafer, S.R., Wei, C.-M. and Horton, S.J. 2000. Carbon dioxide enrichment and nitrogen fertilization effects on cotton (*Gossypium hirsutum* L.) plant residue chemistry and decomposition. *Plant and Soil* **220**: 89-98.

Henning, F.P., Wood, C.W., Rogers, H.H., Runion, G.B. and Prior, S.A. 1996. Composition and decomposition of soybean and sorghum tissues grown under elevated atmospheric carbon dioxide. *Journal of Environmental Quality* **25**: 822-827.

Torbert, H.A., Prior, S.A., Rogers, H.H. and Runion, G.B. 1998. Crop residue decomposition as affected by growth under elevated atmospheric CO_2. *Soil Science* **163**: 412-419.

Van Vuuren, M.M.I., Robinson, D., Scrimgeour, C.M., Raven, J.A. and Fitter, A.H. 2000. Decomposition of ^{13}C-labelled wheat root systems following growth at different CO_2 concentrations. *Soil Biology & Biochemistry* **32**: 403-413.

7.7.2.3. Grassland Species

In the study of Nitschelm *et al.* (1997), white clover exposed to an atmospheric CO_2 concentration of 600 ppm for one growing season channeled 50 percent more newly fixed carbon compounds into the soil than similar plants exposed to ambient air. In addition, the clover's roots decomposed at a rate that was 24 percent slower than that observed for roots of control plants, as has also been reported for white clover by David *et al.* (2001). These observations suggest that soil carbon sequestration under white clover ecosystems will be greatly enhanced as the air's CO_2 content continues to rise, as was also shown for moderately fertile sandstone grasslands (Hu *et al.*, 2001).

Similar results have been observed with mini-ecosystems comprised entirely of perennial ryegrass. Van Ginkel *et al.* (1996), for example, demonstrated that exposing this species to an atmospheric CO_2 concentration of 700 ppm for two months caused a 92 percent increase in root growth and 19 percent and 14 percent decreases in root decomposition rates one and two years, respectively, after incubating ground roots within soils. This work was later followed up by Van Ginkel and Gorissen (1998), who showed a 13 percent reduction in the decomposition rates of CO_2-enriched perennial ryegrass roots in both disturbed and undisturbed root profiles. This and other work led the authors to calculate that CO_2-induced reductions in the decomposition of perennial ryegrass litter, which enhances soil carbon sequestration, could well be large enough to remove over half of the anthropogenic CO_2 emissions that may be released in the next century (Van Ginkel *et al.*, 1999).

In some cases, atmospheric CO_2 enrichment has little or no significant effect on litter quality and subsequent rates of litter decomposition, as was the case in the study of Hirschel *et al.* (1997) for lowland calcareous and high alpine grassland species. Similar non-effects of elevated CO_2 on litter decomposition have also been reported in a California grassland (Dukes and Field, 2000).

In light of these experimental findings, it would appear that as the air's CO_2 concentration increases, litter decomposition rates of grassland species will likely decline, increasing the amount of carbon sequestered in grassland soils. Since this phenomenon is augmented by the aerial fertilization effect of atmospheric CO_2 enrichment, which leads to the production of greater amounts of litter, there is thus a double reason for expecting more carbon to be removed from the atmosphere by earth's grasslands in the future.

Additional information on this topic, including reviews of newer publications as they become available, can be found at http://www.co2science.org/subject/d/decompositiongrass.php.

References

David, J.-F., Malet, N., Couteaux, M.-M. and Roy, J. 2001. Feeding rates of the woodlouse *Armadillidium vulgare* on herb litters produced at two levels of atmospheric CO_2. *Oecologia* **127**: 343-349.

Dukes, J.S. and Field, C.B. 2000. Diverse mechanisms for CO_2 effects on grassland litter decomposition. *Global Change Biology* **6**: 145-154.

Hirschel, G., Korner, C. and Arnone III, J.A. 1997. Will rising atmospheric CO_2 affect leaf litter quality and in situ decomposition rates in native plant communities? *Oecologia* **110**: 387-392.

Hu, S., Chapin III, F.S., Firestone, M.K., Field, C.B. and Chiariello, N.R. 2001. Nitrogen limitation of microbial decomposition in a grassland under elevated CO_2. *Nature* **409**: 188-191.

Nitschelm, J.J., Luscher, A., Hartwig, U.A. and van Kessel, C. 1997. Using stable isotopes to determine soil carbon input differences under ambient and elevated atmospheric CO_2 conditions. *Global Change Biology* **3**: 411-416.

Van Ginkel, J.H. and Gorissen, A. 1998. In situ decomposition of grass roots as affected by elevated atmospheric carbon dioxide. *Soil Science Society of America Journal* **62**: 951-958.

Van Ginkel, J.H., Gorissen, A. and Polci, D. 2000. Elevated atmospheric carbon dioxide concentration: effects of increased carbon input in a *Lolium perenne* soil on microorganisms and decomposition. *Soil Biology & Biochemistry* **32**: 449-456.

Van Ginkel, J.H., Whitmore, A.P. and Gorissen, A. 1999. *Lolium perenne* grasslands may function as a sink for atmospheric carbon dioxide. *Journal of Environmental Quality* **28**: 1580-1584.

Van Ginkel, J.H., Gorissen, A. and van Veen, J.A. 1996. Long-term decomposition of grass roots as affected by elevated atmospheric carbon dioxide. *Journal of Environmental Quality* **25**: 1122-1128.

7.7.2.4. *Woody Plants*

The sequestering of carbon in the soils upon which woody plants grow has the potential to provide a powerful brake on the rate of rise of the air's CO_2 content if the plant litter that is incorporated into those soils does not decompose more rapidly in a CO_2-enriched atmosphere than it does in current ambient air. It is important to determine if this latter constraint is true or false. In this section we review this question with respect to litter produced by conifers and deciduous trees.

Scherzel *et al.* (1998) exposed seedlings of two eastern white pine genotypes to elevated concentrations of atmospheric CO_2 and O_3 in open-top chambers for four full growing seasons, finding no changes in the decomposition rates of the litter of either genotype to the concentration increases of either of these two gases. Likewise, Kainulainen *et al.* (2003) could find no evidence that the litter of 22-year-old Scots pine trees that had been exposed to elevated concentrations of CO_2 and O_3 for three full years decomposed any faster or slower than litter produced in ambient air. In addition, Finzi and Schlesinger (2002) found that the decomposition rate of litter from 13-year-old loblolly pine trees was unaffected by elevated CO_2 concentrations maintained for a period of two full years in a FACE study.

In light of these observations, plus the fact that Saxe *et al.* (1998) have determined that a doubling of the air's CO_2 content leads to more than a doubling of the biomass production of coniferous species, it logically follows that the ongoing rise in the atmosphere's CO_2 concentration is increasing carbon sequestration rates in the soils upon which conifers grow and producing a significant negative feedback phenomenon that slows the rate of rise of the air's CO_2 content.

What about deciduous trees? Scherzel *et al.* (1998) exposed seedlings of yellow poplar trees to elevated concentrations of atmospheric CO_2 and O_3 in open-top chambers for four full growing seasons, finding that rates of litter decomposition were similar for all treatments for the first five months of the study. Thereafter, however, litter produced in the elevated O_3 and CO_2 treatment decomposed at a significantly slower rate, such that after two years had passed, the litter from the elevated O_3 and CO_2 treatment contained approximately 12 percent more biomass than the litter from any other treatment.

Cotrufo *et al.* (1998) grew two-year-old ash and sycamore seedlings for one growing season in closed-top chambers maintained at atmospheric CO_2 concentrations of 350 and 600 ppm. The high-CO_2 air increased lignin contents in the litter produced from both tree species, which likely contributed to the decreased litter decomposition rates observed in the CO_2-enriched chambers. After one year of incubation, for example, litter bags from the CO_2-enriched trees of both species had about 30 percent more dry mass remaining in them than litter bags from the ambient trees. In addition, woodlouse arthropods consumed 16 percent *less* biomass when fed litter generated from seedlings grown at 600 ppm CO_2 than when fed litter generated from seedlings grown in ambient air.

De Angelis *et al.* (2000) constructed large open-top chambers around 30-year-old mixed stands of naturally growing Mediterranean forest species (dominated by *Quercus ilex*, *Phillyrea augustifolia*, and *Pistacia lentiscus*) near the coast of central Italy. Half of the chambers were exposed to ambient air of 350 ppm CO_2, while half were exposed to air of 710 ppm CO_2; and after three years, the lignin and carbon concentrations of the leaf litter of all three species were increased by 18 and 4 percent, respectively, while their nitrogen concentrations were reduced by 13 percent. These changes resulted in a 20 percent CO_2-induced increase in the carbon-to-nitrogen ratio of the leaf litter, which parameter is commonly used to predict decomposition rates, where larger ratios are generally associated with less rapid decomposition than smaller ratios. This case was no exception, with 4 percent less decomposition occurring in the leaf litter gathered from beneath the CO_2-enriched trees than in the litter collected from beneath the trees growing in ambient air.

Cotrufo and Ineson (2000) grew beech seedlings for five years in open-top chambers fumigated with air containing either 350 or 700 ppm CO_2. Subsequently, woody twigs from each CO_2 treatment were collected and incubated in native forest soils for 42 months, after which they determined there was no significant effect of the differential CO_2 exposure during growth on subsequent woody twig decomposition, although the mean decomposition rate of the CO_2-enriched twigs was 5 percent less than that of the ambient-treatment twigs.

Conway *et al.* (2000) grew two-year-old ash tree seedlings in solardomes maintained at atmospheric CO_2 concentrations of 350 and 600 ppm, after which naturally senesced leaves were collected, inoculated with various fungal species, and incubated for 42 days. They found the elevated CO_2 significantly reduced the amount of nitrogen in the senesced leaves, thus giving the CO_2-enriched leaf litter a higher carbon-to-nitrogen ratio than the litter collected from the seedlings growing in ambient air. This change likely contributed to the observed reductions in the amount of fungal colonization present on the senesced leaves from the CO_2-enriched treatment, which would be expected to result in reduced rates of leaf decomposition.

King *et al.* (2001) grew aspen seedlings for five months in open-top chambers receiving atmospheric CO_2 concentrations of 350 and 700 ppm. At the end of this period, naturally senesced leaf litter was collected, analyzed, and allowed to decompose under ambient conditions for 111 days. Although the elevated CO_2 slightly lowered leaf litter nitrogen content, it had no effect on litter sugar, starch, or tannin concentrations. With little to no CO_2-induced effects on leaf litter quality, there was no CO_2-induced effect on litter decomposition.

Dilustro *et al.* (2001) erected open-top chambers around portions of a regenerating oak-palmetto scrub ecosystem in Florida, USA and maintained them at CO_2 concentrations of either 350 or 700 ppm, after which they incubated ambient- and elevated-CO_2-produced fine roots for 2.2 years in the chamber soils, which were nutrient-poor and often water-stressed. They found the elevated CO_2 did not significantly affect the decomposition rates of the fine roots originating from either the ambient or CO_2-enriched environments.

Of these seven studies of deciduous tree species, five are suggestive of slight reductions in litter decomposition rates under CO_2-enriched growth conditions, while two show no effect. With deciduous trees exhibiting large growth enhancements in response to atmospheric CO_2 enrichment, we can expect to see large increases in the amounts of carbon they sequester in the soils on which they grow as the air's CO_2 content continues to rise. And this phenomenon should slow the rate of rise of the atmosphere's CO_2 concentration and thereby reduce the impetus for CO_2-induced global warming.

To summarize, scientific theory and empirical research show that the ongoing rise in the air's CO_2 content will not materially alter the rate of decomposition of the world's soil organic matter. This means the rate at which carbon is sequestered in the world's soils should continue to increase, a joint function of the rate at which the productivity of earth's plants is increased by the aerial fertilization effect of the rising atmospheric CO_2 concentration and the rate of expansion of the planet's vegetation into drier regions of the globe that is made possible by the concomitant CO_2-induced increase in vegetative water use efficiency.

Additional information on this topic, including reviews of newer publications as they become available, can be found at http://www.co2science.org/subject/d/decompconifers.php and http://www.co2science.org/subject/d/decompdeciduous.php.

References

Conway, D.R., Frankland, J.C., Saunders, V.A. and Wilson, D.R. 2000. Effects of elevated atmospheric CO_2 on fungal competition and decomposition of *Fraxinus excelsior* litter in laboratory microcosms. *Mycology Research* **104**: 187-197.

Cotrufo, M.F. and Ineson, P. 2000. Does elevated atmospheric CO_2 concentration affect wood decomposition? *Plant and Soil* **224**: 51-57.

Cotrufo, M.F., Briones, M.J.I. and Ineson, P. 1998. Elevated CO_2 affects field decomposition rate and palatability of tree leaf litter: importance of changes in substrate quality. *Soil Biology and Biochemistry* **30**: 1565-1571.

De Angelis, P., Chigwerewe, K.S. and Mugnozza, G.E.S. 2000. Litter quality and decomposition in a CO_2-enriched Mediterranean forest ecosystem. *Plant and Soil* **224**: 31-41.

Dilustro, J.J., Day, F.P. and Drake, B.G. 2001. Effects of elevated atmospheric CO_2 on root decomposition in a scrub oak ecosystem. *Global Change Biology* **7**: 581-589.

Finzi, A.C. and Schlesinger, W.H. 2002. Species control variation in litter decomposition in a pine forest exposed to elevated CO_2. *Global Change Biology* **8**: 1217-1229.

Kainulainen, P., Holopainen, T. and Holopainen, J.K. 2003. Decomposition of secondary compounds from needle litter of Scots pine grown under elevated CO_2 and O_3. *Global Change Biology* **9**: 295-304.

King, J.S., Pregitzer, K.S., Zak, D.R., Kubiske, M.E., Ashby, J.A. and Holmes, W.E. 2001. Chemistry and decomposition of litter from *Populus tremuloides* Michaux grown at elevated atmospheric CO_2 and varying N availability. *Global Change Biology* **7**: 65-74.

Saxe, H., Ellsworth, D.S. and Heath, J. 1998. Tree and forest functioning in an enriched CO_2 atmosphere. *New Phytologist* **139**: 395-436.

Scherzel, A.J., Rebbeck, J. and Boerner, R.E.J. 1998. Foliar nitrogen dynamics and decomposition of yellow-poplar and eastern white pine during four seasons of exposure to elevated ozone and carbon dioxide. *Forest Ecology and Management* **109**: 355-366.

7.7.3. Temperature and Carbon Sequestration

7.7.3.1. General

It must noted, as stated by Agren and Bosatta (2002), that "global warming has long been assumed to lead to an increase in soil respiration and, hence, decreasing soil carbon stores." Indeed, this dictum was accepted as gospel for many years, for a number of laboratory experiments seemed to suggest that nature would not allow more carbon to be sequestered in the soils of a warming world. As one non-laboratory experiment after another has recently demonstrated, however, such is not the case, and theory has been forced to change to accommodate reality.

The old-school view of things began to unravel in 1999 when two studies presented evidence refuting the long-standing orthodoxy. Abandoning the laboratory for the world of nature, Fitter *et al.* (1999) heated natural grass ecosystems by 3°C and found that the temperature increase had "no direct effect on the soil carbon store." That same year, Liski *et al.* (1999) showed that carbon storage in the soils of both high- and low-productivity boreal forests in Finland actually increased with rising temperatures along a natural temperature gradient.

The following year, Johnson *et al.* (2000) warmed natural Arctic tundra ecosystems by nearly 6°C for eight full years and found no significant effect on ecosystem respiration. Likewise, Giardina and Ryan (2000) analyzed organic carbon decomposition data derived from the forest soils of 82 different sites on five continents, reporting that "despite a 20°C gradient in mean annual temperature, soil carbon mass loss ... was insensitive to temperature."

Thornley and Cannell (2001) ventured forth with what they called "a hypothesis" concerning the matter. Specifically, they proposed the idea that warming may increase the rate of certain physico-chemical processes that transfer organic carbon from less-stable to more-stable soil organic matter pools, thereby enabling the better-protected organic matter to avoid, or more strongly resist, decomposition. Then, they developed a dynamic soil model in which they demonstrated that if their thinking were correct, long-term soil carbon storage would appear to be insensitive to a rise in temperature, even if the respiration rates of all soil carbon pools rose in response to warming, as they indeed do.

Agren and Bosatta's 2002 paper is an independent parallel development of much the same concept, although they describe the core idea in somewhat different terms, and they upgrade the concept from what Thornley and Cannell call a "hypothesis" to what they refer to as the "continuous-quality theory." *Quality*, in this context, refers to the degradability of soil organic matter; and *continuous quality* suggests there is a wide-ranging continuous spectrum of soil organic carbon "mini-pools" that possess differing degrees of resistance to decomposition.

The continuous quality theory states that soils from naturally higher temperature regimes will contain relatively more organic matter in carbon pools that are more resistant to degradation and are consequently characterized by lower rates of decomposition, which has been observed experimentally to be the case by Grisi *et al.* (1998). In addition, it states that this shift in the distribution of soil organic matter qualities—i.e., the higher-temperature-induced creation of more of the more-difficult-to-decompose organic matter—will counteract the decomposition-promoting influence of the higher temperatures, so that the overall decomposition rate of the totality of organic matter in a higher-temperature soil is either unaffected or reduced.

Rising CO_2 levels tend to maintain (Henning *et al.*, 1996) or decrease (Torbert *et al.*, 1998; Nitschelm *et al.*, 1997) CO_2 fluxes from agricultural soils. Consequently, these phenomena tend to increase the carbon contents of most soils in CO_2-enriched atmospheres. However, it is sometimes suggested that rising air temperatures, which can accelerate the breakdown of soil organic matter and increase biological respiration rates, could negate this CO_2-induced enhancement of carbon sequestration,

possibly leading to an even greater release of carbon back to the atmosphere.

Casella and Soussana (1997) grew perennial ryegrass in controlled environments receiving ambient and elevated (700 ppm) atmospheric CO_2 concentrations, two levels of soil nitrogen, and ambient and elevated (+3°C) air temperatures for a period of two years, finding that "a relatively large part of the additional photosynthetic carbon is stored below-ground during the two first growing seasons after exposure to elevated CO_2, thereby increasing significantly the below-ground carbon pool." At the low and high levels of soil nitrogen supply, for example, the elevated CO_2 increased soil carbon storage by 32 and 96 percent, respectively, "with no significant increased temperature effect." The authors thus concluded that in spite of predicted increases in temperature, "this stimulation of the below-ground carbon sequestration in temperate grassland soils could exert a negative feed-back on the current rise of the atmospheric CO_2 concentration."

Much the same conclusion was reached by Van Ginkel *et al.* (1999). After reviewing prior experimental work that established the growth and decomposition responses of perennial ryegrass to both atmospheric CO_2 enrichment and increased temperature, these researchers concluded that, at both low and high soil nitrogen contents, CO_2-induced increases in plant growth and CO_2-induced decreases in plant decomposition rate are more than sufficient to counteract any enhanced soil respiration rate that might be caused by an increase in air temperature. In addition, after reconstructing carbon storage in the terrestrial vegetation of Northern Eurasia as far back as 125,000 years ago, Velichko *et al.* (1999) determined that plants in this part of the world were more productive and efficient in sequestering carbon at higher temperatures than they were at lower temperatures. Similarly, Allen *et al.*, (1999) used sediment cores from a lake in southern Italy and from the Mediterranean Sea to conclude that, over the past 102,000 years, warmer climates have been better for vegetative productivity and carbon sequestration than have cooler climates.

In conclusion, research conducted to date strongly suggests that the CO_2-induced enhancement of vegetative carbon sequestration will not be reduced by any future rise in air temperature, regardless of its cause. The rest of this section looks more closely at research regarding forests and peatlands.

Additional information on this topic, including reviews of newer publications as they become available, can be found at http://www.co2science.org/subject/c/carbonco2xtemp.php.

References

Allen, J.R.M., Brandt, U., Brauer, A., Hubberten, H.-W., Huntley, B., Keller, J., Kraml, M., Mackensen, A., Mingram, J., Negendank, J.F.W., Nowaczyk, N.R., Oberhansli, H., Watts, W.A., Wulf, S. and Zolitschka, B. 1999. Rapid environmental changes in southern Europe during the last glacial period. *Nature* **400**: 740-743.

Casella, E. and Soussana, J-F. 1997. Long-term effects of CO_2 enrichment and temperature increase on the carbon balance of a temperate grass sward. *Journal of Experimental Botany* **48**: 1309-1321.

Fan, S., Gloor, M., Mahlman, J., Pacala, S., Sarmiento, J., Takahashi, T. and Tans, P. 1998. A large terrestrial carbon sink in North America implied by atmospheric and oceanic carbon dioxide data and models. *Science* **282**: 442-446.

Henning, F.P., Wood, C.W., Rogers, H.H., Runion, G.B. and Prior, S.A. 1996. Composition and decomposition of soybean and sorghum tissues grown under elevated atmospheric carbon dioxide. *Journal of Environmental Quality* **25**: 822-827.

Nitschelm, J.J., Luscher, A., Hartwig, U.A. and van Kessel, C. 1997. Using stable isotopes to determine soil carbon input differences under ambient and elevated atmospheric CO_2 conditions. *Global Change Biology* **3**: 411-416.

Torbert, H.A., Prior, S.A., Rogers, H.H. and Runion, G.B. 1998. Crop residue decomposition as affected by growth under elevated atmospheric CO_2. *Soil Science* **163**: 412-419.

Van Ginkel, J.H., Whitmore, A.P. and Gorissen, A. 1999. *Lolium perenne* grasslands may function as a sink for atmospheric carbon dioxide. *Journal of Environmental Quality* **28**: 1580-1584.

Velichko, A.A., Zelikson, E.M. and Borisova, O.K. 1999. Vegetation, phytomass and carbon storage in Northern Eurasia during the last glacial-interglacial cycle and the Holocene. *Chemical Geology* **159**: 191-204.

7.7.3.2. Forests

Liski *et al.* (1999) studied soil carbon storage across a temperature gradient in a modern-day Finnish boreal forest, determining that carbon sequestration in the soil of this forest increased with temperature. In deciduous forests of the eastern United States, White *et al.* (1999) also determined that persistent increases

in growing season length (due to rising air temperatures) may lead to long-term increases in carbon storage, which tend to counterbalance the effects of increasing air temperature on respiration rates.

A data-driven analysis by Fan *et al.* (1998) suggests that the carbon-sequestering abilities of North America's forests between 15 and 51°N latitude are so robust that they can yearly remove from the atmosphere all of the CO_2 annually released to it by fossil fuel consumption in both the United States and Canada (and this calculation was done during a time touted as having the warmest temperatures on record). Moreover, Phillips *et al.* (1998) have shown that carbon sequestration in tropical forests has increased substantially over the past 42 years, in spite of any temperature increases that may have occurred during that time.

Similarly, King *et al.* (1999) showed that aspen seedlings increased their photosynthetic rates and biomass production as temperatures rose from 10 to 29°C, putting to rest the idea that high-temperature-induced increases in respiration rates would cause net losses in carbon fixation. White *et al.* (2000) showed that rising temperatures increased the growing season by about 15 days for 12 sites in deciduous forests located within the United States, causing a 1.6 percent increase in net ecosystem productivity per day. Thus, rather than exerting a negative influence on forest carbon sequestration, if air temperatures rise in the future they will likely have a positive effect on carbon storage in forests and their associated soils.

References

King, J.S., Pregitzer, K.S. and Zak, D.R. 1999. Clonal variation in above- and below-ground responses of *Populus tremuloides* Michaux: Influence of soil warming and nutrient availability. *Plant and Soil* **217**: 119-130.

Liski, J., Ilvesniemi, H., Makela, A. and Westman, C.J. 1999. CO_2 emissions from soil in response to climatic warming are overestimated - The decomposition of old soil organic matter is tolerant of temperature. *Ambio* **28**: 171-174.

Phillips, O.L., Malhi, Y., Higuchi, N., Laurance, W.F., Nunez, P.V., Vasquez, R.M., Laurance, S.G., Ferreira, L.V., Stern, M., Brown, S. and Grace, J. 1998. Changes in the carbon balance of tropical forests: Evidence from long-term plots. *Science* **282**: 439-442.

White, A., Cannell, M.G.R. and Friend, A.D. 2000. The high-latitude terrestrial carbon sink: a model analysis. *Global Change Biology* **6**: 227-245.

White, M.A., Running, S.W. and Thornton, P.E. 1999. The impact of growing-season length variability on carbon assimilation and evapotranspiration over 88 years in the eastern US deciduous forest. *International Journal of Biometeorology* **42**: 139-145.

7.7.3.3. Peatlands

Putative CO_2-induced global warming has long been predicted to turn boreal and tundra biomes into major sources of carbon emissions. Until just a few short years ago it was nearly universally believed that rising air temperatures would lead to the thawing of extensive areas of permafrost and the subsequent decomposition of their vast stores of organic matter, which, it was thought, would release much of the peatlands' tightly held carbon, enabling it to make its way back to the atmosphere as CO_2.

Improved soil drainage and increased aridity were also envisioned to help the process along, possibly freeing enough carbon at a sufficiently rapid rate to rival the amount released to the atmosphere as CO_2 by all anthropogenic sources combined. The end result was claimed to be a tremendous positive feedback to the ongoing rise in the air's CO_2 content, which was envisioned to produce a greatly amplified atmospheric greenhouse effect that would lead to catastrophic global warming.

This scenario was always too bad to be true. Why? Because it did not begin to deal with the incredible complexity of the issue, several important neglected aspects of which have been briefly described by Weintraub and Schimel (2005).

One of the first cracks in the seemingly sound hypothesis was revealed by the study of Oechel *et al.* (2000), wherein long-term measurements of net ecosystem CO_2 exchange rates in wet-sedge and moist-tussock tundra communities of the Alaskan Arctic indicated that these ecosystems were gradually changing from carbon sources to carbon sinks. The transition occurred between 1992 and 1996, at the apex of a regional warming trend that culminated with the highest summer temperature and surface water deficit of the previous four decades.

How did this dramatic and unexpected biological transformation happen? The answer of the scientists who documented the phenomenon was "a previously undemonstrated capacity for ecosystems to metabolically adjust to long-term changes in climate." Just as people can change their behavior in response to environmental stimuli, so can plants. And this ecological acclimation process is only one of several newly recognized phenomena that have caused scientists to radically revise the way they think about global change in Arctic regions.

Camill *et al.* (2001) investigated (1) changes in peat accumulation across a regional gradient of mean annual temperature in Manitoba, Canada, (2) net above-ground primary production and decomposition for major functional plant groups of the region, and (3) soil cores from several frozen and thawed bog sites that were used to determine long-term changes in organic matter accumulation following the thawing of boreal peatlands. In direct contradiction of earlier thinking on the subject, but in confirmation of the more recent findings of Camill (1999a,b), the researchers discovered that above-ground biomass and decomposition "were more strongly controlled by local succession than regional climate." In other words, they determined that over a period of several years, natural changes in plant community composition generally "have stronger effects on carbon sequestration than do simple increases in temperature and aridity." Their core-derived assessments of peat accumulation over the past two centuries demonstrated that rates of biological carbon sequestration can almost double following the melting of permafrost, in harmony with the findings of Robinson and Moore (2000) and Turetsky *et al.* (2000), who found rates of organic matter accumulation in other recently thawed peatlands to have risen by 60-72 percent.

Griffis and Rouse (2001) drew upon the findings of a number of experiments conducted over the past quarter-century at a subarctic sedge fen near Churchill, Manitoba, Canada, in order to develop an empirical model of net ecosystem CO_2 exchange there. The most fundamental finding of this endeavor was that "carbon acquisition is greatest during wet and warm conditions," such as is generally predicted for the world as a whole by today's most advanced climate models. However, since regional climate change predictions are not very dependable, the two scientists investigated the consequences of a $4°C$ increase in temperature accompanied by both a 30 percent increase and decrease in precipitation; and "in all cases," as they put it, "the equilibrium response showed substantial increases in carbon acquisition." One of the reasons behind this finding, as explained by Griffis and Rouse, is that "arctic ecosystems photosynthesize below their temperature optimum over the majority of the growing season," so that increasing temperatures enhance plant growth rates considerably more than they increase plant decay rates.

In summing up their findings, Griffis and Rouse say "warm surface temperatures combined with wet soil conditions in the early growing season increase above-ground biomass and carbon acquisition throughout the summer season." Indeed, they note that "wet spring conditions can lead to greater CO_2 acquisition through much of the growing period even when drier conditions persist." They thus conclude that if climate change plays out as described by current climate models—i.e., if the world becomes warmer and wetter—"northern wetlands should therefore become larger sinks for atmospheric CO_2."

In a somewhat different type of study, Mauquoy et al. (2002) analyzed three cores obtained from a raised peat bog in the UK (Walton Moss) and a single core obtained from a similar bog in Denmark (Lille Vildmose) for macro- and micro-fossils (pollen), bulk density, loss on ignition, carbon/nitrogen ratios, and humification, while they were ^{14}C dated by accelerator mass spectrometry. Among a variety of other things, it was determined, in their words, that "the lowest carbon accumulation values for the Walton Moss monoliths between ca. cal AD 1300 and 1800 and between ca. cal AD 1490 and 1580 for Lille Vildmose occurred during the course of Little Ice Age deteriorations," which finding they describe as being much the same as the observation "made by Oldfield et al. (1997) for a Swedish 'aapa' mire between ca. cal AD 1400 and 1800." They also report that carbon accumulation before this, in the Medieval Warm Period, was higher, as was also the case following the Little Ice Age, as the earth transitioned to the Modern Warm Period. Consequently, whereas the IPCC claims that warming will hasten the release of carbon from ancient peat bogs, these real-world data demonstrate that just the opposite is more likely.

In a somewhat similar study, but one that concentrated more on the role of nitrogen than of temperature, Turunen et al. (2004) derived recent (0-150 years) and long-term (2,000-10,000 years) apparent carbon accumulation rates for several ombrotrophic peatlands in eastern Canada with the help of ^{210}Pb- and ^{14}C-dating of soil-core materials. This work revealed that the average long-term apparent rate of C accumulation at 15 sites was 19 ± 8 g C m^{-2} yr^{-1}, which is comparable to long-term rates observed in Finnish bogs by Tolonen and Turunen

(1996) and Turunen et al. (2002). Recent C accumulation rates at 23 sites, on the other hand, were much higher, averaging 73 ± 17 g C m^{-2} yr^{-1}, which results, in their words, are also "similar to results from Finland (Tolonen and Turunen, 1996; Pitkanen et al., 1999) and for boreal Sphagnum dominated peat deposits in North America (Tolonen et al., 1988; Wieder et al., 1994; Turetsky et al., 2000)." Noting that recent rates of C accumulation are "strikingly higher" than long-term rates, Turunen et al. suggested that increased N deposition "leads to larger rates of C and N accumulation in the bogs, as has been found in European forests (Kauppi et al., 1992; Berg and Matzner, 1997), and could account for some of the missing C sink in the global C budget."

Returning to the role of temperature, Payette et al. (2004) quantified the main patterns of change in a subarctic peatland on the eastern coast of Canada's Hudson Bay, which were caused by permafrost decay between 1957 and 2003, based on detailed surveys conducted in 1973, 1983, 1993 and 2003. This work revealed there was continuous permafrost thawing throughout the period of observation, such that "about 18 percent of the initial frozen peatland surface was melted in 1957," while thereafter "accelerated thawing occurred with only 38 percent, 28 percent and 13 percent of the original frozen surface still remaining in 1983, 1993 and 2003, respectively." This process, in their words, was one of "terrestrialization" via the establishment of fen/bog vegetation, which nearly always results in either no net loss of carbon or actual carbon sequestration. As a result, Payette et al. concluded that "contrary to current expectations, the melting of permafrost caused by recent climate change does not [our italics] transform the peatland to a carbon-source ecosystem." Instead, they say that "rapid terrestrialization exacerbates carbon-sink conditions and tends to balance the local carbon budget."

In a study of experimental warming of Icelandic plant communities designed to see if the warming of high-latitude tundra ecosystems would result in significant losses of species and reduced biodiversity, Jonsdottir et al. (2005) conducted a field experiment to learn how vegetation might respond to moderate warming at the low end of what is predicted by most climate models for a doubling of the air's CO_2 content. Specifically, they studied the effects of 3-5 years of modest surface warming (1°-2°C) on two widespread but contrasting tundra plant communities, one of which was a nutrient-deficient and species-poor moss heath and the other of which was a species-rich dwarf shrub heath. At the conclusion of the study, no changes in community structure were detected in the moss heath. In the dwarf shrub heath, on the other hand, the number of deciduous and evergreen dwarf shrubs increased more than 50 percent, bryophytes decreased by 18 percent, and canopy height increased by 100 percent, but with the researchers reporting they "detected no changes in species richness or other diversity measures in either community and the abundance of lichens did not change." Although Jonsdottir *et al.*'s study was a relatively short-term experiment as far as ecosystem studies go, its results indicate a rise in temperature need not have a negative effect on the species diversity of high-latitude tundra ecosystems and may have a positive influence on plant growth.

In a study that included an entirely new element of complexity, Cole *et al.* (2002) constructed 48 small microcosms from soil and litter they collected near the summit of Great Dun Fell, Cumbria, England. Subsequent to "defaunating" this material by reducing its temperature to -80°C for 24 hours, they thawed and inoculated it with native soil microbes, after which half of the microcosms were incubated in the dark at 12°C and half at 18°C for two weeks, in order to establish near-identical communities of the soils' natural complement of microflora in each microcosm. The former of these temperatures was chosen to represent mean August soil temperature at a depth of 10 cm at the site of soil collection, while the latter was picked to be "close to model predictions for soil warming that might result from a doubling of CO_2 in blanket peat environments."

Next, 10 seedlings of *Festuca ovina*, an indigenous grass of blanket peat, were planted in each of the microcosms, while 100 enchytraeid worms were added to each of half of the mini-ecosystems, producing four experimental treatments: ambient temperature, ambient temperature plus enchytraeid worms, elevated temperature, and elevated temperature plus enchytraeid worms. Then, the 48 microcosms—sufficient to destructively harvest three replicates of each treatment four different times throughout the course of the 64-day experiment—were arranged in a fully randomized design and maintained at either 12° or 18°C with alternating 12-hour light and dark periods, while being given distilled water every two days to maintain their original weights.

So what did the researchers learn? First, they found that elevated temperature reduced the ability of the enchytraeid worms to enhance the loss of carbon from the microcosms. At the normal ambient temperature, the presence of the worms enhanced dissolved organic carbon (DOC) loss by 16 percent, while at the elevated temperature expected for a doubling of the air's CO_2 content they had no effect on DOC. In addition, Cole *et al.* note that "warming may cause drying at the soil surface, forcing enchytraeids to burrow to deeper subsurface horizons;" and since the worms are known to have little influence on soil carbon dynamics below a depth of about 4 cm (Cole *et al.*, 2000), the researchers concluded that this additional consequence of warming would further reduce the ability of enchytraeids to enhance carbon loss from blanket peatlands. In summing up their findings, Cole *et al.* concluded that "the soil biotic response to warming in this study was negative," in that it resulted in a reduced loss of carbon to the atmosphere.

But what about the effects of elevated CO_2 itself on the loss of DOC from soils? Freeman *et al.* (2004) note that riverine transport of DOC has increased markedly in many places throughout the world over the past few decades (Schindler *et al.*, 1997; Freeman *et al.*, 2001; Worrall *et al.*, 2003); they suggest this phenomenon may be related to the historical increase in the air's CO_2 content.

The researchers' first piece of evidence for this conclusion came from a three-year study of monoliths (11-cm diameter x 20-cm deep cores) taken from three Welsh peatlands—a bog that received nutrients solely from rainfall, a *fen* that gained more nutrients from surrounding soils and groundwater, and a riparian peatland that gained even more nutrients from nutrient-laden water transported from other terrestrial ecosystems via drainage streams—which they exposed to either ambient air or air enriched with an extra 235 ppm of CO_2 within a solardome facility. This study revealed that the DOC released by monoliths from the three peatlands was significantly enhanced—by 14 percent in the bog, 49 percent in the fen, and 61 percent in the riparian peatland—by the additional CO_2 to which they were exposed, which is the order of response one would expect from what we know about the stimulation of net primary productivity due to atmospheric CO_2 enrichment, i.e., it is low in the face of low soil nutrients, intermediate when soil nutrient concentrations are intermediate, and high when soil nutrients are present in abundance. Consequently, Freeman *et al.* concluded that the DOC increases they observed "were induced by increased primary production and DOC exudation from plants," which conclusion logically follows from their findings.

Nevertheless, and to further test their hypothesis, they followed the translocation of labeled ^{13}C through the plant-soil systems of the different peat monoliths for about two weeks after exposing them to ~99 percent-pure $^{13}CO_2$ for a period of five hours. This exercise revealed that (1) the plants in the ambient-air and CO_2-enriched treatments assimilated 22.9 and 35.8 mg of ^{13}C from the air, respectively, (2) the amount of DOC that was recovered from the leachate of the CO_2-enriched monoliths was 0.6 percent of that assimilated, or 0.215 mg (35.8 mg x 0.006 = 0.215 mg), and (3) the proportion of DOC in the soil solution of the CO_2-enriched monoliths that was derived from recently assimilated CO_2 (the ^{13}C labeled CO_2) was 10 times higher than that of the control.

This latter observation suggests that the amount of DOC recovered from the leachate of the ambient-air monoliths was only about a tenth as much as that recovered from the leachate of the CO_2-enriched monoliths, which puts the former amount at about 0.022 mg. Hence, what really counts, i.e., the net sequestration of ^{13}C experienced by the peat monoliths over the two-week period (which equals the amount that went into them minus the amount that went out), comes to 22.9 mg minus 0.022 mg = 22.878 mg for the ambient-air monoliths and 35.8 mg minus 0.215 mg = 35.585 mg for the CO_2-enriched monoliths. In the end, therefore, even though the CO_2-enriched monoliths lost 10 times more ^{13}C via root exudation than did the ambient-air monoliths, they still sequestered about 55 percent more ^{13}C overall, primarily in living-plant tissues.

In light of this impressive array of pertinent findings, it would appear that continued increases in the air's CO_2 concentration and temperature would not result in losses of carbon from earth's peatlands. Quite to the contrary, these environmental changes—if they persist—would likely work together to enhance carbon capture by these particular ecosystems.

Additional information on this topic, including reviews of newer publications as they become available, can be found at http://www.co2science.org/subject/c/carbonpeat.php.

References

Berg, B. and Matzner, E. 1997. Effect of N deposition on decomposition of plant litter and soil organic matter in forest systems. *Environmental Reviews* **5**: 1-25.

Camill, P. 1999a. Patterns of boreal permafrost peatland vegetation across environmental gradients sensitive to climate warming. *Canadian Journal of Botany* **77**: 721-733.

Camill, P. 1999b. Peat accumulation and succession following permafrost thaw in the boreal peatlands of Manitoba, Canada. *Ecoscience* **6**: 592-602.

Camill, P., Lynch, J.A., Clark, J.S., Adams, J.B. and Jordan, B. 2001. Changes in biomass, aboveground net primary production, and peat accumulation following permafrost thaw in the boreal peatlands of Manitoba, Canada. *Ecosystems* **4**: 461-478.

Cole, L., Bardgett, R.D. and Ineson, P. 2000. Enchytraeid worms (Oligochaeta) enhance mineralization of carbon in organic upland soils. *European Journal of Soil Science* **51**: 185-192.

Cole, L., Bardgett, R.D., Ineson, P. and Hobbs, P.J. 2002. Enchytraeid worm (Oligochaeta) influences on microbial community structure, nutrient dynamics and plant growth in blanket peat subjected to warming. *Soil Biology & Biochemistry* **34**: 83-92.

Freeman, C., Evans, C.D., Monteith, D.T., Reynolds, B. and Fenner, N. 2002. Export of organic carbon from peat soils. *Nature* **412**: 785.

Freeman, C., Fenner, N., Ostle, N.J., Kang, H., Dowrick, D.J., Reynolds, B., Lock, M.A., Sleep, D., Hughes, S. and Hudson, J. 2004. Export of dissolved organic carbon from peatlands under elevated carbon dioxide levels. *Nature* **430**: 195-198.

Griffis, T.J. and Rouse, W.R. 2001. Modelling the interannual variability of net ecosystem CO_2 exchange at a subarctic sedge fen. *Global Change Biology* **7**: 511-530.

Jonsdottir, I.S., Magnusson, B., Gudmundsson, J., Elmarsdottir, A. and Hjartarson, H. 2005. Variable sensitivity of plant communities in Iceland to experimental warming. *Global Change Biology* **11**: 553-563.

Kauppi, P.E., Mielikainen, K. and Kuusela, K. 1992. Biomass and carbon budget of European forests. *Science* **256**: 70-74.

Mauquoy, D., Engelkes, T., Groot, M.H.M., Markesteijn, F., Oudejans, M.G., van der Plicht, J. and van Geel, B. 2002. High-resolution records of late-Holocene climate change and carbon accumulation in two north-west European ombrotrophic peat bogs. *Palaeogeography, Palaeoclimatology, Palaeoecology* **186**: 275-310.

Oechel, W.C., Vourlitis, G.L., Hastings, S.J., Zulueta, R.C., Hinzman, L. and Kane, D. 2000. Acclimation of ecosystem CO_2 exchange in the Alaskan Arctic in response to decadal climate warming. *Nature* **406**: 978-981.

Payette, S., Delwaide, A., Caccianiga, M. and Beauchemin, M. 2004. Accelerated thawing of subarctic peatland permafrost over the last 50 years. *Geophysical Research Letters* **31**: 10.1029/2004GL020358.

Pitkanen, A., Turunen, J. and Tolonen, K. 1999. The role of fire in the carbon dynamics of a mire, Eastern Finland. *The Holocene* **9**: 453-462.

Robinson, S.D. and Moore, T.R. 2000. The influence of permafrost and fire upon carbon accumulation in high boreal peatlands, Northwest Territories, Canada. *Arctic, Antarctic and Alpine Research* **32**: 155-166.

Schindler, D.W., Curtis, P.J., Bayley, S.E., Parker, B.R., Beaty, K.G. and Stainton, M.P. 1997. Climate-induced changes in the dissolved organic carbon budgets of boreal lakes. *Biogeochemistry* **36**: 9-28.

Tolonen, K., Davis, R.B. and Widoff, L. 1988. Peat accumulation rates in selected Maine peat deposits. *Maine Geological Survey, Department of Conservation Bulletin* **33**: 1-99.

Tolonen, K. and Turunen, J. 1996. Accumulation rates of carbon in mires in Finland and implications for climate change. *The Holocene* **6**: 171-178.

Turetsky, M.R., Wieder, R.K., Williams, C.J, and Vitt, D.H. 2000. Organic matter accumulation, peat chemistry, and permafrost melting in peatlands of boreal Alberta. *Ecoscience* **7**: 379-392.

Turunen, J., Roulet, N.T., Moore, T.R. and Richard, P.J.H. 2004. Nitrogen deposition and increased carbon accumulation in ombrotrophic peatlands in eastern Canada. *Global Biogeochemical Cycles* **18**: 10.1029/2003 GB002154.

Turunen, J., Tomppo, E., Tolonen, K. and Reinikainen, A. 2002. Estimating carbon accumulation rates of undrained mires in Finland: Application to boreal and subarctic regions. *The Holocene* **12**: 69-80.

Weintraub, M.N. and Schimel, J.P. 2005. Nitrogen cycling and the spread of shrubs control changes in the carbon balance of Arctic tundra ecosystems. *BioScience* **55**: 408-415.

Wieder, R.K., Novak, M., Schell, W.R. and Rhodes, T. 1994. Rates of peat accumulation over the past 200 years in five Sphagnum-dominated peatlands in the United States. *Journal of Paleolimnology* **12**: 35-47.

Worrall, F., Burt, T. and Shedden, R. 2003. Long term records of riverine dissolved organic matter. *Biogeochemistry* **64**: 165-178.

7.8. Other Benefits

Other benefits to plants of CO_2 enhancement documented in this section include superior nitrogen-use efficiency, increased nutrient acquisition, greater resistance to pathogens and parasitic plants, greater root development, greater seed and tannin production, and improved performance of transgenic plants. In addition to these benefits to plants, CO_2 enrichment benefits all life on earth by reducing plant emissions of isoprene, a chemical responsible for the production of tropospheric ozone.

7.8.1. Nitrogen-Use Efficiency

Long-term exposure to elevated atmospheric CO_2 concentrations often, but not always, elicits photosynthetic acclimation or down regulation in plants, which is typically accompanied by reduced amounts of rubisco and/or other photosynthetic proteins that are typically present in excess amounts in plants grown in ambient air. As a consequence, foliar nitrogen concentrations often decrease with atmospheric CO_2 enrichment, as nitrogen is mobilized out of leaves and into other areas of the plant to increase its availability for enhancing sink development or stimulating other nutrient-limited processes.

In reviewing the literature in this area, one quickly notices that in spite of the fact that photosynthetic acclimation has occurred, CO_2-enriched plants nearly always display rates of photosynthesis that are greater than those of control plants exposed to ambient air. Consequently, photosynthetic nitrogen-use efficiency, i.e., the amount of carbon converted into sugars during the photosynthetic process per unit of leaf nitrogen, often increases dramatically in CO_2-enriched plants.

In the study of Davey *et al.* (1999), for example, CO_2-induced reductions in foliar nitrogen contents and concomitant increases in photosynthetic rates led to photosynthetic nitrogen-use efficiencies in the CO_2-enriched (to 700 ppm CO_2) grass *Agrostis capillaris* that were 27 and 62 percent greater than those observed in control plants grown at 360 ppm CO_2 under moderate and low soil nutrient conditions, respectively. Similarly, elevated CO_2 enhanced photosynthetic nitrogen-use efficiencies in *Trifolium repens* by 66 and 190 percent under moderate and low soil nutrient conditions, respectively, and in *Lolium perenne* by 50 percent, regardless of soil nutrient status. Other researchers have found comparable CO_2-induced enhancements of photosynthetic nitrogen-use efficiency in wheat (Osborne *et al.*, 1998) and in *Leucadendron* species (Midgley *et al.*, 1999).

In some cases, researchers report nitrogen-use efficiency in terms of the amount of biomass produced per unit of plant nitrogen. Niklaus *et al.* (1998), for example, reported that intact swards of CO_2-enriched calcareous grasslands grown at 600 ppm CO_2 attained total biomass values that were 25 percent greater than those of control swards exposed to ambient air while extracting the same amount of nitrogen from the soil as ambiently grown swards. Similar results have been reported for strawberry by Deng and Woodward (1998), who noted that the growth nitrogen-use efficiencies of plants grown at 560 ppm CO_2 were 23 and 17 percent greater than those of ambiently grown plants simultaneously subjected to high and low soil nitrogen availability, respectively.

In conclusion, the scientific literature indicates that as the air's CO_2 content continues to rise, earth's plants will likely respond by reducing the amount of nitrogen invested in rubisco and other photosynthetic proteins, while still maintaining enhanced rates of photosynthesis, which consequently should increase their photosynthetic nitrogen-use efficiencies. As overall plant nitrogen-use efficiency increases, it is likely plants will grow ever better on soils containing less-than-optimal levels of nitrogen, a point addressed in more detail in Section 7.3.7 of this report.

Additional information on this topic, including reviews of newer publications as they become available, can be found at http://www.co2science.org/subject/n/nitrogenefficiency.php.

References

Davey, P.A., Parsons, A.J., Atkinson, L., Wadge, K. and Long, S.P. 1999. Does photosynthetic acclimation to elevated CO_2 increase photosynthetic nitrogen-use efficiency? A study of three native UK grassland species in open-top chambers. *Functional Ecology* 13: 21-28.

Deng, X. and Woodward, F.I. 1998. The growth and yield responses of *Fragaria ananassa* to elevated CO_2 and N supply. *Annals of Botany* 81: 67-71.

Midgley, G.F., Wand, S.J.E. and Pammenter, N.W. 1999. Nutrient and genotypic effects on CO_2-responsiveness: photosynthetic regulation in *Leucadendron* species of a nutrient-poor environment. *Journal of Experimental Botany* 50: 533-542.

Niklaus, P.A., Leadley, P.W., Stocklin, J. and Korner, C. 1998. Nutrient relations in calcareous grassland under elevated CO_2. *Oecologia* 116: 67-75.

Osborne, C.P., LaRoche, J., Garcia, R.L., Kimball, B.A., Wall, G.W., Pinter, P.J., Jr., LaMorte, R.L., Hendrey, G.R. and Long, S.P. 1998. Does leaf position within a canopy affect acclimation of photosynthesis to elevated CO_2? *Plant Physiology* 117: 1037-1045.

7.8.2. Nutrient Acquisition

Most species of plants respond to increases in the air's CO_2 content by displaying enhanced rates of photosynthesis and biomass production. Oftentimes, the resulting growth stimulation is preferentially expressed belowground, thereby causing significant increases in fine-root numbers and surface area. This phenomenon tends to increase total nutrient uptake under CO_2-enriched conditions, which further stimulates plant growth and development. In this summary, we review how the acquisition of plant nutrients—primarily nitrate and phosphate—is affected by atmospheric CO_2 enrichment. The effects of elevated CO_2 on nitrogen fixation are addressed elsewhere (Sections 7.3.7 and 7.8.1) and more research on the effects of CO_2 enhancement on roots appears in Section 7.8.5.

Smart *et al.* (1998) noted there were no differences on a per-unit-biomass basis in the total amounts of nitrogen within CO_2-enriched and ambiently grown wheat seedlings after three weeks of exposure to atmospheric CO_2 concentrations of 360 and 1,000 ppm. Nevertheless, the CO_2-enriched seedlings exhibited greater rates of soil nitrate extraction than did the ambiently grown plants. Similarly, BassiriRad *et al.* (1998) reported that a doubling of the atmospheric CO_2 concentration doubled the uptake rate of nitrate in the C_4 grass *Bouteloua eriopoda*. However, they also reported that elevated CO_2 had no effect on the rate of nitrate uptake in *Prosopis*, and that it decreased the rate of nitrate uptake by 55 percent in *Larrea*. Nonetheless, atmospheric CO_2 enrichment increased total biomass in these two species by 55 and 69 percent, respectively. Thus, although the uptake rate of this nutrient was depressed under elevated CO_2 conditions in the latter species, the much larger CO_2-enriched plants likely still extracted more total nitrate from the soil than did the ambiently grown plants of the experiment.

Nasholm *et al.* (1998) determined that trees, grasses and shrubs can all absorb significant amounts of organic nitrogen from soils. Thus, plants do not have to wait for the mineralization of organic nitrogen before they extract the nitrogen they need from soils to support their growth and development. Hence, the forms of nitrogen removed from soils by plants (nitrate vs. ammonium) and their abilities to remove different forms may not be as important as was once thought.

With respect to the uptake of phosphate, Staddon *et al.* (1999) reported that *Plantago lanceolata* and *Trifolium repens* plants grown at 650 ppm CO_2 for 2.5 months exhibited total plant phosphorus contents that were much greater than those displayed by plants grown at 400 ppm CO_2, due to the fact that atmospheric CO_2 enrichment significantly enhanced plant biomass. Similarly, Rouhier and Read (1998) reported that enriching the air around *Plantago lanceolata* plants with an extra 190 ppm of CO_2 for a period of three months led to increased uptake of phosphorus and greater tissue phosphorus concentrations than were observed in plants growing in ambient air.

Greater uptake of phosphorus also can occur due to CO_2-induced increases in root absorptive surface area or enhancements in specific enzyme activities. In addressing the first of these phenomena, BassiriRad *et al.* (1998) reported that a doubling of the atmospheric CO_2 concentration significantly increased the belowground biomass of *Bouteloua eriopoda* and doubled its uptake rate of phosphate. However, elevated CO_2 had no effect on uptake rates of phosphate in *Larrea* and *Prosopis*. Because the CO_2-enriched plants grew so much bigger, they still removed more phosphate from the soil on a per-plant basis. With respect to the second phenomenon, phosphatase—the primary enzyme responsible for the conversion of organic phosphate into usable inorganic forms—had its activity increased by 30 to 40 percent in wheat seedlings growing at twice-ambient CO_2 concentrations (Barrett *et al.*, 1998).

In summary, as the CO_2 content of the air increases, experimental data to date suggest that much of earth's vegetation will likely extract enhanced amounts of mineral nutrients from the soils in which they are rooted.

Additional information on this topic, including reviews of newer publications as they become available, can be found at http://www.co2science.org/subject/n/nutrientacquis.php.

References

Barrett, D.J., Richardson, A.E. and Gifford, R.M. 1998. Elevated atmospheric CO_2 concentrations increase wheat root phosphatase activity when growth is limited by phosphorus. *Australian Journal of Plant Physiology* **25**: 87-93.

BassiriRad, H., Reynolds, J.F., Virginia, R.A. and Brunelle, M.H. 1998. Growth and root NO_3^{3-} and PO_4^{3-} uptake capacity of three desert species in response to atmospheric CO_2 enrichment. *Australian Journal of Plant Physiology* **24**: 353-358.

Nasholm, T., Ekblad, A., Nordin, A., Giesler, R., Hogberg, M. and Hogberg, P. 1998. Boreal forest plants take up organic nitrogen. *Nature* **392**: 914-916.

Rouhier, H. and Read, D.J. 1998. The role of mycorrhiza in determining the response of *Plantago lanceolata* to CO_2 enrichment. *New Phytologist* **139**: 367-373.

Smart, D.R., Ritchie, K., Bloom, A.J. and Bugbee, B.B. 1998. Nitrogen balance for wheat canopies (*Triticum aestivum* cv. Veery 10) grown under elevated and ambient CO_2 concentrations. *Plant, Cell and Environment* **21**: 753-763.

Staddon, P.L., Fitter, A.H. and Graves, J.D. 1999. Effect of elevated atmospheric CO_2 on mycorrhizal colonization, external mycorrhizal hyphal production and phosphorus inflow in *Plantago lanceolata* and *Trifolium repens* in association with the arbuscular mycorrhizal fungus *Glomus mosseae*. *Global Change Biology* **5**: 347-358.

7.8.3. Pathogens

As the air's CO_2 content continues to rise, it is natural to wonder—and important to determine—how this phenomenon may impact plant-pathogen interactions. One thing we know about the subject is that atmospheric CO_2 enrichment nearly always enhances photosynthesis, which commonly leads to increased plant production of carbon-based secondary compounds, including lignin and various phenolics, both of which substances tend to increase plant resistance to pathogen attack.

Enlarging upon this topic, Chakraborty and Datta (2003) report that "changes in plant physiology, anatomy and morphology that have been implicated in increased resistance or [that] can potentially enhance host resistance at elevated CO_2 include: increased net photosynthesis allowing mobilization of resources into host resistance (Hibberd *et al.*, 1996a.); reduced stomatal density and conductance (Hibberd *et al.*, 1996b); greater accumulation of carbohydrates in leaves; more waxes, extra layers of epidermal cells and increased fibre content (Owensby, 1994); production of papillae and accumulation of silicon at penetration sites (Hibberd *et al.*, 1996a); greater number of mesophyll cells (Bowes, 1993); and increased biosynthesis of phenolics (Hartley *et al.*, 2000), among others."

Chakraborty and Datta found another way atmospheric CO_2 enrichment may tip the scales in favor of plants in a study of the aggressiveness of the fungal anthracnose pathogen *Colletotrichum gloeosporioides*. They inoculated two isolates of the pathogen onto two cultivars of the tropical pasture legume *Stylosanthes scabra* (Fitzroy, which is susceptible to the fungal pathogen, and Seca, which is more resistant) over the course of 25 sequential infection cycles at ambient (350 ppm) and elevated (700 ppm) atmospheric CO_2 concentrations in controlled environment chambers. This protocol revealed that "at twice-ambient CO_2 the overall level of aggressiveness of the two [pathogen] isolates was significantly reduced on both [host] cultivars." In addition, they say that "as shown previously (Chakraborty *et al.*, 2000), the susceptible Fitzroy develops a level of resistance to anthracnose at elevated CO_2, but resistance in Seca remains largely unchanged." Simultaneously, however, pathogen fecundity was found to increase at twice-ambient CO_2. Of this finding, they report that their results "concur with the handful of studies that have demonstrated increased pathogen fecundity at elevated CO_2 (Hibberd *et al.*, 1996a; Klironomos *et al.*, 1997; Chakraborty *et al.*, 2000)." How this happened in the situation they investigated, as they describe it, is that the overall increase in fecundity at high CO_2 "is a reflection of the altered canopy environment," wherein "the 30 percent larger *S. scabra* plants at high CO_2 (Chakraborty *et al.*, 2000) makes the canopy microclimate more conducive to anthracnose development."

In view of the opposing changes induced in pathogen behavior by elevated levels of atmospheric CO_2 in this specific study—reduced aggressiveness but increased fecundity—it is difficult to know the outcome of atmospheric CO_2 enrichment for the pathogen-host relationship. More research, especially under realistic field conditions, will be needed to clarify the situation; and, of course, different results are likely to be observed for different pathogen-host associations. Results also could differ under different climatic conditions. Nevertheless, the large number of ways in which elevated CO_2 has been demonstrated to increase plant resistance to pathogen attack suggests that plants may well gain the advantage over pathogens as the air's CO_2 content continues to climb in the years ahead.

McElrone *et al.* (2005) "assessed how elevated CO_2 affects a foliar fungal pathogen, *Phyllosticta minima*, of *Acer rubrum* [red maple] growing in the understory at the Duke Forest free-air CO_2 enrichment experiment in Durham, North Carolina, USA ... in the 6th, 7th, and 8th years of the CO_2 exposure." Surveys conducted in those years, in their words, "revealed that elevated CO2 [to 200 ppm above ambient] significantly reduced disease incidence, with 22%, 27% and 8% fewer saplings and 14%, 4%, and 5% fewer leaves infected per plant in the three consecutive years, respectively." In addition, they report that the elevated CO_2 "also significantly reduced disease severity in infected plants in all years (e.g. mean lesion area reduced 35%, 50%, and 10% in 2002, 2003, and 2004, respectively)."

With respect to identifying the underlying mechanism or mechanisms that produced these beneficent consequences, thinking it could have been a direct deleterious effect of elevated CO_2 on the fungal pathogen, McElrone *et al.* performed some side experiments in controlled environment chambers. However, they found that the elevated CO_2 benefited the fungal pathogen as well as the red maple saplings, observing that "exponential growth rates of *P. minima* were 17% greater under elevated CO2." And they obtained similar results when they repeated the *in vitro* growth analysis two additional times in different growth chambers.

Taking another tack when "scanning electron micrographs verified that conidia germ tubes of *P. minima* infect *A. rubrum* leaves by entering through the stomata," the researchers turned their attention to the pathogen's mode of entry into the saplings' foliage. In this investigation they found that both stomatal size and density were unaffected by atmospheric CO_2 enrichment, but that "stomatal conductance was reduced by 21-36% under elevated CO_2, providing smaller openings for infecting germ tubes." They concluded that reduced disease severity under elevated CO_2 was likely due to altered leaf chemistry, as elevated CO_2 increased total leaf phenolic concentrations by 15 percent and tannin concentrations by 14 percent.

Because the phenomena they found to be important in reducing the amount and severity of fungal pathogen infection (leaf spot disease) of red maple have been demonstrated to be operative in most other plants as well, McElrone *et al.* say these CO_2-enhanced leaf defensive mechanisms "may be prevalent in many plant pathosystems where the pathogen targets the stomata." They state their results "provide concrete evidence for a potentially generalizable mechanism to predict disease outcomes in other pathosystems under future climatic conditions."

Malmstrom and Field (1997) grew individual oat plants for two months in pots within phytocells maintained at CO_2 concentrations of 350 and 700 ppm, while a third of each CO_2 treatment's plants were infected with the barley yellow dwarf virus (BYDV), which plagues more than 150 plant species worldwide, including all major cereal crops. They found that the elevated CO_2 stimulated net photosynthesis rates in all plants, but with the greatest increase occurring in diseased individuals (48 percent vs. 34 percent). In addition, atmospheric CO_2 enrichment decreased stomatal conductance by 34 percent in healthy plants, but by 50 percent in infected ones, thus reducing transpirational water losses more in infected plants. Together, these two phenomena contributed to a CO_2-induced doubling of the instantaneous water-use efficiency of healthy control plants, but to a much larger 2.7-fold increase in diseased plants. Thus, although BYDV infection did indeed reduce overall plant biomass production, the growth response to elevated CO_2 was greatest in the diseased plants. After 60 days of CO_2 enrichment, for example, total plant biomass increased by 36 percent in infected plants, while it increased by only 12 percent in healthy plants. In addition, while elevated CO_2 had little effect on root growth in healthy plants, it increased root biomass in infected plants by up to 60 percent. In their concluding remarks, therefore, Malmstrom and Field say that CO_2 enrichment "may reduce losses of infected plants to drought" and "may enable diseased plants to compete better with healthy neighbors."

Tiedemann and Firsching (2000) grew spring wheat from germination to maturity in controlled environment chambers maintained at either ambient (377 ppm) or enriched (612 ppm) atmospheric CO_2 concentrations and either ambient (20 ppb) or enriched (61 ppb) atmospheric ozone (O_3) concentrations, while half of the plants in each of the four resulting treatments were inoculated with a leaf rust-causing pathogen. These procedures revealed that the percent of leaf area infected by rust in inoculated plants was largely unaffected by atmospheric CO_2 enrichment but strongly reduced by elevated O_3. With respect to photosynthesis, elevated CO_2 increased rates in inoculated plants by 20 and 42 percent at ambient and elevated O_3 concentrations, respectively. Although inoculated plants produced lower yields than non-inoculated plants, atmospheric CO_2 enrichment still stimulated yield in infected plants, increasing it by fully 57 percent at high O_3. Consequently, the beneficial effects of elevated CO_2 on wheat photosynthesis and yield continued to be expressed in the presence of both O_3 and pathogenic stresses.

In another joint CO_2/O_3 study, Percy *et al.* (2002) grew the most widely distributed North American tree species—trembling aspen—in twelve 30-m-diameter free-air CO^2 enrichment (FACE) rings near Rhinelander, Wisconsin, USA in air maintained at ambient CO_2 and O_3 concentrations, ambient O_3 and elevated $CO2$ (560 ppm during daylight hours), ambient CO_2 and elevated O_3 (46.4-55.5 ppb during daylight hours), and elevated CO_2 and O_3 over the period of each growing season from 1998 through 2001. Throughout this experiment they assessed a number of the young trees' growth characteristics, as well as their responses to poplar leaf rust (*Melampsora medusae*), which they say "is common on aspen and belongs to the most widely occurring group of foliage diseases." Their work revealed that elevated CO_2 alone did not alter rust occurrence, but that elevated O_3 alone increased it by nearly fourfold. When applied together, however, elevated CO_2 reduced the enhancement of rust development caused by elevated O_3 from nearly fourfold to just over twofold.

Jwa and Walling (2001) grew tomato plants in hydroponic culture for eight weeks in controlled environment chambers maintained at atmospheric CO_2 concentrations of 350 and 700 ppm. At week five of their study, half of the plants growing in each CO_2 concentration were infected with the fungal pathogen *Phytophthora parasitica*, which attacks plant roots and induces water stress that decreases plant growth and yield. This infection procedure reduced total plant biomass by nearly 30 percent at both atmospheric CO_2 concentrations. However, the elevated CO_2 treatment increased the total biomass of healthy and infected plants by the same percentage, so that infected tomato plants grown at 700 ppm CO_2 exhibited biomass values similar to those of healthy tomato plants grown at 350 ppm CO_2. Consequently, atmospheric CO_2 enrichment completely counterbalanced the negative effects of *Phytophthora parasitica* infection on tomato productivity.

Pangga *et al.* (2004) grew well-watered and fertilized pencilflower (cultivar Fitzroy) seedlings—an important legume crop susceptible to anthracnose disease caused by *Colletotrichum gloeosporioides*—within a controlled environment facility maintained at atmospheric CO_2 concentrations of either 350 or 700 ppm, where they inoculated six-, nine- and twelve-week-old plants with conidia of *C. gloeosporioides*. Then, ten days after inoculation, they counted the anthracnose lesions on the plants and classified them as either resistant or susceptible. In doing so, they found that "the mean number of susceptible, resistant, and total lesions per leaf averaged over the three plant ages was significantly (P<0.05) greater at 350 ppm than at 700 ppm CO_2, reflecting the development of a level of resistance in susceptible cv. Fitzroy at high CO_2." With respect to plants inoculated at twelve weeks of age, they say that those grown "at 350 ppm had 60 and 75 percent more susceptible and resistant lesions per leaf, respectively, than those [grown] at 700 ppm CO_2." The Australian scientists say their work "clearly shows that at 350 ppm overall susceptibility of the canopy increases with increasing age because more young leaves are produced on secondary and tertiary branches of the more advanced plants." However, "at 700 ppm CO_2, infection efficiency did not increase with increasing plant age despite the presence of many more young leaves in the enlarged canopy," which finding, in their words, "points to reduced pathogen efficiency or an induced partial resistance to anthracnose in Fitzroy at 700 ppm CO_2."

Finally, according to Plessl *et al.* (2007), "potato late blight caused by the oomycete *Phytophthora infestans* (Mont.) de Bary is the most devastating disease of potato worldwide," adding that "infection occurs through leaves and tubers followed by a rapid spread of the pathogen finally causing destructive necrosis." In an effort to ascertain the effects of atmospheric CO_2 enrichment on this pathogen, Plessl *et al.* grew individual well watered and fertilized plants of the potato cultivar Indira in 3.5-liter pots filled with a 1:2 mixture of soil and "Fruhstorfer T-Erde" in controlled-environment chambers maintained at atmospheric CO_2 concentrations of

either 400 or 700 ppm. Four weeks after the start of the experiment, the first three fully developed pinnate leaves were cut from the plants and inoculated with zoospores of *P. infestans* in Petri dishes containing water-agar, after which their symptoms were evaluated daily via comparison with control leaves that were similarly treated but unexposed to the pathogen.

Results of the German researchers analysis revealed that the 300 ppm increase in CO_2 "dramatically reduced symptom development," including extent of necrosis (down by 44 percent four days after inoculation and 65 percent five days after inoculation), area of sporulation (down by 100 percent four days after inoculation and 61 percent five days after inoculation), and sporulation intensity (down by 73 percent four days after inoculation and 17 percent five days after inoculation). Plessl *et al.* conclude that their results "clearly demonstrated that the potato cultivar Indira, which under normal conditions shows a high susceptibility to *P. infestans*, develops resistance against this pathogen after exposure to 700 ppm CO_2," noting that "this finding agrees with results from Ywa *et al.* (1995), who reported an increased tolerance of tomato plants to *Phytophthora* root rot when grown at elevated CO_2." These similar observations bode well for both potato and tomato cultivation in a CO_2-enriched world of the future.

In conclusion, the balance of evidence obtained to date demonstrates an enhanced ability of plants to withstand pathogen attacks in CO_2-enriched as opposed to ambient-CO_2 air. As the atmosphere's CO_2 concentration continues to rise in the years to come, earth's vegetation should fare ever better in its battle against myriad debilitating plant diseases.

Additional information on this topic, including reviews of newer publications as they become available, can be found at http://www.co2science.org/subject/p/pathogens.php.

References

Bowes, G. 1993. Facing the inevitable: Plants and increasing atmospheric CO_2. *Annual Review of Plant Physiology and Plant Molecular Biology* **44**: 309-332.

Chakraborty, S. and Datta, S. 2003. How will plant pathogens adapt to host plant resistance at elevated CO_2 under a changing climate? *New Phytologist* **159**: 733-742.

Chakraborty, S., Pangga, I.B., Lupton, J., Hart, L., Room, P.M. and Yates, D. 2000. Production and dispersal of

Colletotrichum gloeosporioides spores on *Stylosanthes scabra* under elevated CO_2. *Environmental Pollution* **108**: 381-387.

Hartley, S.E., Jones, C.G. and Couper, G.C. 2000. Biosynthesis of plant phenolic compounds in elevated atmospheric CO_2. *Global Change Biology* **6**: 497-506.

Hibberd, J.M., Whitbread, R. and Farrar, J.F. 1996a. Effect of elevated concentrations of CO_2 on infection of barley by *Erysiphe graminis*. *Physiological and Molecular Plant Pathology* **48**: 37-53.

Hibberd, J.M., Whitbread, R. and Farrar, J.F. 1996b. Effect of 700 μmol per mol CO_2 and infection of powdery mildew on the growth and partitioning of barley. *New Phytologist* **134**: 309-345.

Jwa, N.-S. and Walling, L.L. 2001. Influence of elevated CO_2 concentration on disease development in tomato. *New Phytologist* **149**: 509-518.

Klironomos, J.N., Rillig, M.C., Allen, M.F., Zak, D.R., Kubiske, M. and Pregitzer, K.S. 1997. Soil fungal-arthropod responses to *Populus tremuloides* grown under enriched atmospheric CO_2 under field conditions. *Global Change Biology* **3**: 473-478.

Malmstrom, C.M. and Field, C.B. 1997. Virus-induced differences in the response of oat plants to elevated carbon dioxide. *Plant, Cell and Environment* **20**: 178-188.

McElrone, A.J., Reid, C.D., Hoye, K.A., Hart, E. and Jackson, R.B. 2005. Elevated CO_2 reduces disease incidence and severity of a red maple fungal pathogen via changes in host physiology and leaf chemistry. *Global Change Biology* **11**: 1828-1836.

Owensby, C.E. 1994. Climate change and grasslands: ecosystem-level responses to elevated carbon dioxide. *Proceedings of the XVII International Grassland Congress*. Palmerston North, New Zealand: New Zealand Grassland Association, pp. 1119-1124.

Pangga, I.B., Chakraborty, S. and Yates, D. 2004. Canopy size and induced resistance in *Stylosanthes scabra* determine anthracnose severity at high CO_2. *Phytopathology* **94**: 221-227.

Percy, K.E., Awmack, C.S., Lindroth, R.L., Kubiske, M.E., Kopper, B.J., Isebrands, J.G., Pregitzer, K.S., Hendrey, G.R., Dickson, R.E., Zak, D.R., Oksanen, E., Sober, J., Harrington, R. and Karnosky, D.F. 2002. Altered performance of forest pests under atmospheres enriched by CO_2 and O_3. *Nature* **420**: 403-407.

Plessl, M., Elstner, E.F., Rennenberg, H., Habermeyer, J. and Heiser, I. 2007. Influence of elevated CO_2 and ozone concentrations on late blight resistance and growth of potato plants. *Environmental and Experimental Botany* **60**: 447-457.

Tiedemann, A.V. and Firsching, K.H. 2000. Interactive effects of elevated ozone and carbon dioxide on growth and yield of leaf rust-infected versus non-infected wheat. *Environmental Pollution* **108**: 357-363.

Ywa, N.S., Walling, L. and McCool, P.M. 1995. Influence of elevated CO_2 on disease development and induction of PR proteins in tomato roots by *Phytophthora parasitica*. *Plant Physiolology* **85** (Supplement): 1139.

7.8.4. Parasitic Plants

Parasitic plants obtain energy, water and nutrients from their host plants and cause widespread reductions in harvestable crop yields around the globe. Hence, it is important to understand how rising atmospheric CO_2 levels may impact the growth of parasitic plants and the relationships between them and their hosts.

Matthies and Egli (1999) grew *Rhinanthus alectorolophus* (a widely distributed parasitic plant of Central Europe) for a period of two months on the grass *Lolium perenne* and the legume *Medicago sativa* in pots placed within open-top chambers maintained at atmospheric CO_2 concentrations of 375 and 590 ppm, half of which pots were fertilized to produce an optimal soil nutrient regime and half of which were unfertilized. At low nutrient supply, they found that atmospheric CO_2 enrichment decreased mean parasite biomass by an average of 16 percent, while at high nutrient supply it increased parasite biomass by an average of 123 percent. Nevertheless, the extra 215 ppm of CO_2 increased host plant biomass in both situations: by 29 percent under high soil nutrition and by 18 percent under low soil nutrition.

Dale and Press (1999) infected white clover (*Trifolium repens*) plants with *Orobanche minor* (a parasitic weed that primarily infects leguminous crops in the United Kingdom and the Middle East) and exposed them to atmospheric CO_2 concentrations of either 360 or 550 ppm for 75 days in controlled-environment growth cabinets. The elevated CO_2 in this study had no effect on the total biomass of parasite per host plant, nor did it impact the number of parasites per host plant or the time to parasitic attachment to host roots. On the other hand, whereas infected host plants growing in ambient air produced 47 percent less biomass than uninfected plants growing in ambient air, infected plants growing at 550 ppm CO_2 exhibited final dry weights that were only 20 percent less than those displayed by uninfected plants growing in the CO_2-enriched air, indicative of a significant CO_2-induced partial alleviation of parasite-induced biomass reductions in the white clover host plants.

Watling and Press (1997) infected several C_4 sorghum plants with *Striga hermonthica* and *Striga asiatica* (parasitic C_3 weeds of the semi-arid tropics that infest many grain crops) and grew them, along with uninfected control plants, for approximately two months in controlled-environment cabinets maintained at atmospheric CO_2 concentrations of 350 and 700 ppm. In the absence of parasite infection, the extra 350 ppm of CO_2 increased sorghum biomass by approximately 36 percent. When infected with *S. hermonthica*, however, the sorghum plants grown at ambient and elevated CO_2 concentrations only produced 32 and 43 percent of the biomass displayed by their respective uninfected controls. Infection with *S. asiatica* was somewhat less stressful and led to host biomass production that was about half that of uninfected controls in both ambient and CO_2-enriched air. The end result was that the doubling of the air's CO_2 content employed in this study increased sorghum biomass by 79 percent and 35 percent in the C_4 sorghum plants infected with *S. hermonthica* and *S. asiatica*, respectively.

Hwangbo *et al.* (2003) grew Kentucky Bluegrass (*Poa pratensis* L.) with and without infection by the C_3 chlorophyllous parasitic angiosperm *Rhinanthus minor* L. (a facultative hemiparasite found in natural and semi-natural grasslands throughout Europe) for eight weeks in open-top chambers maintained at ambient and elevated (650 ppm) CO_2 concentrations. At the end of the study, the parasite's biomass (when growing on its host) was 47 percent greater in the CO_2-enriched chambers, while its host exhibited only a 10 percent CO_2-induced increase in biomass in the parasite's absence but a nearly doubled 19 percent increase when infected by it.

Watling and Press (2000) grew upland rice (*Oryza sativa* L.) in pots in controlled-environment chambers maintained at 350 and 700 ppm CO_2 in either the presence or absence of the root parasite *S. hermonthica* for a period of 80 days after sowing, after which time the plants were harvested and weighed. In ambient air, the presence of the parasite reduced the biomass of the rice to only 35 percent of what it was in the absence of the parasite; whereas in air enriched with CO_2 the presence of the parasite reduced the biomass of infected plants to but 73 percent of what it was in the absence of the parasite.

In summary, these several observations suggest that the rising CO_2 content of the air can have wide and variable effects on parasitic plants, ranging from negative to positive growth responses, depending upon soil nutrition and host plant specificity. With respect to the infected host plants, elevated CO_2 generally tends to reduce the negative effects of parasitic infection, so that infected host plants continue to exhibit positive growth responses to elevated CO_2. It is likely that whatever the scenario with regard to parasitic infection, host plants will fare better under higher atmospheric CO_2 conditions than they do currently.

Additional information on this topic, including reviews of newer publications as they become available, can be found at http://www.co2science.org/subject/p/parasites.php.

References

Dale, H. and Press, M.C. 1999. Elevated atmospheric CO_2 influences the interaction between the parasitic angiosperm *Orobanche minor* and its host *Trifolium repens*. *New Phytologist* **140**: 65-73.

Hwangbo, J.-K., Seel, W.E. and Woodin, S.J. 2003. Short-term exposure to elevated atmospheric CO_2 benefits the growth of a facultative annual root hemiparasite, *Rhinanthus minor* (L.), more than that of its host, *Poa pratensis* (L.). *Journal of Experimental Botany* **54**: 1951-1955.

Matthies, D. and Egli, P. 1999. Response of a root hemiparasite to elevated CO_2 depends on host type and soil nutrients. *Oecologia* **120**: 156-161.

Watling, J.R. and Press, M.C. 1997. How is the relationship between the C_4 cereal *Sorghum bicolor* and the C_3 root hemi-parasites *Striga hermonthica* and *Striga asiatica* affected by elevated CO_2? *Plant, Cell and Environment* **20**: 1292-1300.

Watling, J.R. and Press, M.C. 2000. Infection with the parasitic angiosperm *Striga hermonthica* influences the response of the C_3 cereal *Oryza sativa* to elevated CO_2. *Global Change Biology* **6**: 919-930.

7.8.5. Roots

In reviewing the scientific literature pertaining to atmospheric CO_2 enrichment effects on belowground plant growth and development, Weihong *et al.* (2000) briefly summarize what is known about this subject. They report that atmospheric CO_2 enrichment typically enhances the growth rates of roots, especially those of fine roots, and that CO_2-induced increases in root production eventually lead to increased carbon inputs to soils, due to enhanced root turnover and exudation of various organic carbon compounds, which can potentially lead to greater soil carbon sequestration. In addition, they note that increased soil carbon inputs stimulate the growth and activities of soil microorganisms that utilize plant-derived carbon as their primary energy source; and they report that subsequently enhanced activities of fungal and bacterial plant symbionts often lead to increased plant nutrient acquisition.

In a much more narrowly focused study, Crookshanks *et al.* (1998) sprouted seeds of the small and fast-growing *Arabidopsis thaliana* plant on agar medium in Petri dishes and grew the resulting immature plants in controlled environment chambers maintained at atmospheric CO_2 concentrations of either 355 or 700 ppm. Visual assessments of root growth were made after emergence of the roots from the seeds, while microscopic investigations of root cell properties were also conducted. The scientists learned that the CO_2-enriched plants directed a greater proportion of their newly produced biomass into root, as opposed to shoot, growth. In addition, the young plants produced longer primary roots and more and longer lateral roots. These effects were found to be related to the CO_2-induced stimulation of mitotic activity, accelerated cortical cell expansion, and increased cell wall plasticity.

Gouk *et al.* (1999) grew an orchid plantlet, Mokara Yellow, in plastic bags flushed with 350 and 10,000 ppm CO_2 for three months to study the effects of elevated CO_2 on this epiphytic CAM species. They determined that the super-elevated CO_2 of their experiment enhanced the total dry weight of the orchid plantlets by more than two-fold, while increasing the growth of existing roots and stimulating the induction of new roots from internodes located on the orchid stems. Total chlorophyll content was also increased by elevated CO_2—by 64 percent in young leaves and by 118 percent in young roots. This phenomenon permitted greater light harvesting during photosynthesis and likely led to the tissue starch contents of the CO_2-enriched plantlets rising nearly 20-fold higher than those of the control-plantlets. In spite of this large CO_2-induced accumulation of starch, however, no damage or disruption of chloroplasts was evident in the leaves and roots of the CO_2-enriched plants.

A final question that has periodically intrigued researchers is whether plants take up carbon through their roots in addition to through their leaves. Although a definitive answer eludes us, various aspects of the issue have been described by Idso (1989), who we quote as follows.

"Although several investigators have claimed that plants should receive little direct benefit from dissolved CO_2 (Stolwijk *et al.*, 1957; Skok *et al.*, 1962; Splittstoesser, 1966), a number of experiments have produced significant increases in root growth (Erickson, 1946; Leonard and Pinckard, 1946; Geisler, 1963; Yorgalevitch and Janes, 1988), as well as yield itself (Kursanov *et al.*, 1951; Grinfeld, 1954; Nakayama and Bucks, 1980; Baron and Gorski, 1986), with CO_2-enriched irrigation water. Early on, Misra (1951) suggested that this beneficent effect may be related to CO_2-induced changes in soil nutrient availability; and this hypothesis may well be correct. Arteca *et al.* (1979), for example, have observed K, Ca and Mg to be better absorbed by potato roots when the concentration of CO_2 in the soil solution is increased; while Mauney and Hendrix (1988) found Zn and Mn to be better absorbed by cotton under such conditions, and Yurgalevitch and Janes (1988) found an enhancement of the absorption of Rb by tomato roots. In all cases, large increases in either total plant growth or yield accompanied the enhanced uptake of nutrients. Consequently, as it has been suggested that CO_2 concentration plays a major role in determining the porosity, plasticity and charge of cell membranes (Jackson and Coleman, 1959; Mitz, 1979), which could thereby alter ion uptake and organic acid production (Yorgalevitch and Janes, 1988), it is possible that some such suite of mechanisms may well be responsible for the plant productivity increases often observed to result from enhanced concentrations of CO_2 in the soil solution."

In the next two sections we survey the scientific literature on root responses to atmospheric CO_2 enrichment for crops and then trees.

Additional information on this topic, including reviews on roots not discussed here, can be found at http://www.co2science.org/subject/r/subject_r.php under the heading Roots.

References

Arteca, R.N., Pooviah, B.W. and Smith, O.E. 1979. Changes in carbon fixation, tuberization, and growth induced by CO_2 applications to the root zones of potato plants. *Science* **205**: 1279-1280.

Baron, J.J. and Gorski, S.F. 1986. Response of eggplant to a root environment enriched with CO_2. *HortScience* **21**: 495-498.

Crookshanks, M., Taylor, G. and Dolan, L. 1998. A model system to study the effects of elevated CO_2 on the developmental physiology of roots: the use of *Arabidopsis thaliana*. *Journal of Experimental Botany* **49**: 593-597.

Erickson, L.C. 1946. Growth of tomato roots as influenced by oxygen in the nutrient solution. *American Journal of Botany* **33**: 551-556.

Geisler, G. 1963. Morphogenetic influence of (CO_2 + HCO_3^-) on roots. *Plant Physiology* **38**: 77-80.

Gouk, S.S., He, J. and Hew, C.S. 1999. Changes in photosynthetic capability and carbohydrate production in an epiphytic CAM orchid plantlet exposed to super-elevated CO_2. *Environmental and Experimental Botany* **41**: 219-230.

Grinfeld, E.G. 1954. On the nutrition of plants with carbon dioxide through the roots. *Dokl. Akad. Nauk SSSR* **94**: 919-922.

Idso, S.B. 1989. Carbon Dioxide and Global Change: Earth in Transition. IBR Press, Tempe, AZ.

Jackson, W.A. and Coleman, N.T. 1959. Fixation of carbon dioxide by plant roots through phosphoenolpyruvate carboxylase. *Plant and Soil* **11**: 1-16.

Kursanov, A.L., Kuzin, A.M. and Mamul, Y.V. 1951. On the possibility for assimilation by plants of carbonates taken in with the soil solution. *Dokl. Akad. Nauk SSSR* **79**: 685-687.

Leonard, O.A. and Pinckard, J.A. 1946. Effect of various oxygen and carbon dioxide concentrations on cotton root development. *Plant Physiology* **21**: 18-36.

Mauney, J.R. and Hendrix, D.L. 1988. Responses of glasshouse grown cotton to irrigation with carbon dioxide-saturated water. *Crop Science* **28**: 835-838.

Misra, R.K. 1951. Further studies on the carbon dioxide factor in the air and soil layers near the ground. *Indian Journal of Meteorology and Geophysics* **2**: 284-292.

Mitz, M.A. 1979. CO_2 biodynamics: A new concept of cellular control. *Journal of Theoretical Biology* **80**: 537-551.

Nakayama, F.S. and Bucks, D.A. 1980. Using subsurface trickle system for carbon dioxide enrichment. In Jensen, M.H. and Oebker, N.F. (Eds.), *Proceedings of the 15th Agricultural Plastics Congress*, National Agricultural Plastics Association, Manchester, MO, pp. 13-18.

Skok, J., Chorney, W. and Broecker, W.S. 1962. Uptake of CO_2 by roots of Xanthium plants. *Botanical Gazette* **124**: 118-120.

Splittstoesser, W.E. 1966. Dark CO_2 fixation and its role in the growth of plant tissue. *Plant Physiology* **41**: 755-759.

Stolwijk, J.A.J. and Thimann, K.V. 1957. On the uptake of carbon dioxide and bicarbonate by roots and its influence on growth. *Plant Physiology* **32**: 513-520.

Weihong, L., Fusuo, Z. and Kezhi, B. 2000. Responses of plant rhizosphere to atmospheric CO_2 enrichment. *Chinese Science Bulletin* **45**: 97-101.

Yorgalevitch, C.M. and Janes, W.H. 1988. Carbon dioxide enrichment of the root zone of tomato seedlings. *Journal of Horticultural Science* **63**: 265-270.

7.8.5.1. Crops

Hodge and Millard (1998) grew narrowleaf plantain (*Plantago lanceolata*) seedlings for a period of six weeks in controlled environment growth rooms maintained at atmospheric CO_2 concentrations of either 400 or 800 ppm. By the end of this period, the plants in the 800-ppm air exhibited increases in shoot and root dry matter production that were 159 percent and 180 percent greater, respectively, than the corresponding dry matter increases experienced by the plants growing in 400-ppm air, while the amount of plant carbon recovered from the potting medium (sand) was 3.2 times greater in the elevated-CO_2 treatment. Thus, these investigators found that the belowground growth stimulation provided by atmospheric CO_2 enrichment was greater than that experienced above-ground.

Wechsung *et al.* (1999) grew spring wheat (*Triticum aestivum*) in rows in a FACE study employing atmospheric CO_2 concentrations of 370 and 550 ppm and irrigation treatments that periodically replaced either 50 percent or 100 percent of prior potential evapotranspiration in an effort to determine the effects of elevated CO_2 and water stress on root growth. They found that elevated CO_2 increased in-row root dry weight by an average of 22 percent during the growing season under both the wet and dry irrigation regimes. In addition, during the vegetative growth phase, atmospheric CO_2 enrichment increased inter-row root dry weight by 70 percent, indicating that plants grown in elevated CO_2 developed greater lateral root systems than plants grown at ambient CO_2. During the reproductive growth phase, elevated CO_2 stimulated the branching of lateral roots into inter-row areas, but only when water was limiting to growth. In addition, the CO_2-enriched plants tended to display greater root dry weights at a given depth than did ambiently grown plants.

In a comprehensive review of all prior FACE experiments conducted on agricultural crops, Kimball et al. (2002) determined that for a 300-ppm increase in atmospheric CO_2 concentration, the root biomass of wheat, ryegrass and rice experienced an average increase of 70 percent at ample water and nitrogen, 58 percent at low nitrogen and 34 percent at low water, while clover experienced a 38 percent increase at ample water and nitrogen, plus a 32 percent increase at low nitrogen. Outdoing all of the other crops was cotton, which exhibited a 96 percent increase in root biomass at ample water and nitrogen.

Zhao et al. (2000) germinated pea (Pisum sativum) seeds and exposed the young plants to various atmospheric CO_2 concentrations in controlled environment chambers to determine if elevated CO_2 impacts root border cells, which are major contributors of root exudates in this and most other agronomic plants. They found that elevated CO_2 increased the production of root border cells in pea seedlings. In going from ambient air to air enriched to 3,000 and 6,000 ppm CO_2, border-cell numbers increased by over 50 percent and 100 percent, respectively. Hence, as the CO_2 content of the air continues to rise, peas (and possibly many other crop plants) will likely produce greater numbers of root border cells, which should increase the amounts of root exudations occurring in their rhizospheres, which further suggests that associated soil microbial and fungal activities will be stimulated as a result of the increases in plant-derived carbon inputs that these organisms require to meet their energy needs.

Van Ginkel et al. (1996) grew perennial ryegrass (Lolium perenne) plants from seed in two growth chambers for 71 days under continuous $^{14}CO_2$-labeling of the atmosphere at CO_2 concentrations of 350 and 700 ppm at two different soil nitrogen levels. At the conclusion of this part of the experiment, the plants were harvested and their roots dried, pulverized and mixed with soil in a number of one-liter pots that were placed within two wind tunnels in an open field, one of which had ambient air of 361 ppm CO_2 flowing through it, and one of which had air of 706 ppm CO_2 flowing through it. Several of the containers were then seeded with more Lolium perenne, others were similarly seeded the following year, and still others were kept bare for two years. Then, at the ends of the first and second years, the different degrees of decomposition of the original plant roots were assessed.

It was determined, first, that shoot and root growth were enhanced by 13 and 92 percent, respectively, by the extra CO_2 in the initial 71-day portion of the experiment, once again demonstrating the significant benefits that are often conferred upon plant roots by atmospheric CO_2 enrichment. Secondly, it was found that the decomposition of the high-CO_2-grown roots in the high-CO_2 wind tunnel was 19 percent lower than that of the low-CO_2-grown roots in the low-CO_2 wind tunnel at the end of the first year, and that it was 14 percent lower at the end of the second year in the low-nitrogen-grown plants but equivalent in the high-nitrogen-grown plants. It was also determined that the presence of living roots reduced the decomposition rate of dead roots below the dead-root-only decomposition rate observed in the bare soil treatment. Based on these findings, van Ginkel et al. conclude that "the combination of higher root yields at elevated CO_2 combined with a decrease in root decomposition will lead to a longer residence time of C in the soil and probably to a higher C storage."

In conclusion, as the CO_2 content of the air continues to rise, many crops will likely develop larger and more extensively branching root systems that may help them to better cope with periods of reduced soil moisture availability. This chain of events should make the soil environment even more favorable for plant growth and development in a high-CO_2 world of the future.

Additional information on this topic, including reviews of newer publications as they become available, can be found at http://www.co2science.org/subject/r/rootscrops.php.

References

Hodge, A. and Millard, P. 1998. Effect of elevated CO_2 on carbon partitioning and exudate release from Plantago lanceolata seedlings. Physiologia Plantarum 103: 280-286.

Kimball, B.A., Kobayashi, K. and Bindi, M. 2002. Responses of agricultural crops to free-air CO_2 enrichment. Advances in Agronomy 77: 293-368.

Van Ginkel, J.H., Gorissen, A. and van Veen, J.A. 1996. Long-term decomposition of grass roots as affected by elevated atmospheric carbon dioxide. Journal of Environmental Quality 25: 1122-1128.

Wechsung, G., Wechsung, F., Wall, G.W., Adamsen, F.J., Kimball, B.A., Pinter Jr., P.J., LaMorte, R.L., Garcia, R.L. and Kartschall, T. 1999. The effects of free-air CO_2 enrichment and soil water availability on spatial and seasonal patterns of wheat root growth. *Global Change Biology* **5**: 519-529.

Zhao, X., Misaghi, I.J. and Hawes, M.C. 2000. Stimulation of border cell production in response to increased carbon dioxide levels. *Plant Physiology* **122**: 181-188.

7.8.5.2. Trees

Janssens *et al.* (1998) grew three-year-old Scots pine seedlings for a period of six months in open-top chambers maintained at ambient and 700 ppm atmospheric CO_2, finding that the extra CO_2 increased total root length by 122 percent and total root dry mass by 135 percent. In a similar study that employed close to the same degree of enhancement of the air's CO_2 content, Pritchard *et al.* (2001a) grew idealized ecosystems representative of regenerating longleaf pine forests of the southeastern USA for a period of 18 months in large soil bins located within open-top chambers. The above-ground parts of these seedlings experienced a growth enhancement of 20 percent. The root biomass of the trees, however, was increased by more than three times as much (62 percent).

Working with FACE technology, Pritchard *et al.* (2001b) studied 14-year-old loblolly pine trees after a year of exposure to an extra 200 ppm of CO_2, finding that total standing root length and root numbers were 16 and 34 percent greater, respectively, in the CO_2-enriched plots than in the ambient-air plots. In addition, the elevated CO_2 increased the diameter of living and dead roots by 8 and 6 percent, respectively, while annual root production was found to be 26 percent greater in the CO_2-enriched plots. For the degree of CO_2 enrichment employed in the prior two studies, this latter enhancement corresponds to a root biomass increase of approximately 45 percent.

In an open-top chamber study of a model ecosystem composed of a mixture of spruce and beech seedlings, Wiemken *et al.* (2001) investigated the effects of a 200 ppm increase in the air's CO_2 concentration that prevailed for a period of four years. On nutrient-poor soils, the extra CO_2 led to a 30 percent increase in fine-root biomass, while on nutrient-rich soils it led to a 75 percent increase. These numbers correspond to increases of about 52 percent and 130 percent, respectively, for atmospheric

CO_2 enhancements on the order of those employed by Janssens *et al.* (1998) and Pritchard *et al.* (2001a).

Another interesting aspect of the Wiemken *et al.* study was their finding that the extra CO_2 increased the amount of symbiotic fungal biomass associated with the trees' fine roots by 31 percent on nutrient-poor soils and by 100 percent on nutrient-rich soils, which for the degree of atmospheric CO_2 enrichment used in the studies of Janssens *et al.* (1998) and Pritchard *et al.* (2001a) translate into increases of about 52 percent and 175 percent, respectively.

Berntsen and Bazzaz (1998) removed intact chunks of soil from the Hardwood-White Pine-Hemlock forest region of New England and placed them in plastic containers within controlled environment glasshouses maintained at either 375 or 700 ppm CO_2 for a period of two years in order to study the effects of elevated CO_2 on the regeneration of plants from seeds and rhizomes present in the soil. At the conclusion of the study, total mesocosm plant biomass (more than 95 percent of which was supplied by yellow and white birch tree seedlings) was found to be 31 percent higher in the elevated CO_2 treatment than in ambient air, with a mean enhancement of 23 percent above-ground and 62 percent belowground. The extra CO_2 also increased the mycorrhizal colonization of root tips by 45 percent in white birch and 71 percent in yellow birch; and the CO_2-enriched yellow birch seedlings exhibited 322 percent greater root length and 305 percent more root surface area than did the yellow birch seedlings growing in ambient air.

Kubiske *et al.* (1998) grew cuttings of four quaking aspen genotypes in open-top chambers for five months at atmospheric CO_2 concentrations of either 380 or 720 ppm and low or high soil nitrogen concentrations. They found, surprisingly, that the cuttings grown in elevated CO_2 displayed no discernible increases in above-ground growth. However, the extra CO_2 significantly increased fine-root length and root turnover rates at high soil nitrogen by increasing fine-root production, which would logically be expected to produce benefits (not the least of which would be a larger belowground water- and nutrient-gathering system) that would eventually lead to enhanced above-ground growth as well.

Expanding on this study, Pregitzer *et al.* (2000) grew six quaking aspen genotypes for 2.5 growing seasons in open-top chambers maintained at atmospheric CO_2 concentrations of 350 and 700 ppm with both adequate and inadequate supplies of soil nitrogen. This work demonstrated that the trees exposed to elevated CO_2 developed thicker and longer roots than the trees growing in ambient air, and that the fine-root biomass of the CO_2-enriched trees was enhanced by 17 percent in the nitrogen-poor soils and by 65 percent in the nitrogen-rich soils.

Yet another study of quaking aspen conducted by King *et al.* (2001) demonstrated that trees exposed to an atmospheric CO_2 concentration 560 ppm in a FACE experiment produced 133 percent more fine-root biomass than trees grown in ambient air of 360 ppm, which roughly equates to 233 percent more fine-root biomass for the degree of CO_2 enrichment employed in the prior study of Pregitzer *et al.* And when simultaneously exposed to air of 1.5 times the normal ozone concentration, the degree of fine-root biomass stimulation produced by the extra CO_2 was still as great as 66 percent, or roughly 115 percent when extrapolated to the greater CO_2 enrichment employed by Pregitzer *et al.*

In a final quaking aspen study, King *et al.* (1999) grew four clones at two different temperature regimes (separated by 5°C) and two levels of soil nitrogen (N) availability (high and low) for 98 days, while measuring photosynthesis, growth, biomass allocation, and root production and mortality. They found that the higher of the two temperature regimes increased rates of photosynthesis by 65 percent and rates of whole-plant growth by 37 percent, while it simultaneously enhanced root production and turnover. It was thus their conclusion that "trembling aspen has the potential for substantially greater growth and root turnover under conditions of warmer soil at sites of both high and low N-availability" and that "an immediate consequence of this will be greater inputs of C and nutrients to forest soils."

In light of these several findings pertaining to quaking aspen trees, it is evident that increases in atmospheric CO_2 concentration, air temperature and soil nitrogen content all enhance their belowground growth, which positively impacts their above-ground growth.

Turning our attention to other deciduous trees, Gleadow *et al.* (1998) grew eucalyptus seedlings for six months in glasshouses maintained at atmospheric CO_2 concentrations of either 400 or 800 ppm, fertilizing them twice daily with low or high nitrogen solutions. The elevated CO_2 of their experiment increased total plant biomass by 98 and 134 percent relative to plants grown at ambient CO_2 in the high and low nitrogen treatments, respectively. In addition, in the low nitrogen treatment, elevated CO_2

stimulated greater root growth, as indicated by a 33 percent higher root:shoot ratio.

In a more complex study, Day *et al.* (1996) studied the effects of elevated CO_2 on fine-root production in open-top chambers erected over a regenerating oak-palmetto scrub ecosystem in Florida, USA, determining that a 350-ppm increase in the atmosphere's CO_2 concentration increased fine-root length densities by 63 percent while enhancing the distribution of fine roots at both the soil surface (0-12 cm) and at a depth of 50-60 cm. These findings suggest that the ongoing rise in the atmosphere's CO_2 concentration will likely increase the distribution of fine roots near the soil surface, where the greatest concentrations of nutrients are located, and at a depth that coincides with the upper level of the site's water table, both of which phenomena should increase the trees' ability to acquire the nutrients and water they will need to support CO_2-enhanced biomass production in the years ahead.

In another study that employed CO_2, temperature and nitrogen as treatments, Uselman *et al.* (2000) grew seedlings of the nitrogen-fixing black locust tree for 100 days in controlled environments maintained at atmospheric CO_2 concentrations of 350 and 700 ppm and air temperatures of 26°C (ambient) and 30°C, with either some or no additional nitrogen fertilization, finding that the extra CO_2 increased total seedling biomass by 14 percent, the elevated temperature increased it by 55 percent, and nitrogen fertilization increased it by 157 percent. With respect to root exudation, a similar pattern was seen. Plants grown in elevated CO_2 exuded 20 percent more organic carbon compounds than plants grown in ambient air, while elevated temperature and fertilization increased root exudation by 71 and 55 percent, respectively. Hence, as the air's CO_2 content continues to rise, black locust trees will likely exhibit enhanced rates of biomass production and exudation of dissolved organic compounds from their roots. Moreover, if air temperature also rises, even by as much as 4°C, its positive effect on biomass production and root exudation will likely be even greater than that resulting from the increasing atmospheric CO_2 concentration. The same would appear to hold true for anthropogenic nitrogen deposition, reinforcing what was learned about the impacts of these three environmental factors on the growth of quaking aspen trees.

In light of these several experimental findings, it can confidently be concluded that the ongoing rise in the air's CO_2 content, together with possible concurrent increases in air temperature and nitrogen deposition, will likely help earth's woody plants increase their root mass and surface area to become ever more robust and productive.

Additional information on this topic, including reviews of newer publications as they become available, can be found at http://www.co2science.org/subject/r/rootsconifers.php and http://www.co2science.org/subject/r/rootsdeciduous.php.

References

Berntson, G.M. and Bazzaz, F.A. 1998. Regenerating temperate forest mesocosms in elevated CO_2: belowground growth and nitrogen cycling. *Oecologia* **113**: 115-125.

Day, F.P., Weber, E.P., Hinkle, C.R. and Drake, B.G. 1996. Effects of elevated atmospheric CO_2 on fine root length and distribution in an oak-palmetto scrub ecosystem in central Florida. *Global Change Biology* **2**: 143-148.

Gleadow, R.M., Foley, W.J. and Woodrow, I.E. 1998. Enhanced CO_2 alters the relationship between photosynthesis and defense in cyanogenic *Eucalyptus cladocalyx* F. Muell. *Plant, Cell and Environment* : 12-22.

Janssens, I.A., Crookshanks, M., Taylor, G. and Ceulemans, R. 1998. Elevated atmospheric CO_2 increases fine root production, respiration, rhizosphere respiration and soil CO_2 efflux in Scots pine seedlings. *Global Change Biology* **4**: 871-878.

King, J.S., Pregitzer, K.S. and Zak, D.R. 1999. Clonal variation in above- and below-ground growth responses of *Populus tremuloides* Michaux: Influence of soil warming and nutrient availability. *Plant and Soil* **217**: 119-130.

King, J.S., Pregitzer, K.S., Zak, D.R., Sober, J., Isebrands, J.G., Dickson, R.E., Hendrey, G.R. and Karnosky, D.F. 2001. Fine-root biomass and fluxes of soil carbon in young stands of paper birch and trembling aspen as affected by elevated atmospheric CO_2 and tropospheric O_3. *Oecologia* **128**: 237-250.

Kubiske, M.E., Pregitzer, K.S., Zak, D.R. and Mikan, C.J. 1998. Growth and C allocation of *Populus tremuloides* genotypes in response to atmospheric CO_2 and soil N availability. *New Phytologist* **140**: 251-260.

Pregitzer, K.S., Zak, D.R., Maziaasz, J., DeForest, J., Curtis, P.S. and Lussenhop, J. 2000. Interactive effects of atmospheric CO_2 and soil-N availability on fine roots of *Populus tremuloides*. *Ecological Applications* **10**: 18-33.

Pritchard, S.G., Davis, M.A., Mitchell, R.J., Prior, A.S., Boykin, D.L., Rogers, H.H. and Runion, G.B. 2001a. Root dynamics in an artificially constructed regenerating longleaf pine ecosystem are affected by atmospheric CO_2 enrichment. *Environmental and Experimental Botany* **46**: 35-69.

Pritchard, S.G., Rogers, H.H., Davis, M.A., Van Santen, E., Prior, S.A. and Schlesinger, W.H. 2001b. The influence of elevated atmospheric CO_2 on fine root dynamics in an intact temperate forest. *Global Change Biology* **7**: 829-837.

Uselman, S.M., Qualls, R.G. and Thomas, R.B. 2000. Effects of increased atmospheric CO_2, temperature, and soil N availability on root exudation of dissolved organic carbon by a N-fixing tree (*Robinia pseudoacacia* L.). *Plant and Soil* **222**: 191-202.

Wiemken, V., Ineichen, K. and Boller, T. 2001. Development of ectomycorrhizas in model beech-spruce ecosystems on siliceous and calcareous soil: a 4-year experiment with atmospheric CO_2 enrichment and nitrogen fertilization. *Plant and Soil* **234**: 99-108.

7.8.6. Seeds

Elevated CO_2 levels are known to have effects on seeds that are different from their effects on total biomass, roots, and other dimensions examined so far. In this section we survey the scientific literature on this topic regarding crops, grasslands, and trees.

Additional information on this topic, including reviews on seeds not discussed here, can be found at http://www.co2science.org/subject/s/subject_s.php under the heading Seeds.

7.8.6.1. Crops

In a greenhouse study of the various components of seed biomass production, Palta and Ludwig (2000) grew narrow-leafed lupin in pots filled with soil within Mylar-film tunnels maintained at either 355 or 700 ppm CO_2. They found that the extra CO_2 increased (1) the final number of pods and (2) the number of pods that filled large seeds, while it (3) reduced to zero the number of pods that had small seeds, (4) reduced the number of pods with unfilled seeds from 16 to 1 pod per plant, and increased (5) pod set and (6) dry matter accumulation on the developing branches. These several CO_2-induced improvements to key physiological processes resulted in 47 to 56 percent increases in dry matter per plant, which led to increases of 44 to 66 percent in seed yield per plant.

Sanhewe *et al.* (1996) grew winter wheat in polyethylene tunnels maintained at atmospheric CO_2 concentrations of 380 and 680 ppm from the time of seed germination to the time of plant maturity, while maintaining a temperature gradient of approximately 4°C in each tunnel. In addition to the elevated CO_2 increasing seed yield per unit area, they found it also increased seed weight, but not seed survival or germination. Increasing air temperature, on the other hand, increased seed longevity across the entire range of temperatures investigated (14 to 19°C).

Thomas et al. (2003) grew soybean plants to maturity in sunlit controlled-environment chambers under sinusoidally varying day/night-max/min temperatures of 28/18, 32/22, 36/26, 40/30 and 44/34°C and two levels of atmospheric CO_2 concentration (350 and 700 ppm). They determined, in their words, that the effect of temperature on seed composition and gene expression was "pronounced," but that "there was no effect of CO_2." In this regard, however, they note that "Heagle et al. (1998) observed a positive significant effect of CO_2 enrichment on soybean seed oil and oleic acid concentration," the latter of which parameters Thomas et al. found to increase with rising temperature all the way from 28/18 to 44/34°C.

In another soybean study, Ziska et al. (2001) grew one modern and eight ancestral genotypes in glasshouses maintained at atmospheric CO_2 concentrations of 400 and 710 ppm, finding that the extra CO_2 increased photosynthetic rates by an average of 75 percent. This enhancement in photosynthetic sugar production led to increases in seed yield that averaged 40 percent for all cultivars, except for one ancestral variety that exhibited an 80 percent increase in seed yield. Hence, if plant breeders were to utilize the highly CO_2-responsive ancestral cultivar identified in this study in their breeding programs, it is possible that soybean seed yields could be made to rise even faster and higher in the days and years ahead.

Additional information on this topic, including reviews of newer publications as they become available, can be found at http://www.co2science.org/subject/s/seedscrops.php.

References

Heagle, A.S., Miller, J.E. and Pursley, W.A. 1998. Influence of ozone stress on soybean response to carbon dioxide enrichment: III. Yield and seed quality. Crop Science 38: 128-134.

Palta, J.A. and Ludwig, C. 2000. Elevated CO_2 during pod filling increased seed yield but not harvest index in indeterminate narrow-leafed lupin. Australian Journal of Agricultural Research 51: 279-286.

Sanhewe, A.J., Ellis, R.H., Hong, T.D., Wheeler, T.R., Batts, G.R., Hadley, P. and Morison, J.I.L. 1996. The effect of temperature and CO_2 on seed quality development in wheat (Triticum aestivum L.). Journal of Experimental Botany 47: 631-637.

Thomas, J.M.G., Boote, K.J., Allen Jr., L.H., Gallo-Meagher, M. and Davis, J.M. 2003. Elevated temperature and carbon dioxide effects on soybean seed composition and transcript abundance. Crop Science 43: 1548-1557.

Ziska, L.H., Bunce, J.A. and Caulfield, F.A. 2001. Rising atmospheric carbon dioxide and seed yields of soybean genotypes. Crop Science 41: 385-391.

7.8.6.2. Grasslands

Steinger et al. (2000) collected seeds from Bromus erectus plants that had been grown at atmospheric CO_2 concentrations of 360 and 650 ppm and germinated some of both groups of seeds under those same two sets of conditions. They found the elevated CO_2 treatment increased individual seed mass by about 9 percent and increased seed carbon-to-nitrogen ratio by almost 10 percent. However, they also learned that these changes in seed properties had little impact on subsequent seedling growth. In fact, when the seeds produced by ambient or CO_2-enriched plants were germinated and grown in ambient air, there was no significant size difference between the two groups of resultant seedlings after a period of 19 days. Likewise, when the seeds produced from ambient or CO_2-enriched plants were germinated and grown in the high CO_2 treatment, there was also no significant difference between the sizes of the seedlings derived from the two groups of seeds. However, the CO_2-enriched seedlings produced from both groups of seeds were almost 20 percent larger than the seedlings produced from both groups of seeds grown in ambient air, demonstrating that the direct effects of elevated atmospheric CO_2 concentration on seedling growth and development were more important than the differences in seed characteristics produced by the elevated atmospheric CO_2 concentration in which their parent plants grew.

In another study conducted about the same time, Edwards et al. (2001) utilized a FACE experiment where daytime atmospheric CO_2 concentrations above a sheep-grazed pasture in New Zealand were increased by 115 ppm to study the effects of elevated CO_2 on seed production, seedling recruitment and species compositional changes. In the two years of their study, the extra daytime CO_2 increased seed production and dispersal in seven of the eight most abundant species, including the grasses *Anthoxanthum odoratum*, *Lolium perenne* and *Poa pratensis*, the legumes *Trifolium repens* and *T. subterranean*, and the herbs *Hypochaeris radicata* and *Leontodon saxatilis*. In some of these plants, elevated CO_2 increased the number of seeds per reproductive structure, while all of them exhibited CO_2-induced increases in the number of reproductive structures per unit of ground area. In addition, they determined that the CO_2-induced increases in seed production contributed in a major way to the increase in the numbers of species found within the CO_2-enriched plots.

In a five-year study of a nutrient-poor calcareous grassland in Switzerland, Thurig *et al.* (2003) used screen-aided CO_2 control (SACC) technology (Leadley *et al.*, 1997) to enrich the air over half of their experimental plots with an extra 300 ppm of CO_2, finding that "the effect of elevated CO_2 on the number of flowering shoots (+24%) and seeds (+29%) at the community level was similar to above ground biomass response." In terms of species functional groups, there was a 42 percent increase in the mean seed number of graminoids and a 33 percent increase in the mean seed number of forbs, but no change in legume seed numbers. In most species, mean seed weight also tended to be greater in plants grown in CO_2-enriched air (+12 percent); and Thurig *et al.* say it is known from many studies that heavier seeds result in seedlings that "are more robust than seedlings from lighter seeds (Baskin and Baskin, 1998)."

Wang and Griffin (2003) grew dioecious white cockle plants from seed to maturity in sand-filled pots maintained at optimum moisture and fertility conditions in environmentally controlled growth chambers in which the air was continuously maintained at CO_2 concentrations of either 365 or 730 ppm. In response to this doubling of the air's CO_2 content, the vegetative mass of both male and female plants rose by approximately 39 percent. Reproductive mass, on the other hand, rose by 82 percent in male plants and by 97 percent in females. In the female plants, this feat was accomplished, in part, by increases of 36 percent and 44 percent in the number and mass of seeds per plant, and by a 15 percent increase in the mass of individual seeds, in harmony with the findings of Jablonski *et al.* (2002), which they derived from a meta-analysis of the results

of 159 CO_2 enrichment experiments conducted on 79 species of agricultural and wild plants. Because dioecious plants comprise nearly half of all angiosperm families, we may expect to see a greater proportion of plant biomass allocated to reproduction in a high-CO_2 world of the future, which result should bode well for the biodiversity of earth's many ecosystems.

Additional information on this topic, including reviews of newer publications as they become available, can be found at http://www.co2science.org/subject/s/seedsgrasslands.php.

References

Baskin, C.C. and Baskin, J.M. 1998. Seeds: Ecology, Biogeography, and Evolution of Dormancy and Germination. Academic Press, San Diego, CA.

Edwards, G.R., Clark, H. and Newton, P.C.D. 2001. The effects of elevated CO_2 on seed production and seedling recruitment in a sheep-grazed pasture. *Oecologia* **127**: 383-394.

Jablonski, L.M., Wang, X. and Curtis, P.S. 2002. Plant reproduction under elevated CO_2 conditions: a meta-analysis of reports on 79 crop and wild species. *New Phytologist* **156**: 9-26.

Leadley, P.W., Niklaus, P.A., Stocker, R. *et al.* 1997. Screen-aided CO_2 control (SACC): a middle ground between FACE and open-top chambers. *Acta Oecologica* **18**: 39-49.

Steinger, T., Gall, R. and Schmid, B. 2000. Maternal and direct effects of elevated CO_2 on seed provisioning, germination and seedling growth in *Bromus erectus*. *Oecologia* **123**: 475-480.

Thurig, B., Korner, C. and Stocklin, J. 2003. Seed production and seed quality in a calcareous grassland in elevated CO_2. *Global Change Biology* **9**: 873-884.

Wang, X. and Griffin, K.L. 2003. Sex-specific physiological and growth responses to elevated atmospheric CO_2 in *Silene latifolia* Poiret. *Global Change Biology* **9**: 612-618.

7.8.6.3. Trees

How does enriching the air with carbon dioxide impact the reproductive capacity of trees? LaDeau and Clark (2001) determined the reproductive response of loblolly pine trees to atmospheric CO_2 enrichment at Duke Forest in the Piedmont region of

North Carolina, USA, where in August of 1996 three 30-m-diameter FACE rings began to enrich the air around the 13-year-old trees they encircled to 200 ppm above the atmosphere's normal background concentration, while three other FACE rings served as control plots. Because the trees were not mature at the start of the experiment they did not produce any cones until a few rare ones appeared in 1998. By the fall of 1999, however, the two scientists found that, compared to the trees growing in ambient air, the CO_2-enriched trees were twice as likely to be reproductively mature, and they produced three times more cones per tree. Similarly, the trees growing in the CO_2-enriched air produced 2.4 times more cones in the fall of 2000; and from August 1999 through July 2000, they collected three times as many seeds in the CO_2-fertilized FACE rings as in the control rings.

Also working on this aspect of the Duke Forest FACE study were Hussain *et al.* (2001), who report that (1) seeds collected from the CO_2-enriched trees were 91 percent heavier than those collected from the trees growing in ambient air, (2) the CO_2-enriched seeds had a lipid content that was 265 percent greater than that of the seeds produced on the ambient-treatment trees, (3) the germination success for seeds developed under atmospheric CO_2 enrichment was more than three times greater than that observed for control seeds developed at ambient CO_2, regardless of germination CO_2 concentration, (4) seeds from the CO_2-enriched trees germinated approximately five days earlier than their ambiently produced counterparts, again regardless of germination CO_2 concentration, and (5) seedlings developing from seeds collected from CO_2-enriched trees displayed significantly greater root lengths and needle numbers than seedlings developing from trees exposed to ambient air, also regardless of growth CO_2 concentration.

The propensity for elevated levels of atmospheric CO_2 to hasten the production of more plentiful seeds on the trees of this valuable timber species bodes well for naturally regenerating loblolly pine stands of the southeastern United States, where LaDeau and Clark report the trees "are profoundly seed-limited for at least 25 years." As the air's CO_2 content continues to climb, they conclude that "this period of seed limitation may be reduced." In addition, the observations of Hussain *et al.* suggest that loblolly pine trees in a CO_2-enriched world of the future will likely display significant increases in their photosynthetic rates. Enhanced carbohydrate supplies resulting from this phenomenon will likely be used to increase seed weight and lipid content. Such seeds should consequently exhibit significant increases in germination success, and their enhanced lipid supplies will likely lead to greater root lengths and needle numbers in developing seedlings. Consequently, when CO_2-enriched loblolly pine seedlings become photosynthetically active, they will likely produce biomass at greater rates than those exhibited by seedlings growing under current CO_2 concentrations.

Five years later, LaDeau and Clark (2006a) conducted a follow-up study that revealed "carbon dioxide enrichment affected mean cone production both through early maturation and increased fecundity," so that "trees in the elevated CO_2 plots produced twice as many cones between 1998 and 2004 as trees in the ambient plots." They also determined that the trees grown in elevated CO_2 "made the transition to reproductive maturation at smaller [trunk] diameters," and that they "not only reached reproductive maturation at smaller diameters, but also at younger ages." By 2004, for example, they say that "roughly 50% of ambient trees and 75% of fumigated trees [had] produced cones." In addition, they observed that "22% of the trees in high CO_2 produced between 40 and 100 cones during the study, compared with only 9% of ambient trees."

"In this 8-year study," in the words of the two researchers, "we find that previous short-term responses indeed persist." In addition, they note that "*P. taeda* trees that produce large seed crops early in their life span tend to continue to be prolific producers (Schutlz, 1997)," and they conclude that this fact, together with their findings, suggests that "individual responses seen in this young forest may be sustained over their life span."

In a concurrent report, LaDeau and Clark (2006b) analyzed the seed and pollen responses of the loblolly pines to atmospheric CO_2 enrichment, finding that the "trees grown in high-CO_2 plots first began producing pollen while younger and at smaller sizes relative to ambient-grown trees," and that cone pollen and airborne pollen grain abundances were significantly greater in the CO_2-enriched stands. More specifically, they found that "by spring 2005, 63% of all trees growing in high CO_2 had produced both pollen and seeds *vs.* only 36% of trees in the ambient plots." The researchers say precocious pollen production "could enhance the production of viable seeds by increasing the percentage of fertilized ovules," and that "more pollen disseminated from multiple-source trees may also increase rates of gene flow among stands, and could further reduce rates of self-pollination, indirectly enhancing the production of viable seeds." They also say "pine pollen is not a dangerous allergen for the public at large."

Another major study of the reproductive responses of trees to elevated levels of atmospheric CO_2 was conducted at the Kennedy Space Center, Florida, USA, where in 1996 three species of scrub-oak (*Quercus myrtifolia*, *Q. chapmanii*, and *Q. geminata*) were enclosed within sixteen open-top chambers, half of which were maintained at 379 ppm CO_2 and half at 704 ppm. Five years later—in August, September and October of 2001—Stiling *et al.* (2004) counted the numbers of acorns on randomly selected twigs of each species, while in November of that year they counted the numbers of fallen acorns of each species within equal-size quadrates of ground area, additionally evaluating mean acorn weight, acorn germination rate, and degree of acorn infestation by weevils. They found acorn germination rate and degree of predation by weevils were unaffected by elevated CO_2, while acorn size was enhanced by a small amount: 3.6 percent for *Q. myrtifolia*, 7.0 percent for *Q. chapmanii*, and 7.7 percent for *Q. geminata*. Acorn number responses, on the other hand, were enormous, but for only two of the three species, as *Q. geminata* did not register any CO_2-induced increase in reproductive output, in harmony with its unresponsive overall growth rate. For *Q. myrtifolia*, however, Stiling *et al.* report "there were four times as many acorns per 100 twigs in elevated CO_2 as in ambient CO_2 and for *Q. chapmanii* the increase was over threefold." On the ground, the enhancement was greater still, with the researchers reporting that "the number of *Q. myrtifolia* acorns per meter squared in elevated CO_2 was over seven times greater than in ambient CO_2 and for *Q. chapmanii*, the increase was nearly sixfold."

Stiling *et al.* say these results lead them to believe "there will be large increases in seedling production in scrub-oak forests in an atmosphere of elevated CO_2," noting that "this is important because many forest systems are 'recruitment-limited' (Ribbens *et al.*, 1994; Hubbell *et al.*, 1999)," which conclusion echoes that of LaDeau and Clark with respect to loblolly pines.

A third major study of CO_2 effects on seed production in trees has been conducted at the FACE facility near Rhinelander, Wisconsin (USA), where young paper birch (*Betula papyrifera* Marsh.) seedlings were planted in 1997 and have been growing since 1998 in open-top chambers maintained at atmospheric CO_2 concentrations of either 360 or

560 ppm, as well as at atmospheric ozone (O_3) concentrations of either ambient or 1.5 times ambient. There, Darbah *et al.* (2007) collected many types of data pertaining to flowering, seed production, seed germination and new seedling growth and development over the 2004-2006 growing seasons; and as they describe it, "elevated CO_2 had significant positive effect[s] on birch catkin size, weight, and germination success rate." More specifically, they note that "elevated CO_2 increased germination rate of birch by 110%, compared to ambient CO_2 concentrations, decreased seedling mortality by 73%, increased seed weight by 17% [and] increased [new seedling] root length by 59%."

In conclusion, research on a variety of tree species finds that CO_2 enhancement increases the production of viable seeds and seedlings, meaning these species should flourish as CO_2 levels conntinue to rise. Additional information on this topic, including reviews of newer publications as they become available, can be found at http://www.co2science.org/subject/s/seedstrees.php.

References

Darbah, J.N.T., Kubiske, M.E., Nelson, N., Oksanen, E., Vaapavuori, E. and Karnosky, D.F. 2007. Impacts of elevated atmospheric CO_2 and O_3 on paper Birch (*Betula papyrifera*): Reproductive fitness. *The Scientific World JOURNAL* **7**(S1): 240-246.

Hubbell, S.P., Foster, R.B., O'Brien, S.T., Harms, K.E., Condit, R., Wechsler, B., Wright, S.J. and Loo de Lao, S. 1999. Light-gap disturbances, recruitment limitation, and tree diversity in a neotropical forest. *Science* **283**: 554-557.

Hussain, M., Kubiske, M.E. and Connor, K.F. 2001. Germination of CO_2-enriched *Pinus taeda* L. seeds and subsequent seedling growth responses to CO_2 enrichment. *Functional Ecology* **15**: 344-350.

LaDeau, S.L. and Clark, J.S. 2001. Rising CO_2 levels and the fecundity of forest trees. *Science* **292**: 95-98.

LaDeau, S.L. and Clark, J.S. 2006a. Elevated CO_2 and tree fecundity: the role of tree size, interannual variability, and population heterogeneity. *Global Change Biology* **12**: 822-833.

LaDeau, S.L. and Clark, J.S. 2006b. Pollen production by *Pinus taeda* growing in elevated atmospheric CO_2. *Functional Ecology* **20**: 541-547.

Ribbens, E., Silander, J.A. and Pacala, S.W. 1994. Seedling recruitment in forests: calibrating models to predict patterns of tree seedling dispersion. *Ecology* **75**: 1794-1806.

Schutlz, R.P. 1997. *Loblolly Pine—The Ecology and Culture of Loblolly Pine (Pinus taeda L.).* USDA Forest Service Agricultural Handbook 713. USDA Forest Service, Washington, DC, USA.

Stiling, P., Moon, D., Hymus, G. and Drake, B. 2004. Differential effects of elevated CO_2 on acorn density, weight, germination, and predation among three oak species in a scrub-oak forest. *Global Change Biology* **10**: 228-232.

7.8.7. Tannins

Condensed tannins are naturally occurring secondary carbon compounds produced in the leaves of a number of different plants that often act to deter herbivorous insects. How do condensed tannin concentrations in the leaves and roots of trees respond to atmospheric CO_2 enrichment?

Additional information on this topic, including reviews on tannins not discussed here, can be found at http://www.co2science.org/subject/t/subject_t.php under the heading Tannins.

7.8.7.1. Aspen Trees

King *et al.* (2001) grew aspen seedlings for five months in open-top chambers maintained at atmospheric CO_2 concentrations of either 350 or 700 ppm. At the end of this period, naturally senesced leaf litter was collected and analyzed; and it was found that the elevated CO_2 of this particular study had no effect on leaf litter tannin concentration.

A substantially different result was obtained in an earlier study of aspen leaves that was conducted by McDonald *et al.* (1999), who grew aspen seedlings in controlled environment greenhouses that were maintained at either ambient (387 ppm) or elevated (696 ppm) CO_2 concentrations under conditions of either low or high light availability (half and full sunlight, respectively) for 31 days after the mean date of bud break. In this case it was determined that under low light conditions, the CO_2-enriched seedlings exhibited an increase of approximately 15 percent in leaf condensed tannin concentration, while under high light conditions the CO_2-induced increase in leaf condensed tannin concentration was 175 percent.

In a much more complex study than either of the two preceding ones, Agrell *et al.* (2005) examined the effects of ambient and elevated concentrations of atmospheric CO_2 (360 ppm and 560 ppm, respectively) and O_3 (35-60 ppb and 52-90 ppb, respectively) on the foliar chemistry of more mature aspen trees of two different genotypes (216 and 259) growing out-of-doors at the Aspen Free Air CO_2 Enrichment (FACE) facility near Rhinelander, Wisconsin, USA, as well as the impacts of these effects on the host plant preferences of forest tent caterpillar larvae.

In reporting the results of the study, Agrell *et al.* say that "the only chemical component showing a somewhat consistent covariation with larval preferences was condensed tannins," noting that "the tree becoming relatively less preferred as a result of CO_2 or O_3 treatment was in general also the one for which average levels of condensed tannins were most positively (or least negatively) affected by that treatment." The mean condensed tannin concentrations of the aspen 216 and 259 genotypes were 25 percent and 57 percent higher, respectively, under the elevated CO_2 and O_3 combination treatment compared to the ambient CO_2 and O_3 combination treatment.

In light of these findings, it is logical to presume that as atmospheric concentrations of CO_2 and O_3 continue to rise, the increase in condensed tannin concentration likely to occur in the foliage of aspen trees should lead to their leaves becoming less preferred for consumption by the forest tent caterpillar.

Additional information on this topic, including reviews of newer publications as they become available, can be found at http://www.co2science.org/subject/t/tanninsaspen.php.

References

Agrell, J., Kopper, B., McDonald, E.P. and Lindroth, R.L. 2005. CO_2 and O_3 effects on host plant preferences of the forest tent caterpillar (*Malacosoma disstria*). *Global Change Biology* **11**: 588-599.

King, J.S., Pregitzer, K.S., Zak, D.R., Kubiske, M.E., Ashby, J.A. and Holmes, W.E. 2001. Chemistry and decomposition of litter from *Populus tremuloides* Michaux grown at elevated atmospheric CO_2 and varying N availability. *Global Change Biology* **7**: 65-74.

McDonald, E.P., Agrell, J., and Lindroth, R.L. 1999. CO_2 and light effects on deciduous trees: growth, foliar chemistry, and insect performance. *Oecologia* **119**: 389-399.

7.8.7.2. Birch Trees

How do condensed tannin concentrations in the leaves and roots of paper birch (*Betula papyrifera* Marsh.) and silver birch (*Betula pendula* Roth) trees respond to atmospheric CO_2 enrichment with and without concomitant increases in atmospheric temperature and ozone concentrations? We here briefly summarize the findings of several studies that have broached one or more parts of this question.

McDonald *et al.* (1999) grew paper birch seedlings in controlled environment greenhouses that were maintained at either ambient (387 ppm) or elevated (696 ppm) CO_2 concentrations under conditions of either low or high light availability (half and full sunlight, respectively) for 31 days after the mean date of bud break. In doing so, they determined that under low light conditions the CO_2-enriched seedlings exhibited an increase of approximately 15 percent in leaf condensed tannin concentration, while under high light conditions the CO_2-induced tannin increase was a whopping 175 percent.

Peltonen *et al.* (2005) studied the impacts of doubled atmospheric CO_2 and O_3 concentrations on the accumulation of 27 phenolic compounds, including soluble condensed tannins, in the leaves of two European silver birch clones in seven-year-old soil-grown trees that were exposed in open-top chambers for three growing seasons to ambient and twice-ambient atmospheric CO_2 and O_3 concentrations singly and in combination. This work, which was carried out in central Finland, revealed that elevated CO_2 increased the concentration of soluble condensed tannins in the leaves of the trees by 19 percent. In addition, they found that the elevated CO_2 protected the leaves from elevated O_3 because, as they describe it, "all the O_3-derived effects on the leaf phenolics and traits were prevented by elevated CO_2."

Kuokkanen *et al.* (2003) grew two-year-old silver birch seedlings in ambient air of 350 ppm CO_2 or air enriched to a CO_2 concentration of 700 ppm under conditions of either ambient temperature or ambient temperature plus 3°C for one full growing season in the field in closed-top chambers at the Mekrijarvi Research Station of the University of Joensuu in eastern Finland. Then, during the middle of the summer, when carbon-based secondary compounds of birch leaves are fairly stable, they picked several leaves from each tree and determined their condensed tannin concentrations, along with the concentrations of a number of other physiologically important substances. This work revealed that the concentration of total phenolics, condensed tannins and their derivatives significantly increased in the leaves produced in the CO_2-enriched air, as has also been observed by Lavola and Julkunen-Titto (1994), Williams *et al.* (1994), Kinney *et al.* (1997), Bezemer and Jones (1998) and Kuokkanen *et al.* (2001). In fact, the extra 350 ppm of CO_2 nearly tripled condensed tannin concentrations in the ambient-temperature air, while it increased their concentrations in the elevated-temperature air by a factor in excess of 3.5.

In a study of roots, Parsons *et al.* (2003) grew two-year-old paper birch saplings in well-watered and fertilized 16-L pots from early May until late August in glasshouse rooms maintained at either 400 or 700 ppm CO_2. This procedure revealed that the concentration of condensed tannins in the fine roots of the saplings was increased by 27 percent in the CO_2-enriched treatment; and in regard to this finding, the researchers say "the higher condensed tannin concentrations that were present in the birch fine roots may offer these tissues greater protection against soil-borne pathogens and herbivores."

Parsons *et al.* (2004) collected leaf litter samples from early September to mid-October beneath paper birch trees growing in ambient and CO_2-enriched (to 200 ppm above ambient) FACE plots in northern Wisconsin, USA, which were also maintained under ambient and O_3-enriched (to 19 ppb above ambient) conditions, after which the leaf mass produced in each treatment was determined, sub-samples of the leaves were assessed for a number of chemical constituents. The researchers learned that condensed tannin concentrations were 64 percent greater in the CO_2-enriched plots. Under CO_2- and O_3-enriched conditions, condensed tannin concentrations were 99 percent greater.

In conclusion, it appears that elevated concentrations of atmospheric CO_2 tend to increase leaf and fine-root tannin concentrations of birch trees, and that this phenomenon tends to protect the trees' foliage from predation by voracious insect herbivores and protect the trees' roots from soil-borne pathogens and herbivores.

Additional information on this topic, including reviews of newer publications as they become available, can be found at http://www.co2science.org/subject/t/tanninsbirch.php.

References

Bezemer, T.M. and Jones, T.H. 1998. Plant-insect herbivore interactions in elevated atmospheric CO_2, quantitative analyses and guild effects. *Oikos* **82**: 212-222.

Kinney, K.K., Lindroth, R.L., Jung, S.M. and Nordheim, E.V. 1997. Effects of CO_2 and NO_3 availability on deciduous trees, phytochemistry and insect performance. *Ecology* **78**: 215-230.

Kuokkanen, K., Julkunen-Titto, R., Keinanen, M., Niemela, P. and Tahvanainen, J. 2001. The effect of elevated CO_2 and temperature on the secondary chemistry of *Betula pendula* seedlings. *Trees* **15**: 378-384.

Kuokkanen, K., Yan, S. and Niemela, P. 2003. Effects of elevated CO_2 and temperature on the leaf chemistry of birch *Betula pendula* (Roth) and the feeding behavior of the weevil *Phyllobius maculicornis*. *Agricultural and Forest Entomology* **5**: 209-217.

Lavola, A. and Julkunen-Titto, R. 1994. The effect of elevated carbon dioxide and fertilization on primary and secondary metabolites in birch, *Betula pendula* (Roth). *Oecologia* **99**: 315-321.

McDonald, E.P., Agrell, J., and Lindroth, R.L. 1999. CO_2 and light effects on deciduous trees: growth, foliar chemistry, and insect performance. *Oecologia* **119**: 389-399.

Parsons, W.F.J., Kopper, B.J. and Lindroth, R.L. 2003. Altered growth and fine root chemistry of *Betula papyrifera* and *Acer saccharum* under elevated CO_2. *Canadian Journal of Forest Research* **33**: 842-846.

Parsons, W.F.J., Lindroth, R.L. and Bockheim, J.G. 2004. Decomposition of *Betula papyrifera* leaf litter under the independent and interactive effects of elevated CO_2 and O_3. *Global Change Biology* **10**: 1666-1677.

Peltonen, P.A., Vapaavuori, E. and Julkunen-Tiitto, R. 2005. Accumulation of phenolic compounds in birch leaves is changed by elevated carbon dioxide and ozone. *Global Change Biology* **11**: 1305-1324.

Williams, R.S., Lincoln, D.E. and Thomas, R.B. 1994. Loblolly pine grown under elevated CO_2 affects early instar pine sawfly performance. *Oecologia* **98**: 64-71.

7.8.7.3. Oak Trees

Dury *et al.* (1998) grew four-year-old pedunculate oak trees (*Quercus robur* L.) in pots within greenhouses maintained at ambient and twice-ambient atmospheric CO_2 concentrations in combination with ambient and elevated (ambient plus 3°C) air temperatures for approximately one year. This work revealed that elevated CO_2 had only minor and contrasting direct effects on leaf palatability: a temporary increase in foliar phenolic concentrations and decreases in leaf toughness and nitrogen content. The elevated temperature treatment, on the other hand, significantly reduced leaf palatability, because oak leaf toughness increased as a consequence of temperature-induced increases in condensed tannin concentrations. As a result, the researchers concluded that "a 3°C rise in temperature might be expected to result in prolonged larval development, increased food consumption, and reduced growth" for herbivores feeding on oak leaves in a CO_2-enriched and warmer world of the future.

Cornelissen *et al.* (2003) studied fluctuating asymmetry in the leaves of two species of schlerophyllous oaks—myrtle oak (*Quercus myrtifolia*) and sand live oak (*Quercus geminata*)—that dominate a native scrub-oak community at the Kennedy Space Center, Titusville, Florida (USA). Fluctuating asymmetry is the term used to describe small variations from perfect symmetry in otherwise bilaterally symmetrical characters in an organism (Moller and Swaddle, 1997), which asymmetry is believed to arise as a consequence of developmental instabilities experienced during ontogeny that may be caused by various stresses, including both genetic and environmental factors (Moller and Shykoff, 1999).

Based on measurements of (1) distances from the leaf midrib to the left and right edges of the leaf at its widest point and (2) leaf areas on the left and right sides of the leaf midrib, Cornelissen *et al.* determined that "asymmetric leaves were less frequent in elevated CO_2, and, when encountered, they were less asymmetric than leaves growing under ambient CO_2." In addition, they found that "*Q. myrtifolia* leaves under elevated CO_2 were 15.0% larger than in ambient CO_2 and *Q. geminata* leaves were 38.0 percent larger in elevated CO_2 conditions." They also determined that "elevated CO_2 significantly increased tannin concentration for both *Q. myrtifolia* and *Q. geminata* leaves" and that "asymmetric leaves contained significantly lower concentrations of tannins than symmetric leaves for both *Q. geminata* and *Q. myrtifolia*." Specifically, they found induced incfreases in tanning concentrations of approximately 35 percent for *Q. myrtifolia* and 43 percent for *Q. geminata*. In commenting on their primary findings of reduced percentages of leaves experiencing asymmetry in the presence of elevated levels of atmospheric CO_2 and the lesser degree of asymmetry exhibited by affected leaves in the elevated CO_2 treatment, Cornelissen *et al.* say that "a possible explanation for this pattern is the fact that, by contrast to other environmental stresses, which can cause negative effects on plant growth, the predominant effect of elevated CO_2 on plants is to promote growth with consequent reallocation of resources (Docherty *et al.*, 1996)." Another possibility they discuss "is the fact that CO_2 acts as a plant fertilizer," and, as a result, that "elevated CO_2 ameliorates plant stress compared with ambient levels of CO_2."

In a subsequent study conducted at the Kennedy Space Center's scrub-oak community, Hall *et al.* (2005b) evaluated foliar quality and herbivore damage in three oaks (*Q. myrtifolia*, *Q. chapmanii* and *Q. geminata*) plus the nitrogen-fixing legume *Galactia elliottii* at three-month intervals from May 2001 to May 2003, at which times samples of undamaged leaves were removed from each of the four species in all chambers and analyzed for various chemical constituents, while 200 randomly selected leaves of each species in each chamber were scored for the presence of six types of herbivore damage. Analyses of the data thereby obtained indicated that for condensed tannins, hydrolyzable tannins, total phenolics and lignin, in all four species there were always greater concentrations of all four leaf constituents in the CO_2-enriched leaves, with across-species mean increases of 6.8 percent for condensed tannins, 6.1 percent for hydrolyzable tannins, 5.1 percent for total phenolics and 4.3 percent for lignin. In addition, there were large CO_2-induced decreases in all leaf damage categories among all species: chewing (-48 percent), mines (-37 percent), eye spot gall (-45 percent), leaf tier (-52 percent), leaf mite (-23 percent) and leaf gall (-16 percent). Hall *et al.* thus concluded that the changes they observed in leaf chemical constituents and herbivore damage "suggest that damage to plants may decline as atmospheric CO_2 levels continue to rise."

Last, and largely overlapping the investigation of Hall *et al.* (2005b), was the study of Hall *et al.* (2005a), who evaluated the effects of the Kennedy Space Center experiment's extra 350 ppm of CO_2 on litter quality, herbivore activity, and their interactions, over the three-year-period 2000-2002. This endeavor indicated, in their words, that "changes in litter chemistry from year to year were far larger than effects of CO_2 or insect damage, suggesting that these may have only minor effects on litter decomposition." The one exception to this finding was that "condensed tannin concentrations increased under elevated CO_2 regardless of species, herbivore damage, or growing season," rising by 11 percent in 2000, 18 percent in 2001 and 41 percent in 2002 as a result of atmospheric CO_2 enrichment, as best we can determine from the researchers' bar graphs. Also, the five scientists report that "lepidopteran larvae can exhibit slower growth rates when feeding on elevated CO_2 plants (Fajer *et al.*, 1991) and become more susceptible to pathogens, parasitoids, and predators (Lindroth, 1996; Stiling *et al.*, 1999)," noting further that at their field site, "which hosts the longest continuous study of the effects of elevated CO_2 on insects, herbivore populations decline markedly under elevated CO_2 (Stiling *et al.*, 1999, 2002, 2003; Hall *et al.*, 2005b)."

In conclusion, it would appear CO_2-enriched air produces a large and continuous enhancement of condensed tannin concentrations in oak tree foliage, which causes marked declines in herbivore populations observed in CO_2-enriched open-top-chamber studies.

Additional information on this topic, including reviews of newer publications as they become available, can be found at http://www.co2science.org/subject/t/tanninsoak.php.

References

Cornelissen, T., Stiling, P. and Drake, B. 2003. Elevated CO_2 decreases leaf fluctuating asymmetry and herbivory by leaf miners on two oak species. *Global Change Biology* **10**: 27-36.

Docherty, M., Hurst, D.K., Holopainem, J.K. *et al.* 1996. Carbon dioxide-induced changes in beech foliage cause female beech weevil larvae to feed in a compensatory manner. *Global Change Biology* **2**: 335-341.

Dury, S.J., Good, J.E.G., Perrins, C.M., Buse, A. and Kaye, T. 1998. The effects of increasing CO_2 and temperature on oak leaf palatability and the implications for herbivorous insects. *Global Change Biology* **4**: 55-61.

Fajer, E.D., Bowers, M.D. and Bazzaz, F.A. 1991. The effects of enriched CO_2 atmospheres on the buckeye butterfly, *Junonia coenia*. *Ecology* **72**: 751-754.

Hall, M.C., Stiling, P., Hungate, B.A., Drake, B.G. and Hunter, M.D. 2005a. Effects of elevated CO_2 and herbivore damage on litter quality in a scrub oak ecosystem. *Journal of Chemical Ecology* **31**: 2343-2356.

Hall, M.C., Stiling, P., Moon, D.C., Drake, B.G. and Hunter, M.D. 2005b. Effects of elevated CO_2 on foliar quality and herbivore damage in a scrub oak ecosystem. *Journal of Chemical Ecology* **31**: 267-285.

Lindroth, R.L. 1996. CO_2-mediated changes in tree chemistry and tree-Lepidoptera interactions. In: Koch, G.W. and Mooney, H,A. (Eds.). *Carbon Dioxide and Terrestrial Ecosystems*. Academic Press, San Diego, California, USA, pp. 105-120.

Moller, A.P. and Shykoff, P. 1999. Morphological developmental stability in plants: patterns and causes. *International Journal of Plant Sciences* **160**: S135-S146.

Moller, A.P. and Swaddle, J.P. 1997. *Asymmetry, Developmental Stability and Evolution*. Oxford University Press, Oxford, UK.

Stiling, P., Cattell, M., Moon, D.C., Rossi, A., Hungate, B.A., Hymus, G. and Drake, B.G. 2002. Elevated atmospheric CO_2 lowers herbivore abundance, but increases leaf abscission rates. *Global Change Biology* **8**: 658-667.

Stiling, P., Moon, D.C., Hunter, M.D., Colson, J., Rossi, A.M., Hymus, G.J. and Drake, B.G. 2003. Elevated CO_2 lowers relative and absolute herbivore density across all species of a scrub-oak forest. *Oecologia* **134**: 82-87.

Stiling, P., Rossi, A.M., Hungate, B., Dijkstra, P., Hinkle, C.R., Knot III, W.M., and Drake, B. 1999. Decreased leaf-miner abundance in elevated CO_2: Reduced leaf quality and increased parasitoid attack. *Ecological Applications* **9**: 240-244.

7.8.8. Transgenic Plants

Toxins produced by *Bacillus thuringiensis* (Bt) supplied to crops via foliar application have been used as a means of combating crop pests for well over half a century. More recently, the Bt gene for producing the toxin has been artificially inserted in some species of plants, producing transgenic plants that are pest resistant. The effectiveness of this management technique depends primarily upon the amount of Bt-produced toxins that are ingested by targeted insects. Another kind of transgenic plant is wheat that has been made heat resistant by the introduction into its gene code of heat shock protein (HSP) or plastidial EF-Tu (protein synthesis elongation factor). How does atmospheric CO_2 enrichment affect transgenic plants?

If soil nitrogen levels are low, foliar nitrogen concentrations in plants grown in enhanced CO_2 environments are generally reduced from what they are at the current atmospheric CO_2 concentration, which suggests that insects would have to eat more foliage to get their normal requirement of nitrogen for proper growth and development in CO_2-enriched air. But by eating more foliage, the insects would also ingest more Bt-produced toxins, and they would be more severely impacted by those substances.

To test this hypothesis, Coviella and Trumble (2000) grew cotton plants in each of six Teflon-film chambers in a temperature-controlled greenhouse, where three of the chambers were maintained at an atmospheric CO_2 concentration of 370 ppm, and three were maintained at 900 ppm CO_2. In addition, half of the plants in each chamber received high levels of nitrogen (N) fertilization, while half received low levels (30 vs. 130 mg N/kg soil/week). After 45 days of growth under these conditions, leaves were removed from the plants and dipped in a Bt solution, after which known amounts of treated leaf material were fed to *Spodoptera exigua* larvae and their responses measured and analyzed.

By these means, the two researchers determined that the plants grown in the elevated CO_2 chambers did indeed have significantly lower foliar nitrogen concentrations than the plants grown in the ambient CO_2 chambers under the low N fertilization regime; but this was not the case under the high N regime. They also discovered that older larvae fed with foliage grown in elevated CO_2 with low N fertilization consumed significantly more plant material than insects fed with foliage grown in ambient CO_2; but, again, no differences were observed with high N fertilization. Last, and "consistent with the effect of higher Bt toxin intake due to enhanced consumption," they found that "insects fed on low N plants had significantly higher mortality in elevated CO_2." Yet, again, no such effect was evident in the high N treatment. Consequently, with respect to pest management using Bt-produced toxins supplied to crops via foliar application, Coviella and Trumble concluded that "increasing atmospheric CO_2 is making the foliar applications more efficacious."

Coviella *et al.* (2000), in an analogous experiment to that of Coviella and Trumble, grew cotton plants in 12 Teflon-film chambers in a temperature-controlled greenhouse, where six chambers were maintained at an atmospheric CO_2 concentration of 370 ppm and six were maintained at 900 ppm CO_2. Half of the cotton plants in each of these chambers were of a transgenic line containing the Bt gene for the production of the Cry1Ac toxin, which is mildly toxic for *Spodoptera exigua*, while the other half were of a near isogenic line without the Bt gene. In addition, and as before, half of the plants in each chamber received the same low and high levels of N fertilization; and between 40 and 45 days after emergence, leaves were removed from the plants and fed to the *S. exigua* larvae, after which a number of larval responses were measured and analyzed, along with various leaf properties.

This work revealed that the low-N plants in the elevated CO_2 treatment had lower foliar N concentrations than did the low-N plants in the ambient CO_2 treatment, and that the transgenic plants from the low-N, high CO_2 treatment produced lower levels of Bt toxin than did the transgenic plants from the low-N, ambient CO_2 treatment. In addition, the high level of N fertilization only partially compensated for this latter high-CO_2 effect. In the ambient CO_2 treatment there was also a significant increase in days to pupation for insects fed transgenic plants; but this difference was not evident in elevated CO_2. In addition, pupal weight in ambient CO_2 was significantly higher in non-transgenic plants; and, again, this difference was not observed in elevated CO_2.

In discussing their findings, the three researchers wrote that "these results support the hypothesis that the lower N content per unit of plant tissue caused by the elevated CO_2 will result in lower toxin production by transgenic plants when nitrogen supply to the plants is a limiting factor." They also note that "elevated CO_2 appears to eliminate differences between transgenic and non-transgenic plants for some key insect developmental/fitness variables including length of the larval stage and pupal weight."

These findings suggest that in the case of inadvertent Bt gene transference to wild relatives of transgenic crop lines, elevated levels of atmospheric CO_2 will tend to negate certain of the negative effects the wayward genes might otherwise inflict on the natural world. Hence, the ongoing rise in the air's CO_2 content could be said to constitute an "insurance policy" against this potential outcome.

On the other hand, Coviella *et al.*'s results also suggest that transgenic crops designed to produce Bt-type toxins may become less effective in carrying out the objectives of their design as the air's CO_2 content continues to rise. Coupling this possibility with the fact that the *foliar* application of *Bacillus thuringiensis* to crops should become even more effective in a higher-CO_2 world, as found by Coviella and Trumble, one could argue that the implantation of toxin-producing genes in crops is not the way to go in the face of the ongoing rise in the air's CO_2 content, which reduces that technique's effectiveness at the same time that it increases the effectiveness of direct foliar applications.

In a study of three different types of rice—a wild type (WT) and two transgenic varieties, one with 65 percent wild-type Rubisco (AS-77) and one with 40 percent wild-type Rubisco (AS-71)—Makino *et al.* (2000) grew plants from seed for 70 days in growth chambers maintained at 360 and 1000 ppm CO_2, after which they harvested the plants and determined their biomass. In doing so, they found that the mean dry weights of the WT, AS-77 and AS-71 varieties grown in air of 360 ppm were, respectively, 5.75, 3.02 and 0.83 g, while in air of 1000 ppm CO_2, corresponding mean dry weights were 7.90, 7.40 and 5.65 g. Consequently, although the growth rates of the genetically engineered rice plants were inferior to that of the wild type when grown in normal air of 360 ppm CO_2 (with AS-71 producing less than 15 percent as much biomass as the wild type), when grown in air of 1000 ppm CO_2 they experienced greater CO_2-induced increases in growth: a 145 percent increase in the case of AS-77 and a 581 percent increase in the case of AS-71. Hence, whereas the transgenic plants were highly disadvantaged in normal air of 360 ppm CO_2 (with AS-71 plants attaining a mean dry weight of only 0.83 g while the WT plants attained a mean dry weight of 5.75 g), they were found to be pretty much on an equal footing in highly CO_2-enriched air (with AS-71 plants attaining a mean dry weight of 5.65 g while the WT plants attained a mean dry weight of 7.90 g). This finding bodes well for the application of this type of technology to rice crops in a future world of higher atmospheric CO_2 content.

Returning to cotton, Chen *et al.* (2005) grew well watered and fertilized plants of two varieties—one expressing Cry1A (c) genes from *Bacillus thurigiensis* and a non-transgenic cultivar from the same recurrent parent—in pots placed within open-top chambers maintained at either 375 or 750 ppm CO_2 in Sanhe County, Hebei Province, China, from planting in mid-May to harvest in October, throughout which period several immature bolls were collected and analyzed for various chemical characteristics, while others were stored under refrigerated conditions for later feeding to cotton bollworm larvae, whose growth characteristics were closely were monitored. In pursuing this protocol, the five researchers found that the elevated CO_2 treatment increased immature boll concentrations of condensed tannins by approximately 22 percent and 26 percent in transgenic and non-transgenetic cotton, respectively (see Tannins in Section 7.8.7. for a discussion of the significance of this observation). In addition, they found that elevated CO_2 slightly decreased the body biomass of the cotton bollworms and reduced moth fecundity. The Bt treatment was even more effective in this regard, and in the combined Bt-high-CO_2 treatment, the negative cotton bollworm responses were expressed most strongly of all. Chen *et al.* concluded that the expected higher atmospheric CO_2 concentrations of the future will "either not change or only slightly enhance the efficacy of Bt technology against cotton bollworms."

Two years later, Chen *et al.* (2007) reported growing the same two cultivars under the same conditions from the time of planting on 10 May 2004 until the plants were harvested in October, after which egg masses of the cotton bollworms were reared in a growth chamber under ambient-CO_2 conditions, while three successive generations of them were fed either transgenic or non-transgenic cotton bolls from plants grown in either ambient or twice-ambient atmospheric CO_2 concentrations, during which time a number of physiological characteristics of the cotton bollworms were periodically assessed. This work revealed, in the words of Chen *et al.*, that "both elevated CO_2 and transgenic Bt cotton increased larval lifespan," but that they decreased "pupal weight, survival rate, fecundity, frass output, relative and mean relative growth rates, and the efficiency of conversion of ingested and digested food." As a result, they say that "transgenic Bt cotton significantly decreased the population-trend index compared to non-transgenic cotton for the three successive bollworm generations, *especially at elevated CO_2* [our italics]."

Based on these findings, the four researchers concluded that the negative effects of elevated CO_2 on cotton bollworm physiology and population dynamics "may intensify through successive generations," in agreement with the findings of Brooks and Whittaker (1998, 1999) and Wu *et al.* (2006). They additionally concluded that "both elevated CO_2 and transgenic Bt cotton are adverse environmental factors for cotton bollworm long-term population growth," and that the combination of the two factors may intensify their adverse impact on the population performance of the cotton bollworm, which would be good news for cotton growers.

Fu *et al.* (2008) note that "heat stress is a major constraint to wheat production and negatively impacts grain quality, causing tremendous economic losses, and may become a more troublesome factor due to global warming." Consequently, as they describe it, they "introduced into wheat the maize gene coding for plastidal EF-Tu [protein synthesis elongation factor]," in order to assess "the expression of the transgene, and its effect on thermal aggregation of leaf proteins in transgenic plants," as well as "the heat stability of photosynthetic membranes (thylakoids) and the rate of CO_2 fixation in young transgenic plants following exposure to heat stress." These operations led, in their words, "to improved protection of leaf proteins against thermal aggregation, reduced damage to thylakoid membranes and enhanced photosynthetic capability following exposure to heat stress," which results "support the concept that EF-Tu ameliorates negative effects of heat stress by acting as a molecular chaperone."

Fu *et al.* describe their work as "the first demonstration that a gene other than HSP [heat shock protein] gene can be used for improvement of heat tolerance," noting it also indicates that the improvement is possible in a species that has a complex genome, such as hexaploid wheat. They conclude by stating their results "strongly suggest that heat tolerance of wheat, and possibly other crop plants, can be improved by modulating expression of plastidal EF-Tu and/or by selection of genotypes with increased endogenous levels of this protein."

In summary, genetic alterations to crop plants enable them to better withstand the assaults of insects pests or increases in seasonal maximum air temperates. Elevated CO_2 either improves or does not change the effectiveness of genetic alternatives to achieve these objectives, while it simultaneously reduces the severity of possible negative effects that could arise from the escape of transplanted genes into the natural environment.

Additional information on this topic, including reviews of newer publications as they become available, can be found at http://www.co2science.org/subject/t/summaries/transgenicplants.php.

References

Brooks, G.L. and Whittaker, J.B. 1998. Responses of multiple generations of *Gastrophysa viridula*, feeding on *Rumex obtusifolius*, to elevated CO_2. *Global Change Biology* **4**: 63-75.

Brooks, G.L. and Whittaker, J.B. 1999. Responses of three generations of a xylem-feeding insect, *Neophilaenus lineatus* (Homoptera), to elevated CO_2. *Global Change Biology* **5**: 395-401.

Chen, F., Wu, G., Ge, F., Parajulee, M.N. and Shrestha, R.B. 2005. Effects of elevated CO_2 and transgenic Bt cotton on plant chemistry, performance, and feeding of an insect herbivore, the cotton bollworm. *Entomologia Experimentalis et Applicata* **115**: 341-350.

Chen, F., Wu, G., Parajulee, M.N. and Ge, F. 2007. Long-term impacts of elevated carbon dioxide and transgenic Bt cotton on performance and feeding of three generations of cotton bollworm. *Entomologia Experimentalis et Applicata* **124**: 27-35.

Coviella, C.E., Morgan, D.J.W. and Trumble, J.T. 2000. Interactions of elevated CO_2 and nitrogen fertilization: Effects on production of *Bacillus thuringiensis* toxins in transgenic plants. *Environmental Entomology* **29**: 781-787.

Coviella, C.E. and Trumble, J.T. 2000. Effect of elevated atmospheric carbon dioxide on the use of foliar application of *Bacillus thuringiensis*. *BioControl* **45**: 325-336.

Fu, J., Momcilovic, I., Clemente, T.E., Nersesian, N., Trick, H.N. and Ristic, Z. 2008. Heterologous expression of a plastid EF-Tu reduces protein thermal aggregation and enhances CO_2 fixation in wheat (*Triticum aestivum*) following heat stress. *Plant Molecular Biology* **68**: 277-288.

Makino, A., Harada, M., Kaneko, K., Mae, T., Shimada, T. and Yamamoto, N. 2000. Whole-plant growth and N allocation in transgenic rice plants with decreased content of ribulose-1,5-bisphosphate carboxylase under different CO_2 partial pressures. *Australian Journal of Plant Physiology* **27**: 1-12.

Wu, G., Chen, J.F. and Ge, F. 2006. Response of multiple generations of cotton bollworm *Helicoverpa armigera* Hubner, feeding on spring wheat, to elevated CO_2. *Journal of Applied Entomology* **130**: 2-9.

7.8.9. Isoprene

Isoprene (C_5H_8 or 2-methyl-1,3-butadiene) is a highly reactive non-methane hydrocarbon (NMHC) that is emitted in copious quantities by vegetation and is responsible for the production of vast amounts of tropospheric ozone (Chameides *et al.*, 1988; Harley *et al.*, 1999), which is a debilitating scourge of plant and animal life alike. Poisson *et al.* (2000) calculate that current levels of NMHC emissions—the vast majority of which are isoprene, accounting for more than twice as much as all other NMHCs combined—may increase surface ozone concentrations by up to 40 percent in the marine boundary-layer and 50-60 percent over land. They further estimate that the current tropospheric ozone content extends the atmospheric lifetime of methane—one of the world's

most powerful greenhouse gases—by approximately 14 percent. Consequently, it can be appreciated that reducing isoprene emissions from vegetation is to be desired.

Although a few experiments conducted on certain plant species have suggested that elevated concentrations of atmospheric CO_2 have little to no effect on their emissions of isoprene (Buckley, 2001; Baraldi *et al.*, 2004; Rapparini *et al.*, 2004), a much larger number of other experiments are suggestive of substantial CO_2-induced reductions in isoprene emissions, as demonstrated by the work of Monson and Fall (1989), Loreto and Sharkey (1990), Sharkey *et al.* (1991) and Loreto *et al.* (2001).

Rosentiel *et al.* (2003) studied three 50-tree cottonwood plantations growing in separate mesocosms within the forestry section of the Biosphere 2 facility near Oracle, Arizona, USA, one of which mesocosms was maintained at an atmospheric CO_2 concentration of 430 ppm, while the other two were enriched to concentrations of 800 and 1200 ppm for one entire growing season. Integrated over that period, the total above-ground biomass of the trees in the latter two mesocosms was increased by 60 percent and 82 percent, respectively, while their production of isoprene was decreased by 21 percent and 41 percent, respectively.

Scholefield *et al.* (2004) measured isoprene emissions from *Phragmites australis* plants (one of the world's most important natural grasses) growing at different distances from a natural CO_2 spring in central Italy. At the specific locations they chose to make their measurements, atmospheric CO_2 concentrations of approximately 350, 400, 550 and 800 ppm had likely prevailed for the entire lifetimes of the plants. Across this CO_2 gradient, plant isoprene emissions dropped ever lower as the air's CO_2 concentration rose ever higher. Over the first 50-ppm CO_2 increase, isoprene emissions were reduced to approximately 65 percent of what they were at ambient CO_2, while for CO_2 increases of 200 and 450 ppm, they were reduced to only about 30 percent and 7 percent of what they were in the 350-ppm-CO_2 air. The researchers note that these reductions were likely caused by reductions in leaf isoprene synthase, which was observed to be highly correlated with isoprene emissions, leading them to conclude that "elevated CO_2 generally inhibits the expression of isoprenoid synthesis genes and isoprene synthase activity which may, in turn, limit formation of every chloroplast-derived isoprenoid." They state that the "basal emission rate of isoprene is likely to be reduced under future elevated CO_2 levels."

Centritto *et al.* (2004) grew hybrid poplar saplings for one full growing season in a FACE facility located at Rapolano, Italy, where the air's CO_2 concentration was increased by approximately 200 ppm. Their study demonstrated that "isoprene emission is reduced in elevated CO_2, in terms of both maximum values of isoprene emission rate and isoprene emission per unit of leaf area averaged across the total number of leaves per plant," which in their case amounted to a reduction of approximately 34 percent. When isoprene emission was summed over the entire plant profile, however, the reduction was not nearly so great (only 6 percent), because of the greater number of leaves on the CO_2-enriched saplings. "However," as they state, "Centritto *et al.* (1999), in a study with potted cherry seedlings grown in open-top chambers, and Gielen *et al.* (2001), in a study with poplar saplings exposed to FACE, showed that the stimulation of total leaf area in response to elevated CO_2 was a transient effect, because it occurred only during the first year of growth." Hence, they concluded "it may be expected that with similar levels of leaf area, the integrated emission of isoprene would have been much lower in elevated CO_2." Indeed, they say that their data, "as well as that reported by Scholefield *et al.* (2004), in a companion experiment on *Phragmites* growing in a nearby CO_2 spring, mostly confirm that isoprene emission is inversely dependent on CO_2 [concentration] when this is above ambient, and suggests that a lower fraction of C will be re-emitted in the atmosphere as isoprene by single leaves in the future."

Working at another FACE facility, the Aspen FACE facility near Rhinelander, Wisconsin, USA, Calfapietra *et al.* (2008) measured emissions of isoprene from sun-exposed upper-canopy leaves of an O_3-tolerant clone and an O_3-sensitive clone of trembling aspen (*Populus tremuloides* Michx.) trees that were growing in either normal ambient air, air enriched with an extra 190-200 ppm CO_2, air with 1.5 times the normal ozone concentration, or air simultaneously enriched with the identical concentrations of both of these atmospheric trace gases. Results of their analysis showed that for the trees growing in air of ambient ozone concentration, the extra 190 ppm of CO_2 decreased the mean isoprene emission rate by 11.7 percent in the O_3-tolerant aspen clone and by 22.7 percent in the O_3-sensitive clone, while for the trees growing in air with 1.5 times the ambient ozone concentration, the extra

CO_2 also decreased the mean isoprene emission rate by 10.4 percent in the O_3-tolerant clone and by 32.7 percent in the O_3-sensitive clone. At the same time, and in the same order, net photosynthesis rates were increased by 34.9 percent, 47.4 percent, 31.6 percent and 18.9 percent.

Possell *et al.* (2004) grew seedlings of English oak (*Quercus robur*), one to a mesocosm (16 cm diameter, 60 cm deep), in either fertilized or unfertilized soil in *solardomes* maintained at atmospheric CO_2 concentrations of either ambient or ambient plus 300 ppm for one full year, at the conclusion of which period they measured rates of isoprene emissions from the trees' foliage together with their rates of photosynthesis. In the unfertilized trees, this work revealed that the 300-ppm increase in the air's CO_2 concentration reduced isoprene emissions by 63 percent on a leaf area basis and 64 percent on a biomass basis, while in the fertilized trees the extra CO_2 reduced isoprene emissions by 70 percent on a leaf area basis and 74 percent on a biomass basis. In addition, the extra CO_2 boosted leaf photosynthesis rates by 17 percent in the unfertilized trees and 13 percent in the fertilized trees.

Possell *et al.* (2005) performed multiple three-week-long experiments with two known isoprene-emitting herbaceous species (*Mucuna pruriens* and *Arundo donax*), which they grew in controlled environment chambers that were maintained at two different sets of day/night temperatures (29/24°C and 24/18°C) and atmospheric CO_2 concentrations characteristic of glacial (180 ppm), pre-industrial (280 ppm) and current (366 ppm) conditions, where canopy isoprene emission rates were measured on the final day of each experiment. They obtained what they describe as "the first empirical evidence for the enhancement of isoprene production, on a unit leaf area basis, by plants that grew and developed in [a] CO_2-depleted atmosphere," which results, in their words, "support earlier findings from short-term studies with woody species (Monson and Fall, 1989; Loreto and Sharkey, 1990)." Then, combining their emission rate data with those of Rosenstiel *et al.* (2003) for *Populus deltoides*, Centritto *et al.* (2004) for *Populus x euroamericana* and Scholefield *et al.* (2004) for *Phragmites australis*, they developed a single downward-trending isoprene emissions curve that stretches all the way from 180 to 1200 ppm CO_2, where it asymptotically approaches a value that is an order of magnitude less than what it is at 180 ppm.

Working at the Biosphere 2 facility near Oracle, Arizona, USA, in enclosed ultraviolet light-depleted mesocosms (to minimize isoprene depletion by atmospheric oxidative reactions such as those involving OH), Pegoraro *et al.* (2005) studied the effects of atmospheric CO_2 enrichment (1200 ppm compared to an ambient concentration of 430 ppm) and drought on the emission of isoprene from cottonwood (*Populus deltoides* Bartr.) foliage and its absorption by the underlying soil for both well-watered and drought conditions. In doing so, they found that "under well-watered conditions in the agriforest stands, gross isoprene production (i.e., the total production flux minus the soil uptake) was inhibited by elevated CO_2 and the highest emission fluxes of isoprene were attained in the lowest CO_2 treatment." In more quantitative terms, it was determined that the elevated CO_2 treatment resulted in a 46 percent reduction in gross isoprene production. In addition, it was found that drought suppressed the isoprene sink capacity of the soil beneath the trees, but that "the full sink capacity of dry soil was recovered within a few hours upon rewetting."

Putting a slightly negative slant on their findings, Pegoraro *et al.* suggested that "in future, potentially hotter, drier environments, higher CO_2 may not mitigate isoprene emission as much as previously suggested." However, we note that climate models generally predict an intensification of the hydrologic cycle in response to rising atmospheric CO_2 concentrations, and that the anti-transpirant effect of atmospheric CO_2 enrichment typically leads to increases in the moisture contents of soils beneath vegetation. Also, we note that over the latter decades of the twentieth century, when the IPCC claims the earth warmed at a rate and to a level that were unprecedented over the past two millennia, soil moisture data from all around the world tended to display upward trends. Robock *et al.* (2000), for example, developed a massive collection of soil moisture data from over 600 stations spread across a variety of climatic regimes, including the former Soviet Union, China, Mongolia, India and the United States, determining that "In contrast to predictions of summer desiccation with increasing temperatures, for the stations with the longest records, summer soil moisture in the top 1 m has increased while temperatures have risen." And in a subsequent study of "45 years of gravimetrically observed plant available soil moisture for the top 1 m of soil, observed every 10 days for April-October for 141 stations from fields with either winter or spring cereals from the Ukraine for 1958-2002," Robock *et al.* (2005) discovered that these real-world

observations "show a positive soil moisture trend for the entire period of observation," noting that "even though for the entire period there is a small upward trend in temperature and a downward trend in summer precipitation, the soil moisture still has an upward trend for both winter and summer cereals." Consequently, in a CO_2-enriched world of the future, we likely will have the best of both aspects of isoprene activity: less production by vegetation and more consumption by soils.

Finally, we address the issue of how well models predict the response of isoprene emission to future global change. According to Monson *et al.* (2007) such predictions "probably contain large errors," which clearly need to be corrected. The reason for the errors, write the twelve researchers who conducted the study, is that "the fundamental logic of such models is that changes in NPP [net primary production] will produce more or less biomass capable of emitting isoprene, and changes in climate will stimulate or inhibit emissions per unit of biomass." They continue, "these models tend to ignore the discovery that there are direct effects of changes in the atmospheric CO_2 concentration on isoprene emission that tend to work in the opposite direction to that of stimulated NPP," as has been indicated in the research studies described above. Their results showed, in their words, "that growth in an atmosphere of elevated CO_2 inhibited the emission of isoprene at levels that completely compensate for possible increases in emission due to increases in aboveground NPP."

In lamenting this sorry state of global-change modeling, Monson *et al.* say that, "to a large extent, the modeling has 'raced ahead' of our mechanistic understanding of how isoprene emissions will respond to the fundamental drivers of global change," and that "without inclusion of these effects in the current array of models being used to predict changes in atmospheric chemistry due to global change, one has to question the relevance of the predictions."

A year later, Arneth *et al.* (2008) used a mechanistic isoprene-dynamic vegetation model of European woody vegetation to "investigate the interactive effects of climate and CO_2 concentration on forest productivity, species composition, and isoprene emissions for the periods 1981-2000 and 2081-2100," which included a parameterization of the now-well-established direct CO_2-isoprene inhibition phenomenon we have described in the papers above. The study found that "across the model domain," the CO_2-isoprene inhibition effect "has the potential to offset the stimulation of [isoprene] emissions that could be expected from warmer temperatures and from the increased productivity and leaf area of emitting vegetation."

In view of these findings, it appears the ongoing rise in the atmosphere's CO_2 concentration will lead to ever greater reductions in atmospheric isoprene concentrations. As noted in the introductory paragraph of this section, such a consequence would be welcome news for man and nature.

Additional information on this topic, including reviews of newer publications as they become available, can be found at http://www.co2science.org/subject/i/isoprene.php.

References

Arneth, A., Schurgers, G., Hickler, T. and Miller, P.A. 2008. Effects of species composition, land surface cover, CO_2 concentration and climate on isoprene emissions from European forests. *Plant Biology* **10**: 150-162.

Baraldi, R., Rapparini, F., Oechel, W.C., Hastings, S.J., Bryant, P., Cheng, Y. and Miglietta, F. 2004. Monoterpene emission responses to elevated CO_2 in a Mediterranean-type ecosystem. *New Phytologist* **161**: 17-21.

Buckley, P.T. 2001. Isoprene emissions from a Florida scrub oak species grown in ambient and elevated carbon dioxide. *Atmospheric Environment* **35**: 631-634.

Calfapietra, C., Scarascia-Mugnozza, G., Karnosky, D.F., Loreto, F. and Sharkey, T.D. 2008. Isoprene emission rates under elevated CO_2 and O_3 in two field-grown aspen clones differing in their sensitivity to O_3. *New Phytologist* **179**: 55-61.

Centritto, M., Lee, H. and Jarvis, P. 1999. Interactive effects of elevated [CO_2] and water stress on cherry (*Prunus avium*) seedlings. I. Growth, total plant water use efficiency and uptake. *New Phytologist* **141**: 129-140.

Centritto, M., Nascetti, P., Petrilli, L., Raschi, A. and Loreto, F. 2004. Profiles of isoprene emission and photosynthetic parameters in hybrid poplars exposed to free-air CO_2 enrichment. *Plant, Cell and Environment* **27**: 403-412.

Chameides, W.L., Lindsay, R.W., Richardson, J. and Kiang, C.S. 1988. The role of biogenic hydrocarbons in urban photochemical smog: Atlanta as a case study. *Science* **241**: 1473-1475.

Gielen, B., Calfapietra, C., Sabatti, M. and Ceulemans, R. 2001. Leaf area dynamics in a poplar plantation under free-air carbon dioxide enrichment. *Tree Physiology* **21**: 1245-1255.

Harley, P.C., Monson, R.K. and Lerdau, M.T. 1999. Ecological and evolutionary aspects of isoprene emission from plants. *Oecologia* **118**: 109-123.

Loreto, F., Fischbach, R.J., Schnitzler, J.-P., Ciccioli, P., Brancaleoni, E., Calfapietra, C. and Seufert, G. 2001. Monoterpene emission and monoterpene synthase activities in the Mediterranean evergreen oak *Quercus ilex* L. grown at elevated CO_2 concentrations. *Global Change Biology* **7**: 709-717.

Loreto F. and Sharkey, T.D. 1990. A gas exchange study of photosynthesis and isoprene emission in red oak (*Quercus rubra* L.). *Planta* **182**: 523-531.

Monson, R.K. and Fall, R. 1989. Isoprene emission from aspen leaves. *Plant Physiology* **90**: 267-274.

Monson, R.K., Trahan, N., Rosenstiel, T.N., Veres, P., Moore, D., Wilkinson, M., Norby, R.J., Volder, A., Tjoelker, M.G., Briske, D.D., Karnosky, D.F. and Fall, R. 2007. Isoprene emission from terrestrial ecosystems in response to global change: minding the gap between models and observations. *Philosophical Transactions of the Royal Society A* **365**: 1677-1695.

Pegoraro, E., Abrell, L., van Haren, J., Barron-Gafford, G., Grieve, K.A., Malhi, Y., Murthy, R. and Lin, G. 2005. The effect of elevated atmospheric CO_2 and drought on sources and sinks of isoprene in a temperate and tropical rainforest mesocosm. *Global Change Biology* **11**: 1234-1246.

Poisson, N., Kanakidou, M. and Crutzen, P.J. 2000. Impact of non-methane hydrocarbons on tropospheric chemistry and the oxidizing power of the global troposphere: 3-dimensional modeling results. *Journal of Atmospheric Chemistry* **36**: 157-230.

Possell, M., Heath, J., Hewitt, C.N., Ayres, E. and Kerstiens, G. 2004. Interactive effects of elevated CO_2 and soil fertility on isoprene emissions from *Quercus robur*. *Global Change Biology* **10**: 1835-1843.

Possell, M., Hewitt, C.N. and Beerling, D.J. 2005. The effects of glacial atmospheric CO_2 concentrations and climate on isoprene emissions by vascular plants. *Global Change Biology* **11**: 60-69.

Rapparini, F., Baraldi, R., Miglietta, F. and Loreto, F. 2004. Isoprenoid emission in trees of *Quercus pubescens* and *Quercus ilex* with lifetime exposure to naturally high CO_2 environment. *Plant, Cell and Environment* **27**: 381-391.

Robock, A., Mu, M., Vinnikov, K., Trofimova, I.V. and Adamenko, T.I. 2005. Forty-five years of observed soil moisture in the Ukraine: No summer desiccation (yet). *Geophysical Research Letters* **32**: 10.1029/2004GL021914.

Robock, A., Vinnikov, K.Y., Srinivasan, G., Entin, J.K., Hollinger, S.E., Speranskaya, N.A., Liu, S. and Namkhai,

A. 2000. The global soil moisture data bank. *Bulletin of the American Meteorological Society* **81**: 1281-1299.

Rosentiel, T.N., Potosnak, M.J., Griffin, K.L., Fall, R. and Monson, R.K. 2003. Increased CO_2 uncouples growth from isoprene emission in an agriforest ecosystem. *Nature* advance online publication, 5 January 2003 (doi:**10.1038**/nature 01312).

Scholefield, P.A., Doick, K.J., Herbert, B.M.J., Hewitt, C.N.S., Schnitzler, J.-P., Pinelli, P. and Loreto, F. 2004. Impact of rising CO_2 on emissions of volatile organic compounds: isoprene emission from *Phragmites australis* growing at elevated CO_2 in a natural carbon dioxide spring. *Plant, Cell and Environment* **27**: 393-401.

Sharkey, T.D., Loreto, F. and Delwiche, C.F. 1991. High carbon dioxide and sun/shade effect on isoprene emissions from oak and aspen tree leaves. *Plant, Cell and Environment* **14**: 333-338.

7.8.10. Microorganisms

Plants grown in CO_2-enriched atmospheres nearly always exhibit increased photosynthetic rates and biomass production. Due to this productivity enhancement, more plant material is typically added to soils from root growth, turnover and exudation, as well as from leaves and stems following their abscission and falling to the ground during senescence. Such additions of carbon onto and into soils often serve as the only carbon source for supporting the development and growth of microorganisms in terrestrial habitats. Thus, it is important to understand how CO_2-induced increases in plant growth affect microorganisms, a topic omitted from discussion by the IPCC.

Several studies have shown that atmospheric CO_2 enrichment does not significantly impact soil microorganisms. Zak *et al.* (2000), for example, observed no significant differences in soil microbial biomass beneath aspen seedlings grown at 350 and 700 ppm CO_2 after 2.5 years of differential treatment. Likewise, in the cases of Griffiths *et al.* (1998) and Insam *et al.* (1999), neither research team reported any changes in microbial community structure beneath ryegrass and artificial tropical ecosystems, respectively, after subjecting them to atmospheric CO_2 enrichment.

Other studies, however, have found that elevated CO_2 can significantly affect soil microorganisms. Van Ginkel and Gorissen (1998) observed that three months of elevated CO_2 exposure (700 ppm) increased soil microbial biomass beneath ryegrass

plants by 42 percent relative to that produced under ambient CO_2 conditions, as did Van Ginkel *et al.* (2000). Likewise, soil microbial biomass was reported to increase by 15 percent beneath agricultural fields subjected to a two-year wheat-soybean crop rotation (Islam *et al.*, 2000). In a study by Marilley *et al.* (1999), atmospheric CO_2 enrichment significantly increased bacterial numbers in the rhizospheres beneath ryegrass and white clover monocultures. Similarly, Lussenhop *et al.* (1998) reported CO_2-induced increases in the amounts of bacteria, protozoa, and microarthropods in soils that had supported regenerating poplar tree cuttings for five months. In addition, Hungate *et al.* (2000) reported that twice-ambient CO_2 concentrations significantly increased the biomass of active fungal organisms and flagellated protozoa beneath serpentine and sandstone grasslands after four years of treatment exposure.

In taking a closer look at the study of Marilley *et al.* (1999), it is evident that elevated CO_2 caused shifts in soil microbial populations. In soils beneath their leguminous white clover, for example, elevated CO_2 favored shifts towards *Rhizobium* bacterial species, which likely increased nitrogen availability—via nitrogen fixation—to support enhanced plant growth. However, in soils beneath non-leguminous ryegrass monocultures, which do not form symbiotic relationships with *Rhizobium* species, elevated CO_2 favored shifts towards *Pseudomonas* species, which likely acquired nutrients to support enhanced plant growth through mechanisms other than nitrogen fixation. Nonetheless, in both situations, the authors observed CO_2-induced shifts in bacterial populations that would likely optimize nutrient acquisition for specific host plant species.

In an unrelated study, Montealegre *et al.* (2000) reported that elevated CO_2 acted as a selective agent among 120 different isolates of *Rhizobium* growing beneath white clover plants. Specifically, when bacterial strains favored by ambient and elevated CO_2 concentrations were mixed together and grown with white clover at an atmospheric CO_2 concentration of 600 ppm, 17 percent more root nodules were formed by isolates previously determined to be favored by elevated CO_2.

Hu *et al.* (2001) subjected fertile sandstone grasslands to five years of twice-ambient CO_2 concentrations and found they exhibited increased soil microbial biomass while simultaneously enhancing plant nitrogen uptake. The net effect of these phenomena reduced nitrogen availability for microbial use, which consequently decreased microbial respiration and, hence, microbial decomposition. Consequently, these ecosystems displayed CO_2-induced increases in net carbon accumulation. Similarly, Williams *et al.* (2000) reported that microbial biomass carbon increased by 4 percent in a tallgrass prairie after five years exposure to twice-ambient CO_2 concentrations, which contributed to a total soil carbon enhancement of 8 percent.

In summation, as the CO_2 content of the air continues to rise, earth's vegetation will likely respond with increasing photosynthetic rates and biomass production, returning more organic carbon to the soil where it will be utilized by microbial organisms to maintain or increase their population numbers, biomass and heterotrophic activities (Weihong *et al.*, 2000; Arnone and Bohlen, 1998). Shifts in microbial community structure may occur that will favor the intricate relationships that currently exist between leguminous and non-leguminous plants and the specific microorganisms upon which they depend.

Additional information on this topic, including reviews of newer publications as they become available, can be found at http://www.co2science.org/subject/m/microorganisms.php.

References

Arnone, J.A., III and Bohlen, P.J. 1998. Stimulated N_2O flux from intact grassland monoliths after two growing seasons under elevated atmospheric CO_2. *Oecologia* **116**: 331-335.

Griffiths, B.S., Ritz, K., Ebblewhite, N., Paterson, E. and Killham, K. 1998. Ryegrass rhizosphere microbial community structure under elevated carbon dioxide concentrations, with observations on wheat rhizosphere. *Soil Biology and Biochemistry* **30**: 315-321.

Hu, S., Chapin III, F.S., Firestone, M.K., Field, C.B. and Chiariello, N.R. 2001. Nitrogen limitation of microbial decomposition in a grassland under elevated CO_2. *Nature* **409**: 88-191.

Hungate, B.A., Jaeger III, C.H., Gamara, G., Chapin III, F.S. and Field, C.B. 2000. Soil microbiota in two annual grasslands: responses to elevated atmospheric CO_2. *Oecologia* **124**: 589-598.

Insam, H., Baath, E., Berreck, M., Frostegard, A., Gerzabek, M.H., Kraft, A., Schinner, F., Schweiger, P. and Tschuggnall, G. 1999. Responses of the soil microbiota to elevated CO_2 in an artificial tropical ecosystem. *Journal of Microbiological Methods* **36**: 45-54.

Islam, K.R., Mulchi, C.L. and Ali, A.A. 2000. Interactions of tropospheric CO_2 and O_3 enrichments and moisture variations on microbial biomass and respiration in soil. *Global Change Biology* **6**: 255-265.

Lussenhop, J., Treonis, A., Curtis, P.S., Teeri, J.A. and Vogel, C.S. 1998. Response of soil biota to elevated atmospheric CO_2 in poplar model systems. *Oecologia* **113**: 247-251.

Marilley, L., Hartwig, U.A. and Aragno, M. 1999. Influence of an elevated atmospheric CO_2 content on soil and rhizosphere bacterial communities beneath *Lolium perenne* and *Trifolium repens* under field conditions. *Microbial Ecology* **38**: 39-49.

Montealegre, C.M., Van Kessel, C., Blumenthal, J.M., Hur, H.G., Hartwig, U.A. and Sadowsky, M.J. 2000. Elevated atmospheric CO_2 alters microbial population structure in a pasture ecosystem. *Global Change Biology* **6**: 475-482.

Van Ginkel, J.H. and Gorissen, A. 1998. In situ decomposition of grass roots as affected by elevated atmospheric carbon dioxide. *Soil Science Society of America Journal* **62**: 951-958.

Van Ginkel, J.H., Gorissen, A. and Polci, D. 2000. Elevated atmospheric carbon dioxide concentration: effects of increased carbon input in a *Lolium perenne* soil on microorganisms and decomposition. *Soil Biology & Biochemistry* **32**: 449-456.

Weihong, L., Fusuo, Z. and Kezhi, B. 2000. Responses of plant rhizosphere to atmospheric CO_2 enrichment. *Chinese Science Bulletin* **45**: 97-101.

Williams, M.A., Rice, C.W. and Owensby, C.E. 2000. Carbon dynamics and microbial activity in tallgrass prairie exposed to elevated CO_2 for 8 years. *Plant and Soil* **227**: 127-137.

Zak, D.R., Pregitzer, K.S., Curtis, P.S. and Holmes, W.E. 2000. Atmospheric CO_2 and the composition and function of soil microbial communities. *Ecological Applications* **10**: 47-59.

7.8.11. Worms

Perhaps the best known worm in the world is the common earthworm. How will it be affected as the air's CO_2 content continues to climb, and how will its various responses affect the biosphere? What about other worms? How will they fare in a CO_2-enriched world of the future, and what will be the results of their responses?

"Earthworms," in the words of Edwards (1988), "play a major role in improving and maintaining the fertility, structure, aeration and drainage of agricultural soils." As noted by Sharpley *et al.* (1988), "by ingestion and digestion of plant residue and subsequent egestion of cast material, earthworms can redistribute nutrients in a soil and enhance enzyme activity, thereby increasing plant availability of both soil and plant residue nutrients," as others have also demonstrated (Bertsch *et al.*, 1988; McCabe *et al.*, 1988; Zachmann and Molina, 1988). Kemper (1988) describes how "burrows opened to the surface by surface-feeding worms provide drainage for water accumulating on the surface during intense rainfall," noting that "the highly compacted soil surrounding the expanded burrows has low permeability to water which often allows water to flow through these holes for a meter or so before it is sorbed into the surrounding soil."

Hall and Dudas (1988) report that the presence of earthworms appears to mitigate the deleterious effects of certain soil toxins. Logsdon and Linden (1988) describe a number of other beneficial effects of earthworms, including (1) enhancement of soil aeration, since under wet conditions earthworm channels do not swell shut as many soil cracks do, (2) enhancement of soil water uptake, since roots can explore deeper soil layers by following earthworm channels, and (3) enhancement of nutrient uptake, since earthworm casts and channel walls have a more neutral pH and higher available nutrient level than bulk soil. Hence, we should care about what happens to earthworms as the air's CO_2 content rises because of the many important services they provide for earth's plant life.

Edwards (1988) says "the most important factor in maintaining good earthworm populations in agricultural soils is that there be adequate availability of organic matter," while Hendrix *et al.* (1988) and Kladivko (1988) report that greater levels of plant productivity promote greater levels of earthworm activity. Consequently, since the most ubiquitous and powerful effect of atmospheric CO_2 enrichment is its stimulation of plant productivity, which leads to enhanced organic matter delivery to soils, it logically follows that this aerial fertilization effect of the ongoing rise in the air's CO_2 content should increase earthworm populations and amplify the many beneficial services they provide for plants.

The second most significant and common effect of atmospheric CO_2 enrichment on plants is its antitranspirant effect, whereby elevated levels of atmospheric CO_2 reduce leaf stomatal apertures and slow the rate of evaporative water loss from the vast bulk of earth's vegetation. Both growth chamber studies and field experiments that have studied this phenomenon provide voluminous evidence that it often leads to increased soil water contents in many terrestrial ecosystems, which also benefits earthworm populations.

Zaller and Arnone (1997) fumigated open-top and -bottom chambers in a calcareous grassland near Basal, Switzerland with air of either 350 or 600 ppm CO_2 for an entire growing season. They found that the mean annual soil moisture content in the CO_2-enriched chambers was 10 percent greater than that observed in the ambient-air chambers, and because rates of surface cast production by earthworms are typically positively correlated with soil moisture content, they found that cumulative surface cast production after only one year was 35 percent greater in the CO_2-enriched chambers than in the control chambers. In addition, because earthworm casts are rich in organic carbon and nitrogen, the cumulative amount of these important nutrients on a per-land-area basis was found to be 28 percent greater in the CO_2-enriched chambers than it was in the ambient-air chambers. In a subsequent study of the same grassland, Zaller and Arnone (1999) found that plants growing in close proximity to the earthworm casts produced more biomass than similar plants growing further away from them. They also found that the CO_2-induced growth stimulation experienced by the various grasses was also greater for those plants growing nearer the earthworm casts.

These various observations suggest that atmospheric CO_2 enrichment sets in motion a self-enhancing cycle of positive biological phenomena, whereby increases in the air's CO_2 content (1) stimulate plant productivity and (2) reduce plant evaporative water loss, which results in (3) more organic matter entering the soil and (4) a longer soil moisture retention time and/or greater soil water contents, all of which factors lead to (5) the development of larger and more active earthworm populations, which (6) enhance many important soil properties, including fertility, structure, aeration and drainage, which improved properties (7) further enhance the growth of the plants whose CO_2-induced increase in productivity was the factor that started the whole series of processes in the first place.

More earthworms also can increase soil's ability to sequester carbon. As Jongmans *et al.* (2003) point out, "the rate of organic matter decomposition can be decreased in worm casts compared to bulk soil aggregates (Martin, 1991; Haynes and Fraser, 1998)." On the basis of these studies and their own micro-morphological investigation of structural development and organic matter distribution in two calcareous marine loam soils on which pear trees had been grown for 45 years (one of which soils exhibited little to no earthworm activity and one of which exhibited high earthworm activity, due to different levels of heavy metal contamination of the soils as a consequence of the prior use of different amounts of fungicides), they concluded that "earthworms play an important role in the intimate mixing of organic residues and fine mineral soil particles and the formation of organic matter-rich micro-aggregates and can, therefore, contribute to physical protection of organic matter, thereby slowing down organic matter turnover and increasing the soil's potential for carbon sequestration." Put more simply, atmospheric CO_2 enrichment that stimulates the activity of earthworms also leads to more—and more secure—sequestration of carbon in earth's soils.

Don et al. (2008) studied the effects of anecic earthworms—which generally inhabit a single vertical burrow throughout their entire lives that can be as much as five meters in depth, but is generally in the range of one to two meters—on soil carbon stocks and turnover via analyses of enzyme activity, stable isotopes, nuclear magnetic resonance spectroscopy, and the ^{14}C age of their burrow linings. The results of their study indicated that "the carbon distribution in soils is changed by anecic earthworms' activity with more carbon stored in the subsoil where earthworms slightly increase the carbon stocks." In this regard they also state that "the translocation of carbon from [the] organic layer to the subsoil will decrease the carbon vulnerability to mineralization," since "carbon in the organic layer and the surface soil is much more prone to disturbances with rapid carbon loss than subsoil carbon."

Bossuyt et al. (2005) conducted a pair of experiments designed to investigate "at what scale and how quickly earthworms manage to protect SOM [soil organic matter]." In the first experiment, soil aggregate size distribution together with total C and ^{13}C were measured in three treatments—control soil, soil + ^{13}C-labeled sorghum leaf residue, and soil + ^{13}C-labeled residue + earthworms—after a period of 20 days incubation, where earthworms were added after the eighth day. In the second experiment, they determined the protected C and ^{13}C pools inside the newly formed casts and macro- and micro-soil-aggregates. Results indicated that the proportion of large water-stable macroaggregates was on average 3.6 times greater in the soil-residue samples that contained earthworms than in those that lacked earthworms, and that the macroaggregates in the earthworm treatment contained approximately three times more sequestered carbon. What is more, the earthworms were found to form "a significant pool of protected C in microaggregates within large macroaggregates after 12 days of incubation," thereby demonstrating the rapidity with which earthworms perform their vital function of sequestering carbon in soils when plant residues become available to them.

Cole et al. (2002) report that "in the peatlands of northern England, which are classified as blanket peat, it has been suggested that the potential effects of global warming on carbon and nutrient dynamics will be related to the activities of dominant soil fauna, and especially enchytraeid worms." In harmony with these ideas, Cole et al. say they "hypothesized" that warming would lead to increased enchytraeid worm activity, which would lead to higher grazing pressure on microbes in the soil; and since enchytraeid grazing has been observed to enhance microbial activity (Cole et al., 2000), they further hypothesized that more carbon would be liberated in dissolved organic form, "supporting the view that global warming will increase carbon loss from blanket peat ecosystems."

The scientists next describe how they constructed small microcosms from soil and litter they collected near the summit of Great Dun Fell, Cumbria, England. Subsequent to "defaunating" this material by reducing its temperature to -80°C for 24 hours, they thawed and inoculated it with native soil microbes, after which half of the microcosms were incubated in the dark at 12°C and half at 18°C, the former of which temperatures was approximately equal to mean August soil temperature at a depth of 10 cm at the site of soil collection, while the latter was said by them to be "close to model predictions for soil warming that might result from a doubling of CO_2 in blanket peat environments."

Ten seedlings of an indigenous grass of blanket peat were then transplanted into each of the microcosms, while 100 enchytraeid worms were added to each of half of the mini-ecosystems. These procedures resulted in the creation of four experimental treatments: ambient temperature, ambient temperature + enchytraeid worms, elevated temperature, and elevated temperature + enchytraeid worms. The resulting 48 microcosms—sufficient to destructively harvest three replicates of each treatment four different times throughout the course of the 64-day experiment—were arranged in a fully randomized design and maintained at either 12 or 18°C with alternating 12-hour light and dark periods. In addition, throughout the entire course of the study, the microcosms were given distilled water every two days to maintain their original weights.

So what did the researchers find? First, and contrary to their hypothesis, elevated temperature reduced the ability of the enchytraeid worms to enhance the loss of carbon from the microcosms. At the normal ambient temperature, for example, the presence of the worms enhanced dissolved organic carbon (DOC) loss by 16 percent, while at the elevated temperature expected for a doubling of the air's CO_2 content, the worms had no effect at all on DOC. In addition, Cole et al. note that "warming may cause drying at the soil surface, forcing enchytraeids to burrow to deeper subsurface horizons." Hence, since the worms are known to have little influence on soil carbon dynamics below a depth of 4 cm (Cole et al., 2000), they concluded that this additional consequence of warming would further reduce the ability of enchytraeids to enhance carbon loss from blanket peatlands.

In summarizing their findings, Cole et al. say that "the soil biotic response to warming in this study was negative." That is, it was of such a nature that it resulted in a reduced loss of carbon to the atmosphere, which would tend to slow the rate of rise of the air's CO_2 content, just as was suggested by the results of the study of Jongmans et al.

Yeates et al. (2003) report results from a season-long FACE study of a 30-year-old New Zealand pasture, where three experimental plots had been maintained at the ambient atmospheric CO_2 concentration of 360 ppm and three others at a concentration of 475 ppm (a CO_2 enhancement of only 32 percent) for a period of four to five years. The pasture contained about twenty species of plants, including C_3 and C_4 grasses, legumes and forbs. Nematode, or "roundworm," populations increased significantly in response to the 32 percent increase in the air's CO_2 concentration. Of the various feeding groups studied, Yeates et al. report that the relative increase "was lowest in bacterial-feeders (27%), slightly higher in plant (root) feeders (32%), while those with delicate stylets (or narrow lumens; plant-associated, fungal-feeding) increased more (52% and 57%, respectively)." The greatest nematode increases were recorded among omnivores (97 percent) and predators (105 percent). Most dramatic of all, root-

feeding populations of the Longidorus nematode taxon rose by 330 percent. Also increasing in abundance were earthworms: Aporrectodea caliginosa by 25 percent and Lumbricus rubellus by 58 percent. Enchytraeids, on the other hand, decreased in abundance, by approximately 30 percent.

With respect to earthworms, Yeates et al. note that just as was found in the studies cited in the first part of this review, the introduction of lumbricids has been demonstrated to improve soil conditions in New Zealand pastures (Stockdill, 1982), which benefits pasture plants. Hence, the CO_2-induced increase in earthworm numbers observed in Yeates et al.'s study would be expected to do more of the same, while the reduced abundance of enchytraeids they documented in the CO_2-enriched pasture would supposedly lead to less carbon being released to the air from the soil, as per the known ability of enchytraeids to promote carbon loss from British peat lands under current temperatures.

In summary, the lowly earthworm and still lowlier soil nematodes respond to increases in the air's CO_2 content via a number of plant-mediated phenomena in ways that further enhance the positive effects of atmospheric CO_2 enrichment on plant growth and development, while at the same time helping to sequester more carbon more securely in the soil.

Additional information on this topic, including reviews of newer publications as they become available, can be found at http://www.co2science.org/subject/e/earthworms.php.

References

Bertsch, P.M., Peters, R.A., Luce, H.D. and Claude, D. 1988. Comparison of earthworm activity in long-term no-tillage and conventionally tilled corn systems. Agronomy Abstracts 80: 271.

Bossuyt, H., Six, J. and Hendrix, P.F. 2005. Protection of soil carbon by microaggregates within earthworm casts. Soil Biology & Biochemistry 37: 251-258.

Cole, L., Bardgett, R.D. and Ineson, P. 2000. Enchytraeid worms (Oligochaeta) enhance mineralization of carbon in organic upland soils. European Journal of Soil Science 51: 185-192.

Cole, L., Bardgett, R.D., Ineson, P. and Hobbs, P.J. 2002. Enchytraeid worm (Oligochaeta) influences on microbial community structure, nutrient dynamics and plant growth in blanket peat subjected to warming. Soil Biology & Biochemistry 34: 83-92.

Don, A., Steinberg, B., Schoning, I., Pritsch, K., Joschko, M., Gleixner, G. and Schulze, E.-D. 2008. Organic carbon sequestration in earthworm burrows. *Soil Biology & Biochemistry* **40**: 1803-1812.

Edwards, C.A. 1988. Earthworms and agriculture. *Agronomy Abstracts* **80**: 274.

Hall, R.B. and Dudas, M.J. 1988. Effects of chromium loading on earthworms in an amended soil. *Agronomy Abstracts* **80**: 275.

Haynes, R.J. and Fraser, P.M. 1998. A comparison of aggregate stability and biological activity in earthworm casts and uningested soil as affected by amendment with wheat and lucerne straw. *European Journal of Soil Science* **49**: 629-636.

Hendrix, P.F., Mueller, B.R., van Vliet, P., Bruce, R.R. and Langdale, G.W. 1988. Earthworm abundance and distribution in agricultural landscapes of the Georgia piedmont. *Agronomy Abstracts* **80**: 276.

Jongmans, A.G., Pulleman, M.M., Balabane, M., van Oort, F. and Marinissen, J.C.Y. 2003. Soil structure and characteristics of organic matter in two orchards differing in earthworm activity. *Applied Soil Ecology* **24**: 219-232.

Kemper, W.D. 1988. Earthworm burrowing and effects on soil structure and transmissivity. *Agronomy Abstracts* **80**: 278.

Kladivko, E.J. 1988. Soil management effects on earthworm populations and activity. *Agronomy Abstracts* **80**: 278.

Logsdon, S.D. and Linden, D.L. 1988. Earthworm effects on root growth and function, and on crop growth. *Agronomy Abstracts* **80**: 280.

Martin, A. 1991. Short- and long-term effects of the endogenic earthworm *Millsonia anomala* (Omodeo) (Megascolecidae, Oligochaeta) of tropical savannas on soil organic matter. *Biology and Fertility of Soils* **11**: 234-238.

McCabe, D., Protz, R. and Tomlin, A.D. 1988. Earthworm influence on soil quality in native sites of southern Ontario. *Agronomy Abstracts* **80**: 281.

Sharpley, A.N., Syers, J.K. and Springett, J. 1988. Earthworm effects on the cycling of organic matter and nutrients. *Agronomy Abstracts* **80**: 285.

Stockdill, S.M.J. 1982. Effects of introduced earthworms on the productivity of New Zealand pastures. *Pedobiologia* **24**: 29-35.

Yeates, G.W., Newton, P.C.D. and Ross, D.J. 2003. Significant changes in soil microfauna in grazed pasture under elevated carbon dioxide. *Biology and Fertility of Soils* **38**: 319-326.

Zachmann, J.E. and Molina, J.A. 1988. Earthworm-microbe interactions in soil. *Agronomy Abstracts* **80**: 289.

Zaller, J.G. and Arnone III, J.A. 1997. Activity of surface-casting earthworms in a calcareous grassland under elevated atmospheric CO_2. *Oecologia* **111**: 249-254.

Zaller, J.G. and Arnone III, J.A. 1999. Interactions between plant species and earthworm casts in a calcareous grassland under elevated CO_2. *Ecology* **80**: 873-881.

7.9. Greening of the Earth

More than two decades ago, Idso (1986) published a small item in *Nature* advancing the idea that the aerial fertilization effect of the CO_2 that is liberated by the burning of coal, gas and oil was destined to dramatically enhance the productivity of earth's vegetation. In a little book he had published four years earlier (Idso, 1982), he had predicted that "CO_2 effects on both the managed and unmanaged biosphere will be overwhelmingly positive." In a monograph based on a lecture he gave nine years later (Idso, 1995), he said "we appear to be experiencing the initial stages of what could truly be called a *rebirth of the biosphere*, the beginnings of a biological rejuvenation that is without precedent in all of human history."

In light of the fact that Idso's worldview is nearly the exact opposite of the apocalyptic vision promulgated by the IPCC, it is instructive to see what real-world observations reveal about the matter. In this section we review studies that show a CO_2-induced greening of Africa, Asia, Europe, North America, the oceans, and the entire globe.

Additional information on this topic, including reviews not discussed here, can be found at http://www.co2science.org/subject/g/subject_g.php under the heading Greening of the Earth.

References

Idso, S.B. 1982. *Carbon Dioxide: Friend or Foe?* IBR Press, Tempe, AZ.

Idso, S.B. 1986. Industrial age leading to the greening of the Earth? *Nature* **320**: 22.

Idso, S.B. 1995. *CO₂ and the Biosphere: The Incredible Legacy of the Industrial Revolution*. Department of Soil, Water & Climate, University of Minnesota, St. Paul, Minnesota, USA.

7.9.1. Africa

In an article by Fred Pearce that was posted on the website of *New Scientist* magazine on 16 September 2002 titled "Africa's deserts are in 'spectacular' retreat," we were told the story of vegetation reclaiming great tracts of barren land across the entire southern edge of the Sahara. This information likely came as a bit of a surprise to many, since the United Nations Environment Program had reported to the World Summit on Sustainable Development in Johannesburg, South Africa in August of that year that over 45 percent of the continent was experiencing severe desertification. The world of nature, however, told a vastly different story.

Pearce began by stating "the southern Saharan desert is in retreat, making farming viable again in what were some of the most arid parts of Africa," noting that "Burkina Faso, one of the West African countries devastated by drought and advancing deserts 20 years ago, is growing so much greener that families who fled to wetter coastal regions are starting to go home."

The good news was not confined to Burkina Faso. "Vegetation," according to Pearce, "is ousting sand across a swathe of land stretching from Mauritania on the shores of the Atlantic to Eritrea 6000 kilometers away on the Red Sea coast." Besides being widespread in space, the greening was widespread in time, having been happening since at least the mid-1980s.

Quoting Chris Reij of the Free University of Amsterdam, Pearce wrote that "aerial photographs taken in June show 'quite spectacular regeneration of vegetation' in northern Burkina Faso." The data indicated the presence of more trees for firewood and more grassland for livestock. In addition, a survey that Reij was collating showed, according to Pearce, "a 70% increase in yields of local cereals such as sorghum and millet in one province in recent years." Also studying the area was Kjeld Rasmussen of the University of Copenhagen, who reported that since the 1980s there had been a "steady reduction in bare ground" with "vegetation cover, including bushes and trees, on the increase on the dunes."

Pearce also reported on the work of a team of geographers from Britain, Sweden and Denmark that had spent much of the prior summer analyzing archived satellite images of the Sahel. Citing Andrew Warren of University College London as a source of information on this study, he said the results showed "that 'vegetation seems to have increased

significantly' in the past 15 years, with major regrowth in southern Mauritania, northern Burkina Faso, north-western Niger, central Chad, much of Sudan and parts of Eritrea."

Should these findings take us by surprise? Not in the least, as Nicholson *et al.* (1998) reported in a study of a series of satellite images of the Central and Western Sahel that were taken from 1980 to 1995, they could find no evidence of any overall expansion of deserts and no drop in the rainfall use efficiency of native vegetation. In addition, in a satellite study of the entire Sahel from 1982 to 1990, Prince *et al.* (1998) detected a steady rise in rainfall use efficiency, suggesting that plant productivity and coverage of the desert had increased during this period.

That the greening phenomenon has continued apace is borne out by the study of Eklundh and Olsson (2003), who analyzed Normalized Difference Vegetation Index (NDVI) data obtained from the U.S. National Oceanic and Atmospheric Administration's satellite-borne Advanced Very High Resolution Radiometer whenever it passed over the African Sahel for the period 1982-2000. As they describe their findings, "strong positive change in NDVI occurred in about 22% of the area, and weak positive change in 60% of the area," while "weak negative change occurred in 17% of the area, and strong negative change in 0.6% of the area." In addition, they report that "integrated NDVI has increased by about 80% in the areas with strong positive change," while in areas with weak negative change, "integrated NDVI has decreased on average by 13%." The primary story told by these data, therefore, is one of strong positive trends in NDVI for large areas of the African Sahel over the last two decades of the twentieth century. Eklundh and Olsson conclude that the "increased vegetation, as suggested by the observed NDVI trend, could be part of the proposed tropical sink of carbon."

Due to the increase in vegetation over the past quarter-century in the Sahel, the African region was recently featured in a special issue of the *Journal of Arid Environments* titled "The 'Greening' of the Sahel." Therein, Anyamba and Tucker (2005) describe their development of an NDVI history of the region for the period 1981-2003. Comparing this history with the precipitation history of the Sahel developed by Nicholson (2005), they found that "the persistence and spatial coherence of drought conditions during the 1980s is well represented by the NDVI anomaly patterns and corresponds with the documented rainfall anomalies across the region during this time period." In addition, they report that "the prevalence of greener than normal conditions during the 1990s to 2003 follows a similar increase in rainfall over the region during the last decade."

In another analysis of NDVI and rainfall data in the same issue of the *Journal of Arid Environments*, Olsson *et al.* (2005) report finding "a consistent trend of increasing vegetation greenness in much of the region," which they describe as "remarkable." They say increasing rainfall over the last few years "is certainly one reason" for the greening phenomenon. However, they find the increase in rainfall "does not fully explain" the increase in greenness.

For one thing, the three Swedish scientists note that "only eight out of 40 rainfall observations showed a statistically significant increase between 1982-1990 and 1991-1999." In addition, they report that "further analysis of this relationship does not indicate an overall relationship between rainfall increase and vegetation trend." So what else could be driving the increase in greenness?

Olsson *et al.* suggest that "another potential explanation could be improved land management, which has been shown to cause similar changes in vegetation response elsewhere (Runnstrom, 2003)." However, in more detailed analyses of Burkina Faso and Mali, where production of millet rose by 55 percent and 35 percent, respectively, since 1980, they could find "no clear relationship" between agricultural productivity and NDVI, which argues against the land management explanation.

A third speculation of Olsson *et al.* is that the greening of the Sahel could be caused by increasing rural-to-urban migration. In this scenario, widespread increases in vegetation occur as a result of "reduced area under cultivation," due to a shortage of rural laborers, and/or "increasing inputs on cropland," such as seeds, machinery and fertilizers made possible by an increase in money sent home to rural households by family members working in cities. However, Olsson *et al.* note that "more empirical research is needed to verify this [hypothesis]."

About the only thing left is what Idso (1982, 1986, 1995) has suggested, i.e., that the aerial fertilization effect of the ongoing rise in the air's CO_2 concentration (which greatly enhances vegetative productivity) and its anti-transpiration effect (which enhances plant water-use efficiency and enables plants to grow in areas that were once too dry for them) are the major players in the greening phenomenon. Whatever was the reason for the greening of the Sahel over the past quarter-century, it is clear that in spite of what the IPCC claims were

unprecedented increases in anthropogenic CO_2 emissions and global temperatures, the Sahel experienced an increase in vegetative cover that was truly, as Olsson *et al.* write, "remarkable."

Additional information on this topic, including reviews of newer publications as they become available, can be found at http://www.co2science.org/subject/g/africagreen.php.

References

Anyamba, A. and Tucker, C.J. 2005. Analysis of Sahelian vegetation dynamics using NOAA-AVHRR NDVI data from 1981-2003. *Journal of Arid Environments* **63**: 596-614.

Eklundh, L. and Olssson, L. 2003. Vegetation index trends for the African Sahel 1982-1999. *Geophysical Research Letters* **30**: 10.1029/2002GL016772.

Idso, S.B. 1982. *Carbon Dioxide: Friend or Foe?* IBR Press, Tempe, AZ.

Idso, S.B. 1986. Industrial age leading to the greening of the Earth? *Nature* **320**: 22.

Idso, S.B. 1995. *CO₂ and the Biosphere: The Incredible Legacy of the Industrial Revolution*. Department of Soil, Water & Climate, University of Minnesota, St. Paul, Minnesota, USA.

Nicholson, S. 2005. On the question of the 'recovery' of the rains in the West African Sahel. *Journal of Arid Environments* **63**: 615-641.

Nicholson, S.E., Tucker, C.J. and Ba, M.B. 1998. Desertification, drought, and surface vegetation: An example from the West African Sahel. *Bulletin of the American Meteorological Society* **79**: 815-829.

Olsson, L., Eklundh, L. and Ardo, J. 2005. A recent greening of the Sahel—trends, patterns and potential causes. *Journal of Arid Environments* **63**: 556-566.

Prince, S.D., Brown De Colstoun, E. and Kravitz, L.L. 1998. Evidence from rain-use efficiencies does not indicate extensive Sahelian desertification. *Global Change Biology* **4**: 359-374.

Runnstrom, M. 2003. Rangeland development of the Mu Us Sandy Land in semiarid China: an analysis using Landsat and NOAA remote sensing data. *Land Degradation & Development Studies* **14**: 189-202.

7.9.2. Asia

We begin a review of Asia with the modeling work of Liu *et al.* (2004), who derived detailed estimates of the economic impact of predicted climate change on agriculture in China, utilizing county-level agricultural, climate, social, economic and edaphic data for 1275 agriculture-dominated counties for the period 1985-1991, together with the outputs of three general circulation models of the atmosphere that were based on five different scenarios of anthropogenic CO_2-induced climate change that yielded a mean countrywide temperature increase of 3.0°C and a mean precipitation increase of 3.9 percent for the 50-year period ending in AD 2050. In doing so, they determined that "all of China would benefit from climate change in most scenarios." In addition, they state that "the effects of CO_2 fertilization should [also] be included, for some studies indicate that this may produce a significant increase in yield." The significance of these findings is readily grasped when it is realized, in Liu *et al.*'s words, that "China's agriculture has to feed more than one-fifth of the world's population, and, historically, China has been famine prone." They report that "as recently as the late 1950s and early 1960s a great famine claimed about thirty million lives (Ashton *et al.*, 1984; Cambridge History of China, 1987)."

Moving from agro-ecosystems to natural ones, Su *et al.* (2004) used an ecosystem process model to explore the sensitivity of the net primary productivity (NPP) of an oak forest near Beijing (China) to the global climate changes projected to result from a doubling of the atmosphere's CO_2 concentration from 355 to 710 ppm. The results of this work suggested that the aerial fertilization effect of the specified increase in the air's CO_2 content would raise the forest's NPP by 14.0 percent, that a concomitant temperature increase of 2°C would boost the NPP increase to 15.7 percent, and that adding a 20 percent increase in precipitation would push the NPP increase to 25.7 percent. They calculated that a 20 percent increase in precipitation and a 4°C increase in temperature would also boost the forest's NPP by 25.7 percent.

Grunzweig *et al.* (2003) tell the tale of the Yatir forest, a 2,800-hectare stand of Aleppo and other pine trees, that had been planted some 35 years earlier at the edge of the Negev Desert in Israel. An intriguing aspect of this particular forest, which they characterize as growing in poor soil of only 0.2 to 1.0 meter's depth above chalk and limestone, is that

although it is located in an arid part of Asia that receives less annual precipitation than all of the other scores of FluxNet stations in the global network of micrometeorological tower sites that use eddy covariance methods to measure exchanges of CO_2, water vapor and energy between terrestrial ecosystems and the atmosphere (Baldocchi *et al.*, 2001), the forest's annual net ecosystem CO_2 exchange was just as high as that of many high-latitude boreal forests and actually higher than that of most temperate forests. Grunzweig *et al.* note that the increase in atmospheric CO_2 concentration that has occurred since pre-industrial times should have improved water use efficiency (WUE) in most plants by increasing the ratio of CO_2 fixed to water lost via evapotranspiration. They report that "reducing water loss in arid regions improves soil moisture conditions, decreases water stress and extends water availability," which "can indirectly increase carbon sequestration by influencing plant distribution, survival and expansion into water-limited environments."

That this hypothesis is correct has been demonstrated by Leavitt *et al.* (2003) within the context of the long-term atmospheric CO_2 enrichment experiment of Idso and Kimball (2001) on sour orange trees. It has also been confirmed in nature by Fang (2003), who obtained identical (to the study of Leavitt *et al.*) CO_2-induced WUE responses for 23 groups of naturally occurring trees scattered across western North America over the period 1800-1985, which response, Fang concludes, "would have caused natural trees in arid environments to grow more rapidly, acting as a carbon sink for anthropogenic CO_2," which is exactly what Grunzweig *et al.* found to be happening in the Yatir forest on the edge of the Negev Desert.

Based primarily on satellite-derived Normalized Difference Vegetation Index (NDVI) data, Zhou *et al.* (2001) found that from July 1981 to December 1999, between 40 and 70° N latitude, there was a persistent increase in growing season vegetative productivity in excess of 12 percent over a broad contiguous swath of Asia stretching from Europe through Siberia to the Aldan plateau, where almost 58 percent of the land is forested. And in a companion study, Bogaert *et al.* (2002) determined that this productivity increase occurred at a time when this vast Asian region showed an overall warming trend "with negligible occurrence of cooling."

In another study that included a portion of Europe, Lapenis *et al.* (2005) analyzed trends in forest biomass in all 28 ecoregions of the Russian territory, based on data collected from 1953 to 2002 within 3196 sample plots comprised of about 50,000 entries, which database, in their words, "contains all available archived and published data." This work revealed that over the period 1961-1998, as they describe it, "aboveground wood, roots, and green parts increased by 4%, 21%, and 33%, respectively," such that "the total carbon density of the living biomass stock of the Russian forests increased by ~9%." They also report there was a concomitant increase of ~11 percent in the area of Russian forests. In addition, the team of U.S., Austrian and Russian scientists reported that "within the range of 50-65° of latitude [the range of 90 percent of Russian forests], the relationship between biomass density and the area-averaged NDVI is very close to a linear function, with a slope of ~1," citing the work of Myneni *et al.* (2001). Therefore, as they continue, "changes in the carbon density of live biomass in Russian forests occur at about the same rate as the increase in the satellite-based estimate in the seasonally accumulated NDVI," which observation strengthens the findings of *all* satellite-based NDVI studies.

Returning to China for several concluding reports, we begin with the work of Brogaard *et al.* (2005), who studied the dry northern and northwestern regions of the country—including the Inner Mongolia Autonomous Region (IMAR)—which had been thought to have experienced declining vegetative productivity over the past few decades due to "increasing livestock numbers, expansion of cultivated land on erosive soils and the gathering of fuel wood and herb digging," which practices were believed to have been driven by rising living standards, which in combination with a growing population were assumed to have increased the pressure on these marginal lands. In the case of increasing grazing, for example, Brogaard *et al.* note that the total number of livestock in the IMAR increased from approximately 46 million head in 1980 to about 71 million in 1997.

To better assess the seriousness of this supposedly "ongoing land degradation process," as they describe it, the researchers adapted a satellite-driven parametric model, originally developed for Sahelian conditions, to the central Asian steppe region of the IMAR by including "additional stress factors and growth efficiency computations." The applied model, in their words, "uses satellite sensor-acquired reflectance in combination with climate data to generate monthly estimates of gross primary production." To their great surprise, this work

revealed that "despite a rapid increase in grazing animals on the steppes of the IMAR for the 1982-1999 period," their model estimates did "not indicate declining biological production."

Clearly, some strong positive influence compensated for the increased human and animal pressures on the lands of the IMAR over the period of Brogaard *et al.*'s study. In this regard, they mention the possibility of increasing productivity on the agricultural lands of the IMAR, but they note that crops are grown on "only a small proportion of the total land area." Other potential contributing factors they mention are "an increase in precipitation, as well as afforestation projects." Two things they do *not* mention are the aerial fertilization effect and the transpiration-reducing effect of the increase in the air's CO_2 concentration that was experienced over the study period. Applied together, the sum of these positive influences (and possibly others that remain unknown) was demonstrably sufficient to keep plant productivity from declining in the face of greatly increasing animal and human pressures on the lands of the IMAR from 1982 to 1999.

Piao *et al.* (2005a) used a time series of NDVI data from 1982 to 1999, together with precipitation and temperature data, to investigate variations of desert area in China by "identifying the climatic boundaries of arid area and semiarid area, and changes in NDVI in these areas." In doing so, they discovered that "average rainy season NDVI in arid and semiarid regions both increased significantly during the period 1982-1999." Specifically, they found that the NDVI increased for 72.3 percent of total arid regions and for 88.2 percent of total semiarid regions, such that the area of arid regions decreased by 6.9 percent and the area of semiarid regions decreased by 7.9 percent. They also report that by analyzing Thematic Mapper satellite images, "Zhang *et al.* (2003) documented that the process of desertification in the Yulin area, Shannxi Province showed a decreased trend between 1987 and 1999," and that "according to the national monitoring data on desertification in western China (Shi, 2003), the annual desertification rate decreased from 1.2% in the 1950s to -0.2% at present."

Further noting that "variations in the vegetation coverage of these regions partly affect the frequency of sand-dust storm occurrence (Zou and Zhai, 2004)," Piao *et al.* concluded that "increased vegetation coverage in these areas will likely fix soil, enhance its anti-wind-erosion ability, reduce the possibility of released dust, and consequently cause a mitigation of sand-dust storms." They also reported that "recent studies have suggested that the frequencies of strong and extremely strong sand-dust storms in northern China have significantly declined from the early 1980s to the end of the 1990s (Qian *et al.*, 2002; Zhao *et al.*, 2004)."

Piao *et al.* (2006) investigated vegetation net primary production (NPP) derived from a carbon model (Carnegie-Ames-Stanford approach, CASA) and its interannual change in the Qinghai-Xizang (Tibetan) Plateau using 1982-1999 NDVI data and paired ground-based information on vegetation, climate, soil, and solar radiation. This work revealed that over the entire study period, NPP rose at a mean annual rate of 0.7 percent. However, Piao *et al.* report that "the NPP trends in the plateau over the two decades were divided into two distinguished periods: without any clear trend from 1982 to 1990 and significant increase from 1991 to 1999."

The three researchers say their findings suggest that "vegetation growth on the plateau in the 1990s has been much enhanced compared to that in [the] 1980s, consistent with the trend in the northern latitudes indicated by Schimel *et al.* (2001)." In addition, they say that "previous observational and NPP modeling studies have documented substantial evidence that terrestrial photosynthetic activity has increased over the past two to three decades in the middle and high latitudes in the Northern Hemisphere," and that "satellite-based NDVI data sets for the period of 1982-1999 also indicate consistent trends of NDVI increase," citing multiple references in support of each of these statements. Piao *et al.*'s findings, therefore, add to the growing body of evidence that reveals a significant "greening of the earth" is occurring.

Applying the same techniques, Fang *et al.* (2003) looked at the whole of China, finding that its terrestrial NPP increased by 18.7 percent between 1982 and 1999. Referring to this result as "an unexpected aspect of biosphere dynamics," they say that this increase "is much greater than would be expected to result from the fertilization effect of elevated CO_2, and also greater than expected from climate, based on broad geographic patterns." From 1982 to 1999, the atmosphere's CO_2 concentration rose by approximately 27.4 ppm. The aerial fertilization effect of this CO_2 increase could be expected to have increased the NPP of the conglomerate of forest types found in China by about 7.3 percent. (See the procedures and reasoning described in a *CO₂ Science* editorial, September 18,

2002, http://www.co2science.org/articles/V5/N38/EDIT.php). But this increase is only a part of the total NPP increase we could expect, for Fang *et al.* note that "much of the trend in NPP appeared to reflect a change towards an earlier growing season," which was driven by the 1.1°C increase in temperature they found to have occurred in their region of study between 1982 and 1999.

Following this lead, we learn from the study of White *et al.* (1999)—which utilized 88 years of data (1900-1987) that were obtained from 12 different locations within the eastern U.S. deciduous forest that stretches from Charleston, SC (32.8°N latitude) to Burlington, VT (44.5°N latitude)—that a 1°C increase in mean annual air temperature increases the length of the forest's growing season by approximately five days. In addition, White *et al.* determined that a one-day extension in growing season length increased the mean forest NPP of the 12 sites they studied by an average of 1.6 percent. Hence, we could expect an additional NPP increase due to the warming-induced growing season expansion experienced in China from 1982 to 1999 of 1.6 percent/day x 5 days = 8.0 percent, which brings the total CO_2-induced plus warming-induced increase in NPP to 15.3 percent.

Last, we note there is a well-documented positive synergism between increasing air temperature and CO_2 concentration (Idso and Idso, 1994), such that the 1°C increase in temperature experienced in China between 1982 and 1999 could easily boost the initial CO_2-induced 7.3 percent NPP enhancement to the 10.7 percent enhancement that when combined with the 8.0 percent enhancement caused by the warming-induced increase in growing season length would produce the 18.7 percent increase in NPP detected in the satellite data.

In view of these observations, the findings of Fang *et al.* are seen to be right in line with what would be expected to result from the increases in air temperature and atmospheric CO_2 concentration that occurred between 1982 and 1999 in China: a stimulated terrestrial biosphere that is growing ever more productive with each passing year. This is the true observed consequence of rising CO_2 and temperature, and it is about as far removed as one can get from the negative scenarios offered by the IPCC.

Analyzing the same set of data still further, Piao *et al.* (2005b) say their results suggest that "terrestrial NPP in China increased at a rate of 0.015 Pg C yr^{-1} over the period 1982-1999, corresponding to a total increase of 18.5%, or 1.03% annually." They also found that "during the past 2 decades the amplitude of the seasonal curve of NPP has increased and the annual peak NPP has advanced," which they say "may indirectly explain the enhanced amplitude and advanced timing of the seasonal cycle of atmospheric CO_2 concentration (Keeling *et al.*, 1996)," the former of which phenomena they further suggest "was probably due to the rise in atmospheric CO_2 concentration, elevated temperature, and increased atmospheric N and P deposition," while the latter phenomenon they attribute to "advanced spring onset and extended autumn growth owing to climate warming." We are in basic agreement on most of these points, but note that the advanced onset of what may be called biological spring is also fostered by the enhancement of early spring growth that is provided by the ongoing rise in the air's CO_2 concentration.

Citing a total of 20 scientific papers at various places in the following sentence from their research report, Piao *et al.* conclude that "results from observed atmospheric CO_2 and O_2 concentrations, inventory data, remote sensing data, and carbon process models have all suggested that terrestrial vegetation NPP of the Northern Hemisphere has increased over the past 2 decades and, as a result, the northern terrestrial ecosystems have become important sinks for atmospheric CO_2."

In conclusion, the historical increases in the atmosphere's CO_2 concentration and temperature have fostered a significant greening of the earth, including that observed throughout the length and breadth of Asia. It would appear that the climatic change claimed by the IPCC to have been experienced by the globe over the latter part of the twentieth century either did not occur or was dwarfed by opposing phenomena that significantly benefited China, as its lands grew ever greener during this period and its increased vegetative cover helped to stabilize its soils and throw feared desertification into reverse.

Additional information on this topic, including reviews of newer publications as they become available, can be found at http://www.co2science.org/subject/g/asiagreen.php.

References

Ashton, B., Hill, K., Piazza, A. and Zeitz, R. 1984. Famine in China, 1958-1961. *Population and Development Review* **10**: 613-615.

Baldocchi, D., Falge, E., Gu, L.H., Olson, R., Hollinger, D., Running, S., Anthoni, P., Bernhofer, C., Davis, K.,

Evans, R., Fuentes, J., Goldstein, A., Katul, G., Law B., Lee, X.H., Malhi, Y., Meyers, T., Munger, W., Oechel, W., Paw U, K.T., Pilegaard, K., Schmid, H.P., Valentini, R., Verma, S., Vesala, T., Wilson, K. and Wofsy, S. 2001. FLUXNET: A new tool to study the temporal and spatial variability of ecosystem-scale carbon dioxide, water vapor, and energy flux densities. *Bulletin of the American Meteorological Society* **82**: 2415-2434.

Bogaert, J., Zhou, L., Tucker, C.J, Myneni, R.B. and Ceulemans, R. 2002. Evidence for a persistent and extensive greening trend in Eurasia inferred from satellite vegetation index data. *Journal of Geophysical Research* **107**: 10.1029/2001JD001075.

Brogaard, S., Runnstrom, M. and Seaquist, J.W. 2005. Primary production of Inner Mongolia, China, between 1982 and 1999 estimated by a satellite data-driven light use efficiency model. *Global and Planetary Change* **45**: 313-332.

Fang, J., Piao, S., Field, C.B., Pan, Y., Guo, Q., Zhou, L., Peng, C. and Tao, S. 2003. Increasing net primary production in China from 1982 to 1999. *Frontiers in Ecology and the Environment* **1**: 293-297.

Feng, X. 1999. Trends in intrinsic water-use efficiency of natural trees for the past 100-200 years: A response to atmospheric CO_2 concentration. *Geochimica et Cosmochimica Acta* **63**: 1891-1903.

Grunzweig, J.M., Lin, T., Rotenberg, E., Schwartz, A. and Yakir, D. 2003. Carbon sequestration in arid-land forest. *Global Change Biology* **9**: 791-799.

Idso, K.E. and Idso, S.B. 1994. Plant responses to atmospheric CO_2 enrichment in the face of environmental constraints: A review of the past 10 years' research. *Agricultural and Forest Meteorology* **69**: 153-203.

Idso, S.B. and Kimball, B.A. 2001. CO_2 enrichment of sour orange trees: 13 years and counting. *Environmental and Experimental Botany* **46**: 147-153.

Keeling, C.D., Chin, J.F.S. and Whorf, T.P. 1996. Increased activity of northern vegetation inferred from atmospheric CO_2 measurements. *Nature* **382**: 146-149.

Lapenis, A., Shvidenko, A., Shepaschenko, D., Nilsson, S. and Aiyyer, A. 2005. Acclimation of Russian forests to recent changes in climate. *Global Change Biology* **11**: 2090-2102.

Leavitt, S.W., Idso, S.B., Kimball, B.A., Burns, J.M., Sinha, A. and Stott, L. 2003. The effect of long-term atmospheric CO_2 enrichment on the intrinsic water-use efficiency of sour orange trees. *Chemosphere* **50**: 217-222.

Liu, H., Li, X., Fischer, G. and Sun, L. 2004. Study on the impacts of climate change on China's agriculture. *Climatic Change* **65**: 125-148.

Myneni, R.B., Dong, J., Tucker, C.J., Kaufmann, R.K., Kauppi, P.E., Liski, J., Zhou, L., Alexeyev, V. and Hughes, M.K. 2001. A large carbon sink in the woody biomass of Northern forests. *Proceedings of the National Academy of Sciences, USA:* **98**: 14,784-14,789.

Piao, S., Fang, J. and He, J. 2006. Variations in vegetation net primary production in the Qinghai-Xizang Plateau, China, from 1982-1999. *Climatic Change* **74**: 253-267.

Piao, S., Fang, J., Liu, H. and Zhu, B. 2005a. NDVI-indicated decline in desertification in China in the past two decades. *Geophysical Research Letters* **32**: 10.1029/2004GL021764.

Piao, S., Fang, J., Zhou, L., Zhu, B., Tan, K. and Tao, S. 2005b. Changes in vegetation net primary productivity from 1982 to 1999 in China. *Global Biogeochemical Cycles* **19**: 10.1029/2004GB002274.

Qian, Z.A., Song, M.H. and Li, W.Y. 2002. Analysis on distributive variation and forecast of sand-dust storms in recent 50 years in north China. *Journal of Desert Research* **22**: 106-111.

Schimel, D.S., House, J.I., Hibbard, J.I., Bousquet, P., Ciais, P., Peylin, P., Braswell, B.H., Apps, M.J., Baker, D., Bondeau, A., Canadell, J., Churkina, G., Cramer, W., Denning, A.S., Field, C.B., Friedlingstein, P., Goodale, C., Heimann, M., Houghton, R.A., Melillo, J.M., Moore III, B., Murdiyarso, D., Noble, I., Pacala, S.W., Prentice, I.C., Raupach, M.R., Rayner, P.J., Scholes, R.J., Steffen, W.L. and Wirth, C. 2001. Recent patterns and mechanisms of carbon exchange by terrestrial ecosystems. *Nature* **414**: 169-172.

Shi, Y.F., Ed. 2003. An Assessment of the Issues of Climatic Shift from Warm-Dry to Warm-Wet in Northwest China. China Meteorology, Beijing.

Su, H.-X. and Sang, W.-G. 2004. Simulations and analysis of net primary productivity in *Quercus liaotungensis* forest of Donglingshan Mountain Range in response to different climate change scenarios. *Acta Botanica Sinica* **46**: 1281-1291.

White, M.A., Running, S.W. and Thornton, P.E. 1999. The impact of growing-season length variability on carbon assimilation and evapotranspiration over 88 years in the eastern US deciduous forest. *International Journal of Biometeorology* **42**: 139-145.

Zhang, L., Yue, L.P. and Xia, B. 2003. The study of land desertification in transitional zones between the MU US desert and the Loess Plateau using RS and GIS—A case study of the Yulin region. *Environmental Geology* **44**: 530-534.

Zhao, C., Dabu, X. and Li, Y. 2004. Relationship between climatic factors and dust storm frequency in Inner Mongolia of China. *Geophysical Research Letters* **31**: 10.1029/2003GL018351.

Zhou, L., Tucker, C.J., Kaufmann, R.K., Slayback, D., Shabanov, N.V. and Myneni, R.B. 2001. Variations in northern vegetation activity inferred from satellite data of vegetation index during 1981-1999. *Journal of Geophysical Research* **106**: 20,069-20,083.

Zou, X.K. and Zhai P.M. 2004. Relationship between vegetation coverage and spring dust storms over northern China. *Journal of Geophysical Research* **109**: 10.1029/2003JD003913.

7.9.3. Europe

Allen *et al.* (1999) analyzed sediment cores from a lake in southern Italy and from the Mediterranean Sea, developing high-resolution climate and vegetation data sets for this region over the last 102,000 years. These materials indicated that rapid changes in vegetation were well correlated with rapid changes in climate, such that complete shifts in natural ecosystems would sometimes occur over periods of less than 200 years. Over the warmest portion of the record (the Holocene), the total organic carbon content of the vegetation reached its highest level, more than doubling values experienced over the rest of the record, while other proxy indicators revealed that during the more productive woody-plant period of the Holocene, the increased vegetative cover also led to less soil erosion. The results of this study demonstrate that the biosphere can successfully respond to rapid changes in climate. As the 15 researchers involved in the work put it, "the biosphere was a full participant in these rapid fluctuations, contrary to widely held views that vegetation is unable to change with such rapidity." Furthermore, their work revealed that warmer was always better in terms of vegetative productivity.

Osborne *et al.* (2000) used an empirically based mechanistic model of Mediterranean shrub vegetation to address two important questions: (1) Has recent climate change, especially increased drought, negatively impacted Mediterranean shrublands? and (2) Has the historical increase in the air's CO_2 concentration modified this impact? The data-based model they employed suggests that the warming and reduced precipitation experienced in the Mediterranean area over the past century should have had negative impacts on net primary production and leaf area index. When the measured increase in atmospheric CO_2 concentration experienced over the period was factored into the calculation, however, these negative influences were overpowered, with the net effect that both measures of vegetative prowess increased: net primary productivity by 25 percent and leaf area index by 7 percent. These results, in their words, "indicate that the recent rise in atmospheric CO_2 may already have had significant impacts on productivity, structure and water relations of sclerophyllous shrub vegetation, which tended to offset the detrimental effects of climate change in the region."

How can we relate this observation to climate change predictions for the earth as a whole? For a nominal doubling of the air's CO_2 concentration from 300 to 600 ppm, earth's mean surface air temperature is predicted by current climate models to rise by approximately 3°C, which equates to a temperature rise of 0.01°C per ppm CO_2. In the case of the Mediterranean region here described, the temperature rise over the past century was quoted by Osborne *et al.* as being 0.75°C, over which period of time the air's CO_2 concentration rose by approximately 75 ppm, for an analogous climate response of exactly the same value: 0.01°C per ppm CO_2.

With respect to model-predicted changes in earth's precipitation regime, a doubling of the air's CO_2 content is projected to lead to a modest intensification of the planet's hydrologic cycle. In the case of the Mediterranean region over the last century, however, there has been a recent tendency toward drier conditions. Hence, the specific case investigated by Osborne *et al.* represents a much-worse-case scenario than what is predicted by current climate models for the earth as a whole. Nevertheless, the area's vegetation has done even better than it did before the climatic change, thanks to the over-powering beneficial biological effects of the concurrent rise in the air's CO_2 content.

Cheddadi *et al.* (2001) employed a standard biogeochemical model (BIOME3)—which uses monthly temperature and precipitation data, certain soil characteristics, cloudiness, and atmospheric CO_2 concentration as inputs—to simulate the responses of various biomes in the region surrounding the Mediterranean Sea to changes in both climate (temperature and precipitation) and the air's CO_2 content. Their first step was to validate the model for two test periods: the present and 6000 years before present (BP). Recent instrumental records provided actual atmospheric CO_2, temperature and precipitation data for the present period; while pollen data were

used to reconstruct monthly temperature and precipitation values for 6000 years BP, and ice core records were used to determine the atmospheric CO_2 concentration of that earlier epoch. These efforts suggested that winter temperatures 6000 years ago were about 2°C cooler than they are now, that annual rainfall was approximately 200 mm less than today, and that the air's CO_2 concentration averaged 280 ppm, which is considerably less than the value of 345 ppm the researchers used to represent the present, i.e., the mid-point of the period used for calculating 30-year climate normals at the time they wrote their paper. Applying the model to these two sets of conditions, they demonstrated that "BIOME3 can be used to simulate ... the vegetation distribution under ... different climate and [CO_2] conditions than today," where [CO_2] is the abbreviation they use to represent "atmospheric CO_2 concentration."

Cheddadi *et al.*'s next step was to use their validated model to explore the vegetative consequences of an increase in anthropogenic CO_2 emissions that pushes the air's CO_2 concentration to a value of 500 ppm and its mean annual temperature to a value 2°C higher than today's mean value. The basic response of the vegetation to this change in environmental conditions was "a substantial southward shift of Mediterranean vegetation and a spread of evergreen and conifer forests in the northern Mediterranean."

More specifically, in the words of the researchers, "when precipitation is maintained at its present-day level, an evergreen forest spreads in the eastern Mediterranean and a conifer forest in Turkey." Current xerophytic woodlands in this scenario become "restricted to southern Spain and southern Italy and they no longer occur in southern France." In northwest Africa, on the other hand, "Mediterranean xerophytic vegetation occupies a more extensive territory than today and the arid steppe/desert boundary shifts southward," as each vegetation zone becomes significantly more verdant than it is currently.

What is the basis for these positive developments? Cheddadi *et al.* say "the replacement of xerophytic woodlands by evergreen and conifer forests could be explained by the enhancement of photosynthesis due to the increase of [CO_2]." Likewise, they note that "under a high [CO_2] stomata will be much less open which will lead to a reduced evapotranspiration and lower water loss, both for C_3 and C_4 plants," adding that "such mechanisms may help plants to resist long-lasting drought periods that characterize the Mediterranean climate."

Contrary to what is often predicted for much of the world's moisture-challenged lands, therefore, the authors were able to report that "an increase of [CO_2], jointly with an increase of *ca.* 2°C in annual temperature would not lead to desertification on any part of the Mediterranean unless annual precipitation decreased drastically," where they define a drastic decrease as a decline of 30 percent or more. Equally important in this context is the fact that Hennessy *et al.* (1997) have indicated that a doubling of the air's CO_2 content would in all likelihood lead to a 5 to 10 percent increase in annual precipitation at Mediterranean latitudes, which is also what is predicted for most of the rest of the world. Hence, the results of the present study—where precipitation was held constant—may validly be considered to be a worst-case scenario, with the true vegetative response being even better than the good-news results reported by Cheddadi *et al.*, even when utilizing what we believe to be erroneously inflated global warming predictions.

Julien *et al.* (2006) "used land surface temperature (LST) algorithms and NDVI [Normalized Difference Vegetation Index] values to estimate changes in vegetation in the European continent between 1982 and 1999 from the Pathfinder AVHRR [Advanced Very High Resolution Radiometer] Land (PAL) dataset." This program revealed that arid and semi-arid areas (Northern Africa, Southern Spain and the Middle East) have seen their mean LST increase and NDVI decrease, while temperate areas (Western and Central Europe) have suffered a slight decrease in LST but a more substantial increase in NDVI, especially in Germany, the Czech Republic, Poland and Belarus. In addition, parts of continental and Northern Europe have experienced either slight increases or decreases in NDVI while LST values have decreased. Considering the results in their totality, the Dutch and Spanish researchers concluded that, over the last two decades of the twentieth century, "Europe as a whole has a tendency to greening," and much of it is "seeing an increase in its wood land proportion."

Working in the Komi Republic in the northeast European sector of Russia, Lopatin *et al.* (2006) (1) collected discs and cores from 151 Siberian spruce trees and 110 Scots pines from which they developed ring-width chronologies that revealed yearly changes in forest productivity, (2) developed satellite-based time series of NDVI for the months of June, July,

August over the period 1982-2001, (3) correlated their site-specific ring-width-derived productivity histories with same-site NDVI time series, (4) used the resulting relationship to establish six regional forest productivity histories for the period 1982-2001, and (5) compared the six regional productivity trends over this period with corresponding-region temperature and precipitation trends. For all six vegetation zones of the Komi Republic, this work indicated that the 1982-2001 trends of integrated NDVI values from June to August were positive, and that the "increase in productivity reflected in [the] NDVI data [was] maximal on the sites with increased temperature and decreased precipitation."

In discussing their findings, the three scientists state that "several studies (Riebsame *et al.*, 1994; Myneni *et al.*, 1998; Vicente-Serrano *et al.*, 2004) have shown a recent increase in vegetation cover in different world ecosystems." What is special about their study, as they describe it, is that "in Europe, most forests are managed, except for those in northwestern Russia [the location of their work], where old-growth natural forests are dominant (Aksenov *et al.*, 2002)." Consequently, and because of their positive findings, they say we can now conclude that "productivity during recent decades also increased in relatively untouched forests," where non-management-related "climate change with lengthening growing season, increasing CO_2 and nitrogen deposition" are the primary determinants of changes in forest productivity.

In conclusion, this brief review of pertinent studies conducted in Europe strongly contradicts today's obsession with the ongoing rise in the atmosphere's CO_2 content, as well as the many environmental catastrophes it has been predicted to produce. The results of rising CO_2 concentrations and temperatures in the twentieth century were overwhelmingly positive.

Additional information on this topic, including reviews of newer publications as they become available, can be found at http://www.co2science.org/subject/g/europegreen.php.

References

Aksenov, D., Dobrynin, D., Dubinin, M., Egorov, A., Isaev, A., Karpachevskiy, M., Laestadius, L., Potapov, P., Purekhovskiy, P., Turubanova, S. and Yaroshenko, A. 2002. *Atlas of Russia's Intact Forest Landscapes.* Global Forest Watch Russia, Moscow.

Allen, J.R.M., Brandt, U., Brauer, A., Hubberten, H.-W., Huntley, B., Keller, J., Kraml, M., Mackensen, A., Mingram, J., Negendank, J.F.W., Nowaczyk, N.R., Oberhansli, H., Watts, W.A., Wulf, S. and Zolitschka, B. 1999. Rapid environmental changes in southern Europe during the last glacial period. *Nature* **400**: 740-743.

Cheddadi, R., Guiot, J. and Jolly, D. 2001. The Mediterranean vegetation: what if the atmospheric CO_2 increased? *Landscape Ecology* **16**: 667-675.

Hennessy, K.J., Gregory, J.M. and Mitchell, J.F.B. 1997. Changes in daily precipitation under enhanced greenhouse conditions. *Climate Dynamics* **13**: 667-680.

Julien, Y., Sobrino, J.A. and Verhoef, W. 2006. Changes in land surface temperatures and NDVI values over Europe between 1982 and 1999. *Remote Sensing of Environment* **103**: 43-55.

Lopatin, E., Kolstrom, T. and Spiecker, H. 2006. Determination of forest growth trends in Komi Republic (northwestern Russia): combination of tree-ring analysis and remote sensing data. *Boreal Environment Research* **11**: 341-353.

Myneni, R.B., Tucker, C.J., Asrar, G. and Keeling, C.D. 1998. Interannual variations in satellite-sensed vegetation index data from 1981 to 1991. *Journal of Geophysical Research* **103**: 6145-6160.

Osborne, C.P., Mitchell, P.L., Sheehy, J.E. and Woodward, F.I. 2000. Modellng the recent historical impacts of atmospheric CO_2 and climate change on Mediterranean vegetation. *Global Change Biology* **6**: 445-458.

Riebsame, W.E., Meyer, W.B. and Turner, B.L. 1994. Modeling land-use and cover as part of global environmental-change. *Climatic Change* **28**: 45-64.

Vicente-Serrano, S.M., Lasanta, T. and Romo, A. 2004. Analysis of spatial and temporal evolution of vegetation cover in the Spanish central Pyrenees: Role of human management. *Environmental Management* **34**: 802-818.

7.9.4. North America

In a paper titled "Variations in northern vegetation activity inferred from satellite data of vegetation index during 1981-1999," Zhou *et al.* (2001) determined that the magnitude of the satellite-derived normalized difference vegetation index (NDVI) rose by 8.44 percent in North America over this period. Noting that the NDVI "can be used to proxy the vegetation's responses to climate changes because it is well correlated with the fraction of photosynthetically active radiation absorbed by plant canopies and thus leaf area, leaf biomass, and potential photosynthesis," they went on to suggest that the increases in plant growth and vitality implied by their NDVI data were primarily driven by concurrent increases in near-surface air temperature, although temperatures may have actually declined throughout the eastern part of the United States over the period of their study.

Zhou *et al.*'s attribution of this "greening" of the continent to increases in near-surface air temperature was challenged by Ahlbeck (2002), who suggested that the observed upward trend in NDVI was primarily driven by the increase in the air's CO_2 concentration, and that fluctuations in temperature were primarily responsible for variations about the more steady upward trend defined by the increase in CO_2. In replying to this challenge, Kaufmann *et al.* (2002) claimed Ahlbeck was wrong and reaffirmed their initial take on the issue. We believe it was Ahlbeck who was "clearly the 'more correct' of the two camps." (See the discussion in *CO₂ Science* at www.co2science.org/articles/V5/N38/EDIT.php.)

About the same time, Hicke *et al.* (2002) computed net primary productivity (NPP) over North America for the years 1982-1998 using the Carnegie-Ames-Stanford Approach (CASA) carbon cycle model, which was driven by a satellite NDVI record at 8-km spatial resolution. This effort revealed that NPP increases of 30 percent or more occurred across the continent from 1982 to 1998. During this period, the air's CO_2 concentration rose by 25.74 ppm, as calculated from the Mauna Loa data of Keeling and Whorf (1998), which amount is 8.58 percent of the 300 ppm increase often used in experiments on plant growth responses to atmospheric CO_2 enrichment. Consequently, for herbaceous plants that display NPP increases of 30-40 percent in response to a 300-ppm increase in atmospheric CO_2 concentration, the CO_2-induced NPP increase experienced between 1982 and 1998 would be expected to have been 2.6-3.4 percent. Similarly, for woody plants that display NPP increases of 60-80 percent in response to a 300-ppm increase in atmospheric CO_2 (Saxe *et al.*, 1998; Idso and Kimball, 2001), the expected increase in productivity between 1982 and 1998 would have been 5.1-6.9 percent. Since both of these NPP increases are considerably less that the 30 percent or more observed by Hicke *et al.*, additional factors must have helped to stimulate NPP over this period, some of which may have been concomitant increases in precipitation and air temperature, the tendency for warming to lengthen growing seasons and enhance the aerial fertilization effect of rising CO_2 concentrations, increasingly intensive crop and forest management, increasing use of genetically improved plants, the regrowth of forests on abandoned cropland, and improvements in agricultural practices such as irrigation and fertilization. Whatever the mix might have been, one thing is clear: Its effect was overwhelmingly positive.

In a study based on a 48-year record derived from an average of 17 measurements per year, Raymond and Cole (2003) demonstrated that the export of alkalinity, in the form of bicarbonate ions, from the Mississippi River to the Gulf of Mexico had increased by approximately 60 percent since 1953. "This increased export," as they described it, was "in part the result of increased flow resulting from higher rainfall in the Mississippi basin," which had led to a 40 percent increase in annual Mississippi River discharge to the Gulf of Mexico over the same time period. The remainder, however, had to have been due to increased rates of chemical weathering of soil minerals. What factors might have been responsible for this phenomenon? The two researchers noted that potential mechanisms included "an increase in atmospheric CO_2, an increase [in] rainwater throughput, or an increase in plant and microbial production of CO_2 and organic acids in soils due to biological responses to increased rainfall and temperature." Unfortunately, they forgot to mention the increase in terrestrial plant productivity that is produced by the increase in the aerial fertilization effect provided by the historical rise in the air's CO_2 content, which also leads to "an increase in plant and microbial production of CO_2 and organic acids in soils." This phenomenon should have led to an increase in Mississippi River alkalinity equivalent to that which they had observed since 1953.

In a study using data obtained from dominant stands of loblolly pine plantations growing at 94 locations spread across the southeastern United States, Westfall and Amateis (2003) employed mean height measurements made at three-year intervals over a period of 15 years to calculate a site index related to the mean growth rate for each of five three-year periods, which index would be expected to increase monotonically if growth rates were being enhanced above normal by some monotonically increasing factor that promotes growth. This work revealed, in their words, that "mean site index over the 94 plots consistently increased at each remeasurement period," which would suggest, as they further state, that "loblolly pine plantations are

realizing greater than expected growth rates," and, we would add, that the growth rate increases are growing larger and larger with each succeeding three-year period. As to what could be causing the monotonically increasing growth rates of loblolly pine trees over the entire southeastern United States, Westfall and Amateis named increases in temperature and precipitation in addition to rising atmospheric CO_2 concentrations. However, they report that a review of annual precipitation amounts and mean ground surface temperatures showed no trends in these factors over the period of their study. They also suggested that if increased nitrogen deposition were the cause, "such a factor would have to be acting on a regional scale to produce growth increases over the range of study plots." Hence, they tended to favor the ever-increasing aerial fertilization effect of atmospheric CO_2 enrichment as being responsible for the accelerating pine tree growth rates.

Returning to satellite studies, Lim et al. (2004) correlated the monthly rate of relative change in NDVI, which they derived from advanced very high resolution radiometer data, with the rate of change in atmospheric CO_2 concentration during the natural vegetation growing season within three different eco-region zones of North America (Arctic and Sub-Arctic Zone, Humid Temperate Zone, and Dry and Desert Zone, which they further subdivided into 17 regions) over the period 1982-1992, after which they explored the temporal progression of annual minimum NDVI over the period 1982-2001 throughout the eastern humid temperate zone of North America. The result of these operations was that in all of the regions but one, according to the researchers, "δCO_2 was positively correlated with the rate of change in vegetation greenness in the following month, and most correlations were high," which they say is "consistent with a CO_2 fertilization effect" of the type observed in "experimental manipulations of atmospheric CO_2 that report a stimulation of photosynthesis and above-ground productivity at high CO_2." In addition, they determined that the yearly "minimum vegetation greenness increased over the period 1982-2001 for all the regions of the eastern humid temperate zone in North America." As for the cause of this phenomenon, Lim et al. say that rising CO_2 could "increase minimum greenness by stimulating photosynthesis at the beginning of the growing season," citing the work of Idso et al. (2000).

In a somewhat similar study, but one that focused more intensely on climate change, Xiao and Moody (2004) examined the responses of the normalized difference vegetation index integrated over the growing season (gNDVI) to annual and seasonal precipitation, maximum temperature (Tmax) and minimum temperature (Tmin) over an 11-year period (1990-2000) for six biomes in the conterminous United States (Evergreen Needleleaf Forest, Deciduous Broadleaf Forest, Mixed Forest, Open Shrubland, Woody Savanna and Grassland), focusing on within- and across-biome variance in long-term average gNDVI and emphasizing the degree to which this variance is explained by spatial gradients in long-term average seasonal climate. The results of these protocols indicated that the greatest positive climate-change impacts on biome productivity were caused by increases in spring, winter and fall precipitation, as well as increases in fall and spring temperature, especially Tmin, which has historically increased at roughly twice the rate of Tmax in the United States. Hence, "if historical climatic trends and the biotic responses suggested in this analysis continue to hold true, we can anticipate further increases in productivity for both forested and nonforested ecoregions in the conterminous US, with associated implications for carbon budgets and woody proliferation."

Goetz *et al*. (2005) transformed satellite-derived NDVI data obtained across boreal North America (Canada and Alaska) for the period 1982-2003 into photosynthetically active radiation absorbed by green vegetation and treated the result as a proxy for relative June-August gross photosynthesis (Pg), stratifying the results by vegetation type and comparing them with spatially matched concomitant trends in surface air temperature data. Over the course of the study, this work revealed that area-wide tundra experienced a significant increase in Pg in response to a similar increase in air temperature; and Goetz *et al*. say "this observation is supported by a wide and increasing range of local field measurements characterizing elevated net CO_2 uptake (Oechel *et al*., 2000), greater depths of seasonal thaw (Goulden *et al*., 1998), changes in the composition and density of herbaceous vegetation (Chapin *et al*., 2000; Epstein *et al*., 2004), and increased woody encroachment in the tundra areas of North America (Sturm *et al*., 2001)." In the case of interior forest, on the other hand, there was no significant increase in air temperature and essentially no change in Pg, with the last data point of the series being essentially indistinguishable from the first. This latter seemingly aberrant observation is in harmony with the fact that at low temperatures the growth-promoting effects of increasing atmospheric

CO_2 levels are often very small or even non-existent (Idso and Idso, 1994), which is what appears to have been the case with North American boreal forests over the same time period. As a result, Canada's and Alaska's tundra ecosystems exhibited increasing productivity over the past couple of decades, while their boreal forests did not.

Also working in Alaska, Tape *et al*. (2006) analyzed repeat photography data from a photo study of the Colville River conducted between 1945 and 1953, as well as 202 new photos of the same sites that were obtained between 1999 and 2002, to determine the nature of shrub expansion in that region over the past half-century. This approach revealed, in their words, that "large shrubs have increased in size and abundance over the past 50 years, colonizing areas where previously there were no large shrubs." In addition, they say their review of plot and remote sensing studies confirms that "shrubs in Alaska have expanded their range and grown in size" and that "a population of smaller, intertussock shrubs not generally sampled by the repeat photography, is also expanding and growing." Taken together, they conclude that "these three lines of evidence allow us to infer a general increase in tundra shrubs across northern Alaska." Tape *et al*. attribute this to large-scale pan-Arctic warming. From analyses of logistic growth curves, they estimate that the expansion began about 1900, "well before the current warming in Alaska (which started about 1970)." Hence, they conclude that "the expansion predates the most recent warming trend and is perhaps associated with the general warming since the Little Ice Age." These inferences appear reasonable, although we would add that the 80-ppm increase in the atmosphere's CO_2 concentration since 1900 likely played a role in the shrub expansion as well. If continued, the researchers say the transition "will alter the fundamental architecture and function of this ecosystem with important ramifications," the great bulk of which, in our opinion, will be positive.

Working at eight different sites within the Pacific Northwest of the United States, Soule and Knapp (2006) studied ponderosa pine trees to see how they may have responded to the increase in the atmosphere's CO_2 concentration that occurred after 1950. The two geographers say the sites they chose "fit several criteria designed to limit potential confounding influences associated with anthropogenic disturbance." In addition, they selected locations with "a variety of climatic and topo-edaphic conditions, ranging from extremely water-limiting environments ... to areas where soil moisture should be a limiting factor for growth only during extreme drought years." They also say that all sites were located in areas "where ozone concentrations and nitrogen deposition are typically low."

At all eight of the sites that met all of these criteria, Soule and Knapp obtained core samples from about 40 mature trees that included "the potentially oldest trees on each site," so that their results would indicate, as they put it, "the response of mature, naturally occurring ponderosa pine trees that germinated before anthropogenically elevated CO_2 levels, but where growth, particularly post-1950, has occurred under increasing and substantially higher atmospheric CO_2 concentrations." Utilizing meteorological evaluations of the Palmer Drought Severity Index, they thus compared ponderosa pine radial growth rates during matched wet and dry years pre- and post-1950. Overall, the two researchers found a post-1950 radial growth enhancement that was "more pronounced during drought years compared with wet years, and the greatest response occurred at the most stressed site." As for the magnitude of the response, they determined that "the relative change in growth [was] upward at seven of our [eight] sites, ranging from 11 to 133%." With respect to the significance of their observations, Soule and Knapp say their results show that "radial growth has increased in the post-1950s period ... while climatic conditions have generally been unchanged," which further suggests that "nonclimatic driving forces are operative." In addition, they say the "radial growth responses are generally consistent with what has been shown in long-term open-top chamber (Idso and Kimball, 2001) and free-air CO_2 enrichment (FACE) studies (Ainsworth and Long, 2005)." They conclude their findings "suggest that elevated levels of atmospheric CO_2 are acting as a driving force for increased radial growth of ponderosa pine, but that the overall influence of this effect may be enhanced, reduced or obviated by site-specific conditions."

Wang *et al.* (2006) examined ring-width development in cohorts of young and old white spruce trees in a mixed grass-prairie ecosystem in southwestern Manitoba, Canada, where a 1997 wildfire killed most of the older trees growing in high-density spruce islands, but where younger trees slightly removed from the islands escaped the ravages of the flames. There, "within each of a total of 24 burned islands," in the words of the three researchers, "the largest dominant tree (dead) was cut down and a disc was then sampled from the stump height," while

"adjacent to each sampled island, a smaller, younger tree (live) was also cut down, and a disc was sampled from the stump height."

After removing size-, age- and climate-related trends in radial growth from the ring-width histories of the trees, Wang *et al.* plotted the residuals as functions of time for the 30-year periods for which both the old and young trees would have been approximately the same age: 1900-1929 for the old trees and 1970-1999 for the young trees. During the first of these periods, the atmosphere's CO_2 concentration averaged 299 ppm; during the second it averaged 346 ppm. Also, the mean rate-of-rise of the atmosphere's CO_2 concentration was 0.37 ppm/year for first period and 1.43 ppm/year for the second.

The results of this exercise revealed that the slope of the linear regression describing the rate-of-growth of the ring-width residuals for the later period (when the air's CO_2 concentration was 15 percent greater and its rate-of-rise 285 percent greater) was more than twice that of the linear regression describing the rate-of-growth of the ring-width residuals during the earlier period. As the researchers describe it, these results show that "at the same developmental stage, a greater growth response occurred in the late period when atmospheric CO_2 concentration and the rate of atmospheric CO_2 increase were both relatively high," and they say that "these results are consistent with expectations for CO_2-fertilization effects." In fact, they say "the response of the studied young trees can be taken as strong circumstantial evidence for the atmospheric CO_2-fertilization effect."

Another thing Wang *et al.* learned was that "postdrought growth response was much stronger for young trees (1970-1999) compared with old trees at the same development stage (1900-1929)," and they add that "higher atmospheric CO_2 concentration in the period from 1970-1999 may have helped white spruce recover from severe drought." In a similar vein, they also determined that young trees showed a weaker relationship to precipitation than did old trees, noting that "more CO_2 would lead to greater water-use efficiency, which may be dampening the precipitation signal in young trees."

In summary, Wang *et al.*'s unique study provides an exciting real-world example of the benefits the historical increase in the air's CO_2 content has likely conferred on nearly all of earth's plants, and especially its long-lived woody species. Together with the results of the other North American studies we have reviewed, this body of research paints a picture of a significant greening of North America in the twentieth century.

Additional information on this topic, including reviews of newer publications as they become available, can be found at http://www.co2science.org/subject/g/namergreen.php.

References

Ahlbeck, J.R. 2002. Comment on "Variations in northern vegetation activity inferred from satellite data of vegetation index during 1981-1999" by L. Zhou *et al. Journal of Geophysical Research* **107**: 10.1029/2001389.

Ainsworth, E.A. and Long, S.P. 2005. What have we learned from 15 years of free-air CO_2 enrichment (FACE)? A meta-analytic review of the responses of photosynthesis, canopy properties and plant production to rising CO_2. *New Phytologist* **165**: 351-372.

Chapin III, F.S., McGuire, A.D., Randerson, J., Pielke, R., Baldocchi, D., Hobbie, S.E., Roulet, N., Eugster, W., Kasischke, E., Rastetter, E.B., Zimov, S.A., and Running, S.W. 2000. Arctic and boreal ecosystems of western North America as components of the climate system. *Global Change Biology* **6**: 211-223.

Epstein, H.E., Calef, M.P., Walker, M.D., Chapin III, F.S. and Starfield, A.M. 2004. Detecting changes in arctic tundra plant communities in response to warming over decadal time scales. *Global Change Biology* **10**: 1325-1334.

Goetz, S.J., Bunn, A.G., Fiske, G.J. and Houghton, R.A. 2005. Satellite-observed photosynthetic trends across boreal North America associated with climate and fire disturbance. *Proceedings of the National Academy of Sciences* **102**: 13,521-13,525.

Goulden, M.L., Wofsy, S.C., Harden, J.W., Trumbore, S.E., Crill, P.M., Gower, S.T., Fries, T., Daube, B.C., Fan, S.M., Sutton, D.J., Bazzaz, A. and Munger, J.W. 1998. Sensitivity of boreal forest carbon balance to soil thaw. *Science* **279**: 214-217.

Hicke, J.A., Asner, G.P., Randerson, J.T., Tucker, C., Los, S., Birdsey, R., Jenkins, J.C. and Field, C. 2002. Trends in North American net primary productivity derived from satellite observations, 1982-1998. *Global Biogeochemical Cycles* **16**: 10.1029/2001GB001550.

Idso, C.D., Idso, S.B., Kimball, B.A., Park, H., Hoober, J.K. and Balling Jr., R.C. 2000. Ultra-enhanced spring branch growth in CO_2-enriched trees: can it alter the phase of the atmosphere's seasonal CO_2 cycle? *Environmental and Experimental Botany* **43**: 91-100.

Idso, K.E. and Idso, S.B. 1994. Plant responses to atmospheric CO_2 enrichment in the face of environmental constraints: a review of the past 10 years' research. *Agricultural and Forest Meteorology* **69**: 153-203.

Idso, S.B. and Kimball, B.A. 2001. CO_2 enrichment of sour orange trees: 13 years and counting. *Environmental and Experimental Botany* **46**: 147-153.

Kaufmann, R.K., Zhou, L., Tucker, C.J., Slayback, D., Shabanov, N.V. and Myneni, R.B. 2002. Reply to Comment on "Variations in northern vegetation activity inferred from satellite data of vegetation index during 1981-1999: by J.R. Ahlbeck. *Journal of Geophysical Research* **107**: 10.1029/1001JD001516.

Keeling, C.D. and Whorf, T.P. 1998. *Atmospheric CO_2 Concentrations—Mauna Loa Observatory, Hawaii, 1958-1997* (revised August 2000). NDP-001. Carbon Dioxide Information Analysis Center, Oak Ridge National Laboratory, Oak Ridge, Tennessee.

Lim, C., Kafatos, M. and Megonigal, P. 2004. Correlation between atmospheric CO_2 concentration and vegetation greenness in North America: CO_2 fertilization effect. *Climate Research* **28**: 11-22.

Oechel, W.C., Vourlitis, G.L., Verfaillie, J., Crawford, T., Brooks, S., Dumas, E., Hope, A., Stow, D., Boynton, B., Nosov, V. and Zulueta, R. 2000. A scaling approach for quantifying the net CO_2 flux of the Kuparuk River Basin, Alaska. *Global Change Biology* **6**: 160-173.

Raymond, P.A. and Cole, J.J. 2003. Increase in the export of alkalinity from North America's largest river. *Science* **301**: 88-91.

Saxe, H., Ellsworth, D.S. and Heath, J. 1998. Tree and forest functioning in an enriched CO_2 atmosphere. *New Phytologist* **139**: 395-436.

Soule, P.T. and Knapp, P.A. 2006. Radial growth rate increases in naturally occurring ponderosa pine trees: a late-twentieth century CO_2 fertilization effect? *New Phytologist* doi: **10.1111**/j.1469-8137.2006.01746.x.

Sturm, M., Racine, C. and Tape, K. 2001. Increasing shrub abundance in the Arctic. *Nature* **411**: 546-547.

Tape, K., Sturm, M. and Racine, C. 2006. The evidence for shrub expansion in Northern Alaska and the Pan-Arctic. *Global Change Biology* **12**: 686-702.

Wang, G.G., Chhin, S. and Bauerle, W.L. 2006. Effect of natural atmospheric CO_2 fertilization suggested by open-grown white spruce in a dry environment. *Global Change Biology* **12**: 601-610.

Westfall, J.A. and Amateis, R.L. 2003. A model to account for potential correlations between growth of loblolly pine and changing ambient carbon dioxide concentrations. *Southern Journal of Applied Forestry* **27**: 279-284.

Xiao, J. and Moody, A. 2004. Photosynthetic activity of US biomes: responses to the spatial variability and seasonality, of precipitation and temperature. *Global Change Biology* **10**: 437-451.

Zhou, L., Tucker, C.J., Kaufmann, R.K., Slayback, D., Shabanov, N.V. and Myneni, R.B. 2001. Variations in northern vegetation activity inferred from satellite data of vegetation index during 1981-1999. *Journal of Geophysical Research* **106**: 20,069-20,083.

7.9.5. Oceans

Rising air temperatures and atmospheric CO_2 concentrations in the twentieth century also have affected the productivity of earth's seas. We begin with a study that takes a much longer view of the subject.

Elderfield and Rickaby (2000) note that the typically low atmospheric CO_2 concentrations of glacial periods have generally been attributed to an increased oceanic uptake of CO_2, "particularly in the southern oceans." However, the assumption that intensified phytoplanktonic photosynthesis may have stimulated CO_2 uptake rates during glacial periods has always seemed at odds with the observational fact that rates of photosynthesis are generally much reduced in environments of significantly lower-than-current atmospheric CO_2 concentrations, such as typically prevail during glacial periods.

The two scientists provide a new interpretation of Cd/Ca systematics in sea water that helps to resolve this puzzle, as it allows them to more accurately estimate surface water phosphate conditions during glacial times and thereby determine the implications for concomitant atmospheric CO_2 concentrations. What they found, in their words, is that "results from the Last Glacial Maximum [LGM] show similar phosphate utilization in the subantarctic to that of today, but much smaller utilization in the polar Southern Ocean," which implies, according to Delaney (2000), that Antarctic productivity was lower at that time than it is now, but that *sub*antarctic productivity was about the same as it has been in modern times, due perhaps to greater concentrations of bio-available iron compensating for the lower atmospheric CO_2 concentrations of the LGM.

So what caused the much smaller utilization of phosphate in the polar Southern Ocean during the LGM? Noting that the area of sea-ice cover in the Southern Ocean during glacial periods may have been as much as double that of modern times, Elderfield and Rickaby suggest that "by restricting communication between the ocean and atmosphere, sea ice expansion also provides a mechanism for reduced CO_2 release by the Southern Ocean and lower glacial atmospheric CO_2." Hence, it is possible that phytoplanktonic productivity in the glacial Southern Ocean may have been no higher than it is at the present time, notwithstanding the greater supply of bio-available iron typical of glacial epochs.

In the case of the interglacial period in which we currently live, Dupouy *et al.* (2000) say it has long been believed that N_2 fixation in the world's oceans is unduly low, in consequence of the present low supply of wind-blown iron compared to that of glacial periods, and that this state of affairs leads to low phytoplanktonic productivity, even in the presence of higher atmospheric CO_2 concentrations. The evidence they acquired, however, suggests that marine N_2 fixation may be much greater than what has generally been thought to be the case. In particular, they note that several *Trichodesmium* species of N_2-fixing cyanobacteria have "a nearly ubiquitous distribution in the euphotic zone of tropical and subtropical seas and could play a major role in bringing new N to these oligotrophic systems." And this feat, in their words, "could play a significant role in enhancing new production."

The importance of these findings is perhaps best appreciated in light of the findings of Pahlow and Riebesell (2000), who in studying data obtained from 1173 stations in the Atlantic and Pacific Oceans, covering the years 1947 to 1994, detected changes in Northern Hemispheric deep-ocean Redfield ratios that are indicative of increasing nitrogen availability there, which increase was concomitant with an increase in export production that has resulted in ever-increasing oceanic carbon sequestration. These investigators further suggest that the growing supply of nitrogen has its origin in anthropogenic activities that release nitrous oxides to the air. In addition, the increased carbon export may be partly a consequence of the historical increase in the air's CO_2 concentration, which has been demonstrated to have the ability to enhance phytoplanktonic productivity (see Section 6.1.3.4. in this document), analogous to the way in which elevated concentrations of atmospheric CO_2 enhance the productivity of terrestrial plants, including their ability to fix nitrogen.

Further elucidating the productivity-enhancing power of the ongoing rise in the air's CO_2 content is the study of Pasquer *et al.* (2005), who employed a complex model of growth regulation of diatoms, pico/nano phytoplankton, coccolithophorids and *Phaeocystis* spp. by light, temperature and nutrients (based on a comprehensive analysis of literature reviews focusing on these taxonomic groups) to calculate changes in the ocean uptake of carbon in response to a sustained increase in atmospheric CO_2 concentration of 1.2 ppm per year for three marine ecosystems where biogeochemical time-series of the data required for model initialization and comparison of results were readily available. These systems were (1) the ice-free Southern Ocean Time Series station KERFIX (50°40'S, 68°E) for the period 1993-1994 (diatom-dominated), (2) the sea-ice associated Ross Sea domain (76°S, 180°W) of the Antarctic Environment and Southern Ocean Process Study AESOPS in 1996-1997 (*Phaeocystis*-dominated), and (3) the North Atlantic Bloom Experiment NABE (60°N, 20°W) in 1991 (coccolithophorids). Their results, in their words, "show that at all tested latitudes the prescribed increase of atmospheric CO_2 enhances the carbon uptake by the ocean." Indeed, we calculate from their graphical presentations that (1) at the NABE site a sustained atmospheric CO_2 increase of 1.2 ppm per year over a period of eleven years increases the air-sea CO_2 flux in the last year of that period by approximately 17 percent, (2) at the AESOPS site the same protocol applied over a period of six years increases the air-sea CO_2 flux by about 45 percent, and (3) at the KERFIX site it increases the air-sea CO_2 flux after nine years by about 78 percent. Although the results of this interesting study based on the complex SWAMCO model of Lancelot *et al.* (2000), as modified by Hannon *et al.* (2001), seem overly large, they highlight the likelihood that the ongoing rise in the air's CO_2 content may be having a significant positive impact on ocean productivity and the magnitude of the ocean carbon sink.

But what about increasing temperatures? Sarmiento *et al.* (2004) conducted a massive computational study that employed six coupled climate model simulations to determine the biological response of the global ocean to the climate warming they simulated from the beginning of the Industrial Revolution to the year 2050. Based on vertical velocity, maximum winter mixed-layer depth and sea-ice cover, they defined six biomes and calculated how their surface geographies would change in response to their calculated changes in global climate. Next, they used satellite ocean color and climatological observations to develop an empirical model for predicting surface chlorophyll concentrations from the final physical properties of the world's oceans as derived from their global warming simulations, after which they used three primary production algorithms to estimate the response of oceanic primary production to climate warming based on their calculated chlorophyll concentrations. When all was said and done, the thirteen scientists from Australia, France, Germany, Russia, the United Kingdom and the United States arrived at a global warming-induced *increase* in global ocean primary production that ranged from 0.7 to 8.1 percent.

So what do real-world measurements of oceanic productivity reveal? Goes *et al.* (2005) analyzed seven years (1997-2004) of satellite-derived ocean color data pertaining to the Arabian Sea, as well as associated sea surface temperatures (SSTs) and winds. They report that for the region located between 52 to 57°E and 5 to 10°N, "the most conspicuous observation was the consistent year-by-year increase in phytoplankton biomass over the 7-year period." This phenomenon was so dramatic that by the summer of 2003, in their words, "chlorophyll *a* concentrations were >350% higher than those observed in the summer of 1997." They also report that the increase in chlorophyll *a* was "accompanied by an intensification of sea surface winds, in particular of the zonal (east-to-west) component," noting that these "summer monsoon winds are a coupled atmosphere-land-ocean phenomenon, whose strength is significantly correlated with tropical SSTs and Eurasian snow cover anomalies on a year-to-year basis." More specifically, they say that "reduced snow cover over Eurasia strengthens the spring and summer land-sea thermal contrast and is considered to be responsible for the stronger southwest monsoon winds." In addition, they state that "the influence of southwest monsoon winds on phytoplankton in the Arabian Sea is not through their impact on coastal upwelling alone but also via the ability of zonal winds to laterally advect newly upwelled nutrient-rich waters to regions away from the upwelling zone." They conclude that "escalation in the intensity of summer monsoon winds, accompanied by enhanced upwelling and an increase of more than 350 percent in average summertime phytoplankton biomass along the coast and over 300 percent offshore, raises the possibility that the current warming trend of the Eurasian landmass is making the Arabian Sea more productive."

To the north and west on the other side of Eurasia, Marasovic *et al.* (2005) analyzed monthly observations of basic hydrographic, chemical and biological parameters, including primary production, that had been made since the 1960s at two oceanographic stations, one near the coast (Kastela Bay) and one in the open sea. They found that mean annual primary production in Kastela Bay averaged about 430 mg C m^{-2} d^{-1} over the period 1962-72, exceeded 600 mg C m^{-2} d^{-1} over the period 1972-82, and rose to over 700 mg C m^{-2} d^{-1} over the period 1982-96, accompanied by a similar upward trend in percent oxygen saturation of the surface water. The initial value of primary production in the open sea was much less (approximately 150 mg C m^{-2} d^{-1}), but it began to follow the upward trend of the Kastela Bay data after about one decade. Marasovic et al. thus concluded that "even though all the relevant data indicate that the changes in Kastela Bay are closely related to an increase of anthropogenic nutrient loading, similar changes in the open sea suggest that primary production in the Bay might, at least partly, be due to global climatic changes," which, in their words, are "occurring in the Mediterranean and Adriatic Sea open waters" and may be directly related to "global warming of air and ocean," since "higher temperature positively affects photosynthetic processes."

Raitsos et al. (2005) investigated the relationship between Sea-viewing Wide Field-of-view Sensor (SeaWiFS) chlorophyll-a measurements in the Central Northeast Atlantic and North Sea (1997-2002) and simultaneous measurements of the Phytoplankton Color Index (PCI) collected by the Continuous Plankton Recorder survey, which is an upper-layer plankton monitoring program that has operated in the North Sea and North Atlantic Ocean since 1931. By developing a relationship between the two data bases over their five years of overlap, they were able to produce a Chl-a history for the Central Northeast Atlantic and North Sea for the period 1948-2002. Of this record they say that "an increasing trend is apparent in mean Chl-a for the area of study over the period 1948-2002." They also say "there is clear evidence for a stepwise increase after the mid-1980s, with a minimum of 1.3mg m^{-3} in 1950 and a peak annual mean of 2.1 mg m^{-3} in 1989 (62% increase)." Alternatively, it is possible that the data represent a more steady long-term upward trend upon which is superimposed a decadal-scale oscillation. In a final comment on their findings, they note that "changes through time in the PCI are significantly correlated with both sea surface temperature and Northern Hemisphere temperature," citing Beaugrand and Reid (2003).

In a contemporaneous study, Antoine et al. (2005) applied revised data-processing algorithms to two ocean-sensing satellites, the Coastal Zone Color Scanner (CZCS) and SeaWiFS, over the periods 1979-1986 and 1998-2002, respectively, to provide an analysis of the decadal changes in global oceanic phytoplankton biomass. Results of the analysis showed "an overall increase of the world ocean average chlorophyll concentration by about 22%" over the two decades under study.

Dropping down to the Southern Ocean, Hirawake et al. (2005) analyzed chlorophyll a data obtained from Japanese Antarctic Research Expedition cruises made by the Fuji and Shirase ice-breakers between Tokyo and Antarctica from 15 November to 28 December of nearly every year between 1965 and 2002 in a study of interannual variations of phytoplankton biomass, calculating results for the equatorial region between 10°N and 10°S, the Subtropical Front (STF) region between 35°S and 45°S, and the Polar Front (PF) region between 45°S and 55°S. They report that an increase in chl a was "recognized in the waters around the STF and the PF, especially after 1980 around the PF in particular," and that "in the period between 1994 and 1998, the chl a in the three regions exhibited rapid gain simultaneously." They also say "there were significant correlations between chl a and year through all of the period of observation around the STF and PF, and the rates of increase are 0.005 and 0.012 mg chl a m^{-3} y^{-1}, respectively." In addition, they report that the satellite data of Gregg and Conkright (2002) "almost coincide with our results." In commenting on these findings, the Japanese scientists say that "simply considering the significant increase in the chl a in the Southern Ocean, a rise in the primary production as a result of the phytoplankton increase in this area is also expected."

Also working in the Southern Hemisphere, Sepulveda et al. (2005) presented "the first reconstruction of changes in surface primary production during the last century from the Puyuhuapi fjord in southern Chile, using a variety of parameters (diatoms, biogenic silica, total organic carbon, chlorins, and proteins) as productivity proxies." Noting that the fjord is located in "a still-pristine area," they say it is "suitable to study changes in past export production originating from changes in both the paleo-Patagonian ice caps and the globally important Southern Ocean."

The analysis revealed that the productivity of the Puyuhuapi fjord "was characterized by a constant increase from the late 19th century to the early 1980s, then decreased until the late 1990s, and then rose again to present-day values." For the first of these periods (1890-1980), they additionally report that "all proxies were highly correlated (r > 0.8, p < 0.05)," and that "all proxies reveal an increase in accumulation rates." From 1980 to the present, however, the pattern differed among the various proxies; and the researchers say that "considering that the top 5 cm of the sediment column (~10 years) are diagenetically active, and that bioturbation by benthic organisms may have modified and mixed the sedimentary signal, paleo-interpretation of the period 1980-2001 must be taken with caution." Consequently, there is substantial solid evidence that, for the first 90 years of the 111-year record, surface primary production in the Puyuhuapi fjord rose dramatically, while with lesser confidence it appears to have leveled out over the past two decades. In spite of claims that the "unprecedented" increases in mean global air temperature and CO_2 concentration experienced since the inception of the Industrial Revolution have been bad for the biosphere, Sepulveda *et al.* presented yet another case of an ecosystem apparently thriving in such conditions.

Still, claims of impending ocean productivity declines have not ceased, and some commentators single out the study of Behrenfeld *et al.* (2006) in support of their claims. Working with NASA's Sea-viewing Wide Field-of-view Sensor (Sea WiFS), the team of 10 U.S. scientists calculated monthly changes in net primary production (NPP) from similar changes in upper-ocean chlorophyll concentrations detected from space over the past decade. (See Figure 7.9.5.) They report that this period was dominated by an initial NPP increase of 1,930 teragrams of carbon per year (Tg C yr^{-1}), which they attributed to the significant cooling of "the 1997 to 1999 El Niño to La Niña transition," and they note that this increase was "followed by a prolonged decrease averaging 190 Tg C yr^{-1}," which they attributed to subsequent warming.

The means by which changing temperatures were claimed by the researchers to have driven the two sequential linear-fit trends in NPP is based on their presumption that a warming climate increases the density contrast between warmer surface waters and cooler underlying nutrient-rich waters, so that the enhanced stratification that occurs with warming "suppresses nutrient exchange through vertical mixing," which decreases NPP by reducing the supply of nutrients to the surface waters where photosynthesizing phytoplankton predominantly live. By contrast, the ten scientists suggest that "surface cooling favors elevated vertical exchange," which increases NPP by enhancing the supply of nutrients to the ocean's surface waters, which are more frequented by phytoplankton than are under-lying waters, due to light requirements for photosynthesis.

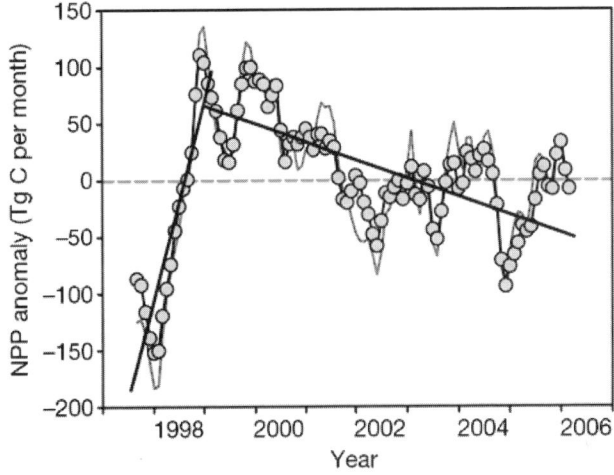

Figure 7.9.5. Monthly anomalies of global NPP (green line) plus similar results for the permanently stratified ocean regions of the world (grey circles and black line), adapted from Behrenfeld *et al.* (2006).

It is informative to note, however, that from approximately the middle of 2001 to the end of the data series in early 2006 (which interval accounts for more than half of the data record), there has been, if anything, a slight increase in global NPP. (See again Figure 7.9.5.) Does this observation mean there has been little to no net global warming since mid-2001? Or does it mean the global ocean's mean surface temperature actually cooled a bit over the last five years? Neither alternative is what one would expect if global warming were a real problem. On the other hand, the relationship between global warming and oceanic productivity may not be nearly as strong as what Behrenfeld *et al.* have suggested; and they themselves say "modeling studies suggest that shifts in ecosystem structure from climate variations may be as [important as] or more important than the alterations in bulk integrated properties reported here," noting that some "susceptible ecosystem characteristics" that might be so shifted include "taxonomic composition, physiological status, and light absorption by colored dissolved organic material." It is possible that given enough time, the types of phenomena Behrenfeld *et al.* describe as possibly resulting in important "shifts in ecosystem structure" could compensate for or even overwhelm what might initially appear to be negative warming-induced consequences.

Another reason for not concluding too much from the oceanic NPP data set of Behrenfeld *et al.* is that it may be of too short a duration to reveal what might be occurring on a much longer timescale throughout the world's oceans, or that its position in time may be such that it does not allow the detection of greater short-term changes of the opposite sign that may have occurred a few years earlier or that might occur in the near future.

Consider, for example, the fact that the central regions of the world's major oceans were long thought to be essentially vast biological deserts (Ryther, 1969), but that several studies of primary photosynthetic production conducted in those regions over the 1980s (Shulenberger and Reid, 1981; Jenkins, 1982; Jenkins and Goldman, 1985; Reid and Shulenberger, 1986; Marra and Heinemann, 1987; Laws *et al.*, 1987; Venrick *et al.*, 1987; Packard *et al.*, 1988) yielded results that suggested marine productivity at that time was twice or more as great as it likely was for a long time prior to 1969, causing many of that day to speculate that "the ocean's deserts are blooming" (Kerr, 1986).

Of even greater interest, perhaps, is the fact that over this particular period of time (1970-1988), the data repository of Jones *et al.* (1999) indicates the earth experienced a (linear-regression-derived) global warming of 0.333°C, while the data base of the Global Historical Climatology Network indicates the planet experienced a similarly calculated global warming of 0.397°C. The mean of these two values (0.365°C) is nearly twice as great as the warming that occurred over the post-1999 period studied by Behrenfeld *et al.*; yet this earlier much larger warming (which according to the ten researchers' way of thinking should have produced a major decline in ocean productivity) was concomitant with a huge increase in ocean productivity. Consequently, it would appear that just the opposite of what Behrenfeld *et al.* suggest about global warming and ocean productivity is likely to be the more correct of the two opposing cause-and-effect relationships.

Moving closer to the present, Levitan *et al.* (2007) published a study of major significance that addresses the future of oceanic productivity under rising atmospheric CO_2 concentrations. In their paper the authors note that "among the principal players contributing to global aquatic primary production, the nitrogen (N)-fixing organisms (diazotrophs) are important providers of new N to the oligotrophic areas of the oceans," and they cite several studies which demonstrate that "cyanobacterial (photo-trophic) diazotrophs in particular fuel primary production and phytoplankton blooms which sustain oceanic food-webs and major economies and impact global carbon (C) and N cycling." These facts compelled them to examine how the ongoing rise in the air's CO_2 content might impact these relationships. They began by exploring the response of the cyanobacterial diazotroph *Trichodesmium* to changes in the atmosphere's CO_2 concentration, choosing this particular diazotroph because it dominates the world's tropical and sub-tropical oceans in this regard, contributing over 50 percent of total marine N fixation.

The eight Israeli and Czech researchers grew *Trichodesmium* IMS101 stock cultures in YBCII medium (Chen *et al.*, 1996) at 25°C and a 12-hour:12-hour light/dark cycle (with the light portion of the cycle in the range of 80-100 μmol photons m^{-2} s^{-1}) in equilibrium with air of three different CO_2 concentrations (250, 400 and 900 ppm, representing low, ambient and high concentrations, respectively), which was accomplished by continuously bubbling air of the three CO_2 concentrations through the appropriate culture vessels throughout various experimental runs, each of which lasted a little over three weeks, during which time they periodically monitored a number of diazotrophic physiological processes and properties.

So what did the scientists learn? Levitan *et al.* report that *Trichodesmium* in the high CO_2 treatment "displayed enhanced N fixation, longer trichomes, higher growth rates and biomass yields." In fact, they write that in the high CO_2 treatment there was "a three- to four-fold increase in N fixation and a doubling of growth rates and biomass," and that the cultures in the low CO_2 treatment reached a stationary growth phase after only five days, "while both ambient and high CO_2 cultures exhibited exponential growth until day 15 before declining."

In discussing possible explanations for what they observed, the researchers suggest that "enhanced N fixation and growth in the high CO_2 cultures occurs due to reallocation of energy and resources from carbon concentrating mechanisms required under low and ambient CO_2." Consequently, they conclude, in their words, that "in oceanic regions, where light and nutrients such as P and Fe are not limiting, we expect the projected concentrations of CO_2 to increase N fixation and growth of *Trichodesmium*," and that "other diazotrophs may be similarly affected, thereby enhancing inputs of new N and increasing primary productivity in the oceans." And to emphasize these points, they write in the concluding sentence of their paper that "*Trichodesmium's* dramatic response to elevated CO_2 may consolidate its dominance in subtropical and tropical regions and its role in C and N cycling, fueling subsequent primary production, phytoplankton blooms, and sustaining oceanic food-webs."

Arrigo *et al.* (2008) introduce their work by writing that "between the late 1970s and the early part of the 21st century, the extent of Arctic Ocean sea ice cover has declined during all months of the year, with the largest declines reported in the boreal summer months, particularly in September (8.6 ± 2.9% per decade)," citing the work of Serreze *et al.* (2007). In an effort to "quantify the change in marine primary productivity in Arctic waters resulting from recent losses of sea ice cover," the authors "implemented a primary productivity algorithm that accounts for variability in sea ice extent, sea surface temperature, sea level winds, downwelling spectral irradiance, and surface chlorophyll *a* concentrations," and that "was parameterized and validated specifically for use in the Arctic (Pabi *et al.*, 2008) and utilizes forcing variables derived either from satellite data or NCEP reanalysis fields."

Arrigo *et al.* determined that "annual primary production in the Arctic increased yearly by an average of 27.5 Tg C per year since 2003 and by 35 Tg C per year between 2006 and 2007," 30 percent of which total increase was attributable to decreased minimum summer ice extent and 70 percent of which was due to a longer phytoplankton growing season. Arrigo *et al.* thus conclude that if the trends they discovered continue, "additional loss of ice during Arctic spring could boost productivity >3-fold above 1998-2002 levels." Hence, they additionally state that if the 26 percent increase in annual net CO_2 fixation in the Arctic Ocean between 2003 and 2007 is maintained, "this would represent a weak negative feedback on climate change."

On the other side of the globe and working in the Southern Ocean, Smith and Comiso (2008) employed phytoplankton pigment assessments, surface temperature estimates, modeled irradiance, and observed sea ice concentrations—all of which parameters were derived from satellite data—and

incorporated them into a vertically integrated production model to estimate primary productivity trends according to the technique of Behrenfeld *et al.* (2002). Of this effort, the two authors say that "the resultant assessment of Southern Ocean productivity is the most exhaustive ever compiled and provides an improvement in the quantitative role of carbon fixation in Antarctic waters." So what did they find? During the nine years (1997-2006) analyzed in the study, "productivity in the entire Southern Ocean showed a substantial and significant increase," which increase can be calculated from the graphical representation of their results as ~17 percent per decade. In commenting on their findings, the two researchers note that "the highly significant increase in the productivity of the entire Southern Ocean over the past decade implies that long-term changes in Antarctic food webs and biogeochemical cycles are presently occurring," which changes we might add are positive.

In light of these several real-world observations, we find no indications of a widespread decline in oceanic productivity over the twentieth century in response to increases in air temperature and CO_2 concentration. In fact, we see evidence that just the opposite is occurring, that environmental changes are occurring that are proving to be beneficial.

Additional information on this topic, including reviews of newer publications as they become available, can be found at http://www.co2science.org/subject/o/oceanproductivity.php.

References

Antoine, D., Morel, A., Gordon, H.R., Banzon, V.J. and Evans, R.H. 2005. Bridging ocean color observations of the 1980s and 2000s in search of long-term trends. *Journal of Geophysical Research* **110**: 10.1029/2004JC002620.

Arrigo, K.R., van Dijken, G. and Pabi, S. 2008. Impact of a shrinking Arctic ice cover on marine primary production. *Geophysical Research Letters* **35**: 10.1029/2008GL035028.

Beaugrand, G. and Reid, P.C. 2003. Long-term changes in phytoplankton, zooplankton and salmon related to climate. *Global Change Biology* **9**: 801-817.

Behrenfeld, M, Maranon, E., Siegel, D.A. and Hooker, S.B. 2002. Photo-acclimation and nutrient-based model of light-saturated photosynthesis for quantifying ocean primary production. *Marine Ecology Progress Series* **228**: 103-117.

Behrenfeld, M.J., O'Malley, R.T., Siegel, D.A., McClain, C.R., Sarmiento, J.L., Feldman, G.C., Milligan, A.J.,

Falkowski, P.G., Letelier, R.M. and Boss, E.S. 2006. Climate-driven trends in contemporary ocean productivity. *Nature* **444**: 752-755.

Chen, Y.B., Zehr, J.P. and Mellon, M. 1996. Growth and nitrogen fixation of the diazotrophic filamentous nonheterocystous cyanobacterium *Trichodesmium* sp IMS101 in defined media: evidence for a circadian rhythm. *Journal of Phycology* **32**: 916-923.

Delaney, P. 2000. Nutrients in the glacial balance. *Nature* **405**: 288-291.

Dupouy, C., Neveux, J., Subramaniam, A., Mulholland, M.R., Montoya, J.P., Campbell, L., Carpenter, E.J. and Capone, D.G. 2000. Satellite captures Trichodesmium blooms in the southwestern tropical Pacific. *EOS: Transactions, American Geophysical Union* **81**: 13, 15-16.

Elderfield, H. and Rickaby, R.E.M. 2000. Oceanic Cd/P ratio and nutrient utilization in the glacial Southern Ocean. *Nature* **405**: 305-310.

Goes, J.I., Thoppil, P.G., Gomes, H. do R. and Fasullo, J.T. 2005. Warming of the Eurasian landmass is making the Arabian Sea more productive. *Science* **308**: 545-547.

Gregg, W.W. and Conkright, M.E. 2002. Decadal changes in global ocean chlorophyll. *Geophysical Research Letters* **29**: 10.1029/2002GL014689.

Hannon, E., Boyd, P.W., Silvoso, M. and Lancelot, C. 2001. Modelling the bloom evolution and carbon flows during SOIREE: implications for future *in situ* iron-experiments in the Southern Ocean. *Deep-Sea Research II* **48**: 2745-2773.

Hirawake, T., Odate, T. and Fukuchi, M. 2005. Long-term variation of surface phytoplankton chlorophyll a in the Southern Ocean during 1965-2002. *Geophysical Research Letters* **32**: 10.1029/2004GL021394.

Jenkins, W.J. 1982. Oxygen utilization rates in North Atlantic subtropical gyre and primary production in oligotrophic systems. *Nature* **300**: 246-248.

Jenkins, W.J. and Goldman, J.C. 1985. Seasonal oxygen cycling and primary production in the Sargasso Sea. *Journal of Marine Research* **43**: 465-491.

Jones, P.D., Parker, D.E., Osborn, T.J. and Briffa, K.R. 1999. Global and hemispheric temperature anomalies— land and marine instrument records. In *Trends: A Compendium of Data on Global Change*. Carbon Dioxide Information Analysis Center, Oak Ridge National Laboratory, U.S. Department of Energy, Oak Ridge, TN, USA.

Kerr, R.A. 1986. The ocean's deserts are blooming. *Science* **232**: 1345.

Lancelot, C., Hannon, E., Becquevort, S., Veth, C. and De Baar, H.J.W. 2000. Modelling phytoplankton blooms and carbon export in the Southern Ocean: dominant controls by light and iron in the Atlantic sector in austral spring 1992. *Deep Sea Research I* **47**: 1621-1662.

Laws, E.A., Di Tullio, G.R. and Redalje, D.G. 1987. High phytoplankton growth and production rates in the North Pacific subtropical gyre. *Limnology and Oceanography* **32**: 905-918.

Levitan, O., Rosenberg, G., Setlik, I., Setlikova, E., Grigel, J., Klepetar, J., Prasil, O. and Berman-Frank, I. 2007. Elevated CO_2 enhances nitrogen fixation and growth in the marine cyanobacterium *Trichodesmium*. *Global Change Biology* **13**: 531-538.

Marasovic, I., Nincevic, Z., Kuspilic, G., Marinovic, S. and Marinov, S. 2005. Long-term changes of basic biological and chemical parameters at two stations in the middle Adriatic. *Journal of Sea Research* **54**: 3-14.

Marra, J. and Heinemann, K.R. 1987. Primary production in the North Pacific central gyre: Some new measurements based on [14]C. *Deep-Sea Research* **34**: 1821-1829.

Pabi, S., van Dijken, G.S. and Arrigo, K.R. 2008. Primary production in the Arctic Ocean, 1998-2006. *Journal of Geophysical Research* **113**: 10.1029/2007JC004578.

Packard, T.T., Denis, M., Rodier, M. and Garfield, P. 1988. Deep-ocean metabolic CO_2 production: Calculations from ETS activity. *Deep-Sea Research* **35**: 371-382.

Pahlow, M. and Riebesell, U. 2000. Temporal trends in deep ocean Redfield ratios. *Science* **287**: 831-833.

Pasquer, B., Laruelle, G., Becquevort, S., Schoemann, V., Goosse, H. and Lancelot, C. 2005. Linking ocean biogeochemical cycles and ecosystem structure and function: results of the complex SWAMCO-4 model. *Journal of Sea Research* **53**: 93-108.

Raitsos, D., Reid, P.C., Lavender, S.J., Edwards, M. and Richardson, A.J. 2005. Extending the SeaWiFS chlorophyll data set back 50 years in the northeast Atlantic. *Geophysical Research Letters* **32**: 10.1029/2005GL022484.

Reid, J.L. and Shulenberger, E. 1986. Oxygen saturation and carbon uptake near 28°N, 155°W. *Deep-Sea Research* **33**: 267-271.

Ryther, J.H. 1969. Photosynthesis and fish production in the sea. *Science* **166**: 72-76.

Sarmiento, J.L., Slater, R., Barber, R., Bopp, L., Doney, S.C., Hirst, A.C., Kleypas, J., Matear, R., Mikolajewicz, U., Monfray, P., Soldatov, V., Spall, S.A. and Stouffer, R. 2004. Response of ocean ecosystems to climate warming. *Global Biogeochemical Cycles* **18**: 10.1029/2003GB002134.

Sepulveda, J., Pantoja, S., Hughen, K., Lange, C., Gonzalez, F., Munoz, P., Rebolledo, L., Castro, R., Contreras, S., Avila, A., Rossel, P., Lorca, G., Salamanca, M. And Silva, N. 2005. Fluctuations in export productivity over the last century from sediments of a southern Chilean fjord (44°S). *Estuarine, Coastal and Shelf Science* **65**: 587-600.

Serreze, M.C., Holland, M.M. and Stroeve, J. 2007. Perspectives on the Arctic's shrinking sea-ice cover. *Science* **315**: 1533-1536.

Shulenberger, E. and Reid, L. 1981. The Pacific shallow oxygen maximum, deep chlorophyll maximum, and primary production, reconsidered. *Deep-Sea Research* **28**: 901-919.

Smith Jr., W.O. and Comiso, J.C. 2008. Influence of sea ice on primary production in the Southern Ocean: A satellite perspective. *Journal of Geophysical Research* **113**: 10.1029/2007JC004251.

Venrick, E.L., McGowan, J.A., Cayan, D.R. and Hayward, T.L. 1987. Climate and chlorophyll a: Long-term trends in the central North Pacific Ocean. *Science* **238**: 70-72.

7.9.6. Global

How have earth's terrestrial plants responded—on average and in their entirety—to the atmospheric temperature and CO_2 increases of the past quarter-century? In this subsection we report the results of studies that have looked at either the world as a whole or groups of more than two continents at the same time.

In one of the earlier studies of the subject, Joos and Bruno (1998) used ice core and direct observations of atmospheric CO_2 and ^{13}C to reconstruct the histories of terrestrial and oceanic uptake of anthropogenic carbon over the past two centuries. This project revealed, in their words, that "the biosphere acted on average as a source [of CO_2] during the last century and the first decades of this century ... Then, the biosphere turned into a [CO_2] sink," which implies a significant increase in global vegetative productivity over the last half of the twentieth century.

More recently, Cao *et al.* (2004) derived net primary production (NPP) values at 8-km and 10-day resolutions for the period 1981-2000 using variables based almost entirely on satellite observations, as described in the Global Production Efficiency Model (GLO-PEM), which consists, in their words, "of linked components that describe the processes of canopy radiation absorption, utilization, autotrophic

respiration, and the regulation of these processes by environmental factors (Prince and Goward, 1995; Goetz *et al.*, 2000)." They learned that over the last two decades of the twentieth century, when temperatures were rising, "there was an increasing trend toward enhanced terrestrial NPP," which they say was "caused mainly by increases in atmospheric carbon dioxide and precipitation."

A year later, Cao *et al.* (2005) used the CEVSA (Carbon Exchanges in the Vegetation-Soil-Atmosphere system) model (Cao and Woodward, 1998; Cao *et al.*, 2002), forced by observed variations in climate and atmospheric CO_2, to quantify changes in NPP, soil heterotrophic respiration (HR) and net ecosystem production (NEP) from 1981 to 1998. As an independent check on the NPP estimate of CEVSA, they also estimated 10-day NPP from 1981-2000 with the GLO-PEM model that uses data almost entirely from remote sensing, including both the normalized difference vegetation index (NDVI) and meteorological variables (Prince and Goward, 1995; Cao *et al.*, 2004). This protocol revealed, in Cao *et al.*'s words, that "global terrestrial temperature increased by 0.21°C from the 1980s to the 1990s, and this alone increased HR more than NPP and hence reduced global annual NEP." *However*, they found that "combined changes in temperature and precipitation increased global NEP significantly," and that "increases in atmospheric CO_2 produced further increases in NPP and NEP." They also discovered that "the CO_2 fertilization effect [was] particularly strong in the tropics, compensating for the negative effect of warming on NPP." Enlarging on this point, they write that "the response of photosynthetic biochemical reactions to increases in atmospheric CO_2 is greater in warmer conditions, so the CO_2 fertilization effect will increase with warming in cool regions and be high in warm environments." The end result of the application of these models and measurements was their finding that global NEP increased "from 0.25 Pg C yr^{-1} in the 1980s to 1.36 Pg C yr^{-1} in the 1990s."

Commenting on their findings, Cao *et al.* note that "the NEP that was induced by CO_2 fertilization and climatic variation accounted for 30 percent of the total terrestrial carbon sink implied by the atmospheric carbon budget (Schimel *et al.*, 2001), and the fraction changed from 13 percent in the 1980s to 49 percent in the 1990s," which indicates the growing importance of the CO_2 fertilization effect. Also, they say "the increase in the terrestrial carbon sink from the 1980s to the 1990s was a continuation of the trend since the middle of the twentieth century, rather than merely a consequence of short-term climate variability," which suggests that as long as the air's CO_2 content continues its upward course, so too will its stimulation of the terrestrial biosphere likely continue its upward course.

Using a newly developed satellite-based vegetation index (Version 3 Pathfinder NDVI) in conjunction with a gridded global climate dataset (global monthly mean temperature and precipitation at 0.5° resolution from New *et al.*, 2000), Xiao and Moody (2005) analyzed trends in global vegetative activity from 1982 to 1998. The greening trends they found exhibited substantial latitudinal and longitudinal variability, with the most intense greening of the globe located in high northern latitudes, portions of the tropics, southeastern North America and eastern China. Temperature was found to correlate strongly with greening trends in Europe, eastern Eurasia and tropical Africa. Precipitation, on the other hand, was *not* found to be a significant driver of increases in greenness, except for isolated and spatially fragmented regions. Some decreases in greenness were also observed, mainly in the Southern Hemisphere in southern Africa, southern South America and central Australia, which trends were associated with concomitant increases in temperature and decreases in precipitation. There were also large regions of the globe that showed no trend in greenness over the 17-year period, as well as large areas that underwent strong greening that showed no association with trends of either temperature or precipitation. These greening trends, as they concluded, must have been the result of other factors, such as "CO_2 fertilization, reforestation, forest regrowth, woody plant proliferation and trends in agricultural practices," about which others will have more to say as we continue.

Working with satellite observations of vegetative activity over the period 1982 to 1999, Nemani *et al.* (2003) discovered that the productivity of earth's terrestrial vegetation rose significantly over this period. More specifically, they determined that terrestrial net primary production (NPP) increased by 6.17 percent, or 3.42 PgC, over the 18 years between 1982 and 1999. What is more, they observed net positive responses over all latitude bands studied: 4.2 percent (47.5-22.5°S), 7.4 percent (22.5°S-22.5°N), 3.7 percent (22.5-47.5°N), and 6.6 percent (47.5-90.0°N).

The eight researchers mention a number of likely contributing factors to these significant NPP increases: nitrogen deposition and forest regrowth in northern mid and high latitudes, wetter rainfall regimes in water-limited regions of Australia, Africa, and the Indian subcontinent, increased solar radiation reception over radiation-limited parts of Western Europe and the equatorial tropics, warming in many parts of the world, and the aerial fertilization effect of rising atmospheric CO_2 concentrations everywhere.

With respect to the latter factor, Nemani *et al.* say "an increase in NPP of only 0.2% per 1-ppm increase in CO_2 could explain all of the estimated global NPP increase of 6.17% over 18 years and is within the range of experimental evidence." However, they report that terrestrial NPP increased by more than 1 percent per year in Amazonia alone, noting that "this result cannot be explained solely by CO_2 fertilization."

We tend to agree with Nemani *et al.* on this point, but also note that the aerial fertilization effect of atmospheric CO_2 enrichment is most pronounced at higher temperatures, rising from next to nothing at a mean temperature of 10°C to a 0.33 percent NPP increase per 1-ppm increase in CO_2 at a mean temperature of 36°C for a mixture of plants comprised predominantly of herbaceous species (Idso and Idso, 1994). For woody plants, we could possibly expect this number to be two (Idso, 1999) or even three (Saxe *et al.*, 1998; Idso and Kimball, 2001; Leavitt *et al.*, 2003) times larger, yielding a 0.7 percent to 1 percent NPP increase per 1-ppm increase in atmospheric CO_2, which would represent the lion's share of the growth stimulation observed by Nemani *et al.* in tropical Amazonia.

The message of Nemani *et al.*'s study is that satellite-derived observations indicate the planet's terrestrial vegetation significantly increased its productivity over the last two decades of the twentieth century, in the face of a host of both real and imagined environmental stresses, chief among the latter of which was what the IPCC claims to be unprecedented CO_2-induced global warming.

Perhaps the most striking evidence for the significant twentieth century growth enhancement of earth's forests by the historical increase in the air's CO_2 concentration was provided by the study of Phillips and Gentry (1994). Noting that turnover rates of mature tropical forests correlate well with measures of net productivity (Weaver and Murphy, 1990), the two scientists assessed the turnover rates of 40 tropical forests from around the world in order to test the hypothesis that global forest productivity was increasing *in situ*. In doing so, they found that the turnover rates of these highly productive forests had indeed been rising ever higher since at least 1960, with an apparent pan-tropical acceleration since 1980. In discussing what might be causing this phenomenon, they stated that "the accelerating increase in turnover coincides with an accelerating buildup of CO_2," and as Pimm and Sugden (1994) stated in a companion article, it was "the consistency and simultaneity of the changes on several continents that lead Phillips and Gentry to their conclusion that enhanced productivity induced by increased CO_2 is the most plausible candidate for the cause of the increased turnover."

Four years later, a group of eleven researchers headed by Phillips (Phillips et al., 1998), working with data on tree basal area (a surrogate for tropical forest biomass) for the period 1958-1996, which they obtained from several hundred plots of mature tropical trees scattered about the world, found that average forest biomass for the tropics as a whole had increased substantially. In fact, they calculated that the increase amounted to approximately 40 percent of the missing terrestrial carbon sink of the entire globe. They suggested that "intact forests may be helping to buffer the rate of increase in atmospheric CO_2, thereby reducing the impacts of global climate change," as Idso (1991a,b) had earlier suggested, and they identified the aerial fertilization effect of the ongoing rise in the air's CO_2 content as one of the factors responsible for this phenomenon. Other contemporary studies also supported their findings (Grace et al., 1995; Malhi et al., 1998), verifying the fact that neotropical forests were indeed accumulating ever more carbon. Phillips et al. (2002) continued to state that this phenomenon was occurring "possibly in response to the increasing atmospheric concentrations of carbon dioxide (Prentice et al., 2001; Malhi and Grace, 2000)."

As time progressed, however, it became less popular (i.e., more "politically incorrect") to report positive biological consequences of the ongoing rise in the air's CO_2 concentration. The conclusions of Phillips and others began to be repeatedly challenged (Sheil, 1995; Sheil and May, 1996; Condit, 1997; Clark, 2002; Clark et al., 2003). In response to those challenges, CO2 Science published an editorial rebuttal (see http://www.co2science.org/articles/V6/N25/EDIT.php), after which Phillips, joined by 17 other researchers (Lewis et al., 2005b), including one who had earlier criticized his and his colleagues' conclusions, published a new analysis that vindicated Phillips et al.'s earlier thoughts on the subject.

One of the primary concerns of the critics of Phillips et al.'s work was that their meta-analyses included sites with a wide range of tree census intervals (2-38 years), which they claimed could be confounding or "perhaps even driving conclusions from comparative studies," as Lewis et al. (2005b) describe it. However, in Lewis et al.'s detailed study of this potential problem, which they concluded was indeed real, they found that re-analysis of Phillips et al.'s published results "shows that the pan-tropical increase in stem turnover rates over the late twentieth century cannot be attributed to combining data with differing census intervals." Or as they state more obtusely in another place, "the conclusion that turnover rates have increased in tropical forests over the late twentieth century is robust to the charge that this is an artifact due to the combination of data that vary in census interval (cf. Sheil, 1995)."

Lewis et al. (2005b) additionally noted that "Sheil's (1995) original critique of the evidence for increasing turnover over the late twentieth century also suggests that the apparent increase could be explained by a single event, the 1982-83 El Niño Southern Oscillation (ENSO), as many of the recent data spanned this event." However, as they continued, "recent analyses from Amazonia have shown that growth, recruitment and mortality rates have simultaneously increased within the same plots over the 1980s and 1990s, as has net above-ground biomass, both in areas largely unaffected, and in those strongly affected, by ENSO events (Baker et al., 2004; Lewis et al., 2004a; Phillips et al., 2004)."

In a satellite study of the world's tropical forests, Ichii et al. (2005) "simulated and analyzed 1982-1999 Amazonian, African, and Asian carbon fluxes using the Biome-BGC prognostic carbon cycle model driven by National Centers for Environmental Prediction reanalysis daily climate data," after which they "calculated trends in gross primary productivity (GPP) and net primary productivity (NPP)." This work revealed that solar radiation variability was the primary factor responsible for interannual variations in GPP, followed by temperature and precipitation variability, while in terms of GPP trends, Ichii et al. report that "recent changes in atmospheric CO_2 and climate promoted terrestrial GPP increases with a significant linear trend in all three tropical regions." In the Amazonian region, the rate of GPP increase was 0.67 PgC year^{-1} decade^{-1}, while in Africa and Asia it was about 0.3 PgC year^{-1} decade^{-1}. Likewise, they report that "CO_2 fertilization effects strongly increased recent NPP trends in regional totals."

In a review of these several global forest studies, as well as many others (which led to their citing 186 scientific journal articles), Boisvenue and Running (2006) examined reams of "documented evidence of the impacts of climate change trends on forest productivity since the middle of the twentieth century." In doing so, they found that "globally, based on both satellite and ground-based data, climatic changes seemed to have a generally positive impact on forest productivity when water was not limiting," which was most of the time, because they report that "less than 7% of forests are in strongly water-limited systems."

Last, Young and Harris (2005) analyzed, for the majority of earth's land surface, a near 20-year time series (1982-1999) of NDVI data, based on measurements obtained from the Advanced Very High Resolution Radiometer (AVHRR) carried aboard U.S. National Oceanic and Atmospheric Administration satellites. In doing so, they employed two different datasets derived from the sensor: the Pathfinder AVHRR Land (PAL) data set and the Global Inventory Modeling and Mapping Studies (GIMMS) dataset. Based on their analysis of the PAL data, the two researchers determined that "globally more than 30% of land pixels increased in annual average NDVI greater than 4% and more than 16% persistently increased greater than 4%," while "during the same period less than 2% of land pixels declined in NDVI and less than 1% persistently declined." With respect to the GIMMS dataset, they report that "even more areas were found to be persistently increasing (greater than 20%) and persistently decreasing (more than 3%)." All in all, they report that "between 1982 and 1999 the general trend of vegetation change throughout the world has been one of increasing photosynthesis."

As for what has been responsible for the worldwide increase in photosynthesis—which is the ultimate food source for nearly all of the biosphere—the researchers mention global warming (perhaps it's not so bad after all), as well as "associated precipitation change and increases in atmospheric carbon dioxide," citing Myneni *et al.* (1997) and Ichii *et al.* (2002). In addition, they say that "many of the areas of decreasing NDVI are the result of human activity," primarily deforestation (Skole and Tucker, 1993; Steininger *et al.*, 2001) and urbanization (Seto *et al.* (2000)).

In conclusion, the results of these many studies demonstrate there has been an increase in plant growth rates throughout the world since the inception of the Industrial Revolution, and that this phenomenon has been gradually accelerating over the years, in concert with the historical increases in the air's CO_2 content and its temperature.

Additional information on this topic, including reviews of newer publications as they become available, can be found at http://www.co2science.org/subject/g/greeningearth.php.

References

Baker, T.R., Phillips, O.L., Malhi, Y., Almeida, S., Arroyo, L., Di Fiore, A., Erwin, T., Higuchi, N., Killeen, T.J., Laurance, S.G., Laurance, W.F., Lewis, S.L., Monteagudo, A., Neill, D.A., Núñez Vargas, P., Pitman, N.C.A., Silva, J.N.M. and Vásquez Martínez, R. 2004. Increasing biomass in Amazonian forest plots. *Philosophical Transactions of the Royal Society of London Series B—Biological Sciences* **359**: 353-365.

Boisvenue, C. and Running, S.W. 2006. Impacts of climate change on natural forest productivity—evidence since the middle of the twentieth century. *Global Change Biology* **12**: 862-882.

Cao, M.K., Prince, S.D. and Shugart, H.H. 2002. Increasing terrestrial carbon uptake from the 1980s to the 1990s with changes in climate and atmospheric CO_2. *Global Biogeochemical Cycles* **16**: 10.1029/2001GB001553.

Cao, M., Prince, S.D., Small, J. and Goetz, S.J. 2004. Remotely sensed interannual variations and trends in terrestrial net primary productivity 1981-2000. *Ecosystems* **7**: 233-242.

Cao, M., Prince, S.D., Tao, B., Small, J. and Kerang, L. 2005. Regional pattern and interannual variations in global terrestrial carbon uptake in response to changes in climate and atmospheric CO_2. *Tellus B* **57**: 210-217.

Cao, M.K. and Woodward, F.I. 1998. Dynamic responses of terrestrial ecosystem carbon cycling to global climate change. *Nature* **393**: 249-252.

Clark, D.A. 2002. Are tropical forests an important carbon sink? Reanalysis of the long-term plot data. *Ecological Applications* **12**: 3-7.

Clark, D.A., Piper, S.C., Keeling, C.D. and Clark, D.B. 2003. Tropical rain forest tree growth and atmospheric carbon dynamics linked to interannual temperature variation during 1984-2000. *Proceedings of the National Academy of Sciences, USA* **100**: 10.1073/pnas.0935903100.

Condit, R. 1997. Forest turnover, density, and CO_2. *Trends in Ecology and Evolution* **12**: 249-250.

Goetz, S.J., Prince, S.D., Small, J., Gleason, A.C.R. and Thawley, M.M. 2000. Interannual variability of global terrestrial primary production: reduction of a model driven with satellite observations. *Journal of Geophysical Research* **105**: 20,007-20,091.

Grace, J., Lloyd, J., McIntyre, J., Miranda, A.C., Meir, P., Miranda, H.S., Nobre, C., Moncrieff, J., Massheder, J., Malhi, Y., Wright, I. andGash, J. 1995. Carbon dioxide uptake by an undisturbed tropical rain-forest in Southwest Amazonia, 1992-1993. *Science* **270**: 778-780.

Graybill, D.A. and Idso, S.B. 1993. Detecting the aerial fertilization effect of atmospheric CO_2 enrichment in tree-ring chronologies. *Global Biogeochemical Cycles* **7**: 81-95.

Hari, P. and Arovaara, H. 1988. Detecting CO_2 induced enhancement in the radial increment of trees. Evidence from the northern timberline. *Scandinavian Journal of Forest Research* **3**: 67-74.

Hari, P., Arovaara, H., Raunemaa, T. And Hautojarvi, A. 1984. Forest growth and the effects of energy production: A method for detecting trends in the growth potential of trees. *Canadian Journal of Forest Research* **14**: 437-440.

Ichii, K., Hashimoto, H., Nemani, R. and White, M. 2005. Modeling the interannual variability and trends in gross and net primary productivity of tropical forests from 1982 to 1999. *Global and Planetary Change* **48**: 274-286.

Ichii, K., Kawabata, A. and Yamaguchi, Y. 2002. Global correlation analysis for NDVI and climatic variables and NDVI trends: 1982-1990. *International Journal of Remote Sensing* **23**: 3873-3878.

Idso, K.E. and Idso, S.B. 1994. Plant responses to atmospheric CO_2 enrichment in the face of environmental constraints: a review of the past 10 year's research. *Agricultural and Forest Meteorology* **69**: 153-203.

Idso, S.B. 1991a. The aerial fertilization effect of CO_2 and its implications for global carbon cycling and maximum greenhouse warming. *Bulletin of the American Meteorological Society* **72**: 962-965.

Idso, S.B. 1991b. Reply to comments of L.D. Danny Harvey, Bert Bolin, and P. Lehmann. *Bulletin of the American Meteorological Society* **72**: 1910-1914.

Idso, S.B. 1995. *CO_2 and the Biosphere: The Incredible Legacy of the Industrial Revolution.* Department of Soil, Water & Climate, University of Minnesota, St. Paul, Minnesota, USA.

Idso, S.B. 1999. The long-term response of trees to atmospheric CO_2 enrichment. *Global Change Biology* **5**: 493-495.

Idso, S.B. and Kimball, B.A. 2001. CO_2 enrichment of sour orange trees: 13 years and counting. *Environmental and Experimental Botany* **46**: 147-153.

Joos, F. and Bruno, M. 1998. Long-term variability of the terrestrial and oceanic carbon sinks and the budgets of the carbon isotopes ^{13}C and ^{14}C. *Global Biogeochemical Cycles* **12**: 277-295.

LaMarche Jr., V.C., Graybill, D.A., Fritts, H.C. and Rose, M.R. 1984. Increasing atmospheric carbon dioxide: Tree ring evidence for growth enhancement in natural vegetation. *Science* **223**: 1019-1021.

Leavitt, S.W., Idso, S.B., Kimball, B.A., Burns, J.M., Sinha, A. and Stott, L. 2003. The effect of long-term atmospheric CO_2 enrichment on the intrinsic water-use efficiency of sour orange trees. *Chemosphere* **50**: 217-222.

Lewis, S.L., Phillips, O.L., Baker, T.R., Lloyd, J., Malhi, Y., Almeida, S., Higuchi, N., Laurance, W.F., Neill, D.A., Silva, J.N.M., Terborgh, J., Lezama, A.T., Vásquez Martinez, R., Brown, S., Chave, J., Kuebler, C., Núñez Vargas, P. and Vinceti, B. 2004a. Concerted changes in tropical forest structure and dynamics: evidence from 50 South American long-term plots. *Philosophical Transactions of the Royal Society of London Series B—Biological Sciences* **359**: 421-436.

Lewis, S.L., Phillips, O.L., Sheil, D., Vinceti, B., Baker, T.R., Brown, S., Graham, A.W., Higuchi, N., Hilbert, D.W., Laurance, W.F., Lejoly, J., Malhi, Y., Monteagudo, A., Vargas, P.N., Sonke, B., Nur Supardi, M.N., Terborgh, J.W. and Vasquez, M.R. 2005b. Tropical forest tree mortality, recruitment and turnover rates: calculation, interpretation and comparison when census intervals vary. *Journal of Ecology* **92**: 929-944.

Malhi Y. and Grace, J. 2000. Tropical forests and atmospheric carbon dioxide. *Trends in Ecology and Evolution* **15**: 332-337.

Malhi, Y., Nobre, A.D., Grace, J., Kruijt, B., Pereira, M.G.P., Culf, A. And Scott, S. 1998. Carbon dioxide transfer over a Central Amazonian rain forest. *Journal of Geophysical Research* **103**: 31,593-31,612.

Myneni, R.C., Keeling, C.D., Tucker, C.J., Asrar, G. and Nemani, R.R. 1997. Increased plant growth in the northern high latitudes from 1981 to 1991. *Nature* **386**: 698-702.

Nemani, R.R., Keeling, C.D., Hashimoto, H., Jolly, W.M., Piper, S.C., Tucker, C.J., Myneni, R.B. and Running. S.W. 2003. Climate-driven increases in global terrestrial net primary production from 1982 to 1999. *Science* **300**: 1560-1563.

New, M., Hulme, M. and Jones, P.D. 2000. Global monthly climatology for the twentieth century (New *et al.*) dataset. Available online (http://www.daac.ornl.gov) from Oak

Ridge National Laboratory Distributed Active Archive Center, Oak Ridge, Tennessee, USA.

Parker, M.L. 1987. Recent abnormal increase in tree-ring widths: A possible effect of elevated atmospheric carbon dioxide. In: Jacoby Jr., G.C. and Hornbeck, J.W. (Eds.), *Proceedings of the International Symposium on Ecological Aspects of Tree-Ring Analysis*. U.S. Department of Energy, Washington, DC, pp. 511-521.

Phillips, O.L., Baker, T.R., Arroyo, L., Higuchi, N., Killeen, T.J., Laurance, W.F., Lewis, S.L., Lloyd, J., Malhi, Y., Monteagudo, A., Neill, D.A., Núñez Vargas, P., Silva, J.N.M., Terborgh, J., Vásquez Martínez, R., Alexiades, M., Almeida, S., Brown, S., Chave, J., Comiskey, J.A., Czimczik, C.I., Di Fiore, A., Erwin, T., Kuebler, C., Laurance, S.G., Nascimento, H.E.M., Olivier, J., Palacios, W., Patiño, S., Pitman, N.C.A., Quesada, C.A., Saldias, M., Torres Lezama, A., B. and Vinceti, B. 2004. Pattern and process in Amazon tree turnover: 1976-2001. *Philosophical Transactions of the Royal Society of London Series B—Biological Sciences* **359**: 381-407.

Phillips, O.L. and Gentry, A.H. 1994. Increasing turnover through time in tropical forests. *Science* **263**: 954-958.

Phillips, O.L., Malhi, Y., Higuchi, N., Laurance, W.F., Nunez, P.V., Vasquez, R.M., Laurance, S.G., Ferreira, L.V., Stern, M., Brown, S. and Grace, J. 1998. Changes in the carbon balance of tropical forests: Evidence from long-term plots. *Science* **282**: 439-442.

Phillips, O.L., Malhi, Y., Vinceti, B., Baker, T., Lewis, S.L., Higuchi, N., Laurance, W.F., Vargas, P.N., Martinez, R.V., Laurance, S., Ferreira, L.V., Stern, M., Brown, S. and Grace, J. 2002. Changes in growth of tropical forests: Evaluating potential biases. *Ecological Applications* **12**: 576-587.

Pimm, S.L. and Sugden, A.M. 1994. Tropical diversity and global change. *Science* **263**: 933-934.

Prentice, I.C., Farquhar, G.D., Fasham, M.J.R., Goulden, M.L., Heimann, M., Jaramillo, V.J., Kheshgi, H.S., Le Quere, C., Scholes, R.J., Wallace, D.W.R., Archer, D., Ashmore, M.R., Aumont, O., Baker, D., Battle, M., Bender, M., Bopp, L.P., Bousquet, P., Caldeira, K., Ciais, P., Cox, P.M., Cramer, W., Dentener, F., Enting, I.G., Field, C.B., Friedlingstein, P., Holland, E.A., Houghton, R.A., House, J.I., Ishida, A., Jain, A.K., Janssens, I.A., Joos, F., Kaminski, T., Keeling, C.D., Keeling, R.F., Kicklighter, D.W., Hohfeld, K.E., Knorr, W., Law, R., Lenton, T., Lindsay, K., Maier-Reimer, E., Manning, A.C., Matear, R.J., McGuire, A.D., Melillo, J.M., Meyer, R., Mund, M., Orr, J.C., Piper, S., Plattner, K., Rayner, P.J., Sitch, S., Slater, R., Taguchi, S., Tans, P.P., Tian, H.Q., Weirig, M.F., Whorf, T. and Yool, A. 2001. The carbon cycle and atmospheric carbon dioxide. Chapter 3 of the Third Assessment Report of the Intergovernmental Panel on Climate Change. *Climate Change 2001: The Scientific Basis*. Cambridge University Press, Cambridge, UK, pp. 183-238.

Prince, S.D. and Goward, S.N. 1995. Global primary production: a remote sensing approach. *Journal of Biogeography* **22**: 815-835.

Saxe, H., Ellsworth, D.S. and Heath, J. 1998. Tree and forest functioning in an enriched CO_2 atmosphere. *New Phytologist* **139**: 395-436.

Schimel, D.S., House, J.I., Hibbard, J.I., Bousquet, P., Ciais, P., Peylin, P., Braswell, B.H., Apps, M.J., Baker, D., Bondeau, A., Canadell, J., Churkina, G., Cramer, W., Denning, A.S., Field, C.B., Friedlingstein, P., Goodale, C., Heimann, M., Houghton, R.A., Melillo, J.M., Moore III, B., Murdiyarso, D., Noble, I., Pacala, S.W., Prentice, I.C., Raupach, M.R., Rayner, P.J., Scholes, R.J., Steffen, W.L. and Wirth, C. 2001. Recent patterns and mechanisms of carbon exchange by terrestrial ecosystems. *Nature* **414**: 169-172.

Seto, K.C., Kaufman, R.K. and Woodcock, C.E. 2000. Landsat reveals China's farmland reserves, but they're vanishing fast. *Nature* **406**: 121.

Sheil, D. 1995. Evaluating turnover in tropical forests. *Science* **268**: 894.

Sheil, D. and May, R.M. 1996. Mortality and recruitment rate evaluations in heterogeneous tropical forests. *Journal of Ecology* **84**: 91-100.

Skole, D. and Tucker, C.J. 1993. Tropical deforestation and habitat fragmentation in the Amazon: satellite data from 1978 to 1988. *Science* **260**: 1905-1909.

Steininger, M.K., Tucker, C.J., Ersts, P., Killen, T.J., Villegas, Z. and Hecht, S.B. 2001. Clearance and fragmentation of tropical deciduous forest in the Tierras Bajas, Santa Cruz, Bolivia. *Conservation Biology* **15**: 856-866.

Weaver, P.L. and Murphy, P.G. 1990. Forest structure and productivity in Puerto Rico's Luquillo Mountains. *Biotropica* **22**: 69-82.

West, D.C. 1988. Detection of forest response to increased atmospheric carbon dioxide. In: Koomanoff, F.A. (Ed.), *Carbon Dioxide and Climate: Summaries of Research in FY 1988*. U.S. Department of Energy, Washington, D.C., p. 57.

Xiao, J. and Moody, A. 2005. Geographical distribution of global greening trends and their climatic correlates: 1982-1998. *International Journal of Remote Sensing* **26**: 2371-2390.

Young, S.S. and Harris, R. 2005. Changing patterns of global-scale vegetation photosynthesis, 1982-1999. *International Journal of Remote Sensing* **26**: 4537-4563.

8

Species Extinction

Introduction

The Intergovernmental Panel on Climate Change (IPCC) claims "new evidence suggests that climate-driven extinctions and range retractions are already widespread" and the "projected impacts on biodiversity are significant and of key relevance, since global losses in biodiversity are irreversible (very high confidence)" (IPCC-II, 2007, p. 213). The IPCC even claims to know that "globally about 20% to 30% of species (global uncertainty range from 10% to 40%, but varying among regional biota from as low as 1% to as high as 80%) will be at increasingly high risk of extinction, possibly by 2100, as global mean temperatures exceed 2 to 3°C above pre-industrial levels" (Ibid.).

These claims and predictions are not based on what is known about the phenomenon of extinction or on real-world data about how species have endured the warming of the twentieth century, which the IPCC claims was unprecedented in the past two millennia. Because we addressed the impact of rising CO_2 concentrations and rising temperatures on plants in detail in the previous chapter, we only briefly recap the evidence concerning terrestrial plants here, finding no evidence of a wave of temperature-driven extinctions, and in fact evidence of just the opposite. We then devote the largest part of the chapter to the effects of global warming on two species to which the IPCC devotes special attention, coral reefs and polar bears.

Additional information on this topic, including reviews of newer publications as they become available, can be found at http://www.co2 science.org/subject/e/extinction.php.

References

IPCC. 2007-II. *Climate Change 2007: Impacts, Adaptation and Vulnerability. Contribution of Working Group II to the Fourth Assessment Report of the Intergovernmental Panel on Climate Change.* M.L. Parry, O.F. Canziani, J.P. Palutikof, P.J. van der Linden and C.D. Hanson. (Eds.) Cambridge University Press, Cambridge, UK.

8.1. Explaining Extinction

There is a large gap between what scientists understand about the definition and causes of extinctions and what is reported in the popular press and even in some headline-seeking scientific journals. We start our analysis by asking what we know about the causes of past extinctions, the shortcomings of popular predictions of pending extinctions due to climate change, and the phenomenon of rapid evolutionary change.

8.1.1. Defining Extinction

Looking at the research papers selected to support the theory of massive warming extinctions, we are struck by many biologists' apparent misunderstanding of extinction. Some biologists seem to believe that effective conservation means every local population of butterflies and mountain flowers must be preserved. This is obviously impossible on a planet with continual, massive climate changes and human impacts. Some biologists try to define more and more local populations as separate species—a subterfuge.

A recent article in *Science* amply illustrates the conflicted feelings and biologists' frantic desire to protect everything. In an article titled "All Downhill from Here," Krajick (2004) laments the supposed danger to pikas (rodents, cousins to rabbits) that live on treeless mountaintops:

> As global temperatures rise, the pika's numbers are nose-diving in far-flung mountain ranges ... researchers fear that if the heat keeps rising, many alpine plants and animals will face quick declines or extinctions ... creatures everywhere are responding to warming, but mountain biota, like cold-loving polar species, have fewer options for coping. ... Comprising just 3% of the vegetated terrestrial surface, these islands of tundra are Noah's ark refuges where whole ecosystems, often left over from glacial times, are now stranded amid un-crossable seas of warm lowlands.

Krajick himself seems to forget the words in his own opening paragraph: pikas "are also some of the world's toughest mammals." As for his "un-crossable seas of warm lowlands," pikas may not be able to thrive in the lowlands competition, but it does seem likely they could find enough vegetation there during their travels to tide them over until they find other, cooler mountaintops.

Parmesan and Yohe (2003) also present a distorted version of the term "extinction." They counted Edith's checkerspot butterflies at 115 North American sites with historical records of harboring the species and then classified the sites as "extinct or intact." They found local checkerspot populations in much-warmer Mexico were four times more likely to be "locally extinct" than those in much-cooler Canada. Similarly, Ian Stirling, an expert on polar bears who is widely quoted in the debate over whether global warming may reduce polar bear populations, was quoted by the World Wide Fund for Nature in 2002 as saying "polar bear numbers will be reduced in the southern portions of their range [and] may even become locally extinct" (WWF, 2002). However, "locally extinct" is not a scientific term. Extinct means "no longer in existence; died out." Gone forever. Parmesan and Yohe are using it in reference to butterfly populations that have simply moved—and even left forwarding addresses farther north. Polar bears, too, are known to migrate to different areas in response to changes in climate and

competition for food. This is not extinction, but extirpation—the loss of a population in a given location. The butterflies are responding effectively to climate change—which is certainly what we would hope a butterfly species would do on a planet with a climate history as variable as Earth's. Parmesan and Yohe found populations of Edith's checkerspot butterflies thriving over most of western North America, but fewer of them at the southernmost extremity of their range—in Baja California and near San Diego—than in the past. However, parts of Canada have been warming into their preferred climate range.

Species resist extinction strongly and they often persist even when humans think they've been wiped out. The supposedly extinct ivory-billed woodpecker was recently found in two forests in eastern Arkansas (Arkansas Game and Fish Commission, 2005). The Nature Conservancy recently found three "extinct" snails in Alabama and California. Botanists have found the Mount Diablo buckwheat plant for the first time since 1936. At least 24 other "extinct" species have been found during natural heritage surveys in North America since 1974 (Holloway, 2005).

References

Arkansas Game and Fish Commission. 2005. Ivory-billed woodpecker found in Arkansas. News release. 5 August.

Holloway, M. 2005. When extinct isn't. *Scientific American,* 9 August.

Krajick, K. 2004. All downhill from here. *Science* **303**: 1600–1602.

Parmesan, C. and G. Yohe. 2003. A globally coherent fingerprint of climate change impacts across natural systems. *Nature* **421**: 37–42.

World Wide Fund for Nature, 2002. *Vanishing Kingdom—The Melting Realm of the Polar Bear.* World Wide Fund for Nature. Gland, Switzerland.

8.1.2. Past Extinctions

Most of the world's major species "body types" were laid down during the Cambrian period 600 million years ago (Levinton, 1992), so we know the major species have dealt successfully through the ages with new pest enemies, new diseases, ice ages, and global warmings greater than today's. Most wild species are at least one million years old, which means they have all been through at least six hundred 1,500-year climate cycles. Not the least of the warmings was the Holocene Climate Optimum, which was warmer than even the predictions of the IPCC for 2100 (IPCC, 2007). That very warm period ended less than 5,000 years ago.

Environmentalists argue that the speed of today's climate change is greater than previous warmings and will overwhelm the adaptive capacities of plants and animals. Yet history and paleontology agree that many of the past global temperature changes arrived very quickly, sometimes in a few decades. For example, 12,000 years ago, the Younger Dryas event suddenly and violently swung from warm temperatures back to Ice Age levels by the shutdown of the Gulf Stream as melting water from the extra trillion tons of ice built up in the glaciers and ice sheets over the previous 90,000 years of frigid climate was released into the oceans. The shutdown of the oceans' Atlantic Conveyor quickly triggered another thousand years of Ice Age. How did wild species deal with Mother Nature's sudden, sharp reversals then? In another example, starting about 1840 a Wyoming glacier went from Little Ice Age cold to near present-day warmth in about a decade (Schuster *et al.,* 2000). There's no evidence of any local species being destroyed by that rapid temperature change.

In contrast to the missing evidence of past climate changes having caused extinctions, we already know how most of the world's extinct species were lost, and in what order of magnitude (Singer and Avery, 2007). The first cause is huge asteroids striking the planet. The web sites of such universities as the University of California–Berkeley, Smith, and North Carolina State are replete with evidence of these "big bangs." Earth's collisions with massive missiles from outer space explode billions of tons of ash and debris into the planet's atmosphere, darken the skies, and virtually eradicate growing seasons for years at a time. There apparently have been more than a dozen such collisions in the Earth's past and they have destroyed millions of species, most of which we know about only through the fossil record.

In 2004, researchers announced that geological evidence suggests an object crashed at the shoreline of what is now Australia's northwestern coast 251 million years ago, creating climate changes and other natural catastrophes. Gugliotta (2004) filed the following report on the discovery for the *Washington Post*:

Scientists said yesterday they have found evidence that a huge meteorite or comet plunged into the coastal waters of the Southern Hemisphere 251 million years ago, possibly triggering the most catastrophic mass extinction in Earth's history. The researchers said that geological evidence suggests that an object about six miles in diameter crashed at the shoreline of what is now Australia's northwestern coast, creating climate changes and other natural catastrophes that wiped out 90 percent of marine species and 70 percent of land species.

The second known cause of extinctions is hunting. For a million years or so, humans along with *Homo erectus* in Southeast Asia and Neanderthals in Europe have hunted whatever they could kill. If it went extinct, we hunted something else. In this sense, the last Ice Age did cause some indirect species extinctions. During that extremely cold period, so much of the world's water was trapped in ice caps and glaciers that sea levels dropped as much as 400 feet below today's levels. Stone Age hunters walked across the Bering Strait from Asia and found hordes of wild birds and mammals that did not fear man. More than 40 edible species were wiped out in a historical eye-blink, including North America's mammoths, mastodons, horses, camels, and ground sloths (Diamond, 1997).

A similar spate of human-hunter extinctions recently has been confirmed in Australia by the discovery of a cave full of fossils beneath the Nullabor Plain. The new fossils disarm the claim that a huge number of Australian species—including its marsupial lion, claw-footed kangaroo, giant wombat, and the Genyornis, the heaviest bird ever known—went extinct 46,400 years ago because of climate change. The discovery team said the species found in the cave died during a dry climate similar to today's, which hadn't changed in 400,000 years. However, the fire-sensitive woodlands, to which the species were adapted, disappeared suddenly about 46,000 years ago—apparently because the newly arrived aboriginal people burned the woods to drive game into the arms of club-wielding hunters. The landscape was reshaped by fire from woods to shrubbery (Western Australian Museum, 2007).

Third: Man learned to farm. Farming for food made us less likely to hunt wild animals and birds to extinction, but eventually we claimed one-third of all the Earth's land area for agriculture. The saving grace was that the best land for farming tended to have few species; instead it had large numbers of a few species, such as bison on the American Great Plains and kangaroo in the Australian grasslands. In contrast, researchers have found as many species in five square miles of the Amazon as in the whole of North America. Fortunately, man tended to farm the best land for the highest sustainable food yields, leaving much of the poorer land (with its diversity) for nature (Avery, 2000).

Fourth: Alien species. Mankind's ships, cars, and planes transport species across natural barriers, enabling them to reproduce and compete with native species. This has made the survival competition among species much more global. Island species, in particular, have found themselves in more intense competition and many have gone extinct.

These four explanations for the major extinctions of the past are each now well documented and understood. The claim that changes in temperature during the twentieth century—which the IPCC calls unprecedented in the past two millennium—are causing extinctions is much more dubious, as we see in the next section.

References

Avery, D.T. 2000. Saving the Planet With Pesticides and Plastics: The Environmental Triumph of High-Yield Farming. Hudson Institute, Indianapolis, IN. 36–37.

Diamond, J. 1997. Guns, Germs and Steel. W. W. Norton & Company, New York. 46–47.

Gugliotta, G. 2004. Impact crater labeled clue to mass extinction. *Washington Post*, 14 May.

IPCC. 2007-II. *Climate Change 2007: Impacts, Adaptation and Vulnerability. Contribution of Working Group II to the Fourth Assessment Report of the Intergovernmental Panel on Climate Change.* Parry, M.L., Canziani, O.F., Palutikof, J.P., van der Linden, P.J. and Hanson, C.E. (Eds.) Cambridge University Press, Cambridge, UK.

Levinton, J. 1992. The big bang of animal evolution. *Scientific American* **267**: 84–91.

Schuster, P.F., *et al.* 2000. Chronological refinement of an ice core record at Upper Fremont Glacier in South Central North America. *Journal of Geophysical Research* **105**: 4657–666.

Singer, S.F. and Avery, D. Unstoppable Global Warming Every 1,500 Years. Rowman & Littlefield Publishers, Inc.

Western Australian Museum. 2007. Ancient Nullabor megafauna thrived in dry climate. Media Alert, 25 January.

8.1.3. Theories of Contemporary Extinctions

In striking contrast to the four known causes of past extinctions, which are backed by extensive fossil and archeological evidence, Thomas *et al.* (2004) simply asserted the *theory* that raising or lowering the Earth's temperature would cause major wildlife extinctions on a linear model. Accordingly, small temperature changes lead to relatively small reductions in species numbers while larger temperature increases drive more species to extinction. The team first defined "survival envelopes" for more than 1,100 wildlife species—in Europe, the Brazilian Amazon, the wet tropics of northeastern Australia, the Mexican desert, and the southern tip of South Africa. Then they used a "power equation" to link loss of habitat area with extinction rates. If the equation showed that a species' potential habitat was projected to decline, it was regarded as threatened; the greater the expected habitat loss, the greater the threat. There was no provision for species adaptation or migration.

One of the Thomas team's "moderate" scenarios was an increase in Earth's temperature of $0.8°$ C in the next 50 years. The researchers said this would cause the extinction of roughly 20 percent of the world's wild species, perhaps one million of them. Fortunately, this prediction can easily be checked. The Earth's temperature has already increased $0.8°$ C over the past 150 years. How many species died out because of that temperature increase? None that we are aware of.

The Thomas paper tells us in its opening sentence: "Climate change over the past 30 years has produced numerous shifts in the distributions and abundances of species, and has been implicated in one species-level extinction." The scientists who are predicting that $0.8°$ C of warming would cause hundreds of thousands of wildlife species extinctions over the next 50 years concede that this level of temperature increase over the past 150 years has resulted in the extinction of *one* species.

Reality takes away even that one extinction claim. Thomas *et al.'s* single cited example of a species driven extinct by the recent warming is the Golden Toad of Costa Rica. That was based on a 1999 paper in *Nature* by J. Alan Pounds and coauthors (Pounds *et al.*, 1999) describing research conducted at the Monteverde Cloud Forest Preserve in Puntarenas, Costa Rica. Pounds *et al.* claimed that, due to rising sea surface temperatures in the equatorial Pacific, 20 of the 50 species of frogs and toads (including the Golden Toad) had disappeared in a cloud forest study area of 30 square kilometers. (Cloud forests are misty habitats found only in the mountains above 1,500 meters, where the trees are enclosed by cool, wet clouds much of the time. The unusual climate serves as a home to thousands of unique plants and animals.) Pounds and a coauthor explained his thesis to a scientific conference in 1999:

> In a cloud forest, moisture is ordinarily plentiful. Even during the dry season ... clouds and mist normally keep the forest wet. Trade winds, blowing in from the Caribbean, carry moisture up the mountain slopes, where it condenses to form a large cloud deck that surrounds the mountains. It is hypothesized that climate warming, particularly since the mid-1970s, has raised the average altitude at which cloud formation begins, thereby reducing the clouds' effectiveness in delivering moisture to the forest. ... Days *without* mist during the dry season ... quadrupled over recent decades (Pounds and Schneider, 1999).

Pounds said at least 22 species of amphibians have disappeared from the cloud forest. Although the other species that disappeared were known to exist in other locations, the Golden Toad lost its only known home. However, two years after Pounds hypothesized that the amphibians lost their cloud forest climate to drying from sea surface warming, another research team demonstrated that it was almost certainly the clearing of lowland forests under the cloud forest of Monteverde that changed the pattern of cloud formation over the Golden Toad's once-mistier home. Lawton *et al.* (2001) noted that trade winds bringing moisture from the Caribbean spend five to 10 hours over the lowlands before they reach the Golden Toad's mountain home in the Cordillera de Tileran. By 1992, only about 18 percent of the original lowland vegetation remained. The deforestation reduced the infiltration of rainfall, increased water runoff, and thus reduced soil moisture. The shift from trees to crops and pasture also reduced the amount of water-holding canopy.

In March 1999, the Lawton team got satellite imagery showing that late-morning dry season cumulus clouds were much less abundant over the deforested parts of Costa Rica than over the nearby still-forested lowlands of Nicaragua. To check their conclusion, the Lawton team simulated the impact of Costa Rican deforestation using Colorado State University's Regional Atmospheric Modeling System. The computer modeling showed that the cloud base over pastured landscape rose above the altitude of the Cordillera peaks (1,800 meters) by late morning. Over forests, the cloud base didn't reach 1,800 meters until early afternoon. Lawton says these values "are in reasonable agreement with observed cloud bases in the area." That puts the blame for the cloud forest dryness squarely on the farmers and ranchers who cleared the lowlands. Pounds' own paper noted deforestation as a major threat to mountain cloud forests. The Lawton study leaves the Thomas team's big computerized study of mass species extinction without any evidence that moderate climate changes—even when abrupt—cause species extinctions.

The two other articles in *Nature* also failed to report any evidence of extinctions caused by the recent warming trend. The closest thing to an extinction threat in the Root *et al.* studies was the expansion of red foxes into the southern former range of arctic foxes in North America and Eurasia. However, this is displacement/ replacement, not extinction. Hersteinsson and Macdonald (1992) concluded that the changes in fox ranges were driven by prey availability. The arctic foxes are found primarily in the treeless regions of the Arctic, where they feed on lemmings and voles in the summer and eat heavily from seal carcasses in the winter. The larger red foxes eat a wider range of prey and fruits and are regarded as stronger competitors in forest and brush land. However, they are less well camouflaged for the winters in the treeless tundra than the arctic foxes in their blue-white winter pelts. Warming temperatures have allowed trees, brush, and red foxes to move farther north in the past 150 years—but they also have allowed arctic foxes to retain enough land and prey to succeed north of the red foxes. We do not know what would have happened if there had been no northern habitat and prey to support the arctic foxes during a red fox expansion, but the foxes already have survived more radical warming than they have faced recently. In earlier parts of the interglacial period, the Arctic temperatures were 2° to 6° C higher than they are now (Taira, 1975; Korotsky *et al.*, 1988).

Returning to the study of Thomas *et al.* one more time, it is interesting to note that an earlier study by Thomas contained findings that completely discredit the thesis on which the 2004 claim rests—that species have readily defined "survival envelopes" outside which they cannot survive. The Thomas team began its 2001 paper by restating the long-believed and broadly held concept that many animals are "relatively sedentary and specialized in marginal parts of their geographical distributions." Thus, creatures are "expected to be slow at colonizing new habitats."

Despite this belief, however, the Thomas team cites its own and many other researchers' studies showing that "the cool margins of many species' distributions have expanded rapidly in association with recent climate warming." This mildly undercuts their thesis. Much worse was to come. The two butterfly species the authors studied "increased the variety of habitat types that they can colonize" and the two species of bush cricket they studied showed "increased fractions of longer-winged (dispersive) individuals in recently founded populations." The longer-winged crickets would be able to fly farther in search of new habitat.

As a consequence of the new adaptations, the Thomas authors report, "Increased habitat breadth and dispersal tendencies have resulted in about 3 to 15-fold increases in [range] expansion rates, allowing these insects to cross habitat disjunctions that would have represented major or complete barriers to dispersal before the expansions started." The changes in the butterfly and cricket populations render Thomas's entire thesis of "survival envelopes" inadequate at best and quite likely irrelevant. Yet this paper was not only written before the Root analysis, it was included in the Root analysis as one of the select few research studies supporting the mega-extinction theory.

Buried in the IPCC report are admissions that the computer models based on the dubious notion of "survival envelopes" that it relies on produce "a picture of potential impacts and risks that is far from perfect, in some instances apparently contradictory" (p. 239) and "climate envelope models do not simulate dynamic population or migration processes, and results are typically constrained to the regional level, so that the implications for biodiversity at the global level are difficult to infer," citing Malcolm *et al.*, 2002 (IPCC, 2007-II, p. 240). We agree, which is why, in the following sections, we present more reliable theories and evidence that paint a much different, and more accurate, picture of the fate of wildlife in a warming world.

References

Hersteinsson, P. and Macdonald, D.W. 1992. Interspecific competition and the geographical distribution of red and Arctic foxes, *Vulpes vulpes* and *Alopex lagopus*. *Oikos* **64**: 505–15.

IPCC. 2007-II. *Climate Change 2007: Impacts, Adaptation and Vulnerability. Contribution of Working Group II to the Fourth Assessment Report of the Intergovernmental Panel on Climate Change.* Parry, M.L., Canziani, O.F., Palutikof, J.P., van der Linden, P.J. and Hanson, C.E. (Eds.) Cambridge University Press, Cambridge, UK.

Korotky, A.M., *et al.* 1988. Development of Natural Environment of the Southern Soviet Far East (Late Pleistocene-Holocene). Kauka, Moscow, Russia.

Root, T., *et al.* 2003. Fingerprints of global warming on wild animals and plants. *Nature* **421**: 57–60.

Lawton, R.O., *et al.* 2001. Climatic impact of tropical lowland deforestation on nearby mountain cloud forests. *Science* **294**: 584–87.

Malcolm, J.R., *et al.* 2002. *Habitats at Risk: Global Warming and Species Loss in Globally Significant Terrestrial Ecosystems*. World Wide Fund for Nature, Gland, Switzerland.

Pounds, J.A., *et al.* 1999. Biological response to climate change on a tropical mountain. *Nature* **398**: 611–15.

Pounds, J.A. and Schneider, S.H. 1999. Present and Future Consequences of Global Warming for Highland Tropical Forests Ecosystems: The Case of Costa Rica. Paper presented at the U.S. Global Change Research Program Seminar, Washington, DC. 29 September.

Taira, K. 1975. Temperature variation of the 'Kuroshio' and crustal movements in eastern and southeastern Asia 700 Years B.P. *Palaeogeography, Palaeoclimatology, Palaeoecology* **17**: 333–338.

Thomas, C.D., *et al.* 2001. Ecological and evolutionary processes at expanding range margins. *Nature* **411**: 577–81.

Thomas, C.D., *et al.* 2004. Extinction risk from climate change. *Nature* **427**: 145–48.

8.1.4. Data on Contemporary Species

What does real-world data say about rates of extinction? In 2002, the United Nations Environmental Program (UNEP) published a new *World Atlas of Biodiversity* (Groombridge and Jenkins, 2002). It reported that the world lost only half as many major wild species in the last three decades of the twentieth century (20 birds, mammals, and fish) as during the last three decades of the nineteenth century (40 extinctions of major species). In fact, UNEP said the rate of extinctions at the end of the twentieth century was the lowest since the sixteenth century—despite 150 years of rising world temperatures, growing populations, and industrialization.

There is a wealth of data to support the fact that

many species have prospered during the twentieth century, starting with research conducted by those who claim rising temperatures have caused a rise in extinctions. Parmesan and Yohe (2003) examined the northern boundaries of 52 butterfly species in northern Europe and the southern boundaries of 40 butterfly species in southern Europe and North Africa over the past century. Given the 0.8° C warming of Europe's temperatures over that period, it is striking to have Parmesan and Yohe tell us that "nearly all northward shifts involved extensions at the northern boundary with the southern boundary remaining stable." Thus, in the researchers own words, "most species effectively expanded the size of their ranges."

Chris Thomas and a coauthor in a study published in 1999 documented changes in the distribution between 1970 and 1990 of many British bird species (Thomas and Lennon, 1999). He found that the northern margins of southerly species shifted northward by an average of 19 km, while the southern margins of northerly species remained unchanged.

On 26 high mountain summits in the middle part of the Alps, a study by Grabherr et al. (1994) of the plant species found "species richness has increased during the past few decades, and is more pronounced at lower altitudes." In other words, the mountaintops show little loss of biodiversity at upper elevations, and increased species richness at lower elevations, where plants from still-lower elevations extended their ranges upward.

Pauli et al. (1996) examined the summit flora on 30 mountains in the European Alps, with species counts that ranged back in history to 1895. They report that mountaintop temperatures have risen by 2° C since 1920, with an increase of 1.2° C in just the last 30 years. Nine of the 30 mountaintops showed no change in species counts, but 11 gained an average of 59 percent more species, and one mountain gained an astounding 143 percent more species. Did historic species get crowded out by the flood of new warmer-zone plants? The 30 mountains showed a mean species loss of 0.68 out of an average of 15.57 species. There was no documentation that any of the species "lost" on particular mountains represented extinctions rather than local disappearances.

Vesperinas et al. (2001) reported that native heat-sensitive plant species have responded to temperature increases in the Iberian Peninsula and the Mediterranean coast over the past 30 years by expanding their ranges "towards colder inland areas where they were previously absent."

Van Herk et al. (2002) reported that the number of lichen species groups present in the Central Netherlands increased from 95 in 1979 to 172 in 2001 as the region warmed. The researchers found the average number of species grouped per site increased from 7.5 to 18.9. Again, more warmth produced increased species richness.

Looking at the distribution of 18 butterfly species widespread and common in the British countryside, "nearly all of the common species have increased in abundance [during the warming], more in the east of Britain than in the west" according to a research team led by Emie Pollard of Britain's Institute of Terrestrial Ecology (Pollard et al., 1995).

Warm-water species of plankton rapidly responded to warming and cooling in the western English Channel, shifting latitudinally by up to 120 miles, and increasing or decreasing their numbers by two to three-fold over 70 years, according to research by A.J. Southward of Britain's Marine Biological Association (Southward, 1995).

In the Antarctic, Adelie penguins need pack ice to thrive, whereas chinstrap penguins prefer ice-free waters. R.C. Smith et al. (1994) found that the Adelie penguins in the West Antarctic Peninsula are declining because the warming on the peninsula favors the chinstraps. Meanwhile the chinstraps in the Ross Sea region are suffering because 97 percent of Antarctica—the part that isn't the peninsula—is getting colder. This can hardly be surprising. Two varieties of a highly mobile species have moved to the sites that favor their respective feeding and breeding requirements while their populations decline in the unfavorable sites.

The Antarctic's only two higher-level plant species have responded to the Antarctic Peninsula's warming by increasing their numbers at two widely separated localities (Smith, 1994). Fortunately, it would take thousands of years of increased warming to melt all the Antarctic ice and a very long period of extended cold to close all the open water around the Antarctic Peninsula. The 1,500-year climate cycle makes it almost certain that Antarctica's plants and penguins will continue to adapt rather than disappear.

Invertebrates in a rocky intertidal site at Pacific Grove, California can't move, but their populations change. The invertebrates were surveyed by Saragin et al. (1999) in 1931–1933 and again in 1993–1996 after a warming of 0.8° C. Ten of the 11 southern species increased in abundance, whereas five of seven northern species decreased.

New photographs were taken by Sturm et al. (2001) to match a set of 1948–1950 photographs of the Brooks Range and the Arctic coast of Alaska. At more than half of the matched locations, researchers found "distinctive and, in some cases, dramatic increases in the height and diameter of individual shrubs ... and expansion of shrubs into previously shrub-free areas."

Western American bird species are pioneering and expanding their ranges over vast areas and huge climatic differences as the climate warms. N.K. Johnson of the University of California–Berkeley compiled records for 24 bird species from *Audubon Field Notes*, *American Birds* and other sources (Johnson, 1994). He found "four northern species have extended their ranges southward, three eastern species have expanded westward, fourteen southwestern or Mexican species have moved northward, one Great Basin-Colorado Plateau species has expanded radially, and two Great Basin-Rocky Mountain subspecies have expanded westward."

Brommer (2004) studied the birds of Finland, which were categorized as either northerly (34 species) or southerly (116 species). Brommer quantified changes in their range margins and distributions from two atlases of breeding birds, one covering the period 1974-79 and one covering the period 1986-89, in an attempt to determine how the two groups of species responded to what he called "the period of the earth's most rapid climate warming in the last 10,000 years." Once again, it was determined that the southerly group of bird species experienced a mean poleward advancement of their northern range boundaries of 18.8 km over the 12-year period of supposedly unprecedented warming. The southern range boundaries of the northerly species, on the other hand, were essentially unmoved, leading once again to range expansions that should have rendered the Finnish birds less subject to extinction than they were before the warming.

Similar results were obtained in a study by Hickling *et al.* (2005) of changes in the northern and southern range boundaries of 37 non-migratory British dragonfly and damselfly species between the two 10-year periods 1960-70 and 1985-95. All but two of the 37 species increased the sizes of their ranges between the two 10-year periods, with the researchers reporting that "species are shifting northwards faster at their northern range margin than at their southern range margin," and concluding that "this could suggest that species at their southern range margins are less constrained by climate than by other factors," which surely appears to be the case.

Chamaille-Jammes *et al.* (2006) studied four unconnected populations of a small live-bearing lizard that lives in peat bogs and heath lands scattered across Europe and Asia, concentrating on a small region near the top of a mountain in southeast France at the southern limit of the species' range. There, from 1984 to 2001, they monitored a number of life-history traits of the populations, including body size, reproductive characteristics, and survival rates, during which time local air temperatures rose by approximately 2.2°C. In doing so, they observed that individual body size increased dramatically in all four populations over the 18-year study period in all age classes and, in the words of the researchers, "appeared related to a concomitant increase in temperature experienced during the first month of life." As a result, since fecundity is strongly dependent on female body size, they found that "clutch size and total reproductive output also increased." In addition, they learned that "adult survival was positively related to May temperature."

In discussing their findings, the French researchers say that since all fitness components investigated responded positively to the increase in temperature, "it might be concluded that the common lizard has been advantaged by the shift in temperature." This finding, as they describe it, stands in stark contrast to what they call the "habitat-based prediction that these populations located close to mountaintops on the southern margin of the species range should be unable to cope with the alteration of their habitat." They conclude that "to achieve a better prediction of a species persistence, one will probably need to combine both habitat and individual-based approaches," noting, however, that individual responses, such as those documented in their study (which were all positive), represent "the ultimate driver of a species response to climate change."

The Audubon Society (2009) released a report in February 2009 calling attention to the correlation between the movement of North American bird populations and an increase in average January temperatures in the lower 48 U.S. states from 1969 to 2005. The report breathlessly recounted the expansion of the northern boundaries of the habitats of 58 percent of observed species of birds over the past four decades and concluded that "we must act decisively to control global warming pollution to curb the worst impacts of climate change, and take immediate steps to help birds and other species weather the changes we cannot avoid." But the study did not ask whether the warming that occurred during this period benefitted or hurt most bird populations by moving poleward the northern edge of their habitats. One might suppose the net effect would be beneficial, and in fact this is what Audubon itself found but failed to mention in the body of the report. In a data table presented in an appendix to the report, one finds that 120 of the 305 species reported in the study (39 percent) showed statistically significant population increases, 128 (42 percent) showed no change, and only 57 (19 percent) showed statistically significant declines. These numbers suggest that North American bird species overall benefited from the modest warming from 1969 to 2005.

Many, and probably most, of the world's species have benefited from rising temperatures in the twentieth century. There is very little evidence of *any* extinctions. What should be plain is that, despite predictions of extinctions based on theories and computer models, real-world observations confirm that a warmer world is more, not less, hospitable to wildlife.

Additional information on this topic, including reviews of newer publications as they become available, can be found at http://www.co2 science.org/subject/r/rangeexpanimals.php and http://www.co2science.org/subject/e/extinction.php.

References

Audubon Society. 2009. *Birds and Climate Change: On the Move — A Briefing for Policymakers and Concerned Citizens.* National Audubon Society, New York, NY. February.

Brommer, J.E. 2004. The range margins of northern birds shift polewards. *Annales Zoologici Fennici* **41**: 391-397.

Chamaille-Jammes, S., Massot, M., Aragon, P. and Clobert, J. 2006. Global warming and positive fitness response in mountain populations of common lizards *Lacerta vivipara. Global Change Biology* **12**: 392-402.

Grabherr, G. *et al.* 1994. Climate effects on mountain plants. *Nature* **369**: 448.

Groombridge, B. and Jenkins, M.D. 2002. *World Atlas of Biodiversity.* United Nations Environmental Program (UNEP) and University of California Press, Berkeley, CA.

Hickling, R., Roy, D.B., Hill, J.K. and Thomas, C.D. 2005. A northward shift of range margins in British Odonata. *Global Change Biology* **11**: 502-506.

Johnson, N.K. 1994. Pioneering and natural expansion of breeding distributions in western North American birds. *Studies in Avian Biology* **15**: 27-44.

Parmesan, C. and Yohe, G. 2003. A globally coherent fingerprint of climate change impacts across natural systems. *Nature* **421**: 37–42.

Pauli, H. *et al.* 1996. Effects of climate change on mountain ecosystems—upward shifting of mountain plants. *World Resource Review* **8**: 382-390.

Pollard, E. *et al.* 1995. Population trends of common British butterflies at monitored sites. *Journal of Applied Ecology* **32**: 9–16.

Saragin, R.D. *et al.* 1999. Climate-related change in an intertidal community over short and long time scales. *Ecological Monographs* **69**: 465–90.

Smith, R.C. *et al.* 1999. Marine ecosystem sensitivity to climate change. *BioScience* **49**: 393–404.

Smith, R.I.L. 1994. Vascular plants as bioindicators of regional warming in Antarctica. *Oecologia* **99**: 322–28.

Southward, A.J. 1995. Seventy years' observations of changes in distribution and abundance of zooplankton and intertidal organisms in the Western English Channel in relation to rising sea temperatures. *Journal of Thermal Biology* **20** (1): 127–55.

Sturm, M. *et al.* 2001. Increasing shrub abundance in the Arctic. *Nature* **411**: 546–47.

Thomas, C.D. and Lennon, J.J. 1999. Birds extend their ranges northwards. *Nature* **399**: 213.

Van Herk, C.M. *et al.* 2002. Long-term monitoring in the Netherlands suggests that lichens respond to global warming. *Lichenologist* **34**: 141–54.

Vesperinas, E.S. *et al.* 2001. The expansion of thermophilic plants in the Iberian Peninsula as a sign of climate change. In G. Walther *et al.* (Eds.) *"Fingerprints" of Climate Change: Adapted Behavior and Shifting Species Ranges.* Kluwer Academic/ Plenum Publishers, New York. 163–84.

8.1.5. Rapid Evolutionary Change

Skelly *et al.* (2007) critiqued the climate-envelope approach to predicting extinctions used by Thomas *et al.* (2004), citing as their primary reason for doing so the fact that this approach "implicitly assumes that species cannot evolve in response to changing climate." As they correctly point out, "many examples of contemporary evolution in response to climate change exist," such as populations of a frog they had studied that had "undergone localized evolution in thermal tolerance (Skelly and Freidenburg, 2000), temperature-specific development rate (Skelly, 2004), and thermal preference (Freidenburg and Skelly, 2004)," in less than 40 years. Similarly, they report, "laboratory studies of insects show that thermal tolerance can change markedly after as few as 10 generations (Good, 1993)."

Adding that "studies of microevolution in plants show substantial trait evolution in response to climate manipulations (Bone and Farres, 2001)," the researchers further noted that "collectively, these findings show that genetic variation for traits related to thermal performance is common and evolutionary response to changing climate has been the typical finding in experimental and observational studies (Hendry and Kinnison, 1999; Kinnison and Hendry, 2001)."

Although evolution will obviously be slower in the cases of long-lived trees and large mammals, where long generation times are the norm, the scientists say the case for rapid evolutionary responses among many other species "has grown much stronger," citing, in this regard, the work of six other groups of researchers comprised of two dozen individuals (Stockwell et al., 2003; Berteaux et al., 2004; Hairston et al., 2005; Bradshaw and Holzapfel, 2006; Schwartz et al., 2006; Urban et al., 2007). As a result, they write, "on the basis of the present knowledge of genetic variation in performance traits and species' capacity for evolutionary response, it can be concluded that evolutionary change will often occur concomitantly with changes in climate as well as other environmental changes (Stockwell et al., 2003; Grant and Grant, 2002; Balanya et al., 2006; Jump et al., 2006; Pelletier et al., 2007)."

Much the same conclusion has been reached by still other groups of scientists. In a study of the field mustard plant, for example, a group of three researchers (Franks et al., 2007) found evidence for what they describe as "a rapid, adaptive evolutionary shift in flowering phenology after a climatic fluctuation," which finding, in their words, "adds to the growing evidence that evolution is not always a slow, gradual process but can occur on contemporary time scales in natural populations."

Likewise, another group of researchers who published in 2007 (Rae et al., 2007)—who worked with hybrids of two *Populus* tree species—obtained results which, as they phrased it, "quantify and identify genetic variation in response to elevated CO_2 and provide an insight into genomic response to the changing environment." The results, they wrote, "should lead to an understanding of microevolutionary response to elevated CO_2 ... and aid future plant breeding and selection," noting that various research groups have already identified numerous genes that appear sensitive to elevated CO_2 (Gupta et al., 2005; Taylor et al., 2005; Ainsworth et al., 2006; Rae et al., 2006).

Life in the sea, in this regard, is no different from life on land. In another study published in 2007, for example, a team of four marine biologists (Van Doorslaer et al., 2007) conducted an experiment with a species of zooplankton in which they say they "were able to demonstrate a rapid microevolutionary response (within 1 year) in survival, age at reproduction and offspring number to elevated temperatures," and they state that "these responses may allow the species to maintain itself under the forecasted global warming scenarios," noting that what they learned "strongly indicates rapid microevolution of the ability to cope with higher temperatures." Many other studies, some of them cited in Section 8.3, have produced analogous results with respect to increases in temperature on corals (Kumaraguru et al., 2003; Willis et al., 2006) and increases in CO_2 on freshwater microalgae (Collins et al., 2006).

In conclusion, many species have shown the ability to adapt rapidly to changes in climate. Claims that global warming threatens large numbers of species with extinction typically rest on a false definition of extinction (the loss of a particular population rather than entire species) and speculation rather than real-world evidence. The world's species have proven to be very resilient, having survived past natural climate cycles that involved much greater warming and higher CO_2 concentrations than exist today or are likely to occur in the coming centuries.

Additional information on this topic, including reviews of newer publications as they become available, can be found at http://www.co2science.org/subject/e/subject_e.php under the subheading Evolution.

References

Ainsworth, E.A., Rogers, A., Vodkin, L.O., Walter, A. and Schurr, U. 2006. The effects of elevated CO_2 concentration on soybean gene expression. An analysis of growing and mature leaves. *Plant Physiology* **142**: 135-147.

Balanya, J., Oller, J.M., Huey, R.B., Gilchrist, G.W. and Serra, L. 2006. Global genetic change tracks global climate warming in *Drosophila subobscura*. *Science* **313**: 1773-1775.

Berteaux, D., Reale, D., McAdam, A.G. and Boutin, S. 2004. Keeping pace with fast climatic change: can arctic life count on evolution? *Integrative and Comparative Biology* **44**: 140-151.

Bone, E. and Farres, A. 2001. Trends and rates of microevolution in plants. *Genetica* **112-113**: 165-182.

Bradshaw, W.E. and Holzapfel, C.M. 2006. Evolutionary response to rapid climate change. *Science* **312**: 1477-1478.

Collins, S., Sultemeyer, D. and Bell, G. 2006. Changes in C uptake in populations of *Chlamydomonas reinhardtii* selected at high CO_2. *Plant, Cell and Environment* **29**: 1812-1819.

Franks, S.J., Sim, S. and Weis, A.E. 2007. Rapid evolution of flowering time by an annual plant in response to a climate fluctuation. *Proceedings of the National Academy of Sciences USA* **104**: 1278-1282.

Freidenburg, L.K. and Skelly, D.K. 2004. Microgeographical variation in thermal preference by an amphibian. *Ecology Letters* **7**: 369-373.

Good, D.S. 1993. Evolution of behaviors in *Drosophila melanogaster* in high-temperatures: genetic and environmental effects. *Journal of Insect Physiology* **39**: 537-544.

Grant, P.R. and Grant, B.R. 2002. Unpredictable evolution in a 30-year study of Darwin's finches. *Science* **296**: 707-711.

Gupta, P., Duplessis, S., White, H., Karnosky, D.F., Martin, F. and Podila, G.K. 2005. Gene expression patterns of trembling aspen trees following long-term exposure to interacting elevated CO_2 and tropospheric O_3. *New Phytologist* **167**: 129-142.

Hairston, N.G., Ellner, S.P., Gerber, M.A., Yoshida, T. and Fox, J.A. 2005. Rapid evolution and the convergence of ecological and evolutionary time. *Ecology Letters* **8**: 1114-1127.

Hendry, A.P. and Kinnison, M.T. 1999. The pace of modern life: measuring rates of contemporary microevolution. *Evolution* **53**: 637-653.

Jump, A.S., Hunt, J.M., Martinez-Izquierdo, J.A. and Penuelas, J. 2006. Natural selection and climate change: temperature-linked spatial and temporal trends in gene frequency in *Fagus sylvatica*. *Molecular Ecology* **15**: 3469-3480.

Kinnison, M.T. and Hendry, A.P. 2001. The pace of modern life II: from rates of contemporary microevolution to pattern and process. *Genetica* **112-113**: 145-164.

Kumaraguru, A.K., Jayakumar, K. and Ramakritinan, C.M. 2003. Coral bleaching 2002 in the Palk Bay, southeast coast of India. *Current Science* **85**: 1787-1793.

Pelletier, F., Clutton-Brock, T., Pemberton, J., Tuljapurkar, S. and Coulson, T. 2007. The evolutionary demography of ecological change: linking trait variation and population growth. *Science* **315**: 1571-1574.

Rae, A.M., Ferris, R., Tallis, M.J. and Taylor, G. 2006. Elucidating genomic regions determining enhanced leaf growth and delayed senescence in elevated CO_2. *Plant, Cell & Environment* **29**: 1730-1741.

Rae, A.M., Tricker, P.J., Bunn, S.M. and Taylor, G. 2007. Adaptation of tree growth to elevated CO_2: quantitative trait loci for biomass in *Populus*. *New Phytologist* **175**: 59-69.

Schwartz, M.W., Iverson, L.R., Prasad, A.M., Matthews, S.N. and O'Connor, R.J. 2006. Predicting extinctions as a result of climate change. *Ecology* **87**: 1611-1615.

Skelly, D.K., Joseph, L.N., Possingham, H.P., Freidenburg, L.K., Farrugia, T.J., Kinnison, M.T. and Hendry, A.P. 2007. Evolutionary responses to climate change. *Conservation Biology* **21**: 1353-1355.

Skelly, D.K. and Freidenburg, L.K. 2000. Effects of beaver on the thermal biology of an amphibian. *Ecology Letters* **3**: 483-486.

Skelly, D.K. 2004. Microgeographic countergradient variation in the wood frog, *Rana sylvatica*. *Evolution* **58**: 160-165.

Stockwell, C.A., Hendry, A.P. and Kinnison, M.T. 2003. Contemporary evolution meets conservation biology. *Trends in Ecology and Evolution* **18**: 94-101.

Taylor, G., Street, N.R., Tricker, P.J., Sjodin, A., Graham, L., Skogstrom, O., Calfapietra, C., Scarascia-Mugnozza, G. and Jansson, S. 2005. The transcriptome of *Populus* in elevated CO_2. *New Phytologist* **167**: 143-154.

Thomas, C.D., *et al.* 2004. Extinction risk from climate change. *Nature* **427**: 145-48.

Urban, M.C., Philips, B., Skelly, D.K. and Shine, R. 2007. The cane toad's (*Chaunus* [*Bufo*] *marinus*) increasing ability to invade Australia is revealed by a dynamically updated range model. *Proceedings of the Royal Society of London B*: 10.1098/rspb.2007.0114.

Van Doorslaer, W., Stoks, R., Jeppesen, E. and De Meester, L. 2007. Adaptive microevolutionary responses to simulated global warming in *Simocephalus vetulus*: a mesocosm study. *Global Change Biology* **13**: 878-886.

Willis, B.L., van Oppen, M.J.H., Miller, D.J., Vollmer, S.V. and Ayre, D.J. 2006. The role of hybridization in the evolution of reef corals. *Annual Review of Ecology, Evolution, and Systematics* **37**: 489-517.

8.2. Terrestrial Plants

The IPCC's global warming extinction scenario is that if it gets "too hot" for a species of plant or animal where it currently lives, individuals of the heat-stressed species would have to move to a cooler location in order to survive. In many cases, however, acclimation can adequately substitute for migration, as has been demonstrated by several studies in which the temperatures at which plants grow best rose substantially (by several degrees Centigrade) in response to increases in the air temperature regimes to which they had long been accustomed (Mooney and West, 1964; Strain *et al.*, 1976; Bjorkman *et al.*, 1978; Seemann *et al.*, 1984; Veres and Williams, 1984; El-Sharkawy *et al.*, 1992, Battaglia *et al.*, 1996).

How does acclimation happen? One possible way is described by Kelly *et al.* (2003). In reference to the view of the IPCC, they note that "models of future ecological change assume that *in situ* populations of plants lack the capacity to adapt quickly to warming and as a consequence will be displaced by species better able to exploit the warmer conditions anticipated from 'global warming'." In contrast to this assumption, they report finding individual trees within a naturally occurring stand of *Betula pendula* (birch) that are genetically adapted to a range of different temperatures. As they describe it, they discovered "the existence of 'pre-adapted' individuals in standing tree populations" that "would reduce temperature-based advantages for invading species," which finding, they say, "bring[s] into question assumptions currently used in models of global climate change."

Another perspective on the adaptation vs. migration theme is provided by the work of Loehle (1998), who notes (using forests as an example) that the CO_2-induced global warming extinction hypothesis rests on the assumption that the growth rates of trees rise from zero at the cold limits of their natural ranges (their northern boundaries in the Northern Hemisphere) to a broad maximum, after which they decline to zero at the warm limits of their natural ranges (their southern boundaries in the Northern Hemisphere). Loehle demonstrates that this

assumption is only half correct. It properly describes tree growth dynamics near a Northern Hemispheric forest's northern boundary, but not its southern boundary.

Loehle notes that in the Northern Hemisphere (to which we will restrict our discussion for purposes of simplicity), trees planted north of their natural ranges' northern boundaries are able to grow to maturity only within 50-100 miles of those boundaries. Trees planted south of their natural ranges' southern boundaries, however, often grow to maturity as much as 1,000 miles further south (Dressler, 1954; Woodward, 1987, 1988). Loehle reports that "many alpine and arctic plants are extremely tolerant of high temperatures, and in general one cannot distinguish between arctic, temperate, and tropical-moist-habitat types on the basis of heat tolerances, with all three types showing damage at 44-52°C (Gauslaa, 1984; Lange and Lange, 1963; Levitt, 1980; Kappen, 1981)."

What Loehle finds from his review of the literature and his experience with various trees in the Unites States is that as temperatures and growing degree days rise from very low values, the growth rates of boreal trees at some point begin to rise from zero and continue increasing until they either plateau at some maximum value or drop only very slowly thereafter, as temperatures rise still higher and growing degree days continue to accumulate. Trees from the Midwest, by comparison, do not begin to grow until a higher temperature or greater accumulation of growing degree days is reached, after which their growth rates rise considerably higher than those of the colder-adapted boreal species, until they too either level out or begin to decline slowly. Lastly, southern species do not begin to grow until even higher temperatures or growing degree day sums are reached, after which their growth rates rise the highest of all before leveling out and exhibiting essentially no decline thereafter, as temperatures and growing degree days continue to climb.

In light of these observations, it is clear that although the northern range limit of a woody species in the Northern Hemisphere is indeed determined by growth-retarding cool growing seasons and frost damage, the southern boundary of a tree's natural range is not determined by temperature, but by competition between the northern species and more southerly adapted species that have inherently greater growth rates.

Whenever significant long-term warming occurs, therefore, earth's coldest-adapted trees are presented with an opportunity to rapidly extend the cold-limited boundaries of their ranges northward in the Northern Hemisphere, as many studies have demonstrated they have done in the past and are doing now. Trees at the southern limits of their ranges, however, are little affected by the extra warmth. As time progresses, they may at some point begin to experience pressure from some of the faster-growing southern species encroaching upon their territory; but this potential challenge is by no means assured of quick success. As Loehle describes it:

> Seedlings of these southern species will not gain much competitive advantage from faster growth in the face of existing stands of northern species, because the existing adult trees have such an advantage due to light interception. Southern types must wait for gap replacement, disturbances, or stand break up to utilize their faster growth to gain a position in the stand. Thus the replacement of species will be delayed at least until the existing trees die, which can be hundreds of years … Furthermore, the faster growing southern species will be initially rare and must spread, perhaps across considerable distances or from initially scattered localities. Thus, the replacement of forest (southern types replacing northern types) will be an inherently slow process (several to many hundreds of years).

In summing up the significance of this situation, Loehle says "forests will not suffer catastrophic dieback due to increased temperatures but will rather be replaced gradually by faster growing types."

Another possibility that must be seriously considered is that northern or high-altitude forests will not be replaced at all by southern or low-altitude forests in a warming world. Rather, the two forest types may merge, creating entirely new forests of greater species diversity, such as those that existed during the warmer Tertiary Period of the Cenozoic Era, when in the western United States many montane taxa regularly grew among mixed conifers and broadleaf schlerophylls (Axelrod 1944a, 1944b, 1956, 1987), creating what could well be called super forest ecosystems, which Axelrod (1988) has described as "much richer than any that exist today."

Possibly helping warmer temperatures to produce this unique biological phenomenon during the Tertiary were the higher atmospheric CO_2 concentrations of that period (Volk, 1987), as has been suggested by Idso (1989). As documented extensively in Chapter 7, elevated concentrations of atmospheric CO_2 significantly stimulate plant growth rates (Kimball, 1983)—especially those of trees (Saxe et al., 1998; Idso and Kimball, 2001)—and they also greatly enhance water use efficiency (Feng, 1999). Even more important, however, is how atmospheric CO_2 enrichment alters plant photosynthetic and growth responses to rising temperatures.

It has long been known that photo-respiration—which can "cannibalize" as much as 40 percent to 50 percent of the recently produced photosynthetic products of C_3 plants (Wittwer, 1988)—becomes increasingly more pronounced as air temperature rises (Hanson and Peterson, 1986). It also has been established that photorespiration is increasingly more inhibited as the air's CO_2 content rises (Grodzinski et al., 1987). Hence, there is a greater potential for rising CO_2 concentrations to benefit C_3 plants at higher temperatures, as was demonstrated by the early experimental work of Idso et al. (1987) and Mortensen (1987), as well as by the theoretical work of Gifford (1992), Kirschbaum (1994) and Wilks et al. (1995). In an analysis of 42 experimental datasets collected by numerous scientists, Idso and Idso (1994) showed that the mean growth enhancement due to a 300-ppm increase in atmospheric CO_2 concentration rises from close to zero at an air temperature of 10°C to 100 percent (doubled growth) at approximately 38°C, while at higher temperatures the growth stimulation rises higher still, as also has been shown by Cannell and Thornley (1998).

Several studies have additionally demonstrated that atmospheric CO_2 enrichment tends to alleviate high-temperature stress in plants (Faria et al., 1996; Nijs and Impens, 1996; Vu et al., 1997); and it has been proven that at temperatures that are high enough to cause plants to die, atmospheric CO_2 enrichment can sometimes preserve their lives (Idso et al., 1989; Idso, 1995; Baker et al., 1992; Rowland-Bamford et al., 1996; Taub et al., 2000), just as it can often stave off their demise in the very dry conditions that typically accompany high air temperatures (Tuba et al., 1998; Hamerlynck, et al., 2000; Polley et al., 2002).

Box 1: The CO$_2$-Temperature-Growth Interaction

The growth-enhancing effects of elevated CO$_2$ typically increase with rising temperature. This phenomenon is illustrated by the data of Jurik *et al.* (1984), who exposed bigtooth aspen leaves to atmospheric CO$_2$ concentrations of 325 and 1,935 ppm and measured their photosynthetic rates at a number of different temperatures. In the figure below, we have reproduced their results and slightly extended the two relationships defined by their data to both warmer and cooler conditions.

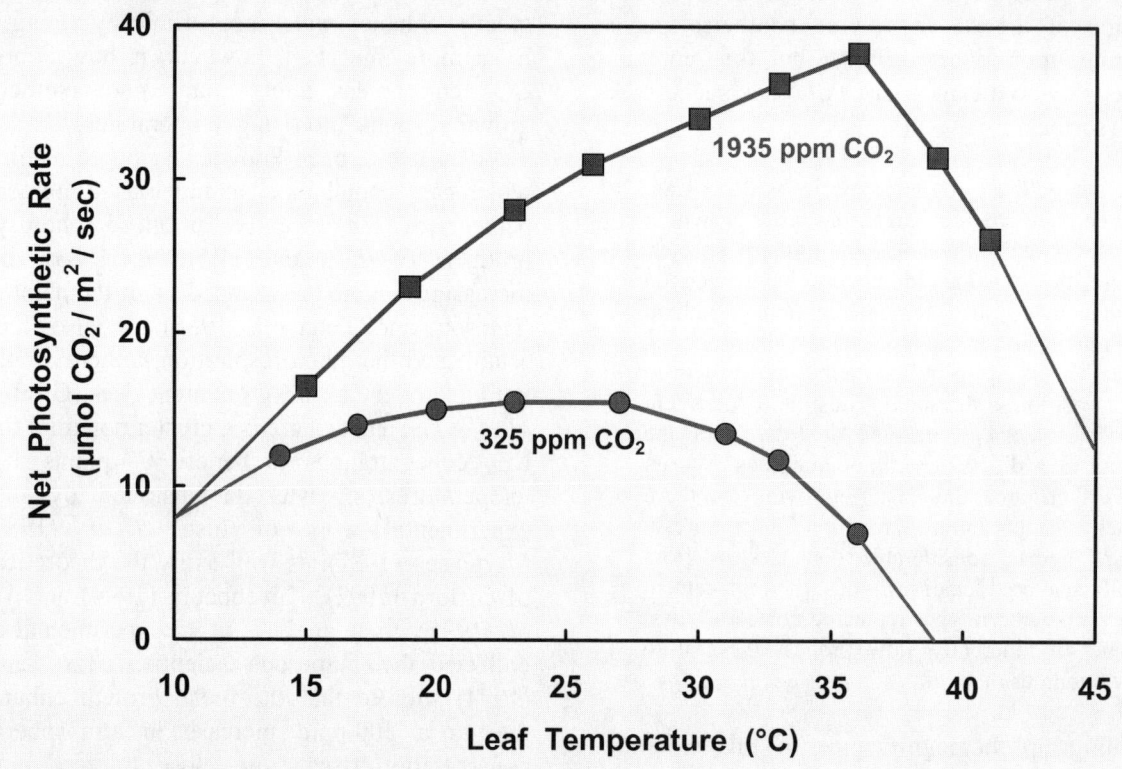

At 10°C, elevated CO$_2$ has essentially no effect on net photosynthesis in this particular species, as Idso and Idso (1994) have demonstrated is characteristic of plants in general. At 25°C, however, where the net photosynthetic rate of the leaves exposed to 325 ppm CO$_2$ is maximal, the extra CO$_2$ of this study boosts the net photosynthetic rate of the foliage by nearly 100 percent; and at 36°C, where the net photosynthetic rate of the leaves exposed to 1,935 ppm CO$_2$ is maximal, the extra CO$_2$ boosts the net photosynthetic rate of the foliage by a whopping 450 percent. The extra CO$_2$ increases the optimum temperature for net photosynthesis in this species by about 11°C, from 25°C in air of 325 ppm CO$_2$ to 36°C in air of 1,935 ppm CO$_2$.

In viewing the warm-temperature projections of the two relationships, it can also be seen that the transition from positive to negative net photosynthesis—which denotes a change from life-sustaining to life-depleting conditions—likely occurs somewhere in the vicinity of 39°C in air of 325 ppm CO$_2$ but somewhere in the vicinity of 50°C in air of 1,935 ppm CO$_2$. Hence, not only was the optimum temperature for the growth of bigtooth aspen greatly increased by the extra CO$_2$ of this experiment, so too was the temperature above which life cannot be sustained increased, and by about the same amount, i.e., 11°C.

A major consequence of these facts is that the optimum temperature (T$_{opt}$) for plant growth—the temperature at which plants photosynthesize and grow best—generally rises with atmospheric CO$_2$ enrichment (Berry and Bjorkman, 1980; Taiz and Zeiger, 1991). An example of this phenomenon is presented in Box 1, where it can be seen that the increase in atmospheric CO$_2$ concentration utilized in this particular study increases the optimum temperature for photosynthesis in this species from a broad maximum centered at 25°C in ambient air to a well-defined peak at about 36°C in CO$_2$-enriched air.

How much is plant optimum temperature typically increased by an extra 300 ppm of CO$_2$? Based largely on theoretical considerations, Long (1991) calculated that such an increase in the air's CO$_2$ concentration should increase T$_{opt}$, in the mean, by about 5°C, while McMurtrie and Wang (1993) calculated the increase at somewhere between 4° and 8°C.

The implication of the finding that plant optimum temperature rises so dramatically in response to increasing atmospheric CO_2 concentration is that if the planet were to warm in response to the ongoing rise in the air's CO_2 content—even to the degree predicted by the worst-case scenario of the IPCC (6.4°C by 2100)—the vast majority of earth's plants would likely not need to migrate towards cooler parts of the globe. Any warming would provide them an opportunity to move into regions that were previously too cold for them, but it would not force them to move, even at the hottest extremes of their ranges; for as the planet warmed, the rising atmospheric CO_2 concentration would significantly increase the temperatures at which most of earth's C_3 plants—which comprise fully 95 percent of the planet's vegetation (Drake, 1992)—function best, creating a situation where earth's plant life would actually prefer warmer conditions.

With respect to the C_4 and CAM plants that make up the remaining 5 percent of earth's vegetative cover, most of them are endemic to the planet's hotter environments (De Jong et al., 1982; Drake, 1989; Johnson et al., 1993), which according to the IPCC are expected to warm much less than the cooler regions of the globe. Hence, the planet's C_4 and CAM plants would not face as great a thermal challenge as earth's C_3 plants in a warming world. Nevertheless, the work of Chen et al. (1994) suggests they too may experience a modest increase in their optimum temperatures as the air's CO_2 content rises (a 1.5°C increase in response to a 350-ppm increase in atmospheric CO_2 concentration). Consequent-ly, and in view of the non-CO_2-related abilities of earth's vegetation to adapt to rising temperatures discussed in the previous section, plants of all photosynthetic persuasions should be able to successfully adapt to any future warming that could be caused by the enhanced greenhouse effect. So obvious is this conclusion, in fact, that Cowling (1999) has bluntly stated, "maybe we should be less concerned about rising CO_2 and rising temperatures and more worried about the possibility that future atmospheric CO_2 will suddenly stop increasing, while global temperatures continue rising."

James Hansen, in testimony to a U.S. House of Representatives committee, has claimed that life in alpine regions is "in danger of being pushed off the planet" because of rising temperatures. In July and August 2003, a team of three researchers investigated this scenario by resurveying the floristic composition of the uppermost 10 meters of 10 mountain summits in the Swiss Alps (Walther et al., 2005), applying the same methodology used in earlier surveys that were conducted there in 1905 (Rubel, 1912) and 1985 (Hofer, 1992). This analysis covered the bulk of the Little Ice Age-to-Current Warm Period transition, and it revealed that plants of many species had indeed marched up the sides of the mountains, as the earth in general—and the Swiss Alps in particular—had warmed. Of even greater significance, however, was the fact that not a single mountaintop species was "pushed off the planet." Between 1905 and 1985 the mean number of species observed on the 10 mountaintops rose by 86 percent, while by 2003 it had risen by 138 percent, providing, in the words of the researchers who conducted the work, "an enrichment of the overall summit plant diversity."

Another research team studied the same phenomenon on 12 mountaintops in the Swiss Alps (Holzinger et al., 2008), making complete inventories of vascular plant species that were present there in 2004, while following—"as accurately as possible," in their words—the same ascension paths used by other researchers in 1885, 1898, 1912, 1913, and 1958, after which they compared their findings with those of the earlier studies. By these means, they detected upward plant migration rates on the order of several meters per decade, which phenomena increased vascular plant species richness at the mountains' summits by 11 percent per decade over the 120-year study period. This finding, in their words, "agrees well with other investigations from the Alps, where similar changes have been detected," and they cited, in this regard, four additional studies (Grabherr et al., 1994; Pauli et al., 2001; Camenisch, 2002; Walther, 2003).

Another pertinent study was conducted by a researcher who analyzed altitudinal shifts in the ranges of alpine and subalpine plants in the mountains of west-central Sweden (Kullman, 2002), where air temperature had risen approximately 1°C over the past hundred years. This work revealed that since the early twentieth century, alpine and subalpine plants had migrated upslope by an average of 200 m. Most importantly, it also indicated, according to the scientist who did the work, that "no species have yet become extinct from the highest elevations," which finding was said by the researcher to "converge with observations in other high-mountain regions worldwide," in support of which statement five more new studies were cited (Keller et al., 2000; Kullman, 2002; Klanderud and Birks, 2003; Virtanen et al., 2003; Lacoul and Freedman, 2006).

In light of these many real-world findings, it is clear that warmer temperatures have not led to an increase in plant extinctions, and in fact have led to the opposite effect: greater plant growth and diversity in larger areas of the globe. Nor did the warming of the past century—which the IPCC claims was unprecedented over the past millennium—push any upward-migrating plants "off the planet" at the tops of its mountains. One of the most highly promoted hypothetical scenarios of the IPCC is contradicted by real-world data.

References

Axelrod, D.I. 1944a. The Oakdale flora (California). *Carnegie Institute of Washington Publication* **553**: 147-166.

Axelrod, D.I. 1944b. The Sonoma flora (California). *Carnegie Institute of Washington Publication* **553**: 167-200.

Axelrod, D.I. 1956. Mio-Pliocene floras from west-central Nevada. *University of California Publications in the Geological Sciences* **33**: 1-316.

Axelrod, D.I. 1987. The Late Oligocene Creede flora, Colorado. *University of California Publications in the Geological Sciences* **130**: 1-235.

Axelrod, D.I. 1988. An interpretation of high montane conifers in western Tertiary floras. *Paleobiology* **14**: 301-306.

Baker, J.T., Allen Jr., L.H. and Boote, K.J. 1992. Response of rice to carbon dioxide and temperature. *Agricultural and Forest Meteorology* **60**: 153-166.

Battaglia, M., Beadle, C. and Loughhead, S. 1996. Photosynthetic temperature responses of *Eucalyptus globulus* and *Eucalyptus nitens*. *Tree Physiology* **16**: 81-89.

Berry, J. and Bjorkman, O. 1980. Photosynthetic response and adaptation to temperature in higher plants. *Annual Review of Plant Physiology* **31**: 491-543.

Bjorkman, O., Badger, M. and Armond, P.A. 1978. Thermal acclimation of photosynthesis: effect of growth temperature on photosynthetic characteristics and components of the photosynthetic apparatus in *Nerium oleander*. *Carnegie Institution of Washington Yearbook* **77**: 262-276.

Camenisch, M. 2002. Veranderungen der Gipfelflora im Bereich des Schweizerischen Nationalparks: Ein Vergleich uber die letzen 80 Jahre. *Jahresber nat forsch Ges Graubunden* **111**: 27-37.

Cannell, M.G.R. and Thornley, J.H.M. 1998. Temperature and CO_2 responses of leaf and canopy photosynthesis: a clarification using the non-rectangular hyperbola model of photosynthesis. *Annals of Botany* **82**: 883-892.

Chen, D-X., Coughenour, M.B., Knapp, A.K. and Owensby, C.E. 1994. Mathematical simulation of C_4 grass photosynthesis in ambient and elevated CO_2. *Ecological Modelling* **73**: 63-80.

Cowling, S.A. 1999. Plants and temperature—CO_2 uncoupling. *Science* **285**: 1500-1501.

De Jong, T.M., Drake, B.G. and Pearcy, R.W. 1982. Gas exchange responses of Chesapeake Bay tidal marsh species under field and laboratory conditions. *Oecologia* **52**: 5-11.

Drake, B.G. 1989. Photosynthesis of salt marsh species. *Aquatic Botany* **34**: 167-180.

Drake, B.G. 1992. Global warming: The positive impact of rising carbon dioxide levels. *Eco-Logic* **1**(3): 20-22.

Dressler, R.L. 1954. Some floristic relationships between Mexico and the United States. *Rhodora* **56**: 81-96.

El-Sharkawy, M.A., De Tafur, S.M. and Cadavid, L.F. 1992. Potential photosynthesis of cassava as affected by growth conditions. *Crop Science* **32**: 1336-1342.

Faria, T., Wilkins, D., Besford, R.T., Vaz, M., Pereira, J.S. and Chaves, M.M. 1996. Growth at elevated CO_2 leads to down-regulation of photosynthesis and altered response to high temperature in *Quercus suber* L. seedlings. *Journal of Experimental Botany* **47**: 1755-1761.

Feng, X. 1999. Trends in intrinsic water-use efficiency of natural trees for the past 100-200 years: A response to atmospheric CO_2 concentration. *Geochimica et Cosmochimica Acta* **63**: 1891-1903.

Gauslaa, Y. 1984. Heat resistance and energy budget in different Scandinavian plants. *Holarctic Ecology* **7**: 1-78.

Gifford, R.M. 1992. Interaction of carbon dioxide with growth-limiting environmental factors in vegetative productivity: Implications for the global carbon cycle. *Advances in Bioclimatology* **1**: 24-58.

Grabherr, G, Gottfried, M. and Pauli, H. 1994. Climate effects on mountain plants. *Nature* **369**: 448.

Grodzinski, B., Madore, M., Shingles, R.A. and Woodrow, L. 1987. Partitioning and metabolism of photorespiratory intermediates. In: Biggins, J. (Ed.) *Progress in Photosynthesis Research*. W. Junk, The Hague, The Netherlands, pp. 645-652.

Hamerlynck, E.P., Huxman, T.E., Loik, M.E. and Smith, S.D. 2000. Effects of extreme high temperature, drought and elevated CO_2 on photosynthesis of the Mojave Desert evergreen shrub, *Larrea tridentata*. *Plant Ecology* **148**: 183-193.

Hanson, K.R. and Peterson, R.B. 1986. Regulation of photorespiration in leaves: Evidence that the fraction of ribulose bisphosphate oxygenated is conserved and stoichiometry fluctuates. *Archives for Biochemistry and Biophysics* **246**: 332-346.

Hofer, H.R. 1992. Veranderungen in der Vegetation von 14 Gipfeln des Berninagebietes zwischen 1905 und 1985. *Ber. Geobot. Inst. Eidgenoss. Tech. Hochsch. Stift. Rubel Zur* **58**: 39-54.

Holzinger, B., Hulber, K., Camenisch, M. and Grabherr, G. 2008. Changes in plant species richness over the last century in the eastern Swiss Alps: elevational gradient, bedrock effects and migration rates. *Plant Ecology* **195**: 179-196.

Idso, K.E. and Idso, S.B. 1994. Plant responses to atmospheric CO_2 enrichment in the face of environmental constraints: A review of the past 10 years' research. *Agricultural and Forest Meteorology* **69**: 153-203.

Idso, S.B. 1989. Carbon Dioxide and Global Change: Earth in Transition. IBR Press, Tempe, AZ.

Idso, S.B. 1995. *CO_2 and the Biosphere: The Incredible Legacy of the Industrial Revolution.* Kuehnast Lecture Series, Special Publication. Department of Soil, Water & Climate, University of Minnesota, St. Paul, MN.

Idso, S.B., Allen, S.G., Anderson, M.G. and Kimball, B.A. 1989. Atmospheric CO_2 enrichment enhances survival of *Azolla* at high temperatures. *Environmental and Experimental Botany* **29**: 337-341.

Idso, S.B. and Kimball, B.A. 2001. CO_2 enrichment of sour orange trees: 13 years and counting. *Environmental and Experimental Botany* **46**: 147-153.

Idso, S.B., Kimball, B.A., Anderson, M.G. and Mauney, J.R. 1987. Effects of atmospheric CO_2 enrichment on plant growth: The interactive role of air temperature. *Agriculture, Ecosystems and Environment* **20**: 1-10.

Johnson, H.B., Polley, H.W. and Mayeux, H.S. 1993. Increasing CO_2 and plant-plant interactions: Effects on natural vegetation. *Vegetatio* **104-105**: 157-170.

Jurik, T.W., Weber, J.A. and Gates, D.M. 1984. Short-term effects of CO_2 on gas exchange of leaves of bigtooth aspen (*Populus grandidentata*) in the field. *Plant Physiology* **75**: 1022-1026.

Kappen, L. 1981. Ecological significance of resistance to high temperature. In: Lange, O.L., Nobel, P.S., Osmond, C.B. and Ziegler, H. (Eds.) *Physiological Plant Ecology. I. Response to the Physical Environment.* Springer-Verlag, New York, NY, pp. 439-474.

Keller, F., Kienast, F. and Beniston, M. 2000. Evidence of response of vegetation to environmental change on high-elevation sites in the Swiss Alps. *Regional Environmental Change* **1**: 70-77.

Kelly, C.K., Chase, M.W., de Bruijn, A., Fay, M.F. and Woodward, F.I. 2003. Temperature-based population segregation in birch. *Ecology Letters* **6**: 87-89.

Kimball, B.A. 1983. Carbon dioxide and agricultural yield: An assemblage and analysis of 430 prior observations. *Agronomy Journal* **75**: 779-788.

Kirschbaum, M.U.F. 1994. The sensitivity of C_3 photosynthesis to increasing CO_2 concentration: a theoretical analysis of its dependence on temperature and background CO_2 concentration. *Plant, Cell and Environment* **17**: 747-754.

Klanderud, K. and Birks, H.J.B. 2003. Recent increases in species richness and shifts in altitudinal distributions of Norwegian mountain plants. *Holocene* **13**: 1-6.

Kullman, L. 2002. Rapid recent range-margin rise of tree and shrub species in the Swedish Scandes. *Journal of Ecology* **90**: 68-77.

Kullman, L. 2007. Long-term geobotanical observations of climate change impacts in the Scandes of West-Central Sweden. *Nordic Journal of Botany* **24**: 445-467.

Lacoul, P. and Freedman, B. 2006. Recent observation of a proliferation of *Ranunculus trichophyllus* Chaix. in high-altitude lakes of Mount Everest Region. *Arctic, Antarctic and Alpine Research* **38**: 394-398.

Lange, O.L. and Lange, R. 1963. Untersuchungen uber Blattemperaturen, Transpiration und Hitzeresistenz an Pflanzen mediterraner Standorte (Costa Brava, Spanien). *Flora* **153**: 387-425.

Levitt, J. 1980. Responses of Plants to Environmental Stresses. Vol.1. Chilling, Freezing, and High Temperature Stresses. Academic Press, New York, NY.

Loehle, C. 1998. Height growth rate tradeoffs determine northern and southern range limits for trees. *Journal of Biogeography* **25**: 735-742.

Long, S.P. 1991. Modification of the response of photosynthetic productivity to rising temperature by atmospheric CO_2 concentrations: Has its importance been underestimated? *Plant, Cell and Environment* **14**: 729-739.

McMurtrie, R.E. and Wang, Y.-P. 1993. Mathematical models of the photosynthetic response of tree stands to rising CO_2 concentrations and temperatures. *Plant, Cell and Environment* **16**: 1-13.

Mooney, H.A. and West, M. 1964. Photosynthetic acclimation of plants of diverse origin. *American Journal of Botany* **51**: 825-827.

Mortensen, L.M. 1987. Review: CO_2 enrichment in greenhouses. Crop responses. *Scientia Horticulturae* **33**: 1-25.

Nijs, I. and Impens, I. 1996. Effects of elevated CO_2 concentration and climate-warming on photosynthesis during winter in *Lolium perenne*. *Journal of Experimental Botany* **47**: 915-924.

Pauli, H., Gottfried, M. and Grabherr, G. 2001. High summits of the Alps in a changing climate. The oldest observation series on high mountain plant diversity in Europe. In: Walther, G.R., Burga, C.A. and Edwards, P.J. (Eds.) *Fingerprints of climate change—Adapted behaviour and shifting species ranges*. Kluwer Academic Publisher, New York, NY, USA, pp. 139-149.

Polley, H.W., Tischler, C.R., Johnson, H.B. and Derner, J.D. 2002. Growth rate and survivorship of drought: CO_2 effects on the presumed tradeoff in seedlings of five woody legumes. *Tree Physiology* **22**: 383-391.

Rowland-Bamford, A.J., Baker, J.T., Allen Jr., L.H. and Bowes, G. 1996. Interactions of CO_2 enrichment and temperature on carbohydrate accumulation and partitioning in rice. *Environmental and Experimental Botany* **36**: 111-124.

Rubel, E. 1912. Pflanzengeographische Monographie des Berninagebietes. Engelmann, Leipzig, DE.

Saxe, H., Ellsworth, D.S. and Heath, J. 1998. Tree and forest functioning in an enriched CO_2 atmosphere. *New Phytologist* **139**: 395-436.

Seemann, J.R., Berry, J.A. and Downton, J.S. 1984. Photosynthetic response and adaptation to high temperature in desert plants. A comparison of gas exchange and fluorescence methods for studies of thermal tolerance. *Plant Physiology* **75**: 364-368.

Strain, B.R., Higginbottom, K.O. and Mulroy, J.C. 1976. Temperature preconditioning and photosynthetic capacity of *Pinus taeda* L. *Photosynthetica* **10**: 47-53.

Taiz, L. and Zeiger, E. 1991. *Plant Physiology*. Benjamin-Cummings, Redwood City, CA.

Taub, D.R., Seeman, J.R. and Coleman, J.S. 2000. Growth in elevated CO_2 protects photosynthesis against high-temperature damage. *Plant, Cell and Environment* **23**: 649-656.

Tuba, Z., Csintalan, Z., Szente, K., Nagy, Z. and Grace, J. 1998. Carbon gains by desiccation-tolerant plants at elevated CO_2. *Functional Ecology* **12**: 39-44.

Veres, J.S. and Williams III, G.J. 1984. Time course of photosynthetic temperature acclimation in *Carex eleocharis* Bailey. *Plant, Cell and Environment* **7**: 545-547.

Virtanen, R., Eskelinen, A. and Gaare, E. 2003. Long-term changes in alpine plant communities in Norway and Finland. In: Nagy, L., Grabherr, G., Korner, C. and Thompson, D.B.A. (Eds.) *Alpine Biodiversity in Europe*. Springer, Berlin, Germany, pp. 411-422.

Volk, T. 1987. Feedbacks between weathering and atmospheric CO2 over the last 100 million years. *American Journal of Science* **287**: 763-779.

Vu, J.C.V., Allen Jr., L.H., Boote, K.J. and Bowes, G. 1997. Effects of elevated CO_2 and temperature on photosynthesis and Rubisco in rice and soybean. *Plant, Cell and Environment* **20**: 68-76.

Walther, G.R. 2003. Plants in a warmer world. *Perspectives in Plant Ecology, Evolution and Systematics* **6**: 169-185.

Walther, G.-R., Beissner, S. and Burga, C.A. 2005. Trends in the upward shift of alpine plants. *Journal of Vegetation Science* **16**: 541-548.

Wilks, D.S., Wolfe, D.W. and Riha, S.J. 1995. Simple carbon assimilation response functions from atmospheric CO_2, and daily temperature and shortwave radiation. *Global Change Biology* **1**: 337-346.

Wittwer, S.H. 1988. *The Greenhouse Effect*. Carolina Biological Supply, Burlington, NC.

Woodward, F.I. 1987. *Climate and Plant Distribution*. Cambridge University Press, Cambridge, UK.

Woodward, F.I. 1988. Temperature and the distribution of plant species and vegetation. In: Long, S.P. and Woodward, F.I. (Eds.), *Plants and Temperature. Vol. 42, The Company of Biologists Limited*, Cambridge, UK, pp. 59-75.

8.3. Coral Reefs

According to the IPCC, "many studies incontrovertibly link coral bleaching to warmer sea surface temperature ... and mass bleaching and coral mortality often results beyond key temperature thresholds" (IPCC 2007-II, p. 235). "Modelling," the IPCC goes on to say, "predicts a phase switch to algal dominance on the Great Barrier Reef and Caribbean reefs in 2030 to 2050." The IPCC further claims that "coral reefs will also be affected by rising atmospheric CO_2 concentrations ... resulting in declining calcification" (Ibid.).

In the following pages we review the scientific literature on coral reefs in an effort to determine if the ongoing rise in the air's CO_2 content, rising temperatures, or rising sea levels pose a threat to these incomparable underwater ecosystems. Because the fate of the earth's corals has become so prominent in the debate over climate change and because our findings are so entirely at odds with those of the IPCC, we present a brief summary of our key findings here:

- There is no simple linkage between high temperatures and coral bleaching.

- As living entities, corals are not only acted upon by the various elements of their environment, they also react or respond to them. And when changes in environmental factors pose a challenge to their continued existence, they sometimes take major defensive or adaptive actions to ensure their survival.

- A particularly ingenious way coral respond to environmental stress is to replace the zooxanthellae expelled by the coral host during a stress-induced bleaching episode by one or more varieties of zooxanthellae that are more tolerant of the stress that caused the bleaching.

- The persistence of coral reefs through geologic time—when temperatures were as much as $10°$-$15°C$ warmer than at present, and atmospheric CO_2 concentrations were two to seven times higher than they are currently—provides substantive evidence that these marine entities can successfully adapt to a dramatically changing global environment. Thus, the recent die-off of many corals cannot be due solely, or even mostly, to global warming or the modest rise in atmospheric CO_2 concentration over the course of the Industrial Revolution.

- The 18- to 59-cm warming-induced sea-level rise that is predicted for the coming century by the IPCC—which could be greatly exaggerated if predictions of CO_2-induced global warming are wrong—falls well within the range (2 to 6 mm per year) of typical coral vertical extension rates, which exhibited a modal value of 7 to 8 mm per year during the Holocene and can be more than double that value in certain branching corals. Rising sea levels should therefore present no difficulties for coral reefs. In fact, rising sea levels may have a positive effect on reefs, permitting increased coral growth in areas that have already reached the upward limit imposed by current sea levels.

- The rising CO_2 content of the atmosphere may induce changes in ocean chemistry (pH) that could slightly reduce coral calcification rates; but potential positive effects of hydrospheric CO_2 enrichment may more than compensate for this modest negative phenomenon.

- Theoretical predictions indicate that coral calcification rates should decline as a result of increasing atmospheric CO_2 concentrations by as much as 40 percent by 2100. However, real-world observations indicate that elevated CO_2 and elevated temperatures are having the opposite effect.

References

IPCC, 2007-II. *Climate Change 2007: Impacts, Adaptation and Vulnerability. Contribution of Working Group II to the Fourth Assessment Report of the Intergovernmental Panel on Climate Change*. M.L. Parry, O.F. Canziani, J.P. Palutikof, P.J. van der Linden and C.E. Hanson. (Eds.) Cambridge University Press, Cambridge, UK.

8.3.1. Indirect Threats

Coral bleaching ranks as probably the most frequently cited indirect negative consequence believed to result from CO_2-induced global warming. It is a phenomenon that is characterized by a loss of color in certain reef-building corals that occurs when the algal symbionts, or *zooxanthellae*, living within the host corals are subjected to various stresses and expelled from them, resulting in a loss of photosynthetic pigments from the coral colony. If the stress is mild, or short in duration, the affected corals often recover and regain their normal complement of zooxanthellae. However, if the stress is prolonged, or extreme, the corals eventually die, being deprived of their primary food source.

We begin our review of the subject by discussing the many suspected causes of coral bleaching, almost all of which have been attributed (often implausibly) to CO_2-induced global warming. Then, we examine the possibility that corals can adapt to the various environmental threats they face, after which we explore whether the widespread bleaching events seen in recent decades are indeed caused by global warming. We conclude our discussion of the major indirect threats facing modern coral reefs by examining the threat of rising sea levels, which the IPCC predicts will occur over the course of the twenty-first century.

8.3.1.1. Coral Bleaching

8.3.1.1.1. Temperature Effects

One of the most frequently cited causes of coral bleaching is anomalously high water temperature (Linden, 1998). The origin of this attribution can be traced to the strong El Niño event of 1982-83, in which widespread bleaching was reported in corals exposed to unusually high surface water temperatures (Glynn, 1988). Since that time, a number of other such observations have been made (Cook *et al.*, 1990; Glynn 1991; Montgomery and Strong, 1994; Brown *et al.*, 1996); and several laboratory studies have demonstrated that elevated seawater temperatures can indeed induce bleaching in corals (Hoegh-Guldberg and Smith, 1989; Jokiel and Coles, 1990; Glynn and D'Croz, 1990).

However, just as anomalously high seawater temperatures have been found to be correlated with coral reef bleaching events, so too have anomalously low seawater temperatures been identified with this phenomenon (Walker *et al.*, 1982; Coles and Fadlallah, 1990; Muscatine *et al.*, 1991; Gates *et al.*, 1992; Saxby *et al.*, 2003; Hoegh-Guldberg and Fine 2004; Yu *et al.*, 2004). These observations suggest that the crucial link between temperature and coral reef bleaching may not reside in the absolute temperature of the water surrounding the corals, but in the rapidity with which the temperature either rises above or falls below the temperature regime to which the corals are normally adapted. Winter *et al.* (1998), for example, studied relationships between coral bleaching and nine different temperature indices, concluding that although "prolonged heat stress may be an important precondition for bleaching to occur," sharp temperature changes act as the "immediate trigger."

In a related study, Jones (1997) reported coral bleaching on a portion of Australia's Great Barrier Reef just after average daily sea water temperature rose by 2.5°C over the brief period of eight days. Likewise, Kobluk and Lysenko (1994) observed severe coral bleaching following an 18-hour *decline* of 3°C in seawater temperature. Because the corals studied by the latter researchers had experienced

massive bleaching two years earlier as a result of an anomalous 4°C *increase* in water temperature, the authors concluded that coral bleaching is more a function the rapidity of a temperature change than it is of the absolute magnitude or sign of the change, i.e., heating or cooling.

Further evidence that high or low seawater temperatures *per se* are not the critical factors in producing coral bleaching is provided by Podesta and Glynn (1997), who examined a number of temperature-related indices of surface waters in the vicinity of Panama over the period 1970-1994. Their analysis revealed that for the two years of highest maximum monthly sea surface temperature, 1972 and 1983, coral bleaching was reported only in 1983, while 1972 produced no bleaching whatsoever, in spite of the fact that water temperatures that year were just as high as they were in 1983.

Additional information on this topic, including reviews of newer publications as they become available, can be found at http://www.co2 science.org/subject/c/bleachingtemp.php.

References

Brown, B.E., Dunne, R.P. and Chansang, H. 1996. Coral bleaching relative to elevated seawater temperature in the Andaman Sea (Indian Ocean) over the last 50 years. *Coral Reefs* **15**: 151-152.

Coles, S.L. and Fadlallah, Y.H. 1990. Reef coral survival and mortality at low temperatures in the Arabian Gulf: New species-specific lower temperature limits. *Coral Reefs* **9**: 231-237.

Cook, C.B., Logan, A., Ward, J., Luckhurst, B. and Berg Jr., C.J. 1990. Elevated temperatures and bleaching on a high latitude coral reef: The 1988 Bermuda event. *Coral Reefs* **9**: 45-49.

Gates, R.D., Baghdasarian, G. and Muscatine, L. 1992. Temperature stress causes host cell detachment in symbiotic cnidarians: Implication for coral bleaching. *Biological Bulletin* **182**: 324-332.

Glynn, P.W. 1988. El Niño-Southern Oscillation 1982-83: Nearshore population, community, and ecosystem responses. *Annual Review of Ecology and Systematics* **19**: 309-345.

Glynn, P.W. 1991. Coral bleaching in the 1980s and possible connections with global warming trends. *Ecology and Evolution* **6**: 175-179.

Glynn, P.W. and D'Croz, L. 1990. Experimental evidence for high temperature stress as the cause of El-Niño-coincident coral mortality. *Coral Reefs* **8**: 181-190.

Hoegh-Guldberg, O. and Fine, M. 2004. Low temperatures cause coral bleaching. *Coral Reefs* **23**: 444.

Hoegh-Guldberg, O. and Smith, G.J. 1989. The effect of sudden changes in temperature, light and salinity on the population density and export of zooxanthellae from the reef corals *Stylophora pistillata* Esper. and *Seriatopora hystrix* Dana. *Journal of Experimental Marine Biology and Ecology* **129**: 279-303.

IPCC 2007. Climate Change 2007: The Physical Science Basis. Contribution of Working Group I to the Fourth Assessment Report of the Intergovernmental Panel on Climate Change. Cambridge University Press.

Jokiel, P.L. and Coles, S.L. 1990. Response of Hawaiian and other Indo-Pacific reef corals to elevated sea temperatures. *Coral Reefs* **8**:155-162.

Jones, R.J. 1997. Changes in zooxanthellar densities and chlorophyll concentrations in corals during and after a bleaching event. *Marine Ecology Progress Series* **158**: 51-59.

Kobluk, D.R. and Lysenko, M.A. 1994. "Ring" bleaching in Southern Caribbean *Agaricia agaricites* during rapid water cooling. *Bulletin of Marine Science* **54**: 142-150.

Linden, O. 1998. Coral mortality in the tropics: Massive causes and effects. *Ambio* **27**: 588.

Montgomery, R.S. and Strong, A.E. 1994. Coral bleaching threatens oceans life. *EOS* **75**: 145-147.

Muscatine, L., Grossman, D. and Doino, J. 1991. Release of symbiotic algae by tropical sea anemones and corals after cold shock. *Marine Ecology Progress Series* **77**: 233-243.

Podesta, G.P. and Glynn, P.W. 1997. Sea surface temperature variability in Panama and Galapagos: Extreme temperatures causing coral bleaching. *Journal of Geophysical Research* **102**: 15,749-15,759.

Saxby, T., Dennison, W.C. and Hoegh-Guldberg, O. 2003. Photosynthetic responses of the coral Montipora digitata to cold temperature stress. *Marine Ecology Progress Series* **248**: 85-97.

Walker, N.D., Roberts, H.H., Rouse Jr., L.J. and Huh, O.K. 1982. Thermal history of reef-associated environments during a record cold-air outbreak event. *Coral Reefs* **1**: 83-87.

Winter, A., Appeldoorn, R.S., Bruckner, A., Williams Jr., E.H. and Goenaga, C. 1998. Sea surface temperatures and coral reef bleaching off La Parguera, Puerto Rico (northeast Caribbean Sea). *Coral Reefs* **17**: 377-382.

Yu, K.-F., Zhao, J.-X., Liu, T.-S., Wei, G.-J., Wang, P.X. and Collerson, K.D. 2004. High-frequency winter cooling and reef coral mortality during the Holocene climatic optimum. *Earth and Planetary Science Letters* **224**: 143-155.

8.3.1.1.2. Solar Radiation Effects

The link between solar radiation and coral reef bleaching goes back more than a century to when MacMunn (1903) postulated that ultraviolet radiation could be potentially damaging to corals. It wasn't until half a century later, however, that scientists began to confirm this suspicion (Catala-Stucki, 1959; Siebeck, 1988; Gleason and Wellington, 1995).

Many investigators of the solar irradiance-coral reef bleaching link have studied the phenomenon by transplanting reef corals from deep to shallow waters. Gleason and Wellington (1993), for example, transplanted samples of the reef-building coral *Montastrea annularis* from a depth of 24 meters to depths of 18 and 12 meters. Using sheets of acrylic plastic to block out ultraviolet radiation on some of the coral samples, they found that the shielded corals experienced less bleaching than the unshielded corals, and that the unshielded corals at the 12-meter depth had significantly lower amounts of zooxanthellae and chlorophyll per square centimeter than all other treatment and control groups. Likewise, Hoegh-Guldberg and Smith (1989) reported bleaching in the corals *Stylophora pistillata* and *Seriatopora hystrix* when they were moved from a depth of 6 meters to 1.2 meters. Vareschi and Fricke (1986) obtained similar results when moving *Plerogyra sinuosa* from a depth of 25 meters to 5 meters. As in the case of temperature stress, however, Glynn (1996) notes that artificially reduced light levels also have been observed to cause coral bleaching.

A number of laboratory studies have provided additional evidence for a link between intense solar irradiance and coral reef bleaching, but identifying a specific wavelength or range of wavelengths as the cause of the phenomenon has been a difficult task. Fitt and Warner (1995), for example, reported that the most significant decline in symbiont photosynthesis in the coral *Montastrea annularis* occurred when it was exposed to ultraviolet and blue light; but other studies have reported coral bleaching to be most severe at shorter ultraviolet wavelengths (Droller *et al.*, 1994; Gleason and Wellington, 1995). Still others have found it to be most strongly expressed at longer photosynthetically active wavelengths (Lesser and Shick, 1989; Lesser *et al.*, 1990; Brown *et al.*, 1994).

As additional studies provided evidence for a solar-induced mechanism of coral reef bleaching (Brown *et al.*, 1994; Williams *et al.*, 1997; Lyons *et al.*, 1998), some also provided evidence for a solar radiation-temperature stress synergism (Gleason and Wellington, 1993; Rowan *et al.*, 1997; Jones *et al.*, 1998). There have been a number of situations, for example, in which corals underwent bleaching when changes in both of these parameters combined to produce particularly stressful conditions (Lesser *et al.*, 1990; Glynn *et al.*, 1992; Brown *et al.*, 1995), such as during periods of low wind velocity and calm seas, which favor the intense heating of shallow waters and concurrent strong penetration of solar radiation.

This two-parameter interaction has much to recommend it as a primary cause of coral bleaching. It is, in fact, the mechanism favored by Hoegh-Guldberg (1999), who claimed—in one of the strongest attempts made to that point in time to portray global warming as the cause of bleaching in corals—that "coral bleaching occurs when the photosynthetic symbionts of corals (zooxanthellae) become increasingly vulnerable to damage by light at higher than normal temperatures." As we shall see, however, the story is considerably more complicated.

Additional information on this topic, including reviews of newer publications as they become available, can be found at http://www.co2science.org/subject/c/bleachingsolar.php.

References

Brown, B.E., Dunne, R.P., Scoffin, T.P. and Le Tissier, M.D.A. 1994. Solar damage in intertidal corals. *Marine Ecology Progress Series* **105**: 219-230.

Brown, B.E., Le Tissier, M.D.A. and Bythell, J.C. 1995. Mechanisms of bleaching deduced from histological studies of reef corals sampled during a natural bleaching event. *Marine Biology* **122**: 655-663.

Brown, B.E., Le Tissier, M.D.A. and Dunne, R.P. 1994. Tissue retraction in the scleractinian coral *Coeloseris mayeri*, its effect upon coral pigmentation, and preliminary implications for heat balance. *Marine Ecology Progress Series* **105**: 209-218.

Catala-Stucki, R. 1959. Fluorescence effects from corals irradiated with ultra-violet rays. *Nature* **183**: 949.

Droller, J.H., Faucon, M., Maritorena, S. and Martin, P.M.V. 1994. A survey of environmental physico-chemical parameters during a minor coral mass bleaching event in Tahiti in 1993. *Australian Journal of Marine and Freshwater Research* **45**: 1149-1156.

Fitt, W.K. and Warner, M.E. 1995. Bleaching patterns of four species of Caribbean reef corals. *Biological Bulletin* **189**: 298-307.

Gleason, D.F. and Wellington, G.M. 1993. Ultraviolet radiation and coral bleaching. *Nature* **365**: 836-838.

Gleason, D.F. and Wellington, G.M. 1995. Variation in UVB sensitivity of planula larvae of the coral *Agaricia agaricites* along a depth gradient. *Marine Biology* **123**: 693-703.

Glynn, P.W. 1996. Coral reef bleaching: facts, hypotheses and implications. *Global Change Biology* **2**: 495-509.

Glynn, P.W., Imai, R., Sakai, K., Nakano, Y. and Yamazato, K. 1992. Experimental responses of Okinawan (Ryukyu Islands, Japan) reef corals to high sea temperature and UV radiation. *Proceedings of the 7th International Coral Reef Symposium* **1**: 27-37.

Hoegh-Guldberg, O. 1999. Climate change, coral bleaching and the future of the world's coral reefs. *Marine and Freshwater Research* **50**: 839-866.

Hoegh-Guldberg, O. and Smith, G.J. 1989. The effect of sudden changes in temperature, light and salinity on the population density and export of zooxanthellae from the reef corals *Stylophora pistillata* Esper. and *Seriatopora hystrix* Dana. *Journal of Experimental Marine Biology and Ecology* **129**: 279-303.

Jones, R.J., Hoegh-Guldberg, O., Larkum, A.W.D. and Schreiber, U. 1998. Temperature-induced bleaching of corals begins with impairment of the CO_2 fixation mechanism in zooxanthellae. *Plant, Cell and Environment* **21**: 1219-1230.

Lesser, M.P. and Shick, J.M. 1989. Effects of irradiance and ultraviolet radiation on photoadaptation in the zooxanthellae of *Aiptasia pallida*: Primary production, photoinhibition, and enzymatic defense against oxygen toxicity. *Marine Biology* **102**: 243-255.

Lesser, M.P., Stochaj, W.R., Tapley, D.W. and Shick, J.M. 1990. Bleaching in coral reef anthozoans: Effects of irradiance, ultraviolet radiation, and temperature on the activities of protective enzymes against active oxygen. *Coral Reefs* **8**: 225-232.

Lyons, M.M., Aas, P., Pakulski, J.D., Van Waasbergen, L., Miller, R.V., Mitchell, D.L. and Jeffrey, W.H. 1998. DNA damage induced by ultraviolet radiation in coral-reef microbial communities. *Marine Biology* **130**: 537-543.

MacMunn, C.A. 1903. On the pigments of certain corals, with a note on the pigment of an asteroid. Gardiner, J.S. (Ed.) *The fauna and geography of the Maldive and Laccadive Archipelagoes*. Cambridge, UK: Cambridge University Press. Vol. 1, pp. 184-190.

Rowan, R., Knowlton, N., Baker, A. and Jara, J. 1997. Landscape ecology of algal symbionts creates variation in episodes of coral bleaching. *Nature* **388**: 265-269.

Siebek, O. 1988. Experimental investigation of UV tolerance in hermatypic corals (Scleractinian). *Marine Ecology Progress Series* **43**: 95-103.

Vareschi, E. and Fricke, H. 1986. Light responses of a scleractinian coral (*Plerogyra sinuosa*). *Marine Biology* **90**: 395-402.

Williams, D.E., Hallock, P., Talge, H.K., Harney, J.N. and McRae, G. 1997. Responses of *Amphistegina gibbosa* populations in the Florida Keys (U.S.A.) to a multi-year stress event (1991-1996). *Journal of Foraminiferal Research* **27**: 264-269.

8.3.1.1.3. Other Causes

In a review of the causes of coral bleaching, Brown (1997) listed (1) elevated seawater temperature, (2) decreased seawater temperature, (3) intense solar radiation, (4) the combination of intense solar radiation and elevated temperature, (5) reduced salinity, and (6) bacterial infections. In a similar review, Meehan and Ostrander (1997) additionally listed (7) increased sedimentation and (8) exposure to toxicants. We have already commented on the four most prominent of these phenomena; we now address the remaining four.

With respect to seawater salinity, Meehan and Ostrander (1997) noted that, as with temperature, both high and low values have been observed to cause coral bleaching. Low values typically occur as a result of seawater dilution caused by high precipitation events or storm runoff; high values are much more rare, typically occurring only in the vicinity of desalinization plants.

A number of studies also have clearly delineated the role of bacterial infections in causing coral reef bleaching (Ritchie and Smith, 1998); this phenomenon, too, may have a connection to high seawater temperatures. In a study of the coral *Oculina patagonica* and the bacterial agent *Vibrio* AK-1, for example, Kushmaro *et al.* (1996, 1997) concluded that bleaching of colonies of this coral along the Mediterranean coast has its origin in bacterial infection, and that warmer temperatures may lower the resistance of the coral to infection and/or increase the virulence of the bacterium. In subsequent studies of the same coral and bacterium, Toren *et al.* (1998) and Kushmaro *et al.* (1998) further demonstrated that this high temperature effect may operate by enhancing the ability of the bacterium to adhere to the coral.

In discussing their findings, Kushmaro *et al.* (1998) commented on the "speculation that increased seawater temperature, resulting from global warming or El Niño events, is the direct cause of coral bleaching." In contradiction of this presumption, they cited several studies of coral bleaching events that were not associated with any major sea surface temperature anomalies, and they explicitly stated, "it is not yet possible to determine conclusively that bleaching episodes and the consequent damage to reefs is due to global climate change." Likewise, Toren *et al.* (1998) noted the extensive bleaching that occurred on the Great Barrier Reef during the summer of 1982 was also not associated with any major sea surface temperature increase; they stated, "several authors have reported on the patchy spatial distribution and spreading nature of coral bleaching," which they correctly noted is inconsistent with the global-warming-induced coral bleaching hypothesis. Instead, they noted, "the progression of observable changes that take place during coral bleaching is reminiscent of that of developing microbial biofilms," a point that will later be seen to be of great significance.

With respect to sedimentation, high rates have been conclusively demonstrated to lead to coral bleaching (Wesseling *et al.*, 1999); most historical increases in sedimentation rates are clearly human-induced. Umar *et al.* (1998), for example, listed such contributing anthropogenic activities as deforestation, agricultural practices, coastal development, construction, mining, drilling, dredging, and tourism. Nowlis *et al.* (1997) also discussed "how land development can increase the risk of severe damage to coral reefs by sediment runoff during storms." But it has been difficult to determine just how much these phenomena have varied over the past few centuries.

Knowledge in this area took a quantum leap forward with the publication of a study by McCulloch *et al.* (2003) that provided a 250-year record of sediment transfer to Havannah Reef—a site on the inner Great Barrier Reef of northern Queensland, Australia—by flood plumes from the Burdekin River. According to the authors of that study, sediments suspended in the Burdekin River contain barium (Ba), which is desorbed from the particles that carry it as they enter the ocean, where growing corals incorporate it into their skeletons along with calcium (Ca). Hence, when more sediments are carried to the sea by periodic flooding and more gradual longer-term changes in land use that lead to enhanced soil erosion, the resultant increases in sediment load are recorded in the Ba/Ca ratio of coral skeleton material. Inspired by these facts, McCulloch *et al.* measured Ba/Ca ratios in a 5.3-meter-long coral core from Havannah Reef that covered the period from about 1750 to 1985, as well as in some shorter cores from Havannah Reef and nearby Pandora Reef that extended the proxy sediment record to 1998.

Results of the analysis revealed that prior to the time of European settlement, which began in the Burdekin catchment in 1862, there was "surprisingly little evidence for flood-plume related activity from the coral Ba/Ca ratios." Soon after, however, land clearance and domestic grazing intensified and the soil became more vulnerable to monsoon-rain-induced erosion. By 1870, baseline Ba/Ca ratios had risen by 30 percent and "within one to two decades after the arrival of European settlers in northern Queensland, there were already massive impacts on the river catchments that were being transmitted to the waters of the inner Great Barrier Reef." During subsequent periods of flooding, in fact, the transport of suspended sediment to the reef increased by fully five- to ten-fold over what had been characteristic of pre-European settlement times.

In a companion article, Cole (2003) reported that corals from East Africa "tell a similar tale of erosion exacerbated by the imposition of colonial agricultural practices in the early decades of the twentieth century." There, similar coral data from Malindi Reef, Kenya, indicate "a low and stable level of barium before about 1910 which rises dramatically by 1920, with a simultaneous increase in variance," a phenomenon that was also evident in the Australian data.

What are the implications of these observations? Cole concludes that "human activity, in the form of changing land use, has added sedimentation to the list of stresses experienced by reefs." Furthermore, as land-use intensification is a widespread phenomenon, she notes that "many reefs close to continents or large islands are likely to have experienced increased delivery of sediment over the past century," which suggests the stress levels produced by this phenomenon are likely to have increased over the past

century as well. In addition, Cole logically concludes that as coastal populations continue to rise, "this phenomenon is likely to expand."

Lastly, a number of poisonous substances are known to have the capacity to induce coral bleaching. Some of them are of human origin, such as herbicides, pesticides, and even excess nutrients that make their way from farmlands to the sea (Simkiss, 1964; Pittock, 1999). Other poisons originate in the sea itself, many the result of metabolic waste products of other creatures (Crossland and Barnes, 1974) and some a by-product of the coral host itself (Yonge, 1968). Each of these toxicants presents the coral community with its own distinct challenge.

Taken together, these findings suggest a number of sources of stress on coral survival and growth that have little or nothing to do with rising CO_2 concentrations or temperatures. It is also clear that human population growth and societal and economic development over the period of the Industrial Revolution have predisposed coral reefs to ever-increasing incidences of bleaching and subsequent mortality via a gradual intensification of near-coastal riverine sediment transport rates.

Additional information on this topic, including reviews of newer publications as they become available, can be found at http://www.co2 science.org/subject/c/subject_c.php under the heading Coral Reefs (Bleaching: Causes).

References

Brown, B.E. 1997. Coral bleaching: Causes and consequences. *Coral Reefs* **16**: S129-S138.

Cole, J. 2003. Dishing the dirt on coral reefs. *Nature* **421**: 705-706.

Crossland, C.J. and Barnes, D.J. 1974. The role of metabolic nitrogen in coral calcification. *Marine Biology* **28**: 325-332.

Kushmaro, A., Rosenberg, E., Fine, M. and Loya, Y. 1997. Bleaching of the coral *Oculina patagonica* by *Vibrio* AK-1. *Marine Ecology Progress Series* **147**: 159-165.

Kushmaro, A., Loya, Y., Fine, M. and Rosenberg, E. 1996. Bacterial infection and coral bleaching. *Nature* **380**: 396.

Kushmaro, A., Rosenberg, E., Fine, M., Ben Haim, Y. and Loya, Y. 1998. Effect of temperature on bleaching of the coral *Oculina patagonica* by *Vibrio* AK-1. *Marine Ecology Progress Series* **171**: 131-137.

McCulloch, M., Fallon, S., Wyndham, T., Hendy, E., Lough, J. and Barnes, D. 2003. Coral record of increased sediment flux to the inner Great Barrier Reef since European settlement. *Nature* **421**: 727-730.

Meehan, W.J. and Ostrander, G.K. 1997. Coral bleaching: A potential biomarker of environmental stress. *Journal of Toxicology and Environmental Health* **50**: 529-552.

Nowlis, J.S., Roberts, C.M., Smith, A.H. and Siirila, E. 1997. Human-enhanced impacts of a tropical storm on nearshore coral reefs. *Ambio* **26**: 515-521.

Pittock, A.B. 1999. Coral reefs and environmental change: Adaptation to what? *American Zoologist* **39**: 10-29.

Ritchie, K.B. and Smith, G.W. 1998. Type II white-band disease. *Revista De Biologia Tropical* **46**: 199-203.

Simkiss, K. 1964. Phosphates as crystal poisons of calcification. *Biological Review* **39**: 487-505.

Toren, A., Landau, L., Kushmaro, A., Loya, Y. and Rosenberg, E. 1998. Effect of temperature on adhesion of *Vibrio* strain AK-1 to *Oculina patagonica* and on coral bleaching. *Applied and Environmental Microbiology* **64**: 1379-1384.

Umar, M.J., McCook, L.J. and Price, I.R. 1998. Effects of sediment deposition on the seaweed Sargassum on a fringing coral reef. *Coral Reefs* **17**: 169-177.

Wesseling, I., Uychiaoco, A.J., Alino, P.M., Aurin, T. and Vermaat, J.E. 1999. Damage and recovery of four Philippine corals from short-term sediment burial. *Marine Ecology Progress Series* **176**: 11-15.

Yonge, C.M. 1968. Living corals. *Proceedings of the Royal Society of London B* **169**: 329-344.

8.3.1.2. Adaptation

Considering the many threats to the health of coral reefs in today's world of extensive socioeconomic and environmental change, how can these incomparable repositories of underwater biodiversity be expected to escape irreversible bleaching and death? In response to this question, Glynn (1996) pointed out that "numerous reef-building coral species have endured three periods of global warming, from the Pliocene optimum (4.3-3.3 million years ago) through the Eemian interglacial (125 thousand years ago) and the mid-Holocene (6000-5000 years ago), when atmospheric CO_2 concentrations and sea temperatures often exceeded those of today." In fact, Glynn observed that "an increase in sea warming of less than 2°C would result in a greatly increased diversity of corals in certain high latitude locations."

How does this happen? Living organisms are resilient. Various lifeforms can tolerate temperatures from below freezing to the boiling point of water; others inhabit niches where light intensity varies from complete darkness to full sunlight. One reason for this great versatility is that, given time to adapt, nearly all living organisms can learn to survive in conditions well outside their normal zones of environmental tolerance. As noted by Gates and Edmunds (1999), results of numerous studies indicate that "corals routinely occupy a physically heterogeneous environment," which "suggests they should possess a high degree of biological flexibility." And indeed they do, as evidenced by their successful responses to the different threats that cause coral bleaching, which are examined in the following subsections.

References

Gates, R.D. and Edmunds, P.J. 1999. The physiological mechanisms of acclimatization in tropical reef corals. *American Zoologist* **39**: 30-43.

Glynn, P.W. 1996. Coral reef bleaching: facts, hypotheses and implications. *Global Change Biology* **2**: 495-509.

8.3.1.2.1. Response to Solar Radiation Stress

One example of adaptation to stress imposed by high solar irradiance comes from studies of corals that exhibit a "zonation" of their symbiont taxa with depth, where symbiont algae that are less tolerant of intense solar radiation grow on corals at greater depths below the ocean surface (Rowan and Knowlton, 1995; Rowan *et al.*, 1997). It has also been demonstrated that zooxanthellae in corals possess a number of light-quenching mechanisms that can be employed to reduce the negative impacts of excess light (Hoegh-Guldberg and Jones, 1999; Ralph *et al.*, 1999). Both the coral host and its symbionts also have the capacity to produce amino acids that act as natural "sunscreens" (Hoegh-Guldberg, 1999); and they can regulate their enzyme activities to enhance internal scavenging systems that remove noxious oxygen radicals produced in coral tissues as a result of high light intensities (Dykens and Shick, 1984; Lesser *et al.*, 1990; Matta and Trench, 1991; Shick *et al.*, 1996).

Another adaptive mechanism to lessen the stress of solar irradiance is coral tissue retraction, according to Brown *et al.* (1994), who studied the phenomenon in the scleractinian coral *Coeloseris mayeri* at coral reefs in Phuket, Thailand by examining the retraction and recovery of coral tissues over a tidal cycle. Results of their analysis showed that extreme tissue retraction was observed approximately 85 minutes after initial sub-aerial coral exposure. Tissue retraction, however, did not involve any reduction in chlorophyll concentration or algae symbiont abundance; the tissues expanded over the coral skeletons to pre-retraction conditions following the return of the tide. The adaptive benefits of tissue retraction, according to the authors, "include increased albedo, leading to a reduction in absorbed solar energy of 10%, ... and possible avoidance of photochemical damage or photoinhibition at high solar irradiance."

Another intriguing idea was proposed by Nakamura and van Woesik (2001), who upon evaluating the bleaching of large and small coral colonies along the western coast of Okinawa, Japan during the summers of 1998 and 2001, argued that small coral colonies should survive thermal and light stress more readily than large coral colonies based on mass transfer theory, which suggests that rates of passive diffusion are more rapid for small colonies than for large colonies. Still another reason why large coral colonies suffer more than small colonies during environmental conditions conducive to bleaching is the fact that small *Acropora* recruits, according to Bena and van Woesik (2004), "contain high concentrations of fluorescent proteins (Papina *et al.*, 2002), which have photoprotective properties (Salih *et al.*, 2000)," and they note that "a high concentration of photoprotective pigments in early life, when planulae are near the surface and as newly settled recruits, may facilitate survival during this phase as well as during stress events involving both high irradiance and thermal anomalies (van Woesik, 2000)."

In addition to the adaptive phenomena described above, the earth appears to possess a natural "heat vent" over the tropics that suppresses the intensity of solar radiation to which corals are exposed whenever dangerously high water temperatures are approached. According to Hoegh-Guldberg (1999), 29.2°C is the threshold water temperature above which significant bleaching can be expected to occur in many tropical corals. However, as Sud *et al.* (1999) have demonstrated, deep atmospheric convection is typically initiated whenever sea surface temperatures (SSTs) reach a value of about 28°C, so that an upper SST on the order of 30°C is rarely exceeded. As SSTs reach 28-29°C, the cloud-base airmass is charged with sufficient moist static energy for the clouds to reach the upper troposphere. At this point, the billowing cloud cover reduces the amount of solar radiation received at the surface of the sea, while cool and dry downdrafts produced by the moist convection tend to promote ocean surface cooling by increasing sensible and latent heat fluxes at the air-sea interface that cause temperatures there to decline. This "thermostat-like control," as Sud *et al.* describe it, tends "to ventilate the tropical ocean efficiently and help contain the SST between 28-30°C," which is essentially a fluctuating temperature band of ±1°C centered on the bleaching threshold temperature of 29.2°C identified by Hoegh-Guldberg.

Some other intriguing observations also point to the existence of a natural phenomenon of this nature. Satheesh and Ramanathan (2000), for example, determined that polluted air from south and southeast Asia absorbs enough solar radiation over the northern Indian Ocean during the dry monsoon season to heat the atmosphere there by 1-3°C per day at solar noon, thereby greatly reducing the intensity of solar radiation received at the surface of the sea. Ackerman *et al.* (2000), however, calculated that this atmospheric heating would decrease cloud-layer relative humidity and reduce boundary-layer mixing, thereby leading to a 25 percent to 50 percent drop in daytime cloud cover relative to that of an aerosol-free atmosphere, which could well negate the surface cooling effect suggested by the findings of Satheesh and Ramanathan. But in a test of this hypothesis based on data obtained from the Extended Edited Cloud Report Archive, Norris (2001) determined that daytime low-level ocean cloud cover (which tends to cool the water surface) not only did not decrease from the 1950s to 1990s, it actually increased ... in both the Northern and Southern Hemispheres and at essentially all hours of the day.

Commenting on this finding, Norris remarked that "the observed all-hours increase in low-level cloud cover over the time period when soot aerosol has presumably greatly increased argues against a dominant effect of soot solar absorption contributing to cloud 'burn-off'." Hence, he says, "other processes must be compensating," one of which, we suggest, could be the one described by Sud *et al.*

Another process is the "adaptive infrared iris" phenomenon described by Lindzen *et al.* (2001).

Working with upper-level cloudiness data obtained from the Japanese Geostationary Meteorological Satellite and SST data obtained from the National Centers for Environmental Prediction, the atmospheric scientists found a strong inverse relationship between upper-level cloud area and the mean SST of cloudy regions, such that the area of cirrus cloud coverage (which tends to warm the planet) normalized by a measure of the area of cumulus coverage (which tends to cool the planet) decreased about 22 percent for each 1°C increase in the SST of the cloudy regions.

"Essentially," in the words of the scientists, "the cloudy-moist region appears to act as an infrared adaptive iris that opens up and closes down the regions free of upper-level clouds, which more effectively permit infrared cooling, in such a manner as to resist changes in tropical surface temperatures." So substantial is this phenomenon, Lindzen *et al.* are confident it could "more than cancel all the positive feedbacks in the more sensitive current climate models," which are routinely used to predict the climatic consequences of projected increases in atmospheric CO_2 concentration.

Is there any real-world evidence the natural thermostat discovered by Sud *et al.* and Lindzen *et al.*. has actually been instrumental in preventing coral bleaching? Mumby *et al.* (2001) examined long-term meteorological records from the vicinity of the Society Islands, which provide what they call "the first empirical evidence that local patterns of cloud cover may influence the susceptibility of reefs to mass bleaching and subsequent coral mortality during periods of anomalously high SST." With respect to the great El Niño of 1998, Mumby and his colleagues determined that SSTs in the Society Islands sector of French Polynesia were above the 29.2°C bleaching threshold for a longer period of time (two months) than in all prior bleaching years of the historical record. However, mass coral bleaching, which was extensive in certain other areas, was found to be "extremely mild in the Society Islands" and "patchy at a scale of 100s of km." What provided the coral relief from extreme sun and heat? As Mumby and his associates describe it, "exceptionally high cloud cover significantly reduced the number of sun hours during the summer of 1998," much as one would have expected earth's natural thermostat to have done in the face of such anomalously high SSTs. The marine scientists also note that extensive spotty patterns of cloud cover, besides saving most of the coral they studied, "may partly account for spatial patchiness in

bleaching intensity and/or bleaching-induced mortality in other areas."

In conclusion, although the natural thermostat cannot protect all of earth's corals from life-threatening bleaching during all periods of anomalously high SSTs, it apparently protects enough of them enough of the time to ensure that sufficiently large numbers of corals survive to perpetuate their existence, since living reefs have persisted over the eons in spite of the continuing recurrence of these ever-present environmental threats. And perhaps that is how it has always been, although there are currently a host of unprecedented anthropogenic forces of site-specific origin that could well be weakening the abilities of some species to tolerate the types of thermal and solar stresses they have successfully "weathered" in the past.

Additional information on this topic, including reviews of newer publications as they become available, can be found at http://www.co2science.org/subject/c/bleachrespsolar.php.

References

Ackerman, A.S., Toon, O.B., Stevens, D.E., Heymsfield, A.J., Ramanathan, V. and Welton, E.J. 2000. Reduction of tropical cloudiness by soot. *Science* **288**: 1042-1047.

Bena, C. and van Woesik, R. 2004. The impact of two bleaching events on the survival of small coral colonies (Okinawa, Japan). *Bulletin of Marine Science* **75**: 115-125.

Brown, B.E., Le Tissier, M.D.A. and Dunne, R.P. 1994. Tissue retraction in the scleractinian coral Coeloseris mayeri, its effect upon coral pigmentation, and preliminary implications for heat balance. *Marine Ecology Progress Series* **105**: 209-218.

Dykens, J.A. and Shick, J.M. 1984. Photobiology of the symbiotic sea anemone *Anthopleura elegantissima*: Defense against photo-dynamic effects, and seasonal photoacclimatization. *Biological Bulletin* **167**: 693-697.

Hoegh-Guldberg, O. 1999. Climate change, coral bleaching and the future of the world's coral reefs. *Marine and Freshwater Research* **50**: 839-866.

Hoegh-Guldberg, O. and Jones, R. 1999. Photoinhibition and photoprotection in symbiotic dinoflagellates from reef-building corals. *Marine Ecology Progress Series* **183**: 73-86.

Lesser, M.P., Stochaj, W.R., Tapley, D.W. and Shick, J.M. 1990. Bleaching in coral reef anthozoans: Effects of irradiance, ultraviolet radiation, and temperature on the activities of protective enzymes against active oxygen. *Coral Reefs* **8**: 225-232.

Lindzen, R.S., Chou, M.-D. and Hou, A.Y. 2001. Does the earth have an adaptive infrared iris? *Bulletin of the American Meteorological Society* **82**: 417-432.

Matta, J.L. and Trench, R.K. 1991. The enzymatic response of the symbiotic dinoflagellate *Symbiodinium microadriaticum* (Freudenthal) to growth under varied oxygen tensions. *Symbiosis* **11**: 31-45.

Mumby, P.J., Chisholm, J.R.M., Edwards, A.J., Andrefouet, S. and Jaubert, J. 2001. *Marine Ecology Progress Series* **222**: 209-216.

Nakamura, T. and van Woesik, R. 2001. Differential survival of corals during the 1998 bleaching event is partially explained by water-flow rates and passive diffusion. *Marine Ecology Progress Series* **212**: 301-304.

Norris, J.R. 2001. Has northern Indian Ocean cloud cover changed due to increasing anthropogenic aerosol? *Geophysical Research Letters* **28**: 3271-3274.

Papina, M., Sakihama, Y., Bena, C., van Woesik, R. and Yamasaki, H. 2002. Separation of highly fluorescent proteins by SDS-PAGE in Acroporidae corals. *Comp. Biochem. Phys.* **131**: 767-774.

Ralph, P.J., Gaddemann, R., Larkum, A.W.E. and Schreiber, U. 1999. In situ underwater measurements of photosynthetic activity of coral-reef dwelling endosymbionts. *Marine Ecology Progress Series* **180**: 139-147.

Rowan, R. and Knowlton, N. 1995. Intraspecific diversity and ecological zonation in coral-algal symbiosis. *Proceeding of the National Academy of Sciences, U.S.A.* **92**: 2850-2853.

Rowan, R., Knowlton, N., Baker, A. and Jara, J. 1997. Landscape ecology of algal symbionts creates variation in episodes of coral bleaching. *Nature* **388**: 265-269.

Salih, A., Larkum, A., Cox, G., Kuhl, M. and Hoegh-Guldberg, O. 2000. Fluorescent pigments in corals are photoprotective. *Nature* **408**: 850-853.

Satheesh, S.K. and Ramanathan, V. 2000. Large differences in tropical aerosol forcing at the top of the atmosphere and Earth's surface. *Nature* **405**: 60-63.

Shick, J.M., Lesser, M.P. and Jokiel, P.L. 1996. Ultraviolet radiation and coral stress. *Global Change Biology* **2**: 527-545.

Sud, Y.C., Walker, G.K. and Lau, K.-M. 1999. Mechanisms regulating sea-surface temperatures and deep convection in the tropics. *Geophysical Research Letters* **26**: 1019-1022.

van Woesik, R. 2000. Modelling processes that generate and maintain coral community diversity. *Biodiversity and Conservation* **9**: 1219-1233.

8.3.1.2.2. Response to Temperature Stress

As living entities, corals are not only acted upon by the various elements of their environment, they also react or respond to them. And when changes in environmental factors pose a challenge to their continued existence, they sometimes take major defensive or adaptive actions to ensure their survival. A simple but pertinent example of one form of this phenomenon is thermal adaptation, which feature has been observed by several researchers to operate in corals.

Fang *et al.* (1997), for example, experimented with samples of the coral *Acropora grandis* that were taken from the hot water outlet of a nuclear power plant near Nanwan Bay, Taiwan. In 1988, the year the power plant began full operation, the coral samples were completely bleached within two days of exposure to a temperature of 33°C. Two years later, however, Fang *et al.* report that "samples taken from the same area did not even start bleaching until six days after exposure to 33°C temperatures."

Similar findings have been reported by Middlebrook *et al.* (2008), who collected multiple upward-growing branch tips of the reef-building coral *Acropora aspera* from three large colonies at the southern end of Australia's Great Barrier Reef and placed them on racks immersed in running seawater within four 750-liter tanks that were maintained at the mean local ambient temperature (27°C) and exposed to natural reef-flat summer daily light levels. Then, two weeks prior to a simulated bleaching event—where water temperature was raised to a value of 34°C for a period of six days—they boosted the water temperature in one of the tanks to 31°C for 48 hours, while in another tank they boosted it to 31°C for 48 hours one week before the simulated bleaching event. In the third tank they had no pre-heating treatment, while in the fourth tank they had no pre-heating nor any simulated bleaching event. And at different points throughout the study, they measured photosystem II efficiency, xanthophyll and chlorophyll *a* concentrations, and *Symbiodinium* densities.

Results of the study indicated that the symbionts of the corals that were exposed to the 48-hour pre-bleaching thermal stress "were found to have more effective photoprotective mechanisms," including "changes in non-photochemical quenching and xanthophyll cycling," and they further determined that "these differences in photoprotection were correlated with decreased loss of symbionts, with those corals that were not pre-stressed performing significantly worse, losing over 40% of their symbionts and having a greater reduction in photosynthetic efficiency," whereas "pre-stressed coral symbiont densities were unchanged at the end of the bleaching." In light of these findings, Middlebrook *et al.* say their study "conclusively demonstrates that thermal stress events two weeks and one week prior to a bleaching event provide significantly increased thermal tolerance to the coral holobiont, suggesting that short time-scale thermal adaptation can have profound effects on coral bleaching."

Moving out of the laboratory and into the real world of nature, Adjeroud *et al.* (2005) initiated a monitoring program on 13 islands (eight atolls and five high volcanic islands) in four of the five archipelagoes of French Polynesia, with the goal of documenting the effects of natural perturbations on coral assemblages. For the period covered by their report (1992-2002), these reefs were subjected to three major coral bleaching events (1994, 1998, 2002) and three cyclones (1997), while prior to this period, the sites had experienced an additional seven bleaching events and 15 cyclones, as well as several *Acanthaster planci* outbreaks.

Results of the monitoring program revealed that the impacts of the bleaching events were variable among the different study locations. In their 10-year survey, for example, they observed three different temporal trends: "(1) ten sites where coral cover decreased in relation to the occurrence of major disturbances; (2) nine sites where coral cover increased, despite the occurrence of disturbances affecting seven of them; and (3) a site where no significant variation in coral cover was found." In addition, they report that "an interannual survey of reef communities at Tiahura, Moorea, showed that the mortality of coral colonies following a bleaching event was decreasing with successive events, even if the latter have the same intensity (Adjeroud *et al.*, 2002)."

Commenting on their and other researchers' observations, the seven French scientists say the "spatial and temporal variability of the impacts observed at several scales during the present and previous surveys may reflect an acclimation and/or adaptation of local populations," such that "coral colonies and/or their endosymbiotic zooxanthellae may be phenotypically (acclimation) and possibly genotypically (adaptation) resistant to bleaching events," citing the work of Rowan *et al.* (1997), Hoegh-Guldberg (1999), Kinzie *et al.* (2001) and Coles and Brown (2003) in support of this conclusion.

Other researchers have also confirmed the phenomenon of thermal adaptation in coral reefs. Guzman and Cortes (2007) studied coral reefs of the eastern Pacific Ocean that "suffered unprecedented mass mortality at a regional scale as a consequence of the anomalous sea warming during the 1982-1983 El Niño." At Cocos Island (5°32'N, 87°04'W), in particular, they found in a survey of three representative reefs, which they conducted in 1987, that remaining live coral cover was only 3 percent of what it had been prior to the occurrence of the great El Niño four years earlier (Guzman and Cortes, 1992). Based on this finding and the similar observations of other scientists at other reefs, they predicted that "the recovery of the reefs' framework would take centuries, and recovery of live coral cover, decades." In 2002, therefore, nearly 20 years after the disastrous coral-killing warming, they returned to see just how prescient they might have been after their initial assessment of the El Niño's damage, quantifying "the live coral cover and species composition of five reefs, including the three previously assessed in 1987." The two researchers report that overall mean live coral cover increased nearly five-fold, from 2.99 percent in 1987 to 14.87 percent in 2002, at the three sites studied during both periods, while the mean live coral cover of all five sites studied in 2002 was 22.7 percent. In addition, they found that "most new recruits and adults belonged to the main reef building species from pre-1982 ENSO, *Porites lobata*, suggesting that a disturbance as outstanding as El Niño was not sufficient to change the role or composition of the dominant species."

With respect to the subject of thermal tolerance, however, the most interesting aspect of the study was the fact that a second major El Niño occurred between the two assessment periods; Guzman and Cortes state that "the 1997-1998 warming event around Cocos Island was more intense than all previous El Niño events," noting that temperature anomalies "above 2°C lasted 4 months in 1997-1998 compared to 1 month in 1982-83." Nevertheless, they report that "the coral communities suffered a lower and more selective mortality in 1997-1998, as was also observed in other areas of the eastern Pacific (Glynn et al., 2001; Cortes and Jimenez, 2003; Zapata and Vargas-Angel, 2003)," which is indicative of some type of thermal adaptation following the 1982-83 El Niño.

One year later in a paper published in *Marine Biology*, Maynard *et al.* (2008) described how they analyzed the bleaching severity of three genera of corals (*Acropora*, *Pocillopora* and *Porites*) via underwater video surveys of five sites in the central section of Australia's Great Barrier Reef in late February and March of 1998 and 2002, while contemporary sea surface temperatures were acquired from satellite-based Advanced Very High Resolution Radiometer data that were calibrated to local ship- and drift buoy-obtained measurements, and surface irradiance data were obtained "using an approach modified from that of Pinker and Laszlo (1991)."

With respect to temperature, the four researchers report that "the amount of accumulated thermal stress (as degree heating days) in 2002 was more than double that in 1998 at four of the five sites," and that "average surface irradiance during the 2002 thermal anomaly was 15.6-18.9% higher than during the 1998 anomaly." Nevertheless, they found that "in 2002, bleaching severity was 30-100% lower than predicted from the relationship between severity and thermal stress in 1998, despite higher solar irradiances during the 2002 thermal event." In addition, they found that the "coral genera most susceptible to thermal stress (*Pocillopora* and *Acropora*) showed the greatest increase in tolerance."

In discussing their findings, Maynard *et al.* write that they are "consistent with previous studies documenting an increase in thermal tolerance between bleaching events (1982-1983 vs. 1997-1998) in the Galapagos Islands (Podesta and Glynn, 2001), the Gulf of Chiriqi, the Gulf of Panama (Glynn *et al.*, 2001), and on Costa Rican reefs (Jimenez *et al.*, 2001)," and they say that "Dunne and Brown (2001) found similar results to [theirs] in the Andaman Sea, in that bleaching severity was far reduced in 1998 compared to 1995 despite sea-temperature and light conditions being more conducive to widespread bleaching in 1998."

As for the significance of these and other observations, the Australian scientists say that "the range in bleaching tolerances among corals inhabiting different thermal realms suggests that at least some coral symbioses have the ability to adapt to much higher temperatures than they currently experience in the central Great Barrier Reef," citing the work of Coles and Brown (2003) and Riegl (1999, 2002). In

addition, they note that "even within reefs there is a significant variability in bleaching susceptibility for many species (Edmunds, 1994; Marshall and Baird, 2000), suggesting some potential for a shift in thermal tolerance based on selective mortality (Glynn *et al.*, 2001; Jimenez *et al.*, 2001) and local population growth alone." Above and beyond that, however, they say their results additionally suggest "a capacity for acclimatization or adaptation."

In concluding their paper, Maynard *et al.* say "there is emerging evidence of high genetic structure within coral species (Ayre and Hughes, 2004)," which suggests, in their words, that "the capacity for adaptation could be greater than is currently recognized." Indeed, as stated by Skelly *et al.* (2007), "on the basis of the present knowledge of genetic variation in performance traits and species' capacity for evolutionary response, it can be concluded that evolutionary change will often occur concomitantly with changes in climate as well as other environmental changes."

One adaptive mechanism that corals have developed to survive the thermal stress of high water temperature is to replace the zooxanthellae expelled by the coral host during a stress-induced bleaching episode by one or more varieties of zooxanthellae that are more heat tolerant, a phenomenon we describe in greater detail in the next section of our report. Another mechanism is to produce heat shock proteins that help repair heat-damaged constituents of their bodies (Black *et al.*, 1995; Hayes and King, 1995; Fang *et al.*, 1997). Sharp *et al.* (1997), for example, demonstrated that sub-tidal specimens of *Goniopora djiboutiensis* typically have much lower constitutive levels of a 70-kD heat shock protein than do their intertidal con-specifics; and they have shown that corals transplanted from sub-tidal to intertidal locations (where temperature extremes are greater and more common) typically increase their expression of this heat shock protein.

Similar results have been reported by Roberts *et al.* (1997) in field work with *Mytilus californianus*. In addition, Gates and Edmunds (1999) have observed an increase in the 70-kD heat shock protein after six hours of exposure of *Montastraea franksi* to a 2-3°C increase in temperature, which is followed by another heat shock protein increase at the 48-hour point of exposure to elevated water temperature. They state that the first of these protein increases "provides strong evidence that changes in protein turnover during the initial exposure to elevated temperature provides this coral with the biological flexibility to acclimatize to the elevation in sea water temperature," and that the second increase "indicates another shift in protein turnover perhaps associated with an attempt to acclimatize to the more chronic level of temperature stress."

How resilient are corals in this regard? No one knows for sure, but they've been around a very long time, during which climatic conditions have changed dramatically, from cold to warm and back again, over multiple glacial and interglacial cycles. Thermal adaptation by coral is a biological response that is overlooked or ignored by the IPCC.

Additional information on this topic, including reviews of newer publications as they become available, can be found at http://www.co2 science.org/subject/c/bleachresptemp.php.

References

Adjeroud, M., Augustin, D., Galzin, R. and Salvat, B. 2002. Natural disturbances and interannual variability of coral reef communities on the outer slope of Tiahura (Moorea, French Polynesia): 1991 to 1997. *Marine Ecology Progress Series* **237**: 121-131.

Adjeroud, M., Chancerelle, Y., Schrimm, M., Perez, T., Lecchini, D., Galzin, R. and Salvat, B. 2005. Detecting the effects of natural disturbances on coral assemblages in French Polynesia: A decade survey at multiple scales. *Aquatic Living Resources* **18**: 111-123.

Ayre, D.J. and Hughes, T.P. 2004. Climate change, genotypic diversity and gene flow in reef-building corals. *Ecology Letters* **7**: 273-278.

Black, N.A., Voellmy, R. and Szmant, A.M. 1995. Heat shock protein induction in *Montastrea faveoluta* and *Aiptasia pallida* exposed to elevated temperature. *Biological Bulletin* **188**: 234-240.

Coles, S.L. and Brown, B.E. 2003. Coral bleaching-capacity for acclimatization and adaptation. *Advances in Marine Biology* **46**: 183-223.

Cortes, J. and Jimenez, C. 2003. Corals and coral reefs of the Pacific of Costa Rica: history, research and status. In: Cortes, J. (Ed.) *Latin American Coral Reefs*. Elsevier, Amsterdam, The Netherlands, pp. 361-385.

Dunne, R.P. and Brown, B.E. 2001. The influence of solar radiation on bleaching of shallow water reef corals in the Andaman Sea, 1993-98. *Coral Reefs* **20**: 201-210.

Edmunds, P.J. 1994. Evidence that reef-wide patterns of coral bleaching may be the result of the distribution of

bleaching susceptible clones. *Marine Biology (Berlin)* **121**: 137-142.

Fang, L.-S., Huang, S.-P. and Lin, K.-L. 1997. High temperature induces the synthesis of heat-shock proteins and the elevation of intracellular calcium in the coral *Acropora grandis*. *Coral Reefs* **16**: 127-131.

Gates, R.D. and Edmunds, P.J. 1999. The physiological mechanisms of acclimatization in tropical reef corals. *American Zoologist* **39**: 30-43.

Glynn, P.W., Mate, J.L., Baker, A.C. and Calderon, M.O. 2001. Coral bleaching and mortality in Panama and Equador during the 1997-1998 El Niño Southern Oscillation event: spatial/temporal patterns and comparisons with the 1982-1983 event. *Bulletin of Marine Science* **69**: 79-109.

Guzman, H.M. and Cortes, J. 1992. Cocos Island (Pacific of Costa Rica) coral reefs after the 1982-83 El Niño disturbance. *Revista de Biologia Tropical* **40**: 309-324.

Guzman, H.M. and Cortes, J. 2007. Reef recovery 20 years after the 1982-1983 El Niño massive mortality. *Marine Biology* **151**: 401-411.

Hayes, R.L. and King, C.M. 1995. Induction of 70-kD heat shock protein in scleractinian corals by elevated temperature: Significance for coral bleaching. *Molecular Marine Biology and Biotechnology* **4**: 36-42.

Hoegh-Guldberg, O. 1999. Climate change, coral bleaching and the future of the world's coral reefs. *Marine and Freshwater Research* **50**: 839-866.

Jimenez, C., Cortes, J., Leon, A. and Ruiz, E. 2001. Coral bleaching and mortality associated with the 1997-1998 El Niño in an upwelling environment in the eastern Pacific (Gulf of Papagayo, Costa Rica). *Bulletin of Marine Science* **69**: 151-169.

Kinzie III, R.A., Takayama, M., Santos, S.C. and Coffroth, M.A. 2001. The adaptive bleaching hypothesis: Experimental tests of critical assumptions. *Biological Bulletin* **200**: 51-58.

Marshall, P.A. and Baird, A.H. 2000. Bleaching of corals on the Great Barrier Reef: differential susceptibilities among taxa. *Coral Reefs* **19**: 155-163.

Maynard, J.A., Anthony, K.R.N., Marshall, P.A. and Masiri, I. 2008. Major bleaching events can lead to increased thermal tolerance in corals. *Marine Biology (Berlin)* **155**: 173-182.

Middlebrook, R., Hoegh-Guldberg, O. and Leggat, W. 2008. The effect of thermal history on the susceptibility of reef-building corals to thermal stress. *The Journal of Experimental Biology* **211**: 1050-1056.

Pinker, R.T. and Laszlo, I. 1991. Modeling surface solar irradiance for satellite applications on a global scale. *Journal of Applied Meteorology* **31**: 194-211.

Podesta, G.P. and Glynn, P.W. 2001. The 1997-98 El Niño event in Panama and Galapagos: an update of thermal stress indices relative to coral bleaching. *Bulletin of Marine Science* **69**: 43-59.

Riegl, B. 1999. Corals in a non-reef setting in the southern Arabian Gulf (Dubai, UAE): fauna and community structure in response to recurring mass mortality. *Coral Reefs* **18**: 63-73.

Riegl, B. 2002. Effects of the 1996 and 1998 positive sea-surface temperature anomalies on corals, coral diseases and fish in the Arabian Gulf (Dubai, UAE). *Marine Biology (Berlin)* **140**: 29-40.

Roberts, D.A., Hofman, G.E. and Somero, G.N. 1997. Heat-shock protein expression in *Mytilus californianus*: Acclimatization (seasonal and tidal height comparisons) and acclimation effects. *Biological Bulletin* **192**: 309-320.

Rowan, R., Knowlton, N., Baker, A. and Jara, J. 1997. Landscape ecology of algal symbionts creates variation in episodes of coral bleaching. *Nature* **388**: 265-269.

Sharp, V.A., Brown, B.E. and Miller, D. 1997. Heat shock protein (hsp 70) expression in the tropical reef coral *Goniopora djiboutiensis*. *Journal of Thermal Biology* **22**: 11-19.

Skelly, D.K., Joseph, L.N., Possingham, H.P., Freidenburg, L.K., Farrugia, T.J., Kinnison, M.T. and Hendry, A.P. 2007. Evolutionary responses to climate change. *Conservation Biology* **21**: 1353-1355.

Zapata, F.A. and Vargas-Angel, B. 2003. Corals and coral reefs of the Pacific coast of Columbia. In: Cortes, J. (Ed.) *Latin American Coral Reefs*. Elsevier, Amsterdam, The Netherlands, pp. 419-447.

8.3.1.2.3. Symbiont Shuffling

Although once considered to be members of the single species *Symbiodinium microadriacticum*, the zooxanthellae that reside within membrane-bound vacuoles in the cells of host corals are highly diverse, comprising perhaps hundreds of species, of which several are typically found in each species of coral (Trench, 1979; Rowan and Powers, 1991; Rowan *et al.*, 1997). One way coral respond to stress is to replace the zooxanthellae expelled by the coral host during a stress-induced bleaching episode with one or more varieties of zooxanthellae that are more tolerant of that particular stress.

Rowan *et al.* (1997) have suggested that this phenomenon occurs in many of the most successful Caribbean corals that act as hosts to dynamic multi-species communities of symbionts, and that "coral communities may adjust to climate change by recombining their existing host and symbiont genetic diversities," thereby reducing the amount of damage that might subsequently be expected from another occurrence of anomalously high temperatures. Buddemeier and Fautin (1993) suggested coral bleaching is an adaptive strategy for "shuffling" symbiont genotypes to create associations better adapted to new environmental conditions. Kinzie (1999) suggested coral bleaching "might not be simply a breakdown of a stable relationship that serves as a symptom of degenerating environmental conditions," but it "may be part of a mutualistic relationship on a larger temporal scale, wherein the identity of algal symbionts changes in response to a changing environment."

This process of replacing less-stress-tolerant symbionts by more-stress-tolerant symbionts is also supported by the investigations of Rowan and Knowlton (1995) and Gates and Edmunds (1999). The strategy seems to be working, for as Glynn (1996) has observed, "despite recent incidences of severe coral reef bleaching and mortality, no species extinctions have yet been documented."

These observations accord well with the experimental findings of Fagoonee *et al.* (1999), who suggest that coral bleaching events "may be frequent and part of the expected cycle." Gates and Edmunds (1999) additionally report that "several of the prerequisites required to support this hypothesis have now been met," and after describing them in some detail, they conclude "there is no doubt that the existence of multiple *Symbiodinium* clades, each potentially exhibiting a different physiological optima, provide corals with the opportunity to attain an expanded range of physiological flexibility which will ultimately be reflected in their response to environmental challenge." In fact, this phenomenon may provide the explanation for the paradox posed by Pandolfi (1999); i.e., that "a large percentage of living coral reefs have been degraded, yet there are no known extinctions of any modern coral reef species." Surely, this result is exactly what would be expected if periods of stress lead to the acquisition of more-stress-resistant zooxanthellae by coral hosts.

In spite of this early raft of compelling evidence for the phenomenon, Hoegh-Guldberg (1999) challenged the symbiont shuffling hypothesis on the basis that the stress-induced replacement of less-stress-tolerant varieties of zooxanthellae by more-stress-tolerant varieties "has never been observed." Although true at the time it was written, a subsequent series of studies has produced the long-sought proof that transforms the hypothesis into fact.

Baker (2001) conducted an experiment in which he transplanted corals of different combinations of host and algal symbiont from shallow (2-4 m) to deep (20-23 m) depths and vice versa. After eight weeks nearly half of the corals transplanted from deep to shallow depths had experienced partial or severe bleaching, whereas none of the corals transplanted from shallow to deep depths bleached. After one year, however, and despite even more bleaching at shallow depths, upward transplants showed no mortality, but nearly 20 percent of downward transplants had died. Why?

The symbiont shuffling hypothesis explains it this way. The corals that were transplanted upwards were presumed to have adjusted their algal symbiont distributions, via bleaching, to favor more-tolerant species, whereas the corals transplanted downward were assumed to not have done so, since they did not bleach. Baker suggested that these findings "support the view that coral bleaching can promote rapid response to environmental change by facilitating compensatory change in algal symbiont communities." Without bleaching, he continued, "suboptimal host-symbiont combinations persist, leading eventually to significant host mortality." Consequently, Baker proposed that coral bleaching may "ultimately help reef corals to survive." And it may also explain why reefs, though depicted by the IPCC as environmentally fragile, have survived the large environmental changes experienced throughout geologic time.

One year later Adjeroud et al. (2002) provided additional evidence for the veracity of the symbiont shuffling hypothesis as a result of their assessment of the interannual variability of coral cover on the outer slope of the Tiahura sector of Moorea Island, French Polynesia, between 1991 and 1997, which focused on the impacts of bleaching events caused by thermal stress when sea surface temperatures rose above 29.2°C. Soon after the start of their study, they observed a severe decline in coral cover following a bleaching event that began in March 1991, which was followed by another bleaching event in March 1994. However, they report that the latter bleaching event "did not have an important impact on coral cover," even though "the proportion of bleached colonies ...

and the order of susceptibility of coral genera were similar in 1991 and 1994 (Gleason, 1993; Hoegh-Guldberg and Salvat, 1995)." They report that between 1991 and 1992 total coral cover dropped from 51.0 percent to 24.2 percent, but that "coral cover did not decrease between 1994 and 1995."

In discussing these observations, Adjeroud et al. write that a "possible explanation of the low mortality following the bleaching event in 1994 is that most of the colonies in place in 1994 were those that survived the 1991 event or were young recruits derived from those colonies," noting that "one may assume that these coral colonies and/or their endosymbiotic zooxanthellae were phenotypically and possibly genotypically resistant to bleaching events," which is exactly what the symbiont shuffling hypothesis would predict. They further state that "this result demonstrates the importance of understanding the ecological history of reefs (i.e., the chronology of disturbances) in interpreting the specific impacts of a particular disturbance."

In the same year, Brown et al. (2002) published the results of an even longer 17-year study of coral reef flats at Ko Phuket, Thailand, in which they assessed coral reef changes in response to elevated water temperatures in 1991, 1995, 1997, and 1998. As they describe it, "many corals bleached during elevated sea temperatures in May 1991 and 1995, but no bleaching was recorded in 1997." In addition, they report that "in May 1998 very limited bleaching occurred although sea temperatures were higher than previous events in 1991 and 1995 (Dunne and Brown, 2001)." What is more, when bleaching did take place, they say "it led only to partial mortality in coral colonies, with most corals recovering their color within 3-5 months of initial paling," once again providing real-world evidence for what is predicted by the symbiont shuffling hypothesis.

The following year, Riegl (2003) reviewed what is known about the responses of real-world coral reefs to high-temperature-induced bleaching, focusing primarily on the Arabian Gulf, which experienced high-frequency recurrences of temperature-related bleaching in 1996, 1998, and 2002. In response to these high-temperature events, Riegl notes that Acropora, which during the 1996 and 1998 events always bleached first and suffered heaviest mortality, bleached less than all other corals in 2002 at Sir Abu Nuair (an offshore island of the United Arab Emirates) and actually recovered along the coast of Dubai between Jebel Ali and Ras Hasyan. As a result, Riegl states that "the unexpected resistance of Sir

Abu Nuair *Acropora* to bleaching in 2002 might indicate support for the hypothesis of Baker (2001) and Baker *et al.* (2002) that the symbiont communities on recovering reefs of the future might indeed be more resistant to subsequent bleaching," and that "the Arabian Gulf perhaps provides us with some aspects which might be described as a 'glimpse into the future,' with ... hopes for at least some level of coral/zooxanthellae adaptation."

In a contemporaneous paper, Kumaraguru *et al.* (2003) reported the results of a study wherein they assessed the degree of damage inflicted upon a number of coral reefs within Palk Bay (located on the southeast coast of India just north of the Gulf of Mannar) by a major warming event that produced monthly mean sea surface temperatures of 29.8 to 32.1°C from April through June 2002, after which they assessed the degree of recovery of the reefs. They determined that "a minimum of at least 50% and a maximum of 60% bleaching were noticed among the six different sites monitored." However, as they continue, "the corals started to recover quickly in August 2002 and as much as 52% recovery could be noticed." By comparison, they note that "recovery of corals after the 1998 bleaching phenomenon in the Gulf of Mannar was very slow, taking as much as one year to achieve similar recovery," i.e., to achieve what was experienced in one *month* in 2002. Consequently, in words descriptive of the concept of symbiont shuffling, the Indian scientists say "the process of natural selection is in operation, with the growth of new coral colonies, and any disturbance in the system is only temporary." Consequently, as they conclude in the final sentence of their paper, "the corals will resurge under the sea."

Writing in *Nature*, Rowan (2004) described how he measured the photosynthetic responses of two zooxanthellae genotypes or clades—*Symbiodinium C* and *Symbiodinium D*—to increasing water temperature, finding that the photosynthetic prowess of the former decreased at higher temperatures while that of the latter increased. He then noted that "adaptation to higher temperature in *Symbiodinium D* can explain why *Pocillopora* spp. hosting them resist warm-water bleaching whereas corals hosting *Symbiodinium C* do not," and that "it can also explain why *Pocillopora* spp. living in frequently warm habitats host only *Symbiodinium D*, and, perhaps, why those living in cooler habitats predominantly host *Symbiodinium C*," concluding that these observations "indicate that symbiosis recombination

may be one mechanism by which corals adapt, in part, to global warming."

Baker *et al.* (2004) "undertook molecular surveys of *Symbiodinium* in shallow scleractinian corals from five locations in the Indo-Pacific that had been differently affected by the 1997-98 El Niño-Southern Oscillation (ENSO) bleaching event." Along the coasts of Panama, they surveyed ecologically dominant corals in the genus *Pocillopora* before, during, and after ENSO bleaching, finding that "colonies containing *Symbiodinium* in clade D were already common (43%) in 1995 and were unaffected by bleaching in 1997, while colonies containing clade C bleached severely." Even more importantly, they found that "by 2001, colonies containing clade D had become dominant (63%) on these reefs."

After describing similar observations in the Persian (Arabian) Gulf and the western Indian Ocean along the coast of Kenya, Baker *et al.* summarized their results by stating they indicate that "corals containing thermally tolerant *Symbiodinium* in clade D are more abundant on reefs after episodes of severe bleaching and mortality, and that surviving coral symbioses on these reefs more closely resemble those found in high-temperature environments," where clade D predominates. They concluded their paper by noting that the symbiont changes they observed "are a common feature of severe bleaching and mortality events," and by predicting that "these adaptive shifts will increase the resistance of these recovering reefs to future bleaching."

Lewis and Coffroth (2004) described a controlled experiment in which they induced bleaching in a Caribbean octocoral (*Briareum* sp.) and then exposed it to exogenous *Symbiodinium* sp. containing rare variants of the chloroplast 23S ribosomal DNA (rDNA) domain V region (cp23S-genotype), after which they documented the symbionts' repopulation of the coral, whose symbiont density had been reduced to less than 1 percent of its original level by the bleaching. Also, in a somewhat analogous study, Little *et al.* (2004) described how they investigated the acquisition of symbionts by juvenile *Acropora tenuis* corals growing on tiles they attached to different portions of reef at Nelly Bay, Magnetic Island (an inshore reef in the central section of Australia's Great Barrier Reef).

Lewis and Coffroth wrote that the results of their study show "the repopulation of the symbiont community involved residual populations within *Briareum* sp., as well as symbionts from the surrounding water," noting that "recovery of coral-algal symbioses after a bleaching event is not solely dependent on the *Symbiodinium* complement initially acquired early in the host's ontogeny," but that "these symbioses also have the flexibility to establish new associations with symbionts from an environmental pool." Similarly, Little *et al.* reported that "initial uptake of zooxanthellae by juvenile corals during natural infection is nonspecific (a potentially adaptive trait)," and "the association is flexible and characterized by a change in (dominant) zooxanthella strains over time." Lewis and Coffroth concluded that "the ability of octocorals to reestablish symbiont populations from multiple sources provides a mechanism for resilience in the face of environmental change." Little *et al.* concluded that the "symbiont shuffling" observed by both groups "represents a mechanism for rapid acclimatization of the holobiont to environmental change."

Writing in the journal *Marine Ecology Progress Series*, Chen *et al.* (2005) reported their study of the seasonal dynamics of *Symbiodinium* algal phylotypes via bimonthly sampling over an 18-month period of *Acropora palifera* coral on a reef flat at Tantzel Bay, Kenting National Park, southern Taiwan, in an attempt to detect real-world symbiont shuffling. Results of the analysis revealed two levels of symbiont shuffling in host corals: (1) between *Symbiodinium* phylotypes C and D, and (2) among different variants within each phylotype. The most significant changes in symbiont composition occurred at times of significant increases in seawater temperature during late spring/early summer, perhaps as a consequence of enhanced stress experienced at that time, leading Chen *et al.* to state their work revealed "the first evidence that the symbiont community within coral colonies is dynamic ... involving changes in *Symbiodinium* phylotypes."

Also in 2005, Van Oppen *et al.* (2005) sampled zooxanthellae from three common species of scleractinian corals at 17 sites along a latitudinal and cross-shelf gradient in the central and southern sections of the Great Barrier Reef some four to five months after the major bleaching event of 2002, recording the health status of each colony at the time of its collection and identifying its zooxanthella genotypes, of which there are eight distinct clades (A-H) with clade D being the most heat-tolerant. Results of the analysis revealed that "there were no simple correlations between symbiont types and either the level of bleaching of individual colonies or indicators of heat stress at individual sites." However, they say "there was a very high post-bleaching abundance of

the heat tolerant symbiont type D in one coral population at the most heat-stressed site."

With respect to the post-bleaching abundance of clade D zooxanthellae at the high heat-stress site, the Australian researchers say they suspect it was due to "a proliferation in the absolute abundance of clade D within existing colonies that were previously dominated by clade C zooxanthellae," and that in the four to five months before sampling them, "mixed C-D colonies that had bleached but survived may have shifted (shuffling) from C-dominance to D-dominance, and/or C-dominated colonies may have suffered higher mortality during the 2002 bleaching event" and subsequently been repopulated by a predominance of clade D genotypes.

Working within Australia's Great Barrier Reef system, Berkelmans and van Oppen (2006) investigated the thermal acclimatization potential of *Acropora millepora* corals to rising temperatures through transplantation and experimental manipulation, finding that the adult corals "are capable of acquiring increased thermal tolerance and that the increased tolerance is a direct result of a change in the symbiont type dominating their tissues from *Symbiodinium* type C to D." Two years later, working with an expanded group of authors (Jones *et al.*, 2008), the same two researchers reported similar findings following the occurrence of a natural bleaching event.

Prior to the bleaching event, Jones *et al.* report that "*A. millepora* at Miall reef associated predominantly with *Symbiodinium* type C2 (93.5%) and to a much lesser extent with *Symbiodinium* clade D (3.5%) or mixtures of C2 and D (3.0%)." During the bleaching event, they report "the relative difference in bleaching susceptibility between corals predominated by C2 and D was clearly evident, with the former bleaching white and the latter normally pigmented," while corals harboring a mix of *Symbiodinium* C2 and D were "mostly pale in appearance." Then, three months after the bleaching event, they observed "a major shift to thermally tolerant type D and C1 symbiont communities ... in the surviving colonies," the latter of which types had not been detected in any of the corals prior to bleaching. They report "this shift resulted partly from a change of symbionts within coral colonies that survived the bleaching event (42%) and partly from selective mortality of the more bleaching-sensitive C2-predominant colonies (37%)." In addition, they report that all of the colonies that harbored low levels of D-type symbionts prior to the bleaching event survived and changed from clade C2 to D predominance.

In conclusion, Jones *et al.* say that "as a direct result of the shift in symbiont community, the Miall Island *A. millepora* population is likely to have become more thermo-tolerant," as they note that "a shift from bleaching-sensitive type C2 to clade D increased the thermal tolerance of this species by 1-1.5°C." They say their results "strongly support the reinterpreted adaptive bleaching hypothesis of Buddemeier *et al.* (2004), which postulates that a continuum of changing environmental states stimulates the loss of bleaching-sensitive symbionts in favor of symbionts that make the new holobiont more thermally tolerant." They state that their observations "provide the first extensive colony-specific documentation and quantification of temporal symbiont community change in the field in response to temperature stress, suggesting a population-wide acclimatization to increased water temperature."

In a much larger geographical study, Lien *et al.* (2007) examined the symbiont diversity in a scleractinian coral, *Oulastrea crispata*, throughout its entire latitudinal distribution range in the West Pacific, i.e., from tropical peninsular Thailand (<10°N) to high-latitudinal outlying coral communities in Japan (>35°N), convincingly demonstrating in the words of the six scientists who conducted the study, "that phylotype D is the dominant *Symbiodinium* in scleractinian corals throughout tropical reefs and marginal outlying non-reefal coral communities." In addition, they learned that this particular symbiont clade "favors 'marginal habitats' where other symbionts are poorly suited to the stresses, such as irradiance, temperature fluctuations, sedimentation, etc." Being a major component of the symbiont repertoire of most scleractinian corals in most places, the apparent near-universal presence of *Symbiodinium* phylotype D thus provides, according to Lien *et al.*, "a flexible means for corals to *routinely cope* [our italics] with environmental heterogeneities and survive the consequences (e.g., recover from coral bleaching)."

Also in 2007, Mieog *et al.* (2007) utilized a newly developed real-time polymerase chain reaction assay, which they say "is able to detect Symbiodinium clades C and D with >100-fold higher sensitivity compared to conventional techniques," to test 82 colonies of four common scleractinian corals (*Acropora millepora*, *Acropora tenuis*, *Stylophora pistillata* and *Turbinaria reniformis*) from eleven different locations on Australia's Great Barrier Reef

for evidence of the presence of background *Symbiodinium* clades. Results of the analysis showed that "ninety-three percent of the colonies tested were dominated by clade C and 76% of these had a D background," the latter of which symbionts, in their words, "are amongst the most thermo-tolerant types known to date," being found "on reefs that chronically experience unusually high temperatures or that have recently been impacted by bleaching events, suggesting that temperature stress can favor clade D." Consequently, Mieog *et al.* concluded that the clade D symbiont backgrounds detected in their study can potentially act as safety-parachutes, "allowing corals to become more thermo-tolerant through symbiont shuffling as seawater temperatures rise due to global warming." As a result, they suggest that symbiont shuffling is likely to play a role in the way "corals cope with global warming conditions," leading to new competitive hierarchies and, ultimately, "the coral community assemblages of the future."

In spite of the hope symbiont shuffling provides—that the world's corals will indeed be able to successfully cope with the possibility of future global warming, be it anthropogenic-induced or natural—some researchers have claimed that few coral symbioses host more than one type of symbiont, which has led some commentators to argue that symbiont shuffling is not an option for most coral species to survive the coming thermal onslaught of global warming. But is this claim correct? Not according to the results of Apprill and Gates (2007).

Working with samples of the widely distributed massive corals *Porites lobata* and *Porites lutea*—which they collected from Kaneohe Bay, Hawaii—Apprill and Gates compared the identity and diversity of *Symbiodinium* symbiont types obtained using cloning and sequencing of internal transcribed spacer region 2 (ITS2) with that obtained using the more commonly applied downstream analytical techniques of denaturing gradient gel electrophoresis (DGGE).

Results of the analysis revealed "a total of 11 ITS2 types in *Porites lobata* and 17 in *Porites lutea* with individual colonies hosting from one to six and three to eight ITS2 types for *P. lobata* and *P. lutea*, respectively." In addition, the two authors report that "of the clones examined, 93% of the *P. lobata* and 83% of the *P. lutea* sequences are not listed in GenBank," noting that they resolved "sixfold to eightfold greater diversity per coral species than previously reported."

In a "perspective" that accompanied Apprill and Gates' important paper, van Oppen (2007) wrote that "the current perception of coral-inhabiting symbiont diversity at nuclear ribosomal DNA is shown [by Apprill and Gates] to be a significant underestimate of the wide diversity that in fact exists." These findings, in her words, "have potentially far-reaching consequences in terms of our understanding of *Symbiodinium* diversity, host-symbiont specificity and the potential of corals to acclimatize to environmental perturbations through changes in the composition of their algal endosymbiont community," which assessment, it is almost unnecessary to say, suggests a greater than previously believed ability to do just that in response to any further global warming that might occur.

In a contemporaneous study, Baird *et al.* (2007) also discount the argument that symbiont shuffling is not an option for most coral species, because, "as they see it," it is the sub-clade that must be considered within this context, citing studies that indicate "there are both heat tolerant and heat susceptible sub-clades within both clades C and D *Symbiodinium*." Hence, the more relevant question becomes: How many coral species can host more than one *sub*-clade? The answer, of course, is that most if not all of them likely do; Baird *et al.* note that "biogeographical data suggest that when species need to respond to novel environments, they have the flexibility to do so."

So how and when might such sub-clade changes occur? Although most prior research in this area has been on adult colonies switching symbionts in response to warming-induced bleaching episodes, Baird *et al.* suggest that "change is more likely to occur between generations," for initial coral infection typically occurs in larvae or early juveniles, which are much more flexible than adults. In this regard, for example, they note that "juveniles of *Acropora tenuis* regularly harbor mixed assemblages of symbionts, whereas adults of the species almost invariably host a single clade," and they indicate that larvae of *Fungia scutaria* ingest symbionts from multiple hosts, although they generally harbor but one symbiont as adults.

Because of these facts, the Australian researchers say there is no need for an acute disturbance, such as bleaching, to induce clade or sub-clade change. Instead, if ocean temperatures rise to new heights in the future, they foresee juveniles naturally hosting more heat-tolerant sub-clades and maintaining them into adulthood.

In a further assessment of the size of the symbiont diversity reservoir, especially among juvenile coral species, Pochon *et al.* (2007) collected more than 1,000 soritid specimens over a depth of 40 meters on a single reef at Gun Beach on the island of Guam, Micronesia, throughout the course of an entire year, which they then studied by means of molecular techniques to identify unique *internal transcribed spacer-2* (ITS-2) types of *ribosomal DNA* (rDNA), in a project self-described as "the most targeted and exhaustive sampling effort ever undertaken for any group of *Symbiodinium*-bearing hosts."

Throughout the course of their analysis, Pochon *et al.* identified 61 unique symbiont types in only three soritid host genera, making the Guam *Symbiodinium* assemblage the most diverse derived to date from a single reef. In addition, they report that "the majority of mixed genotypes observed during this survey were usually harbored by the smallest hosts." As a result, the authors speculate that "juvenile foraminifera may be better able to switch or shuffle heterogeneous symbiont communities than adults," so that as juveniles grow, "their symbiont communities become 'optimized' for the prevailing environmental conditions," suggesting that this phenomenon "may be a key element in the continued evolutionary success of these protests in coral reef ecosystems worldwide."

In support of the above statement, we additionally cite the work of Mumby (1999), who analyzed the population dynamics of juvenile corals in Belize, both prior to, and after, a massive coral bleaching event in 1998. Although 70 percent to 90 percent of adult coral colonies were severely bleached during the event, only 25 percent of coral recruits exhibited signs of bleaching. What is more, one month after the event, it was concluded that "net bleaching-induced mortality of coral recruits ... was insignificant," demonstrating the ability of juvenile corals to successfully weather such bleaching events.

In light of these several observations, earth's corals will likely be able to successfully cope with the possibility of further increases in water temperatures, be they anthropogenic-induced or natural. Corals have survived such warmth—and worse—many times in the past, including the Medieval Warm Period, Roman Warm Period, and Holocene Optimum, as well as throughout numerous similar periods during a number of prior interglacial periods; there is no reason to believe they cannot do it again, if the need arises.

Additional information on this topic, including reviews of newer publications as they become available, can be found at http://www.co2 science.org/subject/c/bleachrespsymb.php.

References

Adjeroud, M., Augustin, D., Galzin, R. and Salvat, B. 2002. Natural disturbances and interannual variability of coral reef communities on the outer slope of Tiahura (Moorea, French Polynesia): 1991 to 1997. *Marine Ecology Progress Series* **237**: 121-131.

Apprill, A.M. and Gates, R.D. 2007. Recognizing diversity in coral symbiotic dinoflagellate communities. *Molecular Ecology* **16**: 1127-1134.

Baird, A.H., Cumbo, V.R., Leggat, W. and Rodriguez-Lanetty, M. 2007. Fidelity and flexibility in coral symbioses. *Marine Ecology Progress Series* **347**: 307-309.

Baker, A.C. 2001. Reef corals bleach to survive change. *Nature* **411**: 765-766.

Baker, A.C., Starger, C.J., McClanahan, T.R. and Glynn, P.W. 2002. Symbiont communities in reef corals following the 1997-98 El Niño—will recovering reefs be more resistant to a subsequent bleaching event? *Proceedings of the International Society of Reef Studies* (Abstract Volume 10: European Meeting, Cambridge, UK, September).

Baker, A.C., Starger, C.J., McClanahan, T.R. and Glynn, P.W. 2004. Corals' adaptive response to climate change. *Nature* **430**: 741.

Berkelmans, R. and van Oppen, M.J.H. 2006. The role of zooxanthellae in the thermal tolerance of corals: a "nugget of hope" for coral reefs in an era of climate change. *Proceedings of the Royal Society B* **273**: 2305-2312.

Brown, B.E., Clarke, K.R. and Warwick, R.M. 2002. Serial patterns of biodiversity change in corals across shallow reef flats in Ko Phuket, Thailand, due to the effects of local (sedimentation) and regional (climatic) perturbations. *Marine Biology* **141**: 24-29.

Buddemeier, R.W., Baker, A.C., Fautin, D.G. and Jacobs, J.R. 2004. The adaptive hypothesis of bleaching. In Rosenberg, E. and Loya, Y. (Eds.) *Coral Health and Disease*, Springer, Berlin, Germany, p. 427-444.

Buddemeier, R.W. and Fautin, D.G. 1993. Coral bleaching as an adaptive mechanism. *BioScience* **43**: 320-326.

Chen, C.A., Wang, J.-T., Fang, L.-S. and Yang, Y.W. 2005. Fluctuating algal symbiont communities in *Acropora palifera* (Schleractinia: Acroporidae) from Taiwan. *Marine Ecology Progress Series* **295**: 113-121.

Dunne, R.P. and Brown, B.E. 2001. The influence of solar radiation on bleaching of shallow water reef corals in the Andaman Sea, 1993-98. *Coral Reefs* **20**: 201-210.

Fagoonee, I., Wilson, H.B., Hassell, M.P. and Turner, J.R. 1999. The dynamics of zooxanthellae populations: A long-term study in the field. *Science* **283**: 843-845.

Gates, R.D. and Edmunds, P.J. 1999. The physiological mechanisms of acclimatization in tropical reef corals. *American Zoologist* **39**: 30-43.

Gleason, M.G. 1993. Effects of disturbance on coral communities: bleaching in Moorea, French Polynesia. *Coral Reefs* **12**: 193-201.

Glynn, P.W. 1996. Coral reef bleaching: facts, hypotheses and implications. *Global Change Biology* **2**: 495-509.

Hoegh-Guldberg, O. 1999. Climate change, coral bleaching and the future of the world's coral reefs. *Marine and Freshwater Research* **50**: 839-866.

Hoegh-Guldberg, O. and Salvat, B. 1995. Periodic mass-bleaching and elevated sea temperatures: bleaching of outer reef slope communities in Moorea, French Polynesia. *Marine Ecology Progress Series* **121**: 181-190.

Jones, A.M., Berkelmans, R., van Oppen, M.J.H., Mieog, J.C. and Sinclair, W. 2008. A community change in the algal endosymbionts of a scleractinian coral following a natural bleaching event: field evidence of acclimatization. *Proceedings of the Royal Society B* **275**: 1359-1365.

Kinzie III, R.A. 1999. Sex, symbiosis and coral reef communities. *American Zoologist* **39**: 80-91.

Kumaraguru, A.K., Jayakumar, K. and Ramakritinan, C.M. 2003. Coral bleaching 2002 in the Palk Bay, southeast coast of India. *Current Science* **85**: 1787-1793.

Lien, Y.-T., Nakano, Y., Plathong, S., Fukami, H., Wang, J.-T. and Chen, C.A. 2007. Occurrence of the putatively heat-tolerant *Symbiodinium* phylotype D in high-latitudinal outlying coral communities. *Coral Reefs* **26**: 35-44.

Lewis, C.L. and Coffroth, M.A. 2004. The acquisition of exogenous algal symbionts by an octocoral after bleaching. *Science* **304**: 1490-1492.

Little, A.F., van Oppen, M.J.H. and Willis, B.L. 2004. Flexibility in algal endosymbioses shapes growth in reef corals. *Science* **304**: 1492-1494.

Mieog, J.C., van Oppen, M.J.H., Cantin, N.E., Stam, W.T. and Olsen, J.L. 2007. Real-time PCR reveals a high incidence of *Symbiodinium* clade D at low levels in four scleractinian corals across the Great Barrier Reef: implications for symbiont shuffling. *Coral Reefs* **26**: 449-457.

Mumby, P.J. 1999. Bleaching and hurricane disturbances to populations of coral recruits in Belize. *Marine Ecology Progress Series* **190**: 27-35.

Pandolfi, J.M. 1999. Response of Pleistocene coral reefs to environmental change over long temporal scales. *American Zoologist* **39**: 113-130.

Pochon, X., Garcia-Cuetos, L., Baker, A.C., Castella, E. and Pawlowski, J. 2007. One-year survey of a single Micronesian reef reveals extraordinarily rich diversity of *Symbiodinium* types in soritid foraminifera. *Coral Reefs* **26**: 867-882.

Riegl, B. 2003. Climate change and coral reefs: different effects in two high-latitude areas (Arabian Gulf, South Africa). *Coral Reefs* **22**: 433-446.

Rowan, R. 2004. Thermal adaptation in reef coral symbionts. *Nature* **430**: 742.

Rowan, R. and Knowlton, N. 1995. Intraspecific diversity and ecological zonation in coral-algal symbiosis. *Proceeding of the National Academy of Sciences, U.S.A.* **92**: 2850-2853.

Rowan, R., Knowlton, N., Baker, A. and Jara, J. 1997. Landscape ecology of algal symbionts creates variation in episodes of coral bleaching. *Nature* **388**: 265-269.

Rowan, R. and Powers, D. 1991. Molecular genetic identification of symbiotic dinoflagellates (zooxanthellae). *Marine Ecology Progress Series* **71**: 65-73.

Trench, R.K. 1979. The cell biology of plant-animal symbiosis. *Annual Review of Plant Physiology* **30**: 485-531.

Van Oppen, M.J.H. 2007. Perspective. *Molecular Ecology* **16**: 1125-1126.

Van Oppen, M.J.H., Mahiny, A.J. and Done, T.J. 2005. Geographic distribution of zooxanthella types in three coral species on the Great Barrier Reef sampled after the 2002 bleaching event. *Coral Reefs* **24**: 482-487.

8.3.1.2.4. Bacterial Shuffling

One final adaptive bleaching mechanism is discussed by Reshef *et al.*, (2006), who developed a case for what they call the coral probiotic hypothesis, and what we call "bacterial shuffling." This concept, in their words, "posits that a dynamic relationship exists between symbiotic microorganisms and environmental conditions which brings about the selection of the most advantageous coral holobiont."

This concept is analogous to the adaptive bleaching hypothesis of Buddemeier and Fautin (1993), or what was referred to in the preceding section as symbiont shuffling, wherein corals exposed to some type of stress—such as that induced by exposure to unusually high water temperatures or solar irradiance—first lose their dinoflagellate symbionts (bleach) and then regain a new mixture of zooxanthellae that are better suited to the stress conditions. In fact, the two phenomena work in precisely the same way, in one case by the corals rearranging their zooxanthellae populations (symbiont shuffling) and in the other case by the corals rearranging their bacterial populations (bacterial shuffling).

In seeking evidence for their hypothesis, the team of Israeli researchers concentrated their efforts on looking for examples of corals developing resistance to emerging diseases. This approach makes sense, because corals lack an adaptive immune system; i.e., they possess no antibodies (Nair *et al.*, 2005), and they therefore can protect themselves against specific diseases in no other way than to adjust the relative sizes of the diverse bacterial populations associated with their mucus and tissues so as to promote the growth of those types of bacteria that tend to mitigate most effectively against the specific disease that happens to be troubling them.

Reshef *et al.* begin by describing the discovery that bleaching of *Oculina patagonica* corals in the Mediterranean Sea was caused by the bacterium *Vibrio shiloi*, together with the finding that bleaching of *Pocillopora damicornis* corals in the Indian Ocean and Red Sea was the result of an infection with *Vibrio coralliilyticus*. But they then report that (1) "during the last two years *O. patagonica* has developed resistance to the infection by *V. shiloi*," (2) "*V. shiloi* can no longer be found on the corals," and (3) "*V. shiloi* that previously infected corals are unable to infect the existing corals." They say "by some unknown mechanism, the coral is now able to lyse the intracellular *V. shiloi* and avoid the disease," and because corals lack the ability to produce antibodies and have no adaptive immune system, the only logical conclusion to be drawn from these observations is that bacterial shuffling must be what produced the welcome results.

With respect to the future of earth's corals in the context of global warming, the Israeli scientists note that "Hoegh-Guldberg (1999, 2004) has predicted that coral reefs will have only remnant populations of reef-building corals by the middle of this century," based on "the assumption that corals cannot adapt rapidly enough to the predicted temperatures in order to survive." However, they report that considerable evidence has been collected in support of the adaptive bleaching hypothesis, and they emphasize that the hundreds of different bacterial species associated with corals "give the coral holobiont an enormous genetic potential to adapt rapidly to changing environmental conditions." They say "it is not unreasonable to predict that under appropriate selection conditions, the change could take place in days or weeks, rather than decades required for classical Darwinian mutation and selection," and that "these rapid changes may allow the coral holobiont to use nutrients more efficiently, prevent colonization by specific pathogens and avoid death during bleaching by providing carbon and energy from photosynthetic prokaryotes," of which they say there is "a metabolically active, diverse pool" in most corals.

Additional information on this topic, including reviews of newer publications as they become available, can be found at http://www.co2 science.org/subject/c/bacterialshuffling.php.

References

Buddemeier, R.W. and Fautin, D.G. 1993. Coral bleaching as an adaptive mechanism. *BioScience* **43**: 320-326.

Hoegh-Guldberg, O. 1999. Climate change, coral bleaching and the future of the world's coral reefs. *Marine and Freshwater Research* **50**: 839-866.

Hoegh-Guldberg, O. 2004. Coral reefs in a century of rapid environmental change. *Symbiosis* **37**: 1-31.

Nair, S.V., Del Valle, H., Gross, P.S., Terwilliger, D.P. and Smith, L.C. 2005. Macroarray analysis of coelomocyte gene expression in response to LPS in the sea urchin. Identification of unexpected immune diversity in an invertebrate. *Physiological Genomics* **22**: 33-47.

Reshef, L., Koren, O., Loya, Y., Zilber-Rosenberg, I. and Rosenberg, E. 2006. The coral probiotic hypothesis. *Environmental Microbiology* **8**: 2068-2073.

8.3.1.3. Widespread Coral Bleaching

8.3.1.3.1. Not Caused by Global Warming

Hoegh-Guldberg (1999) concluded that "coral bleaching is due to warmer than normal temperatures" and that "increased sea temperature is the primary reason for why coral bleaching has occurred with increasing intensity and frequency over the past two decades." As outlined in the preceding sections, there is some evidence that points toward these conclusions, but there is much other evidence that points to alternative possibilities.

Consider, for example, the persistence of coral reefs through geologic time, which provides substantive evidence that these ecological entities can successfully adapt to a dramatically changing global environment (Veron, 1995). What can their history tell us about bleaching and global warming in our day?

The earliest coral reefs date to the Palaeozoic Era, over 450 million years ago (Hill, 1956). The scleractinian corals, which are the major builders of the reefs of today (Achituv and Dubinsky, 1990), appeared in the mid-Triassic some 240 million years later (Hill, 1956), when the earth was considerably warmer than it is currently (Chadwick-Furman, 1996). Although reef-building ceased for a time following the extinctions at the end of the Triassic, the Scleractinia came back with a vengeance during the Jurassic (Newell, 1971; Veron, 1995); they continued to exhibit great robustness throughout the Cretaceous, even when temperatures were as much as 8-15°C higher (Chadwick-Furman, 1996; Veizer *et al.*, 1999), and atmospheric CO_2 concentrations two to seven times higher (Berner and Kothavala, 2001), than present.

At the end of the Cretaceous, 70 percent of the genera and one-third of the families of scleractinian corals disappeared (Veron, 1995) in the greatest biospheric extinction event in geological history, which may possibly have been caused by a large asteroid impact (Alvarez *et al.*, 1980, 1984). They developed again, however, throughout the Cenozoic, particularly the Oligocene and Miocene (Chadwick-Furman, 1996). Finally, throughout the past two million years of the Pleistocene, they survived at least seventeen glacial-interglacial cycles of dramatic climate change and sea-level fluctuation, successfully adapting, over and over again, to these enormous environmental challenges (Pandolfi, 1999). In the words of Benzie (1999), this evidence suggests that "coral reef communities are relatively resilient, have survived previous global climate change, and appear likely to survive future changes." This conclusion leads us to wonder why corals should be succumbing to global warming now.

To answer such an inquiry we must first address the question of what is "normal" for coral reefs in our day. Is it what they look like now? Or what they looked like 30 years ago? Or 300 years ago? Kinzie (1999) has emphatically stated that "it is clear that the definition of a healthy reef as 'what it looked like when I started diving' is fraught not only with hubris but strong temporal bias." Indeed, as Greenstein *et al.* (1998a) have observed, "it must be demonstrated that the classic reef coral zonation pattern described in the early days of coral reef ecology, and upon which 'healthy' versus 'unhealthy' reefs are determined, are themselves representative of reefs that existed prior to any human influence." Only when this criterion is met will we have, in the words of Greenstein *et al.* (1998b), a good replacement for "the temporally myopic view afforded by monitoring studies that rarely span a scientific career." Clearly, therefore, there should be no argument over the key fact that we need a proper understanding of the past to correctly judge the present if we ever are to foretell the future.

In an attempt to obtain a true picture of pristine coral conditions in the western North Atlantic and Caribbean, Greenstein *et al.* (1998a, 1998b) conducted systematic censuses of "life assemblages" and "death assemblages" of corals on healthy modern patch reefs and compared the results with similar censuses they conducted on "fossil assemblages" preserved in Pleistocene limestones in close proximity to the modern reefs. The data revealed a recent decline in thickets of *Acropora cervicornis*, as evidenced by their abundance in the death assemblage, and a concurrent increase in *Porites porites*, as evidenced by their abundance in the life assemblage. In comparing these results with those obtained from the fossil assemblage, they found that the present Caribbean-wide decline of *A. cervicornis* is "without historical precedent" and that it is a dramatic departure from "the long-term persistence of this taxon during Pleistocene and Holocene Optimum time," when "intensifying cycles in climate and sea level" recurred throughout a roughly one-million-year time period.

These observations, along with the similar findings of Jackson (1992) and Aronson and Precht (1997), suggest that if little change in coral community structure occurred throughout the Pleistocene—when it was often warmer than it is now (Petit *et al.*, 1999)—the recent die-off of *A. cervicornis* cannot be due to global warming alone, or even primarily, for this particular coral has clearly weathered several major episodes of global warming and elevated water temperatures in the past with no adverse consequences. Neither can the coral's die-off be due to the CO_2-induced decrease in seawater calcium carbonate saturation state that might possibly be occurring at the present time (see the section later in this chapter on coral calcification); for the air's CO_2 content has not risen sufficiently to have caused this parameter to decline enough to significantly impact reef coral calcification rates (Gattuso *et al.*, 1998, 1999), as is also demonstrated by the opportunistic replacement of *A. cervicornis* by *P. porites*. In addition, in their detailed reconstruction of the history of calcification rates in massive *Porites* colonies from Australia's Great Barrier Reef, Lough and Barnes (1997) report that the mid-twentieth century had the second highest coral growth rate of the past 237 years. Hence, although *A. cervicornis* has indeed suffered an extreme decrease in abundance throughout the Caribbean in recent years (Hughes, 1994), its precipitous decline cannot be attributed to either global warming or the direct effects of rising CO_2.

In light of these data-driven considerations, Greenstein *et al.* (1998a, 1998b) have attributed the increasing coral bleaching of the past two decades to a host of local anthropogenic impacts. This conclusion is accepted in a much wider context as well, as Buddemeier and Smith (1999) have noted that "reviews of the problems facing coral reefs have consistently emphasized that local and regional anthropogenic impacts are a far greater immediate threat to coral reefs than Greenhouse-enhanced climate change."

References

Achituv, Y., and Dubinsky, Z. 1990. Evolution and zoogeography of coral reefs. Dubinsky, Z. (Ed.) *Ecosystems of the world: Coral reefs*. Elsevier, New York, NY.

Alvarez, L.W., Alvarez, W., Asaro, F. and Michel, H.V. 1980. Extraterrestrial cause for the Cretaceous-Tertiary extinction. *Science* **208**: 1095-1108.

Alvarez, W., Alvarez, L.W., Asaro, F. and Michel, H.V. 1984. The end of the Cretaceous: Sharp boundary or gradual transition? *Science* **223**: 1183-1186.

Aronson, R.B. and Precht, W.F. 1997. Stasis, biological disturbance, and community structure of a Holocene coral reef. *Paleobiology* **23**: 326-346.

Benzie, J.A.H. 1999. Genetic structure of coral reef organisms: Ghosts of dispersal past. *American Zoologist* **39**: 131-145.

Berner, R.A. and Kothavala, Z. 2001. GEOCARB III: A revised model of atmospheric CO_2 over phanerozoic time. *American Journal of Science* **301**: 182-204.

Buddemeier, R.W. and Smith, S.V. 1999. Coral adaptation and acclimatization: A most ingenious paradox. *American Zoologist* **39**: 1-9.

Chadwick-Furman, N.E. 1996. Reef coral diversity and global change. *Global Change Biology* **2**: 559-568.

Gattuso, J.-P., Allemand, D. and Frankignoulle, M. 1999. Photosynthesis and calcification at cellular, organismal and community levels in coral reefs: A review on interactions and control by carbonate chemistry. *American Zoologist* **39**: 160-183.

Gattuso, J.-P., Frankignoulle, M., Bourge, I., Romaine, S. and Buddemeier, R.W. 1998. Effect of calcium carbonate saturation of seawater on coral calcification. *Global and Planetary Change* **18**: 37-46.

Greenstein, B.J., Curran, H.A. and Pandolfi, J.M. 1998a. Shifting ecological baselines and the demise of *Acropora cervicornis* in the western North Atlantic and Caribbean Province: A Pleistocene perspective. *Coral Reefs* **17**: 249-261.

Greenstein, B.J., Harris, L.A. and Curran, H.A. 1998b. Comparison of recent coral life and death assemblages to Pleistocene reef communities: Implications for rapid faunal replacement on recent reefs. *Carbonates & Evaporites* **13**: 23-31.

Hill, D. Rugosa and Moore, R.D. 1956. (Eds.) *Treatise on invertebrate paleontology*, Volume F, Coelenterata. Lawrence, KS: Geological Society of America/University of Kansas Press, pp. 233-323.

Hoegh-Guldberg, O. 1999. Climate change, coral bleaching and the future of the world's coral reefs. *Marine and Freshwater Research* **50**: 839-866.

Hughes, T.P. 1994. Catastrophes, phase shifts, and large-scale degradation of a Caribbean coral reef. *Science* **265**: 1547-1551.

Jackson, J.B.C. 1992. Pleistocene perspectives on coral reef community structure. *American Zoologist* **32**: 719-731.

Kinzie III, R.A. 1999. Sex, symbiosis and coral reef communities. *American Zoologist* **39**: 80-91.

Lough, J.M. and Barnes, D.J. 1997. Several centuries of variation in skeletal extension, density and calcification in massive *Porites* colonies from the Great Barrier Reef: A proxy for seawater temperature and a background of variability against which to identify unnatural change. *Journal of Experimental Marine Biology and Ecology* **211**: 29-67.

Newell, N.D. 1971. An outline history of tropical organic reefs. *Novitates* **2465**: 1-37.

Pandolfi, J.M. 1999. Response of Pleistocene coral reefs to environmental change over long temporal scales. *American Zoologist* **39**: 113-130.

Petit, J.R., Jouzel, J., Raynaud, D., Barkov, N.I., Barnola, J.-M., Basile, I., Bender, M., Chappellaz, J., Davis, M.,

Delaygue, G., Delmotte, M., Kotlyakov, V.M., Legrand, M., Lipenkov, V.Y., Lorius, C., Pepin, L., Ritz, C., Saltzman, E. and Stievenard, M. 1999. Climate and atmospheric history of the past 420,000 years from the Vostok ice core, Antarctica. *Nature* **399**: 429-436.

Veizer, J., Ala, D., Azmy, K., Bruckschen, P., Buhl, D., Bruhn, F., Carden, G.A.F., Diener, A., Ebneth, S., Godderis, Y., Jasper, T., Korte, C., Pawellek, F., Podlaha, O. and Strauss, H. 1999. $^{87}Sr/^{86}Sr$, $\delta^{13}C$ and $\delta^{18}O$ evolution of Phanerozoic seawater. *Chemical Geology* **161**, 59-88.

Veron, J.E.N. 1995. *Corals in space and time.* Comstock/Cornell, Ithaca, NY.

8.3.1.3.2. An Alternative Hypothesis

The preceding considerations clearly indicate global warming cannot be the primary cause of the massive coral bleaching the earth has experienced in recent years. However, the IPCC tenaciously clings to this hypothesis because (1) no significant massive and widespread coral bleaching was reported in the 1970s and (2) the global warming hypothesis can account for this observation. Specifically, Hoegh-Guldberg (1999) has suggested that the reason "why mass bleaching events are not seen prior to 1980" is that "increases in sea temperatures have only become critical since in the 1980s, when El Niño disturbances began to exceed the thermal tolerances of corals and their zooxanthellae" as a result of global warming increasing the background temperature to which El Niño thermal effects are added.

This reasoning assumes no other theory is capable of accounting for the fact that modern mass bleaching events did not begin to occur until 1980. On the basis of this *assumption*, Hoegh-Guldberg (1999) concludes that the global warming hypothesis must be correct, even in light of the many problems associated with it. This assumption, however, is not true, for there are other ways of satisfying this critical criterion that do account for the lack of bleaching episodes before 1980, which we describe below.

The North Atlantic Oscillation (NAO) is responsible for multiannual to decadal variability in Northern Hemispheric climate that is numerically represented by the pressure difference between the Azores high and the Icelandic low (Dugam *et al.*, 1997). It has been documented over the past 350 years in Greenland ice core reconstructions (Appenzeller *et al.*, 1998) and explicitly quantified from 1864 through 1994 via actual pressure records

(Hurrell, 1995), which have been updated through 1998 by Uppenbrink (1999).

Plots of these NAO datasets reveal a shift from strong negative index values in the 1950s and 1960s to what Hurrell (1995) describes as "unprecedented strongly positive NAO index values since 1980." This observation is especially important, for during times of high NAO index values, there is a significant reduction in atmospheric moisture transport across southern Europe, the Mediterranean, and north Africa (Hurrell, 1995); and Richardson *et al.* (1999) note that this phenomenon has led to the development of prolonged drought in the Sahel region of Africa since the NAO shift to positive index values in 1980.

One consequence of this drought has been a gradual increase in the dust content of the atmosphere, which in some areas has grown to five-fold what was deemed normal prior to this climatic transition (Richardson *et al.*, 1999). Of particular significance to corals is the fact that this airborne dust carries bacteria, viruses, and fungi that can kill them. Pearce (1999) notes that outbreaks of a number of coral diseases "have coincided with years when the dust load in the atmosphere was highest." In 1983, for example—when the NAO index reached its highest value since 1864 (Hurrell, 1995) and the atmosphere was exceptionally dusty—a soil fungus of the *Aspergillus* genus appeared in the Caribbean, initiating an onslaught of soft coral sea fans that has now destroyed more than 90 percent of them. Pearce (1999) notes there are solid scientific reasons for concluding that "the speed and pattern of the fungus's spread indicates that it could only have arrived on the trade winds from Africa."

In addition to carrying its deadly biological cargo, the positive-NAO-induced airborne dust is rich in iron, which extra supply, in the words of R.T. Barber as quoted by Pearce (1999), "may have spurred the worldwide growth of a variety of invader organisms harmful to coral ecosystems." Such iron-rich dust has the capacity to fertilize algae that compete with zooxanthellae for other scarce nutrients and reef living space. Abram *et al.* (2003), for example, reported that a massive coral bleaching event that killed close to 100 percent of the coral and fish in the reef ecosystem of the Mentawai Islands (located southwest of Sumatra, Indonesia, in the equatorial eastern Indian Ocean) in 1997-1998 was brought about by an anomalous influx of iron provided by atmospheric fallout from the 1997 Indonesian wildfires, which they describe as being "the worst wildfires in the recorded history of southeast Asia."

The enhanced burden of iron, in turn, spawned a large phytoplankton bloom that likely caused the coral and fish death via asphyxiation. In concluding their paper, Abram *et al.* warn that "widespread tropical wildfire is a recent phenomenon, the magnitude and frequency of which are increasing as population rises and terrestrial biomass continues to be disrupted," and "reefs are likely to become increasingly susceptible to large algal blooms triggered by episodic nutrient enrichment from wildfires," which phenomenon, in their words, "may pose a new threat to coastal marine ecosystems that could escalate into the 21st century."

The timeline for the appearance and progression of these several related phenomena matches perfectly with the timeline of the historical buildup of modern coral reef bleaching throughout the 1980s and 1990s. This is not to say, however, that these aggregate phenomena comprise the entire answer to the problem to the exclusion of all other possible causes, even including global warming. We only suggest that they, too, must be seriously considered in attempts to identify the true cause or causes of this development in coral reef history.

Although one can make the case for coral bleaching being caused by global warming, there are too many pieces of evidence that contradict this hypothesis for it to be deemed the sole, or even primary, cause of this modern problem. There is at least one alternative explanation—the unprecedented, strongly positive NAO since 1980—during the past two decades.

References

Abram, N.J., Gagan, M.K., McCulloch, M.T., Chappell, J. and Hantoro, W.S. 2003. Coral reef death during the 1997 Indian Ocean Dipole linked to Indonesian wildfires. *Science* **301**: 952-955.

Appenzeller, C., Stocker, T.F. and Anklin, M. 1998. North Atlantic Oscillation dynamics recorded in Greenland ice cores. *Science* **282**: 446-449.

Dugam, S.S., Kakade, S.B. and Verma, R.K. 1997. Interannual and long-term variability in the North Atlantic Oscillation and Indian summer monsoon rainfall. *Theoretical and Applied Climatology* **58**: 21-29.

Hoegh-Guldberg, O. 1999. Climate change, coral bleaching and the future of the world's coral reefs. *Marine and Freshwater Research* **50**: 839-866.

Hurrell, J.W. 1995. Decadal trends in the North Atlantic Oscillation: Regional temperatures and precipitation. *Science* **269**: 676-679.

Pearce, F. 1999. Coral grief: The cause of reef health problems may be blowing in the wind. *New Scientist* **163** (2193): 22.

Richardson, L.L., Porter, J.W. and Barber, R.T. 1999. Status of the health of coral reefs: An update. U.S. Global Change Research Program Seminar Series, http://www.usgcrp.gov.

Uppenbrink, J. 1999. The North Atlantic Oscillation. *Science* **283**: 948-949.

8.3.1.4. Sea-level Rise

Many people believe rising sea levels, by gradually reducing the amount of life-sustaining light that reaches their algal symbionts, will decimate earth's corals. This assumption is a major concern often expressed in discussions of reef responses to global climate change (Hopley and Kinsey, 1988). But it is probably not valid, for a number of reasons.

First, the 18- to 59-cm warming-induced sea-level rise that is predicted for the coming century (IPCC, 2007)—which could be greatly exaggerated if predictions of CO_2-induced global warming are wrong—falls well within the range (2 to 6 mm per year) of typical coral vertical extension rates, which exhibited a modal value of 7 to 8 mm per year during the Holocene and can be more than double that value in certain branching corals (Hopley and Kinsey, 1988; Done, 1999). Second, most coral reefs are known to have successfully responded to the sea-level rises that occurred between 14,000 and 6,000 years ago—which were accompanied by large changes in "CO_2 concentrations, ... rainfall, cloud cover, storms and currents" (Wilkinson, 1996)—and which were more than twice as rapid as what is being predicted for the coming century (Digerfeldt and Hendry, 1987). Third, earth's oceans have undergone—and their coral reefs have survived (Chadwick-Furman, 1996)—at least 17 major cycles of sea-level rise and fall during the Pleistocene, the most recent low phase of which ended 18,000 years ago with a global sea level some 120-135 meters below where it is now (Grigg and Epp, 1989). Fourth, most coral reefs handle increases in sea level—even *rapid* increases—much better than decreases (White *et al.*, 1998). Yet even if reef vertical growth rates could not keep up with rising sea levels, that would not spell their doom.

One of the important characteristics of essentially all reef cnidarians is their ability to produce free-swimming planulae, spores, or dispersive larval stages. Kinzie (1999) notes that "no matter how quickly sea level might rise, propagules of the species could keep pace and settle at suitable depths each generation," thereby creating what he calls jump-up reefs that "might well contain most of the species present in the original community." Done (1999) notes that "coral communities have a history of tracking their preferred environmental niche which may suggest that as an entity, they will be predisposed to 'adapt' to prospective changes in environment over the next century," citing precedents that clearly demonstrate that "coral communities have historically had a good capacity to track their re-distributed preferred physical niches."

It is not at all surprising, therefore, as Kinzie and Buddemeier (1996) recount, that coral reefs have survived many periods of "massive environmental changes" throughout the geologic record. Reefs are survivors, they state, "because they do not simply tolerate environmental changes" but "exhibit an impressive array of acclimations" that allow them to deal with a variety of challenges to their continued existence in any given area. Hence, it is highly unlikely that anticipated increases in sea level would spell the doom of earth's coral reefs.

Strange as it may seem, rising sea levels may have a *positive* effect on earth's coral reefs (Roberts, 1993). Over the past 6,000 years, relatively stable sea levels have limited upward reef growth, resulting in the development of extensive reef flats; as Buddemeier and Smith (1988) and Wilkinson (1996) have noted, the sea-level rises predicted to result from CO_2-induced global warming should actually be beneficial, permitting increased growth in these growth-restricted areas. In the words of Chadwick-Furman (1996), "many coral reefs have already reached their upward limit of growth at present sea level (Buddemeier, 1992), and may be released from this vertical constraint by a rise in sea level." She also notes that rising sea levels may allow more water to circulate between segregated lagoons and outer reef slopes, which could "increase the exchange of coral propagules between reef habitats and lead to higher coral diversity in inner reef areas." She, too, concludes that "coral reefs are likely to survive predicted rates of global change."

References

Buddemeier, R.W. 1992. Corals, climate and conservation. *Proceedings of the Seventh International Coral Reef Symposium* **1**: 3-10.

Buddemeier, R.W. and Smith, S.V. 1988. Coral-reef growth in an era of rapidly rising sea-level—predictions and suggestions for long-term research. *Coral Reefs* **7**: 51-56.

Chadwick-Furman, N.E. 1996. Reef coral diversity and global change. *Global Change Biology* **2**: 559-568.

Digerfeldt, G. and Hendry, M.D. 1987. An 8000 year Holocene sea-level record from Jamaica: Implications for interpretation of Caribbean reef and coastal history. *Coral Reefs* **5**: 165-170.

Done, T.J. 1999. Coral community adaptability to environmental change at the scales of regions, reefs and reef zones. *American Zoologist* **39**: 66-79.

Grigg, R.W. and Epp, D. 1989. Critical depth for the survival of coral islands: Effects on the Hawaiian Archipelago. *Science* **243**: 638-641.

Hopley, D. and Kinsey, D.W. 1988. The effects of a rapid short-term sea-level rise on the Great Barrier Reef. Pearman, G.I. (Ed.) *Greenhouse: Planning for climate change.* CSIRO Publishing, East Melbourne, Australia, pp. 189-201.

IPCC, 2007-II. Climate Change 2007: Impacts, Adaptation and Vulnerability. Contribution of Working Group II to the Fourth Assessment Report of the Intergovernmental Panel on Climate Change. Parry, M.L., Canziani, O.F., Palutikof, J.P., van der Linden, P.J. and Hanson, C.E. (Eds.) Cambridge University Press, Cambridge, UK.

Kinzie III, R.A. 1999. Sex, symbiosis and coral reef communities. *American Zoologist* **39**: 80-91.

Kinzie III, R.A. and Buddemeier, R.W. 1996. Reefs happen. *Global Change Biology* **2**: 479-494.

Roberts, C.M. 1993. Coral reefs: Health, hazards and history. *Trends in Ecology and Evolution* **8**: 425-427.

White, B., Curran, H.A. and Wilson, M.A. 1998. Bahamian coral reefs yield evidence of a brief sea-level lowstand during the last interglacial. *Carbonates & Evaporites* **13**: 10-22.

Wilkinson, C.R. 1996. Global change and coral reefs: Impacts on reefs, economies and human cultures. *Global Change Biology* **2**: 547-558.

8.3.2. Direct Threats

In addition to the CO_2-induced indirect threats postulated to harm the world's coral reefs, as discussed in Section 8.3.1. of this document, the global increase in the atmosphere's CO_2 content has been hypothesized to possess the potential to harm coral reefs directly. By inducing changes in ocean water chemistry that can lead to reductions in the calcium carbonate saturation state of seawater, it has been predicted that elevated levels of atmospheric CO_2 may reduce rates of coral calcification, possibly leading to slower-growing—and, therefore, weaker—coral skele-tons, and in some cases even death.

We begin this next section by discussing the important role biology plays in driving the physical-chemical process of coral calcification, followed by a discussion of several real-world observations that depict increasing rates of coral calcification in the face of rising temperatures and atmospheric CO_2 concentrations.

8.3.2.1. Ocean Acidification

The rate of deposition of calcium carbonate on coral reefs, or coral calcification rate, is controlled at the cellular level by the saturation state of calcium carbonate in seawater. Oceanic surface waters have likely been saturated or supersaturated in this regard—providing a good environment for coral reef growth—since early Precambrian times (Holland, 1984). Currently, however, as the air's CO_2 content rises in response to anthropogenic CO_2 emissions, more carbon dioxide dissolves in the surface waters of the world's oceans, and the pH values of the planet's oceanic waters are, or should be, gradually dropping, leading to a reduction in the calcium carbonate saturation state of seawater.

This phenomenon has been theorized to be leading to a corresponding reduction in coral calcification rates (Smith and Buddemeier, 1992; Buddemeier, 1994; Buddemeier and Fautin, 1996a,b; Holligan and Robertson, 1996; Gattuso *et al.*, 1998; Buddemeier and Smith, 1999; IPCC, 2007-I, 2007-II; De'ath *et al.*, 2009), which reduction has been hypothesized to be rendering corals more susceptible to a number of other environmental stresses, including "sea-level rise, extreme temperatures, human damage (from mining, dredging, fishing and tourism), and changes in salinity and pollutant concentrations (nutrients, pesticides, herbicides and particulates), and in ocean currents, ENSO, and storm damage" (Pittock, 1999). Kleypas *et al.* (1999), for example, have calculated that calcification rates of tropical corals should already have declined by 6 percent to 11 percent or more since 1880, as a result of the concomitant increase in atmospheric CO_2 concentration, and they predict that the reductions could reach 17 percent to 35 percent by 2100. Likewise, Langdon *et al.* (2000) calculated a decrease in coral calcification rate of up to 40 percent between 1880 and 2065.

The ocean chemistry aspect of this theory is rather straightforward, but it certainly is not as solid as some commentators make it out to be. In evaluating global seawater impacts of model-predicted global warming and direct seawater chemical consequences of a doubling of the air's CO_2 content, Loaiciga (2006) used a mass-balance approach to "estimate the change in average seawater salinity caused by the melting of terrestrial ice and permanent snow in a warming earth." He applied "a chemical equilibrium model for the concentration of carbonate species in seawater open to the atmosphere" in order to "estimate the effect of changes in atmospheric CO_2 on the acidity of seawater." Assuming that the rise in the planet's mean surface air temperature continues unabated, and that it eventually causes the melting of all terrestrial ice and permanent snow, Loaiciga calculated that "the average seawater salinity would be lowered not more than 0.61% from its current 35%." He also reports that across the range of seawater temperature considered (0° to 30°C), "a doubling of CO_2 from 380 ppm to 760 ppm increases the seawater acidity [lowers its pH] approximately 0.19 pH units." He thus concludes that "on a global scale and over the time scales considered (hundreds of years), there would not be accentuated changes in either seawater salinity or acidity from the rising concentration of atmospheric CO_2."

Furthermore, with more CO_2 in the air, additional weathering of terrestrial carbonates is likely to occur, which would increase delivery of Ca^{2+} to the oceans and partly compensate for the CO_2-induced decrease in oceanic calcium carbonate saturation state (Riding, 1996). And as with all phenomena involving living organisms, the introduction of *life* into the ocean acidification picture greatly complicates things. Considerations of a suite of interrelated biological phenomena, for example, also make it much more difficult to draw such sweeping negative conclusions as are currently being discussed. Indeed, as shown in

the next section, they even suggest that the rising CO_2 content of earth's atmosphere may be a positive phenomenon, enhancing the growth rates of coral reefs and helping them to better withstand the many environmental stresses that truly are inimical to their well-being.

References

Buddemeier, R.W. 1994. Symbiosis, calcification, and environmental interactions. *Bulletin Institut Oceanographique*, Monaco, no. special **13**, pp. 119-131.

Buddemeier, R.W. and Fautin, D.G. 1996a. Saturation state and the evolution and biogeography of symbiotic calcification. *Bulletin Institut Oceanographique*, Monaco, no. special **14**, pp. 23-32.

Buddemeier, R.W. and Fautin, D.G. 1996b . Global CO_2 and evolution among the Scleractinia. *Bulletin Institut Oceanographique*, Monaco, no. special **14**, pp. 33-38.

Buddemeier, R.W. and Smith, S.V. 1999. Coral adaptation and acclimatization: A most ingenious paradox. *American Zoologist* **39**: 1-9.

De'ath, G., Lough, J.M. and Fabricius, K.E. 2009. Declining coral calcification on the Great Barrier Reef. *Science* **323**: 116-119.

Gattuso, J.-P., Frankignoulle, M., Bourge, I., Romaine, S. and Buddemeier, R.W. 1998. Effect of calcium carbonate saturation of seawater on coral calcification. *Global and Planetary Change* **18**: 37-46.

Holland, H.D. 1984. *The chemical evolution of the atmosphere and oceans*. Princeton University Press, Princeton, NJ.

Holligan, P.M. and Robertson, J.E. 1996. Significance of ocean carbonate budgets for the global carbon cycle. *Global Change Biology* **2**: 85-95.

IPCC, 2007-I. *Climate Change 2007: The Physical Science Basis. Contribution of Working Group I to the Fourth Assessment Report of the Intergovernmental Panel on Climate Change.* Solomon, S., D. Qin, M. Manning, Z. Chen, M. Marquis, K.B. Averyt, M. Tignor and H.L. Miller. (Eds.) Cambridge University Press, Cambridge, UK.

IPCC, 2007-II. *Climate Change 2007: Impacts, Adaptation and Vulnerability. Contribution of Working Group II to the Fourth Assessment Report of the Intergovernmental Panel on Climate Change.* Parry, M.L., Canziani, O.F., Palutikof, J.P., van der Linden, P.J. and Hanson, C.E. (Eds.) Cambridge University Press, Cambridge, UK.

Kleypas, J.A., Buddemeier, R.W., Archer, D., Gattuso, J-P., Langdon, C. and Opdyke, B.N. 1999. Geochemical consequences of increased atmospheric carbon dioxide on coral reefs. *Science* **284**: 118-120.

Langdon, C., Takahashi, T., Sweeney, C., Chipman, D., Goddard, J., Marubini, F., Aceves, H., Barnett, H. and Atkinson, M.J. 2000. Effect of calcium carbonate saturation state on the calcification rate of an experimental coral reef. *Global Biogeochemical Cycles* **14**: 639-654.

Loaiciga, H.A. 2006. Modern-age buildup of CO_2 and its effects on seawater acidity and salinity. *Geophysical Research Letters* **33**: 10.1029/2006 GL026305.

Pittock, A.B. 1999. Coral reefs and environmental change: Adaptation to what? *American Zoologist* **39**: 10-29.

Riding, R. 1996. Long-term change in marine $CaCO_3$ precipitation. *Mem. Soc. Geol. Fr.* **169**: 157-166.

Smith, S.V. and Buddemeier, R.W. 1992. Global change and coral reef ecosystems. *Annual Review of Ecological Systems* **23**: 89-118.

8.3.2.1.1. Role of Biology

Over half a century ago, Kawaguti and Sakumoto (1948) illustrated the important role played by photosynthesis in the construction of coral reefs. They analyzed numerous datasets recorded in several earlier publications, demonstrating that coral calcification rates are considerably higher in the daylight (when photosynthesis by coral symbionts occurs) than they are in the dark (when the symbionts lose carbon via respiration). A number of more modern studies have also demonstrated that symbiont photosynthesis enhances coral calcification (Barnes and Chalker, 1990; Yamashiro, 1995); and they have further demonstrated that long-term reef calcification rates generally rise in direct proportion to increases in rates of reef primary production (Frankignoulle *et al.*, 1996; Gattuso *et al.*, 1996, 1999). The work of Muscatine (1990) suggests that "the photosynthetic activity of zooxanthellae is the chief source of energy for the energetically expensive process of calcification" (Hoegh-Guldberg, 1999). Consequently, if an anthropogenic-induced increase in the transfer of CO_2 from the atmosphere to the world's oceans, i.e., hydrospheric CO_2 enrichment, were to lead to increases in coral symbiont photosynthesis—as atmospheric CO_2 enrichment does for essentially all terrestrial plants (Kimball, 1983; Idso, 1992)—it is likely that increases in coral calcification rates would occur as well.

There are several reasons for expecting a positive coral calcification response to CO_2-enhanced symbiont photosynthesis. One of the first mechanisms to come to mind is the opposite of the phenomenon that has been proffered as a cause of future declines in coral calcification rates. This reverse phenomenon is the decrease in extracellular CO_2 partial pressure in coral tissues that is driven by the drawdown of aqueous CO_2 caused by the photosynthetic process. With CO_2 being removed from the water in intimate contact with the coral host via its fixation by photosynthesis (which CO_2 drawdown is of greater significance to the coral than the increase in the CO_2 content of the surrounding bulk water that is affected by the ongoing rise in the air's CO_2 content), the pH and calcium carbonate saturation state of the water immediately surrounding the coral host should *rise* (Goreau, 1959), enhancing the coral's calcification rate (Gattuso *et al.*, 1999). If hydrospheric CO_2 enrichment stimulates zooxanthellae photosynthesis to the same degree that atmospheric CO_2 enrichment stimulates photosynthesis in terrestrial plants, i.e., by 30 percent to 50 percent for a 300 ppm increase in CO_2 concentration (Kimball, 1983; Idso 1992, Idso and Idso, 1994), this phenomenon alone would more than compensate for the drop in the calcium carbonate saturation state of the bulk-water of the world's oceans produced by the ongoing rise in the air's CO_2 content, which Gattuso *et al.* (1999) have calculated could lead to a 15 percent reduction in coral calcification rate for a doubling of the pre-industrial atmospheric CO_2 concentration.

Another reason why coral calcification may proceed at a higher rate in the presence of CO_2-stimulated symbiont photosynthesis is that, while growing more robustly, the zooxanthellae may take up more of the metabolic waste products of the coral host, which, if present in too great quantities, can prove detrimental to the health of the host, as well as the health of the entire coral plant-animal assemblage (Yonge, 1968; Crossland and Barnes, 1974). There are also a number of other substances that are known to directly interfere with calcium carbonate precipitation; they too can be actively removed from the water by coral symbionts in much the same way that symbionts remove host waste products (Simkiss, 1964). More importantly, perhaps, a greater amount of symbiont-produced photosynthates may provide more fuel for the active transport processes involved in coral calcification (Chalker and Taylor, 1975), as well as more raw materials for the synthesis of the coral organic matrix (Wainwright, 1963; Muscatine,

1967; Battey and Patton, 1984). Finally, the photosynthetic process helps to maintain a healthy aerobic or oxic environment for the optimal growth of the coral animals (Rinkevich and Loya, 1984; Rands *et al.*, 1992); greater CO_2-induced rates of symbiont photosynthesis would enhance this important "environmental protection activity."

Such observations invoke a number of questions. With ever more CO_2 going into the air, driving ever more CO_2 into the oceans, might we not logically expect to see increasingly greater rates of coral symbiont photosynthesis, due to the photosynthesis-stimulating effect of hydrospheric CO_2 enrichment? Would not this phenomenon, in turn, increasingly enhance all of the many positive photosynthetic-dependent phenomena we have described and thereby increase coral calcification rates? And might it not increase these rates well beyond the point of overpowering the modest negative effect of the purely chemical consequences of elevated dissolved CO_2 on ocean pH and calcium carbonate saturation state?

The answers to these several questions are probably all "yes," but arriving at these conclusions is not as simple as it sounds. For one thing, although many types of marine plant life do indeed respond to hydrospheric CO_2 enrichment (Raven *et al.*, 1985)—including seagrasses (Zimmerman *et al.*, 1997), certain diatoms (Riebesell *et al.*, 1993; Chen and Gao, 2004; Sobrino *et al.*, 2008), macroalgae (Borowitzka and Larkum, 1976; Gao *et al.*, 1993), and microalgae or phytoplankton (Raven, 1991; Nimer and Merrett, 1993)—the photosynthesis of many marine autotrophs is normally not considered to be carbon-limited, because of the large supply of bicarbonate in the world's oceans (Raven, 1997). However, as Gattuso *et al.* (1999) explain, this situation is true only for autotrophs that possess an effective carbon-concentrating mechanism; it is believed that many coral symbionts are of this type (Burris *et al.*, 1983; Al-Moghrabi *et al.*, 1996; Goiran *et al.*, 1996).

Nevertheless, in yet another positive twist to this complex story, Gattuso *et al.* (1999) report that coral zooxanthellae are able to change their mechanism of carbon supply in response to various environmental stimuli. Furthermore, Beardall *et al.* (1998) suggest that an increased concentration of dissolved CO_2, together with an increase in the rate of CO_2 generation by bicarbonate dehydration in host cells, may favor a transition to the diffusional mode of carbon supply, which *is* sensitive to hydrospheric CO_2 concentration. Consequently, if such a change in mode of carbon supply were to occur—prompted,

perhaps, by hydrospheric CO_2 enrichment itself—this shift in CO_2 fixation strategy would indeed allow the several biological mechanisms we have described to operate to enhance reef calcification rates in response to a rise in the air's CO_2 content.

In one final example that demonstrates the importance of biology in driving the physical-chemical process of coral calcification, Muscatine *et al.* (2005) note that the "photosynthetic activity of zooxanthellae is the chief source of energy for the energetically expensive process of calcification," and that long-term reef calcification rates have generally been observed to rise in direct proportion to increases in rates of reef primary production, which they say may well be *enhanced* by increases in the air's CO_2 concentration.

Muscatine *et al.* begin the report of their investigation of the subject by stating much the same thing, i.e., that endosymbiotic algae "release products of photosynthesis to animal cells ... and augment the rate of skeletal calcification." Then, noting that the "natural abundance of stable isotopes ($\delta^{13}C$ and $\delta^{15}N$) has answered paleobiological and modern questions about the effect of photosymbiosis on sources of carbon and oxygen in coral skeletal calcium carbonate," they go on to investigate the natural abundance of these isotopes in another coral skeletal compartment—the skeletal organic matrix (OM)—in 17 species of modern scleractinian corals, after which they compare the results for symbiotic and nonsymbiotic forms to determine the role played by algae in OM development.

Why is this study an important scientific undertaking? It is because, in the words of Muscatine *et al.*, "the scleractinian coral skeleton is a two-phase composite structure consisting of fiber-like crystals of aragonitic calcium carbonate intimately associated with an intrinsic OM," and although the OM generally comprises less than 0.1 percent of the total weight of the coral skeleton, it is, in their words, "believed to initiate nucleation of calcium carbonate and provide a framework for crystallographic orientation and species-specific architecture." In fact, they say inhibition of OM synthesis "brings coral calcification to a halt."

So what did Muscatine *et al.* learn from their experiments? They say their "most striking observation is the significant difference in mean OM $\delta^{15}N$ between symbiotic and nonsymbiotic corals," which makes OM $\delta^{15}N$ an important proxy for photosymbiosis. As an example of its usefulness, they applied the technique to a fossil coral (*Pachythecalis major*) from the Triassic (which prevailed some 240 million years ago), finding that the ancient coral was indeed photosymbiotic. Even more importantly, however, they conclude in the final sentence of their paper that "it now seems that symbiotic algae may control calcification by both modification of physico-chemical parameters within the coral polyps (Gautret *et al.*, 1997; Cuif *et al.*, 1999) and augmenting the synthesis of OM (Allemand *et al.*, 1998)."

Although lacking the research to absolutely identify the "what" and definitively describe the "how" of the hypothesis of hydrospheric CO_2 enhancement of coral calcification, it is clear that something of the nature described above can indeed act to overcome the negative effect of the high-CO_2-induced decrease in calcium carbonate saturation state on coral calcification rate. It has been clearly demonstrated, for example, that corals can grow quite well in aquariums containing water of very high dissolved CO_2 concentration (Atkinson *et al.*, 1995); and Carlson (1999) has stated that the fact that corals often thrive in such water "seems to contradict conclusions ... that high CO_2 may inhibit calcification." There are numerous other examples of such phenomena in the real world of nature, which we examine next.

References

Al-Moghrabi, S., Goiran, C., Allemand, D., Speziale, N. and Jaubert, J. 1996. Inorganic carbon uptake for photosynthesis by the symbiotic coral-dinoflagelate association. 2. Mechanisms for bicarbonate uptake. *Journal of Experimental Marine Biology and Ecology* **199**: 227-248.

Allemand, D., Tambutte, E., Girard, J.-P. and Jaubert, J. 1998. Organic matrix synthesis in the scleractinian coral *stylophora pistillata*: role in biomineralization and potential target of the organotin tributyltin. *Journal of Experimental Biology* **201**: 2001-2009.

Atkinson, M.J., Carlson, B.A. and Crow, G.L. 1995. Coral growth in high-nutrient, low-pH seawater: A case study of corals cultured at the Waikiki Aquarium, Honolulu, Hawaii. *Coral Reefs* **14**: 215-223.

Barnes, D.J. and Chalker, B.E. 1990. *Calcification and photosynthesis in reef-building corals and algae.* Dubinsky, Z. (Ed). Elsevier, Amsterdam, The Netherlands, pp. 109-131.

Battey, J.F. and Patton, J.S. 1984. A reevaluation of the role of glycerol in carbon translocation in zooxanthellae-coelenterate symbiosis. *Marine Biology* **79**: 27-38.

Beardall, J., Beer, S. and Raven, J.A. 1998. Biodiversity of marine plants in an era of climate change: Some predictions based on physiological performance. *Bot. Mar.* **41**: 113-123.

Borowitzka, M.A. and Larkum, W.D. 1976. Calcification in the green alga Halimeda. III. The sources of inorganic carbon for photosynthesis and calcification and a model of the mechanisms of calcification. *Journal of Experimental Botany* **27**: 879-893.

Burris, J.E., Porter, J.W. and Laing, W.A. 1983. Effects of carbon dioxide concentration on coral photosynthesis. *Marine Biology* **75**: 113-116.

Carlson, B.A. 1999. Organism responses to rapid change: What aquaria tell us about nature. *American Zoologist* **39**: 44-55.

Chalker, B.E. and Taylor, D.L. 1975. Light-enhanced calcification and the role of oxidative phosphorylation in calcification of the coral *Acropora cervicornis*. *Proceedings of the Royal Society of London B* **190**: 323-331.

Chen, X. and Gao, K. 2004. Characterization of diurnal photosynthetic rhythms in the marine diatom Skeletonema costatum grown in synchronous culture under ambient and elevated CO_2. *Functional Plant Biology* **31**: 399-404.

Crossland, C.J. and Barnes, D.J. 1974. The role of metabolic nitrogen in coral calcification. *Marine Biology* **28**: 325-332.

Cuif, J.-P., Dauphin, Y., Freiwald, A., Gautret, P. and Zibrowius, H. 1999. Biochemical markers of zooxanthellae symbiosis in soluble matrices of skeleton of 24 *Scleractinia* species. *Comparative Biochemistry and Physiology Part A: Molecular & Integrative Physiology* **123**: 269-278.

Frankignoulle, M., Gattuso, J.-P., Biondo, R., Bourge, I., Copin-Montegut, G. and Pichon, M. 1996. Carbon fluxes in coral reefs. II. Eulerian study of inorganic carbon dynamics and measurement of air-sea CO_2 exchanges. *Marine Ecology Progress Series* **145**: 123-132.

Gao, K., Aruga, Y., Asada, K. and Kiyohara, M. 1993. Influence of enhanced CO_2 on growth and photosynthesis of the red algae *Gracilaria* sp. and *G chilensis*. *Journal of Applied Phycology* **5**: 563-571.

Gattuso, J.-P., Allemand, D. and Frankignoulle, M. 1999. Photosynthesis and calcification at cellular, organismal and community levels in coral reefs: A review on interactions and control by carbonate chemistry. *American Zoologist* **39**: 160-183.

Gattuso, J.-P., Pichon, M., Delesalle, B., Canon, C. and Frankignoulle, M. 1996. Carbon fluxes in coral reefs. I. Lagrangian measurement of community metabolism and resulting air-sea CO_2 disequilibrium. *Marine Ecology Progress Series* **145**: 109-121.

Gautret, P., Cuif, J.-P. and Freiwald, A. 1997. Composition of soluble mineralizing matrices in zooxanthellate and non-zooxanthellate scleractinian corals: Biochemical assessment of photosynthetic metabolism through the study of a skeletal feature. *Facies* **36**: 189-194.

Goiran, C., Al-Moghrabi, S., Allemand, D. and Jaubert, J. 1996. Inorganic carbon uptake for photosynthesis by the symbiotic coral/dinoflagellate association. 1. Photosynthetic performances of symbionts and dependence on sea water bicarbonate. *Journal of Experimental Marine Biology and Ecology* **199**: 207-225.

Goreau, T.F. 1959. The physiology of skeleton formation in corals. I. A method for measuring the rate of calcium deposition by corals under different conditions. *Biological Bulletin* **116**: 59-75.

Hoegh-Guldberg, O. 1999. Climate change, coral bleaching and the future of the world's coral reefs. *Marine and Freshwater Research* **50**: 839-866.

Idso, K.E. 1992. Plant responses to rising levels of atmospheric carbon dioxide: A compilation and analysis of the results of a decade of international research into the direct biological effects of atmospheric CO_2 enrichment. Office of Climatology, Arizona State University, Tempe, AZ.

Idso, K.E. and Idso, S.B. 1994. Plant responses to atmospheric CO_2 enrichment in the face of environmental constraints: A review of the past 10 years' research. *Agricultural and Forest Meteorology* **69**: 153-203.

Kawaguti, S. and Sakumoto, D. 1948. The effects of light on the calcium deposition of corals. *Bulletin of the Oceanographic Institute of Taiwan* **4**: 65-70.

Kimball, B.A. 1983. Carbon dioxide and agricultural yield: An assemblage and analysis of 430 prior observations. *Agronomy Journal* **75**: 779-788.

Muscatine, L. 1967. Glycerol excretion by symbiotic algae from corals and *Tridacna* and its control by the host. *Science* **156**: 516-519.

Muscatine, L. 1990. The role of symbiotic algae in carbon and energy flux in reef corals. *Coral Reefs* **25**: 1-29.

Muscatine, L., Goiran, C., Land, L., Jaubert, J., Cuif, J.-P. and Allemand, D. 2005. Stable isotopes ($\delta^{13}C$ and $\delta^{15}N$) of organic matrix from coral skeleton. *Proceedings of the National Academy of Sciences USA* **102**: 1525-1530.

Nimer, N.A. and Merrett, M.J. 1993. Calcification rate in *Emiliania huxleyi* Lohmann in response to light, nitrate and availability of inorganic carbon. *New Phytologist* **123**: 673-677.

Rands, M.L., Douglas, A.E., Loughman, B.C. and Ratcliffe, R.G. 1992. Avoidance of hypoxia in a cnidarian symbiosis by algal photosynthetic oxygen. *Biological Bulletin* **182**: 159-162.

Raven, J.A. 1991. Physiology of inorganic C acquisition and implications for resource use efficiency by marine phytoplankton: Relation to increased CO_2 and temperature. *Plant, Cell and Environment* **14**: 779-794.

Raven, J.A. 1997. Inorganic carbon acquisition by marine autotrophs. *Advances in Botanical Research* **27**: 85-209.

Raven, J.A., Osborne, B.A. and Johnston, A.M. 1985. Uptake of CO_2 by aquatic vegetation. *Plant, Cell and Environment* **8**: 417-425.

Riebesell, U., Wolf-Gladrow, D.A. and Smetacek, V. 1993. Carbon dioxide limitation of marine phytoplankton growth rates. *Nature* **361**: 249-251.

Rinkevich, B. and Loya, Y. 1984. Does light enhance calcification in hermatypic corals? *Marine Biology* **80**: 1-6.

Simkiss, K. 1964. Phosphates as crystal poisons of calcification. *Biological Review* **39**: 487-505.

Sobrino, C., Ward, M.L. and Neale, P.J. 2008. Acclimation to elevated carbon dioxide and ultraviolet radiation in the diatom Thalassiosira pseudonana: Effects on growth, photosynthesis, and spectral sensitivity of photoinhibition. *Limnology and Oceanography* **53**: 494-505.

Wainwright, S.A. 1963. Skeletal organization in the coral *Pocillopora damicornis*. *Quarterly Journal of Microscopic Science* **104**: 164-183.

Yamashiro, H. 1995. The effects of HEBP, an inhibitor of mineral deposition, upon photosynthesis and calcification in the scleractinian coral, *Stylophora pistillata*. *Journal of Experimental Marine Biology and Ecology* **191**: 57-63.

Yonge, C.M. 1968. Living corals. *Proceedings of the Royal Society of London B* **169**: 329-344.

Zimmerman, R.C., Kohrs, D.G., Steller, D.L. and Alberte, R.S. 1997. Impacts of CO_2 enrichment on productivity and light requirements of eelgrass. *Plant Physiology* **115**: 599-607.

8.3.2.1.2. Coral Calcification

Many people have predicted that rates of coral calcification, as well as the photosynthetic rates of their symbiotic algae, will dramatically decline in response to what they typically refer to as an acidification of the world's oceans, as the atmosphere's CO_2 concentration continues to rise in the years, decades, and centuries to come. As ever more pertinent evidence accumulates, however, the true story appears to be just the opposite.

Herfort *et al.* (2008) note that an increase in atmospheric CO_2 will cause an increase in the abundance of HCO_3^- (bicarbonate) ions and dissolved CO_2, and they also report that several studies on marine plants have observed "increased photosynthesis with higher than ambient DIC [dissolved inorganic carbon] concentrations," citing the works of Gao *et al.* (1993), Weis (1993), Beer and Rehnberg (1997), Marubini and Thake (1998), Mercado *et al.* (2001, 2003), Herfort *et al.* (2002), and Zou *et al.* (2003).

To further explore this subject, and to see what it might imply for coral calcification, the three researchers employed a wide range of bicarbonate concentrations "to monitor the kinetics of bicarbonate use in both photosynthesis and calcification in two reef-building corals, *Porites porites* and *Acropora* sp." This work revealed that additions of HCO_3^- to synthetic seawater continued to increase the calcification rate of *Porites porites* until the bicarbonate concentration exceeded three times that of seawater, while photosynthetic rates of the coral's symbiotic algae were stimulated by HCO_3^- addition until they became saturated at twice the normal HCO_3^- concentration of seawater.

Similar experiments conducted on Indo-Pacific *Acropora* sp. showed that calcification and photosynthetic rates in these corals were enhanced to an even greater extent, with calcification continuing to increase above a quadrupling of the HCO_3^- concentration and photosynthesis saturating at triple the concentration of seawater. In addition, they monitored calcification rates of the *Acropora* sp. in the dark, and, in their words, "although these were lower than in the light for a given HCO_3^- concentration, they still increased dramatically with HCO_3^- addition, showing that calcification in this coral is light stimulated but not light dependent."

In discussing the significance of their findings, Herfort *et al.* suggest that "hermatypic corals incubated in the light achieve high rates of calcification by the synergistic action of photosynthesis," which, as they have shown, is enhanced by elevated concentrations of HCO_3^- ions that come courtesy of the ongoing rise in the air's CO_2 content. As for the real-world implications of their work, the three researchers note that over the next century the predicted increase in atmospheric CO_2 concentration "will result in about a 15%

increase in oceanic HCO_3^-," and they say that this development "could stimulate photosynthesis and calcification in a wide variety of hermatypic corals," a conclusion that stands in stark contrast to the contention of the IPCC.

In another study, Pelejero *et al.* (2005) developed a reconstruction of seawater pH spanning the period 1708-1988, based on the boron isotopic composition ($\delta^{11}B$) of a long-lived massive coral (*Porites*) from Flinders Reef in the western Coral Sea of the southwestern Pacific. Results indicated that "there [was] no notable trend toward lower $\delta^{11}B$ values" over the 300-year period investigated. Instead, they say that "the dominant feature of the coral $\delta^{11}B$ record is a clear interdecadal oscillation of pH, with $\delta^{11}B$ values ranging between 23 and 25 per mil (7.9 and 8.2 pH units)," which "is synchronous with the Interdecadal Pacific Oscillation." Furthermore, they calculated changes in aragonite saturation state from the Flinders pH record that varied between ~3 and 4.5, which values encompass "the lower and upper limits of aragonite saturation state within which corals can survive." Despite this fact, they report that "skeletal extension and calcification rates for the Flinders Reef coral fall within the normal range for *Porites* and are not correlated with aragonite saturation state or pH."

Thus, contrary to claims that historical anthropogenic CO_2 emissions have already resulted in a significant decline in ocean water pH and aragonite saturation state, Pelejero *et al.*'s 300-year record of these parameters (which, in their words, began "well before the start of the Industrial Revolution") provides no evidence of such a decline. What is more, and also contrary to what one would expect from claims of how sensitive coral calcification rate is to changes in pH and aragonite saturation state, they found that huge cyclical changes in these parameters had essentially no detectable effect on either coral calcification or skeletal extension rates.

Moving a little backward in time, in a study of historical calcification rates determined from coral cores retrieved from 35 sites on the Great Barrier Reef, Lough and Barnes (1997) observed a statistically significant correlation between coral calcification rate and local water temperature, such that a 1°C increase in mean annual water temperature increased mean annual coral calcification rate by about 3.5 percent. Nevertheless, they report there were "declines in calcification in *Porites* on the Great Barrier Reef over recent decades." They are quick to point out, however, that their data depict several extended periods of time when coral growth rates were either above or below the long-term mean, cautioning that "it would be unwise to rely on short-term values (say averages over less than 30 years) to assess mean conditions."

As an example of this fact, they report that "a decline in calcification equivalent to the recent decline occurred earlier this century and much greater declines occurred in the 18th and 19th centuries," long before anthropogenic CO_2 emissions made much of an impact on the air's CO_2 concentration. Over the entire expanse of their dataset, Lough and Barnes say "the 20th century has witnessed the second highest period of above average calcification in the past 237 years," which is not exactly what one would expect in light of (1) how dangerous high water temperatures are often said to be for corals, (2) the claim that earth is currently warmer than it has been at any other time during the entire past millennium, and (3) the fact that the air's CO_2 content is currently much higher than it has been for far longer than a mere thousand years.

Similar findings were reported by Bessat and Buigues (2001), who derived a history of coral calcification rates from a core extracted from a massive *Porites* coral head on the French Polynesian island of Moorea that covered the period 1801-1990. They performed this work, they say, because "recent coral-growth models highlight the enhanced greenhouse effect on the decrease of calcification rate," and rather than relying on theoretical calculations, they wanted to work with real-world data, stating that the records preserved in ancient corals "may provide information about long-term variability in the performance of coral reefs, allowing unnatural changes to be distinguished from natural variability."

Bessat and Buigues found that a 1°C increase in water temperature *increased* coral calcification rates at the site they studied by 4.5 percent. Then, they found that "instead of a 6-14% decline in calcification over the past 100 years computed by the Kleypas group, the calcification has increased, in accordance with [the results of] Australian scientists Lough and Barnes." They also observed patterns of "jumps or stages" in the record, which were characterized by an increase in the annual rate of calcification, particularly at the beginning of the past century "and in a more marked way around 1940, 1960 and 1976," stating once again that their results "do not confirm those predicted by the Kleypas *et al.* (1999) model."

Another major blow to the Kleypas *et al.* model was provided by the work of Lough and Barnes

(2000), who assembled and analyzed the calcification characteristics of 245 similar-sized massive colonies of *Porites* corals obtained from 29 reef sites located along the length, and across the breadth, of Australia's Great Barrier Reef (GBR), which data spanned a latitudinal range of approximately 9° and an annual average sea surface temperature (SST) range of 25-27°C. To these data they added other published data from the Hawaiian Archipelago (Grigg, 1981, 1997) and Phuket, Thailand (Scoffin *et al.*, 1992), thereby extending the latitudinal range of the expanded dataset to 20° and the annual average SST range to 23-29°C.

This analysis revealed that the GBR calcification data were linearly related to the average annual SST data, such that "a 1°C rise in average annual SST increased average annual calcification by 0.39 g cm^{-2} year^{-1}." Results were much the same for the extended dataset; Lough and Barnes report that "the regression equation [calcification = 0.33(SST)—7.07] explained 83.6% of the variance in average annual calcification (F = 213.59, p less than 0.00)," noting that "this equation provides for a change in calcification rate of 0.33 g cm^{-2} year^{-1} for each 1°C change in average annual SST."

With respect to the significance of their findings, Lough and Barnes say they "allow assessment of possible impacts of global climate change on coral reef ecosystems," and between the two 50-year periods 1880-1929 and 1930-1979, they calculate a calcification increase of 0.06 g cm^{-2} year^{-1}, noting that "this increase of ~4% in calcification rate conflicts with the estimated decrease in coral calcification rate of 6-14% over the same time period suggested by Kleypas *et al.* (1999) as a response to changes in ocean chemistry." Even more stunning is their observation that between the two 20-year periods 1903-1922 and 1979-1998, "the SST-associated increase in calcification is estimated to be less than 5% in the northern GBR, ~12% in the central GBR, ~20% in the southern GBR and to increase dramatically (up to ~50%) to the south of the GBR."

In light of these real-world observations, and in stark contrast to the doom-and-gloom prognostications of the IPCC, Lough and Barnes conclude that coral calcification rates "may have already significantly increased along the GBR in response to global climate change."

But in *Nature*, Caldeira and Wickett (2003) kept the catastrophe ball rolling. Based on a geochemical model, an ocean general-circulation model, an IPCC CO_2 emissions scenario for the twenty-first century, and a logistic function for the burning of earth's post-twenty-first century fossil-fuel reserves, they calculated the maximum level to which the air's CO_2 concentration might rise, the point in time when that might happen, and the related decline that might be expected to occur in ocean-surface pH. These calculations indicated that earth's atmospheric CO_2 concentration could approach 2,000 ppm around the year 2300, leading to an ocean-surface pH reduction of 0.7 units, a change described by Caldeira and Wickett as being much more rapid and considerably greater "than any experienced in the past 300 million years," which, of course, proves deadly for earth's corals in their scenario.

The following year, similar concerns were aroused by a report prepared for the Pew Center on Global Climate Change, a group advocating immediate action on global warming. In that document, Buddemeier *et al.* (2004) claimed the projected increase in the air's CO_2 content and the simulated decline in ocean-surface pH would dramatically decrease coral calcification rates, which they predicted would lead to "a slow-down or reversal of reef-building and the potential loss of reef structures." Even Buddemeier *et al.* (2004), however, were forced to admit that "calcification rates of large heads of the massive coral *Porites* increased rather than decreased over the latter half of the 20th century," further noting that "temperature and calcification rates are correlated, and these corals have so far responded more to increases in water temperature (growing faster through increased metabolism and the increased photosynthetic rates of their zooxanthellae) than to decreases in carbonate ion concentration."

The most recent claims of impending coral doom derive from the 2009 *Science* study of De'ath *et al.* (2009), who examined coral calcification rates on the Great Barrier Reef over the past 400 years. Results of their analysis indicate that there was a 14 percent decline in *Porites* calcification rate between 1990 and 2005, which observation the authors claim "is unprecedented in at least the past 400 years." As one might expect, the media's coverage of these findings included some ominous declarations. The headline of a BBC News report, for example, proclaimed "coral reef growth is slowest ever," while a Sky News headline read "Barrier Reef coral growth 'will stop'." ABC News went so far as to state *when* it might stop, concluding their report by quoting the research paper's senior author as saying "coral growth could hit zero by 2050."

How correct are such claims? Beginning with the first media claim that "coral reef growth is slowest ever," it can't possibly be right. The scleractinian corals, which are the major builders of the reefs of today, have been around some 200 million years, during most of which time the atmosphere's CO_2 concentration was greater and its temperature was higher than today, which should immediately raise a red flag about the proffered cause of the recent decline in reef growth.

In regard to the recent decline in calcification being unprecedented in the past 400 years, this is a good example of cherry-picking a time frame to make a sensational claim. All one needs to do is follow the published De'ath *et al.* calcification history back in time a mere 33 years more, from 1605 to 1572, to see the coral calcification rate during that earlier time was approximately 23 percent lower than what it was at its twentieth century peak, and the air's CO_2 concentration was more than 100 ppm less than what it is today.

Another way of looking at De'ath *et al.*'s data is to realize that from 1572 to the twentieth century peak, *Porites* calcification rates on the Great Barrier Reef rose by about 29 percent as atmospheric CO_2 concentration and air temperature rose concurrently, after which calcification rates declined, but by a smaller 14 percent. Why would anyone believe that the recent calcification decline implies that *Porites* coral growth "will stop," and that the end will come "by 2050"? They believe it because certain scientists (such as James Hansen) and politicians (such as Al Gore) imply much the same thing, as even De'ath *et al.* do. But when scientists feel compelled to be as correct and as true to their data as possible, such as when writing in *Science*, the three researchers from the Australian Institute of Marine Science clearly state that "the causes for the Great Barrier Reef-wide decline in coral calcification of massive *Porites* remain unknown." And when the causes of the recent decline in coral calcification rate are admitted to be unknown, it seems foolish indeed to predict, not only that the decline will continue, but that it will lead all the way to the demise of the studied coral, and especially at a specified future date, which, we might add, De'ath *et al.* appropriately do not do in their *Science* paper.

A second good reason for not believing that the ongoing rise in the air's CO_2 content will lead to reduced oceanic pH and, therefore, lower calcification rates in the world's coral reefs, is that the same phenomenon that powers the twin processes of coral calcification and phytoplanktonic growth (photosynthesis) tends to increase the pH of marine waters (Gnaiger *et al.*, 1978; Santhanam *et al.*, 1994; Brussaard *et al.*, 1996; Lindholm and Nummelin, 1999; Macedo *et al.*, 2001; Hansen, 2002). This phenomenon has been shown to have the ability to dramatically increase the pH of marine bays, lagoons, and tidal pools (Gnaiger *et al.*, 1978; Macedo *et al.*, 2001; Hansen, 2002) as well as to significantly enhance the surface water pH of areas as large as the North Sea (Brussaard *et al.*, 1996).

In one recent example, Middelboe and Hansen (2007) studied the pH of a wave-exposed boulder reef in Aalsgaarde on the northern coast of Zealand, Denmark, and a sheltered shallow-water area in Kildebakkerne in the estuary Roskilde Fjord, Denmark, reporting that, in line with what one would expect if photosynthesis tends to increase surface-water pH, (1) "daytime pH was significantly higher in spring, summer and autumn than in winter at both study sites," often reaching values of 9 or more during peak summer growth periods vs. 8 or less in winter, (2) "diurnal measurements at the most exposed site showed significantly higher pH during the day than during the night," reaching values that sometimes exceeded 9 during daylight hours but that typically dipped below 8 at night, and (3) that "diurnal variations were largest in the shallow water and decreased with increasing water depth."

In addition to their own findings, Middelboe and Hansen cite those of Pearson *et al.* (1998), who found that pH averaged about 9 during the summer in populations of *Fucus vesiculosus* in the Baltic Sea; Menendez *et al.* (2001), who found that maximum pH was 9 to 9.5 in dense floating macroalgae in a brackish coastal lagoon in the Ebro River Delta; and Bjork *et al.* (2004), who found pH values as high as 9.8 to 10.1 in isolated rock pools in Sweden. Noting that "pH in the sea is usually considered to be stable at around 8 to 8.2," the two Danish researchers thus concluded that "pH is higher in natural shallow-water habitats than previously thought."

With each succeeding year, the physical evidence against the CO_2-reduced calcification theory continues to grow ever more compelling, while support for the positive view promoted here continues to accumulate. Working in the laboratory, for example, Reynaud *et al.* (2004) grew nubbins of the branching zooxanthellate scleractinian coral *Acropora verweyi* in aquariums maintained at 20°, 25°, and 29°C, while weighing them once a week over a period of four weeks. This exercise revealed that coral calcification rates increased in nearly perfect linear fashion with increasing water temperature, yielding values of 0.06 percent, 0.22 percent, and 0.35 percent per day at 20°, 25°, and 29°C, respectively. These data reveal an approximate 480 percent increase in calcification rate in response to a 9°C increase in water temperature and a 160 percent increase in response to a 3°C increase in temperature, the latter of which temperature increases is somewhere in the low to midrange of global warming that the IPCC claims will result from a 300 ppm increase in the air's CO_2 concentration. This positive temperature effect outweighs the negative effect of rising CO_2 concentrations on coral calcification via ocean acidification.

Carricart-Ganivet (2004) developed relationships between coral calcification rate and annual average SST based on data collected from colonies of the reef-building coral *Montastraea annularis* at 12 localities in the Gulf of Mexico and the Caribbean Sea, finding that calcification rate in the Gulf of Mexico increased 0.55 g cm^{-2} year^{-1} for each 1°C increase, while in the Caribbean Sea it increased 0.58 g cm^{-2} year^{-1} for each 1°C increase. Pooling these data with those of *M. annularis* and *M. faveolata* growing to a depth of 10 m at Carrie Bow Cay, Belize, those from reefs at St. Croix in the US Virgin Islands, and those of *M. faveolata* growing to a depth of 10 m at Curacao, Antilles, Carricart-Ganivet reports he obtained a mean increase in calcification rate of ~0.5 g cm^{-2} year^{-1} for each 1°C increase in annual average SST, which is even greater than what was found by Lough and Barnes for *Porites* corals.

In another important study, McNeil *et al.* (2004) used a coupled atmosphere-ice-ocean carbon cycle model to calculate annual mean SST increases within the world's current coral reef habitat from 1995 to 2100 for increases in the air's CO_2 concentration specified by the IPCC's IS92a scenario, after which concomitant changes in coral reef calcification rates were estimated by combining the output of the climate model with empirical relationships between coral calcification rate and (1) aragonite saturation state (the negative CO_2 effect) and (2) annual SST (the positive temperature effect). Their choice for the first of these two relationships was that derived by Langdon *et al.* (2000), which leads to an even greater reduction in calcification than was predicted in the study of Kleypas *et al.* Their choice for the second relationship was that derived by Lough and Barnes (2000), which leads to an increase in calcification that is only half as large as that derived by Carricart-

Ganivet (2004). As a result, it can be appreciated that the net result of the two phenomena was doubly weighted in favor of reduced coral calcification. Nevertheless, McNeil *et al.* found that the increase in coral reef calcification associated with ocean warming outweighed the decrease associated with the CO_2-induced decrease in aragonite saturate state. They calculated that coral calcification in 2100 would be 35 percent higher than what it was in pre-industrial times at the very least. And they found that the area of coral reef habitat expands in association with the projected ocean warming.

Finally, in a study devoted to corals that involves a much longer period of time than all of the others we have discussed, Crabbe *et al.* (2006) determined the original growth rates of long-dead Quaternary corals found in limestone deposits on islands in the Wakatobi Marine National Park of Indonesia, after which they compared them to the growth rates of present-day corals of the same genera living in the same area. This work revealed that the Quaternary corals grew "in a comparable environment to modern reefs"—except, of course, for the air's CO_2 concentration, which is currently higher than it has been at any other time throughout the Quaternary, which spans the past 1.8 million years. Most interestingly, therefore, their measurements indicated that the radial growth rates of the modern corals were 31 percent greater than those of their ancient predecessors in the case of *Porites* species, and 34 percent greater in the case of *Favites* species.

To these papers we could add many others (Clausen and Roth, 1975; Coles and Jokiel, 1977; Kajiwara *et al.*, 1995; Nie *et al.*, 1997; Reynaud-Vaganay *et al.*, 1999; Reynaud *et al.*, 2007) that also depict increasing rates of coral calcification in the face of rising temperatures and atmospheric CO_2 concentrations. Clearly, the net impact of twentieth century increases in atmospheric CO_2 and temperature has not been anywhere near as catastrophically disruptive to earth's corals as the IPCC suggests it should have been. Quite to the contrary, the temperature and CO_2 increases appear to not have been hurtful at all, and in fact appear to have been helpful.

Additional information on this topic, including reviews of newer publications as they become available, can be found at http://www.co2science.org/subject/c/calcification.php.

References

Beer, S. and Rehnberg, J. 1997. The acquisition of inorganic carbon by the sea grass *Zostera marina*. *Aquatic Botany* **56**: 277-283.

Bessat, F. and Buigues, D. 2001. Two centuries of variation in coral growth in a massive *Porites* colony from Moorea (French Polynesia): a response of ocean-atmosphere variability from south central Pacific. *Palaeogeography, Palaeoclimatology, Palaeoecology* **175**: 381-392.

Bjork, M., Axelsson, L. and Beer, S. 2004. Why is Ulva intestinalis the only macroalga inhabiting isolated rockpools along the Swedish Atlantic coast? *Marine Ecology Progress Series* **284**: 109-116.

Brussaard, C.P.D., Gast, G.J., van Duyl, F.C. and Riegman, R. 1996. Impact of phytoplankton bloom magnitude on a pelagic microbial food web. *Marine Ecology Progress Series* **144**: 211-221.

Buddemeier, R.W., Kleypas, J.A. and Aronson, R.B. 2004. *Coral Reefs & Global Climate Change: Potential Contributions of Climate Change to Stresses on Coral Reef Ecosystems*. The Pew Center on Global Climate Change, Arlington, VA, USA.

Caldeira, K. and Wickett, M.E. 2003. Anthropogenic carbon and ocean pH. *Nature* **425**: 365.

Carricart-Ganivet, J.P. 2004. Sea surface temperature and the growth of the West Atlantic reef-building coral *Montastraea annularis*. *Journal of Experimental Marine Biology and Ecology* **302**: 249-260.

Clausen, C.D. and Roth, A.A. 1975. Effect of temperature and temperature adaptation on calcification rate in the hematypic *Pocillopora damicornis*. *Marine Biology* **33**: 93-100.

Coles, S.L. and Jokiel, P.L. 1977. Effects of temperature on photosynthesis and respiration in hermatypic corals. *Marine Biology* **43**: 209-216.

Crabbe, M.J.C., Wilson, M.E.J. and Smith, D.J. 2006. Quaternary corals from reefs in the Wakatobi Marine National Park, SE Sulawesi, Indonesia, show similar growth rates to modern corals from the same area. *Journal of Quaternary Science* **21**: 803-809.

De'ath, G., Lough, J.M. and Fabricius, K.E. 2009. Declining coral calcification on the Great Barrier Reef. *Science* **323**: 116-119.

Gao, K., Aruga, Y., Asada, K., Ishihara, T., Akano, T. and Kiyohara, M. 1993. Calcification in the articulated coralline alga *Corallina pilulifera*, with special reference to the effect of elevated CO_2 concentration. *Marine Biology* **117**: 129-132.

Gnaiger, E., Gluth, G. and Weiser, W. 1978. pH fluctuations in an intertidal beach in Bermuda. *Limnology and Oceanography* **23**: 851-857.

Grigg, R.W. 1981. Coral reef development at high latitudes in Hawaii. In: *Proceedings of the Fourth International Coral Reef Symposium*, Manila, Vol. 1: 687-693.

Grigg, R.W. 1997. Paleoceanography of coral reefs in the Hawaiian-Emperor Chain—revisited. *Coral Reefs* **16**: S33-S38.

Hansen, P.J. 2002. The effect of high pH on the growth and survival of marine phytoplankton: implications for species succession. *Aquatic Microbiology and Ecology* **28**: 279-288.

Herfort, L., Thake, B. and Roberts, J. 2002. Acquisition and use of bicarbonate by *Emiliania huxleyi*. *New Phytologist* **156**: 427—36.

Herfort, L., Thake, B. and Taubner, I. 2008. Bicarbonate stimulation of calcification and photosynthesis in two hermatypic corals. *Journal of Phycology* **44**: 91-98.

Kajiwara, K., Nagai, A. and Ueno, S. 1995. Examination of the effect of temperature, light intensity and zooxanthellae concentration on calcification and photosynthesis of scleractinian coral *Acropora pulchra*. *J. School Mar. Sci. Technol.* **40**: 95-103.

Kleypas, J.A., Buddemeier, R.W., Archer, D., Gattuso, J-P., Langdon, C. and Opdyke, B.N. 1999. Geochemical consequences of increased atmospheric carbon dioxide on coral reefs. *Science* **284**: 118-120.

Langdon, C., Takahashi, T., Sweeney, C., Chipman, D., Goddard, J., Marubini, F., Aceves, H., Barnett, H. and Atkinson, M.J. 2000. Effect of calcium carbonate saturation state on the calcification rate of an experimental coral reef. *Global Biogeochemical Cycles* **14**: 639-654.

Lindholm, T. and Nummelin, C. 1999. Red tide of the dinoflagellate *Heterocapsa triquetra* (Dinophyta) in a ferry-mixed coastal inlet. *Hydrobiologia* **393**: 245-251.

Lough, J.M. and Barnes, D.J. 1997. Several centuries of variation in skeletal extension, density and calcification in massive *Porites* colonies from the Great Barrier Reef: A proxy for seawater temperature and a background of variability against which to identify unnatural change. *Journal of Experimental and Marine Biology and Ecology* **211**: 29-67.

Lough, J.B. and Barnes, D.J. 2000. Environmental controls on growth of the massive coral *Porites*. *Journal of Experimental Marine Biology and Ecology* **245**: 225-243.

Macedo, M.F., Duarte, P., Mendes, P. and Ferreira, G. 2001. Annual variation of environmental variables, phytoplankton species composition and photosynthetic parameters in a coastal lagoon. *Journal of Plankton Research* **23**: 719-732.

Marubini, F. and Thake, B. 1998. Coral calcification and photosynthesis: evidence for carbon limitation. In: *International Society for Reef Studies (ISRS), European Meeting*, Perpignan, September 1-4, 1998, p. 119.

McNeil, B.I., Matear, R.J. and Barnes, D.J. 2004. Coral reef calcification and climate change: The effect of ocean warming. *Geophysical Research Letters* **31**: 10.1029/2004GL021541.

Menendez, M., Martinez, M. and Comin, F.A. 2001. A comparative study of the effect of pH and inorganic carbon resources on the photosynthesis of three floating macroalgae species of a Mediterranean coastal lagoon. *Journal of Experimental Marine Biology and Ecology* **256**: 123-136.

Mercado, J.M., Niell, F.X. and Gil-Rodriguez, M.C. 2001. Photosynthesis might be limited by light, not inorganic carbon availability, in three intertidal Gelidiales species. *New Phytologist* **149**: 431-439.

Mercado, J.M., Niell, F.X., Silva, J. and Santos, R. 2003. Use of light and inorganic carbon acquisition by two morphotypes of *Zostera noltii* Hornem. *Journal of Experimental Marine Biology and Ecology* **297**: 71-84.

Middelboe, A.L. and Hansen, P.J. 2007. High pH in shallow-water macroalgal habitats. *Marine Ecology Progress Series* **338**: 107-117.

Nie, B., Chen, T., Liang, M., Wang, Y., Zhong, J. and Zhu, Y. 1997. Relationship between coral growth rate and sea surface temperature in the northern part of South China Sea. *Sci. China Ser. D* **40**: 173-182.

Pearson, G.A., Serrao, E.A. and Brawley, S.H. 1998. Control of gamete release in fucoid algae: sensing hydrodynamic conditions via carbon acquisition. *Ecology* **79**: 1725-1739.

Pelejero, C., Calvo, E., McCulloch, M.T., Marshall, J.F., Gagan, M.K., Lough, J.M. and Opdyke, B.N. 2005. Preindustrial to modern interdecadal variability in coral reef pH. *Science* **309**: 2204-2207.

Reynaud, S., Ferrier-Pages, C., Boisson, F., Allemand, D. and Fairbanks, R.G. 2004. Effect of light and temperature on calcification and strontium uptake in the scleractinian coral *Acropora verweyi*. *Marine Ecology Progress Series* **279**: 105-112.

Reynaud, S., Ferrier-Pages, C., Meibom, A., Mostefaoui, S., Mortlock, R., Fairbanks, R. and Allemand, D. 2007. Light and temperature effects on Sr/Ca and Mg/Ca ratios in the scleractinian coral *Acropora* sp. *Geochimica et Cosmochimica Acta* **71**: 354-362.

Reynaud-Vaganay, S., Gattuso, J.P., Cuif, J.P., Jaubert, J. and Juillet-Leclerc, A. 1999. A novel culture technique for scleractinian corals: Application to investigate changes in skeletal $\delta^{18}O$ as a function of temperature. *Marine Ecology Progress Series* **180**: 121-130.

Santhanam, R., Srinivasan, A., Ramadhas, V. and Devaraj, M. 1994. Impact of *Trichodesmium* bloom on the plankton and productivity in the Tuticorin bay, southeast coast of India. *Indian Journal of Marine Science* **23**: 27-30.

Scoffin, T.P., Tudhope, A.W., Brown, B.E., Chansang, H. and Cheeney, R.F. 1992. Patterns and possible environmental controls of skeletogenesis of *Porites lutea*, South Thailand. *Coral Reefs* **11**: 1-11.

Weis, V.M. 1993. Effect of dissolved inorganic carbon concentration on the photosynthesis of the symbiotic sea anemone *Aiptasia pulchella* Carlgren: role of carbonic anhydrase. *Journal of Experimental Marine Biology and Ecology* **174**: 209-225.

Zou, D.H., Gao, K.S. and Xia, J.R. 2003. Photosynthetic utilization of inorganic carbon in the economic brown alga, *Hizikia fusiforme* (Sargassaceae) from the South China Sea. *Journal of Phycology* **39**: 1095-1100.

8.3.2.1.3. Other Marine Organisms

In a paper recently published in *Limnology and Oceanography*, Richardson and Gibbons (2008) say there has been a drop of 0.1 pH unit in the global ocean since the start of the Industrial Revolution, and that "such acidification of the ocean may make calcification more difficult for calcareous organisms," resulting in the "opening [of] ecological space for non-calcifying species." In line with this thinking, they report that Attrill *et al.* (2007) have argued that "jellyfish may take advantage of the vacant niches made available by the negative effects of acidification on calcifying plankton," causing jellyfish to become more abundant, and they note that the latter researchers provided some evidence for this effect in the west-central North Sea over the period 1971-1995. Hence, they undertook a study to see if Attrill *et al.*'s findings (which were claimed to be the first of their kind) could be replicated on a much larger scale.

Working with data from a larger portion of the North Sea, as well as throughout most of the much vaster Northeast Atlantic Ocean, Richardson and Gibbons used coelenterate (jellyfish) records from the Continuous Plankton Recorder (CPR) and pH data from the International Council for the Exploration of the Sea (ICES) for the period 1946-2003 to explore the possibility of a relationship between jellyfish abundance and acidic ocean conditions. This work revealed that there were, as they describe it, "no significant relationships between jellyfish abundance and acidic conditions in any of the regions investigated."

In harmony with their findings, the two researchers note that "no observed declines in the abundance of calcifiers with lowering pH have yet been reported." In addition, they write that the "larvae of sea urchins form skeletal parts comprising magnesium-bearing calcite, which is 30 times more soluble than calcite without magnesium," and, therefore, that "lower ocean pH should drastically inhibit the formation of these soluble calcite precursors." Yet they report that "there is no observable negative effect of pH." In fact, they say that echinoderm larvae in the North Sea have actually exhibited "a 10-fold *increase* [our italics] in recent times," which they say has been "linked predominantly to warming (Kirby *et al.*, 2007)." Likewise, they further note that even in the most recent IPCC report, "there was no empirical evidence reported for the effect of acidification on marine biological systems (Rosenzweig *et al.*, 2007)." In light of this body of real-world evidence, Richardson and Gibbons conclude (rather generously, we think) that "the role of pH in structuring zooplankton communities in the North Sea and further afield at present is tenuous."

In another study, Vogt *et al.* (2008) examined the effects of atmospheric CO_2 enrichment on various marine microorganisms in nine marine mesocosms in a fjord adjacent to the Large-Scale Facilities of the Biological Station of the University of Bergen in Espegrend, Norway. Three of the mesocosms were maintained at ambient levels of CO_2 (~375 ppm), three were maintained at levels expected to prevail at the end of the current century (760 ppm or 2x CO_2), and three were maintained at levels predicted for the middle of the next century (1,150 ppm or 3x CO_2), while measurements of numerous ecosystem parameters were made over a period of 24 days.

Results of the analysis showed no significant phytoplankton species shifts between treatments, and that "the ecosystem composition, bacterial and phytoplankton abundances and productivity, grazing rates and total grazer abundance and reproduction were not significantly affected by CO_2 induced effects," citing in support of this statement the work of Riebesell *et al.* (2007), Riebesell *et al.* (2008), Egge *et al.* (2007), Paulino *et al.* (2007), Larsen *et al.*

(2007), Suffrian *et al.* (2008), and Carotenuto *et al.* (2007). With respect to their many findings, the eight researchers say their observations suggest that "the system under study was surprisingly resilient to abrupt and large pH changes."

Expanding the subject of CO_2 effects on other marine organisms, Gutowska *et al.* (2008) studied the cephalopod mollusk *Sepia officinalis* and found that it "is capable of not only maintaining calcification, but also growth rates and metabolism when exposed to elevated partial pressures of carbon dioxide." Over a six-week test period, for example, they found that "juvenile *S. officinalis* maintained calcification under ~4000 and ~6000 ppm CO_2, and grew at the same rate with the same gross growth efficiency as did control animals," gaining approximately 4 percent body mass daily and increasing the mass of their calcified cuttlebone by more than 500 percent. These findings thus led them to specifically conclude that "active cephalopods possess a certain level of pre-adaptation to long-term increments in carbon dioxide levels," and to generally conclude that our "understanding of the mechanistic processes that limit calcification must improve before we can begin to predict what effects future ocean acidification will have on calcifying marine invertebrates."

In another study, Berge *et al.* (2006) continuously supplied five 5-liter aquariums with low-food-content sea water that was extracted from the top meter of the Oslofjord outside the Marine Research Station Solbergstrand in Norway, while CO_2 was continuously added to the waters of the aquaria so as to maintain them at five different pH values (means of 8.1, 7.6, 7.4, 7.1, and 6.7) for a period of 44 days. Prior to the start of the study, blue mussels (*Mytilus edulis*) of two different size classes (mean lengths of either 11 or 21 mm) were collected from the outer part of the Oslofjord, and 50 of each size class were introduced into each aquarium, where they were examined almost daily for any deaths that may have occurred, after which shell lengths at either the time of death or at the end of the study were determined and compared to lengths measured at the start of the study. Simultaneously, water temperature rose slowly from 16° to 19°C during the initial 23 days of the experiment, but then declined slightly to day 31, after which it rose rapidly to attain a maximum value of 24°C on day 39.

A lack of mortality during the first 23 days of the study showed, in the words of the researchers, that "the increased concentration of CO_2 in the water and the correspondingly reduced pH had no acute effects

on the mussels." Thereafter, however, some mortality was observed in the highest CO_2 (lowest pH) treatment from day 23 to day 37, after which deaths could also be observed in some of the other treatments, which mortality Berge *et al.* attributed to the rapid increase in water temperature that occurred between days 31 and 39.

With respect to growth, the Norwegian researchers report that "mean increments of shell length were much lower for the two largest CO_2 additions compared to the values in the controls, while for the two smallest doses the growth [was] about the same as in the control, or in one case even higher (small shells at pH = 7.6)," such that there were "no significant differences between the three aquaria within the pH range 7.4-8.1."

Berge *et al.* say their results "indicate that future reductions in pH caused by increased concentrations of anthropogenic CO_2 in the sea may have an impact on blue mussels," but that "comparison of estimates of future pH reduction in the sea (Caldeira and Wickett, 2003) and the observed threshold for negative effects on growth of blue mussels [which they determined to lie somewhere between a pH of 7.4 and 7.1] do however indicate that this will probably not happen in this century." Indeed, Caldeira and Wickett's calculation of the maximum level to which the air's CO_2 concentration might rise yields a value that approaches 2,000 ppm around the year 2300, representing a surface oceanic pH reduction of 0.7 units, which drops the pH only to the upper limit of the "threshold for negative effects on growth of blue mussels" found by Berge *et al.*, i.e., 7.4. Consequently, blue mussels will likely never be bothered by the tendency for atmospheric CO_2 enrichment to lower oceanic pH values.

In a study of a very different creature, Langer *et al.* (2006) conducted batch-culture experiments on two coccolithophores, *Calcidiscus leptoporus* and *Coccolithus pelagicus*, in which they observed a "deterioration of coccolith production above as well as below present-day CO_2 concentrations in *C. leptoporus*," and a "lack of a CO_2 sensitivity of calcification in *C. pelagicus*" over an atmospheric CO_2 concentration range of 98-915 ppm. Both of these observations, in their words, "refute the notion of a linear relationship of calcification with the carbonate ion concentration and carbonate saturation state." In an apparent negative finding, however, particularly in the case of *C. leptoporus*, Langer *et al.* observed that although their experiments revealed that "at 360 ppm CO_2 most coccoliths show normal morphology," at both "higher and lower CO_2 concentrations the proportion of coccoliths showing incomplete growth and malformation increases notably."

To determine if such deleterious responses might also have occurred in the real world at different times in the past, the researchers studied coccolith morphologies in six sediment cores obtained along a range of latitudes in the Atlantic Ocean. This work revealed that changes in coccolith morphology similar to those "occurring in response to the abrupt CO_2 perturbation applied in experimental treatments are *not* [our italics] mirrored in the sedimentary record." This finding indicates, as they suggest, that "in the natural environment *C. leptoporus* has adjusted to the 80-ppm CO_2 and 180-ppm CO_2 difference between present [and] preindustrial and glacial times, respectively."

In further discussing these observations, Langer *et al.* say "it is reasonable to assume that *C. leptoporus* has adapted its calcification mechanism to the change in carbonate chemistry having occurred since the last glacial maximum," suggesting as a possible explanation for this phenomenon that "the population is genetically diverse, containing strains with diverse physiological and genetic traits, as already demonstrated for *E. huxleyi* (Brand, 1981, 1982, 1984; Conte *et al.*, 1998; Medlin *et al.*, 1996; Paasche, 2002; Stolte *et al.*, 2000)." They also state that this adaptive ability "is not likely to be confined to *C. leptoporus* but can be assumed to play a role in other coccolithophore species as well," which leads them to conclude that such populations "may be able to evolve so that the optimal CO_2 level for calcification of the species tracks the environmental value." With respect to the future, therefore, Langer *et al.* end on a strongly positive note, stating that "genetic diversity, both between and within species, may allow calcifying organisms to prevail in a high CO_2 ocean."

Focusing on another coccolithophore species, Riebesell (2004) notes that "a moderate increase in CO_2 facilitates photosynthetic carbon fixation of some phytoplankton groups," including "the coccolithophorids *Emiliania huxleyi* and *Gephyrocapsa oceanica*." Hence, in a major challenge to the claim that atmospheric CO_2 enrichment will definitely harm such marine organisms, Riebesell suggests that "CO_2-sensitive taxa, such as the calcifying coccolithophorids, should therefore *benefit more* [our italics] from the present increase in atmospheric CO_2 compared to the non-calcifying diatoms."

In support of this suggestion, Riebesell describes the results of some CO_2 perturbation experiments conducted south of Bergen, Norway, where nine 11-m^3 enclosures moored to a floating raft were aerated with CO_2-depleted, normal and CO_2-enriched air to achieve CO_2 levels of 190, 370, and 710 ppm, simulating glacial, present-day, and predicted conditions for the end of the century, respectively. In the course of the study, a bloom consisting of a mixed phytoplankton community developed, and, in Riebesell's words, "significantly higher net community production was observed under elevated CO_2 levels during the build-up of the bloom." He further reports that "CO_2-related differences in primary production continued after nutrient exhaustion, leading to higher production of transparent exopolymer particles under high CO_2 conditions," something that has also been observed by Engel (2002) in a natural plankton assemblage and by Heemann (2002) in monospecific cultures of both diatoms and coccolithophores.

Another important finding of this experiment was that the community that developed under the high CO_2 conditions expected for the end of this century was dominated by *Emiliania huxleyi*. Consequently, Riebesell finds even more reason to believe that "coccolithophores may benefit from the present increase in atmospheric CO_2 and related changes in seawater carbonate chemistry," in contrast to the many negative predictions that have been made about rising atmospheric CO_2 concentrations in this regard. Finally, in further commentary on the topic, Riebesell states that "increasing CO_2 availability may improve the overall resource utilization of *E. huxleyi* and possibly of other fast-growing coccolithophore species," concluding that "if this provides an ecological advantage for coccolithophores, rising atmospheric CO_2 could potentially increase the contribution of calcifying phytoplankton to overall primary production." In fact, noting that "a moderate increase in CO_2 facilitates photosynthetic carbon fixation of some phytoplankton groups," including "the coccolithophorids *Emiliania huxleyi* and *Gephyrocapsa oceanica,*" Riebesell suggests that "CO_2-sensitive taxa, such as the calcifying coccolithophorids, should therefore benefit *more* [our italics] from the present increase in atmospheric CO_2 compared to the non-calcifying diatoms."

Support of Riebesell's findings was recently provided by an international team of 13 researchers

(Iglesias-Rodriguez *et al.*, 2008), who bubbled air of a number of different atmospheric CO_2 concentrations through culture media containing the phytoplanktonic coccolithophore species *Emiliania hyxleyi*, while determining the amounts of particulate organic and inorganic carbon they produced. In addition, they determined the real-world change in average coccolithophore mass over the past 220 years in the subpolar North Atlantic Ocean, based on data obtained from a sediment core, over which period of time the atmosphere's CO_2 concentration rose by approximately 90 ppm and the earth emerged from the frigid depths of the Little Ice Age to experience the supposedly unprecedented high temperatures of the Current Warm Period.

Results of their analysis revealed an approximate doubling of both particulate organic and inorganic carbon between the culture media in equilibrium with air of today's CO_2 concentration and the culture media in equilibrium with air of 750 ppm CO_2. In addition, they say the field evidence they obtained from the deep-ocean sediment core they studied "is consistent with these laboratory conclusions," and that it indicates that "over the past 220 years there has been a 40% increase in average coccolith mass."

Focusing more on the future, a third independent team of seven scientists (Feng *et al.*, 2008) studied *Emiliania huxleyi* coccoliths that they isolated from the Sargasso Sea, and which they grew in semi-continuous culture media at low and high light intensities, low and high temperatures (20 and 24°C), and low and high CO_2 concentrations (375 and 750 ppm). This work revealed that in the low-light environment, the maximum photosynthetic rate was lowest in the low-temperature, low-CO_2 or ambient treatment, but was increased by 55 percent by elevated temperature alone and by 95 percent by elevated CO_2 alone, while in the high-temperature, high-CO_2 or greenhouse treatment it was increased by 150 percent relative to the ambient treatment. Likewise, in the high-light environment, there were maximum photosynthetic rate increases of 58 percent, 67 percent, and 92 percent for the elevated temperature alone, elevated CO_2 alone, and greenhouse treatments, respectively. Consequent-ly, the researchers concluded that "future trends of CO_2 enrichment, sea-surface warming and exposure to higher mean irradiances from intensified stratification will have a large influence on the growth of *Emiliania huxleyi*."

Clearly, claims of impending marine species extinctions due to ocean acidification are not supported by real-world evidence—they are refuted by it.

Additional information on this topic, including reviews of newer publications as they become available, can be found at http://www.co2 science.org/subject/c/calcificationother.php.

References

Attrill, M.J., Wright, J. and Edwards, M. 2007. Climate-related increases in jellyfish frequency suggest a more gelatinous future for the North Sea. *Limnology and Oceanography* **52**: 480-485.

Berge, J.A., Bjerkeng, B., Pettersen, O., Schaanning, M.T. and Oxnevad, S. 2006. Effects of increased sea water concentrations of CO_2 on growth of the bivalve *Mytilus edulis* L. *Chemosphere* **62**: 681-687.

Brand, L.E. 1981. Genetic variability in reproduction rates in marine phytoplankton populations. *Evolution* **38**: 1117-1127.

Brand, L.E. 1982. Genetic variability and spatial patterns of genetic differentiation in the reproductive rates of the marine coccolithophores *Emiliania huxleyi* and *Gephyrocapsa oceanica*. *Limnology and Oceanography* **27**: 236-245.

Brand, L.E. 1984. The salinity tolerance of forty-six marine phytoplankton isolates. *Estuarine and Coastal Shelf Science* **18**: 543-556.

Caldeira, K. and Wickett, M.E. 2003. Anthropogenic carbon and ocean pH. *Nature* **425**: 365.

Carotenuto, Y., Putzeys, S., Simonelli, P., Paulino, A., Meyerhofer, M., Suffrian, K., Antia, A. and Nejstgaard, J.C. 2007. Copepod feeding and reproduction in relation to phytoplankton development during the PeECE III mesocosm experiment. *Biogeosciences Discussions* **4**: 3913-3936.

Conte, M., Thompson, A., Lesley, D. and Harris, R.P. 1998. Genetic and physiological influences on the alkenone/alkenonate versus growth temperature relationship in *Emiliania huxleyi* and *Gephyrocapsa oceanica*. *Geochimica et Cosmochimica Acta* **62**: 51-68.

Egge, J., Thingstad, F., Engel, A., Bellerby, R.G.J. and Riebesell, U. 2007. Primary production at elevated nutrient and pCO_2 levels. *Biogeosciences Discussions* **4**: 4385-4410.

Engel, A. 2002. Direct relationship between CO_2 uptake and transparent exopolymer particles production in natural phytoplankton. *Journal of Plankton Research* **24**: 49-53.

Feng, Y., Warner, M.E., Zhang, Y., Sun, J., Fu, F.-X., Rose, J.M. and Hutchins, A. 2008. Interactive effects of increased pCO_2, temperature and irradiance on the marine coccolithophore *Emiliania huxleyi* (Prymnesiophyceae). *European Journal of Phycology* **43**: 87-98.

Gutowska, M.A., Pörtner, H.O. and Melzner, F. 2008. Growth and calcification in the cephalopod *Sepia officinalis* under elevated seawater pCO_2. *Marine Ecology Progress Series* **373**: 303-309.

Heemann, C. 2002. *Phytoplanktonexsudation in Abhangigkeit der Meerwasserkarbonatchemie*. Diplom. Thesis, ICBM, University of Oldenburg, Germany.

Iglesias-Rodriguez, M.D., Halloran, P.R., Rickaby, R.E.M., Hall, I.R., Colmenero-Hidalgo, E., Gittins, J.R., Green, D.R.H., Tyrrell, T., Gibbs, S.J., von Dassow, P., Rehm, E., Armbrust, E.V. and Boessenkool, K.P. 2008. Phytoplankton calcification in a high-CO_2 world. *Science* **320**: 336-340.

Kirby, R.R., Beaugrand, G., Lindley, J.A., Richardson, A.J., Edwards, M. and Reid, P.C. 2007. Climate effects and benthic-pelagic coupling in the North Sea. *Marine Ecology Progress Series* **330**: 31-38.

Langer, G. and Geisen, M., Baumann, K.-H., Klas, J., Riebesell, U., Thoms, S. and Young, J.R. 2006. Species-specific responses of calcifying algae to changing seawater carbonate chemistry. *Geochemistry, Geophysics, Geosystems* **7**: 10.1029/2005GC001227.

Larsen, J.B., Larsen, A., Thyrhaug, R., Bratbak, G. and Sandaa R.-A. 2007. Marine viral populations detected during a nutrient induced phytoplankton bloom at elevated pCO_2 levels. *Biogeosciences Discussions* **4**: 3961-3985.

Medlin, L.K., Barker, G.L.A., Green, J.C., Hayes, D.E., Marie, D., Wreiden, S. and Vaulot, D. 1996. Genetic characterization of *Emiliania huxleyi* (Haptophyta). *Journal of Marine Systems* **9**: 13-32.

Paasche, E. 2002. A review of the coccolithophorid *Emiliania huxleyi* ((Prymnesiophyceae), with particular reference to growth, coccolith formation, and calcification-photosynthesis interactions. *Phycologia* **40**: 503-529.

Paulino, A.I., Egge, J.K. and Larsen, A. 2007. Effects of increased atmospheric CO_2 on small and intermediate sized osmotrophs during a nutrient induced phytoplankton bloom. *Biogeosciences Discussions* **4**: 4173-4195.

Riebesell, U. 2004. Effects of CO_2 enrichment on marine phytoplankton. *Journal of Oceanography* **60**: 719-729.

Riebesell, U., Bellerby, R.G.J., Grossart, H.-P. and Thingstad, F. 2008. Mesocosm CO_2 perturbation studies: from organism to community level. *Biogeosciences Discussions* **5**: 641-659.

Riebesell, U., Schulz, K., Bellerby, R., Botros, M., Fritsche, P., Meyerhofer, M., Neill, C., Nondal, G., Oschlies, A., Wohlers, J. and Zollner, E. 2007. Enhanced biological carbon consumption in a high CO_2 ocean. *Nature* **450**: 10.1038/nature06267.

Richardson, A.J. and Gibbons, M.J. 2008. Are jellyfish increasing in response to ocean acidification? *Limnology and Oceanography* **53**: 2040-2045.

Rosenzweig, C. *et al.* 2007. Assessment of observed changes and responses in natural and managed systems. In Parry, M.L., Canziani, O.F., Palutikof, J.P., van der Linden, P.J. and Hanson, C.E. (Eds.) *Climate Change 2007: Impacts, Adaptation and Vulnerability. Contribution of Working Group II to the Fourth Assessment Report of the Intergovernmental Panel on Climate Change*. Cambridge University Press, Cambridge, UK, pp. 79-131.

Stolte, W., Kraay, G.W., Noordeloos, A.A.M. and Riegman, R. 2000. Genetic and physiological variation in pigment composition of *Emiliania huxleyi* (Prymnesiophyceae) and the potential use of its pigment ratios as a quantitative physiological marker. *Journal of Phycology* **96**: 529-589.

Suffrian, K., Simonelli, P., Nejstgaard, J.C., Putzeys, S., Carotenuto, Y. and Antia, A.N. 2008. Microzooplankton grazing and phytoplankton growth in marine mesocosms with increased CO_2 levels. *Biogeosciences Discussions* **5**: 411-433.

Vogt, M., Steinke, M., Turner, S., Paulino, A., Meyerhofer, M., Riebesell, U., LeQuere, C. and Liss, P. 2008. Dynamics of dimethylsulphoniopropionate and dimethylsulphide under different CO_2 concentrations during a mesocosm experiment. *Biogeosciences* **5**: 407-419.

8.4. Polar Bears

According to the IPCC, global warming is "inducing declining survival rates, smaller size, and cannibalism among polar bears (Amstrup *et al.*, 2006; Regehr *et al.*, 2006)" (IPCC, 2007-II, p. 88). "Reproductive success in polar bears," the IPCC also claims, "has declined, resulting in a drop in body condition, which in turn is due to melting Arctic Sea ice. Without ice, polar bears cannot hunt seals, their favourite prey (Derocher *et al.*, 2004)" (p. 103). Later in the same report, the IPCC claims to have "very high confidence" that "substantial loss of sea ice will reduce habitat for dependent species (e.g., polar bears)" (p. 213).

As was the case with coral reefs, the IPCC's claims and predictions are based on computer models and untested theories rather than real-world data. They are at odds with much of what is known about sea ice, polar bear populations and behaviors, and the natural ability of wildlife to adapt to climate change.

In this section we review the evidence and conclude polar bears are not endangered by global warming, whether it is caused by human activity or any other causes. Since our findings once again contradict those of the IPCC, we summarize them here:

- There is little or no evidence of global warming-induced reduction in the extent or thickness of Arctic sea ice in the second-half of the twentieth century, particularly during those seasons when polar bears rely on it to reach their favorite food supply (seals), despite what the IPCC calls the "unprecedented warming" of the past century.

- Polar bears have survived changes in climate that exceed those that occurred during the twentieth century or are forecast by the IPCC's computer models.

- Temperatures in Greenland and other Arctic areas exhibit considerable variability described by one group of scientists as "a long term cooling and shorter warming periods."

- Most populations of polar bears are growing, not shrinking, and the biggest influence on polar bear populations is not temperature but hunting by humans, which historically has taken too large a toll on polar bear populations.

- Forecasts of dwindling polar bear populations assume trends in sea ice and temperature that are counterfactual, rely on computer climate models that are known to be unreliable, and violate most of the principles of scientific forecasting.

References

Amstrup, S.C, Stirling, I., Smith, T.S., Perham, C., and Thiemann, G.W. 2006. Recent observations of intraspecific predation and cannibalism among polar bears in the southern Beaufort Sea. *Polar Biology* **29**: 997-1002.

Derocher A.E., Lunn, N.J., and Stirling, I. 2004. Polar bears in a warming climate. *Integrative and Comparative Biology* **44**: 163-176.

IPCC. 2007-II. *Climate Change 2007: Impacts, Adaptation and Vulnerability. Contribution of Working Group II to the Fourth Assessment Report of the Intergovernmental Panel on Climate Change*, M.L. Parry, O.F. Canziani, J.P. Palutikof, P.J. van der Linden and Hanson, C.E. (Eds.) Cambridge University Press, Cambridge, UK.

Regehr, E.V., Amstrup, S.C., and Stirling, I. 2006. Polar bear population status in the Southern Beaufort Sea. U.S. Geological Survey Open-File Report 2006-1337.

8.4.1. Arctic Sea Ice

Global warming is thought to endanger polar bears (*Ursus maritimus*) by melting sea ice. According to Derocher *et al.* (2004), "it is unlikely that polar bears will survive as a species if the sea ice disappears completely as has been predicted by some." (We note that even the IPCC's computer models do not predict the "disappearance" of all sea ice, so this widely cited warning is essentially meaningless.) Amstrup *et al.* (2007) say "Our modeling suggests that realization of the sea ice future which is currently projected, would mean loss of $\approx$ 2/3 of the world's current polar bear population by mid-century."

These dire predictions are based entirely on computer simulations of the effects of warming temperatures on the extent of sea ice in the Arctic region. One way to test the reliability of those simulations is to see what effect the warming of the twentieth century had on the extent and thickness of sea ice. If the warming of the twentieth century hasn't produced trends toward less sea ice extent and thickness, why should we believe computer models that claim future warming would have a different effect? This section summarizes a more comprehensive analysis of Arctic sea ice that appeared in Chapter 4 of this report.

References

Amstrup, S.C., Marcot, B.G., and Douglas, D.C. 2007. Forecasting the rangewide status of polar bears at selected times in the 21st century. USGS Alaska Science Center, Anchorage, Administrative Report.

Derocher A.E., Lunn, N.J., and Stirling, I. 2004. Polar bears in a warming climate. *Integrative and Comparative Biology* **44**: 163-176.

8.4.1.1. Extent

The popular media is filled with reports about the disappearance of sea ice at the North Pole and the possibility of an ice-free summer in the region. The IPCC also focuses on summer (August and September) minimum Arctic sea ice extent (see Figures 4.8, 4.9, and 4.10 on pp. 351-352 of IPCC, 2007-I) and downplays evidence showing relatively little change in sea ice extent at other times of the year (e.g., Figure 4.10 shows relatively stable March sea ice extent but it receives no comment in the text).

Summer sea ice losses are part of an eons-long natural cycle to which polar bears have adapted in order to survive. Polar bears are able to fast for more than four months at a time while fully awake and mobile, without hibernating as other bears do (Watts and Hansen, 1987; Ramsay and Stirling, 1988; Ramsay and Hobson, 1991; Lennox and Goodship, 2008). During their winter feeding season, polar bears rely on sea ice to reach ringed seals (*Phoca hispida*) and in some regions bearded seals (*Erignathus barbatus*), their main sources of food (Derocher *et al.*, 2002; Stirling and Øritsland, 1995). Therefore, the measure of sea ice extent that is most relevant to the survival of polar bears is not summer minimums but multi-year. According to numerous studies described below, multi-year sea ice extent is growing in some parts of the Arctic and declining in others, sometimes changing sign over the course of a single year or two, making it difficult to discern an overall trend. The data show low-frequency oscillations that are not what one would predict based on the greenhouse warming theory and gradually increasing atmospheric CO_2 concentrations.

Johannessen *et al.* (1999) analyzed Arctic sea ice extent over the period 1978-1998 and found it to have decreased by about 14 percent. This finding led them to suggest that "the balance of evidence," as small as it then was, indicates "an ice cover in transition," and that "if this apparent transformation continues, it may lead to a markedly different ice regime in the Arctic," as also was suggested by Vinnikov *et al.* (1999). However, the plots of sea ice area presented by Johannessen *et al.* reveal that essentially all of the drop occurred abruptly over a single period of not more than three years (1987/88-1990/91) and possibly

only one year (1989/90-1990/91). Furthermore, it appears from their data that from 1990/91 onward, sea ice area in the Arctic may have *increased*.

More recently, Kwok (2004) estimated the coverage of Arctic multi-year sea ice at the beginning of each year of the study was 3774×10^3 km^2 in 2000, 3896×10^3 km^2 in 2001, 4475×10^3 km^2 in 2002, and 4122×10^3 km^2 in 2003, representing an *increase* in sea ice coverage of 9 percent over a third of a decade. Belchansky *et al.* (2004) report that from 1988 to 2001, total January multi-year ice area declined at a mean rate of 1.4 percent per year. They note, however, that in the autumn of 1996 "a large multiyear ice recruitment of over 10^6 km^2 fully replenished the previous 8-year decline in total area." They add that the replenishment "was followed by an accelerated and compensatory decline during the subsequent 4 years." In addition, they learned that 75 percent of the interannual variation in January multi-year sea area "was explained by linear regression on two atmospheric parameters: the previous winter's Arctic Oscillation index as a proxy to melt duration and the previous year's average sea level pressure gradient across the Fram Strait as a proxy to annual ice export."

Belchansky *et al.* conclude that their 14-year analysis of multi-year ice dynamics is "insufficient to project long-term trends." They also conclude it is insufficient to reveal "whether recent declines in multiyear ice area and thickness are indicators of anthropogenic exacerbations to positive feedbacks that will lead the Arctic to an unprecedented future of reduced ice cover, or whether they are simply ephemeral expressions of natural low frequency oscillations."

Heide-Jorgensen and Laidre (2004) examined changes in the fraction of open-water found within various pack-ice microhabitats of Foxe Basin, Hudson Bay, Hudson Strait, Baffin Bay-Davis Strait, northern Baffin Bay, and Lancaster Sound over a 23-year interval (1979-2001) using remotely sensed microwave measurements of sea-ice extent, after which the trends they documented were "related to the relative importance of each wintering microhabitat for eight marine indicator species and potential impacts on winter success and survival were examined." Foxe Basin, Hudson Bay, and Hudson Strait showed small increasing trends in the fraction of open water, with the upward trends at all microhabitats studied ranging from 0.2 to 0.7 percent per decade. In Baffin Bay-Davis Straight and northern Baffin Bay, on the other hand, the open-water trend was downward, and at a mean rate for all open-water microhabitats studied of fully 1 percent per decade, while the trend in all Lancaster Sound open-water microhabitats was also downward, in this case at a mean rate of 0.6 percent per decade.

Heide-Jorgensen and Laidre report that "increasing trends in sea ice coverage in Baffin Bay and Davis Strait (resulting in declining open-water) were as high as 7.5 percent per decade between 1979-1999 (Parkinson *et al.*, 1999; Deser *et al.*, 2000; Parkinson, 2000a,b; Parkinson and Cavalieri, 2002) and comparable significant increases have been detected back to 1953 (Stern and Heide-Jorgensen, 2003)." They additionally note that "similar trends in sea ice have also been detected locally along the West Greenland coast, with slightly lower increases of 2.8 percent per decade (Stern and Heide-Jorgensen, 2003)."

Grumet *et al.* (2001) warned that recent trends in Arctic sea ice cover "can be viewed out of context because their brevity does not account for interdecadal variability, nor are the records sufficiently long to clearly establish a climate trend." In an effort to overcome this "short-sightedness," they developed a 1,000-year record of spring sea ice conditions in the Arctic region of Baffin Bay based on sea-salt records from an ice core obtained from the Penny Ice Cap on Baffin Island. They determined that after a period of reduced sea ice during the eleventh through fourteenth centuries, enhanced sea ice conditions prevailed during the following 600 years. For the final (twentieth) century of this period, they report that "despite warmer temperatures during the turn of the century, sea-ice conditions in the Baffin Bay/Labrador Sea region, at least during the last 50 years, are within 'Little Ice Age' variability," suggesting that sea ice extent there has not yet emerged from the range of conditions characteristic of the Little Ice Age.

In an adjacent sector of the Arctic, this latter period of time also was studied by Comiso *et al.* (2001), who used satellite imagery to analyze and quantify a number of attributes of the Odden ice tongue—a winter ice-cover phenomenon that occurs in the Greenland Sea with a length of about 1,300 km and an aerial coverage of as much as 330,000 square kilometers—over the period 1979-1998. By utilizing surface air temperature data from Jan Mayen Island, which is located within the region of study, Comiso and colleagues were able to infer the behavior of this phenomenon over the past 75 years. Trend analyses revealed that the ice tongue has exhibited no statistically significant change in any of the parameters studied over the past 20 years; but the proxy reconstruction of the Odden ice tongue for the past 75 years revealed the ice phenomenon to have been "a relatively smaller feature several decades ago," due to the warmer temperatures that prevailed at that time.

In another study of Arctic climate variability, Omstedt and Chen (2001) obtained a proxy record of the annual maximum extent of sea ice in the region of the Baltic Sea over the period 1720-1997. In analyzing this record, they found that a significant decline in sea ice occurred around 1877. In addition, they reported finding greater variability in sea ice extent in the colder 1720-1877 period than in the warmer 1878-1997 period, suggesting that air temperatures are not the main force at work in determining sea ice extent.

Jevrejeva (2001) reconstructed an even longer record of sea ice duration (and, therefore, extent) in the Baltic Sea region by examining historical data for the observed time of ice break-up between 1529 and 1990 in the northern port of Riga, Latvia. The long date-of-ice-break-up time series was best described by a fifth-order polynomial, which identified four distinct periods of climatic transition: (1) 1530-1640, warming with a tendency toward earlier ice break-up of nine days/century, (2) 1640-1770, cooling with a tendency toward later ice break-up of five days/century, (3) 1770-1920, warming with a tendency toward earlier ice break-up of 15 days/century, and (4) 1920-1990, *cooling* with a tendency toward later ice break-up of 12 days/century.

On the other hand, in a study of the Nordic Seas (the Greenland, Iceland, Norwegian, Barents, and Western Kara Seas), Vinje (2001) determined that "the extent of ice in the Nordic Seas measured in April has decreased by 33% over the past 135 years." He notes, however, that "nearly half of this reduction is observed over the period 1860-1900," during a period, we note, when the atmosphere's CO_2 concentration rose by only 7 ppm, whereas the second half of the sea-ice decline occurred over a period of time when the air's CO_2 concentration rose by more than 70 ppm. If the historical rise in the air's CO_2 content has been responsible for the historical decrease in sea-ice extent, its impact over the last century has declined to less than a tenth of what its impact was over the preceding four decades. This in turn suggests that the increase in the air's CO_2 content over the past 135 years has likely had nothing to do with the concomitant decline in sea-ice cover.

In a similar study of the Kara, Laptev, East Siberian, and Chuckchi Seas, based on newly available long-term Russian observations, Polyakov et al. (2002) found "smaller than expected" trends in sea ice cover that, in their words, "do not support the hypothesized polar amplification of global warming." Likewise, in a study published the following year, Polyakov et al. (2003) report that "over the entire Siberian marginal-ice zone the century-long trend is only -0.5% per decade," while "in the Kara, Laptev, East Siberian, and Chukchi Seas the ice extent trends are not large either: -1.1%, -0.4%, +0.3%, and -1.0% per decade, respectively." Moreover, they say "these trends, except for the Chukchi Sea, are not statistically significant."

Divine and Dick (2006) used historical April through August ice observations made in the Nordic Seas—comprised of the Iceland, Greenland, Norwegian, and Barents Seas, extending from 30°W to 70°E—to construct time series of ice-edge position anomalies spanning the period 1750-2002, which they analyzed for evidence of long-term trend and oscillatory behavior. The authors report that "evidence was found of oscillations in ice cover with periods of about 60 to 80 years and 20 to 30 years, superimposed on a continuous negative trend," which observations are indicative of a "persistent ice retreat since the second half of the 19th century" that began well before anthropogenic CO_2 emissions could have had much effect on earth's climate.

Noting that the last cold period observed in the Arctic occurred at the end of the 1960s, the two Norwegian researchers say their results suggest that "the Arctic ice pack is now at the periodical apogee of the low-frequency variability," and that "this could explain the strong negative trend in ice extent during the last decades as a possible superposition of natural low frequency variability and greenhouse gas induced warming of the last decades." However, as they immediately caution, "a similar shrinkage of ice cover was observed in the 1920s-1930s, during the previous warm phase of the low frequency oscillation, when any anthropogenic influence is believed to have still been negligible." They suggest, therefore, "that during decades to come ... the retreat of ice cover may change to an expansion."

The oscillatory behavior observed in so many of the sea ice studies suggests, in the words of Parkinson (2000b), "the possibility of close connections between the sea ice cover and major oscillatory patterns in the atmosphere and oceans," including connections with: "(1) the North Atlantic Oscillation (e.g., Hurrell and van Loon, 1997; Johannessen et al., 1999; Kwok and Rothrock, 1999; Deser et al., 2000; Kwok, 2000; Vinje, 2001) and the spatially broader Arctic Oscillation (e.g., Deser et al., 2000; Wang and Ikeda, 2000); (2) the Arctic Ocean Oscillation (Polyakov et al., 1999; Proshutinsky et al., 1999); (3) a 'see-saw' in winter temperatures between Greenland and northern Europe (Rogers and van Loon, 1979); and (4) an interdecadal Arctic climate cycle (Mysak et al., 1990; Mysak and Power, 1992)." The likelihood that Arctic sea ice trends are the product of such natural oscillations, Parkinson continues, "provides a strong rationale for considerable caution when extrapolating into the future the widely reported decreases in the Arctic ice cover over the past few decades or when attributing the decreases primarily to global warming."

The latest observations on Greenland's outlet glaciers, reported in January 2009 by a writer for Science who attended a December 2008 meeting of the American Geophysical Union, amply confirm Parkinson's conclusion. They show the acceleration of melting that had occurred between 2003-2005 had "come to an end" and "nearly everywhere around southeast Greenland, outlet glacier flows have returned to the levels of 2000" (Anonymous, 2009). The article quotes a team of researchers led by glacial modeler Faezeh Nick of Durham University in the United Kingdom as saying "our results imply that the recent rates of mass loss in Greenland's outlet glaciers are transient and should not be extrapolated into the future" (Ibid.)

In conclusion, there is little or no evidence of a consistent, global warming-induced reduction in the extent of Arctic sea ice, especially during those seasons when polar bears rely on it to reach their favorite food supply (seals), despite what the IPCC calls the "unprecedented warming" of the past century. This is a key finding that undermines the claim that global warming would have a harmful effect on polar bears in the future.

Additional information on this topic, including reviews of newer publications as they become available, can be found at http://www.co2science.org/subject/s/seaicearctic.php.

References

Anonymous. 2009. Galloping glaciers of Greenland have reined themselves in. Science. **323**: 458.

Belchansky, G.I., Douglas, D.C., Alpatsky, I.V. and Platonov, N.G. 2004. Spatial and temporal multiyear sea ice distributions in the Arctic: A neural network analysis of SSM/I data, 1988-2001. *Journal of Geophysical Research* **109**: 10.1029/2004JC002388.

Comiso, J.C., Wadhams, P., Pedersen, L.T. and Gersten, R.A. 2001. Seasonal and interannual variability of the Odden ice tongue and a study of environmental effects. *Journal of Geophysical Research* **106**: 9093-9116.

Derocher, A.E., Wiig, Ø., and Andersen, M. 2002. Diet composition of polar bears in Svalbard and the western Barents Sea. *Polar Biology* **25**: 448-452.

Deser, C., Walsh, J. and Timlin, M.S. 2000. Arctic sea ice variability in the context of recent atmospheric circulation trends. *Journal of Climate* **13**: 617-633.

Divine, D.V. and Dick, C. 2006. Historical variability of sea ice edge position in the Nordic Seas. *Journal of Geophysical Research* **111**: 10.1029/2004JC002851.

Grumet, N.S., Wake, C.P., Mayewski, P.A., Zielinski, G.A., Whitlow, S.L., Koerner, R.M., Fisher, D.A. and Woollett, J.M. 2001. Variability of sea-ice extent in Baffin Bay over the last millennium. *Climatic Change* **49**: 129-145.

Heide-Jorgensen, M.P. and Laidre, K.L. 2004. Declining extent of open-water refugia for top predators in Baffin Bay and adjacent waters. *Ambio* **33**: 487-494.

Hurrell, J.W. and van Loon, H. 1997. Decadal variations in climate associated with the North Atlantic Oscillation. *Climatic Change* **36**: 301-326.

IPCC 2007-I. *Climate Change 2007: The Physical Science Basis. Contribution of Working Group I to the Fourth Assessment Report of the Intergovernmental Panel on Climate Change.* Solomon, S., Qin, D., Manning, M., Chen, Z., Marquis, M., Averyt, K.B., Tignor, M. and Miller, H.L. (Eds.) Cambridge University Press, Cambridge, United Kingdom and New York, NY.

Jevrejeva, S. 2001. Severity of winter seasons in the northern Baltic Sea between 1529 and 1990: reconstruction and analysis. *Climate Research* **17**: 55-62.

Johannessen, O.M., Shalina, E.V. and Miles, M.W. 1999. Satellite evidence for an Arctic sea ice cover in transformation. *Science* **286**: 1937-1939.

Kwok, R. 2000. Recent changes in Arctic Ocean sea ice motion associated with the North Atlantic Oscillation. *Geophysical Research Letters* **27**: 775-778.

Kwok, R. 2004. Annual cycles of multiyear sea ice coverage of the Arctic Ocean: 1999-2003. *Journal of Geophysical Research* **109**: 10.1029/2003JC002238.

Kwok, R. and Rothrock, D.A. 1999. Variability of Fram Strait ice flux and North Atlantic Oscillation. *Journal of Geophysical Research* **104**: 5177-5189.

Lennox. A.R. and Goodship, A.E. 2008. Polar bears (*Ursus maritimus*), the most evolutionary advanced hibernators, avoid significant bone loss during hibernation. *Comparative Biochemistry and Physiology Part A* **149**: 203-208.

Mysak, L.A., Manak, D.K. and Marsden, R.F. 1990. Sea-ice anomalies observed in the Greenland and Labrador Seas during 1901-1984 and their relation to an interdecadal Arctic climate cycle. *Climate Dynamics* **5**: 111-133.

Mysak, L.A. and Power, S.B. 1992. Sea-ice anomalies in the western Arctic and Greenland-Iceland Sea and their relation to an interdecadal climate cycle. *Climatological Bulletin/Bulletin Climatologique* **26**: 147-176.

Omstedt, A. and Chen, D. 2001. Influence of atmospheric circulation on the maximum ice extent in the Baltic Sea. *Journal of Geophysical Research* **106**: 4493-4500.

Parkinson, C.L. 2000a. Variability of Arctic sea ice: the view from space, and 18-year record. *Arctic* **53**: 341-358.

Parkinson, C.L. 2000b. Recent trend reversals in Arctic sea ice extents: possible connections to the North Atlantic Oscillation. *Polar Geography* **24**: 1-12.

Parkinson, C.L. and Cavalieri, D.J. 2002. A 21-year record of Arctic sea-ice extents and their regional, seasonal and monthly variability and trends. *Annals of Glaciology* **34**: 441-446.

Parkinson, C.L., Cavalieri, D.J., Gloersen, P., Zwally, H.J. and Comiso, J.C. 1999. Arctic sea ice extents, areas, and trends, 1978-1996. *Journal of Geophysical Research* **104**: 20,837-20,856.

Polyakov, I.V., Proshutinsky, A.Y. and Johnson, M.A. 1999. Seasonal cycles in two regimes of Arctic climate. *Journal of Geophysical Research* **104**: 25,761-25,788.

Polyakov, I.V., Alekseev, G.V., Bekryaev, R.V., Bhatt, U., Colony, R.L., Johnson, M.A., Karklin, V.P., Makshtas, A.P., Walsh, D. and Yulin, A.V. 2002. Observationally based assessment of polar amplification of global warming. *Geophysical Research Letters* **29**: 10.1029/2001GL011111.

Polyakov, I.V., Alekseev, G.V., Bekryaev, R.V., Bhatt, U.S., Colony, R., Johnson, M.A., Karklin, V.P., Walsh, D. and Yulin, A.V. 2003. Long-term ice variability in Arctic marginal seas. *Journal of Climate* **16**: 2078-2085.

Proshutinsky, A.Y., Polyakov, I.V. and Johnson, M.A. 1999. Climate states and variability of Arctic ice and water dynamics during 1946-1997. *Polar Research* **18**: 135-142.

Ramsay, M.A. and Hobson, K.A. 1991. Polar bears make little use of terrestrial food webs: Evidence from stable-carbon isotope analysis. *Oecologia* **86**:598-600.

Ramsay, M.A. and Stirling, I. 1988. Reproductive biology and ecology of female polar bears (*Ursus maritimus*). *Journal of Zoology* (London) **214**:601-634.

Rogers, J.C. and van Loon, H. 1979. The seesaw in winter temperatures between Greenland and Northern Europe. Part II: Some oceanic and atmospheric effects in middle and high latitudes. *Monthly Weather Review* **107**: 509-519.

Stern, H.L. and Heide-Jorgensen, M.P. 2003. Trends and variability of sea ice in Baffin Bay and Davis Strait. *Polar Research* **22**: 11-18.

Stirling, I. and Øritsland, N.A. 1995. Relationships between estimates of ringed seal and polar bear populations in the Canadian Arctic. *Canadian Journal of Fisheries and Aquatic Science* **52**:2594-2612.

Vinje, T. 2001. Anomalies and trends of sea ice extent and atmospheric circulation in the Nordic Seas during the period 1864-1998. *Journal of Climate* **14**: 255-267.

Vinnikov, K.Y., Robock, A., Stouffer, R.J., Walsh, J.E., Parkinson, C.L., Cavalieri, D.J., Mitchell, J.F.B., Garrett, D. and Zakharov, V.R. 1999. Global warming and Northern Hemisphere sea ice extent. *Science* **286**: 1934-1937.

Wang, J. and Ikeda, M. 2000. Arctic Oscillation and Arctic Sea-Ice Oscillation. *Geophysical Research Letters* **27**: 1287-1290.

Watts, P.D. and Hansen, S.E. 1987. Cyclic starvation as a reproductive strategy in the polar bear. *Symposium of the Zoological Society of London* **57**:306-318.

8.4.1.2. Thickness

In addition to its extent, the thickness of Arctic sea ice is of concern since it must support the weight of hunting polar bears. Male polar bears can weigh up to 800 kg (1,764 pounds) (DeMaster and Stirling, 1981), and there is anecdotal evidence of bears larger than this (Dowsley, 2005). Thick sea ice can negatively affect populations of marine mammals by reducing the size and number of airholes, thereby suffocating some species of whales and seals and reducing the number of hunting areas for polar bears (Stirling, 2002; Laidre *et al.*, 2008; Harington, 2008). Trends in the thickness of sea ice might also be precursors of changes in sea ice extent.

Based on analyses of submarine sonar data, Rothrock *et al.* (1999) suggested that Arctic sea ice in the mid 1990s had thinned by about 42 percent of the average 1958-1977 thickness. The IPCC reports this finding but also reports that other more recent studies found "the reduction in ice thickness was not gradual, but occurred abruptly before 1991," and acknowledges that "ice thickness varies considerably from year to year at a given location and so the rather sparse temporal sampling provided by submarine data makes inferences regarding long term change difficult" (IPCC 2007, p. 353). Johannessen *et al.* (1999), for example, found that essentially all of the drop occurred rather abruptly over a single period of not more than three years (1987/88-1990/91) and possibly only one year (1989/90-1990/91).

Two years after Johannessen *et al.*, Winsor (2001) analyzed a more comprehensive set of Arctic sea-ice data obtained from six submarine cruises conducted between 1991 and 1997 that had covered the central Arctic Basin from 76°N to 90°N, as well as two areas that had been particularly densely sampled, one centered at the North Pole (>87°N) and one in the central part of the Beaufort Sea (centered at approximately 76°N, 145°W). The transect data across the entire Arctic Basin revealed that the mean Arctic sea-ice thickness had remained "almost constant" over the period of study. Data from the North Pole also showed little variability, and a linear regression of the data revealed a "slight increasing trend for the whole period." As for the Beaufort Sea region, annual variability in sea ice thickness was greater than at the North Pole but once again, in Winsor's words, "no significant trend" in mean sea-ice thickness was found. Combining the North Pole results with the results of an earlier study, Winsor concluded that "mean ice thickness has remained on a near-constant level around the North Pole from 1986 to 1997."

The following year, Holloway and Sou (2002) explored "how observations, theory, and modeling work together to clarify perceived changes to Arctic sea ice," incorporating data from "the atmosphere, rivers, and ocean along with dynamics expressed in an ocean-ice-snow model." On the basis of a number of different data-fed model runs, they found that for the last half of the past century, "no linear trend [in Arctic sea ice volume] over 50 years is appropriate," noting their results indicated "increasing volume to the mid-1960s, decadal variability without significant trend from the mid-1960s to the mid-1980s, then a loss of volume from the mid-1980s to the mid-1990s." The net effect of this behavior, in their words, was that "the volume estimated in 2000 is close to the volume estimated in 1950." They suggest that the initial inferred rapid thinning of Arctic sea ice was, as they put it, "unlikely," due to problems arising from under-sampling. They also report that "varying winds that readily redistribute Arctic ice create a recurring pattern whereby ice shifts between the central Arctic and peripheral regions, especially in the Canadian sector," and that the "timing and tracks of the submarine surveys missed this dominant mode of variability."

In the same year, Polyakov *et al.* (2002) employed newly available long-term Russian landfast-ice data obtained from the Kara, Laptev, East Siberian, and Chuckchi Seas to investigate trends and variability in the Arctic environment poleward of 62°N. This study revealed that fast-ice thickness trends in the different seas were "relatively small, positive or negative in sign at different locations, and not statistically significant at the 95% level." A year later, these results were reconfirmed by Polyakov *et al.* (2003), who reported that the available fast-ice records "do not show a significant trend," while noting that "in the Kara and Chukchi Seas trends are positive, and in the Laptev and East Siberian Seas trends are negative," but stating that "these trends are not statistically significant at the 95% confidence level."

Laxon *et al.* (2003) used an eight-year time series (1993-2001) of Arctic sea-ice thickness data derived from measurements of ice freeboard made by radar altimeters carried aboard ERS-1 and 2 satellites to determine the mean thickness and variability of Arctic sea ice between latitudes 65°N and 81.5°N, which region covers the entire circumference of the Arctic Ocean, including the Beaufort, Chukchi, East Siberian, Kara, Laptev, Barents, and Greenland Seas. These real-world observations (1) revealed "an interannual variability in ice thickness at higher frequency, and of greater amplitude, than simulated by regional Arctic models," (2) undermined "the conclusion from numerical models that changes in ice thickness occur on much longer timescales than changes in ice extent," and (3) showed that "sea ice mass can change by up to 16% within one year," which finding "contrasts with the concept of a slowly dwindling ice pack, produced by greenhouse warming." Laxon *et al.* concluded that "errors are present in current simulations of Arctic sea ice," stating in their closing sentence that "until models properly reproduce the observed high-frequency, and thermodynamically driven, variability in sea ice thickness, simulations of both recent, and future, changes in Arctic ice cover will be open to question."

In a paper on landfast ice in Canada's Hudson Bay, Gagnon and Gough (2006) cite nine different studies of sea-ice cover, duration, and thickness in the Northern Hemisphere, noting that the Hudson Bay region "has been omitted from those studies with the exception of Parkinson *et al.* (1999)." For 13 stations located on the shores of Hudson Bay (seven) and surrounding nearby lakes (six), Gagnon and Gough then analyzed long-term weekly measurements of ice thickness and associated weather conditions that began and ended, in the mean, in 1963 and 1993, respectively. The study revealed that a "statistically significant thickening of the ice cover over time was detected on the western side of Hudson Bay, while a slight thinning lacking statistical significance was observed on the eastern side." This asymmetry, in their words, was "related to the variability of air temperature, snow depth, and the dates of ice freeze-up and break-up," with "increasing maximum ice thickness at a number of stations" being "correlated to earlier freeze-up due to negative temperature trends in autumn," and with high snow accumulation being associated with low ice thickness, "because the snow cover insulates the ice surface, reducing heat conduction and thereby ice growth." Noting that their findings "are in contrast to the projections from general circulation models, and to the reduction in sea-ice extent and thickness observed in other regions of the Arctic," Gagnon and Gough say "this contradiction must be addressed in regional climate change impact assessments."

Finally, the relationship between sea ice thickness and polar bear survival is not so simple that any reduction in the one leads to a reduction in the other. As mentioned previously, research has found that expanding sea ice extent and thickness can negatively affect populations of marine mammals, including polar bears (Stirling, 2002; Laidre *et al.*, 2008; Harington, 2008). Laidre and Heide-Jergensen (2005) report that "cetacean occurrence is generally negatively correlated with dense or complete ice cover due to the need to breathe at the surface," and that "lacking the ability to break holes in the ice," narwhals are vulnerable to reductions in the amount of open water available to them, as has been demonstrated by ice entrapment events "where hundreds of narwhals died during rapid sea ice formation caused by sudden cold periods (Siegastad and Heide-Jorgensen, 1994; Heide-Jorgensen *et al.*, 2002). Such events were becoming ever more likely as temperatures continued to decline and sea ice cover and variability increased." They concluded that "with the evidence of changes in sea ice conditions that could impact foraging, prey availability, and of utmost importance, access to the surface to breathe, it is unclear how narwhal sub-populations will fare in light of changes in the high Arctic."

While the negative effects of too much or too thick ice are well known and documented, the evidence is less clear that thinner ice is a major impediment to hunting by polar bears. The bears are known to hunt on new ice that is less than 30 cm (about 1 foot) thick and to use first-year ice that is greater than 120 cm (about 4 feet) thick for over-wintering and denning (Ferguson *et al.*, 1997; Ferguson *et al.*, 2000). For context, consider that first-year ice in March 2008 was about 160 cm (about 5 feet) thick (NSIDC 2008). Polar bears also are able to migrate to other areas in response to changes in sea ice, as they have been observed to do in response to competition for food, contact with human development, and other environmental impacts (Messier *et al.*, 2001; Dyke *et al.*, 2007).

These observations suggest there has not been a steady or continuing thinning of Arctic sea ice that can be attributed to CO_2-induced global warming. Rather, and as was the case with changes in the extent of Arctic sea ice, changes in sea ice thickness appear to be a consequence of changes in ice dynamics caused by periodic climate oscillations having nothing to do with changes in the air's CO_2 content. Consequently, there is once again no evidence that polar bears are endangered by global warming, whether it results from human activities or other causes.

Additional information on this topic, including reviews of newer publications as they become available, can be found at http://www.co2science.org/subject/s/seaicearcticthick.php.

References

DeMaster, D.P. and Stirling, I. 1981. *Ursus maritimus.* Polar bear. *Mammalian Species* **145**:1-7.

Dowsley, M. 2005. Inuit knowledge regarding climate change and the Baffin Bay polar bear population. Government of Nunavut, Department of Environment, Final Wildlife Report 1, Iqaluit, Nunavut.

Dyck, M.G., Soon, W., Baydack, R.K., Legates, D.R., Baliunas, S., Ball, T.F., and Hancock, L.O. 2007. Polar bears of western Hudson Bay and climate change: Are warming spring air temperatures the "ultimate" survival control factor? *Ecological Complexity* **4**: 73-84.

Ferguson, S.H., Taylor, M.K., and Messier, F. 1997. Space use by polar bears in and around Auyuittuq National Park, Northwest Territories, during the ice-free period. *Canadian Journal of Zoology* **75**: 1585-1594.

Ferguson, S.H., Taylor, M.K., and Messier, F. 2000. Influence of sea ice dynamics on habitat selection by polar bears. *Ecology* **81**: 761-772.

Gagnon, A.S. and Gough, W.A. 2006. East-west asymmetry in long-term trends of landfast ice thickness in the Hudson Bay region, Canada. *Climate Research* **32**: 177-186.

Harington, C.R. 2008. The evolution of Arctic marine mammals. *Ecological Applications* **18** (Suppl.): S23-S40.

Heide-Jorgensen, M.P., Richard, P., Ramsay, M. and Akeeagok, S. 2002. In: *Three Recent Ice Entrapments of Arctic Cetaceans in West Greenland and the Eastern Canadian High Arctic.* Volume 4, NAMMCO Scientific Publications. 143-148.

Holloway, G. and Sou, T. 2002. Has Arctic sea ice rapidly thinned? *Journal of Climate* **15**: 1691-1701.

IPCC. 2007. *Climate Change 2007: The Physical Science Basis. Contribution of Working Group I to the Fourth Assessment Report of the Intergovernmental Panel on Climate Change.* Solomon, S., Qin, D., Manning, M., Chen, Z., Marquis, M., Averyt, K.B., Tignor, M. and Miller, H.L. (Eds.) Cambridge University Press, Cambridge, United Kingdom and New York, NY.

Johannessen, O.M., Shalina, E.V. and Miles, M.W. 1999. Satellite evidence for an Arctic sea ice cover in transformation. *Science* **286**: 1937-1939.

Kwok, R. 2000. Recent changes in Arctic Ocean sea ice motion associated with the North Atlantic Oscillation. *Geophysical Research Letters* **27**: 775-778.

Laxon, S., Peacock, N. and Smith, D. 2003. High interannual variability of sea ice thickness in the Arctic region. *Nature* **425**: 947-950.

Laidre, K.L., Stirling, I., Lowry, L.F., Wiig, Ø., Heide-Jørgensen, M.P., and Ferguson, S.H. 2008. Quantifying the sensitivity of arctic marine mammals to climate-induced habitat change. *Ecological Applications* 18(2, Suppl.): S97-S125.

Laidre, K.L. and Heide-Jorgensen, M.P. 2005. Arctic sea ice trends and narwhal vulnerability. *Biological Conservation* **121**: 509-517.

Messier, F., Taylor, M.K., Plante, A. and Romito, T. 2001. Atlas of polar bear movements in Nunavut, Northwest

Territories, and neighboring areas. Nunavut Wildlife Service and University of Saskatchewan, Saskatoon, SK.

NSIDC 2008. A different pattern of sea ice retreat. (National Snow and Ice Data Center). Web site, last accessed July 17, 2008. http://nsidc.org/arcticsea icenews/index.html

Parkinson, C.L., Cavalieri, D.J., Gloersen, P., Zwally, J., and Comiso, J.C. 1999. Arctic sea ice extent, areas, and trends, 1978-1996. *Journal of Geophysical Research* **104**: 20,837-20,856.

Polyakov, I.V., Alekseev, G.V., Bekryaev, R.V., Bhatt, U., Colony, R.L., Johnson, M.A., Karklin, V.P., Makshtas, A.P., Walsh, D. and Yulin, A.V. 2002. Observationally based assessment of polar amplification of global warming. *Geophysical Research Letters* **29**: 10.1029/2001GL011111.

Polyakov, I.V., Alekseev, G.V., Bekryaev, R.V., Bhatt, U.S., Colony, R., Johnson, M.A., Karklin, V.P., Walsh, D. and Yulin, A.V. 2003. Long-term ice variability in Arctic marginal seas. *Journal of Climate* **16**: 2078-2085.

Rothrock, D.A., Yu, Y. and Maykut, G.A. 1999. Thinning of the Arctic sea ice cover. *Geophysics Research Letters* **26**: 3469-3472.

Siegstad, H. and Heide-Jorgensen, M.P. 1994. Ice entrapments of narwhals (*Monodon monoceros*) and white whales (*Delphinapterus leucas*) in Greenland. *Meddeleser om Gronland Bioscience* **39**: 151-160.

Stirling, I. 2002. Polar bears and seals in the eastern Beaufort Sea and Amundsen Gulf: a synthesis of population trends and ecological relationships over three decades. *Arctic* **55** (Suppl. 1): 59-76.

Winsor, P. 2001. Arctic sea ice thickness remained constant during the 1990s. *Geophysical Research Letters* **28**: 1039-1041.

8.4.2. Temperatures

Polar bears evolved from brown bears (*Ursus arctos*) sometime in the last 400,000 years and probably no more than 200,000 years ago (Amnason *et al.*, 1995; Davis *et al.,* 2008; Harington, 2008). This means they have survived whatever changes in the Arctic climate took place over the course of many millennia, including two major warming periods over the last 11,000 years, the Early Holocene Climatic Optimum and the Medieval Warm Period.

In this section we review evidence of these warm periods first in Greenland and then in the rest of the Arctic. Then we look at more recent temperature trends in Greenland and the rest of the Arctic and find

considerable variability, recently described by one group of scientists as "a long term cooling and shorter warming periods" (Chylek *et al.*, 2006). These topics are addressed in greater detail in Chapter 3 of this report.

References

Arnason, U., Bodin, K., Gullberg, A., Ledje, C., and Mouchaty, S. 1995. A molecular view of pinniped relationships with particular emphasis on the true seals. *Journal of Molecular Evolution* **40**: 78-85.

Chylek, P., Dubey, M.K, and Lesins, G. 2006. Greenland warming of 1920-1930 and 1995-2005. *Geophysical Research Letters* **33**: L11707.

Davis, C.S., Stirling, I., Strobeck, C., and Coltman, D.W. 2008. Population structure of ice-breeding seals. *Molecular Ecology* **17**: 3078-3094.

Harington, C.R. 2008. The evolution of Arctic marine mammals. *Ecological Applications* **18** (Suppl.): S23-S40.

8.4.2.1. Prehistoric Greenland

Dahl-Jensen *et al.* (1998) used data from two ice sheet boreholes to reconstruct the temperature history of Greenland over the past 50,000 years. Their analysis indicated that temperatures on the Greenland Ice Sheet during the Last Glacial Maximum (about 25,000 years ago) were 23 ± 2 °C colder than at present. After the termination of the glacial period, however, temperatures increased steadily to a value that was 2.5°C *warmer* than at present, during the Climatic Optimum of 4,000 to 7,000 years ago. The Medieval Warm Period and Little Ice Age were also evident in the borehole data, with temperatures 1°C warmer and 0.5-0.7°C cooler than at present, respectively. Then, after the Little Ice Age, the scientists report "temperatures reached a maximum around 1930 AD" and that "temperatures have decreased during the last decades."

In another study of Greenland climate, Bard (2002) describes glacial-period millennial-scale episodes of dramatic warming called Dansgaard-Oeschger events (with temperature increases "of more than 10°C"), which are evident in Greenland ice core records, as well as episodes of "drastic cooling" called Heinrich events (with temperature drops "of up to about 5°C"), which are evident in sea surface temperature records derived from the study of North

Atlantic deep-sea sediment cores. In the Greenland record, according to Bard, the progression of these events is such that "the temperature warms abruptly to reach a maximum and then slowly decreases for a few centuries before reaching a threshold, after which it drops back to the cold values that prevailed before the warm event."

Wagner and Melles (2001) retrieved a sediment core from a lake on an island situated just off Liverpool Land on the east coast of Greenland. Analyzing it for a number of properties related to the past presence of seabirds there, they obtained a 10,000-year record that tells us much about the region's climatic history. Key to the study were certain biogeochemical data that reflected variations in seabird breeding colonies in the catchment area of the lake. These data revealed high levels of the various parameters measured by Wagner and Melles between about 1,100 and 700 years before present (BP) that were indicative of the summer presence of significant numbers of seabirds during that "medieval warm period," as they describe it, which had been preceded by a several-hundred-year period of little to no inferred bird presence. Then, after the Medieval Warm Period, the data suggested another absence of birds during what they refer to as "a subsequent Little Ice Age," which they note was "the coldest period since the early Holocene in East Greenland." Their data also showed signs of a "resettlement of seabirds during the last 100 years, indicated by an increase of organic matter in the lake sediment and confirmed by bird observations." However, values of the most recent data were not as great as those obtained from the earlier Medieval Warm Period; and temperatures derived from two Greenland ice cores led to the same conclusion: It was warmer at various times between 1,100 to 700 years BP than it was over the twentieth century.

Kaplan *et al.* (2002) also worked with data obtained from a small lake, this one in southern Greenland, analyzing sediment physical-chemical properties, including magnetic susceptibility, density, water content, and biogenic silica and organic matter concentrations. They discovered that "the interval from 6000 to 3000 cal yr BP was marked by warmth and stability." Thereafter, however, the climate cooled "until its culmination during the Little Ice Age," but from 1,300-900 years BP, there was a partial amelioration of climate (the Medieval Warm Period) that was associated with an approximate 1.5°C rise in temperature.

Following another brief warming between AD 1500 and 1750, the second and more severe portion of the Little Ice Age occurred, which was in turn followed by "naturally initiated post-Little Ice Age warming since AD 1850, which is recorded throughout the Arctic." They report that Viking "colonization around the northwestern North Atlantic occurred during peak Medieval Warm Period conditions that ended in southern Greenland by AD 1100," noting that Norse movements around the region thereafter "occurred at perhaps the worst time in the last 10,000 years, in terms of the overall stability of the environment for sustained plant and animal husbandry."

These many studies of the temperature history of Greenland depict long-term oscillatory cooling ever since the Climatic Optimum of the mid-Holocene, when it was perhaps 2.5°C warmer than it is now, within which cooling trend is included the Medieval Warm Period, when it was about 1°C warmer than it is currently, and the Little Ice Age, when it was 0.5 to 0.7°C cooler than now, after which temperatures rebounded to a new maximum in the 1930s and 1940s, only to fall steadily thereafter.

Polar bears obviously survived these large-scale and often sudden climate changes in Greenland, otherwise none would be found living in Greenland today. This does not mean the population of polar bears remained constant throughout this period; more likely they flourished in some periods and declined in numbers in others. But changes in climate did not lead to their extinction. This history begs the question: Why would temperature changes that are predicted to be of the same scale or less than those that occurred naturally, before there was any human impact on climate, be expected to cause the extinction of polar bears, when early natural cycles did not?

Additional information on this topic, including reviews of newer publications as they become available, can be found at http://www.co2science.org/subject/g/greenland.php.

References

Bard, E. 2002. Climate shock: Abrupt changes over millennial time scales. *Physics Today* **55**: 32-38.

Dahl-Jensen, D., Mosegaard, K., Gundestrup, N., Clow, G.D., Johnsen, S.J., Hansen, A.W., and Balling, N. 1998. Past temperatures directly from the Greenland Ice Sheet. *Science* **282**: 268-271.

Kaplan, M.R., Wolfe, A.P., and Miller, G.H. 2002. Holocene environmental variability in southern Greenland inferred from lake sediments. *Quaternary Research* **58**: 149-159.

Wagner, B. and Melles, M. 2001. A Holocene seabird record from Raffles So sediments, East Greenland, in response to climatic and oceanic changes. *Boreas* **30**: 228-239.

8.4.2.2. Rest of the Prehistoric Arctic

Polar bears faced similar temperature changes in the rest of the Arctic region. Naurzbaev and Vaganov (2000) developed a 2,200-year temperature history using tree-ring data obtained from 118 trees near the upper-timberline in Siberia for the period 212 BC to AD 1996, as well as a similar history covering the period of the Holocene Climatic Optimum (3300 to 2600 BC). They reported that several warm and cool periods prevailed for several multi-century periods throughout the last two millennia: a cool period in the first two centuries AD, a warm period from AD 200 to 600, cooling again from 600 to 800 AD, followed by the Medieval Warm Period from about AD 850 to 1150, the cooling of the Little Ice Age from AD 1200 though 1800, followed by the recovery warming of the twentieth century. In regard to this latter temperature rise, however, the two scientists say it was "not extraordinary" and that "the warming at the border of the first and second millennia [AD 1000] was longer in time and similar in amplitude." In addition, their reconstructed temperatures for the Holocene Climatic Optimum revealed there was an even warmer period about 5,000 years ago, when temperatures averaged 3.3°C more than they did over the past two millennia.

Moore *et al.* (2001) analyzed sediment cores extracted from Donard Lake, Baffin Island, Canada (~66.25°N, 62°W) to produce a 1,240-year record of mean summer temperature for this region that averaged 2.9°C over the period AD 750-1990. Within this period there were several anomalously warm decades with temperatures that were as high as 4°C around AD 1000 and 1100, while at the beginning of the thirteenth century Donard Lake witnessed what they called "one of the largest climatic transitions in over a millennium," as "average summer temperatures rose rapidly by nearly 2°C from AD 1195-1220, ending in the warmest decade in the record," with temperatures near 4.5°C. This latter temperature rise was then followed by a period of extended warmth that lasted until an abrupt cooling

event occurred around AD 1375, resulting in the following decade being one of the coldest in the record and signaling the onset of the Little Ice Age on Baffin Island, which lasted 400 years. At the modern end of the record, a gradual warming trend occurred over the period 1800-1900, followed by a dramatic cooling event that brought temperatures back to levels characteristic of the Little Ice Age, which chilliness lasted until about 1950. Thereafter, temperatures rose once more throughout the 1950s and 1960s, whereupon they trended downwards toward cooler conditions to the end of the record in 1990.

Gedalof and Smith (2001) compiled a transect of six tree ring-width chronologies from stands of mountain hemlock growing near the treeline that extends from southern Oregon to the Kenai Peninsula, Alaska. Over the period of their study (AD 1599-1983), they determined that "much of the pre-instrumental record in the Pacific Northwest region of North America [was] characterized by alternating regimes of relatively warmer and cooler SST [sea surface temperature] in the North Pacific, punctuated by abrupt shifts in the mean background state," which were found to be "relatively common occurrences." They concluded, "regime shifts in the North Pacific have occurred 11 times since 1650." A significant aspect of these findings is the fact that the abrupt 1976-77 shift in this Pacific Decadal Oscillation, as it is generally called, is what was responsible for the vast majority of the past half-century's warming in Alaska, which some commentators wrongly point to as evidence of CO_2-induced global warming.

Kasper and Allard (2001) examined soil deformations caused by ice wedges (a widespread and abundant form of ground ice in permafrost regions that can grow during colder periods and deform and crack the soil). Working near Salluit, northern Quebéc (approx. 62°N, 75.75°W), they found evidence of ice wedge activity prior to AD 140, reflecting cold climatic conditions. Between AD 140 and 1030, however, this activity decreased, reflective of warmer conditions. Then, from AD 1030 to 1500, conditions cooled; and from 1500 to 1900 ice wedge activity was at its peak, when the Little Ice Age ruled, suggesting this climatic interval exhibited the coldest conditions of the past 4,000 years. Thereafter, a warmer period prevailed, from about 1900 to 1946, which was followed by a return to cold conditions during the last five decades of the twentieth century, during which time more than 90 percent of the ice wedges studied reactivated and grew by 20-30 cm, in harmony with a reported temperature decline of 1.1°C observed at the meteorological station in Salluit.

Naurzbaev *et al.* (2002) developed a 2,427-year proxy temperature history for the part of the Taimyr Peninsula, northern Russia, lying between 70°30' and 72°28' North latitude, based on a study of ring-widths of living and preserved larch trees, noting that it has been shown that "the main driver of tree-ring variability at the polar timber-line [where they worked] is temperature (Vaganov *et al.*, 1996; Briffa *et al.*, 1998; Schweingruber and Briffa, 1996)." This work revealed that "the warmest periods over the last two millennia in this region were clearly in the third [Roman Warm Period], tenth to twelfth [Medieval Warm Period] and during the twentieth [Current Warm Period] centuries." With respect to the second of these three periods, they emphasize that "the warmth of the two centuries AD 1058-1157 and 950-1049 attests to the reality of relative mediaeval warmth in this region." Their data also reveal three other important pieces of information: (1) the Roman and Medieval Warm Periods were both warmer than the Current Warm Period has been to date, (2) the beginning of the end of the Little Ice Age was somewhere in the vicinity of 1830, and (3) the Current Warm Period peaked somewhere in the vicinity of 1940.

These studies demonstrate that polar bears throughout the Arctic region experienced periods of warming and cooling in the past that exceed the variability observed in the twentieth century and likely to occur in the twenty-first century. Additional information on this topic, including reviews of newer publications as they become available, can be found at http://www.co2science.org/subject/a/arctictemptrends.php.

References

Briffa, K.R., Schweingruber, F.H., Jones, P.D., Osborn, T.J., Shiyatov, S.G. and Vaganov, E.A. 1998. Reduced sensitivity of recent tree-growth to temperature at high northern latitudes. *Nature* **391**: 678-682.

Gedalof, Z. and Smith, D.J. 2001. Interdecadal climate variability and regime-scale shifts in Pacific North America. *Geophysical Research Letters* **28**: 1515-1518.

Kasper, J.N. and Allard, M. 2001. Late-Holocene climatic changes as detected by the growth and decay of ice wedges on the southern shore of Hudson Strait, northern Québec, Canada. *The Holocene* **11**: 563-577.

Moore, J.J., Hughen, K.A., Miller, G.H. and Overpeck, J.T. 2001. Little Ice Age recorded in summer temperature reconstruction from varved sediments of Donard Lake, Baffin Island, Canada. *Journal of Paleolimnology* **25**: 503-517.

Naurzbaev, M.M. and Vaganov, E.A. 2000. Variation of early summer and annual temperature in east Taymir and Putoran (Siberia) over the last two millennia inferred from tree rings. *Journal of Geophysical Research* **105**: 7317-7326.

Naurzbaev, M.M., Vaganov, E.A., Sidorova, O.V. and Schweingruber, F.H. 2002. Summer temperatures in eastern Taimyr inferred from a 2427-year late-Holocene tree-ring chronology and earlier floating series. *The Holocene* **12**: 727-736.

Schweingruber, F.H. and Briffa, K.R. 1996. Tree-ring density network and climate reconstruction. In: Jones, P.D., Bradley, R.S. and Jouzel, J. (Eds.) *Climatic Variations and Forcing Mechanisms of the Last 2000 Years*, NATO ASI Series 141. 43-66. Springer-Verlag, Berlin, Germany.

Vaganov, E.A., Shiyatov, S.G. and Mazepa, V.S. 1996. *Dendroclimatic Study in Ural-Siberian Subarctic*. Nauka, Novosibirsk, Russia.

8.4.2.3. Twentieth Century

The IPCC theorizes that warming temperatures in the twentieth century have had a harmful effect on polar bears, but it overlooks evidence that temperatures in Greenland and the rest of the Arctic region peaked in the 1930s. We review that research in this section.

Starting in Greenland, Hanna and Cappelen (2003) determined the air temperature history of coastal southern Greenland from 1958-2001, based on data from eight Danish Meteorological Institute stations in coastal and near-coastal southern Greenland, as well as the concomitant sea surface temperature (SST) history of the Labrador Sea off southwest Greenland, based on three previously published and subsequently extended SST datasets (Parker *et al.*, 1995; Rayner *et al.*, 1996; Kalnay *et al.*, 1996). The coastal temperature data showed a *cooling* of 1.29°C over the period of study, while two of the three SST databases also depicted cooling: by 0.44°C in one case and by 0.80°C in the other. Both the land-based air temperature and SST series followed similar patterns and were strongly correlated, but with no obvious lead/lag either way. In addition, it was determined that the cooling was "significantly inversely correlated with an increased phase of the North Atlantic Oscillation (NAO) over the past few decades." The two researchers say this "NAO-temperature link doesn't explain what caused the observed cooling in coastal southern Greenland but it does lend it credibility."

Several other studies also have reported late-twentieth century cooling on Greenland. Based on mean monthly temperatures of 37 Arctic and seven sub-Arctic stations, as well as temperature anomalies of 30 grid-boxes from the updated dataset of Jones, for example, Przybylak (2000) found that "the level of temperature in Greenland in the last 10-20 years is similar to that observed in the 19th century." Likewise, in a study that utilized satellite imagery of the Odden ice tongue (a winter ice cover that occurs in the Greenland Sea with a length of about 1,300 km and an aerial coverage of as much as 330,000 square kilometers) plus surface air temperature data from adjacent Jan Mayen Island, Comiso et al. (2001) determined that the ice phenomenon was "a relatively smaller feature several decades ago," due to the warmer temperatures that were prevalent at that time. In addition, they report that observational evidence from Jan Mayen Island indicates temperatures there cooled at a rate of 0.15 ± 0.03°C per decade during the past 75 years.

Taurisano et al. (2004) examined the temperature history of the Nuuk fjord during the last century, where their analyses of all pertinent regional data led them to conclude that "at all stations in the Nuuk fjord, both the annual mean and the average temperature of the three summer months (June, July and August) exhibit a pattern in agreement with the trends observed at other stations in south and west Greenland (Humlum 1999; Hanna and Cappelen, 2003)." As they describe it, the temperature data "show that a warming trend occurred in the Nuuk fjord during the first 50 years of the 1900s, followed by a cooling over the second part of the century, when the average annual temperatures decreased by approximately 1.5°C." Coincident with this cooling trend there was also what they describe as "a remarkable increase in the number of snowfall days (+59 days)." What is more, they report that "not only did the cooling affect the winter months, as suggested by Hannna and Cappelen (2002), but also the summer mean," noting that "the summer cooling is rather important information for glaciological studies, due to the ablation-temperature relations."

In a study of three coastal stations in southern and central Greenland that possess almost uninterrupted temperature records between 1950 and 2000, Chylek et al. (2004) discovered that "summer temperatures,

which are most relevant to Greenland ice sheet melting rates, do not show any persistent increase during the last fifty years." In fact, working with the two stations with the longest records (both over a century in length), they determined that coastal Greenland's peak temperatures occurred between 1930 and 1940, and that the subsequent decrease in temperature was so substantial and sustained that current coastal temperatures "are about 1°C below their 1940 values." Furthermore, they note that "at the summit of the Greenland ice sheet the summer average temperature has decreased at the rate of 2.2°C per decade since the beginning of the measurements in 1987." Hence, as with the Arctic as a whole, it would appear that Greenland has not experienced any net warming over the most dramatic period of atmospheric CO_2 increase on record. In fact, it has *cooled* during this period.

At the start of the twentieth century, however, Greenland was warming, as it emerged, along with the rest of the world, from the depths of the Little Ice Age. Between 1920 and 1930, when the atmosphere's CO_2 concentration rose by a mere 3 to 4 ppm, there was a phenomenal warming at all five coastal locations for which contemporary temperature records are available. In the words of Chylek et al., "average annual temperature rose between 2 and 4°C [and by as much as 6°C in the winter] in less than ten years." And this warming, as they note, "is also seen in the $^{18}O/^{16}O$ record of the Summit ice core (Steig et al., 1994; Stuiver et al., 1995; White et al., 1997)."

In commenting on this dramatic temperature rise, which they call the "great Greenland warming of the 1920s," Chylek et al. conclude that "since there was no significant increase in the atmospheric greenhouse gas concentration during that time, the Greenland warming of the 1920s demonstrates that a large and rapid temperature increase can occur over Greenland, and perhaps in other regions of the Arctic, due to internal climate variability such as the NAM/NAO [Northern Annular Mode/North Atlantic Oscillation], without a significant anthropogenic influence." These facts led them to speculate that "the NAO may play a crucial role in determining local Greenland climate during the 21st century, resulting in a local climate that may defy the global climate change."

Two years later, Chylek and another team of researchers compared average summer temperatures recorded at Ammassalik, on Greenland's southeast coast, and Godthab Nuuk on the island's southwestern coast, for the period 1905 to 2005 (Chylek et al., 2006). They found "the 1955 to 2005 averages of the summer temperatures and the temperatures of the warmest month at both Godthab Nuuk and Ammassalik are significantly lower than the corresponding averages for the previous 50 years (1905-1955). The summers at both the southwestern and the southeastern coast of Greenland were significantly colder within the 1955-2005 period compared to the 1905-1955."

Chylek *et al.* also compared temperatures for the 10-year periods of 1920-1930 and 1995-2005. They found the average summer temperature for 2003 in Ammassalik was a record high since 1895, but "the years 2004 and 2005 were closer to normal being well below temperatures reached in the 1930s and 1940s." Similarly, the record from Godthab Nuuk showed that while temperatures there "were also increasing during the 1995-2005 period,they stayed generally below the values typical for the 1920-1940 period." The authors conclude that "reports of Greenland temperature changes are .... diverse suggesting a long term cooling and shorter warming periods."

Moving to the rest of the Arctic region, Overpeck *et al.* (1997) combined paleoclimatic records obtained from lake and marine sediments, trees, and glaciers to develop a 400-year history of circum-Arctic surface air temperature. From this record they determined that the most dramatic warming of the last four centuries of the past millennium (1.5°C) occurred between 1840 and 1955. Then, from 1955 to the end of the record (about 1990), the mean circum-Arctic air temperature declined by 0.4°C.

Zeeberg and Forman (2001) analyzed twentieth century changes in glacier terminus positions on north Novaya Zemlya, a Russian island located between the Barents and Kara Seas in the Arctic Ocean, providing in the process a quantitative assessment of the effects of temperature and precipitation on glacial mass balance. This work revealed a significant and accelerated post-Little Ice Age glacial retreat in the first and second decades of the twentieth century; but by 1952, the region's glaciers had experienced between 75 to 100 percent of their net twentieth century retreat. During the next 50 years, the recession of more than half of the glaciers stopped, and many tidewater glaciers actually began to advance. These glacial stabilizations and advances were attributed by the two scientists to observed increases in precipitation and/or decreases in temperature. In the four decades since 1961, for example, weather stations at Novaya Zemlya show summer temperatures to have been 0.3° to 0.5°C colder than they were over the prior 40 years, while winter temperatures were 2.3° to 2.8°C colder than they were over the prior 40-year period. Such observations, in Zeeberg and Forman's words, are "counter to warming of the Eurasian Arctic predicted for the twenty-first century by climate models, particularly for the winter season."

Comiso *et al.* (2000) utilized satellite imagery to analyze and quantify a number of attributes of the Odden ice tongue, including its average concentration, maximum area, and maximum extent over the period 1979-1998. They used surface air temperature data from Jan Mayen Island, located within the region of study, to infer the behavior of the phenomenon over the past 75 years. The Odden ice tongue was found to vary in size, shape, and length of occurrence during the 20-year period, displaying a fair amount of interannual variability. Quantitatively, trend analyses revealed that the ice tongue had exhibited no statistically significant change in any of the parameters studied over the short 20-year period. However, a proxy reconstruction of the Odden ice tongue for the past *75* years revealed the ice phenomenon to have been "a relatively smaller feature several decades ago," due to the significantly warmer temperatures that prevailed at that time. The fact that the Odden ice tongue has persisted, virtually unchanged in the mean during the past 20 years, is in direct contrast with predictions of rapid and increasing warmth in earth's polar regions as a result of CO_2-induced global warming. This observation, along with the observational evidence from Jan Mayen Island that temperatures there actually cooled at a rate of 0.15 ± 0.03°C per decade during the past 75 years, bolsters the view that there has been little to no warming in this part of the Arctic, as well as most of its other parts, over the past seven decades.

Przybylak (2002) conducted a detailed analysis of intraseasonal and interannual variability in maximum, minimum, and average air temperature and diurnal air temperature range for the entire Arctic—as delineated by Treshnikov (1985)—for the period 1951-1990, based on data from 10 stations "representing the majority of the climatic regions in the Arctic." This work indicated that trends in both the intraseasonal and interannual variability of the temperatures studied did not show any significant changes, leading Przybylak to conclude that "this aspect of climate change, as well as trends in average seasonal and annual values of temperature investigated earlier (Przybylak, 1997, 2000), proves that, in the Arctic in the period 1951-90, no tangible manifestations of the greenhouse effect can be identified."

Isaksson *et al.* (2003) retrieved two ice cores (one from Lomonosovfonna and one from Austfonna) far above the Arctic Circle in Svalbard, Norway, after which the 12 cooperating scientists from Norway, Finland, Sweden, Canada, Japan, Estonia, and the Netherlands used $\delta^{18}O$ data to reconstruct a 600-year temperature history of the region. As would be expected—in light of the earth's transition from the Little Ice Age to the Current Warm Period—the international group of scientists reported that "the $\delta^{18}O$ data from both Lomonosovfonna and Austfonna ice cores suggest that the twentieth century was the warmest during at least the past 600 years." However, the warmest decade of the twentieth century was centered on approximately 1930, while the instrumental temperature record at Longyearbyen also shows the decade of the 1930s to have been the warmest. In addition, the authors remark, "as on Svalbard, the 1930s were the warmest decade in the Trondheim record." Consequently, there was no net warming over the last seven decades of the twentieth century in the parts of Norway cited in this study.

In the same year, Polyakov *et al.* (2003) derived a surface air temperature history that stretched from 1875 to 2000, based on measurements carried out at 75 land stations and a number of drifting buoys located poleward of 62°N latitude. From 1875 to about 1917, the team of eight U.S. and Russian scientists found the surface air temperature of the huge northern region rose hardly at all; but then it climbed 1.7°C in just 20 years to reach a peak in 1937 that was not eclipsed over the remainder of the record. During this 20-year period of rapidly rising air temperature, the atmosphere's CO_2 concentration rose by a mere 8 ppm. But then, over the next six decades, when the air's CO_2 concentration rose by approximately 55 ppm, or nearly seven times more than it did throughout the 20-year period of dramatic warming that preceded it, the surface air temperature of the region poleward of 62°N experienced no net warming and, in fact, may have cooled.

Laidre and Heide-Jorgensen (2005), using a combination of long-term satellite tracking data, climate data, and remotely sensed sea ice concentrations to detect localized habitat trends of narwhals—a species of whale that polar bears are known to hunt—in Baffin Bay between Greenland and Canada, home to the largest narwhal population in the world. They found "since 1970, the climate in West Greenland has cooled, reflected in both oceanographic and biological conditions (Hanna and Cappelen, 2003)," with the result that "Baffin Bay and Davis Strait display strong significant increasing trends in ice concentrations and extent, as high as 7.5 percent per decade between 1979 and 1996, with comparable increases detected back to 1953 (Parkinson *et al.*, 1999; Deser *et al.*, 2000; Parkinson, 2000a,b; Parkinson and Cavalieri, 2002; Stern and Heide-Jorgensen, 2003)."

Groisman *et al.* (2006) reported using "a new Global Synoptic Data Network consisting of 2100 stations within the boundaries of the former Soviet Union created jointly by the [U.S.] National Climatic Data Center and Russian Institute for Hydrometeorological Information ... to assess the climatology of snow cover, frozen and unfrozen ground reports, and their temporal variability for the period from 1936 to 2004." They determined that "during the past 69 years (1936-2004 period), an increase in duration of the period with snow on the ground over Russia and the Russian polar region north of the Arctic circle has been documented by 5 days or 3% and 12 days or 5%, respectively," and they note this result "is in agreement with other findings." In commenting on this development, plus the similar findings of others, the five researchers say "changes in snow cover extent during the 1936-2004 period cannot be linked with 'warming' (particularly with the Arctic warming)." Why? Because, as they continue, "in this particular period the Arctic warming was absent."

Karlén (2005), focusing on Svalbard Lufthavn (located at 78°N latitude), which he later shows to be representative of much of the Arctic, reports that "the Svalbard mean annual temperature increased rapidly from the 1910s to the late 1930s," that "the temperature thereafter became lower, and a minimum was reached around 1970," and that "Svalbard thereafter became warmer, but the mean temperature in the late 1990s was still slightly cooler than it was in the late 1930s," indicative of a cooling trend of 0.11°C per decade over the last 70 years of the twentieth century.

In support of his contention that cooling was truly the norm in the Arctic over this period, Karlén goes on to say (1) "the observed warming during the 1930s is supported by data from several stations along the Arctic coasts and on islands in the Arctic, e.g. *Nordklim* data from Bjornoya and Jan Mayen in the north Atlantic, Vardo and Tromso in northern Norway, Sodankylae and Karasjoki in northern Finland, and Stykkisholmur in Iceland," and (2) "there is also [similar] data from other reports; e.g. Godthaab, Jakobshavn, and Egedesmindde in

Greenland, Ostrov Dikson on the north coast of Siberia, Salehard in inland Siberia, and Nome in western Alaska." All of these stations, to quote him further, "indicate the same pattern of changes in annual mean temperature: a warm 1930s, a cooling until around 1970, and thereafter a warming, although the temperature remains slightly below the level of the late 1930s." In addition, he says "many stations with records starting later than the 1930s also indicate cooling, e.g. Vize in the Arctic Sea north of the Siberian coast and Frobisher Bay and Clyde on Baffin Island." Finally, Karlén reports that the 250-year temperature record of Stockholm "shows that the fluctuations of the 1900s are not unique," and that "changes of the same magnitude as in the 1900s occurred between 1770 and 1800, and distinct but smaller fluctuations occurred around 1825."

Karlén notes that "during the 50 years in which the atmospheric concentration of CO_2 has increased considerably, the temperature has decreased," which leads him to conclude that "the Arctic temperature data do not support the models predicting that there will be a critical future warming of the climate because of an increased concentration of CO_2 in the atmosphere." And this is especially important, in Karlén's words, because the model-based prediction "is that changes will be strongest and first noticeable in the Arctic."

All these studies suggest that concern over the effect of the "unprecedented warming" of the twentieth century on polar bears has overlooked a key fact: In the areas where polar bears actually live, there has not been a consistent or unprecedent warming trend in the past 50 years. Consequently, changing temperatures cannot be blamed for changes in polar bear populations, the subject of the next section.

References

Chylek, P., Box, J.E. and Lesins, G. 2004. Global warming and the Greenland ice sheet. *Climatic Change* **63**: 201-221.

Chylek, P., Dubey, M.K, and Lesins, G. 2006. Greenland warming of 1920-1930 and 1995-2005. *Geophysical Research Letters* **33**: L11707.

Comiso, J.C., Wadhams, P., Pedersen, L.T. and Gersten, R.A. 2001. Seasonal and interannual variability of the Odden ice tongue and a study of environmental effects. *Journal of Geophysical Research* **106**: 9093-9116.

Deser, C., Walsh, J.E. and Timlin, M.S. 2000. Arctic sea ice variability in the context of recent atmospheric circulation trends. *Journal of Climatology* **13**: 617-633.

Groisman, P.Ya., Knight, R.W., Razuvaev, V.N., Bulygina, O.N. and Karl, T.R. 2006. State of the ground: Climatology and changes during the past 69 years over northern Eurasia for a rarely used measure of snow cover and frozen land. *Journal of Climate* **19**: 4933-4955.

Hanna, E. and Cappelen, J. 2002. Recent climate of Southern Greenland. *Weather* **57**: 320-328.

Hanna, E. and Cappelen, J. 2003. Recent cooling in coastal southern Greenland and relation with the North Atlantic Oscillation. *Geophysical Research Letters* **30**: 10.1029/2002GL015797.

Humlum, O. 1999. Late-Holocene climate in central West Greenland: meteorological data and rock-glacier isotope evidence. *The Holocene* **9**: 581-594.

Isaksson, E., Hermanson, M., Hicks, S., Igarashi, M., Kamiyama, K., Moore, J., Motoyama, H., Muir, D., Pohjola, V., Vaikmae, R., van de Wal, R.S.W. and Watanabe, O. 2003. Ice cores from Svalbard—useful archives of past climate and pollution history. *Physics and Chemistry of the Earth* **28**: 1217-1228.

Kalnay, E., Kanamitsu, M., Kistler, R., Collins, W., Deaven, D., Gandin, L., Iredell, M., Saha, S., White, G., Woollen, J., Zhu, Y., Chelliah, M., Ebisuzaki, W., Higgins, W., Janowiak, J., Mo, K.C., Ropelewski, C., Wang, J., Leetmaa, A., Reynolds, R., Jenne, R. and Joseph, D. 1996. The NCEP/NCAR 40-year reanalysis project. *Bulletin of the American Meteorological Society* **77**: 437-471.

Karlén, W. 2005. Recent global warming: An artifact of a too-short temperature record? *Ambio* **34**: 263-264.

Laidre, K.L. and Heide-Jorgensen, M.P. 2005. Arctic sea ice trends and narwhal vulnerability. *Biological Conservation* **121**: 509-517.

Overpeck, J., Hughen, K., Hardy, D., Bradley, R., Case, R., Douglas, M., Finney, B., Gajewski, K., Jacoby, G., Jennings, A., Lamoureux, S., Lasca, A., MacDonald, G., Moore, J., Retelle, M., Smith, S., Wolfe, A. and Zielinski, G. 1997. Arctic environmental change of the last four centuries. *Science* **278**: 1251-1256.

Parker, D.E., Folland, C.K. and Jackson, M. 1995. Marine surface temperature: Observed variations and data requirements. *Climatic Change* **31**: 559-600.

Parkinson, C.L. 2000a. Variability of Arctic sea ice: the view from space, and 18-year record. *Arctic* **53**: 341-358.

Parkinson, C.L. 2000b. Recent trend reversals in Arctic Sea ice extents: possible connections to the North Atlantic oscillation. *Polar Geography* **24**: 1-12.

Parkinson, C.L. and Cavalieri, D.J. 2002. A 21-year record of Arctic sea-ice extents and their regional, seasonal and monthly variability and trends. *Annals of Glaciology* **34**: 441-446.

Parkinson, C., Cavalieri, D., Gloersen, D., Zwally, J. and Comiso, J. 1999. Arctic sea ice extents, areas, and trends, 1978-1996. *Journal of Geophysical Research* **104**: 20,837-20,856.

Polyakov, I.V., Bekryaev, R.V., Alekseev, G.V., Bhatt, U.S., Colony, R.L., Johnson, M.A., Maskshtas, A.P. and Walsh, D. 2003. Variability and trends of air temperature and pressure in the maritime Arctic, 1875-2000. *Journal of Climate* **16**: 2067-2077.

Przybylak, R. 1997. Spatial and temporal changes in extreme air temperatures in the Arctic over the period 1951-1990. *International Journal of Climatology* **17**: 615-634.

Przybylak, R. 2000. Temporal and spatial variation of surface air temperature over the period of instrumental observations in the Arctic. *International Journal of Climatology* **20**: 587-614.

Przybylak, R. 2002. Changes in seasonal and annual high-frequency air temperature variability in the Arctic from 1951-1990. *International Journal of Climatology* **22**: 1017-1032.

Rayner, N.A., Horton, E.B., Parker, D.E., Folland, C.K. and Hackett, R.B. 1996. Version 2.2 of the global sea-ice and sea surface temperature data set, 1903-1994. *Climate Research Technical Note 74*, Hadley Centre, U.K. Meteorological Office, Bracknell, Berkshire, UK.

Stern, H.L. and Heide-Jorgensen, M.P. 2003. Trends and variability of sea ice in Baffin Bay and Davis Strait, 1953-2001. *Polar Research* **22**: 11-18.

Steig, E.J., Grootes, P.M. and Stuiver, M. 1994. Seasonal precipitation timing and ice core records. *Science* **266**: 1885-1886.

Stuiver, M., Grootes, P.M. and Braziunas, T.F. 1995. The GISP2 $\delta^{18}O$ climate record of the past 16,500 years and the role of the sun, ocean, and volcanoes. *Quaternary Research* **44**: 341-354.

Taurisano, A., Boggild, C.E. and Karlsen, H.G. 2004. A century of climate variability and climate gradients from coast to ice sheet in West Greenland. *Geografiska Annaler* **86A**: 217-224.

Treshnikov, A.F. (Ed.) 1985. *Atlas Arktiki*. Glavnoye Upravlenye Geodeziy i Kartografiy, Moskva.

White, J.W.C., Barlow, L.K., Fisher, D., Grootes, P.M., Jouzel, J., Johnsen, S.J., Stuiver, M. and Clausen, H.B. 1997. The climate signal in the stable isotopes of snow from Summit, Greenland: Results of comparisons with modern climate observations. *Journal of Geophysical Research* **102**: 26,425-26,439.

Zeeberg, J. and Forman, S.L. 2001. Changes in glacier extent on north Novaya Zemlya in the twentieth century. *Holocene* **11**: 161-175.

8.4.3. Population

The world's polar bear populations live in the wild only in the Northern Hemisphere on land and sea ice in the area surrounding the North Pole. Polar bears tend to stay in, or return to, local areas (Taylor and Lee, 1995; Bethke *et al.*, 1996; Taylor *et al.*, 2001), although some migration is known to occur (Messier *et al.*, 2001; Amstrup *et al.*, 2004). Their range expands and contracts with the accretion and contraction of sea ice with the seasons, with bears moving south during the winter as sea ice advances (Amstrup *et al.*, 2000). In some areas (e.g., Hudson Bay, Foxe Basin, Baffin Bay, and James Bay) polar bears move from sea ice to land for several months during the summer open-water season (Ferguson *et al.*, 1997; Lunn *et al.*, 1997; Taylor *et al.*, 2001, 2005).

The total polar bear population is unknown, since its numbers in the huge central Arctic Basin have never been counted (Aars *et al.*, 2006), although polar bears have been reported there (Van Meurs and Splettstoesser, 2003). A common estimate is of approximately 23,000, with a range of 17,600 to 28,500 (Aars *et al.*, 2006). There is even less certainty regarding the number of polar bears in the 1950s and 1960s, with most estimates around 5,000 to 10,000. Virtually all scientists agree that polar bear populations have grown since the 1970s. For example, Derocher has said "after the signing of the International Agreement on Polar Bears in the 1970s, harvests were controlled and the numbers increased. There is no argument from anyone on this point" (Derocher, 2009).

Even though polar bear populations grew during the second half of the twentieth century, a time when the IPCC claims there was a rapid increase in global temperatures and loss of sea ice, Derocher and others say this population growth is evidence of the effects of hunting bans and quotas and does not contradict their claim that warming temperatures and melting sea ice have hurt polar bear populations. They point, with apparent merit, to negative demographic impacts on polar bear populations identified in the Southern

Beaufort Sea and in Western Hudson Bay, and possible adverse nutritional impacts in the Northern Beaufort Sea and Southern Hudson Bay, due to changes in local sea ice conditions. But this is not evidence that global warming threatens polar bears with extinction. If anthropogenic global warming were a real threat to polar bears, its effects should be observable throughout the Current Warming Period, not just the last few years, and the warming would have to affect more than only a small number of subpopulations, as appears to have been the case (see discussion below). The modified argument—that global warming *only in recent years* is negatively affecting *some subpopulations of* polar bears *but not others*—is not what is being reported in daily newspapers or even what the IPCC claims. The real-world long-term trends in polar bear populations contradict what would be expected if the theory of anthropogenic global warming were true.

The polar bear population is divided into 19 subpopulations for management purposes. According to the IUCN Polar Bear Specialist Group, five subpopulations are declining, five are stable, two are increasing, and seven have insufficient data on which to base a decision. (Aars *et al.*, 2006). Significantly, four of the five subpopulations listed as declining are at risk due to hunting, not reduced sea ice (Aars *et al.*, 2006). This is hardly a picture of a species in steep decline, or even in decline at all. It certainly does not provide an empirical basis for predictions of imminent extinction.

Hunting historically has been the greatest threat to polar bear populations. The arrival of snowmobiles, helicopters, and high-powered rifles led to "harvest" levels that were not sustainable (Taylor *et al.*, 2002; Taylor *et al.*, 2006; Taylor *et al.*, 2008). Hunting was largely unregulated until passage of the 1974 International Agreement for the Conservation of Polar Bears and Their Habitat. Greenland didn't institute a quota for polar bear hunting until 2006 (Polar Bear Technical Committee, 2006). Annual kills for most populations now have been substantially reduced, but it will take at least 20 years for populations to recover.

The range of polar bears is affected by changes in climate but not in a linear fashion with temperature or the extent or thickness of sea ice. As explained previously, sea ice extent and thickness are only indirectly related to polar bear populations. Only two subpopulations—the Western Hudson Bay (WH) and Southern Beaufort Sea (SB) populations—probably have declined due to climate change effects (Ferguson

et al., 2005; Regehr *et al.*, 2006, 2007a,b; Rode *et al.*, 2007; Hunter *et al.*, 2007), and even one of these (WH) is disputed (Dyck *et al.*, 2007).

Taylor and Dowsley (2008) summarized recent population surveys as follows:

> Of six polar bear populations recently evaluated during the climate warming period, two populations appear to have been reduced (WH, SB), 2 populations appear to have remained constant (SH, NB), and one population appears to have increased (DS), and one was abundant but the information was not sufficient to estimate trend (BS). Seven other populations (VM, LS, NW, BB, KB, MC, GB) surveyed during the period of climate warming had vital rates sufficient to sustain substantial rates of harvest [i.e., hunting] at the time they were studied. Information from a Foxe Basin (FB) population survey was sufficient to document that the population had remained abundant although it had been harvested at a relatively high rate, although the survival and recruitment estimates necessary to determine trend were not available. The biological information on the remaining four populations (CS, LS, KS, EG) and the few bears that may inhabit the Arctic Basin is insufficient to suggest anything about current numbers or trend.

Taylor and Dowsley go on to say "the increase to current high numbers of polar bears in the Davis Strait has occurred during the current warming period, and has occurred with declining sea ice conditions that are sometimes less than 40% coverage at winter maximum (Stirling and Parkinson, 2006). Clearly the DS bears do manage to hunt successfully in unconsolidated pack ice." They comment, as do Dyck *et al.* (2007), that polar bears have been observed to successfully hunt seals in tidal flats along shores during ice-free periods.

"Considered together," Taylor and Dowsley conclude, "these demographic data do not suggest that polar bears as a species are headed for extinction in the next three generations (45 years) or the foreseeable future. The demographic data do support increased monitoring, and augmenting periodic population surveys with ecological and behavioral studies." They also observe that "to date, no population has been expatriated due to climate change effects, so the effect of decreased densities, alternative food sources, or behavioral adaptation to less ice on population persistence is not known."

The fact that polar bear populations are not declining should come as no surprise once it is understood that sea ice is not receding (at least not in all areas inhabited by polar bears or during their feeding season) or getting thinner, and that temperatures in the circumpolar region in recent years have not been unusually warm. Polar bears are adapted to the extremes of warming and cooling that can and do occur. Polar bears will move out of affected areas and return when conditions improve and when the sea ice is neither too thick nor too ephemeral. Transition will stress some populations and if warming continues for whatever reason, some populations may be expatriated. However, polar bears as a species are not in danger of depletion, let alone extinction.

References

Aars J., Lunn N.J., and Derocher, A.E. (Eds). 2006. Polar Bears: Proceedings of the 14th Working Meeting of the IUCN/SSC Polar Bear Specialist Group, 20-24 June 2005, Seattle, Washington, USA. Occasional Paper of the IUCN Species Survival Commission. Gland (Switzerland) and Cambridge (UK).

Amstrup, S.C., Durner, G., Stirling, I., Lunn, N.J., and Messier, F. 2000. Movements and distribution of polar bears in the Beaufort Sea. *Canadian Journal of Zoology* **78**: 948-966.

Amstrup, S.C., McDonald, T.L. and Durner, G.M. 2004. Using satellite radio telemetry data to delineate and manage wildlife populations. *Wildlife Society Bulletin* **32**: 661-679.

Bethke, R., Taylor, M.K., Amstrup, S. and Messier, F. 1996. Population delineation of polar bears using satellite collar data. *Ecological Applications* **6**: 311-317.

Derocher, A. 2009. Ask the experts: Are polar bear populations increasing? Polar Bears International. Web site, last accessed April 30, 2009. http://www.polarbears international.org/ask-the-experts/ population/

Dyck, M.G., Soon, W., Baydack, R.K., Legates, D.R., Baliunas, S., Ball, T.F., and Hancock, L.O. 2007. Polar bears of western Hudson Bay and climate change: Are warming spring air temperatures the 'ultimate' survival control factor? *Ecological Complexity* **4**: 73-84.

Ferguson, S.H., Taylor, M.K. and Messier, F. 1997. Space use by polar bears in and around Auyuittuq National Park, Northwest Territories, during the ice-free period. *Canadian Journal of Zoology* **75**: 1585-1594.

Ferguson, S.H., Stirling, I. and P. McLoughlin. 2005. Climate change and ringed seal (Phoca hispida) recruitment in western Hudson Bay. *Marine Mammal Science* **21**: 121-135.

Hunter, C.M., Caswell, H., Runge, M.C., Amstrup, S.C., Regehr, E.V. and Stirling, I. 2007. Polar bears in the southern Beaufort Sea II: Demography and population growth in relation to sea ice conditions. USGS Alaska Science Center, Anchorage, Administrative Report.

Lunn, N.J., Stirling, I., and Nowicki, S.N. 1997. Distribution and abundance of ringed (Phoca hispida) and bearded seals (Erignathus barbatus) in western Hudson Bay. *Canadian Journal of Fisheries and Aquatic Sciences* **54**: 914-921.

Messier, F., Taylor, M.K., Plante, A. and Romito, T. 2001. Atlas of polar bear movements in Nunavut, Northwest Territories, and neighboring areas. Nunavut Wildlife Service and University of Saskatchewan, Saskatoon, SK.

Polar Bear Technical Committee. 2006. Minutes of the 2006 Federal/Provincial/Territorial Polar Bear Technical Committee Meeting, St. John's, Newfoundland and Labrador, 6-8 February 2006. Canadian Wildlife Service, Edmonton, AB.

Regehr, E.V., Amstrup, S.C. and Stirling, I. 2006. Polar bear population status in the Southern Beaufort Sea. U.S. Geological Survey Open-File Report 2006-1337.

Regehr, E.V., Lunn, N.J., Amstrup, S.C., and Stirling, I. 2007a. Survival and population size of polar bears in western Hudson Bay in relation to earlier sea ice breakup. *Journal of Wildlife Management* **71**: 2673-2683.

Regehr, E.V., Hunter, C.M., Caswell, H., Amstrup, S.C., and Stirling, I. 2007b. Polar bears in the southern Beaufort Sea I: survival and breeding in relation to sea ice conditions, 2001-2006. Administrative Report, U.S. Department of the Interior- U.S. Geological Survey, Reston, VA.

Rode, K.D., Amstrup, S.C., and Regehr, E.V. 2007. Polar bears in the southern Beaufort Sea III: stature, mass, and cub recruitment in relationship to time and sea ice extent between 1982 and 2006. Administrative Report, U.S. Department of the Interior-U.S. Geological Survey, Reston, VA.

Stirling, I. and Parkinson, C.L. 2006. Possible effects of climate warming on selected populations of polar bears (Ursus maritimus) in the Canadian Arctic. *Arctic* **59**: 261-275.

Taylor, M.K. and Lee, L.J. 1995. Distribution and abundance of Canadian polar bear populations: a management perspective. *Arctic* **48**: 147–154.

Taylor, M.K., Akeeagok, S., Andriashek, D., Barbour, W., Born, E.W., Calvert, W., Cluff, H.D., Ferguson, S., Laake, J., Rosing-Asvid, A., Stirling, I. and Messier, F. 2001. Delineating Canadian and Greenland polar bear (Ursus maritimus) populations by cluster analysis of movements. *Can. J. Zool.* **79**: 690–709.

Taylor, M.K., Laake, J., Cluff, H.D., Ramsay, M. and Messier, F. 2002. Managing the risk of harvest for the Viscount Melville Sound polar bear population. *Ursus* **13**: 185-202.

Taylor, M.K., Laake, J., McLoughlin, P.D., Born, E.W., Cluff, H.D., Ferguson, S.H., Rosing-Asvid, A., Schweinsburg, R., and Messier, F. 2005. Demography and viability of a hunted population of polar bear. *Arctic* **58**: 203–214.

Taylor, M.K., Laake, J.L., McLoughlin, P.D., Cluff, H.D., and Messier, F. 2006. Demographic parameters and harvest-explicit population viability analysis for polar bears in M'Clintock Channel, Nunavut. *Journal of Wildlife Management* **70**: 1667–1673.

Taylor, M.K., Laake, J., McLoughlin, P.D., Cluff, H.D., Born, E.W., Rosing-Asvid, A., and Messier, F. 2008. Population parameters and harvest risks for polar bears (Ursus maritimus) in Kane Basin, Nunavut and Greenland. *Polar Biology* **31** (4): 491-499.

Taylor, M., and Dowsley, M. 2008. Demographic and ecological perspectives on the status of polar bears. Science & Public Policy Institute, Washington D.C.

Van Meurs, R. and Splettstoesser, J.F. 2003. Farthest North Polar Bear (Letter to the Editor). *Arctic* **56**:309.

8.4.4. Forecasts

Since most polar bear subpopulations are either growing or stable, and since there is little evidence that global warming is causing a loss of sea ice that is affecting most polar bear populations, the IPCC is left only with computer models that predict future declines in polar bear populations. In this section we ask whether those predictions are reliable. The reliability of computer climate models is addressed in greater detail in Chapter 1 of this report.

Green and Armstrong (2007) make the important point that forecasting is a practice and a discipline that is separate from physics, biology, geology, and the other sciences that are often applied to the question of climate change. Physicists, biologists, and other scientists often do not know how to make accurate forecasts, and consequently their predictions of the future are no more reliable than those made by nonexperts (Tetlock, 2005; Ascher, 1978). It is telling that Green and Armstrong's search of the IPCC Working Group I report (2007) "found no references ... to the primary sources of information on forecasting methods" and "the forecasting procedures that were described [in sufficient detail to be evaluated] violated 72 principles. Many of the violations were, by themselves, critical." In other words, the forecasts contained in the IPCC report are unscientific regardless of the scientific qualifications of the report's many contributors.

Scientists working in fields characterized by complexity and uncertainty are apt to confuse the output of *models*—which are nothing more than a statement of how the modeler believes a part of the world works—with real-world trends and forecasts (Bryson, 1993). Computer climate models certainly fall into this class, and they have been severely criticized for their failure to replicate real-world phenomena by many scientists, including Balling (2005), Christy (2005), Frauenfeld (2005), Posmentier and Soon (2005), and Pilkey and Pilkey-Jarvis (2007). Many of these writers observe that computer models can be "tweaked" to reconstruct climate histories after the fact, but this provides no assurance that the new model will do a better job forecasting future climates. Individual climate models often have widely differing assumptions about basic climate mechanisms but are then "tweaked" to produce similar forecasts. This is nothing at all like how real scientific forecasting is done.

Turning to predictions of the possible extinction of polar bears due to anthropogenic global warming, in response to calls to list polar bears as a threatened species under the U.S. Endangered Species Act, the U.S. Geological Survey commissioned nine administrative reports to forecast future polar bear populations. The reports eventually produced were Amstrup *et al.* (2007), Bergen *et al.* (2007), DeWeaver (2007), Durner *et al.* (2007), Hunter *et al.* (2007), Obbard *et al.* (2007), Regehr *et al.* (2007), Rode *et al.* (2007), and Stirling *et al.* (2007). Two of those studies—Anstrup *et al.* (2007) and Hunter *et al.* (2007), thought to give the strongest support for listing polar bears as endangered—were subsequently analyzed by J. Scott Armstrong, Kesten C. Green, and Willie Soon (2008) experts on forecasting at the Wharton School at the University of Pennsylvania and Monash University in Victoria, Australia, and an astrophysicist at the Harvard-Smithsonian Center for Astrophysics in Cambridge, Massachusetts, respectively. The three researchers gave the two studies failing grades for the following reasons:

- Both studies assumed things that are untrue or unknown, such as that "global warming will occur and will reduce the amount of summer sea ice" and "polar bears will not adapt; thus, they

659

will obtain less food than they do now by hunting from the sea ice platform."

- Amstrup *et al.* (2007) "definitely contravened 41 principles and apparently contravened an additional 32 principles," referring to scientific forecasting principles established by the Forecasting Principles Project and set forth in a book titled *Principles of Forecasting* (Armstrong, 2001). "Of the 116 relevant principles, we could find evidence that [Amstrup *et al.* (2007)] properly applied only 17 (14.7 percent)."

- Hunter *et al.* (2007) "clearly contravened 61 principles and probably contravened an additional 19 principles ... the authors properly applied only 10 (9.5 percent) of the 105 relevant principles."

Among the many errors they identified in Amstrup *et al.* (2007) were (1) relying on a single polar bear expert, (2) choosing an extreme forecast rather than a conservative one despite the presence of complex interactions and instability, and (3) failure to include all important variables. Mistakes made by Hunter *et al.* (2007) included (1) heavy reliance on five years of data with unknown measurement errors, (2) failure to include newly available data, (3) failure to give other researchers access to their data, and (4) failure to list possible outcomes and their likelihoods. According to Armstrong, Green, and Soon, the failure of these forecasters to adhere to the principles of scientific forecasting makes their forecasts "of no value to decisionmakers."

More generally, forecasts of the possible extinction of polar bears fail to explain how a decline in population (or some subpopulations) necessarily raises the prospect of species extinction. For example, the IUCN/SSC Polar Bear Specialists Group forecast that climate change might reduce polar bear stocks by as much as 30 percent over the next three generations. Such a reduction, should it occur (and we have given ample reasons to doubt it would), would be unfortunate, but it is not extinction and would not necessarily lead to extinction. It suggests the number of polar bears would decline from approximately 24,000 to approximately 17,000. For the sake of comparison, the National Wildlife Federation supports removing grizzly bears in Yellowstone Park USA from the endangered species list even though that population is estimated to be only 600 (Taylor and Dowsley, 2007).

As Crockford (2008) explains, "even if substantial declines in polar bears and their prey do occur because of anthropogenic global warming ... this does not doom them to extinction: Many species have recovered from far more dramatic declines in population than predicted by even the most pessimistic scenarios conceived of by climate models, including humpback whales (Dalton, 2008), gray whales (Reeves *et al.*, 2002), northern fur seals (Reeves *et al.*, 2002), Atlantic cod (Bigg *et al.*, 2008), and sea otters (Doroff *et al.*, 2003; Estes, 1990), among others. Contrary to common biological assumption, small populations often retain sufficient genetic variation for significant recovery (e.g. Aguilar *et al.*, 2004; Kaeuffer *et al.*, 2007)."

Dyck *et al.* (2008) also point out that the climate models used by the forecasters predict a complete disappearance of sea ice over the central Arctic only for the late summer (September), and sea ice at Hudson Bay during the late winter or early spring is not predicted to completely disappear by the end of this century, even under extreme scenarios. They cite Gagnon and Gough (2005b, p. 291) who concluded that "Hudson Bay is expected to remain completely ice covered in those five models by the end of this century for at least part of the year."

Taylor and Dowsley (2008) report "the majority (60%) of the IPCC models project ice cover in all seasons for the next 40-50 years throughout much of the North American continental shelf where most polar bears reside (Alley *et al.*, 2007; Serreze *et al.*, 2007). The IPCC climate model forecasts for ice reductions in fall, winter, and spring are substantially less than for the summer open water season (Serreze *et al.*, 2007)." This is significant because, as we previously reported, polar bears can fast for four months during the summer when the lack of ice reduces their access to seals (Watts and Hansen, 1987; Ramsay and Stirling, 1988; Ramsay and Hobson, 1991). So even the computer models that predict melting ice, which we've criticized as inherently unreliable and unsuited to forecasting, do not forecast a condition that would make it impossible for polar bears to survive.

Taylor and Dowsley also point out that most forecasts of declining polar bear populations assume a simple linear function between population and habitat availability, an argument similar to the discredited "survival envelopes" theory discussed earlier in this chapter. But the fact that polar bears in most subpopulations are being harvested (hunted) in significant numbers means their populations are not at ecological carrying capacity, which means habitat availability is not the constraining factor affecting population. Populations in some areas (e.g., Viscount Melville Sound, M'Clintock Channel, and Kane Basin) are currently depleted due to over-hunting, so it should be obvious that polar bears there are not constrained by lack of habitat. "There is no evidence that polar bears are at carrying capacity for any population," write Taylor and Dowsley, "and there is no evidence to support any mechanism/s of density dependent population regulation for polar bears."

In conclusion, forecasts of dwindling polar bear populations assume trends in sea ice and temperature that are counterfactual, rely on computer climate models that are known to be unreliable, and violate most of the principles of scientific forecasting. In light of other evidence presented in this section showing no long-term trends toward less sea ice or rising temperatures in the Arctic, plus evidence of rising polar bear populations and their adaptability to climate change and other environmental stresses, we find there is no basis for concern that climate change will ever cause the extinction of polar bears.

References

Aars, J., Lunn, N.J., and Derocher, A.E. (Eds.) 2006. Polar Bears: Proceedings of the 14th Working Meeting of the IUCN/SSC Polar Bear Specialist Group, 20-24 June 2005, Seattle, Washington, USA. Occasional Paper of the IUCN Species Survival Commission 32. Gland (Switzerland) and Cambridge (UK).

Aguilar, A., Roemer, G., Debenham, S., Binns, M., Garcelon, D., and Wayne, R.K. 2004. High MHC diversity maintained by balancing selection in an otherwise genetically monomorphic mammal. *Proceedings of the National Academy of Sciences USA* **101**: 3490-3494.

Alley, R., Berntsen, T., Bindoff, N.L., Chen, Z., Chidthaisong, A., Friedlingstein, P., Gregory, J., Hegerl, G., Heimann, M., Hewitson, B., Hoskins, B., Joos, F., Jouzel, J., Kattsov, V., Lohmann, U., Manning, M., Matsuno, T., Molina, M., Nicholls, N., Overpeck, J., Qin, D., Raga, G., Ramaswamy, V., Ren, J., Rusticucci, M., Solomon, S., Somerville, R., Stocker, T.F., Stott, P., Stouffer, R.J., Whetton, P., Wood, R.A., and Wratt, D. 2007. *Climate Change 2007: The Physical Science Basis, Contribution of Working Group I to the Fourth Assessment Report of the Intergovernmental Panel on Climate Change*. IPCC Secretariat, Geneva Switzerland.

Amstrup, S.C., Marcot, B.G., and Douglas, D.C. 2007. Forecasting the rangewide status of polar bears at selected times in the 21st century. Administrative Report, U.S. Department of the Interior-U.S. Geological Survey, Reston, VA.

Armstrong, J.S. 2001. *Principles of Forecasting – A Handbook for Researchers and Practitioners*. Kluwer Academic Publishers, Norwell, MA.

Armstrong, J.S., Green, K.C., and Soon, W. 2008. Polar bear population forecasts: a public-policy forecasting audit. *Interfaces* **38** (5): 382-405.

Ascher, W. 1978. *Forecasting: An Appraisal for Policy Makers and Planners*. Johns Hopkins University Press. Baltimore, MD.

Balling, R.C. 2005. Observational surface temperature records versus model predictions. In Michaels, P.J. (Ed.) *Shattered Consensus: The True State of Global Warming*. Rowman & Littlefield. Lanham, MD. pp. 50-71.

Bergen, S., Durner, G.M., Douglas, D.C. and Amstrup, S.C. 2007. Predicting movements of female polar bears between summer sea ice foraging habitats and terrestrial denning habitats of Alaska in the 21st century: Proposed methodology and pilot assessment. Administrative Report, USGS Alaska Science Center, Anchorage, AK.

Bigg, G.R., Cunningham, C.W., Ottersen, G., Pogson, G.H., Wadley, M.R., and Williamson, P. 2008. Ice-age survival of Atlantic cod: agreement between palaeoecology models and genetics. *Proceedings of the Royal Society B* **275**: 163-172.

Bryson, R.A. 1993. Environment, environmentalists, and global change: A skeptic's evaluation. *New Literary History* **24**: 783-795.

Christy, J. 2005. Temperature changes in the bulk atmosphere: beyond the IPCC. In Michaels, P.J. (Ed.) *Shattered Consensus: The True State of Global Warming*. Rowman & Littlefield. Lanham, MD. pp. 72-105.

Crockford, S. 2008. *Some things we know—and don't know—about polar bears*. Science and Public Policy Institute, Washington, DC.

Dalton, R. 2008. Whales are on the rise. *Nature* **453**: 433.

DeWeaver, E. 2007. Uncertainty in climate model projections of arctic sea ice decline: An evaluation relevant to polar bears. USGS Alaska Science Center, Anchorage, Administrative Report.

Doroff, A.M., Estes, J.A., Tinker, M.T., Burn, D.M., and Evans, T.J. 2003. Sea otter population declines in the Aleutian Archipelago. *Journal of Mammalogy* **84**: 55-64.

Durner, G.M., Douglas, D.C., Nielson, R.M., Amstrup, S.C., and McDonald, T.L. 2007. Predicting the future distribution of polar bears in the polar basin from resource selection functions applied to 21st century general circulation model projections of sea ice. USGS Alaska Science Center, Anchorage, Administrative Report.

Dyck, M.G., Soon, W., Baydack, R.K., Legates, D.R., Baliunas, S., Ball, T.F., and Hancock, L.O. 2007. Polar bears of western Hudson Bay and climate change: Are warming spring air temperatures the "ultimate" survival control factor? *Ecological Complexity* **4**:73-84.

Estes, J.A. 1990. Growth and equilibrium in sea otter populations. *Journal of Animal Ecology* **59**: 385-401.

Frauenfeld, O.W. 2005. Predictive skill of the El Nino-Southern Oscillation and related atmospheric teleconnections. In Michaels, P.J. (Ed.) *Shattered Consensus: The True State of Global Warming.* Rowman & Littlefield. Lanham, MD. pp. 149-182.

Gagnon, A.S., Gough, W.A., 2005b. Climate change scenarios for the Hudson Bay region: an intermodel comparison. *Climate Change* **69**: 269–297.

Green, K.C. and Armstrong, J.S. 2007. Global warming: forecasts by scientists versus scientific forecasts. E*nergy Environ.* **18**: 997–1021.

Hunter, C.M., Caswell, H., Runge, M.C., Amstrup, S.C., Regehr, E.V. and Stirling, I. 2007. Polar bears in the southern Beaufort Sea II: Demography and population growth in relation to sea ice conditions. USGS Alaska Science Center, Anchorage, Administrative Report.

Kaeuffer, R., Coltman, D.W., Chapius, J.-L., Pontier, D., and Réale, D. 2007. Unexpected heterozygosity in an island mouflon population founded by a single pair of individuals. *Proceedings of the Royal Society B* **274**: 527-533.

Obbard, M.E., McDonald, T.L., Howe, E.J., Regehr, E.V. and Richardson, E.S. 2007. Trends in abundance and survival for polar bears from Southern Hudson Bay, Canada, 1984–2005. Administrative Report, USGS Alaska Science Center, Anchorage, AK.

Pilkey, O.H. and Pilkey-Jarvis, L. 2007. *Useless Arithmetic.* Columbia University Press, New York.

Posmentier, E.S. and Soon, W. 2005. Limitations of computer predictions of the effects of carbon dioxide on global temperature. In Michaels, P.J. (Ed.) *Shattered Consensus: The True State of Global Warming.* Rowman & Littlefield. Lanham, MD. pp. 241-281.

Ramsay, M.A. and Hobson, K.A. 1991. Polar bears make little use of terrestrial food webs: Evidence from stable-carbon isotope analysis. *Oecologia* **86**:598-600.

Ramsay, M.A. and Stirling, I. 1988. Reproductive biology and ecology of female polar bears (Ursus maritimus). *Journal of Zoology* (London) **214**: 601-634.

Reeves, R.R, Stewart, B.S., Clapham, P.J. and Powell, J.A. 2002. *National Audobon Society's Guide to Marine Mammals of the World.* Alfred A. Knopf.

Regehr, E.V., Hunter, C.M., Caswell, H., Amstrup, S.C. and Stirling, I. 2007. Polar bears in the southern Beaufort Sea I: survival and breeding in relation to sea ice conditions, 2001-2006. Administrative Report, U.S. Department of the Interior-U.S. Geological Survey, Reston, VA.

Rode, K.D., Amstrup, S.C. and Regehr, E.V. 2007. Polar bears in the southern Beaufort Sea III: stature, mass, and cub recruitment in relationship to time and sea ice extent between 1982 and 2006. Administrative Report, U.S. Department of the Interior-U.S. Geological Survey, Reston, VA.

Serreze, M.C., Holland, M.M. and Stroeve, J. 2007. Perspectives on the Arctic's shrinking sea-ice cover. *Science* **315** (5818): 1533-1536.

Stirling, I., McDonald, T.L., Richardson, E.S. and Regehr, E.V. 2007. Polar bear population status in the Northern Beaufort Sea. Administrative Report, U.S. Department of the Interior-U.S. Geological Survey, Reston, VA.

Taylor, M.E., and Dowsley, M. 2008. Demographic and ecological perspectives on the status of polar bears. Science and Public Policy Institute, Washington, DC.

Tetlock, P.E. 2005. *Expert Political Judgment—How Good Is It? How Can We Know?* Princeton University Press, Princeton, NJ.

Watts, P.D., and Hansen, S.E., 1987. Cyclic starvation as reproductive strategy in the polar bear. *Symp. Zool. Soc.Lond.* **57**: 305–318.

9

Human Health Effects

9. Human Health Effects
9.1. Diseases
9.2. Nutrition
9.3. Human Longevity
9.4. Food vs. Nature
9.5. Biofuels

The idea that CO_2-induced global warming is harmful to people's health has become entrenched in popular culture, with the reports of the Intergovernmental Panel on Climate Change (IPCC) being the source of much of this concern. In the Working Group II contribution to the Fourth Assessment Report, the authors claim to have "very high confidence" that "climate change currently contributes to the global burden of disease and premature deaths" (IPCC, 2007-II, p. 393). They also claim climate change will "increase malnutrition and consequent disorders ... increase the number of people suffering from death, disease and injury from heatwaves, floods, storms, fires and droughts ... continue to change the range of some infectious disease vectors ... increase the burden of diarrhoeal diseases ... increase cardio-respiratory morbidity and mortality associated with ground-level ozone ... [and] increase the number of people at risk of dengue." The IPCC admits that warming weather would "bring some benefits to health, including fewer deaths from cold," but says those benefits "will be outweighed by the negative effects of rising temperatures worldwide, especially in developing countries" (Ibid.).

Some of these claims have been shown in previous chapters to be counterfactual. For example, research cited in Chapter 6 showed the global warming that occurred in the twentieth century did not cause more "heatwaves, floods, storms, fires and droughts," and that a warmer world is likely to see fewer episodes of these extreme weather events than a cooler world. We will not repeat that analysis in this chapter.

This chapter reviews data on the relationships between temperature and CO_2 and diseases, heat-related mortality, nutrition, and human longevity. We find in each case that global warming is likely to improve human health. Section 9.4 explains how rising CO_2 concentrations in the air will play a positive role in solving the conflict between the need to raise food for a growing population and the need to protect natural ecosystems. Section 9.5 describes the negative role played by increased use of biofuels, which the IPCC advocates in the Working Group III contribution to the Fourth Assessment Report (IPCC, 2007-III), in this same conflict.

References

IPCC. 2007-II. *Climate Change 2007: Impacts, Adaptation and Vulnerability. Contribution of Working Group II to the Fourth Assessment Report of the Intergovernmental Panel on Climate Cha*nge, M.L. Parry, O.F. Canziani, J.P. Palutikof, P.J. van der Linden and C.E. Hanson (Eds.) Cambridge University Press, Cambridge, UK.

IPCC. 2007-III. Climate Change 2007: *Mitigation. Contribution of Working Group III to the Fourth Assessment Report of the Intergovernmental Panel on Climate Change.* B. Metz, O.R. Davidson, P.R. Bosch, R. Dave, L.A. Meyer (Eds.) Cambridge University Press, Cambridge, UK.

9.1. Diseases

Which is more deadly: heat or cold? Rising temperatures or falling temperatures? The IPCC claims warming is the primary danger to be avoided at all costs. Real-world data, however, indicate the opposite.

Systematic research on the relationship between heat and human health dates back to the 1930s (Gover, 1938; Kutschenreuter, 1950; Kutschenreuter, 1960; Oechsli and Buechley, 1970). Early studies by Bull (1973) and Bull and Morton (1975a,b) in England and Wales, for example, demonstrated that normal changes in temperature typically are inversely associated with death rates, especially in older subjects. That is, when temperatures rise, death rates fall; when temperatures fall, death rates rise. Bull and Morton (1978) concluded "there is a close association between temperature and death rates from most diseases at all temperatures," and it is "very likely that changes in external temperature cause changes in death rates."

Since this early research was published, a large number of studies have confirmed the original findings. Contrary to the IPCC's highly selective reading of the literature, the overwhelming majority of researchers in the field have found that warmer weather reduces rather than increases the spread and severity of many diseases and weather-related mortality rates. We review this literature in the following order: cardiovascular diseases, respiratory diseases, malaria, tick-borne diseases, and finally cold- and heat-related mortality from all diseases.

Additional information on this topic, including reviews on the health effects of CO_2 not discussed here, can be found at http://www.co2science.org/subject/h/subject_h.php under the heading Health Effects.

References

Bull, G.M. 1973. Meteorological correlates with myocardial and cerebral infarction and respiratory disease. *British Journal of Preventive and Social Medicine* **27**: 108.

Bull, G.M. and Morton, J. 1975a. Seasonal and short-term relationships of temperature with deaths from myocardial and cerebral infarction. *Age and Ageing* **4**: 19-31.

Bull, G.M. and Morton, J. 1975b. Relationships of temperature with death rates from all causes and from certain respiratory and arteriosclerotic diseases in different age groups. *Age and Ageing* **4**: 232-246.

Bull, G.M. and Morton, J. 1978. Environment, temperature and death rates. *Age and Ageing* **7**: 210-224.

Gover, J. 1938. Mortality during periods of excessive temperatures. *Public Health Rep.* **53**: 1122-1143.

Kutschenreuter, P.H. 1950. Weather does affect mortality. *Amer. Soc. Heat. Refrig. Air-Cond. Eng.* **2**: 39-43.

Kutschenreuter, P.H. 1960. A study of the effect of weather on mortality in New York City. M.S. Thesis. Rutgers University, New Jersey, USA.

Oechsli, F.W. and Buechley, R.W. 1970. Excess mortality associated with three Los Angeles September hot spells. *Environmental Research* **3**: 277-284.

9.1.1. Cardiovascular Diseases

A good place to begin a review of temperature-related mortality is a cold location ... like Siberia. Feigin *et al.* (2000) examined the relationship between stroke occurrence and weather parameters in the Russian city of Novosibirsk, which has one of the highest incidence rates of stroke in the world. Analyzing the health records of 2,208 patients with a sex and age distribution similar to that of the whole of Russia over the period 1982-93, they found a statistically significant association between stroke occurrence and low ambient temperature. For ischemic stroke (IS), which accounted for 87 percent of all strokes recorded, they report that the risk of IS occurrence on days with low ambient temperature is 32 percent higher than on days with high ambient temperature. They recommend implementing "preventive measures ... such as avoiding low temperature."

Hong *et al.* (2003) studied weather-related death rates in Incheon, Korea over the period January 1998 to December 2000, reporting that "decreased ambient temperature was associated with risk of acute ischemic stroke," with the strongest effect being seen on the day after exposure to cold weather. They found that "even a moderate decrease in temperature can increase the risk of ischemic stroke." In addition, "risk estimates associated with decreased temperature were greater in winter than in the summer," suggesting that "low temperatures as well as temperature changes are associated with the onset of ischemic stroke."

Nafstad *et al.* (2001) studied weather-related death rates in Oslo, Norway. Thanks to a Norwegian law requiring all deaths to be examined by a physician who diagnoses cause and reports it on the death certificate, they were able to examine the effects of temperature on mortality due to all forms of cardiovascular disease for citizens of the country's capital over the period 1990 to 1995. They found that the average daily number of cardiovascular-related deaths was 15 percent higher in the winter months (October-March) than in the summer months (April-September), leading them to conclude that "a milder climate would lead to a substantial reduction in average daily number of deaths."

Hajat and Haines (2002) set out to determine if cardiovascular-related doctor visits by the elderly bore a similar relationship to cold temperatures. Based on data obtained for registered patients aged 65 and older from several London, England practices between January 1992 and September 1995, they found the mean number of general practitioner consultations was higher in the cool-season months (October-March) than in the warm-season months (April-September) for all cardiovascular diseases.

Of course, one might say, such findings are only to be expected in cold climates. What about warm climates, where summer maximum temperatures are often extreme, but summer minimum temperatures are typically mild? Research conducted by Green *et al.* (1994) in Israel revealed that between 1976 and 1985, mortality from cardiovascular disease was 50 percent higher in mid-winter than in mid-summer, both in men and women and in different age groups, in spite of the fact that summer temperatures in the Negev, where much of the work was conducted, often exceed 30°C, while winter temperatures typically do not drop below 10°C. These findings were substantiated by other Israeli studies reviewed by Behar (2000), who states that "most of the recent papers on this topic have concluded that a peak of sudden cardiac death, acute myocardial infarction and other cardiovascular conditions is usually observed in low temperature weather during winter."

Evidence of a seasonal variation in cardiac-related mortality has been found in the mild climate of southern California in the United States. In a study of all 222,265 death certificates issued by Los Angeles County for deaths caused by coronary artery disease from 1985 through 1996, Kloner *et al.* (1999) found that death rates in December and January were 33 percent higher than those observed in the period June through September.

Likewise, based on a study of the Hunter region of New South Wales, Australia that covered the period 1 July 1985 to 30 June 1990, Enquselassie *et al.* (1993) determined that "fatal coronary events and non-fatal definite myocardial infarction were 20-40 percent more common in winter and spring than at other times of year." Regarding daily temperature effects, they found that "rate ratios for deaths were significantly higher for low temperatures," noting that "on cold days coronary deaths were up to 40 percent more likely to occur than at moderate temperatures."

In a study of "hot" and "cold" cities in the United States—where Atlanta, Georgia; Birmingham, Alabama; and Houston, Texas comprised the "hot" group, and Canton, Ohio; Chicago, Illinois; Colorado Springs, Colorado; Detroit, Michigan; Minneapolis-St. Paul, Minnesota; New Haven, Connecticut; Pittsburgh, Pennsylvania; and Seattle and Spokane, Washington comprised the "cold" group—Braga *et al.* (2002) determined the acute effects and lagged influence of temperature on cardiovascular-related deaths. They found that in the hot cities, neither hot nor cold temperatures had much impact on mortality related to cardiovascular disease (CVD). In the cold cities, on the other hand, they report that both high and low temperatures were associated with increased CVD deaths, with the effect of cold temperatures persisting for days but the effect of high temperatures restricted to the day of the death or the day before. Of particular interest was the finding that for all CVD deaths the hot-day effect was *five times smaller* than the cold-day effect. In addition, the hot-day effect included some "harvesting," where the authors observed a deficit of deaths a few days later, which they did not observe for the cold-day effect.

Gouveia *et al.* (2003), in a study conducted in Sao Paulo, Brazil using data from 1991-1994, found that the number of cardiovascular-related deaths in adults (15-64 years of age) increased by 2.6 percent for each 1°C decrease in temperature below 20°C, while there was no evidence for any heat-induced deaths due to temperatures rising above 20°C. In the elderly (65 years of age and above), however, a 1°C warming above 20°C led to a 2 percent increase in deaths; but a 1°C cooling below 20°C led to a 6.3 percent increase in deaths, or more than three times as many cardiovascular-related deaths due to cooling than to warming in the elderly.

Similar results have been found in Australia (Enquselassie *et al.*, 1993), Brazil (Sharovsky *et al.*, 2004), England (McGregor, 2005; Carder *et al.*, 2005; McGregor *et al.*, 2004; and Kovats *et al.*, 2004), Greece (Bartzokas *et al.*, 2004), Japan (Nakaji *et al.*, 2004), the United States (Cagle and Hubbard, 2005), and parts of Africa, Asia, Europe, Latin America and the Caribbean (Chang *et al.*, 2004).

These studies demonstrate that global warming *reduces* the incidence of cardiovascular disease related to low temperatures and wintry weather by a much greater degree than it increases the incidence associated with high temperatures and summer heat waves.

Additional information on this topic, including reviews of newer publications as they become available, can be found at http://www.co2 science.org/subject/h/healtheffectscardio.php.

References

Bartzokas, A., Kassomenos, P., Petrakis, M. and Celessides, C. 2004. The effect of meteorological and pollution parameters on the frequency of hospital admissions for cardiovascular and respiratory problems in Athens. *Indoor and Built Environment* **13**: 271-275.

Behar, S. 2000. Out-of-hospital death in Israel - Should we blame the weather? *Israel Medical Association Journal* **2**: 56-57.

Braga, A.L.F., Zanobetti, A. and Schwartz, J. 2002. The effect of weather on respiratory and cardiovascular deaths in 12 U.S. cities. *Environmental Health Perspectives* **110**: 859-863.

Cagle, A. and Hubbard, R. 2005. Cold-related cardiac mortality in King County, Washington, USA 1980-2001. *Annals of Human Biology* **32**: 525-537.

Carder, M., McNamee, R., Beverland, I., Elton, R., Cohen, G.R., Boyd, J. and Agius, R.M. 2005. The lagged effect of cold temperature and wind chill on cardiorespiratory mortality in Scotland. *Occupational and Environmental Medicine* **62**: 702-710.

Chang, C.L., Shipley, M., Marmot, M. and Poulter, N. 2004. Lower ambient temperature was associated with an increased risk of hospitalization for stroke and acute myocardial infarction in young women. *Journal of Clinical Epidemiology* **57**: 749-757.

Enquselassie, F., Dobson, A.J., Alexander, H.M. and Steele, P.L. 1993. Seasons, temperature and coronary disease. *International Journal of Epidemiology* **22**: 632-636.

Feigin, V.L., Nikitin, Yu.P., Bots, M.L., Vinogradova, T.E. and Grobbee, D.E. 2000. A population-based study of the associations of stroke occurrence with weather parameters in Siberia, Russia (1982-92). *European Journal of Neurology* **7**: 171-178.

Gouveia, N., Hajat, S. and Armstrong, B. 2003. Socioeconomic differentials in the temperature-mortality relationship in Sao Paulo, Brazil. *International Journal of Epidemiology* **32**: 390-397.

Green, M.S., Harari, G., Kristal-Boneh, E. 1994. Excess winter mortality from ischaemic heart disease and stroke during colder and warmer years in Israel. *European Journal of Public Health* **4**: 3-11.

Hajat, S. and Haines, A. 2002. Associations of cold temperatures with GP consultations for respiratory and cardiovascular disease amongst the elderly in London. *International Journal of Epidemiology* **31**: 825-830.

Hong, Y-C., Rha, J-H., Lee, J-T., Ha, E-H., Kwon, H-J. and Kim, H. 2003. Ischemic stroke associated with decrease in temperature. *Epidemiology* **14**: 473-478.

Kloner, R.A., Poole, W.K. and Perritt, R.L. 1999. When throughout the year is coronary death most likely to occur? A 12-year population-based analysis of more than 220,000 cases. *Circulation* **100**: 1630-1634.

Kovats, R.S., Hajat, S. and Wilkinson, P. 2004. Contrasting patterns of mortality and hospital admissions during hot weather and heat waves in Greater London, UK. *Occupational and Environmental Medicine* **61**: 893-898.

McGregor, G.R. 2005. Winter North Atlantic Oscillation, temperature and ischaemic heart disease mortality in three English counties. *International Journal of Biometeorology* **49**: 197-204.

McGregor, G.R., Watkin, H.A. and Cox, M. 2004. Relationships between the seasonality of temperature and ischaemic heart disease mortality: implications for climate based health forecasting. *Climate Research* **25**: 253-263.

Nafstad, P., Skrondal, A. and Bjertness, E. 2001. Mortality and temperature in Oslo, Norway. 1990-1995. *European Journal of Epidemiology* **17**: 621-627.

Nakaji, S., Parodi, S., Fontana, V., Umeda, T., Suzuki, K., Sakamoto, J., Fukuda, S., Wada, S. and Sugawara, K. 2004. Seasonal changes in mortality rates from main causes of death in Japan (1970-1999). *European Journal of Epidemiology* **19**: 905-913.

Sharovsky, R., Cesar, L.A.M. and Ramires, J.A.F. 2004. Temperature, air pollution, and mortality from myocardial infarction in Sao Paulo, Brazil. *Brazilian Journal of Medical and Biological Research* **37**: 1651-1657.

9.1.2. Respiratory Diseases

As was true of cardiovascular-related mortality, deaths due to respiratory diseases are more likely to be associated with cold conditions in cold countries. In Oslo, where Nafstad *et al.* (2001) found winter deaths due to cardiovascular problems to be 15 percent more numerous than similar summer deaths, they also determined that deaths due to respiratory diseases were fully 47 percent more numerous in winter than in summer. Likewise, the London study of Hajat and Haines (2002) revealed that the number of doctor visits by the elderly was higher in cool-season than warm-season months for all respiratory diseases. At mean temperatures below 5°C, in fact, the relationship between respiratory disease consultations and temperature was linear, and stronger at a time lag of six to 15 days, such that a 1°C decrease in mean temperature below 5°C was associated with a 10.5 percent increase in all respiratory disease consultations.

Gouveia *et al.* (2003) found that death rates in Sao Paulo, Brazil due to a 1°C cooling were twice as great as death rates due to a 1°C warming in adults, and 2.8 times greater in the elderly. Donaldson (2006) studied the effect of annual mean daily air temperature on the length of the yearly respiratory syncytial virus (RSV) season in England and Wales for 1981-2004 and found "emergency department admissions (for 1990-2004) ended 3.1 and 2.5 weeks earlier, respectively, per 1°C increase in annual central England temperature (P = 0.002 and 0.043, respectively)." He concludes that "the RSV season has become shorter" and "these findings imply a health benefit of global warming in England and Wales associated with a reduction in the duration of the RSV season and its consequent impact on the health service."

The study of hot and cold cities in the United States by Braga *et al.* (2002) found that increased temperature variability is the most significant aspect of climate change with respect to respiratory-related deaths in the U.S. Why is this finding important? Because Robeson (2002) has clearly demonstrated, from a 50-year study of daily temperatures at more than 1,000 U.S. weather stations, that temperature variability *declines* with warming, and at a very substantial rate. The reduced temperature variability in a warmer world would lead to reductions in temperature-related deaths at both the high and low ends of the daily temperature spectrum at all times of the year.

These studies show that a warming world would improve people's health by reducing deaths related to respiratory disease.

Additional information on this topic, including reviews of newer publications as they become available, can be found at http://www.co2 science.org/subject/h/healtheffectsresp.php.

References

Braga, A.L.F., Zanobetti, A. and Schwartz, J. 2002. The effect of weather on respiratory and cardiovascular deaths in 12 U.S. cities. *Environmental Health Perspectives* **110**: 859-863.

Donaldson, G.C. 2006. Climate change and the end of the respiratory syncytial virus season. *Clinical Infectious Diseases* **42**: 677-679.

Gouveia, N., Hajat, S. and Armstrong, B. 2003. Socioeconomic differentials in the temperature-mortality relationship in Sao Paulo, Brazil. *International Journal of Epidemiology* **32**: 390-397.

Hajat, S. and Haines, A. 2002. Associations of cold temperatures with GP consultations for respiratory and cardiovascular disease amongst the elderly in London. *International Journal of Epidemiology* **31**: 825-830.

Nafstad, P., Skrondal, A. and Bjertness, E. 2001. Mortality and temperature in Oslo, Norway. 1990-1995. *European Journal of Epidemiology* **17**: 621-627.

Robeson, S.M. 2002. Relationships between mean and standard deviation of air temperature: implications for global warming. *Climate Research* **22**: 205-213.

9.1.3. Malaria

Rogers and Randolph (2000) note that "predictions of global climate change have stimulated forecasts that vector-borne diseases will spread into regions that are at present too cool for their persistence." Such predictions are a standard part of the narrative of those who believe global warming would have catastrophic effects. However, even the IPCC states "there is still much uncertainty about the potential impact of climate change on malaria in local and global scales" and "further research is warranted" (IPCC, 2007-II, p. 404).

According to Reiter (2000), claims that malaria resurgence is the product of CO_2-induced global warming ignore the disease's history and an extensive literature showing factors other than climate are known to play more important roles in the disease's spread. For example, historical analysis reveals that malaria was an important cause of illness and death in England during the Little Ice Age. Its transmission began to decline only in the nineteenth century, during a warming phase when, according to Reiter, "temperatures were already much higher than in the Little Ice Age."

Why was malaria prevalent in Europe during some of the coldest centuries of the past millennium? And why have we only recently witnessed malaria's widespread decline at a time when temperatures are warming? Other factors are at work, such as the quality of public health services, irrigation and agricultural activities, land use practices, civil strife, natural disasters, ecological change, population change, use of insecticides, and the movement of people (Reiter, 2000; Reiter, 2001; Hay *et al.*, 2002). Models employed by the IPCC predict widespread future increases in malaria because nearly all of the analyses they cite used only one, or at most two, climate variables to make predictions of the future distribution of the disease over the earth, and they generally do not include any non-climatic factors.

In one modeling study that used more than just one or two variables, Rogers and Randolph (2000) employed five climate variables and obtained very different results. Briefly, they used the present-day distribution of malaria to determine the specific climatic constraints that best define that distribution, after which the multivariate relationship they derived from this exercise was applied to future climate scenarios derived from state-of-the-art climate models, in order to map potential future geographical distributions of the disease. They found only a 0.84 percent increase in potential malaria exposure under the "medium-high" scenario of global warming and a 0.92 percent decrease under the "high" scenario. They state that their quantitative model "contradicts prevailing forecasts of global malaria expansion" and "highlights the use [we would say superiority] of multivariate rather than univariate constraints in such applications." This study undercuts the claim that any future warming of the globe will allow malaria to spread into currently malaria-free regions.

Hay *et al.* (2002) investigated long-term trends in meteorological data at four East African highland sites that experienced significant increases in malaria cases over the past couple decades, reporting that "temperature, rainfall, vapour pressure and the number of months suitable for *P. falciparum*

transmission have not changed significantly during the past century or during the period of reported malaria resurgence." Therefore, these factors could not be responsible for the observed increases in malaria cases. Likewise, Shanks *et al.* (2000) examined trends in temperature, precipitation, and malaria rates in western Kenya over the period 1965-1997, finding no linkages among the variables.

Also working in Africa, Small *et al.* (2003) examined trends in a climate-driven model of malaria transmission between 1911 and 1995, using a spatially and temporally extensive gridded climate data set to identify locations where the malaria transmission climate suitability index had changed significantly over this time interval. Then, after determining areas of change, they more closely examined the underlying climate forcing of malaria transmission suitability for those localities. This protocol revealed that malaria transmission suitability did increase because of climate change in specific locations of limited extent, but in Southern Mozambique, which was the only region for which climatic suitability consistently increased, the cause of the increase was increased precipitation, not temperature.

In fact, Small *et al.* say that "climate warming, expressed as a systematic temperature increase over the 85-year period, does *not* appear to be responsible for an increase in malaria suitability over any region in Africa." They concluded that "research on the links between climate change and the recent resurgence of malaria across Africa would be best served through refinements in maps and models of precipitation patterns and through closer examination of the role of nonclimatic influences." The great significance of this has recently been demonstrated by Reiter *et al.* (2003) for dengue fever, another important mosquito-borne disease.

Examining the reemergence of malaria in the East African highlands, Zhou *et al.* (2004) conducted a nonlinear mixed-regression model study that focused on the numbers of monthly malaria outpatients of the past 10-20 years in seven East African highland sites and their relationships to the numbers of malaria outpatients during the previous time period, seasonality and climate variability. They say that "for all seven study sites, we found highly significant nonlinear, synergistic effects of the interaction between rainfall and temperature on malaria incidence, indicating that the use of either temperature or rainfall alone is not sensitive enough for the detection of anomalies that are associated with malaria epidemics." This has also been found by Githeko and Ndegwa (2001), Shanks *et al.* (2002) and Hay *et al.* (2002). Climate variability—not just temperature or warming—contributed less than 20 percent of the temporal variance in the number of malaria outpatients, and at only two of the seven sites studied.

In light of their findings, Zhou *et al.* concluded that "malaria dynamics are largely driven by autoregression and/or seasonality in these sites" and that "the observed large among-site variation in the sensitivity to climate fluctuations may be governed by complex interactions between climate and biological and social factors." This includes "land use, topography, *P. falciparum* genotypes, malaria vector species composition, availability of vector control and healthcare programs, drug resistance, and other socioeconomic factors." Among these are "failure to seek treatment or delayed treatment of malaria patients, and HIV infections in the human population," which they say have "become increasingly prevalent."

Kuhn *et al.* (2003) say "there has been much recent speculation that global warming may allow the reestablishment of malaria transmission in previously endemic areas such as Europe and the United States." To investigate the robustness of this hypothesis, they analyzed the determinants of temporal trends in malaria deaths within England and Wales from 1840-1910. Their analysis found that "a 1°C increase or decrease was responsible for an increase in malaria deaths of 8.3 percent or a decrease of 6.5 percent, respectively," which explains "the malaria epidemics in the 'unusually hot summers' of 1848 and 1859." Nevertheless, the long-term near-linear temporal decline in malaria deaths over the period of study, in the words of the researchers, "was probably driven by nonclimatic factors." Among these they list increasing livestock populations (which tend to divert mosquito biting from humans), decreasing acreages of marsh wetlands (where mosquitoes breed), as well as "improved housing, better access to health care and medication, and improved nutrition, sanitation, and hygiene." They additionally note that the number of secondary cases arising from each primary imported case "is currently minuscule," as demonstrated by the absence of any secondary malaria cases in the UK since 1953.

Although simplistic model simulations may suggest that the increase in temperature predicted for Britain by 2050 is likely to cause an 8-14 percent increase in the potential for malaria transmission, Kuhn *et al.* say "the projected increase in proportional risk is clearly insufficient to lead to the reestablishment of endemicity." Expanding on this statement, they note that "the national health system ensures that imported malaria infections are detected and effectively treated and that gametocytes are cleared from the blood in less than a week." For Britain, they conclude that "a 15 percent rise in risk might have been important in the 19th century, but such a rise is now highly unlikely to lead to the reestablishment of indigenous malaria," since "socioeconomic and agricultural changes" have greatly altered the cause-and-effect relationships of the past.

Zell (2004) states that many people "assume a correlation between increasing disease incidence and global warming." However, "the factors responsible for the emergence/reemergence of vector-borne diseases are complex and mutually influence each other." He cites as an example the fact that "the incidence and spread of parasites and arboviruses are affected by insecticide and drug resistance, deforestation, irrigation systems and dams, changes in public health policy (decreased resources of surveillance, prevention, and vector control), demographic changes (population growth, migration, urbanization), and societal changes (inadequate housing conditions, water deterioration, sewage, waste management)." Therefore, as he continues, "it may be over-simplistic to attribute emergent/re-emergent diseases to climate change and sketch the menace of devastating epidemics in a warmer world." Indeed, Zell states that "variations in public health practices and lifestyle can easily outweigh changes in disease biology," especially those that might be caused by global warming.

Rogers and Randolph (2006) ask if climate change could be responsible for recent upsurges of malaria in Africa. They demonstrate that "evidence for increasing malaria in many parts of Africa is overwhelming, but the more likely causes for most of these changes to date include land-cover and land-use changes and, most importantly, drug resistance rather than any effect of climate," noting that "the recrudescence of malaria in the tea estates near Kericho, Kenya, in East Africa, where temperature has not changed significantly, shows all the signs of a disease that has escaped drug control following the evolution of chloroquine resistance by the malarial parasite." They then go on to explain that "malaria waxes and wanes to the beat of two rhythms: an annual one dominated by local, seasonal weather conditions and a *ca.* 3-yearly one dominated by herd immunity," noting that "effective drugs suppress both cycles before they can be expressed," but that "this produces a population which is mainly or entirely dependent on drug effectiveness, and which suffers the consequence of eventual drug failure, during which the rhythms reestablish themselves, as they appear to have done in Kericho."

Childs *et al.* (2006) present a detailed analysis of malaria incidence in northern Thailand from January 1977 through January 2002 in the country's 13 northern provinces. Over this time period, when the IPCC claims the world warmed at a rate and to a level that were unprecedented over the prior two millennia, Childs *et al.* report a decline in total malaria incidence (from a mean monthly incidence of 41.5 to 6.72 cases per hundred thousand people. Noting "there has been a steady reduction through time of total malaria incidence in northern Thailand, with an average decline of 6.45 percent per year," they say this result "reflects changing agronomic practices and patterns of immigration, as well as the success of interventions such as vector control programs, improved availability of treatment and changing drug policies."

Finally, some researchers have studied the effect of rising CO_2 concentrations on the mosquitos that transmit malaria. Tuchman *et al.* (2003) took leaf litter from *Populus tremuloides* (Michaux) trees that had been grown out-of-doors in open-bottom root boxes located within open-top above-ground chambers maintained at atmospheric CO_2 concentrations of either 360 or 720 ppm for an entire growing season, incubated the leaf litter for 14 days in a nearby stream, and fed the incubated litter to four species of detritivorous mosquito larvae to assess its effect on their development rates and survivorship. This work revealed that larval mortality was 2.2 times higher for *Aedes albopictus* (Skuse) mosquitos that were fed leaf litter that had been produced in the high-CO_2 chambers than it was for those fed litter that had been produced in the ambient-air chambers.

In addition, Tuchman *et al.* found that larval development rates of *Aedes triseriatus* (Say), *Aedes aegypti* (L.), and *Armigeres subalbatus* (Coquillett) were slowed by 78 percent, 25 percent, and 27 percent, respectively, when fed litter produced in the high-CO_2 as opposed to the ambient-CO_2 chambers, so that mosquitoes of these species spent 20, 11, and nine days longer in their respective larval stages when feeding on litter produced in the CO_2-enriched as

compared to the ambient-CO_2 chambers. As for the reason behind these observations, the researchers suggest that "increases in lignin coupled with decreases in leaf nitrogen induced by elevated CO_2 and subsequent lower bacterial productivity [on the leaf litter in the water] were probably responsible for [the] decreases in survivorship and/or development rate of the four species of mosquitoes."

What is the significance of these findings? In the words of Tuchman *et al.*, "the indirect impacts of an elevated CO_2 atmosphere on mosquito larval survivorship and development time could potentially be great," because longer larval development times could result in fewer cohorts of mosquitoes surviving to adulthood; and with fewer mosquitoes around, there should be lower levels of mosquito-borne diseases.

In conclusion, research that takes into account more than one or two variables typically shows little or no relationship between the incidence of malaria and temperature. Many factors are more important than temperature, and those that are subject to human control are being used to steadily reduce the incidence of deaths from this disease. In the words of Dye and Reiter (2000), "given adequate funding, technology, and, above all, commitment, the campaign to 'Roll Back Malaria,' spearheaded by the World Health Organization, will have halved deaths related to [malaria] by 2010" - independent of whatever tack earth's climate might take in the interim.

Additional information on this topic, including reviews of newer publications as they become available, can be found at at http://www.co2 science.org/subject/m/malaria.php

References

Childs, D.Z., Cattadori, I.M., Suwonkerd, W., Prajakwong, S. and Boots, M. 2006. Spatiotemporal patterns of malaria incidence in northern Thailand. *Transactions of the Royal Society of Tropical Medicine and Hygiene* **100**: 623-631.

Dye, C. and Reiter, P. 2000. Temperatures without fevers? *Science* **289**: 1697-1698.

Githeko, A.K. and Ndegwa, W. 2001. Predicting malaria epidemics in the Kenyan highlands using climate data: A tool for decision makers. *Global Change and Human Health* **2**: 54-63.

Hay, S.I., Cox, J., Rogers, D.J., Randolph, S.E., Stern, D.I., Shanks, G.D., Myers, M.F. and Snow, R.W. 2002. Climate change and the resurgence of malaria in the East African highlands. *Nature* **415**: 905-909.

IPCC, 2007-II. *Climate Change 2007: Impacts, Adaptation and Vulnerability. Contribution of Working Group II to the Fourth Assessment Report of the Intergovernmental Panel on Climate Change,* M.L. Parry, O.F. Canziani, J.P. Palutikof, P.J. van der Linden and C.E. Hanson (Eds.) Cambridge University Press, Cambridge, UK.

Kuhn, K.G., Campbell-Lendrum, D.H., Armstrong, B. and Davies, C.R. 2003. Malaria in Britain: Past, present, and future. *Proceedings of the National Academy of Science, USA* **100**: 9997-10001.

Reiter, P. 2000. From Shakespeare to Defoe: Malaria in England in the Little Ice Age. *Emerging Infectious Diseases* **6**: 1-11.

Reiter, P. 2001. Climate change and mosquito-borne disease. *Environmental Health Perspectives* **109**: 141-161.

Reiter, P., Lathrop, S., Bunning, M., Biggerstaff, B., Singer, D., Tiwari, T., Baber, L., Amador, M., Thirion, J., Hayes, J., Seca, C., Mendez, J., Ramirez, B., Robinson, J., Rawlings, J., Vorndam, V., Waterman, S., Gubier, D., Clark, G. and Hayes, E. 2003. Texas lifestyle limits transmission of Dengue virus. *Emerging Infectious Diseases* **9**: 86-89.

Rogers, D.J. and Randolph, S.E. 2000. The global spread of malaria in a future, warmer world. *Science* **289**: 1763-1766.

Rogers, D.J. and Randolph, S.E. 2006. Climate change and vector-borne diseases. *Advances in Parasitology* **62**: 345-381.

Shanks, G.D., Biomndo, K., Hay, S.I. and Snow, R.W. 2000. Changing patterns of clinical malaria since 1965 among a tea estate population located in the Kenyan highlands. *Transactions of the Royal Society of Tropical Medicine and Hygiene* **94**: 253-255.

Shanks, G.D., Hay, S.I., Stern, D.I., Biomndo, K. and Snow, R.W. 2002. Meteorologic influences on *Plasmodium falciparum* malaria in the highland tea estates of Kericho, Western Kenya. *Emerging Infectious Diseases* **8**: 1404-1408.

Small, J., Goetz, S.J. and Hay, S.I. 2003. Climatic suitability for malaria transmission in Africa, 1911-1995. *Proceedings of the National Academy of Sciences USA* **100**: 15,341-15,345.

Tuchman, N.C., Wahtera, K.A., Wetzel, R.G., Russo, N.M., Kilbane, G.M., Sasso, L.M. and Teeri, J.A. 2003. Nutritional quality of leaf detritus altered by elevated atmospheric CO_2: effects on development of mosquito larvae. *Freshwater Biology* **48**: 1432-1439.

Zell, R. 2004. Global climate change and the emergence/re-emergence of infectious diseases. *International Journal of Medical Microbiology* **293** Suppl. 37: 16-26.

Zhou, G., Minakawa, N., Githeko, A.K. and Yan, G. 2004. Association between climate variability and malaria epidemics in the East African highlands. *Proceedings of the National Academy of Sciences, USA* **101**: 2375-2380.

9.1.4. Tick-Borne Diseases

The IPCC claims that one of the likely consequences of the increase in temperature would be expanded geographic ranges of tick-borne diseases, although once again this prediction is highly qualified. "Climate change alone is unlikely to explain recent increases in the incidences of tick-borne disease in Europe or North America," the IPCC admits, and "other explanations cannot be ruled out" (IPCC 2007-II, p. 403).

Randolph and Rogers (2000) reported that tick-borne encephalitis (TBE) "is the most significant vector-borne disease in Europe and Eurasia," having "a case morbidity rate of 10-30 percent and a case mortality rate of typically 1-2 percent but as high as 24 percent in the Far East." The disease is caused by a flavivirus (TBEV), which is maintained in natural rodent-tick cycles; humans may be infected with it if bitten by an infected tick or by drinking untreated milk from infected sheep or goats.

Early writings on the relationship of TBE to global warming predicted it would expand its range and become more of a threat to humans in a warmer world. However, Randolph and Rogers indicate that "like many vector-borne pathogen cycles that depend on the interaction of so many biotic agents with each other and with their abiotic environment, enzootic cycles of TBEV have an inherent fragility," so that "their continuing survival or expansion cannot be predicted from simple univariate correlations." The two researchers decided to explore the subject in greater detail than had ever been done before.

Confining their analysis to Europe, Randolph and Rogers first correlated the present-day distribution of TBEV to the present-day distributions of five climatic variables: monthly mean, maximum, and minimum temperatures, rainfall and saturation vapor pressure, "to provide a multivariate description of present-day areas of disease risk." Then, they applied this understanding to outputs of a general circulation model of the atmosphere that predicted how these five climatic variables may change in the future. The results of these operations indicated that the distribution of TBEV might expand both north and west of Stockholm, Sweden in a warming world. For most other parts of Europe, however, the two researchers say "fears for increased extent of risk from TBEV caused by global climate change appear to be unfounded." They found that "the precise conditions required for enzootic cycles of TBEV are predicted to be disrupted" in response to global warming, and that the new climatic state "appears to be lethal for TBEV." This finding, in their words, "gives the lie to the common perception that a warmer world will necessarily be a world under greater threat from vector-borne diseases." In the case of TBEV, in fact, they report that the predicted change "appears to be to our advantage."

Similarly, Estrada-Peña (2003) evaluated the effects of various abiotic factors on the habitat suitability of four tick species that are major vectors of livestock pathogens in South Africa. This work revealed "year-to-year variations in the forecasted habitat suitability over the period 1983-2000 show a clear decrease in habitat availability, which is attributed primarily to increasing temperature in the region over this period." In addition, when climate variables were projected to the year 2015, Estrada-Peña found that "the simulations show a trend toward the destruction of the habitats of the four tick species." This is the opposite of what is predicted by those who warn of catastrophic consequences from global warming.

Zell (2004) determined that "the factors responsible for the emergence/reemergence of vector-borne diseases are complex and mutually influence each other," citing as an example that "the incidence and spread of parasites and arboviruses are affected by insecticide and drug resistance, deforestation, irrigation systems and dams, changes in public health policy (decreased resources of surveillance, prevention, and vector control), demographic changes (population growth, migration, urbanization), and societal changes (inadequate housing conditions, water deterioration, sewage, waste management)."

In light of these many complicating factors, Zell says "it may be over-simplistic to attribute emergent/re-emergent diseases to climate change and sketch the menace of devastating epidemics in a warmer world." Indeed, he concludes that "variations in public health practices and lifestyle can easily outweigh changes in disease biology," especially those that might be caused by global warming.

References

Estrada-Peña, A. 2003. Climate change decreases habitat suitability for some tick species (Acari: Ixodidae) in South Africa. *Onderstepoort Journal of Veterinary Research* **70**: 79-93.

IPCC, 2007-II. *Climate Change 2007: Impacts, Adaptation and Vulnerability. Contribution of Working Group II to the Fourth Assessment Report of the Intergovernmental Panel on Climate Change*, M.L. Parry, O.F. Canziani, J.P. Palutikof, P.J. van der Linden and C.E. Hanson (Eds.) Cambridge University Press, Cambridge, UK.

Randolph, S.E. and Rogers, D.J. 2000. Fragile transmission cycles of tick-borne encephalitis virus may be disrupted by predicted climate change. *Proceedings of the Royal Society of London Series B* **267**: 1741-1744.

Zell, R. 2004. Global climate change and the emergence/re-emergence of infectious diseases. *International Journal of Medical Microbiology* **293** Suppl. 37: 16-26.

9.1.5. Heat-related Mortality

Keatinge and Donaldson (2001) analyzed the effects of temperature, wind, rain, humidity, and sunshine during high pollution days in the greater London area over the period 1976-1995 to determine what weather and/or pollution factors have the biggest influence on human mortality. Their most prominent finding was that simple plots of mortality rate versus daily air temperature revealed a linear increase in deaths as temperatures fell from 15°C to near 0°C. Mortality rates at temperatures above 15°C were, in the words of the researchers, "grossly alinear," showing no trend. Days with high pollutant concentrations were colder than average, but a multiple regression analysis revealed that no pollutant was associated with a significant increase in mortality among people over 50 years of age. Indeed, only low temperatures were shown to have a significant effect on both immediate (one day after the temperature perturbation) and long-term (up to 24 days after the temperature perturbation) mortality rates.

Keatinge *et al.* (2000) examined heat- and cold-related mortality in north Finland, south Finland, southwest Germany, the Netherlands, Greater London, north Italy, and Athens, Greece in people aged 65-74. For each of these regions, they determined the 3°C temperature interval of lowest mortality and then evaluated mortality deviations from that base level as temperatures rose and fell by 0.1°C increments. The result, according to the researchers, was that "all regions showed more annual cold related mortality than heat related mortality." Over the seven regions studied, annual cold-related deaths were nearly 10 times greater than annual-heat related deaths. The scientists also note that the very successful adjustment of the different populations they studied to widely different summer temperatures "gives grounds for confidence that they would adjust successfully, with little increase in heat related mortality, to the global warming of around 2°C predicted to occur in the next half century." Indeed, they say their data suggest "any increases in mortality due to increased temperatures would be outweighed by much larger short term declines in cold related mortalities." For the population of Europe, therefore, an increase in temperature would appear to be a climate change for the better.

Gouveia *et al.* (2003) conducted a similar study in Sao Paulo, Brazil, where they tabulated the numbers of daily deaths from all causes (excepting violent deaths and deaths of infants up to one month of age), which they obtained from the city's mortality information system for the period 1991-1994. They then analyzed these data for children (less than 15 years of age), adults (ages 15-64), and the elderly (age 65 and above) with respect to the impacts of warming and cooling. For each 1°C increase above the minimum-death temperature of 20°C for a given and prior day's mean temperature, there was a 2.6 percent increase in deaths from all causes in children, a 1.5 percent increase in deaths from all causes in adults, and a 2.5 percent increase in deaths from all causes in the elderly. For each 1°C decrease below the 20°C minimum-death temperature, however, the cold effect was greater, with increases in deaths from all causes in children, adults, and the elderly registering 4.0 percent, 2.6 percent, and 5.5 percent, respectively. These cooling-induced death rates are 54 percent, 73 percent, and 120 percent greater than those attributable to warming.

Kan *et al.* (2003), in a study conducted in Shanghai, China from June 1, 2000 to December 31, 2001, found a V-like relationship between total mortality and temperature that had a minimum mortality risk at 26.7°C. Above this temperature, they note that "total mortality increased by 0.73 percent for each degree Celsius increase; while for temperatures below the optimum value, total mortality decreased by 1.21 percent for each degree Celsius increase." The net effect of a warming of the climate of Shanghai, therefore, would likely be reduced mortality on the order of 0.5 percent per degree Celsius increase in temperature, or perhaps even more, in light of the fact that the warming of the past few decades has been primarily due to increases in daily minimum temperatures.

Goklany and Straja (2000) studied deaths in the United States due to all causes over the period 1979-97. They found deaths due to extreme cold exceeded those due to extreme heat by 80 percent to 125 percent. No trends were found due to either extreme heat or cold in the entire population or, remarkably, in the older, more susceptible, age groups, i.e., those aged 65 and over, 75 and over, and 85 and over. Goklany and Straja say the absence of any trend "suggests that adaptation and technological change may be just as important determinants of such trends as more obvious meteorological and demographic factors."

Davis *et al.* (2003) examined daily mortality rates for 28 major U.S. cities over 29 years between 1964 and 1998. In order to control for changes in the age structure of each city's population that might bias temporal comparisons, they standardized each day's mortality count relative to a hypothetical standard city with a population of one million people, with the demographics of that city based on the age distribution of the entire U.S. population in the year 2000. They found "heat-related mortality rates declined significantly over time in 19 of the 28 cities. For the 28-city average, there were 41.0+/- 4.8 (mean +/- SE) excess heat-related deaths per year (per standard million) in the 1960s and 1970s, 17.3 +/- 2.7 in the 1980s, and 10.5 +/- 2.0 in the 1990s." This 74 percent drop in heat-related deaths occurred despite an average increase in temperature of 1.0°C during the same period. They interpret this to mean that "the U.S. populace has become systematically less affected by hot and humid weather conditions," and they say this "calls into question the utility of efforts linking climate change forecasts to future mortality responses in the United States," something the IPCC explicitly does. The four scientists conclude that "there is no simple association between increased heat wave duration or intensity and higher mortality rates in the United States."

Donaldson *et al.* (2003) determined the mean daily May-August 3°C temperature bands in which deaths of people aged 55 and above were at a minimum for three areas of the world—North Carolina, USA; South Finland; and Southeast England. They then compared heat- and cold-related deaths that occurred at temperatures above and below this optimum temperature interval for each region, after which they determined how heat-related deaths in the three areas changed between 1971 and 1997 in response to: (1) the 1.0°C temperature rise that was experienced in North Carolina over this period (from an initial temperature of 23.5°C), (2) the 2.1°C temperature rise experienced in Southeast England (from an initial temperature of 14.9°C), and (3) the unchanging 13.5°C temperature of South Finland.

First, it was determined that the 3°C temperature band at which mortality was at its local minimum was lowest for the coolest region (South Finland), highest for the warmest region (North Carolina), and intermediate for the region of intermediate temperature (Southeast England). This suggests these three populations were somewhat acclimated to their respective thermal regimes. Second, in each region, cold-related mortality (expressed as excess mortality at temperatures below the region's optimum 3°C temperature band) was greater than heat-related mortality (expressed as excess mortality at temperatures above the region's optimum 3°C temperature band).

Third, the researchers found that in the coldest of the three regions (South Finland, where there was no change in temperature over the study period), heat-related deaths per million inhabitants in the 55-and-above age group declined from 382 to 99. In somewhat warmer Southeast England, where it warmed by 2.1°C over the study period, heat-related deaths declined but much less, from 111 to 108. In the warmest of the three regions (North Carolina, USA, where mean daily May-August temperature rose by 1.0°C over the study period), heat-related deaths fell most dramatically, from 228 to a mere 16 per million.

From these observations we learn that most people can adapt to both warmer and cooler climates and that cooling tends to produce many more deaths than warming, irrespective of the initial temperature regime. As for the reason behind the third observation—the dramatic decline in heat-related deaths in response to warming in the hottest region of the study (North Carolina)—Donaldson et al. attribute it to the increase in the availability of air conditioning in the South Atlantic region of the United States, where they note that the percentage of households with some form of air conditioning rose from 57 percent in 1978 to 72 percent in 1997. With respect to the declining heat-related deaths in the other two areas, they say "the explanation is likely to lie in the fact that both regions shared with North Carolina an increase in prosperity, which could be expected to increase opportunities for avoiding heat stress."

Huynen et al. (2001) analyzed mortality rates in the entire population of Holland. For the 19-year period from January 1979 through December 1997, the group of five scientists compared the numbers of deaths in people of all ages that occurred during well-defined heat waves and cold spells. They found a total excess mortality of 39.8 deaths per day during heat waves and 46.6 deaths per day during cold spells. These numbers indicate that a typical cold-spell day kills at a rate that is 17 percent greater than a typical heat-wave day in the Netherlands.

The researchers note that the heat waves they studied ranged from 6 to 13 days in length, while the cold spells lasted 9 to 17 days, making the average cold spell approximately 37 percent longer than the average heat wave. Adjusting for this duration differential makes the number of deaths per cold spell in the Netherlands fully 60 percent greater than the number of deaths per heat wave. What is more, excess mortality continued during the whole month *after* the cold spells, leading to even more deaths, while there appeared to be mortality deficits in the month following heat waves, suggesting, in the words of the authors, "that some of the heat-induced increase in mortality can be attributed to those whose health was already compromised" or "who would have died in the short term anyway." This same conclusion has been reached in a number of other studies (Kunst et al., 1993; Alberdi et al., 1998; Eng and Mercer, 1998; Rooney et al., 1998). It is highly likely, therefore, that the 60 percent greater death toll we have calculated for cold spells in the Netherlands as compared to heat waves is an underestimate of the true differential killing power of these two extreme weather phenomena.

The Dutch could well ask themselves, therefore, "Will global climate change reduce climate-related mortalities in the Netherlands?" ... which is exactly what the senior and second authors of the Huynen et al. paper did in a letter to the editor of *Epidemiology* (Martens and Huynen, 2001). Based on the predictions of nine different GCMs for an atmospheric CO_2 concentration of 550 ppm in the year 2050—which implied a 50 percent increase in Dutch heat waves and a 67 percent drop in Dutch cold spells—they calculated a total mortality *decrease* for Holland of approximately 1,100 people per year at that point in time.

Data from Germany tell much the same story. Laschewski and Jendritzky (2002) analyzed daily mortality rates of the population of Baden-Wurttemberg, Germany (10.5 million inhabitants) over the 30-year period 1958-1997 to determine the sensitivity of the people living in this moderate climatic zone of southwest Germany to long- and short-term episodes of heat and cold. They found the mortality data "show a marked seasonal pattern with a minimum in summer and a maximum in winter" and "cold spells lead to excess mortality to a relatively small degree, which lasts for weeks," and that "the mortality increase during heat waves is more pronounced, but is followed by lower than average values in subsequent weeks." The authors' data demonstrate that the mean duration of above-normal mortality for the 51 heat episodes that occurred from 1968 to 1997 was 10 days, with a mean increase in mortality of 3.9 percent, after which there was a mean *decrease* in mortality of 2.3 percent for 19 days. Hence, the net effect of the heat waves was a calculated overall decrease in mortality of 0.2 percent over the full 29-day period.

We end with the work of Thomas Gale Moore, an economist at Stanford University USA. In his first publication reviewed here (Moore, 1998), Moore reported the results of two regression analyses he conducted to estimate the effect on the U.S. death rate of a 4.5°F increase in average termperature, the IPCC's "best estimate" at the time (1992) of likely warming over the course of the next century. For the first analysis, Moore "regressed various measures of warmth on deaths in Washington, DC, from January 1987 through December 1989," a period of 36 months, and then extrapolated the results for the entire country. He used Washington, DC because termperatures are recorded for major urban areas, not states, while monthly data on deaths is available from the National Center for Health Statistics only for states, but the center treats the nation's capital as a state. This analysis found a 4.5°F rise "would cut deaths for the country as a whole by about 37,000 annually."

For his second analysis, Moore regressed the death rates in 89 large U.S. counties with various weather variables, including actual average temperatures in 1979, highest summer temperature, lowest winter temperature, number of heating degree days, and number of cooling degree days, and several other variables known to affect death rates (percent of the population over age 65, percent black, percent with 16 years or more of schooling, median household income, per-capita income, air pollution, and health care inputs such as number of hospital beds and physicians per 100,000 population.) He found "the coefficient for average temperature implies that if the United States were enjoying temperatures 4.5 degrees warmer than today, mortality would be 41,000 less. This savings in lives is quite close to the number estimated based on the Washington, DC data, for the period 1987 through 1989." Moore notes that "a warmer climate would reduce mortality by about the magnitude of highway deaths."

Two years later, in a report published by the Hoover Institution, Moore estimated the number of deaths that would be caused by the costs associated with reducing U.S. greenhouse gas emissions (Moore, 2000). "Economists studying the relationship of income and earnings to mortality have found that the loss of $5 million to $10 million in the U.S. GDP [gross domestic product] leads to one extra death," Moore writes. Since the Energy Information Administration (EIA) estimated that meeting the Kyoto Protocol's goal of reducing greenhouse gas emissions to 7 percent below 1990 levels by 2010-2012 would cost $338 billion annually (without emissions trading), "the EIA estimates imply that somewhere between 33,800 and 67,000 more Americans will die annually between 2008 and 2012."

These studies of the effects of temperature on human mortality show that cooling, not warming, kills the largest number of people each year. The number of lives saved by warmer weather, if the IPCC's forecasts of future warming are correct (and we doubt that they are), would far exceed the number of lives lost. The margin in the United States is enormous, with the number of prevented deaths exceeding the number of deaths that occur on the nation's highways each year. Conversely, attempting to stop global warming by reducing emissions would cost lives—between 33,800 and 67,000 a year in the U.S. alone, according to Moore (2000). These staggering numbers leave little doubt that global warming does not pose a threat to human health.

References

Alberdi, J.C., Diaz, J., Montero, J.C. and Miron, I. 1998. Daily mortality in Madrid community 1986-1992: relationship with meteorological variables. *European Journal of Epidemiology* **14**: 571-578.

Davis, D.E., Knappenberger, P.C., Michaels, P.J. and Novicoff, W.M. 2003. Changing heat-related mortality in the United States. *Environmental Health Perspectives* **111** (14): 1712-1718.

Donaldson, G.C., Keatinge, W.R. and Nayha, S. 2003. Changes in summer temperature and heat-related mortality since 1971 in North Carolina, South Finland, and Southeast England. *Environmental Research* **91**: 1-7.

Eng, H. and Mercer, J.B. 1998. Seasonal variations in mortality caused by cardiovascular diseases in Norway and Ireland. *Journal of Cardiovascular Risk* **5**: 89-95.

Goklany, I.M. and Straja, S.R. 2000. U.S. trends in crude death rates due to extreme heat and cold ascribed to weather, 1979-97. *Technology* **7S**: 165-173.

Gouveia, N., Hajat, S. and Armstrong, B. 2003. Socioeconomic differentials in the temperature-mortality relationship in Sao Paulo, Brazil. *International Journal of Epidemiology* **32**: 390-397.

Huynen, M.M.T.E., Martens, P., Schram, D., Weijenberg, M.P. and Kunst, A.E. 2001. The impact of heat waves and cold spells on mortality rates in the Dutch population. *Environmental Health Perspectives* **109**: 463-470.

Kan, H-D., Jia, J. and Chen, B-H. 2003. Temperature and daily mortality in Shanghai: A time-series study. *Biomedical and Environmental Sciences* **16**: 133-139.

Keatinge, W.R. and Donaldson, G.C. 2001. Mortality related to cold and air pollution in London after allowance for effects of associated weather patterns. *Environmental Research* **86A**: 209-216.

Keatinge, W.R., Donaldson, G.C., Cordioli, E., Martinelli, M., Kunst, A.E., Mackenbach, J.P., Nayha, S. and Vuori, I. 2000. Heat related mortality in warm and cold regions of Europe: Observational study. *British Medical Journal* **321**: 670-673.

Kunst, A.E., Looman, W.N.C. and Mackenbach, J.P. 1993. Outdoor temperature and mortality in the Netherlands: a time-series analysis. *American Journal of Epidemiology* **137**: 331-341.

Laschewski, G. and Jendritzky, G. 2002. Effects of the thermal environment on human health: an investigation of 30 years of daily mortality data from SW Germany. *Climate Research* **21**: 91-103.

Martens, P. and Huynen, M. 2001. Will global climate change reduce thermal stress in the Netherlands? *Epidemiology* **12**: 753-754.

Moore, T.G. 1998. "Health and amenity effects of global warming." *Economic Inquiry* **36**: 471–488.

Moore, T.G. 2000. In sickness or in health: The Kyoto protocol vs global warming. *Essays in Public Policy* Hoover Press #104. Stanford, CA.

Rooney, C., McMichael, A.J., Kovats, R.S. and Coleman, M.P. 1998. Excess mortality in England and Wales, and in greater London, during the 1995 heat wave. *Journal of Epidemiology and Community Health* **52**: 482-486.

9.2. Nutrition

Rising concentrations of CO_2 in the atmosphere affect human health indirectly by enhancing plant productivity, a topic examined at length in Chapter 7. In this section we review the scientific literature on CO_2-induced changes to the quantity and quality of food crops—in particular the protein and antioxidants present in grains and fruits—and on the medicinal properties of some plants. We find the overwhelming weight of evidence indicates a positive effect of global warming on human health.

9.2.1. Food Quantity

The concentration of CO_2 in the earth's atmosphere has risen approximately 100 ppm since the inception of the Industrial Revolution. To measure the effect this increase had on wheat, Mayeux *et al.* (1997) grew two cultivars of commercial wheat in a 38-meter-long soil container topped with a transparent tunnel-like polyethylene cover within which a CO_2 gradient was created that varied from approximately 350 ppm at one end of the tunnel to about 200 ppm at the other end. Both wheat cultivars were irrigated weekly over the first half of the 100-day growing season, to maintain soil water contents near optimum conditions. Over the last half of the season, this regimen was maintained on only half of the wheat of each cultivar, in order to create both water-stressed and well-watered treatments.

At the conclusion of the experiment, the scientists determined that the growth response of the wheat was a linear function of atmospheric CO_2 concentration in both cultivars under both adequate and less-than-adequate soil water regimes. Based on the linear regression equations they developed for grain yield in these situations, we calculate that the 100-ppm increase in atmospheric CO_2 concentration experienced over the past century-and-a-half probably increased the mean grain yield of the two wheat cultivars by about 72 percent under well-watered conditions and 48 percent under water-stressed conditions, for a mean yield increase on the order of 60 percent under the full range of moisture conditions likely to have existed in the real world. In other words, the historical rise in CO_2 concentrations may have increased wheat yields by 60 percent, clearly a benefit to a growing population.

This CO_2-induced yield enhancement to wheat production also has been documented by Alexandrov and Hoogenboom, 2000a; Brown and Rosenberg, 1999; Cuculeanu et al., 1999; Dijkstra et al., 1999; Eitzinger et al., 2001; Harrison and Butterfield, 1996; Masle, 2000; Southworth et al., 2002; and van Ittersum et al., 2003. Nor is wheat the only food crop that benefits from CO_2-fertilization. Research reviewed in Chapter 7 showing increased production by other crops exposed to enhanced CO_2, includes the following:

- Alfalfa (De Luis et al., 1999; Luscher et al., 2000; Morgan et al., 2001; Sgherri et al., 1998)

- Cotton (Booker, 2000; Booker et al., 2000; Leavitt et al., 1994; Reddy et al., 1999; Reddy et al., 1998. Tischler et al., 2000)

- Corn (maize) (Baczek-Kwinta and Koscielniak, 2003; Bootsma et al., 2005; Conway and Toenniessen, 2003; Leakey et al., 2004; Magrin et al., 2005; Maroco et al., 1999; Shen et al., 2005; Watling and Press, 1997; Watling and Press, 2000)

- Peanuts (Alexandrov and Hoogenboom, 2000b; Prasad et al., 2003; Stanciel et al., 2000; Vu, 2005)

- Potatoes (Bunce, 2003; Chen and Setter, 2003; Fangmeier and Bender, 2002; Kauder et al., 2000; Lawson et al., 2001; Louche-Tessandier et al., 1999; Ludewig et al., 1998; Magliulo et al., 2003; Miglietta et al., 1998; Olivo et al., 2002; Pruski et al., 2002; Schapendonk, et al., 2000; Sicher and Bunce, 1999; Wolf and van Oijen, 2002; Wolf and van Oijen, 2003)

- Rice (Baker et al., 2000; De Costa et al., 2003a; De Costa et al., 2003b; Gesch et al., 2002; Kim et al., 2003; Kim et al., 2001; Kobayashi et al., 2001; Tako, et al., 2001; Watling and Press, 2000; Weerakoon et al., 2000; Widodo et al., 2003; Ziska et al., 1997)

- Sorgham (Ainsworth and Long, 2005; Ottman et al., 2001; Prior et al., 2005; Watling and Press, 1997)

- Soybeans (Alexandrov and Hoogenboom, 2000b; Allen et al., 1998; Bernacchi et al., 2005; Birt et al., 2001; Bunce, 2005; Caldwell et al., 2005; Ferris et al., 1999; Heagle et al., 1998; Messina, 1999; Nakamura et al., 1999; Rogers et al., 2004; Serraj et al., 1999; Thomas et al., 2003; Wittwer, 1995; Ziska, 1998; Ziska and Bunce, 2000; Ziska et al., 2001a; Ziska et al., 2001b)

- Strawberries (Bunce, 2001; Bushway and Pritts, 2002; Deng and Woodward, 1998)

Based on this voluminous data and much more, Idso and Idso (2000) calculated that the increase in atmospheric CO_2 concentration during the past 150 years probably caused mean yield increases on the order of 70 percent for wheat and other C_3 cereals, 28 percent for C_4 cereals, 33 percent for fruits and melons, 62 percent for legumes, 67 percent for root and tuber crops, and 51 percent for vegetables.

Such major increases in production by important food plants due to the historical increase in the air's CO_2 content have undoubtedly benefitted human health. In fact, it is safe to say that some of the people reading these words would not be alive today were it not for the CO_2 enrichment caused by human industry since the beginning of the Industrial Revolution.

What does the IPCC say about this extraordinary benefit to human health made possible by rising CO_2 levels? Incredibly, it is not mentioned anywhere in the contribution of Working Group I to the Fourth Assessment Report of the IPCC (IPCC 2007-I) or in the chapter on the impact of global warming on human health in the contribution of Working Group II (IPCC 2007-II). It is treated dismissively in the chapter on agriculture in the contribution of Working Group III (IPCC 2007-III), even though the proposals justified in the first two volumes and advanced in the third would *reduce* CO_2 emissions and therefore have a negative impact on crop yields. To call this a gross oversight is to be kind to the authors of these reports.

In view of these observations, it is indisputable that the ongoing rise in the air's CO_2 content has bestowed a huge benefit to human health by expanding the yields of food crops.

Additional information on this topic, including reviews of newer publications as they become available, can be found at http://www.co2science.org/subject/a/agriculture.php.

References

Ainsworth, E.A. and Long, S.P. 2005. What have we learned from 15 years of free-air CO_2 enrichment (FACE)? A meta-analytic review of the responses of photosynthesis, canopy properties and plant production to rising CO_2. *New Phytologist* **165**: 351-372.

Alexandrov, V.A. and Hoogenboom, G. 2000a. The impact of climate variability and change on crop yield in Bulgaria. *Agricultural and Forest Meteorology* **104**: 315-327.

Alexandrov, V.A. and Hoogenboom, G. 2000b. Vulnerability and adaptation assessments of agricultural crops under climate change in the Southeastern USA. *Theoretical and Applied Climatology* **67**: 45-63.

Allen Jr., L.H., Bisbal, E.C. and Boote, K.J. 1998. Nonstructural carbohydrates of soybean plants grown in subambient and superambient levels of CO_2. *Photosynthesis Research* **56**: 143-155.

Baczek-Kwinta, R. and Koscielniak, J. 2003. Anti-oxidative effect of elevated CO_2 concentration in the air on maize hybrids subjected to severe chill. *Photosynthetica* **41**: 161-165.

Baker, J.T., Allen Jr., L.H., Boote, K.J. and Pickering, N.B. 2000. Direct effects of atmospheric carbon dioxide concentration on whole canopy dark respiration of rice. *Global Change Biology* **6**: 275-286.

Bernacchi, C.J., Morgan, P.B., Ort, D.R. and Long, S.P. 2005. The growth of soybean under free air [CO_2] enrichment (FACE) stimulates photosynthesis while decreasing in vivo Rubisco capacity. *Planta* **220**: 434-446.

Birt, D.F., Hendrich, W. and Wang, W. 2001. Dietary agents in cancer prevention: flavonoids and isoflavonoids. *Pharmacology & Therapeutics* **90**: 157-177.

Booker, F.L. 2000. Influence of carbon dioxide enrichment, ozone and nitrogen fertilization on cotton (*Gossypium hirsutum* L.) leaf and root composition. *Plant, Cell and Environment* **23**: 573-583.

Booker, F.L., Shafer, S.R., Wei, C.-M. and Horton, S.J. 2000. Carbon dioxide enrichment and nitrogen fertilization effects on cotton (*Gossypium hirsutum* L.) plant residue chemistry and decomposition. *Plant and Soil* **220**: 89-98.

Bootsma, A., Gameda, S. and McKenney, D.W. 2005. Potential impacts of climate change on corn, soybeans and barley yields in Atlantic Canada. *Canadian Journal of Plant Science* **85**: 345-357.

Brown, R.A. and Rosenberg, N.J. 1999. Climate change impacts on the potential productivity of corn and winter wheat in their primary United States growing regions. *Climatic Change* **41**: 73-107.

Bunce, J.A. 2005. Response of respiration of soybean leaves grown at ambient and elevated carbon dioxide concentrations to day-to-day variation in light and temperature under field conditions. *Annals of Botany* **95**: 1059-1066.

Bunce, J.A. 2003. Effects of water vapor pressure difference on leaf gas exchange in potato and sorghum at ambient and elevated carbon dioxide under field conditions. *Field Crops Research* **82**: 37-47.

Bunce, J.A. 2001. Seasonal patterns of photosynthetic response and acclimation to elevated carbon dioxide in field-grown strawberry. *Photosynthesis Research* **68**: 237-245.

Bushway, L.J. and Pritts, M.P. 2002. Enhancing early spring microclimate to increase carbon resources and productivity in June-bearing strawberry. *Journal of the American Society for Horticultural Science* **127**: 415-422.

Caldwell, C.R., Britz, S.J. and Mirecki, R.M. 2005. Effect of temperature, elevated carbon dioxide, and drought during seed development on the isoflavone content of dwarf soybean [*Glycine max* (L.) Merrill] grown in controlled environments. *Journal of Agricultural and Food Chemistry* **53**: 1125-1129.

Chen, C.-T. and Setter, T.L. 2003. Response of potato tuber cell division and growth to shade and elevated CO_2. *Annals of Botany* **91**: 373-381.

Conway, G. and Toenniessen, G. 2003. Science for African food security. *Science* **299**: 1187-1188.

Cuculeanu, V., Marcia, A. and Simota, C. 1999. Climate change impact on agricultural crops and adaptation options in Romania. *Climate Research* **12**: 153-160.

De Costa, W.A.J.M., Weerakoon, W.M.W., Abeywardena, R.M.I. and Herath, H.M.L.K. 2003a. Response of photosynthesis and water relations of rice (*Oryza sativa*) to elevated atmospheric carbon dioxide in the subhumid zone of Sri Lanka. *Journal of Agronomy and Crop Science* **189**: 71-82.

De Costa, W.A.J.M., Weerakoon, W.M.W., Herath, H.M.L.K. and Abeywardena, R.M.I. 2003b. Response of growth and yield of rice (*Oryza sativa*) to elevated atmospheric carbon dioxide in the subhumid zone of Sri Lanka. *Journal of Agronomy and Crop Science* **189**: 83-95.

De Luis, J., Irigoyen, J.J. and Sanchez-Diaz, M. 1999. Elevated CO_2 enhances plant growth in droughted N_2-fixing alfalfa without improving water stress. *Physiologia Plantarum* **107**: 84-89.

Deng, X. and Woodward, F.I. 1998. The growth and yield responses of *Fragaria ananassa* to elevated CO_2 and N supply. *Annals of Botany* **81**: 67-71.

Dijkstra, P., Schapendonk, A.H.M.C., Groenwold, K., Jansen, M. and Van de Geijn, S.C. 1999. Seasonal changes in the response of winter wheat to elevated atmospheric CO_2 concentration grown in open-top chambers and field tracking enclosures. *Global Change Biology* **5**: 563-576.

Eitzinger, J., Zalud, Z., Alexandrov, V., van Diepen, C.A., Trnka, M., Dubrovsky, M., Semeradova, D. and Oberforster, M. 2001. A local simulation study on the impact of climate change on winter wheat production in north-east Austria. *Ecology and Economics* **52**: 199-212.

Fangmeier, A. and Bender, J. 2002. Air pollutant combinations—Significance for future impact assessments on vegetation. *Phyton* **42**: 65-71.

Ferris, R., Wheeler, T.R., Ellis, R.H. and Hadley, P. 1999. Seed yield after environmental stress in soybean grown under elevated CO_2. *Crop Science* **39**: 710-718.

Gesch, R.W., Vu, J.C., Boote, K.J., Allen Jr., L.H. and Bowes, G. 2002. Sucrose-phosphate synthase activity in mature rice leaves following changes in growth CO_2 is unrelated to sucrose pool size. *New Phytologist* **154**: 77-84.

Harrison, P.A. and Butterfield, R.E. 1996. Effects of climate change on Europe-wide winter wheat and sunflower productivity. *Climate Research* **7**: 225-241.

Heagle, A.S., Miller, J.E. and Pursley, W.A. 1998. Influence of ozone stress on soybean response to carbon dioxide enrichment: III. Yield and seed quality. *Crop Science* **38**: 128-134.

Idso, S.B. 1998. CO_2-induced global warming: a skeptic's view of potential climate change. *Climate Research* **10**: 69-82.

Idso, C.D. and Idso, K.E. 2000. Forecasting world food supplies: The impact of the rising atmospheric CO_2 concentration. *Technology* **7S**: 33-55.

IPCC, 2007-I. *Climate Change 2007: The Physical Science Basis. Contribution of Working Group I to the Fourth Assessment Report of the Intergovernmental Panel on Climate Change.* Solomon, S., D. Quin, M. Manning, Z. Chen, M. Marquis, K.B. Averyt, M. Tingor, and H.L. Miller. (Eds.) Cambridge University Press, Cambridge, UK.

IPCC, 2007-II. *Climate Change 2007: Impacts, Adaptation and Vulnerability. Contribution of Working Group II to the Fourth Assessment Report of the Intergovernmental Panel on Climate Change.* M.L. Parry, O.F. Canziani, J.P. Palutikof, P.J. van der Linden and C.E. Hanson. (Eds.) Cambridge University Press, Cambridge, UK.

IPCC 2007-III. *Climate Change 2007: Mitigation. Contribution of Working Group III to the Fourth Assessment Report of the Intergovernmental Panel on Climate Change.* B. Metz, O.R. Davidson, P.R. Bosch, R.

Dave, and L.A. Meyer. (Eds.) Cambridge University Press, Cambridge, UK.

Kauder, F., Ludewig, F. and Heineke, D. 2000. Ontogenetic changes of potato plants during acclimation to elevated carbon dioxide. *Journal of Experimental Botany* **51**: 429-437.

Kim, H.-Y., Lieffering, M., Kobayashi, K., Okada, M., Mitchell, M.W. and Gumpertz, M. 2003. Effects of free-air CO_2 enrichment and nitrogen supply on the yield of temperate paddy rice crops. *Field Crops Research* **83**: 261-270.

Kim, H.-Y., Lieffering, M., Miura, S., Kobayashi, K. and Okada, M. 2001. Growth and nitrogen uptake of CO_2-enriched rice under field conditions. *New Phytologist* **150**: 223-229.

Kobayashi, K., Lieffering, M. and Kim, H.-Y. 2001. Growth and yield of paddy rice under free-air CO_2 enrichment. In: Shiyomi, M. and Koizumi, H. (Eds.) *Structure and Function in Agroecosystem Design and Management.* CRC Press, Boca Raton, FL, USA, pp. 371-395.

Lawson, T., Craigon, J., Black, C.R., Colls, J.J., Tulloch, A.-M. and Landon, G. 2001. Effects of elevated carbon dioxide and ozone on the growth and yield of potatoes (*Solanum tuberosum*) grown in open-top chambers. *Environmental Pollution* **111**: 479-491.

Leakey, A.D.B., Bernacchi, C.J., Dohleman, F.G., Ort, D.R. and Long, S.P. 2004. Will photosynthesis of maize (*Zea mays*) in the US Corn Belt increase in future [CO_2] rich atmospheres? An analysis of diurnal courses of CO_2 uptake under free-air concentration enrichment (FACE). *Global Change Biology* **10**: 951-962.

Leavitt, S.W., Paul, E.A., Kimball, B.A., Hendrey, G.R., Mauney, J.R., Rauschkolb, R., Rogers, H., Lewin, K.F., Nagy, J., Pinter Jr., P.J. and Johnson, H.B. 1994. Carbon isotope dynamics of free-air CO_2-enriched cotton and soils. *Agricultural and Forest Meteorology* **70**: 87-101.

Ludewig, F., Sonnewald, U., Kauder, F., Heineke, D., Geiger, M., Stitt, M., Muller-Rober, B.T., Gillissen, B., Kuhn, C. and Frommer, W.B. 1998. The role of transient starch in acclimation to elevated atmospheric CO_2. *FEBS Letters* **429**: 147-151.

Luscher, A., Hartwig, U.A., Suter, D. and Nosberger, J. 2000. Direct evidence that symbiotic N_2 fixation in fertile grassland is an important trait for a strong response of plants to elevated atmospheric CO_2. *Global Change Biology* **6**: 655-662.

Louche-Tessandier, D., Samson, G., Hernandez-Sebastia, C., Chagvardieff, P. and Desjardins, Y. 1999. Importance of light and CO_2 on the effects of endomycorrhizal colonization on growth and photosynthesis of potato plantlets (*Solanum tuberosum*) in an *in vitro* tripartite system. *New Phytologist* **142**: 539-550.

Magliulo, V., Bindi, M. and Rana, G. 2003. Water use of irrigated potato (*Solanum tuberosum* L.) grown under free air carbon dioxide enrichment in central Italy. *Agriculture, Ecosystems and Environment* **97**: 65-80.

Magrin, G.O., Travasso, M.I. and Rodriguez, G.R. 2005. Changes in climate and crop production during the twentieth century in Argentina. *Climatic Change* **72**: 229-249.

Maroco, J.P., Edwards, G.E. and Ku, M.S.B. 1999. Photosynthetic acclimation of maize to growth under elevated levels of carbon dioxide. *Planta* **210**: 115-125.

Masle, J. 2000. The effects of elevated CO_2 concentrations on cell division rates, growth patterns, and blade anatomy in young wheat plants are modulated by factors related to leaf position, vernalization, and genotype. *Plant Physiology* **122**: 1399-1415.

Mayeux, H.S., Johnson, H.B., Polley, H.W. and Malone, S.R. 1997. Yield of wheat across a subambient carbon dioxide gradient. *Global Change Biology* **3**: 269-278

Messina, M.J. 1999. Legumes and soybeans: overview of their nutritional profiles and health effects. *American Journal of Clinical Nutrition* **70**(S): 439s-450s.

Miglietta, F., Magliulo, V., Bindi, M., Cerio, L., Vaccari, F.P., Loduca, V. and Peressotti, A. 1998. Free Air CO_2 Enrichment of potato (*Solanum tuberosum* L.): development, growth and yield. *Global Change Biology* **4**: 163-172.

Morgan, J.A., Skinner, R.H. and Hanson, J.D. 2001. Nitrogen and CO_2 affect regrowth and biomass partitioning differently in forages of three functional groups. *Crop Science* **41**: 78-86.

Nakamura, T., Koike, T., Lei, T., Ohashi, K., Shinano, T. and Tadano, T. 1999. The effect of CO_2 enrichment on the growth of nodulated and non-nodulated isogenic types of soybean raised under two nitrogen concentrations. *Photosynthetica* **37**: 61-70.

Olivo, N., Martinez, C.A. and Oliva, M.A. 2002. The photosynthetic response to elevated CO_2 in high altitude potato species (*Solanum curtilobum*). *Photosynthetica* **40**: 309-313.

Ottman, M.J., Kimball, B.A., Pinter Jr., P.J., Wall, G.W., Vanderlip, R.L., Leavitt, S.W., LaMorte, R.L., Matthias, A.D. and Brooks, T.J. 2001. Elevated CO_2 increases sorghum biomass under drought conditions. *New Phytologist* **150**: 261-273.

Prasad, P.V.V., Boote, K.J., Allen Jr., L.H. and Thomas, J.M.G. 2003. Super-optimal temperatures are detrimental to peanut (*Arachis hypogaea* L.) reproductive processes and yield at both ambient and elevated carbon dioxide. *Global Change Biology* **9**: 1775-1787.

Prior, S.A., Runion, G.B., Rogers, H.H., Torbert, H.A. and Reeves, D.W. 2005. Elevated atmospheric CO_2 effects on biomass production and soil carbon in conventional and conservation cropping systems. *Global Change Biology* **11**: 657-665.

Pruski, K., Astatkie, T., Mirza, M. and Nowak, J. 2002. Photoautotrophic micropropagation of Russet Burbank potato. *Plant, Cell and Environment* **69**: 197-200.

Reddy, K.K., Davidonis, G.H., Johnson, A.S. and Vinyard, B.T. 1999. Temperature regime and carbon dioxide enrichment alter cotton boll development and fiber properties. *Agronomy Journal* **91**: 851-858.

Reddy, K.R., Robana, R.R., Hodges, H.F., Liu, X.J. and McKinion, J.M. 1998. Interactions of CO_2 enrichment and temperature on cotton growth and leaf characteristics. *Environmental and Experimental Botany* **39**: 117-129.

Rogers, A., Allen, D.J., Davey, P.A., Morgan, P.B., Ainsworth, E.A., Bernacchi, C.J., Cornic, G., Dermody, O., Dohleman, F.G., Heaton, E.A., Mahoney, J., Zhu, X.-G., DeLucia, E.H., Ort, D.R. and Long, S.P. 2004. Leaf photosynthesis and carbohydrate dynamics of soybeans grown throughout their life-cycle under Free-Air Carbon dioxide Enrichment. *Plant, Cell and Environment* **27**: 449-458.

Schapendonk, A.H.C.M., van Oijen, M., Dijkstra, P., Pot, C.S., Jordi, W.J.R.M. and Stoopen, G.M. 2000. Effects of elevated CO_2 concentration on photosynthetic acclimation and productivity of two potato cultivars grown in open-top chambers. *Australian Journal of Plant Physiology* **27**: 1119-1130.

Serraj, R., Allen Jr., L.H., Sinclair, T.R. 1999. Soybean leaf growth and gas exchange response to drought under carbon dioxide enrichment. *Global Change Biology* **5**: 283-291.

Sgherri, C.L.M., Quartacci, M.F., Menconi, M., Raschi, A. and Navari-Izzo, F. 1998. Interactions between drought and elevated CO_2 on alfalfa plants. *Journal of Plant Physiology* **152**: 118-124.

Shen, S.S.P., Yin, H., Cannon, K., Howard, A., Chetner, S. and Karl, T.R. 2005. Temporal and spatial changes of the agroclimate in Alberta, Canada, from 1901 to 2002. *Journal of Applied Meteorology* **44**: 1090-1105.

Sicher, R.C. and Bunce, J.A. 1999. Photosynthetic enhancement and conductance to water vapor of field-grown *Solanum tuberosum* (L.) in response to CO_2 enrichment. *Photosynthesis Research* **62**: 155-163.

Southworth, J., Pfeifer, R.A., Habeck, M., Randolph, J.C., Doering, O.C. and Rao, D.G. 2002. Sensitivity of winter wheat yields in the Midwestern United States to future changes in climate, climate variability, and CO_2 fertilization. *Climate Research* **22**: 73-86.

Stanciel, K., Mortley, D.G., Hileman, D.R., Loretan, P.A., Bonsi, C.K. and Hill, W.A. 2000. Growth, pod and seed yield, and gas exchange of hydroponically grown peanut in response to CO_2 enrichment. *HortScience* **35**: 49-52.

Tako, Y., Arai, R., Otsubo, K. and Nitta, K. 2001. Application of crop gas exchange and transpiration data obtained with CEEF to global change problem. *Advances in Space Research* **27**: 1541-1545.

Thomas, J.M.G., Boote, K.J., Allen Jr., L.H., Gallo-Meagher, M. and Davis, J.M. 2003. Elevated temperature and carbon dioxide effects on soybean seed composition and transcript abundance. *Crop Science* **43**: 1548-1557.

Tischler, C.R., Polley, H.W., Johnson, H.B. and Pennington, R.E. 2000. Seedling response to elevated CO_2 in five epigeal species. *International Journal of Plant Science* **161**: 779-783.

van Ittersum, M.K., Howden, S.M. and Asseng, S. 2003. Sensitivity of productivity and deep drainage of wheat cropping systems in a Mediterranean environment to changes in CO_2, temperature and precipitation. *Agriculture, Ecosystems and Environment* **97**: 255-273.

Vu, J.C.V. 2005. Acclimation of peanut (*Arachis hypogaea* L.) leaf photosynthesis to elevated growth CO_2 and temperature. *Environmental and Experimental Botany* **53**: 85-95.

Watling, J.R. and Press, M.C. 1997. How is the relationship between the C_4 cereal Sorghum bicolor and the C_3 root hemi-parasites *Striga hermonthica* and *Striga asiatica* affected by elevated CO_2? *Plant, Cell and Environment* **20**: 1292-1300.

Watling, J.R. and Press, M.C. 2000. Infection with the parasitic angiosperm *Striga hermonthica* influences the response of the C_3 cereal *Oryza sativa* to elevated CO_2. *Global Change Biology* **6**: 919-930.

Weerakoon, W.M.W., Ingram, K.T. and Moss, D.D. 2000. Atmospheric carbon dioxide and fertilizer nitrogen effects on radiation interception by rice. *Plant and Soil* **220**: 99-106.

Widodo, W., Vu, J.C.V., Boote, K.J., Baker, J.T. and Allen Jr., L.H. 2003. Elevated growth CO_2 delays drought stress and accelerates recovery of rice leaf photosynthesis. *Environmental and Experimental Botany* **49**: 259-272.

Wittwer, S.H. 1995. Food, Climate, and Carbon Dioxide: The Global Environment and World Food Production. CRC Press, Boca Raton, FL.

Wolf, J. and van Oijen, M. 2002. Modelling the dependence of European potato yields on changes in climate and CO_2. *Agricultural and Forest Meteorology* **112**: 217-231.

Wolf, J. and van Oijen, M. 2003. Model simulation of effects of changes in climate and atmospheric CO_2 and O_3 on tuber yield potential of potato (cv. Bintje) in the European Union. *Agriculture, Ecosystems and Environment* **94**: 141-157.

Ziska, L.H. 1998. The influence of root zone temperature on photosynthetic acclimation to elevated carbon dioxide concentrations. *Annals of Botany* **81**: 717-721.

Ziska, L.W. and Bunce, J.A. 2000. Sensitivity of field-grown soybean to future atmospheric CO_2: selection for improved productivity in the 21st century. *Australian Journal of Plant Physiology* **27**: 979-984.

Ziska, L.H., Bunce, J.A. and Caulfield, F.A. 2001a. Rising atmospheric carbon dioxide and seed yields of soybean genotypes. *Crop Science* **41**: 385-391.

Ziska, L.H., Ghannoum, O., Baker, J.T., Conroy, J., Bunce, J.A., Kobayashi, K. and Okada, M. 2001b. A global perspective of ground level, 'ambient' carbon dioxide for assessing the response of plants to atmospheric CO_2. *Global Change Biology* **7**: 789-796.

Ziska, L.H., Namuco, O., Moya, T. and Quilang, J. 1997. Growth and yield response of field-grown tropical rice to increasing carbon dioxide and air temperature. *Agronomy Journal* **89**: 45-53.

9.2.2. Food Quality

The quantity of food is mankind's primary concern when it comes to survival. But after survival is assured, the *quality* of food rises to the fore. What role does the ongoing rise in the air's CO_2 content play here? In this section we survey the literature on the effects of higher CO_2 air concentration on plant protein and antioxidant content.

9.2.2.1. Protein Content

Idso and Idso (2001) and Idso *et al.* (2001) cited studies where elevated levels of atmospheric CO_2 either increased, decreased, or had no effect on the protein concentrations of various agricultural crops.

The relationship, as we will see, is complex, though in the end it appears that enhanced atmospheric CO_2 has a positive effect on the protein content of most crops.

Pleijel *et al.* (1999) analyzed the results of 16 open-top chamber experiments that had been conducted on spring wheat in Denmark, Finland, Sweden, and Switzerland between 1986 and 1996. In addition to CO_2 enrichment of the air, these experiments included increases and decreases in atmospheric ozone (O_3). The scientists found that while increasing O_3 pollution reduced wheat grain yield it simultaneously increased the protein concentration of the grain. Removing O_3 from the air led to higher grain yield but lower protein concentration. The opposite relationship was found for atmospheric CO_2 enrichment, which increased grain yield but lowered protein concentration. Water stress, which was also a variable in one of the experiments, reduced yield and increased grain protein concentrations.

In an earlier study of CO_2 and O_3 effects on wheat grain yield and quality, Rudorff *et al.* (1996) found that "flour protein contents were increased by enhanced O_3 exposure and reduced by elevated CO_2" but that "the combined effect of these gases was minor." They conclude that "the concomitant increase of CO_2 and O_3 in the troposphere will have no significant impact on wheat grain quality."

Earlier, Evans (1993) had found several other crops to be greatly affected by soil nitrogen availability. Rogers *et al.* (1996) observed CO_2-induced reductions in the protein concentration of flour derived from wheat plants growing at low soil nitrogen concentrations, but no such reductions were evident when the soil nitrogen supply was increased. Pleijel *et al.* concluded that the oft-observed negative impact of atmospheric CO_2 enrichment on grain protein concentration would probably be alleviated by higher applications of nitrogen fertilizers.

The study of Kimball *et al.* (2001) confirmed their hypothesis. Kimball *et al.* studied the effects of a 50 percent increase in atmospheric CO_2 concentration on wheat grain nitrogen concentration and the baking properties of the flour derived from that grain throughout four years of free-air CO_2 enrichment experiments. In the first two years of their study, soil water content was an additional variable; in the last two years, soil nitrogen content was a variable. The most influential factor in reducing grain nitrogen concentration was determined to be low soil nitrogen.

Under this condition, atmospheric CO_2 enrichment further reduced grain nitrogen and protein concentrations, although the change was much less than that caused by low soil nitrogen. When soil nitrogen was not limiting, however, increases in the air's CO_2 concentration did not affect grain nitrogen and protein concentrations; neither did they reduce the baking properties of the flour derived from the grain. Hence, it would appear that given sufficient water and nitrogen, atmospheric CO_2 enrichment can significantly increase wheat grain yield without sacrificing grain protein concentration in the process.

There are some situations where atmospheric CO_2 enrichment has been found to *increase* the protein concentration of wheat. Agrawal and Deepak (2003), for example, grew two cultivars of wheat (*Triticum aestivum* L. cv. Malviya 234 and HP1209) in open-top chambers maintained at atmospheric CO_2 concentrations of 350 and 600 ppm alone and in combination with 60 ppb SO_2 to study the interactive effects of elevated CO_2 and this major air pollutant on crop growth. They found that exposure to the elevated SO_2 caused a 13 percent decrease in foliar protein concentrations in both cultivars; but when the plants were concomitantly exposed to an atmospheric CO_2 concentration of 600 ppm, leaf protein levels decreased only by 3 percent in HP1209, while they actually increased by 4 percent in Malviya 234.

In the case of rice—which according to Wittwer (1995) is "the basic food for more than half the world's population," supplying "more dietary energy than any other single food"—Jablonski *et al.* (2002) conducted a wide-ranging review of the scientific literature, finding that it too appeared to suffer no reduction in grain nitrogen (protein) concentration in response to atmospheric CO_2 enrichment. Likewise, they found no CO_2-induced decrease in seed nitrogen concentration in the studies of legumes they reviewed. This finding is also encouraging, since according to Wittwer (1995) legumes "are a direct food resource providing 20 percent of the world's protein for human consumption," as well as "about two thirds of the world's protein concentrate for livestock feeding." What is more, the biomass of the CO_2-enriched wheat, rice, and legumes was found by Jablonski *et al.* to be significantly increased above that of the same crops grown in normal air. Hence, there will likely be a large increase in the total amount of protein made available to humanity in a future CO_2-enriched world, both directly via food crops and indirectly via livestock.

With respect to the leguminous soybean, Thomas *et al.* (2003) additionally note that "oil and protein comprise ~20 and 40 percent, respectively, of the dry weight of soybean seed," which "unique chemical composition," in their words, "has made it one of the most valuable agronomic crops worldwide." In addition, they say "the intrinsic value of soybean seed is in its supply of essential fatty acids and amino acids in the oil and protein, respectively," and they report that Heagle *et al.* (1998) "observed a positive significant effect of CO_2 enrichment on soybean seed oil and oleic acid concentration."

Legumes and their responses to atmospheric CO_2 enrichment also figure prominently in a number of studies of mixed forage crops. In a study of nitrogen cycling in grazed pastures on the North Island of New Zealand, for example, Allard *et al.* (2003) report that under elevated CO_2, leaves of the individual species exhibited lower nitrogen concentrations but higher water-soluble carbohydrate (WSC) concentrations. They also say "there was a significantly greater proportion of legume in the diet at elevated CO_2," and that this "shift in the botanical composition towards a higher proportion of legumes counterbalanced the nitrogen decrease observed at the single species scale, resulting in a nitrogen concentration of the overall diet that was unaffected by elevated CO_2." They further report that "changes at the species level and at the sward level appeared to combine additively in relation to WSC," and "as there was a significant correlation between WSC and digestibility (as previously observed by Dent and Aldrich, 1963 and Humphreys, 1989), there was also an increase in digestibility of the high CO_2 forage," which result, in their words, "matches that found in a Mini-FACE experiment under cutting (Teyssonneyre, 2002; Picon-Cochard *et al.*, 2004)," where "digestibility also increased in response to CO_2 despite reduced crude protein concentration." These data, plus the strong relationship between soluble sugars (rather than nitrogen) and digestibility, led them to suggest that "the widespread response to CO_2 of increased soluble sugars might lead to an increase in forage digestibility."

Luscher *et al.* (2004) found much the same thing in their review of the subject, which was based primarily on studies conducted at the Swiss FACE facility that hosts what has become the world's longest continuous atmospheric CO_2 enrichment study of a naturally occurring grassland. In response to an approximate two-thirds increase in the air's CO_2 concentration, leaf nitrogen (N) concentrations of

white clover (*Trifolium repens* L.) and perennial ryegrass (*Lolium perenne* L.) were reduced by 7 percent and 18 percent, respectively, when they were grown separately in pure stands. However, as Luscher *et al.* report, "the considerably lower concentration of N under elevated CO_2, observed for *L. perenne* leaves in pure stands, was found to a much lesser extent for *L. perenne* leaves in the bi-species mixture with *T. repens* (Zanetti *et al.*, 1997; Hartwig *et al.*, 2000)." Furthermore, as they continue, "under elevated CO_2 the proportion of N-rich *T. repens* (40 mg N g^{-1} dry matter) increased in the mixture at the expense of the N-poor *L. perenne* (24 mg N g^{-1} dry matter when grown in monoculture)," the end result being that "the concentration of N in the harvested biomass of the mixture showed no significant reduction."

Campbell *et al.* (2000) analyzed research conducted between 1994 and 1999 by a worldwide network of 83 scientists associated with the Global Change and Terrestrial Ecosystems (GCTE) Pastures and Rangelands Core Research Project 1 (CRP1) that resulted in the publication of more than 165 peer-reviewed scientific journal articles. Campbell *et al.* determined from this massive collection of data that the legume content of grass-legume swards was typically increased by approximately 10 percent in response to a doubling of the air's CO_2 content.

Luscher *et al.* (2004) state that "the nutritive value of herbage from intensively managed grassland dominated by *L. perenne* and *T. repens* ... is well above the minimum range of the concentration of crude protein necessary for efficient digestion by ruminants (Barney *et al.*, 1981)." They conclude that "a small decrease in the concentration of crude protein in intensively managed forage production systems [which may never occur, as noted above] is not likely to have a negative effect on the nutritive value or on the intake of forage."

One final forage study is Newman *et al.* (2003), who investigated the effects of two levels of nitrogen fertilization and an approximate doubling of the air's CO_2 content on the growth and chemical composition of tall fescue (*Festuca arundinacea* Schreber cv. KY-31), both when infected and uninfected with a mutualistic fungal endophyte (*Neotyphodium coenophialum* Morgan-Jones and Gams). They found that the elevated CO_2 reduced the crude protein content of the forage by an average of 21 percent in three of the four situations studied: non-endophyte-infected plants in both the low and high nitrogen treatments, and endophyte-infected plants in the high nitrogen treatment. However, there was no protein reduction for endophyte-infected plants in the low nitrogen treatment.

As noted by Newman *et al.*, "the endophyte is present in many native and naturalized populations and the most widely sown cultivars of *F. arundinacea*," so the first two situations in which the CO_2-induced protein reduction occurred (those involving non-endophyte-infected plants) are not typical of the real world. In addition, since the dry-weight biomass yield of the forage was increased by fully 53 percent under the low nitrogen regime, and since the 10-times-greater high nitrogen regime boosted yields only by an additional 8 percent, there would appear to be no need to apply any extra nitrogen to *F. arundinacea* in a CO_2-enriched environment. Consequently, under best management practices in a doubled-CO_2 world of the future, little to no nitrogen would be added to the soil and there would be little to no reduction in the crude protein content of *F. arundinacea*, but there would be more than 50 percent more of it produced on the same amount of land.

With respect to the final plant quality studied by Newman *et al.*—i.e., forage digestibility— increasing soil nitrogen lowered *in vitro* neutral detergent fiber digestibility in both ambient and CO_2-enriched air; this phenomenon was most pronounced in the elevated CO_2 treatment. Again, however, under low nitrogen conditions there was no decline in plant digestibility. Hence, there is a second good reason not to apply extra nitrogen to *F. arundinacea* in a high CO_2 world of the future and, of course, little to no need to do so. Under best management practices in a future CO_2-enriched atmosphere, therefore, the results of this study suggest much greater quantities of good-quality forage could be produced without the addition of any, or very little, extra nitrogen to the soil.

But what about the unmanaged world of nature? Increases in the air's CO_2 content often—but not always (Goverde *et al.*, 1999)—lead to greater decreases in the concentrations of nitrogen and protein in the foliage of C_3 as compared to C_4 grasses (Wand *et al.*, 1999); as a result, in the words of Barbehenn *et al.* (2004a), "it has been predicted that insect herbivores will increase their feeding damage on C_3 plants to a greater extent than on C_4 plants" (Lincoln *et al.*, 1984, 1986; Lambers, 1993).

To test this hypothesis, Barbehenn *et al.* (2004a) grew *Lolium multiflorum* Lam. (Italian ryegrass, a common C_3 pasture grass) and *Bouteloua curtipendula* (Michx.) Torr. (sideoats gramma, a native C_4 rangeland grass) in chambers maintained at either the ambient atmospheric CO_2 concentration of 370 ppm or the doubled CO_2 concentration of 740 ppm for two months, after which newly molted sixth-instar larvae of *Pseudaletia unipuncta* (a grass-specialist noctuid) and *Spodoptera frugiperda* (a generalist noctuid) were allowed to feed upon the grasses. As expected, foliage protein concentration decreased by 20 percent in the C_3 grass, but by only 1 percent in the C_4 grass, when grown in the CO_2-enriched air. However, and "contrary to our expectations," according to Barbehenn *et al.*, "neither caterpillar species significantly increased its consumption rate to compensate for the lower concentration of protein in [the] C_3 grass," noting that "this result does not support the hypothesis that C_3 plants will be subject to greater rates of herbivory relative to C_4 plants in future [high-CO_2] atmospheric conditions (Lincoln *et al.*, 1984)." In addition, and "despite significant changes in the nutritional quality of *L. multiflorum* under elevated CO_2," they report that "no effect on the relative growth rate of either caterpillar species on either grass species resulted" and there were "no significant differences in insect performance between CO_2 levels."

In a similar study with the same two plants, Barbehenn *et al.* (2004b) allowed grasshopper (*Melanoplus sanguinipes*) nymphs that had been reared to the fourth instar stage to feed upon the grasses; once again, "contrary to the hypothesis that insect herbivores will increase their feeding rates disproportionately in C_3 plants under elevated atmospheric CO_2," they found that "*M. sanguinipes* did not significantly increase its consumption rate when feeding on the C_3 grass grown under elevated CO_2," suggesting this observation implies that "post-ingestive mechanisms enable these grasshoppers to compensate for variable nutritional quality in their host plants," and noting that some of these post-ingestive responses may include "changes in gut size, food residence time, digestive enzyme levels, and nutrient metabolism (Simpson and Simpson, 1990; Bernays and Simpson, 1990; Hinks *et al.*, 1991; Zanotto *et al.*, 1993; Yang and Joern, 1994a,b)." In fact, their data indicated that *M. sanguinipes* growth rates may have actually *increased*, perhaps by as much as 12 percent, when feeding upon the C_3 foliage that had been produced in the CO_2-enriched air.

In conclusion, the ongoing rise of the air's CO_2 concentration is not reducing the protein concentration in, or digestibility of, most important plant crops. In cases where protein concentration might by reduced, the addition of nitrogen fertilizer appears to offset the effect.

Additional information on this topic, including reviews of newer publications as they become available, can be found at http://www.co2 science.org/subject/p/ protein.php.

References

Agrawal, M. and Deepak, S.S. 2003. Physiological and biochemical responses of two cultivars of wheat to elevated levels of CO_2 and SO_2, singly and in combination. *Environmental Pollution* **121**: 189-197.

Allard, V., Newton, P.C.D., Lieffering, M., Clark, H., Matthew, C., Soussana, J.-F. and Gray, Y.S. 2003. Nitrogen cycling in grazed pastures at elevated CO_2: N returns by ruminants. *Global Change Biology* **9**: 1731-1742.

Barbehenn, R.V., Karowe, D.N. and Chen, Z. 2004b. Performance of a generalist grasshopper on a C_3 and a C_4 grass: compensation for the effects of elevated CO_2 on plant nutritional quality. *Oecologia* **140**: 96-103.

Barbehenn, R.V., Karowe, D.N. and Spickard, A. 2004a. Effects of elevated atmospheric CO_2 on the nutritional ecology of C_3 and C_4 grass-feeding caterpillars. *Oecologia* **140**: 86-95.

Barney, D.J., Grieve, D.G., Macleod, G.K. and Young, L.G. 1981. Response of cows to a reduction in dietary crude protein from 17 to 13 percent during early lactation. *Journal of Dairy Science* **64**: 25-33.

Bernays, E.A. and Simpson, S.J. 1990. Nutrition. In: Chapman, R.F. and Joern, A. (Eds.) *Biology of Grasshoppers*. Wiley, New York, NY, pp. 105-127.

Campbell, B.D., Stafford Smith, D.M., Ash, A.J., Fuhrer, J., Gifford, R.M., Hiernaux, P., Howden, S.M., Jones, M.B., Ludwig, J.A., Manderscheid, R., Morgan, J.A., Newton, P.C.D., Nosberger, J., Owensby, C.E., Soussana, J.F., Tuba, Z. and ZuoZhong, C. 2000. A synthesis of recent global change research on pasture and rangeland production: reduced uncertainties and their management implications. *Agriculture, Ecosystems and Environment* **82**: 39-55.

Dent, J.W. and Aldrich, D.T.A. 1963. The inter-relationships between heading date, yield, chemical composition and digestibility in varieties of perennial ryegrass, timothy, cooksfoot and meadow fescue. *Journal of the National Institute of Agricultural Botany* **9**: 261-281.

Evans, L.T. 1993. *Crop Evolution, Adaptation and Yield*. Cambridge University Press, Cambridge, UK.

Goverde, M., Bazin, A., Shykoff, J.A. and Erhardt, A. 1999. Influence of leaf chemistry of *Lotus corniculatus* (Fabaceae) on larval development of *Polyommatus icarus* (Lepidoptera, Lycaenidae): effects of elevated CO_2 and plant genotype. *Functional Ecology* **13**: 801-810.

Hartwig, U.A., Luscher, A., Daepp, M., Blum, H., Soussana, J.F. and Nosberger, J. 2000. Due to symbiotic N2 fixation, five years of elevated atmospheric pCO_2 had no effect on litter N concentration in a fertile grassland ecosystem. *Plant and Soil* **224**: 43-50.

Heagle, A.S., Miller, J.E. and Pursley, W.A. 1998. Influence of ozone stress on soybean response to carbon dioxide enrichment: III. Yield and seed quality. *Crop Science* **38**: 128-134.

Hinks, C.R., Cheeseman, M.T., Erlandson, M.A., Olfert, O. and Westcott, N.D. 1991. The effects of kochia, wheat and oats on digestive proteinases and the protein economy of adult grasshoppers, *Malanoplus sanguinipes*. *Journal of Insect Physiology* **37**: 417-430.

Humphreys, M.O. 1989. Water-soluble carbohydrates in perennial ryegrass breeding. III. Relationships with herbage production, digestibility and crude protein content. *Grass and Forage Science* **44**: 423-430.

Idso, C.D. and Idso, K.E. 2000. Forecasting world food supplies: The impact of the rising atmospheric CO_2 concentration. *Technology* **7S**: 33-56.

Idso, K.E., Hoober, J.K., Idso, S.B., Wall, G.W. and Kimball, B.A. 2001. Atmospheric CO_2 enrichment influences the synthesis and mobilization of putative vacuolar storage proteins in sour orange tree leaves. *Environmental and Experimental Botany* **48**: 199-211.

Jablonski, L.M., Wang, X. and Curtis, P.S. 2002. Plant reproduction under elevated CO_2 conditions: a meta-analysis of reports on 79 crop and wild species. *New Phytologist* **156**: 9-26.

Kimball, B.A., Morris, C.F., Pinter Jr., P.J., Wall, G.W., Hunsaker, D.J., Adamsen, F.J., LaMorte, R.L., Leavitt, S.W., Thompson, T.L., Matthias, A.D. and Brooks, T.J. 2001. Elevated CO_2, drought and soil nitrogen effects on wheat grain quality. *New Phytologist* **150**: 295-303.

Lambers, H. 1993. Rising CO_2, secondary plant metabolism, plant-herbivore interactions and litter decomposition. Theoretical considerations. *Vegetatio* **104/105**: 263-271.

Lincoln, D.E., Couvet, D. and Sionit, N. 1986. Responses of an insect herbivore to host plants grown in carbon dioxide enriched atmospheres. *Oecologia* **69**: 556-560.

Lincoln, D.E., Sionit, N. and Strain, B.R. 1984. Growth and feeding response of *Pseudoplusia includens* (Lepidoptera: Noctuidae) to host plants grown in controlled carbon dioxide atmospheres. *Environmental Entomology* **13**: 1527-1530.

Luscher, A., Daepp, M., Blum, H., Hartwig, U.A. and Nosberger, J. 2004. Fertile temperate grassland under elevated atmospheric CO_2—role of feed-back mechanisms and availability of growth resources. *European Journal of Agronomy* **21**: 379-398.

Newman, J.A., Abner, M.L., Dado, R.G., Gibson, D.J., Brookings, A. and Parsons, A.J. 2003. Effects of elevated CO_2, nitrogen and fungal endophyte-infection on tall fescue: growth, photosynthesis, chemical composition and digestibility. *Global Change Biology* **9**: 425-437.

Picon-Cochard, C., Teyssonneyre, F., Besle, J.M. *et al.* 2004. Effects of elevated CO_2 and cutting frequency on the productivity and herbage quality of a semi-natural grassland. *European Journal of Agronomy* **20**: 363-377

Pleijel, H., Mortensen, L., Fuhrer, J., Ojanpera, K. and Danielsson, H. 1999. Grain protein accumulation in relation to grain yield of spring wheat (*Triticum aestivum* L.) grown in open-top chambers with different concentrations of ozone, carbon dioxide and water availability. *Agriculture, Ecosystems and Environment* **72**: 265-270.

Rogers, G.S., Milham, P.J., Gillings, M. and Conroy, J.P. 1996. Sink strength may be the key to growth and nitrogen responses in N-deficient wheat at elevated CO_2. *Australian Journal of Plant Physiology* **23**: 253-264.

Rudorff, B.F.T., Mulchi, C.L., Fenny, P., Lee, E.H., Rowland, R. 1996. Wheat grain quality under enhanced tropospheric CO_2 and O_3 concentrations. *Journal of Environmental Quality* **25**: 1384-1388.

Simpson, S.J. and Simpson, C.L. 1990. The mechanisms of nutritional compensation by phytophagous insects. In: Bernays, E.A. (Ed.) *Insect-Plant Interactions*, Vol. 2. CRC Press, Boca Raton, FL, pp. 111-160.

Teyssonneyre, F. 2002. Effet d'une augmentation de la concentration atmospherique en CO_2 sur la prairie permanete et sur la competition entre especes prairiales associees. Ph.D. thesis, Orsay, Paris XI, France.

Thomas, J.M.G., Boote, K.J., Allen Jr., L.H., Gallo-Meagher, M. and Davis, J.M. 2003. Elevated temperature and carbon dioxide effects on soybean seed composition and transcript abundance. *Crop Science* **43**: 1548-1557.

Wand, S.J.E., Midgley, G.F., Jones, M.H. and Curtis, P.S. 1999. Responses of wild C_4 and C_3 grass (Poaceae) species to elevated atmospheric CO_2 concentration: a meta-analytic test of current theories and perceptions. *Global Change Biology* **5**: 723-741.

Wittwer, S.H. 1995. Food, Climate, and Carbon Dioxide: The Global Environment and World Food Production. CRC Press, Boca Raton, FL.

Yang, Y. and Joern, A. 1994a. Gut size changes in relation to variable food quality and body size in grasshoppers. *Functional Ecology* **8**: 36-45.

Yang, Y. and Joern, A. 1994b. Influence of diet quality, developmental stage, and temperature on food residence time in the grasshopper *Melanoplus differentialis*. *Physiological Zoology* **67**: 598-616.

Zanetti, S., Hartwig, U.A., Van Kessel, C., Luscher, A., Bebeisen, T., Frehner, M., Fischer, B.U., Hendrey, G.R., Blum, G. and Nosberger, J. 1997. Does nitrogen nutrition restrict the CO_2 response of fertile grassland lacking legumes? *Oecologia* **112**: 17-25.

Zanotto, F.P., Simpson, S.J. and Raubenheimer, D. 1993. The regulation of growth by locusts through post-ingestive compensation for variation in the levels of dietary protein and carbohydrate. *Physiological Entomology* **18**: 425-434.

9.2.2.2. Antioxidant Content

Antioxidants are chemical compounds that inhibit oxidation. Some antioxidants found in the human diet, such as vitamin E, vitamin C, and beta carotene, are thought to protect body cells from the damaging effects of oxidation. Scurvy—a condition characterized by general weakness, anemia, gum disease (gingivitis), and skin hemorrhages—is induced by low intake of vitamin C. There is some evidence that the condition may be resurgent in industrial countries, especially among children (Dickinson *et al.*, 1994; Ramar *et al.*, 1993; Gomez-Carrasco *et al.*, 1994). Hampl *et al.* (1999) found that 12-20 percent of 12- to 18-year-old school children in the United States "drastically under-consume" foods that supply vitamin C. Johnston *et al.* (1998) determined that 12-16 percent of U.S. college students have marginal plasma concentrations of vitamin C.

Since vitamin C intake correlates strongly with the consumption of citrus juice (Dennison *et al.*, 1998), and since the only high-vitamin-C juice consumed in any quantity by children is orange juice (Hampl *et al.*, 1999), even a modest role played by the ongoing rise in the air's CO_2 content in increasing the vitamin C concentration of orange juice could prove to be of considerable significance for public health in the United States and elsewhere. Thus, determining if rising CO_2 concentrations increase or hinder the production of antioxidants in human food is relevant to the issue of what effect the historical rise in CO_2 concentrations is having on human health.

Antioxidant concentrations in plants are generally observed to be high when environmental stresses are present, such as exposure to pollutants, drought, intense solar radiation, and high air or water temperatures. Stress generates highly reactive oxygenated compounds that damage plants, and ameliorating these stresses typically involves the production of antioxidant enzymes that scavenge and detoxify the highly reactive oxygenated compounds. In a study of two soybean genotypes, Pritchard *et al.* (2000) reported that three months' exposure to twice-ambient CO_2 concentrations reduced the activities of superoxide dismutase and catalase by an average of 23 and 39 percent, respectively. Likewise, Polle *et al.* (1997) showed that two years of atmospheric CO_2 enrichment reduced the activities of several key antioxidative enzymes, including catalase and superoxide dismutase, in beech seedlings. Moreover, Schwanz and Polle (1998) demonstrated this phenomenon can persist indefinitely, as they discovered similar reductions in these same enzymes in mature oak trees that had been growing near natural CO_2-emitting springs for 30 to 50 years.

The standard interpretation of these results is that the observed reductions in the activities of antioxidative enzymes under CO_2-enriched conditions imply that plants exposed to higher-than-current atmospheric CO_2 concentrations experience less oxidative stress and thus have a reduced need for antioxidant protection. This conclusion further suggests that "CO_2-advantaged" plants will be able to funnel more of their limited resources into the production of other plant tissues or processes essential to their continued growth and development.

On the other hand, when oxidative stresses do occur under high CO_2 conditions, the enhanced rates of photosynthesis and carbohydrate production resulting from atmospheric CO_2 enrichment generally enable plants to better deal with such stresses by providing more of the raw materials needed for antioxidant enzyme synthesis. Thus, when CO_2-enriched sugar maple seedlings were subjected to an additional 200 ppb of ozone, Niewiadomska *et al.* (1999) reported that ascorbate peroxidase, which is the first line of enzymatic defense against ozone, significantly increased. Likewise, Schwanz and Polle (2001) noted that poplar clones grown at 700 ppm CO_2 exhibited a much greater increase in superoxide dismutase activity upon chilling induction than clones

grown in ambient air. In addition, Lin and Wang (2002) observed that activities of superoxide dismutase and catalase were much higher in CO_2-enriched wheat than in ambiently grown wheat following the induction of water stress.

In some cases, the additional carbon fixed during CO_2-enrichment is invested in antioxidative compounds, rather than enzymes. One of the most prominent of these plant products is *ascorbate* or vitamin C. In the early studies of Barbale (1970) and Madsen (1971, 1975), a tripling of the atmospheric CO_2 concentration produced a modest (7 percent) increase in this antioxidant in the fruit of tomato plants. Kimball and Mitchell (1981), however, could find no effect of a similar CO_2 increase on the same species, although the extra CO_2 of their study stimulated the production of vitamin A. In bean sprouts, on the other hand, a mere one-hour-per-day doubling of the atmospheric CO_2 concentration actually *doubled* plant vitamin C contents over a seven-day period (Tajiri, 1985).

Probably the most comprehensive investigation of CO_2 effects on vitamin C production in an agricultural plant—a tree crop (sour orange)—was conducted by Idso *et al.* (2002). In an atmospheric CO_2 enrichment experiment begun in 1987 and still ongoing, a 75 percent increase in the air's CO_2 content was observed to increase sour orange juice vitamin C concentration by approximately 5 percent in run-of-the-mill years when total fruit production was typically enhanced by about 80 percent. In aberrant years when the CO_2-induced increase in fruit production was much greater, however, the increase in fruit vitamin C concentration also was greater, rising to a CO_2-induced enhancement of 15 percent when fruit production on the CO_2-enriched trees was 3.6 times greater than it was on the ambient-treatment trees.

Wang *et al.* (2003) evaluated the effects of elevated CO_2 on the antioxidant activity and flavonoid content of strawberry fruit in field plots at the U.S. Department of Agriculture's Beltsville Agricultural Research Center in Beltsville, Maryland, where they grew strawberry plants (*Fragaria x ananassa* Duchesne cv. Honeoye) in six clear-acrylic open-top chambers, two of which were maintained at the ambient atmospheric CO_2 concentration, two of which were maintained at ambient + 300 ppm CO_2, and two of which were maintained at ambient + 600 ppm CO_2 for a period of 28 months (from early spring of 1998 through June 2000). The scientists harvested the strawberry fruit, in their words, "at the commercially ripe stage" in both 1999 and 2000, after which they analyzed them for a number of different antioxidant properties and flavonol contents.

Before reporting what they found, Wang *et al.* provide some background by noting that "strawberries are good sources of natural antioxidants (Wang *et al.*, 1996; Heinonen *et al.*, 1998)." They further report that "in addition to the usual nutrients, such as vitamins and minerals, strawberries are also rich in anthocyanins, flavonoids, and phenolic acids," and that "strawberries have shown a remarkably high scavenging activity toward chemically generated radicals, thus making them effective in inhibiting oxidation of human low-density lipoproteins (Heinonen *et al.*, 1998)." In this regard, they note that previous studies (Wang and Jiao, 2000; Wang and Lin, 2000) "have shown that strawberries have high oxygen radical absorbance activity against peroxyl radicals, superoxide radicals, hydrogen peroxide, hydroxyl radicals, and singlet oxygen."

They determined, first, that strawberries had higher concentrations of ascorbic acid (AsA) and glutathione (GSH) "when grown under enriched CO_2 environments." In going from ambient to ambient + 300 ppm CO_2 and ambient + 600 ppm CO_2, for example, AsA concentrations increased by 10 and 13 percent, respectively, while GSH concentrations increased by 3 and 171 percent, respectively. They also learned that "an enriched CO_2 environment resulted in an increase in phenolic acid, flavonol, and anthocyanin contents of fruit." For nine different flavonoids, for example, there was a mean concentration increase of 55 ± 23 percent in going from the ambient atmospheric CO_2 concentration to ambient + 300 ppm CO_2, and a mean concentration increase of 112 ± 35 percent in going from ambient to ambient + 600 ppm CO_2. In addition, they report that the "high flavonol content was associated with high antioxidant activity." As for the significance of these findings, Wang *et al.* note that "anthocyanins have been reported to help reduce damage caused by free radical activity, such as low-density lipoprotein oxidation, platelet aggregation, and endothelium-dependent vasodilation of arteries (Heinonen *et al.*, 1998; Rice-Evans and Miller, 1996)."

In summarizing their findings, Wang *et al.* say "strawberry fruit contain flavonoids with potent antioxidant properties, and under CO_2 enrichment conditions, increased the[ir] AsA, GSH, phenolic acid, flavonol, and anthocyanin concentrations," further noting that "plants grown under CO_2

enrichment conditions also had higher oxygen radical absorbance activity against [many types of oxygen] radicals in the fruit."

Deng and Woodward (1998) reported that after growing strawberry plants in air containing an additional 170 ppm of CO_2, total fresh fruit weights were 42 and 17 percent greater than weights displayed by control plants grown at high and low soil nitrogen contents, respectively. Bushway and Pritts (2002) reported that a two- to three-fold increase in the air's CO_2 content boosted strawberry fruit yield by an average of 62 percent. In addition, Campbell and Young (1986), Keutgen *et al.* (1997), and Bunce (2001) reported positive strawberry photosynthetic responses to an extra 300 ppm of CO_2 ranging from 9 percent to 197 percent (mean of 76 percent ± 15 percent); and Desjardins *et al.* (1987) reported a 118 percent increase in photosynthesis in response to a 600 ppm increase in the air's CO_2 concentration.

Other researchers have found similar enhancements of antioxidative compounds under enriched levels of atmospheric CO_2. Estiarte *et al.* (1999), for example, reported that a 180-ppm increase in the air's CO_2 content increased the foliar concentrations of flavonoids, which protect against UV-B radiation damage, in field-grown spring wheat by 11 to 14 percent. Caldwell *et al.* (2005) found that an ~75 percent increase in the air's CO_2 content increased the total isoflavone content of soybean seeds by 8 percent when the air temperature during seed fill was 18°C, by 104 percent when the air temperature during seed fill was 23°C, by 101 percent when the air temperature was 28°C, and by 186 percent and 38 percent, respectively, when a drought-stress treatment was added to the latter two temperature treatments.

Lastly, in an experiment conducted under very high atmospheric CO_2 concentrations, Ali *et al.* (2005) found that CO_2 levels of 10,000 ppm, 25,000 ppm, and 50,000 ppm increased total flavonoid concentrations of ginseng roots by 228 percent, 383 percent, and 232 percent, respectively, total phenolic concentrations by 58 percent, 153 percent, and 105 percent, cysteine contents by 27 percent, 65 percent, and 100 percent, and non-protein thiol contents by 12 percent, 43 percent, and 62 percent, all of which substances are potent antioxidants.

In summary, as the CO_2 content of the air rises, plants typically experience less oxidative stress, and since they thus need fewer antioxidants for protection, antioxidant levels in their leaves decline, which enables them to use more of their valuable resources for other purposes. However, elevated CO_2 also provides more of the raw materials needed for oxidant enzyme synthesis, leading to higher levels of antioxidative compounds—such as *ascorbate,* or vitamin C. Research shows this happens with enough frequency that higher CO_2 levels will lead to higher concentrations of antioxidants, leading to better health.

Additional information on this topic, including reviews of newer publications as they become available, can be found at http://www.co2 science.org/subject/a/antioxidants.php.

References

Ali, M.B., Hahn, E.J. and Paek, K.-Y. 2005. CO_2-induced total phenolics in suspension cultures of *Panax ginseng* C.A. Mayer roots: role of antioxidants and enzymes. *Plant Physiology and Biochemistry* **43**: 449-457.

Barbale, D. 1970. The influence of the carbon dioxide on the yield and quality of cucumber and tomato in the covered areas. *Augsne un Raza (Riga)* **16**: 66-73.

Bunce, J.A. 2001. Seasonal patterns of photosynthetic response and acclimation to elevated carbon dioxide in field-grown strawberry. *Photosynthesis Research* **68**: 237-245.

Bushway, L.J. and Pritts, M.P. 2002. Enhancing early spring microclimate to increase carbon resources and productivity in June-bearing strawberry. *Journal of the American Society for Horticultural Science* **127**: 415-422.

Caldwell, C.R., Britz, S.J. and Mirecki, R.M. 2005. Effect of temperature, elevated carbon dioxide, and drought during seed development on the isoflavone content of dwarf soybean [*Glycine max* (L.) Merrill] grown in controlled environments. *Journal of Agricultural and Food Chemistry* **53**: 1125-1129.

Campbell, D.E. and Young, R. 1986. Short-term CO_2 exchange response to temperature, irradiance, and CO_2 concentration in strawberry. *Photosynthesis Research* **8**: 31-40.

Deng, X. and Woodward, F.I. 1998. The growth and yield responses of *Fragaria ananassa* to elevated CO_2 and N supply. *Annals of Botany* **81**: 67-71.

Dennison, B.A., Rockwell, H.L., Baker, S.L. 1998. Fruit and vegetable intake in young children. *J. Amer. Coll. Nutr.* **17**: 371-378.

Desjardins, Y., Gosselin, A. and Yelle, S. 1987. Acclimatization of ex vitro strawberry plantlets in CO_2-enriched environments and supplementary lighting. *Journal of the American Society for Horticultural Science* **112**: 846-851.

Dickinson, V.A., Block, G., Russek-Cohen, E. 1994. Supplement use, other dietary and demographic variables, and serum vitamin C in NHANES II. *J. Amer. Coll. Nutr.* **13**: 22-32.

Estiarte, M., Penuelas, J., Kimball, B.A., Hendrix, D.L., Pinter Jr., P.J., Wall, G.W., LaMorte, R.L. and Hunsaker, D.J. 1999. Free-air CO_2 enrichment of wheat: leaf flavonoid concentration throughout the growth cycle. *Physiologia Plantarum* **105**: 423-433.

Gomez-Carrasco, J.A., Cid, J.L.-H., de Frutos, C.B., Ripalda-Crespo, M.J., de Frias, J.E.G. 1994. Scurvy in adolescence. *J. Pediatr. Gastroenterol. Nutr.* **19**: 118-120.

Hampl, J.S., Taylor, C.A., Johnston, C.S. 1999. Intakes of vitamin C, vegetables and fruits: Which schoolchildren are at risk? *J. Amer. Coll. Nutr.* **18**: 582-590.

Heinonen, I.M., Meyer, A.S. and Frankel, E.N. 1998. Antioxidant activity of berry phenolics on human low-density lipoprotein and liposome oxidation. *Journal of Agricultural and Food Chemistry* **46**: 4107-4112.

Idso, S.B., Kimball, B.A., Shaw, P.E., Widmer, W., Vanderslice, J.T., Higgs, D.J., Montanari, A. and Clark, W.D. 2002. The effect of elevated atmospheric CO_2 on the vitamin C concentration of (sour) orange juice. *Agriculture, Ecosystems and Environment* **90**: 1-7.

Johnston, C.S., Solomon, R.E., Corte, C. 1998. Vitamin C status of a campus population: College students get a C minus. *J. Amer. Coll. Health* **46**: 209-213.

Keutgen, N., Chen, K. and Lenz, F. 1997. Responses of strawberry leaf photosynthesis, chlorophyll fluorescence and macronutrient contents to elevated CO_2. *Journal of Plant Physiology* **150**: 395-400.

Kimball, B.A., Mitchell, S.T. 1981. Effects of CO_2 enrichment, ventilation, and nutrient concentration on the flavor and vitamin C content of tomato fruit. *HortScience* **16**: 665-666.

Lin, J.-S and Wang, G.-X. 2002. Doubled CO_2 could improve the drought tolerance better in sensitive cultivars than in tolerant cultivars in spring wheat. *Plant Science* **163**: 627-637.

Madsen, E. 1971. The influence of CO_2-concentration on the content of ascorbic acid in tomato leaves. *Ugeskr. Agron.* **116**: 592-594.

Madsen, E. 1975. Effect of CO_2 environment on growth, development, fruit production and fruit quality of tomato from a physiological viewpoint. In: Chouard, P. and de Bilderling, N. (Eds.) *Phytotronics in Agricultural and Horticultural Research*. Bordas, Paris, pp. 318-330.

Niewiadomska, E., Gaucher-Veilleux, C., Chevrier, N., Mauffette, Y. and Dizengremel, P. 1999. Elevated CO$_2$ does not provide protection against ozone considering the activity of several antioxidant enzymes in the leaves of sugar maple. *Journal of Plant Physiology* **155**: 70-77.

Polle, A., Eiblmeier, M., Sheppard, L. and Murray, M. 1997. Responses of antioxidative enzymes to elevated CO$_2$ in leaves of beech (*Fagus sylvatica* L.) seedlings grown under a range of nutrient regimes. *Plant, Cell and Environment* **20**: 1317-1321.

Pritchard, S.G., Ju, Z., van Santen, E., Qiu, J., Weaver, D.B., Prior, S.A. and Rogers, H.H. 2000. The influence of elevated CO$_2$ on the activities of antioxidative enzymes in two soybean genotypes. *Australian Journal of Plant Physiology* **27**: 1061-1068.

Ramar, S., Sivaramakrishman, V., Manoharan, K. 1993. Scurvy—a forgotten disease. *Arch. Phys. Med. Rehabil.* **74**: 92-95.

Rice-Evans, C.A. and Miller, N.J. 1996. Antioxidant activities of flavonoids as bioactive components of food. *Biochemical Society Transactions* **24**: 790-795.

Schwanz, P. and Polle, A. 2001. Growth under elevated CO$_2$ ameliorates defenses against photo-oxidative stress in poplar (*Populus alba x tremula*). *Environmental and Experimental Botany* **45**: 43-53.

Schwanz, P. and Polle, A. 1998. Antioxidative systems, pigment and protein contents in leaves of adult mediterranean oak species (*Quercus pubescens* and *Q. ilex*) with lifetime exposure to elevated CO$_2$. *New Phytologist* **140**: 411-423.

Tajiri, T. 1985. Improvement of bean sprouts production by intermittent treatment with carbon dioxide. *Nippon Shokuhin Kogyo Gakkaishi* **32**(3): 159-169.

Wang, H., Cao, G. and Prior, R.L. 1996. Total antioxidant capacity of fruits. *Journal of Agricultural and Food Chemistry* **44**: 701-705.

Wang, S.Y., Bunce, J.A. and Maas, J.L. 2003. Elevated carbon dioxide increases contents of antioxidant compounds in field-grown strawberries. *Journal of Agricultural and Food Chemistry* **51**: 4315-4320.

Wang, S.Y. and Jiao, H. 2000. Scavenging capacity of berry crops on superoxide radicals, hydrogen peroxide, hydroxyl radicals, and singlet oxygen. *Journal of Agricultural and Food Chemistry* **48**: 5677-5684.

Wang, S.Y. and Lin, H.S. 2000. Antioxidant activity in fruit and leaves of blackberry, raspberry, and strawberry is affected by cultivar and maturity. *Journal of Agricultural and Food Chemistry* **48**: 140-146.

9.2.3. Medicinal Constituents

Primitive medical records indicate that extracts from many species of plants have been used for treating a variety of human health problems for perhaps the past 3,500 years (Machlin, 1992; Pettit *et al.*, 1993, 1995). In modern times the practice has continued, with numerous chemotherapeutic agents being isolated (Gabrielsen *et al.*, 1992a). Until recently, however, no studies had investigated the effects of atmospheric CO$_2$ enrichment on specific plant compounds of direct medicinal value.

Stuhlfauth *et al.* (1987) studied the individual and combined effects of atmospheric CO$_2$ enrichment and water stress on the production of secondary metabolites in the woolly foxglove (*Digitalis lanata* EHRH), which produces the cardiac glycoside *digoxin* that is used in the treatment of cardiac insufficiency. Under controlled well-watered conditions in a phytotron, a near-tripling of the air's CO$_2$ content increased plant dry weight production in this medicinal plant by 63 percent, while under water-stressed conditions the CO$_2$-induced dry weight increase was 83 percent. In addition, the concentration of digoxin within the plant dry mass was enhanced by 11 percent under well-watered conditions and by 14 percent under conditions of water stress.

In a subsequent whole-season field experiment, Stuhlfauth and Fock (1990) obtained similar results. A near-tripling of the air's CO$_2$ concentration led to a 75 percent increase in plant dry weight production per unit land area and a 15 percent increase in digoxin yield per unit dry weight of plant, which combined to produce a doubling of total digoxin yield per hectare of cultivated land.

Idso *et al.* (2000) evaluated the response of the tropical spider lily (*Hymenocallis littoralis* Jacq. Salisb.) to elevated levels of atmospheric CO$_2$ over four growing seasons. This plant has been known since ancient times to possess anti-tumor activity; in modern times it has been shown to contain constituents that are effective against lymphocytic leukemia and ovary sarcoma (Pettit *et al.*, 1986). These same plant constituents also have been proven to be effective against the U.S. National Cancer Institute's panel of 60 human cancer cell lines, demonstrating greatest effectiveness against melanoma, brain, colon, lung, and renal cancers (Pettit *et al.*, 1993). In addition, it exhibits strong anti-viral activity against Japanese encephalitis and yellow, dengue, Punta Tora, and Rift Valley fevers (Gabrielsen *et al.*, 1992a,b).

Idso *et al.* determined that a 75 percent increase in the air's CO_2 concentration produced a 56 percent increase in the spider lily's below-ground bulb biomass, where the disease-fighting substances are found. In addition, for these specific substances, they observed a 6 percent increase in the concentration of a two-constituent (1:1) mixture of 7-deoxynarciclasine and 7-deoxy-trans-dihydronarciclasine, an 8 percent increase in pancratistatin, an 8 percent increase in trans-dihydronarciclasine, and a 28 percent increase in narciclasine. Averaged together and combined with the 56 percent increase in bulb biomass, these percentage concentration increases resulted in a total mean active-ingredient increase of 75 percent for the plants grown in air containing 75 percent more CO_2.

Other plant constituents that perform important functions in maintaining human health include sugars, lipids, oils, fatty acids, and macro- and micro-nutrients. Although concerns have been raised about the availability of certain of the latter elements in plants growing in a CO_2-enriched world (Loladze, 2002), the jury is still out with respect to this subject as a consequence of the paucity of pertinent data.

Literally thousands of studies have assessed the impact of elevated levels of atmospheric CO_2 on the quantity of biomass produced by agricultural crops, but only a tiny fraction of that number have looked at any aspect of food quality. From what has been learned about plant protein, antioxidants, and the few medicinal substances that have been investigated in this regard, there is no reason to believe these other plant constituents would be present in lower concentrations in a CO_2-enriched world and ample evidence that they may be present in significantly higher concentrations and greater absolute amounts.

Additional information on this topic, including reviews of newer publications as they become available, can be found at http://www.co2science.org/subject/h/co2healthpromoting.php.

References

Gabrielsen, B., Monath, T.P., Huggins, J.W., Kefauver, D.F., Pettit, G.R., Groszek, G., Hollingshead, M., Kirsi, J.J., Shannon, W.F., Schubert, E.M., Dare, J., Ugarkar, B., Ussery, M.A., Phelan, M.J. 1992a. Antiviral (RNA) activity of selected Amaryllidaceae isoquinoline constituents and synthesis of related substances. *Journal of Natural Products* **55**: 1569-1581.

Gabrielsen, B., Monath, T.P., Huggins, J.W., Kirsi, J.J., Hollingshead, M., Shannon, W.M., Pettit, G.R. 1992b. Activity of selected Amaryllidaceae constituents and related synthetic substances against medically important RNA viruses. In: Chu, C.K. and Cutler, H.G. (Eds.) *Natural Products as Antiviral Agents*. Plenum Press, New York, NY, pp. 121-35.

Idso, S.B., Kimball, B.A., Pettit III, G.R., Garner, L.C., Pettit, G.R., Backhaus, R.A. 2000. Effects of atmospheric CO_2 enrichment on the growth and development of *Hymenocallis littoralis* (Amaryllidaceae) and the concentrations of several antineoplastic and antiviral constituents of its bulbs. *American Journal of Botany* **87**: 769-773.

Loladze, I. 2002. Rising atmospheric CO_2 and human nutrition: toward globally imbalanced plant stoichiometry? *Trends in Ecology & Evolution* **17**: 457-461.

Machlin, L.G. 1992. Introduction. In: Sauerlich, H.E. and Machlin, L.J. (Eds.) Beyond deficiency: New views on the function and health effects of vitamins. *Annals of the New York Academy of Science* **669**: 1-6.

Pettit, G.R., Pettit III, G.R., Backhaus, R.A., Boyd, M.R., Meerow, A.W. 1993. Antineoplastic agents, 256. Cell growth inhibitory isocarbostyrils from *Hymenocallis*. *Journal of Natural Products* **56**: 1682-1687.

Pettit, G.R., Pettit III, G.R., Groszek, G., Backhaus, R.A., Doubek, D.L., Barr, R.J. 1995. Antineoplastic agents, 301. An investigation of the Amaryllidaceae genus *Hymenocallis*. *Journal of Natural Products* **58**: 756-759.

Stuhlfauth, T. and Fock, H.P. 1990. Effect of whole season CO_2 enrichment on the cultivation of a medicinal plant, *Digitalis lanata*. *Journal of Agronomy and Crop Science* **164**: 168-173.

Stuhlfauth, T., Klug, K. and Fock, H.P. 1987. The production of secondary metabolites by *Digitalis lanata* during CO_2 enrichment and water stress. *Phytochemistry* **26**: 2735-2739.

9.3. Human Longevity

The past two centuries have witnessed a significant degree of global warming as the earth recovered from the Little Ice Age and entered the Current Warm Period. Simultaneously, the planet has seen an increase in its atmospheric CO_2 concentration. What effect have these trends had on human longevity? Although no one can give a precise quantitative answer to this question, it is possible to assess their relative importance by considering the history of human longevity.

Tuljapurkar *et al.* (2000) examined mortality over the period 1950-1994 in Canada, France, Germany (excluding the former East Germany), Italy, Japan, the United Kingdom, and the United States, finding that "in every country over this period, mortality at each age has declined exponentially at a roughly constant rate." In discussing these findings, Horiuchi (2000) notes that the average lifespan of early humans was about 20 years, but that in the major industrialized countries it is now about 80 years, with the bulk of this increase having come in the past 150 years. He then notes that "it was widely expected that as life expectancy became very high and approached the 'biological limit of human longevity,' the rapid 'mortality decline' would slow down and eventually level off," but "such a deceleration has not occurred." "These findings give rise to two interrelated questions," says Horiuchi: (1) "Why has mortality decline not started to slow down?" and (2) "Will it continue into the future?"

Some points to note in attempting to answer these questions are the following. First, in Horiuchi's words, "in the second half of the nineteenth century and the first half of the twentieth century, there were large decreases in the number of deaths from infectious and parasitic diseases, and from poor nutrition and disorders associated with pregnancy and childbirth," which led to large reductions in the deaths of infants, children, and young adults. In the second half of the twentieth century, however, Horiuchi notes that "mortality from degenerative diseases, most notably heart diseases and stroke, started to fall," and the reduction was most pronounced among the elderly. Some suspected this latter drop in mortality might have been achieved "through postponing the deaths of seriously ill people," but data from the United States demonstrate, in his words, that "the health of the elderly greatly improved in the 1980s and 1990s, suggesting that the extended length of life in old age is mainly due to better health rather than prolonged survival in sickness."

Additional support for this conclusion comes from the study of Manton and Gu (2001). With the completion of the latest of the five National Long-Term Care Surveys of disability in U.S. citizens over 65 years of age—which began in 1982 and extended to 1999 at the time of the writing of their paper—these researchers were able to discern two trends: (1) disabilities in this age group decreased over the entire period studied, and (2) disabilities decreased at a rate that grew ever larger with the passing of time. Over the 17-year period of record, the percentage of the group that was disabled dropped 25 percent, from 26.2 percent in 1982 to 19.7 percent in 1999. The percentage disability decline rate per year for the periods 1982-1989, 1989-1994, and 1994-1999 was 0.26, 0.38, and 0.56 percent per year, respectively. Commenting on the accelerating rate of this disability decline, the authors say "it is surprising, given the low level of disability in 1994, that the rate of improvement accelerated" over the most recent five-year interval.

Looking outside the United States, Oeppen and Vaupel (2002) reported that "world life expectancy more than doubled over the past two centuries, from roughly 25 years to about 65 for men and 70 for women." They noted that "for 160 years, best-performance life expectancy has steadily increased by a quarter of a year per year," and they emphasized that this phenomenal trend "is so extraordinarily linear that it may be the most remarkable regularity of mass endeavor ever observed."

These observations demonstrate that if the increases in air temperature and CO_2 concentration of the past two centuries were bad for our health, their combined negative influence was minuscule compared to whatever else was at work in promoting this vast increase in worldwide human longevity. It is that "whatever else" to which we now turn our attention.

It is evident that in developed countries, the elderly are living longer with the passing of time. This phenomenon is likely due to ever-improving health in older people, which in turn is likely the result of continuing improvements in the abilities of their bodies to repair cellular damage caused by degenerative processes associated with old age, i.e., stresses caused by the reactive oxygen species that are generated by normal metabolism (Finkel and Holbrook, 2000).

Wentworth *et al.* (2003) report they found "evidence for the production of ozone in human disease," specifically noting that "signature products unique to cholesterol ozonolysis are present within atherosclerotic tissue at the time of carotid endarterectomy, suggesting that ozone production occurred during lesion development." According to Marx (2003), "researchers think that inflammation of blood vessels is a major instigator of plaque formation," that "ozone contributes to plaque formation by oxidizing cholesterol," and that the new findings "suggest new strategies for preventing atherosclerosis." Also according to Marx, Daniel Steinberg of the University of California, San Diego, says that although it's still too early to definitively state whether ozone production in plaques is a major contributor to atherosclerosis, he expresses his confidence that once we know for sure, we'll know which antioxidants will work in suppressing plaque formation.

Reactive oxygen species (ROS) generated during cellular metabolism or peroxidation of lipids and proteins also play a causative role in the pathogenesis of cancer, along with coronary heart disease (CHD), as demonstrated by Slaga *et al.* (1987), Frenkel (1992), Marnett (2000), Zhao *et al.* (2000) and Wilcox *et al.* (2004). However, as noted by Yu *et al.* (2004), "antioxidant treatments may terminate ROS attacks and reduce the risks of CHD and cancer, as well as other ROS-related diseases such as Parkinson's disease (Neff, 1997; Chung *et al.*, 1999; Wong *et al.*, 1999; Espin *et al.*, 2000; Merken and Beecher, 2000)." As a result, they say that "developing functional foods rich in natural antioxidants may improve human nutrition and reduce the risks of ROS-associated health problems."

Consider, in this regard, the common strawberry. Wang *et al.* (2003) report that strawberries are especially good sources of natural antioxidants. They say that "in addition to the usual nutrients, such as vitamins and minerals, strawberries are also rich in anthocyanins, flavonoids, and phenolic acids," and that "strawberries have shown a remarkably high scavenging activity toward chemically generated radicals, thus making them effective in inhibiting oxidation of human low-density lipoproteins (Heinonen *et al.*, 1998)." They also note that Wang and Jiao (2000) and Wang and Lin (2000) "have shown that strawberries have high oxygen radical absorbance activity against peroxyl radicals, superoxide radicals, hydrogen peroxide, hydroxyl radicals, and singlet oxygen." And they say that "anthocyanins have been reported to help reduce damage caused by free radical activity, such as low-density lipoprotein oxidation, platelet aggregation, and endothelium-dependent vasodilation of arteries (Heinonen *et al.*, 1998; Rice-Evans and Miller, 1996)."

Our reason for citing all of this information is that Wang *et al.* (2003) have recently demonstrated that enriching the air with carbon dioxide increases both the concentrations and activities of many of these helpful substances. They determined, for example, that strawberries had higher concentrations of ascorbic acid and glutathione when grown in CO_2-enriched environments. They also learned that "an enriched CO_2 environment resulted in an increase in phenolic acid, flavonol, and anthocyanin contents of fruit." For nine different flavonoids there was a mean concentration increase of 55 percent in going from the ambient atmospheric CO_2 concentration to ambient + 300 ppm CO_2, and a mean concentration increase of 112 percent in going from ambient to ambient + 600 ppm CO_2. Also, they report that "high flavonol content was associated with high antioxidant activity."

There is little reason to doubt that similar concentration and activity increases in the same and additional important phytochemicals in other food crops would occur in response to the same increases in the air's CO_2 concentration. Indeed, the aerial fertilization effect of atmospheric CO_2 enrichment is a near-universal phenomenon that operates among plants of all types, and it is very powerful (e.g., Mayeux *et al.*, 1997; Idso and Idso, 2000). There must have been significant concomitant increases in the concentrations and activities of the various phytochemicals in these foods that act as described by Wang *et al.* (2003).

Could some part of the rapid lengthening of human longevity reported by Oeppen and Vaupel (2002) be due to enhanced CO_2 in the air putting more antioxidants in our diets? Two recent experiments showing the positive effects of antioxidants on animal lifespan provide some additional evidence that this may be the case.

Melov *et al.* (2000) examined the effects of two superoxide dismutase-/catalase-like mimetics (EUK-8 and EUK-134) on the lifespan of normal and mutant *Caenorhabditis elegans* worms that ingested various concentrations of the mimetics. In all of their experiments, treatment of normal worms with the antioxidant mimetics significantly increased both mean and maximum lifespan. Treatment of normal worms with only 0.05 mM EUK-134, for example, increased their mean lifespan by fully 54 percent. In mutant worms whose lifespan had been genetically shortened by 37 percent, treatment with 0.5 mM EUK-134 restored their lifespan to normal by increasing their mutation-reduced lifespan by 67 percent. It also was determined that these effects were not due to a reduction in worm metabolism, which could have reduced the production of oxygen radicals, but "by augmenting natural antioxidant defenses without having any overt effects on other traits." In the words of the authors, "these results suggest that endogenous oxidative stress is a major determinant of the rate of aging," the significance of which statement resides in the fact that antioxidants tend to reduce such stresses in animals, including man.

Another study addressing the subject was conducted by Larsen and Clarke (2002), who fed diets with and without coenzyme Q to wild-type *Caenorhabditis elegans* and several mutants during the adult phases of their lives, while they recorded the lengths of time they survived. This work revealed that "withdrawal of coenzyme Q from the diet of wild-type nematodes extends adult life-span by ~60 percent." In addition, they found that the lifespans of the four different mutants they studied were extended by a Q-less diet. More detailed experiments led them to conclude that the life-span extensions were due to reduced generation and/or increased scavenging of reactive oxygen species, leading them to conclude in the final sentence of their paper that "the combination of reduced generation and increased scavenging mechanisms are predicted to result in a substantial decrease in the total cellular ROS and thereby allow for an extended life-span."

In light of these many diverse observations of both plants and animals, there is some reason to believe that the historical increase of CO_2 in the air has helped lengthen human lifespans since the advent of the Industrial Revolution, and that its continued upward trend will provide more of the same benefit.

Additional information on this topic, including reviews of newer publications as they become available, can be found at http://www.co2science.org/subject/h/humanlifespan.php.

References

Chung, H.S., Chang, L.C., Lee, S.K., Shamon, L.A., Breemen, R.B.V., Mehta, R.G., Farnsworth, N.R., Pezzuto, J.M. and Kinghorn, A.D. 1999. Flavonoid constituents of chorizanthe diffusa with potential cancer chemopreventive activity. *Journal of Agricultural and Food Chemistry* **47**: 36-41.

Espin, J.C., Soler-Rivas, C. and Wichers, H.J. 2000. Characterization of the total free radical scavenger capacity of vegetable oils and oil fractions using 2,2-diphenyl-1-picryhydrazyl radical. *Journal of Agricultural and Food Chemistry* **48**: 648-656.

Finkel, T. and Holbrook, N.J. 2000. Oxidants, oxidative stress and the biology of ageing. *Nature* **408**: 239-247.

Frenkel, K. 1992. Carcinogen-mediated oxidant formation and oxidative DNA damage. *Pharmacology and Therapeutics* **53**: 127-166.

Heinonen, I.M., Meyer, A.S. and Frankel, E.N. 1998. Antioxidant activity of berry phenolics on human low-density lipoprotein and liposome oxidation. *Journal of Agricultural and Food Chemistry* **46**: 4107-4112.

Horiuchi, S. 2000. Greater lifetime expectations. *Nature* **405**: 744-745.

Idso, C.D. and Idso, K.E. 2000. Forecasting world food supplies: The impact of the rising atmospheric CO_2 concentration. *Technology* **7S**: 33-56.

Larsen, P.L. and Clarke C.F. 2002. Extension of life-span in *Caenorhabditis elegans* by a diet lacking coenzyme Q. *Science* **295**: 120-123.

Manton, K.G. and Gu, X.L. 2001. Changes in the prevalence of chronic disability in the United States black and nonblack population above age 65 from 1982 to 1999. *Proceedings of the National Academy of Science, USA* **98**: 6354-6359.

Marnett, L.J. 2000. Oxyradicals and DNA damage. *Carcinogenesis* **21**: 361-370.

Marx, J. 2003. Ozone may be secret ingredient in plaques' inflammatory stew. *Science* **302**: 965.

Mayeux, H.S., Johnson, H.B., Polley, H.W. and Malone, S.R. 1997. Yield of wheat across a subambient carbon dioxide gradient. *Global Change Biology* **3**: 269-278.

Melov, S., Ravenscroft, J., Malik, S., Gill, M.S., Walker, D.W., Clayton, P.E., Wallace, D.C., Malfroy, B., Doctrow, S.R. and Lithgow, G.J. 2000. Extension of life-span with superoxide dismutase/catalase mimetics. *Science* **289**: 1567-1569.

Merken, H.M. and Beecher, G.R. 2000. Measurement of food flavonoids by high-performance liquid chromatography: a review. *Journal of Agricultural and Food Chemistry* **48**: 577-599.

Neff, J. 1997. Big companies take nutraceuticals to heart. *Food Processing* **58**: 37-42.

Oeppen, J. and Vaupel, J.W. 2002. Broken limits to life expectancy. *Science* **296**: 1029-1030.

Rice-Evans, C.A. and Miller, N.J. 1996. Antioxidant activities of flavonoids as bioactive components of food. *Biochemical Society Transactions* **24**: 790-795.

Slaga, T.J., O'Connell, J., Rotstein, J., Patskan, G., Morris, R., Aldaz, M. and Conti, C. 1987. Critical genetic determinants and molecular events in multistage skin carcinogenesis. *Symposium on Fundamental Cancer Research* **39**: 31-34.

Tuljapurkar, S., Li, N. and Boe, C. 2000. A universal pattern of mortality decline in the G7 countries. *Nature* **405**: 789-792.

Wang, S.Y., Bunce, J.A. and Maas, J.L. 2003. Elevated carbon dioxide increases contents of antioxidant compounds in field-grown strawberries. *Journal of Agricultural and Food Chemistry* **51**: 4315-4320.

Wang, S.Y. and Jiao, H. 2000. Scavenging capacity of berry crops on superoxide radicals, hydrogen peroxide, hydroxyl radicals, and singlet oxygen. *Journal of Agricultural and Food Chemistry* **48**: 5677-5684.

Wang, S.Y. and Lin, H.S. 2000. Antioxidant activity in fruit and leaves of blackberry, raspberry, and strawberry is affected by cultivar and maturity. *Journal of Agricultural and Food Chemistry* **48**: 140-146.

Wang, S.Y. and Zheng, W. 2001. Effect of plant growth temperature on antioxidant capacity in strawberry. *Journal of Agricultural and Food Chemistry* **49**: 4977-4982.

Wentworth Jr., P., Nieva, J., Takeuchi, C., Glave, R., Wentworth, A.D., Dilley, R.B., DeLaria, G.A., Saven, A., Babior, B.M., Janda, K.D., Eschenmoser, A. and Lerner, R.A. 2003. Evidence for ozone formation in human atherosclerotic arteries. *Science* **302**: 1053-1056.

Willcox, J.K., Ash, S.L. and Catignani, G.L. 2004. Antioxidants and prevention of chronic disease. *Critical Reviews in Food Science and Nutrition* **44**: 275-295.

Wong, S.S., Li, R.H.Y. and Stadlin, A. 1999. Oxidative stress induced by MPTP and MPP+: Selective vulnerability of cultured mouse astocytes. *Brain Research* **836**: 237-244.

Yu, L., Haley, S., Perret, J. and Harris, M. 2004. Comparison of wheat flours grown at different locations for their antioxidant properties. *Food Chemistry* **86**: 11-16.

Zhao, J., Lahiri-Chatterjee, M., Sharma, Y. and Agarwal, R. 2000. Inhibitory effect of a flavonoid antioxidant silymarin on benzoyl peroxide-induced tumor promotion, oxidative stress and inflammatory responses in SENCAR mouse skin. *Carcinogenesis* **21**: 811-816.

9.4. Food vs. Nature

Norman Borlaug, the father of the Green Revolution, recently expressed in a *Science* editorial his concern over the challenge of "feeding a hungry world" by noting that "some 800 million people still experience chronic and transitory hunger each year," and that "over the next 50 years, we face the daunting job of feeding 3.5 billion additional people, most of whom will begin life in poverty" (Borlaug, 2007). He described how the scientific and technological innovations he played a major role in discovering and implementing helped reduce the proportion of hungry people in the world "from about 60% in 1960 to 17% in 2000." Had that movement failed, he says, environmentally fragile land would have been brought into agricultural production and the resulting "soil erosion, loss of forests and grasslands, reduction in biodiversity, and extinction of wildlife species would have been disastrous."

Rising CO_2 concentrations in the air helped make it possible to feed a growing global population in the past without devasting nature, but what of the future? The world's poulation in 2008 was estimated to be 6.7 billion and is projected to reach between 9.1 and 9.7 billion by 2050 (United Nations, 2009; U.S. Census Bureau, 2008). There is real concern about our ability to feed the world's population a mere 50 years hence.

Tilman *et al.* (2001) analyzed the global environmental impacts likely to occur if agriculture is to keep pace with population growth. They report that "humans currently appropriate more than a third of the production of terrestrial ecosystems and about half of usable freshwaters." They estimate that the amount of land devoted to agriculture by the year 2050 will have to increase 18 percent to meet the rising demand for food. Because developed countries are expected to withdraw large areas of land from farming over the next 50 years for recreation, open space, and reforestation, the net loss of natural ecosystems to cropland and pasture in developing countries will amount to about half of all potentially suitable remaining land, which would "represent the worldwide loss of natural ecosystems larger than the United States." Similar warnings of a coming food vs. nature conflict have been expressed by other scientists, for example, Wallace (2000) and Raven (2002).

What, if anything, can be done to address this conflict between the need to produce food and the wish to preserve nature? And what role, if any, will climate change play in averting the crisis or making it even worse?

We begin by observing that the fear that there isn't enough land to support a growing population's food needs is a very old one, dating at least to Thomas Malthus (1798) and expressed in our day by popular writers such as Paul Ehrlich (2008) and Al Gore (1992). Predictions of widespread famine have been wrong before, as trends in food production and daily intake of calories per capita, while not linear in the short term, show long-term positive trends that are driven primarily by gains in yields per acre, not expansion of the area under cultivation (Alexandratos, 1995; Goklany, 1999; Waggoner and Ausubel, 2001). Malthusian concerns are misplaced because, as Max Singer once explained, "multiplying food production by five times over the next one hundred or two hundred years will be easier than multiplying it by over seven times as we did in the last two hundred years. No miracles, no scientific breakthroughs, no unknown lands or unexpected new resources, and no reforms of human character or government are required. All that is required is a continuing use of current evolutionary processes in technology and in economic dvelopment, and as much peace as we have had in the last century" (Singer, 1987).

We also agree with the sensible assessment of science writer Gregg Easterbrook that "the whole notion that there is a proper level of population for *Homo sapiens,* or for any species, would be nonsensical to nature" and "there is no reason in principle that the Earth cannot support vastly more human beings than live upon it today, with other species preserved and wild habitats remaining intact" (Easterbrook, 1995). Similar sentiments have been expressed by Waggoner (1995, 1996), Waggoner *et al.* (1996), and Meyer and Ausubel (1999).

Regardless of whether the goal of feeding a growing population while protecting nature is attainable, the question remains about global warming's role in this very real conflict. Tilman and a second set of collaborators, writing a year after their previously cited analysis, said "raising yields on existing farmland is essential for 'saving land for nature'" (Tilman *et al.* (2002). They proposed a three-part strategy: (1) increasing crop yield per unit of land area, (2) increasing crop yield per unit of nutrients applied, and (3) increasing crop yield per unit of water used.

With respect to the first of these efforts— increasing crop yield per unit of land area—the researchers note that in many parts of the world the historical rate-of-increase in crop yield is declining as the genetic ceiling for maximal yield potential is being approached. This "highlights the need for efforts to steadily increase the yield potential ceiling." With respect to the second effort—increasing crop yield per unit of nutrients applied—they note that "without the use of synthetic fertilizers, world food production could not have increased at the rate [that it did in the past] and more natural ecosystems would have been converted to agriculture." Hence, they say the solution "will require significant increases in nutrient use efficiency, that is, in cereal production per unit of added nitrogen." With respect to the third effort—increasing crop yield per unit of water used— Tilman *et al.* note that "water is regionally scarce," and that "many countries in a band from China through India and Pakistan, and the Middle East to North Africa either currently or will soon fail to have adequate water to maintain per capita food production from irrigated land."

The ongoing rise in the atmosphere's CO_2 concentration will help the world's farmers achieve all three parts of the Tilman strategy. First, since atmospheric CO_2 is the basic "food" of nearly all plants, the more of it there is in the air, the better they function and the more productive they become. As discussed in Section 9.2, a 300 ppm increase in the atmosphere's CO_2 concentration would increase the productivity of earth's herbaceous plants by 30 to 50 percent (Kimball, 1983; Idso and Idso, 1994) and the productivity of its woody plants by 50 to 80 percent (Saxe *et al.,* 1998; Idso and Kimball, 2001). These increases will be in addition to whatever yield gains are made possible by advances in plant genetics, pest control, and other agricultural practices. Consequently, as the air's CO_2 content continues to rise, so too will the land-use efficiency and productive capacity of the planet improve.

Regarding the second strategy, of increasing crop yield per unit of nutrients applied, many studies have investigated the effects of an increase in the air's CO_2 content on plants growing in soils with different nitrogen concentrations. (See Chapter 7, Section 7.3.7, for a thorough review of these studies.) These studies found that many plants increase their photosynthetic nitrogen-use efficiency when atmospheric CO_2 concentration is raised. For example, Smart *et al.* (1998) found wheat grown in controlled-environment chambers maintained at an atmospheric CO_2 concentration of 1,000 ppm increased average plant biomass by approximately 15 percent, irrespective of soil nitrogen content.

Zerihun *et al.* (2000) studied the effects of CO_2 enrichment on sunflowers using three different soil nitrogen concentrations and found whole plant biomass values that were 44, 13 and 115 percent greater than those of the plants growing in ambient air at low, medium and high levels of soil nitrogen, respectively. Deng and Woodward (1998) found that strawberries grown in high CO_2 environments produced 17 percent greater fresh fruit weight even when receiving the lowest levels of nitrogen fertilization. Newman *et al.* (2003) investigated the effects of two levels of nitrogen fertilization and an approximate doubling of the air's CO_2 concentration on the growth of tall fescue, an important forage crop. They found the plants grown in the high-CO_2 air and under low N conditions photosynthesized 15 percent more and produced 53 percent more dry matter (DM).

Demmers-Derks *et al.* (1998) grew sugar beets at atmospheric CO_2 concentrations of 360 and 700 ppm and high and low nitrogen treatment levels, and found the extra CO_2 enhanced total plant biomass by 13 percent even in plants receiving the low nitrogen treatments. Also working with sugar beets, Romanova *et al.* (2002) doubled atmospheric CO_2 concentrations while fertilizing plants with three different levels of nitrate-nitrogen. The plants exhibited rates of net photosynthesis that were approximately 50 percent greater than those displayed by the plants grown in ambient air, regardless of soil nitrate availability.

Fangmeier *et al.* (2000) grew barley plants in containers at atmospheric CO_2 concentrations of either 360 or 650 ppm and either a high or low nitrogen fertilization regime. The elevated CO_2 had the greatest relative impact on yield when the plants were grown under the less-than-optimum low-nitrogen regime, i.e., a 48 percent increase vs. 31 percent under high-nitrogen conditions.

Finally, the review and analysis of Kimball *et al.* (2002) of most FACE studies conducted on agricultural crops since the introduction of that technology back in the late 1980s found that in response to a 300-ppm increase in the air's CO_2 concentration, rates of net photosynthesis in several C_3 grasses were enhanced by an average of 46 percent under conditions of ample soil nitrogen supply and by 44 percent when nitrogen was limiting to growth. Clover experienced a 38 percent increase in belowground biomass production at ample soil nitrogen, and a 32 percent increase at low soil nitrogen. Wheat and ryegrass experienced an average increase of 18 percent at ample nitrogen, while wheat experienced only a 10 percent increase at low nitrogen.

Other studies have found that many species of plants respond to increases in the air's CO_2 content by increasing fine-root numbers and surface area, which tends to increase total nutrient uptake under CO_2-enriched conditions (Staddon *et al.*, 1999; Rouhier and Read, 1998; BassiriRad *et al.*, 1998; and Barrett *et al.*, 1998). This once again advances the Tilman strategy of increasing crop yield per unit of available nutrient. (See Chapter 7, Section 7.8.2, for a thorough review of those studies.)

Tilman's third strategy—increasing crop yield per unit of water used—is also advanced by rising levels of CO_2 in the atmosphere. Plants exposed to elevated levels of atmospheric CO_2 generally do not open their leaf stomatal pores—through which they take in carbon dioxide and give off water vapor—as wide as they do at lower CO_2 concentrations and tend to produce fewer of these pores per unit area of leaf surface. Both changes tend to reduce most plants' rates of water loss by transpiration. The amount of carbon they gain per unit of water lost—or water-use efficiency—therefore typically rises, increasing their ability to withstand drought.

In the study of Serraj *et al.* (1999), soybeans grown at 700 ppm CO_2 displayed 10 to 25 percent reductions in total water loss while simultaneously exhibiting increases in dry weight of as much as 33 percent. Likewise, Garcia *et al.* (1998) determined that spring wheat grown at 550 ppm CO_2 exhibited a water-use efficiency that was about one-third greater than that exhibited by plants grown at 370 ppm CO_2. Hakala *et al.* (1999) reported that twice-ambient CO_2 concentrations increased the water-use efficiency of spring wheat by 70 to 100 percent, depending on experimental air temperature.

Hunsaker *et al.* (2000) reported CO_2-induced increases in water-use efficiency for field-grown wheat that were 20 and 10 percent higher than those displayed by ambiently grown wheat subjected to high and low soil nitrogen regimes, respectively. Also, pea plants grown for two months in growth chambers receiving atmospheric CO_2 concentrations of 700 ppm displayed an average water-use efficiency that was 27 percent greater than that exhibited by ambiently grown control plants (Gavito *et al.*, 2000). (See Chapter 7, Section 7.2, for a thorough review of those studies.)

An issue related to water-use efficiency that could become more important in the future is the buildup of soil salinity from repeated irrigations, which can sometimes reduce crop yields. Similarly, in natural ecosystems where exposure to brackish or salty water is commonplace, saline soils can induce growth stress in plants not normally adapted to coping with this problem. The studies reported below show that rising atmospheric CO_2 concentrations also can help to alleviate this problem.

Mavrogianopoulos *et al.* (1999) reported that atmospheric CO_2 concentrations of 800 and 1200 ppm stimulated photosynthesis in parnon melons by 75 and 120 percent, respectively, regardless of soil salinity, which ranged from 0 to 50 mM NaCl. Atmospheric CO_2 enrichment also partially alleviated the negative effects of salinity on melon yield, which increased with elevated CO_2 at all salinity levels.

Maggio *et al.* (2002) grew tomatoes at 400 and 900 ppm in combination with varying degrees of soil salinity and noted that plants grown in elevated CO_2 tolerated an average root-zone salinity threshold value that was about 60 percent greater than that exhibited by plants grown at 400 ppm CO_2 (51 vs. 32 mmol dm$^-$^{3} Cl). The review of Poorter and Perez-Soba (2001) found no changes in the effect of elevated CO_2 on the growth responses of most plants over a wide range of soil salinities, in harmony with the earlier findings of Idso and Idso (1994).

These various studies suggest that elevated CO_2 concentrations will help farmers achieve all three of the strategies Tilman *et al.* say are essential to addressing the conflict between feeding a growing human population and preserving space for nature. The actual degree of crop yield enhancement likely to be provided by the increase in atmospheric CO_2 concentration expected to occur between 2000 and 2050 has been calculated by Idso and Idso (2000) to be sufficient—but just barely—to close the gap between the supply and demand for food some four decades from now. Consequently, letting the evolution of technology take its course—which includes continued emissions of CO_2 into the atmosphere by industry—appears to be the only way we can grow enough food to support ourselves in the year 2050 without taking unconscionable amounts of land and freshwater resources from nature.

In spite of the dilemma described above and the fact that enhanced levels of CO_2, in the air are a necessary part of the solution, the IPCC calls for strict measures to reduce anthropogenic CO_2 emissions—a strategy that, if it has any effect at all on plant and animal life, would lead to lower land-use efficiency, lower nitrogen-use efficiency, and lower plant water-use efficiency, just the opposite of what Tilman *et al.* called for.

One might ask whose predictions are more reliable, the IPCC's computer-model-generated forecasts of catastrophic consequences due to rising temperatures a century or longer from now, or our projections of human population growth and agricultural productivity just four decades into the future? In addition to the obvious time differential between the two sets of predictions, human population growth and agricultural productivity are much better-understood processes than is global climate change, which involves a host of complex phenomena that span a spatial scale of fully 14 orders of magnitude, ranging from the planetary scale of 10^7 meters to the cloud microphysical scale of 10^{-6} meter.

Many of the component processes that comprise today's state-of-the-art climate models are so far from adequately understood (see Chapters 1 and 2) that even the *signs* of their impacts on global temperature change (whether positive or negative) are not yet known. Consequently, in light of the much greater confidence that can realistically be vested in demographic and agricultural production models, it would seem that much greater credence can be placed in our predictions than in the predictions of climate doom.

In conclusion, the aerial fertilization effect of the increase in the air's CO_2 content that is expected to occur by the year 2050 would boost crop yields by the amounts required to prevent mass starvation in many parts of the globe, without a large-scale encroachment on the natural world. Acting prematurely to reduce human CO_2 emissions, as urged by the IPCC, could interrupt this response, resulting in the death by starvation of millions of people, loss of irreplaceable natural ecosystems, or both.

Additional information on this topic, including reviews of newer publications as they become available, can be found at http://www.co2science.org/subject/f/food.php.

References

Alexandratos, N. 1995. *World Agriculture: Towards 2010.* John Wiley & Sons. New York, NY.

Ball, M.C., Cochrane, M.J. and Rawson, H.M. 1997. Growth and water use of the mangroves *Rhizophora apiculata* and *R. stylosa* in response to salinity and humidity under ambient and elevated concentrations of atmospheric CO_2. *Plant, Cell and Environment* **20**: 1158-1166.

Barrett, D.J., Richardson, A.E. and Gifford, R.M. 1998. Elevated atmospheric CO_2 concentrations increase wheat root phosphatase activity when growth is limited by phosphorus. *Australian Journal of Plant Physiology* **25**: 87-93.

BassiriRad, H., Reynolds, J.F., Virginia, R.A. and Brunelle, M.H. 1998. Growth and root NO^{3-} and PO_4^{3-} uptake capacity of three desert species in response to atmospheric CO_2 enrichment. *Australian Journal of Plant Physiology* **24**: 353-358.

Borlaug, N. 2007. Feeding a hungry world. *Science* **318**: 359.

Conley, M.M., Kimball, B.A., Brooks, T.J., Pinter Jr., P.J., Hunsaker, D.J., Wall, G.W., Adams, N.R., LaMorte, R.L., Matthias, A.D., Thompson, T.L., Leavitt, S.W., Ottman, M.J., Cousins, A.B. and Triggs, J.M. 2001. CO_2 enrichment increases water-use efficiency in sorghum. *New Phytologist* **151**: 407-412.

De Luis, J., Irigoyen, J.J. and Sanchez-Diaz, M. 1999. Elevated CO_2 enhances plant growth in droughted N_2-fixing alfalfa without improving water stress. *Physiologia Plantarum* **107**: 84-89.

Demmers-Derks, H., Mitchell, R.A.G., Mitchell, V.J. and Lawlor, D.W. 1998. Response of sugar beet (*Beta vulgaris* L.) yield and biochemical composition to elevated CO_2 and temperature at two nitrogen applications. *Plant, Cell and Environment* **21**: 829-836.

Deng, X. and Woodward, F.I. 1998. The growth and yield responses of *Fragaria ananassa* to elevated CO_2 and N supply. *Annals of Botany* **81**: 67-71.

Easterbrook, G. 1995. *A Moment on the Earth*. Penguin Books, New York, NY.

Ehrlich, P.R. and Ehrlich, A.H. 2008. *The Dominant Animal: Human Evolution and the Environment*. Island Press, Washington, DC.

Fangmeier, A., Chrost, B., Hogy, P. and Krupinska, K. 2000. CO_2 enrichment enhances flag leaf senescence in barley due to greater grain nitrogen sink capacity. *Environmental and Experimental Botany* **44**: 151-164.

Garcia, R.L., Long, S.P., Wall, G.W., Osborne, C.P., Kimball, B.A., Nie, G.Y., Pinter Jr., P.J., LaMorte, R.L. and Wechsung, F. 1998. Photosynthesis and conductance of spring-wheat leaves: field response to continuous free-air atmospheric CO_2 enrichment. *Plant, Cell and Environment* **21**: 659-669.

Gavito, M.E., Curtis, P.S., Mikkelsen, T.N. and Jakobsen, I. 2000. Atmospheric CO_2 and mycorrhiza effects on biomass allocation and nutrient uptake of nodulated pea (*Pisum sativum* L.) plants. *Journal of Experimental Botany* **52**: 1931-1938.

Goklany, I.M. 1999. Meeting global food needs: the environmental trade-offs between increasing land conversion and land productivity. *Technology* **6**: 107-130.

Gore, A. 1992. *Earth in the Balance*. Houghton Mifflin, New York, NY.

Hakala, K., Helio, R., Tuhkanen, E. and Kaukoranta, T. 1999. Photosynthesis and Rubisco kinetics in spring wheat and meadow fescue under conditions of simulated climate change with elevated CO_2 and increased temperatures. *Agricultural and Food Science in Finland* **8**: 441-457.

Huang, J., Pray, C. and Rozelle, S. 2002. Enhancing the crops to feed the poor. *Nature* **418**: 678-684.

Hunsaker, D.J., Kimball. B.A., Pinter Jr., P.J., Wall, G.W., LaMorte, R.L., Adamsen, F.J., Leavitt, S.W., Thompson, T.L., Matthias, A.D. and Brooks, T.J. 2000. CO_2 enrichment and soil nitrogen effects on wheat evapotranspiration and water use efficiency. *Agricultural and Forest Meteorology* **104**: 85-105.

Idso, C.D. and Idso, K.E. 2000. Forecasting world food supplies: The impact of the rising atmospheric CO_2 concentration. *Technology* **7S**: 33-55.

Idso, K.E. and Idso, S.B. 1994. Plant responses to atmospheric CO_2 enrichment in the face of environmental constraints: a review of the past 10 years' research. *Agricultural and Forest Meteorology* **69**: 153-203.

Idso, S.B. and Kimball, B.A. 2001. CO_2 enrichment of sour orange trees: 13 years and counting. *Environmental and Experimental Botany* **46**: 147-153.

Kimball, B.A. 1983. Carbon dioxide and agricultural yield: An assemblage and analysis of 430 prior observations. *Agronomy Journal* **75**: 779-788.

Kimball, B.A., Idso, S.B., Johnson, S. and Rillig, M.C. 2007. Seventeen years of carbon dioxide enrichment of sour orange trees: final results. *Global Change Biology* **13**: 2171-2183.

Kimball, B.A., Kobayashi, K. and Bindi, M. 2002. Responses of agricultural crops to free-air CO_2 enrichment. *Advances in Agronomy* **77**: 293-368.

Maggio, A., Dalton, F.N. and Piccinni, G. 2002. The effects of elevated carbon dioxide on static and dynamic indices for tomato salt tolerance. *European Journal of Agronomy* **16**: 197-206.

Malmstrom, C.M. and Field, C.B. 1997. Virus-induced differences in the response of oat plants to elevated carbon dioxide. *Plant, Cell and Environment* **20**: 178-188.

Malthus, T. 1798. *Essay on the Principles of Population*. Cambridge University Press, Cambridge, UK. (Reprint 1992).

Maroco, J.P., Edwards, G.E. and Ku, M.S.B. 1999. Photosynthetic acclimation of maize to growth under elevated levels of carbon dioxide. *Planta* **210**: 115-125.

Mavrogianopoulos, G.N., Spanakis, J. and Tsikalas, P. 1999. Effect of carbon dioxide enrichment and salinity on photosynthesis and yield in melon. *Scientia Horticulturae* **79**: 51-63.

Nasholm, T., Ekblad, A., Nordin, A., Giesler, R., Hogberg, M. and Hogberg, P. 1998. Boreal forest plants take up organic nitrogen. *Nature* **392**: 914-916.

Newman, J.A., Abner, M.L., Dado, R.G., Gibson, D.J., Brookings, A. and Parsons, A.J. 2003. Effects of elevated CO_2, nitrogen and fungal endophyte-infection on tall fescue: growth, photosynthesis, chemical composition and digestibility. *Global Change Biology* **9**: 425-437.

Meyer, P.S. and Ausubel, J.H. 1999. Carrying capacity: a model with logistically varying limits. *Technological Forecasting & Social Change* **61**(3): 209-214.

Poorter, H. and Perez-Soba, M. 2001. The growth response of plants to elevated CO_2 under non-optimal environmental conditions. *Oecologia* **129**: 1-20.

Raven, P.H. 2002. Science, sustainability, and the human prospect. *Science* **297**: 954-959.

Romanova, A.K., Mudrik, V.A., Novichkova, N.S., Demidova, R.N. and Polyakova, V.A. 2002. Physiological and biochemical characteristics of sugar beet plants grown at an increased carbon dioxide concentration and at various nitrate doses. *Russian Journal of Plant Physiology* **49**: 204-210.

Rouhier, H. and Read, D.J. 1998. The role of mycorrhiza in determining the response of *Plantago lanceolata* to CO_2 enrichment. *New Phytologist* **139**: 367-373.

Saxe, H., Ellsworth, D.S. and Heath, J. 1998. Tree and forest functioning in an enriched CO_2 atmosphere. *New Phytologist* **139**: 395-436.

Serraj, R., Allen Jr., L.H. and Sinclair, T.R. 1999. Soybean leaf growth and gas exchange response to drought under carbon dioxide enrichment. *Global Change Biology* **5**: 283-291.

Singer, M. 1987. *Passage to a Human World.* Hudson Institute, Indianapolis, IN.

Smart, D.R., Ritchie, K., Bloom, A.J. and Bugbee, B.B. 1998. Nitrogen balance for wheat canopies (*Triticum aestivum* cv. Veery 10) grown under elevated and ambient CO_2 concentrations. *Plant, Cell and Environment* **21**: 753-763.

Staddon, P.L., Fitter, A.H. and Graves, J.D. 1999. Effect of elevated atmospheric CO_2 on mycorrhizal colonization, external mycorrhizal hyphal production and phosphorus inflow in *Plantago lanceolata* and *Trifolium repens* in association with the arbuscular mycorrhizal fungus *Glomus mosseae*. *Global Change Biology* **5**: 347-358.

Tilman, D., Cassman, K.G., Matson, P.A., Naylor, R. and Polasky, S. 2002. Agricultural sustainability and intensive production practices. *Nature* **418**: 671-677.

Tilman, D., Fargione, J., Wolff, B., D'Antonio, C., Dobson, A., Howarth, R., Schindler, D., Schlesinger, W.H., Simberloff, D. and Swackhamer, D. 2001. Forecasting agriculturally driven global environmental change. *Science* **292**: 281-284.

United Nations. 2009. World population prospects. The 2008 revision. Department of Economic and Social Affairs, Population Division. ESA/P/WP.210.

U.S. Census Bureau. 2008. Global births increase even as fertility rates decline. [News release] International Population Data Base. 15 Dec.

Waggoner, P.E. 1995. How much land can ten billion people spare for nature? Does technology make a difference? *Technology in Society* **17**: 17-34.

Waggoner, P.E. 1996. Earth's carrying capacity. *Science* **274** (5287): 481-485.

Waggoner, P.E., Ausubel, J.H., and Wernick, I.K. 1996. Lightening the tread of population on the land: American examples. *Population and Development Review* **22** (3): 531-545.

Waggoner, P.E. and Ausubel, P.E. 2001. How much will feeding more and wealthier people encroach on forests? *Population and Development Review* **27**(2): 239–257 .

Wallace, J.S. 2000. Increasing agricultural water use efficiency to meet future food production. *Agriculture, Ecosystems & Environment* **82**: 105-119.

Zerihun, A., Gutschick, V.P. and BassiriRad, H. 2000. Compensatory roles of nitrogen uptake and photosynthetic N-use efficiency in determining plant growth response to elevated CO_2: Evaluation using a functional balance model. *Annals of Botany* **86**: 723-730.

Zhu, J., Goldstein, G. and Bartholomew, D.P. 1999. Gas exchange and carbon isotope composition of *Ananas comosus* in response to elevated CO_2 and temperature. *Plant, Cell and Environment* **22**: 999-1007.

9.5. Biofuels

Biofuels are liquid and gaseous fuels made from organic matter. They include ethanol, biodiesel, and methanol. Biofuels may have some advantages over gasoline and diesel fuels, but they are more expensive to produce and can supply only a small part of the world's total transportation energy needs. Because they compete with food crops and nature for land and nutrients, expanding the use of biofuels could negatively affect human health and natural ecosystems.

The IPCC does not discuss biofuels in the contributions of Group I (Science) or Group II (Impacts, Adaptation and Vulnerability) to the Fourth Assessment Report. When it finally does discuss them, in two sections of the contribution of Group III (Mitigation), it fails to address the likely adverse consequences of increased use of biofuels on human health and the natural environment. We discuss those consequences in this section.

9.5.1. About Biofuels

Biofuels are not new—Henry Ford's first vehicle was fueled by ethanol—and conversion technologies exist or are in development for converting biomass into a wide range of biofuels suitable for heating, electric production, and transportation. For example, residues from agriculture and forestry long have been used by the lumber and papermaking industries to generate heat and power. Methane from animal waste and composting is captured and used locally or sold in commercial markets.

Of particular interest, and the focus of this section, is the biochemical conversion using enzymes of corn, soybeans, sugarcane, and other food crops into ethanol, biodiesel, and other biofuels used mainly for transportation. The country with the most aggressive biofuels program in the world is Brazil. After the country launched its National Alcohol Program in 1975, ethanol production in Brazil rose dramatically and now accounts for approximately 40 percent of total fuel consumption in the country's passenger vehicles (EIA, 2008).

Ethanol became popular as a gasoline supplement in the U.S. during the 1990s, when Congress mandated that oil refiners add oxygenates to their product to reduce some emissions. Congress did not provide liability protection for the makers of methyl tertiary butyl ether (MTBE), ethanol's main competitor in the oxygenate business, so most companies quickly switched from MTBE to ethanol (Lehr, 2006). Some states also began to mandate ethanol use for reasons discussed below.

Most ethanol made in the U.S. comes from corn. Its production consumed 13 percent of the U.S. corn crop (1.43 billion bushels of corn grain) in 2005 and an estimated 20 percent of the 2006 crop. E10 (a blend of 10 percent ethanol and 90 percent gasoline) is widely available. E85 is an alternative fuel (85 percent ethanol and 15 percent gasoline) available mainly in corn-producing states; vehicles must be modified to use this fuel.

The Energy Policy Act of 2005 mandated the use of 4 billion gallons of ethanol in 2006. The 2007 Energy Independence and Security Act (EISA) subsequently mandated the use of 36 billion gallons of renewable fuels by 2022—16 billion gallons of cellulosic ethanol, 15 billion gallons of corn ethanol, and 5 billion gallons of biodiesel and other advanced biofuels (U.S. Congress, 2007).

Federal subsidies to ethanol producers in the U.S. cost taxpayers about $2 billion a year (Dircksen, 2006). Congress protects domestic ethanol producers by imposing a 2.5 percent tariff and 54 cents per gallon duty on imports. Ethanol producers with plants of up to 60 million gallons annual production capacity are eligible to receive a production incentive of 10 cents per gallon on the first 15 million gallons of ethanol produced each year. Ethanol is also subsidized by scores of other countries and by at least 19 U.S. states (Doornbosch and Steenblick, 2007, Annex 1, pp. 45-47).

U.S. ethanol output rose from 3.4 billion gallons from 81 facilities in 2004 to 9 billion gallons from 170 facilities in 2008 (RFA, 2009). According to a forecast by the Energy Information Administration (EIA), "total U.S. biofuel consumption rises from 0.3 quadrillion Btu (3.7 billion gallons) in 2005 to 2.8 quadrillion Btu (29.7 billion gallons) in 2030, when it represents about 11.3 percent of total U.S. motor vehicle fuel on a Btu basis" (EIA, 2008). In 2005 ethanol represented about 2 percent of total gasoline consumption, and biodiesel less than 0.2 percent of diesel consumption, in the U.S.

Doornbosch and Steenblick (2007), in a report produced for the Organization for Economic Cooperation and Development (OECD), reported that "global production of biofuels amounted to 0.8 EJ [exajoule] in 2005, or roughly 1% of total road transport fuel consumption. Technically, up to 20 EJ from conventional ethanol and biodiesel, or 11% of total demand for liquid fuels in the transport sector, has been judged possible by 2050." Also for the world as a whole, EIA predicts "alternative fuels [will] account for only 9 percent of total world liquids use in 2030, despite an average annual increase of 5.6 percent per year, from 2.5 million barrels per day in 2005 to 9.7 million barrels per day in 2030" (EIA, 2008).

References

Dircksen, J. 2006. Ethanol: bumper crop for agribusiness, bitter harvest for taxpayers. *Policy Paper* #121. National Taxpayers Union. July 20.

Doornbosch, R. and Steenblick, R. 2007. Biofuels: is the cure worse than the disease? Organization for Economic Cooperation and Development. Paris.

EIA. 2008. International energy outlook 2008. Energy Information Administration. Report #:DOE/EIA-0484(2008).

Lehr, J.H. 2006. Are the ethanol wars over? *PERC Reports*, Property and Environment Research Center. March.

RFA. 2009. Growing innovation: America's energy future starts at home. 2009 ethanol industry outlook. Renewable Fuels Association. Washington, DC.

U.S. Congress. 2007. Energy Independence and Security Act of 2007.

9.5.2. Costs and Benefits

Proponents of biofuels say their increased production will increase the supply of transportation fuels and therefore lead to lower prices. Critics of biofuels point out that ethanol often costs more, not less, than gasoline, either because of production costs or supplies that can't keep pace with government mandates, and therefore leads to higher prices at least in the short run.

Ethanol has only two-thirds the energy content of gasoline, which makes it a poor value for most consumers. The production cost of ethanol (which is only one component in determining its price) has fallen as a result of technological innovation and economies of scale, but some properties of ethanol continue to make it expensive compared to gasoline. Transportation costs for ethanol, for example, are high because it picks up water if it travels through existing pipelines, diluting the ethanol and corroding the pipelines. Therefore, it is being trucked to the Northeast and along the Gulf Coast. Ethanol must be kept in a different container at the terminal and is blended into the gasoline in the truck on its way to the retailer from the terminal. This has caused regional shortages, further increasing the retail prices in these areas (Dircksen, 2006).

Ethanol also has been promoted as a fuel additive to reduce emissions. It reduces carbon monoxide in older vehicles and dilutes the concentration of aromatics in gasoline, reducing emissions of toxins such as benzene. Because ethanol has only two-thirds the energy content per volume as gasoline, it increases volumetric fuel use (with small increases in energy efficiency.) Ethanol increases air emissions such as aldehydes. In some areas, the use of 10 percent ethanol blends may increase ozone due to local atmospheric conditions (Niven, 2004).

Ethanol also is promoted as a "homegrown" and renewable energy source, so using more of it could help reduce a country's dependency on foreign oil, which in turn might benefit national security and international relations. But ethanol used in the U.S. mostly supplants oil from domestic suppliers, which is more expensive than foreign oil, and leaves the country's dependency on foreign oil the same or even makes it higher (Yacobucci, 2006). Rural communities benefit from the economic boost that comes from higher prices for corn and the jobs created by ethanol plants, but those economic benefits come at a high price in terms of higher food prices and tax breaks financed by government debt or higher taxes on other goods and services.

Finally, biofuels are renewable resources, which advocates say makes them environmentally friendlier than fossil fuels. But the energy consumed to make biofuels—to plant, fertilize, irrigate, and harvest corn and other feedstocks as well as to generate the heat used during the fermentation process and to transport biofuels to markets by train or trucks—is considerable. Fossil fuels (natural gas or coal) are typically the source of that energy. This environmental impact is the focus of the rest of this section.

References

Dircksen, J. 2006. Ethanol: bumper crop for agribusiness, bitter harvest for taxpayers. *Policy Paper* #121. National Taxpayers Union. July 20.

Niven, R.K. 2005. Ethanol in gasoline: environmental impacts and sustainability, review article. *Renewable & Sustainable Energy Reviews* **9**(6): 535-555.

Yacobucci, B.D. 2006. Fuel ethanol: background and public policy issues. Congressional Research Service. March 3.

9.5.3. Net Emissions

The US 2007 Energy Independence and Security Act (EISA) mandates that life-cycle greenhouse gas emissions of corn ethanol, cellulosic ethanol, and advanced biofuels achieve 20 percent, 60 percent, and 50 percent greenhouse gas (GHG) emission reductions relative to gasoline, respectively. But there is considerable controversy over whether these fuels do in fact reduce GHG emissions.

Numerous studies of GHG emissions produced during the life-cycle of ethanol (from the planting of crops to consumption as a fuel) have found them to be less than those of gasoline, with most estimates around 20 percent (Hill *et al.*, 2006; Wang *et al.*, 2007; CBO, 2009). Emissions vary considerably based on the choice of feedstock, production process, type of fossil fuels used, location, and other factors (ICSU, 2009). Liska *et al.* (2009), in their study of life-cycle emissions of corn ethanol systems, found the direct-effect GHG emissions of ethanol (without any offset due to changes in land use) to be "equivalent to a 48% to 59% reduction compared to gasoline, a twofold to threefold greater reduction than reported in previous studies," largely because they incorporate a credit for the commercial use of dry distilled grain (DDG). They report that "in response to the large increase in availability of distillers grains coproduct from ethanol production and the rise in soybean prices, cattle diets now largely exclude soybean meal and include a larger proportion of distillers grains coproduct (Klopfenstein *et al.*, 2008). Thus, the energy and GHG credits attributable to feeding distillers grains must be based on current practices for formulating cattle diets." They give corn ethanol systems DDG credits ranging from 19% to 38% depending on region and type of fossil fuels used.

None of these estimates, however, takes into account the emission increases likely to come about from land-use changes. Righelato and Spracklen (2007) wrote that using ethanol derived from crops as a substitute for gasoline, and vegetable oils in place of diesel fuel, "would require very large areas of land in order to make a significant contribution to mitigation of fossil fuel emissions and would, directly or indirectly, put further pressure on natural forests and grasslands." The two British scientists calculated that a 10 percent substitution of biofuels for gasoline and diesel fuel would require "43% and 38% of current cropland area in the United States and Europe, respectively," and that "even this low substitution level cannot be met from existing arable land."

Righelato and Spracklen add that "forests and grasslands would need to be cleared to enable production of the energy crops," resulting in "the rapid oxidation of carbon stores in the vegetation and soil, creating a large up-front emissions cost that would, in all cases examined, out-weigh the avoided emissions." They report further that individual life-cycle analyses of the conversion of sugar cane, sugar beet, wheat, and corn to ethanol, as well as the conversion of rapeseed and woody biomass to diesel, indicate that "forestation of an equivalent area of land would sequester two to nine times more carbon over a 30-year period than the emissions avoided by the use of the biofuel." They conclude that "the emissions cost of liquid biofuels exceeds that of fossil fuels."

Fargione *et al.* (2008), writing in *Science*, said "increasing energy use, climate change, and carbon dioxide (CO_2) emissions from fossil fuels make switching to low-carbon fuels a high priority. Biofuels are a potential low-carbon energy source, but whether biofuels offer carbon savings depends on how they are produced." They explain that "converting native habitats to cropland releases CO_2 as a result of burning or microbial decomposition of organic carbon stored in plant biomass and soils. After a rapid release from fire used to clear land or from the decomposition of leaves and fine roots, there is a prolonged period of GHG release as coarse roots and branches decay and as wood products decay or burn. We call the amount of CO_2 released during the first 50 years of this process the 'carbon debt' of land conversion. Over time, biofuels from converted land can repay this carbon debt if their production and combustion have net GHG emissions that are less than the life-cycle emissions of the fossil fuels they displace. Until the carbon debt is repaid, biofuels from converted lands have greater GHG impacts than those of the fossil fuels they displace."

Fargione *et al.* calculate the number of years required to repay carbon debts for six areas: Brazilian Amazon (319 years), Brazilian Cerrado wooded (17 years), Brazilian Cerrado grassland (37 years), Indonesian or Malaysian lowland tropical rainforest

(86 years), Indonesian or Malaysian peatland tropical rainforest (423 years), and U.S. central grassland (93 years). They observe that no carbon debt is incurred when abandoned cropland or marginal prairie in the U.S. is used without irrigation to produce ethanol. They conclude that "the net effect of biofuels production via clearing of carbon-rich habitats is to increase CO_2 emissions for decades or centuries relative to the emissions caused by fossil fuel use," and "at least for current or developing biofuels technologies, any strategy to reduce GHG emissions that causes land conversion from native ecosystems to cropland is likely to be counterproductive."

In a companion essay in the same issue of *Science,* Searchinger *et al.* (2008) also describe the carbon debt due to land-use conversion, but measure it as the difference between biofuels and gasoline in GHG emissions measured in grams per MJ (megajoule) of energy. They begin by explaining that "to produce biofuels, farmers can directly plow up more forest or grassland, which releases to the atmosphere much of the carbon previously stored in plants and soils through decomposition or fire. ... Alternatively, farmers can divert existing crops or croplands into biofuels, which causes similar emissions indirectly. The diversion triggers higher crop prices, and farmers around the world respond by clearing more forest and grassland to replace crops for feed and food."

Searchinger *et al.* used the Greenhouse gases Regulated Emissions and Energy use in Transportation (GREET) computer program created by the Center for Transportation Research at Argonne National Laboratory to calculate the GHGs in grams of CO_2 equivalent emissions per MJ of energy consumed over the production and use life-cycles of gasoline, corn ethanol, and biomass ethanol fuels. They observe that "emissions from corn and cellulosic ethanol emissions exceed or match those from fossil fuels, and therefore produce no greehouse benefits," unless biofuels are given a "carbon uptake credit" for the amount of carbon dioxide removed from the air by the growing biofuels feedstocks. When that adjustment is made, they estimate that gasoline (which gets no carbon uptake credit) produces 92g/MJ; corn ethanol, 74g/MJ; and biomass ethanol, 27g/MJ.

Searchinger *et al.* then calculate the amount of land that would be converted from forest and grassland into cropland to support the biofuels and, like Fargione *et al.* (2008), apply the GHG emissions

due to land-use change to each type of fuel. The result is that total net GHG emissions from both kinds of biofuel *exceed* those from gasoline, 177g vs. 92g in the case of corn ethanol and 138g vs. 92g in the case of biomass ethanol. They conclude that "corn-based ethanol, instead of producing a 20% savings, nearly doubles greenhouse emissions over 30 years and increases greenhouse gases for 167 years. Biofuels from switchgrass, if grown on U.S. corn lands, increase emissions by 50%. This result raises concerns about large biofuels mandates and highlights the value of using waste products."

Coming to much the same conclusion, Laurance (2007) observed that "tropical forests, in particular, are crucial for combating global warming, because of their high capacity to store carbon and their ability to promote sunlight-reflecting clouds via large-scale evapotranspiration," which led him to conclude that "such features are key reasons why preserving and restoring tropical forests could be a better strategy for mitigating the effects of carbon dioxide than dramatically expanding global biofuel production."

Doornbosch and Steenblick (2007), while reporting that biofuels could provide up to 11 percent of the total world demand for road transport fuel by 2050, say "an expansion on this scale could not be achieved, however, without significant impacts on the wider global economy. In theory there might be enough land available around the globe to feed an ever increasing world population and produce sufficient biomass feedstock simultaneously, but it is more likely that land-use constraints will limit the amount of new land that can be brought into production leading to a 'food-versus-fuel' debate."

Looking at a different environmental impact of expanded biofuel production, Crutzen and three collaborators calculated the amount of nitrous oxide (N_2O) that would be released to the atmosphere as a result of using nitrogen fertilizer to produce the crops used for biofuels (Crutzen *et al.*, 2007). Their work revealed that "all past studies have severely underestimated the release rates of N_2O to the atmosphere, with great potential impact on climate warming" because, as they report, N_2O "is a 'greenhouse gas' with a 100-year average global warming potential 296 times larger than an equal mass of CO_2." The consequence is that "when the extra N_2O emission from biofuel production is calculated in 'CO_2-equivalent' global warming terms, and compared with the quasi-cooling effect of 'saving' emissions of CO_2 derived from fossil fuel, the outcome is that the production of commonly used biofuels, such as biodiesel from rapeseed and bioethanol from corn, can contribute as much or more to global warming by N_2O emissions than cooling by fossil fuel savings."

Crutzen *et al.* concluded that "on a globally averaged basis the use of agricultural crops for energy production … can readily be detrimental for climate due to the accompanying N_2O emissions." Their concerns were confirmed by a 2009 report from the International Council for Science (ICSU), which found "the increased N_2O flux associated with producing ethanol from corn is likely to more than offset any positive advantage from reduced carbon dioxide fluxes (compared to burning fossil fuels). Even for ethanol from sugar cane or biodiesel from rapeseed, emissions of nitrous oxide probably make these fuels less effective as an approach for reducing global warming than has been previously believed" (ICSU, 2009).

Producing ethanol from crop residues, or stover, is often proposed as a way to avoid carbon emissions arising from land conversion. But as Lal (2007) points out, crop residues perform many vital functions. He reports that "there are severe adverse impacts of residue removal on soil and environmental degradation, and negative carbon sequestration as is documented by the dwindling soil organic carbon reserves." He notes that "the severe and widespread problem of soil degradation, and the attendant agrarian stagnation/deceleration, are caused by indiscriminate removal of crop residues." Lal concludes that "short-term economic gains from using crop residues for biofuel must be objectively assessed in relation to adverse changes in soil quality, negative nutrients and carbon budget, accelerated erosion, increase in non-point source pollution, reduction in agronomic production, and decline in biodiversity."

Finally, while using abandoned or degraded lands to produce biomass, rather than converting existing cropland or forests, is often alleged to reduce carbon emissions (e.g., Fargione *et al.,* 2008), the ICSU report notes that "of course, if the lands have the potential to revert to forests, conversion to biofuels represents a lost opportunity for carbon storage. The environmental consequences of inputs (irrigation water, fertilizer) required to make degraded and marginal lands productive must also be considered" (ICSU, 2009).

In conclusion, the production and use of biofuels frequently does not reduce net GHG emissions relative to gasoline, the fossil fuel they are intended to replace. Therefore, there is no basis from an environmental perspective for preferring them to fossil fuels.

References

CBO. 2009. The impact of ethanol use on food prices and greenhouse-gas emissions. Congressional Budget Office. April.

Crutzen, P.J., Mosier, A.R., Smith, K.A. and Winiwarter, W. 2007. N_2O release from agro-biofuel production negates global warming reduction by replacing fossil fuels. *Atmospheric Chemistry and Physics Discussions* **7**: 11,191-11,205.

Doornbosch, R. and Steenblick, R. 2007. Biofuels: is the cure worse than the disease? Organization for Economic Cooperation and Development. Paris.

Fargione, J., Hill, J., Tilman, D., Polasky, S. and Hawthorne, P. 2008. Land clearing and the biofuels carbon debt. *Science* **319**: 1235-1237.

Hill, J., Nelson, E., Tilman, D., Polasky, S. and Tiffany, D. 2006. Environmental, economic, and energetic costs and benefits of biodiesel and ethanol biofuels. *Proceedings of the National Academy of Sciences* **103** (30): 11206-11210.

ICSU. 2009. Biofuels: environmental consequences and interactions with changing land use. Proceedings of the Scientific Committee on Problems of the Environment (SCOPE) International Biofuels Project Rapid Assessment, International Council for Science (ICSU). 22-25 September 2008, Gummersbach, Germany. R.W. Howarth and S. Bringezu, eds.

Lal, R. 2007. Farming carbon. *Soil & Tillage Research* **96**: 1-5.

Laurance, W.F. 2007. Forests and floods. *Nature* **449**: 409-410.

Liska, A.J, Yang, H.S., Bremer, V.R., Klopfenstein, T.J., Walters, D.T., Galen, E.E. and Cassman, K.G. 2009. Improvements in life cycle energy efficiency and greenhouse gas emissions of corn-ethanol. *Journal of Industrial Ecology* **13** (1).

Righelato, R. and Spracklen, D.V. 2007. Carbon mitigation by biofuels or by saving and restoring forests? *Science* **317**: 902.

Searchinger, T., Heimlich, R., Houghton, R.A., Dong, F., Elobeid, A., Fabiosa, J., Tokgoz, S., Hayes, D. and Yu, T-H. 2008. *Science* **319**: 1238-1239.

Wang, M., Wu, M., and Huo, H. 2007. Life-cycle energy and greenhouse gas emission impacts of different corn ethanol plant types. *Environmental Research Letters* **2** (2): 024001.

9.5.4. Impact on Food Prices

Biofuel refineries compete with livestock growers and food processors for corn, soybeans, and other feedstocks usually used to produce biofuels in the United States, leading to higher animal feed and ingredient costs for farmers, ranchers, and food manufacturers. Some of that cost is eventually passed on to consumers. A study by the Congressional Budget Office (CBO) found "the demand for corn for ethanol production, along with other factors, exerted upward pressure on corn prices, which rose by more than 50 percent between April 2007 and April 2008. Rising demand for corn also increased the demand for cropland and the price of animal feed" (CBO, 2009). The CBO estimated that increased use of ethanol "contributed between 0.5 and 0.8 percentage points of the 5.1 percent increase in food prices measured by the consumer price index (CPI)."

Johansson and Azar (2007) analyzed what they called the "food-fuel competition for bio-productive land," developing in the process "a long-term economic optimization model of the U.S. agricultural and energy system," wherein they found that the competition for land to grow crops for both food and fuel production leads to a situation where "prices for all crops as well as animal products increase substantially." Similarly, Doornbosch and Steenblick (2007) say "any diversion of land from food or feed production to production of energy biomass will influence food prices from the start, as both compete for the same inputs. The effects on farm commodity prices can already be seen today. The rapid growth of the biofuels industry is likely to keep these prices high and rising throughout at least the next decade (OECD/FAO, 2007)."

Runge and Senauer (2007), writing in *Foreign Affairs*, reported that the production of corn-based ethanol in the United States "takes so much supply to keep ethanol production going that the price of corn—and those of other food staples—is shooting up around the world." The rising prices caused food riots to break out in Haiti, Bangladesh, Egypt, and Mozambique in April 2008, prompting Jean Ziegler, the United Nations' "special rapporteur on the right to food," to call using food crops to create ethanol "a crime against humanity" (CNN, 2008). Jeffrey Sachs, director of Columbia University's Earth Institute, said

at the time, "We've been putting our food into the gas tank—this corn-to-ethanol subsidy which our government is doing really makes little sense" (Ibid.). Former U.S. President Bill Clinton was quoted by the press as saying "corn is the single most inefficient way to produce ethanol because it uses a lot of energy and because it drives up the price of food" (Ibid.). Unfortunately, as the CBO report concluded a year later, corn is likely to remain the main source of ethanol for quite some time as "current technologies for producing cellulosic ethanol are not commercially viable" (CBO, 2009).

References

CBO. 2009. The impact of ethanol use on food prices and greenhouse-gas emissions. Congressional Budget Office. April.

CNN. 2008. Riots, instability spread as food prices skyrocket. April 14. http://www.cnn.com/2008/WORLD/americas/04/14/world.food.crisis/ Accessed 4 May 2009.

Doornbosch, R. and Steenblick, R. 2007. Biofuels: is the cure worse than the disease? Organization for Economic Cooperation and Development. Paris.

Johansson, D.J.A. and Azar, C. 2007. A scenario based analysis of land competition between food and bioenergy production in the US. *Climatic Change* **82**: 267-291.

Klopfenstein, T.J., Erickson, G.E. and Bremmer, V.R. 2008. Board-invited review: use of distillers byproducts in the beef cattle feeding industry. *Journal of Animal Science* **86** (5): 1223-1231.

Lehr, J.H. 2006. Are the ethanol wars over? *PERC Reports*, Property and Environment Research Center. March.

Liska, A.J, Yang, H.S., Bremer, V.R., Klopfenstein, T.J., Walters, D.T., Galen, E.E. and Cassman, K.G. 2009. Improvements in life cycle energy efficiency and greenhouse gas emissions of corn-ethanol. *Journal of Industrial Ecology* **13** (1):

OECD/FAO. 2007. *Agricultural Outlook 2007-2016*. Organization for Economic Cooperation and Development/Food and Agriculture Organization (United Nations). Paris, Rome.

Runge, C.F. and Senauer, B. 2007. How biofuels could starve the poor. *Foreign Affairs* **86**: 41-53.

9.5.5. Use of Water

The third strategy proposed by Tilman *et al.* (2002) to address the conflict between growing food and preserving natural ecosystems is finding ways to conserve water. Biofuels, as the following studies demonstrate, fail to advance this objective.

Elcock (2008) projects that 12.9 billion gallons per day of water will be consumed in the manufacture of ethanol by 2030. This "increase accounts for roughly 60% of the total projected nationwide increase in water consumption over the 2005-2030 period, and it is more than double the amount of water projected to be consumed for industrial and commercial use in 2030 by the entire United States."

A 2009 study by Argonne National Laboratory estimated life-cycle water consumption for one gallon of four types of fuel: ethanol, gasoline from domestic conventional crude oil, gasoline from Saudi conventional crude oil, and gasoline from Canadian oil sands (Wu *et al.*, 2009). For ethanol, they estimated an average consumption of 3.0 gallon of water/gallon of corn ethanol during the production process in a corn dry mill, a yield of 2.7 gallons of ethanol per bushel of corn, and the average consumptive use of irrigation water for corn farming in three U.S. Department of Agriculture Regions (5, 6, and 7) representing the vast majority of corn production in the United States. They found "total groundwater and surface water use for corn growing vary significantly across the three regions, producing 1 gallon of corn-based ethanol consumes a net of 10 to 17 gallon of freshwater when the corn is grown in Regions 5 and 6, as compared with 324 gallon when the corn is grown in Region 7." When these figures are adjusted to reflect the lower Btu/gallon of ethanol compared to gasoline (75,700 / 115,000, or .66), the amount of water consumed per gallon of gasoline equivalent ranges from 15.2 to 25.8 gallons in Regions 5 and 6 and 492 gallons in Region 7.

Wu *et al.* (2009) found the amount of water required to create a gallon of gasoline was dramatically less: 3.4-6.6 gallons of water to make one gallon of gasoline from U.S. conventional crude oil, 2.8-5.8 gallons to make one gallon of gasoline from Saudi conventional crude, and 2.6-6.2 gallons to make one gallon of gasoline from Canadian oil sands.

An even more recent review of the literature conducted by the International Council for Science (ICSU) found "the water requirements of biofuel-derived energy are 70 to 400 times larger than other energy sources such as fossil fuels, wind or solar.

Roughly 45 billion cubic meters of irrigation water were used for biofuel production in the [sic] 2007, or some 6 times more water than people drink globally" (ICSU, 2009). The authors also point out that "severe water pollution can result from runoff from agricultural fields and from waste produced during the production of biofuels," and that "the increase in corn [production] to support ethanol goals in the United States is predicted to increase nitrogen inputs to the Mississippi River by 37%."

In light of this evidence, there can be little doubt that biofuels are a much less efficient use of scarce water resources than are fossil fuels. This means increased reliance on fossil fuels would make it more difficult to increase food production per unit of water in the future, one of Tilman *et al.*'s three strategies to solve the food vs. nature conflict.

References

Elcock, D. 2009. Baseline and projected water demand data for energy and competing water use sectors. U.S. Department of Energy, ANL/EUS/TM/08-8 for US DOE/NETL.

Tilman, D., Cassman, K.G., Matson, P.A., Naylor, R. and Polasky, S. 2002. Agricultural sustainability and intensive production practices. *Nature* **418**: 671-677.

ICSU. 2009. Biofuels: environmental consequences and interactions with changing land use. Proceedings of the Scientific Committee on Problems of the Environment (SCOPE) International Biofuels Project Rapid Assessment, International Council for Science (ICSU). 22-25 September 2008, Gummersbach, Germany. R.W. Howarth and S. Bringezu, eds.

Wu, M., Mintz, M., Wang, M. and Arora, S. 2009. Consumptive water use in the production of ethanol and petroleum gasoline. U.S. Department of Energy, Office of Scientific and Technical Information, Center for Transportation Research, Energy Systems Division, Argonne National Laboratory.

9.5.6. Conclusion

The production and use of biofuels has increased dramatically in recent years, due largely to government mandates and taxpayer subsidies. But the alleged environmental benefits of these "renewable fuels" disappear upon close inspection. As Doornbosch and Steenblick (2007) say in their OECD report, "when such impacts as soil acidification, fertilizer use, biodiversity loss and toxicity of agricultural pesticides are taken into account, the overall environmental impacts of ethanol and biodiesel can very easily exceed those of petrol and mineral diesel. The conclusion must be that the potential of the current technologies of choice—ethanol and biodiesel—to deliver a major contribution to the energy demands of the transport sector without compromising food prices and the environment is very limited."

The decision by the IPCC and many environmental groups to embrace ethanol pits energy production against food production, making even worse the conflict between the two that this section has addressed. There can be little doubt that ethanol mandates and subsidies have made both food *and* energy more, not less, expensive, and therefore less

available to a growing population. The extensive damage to natural ecosystems already caused by this poor policy decision, and the much greater destruction yet to come, are a high price to pay for refusing to understand and utilize the true science of climate change.

Additional information on this topic, including reviews of newer publications as they become available, can be found at http://www.co2science.org/subject/b/biofuels.php

References

Doornbosch, R. and Steenblick, R. 2007. Biofuels: is the cure worse than the disease? Organization for Economic Cooperation and Development. Paris.

Appendix 1

Acronyms

ACWT	Atlantic core water temperature
AGAGE	Advanced Global Atmospheric Gases Experiment
AGW	anthropogenic global warming
AMF	*arbuscular mycorrhizal* fungi
AMO	Atlantic Multidecadal Oscillation
AMSR	Advanced Microwave Scanning Radiometer
APSIM	Agricultural Production Systems Simulator
AO/NAO	Arctic Oscillation/North Atlantic Oscillation
AsA	ascorbic acid
ASI	aeolian sand influx
ATLAS	Airborne Thermal and Land Applications Sensor
AVHRR	Advanced Very High Resolution Radiometer
Ba	barium
BATS	Bermuda Atlantic Time-Series Study
BC2	Carlsbad Cavern (New Mexico)
BCC	Buckeye Creek Cave (West Virginia)
BioCON	Biodiversity, Carbon Dioxide, and Nitrogen Effects on Ecosystem Functioning
BIOME3	Biogeochemical Model
BP	before present
BSW	bog surface wetness
Bt	*Bacillus thuringiensis*
BYDV	barley yellow dwarf virus
Ca	calcium
CAM	Crassulacean Acid Metabolism
CASA	Carnegie-Ames-Stanford Approach
CBSC	Carbon-based secondary compounds
CCN	Cloud condensation nuclei

CDC	Canadian Drought Code
CERES	Clouds and the Earth's Radiant Energy System
CEVSA	Carbon Exchanges in the Vegetation-Soil-Atmosphere System
CFC	chlorofluorocarbons
CGCM	Coupled General Circulation Models
CH₂CII	iodocarbon chloroiodomethane
CH₃C₁	methyl chloride
CH₄	methane
CH₂I₂	diiodomethane
CHD	coronary heart disease
CMAP	Climate Prediction Center Merged Analysis of Precipitation
CO₂	carbon dioxide
CPR	Continuous Plankton Recorder
CPY	Climactic Pointer Years
CRF	cosmic ray flux
CRII	cosmic ray-induced ionization
CRP1	Core Research Project 1
CRU	Climate Research Unit
CS₂	carbon disulfide
CSIRO	Commonwealth Scientific and Industrial Research Organization (Australia)
CWP	Current Warm Period
CVD	cardiovascular disease
CZCS	Coastal Zone Color Scanner
DACP	Dark Ages cold period
DDG	dry distilled grain
DGGE	denaturing gradient gel electrophoresis
DM	dry matter
DMS	dimethyl sulfide
DOC	dissolved organic carbon

ECCO	Estimating Circulation and Climate of the Ocean		**IMAR**	Inner Mongolia Autonomous Region
ECMWF	European Centre for Medium-Range Weather Forecasts		**IMR**	Indian Monsoon rainfall
EDC96	European Project for Ice Coring in Antarctica Dome C		**IPCC**	Intergovernmental Panel on Climate Change
EIA	Energy Information Administration (U.S.)		**IPCC 2007-I**	Intergovernmental Panel on Climate Change — Group 1 Contribution
EF-Tu	protein synthesis elongation factor		**IPCC 2007-II**	Intergovernmental Panel on Climate Change — Group II Contribution
ENSO	El Nino-Southern Oscillation		**IPCC 2007-II**	Intergovernmental Panel on Climate Change — Group III Contribution
EQC	eolian quartz content		**IPCC-FAR**	Intergovernmental Panel on Climate Change — First Assessment Report
FACE	Free-air CO_2 Enrichment		**IPCC-SAR**	Intergovernmental Panel on Climate Change — Second Assessment Report
FACTS	Forest Atmosphere Carbon Transfer and Storage		**IPCC-TAR**	Intergovernmental Panel on Climate Change — Third Assessment Report
FB	Foxe Basin		**IPCC-AR4**	Intergovernmental Panel on Climate Change — Fourth Assessment Report
FD	flux data		**IRD**	ice rafted debris
GBR	Great Barrier Reef		**ISCCP**	International Satellite Cloud Climatology Project
GCM	General Circulation Models			
GCR	galactic cosmic rays		**ISM**	Indian Summer Monsoon
GCTE	Global Change and Terrestrial Ecosystems		**ITCZ**	Intertropical Convergence Zone
			ITS2	Internal Transcribed Spacer Region 2
GDP	Gross Domestic Product		**IUCN**	International Union for Conservation of Nature
GEI	Glacier Expansion Index			
GHG	green house gas(es)		**LBM**	larch budmoth
GIMMS	Global Inventory Modeling and Mapping Studies		**LCA**	low cloud amount
			LCLU	land cover and land use
GIS	Greenland Ice Sheet		**LGM**	Last Glacial Maximum
GISS	Goddard Institute of Space Studies		**LIA**	Little Ice Age
GLO-PEM	Global Production Efficiency Model		**LST**	land surface temperature
gNDVI	Normalized Difference Vegetation Index over the Growing Season		**LTM**	long-term mean standardization
			m	meter
GPCP	Global Precipitation Climatology Project		**Ma BP**	million years before present
gr	gram(s)		**MAAT**	mean annual air temperature
GRACE	Gravity Recovery and Climate Experiment		**MBP**	mass balance potential
			MDR	main development region
GREET	Greenhouse gases Regulated Emissions and Energy use in Transportation		**ME**	surface melt
			MJ	mega joule
GSH	glutathione		**MS**	methanesulfonate
GSL	global sea level		**MSA**	methanesulfonic Acid
HC1	Hidden Cave (Guadalupe Mountains)		**MTBE**	methyl tertiary butyl ether
HR	heterotrophic respiration			
HSG	hematite stained grain			
IE	infection efficiency			

MWP	Medieval Warm Period
MXD	maximum latewood density
MY	multiyear
N_2O	nitrous oxide
NABE	North Atlantic Bloom Experiment
NADW	North Atlantic deep water
NAM	Northern Annular Mode
NAO	North Atlantic Oscillation
NAS	National Academy of Sciences
NDVI	Normalized Difference Vegetation Index
NEP	net ecosystem production
NIPCC	Nongovernmental International Panel on Climate Change
NMHC	non-methane hydrocarbon
NPP	net primary production
$nss-SO_4^{2-}$	non-sea-salt sulfate
NWS	National Weather Service
O_3	ozone
OCS	carbonyl sulfide
OLR	outgoing longwave radiation
OM	organic matter
OTC	open-top chambers
P	precipitation
PAL	Pathfinder AVHRR [Advanced Very High Resolution Radiometer] Land
PDO	Pacific Decadal Oscillation
PDSI	Palmer Drought Severity Index
PF	polar front
PGR	post-glacial rebound
PI	potential intensity
PIZ	perennial ice zone
ppb	parts per billion
ppm	parts per million
Ps	solid precipitation
RACM	Regional Atmospheric Climate Model
RCC	rapid climate change
RCS	regional curve standardization
Rd	ratio of diffuse
Rda	area-based dark respiration
Rdm	mass-based dark respiration
rDNA	ribosomal deoxyribonucleic acid
Rg	solar irradiance
ROS	reactive oxygen species
RWP	Roman Warm Period
SACC	Screen-Aided CO_2 Control
SAT	surface air temperature
SB	Southern Beaufort Sea
SCC	Swiss Canopy Crane Project
SCPDSI	Self-Calibrating Palmer Drought Severity Index
SeaWiFS	Sea-Viewing Wide Field-Of-View Sensor
SEPP	Science & Environmental Policy Project
SFP	South Fork Payette
SMB	surface mass balance
SMR	snowmelt runoff
SODA	Simple Ocean Data Assimilation
SOM	soil organic matter
SPAR	Soil-Plant-Atmosphere-Research Facility (Mississippi)
SPCZ	South Pacific Convergence Zone
SPM	Summaries for Policymakers
SPS	sucrose-phosphate synthase
SSM/I	Special Sensor Microwave Imager
SSMR	Scanning Multichannel Microwave Radiometer
SN/SSN	sunspot number
SST	sea surface temperatures
STF	subtropical front
SU	surface sublimation
SWE	snow water equivalent
SWF	shortwave flux
SWM	Southwest Monsoon
TBE	tick-borne encephalitis
TBEV	tick-borne encephalitis virus
TC	tropical cyclones
Tmax	maximum temperature
Tmin	minimum temperature
TMI	Tropical Rainfall Measuring Mission Microwave Imager
T_{opt}	optimum temperature

TP	Tibetan Plateau	**WAIS**	West Antarctic Ice Sheet
TRFO	tropical rainforest	**WH**	Western Hudson Bay
TRMM	Tropical Rainfall Measuring Mission	**WMO**	World Meteorological Organization
TSI	total solar irradiance	**WNP**	Western North Pacific
UHI	urban heat island	**WSC**	water-soluble carbohydrate
UNEP	United Nations Environment Program	**WT**	wild type
UV	ultraviolet	**WUE**	water use efficiency
VS	vertical wind shear		

Appendix 2

Table 7.1.1 – Plant Dry Weight (Biomass) Responses to Atmospheric CO_2 Enrichment

Table 7.1.1 reports the results of peer-reviewed scientific studies indicating the biomass growth response of plants to a 300-ppm increase in atmospheric CO_2 concentration. Plants are listed by both common and/or scientific names, followed by the number of experimental studies conducted on each plant, the mean biomass response to a 300-ppm increase in the air's CO_2 content, and the standard error of that mean. Whenever the CO_2 increase was not exactly 300 ppm, a linear adjustment was computed. For example, if the CO_2 increase was 350 ppm and the growth response was a 60 percent enhancement, the adjusted 300-ppm CO_2 growth response was calculated as (300/350) x 60% = 51%.

The data in this table are printed by permission of the Center for the Study of Carbon Dioxide and Global Change and were taken from its Plant Growth database as it existed on 23 March 2009. Additional data are added to the database at approximately weekly intervals and can be accessed free of charge at the Center's website at http://www.co2science.org/data/plant_growth/dry/dry_subject.php. This online database also archives information pertaining to the experimental conditions under which each plant growth experiment was conducted, as well as the complete reference to the journal article from which the experimental results were obtained. The Center's online database also lists percent increases in plant biomass for 600- and/or 900-ppm increases in the air's CO2 concentration.

Plant Name	# of Studies	Arithmetic Mean	Standard Error
[Phytoplankton]	1	7%	0%
Abelmoschus esculentus [Okra]	1	8%	0%
Abies alba [Silver Fir]	5	33.20%	11.60%
Abies faxoniana [Minjiang Fir]	2	16.50%	1.10%
Absinth Sagewort [Artemisia absinthium]	2	161.50%	44.20%
Abutilon theophrasti [Velvet Leaf]	14	32.10%	13.70%
Acacia aneura [Mulga Acacia]	1	0%	0%
Acacia auriculiformis [Acacia, Earleaf]	1	0%	0%
Acacia catechu [Black Cutch]	6	62.70%	8.50%
Acacia colei [Acacia]	1	89%	0%
Acacia coriacea [Wiry Wattle]	1	50%	0%
Acacia dealbata [Silver Wattle]	1	106%	0%
Acacia implexa	1	77%	0%
Acacia irrorata	1	64%	0%
Acacia magium [Brown Saiwood]	3	19%	6.90%
Acacia mearnsii [Black Wattle]	1	48%	0%

Plant Name	# of Studies	Arithmetic Mean	Standard Error
Acacia melanoxylon [Blackwood]	1	164%	0%
Acacia minuta [Coastal Scrub Wattle]	2	110%	22.60%
Acacia nilotica [Gum Arabic Tree]	8	316.30%	54.10%
Acacia saligna [Orange Wattle]	1	55%	0%
Acacia tetragonophylla [Acacia]	1	55%	0%
Acacia, Earleaf [Acacia auriculiformis]	3	22.70%	7.90%
Acacia, Mulga [Acacia aneura]	1	0%	0%
Acer barbatum [Southern Sugar Maple]	1	95%	0%
Acer pensylvanicum [Striped Maple]	4	28.80%	8.40%
Acer pseudoplatanus [Sycamore]	3	34.30%	12.80%
Acer rubrum [Red Maple]	13	44.20%	13.30%
Acer saccharum [Sugar Maple]	12	48.30%	13.10%
Achillea millefolium [Yarrow]	3	24.70%	12.60%
Aechmea magdalenae [Understory Herb]	1	30%	0%
Agave deserti [Desert Agave]	4	34.80%	10.30%
Agave salmiana [Agave]	1	43%	0%

Plant Name	# of Studies	Arithmetic Mean	Standard Error
Agave vilmoriniana [Leaf Succulent]	2	14%	9.90%
Agave, Desert [Agave deserti]	4	34.80%	10.30%
Agropyron repens [C3 Grass]	1	41%	0%
Agropyron smithii [Western Wheatgrass]	3	50.30%	42.20%
Agrostemma githago [Corncockle]	1	21%	0%
Agrostis canina [Velvet Bentgrass]	1	124%	0%
Agrostis capillaris [Colonial Bentgrass]	9	37.10%	11.70%
Albiza, Tall [Albizia procera]	8	125.60%	16.30%
Albizia procera [Tall Albiza]	8	125.60%	16.30%
Alder, Black [Alnus glutinosa]	3	24.30%	9.10%
Alder, Manchurian [Alnus hirsuta]	5	24%	5.50%
Alder, Mountain [Alnus incana]	1	50%	0%
Alder, Red [Alnus rubra]	6	34.50%	7.70%
Alloteropsis semialata [Cockatoo Grass]	6	41.20%	3.70%
Alnus glutinosa [Black Alder]	3	24.30%	9.10%
Alnus hirsuta [Manchurian Alder]	5	24%	5.50%
Alnus incana [Mountain Alder]	1	50%	0%
Alnus rubra [Red Alder]	6	34.50%	7.70%
Amaranth [Amaranthus tricolor]	1	20%	0%
Amaranth, Grain [Amaranthus hypochondriacus]	2	5.50%	3.90%
Amaranth, Redroot [Amaranthus retroflexus]	16	6.40%	3.50%
Amaranth, Slender [Amaranthus viridis]	4	-2.30%	3.20%
Amaranth, Slim [Amaranthus hybridus]	6	14%	5.60%
Amaranthus hybridus [Slim Amaranth]	6	14%	5.60%
Amaranthus hypochondriacus [Grain Amaranth]	2	5.50%	3.90%
Amaranthus retroflexus [Redroot Amaranth]	16	6.40%	3.50%
Amaranthus tricolor [Amaranth]	1	20%	0%
Amaranthus viridis [Slender Amaranth]	4	-2.30%	3.20%
Ambrosia artemisiifolia [Annual Ragweed]	13	28.80%	7%
Ambrosia dumosa [White Burrobush]	2	164%	36.10%
Ambrosia trifida [Great Ragweed]	3	26%	8.20%
American Pokeweek [Phytolacca americana]	8	-2.40%	13.80%
Amorpha canescens [Leadplant]	3	40.30%	63.30%
Amur Silvergrass [Miscanthus sacchariflorus]	1	-27%	0%
Ananas comosus [Pineapple]	2	5%	9.20%
Andropogon appendiculatus [Vlei Bluegrass]	6	50.50%	6.90%
Andropogon gerardii [Big Bluestem]	12	20.30%	7.20%
Andropogon virginicus [Broomsedge]	2	0%	0%
Anemone cylindrica [Candle Anemone]	1	84%	0%

Plant Name	# of Studies	Arithmetic Mean	Standard Error
Anemone, Candle [Anemone cylindrica]	1	84%	0%
Anthoxanthum odoratum [Sweet Vernal Grass]	1	170%	0%
Anthyllis vulneraria [Common Kidney Vetch]	6	45.30%	17.50%
Arabidopsis thaliana [Thale Cress]	10	219.70%	81.40%
Arachis glabrata [Florigraze Peanut]	2	20.50%	1.10%
Arachis hypogaea [Peanut]	24	84.20%	23.10%
Armeria maritima [Thrift Seapink]	1	-45%	0%
Arrhenatherum elatius [Tall Oatgrass]	7	18.60%	9.60%
Artemisia absinthium [Absinth Sagewort]	2	161.50%	44.20%
Artemisia tridentata [Big Sagebrush]	4	24.30%	6.10%
Asclepias syriaca [Common Milkweed]	1	69%	0%
Asclepias tuberosa [Butterfly Milkweed]	1	108%	0%
Ash, European [Fraxinus excelsior]	6	14.20%	6%
Ash, Green [Fraxinus pennsylvanica]	1	32%	0%
Ash, White [Fraxinus americana]	4	33.80%	7.30%
Aspen, Bigtooth [Populus grandidentata]	1	29%	0%
Aspen, Hybrid [Populus tremula x Populus tremuloides]	6	40.30%	18%
Aspen, Quaking [Populus tremuloides]	32	60.60%	10%
Aster pilosus [White Oldfield Aster]	2	62.50%	26.50%
Aster tripolium [Sea Aster]	4	9.50%	3.90%
Aster, Sea [Aster tripolium]	4	9.50%	3.90%
Aster, White Oldfield [Aster pilosus]	2	62.50%	26.50%
Austrodanthonia caespitosa [Wallaby Grass]	2	78%	4.20%
Avena barbata [Slender Oat]	6	22.80%	5.80%
Avena fatua [Wild Oat]	8	32.90%	8.40%
Avena sativa [Red Oat]	10	23.80%	7.30%
Avicennia germinans [Black Mangrove]	2	22.50%	5.30%
Azolla pinnata [Water Fern]	4	54.30%	27%
Bagpod [Sesbania vesicaria]	1	60%	0%
Bahiagrass [Paspalum notatum]	3	10%	1.90%
Bald Cypress [Taxodium distichum]	4	10.30%	11.90%
Barley [Hordeum vulagare]	15	41.50%	5.70%
Barrelcactus, California [Ferocactus acanthodes]	1	30%	0%
Barrelclover [Medicago truncatula]	6	52.80%	9.60%
Bauhinia variegata [Mountain Ebony]	6	82.80%	9.70%
Bean, Adsuki [Vigna angularis]	4	60%	5.20%
Bean, Broad [Vicia faba]	4	46.30%	13.50%
Bean, Castor [Ricinus communis]	7	53.60%	16.30%
Bean, Garden [Phaseolus vulgaris]	17	64.30%	20.30%

Plant Name	# of Studies	Arithmetic Mean	Standard Error
Bean, Tepary [Phaseolus acutifolius]	2	70%	9.90%
Beech, American [Fagus grandifolia]	1	88%	0%
Beech, European [Fagus sylvatica]	12	30.90%	10.60%
Beet, Common [Beta vulgaris]	24	79.50%	26%
Bellis perennis [Lawn Daisy]	3	95.70%	0.30%
Bentgrass, Colonial [Agrostis capillaris]	9	37.10%	11.70%
Bentgrass, Velvet [Agrostis canina]	1	124%	0%
Berseem [Trifolium alexandrium]	1	41%	0%
Beta vulgaris [Common Beet]	24	79.50%	26%
Betula alleghaniensis [Yellow Birch]	15	34.30%	10.20%
Betula nana [Bog Birch]	1	0%	0%
Betula papyrifera [Paper Birch]	24	72.50%	15.70%
Betula pendula [European White Birch]	27	35.40%	5.40%
Betula platyphylla [Japanese White Birch]	4	15.30%	5%
Betula populifolia [Gray Birch]	4	19.80%	8.30%
Betula pubescens [Downy Birch]	4	25.50%	6.50%
Birch, Bog [Betula nana]	1	0%	0%
Birch, Downy [Betula pubescens]	4	25.50%	6.50%
Birch, European White [Betula pendula]	27	35.40%	5.40%
Birch, Gray [Betula populifolia]	4	19.80%	8.30%
Birch, Japanese White [Betula platyphylla]	4	15.30%	5%
Birch, Paper [Betula papyrifera]	24	72.50%	15.70%
Birch, Yellow [Betula alleghaniensis]	15	34.30%	10.20%
Bittercress, Hairy [Cardamine hirsuta]	2	-0.50%	3.90%
Black Cutch [Acacia catechu]	6	62.70%	8.50%
Blackberry [Rubus]	1	675%	0%
Black-eyed Susan [Rudbeckia hirta]	1	0%	0%
Blackgram [Vigna mungo]	2	87%	19.80%
Blackwood [Acacia melanoxylon]	1	164%	0%
Bluegrass, Alpine [Poa alpina]	2	79.50%	4.60%
Bluegrass, Annual [Poa annua]	10	20.20%	9.10%
Bluegrass, Kentucky [Poa pratensis]	9	113.90%	29.80%
Bluegrass, Rough [Poa trivialis]	1	3%	0%
Bluegrass, Vlei [Andropogon appendiculatus]	6	50.50%	6.90%
Bluestem, Big [Andropogon gerardii]	12	20.30%	7.20%
Bluestem, Little [Schizachyrium scoparium]	9	18.20%	9%
Bottlebrush Squirreltail [Elymus elymoides]	1	24%	0%
Bouteloua curtipendula [Sideoats Grama]	2	7%	22.60%
Bouteloua eriopoda [Black Grama]	1	21%	0%

Plant Name	# of Studies	Arithmetic Mean	Standard Error
Bouteloua gracilis [Blue Grama]	7	2.40%	11.10%
Brachypodium pinnatum [Heath Falsebrome]	1	0%	0%
Bracken [Pteridium aquilinum]	1	0%	0%
Brassica campestris	6	55.80%	8.50%
Brassica carinata [Abyssinian Mustard]	2	28%	1.40%
Brassica juncea [India Mustard]	2	28.50%	1.80%
Brassica kaber [Field Mustard]	1	29%	0%
Brassica napus [Oilseed Rape]	17	52.90%	8.90%
Brassica nigra [Black Mustard]	2	40.50%	10.30%
Brassica oleracea [Broccoli]	5	28.80%	8.50%
Brassica rapa [Mustard]	1	21%	0%
Bristlegrass, Green [Seteria viridis]	1	18%	0%
Bristlegrass, Japanese [Seteria faberi]	5	18.80%	6.70%
Broccoli [Brassica oleracea]	5	28.80%	8.50%
Brome, Compact [Bromus madritensis]	2	7%	14.10%
Brome, Erect [Bromus erectus]	7	34.60%	6.80%
Brome, Foxtail [Bromus rubens]	2	25%	7.80%
Brome, Poverty [Bromus sterilis]	1	0%	0%
Brome, Smooth [Bromus inermis]	1	-13%	0%
Brome, Soft [Bromus hordeaceus]	2	43%	9.90%
Brome, Soft [Bromus mollis]	2	53.50%	2.50%
Bromus erectus [Erect Brome]	7	34.60%	6.80%
Bromus hordeaceus [Soft Brome]	2	43%	9.90%
Bromus inermis [Smooth Brome]	1	-13%	0%
Bromus madritensis [Compact Brome]	2	7%	14.10%
Bromus mollis [Soft Brome]	2	53.50%	2.50%
Bromus rubens [Foxtail Brome]	2	25%	7.80%
Bromus sterilis [Poverty Brome]	1	0%	0%
Bromus tectorum [Cheatgrass]	1	36%	0%
Broomsedge [Andropogon virginicus]	2	0%	0%
Brown Saiwood [Acacia magium]	3	19%	6.90%
Buchloe dactyloides [Buffalo Grass]	1	-5%	0%
Buck Brush [Symphiocarpos orbiculatus]	2	98.50%	42.10%
Buckwheat, Common [Fagopyrum esculentum]	3	17.30%	9.90%
Buffalo Grass [Buchloe dactyloides]	1	-5%	0%
Bunchgrass, Perennial [Oryzopsis hymenoides]	1	0%	0%
Burnet, Small [Sanguisorba minor]	6	81.20%	15.20%
Burr Medick [Medicago minima]	1	-6%	0%
Bushbean, Purple [Macroptilium atropurpureum]	1	43%	0%

Plant Name	# of Studies	Arithmetic Mean	Standard Error
Cabbage [Brassica oleracea]	5	28.80%	8.50%
Calamagrostis epigeios [Chee Reedgrass]	2	98.50%	0.40%
Calluna vulgaris [Heather]	9	17.10%	5.40%
Camphorweed [Heterotheca subaxillaris]	1	20%	0%
Canada Cockleburr [Xanthium strumarium]	7	30.60%	6.40%
Canary Grass [Phalaris arundinacea]	1	49%	0%
Cantaloupe [Cucumis melo]	3	4.70%	0.70%
Caragana intermedia [Deciduous Shrub of Semi-arid Northern China]	6	54.50%	4.50%
Cardamine hirsuta [Hairy Bittercress]	2	-0.50%	3.90%
Carex bigelowii [Bigelow's Sedge]	1	0%	0%
Carex flacca [Heath Sedge]	5	73.80%	18.20%
Carex rostrata [Beaked Sedge]	1	44%	0%
Carob [Ceratonia siliqua]	10	38.10%	9.70%
Carpinus betulus [European Hornbeam]	3	35.30%	26.40%
Carrizo Citrange [Citrus sinensis x Poncirus trifoliata]	2	40.50%	1.80%
Carrot [Daucus carota sativus]	5	77.80%	32.30%
Cassava [Manihot esculenta]	2	73.50%	29.30%
Cassia fasciculata [Sleepingplant]	2	17.50%	27.20%
Cassia nictitans [Partridge Pea]	1	22%	0%
Cassia obtusifolia [Coffeeweed]	2	31.50%	7.40%
Castanea sativa [Sweet Chesnut]	4	12.80%	2.20%
Cattail [Typha latifolia]	1	-2%	0%
Cenchrus ciliaris [Buffel Grass]	1	242%	0%
Cerastium fontanum [Chickweed]	1	51%	0%
Ceratonia siliqua [Carob]	10	38.10%	9.70%
Ceratophytum tetragonolobum	2	60%	42.40%
Chairmaker's Bulrush [Schoenoplectus americanus]	1	27%	0%
Chamaecrista nictitans [Partridge Pea]	1	0%	0%
Chamelaucium uncinatum (Schauer) x Chamelaucium floriferum (MS) [Lady Stephanie]	3	20.70%	3.40%
Chamerion angustifolium	1	228%	0%
Cheatgrass [Bromus tectorum]	1	36%	0%
Chenopodium album [Lambsquarters]	12	31.30%	7.50%
Chenopodium bonus-henricus [Good King Henry]	1	21%	0%
Cherry, Black [Prunus serotina]	2	42%	1.40%
Cherry, Sweet [Prunus avium]	8	59.80%	7.80%
Chesnut, Sweet [Castanea sativa]	4	12.80%	2.20%
Chickweed [Cerastium fontanum]	1	51%	0%
Chlorella pyrenoidosa [Common Freshwater Microalga]	2	8%	0.70%
Cirsium arvense [Canadian Thistle]	3	83.30%	33.50%

Plant Name	# of Studies	Arithmetic Mean	Standard Error
Citrus aurantium [Sour Orange Tree]	3	61.30%	13.30%
Citrus reticulata [Mandarin Orange Tree]	2	29.50%	3.90%
Citrus sinensis [Sweet Orange Tree]	2	38.50%	4.60%
Citrus sinensis x Poncirus trifoliata [Carrizo Citrange]	2	40.50%	1.80%
Clammy Cuphea [Cuphea viscosissima]	1	17%	0%
Clarkia rubicunda [Ruby Chalice Fairyfan]	1	35%	0%
Climbing Nightshade [Solanum dulcamara]	10	46.50%	9.30%
Clover, Crimson [Trifolium incarnatum]	2	21.50%	0.40%
Clover, Japanese [Kummerowia striata]	1	26%	0%
Clover, Purple [Trifolium pratense]	5	16.80%	4.40%
Clover, Silky Prairie [Petalostemum villosum]	3	-9.30%	23.40%
Clover, White [Trifolium repens]	45	59.80%	16.80%
Coastal Scrub Wattle [Acacia minuta]	2	110%	22.60%
Codlins and Cream [Epilobium hirsutum]	1	9%	0%
Coffea arabusta [Coffee]	2	175.50%	76.70%
Coffee [Coffea arabusta]	2	175.50%	76.70%
Coffeeweed [Cassia obtusifolia]	2	31.50%	7.40%
Conium maculatum [Poison Hemlock]	1	21%	0%
Coral Honeysuckle [Lonicera sempervirens]	1	30%	0%
Cordgrass, Common [Spartina anglica]	3	19%	25.10%
Cordgrass, Saltmeadow [Spartina patens]	9	9.70%	4.90%
Cordgrass, Smooth [Spartina alterniflora]	2	0%	0%
Corkscrew Vallisneria [Vallisneria tortifolia]	9	9.70%	4.90%
Corn [Zea mays]	20	21.30%	4.90%
Corncockle [Agrostemma githago]	1	21%	0%
Cornus florida [Eastern Flowering Dogwood]	2	121.50%	37.10%
Correa schlechtendalii	3	9.70%	0.30%
Cotton [Gossypium hirsutum]	32	64.10%	9.40%
Cottongrass, Tussock [Eriophorum vaginatum]	5	105%	71.60%
Cottonwood, Black [Populus trichocarpa]	5	124%	93.20%
Cottonwood, Eastern [Populus deltoides]	5	60.20%	15.80%
Crabgrass, Hairy [Digitaria sanguinalis]	2	39%	12.70%
Crabgrass, Natal [Digitaria natalensis]	6	13%	3.20%
Creosote Bush [Larrea tridentata]	4	105.50%	36.20%
Crested Dogstailgrass [Cynosurus cristatus]	1	30%	0%
Crotalaria juncea [Sunn Hemp]	2	45%	9.90%
Cucumber, Garden [Cucumis sativus]	7	53.10%	7.90%
Cucumis melo [Cantaloupe]	3	4.70%	0.70%
Cucumis sativus [Garden Cucumber]	7	53.10%	7.90%

Plant Name	# of Studies	Arithmetic Mean	Standard Error
Cudweed [Gnaphalium affine]	1	21%	0%
Cuphea viscosissima [Clammy Cuphea]	1	17%	0%
Cynosurus cristatus [Crested Dogstailgrass]	1	30%	0%
Cyperus esculentus [Yellow Nutsedge]	1	16%	0%
Cyperus rotundus [Purple Nutsedge]	1	38%	0%
Dactylis glomerata [Orchardgrass]	11	16.90%	4.60%
Daisy, Lawn [Bellis perennis]	3	95.70%	0.30%
Dalbergia latifolia [Indian Rosewood]	6	51.80%	8.20%
Dallas Grass [Paspalum dilatatum]	1	82%	0%
Danthonia richardsonii Cashmore [Wallaby Grass]	9	25%	8.60%
Datura stramonium [Jimsonweed]	3	36.70%	15.80%
Daucus carota [Carrot]	5	77.80%	32.30%
Deschampsia flexuosa [Wavy Hairgrass]	1	19%	0%
Desmazeria rigida [Ferngrass]	1	26%	0%
Digitalis purpurea [Purple Foxglove]	2	17.50%	2.50%
Digitaria natalensis [Natal Crabgrass]	6	13%	3.20%
Digitaria sanguinalis [Hairy Crabgrass]	2	39%	12.70%
Dock, Bitter [Rumex obtusifolius]	6	18.70%	12.50%
Dogwood, Eastern Flowering [Cornus florida]	2	121.50%	37.10%
Dropseed, Whorled [Sporobolus pyramidalis]	6	33.50%	21.70%
Duckweed, Swollen [Lemna gibba]	2	47%	11.30%
Eastern Purple Coneflower [Echinacea purpurea]	4	191.50%	75.50%
Echinacea purpurea [Eastern Purple Coneflower]	4	191.50%	75.50%
Echinochloa crus-galli [Barnyard Grass]	15	72.10%	16.70%
Echium plantagineum [Salvation Jane]	4	15.80%	6.20%
Ecosystem, 10 Species of Tropical Forest Tree Seedlings	1	-5%	0%
Ecosystem, 12 Species	3	22.70%	5.50%
Ecosystem, 12 Species From Fertile Permanent Grassland	15	67.50%	13.20%
Ecosystem, 7 Species	1	9%	0%
Ecosystem, Pasture	1	12%	0%
Ecosystem, Understory Plants in a Spruce Model Ecosystem	3	33.70%	13.40%
Eelgrass, Common [Zostera marina]	1	24%	0%
Eggplant [Solanum melongena]	1	41%	0%
Eichhornia crassipes [Common Water Hyacinth]	5	51%	19.30%
Eleusine indica [Indian Goosegrass]	6	34%	17.40%
Elm, Winged [Ulmus alata]	1	30%	0%
Elymus athericus	4	46.30%	20%
Elymus elymoides Botlebrush Squirreltail	1	24%	0%
Emblic [Phyllanthus emblica]	8	165%	16%

Plant Name	# of Studies	Arithmetic Mean	Standard Error
Emiliania huxleyi [Marine Coccolithophores]	1	130%	0%
English Holly [Ilex aquifolium]	1	15%	0%
English Laurel [Prunus laurocerasus]	1	56%	0%
Epilobium hirsutum [Codlins and Cream]	1	9%	0%
Eragrostis curvula [Weeping Lovegrass]	6	42.20%	9.80%
Eragrostis orcuttiana [Annual Weed, C4]	1	462%	0%
Eragrostis racemosa [Narrowheart Lovegrass]	6	11.70%	1.80%
Erica tetralix [Crossleaf Heath]	4	23.80%	10.50%
Eriophorum vaginatum [Tussock Cottongrass]	5	105%	71.60%
Eucalyptus cladocalyx [Sugargum]	2	87.50%	9.50%
Eucalyptus miniata [Darwin Woollybutt]	9	-2%	5.40%
Eucalyptus tetrodonta	9	132.80%	19.90%
Euphorbia lathyris [Myrtle Spurge]	2	36.50%	4.60%
European Hornbeam [Carpinus betulus]	3	35.30%	26.40%
European Larch [Larix decidua]	1	142%	0%
Fagopyrum esculentum [Common Buckwheat]	3	17.30%	9.90%
Fagus grandifolia [American Beech]	1	88%	0%
Fagus sylvatica [European Beech]	12	30.90%	10.60%
Fairyfan, Ruby Chalice [Clarkia rubicunda]	1	35%	0%
Falsebrome, Heath [Brachypodium pinnatum]	1	0%	0%
Fenugreek [Trigonella foenum-graecum]	2	91%	33.20%
Fern, Tropical [Pyrrosia piloselloides]	1	78%	0%
Fern, Water [Azolla pinnata var. pinnata]	4	54.30%	27%
Ferngrass [Desmazeria rigida]	1	26%	0%
Ferocactus acanthodes [California Barrelcactus]	1	30%	0%
Fescue, Meadow [Festuca pratensis]	11	20.30%	10%
Fescue, Red [Festuca rubra]	3	30.70%	15.90%
Fescue, Sheep [Festuca ovina]	6	47.70%	16.60%
Fescue, Small [Vulpia microstachys]	2	-4.50%	8.10%
Fescue, Tall [Festuca arundinacea]	11	25.40%	7.20%
Fescue, Tall Meadow [Festuca elatior]	1	40%	0%
Festuca arundinacea [Tall Fescue]	11	25.40%	7.20%
Festuca elatior [Tall Meadow Fescue]	1	40%	0%
Festuca ovina [Sheep Fescue]	6	47.70%	16.60%
Festuca pratensis [Meadow Fescue]	11	20.30%	10%
Festuca rubra [Red Fescue]	3	30.70%	15.90%
Fir, Douglas [Pseudotsuga menziesii]	6	9.70%	3.90%
Fir, Minjiang [Abies faxoniana]	2	16.50%	1.10%
Fir, Silver [Abies alba]	5	33.20%	11.60%

Plant Name	# of Studies	Arithmetic Mean	Standard Error
Flaveria trinervia [Clustered Yellowtops]	1	46%	0%
Flax, Common [Linum usitatissimum]	2	68.50%	18.70%
Flax, Common [Linum usitatissimum] in mixed stands with Silene cretica	2	63.50%	15.90%
Foxglove, Purple [Digitalis purpurea]	2	17.50%	2.50%
Fragaria x ananassa [Hybrid Strawberry]	4	42.80%	13.10%
Fraxinus americana [White Ash]	4	33.80%	7.30%
Fraxinus excelsior [European Ash]	6	14.20%	6%
Fraxinus pennsylvanica [Green Ash]	1	32%	0%
Galactia elliottii [Elliott's Milkpea]	1	110%	0%
Gentian, Dwarf [Gentianella germanica]	1	54%	0%
Gentianella germanica [Dwarf Gentian]	1	54%	0%
Glycine max [Soybean]	162	47.60%	3.10%
Gnaphalium affine [Cudweed]	1	21%	0%
Goldenclub [Orontium aquaticum]	12	19.80%	3%
Goldenrod, Canadian [Solidago canadensis]	2	1750%	1237.40%
Goldenrod, Stiff [Solidago rigida]	1	27%	0%
Goldfields, California [Lasthenia californica]	2	27%	26.90%
Gonolobus cteniophorus	2	3.50%	12.40%
Good King Henry [Chenopodium bonus-henricus]	1	21%	0%
Goosegrass, Indian [Eleusine indica]	6	34%	17.40%
Gossypium hirsutum [Cotton]	32	64.10%	9.40%
Grama, Black [Bouteloua eriopoda]	1	21%	0%
Grama, Blue [Bouteloua gracilis]	7	2.40%	11.10%
Grama, Sideoats [Bouteloua curtipendula]	2	7%	22.60%
Grass, Barnyard [Echinochloa crus-galli]	15	72.10%	16.70%
Grass, Buffel [Cenchrus ciliaris]	1	242%	0%
Grass, C3 [Agropyron repens]	1	41%	0%
Grass, C3 [Koeleria cristata]	3	32%	39.80%
Grass, Cockatoo [Alloteropsis semialata]	6	41.20%	3.70%
Grass, Harding [Phalaris aquatica]	13	24.60%	8.50%
Grass, Kangaroo [Themeda triandra]	8	61.80%	16.20%
Grass, Red Natal [Melinis repens]	6	25.80%	7%
Grass, Sudan [Sorghum sudanense]	1	14%	0%
Grass, Wallaby [Danthonia richardsonii]	9	25%	8.60%
Grassland Community	2	25%	4.20%
Grassland Community	2	19%	2.80%
Grassland Community, Irish Neutral	4	28.50%	6.80%
Grassland, Calcareous, (C3)	11	27.40%	2.80%
Grassland, Calcareous, dominated by Bromus erectus	3	28%	2.90%

Plant Name	# of Studies	Arithmetic Mean	Standard Error
Grassland, California Annual	2	20.50%	2.50%
Grassland, Species-poor on a peaty gley soil	1	50%	0%
Grassland, Species-rich on a brown earth soil over limestone	1	56%	0%
Gray Field Speedwell [Veronica didyma]	1	26%	0%
Groundsel, Common [Senecio vulgaris]	2	66%	25.50%
Guineagrass [Panicum maximum]	1	-2%	0%
Gum Arabic Tree [Acacia nilotica]	8	316.30%	54.10%
Gypsophila paniculata [Babysbreath Gypsophila]	1	23%	0%
Hairgrass, Wavy [Deschampsia flexuosa]	1	19%	0%
Hayfield Tarweed [Hemizonia congesta]	2	48.50%	10.30%
Heath, Crossleaf [Erica tetralix]	4	23.80%	10.50%
Heather [Calluna vulgaris]	9	17.10%	5.40%
Hedera helix [English Ivy]	3	66%	10.60%
Helianthemum nummularium [Sun Rose]	1	0%	0%
Helianthus annus [Sunflower]	11	37.50%	7.80%
Hellroot [Orobanche minor]	1	0%	0%
Hemizonia congesta [Hayfield Tarweed]	2	48.50%	10.30%
Hemlock, Eastern [Tsuga canadensis]	3	34%	4.50%
Hemlock, Poison [Conium maculatum]	1	21%	0%
Heterosigma akashiwo [A Marine Raphidophyte]	2	15%	2.10%
Heterotheca subaxillaris [Camphorweed]	1	20%	0%
Hilograss [Paspalum conjugatum]	1	18%	0%
Hiziki (Brown Seaweed) [Hizikia fusiforme]	1	45%	0%
Hizikia fusiforme [Hiziki (Brown Seaweed)]	1	45%	0%
Holcus lanatus [Common Velvetgrass]	11	38.50%	10.40%
Hordeum vulagare [Barley]	15	41.50%	5.70%
Hydrilla verticillata [Water Thyme]	7	21.90%	3.30%
Hymenocallis littoralis [Spider Lily]	2	52%	2.80%
Hyparrhenia rufa [Jaragua]	2	33.50%	1.80%
Hypericum perforatum [St. John's Wort]	1	72%	0%
Ilex aquifolium [English Holly]	1	15%	0%
Indian Grass, Yellow [Sorghastrum nutans]	3	9.70%	14.40%
Indian Rosewood [Dalbergia latifolia]	6	51.80%	8.20%
Ipomoea batatas [Sweet Potato]	6	33.70%	9.30%
Ipomoea hederacea [Ivyleaf Morningglory]	2	-21%	8.50%
Ipomoea lacunosa [Whitestar]	2	15%	14.80%
Ipomoea purpurea [Tall Morningglory]	2	-29.50%	1.80%
Ivy, English [Hedera helix]	3	66%	10.60%
Japanese Honeysuckle [Lonicera japonica]	4	312.80%	73.70%

Plant Name	# of Studies	Arithmetic Mean	Standard Error
Japanese Knotweed [Polygonum cuspidatum]	3	48%	14.80%
Japanese Larch [Larix kaempferi]	4	26.80%	10.30%
Jaragua [Hyparrhenia rufa]	2	33.50%	1.80%
Jimsonweed [Datura stramonium]	3	36.70%	15.80%
Johnsongrass [Sorghum halepense]	3	-13%	14.30%
Joshua Tree [Yucca brevifolia]	1	65%	0%
Juncus effuses [Soft Rush]	1	-5%	0%
Junegrass, Prairie [Koeleria macrantha]	1	2%	0%
Kielmeyera coriacea [Tropical Savanna Tree]	3	88.70%	55.50%
Knotweed, Curlytop [Polygonum lapathifolium]	5	19.40%	4.20%
Knotweed, Marshpepper [Polygonum hydropiper]	2	40.50%	15.90%
Koeleria cristata [C3 Grass]	3	32%	39.80%
Koeleria macrantha [Prairie Junegrass]	1	2%	0%
Krameria erecta [Littleleaf Ratany]	2	102.50%	1.10%
Kummerowia striata [Japanese Clover]	1	26%	0%
Lactuca sativa [Garden Lettuce]	2	18.50%	6%
Lady Stephanie [Chamelaucium uncinatum (Schauer) x Chamelaucium floriferum]	3	20.70%	3.40%
Ladysthumb, Spotted [Polygonum persicaria]	2	38%	9.20%
Lambsquarters [Chenopodium album]	12	31.30%	7.50%
Lantana [Lantana camara]	2	82.50%	5.30%
Lantana camara [Lantana]	2	82.50%	5.30%
Larix decidua [European Larch]	1	142%	0%
Larix kaempferi [Japanese Larch]	4	26.80%	10.30%
Larix laricina [Tamarack]	1	56%	0%
Larrea tridentata [Creosote Bush]	4	105.50%	36.20%
Lasthenia californica [California Goldfields]	2	27%	26.90%
Layia platyglossa [Coastal Tidytips]	1	12%	0%
Leadplant [Amorpha canescens]	3	40.30%	63.30%
Leaf Succulent [Agave vilmoriniana]	2	14%	9.90%
Ledum palustre [Wild Rosemary]	1	0%	0%
Lemna gibba [Swollen Duckweed]	2	47%	11.30%
Lepidium latifolium [Pepperweed]	2	40%	7.10%
Lespedeza capitata [Roundhead Lespedeza]	3	333.70%	136.10%
Lespedeza cuneata [Chinese Lespedeza]	1	0%	0%
Lespedeza, Roundhead [Lespedeza capitata]	3	333.70%	136.10%
Lettuce, Garden [Lactuca sativa]	2	18.50%	6%
Linum usitatissimum [Common Flax]	2	68.50%	18.70%
Linum usitatissimum [Common Flax] in mixed stands with Silene cretica	2	63.50%	15.90%
Liquidambar styraciflua [Sweetgum]	20	132.40%	23.50%

Plant Name	# of Studies	Arithmetic Mean	Standard Error
Liriodendron tulipifera [Yellow Poplar]	3	34%	4.70%
Littleleaf Ratany [Krameria erecta]	2	102.50%	1.10%
Loblolly Pine (Pinus taeda)	65	61.90%	7.90%
Locust, Black [Robinia pseudoacacia]	4	346%	242.10%
Lolium multiflorum [Italian Ryegrass]	6	11.50%	5.10%
Lolium perenne [Perennial Ryegrass]	74	35.10%	5.30%
Lolium temulentum [Darnel Ryegrass]	3	44.70%	29.20%
Lomentaria articulata [Seaweed]	1	269%	0%
Lonicera japonica [Japanese Honeysuckle]	4	312.80%	73.70%
Lonicera sempervirens [Coral Honeysuckle]	1	30%	0%
Lotus corniculatus [Birdfoot Deer Vetch]	8	49.30%	12.90%
Lotus pedunculatus [Big Trefoil]	6	56%	26.50%
Lovegrass, Narrowheart [Eragrostis racemosa]	6	11.70%	1.80%
Lovegrass, Weeping [Eragrostis curvula]	6	42.20%	9.80%
Lupine, European Yellow [Lupinus luteus]	1	21%	0%
Lupine, Narrowleaf [Lupinus angustifolius]	3	38%	7%
Lupine, Sundial [Lupinus perennis]	11	56.40%	6.60%
Lupine, White [Lupinus albus]	4	10.30%	6.40%
Lupinus albus [White Lupine]	4	10.30%	6.40%
Lupinus angustifolius [Narrowleaf Lupine]	3	38%	7%
Lupinus luteus [European Yellow Lupine]	1	21%	0%
Lupinus perennis [Sundial Lupine]	11	56.40%	6.60%
Lycopersicon esculentum [Garden Tomato]	35	31.90%	5%
Lycopersicon lycopersicum [Tomato]	2	29.50%	5.30%
Macroptilium atropurpureum[Purple Bushbean]	1	43%	0%
Manchurian Wildrice [Zizania latifolia]	1	-5%	0%
Mandarin Orange Tree [Citrus reticulata]	2	29.50%	3.90%
Mangifera indica [Mango]	1	36%	0%
Mango [Mangifera indica]	1	36%	0%
Mangrove, American [Rhizophora mangle]	3	39.30%	6.10%
Mangrove, Black [Avicennia germinans]	2	22.50%	5.30%
Manihot esculenta [Cassava]	2	73.50%	29.30%
Maple, Red [Acer rubrum]	13	44.20%	13.30%
Maple, Southern Sugar [Acer barbatum]	1	95%	0%
Maple, Striped [Acer pensylvanicum]	4	28.80%	8.40%
Maple, Sugar [Acer saccharum]	12	48.30%	13.10%
Maranthes corymbosa	3	69%	14.30%
Marine Coccolithophores [Emiliania huxleyi]	1	130%	0%
Medicago glomerata	1	17%	0%

Plant Name	# of Studies	Arithmetic Mean	Standard Error
Medicago lupulina [Black Medick]	1	68%	0%
Medicago minima [Burr Medick]	1	-6%	0%
Medicago sativa [Alfalfa]	68	33.20%	3.70%
Medicago truncatula [Barrelclover]	6	52.80%	9.60%
Medick, Black [Medicago lupulina]	1	68%	0%
Melinis minutiflora [Molassesgrass]	2	32%	8.50%
Melinis repens [Red Natal Grass]	6	25.80%	7%
Mentha x piperita [Peppermint]	1	29%	0%
Menzies' Baby Blue Eyes [Nemophila menziesii]	1	0%	0%
Mesquite, Honey [Prosopis glandulosa]	13	37.70%	5.20%
Microalga, Common Freshwater [Chlorella pyrenoidosa]	2	8%	0.70%
Microstegium vimineum [Nepalese Browntop]	3	-53.30%	15.60%
Milkpea Elliott's [Galactia elliottii]	1	110%	0%
Milkweed, Butterfly (Asclepias tuberosa)	1	108%	0%
Milkweed, Common [Asclepias syriaca]	1	69%	0%
Millet, Broomcorn [Panicum miliaceum]	1	-13%	0%
Miscanthus sacchariflorus [Amur Silvergrass]	1	-27%	0%
Molassesgrass [Melinis minutiflora]	2	32%	8.50%
Molinia caerulea [Purple Moorgrass]	9	36.60%	9.40%
Moorgrass, Purple [Molinia caerulea]	9	36.60%	9.40%
Morningglory, Ivyleaf [Ipomoea hederacea]	2	-21%	8.50%
Morningglory, Tall [Ipomoea purpurea]	2	-29.50%	1.80%
Mountain Ebony [Bauhinia variegata]	6	82.80%	9.70%
Mungbean [Vigna radiata]	2	31.50%	16.60%
Mustard [Brassica rapa]	1	21%	0%
Mustard, Abyssinian [Brassica carinata]	2	28%	1.40%
Mustard, Black [Brassica nigra]	2	40.50%	10.30%
Mustard, Field [Brassica kaber]	1	29%	0%
Mustard, India [Brassica juncea]	2	28.50%	1.80%
Mustard, Rape Seed [Brassica campestris]	6	55.80%	8.50%
Mustard, White [Sinapis alba]	4	21.30%	4.10%
Myrobalan [Terminalia chebula]	8	442.50%	52.50%
Myrtle Spurge [Euphorbia lathyris]	2	36.50%	4.60%
Nasturtium [Tropaeolum majus]	3	42.30%	10.80%
Nemophila menziesii [Menzies' Baby Blue Eyes]	1	0%	0%
Nepalese Browntop [Microstegium vimineum]	3	-53.30%	15.60%
Nettle, Stinging [Urtica dioica]	1	26%	0%
Nicotiana tabacum [Cultivated Tobacco]	6	60.50%	11%
Night-flowering Catchfly [Silene noctiflora]	6	60.50%	11%

Plant Name	# of Studies	Arithmetic Mean	Standard Error
Nutsedge, Purple [Cyperus rotundus]	1	38%	0%
Nutsedge, Yellow [Cyperus esculentus]	1	16%	0%
Nymphaea marliac [Water Lily]	2	162%	76.40%
Oak, Chapman's [Quercus chapmanii]	2	325%	123.70%
Oak, Cork [Quercus suber]	4	46.80%	13.70%
Oak, Durmast [Quercus petraea]	6	53.20%	18.70%
Oak, Holly [Quercus ilex]	4	38%	4.80%
Oak, Mongolian [Quercus mongolica]	6	54.30%	19.50%
Oak, Northern Red [Quercus rubra]	7	55.30%	25.20%
Oak, Pedunculate [Quercus robur]	8	30.60%	12.80%
Oak, Sand Live [Quercus geminata]	5	9.40%	5.40%
Oak, White [Quercus alba]	6	146.70%	30.10%
Oat, Red [Avena sativa]	10	23.80%	7.30%
Oat, Slender [Avena barbata]	6	22.80%	5.80%
Oat, Wild [Avena fatua]	8	32.90%	8.40%
Oatgrass, Tall [Arrhentherum elatius]	7	18.60%	9.60%
Okra [Abelmoschus esculentus]	1	8%	0%
Olea europaea [Olive Tree]	4	14.50%	9.40%
Olive Tree [Olea europaea]	4	14.50%	9.40%
Opuntia ficus-indica [Prickly Pear]	9	38.20%	13.30%
Orchardgrass [Dactylis glomerata]	11	16.90%	4.60%
Orobanche minor [Hellroot]	1	0%	0%
Orontium aquaticum [Goldenclub]	12	19.80%	3%
Oryza sativa [Rice]	137	34.30%	2.10%
Oryzopsis hymenoides [Perennial Bunchgrass]	1	0%	0%
Panicgrass Blue [Panicum coloratum]	2	1%	15.60%
Panicgrass, Blue [Panicum antidotale]	4	15.50%	6.30%
Panicgrass, Fall (Panicum dichotomiflorum)	1	24%	0%
Panicgrass, Lax [Panicum laxum]	4	30%	11%
Panicum antidotale [Blue Panicgrass]	4	15.50%	6.30%
Panicum coloratum [Blue Panicgrass]	2	1%	15.60%
Panicum dichotomiflorum (Fall Panicgrass)	1	24%	0%
Panicum laxum [Lax Panicgrass]	4	30%	11%
Panicum maximum [Guineagrass]	1	-2%	0%
Panicum miliaceum [Broomcorn Millet]	1	-13%	0%
Panicum virgatum [Switchgrass]	1	-1%	0%
Pansy [Viola x wittrockiana]	1	30%	0%
Papaver setigerum [Dwarf Breadseed Poppy]	1	390%	0%
Pascopyrum smithii [Western Wheatgrass]	4	73.50%	16.80%

Plant Name	# of Studies	Arithmetic Mean	Standard Error
Paspalum conjugatum [Hilograss]	1	18%	0%
Paspalum dilatatum [Dallas Grass]	1	82%	0%
Paspalum notatum [Bahiagrass]	3	10%	1.90%
Paspalum plicatulum [Brownseed Paspalum]	1	9%	0%
Paspalum, Brownseed [Paspalum plicatulum]	1	9%	0%
Pasture	1	9%	0%
Pasture	2	24.50%	13.10%
Pasture in Switzerland [Dactylis glomerata (Orchard Grass) and Trifolium pratense (Red Clover)]	12	29.10%	5.70%
Pea, Blackeyed [Vigna unguiculata]	3	83.70%	12.20%
Pea, Garden [Pisum sativum]	11	33.30%	6.30%
Pea, Partridge [Cassia nictitans]	1	22%	0%
Pea, Partridge [Chamaecrista nictitans]	1	0%	0%
Peach Tree [Prunus persica]	4	27.80%	0.80%
Peanut [Arachis hypogaea]	24	84.20%	23.10%
Peanut, Florigraze [Arachis glabrata]	2	20.50%	1.10%
Peanut, Rhizoma	1	57%	0%
Pencilflower [Stylosanthes scabra]	2	72%	10.60%
Pepino [Solanum muricatum]	4	69.80%	20.50%
Peppermint [Mentha x piperita]	1	29%	0%
Pepperweed [Lepidium latifolium]	2	40%	7.10%
Peruvian Groundcherry [Physalis peruviana]	1	23%	0%
Petalostemum villosum [Silky Prairie-Clover]	3	-9.30%	23.40%
Petunia [Petunia hybrida]	6	55%	9.80%
Petunia hybrida [Petunia]	6	55%	9.80%
Phaeocystis [Phytoplankton]	1	0%	0%
Phalaris aquatica [Harding Grass]	13	24.60%	8.50%
Phalaris arundinacea [Canary Grass]	1	49%	0%
Pharus latifolius [Broad Stalkgrass]	1	144%	0%
Phaseolus acutifolius [Tepary Bean]	2	70%	9.90%
Phaseolus vulgaris [Garden Bean]	17	64.30%	20.30%
Phleum pratense [Timothy]	18	12.20%	3.20%
Phragmites communis [Wetland Reed]	1	-8%	0%
Phragmites japonica [Wetland Reed]	1	-11%	0%
Phyllanthus emblica [Emblic]	8	165%	16%
Physalis peruviana [Peruvian Groundcherry]	1	23%	0%
Phytolacca americana [American Pokeweed]	8	-2.40%	13.80%
Picea abies [Norway Spruce]	11	35.90%	5.50%
Picea glauca [White Spruce]	2	82%	49.50%
Picea koraiensis [Spruce]	7	37.90%	8.70%

Plant Name	# of Studies	Arithmetic Mean	Standard Error
Picea mariana [Black Spruce]	14	22.80%	3%
Picea sitchensis [Sitka Spruce]	7	20.70%	5%
Pine, Black [Pinus nigra]	1	22%	0%
Pine, Eastern White [Pinus strobus]	3	31.70%	9%
Pine, Eldarica [Pinus eldarica]	1	153%	0%
Pine, Jack [Pinus banksiana]	3	18.30%	8.80%
Pine, Japanese Red [Pinus densiflora]	7	15.70%	7.60%
Pine, Korean [Pinus koraiensis]	2	27%	15.60%
Pine, Loblolly [Pinus taeda]	65	61.90%	7.90%
Pine, Longleaf [Pinus palustris]	8	19%	7.40%
Pine, Merkus [Pinus merkusii]	2	200%	43.10%
Pine, Monterey [Pinus radiata]	1	36%	0%
Pine, Mountain [Pinus uncinata]	1	0%	0%
Pine, Ponderosa [Pinus ponderosa]	46	64.20%	11.80%
Pine, Scots [Pinus sylvestris]	46	41%	5.50%
Pineapple [Ananas comosus]	2	5%	9.20%
Pinus banksiana [Jack Pine]	3	18.30%	8.80%
Pinus densiflora [Japanese Red Pine]	7	15.70%	7.60%
Pinus eldarica [Eldarica Pine]	1	153%	0%
Pinus koraiensis [Korean Pine]	2	27%	15.60%
Pinus merkusii [Merkus Pine]	2	200%	43.10%
Pinus nigra [Black Pine]	1	22%	0%
Pinus palustris [Longleaf Pine]	8	19%	7.40%
Pinus ponderosa [Ponderosa Pine]	46	64.20%	11.80%
Pinus radiata [Monterey Pine]	1	36%	0%
Pinus strobus [Eastern White Pine]	3	31.70%	9%
Pinus sylvestris [Scots Pine]	46	41%	5.50%
Pinus taeda [Loblolly Pine]	65	61.90%	7.90%
Pinus uncinata [Mountain Pine]	1	0%	0%
Pisum sativum [Garden Pea]	11	33.30%	6.30%
Plantago erecta [Dwarf Plantain]	2	4.50%	11.70%
Plantago lanceolata [Narrowleaf Plantain]	11	46.20%	15.20%
Plantago major [Common Plantain]	3	31.70%	4.60%
Plantago maritima [Sea Plantain]	3	101.70%	27.90%
Plantago media [Hoary Plantain]	2	26.50%	5.30%
Plantago virginica [Virginia Plantain]	1	45%	0%
Plantain, Common [Plantago major]	3	31.70%	4.60%
Plantain, Dwarf [Plantago erecta]	2	4.50%	11.70%
Plantain, Hoary [Plantago media]	2	26.50%	5.30%

Plant Name	# of Studies	Arithmetic Mean	Standard Error
Plantain, Narrowleaf [Plantago lanceolata]	11	46.20%	15.20%
Plantain, Sea [Plantago maritima]	3	101.70%	27.90%
Plantain, Virginia [Plantago virginica]	1	45%	0%
Poa alpina [Alpine Bluegrass]	2	79.50%	4.60%
Poa annua [Annual Bluegrass]	10	20.20%	9.10%
Poa pratensis [Kentucky Bluegrass]	9	113.90%	29.80%
Poa trivialis [Rough Bluegrass]	1	3%	0%
Poison Ivy [Toxicodendron radicans]	7	75.70%	13.70%
Polygonum cuspidatum [Japanese Knotweed]	3	48%	14.80%
Polygonum hydropiper [Marshpepper Knotweed]	2	40.50%	15.90%
Polygonum lapathifolium [Curlytop Knotweed]	5	19.40%	4.20%
Polygonum pensylvanicum [Pennsylvania Smartweed]	3	55.70%	1.80%
Polygonum persicaria [Spotted Ladysthumb]	2	38%	9.20%
Poplar Clone, Robusta [Populus deltoides x Polulus nigra]	6	63.30%	13.70%
Poplar, Black [Populus nigra]	15	53.90%	12.20%
Poplar, Hybrid [Populus trichocarpa x Populus deltoides]	7	76.70%	23.40%
Poplar, Robusta [Populus euramericana]	13	41.70%	8.80%
Poplar, White [Populus alba]	14	43.90%	4.70%
Poplar, Yellow [Liriodendron tulipifera]	3	34%	4.70%
Poppy, Dwarf Breadseed [Papaver setigerum]	1	390%	0%
Populus alba [White Poplar]	14	43.90%	4.70%
Populus deltoides [Eastern Cottonwood]	5	60.20%	15.80%
Populus deltoides x Polulus nigra [Robusta Poplar Clone]	6	63.30%	13.70%
Populus euramericana [Robusta Poplar]	13	41.70%	8.80%
Populus grandidentata [Bigtooth Aspen]	1	29%	0%
Populus nigra [Black Poplar]	15	53.90%	12.20%
Populus tremula x Populus tremuloides [Hybrid Aspen]	6	40.30%	18%
Populus tremuloides [Quaking Aspen]	32	60.60%	10%
Populus trichocarpa [Black Cottonwood]	5	124%	93.20%
Populus trichocarpa x Populus deltoides [Hybrid Poplar]	7	76.70%	23.40%
Potato, Sweet [Ipomoea batatas]	6	33.70%	9.30%
Potato, White [Solanum tuberosum]	33	29.50%	3.60%
Pothos [Scindapsus aureus]	1	39%	0%
Prairie, Native Tallgrass	3	62%	17.40%
Prairie, Native Tallgrass Dominated by Andropogon gerardii	2	13.50%	9.50%
Prickly Pear [Opuntia ficus-indica]	9	38.20%	13.30%
Prorocentrum minimum [A Marine Dinoflagellate]	2	20%	3.50%
Prosopis flexuosa [Deciduous Tree]	1	42%	0%
Prosopis glandulosa [Honey Mesquite]	13	37.70%	5.20%

Plant Name	# of Studies	Arithmetic Mean	Standard Error
Prunella vulgaris [Selfheal]	2	78%	17.70%
Prunus avium [Sweet Cherry]	8	59.80%	7.80%
Prunus laurocerasus [English Laurel]	1	56%	0%
Prunus persica [Peach Tree]	4	27.80%	0.80%
Prunus serotina [Black Cherry]	2	42%	1.40%
Pseudotsuga menziesii [Douglas Fir]	6	9.70%	3.90%
Pteridium aquilinum [Bracken]	1	0%	0%
Puccinellia maritima [Seaside Alkaligrass]	4	61%	31%
Pueraria lobata (Leguminous Weed)	1	98%	0%
Purple Witchweed [Striga hermonthica]	1	-65%	0%
Pyrrosia piloselloides [Tropical Fern]	1	78%	0%
Quercus alba L. [White Oak]	6	146.70%	30.10%
Quercus cerrioides [Oak]	1	35%	0%
Quercus chapmanii Sargenti [Chapman's Oak]	2	325%	123.70%
Quercus geminata [Sand Live Oak]	5	9.40%	5.40%
Quercus ilex L. [Holly Oak]	4	38%	4.80%
Quercus margaretta [Sand Post Oak]	2	20.50%	14.50%
Quercus mongolica [Mongolian Oak]	6	54.30%	19.50%
Quercus myrtifolia Wild. [Myrtle Oak]	8	140.30%	63.20%
Quercus petraea (Mattuschka) Liebl. [Durmast Oak]	6	53.20%	18.70%
Quercus robur L. [Pedunculate Oak]	8	30.60%	12.80%
Quercus rubra L. [Northern Red Oak]	7	55.30%	25.20%
Quercus suber L. [Cork Oak]	4	46.80%	13.70%
Radish, Wild [Raphanus sativus x raphanistrum]	1	33%	0%
Radish, Wild [Raphanus sativus]	18	75.30%	14.30%
Ragweed, Annual [Ambrosia artemisiifolia]	13	28.80%	7%
Ragweed, Great [Ambrosia trifida]	3	26%	8.20%
Ragwort [Senecio jacobea]	1	21%	0%
Rape, Oilseed [Brassica napus]	17	52.90%	8.90%
Raphanus sativus [Wild Radish]	18	75.30%	14.30%
Raphanus sativus x raphanistrum [Wild Radish]	1	33%	0%
Reedgrass, Chee [Calamagrostis epigeios]	2	98.50%	0.40%
Rhinanthus alectorolophus [European Yellowrattle]	2	75%	68.60%
Rhinanthus minor [Yellow Rattle]	1	50%	0%
Rhizophora mangle [American Mangrove]	3	39.30%	6.10%
Rice [Oryza sativa]	137	34.30%	2.10%
Ricinus communis [Castor Bean]	7	53.60%	16.30%
Robinia pseudoacacia [Black Locust]	4	346%	242.10%
Rosa hybrida [Rose]	4	26.50%	7.90%

Plant Name	# of Studies	Arithmetic Mean	Standard Error
Rose [Rosa hybrida]	4	26.50%	7.90%
Rose, Sun [Helianthemum nummularium]	1	0%	0%
Rosemary, Wild [Ledum palustre]	1	0%	0%
Rubus [Blackberry]	1	675%	0%
Rudbeckia hirta [Black-eyed Susan]	1	0%	0%
Rumex acetosella [Common Sheep Sorrel]	2	24%	2.10%
Rumex obtusifolius [Bitter Dock]	6	18.70%	12.50%
Ryegrass, Darnel [Lolium temulentum]	3	44.70%	29.20%
Ryegrass, Italian [Lolium multiflorum]	6	11.50%	5.10%
Ryegrass, Perennial [Lolium perenne]	74	35.10%	5.30%
Saccharina latissima [Sugar Kelp]	74	35.10%	5.30%
Saccharum officinarum [Sugarcane]	3	25.70%	8.10%
Sage, Pitcher [Salvia pitcheri]	2	25.50%	2.50%
Sagebrush, Big [Artemisia tridentata]	4	24.30%	6.10%
Salvation Jane [Echium plantagineum]	4	15.80%	6.20%
Salvia pitcheri [Pitcher Sage]	2	25.50%	2.50%
Sanguisorba minor [Small Burnet]	6	81.20%	15.20%
Schima superba[Subtropical Tree]	2	27.50%	11%
Schizachyrium scoparium [Little Bluestem]	9	18.20%	9%
Schoenoplectus americanus [Chairmaker's Bulrush]	1	27%	0%
Scindapsus aureus [Pothos]	1	39%	0%
Scirpus lacustris [Softstem Bulrush]	1	-2%	0%
Scirpus olneyi [Salt Marsh Sedge]	10	17.40%	7.70%
Seaweed [Lomentaria articulata]	1	269%	0%
Sedge, Beaked [Carex rostrata]	1	44%	0%
Sedge, Bigelow's [Carex bigelowii]	1	0%	0%
Sedge, Heath [Carex flacca]	5	73.80%	18.20%
Sedge, Salt Marsh [Scirpus olneyi]	10	17.40%	7.70%
Selfheal [Prunella vulgaris]	2	78%	17.70%
Senecio jacobea [Ragwort]	1	21%	0%
Senecio vulgaris [Common Groundsel]	2	66%	25.50%
Sesbania vesicaria [Bagpod]	1	60%	0%
Setaria glauca [Yellow Bristle Grass]	1	16%	0%
Seteria faberi [Japanese Bristlegrass]	5	18.80%	6.70%
Seteria viridis [Green Bristlegrass]	1	18%	0%
Sheep Sorrel, Common [Rumex acetosella]	2	24%	2.10%
Shorea leprosula	2	38.50%	11%
Silene cretica in mixed stands with Linum usitatissimum [Common Flax]	2	98.50%	17.30%
Silene latifolia [White Campion]	6	43.50%	9%

Plant Name	# of Studies	Arithmetic Mean	Standard Error
Silene noctiflora [Night-flowering Catchfly]	1	44%	0%
Sinapis alba [White Mustard]	4	21.30%	4.10%
Sleepingplant [Cassia fasciculata]	2	17.50%	27.20%
Smartweed, Pennsylvania [Polygonum pensylvanicum]	3	55.70%	1.80%
Soft Rush [Juncus effuses]	1	-5%	0%
Softstem Bulrush [Scirpus lacustris]	1	-2%	0%
Solanum curtilobum [Shortlobe Solanum]	2	63%	5.70%
Solanum dulcamara [Climbing Nightshade]	10	46.50%	9.30%
Solanum lycopersicum [Tomato]	2	152.50%	23%
Solanum melongena [Eggplant]	1	41%	0%
Solanum muricatum [Pepino]	4	69.80%	20.50%
Solanum tuberosum [White Potato]	33	29.50%	3.60%
Solanum, Shortlobe [Solanum curtilobum]	2	63%	5.70%
Solidago canadensis [Canadian Goldenrod]	2	1750%	1237.40%
Solidago rigida [Stiff Goldenrod]	1	27%	0%
Sorghastrum nutans [Yellow Indian Grass]	3	9.70%	14.40%
Sorghum [Sorghum bicolor]	24	18.50%	3.60%
Sorghum bicolor [Sorghum]	24	18.50%	3.60%
Sorghum halepense [Johnsongrass]	3	-13%	14.30%
Sorghum sudanense [Sudan Grass]	1	14%	0%
Sour Orange Tree [Citrus aurantium]	3	61.30%	13.30%
Soybean [Glycine max]	162	47.60%	3.10%
Spartina alterniflora [Smooth Cordgrass]	2	0%	0%
Spartina anglica [Common Cordgrass]	3	19%	25.10%
Spartina patens [Saltmeadow Cordgrass]	9	9.70%	4.90%
Spergula arvensis [Corn Spurrey]	2	-20%	7.80%
Sphagnum cuspidatum [Toothed Sphagnum]	1	42%	0%
Sphagnum magellanicum [Magellan's Sphagnum]	1	26%	0%
Sphagnum papillosum [Papillose Sphagnum]	2	57.50%	0.40%
Sphagnum recurvum [Recurved Sphagnum]	4	12.30%	7%
Sphagnum, Magellan's [Sphagnum magellanicum]	1	26%	0%
Sphagnum, Papillose [Sphagnum papillosum]	2	57.50%	0.40%
Sphagnum, Recurved [Sphagnum recurvum]	4	12.30%	7%
Sphagnum, Toothed [Sphagnum cuspidatum]	1	42%	0%
Spider Lily [Hymenocallis littoralis]	2	52%	2.80%
Spinach [Spinacia oleracea]	2	17.50%	1.80%
Spinacia oleracea [Spinach]	2	17.50%	1.80%
Sporobolus pyramidalis [Whorled Dropseed]	6	33.50%	21.70%
Spring Vetch [Vicia lathyroides]	1	114%	0%

Plant Name	# of Studies	Arithmetic Mean	Standard Error
Spruce [Picea koraiensis]	7	37.90%	8.70%
Spruce, Black [Picea mariana]	14	22.80%	3%
Spruce, Norway [Picea abies]	11	35.90%	5.50%
Spruce, Sitka [Picea sitchensis]	7	20.70%	5%
Spruce, White [Picea glauca]	2	82%	49.50%
Spurrey, Corn [Spergula arvensis]	2	-20%	7.80%
St. John's Wort [Hypericum perforatum]	1	72%	0%
Stalkgrass, Broad [Pharus latifolius]	1	144%	0%
Stipa thurberiana [Thurber Needlegrass]	1	11%	0%
Strawberry, Hybrid [Fragaria x ananassa]	4	42.80%	13.10%
Striga hermonthica [Purple Witchweed]	1	-65%	0%
Stylosanthes scabra [Pencilflower]	2	72%	10.60%
Subterranean Clover [Trifolium subterraneum]	1	756%	0%
Sugarcane [Saccharum officinarum]	3	25.70%	8.10%
Sugargum [Eucalyptus cladocalyx]	2	87.50%	9.50%
Sunflower [Helianthus annus]	11	37.50%	7.80%
Sunn Hemp [Crotalaria juncea]	2	45%	9.90%
Sweet Orange Tree [Citrus sinensis]	2	38.50%	4.60%
Sweet Vernal Grass [Anthoxanthum odoratum]	1	170%	0%
Sweetgum [Liquidambar styraciflua]	20	132.40%	23.50%
Switchgrass [Panicum virgatum]	1	-1%	0%
Sycamore [Acer pseudoplatanus]	3	34.30%	12.80%
Symphiocarpos orbiculatus [Buck Brush]	2	98.50%	42.10%
Tamarack [Larix Laricina]	1	56%	0%
Taxodium distichum [Bald Cypress]	5	28%	5.40%
Taxus baccata [English Yew]	5	28%	5.40%
Teak, Common [Tectona grandis]	6	54.20%	5.40%
Tectona grandis [Teak, Common]	6	54.20%	5.40%
Terminalia arjuna [Terminalia]	8	190.60%	35.10%
Terminalia chebula [Myrobalan]	8	442.50%	52.50%
Thale Cress [Arabidopsis thaliana]	10	219.70%	81.40%
Themeda triandra [Kangaroo Grass]	8	61.80%	16.20%
Thinouia tomocarpa Standley	2	94%	28.30%
Thistle, Canadian [Cirsium arvense]	3	83.30%	33.50%
Thrift Seapink Needlegrass [Armeria maritima]	1	-45%	0%
Thurber Needlegrass [Stipa thurberiana]	1	11%	0%
Thyme, Water [Hydrilla verticillata]	7	21.90%	3.30%
Tidytips, Coastal [Layia platyglossa]	1	12%	0%
Timothy [Phleum pratense]	18	12.20%	3.20%

Plant Name	# of Studies	Arithmetic Mean	Standard Error
Tobacco, Cultivated [Nicotiana tabacum]	6	60.50%	11%
Tomato [Lycopersicon lycopersicum]	2	29.50%	5.30%
Tomato [Solanum lycopersicum]	2	152.50%	23%
Tomato, Garden [Lycopersicon esculentum]	35	31.90%	5%
Toxicodendron radicans [Poison Ivy]	7	75.70%	13.70%
Trachypogon plumosus [C4 South American Grass]	2	-28.50%	3.20%
Tree, Tropical Savanna [Kielmeyera coriacea]	3	88.70%	55.50%
Trefoil, Big [Lotus pedunculatus]	6	56%	26.50%
Tridens flavus [Purpletop Tridens]	1	0%	0%
Tridens, Purpletop [Tridens flavus]	1	0%	0%
Trifolium alexandrium [Berseem]	1	41%	0%
Trifolium incarnatum [Crimson Clover]	2	21.50%	0.40%
Trifolium pratense [Purple Clover]	5	16.80%	4.40%
Trifolium repens [White Clover]	45	59.80%	16.80%
Trifolium subterraneum [Subterranean Clover]	1	756%	0%
Trigonella foenum-graecum [Fenugreek]	2	91%	33.20%
Triticum aestivum [Common Wheat]	214	33%	2%
Triticum turgidum [Rivet Wheat]	4	24.50%	2.50%
Tropaeolum majus [Nasturtium]	3	42.30%	10.80%
Tsuga canadensis [Eastern Hemlock]	3	34%	4.50%
Typha latifolia [Cattail]	1	-2%	0%
Ulmus alata [Winged Elm]	1	30%	0%
Understory Herb [Aechmea magdalenae]	1	30%	0%
Urtica dioica [Stinging Nettle]	1	26%	0%
Vaccinium myrtillus [Whortleberry]	6	60.70%	10%
Vallisneria tortifolia [Corkscrew Vallisneria]	2	14%	5.70%
Velvet Leaf [Abutilon theophrasti]	14	32.10%	13.70%
Velvetgrass, Common [Holcus lanatus]	11	38.50%	10.40%
Veronica didyma [Gray Field Speedwell]	1	26%	0%
Vetch, Bird [Vicia cracca]	1	47%	0%
Vetch, Birdfoot Deer [Lotus corniculatus]	8	49.30%	12.90%
Vetch, Common Kidney [Anthyllis vulneraria]	6	45.30%	17.50%
Vicia cracca [Bird Vetch]	1	47%	0%
Vicia faba [Faba Bean]	4	46.30%	13.50%
Vicia lathyroides [Spring Vetch]	1	114%	0%
Vigna angularis [Adsuki Bean]	4	60%	5.20%
Vigna mungo [Blackgram]	2	87%	19.80%
Vigna radiata [Mungbean]	2	31.50%	16.60%
Vigna unguiculata [Blackeyed Pea]	3	83.70%	12.20%

Plant Name	# of Studies	Arithmetic Mean	Standard Error
Viola x wittrockiana [Pansy]	1	30%	0%
Vulpia microstachys [Small Fescue]	2	-4.50%	8.10%
Wallaby Grass [Austrodanthonia caespitosa]	2	78%	4.20%
Water Hyacinth, Common [Eichhornia crassipes]	5	51%	19.30%
Water Lily [Nymphaea marliac]	2	162%	76.40%
Wattle, Black [Acacia mearnsii]	1	48%	0%
Wattle, Orange [Acacia saligna]	1	55%	0%
Wattle, Silver [Acacia dealbata]	1	106%	0%
Wattle, Wiry [Acacia coriacea]	1	50%	0%
Weed, Annual, C4 [Eragrostis orcuttiana]	1	462%	0%
Weed, Leguminous [Pueraria lobata]	1	98%	0%
Weeds, Unspecified	2	0%	0%
Wheat, Common [Triticum aestivum]	214	33%	2%
Wheat, Rivet [Triticum turgidum]	4	24.50%	2.50%
Wheatgrass, Western [Agropyron smithii]	3	50.30%	42.20%
Wheatgrass, Western [Pascopyrum smithii]	4	73.50%	16.80%
White Burrobush [Ambrosia dumosa]	2	164%	36.10%
Whitestar [Ipomoea lacunosa]	2	15%	14.80%
Whortleberry [Vaccinium myrtillus]	6	60.70%	10%
Woollybutt, Darwin [Eucalyptus miniata]	9	-2%	5.40%
Xanthium strumarium var. canadense [Canada Cockleburr]	7	30.60%	6.40%
Yarrow [Achillea millefolium]	3	24.70%	12.60%
Yellow Bristle Grass [Setaria glauca]	1	16%	0%
Yellow Rattle [Rhinanthus minor]	1	50%	0%
Yellowrattle, European [Rhinanthus alectorolophus]	2	75%	68.60%
Yellowtops, Clustered [Flaveria trinervia]	1	46%	0%
Yucca brevifolia [Joshua Tree]	1	65%	0%
Yucca schidigera [Mojave Yucca]	1	86%	0%
Yucca whipplei [Chaparral Yucca]	1	13%	0%
Yucca, Chaparral [Yucca whipplei]	1	13%	0%
Yucca, Mojave [Yucca schidigera]	1	86%	0%
Zea mays [Corn]	20	21.30%	4.90%
Zizania latifolia [Manchurian Wildrice]	1	-5%	0%
Zostera marina [Eelgrass, Common]	1	24%	0%

[this page intentionally blank]

Appendix 3

Table 7.1.2 – Plant Photosynthesis (Net CO₂ Exchange Rate) Responses to Atmospheric CO₂ Enrichment

Table 7.1.2 reports the results of peer-reviewed scientific studies measuring the photosynthetic growth response of plants to a 300-ppm increase in atmospheric CO_2 concentration. Plants are listed by both common and/or scientific names, followed by the number of experimental studies conducted on each plant, the mean photosynthetic response to a 300-ppm increase in the air's CO_2 content, and the standard error of that mean. Whenever the CO_2 increase was not exactly 300 ppm, a linear adjustment was computed. For example, if the CO_2 increase was 350 ppm and the growth response was a 60 percent enhancement, the adjusted 300-ppm CO_2 growth response was calculated as (300/350) x 60% = 51%.

The data in this table appear by permission of the Center for the Study of Carbon Dioxide and Global Change and were taken from its Plant Growth database as it existed on 23 March 2009. Additional data are added to the database at approximately weekly intervals and can be accessed free of charge at the Center's website at http://www.co2science.org/data/plant_growth/dry/dry_subject.php. This online database also archives information pertaining to the experimental conditions under which each plant growth experiment was conducted, as well as the complete reference to the journal article from which the experimental results were obtained. The Center's online database also lists percent increases in plant photosynthetic rate for 600- and/or 900-ppm increases in the air's CO_2 concentration.

Plant Name	# of Studies	Arithmetic Mean	Standard Error
[Tropical Tree] Pseudobombax septenatum	1	68%	0%
Abelmoschus esculentus [Okra]	1	27%	0%
Abies alba [Silver Fir]	2	37.50%	15.20%
Abutilon theophrasti [Velvet Leaf]	6	46.70%	10.30%
Acacia melanoxylon [Blackwood]	1	19%	0%
Acacia minuta [Coastal Scrub Wattle]	1	21%	0%
Acacia nilotica [Gum Arabic Tree]	2	131.50%	14.50%
Acer mono [Shantung Maple]	4	40.80%	15.10%
Acer rubrum [Red Maple]	17	94.80%	22.60%
Acer saccharinum [Silver Maple]	8	18.50%	7.80%
Acer saccharum [Sugar Maple]	3	64%	9%
Achillea millefolium [Yarrow]	3	45.30%	12.50%
Ackama rosaefolia [Small Bushy Tree]	1	-20%	0%
Actinidia deliciosa [Kiwifruit]	1	113%	0%
Agathis microstachya [Semi-Evergreen Rainforest Tree]	1	50%	0%

Plant Name	# of Studies	Arithmetic Mean	Standard Error
Agathis robusta [Queensland Kauri]	1	56%	0%
Agave deserti [Desert Agave]	2	34.50%	3.20%
Agave salmiana [Pulque Agave]	2	39.50%	6.70%
Agave vilmoriniana [Leaf Succulent]	2	37.50%	26.50%
Agave, Desert [Agave deserti]	2	34.50%	3.20%
Agave, Pulque [Agave salmiana]	2	39.50%	6.70%
Agropyron repens [Couch Grass]	1	33%	0%
Agropyron smithii [Western Wheatgrass]	2	-4.50%	6%
Agrostis canina [Velvet Bentgrass]	1	38%	0%
Agrostis capillaris [Colonial Bentgrass]	3	46.30%	20.20%
Albizia procera [Tall Albizia]	2	121%	19.10%
Albizia, Tall [Albizia procera]	2	121%	19.10%
Alder [Alnus firma]	4	83.50%	6.40%
Alder, Black [Alnus glutinosa]	3	51%	11.80%
Alder, Manchurian [Alnus hirsuta]	7	22.10%	17.80%

Plant Name	# of Studies	Arithmetic Mean	Standard Error
Alder, Red [Alnus rubra]	8	73.90%	10.60%
Alfalfa [Medicago sativa]	17	26.20%	4.10%
Alkaligrass, Seaside [Puccinellia maritima]	5	84%	17.80%
Alnus firma [Alder]	4	83.50%	6.40%
Alnus glutinosa [Black Alder]	3	51%	11.80%
Alnus hirsuta [Manchurian Alder]	7	22.10%	17.80%
Alnus rubra [Red Alder]	8	73.90%	10.60%
Alocasia macrorrhiza [Giant Taro]	2	79%	18.40%
Alpine grassland dominated by Carex curvula	2	66.50%	15.20%
Alternanthera crucis [West Indian Joyweed]	1	91%	0%
Amaranth, Grain [Amaranthus hypochondriacus]	1	9%	0%
Amaranth, Redroot [Amaranthus retroflexus]	4	10.80%	5.30%
Amaranth, Slim [Amaranthus hybridus]	2	2.50%	1.80%
Amaranthus hybridus [Slim Amaranth]	2	2.50%	1.80%
Amaranthus hypochondriacus [Grain Amaranth]	1	9%	0%
Amaranthus retroflexus [Redroot Amaranth]	4	10.80%	5.30%
Amate [Ficus obtusifolia]	1	76%	0%
Ambrosia artemisiifolia [Annual Ragweed]	3	37.70%	5.20%
Ambrosia cordifolia [Tuscon Burr Ragweed]	3	45.70%	3.80%
Ambrosia dumosa [White Burrobush]	4	85%	28%
American Pokeweed [Phytolacca americana]	1	56%	0%
Anacardium excelsum	1	19%	0%
Anagallis arvensis [Scarlet Pimpernel]	1	45%	0%
Ananas comosus [Pineapple]	3	168.30%	55.80%
Andropogon gerardii [Big Bluestem]	12	24.30%	6.60%
Andropogon glomeratus [Bushy Bluestem]	2	-1%	3.50%
Anemone [Anemone raddeana]	2	65%	24.70%
Anemone cylindrica [Candle Anemone]	1	100%	0%
Anemone raddeana [Anemone]	2	65%	24.70%
Anthoxanthum odoratum [Sweet Vernal Grass]	1	109%	0%
Anthyllis vulneraria [Common Kidney Vetch]	4	27.80%	10.30%
Antirrhoea trichantha	1	33%	0%
Apple	2	105.50%	18.70%
Apricot [Prusus armeniaca]	4	62.80%	4.20%
Arabidopsis thaliana [Mouse Ear Cress]	8	61.30%	18.30%
Arachis hypogaea [Peanut]	5	36%	7.40%
Arbutus unedo [Strawberry Tree]	4	91.30%	32.10%
Armeria maritima [Thrift Seapink]	1	46%	0%
Arrhenatherum elatius [Tall Oatgrass]	1	39%	0%
Artemisia tridentata [Big Sagebrush]	2	31%	12.70%

Plant Name	# of Studies	Arithmetic Mean	Standard Error
Ash, Green [Fraxinus lanceolata]	1	60%	0%
Ash, Green [Fraxinus pennsylvarica]	5	62.40%	12.90%
Asparagus officinalis [Garden Asparagus]	1	25%	0%
Asparagus, Garden [Asparagus officinalis]	1	25%	0%
Aspen, Bigtooth [Populus grandidentata]	3	181.70%	60.70%
Aspen, Hybrid [Populus tremula x Populus tremuloides]	3	28.70%	3.30%
Aspen, Quaking [Populus tremuloides]	32	59.40%	6.30%
Aster tripolium [Sea Aster]	5	21.40%	8.50%
Aster, Sea [Aster tripolium]	5	21.40%	8.50%
Avena barbata [Slender Oat]	2	52%	5.70%
Avena fatua [Wild Oat]	2	53%	11.30%
Avena sativa [Red Oat]	2	32.50%	3.90%
Avens [Geum reptans]	1	84%	0%
Azolla pinnata [Water Fern]	2	35%	24.70%
Bahiagrass [Paspalum notatum]	1	24%	0%
Bamboo [Fargesia denudata]	1	40%	0%
Barley [Hordeum vulagare]	13	55.20%	12.60%
Barrelcactus, California [Ferocactus acanthodes]	1	30%	0%
Bean, Castor [Ricinus communis]	2	34%	0%
Bean, Faba [Vicia faba]	7	52.30%	10.50%
Bean, Garden [Phaseolus vulgaris]	24	55.80%	10.30%
Beech, American [Fagus grandifolia]	1	96%	0%
Beech, European [Fagus sylvatica]	13	61.80%	9%
Beech, Japanese [Fagus crenata]	5	33.60%	5.20%
Beech, Myrtle [Nothofagus cunninghamii]	1	55%	0%
Beech, Red [Nothofagus fusca]	3	40%	1.70%
Beet, Common [Beta vulgaris]	7	44.70%	4.90%
Begonia [Begonia x hiemalis]	2	70%	21.20%
Begonia x hiemalis [Begonia]	2	70%	21.20%
Beilschmiedia pendula [Slugwood]	4	21.30%	7.70%
Bentgrass, Colonial [Agrostis capillaris]	3	46.30%	20.20%
Bentgrass, Velvet [Agrostis canina]	1	38%	0%
Beta vulgaris [Common Beet]	7	44.70%	4.90%
Blackwood [Acacia melanoxylon]	1	19%	0%
Bluegrass [Poa cookii]	1	55%	0%
Bluegrass, Annual [Poa annua]	6	62%	14%
Bluegrass, Kentucky [Poa pratensis]	2	103%	45.30%
Bluegrass, Rough [Poa trivialis]	1	41%	0%
Bluestem, Big [Andropogon gerardii]	12	24.30%	6.60%
Bluestem, Bushy [Andropogon glomeratus]	2	-1%	3.50%

Plant Name	# of Studies	Arithmetic Mean	Standard Error
Bluestem, Caucasian [Bothriochloa caucasica]	2	15.50%	11%
Bluestem, Little [Schizachyrium scoparium]	3	72.30%	18.10%
Bothriochloa caucasica [Caucasian Bluestem]	2	15.50%	11%
Bottlebrush Squirreltail [Elymus elymoides]	1	21%	0%
Bouteloua curtipendula [Sideoats Grama]	2	18.50%	2.50%
Bouteloua gracilis [Blue Grama]	3	58.70%	28.90%
Brachychiton populneum [Whiteflower Kurrajong]	1	75%	0%
Brachyglottis repanda [Rangiora]	1	120%	0%
Brachypodium pinnatum [Heath Falsebrome]	1	59%	0%
Bracken [Pteridium aquilinum]	8	93.80%	7.30%
Brassica campestris [Rape Seed Mustard]	2	90.50%	37.10%
Brassica carinata [Abyssinian Mustard]	2	34.50%	10.30%
Brassica juncea [India Mustard]	2	50.50%	24.40%
Brassica napus [Canola]	12	63.50%	20.10%
Brassica nigra [Black Mustard]	2	39%	11.30%
Bristlegrass, Green [Setaria viridis]	1	13%	0%
Bristlegrass, Japanese [Setaria faberi]	4	15%	7.10%
Brittlebush, Button [Encelia frutescens]	1	30%	0%
Brome, Compact [Bromus madritensis]	1	12%	0%
Brome, Erect [Bromus erectus]	6	30.20%	9.50%
Brome, Poverty [Bromus sterilis]	1	45%	0%
Brome, Smooth [Bromus inermis]	2	29%	7.80%
Bromus erectus [Erect Brome]	6	30.20%	9.50%
Bromus inermis [Smooth Brome]	2	29%	7.80%
Bromus madritensis [Compact Brome]	1	12%	0%
Bromus sterilis [Poverty Brome]	1	45%	0%
Bromus tectorum [Cheatgrass]	1	56%	0%
Bryum pseudotriquetrum [Common Green Bryum Moss]	2	47.50%	10.30%
Bryum subrotundifolium	2	28.50%	8.10%
Buchloe dactyloides [Buchloe dactyloides]	1	27%	0%
Buck Brush [Symphiocarpos orbiculatus]	2	105%	43.80%
Buffalo Grass [Buchloe dactyloides]	1	27%	0%
Buffel Grass [Cenchrus ciliaris]	1	31%	0%
Bulrush, Seaside [Scirpus maritimus]	1	55%	0%
Bunchgrass, Perennial [Oryzopsis hymenoides]	1	47%	0%
Burnet, Small [Sanguisorba minor]	5	31.60%	4%
Burr Medick [Medicago minima]	1	0%	0%
Burrobush, White [Ambrosia dumosa]	4	85%	28%
Bush, Bellyache [Jatropha gossypiifolia]	1	73%	0%

Plant Name	# of Studies	Arithmetic Mean	Standard Error
Buttercup [Ranunculus]	1	47%	0%
Buttercup, Glacier [Ranunculus glacialis]	3	44%	8.90%
Buttercup, Tall [Ranunculus acris]	1	36%	0%
Cacao [Theobroma cacao]	1	32%	0%
Cactus, Tree [Stenocereus queretaroensis]	1	30%	0%
Cajanus cajan [Pigeonpea]	2	137.50%	26.50%
Calamagrostis epigeios [Chee Reedgrass]	1	27%	0%
Calamondin [Citrus madurensis]	2	195%	74.20%
Calcareous Grassland Community in the Swiss Jura Mountains	2	49.50%	3.20%
California Annual Grassland on Sandstone-Derived Soil	2	21%	3.50%
California Annual Grassland on Serpentine-Derived Soil	2	11%	0.70%
Calluna vulgaris [Heather]	2	49.50%	8.10%
Calophyllum longifolium [Tropical Tree]	1	-15%	0%
Caloplaxa trachyphylla [Lichen]	1	38%	0%
Camphorweed [Heterotheca subaxillaris]	1	17%	0%
Candle Anemone [Anemone cylindrica]	1	100%	0%
Canola [Brassica napus]	12	63.50%	20.10%
Cantaloupe [Cucumis melo]	4	56.80%	0.70%
Capsicum annuum [Bell Pepper]	3	41%	15.50%
Carex bigelowii [Bigelow's Sedge]	1	0%	0%
Carex curvula dominated alpine grassland	2	66.50%	15.20%
Carex flacca [Heath Sedge]	1	55%	0%
Carex paleacea [Chaffy Sedge]	1	34%	0%
Carob [Ceratonia siliqua]	6	40.50%	7.50%
Carpinus betulus [European Hornbeam]	4	14.80%	6.40%
Carrizo citrange [Citrus sinensis x Poncirus trifoliata]	2	76.50%	9.50%
Carya glabra Sweet [Pignut Hickory]	3	57.30%	22.80%
Carya ovata [Shagbark Hickory]	1	35%	0%
Cassava [Manihot esculenta Crantz]	1	56%	0%
Castanea sativa [Sweet Chesnut]	2	66.50%	37.80%
Cecropia longipes	1	17%	0%
Cenchrus ciliaris	1	31%	0%
Ceratonia siliqua [Carob]	6	40.50%	7.50%
Cercis canadensis [Eastern Redbud]	3	127.30%	48%
Chairmaker's Bulrush [Schoenoplectus americanus]	1	27%	0%
Chamaenerion angustifolium [Narrowleaved Fireweed]	1	96%	0%
Cheatgrass [Bromus tectorum]	1	56%	0%
Chenopodium album [Lambsquarters]	3	23.30%	3.40%
Cherry, Black [Prunus serotina]	3	63.70%	5.70%

Plant Name	# of Studies	Arithmetic Mean	Standard Error
Cherry, Sweet [Prunus avium]	1	38%	0%
Chesnut, Sweet [Castanea sativa]	2	66.50%	37.80%
Cinquefoil, Alpine [Potentilla crantzii]	1	25%	0%
Cirsium arvense [Canadian Thistle]	4	56.50%	8%
Cistus [Cistus salviifolius]	1	53%	0%
Cistus salviifolius [Cistus]	1	53%	0%
Citrus aurantium [Sour Orange]	12	111.60%	15.30%
Citrus madurensis [Calamondin]	2	195%	74.20%
Citrus paradisi Macfad. budded to Citrus reticulata Blanco (Cleopatra mandarin) rootstock [Marsh Grapefruit]	2	60%	4.90%
Citrus paradisi Macfad. budded to Poncirus trifoliata (L.) Raf. rootstock [Marsh Grapefruit]	2	70%	14.10%
Citrus reticulata [Mandarin Orange]	2	45.50%	8.80%
Citrus sinensis (L.) Osbeck budded to Citrus reticulata Blanco (Cleopatra manderin) rootstock [Washington Naval Orange]	2	52.50%	10.30%
Citrus sinensis (L.) Osbeck budded to Poncirus trifoliata (L.) Raf. rootstock [Washington Naval Orange]	2	43.50%	8.10%
Citrus sinensis (L.) Osbeck x Citrus reticulata Blanco (Cleoplatra manderin) [Valencia Orange]	2	42.50%	15.90%
Citrus sinensis (L.) Osbeck x Poncirus trifoliata (L.) Raf. [Valencia Orange]	2	54.50%	5.30%
Citrus sinensis [Sweet Orange]	2	57%	4.90%
Citrus sinensis x Poncirus trifoliata [Carrizo Citrange]	2	76.50%	9.50%
Cladonia rangiferina (L.) Wigg. [Fruticose Lichen]	1	31%	0%
Claoxylon sandwicense [Po'ola]	3	100%	12.50%
Climbing Nightshade [Solanum dulcamara]	6	28.80%	3.10%
Clover, Subterranean [Trifolium subterraneum]	1	19%	0%
Clover, White [Trifolium repens]	22	49.40%	7.40%
Coastal Scrub Wattle [Acacia minuta]	1	21%	0%
Cocklebur, Rough [Xanthium strumarium]	4	30.80%	12.90%
Codlins and Cream [Epilobium hirsutum]	1	62%	0%
Coffea arabusta [Coffee]	2	271%	65.10%
Coffee [Coffea arabusta]	2	271%	65.10%
Collema furfuraceum [Jelly Lichen]	1	42%	0%
Conebush, Dune [Leucadendron coniferum]	2	38%	14.10%
Conebush, Golden [Leucadendron laureolum]	2	28.50%	5.30%
Conebush, Limestone [Leucadendron meridianum]	2	18%	10.60%
Conebush, Sickle-leaf [Leucadendron xanthoconus]	2	59%	9.90%
Coolibah Tree [Eucalyptus microtheca]	1	57%	0%
Cordgrass, Common [Spartina anglica]	1	33%	0%
Cordgrass, Saltmeadow [Spartina patens]	5	12.80%	5.10%
Cordgrass, Smooth [Spartina alterniflora]	1	19%	0%
Cordia alliodora [Spanish Elm]	1	41%	0%
Corn [Zea mays]	21	28.50%	11.50%

Plant Name	# of Studies	Arithmetic Mean	Standard Error
Corynocarpus laevigatus [Karaka Nut]	1	79%	0%
Cotton [Gossypium hirsutum]	18	46.40%	5.50%
Cottonwood, Eastern [Populus deltoides]	5	39.20%	9%
Cottonwood, Fremont's [Populus fremontii]	3	30.70%	15.90%
Couch Grass [Agropyron repens]	1	33%	0%
Couch Grass [Elymus repens]	1	176%	0%
Crabgrass, Hairy [Digitaria sanguinalis]	3	75.30%	32.70%
Creosote Bush [Larrea tridentata]	10	103.60%	24.40%
Cress, Mouse Ear [Arabidopsis thaliana]	8	61.30%	18.30%
Cucumber, Garden [Cucumis sativus]	3	155%	107.40%
Cucumis melo [Cantaloupe]	4	56.80%	0.70%
Cucumis sativus [Garden Cucumber]	3	155%	107.40%
Cyperus esculentus [Yellow Nutsedge]	1	-26%	0%
Cyperus rotundus [Purple Nutsedge]	1	6%	0%
Cyprus, Bald [Taxodium distichum]	5	88.60%	14.40%
Dactylis glomerata [Orchardgrass]	11	34.70%	6.20%
Daucus carota [Carrot]	8	105.30%	38%
Dendrosenecio brassica [Afro-alpine Giant Rosette Plant]	1	65%	0%
Dendrosenecio keniodendron [Afro-alpine Giant Rosette Plant]	1	52%	0%
Desert Eveningprimrose [Oenothera primiveris]	1	23%	0%
Digitalis purpurea [Purple Foxglove]	1	78%	0%
Digitaria sanguinalis [Hairy Crabgrass]	3	75.30%	32.70%
Dodonaea viscosa [Florida Hopbush]	1	42%	0%
Dropwort [Filipendula vulgaris]	2	218%	101.80%
Duckweed, Swollen [Lemna gibba]	2	55.50%	3.90%
Echinochloa crus-galli [Barnyard Grass]	7	24.90%	5%
Ecosystem, Ten Species of Tropical Forest Tree Seedlings	1	10%	0%
Ecosystem, Understory Plants in a Spruce Model Ecosystem	2	53%	9.90%
Elatostema repens [Tropical Rain Forest Herb]	1	68%	0%
Eleusine indica [Indian Goosegrass]	1	72%	0%
Elymus elymoides [Bottlebrush Squirreltail]	1	21%	0%
Elymus repens [Couch Grass]	1	176%	0%
Emblic [Phyllanthus emblica]	2	71%	21.90%
Emiliania huxleyi [Marine Coccolithophorid]	5	47.80%	13.10%
Encelia frutescens [Button Brittlebush]	1	30%	0%
Entandrophragma angolense [West African Mahogany]	3	63.30%	7.20%
Epilobium hirsutum [Codlins and Cream]	1	62%	0%
Eragrostis orcuttiana [Annual Weed, C4]	1	-44%	0%
Eriogonum inflatum [Native American Pipeweed]	2	42.50%	15.90%
Eucalyptus and Acacia [Forest Canopy in Lysimeter]	1	53%	0%

Plant Name	# of Studies	Arithmetic Mean	Standard Error
Eucalyptus cladocalyx [Sugargum]	2	47.50%	27.20%
Eucalyptus microtheca [Coolibah Tree]	1	57%	0%
Eucalyptus polyanthemus [Silver Dollar Gum]	1	75%	0%
Eucalyptus tetrodonta	3	26.30%	8.40%
EUROFACE Poplar Plantation	1	82%	0%
European Hornbeam [Carpinus betulus]	4	14.80%	6.40%
European Larch [Larix decidua]	3	62.70%	4.80%
Evernia mesomorpha [Ring Lichen]	1	60%	0%
Fagus crenata [Japanese Beech]	5	33.60%	5.20%
Fagus grandifolia [American Beech]	1	96%	0%
Fagus sylvatica [European Beech]	13	61.80%	9%
Falsebrome, Heath [Brachypodium pinnatum]	1	59%	0%
Fargesia denudata [Bamboo]	1	40%	0%
Feijoa sellowiana [Pineapple Guava]	1	55%	0%
Fern, Tropical [Pyrrosia piloselloides]	1	19%	0%
Fern, Water [Azolla pinnata]	2	35%	24.70%
Ferocactus acanthodes [California Barrelcactus]	1	30%	0%
Fescue [Festuca rupicola]	2	112.50%	54.10%
Fescue, Meadow [Festuca pratensis]	2	34.50%	1.80%
Fescue, Small [Vulpia microstachys]	1	125%	0%
Fescue, Tall [Festuca arundinacea]	2	46.50%	2.50%
Festuca arundinacea [Tall Fescue]	2	46.50%	2.50%
Festuca pratensis [Meadow Fescue]	2	34.50%	1.80%
Festuca rupicola [Fescue]	2	112.50%	54.10%
Ficus insipida	1	47%	0%
Ficus obtusifolia [Amate]	1	76%	0%
Figwort, Desert [Scrophularia desertorum]	3	50%	21%
Filipendula vulgaris [Dropwort]	2	218%	101.80%
Fir, Douglas [Pseudotsuga menziesii]	17	32%	5.30%
Fir, Silver [Abies alba]	2	37.50%	15.20%
Fireweed, Narrowleaved [Chamaenerion angustifolium]	1	96%	0%
Fivespot, Desert [Malvastrum rotundifolium]	3	33%	5.70%
Flaveria floridana [Florida Yellowtops]	1	25%	0%
Flaveria pringlei [Yellowtops]	1	33%	0%
Flaveria trinervia [Clustered Yellowtops]	3	9%	3.70%
Florida Hopbush [Dodonaea viscosa]	1	42%	0%
Fontinalis antipyretica [Antifever Fontinalis Moss]	2	235%	166.20%
Forest Canopy in Lysimeter [Eucalyptus spp. and Acacia spp.]	1	53%	0%
Foxglove, Purple [Digitalis purpurea]	1	78%	0%

Plant Name	# of Studies	Arithmetic Mean	Standard Error
Fragaria x ananassa [Hybrid Strawberry]	14	72.60%	12.50%
Fraxinus lanceolata [Green Ash]	1	60%	0%
Fraxinus pennsylvanica [Green Ash]	5	62.40%	12.90%
Geum reptans [Avens]	1	84%	0%
Geum rivale [Water Avens]	1	31%	0%
Ginkgo biloba [Maidenhair Tree]	4	57.50%	7.60%
Gloria de la Manana [Ipomoea carnea]	1	62%	0%
Glycine max [Soybean]	75	56.20%	9.40%
Goldenclub [Orontium aquaticum]	7	72.10%	9.30%
Goldfields, California [Lasthenia californica]	1	47%	0%
Goosegrass, Indian [Eleusine indica]	1	72%	0%
Gossypium hirsutum [Cotton]	18	46.40%	5.50%
Grama, Blue [Bouteloua gracilis]	3	58.70%	28.90%
Grama, Sideoats [Bouteloua curtipendula]	2	18.50%	2.50%
Grape, California Wild [Vitis californica]	1	57%	0%
Grapefruit, Marsh (Citrus paradisi Macfad. budded to Poncirus trifoliata L. Raf. rootstock)	2	70%	14.10%
Grapefruit, Marsh [Citrus paradisi Macfad. budded to Citrus reticulata Blanco (Cleopatra mandarin) rootstock]	2	60%	4.90%
Grass, Barnyard [Echinochloa crus-galli]	7	24.90%	5%
Grass, C3 [Koeleria cristata]	3	317.30%	237.90%
Grass, Dallas [Paspalum dilatatum]	4	80.80%	18.40%
Grass, Harding [Phalaris aquatica]	2	39.50%	17.30%
Grassland Community, Annual	4	57.30%	22.40%
Grassland dominated by Carex curvula	2	66.50%	15.20%
Guineagrass [Panicum maximum]	2	16%	3.50%
Gum Arabic Tree [Acacia nilotica]	2	131.50%	14.50%
Heather [Calluna vulgaris]	2	49.50%	8.10%
Helianthus annuus [Sunflower]	13	41.50%	6.10%
Helianthus petiolaris [Prairie Sunflower]	2	20%	0%
Herb, Tropical Rain Forest [Elatostema repens]	1	68%	0%
Heterotheca subaxillaris [Camphorweed]	1	17%	0%
Hevea brasiliensis [Rubber Tree]	8	85.90%	3.90%
Hickory, Pignut [Carya glabra]	3	57.30%	22.80%
Hickory, Shagbark [Carya ovata]	1	35%	0%
Holcus lanatus [Common Velvetgrass]	1	85%	0%
Holly-Leaved Daisybush [Olearia ilicifolia]	1	79%	0%
Honduras Mahogany [Swietenia macrophylla]	1	26%	0%
Hordeum vulagare [Barley]	13	55.20%	12.60%
Hylocomium splendens [Splendid Feather Moss]	1	100%	0%
Hymenaea courbaril [Stinkingtoe]	1	70%	0%

Plant Name	# of Studies	Arithmetic Mean	Standard Error
Hyparrhenia rufa [Jaragua]	1	70%	0%
Indian Grass [Sorghastrum nutans]	1	19%	0%
Indian Hawthorn [Rhaphiolepsis indica]	1	79%	0%
Ipomoea batatas [Sweet Potato]	5	39.40%	10.80%
Ipomoea carnea [Gloria de la Manana]	1	62%	0%
Japanese Knotweed [Polygonum cuspidatum]	3	32.70%	3.50%
Japanese Larch [Larix kaempferi]	4	38.80%	16.80%
Jaragua [Hyparrhenia rufa]	2	5%	0%
Jatropha gossypiifolia [Bellyache Bush]	1	73%	0%
Johnsongrass [Sorghum halapense]	1	12%	0%
Joyweed, West Indian [Alternanthera crucis]	1	91%	0%
Juglans nigra [Black Walnut]	1	24%	0%
Kalankoe blossfeldiana	1	47%	0%
Karaka Nut [Corynocarpus laevigatus]	1	79%	0%
Kauri, Queensland [Agathis robusta]	1	56%	0%
Kiwifruit [Actinidia deliciosa]	1	113%	0%
Knotweed, Giant [Polygonum sachalinense]	2	43%	2.10%
Koeleria cristata [C3 Grass]	3	317.30%	237.90%
Krameria erecta[Littleleaf Ratany]	4	115.80%	28.60%
Lactuca serriola [Prickly Lettuce]	1	60%	0%
Lambsquarters [Chenopodium album]	3	23.30%	3.40%
Larix decidua [European Larch]	3	62.70%	4.80%
Larix kaempferi [Japanese Larch]	4	38.80%	16.80%
Larix Laricina [Tamarack]	1	47%	0%
Larrea tridentata [Creosote Bush]	10	103.60%	24.40%
Lasthenia californica [California Goldfields]	1	47%	0%
Leaf Succulent [Agave vilmoriniana]	2	37.50%	26.50%
Lecanora muralis [An Epilithic Lichen]	1	42%	0%
Ledum palustre [Wild Rosemary]	1	0%	0%
Lemna gibba [Swollen Duckweed]	2	55.50%	3.90%
Lettuce, Prickly [Lactuca serriola]	1	60%	0%
Leucadendron coniferum [Dune Conebush]	2	38%	14.10%
Leucadendron laureolum [Golden Conebush]	2	28.50%	5.30%
Leucadendron meridianum [Limestone Conebush]	2	18%	10.60%
Leucadendron xanthoconus [Sickle-leaf Conebush]	2	59%	9.90%
Lichen [Caloplaxa trachyphylla]	1	38%	0%
Lichen, Epilithic [Lecanora muralis]	1	42%	0%
Lichen, Felt [Peltigera canina]	1	24%	0%
Lichen, Felt [Peltigera rufescens]	1	24%	0%
Lichen, Flavopunctelia [Parmelia praesignis]	1	60%	0%

Plant Name	# of Studies	Arithmetic Mean	Standard Error
Lichen, Foliose [Parmelia caperata]	1	16%	0%
Lichen, Foliose [Parmelia kurokawae]	1	30%	0%
Lichen, Foliose [Peltigera polydactyla]	1	30%	0%
Lichen, Fruticose [Cladonia rangiferina]	1	31%	0%
Lichen, Jelly [Collema furfuraceum]	1	42%	0%
Lichen, Membraneous Felt [Peltigera membranacea]	1	83%	0%
Lichen, Menzies' Cartilage [Ramalina menziesii]	1	20%	0%
Lichen, Netted Rimelia [Parmelia reticulata]	1	63%	0%
Lichen, Ring [Evernia mesomorpha]	1	60%	0%
Liquidambar styraciflua [Sweetgum]	18	105.10%	21.80%
Liriodendron tulipifera [Yellow Poplar]	13	64.40%	12.40%
Lobelia telekii [Afro-alpine Giant Rosette Plant]	1	52%	0%
Lolium perenne [Perennial Ryegrass]	28	42.40%	3.80%
Lolium temulentum [Darnel Ryegrass]	2	23.50%	1.80%
Luehea seemannii	2	20%	2.10%
Lupine, Arizona [Lupinus arizonicus]	3	15.70%	6.50%
Lupine, Sundial [Lupinus perennis]	3	71.30%	32.10%
Lupinus arizonicus [Arizona Lupine]	3	15.70%	6.50%
Lupinus perennis [Sundial Lupine]	3	71.30%	32.10%
Lycopersicon esculentum [Garden Tomato]	13	22.60%	5%
Lyonia mariana [Piedmont Staggerbush]	1	27%	0%
Mahogany, West African [Entandrophragma angolense]	3	63.30%	7.20%
Maidenhair Tree [Ginkgo biloba]	4	57.50%	7.60%
Malvastrum rotundifolium [Desert Fivespot]	3	33%	5.70%
Mandarin Orange [Citrus reticulata]	2	45.50%	8.80%
Mangrove, American [Rhizophora mangle]	2	11.50%	0.40%
Manihot esculenta [Cassava]	1	56%	0%
Maple, Red [Acer rubrum]	17	94.80%	22.60%
Maple, Shantung [Acer mono]	4	40.80%	15.10%
Maple, Silver [Acer saccharinum]	8	18.50%	7.80%
Marine Coccolithophorid [Emiliania huxleyi]	5	47.80%	13.10%
Medicago glomerata	1	16%	0%
Medicago minima [Burr Medick]	1	0%	0%
Medicago sativa [Alfalfa]	17	26.20%	4.10%
Melinis minutiflora [Molassesgrass]	2	7.50%	12.40%
Metasequoia glyptostroboides [Dawn Redwood]	1	13%	0%
Millet, Broomcorn [Panicum miliaceum]	2	12%	2.10%
Molassesgrass [Melinis minutiflora]	2	7.50%	12.40%
Moss, Antifever Fontinalis [Fontinalis antipyretica]	2	235%	166.20%
Moss, Rough Goose Neck [Rhytidiadelphus triquetrus]	1	66%	0%

Plant Name	# of Studies	Arithmetic Mean	Standard Error
Moss, Splendid Feather [Hylocomium splendens]	1	100%	0%
Mungbean [Vigna radiata]	2	178%	18.40%
Mustard, Abyssinian [Brassica carinata]	2	34.50%	10.30%
Mustard, Black [Brassica nigra]	2	39%	11.30%
Mustard, India [Brassica juncea]	2	50.50%	24.40%
Mustard, Rape Seed [Brassica campestris]	2	90.50%	37.10%
Myrobalan [Terminalia chebula]	2	147.50%	37.10%
Nauclea diderrichii [Pioneer Tropical Tree]	3	69.70%	2.20%
Nettle, Stinging [Urtica dioica]	1	126%	0%
New Zealand Privet [Griselinia littoralis]	1	96%	0%
Nicotiana sylvestris [South American Tobacco]	1	55%	0%
Nicotiana tabacum [Cultivated Tobacco]	7	83%	14.90%
Night-flowering Catchfly [Silene noctiflora]	1	42%	0%
Nostoc commune	1	5%	0%
Nothofagus cunninghamii [Myrtle Beech]	1	55%	0%
Nothofagus fusca [Red Beech]	3	40%	1.70%
Nutsedge, Purple [Cyperus rotundus]	1	6%	0%
Nutsedge, Yellow [Cyperus esculentus]	1	-26%	0%
Nymphaea marliac [Water Lily]	3	39.70%	12.40%
Oak, Chapman's [Quercus chapmanii]	4	53%	4.10%
Oak, Cork [Quercus suber]	3	28.70%	16.90%
Oak, Downy [Quercus pubescens]	5	136.80%	58.50%
Oak, Durmast [Quercus petraea]	2	23.50%	0.40%
Oak, Holly [Quercus ilex]	11	91.40%	30.10%
Oak, Mongolian [Quercus crispula]	2	19%	3.50%
Oak, Mongolian [Quercus mongolica]	2	19%	3.50%
Oak, Myrtle [Quercus myrtifolia]	9	60.60%	7.70%
Oak, Northern Red [Quercus rubra]	14	65.50%	11.70%
Oak, Pedunculate [Quercus robur]	9	35.80%	6.20%
Oak, Sand Live [Quercus geminata]	6	15.70%	5.90%
Oak, White [Quercus alba]	3	142%	38.30%
Oat, Red [Avena sativa]	2	32.50%	3.90%
Oat, Slender [Avena barbata]	2	52%	5.70%
Oat, Wild [Avena fatua]	2	53%	11.30%
Oatgrass, Tall [Arrhenatherum elatius]	1	39%	0%
Oenothera primiveris [Desert Eveningprimrose]	1	23%	0%
Okra [Abelmoschus esculentus]	1	27%	0%
Olea europaea [Olive]	10	50.20%	12.60%
Olearia ilicifolia [Holly-Leaved Daisybush]	1	79%	0%

Plant Name	# of Studies	Arithmetic Mean	Standard Error
Olive [Olea europaea]	10	50.20%	12.60%
Opuntia ficus-indica [Prickly Pear]	6	22.70%	2.60%
Orange, Sour [Citrus aurantium]	12	111.60%	15.30%
Orange, Sweet, Ridge Pineapple [Citrus sinensis]	2	57%	4.90%
Orange, Valencia [Citrus sinensis (L.) Osbeck x Citrus reticulata Blanco (Cleoplatra manderin)]	2	42.50%	15.90%
Orange, Valencia [Citrus sinensis (L.) Osbeck x Poncirus trifoliata (L.) Raf.]	2	54.50%	5.30%
Orange, Washington Naval [Citrus sinensis (L.) Osbeck budded to Citrus reticulata Blanco (Cleopatra manderin) rootstock]	2	52.50%	10.30%
Orange, Washington Naval [Citrus sinensis (L.) Osbeck budded to Poncirus trifoliata (L.) Raf. rootstock]	2	43.50%	8.10%
Orchardgrass [Dactylis glomerata]	11	34.70%	6.20%
Orontium aquaticum [Goldenclub]	7	72.10%	9.30%
Oryza sativa [Rice]	64	49.70%	5.70%
Oryzopsis hymenoides [Perennial Bunchgrass]	1	47%	0%
Pachysandra terminalis [Japanese Spurge]	1	70%	0%
Panicgrass, Blue [Panicum antidotale]	2	-2.50%	4.60%
Panicgrass, Fall [Panicum dichotomiflorum]	1	18%	0%
Panicgrass, Lax [Panicum laxum]	2	8.50%	5.30%
Panicum antidotale [Blue Panicgrass]	2	-2.50%	4.60%
Panicum dichotomiflorum [Fall Panicgrass]	1	18%	0%
Panicum laxum [Lax Panicgrass]	2	8.50%	5.30%
Panicum maximum [Guineagrass]	2	16%	3.50%
Panicum miliaceum [Broomcorn Millet]	2	12%	2.10%
Panicum virgatum [Switchgrass]	1	64%	0%
Parmelia caperata [Foliose Lichen]	1	16%	0%
Parmelia kurokawae [Foliose Lichen]	1	30%	0%
Parmelia praesignis [Flavopunctelia Lichen]	1	60%	0%
Parmelia reticulata [Netted Rimelia Lichen]	1	63%	0%
Pascopyrum smithii [Western Wheatgrass]	5	57.40%	17.30%
Paspalum dilatatum [Dallas Grass]	4	80.80%	18.40%
Paspalum notatum [Bahiagrass]	1	24%	0%
Pasture	1	29%	0%
Pea, Blackeyed [Vigna unguiculata]	4	67.80%	18.50%
Pea, Garden [Pisum sativum]	4	37.80%	5%
Peach [Prunus persica]	4	37.50%	8.70%
Peanut [Arachis hypogaea]	5	36%	7.40%
Peanut, Rhizoma	1	40%	0%
Peltigera canina [Felt Lichen]	1	24%	0%
Peltigera membranacea [Membraneous Felt Lichen]	1	83%	0%
Peltigera polydactyla [Foliose Lichen]	1	83%	0%

Plant Name	# of Studies	Arithmetic Mean	Standard Error
Peltigera rufescens [Felt Lichen]	1	83%	0%
Pepino [Solanum muricatum]	2	74%	7.10%
Pepper, Bell [Capsicum annuum]	3	41%	15.50%
Pepper, Jamaican [Piper hispidum]	2	77.50%	15.90%
Pepper, Vera Cruz [Piper auritum]	2	67.50%	5.30%
Phalaris aquatica [Harding Grass]	2	39.50%	17.30%
Phaseolus vulgaris [Garden Bean]	24	55.80%	10.30%
Phleum pratense [Timothy]	3	30.30%	3.20%
Phyllanthus emblica [Emblic]	2	71%	21.90%
Phytolacca americana [American Pokeweed]	1	56%	0%
Phytoplankton Communities of the Nutrient-Poor Central Atlantic Ocean	2	10.50%	1.10%
Phytoplankton Community of a Fjord in Southern Norway	1	23%	0%
Picea abies [Norway Spruce]	9	48.60%	8.40%
Picea koraiensis [Spruce]	1	73%	0%
Picea mariana [Black Spruce]	5	33.60%	9.20%
Picea sitchensis [Sitka Spruce]	2	23%	4.20%
Pigeonpea [Cajanus cajan]	2	137.50%	26.50%
Pimpernel, Scarlet [Anagallis arvensis]	1	45%	0%
Pine, Eldarica [Pinus eldarica]	2	133%	0%
Pine, Jack [Pinus baksiana]	4	34.80%	6.20%
Pine, Japanese Red [Pinus densiflora]	5	26.40%	8%
Pine, Korean [Pinus koraiensis]	1	53%	0%
Pine, Loblolly [Pinus taeda]	41	96%	10.80%
Pine, Longleaf	1	41%	0%
Pine, Merkus [Pinus merkusii]	2	44.50%	3.20%
Pine, Mexican Yellow [Pinus patula]	1	47%	0%
Pine, Monterey [Pinus radiata]	13	40.20%	3.80%
Pine, Mountain [Pinus uncinata]	2	44%	7.10%
Pine, Ponderosa [Pinus ponderosa]	10	45.90%	8.80%
Pine, Scots [Pinus sylvestris]	16	77.40%	34.40%
Pineapple [Ananas comosus]	3	168.30%	55.80%
Pineapple Guava [Feijoas sellowiana]	1	55%	0%
Pinus banksiana [Pine, Jack]	4	34.80%	6.20%
Pinus densiflora [Pine, Japonese Red]	5	26.40%	8%
Pinus eldarica [Pine, Eldarica]	2	133%	0%
Pinus koraiensis [Pine, Korean]	1	53%	0%
Pinus merkusii [Pine, Merkus]	2	44.50%	3.20%
Pinus patula [Mexican Yellow Pine]	1	47%	0%
Pinus ponderosa [Ponderosa Pine]	10	45.90%	8.80%
Pinus radiata [Monterey Pine]	13	40.20%	3.80%

Plant Name	# of Studies	Arithmetic Mean	Standard Error
Pinus sylvestris [Scots Pine]	16	77.40%	34.40%
Pinus taeda [Loblolly Pine]	41	96%	10.80%
Pinus uncinata [Mountain Pine]	2	44%	7.10%
Pioneer Tropical Tree [Nauclea diderrichii]	3	69.70%	2.20%
Piper auritum [Vera Cruz Pepper]	2	67.50%	5.30%
Piper hispidum [Jamaican Pepper]	2	77.50%	15.90%
Pipeweed, Native American [Eriogonum inflatum]	2	42.50%	15.90%
Pisum sativum [Garden Pea]	4	37.80%	5%
Plantago asiatica [Plantago Asiatica]	1	45%	0%
Plantago erecta [Dwarf Plantain]	1	43%	0%
Plantago lanceolata [Narrowleaf Plantain]	6	72.50%	24.20%
Plantago maritima [Sea Plantain]	1	-30%	0%
Plantain, Dwarf [Plantago erecta]	1	43%	0%
Plantain, Narrowleaf [Plantago lanceolata]	6	72.50%	24.20%
Plantain, Sea [Plantago maritima]	1	-30%	0%
Platanus occidentalis [American Sycamore]	1	60%	0%
Poa annua [Annual Bluegrass]	6	62%	14%
Poa cookii [Bluegrass]	1	55%	0%
Poa pratensis [Kentucky Bluegrass]	2	103%	45.30%
Poa trivialis [Rough Bluegrass]	1	41%	0%
Poison Ivy [Toxicodendron radicans]	1	116%	0%
Polygonum cuspidatum [Japanese Knotweed]	3	32.70%	3.50%
Polygonum sachalinense [Giant Knotweed]	2	43%	2.10%
Po'ola [Claoxylon sandwicense]	3	100%	12.50%
Poplar, Black [Populus nigra]	6	73.20%	18.30%
Poplar, Hybrid [Populus trichocarpa x Populus deltoides]	2	37.50%	8.80%
Poplar, Robusta [Populus euramericana]	24	79.20%	15.20%
Poplar, White [Populus alba]	6	79.80%	16.40%
Poplar, Yellow [Liriodendron tulipifera]	13	64.40%	12.40%
Populus alba [White Poplar]	6	79.80%	16.40%
Populus cathayanna [Poplar]	1	45%	0%
Populus deltoides [Eastern Cottonwood]	5	39.20%	9%
Populus euramericana [Robusta Poplar]	24	79.20%	15.20%
Populus fremontii [Fremont's Cottonwood]	3	30.70%	15.90%
Populus grandidentata [Bigtooth Aspen]	3	181.70%	60.70%
Populus nigra [Black Poplar]	6	73.20%	18.30%
Populus tremula x Populus tremuloides [Hybrid Aspen]	3	28.70%	3.30%
Populus tremuloides [Quaking Aspen]	32	59.40%	6.30%
Populus trichocarpa x Populus deltoides [Hybrid Poplar]	2	37.50%	8.80%
Potato, Sweet [Ipomoea batatas]	5	39.40%	10.80%

Plant Name	# of Studies	Arithmetic Mean	Standard Error
Potato, White [Solanum tuberosum]	15	33.20%	5.50%
Potentilla crantzii [Alpine Cinquefoil]	1	25%	0%
Prickly Pear [Opuntia ficus-indica]	6	22.70%	2.60%
Prochlorococcus [Marine Picocyanobacterium]	2	5.50%	2.50%
Prunus armeniaca [Apricot]	4	62.80%	4.20%
Prunus avium [Sweet Cherry]	1	38%	0%
Prunus persica [Peach]	4	37.50%	8.70%
Prunus serotina [Black Cherry]	3	63.70%	5.70%
Pseudobombax septenatum [Tropical Tree]	1	68%	0%
Pseudopanax arboreus [Puahou]	1	69%	0%
Pseudotsuga menziesii [Douglas Fir]	17	32%	5.30%
Psychotria limonensis [Forest Shrub]	1	65%	0%
Pteridium aquilinum [Bracken]	8	93.80%	7.30%
Puahou [Pseudopanax arboreus]	1	69%	0%
Puccinellia maritima [Seaside Alkaligrass]	5	84%	17.80%
Pyrrosia piloselloides [Tropical Fern]	1	19%	0%
Quercus alba [White Oak]	3	142%	38.30%
Quercus chapmanii [Chapman's Oak]	4	53%	4.10%
Quercus crispula [Mongolian Oak]	2	19%	3.50%
Quercus geminata [Sand Live Oak]	6	15.70%	5.90%
Quercus ilex [Holly Oak]	11	91.40%	30.10%
Quercus mongolica [Mongolian Oak]	22	74.40%	14.70%
Quercus myrtifolia [Myrtle Oak]	9	60.60%	7.70%
Quercus petraea [Durmast Oak]	2	23.50%	0.40%
Quercus pubescens [Downy Oak]	5	136.80%	58.50%
Quercus robur [Pedunculate Oak]	9	35.80%	6.20%
Quercus rubra [Northern Red Oak]	14	65.50%	11.70%
Quercus suber [Cork Oak]	3	28.70%	16.90%
Radish, Wild [Raphanus raphanistrum]	2	32%	7.10%
Radish, Wild [Raphanus sativus x raphanistrum]	1	109%	0%
Radish, Wild [Raphanus sativus]	8	30.40%	7%
Ragweed, Annual [Ambrosia artemisiifolia]	3	37.70%	5.20%
Ragweed, Tuscon Burr [Ambrosia cordifolia]	3	45.70%	3.80%
Rainforest Tree, Semi-Evergreen [Agathis microstachya]	1	50%	0%
Ramalina menziesii [Menzies' Cartilage Lichen]	1	20%	0%
Rangiora [Brachyglottis repanda]	1	120%	0%
Ranunculus [Buttercup]	1	47%	0%
Ranunculus acris [Tall Buttercup]	1	36%	0%
Ranunculus glacialis [Glacier Buttercup]	3	44%	8.90%

Plant Name	# of Studies	Arithmetic Mean	Standard Error
Raphanus raphanistrum [Wild Radish]	2	32%	7.10%
Raphanus sativus [Wild Radish]	8	30.40%	7%
Raphanus sativus x raphanistrum [Wild Radish]	1	109%	0%
Ratany, Littleleaf [Krameria erecta]	4	115.80%	28.60%
Redbud Tree, Eastern [Cercis canadensis]	3	127.30%	48%
Redwood, Coastal [Sequoia sempervirens]	1	31%	0%
Redwood, Dawn [Metasequoia glyptostroboides]	1	13%	0%
Reedgrass, Chee [Calamagrostis epigeios]	1	27%	0%
Rhaphiolepis indica [Indian Hawthorn]	1	79%	0%
Rhizophora mangle [American Mangrove]	2	11.50%	0.40%
Rhytidiadelphus triquetrus [Rough Goose Neck Moss]	1	66%	0%
Rice [Oryza sativa]	64	49.70%	5.70%
Ricinus communis [Castor Bean]	2	34%	0%
Rosa hybrida [Rose]	1	35%	0%
Rose [Rosa hybrida]	1	35%	0%
Rosemary, Wild [Ledum palustre]	1	0%	0%
Rosette Plant, Giant Afro-alpine [Dendrosenecio brassica]	1	65%	0%
Rosette Plant, Giant Afro-alpine [Dendrosenecio keniodendron]	1	52%	0%
Rosette Plant, Giant Afro-alpine [Lobelia telekii]	1	52%	0%
Ryegrass, Darnel [Lolium temulentum]	2	23.50%	1.80%
Ryegrass, Perennial [Lolium perenne]	28	42.40%	3.80%
Saccharum officinarum [Sugarcane]	6	11%	3.40%
Sage, Pitcher [Salvia pitcheri]	2	26.50%	1.80%
Sage, Introduced [Salvia pratensis]	1	47%	0%
Sagebrush, Big [Artemisia tridentata]	2	31%	12.70%
Salt Marsh Sedge [Scirpus olneyi]	7	61.40%	11.80%
Salvia nemorosa [Woodland Sage]	2	277.50%	136.10%
Salvia pitcheri [Pitcher Sage]	2	26.50%	1.80%
Salvia pratensis [Introduced Sage]	1	47%	0%
Sandspurry, Media [Spergularia maritima]	1	81%	0%
Sanguisorba minor [Small Burnet]	5	31.60%	4%
Schima superba [Sub-tropical Tree]	2	26%	6.40%
Schizachyrium scoparium [Little Bluestem]	3	72.30%	18.10%
Schoenoplectus americanus [Chairmaker's Bulrush]	1	27%	0%
Scirpus maritimus [Seaside Bulrush]	1	55%	0%
Scirpus olneyi [Salt Marsh Sedge]	7	61.40%	11.80%
Scirpus robustus [Sedge]	2	60%	10.60%
Scrophularia desertorum [Desert Figwort]	3	50%	21%
Scrub-Oak Ecosystem	2	98.50%	39.20%

Plant Name	# of Studies	Arithmetic Mean	Standard Error
Sedge [Scirpus robustus]	2	60%	10.60%
Sedge, Bigelow's [Carex bigelowii]	1	0%	0%
Sedge, Chaffy [Carex paleacea]	1	34%	0%
Sedge, Heath [Carex flacca]	1	55%	0%
Sequoia sempervirens [Coastal Redwood]	1	31%	0%
Setaria faberi [Japanese Bristlegrass]	4	15%	7.10%
Setaria viridis [Green Bristlegrass]	1	13%	0%
Shorea leprosula	2	57%	1.40%
Shrub, Forest [Psychotria limonensis]	1	65%	0%
Silene noctiflora [Night-flowering Catchfly]	1	42%	0%
Silver Dollar Gum [Eucalyptus polyanthemus]	1	75%	0%
Six Hardwood Tree Species	2	58%	4.90%
Solanum curtilobum [Shortlobe Solanum]	1	46%	0%
Solanum dulcamara [Climbing Nightshade]	6	28.80%	3.10%
Solanum muricatum [Pepino]	2	74%	7.10%
Solanum tuberosum [White Potato]	15	33.20%	5.50%
Solanum, Shortlobe [Solanum curtilobum]	1	46%	0%
Solidago rigida [Stiff Goldenrod]	1	7%	0%
Sorghastrum nutans [Indian Grass]	1	19%	0%
Sorghum [Sorghum bicolor]	9	26.80%	5.20%
Sorghum bicolor [Sorghum]	9	26.80%	5.20%
Sorghum halapense [Johnsongrass]	1	12%	0%
Southern California Chaparral Ecosystem	3	246.70%	13.40%
Soybean [Glycine max]	75	56.20%	9.40%
Spanish Elm [Cordia alliodora]	1	41%	0%
Spartina alterniflora [Smooth Cordgrass]	1	19%	0%
Spartina anglica [Common Cordgrass]	1	33%	0%
Spartina patens [Saltmeadow Cordgrass]	5	12.80%	5.10%
Spergularia maritima [Media Sandspurry]	1	81%	0%
Sphagnum [Sphagnum fuscum]	3	89.30%	6.60%
Sphagnum fuscum [Sphagnum]	3	89.30%	6.60%
Spinach [Spinacia oleracea]	4	37%	9%
Spinacia oleracea [Spinach]	4	37%	9%
Spruce, Black [Picea mariana]	5	33.60%	9.20%
Spruce, Norway [Picea abies]	9	48.60%	8.40%
Spruce, Sitka [Picea sitchensis]	2	23%	4.20%
Spurge, Japanese [Pachysandra terminalis]	1	70%	0%
Staggerbush, Piedmont [Lyonia mariana]	1	27%	0%
Stenocereus queretaroensis [Tree Cactus]	1	30%	0%
Stiff Goldenrod [Solidago rigida]	1	7%	0%

Plant Name	# of Studies	Arithmetic Mean	Standard Error
Stinkingtoe [Hymenaea courbaril]	1	70%	0%
Stipa thurberiana [Thurber needlegrass]	1	56%	0%
Strawberry Tree [Arbutus unedo]	4	91.30%	32.10%
Strawberry, Hybrid [Fragaria x ananassa]	14	72.60%	12.50%
Sugar Maple [Acer saccharum]	3	64%	9%
Sugarcane [Saccharum officinarum]	6	11%	3.40%
Sugargum [Eucalyptus cladocalyx]	2	47.50%	27.20%
Sunflower [Helianthus annuus]	13	41.50%	6.10%
Sunflower, Prairie [Helainthus petiolaris]	2	20%	0%
Sweetgum [Liquidambar styraciflua]	18	105.10%	21.80%
Swietenia macrophylla [Honduras Mahogany]	1	26%	0%
Swiss Lowland Deciduous Forest	1	100%	0%
Switchgrass [Panicum virgatum]	1	64%	0%
Sycamore, American [Platanus occidentalis]	1	60%	0%
Symphiocarpos orbiculatus [Buch Brush]	2	105%	43.80%
Synechococcus [Unicellular Marine Picocyanobacterium]	1	48%	0%
Talinum triangulare	1	357%	0%
Tamarack [Larix laricina]	1	47%	0%
Taro, Giant [Alocasia macrorrhiza]	2	79%	18.40%
Taxodium distichum [Cyprus, Bald]	5	88.60%	14.40%
Teak [Tectona grandis]	1	37%	0%
Tectona grandis [Teak]	1	37%	0%
Ten Species of Tropical Forest Tree Seedlings	1	10%	0%
Terminalia arjuna [Terminalia]	2	194.50%	3.90%
Terminalia chebula [Myrobalan]	2	147.50%	37.10%
Tetragastris panamensis [Tropical Tree]	1	14%	0%
Themeda triandra [C4 grass]	2	19%	3.50%
Theobroma cacao [Cacao]	1	32%	0%
Thistle, Canadian [Cirsium arvense]	4	56.50%	8%
Thrift Seapink [Armeria maritima]	1	46%	0%
Thurber needlegrass [Stipa thurberiana]	1	56%	0%
Timothy [Phleum pratense]	3	30.30%	3.20%
Tobacco, Cultivated [Nicotiana tabacum]	7	83%	14.90%
Tobacco, South American [Nicotiana sylvestris]	1	55%	0%
Tomato, Garden [Lycopersicon esculentum]	13	22.60%	5%
Toxicodendron radicans [Poison Ivy]	1	116%	0%
Trifolium repens [White Clover]	22	49.40%	7.40%
Trifolium subterraneum [Subterranean Clover]	1	19%	0%
Triticum aestivum [Common Wheat]	83	64.90%	10.20%
Tussock Tundra	2	237.50%	19.40%

Plant Name	# of Studies	Arithmetic Mean	Standard Error
Understory Deciduous Trees in a Pinus taeda L. Plantation	1	70%	0%
Understory Plants in a Spruce Model Ecosystem	2	53%	9.90%
Urtica dioica L. [Stinging Nettle]	1	126%	0%
Vaccinium myrtillus [Whortleberry]	2	47%	9.20%
Velvet Leaf [Abutilon theophrasti]	6	46.70%	10.30%
Velvetgrass, Common [Holcus lanatus]	1	85%	0%
Vernal Grass, Sweet [Anthoxanthum odoratum]	1	109%	0%
Vetch, Common Kidney [Anthyllis vulneraria]	4	27.80%	10.30%
Viburnum marisii [Flowering Shrub]	1	63%	0%
Vicia faba [Faba Bean]	7	52.30%	10.50%
Vigna radiata [Mungbean]	2	178%	18.40%
Vigna unguiculata [Blackeyed Pea]	4	67.80%	18.50%
Virola surinamensis [Tropical Tree]	1	31%	0%
Vitis californica [California Wild Grape]	1	57%	0%
Vulpia microstachys [Small Fescue]	1	125%	0%
Walnut, Black [Juglans nigra]	1	24%	0%
Water Lily [Nymphaea marliac]	3	39.70%	12.40%
Weed, Annual, C4 [Eragrostis orcuttiana]	1	-44%	0%
Wheat, Common [Triticum aestivum]	83	64.90%	10.20%
Wheatgrass, Western [Agropyron smithii]	2	-4.50%	6%
Wheatgrass, Western [Pascopyrum smithii]	5	57.40%	17.30%
Whiteflower Kurrajong [Brachychiton populneum]	1	75%	0%
Whortleberry [Vaccinium myrtillus]	2	47%	9.20%
Woodland Sage [Salvia nemorosa]	2	277.50%	136.10%
Xanthium strumarium L. [Rough Cocklebur]	4	30.80%	12.90%
Yarrow [Achillea millefolium L.]	4	30.80%	12.90%
Yellowtops [Flaveria pringlei Gandoger]	4	30.80%	12.90%
Yellowtops, Clustered [Flaveria trinervia (Spreng.) C. Mohr]	4	30.80%	12.90%
Yellowtops, Florida [Flaveria floridana J.R. Johnston]	4	30.80%	12.90%
Zea mays L. [Corn]	21	28.50%	11.50%

[this page intentionally blank]

Appendix 4

The Petition Project

Appendix 4. The Petition Project

1.1. About the Petition

The petition pictured above has been signed by 31,478 Americans with university degrees in science, including 9,029 with Ph.D.s. The petition reads in part: "There is no convincing scientific evidence that human release of carbon dioxide, methane, or other greenhouse gases is causing or will, in the foreseeable future, cause catastrophic heating of the Earth's atmosphere and disruption of of the Earth's climate. Moreover, there is substantial scientific evidence that increases in atmospheric carbon dioxide produce many beneficial effects upon the natural plant and animal environments of the Earth." A copy of one actual signed petition appears on this page. The majority of the current listed signatories signed or re-signed the petition after October 2007.

The purpose of the Petition Project is to demonstrate that the claim of "settled science" and an overwhelming "consensus" in favor of the hypothesis of human-caused global warming and consequent climatological damage is wrong. No such consensus or settled science exists. As indicated by the petition text and signatory list, a very large number of American scientists reject this hypothesis.

From the clear and strong petition statement that they have signed, it is evident that these 31,478 American scientists are not "skeptics." These scientists are instead convinced that the human-caused global warming hypothesis is without scientific validity and that government action on the basis of this hypothesis would unnecessarily and counterproductively damage both human prosperity and the natural environment of the Earth.

This petition is primarily circulated by U.S. Postal Service mailings to scientists. Included in the mailings are the petition card, a letter from Frederick Seitz (reproduced on the following page), a scientific review article (reproduced on the pages following the directory of petition signers), and a return envelope. If a scientist wishes to sign, he or she completes the petition and mails it to the project by first-class mail. Additionally, many petition signers obtain petition cards from their colleagues, who request these cards from the project. A scientist can also obtain a copy of the petition from www.PetitionProject.org, sign, and mail it. Fewer than 5 percent of the current signatories obtained their petition in this way.

The letter on the following page, from Professor Frederick Seitz, is circulated with the petition. Dr. Seitz, a physicist, was president of the U.S. National Academy of Sciences and of Rockefeller University. He received the National Medal of Science, the Compton Award, the Franklin Medal, and numerous other awards, including honorary doctorates from 32 universities around the world. In August 2007, Dr. Seitz reviewed and approved the article by Dr. Arthur B. Robinson, Dr. Noah E. Robinson, and Dr. Willie Soon that is circulated with the petition and gave his enthusiastic approval to the continuation of the Petition Project. A vigorous supporter of the Petition Project since its inception in 1998, Professor Seitz died on March 2, 2008.

1.2. Qualifications of Signers

Petition project volunteers evaluate each signer's credentials, verify signer identities, and, if appropriate, add the signer's name to the petition list. Signatories are approved for inclusion in the Petition Project list if they have obtained formal educational degrees at the level of Bachelor of Science or higher

in appropriate scientific fields. The petition has been circulated only in the United States.

The current list of petition signers includes 9,029 persons who hold Ph.D.s, 7,153 who hold an MS, 2,585 who hold MDs or DVMs, and 12,711 who hold a BS or equivalent academic degrees. Most of the MD and DVM signers also have underlying degrees in basic science.

All of the listed signers have formal educations in fields of specialization that suitably qualify them to evaluate the research data related to the petition statement. Many of the signers currently work in climatological, meteorological, atmospheric, environmental, geophysical, astronomical, and biological fields directly involved in the climate change controversy. The Petition Project classifies petition signers on the basis of their formal academic training, as summarized below. Scientists often pursue specialized fields of endeavor that are different from their formal education, but their underlying training can be applied to any scientific field in which they become interested.

Outlined below are the numbers of Petition Project signatories, subdivided by educational specialties. These have been combined, as indicated, into seven categories.

1. Atmospheric, environmental, and Earth sciences includes 3,803 scientists trained in specialties directly related to the physical environment of the Earth and the past and current phenomena that affect that environment.

2. Computer and mathematical sciences includes 935 scientists trained in computer and mathematical methods. Since the human-caused global warming hypothesis rests entirely upon mathematical computer projections and not upon experimental observations, these sciences are especially important in evaluating this hypothesis.

3. Physics and aerospace sciences include 5,810 scientists trained in the fundamental physical and molecular properties of gases, liquids, and solids, which are essential to understanding the physical properties of the atmosphere and Earth.

4. Chemistry includes 4,818 scientists trained in the molecular interactions and behaviors of the substances of which the atmosphere and Earth are composed.

5. Biology and agriculture includes 2,964 scientists trained in the functional and environmental requirements of living things on the Earth.

Enclosed is a twelve-page review of information on the subject of "global warming," a petition in the form of a reply card, and a return envelope. Please consider these materials carefully.

The United States is very close to adopting an international agreement that would ration the use of energy and of technologies that depend upon coal, oil, and natural gas and some other organic compounds.

This treaty is, in our opinion, based upon flawed ideas. Research data on climate change do not show that human use of hydrocarbons is harmful. To the contrary, there is good evidence that increased atmospheric carbon dioxide is environmentally helpful.

The proposed agreement would have very negative effects upon the technology of nations throughout the world, especially those that are currently attempting to lift from poverty and provide opportunities to the over 4 billion people in technologically underdeveloped countries.

It is especially important for America to hear from its citizens who have the training necessary to evaluate the relevant data and offer sound advice.

We urge you to sign and return the enclosed petition card. If you would like more cards for use by your colleagues, these will be sent.

Frederick Seitz

Frederick Seitz
Past President, National Academy of Sciences, U.S.A.
President Emeritus, Rockefeller University

6. Medicine includes 3,046 scientists trained in the functional and environmental requirements of human beings on the Earth.

7. Engineering and general science includes 10,102 scientists trained primarily in the many engineering specialties required to maintain modern civilization and the prosperity required for all human actions, including environmental programs.

The outline below gives a more detailed analysis of the signers' educations.

Qualifications of Petition Signers

Atmosphere, Earth, and Environment (3,803)

1. Atmosphere (578)
   a) Atmospheric Science (113)
   b) Climatology (39)
   c) Meteorology (341)
   d) Astronomy (59)
   e) Astrophysics (26)

2. Earth (2,240)
   a) Earth Science (94)
   b) Geochemistry (63)
   c) Geology (1,684)
   d) Geophysics (341)
   e) Geoscience (36)
   f) Hydrology (22)

3. Environment (985)
   a) Environmental Engineering (486)
   b) Environmental Science (253)
   c) Forestry (163)
   d) Oceanography (83)

Computers and Math (935)

1. Computer Science (242)

2. Math (693)
   a) Mathematics (581)
   b) Statistics (112)

Physics and Aerospace (5,810)

1. Physics (5,223)
   a) Physics (2,365)
   b) Nuclear Engineering (223)
   c) Mechanical Engineering (2,635)

2. Aerospace Engineering (587)

Chemistry (4,818)

1. Chemistry (3,126)

2. Chemical Engineering (1,692)

Biochemistry, Biology, and Agriculture (2,964)

1. Biochemistry (744)
   a) Biochemistry (676)
   b) Biophysics (68)

2. Biology (1,437)
   a) Biology (1,048)
   b) Ecology (76)
   c) Entomology (59)
   d) Zoology (149)
   e) Animal Science (105)

3. Agriculture (783)
   a) Agricultural Science (296)
   b) Agricultural Engineering (114)
   c) Plant Science (292)
   d) Food Science (81)

Medicine (3,046)

1. Medical Science (719)

2. Medicine (2,327)

General Engineering and General Science (10,102)

1. General Engineering (9,833)

   a) Engineering (7,280)
   b) Electrical Engineering (2,169)
   c) Metallurgy (384)

2. General Science (269)

1.3. Frequently Asked Questions

1. Is the Petition Project fulfilling the expectations of its organizers?

Yes. In Ph.D. scientist signers alone, the project already includes 15 times more scientists than are seriously involved in the United Nations' IPCC process. The very large number of petition signers demonstrates that, if there is a consensus among American scientists, it is in opposition to the human-caused global warming hypothesis rather than in favor of it. Moreover, the current totals of 31,478 signers, including 9,029 PhDs, are limited only by Petition Project resources. With more funds for printing and postage, those numbers would be much higher.

2. Has the petition project helped to diminish the threat of energy and technology rationing?

The accomplishments of science and engineering have transformed the world. They have markedly increased the quality, quantity, and length of human life and have enabled human beings to make many improvements in the natural environment of the Earth.

Today, scientists are seeing the accomplishments of science demonized and one of the three most important molecular substances that make life possible—atmospheric carbon dioxide (the other two being oxygen and water)—denigrated as an atmospheric "pollutant" in a widely circulated movie. Scientists who have carefully examined the facts know this movie contains numerous falsehoods. This and many other similar misguided propaganda efforts in the media naturally repel men and women who know the truth. The search for truth is the essence of science. When science is misrepresented, scientists are naturally incensed.

There is, therefore, a rapidly growing backlash of opposition among American scientists to this egregious misuse of the reputation and procedures of science. The Petition Project is helping to demonstrate this opposition and, therefore, to reduce the chances of misguided political reductions in science-based technology.

3. Who organized the Petition Project?

The Petition Project was organized by a group of physicists and physical chemists who conduct scientific research at several American scientific institutions. The petition statement and the signatures of its 31,478 signers, however, speak for themselves. The primary relevant role of the organizers is that they are among the 9,029 PhD signers of the petition.

4. Who pays for the Petition Project?

The Petition Project is financed by non-tax deductible donations to the Petition Project from private individuals, many of whom are signers of the petition. The project has no financing whatever from industrial sources. No funds or resources of the Oregon Institute of Science and Medicine are used for the Petition Project. The Oregon Institute of Science and Medicine has never received funds or resources from energy industries, and none of the scientists at the Institute have any funding whatever from corporations or institutions involved in hydrocarbon technology or energy production. Donations to the project are primarily used for printing and postage. Most of the labor for the project has been provided by scientist volunteers.

5. Does the petition list contain names other than those of scientist signers?

Opponents of the Petition Project sometimes submit forged signatures in efforts to discredit the project. Usually, these efforts are eliminated by our verification procedures. On one occasion, a forged signature appeared briefly on the signatory list. It was removed as soon as discovered.

In a group of more than 30,000 people, there are many individuals with names similar or identical to other signatories, or to non-signatories—real or fictional. Opponents of the petition project sometimes use this statistical fact in efforts to discredit the project. For example, Perry Mason and Michael Fox are real scientists who have signed the petition and happen also to have names identical to fictional or real non-scientists.

6. Does the petition project list contain duplicate names?

Thousands of scientists have signed the petition more than once. These duplicates have been carefully removed from the petition list. The list contains many instances of scientists with closely similar and sometimes identical names, as is statistically expected in a list of this size, but these signers are different people, who live at different addresses, and usually have different fields of specialization. Primarily as a result of name and address variants, occasional duplicate names are found in the list. These are immediately removed.

7. Are any of the listed signers dead?

In a group of more than 30,000 people, deaths are a frequent occurrence. The Petition Project has no comprehensive method by which it is notified about deaths of signatories. When we do learn of a death, an "*" is placed beside the name of the signatory. For examples, Edward Teller, Arnold Beckman, Philip Abelson, William Nierenberg, and Martin Kamen are American scientists who signed the Petition and are now deceased.

8. Why is this effort called "Petition Project?"

Signatories to the petition have signed just the petition—which speaks for itself. The organizers—themselves scientists located at several scientific institutions–have designed the project to emphasize this single fact. The use of a post office box mailing address, a generic name– Petition Project—and other institutionally neutral aspects of the project are intended to avoid the impression that the signatories have endorsed the agenda or actions of any institution, group, or other activity. They are simply signers of this petition to the government of the United States.

9. Why was the review article published in the *Journal of American Physicians and Surgeons*?

The authors chose to submit this article for peer-review and publication to the *Journal of American Physicians and Surgeons* because that journal was willing to waive its copyright and permit extensive reproduction and distribution of the article by the Petition Project.

10. Why is the Petition Project necessary?

In December 1997, then U.S. Vice-President Al Gore participated in a meeting in Kyoto, Japan during which he signed a treaty to ration world energy production based upon fear of human-caused global warming. This treaty was not, however, presented to the United States Senate for ratification.

Since before that Kyoto meeting and continuing to the present day, Mr. Gore and his supporters at the United Nations and elsewhere have claimed that the "science is settled" – that an overwhelming "consensus" of scientists agrees with the hypothesis of human-caused global warming, with only a handful of skeptical scientists in disagreement. These proponents of world energy rationing have consistently argued that, in view of this claimed scientific "consensus," no further discussion of the science involved in this issue is warranted before legislative action is taken to heavily tax, regulate, and ration hydrocarbon energy.

Realizing, from discussions with their scientific colleagues, that this claimed "consensus" does not exist, a group of scientists initiated the Petition Project in early 1998. Thousands of signatures were gathered in a campaign during 1998-1999. Between 1999 and 2007, the list of petition signatories grew gradually, without a special campaign. Between October 2007 and March 2008, a new campaign for signatures was initiated. The majority of the current listed signatories signed or re-signed the petition after October 2007. The original review article that accompanied the petition effort in 1998-1999 was replaced in October 2007 with a new review incorporating the research literature up to that date.

The renewed petition campaign in 2007 was prompted by an escalation of the claims of "consensus," release of the movie *An Inconvenient Truth* by Al Gore, and related events. The campaign to severely ration hydrocarbon energy technology has been markedly expanded, and many scientifically invalid claims about impending climate emergencies are being made. Simultaneously, proposed political actions to sharply reduce hydrocarbon use now threaten the prosperity of Americans and the very existence of hundreds of millions of people in poorer countries.

As Professor Seitz states in his Petition Project letter, which speaks of this impending threat to all humanity, "It is especially important for America to hear from its citizens who have the training necessary to evaluate the relevant data and offer sound advice."

The Petition Project is a means by which those citizens are offering that advice.

The Petition Project
P.O. Box 1925
La Jolla, CA 92038
www.PetitionProject.org

1.4. Petition Signers

Alabama

H. William Ahrenholz
Oscar Richard Ainsworth, PhD
John Hvan Aken
Michael L. Alexander
Robert K. Allen
Ronald C. Allison
Barry M. Amyx*
Bernard Jeffrey Anderson, PhD
David W. Anderson
John C. Anderson, PhD
Russell S. Andrews, PhD
Bobby M. Armistead
Ann Askew
Larry N. Atkinson
Alan C. Bailey
Brooks H. Baker, III
Robert Baker
John Wayland Bales, PhD
James Y. Baltar
Ted B. Banner
Robert F. Barfield, PhD
Richard Barnes
Samuel A. Barr
Kenneth A. Barrett
Franklin E. Bates
Ronald G. Baxley
Sidney D. Beckett, PhD
Arthur B. Beindorff, PhD
Victor Bell
Aleksandr A. Belotserkovskiy, PhD
Fred Bender, PhD
William R. Bentley
Reginald H. Benton
M. Bersch, PhD
Raymond E. Bishop
Donald H. Blackwell
Benjie Blair, PhD
Edward S. Blair
Kevin M. Blake
Richard Lee Blanchard, PhD
Anton C. Bogaty, Jr.
Jonathan D. Boland
Desmond H. Bond
James H. Bonds
Terry Bonds
Robert R. Boothe, Jr.
Dale L. Borths
Jerome Boutwell
James L. Box
William D. Boyer, PhD
William C. Bradford
Andrew E. Bradley
Michael W. Bradshaw
Jerome J. Brainerd, PhD
Bradley A. Brasfield
John F. Brass
Claude E. Breed
Thomas H. Brigham
Doyle G. Briscoe
Alfred L. Brown
James Melton Brown, PhD
Robert Alan Brown, PhD
Dushan S. Bukvic
Donald F. Burchfield, PhD
Walter W. Burdin
John E. Burkhalter, PhD
Marshall Burns, PhD

Richard C. Burnside
Eddie C. Burt, PhD
Michael A. Butts
Thomas L. Cain
Arnold E. Carden, PhD
Jason Cassibry, PhD
Darrell W. Chambers
Kenneth E. Chandler
James D. Chesnut, Jr.
Charles Richard Christensen, PhD
Chad P. Christian
Otis M. Clarke
Stan G. Clayton
William Madison Clement, PhD
Barbara B. Clements
James H. Clements, Jr.
David N. Clum
W. Frank Cobb, Jr.
W. A. Cochran, Jr.
Ernst M. Cohn
Robert M. Conry
Robert Bigham Cook, PhD
Franklyn K. Coombs
Kenneth R. Copeland
Clifton E. Couey
Robert W. Coughlin
Sylvere Coussement
Clifton E. Covey
George M. Cox
Justin H. Crain
George S. Crispin
Delton G. Crosier
Delmar N. Crowe, Jr.
Harry Cullinan, PhD
Joseph A. Cunningham
J. F. Cuttino, PhD
Thomas P. Czepiel, PhD
Robert S. Dahlin, PhD
Madge C. Daniel
Thomas W. Daniel, Sr.
Joe S. Darden
Julian Davidson, PhD
Allen S. Davis
Donald Echard Davis, PhD
Gene Davis
Wilfred J. Davis
Michael J. Day, PhD
Charles W. Dean
David Lee Dean, PhD
Edward E. Deason
William M. Decker
James J. Denson
Debra Depiano
Myron G. Deshazo, Jr.
Jerome R. Dickey
Warren D. Dickinson
Wouter W. Dieperink
Allen C. Dittenhoefer, PhD
Wenju Dong, PhD
Francis M. Donovan, PhD
Thomas P. Dooley, PhD
Gilbert F. Douglas, Jr.
James L. Dubard, PhD
Scott A. Dunham
John R. Durant
Paul G. Durr
Z'Bigniew W'Ladyslaw Dybczak, PhD
George Robert Edlin, PhD

Gabriel A. Elgavish, PhD
Rotem Elgavish
Tricia Elgavish
Rush E. Elkins, PhD
Jesse G. Ellard
Howard Clyde Elliott, PhD
Arthur F. Ellis
David A. Elrod, PhD
David J. Elton, PhD
Leonard E. Ensminger, PhD
George Epps
Robert D. Erhardt, Jr.
Edwin C. Ethridge, PhD
Clyde Edsel Evans, PhD
Ken Fann
William S. Farneman
Rodney G. Ferguson
Mason D. Field
Fred H. Fihe
G. L. Fish
Don E. Fitts
Julius D. Fleming
William F. Foreman
William M. Forman
William R. Forrester
John Foshee, Jr
Philip C. Foster
Mark Fowler
Robert Dorl Francis, PhD
Richard M. Franke
Larry D. Franks
Gerald R. Freeman, PhD
Herbert J. Furman
Bob S. Galloway
Bobby R. Ganus
Ronald Gene Garmon, PhD
Henry B. Garrett, Jr.
Norman A. Garrison, PhD
David J. Garvey, PhD
William F. Garvin
Nancy K. Gautier, PhD
W. Welman Gebhart
Gerard Geppert
Thomas A. Gibson
Chris Gilbert, PhD
Ronald E. Giuntini, PhD
Marvin R. Glass, Jr.
John J. Gleysteen
Alexander C. Goforth
Roger L. Golden
Bruce William Gray, PhD
James D. Gregory
Ralph B. Groome
A. M. Guarino, PhD
Wallace K. Gunnells
Leslie A. Gunter
Freddie G. Gwin
Ronald L. Haaland, PhD
Walter Haeussermann, PhD
Leroy M. Hair
Benjamin F. Hajek, PhD
Justin Charles Hamer, PhD
W. Allen Hammack
James W. Handley
Reid R. Hanson, PhD
Richard A. Harkins
Daniel K. Harris, PhD
Gregory A. Harris
Joseph G. Harrison, PhD

John J. Harrity, Jr.
Douglas G. Hayes, PhD
James L. Hayes
Charles D. Haynes, PhD
James Eugene Heath
Paul S. Heck
Bobby Helms
Ron Helms
Robert L. Henderson
John B. Hendricks, PhD
William Henry, Jr.
William D. Herrin
Jerry P. Hethcoat
Mitch Higginbotham
David Higgins, Jr.
Hermon H. Hight
David T. Hill, PhD
Brendall Hinton, PhD
Jerry M. Hobbie
Vincent L. Hodges
Joseph A. Holifield
William A. Hollerman
Frank S. Hollis
David Hood
Joseph E. Hossley
Stephen K. Howard
Keith G. Howell
James F. Howison
James W. Huff, PhD
Dale L. Huffman, PhD
John C. Huggins
Joseph P. Huie
Chih-Cheng Hung, PhD
Hassel E. Hunter
Herbert E. Hunter, PhD
Ray Hunter
Donald J. Ifshin
Victor D. Irby, PhD
Steven K. Irvin
John David Irwin, PhD
Lyman D. Jackson
Holger M. Jaenisch, PhD
Homer C. Jamison, PhD
Donald J. Janes
Kenneth Jarrell
William W. Jemison, Jr.
Penelope Jester
Donald R. Johns
Frank Junior Johnson
Frederic Allan Johnson, PhD
Monroe H. Johnson
Joseph F. Judkins, PhD
Carl D. Jumper
David A. Kallin
James M. Kampfer
Robert Keenum
Lawrence C. Keller
Arthur G. Kelly
Russell R. Kerl
James E. Kingsbury
Earl T. Kinzer, Jr., PhD
Harold A. Kirkland
William Klein
Dorothea A. Klip, PhD
James Knight
William J. Knox
Charles Marion Krutchen, PhD
David E. Labo
Joseph E. Lammon

Philip Elmer Lamoreaux*
John H. Lary
Lloyd H. Lauerman, PhD
Thomas H. Ledford, PhD
William K. Lee
John D. Leffler
Foy K. Lewis
George R. Lewis
Baw-Lin Liu, PhD
Allen Long
James M. Long
Joyce M. Long
John F. Lozowski
Linda C. Lucas, PhD
William R. Lucas, PhD
Brian Luckianow
Randal W. Lycans
Michael A. Macfarlane
George J. Mackinaw
Robert A. Macrae
Frank L. Madarasz, PhD
John F. Maddox
Carl Maltese
I. R. Manasco
Baldev Singh Mangat, PhD
Frank C. Mann
Sven Pit Mannsfeld, PhD
Milton Mantler
Matthew Mariano, PhD
Gordon D. Marsh
David C. Marshal
Arvle E. Marshall, PhD
Paul R. Matthews
Charles R. Mauldin
David Mays, PhD
Van Alfon McAuley
Theresa O. McBride
Mark S. McColl
W. H. McCraney
Geroge M. McCullars, PhD
Phillip I. McCullough
Randall E. McDaniel
Joe A. McEachern
William Baldwin McKnight, PhD
Curtis J. McMinn
Thomas E. McNider
Jasper Lewis McPhail
B. McSpadden
Solomon O. Mester
Joseph P. Michalski
J. G. Micklow, PhD
Arthur J. Milligan
Randall A. Mills
Benjamin K. Miree
Larry S. Monroe, PhD
Dwight L. Moody
Rickie D. Moon
Meg O. Moore
Robert A. Moore, PhD
Wellington Moore, PhD
George S. Morefield
Stephen J. Morisani, PhD
Perry Morton, PhD
Herman A. Nebrig, Jr.
Gary Nelson
Floyd Neth, PhD
Robert W. Neuschaeffer
James Nhool, PhD
Grady B. Nichols, PhD
Billy G. Nippert
Nathan O. Okia, PhD
Byron L. Oliver

J. F. Olivier
Jerry M. Palmer
Edward James Parish, PhD
Michell S. Pate
Roderick J. Patefield
George D. Pattillo
W. Quinn Paulk
David M. Pearsall
Dom Perrotta
Nelson A. Perry
Kenneth F. Persin
Mark R. Pettitt
Tom Pfitzer
John G. Pfrimmer
Kenneth G. Pickett
Sean Piecuch
Donald S. Pierre, Jr.
Charles Thomas Pike
James R. Pike
Peter P. Pincura
Michael Piznar
Morris C. Place
Melvin Price, PhD
Charles W. Prince, PhD
Thaddeus H. Pruett
Jodi Purser
Danny P. Raines
Joseph Lindsay Randall, PhD
Marvin L. Rawls
Michelle B. Ray
Harold M. Raynor
Greg Reardon
Jerry Reaves
Mark Redden
Michael A. Remillard
Robert Ware Reynolds, PhD
Richard G. Rhoades, PhD
William Eugene Ribelin, PhD
Dennis Rich
Martin B. Richardson, PhD
George Richmond
Logan R. Ritchie, Jr.
Alfred Ritter, PhD
Ronnie Rivers, PhD
Hill E. Roberts
Harold Vernon Rodriguez, PhD
Richard B. Rogers, PhD
Thaddeus A. Roppel, PhD
John W. Rouse, PhD
Eladio Ruiz-De-Molina
Leon Y. Sadler, III, PhD
Adel Sakla, PhD
Andreas Salemann, PhD
James Sanford
Ted L. Sartain
D. Satterwhite
Mark Saunders
Carl Schauble, PhD
Bernard Scheiner, PhD
Jason R. Schenk, PhD
Peter Schwartz, PhD
Edmund P. Segner, PhD
William G. Setser
Raymond F. Sewell, PhD
Raymond Lee Shepherd, PhD
Charles H. Shivers, PhD
Don A. Sibley, PhD
Gary D. Sides, PhD
Stephen T. Simpson
Wiliam F. Sims
Norman Frank Six, Jr., PhD
Harold Walter Skalka

Daniel E. Skinner
Peter John Slater, PhD
Donald W. Smaha
Roger J. Smeenge
David A. Smith
Obie Smith
Belton Craig Snyder
Michael Sosebee
Jon J. Spano
D. Paul Sparks, Jr.
Michael P. Spector, PhD
Philip Speir
John T. Spraggins, Jr.
Lethenual Stanfield
Jarel P. Starling
Alfred D. Stevens
Benjamin C. Stevens
Dale M. Stevens
Mike Stewart
J. Stone, PhD
John W. Sumrall
Marvin Laverne Swearingin, PhD
J. M. Tagg
Thomas F. Talbot, PhD
Bruce J. Tatarchuk, PhD
Oscar D. Taunton
Newton L. Taylor
Tommy L. Thompson
Zack Thompson
Eugene Delbert Tidwell
Edward R. Tietel
Timothy C. Tuggle
Charles Tugwell
Ted W. Tyson
James P. Vacik, PhD
James T. Varner
Otha H. Vaughan, Jr.
Phillip G. Vaughan
William W. Vaughan, PhD
Frank L. Vaughn
Thomas Mabry Veazey, PhD
William S. Viall
William Voigt
James E. Waite
Ron C. Waites
William Waldrum Walker, PhD
John Wallace
Edward Hilson Ward, PhD
Wyley D. Ward
Adrian O. Watson
Raymond C. Watson, Jr., PhD
Henry B. Weaver, Jr.
Donald C. Wehner
Lawrence P. Weinberger, PhD
Talmadge P. Weldon
William B. Wells, Jr.
Hans-Helmut Werner, PhD
Francis C. Wessling, PhD
Steven L. Whitfield
Rayburn Harlen Whorton
Leon Otto Wilken, PhD
Charles D. Wilkins
Michael Ledell Williams, PhD
Houston Williamson
Jay C. Willis
Harold J. Wilson
Leighton C. Wilson
Walter W. Wilson
Gregory Scott Windham
Stanley B. Winslow
Edmund W. Winston
Harvey B. Wright

Randy Wynn
Lewis S. Young
Kirk R. Zimmer

Alaska

Ronald Godshall Alderfer, PhD
Donald Ford Amend, PhD
David Anderson
Donald Anderson, PhD
Patrick J. Archey
Randall L. Bachman
Roger L. Baer
Tina L. Baker
Earl R. Barnard
Alex Baskous
Don Bassler
John M. Beitia
Eugene N. Bjornstad
John K. Boarman
William M. Bohon
James F. Boltz
Steven C. Borell
Caroline Bradshaw
Terry T. Brady
Dean C. Brinkman
Mike Briscoe
Jack A. Brockman
Robert Brown
Carrel R. Bryant
Roger C. Burggraf
Duane E. Carson
Glen D. Chambers
Bartly Coiley
Lowell R. Crane
Michael D. Croft
Brent Crowder
Bruce E. Davison
Jonathan Dehn, PhD
Steve Denton
Dave R. Dobberpuhl
Edward M. Dokoozian, PhD
Kathleen Douglas
Robert G. Dragnich
James Drew, PhD
Richard A. Dusenbery, PhD
William E. Eberhardt
Jeffrey D. Eckstein
John Egenholf, PhD
Robert L. Engelbach
William James Ferrell, PhD
Jeffrey Y. Foley
Monique M. Garbowicz
Jon M. Girard
Will E. Godbey
Edward R. Goldmann
Daniel C. Graham
Lawrence G. Griffin
Lenhart T. Grothe
William F. Gunderson
Brian T. Harten
David B. Harvey
Charles C. Hawley, PhD
Tommy G. Heinrich
James R. Hendershot
Steven C. Henslee
Naomi R. Hobbs
Julie K. Holayter
Kurt Hulteen
Lyndon C. Ibele
Steven K. Jones

William W. Kakel
Donald D. Keill
Edward R. Kiddle
Joseph M. Killon
Ted R. Kinney
Christopher C. Klein
George F. Klemmick
Thomas H. Kuckertz, PhD
David W. Lappi
James F. Lebiedz
Harold D. Lee
Harry R. Lee
Erwin L. Long
William E. Long, PhD
Barrie B. Lowe
Everett L. Mabry
Monte D. Mabry
Robert F. Malouf
T. R. Marshall, Jr.
Donald H. Martins, PhD
Jerzy Maselko, PhD
Bruce H. Mattson
Tom E. Maunder
Justin T. Meyring
Jeff R. Michels
J. V. Miesse
William W. Mitchell, PhD
Harold R. Moeser
Jesse R. Mohrbacher
Boyd Morgenthaler
John Mulligan
Erik A. Opstad
Marvin C. Papenfuss, PhD
Pedro E. Perez
Robert A. Perkins, PhD
Greg Peters, PhD
Richard A. Peters
Walter T. Phillips
Kenneth J. Pinard
Bruce Porter
Tom Reed
Richard H. Reiley
Rydell J. Reints
Kermit Dale Reppond
Peter M. Ricca, PhD
Donald R. Rogers
Jimmie C. Rosenbruch
Allan A. Ross
Joe L. Russell
Mortin B. Schierhorn
Lynn W. Schnell
Albert B. Schoffmann
Jeffery M. Scott
Edward M. Sessions
Glenn E. Shaw, PhD
Ernst Siemoneit
Todd A. Sneesby
David K. Soderlund
Damien F. Stella
Michael J. Storer
James W. Styler
Richard C. Swainbank, PhD
Robert C. Tedrick
Tim Terry
Kevin Tomera
Michael D. Travis
Joseph E. Usibelli, Jr.
Duane H. Vaagen
Dominique L. Van Nostrand
Angela C. Vassar
Ross L. Waner
Brenton Watkins, PhD

Jean S. Weingarten
Robert P. Wessels
Michael W. Wheatall
Warrack Willson, PhD
Theron C. Wilson
Frank W. Wince
Marion Yapuncich, PhD
Patrick J. Zettler

Arizona

Richard E. Ackermann
Brook W. Adams
Stanley D. Adams
Stanley P. Aetrewicz
Larry Delmar Agenbroad, PhD
Aida M. Aguirre
Richard Ahern
Edward Ahrens
Lloyd Alaback
Leland C. Albertson
Lee Amoroso, PhD
A. Amr, PhD
Sal A. Anazalone
Anita Teter Anderson
Arthur G. Anderson, PhD
James M. Andrew
John Allen Anthes, PhD
Bruce W. Apland
Ara Arabyan, PhD
Frank G. Arcella, PhD
William W. Archer
Joe R. Arechavaleta
Lew Armer
Mike Assad
David C. Atkins
Jerry C. Atwell
D. Austin
Dirk Den Baars, PhD
A. Terry Bahill, PhD
David A. Bailey
J. Baker, PhD
Jonathan A. Balasa
Roy Jean Barker, PhD
John E. Barkley, PhD
Ross C. Barkley
David A. Barnard
Benny B. Barnes
R. C. Barnett
Charles John Baroczy
Lawrence Dale Barr
Kenneth A. Bartal
John W. Bass
Charles Carpenter Bates, PhD
Roger A. Baumann
Steve Baumann
Frank L. Bazzanella
Randall D. Beck
James R. Beene
W. G. Benjamin
John Bentley
Charles M. Bentzen
Arne K. Bergh, PhD
James Wesley Berry, PhD
Stanley Beus, PhD
Peter F. Bianchetta
William S. Bickel, PhD
Alex F. Bissett
James R. Black
E. Allan Blair, PhD
Marlene Bluestein

Charles W. Boak
Paul L. Boettcher
Bruce Bollermann
Kelsey L. Boltz
Emil J. Bovich
Wendell Bowers
Michael Boxer
Patrick Boyle
John D. Brack
John J. Bradley
W. Newman Bradshaw, PhD
Todd R. Bremner
Harold Brennan
Edward J. Breyere, PhD
Paul Brierley
Fred E. Brinker
James A. Briscoe
Albert Lyle Broadfoot, PhD
Beth M. Brookhouse
Fred B. Brost
John K. Brough
Lansing E. Brown
Raymond F. Brown
Richard E. Brown
S. Kent Brown
Stephen E. Brown
Stephen R. Brown
Will K. Brown
Gerald R. Brunskill
John A. Brunsman
Bruce Bryan
Thomas L. Bryant
Theodore Eugene Bunch, PhD
Laurie B. Burdeaux
Joe Byrne
Ken M. Byrne
Ralph E. Cadger
Roger Cahill
J. B. Caird
Anthony E. Camilli
John D. Camp
Robert J. Campana
John S. Campbell
Robert E. Campbell
M. Durwood Canham
Joe W. Cannon
Bryan J. Carder
Edward H. Carey
Richard B. Carley
E. N. Carlier
Jesus V. Carreon
Charles A. Carroll
Bruce W. Cavender
Robert T. Chapman
Gerald M. Chicoine
Lincoln Chin, PhD
Larry J. Chockie
James R. Civis
Maurice F. Clapp
James G. Clark, PhD
James W. Clark
John Wesley Clayton, PhD
Robert L. Clayton
LaVar Clegg
Jeffrey G. Clevenger
Raymond Otto Clinton, PhD
Elmer Lendell Cockrum, PhD
Theodore L. Cogut
Donald Coleman
Joel E. Colley
Allan W. Collins
Joanne V. Collins*

Ernest W. Colwell
Thomas J. Comi, PhD
Bill J. Conovaloff
Paul Consroe, PhD
Daniel M. Conway
George W. Cook
Glenn C. Cook
Russell M. Corn
Don Corona, Jr.
Roy E. Coulson
Brian Cox
Garland D. Cox
Peter J. Crescenzo
Anne E. Cress, PhD
Richard E. Cribbs
Donald E. Crowell
Gabriel Tibor Csanady, PhD
David C. Cunningham
William H. Curd, PhD
Joseph A. Cusack
William J. Daffron
Charles H. Daggs, Jr.
James F. Dancho
Jerry Danni
Robert Darveaux, PhD
Dan A. Davidson
Lester W. Davis
Francisco Homero De La Moneda, PhD
JoAnn Deakin
Stuart Deakin
Tom E. Deakin
James D. Deatherage
Dirk Den-Baars, PhD
Lemoine J. Derrick
Jack L. Detrick
William Devereux
Michael J. DeWeert, PhD
Jerome P. Dorlac
Harold J. Downey
Gregory J. Dozer
Jean B. Draper
William Dresher, PhD
Patricia Dueck
Gene E. Dufoe
Marie Dugan
Raymond J. Dugandzic
Jonathan DuHamel
J. Durham
Sher M. Durrani
Robert W. Durrenberger, PhD
Ted Earle
John T. Eastlick
Charles N. Emerson
Philip Anthony Emery
Glen B. Engle
T. Enloe
Lawson P. Entwhistle
A. Gordon Everett, PhD
Ted H. Eyde
Steve Fanto
Robert H. Fariss, PhD
Stanley D. Farlin, PhD
Robert P. Farrell
William F. Fathauer
Larry D. Fellows, PhD
John B. Fenger
Chester G. Ferguson
Sam Field
James B. Fink, PhD
Rex G. Finley
William G. Fisher, PhD

Robert T. Fitzgerald
Leon Walter Florschuet, PhD
Branka P. Ford, PhD
Ron Francken
Arnold C. Frautnick
George M. Fritz
Charlotte Frola
F. Ronald Frola
Arthur Atwater Frost, PhD
Nelson M. Funkhouser
Robert Howard Furman
Donald Alan Gall, PhD
Clyde H. Garman
Paddy R. Garver
Roy Lee Gealer, PhD
Robert P. Gervais
James Giles, PhD
Donald K. Gill
Paul C. Gilmour, PhD
Fred Gioglio
Roy L. Givens
Stephen Glacy
William Glass
June Glavin
Robert L. Glick, PhD
Randy Golding, PhD
E. J. Gouvier
Alphonse Peter Granatek
Harold Greenfield, PhD
William M. Greenslade
George Alexander Gries, PhD
Ray Griswold
Joseph F. Gross, PhD
Charles W. Gullikson, PhD
Gordon E. Gumble
Earl S. Gurley
David B. Hackman, PhD
Anton F. Haffer
Ronald J. Hager, Jr.
William R. Hahman
Cathryn E. Hahn
Francis A. Hale
Richard C. Hall
Charles Hallas
Christy N. Hallien
Donald F. Hammer
Raymond E. Hammond
Jerry T. Hanks
Richard W. Hanks, PhD
Joann Brown Hansen, PhD
Deverle Porter Harris, PhD
Ellen Harris
Elmer Otto Hartig, PhD
Scott E. Hartwig
W. L. Harvey
Charles M. Havlik
John W. Hawley
Robert W. Hazlett, PhD
M. W. Heath, Jr.
Lowell H. Heaton
John M. Heermans
Albert K. Heitzmann
Patricia A. Helvenston, PhD
Robert S. Hendricks
Lester E. Hendrickson, PhD
Edward P. Herman
Roy A. Herrington
Tom Hessler
Robin J. Hickson
Douglas W. Hilchie, PhD
Donald C. Hilgers, PhD
Jeffrey H. Hill, PhD

Norman E. Hill
Robert D. Hill
Steven W. Hill
Bruce Hilpert
Daniel L. Hirsch
J. Brent Hiskey, PhD
Corolla K. Hoag
Alfred Joseph Hoehn, PhD
Stuart Alfred Hoenig, PhD
R. N. Holme
G. Edward Holmes
Michael Holtfrerich
Kevin C. Horstman, PhD
Ashley G. House
David C. Howe
William Bogel Hubbard, PhD
Richard O. Huch
Davin L. Huck
Richard R. Huebschman
Robert P. Hughes
Raymond P. Hull
Robert B. Humphrey
Kenneth L. Hunt, Jr.
Charles L. Hunter
Peter P. Hydrean, PhD
Sherwood B. Idso, PhD
Lawrence L. Ingram
G. W. Irvin
Daniel L. Isbell
Kenneth K. Issacson
Teodore F. Izzo
K. A. Jackson, PhD
William A. Jacobs
Marek Jakubowski, PhD
Douglas E. James
Thomas Jancic
Kevin L. Jardine
John H. Jarvis
James Edward Jaskie, PhD
David E. Jeal
Andreas V. Jensen
Harry E. Jensen
Robert M. Jensen
Ralph O. Jewett
David C. Johnson
David Johnston
Billy J. Joplin
Richard Charles Jordan, PhD
Thomas L. T. Jossem
Jim Joyce
George F. Jude
Eric Jungermann, PhD
Richard Spalding Juvet, PhD
Richard J. Kalvaitis
Ronald R. Kamyniski
Justin M. Kapla
Francis Warren Karasek, PhD
Peter P. Kay
Keith M. Keating
Kenneth Lee Keating, PhD
Niles William Keeran
James L. Kelly*
Keith L. Kendall
Shawn B. Kendall
Robert Kendrick
William A. Kennedy, PhD
Kevin M. Kenney
William J. Kerwin, PhD
Harold D. Kessler
Clement Joseph Kevane, PhD
Paul M. Keyser
Paul E. Kienow

William R. Kilpatrick
W. David Kingery, PhD
Dan Kirby
Clayton Ward Kischer, PhD
Roy C. Kiser
R. Jay Kline
William Robert Kneebone, PhD
Vincent E. Knittel
Gregory Dean Knowlton, PhD
Paul Koblas, PhD
Leonard John Koch
Karl F. Kohlhoff
Frederick H. Kohloss
Rudolf Kolaja
Robert R. Koons
William P. Kopp
Melvin J. Kornblatt
Jeff K. Kracht
Henry G. Kreis
Ken James Krolik
Gerald E. Kron, PhD
Edwin H. Krug
David J. Krus, PhD
David L. Kuck
John G. Kuhn
W. R. Kulutachek
Ihor A. Kunasz, PhD
Joseph D. Kutschka
Gary W. Lachappelle
Willard C. Lacy, PhD
Lorin G. Lafoe
Charles F. Lagergren
James L. Lake
Lynn Lalko, PhD
Lionel C. Lancaster
Wayne A. Lattin
Michael J. Lechner
Steven B. Leeland
Randolph S. Lehn
Richard B. Leisure
Tom C. Lepley
Ron L. Levin, PhD
David W. Levinson, PhD
Seymour H. Levy
Ruth L. Leyse, PhD
Hang Ming Liaw, PhD
John M. Liebetreu, PhD
Urban Joseph Linehan, PhD
William G. Lipke, PhD
Ligia B. Lluria
Mark Logan
James D. Loghry
Edward M. Lohman
Armand "A.J." Lombard
David Lorenz
Robert F. Lorenzen
Robert B. Ludden
Roger J. Ludlam
Mark Ludwig, PhD
Ronald J. Lukas, PhD
Robert F. Lundin, PhD
James H. Lundy, Jr.
Donald E. Lynd
Clarence Roger Lynds, PhD
Robert P. Ma, PhD
Richard L. Malafa
Tom Maloney
G. S. Mander
Paul Allen Manera, PhD
James K W Mardian, PhD
Timothy Martin Marsh, PhD
William Michael Marsh

Bijan R. Mashouf
Billy F. Mathews
Thomas W. Mathewson
Donald E. Matlick
Harrison E. Matson
Ralph W. Maughan
J. A. McAllister*
Terry McArthur
Carol Don McBiles
Ron McCallister, PhD
John W. McCracken
Paul McElligott, PhD
Norman E. McFate, PhD
James E. McGaha
Andrew J. McGill
Geogory E. McKelvey
Bruce A. McKinstry
R. L. McPherson
Lorin Post McRae, PhD
John E. McVaugh, Jr.
Wellington Meier, PhD
Arend Meijer, PhD
Norman Anthony Meinhardt, PhD
Gary L. Melvin
Ken Melybe
Charles R. Merigold
Richard E. Merrill
Robert A. Metz
Robert J. Meyers, Jr.
Robert J. Meyers
Harvey D. Michael
Marcus A. Middleton
Donald G. Miller
Glenn Harry Miller, PhD
Kenton D. Miller
Mark A. Miller
Frank R. Milliken
James A. Moeller
Alan R. Mollenkopf
Robert P. Montague
Roger L. Moody
Donald Moon
Jack W. Moore
Ramon A. Morano
Syver W. More
Pamela A. Morford
Phineas K. Morrill
John Ross Mosley, PhD
Robert Zeno Muggli, PhD
Michael C. Mulbarger
Gerald E. Munier
Earl Wesley Murbach, PhD
Fendi R. Murdock, PhD
Robert E. Nabours, PhD
Paula Nadell
Raymond Naumann
John Neff
George E. Nelms, PhD
Frank Nelson
David R. Nielsen
Ronald A. Nielsen
Nyal L. Niemuth
Aundra G. Nix
Robert O. Nixon, PhD
Charles C. Nolan
Eric A. Nordhausen
James J. Novak
Edward A. Nowatzki, PhD
David O. Oakeson
Kevin M. O'Brien, PhD
Ernest L. Ohle, PhD
Francis Oliver

John D. Opalka
Stephen A. Orcutt
Jane M. Orient
Herbert Osborn, PhD
Stephen S. Osder
Charles Lamar Osterberg, PhD
Bernie Ott
Karl J. Palmberg
Paul F. Panebianco
Lloyd M. Parks, PhD
Harvey G. Patterson
Doug S. Pease
E. Howard Pepper
Mike Peralta, PhD
Donald M. Percy
Darlene A. Periconi-Balling
Thomas W. Phillips
S. Pietrewicz
Anthony Pietsch
Fred R. Pitman
Anthony Pitucco, PhD
John A. Plaisted
Wallace Platt
Alan P. Ploesser
James R. Plummer
Gary Pollock
Kent L. Pomeroy
Fernando Ponce, PhD
H. K. Poole
William E. Poole
Jonathan D. Posner, PhD
George J. Potter
Roderick B. Potter
James A. Powell
Arnold Warburton Pratt
Robert F. Prest
Burchard S. Pruett
Robert Putnam
Thomas Pyzdek
Walter J. Quinlan
David D. Rabb
Kenneth D. Rachocki, PhD
Donal M. Ragan, PhD
Steven L. Rauzi
Kyle Rawlings, PhD
Paul R. Reay
K. Redig
Terry A. Reeves
Thomas R. Rehm, PhD
Roger A. Reich
John E. Reichenbach, Jr.
Kenneth G. Renard, PhD
Robert J. Reuss
C. H. Reynolds
Robert R. Reynolds
Harvey H. Rhodes
Terry L. Rice
Warren Rice, PhD
Ralph M. Richard, PhD
James L. Riedl
Timothy R. Robbins
Steve A. Roberts
Timothy Roberts
Chris Robertson
Douglas L. Robinson, PhD
Thomas C. Roche, Jr.
Michael V. Rock
Ralph A. Rockow
John H. Rohrbaugh, PhD
Dwayne A. Rohweder, PhD
Garret K. Ross
Richard H. Rowe

Verald K. Rowe
David Daniel Rubis, PhD
Thomas Rudy
John A. Rupley, PhD
George E. Ryberg
William R. Salzman, PhD
Robert E. Samuelson, PhD
Katherine Sanchez
John B. Sawyer
Thomas E. Scartaccini
Harvey C. Schau, PhD
William G. Scheck
Joseph M. Scherzer
Justin Orvel Schmidt, PhD
Conrad Schneiker
Melvin Herman Schonhorst, PhD
George K. Schuler
Michael H. Schweinsberg
Theodore Thomas Scolman, PhD
Chuck See
Hal Sefton
Mark N. Seidel, PhD
Harner Selvidge, PhD
Robert Shantz, PhD
Steven M. Shaw, PhD
Peter Sherry
Jack L. Shilling
Kirk W. Shubert
Thomas K. Simacek, PhD
William L. Simmermon
James D. Simpson, PhD
Alan M. Sinclair
Mark R. Sinclair, PhD
Ernesto Sirvas
Edward Skibo, PhD
Bruce E. Skippar
W. Roy Slaunwhite, PhD
Bennett Sloan
Thomas L. Sluga
Alice Smith
Delmont K. Smith, PhD
Donald Snyder
John A. Soscia
David M. Spatz, PhD
Earl L. Spieles
John W. Stafford
Daniel Stamps
Royal William Stark, PhD
Travis C. Steele
Frank A. Stephenson, PhD
Lester H. Steward, PhD
Vern S. Strubeck
Robert J. Stuart
Dennis L. Stuhr
Frederick E. Suhm
David Sultana
Soren J. Suver
Paul E. Swain
Gordon Alfred Swann, PhD
John M. Sweet
Moute N. Swetnam
Richard Switzer
Michael Talbot
Charles M. Tarr
Michael Teodori
Marvin W. Teutsch
C. Brent Theurer, PhD
Russell W. Thiele
H. Stephens Thomas
Jess F. Thomas
Glenn Thompson, PhD
David H. Thornton

John R. Thorson
Dennis L. Thrasher
Jacob Timmers
Spencer R. Titley, PhD
George E. Travis
Vernon L. Trimble
Eric Tuch
Roy A. Tucker
Lee A. Tune, Jr.
George C. Tyler
John L. Uhrie, PhD
Bobby Lee Ulich, PhD
Noel W. Urban
Regino B. Urgena
Emmett Van Reed
Willard VanAsdall, PhD
C. Kenneth Vance
Richard V. VanRiper
Edward F. Veverka
Antonio R. Villanueva, PhD
Edward Vizzini, PhD
James W. Vogler, PhD
Peter Vokac
David E. Wahl, Jr., PhD
Robert C. Walish, Jr.
Joe A. Walker
Woodville J. Walker*
Wayne Wallace
Kevin Walsh
Steven I. Walsh
Meredith F. Warner
Steven D. Washburn
Roger L. Waterman
Lee R. Watkins
W. L. Wearly, PhD
Orrin John Webster, PhD
Gene H. Wegner, PhD
Patrick L. Weidner
Jack D. Weiss
Homer William Welch, PhD
William T. Welchert
Herbert C. Wendes
Greg Wenzel
Donald C. White
Clifford K. Whiting
Sue Whitworth
W. Gordon Wieduwilt
John B. Wilburn
George Wilkinson, Jr.
Brandon C. Williams
Steven Kenneth Williams, PhD
Clifford Leon Willis, PhD
Robert E. Willow
Jeff Wilmer
Donald W. Wine
David Wing, PhD
Bruce M. Winn
John B. Winters
James C. Withers, PhD
George K. Wittenberg, PhD
Gerald Woehick
George A. Wolfe
Richard H. Wolters
Gerald L. Wood
David J. Woods
George A. Woods
Will W. Worthington
Glenn Wright, PhD
James A. Yanez
James R. Yingst
Garth L. Young
Robert A. Young, PhD

Itzak Z. Zamir
Roy V. Zeagler, Jr.
Charles Zglenicki
Paul W. Zimmer
Werner G. Zinn

Arkansas

Billy R. Achmbaugh
Alan J. Anderson
James R. Arce
Jerry L. Baber
Harry R. Baker
Phillip A. Barros
Lonnie G. Bassett
Ralph Sherman Becker, PhD
Randolph Armin Becker
Ford Benham
Maximilian Hilmar Bergendahl, PhD
John J. Berky, PhD
C. Dudley Blancke
Robert E. Blanz, PhD
David Bowlin
Robert Edward Bowling, PhD
Delton L. Brown, Jr.
Bryan Burnett
Gordon L. Burr
Stephen Cain
James P. Caldwell
Raymond Cammack, PhD
Samuel Z. Chamberlin
Frank Chimenti, PhD
Thomas Clark
Jeffrey M. Collar
Frederick Clinton Collins, PhD
John D. Commerford, PhD
Reggie A. Corbitt
Kenneth C. Corkum, PhD
Stanley R. Curtis
James Ed Davis
Steven E. Dobbs, PhD
Kenneth L. Duck
James A. Dunlop
Don H. Edington
James E. Erskine
Ronald Everett, PhD
Timsey L. Everett
William Russell Everett, PhD
G. Ferrer, PhD
Karen Ferrer, PhD
Richard Hamilton Forsythe, PhD
Kenneth W. French, PhD
Roland E. Garlinghouse
Walter Thomas Gloor, PhD
William E. Gran
Francis A. Grillot, Jr.
John Louis Hartman, PhD
Roger M. Hawk, PhD
Larry D. Heisserer
Orma L. Henders
Francis M. Henderson
Chester W. Hesselbein
Lewis W. Hirschy
Ed Hiserodt
Edwin J. Hockaday
Charles J. Hoke
Erne Hume
Johnny R. Isbell
Mary E. Jenkins
Charles Jones, PhD
Edward T. Jones

Matti Kaarnakari
Earnest Kavanaugh
Bill W. Keaton
David Wayne Kellogg, PhD
M. K. Kemp
Edward J. Kersey
Donald R. Keys
John W. King, PhD
Tommy C. Kinnaird
David L. Kreider, PhD
Barry Kurth
Sterling S. Lacy
William E. Lanyon
Noel A. Lawson
Wei Li
Dennis D. Longhorn
Gary L. Low
M. David Luneau
Terry L. Macalady
Phillip A. Marak
Mary K. Marks-Wood
Lloyd D. Martin
Richard Harvey Martin
Kenneth L. Mazander
Clark William McCarty, PhD
Hal E. McCloud, PhD
Patrick McGuire
Harlan L. McMillan, PhD
Thomas H. McWilliams
David G. Meador
Allan J. Mesko
Gerald R. Metty
Paul Mixon, PhD
Jeanne Murphy, PhD
X. J. Musacchia, PhD
Bobbie G. Musson
Danny Naegle
Charles A. Nelson, PhD
Lowell Edwin Netherton, PhD
Kelly Hoyet Oliver, PhD*
Vernon L. Pate
John E. Pauly, PhD
Raymond E. Peeples
John D. Pike
Joseph C. Plunkett, PhD
Larry T. Polk
Jack L. Reddin
Douglas A. Rees-Evans
Allan Stanley Rehm, PhD
Al W. Renfroe
Tom E. Richards
J. Herbert Riley
Albert Robinson, PhD
S. Maurice Robinson
Gary W. Russell
William Marion Sandefur, PhD
Boris M. Schein, PhD
Paul H. Schellenberg
Bruce E. Schratz
Mark Shalkowski
B. Sherrill
David E. Sibert
Alfred Silano, PhD
Carroll Ward Smith, PhD
Wayne Smith
Jason Stewart
C. Storm
Mike Robert Strub, PhD
G. Russell Sutherland
John B. Talpas
Aubrey W. Tennille, PhD
Don M. Thomas

Joseph R. Togami
George Toombs
William Walker Trigg, PhD
Daniel L. Turnbow
Alan C. Varner
Patrick D. Walker
Curtis Q. Warner
J. R. Weaver
Melvin Bruce Welch, PhD
Tressa White
Stephen W. Wilson
Mary Wood
James L. Word

California

Earl M. Aagaard, PhD
Charles W. Aami
Ursula K. Abbott, PhD
Janis I. Abele
Robert C. Abrams
Ahmed E. Aburahmah, PhD
Ava V. Ackerman
Lee Actor
Humberto M. Acuna, Jr.
George Baker Adams, PhD
Lewis R. Adams
William John Adams
William H. Addington
Barnet R. Adelman
John H. Adrain
Jack G. Agan
Sven Agerbek
Edward J. Ahmann
John J. Aiello
Arthur W. Akers
Gary L. Akerstrom
Wayne Henry Akeson
John S. Akiyama
Philip R. Akre
G. James Alaback
John A. Alai
Daniel C. Albers
Edward G. Albert
John C. Alden, PhD
Alex F. Alessandrini
Fred Alexander
Ira H. Alexander
Rodolfo Q. Alfonso
R. Allahyari, PhD
Louis John Allamandola, PhD
Levi D. Allen
Robert C. Allen
William Edward Alley, PhD
Charles E. Allman
John J. Allport, PhD
Ronaldo A. Almero
David Altman, PhD
Herbert N. Altneu
Antonio R. Alvarez
Raymond Angelo Alvarez, Jr., PhD
Zaynab Al-Yassin, PhD
Farouk Amanatullah
Carmelo J. Amato
Marvin Earl Ament
Melvin M. Anchell
Torben B. Andersen, PhD
Wilford Hoyt Andersen, PhD
Chris Anderson
Conrad E. Anderson
James Anderson

Jane E. Anderson
Joy R. Anderson, PhD
Orson Lamar Anderson, PhD
Robert E. Anderson
Roscoe B. Anderson
Ross S. Anderson, PhD
Thomas P. Anderson
Warren Ronald Anderson
Karen Andersonnoeck
Lois Andros
Walter S. Andrus
Claude B. Anger
Gregory W. Antal
Achilles P. Anton
Rolando A. Antonio
Arturo Q. Arabe, PhD
John Arcadi
Philip Archibald
Robert L. Archibald
Gary Arithson
Richard W. Armentrout, PhD
Baxter H. Armstrong, PhD
Robert Emile Arnal, PhD
Charles Arney
George V. Aros Chilingarian, PhD
George J. Asanovich
Edward V. Ashburn
Holt Ashley, PhD
Don O. Asquith, PhD
Everett L. Astleford
Greg J. Aten
Robert D. Athey, Jr., PhD
Leonardo D. Attorre
Jerry Y. Au
Mike August
W. David Augustine
Thomas E. Aumock
Henry Spiese Aurand
Kenny Ausmus
Roger J. Austin, PhD
Philip J. Avery
Kenneth Avicola
Luis A. Avila
Theodore C. Awartkruis, PhD
T. G. Ayres
Wesley P. Ayres, PhD
William J. Babalis
Ray M. Bacchi
Gordon R. Bachlund
William E. Backes
Adrian Donald Baer, PhD
Henry P. Baier
Benton B. Bailey
Edmund J. Bailey
Liam P. Bailey
Ronald M. Bailey
Donald W. Baisch
Don Robert Baker, PhD
Mary Ann Baker, PhD
Norman F. Baker, PhD
Roland E. Baker
W. J. Baker
John A. Balboni
Orville Balcom
Barrett S. Baldwin, PhD
David P. Baldwin
Ransom Leland Baldwin, PhD
George Balella
Donald L. Ball
George Ball
Glenn A. Ballard
Martin Balow

John S. Baltutis
John LeRoy Balzer, PhD
Cris C. Banaban
Herman William Bandel, PhD
Richard M. Banister
Ronald E. Banuk
Neil J. Barabas
George Baral
Ronald Barany, PhD
James W. Barcikowski
Norman E. Barclay
Brian S. Barcus
Randolph P. Bardini
Morrie Jay Barembaum
Grigory Isaakovich Barenblatt, PhD
Francis J. Barker
Horace A. Barker, PhD
Richard K. Barksdale
Mary J. Barlow
Albert R. Barnes
Burton B. Barnes
Paul R. Barnes
Russell H. Barnes
James Robert Barnum, PhD
James P. Barrie
Bruce M. Barron
Robert H. Barron
Gary D. Barry
Bruce C. Bartels
Don A. Bartick
Bob W. Bartlett, PhD
Janeth Marie Bartlett, PhD
Sarah H. Bartling
William A. Bartling
W. Clyde Barton, Jr.
Don Bartz
Cecil O. Basenberg
John E. Basinski, PhD
H. Smith Bass
Samuel Burbridge Batdorf, PhD
Barbara Batterson
Edward C. Bauer, PhD
Kurt Baum, PhD
Frank J. Baumann
Hans Peter Bausch
E. Beaton
David Beaucage, PhD
Robert A. Beaudet, PhD
Christine Beavcage, PhD
Horst Huttenhain Bechtel
Robert F. Bechtold
Donald J. Beck
Niels John Beck, PhD
Roy T. Beck
Tom G. Beck
Milton Becker, PhD
William H. Beckley
Arnold Orville Beckman, PhD*
Toni Lynn Beckman, PhD
Gary S. Beckstrom
James P. Beecher
Donald W. Beegle
Mark Beget
Nicholas Anthony Begovich, PhD
Jean E. Beland, PhD
Ralph Belcher, PhD
Charles Vester Bell, PhD
John Bell
Barbara Belli, PhD
Thomas J. Bellon, Jr.
Robert K. Bellue
Francis J. Belmonte

John F. Below, PhD
John W. Ben
David J. Benard, PhD
Robert D. Benbow
Paul F. Bene
Barry P. Benight
Kurt A. Benirschke
John Benjamin, PhD
Istvan S. Benko
William G. Benko
Kenneth W. Benner, PhD
Harold E. Bennett, PhD
Sidney A. Bensen, PhD
Andrew A. Benson, PhD
Herbert H. Benson
John D. Benson
Roger Benson*
Sidney W. Benson, PhD
Margaret W. Benton
Philip H. Benton
John A. Bentsen
John A. Berberet, PhD
Louis Bergdahl
Augustus B. Berger
Lev I. Berger, PhD
Otto Berger
Leo H. Berk
Ami E. Berkowitz, PhD
William I. Berks
Ted Gibbs Berlincourt, PhD
Baruch Berman
Louis Bernath, PhD
Dave Berrier
Lester P. Berriman
Carl E. Berry
Edwin X. Berry, PhD
David J. Berryman
Richard G. Berryman
Georgw J. Bertuccelli
Thomas E. Berty
Bruce A. Berwager
James A. Bethke
John C. Bettinger
Ernest Beutler
Vladislav A. Bevc, PhD
Dimitri Beve, PhD
John H. Beyer, PhD
Ashok K. Bhatnagar
Fred V. Biagini
Carl J. Bianchini
George A. Bicher
Michael D. Bick, PhD
Donald B. Bickler
Donald G. Bickmorc
Lauren K. Bieg
Gregory A. Bierbaum
Richard V. Bierman
Jerry C. Billings
Kenneth William Billman, PhD
Charles J. Billwiller
Paul A. Bilunos
R. L. Binsley
Norman Birch
Kenneth Bird, PhD
James Louden Bischoff, PhD
Clifford R. Bishop
John William Bishop, PhD
Kim Bishop, PhD
Linman O. Bjerken
Lars L. Bjorkman
Paul A. Blacharski
Melvin L. Black

Charles M. Blair, PhD
Francis Louis Blanc
Dean M. Blanchard
Leroy E. Blanchard
Donald W. Blancher
Hiram W. Blanken, Jr.
Michael S. Blankinship, Jr.
Joseph S. Blanton
Dean A. Blatchford
Karl T. Blaufuss
John Blethen, PhD
Zegmund O. Bleviss, PhD
Max R. Blodgett
C. James Blom, PhD
David L. Blomquist
Leonard C. Blomquist
John Bloom, PhD
G. Bluzas
Warren P. Boardman
Carl Bobkoski
Gene Bock
Keith R. Bock
Richard M. Bockhorst
Gene E. Bockmier
William E. Boettger
Harold Bogin
Lawrence P. Bogle
Dale V. Bohnenberger
Eugene Bollay
Ellen D. Bolotin, PhD
Donald H. Boltz
Charles M. Bolus
Joseph C. Bonadiman, PhD
Stephen Alan Book, PhD
E. S. Boorneson
Iris Borg, PhD
W. K. Borgsmiller
Manfred D. Borks, PhD
William R. Bornhorst
Gerald F. Borrmann
Anthony G. Borschneck
Robert B. Bosler, Jr.
Harold O. Boss
Keith A. Bostian, PhD
Danil Botoshanksky
Gerald W. Bottrell
Michel Boudart, PhD
Robert L. Boulware
Kenneth P. Bourke
Robert H. Bourke, PhD
Douglas A. Bourne
Paul K. Bouz
George I. Bovadiieff
Peter F. Bowen
David Bower
Warren H. Bower
John Bowers
William M. Bowers
Doug R. Bowles
Jan Bowman
C. Stuart Bowyer, PhD
Wilson E. Boyce*
Willis Boyd, Sr.
Delbert D. Boyer
William F. Bozich, PhD
Jerry A. Bradshaw
Derek Bradstreet
F. P. Brady, PhD
Matthew E. Brady
William B. Brady
Walton K. Brainerd
J. C. Brakensiek

John W. Bramhall
Francis A. Brandt
Albert Wade Brant, PhD
Thomas E. Braun
Wesley J. Braun
Ben G. Bray, PhD
Warren D. Brayton
James C. Breeding
James D. Brehove
Robert L. Breidenbaugh
Ted Breitmayer
C. H. Breittenfelder
A. C. Breller
Walter B. Brewer
Theodore C. Brice
Alan G. Bridge, PhD
Stephen G. Bridge
Robert M. Bridges*
James E. Briggs, PhD
Robert Briggs
Allan K. Briney
Donald F. Brink, PhD
Tyler Brinker
Francis Everett Broadbent, PhD
Sue Broadston
J. R. Brock
Ivor Brodie, PhD
Woody Brofman, PhD
Ronald J. Bromenschenk
Chistopher Bronny
Charles E. Bronson
Lionel H. Brooks, PhD
Ronald D. Brost
Robert John Brotherton, PhD
David Brown
Hal W. Brown
Howard J. Brown
James R. Brown, PhD
Kenneth Taylor Brown, PhD
Linton A. Brown
Raymond E. Brown
Terrill E. Brown
D. Brownell
Don Brownfield
Peter Brubaker
David Bruce
Gene Bruce
Carl Bruice
Harvey F. Brush
Donald L. Brust
A. Bryan
Glenn H. Bryner
Michael J. Buchan
Steven M. Buchanan
John H. Buchholz, PhD
Smil Buchman
Carl J. Buczek, PhD
Donald R. Buechel
Ronald M. Buehler
Fred W. Bueker
Walter R. Buerger
Robert R. Buettell
Oscar T. Buffalow
Sterling Lowe Bugg
Robert J. Bugiada
Victor E. Buhrke, PhD
Brian J. Bukala
William Murray Bullis, PhD
Ronald Elvin Bullock
Eric Buonassisi
Jacob Burckhard
Harvey Worth Burden, PhD

Herbert S. Burden, Jr.
Willard Burge
Milton N. Burgess
Billy F. Burke
Gary Burke
James E. Burke, PhD
Richard Lerda Burke, PhD
Walter L. Burke
Lawrence H. Burks
James R. Burnett, PhD
Thomas K. Burnham
Leslie L. Burns, PhD
Victor W. Burns, PhD
Roger L. Burtner, PhD
Kittridge R. Burton
Douglas D. Busch
Francis R. Busch
Rick Buschini
Edwin F. Bushman
Mark M. Butier
David V. Butler
Thomas Austin Butterworth, PhD
Sidney Eugene Buttrill, PhD
Gary S. Buxton
Richard G. Byrd
Algyte R. Cabak
Trish Cabral
William P. Cade
Ben Cagle
William M. Cahill
David Stephen Cahn, PhD
Delver R. Cain
Larry Caisuin
Richard E. Cale*
Fred L. Calkins
Gary N. Callihan
Chris Calvert, PhD
Fred D. Campbell
Henry W. Campbell
Malcolm D. Campbell
Nick Campion
Marsha A. Canales
George D. Candella
Robert E. Caniglia
Thomas F. Canning
Arlen E. Cannon
Garry W. Cannon
Peter Cannon, PhD
Harvey L. Canter
Ronald J. Cantoni
Manfred Cantow, PhD
Charles A. Capp Spindt, PhD
Albert J. Cardosa
William Thomas Cardwell
Audrey M. Carlan
Carl E. Carlson
George A. Carlson, PhD
Lloyd G. Carnahan, PhD
Scott Carpenter
Edward Mark Carr
Lester E. Carr, III, PhD
Richard Carr
Gilbert C. Carroll
Jeffery L. Carroll
Walter R. Carrothers
Mary E. Carsten, PhD
David Carta, PhD
Jim Carter
Willie J. Carter, PhD
John G. Carver, PhD
Tony K. Casagranda
Ronald F. Cass

George Cassady
Anthony A. Cassens
Valen E. Castellano
James B. Castles
Kenneth B. Castleton, Jr.
Rick Cataldo
Henry P. Cate, Jr.
Russel K. Catterlin
W. L. Caudry
Thomas Kirk Caughey, PhD
Jerry Caulder, PhD
James E. Cavallin
James A. Cavanah
Chuck J. Cavanaugh
Neal C. Caya
Lee B. Cecil
Carl N. Cederstrand, PhD
Leland H. Celestre
Thomas U. Chace
Pamela Chaffee
Rowand R. Chaffee, PhD
Carlton Chamberlain
Dilworth Woolley Chamberlain, PhD
John S. Chambers
Oliver V. Chamness
Scott O. Chamness
Arthur D. Chan
Sham-Yuen Chan, PhD
Sunney L. Chan, PhD
Steven Chandler
Berken Chang, PhD
Charles S. Chang
Freddy Wilfred L. Chang, PhD
Nicholas D. Change
Mien T. Chao
George Frederick Chapline, Jr., PhD
Ross T. Charest
Bruce R. Charlton
Frank D. Charron
Joseph H. Chasko
E. Cheatham
Boris A. Chechelnitsky
Alwin C. Chen
Donald Chen
Fred Y. Chen
Kun Hua Chen, PhD
Ming K. Chen
Robin S. Chen
Wade Cheng, PhD
J. C. Chernicky
Dallas L. Childress
George V. Chilingar, PhD
Hong Chin, PhD
Jerry L. Chodera
Shary Chotai
Tai-Low Chow, PhD
Emmet H. Christensen
Howard L. Christensen
John D. Christensen
Kent T. Christensen
Steven L. Christenson
George B. Christianson
Kent B. Christianson
Allison L. Christopher
Donald O. Christy, PhD
John E. Chrysler
Daryl Chrzan, PhD
Andy C. Chu
Constantino Chua
Craig P. Chupek
Stanford Church
Steven Ralph Church, PhD

Paul Ciotti
Joseph A. Cipolla
Fernando F. Cisneros
Lawrence P. Clapham
Javier F. Claramunt
Charles R. Clark, PhD
Harold A. Clark
Norman B. Clark
Richard W. Clark
Sharron A. Clark
John Francis Clauser, PhD
Gordon Claycomb
Robert R. Claypool
Bruce Clegg
Carmine Domenic Clemente, PhD
William R. Clevenger
Arnie L. Cliffgard
Watson S. Clifford
H. B. Clingempeel
Mansfield Clinnick
Thomas L. Cloer, Jr.
Joseph F. Cloidt
Ronald E. Clundt
Harold E. Clyde
Paul Jerry Coder
Allen C. Codiroli
C. Robert Coffey
Karl Paley Cohen, PhD
Norman S. Cohen
Sam Cohen
Anthony W. Colacchia
Stefan Colban
Kenneth R. Cole
Miles L. Coleman
Robert G. Coleman, PhD
Joseph D. Coletta
Donald Colgan
John H. Collier
Thomas F. Collier
Dennis R. Collins, PhD
Irene B. Collins
Carlos Adolfo Colmenares, PhD
William B. Colson, PhD
Andre Coltrin
Robert Neil Colwell, PhD
William Tracy Colwell, PhD
Brian Comaskey, PhD
Stephen A. Comfort, PhD
David R. Comish
Jacob C. Compton
Wayne M. Compton
Harry M. Conger
John T. Conlan
Stephen W. Conn
Robert H. Conner
Claud C. Conners
Mahlon C. Connett
Bob A. Conway
Patrick J. Conway
Victoria O. Conway
Brandt Cook
Charles C. Cook
Frank R. Cook, PhD
Karl Cook
Thomas B. Cook, Jr., PhD
James Barry Cooke
James W. Cooksley
Clarence G. Cooper
Martin Cooper
Robert C. Cooper, PhD
Thomas Cooper, PhD
John D. Copley

Claude Coray
Reed S. Coray, PhD
Stephen F. Corcoran
Bruce M. Cordell, PhD
Chris D. Core
John A. Corella
John L. Corl
Joe D. Corless
Roy S. Cornwell, PhD
Nicholas J. Corolis
Wayne T. Corso
Humberto S. Corzo
George J. Cosmides, PhD
Antonio Costa
Harry Cotrill
James R. Coughlin
Danny R. Counihan
George D. Couris
Arnold Court, PhD
Robert E. Covey
William G. Cowdin
Carrol B. Cox
Daniel L. Cox
Kenneth Robert Coyne
Daniel J. Cragin
Kenneth B. Craib
James E. Craig, PhD
Richard F. Craig
Donald James Cram, PhD
Eugene N. Cramer
Leroy L. Crandall
Walter E. Crandall, PhD
Chris L. Craney, PhD
Greg T. Cranham
John D. Craven
Thomas V. Cravy
Dean Crawford
Myron N. Crawford
Dale Creasey
Justin A. Creel, PhD
C. Raymond Cress, PhD
James Creswell
Tom Creswell
Cecil Crews
Phillip O. Crews, PhD
Robert W. Cribbs
Dennis M. Crinnion
Richard G. Crippen
Alipio B. Criste
Luanne S. Crockett
James H. Cronander
Gaines M. Crook, Sr.
James G. Crose, PhD
Kevin P. Cross
Deane L. Crow
Herbert E. Crowhurst
Frank R. Crua
Richard Cruce
Duane Crum, PhD
Jacquelyn Cubre
William K. Culbreth
Arthur G. Cullati, PhD
Donald M. Culler
Floyd Leroy Culler
Peter A. Culley
Murl F. Culp
David Cummings, PhD*
John Cummings
Richard A. Cundiff
A. Cunningham
Edwin L. Currier
Damon R. Curtis

Detlef K. Curtis
Walter E. Curtis
Steven M. Cushman
Kenneth H. Cusick
Donald F. Cuskelly
John M. Cuthbert
Leonard Samuel Cutler, PhD
Robert C. Cutone
Allen F. Dageforde
Himatlal B. Dagli
Daniel P. Dague
Gregory A. Dahlen
Dennis J. Daleiden
Richard Daley, PhD
Lloyd R. Dalton
Robert L. Daly
Philip G. Damask
Marvin V. Damm, PhD
William E. Daniel
Warren Daniels
Raisfeld I. Danse
Moh Daoud
Henry T. Darlington
Gary L. Darnsteadt
Alan D. Dartnell
Renato O. Dato
Clarence Theodore Daub, PhD
Phillip D. Dauben
Arthur A. Daush
Lynn Blair Davidson, PhD
William Davies
Bruce W. Davis, PhD*
H. Turia Davis
James P. Davis
Larry Alan Davis, PhD
Lawrence W. Davis
W. Kenneth Davis*
Don F. Dawson
Paul J. de Fries
Angelo De Min
H. A. De Mirjian
David A. G. Deacon, PhD
John M. Deacon
Willett C. Deady
Douglas L. Dean
Donald Deardorff, PhD
Gerald A. Debeau
Daniel B. DeBra, PhD
Ronald J. Debruin
Robert Joseph Debs, PhD
Paul R. Decker
Curtis Kenneth Deckert
Kent Dedrick, PhD
J. M. Delano, Jr.
Cheryl K. Dell, PhD
Marc Dell'Erba
Charles C. DeMaria
Harold D. Demirjian
Howard D. Denbo
Warren W. Denner, PhD
William J. Denney
Ronald W. Dennison
Wesley M. Densmore
Andrew Denysiak, PhD
Ralph T. Depalma
Joseph George Depp, PhD
Ronny H. Derammelaere
Robert K. Deremer, PhD
Todd C. Derenne
Charles L. Des Brisay
Riccardo DeSalvo, PhD
Don Desborough

Brian J. Deschaine
Harold Desilets
Christopher R. Desley
Alvin M Arden Despain, PhD
Steven A. DeStefano
Robert E. Detrich
Donald P. Detrick
James Edson Devay, PhD
Howard P. Devol
Robert V. Devore, PhD
Edmond M. Devroey, PhD
Thomas Gerry Dewees
Howard F. Dey
Parivash P. Dezham
Arthur S. Diamond
Marian C. Diamond, PhD
Francis P. Diani
James D. Dibdin
Wade Dickinson
Otto W. Dieffenbach
Rodney L. Diehl
Eugene L. Diepholz
Paul A. Diffendaffer
Joseph Brun Digiorgio, PhD
Russell A. Dilley
Ben E. Dillon
John W. Dini
Judy Dirbas
David G. Dirckx
Ray Dirling, Jr.
Kenneth J. Discenza
Byron F. Disselhorst
Kent Diveley
Steven J. Dodds, PhD
Marvin Dodge, PhD
Richard A. Dodge, PhD
Ernest E. Dohner
Roy Hiroshi Doi, PhD
Geoffrey Emerson Dolbear, PhD
Renan G. Dominguez
Chuck Donaldson
Igor Don-Doncon
Armen M. Donian
T. Donnelly, PhD
Kerry L. Donovan, PhD
Brendan P. Dooher, PhD
Robert F. Doolittle, PhD
Timothy B. D'Orazio, PhD
David C. Doreo
Kelly A. Doria
Lowell C. Dorius
Bernhardt L. Dorman, PhD
Ronald J. Dorovi
Weldon B. Dorris
William R. Dotson
Richard L. Double
Steven G. Doulames
Hanania Dover
Douglas B. Dow
Steven Dow
Henry R. Downey
Alexandria Dragan, PhD
Titus H. Drake
William R. Drennen
Daniel D. Drobnis
Richard A. Drossler
Edward A. Drury
Stanley A. Drury
Richard S. Dryden
C. F. Duane
C. Ducoing
Richard D. Dudley

Thomas Dudziak
William T. Duffy, PhD
Paul N. Duggan
Peter Paul Dukes, PhD
William J. Dulude
Arnold N. Dunham
John G. Dunlap
James R. Dunn
Leo P. Dunne
John Ray Dunning, PhD
Thomas G. Dunning
Robin K. Durkee
Gordon B. Durnbaugh
Mark R. Dusbabek
Sophie A. Dutch
John A. Dutton
A. J. Duvall
James G. Duvall, III, PhD
James R. Duvall
Jack Dvorkin, PhD
Paul Dwyer
Denzel Leroy Dyer, PhD
George O. Dyer
J. T. Eagen
Donald G. Eagling
Edwin Toby Earl
Francis J. Eason
Eric G. Easterling
Leslie P. Eastman
A. T. Easton
Kenneth K. Ebel
Richard Eck
Paul R. Edris
Dennis Dean Edwall
David F. Edwards, PhD
Eugene H. Edwards, PhD
J. Gordon Edwards, PhD*
Robert L. Edwards
William R. Edwards
Wilson R. Edwards
Maurice R. Egan, PhD
John P. Ehlen
Kenneth Warren Ehlers, PhD
Walter Eich
Robert E. Eichblatt
Robert Leslie Eichelberger, PhD
Herbert H. Eichhorn, PhD
Donald I. Eidemiller, PhD
David Eitman
Dennis Eland
Wm C. Elhoff
Cindy Eliahu, PhD
Shalom K. Eliahu
Uri Eliahu
Thomas G. Elias
Bert V. Elkins
James C. Elkins
William J. Ellenberger
Jules K. Ellingboe
M. Edmund Ellion, PhD
Robert D. Elliott
Jim E. Ellison
Hugh Ellsaesser, PhD
Ismat E. El-Souki
Gerard W. Elvernum
Cedric B. Emery
Frank E. Emery, PhD
Norman Harry Enenstein, PhD
Rodger K. Engebrethson
Franz Engelmann, PhD
Douglas M. Engh
Harold M. Engle

Richard E. Engle
Grant A. Engstrom
James E. Enstrom, PhD
Bruce Enyeart
Dennis A. Erdman
Wallace J. Erichsen
Alfred A. Erickson
Myriam R. Eriksson
Christine Erkkila
F. H. Ernst, Jr.
Rodney L. Eson
John J. Etchart
Robert H. Eustis, PhD
Charles Andrew Evans, PhD
Charles B. Evans
George W. Evans
Marjorie W. Evans, PhD
Dale Everett
Ray Exley
David S. Fafarman
Robert S. Fagerness
Jack J. Fahey
John C. Fair, PhD
Raymond M. Fairfield
Clay E. Falkner
Steven L. Fallon
Mark W. Fantozzi
Earl L. Farabaugh
W. D. Fargo
Jim O. Farley
Clayton C. Farlow
Emory W. Farr
Gregory L. Farr
James P. Fast
Charles Raymond Faulders, PhD
John R. Favorite
Raymond J. Fazzio
Juergen A. Fehr
John D. Feichtner, PhD
Eugene P. Feist
Betty B. Feldman
W. O. Felsman
David Clarke Fenimore, PhD
John R. Fennell
Paul Roderick Fenske, PhD
Robert B. Fenwick, PhD
Richard K. Fergin, PhD
David B. Ferguson, PhD
Kenneth Edmund Ferguson, PhD
Linn D. Ferguson
Richard L. Ferm, PhD
Louis R. Fermelia
Walter Fetsch
John A. Feyk
William J. Fields
Dale H. Fietz
Donald L. Fife
William Gutierrez Figueroa
Alexandra T. Filer
Mark Filowitz, PhD
Reinald Guy Finke, PhD
Frederick T. Finnigan
William Louis Firestone, PhD
Bryant C. Fischback
Dwayne F. Fischer, PhD
Raymond F. Fish
Donald Fisher
George H. Fisher
Kathleen Mary Flynn Fisher, PhD
Russell L. Fisher
William M. Fishman, PhD
Lanny Fisk, PhD

Jeremy W. Fitch
Richard A. Fitch
Thomas P. Fitzmaurice
James L. Fitzpatrick
William J. Fitzpatrick, PhD
Richard H. Fixler
Loren W. Fizzard
Klaus Werner Flach, PhD
Robert F. Flagg, PhD
Horacio S. Fleischman
Alison A. Fleming, PhD
Donald H. Flowers
Paul H. Floyd
Edward Gotthard Foehr, PhD
Eldon Leroy Foltz
Robert Young Foos*
Michael J. Foote
Irvin H. Forbing
Samuel W. Fordyce
Paul L. Forester
Charles J. Forquer
Robert R. Forsberg
John A. Foster
Robert John Foster, PhD
Henry E. Fourcade
Douglas J. Fouts, PhD
Elliott J. Fowkes, PhD
Brian D. Fox
Wade Hampton Foy, PhD
Donald W. Frames
Dorothea Frames
Dale M. Franchak
Clifford Frank
Robert Frank
Maynard K. Franklin
Randy E. Frazier
Reese L. Freeland
Matt J. Freeman
Reola L. Freeman
Walter J. Freeman
H. Friedemann
Serena M. Friedman
Belmont Frisbee
Tom Frisbee
Earl E. Fritcher
Herman F. Froeb
David Fromson, PhD
Roger J. Froslie
Charles W. Frost
Charles M. Fruey
Si Frumkin
Wilton B. Fryer
George Fryzelka
Frederick A. Fuhrman, PhD
Robert Alexander Fuhrman
Ed D. Fuller
Willard P. Fuller, Jr.
Lawrence W. Funkhouser
Rene G. Fuog
David William Furnas
Ron Gabel
John W. Gabelman, PhD
Jerry D. Gabriel
Michael T. Gabrik, Jr.
Richard A. Gaebel
John Gagliano
Lorenzo A. Gaglio
Gilbert O. Gaines
Russell A. Gaj
John R. Galat
Robert G. Galazin
Darrell L. Gallop, PhD

Yakob V. Galperin, PhD
Kurt Gamara, PhD
Harold D. Gambill
Kurt E. Gamnra, PhD
Luis A. Ganaja
Perry S. Ganas, PhD
Shirish M. Gandhi
Clark W. Gant
Tony S. Gaoiran
Carl W. Garbe
Allen J. Garber
Richard Hammerle Garber, PhD
Alejandro Garcia, PhD
Wayne Scott Gardner, PhD
Walter Garey, PhD
Jack Garfinkel
Joseph F. Garibotti, PhD
Jay M. Garner
Thomas M. Garrett, PhD
William H. Garrison
Thomas D. Gartin
Justine Spring Garvey, PhD
Jerrie W. Gasch
Nicholas D. Gaspar
Robert W. Gassin
Barry Gassner
Robert H. Gassner
Charles F. Gates
Frederick Gates
George L. Gates
Gerald Otis Gates, PhD
Thomas C. Gates
G. R. Gathers, PhD
Phillips L. Gausewitz
Steven D. Gavazza, PhD
Richard L. Gay, PhD
Joseph Gaynor, PhD
Bill A. Gearhart
Colvin V. Gegg, PhD
Dennis Gehri, PhD
Paul Jerome Geiger, PhD
Paul J. Gelger, PhD
Mike Gemmell
Michael S. Genewick
Edward J. George, Sr.
Michael L. Gerber
Renee T. Gerry
Melvin L. Gerst
Paul R. Gerst
Edward C. Gessert
Alex Gezzy
John W. Gibbs
Joseph P. Gibbs
A. Reed Gibby, PhD
Ronald C. Gibson
Warren C. Gibson, PhD
Houghton Gifford
Dominick R. Giglini
Larry L. Gilbert
Lyman F. Gilbert, Sr.
Paul T. Gilbert
Susan Gillen
Wm. R. Gillen
Paul A. Gillespie
William N. Gillespie
Benjamin A. Gillette
Bruce B. Gillies
Robert J. Gilliland
Sherwyn R. Gilliland
Mike Gilmore
Dale P. Gilson
Dezdemona M. Ginosian

William F. Girouard, PhD
Silvio A. Giusti
Dain S. Glad
Dan L. Glasgow
Jerome E. Glass
Stephen M. Glatt
Thomas Glaze
John P. Gleiter
John H. Glenn
Renee V. Glennan
John D. Glesmann
John B. Glode
Vladimir M. Glozman, PhD
Aric Gnesa
Robert W. Goddard
Robert O. Godwin
Susan Godwin
Ludwig Edward Godycki, PhD
Robert W. Goedjen
David Jonathan Goerz
James A. Goethel
Edward Allan Goforth
David J. Goggin, PhD
Alfred Goldberg, PhD
Brian J. Golden
Alan Goldfien
Steven D. Goldfien
Kenneth J. Goldkamp
Bruce Goldman
John Paul Goldsborough, PhD
Norman E. Goldstein, PhD
Vladimir Golovchinsky, PhD
Michel Gondouin, PhD
Dionicio Gonzales
Esteban G. Gonzales
Alexander E. Gonzalez
Ronald Keith Goodman
Roy G. Goodman
Byron Goodrich
James D. Goodridge
James W. Goodspeed
Mikel P. Goodwin
James K. Goodwine, Jr., PhD
Gary E. Gordon
Milton J. Gordon
Rex B. Gordon
Robert Gordon, PhD
William H. Gordon
Nelson G. Gordy
Donald B. Gorshing
Keith E. Gosling
John Ray Goss
Quentin L. Goss
Wilbur Hummon Goss, PhD*
Martin S. Gottlieb
Robert George Gould
Robert D. Gourlay
Darrell Gourley
Lawrence I. Grable
Harald Grabowsky
Ben G. Grady, PhD
Richard W. Graeme
Leroy D. Graff
Kerry M. Gragg
Dee McDonald Graham, PhD
Gary C. Graham, PhD
Alex T. Granik, PhD
Jerry Grant
Edward L. Grau
Walter L. Graves
Clifton H. Gray, Jr.
Lauren H. Grayson

Hue T. Green
Joseph M. Green, PhD
Leon Green, Jr., PhD
Russell H. Green, Jr.
Charles R. Greene, PhD
Eugene Willis Greenfield, PhD
Charles August Greenhall, PhD
Edward C. Greenhood
John Edward Greenleaf, PhD
Jeffrey S. Greenspoon
Peter Gregg
David Tony Gregorich, PhD
Thomas J. Gregory
Gennady H. Grek
Kurt G. Greske
David R. Gress
Donald N. Griffin
Roger D. Griffin
Travis Barton Griffin, PhD
Roy S. Griffiths, PhD
Thomas J. Grifka
Donald Wilburn Grimes, PhD
Leclair Roger Grimes
Charles Groff
Alan B. Gross
Morton Grosser, PhD
James Grote, PhD
Eric Gruenler
Raymond H. Gruetert
Mike A. Grundvig
Mike Gruntman, PhD
Ross R. Grunwald, PhD
Louis E. Grzesiek
Michael R. Guarino, Sr.
Richard Austin Gudmundsen, PhD
Jacques P. Guertin, PhD
Gareth E. Guest, PhD
John O. Guido
Darryl E. Gunderson
Richard R. Gundry
Robert Charles Gunness, PhD
Riji R. Guo
Dwight F. Gustafson
Kermit M. Gustafson
Eugene V. Gustavson
Daniel A. Gutknecht
Steven L. Gutsche
Michael D. Gutterres
John V. Guy-Bray, PhD
Michael A. Guz
Geza Leslie Gyorey, PhD
Bjorn N. Haaberg
Glenn Alfred Hackwell, PhD
D. Haderli
Brian L. Hadley
Arno K. Hagenlocher, PhD
Chuck R. Haggett
Hashem Haghani
Richard B. Hagle
Robert A. Hagn
Pierre Vahe Haig
Samir N. Haji, PhD
Marlund E. Hale, PhD
Paul F. Halfpenny
Kenneth Lynn Hall, PhD
Robert J. Hall
Sylvia C. Hall
Daniel P. Haller
Albert A. Halls, PhD
Herbert H. Halperin
Martin B. Halpern, PhD
Lee Edward Ham

Frank C. Hamann
Ronald O. Hamburger
Edward E. Hamel, PhD
Matt J. Hamilton
Bruce Dupree Hammock, PhD
Stephen Hampton
Anthony James Hance, PhD
Taylor Hancock
Cadet Hand, PhD*
David A. Hand
John W. Hanes
Peter Hangarter
Dale L. Hankins
Gerald M. Hanley
Dean A. Hanquist
Allan G. Hanretta
Ethlyn A. Hansen
L. J. Hansen
Peder M. Hansen, PhD
Robert Clinton Hansen, PhD
Warren K. Hansen
Rowland Curtis Hansford
James C. Hanson
Lloyd K. Hanson
Richard Hanson
John Warvelle Harbaugh, PhD
Ralph Harder
John A. Hardgrove
Edgar Erwin Hardy, PhD
Vernon E. Hardy
Kenneth A. Harkewicz, PhD
Roger N. Harmon
Terry W. Harmon, PhD
John D. Harper, Jr.
John Harper
Kevin J. Harper
Alfred Harral, III
William E. Harries, PhD
David Harriman
Bryan J. Harrington
Kent Harris
Rita D. Harris
Robert O. Harris
S. P. Harris, PhD
Tyler Harris
Burton Harrison
John David Harrison, PhD
Marvin Eugene Harrison
Robin Harrison
Jim Harrower
Philip T. Harsha, PhD
James B. Hart
Darrell W. Hartman
Maurice G. Hartman
George L. Hartmann
William L. Hartrick
Meredith P. Harvan
Jack L. Harvey
Ted F. Harvey, PhD
Dieter F. Haschke
Darr Hashempour, PhD
Hashaliza M. Hashim
Jiri Haskovec, PhD
Robert D. Hass
Steven J. Hassett
George Hathaway, PhD
Mark Hatzilambrou
Warren Hauck
John C. Haugen
Kenneth E. Haughton, PhD
Donald G. Hauser
John E. Hauser

Arthur Herbert Hausman
Alfred H. Hausrath, PhD
Warrnen M. Haussler
Walter B. Havekorst, PhD
Anton J. Havlik, PhD
George E. Hawes
Michael D. Hawkins
Brice C. Hawley
Dale R. Hayden, Sr.
Bill J. Hayes
Robert M. Hayhurst
Carl H. Hayn, PhD
Beth Haynes
William E. Haynes
Gerald Hays
Paul E. Hazelman
R. Nichols Hazelwood, PhD
Dean Head
David L. Heald, PhD
Albert Heaney, PhD
Stephen D. Heath
George E. Heddy, III
Solomon R. Hedges
Larry E. Hedrick
Lee Opert Heflinger, PhD
Frank C. Heggli
Bettina Heinz, PhD
Richard L. Heinze
William D. Heise
Ralph A. Heising
William B. Heitzman
George W. Heller
Denise M. Helm
Carl N. Helmick, Jr.
William F. Helmick
John W. Helphrey
Raymond G. Hemann
F. R. Hemeon
John E. Hench, PhD
Kenneth P. Henderson
Curtis E. Hendrick
Jonathan P. Hendrix
Joseph E. Henn
Carol E. Henneman
Joseph Hennessey
Gary L. Hennings
Donald M. Henrikson
R. R. Heppe
John A. Herb, PhD
Charles A. Herbert
Noel Martin Herbst, PhD
Bruce J. Herdrich
Don D. Herigstad
Elvin Eugene Herman
Ronald C. Herman, PhD
Robert W. Hermsen, PhD
Charles L. Hern
Brian K. Herndon
Leo J. Herrerra, Jr.
James L. Herrick*
David I. Herrington
Christian Herrmann, Jr.
Roger R. Herrscher
John William Baker Hersey, PhD
Robert T. Herzog, PhD
Chris A. Hesse
David A. Hessinger, PhD
Norman E. Hester, PhD
Paul G. Hewitt
Robert E. Heyden
Acle V. Hicks
William B. Hight

Barry J. Hildebrand
Warren W. Hildebrandt
Mahlon M.S. Hile, PhD
Alan T. Hill
John H. Hill
Larry Hill
Robert Hill, PhD
David Hillaker
John R. Hilsabeck
E. B. Hilton
Hubert L. Himes
Carl Hinners
Kenneth A. Hitt
Charles A. Hjerpe
James R. Hoagland
L. C. Hobbs
Peter B. Hobsbawn
Robert A. Hochman
John R. Hoddy
Lee N. Hodge
David Hodgkins
Elizabeth M. Hodgkins
Paul T. Hodiak
Robert S. Hoekstra
William B. Hoenig, Jr.
Robert G. Hoey
C. Hoff
Phil Hoff, PhD
Howard Torrens Hoffman, PhD
Marvin Hoffman, PhD
Carl E. Hoffmeier
Jeffrey E. Hofmann
Roger C. Hofstad
Clarence Lester Hogan, PhD
Roy W. Hogue
Doug Hoiles
Arnold H. Hoines
Franklin K. Holbrook
David Holcberg
Joy R. Holdeman
George R. Holden
Richard Holden
Robert E. Holder
C. H. Holladay, Jr.
Dennis R. Hollars, PhD
Charles L. Hollenbeck
Delbert C. Hollinger
Brent E. Hollingworth
David F. Holman
William H. Holmes
Harold T. Holtom, Jr.
Vincent H. Homer, Jr.
Andrew L. Hon
Richard Churchill Honey, PhD
John D. Honeycutt, PhD
Aaron Hong, PhD
James F. Hood, Jr.
Ronald M. Hopkins, PhD
Gary H. Hoppe
James C. Hoppe
Donald F. Hopps
Thomas G. Horgan
John R. Horn
Lee W. Horn
Erwin William Hornung, PhD
Grant A. Hosack, PhD
Fred L. Hotes
Dale Hotten
David L. Houghton
Leland Richmond House
Loren J. Hov
Max M. Hovaten

Conrad Howan
George O. Howard
S. Dale Howard
Stanley G. Howard
Walter Egner Howard, PhD
George F. Howe, PhD
John P. Howe, PhD
Ward W. Howland
Eric S. Hoy, PhD
C. Hoyt
Chien H. Hsiao, PhD
Hani F. Hskander
Cynthia Hsu
Henry H. Hsu
J. Hsu, PhD
Robert Y. Hsu, PhD
Limin Hsueh, PhD
Kuang J. Huang, PhD
Harmon William Hubbard, PhD
Jason P. Hubbard
Wheeler L. Hubbell
Cyril E. Huber
David Huchital, PhD
Stephen A. Hudson
David E. Hueseman
Hermann F. Huettemeyer
Peter Huetter, III
Jim D. Huff
Thomas J. Huggett
Reginald C. Huggins
Robert A. Huggins
Michael D. Hugh
Larry C. Hughes
Melville P. Hughes
Roxanne C. Hughes
Paul Hull
Charles R. Hulquist
Joseph W. Hultberg
Eric A. Hulteen
Brian Humphrey
John Humphrey
William E. Humphrey, PhD
Steven J. Hunn
George C. Hunt
John P. Hunt, PhD
William J. Hunter
James R. Hurd
Edward T. Hurley
Randy Hurst
Michael C. Husinko
Glen E. Huskey, PhD
Andrew Huszczuk, PhD
Lee Hutchins
Bruce T. Hutchinson
E. S. Hutchison
Alan W. Hyatt, PhD
Eric R. Hyatt, PhD
Charles Hyde
Ralph Hylinski
Umberto C. Iacuaniello
Samuel J. Iarg
Gerald B. Iba
Harold B. Igdaloff
Ronald A. Iltis, PhD
Kenneth T. Ingham
Rodney H. Ingraham, PhD
Donald J. Inman
William Beveridge Innes, PhD
Kaoru Inouye
Dodge Irwin
Donald G. Iselin
Don L. Isenberg, PhD

Robert Isensee, PhD
Byron M. Ishkanian
George Ismael
Farouk T. Ismail, PhD
Guindy Mahmoud Ismail El, PhD
Larry Israel
Olga Ivanilova, PhD
King H. Ives
Lindsay C. Ives
Claude A. Jackman
Dale S. Jacknow
Bruce Jackson
Bruce Jackson
Kingbury Jackson
Robert A. Jackson
Warren B. Jackson, PhD
Sharon Jacobs
John E. Jacobsen
Albert H. Jacobson, Jr., PhD
Fred J. Jacobson
Leslie J. Jacobson
Michael E. Jacobson
Kenneth D. Jacoby
Syed I. Jafri
H. S. James
William L. James
Everett Williams Jameson, Jr., PhD
Robert W. Jamplis
Kenneth S. Jancaitis, PhD
Larry Jang, PhD
Norman C. Janke, PhD
John C. Jaquess
Robert Jastrow, PhD*
Fred J. Jeffers, PhD
George W. Jeffs
David E. Jenkins
Jack D. Jenkins
Sean Jenkins
Richard G. Jenness
David Jennings
Edwin B. Jennings
Frederick A. Jennings
Pete D. Jennings
Gerard M. Jensen, PhD
Paul E. Jensen
Denzel Jenson, PhD
Norman E. Jentz
Herbert C. Jessen
James W. Jeter, Jr., PhD
Nicolai A. Jigalin
Robert M. Jirgal
Zoenek Vaclav Jizba, PhD
Laurie A. Johansen
Craig A. Johanson
Niles W. Johanson
Karl Richard Johansson, PhD
Bertram G. Johnson
Bryce W. Johnson, PhD
Duane Johnson, PhD
Erik Johnson, PhD
G. E. Johnson
Gerald W. Johnson, PhD
H. A. Johnson, PhD
Horace Richard Johnson, PhD
Jim Johnson
Mark A. Johnson
Michael B. Johnson
P. Johnson
Raymond E. Johnson
Richard R. Johnson
Steven M. Johnson
Theodore R. Johnson

Walter F. Johnson
William P. Johnson
William R. Johnson, PhD
George M. Johnston
Viliam Jonec, PhD
Rajinder S. Joneja
Alan B. Jones
Christopher H. Jones
Claris Eugene Jones, Jr., PhD
Edgar J. Jones
Egerton G. Jones
Jeff Jones
Kyle B. Jones
Merrill Jones, PhD
Paul D. Jones
Robert E. Jones
Taylor B. Jones, PhD
Pete Jonghbloed
Peter E. Jonker
Charles Jordan, PhD
L. Jordan
Peter D. Joseph, PhD
Lyman C. Josephs, III
Richard L. Joslin
Ronald F. Joyce
Macario G. Juanola
Robert H. Julian
Bruce B. Junor
George A. Jutila
Charles E. Kaempen, PhD
Robert W. Kafka, PhD
Ron M. Kagan, PhD
Richard L. Kahler
Calvin D. Kalbach
David Kalil
Loren D. Kaller, PhD
Lisa V. Kalman, PhD
Martin D. Kamen, PhD*
Ivan J. Kamezis
Andrew J. Kampe
Sarath C. Kanekal, PhD
Thomas Motomi Kaneko, PhD
Ho H. Kang, PhD
Thomas A. Kanneman, PhD
Paul Thomas Kantz, PhD
Mark S. Kapelke
Alvin A. Kaplan
Hillel R. Kaplan
Eugene J. Karandy
Sid Karin, PhD
Arthur Karp, PhD
Victor N. Karpenko
Ronald A. Kasberger
Brian L. Kash
Daniel E. Kass
Carl J. Kassabian
Jack T. Kassel
William S. Kather
Stanley L. Katten
Fred E. Kattuah
George Bernard Kauffman, PhD
Thomas Kauffman
Alvin Beryl Kaufman
Andrew J. Kay
David W. Kay, PhD
Robert Eugene Kay, PhD
Myron Kayton, PhD
Richard H. Keagy
John R. Keber
Patrick A. Keddington
James Richard Keddy
Ross C. Keeling, Jr.

Walter F. Kelber
William T. Kellermann, Sr.
Charles Thomas Kelley, PhD
Dennis Kelley
Kevin D. Kelley
Patrick R. Kelly
Leroy J. Kemp
Robert E. Kendall
Martin William Kendig, PhD
Peter H. Kendrick
Tom A. Kenfield
Michael T. Kennedy
John M. Kennel, PhD
Brian M. Kennelly, PhD
Clifford Eugene Kent
Kathleen M. Kenyon
Josef Kercso
Clifford Dalton Kern, PhD
Quentin A. Kerns
Anna M. Kerrins
Andrew C. Ketchum
Jesse F. Keville
John D. Keye, Jr.
Tejbir S. Khanna
Simon A. Kheir
Kathleen Kido-Savino
Karl E. Kienow
David C. Kilborn
Kent B. Killian
Jerry Killingstad
David E. Kim
Lawrence K. Kim
D. Kimball
Amon Kimeldorf
William C. Kimpel
Kim S. Kinderman
Chester L. King
Hartley Hugh King, PhD
Joseph E. King
William S. King
William L. Kingston
James W. Kinker
John J. R. Kinney
Gerald Lee Kinnison, PhD
John Kinzell, PhD
E. K. Kirchner, PhD
Tom P. Kirk
Carl S. Kirkconnell, PhD
Richard D. Kirkham
Harry H. Kishineff
Ernest Kiss
Terence M. Kite, PhD
Michael T. Kizer
Kit R. Kjelstrom
Lorentz A. Kjoss
Nicholas Paul Klaas, PhD
Eugene G. Klein
Joseph Klein
Thomas Klein
Aurel Kleinerman, PhD
Robert E. Klenck
Walter Mark Kliewer, PhD
Steve J. Klimowski
Sidney Kline
Thomas J. Kling
Edwin E. Klingman
Gilbert E. Klingman
William Klint
Fred L. Kloepper
Edwin E. Klugman, PhD
Richard M. Klussman
James W. Knapp

Penelope K. Knapp
Richard Hubert Knipe, PhD
Charles R. Knowles
Devin Knowles
Larry P. Knowles
Floyd Marion Knowlton
Stephen A. Kobayashi
Robert D. Kochsiek
Bertram S. Koel
Hans Koellner
Gina L. Koenig
Fred Koester
Robert Cy Koh, PhD
Joe B. Kohen
George O. Kohler, PhD
R. J. Kolodziej
Kazimierz Kolwalski, PhD
Fred W. Koning
B. E. Kopaski
Rudolph William Kopf
John Kordosh
N. Korens
Harrison J. Kornfield
John L. Kortenhoeven
Bart Kosko, PhD
Charles C. Kosky
Mark J. Koslicki
Edward Garrison Kost, PhD
Allen T. Koster
Nagy Hanna Kovacs, PhD
Sankar N. Koyal, PhD
Mitchell M. Kozinski
William R. Krafft
Peter W. Krag, PhD
Jerry Kraim
Roman J. Kramarsic, PhD
Gordon Kramer
Norbert E. Kramer, PhD
William H. Krebs
Ruth E. Kreiss
William T. Kreiss, PhD
Richard M. Kremer, PhD
William B. Krenz
Jeffrey B. Kress
Karl Kretzinger
Joseph Z. Krezandski, PhD
E. Kriva
John Led Kropp, PhD
Loren L. Krueger*
Steven T. Krueger
Gai Krupenkin
E. C. Krupp, PhD
Harvey A. Krygier
Mitsuru Kubota, PhD
Alexander Kucher
Don R. Kuehn, PhD
Frank I. Kuklinski
Eugene M. Kulesza
Kenneth W. Kummerfeld
Joseph Kunc, PhD
Guy Kuncir
Willard D. Kunz
Peter Kurtz, PhD
Alan J. Kushnir, PhD
Paul Kutler, PhD
Leon J. Kutner, PhD
Timothy La Farge, PhD
James La Fleur
Mitchell J. LaBuda, PhD
Leonard L. Lacaze
Kurt D. Ladendorf
Franklin Laemmlen, PhD

Eugene C. Laford, PhD
Bruce Lagasse
Thomas W. LaGrelius
Milton Laikin
Charles L. Laird, III
Cleve Watrous Laird, PhD
John W. Lake
Albert L. Lamarre
Michael A. Lambert, PhD
Ron R. Lambert
Edward W. Lambing, Jr.
Robert A. Lame
H. D. Landahl, PhD
Richard Leon Lander, PhD
William K. Lander
Tom F. Landers
William Charles Landgraf, PhD
Frank L. Landon
Charles S. Landram, PhD
Archie Landry
William G. Landry
F. Lane
Darrell W. Lang
Gregory A. Langan
Thomas H. Lange, PhD
Rolf H. Langland, PhD
Philip G. Langley, PhD
Ward J. Lantier
George R. LaPerle
Gary G. Lapid
Kurt A. Larcher
Lisa W. Larios
Norbert D. Larky
David F. Larochelle
Bruce E. Larock, PhD
Henry Larrucea
David L. Larsen
William E. Larsen
Harry T. Larson
Larry J. Larson
Russell C. Larson
Roderick M. Lashelle
Bill Lee Lasley, PhD
Jason Lau
James Bishop Laudenslager, PhD
Garry E. Laughlin
James W. Laughlin
James H. Laughon
Jim Lauria
Archibald M. Laurie
Michael J. Lavallee
Thomas E. Lavenda
Charles E. Law
Bill Lawler
John John Lawless, Jr., PhD
William H. Lawrence
Edward B. Lawson
Bill R. Lawver
Charles E. Layne
Grant H. Layton
Thomas W. Layton, PhD
Gerry V. Lazzareschi
Julie Leahy
William F. Leahy, PhD
David F. Leake
William D. Leake
James B. Lear
Joseph P. Leaser
Paul Matthew Leavy
Joseph E. Ledbetter, PhD
Robert S. Ledendecker
Bryan D. Lee

Appendix 4: The Petition Project

Bum S. Lee, PhD
Kai Y. Lee
Keun W. Lee
Long Chi Lee, PhD
Min L. Lee, PhD
Paul S. Lee, PhD
Philip R. Lee
Carl B. Leedy
Marjorie B. Leerabhandh, PhD
Robert H. Leerhoff
Franklin E. Lees
Harry A. Leffingwell
Victor E. Leftwich
Walter Lehman
William C. Lehman
James N. Lehmann
Robert F. Lehnen
Edward T. Leidigh
Al G. Leiga, PhD
Jerry S. Leininger
Robert B. Leinster
Larry G. Leiske
George W. Leisz
Ronald J. Lejman
Dan LeMay
Vernon L. Leming
Eric Lemke
Don H. Lenker
John Lenora, PhD
Billy K. Lenser
Robert C. Lentzner
Richard D. Leonard
Peter C. LePort
Robert L. Lessley
James L. Lessman
Max M. Lester, Jr.
Robert W. Lester
Thuston C. LeVay
Lamberto A. Leveriza
Walter Frederick Leverton, PhD
Robert Ernest Le Levier, PhD
Howard Bernard Levine, PhD
Jeff E. Levinger
Roger E. Levoy
J. V. Levy, PhD
Hal Lewis, PhD
William P. Lewis, PhD
Huilin Li, PhD
Shing T. Li, PhD
Thomas T. Liao
Frank Licha
Donald K. Lidster
Kap Lieu
Michael L. Lightstone, PhD
Wayne P. Lill, Jr.
Ray O. Linaweaver
Wilton Howard Lind
Kim R. Lindbery
Robert O. Lindblom, PhD
Charles Alexander Lindley, PhD
Robert R. Lindner
Peter F. Lindquist, PhD
Laurence Lindsay
Robert Lindstrom
Leo C. Linesch
John J. Linker
Paul H. Linstrom
Darrell Linthacum, PhD
Jerome L. Lipin
Arthur L. Lippman
Arthur C. Litheredge
W. M. Liu, PhD

Josep G. Llaurado
Verl B. Lobb
Fred P. Lobban
Timothy A. Lockwood, PhD
Donald O. Lohr
Lewis S. Lohr
Gabriel G. Lombardi, PhD
Robert Ahlberg Loney, PhD
Howard F. Long
James D. Long
Neville S. Long
Paul Alan Longwell, PhD
Clay A. Loomis
Hendricus G. Loos, PhD
David A. Lorenzen
Robert S. Lorusso
Brad M. Losey
Frank H. Lott
Stuart Loucks
Michael E. Lovejoy
C. James Lovelace, PhD
Judith K. Lowe
Alvin Lowi, Jr., PhD
John Kuew Hsiung Lu, PhD
Paul S. Lu
Michael D. Lubin, PhD
Anthony G. Lubowe, PhD
Raymond K. Luci
Samuel R. Lucia
Denise G. Luckhurst
H. Ludwig
Raymond J. Lukens, PhD
William Watt Lumsden, PhD*
Walter F. Lundin
Jorgen V. Lunding
Theodore R. Lundquist, PhD
Pamela G. Lung
Owen Raynal Lunt, PhD
Harold Richard Luxenberg, PhD
Warren M. Lydecker
Dennis Lynch
Laura M. Lynne
Kevin G. Lyons
Richard G. Lyons
Robert S. Lyss
Robert S. MacAlister
Mike S. Macartney
Howard Maccabee, PhD
Alexander Daniel MacDonald, PhD
David V. MacDonald
John W. Mace
Mario A. Machicao
John D. Mack
Donald S. Macko, PhD
John C. Maclay
Lee M. Maclean
Edward H. Macomber
Duane E. Maddux
Joseph T. Maddux
A. Madison
Akhilesh Maewal, PhD
Frank Maga
John L. Magee*
Michael W. Magee
Hans F. Mager
Edward Thomas Maggio, PhD
Tom Alan Magness, PhD
Robert A. Maier
Walter P. Maiersperger*
Douglas J. Malewicki
Michael A. Malgeri
Jim G. Malik, PhD

Calvin Malinka
William Robert Mallett, PhD
Joseph D. Malley, PhD
Albert J. Mallinckrodt, PhD
Kenneth Long Maloney, PhD
George E. Maloof
Neil A. Malpiede
Enrico R. Manaay
Nikola A. Manchev
Isaak Mandelbaum
Nelson L. Mandley
Kent M. Mangold
Anna M. Manley
C. David Mann
George L. Mann
Nancy Robbins Mann, PhD
Philip Mannes
Scheana Mannes
Robert W. Mannon, PhD
William H. Mannon
James Mansdorfer
Greayer Mansfield-Jones, PhD
A. Marasco
Herbert D. Marbach, PhD
George Raymond Marcellino, PhD
Michael Marchese
Stephen C. Marciniec
William C. Marconi
Carol Silber Marcus, PhD
A. J. Mardinly, PhD
Alan Mare
Elwin Marg, PhD
Brian Maridon
Mike J. Marienthal
Michael J. Marinak
William P. Markling
Jack Marling, PhD
Wilbur Joseph Marner, PhD
Patrick M. Maroney
Don Marquis
Marilyn A. Marquis, PhD
Vee L. Marron
Henry L. Marschall
Sullivan Samuel Marsden, PhD
David E. Marshburn
Lorenzo I. Marte
Richard G. Martella
Emil L. Martin
Lincoln A. Martin
Ralph F. Martin
Rebecca Denise Martin
Richard E. Martin
Rodney J. Martin
Stanley Buel Martin
Vurden T. Martin
Ernest A. Martinelli, PhD
Joseph Maserjian, PhD
Arthur J. Mason
George D. Mason
Paul S. Masser, PhD
Alberto G. Masso, PhD
Edgar A. Mastin, PhD
Herbert Franz Matare, PhD
Vincent A. Matera
Charles R. Mathews
Eckart Mathias
Harold C. Mathis
Ronald F. Mathis, PhD
Gene L. Matranga
Kyoko Matsuda, PhD
Jacob P. Matthews
Bill B. May, PhD

Edward H. Mayer
German R. Mayer
David F. Maynard, PhD
G. Mazis
Stephen Albert Mazza
Richard L. McArthy
David M. McCann
William J. McCarter
Edward W. McCauley, PhD
Jon McChesney, PhD
Chester McCloskey, PhD
David L. McClure
R. J. McClure
William Owen McClure, PhD
John M. McCluskey
John V. McColligan
Billy Murray McCormac, PhD
Philip Thomas McCormick, PhD
Thomas E. McCown
Herbert I. McCoy
Ray S. McCoy
Louis Ralph McCreight
Thomas G. McCreless, PhD
James B. McCrumb
Sandra L. McDougald
Barry R. McElmurry
William C. McFadden
Malcolm M. McGawn
William H. McGlasson
Jack F. McGouldrick
Richard J. McGovern
Thomas McGuinness
John P. McGuire
Lawrence McHargue, PhD
Vernon J. McKale
Roger E. McKarus
Charles G. McKay
William Dean McKee, PhD
Donald J. McKenzie
Joe A. McKenzie
William K. McKim
Jerry McKnight
Stephen M. McKown
Charles A. McLean
Steven McLean
Debra McMahan
John D. McMahon
Lester R. McNall, PhD
Gregory R. McNeil
Michael J. McNutt, PhD
Daniel McPherson, Jr.
L. D. McQueen, PhD
Cyril M. McRae
Edgar R. McRae
Gerard J. McVicker
Homer N. Mead
Joseph Meade
Robert C. Meaders
Beverly Meador
Richard A. Meador
Herbert J. Meany
M. G. Mefferd
Merlin Meisner
Stanley Meizel, PhD
Michael A. Melanson
Gloria Melara, PhD
Robert Frederick Meldav
Rodney Melgard
Stanley C. Mellin
Alex S. Meloy
Marvin E. Melton
Bobby J. Melvin

Anthony S. Memeo
Xian-Qin Meng
Steven M. Menkus
William F. Menta
Alan C. Merchant
Leo Mercy
Paul M. Merifield, PhD
Vincent C. Merlin
John Lafayette Merriam
Marshal F. Merriam, PhD
George B. Merrick
Ronald E. Merrill, PhD
William R. Merrill
Seymour Merrin, PhD
Ross A. Merritt
Arthur L. Messinger
Robert Meyearis
Harold F. Meyer
Jewell L. Meyer
Matthew D. Meyer
Rudolf X. Meyer, PhD
Steven J. Meyerhofer
Charles J. Meyers, Jr.
Richard E. Meyers
Gerald James Miatech, PhD
E. Don Michael
Robert C. Michael
Lawrence A. Michel
Lloyd R. Michels, PhD
Michael A. Michelson
Scott J. Mighell
Daniel F. Mika
Nagib T. Mikhael, PhD
Duane Soren Mikkelsen, PhD
Paul G. Mikolaj, PhD
Gerald A. Miles
Melvin H. Miles, PhD
Ralph Franley Miles, PhD
John B. Millard, PhD
Jose B. Millares
Alan D. Miller, PhD
Allan S. Miller, PhD
C. Miller
Cecil Miller, PhD
Daryl D. Miller
Dick Miller, PhD
George Miller
H. L. Miller
James Avery Miller, PhD
John S. Miller
Kenneth J. Miller
Lawrence S. Miller
Robert L. Miller
Sol Miller, PhD
Stanley Leo Miller
Timothy Miller
Wilson N. Miller
Le Edward Millet, PhD
Charles E. Millett
Robert C. Mills
George P. Milton
Thomas Mincer, PhD
Robert E. Minear
Ronald L. Miner
Susie M. Ming
David R. Minor
Tom L. Mintun
John M. Mintz, PhD
Srecko Mirko Mircetich, PhD
Harold Mirels, PhD
Arjang K. Miremadi
Mohammad R. Mirseyedi

Theodore C. Mitchell
Arup P. Mitra
Edward Mittleman
K. L. Moazed, PhD
Julie A. Mobley
Ken L. Moeckel
Randy A. Moehnke
Allen A. Moff
John K. Moffitt
John George Mohler
Daniel E. Mohn
Faramarz Mohtadi
George E. Mohun
Niculae T. Moisidis, PhD
Rogelio A. Molina
Albert James Moll, PhD
John L. Moll, PhD
Terrence V. Molloy
John P. Monahan
Robert E. Moncrieff
Carl L. Monismith
Loren Monroe, Jr.
Myles P. Monroe
Ernesto M. Monteiro
Theodore Ashton Montgomery
Nelson Montoya
Henry L. Moody
Dale W. Mooney
Eugene R. Moore, PhD
Rodney R. Moore
Michael Morcos
William B. Moreland
Richard Leo Moretti, PhD
Dean R. Morford
Robert W. Morford
Donald Earle Morgan, PhD
Leon Frank Morgan
Lucian L. Morgan
Thomas Kenneth Morgan, PhD
Victor G. Morgen
Michael Moroso
Philip J. Morrill
Paul E. Morris
Jeffrey A. Morrish
Allen D. Morrison
Richard J. Morrissey
Dan Morrow
Frances Morse
George A. Morse
Dennis B. Morton
John Robert Morton, PhD
Paul K. Morton
Ray S. Morton
H. David Mosier
Malcolm Mossman
Ronald J. Mosso
Gail F. Moulton, Jr.
Paul Mount, II
Richard C. Much
Jerome Robert Mueller
William L. Mueller
Sig Muessig, PhD
Debasish Mukherjee, PhD
Butch Mulcahey
Fred R. Mulker
Kenneth I. Mullen
L. Frederic Muller
John F. Mulligan
Kary B. Mullis, PhD
Mark B. Mullonbach
T. Munasinge, PhD
David V. Mungcal

Albert G. Munson, PhD
Emil M. Murad
Alexander James Murphy, PhD
Emmett J. Murphy
Gary L. Murphy
Sean Murphy, PhD
William J. Murphy, PhD
Bruce Murray
Frank Murray, PhD
Richard L. Murray
Albert F. Myers
Dale D. Myers
Willard G. Myers, PhD
Donald L. Mykkanen, PhD
Harry E. Nagle
Kenneth A. Nagy, PhD
Yervant M. Nahabedian
Yathi Naidu, PhD
Takuro S. Nakae
Glenn M. Nakaguchi
Dennis B. Nakamoto
M. Nance
A. Naselow, PhD
J. Greg Nash, PhD
Merlin Neff, Jr.
Thomas C. Nehrbas
Donald G. Nelson
Lorin M. Nelson
Richard Douglas Nelson, PhD
Robert L. Nelson
Victor Nelson
Bijan Nemati, PhD
Bruce H. Nesbit
Harold Neufeld
John B. Neuman
Sylvia M. Neumann
Temple W. Neumann
Donald E. Neuschwander
Frank M. Nevarez
James Ryan Neville, PhD
Richard E. Newell
Stanley D. Newell
John M. Newey
Phil W. Newman
Bernard D. Newsom, PhD
H. Newsom, PhD
Kerwin Ng
Thuan V. Nguyen, PhD
D. Paul Nibarger
Edward S. Nicholls
Jack C. Nichols
Mark E. Nichols
Richard A. Nichols, PhD
Richard E. Nicholson
Peter A. Nick
Mathew L. Nickels
Marvin L. Nicola
Albert H. Niden
Tom F. Niedzialek
R. Nieffenegger
Gilbert O. Nielsen
Kurt E. Nielsen
Mark Niemiec
Rodulfo C. Niere
William A. Nierenberg, PhD*
Joseph A. Nieroski
Edwawrd A. Nieto
Danta L. Nieva
James E. Nightingale
Jim B. Nile
George Niles
Soottid Nimitsilpa

Donald A. Nirschl
Kazunori Nishioka
Gilbert A. Nixon
Michael L. Noel
James C. Nofziger, PhD
Richard Nolthenius, PhD
Eugene L. Nooker
Jack D. Norberg
Richard E. Nordan
H. A. Noring
Brent C. Norman
John L. Norris
Tharold E. Northup
Ferdinard J. Nowak
Wesley Raymond Nowell, PhD
Stanley J. Nowicki
Leonard James Nugent, PhD
Maurice Joseph Nugent, Jr., PhD
Erwin R. Null
Robert B. Nungester
James K. Nunn
Amos M. Nur, PhD
Joseph A. Nussbaum
George A. Nyamekye, PhD
Hubert J. Nyser
Michael Nystrom
Harry Alvin Oberhelman
Theodore M. Oberlander, PhD
Rafael H. Obregon
Dale O'Brien
Fabrizio Oddone
Arlo L. Oden*
Dominick Odorizzi
Michal Odyniec, PhD
John D. Oeltman
Jacob J. Offenberger
Naomi Neil Ogimachi, PhD
H. M. Ogle
J. Ogren, PhD
Joon C. Oh
Franklin T. Ohgi
Kurt N. Ohlinger, PhD
Kent A. Ohlson
William Ohm
Arthur F. Okuno
Fred B. Oldham
Arnold N. Oldre
Edward A. Oleata
Edward Eugene Oliphant, PhD
Donald B. Oliver, PhD
Roy R. Oliver
E. Jerry Oliveras
Kenneth L. Olivier, PhD
Bernard J. O'Loughlin, PhD
Sterling B. Olsen
Carl Olson
Jan B. Olson
Larry Olson
Kevin D. Olwell, PhD
Rosalie Omahony, PhD
Willard D. Ommert
Ryan D. O'Neal
Donald E. O'Neill
Anatoli Onopchenko, PhD
Lee B. Opatowsky, PhD
Philip H. Oppenheim
William L. Oppenheim
John J. Oram, PhD
Rolf O. Orchard
Fernando Ore, PhD
David R. Orfant
Mahmoud M. Oriqat

Cornel G. Ormsby
Harold A. Orndorff
Hans I. Orup
Emil A. Osberg
James Osborn
M. Osborn
Bert Osen
James E. Oslund
James R. Oster
Al B. Osterhues
Theodore O. Osucha
Douglas K. Osugi
Arnold Otchin
Michael A. Otnisky
Gary M. Otremba
Wayne Robert Ott, PhD
William M. Otto
Oswald L. Ottolia
John W. Overall, Jr.
Dennis Owen
Scott A. Owen
Thomas W. Owens
Kazimiera J L Paciorek, PhD
Lorenzo M. Padilla
Jeffry Padin, PhD
Seaver T. Page
Coburn Robbins Painter
Thomas Palmer
Patrick E. Pandolfi
Daniel W. Pangburn
Sergio R. Panunzio, PhD
Robert C. Paoluccio
Charles Herach Papas, PhD
Michael L. Pappas
Shashikant V. Parikh
Edward Parilis, PhD
Calvin Alfred Parker, PhD
David W. Parker
Dennis L. Parker
Kenneth D. Parker
Norman F. Parker, PhD
Theodore C. Parker
Robert M. Parkhurst
Malcolm F. Parkman
Merton B. Parlier
Lowell Carr Parode
Christopher M. Parry, PhD
Michael L. Parsons, PhD
Chester R. Partridge
Stan P. Parvanian
Angelo A. Pastorino
James M. Paterson, PhD
Charles M. Patsch
Everett R. Patten
Gaylord Penrod Patten, PhD
Alec M. Patterson
Brenda J. Patterson
Richard M. Patton
James W. Paul
Raymond L. Paulson
Ferene F. Pavlics
Eleftherios B. Pavlis
Hagai Payes
Dalian V. Payne
James Payne, PhD
Andy Peabody
David N. Peacock, PhD
Robert T. Peacock
Gerald F. Pearce
John Pearson
Gerald Pease
Robert A. Pease

Deborah Kerwin Peck
Douglas P. Pedersen
Richard A. Pedersen
Louis E. Pelfini
David Gerard Pelka, PhD
W. S. Penn, Jr., PhD
Paul H. Pennypacker
Richard S. Penska
Jeffery Penta
Allen P. Penton
Linda H. Pequegnat, PhD
Thomas M. Perch
Irma T. Pereira
Robert V. Peringer
Arthur S. Perkins
James Jerome Perrcault
Gerald M. Perry
George Persky, PhD
Alois Peter, Jr.
Marvin Arthur Peters, PhD
Ralph H. Peters
Norman W. Petersen
Patricia J. Petersen
Arthur W. Peterson
Donald J. Peterson
Gary Lee Peterson, PhD
Glenn R. Peterson
Jack E. Peterson, PhD
James D. Peterson
Victor Lowell Peterson
Thomas G. Petrulas
Ray H. Pettit, PhD
Ronald W. Petzoldt, PhD
Bernard L. Pfefer
R. Fred Pfost, PhD
Robert F. Phalen, PhD
Debra Phelps
Lloyd Lewis Philipson, PhD
John P. Phillips
Ronald T. Piccirillo
Benjamin M. Picetti
William Pickett
John H. Pickrell
Bill D. Pierce, PhD
Matthew Lee Pierce, PhD
Terence M. Pierce
Vincent Joseph Pileggi, PhD
Laurence Oscar Pilgeram, PhD
Kurt F. Pilgram
Irwin J. Pincus
Edmund Pinney, PhD
Raymond G. Pinson
Robert G. Piper
Bernard Wallace Pipkin, PhD
Jesse E. Pipkin
Janet C. Piskor
Earl L. Pitkin
Raluca M. Pitts
Michael A. Plakosh
Robert V. Plank
Stephen L. Plett
Joseph S. Plunkett
Gregory E. Polito
Myron Pollycove
Glen D. Polzin
Robert L. Pons
James B. Ponzo
Richard J. Porazynski
David Dixon Porter
Fred C. Porter
Robert Potosnak
Charle E. Pound

Robert D. Pounds
M. L. Powell
James P. Power
Jack Pratt
Gerald O. Priebe
Robert Clay Prim, III, PhD
George B. Primbs
Robert K. Prince
David Prinzing
Lewis W. Pritchett
Winston H. Probert
Richard James Proctor
Thomas Proctor
J. Proffitt
John R. Prosek
Brian S. Prosser
Thomas Prossima, Jr.
Jerry A. Pruett
Vernon L. Pruett
Fernand H. Prussing
Kenneth E. Pruzinsky
Teodor C. Przymusinski, PhD
Timothy G. Psomas
Laurie D. Publicover
Leamon T. Pulley, PhD
Bruce H. Purcell
Everett W. Purcell
Robert G. Purington
Jennifer D. Pursley
Thankamama J. Puthiaparampil
Ramon S. Quesada
Florentino V. Quiaot
John M. Quiel
Louis C. Raburn
Donald Rado
Robert W. Ragen, Jr.
Peter A. Ragusa
James K. Rainforth
James W. Raitt
James A. Ramenofsky
Apolinar Z. Ramiro
Simon Ramo, PhD
Roy E. Ramseier
Lawrence Dewey Ramspott, PhD
Shahida I. Rana
James Rancourt, PhD
Greenfield A. Randall
P. Randall
Thomas J. Rankin
Henry Rapoport, PhD
Daren H. Raskin
Miriam Rasky
Ned S. Rasor, PhD
Howard E. Rast, Jr., PhD
Tom C. Rath
Egan J. Rattin
Michael S. Ratway
Stephen Rawlinson
Monte E. Ray
Everett L. Raymond
Leonard A. Rea
Robert G. Read
Richard L. Reason
Douglas W. Reavie
George J. Rebane, PhD
Jose G. Rebaya
Andreas Buchwald Rechnitzer, PhD
John G. Reddan, III
Damoder P. Reddy, PhD
Allan G. Redeker
Barry Reder
David A. Reed, Jr.

Lester R. Reekers
Harold G. Reeser
Donald F. Reeves
Richard G. Reeves
Steven A. Regis
Charles J. Reich
Kenneth Brooks Reid, PhD
Kyrk D. Reid
Jeff Reimche
Richard D. Rein
Fred W. Reinhart
Mark B. Reinhold, PhD
Marlin E. Remley, PhD
Charles R. Rendall
Daniel F. Renke
Nicholas A. Renzetti, PhD
Josef W. Repsch
Robert Walter Rex, PhD
Arthur A. Reyes, PhD
Diana C. Reyes
Armond G. Rheault
Dennis A. Rhyne
Gary R. Rice
Richard Rice, PhD
Neal A. Richardson, PhD
Ispoone Richlin, PhD
Hannes H. Richter
Harry G. Richter
Philip J. Richter
Corwin Lloyd Rickard, PhD
Joan D. Rickard
Douglas W. Ricks, PhD
R. J. Riddell, PhD
Richard R. Riddell
Michael Riddiford
Fred M. Riddle
John L. Ridell
Adolphus A. Riewe
James W. Riggs, PhD
George P. Rigsby, PhD
G. N. Riley
Gary T. Riley
Lyrad Riley
John A. Rinek
Thomas A. Ring
Robert Ringering
Keith Riordan
Richard L. Ripley
David Ririe, PhD
Martin W. Ritchie, PhD
Arnold P. Ritter
Jack B. Ritter
Robert Brown Ritter
Manuel S. Rivas
R. A. Rivas
James E. Robbins
Roy L. Roberson
Stephen F. Roberts
Donad B. Robertson
Jim D. Robertson
Karen S. Robinson
Richard C. Robinson
Robert B. Roche
Adam Rocke
Leon H. Rockwell, PhD
James W. Rodde
Jonathan P. Rode, PhD
Fredrich H. Rodenbaugh
Bertram J. Rodgers, Jr.
Glenard W. Rodgers
Anthony F. Rodrigues
D. R. Rogers

Dan V. Rogers
Carl A. Rohde
Gerhard Rohringer, PhD
Jack W. Rolston
Ephraim Romesberg
Wendell Hofma Rooks
Steven D. Root
Eugene John Rosa, PhD
Allan B. Rose
Dennis G. Rose
Eric C. Roseen
Alan Rosen, PhD
Dan Yale Rosenberg
Steven Loren Rosenberg, PhD
Donald Edwin Rosenheim
Jack W. Rosenthal
Kermit E. Rosenthal
Rollie D. Rosete
Jonathan P. Rosman
Stephen Ross
Suzi Ross
Ted E. Ross, Jr.
William E. Ross
Mario E. Rossi
A. David Rossin, PhD
Bryant William Rossiter, PhD
Thomas J. Rosten
J. Paul Roston
Adolph Peter Roszkowski, PhD
Ariel A. Roth, PhD
Gerald S. Rothman
Stan A. Rothwell
William Stanley Rothwell, PhD
Nicholas Rott, PhD
Jerome A. Rotter
James E. Roulstone
James M. Rowe, PhD
William R. Rowe
Leroy H. Rowley
George M. Roy
David C. Royer
G. Roysdon
John Rozenbergs, PhD
Balazs F. Rozsnyai, PhD
Leonard Rubenstein
Allen G. Rubin
Benjamin D. Rubin
Efim S. Rudin
Gerard Rudisin
John V. Rudy
Thomas P. Rudy, PhD
W. Ruehle
Edward Rugel
Daniel Ruhkala
Howell Irwin Runion, PhD
Richard A. Runkel, PhD
Jack E. Runnels
Robert C. Rupert
B. Rush, PhD
Andrew Russell
Edmund L. Russell
Lewis B. Russell
Claude Rust, PhD
Paul G. Ruud, PhD
Ed J. Ruzak
Alan S. Ryall, PhD
Philip L. Ryall
Bill Chatten Ryan, PhD
Daberath Ryan
Joe Ryan, PhD
Kevin M. Ryan
Patrick Ryan

Elliott Ryder, PhD
Kathleen Rygiel
Patrick Saatzer, PhD
Frank L. Sabatino
Joseph D. Sabella
William W. Sable
Frank C. Sacco
Marvin H. Sachse
Edgar Albert Sack, PhD
B. Sadri
Frederick M. Sagabiel
Richard A. Sager
William F. Sager
Majid Saghafi, PhD
Kanwar V. Sain
Nirmal S. Sajjan
Roy T. Sakamoto
Eugene Salamin
Robert E. Salfi, PhD
Mikal Endre Saltveit, PhD
Paul K. Salzman, PhD
Larry Sample
Grobert D. Sanborn
William C. Sanborn
Jay C. Sandberg
Ray O. Sandberg
Richard P. Sandell
Thomas L. Sanders, Jr.
Burton B. Sandiford
James K. Sandin
Steven D. Sandkohl
Marvin M. Sando
James S. Sands
Enrique Sanqui
George T. Santamaria
Henry E. Santana
Tom Santillan
Kenneth W. Sapp
A. M. Sam Sarem, PhD
Greg Sarkisian
Raymond Edmund Sarwinski, PhD
Melvin W. Sasse
J. Satko
Richard S. Satkowski
Hugh M. Satterlee, PhD
Tim Saunders
Walt Saunders
Robert E. Saute, PhD
Basil V. Savoy
Austin R. Sawvell
Frederick George Sawyer, PhD
Charles W. Sayles, PhD
Charles R. Saylor
James Scala, PhD
Carolyn A. Scarbrough
Michael P. Scarbrough
Michael P. Scardera
Lido Scardigli
Lawrence A. Schaal
Jeffrey Schaffer
Richard C. Schappert
Edward M. Schaschl
John F. Schatz, PhD
George E. Schauf
Donald E. Scheer
David H. Scheffey
Paul Otto Scheibe, PhD
Thomas J. Scheil
Perry Arron Scheinok, PhD
Michael W. Schell
Deborah S. Schenberger, PhD
Don Van Schenck

Clifton S. Schermerhorn
Don Ralph Scheuch, PhD
Paul G. Scheuerman
Stanley J. Scheurman
Mark Schiller
Ted M. Schiller
Guenter Martin Schindler, PhD
Rudolf A. Schindler
Richard H. Schippers
Hassel E. Schjelderup, PhD
Wilbert H. Schlimmeyer
Evert Irving Schlinger, PhD
Erika M. Schlueter, PhD
John H. Schmedel
Francis R. Schmid
Rudi Schmid, PhD
Alfred C. Schmidt
John W. Schmidt
Kurt C. Schmidt
Richard L. Schmittel
Henry A. Schneider
George L. Schofield, Jr., PhD
Kurt A. Scholz
Martin R. Schotzberger
Robert Schrader
Klaus G. Schroeder, PhD
Ed J. Schryver
Donald A. Schuder
Adolph T. Schulbach
Daniel Herman Schulte, PhD
Robert K. Schultz, PhD
Theodore C. Schultze
Roger W. Schumacher
Joseph F. Schuman
Peter T. Schuyler
Samuel G. Schwab
Benny R. Schwach
Steven C. Schwacofer
Alan B. Schwartz
Gary W. Schwede, PhD
Raymond L. Schwinn
Donald R. Scifres, PhD
Kevin M. Scoggin
Richard A. Scollay
Deborah J. Scott, PhD
Elizabeth Scott
Franklin Robert Scott, PhD
Harrison S. Scott
John F. Scott
Paul Scribner
D. G. Scruggs
Jeffrey Scudder
Christopher L. Seaman, PhD
W. H. Seaman
Paul A. Sease
Robert L. Seat
Bruce E. Seaton
Randall J. Seaver
Leslie I. Sechler
Donald B. Sedgley
Lidia A. Seebeck
Michael Seebeck
Michael L. Seely, PhD
Erwin Seibel, PhD
Edward W. Seigmund
Glenn A. Sels
John R. Selvage
N. T. Selvey
Jospeh Semak
Frederick D. Sena
George W. Sening
James C. Senn

Oscar W. Sepp
Alexander Sesonske, PhD
John D. Severns
Bradley E. Severson
Ordean G. Severud
Archie F. Sexton
James W. Shaffer
Patricia Marie Shaffer, PhD
Jayendra A. Shah
Arsen A. Shahnazarian
Lloyd Stowell Shapley, PhD
Gary Duane Sharp, PhD
Clifford A. Sharpe
Roland L. Sharpe
Clay Marcus Sharts, PhD
Black Shaw
Ian Shaw
Reece F. Shaw
Warren D. Shaw
I. D. Shaylor-Billings
Dennis Shea
Richard Shearer
George F. Sheets
Zubair A. Sheikh
Kent L. Shepherd
John L. Sheport
David A. Sheppard
Russell Sherman
Dalton E. Sherwood
N. Thomas Sherwood, PhD
Wilbert Lee Shevel, PhD
George L. Shillinger
Kaz K. Shintaku
Calvin Shipbaugh, PhD
Michael L. Shira
Ralph E. Shirley
Walter W. Shirley
Arthur W. Shively
Gary Shoemaker, PhD
Robert S. Shoemaker
Oliver B. Sholders
John J. Shore, III
Michael A. Short, PhD
Christopher R. Shubeck
Rex Hawkins Shudde, PhD
Patrick James Shuler, PhD
Dinah O. Shumway
Douglas C. Shumway
Sidney G. Shutt
Raymond E. Sickler
Kurt Sickles
John P. Siegel
Joseph A. Siegel
William Siegfried
Steven M. Siegwein
Brent C. Siemer
Hans R. Sifrig
James Ernest Siggins, PhD
Paul L. Sigwart
Michael Silbergh
Henry W. Silk
Robert T. Silliman
Armando B. Silva
Daniel D. Silva
Gregory P. Silver
Herbert Philip Silverman, PhD
Jacob Silverman, PhD
Herschel W. Silverstone
Craig A. Silvey
William J. Simek
P. John Simic
Michael N. Simidjian

Javid J. Siminou
E. Lee Simmons
Ira B. Simmons
Keith Simmons
William W. Simmons, PhD
Edgar Simons
William H. Simons
Jack Simonton
Donald C. Simpson
Gordon G. Sinclair
John P. Sinek, PhD
Alfredo Hua Sing, PhD
Iqbal Singh
Vernon Leroy Singleton, PhD
Juscelino M. Siongco
Harold T. Sipe
William D. Siuru, PhD
Stanley L. Sizeler
Fritiof S. Sjostrand, PhD
Paul S. Skager
Sidney E. Skarin, PhD
Arlie D. Skelton
Robert E. Skelton, PhD
George I. Skoda
Todd Patrick Slavik
Bernard G. Slavin, PhD
Marina Slepak, PhD
Mikhail E. Slepak, PhD
Richard Slocum, PhD
Robert Gordon Smalley, PhD
Gary Smart
Ronald T. Smedberg
Helen M. Smedbery
Rick G. Smelser
Charlee Smith
Dana L. Smith
Donald A. Smith
Donald C. Smith
Donald R. Smith, PhD
Donna E. Smith
Eric Smith
Gary Smith
Geoffrey R. Smith
James R.E. Smith
Jeff Smith
John R. Smith, PhD
M. R. Smith, PhD
Michael J. Smith, PhD
S. Clarke Smith
Walter J. Smith
Eileen Smithers
Peter Smits
Neil R. Smoots
Walter L. Snell
John P. Snook
William Rossenbrook Snow, PhD
Donald Philip Snowden, PhD
Donald R. Snyder
Stephen J. Snyder
Richard C. Soderholm
James R. Soderman
R. T. Soledberg
Bertram C. Solomon
Ronald C. Sommerfield
John R. Sondeno
Harold S. Song
Loren R. Sorensen
Frank S. Sorrentino
Marco J. Sortillon
Everett R. Southam, PhD
Charles L. Spaegel
Michael E. Spaeth, PhD

William L. Sparks
Russell T. Spears
Aaron B. Speirs
Lawrence C. Spencer
Pierrepont E. Sperry, Jr.
Arlo J. Spiess
Robert Joseph Spinrad, PhD
John Robert Spreiter, PhD
Rodger W. Spriggs
George S. Springer, PhD
Gerard J. Sprokel, PhD
James P. Srebro
Pierre St. Amand, PhD
David St. Armand
Harold Keith Stager
Kenneth E. Stager, PhD
Kim W. Stahnke
Mark A. Stalzer, PhD
Anthony C. Stancik
Clarence H. Stanley
Rosemarie Stanton
Scott R. Stanton
Timothy N. Stanton
Chauncey Starr, PhD
Darrel W. Starr
Edward R. Starr
Mike Starzer
Raymond Stata, PhD
Harrison L. Staub
John F. Steel
Arnold Edward Steele
Thomas C. Steele
Albert J. Stefan
Edward M. Steffani
Richard J. Stegemeier
Michael Steger
Howard Steinberg, PhD
Morris Albert Steinberg, PhD
Richard L. Stennes
Jan Stepek
Ralph L. Stephens
Stuart Stephens, PhD
David A. Stepp
Edward E. Sterling
John A. Stern
Sidney Sternberg
Alvin R. Stetson
David L. Stetzel
Milan R. Steube
Frank Stevens
Lewis A. Stevens
Albert E. Stevenson
Robert E. Stevenson, PhD
Robert Lovell Stevenson, PhD
Gordon Ervin Stewart, PhD
Homer J. Stewart, PhD*
Kathleen M. Stewart
Wayne L. Stewart
Chris Stier
Gerald G. Still, PhD
Howard A. Stine
Daniel P. Stites
Richard P. Stock
Norman Stockdale
Norman D. Stockwell, PhD
Larry J. Stoehr
Donald G. Stoffey, PhD
David Stone
Tabby L. Stone, PhD
W. Ross Stone, PhD
Donald E. Stout
Jay C. Stovall

David J. Stowell
Erwin Otto Strahl, PhD
Robert L. Strand
Edward D. Strassman
Paul M. Straub
Joe M. Straus, PhD
James R. Strawn
Herbert D. Strong, Jr.
Allen Strother, PhD
Wilfred Stroud
Kenneth A. Stroup
Mark W. Strovink, PhD
Harold K. Strunk, PhD
John H. Struthers
Allen Stubberud, PhD
Perry L. Studt, PhD
Justin Stull
Gunther L. Sturm
M. Subramanian, PhD
Marek A. Suchenek, PhD
Mark A. Suden
James Carr Suits, PhD
John F. Sullivan
Joseph H. Sullivan
Robert J. Sultan
Andrew D. Sun
Sally S. Sun
Vane S. Suter
Mark E. Sutherlin
John Svalbe
Curtis Edward Swain, PhD
Daniel Swain, PhD
Robert J. Swain
David Swan
Alan A. Swanson
Linda S. Swanson
Robert Nols Swanson
William Alan Sweeney, PhD
Peter Swerling, PhD
Chauncey Melvin Swinney, PhD
Leif Syrstad
George B. Szabo
Walter S. Szczepanski
Andrew Y. Szeto, PhD
Edwin E. Szymanski, PhD
Leonard Tachner
Spencer L. Tacke
Charles E. Tackels
David Dakin Taft, PhD
Bill H. Taggart
Steve J. Taggart
Samuel Isaac Taimuty, PhD
Girdhari S. Taksali
David B. Talcott
Donald D. Talley
Ralph G. Tamm
Harry H. Tan, PhD
Y. Tang, PhD
James B. Tapp
Waino A. Tapple
Anthony Tate
Rick Tavores, PhD
Richard L. Taw
Donald E. Taylor
Edward Taylor
George F. Taylor
John J. Taylor
Michael K. Taylor
Neil L. Taylor
William G. Taylor
Frank C. Tecca
Edward A. Tellefsen

Edward Teller, PhD*
Orlando V. Telles
Robert Templeton
John T. Tengdin
Jacob Y. Terner
Richard D. Terry, PhD
Steven R. Terwilliger
Henry J. Tevelde
Richard W. Tew, PhD
Edward Teyssier
Michael G. Thalhamer
Joshua M. Tharp, Jr.
Conley S. Thatcher
Gordon H. Theilen
Jerold Howard Theis, PhD
James D. Thissell
Roderick W. Thoits
Charles A. Thomas, Jr., PhD
Dean Thomas
Garfield J. Thomas
Gerald A. Thomas, PhD
James Thomas
Kevin L. Thomas
Russ H. Thomas
Michel A. Thomet, PhD
Dennis P. Thompson
Ken Thompson
William J. Thornhill
Lewis Throop, PhD
Henrik C. Thurfjell
John P. Tibbas
Gerald F. Tice
Karen M. Tierney
William Arthur Tiller, PhD
Edward K. Tipler
Arthur Robert Tobey, PhD
Joseph D. Tobiason, PhD
Brent Tolend
Gerald V. Toler
David C. Toller
Maher B. Toma
Crisanto R. Tomongin
John C. Toomay
Wilfred Earl Toreson, PhD
Eric D. Torguson
Felipe N. Torres
John P. Toth
Charles H. Touton
David C. Tower
Jeffrey G. Towle, PhD
Edward L. Townsend
Norton R. Townsley
Donald Frederick Towse, PhD
Rosalyn Tran
Elbert W. Trantow
Timothy L. Trapp
Mitchell Trauring
William Brailsford Travers, PhD
Frank C. Trayer
B. V. Traynor
Raymond Treder
John D. Trelford
William H. Trent, PhD
Richard L. Trimble
Patrick A. Tripe
Curtis W. Tritchka
Glenn C. Troman
Craig J. Trombly
Tony Troutman
Bretton E. Trowbridge
Eddy S. Tsao
A. N. Tschaeche

Tai Po Tschang
Manuel Tsiang, PhD
Theodore Yao Tsu Wu, PhD
Dean B. Tuft, PhD
Raymond Tulkki
Bryan Tullis
Richard Eugene Tullis, PhD
B. R. Tunai
Andy Tung
Willis E. Tunnell
Richard E. Turk
Robert L. Turk
Robert E. Turner, PhD
Ted H. Tuschka, PhD
Henry A. Tuttle
Robert Tuttobene
Ross W. Tye, PhD
Glenn A. Tyler, PhD
Vincent H. Uhlenkott
Harold B. Uhlig
Joseph J. Unger
Erik Unthank
Robert R. Upp, PhD
James L. Uptegrove
Donadl C. Urfer
Eldon L. Uverne Knuth, PhD
Richard J. Vacherot
Frank H. Vacio
J. O. Vadeboncoeur
John A. Vaillancourt
J. Peter Vajk, PhD
A. Valdos-Meneses
Richard M. Valeriote
Bernard A. Vallerga
Jacob E. Valstar
Job van der Bliek
James R. Van Hise, PhD
Paul M. Van Loenen
Naola Van Orden, PhD*
William W. Van Vorst, PhD
John H. VanAmringe
Peter Vanblarigan, PhD
Mark Vande Pol
Arthur VanDeBrake
Willem Vander Bijl, PhD
Chris J. Vandermaas
Gary J. Vandermolen
Garret N. Vanderplaats, PhD
Larry E. Vanhorn
Walker S. Vaning
Ruth A. vanKnapp
Barry M. Vann
Vito August Vanoni, PhD
Vagarshak V. Vardanyan, PhD
Larry Vardiman, PhD
Perry H. Vartanian, Jr., PhD
Bangalore Seshachalam Vasu, PhD
T. D. Vaughan
Arlie D. Vaughn
King F. Vaughn
Steven W. Vawter
Kenneth S. Vecchio, PhD
Alejandro T. Vega
Edward E. Velarde
Margarita B. Velarde
Daniel W. Velasquez
Wencel J. Velicer
Louis Veltese
Theodore E. Veltfort
Anthony J. Verbiscar, PhD*
Robert J. Verderber
Jared Verner, PhD

Dmitry Vernik, PhD
Daniel J. Vesely
Peter Vessella
Thomas H. Vestal
Louis Vettese
Charles L. Vice
Joan Vickery
Andrew S. Vidikan
Susan L. Vigars
Carlos F. Villalpando
Roberto Villaverde, PhD
Norbert F. Vinatieri
Edgar L. Vincent
Jonathan R. Vinci
R. C. Visser
Richard K. Vitek
Petro Vlahos
Roger Frederick Vogel
Randy L. Vogelgesang
Kevin G. Vogelsang
Edward J. Vollrath
Richard L. Volpe
Robert M. Volpe
Karl Vonderlinden, PhD
Suresh H. Vora
Frederick H. Vorhis
Earl H. Vossenkemper
Frederick C. Vote
Henry P. Voznick
Daniel L. Vrable, PhD
William R. Wachtler, PhD
Don Wade
Glen Wade, PhD
Robert Harold Wade, PhD
William Howard Wade, PhD
Donald Wadsworth, PhD
James K. Wagner
Kenneth E. Wagner
Mark A. Wagner
Lewis D. Wagoner
David P. Wahl
Dennis L. Wahl
Howard W. Wahl
Scott G. Wahl
Richard I. Waite, Jr.
George Wakayama
R. Stephen Waldeck
Jason S. Waldrop
Richard T. Wales
Dennis Kendon Walker, PhD
Jay H. Walker
John C. Walker
Patrick M. Walker
Raymond W. Walker
Verbon P. Walker
Milton B. Walkup
Edward M. Wall
Edwin Garfield Wallace, PhD
Tom S. Wallace
Eleanor A. Wallen
Henry A. Waller
John E. Wallis
Joel D. Walls, PhD
Darrel N. Walter
Herbert G. Walter
Gregg D. Walters
Austin G. Walther
Donald M. Waltz
Bohdan I. Wandzura
Qingqi Wang
T. Wang, PhD
Zhi Jing Wang, PhD

Dennis Wangsness
Casidy A. Ward
Dale Ward
James J. Ward
Jay L. Ward
Paul H. Ward
Charlene L. Wardlow
Ronald S. Wardrop
Richard V. Warnock
Jack S. Warren
Richard Warriner
John P. Waschak
Halbert S. Washburn
Harry L. Washburn
Robert M. Washburn
Claude Guy Wasterlain, PhD
Glenn L. Wasz
Milton N. Watanabe
Dean A. Watkins, PhD
Charlie E. Watson
Gary W. Watson
Guy E. Watson
William W. Watson
R. E. Watts
Walter L. Way
Todd Weatherford, PhD
Robert A. Weatherup
David L. Weaver
Robert D. Weaver
Albert Dinsmoor Webb, PhD
Creighton A. Webb
Jack W. Webb
William P. Webb, PhD
Harry V. Webber
Bruce Warren Webbon, PhD
Barrett H. Weber
Erich C. Weber
William P. Weber, PhD
David Webster
Mark B. Webster
D. J. Wechsler
Saul Wechter
William J. Wechter, PhD
Lloyd Weese
William H. Weese
Robert L. Wehrli
Rudolph W. Weibel
Glen F. Weien, II
Hans Weil-Malherbe, PhD
Carl Martin Weinbaum, PhD
William M. Weinstein
Daniel G. Weis, PhD
Russell J. Weis
Max T. Weiss, PhD
Frank Joseph Welch, PhD
Robin Ivor Welch, PhD
Hugh E. Wells
Lee O. Welter
Gunnar Wennerberg
R. C. Wentworth, PhD
Robert P. Wenzell
Victor H. Werlhof
Robert H. Wertheim
Donald A. Wesley, PhD
A. Wessman
Clinton L. West
Jack H. West
Robert E. West
Rod D. Westfall
Travie J. Westlund
Kenneth Harry Westmacott, PhD
Henry Griggs Weston, Jr., PhD

Duane Westover
Robert J. Wetherall
Mike A. Whatley
Carlos Wheeler
Joseph G. Whelan
Vernon T. Whitaker
David J. White
Gerry W. White
Joel E. White
Richard L. White
Robert Lee White, PhD
Sterling F. White
William R. White
Kenneth E. Whitehead
Kent G. Whitham
Stephen A. Whitlock
David V. Whitmore
Dennis B. Whitney
Robert C. Whitten, PhD
R. Whitting
Derek A. Whitworth, PhD
Eyvind H. Wichmann, PhD
John G. Wichmann
Dave E. Wick
George J. Widly
Arthur F. Widtfeldt
Mark J. Wiechmann, PhD
Daniel W. Wiedman
Francis P. Wiegand
Robert L. Wiegel
Doug Wiens
Don A. Wiggins
John Henry Wiggins, PhD
John S. Wiggins, PhD*
David W. Wilbur, PhD
Corbet E. Wilcox
Michael R. Wilcox
Orland W. Wilcox
Carroll Orville Wilde, PhD
James W. Wilder
Fred R. Wiley
Wilbur F. Wilhelm
Donald W. Wilke
Charles L. Wilkins
Harold E. Wilkins
Ross C. Wilkinson
Bennington J. Willardson
Austin M. Williams
Edgar P. Williams
Forrest R. Williams
Wayne S. Williams
David W. Wilson, PhD
Donald E. Wilson
Garth H. Wilson, PhD
Jack Wilson
Jerome M. Wilson
Linda W. Wilson
Maurice L. Wilson
Melvin N. Wilson
Michael A. Wilson
Royce D. Wilson
William E. Wilson
George D. Wiltchik
Edward J. Wimmer
Jack A. Winchell
Ernest O. Winkler
Robert Winslow
Roger G. Winslow
Guy W. Winston
Harry Winterlin
Wray Laverne Winterlin
Bruce A. Winters

Appendix 4: The Petition Project

Philip Rex Winters
Donald F. Winterstein, PhD
Guy W. Winton, Jr
W. T. Wipke, PhD
Wesley L. Wisdom
Edward Witczak
Eric V. Witt
Kenneth A. Witte, PhD
Robert F. Witters
Lawrence R. Wlezien
James K. Wobser
Virgil O. Wodicka, PhD
Milo M. Wolff, PhD
Paul M. Wolff, PhD
John H. Wolthausen
Fred W. Womble
Ka-Chun Wong
Otto Wong, PhD
Sun Y. Wong
Bruce Wood
David L. Wood
Don E. Wood
James M. Wood, II
Jason N. Wood
Kevin G. Wood
Walter H. Wood
Willis Avery Wood, PhD
Michael L. Woodard
Michael D. Woods
Richard C. Woodward
Robert J. Woodward
Gene A. Worscheck
Edward P. Wosika
Margaret Skillman Woyski, PhD
David G. Wright
Harold V. Wright
Keith A. Wright
Melville T. Wright
Michael E. Wright
William W. Wright
Chris J. Wrigley, Jr.
Jack G. Wulff
Charles R. Wunderlich
Richard A. Wunderlich
Stephen Walker Wunderly, PhD
David E. Wyatt
Jeff Wyatt
Philip J. Wyatt, PhD
Bruce M. Wyckoff
Robert A. Wyckoff
Thomas S. Wyman
Leslie K. Wynston, PhD
William Xenakis
Y. Xie, PhD
Albert R. Yackle
Bohdan M. Yacyshyn
Paul F. Yaggy
Richard N. Yale
Walter M. Yamada, PhD
Jack A. Yamauchi
John S. Yankey, III
John Lee Yarnall, PhD
Anthony Yarosky
Francis Eugene Yates
John L. Yates
Melvin B. Yates
Robert W. Yates
Scott Raymond Yates, PhD
Thomas C. Yaughn
John C. Yeakley
Carlton S. Yee, PhD
Kuo-Tay P. Yeh

Paul Pao Yeh, PhD
Michael W. Yeoman
Ki Jeong Yi
Sherwin D. Yoelin
Denison W. York
G. Young
Jackson Young
Richard Young
Robert D. Young
Stephen G. Young
Wei Young, PhD
Dennis E. Youngdahl
John C. Youngdahl
Mohamad A. Yousef, PhD
Mary Alice Yund, PhD
D. Yundt
Sulhi H. Yungul, PhD
Kirk A. Zabel
David R. Zachary
Michael N. Zaharias
Kamen N. Zakov
Carlos A. Zamano
Alex Zapassoff
J. Edward Zawatson
Jason D. Zeda
Yuan Chung Zee, PhD
Howard C. Zehetner
Ken R. Zeier
Sanford S. Zeller
Kerry Zemp
Robert H. Zettlemoyer
Yi Zhad, PhD
Sigi Zierling, PhD
David L. Zimmerman
Elmer LeRoy Zimmerman, PhD*
Douglas A. Ziprick
Harold Zirin, PhD
Keith Zondervan, PhD
Louis M. Zucker
Yury Zundelevich, PhD
Joe Zupan
Soloman Zwerdling, PhD

Colorado

Wilbur A. Aanes
David M. Abbott, Jr.
Joseph M. Abell
Paul Achmidt
Steven W. Adams
Wayne F. Addy
T. Adkins
Jacques J.P. Adnet
John Aguilar
Alfred Ainsworth
Gary L. Allison
Kevin R. Allison
Bruce W. Allred
Victor Dean Allred, PhD
Robert C. Alson
Greg A. Altberg
Ashton Altieri
Anthony B. Alvarado
Kenneth S. Ammons
Adolph L. Amundson
David Anderson, PhD
Glenn L. Anderson
James P. Anderson
Tom Anderson
Walton O. Anderson
David J. Andes

Mark J. Andorka
George Andreiev
Russell A. Andrews
Stephen P. Antony
James K. Applegate, PhD
Steven B. Aragon
Sidney O. Arola
Charles E. Atchison
William J. Attwooll
Mark Atwood, PhD
James K. August
Kurt L. Austad
Michael N. Austin
Steven G. Axen
Jessica Ayers
Lee R. Bagby
William F. Bagby
Steve G. Bagley
Dana Kavanaugh Bailey
R. V. Bailey
Jack R. Baird, PhD
Daniel Bakker
Alfred Hudson Balch, PhD
Brund Balke
Leslie Ball
G. Arthur Barber
Gerald L. Barbieri
Stephen W. Barnes
Ralph M. Barnett
Charles J. Baroch, PhD
Lance R. Barron
William E. Barton, PhD
Richard L. Bate
Albert Batten, PhD
Bertrand J. Bauer
Martin A. Bauer
Ralph B. Baumer
Eric Paul Baumgartner
David Paul Baumhefner
Robert L. Bayless, Jr.
William H. Bayliff
George W. Bayne
Shelby R. Bear
Dennis E. Beaver
Gayle D. Bechtold
Douglas Beck, PhD
Kenneth A. Beegles
Wayne R. Beeks
Wallace G. Bell, PhD
Donald P. Bellum
Cade L. Benson
Edgar L. Berg
Rodney H. Bergholm
Leonard Bertagnolli
Cornelius E. Berthold
Gregory W. Bertram
Howard H. Bess
Carl E. Beutler
Robert G. Beverly
Wallace J. Bierman
Doug N. Biewitt
D. G. Bills, PhD
Edward R. Binglam
Daniel Bisque
Matthew L. Bisque
Ramon E. Bisque, PhD
John P. Biswurm
Christopher R. Black
Thomas E. Blackman
Barry L. Blair
Robert B. Blakestad
James C. Blankenship

William D. Blankenship
Thomas L. Blanton, PhD
Ronald K. Blatchley
Victor V. Bliden
Duane N. Bloom, PhD
Donald R. Bocast
Jane Haskett Bock, PhD
Bernard L. Bogema
Cody B. Bohall
Richard A. Bohling
Mike Boland
Duane W. Bolling
Robert L. Bolmer
Dudley Bolyard
Lee J. Bongirno
Charles S. Bonnery
Jack N. Boone, PhD
Travis J. Boone
Lawrence Boucher
Larry Q. Bowers
Jean A. Bowles, PhD
Steven C. Boyce, PhD
Lester L. Boyer, Jr., PhD
Brian J. Boyle
Carl F. Branson
Jeff D. Braun
John E. Brehm, Sr.
Michael B. Brewer
Corale Louise Brierley, PhD
James A. Brierley, PhD
Mont J. Bright, Jr.
Michael W. Brinkmeyer
Michael Brittan, PhD
George William Brock
Lawrence R. Brockman
Dean C. Brooks
James R. Brooks
Ben D. Brown
John M. Brown, PhD
Larry F. Brown, PhD
Lynne A. Brown
Charles Robert Bruce, PhD
Thomas J. Bruno, PhD
John L. Brust
Kenton J. Bruxvoort
Donald G. Bryant, PhD
Nathan B. Bryant
Joel W. Buck
James S. Bucks, PhD
Perry Buda
David J. Bufalo
C. James Bulla, Jr.
Stephen D. Bundy
Michael A. Burgess
Leonard F. Burkart, PhD
Dean S. Burkett
Robert H. Bushnell, PhD
Frank J. Buturla
Fredrick G. Calhoun
Mathhew D. Calkins
Terry J. Cammon
H. D. Campbell
Rayford R. Canada
John Canaday
John A. Canning, PhD
Andrew J. Capra
Curtis D. Carey
Charles E. Carlson, III
William A. Carlson
Robert Carpenter
Thomas M. Carroll
Mason Carlton Carter, PhD

763

Lew Casbon
Joseph Cascarelli
Dennis C. Casto
Eugene N. Catalano
P. Ceriani, PhD
Paul D. Chamberlin
Donald R. Chapman
William L. Chenoweth
Robert H. Chesson
Hsien-Hsiang Chiang, PhD
Milton O. Childers, PhD
Sarah J. Chilton
Chris Chisholm
Wiley R. Chitwood
Odin D. Christensen, PhD
Richard L. Christiansen, PhD
Robert Milton Christiansen, PhD
J. W. Christoff
Catherine A. Clark
Ivan L. Clark
Marion D. Clark, PhD
John F. Clarke, Jr.
James S. Classen
J. Clema
Curtis S. Clifton
Peter R. Clute
Leighton Scott Cochran, PhD
William T. Cohan
Lawrence E. Coldren
David R. Cole
Lee Arthur Cole, PhD
Gary W. Collins
Nina T. Collum, PhD
Arthur F. Colombo, PhD
Richard F. Conard
Martin E. Coniglio
Chris M. Conley
Michael S. Connelly
Joe Conway
David F. Coolbaugh, PhD
Harvey L. Coonts
Robert J. Coppin
David Corbin
David B. Corman
Arlen C. Cornett
Garryd D. Cornish
Rodney H. Cornish, PhD
Charles E. Corry, PhD
George E. Cort
Benjamin Costello
Charles K. Cothem
Arthur W. Courtney
Thomas P. Courtney
Thomas R. Couzens
Louis Cox, Jr., PhD
Cecil J. Craft
Bruce D. Craig, PhD
Dexter Hildreth Craig
Don J. Craig
Rex C. Cramer
William Pau Crisler, PhD
David L. Crouse
Alfred John Crowle, PhD
Wolney C. Cunha
Robert L. Curruthers, Jr.
Michael S. Cuskelly
Aaron T. Cvar
Steve D. Dahmer
Carl W. Dalrymple
Kevin M. Daly
Valeria Damiao, PhD
Henry W. Danley

William M. Danley
Eugene A. Darrow
Donald Davidson, PhD
Edward J. Davies, PhD
Charles D. Davis
Daryl W. Davis
James L. Davis
John A. Davis, Jr., PhD
Michael Davison
William Daywitt
John W. Deberard
Peter George Debrunner, PhD
Jeff L. Deeney
Steven L. DeFeyter
Gerald H. Degler
Susan M. Deines
B. J. Delap
Richard A. Denton
William Davis Derbyshire, PhD
Lawrence E. Dernabach
Jeffrey H. Desautels
W. R. Dettmer
John L. Devitt
Rudolph John Dichtl
Douglas Dillon, PhD
Jerry R. Divine
Robert Clyde Dixon
Gene P. Dodd
Lee A. Dodgion
Thomas C. Doe
Eugene Johnson Doering
David J. Doig
Erin R. Dokken
Richard D. Dolecek
Inez G. Dominguez
Michael W. Donley, PhD
Arthur F. Donoho
David R. Donohue
Kiernan O. Donovan
Gerald R. Dooher
Robert A. Doornbos, Jr.
Robert J. Doubek
Edward C. Dowling, Jr., PhD
Mancourt Downing, PhD
William Fredrich Downs, PhD
John E. Dreier, PhD
Jerry D. Droppleman, PhD
Emerson K. Droullard
Murray Dryer, PhD
Harold R. Duke, PhD
Thomas J. Dumull
Joel G. Duncan, PhD
Paul M. Duncan
Harry E. Duprey
Benny R. Dusenbery
William T. Dusterdick
Donald R. East
Dennis M. Eben
William G. Ebersohl
Donald P. Ebright
Anita Eccles
Dana A. Echter
Steven T. Eckmann, PhD
Paul F. Eckstein
Ronald K. Edquist
Thomas Edward
Julie A. Edwards
Thomas B. Egan
Linda L. Ehrlich
William H. Eichelberger
Keith W. Eilers
H. Richard Eisenbeis, PhD

Robert L. Elder
James Eley, PhD
Richard C. Enoch
Alexander Erickson
Christopher Erskine
Thomas M. Erwin
Rudolph Eskra
Ronald L. Estes
Donald Lough Everhart, PhD
Joseph P. Fagan, Jr.
Thomas G. Fails
Frank Farnham
Duane D. Fehringer
Stuart R. Felde
John L. Fennelly
Arthur Thomas Fernald, PhD
Clinton S. Ferris, Jr., PhD
James W. Ferry
Dale A. Fester
Charles J. Fette
Thaddeus C. Fial
Thomas G. Field, PhD
Dennis C. Finn
Carl V. Finocchiaro
Harlan Irwin Firminger
Roland C. Fischer
Werner E. Flacke
Clarence W. Fleming
Peter A. Fleming
Jay E. Foley
Jon R. Ford
Deon T. Fowles
Glenn M. Frank
Andrew P. Franks
Ronald D. Franks
Margo Frantz
Harry D. Frasher
James E. Frazier
Richard J. Frechette
Ryan D. Frederick
William E. Freeman
Tim F. Friday
H. Howard Frisinger, PhD
William R. Fritsch
Norm Froman
L. W. Frowbridge, PhD
Suzanne M. Fujita
Stephen L. Funk
Connie Gabel, PhD
Richard A. Gabel, PhD
Joseph Galeb, PhD
Donald L. Gallaher
Anthony J. Galterio
Patrick S. Galuska
Charles L. Gandy
Paul W. Gard, Jr.
Edward E. Gardner, PhD
Homer J. Gardner
Andrew J. Garner
Thomas R. Gatliffe
Jeffery L. Gay
William M. Gemmell
Milo P. Gerber
Andre A. Gerner
Cynthia A. Gerow, PhD
Douglas O. Gibbs
Elizabeth L. Gibbs
Henry Lee Giclas, PhD
Delbert V. Giddings
Paul N. Gidlund
John M. Giesey
Walter Charles Giffin, PhD

Brian W. Gilbert
Alan D. Gillan
Leland E. Gillan
Rick J. Gillan
Jack D. Gillespie
James R. Gilman
John Gishpert
Dan G. Gleason
Hartley C. Gleason
Earl T. Glenwright
Christine Gloeckler
Robert A. Glover
George G. Goble, PhD
Ronald B. Goldfarb, PhD
Scott L. Gomer
David V. Gonshor
John M. Goodrich
Jeff P. Goodwin
Alfred F. Gort
Mark H. Gosselin
John W. Goth
Donald F. Gottron
Walter Carl Gottschall, PhD
Gary L. Gough
Gordon Gould, PhD
Ronald J. Gould
Thomas L. Gould, PhD
Robert J. Governski
Robert E. Gramera, PhD
Lee B. Grant, Jr.
Lewis O. Grant
Kenneth Donald Granzow
Robert C. Gray
Steven R. Gray
William M. Gray, PhD
Deborah C. Greenwall
Alfred Griebling
James K. Griffin
Paul K. Grogger, PhD
Alfred W. Grohe
Stephen J. Grooters
Michael P. Gross
Eugene L. Grossman
Paul M. Gruzensky, PhD
Ross L. Gubka
Lyle A. Gust
Ronald G. Gutru
Richard Haack, PhD
Donald Haas
Lawrence N. Hadley, PhD*
Richard Frederick Hadley
Frank A. Hadsell, PhD
Gregory A. Hahn
Michael J. Hall
Timothy J. Halopoff
Saheed Hamid
Don D. Hamilton, PhD
Stanley K. Hamilton, PhD
Jack C. Hamm
John R. Hamner
Marvin A. Hamstead, PhD
Joe John Hanan, PhD
Howard W. Hanawalt
Ray A. Hancock
John W. Hand, Sr.
Donald L. Hanlon, Jr.
Barry J. Hansen
Chris E. Hansen
Lowell H. Hansen
Carl D. Hanson
Sergius N. Hanson
Elwood Hardman

W. Henry Harelson
Mark F. Harjes
Stephen T. Harpham
Kevin D. Harrison
Elbert Nelson Harshman, PhD
Arden J. Hartzler
Donald D. Hass
Charles N. Hatcher
Rav P. Hattenbach
Christpher N. Hatton
Niels B. Haubold
Consuelo M. Hauser
Ray L. Hauser, PhD
Michael E. Hayes
Eugene D. Haynie
Eugene C. Head, Jr.
Jenifer Heath, PhD
Robert Bruce Heath
Dale Heermann, PhD
Leonard Heiny
Leslie V. Hekkers
Marvin W. Heller, PhD
Roger A. Heller
William D. Helton
Courtney C. Hemenway
Frank Heming, PhD
Daniel F. Henderson
Phil W. Henderson
Ralph L. Henderson
Raymond L. Henderson
Charles J. Hendricks
Charles S. Henkel, PhD
Raymond P. Henkel, PhD
Ann Henning
Stephen J. Henning
Thomas W. Henry, PhD
Gary A. Herbert
Jeff Herrle
Martin Hertzberg, PhD
Phillip E. Hewlett
Edward D. Hice
Ronald Higgins
Charles L. Hill
Margaret A. Hill
Bill L. Hiltscher
William J. Hine
Lee D. Hinman
John Hinton
John S. Hird
Brian G. Hoal, PhD
Farrel D. Hobbs
Noel Hobbs
Marcus D. Hodges
Marty Hodges
Roy A. Hodson
Arthur P. Hoeft
Harald Hoegberg
Robert J. Hoehn
John Raleigh Hoffman, PhD
Stephen A. Hoffman, PhD
Christopher J. Hogan, PhD
Ronnie E. Hogan
David Clarence Hogg, PhD
Douglas G. Hoisington
Erik Holck
Frank J. Holliday
Victor T. Holm
David A. Holmes
Richard D. Holstad
James C. Holzwarth
Russell M. Honea, PhD
John Hoorer

William E. Hopkinds
Andrea S. Horan
Joseph M. Hornback, PhD
Kim Horsley, PhD
Richard F. Horsnail, PhD
William W. Hoskins
Douglas L. House
James O. Houseweart
Larry B. Howard, PhD
Dennis M. Howe, PhD
Craig A. Howell
Earl L. Huff
Julie Huffman, PhD
Ronald W. Huffman
Timothy J. Hughes
Travis Hughes, PhD
Frederick O. Humke
William R. Humphrey
Gary Hunt
Mack W. Hunt
Leland L. Hurst
Jerome Gerhardt Hust
Robert E. Husted
Robert B. Huston
Eric J. Hutchens
John G. Hutchens
Gary L. Hutchinson
R. W. Hutchinson, PhD
Jay D. Huttenhow
Francis Hutto, PhD
Norman L. Hyndman
Kenneth D. Ibsen
Mark D. Ibsen
Eugene Ignelzi
Duane Imhoff
Robert J. Irish
Larry A. Irons
Scott R. Irvine
William W. Irving
James A. Ives
John B. Ivey
Manly L. Jackson
Stewart A. Jackson, PhD
M. L. Jacobs, PhD
Jay M. Jacobsmeyer
Jimmy Joe Jacobson, PhD
Bahram A. Jafari, PhD
John A. James
Chet H. Jameson, Jr.
Frederick J. Janger
Duane A. Jansen
George J. Jansen
Gustav Richard Jansen, PhD
Karen F. Jass
Seymour Jaye
William F. Jebb
Joe A. Jehn
Eivind B. Jensen
James D. Jessup
Thomas J. Jobin
Carl T. A. Johnk, PhD
Abe W. Johnson
Blane L. Johnson, PhD
Clinton B. Johnson
Curtis L. Johnson
Donal Dabell Johnson, PhD
Michael S. Johnson
Robert Britten Johnson, PhD
Robert C. Johnson
Thomas L. Johnson
Walter E. Johnson
Steve D. Jolly, PhD

G. R. Jones
Paul C. Jones
Robert S. Joy
Ray L. Jukkola
Rodger A. Jump
James E. Junkin
Walter F. Kailey, PhD
Dale C. Kaiser
F. Kamsler
Raymond C. Kane, Jr.
Joe Kaplen
Gretchen A. Kasameyer
Todd R. Kaul
Alvin Kaumeyer
Michael J. Keables, PhD
Philip D. Kearney, PhD
Cresson H. Kearny
Richard A. Keen, PhD
Kenneth L. Keil
Stephen R. Keith
Frederick A. Keller, PhD
Sean P. Kelly
Walter O. Kelm
Graham Elmore Kemp
Robert C. Kendall*
William R. Kendall
Larry Kennedy
Charles L. Kerr
Ronald L. Ketchum
D. F. Kidd
Michael L. Kiefer
Richard E. Kiel
Scott M. Kimble
Thomas G. Kimble
Barry A. King
Thomas A. Kingdom
Monty C. Kingsley, PhD
John Kirkpatrick
Peter H. Kirwin
Hugh M. Kissell
Donald N. Kitchen, PhD
W. Kitdean, PhD
Sharon J. Klipping
Thomas Mathew Kloppel, PhD
Norman C. Knapp
Martin E. Knauss
Duane Van Kniebes
Lisa M. Knobel
John K. Knop
Kirvin L. Knox, PhD
Kenneth Wayne Knutson, PhD
Chuck Koch
Nicholas F. Koch
Donald E. Koenneman
Wilbur L. Koger, Jr.
David L. Kohlman, PhD
William A. Koldwyn, PhD
Arvin L. Kolz
Kenneth D. Kopke
Nicholas C. Kortekaas
Thaddeus S. Kowalik
Jan Krason
Wayne P. Kraus, PhD
George Krauss, PhD
Lee E. Krauth
Robert E. Kribal, PhD
Douglas H. Krohn
Fred J. Kroll
William R. Kroskob
Keith Krugh
Dennis W. Kuhlmann
Ed J. Kurowski

John W. Labadie, PhD
Conrad M. Ladd
John P. Lafollette
Michael A. Laird
Allen B. Lamb
Bruce Landreth
George L. Lane
Alan Lange
Arthur L. Lange
Stephen S. Lange
Carl G. Langner, PhD
Robert K. Lantz, PhD
David D. Larison
Greg A. Larsen
Larry L. Larsen
Byron W. Larson
Roy E. Larson
Thomas A. Larson
Kenneth M. Laura
Linda E. Lautenbach
Leroy D. Lawson
Daryl E. Layne
Jozef Lebiedzik, PhD
Kennon M. Lebsack
Bob L. Lederer
Arthur C. Lee
Robert Allen Lefever, PhD
Gail Legate
William Lehmanu
Michael G. Leidich
Richard W. Lemke
Robert Lentz
Jim Leslie
Barbara B. Lewis
Charles George Liddle
Fred H. Lightner
Richard B. Liming
Peter K. Link, PhD
Kurt O. Linn, PhD
Jeffrey D. Linville
Richard E. Lippoth
Jay Lipson
Donald L. Little
Daniel W. Litwhiler, PhD
George A. Livingston
Thomas O. Livingston
Willem Lodder, PhD
Eric Lodewijk
George O. Lof, PhD
Dave Lofe
Leonard V. Lombardi, PhD
W. Warren Longley, PhD
Jose M. Lopez
Harlie M. Love
Robert W. Lovelace
Delwyn J. Low
T. D. Luckey, PhD
Frank L. Ludeman
Ralph Edward Luebs, PhD
Lilburn H. Lueck
Susan J. Luenser
Kenneth Luff
Forest V. Luke
Larry Lukens, PhD
Robert A. Lunceford
Larry D. Lund
Carl Lyday
Gerald J. Lyons
Robert D. Macdonald
W. Macintyre, PhD
Jerry MacLoughlin, PhD
Logan T. MacMillan

James A. Macrill
M. David Madonna
Kent I. Mahan, PhD
Michael W. Mahoney
Thomas P. Mahoney
M. Maish
A. Hassan Makarechian
Michael A. Malcolm, PhD
George Joseph Maler
Richard Maley
Anthony L. Malgieri
James Leighton Maller, PhD
Michael L. Mallo
James D. Mallorey
Keith N. Malmedal
David E. Malmquist
Jerome J. Malone, Jr.
Leo J. Maloney
Richard H. Mandel, Jr.
Peter A. Mandics, PhD
Robert A. Manhart, PhD
Jay A. Manning
Edward D. Manring
Sam W. Maphis, II
Samuel P. March
George J. Marcoux
Jay G. Marks, PhD
William E. Marlatt, PhD
Anthony D. Marques
Charles F. Marshall
L. K. Marshall
Jack E. Martin, PhD
Randall K. Martin
T. Scott Martin
James A. Martinez
Neal R. Martinez
John B. Maruin
Fredrick J. Marvel
Richard Frederick Marvin
Charles B. Masters
Terry J. Mather, PhD
Michael R. Matheson
J. Paul Mathias
Donald A. Mathison
Bruno A. Mattedi
Weston K. Mauz
Donald E. May
Edwarde R. May
John D. Mayhoffer
James E. Mayrath, PhD
Joseph F. McAleer
Howard S. McAlister
Stewart S. McAlister
Eddie W. McArthur
Lon A. McCaley
John S. McCallie
Shawn W. McCarter
Robert P. McCarthy
Calvin H. McClellan
Tim D. McConnell
Joe M. McCord, PhD
Douglas E. McCormac
Jerry N. McCowan
Dirk W. McDermott
Robert E. McDonald
Larry G. McDonough
Marion Edward McDowell
James D. McFall
Frank E. McGinley
Linda M. McGowan
Stuart D. McGregor
Norman L. McIver, PhD

William J. McKelvey
Quentin McKenna
Floyd McKinnerney
Brenda K. McMillan
Richard B. McMullen
Charles S. McNeil
William McNeill, PhD
Kenneth R. Meisinger, PhD
Eric W. Mende
Keith P. Mendenhall
Michael Mendes
Alan E. Menhennett
Jack M. Merritts
Warren H. Mesloh
James T. Metcalf
James R. Van Meter
Ken Metzer
Harvey J. Meyer, PhD
Troy L. Meyer
Walter D. Meyer, PhD
Ralph Meyertons
Leonard V. Micek
Eugene J. Michal, PhD
Michael C. Mickley, PhD
Theodore W. Middaugh
Matthew J. Mikulich, PhD
William J. Miles, PhD
David T. Miller
George Miller
Glenn E. Miller
Harold W. Miller
Leo J. Miller, PhD
Lisa Marie Miller
Robert J. Miller, PhD
Paul R. Millet
Stephen L. Milller
James H. Mims
Michael J. Minkel
Gregory Minton
Jessalynn Misken
Paul Miskowicz
Bruce S. Mitchell
Gary C. Mitchell
Robert K. Mock
Carroll J. Moench
P. Michael Moffett*
Gunter B. Moldzio, PhD
Frank P. Molli
James R. Mondt
Kenneth W. Monington, PhD
Christopher Monroe, PhD
David Coit Moody, PhD
Richard T. Moolick
Frank D. Moore
Jay P. Moore
John F. Moore
Jonathan C. Moore
Lawrence Y. Moore, PhD
Allan J. Mord, PhD
Duane E. Moredock
Robert C. Morehead
Kenneth O. Morey
W. Lowell Morgan, PhD
Timothy P. Morgen
Jill Moring, PhD
Lawrence C. Morley, PhD
Bruce E. Morrell
Richard L. Morris
Dick Morroni
Jerome Gilbert Morse, PhD
John Jacob Mortvedt, PhD
Vladimir K. Moskver

Larry R. Moyer
Bill L. Mueller
James R. Muhm
James W. Mulholland, PhD
Thomas U. Mullen, Jr.
Kelly J. Murphy
James R. Musick, PhD
Randall L. Musselman, PhD
Jan Mycielski, PhD
Burt S. Myers
Misac Nabighian, PhD
Yoshio Nago
James Nagode
Gene O. Naugle
John D. Nebel
Richard L. Negvesky
Al L. Nelson
David L. Nelson
John Nelson, PhD
Larry K. Nelson
Loren D. Nelson, PhD
Thomas E. Nelson
Toby S. Nelson
Wayne W. Nesbitt
Valmer H. Ness
Michael R. Neumann
Roger C. Neuscheler
Richard E. Newell
Roger Newell, PhD
John B. Newkirk
John Newman
Thomas Newman
James M. Newton
W. W. Newton*
Preston L. Nielsen
Richard L. Nielson, PhD
Gordon Dean Niswender, PhD
James Nolan
Cliff Nolte
Matt R. Nord
Gary Nordlander
C. L. Nordstrom
John D. Norgard, PhD
Stephen N. Norris
Gary Nydegger
Chris M. Nyikos
David Claude O'Bryant
Robert Odien, PhD
Robert T. O'Donnell
Walter G. Oehlkers
Elvert E. Oest, PhD
Eldon L. Ohlen
Ralph L. Ohlmeier
George Ojdrovich
Richard C. Oliver, PhD
Brent Olsen
Daniel Olsen, PhD
Norval E. Olson
James Oltmans, II
Victor C. Oltrogge
Michael T. Orsillo
Joseph T. Osmanski
James R. Ottomans, II
Willard G. Owens
Thomas D. Oxley
Ralph S. Pacini
Michael L. Page
Louis A. Panek, PhD
Arthur J. Pansze, Jr., PhD
Ben H. Parker, Jr., PhD
John M. Parker
Pierce D. Parker, PhD

Neil F. Parrett
Clark D. Parroit
Richard S. Passamaneck, PhD
H. Richard Pate
Gary M. Patton
James Winton Patton, PhD
Thomas R. Paul
Kenneth R. Paulsen
John J. Paulus
Louis A. Pavek, PhD
Mark A. Payne
Mike J. Peacock
Galen L. Pearson
S. Ivar Pearson
John D. Peebles
Michael R. Peelish
M. J. Pellillo
Robert W. Pennak, PhD*
Emo Pentermann
David Peontek
David M. Perkins
Luke A. Perkins
Eric Perry
Michael S. Perry
Reagan J. Perry
Douglas C. Peters
Max S. Peters, PhD
Joseph Claine Petersen, PhD
Alan Herbert Peterson, PhD
George C. Pfaff, Jr.
Ryan A. Phifer
John Phillips
Paul H. Pickard
John J. Pittinger
Byron L. Plumley
Henry Pohs
Charles B. Pollock
Larry Pontaski
George J. Popovich, Jr.
Robert Popovich
Lynn K. Porter, PhD
Bruce D. Portz
Madison J. Post, Sr., PhD
Richard Postma, PhD
Stephan N. Pott
Kathleen M. Power
Kenneth L. Presley
Donald R. Primer
Ronald W. Pritchett
Dick A. Prosence
Russell J. Qualls, PhD
Patrick D. Quinney
Terence Thomas Quirke, PhD
Edward L. Rademacher, Jr.
John G. Raftopoulos
C. Edward Raines
Ted Rains
John Ruel Rairden, III
John M. Rakowski
William Ramer
Owen L. Randall
Donald E. Ranta, PhD
Alan M. Rapaport
David E. Rapley
Keith R. Raschke
Alvin L. Rasmussen
Donald O. Rausch, PhD
Shea B. Rawe
Michael Rawley
David Thomas Read, PhD
James G. Reavis, PhD
David N. Rebol

James A. Reddington
Jack Redmond, PhD
Thomas Binnington Reed, PhD
Bruce B. Reeder
Robert G. Refvem
Scott J. Reiman
Elmar Rudolf Reiter, PhD
Keith A. Rensberger
John J. Reschl, Sr.
Melvin Rettig
Gordon F. Revey
John R. Reynolds
J. J. Richard
James W. Richards
Everett V. Richardson, PhD
Paul Richardson
John Riddle, PhD
Kenton Riggs
Cody Riley
Dan H. Rimmer
David B. Roberts
James L. Robertson, PhD
John Robertson, Jr.
Charles S. Robinson, PhD
Raymond S. Robinson, PhD
Dale W. Rodolff
Brian D. Rodriguez
Scott G. Roen
Raylan H. Roetman
William L. Rogers
John A. Rohr
Charles T. Rombough, PhD
Albert J. Rosa, PhD
John G. Roscoe
Lawrence J. Rose
Sam Rosenblum
Charlie Rossman
Lewis C. Rossman
Ora H. Rostad
David L. Roter
George Rouse, PhD
Raymond L. Ruehle
Donald Demar Runnells, PhD
Susan S. Rupp
William J. Ruppert
Michael T. Rusesky
Cynthia B. Russell
Michael J. Ryan
Barbara Rychlik
Wojciech Rychlik, PhD
Michael A. Rynearson
Burns Roy Sabey, PhD
Harry A. Sabin
Julius Jay Sabo
Alberto Sadun, PhD
Eugene Saghi, PhD
Kenneth C. Saindon
Edward A. Samberson
Justin Sandifer
Mark K. Sarto
Frank Satterlee
Eldon P. Savage, PhD
Larry T. Savard
Robert B. Sawyer
Carl H. Schaftenaar
Karl G. Schakel
John A. Schallenkamp
Michael R. Schardt
Randolph E. Scheffel
Erwin T. Scherich
Jonathan D. Scherschligt
William D. Scheuerman

Andre H. Schlappe
Gerald J. Schlegel
Robert F. Schmal
Gregory S. Schmid
Douglas R. Schmidt
Paul G. Schmidt
Henry J. Schmitt
Robert W. Schrier, PhD
Donald E. Schroeder, Jr.
Russell Schucker
Myron R. Schultz
Steven R. Schurman
Ronald G. Schuyler
J. O. Scott
Richard T. Scott
Samuel A. Scott
Terry A. Scott
Linda M. Searcy
Robert Seklemian
Albert P. Selph
D. W. Sencenbaugh
William H. Sens
Randolph L. Seward
George H. Sewell, Jr.
Alan William Sexton, PhD
Steven R. Shadow
Clarence Q. Sharp
Kenneth Sharp
Alan Dean Shauers
Roy W. Shawcroft, PhD
Daniel R. Shawe, PhD
Eugene M. Shearer
Grant L. Shelton
William M. Sheriff
George M. Sherritt
Kenneth Shonk
William A. Shrode, PhD
Craig E. Shuler, PhD
Edward F. Shumaker
Curtis A. Shuman, PhD
Russell W. Shurts
Jim A. Siano
Herb W. Siddle
Kenneth E. Siegenthaler, PhD
Eugene Glen Siemer, PhD
Nancy Jane Simon, PhD
Brian E. Simpson
Kenneth J. Simpson
Erwin L. Single
Ray L. Sisson
Gary W. Sjolander, PhD
Andrew D. Sleeper, PhD
Darryl Eugene Smika, PhD
Francis J. Smit
Clarence Lavett Smith, PhD
Darryl E. Smith, PhD
Duane H. Smith
Earl W. Smith, PhD
Glenn S. Smith
John Ehrans Smith, PhD
Loren E. Smith
Merritt E. Smith
Robert M. Smith
Scotty A. Smith
Tim L. Smith
Wallace A. Sneddon
Gary D. Snell
James J. Snodgrass
Geoffrey G. Snow, PhD
William H. Snyer
Laurence R. Soderberg
John E. Soma

Ronald W. Southard
Antonio Spagna, PhD
Robert L. Speckman
James W. Spellman
Charles Spencer
John Spezia
Daniel S. Spicer, PhD
Andrew Spiessbach, PhD
Charles W. Spieth
Brian B. Spillane
Derik Spiller
Ralph U. Spinuzzi
Melvin Spira
William J. Spitz
Charles L. Spooner
Douglas S. Sprague
Kirk A. Sprague
Roy D. Sprague
Robert L. Sprinkel, Jr.
Edwin H. Spuhler
Robert P. Spurck, PhD
Douglas S. Stack
Harold W. Stack
Michael Keith Stahl
Joel R. Stahn
Gregory E. Stanbro, Jr.
Robert I. Starr, PhD
Jennifer M. Staszel
Jack B. Stauffer
Robert A. Stears
Karl H. Stefan
Michael H. Steffens
Keith M. Stehman
Walter E. Steige
William D. Steigers, PhD
Richard A. Steineck
Frank A. Stephens
Frank M. Stephens, Jr., PhD
Lou Stephens
Mary Stevenson
Gordon K. Stewart
Richard E. Stiefler
Harold Stienmeier*
Richard Stienmier
Fred B. Stifel, PhD
Julie A. Stiff
Cheryl A. Stiles
Gustav Stolz, Jr.
Brad W. Stone
Bill A. Stormes
Vladimir Straskraba
Bruce A. Straughan
Jimmie J. Straughan
Lawrence V. Strauss
Theron L. Strickland
Arthur W. Struempler, PhD
Howard F. Stup
Ravi Subramanian
Steve J. Sullivan
Wayne Summons
Sherman Archie Sundet, PhD
Joseph D. Sundquist
Harry Surkald
Edward J. Suski
Mark D. Swan
Kevin Swanson
Vernon F. Swanson
Christopher L. Sweeney
Vincent P. Sweeney, Jr.
Jerry L. Sweet, PhD
Henry C. Szymanski
Ronald Dwight Tabler, PhD

George C. Tackels
Kenneth G. Tallman
Joseph U. Tamburini
Charles R. Tate
David A. TenEyck
Ted L. Terrel, PhD
Kendell V. Tholstrom
John P. Thomas
M. Ray Thomasson, PhD
Gerald E. Thompson
Michael B. Thomsen
Curtis G. Thomson
Jack Threet
Gaylen A. Thurston, PhD
Bruce Tiemann, PhD
K. Laus D. Ieter Timmerhaus, PhD
Fred S. D. Toole
Brenndan P. Torres
Terrence J. Toy, PhD
Thomas M. Tracey
Zung Tran, PhD
James M. Treat
John R. Troka
Leslie Walter Trowbridge, PhD
Wade Dakes Troxell, PhD
Harry A. Trueblood, Jr.
Paul A. Tungesvick
Kenneth D. Turnbull
Alistair R. Turner
Alfred H. Uhalt, Jr.
E. H. Ulrich
David J. Ulsh
George M. Upton
Julio E. Valera, PhD
Phillip D. Van Law
James R. Van Meter
Martha Van Seckle
Tim J. Van Wyngarden
Tracy L. Vandaveer
David D. Vanderhoofven
Alwyn J. Vandermerwe, PhD
Gordon H. Vansickle
Kenneth D. Vanzanten
James Vavrina
Roy G. Vawter
Delton E. Veatch
Chris J. Veesaert
Richard Veghte
Roger N. Venables
Stan F. Versaw
Steven G. Vick
Lilly Vigil
Boris L. Vilner
James David Vine
Richard J. Vonbernuth
Tim J. Vrudny
Robert L. Wade
Jerome A. Waegli
David A. Wagie, PhD
Charles F. Wagner
Joseph P. Walker
Cherster A. Wallace, PhD
Wyeth Chad Wallace
Denny M. Wallisch
Kent D. Walpole
W. A. Walther, Jr.
Douglas L. Walton
George G. Walton
Frederick Field Wangaard, PhD
Keith K. Wanklyn
Jeffrey V. Ware
John F. Ware

Edward M. Warner
Darrell R. Warren
Simon P. Waters
Donald K. Wathke
Tim Watson, PhD
Shaunna J. Watterson
Dale Watts
Frank B. Watts
William J. Way
John F. Weaver
Leo Weaver
Rodney L. Weber
Douglas H. Wegener
James A. Wehinger
Irving Weiss, PhD
Kirk R. Weiss
Joseph Leonard Weitz, PhD
John C. Welch
Thomas Wellborn
Lawrence D. Wells
Philip B. Wells
Donald A. Welsh
Robert H. Welton
Harold C. Welz
Anthony J. Wernnan
Lee F. Werth, PhD
Cecil B. Westfall
Ann L. White
Richard L. White
Christopher P. Whitham
Robert W. Whitt
Claude Allan Wiatrowski, PhD
Peter L. Wiebe
Albert H. Wieder
Loren Elwood Wiesner, PhD
Hugh M. Wilbanks
Charles G. Wilber, PhD
Donald Wilde
John D. Wilkes, PhD
Wayne E. Wilkins
Jewel E. Willborn
Ted F. Wille
Ronald M. Willhite
Glen Williams
Richard E. Williamson
Alvin C. Wilson
Laurence M. Wilson
Michael D. Wilson, PhD
Wilmer W. Wilson
Denis J. Winder
Robert C. Winn, PhD
Jack E. Winter
David Wire
David Wirth
David J. Wirtz
Floyd A. Wise
Robert L. Wiswell
Franklin P. Witte, PhD
James D. Wolf
Adrian L. Wolfe, PhD
Bruce R. Wolfe
Laird S. Wolfe, PhD
Charles V. Wolfers
Steven A. Wolpert
Cyrus F. Wood
Robert P. Woodby
Steven Woodcock
Tyler Woolley, PhD
Birl W. Worley
Richard F. Worley
Christopher Wright
Jim M. Wroblewski

Harrison C. Wroton
Richard V. Wyman, PhD
Robert B. Wyman
Dan A. Yaeger
Randy R. Yeager
David R. Yedo, PhD
Gene Yoder
Douglas W. York
Jay L. York, PhD
Masami Yoshimura, PhD
Thomas J. Young
William Yurth
William W. Yust
Stephen G. Zahony
Gerald W. Zander
James G. Zapert
Duane Zavadil
Timothy Dean Ziebarth, PhD
Milton A. Ziegelmeier
Robert Zimmerman
Sally G. Zinke

Connecticut

Robert K. Adair, PhD
Robert P. Aillery
Philip D. Allmendinger
James L. Amarel
David B. Anderson
Janis W. Anderson
Thomas F. Anderson, PhD
Manuel Andrade
Steven M. Andreucci
Angela N. Archon
Philip T. Ashton
James R. Barrante, PhD
Paul Bauer, PhD
Jack W. Beal, PhD
Edward J. Beauchaine, Jr.
Paul D. Bemis
Robert Beringer, PhD
Eugene Berman, PhD
Charles D. Bizilj
R. G. Blain
Alan L. Blake
Steven Allan Boggs, PhD
Karl R. Boldt
Laszlo J. Bollyky, PhD
Walter A. Bork
Harvey B. Boutwell
Norman E. Bowne
Robert S. Bradley
Randolph Henry Bretton
David Allan Bromley, PhD
Andrew B. Burns
Dennis H. Burr
William Patrick Cadogan, PhD
Leonard J. Calbo, PhD
William T. Caldwell
Donald A. Cameron
Zoe N. Canellakis, PhD
Joseph H. Cermola
Robert M. Christman
Boa-Teh Chu, PhD
William H. Church, PhD
Eric P. Cizek
Greg P. Clark
Steven K. Clark
J. David Coakley
Ronald C. Coddington
Henry B. Cole

James N. Colebrook
George F. Collins
William F. Condon, PhD
Ralph D. Conlon
Eugene Anthony Conrad, PhD
Robert Joseph Cornell, PhD
Robert W. Cornell, PhD
Cynthia Coron, PhD
Paul J. Cortesi
William Allen Cowan, PhD
Alan L. Coykendall
Ernest L. Crandall
Edward D. Crosby
Zoltan Csukonyi
Charles C. Cullari
Dwight Hills Damon, PhD*
Rocco N. Dangelo
Michael J. Darre, PhD
Lee Losee Davenport, PhD
Joseph J. De Bartolo
Kenneth A. Decarolis
Eddie Del Valle
Frank Deluca
Anthony J. Dennis, PhD
Hans R. Depold
Raymond J. Dicamillo
J. F. Dinivier
Edward W. Diskavich
Charles H. Doersan, Jr.
Barbara B. Doonan, PhD
Frederick Drasch
Valerie B. Duffy, PhD
K. H. Dufrane
Eliot Knights Easterbrook, PhD
Edwin George Eigel, Jr., PhD
Henry J. Ellenbast, Jr.
Richard A. Eppler, PhD
Paul McKillop Erlandson, PhD
Daniel J. Evans
Jason P. Farren
Robert Feingold
Steve Ferraro
William M. Foley, PhD
Herman J. Fonteyne
John C. Forster
George E. Fournier
G. Sidney Fox
Henry E. Fredericks
Arnold E. Fredericksen
Charles Richard Frink, PhD*
George P. Fulton
Joseph T. Furey
Wayne R. Gahwiller
Louis M. Galie
Ernest B. Gardow, PhD
M. R. Gedge
Albert L. Geetter
Robert Charles Geitz, PhD
Mark Gerstein, PhD
Roger R. Giler
Charles M. Gilman
John W. Glomb, PhD
Efim I. Golub, PhD
Robert Boyd Gordon, PhD
Mario Grippi
Fred J. Gross
Gottfried Haacke, PhD
Robert E. Haag
Sigvard Hallgren
Robert D. Halverstadt
Arthur L. Handman
John Haney

Caryl P. Harkins, PhD
Michael Hawley, PhD
Howard Hayden, PhD
David L. Hedberg
James Heidenreich
David E. Henderson, PhD
William B. Henry
Jack L. Herz, PhD
Donald M. Higgins
Kenneth A. Hiller
Clyde D. Hinman
Jonathan Hoadley
Douglas C. Hoagland
John A. Hofbauer
George Robert Holzman, PhD
John Hoover
Art Hornberger
Donald M. Husted
James C. Hutton
Alfred Ingulli
Gideon E. Isaac
Harry A. Jackson
Eva Vavrousek Jakuba, PhD
Stan Jakuba
Jesse C. Jameson
W. W. Jarowey
Elliot E. Jessen
John William Johnstone, Jr.
Morton R. Kagan, PhD
Frank P. Kalberer
David W. Keefe
Marvin Jerry Kenig, PhD
Dan T. Kinard
Ravi Kiron, PhD
Paul Gustav Klemens, PhD
Bruce G. Koe
Anthony P. Konopka
William M. Kruple
Donald Kuehl
Harold Russell Kunz, PhD
Edward P. Kurdziel
Jai L. Lai, PhD
Robert Carl Lange, PhD*
Robert A. Lapuk
John Larson
John W. Leavitt
S. Benedict Levin, PhD
James J. Licari, PhD
John Liutkus, PhD
Robert W. Loomis
Bernard W. Lovell, PhD
Alvin Higgins Lybeck
Tso-Ping Ma, PhD
James J. Macci
James M. Macdonald, Jr.
Thomas Magee
David J. Mailhot
Harold J. Malone
Richard G. Mansfield
Jack J. Marcinek
William C. Martz
Ceslovas Masaitis
Paul G. Mayer
Robert J. J. Mayerjak, PhD
Donald Mayo, PhD
William Markham McCardell
William Ray McConnell, PhD
J. W. McFarland, PhD
Hugh McKenna
James P. Memery
Robert B. Meny
Miles A. Millbach

Dean and Lois D. Miller
John Milne
Jeffrey N. Mobed
David S. Moelling
Michael Monce, PhD
Richard Moravsik
Timothy A. Morck, PhD
John P. Moschello
Saeid Moslehpour, PhD
Francis Joseph Murray, PhD
Ronald Fenner Myers, PhD
Norman A. Nadel
Peter Nalle
John Nasser
Edward S. Ness, PhD
Charles Henry Nightingale, PhD
Brian O'Brien
Robert L. O'Brien
Ronald R. Oneto
Raymond L. Osborne, Jr.
John Harold Ostrom, PhD
Joseph R. Pagnozzi
Laurine J. Papa
David Frank Paskausky, PhD
Mark R. Pastore
Alfred N. Paul
Zoran Pazameta, PhD
Raul T. Perez
Al Peterson
Ronald Piccoli
Daniel Joseph Pisano, PhD
Thomas Plante
Inge Pope
Robert W. Powitz, PhD
John B. Presti
B. Prokai, PhD
Michael J. Pryor, PhD
Ken J. Pudeler
John A. Raabe, PhD
Arthur L. Rasmussen
John A. Reffner, PhD
Harold Bernard Reisman, PhD
Hans Heinrich Rennhard, PhD
Anna V. Resnansky
William C. Ridgway, III, PhD
Francis J. Riley
Charles A. Rinaldi
Felice P. Rizzo
Frank Roberts, Jr.
Donald Wallace Robinson
J. David Robords
Fred L. Robson, PhD
Robert Henry Roth, PhD
William R. Rotherforth
Robert A. Rubega, PhD
George R. Rumney, PhD
John P. Sachs, PhD
Robert Howell Sammis
Reinhard Sarges, PhD
Jeffrey Satinover, PhD
Emilio A. Savinelli, PhD
John R. Schafer
L. McD. Schetky, PhD
Ronald G. Schlegel
Howard E. Schwiebert
Clive R. Scorey, PhD
Andrew E. Scoville
Thomas Seery, PhD
Ronald Sekellick
Chester J. Sergey, Jr.
Harry Sewell, PhD
Richard C. Sharp

Mark Shlyankevich, PhD
Linton Simeri, PhD
Linton Earl Simerl, PhD
Jaime Simkovitz
E. L. Sinclair
Bolesh Joseph Skutnik, PhD
George N. Smilas
Elwin E. Smith
Bruce Sobol
Lon E. Solomita
Andrzej Stachowiak
Bernard A. Stankevich
William R. Stanley
Steven C. Stanton
Joseph Sternberg, PhD
Charles Lysander Storrs, PhD
Joseph R. Stramondo
Charles J. Szyszko
John Tanaka, PhD
Robert G. Tedeschi
C. Sheldon Thompson, PhD
Robert H. Thompson, PhD
Frederick G. Thumm
Sargent N. Tower
Robert L. Trapasso
Richard F. Tucker
Marvin Roy Turnipseed, PhD
Alan K. Vanags
George Veronis, PhD
Daniel Vesa
John S. Wagner
Terramce J. Walsh
Robert G. Weeles
Edward B. Wenners
Richard Wildermuth
Roger F. Williams
Daniel A. Wisner, Jr., PhD
George C. Wiswell, Jr.
Steven E. Yates
John E. Yocom
Robert L. Yocum
Edward A. Zane
Claude Zeller, PhD
William Arthur Ziegler

Delaware

Earl Arthur Abrahamson, PhD
Albert W. Alsop, PhD
Giacomo Armand, PhD
Joseph Bartholomew Arots, PhD
Charles Hammond Arrington, PhD
Andrejs Baidins, PhD
Lewis Clinton Bancroft
Theo C. Baumeister
Paul Becher, PhD
Scott K. Beegle
Oswald R. Bergmann, PhD
Robert Paul Bigliano
William A. Bizjak
William L. Blackwell
Charles G. Boncelet, PhD
Frank R. Borger
Ernest R. Bosetti
John Harland Boughton, PhD
Charles J. Brown, Jr., PhD
Thomas S. Buchanan, PhD
Bruce M. Buker
Charles B. Buonassisi
Donovan C. Carbaugh, PhD
Louise M. Carter

William B. Carter
Leonard B. Chandler, PhD
Robert John Chorvat, PhD
Emil E. Christofano
Alexander Cisar
Ian Clark
George Rolland Cole, PhD
Albert Z. Conner
Nancy H. Conner
Harry Norma Cripps, PhD
R. D. Crooks
William H. Day, PhD
Daniel M. Dayton
Lawrence De Heer
Pamela J. Delaney
William E. Delaney, III, PhD
Dennis O. Dever
Seshasayi Dharmavaram, PhD
Walter Domorod
Roland G. Downing, PhD
Francis J. Doyle
Arthur Edwin Drake, PhD
C. M. Drummund
Eric James Evain, PhD
Elizabeth W. Fahl, PhD
Stephen R. Fahnestock, PhD
Eric R. Fahnoe
Enrico Thomas Federighi, PhD
Thomas Aven Ford, PhD
John Frederick Gates Clark, PhD
William J. Geimeier
Joseph Edmund Gervay, PhD
Wayne Gibbons, PhD
Roderick J. Gillespie, Jr.
David A. Glenn
William H. Godshall
John E. Greer, Jr.
Alfred A. Gruber
Lachman D. Gupta
Earl T. Hackett, Jr.
R. M. Hagen, PhD
Thomas K. Haldas
Charles W. Hall
Thomas W. Harding, PhD
Charles R. Hartzell, PhD
James R. Hodges
Winfried Thomas Holfeld, PhD
Anthony R. Hollet
Roger L. Hoyt
Chin-Pao Huang, PhD
Robert G. Hunsperger, PhD
Mir Nazrul Islam, PhD
Harold Leonard Jackson, PhD
Vladislav J. Jandasek
Paul R. Jann
George K. Janney
Charles S. Joanedis
John Eric Jolley, PhD
Louise Hinrichsen Jones, PhD
Robert John Kallal, PhD
Robert James Kassal, PhD
John T. Kephart
Charles A. Kettner, PhD
Charles O. King, PhD
Joseph Jack Kirkland, PhD
Henry Kobsa, PhD
Theodore Augur Koch, PhD
Robert F. Kock
Bruce David Korant, PhD
Carl George Krespan, PhD
Palaniappa Krishnan, PhD
Wo Kong Kwok, PhD

Douglas R. Leach, PhD
Bernard Albert Link, PhD
Royce Zeno Lockart, PhD
Francis M. Logullo, PhD
H. Y. Loken, PhD
Rosario Joseph Lombardo, PhD
Ruskin Longworth, PhD
Carl Andrew Lukach, PhD
Jeffrey B. Malick, PhD
Creighton Paul Malone, PhD
Philip Manos, PhD
Charle Eugene Mason, PhD
W. McCormack, PhD
Margaret A. Minkwitz, PhD
Donald M. Mitchell
Edward Francis Moran, PhD
William E. Morris
Calvin Lyle Moyer, PhD
Marcus A. Naylor, Jr., PhD
Howard E. Newcomb
Edmund Luke Niedzielski, PhD
Michael Nollet
Lilburn Lafayette Norton, PhD
Louis P. Olivere
Robert D. Osborne
Alfred Horton Pagano, PhD
Albert A. Pavlic, PhD
Rowan P. Perkins
Michael E. Pilcher
Raymond S. Pusey
Ann Rave, PhD
Richard L. Raymond
Charles Francis Reinhardt, PhD
John Reitsma
Robert D. Ricker, PhD
Louis H. Rombach, PhD
Leonard Rosenbaum
Richard A. Rowe, PhD
Mark H. Russell, PhD
David Frank Ryder, PhD
Bryan B. Sauer, PhD
Robert Andrew Scala, PhD
Jerome C. Sekerke, Sr.
James A. Sinex, III
Robert Smiley, PhD
Jack Austin Snyder, PhD
Frank E. South, PhD
Harold F. Staunton, PhD
Evan R. Steinberger
Douglas C. Stewart
Xavier Stewart, PhD
James W. Stingel
George Joseph Stockburger, PhD
Richard W. Stout
Milton Arthur Taves, PhD
William Russell Thickstun, PhD
James R. Thomen, PhD
Eric J. Trinkle
Siva V. Vallabhaneni
Tina K. Vandyk
J. G. Vermeychuk, PhD
Vernon G. Vernier
Owen W. Webster, PhD
John T. Wellener
Thomas Whisenand, PhD
Arthur Whittemore
Joseph S. Woodhead

District of Columbia

Philip H. Abelson, PhD*
Nicholas A. Alten
Todd Barbosa
James Z. Bedford
Marlet H. Benedick
Donna F. Bethell
John D. Bultman, PhD
Adrian Ramond Chamberlain, PhD
Ronald E. Cohen, PhD
Marguerite Wilton Coomes, PhD
Walter R. Dyer, PhD
Howard L. Egan, PhD
Paul Herman Ernst Meijer, PhD
John B. Fallon
Martin Finerty, Jr.
Donna Fitzpatrick
Bruce Flanders, PhD
Nancy Flournoy, PhD
Edmund H. Fording, Jr.
Robert F. Fudali, PhD
Marjorie A. Garvey
Clarence Joseph Gibbs, PhD
Paul F. Giordano
David John Goodenough, PhD
Kenneth P. Green, PhD
John Carl Harshbarger, Jr., PhD
Niki Hatzilambrou, PhD
Robert L. Hirsch, PhD
John Thomas Holloway, PhD
William S. Hughes
Charles J. Kim, PhD
Thomas Adren Kitchens, PhD
John Cian Knight, PhD
Richard L. Lawson
Frank X. Lee, PhD
Donald R. Lehman, PhD
Devra C. Marcus
Gerald Noah McEwen, PhD
Charles J. Montrose, PhD
F. Ksh Mostofi
Stanley Nesheim
Errol C. Noel, PhD
Frederick J. Pearce, PhD
James S. Potts
Thomas Leonard Reinecke, PhD
Alexander F. Robertson, PhD
Malcolm Ross, PhD
Mark A. Ross
Mark A. Rubin
Joaquin A. Saavedra
Jeffery D. Sabloff
Frank Shalvey Santamour, PhD
Walter Schimmerling, PhD
Nina Scribanu
Frank Senftle, PhD
Anatole Shapiro, PhD
Thomas Elijah Smith, PhD
Thomas R. Stauffer, PhD*
Marjorie R. Townsend
Kyriake V. Valassi, PhD
James D. Watkins
Carl W. Werntz, PhD
Frank Clifford Whitmore, PhD
William Phillips Winter, PhD
William S. Yamamoto
Daniel V. Young
Tomohiro Yuki
Lorenz Eugene Zimmerman

Florida

M. Robert Aaron
Frank D. Abbott
Jose L. Abreu, Jr.
Austin R. Ace
John Adams
Jorge T. Aguinaldo
Tom J. Albert
Evelyn A. Alcantara, PhD
Thomas Alderson, PhD
Samuel Roy Aldrich, PhD
Luis A. Algarra
Mark J. Alkire
Eric R. Allen, PhD
George L. Allgoever
Richard E. Almy
Ramon J. Alonso, PhD
Virgilio E. Alvarez
Raymond J. Anater
Barry D. Anderson
Louis Weston Anderson
Vincent Angelo, PhD
Herbert D. Anton
Henry W. Apfelbach
W. H. Appich, Jr.
Orlando A. Arana
R. Kent Arblaster
Antonio E. Arce
William Bryant Ard, PhD
Vittorio K. Argento, PhD
Ross Harold Arnett, PhD
Jack N. Arnold
James T. Arocho
Rhea T. Van Arsdall
Charles R. Ashford
Victor Asirvatham, PhD
Robert C. Asmus
Lynn A. Atkinson
Robert C. Atwood
Kathi A. Aultman
Alfred Ells Austin, PhD
Andrew B. Avalon
Mark Averett
Donald Avery
Bonnie Jean Wilson Bachman, PhD
Ed Bailey
Travis Bainbridge
Earl W. Baker, PhD
Joseph H. Baker
Theodore Paul Baker
William M. Baldwin
Ferdi M. Baler
Ronald C. Ballard
Lloyd Harold Banning
Peter R. Bannister
Donald E. Barber
Paul D. Barbieri
Andrew M. Bardos
Walter E. Barker, Jr.
Ellis O. Barnes
Kenneth J. Barr
J. H. Barten
Homer E. Bartlett
Niles Bashaw
Mark J. Bassett, PhD
Michael A. Bassford
Sam P. Battista, PhD
Max G. Battle, Sr.
Don J. Bauer
Arthur Nicholas Baumann
Lisa L. Baumbach, PhD

Timothy H. Beacham
James M. Beall
Elroy W. Beans, PhD
James H. Beardall
Mhamdy H. Bechir, PhD
Theodore Wiseman Beiler, PhD
Alfred J. Beljan
Dee J. Bell
Edward Bell
Leonard W. Bell
Randy Bell
Robert J. Bellino
Armando L. Benavides
Deodatta V. Bendre
Carroll H. Bennett
Clayton A. Bennett
Rudy J. Beres
John W. Bergacker
Robert D. Berkebile
Donald C. Berkey
Elliot Berman, PhD
Oran L. Bertsch
Arthur F. Betchart
Ervin F. Bickley, Jr.
David A. Bigler
Robert W. Birckhead
Alfred F. Bischoff
Burt James Bittner
James K. Blaircom
Joel J. Blatt, PhD
Ralph C. Bledsoe
Ernest L. Bliss
Robert R. Blume
Wayne Dean Bock, PhD
Peter Boer, PhD
Colleen H. Boggs
Harry Joseph Boll, PhD
Sam Bolognia
Nicholas H. Booth, PhD
Robert M. Borg
Walter S. Bortko
Joseph F. Boston, PhD
Leroy V. Bovey
Clifford R. Bowers
Gale Clark Boxill, PhD
Joseph C. Bozik
George B. Bradshaw, Jr.
Jerry A. Brady
Walter F. Brander
Richard H. Braunlich
Thomas N. Braxtan
Lloyd J. Bresley
Peter R. Brett
Joe E. Brewer
Greg A. Bridenstine
Michael S. Briesch
Edwin C. Brinker
Anne M. Briscoe, PhD
Joe P. Brittain
Wayne Brittian
Hampton Ralph Brooker, PhD
John Brooks
N. P. Brooks
Bahngrell W. Brown, PhD
Fredrick G. Brown
Gary L. Brown
Kenneth B. Bruckart
Bernard O. Brunegraff
Gerald W. Bruner
Ronald L. Brunk
Richard L. Brutus
Bobby F. Bryan

Frederick T. Bryan, PhD
Gary L. Buckwalter, PhD
David A. Buff
Mark A. Bukhbinder, PhD
Ervin Trowbridge Bullard, PhD
Edward J. Buonopane
Paul Philip Burbutis, PhD
Stanley Burg, PhD
James H. Burkhalter, PhD
Charles Burns
Philip J. Burnstein
Thomas R. Busard
Clair E. Butler
Philip Alan Butler, PhD
John J. Byrnes
Barry M. Bywalec
Charles R. Cabiac
Mary M. Cadieux
Robert Cadieux
Marilee Whitney Caldwell, PhD
Patrick E. Callaghan
Allan B. Callender, PhD
Jeffrey D. Caltrider
James B. Camden, PhD
John R. Cameron, PhD
Clifford A. Campbell
Joseph Campbell, PhD
Jose A. Campoamor
Carlos A. Camps
Jorge A. Camps
Paul Canevaro
Margaret S. Cangro
Hugh N. Cannon
Julian E. Cannon
Daniel James Cantliffe, PhD
John V. Carlson
William J. Carson
Arthur L. Carter
Fred S. Carter
Harvey P. Carter
William J. Caseber, Jr.
Judson A. Cauthen
Wiley M. Cauthen
Robert Cavalleri, PhD
Tito Cavallo
Steven Chabottle
Howard E. Chana
William M. Chandler
Donald G. Chaplin
Gregory W. Chapman
M. J. Charles
Augustus Charos
Fernando G. Chaumont
Robert S. Chauvin, PhD
Mary E. Chavez
Craig F. Cheng
Genady Cherepanov, PhD
Paul A. Chervenick
Roger B. Chewning
Sen Chiao, PhD
R. Chiarenzelli
Henry D. Childs
Craig Chismarick
David Chleck
Mandj B. Chopra, PhD
Wen L. Chow, PhD
Bent Aksel Christensen, PhD
James R. Clapp
Richard Allen Claridge
James R. Clarke
James T. Clay
James D. Cline

Paul L. Clough
Clarence Leroy Coates, PhD
Keith H. Coats, PhD
Leroy M. Coffman
C. Eugene Coke, PhD
Gregory Cole
Samuel Oran Colgate, PhD
Paul G. Colman
Benjamin H. Colmery, PhD
Glenn E. Colvin
Franklin J. Cona
James T. Conklin
Fountain E. Conner
Walter Edmund Conrad, PhD
Alberto Convers
James J. Conway, PhD
Anne M. Cooney
Denise R. Cooper, PhD
Emmett M. Cooper
Ernest B. Cooper
George P. Cooper
Mark S. Cooper
Raymond David Cooper, PhD
Dawson M. Copeland
Eugene Francis Corcoran, PhD
Graydon F. Corn
Richard J. Coston
Leon Worthington Couch, II, PhD
Richard J. Councill
Robert O. Covington
John Cowden
Vincent Frederick Cowling, PhD
Francis M. Coy
Warren S. Craven
James F. Crawford
Buford Creech
G. Kingman Crosby
David G. Cross
William C. Cross
Tom L. Crossman
David Crowe
David L. Crowson
Jose R. Cuarta, Jr.
Matteo A. Cucchiara
Donald A. Cuervo
Carlos J. Cuevas
William M. Cullen
Russell E. Cummings
Gene L. Curen
Peter J. Curry
Fred W. Curtis, Jr.
Rosario E. Cushera
Ron Cusson, PhD
Walter J. Czagas
Donald W. Czubiak
George Clement Dacey, PhD
John A. Dady
Fritz Damveld
Bao D. Dang
Andrew W. Dangelo
Jean H. Darling
Albert N. Darlington
Charlie R. Davenport
Thomas L. Davenport, PhD
Duane M. Davis, PhD
Janice R. Davis
David M. Dawson
Noorbibi K. Day, PhD
Duke Dayton
James W. De Ruiter
Nathan W. Dean, PhD
Stanley Deans, PhD

Wayne Deckert, PhD
Bradford Deflin
Albert M. DeGaeta
James A. deGanahl
Quentin C. Dehaan
Robert T. Dehoff, PhD
Glenn A. DeJong
Gary J. Dellerson
Harold Anderson Denmark
Kenneth Derick
Gary D. Dernlan
Edward Augustine Desloge, PhD
Ronald P. Destefano, PhD
Lawrence E. Deusch
Martha S. Deweese
Charles S. Dexter
Charles M. Dick
Philip A. Dick
John A. Dickerson
James C. Diefenderfer
John R. Dieterman, II
Julie Y. Dieu, PhD
Norman G. Dillman, PhD
Richard H. Dimarco
Robert L. Dimmick
Howard Livingstone Dinsmore, PhD
Roy H. Dippy
James P. Diskin
William L. Dixon
N. Djeu, PhD
Gerald D. Dobie
Jerry L. Dobrovolny, PhD
Robert W. Dobson
Michael Dodane
David A. Dodge
John E. Dolan
William R. Dolbier, PhD
Eugene E. Dolecki
L. Guy Donaruma, PhD
John R. Doner, PhD
Joseph F. Donini
Frederick G. Doran
Daniel A. Doty
Keith L. Doty, PhD
Spencer Douglass
W. Campbell Douglass
Larry J. Doyle, PhD*
Frank J. Dragoun
Edward F. Drass
Neil I. Dreizen
David M. Drenan
F. J. Driggers
Vojtech A. Drlicka
Roger W. Dubble
Joseph E. Duchateau
Loring R. Duff
Edward T. Dugan, PhD
James M. Dunford
Russell F. Dunn, PhD
James L. Dunnie
N. Richard Dunteman
Steven Jon Duranceau, PhD
Fred P. Dwight
Julian Jonathan Dwornik, PhD
Joseph Jackson Eachus, PhD
James B. Earle
Amy E. Eason
Hamel B. Eason
James J. Edmier
Dean Stockett Edmonds, Jr., PhD
Thomas A. Edwards
Warren R. Ehrhardt

Howard George Ehrlich, PhD
Paul Ewing Ehrlich, PhD
Esther B. Eisenstein
Luis R. Elias, PhD
David F. Elliott
John O. Elliott
Scott Ellis
Lamont Eltinge, PhD
Frederick H. Elwell
William A. Engel
John M. England
Jack L. Engleman
John Enns, PhD
Frederick B. Epstein, PhD
Larry R. Erickson
Samuel E. Eubank
Ralph L. Evans
James Legrand Everett
Norm R. Every
John M. Evjen, PhD
Martin Fackler
Atir Fadhli, PhD
Larry E. Fairbrother, III
Peter M. Fallon
Scott D. Farash
Robert L. Farnsworth
Robert C. Faro, PhD
Donald D. Farshing, PhD
Donald Featherman
Henry A. Feddern, PhD
Louis Feinstein, PhD
Donald Fenton, PhD
Emmett B. Ferguson
Patrick J. Ferland
Eugene M. Fetner
Henry T. Fielding
Carl E. Fielland
Joshua A. Fierer*
Norton E. Fincher
Robert D. Finfrock
Ronald L. Finger
Barney W. Finkel
Charles W. Finkl, Jr., PhD
Gerald R. Fishe
Charles H. Fisher
George H. Fisher, PhD
John C. Fisher
William J. Flick
William H. Flowers, PhD
John C. Floyd
Leo J. Flynn
Timothy A. Focht
Peter John Fogg, PhD
James H. Fogleman
Paul D. Folse
Candido F. Font
Anderson M. Foote
Marion Edwin Forsman, PhD
Irving S. Foster, PhD
Norman G. Foster
Robert G. Foster
Don Fournier
James M. Fowler, Jr.
Gerald Fox
Russell E. Frame
Allan J. Frank
Sidney Frank
Anthony D. Frawley, PhD
Stephen Earl Frazier, PhD
Andrew W. Frech
Ronald Harold Freeman, PhD
Doug Freemyer

Thomas V. Freiley
Raymond Friedman, PhD
Dan C. Frodge
Higinio J. Fuentes
Clark W. Furlong
Kenneth E. Fusch
Ronald M. Fussell
George Gabanski
John C. Galen
Robert M. Gallen
Boris Galperin, PhD
Joseph E. Gannon
William Lee Garbrecht, PhD*
Harry E. Gardner, Jr.
Richard Garnache
Carl C. Garner
Charles J. Garrett
Don L. Garrison
Jose R. Garrote
Geoffrey W. Gartner, Jr.
John J. Gaughan
Eugene R. Gaughran, PhD
John F. Gaver
Douglas M. Gebbie
Steve Geci
Anthony F. Gee
James E. Geiger
John Geiger
Paul G. Geiss
Karl R. Geitner
Edmund A. Geller
George T. Genneken
Lawrence D. Gent, Jr.
Larry Gerahiau
Eugene Jordan Gerberg, PhD
Walter M. Gerhold
Umberto Ferdinando Gianola, PhD
Kurt R. Gies
Fred Giflow
Peter Gilbert
Edward R. Gillie
June Gillin
William C. Gilmore, Jr.
George A. Gimber
David N. Girard
W. E. Girton
Johanna M. Glacy-Araos
Richard E. Gladziszewski
Thomas Alexander Gleeson, PhD
Anatol A. Glen
Ronald Gluck
Richard H. Gnaedinger, PhD
C. J. Gober
Louis D. Gold, PhD
Alan S. Goldfarb, PhD
Lionel Solomon Goldring, PhD
Gary J. Goldsberry
Yitzhak Goldstein
Ely Gonick, PhD
Ed E. Gonzalez
Lewis F. Good, Jr.
Brett P. Goodman
John A. Goolsby
Robert Stouffer Gordee, PhD
Theodore B. Gortemoller, Jr.
David Gossett
Henry Gotsch
Charles F. Gottlieb, PhD
Gerald Geza Gould
Karl Gould
Hans C. Graber, PhD
Howard E. Graham

F. R. Grant
Frank Grate
M. R. Grate
Chester W. Graves
Wayne C. Graves
George F. Green, PhD
Willie H. Green
William Greene
Charles W. Gregg
Donald J. Gregg
Frank E. Gregor
Bert L. Grenville
Donald K. Grierson
R. Howard Griffen
Gordon E. Grimes
Chester F. Grimsley
William B. Grinter
Gerry Gruber, PhD
Ivan Grymov
Ernest A. Gudath
Cnelson Guerriere
Robert O. Guhl
Joseph W. Guida
Jean R. Gullahorn
Michael E. Gunger
Phillip W. Gutmann
Thomas J. Gyorog
Mutaz Habal, PhD
James A. Hagans, PhD
Jeffrey S. Haggard
Donald R. Hagge
James T. Haggerty
Theodore W. Hahn
Warren A. Hahn
Oussama Halabi
Douglas K. Hales
John F. Hallahan
Harry Hamburger
Brenton M. Hamil, PhD
Darryl J. Hamilton
Mona S. Hamilton
Douglas K. Hammann
Thomas G. Hammond, Jr.
Larry D. Hamner
M. Hancock
Robert William Hanks, PhD
Donald C. Hanto
Kevin G. Harbin
Jeffrey L. Harmon, PhD
Eric Harms
Joseph J. Harper
Gary G. Harrison
Robert Hartley
William H. Hartmann
Ronald F. Hartung
Eldert Hartwig, Jr.
Ed Harwell
Alden G. Haskins
Elvira F. Hasty, PhD
James H. Hasty, PhD
Charles L. Hattaway
Arthur J. Haug, PhD
Ralph E. Hayden
John A. Hayes
Michael E. Hayes, PhD
Ralph L. Hayes
Jordan M. Haywood
John F. Hazen
Ronald Alan Head, PhD
Robert S. Hearon
George M. Hebbard
Boyd R. Hedrick, Jr.

Roy W. Heffner
Brian E. Heinfeld
David F. Heinrich
Cecil Helfgott, PhD
James F. Helle
Richard B. Hellstrom
Robert L. Helmling
James Brooke Henderson, PhD
Jim Hennessy
Carl H. Henriksen
Jonathan Henry, PhD
James F. Hentges, PhD
Alberto Hernandez
Alvin C. Hernandez, Jr.
Irwin Herman Herskowitz, PhD
David E. Hertel
Richard L. Hester
Richard P. Heuschele
Richard Hevia
Thomas B. Hewton
David K. Hickle
Kathryn Hickman, PhD
Gregory D. Hicks
Thomas M. Hicks
Alberto F. Hidalgo, PhD
Clarence Edward Hieserman
Eric V. Hill, PhD
Richard F. Hill
Robert E. Hillis, PhD
Helmuth E. Hinderer, PhD
Carla J. Hinds
George T. Hinkle, PhD
George M. Hinson
Daniel J. Hirnikel
Thomas James Hirt, PhD
Robert Warren Hisey, PhD
H. Ray Hockaday
John T. Hocker
Edward Hoffmann
James P. Hogan, PhD
Lawrence E. Hoisington, PhD
Wayne Holbrook
David G. Holifield, PhD
Allen B. Hollett
Richard A. Hollmann
Eugene H. Holly
Michael D. Holm
Donald A. Holmer, PhD
H. Duane Holsapple
Larry H. Hooper
E. Erskine Hopkins
Lewis Milton Horger, PhD
Evelio N. Horta, PhD
Paul A. Horton, PhD
Hamid Hosseini
John H. Hotaling
William B. Houghton
Charles M. House
Larry D. Housel
Gordon P. Houston
Louis R. Hovater
James Howze
Sung Lan Hsia, PhD
Robert J. Hudek
George R. Hudson
Russell Henry Huebner, Sr., PhD
Gaylord Huenefeld
Margaret J. M. Huey
Arvel Hatch Hunter, PhD
Jamie Hunter
Ronald L. Hurt
Scotty M. Hutto

Kenneth Hyatt
Carly H. Hyland
Terrence L. Ibbs
Phillip M. Iloff, Jr., PhD
Paul E. Ina
William L. Ingle
David Irvin
Robert M. Ivey
Bruce G. Jackson
J. G. Jackson, III*
Wes S. Jacobs
Jerald O. Jacobson
Michael T. Jaekels
Paul C. Jakob
Clifford H. James
John C. Janus
Samuel E. Jaquinta
Karen C. Jaroch
Adrian Jelenszky
Hanley F. Jenkins
Paul A. Jennings, PhD
Clayton Everett Jensen, PhD
Anthony E. Jernigan
Howard R. Jeter
Eileen D. Johann, PhD
Virgil Ivancich Johannes, PhD
Anthony Johnson
Charles W. Johnson
David A. Johnson
Gerald B. Johnson
H. C. Johnson
J. William Johnson
James Robert Johnson, PhD
Leland B. Johnson
T. Johnson
Frederick W. Johnston, Jr.
Harold E. Johnstone
Barbara C. Jones
Claude D. Jones
Marcia Jones
R. H. Jones, PhD
Roy Carl Jones, PhD
Joseph G. Jordan
Edward Joseph
Edwin A. Joyce, Jr.
Patrick A. Joyce
Hiram Paul Julien, PhD
Brian J. Just, PhD
Martin J. Kaiser
Bernard J. Kane
Kenneth C. Kanige, PhD
Ramanuja Chara Kannan
Robert Karasik
William J. Kardash, Sr.
Delmar W. Karger
Robert R. Karpp, PhD
Jeffery J. Karsonovich
Kenneth Stephen Karsten, PhD
A. J. Kassab, PhD
Michael J. Kaufman, PhD
Michael E. Kazunas
Kar M. Keil
Carroll R. Keim
Richard K. Keimig
Daniel C. Kelley
Whitemore B. Kelley, Jr.
John E. Kemp
Francis J. Kendrick, PhD
Dallas C. Kennedy, II, PhD
Robert W. Kennedy
James A. Kennelley, PhD
Charles W. Kennison

R. D. Kent
James G. Keramas, PhD
Robert Kergai
John E. Kerr
Keith M. Kersch
John M. Kessinger
Ronald J. Kessner
Charles R. Keyser
Suresh K. Khator, PhD
Roosey Khawly
Jennifer L. Kibiger
Richard L. Kiddey
David E. Kiepke
Robert R. Kilgo
Dennis K. Killinger, PhD
Hueston C. King
McLeroy King
Randy L. King
Stan A. Kinmouth
Dalton L. Kinsella
James R. Kircher
Henry W. Kirchner
Michael K. Kirwan
Wendell G. Kish
Abbott Theodore Kissen, PhD
Eva B. Kisvarsanyi
Geza Kisvarsanyi, PhD
Waldemar Klassen, PhD
Steven K. Klecka
Willard R. Kleckner, PhD
Karl M. Klein, Jr.
Karl G. Klinges, PhD
Benny Leroy Klock, PhD
Glenn S. Kloiber
Daniel E. Kludt
Ian G. Koblick
Lee E. Koepke
Lawrence K. Koering
William Francis Kohland, PhD
Paul A. Kohlhepp
Donald H. Komito
George J. Kondelin, Jr.
George E. Konold
Richard J. Kossman
Thomas W. Kotowski
Sidney Kovner
Stanley Krainin
Donald R. Kramer
John Krc, Jr.
Istvan M. Krisko, PhD
John M. Krouse
Walter Hillis Kruschwitz, PhD
Dennis J. Kulonda, PhD
Frank Turner Kurzweg
Bogdan Ognjan Kuzmanovic, PhD
Tung-Sing Kwong
Edward G. Kyle
R. G. Lacallade
J. G. Ladesic, PhD
Charles Lager
Henry J. Lamb
Donald H. Lambing
Stephen Lamer, PhD
Michael C. Lamure
Albert L. Land
B. Edward Lane
Paul J. Langford
JoAnne Larsen, PhD
Reginald Einar Larson
Edwin D. Lasseter
Thomas S. Lastrapes
Mark T. Lautenschlager

Donnie B. Law
David Lawler
Peter K. Lazdins, PhD
Frank W. Leach
George R. Lebo, PhD
John p. Leedy
Irving Leibson, PhD
Gordon F. Leitner
Carl Lentz
Kenneth Allen Leon, PhD
Philip B. Leon
Donald E. Lepic
Michael A. Leuck
Samuel Levin
Stanley S. Levy, PhD
Thomas F. Lewicki
Thomas Lick, PhD
Jon L. Liljequist
Bruce A. Lindblom
Bruce G. Lindsey, PhD
Raul A. Llerena
Vern Lloyd
Salvadore J. Locascio, PhD
Emil P. Loch
Richard F. Lockey
Bruce A. Loeppke
Ernest E. Loft
R. Loftfield
Crispino E. Lombardi
James A. Long
R. W. Long, PhD
Alfredo M. Lopez, PhD
Don D. Lorenc
Robert A. Loscher
Julio G. Loureiro
Richard G. Loverne
Gordon Lovestrand
Robert W. Loyd
Steven A. Lubinski
John W. Luce
Richard R. Ludlam
Bobby R. Ludlum
S. J. Ludwig
Henry C. Luke
Ronald A. Lukert
Arthur Lyall
Victoria S. Lyon
Paul Macchi
Theodore S. Macleod
Robert K. Macmillan
Orville E. Macomber
Guy R. Madden
Robert L. Magann
James G. Magazino
James Mahannah
Raymond C. Mairson
Edmund R. Malinowski, PhD
Paul A. Mallas
Edmund M. Malone
Joe H. Maltby
Eugene H. Man, PhD
Jesse R. Manalo
Robert J. Mandel
Anthony P. Mann
Walter Manning
Joseph Robert Mannino, PhD
W. E. Manry, Jr.*
Robert S. Mansell, PhD
Juliano Maran
Vladimir K. Markoski
Gerardo L. Marquez
Homer L. Marquit

John J. Marra
A. E. Marshall, Jr.
Roger W. Marsters, PhD
Paul R. Martel
Jose L. Martin, PhD
Ralph Harding Martin
Robert W. Martin
William W. Martin
Mason E. Marvel, PhD
Harry L. Mason
Wayne A. Mather
Robb R. Mathias
Harry M. Mathis, PhD
Maria B. Matinchev
Guy C. Mattson, PhD
James W. Maurer
Florentin Maurrasse, PhD
John G. Mavroides, PhD
Vincent E. Mayberry
Marion S. Mayer, PhD
George T. McAllister
Joseph James McBride, PhD
Clarence H. McCall
Edward E. McCallum
Carolyn A. McCann
Ron L. McCartney
R. McCarty
G. J. McCaul
Walter H. McCluskey
W. Phil McConaghey
Daniel F. McConaghy
Daniel G. McCormick
Donald S. McCorquodale, PhD
Scott A. McCoy
Stephen C. McCranie
Robert L. McCroskey, PhD
Victor McDaniel, PhD
Lee Russell McDowell, PhD
Stuart E. McGahee
Anthony McGoron, PhD
David F. McIntosh
Darrell W. McKinley
Michael McKinney, PhD
John McKisson
George Hoagland McLafferty
Charles D. McLelland
Malcolm E. McLouth
Harold R. McMichael
Donald W. McMillan
Donald F. McNeil, PhD
Clyde W. McNew
Edwin R. McNutt
Glenn L. McNutt
Marion L. Meadows
Tonya S. Mellen
Henry Paul Meloche, PhD
Darrell E. Melton
Robert William Menefee, PhD
Gunter Richard Meng
Jerry Merckel, PhD
Daniel N. Mergens
Andrew H. Merritt, PhD
Duane R. Merritt
Henry Neyron Merritt, PhD
Lawrence E. Mertens, PhD
James H. Messer
Aldo J. Messulam, PhD
Alexandru Mezincescu, PhD
Oskar Michejda, PhD
Harold S. Mickley, PhD
Judith A. Milcarsky
Sylvester S. Milewski

John F. Milko
Kenneth J. Miller
Richard H. Miller, PhD
William Knight Miller, PhD
Scott Milroy, PhD
Robert Mitchell Milton, PhD
Malcolm G. Minchin
R. Edward Minchin, Jr., PhD
N. Misconi, PhD
Patricia Mitchell
John B. Mix, PhD
Jeffrey N. Mock
Stanley S. Moles, PhD
Ralph Moradiellos
Christopher A. Morgan, PhD
Robert A. Morrell
Harold N. Morris
James H. Moss
Lee W. Mozes, PhD
Charles L. Mraz
Rosa M. C. Muchovej, PhD
John T. Muller
Robert C. Mumby
Gary L. Murphy
John F. Murphy, PhD
William Parry Murphy
Frank A. Musto
Jay Muza, PhD
Leslie A. Myers, PhD
Mark T. Myers
Daniel A. Myerson
Harry E. Myles
Don R. Myrick
Roger P. Natzke, PhD
Donald K. Nelson
Hugo D. Neubauer, Jr.
James E. Neustaedter
George F. Nevin
Waldo Newcomesz
Jesse R. Newell
McFadden A. Newell
Richard W. Newell
Arthur W. Newett
Carl S. Newman
William A. Newsome
Bill H. Nicholls
David S. Nichols
Kenneth W. Nickerson
Paula W. Nicola
Edward Niespodziany, PhD
Herbert Nicholas Nigg, PhD
Margaret M. Niklas
Edward G. Nisbett
John C. Nobel
Michael J. Noesen
Michael J. Nolan
William G. Norrie
Ronelle C. Norris
Peggy A. Northrop
Albert Notary
Carlos G. Nugent
Robert L. Oatley
Chad J. Oatman
James J. O'Brien, PhD
Thomas W. O'Brien, PhD
T. F. Ochab
Frederic C. Oder, PhD
Norbert W. Ohara, PhD
Daniel J. O'Hara, PhD
J. W. O'Hara, PhD
Kenneth R. Olen, PhD
Oluf E. Olsen

Ronald V. Olsen
Lewis E. Olson
Robert J. Olson
Ray Andrew Olsson
Robert Milton Oman, PhD
Adrain P. O'Neal
Albert E. O'Neall
Gene F. Opdyke
James F. Orofino
Donald Orr
David A. Orser
William Albert Orth, PhD
Harold Osborn
Matthew J. Ossi
Wendall K. Osteen, PhD
John P. Osterberg
Robert Franklin Overmyer
Jerry M. Owen
Terence Cunliffe Owen, PhD
Henry Yoshio Ozaki, PhD
Glenda B. Pace
Thomas J. Padden, PhD
Fred Paetofe
Edward C. Page
Edward Palkovic
Jay W. Palmer, PhD
Ralph Lee Palmer
Gyan Shanker Pande, PhD
Patrick J. Di Paolo
Charles J. Papuchis
Georges Pardo, PhD
Marie J. Parenteau
Charles L. Park
Cyril Parkanyi, PhD
Anthony T. Parker
David H. Parsons
Richard L. Pase
Garland D. Patterson
Ben W. Patz, PhD
Gary W. Pauley
Charles H. Paxton
Robert M. Peart, PhD
Robert Pease
Richard Stark Peckham, PhD
Edward A. Pela
Winston Kent Pendleton, PhD
Robert E. Peppers
Horacio F. Perez
Stanley E. Permowicz
Thomas E. Perrin
James B. Perry
Edward R. Pershe, PhD
Irvin Leslie Peterson
Fred Petito
Wallace M. Philips
Jim M. Phillips
Wayne A. Pickard
Brett H. Pielstick
Frank A. Pierce
James J. Pierotti
Robert F. Pineiro, PhD
Victor J. Piorun
Bruce D. Pisani
R. Clinton Pittman
Ronald Plakke, PhD
Charles D. Plavcan
David A. Pocengal
Frederick J. Pocock
Robert C. Pollard
Richard G. Pollina
Alan Y. Pope
Hugh Popenoe, PhD

Dorian J. Popescu
Richard R. Popham
Ronald R. Porter
Mary B. Portofe
Bonnie W. Posma
Denzil Poston, PhD
Priscilla J. Potter, PhD
Lee A. Powell
P. Mark Powell
Sandra S. Powers
Faustino L. Prado
Kenneth L. Pratt
Steve L. Precourt
Betty Peters Preece
Henri Pregaldin
Irwin M. Prescott
R. Preston
Katherine L. Price
Claude C. Priest
Kermit L. Prime, Jr.
Thomas P. Propert
Lumir C. Proshek
Charles B. Pults
Michael Punicki
Charles W. Putnam
John Ward Putt
Bodo E. Pyko, PhD
Ma Qin
H. Paul Quicksall
Robert H. Quig
Jesus F. Quiles
Donald Joseph Radomski
James A. Ralph
Andrea L. Ramudo
Patricia Ramudo
Charles Addison Randall, PhD
Carlisle Baxter Rathburn, PhD
Gregory K. Ratter
Morgan G. Ray, Sr.
John M. Reardon
Walter T. Rector
William M. Redding
Immo Redeker
John W. Redelfs
Sherman Kennedy Reed, PhD
Gerhard Reethof, PhD
James A. Reger
George Kell Reid, PhD
David E. Reiff
Gerald R. Reiter
Bruce G. Reynolds, PhD
Ernest J. Rice
Bruce A. Richards
William Joseph Richards, PhD
Carl B. Richardson
Richard Wilson Ricker, PhD
Robert R. Righter
Gene A. Rinderknecht
Franklin G. Rinker
Harold L. Riser
George Ritter
James J. Roark
George W. Robbins
George A. Roberts
Kenneth Roberts
Robert S. Roberts
Donald K. Robertson
F. Herbert Robertson
Roger L. Robertson
Gary J. Robinson
Charles L. Robson
Michael W. Rochowiak, PhD

James C. Roden
Billy R. Rodgers, PhD
Jorge A. Rodriguez
Felix Rodriguez-Trelles, PhD
Walter H. Rogers, Jr.
Winston Rogers
Francis C. Rogerson
Richard C. Ronzi
Isaac F. Rooks
William S. Roorda
Peter P. Rosa
Mary C. Roslonowski, PhD
Lenard H. Ross
Jack Rossman
Ronald P. Rowand
Robert Seaman Rowe, PhD
Bob R. Rowland
Joseph A. Roy
Ronald A. Rudolph
Paul W. Runge
Devon S. Rushnell
A. Yvonne Russell, PhD
David E. Russell
Roger L. Russell
Joseph C. Russello
Byron E. Ruth, PhD
Ralph R. Ruyle
Terrell B. Ryan
Frank Sabo
Alfred Saffer, PhD
Alexander A. Sakhnovsky
Israel Salaberrios
Fernando M. Salazar
Richard J. Salk
Arthur S. Salkin
Donald S. Sammis
Ronald E. Samples
Daniel C. Samson
Thomas H. Samter
Oscar A. Sanchez
Kenneth L. Sanders
William E. Sanders
Miguel A. Santos
Herbert P. Sarett, PhD
Philbrook F. Sargent
Noray Sarkisian
Joachim Sasse, PhD
Edward A. Saunders, PhD
Samuel O. Sawyer, III
Eugene D. Schaltenbrand
David O. Scharr
Jay R. Schauer
Zbigniew I. Scheller
Frederick W. Schelm
Richard A. Scheuing, PhD
Blair H. Schlender
Fabiola B. Schlessinger, PhD
Donald J. Schluender
Robert Schmeck
Hubert F. Schmidt
Richard A. Schmidtke, PhD
Karl H. Schmitz
Peter Schroeder, PhD
Robert Schroeder, PhD
Charles Schroeter
John K. Schueller, PhD
Larry E. Schuerman
Robert P. Schuh
John H. Schumertmann, PhD
David J. Schuster, PhD
Eunice C. Schuytema, PhD
Albert Z. Schwartz

John Warner Scott, PhD
Scott E. Scrupski
Walter Tredwell Scudder, PhD
John W. Seabury
Richard A. See
Christopher Seelig
Barry D. Segal
Harvey N. Seiger, PhD
John O. Selby
Luther R. Setzer
Earl William Seugling, PhD
Kenneth N. Sharitz
J. B. Sharp*
M. L. Sharrah, PhD
Bob Sheldon
Preston F. Shelton
Donald M. Shepherd
Walter L. Sherman
Gregory C. Shinn
Aleksey Shipillo
Valentin Shipillo, PhD
John W. Shipley, PhD
Neil Shmunes
Dennis Shoener
James Edward Shottafer, PhD
Eric S. Shroyer
Orren B. Shumaker
John M. Siergiej
Carl Signorelli
Ernest S. Silcox
Paul W. Silver
Fred G. Simmen
Joel M. Simonds
Harry S. Sitren, PhD
Andrew Sivak, PhD
Arthur C. Skinner, Jr.
E. F. Skoczen
Robert P. Skribiski
Dave Skusa
Philip Earl Slade, PhD
William Donald Smart
Jery Smieinski
William H. Smiley
Alexander G. Smith, PhD
Augustine Smith
Bob A. Smith
Bruce W. Smith
Cedric M. Smith
Clifton R. Smith
Dale A. Smith
Diana F. Smith
Lebrun N. Smith
Michael D. Smith
Paul Vergon Smith, PhD
Rudolph C. Smith, II
Samuel W. Smith
Scott H. Smith
Silke G. Smith
T. J Lee Smith, Jr.
Theodore A. Smith
Robert C. Smythe
John T. Snell
James M. Snook
George Snyder, PhD
Warren S. Snyder
Marc P. Sokolay
James E. Solomon
Omelio J. R. Sosa, Jr., PhD
Parks Souther
Donald E. Spade
James A. Spagnola
Richard M. Speer

C. R. Spencer
Richard Jon Sperley, PhD
Norman P. Spofford
Michael D. Sprague
Robert Carl Springborn, PhD
Robert H. Springer, PhD
Donald Platte Squier, PhD
Mark A. Stahmann, PhD
John P. Stancin, PhD
Gregory F. Stanley
Cecil R. Stapleton
Harvey J. Stapleton, PhD
Walter L. Starkey, PhD
Jan D. Steber
Theodore M. Stefanik
Thomas M. Steinert
Osmar P. Steinwald
John E. Stelzer
Russell A. Stenzel
Jesse Jerald Stephens, PhD
Robert D. Stephens
Alan G. Stephenson
Edgar K. Stewart
Ivan Stewart, PhD
John Stewart
William W. Stewart
Werner K. Stiefel
Irvin G. Stiglitz, PhD
Donald Stilwell
Paul R. Stodola
Timothy D. Stoker
Joseph B. Stokes, Jr.
Kenneth J. Stokes
Leonard Stoloff
P. Stransky
Thomas P. Strider
Kelly C. Stroupe
Randy P. Stroupe
Duane H. Strunk
Eugene Curtis Stump, PhD
Frank Milton Sturtevant, PhD
Jacob W. Stutzman, PhD
Spiridon N. Suciu, PhD
Edward J. Sullivan
Chades F. Summer, Jr.
Frederick C. Sumner
John C. Sundermeyer
Peter R. Sushinsky
Howard Kazuro Suzuki, PhD
Laurie A. Swanson
Ronald Lancelot Sweet, PhD
Alan M. Swiercz
Burton L. Sylvern
Walter Symons
James Imre Szabo
Michael Szeliga
John F. Takacs
Gabor J. Tamasy
J. P. Tanner
Robert Tanner
William Francis Tanner, PhD
Donald E. Tannery
Frederick Drach Tappert, PhD
Armen Charles Tarjan, PhD
Amy T. Tatum
Philip Teitelbaum, PhD
Aaron J. Teller, PhD
Jeffrey Tennant, PhD
Harold Aldon Tenney
David S. Teperson
Edward B. Thayer
David T. Therrien

Thomas J. Thiel
Lovic P. Thomas
Richard William Thomas, PhD
David Morton Thomason, PhD
Neal P. Thompson, PhD
Julian C. Thomson, PhD
Bill H. Thrasher, PhD
Lee F. Thurner
Pat T. Tidwell
Jennifer L. Tillman
Andrew D. Tilton
William E. Tipton
Dean H. Tisch
Victor J. Tofany
Robert S. Tolmach
Manolis M. Tomadakis, PhD
Gene T. Tonn
Eric D. Torres
James A. Treadwell
Joseph Trivisonno, Jr., PhD
Byron C. Troutman
Charles H. True
Duane J. Truitt
Steven K. Trusty
Walter Rheinhardt Tschi, PhD
Geoff W. Tucker
R. C. Tucker, Jr., PhD
Stela Tudoran
Leroy L. Turja
Thomas K. Turner
Ralph J. Tursi
Donna Utley, PhD
Tom W. Utley, PhD
Raj B. Uttamchandani, PhD
Augustus Ceniza Uvano, PhD
David P. Vachon
Robert L. Valliere
Harold H. Van Horn, PhD
Raymond Van Pelt
Maria Ventura
William Vernetson, PhD
Daniel F. Vernon
Raymond A. Verville
Shane Vervoort
Fernando Villabona
Mark Villoria, PhD
Thomas R. Visser
Ronald S. Vogelsong, PhD
James Vollmer, PhD
Dale L. Voss
John C. Vredenburgh
William N. Waggener
Laurence F. Wagner, PhD
Nnaette P. Wagner, PhD
Arthur C. Waite
Donald J. Wakely
Stephen D. Waldman
J. David Walker
Harold Dean Wallace, PhD
Joseph M. Walsh
S. Norman Wand
James Wanliss, PhD
Diane Ward
Ralph E. Warmack, PhD
Bertram S. Warshaw
Lawrence K. Wartell
Gerald R. Wartenberg
L. W. Warzecha
Alex C. Waters
Charles Henry Watkins, PhD
Britt Watson
C. Paul Watson

John W. Wayne
Howard W. Webb
Lloyd T. Webb
George W. Webster
Robert D. Webster
David Stover Weddell, PhD
John D. Weesner
Peter J. Weggeman
Stephen L. Wehrmann
John H. Weiler
Mitchell A. Weiner
George A. Weinman
David Weintraub, PhD
Steven Weise
Aaron W. Welch, Jr., PhD
Thomas B. Welch
Buford E. Wells
Daniel R. Wells, PhD
Steven P. Wells
Douglas F. Welpton
Jim C. Welsh
Patrick T. Welsh, PhD
Robert J. Werner
Ronald R. Wesorick
Lewis H. West
Richard S. Westberry
Donald H. Westermann
Edwin F. Weyrauch
George F. Whalen, Jr.
Frank Carlisle Wheeler, PhD
William R. Whidden
Albert C. White
Jacquelyn A. White
John I. White, PhD
George T. Whittle
Alfred A. Wick
Marvin L. Wicker
David Wickham
Donald J. Wickwire
Joseph R. Wiebush, PhD
Kurt L. Wiese
James W. Wiggin
Richard L. Wiker
Dale S. Wiley
Harvey Bradford Willard, PhD
Charles T. Williams
David Williams
George E. Williams
Jonathan Williams
Ralph C. Williams
Tom Vare Williams, PhD
Norman L. Williamson, PhD
Warren P. Williamson, III
Robert Elwood Wilson, PhD
Tun Win
Frank R. Winders
James D. Winefordner, PhD
Lesley Winston
Raymond L. Winterhalter
Hugh E. Wise, Jr., PhD
Robert S. Wiseman, PhD
James T. Wittig
Howard G. Womack
Irwin Boyden Wood, PhD
Richard P. Woodard, PhD
David P. Woodhouse
Roderick F. Woodhouse
Christopher L. Woofter
Douglas Albert Woolley
James A. Woolley
William C. Woolverton
James G. Worth

Donald J. Worthington
David H. Wozab
Dane C. Wren
Floyd D. Wright
William E. Wright
Tien-Shuenn Wu, PhD
C.F. Wynn, PhD
Anton M. Wypych
Theodore E. Yaeger, IV
Chris Yakymyshyn, PhD
David E. Yarrow
Charles Yeh
Wilbur Yellin, PhD
John V. Yelvington
Frank Yi
Richard R. Yindra
Ted S. Yoho
Guy P. York, PhD
Richard A. Yost, PhD
Cindy M. Young
Frank Young, PhD
Douglas J. Yovaish
Terry Zamor
David Aaron Zaukelies, PhD
Fred G. Zauner
Ernest E. Zellmer
Donald M. Zelman
Melanie P. Ziemba
Richard A. Ziemba
Jay Zimmer
Peter H. Zipfel, PhD
Parvin A. Zisman
John Zoltek, PhD
Eli Zonana
Harry David Zook, PhD
Robert M. Zuccaro

Georgia

William G. Adair, Jr.
William P. Adams
L. A. Adkins
Frank Jerrel Akin, PhD
Robert H. Allgood
Mike E. Alligood
Bruce Martin Anderson, Jr.
Byron J. Arceneaux
Robert Arnold, Jr.
Doyle Allen Ashley, PhD
R. Lee Aston, PhD
James Atchison
Keith H. Aufderheide, PhD
C. Mark Aulick, PhD
Paul E. Austin
Dany Ayseur
Charles L. Bachman
Robert Leroy Bailey, PhD
James A. Bain, PhD
George C. Baird
Dana L. Baites
Bill Barks
Larry K. Barnard
Ralph M. Barnes
Gary Bartley
Frank A. Bastidas
Thomas L. Batke
George L. Batten, PhD
Joseph E. Baughman, PhD
Joseph H. Baum, PhD
William Pearson Bebbington, PhD
Gordon Edward Becker, PhD

Wilbur E. Becker
Terrence M. Bedell
James E. Bell
Robert Bell
William A. Bellisle
Steve J. Bennet
Denicke Bennor
Jack C. Bentley
James M. Berge
Gary C. Berliner
Christopher K. Bern
Ray A. Bernard, PhD
Vicky L. Bevilacqua, PhD
Thomas A. Beyerl
Don Black
Mike Blair, PhD
M. Donald Blue, PhD
Barry Bohannon
Jim R. Bone
Arthur S. Booth, Jr.
Charles D. Booth
Edward M. Boothe
Richard E. Boozer
Sorin M. Bota
Spencer R. Bowen
James Ellis Box, PhD
Harry R. Boyd
Paul R. Bradley
George B. Bradshaw
Edward L. Bragg, Sr.
Elizabeth Braham
Greg R. Brandon
J. Allen Brent, PhD
William M. Bretherton, Jr.
Robert N. Brey, PhD
F. S. Broerman
Jerome Bromkowski
Paul C. Broun
Robert E. Brown
Bill D. Browning
William Bruenner
Paul J. Bruner
Sibley Bryan
Richard W. Bunnell
William C. Burnett, PhD
James Lee Butler, PhD
Hubert J. Byrd, III
Sabrina R. Calhoun
Ronnie Wayne Camp
Thomas M. Campbell
James Cecil Cantrell, PhD
Patricia D. Carden, PhD
John L. Carr, III
Laura H. Carreira, PhD
Peter J. Carrillo
Wendell W. Carter
Marshall F. Cartledge
Robert E. Carver, PhD
Billy R. Catherwood
Michael P. Cavanaugh
Matilde N. Chaille
John Champion
Kevin L. Champion
Jack H. Chandler, Jr.
Chung Jan Chang, PhD
Edmund L. Chapman, Jr.
Chellu S. Chetty, PhD
Ken Chiavone
Raghaven M. Chidambaram
George Andrew Christenberry, PhD
Alfred E. Ciarlone, PhD
James A. Claffey, PhD

Mark A. Clements, PhD
Arthur E. Cocco, PhD
Harland E. Cofer, PhD
Charles Erwin Cohn, PhD
Gene Louis Colborn, PhD
James C. Coleman, Jr.
Francis E. Coles, III
P. Jack Collipp
Clair Ivan Colvin, PhD
Leon L. Combs, PhD
Henry P. Conn
David Constans
James H. Cook
Allen Costoff, PhD
Marion Cotton, PhD
Francis E. Courtney
Jack D. Cox
Joe B. Cox
Howard Ross Cramer, PhD
John A. Cramer, PhD
Harry B. Cundiff
Robert A. Cuneo
Randall W. Cunico
Thomas Curin, PhD
Alan G. Czarkowski
Geoffrey Z. Damewood
Ernest F. Daniel
Anne F. Daniels
Charlie E. Daniels
Jagdish C. Das
Shelley C. Davis, PhD
James F. Dawe
Harry F. Dawson
Reuben Alexander Day, PhD*
Juan C. De Cardenas
B. D. Debaryshe
Richard E. Dedels
Johnny T. Deen, PhD
Robert E. Deloach, Jr.
Viral Desai
G. E. Dever, PhD
David Walter Dibb, PhD
John I. Dickinson
John W. Diebold
Herbert V. Dietrich, Jr.
Charles J. Dixon
Hugh F. Dobbins
Harry Donald Dobbs, PhD
Thomas P. Dodson
Clive Wellington Donoho, Jr., PhD
John Dorsey, PhD
Douglas P. Dozier
Thomas E. Driver
Dennis P. Drobny
Earl E. Duckett
Robert L. Dumond
R. E. Dunnells
John W. Duren
James R. Eason
Donald D. Ecker
Teresa Ecker
Stephen W. Edmondson
Marvin E. Edwards, Jr.
Gary L. Elliott
Frampton Ellis
Gary M. Ellis
I. Nolan Etters, PhD
W. E. Evans
Martin Edward Everhard, PhD
Joan H. Facey
Wayne Reynolds Faircloth, PhD
Miguel A. Faria, Jr.

Charles Farley
Christopher J. Farnie
Thomas Farrior
Royal T. Farrow
Michael A. Faten
Mike D. Faulkenberry
John C. Feeley, PhD
Lorie M. Felton
Brent Feske, PhD
Robert Henry Fetner, PhD
Robert Fincher
Burl M. Finkelstein, PhD
Joanna Finkelstein
Reed Edward Fisher
William R. Fisher
J. Ed Fitzgerald, PhD
Carl W. Flammer
Eugene L. Fleeman
K. Fleming, PhD
G. Craig Flowers, PhD
Robert E. Folker
Walter R. Fortner
Gary D. Fowler, Jr.
Steven R. Franco
Thomas Franey
Thomas G. Frangos, PhD
Dean R. Frey
Ralph Fudge
Thomas R. Gagnier
A. Gahr, PhD
Tinsley Powell Gaines
Richard E. Galpin
Gary John Gasche, PhD
George S. Georgalis
Reinhold A. Gerbsch, PhD
Istvan B. Gereben
Lawrence C. Gerow
Georgina Gipson
Charles G. Glenn
Clyde J. Gober
Maria Nelly Golarz de Bourne, PhD
Eugene P. Goldberg, PhD
Earl S. Golightly
Thomas L. Gooch
Walter Waverly Graham, PhD
John B. Gratzek, PhD
James S. Gray, PhD
Robert M. Gray
Clayton Houstoun Griffin
Edward M. Grigsby
Ramon S. Grillo, PhD
William F. Grosser
Erling JR Grovenstein, Jr., PhD
Krishan G. Gupta
Robert E. Hails
Kent W. Hamlin
John R. Haponski
Clyde D. Hardin
James Lombard Harding, PhD
Cliff Hare
Kenneth E. Harper
Joseph Belknap Harris, PhD
Raymond K. Hart, PhD
Lynn J. Harter
Paul V. Hartman
Roger Conant Hatch
Yuichi B. Hattori
Harry T. Haugen
J. Hauger, PhD
William D. Haynes
Thomas D. Hazzard
Charles Jackson Hearn, PhD

Norman L. Heberer
Kenneth F. Hedden, PhD
Walter A. Hedzik
Jeffrey J. Henniger
Gustav J. Henrich
John M. Hester, Jr.
Craig U. Heydon
Donald G. Hicks, PhD
Harold Eugene Hicks
Terry K. Hicks
Thomas J. Hilderbrand
Kendall W. Hill
Warren Ted Hinds, PhD
Roland W. Hinnels
Robert Francis Hochman, PhD
Douglas W. Hochstetler
Dewey H. Hodges, PhD
Kirk Hoefler
Ross Hoffman
William Hogge
Cornelia Ann Hollingsworth, PhD
Leonard Rudolph Howell, Jr., PhD
Chenyi J. Hu
Richard L. Hubbell
James R. Hudson
Bradford Huffines
John Hughes
William C. Humphries
Hugh F. Hunter
Richard Hurd, Jr.
William John Husa, Jr., PhD
James L. Hutchinson, PhD
Al J. Hutko
R. D. Ice, PhD
Walter Herndon Inge, PhD
Donald A. Irwin
James W. Ivey
Dabney C. Jackson
Thomas W. Jackson
David I. Jacob
Stephen B. Jaffe
Vidyasagar Jagadam
Kenneth S. Jago
Jiri Janata, PhD
Robert H. Jarman
Stanley J. Jaworski
Roger D. Jenkins
Wayne Henry Jens, PhD
David R. Jernigan
Ben S. Johnson
Robert H. Johnson, PhD
James P. Jollay
Alan Richard Jones, PhD
Dick L. Jones
E. C. Jones
James K. Jones
Miroslawa Josowicz, PhD
Paul F. Jurgensen
Gerald L. Kaes
Donald C. Kaley
Judith A. Kapp, PhD
M. Katagiri
Ray C. Keause
Raymond J. Keeler
Michael Jon Kell, PhD
William L. Kell, Jr.
Craig Kellogg
Gerald D. Kennett
Warren W. Kent
Joyce E. Kephart
Boris M. Khudenko, PhD
Cengiz M. Kilic

C. Louis Kingsbaker, Jr.
R. C. Kinzie
William H. Kirby, Jr
Adrian F. Kirk
Donald W. Knab
Steven Knittel, PhD
Mark A. Knoderer
David A. Kodl
Juha P. Kokko, PhD
Jenny E. Kopp
Tsu-Kung Ku
Frederick Read Kuc, PhD
Steven B. Kushnick
W. Jack Lackey, PhD
Alexander O. Lacsamana
Myron D. Lair
Trevor G. Lamond, PhD
Willis E. Lanier
Paul H. Laughlin
Leo R. Lavinka
Brian K. Lawrence
Gerald W. Lawson
John C. Leffingwell, PhD
John P. Leffler
Lane P. Lester, PhD
Philip I. Levine
Edward L. Lewis
Kermit L. Likes
Dennis Liotta, PhD
Robert L. Little, PhD
Fred Liu, PhD
Robert Gustav Loewy, PhD
Robert Loffredo, PhD
Stanley Jerome Lokken, PhD
William L. Lomerson, PhD
Earl Ellsworth Long
Justin T. Long, PhD
Larry Lortscher
Jerry Loupee
G. A. Lowerts, PhD
John Lauren Lundberg, PhD
Sarah R. Mack
Joseph Edward MacMillan, PhD
Patrick K. Macy
David A. Madden
Roger Maddocks
James M. Maddry
L. T. Mahaffey
John B. Malcolm
Philip L. Manning
Dale Manos
David E. Marcinko, PhD
Natale A. Marini
John R. Martinec
Arlene R. Martone
Henry Mabbett Mathews, PhD
Walter K. Mathews, PhD
Mike Matis
Curry J. May
Georges S. McCall, II
Neil Justin McCarthy, PhD
Morley Gordon McCartney, PhD
James R. McCord, III, PhD
Burl E. McCosh, Jr.
Sean McCue
Malcolm W. McDonald, PhD
Charles W. McDowell
Larry F. McEver
Ray McKemie
D. K. Mclain, PhD
Archibald A. McNeill
Larry G. McRae, PhD

David Scott McVey, PhD
Thomas R. McWhorter
Andrew J. Medlin, Jr.
Jacob I. Melnik
Michael Menkus
Ronald E. Menze
Allen C. Merritt, PhD
Walter G. Merritt, PhD
Karne Mertins
Glenn A. Middleton
Miran Milkovic, PhD
Jean P. Millen
W. Jack Miller, PhD
R. W. Milling, PhD
Brett A. Mitchell
Carey R. Mitchell
G. C. Mitchell, Jr.
Ralph Monaghan
Carl Douglas Monk, PhD
W. Dupree Moore
Iraj Moradinia, PhD
Stephen T. Moreland
Thomas D. Moreland
Robert G. Morley, PhD
Seaborn T. Moss
Don L. Mueller
Juan A. Mujica
Karl W. Myers
George Starr Nichols, PhD
Jonathan H. Nielsen
Frec C. Nienaber
Ray A. Nixon
Prince M. Niyyar
Stephen R. Noe
Sidney L. Norwood
Susan J. Norwood
James Alan Novitsky, PhD
Wells E. Nutt
Stanley Miles Ohlberg, PhD
Philip D. Olivier, PhD
David R. Olson, PhD
William L. Otwell
Frank B. Oudkirk
Jerry Owen
William E. Owen
Brent W. Owens
Joseph L. Owens
J. Pace, PhD
Ganesh P. Pandya
Les Parker
James Parks
John H. Paterson
William R. Patrick
Daniel D. Payne
Franklin E. Payne
William F. Payne
Terry Peak
Ira Wilson Pence, Jr., PhD
Mel E. Pence
Edgar E. Perrey
Parker H. Petit
Peter J. Petrecca
David W. Petty
Calvin Phillips
John R. Pickett, PhD
Arthur John Pignocco, PhD
Gus Plagianis
Robert V. Plehn
Gayther Lynn Plummer, PhD
Robert A. Pollard, Jr.
Joseph W. Porter
Jerry V. Post

Vijaya L. Pothireddy
Cordell El Prater
Russell Pressey, PhD
Milton E. Purvis
Michael Pustilnik, PhD
Robert E. Rader
Bradford J. Raffensperger
Michael R. Rakestraw
Periasamy Ramalingam, PhD
Stephen C. Raper
Oscar Rayneri
Earnest H. Reade, Jr.
James L. Rhoades, PhD
Robert A. Rhodes, PhD
Peter Rice
Terrence L. Rich
Allan A. Rinzel
Paul W. Risbin
Ken E. Roach
David D. Robertson, PhD
Dirk B. Robertson, PhD
Douglas W. Robertson
John S. Robertson, PhD
David G. Robinson, PhD
Wilbur R. Robinson
Mark E. Robnett
Willie S. Rockward, PhD
Gwenda Rogers
Robert G. Roper, PhD
William J. Rowe
George F. Ruehling
Michael K. Rulison, PhD
James H. Rust, PhD
James Ryan
Marcus Sack
Thomas F. Saffold
George Sambataro
B. Samples
Gary L. Sanford, PhD
Deborah K. Sasser, PhD
Herbert C. Saunders
William E. Sawyer
Kevin Scasny
Kimberly L. Scasny
James R. Schafner
Thomas J. Schermerhorn, PhD
William Schierholz
Robert C. Schlant
Cecil W. Schneider
Michael Charles Schneider, PhD
Terril J. Schneider
Barry P. Schrader
Randy C. Schultz
Robert John Schwartz, PhD
Florian Schwarzkopf, PhD
David Frederick Scott, PhD
Gerald E. Seaburn, PhD
Oro C. Seevers
Robert A. Seitz, PhD
Morton O. Seltzer
Premchard T. Shah
W. A. Sheasin
Andrew J. Shelton
Louis C. Sheppard, PhD
Chung-Shin Shi, PhD
William G. Shira
Joseph John Shonka, PhD
Ronald W. Shonkwiler, PhD
John V. Shutze, PhD
Valery Shver
Steven W. Siegan
Samia M. Siha, PhD

John A. Slaats
Herbert M. Slatton
Earl Ray Sluder, PhD
Joel P. Smith, Jr.
Morris Wade Smith, PhD
Ralph Edward Smith, PhD
Tom Smoot, PhD
Charles C. Somers, Jr.
Richard C. Southerland
Jon A. Spaller
Guy K. Spicer
Firth Spiegel
Charles Hugh Stammer, PhD
Carey T. Stark
Ralph Steger
Fredric Marry Steinberg
John Edward Steinhaus, PhD
Robert J. Stewart
Michael J. Stieferman
Richard C. Stjohn
John C. Stokes
Geo L. Strobel, PhD*
Bill Styer
Joseph H. Summerour
Thomas G. Swanson
Eric Swett
Jose E. Tallet
Darrell G. Tangman
Robert Techo, PhD
Robert L. Terpening
Robert A. Theobald
Peter R. Thomas
M. D. Thompson, PhD
William Oxley Thompson, PhD
Brian Thomson
L. H. Thomson
Francis N. Thorne, PhD
Aloysius Thornton, PhD
Wilfred E. Tinney
Kenneth M. Towe, PhD
Mark Tribby, PhD
Eric Tunison
Nancy Turner
R. W. Turner
Richard Suneson Tuttle, PhD
Noble Ransom Usherwood, PhD
Ahmet Uzer, PhD
James H. Venable
Matthew J. Verbiscer
Jim Vess
Herbert Max Vines, PhD
Terry L. Viness
John Vogel
Carroll E. Voss
Donald L. Voss
Peter Walker
Russell Wagner Walker, PhD
William C. Walker, PhD
Gregory C. Walter
Daniel F. Ward
Fred F. Warden
Raymond Warren
Roland F. Wear
Warren M. Weber
Robert M. Webster
Donald C. Wells
Mike D. Whang, PhD
James Q. Whitaker
Arlon Widder
Richard L. Wilks
Paul T. Willhite
Dansy T. Williams

Carlos A. Wilson
Jeffrey P. Wilson, PhD
Herbert Lynn Windom, PhD
Ward O. Winer, PhD
Alan R. Winn
Robert H. Wise, Jr.
Robert J. Witsell
Monte W. Wolf, PhD
James H. Wood
Scott Wood
Gerald Bruce Woolsey, PhD
Jerry K. Wright, PhD
Neill S. Wyche
Hwa-Ming Yang, PhD
Raymond H. Young, PhD
Andrew T. Zimmerman
Clarence Zimmerman
Glenn A. Zittrauer
D. Zuidema, PhD

Hawaii

John O. Anderson
Jerome K. Bacon
B. Balle
Curtis Beck
Charles Bocage
James Brewbaker, PhD
Richard Brill
Edmond D. Cheng, PhD
Salwyn S. Chinn
John Corboy
Michael Cruickshank, PhD
Charles D. Curtis
Anders P. Daniels, PhD
Harry Davis, PhD
Walter Decker
Guy H. Dority, PhD
Arlo Wade Fast, PhD
Masanobu R. Fujioka
William Gerogi
Susan K. Gingrich
Brian L. Gray
Lloyd Grearson
James A. Griffith
Benjamin C. Hablutzel
Mark Hagadone, PhD
John C. Hamaker, PhD
Kirk Hashimoto
Philip Helfrich, PhD
Philip D. Hellreich
Fred Hertlein, III
Yoshitsugi Hokama, PhD
David R. Howton, PhD
Lester M. Hunkele
James Ingamells, PhD
Glenn Jensen
Richard A. Jurgensen
Takashi Theodore Kadota, PhD
Yong-Soo Kim, PhD
Ronald H. Knapp, PhD
Adelheid R. Kuehnle, PhD
Michael Lane
Joel S. Lawson, PhD
Alan Lloyd
John A. Love
Claus Berthold Ludwig, PhD
Charles Lavern Mader, PhD
Charles K. Matsuda
Edward R. McDowell, PhD
Martin McMorroa

James Mertz
John P. Mihlbauer
Dave Miller
Gabor Mocz, PhD
David R. Moncrief
William F. Moore
Justus A. Muller
Paul D. Nielson
Richard L. O'Connell
Steven Emil Olbrich, PhD
Carlos A. Omphroy
Panagioti Prevedouros, PhD
Edison Walker Putman, PhD
Raymond C. Robeck
Kenneth G. Rohrbach, PhD
Charles G. Rose
James C. Sadler
Arthur A. Sagle, PhD
Narendra K. Saxena, PhD
Jon H. Scarpino
Benjamin R. Schlapak
Steven Seifried, PhD
Rogert Sherman
Shoji Shibata, PhD
Gerry Steiner, PhD
Ramon K. Sy
Michelle Teng, PhD
S. A. Whalen
Gary White
Stephen Wilson
Oliver Wirtki
William K. Wong
Cleveland C. Wu
Alexander Wylly, PhD
Klaus Wyrtki, PhD
Alfred A. Yee

Idaho

Stephen B. Affleck, PhD
Kimbol R. Allen
Wilfred L. Antonson
Elton E. Arensman
Marvin D. Armstrong, PhD
Adrian Arp, PhD
Jim F. Ashworth
Carl Fulton Austin, PhD
George Babits
J. Brent Bagshaw
Craig Riska Baird, PhD
Brent O. Barker
LeRoy N. Barker, PhD
Douglas D. Barman
Justin P. Bastian
Colin J. Basye
Miles F. Beaux, II
Wiley F. Beaux
Matthew A. Beglinger
Donald R. Belville
Glen E. Benedict
David R. Berberick
Julius R. Berreth
Darvil K. Black, PhD
Willis J. Blaine
Wilson Blake, PhD
Richard C. Bland
George Bloomsburg, PhD
Joe Bloomsburg
Carl R. Boehme
John Roy Bower, Jr., PhD
David D. Boyce

Fred W. Brackebusch
Ken N. Brewer
Timothy J. Brewer
Randall A. Broesch
James L. Browne
Shelby H. Brownfield
Merlyn Ardel Brusven, PhD
Brent J. Buescher, PhD
William M. Calhoun
David Lavere Carter, PhD
Thomas Cavaiani, PhD
William R. Chandler
L. F. Cheng, PhD
Grant S. Christenson
Randil L. Clark, PhD
James W. Codding
Joseph A. Coffman
Guy M. Colpron
Michael Cook
Carl Corbit
Ronald C. Crane, PhD
Richard A. Cummings
Raymond C. Daigh
Kim B. Davies
Karl Dejesus, PhD
Robert F. Denkins
Thomas R. Detar
Melvin L. Dewsnup
Mark C. Dooley
Stanley L. Drennan
Larry A. Drew, PhD
Merlyn C. Duerksen
O. Keener Earle
William J. Farrell
Matthew R. Fein
Donald N. Fergusen
Roger Ferguson
Ferol F. Fish, PhD
Dale Fitzsimmons
Richard E. Forrest
Harry Kier Fritchman, PhD
Gary R. Gamble
Robert L. Geddes
Paul L. Glader
Doug J. Glaspey
Donald W. Glenn
V. L. Goltry
Hans D. Gougar, PhD
John G. Haan
Lloyd Conn Haderlie, PhD
Richard Hampton, PhD
Julie J. Hand
Leonid Hanin, PhD
Evan Hansen, PhD
Gerald H. Hanson
Robert N. Hanson
Roger Wehe Harder
Herbert E. Harper
Tim Harper
Herbert J. Hatcher, PhD
John B. Hegsted
Richard Charles Heimsch, PhD
Brian G. Henneman
Robert A. Hibbs, PhD
Richard C. Hill
Robert F. Hill
Larry Hinderager
Kenneth M. Hollenbaugh, PhD
Clifford Holmes
Harold E. Horne
Case J. Houson
John E. Howard

K. B. Howard
Arthur R. Hubscher
Allan S. Humpherys
Richard M. Hydzik
Bert W. Jeffries
C. Thomas Jewell
Janard J. Jobes
Timothy J. Johans
Donald Ralph Johhnson, PhD
James B. Johnson, PhD
Kent R. Johnson, PhD
Lawrence Harding Johnston, PhD
James J. Jones
Jitendra V. Kalyani
Robert W. Kasnitz
Gary J. Kees
Fenton Crosland Kelley, PhD
Paul C. Kowallis
Stephen Kronholm
Philip M. Krueger, PhD
John S. Kundrat
Tim M. Lawton
George J. LeDuc
John W. Leonard
Leroy Crawford Lewis, PhD
William Little, PhD
Richard Luke
Robert A. Luke, PhD
Gary K. Lund, PhD
Patricia A. Maloney
Billy L. Manwill
Jon L. Mason
Jeremiah Mc Carthy, PhD
Kent McGarry
Mark A. McGuire, PhD
Bill McIlvanie
Twylia J. McIlvanie
Michael E. McLean
Brent A. McMillen
Scott G. McNee
Reginald E. Meaker
Donald L. Mecham
Travis W. Mechling
William L. Michalk
Ellis W. Miller, PhD
James A. Miller
Dile J. Monson
John L. Morris, PhD
Jack Edward Mott, PhD
George Murgel, PhD
Thomas J. Muzik, PhD
Michael J. Myhre
Dave Nabbefeld
Roger Donald Nass
Thomas C. Neil, PhD
Doug W. New
Daniel Nogales, PhD
Randy Noriyaki
Mark J. Olson
William Joseph Otting, PhD
Bruce R. Otto
Gerald G. Overly
Aida Patterson
Ed L. Payne, PhD
Raymond M. Petrun
Gene W. Pierson
Harold A. Powers
Vernon Preston
Alan L. Prouty
Thomas R. Rasmussen
Bill Richards
Robert E. Rinker, PhD

Eric P. Robertson
Arthur P. Roeh
Jodie Roletto
Ralph F. Russi
Jerry F. Sagendorf
Charles Sargent
David Laurence Schreiber, PhD
Kevin L. Schroeder
Carl W. Schulte
Sidney J. Scribner
Francis Sharpton, PhD
Glen M. Sheppard
George H. Silkworth
Claude Woodrow Sill
Jay Hamilton Smith, PhD
John Wolfgang Smith, PhD
Thomas H. Smith, PhD
Wesley D. Smith, PhD
Roger F. Sorenson
Ross A. Spackman
John C. Spalding
Barbara Spengler
Thomas Speziale, PhD
Gene E. Start
Raymond J. Stene
Bruce W. Stoddard
Marc Stromberg, PhD
Wilfried J. Struck
Dale Stukenholtz, PhD
Richard A. Suckel
David O. Suhr
Donald C. Teske
Gary K. Thomas
Warren N. Thompson
James P. Todd
Francois D. Trotta
Leland Wayne Tufts
John H. Turkenburg
Edmund Eugene Tylutki, PhD
Joseph James Ulliman, PhD
Randy K. Uranes
Bingham H. Vandyke, Jr., PhD
Chien Moo Wai, PhD
Thornton H. Waite
Leslie M. Walker
Jon H. Warner
Mont M. Warner, PhD
D. T. Westermann, PhD
John E. Wey, Jr.
Jack Weyland, PhD
Robert N. Wiley
George L. Wilhelm
S. Curtis Wilkins
Oliver S. Williams
Roy L. Wise
George D. Wood
Ryan Royce Yee
Austin L. Young
Mark Yuly, PhD
Stephen P. Zollinger, PhD

Illinois

Refaat A. Abdel-Malek, PhD
Eric R. Adolphson
Kenneth Agnes
Sean R. Agnew
Vincent M. Albanese
Rudolph C. Albrecht
Charles W. Allen, PhD
Albert L. Allred, PhD

Kent A. Alms
Duane R. Amlee
Dewey Harold Amos, PhD
Thomas A. Amundsen
David A. Anderson
Joel Anderson
Kevin P. Ankenbrand
Herbert S. Appleman
Douglas E. Applequist, PhD
Anatoly L. Arber, PhD
Ed Arce
Robert W. Arends
Lawrence Ariano
Ralph Elmer Armington, PhD
Aaron J. Arnold
Jaime N. Aruguete
Joseph J. Arx
Warren Cotton Ashley, PhD
Erika J. Atkinson
Keith Atkinson
Luben Atzeff
Darrel D. Auch
Brad August
Arthur J. Avila
Teresita D. Avila
Alison M. Azar
Harold Nordean Baker, PhD
Louis Baker, PhD
Marion John Balcerzak, PhD
Andrew Ateleo Baldoni, PhD
R. M. Bales
Mike A. Banak
Seymour George Bankoff, PhD
Ernest Barenberg, PhD
Charles Barenfanger
John Barney
Lawrence J. Barrows, PhD
Eugene Barth
Douglas B. Bauling
Linda L. Baum, PhD
Richard C. Baylor
Gary Beall, PhD
Rhett Bearmont
Craig A. Beck
Robert P. Becker, PhD
O. Beggs
Albert J. Behn, PhD
Ihor Bekersky, PhD
Gary Lavern Beland, PhD
George Bennett, PhD
Lawrence Uretz Berman
Curtis L. Bermel
Daniel S. Berry
Daniel Best
Herbert W. Beyer
Debanshu Bhattacharya, PhD
John W. Bibber, PhD
Carl Jeffrey Biederstedt
Gregg Bierei
Roger A. Billhardt
M. J. Birck
J. L. Bitner
Thomas E. Blandford
William R. Blew
Robert L. Blood
Loren E. Bode, PhD
Daniel G. Bodine
Robert J. Boehle
Thomas M. Bohn
Mark S. Boley, PhD
Patrick V. Bonsignore, PhD
Lewis D. Book

David L. Booth, PhD
Edgar A. Borda
Harold Joseph Born, PhD
Joseph Carles Bowe, PhD
Robert E. Boyar
R. Braatz, PhD
Robert Giles Brackett, PhD
Dorothy L. Bradbury
Eric M. Bram
Thomas Brandlein
William H. Bray
William Thomas Brazelton, PhD
Manuel Martin Bretscher, PhD
Gregory J. Brewer, PhD
Joseph L. Brewer, PhD
Michael K. Brewer
Mark O. Brien
Steve Briggs, PhD
Robert B. Brigham
Bruce Edwin Briley, PhD
Ed A. Brink
Douglas A. Brockhaus
Mark Bronson
David P. Brown, PhD
William Brown
John S. Brtis
Eldon J. Brunner
Gary G. Bryan
Ronald C. Bryenton
Gregory Buffington
Philip G. Buffinton
Michael Bundra
Ivan L. Burgener
Marty Burke
Peter G. Burnett
Donald S. Burnley
Mike Burnson
William C. Burrows, PhD
Rodney L. Burton, PhD
William H. Busch
Robert Busing
Duane J. Buss, PhD
Margaret K. Butler
Gary J. Butson, PhD
Ralph O. Butz, Jr.
James Albert Buzard, PhD
Roy Byrom
Fernando S. Caburnay
Manual Calzada
Marvin E. Camburn, PhD
Howard S. Cannon, PhD
William L. Carper
Robert C. Casagranda
Phillip M. Caserotti
J. Steven Castleberry
Stanley A. Changnon, Jr.
Michael Andrew Chaszeyka
Robert L. Cheever
A. Cheney
J. A. Cifonelli, PhD
Leroy Clardy
Robert Clark
Russell H. Clark
David G. Clay
James M. Clifford
Alan Clodfelter
Jim Cloud
Benjamin T. Cockrill, Jr.
Fritz Coester, PhD
Allan H. Cohen, PhD
William C. Cohen, PhD
Paul D. Coleman, PhD

Keith A. Collins
Richard A. Comroe, PhD
Paul J. Concepcion
John Conconnan
Patrick Condon
Dennis D. Conner
Gail Rushford Corbett, PhD
Rebecca Corbit
Stephen Watson Cornell, PhD
Thomas Corrigan, PhD
Khalid Cossor
Phillip G. Costantinou
Gordon E. Craigo
Frederick L. Crane, PhD
Donald W. Creger, Sr.
John Edwin Crew, PhD
Gregory A. Crews
George A. Criswell
Michael Summers Crowley, PhD
Andrew A. Cserny, PhD
John Robert Culbert
Peter L. Cumerford
Ronald L. Cutshall, Sr
Russell A. Dahlstrom
Heinz H. Damberger, PhD
Philip Danielson
Morris Juda Danzig, PhD
Kaz Darzinskis
Jeff David
Lyndon L. Dean
Robert H. Dean
Donn Dears
Edward Dale Deboer
Frank Deboer, PhD
Charles S. Dehaan
Robert F. Deibel, Jr.
Manuel J. Delerno
Thomas G. Denton
Edward M. Desrochers, Jr.
Richard DeVries
Donald Dewey, PhD
James Dewey
Justin E. Dewitte
Charles Edward Dickerman, PhD
Bryan J. Dicus
Ray C. Dillon
Timothy W. Dittmer
Omer H. Dix
W. Brent Dodrill
Theodore Charles Doege
Michael F. Dolan
Robert A. Dolehide
Richard J. Dombrowski
Otello P. Domenella
Wayne Donnelly
John E. Dowis
Gerald L. Downing
David R. Doyle
William K. Drake
G. L. Dryder, PhD
John T. Dueker
Carl D. Dufner
John D. Dwyer
Michael Dwyer
Steven Dyer
C. Dykstra, PhD
Philip J. Dziuk, PhD
Philip Eugene Eaton, PhD
Lawrence Thornton Eby, PhD
Charles Ecanow, PhD
Donald E. Eckmann
W. Kent Eden

Paul H. Egbers
Donald A. Eggert, PhD
William W. Elam, PhD
James O. Ellis, Sr.
George Emerle
Louis Emery, PhD
David Engwall, PhD
Jack H. Enloe
Edward Bowman Espenshade, PhD
R. Evans
Don J. Fanslow, PhD
Marjorie Whyte Farnsworth, PhD
Wells E. Farnsworth, PhD
David A. Fehr
Imre M. Fenyes
Joseph Fidley, PhD
Joanne K. Fink, PhD
Patricia Ann Finn, PhD
Ross Francis Firestone, PhD
Robyn Fischer
Steve Fiscor
Emalee G. Flaherty
Donald R. Fletcher
James M. Fox
Stephen Joseph Fraenkel, PhD
Julian Myron Frankenberg, PhD
Raymond Frederici
David M. Frederick
Richard C. Frederick
Allan L. Freedy
Donald Nelson Frey, PhD
Herbert C. Friedmann, PhD
Andrew R. Frierdich
Karl J. Fritz, PhD
Wesley Fritz, PhD
Aaron E. Fundich
Jim Gadwood
John Gaither
Thomas R. Galassi
Charles O. Gallina, PhD
Daniel G. Ganey
Douglas L. Garwood, PhD
Eduardo Gasca
George Gasper, PhD
Roger W. Geiss
William O. Gentry, PhD
Boyd A. George, PhD
George J. Getty
Camillo Ghiron, PhD
Frederick W. Giacobbe, PhD
Larry V. Gibbons, PhD
Jeff Gillespie
John R. Gilmore, PhD
Kenneth J. Ginnard
George Isaac Glauberman, PhD
William D. Glover
William Gong, PhD
William Good
Harold J. Goodman
Joseph Gorsic, PhD
Chris D. Gosling
George Robert Goss, PhD
David G. Gossman
Samuel P. Gotoff
Marcus S. Gottlieb
Robert J. Graham, PhD
William K. Graham
Ted R. Gray
Frank D. Graziano, PhD
Bruce E. Greenfield
D. W. Greger
David Gregg

Gregg T. Greiner
Edward Lawrence Griffin
Harold Lee Griffin
James Edward Griffin, PhD
Ronald E. Groer
Eugene E. Gruber, PhD
Andrew F. Guschwan
Kenneth A. Gustafson
Philip Felix Gustafson, PhD
David Solomon Hacker, PhD
Richard H. Hagedorn
Reino Hakala, PhD
Ken Hall
Suleiman M. Hamway
Daniel F. Hang
Mark E. Hansen
A. O. Hanson, PhD
Robert K. Harbour
Thomas D. Harding
Danny L. Hare
Ben Harrison, PhD
Thomas R. Harwood
John A. Hasdal, PhD
Hans Hasen
Robert Havens
Douglas B. Hayden
Robert J. Heaston, PhD
David Robert Hedin, PhD
Todd Hedlund
Mihaela Hegstrom
Lawrence C. Heidemann
Stan F. Heidemann
Paul K. Heilstedt
Ron Heisner
James Raymond Helbert, PhD
Roy J. Helfinstine
Joseph B. Helms
William C. Helvey
John F. Helwig
Jeannine L. Henderson
Alfred J. Hendron, Jr., PhD
Lester Allan Henning
Richard Henry, PhD
Arthur A. Herm, PhD
Edward Robert Hermann, PhD
Ernest Carl Herrmann
Cynthia Hess, PhD
Karl Hess, PhD
Kenneth E. Hevner
Menard George Heydanek, Jr., PhD
Warren R. Higgins
Raymond M. Hinkle, PhD
Terry L. Hird
Joseph C. Hirschi
Richard A. Hirschmann
Bill J. Hlavaty
Vincent Hodgson, PhD
Timothy J. Holcomb
Gary L. Hollewell
Betsy L. Holli, PhD
Franklin Ivan Honea, PhD
Arie Hoogendoorn, PhD
Doc Horsley, PhD
Craig Horswill, PhD
George E. Hossfeld
Kathleen Ann House, PhD
Robert M. House
Ronald L. Howard
T. E. Hsiu, PhD
Charles Fu Jen Hsu, PhD
Jean Luc Hubert
Craig S. Huff

Dennis Huffaker
Donald C. Huffaker, PhD
Stanley R. Huffman
James A. Huizinga
James Hulvat, PhD
John Gower Hundley, PhD
Stephen A. Hutti
Icko Iben, Jr., PhD
Mac Igbal
Nick Iliadis
Cecil W. Ingmire
Ronald S. Inman
Mitio Inokuti, PhD
Herbert O. Ireland, PhD
Gary J. Isaak
Anthony D. Ivankovich, PhD
David Eugene James, PhD
Lynn M. Janas, PhD
Edwin P. Janus
Frank Henry Jarke
Gerald G. Jelly
William J. Jendzio
Stewart C. Jepson
Glenn Richard Johnson, PhD
Irving Johnson, PhD
Milton Raymond Johnson, PhD
Rodney B. Johnson
Helen S. Johnstone
John Lloyd Jones, PhD
Les Jones
Daniel R. Juliano, PhD
William A. Junk, PhD
George T. Justice
Dennis D. Kaegi
John E. Kaindl
John M. Kalec
Elisabeth M. Kaminsky
Manfred Stephan Kaminsky, PhD
Fred Kampmans
Harvey Sherwin Kantor
George Kapusta, PhD
Dorkie Kasmar
Joseph J. Katz, PhD
R. Gilbert Kaufman, PhD
Michael Kazarinov, PhD
Frederick D. Keady
Clint J. Keifer
Walter R. Keller
Charles D. Kellett
Frank N. Kemmer
Francis S. Kendorski
Albert Joseph Kennedy, PhD
Marina Kennelly, PhD
Robert Kepp, PhD
Naaman Henry Keyser
Gregory B. Kharas, PhD
G. Khelashvili, PhD
John W. Kiedaisch, PhD
John W. Kieken
Myoung D. Kim
Sanford MacCallum King, PhD
Timothy P. Kinsley
John D. Kinsman
Betty W. Kjellstrom
Harold E. Klemptner
James C. Klouda
Edward Andrew Knaggs
Norm Knights
James Otis Knobloch, PhD
James W. Knox
David E. Koehler
James Start Koehler, PhD

Janice H. Koehler
Richard W. Koester
Randall Kok, PhD
Tim Koller
Sam Kongpricha, PhD
Virgil J. Konopinski
Mark J. Konya
Kevin D. Kooistra
Gerald L. Kopischke
Dennis N. Kostic
William Kowal
Kevin Koyle
John J. Krajewski, PhD
John F. Krampien
James J. Krawchuk
Arthur A. Krawetz, PhD
Robert E. Kreutzer, Jr.
Kevin Krist, PhD
Arthur T. Kroll
Moyses Kuchnir, PhD
Eugene James Kuhajek, PhD
Moira Kuhl
Samar K. Kundu, PhD
Walter E. Kunze
Peter Kusel, PhD
Kenneth M. Labas
James R. Lafevers, PhD
Charles Ford Lange, PhD
Ralph Louis Langenheim, PhD
John L. Lapish
Andre G. Lareau
Seymour Larock
Bruce Linder Larson, PhD
Carl S. Larson, PhD
Michael J. Larson
Wayne O. Larson, PhD
Eugene J. Lawrie
James L. Leach, PhD
Dennis R. Lebbin, PhD
Jean-Pierre Leburton, PhD
Richard V. Lechowich, PhD
Charles Lee
Do B. Lee
Gregory E. Lehn
Charles Leland
James D. Lenardson
Terrence A. Leppellere
Dennis Leppin
Ronald G. Leverich
Terrence G. Leverich
Lawrence T. Lewis, PhD
Steven R. Lewis
Ted E. Lewis
Robert D. Libby
David L. Licht
Rebecca A. Lim, PhD
Henry Robert Linden, PhD
William E. Liss
Kenneth J. Little
Chian Liu, PhD
Derong Liu, PhD
Qian Liu, PhD
Stephen R. Lloyd, PhD
John T. Loftus
Roger A. Logeson
John S. Loomis, Jr.
Herb Lopatka
Janet A. Lorenz, PhD
Paul Albert Lottes, PhD
John E. Lovell
Walter S. Lucas
Spomenka M. Luedi

Thomas J. Lukas, PhD
Kenneth P. Lundgren
Channing Harden Lushbough, PhD
Mark R. Lytell, Jr.
Richard W. Lytton
E. Jerome Maas, PhD
Roy P. Mackal, PhD
Arvind J. Madhani, PhD
Alexander B. Magnus, Jr.
James Magnuson
Om Prakash Mahajan, PhD
Eric L. Malaker
Francis M. Mallee, PhD
Leszek L. Malowanski
Christian John Mann, PhD
Carl A. Manthe
Dennis L. Markwell
Donald Paul Martin
Ronald Lavern Martin, PhD
Richard Martinez
Edward L. Marut
Donald Frank Mason, PhD
Richard I. Mateles, PhD
John Mathys, PhD
Michael Matkovich
Robert R. Mazer
Laurence R. Mcaneny, PhD
Ed McCanbe
Ann E. McCombs
Charles R. McConnell, PhD
Sherwood W. McGee
Michael J. McGirr
Randall K. McGivney
Ellen A. McKeon
Peter McKinney
James R. McVicker
Ralph D. Meeker, PhD
Joseph C. Mekel
Kenneth E. Mellendorf, PhD
Edgar L. Mendenhall
Jim Mendenhall
Richard C. Meyer, PhD
Stuart Lloyd Meyer, PhD
Gary L. Miles
Thomas J. Miller
David Mintzer, PhD
Alex Mishulovich, PhD
Vyt Misiulis
Robert H. Mohlenbrock, Jr., PhD
Jeff Monahan, PhD
Raymond W. Monroe
Gregory E. Morris
Robert A. Morris
Gary E. Mosher
Herman E. Muller
Paul E. Mullinax
Dan J. Muno
Alan E. Munter
Bichara B. Muvdi, PhD
John L. Nanak, Jr.
David Ledbetter Nanney, PhD
Thomas F. Nappi
Richard M. Narske, PhD
Joseph J. Natarelli
J. Timothy Naylor
Dale D. Nelson
Harry J. Neumiller, Jr., PhD
Richard William Neuzil
Marcella G. Nichols
Thomas W. Nichols
John Nicklow, PhD
John D. Noble, PhD

Jeffrey J. Nodorft
William L. Nold
Michael L. Norman, PhD
Richard Daviess Norman
Jack D. Noyes
Alan H. Numbers
Bernard C. Ogarek
Fredric C. Olds*
Lawrence Oliver, PhD
Farrel John Olsen, PhD
Larry S. Olthoff
Michael P. O'Mara
William O'Neill, PhD
William C. Orthwein, PhD
Ali M. Oskoorouchi, PhD
Joseph M. O'Toole
Jacques Ovadia, PhD
Carl F. Painter, PhD
Nicholas J. Pappas
Eugene N. Parker, PhD
James D. Parker
Kirit Patel
Natu C. Patel
Clinton P. Patterson
Val E. Peacock, PhD
Barry E. Pelham
David W. Pennington
Roscoe L. Pershing, PhD
Peter Paul Petro, PhD
Dave Pettit
Thomas E. Phipps, PhD
Susan K. Pierce, PhD
Charles Edward Pietri
Glen A. Piland
Dean A. Pilard
Richard A. Pilon
Roger D. Pinc
Mark A. Pinsky, PhD
Chris Piotrowski
Ed Piszynski
Rus Pitch
Philip Andrew Pizzica
Robert J. Podlasek, PhD
Richard Polad
Donald W. Potter
John E. Powell, Jr.
Robert C. Powell, PhD
G. Gary Preston
George W. Price
Michael R. Prisco, PhD
William H. Prokop
David L. Puent
Kay M. Purcell
Kathy Qin
Jim Quandt
Charles E. Quentel
Robert D. Quinn
Carl G. Rako
George Ramatowski
James Rasor, II
Andrew Rathsack
Kent Rausch, PhD
Richard G. Rawlings
Sylvian Richard Ray, PhD
Fred Reader
Craig R. Reckard
Harold Frank Reetz, PhD
Gary K. Regan
Leonard Reiffel, PhD
Vincent Reiling, PhD
Paul G. Remmele
Joseph C. Renn

James M. Richmond, PhD
Bill Tom Ridgeway, PhD
George Roy Ringo, PhD
John V. Roach
Arthur Roberts, PhD
Jene L. Robinson
Jacob Van Roeckel
James A. Roecker
George A. Roegge
Donna L. Rook
Kelly Roos, PhD
Andrew H. Rorick
Albert R. Rosavana
David Rose, PhD
Eric J. Rose
Richard A. Rosenberg, PhD
Peter S. Rosi
Gordon Keith Roskamp, PhD
Timothy Rozycki
George G. Rudawsky
Thomas A. Rudd
Edward Evans Rue
John Ruhl
Timothy J. Rusthoven
Mark Ruttle
Julian G. Ryan
Debra J. Rykoff, PhD
Gregory Ryskin, PhD
George D. Sadler, PhD
M. M. Said, PhD
James A. Samartano
Dennis R. Samuelson
Paul D. Sander, PhD
Mykola Saporoschenko, PhD
Satish Chandra Saxena, PhD
Angelo M. Scanu
Harold E. Scheid
Jay Ruffner Schenck, PhD
Roger M. Schiavoni
Alexander B. Schilling, PhD
Robert Arvel Schluter, PhD
William A. Schmucker
Charles David Schmulbach, PhD
Richard Schockley
Felix Schreiner, PhD
Richard J. Schuerger, PhD
Garmond Gaylord Schurr
Robert W. Schwaner
John J. Sciarra
Paul A. Seaburg, PhD
J. Glenn Seay
Bruce R. Sebree, PhD
Otto F. Seidelman
Lewis S. Seiden, PhD
Richard G. Semonin
Lee H. Sentman, PhD
Gordon E. Sernel
Charles Shabica, PhD
Alan R. Shapiro
John D. Shepherdson
Charles P. Sheridan
Arthur J. Sherman
Douglas B. Sherman
Joseph Cyril Sherrill, PhD
Sanjeev G. Shroff, PhD
James C. Shults
Lawrence G. Sickels
Kent Sickmeyer
Arthur Siegel
Chester Paul Siess, PhD
Gaston N. Siles
Walter Lawrence Silvernail, PhD

Joseph H. Simmons
Ralph O. Simmons, PhD
Gerald Simon
Donald E. Sims
Matthew D. Sink
Amor Sison
John R. Skelley
Gary M. Skirtich
Francis J. Slama, PhD
Arthur Smith
Roger C. Smith*
Ned L. Snider
Richard H. Snow, PhD
Duane Snyder
Milton A. Sobie
Harry J. Soloway
Dennis B. Solt, PhD
Robert E. Sorensen
Peter Sorokin
Chris Soule
Edmund S. Sowa
Brian Spencer
James W. Spires
Mark W. Sprouls
William R. Staats, PhD
Stephen S. Stack
Thomas R. Stack
Paul Stahmann
J. E. Stanhopf
E. J. Stanton
Peter J. Stanul
Stephen E. Stapp
Timothy W. Starck
John J. Staunton
Enrique D. Steider
Frank Jay Stevenson, PhD
D. Scott Stewart, PhD
Robert G. Stewart
Robert Llewellyn Stoffer, PhD
David H. Stone, PhD
Bob Storkman
Glenn E. Stout, PhD
Nathan Stowe
Randy E. Strang
James F. Stratton, PhD
Stephen A. Straub, PhD
Edward A. Streed
Joseph W. Stucki, PhD
Benjamin G. Studebaker
Alan M. Stueber, PhD
John Stueve
Frank C. Suarez
Thomas E. Sullivan
Armando Susmano
Christine Sutton
R. Kent Swedlund
Jerry Sychra, PhD
Steven A. Szambaris
Ajit C. Tamhane, PhD
Rita Tao
Robert E. Teneick, PhD
Lee C. Teng, PhD
Marius C. Teodorescu, PhD
Ram P. Tewari, PhD
W. Tharnish
Carol Tharp
Lawrence Eugene Thielen
Otto G. Thilenius, PhD
Kris Thiruvathukal, PhD
Brian G. Thomas, PhD
John D. Thomas
Don E. Thompson

Stacy T. Thomson
Thomas A. Thornburg
Peter E. Throckmorton, PhD
E. Timson
A. Dudley Tipton
August P. Tiritilli
Cyril M. Tomsic
Kenneth James Trigger
Alkesh N. Trivedi
Edwin S. Troscinski
Alvah Forrest Troyer, PhD
Larry S. Trzupek, PhD
Richard Trzupek
Donald Tuomi, PhD*
Anthony Turkevich, PhD
James J. Ulmes
Ronald T. Urbanik
Larry D. Vail, PhD
Albert P. Van der Kloot
William Van Lue
Donald O. Van Ostenburg, PhD
William Brian Vander Heyden, PhD
Donald D. Vanfossan, PhD
Albert L. Vanness
Edmund Vasiliauskas, PhD
Ira J. Vaziri
John G. Victor, PhD
Larry L. Vieley
Francis J. Vincent
Robert W. Voedisch
Edward D. Vojack
Edward W. Voss, PhD
Mike Wadlington
Howard L. Wakeland
Jack Wakeland
Jeff Walker
John W. Walker
Rodger Walker
Daniel J. Walters
Soo-Young C. Wanda
R. Wanke, PhD
Evelyn Kendrick Wantland, PhD
Brent A. Ward
Toby Ward
Lewis R. Warmington
David D. Warner
James Warner
Bruce Warren
Donald L. Watson
Jason A. Watters
David L. Webb
George E. Webb, Jr.
David Fredrick Weber, PhD
James A. Weber
Olaf L. Weeks
William B. Welch
Eric Welles
Daniel Wenstrup
William A. Weronko
John R. Wesley
Hans U. Wessel, PhD
Douglas Brent West, PhD
Paul J. West
James William Westwater, PhD
Robert Lloyd Wetegrove, PhD
Grady A. White
Laurie C. Wick
Timothy H. Wiley
Arthur R. Williams
Jack M. Williams, PhD
Jack R. Williams
Thomas Williams, PhD

Tom Willibey
Alan P. Wilson
Lorenzo Wilson
R. Wilson, PhD
Carman P. Winarski
Ronald John Wingender, PhD
Robert W. Wingerd
Dale M. Winter
Lester H. Winter
Robert Wisbey
David C. Witkins
Lloyd David Witter, PhD
Dale Wittmer, PhD
Edward Wolf
George D. Wolf
Gene H. Wolfe
Clarke K. Wolfert
Sylvia Wolfson
Sophie M. Worobec
Virgil A. Wortman
Judy L. Wright
Leo A. Wrona
Jaroslav Wurm, PhD*
Peter S. Wyckoff
William P. Yamnitz
X. Terry Yan, PhD
John Herman Yopp, PhD
Don York
Susan D. Younes
Donald E. Young, PhD
Forrest A. Younker
Hany H. Zaghloul
Edward P. Zahora
Zhonggang Zeng, PhD
Arthur J. Zimmer, PhD

Indiana

Bernaard J. Abbott, PhD
Paul Abbott
Chris Adam
Robert Aldridge
Gabriel C. Aldulescu
Joshua C. Allen
Madelyn H. Allen
Jonathan Alley
Lawrence D. Andersen
Eric Anderson
Walter C. Anglemeyer
Morris Herman Aprison, PhD
John Arch
Thomas G. Armbuster
Delano Z. Arvin, PhD
Joe E. Ashby, PhD
Romney A. Ashton
Robert Aten, PhD
Mark D. Atkins
Matthew R. Atkinson
Christian C. Badger
Tim Baer
James E. Bailey
Edward J. Bair, PhD
Dale I. Bales, PhD
David L. Banta
Mark A. Bartlow
Norris J. Bassett
Walter Frank Beineke, PhD
Mark R. Bell, PhD
James Noble Bemiller, PhD
Paraskevi Mavridis Bemiller, PhD
Edward J. Benchik

John M. Bentz
Rod Bergstedt
William P. Best
Robert J. Beyke
Lawrence M. Bienz
Robert Francis Bischoff
Kenneth A. Bisson
Raymond J. Black
Jon Blackburn
Paul V. Blair, PhD
George B. Boder
James M. Bogner
James K. Boomer
W. Bordeaux
Andrew Chester Boston, PhD
Steve Boswell
Edmond Milton Bottorff, PhD
James L. Bowman
Charles A. Boyd
D. Boyd, PhD
Frederick Mervin Boyd, PhD
Raymond D. Boyd
Richard A. Boyd
Redford H. Bradt, PhD
John T. Brandt
Harold W. Bretz, PhD
Tom P. Brignac
Raymond Samuel Brinkmey, PhD
Thomas R. Brinner, PhD
Arlen Brown, PhD
Herbert Brown, PhD
James Brown
John W. Brown
Cornelius Payne Browne, PhD
W. George Brueggemann
Robert F. Bruns, PhD
Russell Allen Buchanan
Stuart Buckmaster
Bradley Burchett, PhD
Kim A. Burke
John Burwell
Tom Busch
Richard W. Butler
John C. Callender
Ernest Edward Campaigne, PhD
Randy S. Cape
Michael E. Caplis, PhD
Eric S. Carlsgaard
Donald D. Carr, PhD
Jerry Allan Caskey, PhD
Stanley W. Chernish
Dhan Chevli, PhD
David L. Christmas
Ellsworth P. Christmas, PhD
John P. Chunga
James M. Clark
Waller S. Clements
David L. Clingman
Roberto Colella, PhD
Christina M. Collester
Don Collier
Andy R. Collins
P. Connolly
Addison G. Cook, PhD
Donald Jack Cook, PhD
Tracy Correa, PhD
Robert J. Corsiglia
Peter A. Costisick
Mike Cox
Donald K. Craft
Tom Crawford, PhD
Robert Crist

David F. Crosley
Marsha L. Culbreth
Stephen L. Cullen
J. William Cupp, PhD
John T. Curran
Theodore Wayne Cutshall, PhD
James M. Czarnik
Marie E. Dasher
Bruce R. Dausman
David G. Davidson
Harold William Davies, PhD
Paul F. Davis
Harry G. Day
Mark E. Deal
David J. Dean
Steve Denner, PhD
Kenneth R. Deremer
Kenneth R. Devoe
Shree S. Dhawale, PhD
Charles V. Di Giovanna, PhD
Tom Dingo
H. Marshall Dixon, PhD
Gerald Ennen Doeden, PhD
Joe Dotzlaf
Steven L. Douglas
David W. Dragoo
Underwood Dudley, PhD
David R. Duffy
Samuel D. Dunbar
John J. Durante
Daniel J. Durbin, PhD
Martin C. Dusel
Robert L. Dyer
James P. Dykes
Jae Ebert
Ronald L. Edwards
Weldon T. Egan
Richard S. Egly, PhD
Calvin W. Emmerson
John L. Emmerson, PhD
Terry L. Endress
Gary L. England
Brian D. Erxleben
Ed Escallon
Edward A. Fabrici
Mitchel D. Fehr
Norman G. Feige
John C. Fenoglio
Steven E. Ferdon
Virginia Rogers Ferris, PhD
Paul A. Feszel
Greg Filkovski
Bruce Fiscus
Elson Fish
Edward W. Fisher
William B. Fisher
John D. Foell
Eugene Joseph Fornefeld, PhD
Dwaine Fowlkes
James W. Frazell
Jim Friederick
Alfred Keith Fritzsche, PhD
Eric Fry
Meredith W. Fry
Philip L. Fuchs, PhD
Forst Donald Fuller, PhD
Stephen E. Funkhauser
Arnold R. Gahlinger
Gustavo E. Galante
Roy E. Gant
Don Gard, PhD
Sudhakar R. Garlapati

David E. Gay
William A. Geary, PhD
Perry J. Gehring, PhD
Gordon H. Geiger, PhD
Doyle Geiselman, PhD
Demosthenes Peter Gelopulos, PhD
Phillip Gerhart, PhD
John S. Gilpin
Robert W. Glasscock
Dan Gleaves
Robert Hamor Gledhill, PhD
Hugh S. Glidewell
Matthew F. Gnezda, PhD
Victor W. Goldschmidt, PhD
William Raymond Gommel, PhD
James P. Graf
Douglas Grahn, PhD
Lawrence W. Grauvogel
Norman W. Gray
Richard Grenne
Susan S. Grenzebach
Steven H. Griffith
Tom U. Grinslade
Larry R. Groves
Joseph C. Gruss
Albert Guilford
C. L. Gunn
Gerald Edward Gutowski, PhD
Don Haack
David L. Hagen, PhD
Richard L. Hahn, PhD
David W. Haines
John E. Haines, Jr.
Todd M. Haley
Harold T. Hammel, PhD
Robert F. Hand
Charles Hantzis
James Hardesty
Robert B. Harnoff
Peter J. Hart, PhD
Clark Hartford
James M. Hartshorne
William H. Hathaway
Bruce A. Hayes
Darrell R. Hazlewood
James C. Heap
Dwight H. Heberer
Thomas J. Helbing
John P. Henaghan
Stanley M. Hendricks, II
Ronald D. Hentz
Wallace E. Hertel
Harold R. Hicks
W. B. Hill, PhD
Donald E. Hilton
Loren John Hoffbeck, PhD
Joseph D. Holder
John E. Holdsworth
John T. Hopwood
Randolph R. Horton
William J. Hosmon
James E. Hough
Karl J. Houghland
Hub Hougland
Don Morgan Huber, PhD
Roger J. Hull
Carl D. Humbarger
Jesse Max Hunter, PhD
Charles C. Huppert, PhD
John M. Igelman
Donald K. Igou, PhD
Gregory J. Ilko

Appendix 4: The Petition Project

E. William Itell
Robert B. Jacko, PhD
Bill G. Jackson, PhD
David E. Jahn
Aloysius A. Jaworski
Robert L. Johnson
Wyatt Johnson
Howard D. Johnston
James E. Jones
John D. Jones
John H. Jones
Ken A. Jones
Eric J. Jumper, PhD
Donald J. Just
Dimitar Kalchev
W. B. Karcher
Gregory A. Katter
Daniel J. Kaufman
Carlos Kemeny
Jim Kent
Steven P. Kepler
B. Charles Kerkhove, Jr.
Edward T. Keve
Dan Kieffner
Winfield S. Kiester
Dennis D. Kilkenny
Jon Kilpinen, PhD
Gary A. Kimberlin
Glen J. Kissel, PhD
Ronald E. Kissell
Peter T. Kissinger, PhD
Lassi A. Kivioja, PhD
Howard Joseph Klein, PhD
John C. Klingler
Steve Klug
Williams G. Knorr
James Paul Kohn, PhD
William A. Koontz
Edmund Carl Kornfeld, PhD
Aaron David Kossoy, PhD
Thomas F. Kowalczyk, Jr.
Michael J. Koyak
Richard Kraft
Bruce L. Lamb
C. Elaine Lane
E. E. Laskowski
Ralph F. Lasley
William S. Laszlo
David Lawson
Marisa S. Leach
Lucky Leavell
Timothy A. Lee
Richard Leidlein
Andrew Lenard, PhD
Peter E. Liley, PhD
Ronald H. Limbach
William C. Link
Donald Linn, PhD
Tim A. Litz
Robert L. Longardner
James S. Lovick
Philip J. Lubensky
Melvin Robert Lund
Neil P. Lynch
Martin Maassen
John C. Mackey
Jeffrey Maguire
John August Manthey
William Markel, PhD
Terry W. Marsh
Bill L. Martz*
Carrie M. Marusek

James A. Marusek
Jojseph B. Materson
Tats Matsuoka
Robert Mayemick
Larry R. Mayton
Allan J. McAllister
Roger D. McClintock
Edwin D. McCoy
James T. McElroy
Orville L. McFadden
Scott McGarvie
John D. McGregor
Edward A. McKaig, PhD
William J. McKenna
James C. McKinstry
Robert A. McKnight, Jr.
Duncan R. McLeish
R. J. McMonigal, PhD
Chuck McPherson
Mark R. Meadows, Sr.
James R. Meier, Sr.
Wilton Newton Melhorn, PhD
Gary Merkis
Mark B. Messmer
Emil W. Meyer
Sandra Miesel
Anthony J. Mihulka
John A. Mikel
Douglas Gene Mikolasek, PhD
Robert Douglas Miles
Dane A. Miller, PhD
George R. Miller
Robert T. Miller
Stephen L. Miller
Bradley G. Mills
Norman T. Mills
Kenneth L. Minett
Thomas J. Miranda, PhD
Michal Misiurewicz, PhD
Donald H. Mohr
John G. Mohr
Kenneth R. Moore, Jr.
Thomas S. Moore, PhD
Alfred L. Morningstar
Raymond L. Morter, PhD
James L. Mottet
John R. Mow
Joel E. Mowatt
Daniel H. Mowrey, PhD
Thomas J. Mueller, PhD
Barry B. Muhoberac, PhD
Jeff K. Munger
Bernard Munos
Haydn Herbert Murray, PhD
Michael L. Nahrwold
Wade L. Neal
David A. Nealy, PhD
Joseph J. Neff
Steven H. Neucks
Charles C. Ney
Kenneth E. Nichols, PhD
George D. Nickas, PhD
Henry Frederick Nolting
Robert C. Novak
John Lewis Occolowitz
George O. P. O'Doherty, PhD
Michael H. O'Donnell
Winston Stowell Ogilvy, PhD
John Olashuk
Timothy C. O'Neill
Jonathan M. Oram
John Robert Osborn, PhD

Albert Overhauser, PhD
Willis M. Overton, IV
C. Subah Packer, PhD
Jim F. Pairitz
Wayne Parke
Charles S. Parmenter, PhD
George A. Payne
Doug Perry
Kenneth D. Perry
Robert G. Phillips
Robert L. Pigott
Ronald C. Pinter
Jeffery J. Poole
Jerry Prescott
Kenneth Price, PhD
Marvin E. Priddy
Salvatore Profeta, Jr., PhD
Gary J. Proksch, PhD
Charles L. Provost
Robert R. Pruse
Phillip G. Przybylinski
Lawrence T. Purcell
David Puzan
Forrest W. Quackenbush, PhD
Beat U. Raess, PhD
Jeffrey T. Rafter
Douglas Ramers, PhD
John E. Ramsey
Phillip Gordon Rand, PhD
Mamunur Rashid, PhD
Lee Raue
Francis Harvey Raven, PhD
James V. Redding
Jack Reed
Roger G. Reed
David W. Reherman
Paul F. Reszel
Brian F. Ricci
Doug A. Richison
David F. Ring
Paul Rivers, PhD
B. D Nageswara Roa, PhD
Larry Roach
Fred E. Robbins
Ira J. Roberts
Charles A. Robinson
Jean-Christophe Rochet, PhD
Nancy Rodgers, PhD
Milton W. Roggenkamp
Thomas K. Rollins
Michael G. Rossmann, PhD
R. E. Rothhaar, PhD
John W. Rothrock, Jr.
James Lincoln Rowe, PhD
Don E. Ruff
Robert G. Rydell
Michael Sain, PhD
Donald G. Scearce
Barry W. Schafer
Donald Joseph Scheiber, PhD
Alfred Ayars Schilt, PhD
Alvin Schmucker
Stephen J. Schneider
George A. Schul
David Schultz, PhD
Bobby J. Sears
George Sears
Mark F. Seifert, PhD
Phillip Sensibaugh
Andrew Sexton
Dennis N. Sheets
Mary A. Sheller, PhD

Steve Shepherd
Kevin M. Sherd
Thomas W. Sherman, Jr.
W. Sherman, II
Ronald W. Shimanek
Vernon Jack Shiner, PhD
Robert L. Shone, PhD
A. Cornwell Shuman, PhD
J. Shung, PhD
Edward Harvey Simon, PhD*
Dennis Skala
Robert H. Slagel
C. B. Slagle
Danny Smith
Roger Smith
William H. Smith, Jr.
Parviz Soleymani
Quentin Francis Soper, PhD
Robert J. Sovacool
Edward Eugene Sowers, PhD
Larry D. Spangler
John M. Spears
John A. Spees
Bill Spindler
William Jacob Stadelman, PhD
Michael C. Stalnecker
S. M. Standish
Alan G. Stanley
Gerald R. Stanley
Donn Starkey
John M. Starkey, PhD
Frank R. Steldt, PhD
Nick C. Steph, PhD
William Stevenson
William R. Stiefel
William H. Stone
Ronald S. Straight
Catherine Strain
Richard N. Streacker
Edgar F. Stresino
Ronald Stroup
Robert M. Struewing
Stanley Julian Strycker, PhD
Jeffrey James Stuart, PhD
Victor Sturm
Carol S. Stwalley, PhD
James S. Sweet
Carey Swihart
John A. Synowiec, PhD
Albin A. Szewczyk, PhD
Virginia Tan Tabib
Elpidio V. Tan
Curtis L. Taylor
Harold L. Taylor, PhD
Harold Mellon Taylor, PhD
James A. Taylor
Scott A. Taylor
Lowell Tensmeyer, PhD
Howard F. Terrill
Nayana B. Thaker
Thomas Delor Thibault, PhD
Jerry A. Thomas
Marlin U. Thomas, PhD
Martin J. Thomas, PhD
Christina Z. Thompson
Larry G. Thompson
Leo Thorbecke
J. Bradley Thurston
Bryan M. Tilley
Glen Cory Todd, PhD
Charles Edward Tomich
William R. Tompkins

Daniel J. Toole
Ruperto V. Trevino
John E. Trok, PhD
Jerry A. True
Lawrence D. Tucker
Robert C. Tucker, PhD
Tim N. Tyler, PhD
John Joseph Uhran, PhD
Tadeusz M. Ulinsnki
Joseph L. Unthank, PhD
Bertram Van Breeman
Richard E. Van Strien, PhD
William Vandemerme, PhD
Vern C. Vanderbilt, PhD
Robert Dahlmeier Vatne, PhD
Spencer Max Vawter
Darrell A. Veach
Arthur A. Verdi
James H. Vernier
Richard Anthony Vierling, PhD
Raymond Viskanta, PhD
Karl J. Vogler, PhD
Donald F. Voros
Vladeta Vuckovic, PhD
Susan M. Waggoner
Norman O. Wagoner
Jerry C. Walker, PhD
James L. Walters
David R. Wanhatalo
Robert W. Weer
John H. Weikel, PhD
George W. Welker, PhD
Lowell Ernest Weller, PhD
Tom Wells
Robert Joseph Werth, PhD
Aubrey L. Wesson
Gordon L. Westergren
Winfield B. Wetherbee, PhD
Bill Wetherton
Alfons J. Wetzel
Dave L. Wheatley
Larry A. Wheelock
Roy Lester Whistler, PhD
Joe Lloyd White, PhD
Lawrence Wiedman, PhD
Charles Eugene Wier, PhD
Gene Muriel Wild, PhD
Jay L. Wile, PhD
James C. William, PhD
Carl J. Williams
Tony D. Wilson
John R. Wingard
Frank A. Witzmann, PhD
William E. Woenker
You-Yeon Won, PhD
Gary K. Woodward
Gregory M. Wotell
Jerry Wright
Lee E. Wright
T. J. Wright
Yao Hua Wu, PhD
Neil Yake
Jason W. Yeager
Richard Yoder
Andres J. Zajac
Ihor Zajac, PhD
Michael W. Zeller
Paul L. Ziemer, PhD
John E. Zimmerman
Steven L. Zirkelback
Earl J. Zwick, PhD

Iowa

William L. Ackerman
Keith L. Amunson
Julia W. Anderson, PhD
Mike E. Anderson
Robert J. Barry
Steven Bateman
Jim Baumer, PhD
Robert L. Beech, PhD
Glen L. Bellows
Lisa A. Beltz, PhD
Dick A. Bergren, PhD
Arnold G. Beukelman
Albert Joseph Bevolo, PhD
Charles H. Black, PhD
Lorraine T. Blanck
Gerald L. Bolingbroke, PhD
Robert M. Bowie, PhD
Bruce A. Braun
Jeb E. Brewer
Alan J. Brinkmann, PhD
Bryan C. Bross
Scott A. Brunsvold
Kenneth D. Bucklin
George V. Burnet, PhD
Wilbur H. Busch
John M. Buss
Stanley E. Buss
Joel Calhoun
Lyle L. Carpenter
C. Clifton Chancey, PhD
James Brackney Christiansen, PhD
Richard H. Cockrum
Thomas W. Conway, PhD
Richard E. Cowart, PhD
Donald J. Cox, PhD
Daniel T. Crawford
Ronald L. Crowley
Abie C. Davis
Bonnie Dawley
Kenneth J. Denault, PhD
William Arna Deskin, PhD
John T. Dolehide
David L. Dooley
Harold E. Doorenbos, PhD*
Dan B. Drahos
Charles W. Dreibelbis
Gary L. Driscoll, PhD
Warren Dunkel
John R. Ebersberger
Charlie R. Edwards
Carl Thomas Egger, PhD
Curt Erickson
Kerry Eubanks
L. C. Evbers
Michael G. Farley
Frank K. Farmer
Leonard Samuel Feldt, PhD
Lavern J. Flage
Gary E. Forristall
Joseph E. Foss
Dale A. Frank
Lloyd R. Frederick, PhD
William A. Gallus, Jr., PhD
Dennis S. Gannon
Greg Gerdes
Frederick S. Gezella
Mohamed Mansour Ghoneim
William H. Gilbert, PhD
Mary Ann Gillbert
Chester Goodrich, PhD

Michael J. Gries
William E. Griffin
Morris Paul Grotheer, PhD
James A. Haigh
Charles Virdus Hall, PhD
Harold Elmore Hammerstrom, PhD
Ronald Scott Harland
Kenneth S. Harris
Nicholas L. Hartwig, PhD
Frederick B. Hembrough, PhD
Herbert Edward Hendriks, PhD
Ken Henrichsen, PhD
Paul W. Herrig
Richard G. Hindman, PhD
Larry L. Hintze
David M. Hodgin
Darrel Barton Hoff, PhD
Palmer Joseph Holden, PhD
Charles Holz
C. D. Van Houweling
Frederick B. Hubler
Roger Huetig
Frank Hummer, PhD
Eric K. Jacobsen
Albert A. Jagnow
Mark Jagnow
Paul G. Jagnow
James Jennison
James L. Johnson
Mark Johnson
Lawrence L. Jones
Rex E. Jorgensen
Richard D. Jorgenson, PhD
Steven R. Junod
John Kammermeyer, PhD
Brian Kenny
Kenneth Kise, Jr., PhD
Thomas Kleen
Thomas Kline, PhD
Tyson Koch
Kris D. Kohl, PhD
Gregory L. Kooker
Frank P. Koontz, PhD
Ann D. Kuenstling
Charles R. Kuhlman
Paul Lancaster
James F. Lardner
Stephen L. Larsen
John H. Lemke, PhD
Wayman Lipsey
Mark G. Lorenz
Kirk A. Macumber
Adeeb Bassili Makar, PhD
Cheryl Marigon
Dennis N. Marple, PhD
Charles R. Marsden
Christopher R. Marshall, PhD
Michael J. Martin, PhD
John C. Mayer
Thomas D. McGee, PhD
Clifford L. Meints, PhD
Charles L. Miksch
Glen E. Miller
Darren R. Moon
Stephen K. Murdock
Richard Neate
Leon R. Nelson
Carl William Niekamp, PhD
John V. Nigro
James L. Nitzschke
Christopher P. Nizzi
Duane A. Nollsch

Albert Stanley Norris
Kevin Charles O'Kane, PhD
Eric Olson
Randy Paap
James C. Parker
Richard S. Parker
Peter A. Pattee, PhD
Ralph E. Patterson III, PhD
Theodore F. Paulson
Larry Preston Pedigo, PhD
David T. Peterson, PhD
John M. Pitt, PhD
Kurt Pontasch, PhD
R. Potratz
David L. Pranger
Richard E. Preston
Michael Pruchnicki
L. L. Pruitt
Keith J. Quanbeck
D. L. Reasons
Theodore Lynn Rebstock, PhD
Peter D. Rens
Winthrop S. Risk
Kevin R. Rogers
Rick P. Salocker
Somnath Sarkar, PhD
George K. Sassmann
Maurice G. Scheider
D. Schmidt
George William Schustek
Vincent A. Schuster
Pamela R. Sessions
Gaylord E. Shaw, PhD
James B. Sheets
Merlin R. Siefker
Leland W. Sims
Gary E. Sindelar
Scott P. Smith, PhD
Daniel E. Sprengeler
David L. Sprunger
Aaron J. Spurr
Robert J. Stalberger
Ole H Viking Stalheim, PhD
Dean Strand
Matt D. Streeter
Steven D. Struble
Kirk Struve
Rexanne Struve
Stephen A. Sundquist
Carl E. Syversen
Eugene W. Taylor
Stephen Tedore
Kent Thompson
Louis M. Thompson, PhD
Richard W. Tock, PhD
Steven A. Tonsfeldt, PhD
Bruce Towne
Jim G. Trettin
Tom B. Ulrickson
Lee E. Vaughan
Jack D. Virtue
Terence C. Virtue
Paul R. Vogt
Dennis F. Waugh
Charles A. Wellman
D. K. Whigman, PhD
Robert A. Wiley, PhD
Grey Woodman
Daniel J. Zaffarano, PhD

Kansas

Dwight L. Adams
Stephen R. Alewine
Donald R. Andersen, Jr.
Harrison Clarke Anderson
Thornton Anderson
William L. Anderson
Robynn Andracsek
Ernest Angino, PhD
Amalia R. Auvigne
James F. Badgett
D. Bahm
Richard C. Bair
Tracy M. Baker
Nathan S. Baldwin
Cliff Bale
Larry Bale
Mark J. Bareta
Neal Barkley
Campbell C. Barnds
Ross W. Barton
Lynn Shannon Bates, PhD
Curtis M. Beecham, PhD
John S. Black
Charles Blatchley, PhD
Robert G. Boling
Alan G. Bosomworth
Lawrence Glenn Bradford, PhD
Don L. Braker
Brian W. Braudaway
Wesley G. Britson
Bruce C. Brooks
Robert A. Brooks, PhD
Ward W. Brown
Dail Bruce
Joan Brunfeldt
Ralph W. Bubeck
William M. Byrne, Jr., PhD
Jon M. Callen
Earl D. Carlson
Craig Caulk
Charles T. Chaffin, PhD
Jon Christensen
Justin Clegg
Calvin E. Coates
Shannon Colbern, PhD
James E. Connor
Max E. Cooper
Ted L. Cooper
Dennis D. Copeland
Murray D. Corbin
Patrick Ivan Coyne, PhD
Kent Craghead
Phillip T. Cross
Robert Hamblett Crowther
Glenn Crumb, PhD
James R. Daniels
Chad Davies, PhD
Eugene William Dehner, PhD*
Ronald A. Dial
J. W. Dohr
Leslie J. Doty
John Doull, PhD
Rod Duke
David W. Dukes
Clinton E. Dunn
A. F. Dyer
Joe R. Eagleman, PhD
Donald R. Eidemiller
Darrel Lee Eklund, PhD
Robert E. Elder

Al Erickson
Michael J. Eslick
Roger W. Evans
Eugene Patrick Farrell
Gregory A. Farrell
Gene Richard Feaster, PhD
Roger Fedde, PhD
Stephen L. Ferry
Theodore C. Finkemeier
Debra Fitzgerald
John R. Floden
Carl A. Fowler
Steve Frankamp
Dale E. Fulcher
David W. Garrett
Dick A. Geis
Donald R. Germann
Clifford Glenn
James N. Glenn
Earl F. Glynn
Bruice Gockel
Kenneth L. Goetz, PhD
David J. Goldak
Peter J. Gorder, PhD
Robert C. Gorman
Albert J. Gotch, PhD
Douglas D. Graver
Lewis L. Gray
Jerry Green
M. Griffin, PhD
William C. Groutas, PhD
John T. Growney
Paul P. Gualtieri
Larry Haffey
Stephen F. Hagan
Quinlan Halbeisen
Wesley H. Hall
William W. Hambleton, PhD
Mark A. Hamilton
Robert M. Hammaker, PhD
Scott E. Hampel
T. Harder, PhD
Geoffrey O. Hartzler
Kirk Hastings
Douglas H. Headley
Dennis Hedke
John Herald
Sandy Herndon
B. Heyen, PhD
Robert J. Hodes
Brett A. Hopkins, PhD
Robert Hopkins
Marta Howard
William C. Hutcheson
Cory Imhoff
Michael D. Jackson
Roscoe G. Jackson, II, PhD
Byron E. Jacobson
Dale P. Jewett
Leland R. Johnson, PhD
David W. Kapple
Kim Karr, PhD
Suzanne Kenton
Michael Kerner
Gerald Francis Kerr
John Kettler
Michael R. Kidwell
Cecil M. Kingsley
Philip G. Kirmser, PhD
Kenneth J. Klabunde, PhD
Robert W. Klee
Burke B. Krueger

Jack K. Krum, PhD
Daniel O. Kuhn
James E. Kullberg
Michael J. Lally
Roger O. Lambson, PhD
David L. Lamp
Vance Lassey
Richard Leeth
Richard Leicht
E. C. Lester
Jingyu Lin, PhD
Edwin D. Lindgren
Glenn Liolios
John B. Loser
James Loving
Leon Lyles, PhD
Keith D. Lynch, PhD
Franklin F. Mackenzie
Susan L. Mann
Robert A. Martinez
Gayle Mason
Keith Mazachek, PhD
Thomas McCaleb
Robert J. McCloud
Richard E. McCoy
James L. McIlroy
George M. McKee, Jr.
Alvin E. Melcher
Dwight F. Metzler
Steven M. Michnick
Frank R. Midkiff
Clinton F. Miller
Richard L. Miller, PhD
Leon J. Mills, PhD
Eldon F. Mockry
Reginald B. Moore
Norman R. Morrow
Fred Moss
Vincent U. Muirhead
E. A. Munyan
Joseph K. Myers
Jay F. Nagori
C. Nobles
Charles Arthur Norwood, PhD
Carl E. Nuzman
Verda Nye
Rocky Nystom
Daniel J. O'Brien, PhD
Steven P. O'Neill
Thomas E. Orr
Dennis D. Ozman
Kyle Parker
Randy C. Parker
Dean Pattison
Charles W. Pauls
David B. Pauly, Jr
Wayne Penrod
Bill Perry
Heide M. Petermann
Leroy Lynn Peters, PhD*
David Pflum
Max E. Pickerill, PhD
Danny L. Piper
Jack J. Polise
Arthur F. Pope
Donald L. Poplin
Robert Pratt
Tom Pronold
John J. Ramm
Thomas E. Ray
Dean D. Reeves
David J. Relihan

James C. Remsberg
Robert L. Reymond, Jr.
Charles A. Reynolds, PhD
Brian K. Richardet
Larry J. Richardson
David Allan Ringle, PhD
Kenneth Roane
Edward H. Roberts
M. John Robinson, PhD
James W. Rockhold
Richard L. Ronning
Karl K. Rozman, PhD
James M. Ryan, PhD
Richard J. Saenger
Frederick Eugene Samson, PhD
L. A. Sankpill
Larry V. Satzler
Michael J. Sauber
Kelly B. Savage
Robert R. Schalles, PhD
Charles Schmidt
J. Richard Schrock, PhD
Robert Samuel Schroeder, PhD
Warren D. Schwabauer, Jr.
Charles M. Schwinger
James E. Sears
Arnold W. Shafer
Dexter Brian Sharp, PhD
Brett E. Sharpe
Brian A. Sheets
Leland M. Shepard
Lloyd W. Sherrill
Kim Shoemaker
Merle Dennis Shogren
Mark A. Shreve
Ray Shultz
Glen L. Shurtz, PhD
James Van Sickle
William E. Simes
Larry Skelton
Edward Lyman Skidmore, PhD
Bruce G. Smith
Michael R. Smith
Randall K. Spare
Feliz A. Spies
Robert P. Spriggs
Jerry L. Stephens
Jeffrey Smith Stevenson, PhD
William T. Stevenson, PhD
Robert T. Stolzle
Richard L. Stoppelmoor
Steven E. Stribling
Doug T. Stueve
Bala Subramaniam, PhD
James W. Suggs
Stuart Endsley Swartz, PhD
Saeed Taherian, PhD
Marcus K. Taylor
Randall Teter
Leslie Thomas
Timothy C. Tredway
Richard S. Troell
James M. Tullis
Richard M. Vaeth
M. Van Swaay, PhD
Gary Vogt
Rosmarie von Rumker, PhD
Bruce L. Wacker
Wilber B. Walton
Dennis L. Wariner
Robert K. Wattson, Jr.
Laurence R. Weatherley, PhD

Allan R. Weide
Kenneth R. Wells
Ronald L. Wells, PhD
Jerry R. Werdel
Steve S. Werner
Tom J. Westerman
Eric R. Westphal, PhD
Carol J. Whitlock
Wendell Keith Whitney, PhD
William Wiener
Jim Wiley
Don K. Wilken
Allen K. Williams
Phillip Brock Williams, PhD
Robert L. Williams, Jr.
Thomas Williams
Lonnie D. Willis
Robert G. Wilson, PhD
Steven J. Wilson
Earl C. Windisch
Elmer J. Wohler
Russell L. Woirhaye
Jim Wolf
Charles Hubert Wright, PhD
Ralph G. Wyss
Shawn Young
Dwight A. Youngberg
Mario K. Yu
Melvin E. Zandler, PhD
Michael T. Zimmer
Adam D. Zorn
Glen W. Zumwalt, PhD

Kentucky

Dell H. Adams
Neil Adams, PhD
Howard W. Althouse
R. Byron Alvey
Michael R. Amick
Robert L. Amster
Donald Applegate
Donald Archer
Raymond A. Ashcraft
A. Ashley, PhD
Richard A. Aurand
Thomas H. Baird
John P. Baker, PhD
Randal Baker
Lee A. Balaklaw
Paul A. Barry
Bonnie L. Bartee
Walter G. Barth
Charles I. Bearse, III
Lloyd Willard Beck
Daniel D. Beineke, PhD
Charles D. Bennett
R. Berson, PhD
Michael Binzer
Mary L. Blair
John G. Bloemer
Richard S. Bonn
Harold Boston
Roscoe C. Bowen, PhD
Jeffrey D. Brock
Fred A. Brooks
William F. Brothers, Jr.
Thomas Dudley Brower
Gerald Richard Brown, PhD
Philip T. Browne
Harry A. Bryan

Barry J. Burchett
Michael A. Burke
Frank C. Campbell, Jr.
Robert E. Campbell
Alan A. Camppli, PhD
Thomas D. Carder
Jeffrey S. Caudill
Robert Lee Caudill
William K. Caylor
James M. Charles
Richard Cheeks
Frederick Chen, PhD
Lijian Chen, PhD
John Albert Clendening, PhD
Thomas R. Coffey
James L. Cole
Donald W. Collier
Stanley W. Collis
Todd Colvin
Shawn W. Combs
Thad F. Connally
Henry E. Cook, Jr.
Maura Corley
Linda G. Corns
James Cramer
Rhonda Creech
Philip G. Crnkovich
William D. Cubbedge
John J. Czarniecki
Frank N. Daniel
Roger A. Daugherty
Aaron R. Davis
Gilber De Cicco
Paul E. Dieterlen
Philip W. Disney
Billy Dobbs
George Charles Doderer
William B. Dougherty
James W. Drye
Donald B. Dupre, PhD
Gregory A. Durffent
Lee G. Durham
Victor E. Duvall
Thomas E. Eaton, PhD
Gary Eberly
Harvey Lee Edmonds, Jr., PhD
William D. Ehringer, PhD
P. Eichenberger
Donald R. Ellison
William K. Elwood, PhD
William G. Emmerling
Jim L. End
Dennis W. Enright
Michael B. Erp
Russell J. Fallon
Lynda R. Farley
Allan George Farman, PhD
David Ray Finnell
Ken W. Fishel
Tommie E. Flora
Burton R. Floyd, Jr.
Paul C. Fogle
David C. Foltz
Gene P. Fouts
Mary Feltner Futrell, PhD
Fletcher Gabbard, PhD
Theodore H. Gaeddert, Jr.
Stanley A. Gall
Theo H. Gammel
J. Steven Gardner
Ronald F. Gariepy, PhD
Daniel L. Garrison

Lawrence K. Gates
Lewis G. Gay
Thomas Edward Geoghegan, PhD
Robert B. George
Earl Robert Gerhard, PhD
Jonathan Gertz
Peter P. Gillis, PhD
Bobby G. Gish
Gordon C. Glass
Peter D. Goodwin
Clifford D. Goss
Kenton J. Graviss, PhD
William Daniel Green, Jr
Ara Hacetoglu
Richard P. Hagan, PhD
Paul F. Haggard
W. Hahn
Ted D. Haley*
Robert Halladay
Harold D. Haller
Charles Edward Hamrin, Jr., PhD
Charles Hardebeck
Byron C. Hardinge
Dean O. Harper, PhD
Kevin W. Harris
Roswell A. Harris, PhD
Dennis R. Hatfield
Aaron Haubner, PhD
John H. Havron
Erv Hegedus
Richard B. Heister, II
Regina C. Henry
Wiley Hix Henson, PhD
Carl W. Hibscher
Don Hise
R. W. Holman, PhD
Kenneth R. Holzknecht
Alan W. Homiak
Ricky Honaker, PhD
Debra House
Kevin E. Houston
David W. Howard
James F. Howard, PhD
J. William Huber
Donald W. Hunter
Hal Hyman
Lorna Hymon
Ralph E. Jackson
Leo B. Jenkins, PhD
Randolph A. Jensen
Elaine L. Jocobson, PhD
Omar M. Johnson
Ray Edwin Johnson, PhD
James H. Justice
Norman E. Karam
Leslie E. Karr
Ann M. Keller
Kenneth F. Keller, PhD
Kurt A. Keller
William W. Kelly, Jr.
John Elmo Kennedy, PhD
Riley Nelson Kinman, PhD
Riley N. Kinners, PhD
W. S. Klein
David Kling
Nancy F. Kloentrup, PhD
Riley N. Kminer, PhD
James E. Krampe
Clarence R. Krebs
Steve Kristoff, PhD
Wasley Sven Krogdahl, PhD
John E. Kuhn, PhD

Garth Kuhnhein
John P. Lambert
Harold Legate
Charles L. Lilly
Merlin D. Lindemann, PhD
Noel W. Lively
Susan Logsdon
James Longo
George E. Love
John Lowbridge, PhD
Charles S. Lown
Gerald J. Lowry, PhD
Phillip Lucas
Robert M. Lukes, PhD
George W. Luxbacher, PhD
William Charles MacQuown, PhD
Dannys Maggard
Alan T. Male, PhD
Robert C. Mania, PhD
Ronald A. Mann, PhD
William J. Mansfield
Dan Marinello
Maurice K. Marshall
Ronald L. Marshall
M. Masthay, PhD
Richard S. Mateer, PhD
Wayne R. McCleese
Terry McCreary, PhD
Bill McGowan, PhD
John Long Meisenheimer, PhD
Phil M. Miles
David B. Miller
Franklin K. Miller
Ronald R. Monson
Dory D. Montzaemi
Duane Moore
Glenn I. Moore
Charle E. Mosgrave
Steven W. Moss
F. Allen Muhl
Fitzhugh Y. Mullins
Michael L. Munday
James K. Neathery, PhD
Jerry S. Newcomb
David A. Newman, PhD
Arthur J. Nitz, PhD
Michael E. Nordloh
Lynn Ogden
Bernie D. Oliver
Steve Ott
Joseph C. Overmann
B. K. Parekh, PhD
Dexter Brian Patton, III
Arthur P. Peel
Douglas W. Peters
Jeffrey L. Peters
Glenn Pfendt
John E. Plumlee
Michael Portman
Ronald T. Presbys
Phil Quire
William Quisenberry
Craig Rabeneck
Manny K. Rafla, PhD
Stephen C. Rapchak
Michael A. Ray
Erick L. Redmon
Frederick C. Rehberg
Phillip J. Reucroft, PhD
Gary Reynolds
J. E. Rhoades
Bart E. Richley

John Thomas Riley, PhD
David Roach, PhD
James O. Roan
Philip J. Robbins
Stewart W. Robinson, PhD
Thomas J. Roe
Jason K. Romain
Lance D. Rowell
Paul B. Rullman
Mary J. Ruwart, PhD
James M. Rynerson
Thomas A. Saladin
Robert Alois Sanford, PhD
Mark S. Sapsford
Thomas A. Saygers
Charles M. Saylor, III
Larry Saylor
Donald J. Scheer, PhD
Frank M. Schuster
George W. Schwert, PhD
Randolph J. Scott
Daniel M. Settles, PhD
Richard W. Sewell
Michael Sherlock
Robin N. Siewert
John A. Sills, PhD
Marek A. Sitarski, PhD
Frank Sizemore
Jon James Smalstig
Avery E. Smith
Hayden Smith
Anne Cameron Snider
James D. Sohl
Gerard T. Sossorg
Joseph Sottile, PhD
Gary L. Southerland
Gary G. Sowards
Donald G. Spaeth, PhD
Werner E. Speer
Robert Wright Squires, PhD
Harry P. Standinger
Douglas E. Stearns
Dennis M. Stephens
Darcy S. Stewart
Richard C. Stocke
Bill Stoeppel
Alan Stone
Richard W. Storey
Vernon Stubblefield, PhD
S. R. Subramani
Tommy L. Sutton
George H. Swearingen, Jr.
Morris A. Talbott
Daniel Tao, PhD
Daniel V. Terrell, III
Gregory R. Thiel
M. P. Thorne
Paul E. Tirey
Albert M. Tsybulevskiy, PhD
Robert S. Tucker
Kot V. Ungrug, PhD
Joseph Van Zee
William York Varney, PhD
Rodney D. Veitschegger, Jr.
Willis G. Vogel
Jim Volz
Konstanty F. von Unrug, PhD
George Vourvopoulos, PhD
William J. Waddell
Bruce E. Waespe
David W. Wallace
Alva C. Ward

Danny R. Ward
Mark T. Warren
Michael Webb
Larry G. Wells, PhD
William Wetherton
Roy E. Whitt
Rudy Wiesemann
Robert O. Wilford
James Cammack Wilhoit, PhD
Bill R. Wimpelberg
Dennis C. Withey
Douglass W. Witt
Earl C. Wood
Farley R. Wood
Alice J. Woosley
Donald R. Yates
Scott Yost, PhD
Ralph S. Young
Peter S. Zanetti
Richard Jerry Van Zee, PhD
Robert J. Zik
David J. Zorn

Louisiana

Joseph P. Accardo
Leonard Caldwell Adams, PhD
Tim Addington
Mark R. Agnew
Steven L. Ainsworth
Christopher G. Allen
Paul W. Allen, PhD
Arthur E. Anderson
James R. Anderson
Roy E. Anderson
Robert E. Angel
Lowell N. Applegate
Lou Armstrong
Daniel J. Aucutt
Bryan Audiffred
William E. Avera
Hugh Bailey
Wallace H. Barrett
Robert J. Bascle
E. Baudoin
Rick Bauman
Edward W. Beall
Joseph A. Beckman, PhD
Charles Bedell
Edward Lee Beeson, Jr., PhD
C. Allen Bell
James M. Bell
E. Paul Bercegeay, PhD
Dan Berry
Craig D. Berteau
R. Dale Biggs, PhD
David L. Billingsley, Jr.
Ron J. Blanton
Virgil L. Boaz, PhD
Larry P. Bodin
Frank R. Boehm, Jr.
Gerald S. Boesch, Jr.
Lawrence G. Bole
Troy N. Book
James E. Boone, PhD
John Boulet
Arnold Heiko Bouma, PhD
Timothy M. Boyd
Samuel B. Brady, IV
Ernest Breaux
Larry Breeding

Karen Brignac
Francis W. Broussard
Mary J. Broussard
Tim Brunson
David W. Bryant
David William Bunch, PhD
Ronald Butler
Augustus George Caldwell, PhD
Gene Callens, PhD
Sidney Campbell
Louis J. Capozzoli, PhD
Bobby L. Caraway, PhD
Edward Carriere, PhD
Oran R. Carter
James Clark Carver, PhD
Thomas E. Catlett
Thomas W. Cendrowski
John N. Cetinich
Bruce Chamberlain
Angelo Chamberlin
John T. Chandler
Harlan H. Chappelle
Emery R. Chauvin
Huimin Chen, PhD
Danny J. Clarke
Chris M. Cobb
J. M. Coffield
R. D. Coles
Ted Z. Collins
James G. Connell
William J. Connick, Jr.
Kenneth R. Copeland
Kevin Corbin
Walter H. Corkern, PhD
Frank A. Cormier
Michael R. Coryn
Judith Coston
Vincent F. Cottone
Ronald A. Coulson, PhD
Ronald Reed Cowden, PhD
Lawson G. Cox
Joel B. Cromartie
Gene Autrey Crowder, PhD
John H. Cunniff
Harold B. Curry
Louis Chopin Cusachs, PhD
Rita Czek
Russell E. Dailey
L. R. Dartez, Jr.
Shawn P. Daugherty
Dennis J. Dautreuil
James H. Davidson
Harry H. Dawson
Robert C. Dawson
Donal Forest Day, PhD
Winston Russel de Monsabert, PhD
John H. Dekker
Anthony J. DeLucca
Charles N. Delzell, PhD
Thomas A. Demars, PhD
Tom DePierri
William G. DePierri, PhD
Marcio Dequeiroz, PhD
Herbert C. Dessauer, PhD
David F. Dibbley
Elizabeth S. Didier, PhD
Paul A. Dieffenthaller
Larry Dorris
Carl L. Douglas, Jr.
Harry Vernon Drushel, PhD
Patrick Dubois
C. J. Duet

Terry M. Duhon
Genet Duke, PhD
Raymond J. Dunn
Henry J. Dupre
Sidney J. Dupuy, III
Eugene F. Earp
D. Elliott, PhD
W. Engle
Thomas L. Eppler, PhD
John R. Eustis
Dabney M. Ewin
Donald C. Faust
Gary Z. Fehrmann
D. Bryan Ferguson
John Ferguson
Jorge M. Ferrer
Aaren Fiedler
Jack Everett Field, PhD
Gary L. Findley, PhD
Robert G. Finkenaur
Marlon R. Fitzgerlad
Robert Wilson Flournoy, PhD
Michael A. Fogarty
Martin M. Fontenot
Rudolph M. Franklin
Benjamin D. Fremming
Jeffrey W. French
Roger D. Fuller
Frank P. Gallagher
Jonathan C. Garrett
Raymond F. Gasser, PhD
Edward William Gassie, PhD
Richard L. Gates
Paul Robert Geissler, PhD
Alexander A. Georgiev, PhD
T. C. Gerhold
Peter John Gerone, PhD
Robert T. Giles
Bob Gillespie
Jeffrey D. Ginnvan
Marvin G. Girod
Nicholas E. Goeders, PhD
John T. Goorley, PhD
V. L. Goppelt
George Gott
Robert J. Gouldie
John A. Grach
Joseph C. Graciana
Eric Greager
Randolph K. Greaves
Donald L. Greer, PhD
Donald A. Griglack
Alan B. Grosbach
Ronald R. Grost
Frank R. Groves, Jr., PhD
Roland J. Guidry, Jr.
William T. Hall
Bruce A. Hallila
John H. Hallman, PhD
Abraham S. Hananel, PhD
D. E. Hansen
M. Wayne Hanson, PhD
Ted N. Harper
Clarence C. Harrey, Jr.
Albert D. Harvey, PhD
Kim L. Harvey
Kerry M. Hawkins
V. R. Hawks, Sr.
William G. Hazen
Phillip C. Hebert
Harold Gilman Hedrick, PhD
Richard E. Heffner

J. Hemstreet, PhD
G. C. Hepburn, Jr.
Stephen R. Herbel
John J. Herbst
Roger N. Hickerson
F. M. Hill, Jr.
Thomas D. Hixson
Paul K. Holt
Lloyd G. Hoover
Sylvester P. Horkowitz
Edwin Dale Hornbaker, PhD
Nancy A. Hosman
Walter A. Hough, PhD
Stephen C. Hourcade
John E. Housiaux
Michael S. Howard, PhD
Fritz Howes
F. Markley Huey
K. Huffman
Woodie D. Huffman
Barry Hugghins
Robert L. Hutchinson, PhD
Terry E. Irwin, II
Edward A. Jeffreys
Dale R. Jensen
Edwin N. Jett
Emil S. Johansen
Adrian Earl Johnson, Jr., PhD
Thomas J. Johnson
Jack E. Jones, PhD
William H. Jones
Douglas L. Jordan
Paul A. Jordan
John Andrew Jung, PhD
Michael Lee Junker, PhD
Robert J. Justice
J. Justiss, Jr.
Richard D. Karkkainen
J. B. Karpa
Kevin Kelly, PhD
Stephen J. Kennedy
Walter Paul Kessinger, Jr., PhD
Michael L. Kidda
Charles A. Killgore
John King
Charles Leo Kingrea, PhD
K. Klingman, PhD
Thomas R. Klopf
Mark A. Knoblach
Daniel E. Knowles
Stanley Phelps Koltun
Anton Albert Komarek
Christopher La Rosa
Dale J. Lampton
Vicki A. Lancaster, PhD
Ricky M. Landreneau
L. Landry
Ronald J. Landry
H. Norbert Lanners, PhD
John LaRochelle, PhD
Dana E. Larson
James A. Latham
Edwin H. Lawson
Lynn L. Leblanc, PhD
Bill Ledford
Grief C. Lee
Joseph M. Leimkuhler
James C. Leisk
Samuel A. Leonard
Gregory Scott Lester
James C. Lin, PhD
Arthur L. Long

William E. Long
William Henry Long, PhD
Jerry A. Louviere
John R. MacGregor
Vincent T. Mallette
Carl W. Mangus
Alvin V. Marks
John Luis Martinez
Barbara L. Matens
Campo Elias Matens
J. L. Mathews
Rex Leon Matlock, PhD
Neil Matthews
Charles M. May, Jr.
David R. Mayfield
Bobby Mayweather
Michael L. McAnelly
Danny W. McCarthy, PhD
Mickey McDonald
John P. McDonnell
C. A. McDowell, Jr.
R. D. McGee
Michael McIntosh
J. T. McQuitty
Alan L. McWhorter, PhD
Tammy S. Meador
Jimmy L. Meier, Sr.
Jerry J. Melton
Robert Merrill
William W. Merrill
Charles R. Merrimen, PhD
Dennis Meyer
Robert A. Meyer
Michael S. Mikkelson
Kelley F. Miles
Harvey I. Miller, PhD
Joseph Henry Miller, PhD
Robert H. Miller, PhD
Arnout L. Mols
Christy A. Montegur
Douglas Moore
J. Stephen Moore
Richard Newton Moore
Paul H. Morphy
Harry Edward Moseley, PhD
Peter V. Moulder
Clifford Mugnier
Douglas V. Mumm
Donald D. Muncy
Jack P. Murphy
John D. Murry
Adil Mustafa
Rocco J. Musumeche
Peter C. Mutty
Lynn J. Myers
Billy J. Neal
William J. Needham
Joseph Navin Neucere
Robert H. Newll
Charles H. Newman, PhD
Conrad E. Newman
Rogers J. Newman, PhD
James R. Nichols
Martin R. Nolan
Richard Dale Obarr, PhD
Liang S. Ocy, PhD
William B. Oliver, III
Donald A. Olivier
Dean S. Oneal
Addison Davis Owings, PhD
Richard Milton Paddison
Muriel Signe Palmgren, PhD

Richard Parish, PhD
Kirt S. Patel
Frank M. Pattee
John W. Paxton
Julian Payne
Armand Bennett Pepperman, PhD
John H. Pere
James L. Peterson
George E. Petrosky
Chester Arthur Peyronnin, Jr.
James E. Pfeffer
John R. Pimlott
Richard M. Pitcher
Larry Plonsker, PhD
Timothy M. Power
Irving J. Prentice
Charles W. Pressley
R. L. Prichard
William A. Pryor, PhD
Elliott Raisen, PhD
Ganesier Ramachandran
Charles L. Rand
Clinton M. Rayes
William Rehmann
Alfred Douglas Reichle, PhD
Robert T. Reimers, PhD
Leslie H. Reynolds
Lee J. Richard
Leonard Frederick Richardson, PhD
Neil Richter
Herman H. Rieke, PhD
Jim Rike
Kent R. Rinehart
Terry D. Rings
Bruce A. Rogers
Gerard A. Romaguera
Richard Rosenroetter
M. Rosenzweig
Lawrence James Rouse, PhD
Steve Rowland
Gillian Rudd, PhD
Kelli Runnels, PhD
Louis J. Rusoff, PhD
Richard A. Sachitano, PhD
Charles M. Sampson
Ronald G. Sarrat
Ralph J. Saucier
Robert Joseph Schramel
Roy W. Schubert, PhD
Jeffrey A. Schwarz
Otto R. Schweitzer, PhD
Albert Edward Schweizer, PhD
Michael J. Screpetis
Richard D. Seba, PhD
Joseph E. Sedberry, Jr., PhD
Rowdy L. Shaddox
Gordon Shaw
Lawrence H. Shaw
David Preston Shepherd, PhD
Frederick F. Shih, PhD
Muhammad D. Shuja
Karen L. Shuler
Mark Sibille
Harold D. Siegel
William Carl Siegel
Larry Sifton
Lucila Silva
William E. Simon, PhD
A. Craig Simpson
Howard F. Sklar
Denny L. Small
James M. Smith

Larry Smith
Thomas M. Snow
Kepner D. Southerland
Elmer N. Spence
Donald Sprowl, PhD
Burt L. St. Cyr
Paul A. Stagg
Kevin Stanley
Eric States
G. Sterling
Arthur E. Stevenson
Rune Leonard Stjernholm, PhD
Edward H. Stobart
Theresa D. Stokeld
Robert C. Stone, Jr.
James P. Storey
William T. Straughan, PhD
Roland A. Sturdivant
Alberto Suarez
Ruth Sundeen
Frank J. Sunseri
Tony E. Swisher
Katherine Talluto, PhD
Orey Tanner, Jr.
Eric R. Taylor, PhD
Erik D. Taylor
Paul R. Tennis
Frank J. Tipler, PhD
Wallace K. Tomlinson
James E. Toups
Ed Trahan
Ann F. Trappey
John H. Traus
Melvin L. Triay
Peter J. Van Slyke
Anthony R. Venson, PhD
John R. Vercellotti, PhD
Robert C. Vestal
Marc Vezeau
Clemens J. Voelkel
Eugene von Rosenberg
Coerte A. Voorhies
Charles Henry Voss, Jr., PhD
Harold V. Wait
John Wakeman, PhD
Nell Pape Waring
F. J. Warren, Jr.
Edis W. Warrif
W. P. Weatherby
Luke T. Webb
Anthony J. Weber
Donald J. Weintritt
Shane Wells
Juergen Wesselhoeft
Richard M. West
Tom West
Albert J. Wetzel
James Henry Wharton, PhD
Ralph E. Wharton
Paul R. Whetsell
Tadeusz Karol Wiewiorowski, PhD
Monty Wilkins
Louis E. Willhoit, Jr., PhD
Oneil J. Williams, Jr.
Ron Williams
Chester G. Wilmot, PhD
Leslie C. Wilson, PhD
Kathleen S. Wiltenmuth
William J. Woessner
William H. Wohler
Andrzej Wojtanowicz, PhD
Laurence Wong, PhD

J. Stuart Wood, PhD
Mike Wood
Adonna M. Works
Roger M. Wright
David B. Wurm, PhD
Tom Wyche
Bruce C. Wyman, PhD
David Allou Yeadon
Michael C. Yohe
L. L. Yoho
Gerald M. Young
Ronnie M. Youngblood
Chester E. Zawadzki
Juan C. Zeik
Fred C. Zeile
Edward J. Zisk
Marvin L. Zochert

Maine

Clayton H. Allen, PhD
Roger B. Allen, PhD
Charles G. Beudette
James V. Bitner
Henry H. Blau, PhD
Robert Perry Bosshart, PhD
Peter B. Brand
John Bridge
Forest D. Briggs
R. S. Chamberlin
Andrew Jackson Chase
Joseph M. Chirnitch
Michael S. Coffman, PhD
H. Douglas Collins
Carl Cowan
Mark R. Daigle
Lee C. Devito
Donald Dobbin
Herbert W. Dobbins
William L. Donnellan, PhD
Stephen Doye
Kenneth M. Eldred
Stephen G. Eldridge
Robert C. Ender
John W. Fitzpatrick
Randall C. Foster
Robert C. Frederich
Joseph M. Genco, PhD
Edward T. Gerry, PhD
Clark A. Granger, PhD
Steven R. Grant
Douglas Griffin
Raymond E. Hammond
R. W. Hannemann
Wilkes B. Harper
Earnest M. Hayes
Grove E. Herrick, PhD
Alan E. Hitchcock
Aaron L. Hoke
William G. Housley, PhD
H. Blane Howell
Terence J. Hughes, PhD
Herbert C. Jurgeleit
Kathy A. Kaake
Charles Kahn
Roger F. Karl
Roland Lebel, PhD
Richard R. Lecompte
Dulcie Lishness
Alan B. Livingstone
Albert F. Lopez

Jerry D. Lowry, PhD
Sylvia B. Lowry
John P. Lynam
Peter L. Madaffari
Charles E. Maguire
Robert Lawrence Martin, PhD
Eithne McCann
Harold K. McCard
Ivan Noel McDaniel, PhD
Ralph R. McDonough
Robert J. McNally
Clark Nichols
Jackson Nichols
David Sanborne Page, PhD
Richard A. Parker
Doris S. Pennoyer
Burce E. Philbrick
Carl R. Poirier
Waldo C. Preble
Glenn Carleton Prescott
Robert W. Rache
Malcolm H. Ray, PhD
Charles Davis Richards, PhD
Thomas L. Richardson
Harold A. Rosene
Robert B. Russell, Jr.
Rudolph P. Sarna
Razi Saydjari
Peter Schoonmaker
Mark D. Semon, PhD
Timothy G. Shelley
Paul L. Sherman
Thomas F. Shields
Charles William Shipman, PhD*
Donald M. Stover
James H. Stuart, PhD
David B. Thurston
John A. Tibbetts
Frank Trask
Erik J. Wiberg
Arthur F. Wilkinson
H. Hugh Woodbury, PhD
Robert E. Zawistowski
Jay Zeamer

Maryland

Roger L. Aamodt, PhD
Donald W. Abbott
Wayne L. Adamson
Winford R. Addison
Perry Baldwin Alers, PhD
Kelsey Alexander
Frederick C. Althaus
Melvyn R. Altman, PhD
Edward J. Ames, III
Martin R. Ames
Kenneth J. Anderer
Charles R. Anderson, PhD
Leif H. Anderson
Ivan J. Andrasik
J. B. Aquilla
Casper J. Aronson
Goro G. Asaki
Raymond J. Astor, Sr.
M. Friedman Axler, PhD
Azizollah Azhdam
James A. Ball
George P. Bancroft
Carroll Marlin Barrack, PhD
Stephen S. Barranco

Carol J. Bartnick, PhD
Warren J. Bayne
Jeffrey P. Beale
Rowland Bedell, Jr.
Harold E. Beegle
Roger A. Bell, PhD
Robert C. Beller
Warren Walt Berning
Joseph M. Bero
Charles Frank Bersch
Arlen D. Besel
Harry E. Betsill
Brooke A. Beyer
Andrew L. Biagioni
William Elbert Bickley, PhD
Barbara O. Black
Thomas T. Blanco
Gilbert Sanders Blevins, PhD
Ralph N. Blomster, PhD
Samuel Blum
Paul W. Boldt
Morris Reiner Bonde, PhD
Michael C. Bonsteel
Warren J. Boord
Wayne E. Booth
Alexej B. Borkovec, PhD
George Bosmajian, PhD
Robert Clarence Bowers
Frank J. Bowery, Jr.
Emanuel L. Brancato
Allan R. Brause, PhD
John P. Brennan
Owen P. Bricker, III, PhD
Philander B. Briscoe
John D. Bruno, PhD
Jeffrey W. Bullard, PhD
Philip Gary Burkhalter
Richard Burkhart
Thomas Butler
Paul M. Butman
William L. Cameron
Frederick W. Camp, PhD
Mark L. Campbell, PhD
Robert J. Cangelosi
Robert N. Carhart
Donald E. Carlson, PhD
C. Jelleff Carr, PhD
David G. Carta, PhD
Lynn K. Carta, PhD
John Paul Carter
Vincent O. Casibang, PhD
Peter Castruccio, PhD
Joel V. Caudill
Nancy L. Centofante, PhD
David Leroy Chamberlain, PhD
John T. Chambers
Yung Feng Chang, PhD
Lloyd W. Charles
Zachary L. Chattler
David W. Christensen
Richard G. Christensen
Ross D. Christman
Alfred Lawrence Christy, PhD
Paul Chung
Bill Pat Clark, PhD
John B. Clark
Leo J. Clark, Jr.
Charles Frederick Cleland, PhD
David A. Cline
Gideon Marius Clore, PhD
Charles A. Clough
John B. Coble

Robert Cohn
James D. Coleman
Todd A. Collins
William Henry Collins
Marco Colombini, PhD
Charles D. Conner, PhD
Lawrence Cooper
Robert S. Cooper, PhD
Jules M. Coppel
William Sydney Corak, PhD
Rebecca B. Costello, PhD
Ross Couwenhoven, PhD
William Crawford
Joseph Presley Crisler, PhD
Thomas Benjamin Criss, PhD
John Cullom
Robert D. Culver
Lawrence E. Cunnick
Ralph L. Cunningham
Dalcio Kisling Dacol, PhD
Charles Dwelle Daniel, PhD
Jayant J. Darji
Wes R. Daub
Leroy Edward Day
Forest C. Deal, Jr., PhD
Thakor M. Desai
Kamalinee V. Deshpande
Armand J. Desmarais
Michael Despines, PhD
Ronald Dewit, PhD
Edmund Armand Dimarzio, PhD
Cecil Malmberg DiNnuno
Kent C. Dixon
Thomas O. Dixon
David B. Doan
David B. Dobson
Abel M. Dominguez, PhD
James E. Donaghy
James E. Drummond, PhD
Timothy P. Dudenhoefer
Joseph J. Dudis, PhD
Thomas E. Dumm
Frederick C. Durant, III
Sajiad H. Durrani, PhD
Assaf J. Dvir
Donald Eccleshall, PhD
Joseph D. Eckard, Jr., PhD
Charles N. Edwards
William Frederick Egelhoff, PhD
Gerald I. Eidenberg
Jokomo A. Ekpaha-Mensah, PhD
Stephen D. Elgin
Thomas W. Eliaren, Jr.
Emil Elinsky
Bruce L. Elliott, PhD
John Elsen
George F. Emch
Philip H. Emery, Jr., PhD
Sarah Eno, PhD
Duane G. Erickson, PhD
Charles M. Ernst
Michael J. Ertel
Neal C. Estand
Norman Frederick Estrin, PhD
Albert Edwin Evans, PhD
William Evans
Phyllis B. Eveleth, PhD
Mostafa A. Fahmy, PhD
Alva R. Faulkner
Maria Anna Faust, PhD
Wayne D. Fegely
David M. Feit

Mary K. Felsted
Ronald L. Felsted, PhD
Charles M. Ferrell
Aldo Ferretti, PhD
Malcolm S. Field, PhD
David Fischel, PhD
Eugene Charles Fischer, PhD
William W. Fitchett, Jr.
Michael M. Fitelson, PhD
Hugh Michael Fitzpatrick
Lewis W. Fleischmann
Alvin R. Flesher
Jeffrey S. Flesher
Joseph H. Flesher
John E. Folk, PhD
Allan L. Forbes
Vaughn M. Foxwell, Jr.
William E. Franswick
Joseph C. Fratantoni
Bernard. A. Free
Ernest Robert Freeman
David J. Fry, PhD
Ronald S. Fry
Sabit Gabay, PhD
William George Galetto, PhD
Bernard William Gamson, PhD
Robert Vernon Garver
Godfrey R. Gauld
Joseph C. Gerstner
Thomas George Gibian, PhD
Eric Giosa
James A. Given
George Gloeckler, PhD
Constance Glover, PhD
Dale P. Glover
James Glynn, PhD
William R. Godwin
Mark J. Gogol
Mikhail Goloubev, PhD
John Roland Gonand, PhD
Donald A. Gooss, Sr.
Stephen Gordey
William B. Gordon, PhD
John W. Gore, Jr.
Willis C. Gore, PhD
Gio Batta Gori, PhD
Owen P. Gormley
William M. Gould
C. Grabowski
Harvey W. Graves
Lucy M. Gray
George Linden Greene, PhD
Gerald Alan Greenhouse, PhD
Erik Gregg
Claude W. Gregory
Joseph R. Greig, PhD
Linda M. Gressitt
Tom Griffiths
Peter L. Grimm
Sigmund S. Grollman, PhD
Philip K. Grotheer
H. Guenterberg
Mustafa Guldu
Alvin Guttag
Rooney Neal Hader
Albert Felix Hadermann, PhD
John Haffner
Richard David Hahn
Mary B. Hakim
Martha L. Hale, PhD
Timothy A. Hall, PhD
Diane K. Hancock, PhD

Patrick J. Hannan
E. M. Hansen
Lawrence R. Harding
George E. Harmening
Bruce W. Harrington
Benjamin L. Harris, PhD
Edward R. Harris
Greg A. Harrison, PhD
Michael H. Hart, PhD
Lumbert U. Hartle
Robert A. Hartley
Todd Hathaway
Francis Hauf
Frank L. Haynes, PhD
Earl Larry Heacock
Donald J. Healy, PhD
John B. Hearn
Robert E. Hedden
Joseph Mark Heimerl, PhD
Philip B. Hemmig
Ronald Hencin, PhD
Andrew L. Henni
Mark A. Hermeling
Richard Allison Herrett, PhD
Frank F. Hertsch, III
Jeffrey L. Hess, PhD
Sol Hirsch
Michael Ho
Viet Hoang
Robert W. Hobbs, PhD
Stephen Hochman
Herbert Hochstein, PhD
Frederick A. Hodge, PhD
Paul Hogroian
Russell Holland, PhD
Jan M. Hollis, PhD
Otto A. Homberg, PhD
Carl F. Hornig
Donald Paul Hoster, PhD
Riley Dee Housewright, PhD
Dwight D. Howard
Franklin P. Howard
Russell D. Howardgully, PhD
Mark T. Hubbard
Sue Huck, PhD
John Stephen Huebner, PhD
Sean P. Hunt
David L. Hursh
Anthony L. Imbembo
John Edwin Innes, PhD
Peter D. Inskip
Patricia A. Irwin
Terri Isakson
Bruce Jacobs
James J. Jaklitsch
Neldon Lynn Jarvis, PhD
John P. Jastrzembski
Kuan-Teh Jeang, PhD
Joseph Victor Jemski, PhD
Donald A. Jennings
Arthur Seigfried Jensen, PhD
LeeAnn Jensen, PhD
Stephen Johanson
Robert V. Johnson, PhD
Harold B. Johnsson, III
Gerald S. Johnston
Alan D. Jones
Charlie E. Jones, PhD
Emmet Jones, PhD
David Joseph, PhD
Stanley Robert Joseph, PhD
John Juliano

Peter V. Juvan
Joseph A. Kaiser, PhD
B. Michael Kanack, PhD
Stanley Martin Kanarowski
Gerson N. Kaplan
Olgerts Longins Karklins, PhD
Thomas J. Karr, PhD
Herman F. Kaybill, PhD
Patrick Norman Keating, PhD
Robert Kelly
Ronald G. Kelsey
Dennis A. Kennedy
James G. Kester
Harry W. Ketchun, Jr.
David R. Keyser, PhD
Chul Kim, PhD
Ralph C. Kirby
Henry Klein
Karl C. Klontz
Frederick R. Knoop
William F. Kobett
Alex Kokolios
William A. Korvin
Robert J. Kostelniki, PhD
Eugene George Kovach, PhD
Allen A. Kowarski
Kenneth Koziol
Kenyon K. Kramer
Robert W. Krauss, PhD
Joseph Henry Kravitz, PhD
Lawrence C. Kravitz, PhD
Ron Kreis, PhD
Peter W. Kremers
Edward A. Kriege
Jacob K. Krispin, PhD
John R. Krouse
Donald M. Krtanjek
Ann Marie Krumenacker
Lorin Ronald Krusberg, PhD
Paul R. Kuraguntla
Chris E. Kuyatt, PhD
Andrew J. Kuzmission
Charles T. Lacy, Sr.
Jaynarayan H. Lala, PhD
Robert J. Lambird
John P. Lambooy, PhD
Malcolm Daniel Lane, PhD
Mike Lauriente, PhD
John F. Lawyer
James A. Lechner, PhD
Matthew A. Lechowicz
Charles H. Lee
Clarence E. Lee
James Lewis Leef, PhD
Herwig Lehmann, PhD
Gregory S. Leppert, PhD
James D. Lesikar, PhD
Howard Lessoff
Gilbert V. Levin, PhD
Alexander D. Leyderman, PhD
Milton Joshua Linevsky, PhD
Thomas R. Lipka
Robert W. Lisle
G. D. Little, Jr.
Donlin Martin Long, PhD
Gerald W. Longanecker
David Loninotti
Philip E. Luft, PhD
William Hamilton Lupton, PhD
Franklin D. Maddox
Carl W. Magee
B. Maghami

Richard Magno, PhD
Robert Maher, PhD
Andrew E. Mance
Naga Bhushan Mandava, PhD
Stephen W. Manetto
Phillip Warren Mange, PhD
Lawrence Raymond Mangen
Martin Marcus
Edward J. Martin
Steven M. Martin
Wayne Martin
Donald Massey
Richard E. Matz
Jean Mayers
David J. Mayonado, PhD
Paul Henry Mazzocchi, PhD
Bruce E. McArtor, PhD*
Jeffrey P. McBride
Maclyn McCarty, Jr., PhD
Frank Xavier McCawley
Daniel G. McChesney, PhD
David H. McCombe
Samuel R. McConoughey
John Dennis McCurdy, PhD
Wilbur Renfrew McElroy, PhD
Edward P. McMahon, PhD
Bryce H. McMullen
Richard B. McMullen
William L. Mehaffey
Robert S. Melville, PhD
Robert J. Melvin
John. N. Menard
Peter L. Metcalf, PhD
M. Kent Mewha
Frederick G. Meyer, PhD
George B. Michel
Wyndham Davies Miles, PhD
Brian C. Miller
Kenneth R. Miller
Michael H. Mirensky
Wayne Mitzner, PhD
John A. Moeller
Matthew H. Moles
W. Bryan Monosmith, PhD
Marcia M. Moody, PhD
Henry F. Moomau
Peter T. Mora, PhD
Alice B. Moran, PhD
Leonard Mordfin, PhD
Jim Morentz, PhD
Joe W. Moschler, Jr.
Arthur B. Moulton
Gary M. Mower
Barbara Mroczkowski, PhD
Joseph J. Mueller
Charles Lee Mulchi, PhD
Jim Mullen
John Irvin Munn, PhD
Wayne K. Murphey, PhD
John Cornelius Murphy, PhD
Scott Myers
Norbert Raymond Myslins, PhD
Zane Elvin Naibert, PhD
Robert A. Neff
George D. Nesbitt
Nicolas G. Nesteruk
William J. Neville
Richard Nieporent, PhD
David T. Nies
Danny E. Niner
Mark A. Noblett
Jim K. Noffsinger, PhD

Thomas E. Nolan
Richard B. North
Nicholas J. Nucci
Robert W. Olwine
Roger P. Orcutt, PhD
Johathan H. Orloff, PhD
Edward T. O'Toole, PhD
William G. Palace
Timothy T. Palmer, PhD
Spyridon G. Papadopoulos
Dale Wayne Parrish, PhD
Marshall F. Parsons
Malcolm Patterson
David Chase Peaslee, PhD
Samuel Penner, PhD
Robert Edward Perdue, Jr., PhD
Russel G. Perkins
Charles A. Pessagno
Stephen G. Petersen
George D. Peterson, PhD
Frank C. Pethel
William Leo Petrie
Edward M. Piechowiak
Kirvan H. Pierson, Jr.
Arthurs H. Piksis, PhD
Lester Y. Pilcher
Joseph T. Pitman
Karl W. Plumlee, PhD
George Pollock
Robert F. Poremski
Richard F. Porter
Louis Potash, PhD
Conrad P. Potemra
James Edward Potzick
Kendall Gardner Powers, PhD
Hullahalli Rangaswamy Prasan, PhD
H. Price
Donald William Pritchard, PhD
Leon Prosky, PhD
William M. Pulford
Herman Pusin
Earl Raymond Quandt, PhD
Bernard Raab, PhD
Mohsen Khatib Rahbar, PhD
John W. Ranocchia, Sr.
Harris E. Reavin
Richard L. Rebbert, PhD
Sterlin M. Rebuck
Charles W. Rector, PhD
James D. Redding
Frank J. Regan
William G. Regotti
Fred M. Reiff
Hugh T. Reilly
Ralph E. Reisler
Chad Reiter, PhD
L. Jan Renfro
Hee M. Rhee, PhD
James K. Rice
Joseph Dudley Richards
Nancy Dembeck Richert, PhD
Michael Richmond
Melvin W. Richter
Claude Frank Riley, Jr.
Lauro P. Rochino
Theodore Rockwell
David F. Rogers, PhD
Frederick A. Rosell
David A. Rossi
Anthony Richard Ruffa, PhD
Kevin C. Ruoff
Jeffrey W. Ryan

M. Kirkien Rzeszotarsk, PhD
Waclaw Janusz Rzeszotarski, PhD
Frank D. Sams
Jerrell L. Sanders
Paul Santiago
Badri N. Satwah
David R. Savello, PhD
William A. Scanga
Mark Thomas Schaefer
Charles W. Scheck
I. Morton Schindler
Joseph J. Schmidt
Philip L. Schmitz
George J. Schonholtz
Ethan Joshua Schreier, PhD
W. L. Schweisberger
Robert G. Sebastian
Frank David Seydel, PhD
Thomas P. Sheahen, PhD
Gennady M. Sherman
Howard S. Shieh
Kevin J. Shields
Kandiah Shivanandan, PhD
John B. Shumaker, PhD
Paul T. Shupert
Edwin B. Shykind, PhD
Ronald E. Siatkowski, PhD
Payson U. Sierer
Jerry Silhan
Joseph Silverman, PhD
Joseph D. Silverstein, PhD
William Simmons
Harjit Singh
Dale Dean Skinner
Paul Slepian, PhD
Bruce H. Smith
David A. Smith, PhD
Gilbert Howlett Smith, PhD
Joseph Collins Smith
Richard L. Smith, PhD
Robert J. Snoddy
Michael R. Snyder
Edward D. Soma
Richard M. Sommerfield
John N. Sorensen
Leo E. Soucek
W. B. Spivey
Pete R. Spuler
Gregory A. Staggers
Charles J. Stahl, III
John Starkenberg, PhD
Jerry E. Steele
William Henry Steigelmann
Louis E. Stein
Howard Odell Stevens
Peter Beekman Stifel, PhD
Scott R. Stiger
Hamilton W. Stiles, Jr.
Thomas L. Stinchcomb
Vojislav Stojkovic, PhD
Robert R. Strauss, PhD
Edward Strickling, PhD
Alan E. Strong, PhD
Rolland E. Stroup
Gene Strull, PhD
Harry H. Suber, PhD
James E. Sundergill
John Supp
Martin Surabian
Hilmar W. Swenson, PhD
Barbara J. Syska
Donald Roy Talbot

John C. Tankersley
Robert Ernest Tarone, PhD
Tom Tassermyer
Tyler Tate
Hari P. Tayal
Louis F. Terenzoni
Nick van Terheyden
Glenn A. Terry, PhD
Charles N. Tesitor
John Aloysius Tesk, PhD
Alan Thornton
Rod S. Thornton
Terry N. Thrasher, PhD
Glenn E. Tisdale, PhD
James G. Topper
Dan E. Tourgee
Edmund C. Tramont
M. J. Travers
Joseph M. Tropp
Nadine Tuaillon, PhD
Herbert M S Uberall, PhD
Kenneth S. Unruh, PhD
George F. Urbancik, Jr.
O. Manuel Uy, PhD
Walter Robert Van Antwerp*
Andrew Van Echo
Lawrence J. Vande Kieft, PhD
James G. Vap
Lyuba Varticouski, PhD
Marc Vatin
James Ira Vette, PhD
James Hudson Vickers
Arthur Viterito, PhD
Clifton O. Wallace, Jr.
William E. Wallace
Robert Jerome Walsh
Charles E. Walter
Donald K. Walter
John Fitler Walter, PhD
John S. Walter
Rudolf R. Walter
Barbara S. Walters
William Walters, PhD
Paul John Waltrup, PhD
John Huber Wasilik, PhD
Thomas L. Watchinsky, Sr.
David J. Waters, PhD
Richard Milton Waterstrat, PhD
Frank M. Watkins
George H. Way
Howard H. Weetall, PhD
Eugene C. Weinbach, PhD
Howard L. Weinert, PhD
Edward Earl Weir, PhD
Gerald C. Wendt
Richard P. Wesenberg
Larry L. West
James C. Wharton
Philip J. Whelan
Edmund William White, PhD
Reed B. Wickner
Dennis A. Wilkie
Scott R. Williams
Vernon L. Williams
Steven P. Willing
James E. Wilson
Clarence A. Wingate
Ernest Winter
John M. Winter, Jr., PhD
Thomas Wisniewski
John B. Wolff, PhD
Winnie Wong Ng, PhD

Keith Woodard
John N. Woodfield
Walter E. Woodford, Jr.
Rita J. Wren
Carol G. Wright
George M. Wright, PhD
Sidney Yaverbaum, PhD
Robert Gilbert Yeck, PhD
Hubert Palmer Yockey, PhD
George. A. Young, PhD
William Nathaniel Zeiger, PhD
Robert George Zimbelman, PhD
John Gordon Zimmerman, PhD
Paul Carl Zmola, PhD
Edwin Zucker
Howard A. Zwemer

Massachusetts

Ann S. Adams
Albert H. Adelman, PhD
Norman Adler, PhD
Vincent Agnello
Mumtaz Ahmed, PhD
Sol Aisenberg, PhD
Harl P. Aldrich, PhD
Charles D. Alexson
Richard Alan Alliegro
Edward E. Altshuler, PhD
Wayne P. Amico
Dean P. Amidon
Jonathan Arata, PhD
Neil N. Ault, PhD
Bob J. Aumueller
Joseph Avruch
Frederic S. Bacon, Jr.
Joseph Anthony Baglio, PhD
James L. Baird
Andrew D. Baker
Sallie Baliunas, PhD
William Warren Bannister, PhD
Fioravante A. Bares, PhD
Allen Vaughan Barker, PhD
Debra Barngrover, PhD
Wylie W. Barrow, Jr.
Samuel M. Barsky
Peter J. Barthuly
F. R. Barys
Carl H. Bates, PhD
Herbert Beall, PhD
Timothy L. Beauchemin
Richard J. Becherer, PhD
Harry Carroll Becker, PhD
John C. Becker
John J. Benincasa
Sharon Benoit
Edward J. Benz
Daniel Berkman
Robert L. Bertram
Richard J. Bertrand
Robert E. Beshara
Peter R. Beythen
Georgy Bezkorovainy
Maneesh Bhatnagar
Joseph D. Bianchi
Conrad Biber, PhD
Yan Bielek
Klaus Biemann, PhD
Seymour J. Birstein
Aaron Led Bluhm, PhD
Bruce Plympton Bogert, PhD

Norman Robert Boisse, PhD
John A. Bologna
Gregory M. Bonaguide
Laszlo Joseph Bonis, PhD
Robert S. Borden
Kenneth Jay Boss, PhD
Edward A. Boulter
Kevin Bourque
John F. Bovenzi, PhD
H. Kent Bowen, PhD
Clifford W. Bowers
Sidney A. Bowhill, PhD
Colin Bowness, PhD
Karl W. Boyer
John J. Brady
Robert J. Breen
Sean P. Brennan
Lawrence S. Bright
Ernest F. Brockelbank
Ronald F. Brodrick
Delos B. Brown
Richard D. Brown, PhD
Robert G. Brown
Walter Redvers John Brown
Wilson L. Brown
Alex A. Brudno, PhD
Barry W. Bryant, PhD
Arnold R. Buckman
Avraam Budman
H. Franklin Bunn
Michael D. Burday
Charles Burger
E. A. Burke
Stephen E. Burke
Ronald T. Burkman
George H. Burnham
James H. Burrill, Jr.
Jennifer Burrill
Joyce Burrill
Richard S. Burwen
Inez L. Busch
Scott E. Butler, PhD
E. B. Buxton
Valery K. Bykhovsky, PhD
Frank Cabral, Jr.
Michael S. Cafferty
Daniel L. Cahalan, PhD
Daniel E. Caless
Wendell J. Caley, Jr., PhD
John M. Calligeros
Allan Dana Callow, PhD
Nicholas A. Campagna, Jr.
Harry F. Campbell
Donald A. Carignan
Arthur Carpenito
Jack William Carpenter, PhD
Jerome B. Carr, PhD
James C. Carroll
Jeffrey A. Casey, PhD
Renata E. Cathou, PhD
Larry Cecchi
Joseph F. Chabot
Antohny J. Chamay
Kenneth S.W. Champion, PhD
Charles H. Chandler
Robin Chang, PhD
Robert A. Charpie, PhD
Dennis William Cheek, PhD
Zafarullah K. Cheema, PhD
Chi-Hau Chen, PhD
Mark R. Chenard
Ronald B. Child

Alm Christensen
Robert T. Church
Donald John Ciappenelli, PhD
Daniel E. Clapp
Richard H. Clarke, PhD
John T. Clarno
James J. Cleary
John P. Cocchiarella
Michael D. Coffey
Lawrence B. Cohen
Jonathan Cohler
Jerry M. Cohn
Jennifer Cole, PhD
Martin Cole
Anthony J. Colella
Sumner Colgan
Robert J. Conrad
Robert F. Cooney
Robert D. Cotell
Eugene E. Covert, PhD
John Merrill Craig
John B. Creeden
Peter Crimi, PhD
Edward Crosby
Robert P. Cunkelman
Maria A. Curtin, PhD
Cameron H. Daley
George F. Dalrymple
Anthony D. D'Ambrosio
Balkrishna S. Dandekar, PhD
John T. Danielson, PhD
Joseph Branch Darby, Jr., PhD
Leo G. Darian
Eugene Merrill Darling, Jr.
Thonet Charles Dauphine
Donald J. David, PhD
Charles F. de Ganahl
Ronald K. Dean
Frank D. Defalco, PhD
John A. Defalco
Carlo J. DeLuca, PhD
Kevin A. Demartino, PhD
Don W. Deno, PhD
Eolo D. Derosa
Martin A. Desantis
David P. Desrosiers
Marshall Emanuel Deutsch, PhD
Robert M. Devlin, PhD
Pedro Diaz
Robert A. Donahue, PhD
Alejandro M. Dopico, PhD
Paul S. Doucette
James Merrill Douglas, PhD
Westmoreland J. Douglas, PhD
George T. Dowd
Margaret Driscoll
Robert C. Dubois
Harold Fisher Dvorak
Stephen M. Dyer
Bruce P. Eaton
Eric B. Eby
George F. Edmonds, Jr.
Andrew E. Ellis
Sherif Elwakil, PhD
L. R. Engelking, PhD
John E. Engelsted, PhD
Richard A. Enos
Nancy Enright
Klaas Eriks, PhD
Alvin Essig
T. Farber
Richard J. Farris, PhD

Ernest H. Faust, Jr.
James T. Fearnside
Robert D. Feeney, PhD
Zdenek F. Fencl
William Andrew Fessler, PhD
Edward Mackay Fettes, PhD
Kenneth G. Fettig
Fred F. Feyling
John Fierke
Melvin William First, PhD
Fred Fisher
Edward L. Fitezmeyer, Jr.
Maurice E. Fitzgerald, PhD
John Ferard Flagg, PhD
Carl E. Flinkstrom
John L. Flint
Harold W. Flood
David Flowers
Timothy Fohl, PhD
John H. Folliott
Clifford J. Forster
Christopher H. Fowler
Irving H. Fox
Myles A. Franklin
Martin S. Frant, PhD
Al A. Freeman
Hanna Friedenstein
Melvin Friedman
Peter Friedman, PhD
Francis S. Furbish, PhD
Neil A. Gaeta
Christopher P. Gagne
John P. Gallien
Ethan Charles Galloway, PhD
Richmond Gardner
Joseph M. Gately
Robert F. Gately
George F. Gehrig
Grantland W. Gelette
Vlasios Georgian, PhD
Elliot L. Gershtein
Adolf J. Giger, PhD
Steven D. Gioiosa
Horst E. Glagowski
Homer Hopson Glascock, II, PhD
Julie Glowacki, PhD
Valery Anton Godyak, PhD
Jeff M. Goldsmith
Leonid Goloshinsky
Maya Goloshinsky
Alex Gonik
Brian G. Goodness
Scharita Gopal
Marina J. Gorbunoff, PhD
Alan B. Goulet
James T. Gourzis, PhD
Nicholas John Grant, PhD
Jurgen Michael Grasshoff, PhD
John F. Gray
Howard Green
Anton C. Greenland, PhD
William T. Greer
James B. Gregg
Joseph W. Griffin, Jr., PhD
Reginald M. Griffin, PhD
Linda Griffith, PhD
Rosa Grinberg, PhD
David P. Gruber
Nicholas J. Guarino
John E. Guarnieri
Peter H. Guldberg
Joseph M. Gwinn

Douglas C. Hahn
Robert Simspon Hahn, PhD
Charles W. Haldeman, III, PhD
John R. Hall, PhD
Arnold D. Halporn, Jr.
Priscilla O. Hamilton, PhD
John C. Hampe
Linda C. Hamphill
Bruce E. Hanna
Bruce L. Hanson
Charles S. Hatch
James W. Havens
Robert L. Hawkins, PhD
John Heard
Milton W. Heath, Jr.
Thomas S. Hemphill
Richard G. Hendl, PhD
Jozef H. Hendriks
Brian B. Hennessey
Karl P. Hentz
Hans Heinrich Wolfgang Herda, PhD
David J. Herlihy
Thomas F. Herring
Jeffrey L. Herz
George Herzlinger, PhD
Carl E. Hewitt, PhD
Richard E. Hillger, PhD
Charles H. Hillman
Charles J. Hinckley
Harold Hindman
John C. Hoagland
J. D. Hobbs
Melvin Clay Hobson, PhD
Lon Hocker, PhD
George Raymond Hodgins
Mark A. Hoffman, PhD
Donald J. Hoft
Joseph L. Holik, PhD
Lawrence R. Holland
F. Sheppard Holt, PhD
Lowell Hoyt Holway, PhD
Francis J. Hopcroft
Harold H. Hopfe
Donald P. Horgan
Ken Hori
Mark Lee Horn, PhD
John E. Huguenin, PhD
George William Ingle
Lucy Ionas
Martin Isaks, PhD
Gerald William Iseler, PhD
Max Ito, PhD
Kenneth M. Izumi, PhD
James G. Jacobs
Lenard Jacobs
Joseph H. Jessop
H. William Johansen, PhD
Alan W. Johnson
Richard B. Johnson
H. M. Jones, PhD
Harry B. Jones
Thomas A. Jordan
George J. Kacek
Robert A. Kalasinsky
Antoine Kaldany
James B. Kalloch
Matthew S. Kaminske
Paul D. Kearns
James W. Keating
Maura E. Keene
Charles Kelley, PhD
Mason E. King

Edwin G. Kispert, PhD
John F. Kitchin, PhD
Michael J. Kjelgaard
Donald J. Kluberdanz
A. G. Kniazzeh, PhD
Bryan S. Kocher, PhD
Robert D. Kodis
F. Theodore Koehler
Edward C. Koeninger
Janos G. Komaromi
Dennis J. Kopaczynski
Richard E. Koppe
Harold J. Kosasky
Michael F. Koskinen
Kerry L. Kotar, PhD
Paul Krapivsky, PhD
Christopher P. Krebs
Henry C. Kreide
John Gene Kreifeldt, PhD
Jacqueline Krim, PhD
Boris B. Krivopal, PhD
Louis Kuchnir
Amarendhra M. Kumar, PhD
Thomas Henry Kunz, PhD
Jing-Wen Kuo, PhD
Glenn H. Kutz
Robert W. Kwiatkowski
Ludmila K. Kyn
Melvin Labitt
Martin R. Lackoff, PhD
John Lahoud
Richard R. Langhoff
Augustin F. Lanteigne
L. Lasagna
Eric J. Lastowka
Robert L. Laurence, PhD
Robert A. Lavache
Jerome M. Lavine, PhD
Gilbert R. Lavoie, PhD
Paul D. Lazay, PhD
Louis A. Ledoux
Kenneth M. Leet, PhD
David Leibman
Andrew Z. Lemnios, PhD
John F. Lescher
Karen N. Levin
Robert E. Levin, PhD
Yuri Levin
Linda J. Levine
Howard L. Levingston
Phil Levy
Robert P. Leyden
Barry J. Liles, PhD
William Tenney Lindsay, PhD*
Richard S. Lindzen, PhD
Stuart J. Lipoff
Janet M. Lo
John H. Lok
Morgan J. Long
Albert W. Lowe
J. Lowenstein, PhD
Henry E. Lowey
Howard W. Lowy
Richard E. Lundberg
David Lunger
Gordon F. Lupien
Sidney John Lyford, PhD
Leah Lyle
Scott Lyle
Jenner Maas
Brian MacDonald
Charles L. Mack, Jr.

Charle Aram Magarian
John S. Magyar
Joseph D. Malek
David H. Mallalieu
Nicholas J. Mandonas
Harold J. Manley
J. Mannion
John A. Manzo
Richard W. Marble
Robert W. Marculewicz
Mikhail D. Margolin
Richard M. Marino, PhD
William A. Marr, PhD
Peter D. Martelly
Eric Martz, PhD
James V. Masi, PhD
Mihkel Mathiesen, PhD
Charles F. Mattina, PhD
Thomas E. Mattus
Robert S. Mausel
John Elliott May, PhD
Ernst Mayr, PhD
Siegfried T. Mayr, PhD
Janes W. McArthur
David W. McCabe
Russell F. McCann
John Joseph Gerald McCue, PhD
Kilmer S. McCully
Sean D. McEnroe
Leonard J. McGlynn, Jr.
James B. McGown
Gary H. McGrath
Peter D. McGurk
Edward J. McNiff
Kenneth Earl McVicar
Donald K. Medeiros, Sr.
Frank J. Meiners
Ivars Melingailis, PhD
Roland W. Michaud
Joseph J. Mickiewicz
Klaus A. Miczek, PhD
Joseph J. Miliano
Fanny I. Milinarsky
Richard James Millard
Robert C. Miller, PhD
Philip G. Milone
John R. Mirabito, Jr.
Harold W. Moore
David A. Morano, PhD
Richard D. Morel
Raymond C. Morgan
Michael F. Morris
Walter J. Mozgala
Michael J. Mufson
George F. Mulcahy
Richard F. Mullin
John D. Munno
Luba Mushkat
Rafael I. Mushkat
Paul J. Nagy
Romen M. Nakhtigal
Francis J. Nash, Jr.
Robert R. Nersasian, PhD
Paul Nesbeda, PhD
Walter P. Neumann
Phillip E. Neveu
James L. Nevins
Paul M. Newberne, PhD
Sydney I. Newburg
Walter R. Niessen
Donald W. Noble
James A. Nollet

Carole M. Noon
John C. Ocallahan, PhD
Daniel J. O'Connor
Kenneth C. Ogle
Bernard X. Ohnemus
Thomas O'Leapy, Jr.
William P. Oliver, PhD
Curtis R. Olsen, PhD
Alexander Olshan
Ludmica Olshan
Dexter A. Olsson
Austin S. O'Malley
Peg Opolski
Robert I. Orenstein
Charles Ormsby, PhD
Demetrius George Orphands, PhD
Curtis M. Osgood
Robert Osthues
John D. Palmer, PhD
M. M. Palmer
Robert T. Palumbo
Dana M. Pantano, PhD
Leonard Parad
W. N. Parsons, PhD
Bhupendra C. Patel
Mangal Patel
Brian Patten, PhD
Charles J. Patti
Robert J. Pawlak
Raymond Andrew Paynter, PhD
Edward J. Peik
Ronald Peik
Steven J. Pericola
Stig Persson
Mikhail I. Petaev, PhD
Paul R. Petcen
Roland O. Peterson
Samuel Petrecca
Louis John Petrovic, PhD
Eleanor Phillips, PhD
Richard Arlan Phillips, PhD
Sanborn F. Philp, PhD
Charles A. Pickering
Allan Pierce, PhD
William J. Pietrusiak
Joseph L. Polidoro
Jay M. Portnow, PhD
David L. Post
Pranav M. Prakash
Susan M. Prattis, PhD
David S. Prerau, PhD
George D. Price
Ronald L. Prior, PhD
Daniel Puffer
Evan R. Pugh, PhD
James R. Qualey, PhD
Ryan D. Quam
Prosper E. Quashie
Arthur Robert Quinton, PhD
Steve Rabe
Harold R. Raemer, PhD
Keen A. Rafferty, PhD
Nicholas D. Raftopoulos
Alvin O. Ramsley
David Rancour, PhD
Thomas Raphael, PhD
Amram Rasiel, PhD
Dennis Rathman, PhD
John A. Recks
Robert Harmon Rediker, PhD
David W. Rego
Arnold E. Reif, PhD

Oded A. Rencus
Harris Burt Renfroe, PhD
Charles T. Reynolds
John R. Rezendes
Allyn St. Clair Richardson, PhD
Edward J. Rickley
Donald E. Ridley
Stephen Joseph Riggi, PhD
Paul B. Rizzoli
Robert S. Rizzotto
Charles D. Robbins
Francis Donald Roberts, PhD
Abraham P. Rockwood
Robert W. Rodier
Peter Rogers, PhD
Randy Rogers
Peter D. Roman
Thomas Ronay, PhD
Peter E. Roos
Anthony V. Rosa
Byron Roscoe
Philip Rosenblatt
Stuart Rosenthal
James N. Ross, Jr., PhD
Sanford Irwin Roth
William Clinton Roth
Raymond A. Rousseau
Robert T. Rubano
Alan Rubin, PhD
Paula A. Ruel
Richard F. Russo
Jean L. Ryan, PhD
Carroll J. Ryskamp
Faina Ryvkin, PhD
Richard J. Salmon
Erdjan Salth, PhD
Alan M. Salus
Dominick A. Sama, PhD
Sarmad Saman, PhD
George W. Sampson
Oli A. Sandven, PhD
Joseph B. Sangiolo
Michele S. Sapuppo
Richard Sasiela, PhD
Paul S. Satkevich
Daniel R. Saunders
Carol L. Savage
Samuel Paul Sawan, PhD
Charles J. Scanio, PhD
Jeffrey Schenkel
Allan L. Scherr, PhD
Martin Schetzen
Elliot R. Schildkraut
Mikhail Schiller
Warner H. Schmidt
William Harris Schoendorf, PhD
Bradford R. Schofield
Nick R. Schott, PhD
Stephen Schreiner, PhD
Hartmut Schroeder, PhD
Paul K. Schubert
Manfred B. Schulz, PhD
Yael A. Schwartz, PhD
Henry G. Schwartzberg, PhD
Bruce W. Schwoegler
Stylianos P. Scordilis, PhD
Ryan S. Searle
Norman E. Sears
George Seaver, PhD
Walter Sehuchard
Sal J. Seminatore
Edward C. Shaffer, PhD

Frank G. Shanklin
Edward Kedik Shapiro, PhD
Ralph Shapiro, PhD
S. Shayan
Dov J. Shazeer, PhD
Vedat Shehu, PhD
Daniel H. Sheingold
Paul S. Shelton
Hao Ming Shen, PhD
Thomas B. Shen, PhD
Vivian Ean Shih
George A. Shirn, PhD
Steven E. Shoelson, PhD
Earl Henry Sholley
Werner E. Sievers
Maya Simanovsky
Leo Siminavosky, PhD
Christpher Simpson
Robert I. Sinclair
Joseph J. Singer
William M. Singleton
Richard Henry Sioui, PhD
Norman Sissenwine
Cary Skinner
Kenneth J. Skrobis, PhD
Gilbert Small, Jr., PhD
Douglas E. Smiley
Nathan Lewis Smith, PhD
Verity C. Smith
Edna H. Sobel
Monique S. Spaepen
Howard C K Spears
Timothy Alan Springer, PhD
Archibald D. Standley, Jr.
Mark A. Staples, PhD
James C. Stark, PhD
Frank A. Stasi
Hermann Statz, PhD
Thomas N. Stein
Peter Stelos, PhD
James Irwin Stevens
Harold E. Stinehelfer
James H. Stoddard, PhD
Donald B. Strang
E. Whitman Strecker
Jerrold H. Streckert
Peter Sultan, PhD
Paul J. Suprenant
Laurence R. Swain
Oscar P. Swecker
George J. Szecsei
Geza Szonyi, PhD
John R. Taft
Daniel Joseph Tambasco, PhD
Jeffery P. Tannebaum
Peter E. Tannewald, PhD
Allan F. Taubert, Jr.
Kenneth J. Tauer, PhD
Kenneth E. Taylor
John W. Telford, PhD
Joseph Ignatius Tenca
Robert J. Thompson
George W. Thorn
Serge Nicholas Timasheff, PhD
Nancy S. Timmerman
Dean Tolan, PhD
Kurt Toman, PhD
Vladimir Petrovich Torchilin, PhD
Diogenes R. Torres, PhD
Ildiko Toth, PhD
Paul S. Tower
C. David Trader, PhD

Paul J. Trafton, PhD
T. C. Trane
James W. Trenz
Edwin P. Tripp, III
Donald E. Troxel, PhD
Brian Edward Tucholke, PhD
Ben Tuval
Yuri Tuvim, PhD
David L. Tweedy
James F. Twohy
George W. Ullrich
Eric Edward Ungar, PhD
Robert O. Valerio
Robert P. Vallee, PhD
W. Van De Stadt
Larry Van Heerdan
Thomas J. Vaughan
Eugene D. Veilleux
Erich Veynl
Alfred Viola, PhD
Otto Vogl, PhD
Michael A. Volk
Edward A. Vrablik
Edward R. Wagner
Janet M. Walrod
Myles A. Walsh, PhD
Peter R. Walsh
Kurt D. Walter
Michael D. Walters, PhD
Roy F. Walters
John F. Wardle, PhD
Alan A. Wartenberg
P. C. Waterman, PhD
James L. Waters
John F. Waymouth, PhD
Raymond N. Wear
William P. Webb
Bruce Daniels Wedlock, PhD
Roger P. Wessel
Paul Westhaver
Burton Wheelock
Kenneth Steven Wheelock, PhD
Douglas White, PhD
William R. White
Richard H. Whittle
Mel Wiener
George Friederich Wilgram, PhD
Leonard Stephen Wilk
Denise Williams
Byron H. Willis, PhD
David J. Wilson, PhD
James W. Winkelman
Dave Withy
George M. Wolfe
James A. Wolstenholme
Kenneth S. Woodard
Leonard W. Woronoff
Noga K. Woronoff
Donald Prior Wrathall, PhD
Robert D. Wright
John M. Wuerth
Geert J. Wyntjes
John C. Yarmac, III
Marvel J. Yoder, PhD
William C. Young
Chester W. Zamoch
Gregory G. Zarakhovich, PhD
Inna Y. Zayas
Harold Zeckel, PhD
Dong Er Zhang, PhD
Ming Zhang, PhD
Alexander Zhukovsky, PhD

Thomas E. Zipoli
Martin Vincent Zombeck, PhD
Vincent M. Zuffante

Michigan

Douglas R. Abbott
V. Harry Adrounie, PhD
Henry Albaugh
Dave Alexander
Amos Robert Anderson, PhD
Craig A. Anderson
James R. Anderson
Jon C. Anderson
Mitchell Anderson
R. L. Anderson
Rodney C. Anderson, PhD
Peter R. Andreana, PhD
Mohammed R. Ansari
Clifford B. Armstrong, Jr.
Robert Lee Armstrong, PhD
Charles G. Artinian
Tom Asmas, PhD
Donald W. Autio
Charles R. Bacon, PhD
Lloyal O. Bacon
Mary Bacon
Terry Baker
George Bakopoulos
Jeffrey R. Bal
Carr W. Baldwin
Ronald L. Ballast
Dan W. Bancroft
Rudolph Neal Band, PhD
Richard A. Barca
Thomas Barfknecht, PhD
Michael J. Barjaktarovich
Daniel D. Barnard
Tom Barnard
Josh T. Barnes
Donald J. Barron
Ceo E. Bauer
Frederick Bauer, PhD
Thomas Bauer
Barry A. Bauermeister
Wallace E. Beaber
Bruce A. Beachnau
Joseph M. Beals
William Boone Beardmore, PhD
Wayne D. Beasley
William W. Beaton
Carl G. Becker
Bob W. Belfit
Dwight A. Bell
Thomas G. Bell, PhD
Daniel Thomas Belmont, PhD
John V. Bergh
Ernest Bergman
Jeffrey J. Best
Robert Bilski
James R. Bishop
Charles E. Black
Tomas H. Black
Henry S. Blair
Luther L. Blair
John R. Blaisdell
Dav Blakenhagen
Barbara Blass, PhD
Brian Bliss
Larry D. Blumer
John G. Bobak

Lawrence C. Boczar
David Edwin Boddy, PhD
Willard A. Bodwell
Wladimir E. Boldyreff
Elizabeth Bolen, PhD
Scott Boman
William Bond
Lawrence P. Bonicatto
Philip Borgending
David P. Borgeson
Dwight D. Bornemeier, PhD
Steven L. Bouws
Alan D. Boyer
David Brackney
Gary L. Bradley
Peggy A. Brady, PhD
Mark S. Braekevelt
Alan David Brailsford, PhD
James A. Brandt
Albert J. Brant
Webb Emmett Braselton, PhD
Harvey H. Braun
James I. Breckenfeld
Ross J. Bremer
Bart J. Bremmer
Michael J. Brennan
William A. Brett
Darlene Rita Brezinski, PhD
Blaschke Briggs
Dale Edward Briggs, PhD
Phil N. Brink
Paul Brittain
B. J. Broad
Tammy A. Brodie
William J. Broene
Richard K. Bronder
James S. Brooks
John W. Broviac
J. Brower
Richard B. Brown, PhD
Richard R. Brown
Roy T. Browning
Burton Dale Brubaker, PhD
Douglas B. Brumm, PhD
Paul O. Buchko
Charles D. Bucska
Stephen G. Buda
David E. Bullock
David Richard Buss, PhD
Stephen L. Bussa
Ceil Bussiere
Frank L. Butts
Forrest K. Byers
David Byrd
Derek Byrd
Charles Cain, PhD
Louis Capellari
Darrel E. Cardy
F. L. Carlsen, Jr.
Peter E. Carmody
Paul Carolan
E. Louis Caron
Glenn S. Carter
Steven K. Casey
William J. Cauley
Robert Champlin, PhD
Daniel W. Chapman
C. Robert Charles
Benjamin Frederic Cheydleur
Lawrence O. Chick
Dale J. Chodos
Sue C. Church

Gene S. Churgin
Joseph R. Cissell
Robert Malden Claflin, PhD
William Clank
Tim Clarey, PhD
John Alden Clark, PhD
John C. Clark
John R. Clark
Kent Clark
Richard P. Clarke, PhD
Bill Cleary
Paul W. Clemo
William G. Clemons
M. Gerald Cloherty
Samuel W. Coates, PhD
James L. Coburn, PhD
Chester Coccia
Flossie Cohen
Paul L. Cole
A. Collard
Ralph O. Collter
William B. Comai
John F. Conroy
Robert C. Cook
Daniel M. Cooper
Bahne C. Cornilsen, PhD
John D. Cowlishaw, PhD
R. E. Craigie, Jr.
Ronald Cresswell, PhD
Dale Scott Cromez
Dean A. Cross
C. Richard Crowther, PhD
Warren B. Crummett, PhD
Kenneth D. Cummins
Merlyn Curtis
Charles Eugene Cutts, PhD
Robert C. Cyman
Werner J. Dahm, PhD
Paul D. Daly
Nicholas Darby, PhD
Forrester B. Darling
Kent R. Davis
Ralph Anderson Davis
H. M. DeBoe
Daniel DeBoer
Dennis B. Decator
James L. Delahanty
E. F. Delitala
John A. DeMattia
Frank Willis Denison, PhD
Robert Dennett
Glenn O. Dentel
David R. Derenzo
Larry D. Dersheid
Maruthi N. Devarakonda, PhD
Jaap B. Devevie
Marco J. Di Biase
Aaron M. Dick
Macdonald Dick, II
Reynold J. Diegel, PhD
Jerry A. Dieter, PhD
Alma Dietz
David Ross Dilley, PhD
Pryia C. Dimantha
Jim V. Dirkes
James D. Dixon, PhD
Thomas A. Doane
Herbert Dobbs, PhD
Timothy F. Dobson
Glen R. Dodd
Dale E. Doepker
John Domagala, PhD

Thomas Donahue
B. Donin
Thomas MC Donnell
Michael Cameron Drake, PhD
Gary M. Drayton
Michael T. Drewyor
James F. Driscoll, PhD
S. C. Dubois
LeRoy Dugan, PhD
Gary Rinehart Dukes, PhD
L. Jean Dunegan
Duane F. Dunlap
Harold Dunn
William A. Dupree
James A. Durr
Beecher C. Eaves
Earl A. Ebach, PhD
Gordon Eballardyce
Floyd S. Eberts, PhD
Dale P. Eddy
Robert Egbers
Val L. Eichenlaub, PhD
Jacob Eichhorn, PhD
Paul J. Eisele, PhD
Jack R. Elenbaas
Hans George Elias, PhD
Richard I. Ellenbogen
Duane F. Ellis
Robert W. Ellis, PhD
Rodney Elmore
Mark Elwell, PhD
Gordon H. Enderle
Don J. Erickson
Lars Eriksson
Shanonn E. Etter
De James J. Evanoff
Bradley Evans
David Hunden Evans, PhD
Leonard Evans, PhD
Eugene C. Fadler
Robert Feisel
William John Felmlee
Robert Fenech
Howard Ferguson
James H. Fernandes
Verno Fernandez, PhD
Marion S. Ferszt
Terence M. Filipiak
Charles Richard Finch, PhD
John Bergeman Fink, PhD
Donald W. Finney, Jr.
Henry P. Fleischer
Gary R. Foerster
Jay Ernest Folkert, PhD
Dwain Ford, PhD
Gary L. Foreman
Mary R. Forintos
Ingemar Bjorn Forsblad, PhD
Guy J Del Franco
Bruce A. Frandsen
Douglas Frantz
Lewis G. Frasch
Nile Nelson Frawley, PhD
James L. Frey
Wilfred Lawson Freyberger
Robert Edsel Friar, PhD
Thomas W. Fritchek
Edwin H. Frowbieter, Jr.
James H. Frye
Lyle E. Funkhouser
James A. Gallagher, PhD
Harendra Sakarlal Gandhi, PhD

Kent Gardner
H. Richard Garner
Jay B. Gassel
Geoffrey C. Geisz
Charles B. Gentry
John W. Gesink, PhD
Lawrence J. Giacoletto, PhD
Randy Gibbs
Kernon M. Gibes
Timothy P. Gilberg
Warren D. Gilbert
Tony Gill
Gary Gillespie
Thomas D. Gillespie, PhD
Tim Gilson
Dan Glazier
Robert I. Goldsmith
Herbert H. Goodwin
Sanford Jay Gorney
Gary J. Gorsalitz
Stephen R. Gorsuch
Vincent J. Granowicz
Geraldine Green, PhD
Marvin L. Green
Charles W. Greening
Henry V. Greenwood
David Henry Gregg, PhD
John D. Grier
Robert F. Gurchiek
Robert J. Gustin
Robert A. Haapala
Herald A. Habenicht
William C. Haefner
William Carl Hagel, PhD
Russell H. Hahn
Forest E. Haines, PhD
Raymond M. Halonen
George J. Hambalgo
Ralph E. Hamilton
Philip Hampton
Martin J. Hanninen
Daniel L. Hanson
Stewart T. Harman
Timothy A. Harmsen
Arthur B. Harris
B. Harris
Bruce V. Harris
James Edward Harris
Keith N. Harris
Kenneth Harris
William A. Harrity
Albert Hartman
William Louis Hase, PhD
Irwin O. Hasenwinkle
Richard E. Haskell, PhD
Bruce D. Hassen
Gregory J. Havers
James P. Hebbard
Donald B. Heckenlively, PhD
John Heckman, PhD
James B. Heikkinen
Kevin G. Heil
Richard L. Heiny, PhD
Ed W. Hekman
Alan K. Hendra
Owen A. Heng, PhD
William A. Hennigan
Raymond L. Henry, PhD*
William J. Henry
Neal Hepner
Jerry L. Herrendeen
Luis F. Herrera

Todd W. Herrick
Robert A. Herzog
John C. Hill, PhD
Raymond Hill, Jr.
Gregory D. Hines
John R. Hines
Marshall Hines
W. G. Hines, PhD
Jack T. Hinkle, PhD
Jack Wiley Hinman, PhD
Jivig Hinman, PhD
Roger R. Hinshaw
Kenneth Hinze
Ralph J. Hodek, PhD
Charles P. Hodgkinson
John P. Hoehn, PhD
Kenneth M. Hoelst
Richard Hofsess
David V. Holli
Don H. Holzhei, PhD
William John Horvath, PhD*
Robert L. Hosley
Eric C. Houze
Scott Hover
Robert George Howe, PhD
Thomas J. Hrubovsky
Eugene Yuching Huang
John A. Hudak
Larry Hudson
Harold L. Hughes
John R. Hughes
Bradley Eugene Huitema, PhD
Calvin T. Hulls
Mary Anne Hunter
Frank Hussey
John M. Hutto
David Peter Hylander
Bradley S. Hynes
Marvin H. Ihnen
Raymond Ingles
David C. Irish
Donald Richard Isleib, PhD
Asaad Istephan, PhD
John F. Itnyre
Thomas R. Jackson
Richard G. Jacobs
Ray Jaglowski
Arnold Jagt
C. C. Jakimowicz
Daryl N. James
Borek Janik, PhD
Edward S. Jankowski
Dennis C. Jans
Steven M. Japar, PhD
John A. Jaszczak, PhD
Bernard Jerome
Guy C. Jeske
Gerald E. Johnson
Gordon E. Johnson
James S. Johnson
Frank Norton Jones, PhD
Bernard William Joseph
Raymond P. Joseph
Richard Harvey Kabel
Michael G. Kalinowski
Jeremy M. Kallenbach, PhD
Brian D. Kaluzny
Margaret A. Kaluzny
Joseph M. Kanamueller, PhD
Alexander A. Kargilis
Neil R. Karl
Lawrence A. Kasik

Raymond J. Kastura
Jack H. Kaufman
Thomas H. Kavanagh
Charles Kelly
Peter Kelly
Mike Kendall
Wilbur W. Kennett
William G. Kern
Majiid Khalatbari
J. S. King, PhD
Horst Kissel
Karel Kithier, PhD
Edgar W. Kivela, PhD
Bruce Holmes Klanderman, PhD
Rolf E. Kleinau
Gerald J. Kloock
Charles Philip Knop, PhD
Wayne Knoth*
Chung-Yu Ko
Ludovik F. Koci
William P. Koelsch
John C. Koepele
Adam J. Kollin
Michael K. Kondogiani
Joseph P. Konwinski
Dale W. Koop
Raoul Kopelman, PhD
R. Kosarski
Robert Allen Koster, PhD
Louis C. Kovach
James E. Kowalczyk
Adam Kozma, PhD
Ted Kozowyk
Kenneth Kramer
Tom J. Krasovec
Clyde Harding Kratochvil, PhD
David J. Krause, PhD
John Belshaw Kreer, PhD
James W. Kress, PhD
Matthew P. Kriss
Kenneth S. Kube
David J. Kubicek
Edward P. Kubiske
Richard E. Kuelske
Andrew Kujawiak
Gerard J. Kulbieda
Norman R. Kurtycz
Martin D. Kurtz
Clayton La Pointe, PhD
Edward R. Lady, PhD
Thomas J. Laginess
Kenneth Lagrand
Donald Joseph Lamb, PhD
Jeffrey C. Lamb
Richard W. Lambrecht, Jr.
Harold H. Lang, PhD
Ronald A. Lang
George R. Lange
William R. Langolf
Richard L. Lanier
James J. Laporte
Vernon L. Larrowe, PhD
Elisabeth Larsen
Eric R. Larsen, PhD
Kenneth Larson
Michael G. Last
Ruth C. Laugal
Daniel G. Laviolette
Walter E. Lawrence
Edward J. Lays
Ruben C. Legaspi
John M. Leinonen

F. Lemke, PhD
Geoffrey Lenters, PhD
Dennis J. Leonard, PhD
Joseph W. Leone
Bruno Leonelli
Frederick C. Levantrosser
Roy W. Linenberg
John C. Linton
Michael Forrester Lipton, PhD
Kurt R. List
Georgetta S. Livingstone, PhD
Edward T. Lock
James E. Lodge
Steve Loduca, PhD
Lawrence Hua Hsien Louis, PhD
Don L. Loveless
Cole Lovett, PhD
Marvin Lubbers
Frederick E. Lueck
Mike Lueck
Donald R. Lueking, PhD
Robert Luetje
Randolph M. Luke
James R. Lumley
Dwight E. Lutsko
Dick J. Macadams
John F. Macgregor
John T. Madl
Wiliam Thomas Magee, PhD
Edward W. Maki
Eugene R. Maki
Ernest W. Malkewitz
David J. Maness
L.B. Mann
Nasrat George Mansoor
George E. Marks
George E. Maroney, PhD
Dennis Marshall
Norman B. Marshall, PhD
Richard Marshall
George H. Martin, PhD
Ronald H. Martineau
Conrad Jerome Mason, PhD
John L. Massingill, Jr., PhD
Mark L. Matchynski
Mundanilkuna A. Mathew, PhD
David C. Matzke
Augustin D. Matzo
Daniel W. May
S. Mazil, PhD
William C. McAllister
Shaun Leaf McCarthy, PhD
Leslie Paul McCarty, PhD
Neil McClellan
Edward L. McConnell
Donald Alan McCrimmon, PhD
Michael C. McDermit
William J. McDonough
Robert D. McElhaney
Jeffrey A. McErlean
Daniel R. McGuire
Richard J. McMurray
Ruth D. McNair, PhD
Dennis McNeal
Dan B. McVickar, PhD
Ken Mead
Rodney Y. Meade
Loida S. Medina
Dale J. Meier, PhD
Robert J. Meier, Jr., PhD
Peter D. Meister, PhD
Joseph Meites, PhD

George W. Melchior, PhD
Bohdan Melnyk
Albert R. Menard, PhD
George D. Mendenhall, PhD
Roberto Merlin, PhD
James S. Merlino
Herman Merte, PhD
Donald Irwin Meyer, PhD
Heinz Friedrich Meyer, PhD
Howard J. Meyer
Robert F. Meyer, PhD
Eugene Mezger
Joseph M. Michalsky
David Michelson
Carol J. Miller, PhD
Herman L. Miller
John W. Miller, PhD
Steven J. Miller
Francis J. Mills, III
Jack F. Mills, PhD
James A. Mills
Mark M. Miorelli
William J. Mitchell
Philip V. Mohan
Lydia Elizabeth Moissides-Hines, PhD
Lawrence J. Moloney
David G. Mooberry
Leonard O. Moore, PhD
Richard Anthony Moore, PhD
Ronald Morash, PhD
Eduardo A. Moreira
Gene E. Morgan
Mark A. Moriset
Charles R. Morrison
Timothy B. Mostowy
Bob Mottice
Peter Roy Mould, PhD
Craig Brian Murchison, PhD
Pamela W. Murchison, PhD
Raymond Harold Murray
Donald Louis Musinski, PhD
James G. Musser
Wesley F. Muthig
Marvin Myers
Michael J. Nadeau
Edward M. Nadgorny, PhD
Robert F. Nagaj
Champa Nagappa, PhD
Ray Nalepka
Robert J. Nankee
Mark R. Napolitan
Terry T. Neering
Gary R. Neithammer
Jan G. Nelson
Keith Nelson
A. David Nesbitt
Carl C. Nesbitt, PhD
Martin Newcomb, PhD
Russell D. Newhouse
Joseph P. Newton
Jerome C. Neyer
Roberta J. Nichols, PhD
Kathleen M. Nicholson
Alfred Otto Niemi, PhD
Bob Niemi
James M. Nieters
Harmon S. Nine, PhD
Ivan Conrad Nordin, PhD
J. Nordin
Jim T. Nordlund
David C. Norton

Raymond Francis Novak, PhD
William R. Nummy, PhD
Richard R. Nunez
Richard Allen Nyquist, PhD
Le Roy T. Oehler
Walter K. Ogorek
Stanley P. Oleksy*
Earl D. Ollila
Duane A. Olson
Gary Orvis
George F. Osterhout
Gary Ovans
Thomas L. Paas
Jorge A. Pacheco
Edward D. Pachota
Karur R. Padmanabhan, PhD
Glenn Palmbos
William Pals
Jerome C. Pando, PhD
Jack S. Panzer
Jack Parker
Gary Pashaian
Bharat K. Patel
Michael K. Paul
James Marvin Paulson, PhD
Alfred A. Pease
Roger Peck
Dana Pelletier
William T. Pelletier, PhD
Paul S. Pender
J. Percha
Calvin R. Peters
James B. Peters
William J. Peters
Edward Charles Peterson
Valeri Petkov, PhD
Frederick Martin Phelps, PhD
Perry T. Piccard
Anton J. Pintar, PhD
James C. Plagge, PhD
Alan Edward Platt, PhD
Daniel E. Pless
Susan Pless
Charles E. Plessner
Robert E. Plumley
Howard K. Plummer
James W. Pollack
Joseph M. Post
Darrell L. Potter
Harold Anthony Price, PhD
Kenneth Prox
Andrzej Przyjazny, PhD
Siegfried Pudell
E. Dale Purkhiser, PhD
Rena H. Quinn
Todd L. Rachel
Michael T. Radvaw
Charles F. Raley, PhD
Dario Ramazzotti
Sonia R. Ramirez
Eero Ranta
Robert E. Rapp
John W. Rebuck, PhD
Foster Kinyon Redding, PhD
William M. Redfield
Bart J. Reed
C. Reed, PhD
Robert R. Reiner
Richard F. Reising, PhD
Robert F. Reynolds
Kathy A. Rheaume
Anthony J. Rhein

William Bennett Ribben, PhD
Allan Richards
Jeffrey E. Richards
Gregory B. Rickmar
Otto K. Riegger, PhD
Paul E. Rieke, PhD
Stanley K. Ries, PhD
James Rigby, PhD
James H. Rillings, PhD
Beverly Riordan
Harlan Ritchie, PhD
Glen A. Roberts
Bernard I. Robertson
George Henry Robinson
Mary T. Rodgers, PhD
Karen J. Roelofs
Ignatius A. Rohrig
David J. Romenesko
Norman L. Root
Douglas N. Rose, PhD
Leonard Rosenfeld
Paul E. Rosten
Jean E. Russell, PhD
Michele M. Ryan
Timothy M. Ryan
Charles Joseph Ryant, Jr., PhD
Carol Saari
Steven T. Salli
Vernon Ralph Sandel, PhD
John F. Sandell, PhD
Peter P. Sandretto
Greg Savasky
Todd S. Schaedig
Carl Schafer
Albert L. Schaller
Carl Alfred Scheel, PhD
Francis Matthew Scheidt, PhD
Alexander W. Schenck
Edgar H. Schlaps
Harold E. Schlichting, PhD
Robert A. Schmall, PhD
Curtis S. Schmaltz
Earl A. Schmidt
Helen Schols
Keith Schreck
Paul E. Schroeder
Arthur W. Schubert
Norman W. Schubring
Robert W. Schubring
Albert Barry Schultz, PhD
Donald D. Schuster
Richard C. Schwing, PhD
Cathie Jo Seamon
Philip A. Seamon
Martin F. Seitz
Richard R. Seleno
Arthur Seltmann
Robert Seng
David Shah
Shirish A. Shah, PhD
Donald H. Shaver
Mark C. Shaw
Stephen R. Sheets
Dan Shereda
Richard D. Show
Martin Sichel, PhD
Charles H. Sidor
Larry D. Siebert
Kent S. Siegel
Michael D. Singer, PhD
Lal Pratap S. Singh, PhD
Julie A. Sinnott

Fred Z. Sitkins, Sr.
John E. Skidmore
Harold T. Slaby, PhD
David F. Slater
Nelson S. Slavik, PhD
Joh Slaybaugh
T. Wentworth Slitor
George Slomp, PhD
James Blair Smart, PhD
Herbert J. Smedley
Hadley James Smith, PhD
Louis G. Smith
Paul Dennis Smith, PhD
Rick Smith
Ron Smith
C. R. Snider
Robert James Sokol
Don V. Somers
Richard E. Sonntag, PhD
Nick Sorko-Ram
James H. Sosinski
Gary W. Sova
Chris J. Spaseff
Steven Tremble Spees, PhD
John Leo Speier, PhD
Robert F. Spink
Henry J. Spiro
John G. Spitzley
Kenneth E. Spray
Peter Springsteen
Gordon W. Squires
Dave Stamy
Charles Stanich
Reed R. Stauffer
David A. Steen, PhD
Kevin J. Steen
J. Stephanoff
George L. Stern
William C. Stettler, PhD
Al Worth Stinson
Donald Lewis Stivender
Richard C. Stouffer
James Shive Strickland, PhD
Edwin Joseph Strojny, PhD
John H. Stunz, Jr.
Darrel G. Suhre
James A. Surrell
Todd W. Sutton
Jon R. Swanson, PhD
J. A. Sweet
LeRoy Swenson
Don Eugene Swets
Richard Swiatek
Ronald J. Swinko
Joseph Vincent Swisher, PhD
B. J. Szappanyos
Greg A. Szyperski, PhD
Edward F. Tallant
Carl I. Tarum
Frank J. Taverna
B. Ross Terry
Raymond Thacker, PhD
Dilli J. Thapa
Alan W. Thebert
John Thebo
Don L. Theis
Sean Theisen
Joe C. Thomas
Steve Thomas
E. A. Thompson
Ted Thompson
Thomas Thornton

Richard Perry Tison
Stephen Winter Tobey, PhD
Mary Ann Tolker
David W. Tongue
Calvin Douglas Tormanen, PhD
Pascual E. Tormen
William N. Torp
Paul Eugene Toth
Mike Touchinski
Donald J. Treder
Tom Troester
James Trow, PhD
Judith Lucille Truden, PhD
Michael A. Tubbs
William A. Turcotte
Richard R. Turkiewicz
Roy C. Turnbull
Almon George Turner, Jr., PhD
Joan L. Tyrrell
Peter I. Ulan
Charles R. Ulmer
Robin J. Ungar
Kurt E. Utley, PhD
Richard L. Van De Polder
Richard Michael Van Effen, PhD
Clayton Edward Van Hall, PhD
Keith D. Van Maanen
Verlan H. Van Rheenen, PhD
Kamala J. Vanaharam
Thomas L. Vanmassenhove
William Vecere
Roger F. Verhey, PhD
George E. Vogel
Paul A. Volz, PhD
John P. Von Plonski
Howard Voorhees, PhD
Jaroslav J. Vostal
A. R. Wagner
Richard E. Wainwright
Donald E. Waite
F. A. Walas
Rick S. Wallace
Bruce W. Walters
John Walther
Jan A. Wampuszyc
Brian Warner
Peter O. Warner, PhD
Alfred B. Warwick
Frederick H. Wasserman
Michael T. Wattai
Gary S. Way
Michael R. Weber
Walter J. Weber, Jr., PhD
Alfred R. Webster
Gordon J. Webster
James W. Weems
Warren O. Weingarten
William J. Weinmann
Allan Orens Wennerberg
Stanley J. Wenzlick
Leslie Morton Werbel, PhD
James L. Wesselman
Edgar Francis Westrum, PhD
Norris C. Wetters
Joseph W. Whalen, PhD
Thomas J. Whalen, PhD
Russell P. Whitaker
Alan B. Whitman, PhD
Ronald W. Whiton
Gregg B. Wickstrom
Anthony Widenmann, III
Thomas D. Wiegman

Tom V. Wilder
Gary J. Wilkins
Robert A. Wilkins, PhD
Peirre A. Willermet, PhD
Donald H. Williams, PhD
Jefrey F. Williams, PhD
William W. Willmarth, PhD
Douglas J. Wing
Martin J. Wininger
Edward J. Wolfrum, PhD
Susan E. Wolfrum
Christoper S. Wood
Greg J. Woods
Richard D. Woods, PhD
Clark D. Woolpert
K. Worden
Gordon Wright, PhD
Weimin Wu, PhD
Walter P. Wynbelt
Ning Xi, PhD
Stuart Yntema, Sr.
Brenda Young
David Caldwell Young, PhD
Joseph A. Zagar
Eugene F. Zeimet
Albert Fontenot Zellar, PhD
Jiri Zemlicka, PhD
Roger Ziemba
Richard J. Zimmer

Minnesota

Terry D. Ackman
Bryan C. Adams
Paul Bradley Addis, PhD
Ronald R. Adkins, PhD
Gene P. Andersen
Ingrid Anderson, PhD
Ken Anderson
Nathan Anderson
B. M. Anose, PhD
Dana Arndt
Orv B. Askeland
Bryan Baab
Ronald R. Bach, PhD
A. Richard Baldwin, PhD
Keith P. Barnes
Douglas W. Barr
Blaine W. Bartz
Milton Bauer
Wolfgang J. Baumann, PhD
Brian P. Beecken, PhD
Richard Behrens, PhD
Andrew H. Bekkala, PhD
Delfin J. Beltran
David M. Benforado
Jody A. Berquist
Ralph H. Bertz
Jack N. Birk
Rolland L. Blake, PhD
Rodney L. Bleifull, PhD
William R. Block
Todd E. Boehne
Clay B. Bollin
Maurice M. Bowers
Susan H. Bowers
Robert B. Bradley
Raymond J. Brandt
Arvid J. Braun
E. J. Bregmann
Charles W. Bretzman

Allan D. Brown
Roderick B. Brown
Stephen M. Brzica
Richard L. Buchheit
Donald Burke
R. Bruce Burton
Mary E. Butchert
Denise Butler
James Calcamuggio
Elwood F. Caldwell, PhD
Herbert L. Cantrill
David J. Carlson
Orwin Lee Carter, PhD
Victor M. Castro
Jim Caton
Eugene Chao
John W. Chester
Terry R. Christensen, PhD
Arnold A. Cohen, PhD
Mark W. Colchin
Mariette Cole, PhD
James A. Collinge
Kent W. Conway
Robert Kent Crookston, PhD
Donald D. Dahlstrom
Moses M. David, PhD
Thomas Jonathan Delberg, PhD
Fletcher G. Driscoll, PhD
Terrance W. Duffy
Wayne K. Dunshee
Dedi Ekasa
Wayne G. Eklund
James H. Elleson
Paul John Ellis, PhD*
Richard F. Emslander
Arthur E. Englund
Douglas J. Erbeck, PhD
John Gerhard Erickson, PhD
John A. Eriksen
Lee M. Espelan
Robert W. Everett, PhD
Craig T. Evers, PhD
Eric E. Fallstrom
Homer David Fausch, PhD
Daniel A. Feeney
Keith Fellbaum
Herbert John Fick
Stephen D. Fisher
Eugene Flaumenhaft, PhD
Carolyn R. Fletcher
Dean G. Fletcher
Terrence F. Flower, PhD
William R. Forder
David William Fox, PhD
Melvin Frenzel
Melchior Freund*
Frank D. Fryer
Mark A. Fryling, PhD
John Gaffrey
Mary Carol Gannon, PhD
Frank Germann
Harold E. Goetzman
Lawrence Eugene Goodman, PhD
William Grabski
Max Green
Gregory Greer
Troy Gregory
Carl L. Gruber, PhD
Sam Gullickson
Kelleen M. Gutzmann
Jeff Hallerman
Arthur Hamburgen

George Charles Hann
Steven Hanson
Jonathan Hartzler, PhD
Peter Havanac
Dean J. Hawkinson
Jerry W. Helander
Neil R. Helming
Ryan Henrichsen
Tara Henrichsen
Paul Henriksen
Donald W. Herrick, PhD
Frederick George Hewitt, PhD
David A. Himmerich
Al Hoidal
Charles D. Hoyle, PhD
Mark D. Huschke
Valentin M. Izraelev, PhD
Wayne D. Jacobson
Rodney Jasmer
Mark T. Jaster
Jean Jenderlco
Timothy Berg Jensen, PhD
Robert P. Jeub
James R. Johnson, PhD
Scott Johnson
Richard W. Joos, PhD
William P. Kamp, PhD
Frank D. Kapps
Robert J. Kaszynski
Michael Peter Kaye
Charles Keal
Allan H. Kehr
Patrick L. Kelly
Paul T. Kelly
Frank W. Kemp
James L. Kennedy
Bridget R. King
Donald W. Klass
William P. Klinzing, PhD
Roger C. Klockziem, PhD
Charles Kenneth Knox, PhD
David Kohlstedt, PhD
Richard A. Kowalsky
Michael S. Kuhlmann
Joseph M. Kuphal
John J. Lacey, Jr.
Richard Lacher
Robert F. Lark
Ashley V. Larson
Allen Latham
Wayne Adair Lea, PhD
R. Douglas Learmon
Scott A. Lechtenberg
Brian W. Lee, PhD
Bruce Legan, PhD
Mike Lehman
Ernest K. Lehmann
Wendell L. Leno
Roland E. Lentz
Donald A. Letourneau
Benjamin Shuet Kin Leung, PhD
Wyne R. Long
Donald Hurrell Lucast, PhD
Mariann Lukan
Rufus Lumry, PhD
Richard G. Lunzer
Mac Macalla
James D. MacGibbon
Jay Mackie
John Maclennan
John R. Manspeaker
Jean A. Marcy-Jenderko

William N. Marr
W. N. Mayer
W. T. McCalla
John McCauley
Tom McNamara
William H. McNeil
Igor Melamed, Sr.
Robert C. Melchior, PhD
David L. Mellen
Frank Henry Meyer
Maurice W. Meyer, PhD
Daniel W. Mike
K. Milani
David W. Miller
Stephen A. Miller
Robert Moe
Robert Leon Moison
Glenn D. Moore, PhD
David L. Mork, PhD
Howard Arthur Morris, PhD*
Dave Mueller
Edward S. Murdock, PhD
Richard C. Navratil
Kenneth H. Nebel
Kevin F. Nigon
Wayland E. Noland, PhD
Frank Q. Nuttall, PhD
Richard P. Nyberg
August J. Olinger
Leonard G. Olson
Mark G. Olson, PhD
Joseph Wendell Opie, PhD
Charlotte Ovechka, PhD
John S. Owens
Gordon Squires Oxborrow
Richard Palmer
Robert E. Palmquist
Guy R. Paton
Timothy A. Patterson
John W. Paulsen
Alfred Pekarek, PhD
Michael Pestes
Donald G. Peterson
Michael R. Peterson
Steven F. Peterson
Douglas D. Pfaff
John N. Pflugi
Frederic Edwin Porter, PhD
Russell C. Powers
Thomas F. Prehoda
Randy M. Puchet
Steven M. Quinlan
Byron K. Randall
Steven T. Ratliff, PhD
Nancy C. Raven
Richard J. Reilly
Timothy J. Reilly
Joseph E. Richter
Kristin Riker-Coleman
David Joel Rislove, PhD
Janis Robins, PhD
Glen M. Robinson, PhD
Robert G. Robinson, PhD
Robert Rosene
Janet M. Roshar, PhD
Olaf Runquist, PhD
Peter A. Rzepecki, PhD
Albert L. Saari, PhD
Wilmar Lawrence Salo, PhD
Wade D. Samson
Richard M. Sanders, PhD
Peter K. Sappanos

Jay Howard Sautter, PhD
Paul Savaryn
Curt C. Schmidt
Tony Schmitz
Thomas W. Schmucker
Oscar A. Schott
Gerald Schramm
Merle D. Schule
Anthony Schulz
Jeanette Schulz
Kevin Schulz, PhD
James W. Seaberg
Dave Seibel
James M. Sellner
James C. Sentz, PhD
James B. Serrin, PhD
Arlen Raynold Severson, PhD
Dennis F. Shackleton
George P. Shaffner
Mark W. Siefken, PhD
William E. Skagerberg
Frank J. Skalko, PhD
Neil A. Skogerboe
Ivan Hooglund Skoog, PhD
Norman Elmer Sladek, PhD
Kenneth Sletten
Aivars Slucis
Chad J. Smith
Bryan Smithee
David Perry Sorensen, PhD
Harold G. Sowmam, PhD
Dale R. Sparling, PhD
D. Dean Spatz
Edward Joseph Stadelmann, PhD*
Leon Stadtherr, PhD
Larry A. Stein
Truman M. Stickney
Sandy Stone
Bart A. Strobel
Patrick Suiter
Bruce M. Sullivan
Arlin B. Super, PhD
Frederick Morrill Swain, PhD
David R. Swanberg
Brian M. Swanson
Robert M. Swanson
Brion P. Szwed
Robert T. Tambornino
Gerald T. Tasa
Gregory D. Taylor
Henry William Tenbroek
Walter Eugene Thatcher, PhD
Brenda J. Theis
James A. Thelen
Mark Thoma
Herbert Bradford Thompson, PhD
Mary E. Thompson, PhD
Richard David Thompson
Arnold William Thornton, PhD
Edward A. Timm
Patrick A. Tuzinski
John R. Tweedy
Oriol Tomas Valls, PhD
William R. Vansloun
Gloria E. Verrecchio
George M. Waldow
David R. Wallace
James R. Waller, PhD
E. C. Ward
Robert Wardin, PhD
Douglas Eugene Weiss, PhD
Rod Wells

James E. Wennen
James F. Werler
Clarence L. Wesenberg
Darrell J. Westrum
Robert W. Whitmyer
John F. Wilkinson
Daryl P. Williamson
Richard D. Williamson
Ronn A. Winkler
Jerry Witt, PhD
Mark Wolf
Bruce Frederick Wollenb, PhD
Eric Woller
John Woods, PhD
Nancy Zeigler
Richard R. Zeigler
William J. Zerull
Daryl E. Zuelke

Mississippi

Hamed K. Abbas, PhD
William W. Adams
Phillip S. Ahlberg
Richard G. Ahlvin
Timothy A. Albers
Robert T. Van Aller, PhD
William M. Ashley
Charles Edward Bardsley, PhD
Lawrence R. Baria
Rolon Barnes
F. Scott Bauer
Fred H. Bayley
James T. Bayot, PhD
Henry Joe Bearden, PhD
Jared M. Becker
Fred E. Beckett, PhD
Albert G. Bennett, PhD
Roy A. Berry, PhD
Curt G. Beyer
Harold L. Blackledge
Michael Blackwell
Hamid Borazjani, PhD
Steven T. Boswell
John C. Bourgeois
William E. Bowlus
Cynthia R. Branch
John R. Brinson
Ronald A. Brown, PhD
George A. Brunner
Stanley F. Bullington, PhD
Harry Dean Bunch, PhD
Rufus T. Burges
William P. Burke
Charles C. Bush
Carolyn M. Buttross
Paul D. Bybee, Jr.
Charles E. Cain, PhD*
Curtis W. Caine
S. Pittman Calhoun
James E. Calloway, PhD
Christopher P. Cameron, PhD
John J. Carley
Kenneth G. Carter
Garland L. Cary
Sidney Castle
Paul David Cate
David L. Chastain
J. Mike Cheeseman
Alfred P. Chestnut, PhD
Le Roy H. Clem, PhD

Charles B. Cliett
Avean Wayne Cole, PhD
Billy E. Colson
George E. Colvin
F. Dee Conerly, Jr.
Sara J. Cooper
Gordon E. Cordes
Sidney G. Cox
Robert M. Crout
Kay Crow
David B. Crump
Verne L. Culbertson
Robert William Cunny
Marion E. Davenport
Donald D. Davidson
Don Davis
Sandra Davis
Paul Day
Ken Dillard
Wanda L. Dodson, PhD
Donald Warren Emerich, PhD
Rick L. Ericksen
Edward Estalote, PhD
Henry P. Ewing, PhD
Thomas C. Ewing
Clarence S. Farmer
Robert William Farwell, PhD
Harold D. Fikes
William E. Finne
Robert Rodney Foil, PhD
William R. Ford
Marcial D. Forester
Arley G. Frank, PhD
Melvin H. Franzen
Richard T. Furr
James B. Furrh, Jr.
Lanelle Guyton Gafford, PhD
Robert S. Gaston
B. J. Gilbert
Roy Glenn
Allan Guymon, PhD
John D. Haien
Duane E. Haines, PhD
David Hancek
Carey F. Hardin
Julian C. Henderson
Clifton L. Hester, Jr.
John Pittman Hey, PhD
Robert M. Hildenbrand
William G. Hillman
Jack D. Hilton
Gerald R. Hodge
James R. House, III
Gary A. Huffman
Dudley J. Hughes
Arthur Scott Hume, PhD
Harold R. Hurst, PhD
Stephen L. Ingram, Sr.
Jonathan Janus, PhD
Charles G. Johnson
Christopher W. Jones
Johnny E. Jones
Paul Jurik
Harold E. Karges
Elizabeth A. Keene, PhD
Gerald W. Kinsley
Wilbur Hall Knight
Richard S. Kuebler
Bobby D. Lagrone
Willem Lamartine Lamar
Elmer W. Landess
Tommy Leavelle, PhD

Leslie Leigh
George L. Leonard
Nelson J. Letourneau, PhD
Clarence D. Lipscombe, III, PhD
Henry R. Logan
Mark J. Losset
Allen Lowrie
Dennis L. Lundberg, PhD
Danny L. Magee
Kenneth R. Magee
Angel Kroum Markov
W. Maret Maxwell
Phillip W. McDill
Calvin P. McElreath
John McGowan
Thomas D. McKewen
Louis D. Meegehee
Maurice A. Meylan, PhD
Cindy Miller
Dalton L. Miller
Donald Piguet Miller, PhD
Richard T. Miller
Dale Mitchell
Ray E. Mitchell
John L. Montz
Thomas C. Moore
Kenneth W. Moss
Vernon B. Muirhead
Robert Muller
William L. Nail
Grey L. Neely, Jr.
Isaac A. Newton, Jr.
Stephen Oivanki
Marvin L. Oxley
Henry L. Parker
Blakeslee A. Partridge
Bernard Patrick
Richard Patton, PhD
Emil H. Pawlik
Tom L. Pederson
Biuy R. Powell
James W. Pressler
Edward W. Puffer
John A. Pybass
Frank A. Raila
Charles C. Randall
James W. Rawlins, PhD
James H. Rawls
Armando Ricci, Jr.
Karl A. Riggs, PhD
Robert Wayne Rogers, PhD
A. Kirk Rosenhan
Thomas E. Rosser
Stephen P. Rowell
Ken D. Ruckstuhl
Ernest Everett Russell, PhD
Jacob B. Russell
Edward C. Schaub
Amy M. Schmidt
Robert Schneeflack
Thomas Schwager
Lawrence L. Seal
Francis G. Serio
John Herbert Shamburger
Bernard L. Shipp
Joseph A. Shirer
Gary Shupp
Harold A. Simmons, PhD
Carl Simms
Donald L. Smith
James Winfred Smith, PhD
Gary Sneed

Harry V. Spooner, Jr.
Vance Glover Sprague, Jr.
Maitland A. Steele
James O. Stephens
Darryl Stewart
William A. Stewart
Robson F. Storey, PhD
Jeffrey Summers
Ronald J. Tarbutton
Thomas G. Tepas
James A. Thornhill
H. Thornton
Michael M. Tierney
A. C. Tipton, Jr.
Roger Townsend
Glover Brown Triplett, PhD
E. H. Tumlinson
Manson Don Turner, PhD
Michael M. Turner
Waheed Uddin, PhD
John M. Usher, PhD
Robert T. Van Aller, PhD
George B. Vockroth
Thurman C. Waller
Joe D. Warrington
Zahir U. Warsi, PhD
Charles L. Wax, PhD
Charles W. Werner, PhD
Lyle D. Wescott, PhD
Raymond Westra, PhD
Randall H. White
Stacy White
Charles E. Williams
Gene D. Wills, PhD
Wilbur William Wilson, PhD
Robert Womack, Jr.
E. Greg Wood, III
Donald E. Yule

Missouri

Anthony W. Adams
Richard L. Adams
C. William Ade
Richard John Aldrich, PhD
Robert W. Allgood
Robert W. Andersohn
Mary P. Anderson
Richard Alan Anderson, PhD
Richard M. Andres, PhD
Mel Andrews
Keith Angle
Charles Apter, PhD
Marcia L. Arnold
John H. Arns, Jr
John P. Atkinson
Robert J. Bachta
John J. Bailey, Jr.
Lee J. Bain, PhD
Robert D. Baker, PhD
Connie L. Baldwin
Lorry T. Bannes
Charles L. Barry
Rose Ann Bast, PhD
Dane G. Battiest
John G. Bayless
Carl L. Beardeh
David A. Beerbower
Donald T. Behrens
Abdeldjelil Belarbi, PhD
Max Ewart Bell, PhD

799

David E. Bemath
Roger Berger
Robert C. Beste
Harold Victor Biellier, PhD
Daniel R. Billbrey
Charles F. Bird
Wayne Edward Black, PhD
Michael M. Blaine
J. E. Blanke
Joseph Terril Bohanon, PhD
Carl D. Bohl
Craig H. Boland
Dean Bolick
J. B. Boren, PhD
Thomas G. Borowiak
Louis Percival Bosanquet, PhD
John W. Botts
Gary Boyer, PhD
Denis Boyle, PhD
Erwin F. Branahl
Anne F. Bransoum
Brian L. Brau
Stanton H. Braude, PhD
Robert L. Breckenridge
Monroe F. Brewer
William J. Bridgeforth
Steven R. Brody
Olen Ray Brown, PhD
Robert E. Brown, PhD
Jeff E. Browning
Michael A. Brueckmann
John J. Bruegging
Don D. Brumm, II
Billy L. Bruns
Lester G. Bruns, PhD
David A. Bryan, PhD
Michael R. Buckheit
Jim E. Buckley
Eugene B. Buerke
Matt J. Bujewski
Martin Bunton
Larry Burch
Joe T. Burden
Joel G. Burken, PhD
George F. Burnett
Joe D. Burroughs
John Frederick Burst, PhD
Stanley J. Butt
Eugene C. Bybee
Christopher I. Byrnes, PhD
Raymond J. Caffrey
Lynn B. Calton
Paul Cameron
Martin Capages, Jr., PhD
Will Dockery Carpenter, PhD
Hugh Carr
Marino Martinez Carrion, PhD
John Carroll, PhD
Barry D. Carter
J. K. Cassil
John Arthur Caughlan
Harold Chambers, PhD
Kenneth P. Chase
Kuang Lu Cheng, PhD
Lou Chenot
Henry H. Chien, PhD
Larry B. Childress
David T. Chin, PhD
Troy L. Chockley
Baek-Young Choi, PhD
Rich Christianson
Rudy Clark

Sally Clark
Kent H. Clotfelter
David K. Cochran, PhD
Jane E. Collins
Keith R. Conklin
Jack L. Conlee
Gheorghe M. Constantinescu, PhD
Alfred A. Cook
Creighton N. Cornell
Raymond C. Cousins, PhD
Micheal A. Cowan
Braden S. Cox
John M. Cragin, PhD
Ronald W. Craven
Clara D. Craver
Bradford R. Crews
Thomas Bernard Croat, PhD
Stephen Crouse
Frazier Curt
D. Dankel
Kenton T. Davidson
Brent Davis
James A. Davis, PhD
Bruce Dawson
John A. Deal
Jack Dederich
Lynn I. Demarco
Randall G. Dempsey
Jeffery R. Derrick
Kevin A. Deschler
R. Deufel, PhD
Joseph E. Devine, PhD
James D. Dexter
Shawn M. Diederich
Charles Dietrick
Nicholas T. Dimercurio
Mark A. Ditch
Donald A. Ditzler
Marvin P. Dixon, PhD
Jim Doehla
William Waldie Donald, PhD
Nirunjan K. Doshi
Eric Doskocil, PhD
Robert R. Dossett
Michael Gilbert Douglas, PhD
William L. Drake, Jr.
Matthew G. Dreifke
Randall G. Dreiling
Raymond L. Driscoll
Gil L. Dryden, PhD
Richard Edward Dubroff, PhD
Paul J. Dumser
B. B. Dunagan
Henry H. Dunham, PhD
Robert G. Dunn, Jr.
Jamese Durig, PhD
Reginald W. Dusing
Roger W. Dutton
Ronald R. Dutton
Thomas A. Dvorak
Donovan A. Eberle
Tom M. Edelman
Jozef W. Eerkens, PhD
William C. Eggers
Barry R. Eikmann
Michael J. Elli
Edward Mortimer Emer, PhD
Edgar G. H. Emery
Adolph M. Engebretson, PhD
Ron A. Engelken
Charles P. Etling
James L. Fallert

Mark E. Farnham
Milo M. Farnham
Carl H. Favre
A. N. Feelem
Walter C. Feld
David M. Felkner
Dean R. Felton
Frederic L. Fenier
Kenneth P. Ferguson
Byron Dustin Field
Quay G. Finefrock
James L. Fletcher, Jr.
Steven T. Fogel
Denis Forster, PhD
Frank Cavan Fowler, PhD
Todd L. Fraizer
David M. Fraley, PhD
Harold J. Frank
George E. Franke
Gordon R. Franke, PhD
Michael J. Frayne
Karen S. Frederich
Max Freeland, PhD
Gary W. Friend
Clark H. Fuhlage
Wm. Terry Fuldner
Michael E. Gaddis
Robert Gene Garrison, PhD
Ray A. Gawlik
Daryl D. George, PhD
James C. Gerdeen, PhD
Damian J. Gerstner
Jerry D. Gibson
Ken Giebe
Leo Giegzelmann
James W. Gieselmann
Melvin E. Giles
Howard D. Gillin
Steven P. Gilmore
Homer F. Gilzow
Richard Joel Gimpelson
Mark Ginoling
Conway Goehman
Royal D. Goerz
Terry L. Gooding
Norman E. Goth
Todor K. Gounev, PhD
Gary Graeser
George Stephen Graff
Tom R. Grass
Mary S. Graves
James Greenhaw
Mary A. Grelinger
Kurt L. Gremmler
Don Griese
Mike Griese
Brian S. Griffen
Virgil Vernon Griffith, PhD
Barbara Grimm
Louis John Grimm, PhD
Arthur N. Griswold
Herb W. Gronomeyer
Fred H. Groves, PhD
Elaine Guenther
Paul E. Guidry
Brian T. Hackett
Earl R. Hackett
Jacqueline Hackett
R. C. Hafner
Henry E. Haggard
Eugene W. Hall
Johnnie E. Hall

James W. Hamilton, PhD
Adrian J. Hampshire
Lewis F. Hancock
John P. Hansen, PhD
Ronald A. Hansen
Willis Dale Hanson
Lawrence R. Happ
Nicholas Konstan Harakas, PhD
William B. Hardin
Christopher Harman
Brian Harrington
Melissa Hart
Pam C. Hart
Randall E. Hart
Clay E. Haynes
Ross Melvin Hedrick, PhD
Roger W. Heitland
Duane Helderlein
James C. Hemeyer
Bruce C. Hemming, PhD
Steve D. Hencey
Stephen J. Henderson
Marty C. Henson
John Frederick Herber, PhD
Carl K. Herkstroeter
Troy L. Hicks, PhD
Gene H. Hilgenberg
Jack Filson Hill, PhD
Jerry L. Hofman
K. Hogan
Troy J. Hogsett
David Eldon Hoiness, PhD
Tammy Holder
Robert L. Hormell
Edward E. Hornsey, PhD
William Burtner House, PhD
Jack E. Howard
James P. Howard
Tina M. Howerton
Robert B. Hudson
Wayne Huebner, PhD
James B. Hufham, PhD
Randy L. Hughes
Michael D. Hurd
Victor E. Hutchison
Frank D. Illingworth
Jeffrey A. Imhoff
Vincent J. Imhoff
Walter L. Imm
Hobart L. Jacobs
Jerome S. Jacobsmeyer
Leon U. Jameton
Clifford S. Jarvis
Theresa Jeevanjee, PhD
Max Jellinek, PhD
Guy E. Jester, PhD
George M. Johnson, IV
James W. Johnson, PhD
Peter F. Johnson
Richard Dean Johnson, PhD
Robert G. Johnson
George Johnston
Robert W. Jonasen
Phillip A. Jozwiak
William Jud
Frederick J. Julyan, PhD
Hemendra A. Kalbamna
Doris Kalita
Salma B. Kamal, PhD
Joe C. Kamalay, PhD
Alan A. Kamp
Earl J. Kane

Arnold F. Kansteiner, PhD
Raymond Kastendieck
Brian C. Kaul, PhD
John D. Keich
Ed Kekec
Roy Fred Keller, PhD
Amy L. Kelly
Marion Alvah Keyes, IV
Ali M. Khojasteh
John B. Kidney
Charles N. Kilo
Justin A. Kindt
Philip A. King
Jerome W. Kinnison
Jeffrey Kiviat
Raymond Kluczny, PhD
Charles E. Knight
Ronald A. Knight, PhD
Bedford F. Knipschild
Donald J. Kocian
David M. Koenig
Richard W. Koenig
Derek B. Koestel
Ernest J. Koestering
Ronald A. Kohser, PhD
Alois J. Koller, Jr.
Kenneth J. Kopp, PhD
Otto R. Kosfeld
Leslie R. Koval, PhD
Vincent J. Kovarik
Raymond J. Kowalik
Bob Kraemer
Lawrence Krebaum, PhD
Roy Krill
Joseph F. Krispin
Frank G. Kriz
Lester H. Krone
Joseph T. Krussel
Donald E. Kuenzi
Lowell L. Kuhn
Stanley P. Kuncaitis
David W. Kuneman
Karen M. Kurosz
Vicent E. Kurtz, PhD
Henry Kurusz
Tom Lacey
David H. Lah
William M. Landau
Stacy O. Landers
Brownell W. Landes
Stephen R. Langford
Edward Lanser
Michael A. Lawson
Christopher B. Lee
Paula Lee
Robert Lee
Randy W. Lehnhoff
Joshua M. Leise, PhD
Leslie Roy Lemon
Donald W. Lett
Steven K. Lett
Marc S. Levin
Kenneth L. Light
John D. Lilly
Arthur Charles Lind, PhD
John F. Lindeman
Jesse B. Lininger
Ellen Kern Lissant, PhD
Kenneth Jordan Lissant, PhD
Henry Liu, PhD
Robert S. Livingston, PhD
Peter F. Lott, PhD

Allan R. Louiselle
Forrest G. Lowe
Anthony Lupo, PhD
Robert J. Macher
Alexander E. Macias
Michael A. Madonna
Robert E. Mady
Dillon L. Magers
Philip W. Majerus
Bob Mallory, PhD
Carole J. Maltby, PhD
Warren O. Manley
Lena M. Mantle
Oliver K. Manuel, PhD
Ronald E. Markland
Clifford Marks
Thomas R. Marrero, PhD
Thomas E. Marshall, PhD
Wm. N. Marshall
Byron L. Martin
Edwin J. Martin, PhD
Stanley E. Massey
Balakrishnan Mathiprakasam, PhD
Charles R. Mattox, Jr.
Joe May
James L. McAdamis
Joseph C. McBryan
Edwin B. McBurney, Jr.
Martin H. McClurken
M. McCorcle, PhD
Hulon D. McDaniel
Hector O. McDonald, PhD
Kerry McDonald, PhD
William S. McDowell
Kenneth Laurence McHugh, PhD
William M. McLaurine
Joseph U. McMillan
Kathleen A. McNelis
Leo J. Menz, PhD
Roger J. Mercer
Dean Edward Metter, PhD
D. Joseph Meyer
John A. Mihulka
Dennis Miller
Carol R. Mischnick
Eugene L. Mleczko
Gary L. Montgomery
D. G. Moore, PhD
Francis E. Moore, Jr.
Jason T. Moore
Robert D. Moore, Jr.
Bernard J. Moormann
William T. Morgan, PhD
Randy S. Morris
Brad Moseley
Melvyn Wayne Mosher, PhD
William H. Mount
Kevin M. Mowery
Wayne K. Mueller
John E. Mullins
Sarah M. Mundy
Archie Lee Murdock, PhD
Orin L. Murray
Scott H. Muskopf
Vivek Narayanan, PhD
Robert Nash
Robert Overman Nellums
David J. Nelsen
G. Barry Nelson
George Douglas Nelson
Gregory J. Neuner
Robert E. Nevett

Eugene Haines Nicholson, PhD*
Paul J. Nord, PhD
Willard Norton
Robert Nothdurft, PhD
Darwin A. Novak, Jr., PhD
Laurence M. Nuelle
Deborah O'Bannon, PhD
Marvin Loren Oftedahl, PhD
Richard B. Oglesby
Andrew Oldroyd, PhD
Mark A. Ousnamer
Donald S. Overend
Sastry V. Pappu, PhD
Robert B. Park
Joseph Gilbert Patterson
Garry W. Perrey
Elroy J. Peters, PhD
Philip C. Peterson
Randall P. Petresh
George W. Petri
Larry T. Pettus
Donald L. Pfost, PhD
Harry M. Philip
Bruce B. Phillips, PhD
Lawrence Pilgram
Jerry P. Place, PhD
Ronald J. Poehlmann
M. E. Postlethwaite
Neil E. Prange
Daphne Press
R. H. Prewitt, Jr.
Melvin H. Proctor
John E. Rabbitt
Dennis P. Rabin, PhD
Jerome D. Radar
Rodney Owen Radke, PhD
Donald L. Rapp
H. C. Rechtien, Jr.
Norman Van Rees
James A. Reese
Richard B. Renz
Walter L. Reyland
George W. Reynolds
Vic Reznack
Gene M. Rice
Michael K. Rice
Graydon Edward Richards, PhD
Bill M. Rinehart
Frank S. Riordan, Jr., PhD
Ernest Aleck Robbins, PhD
Warren A. Roberts
Rodney A. Robertson
Daniel Rode, PhD
Michael Rodgers, PhD
Dustin J. Rogge
Dan Rohr
Dallas W. Rollins
Marston Val Roloff, PhD
Max E. Roper
Bruce David Rosenquist, PhD
Harold Leslie Rosenthal, PhD
Donald Kenneth Ross, PhD
James A. Ross
John E. Roush, Jr.
Charles F. Row
David M. Rucker
Gerald Bruce Rupert, PhD
Richard E. Ruppert
Theodore A. Ruppert
Paul J. Russ
Robert B. Russell
Robert Raymond Russell, PhD

Paul H. Rydlund
Edwin L. Ryser
J. Evan Sadler, PhD
Daniel K. Salisbury
Mohammad Samiullah, PhD
Leslie S. Samuels
Fred J. Sanders
Richard D. Sands
Thomas J. Sawarynski
Gutherie Scaggs
Barrett Lerner Scallet, PhD
Kenneth B. Schechtman, PhD
Carl R. Scheske
Edward P. Schneider
Lloyd E. Schultz
Elroy E. Schumacher
Ignatius Schumacher, PhD
Gerald V. Schwalbe
Gary W. Schwartz
Kirk G. Schweiss
Kerry G. Scott
Chuck Seger
Jan Segert, PhD
David L. Seidel
Jay Shah
Robert Ely Shank, PhD
Franklin H. Sharp
Bob D. Shaw
Ralph Waldo Sheets, PhD
Harvey D. Shell
George Calvin Shelton, PhD
Michael F. Shepard
Ronnie C. Shy
Daniel Robert Sidoti
Oliver W. Siebert
Robert H. Sieckhaus
Dennis J. Sieg
Marc S. Silverman
Virgil Simpson
Joseph J. Sind
Charles Carleton Sisler
Harvey D. Small
Haldon E. Smith
Wes Sorenson, PhD
Jack P. Spenard
Wayne K. Spillner
Alfred Carl Spreng, PhD
Michael E. Spurlock, PhD
Paul F. Stablein, PhD
Roger D. Steenbergen
Theo J. Steinmeyer
Robert E. Stevens
Raymond Fred Stevenson
Norman D. Stickler
Charles A. Stiefermann
James Osber Stoffer, PhD
Mel J. Stohl
Daniel F. Stokes
William Stoneman
Don B. Storment
Keith A. Stuckmeyer
Gary Stumbaugh
George Sturmon
Amy D. Sullivan
Dan T. Sullivan
Daniel T. Sullivan
Xingping Sun, PhD
Salvatore P. Sutera, PhD
J. S. Sutterfield, PhD
Mark E. Swanson
Peter H. Sweeney
Jason R. Szachnieski

David Tan
Tzyh-Jong Tarn, PhD
Jerry W. Tastad
Joseph R. Tauser
Michael J. Tautphaeus
Dugald A. Taylor
Kenneth R. Teater
Mark W. Tesar
Guy G. Thacker
Vernon James Theilmann, PhD
Richard C. Theuer, PhD
Marc G. Thomas
Rhys N. Thomas, PhD
Granville Berry Thompson, PhD
Kenneth C. Thomson, PhD
Stephen D. Tiesing
Dudley Seymour Titus, PhD
Douglas M. Tollefsen, PhD
Justin W. Tomac
Merrill M. Townley
Clarence A. Trachsel
Frank E. Tripp
Sheryl Tucker, PhD
John E. Turnage
Donald J. Turner
Zbylut Jozef Twardowski, PhD
Blaine A. Ulmer
Gary S. Urich
Frank D. Utterback
William P. Vale
Cornelius Van Dyke
Michael E. Vandas
Walter F. Vandelicht
Jerry G. VanderTuig, PhD
Allen L. Vandeusen
J. J. Ventura
Stanley Viglione
David N. Visnich
C. L. Voellinger
Michae F. Vogt
Frederick W. Vonkaenel
James R. Waddell
Billy C. Wade
Elizabeth L. Walker, PhD
Russell A. Walker
Timothy J. Walsh
Richard C. Waring
Don Warner, PhD
James A. Warnhoff
Jay Weaver
Charles W. Wehking, Jr.
Mark C. Wehking
A. R. Weide
Frank G. Weis
Richard E. Wendt, PhD
Roger Michael Weppelman, PhD
Samuel A. Werner, PhD
Terry L. Werth
John P. Werthman
James F. Westcott
Dallas G. Wetzler
Howard Martin Whitcraft, PhD
Charles K. Whittaker
Jon H. Wiggins, PhD
James M. Williams
Roger W. Williams*
Clyde Livingston Wilson, PhD
David Wilson
J. W. Wilson, PhD
Royce A. Wilson
Robert Winder
David F. Winter

James J. Wissman
Michael Witt
Edward W. Wolfe
William A. Wolff
Jared J. Wolters
Randall Wood, PhD
Marvin S. Wool
George T. Wootten
Paul N. Worsey, PhD
George G. Wright
John Scott Yaeger
Libby Yunger, PhD
Marvin Leon Zatzman, PhD
Chad E. Zickefoose
Thomas W. Ziegler
Ferdinand B. Zienty, PhD
Gary Zimmermann
Joseph E. Zweifel

Montana

David J. Abbott
Frank Addison Albini, PhD
Corby G. Anderson, PhD
David W. Ballard
William Ballard, PhD
Allen Barr
Wayne Bauer
George L. Baughman
Monte P. Bawden, PhD
Haley Beaudry
Pat Behm
Wade L. Bellon
Robert A. Bellows, PhD
Douglas N. Betts
Albert R. Blair
William V. Bluck
Jason W. Boeckel
Harold L. Bolnick
Fred M. Bonnett
Brian J. Bozlee, PhD
Jeff Briggs
Daniel Brimhall
Lawrence E. Bronec
Garry J. Carlson
A. Lee Child
James W. Crichton
Jerry Croft
Frank Patrick Crowley
Chris A. Cull
Gregory N. Cunniff
Thomas A. Dale
Nigel L. Davis
Ward S. Dewitt
Wade T. Diehl
Dennis Dietrich
Patrick J. Downey
Edward A. Dratz, PhD
J. Robert Dynes, PhD
Fred N. Earll, PhD
Wayne G. Farley
H. Richard Fevold, PhD
Lee H. Fisher
Fess Foster, PhD
Bob W. Fulton
Larry W. Geisler
David F. Gibson
John E. Grauman, Jr.
Mike Greger
Leland Griffin
Evan L. Griffith

Robert J. Guditis
James F. Hammill
Lowell C. Hanson
James C. Hansz
Earl Ray Harrison, Jr.
Charles M. Hauptman
Leo A. Heath
Alfred Hendrickson
Conrad R. Hilpert, PhD
Larry C. Hoffman
C. R. Hunkins
Ronald D. Isackson
Charlie M. Jackson, Jr.
William C. Jenkins
B. Johnke
John P. Jones
John M. Jurist, PhD
John H. Kennah
Jarvis R. Klem
Stephen J. Knapp
Stephan T. Kujawa, PhD
Michael L. Laird
Gerald J. Lapeyre, PhD
Wayne E. Lee
Jerry G. Lockie
Robert S. Macdonald
William C. Maehl
Daniel E. March
Ira Kelly Mills, PhD
Lyle Leslie Myers, PhD
Kenneth Nordtvedt, PhD
Michael J. Oard
Thomas M. Olson
Patricia H. Paulson-Ehrhardt
John P. Pavsek
Mark Pfau
Ronald Pifer
George R. Powe
Frank Radella
Keith A. Rae
Charles Rawlins, PhD
Clark S. Robinson, PhD*
Frank Rosenzweig, PhD
John H. Rumely, PhD
William R. Sacrison
Philip K. Salitros
James R. Scarlett
Darrell Scharf
Fred P. Schilling, PhD
Robert E. Schumacher
Mike Schweitzer
Robert M. Schweitzer
Fred Schwendeman
Larry R. Seibel
Robert K. Sengl
Ray W. Sheldon
Steven L. Shelton
James C. Simmons
Earl O. Skogley, PhD
Gordon E. Sorenson
Gilbert Franklin Stallknecht, PhD
George D. Stanley, PhD
Peter F. Stanley
Michael J. Stevermer
Ralph F. Stockstad
Arthur L. Story
Kyle Strode, PhD
James E. Taylor
John Edgar Taylor, PhD
Trevor Taylor
Deniz Tek
Donald L. Tennant

Forrest D. Thomas, II, PhD
Melinda Tichenor
Bradley S. Turnbull
Charles E. Umhey, Jr.
Saralee Neumann Visscher, PhD
Cory Vollmer
Scott J. Wagner
Russ Wahl
Kevin K. Walsh
J. Michael Wentzell
Wesley R. Womack
David Eugene Worley, PhD
Thomas J. Worring
Randi W. Wytcherley
Courtney Young, PhD
Greg D. Zeihen
Gordon F. Ziesing
A. L. Zimmerman
Donald G. Zink, PhD

Nebraska

Robert J. Anderson
R. Arnold, PhD
Donald R. Arrington
Leroy E. Baker
Ronald Bartzatt, PhD
Don Becker
Tiffany Bendorf
Lloyd Benjamin
Ray Bentall*
William M. Berton
John R. Brady
Albert L. Brown, PhD
Lloyd B. Bullerman, PhD
John Bryan Campbell, PhD
James A. Carroll, PhD
Raymond W. Conant
Sean M. Corday, PhD
C. Michael Cowan, PhD
Clifford H. Cox, PhD
Jon P. Dalton
Douglas A. Delhay, PhD
David A. Demick
Jerry A. Doctor
James F. Drake, PhD
Vincent Harold Dreeszen
William N. Durand
Allen Ray Edison
Jeffrey A. Ehler
James G. Ekstrand
Jeff H. Erquiaga, PhD
Merlin Feikert
Gerald C. Felt
Arthur F. Fishkin, PhD
Glenn Wesley Froning, PhD
Elisha J. Fuller
David Gambal, PhD
James Garretson
Fred M. Gawecki
Ron Gawer
M. Glasser
Bradley A. Gloystein
Robert Dwight Grisso, PhD
Terry Grubaugh
Myron E. Haines
Ray J. Hajek
John F. Hartwell
Dwight Haworth, PhD
Robert Mathew Hill, PhD
Ronald D. Hoback

Loren G. Hoekema
Steve N. Hoody
Ken Hucke
Kermit Jacobsen
Sherwin W. Jamison
Bryce G. Johnson
William W. Johnson
Fritz E. Kain
Paul Karr, PhD
Dan L. Keiner
Bradley N. Keller
Floyd C. Knoop, PhD
Duane G. Koenig
James G. Krist
Phillip Laroe
Dwight L. Larson
Leon D. Leishman
Louis I. Leviticus, PhD
Max W. Linder
Daryl C. Long, PhD
William W. Longley, Jr., PhD*
William F. Lorenz
Lawrence H. Luehr
Douglas E. Lund, PhD
Jeffrey E. Malan
Douglas W. Mallenby, PhD
John P. Maloney, PhD
Charles Wayne Martin, PhD
Bob C. McFarland
Thomas R. McMillan
David B. Mead
James Mertz
Gary D. Michels, PhD
Therese Michels, PhD
James P. Millard
Gail L. Miller
Larry E. Moenning
H. E. Moller
Chanin Monestero
Michael C. Mulder, PhD
Lyle W. Nilson
Richard A. Onken
James T. O'Rourke, PhD
David R. Paik
George Phelps
William Poley
William J. Provaznik
Kent W. Pulfer, PhD
Henry J. Quiring
John W. Reinert
George Remmenga
Keith Riese
Neil Wilson Rowland, PhD
Melvin Dale Rumbaugh, PhD
Constance B. Ryan
Wayne L. Ryan, PhD
Morris Henry Schneider, PhD
J. Shield, PhD
James B. Siebken
Eugene W. Skluzacek, PhD
Tim J. Sobotka
Mike J. Spoto
Franklin L. Stebbing
Stephen Stehlik
Herschel E. Stoller
Noble L. Swanson
Thomas Tibbels
John Toney
Rodney J. Tremblay
David George Tunnicliff, PhD
Norman Russell Underdahl
Glenn Underhill, PhD

Fritz Van Wyngaarden
Ronald E. Waggener, PhD
Frederick William Wagner, PhD
Laurie Walrod
Dean D. Watt, PhD
John Robert Webster
Dwight J. Williams
Richard Barr Wilson
Jake Winemiller
Gene Wubbels, PhD
Sabastian A. Zarbano
Thomas Herman Zepf, PhD

Nevada

Howard J. Adams
Opal Adams
Amy L. Anderson
Leslie Anderson, PhD
Joseph S. Armijo, PhD
Timothy D. Arnold
Orazio J. Astarita
C. David Baer
Matthew P. Bailey, PhD
Leland B. Baldrick
Richard L. Balogh, Jr.
William M. Bannister
Ted Barben
Jon Barth
Scott Beckstrand
Richard L. Bedell, Jr.
Denis Beller, PhD
Kenton E. Bentley, PhD
Bernard D. Benz
Frank Bergwall
Rahul S. Bhaduri
Arden E. Bicker
Arthur C. Bigley, Jr.
E. Boehmer
Raul H. Borrastero
Richard P. Bowen
Charles Arthur Bower, PhD
Sydney D. Bowers
William J. Brady
Alan D. Branham
Timothy J. Bray
Doug Brewer
Steve H. Brigman
James A. Brinton
Robert C. Broadbent
Russell K. Brunner
Donald H. Buchholz
Paul Buller, Jr.
Richard A. Calabrese
Roy Eugene Cameron, PhD
Richard S. Carr, III
Michael R. Cartwright
John W. Catledge
Marc Cave
Bob A. Chambers
Ken Chambers
Alan R. Charette
Fred Charette, PhD
Henry Chessin
Bertrand Chiasson, PhD
Randy Chitwood
Lawrence S. Chun
Edward M. Cikanek
Paul B. Clark
Peter J. Clarke, PhD
Ronald W. Clayton

John G. Cleary
James W. Cole
J. R. Colgan
James Collier, PhD
Sharon L. Commander
Teresa A. Conner
Neil D. Cos, PhD
Sheldon F. Craddock
Dale E. Crane
Michel W. Creek
Kurt E. Criss
Marla A. Criss
Steven R. Custer
Jaak J. K. Daemen, PhD
Fred J. Daniels, Jr.
Jeanette Daniels
John H. DeTar, PhD
Howard W. Dickson
Steve Dixon
Donald C. Dobson, PhD
Richard M. Dombrosky
Patrick J. Dougherty
Bruce M. Douglas, PhD
Ralph C. Dow
Stephen D. Dow
Doyle J. Dugan
W. P. Duyvesteyn, PhD
Carrie M. Eddy
J. James Eidel
Howard Lyman Ellinwood, PhD
Edi K. Engeln
Larry R. Engle
Rex L. Evatt, III
Karl C. Fazekas
Carmen Fimiani, PhD
Robert D. Fisher
David C. Fitch
W. Darrell Foote, PhD
Robert T. Forest
Helen L. Foster, PhD
Corri A. Fox
Forrest L. Fox
Neil Stewart Fox, PhD
Gary Fullington
Steven Alexander Gaal, PhD
Roy M. Gale
Peter E. Galli
Charles R. Galloway
Larry Joe Garside
James P. Gerner
Thomas E. Gesick
R. L. Giacomazzi
Peter F. Giddings
Dewayne Everett Gilbert, PhD
James M. Gills
James S. Goft
Patrick Goldstrand, PhD
Ernesto A. Gonzaga
Robert E. Gordon
C. Thomas Gott, PhD
William P. Graebel, PhD
Mihail I. Grigore
Robert E. Gross
Kathleen J. Gundy
Lewis B. Gustafson, PhD
M. Craig Haase
Jeffrey M. Haeberlin
C. Troy Haggard
Philip B. Halverson, PhD
Robert W. Handford
Daniel Harms
Elsie L. Harms

Anne Marie Harris
Floyd T. Harris
Larry W. Hatcher
Mark G. Hayden, PhD
William R. Henkle, Jr.
Michael S. Henson
James J. Hodos
K. E. Hodson
Thomas Edward Hoffer, PhD
Lee E. Hoffman
James B. Holder
Robert C. Horton
Dan Howard
Joseph E. Howland, PhD
Liang Chi Hsu, PhD
Ernest L. Hunsaker, III
John H. Huston
Craig Hutchens
E. Carl Hylin, PhD
John Walter Hylin, PhD
Arshad Iqbal
Jeffrey Janakus
Donald K. Jennings
Marc Z. Jeric, PhD
Scott Jimmerson
William P. Johnston, PhD
Doug Jones
Robert A. Jones, PhD
Edward P. Jucevic
Kirk K. Kaiser
Thomas R. Kalk
William J. Kane
Charles P. Kelly
Kathryn E. Kelly, PhD
Raymond C. Kelly, PhD
Robert Ray Kinnison, PhD
Stephene C. Kinsky, PhD
Herbert A. Klemme
David C. Knight
Paul W. Knoop
Robert J. Kopp, PhD
Larry D. Kornze
Mary Korpi
Winnie Kortemeier
Charle Kotulski
Robert M. Koval
Stephen L. Kozarich, PhD
Peter A. Krenkel, PhD
Dale D. Kulm
Charles A. Lacugna
Jan B. Lamb
Debbie Laney
David J. Langston
Richard G. Laprairie
Judd Larowe
Clark D. Leedy, PhD
Peter E. Lenz
Thomas M. Leonard
Alfred Letcher
Carl R. Leviseur
Kenton M. Loda
J. Vance Longley
Richard B. Loring
Kenneth A. Lucas
Paul Lumos
Farrel Wayne Lytle
Mel G. Maalouf
Joseph M. Maher
Robert L. Mann
John Craige Manning, PhD
Lindley Manning
Martin J. Manning

Carl R. Manthei
Edward F. Martin
William C. Mason
William H. Matchett, PhD
David C. Mathewson
Lauren H. Mattingly
Perry McCart
Robert J. McConkie
Lee B. McConville
Karl W. McCrea
Dayton T. McDonald
John C. McHaffie
Harold P. Meabon
Robert F. Merchant, Jr.
George Clement Messenger, PhD
George J. Miel, PhD
Greg Millspaugh
M. Myd Min
Dennis M. Molte, PhD
Cynthia R. Moore
Paul V. Morgan
Neil J. Mortensen
Kenneth L. Moss
Ralph D. Mulhollen
David L. Mumford
Ralph G. Musick, Jr.
Richard F. Nanna
John Neerhout, Jr.
Michael L. Neeser
Genne M. Nelson
Owen N. Nelson
R. William Nelson
Daniel P. Neubauer
William N. Neumann
Joe D. Newton
Donald R. Nichols
Thomas L. Nimsic
Elray S. Nixon, PhD
Ann T. Nunnemaker
Samuel G. Nunnemeker
Mark A. Odell
Frederick Kirk Odencrantz, PhD
Gary L. Oppliger, PhD
William V. Orr
Marla A. Osborne
Thomas Paskowski
Sergio Pastor
Gilbert W. Patterson
Robert Paul
Durk Pearson
Alan Embree Peckham
Alan C. Peitsch
Frank M. Pelteson
Penio D. Penev
Robert A. Perkins
Richard M. Perry
Mitchel Phillips
David M. Pike
John Polish
John Porterfield
Alan T. Power
Neal J. Prendergast
Robert F. Prichett
Victor H. Prodehl
Ralph Quosig
Charles G. Ranstrom
Richard R. Redfern
Kent Redwine, PhD
Kenneth M. Reim
Fredrick K. Retzlaff
Brian Ridpath
Kevin J. Riley

Bradley R. Rising
Daniel E. Robertson
Raymond F. Robinson
R. S. Rodda
Frank A. Rogers
Franklin J. Ross
Zsolt F. Rosta
Edwin S. Rousseau
Kim J. Runk
Myrl J. Saarem
Ralph R. Sacrison
Stanley F. Schmidt, PhD
Thom J. Seal
David W. Seeley
Wen Shen
Tyler Shepherd
Bernie J. Sherin, Jr.
Lisa Shevenell, PhD
Michael E. Silic
Michael B. Simpson
Doug J. Siple
Colin A. Smith
Olin K. Smith
Charles D. Snow
John Snow
Kenneth D. Snyder, PhD
M. Bradford Snyder, PhD
Scott E. Soderstrom
James P. Solaro
Lyle C. Southworth
Robert M. St. Louis
William R. Stanley
David J. Starbuck
Kenneth L. Steffan
Don W. Stevens
Harry B. Stoehr
John Grover Stone, PhD
Kevin W. Sur
Edward J. Sutich
Charles A. Sweningsen
Edgar R. Terlau
David B. Thompson, PhD
Tommy Thompson, PhD
Rory Tibbals
George P. Timinskas
Robert W. Titus
Luis A. Topete
Richard R. Tracy, PhD
Robert J. Trauger
John G. Trulio, PhD
Glenn W. Tueller
Paul T. Tueller, PhD
Michael G. Turek
Nancy Vardiman-Hall
John Vivier
Darrell E. Wagner
John D. Walter, PhD
Quinten E. Ward
Chuck Weber
W. Layne Weber
James M. Wehrman
Carl M. Welch
Herbert C. Wells
John D. Welsh
Henri Wetselaar
Keni Whalen
Gordon R. Wicker
John B. Wigglesworth
Scott W. Wiljanen
Clavin E. Willoughby
Sylvan Harold Wittwer, PhD
Daniel J. Wonders

Morris T. Worley
Peter F. Young
Ken Zaike
Danny Zampirro
Robert L. Zerga

New Hampshire

Henry J. Adams
Arthur Edward Albert, PhD
Lynn C. Anderson
Ernest F. Angelicola
Ralph H. Baer
Daniel L. Banowetz
Walter Barquist
Kenneth H. Barratt
Hans D. Baumann, PhD
Wayne Machon Beasley
Paul R. Beswick
Peter Bird
Harvey J. Bloom
Frank T. Bodurtha, PhD
Philip D. Bogdonoff, PhD
Ray Book
Thomas A. Bover
Andrew J. Breuder
Ray O. Bristol
Bruce L. Brown
James A. Browning
Lawrence F. Buckland
Charles Rogers Buffler, PhD
John C. Burgeson
Roland Polk Carreker, PhD
Robert M. Chervincky
John A. Clements
Dennis E. Coburn
Norm G. Cote
Jesse F. Crump
Thomas John Curphey, PhD
Benoit Cushman-Roisin, PhD
Joseph S. D'Aleo
Robert Desmarais
George Dunnavan
Andrew L. Eastman
Richard N. Erickson
George Fallet
Daniel J. Filicicchia
Thomas L. Fowke
Roy R. Frampton, PhD
Edwin Francis Fricke, PhD
Joe Gagliardi
Leon H. Geil
Glen C. Gerhard, PhD
Katherine J. Getchell
Wayne J. Getchell
Michael D. Gilbert, PhD
Thomas J. Gilligan
William T. Golding
Robert V. Gould*
Kevin J. Gray, PhD
David L. Gress, PhD
Gordon W. Gribble, PhD
Carlton W. Griffin
Jay S. Grumbling
David R. Hall
William A. Hamill
Stephen P. Hansen
Robert Duane Harter, PhD
Norman H. Heinze
Rick Hickerson
Lawrence Hoagland, PhD

Scott D. Hobson
Tim Horton
John S. Howland
Fred R. Huber
Everett Clair Hunt, PhD
Albert H. Huntoon
Karl Uno Ingard, PhD
Lawrence R. Jeffery
Carl N. Johnson, PhD
Arthur Robert Kantrowitz, PhD
Gerson Kegeles, PhD
Anthony J. Kelley
James F. Kerivan
Michael Klingler
Louis H. Klotz, PhD
Robert G. Koch
Dennis A. Lauer
Marcel Elphege LaVoie, PhD
John W. Leech, PhD
Semon M. Lilienfeld
Russell W. Maccabe
William G. Machell
C. J. Manning
Albert V. Martore
James Maynard
Kenneth E. Mayo
Kevin M. McEneaney
Eugene D. Mellish
Edward J. Merrell
David G. Miller
David O. Mulgrew
John H. Muller, PhD
George A. Oldham
Karle Sanborn Packard
R. Palmer
Radi Pejouhy
Howard G. Pritham
Jonathan S. Remillard
Harold C. Ripley
Gilbert S. Rogers
Barbara Rucinska, PhD
Jason J. Saunderson, PhD
Ronald E. Scott, PhD
Robert Seaman
John J. Segedy, PhD
Faye Slift
Bernard G. E. Stiff
David E. Strang
Augustine R. Stratoti
Robinson Marden Swift, PhD
Morgan Chuan Yuan Sze, PhD
Donald M. Taylor
Lawrence J. Varnerin, PhD
Charles Wagner
Richard W. Waldron, PhD
Maynard C. Waltz
William T. West, PhD
James R. Weth
George R. Winkler
Stephen S. Wolf
Yin Chao Yen, PhD

New Jersey

Daniel T. Achord, PhD
Phillip Adams, PhD
Richard Ernest Adams
Ben J. Addiego
Anthony J. Adrignolo, PhD
Robert Aharonov
Andrew J. Alessi

Roger C. Alig, PhD
Emma Allen, PhD
A. Frank Alsobrook
Martin E. Altis
Pushpavati S. Amin
G. Anderle, PhD
George H. Andersen
Eva Andrei, PhD
Frederick T. Andrews
Charles H. Antinori, PhD
P. J. Apice
Alan Appleby, PhD
Earl F. Arbuckle
Zaven S. Ariyan, PhD
John V. Artale
Robert Artz
Robert H. Austin, PhD
Frederick Addison Bacher
Alan G. Backman
James H. Bailey, Jr.
Robert J. Baker
William Oliver Baker, PhD
Benjamin H. Bakerjian
Nicholas N. Baldo, Jr.
Arthur Ballato, PhD
Zinovy I. Barch, PhD
Jeffrey D. Barczak
Yeheskel Bar-Ness, PhD
William R. Barretti
Douglas G. Bartels
Peter Bartner
Alton H. Bassett, PhD
Nayan K. Basu
Richard Baubles
Eric Baum, PhD
John W. Bauman, PhD
Robert M. Bean, PhD
Herbert Ernest Behrens
Raymond L. Bendure, PhD
James A. Benjamin, PhD
David R. Benner
Gordon D. Benson
Paul R. Berkowite
Daniel D. Bertin
Harry M. Betzig
Rowland Scott Bevans, PhD
Joseph W. Bitsock
David Blewett
Robert W. Blocker
Rodney J. Blouch
Claire Bluestein, PhD
Joseph A. Bluish
Sheri L. Blystone, PhD
Anthony Vincent Boccabella, PhD
Richard F. Bock
Nicholas Bohensky
Kees Boi, PhD
Gaston A. Boissard
Theodore Edward Bolden
William V. Booream
William G. Borghard, PhD
William K. Boss, Jr.
Robert W. Bosse
Henry Bovin
Roger L. Boyell
Paul S. Boyer, PhD
Matthew A. Braccio
Robert Brady
Robert H. Brakman
William Brennan
Kurt Brenner
Charles O. Brostrom, PhD

James Brown
Richard L. Brown
William M. Brown, PhD
Paul Eugene Brubaker, PhD
Gregory J. Brunetta
James A. Bruni
Charles Frank Bruno, PhD
Ralph Bruzzichesi
Daniel M. Bubb, PhD
Carl J. Buck, PhD
David Richard Burley, PhD
Samuel R. M. Burton
Richard S. Butryn
Michael J. Byrne
Edwin Caballero
James L. Calderella
John D. Calvert
Douglas J. Campbell
Daniel T. Canavan
Peter J. Canterino
Walter J. Canzonier
Timothy J. Carlsen
Adam S. Carney
Michelle Carney
David C. Carter
Lilana I. Cecan, PhD
Bernard H. Chaiken
Partitosh M. Chakrabarti, PhD
Chun K. Chan, PhD
Donald Jones Channin
Victor S. Cheng
Roy T. Cherris
Arthur Warren Chester, PhD
Thomas Chestone
Ye C. Choi, PhD
Chang H. Chung
Eugene L. Church, PhD
Anthony J. Cirillo, Jr.
J. Clemente
Otto M. Clemente
Mary P. Coakley, PhD
Robert Coffee
Harvey H. Cohen
Paul H. Cohen
Francis Colace
Roger T. Cole
Donald E. Collins
Robert B. Comizzoli, PhD
George J. Conrad
Alicia S. Conte
Joseph F. Conte
Maryann A. Conte
Edward M. Cooney
Thomas E. Cope, Jr.
Lois J. Copeland
Deborah A. Corridon
Bernard J. Costello
Val Francis Cotty, PhD
James C. Coyne, PhD
Frank Cozzarelli
Carolyn S. Crawford
James Glen Crawford
Kenneth Alan Crossner, PhD
Edwin Patrick Crowell
William J. Cruice
Donald C. Cuccia
Richard Williamson Cummins, PhD
Michael A. Cuocolo, Jr.
Janet C. Curry
Eugene J. Daly
Theo C. Damen
Wayne E. Daniel

Henry D. Dankenbring
Matthew A. Danza
Alexandra D'Arcangelis, PhD
Samuel D'Arcangelis, PhD
Robert F. Dauer
Edward Emil David, Jr., PhD
Thomas A. Davis, PhD
William F. Davis
Arthur Donovan Dawson, PhD
Albert Dazzo
Stephen Dazzo
Tom V. Debrock
Donald R. Degrave
James R. Deland, Jr.
Joseph A. Delvers
Michael R. Demcsak
David T. Denhardt, PhD
Robert A. Dennin, Jr
Felix M. Depinies, PhD
Dimitris Dermatis, PhD
Vijaya S. Desai
Salvatore J. Desalva, PhD
R. Detig, PhD
Anthony J. Devivo
Thomas F. Devlin, PhD
Franklin D. Dexter
William J. Dieal
Ray DiMartini, PhD
Earl M. Dipirro
William A. Dittrich
Richard W. Dixon, PhD
George F. Doby
Emerson B. Donnell
Sandra Steranka Donovan, PhD
Patrick Dorgan
Milos A. Dostal, PhD
Nancy M. Doty
Nicholas J. Driscoll
Donald C. Dunn
James A. Dusenbury
Dushan Dvornik, PhD
Freeman John Dyson
Wayne F. Eakins
Harold W. Earle
Alieta Eck
Peter Egli
I. Robert Ehrlich, PhD
Saundra M. Ehrlich
Richard Edward Elden
Shaker H. El-Sherbini, PhD
Youichi Endo, PhD
William J. Engelhard
Seymour Epstein, PhD
Joshua A. Erickson
John G. Esock
Ramon L. Espino, PhD
Michael A. Esposito
Glen A. Evans
Alexander Fadeev, PhD
Peter W. Failla
Robert J. Farquharson
Michael L. Fasnacht
Stephen E. Fauer
Charles E. Fegley
Edwin L. Fehre
Abraham S. Feigenbaum, PhD
Paul M. Feinberg, PhD
Frederick P. Fendt
Eric V. Fernstrom
Robert M. Feuss
Paul Finkelstein, PhD
Gordon H. Fish

T. Burnet Fisher
Richard E. Foerster
Dennis Fost, PhD
Benjamin T. Fought
Philip Franco
Peter M. Franzese
Daniel W. Frascella, PhD
Inge Friedrich
Jack A. Frohbieter
Gerhard J. Frohlich, PhD
Roland Charles Fulde, PhD
Shun Chong Fung, PhD
John C. Gabriel
L. John Gagliardi, PhD
George Gal, PhD
Anthony Gallo
Paul H. Garnier
Douglas G. Gehring
Thomas L. Geib
Celine Gelinas, PhD
Gary Gerardi, PhD
Norbert Gerlach
Raymond P. Germann
Theodore A. Giannechini
Anne M. Giedlinski
Albert M. Giudice
Sivert Herth Glarum, PhD
Ronald N. Glass
Werner B. Glass
Frederick G. Glatter
Robert A. Gleason
Douglas J. Glorie
Stephen Gold, PhD
Xavier F. Gonzalez
Lewis S. Goodfriend
Jerome D. Goodman, PhD
Martin Leo Gorbaty, PhD
Eugene Gordon, PhD
Edward A. Gorka
Jerome J. Gramsammer
Jeffrey E. Grant
Ashby A. Grantham
Wayne P. Grau
Martin Laurence Green, PhD
Marvin I. Greene
Frederic C. Grigg
Nicholas A. Grippi
Richard G. Grisky, PhD
Don J. Grove, PhD
Y. Guberer, PhD
James G. Guider
Vajira K. Gunawardana
Stuart S. Gut
Berwin A. Guttormsen
George Hacken, PhD
Stephanos S. Hadjiyannis
Erich L. Hafner, PhD
James J. Hagan, PhD
Robert E. Hague, PhD
Leslie Hamilton
James M. Hammell
Sunil P. Hangal, PhD
I. Hapij, PhD
William Happer, PhD
John R. Harding
Wayne A. Harmening
Clifford Harvey, PhD
Frank C. Hawk, PhD
Philip F. Healy
Sylvester B. Heberlig
James B. Heller
Edwin Hendler, PhD

Tom A. Hendrickson
Kenneth H. Hershey
Uri Herzberg, PhD
Jack Hickey
Thomas J. Hicks
Mark J. Higgins
Walter L. Hinman, Jr.
Arthur Hirsch, PhD
Michael J. Hnat, PhD
Frederick J. Hofmann
Austin W. Hogan, PhD
Thomas C. Holcombe
Charles B. Hopkins, Jr.
Harry M. Horn, PhD
Alan J. Horowitz
Zdenek Hruza, PhD
Taras P. Hrycyshyn
Roy E. Hunt
William E. Hymans, PhD
Arnold Gene Hyndman, PhD
K. Hyun, PhD
Bradley Ingebrethsen, PhD
Basil J. Ingemi
Criton George S. Inglessis, PhD
Nicholas A. Ingoglia, PhD
Shabana S. Insaf, PhD
Shahid S. Insaf
Michelle L. Jablons
Johan Jaffe, PhD
Kenneth Irwin Jagel, PhD
Paul J. Jakubicki
Robert R. Jamieson
Akshay Javeri
Hugh F. Jenkins
Max F. Jenne
Luis E. Jimenez, PhD
Amos E. Joe
Leo Francis Johnson, PhD
Scott R. Johnston
John G. Johnstone
Leonard Juros
Liutas K. Jurskis
Raymond F. Kaczynski
Michael Z. Kagan, PhD
Sergei L. Kanevsky
Stanley F. Kantor
Guido George Karcher
Bennett C. Karp, PhD
Frederick C. Kauffman, PhD
Yuri Kazakevich, PhD
John H. Kearney
Horst H. Keasdy, PhD
Craig T. Keeley
David B. Keller
Joseph J. Keller
Joseph Kelley, PhD
John Kellgren, PhD
George Cantine Kent, PhD
Michael John Keogh, PhD
Thomas J. Kesolits
Mark Otto Kestner, PhD
Z. Khan
Charles W. Kilgore, II
Snezana Kili-Dalafave, PhD
Aejin Kim
Robert M. Kipperman, PhD
Hartmann J. Kircher
Steven G. Kisch
Walter B. Kleiner, PhD
Lawrence Paul Klemann, PhD
Robert C. Kluthe
Richard J. Knopf

William J. Koehl, Jr., PhD
William J. Koenecke
Norm C. Koller
Catherine Koo, PhD
George Eugene Kosel
John Norman Kraeuter, PhD
Walter H. Kraft
Ralph H. Kramer
Paul J. Kraus
Michael S. Krepky
Robert Kricks, PhD
Geddy J. Krul
Stephen A. Kubow, PhD
James Edgar Kuder, PhD
Stephen J. Kuritz, PhD
Frank Michael Labianca, PhD
Paul Albert Lachance, PhD
John F. Laidig
James C. Lane
Robert L. Langerhans
Robert Charles Langley
Ralph Lanni
Geoffrey R. Lanza
John P. Larocco
Jack Samuel Lasky, PhD
Alan R. Latham, PhD
Robert J. Laufer, PhD
Ralph John Leary, PhD
Robert Lefelar
Richard O. Leinbach
Mario Leone
Thomas E. Leonik
Allan L. Levey
Sidney B. Levinson
J. Levy
Jacek K. Leznicki, PhD
Wayne W. Lippincott
John E. Littlefield
John M. Lix
William Lockett, Jr.
G. M. Loiacono
Carlos A. Londono
Donald H. Lorenz, PhD
R. W. Love
Jan Lovy, PhD
John J. Lowney
Brian Lubbert
Peter J. Lucchesi, PhD
Frank P. Lunn, III
Luigi Lupinacci
Patrick B. Lynam
Charles S. Lyon
Margaret J. Lyons
Wilfred J. Mabee
Douglas J. MacNeil, PhD
Alfred Urquhart MacRae, PhD
Chandra P. Mahadass
George F. Mahe
John Philip Maher
Robert Maines
Robert C. Maleeny
Joseph T. Maloy, PhD
James Manfreda
George J. Manik
Louis J. Mankowski
Bryan C. Mann
August Frederick Manz
Betty L. Marchant
Stanley S. Marcus
Herman Lowell Marder, PhD
Andrew C. Marinucci, PhD
James W. Martin

Ramon Martinelli, PhD
Albert Masetti
Harry L. Masten
R. Mastracchio
Norman J. Matchett
Melvin K. Mathisen
Richard A. Matula, PhD
Bryce Maxwell
Kenneth Maxwell
Robert W. McAdams
Robert J. McCabe
Francis D. McCann
Kristine McCool
John Price McCullough, PhD
Richard J. McDermott
Wilbur Benedict McDowell, PhD
Edward R. McFarlan
Peter McGaughran
Kenneth J. McGuckin
George Francis McLane, PhD
Pat McMahon
William A. McMahon, Jr.
H. McVeigh, PhD
Sidney S. Medley, PhD
George Megerle, PhD
Charles T. Melick
James Richard Mellberg
Neri Merlini, PhD
Frank J. Mescall
George F. Meshia
Joseph N. Miale
Wayne Michaelchuck
Paul Charles Michaelis
William J. Mikula
Dick M. Miller
Dale F. Milsark
William B. Mims, PhD
Robert S. Miner, Jr., PhD
Timothy R. Minnich
Jerry B. Minter
Charles W. Moehringer
H. Walter Moeller
Edward J. Moller
Mark D. Morgan, PhD
Antonio Moroni, PhD
Lester D. Morris
M. E. Morrison, PhD
Edward G. Moss
Helmut Mrozik, PhD
Charles Norment Muldrow, Jr., PhD
Jeffrey S. Mumma
Richard C. Murgittroyd
Charles H. Murphy, PhD
Joseph A. Musci
Ernest M. Myhren
Joseph O. Myslinski
Michael J. Napoliello
John T. Nardone
Franklin Richard Nash, PhD
Errol S. Navata, Sr.
Edward G. Nawy, PhD
John P. Neglia
Lasalle L. Nolin
Frank R. Nora
Wayne J. Norman
Jerry J. Notte
William J. Novick, PhD
Susan R. Obaditch
Dennis O. Odea
Hans Oesau
Edward A. Ohm, PhD
Vilma E. Oleri

Horace G. Oliver, Jr.
Gregory L. Olson
Charles S. Opalek
Anthony M. Ordile
Gerald J. Orehostky
Donald H. Ort
Paul W. Osborne, PhD
Tom J. Osborne, Jr.
Donald O'Shea
Dan A. Ostlind, PhD
Richard C. Otterbein
Sally Bulpitt Padhi, PhD
M. Robert Paglee
Felice Charles Palermo
Richard Walter Palladino
Lawrence G. Palmer
Virendra N. Pandey, PhD
Gabriel J. Paoletti
Pat A. Papa
Laura Z. Papazian
Robert P. Parisi
Arthur Parks, Sr.
Robert A. Pascal, PhD
Raymond A. Patnode
Rolf Paul, PhD
Walter B. Paul, PhD
John D. Pearson
Edward R. Pennell
Samuel C. Perks, PhD
Anthony James Petro, PhD
Raymond J. Pfeiffer, PhD
Francis A. Pflum
John Philipp
Wendell Francis Phillips
Joseph F. Pilaro
August W. Pingpank
Steven S. Plemenos
John Polcer
Robert D. Popper
Reddeppa N. Pothuri, PhD
Robert C. Potter, PhD
Kurt Pralle
David Pramer, PhD
Bernard L. Prince
James R. Prisco
Charles John Prizer
Eugene M. J. Pugatch, PhD
Peter T. Queenan
Jack G. Rabinowitz
Attilio Joseph Rainal, PhD
Karel Frantiscek Raska, Jr., PhD
Susanne Raynor, PhD
Horace A. Reeves
Franklin G. Reick
Kenneth E. Resztak
George Henry Reussner
Edward T. Rhode
Bruce C. Rickard
Martin Max Rieger, PhD
James Louis Rigassio
R. Rivero
George H. Robertson
Michael J. Rohal
Louis D. Rollmann, PhD
Lee Rosenthal, PhD
John A. Rotondo, PhD
Arthur Israel Rubin
Ralph A. Runge
Irving I. Rusoff, PhD
George M. Rynkiewicz
John W. Ryon, PhD
Edward Stephen Sabisky, PhD

Frithiof N. Sagerholm, Jr.
Raymond J. Salani
Solomon J. Salat
Joseph J. Salvatorelli
Robert H. Salvesen, PhD
R. E. Sameth
Edward W. Sapp
John W. Sarappo
Eric F. Sarshad
Louis J. Sas
Charles M. Sather
Matthew J. Schaeffer
Garrett C. Schanck
Harold C. Schanck
Charles T. Schenck
Andrew L. Schlafly
Steven I. Schneider
Ken Schoene
Robert Edwin Schofield, PhD
Gervasia M. Schreckenberg, PhD
William Lewis Schreiber, PhD
Peter Schroder
Eric Schuler
Donald Frank Schutz, PhD
Fred A. Schwizer
Robert L. Scotto
Thomas J. Scully
Fred Seeber, PhD
John M. Segelken
Glennda Koon Selah
Roger C. Selah
Bernard G. Senger, III
Gray F. Sensenich, Jr.
Nicholas M. Setteducato
Mahmoud P. Shahangian
Junaid R. Shaikh
Michael L. Shand, PhD
Thomas J. Sharp
Richard H. Sharrett
Milton C. Shedd
Everett O. Sheets
Joshua Shefer, PhD
James Churchill Shelton, PhD
Mark V. Sheptock
Peter B. Sheridan
Dennis Shevlin, PhD
Kathleen M. Shilling
Frank Shinneman
Taposh K. Shome
David A. Shriver, PhD
Donald Bruce Siano, PhD
John Philip Sibilia, PhD
Hsien Gieh Sie, PhD
Daniel J. Silagi
Frederick Howard Silver, PhD
Karl Leroy Simkins, PhD
Frederick H. Singer
Gainmattie Gail Singh
Roy J. Sippel, PhD
Leon J. Skvirsky
Jim R. Slate
Alvin Slomowitz
Patrick L. Smiley
Albert Smith
Bryan J. Smith
Eric S. Smith
Mariah Jose Smith, PhD
Marvin F. Smith, Jr.
Patrick Sean Smith
Roy C. Smith
William Smith
Joseph Smock

Angela Snow, PhD
Harold J. Sobel
Frank B. Sorgie
George F. Spagnola
J. C. Spitsbergen, PhD
Marie T. Spoerlein, PhD
Ralph Joseph Spohn, PhD
Andrew P. Srodin
Gilbert Steiner, PhD
Timothy D. Steinhiber
William D. Stevens, PhD
N. Stiglich
Bruce D. Stine
Tom Stinnett
Ronald S. Strange, PhD
Anthony W. Strano
Szymon Suckewer, PhD
Norman Sutherland
Jan O. Svoboda
Karl P. Svoboda
James R. Sweeney
Joseph M. Tak
Alan Wayne Tamarelli, PhD
George P. Tanczos
Robert J. Tarantino
Richard M. Taylor
Patrick H. Terry
Sant S. Tewari, PhD
Warren Alan Thaler, PhD
John L. Thein
Lowell Charles Thelin
William L. Thoden
Gordon A. Thomas, PhD
Ellen K. Thompson
Jeffery Thompson
William Baldwin Thompson, PhD
Henry C. Tillson, PhD
John H. Tinker
William C. Tintle
John M. Toto
Donald R. Tourville, PhD
Ralph Trambarulo, PhD
Ralph A. Treder, PhD
Jeffrey Charles Trewella, PhD
Jane Tsai
Stephen Tsingas
Stephen P. Tyrpak
P. Roy Vagelos
James P. Van Hook, PhD
Steven Vanata
Thomas Henry Vanderspurt, PhD
William Vandersteel
Christina VanderWende, PhD
Andreas H. Vassiliou, PhD
Philip J. Vecere, Sr.
Patrick G. Vemacchia
Karl G. Verebey, PhD
John A. Vigelis
Francis V. Villani
Frank A. Vinciguerra
Luciano Virgili, PhD
Dale E. Vitale, PhD
George Viveiros
Gert Volpp, PhD
Paul Justus Volpp, PhD
Gerald N. Wachs
Walter M. Walsh, Jr., PhD
Arthur Ernest Waltking
Stanley Frank Wanat, PhD
Hsueh Hwa Wang
W. Steven Ward, PhD
Clifford A. Warren

Frank Warren, Jr.
Frank R. Warshany
Madeline S. Wasilas
Stuart L. Weg
Walter F. Wegst, PhD
Roy W. Weiland
George Austin Weisgerber, PhD
Barry R. Weissman
William G. Weissman
Warren D. Wells
Jack Harry Wernick, PhD
George M. Wheeler
John Michael Whelan, PhD
Joseph Marion Wier, PhD
Kenneth C. Wilding, Jr.
Jos B. Wiley, Jr.
Karla E. Williams
Neal Thomas Williams
W. H. Williams, PhD
Frederick Willis
Robert K. Willmot, Jr.
Lloyd Winsor
Anthony E. Winston
Donald F. Wiseman
Peter Joseph Wojtowicz, PhD
Chi-Yin Wong
Joseph Eliot Woodbridge, PhD
Patrick J. Wooding
John R. Woodruff
Von Worthington
Paul E. Wyszkowski
Jim Yarwood
Dennis K. Yoder
Richard H. Young
Stephen A. Yuhas
Robert F. Yuro
Egon J. Zadina
A. Zamolodchikov, PhD
Vlasta K. Zbuzek, PhD
Debi Y. Zeno
Shumin Zhuang
John Benjamin Ziegler, PhD
David Zudkevitch, PhD
Melvin L. Zwillenberg, PhD

New Mexico

Carl M. Abrams
Donald Adams
Stuart R. Akers
Frederick I. Akiya
James B. Alexander
M. Dale Alexander, PhD
Charles C. Allen
Lynford L. Ames, PhD
James F. Andrew, PhD
Harold V. Argo, PhD
Robert L. Armstrong, PhD
Charles Arnold, PhD
Walter R. Ashwill
Erik Aspelin
Dustin M. Aughenbaugh
Gerald V. Babigian
Lester Marchant Baggett, PhD
Lara H. Baker, PhD
Arden Albert Baltensperger, PhD
James A. Baran, PhD
Franklin Brett Barker, PhD
Scott L. Beardemphil
David F. Beck
A. D. Beckerdite

Glenn W. Bedell, PhD
Rettig Benedict, PhD
H. W. Benischek
Kaspar G. Berget
M. Berkey
Marshall Berman, PhD
Theodore F. Berthelote
Frances M. Berting, PhD
Andrew J. Betts
John Beyeler
William Edward Blasé
Clay Booker, PhD
Kevin W. Boyack, PhD
Ben M. Boykin
Richard J. Boyle
Ben J. Brabb, PhD
Charles D. Braden
Martin Daniel Bradshaw, PhD
Frederick S. Breslin, PhD
Robert E. Bretz, PhD
Clifton Briner
Kay Robert Brower, PhD
Donnie E. Brown
James S. Brown
Lowell S. Brown, PhD
James R. Buchanan
J. F. Buffington
Merle E. Bunker, PhD
Frank Bernard Burns, PhD
Robert C. Byrd
Suzanne C. Byrd
Fernando Cadena, PhD
Howard Hamilton Cady, PhD
Dennis R. Cahill
George Melvin Campbell, PhD
Jeff Campbell
Gregory Cannon
Frederick Herman Carl Schultz, PhD
Gunnar C. Carlson
Peter R. Carlson
John Granville Castle, PhD
Roy Dudley Caton, Jr., PhD
Charles W. Causey
Philip Ceriani
Peter E. Christensen
Petr Chylek, PhD
Kenneth W. Ciriacks, PhD
William F. Clark
Frank Welch Clinard, PhD
Michael R. Clover, PhD
Richard Lee Coats, PhD
Allen H. Cogbill, PhD
Michael W. Coleman
Kenneth B. Collins
Albert D. Corbin
Samuel Douglas Cornell, PhD
Maynard Cowan, Jr.
Jerry Ferdinand Cuderman, PhD
John E. Cunningham, PhD
Dirk A. Dahlgren, PhD
James P. Daley
John C. Dallman, PhD
Michael Daly
Richard Aasen Damerow, PhD
Edward George Damon, PhD
James M. Davidson, PhD
Frank W. Davies
Herbert T. Davis, PhD
Robert D. Day, PhD
Michael J. Dietz, Sr.
Malcolm Dillon
Robert Hudson Dinegar, PhD*

Steven K. Dingman
Eugene H. Dirk, PhD
Paul Dobak
D. Donaldson, PhD
Gary B. Donart, PhD
Rae A. Dougherty
John R. Doughty, PhD
Charles O. Dowell
Richard T. Downing
Clifton Russell Drumm, PhD
Sherman Dugan
Thomas A. Dugan
Herbert M. Dumas
C. M. Duncan, III
Thomas G. Dunlap
Jed E. Durrenberger
Lawrence G. Dykers
Robert John Eagan, PhD
Tim E. Eastep
George W. Eaton, Jr.
Keith Ehlert
Chris S. Ellefson
Robert W. Endlich
Albert George Engelhardt, PhD
Raymond Engelke, PhD
Steve R. England
Roger Charles Entringer, PhD
Andy C. Erickson
Jay Erikson, PhD
Mary Jane Erikson
R. W. Erwin
William R. Espander, PhD
Eugene Henderson Eyster, PhD
Ron Farmer
Douglas Fields, PhD
Morris Dale Finkner, PhD
Philip C. Fisher, PhD
Fred F. Fleming, Jr.
Stuart Flicker
K. Randy Foote
Clarence Q. Ford, PhD
Robert F. Ford, PhD
David Wallace Forslund, PhD
Eric Beaumont Fowler, PhD
Harry James Fox
Julian R. Franklin
Bruce L. Freeman, PhD
Susan H. Freeman
Michael Frese, PhD
J. M. Fritschy
Ian J. Fritz, PhD
Paul Fuierer, PhD
Julie Fuller
Michael S. Fulp
Irwin M. Gabay
Bruce E. Gaither
Quentin Galbraith
Ri Garrett
John Eric Garst, PhD
J. C. Garth, PhD
Jackson Gilbert Gay
Arthur Gebean
Raymond S. George, PhD
Timothy Gordon George
Claude Milton Gillespie, PhD
David L. Gilmer
H. Scott Gingrich
Thomas H. Giordano, PhD
Earl R. Godwin
Terry Jack Goldman, PhD
James Wylie Gordon, PhD
Walter L. Gould, PhD

David G. Grabiel
Mathew O. Grady
Charles Thornton Gregg, PhD
Clifton E. Gremillion
Reid Grigg, PhD
A. G. Griswold
Walter Wilhelm Gustav Lwowsk, PhD
Eero Arnold Hakkila, PhD
Frank C. Halestead
Charles Ainsley Hall, PhD
V. A. Hamilton
James T. Hanlon, PhD
Harry C. Hardee, PhD
Daniel E. Harmeyer
Gerald E. Harrington
Richard A. Harris, PhD
E. Frederick Hartman
Eric W. Hatfield
Robert F. Hausman, Jr., PhD
K. Havernor, PhD
Russell Haworth
Dennis B. Hayes, PhD
Dale B. Henderson, PhD
Jon D. Hicks
Mary Jane Hicks, PhD
Michael Hicks
Dwight S. Hill
Fred L. Hinker
Terry D. Hinnerichs, PhD
Charles Henry Hobbs
Stephen M. Hodgson
David K. Hogan
Gregory A. Hosler
Michael G. Houts, PhD
Volney Ward Howard, Jr., PhD
James M. Hylko
J. Charles Ingraham, PhD
Mehraboon S. Irani
David H. Jagnow
Andrew John Jason, PhD
Ronald L. Jepsen
Robert Johannes, PhD
Ken L. Johnson
Michael L. Johnson, PhD
Wesley Morris Jones, PhD
Richard D. Juel
Jon P. Kahler
C. Keenan, PhD
William E. Keller, PhD
John Daniel Kemp, PhD
David R. Kendall
Claude Larry Kennan, PhD
H. Grant Kinzer, PhD
David I. Knapp
John H. Kolessar, III
Fleetwood R. Koutz
Gary R. Kramer, PhD
Lillian A. Kroenke
William Joseph Kroenke, PhD
Paul Kuenstler, PhD
Roger W. Lamb
Leo J. Lammers
James G. Lareau
Signa Larralde, PhD
Tom B. Larsen, PhD
Laurence Lattman, PhD
Roger X. Lenard
Lary R. Lenke
Frank L. Lichousky
William C. Lindemann, PhD
Stanley E. Logan, PhD

H. Jerry Longley, PhD
Gabriel P. Lopez, PhD
Radon B. Loveland
Bruce P. Lovett
Kenneth Ludeke, PhD
Mark J. Ludwig
Jesse Lunsford
Raymond A. Madson
Jeffrey Mahn
Norman Malm, PhD
Bob Malone
Gary M. Malvin, PhD
Dorinda Mancini
Joseph Bird Mann, PhD
Charles Robert Mansfield, PhD
Greg Mansfield
Jose Eleazar Martinez, Jr.
David S. Masterman
Mark A. Mathews, PhD
James W. Mayo
Douglas K. McCullough
Kirk McDaniel
Patrick D. McDaniel, PhD
Luther F. McDougal
Thomas D. McDowell, PhD
Heath F. McLaughlin
Virginia McLemore, PhD
Reed H. Meek
Jon J. Mercurio, PhD
James H. Metcalf
Andre F. Michaudon, PhD
Edmund Kenneth Miller, PhD
Peter W. Milonni, PhD
James J. Mizera*
Raymond C. Mjolsness, PhD
Charles E. Moeller
Joe P. Moore
Michael Stanley Moore, PhD
Tamara K. Morgan
Henry Thomas Motz, PhD
Joseph Mraz
Arthur Wayne Mullendore
Darrell Eugene Munson, PhD
Daniel Neal, PhD
Leland K. Neher, PhD
Burke E. Nelson, Sr., PhD
R. T. Nelson
James P. Niebaum
Clair W. Nielson, PhD
Clyde John Marshall Northrup, PhD
Bob Norton
Henry C. Nowail
Sean G. O'Brien, PhD
Connon R. Odom
Edward E. O'Donnell, PhD
William D. Ohmstede
Paul J. Ortwerth, PhD
Robert C. O'Shields
David K. Overmier
Thomas G. Parker
William L. Partain, PhD
Stanley J. Patchet, PhD
James Howard Patterson, PhD
Randy Patterson
Robert H. Paul, PhD
David C. Pavlich
Tamara E. Payne, PhD
Daniel N. Payton, III, PhD
Jeffrey H. Peace
Jerry L. Peace
Lester D. Peck
Ralph B. Peck, PhD

James A. Phillips, PhD
Rex D. Pieper, PhD
Gordon E. Pike, PhD
Pamela D. Pinson
James D. Plimpton, PhD
Paul Pompeo
William Morgan Porch, PhD
Nolan Probst
Irving Rapaport
J. R. Ratcliff
Leonard Raymond
Gary D. Rayson, PhD
George R. Reddy
Joe W. Reese
Paul A. Rehme
John Douglas Reichert, PhD
Mark M. Reif, PhD
William E. Reifsnyder
Anita S. Reiser
Norbert T. Rempe
Elliott A. Riggs, PhD
Arnold Robinson
David A. Rockstraw, PhD
John E. Rockwell, III
Larry D. Rodolph
John D. Roe
G. Rollet
Gary R. Rollstin
James M. Romero
Benny H. Rose, PhD
Daniel G. Rossbach
B. A. Rosser
Darryl D. Ruehle
Lawrence D. Rutherford
Kenneth Sabo, PhD
Richard John Salzbrenner, PhD
George Albert Samara, PhD*
Ian A. Sanders
Donald James Sandstrom
Alcide Santilli
David P. Sauter
Frank R. Schenbel
Scott A. Schilling
Jeffrey S. Schleher
Glen L. Schmidt, PhD
Theodore R. Schmidt, PhD
David A. Schoderbek
David W. Scholfield, PhD
Charles R. Schuch
Joseph Albert Schufle, PhD
L. Schuster
John Seagrave, PhD
Robert D. Sears
Gary L. Seawright, PhD
Jack Behrent Secor, PhD
Dwight Sederholm
Thomas J. Seed, PhD
David W. Seegmiller, PhD
Fritz A. Seiler, PhD
James Shaffner
Thomas L. Shelley
Lawrence Sher
Arnold John Sierk, PhD
Randal S. Simpson
Wilbur A. Sitze
Joseph L. Skibba, PhD
Florentine Smarandache, PhD
Clay T. Smith, PhD
Garmond Stanley Smith, PhD
Harvey M. Smith
Maynard E. Smith, PhD
Jon E. Sollid, PhD

Appendix 4: The Petition Project

Michael Scott Sorem, PhD
Morgan Sparks, PhD
David J. Sperling
James Kent Sprinkle
Vincent W. Steffen*
Stephen M. Sterbenz, PhD
Philip J. Sterling
Billy Stevens
Regan W. Stinnett, PhD
Leo W. Stockham, PhD
James William Straight, PhD
Thomas F. Stratton, PhD
William R. Stratton, PhD
Leonard Richard Sugerman
Donald Lee Summers
A. Einar Swanson
Donald M. Swingle, PhD
Mariano Taglialegami
Willard Lindley Talbert, PhD
David R. Tallant, PhD
Glen M. Tarleton
Gene W. Taylor, PhD
Javin M. Taylor, PhD
Troy Lynn Teague
Michael M. Thacker
Edward Frederic Thode, PhD
Angela Thomas
James L. Thomas, PhD
John R. Thompson, PhD
Seth J. Thompson
Billy Joe Thorne, PhD
Gary Tietjen
Robert Allen Tobey, PhD
Leonard A. Traina, PhD
Anthony J. Trennel
Charles V. Troutman
Patricio Eduardo Trujillo
John D. Underwood
Jurgen H. Upplegger, PhD
Bruce Harold Van Domelen, PhD*
Ron Van Valkenburg
Valentine W. Vaughn, Jr.
Don R. Veazey
Charles G. Vivion
George Leo Voelz
James Von Husen
Frederick Ludwig Vook, PhD
Teodor Vulcan, PhD
Francis J. Wall, PhD
Robert F. Walter, PhD
Roddy Walton, PhD
Norman R. Warpinski, PhD
Mial Warren, PhD
James D. Waters
William W. Weiss
William J. Whaley
Gerald W. Whatley
David Wheeler
H. E. Whitfacre
Larry K. Whitmer
Raymond V. Wick, PhD
James Wifall
Gary R. Williams
Gearld F. Willis
Donald E. Wilson
Norman G. Wilson
Bob Winn
Kenelm C. Winslow*
Donald Wolberg, PhD
Robert O. Woods, PhD
Eric J. Wrage
Harlow Wright

Ely Yao, PhD
John L. Yarnell, PhD
Frank Yates, Jr.
George W. York, Jr., PhD
Jeffrey L. Young
Lloyd M. Young, PhD
Phillip L. Youngblood

New York

Ralph F. Abate
Alan V. Abrams
Paul B. Abramson, PhD
John K. Addy, PhD
Siegfried Aftergut, PhD
M. C. Agress
Robert J. Alaimo, PhD
Rogelio N. Alama
Jay Donald Albright, PhD
Zeki Al-Saigh, PhD
Randy J. Alstadt
Vincent O. Altemose
Peter Christian Altner
Burton Myron Altura, PhD
Ronald F. Amberger, PhD
Leonard Amborski, PhD
Richard Amerling
Moris Amon, PhD
H. C. Anderson
Felixe A. Andrews
Dan Antonescu-Wolf
John L. Archie
William J. Arion, PhD
Alfred Arkell, PhD
Gertrude D. Armbruster, PhD
Seymour Aronson, PhD
Clement R. Arrison
James S. Arthur, PhD
Alvin Ashman
Charles W. Askins
Winifred Alice Asprey, PhD
Walter Auclair, PhD
Louis A. Auerbach
Rosario D. Averion
John M. Babli
Christopher J. Bablin
Henry Lee Bachman
Reid Bader
Richard G. Badger
Lcdr Paul Baham
Betty L. Bailey
Randall L. Bakel
Joseph M. Ballantyne, PhD
Debendranath Banerjee, PhD
Donald E. Banzhaf
Octavian M. Barbu
Charleton C. Bard, PhD
Robert Barish, PhD
Jeremiah A. Barondess
Louis Joseph Barone
Benjamin Austin Barry
Thomas G. Bassett
Robert William Batey
Frances Brand Bauer, PhD
Henry H. Baxter
Mark K. Beal
David Beckner
David F. Bedey, PhD
Allan Lloyd Bednowitz, PhD
Dan Beldy
Robert G. Bellinger

Douglas A. Bennett
Gerald William Bennett, PhD
Jay Manton Berger, PhD
Gerald Berkowitz, PhD
Herbert Weaver Berry, PhD
Kenneth Del Bianco
Jean M. Bidlack, PhD
Kenneth L. Bielat, PhD
Joseph F. Bieron, PhD
Jean M. Bigaouette
John Edward Bigelow
Rodney Errol Bigler, PhD
William Bihrle
Arthur Bing, PhD
George Birman
David L. Black, PhD
Donald E. Black
J. Warren Blaker, PhD
Nik Blaser
Sheldon Joseph Bleicher
Marvin E. Blumberg
M. Blumenberg, PhD
Fred G. Bock, PhD
Mary M. Bockus
J. Neil Boger
Haig E. Bohigian, PhD
Donald A. Bolon, PhD
Shirley C. Bone
Robert M. Booth
Sunder S. Bora, PhD
David Borkholder, PhD
George Boulter
Mohamed Boutjdir, PhD
Leo J. Brancato
Mike Brannen
Sydney Salisbury Breese
Jan Leslie Breslow
Rainer H. Brocke, PhD
John W. Brook, PhD
Robert R. Brooks, PhD
Walter Brouillette, PhD
Allen S. Brower
Melvin H. Brown
Samuel S. Browser, PhD
Eugene John Brunelle
Howard Bryan
Edward Harland Buckley, PhD
Herbert Budd, PhD
Richard F. Burchill
James M. Burlitch, PhD
David W. Burnett
Lawrence E. Burns
Robert A. Buroker
Steve Burrell
Tracee A. Burzycki
Timothy P. Bushwell, Jr., PhD
K. B. Cady, PhD
Angelo Cagliostro
Stephen R. Cain, PhD
Robert Ellsworth Calhoo, PhD
Gerald Allan Campbell, PhD
Antonio Mogro Campero, PhD
Enrico M. Camporesi
A. Van Caneghem
Thomas J. Carpenter
John L. Carroll
Robert B. Case
George S. Case, Jr
Anthony S. Caserta
Chris Cash
Edward Cassidy
Joseph Cassillo

Alan William Castellion, PhD
Lawrence M. Cathles, PhD
Robert E. Cech, PhD
Lawrence J. Cervellino
Charles C. Chamberlain, PhD
Shu Fun Chan, PhD
Richard Joseph Charles, PhD
Jacob Chass
Stuart S. Chen, PhD
Tak-Ming Chen
Relly Chern
A. Cheung, PhD
Joseph F. Chiang, PhD
Yuen Sheng Chiang, PhD
Robert L. Christensen, PhD
Kenneth G. Christy
Phillip S. Cimino
Emil T. Cipolla
Michael Circosca
Stephen F. Claeys
Earlin N. Clarke
Nicholas L. Clesceri, PhD
Theodore James Coburn, PhD
Paul Cochran
William D. Coe
Anthony Cofrancesco, PhD
Ezechiel Godert David Cohen, PhD
Randall Knight Cole, PhD
Philip J. Colella
Lawrence Colin
Hans Coll, PhD
George H. Collins
Robert J. Conciatori
Norman I. Condit
John P. Coniglio
Dennis Conklin, PhD
John Joseph Connelly, PhD
John A. Constance
William S. Cooley
Harry B. Copelin
Thomas M. Cornell
Arthur J. Cosgrove
Wilfred Arthur Cote, PhD
Eric W. Cowley
Robert Stephen Craxton, PhD
Robert J. Cresci, PhD
Nicholas G. Cristy
Alfred G. Cuffe
John S. Cullen
Edoardo S. Cuniberti
Ira B. Current
Darisuz Czarkowski, PhD
Lawrence B. Czech
Vincent F. D'Agostino
Barton Eugene Dahneke, PhD
William Dalton
Charles A. Daniel
Wayne Daniels
Zafar I. Dar
Ray L. Darling, Jr.
Baldev B. Das
Ellen S. Dashefsky
Edward Daskam
John R. Davis, Jr.
Raymond Davis, PhD
Ivana P. Day, PhD
Alain de La Chapelle
David J. De Marle
Nixon de Tamowsky
Vincent T. De Vita
Robert E. DeBrecht
Erwin Delano, PhD

John W. Delano, PhD
William M. Delaware
Jon Delong
Albert M. Demont
Michael C. Detraglia, PhD
Gregory J. Devlin
Charles DeVoe, Jr.
Robert C. Devries, PhD
William B. Dickinson, PhD
Joseph DiSisto
Carlos M. Dobryn
Edwin S. Dojka
Bruce J. Dolnick, PhD
Robert M. Domke
Robert W. Doty, PhD
Lawrence G. Doucet
Patricia Doxtader, PhD
John Droz, Jr.
Bernard S. Dudock, PhD
James R. Dunn, PhD
George J. Dvorak, PhD
Robert H. Easton
Constantino Economos, PhD
James P. Edasery, PhD
Robert Flint Edgerton, PhD
John A. Edward
Kurt F. Eise
Richard J. Engel
Charles Edward Engelke, PhD
Louis James Esch, PhD
Theodore Walter Esders, PhD
Levent Eskicakit
James Estep, PhD
John H. Estes, PhD
Howard Edward Evans, PhD
Woodrow W. Everett
Abraham Eviatar
Carl Fahrenkrug, PhD
Susan Fahrenkrug, PhD
Amedeo Alexander Fantini, PhD
Gerald N. Fassell
Anthony John Favale
Ralph A. Favale
Jay M. Feder
Tom Feil
George Feinbaum
Jean F. Feldman
Martin Felicita
Archie D. Fellenzer, Jr.
Charles P. Felton
John W. Fenton, II, PhD
Maj Virgil F. Ficarra
Robert H. Fickies
Bradley J. Field
James R. Fienup, PhD
Anthony Salvatore Filandro
Peter Stevenson Finlay, PhD
Arnold G. Fisch
Grahme P. Fischer
Nancy Deloye Fitzroy
Delbert Dale Fix, PhD
John Edward Fletcher, PhD
Irving M. Fogel
Charles E. Frank
Dorothea Zucker Franklin
Andrew Gibson Frantz
Donald M. Fraser
Gilbert Friedenreich
Gerald M. Friedman, PhD
Robert Louis Fullman, PhD
Maria I. Galaiko
Eugene H. Galanter, PhD

Gregory W. Gallagher
Richard D. Galli
Franics M. Gallo
Rosa M. Gambier, PhD
Gerard C. Gambs
Donald D. Garman
Robert Harper Garmezy
Howard B. Gates
Raymond E. Gaus
Robert Frank Gehrig, PhD
Gunther Richard Geiss, PhD
James G. Geistfeld, PhD
Herbert Leo Gelernter, PhD
Angelos S. Georgopoulos
J. P. Gergen, PhD
Anne A. Gershon
Eric B. Gertner
Arthur Gerunda
Mellor A. Gill
Robert Gillcrist
Lawrence B. Gillett, PhD
Ernest Rich Gilmont, PhD
Helen S. Ginzburg
Ronald M. Gluck
Charles J. Godreau
Leslie I. Gold, PhD
Malcolm Goldberg, PhD
Michelle Goldschneider
Edward B. Goldstein, PhD
Paul N. Goodwin, PhD
Leonard Gordon
Steven Grassl, PhD
Dennis J. Graves
Saul Green, PhD
Alex Gringauz, PhD
Herbert Grossman
C. Edward Grove, PhD
Timothy R. Groves, PhD
Johathan R. Gruchala
Bruce B. Grynbaum*
Robert F. Guardino
Heinrich W. Guggenheiher, PhD
Paul F. Gugliotta, PhD
Michael A. Guida
Raj K. Gupta, PhD
Susan Gurney
James Guyker, PhD
Daniel Habib, PhD
Albert Haim, PhD
Mike Hale, PhD
Norm Hall
Joseph Halpern, PhD
Jack A. Hammond
Stanley E. Handman
Sultan Haneed, PhD
Leon Dudley Hankoff
Kevin Joseph Hanley
George E. Hansen
Gordon Charles Hard, PhD
George Bigelow Hares, PhD
Henry Gilbert Harlow
Dean Butler Harrington
Bruce Harris
Donald R. Harris, PhD
Ramon P. Harris
Delmer L. Hart
Sylvan Hart
Hermen Hartjens
Richard W. Hartmann, PhD
Richard M. Harwell
Frederick C. Hass, PhD
Floyd N. Hasselrlis

Maximilian E. Hauser
Wayne Haushnecht
William W. Havens, Jr., PhD
Terrance Haviland
G. Wayne Hawk
Daniel P. Hayes, PhD
F. Terry Hearne
Ira Grant Hedrick
Leslie B. Hegeman
John C. Heiman
Charles Gladstone Heisig, PhD
Mark Heitz, PhD
David Helfand
Harold R. Hellstrom
Allan Hennings
Lloyd Emerson Herdle, PhD
Zvi Herschman
Ralph Allan Hewes, PhD
Patrick Hickey
Walter A. Hickox
Robert E. Hileman, PhD
James O. Hillis
H. O. Hoadley
Stanley H. Hobish
Merrill K. Hoffman
Roger Alan Hoffman, PhD
Theodore P. Hoffman, PhD
Walter H. Hoffmann
Gary Hohenstein
John C. Holko
Alfred C. Hollander
Harold A. Holz
Kenneth E. Hoogs
Robert H. Hopf
Harvey Allen Horowitz
Mark G. Horschel
Malcolm D. Horton
Barbara J. Howell, PhD
Jan Hrabe, PhD
Lawrence R. Huntoon, PhD
Michael John Iatropoulos, PhD
Marvin L. Illingsworth, PhD
A. Rauf Imam, PhD
Charles G. Inman
A. M. Irving
Erich Isaac, PhD
Zafar A. Ismail, PhD
Patrick Thomas Izzo, PhD
Joseph P. Jackson
James Jacob, PhD
Theodore G. Jacobs
Brian Jacot
Jules Jacquin
William Julian Jaffe, PhD
Lauren E. Jaquin
George A. Jensen
Thomas E. Jensen, PhD
Henry Louis Jicha, Jr., PhD
M. Jin, PhD
Karel Jindrak, PhD
Henry M. Joerz, Jr.
Sune Johansson
Francis Johnson, PhD
Horton A. Johnson
Thomas M. Johnson
Andrew E. Jones, PhD
Larry Josbeno
Robert O. Kalbach, PhD
Norman Wayne Kalenda, PhD
Moses K. Kaloustian, PhD
Robert M. Kamp
Walter Reilly Kane, PhD

Walter T. Kane
Munawar Karim, PhD
Yefim Kashler
Bertil J. Kaudern
George Kaufman
Kevin J. Kearney
Donald Bruce Keck, PhD
Zvi Kedem, PhD
Robert Adolph Keeler
Jerome Baird Keister, PhD
D. Steven Keller, PhD
Spencer II Kellogg, II
J. Kelly
Margaret M. Kelly
Donald Laurens Kerr, PhD
S. F. Kiersznowski
Thomas Kileer
Robert William Kilpper, PhD
Kenneth M. King
Paul I. Kingsbury, PhD
George Stanley Klaiber, PhD
Kevin J. Klees
Vasilios Kleftis
Ludwig Klein, PhD
Eugene M. Klimko, PhD
Sylvia K. Knowlton
Elisa Konofagou, PhD
Howard J. Kordes
Karl F. Korfmacher, PhD
Jack I. Kornfield, PhD
Lee Ming Kow, PhD
Russell Kraft, PhD
Robert W. Kress
Joel Kronfeld, PhD
Valentina R. Kulick, PhD
Max Kurtz
Theodore W. Kury, PhD
Peter La Celle, PhD
Paul L. Lacelle, PhD
Joann Lachut
Sanford Lacks, PhD
Robert J. Laffin, PhD
Peter L. Laino
David E. Landers
Richard L. Lane, PhD
Arthur M. Langer, PhD
Frank Lanzafame, PhD
Joseph M. Lanzafame, PhD
Peter Lanzetta, PhD
Evelyn P. Lapin, PhD
Charles Conrad Larson, PhD
Ernest T. Larson
Keith F. Lashway
Kimberly F. Lavery
Kent P. Lawson, PhD
Jeffrey C. Lawyer
Norman Lazaroff, PhD
Robert J. Lecat, PhD
Andre Leclair, PhD
N. A. Legatos
August Ferdinand Lehman
Nathaniel S. Lehrman
Herbert August Leupold, PhD
Richard D. Levere
Jane R. Levine
Herbert August Levpold, PhD
Harold W. Lewis, PhD
Emanuel Y. Li
Yong J. Li, PhD
Irwin A. Lichtman, PhD
James A. Liggett, PhD
Ralph Linsker, PhD

Appendix 4: The Petition Project

Arnold A. Lipton
Clayton Lee Liscom
Bruce A. Lister
Richard G. Livingston
Clyde R. Locke
Joseph Carl Logue
Anna K. Longobardo
Jerome J. Looker, PhD
John Looney
Ronald J. Lorenz
David Lorenzen
Jerry A. Lorenzen, PhD
Thomas Lowinger, PhD
Gerald Luck
Frederick J. Luckey
George W. Luckey, PhD
Carol J. Lusty, PhD
Bruce MacDonald
John Fraser Macdowell
Elbert F. Macfadden, Jr.
Robert E. Madden
Robert W. Madey, PhD
Edward F. Magnusson
Leathem Mahaffey, PhD
Hug David Maille, PhD
Jay H. Maioli
Michael Maller, PhD
Timothy M. Maloney
William Muir Manger, PhD
Stephen H. Manglos, PhD
Ruben Manuelyan
Karl Maramorosch, PhD
Morris J. Markovitz
Paul Marnell, PhD
Robert E. Marrin
Vincent F. Marrone, PhD
Elliott R. Marsh
James A. Marsh, PhD
L. Gerald Marshall
Irving Marvin
John J. Mastrangelo
Donald E. Mathewson
David Matthew
Paul R. Mattinen
Paul B. Mauer
Robert E. Maurer, PhD
A. Frank Mayadas, PhD
Theodor Mayer, PhD
Denis G. Mazeika
John R. Mcbride, PhD
Thomas A. McClelland, PhD
J. R. D. McCormick, PhD
John P. McGuire
William R. McLean
William McNamara, PhD
Paul F. McTigue
Winthrop D. Means, PhD
Allen G. Meek
Karin W. Megerle
Karl R. Megerle
Otto Meier, Jr.
Daniel T. Meloon, PhD
Ernest Menaldino
Paul Barrett Merkel, PhD
David L. Messenger
John D. Metzger, PhD
Richard H. Meyer, Jr.
Robert P. Meyer, PhD
John C. Midavaine
Kevin J. Miley
Allan F. Miller
Gary S. Mirkin

Gerald Jude Mizejewski, PhD
Robert P. Moehring
Henry W. Moeller, PhD
Joseph C. Mollendorf, PhD
Leo F. Monaghan
Salvatore J. Monte
Richard T. Mooney
John Moran
Thomas C. Morgan
Paul W. Morris, PhD
William Guy Morris, PhD
Roger Alfred Morse, PhD
Kenneth Ernest Mortenson, PhD
Edward P. Mortimer
Wayne Morton
Lloyd Motz, PhD
Eldridge M. Mount, III, PhD
Charles F. Mowry
Shakir P. Mukhi
L. Muller
Charles Munsch
Shyam Prasad Murarka, PhD
Daniel B. Murphy, PhD
Victor L. Mylroie
Eric Nary
Hugh M. Neeson
Jarda D. Nehybka
Charles A. Nelson, PhD
Sven Nelson
Wendell Neugebauer
Hans P. Nichols
Richard Nickerson
Sissy A. Nikolaou, PhD
Joseph F. Noon
R. A. Norling
William T. Norton, PhD
Julian Nott
Jerzy Nowakowski, PhD
Tom O'Brien, PhD
Matthew F. O'Connell
G. T. O'Connor
Robert T. Ohea
Tonis Oja, PhD
Joanne M. Ondrako, PhD
Dino F. Oranges
Brian P. O'Rourke
Michael J. O'Rourke, PhD
Ernest Vinicio Orsi, PhD
Harold R. Ortman
Paul C. Oscanyan, PhD
Maurice J. Osman
C. M. Oster
Hans Walter Osterhoudt, PhD
Murali Krishna Pagala, PhD
Donald G. Paish
Christopher R. Palma, PhD
Giorgio Pannella, PhD
Lawrence D. Papsidero, PhD
Hemant Bhupendra Parikh
David J. Parker
Zohreh Parsa, PhD
Richard E. Partch, PhD
Robert William Pasco, PhD
Kenneth Pass, PhD
Harini Patel, PhD
Sandra Patrick
Gregory J. Patterson
Michael V. Pavlov
George John Pearson, PhD
Erik Mauritz Pell, PhD
Joseph C. Pender
John P. Pensiero

Theodore Peters, Jr., PhD
John H. Peterson
Dana M. Pezzimenti
Tuan Duc Pham, PhD
Kevin Phelan
Neal Phillip, PhD
Ernest E. Phillips
Erling Phytte, PhD
Carlson Chao Ping Pian
Richard N. Pierson
Valter Ennis Pilcher, PhD
Vernor A. Pilon
Walter Thornton Pimbley, PhD
David Pimentel, PhD
William R. Pioli
Anthony W. Pircio, PhD
Rahn G. Pitzer
Wendy K. Pogozelski, PhD
V. Pogrebnyak, PhD
Wilson Gideon Pond, PhD
Om P. Popli
Gerhard Popp, PhD
Raymond Porzio
Maria A. Pospischil
George Henry Potter, PhD
R. Priefer, PhD
Svante Prochazka, PhD
John James Prucha, PhD
Richard P. Puelle
Donald H. Puretz, PhD
D.C. Putman
Orrea F. Pye, PhD
James Roger Quinan, PhD
Richard L. Racette
Joseph Wolfe Rachlin, PhD
Stanislaw P. Radziszowski, PhD
Alan D. Raisanen, PhD
Michael Ram, PhD
Madeline K. Ramsey
Paul Ellertson Ramstad, PhD
Hugh Rance, PhD
Navagund Rao, PhD
Arthur A. Rauth
Stanley R. Rawn, Jr.
Geoffrey Recktenwald, PhD
Michael L. Reichard
Jerome A. Reid
Anthony A. Reidlinger, PhD
Norman E. Reinertsen
William E. Rew
William T. Rhea
John T. Rice, PhD
Guercio Louis Richard
Elmer A. Richards
C. II. B. Richardson
Charles M. Richardson
Selma Richman
Malcolm M. Riggs
James Lewis Ringo, PhD
Arnold Risman
Walter Lee Robb, PhD
Leslie Earl Robertson
Terence Lee Robinson, PhD
Maxine Lieberman Rockoff, PhD
James D. Rodems
Robert M. Roecker, PhD
Lloyd Sloan Rogers
Robert C. Rohr, PhD
Mario F. Romagnoli
David M. Rooney, PhD
Robert L. Rooney
Milton Jacques Rosen, PhD

Arthur Rosenshein
Joseph Rosi
John Rostand
R. A. Roston
Ronald D. Rothchild, PhD
Elliot M. Rothkopf, PhD
Lewis P. Rotkowitz
William N. Rotton
Gaylord Earl Rough, PhD
Ibrahim Rubeiz, PhD
Leonard Rubenstein
Mario J. Rufino
Arthur L. Ruoff, PhD
William F. Ryan
James A. Sacco
Frederick Sachs, PhD
Robert A. Saia
John Salvador, Jr.
Jay Salwen
Robert S. Samson
Jose Sanchez, PhD
Juan Sanchez
Jon Henry Sanders, PhD
Andrew M. J. Sandorfi, PhD
Edward James Sarcione, PhD
Balu Sarma, PhD
Frank Scalia, PhD
Marc A. Scarchilli
Len Schaer
Simon Schaffel, PhD
Anthony J. Schaut
Alan J. Schecter, PhD
Edward J. Schmeltz
Frederick W. Schmidlin, PhD
Richard F. Schneeberger
Samuel Ray Scholes, Jr., PhD
James J. Schultheis
Gerald Schultz
W. E. Schwabe
John Schwaninger
Frederick A. Schweizer
Daniel Sciarra
Wright H. Scidmore
Terrence R. Scott
Brian E. Scully
David R. Seamen
Boris M. Segal
Robert Seider
Sharon Seigmeister
Frederick Seitz, PhD*
M. Seleem, PhD
Domenick A. Serrano
Bryon R. Sever, PhD
Wayne K. Shaffer
R. Shapiro
John B. Sharkey, PhD
Clifford J. Shaver
Richard Gregg Shaw, PhD
Robert Shaw
Chester Stephen Sheppard, PhD
Malcolm J. Sherman, PhD
Robert M. Shields, PhD
Avner Shilman, PhD
Steve Shoecraft
John O. Shotwell
Avanti C. Shroff, PhD
Eugene E. Shube
Barbara Shykoff, PhD
Richard W. Siegel, PhD
John Siegenthaler
Richard M. Sills
Albert Simon, PhD

John M. Simpson
Emil D. Sjoholm
Victor Skormin, PhD
Jane Ann Slezak, PhD
Harry Kim Slocum, PhD
Albert A. Smith
Frederick W. Smith, PhD
Neil M. Smith
Bronislaw Smura, PhD
Herman Soifer
Robert Soley
Chester P. Soliz
John Robert Sowa, PhD
Paul William Spear
William E. Spears, Jr.
Abraham Spector, PhD
Stanley B. Spellman
William J. Spezzano
Michael J. Spinelli
John M. Spritzer
Alfred E. Stahl
Laddie L. Stahl
Zygmunt Staszewski
D. E. Steeper
Daniel T. Stein
Martin F. Stein
Harry Stevens, Jr.*
Carleton C. Stewart
William Andrew Stewart
Gerhard Stohrer, PhD
Norman Stanley Stoloff, PhD
John S. Stores, PhD
Stewart P. Stover
Fred Strnisa, PhD
Forrest C. Strome, PhD
Raymond G. Stross, PhD
Henry Stry
Leon S. Suchard
Michael Surgeary
Bob Sutaria
Paris D. Svoronos, PhD
John M. Swab
Charles Swain
Donald Percy Swartz
Donald Sykes, PhD
Joseph Szczesniak
Robert Szego, PhD
Boleslaw K. Szymanski, PhD
Charles Peter Talley, PhD
Mehmet Y. Tarhan
Pamela A. Tarkington
Nixon De Tarnowsky
Aaron M. Taub, PhD
James E. Taylor, PhD
Kenneth J. Teegarden, PhD
Stephen Whittier Tehon, PhD
Aram V. Terchwian
George Terwilliger, PhD
David K. Thom
Edward Sandusky Thomas, PhD
G. J. Thorbecke, PhD
Craig E. Thorn, PhD
Larry Tilis
Tore Erik Timell, PhD
William Timm
Lu Ting, PhD
Sathyan C. Tivakaran
Thomas B. Tomasi, PhD
Joseph Torosian
Steven M. Trader
Joseph M. Tretter
Robert Bogue Trimble, PhD

Anthony Trippe, PhD*
Donald Troupe, PhD
Min-Fu Tsan, PhD
Herbert F. Tschudi
Robert Joseph Tuite, PhD
James B. Turchir
William Joseph Turner, PhD
Arthur Glenn Tweet, PhD
George A. Tyers
John Urdea
Charles F. Vachris
Marius P. Valsamis
Vida K. Vambutas, PhD
Ruth VanDeusen
R. V. Vanhoweistine, PhD
Michael O. Varner
Thomas S. Velz
David Augustus Vermilyea, PhD
Morrill Thayer Vittum, PhD
Kalman N. Vizy, PhD
Alexander V. Vologodskii, PhD
Richard W. Vook, PhD
Leonid Vulakh, PhD
Ernest N. Waddell
Thomas J. Wade, PhD
John M. Wadsworth
Salome G. Waelsch, PhD
Anne M. Wagner
Roy C. Wagner, PhD
Christopher Walcek, PhD
Joseph Walters
Robert L. Walton
James R. Wanamaker
Howard C. Ward, Jr.
Robert S. Ward
Barbara E. Warkentine, PhD
Arthur P. Weber, PhD
Julius Weber
Brent M. Wedding, PhD
Paul M. Wegman
Leonard Harlan Weinstei, PhD
Herbert Weinstein
Leonard Weisler, PhD
Harvey Jerome Weiss
R. B. Wenger
Jack Hall Westbrook, PhD
Calvin E. Weyers
Elizabeth M. Whelan, PhD
Donald Robertson White, PhD
Carl F. Whitehead
David C. Widrig
James Wiebe
Grace M. Wieder, PhD
Kevin Williams
William H. Williams, PhD
Nicolaos Daniel Willmore, PhD
Jack Belmont Wilson, PhD
Janet W. Wilson
Frank W. Wise, PhD
Charles E. Wisor
Paul C. Witbeck
Mark Witowski
Harold Ludwig Witting, PhD
Robert C. Wohl, PhD
Paul Wojnicz
Arthur J. Wolf
Peter Wolle
John T. Wozer
Robert W. Wright, PhD
Konran T. Wu, PhD
Klaus Wuersig
John E. Wylie

Dmeter Yablonsky, PhD
Kenneth R. Yager
Scott Yanuck
William D. Young, Jr.
Wen Shi Yu, PhD
Chia Liu Yuan, PhD
Anthony M. Yurchak
Richard Yuster
Marco Zaider, PhD
Ethel Suzanne Zalay, PhD
Andreas Athanasios Zavitsas, PhD
Allen Zelman, PhD
Frank F. Zhang, PhD
Herman Ernst Zieger, PhD
Edward N. Ziegler, PhD
Hirsch J. Ziegler
Jehuda Ziegler, PhD
Ronald F. Ziolo, PhD
Richard J. Ziomkowski
Thomas A. Zitter, PhD
David Zuckerberg
Carl William Zuehlke, PhD
Petr Zuman, PhD

North Carolina

Eugene Adams
Rolland W. Ahrens, PhD
Allan J. Albrecht
William D. Albright
Moorad Alexanian, PhD
Charles M. Allen, PhD
Gary L. Allen, PhD
Reevis Stancil Alphin, PhD
John Pruyn Van Alstyne
Robert H. Appleby
Webster J. Arceneaux, Jr.
Gregory S. Augspurger
Robert L. Austin
Max Azevedo
Rodger W. Baier, PhD
Jim Bailey
Lloyd W. Bailey
Richard J. Baird
John L. Bakane
Charles R. Baker, PhD
D. K. Baker, PhD
Walter E. Ballinger, PhD
Steven Bardwell, PhD
Stanley O. Barefoot
James A. Bass
Norman J. Bedwell
Carl Adolph Beiser, PhD
John C. Bemath
Edward George Bilpuch, PhD
Jack C. Binford
Larry G. Blackburn
Stephen T. Blanchard
Catherine E. Blat
Nathan Block
Arthur Palfrey Bode, PhD
Charles Edward Boklage, PhD
E. Arthur Bolz
William George Bottjer, PhD
Daniel P. Boutross
Richard D. Bowen
Stephen G. Boyce, PhD
Dale W. Boyd, Sr.
Lyle A. Branagan, PhD
David K. Brese
Charles B. Breuer, PhD

Richard L. Brewer
Philip M. Bridges
David W. Bristol
John Oliver Brittain, PhD
James E. Brookshire
Ben Brown
David Brown
Henry S. Brown, PhD
James E. Brown
M. Frank Brown
Robert C. Brown
Brett H. Bruton
Donald Buckner
Francis P. Buiting
Leonard Seth Bull, PhD
Paul Leslie Bunce
James J. Burchall, PhD
Hugh J. Burford, PhD
Steven V. Burgess
Rollin S. Burhans
John Nicholas Burnett, PhD
Loy R. Burris
David A. Burton
Neal E. Busch, PhD
William J. Bushkie, Sr.
Domingo E. Cabinum
Brett Callaway, PhD
Dixon Callihan, PhD
Harry B. Cannon, Jr.
J. David Carlson, PhD
David W. Carnell
Benjamin Harrison Carpenter, PhD
Randall L. Carver
Charles D. Case
Anthony J. Castiglia
Frank P. Catena
Gregory P. Chacos
Raymond F. Chandler
John Judson Chapman, PhD
Richard A. Chase
Brad N. Chazotte, PhD
John Garland Clapp, PhD
Howard G. Clark, PhD
James William Clark
William M. Clark
Larry Clarke
Raymond E. Clawson
Kenneth W. Clayton
Frank Clements
Warren Kent Cline, PhD
Todd Cloninger, PhD
Donald Gordon Cochran, PhD
Bertram W. Coffer
William L. Cogger, Jr.
Chris Coggins, PhD
Richard Paul Colaianni
Stephen K. Cole, PhD
Elwood B. Coley
Clifford B. Collins
John R. Cone
Paul Kohler Conn, PhD
John M. Connor
Maurice Gayle Cook, PhD
Anson Richard Cooke, PhD
Jackie B. Cooper
Daniel S. Corcoran
Alfred R. Cordell
John W. Costlow, PhD
Frederick Russell Cox
John T. Coxey
Terry A. Cragle
Ely J. Crary

David F. Creasy
Ronald P. Cronogue
John M. Crowley*
Alan L. Csontos
Bradford C. Cummings
William S. Currie
George B. Cvijanovich, PhD
Anthony A. Dale
Walter E. Daniel, PhD
Jose Joaquim Darruda, PhD
Dean Daryani
Tony Dau
Howard C. Davenport
Clayton L. Davidson
W. H. Davidson
Harold Davies
Dana E. Davis
Warren B. Davis
William Robert Davis, PhD
Eugene De Rose, PhD
Robert J. Dean
Nancy E. Dechant
Donald W. Dejong, PhD
James B. Delpapa
Florian E. Deltgen, PhD
Adam B. Denison
Gregory W. Dickey
Robert A. DiLorenzo
Raymer G. Dilworth
Michael Dion
Nama Doddi, PhD
Hugh J. Donohue
G. Double, PhD
E. Douglass
Donald A. Dowling
Thomas R. Drews
Earl G. Droessler
William DuBroff, PhD
Joseph M. Ducar
Donald John Duoziak, PhD
Kevin D. Durs
Paul B. Duvall
Robert S. Eckles
Jim Edgar
Steve S. Edgerton
Linda C. Ehrlich, PhD
Douglas F. Eilerston
Ademola Ejire, PhD
Warner A. Eliot
Norma S. Elliott, Jr.
William H. Elliott, Jr.
John Joseph Ennever
George E. Erdman
Harold P. Erickson, PhD
Paul S. Ervin
M. Frank Erwin
Joseph J. Estwanik
David P. Ethier
Marshall L. Evans
Ralph Aiken Evans, PhD
John O. Everhart
Herman A. Fabert, Jr.
Alan L. Falk
David W. Fansler
Michael Farona, PhD
William Joseph Farrissey, Jr., PhD
Robert Y. Felt
John James Felten, PhD
V. Hayes Fenstermacher
Courtney S. Ferguson, PhD
Herman White Ferguson
Carl F. Fetterman

James Fiordalisi, PhD
Joseph W. Fischer
David C. Fischetti
Timothy J.R. Fletcher
Ken Flurchick, PhD
Thomas P. Foley
Roger A. Foote
Michael J. Forster, PhD
Mac Foster, PhD
Robert Middleton Foster
Neal Edward Franks, PhD
Scott French
Pete G. Friedman
Robert S. Fulghum, PhD
Louis Galan
Dan Galloway
Clifton A. Gardner, Jr.
Donald M. Garland
Watson M. Garrison
Daniel S. Garriss
Mark P. Gasque
Edward F. Gehringer, PhD
John C. Geib
William T. Geissinger
Joseph A. Gerardi
Hugh Gerringer
Forrest E. Getzen, PhD
Harry F. Giberson
Gerald W. Gibson
Michael J. Gibson, PhD
Gray T. Gilbert
William C. Gilbert
Nicholas W. Gillham, PhD
John H. Gilliam
Thomas R. Gilliam
Elizabeth H. Gillikin
William Budd Glenn, PhD
Lucinda H. Glover
Robert A. Goetz
Eric Charles Gonder, PhD
Tim Good
Robert Goodman
Donald Gorgei
Bill Grabb
Tommy Grady
Robert B. Gregory
Tim A. Griffin
Eugene W. Griner
James J. Grovenstein
Robert M. Gruninger
Frederic C. Gryckiewicz
Harold A. Guice
Robert J. Gussmann
Joseph A. Gutierrez
Kevin Hackenbruch
Kenneth Doyle Hadeen, PhD
John V. Hamme, PhD
Charles Edward Hamner, PhD
N. Bruce Hanes, PhD
Steve Hansen, PhD
James E. Hardzinski
Charles M. Harman, PhD
Michael A. Harpold, PhD
Bernard C. Harris
Joe Harris, PhD
Heath Harrower
Paul J. Harry
James Ronald Hass, PhD
Thomas W. Hauch
John R. Hauser, PhD
Gerald B. Havenstein, PhD
Richard J. Hawkanson

Leland E. Hayden
John Lenneis Haynes
George L. Hazelton
Jane K. Hearn
Ralph C. Heath
Jonathan Daniel Heck, PhD
Edward M. Hedgpeth
Luther W. Hedspeth
Carl John Heffelfinger, PhD
John A. Heitmann, PhD
Henry Hellmers, PhD
William T. Helms
Miriam M. Henson, PhD
O'Dell W. Henson, PhD
Teddy T. Herbert, PhD
John Key Herdklotz, PhD
James E. Hester
Thomas R. Hewitt
William J. Hindman, Jr.
James H. Hines, Jr.
William B. Hinshaw, PhD
Willie L. Hinze, PhD
Maureane R. Hoffman, PhD
Kenneth W. Hoffner
Anatoli T. Hofle
Charles E. Holbrook
Richard Thornton Hood, Jr.
Philip J. Hopkinson
Charles R. Hosler
Dennis E. House
John L. Hubisz, PhD
Colin M. Hudson, PhD
Allen S. Hudspeth
B. Huggins
C. G. Hughes
Harold Judson Humm, PhD
Francis X. Hurley, PhD
William R. Hutchins
William Hutchins
Richard Ilson
Albert M. Iosue
James Bosworth Irvine
Rosemary B. Ison
F. K. Iverson
Robert A. Izydore, PhD
Hank Z. Jackson, Jr.
Julius A. Jackson
Mordecai J. Jaffe, PhD
Robert Janowitz
Chueng R. Ji, PhD
Friedlen B. Jones
Melvin C. Jones
Edward Daniel Jordan, PhD
Donald G. Joyce
Paul Judge
Karl Kachadoorian
Thomas R. Kagarise
Antone J. Kajs
Victor V. Kaminski, PhD
A. J. Karalis
George Keller, PhD
Brad A. Kemmerer
Donald L. Kimmel, Jr., PhD
Emmett S. King
Oscar H. Kirsch
Duane L. Kirschenman
Sam D. Kiser
R. Klein, PhD
Thomas Rhinehart Konsler, PhD
John T. Kopfle
Jeffrey L. Kornegay
Lemuel W. Kornegay

John M. Kramer
Howard R. Kratz, PhD
Steven Kreisman
Frederick William Kremkau, PhD
Lance Whitaker Kress, PhD
George James Kriz, PhD
Raymond Eugene Kuhn, PhD
James Richard Kuppers, PhD
Evan Dean Laganis, PhD
Jack W. Laney
Dennis J. Lapoint, PhD
Robert E. Lasater
William E. Laupus
Douglas M. Lay, PhD
Suzanne M. Lea, PhD
Harvey Don Ledbetter, PhD
Richard A. Ledford, PhD
Keith E. Leese
Dennis M. Leibold
Michael L. Leming, PhD
Robert Murdoch Lewert, PhD
George W. Lewis
Gordon Depew Lewis, PhD
Chia-yu Li, PhD
John B. Link
Mark D. Lister, PhD
Harold W. Lloyd, Jr.
Charles H. Lochmuller, PhD
Harry G. Lockaby
Richard A. Loftis
John B. Longenecker, PhD
Ian S. Longmuir
H. M. Losee
Kyle W. Loseke
Jeff S. Lospinoso
Doug Loudin
Sadler Love
Jeffery D. Loven
Holly E. Lownes
David L. Lucht
Arthur Lutz
David S. Lutz
John Henry MacMillan
Cecil G. Madden
Rodney K. Madsen
Bohdan E. Malyk
Jethro Oates Manly, PhD
William Robert Mann, PhD
P. K. Marcon
Gerald R. Marschke
Bryon E. Martin
Douglas Lee Martin, PhD
Harold L. Martin
J. C. Martin
Leila Martin
Robert C. Matejka
Joe K. Matheson
Nils E. Matthews
Eric J. Matzke
Kenneth Nathaniel May, PhD
Michael S. Maynard, PhD
Robin S. McCombs
Philip Glen McCracken, PhD
Jennifer McDuffie, PhD
Dale A. McFarland
Kenneth G. McKay, PhD
Hugh M. McKnight
Nathan H. McLamb
Nicole B. McLamb
Beverly N. McManaway
M. S. Medeiros, Jr.
Richard Young Meelheim, PhD

Bill R. Merritt
Genevieve T. Michelsen
Donald V. Micklos
Warren E. Milbrandt
Alan Millen
Daniel Newton Miller, PhD
James R. Miller
Robert James Miller, PhD
James F. Mills
Jesse R. Mills
James W. Mink, PhD
Gary N. Mock, PhD
Masood Mohiuddin, PhD
Paul Richard Moran, PhD
Edward G. Morris
Willard L. Morrison, Jr.
Thomas M. Morse
Don R. Morton
Kenneth L. Morton
Susan R. Morton
Parks D. Moss
Christopher B. Mullen
Jamie P. Murphy
Raymond L. Murray, PhD
Robert D. Neal, PhD
Kenneth E. Neff
Hesam O. Nekooasl
A. Carl Nelson, Jr.
Richard D. Nelson
John T. Newell, PhD
Alan H. Nielsen
John Merle Nielsen, PhD
Larry Nixon
D. B. Nothdurft
Donald E. Novy
Robert A. Novy, PhD
Joe Allen Nuckolls
Sylvanus W. Nye
Michael J. O'Connell
Calvin M. Ogburn
Jerry A. Orr
Mitchell M. Osteen
Robert J. Owens
Virgeon A. Pace
Teresa W. Page
Bert M. Parker, PhD
George W. Parker
Bernard L. Patterson
Orus F. Patterson, III
Dennis R. Paulson
Ira W. Pearce
George Wilbur Pearsall
Mary Alice Penland
Kim Percell
Ralph Matthew Perhac, PhD
Derrick O. Perkins
Peter Petrusz, PhD
DeWitt R. Petterson, PhD
Dwayne Phillips
Cu Phung, PhD
Mark S. Pierce
William C. Piver
Harvey Pobiner
Mark A. Pope
Thomas D. Pope
Kevin T. Potts, PhD
Roger Allen Powell, PhD
Eugene A. Praschan
Gary C. Prechter
Mark L. Prendergast
William Pulyer
Peter M. Quinn

Karl Spangler Quisenberry, PhD
Charles W. Raczkowski, PhD
Jay M. Railey
James Arthur Raleigh, PhD
Colin Stokes Ramage
Walter B. Ramsey
James A. Reagan
Jamie Redenbaugh
Bernice E. Redmond
Michael R. Reesal, PhD
William O. Reeside
G. George Reeves, PhD
Nancy Reeves
Stanley E. Reinhart, PhD
Stephen G. Richardson, PhD
Jerry F. Rimmer
Benjamin L. Roach
Paul J. Robert
C. Gordon Roberts, PhD
Joe L. Robertson
Norman Glenn Robison, PhD
John M. Roblin, PhD
Neill E. Rochelle
John M. Rodlin, PhD
Neil Roeth, PhD
Michael E. Rogers
John H. Rohde, PhD
Joseph W. Rose
Ollie Rose, PhD
John P. Ross, PhD
Thomas M. Royal
Sohindar S. Sachdev, PhD
Robert L. Sadler
Walter Carl Saeman
Joseph R. Salem, Sr.
Francis L. Sammt
William August Sander, PhD
Foster J. Sanders
Jerry Satterwhite
Vinod Kumar Saxena, PhD
Howard John Schaeffer, PhD
J. T. Scheick, PhD
Keith Schimmel, PhD
Laurence W. Schlanger
Douglas G. Schneider
Howard A. Schneider, PhD
William S. Schwartz
James Alan Scott
Allen J. Senzel, PhD
Charlie C. Sessoms, Jr.
Stephen T. Sharar
Lewis J. Shaw
Joe H. Sheard
Donald Lewis Shell, PhD
Kevin A. Shelton
Bruce A. Shepherd
Lamar Sheppard, PhD
Moses Maurice Sheppard, PhD
William C. Sides
Charles N. Sigmon
Dorothy Martin Simon, PhD
Anthony D. Simone
Richard E. Slyfield
George Smith
James R. Smith
Norman Cutler Smith
Warren A. Smith
Robert L. Smyre
Stanley Paul Sobol
Ken Soderstrom, PhD
Ronald G. Soltis
Jan J. Spitzer, PhD

James B. Springer, PhD
James L. Spruill
Alexander Squire
Hans H. Stadelmaier, PhD
Sanford L. Steelman, PhD
Bern Steiner
Charles T. Steinman
Paul B. Stelzner
Gene Stephenson
Richard M. Stepp
Richard H. Stickney
Henry A. Stikes
William T. Stockhausen
Waldemar D. Stopkey
Ralph D. Stout
William R. Stowasser
Richard Strachan, PhD
Robert F. Strauss
Donald W. Strickland
Roger D. Stuck
William Alfred Suk, PhD
Lyman E. Summerlin
Darrell D. Sumner, PhD
Erick D. Swanger
Kim H. Tan, PhD*
Marshall C. Taylor
John Pelham Thomas, PhD
Stanley J. Thomas, PhD
Vernard Ray Thomas, Jr., PhD
Steven G. Thomasson
Thomas Tighe
Ronald W. Timm
Daniel L. Timmons
William C. Timmons
Samuel Weaver Tinsley, PhD
Arthur Toompas
Alvis Greely Turner, PhD
William J. Turpish
Frank M. Tuttle
Robert K. Tyson, PhD
David L. Uhland
Laura Valdes
John A. VanOrder
Paul Varlashkin, PhD
Chadwick D. Virgil
Denniss E. Voelker
Robert A. Vogler
H. Von Amsberg, PhD
Garry W. Voncannon
Nicholas C. Vrettos
William Delany Walker, PhD
Jie Wang, PhD
Mansukhlal C. Wani, PhD
W. C. Warlick
Mark R. Warnock
David B. Waters
Jack Watson
Robert B. Watson
James D. Waugh
Norman L. Weaver
Jerome Bernard Weber, PhD
David Weggel, PhD
James T. Welborn
Gregory V. Welch
Frank Welsch, PhD
Edward L. Wentz
Kenneth T. Wheeler, PhD
Russell D. White
Stuart R. White
Thomas W. Whitehead, Jr., PhD
Brooks M. Whitehurst
James N. Wilkes

Louis A. Williams
Roberts C. Williams, Jr.
Candler A. Willis
Cody L. Wilson, PhD
Samuel Winchester, PhD
Perry S. Windsor
Gary M. Wisniewski
Richard Lou Witcofski, PhD
Peter Witherell, PhD
Karol Wolicki
G. C. Wolters
J. Lamar Worzel, PhD
Charles A. Yost
Gregory C. D. Young, PhD
Stanley Young, PhD
Thomas R. Zimmerman
Bruce John Zobel, PhD

North Dakota

Joseph E. Adducci
Richard A. Adsero
Robert F. Agnew
James L. Alberta
Donald A. Andersen, PhD
Timothy W. Bartel
Robert J. Bartosh, Jr.
Armand Bauer, PhD
James H.M. Beaudry
Duane R. Bentz
George Bibel, PhD
Timothy A. Bigelow, PhD
Shawn Bjerke
Joseph E. Bowling
Mike Briggs
Richard E. Broschat
Jeremy J. Brown
Justin Burggroff
Edward C. Carlson, PhD
Jon M. Carroll
Greg D. Dehne
Alan Dexter, PhD
Mark Dihle
Scott D. Dihle
William Erling Dinusson, PhD
John C. Duffey
Mary A. Durick
John W. Enz, PhD
Joe T. Fell
Joe Friedlander
Mark A. Gonzalez, PhD
Robert S. Granlund
James J. Gress
Harvey Gullicks, PhD
Wayne M. Haidle
Steven H. Harris
Harry Lee Holloway, PhD
Bruce Imsdahl
Francis Albin Jacobs, PhD
Dennis R. James
Larry C. Katcher
James Kerian
John Kerian, PhD
Ronald C. Koehler
Mark S. Kristy
Troy J. Leingay
O. Victor Lindelow
Kathleen M. Lucas
Rodney G. Lyn, PhD
Calvin G. Messersmith, PhD
Cameron B. Mikkelsen

Donald A. Moen, PhD
Duane R. Myron, PhD
John Dennis Nalewaja, PhD
Joel K. Ness, PhD
Randal Ness
James R. Olson
Jim L. Ozbun, PhD
Kellan Pauly
Terry L. Rowland
Casey J. Ryan
David P. Schaaf
E. Brett Schafer
Timothy A. Schmidt
Bruce D. Seelig, PhD
Garry Austin Smith, PhD
Leroy A. Spilde, PhD
Dean R. Strinden
Harold Sundgren
Earl T. Torgerson
Ward Uggernd
Robert J. Werkhoven
John F. Wheeler
Alicia A. Wisnewski

Ohio

Robert E. Abell
Gene H. Abels
Wayne Aben
Harold Elwood Adams, PhD
Richard W. Adams
Verne E. Adamson
Kevin Ahlborg
Roy A. Albertson
Steven J. Alessandro
George C. Alexander
Jennifer M. Alford
Mary E. Alford
Ben C. Allen, PhD
Robert G. Allen
Carl J. Allesandro
Timothy L. Altier
Patrick Aluotto, PhD
Charles David Amata, PhD
Paul Gerard Amazeen, PhD
Ernest J. Andberg
John P. Anders
Doug E. Andersen
Christopher Anderson
Randall H. Anderson
Charles S. Andes
William D. Ankney
Stuart H. Anness
Robert D. Anthony
Clarence R. Apel
Carl Apicella
Kenneth P. Apperson
Richard Apuzzo, II
Howard Arbaugh
Raymond D. Arkwright
Mark Armstrong
Harold H. Arndt
Bob Ashworth
Otilia J. Asuncion
William J. Atherton, PhD
Lester C. Auble
Frederick N. Aukeman
Neil C. Babb
Kenneth A. Bachmann, PhD
Frank Rider Bacon
Maurice F. Baddour

Keith J. Bagarus
Donald R. Bahnfleth
Benny H. Baker
Gary E. Baker
Robert L. Baker
Roger P. Baker
Fredrick B. Ball
D. Ballal, PhD
William L. Ballis
Evan S. Baltazzi, PhD
William R. Banman
Elmer Alexander Bannan
Zot Barazzotto
Kurt L. Barbee
Donald O. Barici
Anthony Barlow, PhD
Steven J. Barlow
Herbert G. Barth
Robert F. Bartholomew
Benjamin M. Bartilson
James W. Bartos
David L. Bartsch
Samuel J. Basham, Jr.
Roland E. Bashore
Richard Basinski
Robert Dean Battershell, PhD
Robert Bauman
Daniel E. Beasley
J. Donald Beasley, PhD
Tom B. Bechtel, PhD
Giselle Beeker
Stephen W. Bell
David F. Bellan
John F. Beltz
John Wilfrid Benard*
James Harold Benedict, PhD
Ronald G. Benneman
William W. Bennett
George Benzing
Kenneth W. Berchak
Hugh Berckmueller
Alexander T. Berkulis
Andre Bernier
Albert C. Bersticker
Kevin J. Bertermann
James H. Bertke
Larry W. Best
Dennis W. Bethel
William F. Beuth
Vincent Darell Bevill
Lawrence Romaine Bidwell, PhD
Daniel C. Bieberitz
Bill J. Bielek
Donald Kent Bierley
Donald Bigg, PhD
Ted F. Billington
George Edward Billman, PhD
Layton C. Binon
Matthew C. Birch
Dennis W. Bishop
Michael G. Bissell
James J. Bjaloncik
John F. Blackmer
Marc E. Blankenship
John R. Blasing
Charles William Blewett, PhD
Raymond Blinn
Dexter W. Blome, PhD
James W. Blue, PhD
Paul J. Boczek
Howard Boeing
John E. Bohl

Edward H. Bollinger, PhD
Irwin H. Bollinger
Michael T. Bonnoitt
Peter Frank Bonventre, PhD
Samia Borchers
William Borchers
Dan Borgnaes, PhD
William Earle Bosken
Kenwood A. Botzner
Jim D. Boucher
Gerardus D. Bouw, PhD
Richard O. Bowerman
Scott C. Boyer
John H. Boyles, Jr.
Lincoln L. Braden
Thomas Brady, PhD
Robert E. Brandt
Arthur M. Brate
David Braun
Jennifer L. Braun
Johnny A. Brawner
Bruce H. Brazelton
Merlin Breen, PhD
Jerald R. Brevick, PhD
Joseph Brinck
Thomas A. Brisken
Gloria C. Britton
Robert Walter Broge, PhD
John M. Bronstein, PhD
Gordon Brooks
Lawrence U. Brough
Karen Brown
Larry D. Brown
P. T. Brown
Robert Brown
Roy W. Brown
Thane A. Brown
Thomas V. Brown
Timothy H. Brown
John J. Brumbaugh
John B. Brush
Kenneth K. Brush
Mark R. Bruss
Tom Bryan
Kenneth J. Brzozowski, PhD
Gregory J. Brzytwa
Tom Buckleitner
Ronald E. Buckley
Kevin Buddie
Edwin V. Buehler, PhD
Jim E. Bunds
David C. Burgess
Amy S. Burke
Martin P. Burke, Jr.
Beth Burns
Jerry W. Burns
John R. Burns
Robert David Burns, PhD
Timothy Burns
Richard S. Burrows, PhD
B. Burton, PhD
Alan G. Burwinkel
Robert L. Bush
Jeff N. Butler
Richard H. Byers
Garth Arthur Cahoon, PhD
Mike Cameron
Larry Campbell
Leo V. Campbell
David R. Canterbury
Ronald E. Cantor
Robert S. Carbonara, PhD

Brenda L. Carr
Chuck Carson
Gwendowyn B. Carson, PhD
Richard M. Carson
Richard Cartnal, PhD
William O. Cass
Thomas J. Castor
Michael Caudill, PhD
Rosemary Louise Centner
Edwin B. Champagne, PhD
Chiou S. Chen, PhD
Peter Chesney
Paul W. Chester
Joseph S. Chirico
Patricia S. Choban
Frank W. Chorpenning, PhD
Kenneth D. Christman
Ryan A. Chrysler
John E. Cicillian
Kimball Clark, PhD
Robert L. Clark
Shane F. Clark
Peter Clarke
James F. Clements
Douglas J. Clow
Steven D. Coder
Steven J. Cohen
Edward Eugene Colby
Christopher J. Cole
Clarence Cole, PhD
John Cole
Edward A. Collins, PhD
Horace Rutter Collins
E. Keith Colyer
David M. Combs
Floyd Sanford Conant
Paul D. Conkel
James K. Conlee
William E. Conway
Wayne L. Cook, PhD
William Edward Cooley, PhD
Jack L. Coomer
Harry C. Cooper
Christopher T. Cordle, PhD
Robert W. Core
Billy D. Cornelius
Dale R. Cornett
David George Cornwell, PhD
Norman R. Cox
Aaron S. Coyan
Richard A. Crago
Richard S. Cremisio, PhD
Larry G. Criswell
Eugene E. Crosby
Robert Franklin Cross, PhD
D. B. Crouch
Frank Cutshaw Croxton, PhD
Ralph G. Crum, PhD
Richard Lee Cryberg, PhD
Tom E. Ctvrtnicek
Hongjuan Cui, PhD
Bruce A. Currie
Tom P. Currie
Mary J. Custer
James R. Cutre
Joseph M. Czajka
James Thomas Dakin, PhD
James J. Dale
Raphael D'Alonzo, PhD
Henry J. Dammeyer
Richard John Danke, PhD
S. Dantiki, PhD

Rodney C. Darrah
Charles R. Daub
Pierre H. Dauby, PhD
Julian Anthony Davies, PhD
James H. Davis
Mark Davis
William W. Dawson
Mike P. Day
Robert E. Dean
Charles W. Dearmon
Ronald W. Decamp
Anthony A. Dechiara
Arthur John Decker, PhD
Clarence Ferdinand Decker, PhD
Dan Deckler, PhD
Rowan B. Decoster
Charles D. Degraff
Everett DeJager
John Dejager
Ken Dejager
Robert Mitchell Delcamp, PhD
Joseph F. Denk, Sr.
Cecil F. Desai
Ravi S. Devara
Paul D. Deverteuil
Edward J. Devillez, PhD
James K. Devoe
Thomas E. Devoe
Richard J. Dick
Winifred Dickinson, PhD
Henry A. Diederichs
John D. Dietsch
Harold L. Dimond, PhD
James S. Dittoe
Roy Richard Divis
Brown M. Dobyns, PhD
Robert Dodd
Stephen R. Doe
Erich D. Dominik
Charles P. Dorian
Thomas M. Dorr
David H. Douglas
Robert W. Douglass, PhD
Jeffrey D. Downing
Michael L. Dransman
Harry J. Driedger
Ned E. Druckenbrod
Earl W. Duck
Thomas Joseph Dudek, PhD
Richard E. Dugan
John Durig
D. Durkee, PhD
Michael W. Durner
Somnath Dutta, PhD
Leonard A. Duval
T. Dzik, PhD
Jimmy P. Easley
Jared C. Ebbing
Arthur C. Eberle
Tom Edgar
Paul Edmiston, PhD
Brian Edwards
Gary L. Eilrich, PhD
William Anthony Eisenhardt, Jr.,
PhD
J. David Ekstrum
Klaus Emil Eldridge, PhD
Sylvan D. Eller
W. S. Elliott
Frank E. Ellis
James C. Ellsworth
Richard H. Engelmann

Dale Ensminger
John K. Erbacher, PhD
Robert H. Erdmann
Ted Erickson
Carl Eriksson
Steven J. Erlsten, PhD
George Ernst, PhD
John T. Eschbaugh
Robert H. Essenhigh, PhD
Gustave Alfred Essig
Daniel Esterline, PhD
Frances C. Esteve
Brian K. Fabel
W. J. Fabish
James P. Fadden
David M. Faehnle
William O. Fahrenbruck
Howard Fan, PhD
John M. Farley
David E. Farrell, PhD
David N. Fashimpaur
Sherwood Luther Fawcett, PhD
Leonid Z. Fedner
Robert William Fenn, PhD
John W. Fenton
Jeffrey J. Ferguson
Larry Fernandez
Rene Fernandez
Richard J. Fidsihum
Conrad William De Fiebre, PhD
Rex N. Figy
Richard P. Fink
John Martin Finn, PhD
John Fischley
Ira B. Fiscus
David L. Flannery, PhD
Raymond J. Flasck
Russell W. Flax
Wolfgang H. Fleck
James R. Fletcher, Jr.
M. R. Flitcraft
Richard Kirby Flitcraft, II
Gary Flood
John Flynn
Dale Force
John Forman
Eric N. Forsberg
Norman A. Fox
Walter J. Frajola, PhD
Charles E. Frank, PhD
Douglas G. Frank, PhD
Michael Frank
R. Thomas Frazee
Andrew M. Freborg
Anton Freihofner
Billy W. Friar, PhD
John Albert Fridy, PhD
Alfred E. Friedl
Miriam F. Friedlander
Dale J. Friemoth
Hilda K. Frigic
Bret A. Fry
Thomas C. Furnas, PhD
Dennis J. Fuster
Ronald M. Gabel
Frederick Worthington Gage
John H. Galla
Heather Gallacher, PhD
Joan S. Gallagher, PhD
Rodger L. Gamblin, PhD
Stephen J. Ganocy
Harry Richard Garner

Jeanette M. Garr, PhD
Daniel L. Garver
A. D. Gate
Carl Gausewitz
David M. Gausman
Peter D. Gayer
Stephen A. Geers, Jr.
Alexander Salim Geha
Walter B. Geho, PhD
Philip A. Geis, PhD
Elton W. Geist
Nick Genis
Jerome J. Gerda
Ulrich H. Gerlach, PhD
Richard Paul Germann, PhD
Glenn R. Gersch
John E. Getz
Louis Charles Gibbons, PhD
Thomas E. Giffels
Stephen Warner Gilby, PhD
T. J. Gilligan, PhD
W. J. Girdwood
Norbert S. Gizinski
Travis W. Glaze
Daniel Glover, PhD
Myra A. Gold
Richard Goldthwait, PhD
John C. Goon
Manuel E. Gordillo
E. D. Gordon
Scott Gordon
John R. Gorham, PhD
Stephen M. Goss
Charles E. Gottschalk
Edward J. Gouvier
Lawrence F. Gradwell
Bob Graham
Otto G. Gramp
Bruce Howard Grasser
Mark A. Gray
Frank W. Green
David C. Greene
John J. Greene
Dane W. Gregg
N. E. Grell
Frank E. Gren
Daniel R. Grieser
Walter M. Griffith, Jr., PhD
Otto Grill
Martin J. Grimm, Jr.
Raymond L. Grismer
Thomas Albin Grooms, PhD
Percy J. Gros, Jr.
Steven L. Grose
William C. Grosel
Leo C. Grosser
Henry M. Grotta, PhD
Andrew P. Grow
James B. Grow
Shannon Grubb
William J. Grunenwald
Paul W. Gruner
Douglas A. Gruver
Terry Guckes, PhD
Robert F. Guenther
Yann G. Guezennec, PhD
Bruce P. Guilliams
Carl P. Gulla
Rand A. Gulvas
Don R. Gum
Atul A. Gupte
Donald T. Guthrie

Ronald Gutwiller
George J. Guzauskas
William Haas
Robert Morton Haber, PhD
Paul S. Hadorn, PhD
Ron Haeffer
Don C. Haeske
G. Richard Hagee, PhD
Thomas J. Hager
G. P. Hall
Joseph L. Hall
Judd Lewis Hall, PhD
Kent Hall
Matthew R. Halter
Bart P. Hamilton
Jeffrey V. Handorf
David E. Haney
Wilbur Leason Hankey, Jr., PhD
John J. Hannan
Sandra Hardy
Richard W. Harkins
Ronald O. Harma
Steven Harmath
Mary L. Harmon
W. T. Harmon
Andrea Harpen
B. Harper, PhD
Richard E. Hartle
Frederick A. Hartman, PhD
Robert R. Hartsough
Robert Rice Haubrich, PhD
Andrew F. Haumesser
Hans Hauser
Arthur B. Havens
Paul W. Haverkos
Thomas J. Hawk
Robert D. Hawkins
Robert Hawthorne
Jamie Haycox
Fred Hayduk
U. P. Hayduk
Peter C. Hazel
Gregory S. Hazlett
Paul L. Heater
George L. Heath
Lynn C. Hebebrand Edgar
Art L. Hecker, PhD
William H. Hedley, PhD
Edward H. Heineman
Robert J. Heintz, Jr.
J. Heisey
Kenneth J. Hemmelgarn
Klaus H. Hemsath, PhD
Jerry S. Henderson
Shawn P. Heneghan, PhD
Hugh M. Henneberry
Paul M. Henry
Al Hepp, PhD
Charles E. Herdendorf, III, PhD
William Lee Hergenrother, PhD
Vernon J. Hershberger
John A. Hersman
Evelyn V. Hess
Mary Lee Hess
Don J. Heuer
Chares R. Hewitt, Jr., PhD
Howard Minor Hickman
E. S. Hicks
Bob Hill
Dexter W. Hill
Mark S. Hill
Don R. Hiltner

Ashley Stewart Hilton, PhD
Bradley T. Hina
R. J. Hipple
Robert T. Hissong
Gary E. Hoam
Richard T. Hoback
Michael Hoch, PhD
David A. Hoecke
Harry R. Hoerr
Bob Hoffee
Ralph B. Hoffman
Thomas Hoffman
Scott B. Hoffmann, Jr.
William A. Hohenstein
Charles M. Hohman
Allen Hoilanson
Thomas J. Holderread
John H. Hollis
Thomas P. Holloran
Edward Hopkins
Ninnian E. Hopson, PhD
Howard Horne
Myer George Horowitz, PhD
Lloyd A. Horrocks, PhD*
Ralph L. Hough
David D. Houston
Edgar L. Howard
Edward J. Hren
Wilbert N. Hubin, PhD
Norman Thomas Huff, PhD
Dave Huffman
Donald H. Hughes, PhD
Walter P. Huhn
Bruce Lansing Hull
John H. Hull
Ronald E. Hulsey
Steve A. Hunter
Frank D. Huntley
John O. Hurd
Shakir Husain, PhD
Ira B. Husky
Keith Arthur Huston, PhD
Darrell A. Hutchinson
Ahmed A. Ibrahim
Douglas K. Iles
Milton S. Isaacson
R. Jackman
Eunice F. Jackson
John W. Jacob
George D. Jacobs
John W. Jacox
Ralph R. Jaeger, PhD
Alexander A. Jakubowycz
Richard J. Jambor
Jerry J. James
Blair F. Janson, PhD
F. Jarka
Charles St. Jean
Lisa A. Jeffrey
Bernhart Jepson, PhD
Peter Jerabek
George G. Jetter
Vinton H. Jewell
Stanley W. Joehlin
Dennis M. Johns, PhD
Lars R. Johnson
Robert Oscar Johnson, PhD
Ronald Gene Johnson, PhD
Wendell Johnson
William E. Johnson
F. Thomas Johnston
Matt Johnston

Kenneth L. Jonas
Don Jones
Daniel J. Joseph
Blaine R. Joyce
R. M. Judy
Frank E. Kalivoda, Jr.
Harold Kalter, PhD
Antony Kanakkanatt, PhD
Victor G. Karpov, PhD
Lewis P. Kasper
Rosalind Kasunick
Vladimir Katovic, PhD
Robert L. Katz
Christopher W. Keefer
Terrence P. Keenan
Harold Marion Keener, PhD
Calvin W. Keeran
Hubert Joseph Keily, PhD
John Eugene Keim, PhD
George E. Kelbly, Jr.
Joseph H. Keller, III
Joseph H. Keller
Parry J. Keller
Albert Kelly
Thomas A. Kenat, PhD
Jeffrey S. Kennedy
Thomas C. Kennicott
Harold Benton Kepler
Harold E. Kerber, PhD
James Gus Kereiakes, PhD
Vincent J. Kerscher, Jr.
Jack D. Kerstetter
Ned W. Kerstetter
Ricky D. Kesig
Tom Kesler
Mel Keyes, PhD
John E. Kiefer, Jr.
Richard W. Kieffer
Donald W. Kifer
Arthur M. Killin
Hyun W. Kim, PhD
Harry S. King, PhD
John Frederick Kircher, PhD
James W. Kirchner
Chester Joseph Kishel, PhD
Craig A. Kister
Ronald W. Klenk
Robert Kline, PhD
Lee A. Klopfenstein
David Kloppenburg
Albert Leonard Klosterman, PhD
Kent Knaebel, PhD
Matthew D. Knecht
Slavko Knezevic
Dennis D. Knowles
Arthur T. Koch, PhD
Kevin J. Kock
Ken E. Kodger
Stan Koehlinger
Clint Kohl, PhD
Bernard R. Kokenge, PhD
Frank Komorowski
Daniel R. Kory, PhD
Joseph J. Kotlin
Frank Louis Koucky, PhD
Peter Kovac
J. L. Kovach, PhD
Kenneth W. Krabacher
Deborah J. Krajicek
Eric Russell Kreidler, PhD
Julius Peter Kreier, PhD
Mark L. Kreinbihl

Warren C. Kreye, PhD
William G. Krochta, PhD
David J. Krupp
James R. Krusling
James Kulick
Gordon Kuntz, PhD
Donald Kunz, PhD
Frederick L. Kuonen
David W. Kurtz, PhD
Leonard G. Kutney
Rosemary Lacher
James Walter Lacksonen, PhD
Jerry P. Lahmers
Russell J. Lahut
Alan Van Lair, PhD
George W. Lambrott
Norman J. Lammers, PhD
Scott A. Landgraf
Jerome B. Lando, PhD
Robt Lane, PhD
Albert Langley, PhD
Robert Larson, PhD
Robert P. Lattimer, PhD
Duane E. Lau
Robert J. Lauer
Elbert D. Lawrence
Anne M. Lawson, PhD
Hugh J. Leach
Stephen J. Lebrie, PhD
Lawrence M. Lechko
Michael T. Lee
James Frederick Leetch, PhD
Kevin Lefler
Jay H. Lehr, PhD
Mark Leifer
Jon Leist
Bernard J. Leite
John C. Leite
Paul W. Leithart
Terry L. Lemley, PhD
John A. Lengel, Jr.
Mary A. Lenkay
Dean J. Lennard
Ralph Lennerth
Gary A. Lensch
John A. Leonatti
Joseph Leonelli, PhD
Arkady I. Leonov, PhD
Fred E. Leupp
Stanley M. Levers
Hudnall J. Lewis
Peter A. Lewis, PhD
Robert B. Licht
Thomas E. Lieser
Terrance R. Liette
Charles G. Lindeman
Larry Linder
Jonathon P. Llewellyn
Harold R. Lloyd
Shane A. Locke
David A. Lockmeyer
Dwight N. Lockwood
John R. Long, PhD
Lawrence E. Long
Roland V. Long
Daniel Longo
Dan C. Lostoski
Martha Loughlin
Gerard A. Loughran
Rodney R. Louke
Donald F. Lowry
Alan H. Lubell

Hugh H. Lucas
Caroline N. Luhta
Thomas Elwin Lynch
Richard H. Lyndes
J. Machen
Alex L. MacInnis
Peter Mackenzie
James B. Macknight
Leo R. Maier, PhD
Edward F. Maleszewski
Eugene Malinowksi, PhD
John Malivuk
Irving Malkin
Harry N. Malone
Lancing P. Malusky, PhD
Neil W. Mann
Ronald W. Manus, PhD
David R. Marcus
Joseph Marhefka
Daniel F. Marinucci
Philip C. Marriott
Jack Marshall
Martin D. Marsic
Daniel William Martin, PhD
Frank P. Martin
Jay W. Martin
Joseph P. Martino, PhD
Andrew J. Maslowski
Lester George Massay, PhD
Jim E. Masten
Joseph B. Masternick
B. Matty
John F. Maxfield
Michael Maximovich, PhD
Robert Mayhan, PhD
Terry J. Mazanec, PhD
August C. Mazza
Glenn W. McArdle
Emerson O. McArthur, III
Gregory W. McCall
Joseph F. McCarthy
Brenda C. McCune
Tim W. McDaniel
James N. McDougal, PhD
Kevin C. McGann
James A. McGarry
Robert P. McGough
Philip McGuinness
Gene McKenzie
Irven J. McMahon
Robert H. McMaster
Paul E. McNichol
Pearson D. McWane, PhD
Earle M. Mead
Dean Meadows
Ronald G. Meadows
John C. Medici, Jr.
Mitchell I. Meerbaum
Carl P. Meglan
W. Meilander
Walter Meinert
Paul W. Meisel
Steve Meisel
Ame T. Melby
C. Benjamin Meleca, PhD
Todd Melick
Mike Melnick
Max G. Menefee, PhD
Harry Winston Mergler, PhD
Robert Edward Merritt
Robert Mertens, PhD
William A. Metcalf

Kenneth A. Metz
A. R. Metzer
Dave Meyer, PhD
Thomas G. Meyer
Andrew J. Meyers
Fritz D. Meyers
Dwight W. Michener
M. B. Mick
Tim K. Mickey
Gerald P. Migletz
George F. Mihalich
Louis C. Mihaly
Frederick John Milford, PhD
C. Miller, PhD
David Miller, PhD
Jeffrey Miller
Kurtz K. Miller
Laura S. Miller
Laverne L. Miller
Marlen L. Miller
Matthew Miller
Raymond E. Miller, PhD
Timothy P. Miller
George A. Minges
Flora Miraldi, PhD
David A. Miskimen
Ben G. Mitchell
David J. Mitchell
Howard R. Mitchell, Jr.
Jay P. Mitchell
Pradip K. Mitra
Marvin Joseph Mohlenkamp, PhD
Lars E. Molander
Karl H. Moller
Stephen P. Molnar, PhD
Michael W. Monk
Michael M. Montgomery
Scott M. Moody, PhD
Young I. Moon, PhD
Robert T. Moore
John A. Moraites
John P. Morelock
Dave Morgan
Stanley L. Morgan
P. J. Mormile
Kenneth P. Morrell
Edward C. Morris
Lynn A. Morrison
Brad G. Morse
Richard F. Mortensen, PhD
T. R. Morton
Stewart Wayne Moser
Thomas E. Moskal, PhD
Pual W. Mossey
John D. Mowery
Jeffery L. Moyer
Wayne E. Moyer
Gordon Mark Muchow, PhD
Robert L. Mullen, PhD
David Munn, PhD
Philip C. Munro, PhD
Richard August Muntean, PhD
Robert L. Murdoch
Daniel B. Murphy
Joel Muse, Jr., PhD
Dale W. Musolf
Chrisy M. Myers
Sam A. Myers
Richard R. Nadalin
Sonia Najjar, PhD
Sang Boo Nam, PhD
John Christopher Nardi

Bhaskara R. Narra
Harry C. Nash, PhD
Ronald O. Neaffer, PhD
John E. Nebergall
Daniel W. Nebert
Glen Ray Needham, PhD
John T. Nenni
Neal M. Nesbitt
James F. Neuenschwandor
Jesse V. Newburn
Mark A. Newman
Simon L. Newman, PhD
Richard S. Newrock, PhD
Thomas Nichols
Jerry L. Nickol
Dennis A. Nie
John A. Niebauer
Harry D. Niemczyk, PhD
Alan J. Niemira
Samuel Nigro
Henry Nikkel
Paul John Nikolai, PhD
William E. Noble
Hallan Costello Noltimier, PhD
David B. Norby
Peter J. Nord, PhD
Tim L. Norris
Gregory J. Nortz
Thomas E. Noseworthy
A. Novak, PhD
Edward F. Novak
Marwan Nusair, PhD
Terresa L. Nusair, PhD
William E. Nutter, PhD
Elmer J. Obermeyer
Allan O'Brien
Wayne M. O'Connor
Herbert A. Odle
Richard W. Oehler, Jr.
Robert A. Oetjen, PhD
Pearl Rexford Ogle, PhD
Jack M. Ohmart
Jerry C. Olds
George L. Opdycke
Richard C. Organ
James E. Orwig
Dave Osterhout
Robert R. Osterhout
Mike L. Oxner
James Oziomek, PhD
Tony Pace, PhD
Gilbert E. Pacey, PhD
Frank Patrick Palopoli, PhD
Bob J. Palte
Richard E. Panek
Salvatore R. Pansino, PhD
H. L. Parks
Alice B. Parsons
Stuart Parsons
Charlie Patchett
Geraldine C. Patman
Bruce H. Pauly
Charles W. Pavey
Anthony J. Pearson, PhD
Michael J. Pechan, PhD
Lyman Colt Peck, PhD
Bruce Pedersen
Robert G. Peed
Larry J. Pemberton
Earl R. Pennell
Joel W. Percival
Joseph C. Perin, Jr.

Michael D. Perrea
George Perry, PhD
Steven P. Petrosino, PhD
James C. Petrosky, PhD
J. A. Petrusha
Robert W. Pfaff
William H. Pfaff
Douglas R. Pfeiffer, PhD
John R. Pfieffer
Ronald John Pfohl, PhD
Keith Phillips
Clarence R. Piatt
Kathi A. Piergies
Richard E. Pietch
Jeffrey T. Pietz
Charles F. Pilati, PhD
Richard W. Pine
David G. Pinney
Ervin L. Piper
John A. Plenzler
B. Poling, PhD
Donald S. Poole
Herbert G. Popp
Terry Potter
Louis A. Povinelli, PhD
Jean D. Powers, PhD
William G. Preston
William H. Prior
David G. Proctor, PhD
Richard A. Proeschel
Zenon C. Prusas, PhD
Jamie Przybylski
T. O. Purcell, PhD
Abbott Allen Putnam
James C. Putt
Leon L. Pyzik
James F. Quigley
Kristine J. Raab
M. J. Rabinowitz, PhD
Roger L. Radeloff
Thomas S. Raderstorf
Marvin C. Raether
David E. Rainer
Robert P. Raker
John Ramus
Robert T. Randall
Roberta B. Randall
Oscar Davis Ratnoff
James P. Rauf
Joseph R. Raurora
Park Rayfield, PhD
Harold H. Reader, Jr.
Larry K. Reddig
Richard L. Reddy
Robert F. Redmond, PhD
Roy Franklin Reeves, PhD
Dean J. Reichenbach
Charles Bernard Reilly, PhD
Richard H. Reinhardt
Eugene Reiss
David A. Retterer
R. H. Reuter, PhD
David Stephen Reynolds
Young W. Rhee
Daniel Rhodes
Perry J. Ricciardi
Bernard L. Richards, PhD
Aimee M. Richmond
Thomas Ricketts
N. R. Ricks, Jr.
Ralph E. Ricksecker
Harold P. Rieck

John Samuel Rieske, PhD
Frederick J. Riess
Ali Rihan
William M. Rike
Boyd T. Riley, Jr., PhD
Michelle M. Rising
Nancy Risner
Charles I. Ritchie
Malcolm L. Ritchie, PhD
Edgardo H. Rivera
Jerald Robertson
Lee A. Robinson
Robert L. Robke
Alex F. Roche, PhD
Steve Rockow
Walter W. Rodenburg, PhD
Boley J. Rog
Thomas R. Rolfes
Floyd Eugene Romesberg, PhD
Eldon E. Ronning, PhD
I. C. Rorquist
Frederick R. Rose, Jr.
William E. Rosenberg
Melvin S. Rosenthal
Mark A. Rosenzweig
Arthur L. Ross
Charles J. Rostek, PhD
Lloyd B. Roth
John Rottenborn
David L. Rouch, PhD
William R. Rousseau
Melinda Rowe
Trent Rowe
Brian Rowles
William C. Rowles
Csaba Rozsa, PhD
Neil S. Rubin
Stephen A. Ruehlman
Ed Ruff
Dan J. Rundell
Charles Runyon
Mark V. Runyon
Nick P. Rusanowsky
Harry C. Russell
George B. Rutkowski
Willis C. Ryan
Edward Sabo
Elmer T. Sabo
David Sachs, PhD
George S. Sajner
Stephen G. Salay
Seppo Ossian Salminen, PhD
Stuart C. Salmon, PhD
S. C. Sandusky
Sam Sangregory
John R. Sans, PhD
Richard L. Sanson
Michael D. Santry
Fred Saris, PhD
Frank J. Sattler
Rudolph A. Sattler
Lee A. Scabeck
Patrick J. Scarpitti
Rudolph J. Scavuzzo, PhD
Gary A. Schackow
James K. Schaffer
Daniel Schaper
Daniel Scheffel
Timothy William Schenz, PhD
Roger Scherer, PhD
Philip Eowin Schick, PhD
Charles T. Schieman

Richard Schiewetz
Robert E. Schilling
John Schlaechter
Don L. Schlegel
James A. Schlunt
Donald J. Schmitt
Gordon T. Schmucker
Karl. A. Schnapp
David Schneider
John Matthew Schneider, PhD
John Schneider
Philip Schneider
Ronald E. Schneider, PhD
Steven M. Schneider, PhD
Erwin Schnetzer
Albert William Schreiner
Leo Schrider
Charles R. Schroeder
James W. Schroeder
Franklin D. Schrum, Jr.
Steve Schulte
Frank Schultz
Donalo R. Schuster
Robert R. Schutte
Robert C. Schwendenman
James A. Schwickert
John Scofield, PhD
James L. Seals
Robert Secor
Jorge F. Seda
Warren George Segelken, PhD
Robert Seigneur
Maurie Semel, PhD
Mark Setele
Christopher E. Shadewald
David A. Shadoan
William Shafer
Jane J. Shaffer
Jeffrey S. Shane
Fred J. Sharn
Fredrick G. Sharp
Elwood R. Shaw
Robert Shaw
Lyle D. Sheckler
Steven A. Shedroff
Paul Sheil
Nadia M. Shenouda, PhD
Douglas A. Sheridan
Paul Griffith Shewmon, PhD
Lee R. Shiozawa, PhD
Leila Shiozawa
Larry Shivak
Andrew Shkolaik
Ted H. Short, PhD
Doug L. Shrake
Glenn Sickles
Jerome M. Siekierski
Michael Peter Siklosi, PhD
Arthur Dewitt Sill, PhD
Lester C. Simmons
Hugh W. Simpson
Robert S. Simpson
Allen W. Singer
Stephen E. Singleton
Paul E. Sisco
Michael L. Sitko, PhD
Ron Sizemore
Hugh B. Skees
Marek J. Skoskiewicz, PhD
Robert B. Skromme
Alan Skrzyniecki, PhD
Jerry L. Slama

Raymond J. Slattery
Francis Anthony Sliemers, Jr.
James L. Smialek, PhD
Allan M. Smillie, PhD
Davey L. Smith
David K. Smith, PhD
Ivan K. Smith, PhD
Kevin T. Smith
Mark A. Smith, PhD
Myron R. Smith
Raymond L. Smith
Rick A. Smith
Watson C. Smith
David Smithrud, PhD
George Ray Smithson, Jr.
Cloyd Arten Snavely, PhD
David T. Snelting
Joe Snyder
Rodney K. Snyder
Gifford G. Solem
Thomas A. Somerfield
Thomas M. Sopko
William T. Southards
Larry M. Southwick
Philip Andrew Spanninger, PhD
Richard Clegg Spector
Robert R. Speers, PhD
Hermann A. Spicker
Ernest G. Spittler, PhD
Robert Bruce Spokane, PhD
Richard M. Sprang
Allan Springer, PhD
Lindsay E. Spry
Kevin J. Sroub
Thomas M. Stabler
Murray D. Stafford
Larry C. Stalter
Ronald L. Stamm
Norman Weston Standish, PhD
Dann Stapp
Fred J. Starheim, PhD
Peter Staudhammer, PhD
D. Stefanescu, PhD
John W. Steinberger
John C. Steiner
David George Stephan, PhD
James P. Stephens
Henry C. Stevens, PhD
Mark Stewart
Ron Stieger
C. Chester Stock, PhD*
Brent T. Stojkov
Michael R. Stone
Robert A. Straker
Randall L. Stratton
Carl Richard Strauss, PhD
Gary D. Streby
Richard A. Strong
Warren Stubblebine, PhD
John Stubbles, PhD
C. D. Stuber, PhD
Mark A. Stuever
Richard Sutera
George E. Sutton, PhD
James Lowell Sutton
Arden W. Swanson
Ronald E. Symens
Widen Tabakoff, PhD
Mowafak M. Taha
Sherwood G. Talbert
Dorrence C. Talbut
Richard F. Tallini

Louis Anthony Tamburino, PhD
Richard R. Tattoli
Thomas L. Taubken
Douglas Hiram Taylor, PhD
David M. Tennent, PhD
Martha Melay Tennent
Lou Testa
Nicholas J. Teteris
James R. Theaker
Lawrence E. Thieben
Dudley G. Thomas
Thomas Thompson, PhD
Roger D. Thornton
Anthony D. Ticknor
Clyde Raymond Tipton, Jr.
Kenneth M. Tischler
Sarah A. Tjioe, PhD
Michael J. Toffan
Paul E. Tomes
Terry L. Tomlinson
Peter J. Torvik, PhD
Daniel Town, PhD
D. A. Towner
William L. Trainor
John H. Trapp
Leora A. Traynor
Michael J. Trimeloni
Charles F. Trivisonno
William Trommer
David V. Trostyanetsky
Harris C. True
Igor Tsukerman, PhD
Timothy R. Tuinstra
Michael S. Tullis
Richard C. Tumbleson
David W. Turner
Robert C. Turner
Joseph Tutak
Tony Tye
Russell A. Ulmer
Thomas J. Ulrich
Israel Urieli, PhD
Dave Petrovich Uscheek
Johannes M. Uys, PhD
Stephen J. Vamosi
David Van Sice
Sam Vandivort
Brian S. Vandrak
John A. Varhola
Joseph M. Varvaro
Alex Vary
Gregory S. Vassell
Steve Vatovec
Jeff Vaughn
Richard P. Veres
William J. Veroski
Ronald James Versic, PhD
Kent Vickie
Thomas P. Viggiano
Keith E. Vilseck
Robert L. Vitek, Sr.
James W. Voegele
Joseph C. Vogel
Walter E. Vonau
Lois B. Wachtman
Jeffrey K. Wagner, PhD
Raymond S. Wagner*
William M. Wagner, PhD
W. B. Walcott
John J. Waldron
Brent Wallace
Michael Wallace

Richard F. Wallin, PhD
Eugene J. Walsh
John K. Walsh
Edward P. Waltz
Lance G. Wanamaker
Graham T. Wand
Douglas Eric Ward, PhD
Roscoe Fredrick Ward
Donald E. Washkewicz
Chatrchai Watanakunakorn
Maurice E. Watson, PhD
Timothy E. Weaver, PhD
William M. Weaver, PhD
Allen E. Webster
Donald K. Wedding
Thomas A. Weedon
Brenda Wehner
Brenda Wehnermuck
Carl S. Wehri
Edward David Weil, PhD
Frank Carlin Weimer, PhD
Daniel Weiss
Michael Welch
Elbert J. Weller
John Fengpinct Wen, PhD
Marlin L. Wengerd
Carl Werhi
John C. Werner
Clark D. West
David B. West
Tammy Westerman
Fred Ernst Westermann, PhD
Harry Whitney Wharton, PhD
Robert E. Wheeler
R. E. Whitam
Douglas S. White
Eugene A. White
Joseph E. White
Thomas J. Whitney, PhD
Charles E. Wickersham, PhD
William T. Wickham, PhD
Lawrence H. Wickstrom
James A. Wiechart
Edward William Wiederhold
Robert Wieneke
Robert George Wiese, PhD
Jesse H. Wilder
Dwight R. Wiles
L. B. Willett, PhD
Craig F. Williams
Daniel G. Williams
Paul Williams
Louis A. Williams, Jr
Thomas A. Willke, PhD
Alan J. Willoughby
Robert W. Wilmott
Daniel L. Wilson
Edward A. Winsa
Chester Caldwell Winter
Frederick S. Wintzer
Herbert A. Wise, Jr.
Richard Melvin Wise, PhD
James M. Withrow
James K. Woessner
Andrew W. Wofford
Donald R. Wogaman
S. Raymond Woll
Jackie Dale Wood, PhD
Malcolm Wood
Jim Woodford
Ted Woods
Jim Woodworth

819

David R. Worner, PhD
Gene E. Wright
Nathan A. Wright
Milton Wyman
R. A. Wynveen, PhD
Kefu Xue, PhD
Edward Yachimiak, Jr.
Riad Yammine
Gilbert Yan
Anthony J. Yankel, Jr.
William Harry Yanko, PhD
Donald J. Yark
Ronald A. Yates
Fred Young
Terrill Young
Edward A. Yu
Stanley Zager, PhD
Joseph Zappia
Lawrence E. Zeeb
James W. Zehnder, II
R. A. Zeuner
Perry H. Ziel
Steven A. Zilber
Tommy L. Zimmerman, PhD
Steven P. Zody
Raymond N. Zoerkler
Gary B. Zulauf
John F. Zurawka

Oklahoma

Ernest R. Achterberg
Brian D. Adam, PhD
Albert W. Addington
Robert M. Ahring, PhD
Brian R. Ainley
William L. Albert
James C. Albright, PhD
Ben Alderton
John C. Alexander
Kenneth L. Allen
Craig Allison
Ivan D. Allred
Ronny G. Altman
Bradley A. Aman
Terrell Neils Andersen, PhD
James K. Anderson
Jack R. Anthony
William H. Audley
John B. Aultmann, Jr.
Gerald E. Baehler
Dennis D. Baggett
Daniel M. Baker, PhD
Newton C. Baker
Richard B. Banks
Lindsey B. Barnes
David T. Basden
Haskell H. Bass
Robert Batay
David George Batchelder
T. Mack Baugh
Brian Bean
Edward A. Beaumont
Bennett E. Bechtol
Bruce M. Bell, PhD
Thomas C. Bennett
Glen L. Berkenbile
D. E. Berry
Charles W. Bert, PhD
James M. Beyl
Jeffrey M. Bigelow, PhD

James Biggs, PhD
John Lawrence Bills, PhD
Michael A. Birch
Marshall D. Bishop
Brian H. Blackwell
Larry J. Bledsoe
Robert Blevins
Edward Forrest Blick, PhD
Jack E. Bobek
Terry Bobo
John M. Bohannon
Victor Bond, PhD
Howard William Bost, PhD
John N. Botkin
Clarence Bowers
B. W. Box
Barth Bracken
Shawn Braden
Edward Newman Brandt, PhD
Lindell C. Bridges
David A. Brierley
James P. Brill, PhD
Bob Diggs Brown
C. A. Brown
Hal Brown
Keith H. Browne
John H. Bryant
Brad D. Burks
Jay B. Burner
Vaud A. Burton
James Robert Burwell, PhD
Gary Buser
P. Edward Byerly, PhD
W. D. Byrd
Warren D. Cadwell
William S. Cagle
Peter K. Camfield, PhD
Jerry L. Camp
Jesse Campbell
John Morgan Campbell, PhD
Philip Jan Cannon, PhD
Alan Carlton
Bruce N. Carpenter
Robert K. Carpenter
John Cassidy, Jr.
Gene C. Cates
Kenneth F. Cathey
John M. Cegielski, Jr., PhD
Raymond Eugene Chapel
Robert C. Charles
Jim C. Chase
Ken K. Clark
Kenneth M. Clark
Merlin Clark
Harris A. Clay
J. C. Clemmer
Terry Cluck, PhD
James H. Cobbs
Joseph B. Cole
William B. Collier, PhD
Kenneth Edward Conway, PhD
Michael C. Cook
Lowell L. Coon
William L. Coppoc
John T. Corson
Donald L. Crain, PhD
Kenneth S. Cressman
Curtis W. Crowe
Billy Crynes, PhD
Duane C. Dahlgren
Glenn Hilburn Dale
Rodnoosh R. Davari

Dale E. Dawson
John R. Dean
Roy Dennis
Lawrence A. Denny
Henry N. Deshazo, II
John Robert Dille
Ross J. Dixon
Elliott P. Doane, PhD
Kenneth J. Dormer, PhD
William F. Dost, Jr.
Jan Frederic Dudt, PhD
Donald T. Duke
Larry Dunn
Marcus Durham, PhD
D. Dutton
Jeffrey L. Dwiggins
R. C. Earlougher
Jason Egelston
Abdalla M. Elias
Lloyd S. Elliott
David Allen Ellis, PhD
Robert S. Ellis
Charles F. Engles
Jack F. Epley
Ralph H. Espach
Craig R. Evans
F. Monte Evens, PhD
Arthur C. Falkler
Michael K. Farney
Charles H. Farr, PhD
S. D. Farrar
Clarence Robert Fast
Robert O. Fay, PhD
James K. Fayard
J. Roberto Feige
Robert Fergusen
Louis G. Fernbaugh
Craig Ferris
Franklin E. Fields
Warren V. Filley
Leon Fischer, PhD
Wayne W. Fish, PhD
J. W. Fishback, II
John Berton Fisher, PhD
John I. Fisher
Raymond E. Fletcher
Jack Fly
Melton L. Fly
James M. Forgotson, PhD
Bill E. Forney
Lee C. Francis
James M. Franklin
Joes R. Friedman
Profeesor Friess
H. Robert Froning, PhD
S. W. Fruehling
Maryann C. Fuchs
John Gallagher, PhD
Ornald L. Gambrell
Thomas P. Gamwell
Charles R. Garner
Harvey L. Gaspar
George Hiram Gass, PhD
Rick D. Gdanski, PhD
William C. Geary
Robert L. Geyer, PhD
William A. Giffhorn
Bertis Lamon Glenn, PhD
Kent J. Glesener
Wilmer E. Goad
Leroy Goodman
Robert G. Graf

Ashley Q. Graham
Dick Greenly
Billy D. Griffin
Ivan Vincent Griffin
Brandon Griffith
Mark Grigsby, PhD
James I. Grillot
Richard S. Grisham
Fred Grosz, PhD
Dane Gruben
Elard L. Haden
Pablo Hadzeriga
M. H. Halderson
John Hales
Thomas G. Halfast
Phillip M. Hall
Rick Hall
Justin Hamlin
Daniel L. Hansen, PhD
Harold Cecil Harder, PhD
Barry A. Hardy
Bryant J. Hardy
Gerald R. Hare
Charles E. Harmon
Charles R. Harvey
Charles N. Haskell
Philip J. Hawkins
Anthony V. Hayes
Kendall Hays
Kenneth Heacock
George B. Heckler
James H. Hedges, PhD
Richard Warren Hedlund, PhD
Robert A. Hefner, IV
Stuart W. Henderson
Vinson D. Henderson
Henry W. Hennigan
Timothy L. Hermann
Chesley C. Herndon, Jr.
Barry E. Herr
John P. Herzog
Neil C. Hightower
Irving A. Hill
George W. Hillman
Donald O. Hitzman
Tong Yun Ho, PhD
Joe M. Hodgson
Carl C. Holloway, II, PhD
Harvey H. Holman
James A. Holt
David M. Holy
Wendell A Van Hook
Gary M. Hoover, PhD
Rodney T. Houlihan, PhD
Ray A. Howell
George M. Hudak
Steven P. Huddleston
David A. Hudgins
Deanne D. Hughes
Jerry Hughes
Kenneth James Hughes
W. Bryan Hughes
Ronald Hull
Hubert B. Hunt
Luverne A. Husen
Michael D. Huston
Walter E. Hyde
Rodney D. Ice, PhD
John R. Imel
Lisa Funkhouser Ingle
Donald C. Jacks
Vincent Francis Jennemann, PhD

Appendix 4: The Petition Project

Michael W. Jezercak, PhD
Terry E. Johnson
Bill Johnston
Harlin Dee Johnston, PhD
Ralph S. Jones
Rhea A. Jones
Pat Joyce
Richard L. Jueschke
Brian O. Keefer
Dan F. Keller
Arthur L. Kelley
Jim M. Kelly
W. J. Kelly, PhD
Tim Kennemer
John F. Kerr
Francis W. King
John G. Kinsey
H. Kleemeier
Gary C. Knight
Jeffrey R. Knoles
Katherine M. Kocan, PhD
Eunsook Tak Koh, PhD
Kurtis Koll, PhD
John J. Kueser, Jr.
J. Michael Lacey
J. A. Landrum
Robert B. Lange
Jon LaRue, PhD
Reginald M. Lasater
Kenneth E. Laughrey
Richard Lawson
J. T. Lee
Leo V. Legg
Robert E. Lemmon, PhD
James H. Lieber
James B. Lockhart
Lyle G. Love
Dale D. Lovely
William R. Low, III
Samuel E. Loy, III
Robert C. Lucas
William Glenn Luce, PhD
Frank Lucenta
C. Luger
Scott E. Lugibihl
Mike S. Mabry
Scott E. Maddox
Bruce H. Mailey
Martin D. Malahy
Scott Maley
Phillip Gordon Manke, PhD
Francis S. Manning, PhD
Sam S. Marbry, Jr.
Douglas L. Marr
Jerry Martin, PhD
Richard Martin, PhD
Samuel T. Martner, PhD
Ted P. Matson
Wallace I. May
Kevin M. Mayes
David L. McCarley
Douglass A. McClure
Bob M. McCraw
Max P. McDaniel, PhD
Leslie Ernest McDonald, PhD
M. McElroy, PhD
Gregory B. McGowen
Stephen Edward McGuire, PhD
W. D. McIntosh
Cameron R. McLain
Wilfred E. McMurphy, PhD
James M. McUsic

Ralph M. McVay
Douglas W. Meeks
Allen D. Meese
Verlin G. Meier
Neal E. Mercer
Dean H. Mikkelson
Gary Miller
Larry O. Miller
Michael G. Mills
Claude E. Mised
Paul L. Mitchell
Bj J. Moon
Donald N. Mooney
David K. Moore
Jon L. Morgan
Oscar P. Morgan, Jr.
Ronnie G. Morgan, PhD
Teruo Morishige, PhD
O. Charles Morrison
W. Mike Morrison
W. A. Morse
Ora M. Moten
Michael G. Mount
Joseph F. Mueller
Bryan Newell
John Nichols, PhD
Leo A. Noll, PhD
Robert A. Northcutt
Doug Norton
John E. Norvell, PhD
Evan P. Obannon
Richard C. Obee
Randy M. Offenberger
Kenneth D. Oglesby
John C. Orloski
S. J. Orloski
Jim D. Outhier
Bill G. Owen
James Robert Owen, PhD
Fredric Newell Owens, PhD
William W. Owens
Bruce Robert Palmer, PhD
Jerald Parker, PhD
Gary L. Parks
Stephen Laurent Parrott, PhD
Peter C. Patton, PhD
Charles A. Paulson
R. Payne
H. Peace
Herbert E. Pearson
Ray E. Penick
Frederic J. Penner
Charles C. Perry, Jr., PhD
Raj Phansalkar, PhD
Don A. Phillips
Wallace C. Philoon, PhD
George Pierson
Gerald Pierson
Raymond E. Pletcher
Dennis L. Poindexter
William Jerry Polson, PhD
David E. Powley
Victoria Prevatt, PhD
Geoffrey L. Price, PhD
James E. Pritchard, PhD
Barry Profeta
Richard W. Radeka
Benjamin F. Ramsey
Spencer G. Randolph
Gujar N. Rao, PhD
J. B. Red
Harold J. Reddy

Thomas R. Redman
Larry Redmon, PhD
Max E. Reed, PhD
Philip W. Reed
Harold J. Reedy
Homer Eugene Reeves, PhD
Carl J. Regone
B. J. Reid
Don J. Remson
Thomas W. Reynolds
Ken J. Richards, PhD
J. Mark Richardson, PhD
Verlin Richardson, PhD
Greg A. Riepl
Olen L. Riggs, Jr., PhD
David Rippee
Jerry Lewis Robertson, PhD
Stanley L. Robertson, PhD
Wilbert Joseph Robertson, PhD
Robert L. Rorschach
Jeffrey S. Ross
Dighton Francis Rowan, PhD
Donal E. Ruminer
Larry J. Rummerfield
Geo Rushton
Mamdouh M. Salama, PhD
John T. Sanner
Charles Gordon Sanny, PhD
James E. Schammerhorn
Frank W. Schemm
Dwayne A. Schmidt
Carl F. Schneider, Jr.
James G. Schofield
Robert G. Schroeder
Roland Schultz, PhD
Charles R. Schwab
Radny Schwab
Charles N. Scott
Robert L. Scott
John H. Seader
Walter E. Seideman, PhD
Lawrence R. Seng
Charles Shackelford
Jack A. Shannon
Brian S. Shaw
William W. Sheehan
Jon P. Sheridan
Adolph Calveran Shotts
John P. Siedle, PhD
Jay K. Smith
Kevin L. Smith
Sherman E. Smith
Sofner Smith
Kenneth Snavely
Theodore Snider, PhD
Wallace W. Souder, PhD
Thomas C. Spear
Glenn E. Speck
Harry Spring
Robert M. St. John, PhD
John Stark
Gary Franklin Stewart, PhD
Edwin Tanner Still
P. Doug Storts
J. Story
Dennis E. Stover, PhD
George G. Strella
Raymond Walter Suhm, PhD
Gary D. Sump, PhD
Lyll S. Surtees
Earl W. Sutton
William D. Sutton

James A. Svetgoff
Shaun H. Sweiger
Allen G. Talley
Terry F. Tandy
James R. Taylor
William N. Thams
Barbara J. Thomas
Cullen Thomas
William Hugh Thomason, PhD
Robert R. Thompson, PhD
Timothy R. Thompson
W. H. Thompson, Jr.
Jimmy L. Thornton
Michael B. Tibbits
Buck J. Titsworth
Robert Totusek, PhD
Lawrence H. Towell
Harry P. Trivedi
Joe L. Troska
Steven M. Trost, PhD
Billy B. Tucker, PhD
James M. Tully
Brian Turner
Lynn D. Tyler, PhD
Mark W. Valentine
John Warren Vanderveen, PhD
James Milton Vanderwiele
Gary V. Vanmeter
James R. Vaughan
James P. Vaughn
Joseph C. Vaughn
Laval M. Verhalen, PhD
Suzanne Vincent, PhD
Ed Wagner
William D. Wakley, PhD
Charles Wall
Kenneth K. Warlick
William E. Warnock, Jr.
Lawrence A. Warzel, PhD
George A. Waters
J. R. Webb
Curtis W. Weittenhiller
Irving P. West
Delmar G. Westover
Gene L. Whitaker
James D. White
James R. Whiteley, PhD
Joe V. Whiteman, PhD
Thomas L. Whitsett
Michael L. Wiggins, PhD
Jack A. Williams, PhD
Keith P. Willson
David B. Wilson
David M. Wilson
Timothy M. Wilson, PhD
Weldon J. Wilson, PhD
Donn Braden Wimmer
William K. Winter, PhD
Richard B. Winters
Thomas H. Wintle
Kenneth Wolgemuth, PhD
I. K. Wong, PhD
C. Wootton
Wm. P. Wortman
Bill J. Wright
John L. Wright
Paul McCoy Wright, PhD
Demetrios V. Yannimaras, PhD
Donald Yaw
Marvin E. Yost, PhD
Donald Frank Zetjk, PhD
Charles W. Ziemer, PhD

John C. Zink, PhD

Oregon

Gail D. Adams, PhD
Janet Alanko
Randy A. Alanko
Fred A. Allehoff
Arthur W. Allsop
Allen A. Alsing
Elmer A. Anderson, PhD
Kenneth E. Anderson
Micheal J. Anhorn
Stig A. Annestrand
Lynn Apple
Edward E. Ashley
Owen H. Auger
Cyril G. Bader
Eric S. Ball
Harold Ball
Jeffrey L. Ball
Brian F. Barnett
Cary N. Barrett, PhD
Patrick H. Barrett
Earl M. Bates
Bruce L. Bayley
Matthew Beals
Thomas Erwin Bedell, PhD
Sheeny Behmard
Frederick H. Belz
Ronald Berg
Richard A. Bernhard
Siegried R. Berthelsford
James J. Besemer
Stephen E. Binney, PhD
Guy William Bishop, PhD
Warren M. Bliss
Vern Allen Blumhagen
Dennis B. Bokovoy
S. Bonar
Daniel C. Boteler
Joseph K. Bozievich
William J. Brady
William Henry Brandt, PhD
Gregory Brenner, PhD
David Briggs
Karel M. Broda
James R. Brunner
John D. Bryan
James L. Buchal
Charles M. Buchzik
John P. Buckinger
Hallie Flowers Bundy, PhD
Edgar M. Burton, Jr.
Robert E. Bynum
Donald D. Calvin
A. Eugene Carlson
Glenn Elwin Carman, PhD
Richard A. Carpenter
Richard N. Carter
David Owen Chilcote, PhD
Donald Ernest Chittick, PhD
Duane Christensen
Jason J. Churchill, PhD
David Thurmond Clark, PhD
Cecil K. Claycomb, PhD
Vyvyan S. Clift
Ray W. Clough
John D. Coates
D. W. Coleman
Jerry Colombo

Erik E. Colville
F. N. Cooke, PhD
Roc A. Cordes
Rebecca A. Coulsey
Edward A. Crane
Michael A. Creager
Harry G. Cretin
Arthur D. Crino
William Craig Cullen, PhD
Stanley E. Cutrer
Michael Daly, PhD
Harold Datfield
Jay C. Davenport
Kirk C. Davis
Fred W. Decker, PhD
Martin J. Dehaas
V. James DeMeo, PhD
William David Detlefsen, PhD
William J. DeVey
Lee G. Dickinson
Edward Joseph Dowdy, PhD
Nicholas Drapela, PhD
Arthur B. Drescher
Robert W. Du Priest
Dennis Dunning
John Durbetaki
Owen W. Dykema
David C. Ell
Rodney D. Elmore
David M. Enloe
Richard F. Erpelding, PhD
Steven P. Eudaly
Clinton Faber
Fred W. Fahner
Gary Farmer
William Farnham
David G. Farthing, PhD
Andrew J. Fergus
Cyrus West Field, PhD
D. Findorff
Robert Findorff
James W. Fitzsimmons
Richard H. Fixott
Bruce Ingram Fleming, PhD
J. Ford, PhD
Steven Robert Fordyce
Dwayne T. Friesen, PhD
Herbert Farley Frolander, PhD
Gordon J. Fulks, PhD
Larry D. Gage, PhD
Herbert G. Gascon
Otto M. Gobina, PhD
Jay A. Goin
Jerry J. Gray*
Thomas B. Gray
Garrison Greenwood, PhD
James Robb Grover, PhD
M. S. Gurney
Christopher T. Haffner
Paul Ellsworth Hammond, PhD
Raymond J. Hardiman
Floyd A. Harrington
Thomas S. Harrison, IV
John P. Harville, PhD
Paul L. Hathaway
Harold Franklin Heady, PhD
William M. Heerdt
Stanley O. Heinemann
Robert H. Heitmanek
Roger A. Helvey
Charles A. Hen
Robert R. Henderson, II

Tom R. Herrmann, PhD
Dennis E. Hess
Clayton Hill
John Donald Hill
John Holbrook
Norman B. Holman
James Frederick Holmes, PhD
Ronald G. Holscher
John L. Hult, PhD
Robert D. Hunsucker, PhD
Douglas Hunt
Brad L. Hupy
W. E. Hutley
John F. Hutzenlaub, PhD
Brian J. Imel
Merlyn Isaak
Robert E. Jamison
Ryan M. Jefferis
Yih-Chyun Jenq, PhD
Karl J. Johnson
Terrance Johnson, PhD
Jenri B. Joyaux
Edward Lynn Kaplan, PhD
Larry Kapustka, PhD
Dennis J. Kavanaugh
Ronald W. Kelleher
Robert C. Kieffer
Tatyana P. Kilchugina, PhD
Marion E. Kintner
Donald Kinzer
Gregory F. Kohn
Loren D. Koller, PhD
Carl R. Kostol
Robert W. Koza
David A. Kribs
Leonard J. Krombein
Travis A. Kruger
Donald L. Lamar, PhD
Andrew Lambie
Don LaMunyon, PhD
Darrin L. Lane
Robert Larsell
Eric Larsen, PhD
Bryan C. Larson
Harold M. Lathrop
William Orvid Lee, PhD
Guy L. Letourneau
Kenneth W. Lewin
Kurt D. Liebe
Mark A. Liebe, PhD
W. J. Lindblad
Garry N. Link
Robert K. Linn
Adam W. Lis, PhD
Daniel C. Lockwood
Michael Love
Herbert F. Lowe
R. Kent Lundergan
J. Malcom
Aaron Manley
Niels L. Martin, PhD
Philip O. Martinson
Lorrin L. Marvin
Richard R. Mason, PhD
Curtis R. Matteson
Mark W. Matthes
Jerome A. Maurseth
Michael D. McCaffery
Mary McCarthy
Henry F. McKenney
Harold A. McSaaden
Harry H. Mejdell, PhD

Deering D. Melin
Daniel Merfeld, PhD
Paul Louis Merz, PhD
Sabine Meyer, PhD
Cameron D. Mierau
Kevin L. Miles
Larry D. Miller
Leonid Minkin, PhD
J. Minoff, PhD
Johnny D. Mitchell
Robert H. Mitchell
William A. Mittelstadt
Donovan B. Mogstad
Javid Mohtasham, PhD
Darrell W. Monroe
Larry Wallace Moore, PhD
Joseph T. Morgan
Emile C. Mortier
Jane E. Mossberg
Clifford John Murino, PhD
Jeffrey M. Murphy
Ruth Naser
Steven L. Neal
Victor Thomas Neal, PhD
David Michael Nelson, PhD
Harold W. Nichols
William Nisbet
Richard E. Noble
Ella Mae Noffsinger
Maya M. Nowakowski
R. L. Nowlin
Gerald O'Bannon, PhD
Laurence B. Oeth
Roger Dean Olleman
Loren K. Olson
Cherri Pancake, PhD
Donald H. Parcel
Albert C. Parr, PhD
Albert Marchant Pearson, PhD
Joseph C. Peden
Wendell Pepperdine, PhD
Heriberto Petschek
Richard F. Pfeifer
Lincoln Phillippi
Dennis Phillips, PhD
Brian Phipps
David N. Pocock
Leroy D. Poindexter
Serge Preston, PhD
Casey Jo Price
J. J. Price, III
William E. Purnell
William J. Randall, PhD
Bob Raser
Alfred Ratz, PhD
Jerry Re
Joseph Arthur Reed
Kenneth R. Reed
Michael E. Rehmer
James W. Richards
Jerry O. Richartz
William J. Robinet
Arthur B. Robinson, PhD
Arynne L. Robinson
Bethany R. Robinson
Joshua A. Robinson
Matthew Loren Robinson
Noah E. Robinson, PhD
Zachary W. Robinson
James D. Rodine, PhD
Francis Rogan
Charles Raymond Rohde, PhD

K. Rudestam
Lee Chester Ryker, PhD
Michael A. Sandquist
Dennis A. Schantzen
Rolf Schaumann, PhD
Larry Scheckter, PhD
Chester O. Schink, PhD
Dyrk H. Schlingman
Clifford Leroy Schmidt, PhD
Roman A. Schmitt, PhD
Robert M. Schwarze
Doug P. Schwin
Richard N. Seemel
Eric Sharpnack
Donald K. Shields
Terry Shortridge
Larry D. Shuttlesworth
Jerry A. Siemens
Richard Ernest Siemens
Bruce Siepak
Ronald Smelt, PhD
Deboyd Smith
Troy L. Smith, Jr.
Ronald P. Staehlin
Paul Edward Starkey
Scott M. Steckley
Robert C. Stones, PhD
William L. Strangio, PhD
Robert Leonard Straube, PhD
Charles Strohkirch
Ralph C. Swanson
Danuta Szalecka
Wojciech Szalecki, PhD
Carson William Taylor
Richard W. Thoresen
Jerry Tobin
William L. Toffler
Albert L. Tormey
John D. Trudel
Robert W. Trumbull
Dah Weh Tsang, PhD
David B. Tyson
Henry J. Van Hassel, PhD
Philip A. Volker
R. A. Vollar
James Arthur Vomocil, PhD
Mike Vossen
Orvin Edson Wagner, PhD
Neil E. Warren
A. D. Watt
Susan L. Wechsler
Dennis H. Wessels
Jack A. Weyandt
Richard V. Whiteley, Jr., PhD
James Whiting
Chuck F. Wiese
Robert Stanley Winniford, PhD*
C. N. Winningstad
Mark C. Wirfs
George F. Wittkopp
Peter Wolmut
Jeffry T. Yake
Elisabeth S. Yearick, PhD
Nicholas J. Yonker
Donald J. Zarosinski

Pennsylvania

Joe L. Abriola, Jr.
David A. Acerni
John R. Ackerman

Roy Melville Adams, PhD
Lionel Paul Adda, PhD
George Aggen, PhD
Thomas I. Agnew, PhD
David J. Akers
Munawwar M. Akhtar
Greg Alan
Eric K. Albert, PhD
James T. Albert
Fred Ronald Albright, PhD
Robert Lee Albright, PhD
Reuben J. Aldrich
Joseph J. Alex
Danrick W. Alexander
Harold R. Alexander
Kevin Alexander
George L. Allerton
Robert Q. Alleva
Gary W. Allshouse
R. A. Allwein
James A. Aloye
Joseph R. Ambruster
Richard D. Amori
Fred Amsler
Arvid Anderson
David Robert Anderson, PhD
J. Hilbert Anderson
Harry N. Andrews
Raynal W. Andrews
Timothy Andreychek
Edward A. Andrus
Francis M. Angeloni, PhD
Howard P. Angstadt, PhD
Edward J. Annick
Bradley C. Antanaitis, PhD
David R. Appel
Diane M. Archer
Desiree A. Armstrong, PhD
James R. Armstrong
Charles W. Arnold
Randall W. Arnold
Jerome P. Ashman
Abhay Ashtekar, PhD
Philip R. Askman
Andrew P. Assenmacher
Victor Hugo Auerbach, PhD
Dale A. Augenstein, PhD
Joe Augspurger, PhD
J. Todd Aukerman
Kent P. Bachmann
Klemens C. Baczewski
Walter J. Bagdon, PhD
David George Bailey, PhD
Egon Balas, PhD
W. Lloyd Balderston, PhD
Paul F. Balint
Antonio P. Ballestero, Jr.
Kent F. Balls
Chad J. Bardone
Philip A. Barilla
Barry L. Barkley
Larry E. Barnes
Robert M. Barnoff, PhD
James E. Barrick, PhD
Paul Barton, PhD
Patrica W. Bartusik
Jonathan Langer Bass, PhD
Robert L. Batdorf, PhD
Joe Battista
Carl August Bauer, PhD
Kerry E. Bauer
Michael H. Bauer, Jr.

Wolfgang Baum, PhD
Kenneth L. Baumert, PhD
A. W. Baumgartner
Richard W. Bayer
Theodore H. Bayer
Buddy A. Beach
James Monroe Beattie, PhD
Frank V. Beaumont
Anne C. Beavers
Ellington McHenry Beavers, PhD
Thomas F. Beck
Robert S. Becker, PhD
Robert B. Beelman, PhD
Cari R. Beenenga
David B. Behar
John Behun, PhD
Alan R. Beiderman
William W. Bele
George C. Bell
Stanley C. Bell, PhD
John A. Bellak
James R. Bellenoit
Erwin Belohoubek, PhD
Daniel W. Bender
David J. Bender, PhD
Ashley Bengali, PhD
Allen William Benton, PhD
Charles R. Bepler
Santo C. Berasi
Alan Berger
Lee E. Berkebile
William G. Berlinger
David Berry, PhD
Robert Walter Berry, PhD
Christopher Anthony Bertelo, PhD
Joseph M. Beskit
Tirlochan S. Bhat
Paul D. Bianculli
Stanley J. Biel, Jr.
Richard J. Bielicki
George J. Bierker
Keith L. Bildstein, PhD
Dan A. Billman
Terrence L. Bimle
Lisa J. Binkley
William W. Bintzer*
James A. Bishop, Jr.
Richard E. Bishop
Curtis S. Bixler
Jerry M. Blackmon
Douglas R. Blais
Paul R. Blankenhorn, PhD
James R. Blankenship, PhD
William Blanset
Paul J. Blastos
Robert C. Blatz
Harold Frederick Bluhm
Edward S. Bober
David Bochnowich
Bohdan K. Boczkaj, PhD
Kenneth E. Bodek
Gerald R. Bodman
Arthur W. Boesler
William Emmerson Boggs
N. Charles Bolgiano, PhD
James C. BolliBon
Jeffery L. Bolze
James M. Bondi
Oliver P. Bonesteel
Charles M. Bonner
Bruce W. Booth, PhD
Frederick Bopp, III, PhD

Stuart Boreen
Theodore T. Borek, II
Robert Bores, PhD
Paul N. Bossart, Jr.
Thomas H. Bossler
Aksel Arnold Bothner, PhD
Paul Andre Bouis, PhD
William C. Bowden
A'Delbert Bowen
Mark Bowen
Dennis L. Bowers
Paul R. Bowers
Ed Bowersox
Arthur F. Bowman
Gerald L. Bowman
Lewis Wilmer Bowman, PhD
Robert S. Bowman, PhD
Carl A. Bowser
Robert Allen Boyer, PhD
John J. Boyle
Herbert H. Braden
William Earle Bradley
Keneth A. Brame
Bruce Livingston Bramfitt, PhD
Patrick G. Brannac
Frank J. Braun
James P. Brazel
Darwin Brendlinger
Joseph N. Breston, PhD
Robert N. Brey
Pierre Leonce Thibaut Brian, PhD
C. Carroll Brice
Paul P. Bricknell
Kevin V. Bridge
Randall A. Brink
R. Brisbin
Frederick N. Broberg
E. J. Brock, PhD
Edward C. Broge, PhD
Richard P. Brookman
James Brubaker
Daniel E. Bruce
Linda Bruce-Gerz
Glenn G. Bruckno
William H. Bruhns, PhD
Roger R. Brumbaugh
William M. Brummett
Jacomina A. Bruno
Jeffrey W. Brunson
Frederick Vincent Brutcher, Jr., PhD
Walter T. Bubern
Ralph S. Bucci
Robert C. Buckingham
Daniel E. Bullard, PhD
Jacob Burbea, PhD
Herbert A. Burgman
David M. Burkett
Cynthia A. Burr
John F. Burr
Edward G. Busch
Elsworth R. Buskirk, PhD
James S. Butcofski
James J. Butler
Phillip E. Butler
Douglas S. Byers
R. Lee Byers, PhD
Mike Cabirac
John L. Caddell
Craig R. Calabria, PhD
David R. Calderone
Lawrence A. Caliguiri
David S. Callaghan

William R. Camerer
Robert Camero
Richard W. Campbell, PhD
Robert Campo, PhD
James N. Canfield
Joseph M. Cannella
Frederic S. Capitosti
Anthony P. Caprioli
Steven Carabello
Mario Domenico Carelli, PhD
John Carey
George F. Carini, PhD
Harold Edwin Carley, PhD
Burnette J. Carlson
Douglas E. Carlson
Ralph Carlson, PhD
Brad H. Carpenter
Charles Patten Carpenter, PhD
Michael J. Carrera
Roger E. Carrier, PhD
Ron A. Carrola
J. Randall Carroll
Robert J. Carroll
Stanley J. Carroll, Jr
Thomas C. Carson
Joanne M. Caruno
Floyd L. Cassidy
Ronald E. Castello
Anthony J. Castellone
Robert L. Catherman
David Cattell, PhD
Joseph J. Catto
William J. Cauffman
Edward L. Caulkins
Paul Cazier
Gilbert J. Celedonia
Chris Cellucci, PhD
Nick Cemprola
Zoltan Joseph Cendes, PhD
John J. Cenkner
Samuel Cerni, PhD
George G. Cervenka
John Chadbourne, PhD
Ronald A. Chadderton, PhD
John Chappell
Kenneth Chaquette
Stanley H. Charap, PhD
Edward M. Charney
Edward Chastka
Chi-Yang Cheng, PhD
Kuang Liu Cheng, PhD
John M. Chenosky
Kenneth P. Chepenik, PhD
Joseph R. Chessario
Peter Cheung, PhD
V. Chiappardi
James P. Childress
D. S. Choi, PhD
Janet C. Christiansen
Alan Christman, PhD
Kyung Y. Chung
Jon A. Ciotti
Michele A. Cipollone
John S. Clark
Duane Grookett Clarke, PhD
Robert L. Clouse
David A. Cobaugh
Carl Cochrane
Charles W. Coe, II
David B. Coghlan
Jennifer M. Cohen, PhD
Merrill Cohen

Michael R. Cohen, PhD
James M. Coleman, PhD
Robert E. Coleman, PhD
William E. Coleman
David B. Collins
Christopher J. Commans
Joseph R. Compton
Leonard Conapinski
James J. Concilla
Harold Conder, PhD
Albert Carman Condo, Jr.
Rex B. Conn
Garret H. Conner
Paul W. Conrad
Donald R. Conte
John Cooper, PhD
Nicholas N. Coppage
Robert B. Corbett, PhD
John J. Corcoran
Vincent F. Cordova
William E. Cormier
John Cornell
Kevin Cotchen
Larry Coudriet
Robert A. Cousins
Fred Covelli
Brent W. Cowan
Jeffrey L. Coward
Glenn E. Cowher
Michael F. Cox
Robert L. Cragg
Jonathan Crawford
Michael J. Cribbins
Robert D. Cryer
Dezso L. Csizmadia
Kevin Cumblidge
Donald E. Cummings
Howard Cunningham, PhD
Robert Cunningham
Francis M. Curran
Ken E. Cutler
Glenn Czulada
George F. Daebeler
W. V. Dailey
James M. Daley
Roy W. Danielsen
Victro M. Danushevsky, PhD
D. K. Davies, PhD
Mark Davis
Frank Robert Dax
Sheldon W. Dean, Jr., PhD
Joseph Deblassio, PhD
David R. Debo
Paul E. DeCusati, PhD
Emil W. Deeg, PhD
Robert W. DeFayette
Jonh D. Defelice
Vincent A. Degiusto
Andrew Dekker
Paul Dellevigne
John J. DeLuccia, PhD
Paul F. Demmert
George T. Demoss
Louis J. Denes, PhD
William E. Dennis, PhD
Norman C. Deno, PhD
Fred E. Derks
T. Derossett, PhD
Bhasker C. Desai
Daniel Frank Desanto, PhD
George R. Desko
Ryan Desko

Neal D. Desruisseaux, PhD
Maurice Deul
Al Deurbrouck
Frederick William Deuries
Robert Dewitt, PhD
Ronald K. Dickey
Rodger L. Diehl
Frederick M. Dietrich
Edward J. Diggs, III
Jeffrey A. Dille
Frederick R. Dimasi
Gene Dipasquale, PhD
Edward R. Dobson
Ronald P. Dodson, Jr.
James B. Doe
Ralph O. Doederlein
Robert C. Doerr
Michael J. Doherty
Gregory Dolise
Donald E. Dorney
Frank A. Dottore
W. D'Orville Doty, PhD
Charles B. Dougherty
Patrick A. Dougherty
Thomas J. Dowling
Bruce D. Drake, PhD
David F. Drinkhouse
Richard M. Drisko, PhD
Guy A. Duboice
Ronald H. Duckstein, Jr.
Ken Dudeck
Michael J. Duer
William O. Dulling
Leon B. Duminiak
Charles Lee Duncan, PhD
Lawrence A. Dunegan
David P. Dunlap
Jack M. Durfee
Henry W. Durrwachter
Dennis J. Duryea
John B. Duryea
Jean F. Duvivier, PhD
James Philip Dux, PhD
Arthur H. Dvinoff, PhD
Judith M. Dvorsky
Francis Gerard Dwyer, PhD
Alan C. Dyar
John C. Dzurino
Russell E. Earle
Douglas A. East, PhD
Richard J. Ebert
Charles Eckman
Charles K. Edge, PhD
Dion R. Ehrlich
H. Paul Ehrlich, PhD
Michael John Elkind
Richard J. Ellis, PhD
Don C. Ellsworth
K. Donald Ellsworth
Franklin T. Emery, PhD
Thomas J. Emory
Robert J. Ennings Heinsohn, PhD
J. Michael Enyedy
Robert Allan Erb, PhD
Richard R. Erickson, PhD
Jan Erikson, PhD
Terry Ess
Ben Edward Evans, PhD
George H. Evans
James M. Eways
Michael William Fabian, PhD
Thomas John Fabish, PhD

Leonard J. Facciani
David R. Fair
Frank G. Falco
Joe Falcone, PhD
Phillip A. Farber, PhD
Robert L. Fark
Raymond G. Fasiczka, PhD
Dennis A. Faust
Jerry W. Faust
Donald P. Featherstone
Thomas V. Febbo
Peter W. Fedum
James F. Feeman, PhD
Michael L. Fenger
Ronald W. Fenstermaker
Albert B. Ferguson
Michael W. Ferrelli, PhD
David Field
Albert J. Fill
James W. Finlay
Frances M. Finn, PhD
Leonard W. Finnell
Grace Mae Fischer
Robert P. Fischer
Douglas A. Fisher, PhD
Wade L. Fite, PhD
Henry J. Fix
R. M. Fleig
Charles W. Fleischmann, PhD
Harry Fleming
Scott J. Fletcher
Robert Flicker
Anthony Foderaro, PhD
James F. Foley
James S. Fontaine, Jr.
Robert T. Forbes
Albert J. Forney
James L. Foster, Jr.
Bennett R. Fox
J. Fox
Timothy J. Fox
Sandra W. Francis, PhD
Dimitrios N. Frangiadis
William Frankl
Norbert W. Franklin
August R. Freda, PhD
Glenn C. Frederick
Wallace L. Freeman, PhD
L. L. French
Glenn Frey
Robert L. Frey, Sr.
Larry S. Friedline
S. Scott Frielander
Alfred E. Fuchs
Dennis Light Funck, PhD
William R. Funk
Rande P. Funkhauser
Harold R. Fuquay
Samuel T. Furr
Gabriel C. Fusco, PhD
P. S. Gaal, PhD
Peter Gaal
James R. Gage
Steven R. Gage
William A. Gallus
James J. Gambino
Donald J. Gandenberger
Joseph E. Gantz
Richard J. Garbacik
George F. Garcelon
Celso-Ramon Garcia
Albert S. Garofalo

Jennifer J. Garrett
Donald W. Gauntlett
James Gaynor
Donald Gebbie
David Geeza
Roland P. Gehman
Evelyn M. Gemperle
Alfonso R. Gennaro, PhD
Richard J. Genova
F. Jay Gerchow
Ronald E. Gerdeman
J. Calvin Gerhard
George William Gerhardt, PhD
Henry Dietrich Gerhold, PhD
William P. Gibbons
Donald G. Gibson
C. Burroughs Gill, PhD
Michael S. Gillan
Daniel Gillen
Harry K. Gillespie
Thomas E. Gillingham, PhD
Jacques Gilloteaux, PhD
Joseph A. Giordano
Ernie Giovannitti
Melvin H. Gitlitz, PhD
James S. Glessner
Barbara W. Glockson, PhD
Jonathan C. Goble
James R. Gockley
Frederick A. Goellner
Thomas W. Goettge
Theodore Philip Goldstein, PhD
Walter J. Goldsworthy
Jonathan M. Golli
Ralph Golta, PhD
Michael S. Goodman
John H. Goodworth
Robert B. Gordon
Henry B. Gorman
Scarlette Z. Gotwals
Fred L. Graf
Jamieson Graf
Jim Graham
Michael W. Graham
Norman Howard Grant, PhD
Richard G. Graven
D. Gray, PhD
Richard E. Gray
Ronald D. Gray
Kim W. Graybill
Claudius A. Greco
J. D. Green
Patricia A. Green
George M. Greene, PhD
Miles G. Greenland
Ronald H. Greenwood
Howard M. Greger
Richard A. Gress
J. Tyler Griffin
Jean Griffin
Jonathan M. Grohsman
William G. Gross
Timoth Grotzinger
Joseph S. Grubbs
Gerald William Gruber, PhD
Geza Gruenwald, PhD
Reed H. Grundy
Gavin G. Guarino
William G. Gunderman
Hans Heinrich Gunth, PhD
Ropert E. Gusciora
William A. Gustin

David Lawrence Gustine, PhD
William C. Guyker, Jr., PhD
James T. Gwinn
George J. Haberman
John A. Habrle
Ross A. Hackel
Richard S. Hackman
Andree H. Hadden
Carl W. Haeseler, PhD
Joseph H. Hafkenschiel
Oskar Hagen, PhD
Jack W. Hager
John A. Haiko
David A. Hair
Michael S. Haladay
James L. Hales
John P. Halferty
Kenneth R. Ham
Erwin F. Hamel
James V. Hamel, PhD
Bruce S. Hamilton
Gordon Andrew Hamilton, PhD
James G. Haney
Jerry E. Hankey
Michael L. Haraczy
Michael P. Harasym
Henry Reginald Hardy, PhD
Robert Ian Harker, PhD
George E. Harper
John A. Harris, PhD
Clark D. Harrison
Don E. Harrison, PhD
Ernest A. Harrison, PhD
T. W. Harshall
George R. Hart
Edward L. Hartmann
Nathan L. Hartwig, PhD
W. P. Hartzell
Richard Harwood, PhD
Michael K. Hatem
Peter L. Hatgelakas
James N. Hauff
Helga Francis Havas, PhD
Eugene J. Hebert
Neil Hedrick
James J. Heger
Donald L. Heim
Walter N. Heine
Paul E. Heise
Wesley E. Heisley
Herbert Heller
Mike D. Helmlinger
Robert P. Helwick
James C. Henderson
William R. Henning, PhD
David S. Hepler
Phillip D. Herman
Roger Myers Herman, PhD
J. Wilson Hershey, PhD
John E. Herweh
Robert P. Heslop
Eugene L. Hess, PhD
Marilyn E. Hess, PhD
Thomas E. Hess
William John Adrian Vanden
Heuvel, III, PhD
Kenneth D. Hickey, PhD
Richard T. Hilboky
Hilton F. Hinderliter, PhD
Bruce J. Hinkle
Jan D. Hinrichsen
Millard L. Hinton

Raymond Price Hinton
Raymond G. Hobson
J. Hochendoner
Paul G. Hofbauer
Richard J. Hoffert
Charles F. Hoffman
Thomas F. Hoffman
Joseph L. Hoffmann
William R. Hogan
Deborah Hokien, PhD
Richard M. Holcombe
Rodney E. Holderbaum
Philip M. Holladay, PhD
Werner J. Hollendonner
Robert Holleran
Ronald Holmes
Ken A. Holtz
Jerry E. Holtzer
Richard A. Hoover
David A. Hopkins
Edward H. Hoppe, IV
Richard A. Horenburger
James W. Howard
Jason Howe
Benjamin F. Howell, Jr., PhD
Paul J. Hoyer, PhD
Nicholas C. Hrnjez
Dale A. Hudson, PhD
John Laurence Hull
John Kenneth Hulm, PhD
Frank Hultgren, PhD
Philip Wilson Humer, PhD
Kenneth Hunt
Philip R. Hunt
Richard Hurley, PhD
Jason S. Hustus
J. J. Huth, III
Andre M. Iezzi
Bruce E. Ilgen
Giovanni C. Ingegneri
Alvin Richard Ingram, PhD
Patricia A. Innis
Robert A. Irvin
Philip George Irwin, PhD
Richard W. Irwin
George Isajiw
Sheldon Erwin Isakoff, PhD
Gordon A. Israelson
William M. Jacobi, PhD
Gail L. Jacobs
Donald Jaffe, PhD
Douglas E. Jakim
Will B. Jamison
Mark Jancin
Ken M. Janke
Samuel M. Jenkins
Raymond W. Jensen
William A. Jester, PhD
Benjamin L. Jezovnik
Spurgeon S. Johns
Alan W. Johnson, PhD
Dale L. Johnson
Melvin Walter Johnson, PhD
Robert M. Johnson
William Dwight Johnston, PhD
David H. Jones, PhD
Hadley H. Jones
Richard Patterson Jones
Roger F. Jones
Robert Kenneth Jordan
Thomas E. Jordan
Pisica H. Joseph

Donald Joye, PhD
Joseph Malcahi Judge, PhD
Thomas J. Junker
Walter J. Jurasinski
Stephen A. Justham, PhD
Robert Kabel, PhD
George C. Kagcer, PhD
Joh R. Kalafut
Beth Kalmes
James Herbert Kanzelmeyer, PhD
Norman G. Kapko
Naum M. Kaplan
M. Kapolka, PhD
George A. Kappenhagen
Jason Karasinski, PhD
Eskil Leannart Karlson, PhD
John J. Karpowicz
Thomas W. Karras, PhD
Ronald L. Kasaback
Thomas Kasuba
Michael J. Kaszyski
Joel Kauffman, PhD
Leo F. Kaufhold
Robert L. Kay, PhD
J. M. Kazan
Donald E. Keefer
James D. Keith
Jay H. Kelley, PhD
Ian A. Kellman
James M. Kelly
George R. Kemp, Jr.
Robert J. Kendra
Edward J. Kenna
Phil L. Kershner
Donald L. Kettering
Donald E. Keyt
Z. Khatchadourian
Andrew Kicinski
Anthony Stanley Kidawa
Joseph C. King
Norma A. Kinsel, PhD
Maclean Kirkwood
Pamela D. Kistler, PhD
Amy G. Klann, PhD
Louis T. Klauder, PhD
John Klein
Donald R. Kleintop
Stanley J. Kletch
Irvin C. Klimas
Gary P. Klipe
Kenneth A. Kloes
James J. Knapik
Dale B. Knepper, Jr.
Reinhard H. Knerr, PhD
F. James Knight
Thomas E. Kobrick
Donald J. Koestler
Robert N. Kohler
William C. Kohnle
Paul M. Kolich
Paul J. Konkol
Robert Philip Koob, PhD
Joseph Korch
Edward Korostoff, PhD
Randolph M. Kosky
Kishen Koul, PhD
William P. Kovacik, PhD
Edward R. Kovanic
Theodore J. Kowalyshyn
John G. Kraemer
Drew A. Kramer
Douglas J. Kramm

Kenneth G. Krauss, PhD
Thurman R. Kremser, Jr.
Thurman R. Kremser, Sr., PhD
Arthur F. Krieg
Scott Kriner
Thomas S. Krivak, PhD
Joseph F. Kroker
Paul H. Krumrine, PhD
Regis W. Kubit
George Kugler, PhD
Casimir A. Kukielka
Andrei Kukushkin
Francis R. Kull
Jeff Kumer
Robert F. Kumpf
Irving B. Kun
Sean T. Kuplean
Scott T. Kupper
Robert John Kurland, PhD
Andy Kuzma, Jr.
Peter Labosky, PhD
Joseph Thomas Laemmle, PhD
James A. Lageman
Dale B. Lancaster, PhD
Ronald E. Land
George P. Lang
Ruth E. Lang
John M. Langloss, PhD
Thomas S. Larko
Stanley B. Lasday
Joseph Lasecki
John C. Lathrop
Gordon Fyfe Law
Joethel T. Lawler
Robert W. Lawson
Richard D. Leamer, PhD
Paul W. Lebarron
David C. Leber
Donald D. Lee, PhD
Kuo Hom Lee Hsu, PhD
E. J. Leffler
Edward F. Leh
Joseph M. Lehman
Paul H. Lehman, PhD
Richard K. Lehman
Ronald G. Lehman
Harry J. Lehr
Daniel Leiss
David Leiss
Stanley L. Leitsch
Craig Lello
George W. Leney
Jim W. Lennen
Ronald S. Lenox, PhD
Stephen Lentz, PhD
George W. Leroux
David M. Lesak
Richard E. Lesher
Hans E. Leumann
Mark Levi, PhD
William A. Levinson, PhD
D. Lewis
John A. Lewis, PhD
David Leyshon
William T. Liddle
L. Charles Lightfoot
Robert E. Liljestrand
Wayne B. Lingle
Donna J. Link
Jeffrey M. Linn
David E. Lipkin
Jack Lipman

Bryan L. Lipp
Mitchell Litt, PhD
Ed Liwerant
John E. Logan
John P. Logan, Jr.
David M. Lolley, PhD
Paul S. Lombardo
Bruce Long, PhD
Richard K. Lordo
Dale R. Lostetter
Harold Lemuel Lovell, PhD
John Robert Lovett, PhD
Mark Lowell, PhD
Douglas E. Lowenhaupt
Jerald Frank Lowry
Ramon G. Lozano
Louis A. Lucchetti
John J. Luciani
Theodore M. Lucidi
Dennis E. Ludwig
Oliver G. Ludwig, PhD
Gene S. Luff
Frank L. Luisi
James W. Lundy
R. Dwayne Lunsford, PhD
Peter J. Lusardi
Arthur P. Luthy
John R. Lydic
W. R. Lyman
Bernard J. Lynch
John J. Lynch, PhD
Roland H. Lynch
Digby D. MacDonald, PhD
Robert A. MacFarlane
Lawrence A. Mack
Ronald B. Madison, PhD
Quirico R. Magbojos
Robert Dixon Mair, PhD
Daniel D. Malinowski
Frank Bryant Mallory, PhD
Frederick J. Managhan
James M. Mandera
Lucia Manea
Samuel John Manganello
Kristeen Maniscalco
John H. Manley, PhD
Frederick J. Mannion, PhD
Mark S. Manzanek
Zhi-Hong Mao, PhD
Bernard J. Maopolski
Arnold Marder, PhD
Anthony B. Marino
Constance F. Marsh
Harold Gene Marshall, PhD
Terrel Marshall, Jr., PhD
Edward Shaffer Martin, PhD
Peter M. Martin
William F. Martinek
Albert A. Martucci
John T. Maruschak, Jr.
Putinas V. Masalaitis
Michael P. Mascaro
Philip X. Masciantonio, PhD
Carmime C. Mascoli, PhD
Michael John Mash
Robert J. Mathieu
Dilip Mathur, PhD
John Matolyak, PhD
Kenneth D. Matthews
Roger L. Mauchline
Frank D. Mawson, Jr.
William D. Mayercheck

Anthony E. Mayne
Eugene A. Mazza
T. Ashley McAfoose
Darrell McAlister
Bruce Ronald McAvoy
Richard J. McCarthy
Earl W. McCleerey, Sr.
Thomas A. McClelland
Robert McCollum
Robert A. McConnell, Jr.
John E. McCool
Clifford W. McCoy, PhD
Thomas W. McCreary
Dennis E. McCullough, PhD
Wallace G. McCune
David A. McDevitt
Robert H. McDonald
Ronald W. McDonel, Jr.
Omer H. McGee
Michael F. McGuire
John Philip McKelvey, PhD
Maribel McKelvy
James Alan McLennan, PhD
Clyde S. McMillen
James I. McMillen
Kenneth M. McNeil, PhD
Robert H. McNeill
Paul V. McQuade
Jack L. McSherry
Jeffrey L. Meadows
Paul Meakin, PhD
Rick Meckley
Thrygve Richard Meeker, PhD
John Mehltretter
John T. Melick
A. Meliksetian, PhD
Norman D. Melling
Stephen D. Meriney, PhD
Daniel Merkovsky
Wendle P. Mertz
Robert E. Metzger
William E. Michael
Andrew F. Michanowicz
John J. Michlovic
Raymond D. Mikesell
Dale Lloyd Miller
Foil A. Miller, PhD
George J. Miller
James R. Miller
Joseph H. Miller
Lewis E. Miller
Richard Miller
Robert C. Miller, PhD
Tad W. Miller
Michael R. Millhouse
Howard Minckler
Michael J. Minckler
Tom J. Minock
Richard Edward Mistler, PhD
Charles J. Mode, PhD
Arthur F. Moeller
Elwood R. Mohl
C. John Mole
Gregory C. Moll
Harold Gene Monsimer, PhD
Robert H. Montgomery, Jr.
Michael R. Moore, PhD
Walter Calvin Moore
George M. Morley
Douglas R. Morrissey
Anthony F. Morrocco
David L. Morton

Thomas E. Morton
Lester R. Moskowitz, PhD
Michael T. Moss
Faith E. Moyer
Walter A. Mroziak
Francis Mulcahy, PhD
Thomas G. Mulcavage
James Alan Mullendore, PhD
Karl F. Muller, PhD
Stanley A. Mumma, PhD
Mark C. Mummert, PhD
Michael F. Murphy
William James Murphy, PhD
Steven R. Musial, II
R. Musselman, PhD
Glenn A. Musser
Earl Eugene Myers, PhD
Glenn L. Myers
Donna L. Naples, PhD
Michael Nawrocki
Frank H. Neely
Roy H. Neer
Stuart Edmund Neff, PhD
Myron P. Nehrebecki
Jeremy A. Nelson
Richard Lloyd Nelson
Bradley G. Neubert
Frederick C. Newton
Gabriel P. Niedziela
Larry A. Niemond
Thomas George Nilan, PhD
Richard M. Ninesteel
David L. Nirschl
George Niznik, PhD
Edward J. Nolan, PhD
James Noss
James P. Novacek
Timothy K. Nytra
Charles J. Odgers
Orin L. Odonel
Sakir S. Oguz
Wm S. Oleyar
David A. Oliver
Andrew M. Ondish
John Allyn Ord
William Orlandi
Robert J. Ormesher
Gus G. Orphanides, PhD
Andrew A. Orr
William D. Orville, PhD
William J. Osborne
Richard H. Osman
Gerald S. Osmanski
Carl W. Ott
Michael J. Otto, PhD
Mark Owens
William E. Paff, II
Gary Page
George A. Paleos
Vincent Palmer, PhD
Richard R. Palski
John A. Pantelis
Stephen A. Pany
Charles A. Papacostas, PhD
Spiro J. Pappas
Robert E. Park, PhD
Donald Parker, PhD
Wanda J. Parker
William H. Parker
Frederick L. Parks
A. Paterno Parsi, PhD
Ralph Edward Pasceri, PhD

Robert Patarcity
Edward W. Patchell, Jr.
Mina G. Patel
Christopher J. Patitsas
Marcia L. Patrick
Richard G. Patrick
Donald Patten
R. Dean Patterson
Anton D. Paul, PhD
Timothy P. Pawlak
Charles W. Pearce, PhD
F. Gardimes Pearson, PhD
James B. Pease
Richard G. Pedersen
Donald F. Pegnataro, Jr.
Stanley J. Penkala, PhD
Charles R. Penquite
James M. Perel, PhD
Victor M. Perez
Phillip J. Persichini, PhD
R. Peters, PhD
Raymond W. Peterson
Garold W. Petrie
Howard L. Petrie, PhD
Francis X. Petrus
Herhsel L. Phares
Edward J. Phillips
Steven J. Phillips
Louis Pianetti, Jr.
George J. Piazza, PhD
Earl F. Pickel
John E. Pierce
William Schuler Pierce
Edwin Thomas Pieski, PhD
Philip A. Pilato, PhD
James W. Pillsbury
Kenneth Elmer Pinnow, PhD
Peter E. Pityk, Jr.
Michael Pizolato
Lucian B. Platt, PhD
Richard P. Platt
Andrew Walter Plonsky
Tom R. Plowright
James H. Poellot, PhD
Max E. Pohl
James M. Policelli
Frank C. Polidora
James R. Polski
Vincent J. Pongia
John T. Popp
Thomas J. Porsch
Edward S. Porter
Thomas J. Porterfield
Kenneth A. Potter, Jr.
Wayne G. Pottmeyer
Daniel B. Pourreau, PhD
Joseph J. Poveromo, PhD
Derek Powell
Dwight R. Powell
Donald L. Pratt, PhD
Frank M. Precopio, PhD
George C. Prendergast, PhD
Richard D. Prescott
Henry L. Price, PhD
Joseph W. Proske, PhD
Leonard J. Pruckner
John Pusatari
Thomas H. Putman, PhD
Albert E. Pye, PhD
Louis L. Pytlewski, PhD
Jose J. Quintero, Sr.
Sidney C. Rabin

John R. Ragase
James J. Rahn, PhD
William W. Ramage
Robert A. Ramser
Ann Randolph, PhD
Vinod E. Rao
Richard J. Ravas, PhD
Martin A. Rawhouser
Stuart Raynolds, PhD
Nelson Henry Reavey-Cantwell, PhD
Kenneth R. Reber, Jr.
John G. Redic
Issac R. Reed
O. W. Reen, PhD
Eber O. Rees, PhD
John L. Reese
Floyd K. Reeser
Claude Virgil Reich, PhD
Paul H. Reimer
John B. Reinoehl
Phyllis A. Reis
Robert G. Reis
David Reiser, PhD
M. Paul Reiter
Joseph Francis Remar, PhD
Mack Remlinger
Todd Renney
Robert Kenneth Resnik, PhD
William B. Retallick, PhD
William R. Rex
Douglass C. Reynolds
John C. Reynolds
James D. Rhoads
Gary Rhodes
Kevin Rhodes
James P. Rice
Ralph J. Richardson, PhD
Ed Richman, PhD
Tomas Richter
James V. Ricigliano
Russell Kenneth Rickert
Thomas S. Riddle
Karen A. Riley, PhD
William R. Rininger
John Roberts
William D. Roberts
Charles A. Robinson, PhD
John B. Robinson
Vincent P. Rocco
Delbert W. Rockwell
Thomas E. Rodwick
John Roede
Charles C. Rogala
Daniel E. Rogers
Robert Romancheck, PhD
John A. Romberger, PhD
Ivor C. Rorquist
Cloyd J. Rose
Kenneth L. Rose, PhD
Ralph Rose
Lawrence W. Rosendalc
Charles Thomas Rosenmayer, PhD
Arthur B. Ross
Arthur Leonard Ross, PhD
John W. Ross
Lawrence Ross
Hansjakob Rothenbacher, PhD
Robert J. Roudabush
Thomas S. Rovnak
Blake A. Rowe
Bruce F. Rowell, PhD
Robert Roy

Arthur M. Royce
Mae K. Rubin, PhD
Douglas Rudenko
Charles G. Rudershausen, PhD
Thomas W. Ruffner
Brian Mandel Rushton, PhD
Harold D. Rutenberg
Sigmond S. Rutkowski III
John P. Rutter
Frederick M. Ryan, PhD
Robert P. Rynecki
Richard E. Sacks
Nicholas Sagliano, Jr., PhD
James A. Salsigiver
Douglas H. Sampson, PhD
James B. Sandford
Vincent J. Sands
Foster C. Sankey
Gordon F. Santee
John F. Sarnicola, PhD
Daryl Sas, PhD
Sreela Sasi, PhD
Joshua Sasmor, PhD
Ronald Wayne Satz, PhD
Leonard M. Saunders, PhD
Paul R. Saunders
Gayle E. Sauselein
Theodore B. Sauselein
John J. Sayen
John B. Schaefer
Richard W. Scharf
William J. Scharle
Gary Scheeser
Howard Ansel Scheetz
Frank Schell
Frederick H. Schellenberg
Guy R. Schenker
Robert H. Schiesser, PhD
Lee Charles Schisler, PhD
George J. Schlagnhaufer
James S. Schlonski
Robert W. Schlosser
David J. Schmidle
Terry W. Schmidt
Frederick R. Schneeman
Charles F. Schneider, Jr.
M. C. Schneider, PhD
Robert L. Schnupp
Robert J. Schoenberger, PhD
Richard Alan Schofield
Giles M. Schonour
Benjamin A. Schramze
Edward C. Schrom
Steven E. Schroth
Herbert A. Schueltz
Charles H. Schultz, PhD
Earl D. Schultz
Edward M. Schulz, PhD
August H. Schwab
Carl W. Schweiger
Donna P. Scott
William J. Seaman, PhD
Paul J. Sebastian
James L. Seibert
Jack E. Seitner
Donald M. Self
James Edward Seltzer, PhD
John J. Serdy
Frederick Hamilton Sexsmith, PhD
Charles Lewis Shackelford
Alexander C. Shafer, Jr.
Brian N. Shaffer

John L. Shambaugh
Brett D. Shamory
Terrry Shannon
Cyrus J. Sharer, PhD
William David Sheasley, PhD
Thomas R. Shenenbarger
Henry A. Shenkin
Anthony M. Sherman, PhD
Malcolm H. Sherwood
Bruce M. Shields
Bruce S. Shipley
Glenn A. Shirey
John G. Shively
Joseph Shott
Scott A. Shoup
David P. Shreiner
Mark E. Shuman, PhD
Michael L. Sidor
Charles Signorino, PhD
Leonard Stanton Silbert, PhD
Dan E. Simanek, PhD
David L. Simon
Craig Simpson
Stephen G. Simpson, PhD
Samuel J. Sims, PhD
Dom C. Sisti
Lawrence H. Skromme
Gary E. Slagel
Alfred A. Slowik
Joseph Slusser
Ralph M. Smailer
Herbert L. Smith, PhD
James A. Smith
James L. Smith, Jr.
James L. Smith, PhD
K. Smith
Richard Smith, PhD
Robert L. Smith
Robert Lawrence Smith, PhD
Susan M. Smith, PhD
Barry R. Smoger
Trent Snider, PhD
Dudley C. Snyder, PhD
Robert K. Soberman, PhD
Donald W. Soderlund
Pierce S. Soffronoff
Joseph C. Sofianek
Walter C. Somski
P. Sonnet, PhD
David E. Sonon
Paul T. Spada
William C. Spangenberg
Richard F. Speaker
Paul K. Sperber
Lawrence G. Spielvogel
Joel S. Spira
David T. Springer
W. H. Springer
William G. Stahl
Randall K. Stahlman
James J. Stango
Frank X. Stanish
Edward A. Stanley, PhD
Francis J. Stanton
Walter F. Staret
Keith E. Starner
Stephen J. Stasko
Ronald Staut, PhD
Kathleen P. Steele, PhD
Bruce G. Stegman
Thomas H. Stehle
Timothy J. Stehle

Roger Arthur Steiger, PhD
Bob Stein
Michael A. Steinberg
Kenneth Brian Steinbruegge
Edwin G. Steiner
Herbert E. Steinman
Margery L. Stem
Melinda Stephens, PhD
James W. Sterenberg
Linda T. Stern
Chris G. Stewart
Glenn Alexander Stewart, PhD
Ronald H. Stigler
Richard X. Stimpfle
Henry Joseph Stinger
Richard Floyd Stinson, PhD*
Donna J. Stockton
E. F. Stockwell, Jr.
Paul A. Stoerker
Sam Stone
William C. Stone
William T. Storey
Edward J. Stoves
Carole Stowell
David H. Strome, PhD
David B. Strong, PhD
Dick Strusz
Kenneth Stuccio
William Lyon Studt, PhD
Jeffrey C. Sturm
Wen Su, PhD
John L. Suhrie
Frederick H. Suydam, PhD
Rens H. Swan
Robert W. Swartley
Robert V. Swegel
Don Sweitzer
Joseph W. Swenn
Robert A. Sylvester, Jr.
Leonard M. Syverson
Johann F. Szautner
Joseph R. Szurek
Allen C. Taffel
William R. Taft
Jim Tallon
Richard C. Tannahill
Timothy R. Tate
Tom L. Tawoda
Larry R. Taylor, PhD
Joseph George Teplick
Dixon Teter, PhD
Vassilios John Theodorides
Richard B. Thomas, PhD
Ralph N. Thompson
Ramie Herbert Thompson
Richard G. Thornhill
Ruth D. Thornton, PhD
Norman E. Tilden
Walter J. Till
Richard H. Tinsman
John Anthony Tirpak, PhD
Ronald M. Tirpak
Peter G. Titlow
Charles H. Titus
Richard H. Toland, PhD
Lazar Trachtenberg, PhD
Charles W. Tragesser
Dale B. Traupman
Charles L. Tremel, Sr.
Karl P. Trout
Richard A. Trudel
Robert C. Tucker

Natalia Turaki, PhD
Michael A. Turco
Richard Turiczek
John C. Tverberg
Somdev Tyagi, PhD
Michael J. Uhren
Alexander Ulmer
Robert Urban
William Andrew Uricchio, PhD
David H. Vahlsing
Jerome W. Valenti
Thomas J. Valiknac
Clarence Gordon Van Arman, PhD
Desiree A. J. Van Arman, PhD
Bingham H. Van Dyke
Drew R. Van Orden
Richard J. Van Pelt
Samuel W. Vance
Richard A. Vandame, Jr.
Edward F. Vanderbush
Todd P. Vandyke
David H. Vanhise
Henricus F. Vankessel
Albert Vannice, PhD
Catherine L. Vanzyl
John Vare
Gabriella Anne Varga, PhD
Francis J. Vassalluzzo
Jaroslav V. Vaverka, PhD
Paul B. Venuto, PhD
Deepak Kumar Verma, PhD
Glenn E. Vest
Edward J. Vidt
Richard J. Virshup
James A. Vitek
James F. Vogel
Jason Vohs, PhD
Philip B. Vollmer
Norman Vorchheimer, PhD
Emil J. Vyskocil
Stephen C. Wagner
William G. Wagner
Haney N. Wahba
David Allen Walker, PhD
Rex A. Waller
James C. Walter
Jeremy L. Walter, PhD
W. Timothy Walter
Robert A. Walters
John R. Warrington
Stephen Shepard Washburne, PhD
George Wasko
Andrew P. Watson
Elwood G. Watson
Jeff Watson
Lin K. Watson
Donald J. Weaver
James P. Webb
Alfred C. Webber
Cynthia E. Weber
David Weber
Gary Weber, PhD
Jan R. Weber
Oliver J. Weber
Robert C. Weber
George G. Weddell
Charles P. Weeks
Loraine Hubert Weeks
Dwight Wegman
Robert P. Wei, PhD
T. Craig Weidensaul, PhD
Eugene P. Weisser

Irving Wender, PhD
Michael G. Wendling
Kimberly A. Werner
Laurence N. Wesson
Thomas E. Weyand, PhD
Joe A. Whalen, Jr.
Jack Whelan
John P. Whiston
Chuck White
Malcolm Lunt White, PhD
James R. Whitley
Alan M. Whitman, PhD
Stanley L. Whitman
Joseph C. Wilcox
Robert A. Wilks
Jack P. Willard
Jack H. Willenbrock, PhD
David G. Willey
Bernard J. Williams
Brian Williams, PhD
Gregory S. Willis
Armin Guschel Wilson, PhD
Harry W Alton Wilson, Jr.
Lowell L. Wilson, PhD*
Thomas L. Wilson
Richard Wiltanger
Gregory A. Wimmer
Gilbert Witschard, PhD
Warren F. Witzig, PhD
David Alan Wohleber, PhD
Kenneth Joseph Wohlpart
Bradley D. Wolf
Mark S. Wolfgang
Tammy L. Wolfram
Thomas Hamil Wood, PhD
William A. Wood, PhD
William T. Wood, II
C. L. Woodbridge, PhD
Clifford A. Woodbury, PhD
Kenneth L. Woodruff
David W. Woodward, PhD
Mark J. Woodward
James H. Workley
Dexter V. Wright
Thomas Wynn, PhD
Richard F. Wyse
William C. Yager
Richard Howard Yahner, PhD
Michael M. Yanak
Howard W. Yant
Campbell C. Yates
Paul Yesconis
Thomas Lester Yohe, PhD
Charles L. Yordy
Leland L. Young
Scott G. Young, PhD
Robert F. Youngblood
Larry M. Younkin, PhD
M. Zagorski
John S. Zavacki, PhD
John Zebrowski
Arthur F. Zeglen
Richard F. Zehner
Bruce Allen Zeitlin
Paul S. Zell
James V. Zello
Michael Alan Zemaitis, PhD
Nicholas A. Zemyan
Mark L. Ziegler
William N. Zilch
Richard Eugene Zindler, PhD
Donald W. Zipse

William C. Zollars
Richard C. Zolper
Charles W. Zubritsky

Puerto Rico

Karlis Adamsons, Jr., PhD
Jaime Arbona-Fazzi, PhD
Bernard V Illars Baus, PhD
Robert Fuhrer
Francisco J. Martinez
Paul Retter, PhD

Rhode Island

Roger L. Allard
John Robert Andrade, PhD
Carl Baer, PhD
David Bainer
Geoffrey Boothroyd, PhD
Anthony Bruzzese
Bruce Caswell, PhD
Joseph K. Cherkes
Deborah M. Ciombor, PhD
Alan Cutting
Jerry Daniels, PhD
Cynthia J. Desjardins
Peter Dewhurst, PhD
David A. Dimeo
Harry F. Dizoglio
William Theodore Ellison, PhD
Earl T. Faria, Jr.
John H. Fernandes, PhD
Peter J. Freeburg
Walter F. Freiberger, PhD
Daniel S. Freitas
Russell F. Geisser
Peter Glanz, PhD
William H. Greenberg
David Steven Greer
Ram S. Gupta, PhD
William J. Hemmerle, PhD
G. Hradil, PhD
Linda Ann Hufnagel, PhD
William Howard Jaco, PhD
Philip N. Johnson, PhD
Edward Samuel Josephson, PhD
Herbert Katz
Mark B. Keene
Harris J. Kenner
Robert G. LaMontagne, PhD
Royal Laurent
Mark Lord
JoAnn D. MacMillan
Carleton A. Maine
Jean M. Marshall, PhD
Robley Knight Matthews, PhD
James P. McGuire, PhD
John P. Mycroft, PhD
John Tse Tso Ning, PhD
Stanley J. Olson
Anthony D. Palombo
Jay Pike, PhD
Allan Hubert Reed, PhD
August J. Saccoccio
Saul Bernhard Saila, PhD
John A. Shaw
Theodore Smayda, PhD
Homer P. Smith, PhD
Henry R. Sonderegger

Paul D. Spivey
Eugene F. Spring
Barbara S. Stonestreet
Benjamin Y. Sun
Paul A. Sylvia
Joseph Charle Tracy, PhD
Kenneth M. Walsh
Michael F. Wilbur
South Carolina

Daniel Otis Adams, PhD
Louis W. Adams, PhD
Gary N. Adkins
Donald R. Amett
Albert S. Anderson
Douglas J. Anderson
Gregory B. Arnold
Luther Aull, PhD
Ricardo O. Bach, PhD
Carl Williams Bailey, PhD
Frederic M. Ball
J. Austin Ball
Tom Bane
Seymour Baron, PhD
Charles William Bauknight, Jr., PhD
Norman Paul Baumann, PhD
Randy C. Baxter
Richard C. Beckert
Wallace B. Behnke, Jr.
Warrem C. Bergmann
William H. Berry
Robert G. Best, PhD
Minnerd A. Blegen
William Bleimeister
James A. Bloom
Elton T. Booth
Dwight L. Bowen
Robert Foster Bradley, PhD
Richard G. Braham
Robert E. Brown
Robert S. Brown, PhD
Jay B. Bruggeman
Carl Brunson
Philip Barnes Burt, PhD
Gregory Bussell
Jake D. Butts
Ed F. Byars, PhD
Vincent J. Byrne
Edward C. Cahaly
Michael J. Cain
Thomas L. Campisi
Howard L. Carlson
Nelson W. Carmichael
George R. Caskey, PhD
Rex L. Cassidy
Curtis A. Castell
Frederick J. Chapman
Ken A. Chatto
Thomas Clement Cheng, PhD
Emil A. Chiarito
J. D. Chism
Frank M. Cholewinski, PhD
Allen D. Clabo, PhD
Hugh Kidder Clark, PhD
Tommy Clawson
Fred R. Clayton, PhD
Paul Cloessner, PhD
Daniel Codespoti, PhD
William Weber Coffeen, PhD
Benjamin Theodore Cole, PhD
Floyd L. Combs
J. J. Connelly

Bingham M. Cool, PhD
Horace Corbett
Leon W. Corley
Andrew K. Courtney
William H. Courtney, PhD
William G. Cousins
Norman J. Cowen
Garnet Roy Craddock, PhD
Irving M. Craig
Van T. Cribb
Carl J. Croft
George T. Croft, PhD
Donald C. Cronemeyer, PhD
Peter M. Cumbie, PhD
James L. Current
Bobby L. Curry
William C. Daley
Robert F. Dalpiaz
Colgate W. Darden, PhD
William E. Dauksch
Francis K. Davis, PhD
Richard C. Davis
Murray Daw, PhD
John D. Deden
Robert W. Degenhart
Joseph V. Depalma
George T. Deschamps
Jerry B. Dickey
Charles R. Dietz
William R. Dill
Dennis R. Dinger, PhD
Thomas J. Dipressi
Louis Dischler
James H. Diven
Barry K. Dodson
Richard M. Dom, PhD
Henry Donato, PhD
John J. Donegan, PhD
Theron W. Downes, PhD
Frank E. Driggers, PhD
William W. Duke
Travis K. Durham
Martin D. Egan
Eric P. Erlanson
Laurie N. Ervin
Henry S. Famy
Pauline Farr
Charles J. Fehlig
Jeff S. Feinstein
Oren S. Fletcher
Peter J. Foley
Dennis Martin Forsythe, PhD
R. S. Frank
David E. Franklin
William R. Frazier
Donley D. Freshwater
Carlos Wayne Frick
Jerry F. Friedner
S. Funkhouser, PhD
Joseph E. Furman
Alfred J. Garrett, PhD
Alan A. Genosi
Eric Gerstenberger
George F. Gibbons
John H. Gilliland
John H. Gilmore
Paul M. Glick
Lewis W. Goldstein
Charles Gooding, PhD
Stephen H. Goodwin
Nicholas W. Gothard
Christian M. Graf

Sutton L. Graham
Michael Gray, PhD
Ted W. Gray
John T. Gressette
Edwin N. Griffin
Paul P. Griffin
Fredrick Grizmala
Peter P. Hadfalvi
Gerhard Haimberger
Christian M. Hansen
Henry F. Harling
Don Harris
Eric W. Harris
Jim L. Harrison
Grover C. Haynes
Thomas P. Henry
Walter J. Henson
Don Herriott
Arnold L. Hill
Theodore C. Hilton
Kearn H. Hinchman
John C. Hindersman
Stanley H. Hix
Robert Charles Hochel, PhD
Louis J. Hooffstetter
Michael L. Hoover
Jeffrey Hopkins
Alvin J. Hotz, Jr.
Vince Howel
Monte Lee Hyder, PhD
Robert Hiteshew Insley
Walter F. Ivanjack
Wade Jackson
Doyle J. Jaco
William John Jacober, PhD
Jim J. Jacques
S. Tonebraker Jennings, PhD
Wayne H. Jeus, PhD
Roger E. Jinkins
Jeremy P. Johnson
Edwin R. Jones, PhD
Jay H. Jones
Henry S. Jordan
John W. Kalinowski
Noel Andrew Patrick Kane-Maguise, PhD
Randy F. Keenan
Douglas E. Kennemore
James F. Knight
Jere Donald Knight, PhD
Charles F. Knobeloch
Harry E. Knox
Jacob John Koch
Lawrence A. Kolze
Norman L. Kotraba
Karl Walter Krantz, PhD
Lois B. Krause, PhD
Darryl E. Krueger
Thomas A. Krueger
Kenneth F. Krzyzaniak
Donald Gene Kubler, PhD
Jerry Roy Lambert, PhD
Carl Leaton Lane, PhD
John Lavinskas
Louis Leblanc
Burtrand Insung Lee, PhD
Duane G. Leet, PhD
Victoria Mary Leitz, PhD
W. Leonard
Lionel E. Lester
Donald F. Looney
Robert Irving Louttit, PhD

Thomas G. MacDonald
D. E. Maguire
Robert E. Malpass
Robert Arthur Malstrom, PhD
Wayne Maltry
Gregory J. Mancini, PhD
Donald G. Manly, PhD
Ronald Marks, PhD
Leif E. Maseng
George T. Matzko, PhD
James Thomas McCarter
Kevin F. McCrory
F. Joseph McCrosson, PhD
David K. McCurry
James D. McDowall
Roger D. Meadows
Tarian M. Mendes
Paul J. Miklo
John D. Miles
Nawin C. Mishra, PhD
Michael I. Mitelman
Adrian A. Molinaroli
Al Montgomery
Richard Morrison, PhD
William S. Morrow, PhD
Duane Mummett
Edward A. Munns
Julian M. Murrin
John D. Myers
Michael J. Myers
George D. Neale
Robert Nerbun, PhD
Chester Nichols, PhD
Richard F. Nickel
K. Okafor, PhD
Thomas E. O'Mara
George Franci O'Neill, PhD
Errol Glen Orebaugh, PhD
Terry L. Ortner
Joseph E. Owensby
John Michael Palms, PhD
Chanseok Park, PhD
Roy A. Pastorek
Dewey H. Pearson
David W. Peltola
Larry Pensak
Mark Perry
Edward A. Pires
Dean Pitts
Branko N. Popov, PhD
William F. Prior
William A. Quarles
Robert C. Ranew
Margene G. Ranieri, PhD
Charles H. Ratterree
Robert H. Reck
Chester Q. Reeves
James L. Reid, PhD
William J. Reid, III, PhD
George M. Rentz
Harry E. Ricker, PhD
Elizabeth S. Riley
John S. Riley, PhD
Samuel E. Roberts
Robert E. Robinson, PhD
Neil E. Rogen, PhD
Emil Roman
George Moore Rothrock, PhD
Carl Sidney Rudisill, PhD
Harvey R. Ruggles
Raymond J. Rulis
David J. Salisbury

Hubert D. Sammons
Robert M. Sandifer
Marshall C. Sasser
Frederick Charles Schaefer, PhD
Stephen Schaub
Robert C. Schnabel
James F. Scoggin, Jr., PhD
Doug P. Seif
Joe R. Selig
Taze Leonard Senn, PhD
Jerry W. Shaw
Thomas A. Sherard
Phillip A. Shipp
John E. Shippey
Jonathan R. Shong
Jack S. Sigaloff
Michael R. Simac
Edwin K. Simpson
Charles O. Skipper, PhD
Thomas E. Slusher
Glendon C. Smith
Levi Smith, III
Paul D. Smith
Robert W. Smithem
William H. Spence, PhD
Roland B. Spressart
Nesan Sriskanda, PhD
Robert G. Stabrylla
Laurie Stacey
Paul R. Staley
Edwin H. Stein, PhD
Neil I. Steinberg, PhD
Jeff Stevens
Keith C. Stevenson
G. David Stewart
William Hogue Stewart, PhD
Robert M. Stone, PhD
Roger William Strauss, PhD
Terry A. Strout, PhD
Ronald K. Stupar
John Sulkowski
Arthur B. Swett
William B. Sykes
John L. Tate
Victor Terrana, PhD
Wayne Therrell
Hugh S. Thompson
Jerry R. Tindal
Reuben R. Tipton
Laura L. Tovo, PhD
Dean B. Traxler
James S. Trowbridge
Roy R. Turner
Jerry A. Underwood
Ivan S. Valdes
Neil George VanderLinden, PhD
George C. Varner, PhD
Henry E. Vogel, PhD
Peter John Vogelberger
J. Rex Voorhees
Stanley K. Wagher
Earl K. Wallace
Michael F. Warren
Linda A. Weatherell
George M. Webb
James W. Weeg
Joel D. Welch
William B. Wells
William J. Westerman, PhD
Richard L. Westmore
Edward Wheaton
Douglas R. White

Robert L. Whitehead
Lisa C. Williams
James C. Williamson
Nigel F. Wills
Charles A. Wilson, PhD
David N. Wilson
John E. Wilson
Rick L. Wilson
Robert W. Wise
Walter P. Witherspoon, Jr.
Herbert S. Wood
Roberta L. Wood
Bruce D. Woods
John J. Woods
James M. Woodward
Walt Worsham, PhD
Franklin Arden Young, PhD
William M. Zobel

South Dakota

Jerrold Abernathy
Dale E. Arrington, PhD
B. Askildsen
Michael J. Barr
Larry G. Bauer, PhD
Jan Berkhout, PhD
Jerrold L. Brown
Lewis F. Brown, PhD
Kelly W. Bruns
Bruce W. Card
Harold E. Carda
Robert L. Corey, PhD
Cyrus William Cox
C. H. Cracauer
John T. Deniger
Norm Dewit
Raymond Donald Dillon, PhD
Clifford Dean Dybing, PhD
Jose Flores, PhD
A. Martin Gerdes, PhD
Joseph W. Gillen
Charles F. Gritzner, PhD
Thomas F. Guetzloff, PhD
Robert E. Hayes
Clark E. Hendrickson
Steve Hietpas, PhD
Samuel D. Hohn
Stanley M. Howard, PhD
Dean A. Hyde
Dennis V. Johnson
L. R. Johnson
Benjamin H. Kantack, PhD
Jon J. Kellar, PhD
Martin J. Kiousis
Clyde A. Kirkbride
Kenneth F. Klenk, PhD
Paul L. Koepsell, PhD
Robert Kohl, PhD
Dave F. Konechne
Jerrod N. Krancler
Robert J. Lacher, PhD
Richard John Landberg, PhD
Wayne K. Larsen
Jaime Aquiles Lescarboura, PhD
David R. Lingenfelter
Willard John Martin, PhD
Stewart L. Moore
Darrell M. Nielsen
Gary Novak
Ronald R. O'Connell

Thomas K. Oliver, PhD
Gary Wilson Omodt, PhD
Paul M. Orecchia
Recayi Pecen, PhD
Andrey Petukhov, PhD
Terje Preber, PhD
John Prodan
Linda E. Rabe
Perry H. Rahn, PhD
Dale Leslie Reeves, PhD
Charles P. Remund, PhD
D. Roark, PhD
Paul D. Roberts
Thomas M. Robertson
Dale M. Rognlie, PhD
Mark A. Rolfes
Michael Harris Roller, PhD
Rolland R. Rue, PhD
Frank W. Rust
Randall J. Sartorius
Anthony V. Schwan
Jayna L. Shewak
Sung Y. Shin, PhD
Arthur Lowell Slyter, PhD
Thane Smith, PhD
Leo H. Spinar, PhD
David A. Stadheim
Larry D. Stetler, PhD
Glen A. Stone, PhD
Bob Suga
Don Talsma, PhD
Chad Timmer
Merlin J. Tipton
E. Brent Turnipseed, PhD
James H. Unruh
Theodore Van Bruggen, PhD
L. L. Vigoren
Larry L. Wehrkamp
Carl A. Westby, PhD
Dana Windhorst
Thomas J. Wing

Tennessee

B. Steven Absher
Sally Absher
Anthony F. Adamo
George F. Adams
Rusty Adcock
Robert A. Ahokas, PhD
Igor Alexeff, PhD
Ernest R. Alley
Sally S. Alston
John Stevens Andrews, PhD
Nick J. Antonas
Stephen Arnold
William Archibald Arnold, PhD
Michael W. Ashberry
Larry P. Atkins
Richard Attig
Frank Averill, PhD
Drury L. Bacon
Allen J. Baker, PhD
Herbert H. Baker
Ollie H. Ballard, IV
Joseph S. Ballassai
Derek M. Balskin
John Paul Barach, PhD
Eric N. Barger
Paul Edward Barnett, PhD
A. V. Barnhill, Jr.

George H. Barrett
Bruce E. Bates
W. Z. Baumgartner
Samuel E. Beall, Jr.
Harold R. Beaver
Rogers H. Beckham
William Clarence Bedoit, PhD
Whitten R. Bell
William D. Bernardi
William Robert Bibb, PhD
Alan P. Biddle, PhD
Jerry N. Biery
Charles Biggers, PhD
Chris M. Bird
Alan R. Bishop
David L. Black, PhD
Kurt E. Blum, PhD
Charles Nelson Boehms, PhD
Glenn C. Boerke
A. Bollmeier
Walter D. Bond, PhD
Joseph N. Bosnak
Marion R. Bowman
Randal J. Braker
Charles T. Brannon
George R. Bratton, Jr., PhD
David M. Breen
Marcus N. Bressler
W. R. Briley, PhD
Frank Broome
Jim Brown
Lloyd L. Brown
Marvin K. Brown
Charles Bryson
E. Clark Buchi
Edsel Tony Bucovaz, PhD
Larry D. Buess
George Jule Buntley, PhD
Edward Walter Burke, PhD
Calvin C. Burwell
T. Bush
Rafael B. Bustamante, PhD
Joel T. Cademartori, Sr.
Edward George Caflisch, PhD
George Barret Caflisch, PhD
Steven G. Caldwell, PhD
Colin Campbell
Susan B. Campbell
Warren Elwood Campbell, PhD
Jay Caplan
James D. Carrier
James B. Carson
Randy R. Casey
Don S. Cassell
Mural F. Castleberry
Carroll W. Chambliss, PhD
James V. Christian
Warner Howard Christie, PhD
Habte G. Churnet, PhD
Charles D. Cissna
Terry A. Clark
Grady Wayne Clarke, PhD
William L. Cleland
David Matzen Close, PhD
Walter G. Cochran
Frederick L. Cole
Jack A. Coleman, PhD
Jimmie Lee Collins, PhD
George T. Condo, PhD
Gregory J. Condon
Ronald A. Conley
Nelson F. Consumo

Appendix 4: The Petition Project

David Franklin Cope, PhD
A. D. Copeland
Donald Copinger
Francis Merritt Cordell, PhD
Patrick Core
Eugene Costa
George S. Cowan
Michael A. Crabb
John Craft
Nicholas T. Crafton
Sterling R. Craig
Harvey A. Crouch
James J. Cudahy
John W. Dabbs, PhD
John Irvin Dale, III, PhD
J. Marshall Davenport
Annita Davis
Montie G. Davis, PhD
Wallace Davis, Jr., PhD
Philip M. Deatherage
James H. Deaver, PhD
W. Edward Deeds, PhD
Roy L. Dehart
Joe DeLay, PhD
Mark F. Dewitt
Albert W. Diddle
John Willard Dix
Mathilda Doorley
Nicholas M. Dorko
Wesley C. Dorothy
Robert S. Dotson
Thomas C. Dougherty
Michael A. Dudas
Peter L. Duncanson
Linwood B. Dunn
Joseph F. Edwards
Kenneth J. Eger, PhD
Kevin S. Elam
Alice Elliott, PhD
George B. Ellis
J. Michael Epps
Kenneth Eric Estes
David Richard Fagerburg, PhD
Gregory Farmer
James Rodney Field
Scot N. Field
Theodore H. Filer, PhD
Patrick Carl Fischer, PhD
Gary A. Flandro, PhD
Dennis D. Flatt
Darryl Fontaine
Bobby Forrester
Edwin P. Foster, PhD
Samuel E. French, PhD
Walter L. Frisbie
Charles S. Fullgraf, PhD
Charles P. Furney, Jr.
Nelson Fuson, PhD
John F. Fuzek, PhD
Michael L. Gamblin
Floyd Wayne Garber, PhD
Ted D. Garretson
Paul C. Gate
Dale Gedcke, PhD
Larry Gerdom, PhD
Richard L. Giannelli
Edward Hinsdale Gleason, Jr., PhD
David L. Gleaves
Virginie M. Goffaux
Clifford M. Gordon
Lee M. Gossick
Nicholas J. Gotten

John L. Grady
Peter Michael Gresshoff
Ben O. Griffith
William E. Griggs
Frederick Peter Guengerich, PhD
Kenneth H. Gum
B. K. Gupta
Donald B. Habersberger
Robert M. Hackett, PhD
Ronald D. Hackett
Werner B. Halbeck
Alvin V. Hall
Frank J. Hall, Jr.
Robert C. Hall
William H. Hall
Howard L. Hamilton
Zhihua Han, PhD
Samuel Leroy Hansard, PhD
William N. Hansard
William Otto Harms, PhD
Delbert D. Harper
Joseph J. Hasper
Stephen G. Hauser
Robert M. Hayes, PhD
James Leslie Heaberlin
Rita W. Heckrotte
Barry Henderson
Gregory W. Henry
Zachary Adolphus Henry, PhD
Rudolph W. Hensel
Paul E. Hensley
Robert B. Hern
Mark A. Herrmann
Jesse L. Hicks
Willena M. Highsmith
Lloyd L. Hiler, PhD
Michael J. Hill
Milbrey D. Hinrichs
Shean J. Hobbs
Mary E. Hohs, PhD
Kenneth L. Holberg
David H. Holcomb
Ray Walter Holland, PhD
Richard M. Holland
James N. Holliday, PhD
Kenneth A. Hollowell
Howard F. Holmes, PhD
Johnny B. Holmes, PhD
Roy E. Holmes, Jr
Moon W. Hong
Fred E. Hossler, PhD
Mark Hostetter
W. Andes Hoyt
Charlie Huddleston
James M. Hudgins
Phillip Montgomery Hudnall, PhD
B. Jerry L. Huff, PhD
Ericq Huffine
Gene E. Huffine
Ray J. Hufft
Michael R. Hunter
Jack A. Hurlbert
Randall L. Hurt
Paul Henry Hutcheson, PhD
James H. Hutchinson, PhD
Bradley J. Huttenhoff
John Anthony Hyatt, PhD
Bradley G. Hyde
Tony Indriolo
John B. Irwin
Donathan M. Ivey
Virgil E. James, PhD

King W. Jamison, PhD
Carol C. Jenkins
Gordon Jonas
Larry S. Jones
Frank E. Jordan
D. Junker, PhD
Karl Philip Kammann, Jr., PhD
Jon R. Katze, PhD
Lee Keel, PhD
Robert S. Kehrman
Shane Kelley, PhD
Minton J. Kelly, PhD
Francis T. Kenney, PhD
William Kerrigan
Elbert R. Key
Clyde E. Kidd
Vincent J. Kidd, PhD
Earl T. Kilbourne, Jr.
Anthony W. Kilroy
Keith McCoy Kimbrell
Brice W. Kinyon
Christine L. Klitzing
Ronald Lloyd Klueh, PhD
Terence E. C. Knee, PhD
Roger Lee Kroosma, PhD
Kenneth A. Kuiken, PhD
Nipha Kumar
Robert E. Laine
Paris Lee Lambdin, PhD
Scott Lane
Paul B. Langford, PhD
William D. Lansford
Andrew Lasslo, PhD
Steven Lay, PhD
John Kennon Ledbetter
Larry H. Lee
Mark P. Lee
Paul D. Lee, PhD
Carl F. Leitten, Jr.
Gary Lynn Lentz, PhD
James P. Lester
Edward Lewis, PhD
Pamela A. Lewis
R. Lewis
Russell J. Lewis, PhD
Alex Lim
Robert Simpson Livingston, PhD
Davie L. Lloyd
Bryan H. Long
Martin Shelling Longmire, PhD
Allen Loy
Martin S. Lubell
C. T. Ludwig
Alvin L. Luttrell
James E. Lyne, PhD
William Southern Lyon
Harold Lyons, PhD
P. MacDougall, PhD
Thomas M. Mahany, PhD
Arthur A. Manthey, PhD
W. Marking, PhD
Daniel E. Martin
Eric Martin
Mario Martini, PhD
Michael R. Mason
Frank W. McBride
W. E. McBrown
D. A. McCauley
Jack R. McCormick
David L. McCullev
Ted Painter McDonald, PhD
Robert McElroy

Alfred Frank McFee, PhD
Charles Robert McGhee, PhD
Jack W. McGill
Michael B. McGinn
Robert R. McGinnis
Winford R. McGowan
Mark E. McGuire
Carl McHargue, PhD
Sam L. McIllwain
Steve A. McNeely, PhD
David W. McNeil
Fred E. Meadows
Thomas Meek, PhD
Norman L. Meyer, PhD
James L. Miller, PhD
Joseph A. Miller
Truesdell C. Miller
Benjamin J. Mills
Hugh H. Mills, PhD
Allyn K. Milojevich
A. Wallace Mitchell
Standley L. Moore
John R. Morgan
Relbue Marvin Morgan, PhD
Robert L. Morgan, PhD
James R. Morris
Stephen Morrow
Gordon Moughon
James Thomas Murrell, PhD
Michael G. Muse
Robert F. Naegele, PhD
Roy A. Nance
Roland W. Nash
John Neblett
James Warwick Nehls, PhD
John Henry Neiler, PhD
Matthew E. Nemeth
Billy C. Nesbett
Marlyn Newhouse, PhD
John A. Niemeyer
Eugene Nix
Larry L. Nixon
Thomas J. Nolan
Edward David North, PhD
Scott H. Northrup, PhD
B. G. Norwood
Clinton B. Nosh, PhD
James J. Oakes
Lester C. Oakes
Ordean Silas Oen, PhD
David Q. Offutt, PhD
Dom W. Oliver
Donald R. Overdorf
Donald F. Owen
Oladayo Oyelola, PhD
Andreas M. Papas, PhD
William M. Pardue
Robert Paret
Clifford R. Parten, PhD
Arvid Pasto, PhD
James Wynford Pate
James Holton Patric
Billy Patterson
Wayne A. Pavinich
Dewitt Allen Payne, PhD
Earl Pearson, PhD
Jesse T. Perry
Raymond A. Peters
Lawrence Marc Pfeffer, PhD
Brian M. Phillipi
Bob Pinner
Jeffrey L. Pintenich

Russell Pitzer
David Martin Pond, PhD
M. G. Popik
Glenn W. Prager
William C. Presley
Lee Puckett
T. Linn Pugh
Jasmine Pui
John Joseph Quinn, PhD
David R. Raden
William T. Rainey, PhD
Harmon Hobson Ramey, PhD
Ronald D. Reasonover
Ronnie G. Reece
Larry D. Reed
Daniel P. Renfroe
David L. Rex
Joseph E. Reynolds
Mary Rhyne
Thomas E. Richardson
John R. Rickard
Tewfik E. Rizk
Kelly R. Robbins, PhD
Russel H. Robbins
Dewayne Roberts, PhD
William C. Roberts
Ben C. Rogers, PhD
Robert J. Roselli, PhD
Ross F. Russell, PhD
Nicholas Charles Russin, PhD
Edmond P. Ryan, PhD
Michael Salazar, PhD
Jacqueline B. Sales
Norman R. Saliba
L. C. Sammons, Jr.
Ann G. Samples
John Paul Sanders, PhD
Kenneth M. Sanders
Greg J. Sandlin
Michael Santella, PhD
Royce O. Sayer, PhD
Fred Martin Schell, PhD
Glenn C. Schmidt
Marcel R. Schmorak, PhD
Edward G. Schneider, PhD
John D. Schoolfield
Theodore A. Schuler
Stephen J. Schultenover
Marc G. Schurger
Herbert Andrew Scott
Todd F. Scott
Paul Bruce Selby, PhD
John R. Sellars
Kirk L. Sessions, PhD
Nelson Severinghaus, Jr.
Allan Shack
James R. Shackleford, III, PhD
Kenneth Shappely
Lydia F. Sharpe
John J. Shea, Jr.
Gerald L. Shell
Gordon Sherman, PhD
Frederick A. Sherrin
Glenn L. Shoaf, PhD
Carol Shreve
Thomas F. Shultz
Roland G. Siegfried
Priyadarshi Narain Sin, PhD
James A. Sitter
W. Keith Sloan
Nathan H. Sloane, PhD
Tim K. Slone

Glendon William Smalley, PhD
Dan R. Smith
Richard Smith
Rocky W. Smith
Robert Edwin Smylie
Alan Solomon
Reza E. Soltani
John V. Spencer
William Kenneth Stair, PhD
William P. Stallworth
George N. Starr
Richard Stauffer
William David Stewart
F. D. Sticht
Thomas C. Stillwell
Norton Duane Strommen, PhD
Victor W. Stuyvesant
James A. Sumner
Skip Swanner
Don E. Syler
Donald W. Tansil
Gordon J. Taras
Douglas E. Tate
Byron Taylor
Robert C. Taylor
Donald C. Thompson
Elton R. Thompson
John T. Thompson
Leslie Thompson
Sherle R. Thompson
Ker C. Thomson
Eyvind Thor, PhD
John G. Thweatt, PhD
John L. Tilstra
R. Lane Tippens
Norman Henry Tolk, PhD
Harmon L. Towne
Anthony Trabue
Rex Trammell
Jeffrey T. Travis
Nicholas Tsambassis
David Turner
Ella Victoria Turner, PhD
Ed Tyczkowski, PhD
James E. Vath
Robert D. Vick
Richard A. Vognild
Bruce E. Walker, PhD
Johnny K. Walker
Ralph E. Walker
Samuel J. Walley, PhD
F. John Walter, PhD
William D. Watkins
Dianne B. Watson
Marshall Tredway Watson, PhD
Michael F. Watson
Garry K. Weakley, PhD
David J. Wehrly
Marion R. Wells, PhD
Theodore Allen Welton, PhD
James J. Wert, PhD
Alan Wayne White, PhD
John L. White, PhD
Jerry M. Whiting, PhD
Alan L. Whiton
Phil P. Wilbourn
Alan Wilks
Doug Williams
Horace E. Williams, PhD
Robert L. Williams
Giles W. Willis, III
Al Wilson

James E. Wilson
Craig Wise
Cyrus Wymer Wiser, PhD
Robert C. Wolfe
John Lewis Wood, PhD
Robert Wood
B. D. Wyse, PhD
Steve Wyse
Stephen L. Wythe, PhD
Ying C. Yee, PhD
Craig S. Young
Curt Zimmerman

Texas

Wyatt E. Abbitt, III
Eugene Abbott
Albert S. Abdullah
Alan E. Abel
Marshall W. Abernathy
Grady L. Ables
John W. Achee, Sr.
Gene L. Ackerman
Donald O. Acrey
Audrey W. Adams
Daniel B. Adams, Jr.
Gerald J. Adams, PhD
Gregory A. Adams
James D. Adams
Jim D. Adams
John Edgar Adams, PhD
Kent A. Adams
Lee A. Adams, Jr.
N. Adams
Roy B. Adams
Steve W. Adams
William D. Adams
Wilton T. Adams, PhD
George Adcock
Robert E. Adcock
Marshall B. Addison, PhD
Michael F. Adkins
Wilder Adkins
Perry Lee Adkisson, PhD
Steve E. Aeschbach
Frederick A. Agdern
David Agerton, PhD
Nathan Agnew
Mark Ahlert
Thane Akins
Kenneth O. Albers
Robert M. Albrecht
Dennis J. Alexander
Rex Alford
Robert L. Alford
William David Alldredge, Jr.
David J. Allen, PhD
James L. Allen, PhD
John L. Allen
Randall Allen
Stewart J. Allen
Thomas Hunter Allen, PhD
Randall W. Allison
Terry G. Allison
George J. Allman
Anthony H. Almond
Jorge L. Alonso
Ali Yulmaz Alper
John Henry Alsop, PhD
George A. Alther
Larry W. Altman

Vern J. Always
James I. Alyea
Bonnie B. Amos, PhD
James P. Amy
Kenneth L. Ancell
C. M. Anderson, Jr.
David O. Anderson, PhD
Fred G. Anderson
Greg J. Anderson
Keith R. Anderson
Larry Anderson, PhD
Mark Anderson
P. Jennings Anderson
Reece B. Anderson
Richard C. Anderson
Gilbert M. Andreen
Douglas R. Andress
T. Angelosaute
Robert H. Angevine
Elizabeth Y. Anthony, PhD
John K. Applegath
Harry D. Arber
Leon M. Arceneaux
William Ard
Christopher Arend
John W. Argue
Bradley Armentrout
Robert L. Arms
Glenn M. Armstrong
James E. Armstrong
Lowell Todd Armstrong
David Arnold
Edwin L. Arnold
Herbert K. Arnold
Lester C. Arnwine
Charles H. Asbill
Maynard B. Ashley
Wayne A. Ashley
Monroe Ashworth
Robert A. Ashworth
Robert S. Ashworth
Curtis L. Atchley
James Athanasion
Arthur C. Atkins
D. O. Atkinson
Stanley L. Atnipp
William R. Aufricht
Richard Aurisano, PhD
Joeseph D. Aurizio
Brian E. Ausburn
Harold T. Austin
Ward H. Austin
Jon R. Averhoff
Nathan M. Avery
William P. Aycock
Robert C. Ayers, Jr., PhD
Bill E. Babyak
Bruce E. Bachman
Patrick J. Back
J. Robert Bacon
David Badal
Tanwir A. Badar
T. Dale Badgwell
Jay K. Baggs
Vincent P. Baglioni
Georgw C. Bagnall
R. A. Baile
C. Bailey
Dane E. Bailey
John A. Bailleu
Robert M. Bailliet
Gerald S. Baker

Gilbert Baker
Howard T. Baker
Lee Edward Baker, PhD
Roger G. Baker
Francis J. Balash
Brent P. Balcer
Edgar E. Baldridge
Steven J. Baldwin
Amir Balfas
P. Balis
Craig Balistrire
Jerry C. Ball
Harold N. Ballard
Ashok M. Balsaver
Jerome E. Banasik, Sr.
John J. Banchetti
Robert D. Bankhead
Attila D. Banki
James Noel Baptist, PhD
Bill Barbee
Anslem H. Barber, Jr.
Paul Barber
Robert A. Bardo
O'Gene W. Barkemeyer
Eli F. Barker
Theodore S. Barker
William M. Barlow
Alex Barlowen
Allen L. Barnes, PhD
Mike Barnett
William J. Barnett
Linard T. Baron
Clem A. Barrere, PhD
Christopher M. Barrett
Damian G. Barrett
Timothy M. Barrett
Leland L. Barrington
Allen C. Barron
Oscar N. Barron
Thomas D. Barrow, PhD
Paul W. Barrows, PhD
John Bartel, PhD
B. Bartley, PhD
Hugh B. Barton
Jerry G. Bartos
William P. Bartow
William L. Basham, PhD
George M. Baskin
Andy H. Batey, PhD
Dara Batki
Stuart L. Battarbee
Jack L. Battle
Mark S. Bauer
Frederick C. Bauhof
Bernard D. Bauman, PhD
Max Baumeister
John E. Baures
Steve Bayless
Robert Baylis
Stephen L. Baylock
Melvin A. Bayne
Charles Beach
Clifton W. Beach
Dwight Beach, Jr.
Bobby Joe Beakley
Bobby D. Beall
James F. Beall
Paula Thornton Beall, PhD
Terry W. Beall
James B. Beard, PhD
Thomas L. Beard
Reginald H. Bearsley

David C. Beaty
Edward G. Beauchamp
Weldon H. Beauchamp, PhD
Curt B. Beck
Robert Beck
William B. Beck
Bill Becker
Thomas G. Becnel
Rudolph J. Bednarz
Kenneth E. Beeney
Howard Dale Beggs, PhD
Francis Joseph Behal, PhD
Anthony J. Beilman
Deanna K. Belanger
Richard J. Belanger
Howard F. Bell
Jeffrey Bell
Thomas H. Belter
Randy L. Bena
Gilbert G. Bending
Robert G. Bening
Robert I. Benner
Alan Bennett
Judith Bennett
R. S. Bennett
William F. Bennett, PhD
Kimberly R. Bennetts
Faycelo L. Bensaid
Fred C. Benson, PhD
Harold E. Benson
Robert Lloyd Benson
Kenneth E. Benton
Leonidas A. Berdugo
Brian Berger
T. F. Berger
C. B. Bergin
Eric N. Berkhimer
David T. Berlin
Jerry D. Berlin, PhD
Franklin Sogandares Bernal, PhD
Charles A. Berry
John R. Berryhill, PhD
James E. Berryman
Allen J. Bertagne
Robert G. Bertagne
John T. Bertva
Robert K. Bethel
G. W. Bettge
Austin Wortham Betts*
Richard O. Beyea
Swapan K. Bhattacharlee, PhD
John M. Biancardi
Peter W. Bickers
Lewis J. Bicking, Jr.
Edward R. Biehl, PhD
Donald N. Bigbie
Tim Bigham
Dean M. Bilden
Wallace David Billingsley
Charles R. Bills
Gene R. Birdwell
John R. Birdwell
Craig E. Bisceglia
James Merlin Bisett
Amin Bishara
Joe O. Bishop
Richard E. Bishop, PhD
Richard H. Bishop, PhD
Robert Bittle, PhD
Benny Bixenman
Sidney C. Bjorlie
Dean Black, PhD

Dennis Black
Sammy M. Black
Victoria V. Black
William L. Black
Brian D. Blackburn
Randolph Russell Blackburn
James C. Blackmon
Steve Blaglock
Bruce A. Blake
William M. Bland, Jr.
Linda D. Blankenship
Scott C. Blanton, PhD
George Albert Blay, PhD
G. Blaylock
W. N. Bledsoe
Donald J. Blickwede, PhD
Roger P. Bligh
Tedde R. Blunck
Don A. Boatman
Lawrence T. Boatman
Joshua T. Boatwright
Robert S. Bobbitt
Thomas C. Boberg, PhD
Stephen J. Bodnar, PhD
Hollis Boehme, PhD
Jack E. Boers, PhD
David H. Boes
Johnny F. Boggs
Leslie K. Bogle
Kevin Michael Bohacs, PhD
Mark Bohm
Jack C. Bokros, PhD
Regnald A. Boles
Robert B. Boley, PhD
Charles Bollfrass
Frank R. Bollig
Gerald J. Bologna
John Joseph Boltralik
Sam T. Boltz, Jr.
Patrick L. Bond
Arnold B. Booker
R. J. Boomer
Paul M. Boonen, PhD
Lawrence E. Boos, Jr., PhD
Edward J. Booth
James H. Bordelon
Jerry Borges
J. Borin
Annette H. Borkowski
Sebastian R. Borrello
Randal S. Bose
David Boss
Robert Bossung
George M. Boswell
Kathleen B. Bottroff
Harry Elmo Bovay, Jr.
Roger L. Bowers
Randy Bowie
Christopher Bowland
David L. Bowler
Gary R. Bowles
Lamar D. Bowles
John T. Boyce
Jimmy W. Boyd
Phillip A. Boyd, PhD
Robert Boyd, III
Brad B. Boyer
Robert Ernst Boyer, PhD
Sherry L. Brackeen
Brett K. Bracken
James C. Brackmor
Vincent H. Bradley

Charles M. Bradshaw
Bruce B. Brand
Richard Brand
Stanley George Brandenberger, PhD
Robert E. Brandt
Thomas Brandt
Michael S. Brannan
Raymond E. Brannan, PhD
Ross E. Brannian
Glenn S. Brant
William H. Branum
James E. Brasher
Jeffrey M. Braun
Bruce G. Bray, PhD
Carl W. Bray
James F. Brayton
William Breach
Louis Breaux, PhD
Theodore M. Breaux
Jimmy L. Breazeale
Kenneth J. Breazeale
Bob Breeze
Martin L. Bregman, PhD
Berryman M. Breining
Harry L. Brendgen
John H. Bress
Herbert L. Brewer
J. R. Brewer
James H. Bridges
Claudia Briell
Arthur R. Briggs
W. Briggs, PhD
Robert A. Brimmer
Charles D. Briner
Niz Brissette
Earl Bristow
Michael W. Britton
Morris L. Britton
Ronald A. Britton
Paul R. Brochu
H. Kent Brock
Howard M. Brock
James P. Brock
Russell G. Broecklemann
Harry J. Broiscoe
Phiilip F. Bronowitz
John D. Bronson
Mark A. Bronston, PhD
Britt E. Brooks
James Elwood Brooks, PhD
Royce G. Brooks
John R. Brose
John W. Brosnahan
Thomas S. Brough
Jack F. Browder, Jr.
Rick A. Brower
Brenda E. Brown
Byron L. Brown
C. Douglas Brown
Charles R. Brown
Glenn Lamar Brown, PhD
Jim Brown
John T. Brown
Larry P. Brown, PhD
Leonard F. Brown, Jr., PhD
Murray Allison Brown, PhD
Robert G. Brown, PhD
Scott A. Brown
Steven Brown
Jimmy D. Browning
Thomas Broyles
David L. Bruce

George H. Bruce
Paul L. Bruce
Sandra Bruce
Herman M. Bruechner
Robert L. Brueck
Charles Brueggerhoff
Larry W. Bruestle
Tim D. Brumit
Lowell D. Brumley
Harrison T. Brundage
Allen Brune
Scott R. Bryan
John M. Bryant
Michael David Bryant, PhD
Thomas L. Bryant
Charles B. Bucek
Chris R. Buchwald
Charles G. Buckingham
Ellis P. Bucklen
Jack B. Buckley
J. Fred Bucy, Jr., PhD
Roy S. Bucy
Travis L. Budlong
Kenneth D. Bulin
Rex G. Bullock
Richard D. Bullock
John M. Bunch
William H. Bunch, Jr.
Douglas A. Buol
James Burckhard
Roy A. Burgess
David T. Burke
David S. Burkhalter
James F. Burkholder
Jeffrey C. Burkman
Ralph D. Burks
D. Burleson, PhD
Ned K. Burleson, PhD
Thomas M. Burnett, PhD
James E. Burnham
Robert B. Burnham
Robert W. Burnop
John D. Burns
Stanley S. Burns
William P. Burpeau, Jr.
W. F. Burroughs
Arthur B. Busbey, PhD
Donald L. Buscarello
Harry H. Bush, Jr.
Russel L. Bush
Houston D. Butcher
Arthur P. Buthod
Dennis L. Butler
Don W. Butler
O. Doyle Butler
Bruce L. Butterfield
Charles J. Butterick
Joe W. Button
Byron R. Byars, Jr.
Nelson Byman
Richard Dowell Byrd, PhD
Fred V. Byther
Edwin D. Cable
C. Cadenhead, PhD
Stephen A. Cady
James E. Caffey, PhD
Harry J. Cain, III
Gregory S. Caine
James B. Caldwell, PhD
Brian S. Calhoun
Tom Calhoun
Ray L. Calkins, PhD

Michael J. Callaghan, PhD
Mike Callahan
Mickey J. Callanan
Lester H. Callaway
David D. Calvert
Richardo E. Calvo, PhD
Lee Cambre
C. David Campbell
Harvey A. Campbell
Lynn D. Campbell
R. L. Campbell
Steve K. Campbell
Richard J. Cancemi
Joe A. Cantlon
Hengshu Cao, PhD
Juan J. Capello
Ronald W. Capps, PhD
David V. Cardner, PhD
Philip H. Carlisle
David R. Carlson
Lawrence O. Carlson
Roy F. Carlson
Thomas Carlson
David C. Carlyle, PhD
Walter J. Carmoney
B. Ronadl Carnes
Peyton S. Carnes, Jr.
Spencer Carnes
David L. Carpenter
Leo C. Carr
Brent Carruth
Charles D. Carson, PhD
E. Forrest Carter
Greg Carter
Aubrey Lee Cartwright, Jr., PhD
Ralph V. Caruth
Antone V. Carvalho
Audra B. Cary
Eddie Case
Jeffrey J. Casey
John R. Cassata
Patrick Cassidy, PhD
Ralph J. Castille
Joseph L. Castillo
Ron Castleton
Raphael A. Castro
Xiaolong Cat, PhD
Ken D. Caughron
James C. Causey
P. G. Cavazos
Bob C. Cavender
Arthur Vorce Chadwick, PhD
John Chadwick
Barry Chamberlain
William S. Chambless
Doyle Chamgers, PhD
Michael M. Chan
Clarence R. Chandler
Tyne-Hsien Chang, PhD
David P. Chapman
Bert R. Charan
William Charowhas
Vanieca L. Charvat
Thomas J. Chastant
Ashok K. Chatterjee
Lynn Chcoran
R. S. Cheaney
Curtis A. Cheatham
Paul S. Check
Robert F. Chesnik
William J. Chewning
Russell Chianelli, PhD

Thomas W. Childers
James H. Childres
James N. Childs
William M. Chop, Jr.
Ceasar C. Chopp
Peter S. Chrapliwy, PhD
Don Christensen
Ron V. Christensen
Allen L. Christenson
David R. Christiansen
Wiley Christie, Jr.
James P. Chudleigh
J. Joseph Ciavarra, Jr.
Charles J. Cilfone
Atlam M. Citzler, Jr.
Edwin J. Claassen, Jr., PhD
Gary D. Clack
John D. Clader, PhD
Gene W. Clark
J. Donald Clark
James B. Clark, PhD
Randall D. Clark
Richard T. Clark
Robert Clark
Ewell A. Clarke
Calvin Class, PhD
Nick W. Classen
Don Clauson
David B. Clayton, PhD
Arthur M. Clendenin
W. J. Clift
Fred W. Clinard
James T. Cline
Robert E. Cliver
Aaron Close
Kelton W. Cloud
Thomas F. Clough
Mike Clumper
Thomas John Clunie, PhD
Howard L. Cobb
Jon F. Cobb
William Cobb
John H. Cochrane
Jerry R. Cockerham
William H. Cockerham
Joseph W. Coddou
Charles A. Cody
Curtis Coe
Charles R. Cofer
William R. Coffelt
James D. Coffman
John R. Cogdell, PhD
Clayton Coker
Glenn P. Coker
Ralph Coker
Tom B. Coker
Joseph F. Colangelo
W. B. Colburn
Douglas E. Cole
Frank W. Cole
John F. Cole, PhD
Forrest Donald Colegrove, PhD
Bobby Coleman
Richard A. Coleman
Woodrow W. Coleman
Don S. Collida
Charles A. Collins
Glynn C. Collins
Stephen L. Collins
Ted Collins, Jr.
William P. Collins
Bruce G. Collippns, Jr.

Nicholas E. Combs
John S. Comeaux
William T. Comiskey
Hanson Cone
Carter B. Conlin
Jack D. Connally
H. E. Connell, Jr.
Ralph L. Conrad
Jesus Constante
Ann S. Cook
Arthur B. Cook
Douglas H. Cook
Preston K. Cook
Steven C. Cook
Wendell C. Cook
Edward F. Cooke
Willis R. Cooke
James M. Cooksey
Daniel F. Cooley
Denton Arthur Cooley
Bruce Cooper
Gordon Cooper, PhD
John C. Cooper
Thomas B. Coopwood
Kevin Copeland
M. Yavuz Corapcioglu, PhD
Jim Corbit
David L. Corder
Eugene H. Core
Wayland Corgill
Henry E. Corke, PhD
Jimmie A. Corley
Robert L. Cornelius
Holley M. Cornette
John Corrigan, III
Suzanne L. Corrigan
Kenneth M. Cory
J. Paul Costa
Alan T. Costello
Mark J. Costello
Steven A. Costello
Richard O. Cottie
William R. Cotton, PhD*
Marcus L. Countiss
Gary R. Countryman
Galen L. Coupe
Lonzo C. Courtney, Jr.
Chris E. Covert
Kenneth W. Covey
Daniel Francis Cowan
Tracy Cowan
Robert S. Cowperthwait
Alice D. Cox
Bruce W. Cox
Dan M. Cox
David L. Cox
Hiram M. Cox
James A. Cox
Jerry D. Cox
Leon W. Cox
Edward Jethro Cragoe, Jr., PhD
John A. Craig
C. Ryan Crain
Glenn D. Crain
Richard D. Cramer
John R. Crandall
David C. Crane, PhD
James F. Cravens
Doyle Craver
Ken Craver
Lionel W. Craver, PhD
Russell S. Cravey

Carter D. Crawford
Don L. Crawford
Duane Austin Crawford
James D. Crawford
Ron Creamer
Wendell R. Creech
Prentice G. Creel
T. C. Creese
Sondra L. Creighton
Harold W. Criswell
J. L. Crittenden
Dan W. Crofts
Thomas J. Crosier
Don W. Cross
George A. Cross
Bill B. Crow
Morgan L. Crow
Carroll M. Crull
Harry A. Crumbling
E. L. Crump
Eligio D. Cruz, Jr.
Tihamer Zoltan Csaky
Donald Cudmore
Samuel F. Culberson
A. S. Cullick, PhD
Reid M. Cuming
Tom Cundiff
Debra Cunningham
R. Walter Cunningham, PhD
Peter A. Curka
Clarence L. Curl
Gene Curry
Timmy F. Curry
Rankin A. Curtis
Stuart C. Curtis
Charles E. Cusack, Jr.
Herman C. Custard, PhD
Hugo C. da Silva, PhD
Calvin Daetwyler, PhD
Harry Martin Dahl, PhD
Keith F. Dahnke, PhD
Peter C. Daigle
Joseph W. Dalley, PhD
Jesse Leroy Dally, PhD
John J. Dalnoky
Charles Dalton, PhD
James U. Daly
Jerome Samuel Danburg, PhD
Richard M. Dangelo
Adam S. Daniec
Scott Daniel
Stephen R. Daniel
Ned R. Daniels
William R. Dannels
Ronald Darby, PhD
David H. Darling
Russ C. Darr
Joseph E. Darsey
Hriday K. Das, PhD
George F. Davenport
Gregory W. Daves
Cecil W. Davidson
Thomas Davidson
Emlyn B. Davies, PhD
Alfred Davis
D. K. Davis
Donald J. Davis
Frances M. Davis, PhD
Jay F. Davis
Joseph R. Davis, PhD
Michael J. Davis
Theodore R. Davis

Wendell Davis, PhD
H. R. Dawson, PhD
Mike Dawson
Wyatt W. Dawson, Jr.
Raul J. De Los Reyes
Frederik Willem De Wette, PhD
Don A. Dean
Robert A. Dean
William D. Dean
Roy F. Dearmore
Michael L. Deason
Bobby Charles Deaton, PhD
Charles Deboisblanc
Francis E. Debons, PhD
Howard E. Decker
David G. Deeken
Philip E. Deering
John M. Dees
John A. Deffner
William L. Deginder
Charles F. Deiterich
Phillip T. DeLassus, PhD
Kirk F. Delaune
Nicholas Delillo, PhD
Mark T. Delinger
Lyman D. Demand
J. Demarest, PhD
David C. DeMartini, PhD
Scott C. Denison, Jr.
David R. Denley, PhD
Richard G. Denney
Richard S. Dennis
Al M. Denson
James D. Denton
Newton B. Derby
Edward B. Derry
Peter A. Desantis
Rodney F. Deschamps
Cheryl E. Desforges
John J. Deshazo
John W. Devine
Aaron R. Dewispelare, PhD
Jim Dews
Lloyd Dhren
Paul J. Dial
John K. Dibitz
Marvin R. Dietel
Ronald L. Diggs
Robert Garling Dillard
Roger McCormick Dille
Howard Dingman
R. W. Dirks
James P. Disiena
Albert K. Dittmer
Charles J. Diver
Howard E. Dixon
Richard C. Doane
Gerard R. Dobson, PhD
Earl S. Doderer, PhD
Charles Fremont Dodge, PhD
Stephen G. Dodwell
Matthew A. Doffer
Harold H. Doiron, PhD
Michael J. Doiron
Ronald U. Dolfi
W. W. Dollison
Calvin W. Donaghey
Bob L. Donald
Allen Donaldson
James E. Donham
George Donnelly
Billie G. Dopstauf

James H. Dorman
John B. Dorsey
Floyd Doughty
Ralph D. Doughty
Charles E. Douglas, PhD
Arthur Constant Doumas, PhD
Gary L. Douthitt
Calhoun Dove
William Louis Dowdy
James D. Dowell
Jack D. Downing
Terrell Downing
Stanford L. Downs
Tom D. Downs, PhD
B. J. Doyle
Charles W. Drake
Rex A. Drake
George L. Drenner, Jr.
Carl S. Droste, PhD
William C. Drow
John A. Drozd
A. J. Druce, Jr.
Del Rose M. Dubbs, PhD
Louis H. Dubois
Thomas Dudley
Roy L. Dudman
Robert J. Duenckel
Taylor Duke
Thomas L. Dumler
Dean D. Duncan, PhD
Raynor Duncombe, PhD
James G. Dunkelberg
Cleo C. Dunlap
Henry Francis Dunlap, PhD
Johnny L. Dunlap
R. C. Dunlap, Jr.
Roy L. Dunlap
Dale E. Dunn
Francis P. Dunn
James F. Dunn
John Dale Dunn
Neil M. Dunn
Wade H. Dunn, PhD
Nora E. Dunnell
James D. Duppstadt
Dermot J. Durcan
Jack D. Duren
Ray R. Durrett
Allen L. Dutt
Chizuko M. Dutta, PhD
Granville Dutton
Julie A. Duty, PhD
Neil T. DuVernay
Roger L. Duyne, PhD
Isaac Dvoretzky, PhD
Lawrence D. Dyer, PhD
Bertram E. Eakin, PhD
R. C. Earlougher, Jr., PhD
Richard L. Easterwood
James H. Eaton
John R. Eaton
H. Eberspacher
Stanley R. Eckert
Dominic Gardiner Bowlin Edelen, PhD
Richard Carl Eden, PhD
John M. Edgington
Peter J. Edquist
Leo T. Effenberger
Darryl J. Egbert
James P. Egger
Christine Ehlig-Economides, PhD

Steven M. Ehlinger
C. D. Ehrhardt, Jr.
Paul W. Ehrhardt
Peter B. Eichelberger
Peter M. Eick
Dean L. Eiland
Jack Gordon Elam, PhD
Bill F. Eldridge
Jack G. Elen, PhD
Lloyd E. Elkins, Sr.
Joseph A. Ellerbrock
David E. Ellermann
Douglas G. Elliot, PhD
John G. Elliott, Jr.
Mark H. Elliott
Rodger L. Elliott
Edison M. Ellis
James L. Ellis
Robert L. Ellis
Walter D. Ellis
Grover C. Ellisor
Kevin L. Elm
Andy Elms
Sandy Elms
Howard P. Elton
Gary M. Emanuel
Barry E. Engel
James J. Engelmann
Newton England, Jr.
Donald D. Engle*
Mike S. Engle
Michael Engler, PhD
Kenneth C. English
Gilbert K. Eppich
Jay M. Eppink
Russ Eppright, PhD
John F. Erdmann
Alvin J. Erickson, Jr.
Roger Erickstad
Martin J. Erne
James Lorenzo Erskine, PhD
Don R. Erwin
Brenda J. Eskelson
E. Esparza
Jimmie L. Estill
Beverly A. Evans
Charles R. Evans
Claudia T. Evans, PhD
James A. Evans
Kenneth R. Evans
Steven Evans, PhD
Harmon Edwin Eveland, PhD
Thomas M. Even
Keith A. Everett
Robert H. Everett
Bryan J. Ewing
Richard E. Eyler
William A. Fader
Dennis E. Fagerstone
K. Marshall Fagin
James R. Fair, PhD
Jack Fairchild, PhD
Ken Paul Fairchild
Leonard S. Falsone
Daniel J. Faltermeier
Billy Don Fanbien, PhD
Robert S. Fant
Timothy Farage
Billy L. Farmer
Jim Farr
Charles P.O. Farrell, PhD
Elnora A. Farrell

P. A. Farrell
Billie R. Farris
David Faulkinberry
Claude Marie Faust, PhD*
C. Featherston
S. Matthew Feil
Hank Feldstein, PhD
Thomas Felkai
Steve Fenderson
Edward L. Fennell
Felix West Fenter, PhD
Dave D. Ferguson
Keith Ferguson
Dino J. Ferralli
John M. Ferrell
Robert J. Ferry, PhD
Thomas H. Fett
Mark E. Fey
Carl L. Fick
Bruce Ficken
Kenneth J. Fiedler
Douglas B. Field
Paul L. Figel
Tom Filesi
Marion F. Filippone
Reggie Finch
Robert D. Finch, PhD
Donald F. Fincher
William E. Findley
Jerry Finkelstein
Robert E. Finken
John C. Finneran, Jr.
Roger W. Fish
James Fisher
John D. Fisk
Glenn D. Fisseler
Travis G. Fitts, Jr.
Gregory N. Fitzgerald
Mark R. Fitzgerald
Michael J. Flanigan
Adrian Ede Flatt, PhD
C. D. Flatt
William T. Flis
John A. Flores
Dagne Lu Florine, PhD
Daniel Fort Flowers, PhD
Joseph Calvin Floyd, PhD
Monroe H. Floyd
Lowell R. Flud
David A. Flusche
George E. Fodor, PhD
Gerald W. Foess, PhD
James L. Folcik
Aileen M. Foley
Charles T. Folsom
Marc F. Fontaine, PhD
Jack G. Foote
R. S. Foote
Phillip Forbes
Donald P. Ford
George H. Ford
Kenneth B. Ford
Roger G. Ford, PhD
John Wiley Forsyth, PhD*
A. Gerald Foster, PhD
Charles R. Foster
Donald M. Foster, PhD
J. S. Foster
Randall R. Foster
Walter E. Foster, PhD
Doyle F. Fouquet
James L. Fowler

John Greg Fowler
Leon L. Fowler
Louis H. Fowler
Mike Fowler
Donald W. Fox
Michael S. Francisco
Anne R. Frank
Donald A. Frank
Milton C. Franke
Homer Franklin, Jr.
Edwin R. Franks
Bruce Frantz
Warren L. Franz, PhD
Frank E. Frawley, Jr.
Charles W. Frazell
Marshall Everett Frazer, PhD
Don W. Frazier
Richard R. Frazier
Robert S. Frederick, PhD
Chris Frederickson, PhD
Jack C. Freeman
Johnny A. Freeman
Burlin E. Freeze*
Emil J. Freireich
Stephen M. Fremgen
Louisa J. Fremming
Kenneth A. French, PhD
William S. French, PhD
David C. Fresch
Marion T. Friday
Gerald E. Fritts
Charles W. Frobese
Charles W. Fromen
U. George Frondorf
John E. Frost, PhD
Wade J. Frost
David H. Fruhling
William K. Fry
Rodney G. Fuchs
Charles L. Fuller
Thomas R. Fuller
Terry L. Furgiuele
Dionel Fuselier
Morris L. Gabel
William A. Gabig
Randall Gabrel
Leo H. Gabro
Dean E. Gaddy
Robert A. Gahl
Will Gaines
Douglas W. Gaither
Joseph W. Galate
R. Gale, PhD
Chrisitna A. Galindo
Marcus J. Galvan
Gary L. Galyardt
B. M. Gamble
Harvey M. Gandy
S. Paul Garber
Gerard G. Garcia
Hector D. Garcia, PhD
Rafael Garcia, PhD
Thomas L. Gardner
Tony G. Gardner
D. Garett
Fred S. Garia
Norman E. Garner, PhD
R. Dale Garner
Claude H. Garrett
Frederick Garver, Jr.
Robert J. Gary
Ron Gasser

Carolyn J. Gaston, PhD
Stephen L. Gates
Thea B. Gates
Anthony Roger Gatti, PhD
John Gatti
James R. Gattis
Herbert Y. Gayle
Al Gaylord
Michael J. Gaynor
Bob J. Gebert
William E. Gee
William F. Geisler
R. Genge
Ronald L. Genter
Joseph C. Gentry
Charles J. George
Robert E. Gerald
Thomas G. Gerding, PhD
George S. Gerlach
Richard J. Geshay
Richard G. Ghiselin
Gordon A. Gibbs
Bobby W. Gibson
Daniel M. Gibson, PhD
George R. Gibson, PhD
Lee B. Gibson, PhD
Byron J. Gierhart
Donald C. Gifford
Hugh W. Gifford
Joel Gilbert
W. Allen Gilchrist, Jr., PhD
William H. Gilchrist
John Giles
Amarjit S. Gill
Jimmy D. Gillard
Robert Gillespie, PhD
Trauvis H. Gillham
Clarence F. Gilmore
Merrill Stuart Ginsburg, PhD
Walter Glasgow
Bryan Glass
Robert W. Gleeson
Mark E. Glover, PhD
Robert R. Glovna
Earnest Frederick Gloyna, PhD
John M. Glynn
Barry G. Goar
Forrest Gober
Ferd S. Godbold, III
Charles B. Godfrey
Frederick W. Goff
William C. Goins
Larry O. Goldbeck
Michael H. Golden
Charles Goldenzopf, PhD
Fred L. Goldsberry, PhD
R. K. Golemon
Ruth Gonzalez, PhD
Richard L. Good
John Bannister Goodenou, PhD
Loy B. Goodheart
Charles Thomas Goodhue, PhD
Kent Goodloe
Billy J. Goodrich, Jr.
William A. Goodrich, Jr.
Korwin J. Goodwin
Philip W. Goodwin, Jr.
Jack Ray Goodwyn, PhD
Terry L. Goosey
Edward F. Gordon
Scott D. Gordon
D. Gorham, PhD

Cheryl Goris
Waldemar Gorski, PhD
Paul L. Gorsuch
Gary R. Gosdin
Darrell L. Goss
Nicolas Goutchkoff
Robert D. Grace
G. Robert Graf
Bill D. Graham
Charles Richard Graham
Harold P. Graham, PhD
Irwin Patton Graham
Jon Graham
Marie M. Graham
Seldon B. Graham, Jr.
Richard G. Grammens
Curtis E. Granberry
Milton J. Grant, II
R. R. Grant
William F. Grauten
Charles B. Graves
Gregory K. Graves
Leonard M. Graves
Gordon H. Gray
Joe C. Gray
William W. Gray
Gerald T. Greak
Michael G. Grecco
David F. Greeley
Andrew Green
Bobby L. Green
Hubert G. Green
John W. Green
Milton Green
Taylor C. Green
Tom F. Green
Donald M. Greene, PhD
Vance T. Greene, Jr.
John F. Greenip, Jr.
Donald R. Greenlee
Jack D. Greenwade
Howard E. Greenwell
B. Marcum Greenwood
Gary C. Greer
Ralph D. Greer
William A. Gregorcyk
Ted M. Gregson
David M. Gresko
Francis J. Greytok
Doreen V. Grieve
B. W. Griffith, Jr.
M. Shan Griffith
Paul G. Griffith, PhD
Robert W. Griffith
Edward T. Grimes
Jim T. Grinnan
Ricahrd T. Grinstead
Larry Grise
Strother Grisham
Donald L. Griswold, Jr.
Norman C. Griswold, PhD
Fred R. Grote
Barney Groten, PhD
William S. Groves
S. D. Grubb
Frank C. Gruszynski
Harold Julian Gryting, PhD
Johnnie F. Guelker
Gary Guerrieri
Charles G. Guffey, PhD
William R. Guffey
Eugene P. Gugel

Stephen N. Guillot
David C. Guinn
J. Arvis Gully, PhD
Mark T. Gully
Alberto F. Gutierrez
Jeng Yih Guu, PhD
Necip Guven, PhD
Frank W. Guy, PhD
Joe M. Haas
Merrill Wilber Haas
James M. Hackedorn
Larry G. Hada
Chuck H. Hadley
Frederick R. Hafner
Robert Hage
Cecil W. Hagen
Robert O. Hagerty, PhD
Ismail B. Haggag, PhD
Gerow Richard Hagstrom, PhD
Wayne A. Hahne
Anthony Haines
Stephen W. Haines
Delilah B. Hainey
William D. Hake
Michel Thomas Halbouty, PhD
Gary D. Halepeska
R. A. Haley
Richard L. Haley, PhD
Billy R. Hall
Douglas Lee Hall, PhD
Gary R. Hall
George Hall
John Hall
Kenneth R. Hall, PhD
R. W. Hall, Jr., PhD
Tommy G. Hall
Robert T. Halpin
Barry Halvorsen
Robert L. Halvorsen
Dixie G. Hamilton
Robert Hamilton, PhD
Stephen L. Hamilton
Thomas M. Hamilton, PhD
David J. Hammel
Matthew M. Hammer
Andrew W. Hampf
Bernold M. Handon
Eugene Hanegan
Arthur D. Hanna, PhD
Thomas L. Hanna
Susan K. Hannaman
Douglas B. Hansen
Hugh T. Hansen
Judd A. Hansen
Robin Hansen
Harry R. Hanson
E. W. Hanszen
Richard A. Haralson
Allan Wilson Harbaugh, PhD
Bob C. Harbert
N. J. Hard, PhD
Robert Hard
Stephen D. Hard
James Edward Hardcastle, PhD
Glenn L. Hardin
Andrew T. Harding
Henry W. Harding, Jr.
William H. Harding
Hugh W. Hardy
Robert E. Hardy
Jesse W. Hargis, Jr.
Mary W. Hargrove

Wendell N. Harkey
Leslie Parsons Harman
Scott I. Harmon
John R. Harmonson
Ralph K. Haroldson
Jordan Harp
Laddie J. Harp
Ronald Harp
Henry S. Harper
Charles D. Harr, PhD
Ben Gerald Harris, PhD
Billy W. Harris
David C. Harris
Dennis J. Harris
Jimmy D. Harris
John C. Harris
L. Harris
William E. Harris
Mathew R. Harrison
Maxwell M. Hart
Paul Robert Hart
Louis W. Hartman
Glenn A. Hartsell
Charles M. Hartwell
Dan E. Hartzell
F. Reese Harvey, PhD
Kenneth C. Harvey, PhD
Meldrum J. Harvey
Robert E. Harvey
William H. Harwood, PhD
Syed M. Hasan
Steven R. Haskin
Jill Hasling
Jay Hassell, PhD
Pat Hastings, III
Turner Elilah Hasty, PhD
William B. Hataway
Tim Hatch
Philip R. Haught
S. Mark Haugland
Victor LaVern Hauser, PhD
Rudolf H. Hausler, PhD
William K. Hawkins
Y. Hawkins
Edward F. Haye
James O. Haynes
Richard D. Haynes
J. Ross Hays
James E. Hays
James T. Hays, PhD
Stephen M. Hazlewood
Klyne Headley
Forrest Dale Heath
James Edward Heath, PhD
Maxine S. Heath, PhD
Milton Heath, Jr.
James E. Heavner, PhD
James B. Hebel
Bobby D. Hebert
Donald Hecker
Bobbie Hediger
John A. Hefti
Daniel J. Heilman
William Joseph Heilman, PhD
James R. Heinze
Peter J. Heinze
William Daniel Heinze, PhD
Mark B. Heironimus
Robert W. Helbing
Stephen C. Helbing, Sr.
Jerry Helfand
Paul E. Helfer

James H. Helland
Ronald A. Hellstern, PhD
Langley Roberts Hellwig, PhD
Donald C. Helm
Stephen A. Helmberger
Robert E. Helmkamp, PhD
Maher L. Helmy
Bill D. Helton
Glenn R. Hemann, Jr.
Douglas J. Henderson
Gerald J. Henderson, PhD
Loren L. Henderson
Robert J. Henderson
Dennis L. Hendrix
Malcolm Hendry, PhD
Erwin M. Hengst, Jr.
Don E. Henley, PhD
Ernest J. Henley, PhD
Gary E. Henry
Terry L. Henshaw
Roger P. Herbert
John F. Herbig
John F. Herbig, Jr.
Stevens Herbst
Milton B. Herndon
Robert K. Heron
Joe S. Herring
Randy B. Herring
Robert P. Herrmann
Charles H. Herty, PhD
Henry J. Hervol
Donald E. Herzberg
Allen H. Hess
Barry G. Hexton
Carlos W. Hickman
James L. Hickman
Karen Hickman, PhD
James M. Hicks
Willie K. Hicks
Joseph H. Higginbotham, PhD
Edward D. Higgins
Michael J. Higgins
Margaret A. Hight
William K. Hilarides
Thomas W.C. Hilde, PhD
Donald L. Hildebrand
Andrew T. Hill
Harvey F. Hill
John V. Hilliard
Lee Hilliard
William D. Hillis
John Hills
Philip M. Hilton
Ray Hilton, PhD
Mark Fletcher Hines
Howard H. Hinson
Tod Hinton
Mark A. Hitchcock
Rankin V. Hitt
Bryan E. Hivnor
Grace K. Hivnor
Richard E. Hixon
Sumner B. Hixon, PhD
Wai Ching Ho, PhD
George H. Hobbs
Michael W. Hoblet
Sidney Edward Hodges, PhD
Albert Bernard Hoefelmeyer, PhD
Gustave Leo Hoehn, PhD
Earl C. Hoffer
Charles L. Chuck Hoffheiser
Paul F. Hoffman

Gerald P. Hoffmann
James E. Hoffmann
Timothy J. Hogan
Laurence H. Hogue
David L. Holcomb
Robert Marion Holcomb, PhD
Howard D. Holden, PhD
Stephen A. Holdith, PhD
John M. Holladay, Jr
Howard Holland
Henry B. Holle
Clifford R. Holliday
George Hayes Holliday, PhD
John C. Holliman
Frank Joseph Holly, PhD
Anchor E. Holm
Eldon Holm
Hewett E. Holman
Russell Holman
Peter F. Holmes
Ken D. Holt
Robert J. Holt, Sr.
Bud Holzman
Harry T. Holzman, Jr.
Frank D. Holzmann
Baxter D. Honeycutt
Russell H. Hoopes
Harry A. Hope
George H. Hopkins, Jr., PhD
Richard B. Hopper
Carlos L. Horler, Sr.
Frank A. Hormann
Jerrold S. Horne
R. Dwight Horne, Jr.
Michael A. Hornung
John Horrenstine
Randy D. Horsak
Carl W. Horst
Douglas J. Horton
Edward Horton
Howard T. Horton
Roger Horton
Friedrich Horz, PhD
B. Wayne Hoskins
Richard F. Houde
Albin Houdek
Joel S. Hougen, PhD
Jace Houston
James P. Howalt
Ben K. Howard, PhD
Bruce Howard
Charles H. Howard
Robert L. Howard
Wendell E. Howard
William L. Howard, Sr., PhD
John Howatson, PhD
Terry Allen Howel, PhD
John C. Howell
Randall L. Howell
John E. Howland, PhD
Donald L. Howlett
Stan J. Hruska
Bartholomew P. Hsi, PhD
Yen T. Huang, PhD
Bradford Hubbard
Russell H. Hubbard
Terry K. Hubele
Douglas W. Huber
Frank A. Hudson, PhD
Hank M. Hudson
Stephen H. Hudson
Fred B. Hudspeth

Harry J. Huebner
Fred Robert Huege, PhD
Clark K. Huff
James Douglas Huggard
Dan A. Hughes
Michael Hughes, PhD
Robert G. Hughes
Mark O. Hughston
Robert C. Hulse
David Hulslander
Charles A. Hummel
Ray Eicken Humphrey, PhD
Cathi D. Humphries, PhD
W. Houston Humphries
Cecill Hunt, Jr.
Dale L. Hunt
William H. Hunten
Hassell E. Hunter
Douglass M. Hurt
Vivian K. Hussey
David Hutcheson, PhD
Carl M. Hutson
James H. Hyatt
Mike Ibarguen
Alex Ignatiev, PhD
Piloo Eruchshaw Ilavia
Cecil M. Inglehart
Robert E. Irelan
Charles F. Irwin
James M. Irwin
Jerry K. Irwin
E. Burke Isbell
Cary W. Iverson
Samuel E. Ives, PhD
Edwin Harry Ivey
Turner W. Ivey, Sr.
Jerry W. Ivie
Cleon L. Ivy
David Ivy
Gerald Jacknow
George R. Jackson
James R. Jackson
Jeral Jackson
Nichole Jackson
Scott L. Jackson
William Jackson
Garry H. Jacobs
J. C. Jacobs, PhD
Roy P. Jacobs
Donald F. Jacques, PhD
Robert B. Jacques
Calvin R. James
John E. James
Patrick H. Jameson
Harwin B. Jamison
Marilyn R. Janke
Maximo J. Jante, Jr.
Rupert Jarboe
Richard P. Jares
Kenneth E. Jarosz
Herbert F. Jarrell, PhD
Douglas E. Jasek
Buford R. Jean, PhD
Thomas T. Jeffries, III
Hugh Jeffus, PhD
Michael J. Jellison
Veronon Kelly Jenkins, PhD
Jack H. Jennings
Randall D. Jensen
Stanley M. Jensen
Walter P. Jensen, Jr.
Bill E. Jessup

Richard L. Jodry, PhD
Knut A. Johanson, Jr.
Charles W. Johnson
Delmer R. Johnson
Doug Johnson
Duane J. Johnson, PhD
Earl F. Johnson
Fred Lowery Johnson, PhD
Howard R. Johnson
Jeffrey L. Johnson
Keith Johnson
Paul D. Johnson
Raynard J. Johnson
Stuart G. Johnson, PhD
Thomas P. Johnson
Bonnie Johnston
D. W. Johnston
Daniel Johnston, PhD
La Verne Albert Johnston, PhD
Marshall C. Johnston, PhD
Stephan E. Johnston
Stephen Albert Johnston, PhD
William R. Johnston
Andrew L. Jonardi
Kevin W. Jonas
B. C. Jones
Bill Jones, PhD
Granby D. Jones
James Ogden Jones, PhD
James T. Jones
Mitchell Jones
Richard A. Jones, PhD
Robert E. Jones, PhD
Victor J. Jones, III, PhD
George F. Jonke
Duane P. Jordan, PhD
Jonathan D. Jordan
James R. Jorden
Michael J. Joyner, PhD
Paul G. Judas
Cynthia K. Jungman
Terence B. Jupp
Joseph Jurlina
James Horace Justice, PhD
Richard J. Kabat
Joan D. Kailey
Hugh D. Kaiser
Klem Kalberer
Robert R. Kallman, PhD
Ronald P. Kaltenbaugh
Medhat H. Kamal, PhD
William W. Kaminer
Mark P. Kaminsky, PhD
John Kapson
Dennis R. Karns
Lester Karotkin, PhD
Ross L. Kastor
Ira Katz, PhD
Marvin L. Katz, PhD
Bill Kaufman
Robert R. Kautzman
Don M. Kay
William Kaylor, Jr.
Marvin D. Kays, PhD
William Kazmann
Gene Keating
Lawrence C. Keaton, PhD
Iris Keeling
Joseph Aloysius Keenan, PhD
Ralph O. Kehle, PhD
Byron L. Keil
Kenwood K. Keil

John W. Kelcher
Raymond A. Kelinske
Herman Keller
Michael Keller
Paul W. Keller
R. Keller
Glen E. Kellerhals, PhD
Bob Kelley
Jim D. Kelley
Monica W. Kelley
Dan A. Kellogg
Colin M. Kelly
Charles H. Kelm
Marc Keltner
Arthur H. Kemp
L. N. Kendrick
Anthony Drew Kennard
John W. Kenneday
Howard V. Kennedy, PhD
William Kennedy
J. F. Kenney, PhD
Theodore R. Keprta
Odis D. Kerbow
T. Lamar Kerley, PhD
Linda Kerr
Thomas W. Kershner
James B. Ketchersid
Frank Key
Joe W. Key
Donald Arthur Keyworth, PhD
Daniel E. Kiburz
Harold J. Kidd, PhD
Thornton L. Kidd
Albert Laws Kidwell, PhD
Nat Kieffer, PhD
Robert Mitchell Kiehn, PhD
Marvin E. Kiel
Rodney A. Kiel
William H. Kielhorn
Charles H. Kilgore
Marion D. Kilgore
John E. Kimberly
Kenneth B. Kimble
David R. Kimes
Thomas Fredric Kimes, PhD
Edwin B. King
F. King
Gerald S. King
Kathryn E. King
Roy L. King
Thomas J. King, Jr.
Karl Kinley
Roy H. Kinslow, PhD
Thyl E. Kint
Ken K. Kirby
Fred Kirchhoff, Jr.
Earl Kirk, Jr.
James E. Kirkham
Matthew L. Kirkland
Dennis Kirkpatrick, PhD
Haskell M. Kirkpatrick
Hugh R. Kirkpatrick
Pamela K. Kirschner
Julianne J. Kisselburgh
Mark C. Kittridge
Thomas R. Kitts
Alfred Klaar
Miroslav Ezidor Klecka, PhD
Gordon Leslie Klein
Richard G. Klempnauer
David L. Klenk
Melvin Klotzman

John L. Knapp
Barry Kneeland
Peter Knightes, PhD
William Knighton
Christian W. Knudsen, PhD
Brody Knudtson
Buford R. Koehler
Wellington W. Koepsel, PhD
Moses M. Koeroghlian
Charles A. Kohlhaas, PhD
Dusan Konrad, PhD
Kenneth K. Konrad
Walter R. Konzen
Kenneth T. Koonce, PhD
Charles B. Koons, PhD
Micah S. Koons
Johnny A. Kopecky, PhD
David E. Kosanda
Charlie Kosarek
Merwyn Mortimer Kothmann, PhD
Christopher C. Kraft, Jr., PhD
Paul M. Krail, PhD
Stephen G. Kramar
George G. Krapfel
Garry D. Kraus
Mark Krause
Dennis A. Krawietz
Robert F. Kraye
M. Fred Krch
William F. Krebethe
K. Kreckel
Rodger L. Kret
Robert W. Kretzler
Charles R. Kreuz
Victor M. Kriechbaum
Daniel R. Krieg, PhD
Magne Kristiansen, PhD
Tony R. Kroeker
James C. Kromer
Paul H. Kronfield
Tim C. Kropp
Julius Richard Kroschewsky, PhD
Glenn L. Krum
Karen Sidwell Kubena, PhD
Marc S. Kudla
Antonin J. Kudrna
Stanley E. Kuenstler
Bernard J. Kuhn
Janice Oseth Kuhn, PhD
Carl W. Kuhnen, Jr.
Kenneth K. Kulik
Arun D. Kulkarni, PhD
Edgar V. Kunkel
George William Kunze, PhD
Peter T. Kuo
Joseph M. Kupper
M. Kurz, PhD
Charles R. Kuykendall
Edmond R. La Belle
Ronnie R. Labaume
John M. Lagrone
William E. Laing
Dusan J. Lajda, PhD
Jeffrey J. Lamarca, Sr.
George W. Lamb
W. E. Lamoreaux
S. S. Lan
Malcolm Lancaster
John R. Landreth
William R. Landrum
Alis C. Lane
Clifford Lane

Robert Lane, PhD
Newton L. Lang
William R. Lang
Normal L. Langham
Gerald B. Langille, PhD
Hans A. Langsjoen
Peter H. Langsjoen
Richard M. Lannin
Robert G. Lannon
Vince L. Lara
William E. LaRoche
Donald E. Larson
Ronald Larson, PhD
Lawrence B. Laskoskie
Stanley J. Laster, PhD
Benny L. Latham
Gail Latimer
Pierre Richard Latour, PhD
William H. Laub
Charles E. Lauderdale
Walter A. Laufer
David D. Laughlin
William R. Laughlin
Dallas D. Laumbach, PhD
Bob J. Lavender
Delman Law
James H. L. Lawler, PhD
Harold B. Lawley
Joseph D. Lawrence
Philip Linwood Lawrence
Donald K. Lawrenz
Bill Lawson
Royce E. Lawson, Jr.
Marsha A. Layton
Robert E. Layton, Jr.
Lawrence E. Leach
William D. Leachman
Joanne H. Leatherwood
Floyd L. Leavelle, PhD
Jacob M. Lebeaux, PhD
Rebel J. Leboeuf
Thomas K. Ledbetter
Frank F. Ledford
Daniel V. Lee
Harry A. Lee
Karl E. Lee
Larry E. Lee
Lindsey D. Lee
Richard F. Lee
Steve W. Lee
Gil L. Legate
Thomas H. Legg
Robert B. Leggett
Ruw W. Lehde
Barbara T. Lehmann
Elroy Paul Lehmann, PhD
William L. Lehmann, PhD
David J. Lehmiller
Roger D. Leick
W. L. Lemon
David V. Lemone, PhD
Roland D. Leon
Albert O. Leonard
Paul G. Leroy
Ming-Biu Leung, PhD
Emil M. Leutwyler
Donn W. Leva
Stewart A. Levin, PhD
Catherine Lewis, PhD
Philip E. Lewis
Robert A. Lewis
Timothy E. Lien

David W. Light
Curt Lightle
Steven C. Limke
Philip V. Lindblade
R. Lindemann, PhD
Gerald S. Lindenmoen
Jay T. Lindholm
Cecil H. Link
Andrew Franklyn Linville, Jr.
Anthony Pasquale Lioi, PhD
Eugene G. Lipnicky
Merrill I. Lipton
Jim Litchfield
David Litowsky
Frank K. Little
Jack E. Little, PhD
Larry E. Little
Hung-Wen (Ben) Liu, PhD
William H. Livingston
Barry H. Lloyd
David M. Lloyd
James R. Lloyd, PhD
Sheldon G. Lloyd
James R. Lobb
Gene M. Lobrecht
Brent Lockhart
Paul Lockhart, Jr.
Robert M. Lockhart
Travis W. Locklear
William R. Locklear
Paul L. Lockwood
Alfred R. Loeblich, III, PhD
Larry K. Lofton
R. H. Lofton
Jerry L. Logan
Robert J. Logan
Thomas L. Logan
Charles B. Loggie
Nicolas Logvinoff, Jr.
H. William Lollar
Henry R. Longcrier
Kenneth C. Longley
Guy J. Lookabaugh
Ronald L. Loper
H. C. Lott, Jr.
L. Richard Louden, PhD
Ralph W. Love
Richard M. Love
Ben A. Lovell
Wayne T. Lovett
Robert F. Loving
Glen R. Lowe
Robert M. Lowe
John D. Lowery
Charles B. Lowrey, PhD
Justin Lowry
D. Mark Loyd
Dale R. Lucas
Timothy W. Lucas
Michael W. Luck
Gerald S. Ludwig
William P. Ludwig
Donald L. Luffell
Gordon D. Luk
Richard Lumpkin, PhD
J. Lund
Brenda Lunsford
Mark J. Lupo, PhD
Francis M. Lynch
Michael S. Lynch
Malcolm K. Lyon
James I. Lyons

Robert Leonard Lytton, PhD
Robert L. Maby, Jr.
Charles B. Macaul
Robert N. MacCallum, PhD
Richard E. Macchi
Alexander MacDonald
Charles E. Mace
Allan Macfarland
Marc Machbitz
Thomas J. Machin
Edward A. Maciula
John B. Mack
Bruce C. Macke
Henry James Mackey, PhD
Patrick E. Mackey
Steve D. Maddox
Randall N. Maddux
Hulon Madeley, PhD
Jack T. Madeley
Neil S. Madeley
James E. Madget, II
Kenneth Olaf Madsen, PhD
John M. Maerker, PhD
John J. Magee
Ronald Magel
Robert E. Magers
Allen H. Magnuson, PhD
John F. Maguire, PhD
William Mahavier, PhD
Salah E. Mahmoud
Laurie A. Main
Charles F. Maitland
Lewis R. Malinak, PhD
James Jo Malley, PhD
Meredith Mallory, Jr.
William H. Malone
Glenn Maloney
Harold R. Mancusi-Ungaro, Jr.
Sharad V. Mane
John L. Maneval
J. David Manley, IV
T. Mann
A. Bryant Mannin
David T. Manning
Lewis A. Manning
Robert A. Manning
Stephen M. Manning
Terry Manning
Robert A. Marburger
Karla Wade Marchell
Kirk A. Marchell
Matthew C. Marcontell
Ronald Marcotte, PhD
Roger W. Marcum
Louis F. Marczynski
Todd A. Marek
George Marklin, PhD
Daniel B. Marks
Stuart A. Markussen
James L. Marsh, PhD
David W. Marshall, Jr.
William Marshall
Antonio S. Martin
Benjamin F. Martin, Jr.
Brent H. Martin
John L. Martin
Monty G. Martin
Rex I. Martin
George M. Martinez
Jack M. Martt
Vernon J. Maruska
Perry S. Mason, PhD

Riaz H. Masrour
Wulf F. Massell, PhD
Lenita C. Massey
John A. Massoth
John R. Masters
Robert Mastin
T. Mather
John R. Mathias
Francis J. Mathieu
Rickey L. Mathis
Hudson Matlock
Ron J. Matlock
Roy E. Matthews
Charles Matusek
Samuel Adam Matz, PhD
A. Bruce Maunder, PhD
Roger A. Maupin
Margaret N. Maxey, PhD
Arthur R. Maxwell, PhD
John Crawford Maxwell, PhD
Henry K. May
Jimmy E. May
Greg L. Mayes
Harry L. Mayes
J. Mayfield
Alfred M. Mayo
Russ K. Mayors
Joel T. Mays
C. Gordon McAdams
William N. McAnulty, Jr., PhD
Melinda S. McBee
William D. McCain, PhD
David McCalla
John R. McCalmont
Michael F. McCardle
A. T. McCarroll
Andrew C. McCarthy, PhD
Glenn J. McCarthy
Keith F. McCarthy
Michael J. McCarthy
Bramlette McClelland
James O. McClimans
Jack L. McClure
Lawrence McClure
Jalmer R. McConathy
Stewart McConnell
Don L. McCord
James W. McCown
Jeff McCoy
Lawrence T. McCoy
Rayford L. McCoy
Michael L. McCrary
Kem E. McCready
Lynn McCuan
Robert G. McCuistion
David W. McCulloch
Thane H. McCulloh, PhD
Dennis W. McCullough, PhD
Keith D. McCullough
Michael C. McCullough
Martin K. McCune
Dolan McDaniel
Douglas McDaniel
William D. McDaniel
C. McDaniels
Floyd McDonald
Lynn Dale McDonald, PhD
Robert I. McDougall, PhD
Edward R. H. McDowell, Jr.
John C. McDuffie, Jr.
Paul M. McElfresh, PhD
Raymond E. McFarlane

Michael A. McFerrin
Robert G. McGalmery
Mickey McGaugh
James M. McGee
Richard Heath McGirk, PhD
Donald Paul McGookey, PhD
Ron McGregor
Brady J. McGuire
Gary McHale
Gary B. McHalle, PhD
Timothy B. McIlwain
Roy L. McKay
Fount E. McKee
Bishop D. McKendree
Michael G. McKenna
William R. McKenna
Samuel R. McKenney
John McKetta, Jr., PhD
Charles W. McKibben
Elvie F. McKinley
Ron J. McKinley
Paul D. McKinney
Samuel J. McKinney
J. D. McLaughlin
L. A. McLaurin
James C. McLellan, PhD
Jerry D. McMahon
Perry R. McNeil, PhD
Roger J. McNichols, PhD
Allen F. McNight
Clara McPherson
Clinton M. McPherson, PhD
D. Sean McPherson
Richard C. McPherson
P. S. McReynolds
C. L. McSpadden
Kevin D. Mcvey
Michael R. McWiliams
Sally J. Meader-Roberts
Samantha S. Meador
Alan J. Mechtenberg
James M. Medlin
William Louis Medlin, PhD
Russel P. Meduna
Donald N. Meehan, PhD
Preston L. Meeks
Paul E. Melancon
James C. Melear
Michael Brendan Melia, PhD
Jim Mellott
B. Melton
James Ray Melton, PhD
Arnold Mendez
William Menger
Samuel H. Mentemeier
Erhard Roland Menzel, PhD
Carl Stephen Menzies, PhD
J. Steven Mercer
Sherrel A. Mercer
James B. Merkel
G. Merkle
John C. Merritt, Jr.
Mark B. Merritt
Frederick Paul Mertens, PhD
Carl W. Mertz
Charles L. Messler
Ron G. Metcalf
Michael W. Metza
Leslie P. Metzgar
John Wesley Meux, PhD
Thomas R. Mewhinney
Frank E. Meyers

Nicholas Michaels, PhD
F. Curtis Michel, PhD
W. M. Midgett
C. R. Miertschin
Edward Miesch, PhD
Alton Migl
Nelson F. Mikeska
John A. Mikus, PhD
Otto J. Mileti
Paul B. Milios
Charleton J. Miller
Clifford A. Miller
David Todd Miller
Gary D. Miller
Jeffrey A. Miller
John C. Miller
Kerry C. Miller
Leland H. Miller
Leslie T. Miller
Loyle P. Miller
Matt C. Miller
P. H. Miller
Philip Dixon Miller, PhD
Richard T. Miller, PhD
Fay E. Millett
Spencer Rankin Milliken, PhD
Curtis R. Mills
John James Mills, PhD
Thomas H. Milstead
Elmer A. Milz
James F. Minter
A. E. Minyard
Raymond W. Mires, PhD
John P. Mireur
Conarad Mirochna
Gustave A. Mistrot, III
Billy F. Mitcham
Edward H. Mitchell
Robert J. Mitchell
Thomas J. Mittler
Perry J. Mixon
Jack Pitts Mize, PhD
Glen Lavell Mizer
Henry A. Mlcak
Richard G. Mocksfield
Jerry L. Modisette, PhD
Fersheed K. Mody, PhD
William R. Moeller
Camilla M. Moga
Traian C. Moga
Mohammad M. Mokri
John Molloy
Thomas D. Molzahn
D. Mommsen
Thomas M. Momsen
Peter A. Monteleone
Edward Harry Montgomery, PhD
Gerald T. Montgomery
Joseph S. Montgomery
Monty Montgomery
Harley Moody
Wendell B. Moody
Elizabeth A. Mooney
B. L. Moore, Jr.
Craig Moore
Daniel C. Moore
George E. Moore
Jimmie Moore, PhD
Kenneth L. Moore
Walter L. Moore, PhD
William Moorhead, PhD
George A. Moran

Thomas M. Moran
Hubert L. Morehead, PhD
Clyde N. Morgan
Ronald B. Morgan, PhD
Jerry I. Moritzq
John C. Morrill, PhD
Ben L. Morris
Brock A. Morris
Charles W. Morris, PhD
Perry B. Morris
Robert A. Morris
Dennis Moseley
Ron R. Moser
David A. Mosig
Jerry E. Mount
R. Mowell
Dan L. Mueller
Philip M. Mueller
William B. Mueller
J. Mulcahey
Bertram S. Mullan
James C. Mullen
Dennis Mulvey, PhD
Michael E. Munoz
Rodney D. Munsell
David M. Munson
Jack G. Munson
Ronald E. Munson
Henry W. Murdoch
Karl Muriby
Michael M. Murkes, PhD
John R. Murphey, Jr.
Lawrence Eugene Murr, PhD
Clarence Murray, Jr.
Glenn Murray
James W. Murray
Daniel M. Musher
Jack Thompson Musick
Harry C. Mussman, PhD
Stuart Creighton Mut
Walter F. Muzacz
Punam S. Myer
George M. Myers, PhD
Glen Myska
Larry D. Nace
Carl M. Naehritz, Jr.
Harry E. Nagel
Joseph Nagyvary, PhD
B. J. Nailon
Marie M. Naklie
Julian Nalley
Jerome Nicholas Namy, PhD
Howard L. Nance
Michael L. Nance
Robert K. Nance
Cynthia Y. Naples, PhD
David Naples
Kenneth P. Naquin
Karen M. Needham
Thomas H. Neel
Joe C. Neeley
Dana Neely
James H. Nelland
David L. Nelson
J. Robert Nelson
James Arly Nelson, PhD
John J. Nelson
Mart D. Nelson
Ronald G. Nelson
Scott R. Nelson
Stephen Nelson
Charles E. Nemir

Jerry E. Nenoon
R. B. Nesbitt
Dan R. Neskora, PhD
Michael E. Nevill
Daniel B. Nevin
Roman N. Newald
Jean-Jaques Newey
Robert M. Newman
Sidney B. Nice
Clifford E. Nichols
James R. Nichols
Richard E. Nichols
James H. Nicholson
Robert L. Nickell, Jr.
Thomas D. Nickerson
Michael J. Nicol
Billy W. Nievar
James A. Nitsch
James W. Nixon
Paul Robert Nixon
Marion D. Noble
Bertram Nolte, PhD
Philip A. Norby
James C. Nordgren
Carl H. Nordstrand
Gregg A. Norman
Billy N. Norris
Hughie C. Norris
Douglas F. North
L. D. Northcott
Michael W. Norton
S. J. Norton, PhD
H. Scott Norville, PhD
Robert J. Noteboom
J. D. Novotny
Shirley E. Nowicki
Gary P. Noyes, PhD
Ronnie L. Nye
James Eugene Nymann, PhD
Edward L. Oakes
Victor D. Obadiah
Douglas K. Obeck
Donald Oberleas, PhD
Mark J. O'Brien
Carlos Ochoa
George S. Ochsner
John D. Ochsner
J. Oden, PhD
Charles B. Odom
Vencil E. O'Donnell
Carl O. Oelze
Charles Patrick Ofarrel, PhD
John H. Ognibene
David John Ogren
Robert E. Old
Daniel R. Olds
Enrique A. Olivas, PhD
Arnold W. Oliver
Ray E. Olsen
Christopher E. Olson
Dale C. Olson
Frank I. Ondrovik
Charles L. Oney
Curtis H. Orear
John A. Oren
J. Dale Ortego, PhD
Ramon E. Ortiz-Juan
William F. Ortloff
James Wilbur Osborn
Zofia M. Oshea
Jeffrey S. Oslund
William T. Osterloh, PhD

David P. Osterlund
Osamudiamen M. Otabor
Leo E. Ott, PhD
Jack M. Otto
Carroll Oubre, PhD
Steve H. Ousley
Susan B. Ouzts
Robert G. Ovellette
Lanny M. Overby
Rufus M. Overlander
F. A. Overly
J. David Overton
Scott C. Overton
Donald E. Owen, PhD
Prentice R. Owen
Richard Owen
William C. Owens
R. M. Ownesby
T. M. Ozymy
David R. Paddock
Susan Paddock
Carey P. Page
John H. Painter, PhD
Lorn C. Painter
Jim R. Palmore
Shambhu A. Pai Panandiker
James M. Pappas
Ronald E. Paque, PhD
Howell W. Pardue
Manuel Paredes
William Park
Billy Y. Parker
H. William Parker, PhD
Michael D. Parker
Raymond G. Parker
Sidney G. Parker, PhD
Harold R. Parkinson
David B. Parks
Edward M. Parma
Sheldon C. Parmer
Charles H. Parr
W. M. Parr
Clinton Parsons, PhD
Donald A. Parsons
William M. Pate
Pradeep V. Patel, PhD
Margaret D. Patin
Wesley Clare Patrick, PhD
Ben M. Patterson, Jr.
Calvin C. Patterson, PhD
Don R. Patterson
James C. Patterson
Jeff D. Patterson
Robert W. Patterson
Sharon Patterson
Wayne R. Patterson, PhD
Donald R. Pattie
Robert J. Patton
Cynthia M. Patty
James T. Paul
Ralph L. Pauls
Bernard A. Paulson
Robert C. Paulson
Darrell G. Pausky
Max L. Paustian
Michael A. Pauwels
Barry Payne
Cyril J. Payne
D. Payne
F. R. Payne, PhD
James R. Payne
Michael A. Payne, PhD

Ralph Payne
Charles Pearson
Daniel Bester Pearson, III, PhD
Stephen I. Pearson
Rod W. Pease
Patrick A. Peck
Christopher B. Peek
Vernon L. Peipelman
Gary R. Pekarek
John L. Pellet
Richard R. Pemper, PhD
Yarami Pena
Pawel Penczek, PhD
Jim R. Pendergrass
David Pendery
A. Penley
B. F. Pennington
Victor H. Peralta
Luzviminda K. Peredo
James D. Periman
F. M. Perkins, Jr.
Thomas Perkins, PhD
Dean H. Perry
Swan D. Person
Cyril J. Peruskek
David W. Peters, PhD
Christopher K. Peterson
D. Peterson
Emmett M. Peterson
Floyd M. Peterson
Gerald L. Peterson
Joel E. Peterson
Mark A. Peterson, PhD
Arthur K. Petraske
Chester A. Peyton, Jr.
William D. Pezzulich
Edward Phillips, PhD
Felton Ray Phillips
Karen S. Phillips, PhD
Perry Edward Phillips, PhD
Kenneth Piel, PhD
Walter H. Pierce, PhD
D. Pigott, PhD*
Jeffrey A. Pike
Paul E. Pilkington
John H. Pimm
R. E. Pine
Carl O. Pingry
Michael R. Piotrowski, PhD
James J. Pirrung
Shane W. Pirtle
John Robert Piskura
David J. Pitts
Gerald S. Pitts
Gregory S. Pitts
James J. Piwetz, Jr.
Frank J. Pizzitola
Joe Pizzo, PhD
John E. Plapp, PhD
E. Robert Plasko
Carl W. Ploeger
Daniel T. Plume
John F. Podhaisky
Richard D. Poe, PhD
Steve C. Poe
Ronald F. Pohler
Craig Poindexter
Adrian M. Polit
James K. Polk
R. Scott Pollard
William D. Pollard
Walter L. Pondrom, PhD

Philip C. Pongetti
Sam L. Pool
Russel Poole
Lee Pope
Richard Porter
Vernon Ray Porter, PhD
Mike Posson
Keith M. Potter
Rainer Potthast, PhD
Michael Robert Powell, PhD
Darden Powers, PhD
Don Gary Powers
Adam A. Praisnar, Jr.
Samuel F. Pratt, Jr.
Scott H. Prengle
Richard S. Prentice
Tony M. Preslar
Allister L. Presnal
Jackie L. Preston
Charles R. Price
Donald G. Price, Jr.
R. E. Price
Jerry F. Priddy
Bill Priebe
J. Prieditis, PhD
Guy T. Priestly
Robert Prince
James W. Pringle
Vernon Pringle, Jr.
Alan N. Pritchard
Charles A. Proshek
Jesus Roberto Provencio, PhD
Basil A. Pruitt, Jr.
Jay P. Pruitt
Charles J. Pruszynski
Shane Pryor
Stanley E. Ptaszkowski
Juan R. Puerto
Richard R. Pugh
Viswanadha Puligandla, PhD
Layne L. Puls
Cary C. Purdy
Paul E. Purser
George T. Pyndus
Jeffrey P. Quaratino
Miller W. Quarles
John P. Quinn
Perry C. O. Quinn
Richard H. Quinn, PhD
Shirley J. Quinn
J. R. Quisenberry
Patrick W. Quist
Ashley Rabalais
Brian T. Raber
Thomas A. Rabson, PhD
Robert A. Rademaket
Nancy R. Radercic, PhD
Norman D. Radford, Jr.
Steven R. Radford
Lewis E. Radicke, PhD
Donald E. Radtke
Ralph D. Ragsdale
Joe M. Rainey
Robert N. Rainey
DeJan Rajcic
C. L. Rambo
Frederick H. Rambow, PhD
Guadalupe Ramirez
Jerry Dwain Ramsey, PhD
Milton H. Ramsey
Thomas E. Ramsey
Jack E. Randorff, PhD

Jesus J. Rangel, Jr.
C. J. Ransom, PhD
W. R. Ransone, PhD
Marco Rasi, PhD
Beatrice K. Rasmussen
Mark Rasmussen
Lee Ratcliff
Jason Rathweg
Carroll J. Rawley
Brad Ray
Clifford H. Ray, PhD
Donald R. Ray
Michael B. Ray, PhD
Herbert Raymond
Bill E. Raywinkle
Robert C. Reach
W. Rector
Brian Redlin
Cynthia A. Reece
Steven Reeder
Richard W. Rees
Dale O. Reese
David D. Reese
W. R. Reeves, Jr.
M. Loren Regier
Larry M. Rehg
Frank E. Reid
William M. Reid, PhD
Erwin A. Reinhard, PhD
James A. Reinhert, PhD
Samuel Reiser
Russel J. Reiter, PhD
Jimmie J. Renfro
Kevin D. Renfro
Edward G. Rennels, PhD
Kenneth L. Rergman
F. E. Resch
Lynn A. Revak
William Revelt
Randall H. Reviere, PhD
Bill Rhoades
John E. Rhoads, PhD
James D. Rhodes
John R. Rhodes
R. David Rhodes
Charles N. Rice
Patrick Rice
Robert M. Rice
David A. Rich
Otis H. Richards
Russell Richards
Thomas C. Richards, PhD
Albert T. Richardson
Clarence W. Richardson, PhD
George A. Richardson
J. C. Richardson, Jr.
Louis A. Rickert
Richard L. Ricks
David C. Riddle, PhD
John W. Riddle
Napoleon B. Riddle
Kerry J. Ridgeway
Susan J. Riebe
Edward Richard Ries, PhD
Glen A. Ries
Noel D. Rietman
Charles M. Riley, PhD
J. W. Rimes
Gregory S. Rinaca
Charles E. Rinehart, Jr., PhD
Stephen J. Ringel
David P. Ringhausen

David Rios-Aleman
George R. Ripley
Don L. Risinger
Carolyn J. Ritchie
Steven J. Ritter, PhD
Russell E. Ritz
George A. Rizk
Patrick C. Roark
Norman B. Robbins
Alvis D. Roberts, Jr.
Arthur R. Roberts
Henry E. Roberts
James W. Roberts
Larry C. Roberts
Michael A. Roberts, Jr.
Robert H. Roberts
William E. Roberts
David R. Robertson
Gordon W. Robertstad, PhD
Elizabeth N. Robinson
J. Michael Robinson, PhD
M. Robinson, PhD
Stephen L. Robinson
Ken Robirds
E. Douglas Robison
G. Alan Robison, PhD
Jackie L. Robison
Mary Drummond Roby, PhD
James C. Rock, PhD
Cary O. Roddy
Jack W. Rode
Rocky R. Roden
Charles Alvard Rodenberger, PhD
Robert B. Rodriguez
Weston A. Roe
Isaac F. Roebuck
Robert C. Roeder, PhD
Anthony C. Rogers
Dave E. Rogers
Glenn M. Rogers
Marion Alan Rogers, PhD
Tom Rogers, PhD
Howard J. Rohde
Don A. Rohrenbach
Ronald Rolando
Jude R. Rolfes
Albert W. Rollins
Thomas W. Rollins
Joe T. Romine
Jeff H. Ronk
Charles H. Roos
Paul James Roper, PhD
G. Rorschach
Paul N. Roschke, PhD
Ward F. Rosen
Charles R. Rosenfeld
Joshua H. Rosenfeld, PhD
Betty J. Rosenthal
Randy E. Rosiere, PhD
Hayes E. Ross, Jr., PhD
Nealie E. Ross
Norman B. Ross
Randall R. Ross
William F. Ross
Audrey Rossi
Bela P. Roth, PhD
Charles H. Roth, PhD
Thomas P. Roth
Ronald C. Rothe
Billy J. Rouser
Lee J. Rousselot
Martin D. Rowe, Jr.

Rex L. Rowell
David A. Rowland, PhD
Lenton O. Rowland, PhD
Stephen H. Rowley
Richard L. Royal
Tad T. Rozycki
Carter F. Rubane
Jim R. Rucker
Kenneth E. Ruddy
Douglas L. Rue
Walter E. Ruff
T. H. Ruland
Robert E. Rundle
Edward E. Runyan
Andrew Rusiwko, PhD
Charles L. Russell
Carri A. Rustad
James D. Rutherford
Paul J. Rutherford
Robert S. Rutherford
Durward E. Rutledge
Donald Rux
Lester M. Ryol, PhD
Deborah K. Sacrey
Tom R. Sadler
Harry C. Sager
Winston Martin Sahinen
Nelton O. Salch
Belinda Salinas
Joe G. Saltamachia
Richard W. Sammons
Michael T. Sample
Cheryl K. Sampson
Freddy J. Sanches
Isaac C. Sanchez, PhD
Robert M. Sanford
Thomas G. Sarek
Cornel Sarosdy
Robert Sartain, PhD
Ray N. Sauer
Richard L. Sauer
Lloyd Saunders, PhD
John D. Savage
George P. Saxon, PhD
Hugh A. Scanlon
Paul Scardaville
J. F. Scego
Brett Schaffer
Raymond A. Schakel
Thomas S. Schalk
Laird F. Schaller
Richard Allan Schapery, PhD
Richard M. Scharlach
Marvin W. Schindler
Raymond C. Schindler
Brent D. Schkade
Richard D. Schlomach
Lester E. Schmaltz
Brian F. Schmidt
Ralph Schmidt
Allan M. Schneider
Louis I. Schneider, Jr.
Martin L. Schneider
William P. Schneider
Lewes B. Schnitz
Arthur Wallace Schnizer, PhD
William R. Schoen
John L. Schoenthaler
Stephen L. Schrader
Martin William Schramm, PhD
Wilburn R. Schrank, PhD
Robert Alvin Schreiber, PhD

Max P. Schreiner
R. J. Schrior
Steve A. Schroeder
William Schrom
Dale R. Schueler
Frank J. Schuh
Robert E. Schuhmann, PhD
P. Schulle
Michael Schuller, PhD
Norrell D. Schulte
Lowell E. Schultz
Thomas A. Schultz
John E. Schumann
Max A. Schumann, Jr.
Gerard Majella Schuppert, Jr.
Mark J. Schusler
Michael Frank Schuster, PhD
Gilbert C. Schutza
Jeffry A. Schwarz
Colman P. Schweikhardt
Thomas R. Schwerdt
Bill H. Scott
C. J. Scott
Linda Scott
Marion Scott, PhD
Oscar T. Scott, IV, PhD
R. W. Scott
Wayne S. Scott
Michael E. Scribner, PhD
Richard Burkhart Sculz
Harry G. Scurlock
Daniel R. Seal
Ryan B. Seals
Brian E. Sealy
Michael T. Searfass
William J. Sears, PhD
Thomas M. Seay
Mary Ann Sechrest
Paul R. Seelye
Philip A. Seibert
James A. Seibt
William Edgard Seifert, PhD
Jerold Alan Seitchik, PhD
Joseph J. Sekerka
Joe Selle
John S. Sellmeyer
Scott H. Semlinger
Edwin T. Sewall
James W. Sewell
Franklin W. Shadwell
W. A. Shaeffer, III*
William L. Shaffer
Albert M. Shannon
Edward M. Shapiro
James W. Sharp
Stephen L. Shaw
David R. Sheahan
Robert E. Sheffield
Joseph Sheldon
Robert C. Shellberg
William T. Sheman
Mark B. Shepherd, Jr.
Kalapi D. Sheth
Scott Shifflett
Don I. Shimmon
Ross L. Shipman
Gene M. Shirley
Richard R. Shirley
William C. Shockley
Dale M. Short
Allen Shotts
Michael Shouret

Loy William Shreve, PhD
Robert E. Shwager
David E. Sibley, Jr.
Willis P. Sibley
A. J. Siefker, PhD
Harold L. Siegele
Wayne L. Sievers, PhD
Curtis J. Sievert
Henno Siismets
John William Sij, PhD
I. J. Silberberg, PhD
Miles Silberman, PhD
Jerome L. Silverman
John Simion
Larry B. Simmers
Charles M. Simmons
Kelly V. Simmons
Lynn A. Simpson
Rau E. Simpson
David C. Sims
John C. Sinclair
Robet B. Singer
Raj N. Singh
Joseph R. Sinner
Melvin M. Sinquefield
Lou Di Sioudi
J. L. Sipes, Jr.
William Allen Sistrunk, PhD
Richard L. Sitton
Michael Sivertsen
Phillip S. Sizer
Damir S. Skerl
Allen C. Skiles
Leslie D. Skinner
Marvin R. Skinner
Nicholas R. Skinner
Ronald N. Skufca
Bill E. Slade
Clifford V. Slagle
Mark E. Slezak
Jim H. Slim
William M. Sliva, PhD
Harold S. Slusher, PhD
Jeffrey S. Small
Samuel W. Small
G. A. Smalley, Jr.
T. Smiley, Jr.
Al H. Smith
Alan Lyle Smith, PhD
Daniel L. Smith
David L. Smith, Jr.
David Smith
Derek L. Smith, PhD
Ead C. Smith
Floyd A. Smith
Louis C. Smith, PhD
Louis J. Smith
Mark Thomas Smith
Milton Louis Smith, PhD
O. Lewis Smith, Jr.
Raymond H. Smith
Robert P. Smith, PhD
Thomas L. Smith
Thomas M. Smith
Tina Smith
Todd G. Smith
Tommy L. Smith
Eugene A. Smitherman
Billy J. Sneed
Dennis R. Sneed
Gilbert W. Snell
Glenn C. Snell

Jon S. Snell
R. Larry Snider
Paul R. Snow
Blaine Snyder
Charles F. Snyder
David A. Snyder, PhD
Fred F. C. Snyder
Nicholas Snyder
Robert B. Snyder
Arthur J. Socha, Sr.
E. H. Soderberg
Dean E. Soderstrom
Harold K. Sohner
Robert E. Sokoll
Dale M. Solaas
Turner Solari
Albert K. Solcher
Eugene A. Soltero
Siva Somasundaram, PhD
Xiaohui M. Song, PhD
Philip M. Sonleitner
Danny C. Sorrells
Gordon Guthrey Sorrells, PhD
Armand Max Souby
Robert G. Spangler, PhD
Aubrey A. Spear
Tim R. Speer
Frederick Speigelberg
William A. Spencer
George V. Spires
Marion E. Spitler
William J. Spray
Maurice L. Sproul
Eve S. Sprunt, PhD
Hugh H. Sprunt
Charles F. Squire, PhD
Douglas E. Stafford
Frank B. Stahl
Fred Stalder
Dennis D. Stalmach
Michael R. Stamatedes
Albert L. Stanford
Drago Stankovic
John L. Stanley
Steven Stanley
Robin L. Stansell
Jerry Staples
John Starck
Kelly L. Stark
Richard R. Statton
Eric A. Stauch
Frederick L. Stead
Werner W. Stebner
Jame Harlan Steel
George M. Steele
Donald N. Steer
Mark C. Stefanov
Tom J. Steffens
Neil A. Stegent
David L.R. Stein
William H. Steiner
Ray L. Steinmetz
Michael F. Stell
William K. Stenzel
Pamela A. Stephens, PhD
William C. Stephens
Michael Steppe
Winfield Sterling, PhD
Charles A. Sternbach
James Stevens
H. Leighton Steward
William C. Stewart

William L. Stewart, Jr.
David W. Stilson
Louis Stipp
Dan Stoelzel
Kim R. Stoker
Dana D. Stokes
Michael F. Stolle
Jay D. Stone, PhD
Michael E. Stones
Charles L. Strain
David A. Strand
Kurt A. Stratmann
William L. Straub
Roger A. Streater
Robert Lewis Street, PhD
John C. Strickland
Sarah Taylor Strinden, PhD
John J. Strojek
James E. Strozier
Glenn A. Strube
Malcolm K. Strubhar
John B. Stuart
Roger G. Stuart, Jr.
Telton Stubblefield, PhD
Harry T. Stucker
Donald P. Stuckey
David Owen Stuebner
Robert J. Stupp
Jerry D. Sturdivant
G. Sturges
William J. Sturtz
K. Stutler
Ralph E. Styring
H. Su, PhD
Felipe J. Suarez
Daniel J. Subaen, PhD
Jonathan D. Such
John Suggs
Bruce M. Sullivan
Carl Summers
Charles Summers
Norman A. Sunderlin
A. J. Sustek, Jr., PhD
Charles Suter
William R. Suter
James A. Sutphen
Paul A. Svejkovsky
Robert S. Svoboda
William C. Swart, Jr.
John H. Swendig
James E. Swenke
Ronald F. Swenson
Gary S. Swindell
Rita D. Swinford
Charles J. Swize
Paul M. Swoboda
James R. Sydow
Elwood Sylvanus
Robert E. Sylvanus
Ronald E. Symecko
Edward J. Szymczak
Jerrry L. Tabb
Kelley A. Tafel, PhD
Gerald W. Tait
Emil R. Talamo
Alvin W. Talash, PhD
John S. Talbot
Larry D. Talley, PhD
Russell Talley
James L. Tally
Edward A. Talmage
Ashley C. Tanner

Sean Tanner
Joann H. Tannich
John Tarpley
Jeffrey J. Tarrand
Alan R. Tarrant
Michael J. Tarrillion
Gordon E. Tate
Nahum A. Tate
Ben E. Taylor
Gordon E. Taylor
Lyndon Taylor, PhD
Richard Melvin Taylor, PhD
Steve D. Taylor
William Charles Taylor-Chevron
Neal Teague
Edward Teasdale
Todd X. Teitell
Robert W. Temple
Robert J. Templin
Peter W. Ten Eyck
Richard N. Tennille
Jamie B. Terrell
Robert E. Terrill
Earl Tessem
Norman Jay Tetlow, PhD
W. G. Teubner
Raynold J. Thibodeaux
Kenneth A. Thoma
A. D. Thomas
Charles Thomas, PhD
J. Todd Thomas
Jeremy Miles Thomas
Leslie M. Thomas
Blair D. Thompson
Evan C. Thompson
Greg Thompson
Guy Thompson
J. Christy Thompson
James R. Thompson, PhD
Richard J. Thompson, PhD
Richard W. Thompson, Jr.
Roger Thompson
Ronald E. Thompson
Tommy L. Thompson
William S. Thompson III
Walter Thomsen
J. T. Thorleifson
Robert L. Thornton
Purvis J. Thrash, Jr.
Robert F. Thrash
Billy H. Thrasher, PhD
Ben H. Thurman
Kenneth Shane Tierling
Richard B. Timmons, PhD
Clarence N. Tinker
David R. Tinney
William R. Tioton
George R. Tippett
Craig A. Tips
William R. Tipton
Herbert G. Tiras
B. H. Tjrasjer, PhD
Robert M. Todor
James L. Tomberlin
Jocelyn Tomkin, PhD
William Harry Tonking, PhD
George Tope
Mary E. Totard
Mills Tourtellotte
Rusty Towell, PhD
Lawrence E. Townley
Phinn W. Townsend

Charles D. Towry
Kenneth L. Trachte, PhD
Laurence Munro Trafton, PhD
George Thomas Trammell, PhD
O. E. Trechter
Anthony D. Trevelstead
J. Trevino
David C. Triana
Leland Floyd Tribble, PhD
R. G. Tribble
George A. Trimble
Jay H. Troell
John S. Troschinetz
J. Michael Trotter
Leonard L. Trout
C. E. Trowbridge
John Parks Trowbridge
G. I. Troyer
Robert E. Truly
Thomas H. Tsai
Julio C. Tuberquia
Bill C. Tucker
Jasper M. Tucker
W. David Tucker
William E. Tucker
Daniel S. Tudor, PhD
John O. Tugwell
Frank Tull
Alton L. Tupa
Peter L. Turbett
D. Turbeville
Charles Paul Turco, PhD
Albert J. Turk
James H. Turk
Morris Turman
Dwight J. Turner
M. O. Turner
Michael D. Turner
Michael S. Turner
Rick Turner
E. W. Tuthill
James O. Tyler
Ronnald P. Tyler
John T. Tysseling
William S. Ullom
L. T. Umfleet
Adelbert C. Underwood
John R. Underwood, PhD
Paul Walter Unger, PhD
Simon Upfill-Brown
Chester R. Upham, Jr.
James Upton
Randal W. Utech
Alfred L. Utesch
Raymond E. Vache
Ashokkumar N. Vachhani
Claudio Valenzuela
William Pennington Van, PhD
Jack C. Van Horn
Earl D. Van Reenan
Rick Van Surksum
William E. Vanarsdale, PhD
Jerry L. Vanden Boom, PhD
Rainer A. Vanoni
William D. VanScoy
George G. Vanslyke
Kenneth L. Vantine
James E. Varnon, PhD
Mihal A. Vasilache
Joe E. Vaughan, PhD
David D. Vause
William Austin Veech, PhD

Felix J. Vega
Steve Venner
Bruce L. Veralli
G. Verbeck, PhD
John D. Vermaelen
Billy E. Vernon
Lonnie W. Vernon, PhD
Richard R. Vernotzy, Sr.
Ralph W. Vertrees
Harry A. Vest
Dan E. Vickers
Keith V. Vickers
Fredrick L. Vieck
Rodolfo L. Villarreal
Mikhail M. Vishik, PhD
Paul W. Visser
Russel C. Vlack
Rodney D. Vlotho
Jason P. Volk
George D. Volkel
Walter W. von Nimitz, PhD
Max R. Vordenbaum
Steve C. Voss
Bill S. Vowell
Don C. Vu
Elmo B. Wade
James W. Wade, PhD
John C. Wade
William M. Wadsworth
Don B. Wafer
Alfred Wagner, Jr.
David P. Wagner
Ray F. Wagner
Ross E. Wagner
Marvin L. Wagoner
Richard B. Waina, PhD
Charles Wakefield, PhD
Brian E. Waldecker, PhD
C. Ernest Wales
Harry M. Walker, PhD
John A. Walker
P. David Walker
Rodger L. Walker
William F. Walker
James P. Wallace
Lynn H. Wallace
Mark A. Wallace
Anthony Waller
William E. Wallis
William Walls
William N. Wally
Bruce W. Walter
James C. Walter
George Alan Walton
Donald C. Wambaugh
David Y. Wang, PhD
Gordon P. Ward
T. G. Ward, Jr.
John B. Wardell
Ray W. Ware
Darrell G. Warner
Robert A. Warner
Donald R. Warr
Bruce A. Warren
Cedric D. Warren
Joseph F. Warren
Robert L. Warters
G. S. Wassum
Damon L. Watkins
Michael J. Watkins
Mildred E. Watson
Randy K. Watson

Richard L. Watson, PhD
George Watters
Jack T. Watzke
Darel A. Wayhan
William Dewey Weatherford, Jr.,
PhD
Mark Weatherston
George L. Weaver
John M. Weaver
Ronald R. Weaver
Samuel R. Weaver
James F. Webb
Theodore Stratton Webb, PhD
David J. Weber
Monty Weddell
Curtis E. Weddle
Timothy R. Weddle
Olan B. Weeks
Scott C. Wehner
Bernard William Wehring, PhD
Jorge S. Weibel
Felicia K. Weidman
Donald G. Weilbaecher
Richard D. Weilburg
Lawrence Weinman
Lee Weis
R. Stephen Weis, PhD
Kent M. Weissling
M. E. Welbourn
Thomas L. Welch
Joe H. Wellborn
Howard T. Wells
Lewis W. Wells
Roger Murray Wells
Bruce Welsh
Clayton H. Wene
Richard A. Werner
Kenneth G. Wernicke
David E. West
J. C. West
Sidney E. West
Glen R. Westall
Gerald T. Westbrook
Mark E. Westcott
Jim P. Westerheid
William M. Westhoff
Harry Westmoreland
Thomas D. Westmoreland, Jr., PhD
Scott A. Westveer
Michael L. Wharton
Merle M. Whatley, PhD
Mitchell R. Whatley
C. Wheeler
Clyde C. Wheeler
Douglas H. Wheeler
Thomas J. Wheeler
William L. Wheeler
Lloyd Leon Whetzel
Richard M. Whiddon
Christian L. Whigham
B. Lee White, Jr.
D. J. White
Jerry L. White
Charles Hugh Whiteside, PhD
Jack Whiteside
Montague Whiting, PhD
Raymond B. Whitley
Richard E. Whitmire
Philip Whitsitt
Christopher J. Whitten
James C. Whitten
Lester B. Whitton

Charles D. Whitwill
Paul A. Wichmann
S. A. Wickstrom*
Steven G. Widen, PhD
Vernon R. Widerquist
Robert H. Widmer
J. A. Wiebelt, PhD
G. Wieland
John Herbert Wiese, PhD
Alan Wiggins
Larry Wiginton
Kenneth A. Wigner
Herman S. Wigodsky, PhD
Bill O. Wilbanks
Joe A. Wilbanks
Kenneth Alfred Wilde, PhD
James Wilder
Charles B. Wiley
Richard B. Wilkens
James Wilkes
Lambert H. Wilkes
Monte G. Wilkes
Stella H. Wilkes
Eugene M. Wilkins, PhD
Kevin L. Wilkins
Hector M. Willars
Al V. Williams
C. H. Williams, PhD
Danny E. Williams
Douglas B. Williams
Gary L. Williams
H. Hr. Williams
Jim Williams
Reginald D. Williams
Richard D. Williams, PhD
Roy D. Williams
Talmage T. Williams
Joe W. Williamson
Robert A. Williamson
Glen R. Willie
Allen H. Williford
H. Earl Willis
Gary L. Willmann
Charles L. Willshire
Bobby Wilson, PhD
David A. Wilson
Owen D. Wilson
Rick Wilson
Willie J. Wilson
D. Winans
Lars I. Wind
Edgar C. Winegartner
Christopher L. Wingert
David M. Winkeljohn
Weldon O. Winsauer
Donald K. Winsor
Frederic J. Winter
Joe E. Wirsching
Michael B. Wisebaker
Del N. Wisler
Eugene Harley Wissler, PhD
Wojciech S. Witkowski
Donald A. Witt
A. J. Wittenbach
Thomas R. Woehler
H. B. Wofford
Craig A. Wohlers
Earl G. Wolf
John C. Wolfe, PhD
Gary L. Womack
Robert Womack
Bob G. Wonish

David R. Wood
Scott Emerson Wood, PhD
Steve Wood
Charles R. Woodbury
Edward G. Woods, PhD
John P. Woods, PhD
Robert F. Woods
Herbert H. Woodson
Gary R. Wooley, PhD
Stephen T. Woolley
Bobby D. Woosley
Rodney L. Wooster
Darrell R. Word, PhD
Joe Max Word
Roy A. Worrell
Jonathan H. Worstell, PhD
Conrad H. Wright
Harold E. Wright
Thomas R. Wright, Jr.
Keith H. Wrolstad, PhD
Gang M. Wu, PhD
Peter T. Wu, PhD
W. H. Wurth
Reece E. Wyant
Philip R. Wyde, PhD
James M. Wylie
E. Staten Wynne, PhD
Will Yanke
Harold L. Yarger, PhD
Tommie E. Yates
Curtis Yawn
J. C. Yeager
Mark Yeager
Wilbur A. Yeager, Jr.
Jim P. Yocham
B. Biff Yochum
John G. Yonkers
Clifton L. Young
Farrile S. Young, PhD
Joe A. Young
Ruby D. Young
Susan W. Young
James L. Youngblood, PhD
James D. Younger
William M. Zahn, Jr.
Joseph E. Zanoni
Hua-Wei Zhou, PhD
Lois A. Ziler
James S. Zimmerman
John Zimmerman, PhD
Ralph Anthony Zingaro, PhD
Richard J. Zinno
Jerome Zorinksy, PhD
Alfonso J. Zuniga
Michael J. Zuvanich
Philip L. Zuvanich

Utah

Jason Abel
Terry Ahlquist
Marvin E. Allen
Merrill P. Allen
Raul C. Alva
D. Andersen
Russell Anderson
Wilbert C. Anderson
M. B. Andrus, PhD
Melvin B. Armstrong
Kenny Ashby
John R. Atkinson

D. Austin, PhD
Lloyd H. Austin
Robert J. Baczuk
Jay M. Bagley, PhD
R. E. Bailey
Ramon Condie Baird, PhD
Roger E. Banner, PhD
Mark Owens Barlow
M. D. Barnes, PhD
Howard F. Bartlett
Mark R. Bartlett, PhD
T. David Bastian
Joseph Clair Batty, PhD
Ken L. Beatty
Peggy Beatty
Albert I. Beazer
Boyd R. Beck, PhD
Merrill W. Beckstead, PhD
Blake D. Beckstrom
Gregory J. Bell
Alvin Benson, PhD
Wesley George Bentrude, PhD
David Alan Berges, PhD
John A. Bianucci
Kip A. Billings
Robert F. Bitner
Jan Bjernfalk
Clyde Blauer
Douglas Bledsoe
Lori Blum
Jerry C. Bommer, PhD
Blair E. Bona, PhD
Jay Frank Bonell
P. R. Borgmeier, PhD
David Borronam
John F. Boyle
Bruce Bradley
Ernest T. Bramwell
William W. Brant
Kennethe Martin Brauner, PhD
Thomas O. Breitling
Lynn H. Bringhurst
H. Smith Broadbent, PhD
Boyd C. Bronson
Billings Brown, PhD
John N. Brown
Arthur D. Bruno
Shawn P. Buchanan
Wallace Don Budge, PhD
Mark D. Bunnell
David D. Busath
Clinton Bybee
Charles H. Cameron
Earl E. Castner
Norman L. Challburg
Kenneth W. Chase, PhD
Norman Jerry Chatterton, PhD
Joseph A. Chenworth
Allen D. Child
Joseph G. Christensen
Jay P. Christofferson, PhD
John C. Clegg, PhD
Calvin Geary Clyde, PhD
Ralph L. Coates, PhD
Richard M. Cohen, PhD
Don J. Colton
Jack Comeford, PhD
Melvin Alonzo Cook, PhD
Michael L. Cosgrave
Robert E. Covington
Mac Crosby
Larry D. Culberson, PhD

Gary Monroe Cupp
Lowell C. Dahl
Donald Albert Dahlstrom, PhD
Jonathan H. Daines
Leland P. Davis
Robert Elliott Davis, PhD
Tom Dawson
Karl Devenport
C. Nelson Dorny, PhD
Andrew Drake
John J. Drammis
Gary L. Draper
Bryan B. Drennan
David A. Duncan
William S. Dunford
William R. Edgel, PhD
Allan C. Edson
George Eisenman
George L. Eitel
David A. Eldredge, Jr.
Mark T. Ellis
Robert W. English
Norman T. Erekson, PhD
Jerry A. Erickson, PhD
Creed M. Evans
Jeff Farrer, PhD
W. Dale Felix, PhD
David Feller, PhD
David Allan Firmage
Thomas H. Fletcher, PhD
Paul V. Fonnesbeck, PhD
Grant Robert Fowles, PhD
Richard L. Frost, PhD
Norihiko Fukuta, PhD
Ron Fuller
Blair O. Furner
Michael Aaron Gabrielson
Charles O. Gale
Lynn E. Garner, PhD
S. Parkes Gay, Jr.
Mark G. Gessel
Thon T. Gin, PhD
George P. Gowans
Fred L. Greer
K. Greer
Glen Griffin
Robert W. Griffiths
Richard W. Grow, PhD
Pamela Grubaugh-Littig
Michel B. Guinn
Robert L. Haffner
Peter H. Hahn
Edward H. Hahne
David R. Hall
Bill A. Hancock
Dallas Hanks
Darrell G. Hansen
Wayne R. Hansen
Larry R. Hardy
Ellis D. Harris
Franklin S. Harris, Jr., PhD
Ronald L. Harris
William Joshua Haslem
Scott E. Hassett
Stuart Havenstrite
Ralph W. Hayes
Creed Haymond
Dale G. Heaton
Lavell Merl Henderson, PhD
Deloy G. Hendricks, PhD
R. Dahl Hendrickson
James O. Henrie

Lyle G. Hereth
Charles Selby Herrin, PhD
John C. Higgins, PhD
Archie Clyde Hill, PhD
Allen E. Hilton, II
Clifford M. Hinckley, PhD
Robert L. Hobson, PhD
Grant Holdaway
David E. Holt
M. I. Horton, PhD
Ralph M. Horton, PhD
Frederick A. Hottes
Clayton Shirl Huber, PhD
J. Poulson Hunter
Harvey L. Hutchinson
Kyle Jackson
Daniel J. Janney
August Wilhelm Jaussi, PhD
Brian Jensen
L. Carl Jensen
Layne D. Jensen
Leland V. Jensen
Marcus Martin Jensen, PhD
Arlo F. Johnson, PhD
Owen W. Johnson, PhD
Todd Johnson, PhD
Von D. Jolley, PhD
Percy C. Jonat
Merrell Robert Jones, PhD
Walter V. Jones
John Jonkman
Royce A. Jorgensen
Mark Kaschmitter
Stephen Jorgensen Kelsey, PhD
Boyd M. Kitchen, PhD
Kenneth T. Klebba
Frederick Charles Kohout, PhD
David L. Kooyman, PhD
Steven K. Koski
William B. Laker
Burton B. Lamborn
Dale G. Larsen
Lloyd D. Larsen, PhD
Robert Paul Larsen, PhD
Robert W. Leake
Charles Lear
Gregory L. LeClaire
David J. Lee*
James Norman Lee, PhD
Lawrence Guy Lewis, PhD
Stephen Liddle, PhD
Thomas W. Lloyd
Wilburn L. Luna
Mark Wylie Lund, PhD
Charles P. Lush
Joseph L. Lyon
William E. Maddex
Robert K. Maddock
Jules J. Magda, PhD
Kenneth A. Mangelson, PhD
Brent P. Marchant
Eugene Marshall
John May
Vern T. May
D. Maynes, PhD
Robert Eli McAdams, PhD
John McDonald, PhD
Susan O. McDonald
Timothy W. McGuire
Jeffrey D. Mckay
Cyrus Milo McKell, PhD
Jefferson D. McKenzie

Hugh A. McLean
Dick N. Mechem
Vincent J. Memmott
John M. Mercier
LaVere B. Merritt, PhD
Daniel W. Miles, PhD
John G. Miles
Gene W. Miller, PhD
Wade Elliott Miller, II, PhD
Paul Mills
Bryant A. Miner, PhD
Terrence Mischler, PhD
Robert Alan Mole
J. Ward Moody, PhD
M. Lyman Moody
Karl G. Morey
Howard C. Morrill
Kay S. Mortensen, PhD
William R. Mosby
Thomas C. Moseley
Loren Cameron Mosher, PhD
Alfred R. Motzkus
Glenn L. Mower
Harold W. Mulvany
Grant R. Murdoch
Donald L. Myrup
Dave Neale
Kenneth W. Nelson
Manfred R. Nelson
Michael G. Nelson, PhD
Pamela Nelson
Gregg Nickel
Barry C. Nielsen
Dennis M. Nielsen
Donald R. Nielsen, PhD
Edwin Clyde Nordquist
Robert Quincy Oaks, Jr., PhD
Hayward B. Oblad, PhD
Leonard Olds
Larry S. Olsen
Merrill H. Olson
Ted Olson
Robert E. O'Neill
Alfred M. Osborne
Robert Pack, PhD
Lowell L. Palm
Tom T. Paluso
Joseph M. Papenfuss, PhD
Gerald M. Park
Robert Lynn Park, PhD
Forest Parry
Robert Walter Parry, PhD
Ernest B. Paxson, PhD
Clifford C. Payton
Justin B. Peatross, PhD
Leslie H. Pennington
Danial L. Perry
Jay D. Peterson
Randall D. Peterson
Jay W. Phippen, PhD
David Pierson
Paul E. Pilgram
Michael L. Pinnell
Charles H. Pitt, PhD
William G. Pitt, PhD
Gary F. Player
Steven J. Postma
Bradley M. Powell
John D. Prater
Jim S. Priskos
Frederick Dan Provenza, PhD
Kenneth Puchlik

Ronald J. Pugmire, PhD
Samuel C. Quigley
Francis Rakow
Carole A. Ray
Jory Ty Redd, PhD
Donal J. Reed, PhD
Alvin C. Rencher, PhD
Brian W. Rex
Max J. Reynolds
Michael D. Rice, PhD
Lynn S. Richards, PhD
Gary Haight Richardson, PhD
Harold W. Ritchey, PhD
Samuel P. Roberts
Alden C. Robinson
Clay W. Robinson
T. F. Robinson, PhD
Lee E. Rogers
William C. Romney
Simpson Roper
Harold L. Rosenhan
Charles Ross
George W. Sandberg
LeRoy J. Sauter
Keith Jerome Schaiger, PhD
Ernest F. Scharkan
Edward B. Schoppe
Spencer L. Seager, PhD
Richard C. Sharp
W. Kenneth Sherk
Jay G. Shields
Charles W. Shoun
Richard Phil Shumway, PhD
Val E. Simmons, PhD
David Michael Sipe, PhD
Duane Sjoberg
Ross A. Smart
Robert B. Smith
Tracy W. Smith
Leonard W. Snellman, PhD
Richard L. Snow, PhD
Phillip T. Solomon
Carl D. Sorensen, PhD
Wesley K. Sorensen
Brent A. Sorenson
Forrest L. Staffanson, PhD
Kent C. Staheli
Norman E. Stauffer, PhD
K. S. Stevens, PhD
Gene M. Stevenson
Leo M. Stevenson
Kenneth J. Stracke
Glen Evan Stringham, PhD
William J. Strong, PhD
Calvin K. Sudweeks
Robert Summers, PhD
Richard Swenson
B. A. Tait, PhD
Carlos Tavares
Doug Taylor
James S. Taylor, PhD
Lester Taylor
Randall L. Taylor
Richard Tebbs
Abraham Teng, PhD
Lawrence E. Thatcher
John D. Thirkill
Don Wylie Thomas, PhD
James H. Thomas, PhD
Grant Thompson, PhD
Philip Thompson
Richard M. Thorup, PhD

Richard N. Thwaitis, PhD
Rudy W. Tietze
Terry L. Tippets
David O. Topham
Anthony R. Torres
Del Traveller
J. Paul Tullis, PhD
George Uhlig, PhD
Richard M. Vennett, PhD
John M. Viehweg
Charles T. Vono
William Vorkink, PhD
John W. Vosskuhler
Kathleen L. Wegener
Joseph C. Wells, Jr.
Lewis F. Wells
Eleroy H. West
D. R. Westenskow, PhD
Edward B. Whicker
Brent L. White
Lowell D. White, PhD
Thomas G. White
Leslie Whitton, PhD
Harry E. Wickes, PhD
Clifford Lavar Wilcox, PhD
David P. Wilding
W. Vincent Wilding, PhD
Bill Beauford Wiley, PhD
Charles Ronald Willden, PhD
Arnold Wilson, PhD
Byron J. Wilson, PhD
Stephen E. Wilson
Robert J. Wise
John K. Wood, PhD
Jack L. Woods
Richard E. Wyman
Thomas B. Yeager
Glen A. Zumwalt

Vermont

Paul A. Bailly, PhD
Robert A. Baum
Bruce Irving Bertelsen
Lawrence Bilodeau
James H. Burbo
Bernard Byrne, PhD
Robert E. Chiabrandy
Donald Lyndon Clark, PhD
William Wesley Currier, PhD
Barry S. Deliduka
Blair J. Enman
Nils C. Ericksen
James Evans
Paul D. Fear
Daniel P. Foty, PhD
Paul D. Gardner
William Halpern, PhD
Ray A. Hango
Grant Hopkins Harnest, PhD*
David Japikse, PhD
David J. Kuhnke
Thomas J. Laplaca, PhD
Anthony G. Lauck
Peter P. Lawlor
George Butterick Maccollom, PhD
Kenneth Gerard Mann, PhD
William W. Martinez
John McClaughry
Jack Noll
Duncan G. Ogden

Donald C. Oliveau
Frank R. Paolini, PhD
Henry B. Patterson
Malcolm J. Paulsen
James A. Peden
Hans J. Pfister
Kerri L. Polli
Olin E. Potter
Frederick H. Raab, PhD
Benjamin F. Richards, Jr., PhD
Albert L. Robitaille
Thomas Dudley Sachs, PhD
Phillip J. Savoy
Christian T. Scheindel
Josh Schultz
Byron G. Sherman
Thomas Smith, PhD
Robert F. Stuart
Frank J. Thornton
Waclaw Timoszyk, PhD
Ralph Wallace, PhD

Virginia

Allwyn Albuquerque
Ernest Charles Alcaraz, PhD
James F. Alexander, Jr.
David M. Allen
Kristin L. Allen
Frank Murray Almeter, PhD
Andrew S. Anderson, PhD
Gerald L. Anderson
John Andrako, PhD
Scott Andrews, PhD
Steven T. Angely
Lee Saunders Anthony, PhD
Richard J. Ardine Arthur
William E. Armour
Randolph W. Ashby, PhD
Michael J. Atherton, PhD
James C. Auckland
Brian Augustine, PhD
Alex Avery
Vernon H. Baker, PhD
David A. Baldwin, PhD
Mark A. Ball
Greg Ballengee
George R. Barber, Jr.
Theodore K. Barna, PhD
Scott A. Barnhill
Maria V. Bartha
John Wesley Bartlett, PhD
Thomas A. Bartlett
Mark D. Basile
James Bateman
Alan P. Batson, PhD
Kirk Battleson, PhD
Bradley M. Bauer
Peyton B. Baughan
Lester F. Baum
Mike B. Bealey
George A. Beasley
Kenneth L. Bedford, PhD
Eric Bellows
Paul J. Berenson, PhD
James R. Berger
Marion Joseph Bergin
Peter F. Bermel
Simon S. Bernstein, PhD
Robert W. Berry
Eguene R. Biagi

Rudolf G. Bickel
Charles F. Bieber
Billy Elias Bingham, PhD
Harlan Fletcher Bittner, PhD
Rebecca B. Bittner
Thomas E. Bjerke
Albert W. Blackburn
Paul M. Blaiklock, PhD
Robert V. Blanke, PhD
George K. Bliss
John C. Bloxham
Randy L. Boice
Arthur F. Boland
Robert G. Boldue
W. Kline Bolton
Robert A. Book, PhD
John W. Boring, PhD
Robert Clare Bornmann
Benjamin A. Bosher*
Tye W. Botting, PhD
Truman Arthur Botts, PhD
Donald C. Bowman
Ricky L. Bowman
Joseph V. Braddock, PhD
James K. Brandau
William L. Brandt
Townsend D. Breeden
William B. Brent, PhD
Philip W. Brewer
Frank Brimelow
Robert Britton
John A. Brockwell, Sr.
Kevin J. Brogan, PhD
Kent Brooks, PhD
Steven D. Brooks
Paul Wallace Broome, PhD
Christopher J. Brown
Doug L. Brown
Henry B. R. Brown
J. Paul Brown
James A. Brown
Jeffrey P. Brown
Jerry W. Brown, PhD
Charles S. Bryant
Robert D. Buchy
Joseph A. Buder
Merton W. Bunker, Jr.
Kenneth Burbank, PhD
Annette Burberry
Thomas J. Burbey, PhD
Robert J. Burger
John P. Burkett
Harold Eugene Burkhart, PhD
Mahlon Admire Burnette, III, PhD
Rhendal Butler
Robert Campbell
William H. Campbell, PhD
Homer Walter Carhart, PhD
David R. Carlson
Kenneth G. Carlson
Aubrey M. Cary, Jr.
Douglas R. Case
Davis S. Caskey
Armand Ralph Casola, PhD
Neal Castagnoli, PhD
Derek Cate
James R. Cavalet
Steven L. Cazad
Charles Mackay Chambers, PhD
Sanjay Chandra
Fred R. Chapman
Rees C. Chapman

Telesphore L. Charland
Joseph D. Chase, PhD
John T. Chehaske
Wayne P. Chepren, PhD
Nancy A. Chiafulio
Melvin C. Ching, PhD
Alfred N. Christiansen
John A. Cifala
Donald T. Clark
Randall A. Clark
R. Clay
Dennis Cline
James C. Clow
Daniel G. Coake
Walter Cobbs, PhD
James F. Coble, PhD
Benjamin R. Cofer
Murry J. Cohen
Albert R. Colan
James E. Collins
William F. Collins, PhD
Timothy B. Compton
Bryan Keith Condrey
David A. Conley
John Irving Connolly, Jr., PhD
Frederick E. Conrad
Terence G. Cooper
Richard H. Coppo
William P. Corish
Frank Corwin
Roscoe Roy Cosby, Jr.
David G. Cotts
Edwin Cox
Forrest E. Coyle
Barry D. Crane, PhD
Steven P. Crawford
Stephen J. Creane
Walter B. Crickman
A. Scott Crossfield
Lloyd R. Crowther
John C. Crump, III, PhD
Daniel J. Cunningham, III
Thomas Fortsen Curry, PhD
James M. Cutts
James Frederick Dalby, PhD
Harry P. Dalton, PhD
Thomas J. Dalzell
David A. Danello
Joe E. Darby
Ralph E. Darling
Bradley Darr
John Siv Davis
Philip E. Davis
Barry G. Dawkins
James Carroll Deddens
Gary C. Defotis, PhD
Rosalie F. Degiovanni-Donnelly, PhD
Stephen Degray
Michael P. Deisenroth, PhD
Joseph F. Delahanty
Barbu J. Demian, PhD
Alan M. Dennis
Edmund G. Dennis
Raymond Edwin Dessy, PhD
Raymond A. DeStephen
G. W. Detty
John F. Devaney
James A. Deye, PhD
John Dickmann
Robert B. Dillaway, PhD
John K. Dixon, PhD

Wayne A. Dixon
Robert Dobessette
D. C. Doe
Robert C. Dolecki
Jocelyn Fielding Douglas, PhD
Rose T. Douglas
Basil C. Doumas, PhD
Frederick Joseph Doyle, PhD
David J. Drahos, PhD
Bill Drennan
P. Du
Michael Serge Duesbery, PhD
Samuel M. Dunaway, Jr.
Charles F. Dunkl, PhD
Kevin M. Dunn, PhD
James H. Durham
Thomas Earles
W. S. Easterling, PhD
Charles J. Eby, PhD
Michael T. Eckert
Arthur L. Eller, PhD
Stanley M. Elmore
R. Elswick, PhD
Thomas C. Elwell
Thomas H. Emsley
John J. Engel
John R. English
Gerard J. Enigk
James W. Enochs, Jr.
Jeffrey F. Eppink
James L. Erb
Doug Erickson
John Erickson
Manuela Erickson
John R. Esser
Edward R. Estes, Jr.
Juliette A. Estevez
Pamel F. Faggert
Dennis A. Falgout, PhD
Michael A. Farge
Anthony L. Farinola
W. Michael Farmer, PhD
Ollie J. Farnam
Frank Hunter Featherston, PhD
James Love Fedrick, PhD
Michael B. Feil
Harvey L. Fein, PhD
Michael D. Fenton, PhD
Joseph L. Ferguson
William P. Fertall
David C. Finch
Alvin W. Finestone
Martin C. Fisher
Richard Harding Fisher
Donald E. Fleming
Donald K. Forbes
Michael Shepard Forbes, PhD
R. C. Forrester, PhD
Mark A. Forte
Eugene K. Fox
Russell Elwell Fox, PhD
Roger Frampton, PhD
Daniel Frederick
Herbert Friedman, PhD
Charles Sherman Frye
Keith Frye, PhD
Michael L. Fudge
Harold Walter Gale, PhD
Mario R. Gamboa
A. K. Ganguly, PhD
Hessle Filmore Garner, PhD
Jerry H. Gass

Stephen Jerome Gawarecki, PhD
Henry T. Gawrylowicz
James J. Gehrig
Richard B. Geiger
Daniel F. Geldermann
Jerauld Gentry
George C. Gerber
David P. Gianettino
Michael G. Gibby, PhD
Jonathan Gifford, PhD
George Thomas Gillies, PhD
Mark S. Goff
Arthur A. Goodier
Hubert C. Goodman, Jr.
Louie Aubrey Goodson
Heinz August Gorges, PhD
Paul Gorman
Steven D. Goughnor
Roy R. Goughnour, PhD
S. William Gouse, Jr., PhD
Martha S. Granger
George A. Gray, PhD
Michael N. Greeley
Steve R. Green
John P. Greenaway
Everette D. Greinke
Norman J. Griest
Jerry L. Grimes
Ronald J. Gripshover, PhD
Sharon L. Gripshover
Ernest E. Grisdale
Bernardo F. Grossling, PhD
Daniel G. Hadacek
Henry A. Haefele
Gary J. Hagan
Lars B. Hagen
Harold B. Haley
Brian J. Hall
Brian Hall
Harvey Hall, PhD
Christian L. Haller
Gordon S. Hamilton
Stevan T. Hanna
Edward Hanrahan, PhD
Robert H. Hansen
Richard C. Hard
Mary K. Harding
P. L. Hardy, Jr.
Robin C. Hardy
Thomas J. Harkins
David W. Harned
Russell S. Harris, Jr.
Hetzal Hartley
Homer A. Hartung, PhD
Ken Haught, PhD
Dexter Stearns Haven
Harold E. Headlee
Sean M. Headrick
Chase P. Hearn
Christian W. Hearn
David R. Heebner
Glenn R. Heidbreder, PhD
Robert Virgil Hemm, PhD
Frank D. Henderson
Ronald J. Hendrickson
George Burns Hess, PhD
William R. Hestand
Dennis W. Heuer
James Veith Hewett, PhD
Richard M. Hicks
Ronald Earl Highsmith, PhD
Peter J. Hill

Edward W. Hiltebeitel
Barton Leslie Hinkle, PhD
Jeffery G. Hinson
Donald A. Hirst, PhD
Charles C. Hogge
Thurman Holcomb
Michael D. Holder
Edward J. Hollos
W. J. Hooker, PhD
Carroll B. Hooper
Philip E. Hoover
T. E. Hopkins
William J. Horvath
Eric W. Howard
Kevin B. Howerton
James Secord Howland, PhD
Tsuying Hsieh, PhD
Juergen Hubert, PhD
William L. Hudgins
Francis David Hufford
Greg R. Hughey
David M. Human
Jeffrey E. Humphrey, PhD
Joseph W. Hundley, Jr.
Lee McCaa Hunt
David Hunter
Thomas C. Hunter
Samir Hussamy, PhD
James B. Hutt, Jr.
Michael B. Hydorn, PhD
Robert Mitsuru Ikeda, PhD
Anton Louis Inderbitzen, PhD
Jeffrey Ingram
Carl T. Jackson
George A. Jahnigen
John A. Jane, Sr., PhD
Dale R. Jensen
Mark Jerussi
William A. Jesser, PhD
Tedd H. Jett
Charles Minor Johnson, PhD
Robert S. Johnson, PhD
Gerald L. Jones
Oliver Jones
Richard D. Jones
Wellington H. Jones
C. R. Judy
H. Julich
Mario George Karfakis, PhD
Jeffrey M. Kauffman
Peter W. Kauffman
Saul Kay
John Kckert
Ray N. Keffer
Seth R. Kegan
Karl S. Keim
Bernhard Edward Keiser, PhD
Claud Marvin Kellett
Timothy E. Kendall
Jack Martin Kerr
John C. Kershenstein, PhD
Donald J. Kienast
Theo Daniel Kimbrough, PhD
Bessie H. Kirkwood, PhD
Ahto Kivi
George J. Klett
Denny L. Kline
S. J. Kloda
John M. Knew
Maurice Kniceley, Jr.
C. P. Koelling, PhD
Gary L. Koerner

Fred Koester, Jr.
Joshua O. Kolawole, PhD
Andrew Kolonay
Brent D. Kornman
Theresa M. Koschny
E. H. Krafft
Geoffrey A. Krafft, PhD
David E. Kranbuehl, PhD
Robert K. Krout
Peter George Krueger, PhD
Eric M. Krupacs
John M. Lafferty, Jr.
William A. Laing
Stephen A. Lamanna, PhD
Roland J. Land
Richard A. Lander
Thomas Larry, PhD
Calvert T. Larsen, PhD
Jeffrey G. Larson
Thomas J. Lauterio, PhD
William F. Lavecchia
Armistead M. Leggett
Daniel D. Lelong
Stanley H. Levinson, PhD
Michael D. Lewes
David W. Lewis, PhD
Peter E. Lewis
Scott B. Lewis, PhD
Verna M. Lewis
Gabriel Lifschitz
Philip D. Lindahl
Charles J. Lindemann, II
Robert B. Lindemann
Lenny W. Lipscomb
Clement M. Llewellyn, Jr.
Krystyna Kopaczyk Locke, PhD
Raymond Kenneth Locke
David Loduca
Edward R. Long, PhD
Jean E. Loonam, PhD
William Farrand Loranger, PhD
Thomas A. Loredo
James L. Loveland
James A. Lovett
David N. Lucas
Daniel D. Ludwig, PhD
John Henry Lupinski, PhD
Joseph W. Luquire, PhD
Travis O. Lyday
Michael E. Lyle
William G. Maclaren, Jr., PhD
Michael T. Madden
Dave N. Mahony
Thomas H. Maichak
Albert C. Malacarne
Martin J. Mangino, PhD
John E. Mansfield, PhD
Robert R. Mantz
George E. Marak
Glenn R. Marchione
Thomas Ewing Margrave, PhD
John L. Marocchi
Robert L. Marovelli
Patrick D. Marsden
Sewell R. Marsh
James A. Marshall, PhD
Allen R. Martin
Thomas D. Mattesonn
Robert D. Matulka, PhD
Robert L. Maynard, Jr.
Steven E. Mays
John Hart McAdoo, PhD

William C. McCarthy
Robert F. McCarty
James K. McClanahan
Tracy McCombs
Fred Campbell McCormick, PhD
Irwin H. McCumber
Daniel F. McDonald, PhD
Diane McDonaugh
David S. McDougal
Tomas T. McFie, PhD
Charles Wesley McGary, Jr., PhD
Thomas J. McGean
Henry Alexander McGee, Jr., PhD
James G. McGee
Thomas H. McGorry
Lea B. McGovern
James E. McGrath, PhD
Robert J. McKay
Thomas E. McKiernan
George McKinney
Robert L. McMasters, PhD
John W. McNair
Harry W. Meador
Walter S. Medding
Barbara Mella
Arthur Mendonsa
T. Metzgar, PhD
William F. Meyer
Earl Lawrence Meyers, PhD
Patrick J. Michael, PhD
David John Michel, PhD
Robert Michel
Brooks A. Mick
Ronald J. Mickiewicz
Charles R. Miller, III
Marshall Miller
Larry L. Misenheimer
Alfred F. Mistr, Jr.
John E. Mock, PhD
Jacob Moll
Thomas W. Moodie
Thomas K. Moore, PhD
Matthew Moriarty
Cecil Arthur Morris, PhD
Leonard Morrow, PhD
Melody A. Morrow
Vivekanand Mukhopadhyay, PhD
Ragnwald Muller
Robert Murdock
Roy K. Murdock
Robert Stafford Murphey, PhD
Wayne K. Murphy, PhD
Clifford A. Myers
Larry K. Myrick
Fred W. Nagle, Jr.
Brett Newman, PhD
Nancy Newman
K. D. Ngo, PhD
Lee Lockhead Nichols, PhD
Alvin E. Nieder
Mark J. Norden
Pamela Norris, PhD
Thomas Nyman
E. J. O'Brien
Gordon Oehler, PhD
Walter Clayton Ormsby, PhD
Peter P. Ostrowski, PhD
Wesley Emory Pace
Jack Paden
Robert R. Palik
Nina Parker, PhD
K. W. Parrish

H. D. Parry
Joseph A. Paschall
Thomas Joseph Pasko
Marvin P. Pastel, PhD
Kerry B. Patterson
Wilbur I. Patterson, PhD
Harold J. Peake
Edward F. Pearsall, PhD
Eugene Lincoln Peck, PhD
Thomas R. Pepitone, PhD
Paul E. Perrone
John Edward Perry
Rafayel A. Petrossian
David Van Petten
Gerard David Pfeifer, PhD
Arun G. Phadke, PhD
John J. Phillips, Jr.
Roland W. Phillips, Jr.
Timothy Pickenheim
Rhonda L. Pierce
O. Pierrakos, PhD
Vincent A. Pierro
John B. Pietron
H. Alan Pike, PhD
Herman H. Pinkerton, PhD
Ricker R. Polsdorfer
Benjamin W. Pope
H. William Porterfield
William W. Porterfield, PhD
John T. Postak
Donald H. Potter
Edward T. Powell, PhD
Arel W. Powers
Steven Price, PhD
Fred H. Proctor, PhD
Anthony J. Provenzano, PhD
Charles S. Pugh
Joan L. Purdy
Walter V. Purnell, Jr.
Robert H. Pusey, PhD
Richard Lee Puster
William P. Raney, PhD
Bhakta Bhusan Rath, PhD
Richard Travis Rauch, PhD
Tomas Reamsnyder
Kay P. Reasor
John G. Reed
Barry G. Reid
Keith W. Reiss, PhD
David E. Reubush
James M. Reynolds
Fred R. Riddell, PhD
Harold E. Ritter
Joseph Peter Riva
Joseph A. Robinson
Samuel A. Rodgers
David Rogers
Henry M. Rogers
Thomas M. Rohrbaugh
Oscar A. Rondon-Aramayo, PhD
Thomas J. Rosener, PhD
Fred Rosi, PhD
Tom L. Rourk
John Rowland, PhD
Charlotte E. Roydhouse
John Rudolf
John R. Ruff
Larry E. Ruggiero
Clifford D. Russell
Todd M. Rutherford
John Ruvalds, PhD
Fredrick J. Ryan

Robert R. Ryland
Aubrey E. Sadler
Donald L. Sage
Gerald O. Sakats
Donald R. Salyer
W. J. Samford
Vernon C. Sanders
G. S. Santi
David P. Sargent
Peter Thomson Sarjeant, PhD
James M. Sawhill, Jr., PhD
Francis T. Schaeffer
Joseph T. Schatzman
Frank R. Scheid, Jr.
Karl A. Schellenberg, PhD
Joseph A. Schetz, PhD
Robert J. Scheuplein, PhD
Raymond Schneider, PhD
Robert W. Schneiter, PhD
Matthew J. Schor
Dunell V. Schull
Eugene H. Scott, PhD
James Byrd Seaborn, PhD
Paul Sebula
John H. Seipel, PhD
J. Michael Selnick
Karl A. Sense
John L. Sewall
James Shaeffer, PhD
Donald G. Shaheen
Paul Shahinian, PhD
Donald E. Shaneberger
John A. Shannon
Anthony R. Sharkey
James J. Shea
Gary A. Sheehan
Jennifer P. Sheets
Scott T. Shipley, PhD
Gilbert R. Shockley, PhD
Stephen A. Shoop
George M. Shrieves
George John Simeon, PhD
Gordon B. Sims, Jr.
S. Fred Singer, PhD
Herbert James Sipe, Jr., PhD
Jerome P. Skelly, PhD
Alfred Skolnick, PhD
J. L. Slaughter
Daniel F. Sloan
Gregory E. Slominski
Anthony R. Slotwinksi
Blanchard D. Smith
Bryan W. Smith
Dale Smith, PhD
Emerson W. Smith
Ray M. Smith
Lyle C. Snidow, III
Glenn J. Snyder
Jaroslaw Sobieski, PhD
Joseph E. Sohoski
Andrew Paul Somlyo
Donald E. Sours, Sr.
Samuel V. Spagnolo
David W. Speer
Gilbert Neil Spokes, PhD
Lucian Matthew Sprague, PhD
Edward James Spyhalski, PhD
William S. Stambaugh, PhD
Raymond M. Standahar
Richard G. Starck, II
Lendell Eugene Steele
Bland A. Stein

Kenneth Stepka
Scott M. Stevens
Robert R. Stimpson
Jan Stofberg, PhD
Evangelos P. Stoyas
Sharon R. Streight, PhD
Walter Strohecker
Harry M. Strong
Erick Stusnick, PhD
Jack A. Suddreth
Ronald M. Sundelin, PhD
William G. Sutton, PhD
Waller C. Tabb
Keith A. Taggart, PhD
Sandra R. Tall
Lukas Tamm, PhD
Suren A. Tatulian, PhD
Joseph D. Taylor, PhD
Gilbert A. Tellefsen
Wilton R. Tenney, PhD
Milton D. Tetterton, Jr.
John S. Theon, PhD
Matthew A. Thexton
Melton Sherwood Thoele
Thomas Thorpe
Albert William Tiedemann, PhD
Samuel C. Tignor, PhD
Calvin Omah Tiller
James F. Tobey, Jr.
Joseph J. Tobias
Gregory A. Topasna, PhD
Edward Thomas Toton, PhD
Priestley Toulmin, PhD
Richmond S. Trotter
Walter L. Turnage
John Charles Turner, PhD
Ian Charles Twilley, PhD
Russell K. Tyson
Brian G. Valentine, PhD
George J. Vames
Richard Van Blaricom, PhD
Jim VanEaton, PhD
Dirk G. Vanvort
Frederick Vastine, PhD
Sidney Edwin Veazey, PhD
Karl F. Veith, PhD
Willard A. Wade, II
Jon A. Wadley
Ann S. Walker
Richard D. Walker, PhD
Sharyl E. Walker, PhD
J. L. Wallace
Raymond Howard Wallace
Richard C. Walsh
Theodore Ross Walton, PhD
John W. Ward, PhD
Holland Douglas Warren, PhD
David T. Warriner
William A. Watson, III, PhD
Harry Weaver
Robert C. Weaver
Randy L. Weingart
Brian L. Weinstein
Martin Weiss, PhD
Elizabeth Fortson Wells, PhD
John L. Welsh
Roger L. Wesley
Alfred L. West, II
Warwick R. West, PhD
Kurt O. Westerman
Curt Wexel
David E. Wheeler

Jerry Whidby, PhD
Oliver B. White
Richard P. White
Dale Whitman
Robert R. Wiederkehr, PhD
King W. Wieman
Donald Richard Wiesnet
Gaylord S. Williams
Stafford H. Williams
Sheldon P. Wimpfen
Russell Lowell Wine, PhD
Billy K. Wingfield
Reed H. Winslow
Eleanor Stiegler Wintercorn, PhD
Dermot M. Winters
James C. Wirt
Lawrence Arndt Wishner, PhD
Abund O. Wist, PhD
Norman C. Witbeck, PhD
James M. Witting, PhD
Marcus M. Wolf
Eligius A. Wolicki, PhD
Walter Ralph Woods, PhD
Eugene G. Wrieden
Ed Wright
M. R. J. Wyllie, PhD
Yijing Xu, PhD
Edward Carson Yates, Jr., PhD
Philip R. Young, PhD
Joseph A. Zak, PhD
Laura Zischke
Todd Zischke
Warren Zitzmann
Ernst G. Zurflueh, PhD

Washington

Bryant L. Adams, PhD
William M. Adams, PhD
Grant Alberich
Marcus Albro
Garrett D. Alcorn
Franklin Dalton Aldrich, PhD
Joel W. Alldredge
William Allen, Jr.
Sidney J. Altschuler
Dayton L. Alverson, PhD
Larry C. Amans
Donald Heruin Anderson, PhD
Mark A. Anderson
Theodore D. Anderson
Tom P. Anderson
William L. Anliker
Norman Apperson
William S. Arnett
Bob J. Ascherl
Alvin G. Ash
Eugene Roy Astley
Kent E. Ausburn, PhD
William Dudson Bacon, PhD
Quincey Lamar Baird, PhD
Steven M. Baker, PhD
George M. Barrington, PhD
William G. Bartel
Edward J. Barvick
Dale E. Bean
Richard W. Becker
Wendland Beezhold, PhD
John A. Bell, PhD
James C. Bellamy
Randy Belstad

Dean A. Bender
John A. Bennett
Andrew W. Berg
John C. Berg, PhD
Richard A. Bergquist
Bertram Rodney Bertramson, PhD
Bob Betcher
Nedavia Bethlahmy, PhD
Jack S. Bevan
Joseph John Bevelacqua, PhD
Hans Bichsel, PhD
Darrel Rudolph Bienz, PhD
Theo Karl Bierlein, PhD
George A. Block
Leo R. Bohanick
Michael A. Bohls
David Boleneus
Mary Ann Bolte
Glenn G. Bonacum
L. P. Bonifaci
John Franklin Bonner, PhD
Peter B. Bosserman
Kaydell C. Bowles
Dale Boyce
James Boyd
Kim R. Boyer
James A. Boyes
James A. Bradbury
Douglas G. Braid
Grigore Braileanu, PhD
William C. Breneman
Henry W. Brenniman, Jr.
Larry K. Brinkman
Richard K. Brinton
Alfred Carter Broad, PhD
Edwin C. Brockenbrough
Casey Brooks
Gordon W. Bruchner
Hayes R. Bryan, PhD
Ben S. Bryant, PhD
John P. Buchanan, PhD
Fred C. Budinger
Robert Joseph Buker, PhD
Wilbur Lyle Bunch
Sherman K. Burd
Robert S. Burdick
Everett C. Burts, PhD
Robert C. Buto
Jerry R. Cahan
Dennis Caird
John M. Calhoun
Brian K. Callagan
Martin J. Campbell
Milton Hugh Campbell
Carl M. Canfield
Frank Caplan
Frederick Paul Carlson, PhD
William D. Carson
Joseph W. Cartwright
Robert L. Cashion
Stan J. Casswell
Dominic Anthony Cataloo, PhD
Howard E. Ceion
Kyle W. Chapman
Douglas E. Chappelle
William R. Chastain
Thomas David Chikalla, PhD
John G. Chittick
Darrel L. Choate
Mahlon F. Christensen
Robert N. Christenson
Robert J. Cihak

Marc Cimolino, PhD
Arlen B. Clark
Kenneth Clark
Robert Clark
Gerhardt C. Clementson
William A. Clevenger
Gary S. Collins, PhD
Ralph J. Comstock
Michael W. Conaboy
John P. Conway
Mary W. Conway
Gary A. Cooke
Joan M. Cooke
Peter Cooper
Charles J. Cordalis
Jeffrey A. Corkill, PhD
Neil Cornia
Matthew Corulli
Kirk D. Cowan
Brian J. Coyne, PhD
Dan Crackel
Elmer M. Cranton
Richard A. Craven
Bruce C. Crist
William S. Croft
Harvey D. Curet
Walter E. Currah
Roy E. Dahl, PhD
Hubert A. Dame
Robert Arthur Damon, PhD
Jess Donald Daniels, PhD
Edwin M. Danks
William C. Davis
D. M. Day
Edgar Dale De Remer, PhD
Gary M. De Winkle
Douglas A. Dean
Samuel J. Dechter
Hans G. Dehmelt, PhD
Robert O. Dempcy, PhD
Douglas C. Dennison
Herman A. Dethman
Roy J. Dewey
Ian Andrew Dick
Ronald John Dinus, PhD
Arnold D. Ditmar
Carroll Dobratz, PhD
Ralph J. Domsife
Bruce R. Donaldson
Roark M. Doubt
William B. Dress, PhD
Philip Gordon Dunbar, PhD
Richard G. Duncan
Michael J. Dunn
Kenneth Laverne Dunning, PhD
Mark W. Durrell
Frank Arthur Dvorak, PhD
David L. Dye, PhD*
Eric Dysland
Don J. Easterbrook, PhD
Fred Edelman
Donald Carl Elfving, PhD
Everett L. Ellis, PhD
Mark S. Erdman
Dale L. Erickson
Roger Eriekson
Peggy R. Erskine
Bruce V. Ettling, PhD
William Henry Eustis, PhD
Thomas Walter Evans
F. R. Fahland
Robert E. Fankhauser

Marc W. Fariss, PhD
David Farrell
William D. Farrington
George W. Farwell, PhD
Felix Favorite, PhD*
V. Robert Feltin
David L. Feray
Robert B. Ferguson
Kenneth A. Feucht, PhD
Tim Figgie
Roy H. Filby, PhD
M. Firpo
Mel Fiskum
Trevor Fitzgerald
H. Dean Fitzsimmons
Jane A. Fore
Harold K. Forsen, PhD
Mark B. Fosberg
Norman Charles Foster, PhD
Earl Fox
Michael R. Fox, PhD
E. L. Freddolino
James Harold Freisheim, PhD
George M. Frese
Roman D. Fresnedo, PhD
Gary S. Friehauf
Robert O. Frink
Paul F. Fuglevand
Charles V. Fulmer, PhD
Alan D. Fure
Charles Gades
Roger W. Gallington, PhD
Scott Garbarino
Maurice David Gardner
Todd Garvin
LeRoy Gebhardt, PhD
Michael Gelotte
William W. Gill, PhD
George A. Gillies, PhD
Victor C. Gilliland
John Glissmeyer
Edward L. Godsey
Patrick Goldsworthy, PhD
Guillerme Gonzalez, PhD
Torben Goodhope
Howard Gorsuch
John C. Graham
Arvid P. Grant, PhD
Arthur R. Green
J. T. Gregor
Lawrence E. Gulberg, PhD
Rentz Gullick
Alankar Gupta
Stephen A. Habener
Nathan Albert Hall, PhD
John Emil Halver, PhD
John K. Hampton, PhD
Barry Hankins
William Hankins
Jeff Hanna
Douglas E. Hansen
Bill H. Hanson
Mervyn Gilbert Hardinge, PhD
Stanley Harris
Francis L. Harrison
Douglas R. Hartsock
Reese P. Hastings
James L. Hatch
Gerald D. Hauxwell, PhD
Gordon W. Hayslip, Jr.
Louis W. Heaton
E. Helvoqt

Charles H. Henager
Gary W. Henderson
Wayne A. Hendricksen
John P. Hendrickson
Robert J. Hennig
Malcolm Hepworth, PhD
G. Herbert
Nicholas A. Hertelendy
Donald E. Hetzel, PhD
Harry V. Hider
Gene Higgins
William Higham
Eugene R. Hill
Dell E. Hillger
Richard M. Hilliard
John E. Hiner
Jack W. Hitchman
John B. Hite
Dean Hobson
Lawrence D. Hokanson, PhD
Eric H. Holmes, PhD
David B. Holsington
Stanley P. Horack
Terry Horn
J. R. Hoskins, PhD
Robert D. Hoss
Ric Howard
Gerald R. Howe
Warren B. Howe
William P. Hoychuk
Pavel Hrma, PhD
Jeff T. Hubbell
John R. Huberty, PhD
Glen W. Hufschmidt
John F. Hultquist, PhD
Nancy B. Hultquist, PhD
Rodney William Hunziker, PhD
Charles W. Hurter
David Hylton
Roar L. Irgens
James Ivers
Rick L. Jackson
Glen O. Jacobson
Arthur Dean Jacot
Neal M. Jacques
Cole Janicki
James J. Jarvis
Keith Bartlett Jefferts, PhD
Anchor D. Jensen
Creighton Randall Jensen, PhD
Daryl T. Johns
Donald Curtis Johnson, PhD
Howard W. Johnson, PhD
James R. Johnson, PhD
Sara Johnson
Skyler Johnson
Stephen Johnson
William Everett Johnson, PhD
Sheridan Johnston, PhD
Wright H. Jonathan
Kay H. Jones, PhD
Robert F. Jones
Gary Jordan
John D. Kaser, PhD
Yale H. Katz
Armen Roupen Kazanjian, PhD
Bruce J. Kell
Bruce J. Kelman, PhD
John R. Kennedy
Kenneth H. Kinard
Alan O. King
Charles J. Kippenhan, PhD

Roderick R. Kirkwood
James D. Klein
John G. Kleyn, PhD
Max E. Kliewer
Gary L. Kloster
Barry S. Knight
George F. Koehler*
Milan Rastislav Kokta, PhD
Lee R. Koller, PhD
Roger W. Kolvoord, PhD
David V. Konsa
Edward F. Krohn
James W. Kross
James D. Krueger
Jerry Krumlik
Peter M. Kruse
Mathew H. Kulawiec
Karl G. Kunkle
Robert W. Kunkle, PhD
Edgars A. Kupcis
Lydiane Kyte
Rick Lally
Michael M. Landon
Charles J. Lang
James C. Langford
Stephen C. Langford, PhD
Stanley G. Langland, Jr.
John Clifford Lasnetske
Albert Giles Law, PhD
Girvis E. Ledbetter
John Leicester
Jennifer B. Leinart
Robert W. Lemon
William F. Lenihan
Stan Lennard
Gary R. Letsinger
Richard H. Levy, PhD
Milton Lewis, PhD
Donald H. Lindeman
Lawrence S. Lindgren
Wesley Earl Lingren, PhD
Winston Woodard Little, Jr., PhD
Thomas J. Lockhart
Robert L. Loofbourow
John V. Loscheider
Heinz Lycklama, PhD
Charles R. Lyons
Douglas M. Lyons
Harry Michael Macksey, PhD
Elwin A. Magill
H. W. Maier
Michael F. Maikowski
Stephen C. Maloney
L. Frank Maranville, PhD
Steven C. Marble
John D. Marioni
Michael Markels, Jr., PhD
Franz J. Markowski
Elbert L. Marks
Todd H. Massidda
Charles E. Mathieu
Donald W. Matter
Nicholas C. Maximovich
Roy T. Mayfield
Lawrence L. Mayhew
Donald R. McAlister, PhD
William Bruce McAlister, PhD
Donald Lee McCabe
Andrew McComas
Richard P. McCullough
Kevin McCurry
Leslie Marvin McDonough, PhD

William McElroy, PhD
Dean Earl McFeron, PhD
Sam McIlvanie
Dwaine McIntosh
Doug McKeever
Ernest J. McKeever
Stephen M. McKinney
John C. McMillan
Natalie C. Meckle
William K. Mellor
James A. Mercer, PhD
Robert Mermelstein, PhD
Norbert E. Methven
Leroy R. Middleton
M. Howard Miles, PhD
Claude P. Miller
Howard Miller
James A. Miller
Ralph Hugh Minor
Paul M. Mockett, PhD
Mahmoud Mohamed Abdel Monem,
PhD
Jon F. Monsen, PhD
Fred Moore, PhD
Kou-Ying Y. Moravan, PhD
John Rolph Morrey, PhD
David J. Morris
Arne Mortensen, PhD
Charles S. Mortimer
Lonnie E. Moss
Karl W. Mote
George Mueller, PhD
Mark P. Mullen
Peter J. Murray, PhD
Harold E. Musgrove
Robert J. Musgrove
Gerald B. Myers
Cliff R. Naser
Thomas J. Neidlinger
Carolyn R. Nelson
Charles L. Nelson
Errol Nelson
Jay C. Nelson
Arthur Hobart Nethercot, PhD
Edward F. Neuzil, PhD
R. Newhouse
Darrell Francis Newman
Walter F. Nicaise, PhD
Davis Betz Nichols, PhD
Robert Fletcher Nickerson, PhD
Ronald D. Nielson
Kerry M. Noble
Sherman Berdeen Nornes, PhD
Bruce W. Novark
Jack H. Nuszbaum
Hugh Nutley, PhD
Francisco Ochoa
M. G. Oliver
Bertel O. Olson
Edwin A. Olson, PhD
Jerry Olson
Joe R. O'Neal
W. E. Osborne
Jay Palmer
Steven P. Pappajohn
James Lemuel Park, PhD
James Floyd Parr, PhD
Jerry Alvin Partridge, PhD
Ian J. Patterson
Ronald Stanley Paul, PhD
Craig L. Paulsen
Russell D. Paxson

Edson Ruther Peck, PhD
John L. Perreault
David L. Perry
Donald Peter
Kenneth M. Phillips
Rich Phillips
Michael F. Pickett
Gary Plyler
Karl Hallman Pool, PhD
Gary R. Porter
Gerald J. Posakony
Dale E. Potter, PhD
David T. Powell, Jr.
R. C. Powell
John D. Power
Ekkehard Preikschat, PhD
James H. Price
Robert L. Pritchett
Allan E. Query
Jennifer L. Rademacher
Dean C. Ralphs
David C. Ransower
Lowell W. Rasmussen, PhD
William R. Rauch
Glen Reid
Ronald J. Reiland
Susan Rempe, PhD
Jeffrey M. Reynolds, PhD
Nick Rezek
Brian Rhodes
Hollis E. Rice
Gerald L. Richardson
Jeffrey Ries
William T. Riley
Gary A. Ritchie, PhD
Jonathan Ritson, PhD
John Rivers
Lamont G. Robbins
Gregory F. Robel, PhD
Joseph L. Roberge
Norman Hailstone Roberts, PhD
Daniel B. Robertson
Gary W. Robertson
Roger Robinett
Suzanne Robinson
Steven P. Roblyer
G. J. Rogers
Gordon W. Rogers
Steven B. Rogers
Duane I. Roodzant
Erik J. Rorem
John J. Rosene
Thomas M. Rothacker
Sue Ruff
Susan E. Rutherford
Thomas B. Salzano
William Henry Sandmann, PhD
Wayne Mark Sandstrom, PhD
Lynn Redmon Sarles, PhD
Lester Rosaire Sauvage
Martin W. Savinsky
Eric A. Schadler
Ed R. Schild
Lucien A. Schmit, Jr.
D. J. Schoenberg
Michael J. Scholtens
Roderic Schreiner
Charles M. Schultz
Paul M. Schumacher
Kevin N. Schwinkendorf, PhD
Marvin J. Scotvold
Joel K. Sears

Timothy A. Sestak
James M. Severance
William L. Shackleford, PhD
Richard U. Shafer
Grant William Sharpe, PhD
Robert G. Sheehan
Matthew M. Sherman
Frank Connard Shirley, PhD
Dalton Shotwell
Gail K. Shuler
Harry C. Simonton
John C. Sindorf
Philip Skehan, PhD
Thomas H. Skrinar
Rolf T. Skrinde, PhD
John R. Sladek
Robert E. Slalen
Frederick Watford Slee, PhD
Brenton N. Smith
Francis Marion Smith
Maurice Smith
S. T. Smith
Sam Corry Smith, PhD*
John Sommer
Norman D. Sossong, PhD
Darrel H. Spackman, PhD
Walter Spangenberg
Kemet Dean Spence, PhD
Mike Stansbury
George J. Stead
Robert B. Stewart
Robert J. Stewart, PhD
Arthur L. Storbo
J. Michael Summa, PhD
Earl C. Sutherland, PhD
Jordan L. Sutton
Thomas B. Swearingen, PhD
R. Paul Swift, PhD
David M. Sypher
Gerald V. Tangren
William G. Tank
Ronald Lee Terrel, PhD
Andrew H. Thatcher
Paul Tholfsen, PhD
L. T. Thomasson
Joseph J. Thompson
Lael J. Thompson
A. E. Thornton
Robert Kim Thornton, PhD
Richard L. Threet, PhD
Patricia Tigges
Darrol Holt Timmons, PhD
Frederic E. Titus
Michael L. Tjoelker
Victor E. Tjulander
Harvey C. Trengove
James Ray Tretter, PhD
D. H. Treusdell
Willard Tribble
Gary L. Troyer
Gerald P. Trygstad
Kenneth R. Tucker
William L. Ullom
Dana M. Updegrove
Juris Vagners, PhD
Gregory Van Doren
Wayne Paul Van Meter, PhD
Patrick R. Vasicek
John R. Vasko, Jr.
Thomas E. Vaught
Gerard Verkaik
Joe Vierra

M. Douglas Vliet
Erich R. Vorpagel, PhD
Bruce G. Wakeham
Mark R. Waldron
C. B. Ward
Vernon D. Warmbo
Bruce P. Warren
Maurice Wassmann
Thomas L. Webber
Alfred Peter Wehner
Marvin Wehrman
Ben Weigandt
Selfton R. Wellings, PhD
Howard E. Wellsfry
John A. Westland, PhD
John R. Wheatley
Kerry M. Whitaker
W. E. Whitney, Jr.
Jackson E. Wignot
Gary W. Wilcox, PhD
Larry Wilhelmsen
Richard S. Wilkinson
William H. Will
Bradley R. Williams
Jerry Williams
Max W. Williams, PhD
C. Ray Windmueller
Sam R. Windsor
Philip C. Wingrove
Holden W. Withington
Casey W. Witt
Charles Edward Woelke, PhD
R. P. Wollenberg
Frank William Woodfield
David Dilley Wooldridge, PhD
Mark D. Workman
Jonathan V. Wright
Milo Wright
Robert C. Wright
Peter Wyzinski
Edwin N. York
William B. Youmans, PhD
Tim A. Younker
Ralph Yount, PhD
Ivan Zamora
Waldo S. Zaugg, PhD
Tianchi Zhao
Gerald A. Zieg
Bernard M. Ziemianek
Douglas Jerome Zimmer
Edward E. Zinser
Steven C. Zylkowski

West Virginia

M. H. Akram, PhD
Charles Alt
Stephen Edward Always, PhD
Robert Anderson
Guvenc Argon
Jerome C. Arnett, Jr.
Jay K. Badenhoop, PhD
Carlos T. Bailey
Ed Bailey
Barton Scofield Baker, PhD
Ever Barbero, PhD
Kevin P. Batt
Thomas P. Baumgarth
Franklin L. Binder, PhD
Samuel Oscar Bird, II, PhD
Samuel G. Bonasso

Stephon Thomas Bond, PhD
Anthony John Bowdler, PhD
S. D. Brady, III
Donald M. Brafford
William E. Broshears
Robert William Bryant
Warren Burch
Michael J. Burrell
Darrell K. Carr
Richard H. Clapp
Alan R. Collins, PhD
Sabin W. Colton, PhD
Roddy Conrad, PhD
Ronald E. Cordell
Richard L. Coty
Rex R. Craft
Edward Hibbert Crum, PhD
Ricky E. Curry
H. Douglas Dahl, PhD
Richard M. Dean
Warren Edgell Dean, PhD
Ronnie L. Delph
Edmond Joseph Derderian, PhD
Albert John Driesch
Norwood F. Dulin
Herbert Louis Eiselstein
Richard W. Eller
Nick Endrizzi
Alvin L. Engelke
Philip L. Evans
Gary D. Facemyer
Joseph A. Farinelli
David F. Finch
Gunter Norbert Franz, PhD
Arthur L. Fricke, PhD
Frank L. Gaddy
Joyce A. Gentry
Mark W. Getscher
Lawrence J. Gilbert
William Harry Gillespie
Donald Wood Gillmore, PhD
Wilbert Eugene Gladfelter, PhD
John Gottschling
Joseph W. Grahame
Teddy Hodge Grindstaff, PhD
James Hansen, PhD
R. Thomas Hansen, PhD
William W. Hartman
Timothy M. Hawkes
H. G. Henderson, PhD
Roger L. Henderson
Gregory E. Hinshaw
Jay W. Honse
Charles G. Howard
Clayton L. Huber
J. David Hurt
George S. Jones, II
Wilber C. Jones, PhD
Mitchell M. Kalos
Everett Lee Keener
Alan F. Keiser
Lynette Kell
Richard S. Kerr
A. Wahab Khair, PhD
Gregory A. Kozera
Scott Kuhn
Andrew G. Kulchar
William C. Kuryla, PhD
Richard A. Lambey
Thomas Lee Lantz
Howland Aikens Larsen, PhD
Dickey L. Laughlin

Layle Lawrence, PhD
Gerald Roger Leather, PhD
D. M. Lee
Paul Lee, PhD
H. William Leech, PhD
Arnold David Levine, PhD
Antonio S. Licata
Andrew J. Lino
Ashby B. Lynch, Jr.
Philip L. Martin
Louise Tull Mashburn, PhD
Thompson Arthur Mashburn, PhD
Marvin Richard McClung, PhD*
Russell K. McFarland, Jr
Douglas E. McKinney
Scott D. Melville
Andrew J. Mesaros
Edwin Daryl Michael, PhD
Michael H. Mills
Robert E. Mitch
Raymond F. Mize
Richard W. Morrison, PhD
Kenneth L. Murphy, PhD
Larry T. Murray
Hannah Murrell
Bruce V. Mutter, PhD
Charles H. Myslinsky
Jeffrey O. Nelson
Roy Sterling Nutter, PhD
Richard C. Oldack
Michele O'Neil
Brian M. Osborn
Joseph R. Overbay, Jr.
Gary A. Patterson
Arthur Stephen Pavlovic, PhD
Henry E. Payne, III, PhD
Salvatore Pecoraro
G. Albert Popson, PhD
John M. Praskwiecz
Charles D. Preston
Jacky Prucz, PhD
Robert L. Raines
Lyn B. Ratcliff, PhD
Herman Henry Rieki, PhD
James Ritscher, PhD
Satya P. Rochlani
Robert M. Rohaly, Jr.
Norman H. Roush
Daniel Rubinstein, PhD
Tim A. Salvati
Robert Lee Sandridge, PhD
Dalip K. Sarin
Todd V. Sauble
John L. Schroder
Gregory L. Schumacher
Edward L. Scram
Robert G. Shirey
H. Drexel Short
Paul E. Smalley
Daniel D. Smith
Gregory C. Smith
Percy Leighton Smith
Nicholas M. Spands
Don S. Starcher
William K. Stees
Walter M. Stewart
Dan L. Stickel
Calvin L. Strader
Thomas B. Styer
Francis Suarez
Elizabeth Davis Swiger, PhD
Thomas E. Tabor

Jasmin A. Tamboli
Richard L. Terry
Lester A. Tinkham
Michael W. Tobrey
Kathleen Umstead
Thomas E. Urquhart
Jess H. Wadhams
Charles R. Waine
Kim A. Walbe
George Paul Walker
Haven N. Wall, Jr.
Glen C. Warden
Royal J. Watts
Lewis Wertz
Ronald Kevin Whipkey
Harry V. Wiant, Jr., PhD
Steve Williams
W. Randy Williams, PhD
Walter E. Williamson
Dayne A. Willis
Henry N. Wood, PhD
Gregory Wooten
Charles S. Workman
Paul Zinner

Wisconsin

Albert H. Adams
James William Adams
Rafique Ahmed, PhD
John F. Alleman
Jason D. Allen
Roger O. Anderson
Mary J. Anzia, PhD
Eric C. Araneta
Philip J. Arnholt, PhD
John K. Arnold
Jacob F. Asti
Carlton L. Austin
Edward T. Auth
Bruce D. Ayres, PhD
Edward A. Baetke
A. D. Baggerley
Jeffrey M. Baker
Robert R. Baker
Robert W. Baker, PhD
Gilbert B. Bakke
James R. Bakken
Cathy Bandyk, PhD
Robert F. Barnes, PhD
Jack Bartholmai
Joseph H. Battocletti, PhD
Roy M. Bednarski
Ralph J. Beiermeister
Robert W. Bellin
J. C. Belling
Thomas E. Bellissimo
John P. Benecke
Ralph E. Benz
Laurence C. Berg
Joseph S. Bernstein
Floyd Hilbert Beyerlein, PhD
Katherine Bichler, PhD
John A. Bieberitz
Harvey A. Biesel
John Merlyn Bilhorn
Russell A. Billings
William P. Birkemeier, PhD
Gerald E. Bisgard, PhD
Martin T. Bishop

Clarence L. Blahnik
Brooks K. Blanchard
Philip J. Blank, PhD
Gilbert K. Bleck
Fred G. Blum
Gregory E. Bolin
Jack M. Bostrack, PhD
R. K. Boutwell, PhD
Francis J. Bradford
Daniel C. Bradish
Michael L. Bradley
William M. Breene, PhD
Charles F. Bremigan
Gunnars J. Briedis
Robert P. Bringer, PhD
Dean A. Brink
Timothy C. Bronn
Charles E. Brown
Carl W. Bruch, PhD
Eugene S. Brusky
Reid Allen Bryson, PhD
Ellen F. Buck
Robert O. Buss
Anthony F. Cafaro
Bruce J. Calkins
James B. Calkins, PhD
Hugh W. Carver
Russell J. Cascio
Michael R. Cesarz
David T. Champion
Peter D. Chapman
Lyle E. Cherney
Julie L. Chichlowski
Paul I. Chichlowski
Edward F. Choulnard, PhD
Joyce R. Christensen
Richard W. Christensen, PhD
Dennis Christopherson
Patrick E. Ciriacks
James L. Clapp, PhD
William B. Clayton
W. Wallace Cleland, PhD
John E. Clemons
Mark A. Clemons
Peter G. Cleveland, PhD
Donald T. Clickner
Edward D. Coen
Robert E. Coffey
Mark V. Connelly
Kenneth Cook
Mark Eric Cook, PhD
Tanya M. C. Cook
Frederick Albert Copes, PhD
Jerry L. Corkran
Dennis M. Corrigan
Robert P. Cox
James E. Craft
David R. Cress
Chris Cruz
Ronald E. Cunzenheim
Frederick D. Curkeet
John W. Curtis
M. L. Curtis
Shannon Czysz
Frank J. Dadam
Kenneth R. Dahlstrom
Ruth M. Dalton
Heather M. Daniels
Eugene F. Dannecker
Paul R. Danyluk
Anthony C. De Vuono, PhD
Roger J. Debaker

Mark E. DeCheck
Randall L. Degier
Raymond A. Delponte
Gerald L. Demers
J. Dennitt
Edward P. DePreter
Jerry J. Dewulf
Charlesworth Lee Dickerson, PhD
John M. Dierberger
Daniel M. Dillon
Michael P. Doble
William V. Dohr
Joseph Domblesky, PhD
Edwin Dommisse
John J. Donnell
Lloyd H. Dreger, PhD
Jack C. Dudley
Thomas D. Durand
Ernest John Duwell, PhD
Dwight Buchanan Easty, PhD
Dean W. Einspahr, PhD
David A. Ellis
Mark Engebretson
Ron J. Engel
Leon R. Engler
Richard C. Entwistle
William A. Erby, PhD
Karl L. Erickson, PhD
Robert H. Erskine
Thomas Alan Fadner, Sr., PhD
Gary A. Fahl
Donald Fancher, PhD
Dennis C. Fater, PhD
Jeanene L. Feder
Robert J. Felbinger
William M. Fesenmaier
Brandon M. Fetterly, PhD
Peter A. Fitzgerald
Stephen Wayne Flax, PhD
Eugene J. Fohr
Brandon A. Foss
Raymond A. Fournelle, PhD
James Foy, PhD
Lowell Frank
Margaret Shirley Fraser, PhD
Robert Watt Fulton, PhD
Lloyd C. Furer, PhD
Richard A. Gaggioli, PhD
Cheryl L. Gardner
Tom A. Gardner
Jerry Gargulak, PhD
Ronald L. Garrison
James C. Gaskell
Robert C. Gassert
Dave Gaylord
Jolene K. Geary
R. Geerey, PhD
Richard L. Geesey, PhD
Charles E. Geisel
Barbara Geldner
Robert R. Gerger
Kenneth H. Gersege
Frederic Arthur Giere, PhD
Rodney L. Goldhahn
Russell S. Gonnering
Tom O. Goodney
G. Robert Goss, PhD
Larry F. Gotham
Zech Gotham
Charles Thomas Govin, Jr.
Kim A. Gowey
John Graf

Ray H. Griesbach
Lynn M. Griffin
Edward Griggs
Karl P. Grill
Paul J. Grosskreuz
Leonard Charles Grotz, PhD
James A. Gundry
Paul B. Gutting
Earl Hacking
James E. Hagley
Albert C. Hajny
Donal W. Halloran
Perry T. Handziak
Robert B. Hanna
Brad S. Harlander
Jerome Scott Harms, PhD
Lynn Harper
David Harpster, PhD
Heather Harrington
Leon Harris
James P. Harsh
John Michael Hassler
Gregory Dexter Hedden, PhD
Donald J. Helfrecht
Roger W. Helm
Michael R. Henderson
Philip S. Henkel
David Henzig
Mike S. Hess
William R. Hiatt
Patrick J. Hickey
Robert C. Hikade
Duane F. Hinz
Lawrence H. Hodges
Gary L. Hoerth
Gregory J. Hofmeister
Samuel E. Hoke
David Henry Hollenberg, PhD
Donald E. Honaker
Earl W. Horntvedt
David R. Horsefield
Steve Horton
Lowell A. Howard
John L. Hussey
George Keating Hutchinson, PhD
Reinhold J. Hutz, PhD
Erik J. Hyland
Steven J. Hynek
Donald V. Hyzer
Charles Albert Ihrke, PhD
Donald D. Ingalls
Raymond W. Ingold
Jerome M. Iverson
George Thomas Jacobi
Thomas J. Jacobi
Lois Jean Jacobs, PhD
Robert L. Jeanmaire
Allen J. Jenquin
Roy J. Jensen
Mary A. Johanning, PhD
Eric W. Johnson
S. Johnston, PhD
David Jowett, PhD
Joe Kahm
Dean M. Kaja
Robert J. Kalscheur
Gabor Karadi, PhD
Vernon D. Karman
Mack A. Karnes
E. A. Kasel
Charles B. Kasper, PhD
Michael J. Kasten

Richard Keller
Andrew Augustine Kenny
Chris E. Kensel
Patrick J. Keyser
Todd D. Kile
Bruce F. Kimball
Peter O. Kirchhoff
Arthur S. Klein
Richard G. Knight, PhD
Steven G. Knorr
Henry Koch, Jr.
Eldo Clyde Koenig, PhD
Robert L. Kogan, PhD
Lewis E. Kohn
Van E. Komurka
Kirk A. Konkel
Jeffrey A. Konop
Mary K. Kopmeier
Joseph Edward Kovacic
Joseph D. Kovacich
Jeffrey G. Kramer
Melvin W. Kraschnewski
Kenneth W. Kriesel
Alvin C. Krings
Larry A. Kroger, PhD
Gary Krogstad
Gary S. Krol
David P. Krum
Oliver M. Kruse
Peter Mouat Kuhn, PhD
Bernard P. Lakus
Conrad Marvin Lang, PhD
Ivan M. Lang, PhD
Richard C. Lange
Albert U. Langenegger
Elma Lanterman, PhD
David L. Larson
Philip Rodney Larson, PhD
Sherman D. Laviolette
Clifton E. Lawson
Bruce W. Leberecht
Davin R. Lee
Kenneth B. Lehner
John P. Leifer
Paul B. Lemens
Ron Lenaker
Richard J. Leonard
Thomas J. Leonard
Bohumir Lepeska, PhD
Larry A. Lesch
Robert J. Lex
John Lien
Mark D. Lillegard
Keith R. Lindsley
Emil Patrick Lira, PhD
Gregory R. Lochen
Donald M. Lochner, PhD
A. Locke
Leo J. Lofland
Donald G. Logan
Joan Logan
Robert S. Logan
Julian H. Lombard, PhD
Steven Paul Lorton, PhD
Edward Thomas Losin, PhD
Scott T. Luckiesh
Rita M. Luedke
Steven H. Lunde
Michael J. MacDonald
Paul Machmeier, PhD
Frederick Mackie
Thomas J. Madden

Jim Maerzke
Greg J. Magyera
Gregory H. Mahairas
Guenther H. Mahn
Ingo G. Mahn, PhD
Frank Mahuta, Jr., PhD
James D. Maki
Randall L. Maller
Steven J. Mamerow
Swarnambal Mani
Richard F. Manthe
Marian Manyo
Anthony Maresca
Jeffery P. Martell
David K. Mathews
Johan A. Mathison
Timothy V. Mattson
Stephen A. Maus
Deanna J. May
Jerome S. Mayersak
Percy M. Mayersak
Roger D. Mayhew
Don A. McAllister
Greg W. McCormack
James C. McDade, PhD
George W. Mead
Edward C. Mealy, PhD
Lynn A. Mellenthin
Michael D. Mentzel
Steven L. Merry
Cynthia Meyer
Lynn D. Meyer
Steven L. Meyer
Victor L. Michels, PhD
J. R. Miller
Kevin J. Miller
Lee Miller
Roger G. Miller
Bruce Mills
Mark J. Mobley
Kennth G. Molly
Randy Montgomery
John Duain Moore, PhD
David L. Morris
Marion Clyde Morris, PhD
Michael C. Mortl
Carl Mueller
Tim Mueller
Lee J. Murray
Mark L. Nagurka, PhD
Gangadharan V. M. Nair, PhD
Phil Naunann
Daniel J. Naze
Percy L. Neel, PhD
Amy L. Nehls
Kenneth R. Neil
David M. Nelson
Elwyn F. Nelson
Sherry L. Nesswenum
Arthur Nethery, PhD
Joachim Peter Neumann, PhD
Manfred E. Neumann
Paul Nevins
Carroll Raymond Norden, PhD
Dexter B. Northrop, PhD
Gerald Novotny
Lynn E. Nowak
Richard O'Connor
T. A. O'Connor
David P. Ohrmundt
Andre Olivas
Randy Olson, PhD

Marshall F. Onellion, PhD
Edward Scott Oplinger, PhD
Apolonio M. Ortiguera
Joseph C. Ostrander
David Ozsvath, PhD
Dale R. Paczkowski
Edward H. Pahl
John Charles Palmquist, PhD
William C. Pappathopoulos
Andrew M. Paretti
Hyunjaz Park, PhD
Robert O. Parmley
J. Parrish
Marshall E. Parry
Scott A. Patulski
Kermit E. Paulson
David M. Paulus, Jr., PhD
Paul E. Pawlisch, PhD
Thomas E. Pelt
Lynn A. Peterson
Roy L. Peterson
Terrance A. Peterson
John Phillips
Keith M. Picker
Adam J. Pientka
William G. Piper
Jeff Plass
Huberto R. Platz
William A. Polley
Harvey Walter Posvic, PhD
Rial S. Potter
Richard H. Preu
Timothy B. Pryor
Mary J. Purvis
Arnold J. Quakkelaar
Daryl Raabe
Dennis S. Rackers
Richard E. Radak
Hershel Raff, PhD
Gregory J. Rajala
Tom Rank, PhD
Mike Ratcliff
Hukum Singh Rathor, PhD
Michael P. Rau
Gerald W. Rausch, PhD
Paul Reesman
David A. Reid
William Stanley Rhode, PhD
Karl W. Richter
Michael Riley
Terri A. Rinke
William Robisch
Michael J. Robrecht
William H. Roche
Randall J. Roeder
Martin A. Rognlien
Ellen A. Roles
John A. Roth
Carl B. Rowe
Paul Allen Roys, PhD
Charles A. Rucks
John A. Rybicki
Stephen L. Saeger
Eric M. Sanden, PhD
Robert A. Sanders
Louis W. Sandow
Donald C. Sassor
Suzanne Sawyer
Umesh K. Saxena, PhD
Fred W. Schaejbe
William J. Schapfel
Leo J. Scherer

Michael R. Schindhelm
Jeffrey D. Schleif
Donald A. Schlernitzauer
Charles A. Schmitt
David L. Schmutzler
Timothy D. Schnell, PhD
Carl P. Schoen, PhD
Randall R. Schoenwetter
John L. Schrag, PhD
Kurt Schrampfer
Hans Schroeder
Arthur R. Schuh
Clemens J. Schultz
Frederick H. C. Schultz, PhD
Lynn V. Schultz
Steve A. Schultz
Michael A. Schulz
Glen R. Schwalbach
James Schweider
Carol D. Seaborn, PhD
Ralph Walter Seelke, PhD
Steven Seer
Daniel F. Senf
Liza Servais
Paul E. Seul
Keith R. Seward
Richard A. Shellmer
Alan D. Shroeder
Larry R. Shultis
Barteld Richard Siebring, PhD
Terry V. Sieker
Jeff P. Simkowski
John C. Simms, PhD
Doug Sitko
Thomas G. Sladek
Lisa A. Soderberg
Dave Allen Soerens, PhD
Alan R. Sorenson
James Alfred Sorenson, PhD
William E. Southern, PhD
Dale A. Spires
Gary A. Splitter, PhD
Julien C. Sprott, PhD
Joseph Staley
Glenn Stanwick
Gary M. Stanwood
Richard Steiner
Shawn D. Steinhoff
William John Stekiel, PhD
Richard A. Steliga
Michael Fred Stevens, PhD
Patrick J. Stiennon
James R. Stieve
E. N. Stillings
John Stoffel
Lawrence W. Strelow
D. H. Strobel
John R. Stubbles, PhD
Warren Robert Stumpe
Harry W. Stumpf
Glenn Richard Svoboda, PhD
Stacy L. Swantz
Neal L. Sykora
Steve Stanley Szabo, PhD
Al P. Szews, PhD
Lars J.S. Taavola
Kuni Takayama, PhD
Louis J. Tarantino
Paul J. Taylor, PhD
Kenneth J. Terbeek, PhD
Steven R. Tesch
Roger Thiel

Theodore William Tibbitts, PhD
Jon Tiger
Arthur T. Tiller
Brian P. Tonner, PhD
John W. Troglia
Vladimir M. Uhri
George Uriniuk
Mary P. Utzerath
S. Verrall, PhD
Thomas E. Vik
James F. Voborsky
Randall Vodnik
Bill Vogel
Joseph P. Vosters
Allan G. Waelchli
Richard F. Wagner
James E. Waldenberger
John L. Walkowicz
Gordon David Waller, PhD
Robert R. Wallis
Roger C. Wargin
Frank E. Weber, PhD
Mary Weber
Randall James Weege
Mark J. Wegner
Eugene Allen Weipert, PhD
Allen C. Wendorf
John Westfall
Donald Edward Whyte, PhD
Robert C. Wiczynski
Everette C. Wilson
Richard D. Winter
Ronald Wolf
Carl E. Wulff
Clyde W. Yellick
Sharon H. Young
Nathaniel K. Zelazo
Brad W. Zenko
Carl Zietlow
Bob Ziller
Myron E. Zoborowski
Edward John Zuperku, PhD

Wyoming

Charles K. Adams
John W. Ake
Lowell Ray Anderson
Percy G. Anderson, Jr.
Marion L. Andrews
Lance L. Arnold
Joseph P. Bacho
Richard E. Bacon
Donald T. Bailey
Robert C. Balsam, Jr.
Charles J. Barto, Jr.
Curtis M. Belden
John C. Bellamy, PhD
Donald W. Bennett
Joseph C. Bennett
David Bishop
Floyd A. Bishop
Lloyd P. Blackburn
Richard G. Boyd
Dennis Brabec
Lee Brecher, PhD
Francis M. Brislawn
Paul Brutsman
J. B. Bummer
Roger T. Bush
John D. Campanella

Scott P. Carlisle
Francis Carlson
Carl D. Carmichael
Keith T. Carron, PhD
Paul C. Casperson
Robert H. Chase
Robert G. Church
E. James Comer
Daniel P. Coolidge
Robert F. Cross, PhD
John J. Culhane
Ray Culver
William Hirst Curry, PhD
Jan Curtis
Steve L. Dacus
Mark A. Davidson
Dick Davis, PhD
Robert Joseph Dellenback, PhD
Hugh D. Depaolo
Martin L. Dobson
Scott N. Durgin
John C. Elkin
Fred M. Emerich
Paul M. Fahlsing
Susan J. Fahlsing
Kenneth K. Farmer
S. R. Faust, Jr.
Ray A. Field, PhD
Stuart C. Fischbeck
James P. Ginther
Jimmy Goolsby, PhD
John David Green
James E. Greer
Richard C. Greig
Durward R. Griffith
Falinda F. Hall
Daniel J. Haman
Homer R. Hamilton, PhD
C. J. Hanan
Paul Michael Harnsberger
Robert A. Harrower
William D. Hausel
Mike C. Hawks
Henry William Haynes, Jr., PhD
Scott J. Hecht
Bill K. Heinbaugh
John H. Hibsman, Jr.
Paul R. Hildenbrand
Mark Hladik
Michael S. Holland
Kenneth O. Huff
G. William Hurley
Irven Allan Jacobson
Archie C. Johnson
Frederick S. Johnson
Terrell K. Johnson
Jack B. Joyce
Victor R. Judd
William T. Kane, PhD
J. Kauchich
William Keil
Hugh Kendrick, PhD
Leroy Kingery
Robert W. Kirkwood
Duane D. Klamm
Paul Koch, PhD
Robert A. Koenig
Bernard J. Kolp, PhD
Joe E. Kub, Jr.
Harry C. La Bonde, Jr.
Richard Laidlaw, PhD
Jerry T. Laman

Donald Roy Lamb, PhD
William Anthony Laycock, PhD
Lewis A. Leake
Howard R. Leeper
James G. Lemmers
Dayton A. Lewis, Jr.
Don J. Likwartz
Michael C. Lock
C. Barton Loundagin
Thomas L. Lyle
Richard E. Mabie
Don Madden
Rick D. Magstadt
Leland C. Marchant
John F. McKay, PhD
Robert E. "Bob" McKee
Edmond Gerald Meyer, PhD
John W. Miller
Kenneth R. Miller
Reid J. Miller
Terry S. Miller
Tom D. Moore
Harold C. Mosher
Roger B. Mourich
Lance T. Moxey
Evart E. Mulholland
Ralph W. Myers
Kevin J. Negus, PhD
Lance Neiberger
Judith E. Nelson
Daniel Anthony Netzel, PhD
Keith A. Neustel
Mark A. Newman
Despina I. Nikolova
Edmond L. Nugent
Dale Nuttall
Robert D. Odell
Paul O. Padgett
Thomas Parker
Leonard Payne
Bruce H. Perryman
Paul T. Peterson
Donald Polson
Larry C. Raymond
Wallace K. Reaves
Paul Albert Rechard
Kenneth F. Reighard
Steven Y. Rennell
Timothy C. Richmond
Robert W. Riedel
Terry P. Roark, PhD
W. F. Robertson
Ted Roes
Arthur R. Rogers
Robert G. Rohwer
Larry J. Roop
Richard D. Rosencrans
Todd Schmidt
David H. Scriven
Leslie E. Shader, PhD
Jim Shriver
Riley Skeen
Terry K. Skinner
Ralph Smalley, Jr.
Ray C. Smith
Stan Smith
Thomas B. Smith
James P. Spurrier
Kimberly J. Starkey
Robert J. Starkey, Jr., PhD
Larry R. Stewart
Donald Leo Stinson, PhD

Tony C. Stone
Eldon D. Strid
K. Sundell, PhD
Archer D. Swank
Tim L. Thamm
S. Thompson

Keith A. Trimels
John F. Trotter
Ron M. Tucker
Kenneth F. Tyler
Ronald D. Wagner
B. Watne, PhD

Larry Weatherford, PhD
Michael Wendorf
Frank D. Werner, PhD
Robert J. Whisonant
Douglas C. White
Donna L. Wichers

James Williams
Larry Dean Hayden Wing, PhD
Robert D. Winland
Bret H. Wolz
Charles K. Wolz
Marcelyn E. Wood
Thomas R. Zachar

[this page intentionally blank]

Environmental Effects of Increased Atmospheric Carbon Dioxide

Arthur B. Robinson, Noah E. Robinson, and Willie Soon

Oregon Institute of Science and Medicine, 2251 Dick George Road, Cave Junction, Oregon 97523 [artr@oism.org]

ABSTRACT   A review of the research literature concerning the environmental consequences of increased levels of atmospheric carbon dioxide leads to the conclusion that increases during the 20th and early 21st centuries have produced no deleterious effects upon Earth's weather and climate. Increased carbon dioxide has, however, markedly increased plant growth. Predictions of harmful climatic effects due to future increases in hydrocarbon use and minor greenhouse gases like CO_2 do not conform to current experimental knowledge. The environmental effects of rapid expansion of the nuclear and hydrocarbon energy industries are discussed.

SUMMARY

Political leaders gathered in Kyoto, Japan, in December 1997 to consider a world treaty restricting human production of "greenhouse gases," chiefly carbon dioxide (CO_2). They feared that CO_2 would result in "human-caused global warming" – hypothetical severe increases in Earth's temperatures, with disastrous environmental consequences. During the past 10 years, many political efforts have been made to force worldwide agreement to the Kyoto treaty.

When we reviewed this subject in 1998 (1,2), existing satellite records were short and were centered on a period of changing intermediate temperature trends. Additional experimental data have now been obtained, so better answers to the questions raised by the hypothesis of "human-caused global warming" are now available.

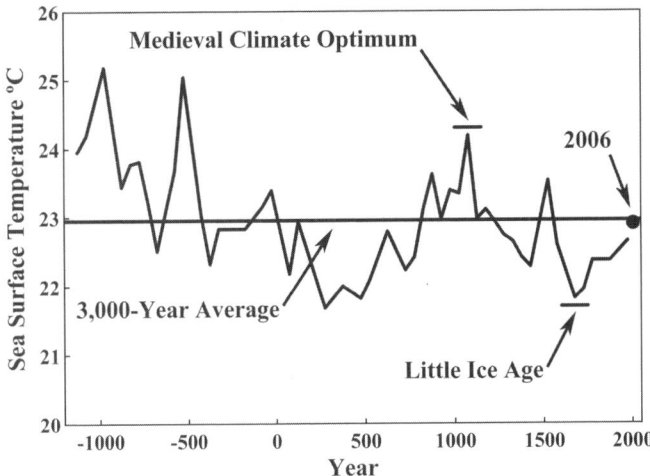

Figure 1: Surface temperatures in the Sargasso Sea, a 2 million square mile region of the Atlantic Ocean, with time resolution of 50 to 100 years and ending in 1975, as determined by isotope ratios of marine organism remains in sediment at the bottom of the sea (3). The horizontal line is the average temperature for this 3,000-year period. The Little Ice Age and Medieval Climate Optimum were naturally occurring, extended intervals of climate departures from the mean. A value of 0.25 °C, which is the change in Sargasso Sea temperature between 1975 and 2006, has been added to the 1975 data in order to provide a 2006 temperature value.

The average temperature of the Earth has varied within a range of about 3°C during the past 3,000 years. It is currently increasing as the Earth recovers from a period that is known as the Little Ice Age, as shown in Figure 1. George Washington and his army were at Valley Forge during the coldest era in 1,500 years, but even then the temperature was only about 1° Centigrade below the 3,000-year average.

The most recent part of this warming period is reflected by short-

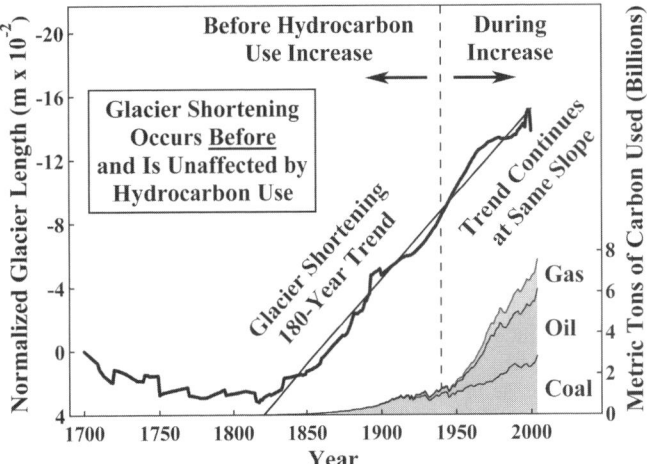

Figure 2: Average length of 169 glaciers from 1700 to 2000 (4). The principal source of melt energy is solar radiation. Variations in glacier mass and length are primarily due to temperature and precipitation (5,6). This melting trend lags the temperature increase by about 20 years, so it predates the 6-fold increase in hydrocarbon use (7) even more than shown in the figure. Hydrocarbon use could not have caused this shortening trend.

ening of world glaciers, as shown in Figure 2. Glaciers regularly lengthen and shorten in delayed correlation with cooling and warming trends. Shortening lags temperature by about 20 years, so the current warming trend began in about 1800.

Atmospheric temperature is regulated by the sun, which fluctuates in activity as shown in Figure 3; by the greenhouse effect, largely caused by atmospheric water vapor (H_2O); and by other phenomena that are more poorly understood. While major greenhouse gas H_2O substantially warms the Earth, minor greenhouse gases such as CO_2

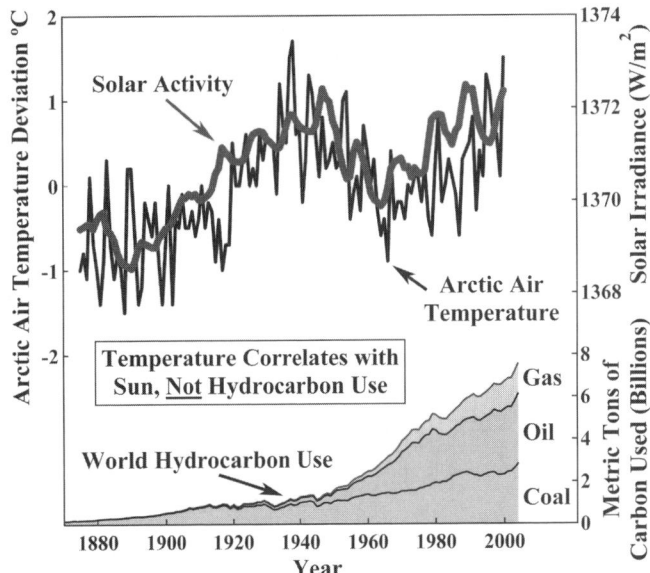

Figure 3: Arctic surface air temperature compared with total solar irradiance as measured by sunspot cycle amplitude, sunspot cycle length, solar equatorial rotation rate, fraction of penumbral spots, and decay rate of the 11-year sunspot cycle (8,9). Solar irradiance correlates well with Arctic temperature, while hydrocarbon use (7) does not correlate.

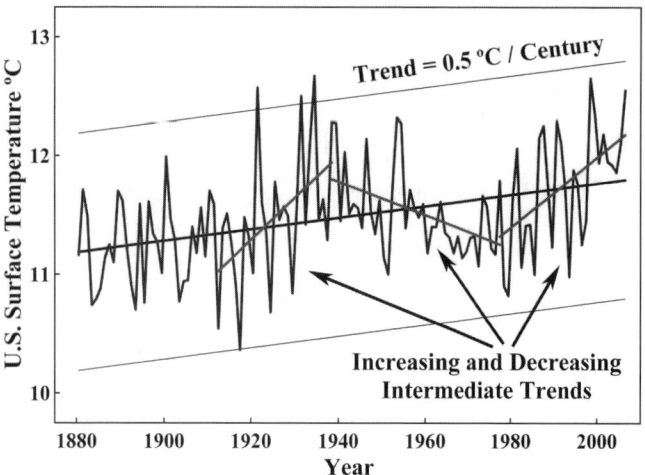

Figure 4: Annual mean surface temperatures in the contiguous United States between 1880 and 2006 (10). The slope of the least-squares trend line for this 127-year record is 0.5 °C per century.

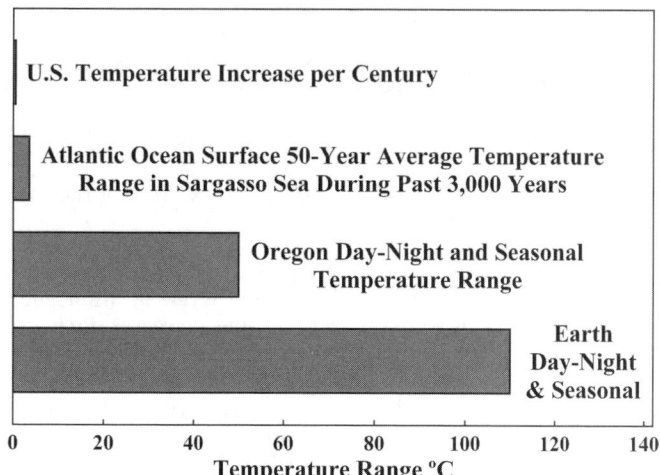

Figure 6: Comparison between the current U.S. temperature change per century, the 3,000-year temperature range in Figure 1, seasonal and diurnal range in Oregon, and seasonal and diurnal range throughout the Earth.

have little effect, as shown in Figures 2 and 3. The 6-fold increase in hydrocarbon use since 1940 has had no noticeable effect on atmospheric temperature or on the trend in glacier length.

While Figure 1 is illustrative of most geographical locations, there is great variability of temperature records with location and regional climate. Comprehensive surveys of published temperature records confirm the principal features of Figure 1, including the fact that the current Earth temperature is approximately 1 °C lower than that during the Medieval Climate Optimum 1,000 years ago (11,12).

Surface temperatures in the United States during the past century reflect this natural warming trend and its correlation with solar activity, as shown in Figures 4 and 5. Compiled U.S. surface temperatures have increased about 0.5 °C per century, which is consistent with other historical values of 0.4 to 0.5 °C per century during the recovery from the Little Ice Age (13-17). This temperature change is slight as compared with other natural variations, as shown in Figure 6. Three intermediate trends are evident, including the decreasing trend used to justify fears of "global cooling" in the 1970s.

Between 1900 and 2000, on absolute scales of solar irradiance and degrees Kelvin, solar activity increased 0.19%, while a 0.5 °C temperature change is 0.21%. This is in good agreement with estimates that Earth's temperature would be reduced by 0.6 °C through particulate blocking of the sun by 0.2% (18).

Solar activity and U.S. surface temperature are closely correlated, as shown in Figure 5, but U.S. surface temperature and world hydrocarbon use are not correlated, as shown in Figure 13.

The U.S. temperature trend is so slight that, were the temperature

change which has taken place during the 20th and 21st centuries to occur in an ordinary room, most of the people in the room would be unaware of it.

During the current period of recovery from the Little Ice Age, the U.S. climate has improved somewhat, with more rainfall, fewer tornados, and no increase in hurricane activity, as illustrated in Figures 7 to 10. Sea level has trended upward for the past 150 years at a rate of 7 inches per century, with 3 intermediate uptrends and 2 periods of no increase as shown in Figure 11. These features are confirmed by the glacier record as shown in Figure 12. If this trend continues as

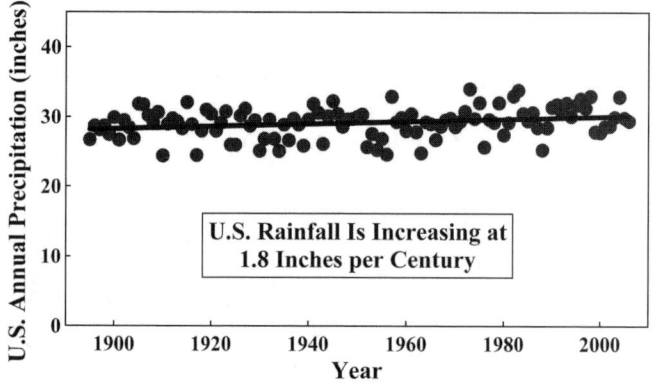

Figure 7: Annual precipitation in the contiguous 48 United States between 1895 and 2006. U.S. National Climatic Data Center, U.S. Department of Commerce 2006 Climate Review (20). The trend shows an increase in rainfall of 1.8 inches per century – approximately 6% per century.

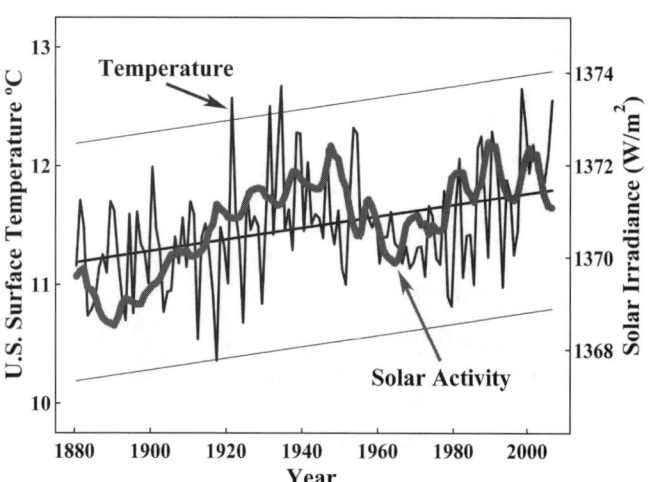

Figure 5: U.S. surface temperature from Figure 4 as compared with total solar irradiance (19) from Figure 3.

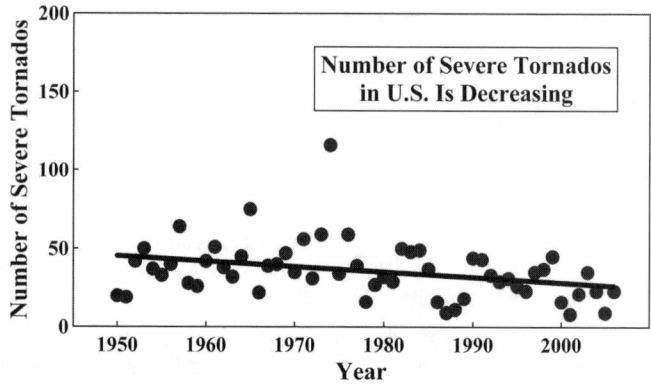

Figure 8: Annual number of strong-to-violent category F3 to F5 tornados during the March-to-August tornado season in the U.S. between 1950 and 2006. U.S. National Climatic Data Center, U.S. Department of Commerce 2006 Climate Review (20). During this period, world hydrocarbon use increased 6-fold, while violent tornado frequency decreased by 43%.

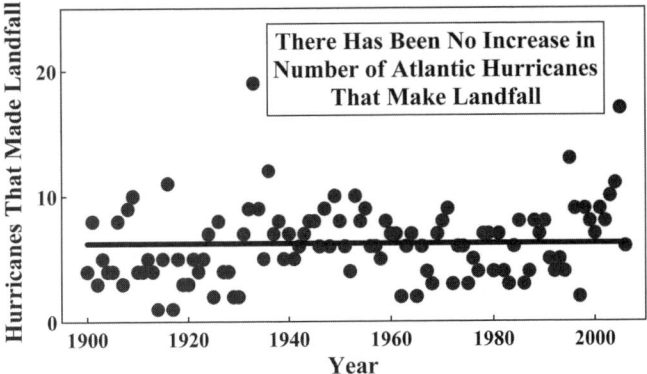

Figure 9: Annual number of Atlantic hurricanes that made landfall between 1900 and 2006 (21). Line is drawn at mean value.

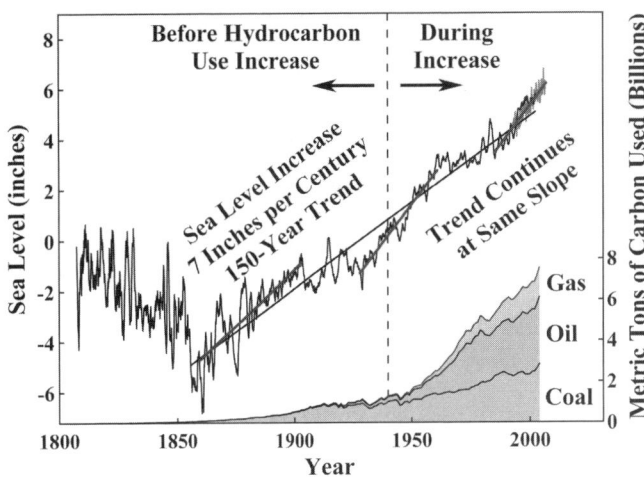

Figure 11: Global sea level measured by surface gauges between 1807 and 2002 (24) and by satellite between 1993 and 2006 (25). Satellite measurements are shown in gray and agree with tide gauge measurements. The overall trend is an increase of 7 inches per century. Intermediate trends are 9, 0, 12, 0, and 12 inches per century, respectively. This trend lags the temperature increase, so it predates the increase in hydrocarbon use even more than is shown. It is unaffected by the very large increase in hydrocarbon use.

did that prior to the Medieval Climate Optimum, sea level would be expected to rise about 1 foot during the next 200 years.

As shown in Figures 2, 11, and 12, the trends in glacier shortening and sea level rise began a century *before* the 60-year 6-fold increase in hydrocarbon use, and have not changed during that increase. Hydrocarbon use could not have caused these trends.

During the past 50 years, atmospheric CO_2 has increased by 22%. Much of that CO_2 increase is attributable to the 6-fold increase in human use of hydrocarbon energy. Figures 2, 3, 11, 12, and 13 show, however, that human use of hydrocarbons has not caused the observed increases in temperature.

The increase in atmospheric carbon dioxide has, however, had a substantial environmental effect. Atmospheric CO_2 fertilizes plants. Higher CO_2 enables plants to grow faster and larger and to live in drier climates. Plants provide food for animals, which are thereby also enhanced. The extent and diversity of plant and animal life have both increased substantially during the past half-century. Increased temperature has also mildly stimulated plant growth.

Does a catastrophic amplification of these trends with damaging climatological consequences lie ahead? There are no experimental data that suggest this. There is also no experimentally validated theoretical evidence of such an amplification.

Predictions of catastrophic global warming are based on computer climate modeling, a branch of science still in its infancy. The empirical evidence – actual measurements of Earth's temperature and climate – shows no man-made warming trend. Indeed, during four of the seven decades since 1940 when average CO_2 levels steadily increased, U.S. average temperatures were actually decreasing.

While CO_2 levels have increased substantially and are expected to continue doing so and humans have been responsible for part of this increase, the effect on the environment has been benign.

There is, however, one very dangerous possibility.

Our industrial and technological civilization depends upon abundant, low-cost energy. This civilization has already brought unprecedented prosperity to the people of the more developed nations. Billions of people in the less developed nations are now lifting themselves from poverty by adopting this technology.

Hydrocarbons are essential sources of energy to sustain and extend prosperity. This is especially true of the developing nations, where available capital and technology are insufficient to meet rapidly increasing energy needs without extensive use of hydrocarbon fuels. If, through misunderstanding of the underlying science and through misguided public fear and hysteria, mankind significantly rations and restricts the use of hydrocarbons, the worldwide increase in prosperity will stop. The result would be vast human suffering and the loss of hundreds of millions of human lives. Moreover, the prosperity of those in the developed countries would be greatly reduced.

Mild ordinary natural increases in the Earth's temperature have occurred during the past two to three centuries. These have resulted in some improvements in overall climate and also some changes in

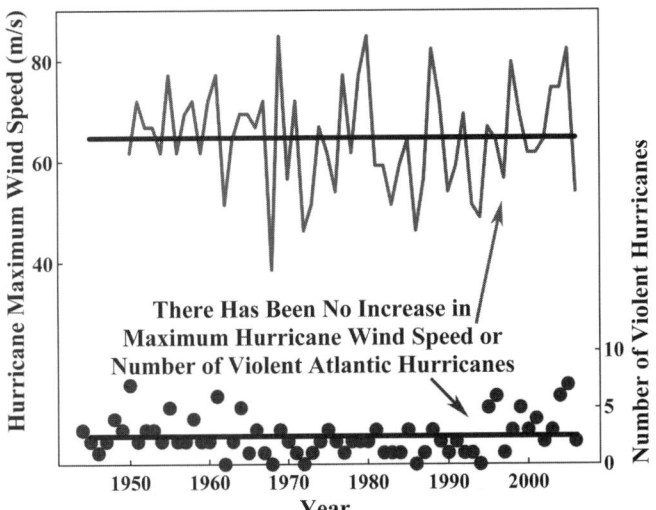

Figure 10: Annual number of violent hurricanes and maximum attained wind speed during those hurricanes in the Atlantic Ocean between 1944 and 2006 (22,23). There is no upward trend in either of these records. During this period, world hydrocarbon use increased 6-fold. Lines are mean values.

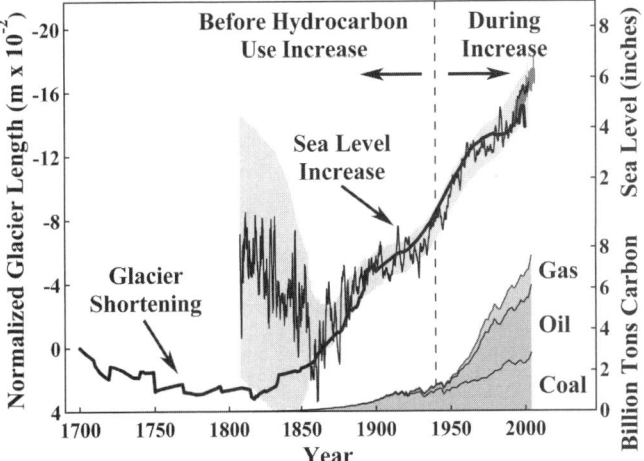

Figure 12: Glacier shortening (4) and sea level rise (24,25). Gray area designates estimated range of error in the sea level record. These measurements lag air temperature increases by about 20 years. So, the trends began more than a century before increases in hydrocarbon use.

the landscape, such as a reduction in glacier lengths and increased vegetation in colder areas. Far greater changes have occurred during the time that all current species of animals and plants have been on the Earth. The relative population sizes of the species and their geographical distributions vary as they adapt to changing conditions.

The temperature of the Earth is continuing its process of fluctuation in correlation with variations in natural phenomena. Mankind, meanwhile, is moving some of the carbon in coal, oil, and natural gas from below ground to the atmosphere and surface, where it is available for conversion into living things. We are living in an increasingly lush environment of plants and animals as a result. This is an unexpected and wonderful gift from the Industrial Revolution.

ATMOSPHERIC AND SURFACE TEMPERATURES

Atmospheric and surface temperatures have been recovering from an unusually cold period. During the time between 200 and 500 years ago, the Earth was experiencing the "Little Ice Age." It had descended into this relatively cool period from a warm interval about 1,000 years ago known as the "Medieval Climate Optimum." This is shown in Figure 1 for the Sargasso Sea.

During the Medieval Climate Optimum, temperatures were warm enough to allow the colonization of Greenland. These colonies were abandoned after the onset of colder temperatures. For the past 200 to 300 years, Earth temperatures have been gradually recovering (26). Sargasso Sea temperatures are now approximately equal to the average for the previous 3,000 years.

The historical record does not contain any report of "global warming" catastrophes, even though temperatures have been higher than they are now during much of the last three millennia.

The 3,000-year range of temperatures in the Sargasso Sea is typical of most places. Temperature records vary widely with geographical location as a result of climatological characteristics unique to those specific regions, so an "average" Earth temperature is less meaningful than individual records (27). So called "global" or "hemispheric" averages contain errors created by averaging systematically different aspects of unique geographical regions and by inclusion of regions where temperature records are unreliable.

Three key features of the temperature record – the Medieval Climate Optimum, the Little Ice Age, and the Not-Unusual-Temperature of the 20th century – have been verified by a review of local temperature and temperature-correlated records throughout the world (11), as summarized in Table 1. Each record was scored with respect to those queries to which it applied. The experimental and historical literature definitively confirms the primary features of Figure 1.

Most geographical locations experienced both the Medieval Climate Optimum and the Little Ice Age – and most locations did not

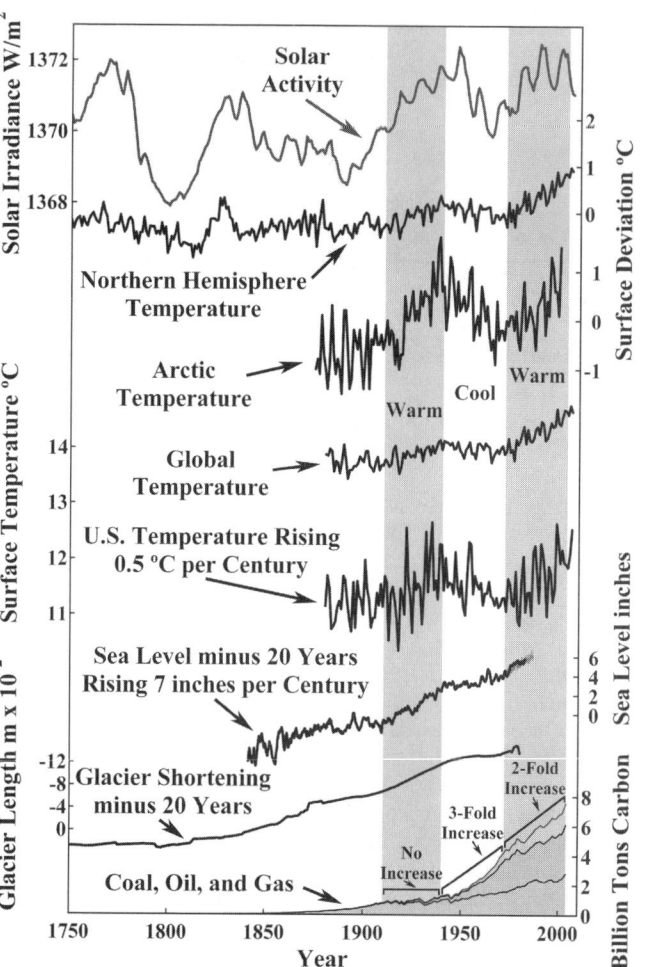

Figure 13: Seven independent records – solar activity (9); Northern Hemisphere, (13), Arctic (28), global (10), and U.S. (10) annual surface air temperatures; sea level (24,25); and glacier length (4) – all qualitatively confirm each other by exhibiting three intermediate trends – warmer, cooler, and warmer. Sea level and glacier length are shown minus 20 years, correcting for their 20-year lag of atmospheric temperature. Solar activity, Northern Hemisphere temperature, and glacier lengths show a low in about 1800.

Hydrocarbon use (7) is uncorrelated with temperature. Temperature rose for a century before significant hydrocarbon use. Temperature rose between 1910 and 1940, while hydrocarbon use was almost unchanged. Temperature then fell between 1940 and 1972, while hydrocarbon use rose by 330%. Also, the 150 to 200-year slopes of the sea level and glacier trends were unchanged by the very large increase in hydrocarbon use after 1940.

Table 1: Query	Yes	No	Yes/No	Two-Tailed Probability
Warm Climatic Anomaly 800-1300 A.D.?	**88**	2	7	> 99.99
Cold Climatic Anomaly 1300-1900 A.D.?	**105**	2	2	> 99.99
20th Century Warmest in Individual Record?	7	**64**	14	< 0.0001

Table 1: Comprehensive review of all instances in which temperature or temperature-correlated records from localities throughout the world permit answers to queries concerning the existence of the Medieval Climate Optimum, the Little Ice Age, and an unusually warm anomaly in the 20th century (11). The compiled and tabulated answers confirm the three principal features of the Sargasso Sea record shown in Figure 1. The probability that the answer to the query in column 1 is "yes" is given in column 5.

experience temperatures that were unusually warm during the 20th century. A review of 23 quantitative records has demonstrated that mean and median world temperatures in 2006 were, on average, approximately 1 °C or 2 °F cooler than in the Medieval Period (12).

World glacier length (4) and world sea level (24,25) measurements provide records of the recent cycle of recovery. Warmer temperatures diminish glaciers and cause sea level to rise because of decreased ocean water density and other factors.

These measurements show that the trend of 7 inches per century increase in sea level and the shortening trend in average glacier length both began a century before 1940, yet 84% of total human annual hydrocarbon use occurred only after 1940. Moreover, neither of these trends has accelerated during the period between 1940 and 2007, while hydrocarbon use increased 6-fold. Sea level and glacier records are offset by about 20 years because of the delay between temperature rise and glacier and sea level change.

If the natural trend in sea level increase continues for another two centuries as did the temperature rise in the Sargasso Sea as the Earth entered the Medieval Warm Period, sea level would be expected to rise about 1 foot between the years 2000 and 2200. Both the sea level and glacier trends – and the temperature trend that they reflect – are

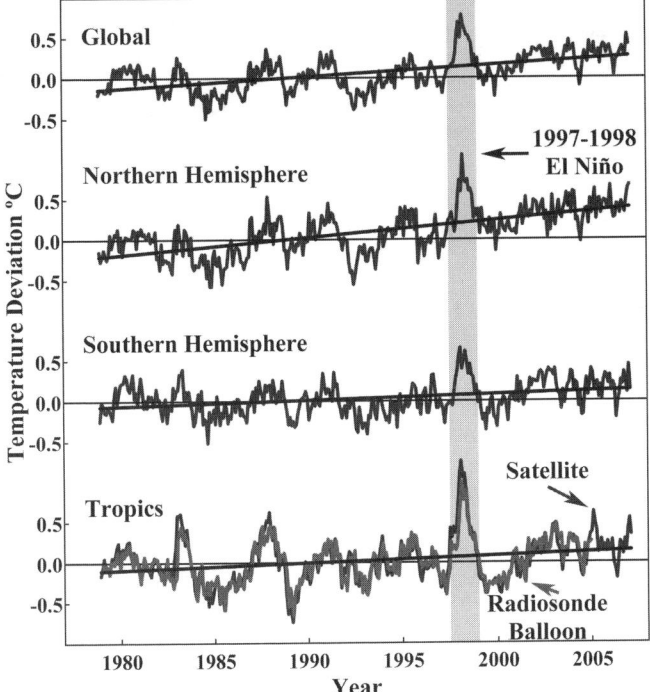

Figure 14: Satellite microwave sounding unit (blue) measurements of tropospheric temperatures in the Northern Hemisphere between 0 and 82.5 N, Southern Hemisphere between 0 and 82.5 S, tropics between 20S and 20N, and the globe between 82.5N and 82.5S between 1979 and 2007 (29), and radiosonde balloon (red) measurements in the tropics (29). The balloon measurements confirm the satellite technique (29-31). The warming anomaly in 1997-1998 (gray) was caused by El Niño, which, like the overall trends, is unrelated to CO_2 (32).

unrelated to hydrocarbon use. A further doubling of world hydrocarbon use would not change these trends.

Figure 12 shows the close correlation between the sea level and glacier records, which further validates both records and the duration and character of the temperature change that gave rise to them.

Figure 4 shows the annual temperature in the United States during the past 127 years. This record has an upward trend of 0.5 °C per century. Global and Northern Hemisphere surface temperature records shown in Figure 13 trend upward at 0.6 °C per century. These records are, however, biased toward higher temperatures in several ways. For example, they preferentially use data near populated areas (33), where heat island effects are prevalent, as illustrated in Figure 15. A trend of 0.5 °C per century is more representative (13-17).

The U.S. temperature record has two intermediate uptrends of comparable magnitude, one occurring before the 6-fold increase in hydrocarbon use and one during it. Between these two is an intermediate temperature downtrend, which led in the 1970s to fears of an impending new ice age. This decrease in temperature occurred during a period in which hydrocarbon use increased 3-fold.

Seven independent records – solar irradiance; Arctic, Northern Hemisphere, global, and U.S. annual average surface air temperatures; sea level; and glacier length – all exhibit these three intermediate trends, as shown in Figure 13. These trends confirm one another. Solar irradiance correlates with them. Hydrocarbon use does not.

The intermediate uptrend in temperature between 1980 and 2006 shown in Figure 13 is similar to that shown in Figure 14 for balloon and satellite tropospheric measurements. This trend is more pronounced in the Northern Hemisphere than in the Southern. Contrary to the CO_2 warming climate models, however, tropospheric temperatures are not rising faster than surface temperatures.

Figure 6 illustrates the magnitudes of these temperature changes by comparing the 0.5 °C per century temperature change as the Earth recovers from the Little Ice Age, the range of 50-year averaged Atlantic ocean surface temperatures in the Sargasso Sea over the past 3,000 years, the range of day-night and seasonal variation on average

in Oregon, and the range of day-night and seasonal variation over the whole Earth. The two-century-long temperature change is small.

Tropospheric temperatures measured by satellite give comprehensive geographic coverage. Even the satellite measurements, however, contain short and medium-term fluctuations greater than the slight warming trends calculated from them. The calculated trends vary significantly as a function of the most recent fluctuations and the lengths of the data sets, which are short.

Figure 3 shows the latter part of the period of warming from the Little Ice Age in greater detail by means of Arctic air temperature as compared with solar irradiance, as does Figure 5 for U.S. surface temperature. There is a close correlation between solar activity and temperature and none between hydrocarbon use and temperature. Several other studies over a wide variety of time intervals have found similar correlations between climate and solar activity (15, 34-39).

Figure 3 also illustrates the uncertainties introduced by limited time records. If the Arctic air temperature data before 1920 were not available, essentially no uptrend would be observed.

This observed variation in solar activity is typical of stars close in size and age to the sun (40). The current warming trends on Mars (41), Jupiter (42), Neptune (43,44), Neptune's moon Triton (45), and Pluto (46-48) may result, in part, from similar relations to the sun and its activity – like those that are warming the Earth.

Hydrocarbon use and atmospheric CO_2 do not correlate with the observed temperatures. Solar activity correlates quite well. Correlation does not prove causality, but non-correlation proves non-causality. Human hydrocarbon use is not measurably warming the earth. Moreover, there is a robust theoretical and empirical model for solar warming and cooling of the Earth (8,19,49,50). The experimental data do not prove that solar activity is the only phenomenon responsible for substantial Earth temperature fluctuations, but they do show that human hydrocarbon use is not among those phenomena.

The overall experimental record is self-consistent. The Earth has been warming as it recovers from the Little Ice Age at an average rate of about 0.5 °C per century. Fluctuations within this temperature trend include periods of more rapid increase and also periods of temperature decrease. These fluctuations correlate well with concomitant fluctuations in the activity of the sun. Neither the trends nor the fluctuations within the trends correlate with hydrocarbon use. Sea level and glacier length reveal three intermediate uptrends and two downtrends since 1800, as does solar activity. These trends are climatically benign and result from natural processes.

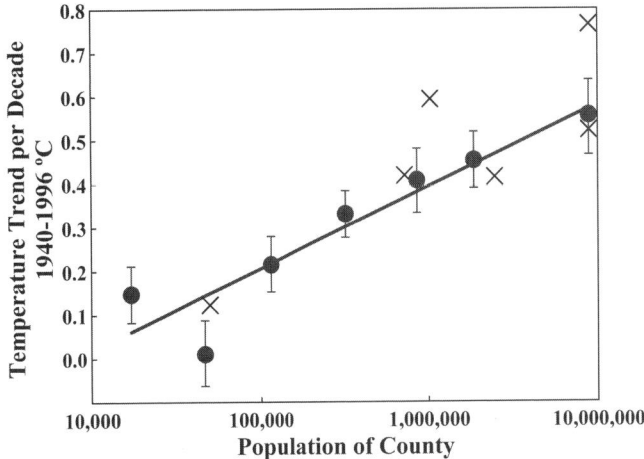

Figure 15: Surface temperature trends for 1940 to 1996 from 107 measuring stations in 49 California counties (51,52). The trends were combined for counties of similar population and plotted with the standard errors of their means. The six measuring stations in Los Angeles County were used to calculate the standard error of that county, which is plotted at a population of 8.9 million. The "urban heat island effect" on surface measurements is evident. The straight line is a least-squares fit to the closed circles. The points marked "X" are the six unadjusted station records selected by NASA GISS (53-55) for use in their estimate of global surface temperatures. Such selections make NASA GISS temperatures too high.

-5-

ATMOSPHERIC CARBON DIOXIDE

The concentration of CO_2 in Earth's atmosphere has increased during the past century, as shown in Figure 17. The magnitude of this atmospheric increase is currently about 4 gigatons (Gt C) of carbon per year. Total human industrial CO_2 production, primarily from use of coal, oil, and natural gas and the production of cement, is currently about 8 Gt C per year (7,56,57). Humans also exhale about 0.6 Gt C per year, which has been sequestered by plants from atmospheric CO_2. Office air concentrations often exceed 1,000 ppm CO_2.

To put these figures in perspective, it is estimated that the atmosphere contains 780 Gt C; the surface ocean contains 1,000 Gt C; vegetation, soils, and detritus contain 2,000 Gt C; and the intermediate and deep oceans contain 38,000 Gt C, as CO_2 or CO_2 hydration products. Each year, the surface ocean and atmosphere exchange an estimated 90 Gt C; vegetation and the atmosphere, 100 Gt C; marine biota and the surface ocean, 50 Gt C; and the surface ocean and the intermediate and deep oceans, 40 Gt C (56,57).

So great are the magnitudes of these reservoirs, the rates of exchange between them, and the uncertainties of these estimated numbers that the sources of the recent rise in atmospheric CO_2 have not been determined with certainty (58,59). Atmospheric concentrations of CO_2 are reported to have varied widely over geological time, with peaks, according to some estimates, some 20-fold higher than at present and lows at approximately 200 ppm (60-62).

Ice-core records are reported to show seven extended periods during 650,000 years in which CO_2, methane (CH_4), and temperature increased and then decreased (63-65). Ice-core records contain substantial uncertainties (58), so these correlations are imprecise.

In all seven glacial and interglacial cycles, the reported changes in CO_2 and CH_4 lagged the temperature changes and could not, therefore, have caused them (66). These fluctuations probably involved temperature-caused changes in oceanic and terrestrial CO_2 and CH_4 content. More recent CO_2 fluctuations also lag temperature (67,68).

In 1957, Revelle and Seuss (69) estimated that temperature-caused out-gassing of ocean CO_2 would increase atmospheric

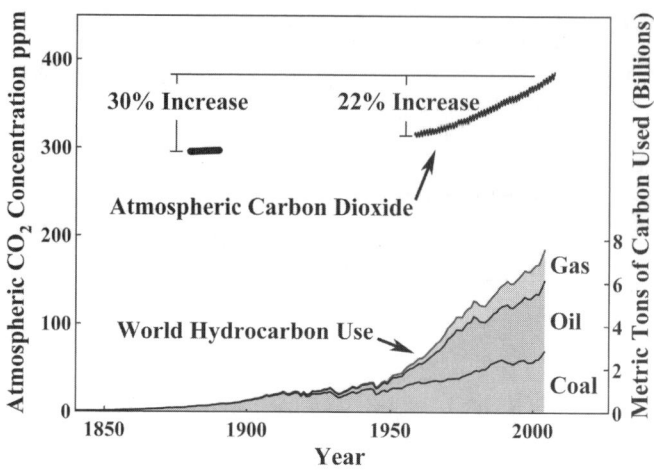

Figure 17: Atmospheric CO_2 concentrations in parts per million by volume, ppm, measured spectrophotometrically at Mauna Loa, Hawaii, between 1958 and 2007. These measurements agree well with those at other locations (71). Data before 1958 are from ice cores and chemical analyses, which have substantial experimental uncertainties. We have used 295 ppm for the period 1880 to 1890, which is an average of the available estimates. About 0.6 Gt C of CO_2 is produced annually by human respiration and often leads to concentrations exceeding 1,000 ppm in public buildings. Atmospheric CO_2 has increased 22% since 1958 and about 30% since 1880.

CO_2 by about 7% per °C temperature rise. The reported change during the seven interglacials of the 650,000-year ice core record is about 5% per °C (63), which agrees with the out-gassing calculation.

Between 1900 and 2006, Antarctic CO_2 increased 30% per 0.1 °C temperature change (72), and world CO_2 increased 30% per 0.5 °C. In addition to ocean out-gassing, CO_2 from human use of hydrocarbons is a new source. Neither this new source nor the older natural CO_2 sources are causing atmospheric temperature to change.

The hypothesis that the CO_2 rise during the interglacials caused the temperature to rise requires an increase of about 6 °C per 30% rise in CO_2 as seen in the ice core record. If this hypothesis were correct, Earth temperatures would have risen about 6 °C between 1900 and 2006, rather than the rise of between 0.1 °C and 0.5 °C, which actually occurred. This difference is illustrated in Figure 16.

The 650,000-year ice-core record does not, therefore, agree with the hypothesis of "human-caused global warming," and, in fact, provides empirical evidence that invalidates this hypothesis.

Carbon dioxide has a very short residence time in the atmosphere. Beginning with the 7 to 10-year half-time of CO_2 in the atmosphere estimated by Revelle and Seuss (69), there were 36 estimates of the atmospheric CO_2 half-time based upon experimental measurements published between 1957 and 1992 (59). These range between 2 and 25 years, with a mean of 7.5, a median of 7.6, and an upper range average of about 10. Of the 36 values, 33 are 10 years or less.

Many of these estimates are from the decrease in atmospheric carbon 14 after cessation of atmospheric nuclear weapons testing, which provides a reliable half-time. There is no experimental evidence to support computer model estimates (73) of a CO_2 atmospheric "lifetime" of 300 years or more.

Human production of 8 Gt C per year of CO_2 is negligible as compared with the 40,000 Gt C residing in the oceans and biosphere. At ultimate equilibrium, human-produced CO_2 will have an insignificant effect on the amounts in the various reservoirs. The rates of approach to equilibrium are, however, slow enough that human use creates a transient atmospheric increase.

In any case, the sources and amounts of CO_2 in the atmosphere are of secondary importance to the hypothesis of "human-caused global warming." It is human burning of coal, oil, and natural gas that is at issue. CO_2 is merely an intermediate in a hypothetical mechanism by which this "human-caused global warming" is said to take place. The amount of atmospheric CO_2 does have profound environmental effects on plant and animal populations (74) and diversity, as is discussed below.

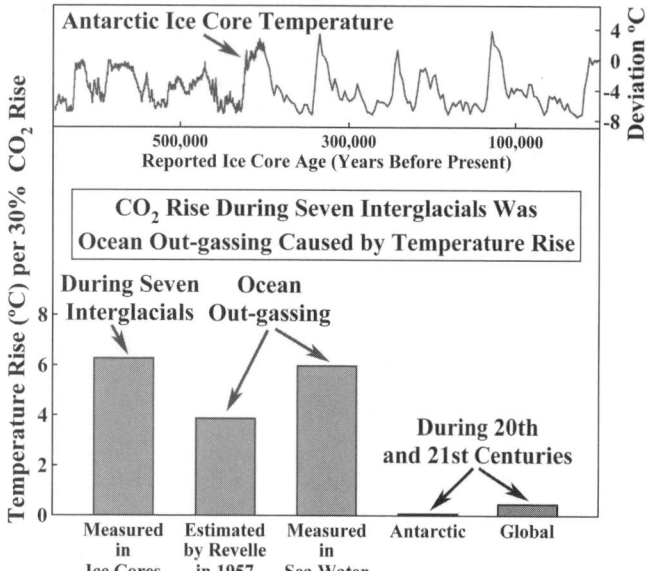

Figure 16: Temperature rise versus CO_2 rise from seven ice-core measured interglacial periods (63-65); from calculations (69) and measurements (70) of sea water out-gassing; and as measured during the 20th and 21st centuries (10,72). The interglacial temperature increases caused the CO_2 rises through release of ocean CO_2. The CO_2 rises did not cause the temperature rises.

In addition to the agreement between the out-gassing estimates and measurements, this conclusion is also verified by the small temperature rise during the 20th and 21st centuries. If the CO_2 versus temperature correlation during the seven interglacials had been caused by CO_2 greenhouse warming, then the temperature rise per CO_2 rise would have been as high during the 20th and 21st centuries as it was during the seven interglacial periods.

CLIMATE CHANGE

While the average temperature change taking place as the Earth recovers from the Little Ice Age is so slight that it is difficult to discern, its environmental effects are measurable. Glacier shortening and the 7 inches per century rise in sea level are examples. There are additional climate changes that are correlated with this rise in temperature and may be caused by it.

Greenland, for example, is beginning to turn green again, as it was 1,000 years ago during the Medieval Climate Optimum (11). Arctic sea ice is decreasing somewhat (75), but Antarctic ice is not decreasing and may be increasing, due to increased snow (76-79).

In the United States, rainfall is increasing at about 1.8 inches per century, and the number of severe tornados is decreasing, as shown in Figures 7 and 8. If world temperatures continue to rise at the current rate, they will reach those of the Medieval Climate Optimum about 2 centuries from now. Historical reports of that period record the growing of warm weather crops in localities too cold for that purpose today, so it is to be expected that the area of more temperate climate will expand as it did then. This is already being observed, as studies at higher altitudes have reported increases in amount and diversity of plant and animal life by more than 50% (12,80).

Atmospheric temperature is increasing more in the Northern Hemisphere than in the Southern, with intermediate periods of increase and decrease in the overall trends.

There has been no increase in frequency or severity of Atlantic hurricanes during the period of 6-fold increase in hydrocarbon use, as is illustrated in Figures 9 and 10. Numbers of violent hurricanes vary greatly from year to year and are no greater now than they were 50 years ago. Similarly, maximum wind speeds have not increased.

All of the observed climate changes are gradual, moderate, and entirely within the bounds of ordinary natural changes that have occurred during the benign period of the past few thousand years.

There is no indication whatever in the experimental data that an abrupt or remarkable change in any of the ordinary natural climate variables is beginning or will begin to take place.

GLOBAL WARMING HYPOTHESIS

The greenhouse effect amplifies solar warming of the earth. Greenhouse gases such as H_2O, CO_2, and CH_4 in the Earth's atmosphere, through combined convective readjustments and the radiative blanketing effect, essentially decrease the net escape of terrestrial thermal infrared radiation. Increasing CO_2, therefore, effectively increases radiative energy input to the Earth's atmosphere. The path of this radiative input is complex. It is redistributed, both vertically and horizontally, by various physical processes, including advection, convection, and diffusion in the atmosphere and ocean.

When an increase in CO_2 increases the radiative input to the atmosphere, how and in which direction does the atmosphere respond? Hypotheses about this response differ and are schematically shown in Figure 18. Without the water-vapor greenhouse effect, the Earth would be about 14 ºC cooler (81). The radiative contribution of doubling atmospheric CO_2 is minor, but this radiative greenhouse effect is treated quite differently by different climate hypotheses. The hypotheses that the IPCC (82,83) has chosen to adopt predict that the effect of CO_2 is amplified by the atmosphere, especially by water vapor, to produce a large temperature increase. Other hypotheses, shown as hypothesis 2, predict the opposite – that the atmospheric response will counteract the CO_2 increase and result in insignificant changes in global temperature (81,84,85,91,92). The experimental evidence, as described above, favors hypothesis 2. While CO_2 has increased substantially, its effect on temperature has been so slight that it has not been experimentally detected.

The computer climate models upon which "human-caused global warming" is based have substantial uncertainties and are markedly unreliable. This is not surprising, since the climate is a coupled,

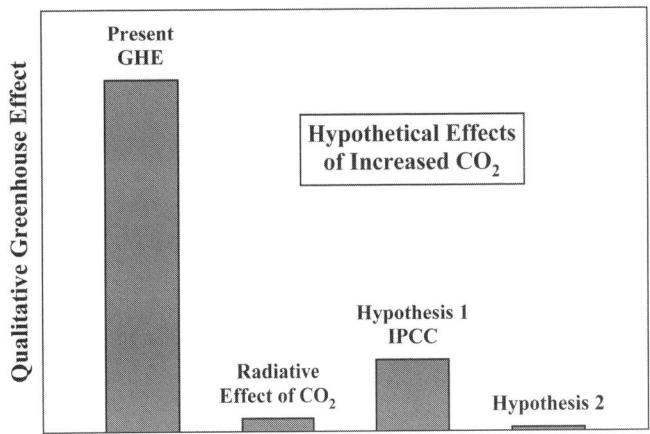

Figure 18: Qualitative illustration of greenhouse warming. "Present GHE" is the current greenhouse effect from all atmospheric phenomena. "Radiative effect of CO_2" is the added greenhouse radiative effect from doubling CO_2 without consideration of other atmospheric components. "Hypothesis 1 IPCC" is the hypothetical amplification effect assumed by IPCC. "Hypothesis 2" is the hypothetical moderation effect.

non-linear dynamical system. It is very complex. Figure 19 illustrates the difficulties by comparing the radiative CO_2 greenhouse effect with correction factors and uncertainties in some of the parameters in the computer climate calculations. Other factors, too, such as the chemical and climatic influence of volcanoes, cannot now be reliably computer modeled.

In effect, an experiment has been performed on the Earth during the past half-century – an experiment that includes all of the complex factors and feedback effects that determine the Earth's temperature and climate. Since 1940, hydrocarbon use has risen 6-fold. Yet, this rise has had no effect on the temperature trends, which have continued their cycle of recovery from the Little Ice Age in close correlation with increasing solar activity.

Not only has the global warming hypothesis failed experimental tests, it is theoretically flawed as well. It can reasonably be argued that cooling from negative physical and biological feedbacks to greenhouse gases nullifies the slight initial temperature rise (84,86).

The reasons for this failure of the computer climate models are subjects of scientific debate (87). For example, water vapor is the largest contributor to the overall greenhouse effect (88). It has been suggested that the climate models treat feedbacks from clouds, water vapor, and related hydrology incorrectly (85,89-92).

The global warming hypothesis with respect to CO_2 is not based upon the radiative properties of CO_2 itself, which is a very weak greenhouse gas. It is based upon a small initial increase in temperature caused by CO_2 and a large theoretical amplification of that temperature increase, primarily through increased evaporation of H_2O, a

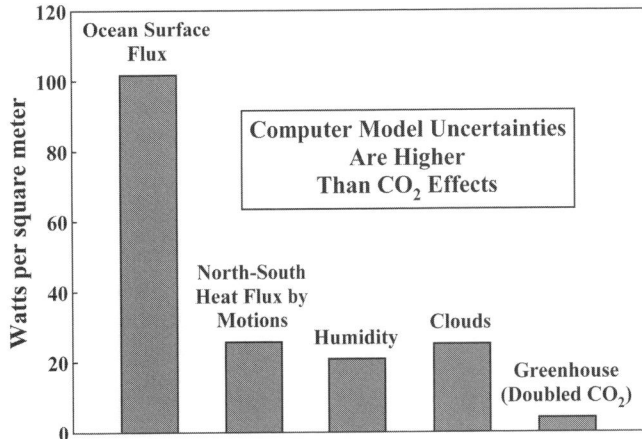

Figure 19: The radiative greenhouse effect of doubling the concentration of atmospheric CO_2 (right bar) as compared with four of the uncertainties in the computer climate models (87,93).

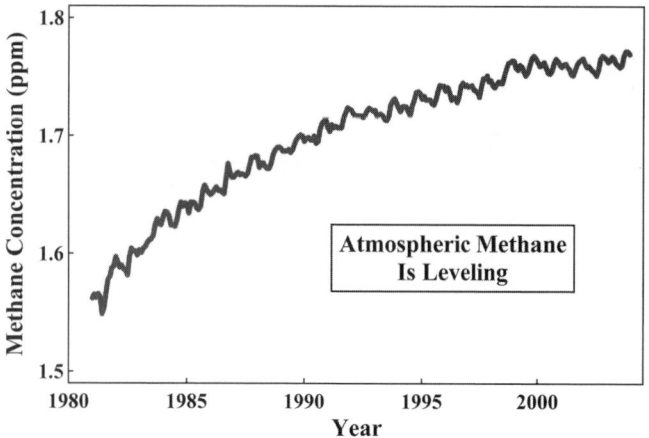

Figure 20: Global atmospheric methane concentration in parts per million between 1982 and 2004 (94).

strong greenhouse gas. Any comparable temperature increase from another cause would produce the same calculated outcome.

Thus, the 3,000-year temperature record illustrated in Figure 1 also provides a test of the computer models. The historical temperature record shows that the Earth has previously warmed far more than could be caused by CO_2 itself. Since these past warming cycles have not initiated water-vapor-mediated atmospheric warming catastrophes, it is evident that weaker effects from CO_2 cannot do so.

Methane is also a minor greenhouse gas. World CH_4 levels are, as shown in Figure 20, leveling off. In the U.S. in 2005, 42% of human-produced methane was from hydrocarbon energy production, 28% from waste management, and 30% from agriculture (95). The total amount of CH_4 produced from these U.S. sources decreased 7% between 1980 and 2005. Moreover, the record shows that, even while methane was increasing, temperature trends were benign.

The "human-caused global warming" – often called the "global warming" – hypothesis depends entirely upon computer model-generated scenarios of the future. There are no empirical records that verify either these models or their flawed predictions (96).

Claims (97) of an epidemic of insect-borne diseases, extensive species extinction, catastrophic flooding of Pacific islands, ocean acidification, increased numbers and severities of hurricanes and tornados, and increased human heat deaths from the 0.5 °C per century temperature rise are not consistent with actual observations. The "human-caused global warming" hypothesis and the computer calculations that support it are in error. They have no empirical support and are invalidated by numerous observations.

WORLD TEMPERATURE CONTROL

World temperature is controlled by natural phenomena. What steps could mankind take if solar activity or other effects began to shift the Earth toward temperatures too cold or too warm for optimum human life?

First, it would be necessary to determine what temperature humans feel is optimum. It is unlikely that the chosen temperature would be exactly that which we have today. Second, we would be fortunate if natural forces were to make the Earth too warm rather than too cold because we can cool the Earth with relative ease. We have no means by which to warm it. Attempting to warm the Earth with addition of CO_2 or to cool the Earth by restrictions of CO_2 and hydrocarbon use would, however, be futile. Neither would work.

Inexpensively blocking the sun by means of particles in the upper atmosphere would be effective. S.S. Penner, A.M. Schneider, and E. M. Kennedy have proposed (98) that the exhaust systems of commercial airliners could be tuned in such a way as to eject particulate sun-blocking material into the upper atmosphere. Later, Edward Teller similarly suggested (18) that particles could be injected into

the atmosphere in order to reduce solar heating and cool the Earth. Teller estimated a cost of between $500 million and $1 billion per year for between 1 °C and 3 °C of cooling. Both methods use particles so small that they would be invisible from the Earth.

These methods would be effective and economical in blocking solar radiation and reducing atmospheric and surface temperatures. There are other similar proposals (99). World energy rationing, on the other hand, would not work.

The climate of the Earth is now benign. If temperatures become too warm, this can easily be corrected. If they become too cold, we have no means of response – except to maximize nuclear and hydrocarbon energy production and technological advance. This would help humanity adapt and might lead to new mitigation technology.

FERTILIZATION OF PLANTS BY CO_2

How high will the CO_2 concentration of the atmosphere ultimately rise if mankind continues to increase the use of coal, oil, and natural gas? At ultimate equilibrium with the ocean and other reservoirs there will probably be very little increase. The current rise is a non-equilibrium result of the rate of approach to equilibrium.

One reservoir that would moderate the increase is especially important. Plant life provides a large sink for CO_2. Using current knowledge about the increased growth rates of plants and assuming increased CO_2 release as compared to current emissions, it has been estimated that atmospheric CO_2 levels may rise to about 600 ppm before leveling off. At that level, CO_2 absorption by increased Earth biomass is able to absorb about 10 Gt C per year (100). At present, this absorption is estimated to be about 3 Gt C per year (57).

About 30% of this projected rise from 295 to 600 ppm has already taken place, without causing unfavorable climate changes. Moreover, the radiative effects of CO_2 are logarithmic (101,102), so more than 40% of any climatic influences have already occurred.

As atmospheric CO_2 increases, plant growth rates increase. Also, leaves transpire less and lose less water as CO_2 increases, so that plants are able to grow under drier conditions. Animal life, which depends upon plant life for food, increases proportionally.

Figures 21 to 24 show examples of experimentally measured increases in the growth of plants. These examples are representative of a very large research literature on this subject (103-109). As Figure 21 shows, long-lived 1,000- to 2,000-year-old pine trees have shown a sharp increase in growth during the past half-century. Figure 22 shows the 40% increase in the forests of the United States that has

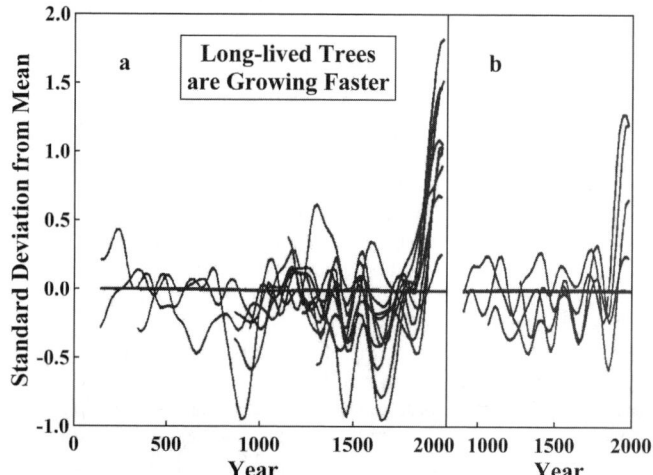

Figure 21: Standard deviation from the mean of tree ring widths for (a) bristlecone pine, limber pine, and fox tail pine in the Great Basin of California, Nevada, and Arizona and (b) bristlecone pine in Colorado (110). Tree ring widths were averaged in 20-year segments and then normalized so that the means of prior tree growth were zero. The deviations from the means are shown in units of standard deviations of those means.

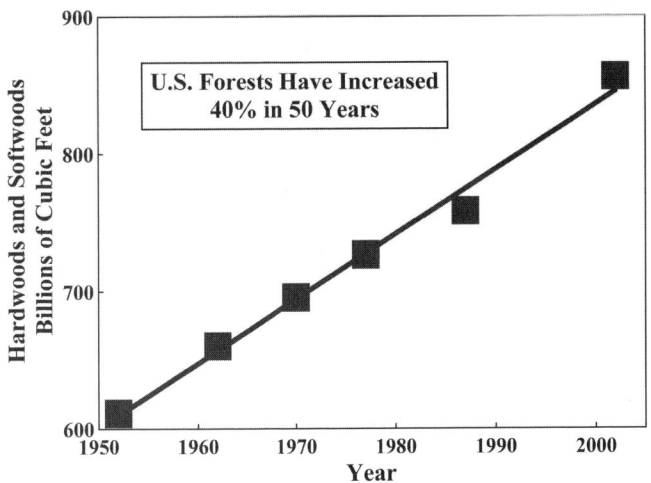

Figure 22: Inventories of standing hardwood and softwood timber in the United States compiled in *Forest Resources of the United States, 2002,* U.S. Department of Agriculture Forest Service (111,112). The linear trend cited in 1998 (1) with an increase of 30% has continued. The increase is now 40%. The amount of U.S. timber is rising almost 1% per year.

taken place since 1950. Much of this increase is due to the increase in atmospheric CO_2 that has already occurred. In addition, it has been reported that Amazonian rain forests are increasing their vegetation by about 900 pounds of carbon per acre per year (113), or approximately 2 tons of biomass per acre per year. Trees respond to CO_2 fertilization more strongly than do most other plants, but all plants respond to some extent.

Since plant response to CO_2 fertilization is nearly linear with respect to CO_2 concentration over the range from 300 to 600 ppm, as seen in Figure 23, experimental measurements at different levels of CO_2 enrichment can be extrapolated. This has been done in Figure 24 in order to illustrate CO_2 growth enhancements calculated for the atmospheric increase of about 88 ppm that has already taken place and those expected from a projected total increase of 305 ppm.

Wheat growth is accelerated by increased atmospheric CO_2, especially under dry conditions. Figure 24 shows the response of wheat grown under wet conditions versus that of wheat stressed by lack of water. The underlying data is from open-field experiments. Wheat was grown in the usual way, but the atmospheric CO_2 concentrations of circular sections of the fields were increased by arrays of com-

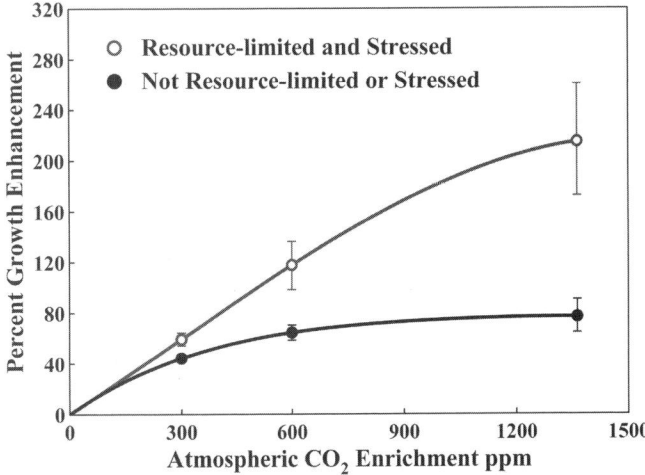

Figure 23: Summary data from 279 published experiments in which plants of all types were grown under paired stressed (open red circles) and unstressed (closed blue circles) conditions (114). There were 208, 50, and 21 sets at 300, 600, and an average of about 1350 ppm CO_2, respectively. The plant mixture in the 279 studies was slightly biased toward plant types that respond less to CO_2 fertilization than does the actual global mixture. Therefore, the figure underestimates the expected global response. CO_2 enrichment also allows plants to grow in drier regions, further increasing the response.

puter-controlled equipment that released CO_2 into the air to hold the levels as specified (115,116). Orange and young pine tree growth enhancement (117-119) with two atmospheric CO_2 increases – that which has already occurred since 1885 and that projected for the next two centuries – is also shown. The relative growth enhancement of trees by CO_2 diminishes with age. Figure 24 shows young trees.

Figure 23 summarizes 279 experiments in which plants of various types were raised under CO_2-enhanced conditions. Plants under stress from less-than-ideal conditions – a common occurrence in nature – respond more to CO_2 fertilization. The selections of species in Figure 23 were biased toward plants that respond less to CO_2 fertilization than does the mixture actually covering the Earth, so Figure 23 underestimates the effects of global CO_2 enhancement.

Clearly, the green revolution in agriculture has already benefitted from CO_2 fertilization, and benefits in the future will be even greater. Animal life is increasing proportionally, as shown by studies of 51 terrestrial (120) and 22 aquatic ecosystems (121). Moreover, as shown by a study of 94 terrestrial ecosystems on all continents ex-

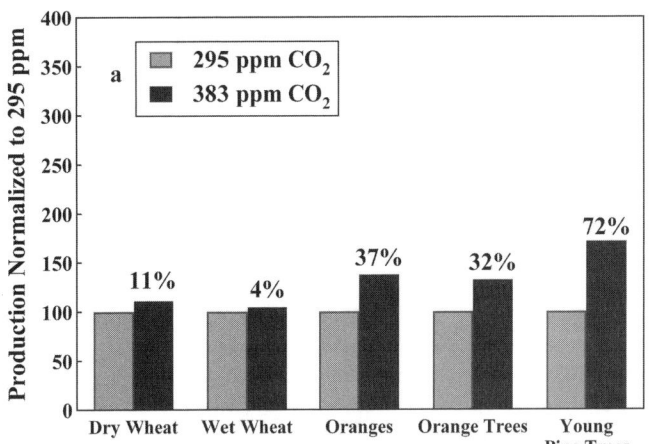

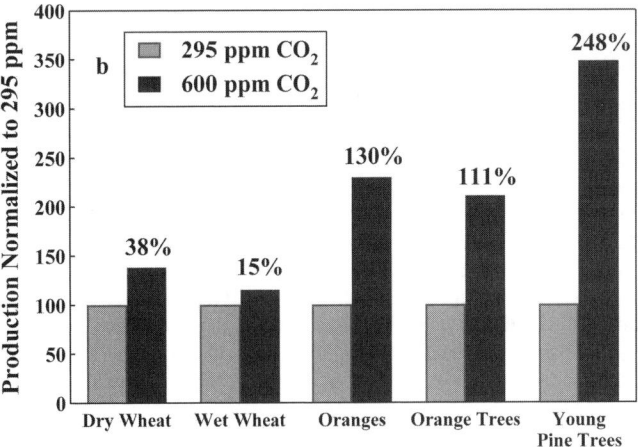

Figure 24: Calculated (1,2) growth rate enhancement of wheat, young orange trees, and very young pine trees already taking place as a result of atmospheric enrichment by CO_2 from 1885 to 2007 (a), and expected as a result of atmospheric enrichment by CO_2 to a level of 600 ppm (b).

cept Antarctica (122), species richness – biodiversity – is more positively correlated with productivity – the total quantity of plant life per acre – than with anything else.

Atmospheric CO_2 is required for life by both plants and animals. It is the sole source of carbon in all of the protein, carbohydrate, fat, and other organic molecules of which living things are constructed.

Plants extract carbon from atmospheric CO_2 and are thereby fertilized. Animals obtain their carbon from plants. Without atmospheric CO_2, none of the life we see on Earth would exist.

Water, oxygen, and carbon dioxide are the three most important substances that make life possible.

They are surely not environmental pollutants.

ENVIRONMENT AND ENERGY

The single most important human component in the preservation of the Earth's environment is energy. Industrial conversion of energy into forms that are useful for human activities is the most important aspect of technology. Abundant inexpensive energy is required for the prosperous maintenance of human life and the continued advance of life-enriching technology. People who are prosperous have the wealth required to protect and enhance their natural environment.

Currently, the United States is a net importer of energy as shown in Figure 25. Americans spend about $300 billion per year for imported oil and gas – and an additional amount for military expenses related to those imports.

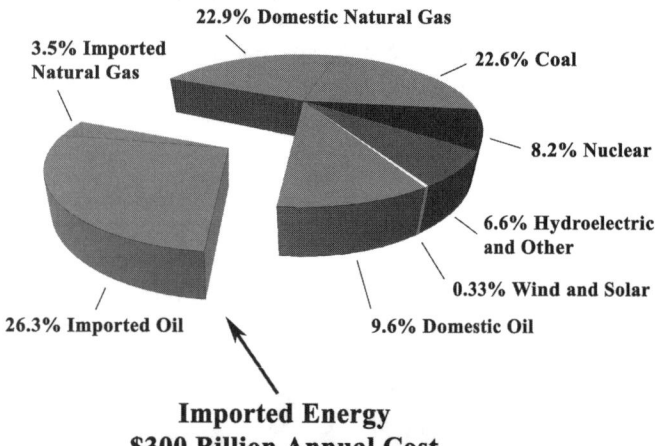

Imported Energy
$300 Billion Annual Cost

Figure 25: In 2006, the United States obtained 84.9% of its energy from hydrocarbons, 8.2% from nuclear fuels, 2.9% from hydroelectric dams, 2.1% from wood, 0.8% from biofuels, 0.4% from waste, 0.3% from geothermal, and 0.3% from wind and solar radiation. The U.S. uses 21 million barrels of oil per day – 27% from OPEC, 17% from Canada and Mexico, 16% from others, and 40% produced in the U.S. (95). The cost of imported oil and gas at $60 per barrel and $7 per 1,000 ft³ in 2007 is about $300 billion per year.

Political calls for a reduction of U.S. hydrocarbon use by 90% (123), thereby eliminating 75% of America's energy supply, are obviously impractical. Nor can this 75% of U.S. energy be replaced by alternative "green" sources. Despite enormous tax subsidies over the past 30 years, green sources still provide only 0.3% of U.S. energy.

Yet, the U.S. clearly cannot continue to be a large net importer of energy without losing its economic and industrial strength and its political independence. It should, instead, be a net exporter of energy.

There are three realistic technological paths to American energy independence – increased use of hydrocarbon energy, nuclear energy, or both. There are no climatological impediments to increased use of hydrocarbons, although local environmental effects can and must be accommodated. Nuclear energy is, in fact, less expensive and more environmentally benign than hydrocarbon energy, but it too has been the victim of the politics of fear and claimed disadvantages and dangers that are actually negligible.

For example, the "problem" of high-level "nuclear waste" has been given much attention, but this problem has been politically created by U.S. government barriers to American fuel breeding and reprocessing. Spent nuclear fuel can be recycled into new nuclear fuel. It need not be stored in expensive repositories.

Reactor accidents are also much publicized, but there has never been even one human death associated with an American nuclear reactor incident. By contrast, American dependence on automobiles results in more than 40,000 human deaths per year.

All forms of energy generation, including "green" methods, entail industrial deaths in the mining, manufacture, and transport of resources they require. Nuclear energy requires the smallest amount of such resources (124) and therefore has the lowest risk of deaths.

Estimated relative costs of electrical energy production vary with geographical location and underlying assumptions. Figure 26 shows a recent British study, which is typical. At present, 43% of U.S. energy consumption is used for electricity production.

To be sure, future inventions in energy technology may alter the relative economics of nuclear, hydrocarbon, solar, wind, and other methods of energy generation. These inventions cannot, however, be forced by political fiat, nor can they be wished into existence. Alternatively, "conservation," if practiced so extensively as to be an alternative to hydrocarbon and nuclear power, is merely a politically correct word for "poverty."

The current untenable situation in which the United States is losing $300 billion per year to pay for foreign oil and gas is not the result of failures of government energy production efforts. The U.S. government does not produce energy. Energy is produced by private industry. Why then has energy production thrived abroad while domestic production has stagnated?

This stagnation has been caused by United States government taxation, regulation, and sponsorship of litigation, which has made the U.S. a very unfavorable place to produce energy. In addition, the U.S. government has spent vast sums of tax money subsidizing inferior energy technologies for political purposes.

It is not necessary to discern in advance the best course to follow. Legislative repeal of taxation, regulation, incentives to litigation, and repeal of all subsidies of energy generation industries would stimulate industrial development, wherein competition could then automatically determine the best paths.

Nuclear power is safer, less expensive, and more environmentally benign than hydrocarbon power, so it is probably the better choice for increased energy production. Solid, liquid and gaseous hydrocarbon fuels provide, however, many conveniences, and a national infrastructure to use them is already in place. Oil from shale or coal liquefaction is less expensive than crude oil at current prices, but its ongoing production costs are higher than those for already developed oil fields. There is, therefore, an investment risk that crude oil prices could drop so low that liquefaction plants could not compete. Nuclear energy does not have this disadvantage, since the operating costs of nuclear power plants are very low.

Figure 27 illustrates, as an example, one practical and environmentally sound path to U.S. energy independence. At present 19% of U.S. electricity is produced by 104 nuclear power reactors with an average generating output in 2006 of 870 megawatts per reactor, for a total of about 90 GWe (gigawatts) (125). If this were increased by 560 GWe, nuclear power could fill all current U.S. electricity requirements and have 230 GWe left over for export as electricity or as hydrocarbon fuels replaced or manufactured.

Thus, rather than a $300 billion trade loss, the U.S. would have a $200 billion trade surplus – and installed capacity for future U.S. re-

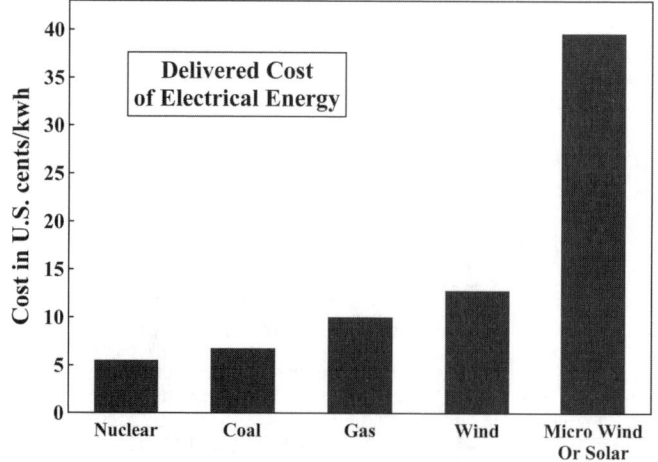

Figure 26: Delivered cost per kilowatt hour of electrical energy in Great Britain in 2006, without CO_2 controls (126). These estimates include all capital and operational expenses for a period of 50 years. Micro wind or solar are units installed for individual homes.

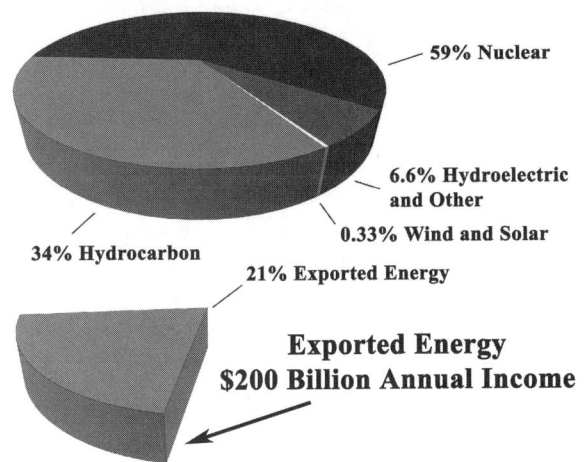

59% Nuclear

6.6% Hydroelectric and Other

0.33% Wind and Solar

34% Hydrocarbon

21% Exported Energy

Exported Energy
$200 Billion Annual Income

Figure 27: Construction of one Palo Verde installation with 10 reactors in each of the 50 states. Energy trade deficit is reversed by $500 billion per year, resulting in a $200 billion annual surplus. Currently, this solution is not possible owing to misguided government policies, regulations, and taxation and to legal maneuvers available to anti-nuclear activists. These impediments should be legislatively repealed.

quirements. Moreover, if heat from additional nuclear reactors were used for coal liquefaction and gasification, the U.S. would not even need to use its oil resources. The U.S. has about 25% of the world's coal reserves. This heat could also liquify biomass, trash, or other sources of hydrocarbons that might eventually prove practical.

The Palo Verde nuclear power station near Phoenix, Arizona, was originally intended to have 10 nuclear reactors with a generating capacity of 1,243 megawatts each. As a result of public hysteria caused by false information – very similar to the human-caused global warming hysteria being spread today, construction at Palo Verde was stopped with only three operating reactors completed. This installation is sited on 4,000 acres of land and is cooled by waste water from the city of Phoenix, which is a few miles away. An area of 4,000 acres is 6.25 square miles or 2.5 miles square. The power station itself occupies only a small part of this total area.

If just one station like Palo Verde were built in each of the 50 states and each installation included 10 reactors as originally planned for Palo Verde, these plants, operating at the current 90% of design capacity, would produce 560 GWe of electricity. Nuclear technology has advanced substantially since Palo Verde was built, so plants constructed today would be even more reliable and efficient.

Assuming a construction cost of $2.3 billion per 1,200 MWe reactor (127) and 15% economies of scale, the total cost of this entire project would be $1 trillion, or 4 months of the current U.S. federal budget. This is 8% of the annual U.S. gross domestic product. Construction costs could be repaid in just a few years by the capital now spent by the people of the United States for foreign oil and by the change from U.S. import to export of energy.

The 50 nuclear installations might be sited on a population basis. If so, California would have six, while Oregon and Idaho together would have one. In view of the great economic value of these facilities, there would be vigorous competition for them.

In addition to these power plants, the U.S. should build fuel reprocessing capability, so that spent nuclear fuel can be reused. This would lower fuel cost and eliminate the storage of high-level nuclear waste. Fuel for the reactors can be assured for 1,000 years (128) by using both ordinary reactors with high breeding ratios and specific breeder reactors, so that more fuel is produced than consumed.

About 33% of the thermal energy in an ordinary nuclear reactor is converted to electricity. Some new designs are as high as 48%. The heat from a 1,243 MWe reactor can produce 38,000 barrels of coal-derived oil per day (129). With one additional Palo Verde installation in each state for oil production, the yearly output would be at least 7 billion barrels per year with a value, at $60 per barrel, of

more than $400 billion per year. This is twice the oil production of Saudi Arabia. Current proven coal reserves of the United States are sufficient to sustain this production for 200 years (128). This liquified coal exceeds the proven oil reserves of the entire world. The reactors could produce gaseous hydrocarbons from coal, too.

The remaining heat from nuclear power plants could warm air or water for use in indoor climate control and other purposes.

Nuclear reactors can also be used to produce hydrogen, instead of oil and gas (130,131). The current cost of production and infrastructure is, however, much higher for hydrogen than for oil and gas. Technological advance reduces cost, but usually not abruptly. A prescient call in 1800 for the world to change from wood to methane would have been impracticably ahead of its time, as may be a call today for an abrupt change from oil and gas to hydrogen. In distinguishing the practical from the futuristic, a free market in energy is absolutely essential.

Surely these are better outcomes than are available through international rationing and taxation of energy as has been recently proposed (82,83,97,123). This nuclear energy example demonstrates that current technology can produce abundant inexpensive energy if it is not politically suppressed.

There need be no vast government program to achieve this goal. It could be reached simply by legislatively removing all taxation, most regulation and litigation, and all subsidies from all forms of energy production in the U.S., thereby allowing the free market to build the most practical mixture of methods of energy generation.

With abundant and inexpensive energy, American industry could be revitalized, and the capital and energy required for further industrial and technological advance could be assured. Also assured would be the continued and increased prosperity of all Americans.

The people of the United States need more low-cost energy, not less. If this energy is produced in the United States, it can not only become a very valuable export, but it can also ensure that American industry remains competitive in world markets and that hoped-for American prosperity continues and grows.

In this hope, Americans are not alone. Across the globe, billions of people in poorer nations are struggling to improve their lives. These people need abundant low-cost energy, which is the currency of technological progress.

In newly developing countries, that energy must come largely from the less technologically complicated hydrocarbon sources. It is a moral imperative that this energy be available. Otherwise, the efforts of these peoples will be in vain, and they will slip backwards into lives of poverty, suffering, and early death.

Energy is the foundation of wealth. Inexpensive energy allows people to do wonderful things. For example, there is concern that it may become difficult to grow sufficient food on the available land. Crops grow more abundantly in a warmer, higher CO_2 environment, so this can mitigate future problems that may arise (12).

Energy provides, however, an even better food insurance plan. Energy-intensive hydroponic greenhouses are 2,000 times more productive per unit land area than are modern American farming methods (132). Therefore, if energy is abundant and inexpensive, there is no practical limit to world food production.

Fresh water is also believed to be in short supply. With plentiful inexpensive energy, sea water desalination can provide essentially unlimited supplies of fresh water.

During the past 200 years, human ingenuity in the use of energy has produced many technological miracles. These advances have markedly increased the quality, quantity, and length of human life. Technologists of the 21st century need abundant, inexpensive energy with which to continue this advance.

Were this bright future to be prevented by world energy rationing, the result would be tragic indeed. In addition to human loss, the Earth's environment would be a major victim of such a mistake. Inexpensive energy is essential to environmental health. Prosperous people have the wealth to spare for environmental preservation and enhancement. Poor, impoverished people do not.

CONCLUSIONS

There are no experimental data to support the hypothesis that increases in human hydrocarbon use or in atmospheric carbon dioxide and other greenhouse gases are causing or can be expected to cause unfavorable changes in global temperatures, weather, or landscape. There is no reason to limit human production of CO_2, CH_4, and other minor greenhouse gases as has been proposed (82,83,97,123).

We also need not worry about environmental calamities even if the current natural warming trend continues. The Earth has been much warmer during the past 3,000 years without catastrophic effects. Warmer weather extends growing seasons and generally improves the habitability of colder regions.

As coal, oil, and natural gas are used to feed and lift from poverty vast numbers of people across the globe, more CO_2 will be released into the atmosphere. This will help to maintain and improve the health, longevity, prosperity, and productivity of all people.

The United States and other countries need to produce more energy, not less. The most practical, economical, and environmentally sound methods available are hydrocarbon and nuclear technologies.

Human use of coal, oil, and natural gas has not harmfully warmed the Earth, and the extrapolation of current trends shows that it will not do so in the foreseeable future. The CO_2 produced does, however, accelerate the growth rates of plants and also permits plants to grow in drier regions. Animal life, which depends upon plants, also flourishes, and the diversity of plant and animal life is increased.

Human activities are producing part of the rise in CO_2 in the atmosphere. Mankind is moving the carbon in coal, oil, and natural gas from below ground to the atmosphere, where it is available for conversion into living things. We are living in an increasingly lush environment of plants and animals as a result of this CO_2 increase. Our children will therefore enjoy an Earth with far more plant and animal life than that with which we now are blessed.

REFERENCES

1. Robinson, A. B., Baliunas, S. L., Soon, W., and Robinson, Z. W. (1998) *Journal of American Physicians and Surgeons* **3**, 171-178.
2. Soon, W., Baliunas, S. L., Robinson, A. B., and Robinson, Z. W. (1999) *Climate Res.* **13**, 149-164.
3. Keigwin, L. D. (1996) *Science* **274**, 1504-1508. ftp://ftp.ncdc.noaa.gov/pub/data/paleo/contributions_by_author/keigwin1996/
4. Oerlemanns, J. (2005) *Science* **308**, 675-677.
5. Oerlemanns, J., Björnsson, H., Kuhn, M., Obleitner, F., Palsson, F., Smeets, C. J. P. P., Vugts, H. F., and De Wolde, J. (1999) *Boundary-Layer Meteorology* **92**, 3-26.
6. Greuell, W. and Smeets, P. (2001) *J. Geophysical Res.* **106**, 31717-31727.
7. Marland, G., Boden, T. A., and Andres, R. J. (2007) Global, Regional, and National CO2 Emissions. In Trends: *A Compendium of Data on Global Change*. Carbon Dioxide Information Analysis Center, Oak Ridge National Laboratory, U.S. Department of Energy, Oak Ridge, TN, USA, http://cdiac.ornl.gov/trends/emis/tre_glob.htm
8. Soon, W. (2005) *Geophysical Research Letters* **32**, 2005GL023429.
9. Hoyt, D. V. and Schatten, K. H. (1993) *J. Geophysical Res.* **98**, 18895-18906.
10. National Climatic Data Center, *Global Surface Temperature Anomalies* (2007) http://www.ncdc.noaa.gov/oa/climate/research/anomalies/anomalies.html and NASA GISS http://data.giss.nasa.gov/gistemp/graphs/Fig.D.txt.
11. Soon, W., Baliunas, S., Idso, C., Idso, S., and Legates, D. R. (2003) *Energy & Env.* **14**, 233-296.
12. Idso, S. B. and Idso, C. D. (2007) Center for Study of Carbon Dioxide and Global Change http://www.co2science.org/scripts/CO2ScienceB2C/education/reports/hansen/hansencritique.jsp.
13. Groveman, B. S. and Landsberg, H. E. (1979) *Geophysical Research Letters* **6**, 767-769.
14. Esper, J., Cook, E. R., and Schweingruber, F. H. (2002) *Science* **295**, 2250-2253.
15. Tan, M., Hou, J., and Liu, T. (2004) *Geophysical Research Letters* **31**, 2003GL019085.
16. Newton, A., Thunell, R., and Stott, L. (2006) *Geophysical Research Letters* **33**, 2006GL027234.
17. Akasofu, S.-I. (2007) International Arctic Research Center, Univ. of Alaska, Fairbanks http://www.iarc.uaf.edu/highlights/2007/akasofu_3_07/Earth_recovering_from_LIA_R.pdf
18. Teller, E., Wood, L., and Hyde, R. (1997) 22nd International Seminar on Planetary Emergencies, Erice, Italy, Lawrence Livermore National Laboratory, UCRL-JC-128715, 1-18.
19. Soon, W. (2007) private communication.
20. U.S. National Climatic Data Center, U.S. Department of Commerce 2006 Climate Review. http://lwf.ncdc.noaa.gov/oa/climate/research/cag3/na.html
21. Landsea, C. W. (2007) *EOS* **88** No. 18, 197, 208.
22. Landsea, C. W., Nicholls, N., Gray, W. M., and Avila, L. A. (1996) *Geophysical Research Letters* **23**, 1697-1700.
23. Goldenberg, S. B., Landsea, C. W., Mesta-Nuñez, A. M., and Gray, W. M. (2001) *Science* **293**, 474-479.
24. Jevrejeva, S., Grinsted, A., Moore, J. C., and Holgate, S. (2006) *J. Geophysical Res.* **111**, 2005JC003229. http://www.pol.ac.uk/psmsl/author_archive/jevrejeva_etal_gsl/
25. Leuliette, E. W., Nerem, R. S., and Mitchum, G. T. (2004) *Marine Geodesy* **27**, No. 1-2, 79-94. http://sealevel.colorado.edu/
26. Lamb, H. (1982) *Climate, History, and the Modern World*, Methuen, New York.
27. Essex, C., McKitrick, R., and Andresen, B. (2007) *J. Non-Equilibrium Therm.* **32**, 1-27.
28. Polyakov, I. V., Bekryaev, R. V., Alekseev, G. V., Bhatt, U. S., Colony, R. L., Johnson, M. A., Maskshtas, A. P., and Walsh, D. (2003) *Journal of Climate* **16**, 2067-2077.
29. Christy, J. R., Norris, W. B., Spencer, R. W., and Hnilo, J. J. (2007) *J. Geophysical Res.* **112**, 2005JD006881. http://vortex.nsstc.uah.edu/data/msu/t2lt/uahncdc.
30. Spencer, R. W. and Christy, J. R. (1992) *Journal of Climate* **5**, 847-866.
31. Christy, J. R. (1995) *Climatic Change* **31**, 455-474.
32. Zhu, P., Hack, J. J., Kiehl, J. T., and Berthorton, C. S. (2007) *J. Geophysical Res.*, in press.
33. Balling, Jr., R. C. (1992) *The Heated Debate*, Pacific Research Institute.
34. Friis-Christensen, E. and Lassen, K. (1991) *Science* **254**, 698-700.
35. Baliunas, S. and Soon, W. (1995) *Astrophysical Journal* **450**, 896-901.
36. Neff, U., Burns, S. J., Mangini, A., Mudelsee, M., Fleitmann, D., and Matter, A. (2001) *Nature* **411**, 290-293.
37. Jiang, H., Eiriksson, J., Schulz, M., Knudsen, K., and Seidenkrantz, M. (2005) *Geology* **33**, 73-76.
38. Maasch, K. A., et. al. (2005) *Geografiska Annaler* **87A**, 7-15.
39. Wang, Y., Cheng, H., Edwards, R. L., He, Y., Kong, X., An, Z., Wu, J., Kelly, M. J., Dykoski, C. A., and Li, X. (2005) *Science* **308**, 854-857.
40. Baliunas, S. L. et al. (1995) *Astrophysical Journal* **438**, 269-287.
41. Fenton, L. K., Geissler, P. E., and Haberle, R. M. (2007) *Nature* **446**, 646-649.
42. Marcus, P. S. (2004) *Nature* **428**, 828-831.
43. Hammel, H. B., Lynch, D. K., Russell, R. W., Sitko, M. L., Bernstein, L. S., and Hewagama, T. (2006) *Astrophysical Journal* **644**, 1326-1333.
44. Hammel, H. B., and Lockwood, G. W. (2007) *Geophysical Research Letters* **34**, 2006GL028764.
45. Elliot, J. L. et al. (1998) *Nature* **393**, 765-767.
46. Elliot, J. L., et. al. (2003) *Nature* **424**, 165-168.
47. Sicardy, B., et. al. (2003) *Nature* **424**, 168-170.
48. Elliot, J. L., et. al. (2007) *Astronomical Journal* **134**, 1-13.
49. Camp, C. D. and Tung, K. K. (2007) *Geophysical Research Letters* **34**, 2007GL030207.
50. Scafetta, N. and West, B. J. (2006) *Geophysical Research Letters* **33**, 2006GL027142.
51. Goodridge, J. D. (1996) *Bull. Amer. Meteor. Soc.* **77**, 3-4; Goodridge, J. D. (1998) private comm.
52. Christy, J. R. and Goodridge, J. D. (1995) *Atm. Environ.* **29**, 1957-1961.
53. Hansen, J. and Lebedeff, S. (1987) *J. Geophysical Res.* **92**, 13345-13372.
54. Hansen, J. and Lebedeff, S. (1988) *Geophysical Research Letters* **15**, 323-326.
55. Hansen, J., Ruedy, R., and Sato, M. (1996) *Geophysical Research Letters* **23**, 1665-1668; http://www.giss.nasa.gov/data/gistemp/
56. Schimel, D. S. (1995) *Global Change Biology* **1**, 77-91.
57. Houghton, R. A. (2007) *Annual Review of Earth and Planetary Sciences* **35**, 313-347.
58. Jaworowski, Z., Segalstad, T. V., and Ono, N. (1992) *Science of the Total Environ.* **114**, 227-284.
59. Segalstad, T. V. (1998) *Global Warming the Continuing Debate*, Cambridge UK: European Science and Environment Forum, ed. R. Bate, 184-218.
60. Berner, R. A. (1997) *Science* **276**, 544-545.
61. Retallack, G. J. (2001) *Nature* **411**, 287-290.
62. Rothman, D. H. (2002) *Proc. Natl. Acad. Sci. USA* **99**, 4167-4171.
63. Petit et. al., (1999) *Nature* **399**, 429-436.
64. Siegenthaler, U., et. al. (2005) *Science* **310**, 1313-1317.
65. Spahni, R., et. al. (2005) *Science* **310**, 1317-1321.
66. Soon, W. (2007) *Physical Geography*, in press.
67. Dettinger, M. D. and Ghill, M. (1998) *Tellus*, **50B**, 1-24.
68. Kuo, C., Lindberg, C. R., and Thornson, D. J. (1990) *Nature* **343**, 709-714.
69. Revelle, R. and Suess, H. E. (1957) *Tellus* **9**, 18-27.
70. Yamashita, E., Fujiwara, F., Liu, X., and Ohtaki, E. (1993) *J. Oceanography* **49**, 559-569.
71. Keeling, C. D. and Whorf, T. P. (1997) Trends Online: *A Compendium of Data on Global Change*, Carbon Dioxide Information Analysis Center, Oak Ridge National Laboratory; http://cdiac.ornl.gov/trends/co2/sio-mlo.htm http://www.esrl.noaa.gov/gmd/ccgg/trends/co2_data_mlo.html
72. Schneider, D. P. et. al. (2006) *Geophysical Research Letters* **33**, 2006GL027057.
73. Archer, D. (2005) *J. Geophysical Res.* **110**, 2004JC002625.
74. Faraday, M. (1860) *The Chemical History of a Candle*, Christmas Lectures, Royal Institution, London.
75. Serreze, M. C., Holland, M. M., and Stroeve, J. (2007) *Science* **315**, 1533-1536.
76. Bentley, C. R. (1997) *Science* **275**, 1077-1078.
77. Nicholls, K. W. (1997) *Nature* **388**, 460-462.
78. Davis, C. H., Li, Y., McConnell, J. R., Frey, M. M., and Hanna, E. (2005) *Science* **308**, 1898-1901.
79. Monaghan, A. J., et. al. (2006) *Science* **313**, 827-831.
80. Kullman, L. (2007) *Nordic Journal of Botany* **24**, 445-467.
81. Lindzen, R. S. (1994) *Ann. Review Fluid Mech.* **26**, 353-379.
82. IPCC Fourth Assessment Report (AR4), Working Group I Report (2007).
83. Kyoto Protocol to the United Nations Framework Convention on Climate Change (1997).
84. Sun, D. Z. and Lindzen, R. S. (1993) *Ann. Geophysicae* **11**, 204-215.
85. Spencer, R. W. and Braswell, W. D. (1997) *Bull. Amer. Meteorological Soc.* **78**, 1097-1106.
86. Idso, S. B. (1998) *Climate Res.* **10**, 69-82.
87. Soon, W., Baliunas, S., Idso, S. B., Kondratyev, K. Ya., and Posmentier, E. S. (2001) *Climate Res.* **18**, 259-275.
88. Lindzen, R. S. (1996) *Climate Sensitivity of Radiative Perturbations: Physical Mechanisms and Their Validation*, NATO ASI Series 134, ed. H. Le Treut, Berlin: Springer-Verlag, 51-66.
89. Renno, N. O., Emanuel, K. A., and Stone, P. H. (1994) *J. Geophysical Res.* **99**, 14429-14441.
90. Soden, B. J. (2000) *Journal of Climate* **13**, 538-549.
91. Lindzen, R. S., Chou, M., and Hou, A. Y. (2001) *Bull. Amer. Meteorlogical Soc.* **82**, 417-432.
92. Spencer, R. W., Braswell, W. D., Christy, J. R., and Hnilo, J. (2007) *Geophysical Research Letters* **34**, 2007GL029698.
93. Lindzen, R. S. (1995), personal communication.
94. Khalil, M. A. K., Butenhoff, C. L., and Rasmussen, R. A. (2007) *Environmental Science and Technology* **41**, 2131-2137.
95. Annual Energy Review, U.S. Energy Information Admin., Report No. DOE/EIA-0384 (2006).
96. Essex, C., Ilie, S., and Corless, R. M. (2007) *J. Geophysical Res.*, in press.
97. Gore, A. (2006) *An Inconvenient Truth*, Rodale, NY.
98. Penner, S S., Schneider, A. M., and Kennedy, E. M. (1984) *Acta Astronautica* **11**, 345-348.
99. Crutzen, P. J. (2006) *Climatic Change* **77**, 211-219.
100. Idso, S. B. (1989) *Carbon Dioxide and Global Change: Earth in Transition*, IBR Press.
101. Lam, S. H. (2007) *Logarithmic Response and Climate Sensitivity of Atmospheric CO_2*, 1-15, www.princeton.edu/~lam/documents/LamAug07bs.pdf.
102. Lindzen, R. S. (2005) *Proc. 34th Int. Sem. Nuclear War and Planetary Emergencies*, ed. R. Raigaina, World Scientific Publishing, Singapore, 189-210.
103. Kimball, B. A. (1983) *Agron. J.* **75**, 779-788.
104. Cure, J. D. and Acock, B. (1986) *Agr. Forest Meteorol.* **8**, 127-145.
105. Mortensen, L. M. (1987) *Sci. Hort.* **33**, 1-25.
106. Lawlor, D. W. and Mitchell, R. A. C. (1991) *Plant, Cell, and Environ.* **14**, 807-818.
107. Drake, B. G. and Leadley, P. W. (1991) *Plant, Cell, and Environ.* **14**, 853-860.
108. Gifford, R. M. (1992) *Adv. Bioclim.* **1**, 24-58.
109. Poorter, H. (1993) *Vegetatio* **104-105**, 77-97.
110. Graybill, D. A. and Idso, S. B. (1993) *Global Biogeochem. Cyc.* **7**, 81-95.
111. Waddell, K. L., Oswald, D. D., and Powell D. S. (1987) *Forest Statistics of the United States*, U.S. Forest Service and Dept. of Agriculture.
112. Smith, W. B., Miles, P. D., Vissage, J. S., and Pugh, S. A. (2002) *Forest Resources of the United States*, U.S. Forest Service and Dept. of Agriculture.
113. Grace, J., Lloyd, J., McIntyre, J., Miranda, A. C., Meir, P., Miranda, H. S., Nobre, C., Moncrieff, J., Massheder, J., Malhi, Y., Wright, I., and Gash, J. (1995) *Science* **270**, 778-780.
114. Idso, K. E. and Idso, S. (1974) *Agr. Forest Meteor.* **69**, 153-203.
115. Kimball, B.A., Pinter Jr., P. J., Hunsaker, D. J., Wall, G. W. G., LaMorte, R. L., Wechsung, G., Wechsung, F., and Kartschall, T. (1995) *Global Change Biology* **1**, 429-442.
116. Pinter, J. P., Kimball, B. A., Garcia, R. L., Wall, G. W., Hunsaker, D. J., and LaMorte, R. L. (1996) *Carbon Dioxide and Terrestrial Ecosystems* 215-250, Koch and Mooney, Acad. Press.
117. Idso, S. B. and Kimball, B. A. (1991) *Agr. Forest Meteor.* **55**, 345-349.
118. Idso, S. B. and Kimball, B. A. (1994) *J. Exper. Botany* **45**, 1669-1692.
119. Idso, S. B. and Kimball, B. A. (1997) *Global Change Biol.* **3**, 89-96.
120. McNaughton, S. J., Oesterhold, M., Frank, D. A., and Williams, K. J. (1989) *Nature* **341**, 142-144.
121. Cyr, H. and Pace, M. L. (1993) *Nature* **361**, 148-150.
122. Scheiner, S. M. and Rey-Benayas, J. M. (1994) *Evol. Ecol.* **8**, 331-347.
123. Gore, A., Pelosi, N., and Reid, H. (June 29, 2007) *The Seven Point Live Earth Pledge*. Speaker of the House Website, www.speaker.gov. and www.liveearth.org.
124. Beckmann, P. (1985) *The Health Hazards of NOT Going Nuclear*, Golem, Boulder, Colorado.
125. American Nuclear Society, *Nuclear News* (2007) March, 46-48.
126. McNamara, B. (2006) Leabrook Computing, Bournemouth, England.
127. *Projected Costs of Generating Electricity: 2005 Update* (2005), Paris: Nuclear Energy Agency, OECD Publication No. 53955 2005, Paris.
128. Penner, S. S. (1998) *Energy* **23**, 71-78.
129. Posma, B. (2007) *Liquid Coal*, Fort Myers, Fl, www.liquidcoal.com.
130. Ausubel, J. H. (2007) *Int. J. Nuclear Governance, Economy and Ecology* **1**, 229-243.
131. Penner, S. S. (2006) *Energy* **31**, 33-43.
132. Simon, J. L. (1996) *The Ultimate Resource 2*, Princeton Univ. Press, Princeton, New Jersey.